1 MONTH OF
FREE
READING

at

www.ForgottenBooks.com

By purchasing this book you are eligible for one month membership to ForgottenBooks.com, giving you unlimited access to our entire collection of over 700,000 titles via our web site and mobile apps.

To claim your free month visit:

www.forgottenbooks.com/free207483

ISBN 978-0-265-20405-4
PIBN 10207483

FOREST PHYSIOGRAPHY

PHYSIOGRAPHY OF THE UNITED STATES
AND PRINCIPLES OF SOILS IN
RELATION TO FORESTRY

BY

ISAIAH BOWMAN, Ph. D.

Assistant Professor of Geography in Yale University

FIRST EDITION
FIRST THOUSAND

NEW YORK
JOHN WILEY & SONS
LONDON: CHAPMAN & HALL, LIMITED
1911

Stanhope Press
F. H. GILSON COMPANY
BOSTON, U.S.A.

To

Eugene Woldemar Hilgard

LEADER IN AGROGEOLOGY

PREFACE

STUDENTS of forestry in the United States are constantly demanding a guide to the topography, drainage, soils, and climatic features of the country. The need for such a book is keenly felt, since students in the professional schools of forestry have little time for the study of original sources of material in a subject which is not forestry but the basis of forestry. On the other hand, such general sources as are available are both too brief and too elementary for the somewhat comprehensive requirements of the American forester.

In preparing this book I have attempted to steer a middle course between that purely descriptive writing with which the forester is too often satisfied, and that altogether explanatory writing which the technical physiographer is inclined to regard as the real substance of a scientific book. The descriptive portions are intentionally rather comprehensive, for the relief and not the explanation of it is of immediate value to the forester, no matter how important the explanation may be in assisting him to appreciate and remember the relief. A chief concern has been the reduction of geologic data to a minimum. A geologic statement frequently runs off into so many consequences that the most important of these almost force one to a more extended and complex discussion than the forester, a lay student of geologic and geographic science, can assimilate. Emphatically, some geologic data are essential, but only in so far as they have an immediate physiographic bearing.

A further point concerning the organization of the material of this book requires statement here. It seems so clear that one can not know forestry without knowing under what physical conditions trees grow, that one finds it impossible to see how even the least philosophical view of the subject can exclude a knowledge of physiography. It would seem that one should pay a great deal of attention to lumbering as related to drainage and relief, to silviculture as related to soils, climate, and water supply, and in general that one should emphasize the forester's dependence upon physical conditions. This would appear to be so plain a doctrine as not to require restatement here, were it not for the fact that some students of forestry and even some schools of forestry still

vii

pay too little attention to the subject. If the forest is accepted merely as a fact, and the chief concern is its immediate and thoughtless exploitation, physiography may indeed be the fifth wheel to the coach, although even so practical a view as the lumberman's must include some knowledge of topography and drainage if merely to put forest products upon the market. But forestry is more than lumbering, and if forests are to be conserved, if they are to be improved and extended, every direct relation of the tree to its physiographic environment is vital.

The title, "Forest Physiography," does not imply a book on forestry but rather a book on physiography for students of forestry; and, as nearly as has seemed advisable from the nature of the subject, it has been prepared for their special needs. It is hoped, however, that the book may be of service to historians also, and to economists, since a knowledge of the physiography of the United States has heretofore depended upon one or two short and general chapters on the subject, or upon a study of hundreds of original papers and monographs.

No attempt has been made to show the connection between soils and agriculture, which is the general theme of text-books on soils. Neither has it been attempted to make a complete classification of all the types of rocks and soils found in nature. The distinctions which such a classification implies may be serviceable to the geologist, the farmer, and the gardener, but they are too finely drawn for the forester, whose needs are met by a broader classification based upon qualities of wider application. It is our purpose to discuss the origin of soil, and the physical and chemical transformations it undergoes in the process of gradual decay and of interaction with the plants that thrive upon it. Soil water, drainage, plant foods in the soil, soil warmth, and soil improvement are additional topics; the most important point of all concerns the actual preservation of soils that occur upon forested lands. In a broad way all soils are an inheritance from a geologic past; they are slow of accumulation, precious, and vital to man's welfare, for agriculture is the basis of our modern organized life, and soil is the basis of agriculture. Imprudent forest cutting and thoughtless land tillage tend to disturb a natural balance between great forces. That they may be a sowing to the wind is amply shown by the whirlwind of destruction which man is reaping in extensive tracts of deforested, soilless uplands of America, Europe, and Asia. Forestry affects not only trees but also soils, one of the greatest of man's geologic inheritances.

I am under obligations to many for advice and assistance. First of all to Prof. H. E. Gregory, who has given most generously of his time in reading both manuscript and proof. Prof. J. Barrell gave helpful advice in the preparation of certain chapters in Part One, and Prof. J. W. Toumey of the Yale Forest School supplied important criticisms. Dr. G. E. Nichols of the Sheffield Scientific School made a large number of corrections and alterations in the botanical descriptions. Prof. E. W. Hilgard of the University of California, and his colleague, Prof. R. H. Loughridge, have read Part One with great care, and the benefit of their searching criticisms lays me under deep obligations. My obligations to Professor Hilgard extend beyond this, however, for his great work on soils has been an invaluable source of experimental data.

The United States Geological Survey has followed its usual generous policy and supplied many of the illustrations. Prof. R. DeC. Ward has kindly allowed me to use the expensive original drawings of two climatic maps after Köppen. A number of publishers, acknowledged in a separate list, have permitted the reproduction of illustrative material. I have also obtained illustrations from the Canadian Geological Survey, Prof. C. R. Gould of Oklahoma University, Prof. C. A. Reeds of Bryn Mawr, Mr. H. Brigham, Jr., of Cortez, Colorado, and a number of my students in forest physiography.

ISAIAH BOWMAN.

YALE UNIVERSITY,
June 10, 1911.

ACKNOWLEDGMENT OF ILLUSTRATIONS

Fig. 10, p. 91, from H. W. Wiley, *Soils*, Chemical Publishing Company, Easton, Penn.

Fig. 11, p. 99, from E. W. Hilgard, *Soils*, Macmillan Company, N. Y.

Fig. 13, p. 111, and Fig. 14, p. 113, from R. DeC. Ward, *Climate*, G. P. Putnam's Sons, N. Y. Based on Köppen.

Fig. 29, p. 146; Fig. 48, p. 189; Fig. 61, p. 230; Fig. 62, p. 231, from F. H. Newell, *Irrigation in the United States*.

Fig. 80, p. 285, from Gilbert and Brigham, *Physical Geography*, D. Appleton & Co.

Fig. 271, p. 676. From E. de Martonne, *Traité de Geographie Physique*, Armand Colin, Paris, France.

The largest number of illustrations are from the *Collections of the United States Geological Survey*.

A number of the block diagrams of the Basin Ranges are from W. M. Davis, *Mountain Ranges of the Great Basin*, Bull. Mus. Comp. Zool.

CONTENTS

PART ONE—THE SOIL

PART TWO—PHYSIOGRAPHY OF THE UNITED STATES

LIST OF ILLUSTRATIONS

xvii

• LIST OF PLATES

FOREST PHYSIOGRAPHY

PART ONE

THE SOIL

CHAPTER I

THE IMPORTANCE, ORIGIN, AND DIVERSITY OF SOILS

THE SOIL IN RELATION TO LIFE

MEN whose work brings them into touch with the soil and its relation
to life do not employ the phrase "mother earth" in a casual sense.
The great hosts of plant and animal life that people the lands in large
part have their origin in or draw their support from the cover of land
waste whose upper layers are the soil. They are, directly or indirectly,
the dependent children of the earth. Viewed from such a standpoint
the soil is not mere dirt, a substance to be despised, a synonym for
filth, but a great storehouse of energy, a great home, a bountiful mother.
Countless billions of micro-organisms — the bacterial flora — throng its
dark passageways while the roots of countless higher plants ramify
through it in eager quest for food and water. Only less numerous are
the earthworms, insects, and burrowing animals that delve into it for
food as well as for shelter. To supply all these needs is no mean function; it is probable that no other planet in our solar system has so large
an endowment of life-giving, life-supporting soil; the evolution of the
life of the earth would have been on far lower levels if the endowment
had not been so generous.

THE SOIL AND THE FOREST

From the standpoint of the forest the soil is a factor of great importance. The home of the tree is the soil and the air; and a forester, whose
chief concern is the tree, requires a somewhat comprehensive knowledge
of these two elements of the environment of every forest. Without

soil in some amount tree growth of any kind is impossible, although the amount required to produce a poor growth may be very small. Low forms of vegetation, such as lichens, mosses, and shrubs of many varieties, may find life possible in a region where there appears to be practically no soil; but a careful examination will usually disclose rock pockets partially filled with small quantities of soil, tiny crevices that contain particles of dust, and joints in greater or lesser number that have caught soil fragments washed from the adjacent surfaces almost as fast as formed. These accumulations afford a foothold for the lower forms of vegetable life which tend to disintegrate both soil and rock and further to increase the amount of available soil. Ordinary weathering will tend toward the same result, and if the climatic conditions, the relief, etc., are favorable, a soil cover capable of sustaining a denser growth of vegetation of a higher order will eventually be formed. In time and through the gradual development of a soil cover a dense forest may grow on what was in a preceding portion of a geologic period a bare rock terrane.

While the broad relation of the soil to the forest is thus readily distinguishable, the finer relations are often difficult of determination and there are many physical conditions that evoke no recognizable response in the forest world. A study of the maps, Fig. 23 and Plates IV and V, representing respectively the forest regions of the United States, the physiographic provinces and the geologic formations, will enable the student of forest physiography to appreciate at once that the physiographic features and related soil types are of more local development than the broad forest types which they support; and that the finer distinctions between soil types are of little value in understanding the range of a given forest type, however directly they may affect the welfare of the individual tree by modifying its habitat. In short, it may be said that the conditions which limit either the growth or the distribution of most forest species are so extreme that they embrace or overlap a large number of physical subdivisions.

In general plants are rather impartial as to soil unless the soil characters are of an extreme type; relatively few have absolute soil requirements. Competition among forest trees may drive out some species, in which case the unsuccessful species can not be described as incapable of growth on the soil from which they have been driven; simply, their competitors are able better to use the given resources of soil and air. It often happens that a given species is markedly tolerant of soil and climate except at the limits of its range where competition begets an apparent intolerance.

It is concluded that plants possess a peculiar inherent force by the exercise of which they directly adapt themselves to new conditions and become fitted for existence in accordance with new surroundings. Thus plants are thought to have a certain physiologic plasticity or power of self-regulation that tends to adjust them to a new environment, a feature that goes far in explaining the absence of a rigid control of physiographic conditions over forest distributions although an approximate control is often manifested.[1] The forester, then, requires a scientific knowledge of soils and climate, but in the final application of his knowledge to the distribution and growth of forests it is often necessary for him to employ somewhat broader generalizations than those employed by the geographer and the botanist for the special purposes of their sciences.

The Maintenance of a Soil Cover

Everywhere on the land we find at work the two forces of soil making and soil removal. In regions of aggradation the two tend in the same direction; in regions of degradation the soil may be removed as fast as formed and bare rock everywhere exposed at the surface; or there may be established so delicate a balance between soil formation and soil removal that though the soil cover is continually wasted the process of soil formation takes place at an equivalent rate and a covering of soil is perpetually maintained.

The matter of soil formation is of special importance to the people of North America, for not only is denudation (chiefly glacial) responsible for an area of bare, denuded country almost twice as great as that of the continent next in order in this respect, Africa, but denudation is probably proceeding at a faster rate on our continent than on any of the other five. The saving quality of glacial denudation in the past has been its occurrence chiefly in mountain regions of the United States and in the upper boreal and the arctic regions of Canada where extreme climatic conditions would largely offset the advantages of soil.[2] A physiographic map of the United States, Plate IV, appears to show that approximately one-fifth to one-sixth of the total area is now undergoing alluviation; everywhere else, no matter what process has been active in the immediate geologic past, the surface is now being eroded. The action is so slow in some places as to be exceeded by the rate of rock decay, and in such cases no fear need be entertained for the safety of

[1] See especially, V. M. Spalding, Distribution and Movements of Desert Plants, Carnegie Inst. Pub. No. 113, 1909; also E. Warming, Œcology of Plants, Ox. ed., 1904, pp. 370–372.

[2] For an outline presentation of the balance of soil-making and soil-destroying forces that have produced the main soil types now at the surface of the earth see the table, p. 25.

the soil; in other localities the action is rapid and disastrous and its checking should be a matter of the gravest concern. As early as 1890, Shaler[1] estimated that the soils of about 4000 square miles of country had been impoverished through wasteful agricultural methods, representing a loss of food resources sufficient to support a million people; and that at least 5% of the soils which at one time proved fertile under tillage "are now unfit to produce anything more valuable than scanty pasturage." To us of a later generation this figure appears gratifyingly small beside the figure that would express the deplorable ruin of the past quarter century of reckless timber cutting and baneful neglect of fields abandoned by the upland farmer.

The forest is an important factor in soil erosion because it plays a considerable part in the flow of water by which such erosion is effected. The inequalities of the forest floor offer many mechanical obstacles to the flow of surface waters. Innumerable pools of water collect in hollows and are gradually absorbed by the underlying soil instead of running off at the surface. The leaf canopy catches and reëvaporates about 12% of the rainfall, while 10% of it runs along the tree trunks and reaches the ground by a circuitous course. The forest litter, the moss-covered and leaf-strewn ground, is capable of absorbing water at the rate of from 40,000,000 to 50,000,000 cubic feet per square mile in 10 minutes, — water whose progress is delayed by some 12 to 15 hours after the first effects of a heavy freshet have passed.[2]

While the forest thus plays an important part in the maintenance of a soil cover and in the better equalization of the run-off of streams, it would be a mistake to assume that it is the only agent which accomplishes these highly beneficial results. It is scarcely more necessary to know that deforestation may permit a precious soil cover to be wasted than it is to understand that many other types of vegetal covering besides the forest effect these desirable results. Of the same order of importance is the fact that the effects that follow upon deforestation are not equally harmful upon all types of topography and soil. If these conclusions are true it is necessary that the soil, the topography, and the secondary vegetative forces of a given region be evaluated before the statement is made that excessive soil erosion is the necessary correlative of deforestation.

That the soil is protected by many vegetal types other than the forest is now a well-established fact. The high alpine meadows of the Pacific

[1] N. S. Shaler, The Origin and Nature of Soils, 12th Ann. Rept. U. S. Geol. Surv., pt. 1, 1890–1891, p. 333.
[2] B. E. Fernow, Relation of Forest to Water Supplies, in Forest Influences, Bull. U. S. Dept. Agri., Forestry Division, No. 7, 1902, p. 158.

Cordillera consist in many cases of a thick turf which supports natural grasses of luxuriant growth. The interlacing roots of the grasses in many situations bind the soil past all reasonable possibility of excessive erosion, the stems of the grasses impede the run-off by breaking up in one locality any incipient streams formed in another locality exceptionally favorable to concentration of run-off, the whole grass cover breaks the force of the falling rain and prevents erosion. Added to these is the influence of ponds and lakes of glacial origin in the higher situations. The retention of the soil above timber line on the Beartooth Plateau, southwestern Montana, is attributed to such a combination of grass cover and storage basins;[1] it was as natural a result that gullying should follow overgrazing of these meadows as that evil results should follow deforestation in the well-known case of the Southern Appalachians. The binding of the soil and the checking of erosion are also effected by brush and vines which in moist regions may spring up in a few years following deforestation and form an almost impenetrable covering. Such a tangle of vegetation offers even greater resistance to the surface flow of water than does the vegetation of a forest, besides permitting the formation of snowdrifts, one of the most important forms of surface water storage.[2]

The retention of rainfall by the mosses that cover the hill slopes of the Laurentian Plateau, a feature especially well developed in the Labrador peninsula, diminishes the run-off and equalizes it to a degree far exceeding that of the thin spruce forests of the region. Even the steepest slopes are slippery with loose dripping Sphagnum moss, whose effects obviously exceed those of the most porous forest floor. The very existence of a steep hillside bog is in itself proof of an unusually powerful retentive effect of the moss cover upon both soil and run-off.

All of these consequences are subject to changes in degree depending upon variations in soil and topography. If the soil is very porous the imbibition of rain water is rapid and run-off and erosion are correspondingly lessened; if the soil is compact there is little absorption, and run-off and soil erosion are more active. A hill-and-valley country, one consisting entirely of slopes of strong gradient, such as the well-dissected Allegheny Plateau, has a high percentage of run-off and soil erosion, for almost every drop of water falls upon a slope and begins a downhill movement the moment it strikes the surface. A flat surface like the till plains of central Indiana or the outer part of the coastal

[1] J. B. Leiberg, Forest Conditions in the Absaroka Division of the Yellowstone Forest Reserve, Montana, Prof. Paper U. S. Geol. Surv. No. 29, 1904, pp. 15-19.

[2] J. C. Stevens, Water Powers of the Cascade Range, pt. 1, Southern Washington, Water-Supply Paper U. S. Geol. Surv. No. 253, 1910, p. 16.

plain of South Carolina absorbs a high proportion of the rainfall, and soil erosion is of trifling importance. Combinations of these factors are both numerous and variable. A part of the southern slope of Long Island is a natural prairie unforested even before the coming of the whites. It bears almost no signs of erosion, and such as occur had in most cases a very special origin. The absence of erosional features is not surprising when the low gradient of the plain, 10 feet per mile, and the high porosity of the sand are taken into account. These flat-lying porous sands absorb from 60% to 75% of the rainfall, perhaps as great a value as that found on any other area of equal size in the eastern half of the United States.

In New England it has been noted that the quick-growing brush and the special qualities of the glacial soil prevent the undue erosion of de-forested hill slopes in the Berkshires where the relief is so strong that landslides sometimes occur. The pebbles and bowlders of the till constantly divert the run-off and lessen its velocity, while the bottoms of ravines sunk into the till are in a measure protected from erosion by a pavement of stones derived from the till. In many cases in western Massachusetts and Vermont and New Hampshire steep mountain slopes "have been several times stripped of their forest growth with little, though doubtless some, injury to the soil," and "the mountain streams are beautifully clear except immediately after a heavy rain."[1]

The large number of rock ledges that occur in this region contribute to the same effect. The soil is thin, and irregularities of the underlying bed-rock assist in holding it in place not only by physically retaining it but also by preventing the streams held upon the projecting rock ledges from expending the whole of their erosive energy upon its loose material.

The effect of the forest upon the run-off alone is extremely difficult of determination, for soil and topography are in this respect of much greater importance. That forests tend to conserve the run-off is clear; their effects in individual cases, however, may be so small as compared with the effects of soil and topography as to be overshadowed by the latter.

"Donner und Blitzen River, in central Oregon, is a very uniform stream with a well-main-tained summer flow, but its area does not produce a tree, except here and there a juniper. On the other hand, Silvies River, which exists under the same climatic conditions as Donner und Blitzen River and discharges its waters into the same lake, is anything but uniform in its flow, although its drainage area is heavily forested. Niobrara and Loup rivers, in Nebraska, are very uniform in flow, but there is hardly enough timber on both areas to build a cabin. Nearly

[1] H. F. Cleland, The Effects of Deforestation in New England, Science, n. s., vol. 32, 1910, pp. 82–83.

all the streams of western Oregon and Washington are subject to enormous floods, and all run comparatively low in summer, yet no streams in the world have more densely forested drainage areas." [1]

The conclusion that forests are not a guaranty of uniform stream flow, in spite of the fact that they tend in the direction of uniformity, does not diminish the interest of students in such influence as forests do exert, since theirs is a *controllable* influence. Man can not greatly modify the porosity of the soil or the slopes of the land, and the effects that follow upon these causes are therefore irremediable; but man may save a forest or plant one and thus mitigate effects which he can not wholly prevent. In precisely those regions where run-off and soil erosion are most extreme through unfavorable topographic and soil conditions, man may find it possible to preserve a tolerable state of affairs by saving the forest from destruction. In regions where the conditions are critical the destruction of the forest may mean the quick destruction of the soil, its preservation the preservation of the soil and the indefinite occupation of the region by man.

The retaining influence of the forest on the soil is most strikingly exhibited where the balance between soil formation and soil removal is delicately established and may be easily destroyed. An extreme instance is Kanab Creek, Utah, where the burning of the forest and the overgrazing of the pastures have resulted in torrent conditions. The tributaries have become deep washes, many new and deep gulches have been formed, dams and bridges have been destroyed by the floods and coarse gravel deposited on formerly arable valley lands. [2]

It is of importance, then, to examine at the outset the relations of soil denudation and soil accumulation, that we may be the better prepared to study those forces which tend to bind and partially to retain the covering of soil; not only that forests themselves may be perpetuated, but also that the flood-plain soils on the borders of the forests may be adequately preserved and the natural advantages of the waterways retained.

SOIL-MAKING FORCES [3]

The complex cover of rock waste which we call the soil is the product of a great number and variety of forces. Only the principal ones are here outlined.

[1] J. C. Stevens, Water Powers of the Cascade Range, Southern Washington, Water-Supply Paper U. S. Geol. Surv. No. 253, 1910, p. 17.

[2] H. S. Graves, The Forest and the Nation, American Forestry, vol. 16, 1910, p. 608.

[3] The section on soils is necessarily brief and somewhat technical and assumes on the part of the student a knowledge of ordinary rocks and rock-making minerals as well as an elementary knowledge of chemistry. Those students who are deficient in such knowledge should consult

OXIDATION.

CARBONATION.

HYDRATION.

SOLUTION.

MECHANICAL ACTION OF WATER AND ICE.

TEMPERATURE EFFECTS.

WIND.

BACTERIA.

ANIMALS AND THE HIGHER PLANTS.

OXIDATION

Oxygen is the most active element of the air, and the process of oxidation is of great importance in reducing rock masses to soils. The action is perceptibly manifested only in rocks containing iron either as a sulphide, a carbonate, or a silicate. Of these the sulphides are changed to sulphates which are soluble and may be removed in solution. The most common minerals attacked are ferrous carbonate associated with the carbonates of lime and magnesia and the silicates of mica, amphibole, and pyroxene. The minerals become gradually decomposed through oxidation and disintegrate into unrecognizable forms. The oxidation of the iron-making minerals of a rock is always attended by increase in bulk, and when this takes place in cracks and crevices it tends, like the freezing of water, to widen the cracks and to increase the surface exposed to attack. In general the action of the air in soil formation is of secondary importance and depends chiefly on the oxidation of the lower to the higher basic forms. The ferrous and ferroso-ferric oxides are converted into ferric oxide or its hydrate limonite, iron rust, which gives to soils containing much iron their characteristic reddish or yellowish colors.

The presence of ozone[1] in air without doubt causes it to have a more active oxidizing effect. Ozone is present in considerable quantity in the air only when the air is free from organic impurities and products of decay. The average amount of ozone in a hundred cubic meters of air is 1.4 mg., but the amount may be doubled after thunderstorms.[2]

Merrill, Rocks, Rock-weathering and Soils, 1897. For purposes of brief inspection of the mineralogical composition of ordinary rocks and the chemical composition of rock-making minerals the tables in Appendix C in this book should be consulted, and a text-book of Lithology such as Pirsson, Rocks and Rock-forming Minerals, 1909.

[1] Ozone is a very active form of oxygen in which the molecules consist of three atoms of oxygen instead of two atoms as in a molecule of ordinary oxygen. It is formed by silent electrical discharges, and is chemically unstable, readily parting with one of its atoms, hence chemically active.

[2] J. Hann, Handbook of Climatology, 1903, pp. 80–81.

The amount of ozone in the air is determined by the rate of change in an easily oxidized substance — not a very accurate method.[1] Ebermayer emphasizes the more powerful oxidizing effects of ozone in the air and its formation in the forest in unusual amounts.[2]

CARBONATION

The oxidation of organic materials (both plant and animal remains) by bacteria and oxygen in the zone of weathering produces a concentration of carbon dioxide near the surface. The degree of concentration of this gas in the zone of weathering is appreciated by comparison of the soil air with the atmosphere. The amount of carbon dioxide in the atmosphere is 45 parts in 10,000 by weight; the amount in soil air or soil gases is represented in the following table:

AMOUNT OF CARBON DIOXIDE IN SOIL AIR[3]

Derivation	Parts by weight in 10,000
Air from sandy subsoil of forest..........	38
Air from loamy subsoil of forest..........	124
Air from surface soil of forest.............	130
Air from pasture soil.....................	270
Air from soil rich in humus...............	543

The carbon dioxide in the soil is the agent in the important weathering process known as carbonation, by which is meant the union of carbonic acid with bases in the formation of carbonates. It is dominantly accomplished by the substitution of carbonic for silicic acid. To some extent carbonates are also formed (1) by the substitution of carbonic acid for other acids, e.g., phosphoric acid, and (2) by the union of carbon dioxide with oxides not united with other acids, e.g., ferrous oxide in magnetite.[4]

The process of carbonation takes place on a vast scale. It is most rapid in the tropics, takes place at a moderate rate in temperate lands, and is least important in the frigid zones and in arid regions. It has a direct relation to the amount of vegetation, since it is chiefly through the decay of the vegetation that carbon dioxide is supplied for the reaction involved in carbonation. On the other hand a soil containing carbon-

[1] W. M. Davis, Elementary Meteorology, 1898, p. 5.
[2] E. Ebermayer, Lehre der Waldstreu, etc., 1876, p. 202.
[3] Boussingault and Levy, quoted by G. P. Merrill, Rocks, Rock-weathering and Soils, 1897, p. 178.
[4] C. R. Van Hise, A Treatise on Metamorphism, Mon. U. S. Geol. Surv., vol. 47, 1904, p. 475.

ates is ordinarily fertile and supports an abundant vegetation. The surface vegetation and the soil carbonates are therefore mutually inter-active and helpful. The cumulative effect of the act of carbonation would therefore appear to be constantly increasing amounts of carbon-ate substances. But this tendency is offset or matched by the libera-tion of silica in the process of carbonation; about one and one-third times as much silica is released from the silicates as there is carbon dioxide combined in the carbonates.[1]

HYDRATION

The action of hydration is the union of water with chemical com-pounds in the production of hydrous minerals. It is the most extensive reaction in the zone of weathering and next to carbonation the most important. It affects practically all of the anhydrous silicate minerals of the igneous, sedimentary, and metamorphic rocks to some degree, and many of them to a great degree. The decomposition products of the rock minerals are almost all strongly hydrated, such as the zeolites, chlorites, and kaolin in the silicate class and aluminum and iron among the oxides.

The action of hydration is always accompanied by the liberation of great quantities of heat and by increase in bulk. It is calculated that the transition of a granite rock into arable soil, provided such transition takes place without loss of material, is attended by an increase in bulk of 88%. In rocks as a class hydration effects volume increases which range from a very small per cent to 160% (corundum to gibbsite). In general the increase is less than 50%. Such volume increases prevent complete hydration at any great depth below the surface; partly hydrated rock when artificially exposed at the surface, as in railway cuttings, may become completely hydrated at so rapid a rate as to expand greatly in volume and soon disintegrate. Notable in-crease in bulk does not follow if the pore space is ample; if the pore space is small and the rock dense the action is either incomplete or involves great increase in bulk.

Commonly hydration takes place in connection with carbonation and solution. In so far as soil water is consumed in the formation of new (hydrous) minerals it can not be used in the process of solution; the amount so consumed is, however, but a small part of the whole, and solution is therefore a companion process of hydration.

[1] C. R. Van Hise, A Treatise on Metamorphism, Mon. U. S. Geol. Surv., vol. 47, 1904, p. 480.

SOLUTION

WATER AS AN AID TO CHEMICAL ACTION

Water is the most important weathering agent, not only because of its direct effects but also because its presence conditions all phases of weathering, such as hydration, etc. It has so great a dissolving power that few substances found in rocks are wholly insoluble in it, while in water charged with acids of various sorts many rocks are readily soluble and all are somewhat soluble. The number of such acids is small, but their action is so general that they powerfully aid solution in reducing rocks to soil. Nitric acid is present in some amount in rainfall, in surface waters, and in the soil water, in which it may be supplied in small quantities by the action of bacteria. Sulphuric acid may be derived in somewhat the same manner; an important source in some regions is iron pyrites, which on decomposition may yield free sulphuric acid. It is altogether probable that many if not all soils constantly receive small amounts of sulphuric acid, and it is possible that in some cases the solvent action of this acid on the mineral constituents of the soil may become important.[1] Among these substances is chlorine the amount of which in the air varies with the distance from the sea and is greatest at the seashore. On the island of Barbados 116 pounds of chlorine are contributed annually to each acre.[2] Two or three extractions of soils, however, seldom show the presence of any free acid other than carbonic acid.

Carbon dioxide, which is the basis of carbonic acid, is contained in all natural water and in rainfall so that all percolating waters are real acid solvents and exercise a far-reaching effect, a fact now universally recognized. That dissolved carbon dioxide may act directly as an acid, thus increasing the solvent power of the water in which it is contained, is probable.[3] The destructive action of water charged with carbonic acid is most strikingly exhibited in limestone but it is not confined to this type of rock; even quartzose rocks of the ordinary kinds are attacked by it and granite and related rocks are rather quickly affected. Its effect both in the soil and in the zone of weathering [4] generally is due largely to a reduction of the mass of the hydrates of the hydrolyzed bases by the formation of bicarbonates. The result of its action upon the feldspars is the forma-

[1] C. R. Van Hise, A Treatise on Metamorphism, Mon. U. S. Geol. Surv., vol. 47, 1904, p. 205 et al.

[2] Harrison and Williams, Jour. Am. Chem. Soc., vol. 19, 1897, p. 1.

[3] Carbon dioxide is soluble in water to the extent of equal volumes at ordinary temperature and barometric pressure.

[4] The zone of weathering extends from the surface to the ground water (Van Hise, A Treatise on Metamorphism, Mon. U. S. Geol. Surv., vol. 47, 1904).

tion of clay, a most essential element of soils from the physical stand-point, and the freeing of potash, one of the most essential plant foods. In the case of granite rocks the silica set free by the carbonic acid remains partially or wholly in the resulting soils; in the case of limestones the lime at first remains in the form of a carbonate, but potash and soda compounds, which are readily soluble in water, are largely carried away either by percolating water or absorbed by plants. The action of carbon dioxide is also manifest in the formation of carbonates of iron and magnesium.

In certain experiments carried on in the laboratory of the U. S. Bureau of Soils some powdered minerals, among which were muscovite and albite, were kept in contact for fourteen months with water and certain solutions in paraffin cylinders. Excepting the results obtained with albite, those obtained by treatment with water saturated with carbon dioxide are so much greater than the corresponding results obtained with pure water that no reasonable doubt can exist that the effect of the carbon dioxide is not only to hasten the rate of solution, but actually to increase the absolute solvent action of the water.[1]

In nature all the elements in the rock and the soil minerals in the zone of weathering are being dissolved all the time, but at variable rates depending (1) upon the strength and abundance of the active compounds in solution and (2) upon the solubility of the constituent minerals.

WATER AS A CARRIER

In addition to being the substance necessary for the chemical decom-position of nearly all kinds of rocks and soil, water has an important influence in removing large amounts of soluble plant food from the soil. Nearly five billions of tons of mineral matter are annually carried away in solution from the land into the sea, while the amount of sediment is many times greater.[2]

The amount of nitric acid found in drain water (water that runs off through drains, i.e., tiles, etc.) sometimes shows a heavy depletion of the land by the leaching out of this highly important substance. In all drain water lime is found to be leached out most abundantly, mainly in the form of bicarbonate. Magnesia is next in order, then soda and other substances of minor value. Potash is present in drain water in small amounts. Carbonic acid is the most abundant of the acids found in such water, and chlorine and silicic acids are next in order.

MECHANICAL ACTION OF WATER AND ICE

An important mechanical effect of water is exhibited during rain storms when the erosion of soil on all slopes and its rapid erosion on unprotected steep slopes occur and may lay bare the rock surface

[1] Cameron and Bell, The Mineral Constituents of the Soil Solution, Bull. U. S. Bur. Soils No. 30, 1905.
[2] E. W. Hilgard, Soils, 1906, p. 24.

and enable other soil-making forces again to act upon the exposed rock. Falling raindrops also beat upon and jostle the soil grains or move them about upon the rock surface in such manner as to break off smaller particles, an action which on flat surfaces may tend to increase the amount of soil.

It has been computed with reasonable accuracy that 783 million tons of earth and rock measured as soil are removed each year by erosion from the surface of the United States. The amount removed from different watersheds varies greatly, not only on account of differences in the sizes of the drainage areas but also on account of differences in the depth and porosity of the soil, the extent and nature of the vegetable cover, the lengths and declivities of the slope, the rainfall, the temperature, the extent of lakes, etc. In the north Atlantic basin the rate is 130 tons per square mile per year; the rate in the Hudson Bay basin is 28 tons and is the lowest on the continent; the southern Pacific basin heads the list with 177 tons. Individually the Colorado River brings down the greatest amount of suspended matter; it delivers 387 tons per year for every square mile of its drainage basin. Practically no suspended matter is transported by the St. Lawrence River, since the water is cleared by sedimentation in the Great Lakes. In general the northern streams carry much less suspended matter than the southern streams, a result due probably to the large number of lakes in the drainage basins of the northern streams, the large extent of bare rock outcrop, and the hindrances to erosion imposed by soil frozen during a large part of the year.[1]

The action of freezing water is due mainly to the expansive force it manifests as it passes from water to ice, and has been described as equivalent to the pressure of a column of ice a mile high, or about 150 tons to the square foot. If a given rock contains much water in its pore space and is repeatedly subjected to freezing temperatures, the rock will in time be disrupted by heavy internal strains. The extent of the strain effect depends (1) upon the climate, (2) upon the weather conditions, whether uniform or variable, and (3) upon the amount of water contained by the various kinds of rock, which in turn differs with the nature of the minerals and their state of aggregation.

All rocks when freshly exposed hold by capillary attraction a certain amount of water, which occurs largely as interstitial water. The amounts that may be contained are expressed roughly as follows: granite, 0.37% by weight; chalk, 20%; ordinary compact limestone, 0.5% to 5%; and sandstones from 10% to 12%; while clay may contain nearly one-fourth its weight

[1] These computations show that the surface of the United States is being removed at the average rate of .0013 of an inch per year, or 1 inch in 760 years. Dole and Stabler, Water-Supply Paper U. S. Geol. Surv. No. 234, 1910 (Denudation), pp. 82–83.

in water. The amount in white chalk is as much as 19% and a piece of such chalk may be shattered into fragments by a single night's frost. The freezing of absorbed water is one of the most general sources of disintegration of building stones.[1]

In addition to the expansive force of interstitial water when frozen is the action of freezing water in the joints of the rock, which tends to disrupt large masses from the faces of cliffs and other bare rock surfaces.

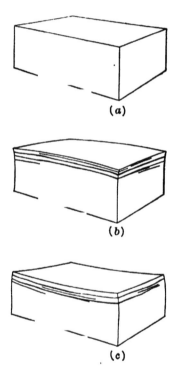

(a)

(b)

(c)

Fig. 1.—Effect of unequal heating of the surface of a rock. (a) shows the condition of a block at uniform temperature; (b) the manner in which the upper portion expands when heated above the average temperature; (c) the contraction of the upper portion by cooling below the average temperature. When contraction and expansion are sufficiently great they result in the splitting of the surface layers. (Van Hise, U. S. Geol. Surv.)

The effect is heightened if freezing and thawing alternate in periods of short duration. Alternate freezings and thawings may be beneficial to the soil after formation, because the freezing of the water in the pore spaces increases the bulk of the whole mass of frozen soil, — an increase which is not immediately compensated on thawing, so that aeration and root penetration derive a certain benefit from the process.

TEMPERATURE EFFECTS

While the agencies we have so far enumerated are active in nature in breaking down rock, soil would be formed without such agencies, though at a slower rate, through the inherent instability of the rock under changing temperatures. The breaking of a hot glass plunged suddenly into cold water is a manifestation of the same force that in humid regions to a lesser extent, in arid regions to a greater extent, tends to disrupt rock masses and to form soil. The temperature effects upon rock are of several kinds: (a) the breaking apart of large rocks into smaller masses through expansion and contraction of the whole mass at an unequal rate; (b) the peeling off of rock layers and chips from a rock surface or from the surfaces of bowlders through unequal expansion and contrac-

[1] A. D. Hall, The Soil, 1907, p. 11.

tion between the surface layer and the layer immediately beneath, a process known as exfoliation; and (c) temperature changes which expand different minerals at different rates and cause an internal strain to which the rock may finally yield. The first process (a) is so familiar as to require no description. The second process (b) may be understood by the recognition of the high temperatures which bare rock surfaces attain when exposed to the summer sun. On a hot day they may attain a temperature of 150° or 160°, a temperature so high that the hand can not be held on exposed surfaces without being burned. Between the highly heated surface particles and the particles some distance beneath the surface there must be a zone of shear, for at these high temperatures the surface rock will expand greatly, while the cool rock only a short distance beneath the surface is so much lower in temperature as to occupy smaller bulk, Fig. 1. From observations made near Edinburgh, Scotland, during 1841–42, the range of earth temperatures at varying depths in soil, sandstone, and trap rock was determined to be as follows:

VARIATION OF TEMPERATURE WITH DEPTH [1]

Depth	Trap Rock			Sand of Garden			Craigleith Sandstone		
	Max.	Min.	Range	Max.	Min.	Range	Max.	Min.	Range
3 feet	52.85°	38.88°	13.97°	54.50°	37.85°	17.65°	53.15°	38.25°	14.90°
6 feet	51.07	40.78	10.29	52.95	39.55	13.40	51.90	38.95	12.95
12 feet	49.00	44.20	4.80	50.40	43.50	6.90	50.30	41.60	8.70
24 feet	47.50	46.12	1.38	48.10	46.10	2.00	48.25	44.35	3.90

Of course the surface inch or two or three inches show much greater ranges; and between the first and twelfth inches the differences may be extreme on hot summer days. The author has noted as the result of temperature observations on loose dry soil during several summer months a maximum difference of 35° to 50° between the first and fifteenth inches, which was reduced to 5° or 10° before the following sunrise.

Translating differences of temperature into units of expansion we have the rate of horizontal expansion varying from .000004825 inch per foot for each degree Fahrenheit in granite to .000009532 inch in sandstone.[2] A change of temperature of 150° in a sheet of granite 100 feet in diameter would thus produce a lateral expansion of about 1 inch, an expansion that tends to lessen the cohesion of the rock and to cause a shearing of

[1] Trans. Royal Society of Edinburgh, vol. 16, 1849. From G. P. Merrill, Rocks, Rock-weathering and Soils, 1897, p. 184.

[2] W. H. Bartlett, Experiments on the Expansion and Contraction of Building Stones, etc. Amer. Jour. Sci., vol. 22, 1832, p. 136.

the upper over the lower layers. Although these movements seem slight they are sufficient to produce in time a weakening and breaking of the rock, thus affording a better opportunity for the action of other physical and chemical agencies such as freezing water, bacteria, plant roots, etc.[1]

This form of rock disintegration is most pronounced in massive coarse-grained rocks located in regions of great extremes of daily temperature such as occur in the arid and semiarid portions of the West. When it occurs in homogeneous massive rock it may produce rounded forms or bosses of very characteristic appearance.

Stone Mountain, Georgia, 650 feet high, 2 miles long, and 1¼ miles wide, owes its elliptical dome-like form to such exfoliation along preëxisting lines of weakness in the form of joints. The surface sheets are buckled up in very characteristic forms. They are rarely more than 10 inches thick, but are 10 or 20 feet in diameter. In a few instances small avalanches have been caused by the giving way of such sheets.[2]

The third process (c) of rock disintegration through temperature change, that of crystal crowding, may be understood from the fact that in all rock composed of crystals there is an internal strain due to the unequal rates at which the component minerals expand upon increase of temperature. Such expansion has two forms of inequality: (a) the cubical expansion varies with the kind of mineral and (b) the rates along the various crystallographic axes of the constituent minerals are unequal.[3]

It is self-evident that a coarsely crystalline rock under these conditions will disintegrate more rapidly than a rock of finer grain. Rocks of granular structure, all other things being equal, undergo disintegration much more quickly than those in which the individual minerals are closely compacted, as a diabase or a quartzite. It is believed that dark-colored basic rocks tend to respond to the forces manifested by changes of temperature somewhat more readily than do light-colored acidic rocks, because of the more rapid absorption of heat by dark-colored objects in sunlight. It has also been shown that the thermal conductivity of rocks varies in direction according to the structure, being greatest in the direction of the schistosity, where this feature is developed. The result is that in massive, homogeneous rocks the conductivity is the same in all directions, while in finely fissile rocks it may be four times as great in the direction of the fissility as at right angles thereto.[4]

WIND AS AN AGENT IN SOIL FORMATION

While the action of wind is most clearly seen on the surfaces of land waste where loose dry material may be shifted about in the form of dunes and sand drifts of variable size and shape, wind may also be an important agent in the actual production of soil. Loose particles of rock may be driven through the air against projecting rock ledges, and not only do they themselves tend to become finer through attrition in

---

[1] G. P. Merrill, Rocks, Rock-weathering and Soils, 1897, p. 181.
[2] Idem, pp. 245-246.
[3] G. P. Merrill, Stones for Building and Decoration, p. 419.
[4] G. P. Merrill, Rocks, Rock-weathering and Soils, 1897, p. 184.

the air, but they also abrade obstructing ledges. This action takes place on a considerable scale in dry regions and may become one of the most important agents in the denudation of desert lands.[1] It has been estimated that the dust in a cubic mile of lower air during a dry storm weighs not less than 225 tons, while the amount of dust in the same volume of air during a severe storm may reach 126,000 tons.[2] The great importance of the wind as a soil builder is shown by the wide distribution of the loess deposits of the world. "It would perhaps be an exaggeration to say that every square mile of land surface contains particles of dust brought to it by the wind from every other square mile, but such a statement would probably involve much less exaggeration than might at first be supposed."[3] Dust transportation is not confined to desert regions, but takes place almost everywhere on some scale.

Not the least important of the effects of wind on soil has been the wide distribution of volcanic dust as in Oregon, southern Idaho, etc. In the Tertiary period volcanic eruptions took place on a vast scale in many portions of the West. Great quantities of volcanic dust were raised aloft and then deposited at varying distances from the volcanic vents. In many instances such bodies of dust were swept by the wind hundreds of miles from their points of origin, and finally deposited as a sheet of loose waste. Since their original deposition the wind has played upon the surface layers, shifting them about, reworking and redepositing the material, and by these means modifying the qualities of the surface soil of many great tracts.

BACTERIA

Certain bacteria are able to draw their nourishment directly from the air in a purely mineral environment such as that found upon the surfaces of bare rock, so that even the denuded rocks of high mountains may be populated by these minute organisms. The ragged rocks of high altitudes and steep slopes in the Alps, the Pyrenees, the Vosges, etc., are composed of minerals of the most varied nature, all of which have been found to be coated with a "nitrifying ferment." These bacteria develop by absorbing carbonate of ammonia and vapors of alcohol from the air and they even assimilate carbon dioxide. The wide distribution of these organisms, their great number, and the manner in which

---

[1] W. Cross, Wind Erosion in the Plateau Country, Bull. Geol. Soc. Am., vol. 19, 1908, pp. 53–62; S. Passarge, Die Kalahari, 1905; W. M. Davis, The Geographical Cycle in an Arid Climate, Jour. Geol., vol. 13, 1905, pp. 381–407.

[2] J. A. Udden, A Geological Romance, Pop. Sci. Mo., vol. 44, 1898, pp. 222–229.

[3] Chamberlin and Salisbury, Geology, vol. 1, 1904, p. 22.

Fig. 2.—Underground burrows and chambers of earthworm population, in well-drained sand bank near New Haven. The vertical burrows connect with underground chambers showing as dark lenses within an inch and an inch and a half of the bottom of the photograph. Seven or eight can easily be identified. At the time the photograph was taken many of the chambers and some of the burrows were completely filled with dark humus collected by the earthworms. When the humus was scraped away it was seen that each chamber had a smooth and generally flat floor and an irregularly arched roof. Such of the openings as had been filled with humus had been abandoned. The open chambers and burrows were teeming with earthworms and with insects of many genera. The upper edge of the bank was broken down during a dry spell and at a time when the ground was fully stocked. The scale in the upper right hand corner is three inches long. (Photograph by Mason.)

they prepare the rock for the microscopic vegetation which is usually found in the form of a layer covering rock surfaces and soil particles, place them among the important geologic agencies that have effected the disintegration of the rock and the formation of soils.[1]

Their action is carried on on a small scale in cold temperatures and on a very large scale under normal temperatures where the rock is but thinly covered with earth. The action is also carried on among rock fragments, tending gradually to reduce them. The bacterial organisms penetrate every cleft or crevice, and by the chemical action of their secretions continually reduce the rock to smaller and smaller sizes, acting even on the most minute fragments. Each rock particle is found covered with a film of organic matter accumulated by these bacteria and the plants of a higher order which secure through them a food supply.

ANIMALS AND HIGHER PLANTS AS SOIL MAKERS

The most active animal agencies in producing and modifying soils are earthworms, which influence the physical state of soils by making them more porous and open. Darwin's studies showed that the intestinal content of worms has an acid reaction which has an effect on the soils passing through the alimentary canal. They still further modify soils by drawing into their holes leaves and other organic materials which gradually decay and are converted into humus, Fig. 2. Darwin estimated that an average of about 11 tons of organic matter is in this manner annually added to each acre of soil in regions where earthworms abound. Earthworms tend also to increase the amount of ammonia in the soil, to make the soil finer by attrition during the process of digestion, and to mix the soil by bringing to the surface portions of the subsoil. They thus play a notable part in maintaining soil fertility. The excreta of earthworms contains more nitrogen and other easily oxidized compounds than the original soil, and after excretion by worms soil contains phosphoric acid and calcium carbonate in more readily soluble forms.[2] It has been observed that when earthworms are drowned out the surface soil layer remains compacted and vegetation grows very feebly until earthworm immigration has restocked the soil.[3] The action of burrowing animals, worms, insects, etc., in the soil, allows freer access of rain water and drainage water and increases the depth and the rate of rock and soil decay. Ants sometimes produce

[1] H. W. Wiley, Principles and Practice of Agri. Analysis: Soils, vol. 1, 1906, pp. 39–40.
[2] Cameron and Bell, Mineral Constituents of the Soil Solution, Bull. U. S. Bureau of Soils No. 30, 1905, p. 41.
[3] A. D. Hall, The Soil, 1904, p. 159.

highly important effects upon the soil, especially in tropical countries as in Brazil, where they often build tunnels hundreds of yards long which give the soil-making forces access to the subsoil. They carry great quantities of leaves into their nests and by this means and by their excreta contribute vegetable acids to the soil and thus promote rock and soil decay.[1]

More general in their action upon both rock and soil are plants and plant roots. Roots force themselves into the crevices of rocks and minerals, they wedge apart rock masses, and thus expose new surfaces and a larger extent of surface to the action of other forces. The mechanical force exerted by growing roots is very great. The root of the garden pea has a wedging force equal to 200 or 300 pounds per square inch, a force that is exerted without harmful effects upon the small roots because of the protection afforded by the corky layer of the root tips.[2] Vegetation also assists in rock decay by shading the surface and permitting the retention of a larger amount of water upon the immediate surface where it exercises a dissolving action as already described. The organic matter contributed to the soil by decaying vegetation promotes rock decay by furnishing carbon dioxide or carbonic acid. Rootlets of plants in contact with limestone dissolve large portions of this rock through the solvent action of root moisture. The action of rock disintegration is also extensively carried on by the lower forms of vegetation, among which the lichens produce the most important effects. The rock surface is corroded and a soil cover originated. A prominent ingredient of the lichens is oxalic acid, an acid that compares in strength with hydrochloric and nitric acids, and that powerfully aids rock decay.

Plant roots that permeate the soil are agencies of oxidation which have a very appreciable effect in altering some soil constituents and influencing soil fertility. Active extracellular oxidation is carried on by the plant roots chiefly by means of the enzymes (the name applied to any unorganized chemical ferment such as diastase, pepsin, etc.) which they excrete, and not to organic (carbonic acid alone excepted) and inorganic acids which they were formerly supposed to excrete. It is believed that the effect of the roots of growing plants in dissolving the organic substances of the soil is due chiefly to this action. Among phanerogams extracellular oxidation is strongly localized and limited to the absorbing surface of the root; the most intense oxidation is effected by the root hairs. Oxidation is more marked when an optimum water

[1] J. C. Branner, Ants as Geological Agents in the Tropics, Bull. Geol. Soc. Am., vol. 7, 1896, p. 255; idem, Geologic Work of Ants in Tropical America, Bull. Geol. Soc. Am., vol. 21, 1910, pp. 449–496.
[2] E. W. Hilgard, Soils, 1906, p. 19.

content is maintained in the soil, while saturation of the soil produces a decided depression in the rate of oxidation. · Oxidation by plant roots is increased by the presence of calcium salts, potassium salts, phosphates, nitrates, etc.[1]

The action of plants in promoting the formation of soil is well illustrated by the manner in which plant associations succeed each other upon extensive areas of bare rock, a succession that is dependent upon the ability of each plant group to live under the hard conditions which exclude the next higher group. Crustose lichens are the only plants which are able to establish themselves on a bare rock face. Upon the thin soil formed by these, other lichens are able to secure a foothold. Then appear the mosses, for example Hedwigia, Grimmia, etc., which eventually eliminate all but the erect (fructicose) lichens. The mosses still further increase the soil layer, both through the accumulation of mineral particles and by their own decay, and are in turn wholly or partly displaced by the more xerophytic species of ferns such as the spleenworts. With the ferns appear many herbaceous flowering plants, notably the stonecrops and saxifrages, and certain grasses. In addition to the carbonic acid which is excreted by the roots of all plants many of them secrete vegetable acids such as oxalic and citric acids which assist in soil formation. The herbaceous plants in turn prepare the way for, and are eventually succeeded by, shrubs and trees.

Senft describes the vegetation that takes hold of landslips and coarse terrace deposits (near Eisenach) and shows how it undergoes great changes in type due to soil changes brought about chiefly by the vegetation itself. In the locality examined the bare stony heaps were first clothed with mosses, then xerophytic grasses; later other xerophytic herbs came in and also shrubs like the juniper which gave rise to dense, bushy growths. In twelve years an impenetrable bush land had arisen and finally Sorbus, Fagus, and other trees appeared and a forest arose. During this change in vegetation type the soil had constantly changed and improved. Each kind of vegetation suppressed another — bush land vanquished xerophytic grasses, and forest vanquished bush land.[2]

The effects of minor rock structures upon plants is extremely interesting. Peculiar and variable rock habitats induce variations in plant societies which are due chiefly to local differences in the nature of crevices — joints, fissures, etc. — in the rock.

[1] Schreiner and Reed, The Rôle of Oxidation in Soil Fertility, Bull. U. S. Bureau of Soils No. 56, 1907, pp. 7–9 et al.

[2] Quoted by Warming, Œcology of Plants, 1909, Ox. ed., pp. 352–353.

"Some of these receive water percolating from higher parts of the mountain, and may remain moist throughout prolonged periods of drought; other crevices obtain their water exclusively from the strictly local rain.   Some crevices contain abundant detritus and are therefore endowed with a greater power of storing water; others are poor in detritus and allow the water to pass away.   The chemical composition of the detritus also varies, as some crevices contain abundant humus, in which numerous earthworms may lurk, whereas others are poor in humus.   Cracks in rocks supply an endless variety of habitats, each of which forms a special kind of environment." [1]

The point is of exceptional importance in regions of thin soil where the lower roots of all plants and all the roots of some plants are in intimate association with the rock.   Talus slopes frequently show similar variations.   Their lower slopes or lower margins are commonly wooded because of the finer rock waste and the greater amount of water which here reappears.   The loose upper slopes are treeless except where the talus has become temporarily stable and clogged with finer waste or where a change of geologic formation or the arrangement of crevices cause seepage, a common condition at the upper edge of talus slopes.

THE CAUSES OF SOIL DIVERSITY

The preceding discussion will enable us to see that the soil is an extremely complex mixture.   There are present mineral débris from rock degradation and decomposition; organic matter, the partly decomposed remnants of former plant and animal tissues; the soil atmosphere, always richer in carbon dioxide and generally in water vapor than the atmosphere above the soil; living organisms, such as various kinds of bacteria, and often molds, ferments, and enzymes; and finally the soil water, a solution of products yielded by other substances.   This enumeration is sufficient to show how diverse are the origins of the different components, and to suggest how varied are the reactions that take place in the soil even after its formation.   These facts need emphasis because of the general view that soil is mere dirt or rock waste, or that it is everywhere the same, whereas the truth lies nearer the other extreme. The soil is a great complex of varied elements, formed in many ways, and subject to the most widely diverse changes after its formation. Nor have we exhausted the list of diversifying forces.   We have yet to consider briefly the transportation of rock waste after formation, the various agencies concerned in it, and the results upon the soil texture and fertility.

All soils are subject to some movement at the earth's surface, and since the soil particles are of different sizes, weights, and compositions, they must respond in different ways to the forces that tend to move

[1] E. Warming, Œcology of Plants, Ox. ed., 1909, p. 245

them. The simplest case is that of deposition by running water, with gradually diminishing velocity, where on the whole the coarser particles are deposited first and the finer last. The action and the result are so familiar to all as to require no extended discussion. The distribution of material in an alluvial fan or a delta or a flood plain always follows this well-recognized law. The effects of creep and rainwash are not so simple. Under the influence of constantly changing temperature all hillside or *colluvial* soils tend to move down slope. Contributing toward the same result is the action of percolating soil water and ground water, cultivation, etc. In all these cases the fundamental and ultimate force is gravity, but because gravity is manifested in so large a number of forms it is clearer to consider the forms themselves and not the basic force on which they depend. In such cases of creep there is not that suspension of particles in water that permits thorough stratification or sorting, consequently coarse and fine are left mixed together, and the rate of movement may be so slow as scarcely to be perceptible.

Where rock formations succeed each other in short distances soil creep may cause an important lack of sympathy between the underlying rock and the overlying soil. The soils of the higher may come to rest over the lower formations, and the rock character give little clue to the nature of the overlying soil. Cases of this kind are frequent in the Appalachian Mountains and the ridges of the Great Appalachian Valley, and are especially well marked where the boundary between two unlike formations occurs on a hill slope. If the slopes are quite steep and the rock formations numerous on a given slope, such overplacement of soils may produce extreme effects and the waste from the different rocks become so mixed as to show at the foot of a slope or on the inner border of a foreland plain but little relation to any particular rock. On broad plateaus where the boundaries between rock formations are far apart mixtures of soil types take place to an important degree only near the boundaries of the formations; over the greater part of the outcrop of a given formation there is a close relation between the underlying rock and the soil. The Colorado Plateaus of the Southwest and portions of the Cumberland Plateau furnish excellent examples of this law.

The rate of movement has been made the basis of a classification of soils according to origin that deserves a word of explanation. While all soils are subject to some movement, the movement may be so slow as to be of no importance, as on portions of flat tablelands or base-leveled areas like the Piedmont Plateau of Georgia and Maryland. Such cases of no movement or of little movement will tend to cause a certain sympathy between soil and rock in a given locality, and the minerals that occur in the rock are found in the soil or at least their decomposition products are found there. Such soils are sedentary or residual soils, and the term generally implies a fundamental relation between a rock terrane and the soil covering it. On the other hand, if the soil has once been in the grip of a transporting agent such as the wind, running water in the form of river or brook, or a continental ice sheet, or a glacier, it is considered a transported soil; and the term *transported* always connotes a mixture of soil ingredients in the case of ice, and sorting in the case of water and wind. By these agencies soils may come to rest far from their place of origin. The alluvial deposits of the Mississippi flood plain are derived from fully one-fourth the whole United States. The underlying rock now perhaps deeply buried in a given locality may be limestone, but the alluvial soil overlying it may be deficient in lime and in still other ways less fundamental bear little or no relation to it because it was not derived from it to any important degree, or perhaps to any degree at all. No less true are these statements when applied to wind-borne soils. In every dry region whirlwinds raise aloft great clouds of dust that settle down near by or become more or less permanently lodged perhaps hundreds of miles away in extra-desert regions, where their relation to the rock or soil they overlie is purely fortuitous. The great loess deposits of western China, the dust soils of Oregon and Idaho, the loess deposits of the Mississippi Valley, all are illustrations of wind-borne soils, though they are not all fundamentally related to the extremes of arid conditions.

Glacially transported material has or may have the same discordance with respect to the underlying rock. Over the sandstone and limestone areas of the Great Lake region has been swept glacial detritus in vast amounts, and although the material reflects the character of the underlying rock to a notable degree (perhaps on the average about 80% of it is locally derived), yet an important share is also derived from northern localities. Bowlders of granite, gneiss, greenstone, slate, and basalt may be found scores and even hundreds of miles from their nearest outcrop, and everywhere in the glacial till are found important amounts of clay which were derived at least in small part from northern localities. The effects of the continental ice cap on the soil are discussed in greater detail in succeeding pages, and need not be further described here. Alpine glaciers have had less important effects upon soil because of their relatively slight development and because they have produced their effects chiefly in mountain valleys where their deposits are so restricted that though they may be important to the farmer they are relatively of less importance to the forester.

The part that the various agencies concerned in soil formation have played in the making of the soil of North America is brought out in the following table.[1]

DISTRIBUTION OF SOIL TYPES BY REGIONS [2]

(The surface of each continent is taken as 100)

	North America	South America	Europe	Africa	Asia	Oceania	New World	Old World	Whole Land Surface
I. Alluvial regions:									
Loam predominating	17	2	22	1	37	15	10	21	18
Mountain débris (gravel, etc.)	....	....	....	....	....	1	....	1	0
Laterite (red ferruginous residual clay characteristic of the tropics)	9	43	....	49	16	16	24	25	25
II. Equality of destruction and transportation	4	9	8	3	3	0	6	3	4
III. Denudation preponderating:									
Eolian denudation	2	....	....	14	7	2	1	8	6
Glacial denudation	25	1	9	....	0	....	14	1	5
IV. Accumulation preponderating:									
Glacial accumulations	23	4	36	....	1	....	15	4	8
Marine accumulations	....	0	....	0	....	....	0	0	0
Stream and lake accumulations	1	27	5	2	3	....	13	2	5
Shifting sand	0	1	0	13	8	19	1	10	7
Fine eolian accumulations (steppe soils)	13	1	13	18	20	41	7	21	17
Volcanic accumulations	1	2	0	0	1	2	2	1	1
V. Dissected loess deposits	5	10	7	....	3	0	7	2	4
VI. Coral Islands	....	....	....	0	0	5	....	1	0
Total	100	100	100	100	100	100	100	100	100

The kinds of residual soil that result from the decomposition of the various kinds of rock at the earth's surface are both numerous and variable. For a number of typical illustrations the student is referred to the convenient tables from Merrill in Appendix B. These tables present in summary fashion the chief facts with which he should be acquainted. They may well be supplemented by readings in Merrill[3] and Pirsson,[4] and by the chapter on Land Waste by Davis.[5] In the interpretation of these data it is well to bear in mind that not always is the rock character revealed in the soil character even in the case of residual soils. Limestone soils usually contain adequate amounts of lime, but sometimes they are so deficient in this respect that artificial liming is one of the chief neces-

[1] Compiled by A. von Tillo from Sheet 4 of Berghaus' Physikalischer Atlas and from Richthofen's Führer für Forschungsreisende, p. 498. Original table occurs in Die Geographische Verteilung von Grund und Boden, Petr. Mittheil., vol. 39, 1893, pp. 17–19.

[2] Translation from von Tillo with modifications.

[3] Rocks, Rock-weathering and Soils, 1896.

[4] Rocks and Rock-making Minerals, 1909.

[5] Physical Geography, 1899, pp. 263–296.

sities to bring them up to normal fertility. Likewise the soil resulting from the decay of rocks such as certain basalts of Idaho that contain a great deal of apatite (phosphate of lime), a mineral which generally yields phosphoric acid to the soil, may hold the phosphorus in an insoluble form and make the addition of this ingredient one of the first necessities. To some extent also the geologic history of a region is important in the interpretation of the soils, for a cherty dolomite overlying a shale may yield its insoluble elements to the shale surface long after the soluble part of the dolomite has been removed. Some regions have been dry that now are moist, some moist that now are dry, and each change has effected a change in the soil. Many similar geologic inheritances are known that produce soil effects of fundamental importance.

CHAPTER II

PHYSICAL FEATURES OF SOILS

SIZE AND WEIGHT OF SOIL PARTICLES

WE have now seen that the soil is a complex mixture of mineral particles, soluble and insoluble chemical substances, gases, liquids, living organisms and dead organic matter, various kinds of bacteria, and often molds and enzymes. But the chief ingredients of most soils and important ingredients in all soils are the mineral particles originally derived from rock. A soil may be so coarse that it consists of little more than huge stones and bowlders with which are intermingled small quantities of rock fragments; or it may be so fine that, as in the case of the finest clays, the diameter of the individual particles is only one-thousandth of a millimeter. A pound of such material would be composed of grains whose aggregate surface extent would be 110,538 square feet, or more than 2½ acres. The number of grains in a gram of soil of such fineness would be 720,000 billion. In ordinary soils the number of grains in a single gram varies from about 2 to about 5 million.[1]

The average specific gravity of soils with an ordinary amount of humus will range between 2.55 and 2.75. The lightest constituent is kaolinite, 2.60, and the heaviest are mineral particles containing much iron such as mica and hornblende, which may range over 3.00. The specific gravity is, however, of less importance than the volume weight, or the weight of the natural soil in terms of an equal volume of water. A cubic foot of water weighs 62½ pounds, while the average weight of an equal volume of soil is about 75 or 80 pounds. The extremes are represented by calcareous sand, 110, and peaty and swampy soils, 30.[2] It is important to see at once that what are known as light and heavy soils in agriculture and forestry are not light and heavy in terms of either gravity or volume weight but in terms of tillage and root penetration. Clay is a light soil (70–75 pounds) as to volume weight; pure or moderately pure clay soils are among the heaviest known to agriculture. In general the greater the amount of humus in the soil the lighter it is.[3]

[1] Milton Whitney, Bull. U. S. Weath. Bur., No. 4; F. H. King, Physics of Agriculture, p. 117.

[2] E. W. Hilgard, Soils, 1907, p. 107.

[3] For standard methods of determining the specific gravity of soil see H. W. Wiley, Prin. and Prac. of Agri. Anal., vol. 1, 1906, pp. 96–97.

PORE SPACE AND TILTH

The physical organization of the soil is extremely varied from place to place. Certain sandstones are composed of grains of very uniform size, and weather into a soil of unusually uniform texture. The relations

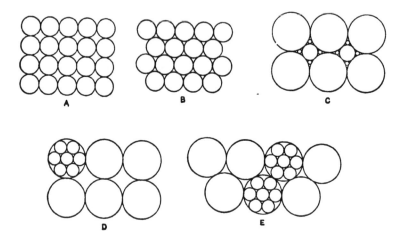

Fig. 3. — Diagram to illustrate pore space in soils

of the individual grains of such a case are represented in Fig. 3 B, where a high percentage of pore space is afforded. If, however, the grains vary in size, the smaller will occupy the interstices between the larger, the

Fig. 4. — The most compact packing of a mass of spheres (left), and unit element of the pore space in such a mass (right). (Slichter, U. S. Geol. Surv.)

volume weight will be increased and the pore space will be diminished, Fig. 3 C. The extent of the pore space may be determined by finding the difference between the specific gravity and the volume weight of the soil. For soils composed of particles of uniform sizes the smaller the size of the grain the smaller the unit pore space, the larger the size of the grain the larger the unit pore space; under all circumstances the amount of pore space is decreased by increase of variability in the size of the individual particles. The value of the porosity is independent of the size of the grains where these are uniform in a given mass; it is dependent merely upon the manner of packing. The minimum porosity of a mass of uniform spheres is 25.95% of the whole mass occupied by them, Fig. 4; the maximum porosity is 47.64% of the whole space.[1] The pore space in most cultivated soils ranges from 35% to 50%, and theoretically may be as high as 74.05%, if the soil is flocculated and the floccules or crumbs arranged as in Fig. 3 D; in sandy soils it may be only 20%.

The actual field condition of a soil as to pore space will depend upon the rainfall, methods of tillage, types of vegetal growths, and the chemical composition of the soil particles. Cultivation results in the formation of soil crumbs or soil aggregates which give the soil a loose condition known as tilth. The volume weight is decreased and the pore space increased by this condition. Wollny[2] estimates that the increase of soil volume through such flocculation may amount in the case of consolidated clay to as much as 41.9%; on moistening, dry clay increases in bulk 30% to 40%. Flocculation may be caused by lime carbonate, and to the action of this substance are due the easy tillage of limestone soils and the loose condition of the loess deposits of the western part of the country. The aggregates may range in size from 10 or more inches to those of microscopic size. The steep bluffs characteristic of loess are explained by the gripping together of the rough soil aggregates which compose the loess. The most common cement that aids crumb structure or soil flocculation is clay, and the action is one of the most important attributed to this soil substance. The fresh colloidal humates of lime, magnesia, and iron act in the same manner, while silica, silicates, and ferric hydrates have the same action but to a less important extent. Tilth is maintained in nature by humus (one of its most important functions) and by root penetration, while the deflocculating action of beating rains is prevented in nature by a cover of forest litter or of grass and herbage. Saturated soils have many defects, such as a high percentage of acids, lack of aeration, etc., but one of their chief defects is the absence of crumb structure, the compact subsoil to which this leads, and the difficulty of root penetration.

[1] C. S. Slichter, The Motions of Underground Waters, Water-Supply Paper U. S. Geol. Surv. No. 67, 1902, p. 20.
[2] Quoted by E. W. Hilgard, Soils, 1907, p. 110.

SPECIAL ACTION OF CLAY

When soils contain too much clay they may act almost like pure clay and, like certain of the prairie soils of the western part of the United States, develop wide cracks during the dry summer months. This action has been described as contributing greatly to the drying out of the soil to some depth, the mechanical tearing of the delicate roots, and sometimes the total destruction of vegetation. In some clay soils the shrinkage after rain or irrigation is sufficient to cause the surface crust to contract about the stem and so injure the bark as to interfere with the proper growth of the plant.[1] The amount of shrinkage in the case of heavy clay soils has been measured; it ranges from 28% to 40% of the original volume. When the content of colloidal clay falls below 15% the shrinkage in drying is so slight as to produce no damage, and in sandy soils no perceptible volume change occurs. In the case of certain alkali soils of California the alkaline carbonates prevent floc-culation and render tillage difficult or impossible, a result that is effec-tively offset by the use of gypsum. Such action is not, however, wholly destructive, for the falling into the cracks of the surface material at the next wetting causes a sort of inversion of the soil, to which is thought to be due in part the long duration of fertility in the case of many clay soils.

SOIL AND SUBSOIL

No sharp distinction has yet been generally accepted between soil and subsoil. A change in color from the darker surface layer (due to the presence of humus or ferric hydrate) to the lighter subsurface layer is the usual basis of the distinction, but this breaks down in the arid region where humus is either present in very small quantities or is wholly absent. Some would base the distinction upon change from rock to rock débris, and while this works well in the case of a shallow cover of land waste it is unsatisfactory in the coastal plains of the world and in a number of other types of places where the loose material of geologi-cally recent deposition may reach down hundreds of feet. The color and humus distinctions harmonize with a number of other qualities, such as absence of structure, in setting off the surface layer of a few inches to a few feet as the soil and the material below that as the subsoil, and we shall regard this distinction as the most valid and acceptable.

Among the distinctions noted between soil and subsoil one of the most important is the humus content. The depth to which humus is found varies somewhat with the nature of the plant growth and is often

[1] E. W. Hilgard, Soils, 1907, p. 112.

found to extend in notable quantities to the lower limit of development of the roots of annual plants. Variation in the root habit of different kinds of plants therefore brings about a variation in the depth of the soil as determined by the humus content. Fertilization tends to change the structure, chemical composition, and degree of compactness. In swamps and marshes the humus tint may reach to such depth as to invalidate the distinctions based upon humus content.

Since humus is porous and has a high water-absorbing and water-holding capacity, and since the surface layer of soil may consist of much finer particles than the subsoil, it is clear that the water content of soil and subsoil may be very unlike, both in time of drought and in time of abundant rain. Aeration is also more nearly perfect in the surface soil than in the subsoil and perpetually continues many chemical processes of great importance in transforming soil substances into available plant food.

Further differences between soil and subsoil deserve consideration. As a rule the subsoil is more clayey than the surface soil, hence subsoils of residual origin are generally less pervious and more retentive of moisture and plant foods in solution than the surface soil. Since the finest particles of soil are usually those richest in plant foods, subsoils as a rule would tend to accumulate larger potential supplies of plant food than the surface soil. These results are due to the penetration of the soil by rain water, which carries the finer particles down with it into the subsoil. The steady depletion of the surface soil in the humid regions by downward percolation would tend to produce far-reaching contrasts between soil and subsoil were it not for the fact that between rains evaporation progresses steadily and capillary action tends constantly to bring soluble salts nearer the surface, where they are deposited through evaporation of the water in which they are held, thus periodically increasing the amount of available plant food in the surface soil. The process is always beneficial in a humid region, but in an arid region may result in so large a surface accumulation of salts (p. 97) as to be injurious to plant growth.

Subsoil is usually more calcareous than the overlying surface soil, and this difference is so marked in some cases that the surface soil requires lime replacement when the subsoil contains a relatively large amount of lime carbonate. It may accumulate to such an extent as to form a solid subsoil mass or hardpan. It is noteworthy that the minerals of the subsoil are in a less weathered condition than those of the surface soil — a condition due largely to the absence of humus and associated carbonic and other acids, so that the subsoil is often spoken of as "raw." In arid regions the characteristic differences between soil and subsoil as developed in humid regions disappear to a large extent. The slight percolation of water does not greatly favor the accumulation of fine colloidal clay in the subsoil, so that both soil and subsoil are of the sandy type and air penetrates to a great distance. Extreme figures for aeration in arid subsoils are a few hundred to a thousand feet, as shown by oxidation of ore in rock to that depth. The distinction between soil and subsoil in humid regions as based on the humus content likewise disappears in arid regions partly because the amount of vegetable matter contributed to the soil is so small, partly because oxidation is so active that vegetation is in some cases completely "burnt up" and is not incorporated in the soil, and partly because the long roots of arid region plants are widely distributed through the deeper layers and largely absent from the surface layer.

The porosity of arid soils and subsoils and their dryness result in an extraordinary root penetration in trees, shrubs, and taproot herbs whose fibrous feeding roots are found deep in the subsoil and sometimes wholly absent from the surface soil. The roots of grapevines have been found 22 feet below the surface, and in the loess of Nebraska the roots of the native Shepherdia have been found at a depth of 50 feet. It is quite otherwise in the case of humid soils. The greater fertility of the surface layer and the abundance of air and other desirable substances, including humus, result in a great development of small feeding roots at the surface, where the largest and most active portion of the root system is found; while the water supply is derived either through a long taproot or through deeply penetrating water roots having the same function, or the whole root system of the plant has been modified to suit moist surface or subsoil conditions.

Sometimes the geologic mode of origin of the land waste of a locality is responsible for an extremely coarse subsoil overlain by thin layers of fine wind-blown or stream-deposited material. The coarseness of the subsoil of such a locality tends not only to depress the water table (p. 44) but also to repress capillary action. The result is often disastrous to all but the most deep-rooted vegetation. Indeed some localities exhibit extreme conditions; the subsoil may be almost perfectly dry during a dry season, while the surface soil is perceptibly moist and supports growing plants. This condition is seen on many outwash plains of the glaciated region and is one of the most serious drawbacks in the development of agriculture on the outwash plains of Long Island, as well as in the larger valleys of the arid region. The possibility of the condition should always be borne in mind, for it is always a reasonable expectation in transported and water-laid soils. This class of soils may offer con- ditions of soil and subsoil quite unlike those outlined above. When they are of recent origin almost any contrast between soil and subsoil may occur. The wandering of a waste-laden stream over an aggrading flood plain may bring about a covering of fine silt over clay or gravel; on the other hand the same stream in a near-by locality may cause the formation of clayey deposits where it formerly deposited the coarsest material. On the seaward margin of a coastal plain it is also common to find wide variations between soil and subsoil, but the variations are not always of the same kind or to the same degree, and will depend to some extent upon the shore conditions at the time of deposition of the coastal plain sediments of a given locality, and to a great extent upon the drainage conditions and degree of dissection since uplift of the coastal plain.

Soil Air

Plants may be drowned through lack of free oxygen when their roots are submerged. Besides this harmful effect of submersion is the lack of continued oxidation of the soil particles and the formation of plant foods. Furthermore, nitrification ceases and denitrification sets in. Aeration is therefore an essential process for the best plant growth, and by this is meant the admission of air not merely into the surface layers but deeply into the root zone so that the decaying organic matter in the form of roots and leaves carried down by earthworms may be formed into nitrogen by the carrying forward of the nitrifying process, which is dependent upon a supply of oxygen. In addition it should be noted that many purely chemical reactions essential to soil fertility require a certain amount of oxygen and carbon dioxide for their continuance.

The amount of air space in soils is from 35% to 50% of their volume, and when soils are in their best condition for the support of vegetation about one-half of this space is filled with water, the other half with air. A number of investigators agree in assigning to unculti- vated forest soil only about half as much air per acre foot as in the case of a well-cultivated garden soil, or from 4000 to 6000 cubic feet. The com- position of soil air is different from that of the atmosphere in that soil air usually contains a larger amount of water vapor, a higher nitrogen content, a lower oxygen content, a larger amount of carbon dioxide, etc.

One of the chief objects in draining a soil is to facilitate aeration. For while soil water is composed in part of oxygen it is not in a free state and requires chemical alteration to be suitable for the transformations in which oxygen plays a part. The various bad effects of lack of aeration are, chiefly, a stoppage of the important process of nitrification, the but partial decomposition of organic matter in the soil, a drowning out of earthworms, insects, etc., whose effects in maintaining soil fertility are important, and a reduction of the soil temperature. Aeration denotes good drainage; lack of aeration poor drainage, except in the case of stiff clay, where air may be excluded to a certain extent even when the clay is relatively dry. Clay soils are in general poorly aerated because their fineness of texture causes a large area of grain surface and this in turn a large water-retaining capacity. On certain areas of the Oxford clay and London clay of England the pastures degenerate in a few years into masses of creeping plants and the land must be cultivated afresh in order to aerate it. The clays are so fine-textured that they become water-logged when allowed to stand without cultivation. In order to aerate the soil the Dutch farmer of the lowlands causes the water table to sink to a depth of a meter during autumn and winter, but during the remaining months only to a depth of a half-meter; and a similar practice is followed in certain meadows in Denmark. A wet, badly aerated soil, poor in oxygen, obstructs plant respiration and represses the functional activity of the roots.

The amount of air in the soil affects the internal structure of the plant so that in very wet soil the plants that thrive frequently have very large internal air spaces which are in communication with one another throughout the whole plant and can even convey air from the atmosphere itself to the most distant root tips and parts of the rhizomes.[1]

Loam, Silt, and Clay

Loam, silt, and clay are of exceptional importance in the soil. The fineness of their constituents causes notable effects upon the water content of soils, upon the solubility of the soil substances, and upon the facility of root penetration, so that it may be taken as a general principle that the fine material of a soil has an importance quite out of proportion to its relative volume or volume weight, and the determination of its existence and amount are of the greatest importance in a mechanical analysis of the soil.

LOAM

The term "loam" is perhaps one of the most indefinite words in the vocabulary of the layman who undertakes to speak about soils. It is used in a very loose and sometimes wholly indefinite way even by some soil physicists and it is therefore necessary to note its features in a special manner. From the table on p. 722 it is easy to derive the empirical formula for loam. As there defined it is a soil that contains less

---

[1] E. Warming, Œcology of Plants, Ox. ed., 1909, p. 44.

than 55% of silt and more than 50% of silt and clay; but the essentials of that formula are not brought out by its mere statement. The essential feature of a loam is that it is a mixture in certain proportions of fine with less fine material. That mixture may represent the widest extremes of soil material, such as stones and bowlders mixed with silt and clay in right proportions, and would then be called a stony loam; or it may be either a coarse or fine or medium sand that is mixed with clay or silt or a mixture of these two substances, in which case it would be called a sandy loam. *Loam is therefore not to be thought of as a certain soil ingredient such as sand or silt or stones, but as a condition of mixture which makes it desirable to designate the mixture and not the individual components and to express such designation by a specific name.* This explanation can not be stated too emphatically, because a great many writers loosely consider a soil to be a loam when "loam" predominates, and call it a sandy loam if a certain important amount of sand occurs in the soil. If on the other hand the sand predominates they call the sample a sandy loam instead of a loamy sand. Such a designation makes the erroneous assumption that "loam" is a substance instead of a condition of mixture. It is scarcely necessary to add that humus added to clay or sand or silt in right proportions makes a loam or that clay added to gravel in certain proportions makes a loam, etc.

SILT

The term "silt" is commonly employed to denote the finest material of the soil above clay. In the mechanical analysis of soils only differences in size of grain are determined and it might therefore be assumed that silt resembles sand in chemical constitution. They are on the contrary unlike in their chemical nature but the differences are not radical. In sand, quartz is the principal mineral; in silt the hydroxides and hydrous silicates predominate. Neither product is wholly free from the other and neither is uniform in character since the character of the sediments everywhere reflects to some degree the character of the rock from which the sediments were derived. In regions of basic rock, for example, the sediments are rich in iron; in granite regions the sediments are composed largely of aluminous residues.[1]

CLAY

Residual clays originate through the decomposition of crystalline rocks. They are pure or of high grade when they are derived from rocks which contain only silicates of alumina or when the movement of the

[1] F. W. Clarke, The Data of Geochemistry, Bull. U. S. Geol. Surv. No. 330, 1908, p. 428.

ground water is thorough enough to remove other more soluble salts as fast as formed. Since these conditions are rarely fulfilled it follows that even the purest deposits of clay usually contain crystals of quartz and other types of resistant minerals. The acids of ground water have far less effect upon aluminum than upon the other bases, so that the greater part of this base remains in the soil and collects in such amounts as to form deposits containing large percentages of clay (silicate of aluminum) or kaolin.

"Clays may be defined as mixtures of minerals of which the representative members are silicates of aluminum, iron, the alkalies, and the alkaline earths. The hydrated aluminum silicate, kaolin ($Al_2O_3.2SiO_2.2H_2O$), is the most characteristic of these. Some feldspar is usually present. The grains of these minerals may show crystal faces (especially in the case of kaolins), but more commonly they are of irregular shapes. Upon most of these grains is an enveloping colloid coating. This is mainly of silicate constitution, but may consist partly of organic colloids, of iron, manganese, and aluminum hydroxides, and of hydrated silicic acid. Quartz grains, which are generally present, and mica, which is frequently present, do not have the colloid coating or have it in much less degree. Almost any mineral may be present in clays and modify the properties somewhat. The combination of granular materials and colloids is in such proportion that, when reduced to sufficiently fine size (by crushing, sifting, washing, or other means) and properly moistened with an appropriate amount of water, plasticity is developed. If the colloid matter is in excess the clay is considered very plastic, fat, or sticky, but if the granular matter is in excess it is called sandy, weak, or nonplastic."[1]

Transported clays owe their existence to the sorting action of water. The deposition of the transported material according to size leaves the fine clay to be deposited last. Hence clay deposits may be found on lowlands where more or less regular inundations permit of the subsidence of clay as the end of a series of quiet water depositions in back swamps and flood plains. Clay may also be formed as a marine or lacustrine deposit to become dry land either through the elevation of the deposits by crustal movement or through the draining of a lake or partial draining by the cutting down of the outlet, the silting up of the lake floor, the tilting of the land, or the growth of shore vegetation. The formation of pure clay can take place only under exceptional conditions, since deposition is usually in the form of floccules of coarser material which carry down the finer particles with them even when the water is undisturbed.

COLLOID CLAY

The name "colloid" (resembling glue or jelly) is applied to the clay particles that remain suspended in water for 24 hours or longer. The presence of an electrolyte such as a soluble salt causes the discharge of the static electrical charges either positive or negative to which the suspension is thought to be due and the subsidence of the colloid follows.

[1] H. E. Ashley, The Colloid Matter of Clay and its Measurement, Bull. U. S. Geol. Surv. No. 388, 1909, p. 7.

This explains the hastened subsidence of colloidal matter in river water when it is discharged into a salty sea. Besides the true clay of the extremely fine substance of soils there are present, usually in all soils, substances such as silicic, aluminic, and zeolitic hydrates, which are all nonplastic and yet fine enough to form part of the clay substance as usually described.[1] The plasticity appears to be due solely to those particles of the clay substance which do not settle in the course of 24 hours through a column of pure water 8 inches high. Colloid clay is a jelly-like substance which shrinks greatly on drying and when dry appears like glue. It adheres to the tongue with great tenacity, swells rapidly when wetted, and is highly adhesive and plastic. It may also be separated from water by evaporation of the water or by the use of lime which flocculates the clay particles and causes them to subside.[2]

CLASSIFICATION OF CLAYS

Clays are designated according to the predominance of certain constituents: some are calcareous and are called marls; some contain a great deal of fine quartz and are known as siliceous clays; while others are rich in iron oxides and are called ferruginous clays or ochers, etc.[3] The brickmaker, the ceramist, and others have refined classifications based upon the special qualities clays exhibit when used for special purposes and under special conditions. These are, however, not natural features of clay in the soil and therefore lie outside our field of study.

Of all the mineral constituents of the soil clay is without doubt the most important because of its peculiar rôle in the physical structure of the soil, whereby it affects root penetration, drainage, fertility, etc. Its fineness and plasticity cause it to fill the soil spaces to the degree to which it is present and to cause the soil particles to adhere and to form soil crumbs, or floccules, which in turn results in a more open structure and therefore better aeration, better drainage, more rapid humification of organic matter, etc. Without clay, sand flocculates only when moist; when thoroughly wet or thoroughly dry the particles collapse and the soil assumes a single grain structure instead of a crumb structure. The whole soil mass then becomes densely packed and its fertility reduced

[1] H. W. Wiley, Prin. and Prac. of Agri. Anal.: The Soil, 1906, p. 182; and E. W. Hilgard, Soils, 1907, 59-85.

[2] For further data concerning the various theories of plasticity and the composition of clays see the following references:

Th. Schloesing, The Constitution of the Clays, Compt. Rend., vol. 79, 1874, pp. 386-390, 473-477.

A. S. Cushman, The Colloid Theory of Plasticity, Trans. Am. Ceram. Soc., vol. 6, pp. 65-78.

[3] L. V. Pirsson, Rocks and Rock-forming Minerals, 1909, p. 281.

both because of the exclusion of adequate amounts of water and because roots are not able to penetrate it. The compacting is furthered by the presence of grains of many sizes in somewhat equal proportions; under these circumstances the smaller grains fit into the interstices of the larger and give the soil an imperviousness that makes it very difficult of cultivation. The same result may be achieved in a soil with a high percentage of fine sediments and an equally high percentage of large grains. This combination in the absence of either intermediate grains or clay will effect a high degree of impermeability. The comparable influence of forest litter, humus, etc., on soil tilth, is discussed in Chapter VI and will not be treated here.

It is to the tendency of clay to bind the particles of the soil and give it tilth or open texture that the loaminess of soils is due when their chief constituent is sand. The small percentage of clay required to produce important effects is shown in the following table,[1] but in interpreting it the reader should keep in mind that by clay is meant the colloidal clay as noted above and not alone the fine substance separated by elutriation and of different character both physically and chemically from a colloid.

Very sandy soils	.5% to 3% clay
Ordinary sandy lands	3.0% to 10% clay
Sandy loams	10.0% to 15% clay
Clay loams	15.0% to 25% clay
Clay soils	25.0% to 35% clay
Heavy clay soils	35.0% to 45% and over

Like humus, clay is very retentive of water and soil gases as well as of solids dissolved in water, qualities so markedly absent in certain coarse soils as to render them almost useless for agriculture in spite of the presence of a rather large amount of plant food as shown on chemical analysis. Furthermore the clay substance in the soil while it itself contains nothing of value to the plant (silicate of aluminum in its pure state being of no importance whatever in nutrition) yet contains within its mass in a fine, easily dissolved, and highly decomposed condition other soil minerals or substances of great importance. Among the most important of these are potash, lime, soda, etc. As an illustration of the origin of such substances may be mentioned the soda-lime and potash feldspars. Those containing lime are more readily attacked than those containing potash. All clays arise from the decomposition of the feldspars, augite, hornblende, etc., and as these minerals all gen-

[1] According to E. W. Hilgard, Soils, 1907, p. 84.

erally contain potash the clays are the source of the available potash in the soil; therefore the amount of potash in the soil usually varies with the amount of clay.[1] In many cases also zeolitic compounds are associated with clay. These are hydrous silicates of lime or alumina which in the presence of a solution containing a stronger base such as potash or soda may yield the displaced base to the soil as a soluble substance of great potential value to plants.

The insolubility of clay, suggested by the fact that it is an ultimate product of rock decomposition, is one of its chief defects, though the defect is generally not apparent in nature because clay has a strong affinity for many soluble salts of great importance as plant food. The manner in which the soluble material of a soil rapidly increases with increase in fineness and the importance of clay in this respect are well brought out in the following table modified from the table by R. H. Loughridge.[2]

RELATION OF SOLUBLE MATTER TO SOIL CLASS

Conventional Name	Clay	Finest Silt	Fine Silt	Medium Silt	Coarse Silt
Per Cent in Soil	21.64%	23.56%	12.54%	13.67%	13.11%
Diameter of Particles	?	mm. .005–.011	mm. .013–.016	mm. .022–.027	mm. .033–.038
Constituents	%	%	%	%,	%
Insoluble residue...................	15.96	73.17	87.96	94.13	96.52
Soluble silica......................	33.10	9.95	4.27	2.35	....
Potash (K₂O)......................	1.47	0.53	0.29	0.12	....
Soda (Na₂O).......................	(1.70)	0.24	0.28	0.21	....
Lime (CaO)........................	0.09	0.13	0.18	0.09	....
Magnesia (MgO)...................	1.33	0.46	0.26	0.10	....
Manganese (MnO₂)................	0.30	0.00	0.00	0.00	....
Iron sesquioxide (Fe₂O₃)...........	18.76	4.76	2.34	1.03	....
Alumina (Al₂O₃)...................	18.19	4.32	2.64	1.21	....
Phosphoric acid (P₂O₅)............	0.18	0.11	0.03	0.02	....
Sulphuric acid (SO₂)...............	0.06	0.02	0.03	0.03	....
Volatile matter....................	9.00	5.61	1.72	0.92	....
Totals........................	100.14	99.30	100.00	100.21	....
Total soluble constituents..........	75.18	20.52	10.32	5.16	....

The table shows that clay contains about 33% of soluble silica, finest silt about 10%, and medium silt about 2½%. The total soluble in-

[1] A. D. Hall, The Soil, 1907, p. 19.

[2] On the Distribution of Soil Ingredients among the Sediments obtained in Silt Analysis, Am. Jour. Sci, vol. 7, 1874, p. 18. Analysis based on a yellow loam from Mississippi. Designations of soil classes do not follow present conventions.

gredients in the same order are 75%, 20½%, and 5%. The clay is by far the richest in mineral ingredients, the amount being more than twice as great as that contained by all the other soil substances combined. Its insoluble residue is very small, its volatile matter is the largest, it contains more soda and manganese, and it heads the list in the amount of free silica it contains. The availability of the soluble material, however, depends on the tilth and the water supply to a large degree, and a fine soil must have a proportionately greater water supply than a coarse one or its otherwise more favorable qualities will be counterbalanced by excessively slow transference of plant food.

CHAPTER III

WATER SUPPLY OF SOILS

Relation to Plant Growth and Distribution

WATER is of fundamental importance in ecology. It constitutes from 65% to over 95% of the tissues of plants, is a necessary part of all cell walls and of protoplasm, is vital to all transference of plant food and even to the forming of plant food in the soil, is the agent of respiration, in general is the factor that most frequently conditions life and death, and hence has a predominating influence upon both the internal and external structures of the plant.[1] Not only does the rainfall determine the great regional types of vegetation; it determines also the finer shades of detailed distribution where topographic differences occasion great variability in the rainfall distribution from point to point. It is of even more importance than heat, for it is of more irregular distribution. Its importance is reflected in a number of indirect ways as well as in the more familiar direct ways. For example, a windy region is likely to be a dry region for plants, and if not dry in a physical sense is almost bound to be dry in a physiological sense.[2] Wind dries the soil and increases transpiration in the plant to such a degree that places most exposed to it have a relatively xerophilous vegetation. The eastern protected hill slopes of central Jutland are clothed with forest; the western exposed hill slopes are covered with heath. On the northern border of the subarctic forest, bands of trees extend down the sheltered valleys far beyond the continental timber line. The most remarkable case is that of the Ark-i-linik, a tributary of Hudson Bay, which is bordered for 200 miles (lat. $62\frac{1}{2}°$ to $64\frac{1}{2}°$) by a nearly continuous belt of spruce, although the stream flows in the midst of the Barren Grounds.[3] Undoubtedly in the last-named case the distribution is favored also by the higher temperature of the seepage water on the lower slopes and valley floors during the autumn, a condition that prolongs

[1] E. Warming, Œcology of Plants, Ox. ed., 1909, pp. 28–29.

[2] For definition of physiological dryness see Schimper, Plant Geography upon a Physiological Basis, Ox. ed., 1903, p. 2.

[3] E. A. Preble, A Biological Investigation of the Athabaska-Mackenzie Region, No. Am. Fauna, U. S. Dept. Agri., No. 27, 1908, p. 48. Excellent for exact delineation of the continental tree line in northern Canada.

the growing season and mitigates the effects of extremely low air temperatures. Some part of the effect may be attributed also to the deeper snows of the valleys which prevent extremely low ground temperatures.

It is found that each species of plant requires its own specific water supply for most favorable conditions of growth, and that the quantity of water in the soil has a greater influence than any other condition on the distribution of plant species. To illustrate adaptations within a single genus the larches may be taken. *Larix decidua* prefers loose, well-drained soil and hence flourishes in dry situations where many other species die.[1] It is partly to similar adaptations with respect to physiologic dryness that *Larix sibirica* owes its northerly range in Siberia where there is an extremely short growing season. The tamarack (*Larix laricina*) prefers a swamp habitat, though it will endure a hillside situation; it often occupies shallow lake basins recently reclaimed or partially reclaimed by lower forms of vegetation.[2] With the ascent or descent of the ground water new species may come in and old ones die out, so that changes in the level of the ground water have been found gravely to affect the character of the grasses and shrubs and even the trees of a region.[3]

The amount of water required by growing plants is large in proportion to the amount of dry vegetable substance produced. It varies according to the extent and structure of the leaf surface, the number and size of the stomata of the leaves, and the climatic conditions, especially the wind, which when strong and continuous has so intense a drying action on plants as sometimes to lead to special modifications of structure even when the ground is well supplied with moisture, a feature well developed in vegetation that occurs on windy mountain slopes. It has been found that the same plants use more water in humid than in dry climates, as if physiologic adjustment had been made in the latter case in response to a lessened water supply before the development of special structural adaptations.

In general the amount of water required by a growing plant varies from 50 to 900 times the weight of the dry substance. Birch and linden transpire 600 to 700 pounds of water for every pound of dry matter fixed in the plant; oak, 200 to 300 pounds; spruce, fir, and pine, 30 to 70 pounds; European evergreen oak, 500 pounds. What this means in terms of rainfall may be estimated from the last-named case, the European evergreen oak, which, with the water requirement indicated, and with 250 trees to the acre, and 40 pounds of dry matter per season to the tree, would require a rainfall of 22½ inches per year. In general, about 15% of the rainfall is lost through plant transpiration.

[1] H. L. Keeler, Our Native Trees, 1905, p. 480.
[2] Idem.
[3] P. Feilberg, Om Enge og vedvarende Græsmarker. Tidsskr. Landökon. Kjøbenhavn, p. 270. Quoted by Warming, p. 46.

Naturally the required amount of rainfall varies with the kind of soil, whether porous and nonretentive or compact and retentive, with the topography, and with the seasonal and yearly fluctuations in cloudiness, insolation, etc. The amount of rainfall necessary to the growth of forests is about the same as the amount necessary for agriculture without irrigation; that is, from 20 to 40 inches. Timber growth in regions having a mean annual rainfall less than the minimum amount is of so stunted a character as to be of little value. Furthermore the growth is so slow that once the timber has been removed by fire or lumbermen, the time necessary for a new growth is very great.[1] In studying the water supply of a region the lowest rainfall is of quite as much if not more value than the mean rainfall, for it is in a season of unusual drought that the growing trees may be most affected, so that in dry climates it is difficult to establish a forest without prohibitive expense.[2] It is suggested that the cyclic changes in climate which appear to affect the entire earth might be studied to the benefit of the forester in planting forest seedlings, by enabling him to plant during the time of greatest rainfall so that an adequate root system shall have been developed before the advent of the driest years.

The high water content of the soil may in part make up for the dryness of the air, as on the banks of streams in tropical savannas where lines of forest occur, or on steppes and deserts where trees are found near running water or where the ground water approaches the surface.[3] This is, however, not a universal condition, for some plants flourish on very dry soil and in a humid air but are excluded from places with dry air. The heads of alluvial fans and cones where rivers leave mountain canyons at the common border of mountains and piedmont plain are often covered with small patches of forest. The forest vegetation is maintained by a kind of subirrigation or seepage through the porous sands and gravels of the fans. It often happens that this natural watering is too deep for agriculture and that forests in such cases grow where agriculture without irrigation is not possible.[4]

As an instance of the effect of soil character upon the amount of soil water available for plants and hence upon the specific character of the vegetation may be mentioned the Steilacoom Plains, south of Tacoma,

[1] J. W. Powell, The Lands of the Arid Region of the United States, U. S. Geog. and Geol. Surv. of the Rocky Mountain Region, 1879, pp. 14–20.

[2] J. Wilson, The Modern Alchemist, Nat. Geog. Mag., vol. 18, 1907, p. 791; see also Rept. of the Sec'y of Agri. for 1907.

[3] J. W. Powell, The Lands of the Arid Region of the United States, U. S. Geog. and Geol. Surv. of the Rocky Mountain Region, 1879, pp. 15–16.

[4] Idem.

Wash., which have such an extremely porous soil of coarse gravel, with only a thin veneer of silt, that they constitute a locally semi-arid district in what is otherwise a humid region. The rainfall is about 44 inches per year, but percolation is so rapid in the loose stony ground that the district is a barren island surrounded by dense forests characteristic of the region. Instead of the Douglas spruce (*Pseudotsuga taxifolia*), the white fir (*Abies grandis*), the tideland spruce (*Picea sitchensis*), and the western hemlock (*Tsuga mertensiana*), the district bears the yellow pine (*Pinus ponderosa*); and species of gophers and the desert horned lark, which are at home in the dry districts east of the Cascades, are also at home in this restricted and peculiar area.[1]

The Coalinga district of California exhibits a plant distribution intimately related to water conditions. Certain gravelly and sandy beds of the district have superior absorptive capacity, while the adjacent

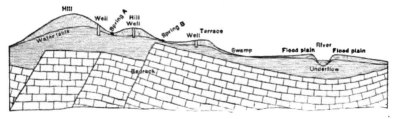

Fig. 5. — Ideal section representing the ground water in relation to the surface and the bed rock. (Slichter, U. S. Geol. Surv.)

clay beds have but little power of absorption.[2] The sudden rains run off the clay beds without wetting them notably. The coarse beds are therefore marked by a varying abundance of vegetation; the clay beds do not support vegetation at all. The result is a marked parallelism and alternation of belts of vegetation and belts of barren country in sympathy with the belted outcrop of the strata.

FORMS OF OCCURRENCE

Water is contained in the soil in three different ways — as ground water, as capillary water, and as hygroscopic water.

GROUND WATER

Ground water is the name applied to the water in the saturated zone of soil or rock; it occurs from a few feet to a few hundred feet below the surface, Fig. 5; in humid regions it is found usually from five to fifty feet

[1] Willis and Smith, Tacoma Folio, Wash., U. S. Geol. Surv. No. 54, 1899, p. 2.

[2] Arnold and Anderson, Geology and Oil Resources of the Coalinga District, Cal., Bull. U. S. Geol. Surv. No. 398, 1910, p. 33.

below the surface. The surface of the saturated zone or of the ground water is known as the water table or water plane. The depth of the water table below the surface varies in a striking way not only as between arid and humid regions, but also from place to place in a given region as shown in Fig. 6, because of topographic, soil, and other variations.[1] The available pore space of the surface rocks occupied by water or moisture is generally about 10% of their total volume.

The water contained in porous soils and rocks as ground water is not stationary but possesses a very slow although perfectly definite

Fig. 6. — Contour map of water table (continuous lines), showing direction of motion of ground water (arrows) and drainage lines (heavy lines). (Slichter, U. S. Geol. Surv.)

motion as shown by geologic data which indicate very important chemical and physical effects due to permeating waters and by direct measurement of the rate of movement. The cause of the movement of ground water is gravity alone, and the rate depends upon the size of the pores, the total porosity, the pressure gradient in the direction of flow, and the temperature of the water, being more rapid the larger the size of the pores, the greater the porosity, and the higher the gradient and the temperature. The motion of the ground water as a whole is somewhat like the slow motion of a viscous substance, but is not generally in the nature of an underground stream as is ordinarily supposed. Underground streams may exist in limestone regions in great numbers, but they are on the whole exceptional hydrologic

[1] C. S. Slichter, The Motions of Underground Waters, Water-Supply Paper U. S. Geol. Surv. No. 67, 1902, p. 33.

features. The general trend of moving ground water is into neighboring streams and lakes; and the marshy zone on the borders of a valley flat, or on the bluffs of an intrenched valley, or on the shore of a lake, is a manifestation of the ground water appearing at the surface. In many dry western localities the ground water does not find its way immediately into the channel of the river, but takes a general course down the valley within the porous material of the valley floor; this movement, called the underflow, may often be utilized by constructing across the valley a subsurface dam which causes the ground water to rise to the surface or so near it as to become available to plants. If a natural dam crosses the valley the effect may be a similar raising of the underflow and of the ground water, as is the case at the Bunker Hill dike near San Bernardino in southern California.[1] A convergence of canyon walls will produce a similar augmentation of the water in a river because the underflow is forced to the surface. The debouchures of the rivers in such dry regions are usually marked by huge alluvial fans, as along the western base of the Sierra Nevada. In such cases both the underflow and the surface flow are distributed by a set of complex and anastomosing[2] distributaries through gravelly fan deposits and so are gradually dissipated. Sometimes the underflow may be quite independent of the water flowing in the surface channel.

The surface of the water table is seldom level; the nearest approach to this condition is found in the case of a flat topography such as local areas of a coastal plain and of alluvial bottom lands. The surface of the water table shows a close sympathy with the surface contours of the land, although geologic conditions may greatly modify this general fact. Subsurface layers of impervious material may cause a rise and fall of the water table quite out of harmony with the surface contours, Fig. 5. The surface of the ground water is never fixed, for its level is responding continually to changes in rainfall, in barometric pressure, and in temperature by such important amounts as notably to affect the strength of flow of springs and flowing wells.[3]

After a rainy period has passed, the surface zone of saturation gradually descends through the flowing off of water from the surface of the saturated zone, and continues to sink at a constantly decreasing rate

[1] W. C. Mendenhall, The Hydrology of San Bernardino Valley, Cal., Water-Supply Paper U. S. Geol. Surv. No. 142, 1905, p. 26, and plate 11, p. 72.

[2] A term applied to the characteristic branching and reuniting pattern exhibited by streams that terminate upon piedmont alluvial plains.

[3] A. C. Veatch, and others, The Underground Water Resources of Long Island, N. Y., Prof. Paper U. S. Geol. Surv. No. 44, 1906; also idem, Fluctuations of the water level in wells, with special reference to Long Island, N. Y., Water-Supply Paper U. S. Geol. Surv. No. 155, 1906.

until another rainy season causes it to rise again toward the surface. The responses of the ground water to rainfall are not immediate, and depend upon the depth of the water table and the duration of both the rainy and the rainless period. It may happen that the water table is actually rising between rains and falling during a rainy period. The amount of such lag is usually rather small, however. In arid regions, where the ground water is far below the surface, say 100 feet, a rain of considerable magnitude may be absorbed by the dry surface layer and again reëvaporated without replenishing the ground water at all. In humid regions light rains may be evaporated in the same way, but the water of prolonged rains is contributed to the ground water to the extent of 35% to 60%; the remainder is disposed of in the immediate run-off and by evaporation.

Plants growing upon the soil rob it of moisture during the growing season to a degree that varies with the temperature, the kind of plant, and the texture of the soil, so that the amount of water in the soil diminishes from spring to autumn, at which time the water table is at its lowest and may be from five to seven feet lower than in the spring. In the forest various species of plants act as weeds because they consume water before it reaches the roots of the trees. Shallow-rooted plants in the forest have on the whole a relatively small effect, however, because the greater supply of moisture for trees is derived from deeper lying sources.

The level of the ground water is invariably lower in a forested tract, for the forest consumes water in exceptional quantities from the subsoil; and this in spite of the fact that the surface soil of forested regions is as a rule moister than the surface soil of unforested regions. Many trees assume a peculiar shape or can not grow to normal height in a soil in which the ground water is near the surface.[1] The forest sometimes has an important power in maintaining the soil water (not ground water) near the surface, i.e., not only the immediate surface but the whole surface zone in which the plant roots are found. The removal of the forest cover may in delicate cases destroy the capacity of the surface soil for water. Forest litter and particularly humus have high capacities not only for water but also for water vapor, and their destruction by leaching, burning out through excessive oxidation, and the absence of any additions through fallen foliage, trees, twigs, etc., may cause a region that was once fairly moist to become dry. A concrete instance is furnished by the Karst of Austria. This region lies

---

[1] For the rate of evaporation from a free water surface, from bare soil, and from soil covered with vegetation see R. Warington, Lectures on Some of the Physical Properties of the Soil, Oxford, 1900, pp. 107–126, where the results of Ebermayer, King, and others are discussed in detail.

along the Austrian shores of the Adriatic and is composed of porous, fissured limestone. For centuries it furnished the ship timber and wood supplies of Venice, but excessive cutting, burning, and pasturing left it almost a desert waste, not only by decreasing the amount of soil water but also by allowing excessive soil erosion. Through government assistance 400,000 acres of the karst were placed under forest, beginning in 1865, and the government has also backed up planting efforts by passing (1884) a reforestation law to control torrents.[1]

The beneficial effect of the forest in maintaining a soil cover by decreasing the delivery of ground water and retarding the immediate run-off is easily appreciated by recalling the fact that each of the waterways in the forest is occupied by a perennial brook fed from the spongy soil, while the small stream beds of tilled land are dry except when rain is actually falling. The difference in the amount of erosive energy applied by the rain to the earth in these two contrasted conditions is very great. In the forest the rain creeps through the openings in the vegetal coating and moves so slowly that it does not expend any sensible energy upon the soil cover, while, if the surface is deprived of vegetation, the water may have a swift motion and an intense erosive force may be expended upon the incoherent soil.[2]

The fact and rate of movement of the ground water may be determined (a) by the electrical method of Slichter,[3] in which the gradual motion of the ground water from one point to another is determined by the use of an electrolyte which passes with the ground water in its general direction of movement; the method is very accurate, and is the standard one in use by the United States Geological Survey to-day; (b) by the use of the lysimeter, which is a receptacle inserted into the ground in such a manner that one side has an outlet discharging into a measuring gauge. It has been found by the Slichter method of measurement that the rate of movement is on the average from 2 to 10 feet per day for areas of moderate relief. A closer average figure is not possible because of the effects of variable soil texture, variable topographic and structural gradients, differences in the amount and time of occurrence of rainfall, etc.

Since the amount of soil moisture most favorable to plants (the optimum water content) is about half the maximum it can hold, it is clear that the saturated zone does not supply the most favorable conditions

---

[1] European Countries' Reclamation of Waste Land, Forestry Bull., Dec. 12, 1909, p. 2.

[2] N. S. Shaler, The Origin and Nature of Soils, 12th Ann. Rept. U. S. Geol. Surv., pt. 1, 1890–91, p. 254.

[3] C. S. Slichter, The Motions of Underground Waters, Water-Supply Paper U. S. Geol. Surv. No. 67, 1902, pp. 48–51.

for plant growth. These conditions are supplied only in the zone immediately above the mean position of the ground water, where occasional immersion takes place through the raising of the water table but where opportunity is afforded for aeration and for the formation of plant foods without their being swept away immediately by movements of the soil water. Chemical decay and the formation of soluble plant foods take place in the soil only when a certain amount of water is present, but their most rapid rate is attained above the ground water not in it; with an excess of water the soil chemicals are too widely diffused upon their formation to effect soil changes of sufficient importance for the immediate needs of a plant. The most favorable conditions for the full utilization of the advantages of ground water are to be found in those places where the water table is from 5 to 10 feet beneath the surface, is relatively constant in position, the rainfall evenly distributed throughout the year, and where the vegetation is in the form of trees.

It is noteworthy that the root systems of trees are very responsive to the ground water. A layer of feeding roots occurs in the surface soil where there is the greatest amount of soluble plant food immediately available, while the roots supplying moisture to the tree will be found to descend almost vertically to a point a little above the surface of the ground water, where a broad extension of the terminal roots may be found. Serious disturbance in the life of the tree is occasioned by sudden and unusually large changes in the level of the ground water, as through irrigation, which may raise the level of the ground water and cause the root terminals to suffer from want of aeration; or by too thorough underdrainage, either by tile or pumping, which may more or less permanently depress the water table and move it out of reach of the deep-lying roots adapted to a certain normal position. It is important, whatever the position of the ground water, that it be maintained at a relatively fixed position. Even short periods of immersion may work great injury to roots accustomed to perform their functions in an aerated soil.

In an undrained soil the roots are confined to a shallow layer from which they quickly abstract the available moisture, and if the subsoil is clay the plant may suffer through the inability of capillarity to supply the needed amount of water. In a drained soil the roots traverse the whole three feet or more into which air is admitted, and this mass of soil holds a very large quantity of capillary water. A water-logged soil is one in which the harm is not confined to the above results but extends to the solution of plant foods in superabundance and their removal in the water. The same condition also leads to the setting free of a large part of the nitrogen as nitrogen gas instead of its accumulation as a nitrate and to the breaking down of nitrates in the soil. Vegetable acids in exceptional amounts occur in wet soils, and in their presence bacteria can not thrive. Earthworms and insects and the benefits derived from their action are excluded from all saturated soils.

CAPILLARY WATER

Capillary water is the most important form of water in the soil, since it is the normal means by which plants absorb food and sustain the rapid evaporation of the hot summer season. Furthermore, few plants have roots adapted to normal action in the saturated zone where free oxygen is not available. There is for all land plants a definite time limit beyond which their roots can not live or at least remain healthy in a submerged state. The period is about three weeks for deciduous orchards when in their winter condition.

In most cultivated soils the pore space is about 25% to 50% of their volume and this is known as their maximum water capacity or saturation point. The amount of this space occupied by water and required for the best development of plants is generally not more than 50% nor less than 40%, which means that the pore space must be about half filled with air for best results. All of these figures are subject to considerable variation in individual cases. For example, the maximum water content for lodgepole pine as it occurs in the dry hills about Sulphur Springs in Middle Park, Colorado, is 35% in loam and about half as much in sand and gravel. The optimum water content is between 12% and 15%, rising to 20% where the rapid decay of the needle cover decreases the amount of available water. The minimum water content may fall below 5% in gravel without injury to the tree except in decreasing its rate of growth.[1]

Loblolly pine also illustrates the great variation among trees in the amount of water they can endure. While this tree is adapted to a wide range of soils and can grow almost equally well on poor sandy upland soils and low rich bottom lands, it everywhere requires an abundant supply of water. When the soil becomes dry the loblolly pine of Texas and North Carolina gives way ultimately to the longleaf and shortleaf pines. Its immediate occurrence in the zone of contest between it and the other pines sometimes follows, not because the water supply is at an optimum for it but because its prolific seeding and rapid early growth cause it to come in more readily on land made vacant by fire or by lumbermen.[2]

[1] F. E. Clements, The Life History of Lodgepole Burn Forests, Bull. U. S. Forest Service No. 79, 1910, p. 52.
[2] R. Zon, Loblolly Pine in Eastern Texas, Bull. U. S. Forest Service No. 64, 1905, p. 8.

NATURE OF CAPILLARY ACTION

The phenomena of capillarity depend upon the well-known fact of *surface tension.* If the molecular attraction of the particles of a solid for those of a liquid exceeds the attraction of the liquid molecules for each other, the liquid adheres to or wets the solid and the water rises until the pressure of the raised water column equals the pull (molecular attraction) of the solid upon the liquid.[1]

If a soil is saturated with water the whole pore space is filled, and, when this is allowed to drain away, some of the water is pulled down by gravity, but much remains clinging to the particles in a state of tension which just balances gravity. Reversing the process we find that water will always pass into the soil from a wet to a dry place until the film surrounding the particles is evenly stretched throughout. The capillary rise of water in soil materials is well shown in the accompanying illustration from Johnson, Fig. 7.

The nature of capillary action is easily illustrated by the immersion in a basin of water of an open

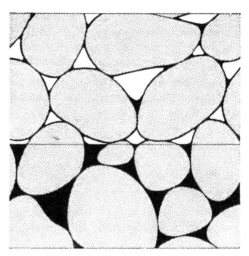

Fig. 7. — Capillary water about soil grains. The horizontal line is the water table.

tube filled with soil. The water will rise in the soil of the tube to a height depending upon the temperature of soil and water, upon the amount of pore space in the soil, and upon the size of the capillary tubes or pores. The finer the particles of a soil the greater its water-holding capacity, the slower the capillary movement in a given unit of time, and the higher the ultimate capillary rise. The maximum height of capillary rise thus far observed is 10.17 feet in material whose particles range from .0005 mm. to .016 mm. in diameter, although eighteen months are required to obtain the maximum. The rise of water in capillary tubes is at first rapid, but soon becomes

[1] For a complete statement of the laws of capillary action see any standard text-book of physics.

slower and after a few months usually reaches a maximum height beyond which it can not rise. The most rapid continuous rise and the longest and ultimately the highest rise usually occurs in salty soils containing a small percentage of clay. Capillary movement takes place in moist soils much more rapidly than in dry soils, but the final adjustment as to height and water content will be the same. Wetting the surface layer or cooling it as in a cold rain tends to raise additional supplies from below, but in the latter case the action is probably to be attributed in part to the condensation of the evaporated subsoil moisture on coming into contact with the cool surface zone.

Water will rise in the capillary tubes of the soil to a greater distance than will any other soil liquid. It is thought to be the limit of capillarity in trees which determines to a large extent the limit of their height.

REGULATING ACTION OF THE SUBSOIL

The subsoil acts as a regulator of the amount of water contained in the surface soil. It absorbs the water which percolates through the surface during the rainy seasons and yields water to the soil during the dry periods by capillary action. This is well illustrated in the following table where the gain and loss of the surface layer are shown.

RELATION OF THE SUBSOIL TO THE WATER CONTENT OF THE SURFACE SOIL

Water in inches	Months and Days			
	30/iv to 30/v	30/v to 9/vii	9/vii to 7/ix	7/ix to 27/x
Rainfall....................................	0.18	4.53	3.47	5.65
Evaporation.................................	3.45	2.96	5.71	1.83
Gain or loss of water in top foot..............	−1.0	+1.4	−0.24	+0.61
Water furnished by (−), or passed on to (+) subsoil....................................	−2.27	+0.17	−2.0	+3.21

Since the changes in the water content of the surface layer are not represented by the difference between the rainfall and the evaporation, some water must in the one case have passed to the subsoil and in the other case have been lifted from it by capillary action.[1]

The studies of the Dnieper above Kiev during a twenty-nine-year period (1876–1905) appear to show that there is in certain periods an overconsumption by evaporation of the moisture stored in the soil and made available by capillarity. This overconsumption has to be supplied during the first wet year which follows one or more dry years, so that the amount

[1] A. D. Hall, The Soil, 1907, p. 95.

of evaporation may not in a given year be strictly the difference between the rainfall and the discharge, being less than this difference in wet years and greater in dry years. The influence of forests and marshes on the discharge in the summer of a dry year is to diminish the discharge; but in wet years the forest stores water.[1]

The steady movement of capillary water toward the surface and its evaporation in the surface layer of soil results in highly beneficial effects. The upward-moving water holds in solution the soluble products of soil or rock decay, or both, and as it approaches the surface and is steadily evaporated it becomes more and more concentrated and may finally deposit its contained salts upon and among the soil grains. In either case enrichment of the surface soil or of the soil solution is the result. To be sure, the downward percolation of water, as after rains, neutralizes these effects to some degree, but since percolation is most active, almost wholly active, in the larger openings among the soil grains, and since downward percolation in the surface soil is an exceedingly temporary phenomenon, the contributions of plant food from below upward are more abundant than the losses by downward percolation.[2] A slow concentration of plant food thus goes on in the surface soil, and it is therefore in the surface soil that the feeding roots of plants are chiefly disposed. It is from the surface layer or layers that they derive their chief nutrient substances; deeper-lying roots are mainly for supplies of water. In arid regions, where the ground water is far below the surface and the zone of weathering correspondingly deeper, and plant food more widely disseminated, plant roots are of course not confined to the surface soil.

LIMITS OF ADEQUACY

The entire amount of capillary water in the soil is not available to plants, for their roots are not in contact with all soil particles of the mass of earth they permeate, and, long before the mass as a whole has become dry, the particles in contact with the roots may be robbed of their moisture to such an extent that the soil may be said to be physiologically dry. In the case of certain apple trees in the arid region of California, 8.3% of water was sufficient to keep the trees in excellent condition on a loam soil, while on a clay soil 12.3% was too small an amount for proper growth.[3] It is thus seen that the welfare of the plant is determined not by the total moisture content but by the free moisture held as capillary water by the capillary tubes.

[1] E. V. Oppokov, 11th International Navigation Congress, St. Petersburg, Russia, 1905.
[2] Cameron and Bell, The Mineral Constituents of the Soil Solution, Bull. U. S. Bur. Soils No. 30, 1905, p. 68.
[3] R. H. Loughridge, Rept. Cal. Exp. Sta., 1897-98, pp. 65-96.

HYGROSCOPIC WATER

Dry soil if exposed to moist air absorbs water vapor, the rate and amount of absorption varying greatly with the character of the soil and the degree of saturation of the air to which the soil is exposed. The finer the particles the greater the capacity of the soil to absorb water vapor. Humus and finely divided ferric hydrates are substances with exceptional capacities for such absorption. Sachs has shown by experiment that the amount of moisture absorbed by dry soil as aqueous vapor may be so high in the presence of saturated atmosphere as to supply distinct portions of the normal vegetal demands. For example, in the arid regions the chief supply of water is derived through the deeply penetrating main roots; on the other hand the feeding roots of the plant, which are nearer the surface, are surrounded by soil that is almost air dry and yet slow growth and nutrition are possible. In such cases the water made available through the absorption of aqueous vapor is thought to be sufficient to have an effect upon vegetation especially in coast regions of summer fog, e.g., the coasts of California, northern Chile, and Peru. In the last-named cases a fog bank hangs over the edge of the land almost every night and frequently during the day at an altitude of 1500 to 2000 feet, and a band of vegetation thrives at this elevation, whereas at lower and higher elevations the natural vegetation is much inferior or wholly lacking.

High moisture absorption at night and its evaporation by day are also thought to prevent the rapid and undue heating of the soil and thus to improve the condition of plants under extreme temperature conditions. In humid regions where plants have become adapted to a higher water content hygroscopic water can not maintain plant growth, for wilting begins some time before even the capillary water is exhausted. Sachs[1] found that a young plant began to wither when the dark humus soil in which it grew still contained water equivalent to 12.3% of its dry weight; and plants were found to wither on loam and sand when the percentages of water fell to 8% and 1.5% respectively. A conservative estimate of the value of hygroscopic water would be that ordinarily it has little if any value to plants directly; but, by increasing the amount of water in the soil that will be evaporated the following day, it lowers the temperature and thus indirectly increases the amount of available water by decreasing the rate of evaporation.[2]

[1] J. von Sachs, Handbuch der Experimental-Physiologie der Pflanzen, 1865, p. 173.

[2] For velocity of flow of aqueous vapor through soil and its control by the dimensions of the apertures between the soil grains see Brown and Escombe, Static Diffusion of Gases and Liquids in Relation to the Assimilation of Carbon and Translocation in Plants, Phil. Trans., vol. 193, 1900, pp. 283–291. Abstract in Annals of Botany, London, vol. 14, 1900, pp. 537–542.

CHAPTER IV

SOIL TEMPERATURE

ECOLOGIC RELATIONS

THERE are for each plant certain air and soil temperatures most favorable to development, known as optimum temperatures. The red birch (*Betula nigra*), peculiar among the birches in preferring a warm habitat, will grow in situations where important temperature changes occur, but it reaches its greatest size in the damp misty lowlands of Texas and the bayous and swamps of Florida and Louisiana. For most plants it is true that if the temperature at any time varies widely from the mean the activity of the vegetative functions is diminished or stopped, or the plant enters into a pathologic condition or dies. Beech, oak, and ash can survive in an air temperature of − 9.4° F.; their finer roots succumb to cold at from 8.6° to 3.2° F.[1]

Were the harm confined to mere stoppage of growth it would not be great, for a return to favorable temperatures would mean a revival of plant growth. But when during the summer season either seeds or plants remain in the soil at a temperature but little above the freezing point, bacteria and fungi of many varieties which are able to live at low temperatures may attack and destroy the vegetation. The limit below which most cultivated plants are practically inactive lies between 40° and 45° F. Tropical plants usually germinate at a temperature of about 95° F. Even when seed germination progresses at a low temperature the rate is very greatly hastened by a higher temperature, and the same holds true of the normal growth of the plant. The temperature most favorable to germination and growth, and the degree of tolerance of high or low temperatures vary greatly with different plants; apparently each plant has become adapted to a certain mean temperature as well as to a certain range of temperature. Seeds and seedling plants should be put into the ground at a time when the temperature is most favorable for active growth, otherwise they may be destroyed by the micro-organisms of the soil.

[1] C. von Mohl, Über das Erfrieren der Zweigspitzen mancher gewisser Phycochromaceon. Bot. Zeitg., vol. 41, 1848.

The degree of adaptation to cold made by some plants is quite remarkable. In the Arctic the shallow-rooted flora develops rapidly under the influence of continuous sunshine in the course of five to eight weeks. The seeds are capable of germination at very low temperatures, so that a mass of flowers may be found growing only a few feet from a snow bank or a glacier. The extreme conditions of development are easily appreciated. The ground is soaked with water nearly ice cold, and at a depth of only a few inches, and at the most but a few feet, ground ice may occur in large masses. But insolation during the period of continuous sunshine is very great and on June 21st surface insolation at the pole is almost as great as at the equator.[1] The conditions under which Arctic plants live during their short cycle of growth in the Arctic summer have been described in a number of records of experiments and observations among which are those noted below.[2]

Soil temperature is of further importance in plant growth because of the stimulation which relatively high temperatures give to the useful bacteria of the soil, — bacteria which increase the supply of available nitrogen. It has been found that bacteria cease to develop nitric acid from humus when the temperature drops below 41° F., their action is of trifling importance when the temperature is at 54° F., it becomes vigorous at 58° F., but at extremely high temperatures the activity is reduced to a degree as unimportant as when too low temperatures prevail.[3] The influence of high temperatures in promoting rapid chemical action in the soil is shown by the sharp contrast between the highly decomposed soils of wet tropical regions and the moderately decomposed soils of polar regions. The contrast is of course heightened by the greater rainfall in the tropics.

INFLUENCE OF WATER ON SOIL TEMPERATURE

Water has a predominating influence upon the temperature conditions of soils in humid regions. This is because the capacity of water for heat is about four or five times as great as the heat capacity of the average soil, weight for weight; so that while one unit of heat is required to raise one pound of water 1°, the same change of temperature is produced in a pound of dry sand by the expenditure of .19 unit, and a pound of pure clay requires about .224 unit. Indeed water has the greatest capacity for heat or the greatest *specific heat* among known substances. This means that when the sun shines upon moist sand or clay a large amount of heat is expended in evaporating the water in it, while a relatively small amount is expended in raising the temperature of the soil particles. A well-drained field is therefore warmer on the whole than a poorly

[1] J. Hann, Handbook of Climatology, 1903, p. 93; R. DeC. Ward, Climate, 1908, p. 15.

[2] M. Smith, Gardening in Northern Alaska, Nat. Geog. Mag., vol. 14, 1903, pp. 355–357; Raising Crops in the Far North, Geog. Notes, Jour. Geog., vol. 3, 1904, p. 91; Agriculture and Grazing in Alaska, Geog. Notes, Jour. Geog., vol. 11, 1903, pp. 528–529.

[3] F. H. King, The Soil, 1905, p. 224.

drained field and a dry soil warmer than a wet soil.[1] It also follows that a fine-grained soil like clay will have a lower temperature than a coarse-grained and easily drained soil like gravel or sand. Hence clay soils are "cold" and sand soils are "warm." Were the clay and the sand air dry, the clay soil would be the warmer because its volume weight is less than the volume weight of sand. Since, however, few soils in the humid region contain no water, it is clear that clay will always be relatively cold and sand relatively warm under comparable conditions of water content. The following table summarizes the temperature differences between clayey and sandy soils, the table representing observations on a well-drained clay loam and a well-drained sandy loam.

TEMPERATURE CONTRASTS BETWEEN SANDY AND CLAYEY SOILS[1]

	First Foot	Second Foot	Third Foot
Sandy loam......	76.5° F.	74.7° F.°	72.1° F.°
Clay loam.......	69.5°	69.3°	67.0°
Difference....	7.0°	5.4°	5.1°

SOIL TEMPERATURE AND CHEMICAL ACTION

One of the first functions of soil water is to take into solution from the soil mineral substances which dissolve under all conditions with extreme slowness. It is here that the influence of soil temperature is perhaps as marked as in the beginnings of seed germination and plant growth in the spring. With a rise in the temperature of the soil chemical action becomes more effective, the supply of plant food in the soil rapidly increases, and osmosis and the diffusion of the dissolved material away from the soil and through the roots and other tissues of the plant are hastened. When we recall the fact that the soil air can occur in favorable quantities only in a well-drained soil and that both high temperature and an abundant supply of air are necessary to many chemical reactions in the soil, it is clear that the conditions favoring a high temperature favor the disintegration of the soil minerals and the formation of available plant food.

[1] The exception to this condition may be noted in the autumn when the warmer soils are those containing the more water on account of the slow radiation of heat by water. The condition may be compared to that of a lake in the temperate zones, which is colder in summer but warmer in autumn and winter than the adjacent land.

[1] F. H. King, The Soil, 1905, p. 228.

INFLUENCE OF SLOPE EXPOSURE, SOIL COLOR, RAINFALL, AND VEGETATION

A rough surface will be colder than a smooth surface, other things being equal, because a larger surface of soil grains is exposed when the ground is rough than when it is smooth, and while the slopes exposed directly to the sun receive more heat than the sheltered slopes they lose more than they gain, by radiation and by contact with the air. This is overcome in agriculture by leveling the land or "rolling" it. The slope of the whole surface with respect to the sun's rays also has an important influence on the temperature of the soil. This principle is illustrated by Fig. 8. It will be seen that the slopes *ad* and *db* are equal in gradient and length. The difference in the amounts of

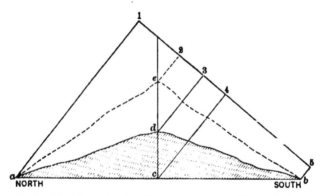

Fig. 8. — The influence of surface slope upon the amount of heat received per unit area.

insolation is shown by the difference between 1–3 and 3–5, 1–*a*, 2–*e*, etc., representing the sun's rays. Upon a flat surface, as *acb*, the amount of heat received in the two cases is the same, that is, 1–4 equals 4–5. It also follows from the diagram that the greater the relief the sharper the contrasts between the slopes directed toward the sun and those directed away from the sun, i.e., 1–2 is much smaller than 2–5. Southern slopes in the northern hemisphere are dry and warm, northern slopes are cool and moist, and the two often bear markedly different types of vegetation. Near Findelen, Switzerland, one may observe patches of snow on northern mountain slopes at elevations *lower* than the barley and rye fields on the southern slopes at 6900 feet.[1] This difference is most marked in high latitudes and high altitudes combined

[1] H. E. Gregory (Gregory, Keller, and Bishop), Physical and Commercial Geography 1910, p. 105.

with strong local relief. It would be shown in the diagram by causing 1a — a ray of sunshine — to approach and finally to fall below ad. It is shown even in tropical situations close to the equator, though its expression in such a case is likely to be exaggerated by contrasts in rainfall derived from the trade winds. North of the equator these blow from the northeast and water the north or shadier (for most of the year) slopes copiously and leave the southern slopes dry; south of the equator the southeast trades produce a similar effect upon the southern or shady slopes.

The temperature of the soil will be affected also by its color. Sandy soils are light in color and to that extent are cool because they reflect a great deal of sunlight from their white or light yellow surfaces. Dark-colored soils like dark loams and humus, and dark-colored rocks such as basalt and other basic rocks are raised to a higher temperature under comparable conditions of water supply and insolation. Humboldt found that a black basalt sand on the island of Graciosa reached a temperature of 147° F. while white quartz sand in the same situation reached only 122° F. Dark soils cool more rapidly at night than light-colored soils, but do not become colder than the latter. Among all known substances charcoal absorbs and radiates the sun's heat rays most powerfully, so that its absorptive power is taken as 100. Gardeners and vine growers in the colder parts of Europe sometimes take advantage of the great absorptive power of carbon by spreading charcoal over the surface of the soil when early maturity is desired. The peasants of Chamouni hasten the melting of the snow by sprinkling slate powder over it.[1] A similar practice has been observed among the Ladakhis in the upper Indus valley, where the peasants dig up earth in the fall, store it in the stables and houses all winter, and in the spring hasten the late melting of the deep snow by scattering the stored earth over it.[2]

One of the most effective means of increasing the soil temperature is through warm percolating rains which displace the cold soil water below the root zone. Its effect may be understood by recalling that the specific heat of water is higher than that of any soil by four or five times and that if a pound of rain water at 60° F. carry ten heat units into the ground each heat unit raises the temperature of a pound of sand, not one degree as would be the case with water, but four or five degrees. Cold rains produce the opposite result, and one of the most important beneficial effects produced by proper drainage of a forest area is the

[1] E. W. Hilgard, Soils, 1907, p. 304.
[2] E. Huntington, The Pulse of Asia, 1907, pp. 50-51.

removal of cold water from the soil during the spring. If the region is one which is subject to summer droughts, such rapid removal of even cold water may be harmful, for the cold rains of spring would have less effect on the vegetation in delaying the beginnings of seasonal growth than would the lack of moisture in the dry season.

A covering of vegetation, either living or dead, diminishes the soil temperature below that of bare fields. During different parts of the day these differences may rise to 4° or 5° F. a few inches below the surface and to far greater values at the immediate surface. The monthly averages of two localities rarely exceed 1½° F. The differences are greatest in the summer season, and when the covering of vegetation is thick the effect is more marked than when it is thin, so that forests exert a cooling influence and on the whole tend to diminish direct evaporation from the soil.

TEMPERATURE VARIATIONS WITH DEPTH

The effect of either extremely high or extremely low temperatures is felt in a surprisingly shallow layer at the earth's surface. In temperate regions the daily temperature variations affect only the surface two or three feet of soil or rock and vary according to the nature and condition of the soil material.

The monthly variations reach to greater depths and the annual variations affect a layer from 35 to 75 feet in thickness. Below the zone of change the same temperature is found year after year, and though there are many exceptions to the rule, yet it is in general true that the deeper the point of observation the higher the temperature.[1] In the Arctic regions the level of no variation in temperature is but little below the surface in spite of surface variations between − 40° and − 60° on the one hand and 80° and 90° on the other. This is due to the presence of ice a short distance below the surface, and ice is a very poor conductor of heat. In the tropics the annual temperature variation affects the surface layer to a depth less than 2 feet because of the very slight seasonal changes in temperature.

The temperature of the soil during the cold season generally exceeds that of the air; and the absolute minimum temperature of the soil is always higher than the absolute minimum temperature of the air during the cold season. During the warm season the soil temperatures in general exceed the air temperatures. In summer sandy soils are warmest, loam soils next in order, and clay coldest; in winter these conditions are reversed. It is to the fact that soil temperatures are higher than air temperatures in cold situations and in all situations in cold seasons that render inhospitable localities possible to plants.

Dew is sometimes derived from the soil owing to differences of temperature between soil and air. This occurs when the soil is warmer than

---

[1] The mean temperature of the earth's soil is estimated by Tabert to be raised by conduction from the internal heat of the earth by the trifling amount of 0.225° F.

the air above it during a summer night. When the soil temperature falls to the proper figure dew is deposited within the soil to a depth at which the critical temperature is found. The daily repetition of this process at different depths exerts a considerable influence upon the distribution of moisture in the soil. It is probable that the formation of dew within the soil materially assists capillarity in distributing the soil moisture more uniformly.

Recently an investigation has been made of the rate of flow of heat through the soil under certain standard conditions of moisture content, specific volume, and effective specific heat.[1] The practical value of the work lies in indicating the nature of the soil control which should be exercised in order to secure a warm seed bed and good germination in the preparation of forest seedlings, the handling of cranberry marshes, etc. It has been found that heat will pass from a soil grain to soil water 150 times easier than from a soil grain to soil atmosphere, and this points to one reason why an air-dry soil shows such low heat conductivity.[2] Increase in heat conductivity due to the wetting of a soil is caused by a better contact between the soil grains. Coarse-grained soil has a lower heat conductivity than a fine-grained soil. A crumb structure in the soil causes the formation of air spaces, and the air acts as insulation against the passage of heat. When the soil crumbs are destroyed heat is conducted more rapidly and there is a more rapid rise of temperature.[3]

[1] H. E. Patten, Heat Transference in Soils, Bull. U. S. Bur. Soils No. 259, 1909, pp. 1-54.
[2] Idem, p. 49.
[3] Idem, p. 51.

CHAPTER V

CHEMICAL FEATURES OF SOILS

RELATIVE VALUE OF CHEMICAL QUALITIES

A SOIL can be considered fertile only when it possesses certain necessary physical qualities. If the physical condition excludes water, air, and plant roots, its stores of chemical substances remain locked within it, as useless as if they did not exist. It is also true that a soil poor in plant food and yet richly endowed with favorable physical properties may support an abundant vegetal growth. It is for these reasons chiefly that many investigators consider physical properties as paramount in soil fertility. The chief objection to any rigid claims for either side of the contention regarding the relative value of physical and chemical properties is founded on the fact that physical and chemical conditions are often evenly balanced, and when the balance is destroyed it is as often because of unfavorable chemical as of unfavorable physical conditions. An ecologist would see in the distribution of vegetation in the Coalinga district, Cal. (p. 44), or in the Steilacoom Plains, Wash. (P. 43), strong physical control of both plant distribution and plant species. On the other hand the strong chemical contrasts afforded by the various types of igneous and metamorphic rocks in many parts of the Laurentian area of Canada have very close counterparts in the vegetative contrasts,[1] for the glacial and postglacial soils of the Laurentian area are thin and the rock character has equal opportunity with the soil character for vegetative expression. The almost universal abundance of moisture in the eastern part of the United States permits many variations in plant distributions based on chemical differences in soil and rock; in the West the general scarcity of water causes it to be as a rule the dominant factor in distribution. Chemical differences in soil have probably resulted in some cases in the development of new species of plants. *Viola calaminaria*, for example, is thought to have arisen from *Viola lutea* by the action of zinc in the soil.[2] In the Alps there appears to be a wide difference between parallel species occupying mountains of limeless slate on the one hand and mountains of limestone on the other.[3]

[1] M. L. Fernald, The Soil Preferences of Certain Alpine and Subalpine Plants, Rhodora, vol. 9, 1907, pp. 149–193.

[2] A. F. W. Schimper, Plant Geography upon a Physiological Basis, Ox. ed., 1903.

[3] A. Kerner von Marilaun, Die Abhängigkeit der Pflanzengestalt von Klima und Boden, 1858. Quoted by Warming.

Each plant distribution is essentially the result of three groups of variable factors — physical factors, chemical factors, and biotic factors. All three groups of factors must be comprehended. Only exceptionally is a single factor a determining factor in plant growth or distribution. It is much more common to find these results controlled by combinations of factors, some physical, some chemical, some biologic. By physiologic adjustment and structural adaptation plants still further diversify their character, extend their distribution, and complicate the problem of tracing a given effect back to its fundamental causes.

SOIL MINERALS

All soils are based to a greater or less extent upon the existence and destruction of fragments of rock-making minerals. Practically all the common rock-forming minerals are to be found in any ordinary arable soil, but the relative amounts may vary widely from those in the rock from which the soil was formed and from each other. This conclusion is true even in such apparently homogeneous substances as brick clays.[1] Retgers found 23 different kinds of minerals in the dune sands of Holland.[2] Examinations by the U. S. Bureau of Soils show that some of the mineral species of the soil present clear and unaltered faces, although frequently, and especially with the feldspars, alteration products may be observed on the mineral fragments.

The composition of the chief minerals in the solid crust of the earth is shown in the following table.[3]

	Silica	Potash	Soda	Magnesia	Lime	Alumina	Ferrous Oxide	Ferric Oxide	Water
Quartz	100								
Feldspar { Orthoclase	64.2	17				18.4			
Feldspar { Albite	68.6		11.8			19.6			
Feldspar { Anorthite	43.1				20	36.9			
Mica	45 to 50	6 to 10	0 to 1.5			26 to 36			1 to 4.7
Hornblende { Augite	39 to 49			10 to 27	10 to 15	3 to 15	3 to 20		
Olivine	41			49.2			9.8		
Talc	63.5			31.7					4.8

The soil minerals are all soluble to a certain extent in water, although the rate of solution may be quite slow and the actual amount dissolved

[1] Cameron and Bell, Mineral Constituents of the Soil Solution, Bull. U. S. Bur. Soils No. 30, 1905, p. 9.

[2] L. V. Pirsson, Rocks and Rock Minerals, 1908, p. 280.

[3] A. D. Hall, The Soil, 1907, p. 16.

at any one time small. Many of the soil minerals are in a very finely divided or pulverulent condition and are therefore easily attacked by chemical agencies such as water, oxygen, and the various soil acids. Among the important chemical substances in the soil that may be classed as mineral are the zeolitic compounds (the hydrosilicates of alkaline earths and alkalies) easily decomposable by acids and capable of exchanging a part or the whole of their basic ingredients with solutions of other bases that enter the soil. The zeolitic compounds readily yield up a part of their ingredients to acid solvents and tend to fix a part or all of the soluble compounds that may be set free in the soil. The yielding up of their ingredients to acids is of great importance in that it enables plants to draw upon the reserve stores of food within the soil, the active solvent surrounding plant roots being water impregnated more or less with carbonic acid and possibly other acids.

Perhaps nowhere outside the arid regions are the mineral and rock particles of the soil in such a fresh condition as in glacial soils; and to the presence of large quantities of undecomposed material in such soils may be attributed their prolonged fertility, although their immediate fertility may be far below that of rocks in a stage of more complete decomposition.

ELEMENTS OF THE SOIL

Of the nearly 80 elements that have been identified in the earth's crust but 18 occur in important amounts. These arranged roughly in order of abundance are as follows:

AVERAGE COMPOSITION OF LITHOSPHERE [1]

Oxygen	47.07 %
Silicon	28.06
Aluminum	7.90
Iron	4.43
Calcium	3.44
Magnesium	2.40
Sodium	2.43
Potassium	2.45
Hydrogen	.22
Titanium	.40
Carbon	.20
Chlorine	.07
Phosphorus	.11
Sulphur	.11
Barium	.09
Manganese	.07
Strontium	.03
Fluorine	.02
All other elements	.50
	100.00 %

[1] F. W. Clarke, The Data of Geochemistry, Bull. U. S. Geol. Surv. No. 330, 1908, p. 32.

In the higher plants that have been investigated up to the present time the elements indispensable to normal development are invariably ten in number: oxygen, hydrogen, carbon, nitrogen, phosphorus, sulphur, iron, potassium, calcium, magnesium. If a single one of those substances is in a chemical form unavailable to the plant the plant enters into a pathologic condition or refuses to grow. Besides these ten substances all plants absorb various other substances whose utility is unknown.[1]

RELATIONS OF SOIL ELEMENTS TO PLANTS

While plants can not thrive in a soil that is without the substances noted above, the amount of each substance is found to vary according to the amounts of the others present; when a large percentage of lime is present, smaller percentages of potash, nitrogen, and phosphorus are required. But a soil entirely lacking in any one of these ingredients is an infertile soil. If the material of plants is burned the mineral residue contains potassium, calcium, magnesium, and a little iron among bases, and phosphorus, chlorine, sulphur, and silicon among non-metallic elements. Nearly all plants contain very small quantities of sodium and manganese; and radium, zinc, and copper have been found in traces in some plants.

Among these various elements carbon, hydrogen, and oxygen are drawn from the air or the water, while the other substances, equally indispensable to the plant, are derived chiefly by way of the roots from the soil. This fact makes it unnecessary to have a complete chemical analysis of the soil to understand certain aspects of its fertility, since the non-essential substances in the soil are simply the physical media in which plants grow; they therefore have no chemical importance. A chemical analysis of a soil is required to show the amount of nitrogen, phosphorus, potassium, and calcium in the soil besides other substances of less importance. The determination of the carbon compounds of the soil is of importance, and of the carbonates of calcium and magnesium which in most soils are active in neutralizing the acids that are harmful to bacteria.[2]

The degree of fertility of the soil, that is, the degree to which all the essential foods are present in available form, markedly affects plants in many ways, as, for example, the root habit of trees.[3] The more

---

[1] E. Warming, Œcology of Plants, Ox. ed., 1909, p. 55.

[2] A. D. Hall, The Soil, 1907, pp. 128–129.

[3] J. von Sachs, Über den Einfluss der chemischen und der physikalischen Beschaffenheit des Bodens auf die Transpiration. Landw. Ver.-Sta., vol. 1, 1859, p. 179.

concentrated the nutrient solution of the soil the shorter the roots; and the poorer the soil the longer and more feebly branched are the roots; roots branch very copiously and form dense clumps in rich soil. In case of changing fertility with changing strata the roots display very marked contrasts in the degree of ramification within the different strata. In all such cases an unfavorable water supply in the otherwise richer medium will prohibit an exceptional root development in it.

The same species will absorb various substances from different soils in different proportions. Individuals of the same species contain much silica if they grow in granite soil and much lime if they grow in calcareous soil. Certain plants have the power of making both quantitative and qualitative selections of soil substances, and even when a desired substance is distributed in the soil in very small quantities such plants can in time absorb it in surprisingly large amounts. It is this selective power of root action combined with the fact that the substances indispensable to plants occur in nearly all soils in quantities so considerable that almost every plant may extract more than the minimum amount, that results in the distribution of a given kind of plant upon soils of very widely different character. Furthermore each plant community and each plant species has its own peculiar economy, its own peculiar root system and general demands, which make it possible sometimes for many species to live side by side on the same soil without competing for food.[1]

CHARACTERISTICS AND FUNCTIONS OF THE PRINCIPAL SOIL ELEMENTS

OXYGEN

Oxygen is the most abundant of the elements, forming about one-half of all known terrestrial matter. It is a constituent of nearly all minerals in both soil and rock, and occurs in most rocks in amounts ranging between 45% and 50%. It is present in the soil as air, has an important effect as free oxygen in the oxidizing of various soil minerals and substances to form oxygen compounds, and plays a very prominent part in the many chemical changes which take place in both soil and vegetation. Without oxygen the nitrifying bacteria, which are of the utmost importance in maintaining the supply of soil nitrogen, can not live, earthworms are excluded, plants die, and some of the chemical changes important in maintaining the supply of plant food suffer a diminished rate of action.

[1] E. Warming, Œcology of Plants, Ox. ed., 1909, pp. 56–58.

SILICON

Silicon occurs in both soil and rock in the form of silica (SiO_2) or quartz. It is one of the most indestructible of natural compounds and is the prevailing constituent in nearly all sands and soils because of its great resistance, so that soils from whatever source derived will differ from each other mainly in the relative proportions of the siliceous and clayey constituents.[1] Silica requires even in its most soluble form ten thousand times its weight of water for solution. Its relative indestructibility causes it to accumulate to such an extent in practically all soils that it is a matter of no concern. While it is not an essential plant food it is important to plants as a medium in which their roots are disposed and anchored. The amount of silica (SiO_2) in the parent rock determines to a large extent the rate of weathering, the rate decreasing with increasing quantities of this substance. Soil is derived from quartz schist, for example, with extreme slowness because of the resistance of quartz to weathering, and since this rock is composed chiefly of quartz its soil supplies but little plant food. Such a soil would be almost absolutely barren but for the frequent occurrence in the parent rock of accessory minerals that on decomposition yield important plant foods. Sandstone and sandy soils are usually poor because the sand almost always consists chiefly of quartz grains and the finer portions alone are of importance in plant nutrition. The oxide of silicon is the principal constituent of quartz sand; with it are usually associated however particles of other substances so that even a beach sand may have all the necessary elements for the limited growth of plants.

ALUMINUM

Next to oxygen and silicon the most important element is aluminum. It is the most abundant of all the metals and occurs chiefly in combination with silicon and oxygen, forming an important series of minerals known as aluminous silicates, such as feldspars, micas, zeolites, etc. Aluminum is so easily oxidized that (with the exception of the fluorides) only oxidized compounds of aluminum occur in nature. As a silicate, aluminum occurs as the principal constituent of all clays, and while insoluble in this form, soluble potash, lime, etc., are usually associated with it. Chemically pure clay is very insoluble and of little importance for it is the end product of the chemical decomposition of the soil minerals enumerated above, but it is very important in its capacity to retain soluble salts. The physical action of clay in producing flocculation and tilth has already been described (p. 37) and is of the highest importance.

[1] Smith and Macauley, The Mineral Resources of Alabama, Geol. Surv. Ala., 1904, p. 74.

IRON

Iron gives a characteristic reddish or yellowish color to soils, occurs on the surface of the earth as an oxide and at greater depths or on fresh rock surfaces as a carbonate, sulphide, or silicate. Its principal forms are hematite, limonite, magnetite, pyrite, etc. It is essential to plants in the development of chlorophyll, without which there is improper nutrition. On account of its almost universal distribution in some form or other it is not a matter of concern in soil fertility.

CALCIUM

Calcium occurs in combination with carbon dioxide in great abundance in limestone; in the form of calcium carbonate it is slightly soluble in water containing carbonic acid and hence is an almost universal ingredient of all natural water. It is an important constituent of the principal silicates. It is one of the most essential substances in the soil because of its physical effects (p. 29) and because of its direct use in the formation of plant substances.

We have already noted the effect of lime, the carbonate of calcium, in promoting tilth, though it should be observed that an excess of 2% lime does not increase the tilth of the soil. Besides this it favors the important process of nitrification by prohibiting that acidity which excludes the nitrifying bacteria (p. 88). It seems to be an established fact that about 1% of lime is a high percentage of this ingredient in virgin soils.[1] In this connection it is important to see that when a large proportion of lime carbonate is present in the soil lower percentages of potash, phosphoric acid, and nitrogen are adequate. Among the other influences of lime in the soil are the rapid conversion of vegetable matter into black neutral humus and an increase in the nitrogen supply, an acceleration of the oxidation of carbon and hydrogen, a counteracting of the injurious effects of an excess of magnesia and of soluble salts in alkali lands, and a liberating effect upon the potash held in zeolitic compounds. An excess of lime, from 8% or more, disturbs nutrition, suppresses or diminishes the formation of chlorophyll and starch, and is in general deleterious to plant growth.[2]

The longleaf and shortleaf pine regions of the United States are poor in lime and have long remained almost uncultivated; the excess of lime in many of the chalk lands of Europe causes them to be equally infertile. The maritime pine and chestnut tree are both antagonistic to lime and any considerable amount of it will cause them to die or to

[1] E. W. Hilgard, Soils, 1907, p. 346.
[2] For a discussion of lime effects see E. W. Hilgard, Soils, 1907, pp. 378–381 et al.

deteriorate, in contrast to the Corsican pine, which is a lime-loving tree.

The .higher the clay percentage of a soil the more lime carbonate it must contain in order to exhibit the advantages of a calcareous soil. In sandy lands a characteristic lime growth may reflect the presence of only 0.10% of lime, while in heavy clay soils 0.6% is required to produce the same result. This explains the prominent color of dark-tinted humus in sand soils when very small amounts such as 0.2% of lime occur, whereas a comparable effect is produced in clay soils only when the percentage rises to 1%. A soil that effervesces with acids contains at least 5% of carbonate of lime, and percentages so small as to make the soil distinctly calcareous are not distinguishable by this mode of analysis.

In the study of the effect of lime upon vegetation a difference of opinion has arisen, probably due to the very different methods pursued and the different regions in which the students have worked. Hilgard, who has done the most extensive work of this character in America, concludes that the moisture of the soil is the point of first importance in the distribution and welfare of vegetation but that the condition next in importance is the amount of lime present. He grants, of course, that certain species are indifferent to lime, but holds that most species respond to lime in a marked manner and that on the whole the presence of lime tends to greater fertility except in the obvious case where it is present in excess. He finds that in Mississippi and Alabama there is a marked correspondence between the growth habit and types of trees and the geologic formations, so that a geologic map of the region would also be to a large extent a map of the various tree zones. The conclusions of Hilgard are of great value because they were formulated as early as 1860 after extensive study of native vegetation which grew in a soil that was almost undisturbed by man, hence a vegetation that represented a long term of adjustment to the soil. This is obviously an advantage over the conditions in Europe, where the observers of the vegetation have quite constantly to eliminate the influence of cultivation and other disturbing influences due to the long occupation of the land by man. It would not be fair, however, to dispose of the matter by merely stating the ground of Hilgard's conclusions. The contentions of the European students are cogent and interesting and deserve equal attention here.

Warming says:

"Although the characteristics of the lime-flora are clear and distinct, yet in the past the influence of lime upon vegetation has been overestimated. Indeed, a distinction has been made between calciphilous and calciphobous plants. Recently it has been definitely established that the amount of lime in itself, in so far as it does not operate physically, can not be the cause of differences in the flora, for not only can calcicolous plants be cultivated in soil that is poor in lime, but silicicolous plants, and even bog-mosses, which are regarded as preëminently calciphobous, can grow vigorously in pure limewater if the aqueous solution be otherwise poor in dissolved salts. It has been overlooked that nearly all lime soils are rich in soluble mineral substances, and this wealth excludes plants belonging to poorer soils; beyond this the important physical characters of calcareous soil, compared with granite soil, come into play."[1]

In general the disagreement of the conclusions as to the power of lime to control plant distribution appears to be due to the absence of a standard conception as to what constitutes a lime soil. By some a soil is regarded limey only when it effervesces with acids, yet not less than 4% of calcium carbonate in a soil will respond to the acid test, whereas 0.1% will have important effects on plants provided the soil is sandy. So far, most of the conclusions have been mere *obiter dicta*. The conclusions of American investigators, based on native and practically undisturbed vegetation, are essentially sound, though they require important modifications where the physical conditions become extreme.

MAGNESIUM

Magnesium forms an essential part of the rock known as dolomite and may exist as a silicate in such rocks as serpentine, talc, etc. In igneous rocks it occurs in the minerals pyroxene, mica, olivine, etc. Magnesia is invariably and rather abundantly found in the seeds of plants and is a very important plant-food ingredient, but must not occur in excess or it will cause, through chemical action, a pronounced change in the capacity for imbibition, and thus particularly disturb the functions of the plant. Magnesia is especially concerned in the transfer of phosphoric acid through the plant tissues; while magnesia predominates in the fruit of a number of crop plants lime predominates in the leaves, so that there is apparently a connection between the extension of leaf surface and the lime requirement.

SODIUM

Sodium occurs in largest percentage in the igneous rocks as a constituent of the soda-lime feldspars, amounting on the average to $2\frac{1}{2}\%$ of the igneous rocks. These feldspars are more readily attacked by water and carbon dioxide than are the other common minerals save certain basic silicates, so that the whitening and the softening of feldspar is one of the first signs of rock decay. The sodium salts are so soluble, however, that they are leached away almost as rapidly as formed, with the result that soils are normally rich in potash but poor in soda. In poorly drained soils of arid lands, however, these qualities result in

[1] E. Warming, Œcology of Plants, Ox. ed., 1909, p. 58.

a concentration of soda through the continued evaporation of ground water, where as carbonate, sulphate, and chloride it gives rise to alkali tracts poisonous to all but specially adapted species of plants. In northern Chile, one of the most arid tracts of the world, the sodium has become concentrated in nitrate deposits. These on account of the fixed nitrogen which they contain are extensively mined·and shipped to many agricultural regions in humid lands as a fertilizer. Thus through its chemical properties — the extreme basicity of the element and the solubility of its salts — sodium causes the most arid deserts to add to the fertility of the garden spots of the world. Sedimentary rocks, the accumulations of the constituents of soils of former ages, are usually deficient in sodium and may be almost free from that element. Their soils, the result of a second cycle of leaching, tend to be still more barren in sodium. The use by land plants of potassium is doubtless an adjustment to the prevailing composition of soils; marine plants, living in an environment where sodium is dominant, show a parallel use of sodium in their tissues though they use potassium also, to some extent.

POTASSIUM

One of the most important elements of the earth's surface from the standpoint of plant growth is potassium, which in the form of a nitrate is found in nature (saltpeter). It occurs in small and large quantities in a great variety of igneous and metamorphic rocks, but may be absent from sedimentary rocks. It is present in mica, amphibole, and pyroxene, and when combined with silica is an important member of orthoclase and other minerals. Granite soils generally contain a good supply of potash on account of the common occurrence of potash feldspar in them. Granite soils may be deficient in lime, however, unless hornblende is present, since lime-feldspar is not likely to occur as an accessory ingredient of granite.

The amount of potash necessary for high soil productivity is about 0.5% and at this figure the addition of potash has but little effect upon the fertility. At 0.25% there is a deficiency that must be made up by fertilization. These figures do not apply, however, in arid or tropic lands. In tropic lands the prevailingly high temperature, the great rainfall, and the continuous leaching, cause a very rapid liberation of potash from its insoluble form as well as its rapid removal; smaller amounts are therefore necessary at any given moment. In arid regions the absence of rapid leaching allows the accumulation of earthy salts of many kinds, among which potash is prominent.[1]

[1] E. W. Hilgard, Soils, 1907, pp. 354-355.

PHOSPHORUS

Of very high importance in soil fertility is phosphorus, found in the minerals vivianite and apatite, in the bones of animals, and in the seeds of plants. Apatite (phosphate of lime) is an almost universal constituent of granitic rocks, but occurs in very small quantities.[1] The amount of phosphoric acid (P_2O_5) contained in granitic rocks rarely exceeds 0.2% and may fall as low as 0.05%; but small as the amount is it probably is the main source of supply of phosphates existing in the soil. Phosphorus is most abundant in the basic eruptive rocks such as diorites and gabbros and deficient in such rocks as sandstone and slate.

Where the minerals vivianite and apatite are abundant in the country rock, as in the basaltic lavas of Hawaii, phosphoric acid may be present in the soil in exceptional amounts, — nearly 2%. Unfortunately in this particular case it occurs in the form of an insoluble basic iron compound, ferric phosphate, which is dissolved with such difficulty that it is wholly unavailable to vegetation and the soil containing it is actually phosphate poor. The same is probably true of certain ferruginous soils in California and the South. The lower limit of adequacy of phosphoric acid in the soil is about .05%. Exceptionally soils may contain as much as .30%, while .15% is regarded as adequate. In non-ferruginous lands the amount required is smaller than in the case of ferruginous lands because the iron renders phosphoric acid inert by forming ferric phosphate, an insoluble substance.[2]

Phosphate deposits are derived chiefly from animal remains, but animals derive it from plants, which in turn depend for their supply upon the alteration products of apatite. Commercially important deposits of apatite occur in Spain, Canada, and Norway. From the Norwegian deposits a commercial fertilizer is now manufactured, a phosphoric-chalk manure. Phosphorus is one of the rarest essential plant foods, and its conservation should be a matter of great concern. Sewage contains relatively high percentages of it, and the application of sewage to the land instead of its wastage in rivers and the sea is one of the most important though as yet limited uses of this neglected fertilizer.

SULPHUR

Sulphur plays an important part in the nourishment of plants, since it is an essential constituent of vegetable albumen and allied compounds,

[1] The pure crystalline mineral apatite rarely occurs in large masses. Minute crystals of it are found widely scattered in many rocks, granite, basalts, etc. The largest deposits occur in connection with carbonate of lime in rocks known as phosphorites which closely resemble limestone. Extensive phosphate deposits are found in southern California, Florida, Alabama, Tennessee, Kentucky, Wyoming, Utah, Idaho, Montana, etc.

[2] E. W. Hilgard, Soils, 1907, pp. 393 et al.

hence a soil to be fertile should always contain sulphates in available form. The usual form of occurrence is in combination with the metals to form sulphides or with oxygen and a metal to form sulphates. It is an essential part of the mineral pyrite, and when combined with oxygen and calcium forms the valuable fertilizer gypsum. The amount of available sulphur existing in any soil is usually very small. The relative available amount in ordinary soils is indicated in the following table, in which it is assumed (1) that the mean dry weight of a surface foot of soil is 80 pounds, and (2) that the amount of soluble material is accurately represented by the results of a large number of analyses collected in the Tenth Census Reports.

RELATIVE AMOUNTS OF CERTAIN ESSENTIAL PLANT FOODS IN AN ACRE-FOOT [1]

Potash in surface foot, per acre.................................... 3.76 tons
Soda in surface foot, per acre..................................... 1.58 "
Magnesia in surface foot, per acre .... 3.92 "
Lime in surface foot, per acre .... 1.88 "
Phosphoric acid in surface foot, per acre.......................... 1.97 "
Sulphuric acid in surface foot, per acre........................... .91 "
Soluble silica in surface foot, per acre .... 73.40 "

In the decomposition of the various compounds of iron and sulphur, oxidation affects either or both the iron and the sulphur; when the iron alone is oxidized the sulphur or some part of it separates as hydrosulphuric, sulphurous, or sulphuric acid. The sulphur of the hydrosulphuric acid may be later oxidized to sulphurous or sulphuric acid through the action of water and oxygen with or without the assistance of bacteria, though bacteria are often the inciting cause of the oxidation. In the form of sulphuric acid sulphur is immediately available to plants, indeed sulphuric acid itself is found in the tissues of plants in small quantities. It is subject to steady depletion in this form by percolating water and by combination with bases in the formation of sulphates, in addition to the demands upon it by growing vegetation.[2]

TOTAL PLANT FOOD; AVAILABLE PLANT FOOD

The data concerning plant food are to be distinguished from those derived by mere chemical analyses of soils, which in themselves are of small value in understanding ecologic conditions. Almost all soils show on ultimate chemical analysis an abundance of the elements required for almost any given crop, but there is the widest difference between the forms in which the elements occur, so that it is the available plant food, and not the ultimate amount of plant food that may be produced on complete decomposition, that is a matter of chief interest

[1] F. H. King, The Soil, 1905, p. 102.
[2] C. R. Van Hise, A Treatise on Metamorphism, Mon. U. S. Geol. Surv., vol. 47, 1904, p. 468.

in the study of the chemistry of soils. The soil must be regarded as possessing most of its plant food in such a state of combination that it can not be utilized by the plant directly, but must by weathering slowly pass into the soluble, i.e., the available, form.

DETERMINATION OF SOIL FERTILITY

The approximate chemical nature of an ordinary soil may be ascertained in a direct manner by the determination of both the decomposed and undecomposed minerals present in it. The determination of its fundamental nature requires an examination of the undecomposed minerals only, since it is presumed that the decomposed part of the soil has been derived from the constituent minerals and since the undecomposed material forms by far the greater bulk of the soil. But such an analysis is of less value than direct qualitative and quantitative chemical analyses of the soil character. Even the latter analysis does not always furnish a reliable guide as to the productivity of the soil. The previous history of the land and the physical characters of the soil may be predominating factors.

In attempting to ascertain the nature and amount of the decomposed portion of the soil various working plans, which attempt to imitate plant action, have been tried. The water-soluble ingredients of the soil are only a portion of the total number of substances upon which plants may draw for food, because the plant roots act not alone through the medium of water but also through water charged with carbonic and possibly other acids. Clearly, then, the action of plant roots may be imitated more closely by employing in the analysis a weak acid solvent that will act upon the soil in a manner similar to the soil acids. The weak acid solvent employed is empirically determined, for no one has yet analyzed the soil about a growing plant in such a way as to ascertain under precisely what conditions the various soil acids act. The results of soil analyses by means of weak acid solvents must be compared with cultural experience and observations on natural plant growth; the results are thus empirical approximations, but they are the best that have been achieved. It has been found by such observations that all soils are continuously soluble to some extent but that the differences between the solutions derived from soils of low and of high productivity are very striking.

Plants differ very greatly in the energy and quality of their action upon reserve soil ingredients, so that no single solvent used in an analysis could properly represent the action of plant roots in general. Among the many solvents employed for the purposes of soil extraction and the determination of immediate soil productivity are citric acid (a 1% solution is most commonly employed) and aspartic acid, among the weak acids; and hydrochloric, nitric, and a few others, among strong acids. The hydrochloric acid is employed in densities ranging from 1.1 to 1.16, while a density of 1.115 has been found most convenient and satisfactory of all. From experience with acids of different strength it has been found that a five-day period of digestion with hydrochloric acid (density, 1.115) is sufficiently effective in showing what plant-food ingredients of the soil maintain its permanent productive capacity. This appears to be the natural limit of the action of the acid upon the soil and produces a maximum effect. There is much to justify the contention that the only legitimate solvent in determinations of soil fertility is carbonic acid, the commonest, the natural, and the most abundant acid in the soil.[1] The immediate soil requirements may also be empirically determined with a fair degree of accuracy by analyzing the ash of the vegetable growth and establishing a ratio between the normal ash ingredient and the actual soil ingredients. It has been found that in the case of a deficiency of certain kinds of plant food there is a disturbance of nutrition that manifests itself in the form or in the development of the plant and affords a direct basis for future determinations of a similar sort without the repetition of the full chemical analysis of the ash.

Unless extreme physical characters interfere with normal plant growth virgin soils showing high percentages of plant food as determined by extraction with acids[2] (hydrochloric, nitric, etc.) invariably prove highly productive. Hilgard states the law as follows: "The actual amounts of soil ingredients . . . rendered accessible to plants are . . . more or less proportional to the totals of acid-soluble plant-food ingredients present."[3] The natural condition which this result indicates may be stated thus: the larger the total amount of plant food in the soil and ultimately available the greater will be the immediate effect of the weathering agencies that produce available plant food. This conclusion appears to be rather well established and indicates how high a degree of importance should be attached to the chemical composition of soils despite the relatively higher importance of physical qualities in general. Neither group of qualities can be adequately applied to a soil type to the exclusion of the other group. All the field conditions must be evaluated before a soil analysis can be regarded as complete. It is in the highest degree unscientific longer to advocate the supreme importance under all circumstances of a single group of soil qualities.

[1] A. D. Hall, The Fertility of the Soil, Science, n. s., vol. 32, 1910, p. 366.
[2] E. W. Hilgard, Soils, 1907, pp. 343–353, etc.
[3] Idem, p. 346.

Harmful Organic Constituents of the Soil

In concluding this discussion of some of the chemical properties of soils it seems desirable to note an important recent result in one of the most complicated branches of soil chemistry, the identification of harmful organic substances that occur in the soil, the determination of their properties and the formulation of remedies to offset their effects. It has been found that many plants can possibly excrete, as the result of growth, organic compounds which are poisonous to the plants producing them. It has been found that these organic substances while inhibitory to the plants which produce them have little or no effect upon other plants. It is therefore concluded that they may be positive forces in producing a natural succession or rotation of wild vegetation not explained by any change of soil or climate.[1]

These harmful substances are constantly being added to the soil, and some of them are injurious in quite small amounts and are a cause of infertility even when the amount of plant food in the soil is abundant.[2]

The most markedly harmful bodies are found within the plant but are not apparently an essential part of the life of the plant. Of this class are arbutin, vanillin, heliotropine, terpenes, etc., all of which are very injurious. Tyrosine is injurious in quite small amounts, while some of the protein decomposition products are not only harmless but appear to act as plant nutrients; such for instance are asparagine and leucine.[3] The harmful substances have a toxic effect on the plants, but are destroyed or rendered harmless by other substances like soda and lime in coöperation with plant roots. Oxidation in a high degree converts some of these bodies into harmless substances, but a low rate of oxidation causes the organic matter to decompose incompletely and gives rise to organic compounds unfavorable to plant growth.[4]

Hall suggests that these so-called toxic substances may be normal products of bacterial action upon organic residues in the soil and that as such they may be as abundant in fertile soils rich in organic matter as in the sterile soils from which they were extracted. He points to the great power of the soil in precipitating soluble materials within it as a possible natural remedy for such toxic substances.[5]

It has also been found that tree roots have a toxic effect on the growth of wheat. The absence of grasses about certain trees is thus attributed not to depletion of plant food alone but to the toxic effects of the tree roots heightened by shade and the well-known injurious effects of washings from trunk and leaves.[6]

[1] Schreiner and Shorey, The Isolation of Harmful Organic Substances from Soils, Bull. U. S. Bur. Soils No. 53, 1909, pp. 1–3.

[2] Idem, p. 29.

[3] Idem, pp. 12–13.

[4] Schreiner and Reed, Certain Organic Constituents of Soils in Relation to Soil Fertility, Bull. U. S. Bur. Soils No. 47, 1907, p. 13.

[5] A. D. Hall, the Fertility of the Soil, Science, n. s., vol. 32, 1910, p. 367.

[6] C. A. Jensen, Some Mutual Effects of Tree Roots and Grasses on Soils, Science, n. s., vol. 25, 1907, pp. 871–874.

CHAPTER VI

HUMUS AND THE NITROGEN SUPPLY OF SOILS

Sources and Plant Relations

THE nitrogen of the soil is a matter of paramount importance in forestry, especially in the maintenance of a proper forest growth and in efforts at reforestation, for as we have seen a supply of nitrogen in some form or other is absolutely indispensable to plants.[1] We shall therefore discuss it somewhat more fully than the other plant foods of the soil. To be available to plants nitrogen must be in soluble form, and it is therefore as a nitrate that it is used by the plant. The main source of nitrogen is humus, whence it is derived chiefly by bacterial action; although nitrogen exists in unhumified organic matter it is not in an available form. Other sources of nitrogen are (a) nitrogen-fixing bacteria that live in a free state in the soil and derive their nitrogen from the soil air, (b) nitrogen-fixing bacteria that live in symbiotic association with legumes and other plants, and (c) rain water. The amount of nitrogen contributed to the land in the last-named manner amounts to a half pound or a pound or more per acre with a rainfall of about 30 inches per annum.[2]

NITROGEN BROUGHT TO THE SURFACE OF THE EARTH BY RAIN

(Pounds per acre per annum.)

Locality	Nitrogen			Remarks
	Ammoni-acal	Nitric	Total	
Rothamsted, England................	2.823	0.917	3.74	In 1888–89
Barbados...........................	1.009	2.443	3.452	5 years' average
British Guiana......................	1.351	2.190	3.541	7 years' average
Kansas.............................	2.63	1.06	3.69	3 years' average
Utah...............................	5.06	.356	5.42	Do.

Nitrogen is usually the first element to become exhausted in the soil because the nitrates are exceedingly soluble and no part of the soil has any special power of holding back nitric acid when it passes in aqueous solution through its pores, so that the nitric acid produced in the soil

[1] A. D. Hall, The Fertility of the Soil, Science, n. s., vol. 32, 1910, p. 368.
[2] H. W. Wiley, Prin. and Prac. of Agri. Anal.: Soils, vol. 2, 1906, p. 448.

passes at once into the vegetation, or remains in store in dry periods, or passes into the drainage water and is lost.[1]

It is important that the forester retain the humus of the forest soil and increase it or make the amounts already there more useful, since the nitrogen which it yields is one of the rarest of the essential plant foods in the soil. The maintenance of humus in the soil requires a forester's constant attention to renewal of growth after cutting, proper drainage, a shaded surface, etc.[2] Different species of plants demand very different amounts of humus: some plants appear to require none at all, as those that develop on bare rock; some require a moderate amount, as is the case with certain grains; and others, notably the moorland plants, thrive only in rich humus, and have special methods of nutrition dependent upon the kind of soil in which they live.[3]

The great value of the birch and the aspen lies in their power of rapid dispersion and quick germination and growth in sterile soil or soil robbed of humus by repeated fires. They thus prevent excessive erosion, which is usually so destructive of the soil after fire has destroyed either or both the forest cover and the soil humus. In this manner and by their rapid growth they often afford an opportunity for the seedlings of longer-lived and more valuable trees to come in under conditions that insure their successful growth.[4]

As an illustration of the importance of humus in maintaining an original growth of vegetation may be cited the fact that a change in the forest vegetation is known to have occurred in Denmark during past millennia, owing in part to the action of the wind, which by blowing leaves out of the forest has reduced the amount of humus.

"When a forest soil is exposed to desiccation and the fallen leaves are carried away by wind, the earthworms vanish, the soil becomes dry and hard, and the vegetation suffers. In acid soil (bog, heath) and dune, earthworms are wanting. Upon their presence or absence depends the occurrence of a humus soil or a raw humus soil in north temperate forest and heath. Conversely they disappear upon the production of raw humus and humous acids. Even upon the growth of rhizomatous plants in the forest do they exert an action; their presence or absence causes a series of variations in the kinds of soil that corresponds to a series of variations in the plants clothing it."[5]

It is noted also that a covering of leaves in a beech forest has a great influence upon ground vegetation in that it supports mosses and other plants, produces humus, and provides food for animals living in the

[1] W. J. Spillman, Renovation of Worn-out Soils, Farmers' Bull. U. S. Dept. Agri. No. 245, 1906, p. 5.
[2] E. Ebermayer, Lehre der Waldstreu, etc., 1876, p. 206.
[3] E. Warming, Œcology of Plants, Ox. ed., 1909, pp. 62–64.
[4] C. S. Sargent, Manual of the Trees of North America, 1905, pp. 155–201.
[5] E. Warming, Œcology of Plants, Ox. ed. 1909, p. 78.

forest soil, among which earthworms are considered to be the most important.[1]

One of the most important qualities of humus that affects its value to the soil is its natural porosity, which renders it very absorptive of gases, especially aqueous vapor. Dry humus swells up when wetted and the volume of weight increases in the ratio of 2 or 8 to 1; in fact humus stands first in this respect among the soil ingredients.

Although in general the presence of the organic matter of plants increases the power of soils to hold moisture, some kinds of organic matter are known to cause a low water-holding power, as in certain California soils, a condition due to the peculiar and special qualities of the organic matter, which when extracted is found to have the properties of a varnish, repelling water to an extreme degree.[2]

The density of natural humus as compared with ordinary soil is about 1:4. It is the lightest soil material and greatly promotes tilth, aeration, water supply, etc. Besides nitrogen, humus contains mineral plant food ingredients which are capable of nourishing plant growth. These mineral ingredients are probably made available to plants largely through the direct and indirect action of the humus compounds; for it has been shown that the richer the soil is in humified organic matter the more rapidly the mineral matter of the soil is made available for plant nutrition. With an increase of soil humus there is a corresponding increase in the amount of mineral plant food extracted from the soil by a 4% solution of ammonia such as is employed in the Grandeau method of humus determination (p. 82).[3]

ORGANIC MATTER IN THE SOIL

The difference between soil and a mere mass of sand or disintegrated rock is that soil contains some organic matter. Soil becomes arable and furnishes a medium suitable for the growth of higher plants when a certain amount of organic material has been accumulated in it from the growth and decay of lower plants. Animal remains, such as insects and worms, also have a prominent place as a source of soil organic matter. The accumulation of the remains of micro-organisms and of the vegetation which they modify is an important factor in the transformation of land waste into soil. The final product of these and other processes is a mixture of stuff fully as complex as the processes to which it owes its origin.

[1] E. Warming, Œcology of Plants, Ox. ed. 1909, p. 74. For a very clear and comprehensive discussion of the chemical modification of forest litter and its value to the soil, see E. Ebermayer, Die gesammte Lehre der Waldstreu mit Rücksicht auf die chemische Statik des Waldbaues, 1876.

[2] Schreiner and Shorey, Chemical Nature of Soil Organic Matter, Bull. U. S. Bur. Soils No. 74, 1910, pp. 8–9.

[3] E. F. Ladd, Bull. So. Dakota Agri. Exp. Station, Nos. 24–32, 35, 47. Quoted by Hilgard, pp. 140–141.

The organic matter of both plants and animals is made up of protein, fats, and carbohydrates principally, but besides these substances there is a host of other compounds such as resins, hydrocarbons, and derivatives, that is, alkaloids, acids, etc. Furthermore, all living matter has as essential components some organic compounds that contain nitrogen, compounds which for the most part are those related to protein. When the complex molecules of proteins, fats, and carbohydrates break down into simpler bodies they pass through the same changes whether these changes are brought about through the agency of micro-organisms as in decay or through the agency of acids; and they are subject to still further decomposition through the same or other agencies. The secondary products are very numerous and of widely varying composition and structure, as well as chemical and physical composition and properties, so that the final products of decay are very different under different conditions.

When organic matter is contributed to the soil there is a continuous "building-down" process from the original complex molecule to simpler ones and these again to still simpler molecules, until in some instances the substances are reduced to their most elementary constituents.[1] All these chemical changes are affected by the temperature of the soil, the amount of air it contains and the amount of water, etc. Dry leaves, wood, and litter generally do not change in dry air or change very slowly; but if the ground is moist the process goes on rapidly.[2] In the absence of air bacterial transformations cease and the bacteria die. Extremely high or low temperatures likewise limit their activities (p. 86).

SOIL HUMUS

AMOUNT AND DERIVATION

The chief source of soil nitrogen, as we have already noted, is the organic matter in the soil, and this occurs to a large extent in the form of "humus." The total amount of organic matter in ordinary soils is 2.06% for the soil and 0.83% for the subsoil, figures based on the analysis of thousands of samples from many different portions of the United States.[3] There is an average of 28 tons of organic matter in the soil per acre taking it to a depth of 8 inches, and 50 tons of soil and subsoil taking it to a depth

[1] Schreiner and Shorey, The Isolation of Harmful Organic Substances from Soils, Bull. U. S. Bur. of Soils No. 53, 1909, pp. 9–11.

[2] E. Ebermayer, Die gesammte Lehre der Waldstreu mit Rücksicht auf die chemische Statik des Waldbaues, 1876, p. 202.

[3] Schreiner and Reed, Certain Organic Constituents of Soils in Relation to Soil Fertility, Bull. U. S. Bur. Soils No. 47, 1907, p. 10.

of 2 feet.[1] The amount of nitrogen is generally not far from 0.1%. In arid soils the average amount of humus is much lower, rarely exceeding 1% and frequently falling to or below 0.30%; but the nitrogen content of the humus of arid soils is very much higher than in the case of humid soils. Woodlands and old meadows as a rule show a high humus content in their surface soils. The humus content of peat and marshland is also high.

While figures for the amount of humus which may be derived from a given quantity of vegetable matter must vary greatly according to the conditions under which humification takes place, it may be said that in the humid regions roughly one part of normal soil humus may be formed from five to six parts of dry plant débris. This ratio is based upon the assumption that the average nitrogen content of plant débris is 1%. The ratio will vary according to the temperature, the degree of access of air and moisture, etc. In hot arid regions all vegetation may wholly disappear by oxidation at the surface of the ground, and the proportion of humus derived from the decaying vegetation of arid regions is in general very much smaller than that of humid regions where it is rather rapidly incorporated in the surface soil.[2]

The absolute amount of humus decreases rather regularly downward except in the case of depths that represent the maximum root development, at which level there is always a slight and often a notable increase in the humus content. Below this level decrease again takes place. The nitrogen percentage in the humus (which is to be distinguished from the total nitrogen content of the soil) exhibits a general decrease in the same direction, probably due to decrease in the amount of oxygen and a diminished rate of oxidation with increase in depth.

The humus content of soils has a very close correspondence in some cases with the root development and is large in the surface layer and small in the subsurface layer. In other cases, while the decay of organic material takes place chiefly at the surface, active animal agencies may carry the organic remnants downward into the soil and effect a somewhat uniform distribution. If both root penetration and animal agencies are restricted by the compactness of the subsoil only a light surface layer of mold will be formed and what little humus is found in the lower soil layers will be derived entirely from the decay of a very limited number of roots. Cultivation or timber cutting when followed by excessive erosion prevents the accumulation of vegetable matter, from which

[1] Schreiner and Shorey, The Isolation of Harmful Organic Substances from Soils, Bull. U. S. Bur. Soils No. 53, 1909, p. 26.
[2] E. W. Hilgard, Soils, 1907, p. 128.

humus is chiefly derived, and allows the too rapid aeration and destruction of the humus.

In general it seems necessary to keep the nitrogen percentage of soil humus at about 4% to insure satisfactory results, and for the growth of grasses a nitrogen percentage in the humus of 1.7% is quite inadequate, no matter how much humus may be present. Different plants will accept this minimum, as might be expected from the differences of root habit, water supply, lime percentage, etc., which have an influence upon the rate of nitrification and of the leaching of nitrogen from the soil.

If a moderate amount of moisture is present in the soil and there is in consequence a relatively free circulation of the air, and if earthy carbonates are present, especially lime, so as to neutralize the acids of the soil as fast as formed, fungoid and bacterial growths effect the steady humification of organic matter. Oxygen and hydrogen are eliminated in the form of water and carbon dioxide, and there is an increase in the percentage of carbon and generally of nitrogen. When once humification is completed oxidation affects mainly the carbon and the hydrogen, so that the nitrogen content may for a time rise steadily and reach very high figures. Under unfavorable conditions the conversion of organic material to soluble nitrogen may be so slow that a soil containing as much as 40% of unhumified organic matter may contribute during a single year but a small quantity of available nitrate.

HUMUS DEFINED

Originally the term *humus* had no special significance and was only a name for dark-colored vegetable mold; later it came to be applied to this mold material when incorporated to a greater or lesser degree in soils. The term now has a more restricted meaning, at least among soil chemists, and is applied exclusively to the dark-colored organic matter extracted from soils by dilute solutions (usually 5% solutions) of sodium or ammonium hydrate.[1] The method of determination is purely empirical and ascertains the existence of only a portion of the organic matter of the soil.

Among agricultural folk and in general among foresters the term *humus* still retains its early meaning — partly decomposed organic material, dark in color, light in weight, and mixed with more or less mineral matter.

The amount of humus in the soil is sometimes determined by means of dry or wet combustion in which process the humus is calculated from the carbon dioxide formed and the nitrogen gas is measured directly. Obviously this measurement includes the entire organic matter of

[1] Methods of Analysis, Humus in Soil, Bull. U. S. Bur. Chem. No. 107, 1907, p. 19; see also C. A. Davis, Peat, 8th Ann. Rept. Mich. Geol. Surv., 1906.

the soil whether humified or unhumified. To obtain the amount of actual functional humus in the soil (only humified organic matter can be directly nitrified) a solvent must be employed and discrimination between humified and unhumified material made. The accepted process is the method of Grandeau. By this method the soil is first extracted with dilute acid in order to set the humic substances free from their combinations of lime and magnesia and a subsequent extraction is made with moderately dilute solutions of ammonia. After the evaporation of the ammonia solution the humus is left behind in the form of a black substance somewhat resembling soot. This is then weighed and afterwards burned and the ash weighed. The amount of functional humus is considered to be the difference between these two weights. The nitrogen content of the humus may be determined directly by substituting in this process potash or soda lye for ammonia water and determining the nitrogen by the Kjeldahl method in the filtrate.[1]

The chemical composition of humus is indefinite because it itself is a variable mixture of substances of complex composition. It always contains more carbon and nitrogen and less oxygen and hydrogen than the substances from which it was formed. The following table shows the chemical composition of grass and of the top brown layer of turf in a peat bog, also the composition of peat of greater age, at 7 feet and 14 feet respectively.

CHEMICAL COMPOSITION OF VARIOUS TYPES OF ORGANIC MATTER.[2]

	Grass	Top Turf	Peat at 7'	Peat at 14'
Carbon........	50.3	57.8	62	64
Hydrogen......	5.5	5.4	5.2	5
Oxygen........	42.3	36	30.7	26.8
Nitrogen......	1.8	0.8	2.1	4.1

It should be remembered in the inspection of this table as well as in the general consideration of humus that vegetable matter consists, in addition to carbohydrates, of other carbon compounds containing nitrogen, and in some cases both nitrogen and phosphorus, which all break down under bacterial and acid action into dark-colored substances called humus. In the process of humus formation the nitrogen-containing bodies resist the action of bacteria longer than the carbohydrates, so that the later the stage of decay the greater the proportion of nitrogen the humus will carry. It follows that during the gradual depletion of the humus higher and higher percentages of nitrogen will be developed in that part not removed.[3]

Humin and *humic acid* are terms used in designating vegetable acids resulting from the decay of vegetable matter in the formation of peat, etc. The descriptions and formulæ for humic acid and related bodies

[1] For a description of the Kjeldahl method see Wiley, Soils, 1906, p. 491 et seq.

[2] A. D. Hall, The Soil, 1907, p. 42.

[3] Idem, pp. 41–44.

are about as numerous as the number of investigators.[1] Certain elaborate experiments have resulted in the attempt to show that humic acid consists for the most part of an insoluble body of a protein nature, but this proves only that the humic acid examined by this investigator was either of a protein nature, a mixture of protein decomposition products, or probably both, together with some unknown body.[2]

Humic acid, ulmic acid, ulmin, etc., are commonly used as if they were definite bodies of well-established composition, but this is not the case, as their very existence has never been satisfactorily demonstrated. No attempt should be made to ascribe formulæ to acids of so complex and variable a nature. As commonly written the formulæ for the group are as follows:[3]

ULMIN AND ULMIC ACID

Carbon	67.1%	
Hydrogen	4.2	Corresponding to $C_{40}H_{28}O_{12} + H_2O$
Oxygen	8.7	

HUMIN AND HUMIC ACID

Carbon	64.4%	
Hydrogen	4.3	Corresponding to $C_{21}H_{24}O_{12} + 3 H_2O$
Oxygen	31.3	

CRENIC ACID

Carbon	44.0%	
Hydrogen	5.5	Corresponding to $C_{12}H_{12}O_8$?
Nitrogen	3.9	
Oxygen	46.6	

APOCRENIC ACID

Carbon	34.4%	
Hydrogen	3.5	Corresponding to $C_{24}H_{24}O_{12}$?
Nitrogen	3.0	
Oxygen	39.1	

The ulmic, crenic, and apocrenic acid groups are therefore names for exceedingly complex and unstable compounds as yet but very little understood. Although they have never been isolated and their character definitely determined it should be remembered that there is reason for believing that they have at least a very short-lived existence in the soil. If they exist at all they pass quickly into the higher stages of oxidation and their final condition is CO_2; yet it is believed that during their supposed brief existence they may not only attack alkalies and alkali earths but also dissolve even silica.[4]

Clarke asserts that they have a very appreciable solvent power and advance the decomposition of rocks, but notes that their constitution is but little understood.[5]

[1] Schreiner and Reed, Certain Organic Constituents of Soils in Relation to Soil Fertility, Bull. U. S. Bur. of Soils No. 47, 1907, pp. 14–16.

[2] Suzuki, Bull. Col. Agri. Tokio, vol. 7, 1907, p. 513. Quoted by Schreiner and Shorey in Bull. U. S. Bur. of Soils No. 53, p. 19.

[3] G. P. Merrill, Rocks, Rock-weathering and Soils, 1896, pp. 189–199.

[4] Sir Archibald Geikie, Text-book of Geology, 3d. ed., 1886, p. 472.

[5] F. W. Clarke, The Data of Geochemistry, Bull. U. S. Geol. Surv. No. 330, 1908, p. 409.

Hilgard says that an acid reaction is characteristic of the soils of many woodlands, as is notable of the soils of the long leaf pine region of the South as well as of many deciduous forests in northern climates.   He believes that in the course of time oxidation converts the natural neutral humin and ulmin into humic and ulmic acids capable of combining with the bases. Still further action is thought to result in the formation of crenic and apocrenic acids, which are readily soluble in water and in part form soluble salts with lime, magnesia, and other bases.

"These acids act strongly upon the more readily decomposable silicates of the soil, and in the course of time may dissolve out, and aid in the removal, by leaching, of most of the plant-food ingredients . . . of the soil."[1]

It is conceived that this agency may be responsible for the almost complete absence of mineral plant food in the lower portions of peat beds and the subclays of coal beds.

Besides the uncertain humus compounds of ulmic, humic, crenic, and apocrenic acids, there are others of similar derivation the action of which is not yet understood; among them are xylic acid, saccharic acid, and glucinic acid, and a black humus acid containing 71.5% of carbon and 5.8% of hydrogen.

In the old forests of northern climates there is sometimes formed at the surface a partially decomposed peaty layer or vegetable remains which retards the full production of the land both while occupied by forests and for some time after being cleared.[2]   In the case of such arrested humus development the soil becomes sour.   The sourness is thought to be due to the presence of the above-named acids.

Raw humus or unhumified organic material consists of incompletely decomposed plant remains, and may be found so rich in such remains from 50% to 60% organic matter as to be employed as fuel.   It has free vegetable acids in abundance, earthworms can not penetrate it, and by itself it is of no value to plants.   Raw humus appears in the forest in poorly drained places or in places exposed to wind, while ordinary humus with its earthworms, insects, etc., occurs in fresh, sheltered places. Ordinary humus or vegetable mold contains many fungi, besides earthworms and insects, and is an excellent nutritive substance for plants: the rapid production of humus in the forest is therefore a kind of natural manuring.   When ordinary humus in the beech forests of Europe is displaced by raw humus because of timber falls, etc., the beech is no longer capable of regeneration and disappears, being replaced by a heath.[3]

Unhumified organic material has a potential value through the possibility of its nitrification into active humus.   It also lightens the soil by rendering it more pervious to air and water, and by progressive decay gives off carbon dioxide, which is the basis of carbonic acid so important in soil decomposition.

[1] E. W. Hilgard, Soils, 1907, p. 126.
[2] Muller, Natürliche Humusformen.
[3] E. Warming, Œcology of Plants, Ox. ed., 1909, pp. 62–63.

ACTION OF BACTERIA

We have as yet only briefly mentioned the action of bacteria in relation to soil nitrogen. Further discussion is required in order that the conclusions concerning the control of bacterial action may be adequately understood. Since the growth of bacteria is a large factor in maintaining the supply of some of the most important plant foods, the conditions which promote the activity of such bacteria are matters of serious interest to those engaged in the production and care of plants.

Bacteria are plants that form the simplest group of fungi and are lacking in chlorophyll. They are very minute. The largest forms may reach a diameter of 0.008 mm. (0.000352 inch), and the majority are not more than 0.005 mm. (0.000197 inch) in diameter. It is believed that some bacteria are too small to be seen even with the most powerful microscope. Though they are of small size they are concerned with almost every phase of our daily life and overcome their apparent insignificance by their incredible numbers and ceaseless activity. A fertile soil has from 500,000 to 10 million bacteria to the gram, or from 15 million to 300 million to the ounce.[1] In a drop culture of one cubic millimeter an experimenter found that one-tenth the total volume was composed of bacteria. In 24 hours, 48 generations will produce 281,500 billions of organisms.[2] They are most numerous near though generally not at the surface, decrease in numbers rapidly downward, and generally vanish at seven or eight feet. Water drawn from deep wells does not show any bacterial growth.

The functions and value of soil bacteria are variable, but the kinds that thrive in the soil are chiefly beneficial. Their action in the soil is affected by moisture, temperature, degree of comminution of soil particles, aeration, drainage, etc., besides which bacteria have associative relations with each other whose reactions may be either beneficial or harmful. Some of them decompose dead plant and animal matter into simpler compounds, reconstruct various inert materials, and thus constantly renew certain elements in the soil and maintain its fertility. It should be remembered, however, that if the conditions of food supply and environment are unfavorable harmful groups of bacteria may destroy the fertility of the soil.[3]

Humus is essentially a product of bacterial action, but this action should be carefully distinguished from the later action, called nitrification, which transforms the nitrogen of the humus into available nitrates. The formation of humus is accomplished mainly by the breaking down of carbon compounds, especially the carbohydrates of plant tissues, with the production of marsh gas or hydrogen, carbonic acid, and humus. With a surplus of air the humus-forming fermenta-

[1] K. F. Kellermann, The Functions and Value of Soil Bacteria, Yearbook U. S. Dept. Agri., 1909, pp. 219–226.
[2] Grandeau, Ann. Sci. Agr., vol. 1905, p. 456.
[3] For Winogradsky's classification of nitrifying organisms see H. W. Wiley, Prin. and Prac. of Agri. Anal.: Soils, vol. 1, 1906, p. 557.

tion is replaced by one which results in the complete combustion of the organic matter to carbonic acid. It is largely for this reason that more humus is found in a pasture or a forest than in a continuously tilled field.

Most aerobic (oxygen-consuming) bacteria require for their well-being, or even in some cases their existence, some carbon compounds of nitrogen, and will begin to break down proteids and other nitrogen-containing materials. The products of their action, Fig. 9, are successively peptones (like leucin and tyrosin), eventually ammonia, and probably free nitrogen, but the formation of ammonia is the most characteristic effect of the bacterial fermentation of a proteid. When organic matter has decayed to the stage of such simple compounds as ammonia, nitric acid, carbonic acid, etc., it is no longer organic matter and much of it may escape from the soil altogether.[1]

Ammonia is not directly assimilable in the soil when delivered to it by the air or when occurring as the product of plant decay. The ammonia of air and soil is converted into nitrous acid by bacteria or by the oxidation produced under the influence of the catalytic activity of ferric hydroxide. The latter process takes place at a temperature of $15°$ to $25°$ C., and under its influence a certain amount of available nitrogen is developed in the soil independently of the activity of the nitrifying ferments. The conversion of ammonia into nitrates is, however, chiefly accomplished by two groups of organisms, (1) nitrosomonas or nitrosococcus and (2) nitrobacteria. The action of the first group is limited to the formation of a nitrite; the action of the second group is the oxidation of the nitrite to a nitrate. Both groups are active only upon humus and its products, and are to be distinguished from the ammoniacal bacteria (such as *Bacillus mycoides*) that effect the reduction of the carbohydrates and the oxidation of proteid compounds to humus and ammonia. The latter have wholly different habits. In general the nitrifying organisms require both organic and mineral substances for proper growth. Indeed some forms of nitrifying organisms have the power of subsisting wholly upon mineral substances.[2]

The proper conditions for both groups of organisms are somewhat definite. A fairly high temperature, $75°$ F., is most favorable, and there must be a certain amount but not an excess of moisture present in the soil. If the temperature is low and water is present in excess, bacterial action may be incomplete in its effects or cease altogether. This is illustrated by the preservation of leaves, stems, and seeds for thousands of

---

[1] Schreiner and Shorey, Chemical Nature of Soil Organic Matter, Bull. U. S. Bur. Soils No. 74, 1910, p. 45.

[2] H. W. Wiley, Prin. and Prac. of Agri. Anal.: Soils, 1906, pp. 522–526 et al.

years in peat bogs, and by the well-known antiseptic properties of sour humus or peat. Free oxygen in large quantities is required, and there must be present a base or its carbonate, such as lime, with which the acids due to oxidation immediately unite. Sour soils exclude nitrifying bacteria through the action of the organic acids that have not been neutralized.

The neutralizing salts favorable to bacterial development are not restricted to the carbonates. Potassium chloride acts favorably up to 0.3% but suppresses nitrification at 0.8%. Earthy and alkaline sulphates act favorably up to 0.5%. Among the latter gypsum is most beneficial and accelerates the process more than any other substance known. Common salt inhibits nitrification to a notable degree, and to this fact is due in great part the absence of nitrates in low-lying seacoast tracts. Arranged in the order of their value to the nitrifying process the various substances stand as follows, taking gypsum at 100% (after Pichard).

Gypsum... 100%
Sodic sulphate.. 47.9
Potassic sulphate... 35.8
Calcic carbonate.. 13.3
Magnesic carbonate... 12.5

The value of thorough aeration in nitrification is shown by the experiments of Deherain in which a cubic meter of soil was left undisturbed for several months while a similar mass was agitated in air once a week for the same period. At the end of the experiment it was found that the ratio of nitrogen content in the two samples was 1: 70 respectively.[1] It has also been shown[2] that the brown substances of humus and analogous compounds are directly oxidized to some extent under the influence of air and sunlight, thus forming carbonic acid. The process is purely chemical, has no relation whatever to bacteria, and is rendered more effective by cultivation, principally through the better aeration thus effected. The succession of changes through which organic matter passes in the processes of nitrification and denitrification is shown in the accompanying diagram, Fig. 9. It presents in summary form the principal changes that have thus far been described.

The immensely important conclusion has recently been established that all soils contain groups of protozoa which feed upon living bacteria and restrict their numbers, thus acting as beasts of prey.[3] Their predatory activities seriously restrict the limits of nitrogen production even when the amount of organic matter in the soil is greatly increased. Happily a remedy has been found in heating or in treatment with antiseptics. Crop increases of 30% have been effected by a 48-hour treat-

[1] E. W. Hilgard, Soils, 1907, p. 147.
[2] Berthelot and Andre, Comptes Rendus Academie de Paris, 114, 1892, pp. 41-43.
[3] A. D. Hall, The Fertility of the Soil, Science, n. s., vol. 32, 1910, pp. 370-371.

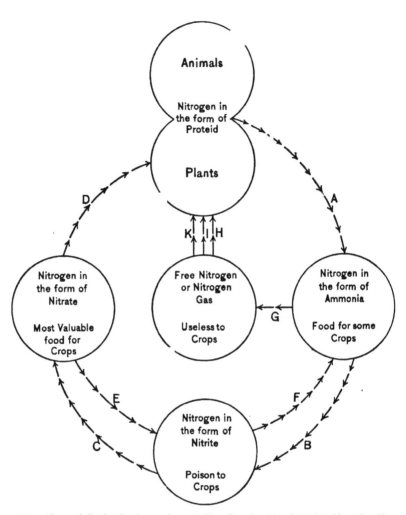

Fig. 9. — Diagram indicating the nitrogen changes in the soil produced by the action of bacteria. The arrows indicate the course of the changes which various groups of bacteria may produce in the nitrogen compounds of the soil. A, action of ammonifying bacteria which change organic nitrogen to ammonia; B, action of nitrifying bacteria which change ammonia to nitrite; C, action of nitrifying bacteria which change nitrite to nitrate; D, assimilation of nitrate by green plants; E, action of denitrifying bacteria which change nitrate to nitrite; F, action of denitrifying bacteria which change nitrite to ammonia; G, action of denitrifying bacteria which change ammonia to nitrogen gas; H, action of bacteria which change nitrogen gas into proteid nitrogen; I, action of bacteria which in symbiosis with leguminous plants change nitrogen gas into proteid nitrogen; K, action of bacteria which in symbiosis with certain non-leguminous plants change nitrogen gas into proteid nitrogen.[1]

[1] Yearbook, Dept. Agri., 1909, p. 222.

ment of the soil with vapors of toluene, chloroform, etc., followed by complete volatilization of these antiseptics. Analyses of the plant material so produced shows an assimilation of greater amounts of other plant foods as well as of nitrogen. It follows that the extra growth does not represent mere temporary stimulation but an absolute increase in the available stores of plant food. While great numbers of bacteria also succumb in the application of the remedial measures, some of them escape, and these, immune from attack, increase at a prodigious rate and almost at once increase the soil fertility.[1]

Several species of bacteria have the power of direct fixation of nitrogen from the soil air. Some of the most important of these bacteria are *Clostridium pasteurianum*, *Bacillus alcaligenes*, *Bacillus tumescens*, *Pseudomonas radicicola*, Granulobacter and several species of Azotobacter.[2] The abundance of these bacteria in the soil seems to indicate the measure of the natural nitrogen-recuperative power of the soil. In the coastal-plain soils the genus Azotobacter occurs only in the surface layer of a few inches; in the deep and almost exhaustless soils of certain sections of the West the same genus is found in active condition even down in the fifth foot below the surface. While most investigators attach considerable importance to the direct fixation of nitrogen, Hall considers that it is yet to be ascertained if the direct fixation of nitrogen by bacteria has any very important part in re-creating the store of uncombined nitrogen in the soil.[3]

An interesting and elaborate series of experiments by Lipman[4] makes it seem likely that *Azotobacter chroococcum* has the power of fixing atmospheric nitrogen when in symbiotic association with certain green algæ with which it is commonly found and which develop with great rapidity upon limestone soils. It has also been shown by Lipman that *Azotobacter vinelandii* does not require symbiosis with algæ to fix atmospheric nitrogen, but that a mixture of it and another bacillus (No. 30) caused a doubling of the nitrogen development.

Certain bacteria have nitrogen-fixing power when in symbiotic association with various legumes such as clover, etc., and accumulate large amounts of soluble nitrates which are ultimately used by the host plant.

[1] There are also present in the soil anaerobic or denitrifying bacteria whose life functions are not dependent upon the presence of air since they are able to avail themselves of combined oxygen by reducing the oxides present in the soils. Some of them cause the reduction of nitrates to nitrites and finally to ammonia as shown in Fig. 9. Such reductive processes are carried on by *bacillus denitrificans* I, and occur chiefly in soils rich in organic matter or badly aerated. The result is a loss of nitrogen and a depletion of the plant food in the soil.

[2] Yearbook Dept. Agri., 1909, p. 225.

[3] A. D. Hall, The Soil, 1907, p. 171.

[4] Rept. Agri. Exp. Station, New Jersey, 1903–1904.

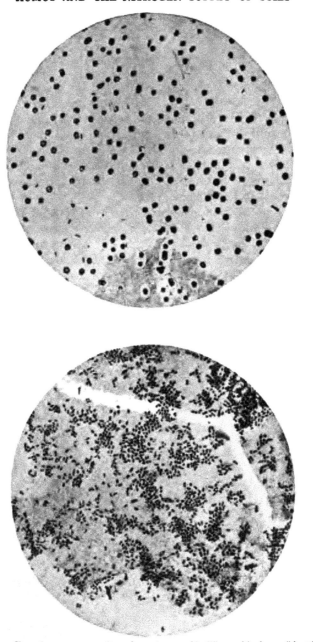

Fig. 10. — Upper figure represents nitrous ferment prepared by Winogvodsky from soil from Cito
Lower figure represents nitric ferment from the same source. (Wiley Soils.)

The bacteria live as parasites within the roots of the host plant from which they derive their carbohydrate food supply. The presence of the parasites stimulates the development, by the root, of the peculiar swellings known as root nodules or root tubercles. These become filled with a bacterial mass consisting principally of swollen and abnormal (hypertrophied) "Bacteroids" having forked outlines, but in part also of bacteria which remained in their normal condition.[1] For a time the nodule increases in size and the bacteria continue to furnish a steady supply of nitrogenous material to the host plant. Ultimately the nodules cease growing, the Bacteroids degenerate, and their substance is absorbed by the host. Normal bacteria, however, remain and provide for future reproductions.

When the root nodules cease growing the larger group of bacteria, called also bacteroids, gradually collapse and are then depleted of their nitrogenous substance. The bacteria capable of this action draw their supply of food largely from the plant with which they grow in symbiotic relationship. It follows also that when the plant becomes seasonably inactive the bacteroids also become inactive for a season; they are found only upon actively growing roots and not upon the roots of dead legumes.

A fourth process of nitrogen fixation is by symbiosis, in which the plant involved is one other than a legume. Whether this process has any practical value remains to be determined, but bacterial nodules have been found upon at least one species of Alnus (alder), upon *Ceanothus americanus* (red root, New Jersey tea), *Ceanothus velutinus, Elaeognus argentea* (silvery berry), *Shepherdia argentea* (buffalo berry or rabbit berry), *Podocarpus macrophylla*, and several genera of the cycads, though the nodules of the latter group of plants are quite different in some ways from the nodules of legumes.[1]

One investigator concludes that the property of utilizing atmospheric nitrogen belongs to many other plants than the legumes.[3] He contends that in all the plants examined by him structures were found which are capable of absorbing it from the air and transforming it into the organic state. The chlorophyll cell seems to possess this power to a high degree. It absorbs free nitrogen and transforms it into organic compounds. Certain organs, called producers of nitrogen, occur in the tender parts of very young leaves or their petioles. But the whole matter of such utilization by non-leguminous plants requires much further work before these broad conclusions can be accepted.

[1] Sharburger, Noll, Shenck, and Karsten. A Text-book of Botany, 3d Engl. ed., 1908, p. 232.
[1] K. F. Kellermann, The Functions and Value of Soil Bacteria, Yearbook Dept. Agri., 1909, p. 226.
[3] Jamieson, Rept. Agri. Research Assn. of the Northeastern Counties of Scotland, 1905, p. 16.

It has been fairly well demonstrated that forests are able to appropriate free nitrogen by means of their trichomes.[1] In a large number of chemical tests conducted in such a manner as, it is believed, to exclude all other sources of nitrogen than the atmosphere it was found that nitrogen is always present in the trichomes. Among the forest trees that have been tested are *Acer campestre, Tilia europaea, Ulmus campestris, Sorbus Aucuparia, Fagus silvatica,* and *Abies concolor.* The process is one of unusual interest, for it adds one more to the very brief list of means by which nitrogen is made available as a plant food. This statement should not be taken to indicate that plants in general are capable of fixing nitrogen. The experiments of two Rothamsted (England) investigators, Lawes and Gilbert, and of Boussingault in the middle of the nineteenth century are very conclusive in their determination of the inability of cultivated plants to fix atmospheric nitrogen except in symbiosis with nitrogen-fixing bacteria.[2] The decay of forest litter and its transformation into nitrates by bacterial means are the chief processes that supply nitrogen in forested areas.[3]

ACTION OF FUNGI

Contributing to the general store of humus in the soil are a large number of fungi among which are Penicillium, Mucor, Aspergillus, Oidium, etc.[4] They take a prominent part in the conversion of vegetable matter into black neutral insoluble humus, a process that is always marked by the formation of carbon dioxide. It is always found that both the fungous tissues and the humus resulting from their decay are notably richer in nitrogen and carbon than the higher plants. These organisms are found about decaying roots and plants, though they are not confined to this occurrence. Their tiny fibrils or hyphæ are scattered through the ground much more thoroughly than even the tiniest rootlets with which they are associated. It is this thorough distribution that makes their presence so important in soil fertility, for it results in a correspondingly wide dissemination of the humus to which the fibrils contribute. Their life functions are dependent upon air, so that thorough aeration promotes their activity and increases their number; it follows that they will be most numerous near the surface and that

[1] T. Cleveland, Jr., Forests as Gatherers of Nitrogen, Science, n. s., vol. 31, 1910, p. 908.
[2] A. D. Hall, The Soil, 1907, p. 162.
[3] Warming suggests mutual reactions between plants as possible sources of nitrogen. "Experience has shown that where *Picea excelsa* has been planted in Jutland it flourishes better in company with *Pinus montana* than without it. It is probable that in this case the mountain pine provides the spruce with nitrogen."
[4] A. D. Hall, The Soil, 1907, p. 182.

the nature of the soil drainage will determine their maximum depth or even their very existence.

From the standpoint of the forest it is fortunate that these growths are not confined to decaying vegetation. They infest the roots of a large number of trees and shrubs and enable the latter to assimilate indirectly the decaying organic and inorganic matter that would otherwise be unavailable. They are in symbiotic association or coöperation with pines, firs, beeches, aspens, and many other forms, all of which appear to depend very largely for their healthy development upon such association.[1] The degree of dependence of the host plant upon the fungoids that inhabit it is emphasized by Frank, who says that some host plants are so dependent upon this relation to obtain the necessary carbon from the humus that they possess no green carbon-assimilating foliage, a condition illustrated by the *Neottia nidusavis* or Bird's-Nest Orchid found among beech underwood in England. The action requires close association of host plant and fungus, and in some cases the fungus penetrates the cortical tissue of the root or forms a sort of cap on the ends of the smallest and shortest rootlets.[2] The association, when it results in the supplying of the host plant with mineral substances, is especially characteristic of plants that grow in soils subject to drought or poor in mineral salts or rich in humus. In general those plants that have feeble transpiration capacity are well supplied about their roots with fungus through which they obtain food of many kinds from the soil. Such plants have also a very limited starch development in the leaf. Fungi may be harmful in their action when they rapidly transform the plant food into available form and produce such large amounts as to cause the excess to be removed by solution, so that a luxuriant but short-lived growth occurs, as in the "faery rings" common in poor pastures.[3]

[1] E. W. Hilgard, Soils, 1907, p. 157.
[2] A. D. Hall, The Soil, 1907, p. 183.
[3] Idem, pp. 184–185 and 219.

CHAPTER VII

SOILS OF ARID REGIONS

GENERAL QUALITIES

THE rainfall of arid regions is insufficient to leach out of the soil the salts that tend to form in it during the progressive weathering of the rock. In humid regions where the rainfall is abundant these salts are leached out and pass in very dilute form through springs and streams into the sea, and relatively small portions of the salts formed by weathering are retained in the soil and utilized by plants. It follows that when the degree of aridity is variable the salt of the soil will be variable in kind and amount and that in extremely arid regions practically the entire mass of salts contained in the rocks remains in the soil.

We have already seen that arid soils are potentially more fertile than humid soils, for abundant rainfall in the latter case leaches the soil of valuable fertilizing elements that are easily soluble. In arid regions these elements are retained in the soil and many of them are very productive. To this advantage is added the greater number of bright clear days in arid regions and a maximum of insolation. The favorable features of arid soils are illustrated by the bench lands about San Bernardino, which are fertile, warm, equable in temperature, free from alkalinity, and, where water can be directed upon them for irrigation, among the most valuable agricultural lands in the United States.[1]

The most striking characteristic of the soils of arid regions is the uniformly high percentage of lime and, as a rule, of magnesia, no matter what the underlying geologic formations happen to be. This condition is all the more apparent in the United States because of the general absence of limestone formations in the more arid portion of the country west of the Rocky Mountains and the wide distribution of such formations east of the Rocky Mountains. No matter what the formation is, whether it is the granite of the Sierra Nevada foothills, the eruptives of the coast ranges of California and Oregon, or the great basalt sheets of Idaho and Washington, they are all alike in producing soils with a high lime content.

It is largely to the high percentage of lime that the flocculated condition of arid soils is due. This is one of their notable characteristics,

[1] W. C. Mendenhall, The Hydrology of San Bernardino Valley, Cal., Water-Supply Paper, U. S. Geol. Surv. No. 142, 1905, p. 17.

and is related to the great depth of root penetration and easy tillage in arid lands.

While in humid regions the average nitrogen content of soil humus is less than 5%, in the upland soils of arid regions the percentage rises as high as 22%, with a general average between 15% and 16%. This is probably due both to the presence of the lime which keeps the acids that tend to form in the soil completely neutralized, and to the deficiency of vegetable matter which allows more complete nitrification than when an excess of such matter is present. The absence of an abundance of vegetable matter in the surface soil and of other qualities that differentiate it from the subsoil is a distinguishing character of arid-region soils. Subsoil and surface soil have no sharp line of separation either in fertility or general appearance or composition. The decomposed state of arid subsoils, the absence of that rawness that characterizes humid subsoils, and the possibilities of their utilization are shown by the manner in which quick growths of yellow pine (*Pinus ponderosa*) appear in the placer mines of the foothills of the Sierra Nevada of California where the subsoil was thrown out years ago and became the surface soil. The timber growth is now of sufficient size to be used for mine timber and a second young forest is springing up on the red earth which once appeared as barren as the desert.[1]

In the case of the relatively insoluble ingredients of the soil such as quartz or silica, the substance making up the greater part of sand, the humid region contains the larger amount, 84%, as compared with the 69% of the arid region where other substances exist in greater proportions. So that while sand of humid regions ordinarily consists of a collection of quartz grains with relatively clean surfaces, in arid regions it consists of a great variety of minerals in a partially decomposed condition. Mixtures of fine and coarse particles at the surface are also more common in arid than in humid soils, due to the absence of thorough sorting in water. Thus the soils of arid regions often have uniform physical and chemical characters to a great depth.

Scantiness of rainfall in arid regions effects a great retardation in the rate at which clay forms from feldspathic rocks and the sediments derived from them. This is shown in the distribution of two broadly different soil types in the eastern wet and western dry portions of the country. The soils of the Atlantic slope are prevalently loams and contain considerable clay; the soils of the arid region are generally sandy or silty with a small amount of clay unless derived directly or indirectly from clay or clay shales.

[1] E. W. Hilgard, Soils, 1907.

The amount of phosphoric acid contained in the soils of arid regions is not different from that contained in the soils of humid regions. Phosphorus is a substance tenaciously retained by all soils and appears to be independent of leaching. On the other hand the leaching process has such a marked influence upon the compounds of the alkaline metals, potassium and sodium, which are readily soluble in water, that the ratios of their percentages show a marked difference in humid and in arid regions. The average ratio for potash is .216% to .672%, the ratio for soda is .140% to .420%. Potash occurs in greater abundance than soda in all soils because it is tenaciously held through reactions effected by zeolitic compounds in which soda is wholly or partially displaced when brought into solution with a potash compound, so that soda accumulates only where the rainfall or drainage is insufficient to effect proper leaching; in such places it results in what is generally known as an alkali soil. Potash is therefore relatively abundant in arid soils and is one of the last substances added to them for increasing their productivity. They rarely contain much less than 1% of acid-soluble potash, occasionally rising as high as 1.8%.

Among the most notable differences in the composition of arid and humid soils is not only the smaller amount of humus found in arid regions but also the relatively higher nitrogen content of the arid-soil humus. On the average, the humus of arid soils contains about $3\frac{1}{2}$ times as much nitrogen as the humus of normal soils and in extreme cases the amount may be 6 times as great, in which case the nitrogen percentage in the arid humus considerably exceeds that of the albuminoid group, so that in arid regions a humus percentage which in humid regions would be considered quite inadequate may be considered entirely sufficient for all crop demands.[1] In arid regions the substance that first requires replacement is phosphoric acid, the second is nitrogen.

ALKALI SOILS

For reasons already stated in connection with the higher percentage of soluble salts in arid regions, certain tracts may under special topographic and drainage conditions develop alkalinity, a condition due to the presence of three compounds which usually form the main mass of the salts — the sulphate, chloride, and carbonate of sodium. Among these, calcium sulphate is nearly always present, magnesium sulphate (Epsom salt) is in many cases very abundant, and calcium chloride is present occasionally. The composition of a more or less typical alkali soil in California is shown in the subjoined table.[2]

[1] E. W. Hilgard, Soils, 1907, p. 138.
[2] Hilgard and Weber, Bull. Cal. Agri. Ex. Sta. No. 82, p. 4. Quoted by Wiley.

TABLE SHOWING COMPOSITION OF ALKALI SALTS IN SAN JOAQUIN VALLEY

	Tulare County					
	Goshen	People's Ditch	Near Lake Tulare	Visalia	Lemoore	Tulare Expm't Station
	Surface Soil	Alkali Crust	Surface Soil	Surface Soil	Alkali Crust	Alkali Crust
Soluble salts in 100 parts soil............	1.40	....	0.83	1.26	....	......
Potassium sulphate......................	....	....	....	....	....	small
a Potassium nitrate.......................	....	....	....	....	....	small
Potassium carbonate (saleratus).........	....	....	....	18.80	....	......
Sodium sulphate (Glauber's salt)........	44.24	1.22	31.30(n)	13.4	chiefly	32.8
Sodium carbonate (sal soda)............	32.98	88.09	18.2	45.3	....	13.16
Sodium chloride (common salt).........	16.74	1.00	....	4.4	little	31.16
a Sodium phosphate.....................	1.97	....	0.22	10.4	....	......
Calcium sulphate (gypsum).............	....	....	....	....	little	......
Magnesium sulphate (Epsom salt).......	....	....	....	8.1	moderate	......
Organic matter.........................	1.97	9.21	7.5	....	....	5.37

a Very generally present, but not always in quantities sufficient for determination.

Alkali lands are so widely distributed in the desert regions of the world[1] that the problem of their improvement and utilization for agriculture is of great importance. These lands when properly treated have great fertility on account of the many soluble plant foods in them, and they may in many places be turned from waste lands to fertile oases. Their natural vegetation is of little value except in a few cases, as the salt bushes and wild clover of South America and Australia, which form valuable pasture. Considerable areas of alkali lands are either destitute of vegetation or bear resistant growths of little value as forage. The effects of sodic carbonate on plants grown in alkaline regions are seen in a scant leafage, short growth of shoots, and a deadening of the roots. The cortex assumes a brownish tinge just above the surface in the case of green herbaceous stems, and, in the case of trees, the outer bark assumes an almost black tint and the green layer underneath turns brown. The maximum injury is usually at or near the surface, where there is a maximum accumulation of salts, due to evaporation at the surface. The vertical distribution of the alkali salts in a California soil is shown in the diagram below, Fig. 11.[2]

Certain native plants that live upon alkali soils have adapted their root systems to a very interesting condition. Figure 11 shows that down to a depth of 15 inches there is practically no alkaline content; and it is within these 15 inches of soil that the native plants develop their roots

[1] The total area of the arid lands of the world computed from the total area of interior basin drainage is given by Sir John Murray as 11½ million square miles, or one-fifth of the total land surface of the globe — The Origin and Character of the Sahara, Science, vol. 16, 1890, p. 106.

[2] E. W. Hilgard, Soils, 1907, p. 432.

and develop their growth. The bulk of the salts accumulate at the greatest depth to which the annual rainfall of seven inches reaches, where it forms a hardpan. It is above this hardpan that the seeds of the shallow-rooted plants germinate and extend their roots. The soil moisture of the surface layer is so thoroughly consumed by the plants that no alkali is brought up from below by evaporation and the life cycle begins the following season. It is in this manner that the luxuriant vegetation of the San Joaquin plains is maintained except where occasional alkaline spots occur. The horizontal distribution of alkali is variable and the location of the salts changes from year to year, especially in irrigated lands, so that the cultivation of alkali lands, the determination of their position, etc., must be carried on with great care.

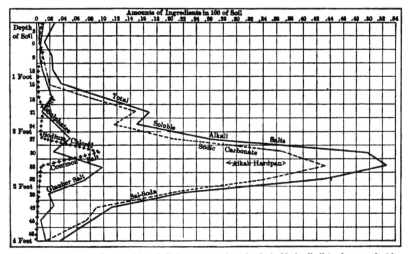

Fig. 11. — Amounts and composition of alkali salts at various depths in black alkali lands covered with native vegetation. (Hilgard, Soils.)

Two types of alkali soils are noticeable: the one is due to the presence of carbonate of soda and is called black alkali; the other is due to the presence of the sulphates and chlorides of sodium, and is called white alkali. The latter is much milder in its effect on plants. In California, outside the main valleys no important amounts of alkali salts are found at depths exceeding four feet. The total amount found in alkali lands which show saline efflorescences at the surface in the dry season is from one-tenth of one per cent to as much as three per cent of the weight of the soil taken to a depth of four feet. Alkali lands also have a high lime content and a high potash content — higher than the average amount of phosphates; while nitrates are usually scarce or altogether absent, though nitrates may occur along the margins of the alkali spots.[1]

In sloping valleys or basins where the central lowest portion receives the salts leached out of the soils of the adjacent slopes, occur belts of variable width in which the alkali impregnation may reach to a depth of 10 or 12 feet. Such areas are, however, quite limited and irreclaimable, and the predominating ingredient is usually common salt, as is illustrated in the Great Salt Lake basin of Utah, in the Antelope and Perris valleys of southern California, and the Yellowstone Valley near Billings, Montana.[2] The salts of alkali lands are not permanent in their vertical position but follow the movement of the moisture, descending in the rainy season to the lower limit of the absorbed rainfall, and re-ascending in response to surface evaporation, so that at the end of the dry season a saline efflorescence may occur or the entire mass may be found within a few inches of the surface.

Carbonate of soda exercises a puddling action on the soil, destroys its crumb structure, and renders it almost untillable and impervious. It also tends to form a tough impervious hardpan as resistant to roots as to implements. Hilgard has shown that the proper treatment of alkaline soils is leaching (after treatment with gypsum in the case of black alkali), together with subdrainage. Flooding alone is not sufficient, for if the process is carried on generally the alkali spots grow larger to the destruction of adjacent lands. If coöperative subdrainage is carried out the salts may be entirely removed by an excess of irrigation water or be carried down to so low a level as to have no injurious effect upon plants. The amount of alkali in the soil may be diminished also by cultivating plants that take up considerable amounts of salt, — a notable property of the greasewoods (Sarcobatus, Allenrolfea), which contain from 12% to 20% of alkaline ash. When grown upon the land and then cut and removed, such plants will markedly diminish the amount of alkali in the soil. A few such salt-consuming plants are fit for pasture, such as the Argentine plant (*Atriplex chachiyuyun*) and the Australian salt bushes (*Atriplex halimoides*), Vesicaria and Lepto-carpus, and a Chilean plant (*Modiola procumbens*). The results of the reclamation experiments of the U. S. Bureau of Soils are surprisingly good and indicate the range of possibilities in regard to the use of alkali lands.[3] Tracts originally covered with a white crust of alkali and

[1] E. W. Hilgard, Soils, 1907, pp. 439, 444, 448.

[2] Farmers' Bull. U. S. Dept. Agri. No. 88, 1899.

[3] For specific descriptions of the localities where alkali soils have been experimentally improved and reclaimed, see C. W. Dorsey, Reclamation of Alkali Soils, 1907, and Reclamation of Alkali Land in Salt Lake Valley, Utah, Bull. U. S. Bur. Soils No. 43, 1907.

supporting a scanty growth of greasewood were sweetened by flooding and drainage and then sown to alfalfa, various vegetables, grains, etc., with very beneficial results.

The accumulation of mineral salts at or near the surface has given rise to the formation in most arid regions of a characteristic deposit known as "caliche." It often consists of a variety of substances, but the most common constituent is nitrate of soda. The greatest deposit of this sort is found in Chile in the province of Tarapacá, where nitrate of soda is produced on a very large scale, practically the whole of the world's supply being derived from this desert region. Smaller tracts are found in many other deserts, notably in the Southwest as in Death Valley, California, but the scale of production in all these cases is decidedly limited. In general the layer of caliche is covered with a deposit of earth from a few inches to a few feet in thickness. It may consist in part of wind-blown material, in part of water-laid material deposited since the caliche was formed. The largest beds of caliche are probably due to the crystallization of mineral salts from bodies of water which have disappeared either through a change of climate or a change in the level of the land or both. The origin of this class of material is, however, still in doubt, and although the result is closely allied to aridity it is not yet clear what combination of arid conditions with topography, drainage, and chemical character of rock and vegetation brings about its existence.

CHAPTER VIII

SOIL CLASSIFICATION

It has been generally agreed among soil investigators that because of the predominating influence of physical characteristics in soil fertility, physical and not chemical qualities shall be made the basis of soil classification. This decision is strengthened by the immemorial custom among agricultural folk of designating soils principally by their physical character. The terms *sand, gravel, clay,* etc. (or their equivalents), are common non-technical words which convey a fairly definite meaning the world over. When, however, a soil is to be scientifically investigated, its characters strictly defined, and its value and its needs formulated, somewhat precise terms must be employed, careful experiments conducted, and conventional symbols devised which have stricter meanings than the colloquialisms of the farmer. Hence a relatively refined classification has been elaborated, based primarily upon physical character. In examining the accepted classification we should not lose sight of the fact that while the forester must acquaint himself with it in order to make the literature of soil investigators serve his purpose, he generally requires for ordinary field work a somewhat rougher scheme of classification. Gravel, sand, silt, clay, and peat or muck are the main types he is required to recognize, modifying his choice of terms by mention of such secondary qualities as the soil of a particular locality exhibits. He will also be required to distinguish between various grades of gravel, sand, etc., as coarse, medium, and fine, and it is obviously to his advantage to employ for these subdivisions the basis employed by soil specialists, in so far as this is possible.

In determining the sizes of soil grains a number of methods may be employed; the three principal ones are (*a*) sieving the soil samples, (*b*) elutriating them, and (*c*) separating the various grades by the subsidence method.

(*a*) The sieves used for soil analyses have round holes of carefully determined diameter. The unit employed may be fractions of an inch, but the smaller units of the metric scale (millimeters) are decidedly preferable both because of their international acceptance and their easy use in computations. The soil samples are sifted after being rubbed so as to destroy the lumps or soil crumbs composed of both fine and coarse material that behave in some respects as large individual particles the size of the lumps. Separation may be more easily accomplished by playing water on the sieve; without it the clay particles and even the silt particles tend to cling to the sand as soon as the grain sizes in the latter fall much below ½ mm.

(b) The elutriator, Fig. 12, is an instrument employed to separate soil grains of different sizes (after removal of the clay by subsidence) by an ascending current of water at various fixed velocities. The soil grains are carried off in exact conformity to their several sizes or volume weights. The maintenance of the current at a fixed velocity for a long enough period will result in the practically complete removal of all grains below a certain size. The different velocities are adapted to certain desired grain sizes,

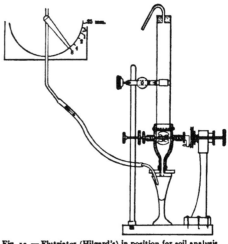

Fig. 12.— Elutriator (Hilgard's) in position for soil analysis.

and the volumetric determinations that follow elutriation form the basis for classification.

(c) The subsidence method is based upon the assumption that if a soil sample is thoroughly mixed in water the different grain sizes will settle according to their weights. A successful outcome requires the removal of the clay by repeated sedimentations of the non-clay material, the decantation of the water in which the suspended clay particles are held, and the final sedimentation of the coarser grades. The necessity for the removal of the clay is due to the greater viscosity of the water in which the clay is suspended and its interference with normal and accurate sedimentation. The clay itself is determined from the several clay waters by evaporation of the water. Precipitation will not suffice, for the finest colloidal clay will not subside for years, — a condition thought to be due to a change in its physical and possibly in its chemical nature.[1] A defect of the subsidence method is the impossibility of abstracting all the clay, a defect the more serious because of the high importance of the

[1] W. H. Brewer, On the Suspension and Sedimentation of Clays, Am. Jour. Sci., vol. 29, 1885, p. 1.

clay constituent of soils (p. 35). Some clay particles are invariably carried down by heavier constituents and deposited with them. A similar result does not occur in elutriation because of (1) the agitation (which prevents flocculation into heavy aggregates) of the ascending current and (2) the grain sizes are expelled in reverse order, the finest first, and so on.

Purpose of a Soil Analysis

The physical analysis of a soil is not alone for the purpose of finding a name for it in the series proposed in the table, Appendix A. This is the least of its purposes. It aims, in addition, clearly to present the controlling constituent among the soil grains. Soils are not generally composed of grains rather equally distributed among the several sizes. Some particular size usually predominates, and gives the soil its strongest individual character. This is clearly illustrated by the Volusia soils spread over some 10,000,000 acres of western Pennsylvania, southern New York, Ohio, and West Virginia. They are poor in their present condition, yet nothing in their chemical nature suggests poverty. Their unproductiveness is due largely to improper drainage, for their physical composition is such that the natural drainage is not adequate; acid therefore forms and accumulates and makes the soil sour. Their improvement is not to be sought in the use of fertilizers alone, for the fine-grained fertilizers by themselves would still further clog the soil pores. It is suggested that adequate drainage and deep aeration would prove remedial.[1]

The remedies proposed are particularly interesting because they are among the relatively small number which the forester finds it possible to apply to economic advantage over large areas. The forester can not fertilize the ground by the application of manures or mineral fertilizers; the very scale of his work makes it impossible to do more than enable a forest soil to improve itself by encouraging processes already in action. This he is able to do in many ways. Proper cutting and seeding or replanting are almost always possible in both the physical and the commercial sense of the term, and by maintaining shade the drying of the surface soil, in which seedlings make their first growth, is prevented. Proper drainage is also feasible and is at once a means of partly controlling the kind of growth and the rate of growth.

It is fundamental in drainage to ascertain the nature of the subsoil as well as the soil. If the subsoil is open and porous it is likely to be dry; if clay forms the subsoil it will probably be wet and by capillarity

---

[1] M. E. Carr, The Volusia Soils, Bull. U. S. Bur. of Soils No. 60, 1909, pp. 21–22.

supply the surface soil with water in the dry season. Not only for this reason, but also because so important a part of the root systems of trees occurs in the subsoil, is it necessary to secure data as to its nature and its effects upon the surface soil in different seasons.

DIFFERENT BASES OF SOIL CLASSIFICATIONS

It is unfortunate that no general scheme of classification has been universally adopted by soil physicists. In Appendix Å the scheme of the U. S. Bureau of Soils is described because the literature of this Bureau is now both extensive and valuable and reference to it indispensable on the part of any one beginning the study of the soils of a given region. Nevertheless it should be noted that many other bases of subdivision of soil types are in vogue. It has been proposed by Hilgard and others, that all constituents of soils that are too large to pass through a sieve with meshes 0.5 millimeter in width should be called the "soil skeleton" and that the remaining constituents that pass through the sieve should be called "fine earth." He regards the fine earth as having a special relation to plant life as food material and through its physical attributes. For the purposes of ecologic studies Warming distinguishes six different kinds of soil as follows [1]:

Rock soil.	Clay soil.
Sand soil.	Humus soil.
Lime soil.	Saline soil.

Soils may be distinguished also by classes, as rigid, stiff, mellow, lax, loose, and shifting, in order to express various grades of compactness. Hilgard suggests the use of broad types such as sand, clay, and humus.

Considered in reference to their origin, soils may be classified as sedentary and transported. A sedentary soil may then be either residual where it remains upon the rock from which it was derived or colluvial where it is subject to slow down-hill movement on hill slopes. Transported soils may also be divided into alluvial where the soil is deposited on bench lands or flood plains, and eolian where it is deposited by the wind. The most common classification is based on texture, as gravel, sand, silt, clay, and their subdivisions and derivatives. Again, soils may be called humid or arid according as they are formed in one or the other of these climates. Or a soil may be classified according to its chemical properties. Thus, we have a lime soil with its lime-loving vegetation, or a magnesium soil, or a soil exceptionally

[1] Œcology of Plants, Ox. ed., 1909, p. 60.

rich in potash or gypsum.   A classification based on the distribution of vegetation is often helpful but the soil requirements of plants are not rigid except in the case of extreme types of soils and plants. Doubtless the physical classification has the widest practical importance since physical features most commonly have a fundamental control over vegetation and are the most unchangeable.   The ideal classification would be adaptable to, and would take cognizance of, all important soil characters.   It is at least certain that no single scheme is applicable to all kinds of soils in all kinds of climates.[1]

With this brief suggestion of the nature of other classifications in mind the student is referred to Appendix A for a description of a suggested outline of a soil survey with such description of the field methods required as bear on soil studies practicable in forestry.

[1] Hilgard and Loughridge, The Classification of Soils, Verh. der II Int. Agrogeologen-conferenz, Stockholm, 1911.

PART TWO

PHYSIOGRAPHY OF THE UNITED STATES

CHAPTER IX

PHYSIOGRAPHIC REGIONS, CLIMATIC REGIONS, FOREST REGIONS

INTRODUCTION

PHYSIOGRAPHY is indispensable to the environmental study of organisms of every kind, whether trees, or men, or bacteria. Soil, topography, and climate are positive forces in the development of forests and the harvesting of forest products. They underline the main possibilities as well as the main limitations of nature. We have already seen that soil in at least small amounts is a necessary condition of tree growth; of the same order of importance are the facts that the broader forest distributions depend upon climate, while the accessibility of forests depends to a large extent upon topography. It is doubtful, for example, whether some of the best timber of the Sierra Nevada will ever be harvested because of topographic difficulties — steep canyon walls, sharp spurs, and remoteness from transportation lines. Climatic conditions exclude trees from the larger part of the arid and semi-arid West and from the higher, colder, and windier parts of the western mountains. Forests are excluded also from great areas of bare rock outcrop in regions of glacial denudation or from soils rendered infertile by extreme physical or chemical properties. The forester must take account not only of these relations but also of the larger relations of forests to stream flow and soil erosion. Each soil type has its own peculiar water-holding or water-shedding capacity, each topographic province has certain slopes upon which either agriculture or forestry can or can not be conducted, each natural region has its own climatic possibilities and restrictions, a study of which enables the forester to improve natural conditions and repress harmful forces to the benefit of mankind.

It is our purpose in these pages to present the physical basis of forestry in the United States. No attempt is made, however, to discuss either regional ecology or the principles of ecology. The ecology of the forest is regarded as a subject which possesses a body of facts and laws of its own. The single object of this book is to acquaint the forester with the geographic basis of his work, with such references to the forest as point the direction of his more special subjects.

This attitude should be appreciated here, lest the organization of the following chapters be misunderstood. For example, a forester requires a certain group of physical data in developing, let us say, the forests of the Black Hills. It is our object to discuss not the silvicultural or lumbering methods best adapted to the Black Hills and the relations of these to the physical geography of the region, but to make the student so familiar with the geography that a knowledge of it may be assumed when he begins a study of the forestry.

Physiographic Regions

The description and explanation of any large and varied portion of the lands proceed naturally by subdivisions, each subdivision embracing a tract in which the topographic expression is in the main uniform. Since uniformity of topographic expression is achieved only when geologic structure, physiographic process, and stage of development in the geographic cycle are also uniform, each subdivision has an essential uniformity or unity of geologic and physiographic conditions. It is customary to speak of each subdivision as a natural region or province, and to bound each region by lines which represent the limits of unity. In some cases the boundary lines are very precisely located, as along the eastern edge of the Allegheny Plateau in Pennsylvania, or the western edge of the Colorado Plateaus in Arizona, where great scarps mark out sharply defined borders; in other cases the transition zone between provinces is relatively wide and has characteristics which partake of the nature of both adjoining provinces, as between the Columbia Plateau of southern Oregon and the Great Basin south of it.

It must not be supposed that the idea of physiographic unity is applied in an absolutely rigid manner. Some of the physiographic provinces appear to have great topographic variation and but little unity. In such cases it appears at first sight impossible to group the forms in a rational manner until soil, climate, topography, and geologic structure are all examined, when prevailing or group characters always become apparent. Exceptions to group qualities are often observed, and in some cases these are of great importance, but on the whole they affect only the minor physiographic features. Thus the northern Rockies of Montana and Idaho are in striking contrast to the high plains on either side, — the Columbia lava plain of Washington on the west and the Great Plains of Montana on the east. This contrast between a belt of high, rugged mountains and gently rolling, bordering plains forms a primary basis for subdivision. Nevertheless the traveler in crossing the northern Rockies finds the landscape changing continually. Everywhere the ranges trend in the same general direction, but the kind of rock, the structure and hardness of the rock, and the kinds of dissection affecting the range masses vary from point to point; everywhere the major valleys have a distinctive trough-like cross section, but the minor valleys are of many varieties. No two ranges, therefore, are alike in detail, but all are alike in being rugged and high, while the bordering lands are smooth and low (relatively). Some of the mountains are dissected plateaus, as the Clearwater Mountains of Idaho; others are dissected anticlines and synclines, as the Lewis and Livingston Mountains of Montana; yet they are all alike in being *deeply* dissected.

The broad *similarities* among the features of a physiographic province are frequently expressed in the name of the province, as "Prairie *Plains,*" or "Colorado *Plateaus,*" or "Great *Basin*"; yet the Prairie Plains are locally rough, as where high and irregular morainic belts cross them, the Colorado Plateaus are diversified by volcanic mountains like Mt. Taylor and San Francisco Mountains, whose summits reach above timber line, and the Great Basin consists of many independent basins broken by, and mutually separated by, mountains of marked height and ruggedness. The *dissimilarities* among the features of a region may be classified and grouped by subregions. The Older Appalachians (Chap. XXVIII) for example have certain very distinctive and uniform features over a great belt of country from Maine to Alabama, such as their great geologic age, their highly complex structure, their prevailingly crystalline character, and the tremendous erosion which they have suffered. But throughout the region marked dissimilarities also occur which require recognition. The scenery about Asheville, North Carolina, is quite unlike that about New Haven; the mountain basin in the first case falls in the Appalachian Mountains, a subdivision of the Older Appalachians, the valley lowland of the second case lies in the glaciated New England subdivision of the same province.

Many of the physiographic regions of the United States are of great size. The Great Plains are as extensive as European Russia; the Lower Alluvial Valley of the Mississippi is at least half as large as Italy; the Alps are less extensive than that part of the Rockies south of the international boundary. Many individual topographic features are developed on a large scale. Hurricane Ledge, an eroded fault

scarp in the Colorado Plateaus, is 700 miles long, the Great Valley
of the Newer Appalachian region extends with but local interruptions
from Alabama to Quebec, the Grand Hogback of western Colorado
is a bold escarpment of erosion which extends unbroken for over 200
miles.

In addition to the great size of individual provinces and features is
the great variety of physiographic features which the country exhibits.
The physiography is also in many respects unique. Exploration has not
revealed anywhere else on the earth structures and forms like those
of the Colorado Plateaus, either in respect of scale or perfection of development. The Newer Appalachian ridges and valleys are so perfectly
developed that "Appalachian structure and topography" has become a
technical phrase. The till sheets of the upper basin of the Mississippi,
in the heart of the continent, are so extensive and their succession so
complete that the history of the glacial period was first worked out to
partial completeness in America. The Columbia lava flows constitute
one of the few really great basalt fields of the world.

We should ascribe to the great variety of physical conditions in the
United States no small share of the general interest in American forestry,
for the forests and forest problems are almost as varied as the relief
upon which, either directly or indirectly, they so commonly depend.
The vital relation of the forests of the arid West to the general welfare
of the people has given western forestry a scientific interest not exceeded
elsewhere. Man's control of the desert had its beginning in his control
of water. At first that control was in the nature of art; now it is in the
nature of science. As much care was at first bestowed upon the ritual
of rain as upon the construction of a dam; we now study the laws of
rainfall, measure its several dispositions, exercise control over it in measur-
able degree, relate cause and effect, and know the limits of irrigation
enterprises before they are begun. Throughout the West the problem
of water control has been found to be also a problem in forest control
and grass control. Irrigation, forestry, and grazing are parts of one
scientific problem. Of lesser but still of great importance are the relations of run-off and forests in the humid East, where over large areas a
mountainous relief diminishes the value of the land for agriculture and
increases its value in relation to forests and stream flow. Mountain
influences are extended out upon the plains, where resides a dense agricultural population whose commerce and towns and fertile valley lands
are often largely dependent upon the behavior of the mountain-born
streams.

CLIMATIC REGIONS

The determining *factors* of climate are chiefly latitude, the relative distribution of land and water, the elevation of the land above the sea, and the prevailing winds; the most important climatic *elements* are temperature, moisture, wind, pressure, and evaporation. The climatic elements are shown graphically on the accompanying series of maps for North America. Among them temperature and rainfall are of most importance in relation to forests, and a brief discussion of these elements of our climate is therefore presented in this chapter. In Plate I, representing the climatic provinces which correspond with the life regions of Merriam,[1] they are combined in such a way as to show their effect upon the life of the continent.

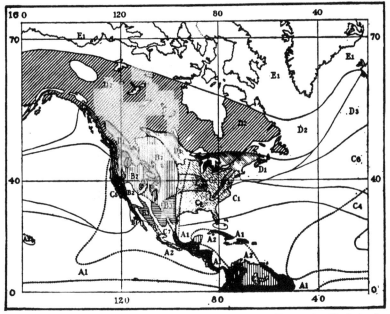

Fig. 13. — Köppen's classification of climates in relation to vegetation. Tracts enclosed by broken line have distinct dry seasons (rain probability < 0.20). A₁, Liana; A₃, Baobab; B₁, Garva; B₃, Simoon; B₅, Mesquite; B₇, Prairie; C₁, Hickory; C₃, Corn; C₄, Olive; C₅, Heath; C₇, High Savanna; D₁, Oak; D₃, Birch; E₁, Arctic Fox.   (Ward, Climate.)

TEMPERATURE

The temperature map, after Köppen, Fig. 14, is peculiarly useful to a forester, since on it temperatures are not reduced to sea level and the

[1] C. H. Merriam, Life Zones and Crop Zones of the United States, Bull. Div. Biol., U. S. Dept. Agri., No. 10, 1898. The fourth edition of this map forms the basis for Plate I.

boundaries of the various belts have certain definite relations to tree growth.

"A normal duration of a temperature of 50° for less than a month fixes very well the polar limit of trees and the limits of agriculture. Near this line are found the last groups of trees in the tundras. A temperature of 50° for four months marks the limit of the oak, and also closely coincides with the limits of wheat cultivation."[1]

The greater part of the United States lies in the belt of westerly winds, hence we should expect marine temperature influences to be felt farther inland on the Pacific or windward coast than on the Atlantic or leeward coast. However, the Pacific coastal tract has a relief including high mountain ranges trending at right angles to the prevailing winds, hence marine influences affect a belt of country sharply limited on the east. They do not extend farther inland than the Sierra Nevada in California and the Cascades in Oregon and Washington, except along the valley of the Columbia (which cuts across the Pacific mountains), where unusually high temperatures prevail for some distance east of the Cascades. Washington, Oregon, and California have strikingly equable temperature conditions. On the Atlantic coast marine influences do not extend so far inland as a rule nor are they so pronounced. They are, however, distinct, as is shown both by the lower absolute and mean monthly temperatures at a shore station like New London as compared with a station like Middletown, Connecticut, 20 miles inland. Southern species of birds and plants follow the Atlantic coast in narrowing belts surprisingly far toward the north. The northernmost occurrence of persimmon trees, a southern species, is at Lighthouse Point, New Haven. The temperature contrasts between the two coasts are shown in the following table.

CONTRASTS BETWEEN ATLANTIC AND PACIFIC COAST TEMPERATURES[1]

Stations	Latitude		Annual Mean	Winter Mean	Summer Mean
Savannah, Ga.............	32°	5′	66° F.	52° F.	81° F.
San Diego, Cal............	32	43	61	55	68
Cape May, N. J...........	38	50	54	36	72
San Francisco, Cal.........	37	18	56	51	59
Nantucket, Mass..........	41	17	49	33	65
Eureka, Cal..............	40	48	52	47	56
Chatham, N. B...........	47	3	39	13	63
Fort Canby, Wash.........	46	17	50	42	58

[1] R. DeC. Ward, Climate, 1908, p. 28.
[1] A. J. Henry, Climatology of the United States, Bull. Q, U. S. Weather Bureau, 1906, p. 26. The statistical data in the section on climate have been derived chiefly from this source.

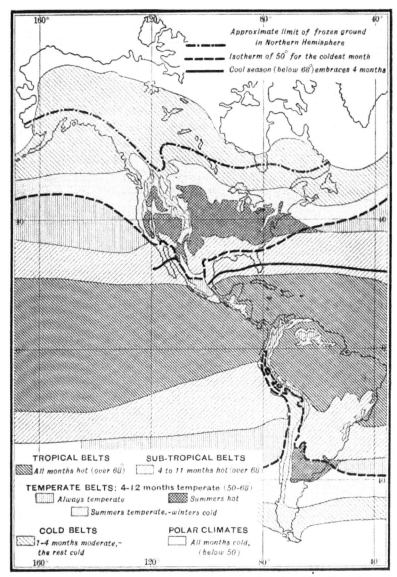

Fig. 14.—Temperature zones of the western hemisphere. (Ward, adapted from Köppen.)

It will be noted from the table that the winter means on the Atlantic coast are regularly lower than those on the Pacific, since, being on the leeward side of the continent, continental influences are more strongly marked than marine. The summer means are all higher for the same reason: the land is always warmer in summer and colder in winter than the sea.

The greatest extremes of temperature are experienced in the interior of the country far from the influence of the oceans, where the continental type of climate prevails. The valleys of eastern Montana experience

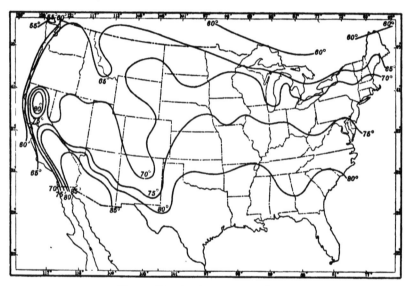

Fig. 15. — Normal surface temperatures for July.

the lowest absolute temperatures; —65° below zero was recorded at Miles City, Mont., January, 1888. The whole plains region in the latitude of the international boundary is subject to great and sudden variations of temperature, since its open and vast expanses are exposed both to the cold winds from the mountains and the north, and to hot winds from the south. The 100° maximum at times extends into the Canadian Northwest. The northerly winds are sometimes of great velocity and in winter are often attended by light, dry snow, conditions which reach their culmination during a blizzard, when the wind may attain a velocity of 60 miles an hour. High winds in summer are often attended by dust and give rise to the "dust storms" of the plains.

When they are marked by high temperatures they may be even more harmful to crops than the winter blizzards are to livestock.

Maximum temperatures of 100° and over are experienced in all parts of the United States except in the higher portions of the Atlantic and Pacific Cordilleras, the immediate coasts of both oceans north of 40°, in the peninsula of Florida, along the Gulf coast, and in portions of the Great Lake region. The highest recorded temperature in the entire country is 130°, recorded in the Colorado Desert in southern California. Maximum temperatures of 112° to 115° are frequent in southwestern Arizona and southern California. The only Weather Bureau stations

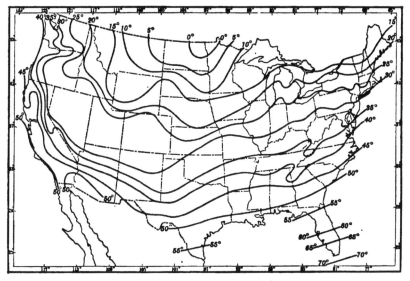

Fig. 16. — Normal surface temperatures for January.

in the United States where a minimum temperature below freezing has not been experienced are Key West, Fla., and San Diego, Cal., with absolute minima of 41° and 32° respectively. South of the mouth of Chesapeake Bay the Atlantic coast has never experienced a temperature below zero, nor have zero temperatures ever been recorded along the Gulf coast, at any point on the Pacific coast, or in the Great Valley of California. The mountain summits of both the Atlantic and the Pacific Cordilleras have minima comparable to those experienced on the north-central Great Plains and in the Arctic regions. The lowest recorded temperature on Mount Washington, N. H. (6293 feet), is — 50°, the lowest on Pikes Peak, Col. (14,134 feet), is — 37°.

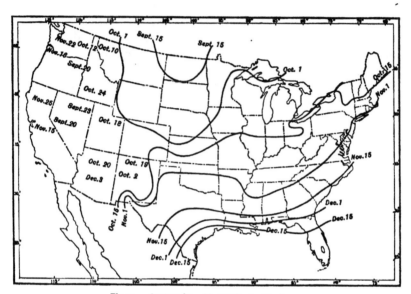

Fig. 17. — Average date of last killing frost in Autumn.

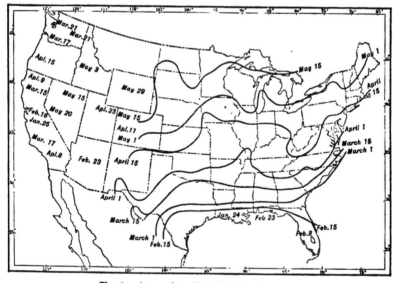

Fig. 18. — Average date of first killing frost in Spring.

The mean annual range of temperature is about 30° in the states of the Gulf and Atlantic Coastal Plain, 40° to 50° in the interior valleys, Rocky Mountain region, and Middle Atlantic States, and from 55° to 65° over the northeastern Rocky Mountain slope and eastward to Lake Superior. The greatest daily range occurs in the arid and semi-arid Southwest on account of the prevailingly clear skies and the lack of vegetation; the greatest mean daily range (30° to 35°) is in the Plateau region, the least (8° to 12°) is along the Pacific and Gulf coasts. East of the Mississippi Valley the mean daily range is generally less than 20°.

PRECIPITATION

The chief causes of an abundant rainfall are (1) nearness to the ocean or other large body of water such as the Gulf of Mexico or the Great Lakes in the United States, (2) location within or near the track of cyclonic storms, and (3) mountain ranges athwart the rain-bearing winds. The western slopes of the Coast Ranges of Oregon face the ocean and run at right angles to the westerly winds, and their rainfall exceeds 100 inches a year; the Ohio Valley lies in the track of the more or less regular cyclonic storms that move northwestward from the Gulf, and receives a rainfall of 40 to 50 inches a year; nearness to the sea gives the greater part of the Atlantic and Gulf coasts a higher rainfall, 50 to 60 inches, than is enjoyed by any portion of the eastern half of the country except the mountains of western North Carolina. By contrast the mountain-rimmed parks of Colorado, and the Great Basin of Nevada, are regions of diminished rainfall; the coast of southern California owes its dryness chiefly to its position outside the belt of cyclonic storms; the dryness of North Dakota is chargeable chiefly to remoteness from the sea, although in this, as in the other cases cited, the rain-producing or rain-resisting forces commonly operate in combination with other forces, so that the influence cited should be understood to be the predominating and not the sole influence.

Rainfall is always due to the cooling of the air to and below the point of saturation. This may be accomplished (1) by the rise of air on a mountain flank—the air expands on rising and since the heat of the air supplies the energy for expansion, the air is cooled to and beyond the point of saturation and rain falls; (2) by convectional air currents produced by a local overturning of the lower air as during a summer thunder shower; and (3) by radial inflow and ascensional movement, as in cyclonic storms, with expansion and cooling to the point where rain falls.

The seaward slopes of the Coast Ranges of Oregon and Washington receive the heaviest rainfall in the United States, from 60 to 150 inches a year. Rains are frequent during the entire year, but most frequent during the winter season, from November to May. The rain-bearing winds change from southeasterly to westerly with the approach and passage of cyclonic storms. The rain begins with the southeast wind and ends with the westerly wind. The result is that the leeward or eastern slopes of the mountains are also well watered, though the fall is lighter than that on the windward or western slopes. Northerly winds bring fair weather at all seasons. Southward from the well-watered strip along the northern part of the Pacific coast the rainfall decreases rapidly, falling from 67 to 22 inches between the northern boundary of

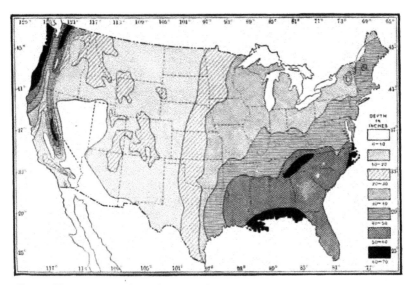

Fig. 19.—Mean annual precipitation in the United States reduced to inches of rainfall. Note the Adirondacks, White Mountains, and Black Hills islands. (U. S. Geol. Surv.)

California and San Francisco. Toward the south it continues to decrease and falls to a minimum of less than 10 inches at San Diego, in the horse latitude belt of light uncertain winds between the westerlies and the trades. The Pacific coast thus exhibits a range in rainfall of about 100 inches.

The main Pacific coast valleys, embracing the Great Valley of California, Salton Sink, the Willamette Valley, and the Puget Sound depression, have a much lighter rainfall than the rain-obstructing Coast Ranges, since they lie in the lee of the latter. The valley of southern California is an extremely dry desert. In the Great Valley the rainfall varies from about 10 inches at Fresno in the south to 25 inches in the north; the

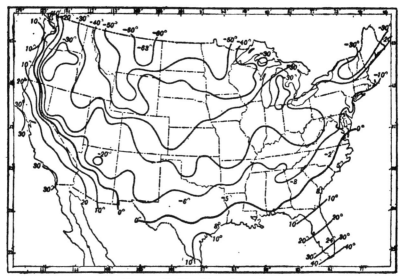

Fig. 20. — Absolute minimum temperatures.

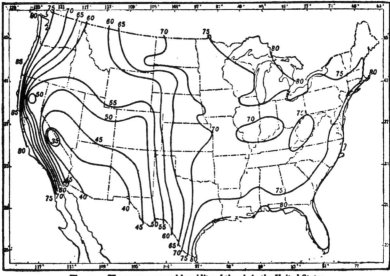

Fig. 21. — The average annual humidity of the air in the United States.

rainfall of the Willamette Valley varies from 25 inches in the south to 45 inches in the north; while the Puget Sound region has an average rainfall of about 45 inches. The precipitation increases rapidly eastward as the winds ascend the western slopes of the Sierra Nevada and the Cascades. It reaches a maximum of about 100 inches in Washington and Oregon, and from 40 to 80 inches in California at elevations between 3500 and 5000 feet. Beyond this point the precipitation

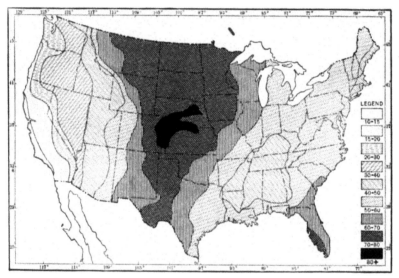

Fig. 22. — Percentage of annual rainfall received in the six warmer months, April to September inclusive.

diminishes again toward the summit and becomes insignificant at the eastern base of the mountains.

The great height and continuity of the Sierra Nevada and the Cascades cause these mountains thoroughly to obstruct the westerly winds in respect of moisture, with the result that great stretches of country east of them are arid wastes. Where ranges of exceptional height occur in the country east of the Cascades and the Sierra Nevada the rainfall may exceed 25 inches a year, but by far the greater part of the region has less than 12 inches a year. Southwestern Arizona and southern California are the driest regions in the United States, and the rainfall of the lowlands is almost wholly confined to the winter months. In the rain shadow of the Sierra Nevada the mean annual rainfall is between 5 and 6 inches and locally as low as 3 inches. It is characteristic of the region that occasionally some portion of it receives a rather abundant rainfall. About once in six years the greater part of the rainfall of a locality comes in a single month. In 1891, for example, 2.5 inches or 93% of the annual rainfall of the lower Colorado Basin fell in January. In January, February, and March, 1905, 8 inches of rain fell at Yuma, Arizona, whereas the average annual rainfall of that place is but 2.7 inches.

The Rocky Mountain region rises to such a height as to provoke a heavier rainfall than the plateaus and basins on the west, though the remoteness of the region from the sea gives it a much lighter rainfall than occurs in the southern Appalachians of far lower elevation, or the relatively low Coast Ranges of Oregon. During the winter the western windward slopes of the Rockies are more heavily watered than the eastern, but the reverse is true for portions of the system during the spring and summer. In New Mexico and western Texas the mountains of the Trans-Pecos region have a greater rainfall than the surrounding plains and basins, but it is never heavy in an absolute sense. It comes in July and August and appears to be rather evenly distributed on all sides as if due to local updraft of air from plains to mountains. Nowhere does more than 50 inches of rain occur in the Rockies and the average amount is far smaller. The rainfall is very unevenly distributed, as might be expected owing to the irregular trends of the intermont basins and mountain ranges and the variable dispositions and heights of the mountains. The maximum precipitation occurs probably in northern Idaho; the minimum is between 6 and 8 inches and falls in San Luis Park in south-central Colorado.

Eastward of the Rockies as far as the Atlantic coast the topographic features lack great height, hence the rainfall distribution is controlled chiefly by the frequency and direction of movement of the rain-bearing cyclonic storms. The greater height of the Unakas, Great Smokies, and associated ranges in western North Carolina and South Carolina and northern Georgia cause their rainfall to exceed that of any other region east of the Pacific mountains. It is more than 70 inches a year. The rugged and high eastern portion of West Virginia, the Adirondacks, and the White and the Green mountains are other centers of heavy rainfall that owe their influence upon climate to their greater height. A heavier rainfall depending not on elevation but on nearness to the sea and to position within the track of frequent cyclones occurs in southern Louisiana and Alabama, — 60 to 70 inches.

The Great Plains region has a diminished rainfall owing to its position in the rain shadow of the Rockies and its remoteness from the sea.

Fortunately such rain as falls comes chiefly in the summer or growing season. From the 101st meridian to the Rockies the rainfall is from 10 to 15 inches. Eastern Colorado is a region of small precipitation, with an average fall of about 12 inches and a maximum yearly fall rarely in excess of 20 inches. In western Kansas the precipitation of the driest year was 9.9 inches; of the wettest, 33.7 inches. The last-named illustration is typical of the wide differences between the extremes of rainfall in the arid and semi-arid portions of the West. The wettest years have a rainfall sufficiently great for agriculture; the means and minima are far below the necessary amount.

Of the seven climatic and life provinces of North America, Plate I, but four fall within the limits of the United States, except that the southernmost or tropical province touches southern Florida and the lower valley of the Colorado, and the northernmost or Arctic province is developed on a few of the highest summits like Shasta in northern California and Blackfoot Mountain, Montana.

The Boreal province is developed in the southern Appalachians of western North Carolina, eastern West Virginia, the Catskills, the Adirondacks, the White and the Green Mountains, and in the Superior Highlands of northern Michigan and Wisconsin and on all the main divisions of the Pacific Cordillera, where its upper limit coincides with the timber line. It is marked by a low mean annual temperature, generally between 32° and 40°, by long, cold winters, and in general by a heavy snowfall. Its forest growth is spruce and pine in New England, spruce and balsam in North Carolina, chiefly white pine in Michigan and Wisconsin, and spruce, fir, and cedar in the Pacific Cordillera.

The greater part of the mountain forests of the United States is found in the Transition province which includes the cool temperate portions of the country with a generally high mean annual precipitation. The mean annual temperature is about 45°, but the temperature is in general marked by frequent and sudden changes. Snow falls throughout the entire province, though it is variable in amount owing to differences·of elevation, exposure, etc.

In the northern portion of this province both broad-leaved deciduous trees and conifers grow; similarly in the west, scattered growths of oak, piñon pine, and sycamore of the lower mountain slopes mingle or shade into the spruce and yellow and white pines of the upper slopes. The province as a whole has few distinctive plants; it is marked rather by the mingling of southern species that here find their northern (on the mountains their upper) limit and of northern species which find their southern (or lower) limit of occurrence.

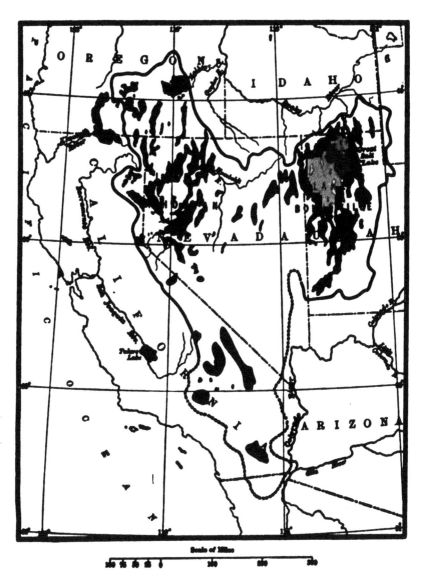

Scale of Miles

100 75 50 25 0   100   200   300

Plate I. — Climatic and Life Provinces of North America.

The western division of the Upper Austral province is so dry that agriculture without irrigation is impossible and tree growth is limited to the heads of the better watered alluvial fans and the banks of streams. The eastern portion of the province was originally covered with a varied and dense forest of hickory, maple, oak, and chestnut, as in the Ohio Valley and the Piedmont and the Appalachian Plateaus. The climate is warm temperate, with a long summer season.

The Lower Austral province resembles the Upper in the western part of the country in its general treelessness except along the streams. East of the 100th meridian the climate is wetter, and throughout the entire Gulf and Atlantic Coastal Plain the temperature and rainfall are favorable to the growth of great forests of southern species of pine and of cypress. The winters are very mild, snowfall is absent, and the long, hot summers have an abundant rainfall. The mean winter temperature is 40° to 52°, the summer temperature from 75° to 80°.

FOREST REGIONS

A single unbroken forest belt extends across North America, the spruce forest of Canada. Its northern border, a timber line determined by cold and physiological dryness, extends from Hudson Bay northwestward to the head of the Mackenzie delta, thence westward and southwestward across Alaska. Its southern margin is the 60th parallel in the Canadian Northwest and the 50th parallel in the Great Lake region. Black and white spruces, poplar, canoe birch, aspen, and tamarack are typical growths; the presence of only a few species of trees is characteristic. The spruce forest includes a large part of the lake region of North America with an abundance of lakes and swamps (p. 565). The spruces and the gray pine grow on the uplands between lakes and swamps; poplar, dwarf birch, willow, and alder occupy the cold wet bottom lands. While the trees attain fair size on the southern portions of the belt in which they occur, they are never large, and decrease notably in size toward the north, where they finally become so stunted as to be of little economic importance.

Southward from the broad transcontinental forest belt are an Atlantic forest, a Pacific forest, and a Rocky Mountain forest. The two intervening belts of country — the Great Plains and the Great Basin — are forestless though not treeless. This distribution is controlled largely by rainfall, though the distribution of species within each region is also controlled by insolation, temperature, wind velocity, water supply, and geographic relation to postglacial centers of dispersal. By the same token the forests are not distributed evenly over a given region,

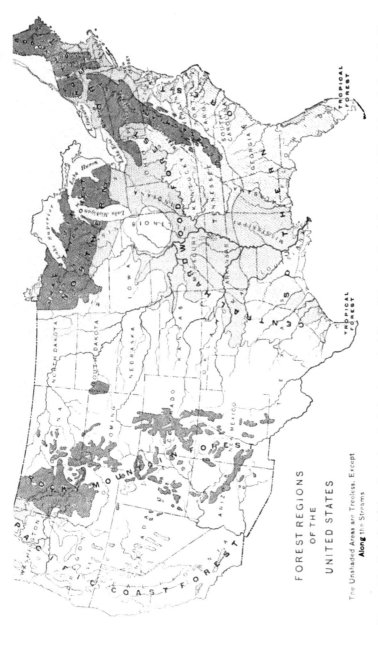

FOREST REGIONS
OF THE
UNITED STATES

The Unshaded Areas are Treeless, Except
Along the Streams

Fig. 23. — The effects of rainfall are clearly shown in the forest contrasts of East and West; the effect of elevation on rainfall and forests is shown in the Blue Mountains of Oregon, the Black Hills of South Dakota, the Trans-Pecos ranges, etc.; the effect of elevation on temperatures and forests is shown in the long arm of northern forest that extends south along the Appalachian system; the effects of high temperatures in low latitudes is shown in the tropical forest of southern Florida and southern Texas; the effects of ground water are most apparent on the western border of the central hardwoods area where finger-like extensions project westward along the river bottoms; the exclusion of the forest (1) by an excess of water is shown in southern Louisiana, (2) by a deficiency of water, in the unshaded portions of the West.   (Map from U. S. Dept. of Agri., Bur. of For.)

but vary from windward to leeward slopes, from warm southern to cool northern slopes, and from dry to wet situations as controlled by more local conditions.

The eastern or Atlantic forest tract consists largely of hardwoods, though large tracts of conifers are also found; the western forests are principally conifers of unrivaled size and beauty and hardwoods are comparatively rare. The great variety and in some cases the great size of the trees of the Atlantic forest are distinctive features. The primary divisions of the Atlantic forest belt are (1) a belt of conifers, (2) a white pine belt, and (3) a belt of southern pines.

The belt of conifers in which the white pine is the most important species extends from southeastern Canada and Massachusetts westward to northern Michigan, Wisconsin, and Minnesota. It reaches its best development in the light, dry, sandy soil of the glacial drift in the northern part of the southern peninsula of Michigan. White cedar, hemlock, fir, larch, and spruce are other conifers of minor importance in the tract.

South of the white pine belt is a belt of hardwoods which extends through the Great Lake states and the larger part of the great Appalachian region. The most notable types of trees are many species of oaks, several kinds of hickory, the chestnut, basswood, magnolia, tulip tree, and cottonwood. In this belt the hickories reach their greatest size in the Ozark region, the oaks in the central and southern portions of the eastern United States, the tulip tree in Kentucky. Among the exceptions to the characteristic growths of this division may be mentioned the spruce and hemlock forests on the summits of the Pisgah and other ranges in western North Carolina, where boreal conditions prevail.

The Atlantic and Gulf Coastal Plain is occupied by longleaf, shortleaf, loblolly, and slash pines. The first two occupy dry sandy uplands for the most part; the loblolly pine grows best on the drier portions of the moist lowlands of eastern Texas.

Among the trees of the Pacific coast forests in the United States the red or Douglas fir is the most important. It has its best development in the wet Puget Sound region, where it grows to a height of several hundred feet. It is associated with the tide-land spruce, hemlock, and red cedar. This portion of the Pacific forest is one of the densest and commercially one of the most valuable in the country. Southward and eastward the forest changes in character. In northern California, where the rainfall is heavy, the redwood is the most important forest tree, and between it and the fir forest on the north is a tract occupied by the Port Orford cedar. The dry and nearly treeless Great Valley of California divides the Pacific forest into an eastern and a western section. The well-watered Sierra Nevada in the east has a heavy forest growth of sugar pine, red fir, yellow pine, and hemlock besides the famous *Sequoia gigantea*. The dry Columbia Plateaus have a discontinuous forest in which pine and larch are the most important types. The Great Basin region is still drier and supports an even more limited growth of pine and juniper at the lower elevations, and a scanty growth of fir and spruce at the higher elevations.

The Rocky Mountains forest extends from the borders of the surrounding plains and the intermont basins up to elevations of 9000 to 11,000 feet and is much broken into forest islands by the restricted areas of mountain land which are sufficiently high to provoke an adequate rainfall from the prevailing westerly winds. Spruce grows luxuriantly at elevations ranging from 8000 to 10,000 feet, and at lower altitudes yellow pine, red fir, and white fir are abundant. This growth is characteristic of the Rockies as far south as New Mexico, where the lower elevations of the mountains of the Trans-Pecos region cause the forest to disappear or to become restricted to a few higher ranges such as the Sacramento and Davis Mountains of western Texas and eastern New Mexico.

Two small forest tracts of exceptional character deserve mention in even this brief description. In Florida the subtropical climate has given rise to an Antillean type of flora among which mahogany, royal palm, and mangroves are of chief interest. The second tract is in southernmost Texas, where vegetation occurs whose affinities are with the Mexican flora.

CHAPTER X

COAST RANGES

THE relief of the relatively dry western half of the United States has a high importance in the distribution of the forests because of the effects of relief upon two of the controlling factors of forest growth — temperature and rainfall. We shall therefore begin our consideration of the physiography of the United States by a study of the West. Of the 190 million acres of national forests, more than 185 million occur west of the eastern front of the Rockies, a fact that further increases the forester's interest in the topography, drainage, soils, and rainfall of this vast region.

SUBDIVISIONS

The Coast Range system of mountains in the United States extends from southern California to the Straits of Juan de Fuca. The mountains of the system do not have uniform topographic qualities throughout but consist of four somewhat dissimilar sections. The southern section extends from southern California to the 40th parallel; farther north are the Klamath Mountains, which extend from the 40th to the 43d parallel; the third section embraces the low Coast Ranges of western and northwestern Oregon; the fourth includes the Olympic Mountains, which rise to heights of over 8000 feet and are the highest mountains of the system next to those in southern California.

COAST RANGES OF CALIFORNIA

The Coast Ranges of California terminate on the northern margin of Humboldt County (p. 141); beyond them to the northeast lie the Klamath Mountains, which are more closely allied to the Sierra Nevada Mountains than to the Coast Ranges in rock character and geologic history though not in geographic position. The Coast Ranges of California are sometimes regarded as ending on the south in Santa Barbara County, there giving way to the mountains of southern California. We shall here include the mountains of southern California with those of the coast of California to the 40th parallel in a single coast group because of (1) the extension of the tectonic lines of the northern mountains into the mountains of southern California and (2) the closely related fact that the movements along these lines — movements to which the larger topographic features are due — date in both cases from the close of the Tertiary. A unity of both structural and topographic characters is thus given the entire group of ranges. The group is divided, however (largely on the basis of trend), into three subgroups: (1) the Coast Ranges proper; (2) a broad chain extending from Santa Barbara County to the eastern and southeastern side of the Colorado desert, with a general trend westnorthwest, and including the San Rafael, Santa Ynez, Santa Susannah, Santa Monica, and San Gabriel ranges, a chain parts of which are known locally as the Sierra Madre, though the application is neither uniform nor clear; and (3) the mountainous country of the Valley of southern California. The principal ranges of this group are the northwestwardtrending Santa Ana and San Jacinto mountains, sometimes called the Peninsular chain.[1]

On both the north and the south the Coast Ranges are from 5000 to 8000 feet high; the elevations of the central portions are from 3000 to 4000 feet. San Lucia Peak, the highest peak of the central Coast Ranges, is less than 6000 feet high. In general the crests range from 2000 to 4000 feet.

The eastern margin of the Coast Ranges of California rises abruptly from the floor of the great central valley of California as a well-marked continuous mountain front. At its southern end it is a dissected fault scarp; elsewhere a smaller amount of faulting has taken place, but everywhere the eastern border represents a line of strong deformation. In a broad view the western margin of the Coast Ranges is not at the shore line but at the edge of the continental platform on the 600-foot submarine contour, where the sea bottom changes its slope abruptly from a previously gentle incline and plunges steeply down to depths of 8000 feet and more.[2] At the foot of this steep decline the sea bottom again assumes low gradients. The slope constitutes a notable mountain front rising from the floor of the Pacific and forming the natural western boundary of the coast system of mountains. It is interpreted as a great submarine fault scarp or series of fault scarps comparable to those that form the eastern front not only of the Coast Ranges of California but also of the Sierra Nevada. At the base of the steep submarine scarp, dredgings (Tuscarora explorations) at the depth of 12,000 feet have brought up fragments of bituminous shale which are con-

[1] A. C. Lawson and others, Section on Geology, The California Earthquake of April 18, 1906, Carnegie Inst., vol. 1, pt. 1, pp. 2, 3 et al.

[2] Andree's Handatlas, bathymetric chart, No. 157.

GEOMORPHIC MAP.

OF

CALIFORNIA

SHOWING THE DIASTROPHIC
CHARACTER OF THE RELIEF
AND THE MOST IMPORTANT
KNOWN FAULTS

Scale of Miles

Fig. 24. — Map of California. The heavy lines indicate the principal faults.

sidered to be talus débris of so recent origin as not yet to have been buried by oceanic sediments.

The coastal scarp above sea level is not everywhere regular in development; the most noticeable interruption is the Bay of Monterey and adjacent slopes, which form parts of a synclinal trough whose axis is at right angles to the trend of the coast and of the Coast Ranges as a belt. A second interruption of the continuity of the coastal scarp is at the Golden Gate and is due to a depression of the Coast Ranges which resulted in the drowning of the lower portions of land valleys that formerly crossed the coastal mountains. The Point Rees peninsula is a third important break in the continuity of the coast line of California and is due to the manner in which the depression east of the ridge forming the peninsula has been drowned; the northern end of the valley is occupied by Tomales Bay, the southern by Bolinas lagoon.[1]

The Coast Ranges of California consist on the whole of a series of parallel ridges composed of sedimentary strata (Cretaceous[2] and later) that have been deformed on broad lines by deep-seated causes. There has been crumpling of the strata besides a certain amount of igneous eruption, but the major features are due to the effects of great dissection and later block faulting on a large scale, attended and followed by erosion. The character of the relief is in many cases markedly diastrophic and the relief features commonly have the rectilinear quality associated with pronounced faulting, which also explains to a large degree the parallelism of the ridges. Examples are Castle Rock Ridge, Cavilan range, the Santa Cruz range, and many others whose borders are marked by the San Andreas fault, the Castle Rock fault, etc., which during the California earthquake of 1906 were the loci of maximum earthquake intensity.[3]

The most important line of faulting in the Coast Ranges of California is the Rift, as it has been termed, or the San Andreas Rift, a name taken from the San Andreas valley of the peninsula of San Francisco. The Rift is a continuous topographic depression for at least 190 miles from Point Arena to San Juan, and in this part of its course is nearly straight, following an old line of seismic disturbance which has a much greater extent — that is to say, from southwest of the Point to southern California, or about 600 to 700 miles. Indeed the Rift may extend much farther to the south and may be associated with the origin of the

[1] Lawson, loc. cit., pp. 12–15.

[2] For geologic time names consult Appendix D.

[3] Atlas of maps and seismographs accompanying the Report of the State Earthquake Investigation Commission upon the California Earthquake of April 8, 1906, Maps 1, 22, and 23. Also the Santa Cruz quadrangle, U. S. Geol. Surv.

Colorado desert and the Gulf of California. The physical habit of the Rift valley, for example in the Bolinas-Tomales section, is that of a remarkably straight depression, with the southwestern wall steep, the northeastern wall gentle. The character of a pronounced topographic depression, however, is not everywhere sustained.

The southern end of the great Rift may be traced for an unknown distance along the base of the mountains bordering the Salton Basin upon the northeast, where it probably dies out gradually. It is coincident with long and narrow valleys whose orientation is controlled by faulting along the Rift but whose detailed features are in large measure determined by erosion upon the exposed edges of formations of varying hardness. The depressions which constitute the major Rift along the southern margin of the Mohave desert appear to be almost wholly diastrophic. The steep northern flank of the San Rafael and San Gabriel ranges on the south side of the Mohave desert are degraded fault scarps, the walls of the great Rift valley. The exact share in all these various sections of the Rift valley that may be ascribed on the one hand to crustal deformation of the fault block type and on the other to erosion has not been determined. For miles at a stretch the earth on one side or the other of the fault in the southern part of the Coast Ranges has sunk in such manner as to give rise to basins and cliffs measured in terms of several hundred feet.

The individual ridges of the Coast Ranges of California have a pronounced parallelism in a direction somewhat oblique to the main trend of the coast, so that they tend constantly to emerge upon the coast in the form of northwestward-trending peninsulas. The courses of the longitudinal valleys correspond either to the strike of the rocks or the trend of the fault lines and are oblique to the general trend of the coast range belt. The general drainage is therefore termed subsequent, for the streams have extended themselves along belts of weak rock or along fault depressions at the expense of an earlier drainage crossing the region in a westerly direction or transverse to the structure. Short sections of the streams cross the ridges in steep-sided valleys or gorges, and these only may be termed antecedent.[1]

The tops of the ridges in some respects are more or less flat and present the character of a rolling, mature upland; but more commonly they are determined by the intersection of the slopes of adjacent valleys; even in the latter case, however, it is generally true that the ridge crests over wide areas reach about the same altitude and in a broad view give the impression of an upland with fairly uniform elevations and gentle

[1] Lawson, loc. cit., p. 20.

slopes. The stream valleys, cut below the level of the dissected upland, are usually wide-bottomed in the softer and narrow-bottomed in the harder rocks.[1]

The Santa Lucia Range illustrates many of the general features of the region. It is the dominant mountain range of the coast of California for over 100 miles between latitude 35° and 36°30′ N., Fig. 24. For much of this distance it rises boldly from the Pacific Ocean and forms the most picturesque portion of the California coast. In places the spurs of the range terminate in cliffs several hundred feet high; in other places the range is bordered on the seaward side by a gently sloping platform or terrace which is from 40 to 80 feet high on its cliffed outer margin and 100 feet high on its inner margin. This platform is primarily a wave-cut terrace, though its surface is thinly covered with wash from the bordering hills.[2] The Santa Lucia Range has an even sky line many miles long, and a summit from 2 to 4 miles wide. Its regular front is a bold, compound, fault scarp. The range is traversed by narrow canyons which open out headward into broad valleys in an advanced stage of topographic development. In this respect the range resembles many others among the Pacific mountains. An earlier surface, in some places softened and subdued with moderate waste-covered slopes, in other places a true peneplain, was deformed by faulting. The summit levels of the uplifted fault-blocks (the present ranges) display remnants of the ancient smoothly-contoured surface in strong contrast to the steep borders of the ranges sharply outlined by more recent faulting and now in process of vigorous dissection.

Large portions of the Coast Ranges are unknown even through reconnaissance surveys. Among the known portions the Santa Cruz section between San Francisco Bay and the Bay of Monterey presents features of special interest. Here the parallelism of the valleys and ridges is apparent in the larger features of the topography but is less marked or absent in the minor relief. The main ridges have a steep northeast slope bordered by a series of valleys lying along the San Andreas Rift. The lines of the major folds of the Santa Cruz region are marked by more or less continuous valleys, and in the case of both these larger valleys and the main ridges the topographic and geologic features are in sympathetic relation.[3] The hillsides of the region are generally covered with a deep coating of soil, and cliffs are rare, owing both to the friability of most of the rocks and to the advanced state

---

[1] Lawson, loc. cit., p. 20.

[2] H. W. Fairbanks, San Luis Folio U. S. Geol. Surv. No. 101, 1904, p. 1.

[3] Branner, Newsom, and Arnold, Santa Cruz Folio U. S. Geol. Surv. No. 163, 1906, p. 1.

of topographic development which was reached before the last and recent uplift. An unusual feature of the topography of the Santa Cruz region is the occurrence of very steep yet soil-covered hillsides; 35° to 40° slopes are not uncommon, and in one place is found a soil- and vegetation-covered hillside with a slope of 50° from the horizontal. There is a dense growth of timber and underbrush over much of the area, which does not prevent the thick covering of soil from being frequently involved in landslides in the belts of greatest faulting and folding.[1]

The Coast Ranges of northern California include, besides the mountains proper, a coastal tract which was eroded (Pliocene) to the form of a peneplain. The coastal peneplain was then uplifted and its streams intrenched; it now forms a dissected plateau with long and roughly level-topped ridges separated by equally long, narrow valleys. The ridges are remarkably constant in general altitude, and the sky line is essentially level. In a general perspective the view is that of a plain or sloping plateau of low relief. The peneplain was uplifted to an elevation of 1600 feet above the sea on the seaward margin, and to 2100 feet on the inner margin. The mountainous tract adjacent on the east participated in the same movement. In Humboldt County several sharp peaks rise abruptly above the general level of the dissected plateau to 4000 or 5000 feet, but they are clearly encircled by remnants of the plateau which give to the mid-slopes of the peaks a distinctly terraced aspect. The peneplain may be followed in among clusters of mountain peaks and ridges and extends at least as far as the Bear River ridge. That the present dissected plateau was once a peneplain is inferred from the facts that the rocks composing it are of varying ages and of varying degrees of hardness, and that the general surface of the region bevels rather evenly across the deformed strata. On the summit of some of the ridges of the plateau numerous water-worn pebbles have been found, at 1600 feet, which are reasonably interpreted as remnants of larger bodies of stream gravels formed upon an erosion surface.[2]

The coastal peneplain grades into a region of stronger relief on the east where the stream courses were still completely under the control of geologic structure at the end of the first erosion cycle and flowed in mature subsequent valleys which were inherited by the streams of the second or present cycle of erosion. The abrupt coastal margin of the uplifted peneplain of northern California has given rise to a youthful

[1] Branner, Newsom, and Arnold, Santa Cruz Folio U. S. Geol. Surv. No. 163, 1906, p. 10.
[2] A. C. Lawson, The Geomorphogeny of the Coast of Northern California, Univ. Cal. Bull., Dept. Geol., vol. 1, pp. 242-244.

topography along the coast; the coastal canyons are narrow and precipitous, and V-shaped profiles predominate.   In the middle stretches of
the streams degradation is less intense and the topography appears
somewhat less rugged.

Recent events following the uplift and dissection of the coastal peneplain of California are a subsidence of at least 370 feet at the mouth of
the Sacramento River which flooded the lower portions of that valley
and gave rise to the magnificent harbor of San Francisco.   The drowned
mouth of the river once discharging across the Coast Ranges at this
point is known as the Golden Gate.   The last episode in the region
has been a slight uplift in the vicinity of the Straits of Carquinez.[1]

Fig. 25. — Coastal terraces produced by wave erosion, west of Santa Cruz, California. (U. S. Geol. Surv.)

The uplift of the coastal peneplain of northern California was not
accomplished in a single continuous movement but was interrupted
by many halts.   During these periods of relative stability there were
formed well-developed ocean terraces which are among the most prominent features of the coastal topography.   Such terraces were always
involved in later uplifts and now stand at high levels, the highest
representing the algebraic sum of all coastal changes whether of uplift
or depression since the beginning of the last series of changes in the
level of the land.   The highest terrace of northern California is about

[1] A. C. Lawson, The Geomorphogeny of the Coast of Northern California, Univ. Cal. Bull.,
Dept. Geol., vol. 1, pp. 270–271.

1500 feet above sea level. Below this are prominent terraces at 1400, 1180, 760, 440, 350, and 280 feet, respectively, above sea level, with many less prominent terraces at intermediate levels. The lower terraces have all the characters associated with wave and current origin, such as a rather regular seaward slope, upturned strata smoothly planed off, residual stacks, beach bowlders, and sea cliffs with horizontal base lines. The higher ones are usually not so clear, though even the highest have sufficient definition in the form of sea cliff, sloping terrace, and bowlder beach to make its character certain.[1]

MOUNTAINS OF SOUTHERN CALIFORNIA

In the southernmost division of the coastal mountains of California (see p. 127) faults have also played a very important part in the topography. Both the northern and southern sides of the San Gabriel range are determined by a profound fault; the range may be interpreted as a horst thrust up between two bounding faults. Since uplift the range has been thoroughly dissected and older surfaces of erosion destroyed.

"One seeks in vain for horizontal lines along the San Gabriel tops; a confusion of peaks and ridges of discordant and seemingly unrelated heights makes up the mountain mass. . . . [They] present a labyrinth of canyons and ridges and peaks, with no level areas of any size. The ridges have narrow summits; the peaks are sharp; the streams are all evenly graded from source to mouth."[2]

The Santa Ana Mountains are a tilted, seaward-sloping mountain block with a very straight and abrupt fault scarp that faces the northeast and overlooks the Perris plain. The block is an elevated and as yet but little dissected peneplain (Cretaceous) with remnants of younger (Tertiary) deposits upon it, indicating that it has in part at least been resurrected in recent times from a buried condition. It is thought that the same tilted block structure extends beyond the Santa Ana Mountains southward to the international boundary and even beyond. Both sides of the San Jacinto Mountains are precipitous and probably determined by faults, so that the ridge has very bold margins.

Among the drainage features of these mountains are the interesting valleys of the Santa Ana and Santa Margherita rivers which are antecedent to the tilting of the region; they persisted in their southwestward courses during the development of the fault scarps, and now cut squarely across the range, draining the valley lands on the northeast.[3]

[1] A. C. Lawson, The Geomorphogeny of the Coast of Northern California, Univ. Cal. Bull., Dept. Geol., vol. 1, pp. 246–247.

[2] W. C. Mendenhall, Ground Waters and Irrigation Enterprises in the Foothill Belt, Southern California, Water-Supply Paper U. S. Geol. Surv. No. 219, 1908, p. 17.

[3] A. C. Lawson and others, The California Earthquake of April 18, 1906, Carnegie Inst., vol. 1, pt. 1, pp. 23–24.

SAN BERNARDINO RANGE

The San Bernardino range of southern California is a distinct topographic unit and does not have a close genetic relationship with the other members of the Coast Range System in southern California. It is much younger than the San Gabriel range and appears to have had a history different from that of the San Jacinto range south of it. The relief of the mountains is outlined upon an uplifted fault block once of somewhat more regular development than at present. Remnants of an old surface of moderate relief, broad elevated valleys,

Fig. 26. — Redlands and San Bernardino and San Gorgonio Peaks, San Bernardino Mountains, California. (Mendenhall, U. S. Geol. Surv.)

plateau-like ridges, and several interior basins like those in the Mohave desert on the north, are the principal secondary topographic elements. At its western end is displayed a long even sky line at elevations between 5000 and 6000 feet above the sea.

n.   " . . . there are many wide upland valleys, forested and grassy glades, and lakes or playas
.   "ke Bear Lake and Baldwin Lake. Where these upland levels are attained it is difficult to
invc.lize that one is actually in the high mountains. The surrounding topographic forms are
repreaded and gentle, the level areas are extensive, the streams meander placidly through broad
or dep ws, and the topographic type is that of a rolling country of moderate elevation. But
level ol edge of these interior uplands is approached the streams plunge into precipitous canyons,
        bes are as steep as earth and rock can stand, the roads and trails twist and turn and
     ¹ A. C. to find a devious and precarious way to the valleys below." [1]
Dept. Geo C. Mendenhall, Ground Waters and Irrigation Enterprises in the Foothill Belt, South-
fornia, Water-Supply Paper U. S. Geol. Surv. No. 219, 1908, p. 17.

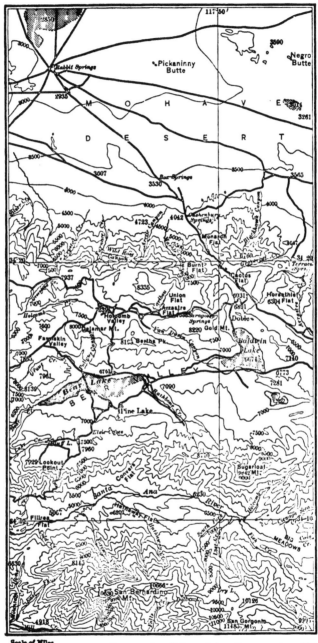

Scale of Miles

0 ½ 1   2 - 3   4   5

Contour interval 500 feet

Fig. 27. — Bear Valley and the adjacent country exhibit the subdued relief characteristic of the interior of the San Bernardino
Mountains.   The parallel of 34° 20′ coincides with a fault
scarp on the northern border of the range; the southern edge of the map represents a part of the southern
border of similar origin.   Both borders are deeply dissected.   Note the withering streams, foreland
plain, and playa of the Mohave desert.   (San Gorgonio quadrangle, U. S. Geol. Surv.)

The mountain mass was blocked out of a portion of the earth's crust that was at one time continuous with the Mohave desert region and the San Bernardíno Valley. The highest portion of the range is a rather sharp ridge about six miles long culminating in San Bernardino Mountain (10,630 feet) on the west and San Gorgonio Mountain (11,480 feet) on the east.

A surprising feature of the topography of the range is the occurrence of glacial cirques, moraines, and basin-like depressions. The southern limit of glaciation in this longitude has until lately been thought to be somewhat farther north, so that this occurrence of glacial features is one of the southernmost in the country.[1] Both the northeast and the northwest slopes of the main ridge appear to have been glaciated, the snowy accumulations having been formed in alcoves near the summit where drifted snows still gather. At the head of Hathaway Creek are five semicircular terminal moraines a mile and a half below the cirque-like basin close under the crest. Of interest in this connection is the fact that the summit of the range still contains a distinctly boreal fauna and flora.[2]

THE KLAMATH MOUNTAINS

In the second major division of the Coast Ranges dominated by the Klamath Mountains and designated the Klamath sub-province there are three well-marked subdivisions: (1) a narrow coastal plain, (2) high marine terraces, and (3) a well-dissected plateau.

The coastal plain is from one to five miles wide, and its inner margin stands several thousand feet above the sea. Swamps border the expanded lower courses of the streams where lagoons have been formed back of the sand reefs that fringe the coast. An interesting feature of portions of the coast is the variable position of the stream mouths through the year; the south and southwest storms of winter produce a coastal drift northward and the inlets through which the rivers discharge are moved in this direction; but when the northwest winds of summer prevail the movement is southward. In many places the winds have blown the reef sands into dunes whose shifting character may long prevent tree growth. There appears to be a natural limit to this action, however, for each locality, so that ultimately lower forms of vegetation take hold of the sand and bind it, allowing the trees to come in. The

[1] Other southerly localities where glacial features have been found are (a) near Santa Fe, (b) on San Francisco Mountain, (c) near Nogales, etc. For a résumé of these occurrences see D. W. Johnson, The Southernmost Glaciation in the United States, Science, n. s., vol. 31, 1910, pp. 218-220.

[2] Fairbanks and Carey, Glaciation in the San Bernardino Range, California, Science, n. s., vol. 31, 1910, pp. 32-33.

action is well illustrated along the inner margin of some of the dunes near Coos Bay, Oregon. Locally dunes have been driven inland so far from the source of sand supply, a mile or more, as at last to make little progress and to become covered with a forest growth.

On its inner margin the coastal plain has been moderately dissected; the outer margin of the plain still bears marks of extreme youthfulness in the form of coastal lagoons and recent marine sediments. Occasional rock stacks persist, of which Tupper Rock is a conspicuous illustration; they represent harder or more favorably located rock masses that withstood the wave erosion which carried away the softer surrounding rocks. Although the coastal plain of this part of Oregon is narrow it contains by far the greater part of the people of the region, a fact due to its flat tillable surface and the dark, rich loam which favors the interests of agricultural people.

The ascent from coastal plain to high-level plateau is made by a series of terraces sculptured upon the prominent spurs that define the interfluves. Ancient sea cliffs with ancient beaches at their foot alternate with long gentle slopes marking the wave-cut terraces that once extended seaward from the cliff as a submarine platform. The terraces range in height from 500 to 1500 feet. At the latter elevation is a well-marked, though discontinuous, sea cliff which has been traced for many miles along the coast. The preservation of these old cliffs and benches of a former shore line at such high elevations above the sea are suggestive of the rapidity that characterizes uplift on these shores.

The Klamath Mountains proper embrace all those peaks and ridges lying between the 40th and 43d parallels. Some of their most conspicuous members are the Salmon, Trinity, and Scott mountains of California and the Siskiyou and Rogue River mountains of Oregon. The mountains are composed in large part of rocks similar to those found in the Sierra Nevada, — limestone, sandstone, shale, schist, diabase, etc., — with traces here and there of lavas having a close relationship to those of the Cascade Mountains; in late physiographic history and in geographic position, however, they are related to the Coast Ranges.[1]

The dominating physiographic feature of the Klamath sub-region is the Klamath plateau. From one of the higher summits a general view of the landscape may be obtained which shows that while there are many small irregularities, the summit levels approximate a general plane with moderate inclination toward the sea. The elevation of the plateau is from 2000 feet on the west to 4000 and 5000 feet and more on the east. In many places decidedly flat summits may be noted, so that in a

[1] J. S. Diller, Roseburg Folio U. S. Geol. Surv. No. 49, 1898, p. 1.

general view the surface appears to be a practically level-topped plateau deeply trenched by streams. The South Fork range in Trinity County, Oregon, has an even sky line more than 40 miles long at an elevation exceeding 5000 feet in spite of its variable structure. Such a relation of surface to structure is indicative of a long erosion period in which rocks of diverse altitudes, hardnesses, etc., were brought to essentially the same level; in short, that the region was peneplaned, that is, reduced by long-continued erosion at one level to the form of an almost featureless plain. Uplift is indicated not only by the relatively high level at which the plain, once formed at sea level, now stands, but also by deep dissection.

The fact of early peneplanation and later dissection is also well shown by a comparison of the upper and lower valley slopes. The lower portions of the valleys are in general narrow and canyon-like, with prevailingly steep descents, while the upper portions of the valleys are wide and the slopes gentle. The upper gentle slopes are the slopes of an early valley system which is now being destroyed by the present drainage cut far below the old level since the uplift of the region. One of the best preserved of the early valleys is the Pitt River valley. The level of the broad, shallow, old valley of the Pitt is but 500 feet below the flat backbone of the ridges across which its course is directed, and is in very strong contrast to the deep, narrow, canyon-like valley of the present river. Traces of earlier valleys may also be found on the uplands along the McCloud and Little Sacramento valleys.

An interesting fact which bears upon the origin of the older valleys and the former existence of a peneplain is the occurrence at Potters Creek cave of the bones of some forty species of animals of which at least seventeen, including the mastodon, elephant, and tapir, are extinct. The character of the fauna indicates low relief and a condition quite out of harmony with the present topography.[1] The low relief that must have existed here when the peneplain was nearing its latest stages of development is also indicated by the fine character of the corresponding sediments (Ione formation) which like the characters of the fossil flora and fauna suggests a flat coastal region whose climate was not notably different from that of Florida to-day.[2]

The Klamath peneplain in an uplifted and deeply dissected state has been traced southwestward to the head of the Sacramento Valley, California, where the slopes of the mountains become gentler as they approach the

---

[1] J. S. Diller, A Preliminary Account of the Exploration of the Potters Creek Cave, Shasta County, California, Science, n. s., vol. 17, 1903, pp. 708–712.

[2] J. S. Diller, Redding Folio, Cal. U. S. Geol. Surv. No. 138, 1906, p. 10.

highest summits. These flattish crests approximate a general plain and indicate that the region was one of gentle relief before the last uplift.

Turning now to the more rugged interior portions of the Klamath district we find that the main ranges fall into two rather well-defined systems which cross each other nearly at right angles. The most

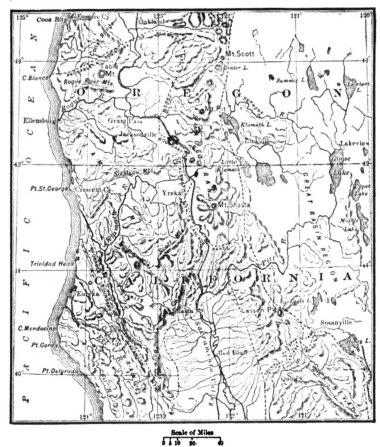

Fig. 28. — Boundaries between the Sierra Nevada, Cascades, Coast Ranges, and the Klamath Mountains. The Lassen Peak volcanic ridge extends from the Pitt River on the north to the North Fork of the Feather River on the south. It is the southern part of the Cascade Range. Goose Lake discharges into Pitt River only at long intervals. (After Diller, U. S. Geol. Surv.)

prominent ranges have an east-west trend, as the Rogue River, Siskiyou, Scott, and Trinity; on the other hand the Yallo Bally, Bally Choop, South Fork, and Salmon River mountains and many less important ridges run approximately in a northerly direction. Even in the central portion of the group, as between the Rogue River and the Trinity valleys, the north-south trends are apparent.

The mountain-rimmed basins of the region have two characteristic features; all receive the drainage of comparatively large areas, and each is drained by a main stream that leaves the basin to cross a bordering range through a deep canyon. Scott Valley, for example, 8 by 25 miles in extent, has a nearly level floor, and an extensive system of centripetal tributaries; it drains through an almost impassable canyon more than 20 miles long. Other interior valleys of this type are Hay Fork, Trinity, and Illinois.

It seems clear from the persistent manner in which the ranges lie athwart the main drainage lines that the ranges were developed after the drainage had become well established. It is conceived that the eastward-trending ranges had been developed, peneplanation had been accomplished, and the streams by gradual development had gained courses westward to the sea when elevation along north-south axial lines deformed the peneplain and gave rise to a system of cross ranges with intervening structural valleys and valley basins.[1]

The deformations of the Klamath peneplain were sufficiently acute to cause the north-south ranges to stand well above the general level of the broadly uplifted portions of the peneplain. The east-west ranges in the higher and more rugged portions of the Klamath group have a residual relation to the uplifted and dissected peneplain about them. They represent unreduced elevations and display a boldness of form and an irregularity of relief in sharp contrast to the plateau character of the marginal tracts of the region. They are nowhere lofty, however, nor does their ruggedness in any place have alpine characteristics. In general their elevation exceeds the elevation of the bordering peneplain from 2000 to 4000 feet.

COAST RANGES OF OREGON

The Coast Ranges of Oregon constitute the third member of the Coast Range System of mountains. Their geology and geography have not yet been studied in sufficient detail to make generalization very profitable. It is known that they consist in part of sandstones (Eocene) and in part of volcanic rocks, the latter type constituting a considerable part of the ranges south of the Columbia River.[2] In the Coos Bay region a portion of the Coast Ranges of Oregon has been described as

---

[1] F. M. Anderson, The Physiographic Features of the Klamath Mountains, Jour. Geol., vol. 10, 1902, pp. 144-159.

[2] Willis and Smith, Tacoma Folio U. S. Geol. Surv. No. 54, 1899, p. 1.

exhibiting somewhat flat though narrow hill and ridge crests from which steep slopes descend to the valley floors.[1] Farther north similar qualities are exhibited, and in addition there are a number of rather flat-topped tablelands which represent remnants of an elevated peneplain. The even summits rise to maximum heights from 1200 to 1700 feet above the sea south of the Columbia River. Above them are upper mountain slopes and a considerable number of peaks against which the plain breaks abruptly. The peaks form true monadnocks among which Saddle Mountain displays typical features and relations.[2]

While there is great variability in rock hardness from point to point it is notable that in general the rocks are so soft as to permit rapid erosion wherever the forest is removed. Under natural conditions erosion is prevented by an extremely dense vegetal covering which not only breaks the force of the heavy rains but also binds the soil and delays run-off in other familiar ways.

The Coast Ranges of southwestern Oregon almost meet the western spurs of the Cascades. The Willamette Valley narrows toward its head, and beyond it and to the south are other streams of still more restricted valley development. At Roseburg (43° 10′), the depression between the ranges narrows to but fifteen miles. The foothills of the Cascades form a prominent though not a precipitous mountain border. The streams descend from the long, sloping western flank of the volcanic tableland of the Cascades and emerge from their rugged canyons to enter the more open valley stretches of the Umpqua, or Rogue, or Klamath rivers. Among these valleys the Umpqua alone lies north of the Klamath Mountains. It maintains an open character for but a short distance, however, then strikes boldly into and across the Coast Ranges, where its valley becomes a canyon. The most remarkable feature of the canyon is its winding course, which appears to represent the meanderings of its stream in an earlier topographic cycle when it flowed upon the surface of a peneplain now represented by the even and accordant crest lines of the Coast Ranges. The courses of the master streams, as the Nehalem in northwestern Oregon, resemble the Umpqua in the manner in which they cut across the mountains. They gained their courses on the coastal peneplain and since its uplift to form the Coast Ranges they have persisted in their courses. The smaller streams all show a sympathetic relation to the structure; their valleys in general follow the outcrop of the softer rocks.

[1] J. S. Diller, Coos Bay Folio U. S. Geol. Surv. No. 73, 1901, p. 1.
[2] J. S. Diller, A Geological Reconnaissance in Northwestern Oregon, 17th Ann. Rept. U.S Geol. Surv., pt. 1, 1895-96, pp. 449, 488.

The eastern front of the Coast Ranges of Oregon is a bold, partly dissected fault scarp, about 2000 feet high, formed of massive sandstone which stands above the lowland developed upon the shales and thin-bedded sandstones east of it. The mountain spurs running westward to the sea have a longer and gentler descent than those extending eastward to the Willamette Valley. The western spurs terminate on the coast as prominent and cliffed headlands connected by stretches of sand beach covered with a dense growth of grass and ferns.

Great terraces have been developed on the coastal margin of the Coast Ranges of Oregon just as on the western borders of the Klamath Mountains and the Coast Ranges of California. Since their development the terraces have been uplifted to heights of hundreds of feet, the highest attaining an elevation of 1500 feet. Above this elevation uniformity of level is less marked but still sufficiently marked to indicate the existence before the last uplift of an extensive plain of erosion now maturely and deeply dissected by the rejuvenated streams.[1]

OLYMPIC MOUNTAINS

The Olympic Mountains are the most conspicuous member of the northernmost section of the Coast Range System. They lie north of the Columbia River and west of Puget Sound. Like the Cascades the dominant peaks are volcanoes that rest upon a much older schistose rock. The highest peak of the Olympics, Mount Olympus, rises 8200 feet above the sea, and crowns a magnificent range in full view from the eastern side of the Sound. The higher mountains are alpine with sharp spires and serrate ridges from 6000 to 8000 feet high. The mountains have a roughly circular form and are about 40 miles across. The drainage of the region is radial, the streams being arranged much like the spokes of a wheel of which the region of high mountains is the hub; it has been suggested that this feature is due to the domed warping of a former flattish surface of erosion.[2]

The uplift of the mountains is still progressing, or at least uplift has occurred in postglacial time, as shown by the gently folded and tilted glacial clays, sands, and gravels in the vicinity of Port Angeles. The range is one but little known to-day on account of the ruggedness of the country, the fallen trees, the lichen-covered rock slopes, and the extreme density of the tangled underbrush. Because of the high degree of humidity, the great rainfall, and the equable and moderate temperature

---

[1] J. S. Diller, Roseburg Folio U. S. Geol. Surv. No. 49, 1898, p. 4.
[2] Ralph Arnold, Geological Reconnaissance of the Coast of the Olympic Peninsula, Washington, Bull. Geol. Soc. Am., vol. 17, 1906, pp. 451–468.

the mountain slopes are clothed with an almost impenetrable forest up to an altitude of 7000 feet. The Olympic forest is the densest in Washington and with few exceptions the densest in the country. It consists chiefly of red fir and hemlock.

CLIMATE, SOIL, AND FORESTS

The Coast Ranges extend through 20 degrees of latitude and lie in two distinct climatic belts, the belt of the westerly winds and the horse latitude belt. Southern California lies wholly in the latter belt, where the rainfall is scant on the lowlands and limited on the highlands or mountains, yet sufficient on the higher exposed slopes to support a forest growth. Its forests are in general of small extent, although a number of districts in the coastal ridges and the San Bernardino Mountains have good stands of timber. The Forest Service estimates the total standing live timber of merchantable size in this district, at approximately 1 % of the total for the State. At present there is in some localities a tendency toward eucalyptus culture, which may eventually have a beneficial effect upon the run-off of the region,[1] besides supplying the demand for a hardwood, one of the great defects of the Pacific forests.[2] It is, however, particularly sensitive to cold and especially will not endure frost, hence the range of its culture will be distinctly limited.

The Coast Ranges south of San Francisco lie in the belt of winter rains and are almost rainless in summer months. Nearness to the sea, however, brings climatic responses of great importance to vegetation, even in summer. The regular northwest winds of summer blow from the sea and for several months are accompanied by cool, damp fogs which sweep inland forty to fifty miles. They temper the hot summer weather, depress the rate of evaporation, and in the lands they overlie they make possible the production of certain crops without irrigation.

The larger part of the rainfall occurs on the western slopes of the westernmost ranges, decreasing on each range in eastward succession. It is nowhere sufficient to support a true forest vegetation, except immediately south of San Francisco where the Coast Ranges are covered with a heavy growth of timber and underbrush. At the heads of the valleys which drain the higher portions of the Santa Cruz Range

1 Van Winkle and Eaton, Quality of the Surface Waters of California, Water-Supply Paper, U. S. Geol. Surv. No. 237, 1910, p. 65.
2 Betts and Smith, Utilization of California Eucalypts, Circular U. S. Forest Service, No. 179, 1910, p. 6.

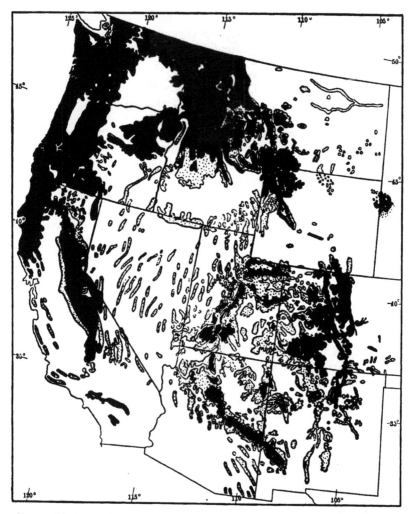

Fig. 29. — Distribution of western forests and woodlands. Solid black represents continuous forests; dotted areas represent woodland, that is, a thin scattered growth of forest vegetation. (Newell.)

the rainfall is heavy and originally supported a dense growth of redwoods.[1]

Farther south, at San Luis Obispo (lat. 35° 20′) the rainfall is 21 inches, though it has the variable quality of the true desert type of rainfall, ranging in different years from 5 to 40 inches, a feature which

[1] Branner, Newsom, and Arnold, Santa Cruz Folio U. S. Geol. Surv. No. 163, 1909, pp. 9, 11.

greatly limits the forest growth since in dry years only the most favored situations supply trees with the necessary moisture. The higher and steeper mountain ridges are generally covered with a dense growth of low shrubs or chaparral, among which are the manzanita, scrub oak, and California lilac. The sycamore follows the watercourses and is grouped about springs, in places forming dense groves. The canyons and marshy tracts have a growth of willow and laurel. In the higher valleys white oak is abundant, while the Digger pine is common on the ranges east of those on the immediate coast.

There is great variation in the soils throughout the Coast Ranges of California and related variations are clearly traceable in many of the plant distributions. At San Luis Obispo heavy and rich soils support grasses and wild oats which replace the shrubby vegetation even on steep slopes. Live oak and laurel are found on areas where a rich soil and a good water supply are combined. Vegetation is most scanty where soils of poor quality occur, for example over areas underlain by serpentine rocks. The best growth of grasses is found on soil derived from an earthy sandstone (San Luis formation) intruded by basic rocks; these yield on decay a deep, residual soil of great fertility even on steep hillsides.[1]

North of San Francisco the rainfall increases rapidly as one enters the belt of permanent westerly winds. While the rains are more frequent and heavy during the winter season they do not fail in summer as is the case farther south. In Oregon and Washington the Coast Ranges are heavily watered and receive more precipitation per unit area than any other tract in the country. At 2000 feet the Coast Ranges of northwestern Oregon enjoy a total precipitation of 138 inches[2] and at higher elevations the precipitation is estimated at 150 inches.[3] While this enormous precipitation is not evenly distributed throughout the year the rainfall is heavy even in the relatively drier season, hence the forest growth is extremely dense and the trees of great size. Because of increasing temperature there is a marked increase in size with decreasing elevation in the well-watered portions of the mountains.

These two primary controls of forest distribution, precipitation and temperature, have important variations in altitude and latitude that are well expressed in the Coast Range System of North America as a

[1] H. W. Fairbanks, San Luis Folio U. S. Geol. Surv. No. 101, 1904, pp. 1, 2, 14.

[2] A. J. Henry, Climatology of the United States, Bull. Q, U. S. Weather Bureau, 1906, pp. 948–949.

[3] J. C. Stevens, Water Powers of the Cascade Range, pt. 1, Southern Washington, Water-Supply Paper U. S. Geol. Surv. No. 253, 1910, p. 4.

whole. In Alaska the greater portions of the mountains are bare or covered with snow fields and glaciers; in the Coast Ranges the forest is restricted to a belt between 2500 feet and sea level. In the interior mountains of Alaska, such as the Endicott Mountains, no forests at all occur because of cold and at least physiological dryness. Farther south the cold timber line has a greater elevation and the forest belt is wider, attaining its maximum development in Washington, Oregon, and northern California. Still farther south the dryness of the lower slopes causes the forest growth to be restricted at lower elevations on account of drought, as it is restricted at higher elevations because of cold. Like the cold timber line the dry timber line lies at progressively higher elevations with decreasing latitude and increasing temperature, so that in southern California the forest growth is restricted to the upper slopes and summits in the zone of maximum rainfall.[1]

The higher mountain slopes and mountain summits of Alaska are without forests; in southern California the plains and lower valleys are without forests. The mountain summits in the latter case are sufficiently warm to support forests, but the lower slopes are too dry. The intermediate tract has neither the great cold of Alaska nor the great dryness of southern California. Its forests of fir and cedar in Oregon and Washington and of redwood in northern California are among the most magnificent in the world. Broad-leaved trees are rare however. A few specimens are found along the streams and in the lower valleys, as the maple, cottonwood, ash, and alder in Washington and the oaks in California.[2] The forest is composed chiefly of conifers, but within it there is considerable variation in the distribution of species on account of differences of soil, climate, and topography. The cedars thrive best in the moister valleys and along the watercourses, though their range includes a large extent of higher mountain slopes. The firs thrive on the drier (though in an absolute sense wet) uplands, ridges, and steep mountain slopes, but they are also tolerant of wetter situations. On the steep declivities and sharp ridges, between 4000 and 8000 feet, as well as in more favorable situations, the sugar pine is found; in the middle of its range it attains great size and remarkable symmetry of form.

[1] J. Hann, Handbook of Climatology, 1903, pp. 305-308, presents an important summary of knowledge concerning increase and decrease of rainfall with increasing altitude and discusses the altitude of the zone of maximum rainfall. Scarcely any observations of the elevation of the zone of maximum rainfall have been made on mountains in middle latitudes, but it is probably between 3000 and 6000 feet, varying with the exposure, the temperature, etc.

[2] I. C. Russell, North America, 1904, pp. 238-239.

CHAPTER XI

CASCADE AND SIERRA NEVADA MOUNTAINS

CASCADE MOUNTAINS

CENTRAL CASCADES

THE Cascade Mountains form a separate physiographic province not only because of the distinct manner in which their borders lie above the surrounding country but also because of their characteristic interior features. The province is set off from the Sierra Nevada Mountains on the south, Fig. 28, by the valley of the North Fork of Feather River in northeastern California; the northern Cascades terminate quite as sharply immediately north of the international boundary line and south of the Frazer River Valley. The western and eastern borders of the Cascades descend steeply to the Sound Valley and the Columbia Plateaus respectively. The eastern border of the Cascades is particularly well marked in central Washington and in Oregon immediately south of the Columbia River, where the steepness of the slopes suggests an origin through either very sharp folding or faulting, but faults have not been actually observed, for lava flows to a large extent conceal the underlying structure.

Except for the breaks of the Columbia, Klamath, and Pitt valleys the Cascades possess marked continuity, and roads have been built across them only with great difficulty. Three railways cross the mountains to Tacoma and Seattle, but the grades are very steep and each line at the highest point requires a tunnel about a mile long.

In the earlier descriptions of the Cascade Mountains great attention was paid to the line of lofty volcanoes that are dominating elements in almost every view. Conspicuous among these elevations are Mount Rainier (14,500), Mount St. Helens (9700), Mount Baker (10,800), and Mount Adams (9500) in Washington, and Mount Hood (11,200), Mount Jefferson (10,200), and Mount Pitt (9700) in Oregon. The early explanation of the Cascades, suggested by the fact that a large amount of volcanic material occurs in the region, ascribed their forms wholly to volcanic processes, but later studies[1] show that the Cascades, at least in

[1] I. C. Russell, A Geological Reconnaissance in Central Washington, Bull. U. S. Geol. Surv. No. 108, p. 30.

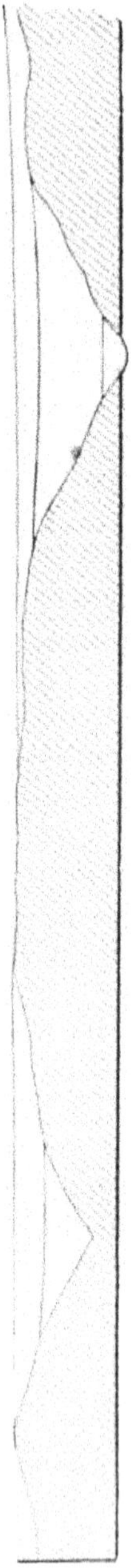

Washington and north-central Oregon, have been formed not mainly by the piling up of volcanic material but by the broad uplift and deformation of lava sheets, granites, and sedimentary strata. The great volcanoes that appear to be such prominent features of the range are secondary to the main mountain forms which consist of deeply intrenched valleys and sharp ridge crests with accordant altitudes. It has also been determined that the structure of the range is highly complex and that the conception of a warped monoclinal fault block sculptured by erosion such as is properly applied to the Sierra Nevada requires considerable modification here.[1] On the basis of truncated folds and a general lack of sympathy between surface and structure over broad areas, it is concluded that the Cascades may be termed mountains of the second generation; that is to say they are mountains which have been formed by the broad uplift and deep erosion of an almost base-leveled or peneplaned surface.[2]

Perhaps the two best localities from which to observe that uniformity of summit levels which is an inheritance from the period of peneplanation are at the head of Cold Creek, Washington, or near Cascade Pass,[3] although in many localities within the province, accordance of summit levels can not be observed because of (1) the complex nature of the later deformation that affected the ancient surface, (2) the volcanic outpourings that have in many places obliterated the old relief, or (3) the presence of unreduced or residual masses. This is true especially of the more elevated portions of the range where no recognizable flat-topped remnants of the original plateau are to be found.[4]

Fig. 30. — Profile across the Chelan Range (Cascades), showing the peneplaned surface uplifted and dissected. Two pat ial cycles are shown (1) in the lower dotted line and associated slopes, (2) in the steeper valley slopes at the lower level. (Smith and Willis, U. S. Geol. Surv.)

[Labels on figure: Mount Tyee. — Entiat Valley. — Chelan Range. — Lake Chelan.]

[1] I. C. Russell, A Preliminary Paper on the Geology of the Cascade Mountains in Northern Washington, 20th Ann. Rept. U. S. Geol. Surv., pt. 2, 1899, p. 137.

[2] Idem, p. 140.

[3] Smith and Willis, The Physiography of the Cascades in Central Washington, Prof. Paper U. S. Geol. Surv. No. 19, Plates 9 and 10, pp. 53–54.

[4] I. C. Russell, A Preliminary Paper on the Geology of the Cascade Mountains in Northern Washington, 20th Ann. Rept. U. S. Geol. Surv., pt. 2, p. 141.

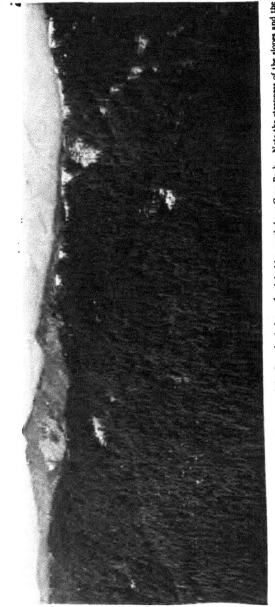

Fig. 31. — General view of the accordant ridge crests of the Cascades in lat. 45° 15′, looking south from Cone Peak.   Note the steepness of the slopes and the heavy timber covering.   (U. S. Geol. Surv.)

Fig. 32. — Details of Cascade topography in lat. 44° 15', looking southeast from Cone Peak to the Three Sisters.   Fig. 31 represents a more distant and general view, looking south.   (U. S. Geol. Surv.)

Fig. 33.— View of High Cascades from near Cascade Pass, showing uniformity of summit levels.   (Smith & Willis, U. S. Geol. Surv.)

The uniformity of summit level has been found to extend over the greater part of the Cascades, but it is clearly visible only from selected viewpoints and is generally expressed in the ideal plane of the ridge crests and the hilltops rather than in the few undissected remnants of the former plateau now remaining. It occurs at elevations varying from 4000 to 8000 feet; the maximum of 8000 feet is attained north of the 47th parallel and continues to the 49th parallel. In the vicinity of Mount Rainier the plateau remnants, at about 7500 feet, form the platform on which the cone of Mount Rainier stands. Toward the south the altitude of the plateau decreases and becomes about 4000 feet in southern Oregon where the width of the province at the plateau level is from 60 to 75 miles. Between latitude 44° and 45° N. in the Mount Washington-Mount Jefferson country the accordance of summit levels among the hill and ridge crests is quite remarkable, the broad bases of the snow-capped peaks resting upon the general summit of the plateau as shown in Fig. 32. The plateau is at an elevation of 5000 feet above Mount Baker, in Washington, and thence descends westward to the Sound level in the form of a broad and somewhat regular slope.

As a whole, then, the Cascade range has a broad though greatly dissected summit about 75 miles wide, upon which is a remarkably straight north-south line of several score volcanic peaks. These lie not along the central portion of the range but near the eastern margin, so that the greater mass of the range lies west of the geographic summit and watershed. From the crowning line of peaks long, broad, flat-topped spurs, some of them having almost the magnitude of distinct ranges, descend eastward to the Columbia Plateaus and Great Basin; that part of the range west of the summit consists of long massive mountain spurs (generally with accordant altitudes on north-south lines) and intervening deep canyons. The lava flows extending outward from the line of the now extinct volcanoes have altered the drainage courses greatly by damming the streams, thus causing lakes, marshes, and new outlets. These changes have taken place so recently as to give the drainage lines many youthful features.

The eastern margin of the Cascades extends southward from the vicinity of Osoyoos lake on the 49th parallel to Ellensberg and thence down the west side of the Yakima Valley and across the Columbia into Oregon. On this side the slope has marginal flutings which extend nearly at right angles to the major uplift some distance into the lower Columbia Plateaus on the east and give the margin in places an extremely irregular outline.[1]

[1] Smith and Willis, The Physiography of the Cascades in Central Washington, Prof. Paper U. S. Geol. Surv. No. 19, 1900, p. 25.

Long, gentle slopes along the flanks of the ridges descend to the valley floors. The dip of the rock and the inclination of the slopes of the ridges agree in direction but differ in amount, the dip of the rock commonly exceeding the inclination of the slope. Such a relation of form to structure requires the assumption of an erosion period in which was developed a topography moderately discordant with respect to structure and later deformed into the attitude in which we find it to-day.

The slopes of the older surface have a remarkably smooth development and are so regularly coördinated that the marks of recent dissection are not readily distinguishable in a view along the border. There appears to be only a gently inclined surface extending without perceptible break from the even-crested ridges to the valley floor. As a matter of fact, narrow gulches alternate with the ridges.[1]

Deformations of the ancient surface occurred in many places, and all show that the uplift of the Cascade lowland to form the Cascade plateau or mountains was not a simple broad anticlinal uplift or fault block deformation but a deformation of complex character.[2] There was at least one important halt in the uplift during which a mature topography was developed in places. One of the most important facts relating to the broad deformation of the formerly peneplaned surface now uplifted and dissected into the forms of the Cascades is the antecedent course of the Columbia River. After flowing for several hundred miles along the eastern front of the Cascades in Washington this trunk

Fig. 34. — Section showing relations of former and present forest near Prospect Peak, Cal.
1. Original soil.   2. Volcanic ashes and lapilli.   3. Tree of former forest, killed by shower of volcanic ashes.   4. Pit formed by decay of old stump.   5. Tree of present forest. (Diller, U. S. Geol. Surv.)

stream turns nearly at right angles and strikes boldly across the very heart of the Cascades. From a width of 2000 feet at the point where the Snake River enters it, the Columbia narrows to from 130 to 200 feet at "the Dalles," where it is bordered by high basaltic cliffs.[3] The river thus bears an antecedent relation to the range, having maintained a course outlined upon the Tertiary (Pliocene) peneplain.

[1] Smith and Willis, The Physiography of the Cascades in Central Washington, Prof. Paper U. S. Geol. Surv. No. 19, 1900, p. 26.

[2] Idem.

[3] The narrow portion of the channel terminates at the foot of a line of falls and rapids called "The Cascades," whence the name of the mountains (see especially George Gibbs, Physical Geography of the North-Western Boundary of the United States, Jour. Am. Geog. Soc., vol. 3, 1870–71, pp. 144, 147, 148).

SOUTHERN CASCADES

The southern end of the Cascades is the Lassen Peak volcanic ridge which extends southeast from the Pitt River to the North Fork of the Feather River. The ridge is about 25 miles wide and 50 miles long. It was built up by eruptions from more than 120 volcanic vents. A few of the craters are over a mile in diameter and were centers of enormous eruptions. All the prominent peaks of the ridge are volcanic cones. The last of the eruptions occurred very recently, a number of them taking place probably not more than 200 years ago; some of the trees killed at the time are still standing. Large portions of the original pine forest were covered with a mantle of volcanic sand or overwhelmed by lava during the more recent eruptions. In places the trees of the older forest project above the volcanic sand, their bare trunks forming a striking contrast to the new green forest developed at a higher level, Fig. 34.

The western slope of the volcanic ridge is relatively gentle and is underlain by volcanic material in the form of lava flows or agglomerate tuff. It is dry and sterile, and the larger part is strewn with rough lava fragments. The eastern slopes are in general bold.

The Lassen Peak volcanic ridge is from 5000 to 9000 feet high and about 4000 feet above the Great Valley of California on the southwest and the Great Basin on the east. Its highest point, Lassen Peak, is 10,437 feet above the sea. The ridge receives a sufficient rainfall to support an open forest of pines.

NORTHERN CASCADES

Immediately north of the 49th parallel the Cascades terminate abruptly and descend to a plateau several thousand feet lower; immediately south of the 49th parallel the Cascades have their greatest development. The distance from Mount Chopaka on the eastern to Mount Baker on the western side of the northern Cascades is about 90 miles. Thus by a pure coincidence the international boundary is also a physiographic boundary although it follows a parallel.

Like the Sierra Nevada Mountains, the Cascades terminate on the north in a triple set of subranges, the Okanogan, Hozomeen, and Skagit mountains. The Okanogan Mountains extend from Mount Chopaka to the valley of the Pasayten River. On the east the Okanogan Mountains terminate abruptly in a narrow foothill belt. Mount Chopaka here rises as a steep wall over 7000 feet high on the border of the Similkameen Valley. Between the Pasayten River and the Skagit River is the Hozomeen range; west of the Skagit River are the Skagit Mountains.

The north-south valleys of the Pasayten and Skagit thus form the dividing lines between the three subranges of the northern Cascades. We shall now briefly examine the detailed characteristics of each of these ranges.

The Okanogan Mountains consist of a great batholith of granitic rocks and are perhaps the most important igneous member of the Cas-

Fig. 35. — Cathedral Peak, Okanogan Mountains, Washington, showing glaciated summit of the matterhorn type. Nearly vertical jointing in granite rock has assisted glacial erosion in producing rugged forms. (Smith & Calkins, U. S. Geol. Surv.)

cades. The Skagit and Hozomeen ranges are composed chiefly of sedimentary rock, with a large amount of conglomerate, slate, and schist, the latter having structures with a north-south trend.[1] The Okanogan Mountains have a number of high peaks such as Chopaka, Cathedral, Remmel, and Bighorn, with a nearly uniform elevation of 8000 to 8500 feet. Almost all the mountain peaks are above the 7000-foot level. The highest peaks are extremely rugged, and are bordered by deep glaciated valleys. Glaciers still persist on the north sides of a few peaks, but they are of small size. The evidences of former more extensive glaciation are particularly well shown in the deeply carved northern aspects of spurs and ridges where steep-sided cirques and gulches

[1] Smith and Calkins, A Geological Reconnaissance Across the Cascade Range near the 49th Parallel, Bull. U. S. Geol. Surv. No. 235, 1904, p. 84.

abound. The southern slopes are more regular and without glacial modification. Small lakes with bordering snow banks occupy almost all the glacial amphitheaters that are tributary to the Similkameen.

Hozomeen range includes the central and main crest of the northern Cascades. Its western flank is scored by a number of remarkably narrow canyons, among which is the Skagit, whose mouth is so narrow as barely to permit the passage of the stream. The divide of the range has an elevation of 7000 to 8000 feet, and about ten miles south of the 49th parallel consists of sharp peaks with rugged outlines due to the irregular, coarse, and resistant conglomerate of which they are composed and the deep dissection of the range of which they form a part. Numerous glaciers occur and glacial cirques have been cut back into the main mass of the mountain to such an extent that the peaks are largely of the pyramidal or matterhorn type. North of the boundary the topography of this subrange becomes much less bold.

Skagit Mountains form the western subrange of the Cascades and include the wildest and most rugged country of the entire section. High peaks with precipitous sides abound and the scenery is extraordinarily picturesque. The mountain slopes are so steep that much of the country is practically inaccessible and unknown even to prospectors. Sharp, glaciated, pyramidal peaks are characteristic of the range about the headwaters of the Nooksak and Chilliwhack rivers. Some of the higher peaks are still flanked by glaciers, the feeble descendants of the large Pleistocene glaciers that are responsible for the development of amphitheaters with extremely steep walls and for the pyramidal peaks which occur throughout the range. The western portion of the Skagit Mountains consists of broader ridges, essentially flat-crested, smooth, grass-covered, and separated by broad steep-walled canyons. The western border of the northern Cascades is marked by an abrupt descent to the gravel-covered plain that extends west to the coast.[1]

The flat-topped ridges of the Skagit range west of the Mount Baker district have a somewhat uniform elevation of 5000 to 6000 feet and a gentle westerly inclination. The flat tops are interpreted as the remnants of a preëxisting topography which once occurred in the form of a fairly perfect lowland with few residuals rising above the general surface. Later uplift supplied an opportunity for vigorous dissection. Where the uplift was from 5000 to 6000 feet remnants of the old surface remain on the divides. Where the uplift was 7000 feet and more no traces of the old lowland persist, the peaks are acute pinnacles, the

---

[1] Smith and Calkins, A Geological Reconnaissance Across the Cascade Range near the 49th Parallel, Bull. U. S. Geol. Surv. No. 235, 1904, pp. 13-17.

divides mere knife-edges. In the Okanogan Mountains the peaks at 8000 feet are approximately uniform in altitude and merely suggest an older topography. At the time of uplift of the northern Cascades from base level there appears to have been not a single broad upwarp; the three subranges composing the northern Cascades represent distinct upwarps separated from each other by downwarps which have determined the positions of Pasayten and Skagit valleys.[1]

The abrupt termination of the Cascade Mountains at the international boundary line has been shown to be due to a difference in

Fig. 36. — Plateau of the Cascades representing uplifted and dissected lowland surface. Western portion of Skagit Mountains, Mount Baker in left background. Looking west from Bear Mountain. (Smith and Calkins, U. S. Geol Surv.)

degree of uplift between the Cascade country and the Interior Plateau of British Columbia since the last erosion cycle common to both mountains and plateau. The earlier structures so far as known appear to extend from one province to the other, denoting a common geologic history down to the Eocene.[2] In the latter part of the Pliocene, uplift of the base-leveled surface common to both took place, but the uplift was differential and amounted to about 4000 feet in the Interior Plateau and about 8000 feet in the Cascades. The uplift of 8000 feet was, however, not uniform throughout the Cascades; it was an uplift in the form of upwarps and downwarps, the upwarps being represented to-day by the three parallel ranges constituting the northern end of the Cascades, and

[1] Smith and Calkins, A Geological Reconnaissance Across the Cascade Range near the 49th Parallel, Bull. U. S. Geol. Surv. No. 235, 1904, p. 89.

[2] G. M. Dawson, Trans. Royal Soc. Canada, sec. 4, 1890, p. 16.

the downwarps by the intervening valleys. During the upwarping certain streams persisted across the structures upraised in their paths, such as the Skagit across the Skagit Range, the Columbia across the Cascades, and the Frazer across the Interior Plateau. All these streams are therefore antecedent in portions of their courses. The differences between the Cascade ranges on the one hand and the Interior Plateau and Columbia Plateaus on the other are therefore due fundamentally to differences of elevation and degree of dissection conditioned by uplift. The eastern margin of the peneplain of the Cascades descends gradually to the plateaus of the Columbia apparently without a break; a similar but more sudden descent marks the northern end of the Cascades where they descend to the level of the Interior Plateau of British Columbia.

The rather general occurrence of the Columbia River lavas over a large part of the northern and central Cascades is accounted for by the fact that the extrusion of the lavas took place mainly at a time when the low relief of the land allowed their widespread distribution. These great basaltic inundations are to be distinguished from the much later volcanic outpourings that formed the cones now surmounting the Cascades. Mount Baker, for example, like most of the highest peaks of the Cascades, is an extinct volcano, built up of andesitic lavas of rather recent age poured out upon a topography as rough as the present; and the main drainage feature of the region, the Nooksak, was, at the time of the eruption of the lavas, practically at the present level and in the present position.[1]

All the higher peaks of the Cascades, as well as the lower country, were glaciated during the Pleistocene period, the lower limit of glaciation ranging from sea level in the Sound Valley to heights of several thousand feet, its position depending on exposure, latitude, etc. The evidences of past glaciation are of the familiar sort and consist of terminal and lateral moraines, glacially modified valleys, and aggraded stream courses at lower altitudes. Probably the most important topographic and drainage effects in the Cascades occurred in Washington in the vicinity of Lake Chelan.

"Lake Chelan is a splendid body of water 65 miles long whose southeastern end lies open to the sky between the grass-grown hills of the outer Columbia Valley, while its northwestern end lies in shadow between precipitous mountains in the heart of the Cascade range. There are sandy shallows near its outlet, but beneath the cliffs of its upper course the water is profoundly deep."[2]

[1] Smith and Calkins, A Geological Reconnaissance Across the Cascade Range near the 49th Parallel, Bull. U. S. Geol. Surv. No. 235, 1904, p. 35.

[2] Smith and Willis, The Physiography of the Cascades in Central Washington, Prof. Paper U. S. Geol. Surv. No. 19, 1900, p. 58.

The lake lies in the canyon of the Stehekin-Chelan River and 32 miles of its length lies within the Cascade Mountains.   The depth of the lake varies from 1000 to 1400 feet, and as the water's surface is but 1079 feet above the sea, the bottom of the lake is at one place 300 feet below sea level.   The water is partially retained at its present level by a dam of sand and gravel.   It appears that the valley now partly filled by Lake Chelan was occupied by a great mountain glacier that deepened and widened the preglacial valley and steepened the valley walls, giving them their present precipitous character.

The glacial features of the northern Cascades are markedly asymmetric.   For example, the U-shaped canyons that drain eastward to

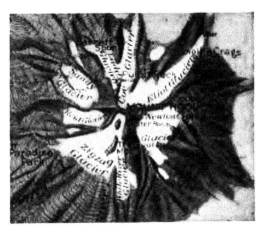

Fig. 37. — Relief map of Mount Hood, Oregon, showing the eroded condition of the volcano and the extent of its glacier systems.   (U. S. Geol. Surv.)

the Pasayten have northern walls of great simplicity and southern or shady walls carved into niches or hanging cirques of glacial origin. The average or prevailing aspect of the glacial cirques of the entire region is about due northeast, a feature probably due to the preglacial topography, lesser insolation on that aspect, and a certain excess of snow accumulated by drifting across the divide.   The more favorable easterly aspect is well illustrated by the glacial erosion of the Hozomeen range, the glaciers flowing eastward having eaten back into the heart of the range much more than their rivals in the western valleys.   So marked is this feature that many of the higher peaks have a degree of asymmetry suggesting a breaking wave.[1]

[1] Smith and Calkins, A Geological Reconnaissance Across the Cascade Range near the 49th Parallel, Bull. U. S. Geol. Surv. No. 235, 1904, pp. 90–91.

Glacier systems are still prominent features of the higher peaks. Mount Rainier, Mount Hood, and others are flanked by short glaciers above the 8000-foot level, Fig. 37. The snow and ice fields not only enhance the beauty of these splendid volcanoes but also serve to steady the discharge of the rivers which they feed.

SOIL, CLIMATE, AND FORESTS

The almost endless variety of rocks which form the Cascades causes the valley soils to be formed of a great variety of mineral fragments. The valley soils do not therefore reflect the nature of the nearest rock exposures to an important degree. In the higher valleys and basins which lie parallel to the trends of the structures the soil retains to some degree the mineral characteristics of the rock (chiefly granite) from which it was derived. So far as surveys have been made the granitic rocks appear to be sandier and drier because of the mineral composition of the parent rock. They are therefore covered with a shrubby, grassy vegetation; the schist and slate rocks furnish a finer soil with a larger proportion of clay, and appear to be heavily wooded. At the higher elevations, however, the one exhibits rounded forms, the other is developed into a sharply serrate topography unfavorable to heavy tree growth. Toward the valley heads the land waste is coarse and bowldery; down valley a gradation in size is effected through transportation and deposition by water; the soils of the lower valleys are fine.

The total precipitation of the central Cascades varies from 60 to 100 inches a year. Three-fourths of it occurs in the winter season from November to May. In the higher portions of the range the snowfall is heavy, varying from 4 to 10 feet in depth. In these situations it remains throughout the winter, and, in restricted summit areas, throughout the year. On the highest peaks precipitation is almost wholly in the form of snow and well-defined glaciers, from a fraction of a mile to several miles in length, flank Mount Hood, Mount Rainier, and other peaks. The more important streams as a rule have their headwater sources in fields of perpetual ice and snow. Snowstorms in the lower and middle portions of the Cascades (Mount Hood region) are usually followed by rains and warm chinook winds which dissolve the snows and cause extremely heavy freshets in all the streams.[1]

The forests of Washington cover the state as a thick mantle from high elevations on the Cascade range westward to the Pacific. In this

[1] H. D. Langville and others, Forest Conditions in the Cascade Range Forest Reserve, Oregon, Prof. Paper U. S. Geol. Surv. No. 9, 1903, pp. 29, 30.

great region the only mountains that reach above timber line are the Olympics in the Coast Range System and a limited number of peaks in the Cascades. The forests are the densest, heaviest, and most continuous in the United States except for the redwood area in northwestern California, and are marked by a dense and tangled undergrowth. The largest trees are from 12 to 15 feet in diameter and 250 feet in height. Red or

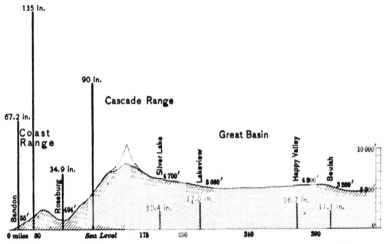

Fig. 38. — Topographic profile in relation to rainfall in the Coast Ranges and the Cascades of Oregon.

yellow fir, in the zone of heaviest rainfall and where there is at least a moderate soil cover, constitutes the larger part of the forest, with an intermingling of spruce, hemlock, and cedar.[1]

West of the Cascade range the country is occupied mainly by four species, red fir, cedar, hemlock, and spruce. The percentages of composition arranged in the same order are as follows: 64, 16, 14, and 6, with the proportions of cedar and spruce increasing toward the coast. At the highest elevations the fir disappears and hemlock and cedar come in. East of the Cascades the climate becomes rapidly drier and the timber consists almost entirely of lodgepole and yellow pine.[2]

In Oregon the timber consists of about the same species as in Washington, with the addition in the southwestern part of the state of sugar pine, noble fir, and yellow pine. The red fir constitutes by far the larger part of all the timber in the state. Cedar, hemlock, and spruce are comparatively unimportant, except along the coast. The fir occupies the entire timbered portion of the western slope of the Cascades, the

1 Henry Gannett, 19th Ann. Rept. U. S. Geol. Surv., 1897–98, p. 26.
2 Idem, p. 27.

eastern slope of the Coast Ranges, and the depression between these mountains where it forms more than three-fourths of the forest. The cedar occurs mainly at mid-altitudes upon the Coast Ranges and the Cascades but forms a small proportion of the forest. Hemlock occurs notably upon the western slope of the Cascade range at mid-altitudes. East of the Cascade range in Oregon the forests are largely of yellow pine. Sugar pine extends over the entire breadth of the Cascades and from California northward to the Columbia and westward to the coast.

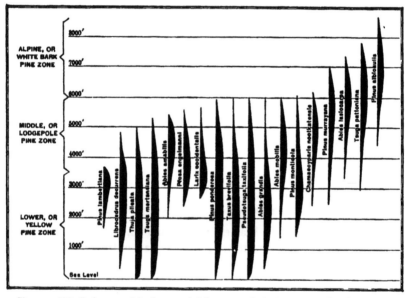

Fig. 39. — Altitudinal range and development of timber-tree species in the central portion of the Cascade Mountains. (U. S. Geol. Surv.)

On account of the exceptional height, breadth, and topographic boldness of the northern Cascades their climatic and forest characters deserve special discussion. The northern end of the Cascades is a region of short summers, comparatively free from rains. The winter snowfall, however, is extremely heavy. Besides its important relation to irrigation through its tendency to equalize stream flow, the heavy snowfall is of ecologic importance, for it determines the distribution of certain definite plant communities, influences the forms of plants as along the cold timber line where dwarfed or gnarled forms occur, and prevents excessive temperatures and too rapid transpiration during winter storms when the roots are incapable of absorbing water. In the

Okanogan Mountains the snow does not disappear until the middle of July, and several miles south of the 49th parallel deep snow remains throughout the summer and snow squalls occur even in July and August.

Similar climatic conditions occur in the eastern part of the Hozomeen range, but the western slope of this range has a much greater annual precipitation. The forests become extensive, large trees with dense underbrush cover the valley bottoms and extend well up the slopes, and grass is not plentiful except on a few ridges. Banks of snow persist throughout the year and glaciers occur on all the higher peaks. In the Skagit Mountains the summer is very short and July and August are the only months in which snow does not fall in considerable amounts. At no time in the year are the passes in the Skagit Mountains free from snow.

The climatic contrasts between the east and west slopes of the Cascades are well shown by the conditions of the old vistas cut on portions of the boundary line. In the Pasayten Valley the cuttings can be found easily, the stumps and logs are sound, and camp stools remain as they were left forty years ago. On the western slopes and in the Skagit Valley the old stumps are so decayed as to be barely recognizable, and the vista is here occupied by trees 75 feet high and 14 inches in diameter, which have grown up in the old cuttings in that short time.[1]

On the western slopes of the northern Cascades there is a heavy precipitation which supports a dense forest growth. Timber line is at 6000 feet; above this elevation trees occur only in groves or singly and most of the mountain summits are treeless. Grasses, sedges, and heather are the common growths above the forest. The best forest growth is below 4000 feet. Above that elevation the trees are apt to be shorter and more branched, and the trunks twisted or otherwise defective. Some of the basins at rather low altitudes are without forests on account of the accumulated snows which melt out too late in the spring to favor anything but a growth of grass. The red fir belt is the lowest of the three forest belts recognized here, and has associated with it the Sitka spruce, the silver fir, the hemlock, and other species. The main slopes of the mountain spurs are covered with hemlock and white fir, which constitute the second forest belt. In the third or alpine belt on the summits of the principal spurs and on the divides the growth is sparse or absent. The principal trees of the alpine zone are the mountain hemlock, alpine fir, and Engelmann spruce.[2]

The heavy rainfall of the western slopes slightly overlaps the upper part of the eastern slope of the Cascades and is accompanied by trees

[1] Smith and Calkins, A Geological Reconnaissance across the Cascade Range near the 49th Parallel, Bull. U. S. Geol. Surv. No. 235, 1904, p. 18.

[2] H. B. Ayres, Washington Forest Reserve, 19th Ann. Rept. U. S. Geol. Surv., pt. 5, 1897–98, pp. 283–293.

characteristic of the western zone. These extend eastward over the geographical summit of the range, a feature less marked on the higher saddles, ridges, and peaks and very strongly marked in the low passes.

SIERRA NEVADA MOUNTAINS

The Sierra Nevada Mountains on the eastern border of California are a bold, continuous range about 75 miles in width. They have many well-defined peaks; the larger number occur in a line west of Lake Tahoe, Owen's Lake, etc., and constitute what is generally known as the High Sierra, the crest line of the Sierra Nevada Mountains over 11,000 feet in height. By reason of the dominating character of the Sierra Nevada and its high degree of effectiveness in barring the rains and snows of the westerly winds, it has an abundant water supply and is well clothed with forests of pine, fir, hemlock, etc.; in both these respects it is strikingly different from the minor north-south ranges on the east that comprise so large a portion of the Great Basin, and from the nearly treeless central valley of California on the west.

The Sierra Nevada is a notable example of a mountain range of great geologic complexity whose general physiography is of a rather simple type. In order that this may be realized a few geologic details may be given. It is probable that the Sierra Nevada rock formations range in age from Archæan or Algonquin to Recent.[1] The rocks have been profoundly affected by crustal compression accompanied by close faulting and schistosity, in many places carried to the point where the original nature of the sediments has been completely altered. These statements are sufficient to show that a complete geologic study of the mountains would include a wide range of facts related to almost every department of geologic science. Compression and folding occurred at the end of the Paleozoic, as well as a certain amount of igneous intrusion, and later the mountains were greatly eroded. At the close of the Jurassic the Sierra Nevada region was again compressed and folded as well as uplifted into the form of a prominent mountain range. Great batholithic intrusions also took place at this time. In connection with these intrusions the sedimentary rocks were largely metamorphosed and rendered schistose and platy.[2] These two facts, as we shall see in succeeding pages, are of first importance in the interpretation of the canyon forms associated with the Yosemite, Merced, Tuolumne, and other mountain streams.

In spite of the great geologic complexity of the Sierra Nevada Mountains their broader physiographic features are somewhat simple. Whatever the original relief, now lost, may have been, and however complex the structural changes that have taken place, these are on the whole of lesser importance geographically than peneplanation which brought about the existence of a topography of little relief. The highest mountains were reduced to residual mountains, the valleys were broadened out to great width, and the streams flowed in courses of slight gradient. As in so many other instances of peneplanation, the

1 H. W. Turner, 14th Ann. Rept. U. S. Geol. Surv., pt. 2, p. 445.
1 J. S. Diller, Bull. U. S. Geol. Surv. No. 353, 1909, pp. 8–9.

structure of the country was practically unexpressed in its topography: high and low masses, hard and soft rocks, were as a rule brought down to a common topographic expression. Upon the floors of the ancient valleys that but slightly diversified the relief of the ancient peneplain (completed in the Miocene) auriferous gravels were deposited, and among the finer sediments are fossil leaves of the fig, oak, and other plants indicative of a low coastal country somewhat like Florida. The uplift of the peneplain was accompanied by volcanic activity. From volcanic vents near the low crests streams of lava issued and followed the water-courses, covering the auriferous valley gravels and displacing many of the streams.[1]

The Sierra Nevada Mountains, as we know them to-day, are among the major relief features of the continent, and the contrast between their former peneplaned and their present mountainous condition can be understood from the fact that block faulting on a large scale has taken place, resulting in both the bodily uplift and the tilting of a large crust block. The eastern face of the Sierra Nevada over a distance of several hundred miles is exceedingly steep and forms a fault scarp which is to be compared in steepness and continuity only with the eastern face of the Lewis Mountains in western Montana (p. 307). Among the facts supporting the hypothesis of faulting are the stream gravels on the very summits of the mountains above the steep eastern scarp where in places they are displaced through 3000 feet of vertical distance.[2]

Evidence of recent faulting has been found along the eastern base of the mountains, near Genoa, where alluvial deposits (Pleistocene) have been displaced some 40 feet; it has also been found that the Carson River on emerging from the mountains increases its grade abruptly toward the east, suggesting recent dislocation of its valley. It is concluded that the first dislocation along the eastern face of the Sierra Nevada Mountains took place at the close of the Cretaceous and that it has continued down to the present day, thus making the faulting complex. A number of more or less parallel faults have been identified within a belt 25 miles wide.[3]

Along most of the range the rocks of the Sierra Nevada scarp do not end finally, but occur in the ranges to eastward, a feature explained by a system of compound faults parallel with the eastern front of the major range. It is in the depressions between the main Sierra block and the subsidiary blocks that the chief lakes of the region occur, as Owen's Lake, Mono Lake, Tahoe Lake.[4] The Carson topographic sheet well represents

[1] J. S. Diller, Bull. U. S. Geol. Surv. No. 353, 1909, p. 9.

[2] J. S. Diller, 14th Ann. Rept. U. S. Geol. Surv., pt. 2, p. 432; H. W. Turner, 14th Ann. Rept. U. S. Geol. Surv., pt. 2, p. 442; I. C. Russell, 8th Ann. Rept. U. S. Geol. Surv., pt. 1, p. 322.

[3] Auriferous Gravels of the Sierra Nevada, Jour. Geol., vol. 4, quoted by Spurr in Descriptive Geology of Nevada South of the 40th Parallel, and Adjacent Portions of California, Second Edition, Bull. U. S. Geol. Surv. No. 208, p. 222.

[4] See Contour Map of the United States, scale 111 miles to the inch, U. S. Geol. Surv.

this feature so common in the Great Basin region and marked out on very strong lines along the eastern border of the Sierra Nevada. The main intermont valley is broken by secondary blocks into a series of subordinate valleys. The secondary blocks are generally discontinuous longitudinally and the greater number of them are roughly parallel to the primary fault plane. The eastern slope of the Sierra Nevada Mountains thus exhibits a multiple scarp.

As an illustration of a depression due to deformation of the crust block type may be cited Owen's Valley, a V-shaped depression between the Sierra Nevada and the White Mountain blocks. This conclusion is based upon the fact that faulting and associated phenomena have been observed in many places along the eastern margin of the Sierra Nevada and also on the eastern margin of the White Mountains. Furthermore, hot springs occur in the marginal zone from the midst of Owen's Valley to Mono Lake, and a mud geyser is known at Casa Diablo, two features whose occurrence is commonly associated with faulting. The sharp truncation along the eastern border of the mountains of the inclined peneplain of the western slope of the Sierra Nevada, a feature well developed on the steep eastern face of the White Mountains in sharp contrast to the gentle western slope, points to the same conclusion. Finally, faulting and crustal movements of considerable magnitude and accompanied by earthquake shocks have taken place in Owen's Valley in historic time.[1]

In contrast to the steep eastern front of the Sierras is the gentle westward slope that descends from 9000 feet on the east to the 1000-foot level on the west at the border of the central valley of California. From any commanding view between the crest of the range and the valley of California one looks out upon a plateau whose general accordance of summit levels is striking and significant. This was once a broad plain near sea level, now uplifted to a great height.[2] Though peneplanation may be safely inferred from the discordance between the plane of the sublevel hilltops and the structure, the region is nevertheless deeply dissected by the rejuvenated streams. Canyons have been formed of such

---

[1] W. T. Lee, Geology and Water Resources of Owen's Valley, California, Water-Supply Paper U. S. Geol. Surv. No. 181, 1906, p. 25.

[2] The peneplain of the western slope of the Sierra Nevada was first recognized by Gilbert (G. K. Gilbert, Science, vol. 1, 1883, pp. 194–195). Diller showed that the planation was probably accomplished during Miocene time (J. S. Diller, Tertiary Revolution in Topography of the Pacific Coast, 14th Ann. Rept. U. S. Geol. Surv., pt. 2, 1894, pp. 404–411). He also found that gravel deposited upon this peneplain had been elevated, faulted, and tilted, the degree of vertical displacement along the eastern face of the range being 3000 feet at the northern end.

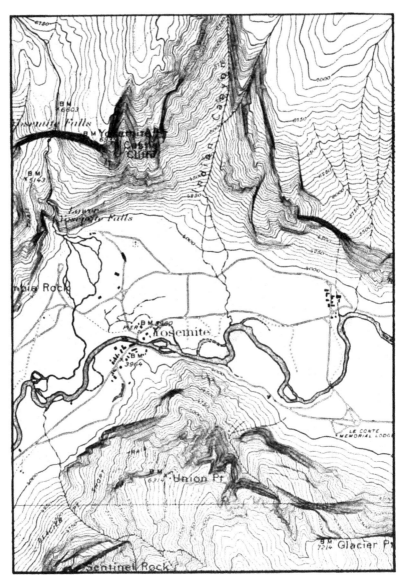

Fig. 40. — Typical portion of Yosemite Valley, showing cliffed margins, waterfalls, flat floor, and meandering stream.   Partly stream eroded, partly ice eroded canyon whose details of cliff structure are
due to geologic structure.   (Matthes, Yosemite Valley Map, U. S. Geol. Surv.)

profound depth and steepness as to constitute the chief scenic feature of the range outside the snow-capped High Sierra.

The main streams such as the Merced, Tuolumne, Feather, and Hetch-Hetchy flow in canyons from a half mile to a mile deep. The encanyoned portions of the streams are but fractions of the total lengths of the streams, since the canyons do not appear either at the headwaters or in the foothill belt. They are limited to the intermediate levels where the alpine glaciers once occupying the largest valleys produced their most important topographic effects. All the canyons have exceedingly steep walls over which tributary streams form waterfalls of celebrated beauty. Rarely does flat land of any extent occur on the canyon floors, though this feature is well developed in the Yosemite and the Hetch-Hetchy where open grassy glades diversify the forest cover and possess a charm possessed in equal degree by few regions in the world. The canyon of the Yosemite, the most celebrated in the region, has long been regarded as due wholly to glacial erosion, but recently it has been more safely determined to be not solely a normal product of ordinary stream or ice erosion but equally a function of the structure of the country rock. The granites of the Yosemite region consist of many huge monolithic masses embedded in a matrix of more or less strongly fissured rock. There is in consequence extreme inequality of resistance to dissection and the landscape reflects very faithfully each structural character of the material from which it was carved. The dominating heights, such as Half Dome, consist invariably of resistant monoliths; the canyons and gorges are due to the erosion of zones of nonresistant fissile rocks; the courses of the rivers and lake basins in the valley floors and the scarp-like rock walls were in all cases evolved in obedience to local structural conditions. The trend and profile of each cliff are an expression of its associated structure.[1]

The Sierra Nevada block south of latitude 38° 30′ is displaced along its eastern face, chiefly along a single fault line, and the range is here much higher than to the north, and bears upon its summit a number of exceptionally high peaks, among them Mount Whitney, the highest mountain in the United States (14,800 feet).

This portion of the mountains, the High Sierra, should not be considered as a part of the ancient peneplain, since it stands several thousand feet above the general summit level of the Sierra Nevada. It is an unreduced or residual portion of the region. The main range of the Sierra is thus a belt of extremely high relief about 25 miles wide,

---

[1] F. E. Matthes, The Cliff Sculpture of the Yosemite Valley, Paper before the Geol. Soc. Am., Dec., 1909, p. 4.

with numerous sharp spurs and ridges among which there are few tracts of valley land. Level tracts of limited extent, however, occur at the heads of most of the larger streams, and a few large streams have head-waters draining big valleys or lakes. This is true of the northern and lower Sierra Nevada especially. Thus the Middle Fork of the Feather River issues from a large level tract known as Sierra Valley, about 60,000 acres in extent, and Truckee River flows out of Lake Tahoe, 190 square miles in extent. Carson Valley and Little Valley west of Washoe Lake are further illustrations. The smaller streams drain small glades, ponds, or lakelets or rise directly from steep mountain flanks.[1]

In a broad view of the Sierra Nevada, five categories of form are thus distinguishable: (1) a summit zone of residual elevations, including the High Sierra and many lesser elevations, (2) a plateau zone representing an uplifted, tilted, and partly dissected peneplain, (3) a fault zone on the eastern border of the range, including steep, recently formed, and but little dissected fault scarps which are in almost startling contrast to the flat ridge tops that represent peneplain remnants, (4) a narrow, structural valley zone on the east in which the valleys represent fault depressions, (5) a broader valley zone on the west, in which the valleys are canyons of erosional origin whose architectural details alone are responses to structure.[2]

The uplift of the Sierra Nevada block rearranged the drainage of the region in many localities. In a general view of the drainage prominent features are the short and steep streams that descend the eastern scarp of the Sierra Nevada block and the generally direct courses of the streams draining the gentler western slope. It is noteworthy that the westward-flowing streams have a certain axial directness down the incline of the block, courses which appear to have been gained during or as a result of the block deformation, for their inharmonious relations to the structure indicate that they are formed upon a peneplaned surface deeply covered with land waste and at a time when hard and soft rocks did not show any important topographic differences. Several streams appear to have maintained their courses across the rising Sierra Nevada block in spite of the uplift. The forks of the Feather River persisted in their courses and cut deep canyons directly across the rising crests of some of the individual blocks of the range. The effect of the deformation

[1] J. B. Leiberg, Forest Conditions in the Northern Sierra Nevada, California, Prof. Paper U. S. Geol. Surv. No. 8, 1902, p. 17.

[2] For a detailed description of the first four zones as developed in the Carson Lake district northeast of Lake Tahoe see J. A. Reid, The Geomorphogeny of the Sierra Nevada Northeast of Lake Tahoe, Bull. Dept. Geol., Univ. Cal. Pub., vol. 6, 1911, p. 108.

was to bring the ancient bed of the Jura River to unequal elevations above the present bed, giving the ancient bed an abnormal profile, so that the old and now lithified river gravels arch over a secondary block of the Sierra.

The northern end of the Sierra Nevada is more complex than other portions of the range. Beyond Lake Tahoe are three main crests, Clermont Hill Ridge, Grizzly Mountains, and Diamond Mountain, and each crest has a valley at its northeastern base. Each crest represents the edge of a tilted crust block whose northeastern face is a fault scarp and whose southwestern descent is a dissected remnant of a former peneplain. The western crest (Clermont) is continuous with the main crest of the range north of Lake Tahoe. The Diamond Mountain block of the northern Sierra Nevada has a long gentle slope towards the southwest, which gives that side the appearance of a plateau. Toward the northeast the mountain presents an escarpment over 200 feet high in a short steep slope to Honey Lake, Fig. 28. This escarpment is remarkably regular, with few prominent spurs and reëntrants and no important stream features which point to a recent origin through faulting. The upper courses of the westward-flowing streams are in broad shallow valleys developed upon the regional peneplain before uplift and deformation took place, but as they continue to the southwest their valleys become progressively deeper until they have true canyon profiles. All of them open into broad alluvial valleys at the foot of the mountain slope. Lights Canyon, Cooks Canyon, and the canyons of Indian and Squaw creeks are illustrations. The Grizzly Mountain crust block repeats the essential features of the Diamond Mountain block except that its crest line is less regular; a broad gap has been developed across the range and no definite traces of an inclined peneplain may be found upon the southwestern slope of the mountain.[1]

SOILS, PRECIPITATION, AND FOREST BELTS

The soils of the western forested slopes of the Sierra Nevada are chiefly residual and have been derived from the weathering of granitic rocks, diabase, amphibolites, slates, serpentine, and volcanic material. They are prevailingly of light-red to deep-red color, and generally of somewhat compact structure. The soils are sometimes separated from the underlying parent rock by a thin stratum of adobe-like material. They are frequently very shallow and marked by abundant rock outcrops, bowlders, and rough, rocky areas.

[1] J. S. Diller, Geology of the Taylorsville Region, California, Bull. U. S. Geol. Surv. No. 353, 1909, pp. 9–12.

The slopes of the canyons are generally rocky and almost denuded of soil, but deep residual soils are found along the summits of the ridges below 4000 feet.   Deep-red soils are as a rule found on the andesite, gabbro, and diabase-porphyry rocks, while the sedimentary rock is usually covered with a poor shallow soil.[1]

The foothill soils are almost entirely residual and vary in character with the nature of the underlying rock.   The poorest soil, light colored and shallow, is found upon slate.   A deeper and warmer soil is found upon granite, and the best soil of all — the one richest in plant food — is the so-called "red soil," derived principally from the disintegration of diabase and amphibolite.[2]

Above 5500 feet on the north and 6500 feet on the south the Sierra Nevada has been glaciated and abounds in rocky slopes, tracts of bare rock from which the soil has been swept.   The present glaciers of the High Sierra are very small and occupy only the headwater amphitheaters

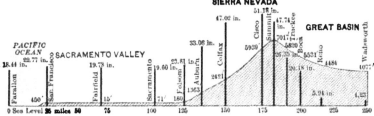

Fig. 41. — Relation of topography to rainfall, Sierra Nevada Mountains.

formed by the larger ancestral glaciers.   The extensive glaciers and snow fields of the past formed a large part of the existing soil; the present glaciers are above tree line, occupy but an insignificant fraction of the total surface, and have no important relation to soils or forests.

The annual precipitation over the Sierra Nevada, Fig. 41, ranges from 40 inches at elevations of about 3000 feet to a maximum of 70 inches at 7000 feet, with less precipitation on the eastern slopes than at corresponding elevations on the western slopes.[3]   The winter snowfall at elevations above 4500 to 5000 feet is as a rule heavy, the largest banks in the deepest woods persisting until late summer.   In the valleys and foothills below 1500 feet snow seldom occurs and lasts for only a few hours.

On the western slope of the Sierra Nevada three well-defined zones of vegetation may be distinguished: (1) the well-watered, heavily for-

[1] W. Lindgren, Colfax Folio U. S. Geol. Surv. No. 66, 1900, p. 10.
[2] W. Lindgren, Sacramento Folio U. S. Geol. Surv. No. 5, 1894, p. 3.
[3] W. Lindgren, Pyramid Peak Folio U. S. Geol. Surv. No. 31, 1896, p. 1.

ested zone between 3000 and 6000 feet, known as the "timber belt," (2) the drier transition belt of thin forest below 3000 feet, and (3) the nearly treeless rolling grassy hills which occur between the floor of the Great Valley and the slope of the Sierra up to 1500 feet.

The timber belt contains magnificent forests among which the conifers predominate both in size and number. They include the yellow pine, the sugar pine, and the famous "big trees." The last-named grow in quiet hollows protected from the winter storms by the bordering ridges and surrounding forests of pines. All the larger conifers likewise flourish best in sheltered areas, though they are also found in diminished numbers on the ridges. Undergrowth is usually lacking except near springs or small streams, and this condition together with the open stand of the trees gives the forests a pleasant, park-like character. Within the timber belt are also found the spruce, fir, tamarack pine, and silver fir, the last-named generally clinging to the ridges and higher slopes, while the tamarack associated with willows and poplars is found in the low, marshy places. In the upper glaciated portion of the timber belt the rock has been swept bare of soil and the trees grow under hard conditions, their roots penetrating soil that fills cracks and joints in the granite.

The higher elevations from 6000 to 9000 feet have been denuded of their soil by glacial action and are characterized by various species of firs, spruce, and tamarack; the silver fir, for example, grows chiefly above 8000 feet. All the timber of the higher belt is sparse and of poorer quality than that found on the lower elevations. The highest slopes are rocky and inaccessible and without vegetation.

The ranges of a number of characteristic species of the northern Sierra is shown in Fig. 42. The Patton hemlock has a restricted and uneven distribution following a granite axis on the summit of the range over which its distribution corresponds with the belt of heaviest precipitation. Its continuity is broken by deep, low valleys. The Shasta fir is another essentially mountain species. It appears to require a precipitation of at least 50 inches and hence does not occur below elevations of 4800 feet. Since it is restricted to limited areas because of its temperature and moisture requirements, it is not distributed in a continuous belt but like the Patton hemlock is broken by deep valleys and canyons. The sugar pine has a wider and more continuous distribution. It is absent below 2000 feet in the western foothills and occurs on the eastern and upper margin of the belt in detached fragmentary bodies and east of the mountains almost not at all. The yellow pine has the widest range of any of the Sierra species. Its lower limit of distribution is about 1500 feet, its upper from 6500 to 7000 feet. The high precipitation and low temperature of the summit of the Sierra prevent the yellow pine from occupying the higher ridges, just as dryness prevents it from occupying the Sacramento Valley.

Some of the high ridges covered with andesitic breccia are very dry and support either no trees or shrubs or those types normally found at a lower altitude under drier conditions.[1]

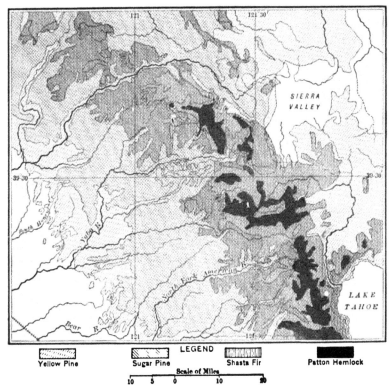

LEGEND

Yellow Pine   Sugar Pine   Shasta Fir   Patton Hemlock

Scale of Miles
10   5   0   10   20

Fig. 42. — Ranges of four characteristic species in the northern Sierra Nevada.
(Compiled from U. S. Geol. Surv. maps.)

The trees of the thinner forests below 3000 feet grow in thin groves or are scattered over grassy slopes. Yellow pine occurs on the higher ridges as an outlying fringe of the great forests of the timber belt, while oaks are particularly abundant on dry areas underlain by the older volcanic breccias and tuffs. Many hills are thickly covered with an evergreen shrub (greasewood), and in the lower part of the belt the Digger pine

[1] Turner and Ransome, Big Trees Folio U. S. Geol. Surv. No. 51, 1898, p. 3.

comes in, besides stunted oaks, and a number of shrubs. The lower-most zone of vegetation in the foothill belt also contains the Digger pine and stunted oaks, and shrubs characteristic of greater dryness, such as the manzanita.[1]

Although the forests of the Sierra Nevada are more restricted in vertical range toward the south owing to increasing aridity in this direction, they persist as far as the 36th parallel, where the elevation of the mountains begins to decrease. With lower elevation the rainfall diminishes to the point where a stunted vegetation appears, and at the extreme south a true desert flora is developed. By contrast, the Coast Ranges exhibit a desert vegetation as far north as the Bay of Monterey (lat. 37°), and nowhere bear extensive forests south of San Francisco. The lesser elevation of the Coast Ranges accounts for a large part of this deficiency, though a part also must be attributed to their position on the margin of the belt of westerly winds. They receive their rainfall chiefly during the winter season, when the westerly wind belt with its cyclonic storms has migrated southward over them. But the latter deficiency is shared by the Sierra Nevada. We may therefore say that were the altitudes of the two ranges reversed the Coast Ranges would be densely wooded at least as far south as the Bay of Monterey while the Sierra Nevada would be practically without forests.

It seems clear from the topographic character of the Sierra Nevada as well as the Cascades that large portions of the existing forests grow under conditions that must for a long time to come, perhaps forever, prevent the utilization of their products. Especially is this true of detached areas of forest on high mountains and steep slopes separated from the main body of the forest by deep and almost unscalable canyons and gorges, and far from centers of population upon which the lumberman must depend for a labor supply as well as for the consumption of the forest products. The steep, cliff-like walls of the larger valleys themselves offer difficulties of the highest order. Ordinary lumbering methods are useless, extraordinary methods are so expensive as to make the development of the more difficult forest areas highly improbable. The potential value of the forests in the more difficult tracts is therefore very limited in relation to the lumber supply, but their practical value is almost unlimited in relation to stream flow. The retardation of the run-off on steep slopes is a matter of the greatest concern where both the rainfall and the snowfall are heavy and the topography extraordinarily rugged, and this is everywhere to some extent and in some places to a large extent effected by the forest cover.

---

[1] F. L. Ransome, Mother Lode District Folio U. S. Geol. Surv. No: 63, 1900, p. 1.

CHAPTER XII

PACIFIC COAST VALLEYS

GENERAL GEOGRAPHY

ONE of the larger features of the continent of North America is the discontinuous line of valleys known as the Pacific coast downfold, a structural and topographic depression between the Coast Ranges on the west and the Sierra Nevada and Cascade mountains on the east. Its most striking expression is in the valleys of the San Joaquin and the Sacramento in central California and in the Willamette Valley in northwestern Oregon, where the Coast Ranges descend with marked abruptness to the level of the alluvium-filled valley floors. The width of the downfold is about 100 miles from crest to crest of the bordering ranges and from 50 to 75 miles from mountain front to mountain front; the total length is about 2500 miles. The northern and southern ends of the downfold are submerged, forming Puget Sound and the Gulf of California respectively; the higher unsubmerged sections extend through northern California, Oregon, and Washington and are separated by two mountain groups. The divisions are as follows: (1) a southern division extends from Gulf of California to Los Angeles, (2) a central division constitutes the Great Valley of California, and (3) a northern division comprises the Willamette Valley in Oregon, the Cowlitz Valley in Washington, and the broad depression at the head of Puget Sound. The mountains separating the southernmost depression from the Great Valley of California are the San Gabriel, San Rafael, San Bernardino, and others; those separating the Great Valley of California from the Willamette Valley are the Klamath, a group rather than a range of mountains, consisting of a number of secondary ranges among which are the Siskiyou, Rogue River, and others.

The Pacific coast downfold has been a feature of the western coast since the Cretaceous period, and during several geologic periods was so deeply depressed as to lie beneath sea level and receive a considerable body of sediments. The later phases of alluviation are due to the action of the tributary streams which descend in steep courses from the flanks of the high mountains near by and contribute a vast body of detrital material to the upbuilding of the valley floors.

177

WILLAMETTE, COWLITZ, AND PUGET SOUND VALLEYS

The northernmost section of the Pacific coast downfold consists chiefly of the Willamette and Puget Sound valleys. The alluvial portion of the Willamette Valley heads near Eugene, Oregon, and extends north to the Columbia. North of that river the depression is continued by the Cowlitz Valley and lesser valleys tributary to the southern end of Puget Sound. The depression is important climatically, since on the north it lets in the sea in the form of a great mediterranean that extends 150 miles inland; and in the Willamette Valley on the south it results in a much greater seasonal range of temperature than occurs in the coastal section near by from which it is separated by the Coast Ranges. Everywhere in the northern depression the rainfall is markedly less than on the windward (western) slopes of the bordering Coast Ranges and Cascades.

In Washington the greater part of the depression is composed of alluvium and glacial or fluvio-glacial deposits. These consist of till, sand, and gravel, and were formed during and at the close of the glacial period when piedmont and valley glaciers descended from the bordering mountains and discharged into the waters of Puget Sound. The surficial deposits overlie and partly obscure an older topography, a well-developed valley system coördinated with the present system of converging sounds and bays so suggestive of the drowning which occurred here. The postglacial changes are due chiefly to the extension of the deltas at the mouths of the streams. These advance into the bays, reclaim their heads, and thus greatly modify both valley and shore.[1] The irregularities of the shore line of Puget Sound are not attributable to depression alone. The glaciation of the sound deepened and widened the depressions that were the lines of glacial movement, but the deepening was so much more important than the widening that the channels are deep and narrow. The low water-parting between the Cowlitz Valley and the valleys tributary to Puget Sound is due to the deposition of alluvium and glacial deposits upon a previously nearly level-floored intermontane depression.[2]

The Willamette Valley south of the Columbia River is to the Cowlitz north of the Columbia what the San Joaquin is to the Sacramento. It is intensively farmed, relatively, and is almost unforested in contrast to the densely forested, because better-watered, hills and mountains on

[1] Willis and Smith, Tacoma Folio U. S. Geol. Surv. No. 54, 1899, p. 2.
[2] I. C. Russell, North America, 1904, p. 160.

either side. Its deposits are likewise of glacial and fluvial origin, deposits of the latter kind predominating.[1] The valley is 150 miles long and at present constitutes the most important single tract of arable land in the state.

GREAT VALLEY OF CALIFORNIA

CLIMATIC FEATURES

In the study of the Great Valley of California, and indeed of the physiography of the California district as a whole, one must keep in mind the great range in latitude between its northern and southern ends. The state is 800 miles long, and if it were transposed to the Atlantic seaboard with its southern end placed on Charleston, South Carolina, its northern end would lie approximately on New Haven, Connecticut. The southern end of California is a region of deserts, desert mountains, salt lakes, a sparse and specialized vegetation, and other features associated with pronounced aridity. The northern end of California lies on the whole in the belt of adequate rains; on the windward slopes of the mountains in the northwest corner of the state there is a mean annual rainfall of 81 inches,[2] and dense forests of redwood clothe the mountain slopes.

These great differences in precipitation are due to two conditions: (1) the state is of unusual size and has a wide range of latitude; (2) it lies partly within two climatic belts, the belt of westerly winds, and the horse latitudes. The mean annual rainfall varies from 1 inch to 81 inches. In the extreme southern part of California there live many people who have never seen snow in any form. At Summit, near Donner, in northern California, an annual snowfall of 697 inches, or nearly 60 feet, has been reported. Farming is conducted in an ordinary manner in large sections of northern California, though some irrigation is practiced; irrigation is the indispensable condition of the agriculture and the horticulture of southern California. Little wonder is it that under these circumstances Californians should speak of northern California and southern California as two very unlike regions. The degree of unlikeness is so extreme, the different interests so divergent in many respects, that the idea is quite widely entertained that California should be separated into two states for the better safeguarding of local interests.

In addition to the climatic differences between northern and southern California are east-west differences of climate dependent upon strong

[1] For the character of the drainage and the topography of the upper Willamette Valley see the Eugene quadrangle, U. S. Geol. Surv.

[2] A. J. Henry, Climatology of the United States, Bull. Q. U. S. Weather Bureau, 1906, pp. 9–72.

contrasts in the different north-south belts of relief. These are summarized by Hilgard[1] as follows:

(1) Bay and coast region characteristics: Small range of temperature, the extremes being
only 53° apart. Means of summer and winter are only 6° apart. There is no intense heat
and frosts are very rare. Fogs from the sea are quite common on summer afternoons. Rainfall averages 27.3 inches, about 25 inches of which falls between December and May.

(2) Great Valley characteristics: Average winter temperatures lower than those of the
coast, though minimum is about the same. Frosts are rare. Summer heat is very intense,
often above 100°. The nights are warm but dry, and are therefore less oppressive. Extreme
range of temperature 76°, mean range 23.6°. Rainfall averages about 21.5 inches, of which
19.8 inches fall between December and May.

(3) Sierra slope characteristics: Cool summers with frequent thunderstorms. The winters
are often severe, with much rain and snow. Mean summer temperatures, 57.5°, with a mean
range of 14° between that and the winter temperature of 43.5°. Rainfall averages 57.24 inches,
fairly well distributed throughout the season.

GENERAL GEOGRAPHIC AND GEOLOGIC FEATURES

The Great Valley of California, the largest unit of the great Pacific
coast downfold, lies between the two main chains of that state, the
Sierra Nevada on the east and the Coast Ranges on the west. It is
about 400 miles long, has an average width of about 50 miles, and
contains about 20,000 square miles. It consists chiefly of two long and
relatively narrow piedmont alluvial plains with a monotonously level
surface and a marked parallelism with all the main physiographic
features of the state lying north of the 35th parallel.

The drier southern end of the Great Valley is a region of large wheat
ranches, but in later years fruit raising has begun to supplant this
industry. Grazing is also a principal resource. The better-watered
northern end of the valley produces lumber, dairy products, fruits, and
vegetables; and the greater rainfall of the Sacramento Valley and bordering ranges so well maintains the level of the Sacramento River that a
navigable depth of seven feet from Sacramento to the river's mouth is
maintained at slight expense.[2]

The history of the Great Valley dates from the great orogenic disturbance at the close of the
Miocene which gave birth to the Coast Ranges as a connected mountain chain. Later still
(at the close of the Pliocene) the Sierra Nevada block was further uplifted and the Coast Ranges
increased in height, an increase which has continued down to the present time.[3] During the
post-Pliocene elevation of the crest of the Sierra and of the Coast Ranges and also in the
Pleistocene period the Great Valley was gradually and finally cut off from the sea, closed in
by mountains, and changed to a definite well-bounded area of sediments, upon which stream

[1] E. W. Hilgard, quoted by Van Winkle and Eaton, Quality of the Surface Waters of California, Water-Supply Paper U. S. Geol. Surv. No. 237, 1910, pp. 10–11.

[2] Document No. 1123, 60th Congress, 1909.

[3] F. L. Ransome, The Great Valley of California, Bull. Dept. Geol., Univ. Cal., vol. 1,
1896, p. 387.

deposits began to form. The whole Great Valley is now completely walled in by mountains except where the Sacramento and San Joaquin unite to flow through the straits of Carquinez into San Francisco Bay.

The Great Valley is divided into three parts: (1) the Sacramento Valley, drained by the Sacramento River; (2) the San Joaquin Valley, drained by the San Joaquin River; and (3) the Tulare Valley, which might be considered a subdivision of the San Joaquin Valley, for it is sometimes tributary to it.

SACRAMENTO AND SAN JOAQUIN VALLEYS

The Sacramento Valley is a broad and nearly flat alluvial plain. It gradually diminishes in breadth northward and terminates near Red Bluff at an altitude about 300 feet above the sea. At this point the valley is composed of low alluvial fans which have developed to the point of confluence. On the western side of the valley the mountain slopes rise abruptly from the plain; on the east the slopes of the Sierra Nevada rise in a more regular and even manner. All the streams from the Sierra Nevada that enter the Great Valley carry large amounts of rock waste and those that drain the largest basins carry exceptionally large amounts. Their gradients are greatly decreased as they emerge from their deep mountain canyons, and a part, sometimes a large part, of their water is absorbed or evaporated. Thus the carrying power of the streams diminishes rapidly, and eventually a part of the load of land waste is dropped in the form of alluvial fans some of which are 40 to 50 miles in radius. The alluvial fans are composed of coarse waste near the mountain foot, — rough, bowldery material, very pervious, and therefore very dry. Farther from the mountains the material becomes finer; on the lower valley flats it is chiefly fine silt.

Across the broad plain of gravel and sand which forms the northern end of the Sacramento Valley the river and its tributaries have cut valleys from one-fourth mile to four miles in width and to depths sometimes reaching 100 feet. The floors of the valleys are generally flat and may be called the valley flats and flood plains of the adjacent streams. They are covered with fine alluvial soil which when well watered is excellent for agricultural purposes.[1]

The tributaries of the Sacramento before reaching the main stream turn aside and discharge in stagnant sloughs which expand and overflow large areas during the wet season. Broad belts of swamp land and lake therefore occur on both sides of the Sacramento River and are usually covered by a dense growth of tule (*Scirpus lacustris*). The plains and the lowest rolling foothills are on the whole without arboreal

[1] J. S. Diller, Redding Folio, U. S. Geol. Surv. No. 138, 1906, p. 1.

vegetation save for scattered oak trees which give a park-like character
to the landscape. The river is usually lined by tule swamps and the
banks support a dense vegetation of brush and willows.[1]   Farther south
the Sacramento River and its principal tributary, the Feather River,
flow in channels well above the general level of the flood plain.   The
case of the Yuba River is of peculiar interest.   Mining operations in its
valley have caused the delivery to the stream of an exceptional amount
of alluvium, so that the town of Marysville, which was formerly well
above the river, is now considerably below it at high water.

The streams draining the western slopes of the Sierra Nevada constitute the larger part of the drainage of the Sacramento Valley and have
a relatively constant flow reaching the Sacramento through definite
channels; the smaller streams draining the eastern slopes of the Coast
Ranges seldom reach the Sacramento River at the surface but are lost
in the intricacies of the sloughs which meander through the bordering tule lands.   This difference in the amount of water and therefore of
alluvium contributed to the Sacramento and the San Joaquin valleys
on the east and west sides has resulted in a marked asymmetry of valley
form, both rivers lying on the western sides of the plain.   The San
Joaquin, especially, flows close to the base of the Coast Ranges, having
been pushed farther and farther west by the building up of low confluent alluvial fans at the mouths of the Sierra streams.   In a similar way
Lake Tulare lies near the western edge of the valley on account of the
encroachment of the extensive fan of the Kaweah River combined with
the deltas of the Kings River and other streams.

On the north the alluvial portion of the Sacramento Valley is bordered
by a well-marked plain of erosion (Pliocene) which passes under the

SECTION NEAR ELDER CREEK, TEHAMA CO.
Fall—130 feet per mile

Fig. 43. — Base-leveled plain on the northern border of the Great Valley of California.
(Diller, U. S. Geol. Surv.)

lavas of the Lassen Peak district and in some places changes gradually
and in others abruptly into the mountain slopes of the Klamath region.
It is considered to be a continuation of the peneplain recognized within
that region.   The width of the base-leveled plain at the northern end
of the valley varies from one to fourteen miles, and was once part of
an extensive erosion plain which formerly included middle and northern

[1] Lindgren and Turner, Marysville Folio U. S. Geol. Surv. No. 17, 1895, p. 1.

California, southern Oregon, and possibly an even greater area. The plain cuts across Cretaceous and Tertiary strata along the eastern border of the northern part of the Sacramento Valley. These strata pass by gentle dips westward beneath the valley and rise again to the surface along the western border, thus outlining the northern part of the valley as a broad shallow geosyncline filled with deposits of late geologic age.[1]

TULARE VALLEY AND LAKE

Tulare Valley, near the southern end of the Great Valley of California, contains Tulare Lake, which has no regular surface drainage to the sea. The waters of the lake are separated from the San Joaquin system by a gentle swell of alluvium, so that in seasons of unusual rainfall a surface connection is established between the lake and the river.

The combination of drainage conditions about Lake Tulare reminds one very strikingly of those in the Salton Sink region. The basin of Lake Tulare is due chiefly to the building up of alluvial fans across the San Joaquin Valley north of it, especially between Kings River on the east and Los Gatos and other creeks in the northern part of the Coalinga district, the latter having formed exceptionally large alluvial fans for Coast Range streams. Lake Tulare is therefore a broad shallow water body developed upon an almost level floor of alluvium and represents an expanse of water above an obstructing dam formed by alluvial fans. It derives its water supply from several streams that descend from the Sierra Nevada and spread numerous distributaries over the valley floor. Practically no surface water reaches the lake from the mountains on the west side in spite of their closer proximity. On all sides the lake is bordered by broad tule-covered swamps, hence its name.

Lake Tulare has no regular surface outlet; the water level is controlled by seepage and evaporation. It is therefore subject to great fluctuations and in periods of high water the whole central portion of the valley becomes flooded and marshy. In earlier years the lake was one of the largest bodies of fresh water in California; in later years it has been gradually declining in size owing largely to decreased rainfall, to the use of the water of its tributary streams for irrigation, and to the reclamation by dikes of the land formerly covered by it. In 1880 it overspread an area about 27 miles long and 20 miles wide; in 1889 it was 20 miles long and 15 miles broad. Still more recently successive dikes have been constructed, the lake has almost dried up, and most of the former lake bottom has been cultivated. In 1907 the precipitation

---

[1] A. C. Lawson, Bull. Dept. Geol., Univ. Cal., vol. 1, 1896, p. 271.

was unusually large and the whole central portion of the valley was again inundated, the lake extending almost to its old shore line of 1880 near the base of the Kettleman Hills bordering the Coast Ranges.

During late Quaternary time Tulare was much greater than at present, as shown by an old beach a hundred feet above it. The beach is in the form of a ridge of sand about 20 feet wide, 6 to 8 feet high, and somewhat eroded and covered with vegetation. A line of depression across the middle of the main valley connects Tulare Lake by a low marshy tract with two smaller lakes, Kearn and Buena Vista, 50 miles to the south, which owe their position near the base of the Coast Ranges to the westward growth of the large delta of the Kearn River.[1]

VALLEY OF SOUTHERN CALIFORNIA

LOCATION AND CLIMATIC FEATURES

The valley of southern California is a lowland area limited on the north by the San Gabriel range and separated from the Mohave and Colorado deserts on the east by the San Bernardino and San Jacinto mountains. On the west the plain extends to the Pacific; on the south its limits are irregular and indefinite, a broad transitional belt occurring between it and the heights of the Sierra Madre of Mexico. It is not a part of the great Pacific downfold but a separate lowland unit. The valley of southern California is more populous, more intensely cultivated, and has more concentrated wealth than any similar area in the Southwest. These unique features depend upon its climate, its fertile soil, and its valuable products. Its southerly position gives it a moderately high mean annual temperature of about 62°. The open exposure of its surrounding mountains to the Pacific results in a marked rainfall and hence they supply more water for irrigation than is commonly supplied to the alluvial plains of the West. The streams that descend form the seaward slopes of these mountains and water the alluvial plain between them and the sea are, as such streams go, of great size and permanence of flow. The soils are generally well disintegrated arid land soils with a high percentage of soluble plant food.

TOPOGRAPHY AND DRAINAGE

The valley of southern California, Fig. 44, is divided by the Santa Ana Mountains and the Puenta Hills into two parts: (1) a coastal portion, the coastal plain, and (2) an interior portion, the interior valley. The coastal plain of southern California is about 50 miles long and 15 to 20 miles wide. Its relief is in general low and the regional slope is seaward

[1] Arnold and Anderson, Geology and Oil Resources of the Coalinga District, California, Bull. U. S. Geol. Surv. No. 398, 1910, pp. 39–382.

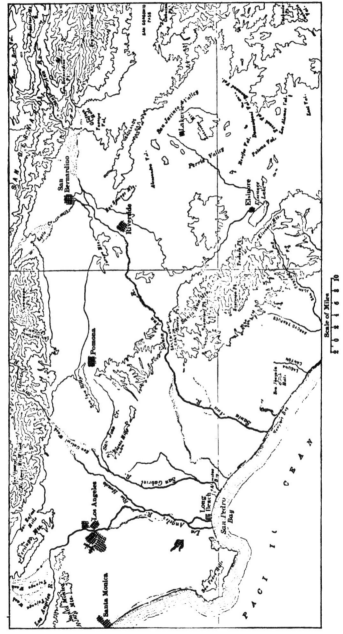

Fig. 44. — Bench lands, coastal plains, and mountains of southern California.

from an elevation of 200 to 300 feet along the inner margin to the salt marshes and sand dunes of the coast. The chief interruptions of its level surface are San Pedro Hill and a low ridge that extends southeastward from the vicinity of Palms. The inner edge of the coastal plain forms a fringe of bench land which contours the higher mountains back of it and forms bluffs except where the mountains approach the coast. It is somewhat dissected by the canyons of the small streams that cross it. These gullied benches are conspicuous back of Santa Monica and along the southern base of the Santa Monica Mountains, where they are composed of stream-deposited sands, gravels, and clays, in contrast to

Fig. 45. — Inner edge of the coastal plain of southern California near Whittier.
(Mendenhall, U. S. Geol. Surv.)

the marine deposits which make up a large part of the coastal plain.[1] The coastal plain of southern California is regarded as a former broad embayment of the Pacific in which débris brought down by the mountain streams accumulated; it was finally exposed by uplift and slightly modified by erosion.[2]

FORESTS AND WATER SUPPLY

In southern California, where water supply is a dominating economic necessity, a very intimate relation has been found to exist between the amount of forest cover in the mountains and the water supply

[1] W. C. Mendenhall, Development of Underground Water in the Western Coastal Plain Region of Southern California, Water-Supply Paper U. S. Geol. Surv. No. 138, 1905, pp. 9-11.
[2] Idem, p. 11.

Fig. 46. — West Riverside district, California, representing typical relations of mountains and bordering alluvial plains, valley of southern California. (Mendenhall, U. S. Geol. Surv.)

Fig. 47. — West Riverside district, California. This view is panoramic with the one above. The mountain notch on the right is the notch on the left in Fig. 46.

of the bordering plains. The matter is of great importance to horti-
cultural interests in southern California because of the nearly rainless summer, the precipitation occurring almost wholly in the winter
months from November to April. Practically all of the rain that falls
upon the flat porous valley lands is immediately absorbed by the soil
or evaporated into the air. On the mountain slopes a large proportion is absorbed by the soil and humus where vegetation has not been
destroyed by fire and the unprotected soil swept off. It is the water
absorbed by the soil and forest litter of the mountain slopes that is the
source of the important summer flow of the mountain streams. It has
been found that the denser the vegetal growth and the thicker the soil
on the mountain slopes, the more effectively are the winter rains stored
and the more uniform is the summer flow. The effect of the forest in
decreasing surface flow during the rainy season is enormous, the average
of four different basins showing an absorption of 95% on the forest-
covered areas and only 60% on the nonforested areas, where the
rainfall is much less. A comparison of three other areas gave the
following result:

"The three forested catchment areas, which, during December, experienced a run-off of but
5% of the heavy precipitation for that month and which during January, February, and March
of the following year had a run-off of approximately 37% of the total precipitation, experienced
a well-sustained stream flow three months after the close of the rainy season. The nonforested
catchment areas, which during December experienced a run-off of 40% of the rainfall and which
during the three following months had a run-off of 95% of the precipitation, experienced a run-
off in April (per square mile) of less than one-third of that from the forested catchment areas,
and in June the flow from the nonforested area had ceased altogether."[1]

The disastrous results that follow deforestation are of great concern in
this thinly forested region, for the private lands are being rapidly deforested by large lumber camps, whose operations cause ever-increasing
danger from forest fires, floods, and summer droughts.[2]

Soils of the Pacific Coast Valleys

The soils of the Pacific coast valleys range from residual and colluvial
soils of the mountain foothills to deep and extensive river flood-plain
and delta sediments and ancient and modern marine and lacustrine
shore deposits. The wide range in value of these soils and their adapta-

[1] J. W. Toumey, The Relation of Forests to Stream Flow, Yearbook, Dept. of Agri., 1903,
pp. 286–287.
[2] Van Winkle and Eaton, Quality of the Surface Waters of California, Water-Supply Paper
U. S. Geol. Surv. No. 237, 1910, p. 17.

tion to crops is dependent largely upon the possibilities of irrigation and upon local climatic conditions of rainfall and temperature.

The soils of the alluvial fan deposits, colluvial and alluvial wash from foothills and higher adjacent soil bodies, and occasional small areas of residual material are derived mainly from sandstones, shaly sandstones,

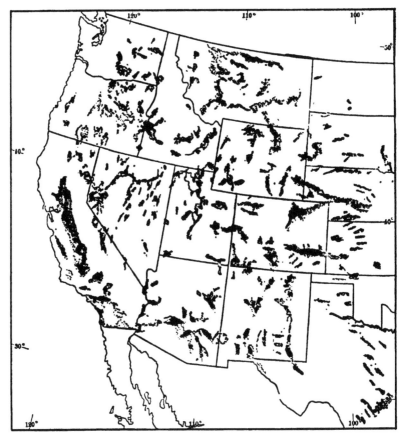

Fig. 48. — Irrigation map of the West. Irrigated areas solid black, irrigable areas dotted.
(Newell, Irrigation in the U. S.)

and shales (Cretaceous and Tertiary), and occur upon rolling marginal hills, sloping, elevated, and dissected mesa or bench lands, and in some places on the margins of lower nearly level valley plains.

The soils composed of recent alluvial materials derived from a great variety of rocks and deposited as river and delta plains generally occupy a lower topographic position, are of more recent origin, are subject to more frequent overflow than other soils of the region, and often support a growth of swamp vegetation, brush and willow thickets, and timber in the river bottoms and lower valley plains. The surface is generally level, slightly sloping or sometimes uneven, and is frequently marked by sloughs or the interlacing channels of streams many of which carry water only in times of flood and disappear in sandy washes. The heavier members are frequently marked by adobe structure and the soils are generally free from gravel or bowlders.

The lower-lying areas are frequently poorly drained, subject to the influence of seepage water from irrigation, and contain alkali. They are generally underlain by subsoils of fine ashy texture, light color, and compact, close structure, usually separated from the overlying soil by an alkali carbonate hardpan of white or light-gray color. The hardpan softens slowly upon the application of irrigation water, but is normally impenetrable to the roots of growing plants.

An important group of soils is composed of alluvial deposits formed by shifting streams and mountain torrents and occurring as broad, low, alluvial cones occupying gently sloping plains or slightly rolling valley slopes, generally treeless and lying above present stream flood plains. The soils of this series are derived from a variety of rocks, but generally from those of granitic and volcanic character, or from sandstones carrying large amounts of granitic material. They are generally treeless and support only a desert vegetation except where they are irrigated. They are frequently cut by arroyos, and are usually gravelly and often strewn with bowlders. These soil bodies vary from small areas of irregular outline to broad, extensive, uniform sheets. They are generally dark-colored, open-textured, well drained, and free from alkali.

The Great Valley contains no true forest growths, only lines of poplars and sycamores along the rivers. The lower foothills on the borders of the valley and up to elevations of 1000 feet are dotted with Douglas oak and live oak and occasional patches of a thorny shrub (*Ceanothus cuneatus*). During the dry summer season the main expanse of the valley presents a rather monotonous and cheerless view since it is practically treeless and covered with a dry gravelly soil or a parched growth of sparse grasses that spring up in the wet season. Along the stream courses in the axis of the valley trees are sometimes found, though willow and alder shrubs are more common growths.[1]

The soils of the valley of southern California are similar in origin, topographic position, and texture to those of the Great Valley. The

---

[1] Turner and Ransome, Sonora Folio U. S. Geol. Surv. No. 41, 1897, p. 3.

most important departure from the general character is made on the coastal plain. Its outer margin has soils of distinctive character derived from marine sediments. Except along stream courses and at the heads of some of the alluvial fans arboreal vegetation is entirely wanting.

In the Tacoma region at the northern end of the series of coast valleys the bottom-land soils of the valley floors are stream-laid and vary in texture from gravel to fine silt. Silt is most common, gravel and sand have a more restricted development and are generally found near the mouths of the canyons and at the heads of the alluvial fans of tributary streams.

The soils of the uplands are disposed on steep sharp ridges and washed to such a degree as to be unfit for tillage, except in limited areas on rounded hills. Locally the soil is open in texture and very sterile, as in the case of the Steilacoom plains south of Tacoma. Although these plains receive about 44 inches of rain per year, the water drains away so rapidly in the loose, stony ground that they are an arid tract in the midst of a humid region.[1]

The primeval growth of the valley soils was fir and hemlock with cedar and maple predominating in the wet lands along the hollows and tributary gullies.

[1] Willis and Smith, Tacoma Folio U. S. Geol. Surv. No. 54, 1899, p. 7.

CHAPTER XIII

COLUMBIA PLATEAUS AND BLUE MOUNTAINS

COLUMBIA PLATEAUS

EXTENT AND ORIGIN

THE Columbia physiographic province includes a large variety of physical features whose basis of unity lies in their common association with those widespread sheets of basaltic lava that form the larger part of the region. It is important therefore to have at the outset a clear conception of the geologic origin and nature, areal extent, and physiographic features of the Columbia River and Snake River lavas.

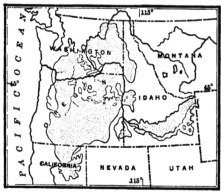

Fig. 49. — Lava fields of the Northwest. (Data from Geologic Map of North America, by Willis, U. S. Geol. Surv.)

In sharp contrast to the mountainous borders of the province are the flat or apparently flat basalt plains which cover an area of over 250,000 square miles and form probably the most extensive single field of sensibly flat basalt in the world. Their development involved the transference of about 100,000 cubic miles of dense rock from the earth's interior to the surface. For many years all these basaltic flows were believed to have originated from fissures concealed by the lava that issued from them. This was the view of Sir Archibald Geikie, for example, when in 1882 he described the Snake River basalt plains.[1] But the studies of Russell have shown conclusively that no direct evidence has been found in support of this explanation in the greater part of the Snake River lavas, while the region abounds in evidence pointing to the local accumulation of material from a large number of vents of several varieties. In many localities low mounds may be seen rising by gentle

[1] Sir Archibald Geikie, Geological Sketches at Home and Abroad, 1882, pp. 237-242

gradients above the general level of the country.   Their summits are flat, and from them may be traced lava flows which extend outward in all directions until they merge into the flat expanses of the plain.   Here and there on the borders of the region flows may be traced down tributary valleys at whose mouths they expand to form part of the general surface.[1]

To produce so flat and so extensive a plain, lava must have great fluidity and must originate from separate vents thickly strewn over a given region.   Both these conditions are fulfilled here.   Vents are numerous and their flows have been traced into each other to such an extent that the region may be described as consisting of a large number of low gradient plains merging inperceptibly into each other.   Some of the constituent minerals of the basalt (augite, etc.) fuse at such low temperatures that a high degree of fluidity had been attained by the lava. It ran readily over the gentle slopes and spread far and wide before it was congealed.   Were the flows less fluid they would have gathered closer about the volcanic vents and would now show far steeper gradients, as in the Cascades.

While the Snake River lava was thus formed by volcanic extravasation of highly fluid material from many different vents, the basalt of a large portion of eastern Oregon and Washington is equally well known to have originated from a vast number of fissures.   On the eastern border of the Cascades there is a great system of feeding dikes in the sandstone beneath the basalt; similar relations have been observed in the Blue Mountains, where the once overlying basalt has been removed by erosion.[2]

In central Idaho on the eastern edge of the basalt plain east of Lewiston no ash cones or tuff volcanoes are found nor are any dikes of basalt observed in the foothills of the Clearwater Mountains adjacent to the basalt.   The eruption of the fluid rock must have taken place in this locality without explosive action and from fissures.[3]   In the Eagle Greek range, Oregon, a local center of eruption has been discovered near Cornucopia, where at elevations of 7000 to 8000 feet there occurs a perfect network of basalt dikes intersecting the schists and granites immediately above the lava plateau.[4]

The lavas comprising the Snake River and Columbia River plains are of two general sorts, basalt and rhyolite.   The basalt, as we have already seen, was spread over the surface in great flows and in a highly

[1] I. C. Russell, Geology and Water Resources of the Snake River Plains of Idaho, Bull. U. S. Geol. Surv. No. 199, p. 66.

[2] F. C. Calkins, Geology and Water Resources of a Portion of East-Central Washington, Water-Supply Paper U. S. Geol. Surv. No. 118, 1905, p. 19.

[3] W. Lindgren, A Geological Reconnaissance Across the Bitterroot Range and Clearwater Mountains in Montana and Idaho, Prof. Paper U. S. Geol. Surv. No. 27, 1904.

[4] W. Lindgren, The Gold Belt of the Blue Mountains of Oregon, 22nd Ann. Rept. U. S. Geol. Surv., pt. 3, 1902, pp. 740–745.

fluid condition. The rhyolite, on the other hand, was extruded in part in the form of sheets or flows and part as ejectamenta — volcanic dust, gravel, lapilli, and other angular fragments. These two kinds of eruptions occurred at different periods; the greatest inundations of basaltic lava in eastern Oregon took place before the rhyolitic eruption. After the rhyolite had been extruded there came a renewal of the basaltic eruptions, which continued from time to time almost to the present day but on a far less extensive scale. The latest basaltic eruptions of the Snake River plains are so fresh as to appear to have been formed within historic times, probably not more than several hundred years ago; the earliest eruptions date back to the Tertiary (Miocene). The total number of flows varies from place to place. but in a number of cases at least 100 are known to have occurred.

There is considerable difference in the age of the Columbia and the Snake River lavas. The Columbia River lava is deeply decayed and over large areas has been changed to a soft clay-like soil having a depth of sixty feet or more, while the Snake River lavas are still fresh and even the exposed portions of older sheets show but slight changes.[1]

The volcanic outpourings of the region were not limited to the Columbia and the Snake River valleys; they extend into the Cascades, which are in large part formed of volcanic material; the southern Cascades are almost exclusively volcanic. The greater part of the eastern border of these mountains is completely buried beneath recent flows. The basaltic flows also extend northward into Canada, where they form a considerable portion of the Interior Plateau of British Columbia. Lava flows of the same kind and approximately the same age are found in many other localities in the Pacific Cordillera, as in the Great Basin and the Colorado Plateaus, where they form highly important elements of the relief.

BURIED TOPOGRAPHY BENEATH THE BASALT

To understand the present distribution of the lava of the Columbia Plateaus and the detailed character of the surface one must know that before the lava was extruded the region had considerable relief; canyons and gorges had been cut, and the whole surface was in a state of vigorous dissection. The country rock consisted of old volcanic and sedimentary formations, mainly granite, rhyolite, quartzite, and limestone.[2]

The effect of the basaltic inundations was to fill the valleys and, to a large extent, to bury the older topography. In some cases hills and mountains of older material project through the basalt as islands project above the surface of the sea, for example, Big, Middle, and

[1] I. C. Russell, Geology and Water Resources of the Snake River Plains of Idaho, Bull. U. S. Geol. Surv. No. 199, 1902, p. 61.

[2] Idem, p. 15.

East Butte, Idaho.[1] In other cases the old divides extend for long distances into the lava fields as capes and promontories against which the basalt came to rest.

An excellent locality for the study of the relations of the present to

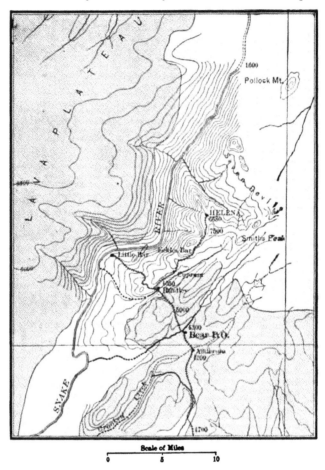

Scale of Miles

Fig. 50. — Canyon of the Snake River at the Seven Devils. Sketch contours, interval approximately 500 feet. The cross-lined areas represent basalt-covered surface; the blank areas represent older slates, schists, etc. (After Lindgren, U. S. Geol. Surv.)

the buried topography is northeast of the Blue Mountains, Oregon, where the Snake River has cut a gorge across a great structural arch in the basalt and has exposed successive flows with their interstratified

[1] I. C. Russell, Geology and Water Resources of the Snake River Plains of Idaho, Bull. U. S. Geol. Surv. No. 199, 1902, p. 62.

beds of dust and lake deposits. Near the Seven Devils, Fig. 50, the gorge is 4000 to 6000 feet deep, and is one of the most remarkable erosion features in the United States. It extends northward for 125 miles as far as Asotin, a few miles above Lewiston, and is comparable in grandeur and depth to the canyon of the Colorado. Above Asotin the canyon of the Snake River becomes deeper and furnishes many striking illustrations of columnar basalt which rises tier on tier more than 3000 feet. Near Buffalo Rock may be seen to best advantage the relations of the old and now buried topography to the basalt cover. The metamorphic rocks which formed the surface of the country before the basalt inundated it rise at least 2000 feet into the horizontally bedded flows. The river has thus cut its gorge across a buried mountain and has exposed the rocks composing it for a stretch of about a mile. The undisturbed horizontal layers of basalt abut sharply against the steep waste-free slopes of the old mountain which descend to the lowest level of the Snake River and have ancient valleys coördinated with them deeper than the present deep canyon of the Snake.[1] The crest line of the buried mountain is rugged and serrate; the gorges between the higher crests and spurs are filled with horizontal sheets of lava, showing that the flood of basalt flowed about the highest peaks and for a time left them as islands in a molten sea of rock, then overtopped their summits and completely buried them.[2] The topographies of the canyon wall above and below the contact of these two rock types are very dissimilar. The buried surface developed upon the older rocks is exceedingly irregular and steep, the spurs ending in precipices that are sometimes almost a thousand feet sheer. On the other hand the dull-brown and relatively flat-lying basalt is weathered into cliff and talus and a more regular type of topographic architecture.

DRAINAGE EFFECTS OF THE BASALT FLOODS

SNAKE RIVER VALLEY

The repeated and extensive outpourings of lava resulted in widespread hydrographic changes. The ancestral Snake River was dammed by lava flows to such an extent that a large lake, so-called Lake Payette, was formed, upon whose floor sediments were laid down. The lake appears to have been invaded time and again by contempo-

---

[1] I. C. Russell, A Reconnaissance in Southeastern Washington, Water-Supply Paper U. S. Geol. Surv. No. 4, 1897, p. 31.

[2] Idem, p. 35.

raneous lava flows, and it is with these flows and with the widespread sheets of volcanic sand and dust that the sediments are associated. The lake beds filled the valley to a great depth, and are divided into an older (Miocene) and a younger (Pleistocene) division which carry mammalian remains and fresh-water mollusks.[1]

The waters of Payette Lake entirely surrounded the Owyhee range of Idaho, as is shown on the west side of this range where the nearly horizontal soft sandstones and shales of lacustral origin rest against eruptives (Miocene) and display a well-defined shore line from 5400 to 5500 feet above sea level. This shore line is also identified along the Boise River but at 1000 feet higher elevation, and indicates by its variable elevation at many points in the Snake River Valley that notable deformation has taken place since deposition of the lake beds.

In addition to the main Miocene lake there were formed many small lake basins caused by lava dams at the valley mouths, as in Long Valley in the northern part of Boise County.[2]

The present valley of Snake River extends across Idaho in a semicircular course about 80 miles wide. Its course is underlain by lake

Fig. 51. — South shore of Malheur Lake, Oregon. Salacornia growing in alkali.
(Russell, U. S. Geol. Surv.)

beds and intercalated flows of basalt that slope gently from the bordering mountains to the axis of the valley. Into these beds the river has cut a sharp canyon 400 to 1000 feet deep, thus exposing the structure. The average elevation of the Snake River plains of Idaho is from

[1] W. Lindgren, The Gold and Silver Veins of Silver City, De Lamar, and Other Mining Districts, Idaho, 20th Ann. Rept. U. S. Geol. Surv., pt. 3, 1898–99, p. 80.
[2] Idem, pp. 95–96.

3975 feet at Shawnee to 2125 feet at Weiser. Alluvial and to some extent irrigated bottom lands occur at a number of places along the river, and they also exist along the lower courses of the Boise and Payette tributaries.[1]

HARNEY-MALHEUR SYSTEM

In Oregon the course of the Malheur River has been profoundly modified so as to bring about a loss of fully one-third of its former drainage basin. A lava flow dammed its channel in the vicinity of Mule River and caused the formation of the basin of Harney and Malheur lakes, Fig. 52. The surface of the lava is only about 10 or 15 feet above the normal level of Malheur Lake, and so delicately are the topographic and climatic conditions balanced that a very slight increase in rainfall would result in the discharge of Malheur River down the line of its old valley. The occurrence has additional interest in that the ponding of the water above the lava dam causes the entire region now draining into Harney and Malheur lakes, about 4500 square miles in area, to be removed from the Pacific slope drainage and added to the drainage of the Great Basin.

A further modification of the Harney-Malheur drainage has been brought about by the formation of hills of wind-drifted sand which invaded what was once a single basin, making the two basins now occupied by Malheur and Harvey lakes. The hills are in part grass-covered, with steep-sided basins among them, and furnish a barrier between the two lakes which is only crossed by the water of Malheur Lake during high-water stages.

Deformations of the Basalt Cover

Although the basalt plains of the Columbia region were formed in a nearly horizontal position and although these plains appear to be approximately horizontal to-day, there are in reality many important departures from horizontality. The Snake River plains are now in the form of a broad trough or downfold reaching from Lost River and Sawtooth Mountains on the north to Goose Creek and Bear River Mountains on the south. Many minor irregularities of structure have been noted. In southwestern Idaho the lavas and intercalated lake

---

[1] W. Lindgren, The Gold and Silver Veins of Silver City, De Lamar, and other Mining Districts in Idaho, 20th Ann. Rept. U. S. Geol. Surv., pt. 3, 1898–99, p. 77. See this author's geologic map, Plate 8, p. 76, for the distribution of the various types of rocks found in a little-known section of central Idaho south of the National Forest that lies east of Mount Idaho and northeast of the Snake River Valley.

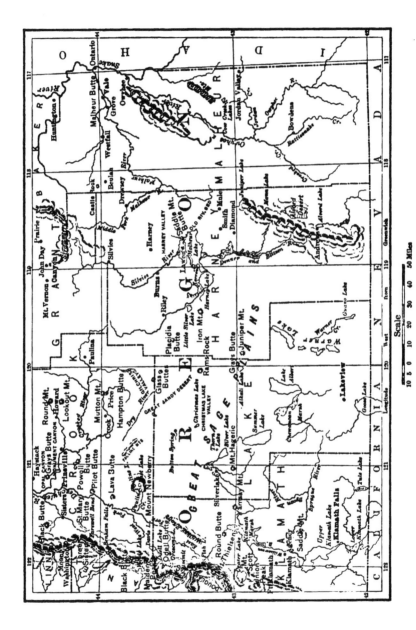

Fig. 52. — Sketch map of Southeastern Oregon, showing drainage features. (Modified from Russell, U. S. Geol. Surv.)

and river sediments have been gently flexed and, near the bases of the bordering mountains, broken and faulted.[1] These structural irregularities are of considerable economic importance, for it is upon the trough-like arrangement of the beds that the artesian condition of the deeper waters depends. The economic development of the region has been accomplished to a notable extent by the use of artesian waters for domestic purposes and for supplying stock as well as for a small amount of irrigation.

The existing relation of drainage to relief implies antecedent conditions on the part of the streams. The lava, originally disposed in an essentially horizontal position, has been deformed from this position but the deformation has not proceeded so rapidly as to rearrange the drainage courses. Stream courses laid out upon the nearly flat lava sheets in response to the initial slopes have persisted in their courses, and where there have been great uplifts athwart the streams we now find great canyons. The explanation is similar to the one applied to the present course of the Columbia across the Cascades except that in the Cascades it was a base-leveled and to some extent a lava-covered surface and not exclusively a sheet of lava that was uplifted across the path of the river. Had the lava been in its present attitude when the Snake River first gained its course the river would now run in an opposite direction for some distance south of the great canyon.

Coulees of the Columbia Plateaus

Among the striking physiographic features of the Columbia River region are the coulees that occur in Washington from Patterson to Connell east of the Cascades. Many of them are scores of miles in length and of notable width and depth. They are dry or contain only small streams in spite of their great size. They have all the topographic qualities of river valleys in addition to tributary systems of branching valleys with stream-carved bluffs. Some of them represent an earlier period of more abundant rainfall during which wide, flat-bottomed, and cliff-walled canyons were formed. Others are related to the bodily diversion of the streams from their preglacial courses and the post-glacial abandonment of the temporary courses.

The largest and most rambling of the coulees is Grand Coulee. It is not only of greater interest scenically but its geologic history has been unusually interesting. This great canyon was cut by the Columbia River at a time when the profound gorge Columbia was temporarily

[1] I. C. Russell, Geology and Water Resources of the Snake River Plains of Idaho, Bull. U. S. Geol. Surv. No. 199, 1902, p. 16.

dammed by a great glacier that came down the Okanogan Valley, filling it to a depth of several thousand feet. Upon reaching the Columbia River Valley the glacier expanded and spread out over the plateau for about 35 miles. The displaced Columbia flowed along the eastern face of the glacier for two miles below Coulee City, where the bottom of the canyon drops abruptly and where the Columbia River once poured over a mighty cataract 400 feet high. The gorge below the cataract extends southward for 15 miles.

The lakes of the plain of the Columbia are situated in the coulees. They occupy basins that represent the former irregularities of channel bottom, or that have been formed by the irregular distribution of wind-blown material. They are elongate in form and often disposed in chains. This is the character of Moses Lake, the largest body of water in the district. The lakes in the northern part of the Grand Coulee occasionally flow to the south at seasons of high water and are therefore comparatively fresh and palatable. On the other hand the waters of the southern lakes become successively more alkaline, and Soap Lake, the most southern, is extremely alkaline. On stormy days it is beaten into great masses of white foam on the exposed shore. The substances in solution consist essentially of carbonates, sulphates, and bicarbonates of soda.

STREAM TERRACES

Later episodes in the history of the Columbia Plateaus are inferred from the gravel deposits associated with the streams in the form of terraces. They indicate that after the erosion of the deep, rock canyons there followed a period during which the streams aggraded their valleys to depths of many feet, — in the case of the Columbia at least several hundred feet. Later still, a second period of down-cutting was inaugurated in which the streams degraded their channel bottoms, in some cases to bedrock. There were temporary halts in the down-cutting which are now expressed by the lower terraces.[1] In many cases the streams have again begun the aggradation of their valley floors.[2]

The canyon walls of the Columbia River, for example, are marked by a succession of terraces in part carved out of bedrock and in part composed of stream-laid gravels. The largest of these terraces is known as the "Great Terrace of the Columbia" and is distinguished from its neighbors by its great extent and perfection of development. The terraces above the Great Terrace were built by streams along the sides of

---

[1] F. C. Calkins, Water-Supply Paper U. S. Geol. Surv. No. 118, 1905, p. 43.
[2] Idem, p. 45.

the Okanogan glacier during the period of its decline. These terraces are less regular than the lower ones and are characterized by more numerous deep pits or kettle holes, while the material composing the terraces suggests glacial débris slightly modified by water. On the other hand the Great Terrace is of postglacial age and is due to the fact that the canyon of the Columbia was once filled with fluviatile deposits from 300 to 500 feet above the bedrock floor and to the level of the Great Terrace. The terraces below the Great Terrace simply mark halts in the process of dissection in postglacial time of the original gravel filling.[1]

Gravel plains of large extent were also developed along the international boundary line in the form of low delta deposits and terraces, the variety of minor features being due to the interaction of ice and stream work.

CLIMATE, SOIL, AND VEGETATION

CLIMATE

The Columbia Plateaus everywhere receive a deficient rainfall, since the province is practically surrounded by mountains of notable height and continuity. On the west are the Cascades, which so thoroughly intercept the rain-bearing westerly winds as to leave even the higher eastern slopes of these mountains rather dry and the lower slopes very dry. The lower and flatter country to eastward is also very dry and in general receives from 8 to 25 inches of rainfall; the lesser amount falls on the Great Sage Plains of central Oregon and the lower Snake River Valley, the greater amount being restricted to the higher relief features such as the Blue Mountains of northeastern Oregon and other ranges of lesser height. Sufficient rain falls upon these mountains to support a forest growth, but the amount in even the most favored situations is nowhere large, and none of the forests is dense. The Blue Mountains shut in the plain of the Columbia on the south in eastern Washington, just as the Owyhee, Goose Creek, and Bear River ranges shut in the Snake River Valley on the south. On the north the plain of the Columbia is sheltered from cold winter winds by the Okanogan and Columbia ranges, on the east by the Cœur d'Alene; the valley of the Snake is similarly sheltered in these directions by the great mountain mass of the Clearwater and Salmon River mountains of central Idaho and the Bitterroot Mountains on the Idaho-Montana line. In both cases, however, the bordering ranges exact their toll of rainfall and leave the plains relatively dry. In eastern Washington the rainfall increases

---

[1] Smith and Calkins, A Geological Reconnaissance Across the Cascade Range near the 49th Parallel, Bull. U. S. Geol. Surv. No. 235, 1904, pp. 38–41.

somewhat because of increasing elevation, though this increase should not be confounded with increases on the mountainous elevations farther east. It is sufficient over a restricted tract to make possible farming without irrigation.

The partly enclosed character of the Columbia region causes its mean annual temperature to be much higher than one would expect from its latitude and altitude. Very low winter temperatures prevail, however, in the mountain tracts, especially in the mountainous area of central Idaho, with the result that the snowy precipitation is heavy and the snow melts slowly and remains late the following spring. The effect is to give the streams draining across the bordering valleys and plains access to a natural storage of precipitation of the greatest value in maintaining a proper flow. Since forest trees normally require only about 100 days a year free from snow, the slow melting in late spring of the heavy snowfall of winter greatly increases the value of the total yearly precipitation available to the forests of the region without interfering with their growth. In the Columbia region grazing must always constitute the most important industry from the standpoint of the extent of territory involved. Irrigation by means of mountain streams may, however, become the most important industry as regards the value of the products.

SOILS

The desert quality of the climate of the region is emphasized by the presence of an extensive layer of pumiceous gravel, sand, and dust. It covers tracks aggregating several thousand square miles in extent. It is particularly abundant east of the Cascades, and also extends westward over portions of the mountains and down their slopes for twenty miles or more. Both the Great Sandy Desert (see Fig. 52) and the country west of it display this feature. In many places the layer of volcanic dust is fully 50 feet thick and a maximum thickness of 70 feet has been recorded.[1] It is extremely porous, permits the quick descent of the ground water, and invariably accentuates the dryness of the climate east of the Cascades.

On the great plain of the Columbia in eastern Washington and in many other localities the soils have been described as residual, as having been formed in place by the decay of the basalt. Calkins shows conclusively,[2] however, that in many portions of Washington the soils

[1] I. C. Russell, Geology and Water Resources of Central Oregon, Bull. U. S. Geol. Surv. No. 252, 1905, p. 16.

[2] F. C. Calkins, Geology and Water Resources of a Portion of East-Central Washington, Water-Supply Paper U. S. Geol. Surv. No. 118, 1905, p. 45.

have been formed by wind action. The conclusion is based upon the facts that there is an absence of lamination in the soil, that it is extremely fine in texture, that there is a remarkably sharp definition between the soil and basalt, and that comparative chemical analyses indicate soils not of the character naturally to be expected from the decomposition of basalt in this climatic province. The principal source of the material appears to be the soft sedimentary beds (Ellenberg formation) in the southwestern portion of the Columbia plains.

The soils are fine loams, very light, open, and friable, with a light, tawny, brown color. The thickness of the soils varies according to location from 25 to 50 feet, the greatest thickness being on the brows of slopes where there is the most favorable opportunity for the accumulation of wind-blown material.

The character of the so-called "dust soils" of Oregon and Washington is typically represented by the following analysis. They are light, dry soils raised into clouds under natural conditions at the slightest wind and probably originated entirely through wind action akin to that which resulted in the formation of loess.[1]

CHEMICAL ANALYSES OF DUST SOILS [2]

Constituents	I Atathnam Prairie, Yakima County, Washington	II Rattlesnake Creek, Kittitas County, Washington	III Plateau on Willow Creek, Morrow County, Oregon
	%	%	%
Insoluble matter	71.67 } 76.78	78.33 } 80.53	79.21 } 81.51
Soluble silica	5.11	2.20	2.30
Potash (K_2O)	1.07	0.70	0.89
Soda (Na_2O)	0.35	0.24	0.05
Lime (CaO)	2.00	2.08	1.37
Magnesia (MgO)	1.34	1.47	1.08
Brown oxide of manganese (Mn_3O_4)	0.04	0.07	0.06
Peroxide of iron (Fe_2O_3)	6.88	6.13	5.63
Alumina (Al_2O_3)	7.91	6.12	6.02
Phosphoric acid (P_2O_5)	0.13	0.18	0.18
Sulphuric acid (SO_3)	0.02	0.02	0.03
Water and organic matter	2.82	2.35	2.55
Total	99.33%	99.90%	99.35%
Humus	4.10	....	0.44
Hygroscopic moisture	4.98	3.20	4.92

The most extensive and uniform soil types of the Columbia Plateaus consist of residual materials overlying extensive basaltic lava plains. In

---

[1] G. P. Merrill, Rocks, Rock-weathering and Soils, 1896, p. 345.
[2] Bull. U. S. Weather Bureau, No. 3, 1892.

some cases the soils have been derived from granitic rocks or from ancient lacustrine sediments or extensive lake beds now more or less modified by erosion or æolian agencies. The margins of the lacustrine or residual deposits are covered by sloping plains and fans of colluvial wash from the adjacent mountain borders, while in the vicinity of the larger streams, which have carved and terraced the lacustrine beds and residual soils, are recent alluvial stream sediments derived from reworked materials of the lake beds or from the weathered products of the mountains. Soils of this type constitute a large portion of the great grain-producing lands of the Northwest. Everywhere the soils are treeless or sparsely timbered, except in the vicinity of streams.

The higher lying areas are often rough and hilly, marked by rock outcrop, bowlders, or morainic débris, and deeply cut by stream channels. Their soils "are generally of dark color, and are underlain by sticky subsoils of light-gray or yellow color. The soils and subsoils are generally gravelly, the gravel varying from fine angular chips to large, well-rounded or angular blocks and cobbles. The soils are dry farmed to grains or, when not occupying too high a position, are irrigated and devoted to grains, alfalfa, clover, and fruits." [1]

Recent flood-plain deposits are underlain by beds of gravel and cobbles, usually from a few inches to a few feet thick, sometimes partially cemented by lime. They are often marked by shallow beds or channels of meandering streams, and are frequently timbered or covered with willow or brush thickets in the vicinity of streams. They usually occur as small irregular to extensive areas, often subject to overflow. The flood-plain soils are generally rich in organic matter and of a mucky consistency, except in the lighter, higher lying members, and sometimes contain alkali. [2] The basalt of the Columbia plain weathers easily and has been decomposed to form a rich residual soil which mantles the surface and gives its slopes characteristic soft, rounded, flowing outlines. [3]

The greatest difficulty in the utilization of the water of the Snake and the Columbia arises from the fact that large stretches of these rivers occur at great depths below the general level of the country. The Columbia flows in a canyon with fairly abrupt walls sunk from 100 to 1000 feet below the broad stretches east of it in Washington. Furthermore, its gradient is much lower than that of much of the land along

[1] Soil Survey Field Book, U. S. Bur. of Soils, 1906.
[2] Idem.
[3] I. C. Russell, A Reconnaissance in Southeastern Washington, Water-Supply Paper U. S. Geol. Surv. No. 4, 1897.

it and the application of its water to the soil is therefore exceedingly difficult. The most important streams, from the standpoint of irrigation, are those which drain the eastern flanks of the Cascades, as the Yakima, Kittitas, and others.[1]

VEGETATION

Throughout southern Idaho and over the greater portion of Oregon east of the Cascade Mountains the sage-bush is the characteristic plant. It is nowhere absent save on the barren mud plains left by the drying up of the ephemeral lakes, or upon the summits of the mountains. While not so plentiful as the sage-bush the bunch grass is dispersed almost as widely. The fresh-water ponds of the coulee bottoms are bordered by tule, while the meadows in the same situations are covered with wild grasses. The small streams are fringed with a scattered growth of willow, birch, and wild cherry.[2] With increase in elevation the juniper makes its appearance, and beyond the lower limit of the juniper are thickets and groves of mountain mahogany. At still higher elevations, yet within the range of the mountain mahogany, the pine appears and reaches up to an elevation of 8000 to 10,000 feet. However in only two areas in the whole southeastern quarter of Oregon do forests of any considerable extent occur. Castle Rock in Malheur County, Oregon, is on the border of an extensive forest of pines, as well as juniper, mountain mahogany, and many other trees, and a splendid forest exists on the mountains northwest of Harney and Burns in which the Silvies River has its sources.

Above 7500 feet the peaks of the Seven Devils and of the Eagle Creek range in Oregon are flecked with snowdrifts and scored by rock slides and a forest cover is wanting. A forest zone consisting of black pine above and yellow pine below extends from 4000 feet to 7500 feet. Below 4000 feet the canyon walls of the Snake are again almost bare.[3] The bottom of the canyon is at an elevation of about 1600 feet; snow rarely falls in it, and the rainfall is almost equally scant, so that only desert types of vegetation grow upon it.

The occurrence of forests in the region depends upon many factors, chief of which is the moisture supply. The whole tract is extremely dry and one must always ascend to a considerable elevation to find a forest growth. The lower edge of the forest growth is called the "dry

[1] F. C. Calkins, Water-Supply Paper U. S. Geol. Surv. No. 118, 1905, p. 18.
[2] Idem, p. 22.
[3] W. Lindgren, The Gold and Silver Veins of Silver City, De Lamar, and Other Mining Districts in Idaho, 20th Ann. Rept. U. S. Geol. Surv., 1898–1899, pt. 3, 1899, p. 92.

timber line." On ascending the forest-clothed mountains one reaches also an upper limit of tree growth, the "cold timber line," beyond which only shrubs, stunted trees, and alpine flowers exist. If the aridity is intense the dry timber line will be high and the forest belt correspondingly narrow; if the aridity is not extreme the forest belt will be wide. In many cases the aridity elevates the dry timber line until it coincides with the cold timber line and no forest exists in such cases even though the mountains have a great elevation. It is for this reason the prominent Stein Mountains of southeastern Oregon have no forests. It may readily be seen that mountains which project above the dry timber line but which do not reach the cold timber line have forest-clothed summits, while those whose summits reach to elevations below the dry timber line or above the cold timber line are bare.[1] On the Cascade Mountains in central Oregon the cold timber line has an elevation of about 8000 feet, while the dry timber line, marked by the cessation of the yellow pine, may be taken at approximately 4000 feet.

In the use of the vegetation the scarcity of surface water upon areas underlain by volcanic ash is apparent in two main ways. Water is not available for fighting forest fires even though there is sufficient ground water to support a forest growth; and over large areas there is an excellent growth of grass untouched by sheep or cattle because of the absence of drinking water in the form of springs or running streams. The ash cover is a rapid absorbent and rain water that falls upon it almost immediately becomes ground water. A heavy restriction is thus laid upon the use of the land and its products, though at least one beneficial result follows — the larger streams of the region are kept in more even flow.

BLUE MOUNTAINS

The Blue Mountains of northeastern Oregon have received far less attention than they deserve from physiographers and geologists as well as from foresters. They lie in a very interesting position midway between the mountain complex of central Idaho and the Cascades, and are a projecting spur of the great crust-block composed of the Lost River, Bitterroot, Clearwater, and Salmon River mountains. While they extend well across the basin and plateau region between the Rocky Mountain system and the Pacific Mountains, they do not connect directly with the Sierra Nevada and the Cascades. From the standpoint of rainfall and forests the Blue Mountains (8500 feet) are the

---

[1] I. C. Russell, Notes on the Geology of Southwestern Idaho and Southeastern Oregon, Bull. U. S. Geol. Surv. No. 217, 1903, p. 11.

most important relief feature in the entire region between the Cascades and the Rockies since they cause a local rainfall (15 to 25 inches) that waters the fertile valley flats and a belt of peripheral country of considerable extent.

Topographically the Blue Mountains consist of all that group of complex ranges that constitute the country between the Deschutes Valley on the west, the Malheur and Harney deserts on the south, and the Snake and Columbia rivers to the east and north. Within the mountain knot thus outlined are a number of ranges with such specific qualities that they have received separate designations. Conspicuous among these are the Eagle Creek Mountains, the Elkhorn Range, the Greenhorn Ridge, the Strawberry Range, and others. The mountain group thus defined stands out prominently above the surrounding plain, which lies from 4000 to 6000 feet above sea level. The highest peaks of the Blue Mountains rise to heights over 9000 feet above the sea and many exceed 8000 feet.

The geologic features necessary to an understanding of the physiography of these mountains may be briefly stated. The sedimentary rocks of which they are chiefly composed have been not only extensively and rather generally folded but also quite thoroughly intruded by granodiorite, diorite, gabbro, and peridotites. The intrusions were accompanied by uplift, — the net result of folding, intrusion, and uplift being the formation (Jurassic and early Cretaceous) of a mountain knot of impressive height. Upon this complex mass erosion produced profound effects, stripping off a great mass of material and laying bare the heart of the mountains. While they were thus deeply dissected they were never reduced to an old-age condition; erosion was carried only so far as to make them very rugged; so that were one to take away the lava flows about the Blue Mountains they would stand out as imposing heights. Lava flows (Miocene) derived from numberless fissures on the flanks of the mountains were then spread far and wide over the surrounding country, burying the older topography, subduing the relief, and separating the Blue Mountains from the Rockies, causing them to stand out as islands in a basaltic sea. The first flows were rhyolites and andesites, the later flows were basalts in increasing volume.

The effects of the great lava flows were not confined to topographic and drainage changes in the valleys of the Columbia and the Snake, but were exhibited as well in the marginal tracts of the Blue Mountains. The effects are all the more striking because of the former well-developed character of the drainage. The present river courses and the sediment-filled upper basins that are the products of volcanic flows are among the most difficult physiographic problems of the region. The lower parts of the watercourses became filled with basalt, damming the headwaters and creating lakes that were afterwards drained to a large extent by the downcutting at their outlets, thus producing physiographic effects of puzzling complexity.[1]

The precipitation of the Blue Mountains is heavy enough to produce a forest of yellow pine which shades off to mountain mahogany, juniper, and other types, in lower and therefore drier situations. The greater part of the tree cover consists of an open woodland growth since even the best watered areas receive a limited rainfall and snowfall. Small summit areas on the higher mountains, such as the Strawberry Range, Fig. 52, and the Powder River Range north of it, are without a tree cover since their elevations exceed that of the cold timber line, about 8000 feet. The alluvium-filled mountain basins, noted above, and the forest cover combine to keep the streams in more even flow than would otherwise be the case, thus making possible an important amount of agriculture in the lower irrigated valleys. The result is a fringe of population about the flanks of the mountains with finger-like extensions down the major depressions.

[1] W. Lindgren, The Gold Belt of the Blue Mountains, 22d Ann. Rept. U. S. Geol. Surv., pt. 2, pp. 574, 575, 594, 597, et al.

CHAPTER XIV

GREST BASIN

Arid Region Characteristics; Hydrographic Features

In spite of the prevailingly arid character of the western half of the United States its streams are in large part through-flowing, a feature due chiefly to the loftiness and position with respect to rain-bearing winds of the mountain groups and ranges in which their sources lie. The Columbia and the Colorado, for example, although they lose a part of their water by evaporation and absorption, yet maintain a considerable volume up to the point of discharge; and in other similar cases the large headwater contributions maintain a perennial flow. Therefore, in respect of drainage a relatively small portion of the arid West has the characteristics usually associated with pronounced aridity — interior basins, large streams which disappear on piedmont slopes, and salt lakes such as those that characterize the deserts of Asia, or the Sahara desert in Africa.

Two large physiographic provinces of the United States are exceptions to this general rule — the Great Basin and the basin of the Lower Colorado. Of the two the Great Basin has the more remarkable development of those drainage features that are an index of extreme aridity. The drainage of the entire Great Basin is of the interior-basin variety, no part of the water that falls within it reaching the ocean by surface drainage. Everywhere the streams descend from the better-watered mountain ranges to waste-floored forelands, where a large part of their water — sometimes the whole — is lost by evaporation and absorption. Such excess of water as locally fails to be absorbed by the porous sands and gravels of the piedmont regions is gathered upon the floors of depressions between mountain ranges in the form of salt lakes or lakes that are strongly alkaline.

Salt Lakes of the Great Basin

Chief among the salt lakes of the Great Basin are Great Salt Lake, Lake Humboldt, Carson Lake, and the group of saline lakes that occupy the Sage Plains of central Oregon, Fig. 52. Some of the lakes of the Great

Fig. 53. — Illustrates the small size and independent character of the drainage systems of the Great Basin region. (Russell, U. S. Geol. Surv.)

Basin are composed of very dense brine of common salt, sodium sulphate, and other substances. It has been estimated that Great Salt Lake alone contains 400,000,000 tons of common salt.[1] Not all the lakes tributary to or in the Great Basin are of this character however. Some of them consist of pure wholesome water, such as Utah Lake, Bear Lake, and Lake Tahoe on the western edge of the Great Basin and near the forested heights of the Sierra Nevada. Wherever a constant outlet is assured, the lakes consist of sweet water, but an interrupted outflow is always indicated by an increase of salinity, and the absence of an outflow results in the concentration of chemical salt to such an extent as to result in dense brines.

Another important feature of many lakes of the Great Basin is their ephemeral character. In a large number of instances lakes exist on the basin floors only during periods of high water in the feeding rivers; in dry seasons such lakes evaporate and expose broad flat expanses of mud which soon become dry and sun-cracked and present on the whole a monotonous and forbidding appearance. These are called *playas* and are desert features as characteristic as sand dunes or lost rivers.

The rapid manner in which some playas are transformed into shallow lakes is almost incredible to one unacquainted with rainfall conditions in desert regions. A single storm will sometimes form a shallow lake whose waters are spread far and wide over a basin floor. The disappearance of such a lake may be almost as rapid as its formation, for the mud cracks, at least for a time, allow easy passage of the water to lower levels, and with the high temperature and clear skies characteristic of arid regions, surface evaporation is rapid, and often within a few days, sometimes even in a few hours, after the rain has ceased, the smaller lakes have disappeared.

It is easily seen that the sudden appearance and disappearance of lakes in the Great Basin region are to a large extent functioned by the flatness of the basin floors on which the waters rest. Sudden and great differences in topographic level are not characteristic of regions whose surface forms are products of alluviation. The graded and gentle waste slopes of piedmont forelands and aggrading basin floors are surfaces of such slight relief that waters which come to rest upon them may be spread over great areas and yet nowhere be of great depth. Great Salt Lake is an illustration in point. Its average depth is from 15 to 18 feet; its maximum depth is less than 50 feet. Its area in 1850 was 1750 square miles, but by 1869 its volume had increased to 2170

[1] I. C. Russell, North America, 1904, p. 142.

square miles.[1] From 1900 to 1904 it was feared it would disappear entirely and its bed become a salt desert. Since that time the level has risen to such an extent that large engineering works, like the Lucin cut-off of the Southern Pacific and the roadbed of the Western Pacific are endangered and may have to be abandoned. These changes are mainly in response to changes of rainfall, 1910 having been abnormally wet.[2] The changes that have taken place in the past half century are to be regarded as changes that may be repeated at any time in the future.

The rate of supply of water to the lake is constantly undergoing changes in sympathetic response to changes in precipitation in the drainage basin. Likewise, the rate of evaporation at the lake surface is subject to considerable fluctuation. A third variable factor is the rate at which water is diverted from tributary streams for irrigation purposes. All of the natural changes are subject to periodic fluctuations. Climatologic studies have shown that rather definite changes are characteristic of the climatic elements in all portions of the earth. These changes are cyclic in character and occur in periods of 11 and 35 years and undoubtedly in even longer periods of a hundred, several hundred, and even many thousands of years. The effects of the cyclic return of wetter conditions have been noted in the case of many lakes of humid regions and in their discharging streams; but the maximum effects of such climatic changes are felt in regions in which the drainage is of the interior-basin variety. The water of an interior basin must rise in response to wetter conditions until evaporation from the expanded surface just equals the rate of supply. The amount of expansion would represent, roughly, the increase of rainfall. If the changes from wet to dry and from dry to wet are not only cyclic but also progressive in one direction, as appears to have been the case at least during and since the glacial period, the repeated rise and fall of the lake surface by large amounts would be recorded in the form of shore features of familiar kinds. Thus there are small changes of lake level that affect all lakes in all climates; also changes of greater amplitude and of far greater physiographic importance, that may be designated as geologic. Such changes are best recorded in the basins of desert regions, for these are, on the whole, without outlet, and the surplus waters are confined and must faithfully record upon their margins the manner in which the changes occur.

[1] G. K. Gilbert, Lake Bonneville, Mon. U. S. Geol. Surv, vol. 1, 1890.
[2] Ebauch and Macfarlane, Comparative Analyses of Water from the Great Salt Lake, Science, n. s., vol. 32, 1910, p. 568.

Since the Great Basin is the only large physiographic province in the United States in which interior-basin drainage predominates, it follows that the clearest drainage records of climatic change are to be found there. During the glacial period a wetter climate prevailed in extra-glacial regions such as the Great Basin, and each lake in this province expanded until the rate of evaporation or evaporation and discharge just equaled the supply. The two largest of these lakes were in Utah and Nevada. The ancient lake that once existed in Utah has been called Lake Bonneville; its counterpart in Nevada is known as Lake Lahonton. Lake Bonneville has all but disappeared, its descendant being Great Salt Lake; and the discontinuous water bodies called Lake Lahonton have shrunk to such an extent that the lowest parts of the various basins are to-day occupied by a few salt- and brackish-water lakes such as Carson Lake, Humboldt Lake, etc. During its maximum development Lake Lahonton contained salt water, as its shrunken remnants do to-day, though it was undoubtedly less salt than they; on the other hand, Lake Bonneville was fresh or only slightly alkaline when it stood at its highest level, for it rose so high as to discharge for a time over the col on the divide between its basin and the Snake River Valley.

About the borders of the basin of Great Salt Lake may be seen shore features associated with the ancient lake levels and still in an almost perfect state of preservation. Upon the surrounding slopes and up to elevations of 1500 feet above the surface of Great Salt Lake are well-defined deltas, bars, beaches, spits, capes, cliffed promontories, and bottom deposits, all formed by or associated with the ancient lakes whose waters once stood at these high levels.[1]

Some of the most interesting and important lake features of the Great Basin are associated with the lakes of southern Oregon. For example, the water bodies in Lake County, southern Oregon, are all shallow. None exceeds 25 feet in depth. The size of these shallow water bodies depends on the seasonal rainfall, and changes in size are characteristic features in the absence of an outlet to the sea which might permit the maintenance of a more or less constant level, the level of the point of discharge. Important changes in the outline of some of these lakes have taken place since the settlement of the country.

In the early days of Lake View the town was on the edge of Goose Lake, but it is now six miles from it. In 1869 the lake overflowed for a short time southward into Pitt River. In 1881 it also overflowed for two hours during a severe gale from the north.[2] The fluctuation in

---

[1] G. K. Gilbert, Lake Bonneville, Mon. U. S. Geol. Surv., vol. 1, 1890; I. C. Russell, Geological History of Lake Lahonton, Mon. U. S. Geol. Surv., vol. 11, 1885.

[2] I. C. Russell, 4th Ann. Rept. U. S. Geol. Surv., 1884, pp. 456–457.

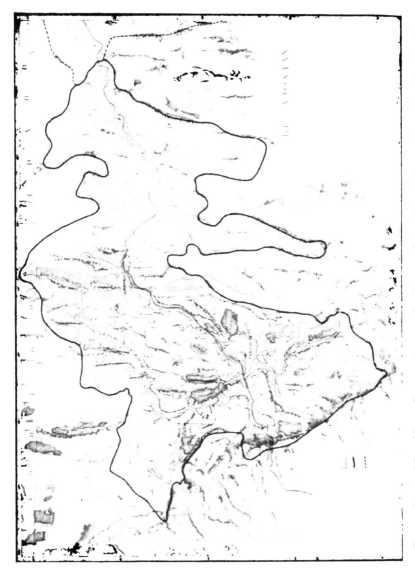

Fig. 54. — The dotted lines of intermediate strength represent Post-Quaternary faults. Scale, 1 inch = 70 miles. (Adapted from Russell, Mon. U. S. Geol. Surv., vol. II, Plates 4 and 44.)

the level of Goose Lake is also marked by the fact that in the early emigrant days the trail crossed Goose Lake Valley at a point where now the floor of the valley is under several feet of water.[1] The fluctuations in lake level have brought about a certain amount of litigation in relation to lands in the valley of Warner Lake, the decision turning upon the question whether some 4000 or 5000 acres now dry was swamp land or part of the bed of the lake at the time of the passage of the Swamp Land Act of 1860. After the exceptionally dry season of 1887–88, Silver Lake dried up, its bed was cultivated by farmers and one season's crops were gathered before the lake again occupied its old floor.[2]

The degree of alkalinity of the lakes of southern Oregon may be appreciated by comparison with the average alkaline content of the fresh-water lakes of North America as given by Russell,[3] who notes that the amount of alkaline material in the fresh-water lakes of this continent is between 15 and 18 parts in 100,000. The water of Summer Lake, Oregon, on the other hand, has 500 or more parts of salt (sulphate of soda, carbonate of soda, and bicarbonate of soda) in 100,000.[4] The practical bearing of these facts may be appreciated when it is known that the limit of alkalinity for domestic or irrigation purposes is about 400 parts to 100,000, although this limit depends largely on the character of the salt in each particular instance. The water of Lake Albert, southern Oregon, has a content of 3.9 % of salt, showing that the water is more strongly impregnated than ocean water, which contains about 3.5 % of mineral salt.[5]

RIVERS OF THE GREAT BASIN; PRECIPITATION

The chief rivers of the Great Basin receive their principal supply of water not from rainfall in their middle or lower courses but from melting snows in the mountains, as on the eastern and western borders of the basin. The stream discharge of the region is characteristic, the maximum occurring in the late spring or early summer, with a decrease of flow during the summer and a minimum during the winter months. The streams receive little or no additions after leaving the mountains, and diminish in size and often cease to flow at the surface.[6] The streams which discharge eastward from the Sierra Nevada, such as the Carson and the Truckee, have an immense run-off during the late spring, although the snows accumulate to a great depth in their thickly forested head-

[1] G. A. Waring, Geology and Water Resources of a Portion of South-Central Oregon, Water-Supply Paper U. S. Geol. Surv. No. 220, 1908, p. 12.

[2] Idem, p. 12.

[3] I. C. Russell, Lakes of North America, 1897, p. 55.

[4] G. A. Waring, Geology and Water Resources of a Portion of South-Central Oregon, Water-Supply Paper U. S. Geol. Surv. No. 220, 1908, p. 14.

[5] Idem, p. 13.

[6] La Rue and Henshaw, Surface Water Supply of the United States, 1907–08, pt. 10, The Great Basin, Water-Supply Paper U. S. Geol. Surv. No. 250, 1910, p. 28.

water regions and a considerable quantity is stored in natural lakes which supply water gradually to the streams into which they discharge.[1]

The streams that descend from the basin ranges are variable as to length and discharge, the latter feature depending ultimately upon the great variations in the height of the mountains (Fig. 55) with which the rainfall and snowfall are inevitably associated. Most of them disappear in the loose material of piedmont slopes, some of them discharge into lakes, all of them are subject to considerable variation in volume. These variations are immediately related to the forest and soil cover of the mountains in which the sources lie and to the rapid melting of the winter snows provided the mountains are of sufficient height to receive their precipitation in this form during the winter months. Only a few of the highest ranges were ever glaciated, hence few lakes occur in the regions of snowy precipitation, and natural storage of mountain waters is markedly absent. Those streams that are supplied by melting snows have great changes in volume from season to season, especially if the supply from springs is exceptionally deficient.[2] Thus the Humboldt River derives its supply from the melting of snows in headwater regions, and the run-off during the spring and summer months is very heavy; but as soon as the snow is gone the rivers are left practically without a source of supply and their channels gradually become dry.

Few better illustrations can be found of a stream not subject to excessive changes of volume and yet without forests or extensive meadows than Bear River, which drains the northern slope of the Uinta Mountains and discharges its waters into Great Salt Lake. The basin contains no marshes and but few small lakes near the head of the river, but the greater part of the precipitation is in the form of snow and the chief sources of supply are from melting snow and from numerous small springs. The latter form so steady a supply that after the annual high-water period during May and June the stream although diminished in volume does not cease to flow.[3]

Special Topographic Features

Having no outlet to the sea, the streams of the Great Basin sink into the alluvium of the basin slopes and floors or evaporate from the surfaces of salinas and salt lakes. The surface evaporation of the drainage waters also halts the waste which the drainage water carries and it

[1] La Rue and Henshaw, Surface Water Supply of the United States, 1907–08, pt. 10, The Great Basin, Water-Supply Paper U. S. Geol. Surv. No. 250, 1910, p. 100.

[2] Idem, p. 56.

[3] La Rue and Henshaw, Surface Water Supply of the United States, pt. 10, 1907–08, The Great Basin, Water-Supply Paper U. S. Geol. Surv. No. 250, 1910, p. 29.

therefore accumulates upon the land and tends to aggrade it. The disposition of the waste brought down by desert streams to lower levels is thus of special interest. Each basin floor becomes a local base level if there is no outlet to a lower basin. In the case of normal topographic development the surface of the land, though it may be built up temporarily, tends ultimately to be worn down to base level, which is almost sea level. In the development of the topography of a desert tract with interior-basin drainage, like the Great Basin, the elevations are worn down but the depressions are built up. The plane of reference, the common plane to which these forces will ultimately bring both depressions and elevations, is then not sea level but some higher level of indeterminate position. When this level has been attained further degradation of the surface will be chiefly through the exportation of dust by the wind to extra-desert regions. The drainage systems, at first independent and local, will become more and more interdependent and general, as one basin after another becomes filled or captured and enters into tributary relations with some lower neighboring depression. When the heights have been reduced and the basins all filled to a common level the rainfall is lessened by the decrease of relief and the streams will become enfeebled and disorganized or over large tracts cease to flow altogether.

The Great Basin streams are still in the early stages of basin filling. The region was deformed so recently (Miocene) that the topographic forms produced by the last deformation are still of mountainous proportions and the basins are only partly filled with land waste. Whether an ultimate level will be reached will depend (1) on the stability of the present climate: at one time (Pleistocene) a Great Basin lake (Bonneville) overflowed to the sea because of a wetter climate, and the same occurrence may be repeated. (2) It will also depend on crustal stability: some crustal movements have occurred in late geologic and even in historic time; if they are repeated and extensive they may offset the tendency toward leveling on the part of the streams.

BASIN RANGES

The most important topographic elements of the Great Basin are the roughly parallel mountain ranges that cross it from south to north and that diversify its surface to a greater degree in Nevada than elsewhere. The mountains are long and narrow and frequently have sharp crests with steep slopes on one side and relatively gentle slopes on the other. The mountain ranges of the Great Basin are explained by block faulting. In general each range may be considered as the upturned edge of a

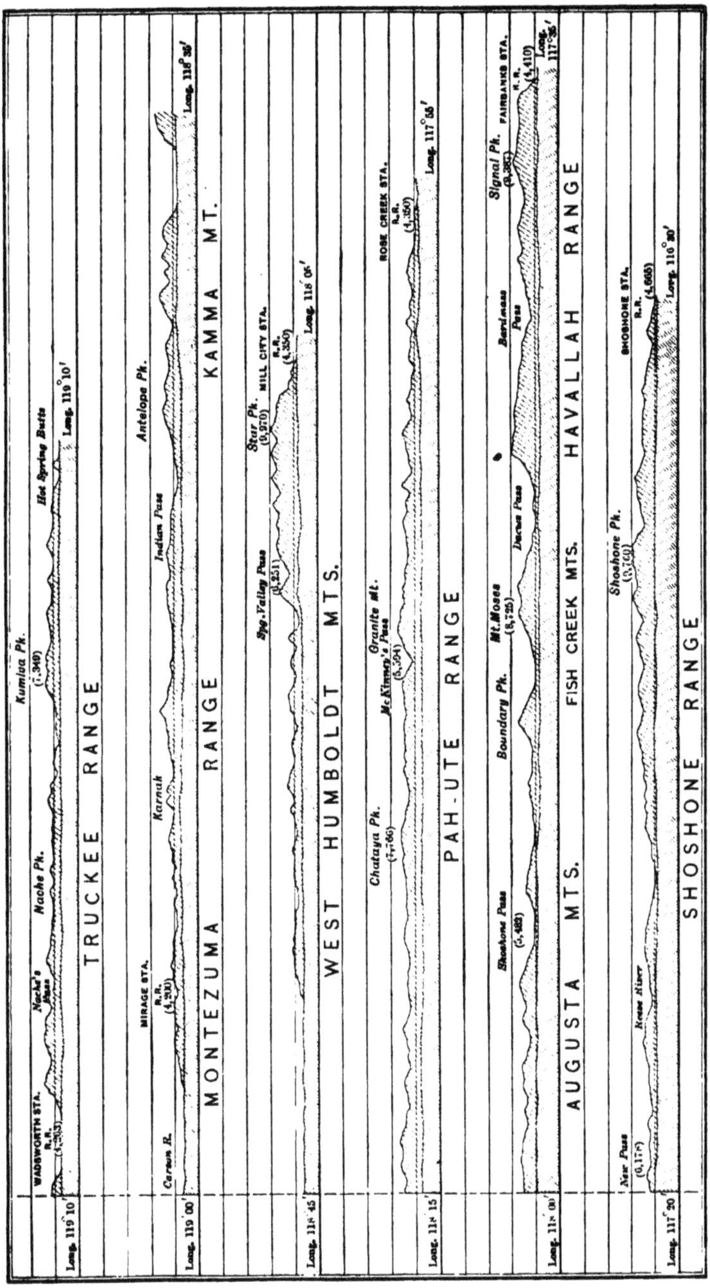

Fig. 55. — Longitudinal profiles of a number of the Basin Ranges. Lighter shading represents elevation of ranges above surrounding country. Horizontal scale, 25 miles to the inch; vertical scale, 37,500 feet to the inch. (King Surveys.)

block of the earth's crust, the faulted edge of the block forming a steep scarp, the tilted back of the block forming a long and relatively gentle declivity that merges imperceptibly into the bordering plain.

In some instances the rocks of the basin ranges appear to have been so complexly faulted that the relief has a peculiarly irregular quality with an absence of parallel ridges and valleys such as characterize the

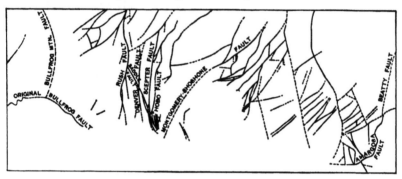

Fig. 56. — Plan of the principal faults in the Bullfrog district, Nevada.   (Emmons, U. S. Geol. Surv.)

greater number of the basin ranges.   The result has been described as a fault mosaic.   The complexity of the main fault systems in the Bullfrog district, Nev., is indicated in plan in Fig. 56; the vertical displacements are shown in Fig. 57.   Extensive mining development in Nevada in recent years has led to a much more detailed study of actual faults

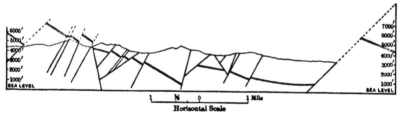

Fig. 57. — Diagram illustrating fault-block displacements in the Bullfrog district, Nevada.
(Emmons, U. S. Geol. Surv.)

than has heretofore been possible and has removed from the realm of theory to that of well-determined fact the question of the existence of faults that affect the topography.[1]

GEOLOGIC DATA

The structure and geologic history of the Basin Ranges are so directly related to the physiographic features of the Great Basin that we

[1] In this connection see the paper by W. H. Emmons, A Reconnaissance of Some Mining Camps in Central Nevada, Bull. U. S. Geol. Surv. No. 408, 1910, pp. 76–81 et al.

may well consider for a moment the later geologic events of the region in so far as they serve as a guide to the interpretation of the topographic forms.

The internal structures of the Basin Ranges indicate an early period of mountain making in which were formed anticlines and synclines whose surface expression was probably not markedly unlike that of the Appalachian Mountains of to-day. These beginnings of physiographic history took place in the Jurassic period and resulted from the extensive folding to which the region was subjected at that time and from its gradual uplift above the sea. It is inferred from the internal structures of the ranges that these early deformations produced mountains of great size and height.[1]

The second period of folding (post-Jurassic) took place almost entirely on north-south lines or on lines very closely approximating this direction. East-west lines of folding are extremely rare, and the ranges resulting from such folding were neither long nor important. It should be carefully noted that the axes of folding trend somewhat more east of north than the ridge lines of the present ranges. The formation of these early mountain ranges was followed by a long period of erosion and the development of a topography of very low relief.

The next important geologic event was the beginning of crustal warping and the production of fresh-water lakes in considerable numbers and of large size. During the period of crustal warping explosive eruptions took place from a number of volcanic centers, and rhyolitic outpourings covered large tracts of land. Great faults were then developed and a period of active differential elevation inaugurated. It was the differential movement of large crust blocks which produced the present mountain ranges and broad intermontane basins. Adjacent blocks rose or sank as units, though they did not move as absolutely rigid units, for there is evidence of internal deformations of both folding and warping on a limited scale. The result of the differential movements of crust blocks was the formation of the Great Basin as an interior basin, the formation of mountain ranges by block faulting, and the development of intermontane valleys or great structural troughs whose broad characters do not rest upon stream erosion but upon faulting and stream aggradation. Since the period of principal faulting dissection has progressed to the point where alluvial fans and cones of great size have been formed.

[1] G. D. Lauderback, Basin Range Structure of the Humboldt Region, Bull. Geol. Soc. Am., vol. 15, 1904, p. 336.

VARIATION IN TOPOGRAPHIC DEVELOPMENT

The Basin Ranges were not all formed at the same time nor of the same material; they therefore possess distinctly variable topographic qualities. In the northwestern corner of the basin (southeastern Oregon) the ridges appear to be of very recent origin, and their forms have been but slightly changed from the original outlines of the tilted blocks.

Their youthful condition is indicated by the remarkably straight and regular character of their frontal scarps and crest lines and by the small amount of dissection which they have suffered. Another indication of youth is the frequent occurrence of landslides upon their steep faces where gradation has not yet produced a surface over which land waste is transported in an orderly manner.

A conspicuous instance is found in the Satas ridge in southern Washington, where the face of the uplifted block is so steep that huge landslides have occurred, one of which affected the cliff face over a distance of half a mile and at a height of about 2500 feet above the adjacent plain. It has a very irregular surface and its margin is circled by a line of hills 200 feet high where the material of the plain was pushed up ahead of the slide.[1]

The ranges occupying the central portion of the Great Basin are of later origin. Like the ridges of Oregon they appear to have been formed by the uplifting and tilting of long narrow blocks, but the blocks are larger in Nevada than in Oregon, and the displacements are of greater value. Also the dissection of the block mountains of Nevada has progressed much further, so that they may be described as maturely dissected. Each range has one relatively short steep slope and one long gentle slope, but the outlines of the original block are scarcely discernible and the crest lines of the ranges are minutely irregular.

In southeastern California and southwestern Arizona many of the fault-block mountains are in a still older stage of development and indicate the condition which the young block mountains of Oregon and the maturely dissected block mountains of Nevada will ultimately reach. They have often been described as presenting the appearance of buried mountains, for they are surrounded on all sides by long gentle slopes of gravelly waste sometimes overlying a smooth rocky floor as a veneer. The original outlines of the blocks have been completely lost; only the alignment of the ranges has been maintained. The opposite slopes of each range have become nearly equal in both gradient and length. There are no pronounced spurs or deep, profound valleys. The mountains have rather gentle relief and occupy but a small portion of the total surface.

[1] W. M. Davis, Physical Geography, 1899, p. 162.

EVIDENCES OF FAULT-BLOCK ORIGIN

One of the most interesting features concerning the fault-block mountains of the Great Basin is the physiographic evidence of faulting that is there displayed; indeed this class of evidence is of the utmost importance in an interpretation of the Basin Ranges, for the prodigious quantity of waste accumulated about the borders of the mountains and in the intermontane basins makes it impossible always to observe the structure in detail and to establish by direct observation the fact of faulting.[1]

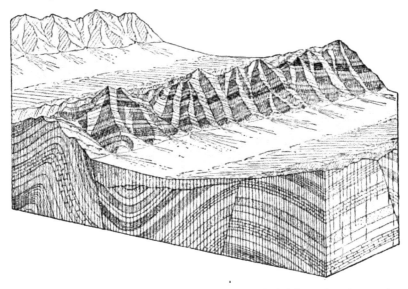

Fig. 58. — Diagram to illustrate the manner in which the strike of the bed diverges from the mountain front in the fault-block mountains of the Great Basin. (Davis, Science, 1901.)

MOUNTAIN BORDER AND INTERNAL STRUCTURE

A condition of first importance in determining the origin and physiographic aspects of the Basin Ranges is the manner in which the border

[1] In an attempt to formulate the principles that should guide the interpretation of such forms, Davis has made an elaborate study of the theoretic problem of the explanation of fault-block mountains. The block diagrams from that paper (W. M. Davis, The Mountain Ranges of the Great Basin, Bull. Mus. Comp. Zoöl., vol. 42, Geol. Surv., vol. 6, 1906), reproduced here, very clearly present the main problems associated with the three elements necessarily involved in the problem of block faulting — the pre-fault topography, the topographic effect of the faulting, and the degree of dissection of the faulted blocks. The paper includes a comparison of the base line of residual mountains with that of fault-block mountains; a description of the canyons and ravines of block mountains; and the order of development and means of determination of spurs and terminal facets.

of each range cuts across the internal or primary structure so that the structure is very commonly oblique to the mountain front, Fig. 58. It is a safe inference that there can be in such a case no direct relation between the trend of the range and its primary structure, otherwise there should be a certain harmony between them, such as exists for example between the strike of resistant strata in the Appalachian Mountains and the trend of the Appalachian ridges.

CONTINUITY OF RANGE CREST

One of the most important indications of the origin of the Basin Ranges through block faulting of later date than the internal structure of the ranges is the fact that the body of each range is usually continuous although incised by sharply cut valleys. If the mountain ranges were residuals of a period of long-continued and undisturbed erosion such as must be postulated if the broad intermont valleys are explained by erosion, each range should be dissected into a number of short mountain ranges and isolated peaks. Stream action profound enough to excavate the broad intermont valleys would produce in the same time equally profound differential erosion in the body of the ranges. On the other hand, had the region been reduced at one time to a lowland of faint relief and the relief still further reduced by lava outpourings, the formation of great fault blocks would result in broad and flat intermont valleys as conspicuous as the ranges between which they lie, and the ranges themselves, if not too maturely dissected, would possess a marked rectilinear quality.[1]

THE OREGON RANGES EXHIBIT CRITICAL FEATURES

The indifference of mountain border to internal structure, and the continuity of the individual ranges, are both seen to best advantage in southern Oregon. In most parts of the Great Basin the typical basin-range structure produced by the faulting and tilting of long narrow crust blocks is largely obscured by erosion or by the topographic effects of complex internal folds and faults. In south-central Oregon, however, the crust blocks have been deformed so recently that erosion has but slightly modified them, and no internal deformation in the body of the blocks preceded the faulting.[2]

In the bedded lavas of Lake County, Oregon, topographic features[3] occur which seem closely related to a great upward fold or anticline

[1] W. M. Davis, Current Notes in Physiography, Science, n. s., vol. 14, 1901.
[2] Idem.
[3] G. A. Waring, Geology and Water Resources of a Portion of South-Central Oregon, Water-Supply Paper U. S. Geol. Surv. No. 220, 1908, p. 25.

which has been extensively faulted in places. Chewancan River has cut its channel along the axis of the fold for a number of miles. In Summer Lake Valley the anticline is broken down, the western side remaining in place to form Winter River Valley, while the eastern is buried beneath lake deposits. Goose Lake Valley lies on the dropped keystone of the anticlinal arch, its eastern side being marked by a steep slope, its western side by a longer monoclinal slope.[1]

An immediate result of the earth movements by which the ridges were formed was the formation of a large number of enclosed basins whose floors are now occupied, to some extent, by lakes. There are all gradations between basins so small and poorly supplied with water that none whatever accumulates on the basin floor, and basins in which the rainfall is sufficient to maintain either temporary or permanent alkaline or saline lakes; or, as in the case of Goose Lake basin, a water supply sufficient to cause the basin occasionally to drain into the sea. Distinct shore terraces indicating the level of ancient Quaternary lakes that once existed here are to be found on the slopes of many of the basins.

EVIDENCES OF PROGRESSIVE AND RECENT FAULTING

BROKEN WASTE SLOPES

Not only have the Basin Ranges been blocked out by great faults, but faulting has continued down to the present. It is expressed among other ways in broken fans at the foot of the scarps which form the range fronts. Such broken fans have been noted repeatedly in glacial and postglacial deltas and fans along the base of the Wasatch Mountains, as in the delta of Rock Canyon Creek near Provost.[2] Low escarpments in lacustral beds and alluvial slopes in places form irregular lines along the bases of the mountains, and at times cross the valleys. They present a small cliff or steep ascent between two nearly horizontal plains.[3] The crests of the scarps are always irregular and sometimes form zigzag lines that may be followed for miles; they are fault scarps of very late origin. In many cases it is believed that they could not have existed in their present condition more than a few years. Sometimes they are more than a hundred miles long and vary from a few feet to more than a hundred feet in height.[4] That they are recent fault scarps is shown by the fact that they commonly occur in Quaternary lake deposits and recent alluvial slopes but little modified by erosion; and in many instances they are without vegetation. Similar scarps have been observed at the eastern base of the Sierra Nevada and at the foot

[1] G. A. Waring, Geology and Water Resources of a Portion of South-Central Oregon, Water-Supply Paper U. S. Geol. Surv. No. 220, 1908, p. 26.

[2] W. M. Davis, The Mountain Ranges of the Great Basin, Bull. Mus. Com. Zoöl., vol. 42, p. 160.

[3] I. C. Russell, Geological History of Lake Lahonton, a Quaternary Lake of Northwestern Nevada, Mon. U. S. Geol. Surv., vol. 11, p. 274.

[4] Idem, p. 375.

of the slopes of many of the Basin Ranges. In the Lahonton area recent fault scarps are a common feature in the topography of the valleys, Fig. 54. Scarps of a similar nature were first observed in the Great

Fig. 59. — Post-Quaternary fault on the south shore of Humboldt Lake. (Russell, U. S. Geol. Surv.)

Basin by Gilbert and were recognized as the result of recent crustal movements.[1]

The recent faults of the Basin Ranges occur most commonly on the steeper sides of the mountains and invariably the throw is toward the valley. Occasionally they cross stream channels and cause rapids, as in the case of the American Fork, Utah, where it crosses the Wasatch fault. The distribution of recent faults is in marked sympathy with the ancient lines of displacement as determined by evidences of a topographic character such as have just been outlined. But it should be remembered that the recent faults are but a small fraction of the entire displacement.

STREAM PROFILES AND RECENT FAULTING

Among the significant elements of topographic form indicative of recent faulting are the abnormal profiles of many stream channels crossing the fronts of the fault blocks. Prolonged erosion of a stable block

[1] Second Ann. Rept. U. S. Geol. Surv., 1880–1881, p. 192.

mountain would result in the development of stream gradients of the normal type whose descent from the headwaters of the region would be progressively more and more gentle. In the Basin Ranges, however, it has been frequently noted that the stream gradients are distinctly abnormal, and that they are notably peculiar in that a V section persists down to the mountain base where the steep-sided ravines suddenly open upon gravel fans that form parts of wide piedmont alluvial plains. Typical examples occur in the Pueblo range and in the Weber and Ogden canyons of the Spanish Wasatch. It is noteworthy that the steep-walled canyons that appear near the base of the range are in contrast with the upper portions of the valleys where flatter gradients occur. It appears that the progressive elevation of the fault-block mountains causes progressive down-cutting on the part of the draining streams. This lack of stability in the mountain mass and constant rejuvenation of the streams by repeated uplift prevent the streams from widening their valleys to the normal form.

TERMINAL FACETS OF THE MOUNTAIN SPURS

Another feature indicative of progressive faulting is the occurrence of terminal facets of peculiar and significant character and of very per-

Fig. 60. — Ravines, spurs, and terminal facets of the Spanish Wasatch, looking east. Note the even base line developed on rocks of varying resistance. (Davis, Bull. Mus. Comp. Zoöl.)

fect development. In the case of the Spanish Wasatch the facets slope at an angle of 38° or 40°, are of remarkably regular occurrence and form, and are set off from each other by deep ravines that diversify the mountain front. The base line of this range is almost rectilinear, a feature in itself of the greatest importance in the interpretation of the morphology of a mountain mass whose structure possesses sufficient diversity to occasion under ordinary conditions topographic irregularities of considerable degree.

Throughout the entire Great Basin there is a rather intimate association between thermal springs and lines of recent faulting; and the
hottest springs almost invariably occur on the lines of displacement
that have suffered most recent movement.[1] This relation of thermal
springs to recent faults along the bases of the frontal scarps is almost
constant and is in entire sympathy with the explanation of continued
block faulting in the Great Basin region.[2]

FEATURES OF THE DEATH VALLEY REGION

If we now turn to other ranges in the Great Basin than those we
have thus far noted we shall find a striking persistence of the structural
and physiographic features already examined. In the Death Valley
region the fault-block type of structure and topography has been clearly
identified. The strata of this region suffered deformation (Eocene)
in which faulting and tilting took place and parallel mountains and
valleys formed that trend northwest in the general direction of the
Sierra Nevada. Deformation of any type tends to produce enclosed
basins and in an arid climate this tendency is usually realized in a pronounced way. In the Death Valley region faulting and tilting produced
enclosed basins in which lakes were formed and lake sediments deposited to a thickness of several thousand feet. These lake sediments
include great deposits of salt, gypsum, soda, and borax. Later still
(Miocene) another period of deformation set in. It was characterized
by faulting and tilting as in the earlier period of deformation but along
lines more nearly north than before and parallel with the basin ranges.
Immense mountain ranges were the result, such as the Funeral and
Panamint ranges, with the Panamint, Detah, and Amaragosa valleys
between them. In the enclosed basins that were thus formed lakes
existed for a time, and on their floors and about their borders were
deposited sediments similar to those of the earlier lake period.[3]

---

[1] I. C. Russell, Mon. U. S. Geol. Surv., vol. 11, 1885, p. 276.

[2] The relation of hot springs to recent faulting is brought out clearly in a map of the
United States published in 1875: Report upon Geographical and Geological Exploration and
Surveys West of 100th Meridian, Wheeler Surveys, vol. 3, Geology, 1875, pp. 148–150. Sixty-
seven springs occur in the western region and but 15 in the eastern. Forty-seven of the first
group have a temperature as high as 100° F.; only 2 in the latter group reach this temperature.
The areas are in the ratio of 13 to 3. If the country were better known the ratio would show
an even greater preponderance of springs in the western region.

[3] M. R. Campbell, Basin Range Structure in the Death Valley Region of Southeastern
California (Abstract), Bull. Am. Geol. Soc., vol. 14, 1903, pp. 551–552.

Soils of the Great Basin

The soils of the Great Basin are derived from a great variety of rocks, and consist of colluvial wash of the mountain slopes, thick lacustrine and shore deposits associated with ancient Lake Bonneville, and recent stream-valley sediments and river-delta deposits. When not situated above or outside the limits of irrigation, or rendered unfit for cultivation by accumulations of alkali or seepage waters, they are of great agricultural importance.

The soils of alluvial cone deposits are usually gravelly and very dry, and therefore treeless, except in the immediate vicinity of stream courses. The more elevated areas are frequently rough and hilly and marked by the presence of rock outcrop and bowlders. They are frequently cut by washes or intermittent stream channels, and are well drained, except in the lower-lying areas occupying depressions.

The soils of lacustrine sediments and material derived from stream deltas occur upon low, level plains, marking the site of recent lake bottoms. They are generally barren, deficient in drainage, and heavily impregnated with alkali salts. They are derived from eruptive, sedimentary, and altered rocks of various ages and are without gravel. They cover extensive areas, are usually dark in color, and in general have little or no agricultural importance.

The soils formed of colluvial mountain wash or of residual material mingled with alluvial deposits of intermittent or torrential streams are often gravelly, sometimes marked by rock outcrop, and frequently cut by washes and intermittent stream channels, and generally treeless. The soils are derived primarily from red sandstone, modified in places by an admixture of material derived from shales, slates, eruptive rocks etc., and are typically of vermilion or bright-red color. They occur generally as extensive areas. The lower-lying and heavier soils are often poorly drained and alkaline.

Along valley troughs and in the vicinity of river flood plains, stream sediments of recent origin or in process of formation form an important group of soils. They occupy low or slightly elevated valley plains, have a smooth, nearly level surface, and are frequently marked by the presence of stream channels or sloughs. They are derived mainly from eruptive, early sedimentary, and altered sedimentary rocks, are generally dark in color, and are underlain by light-colored sands or sandy loams or by heavy red subsoils.

FORESTS AND TIMBER LINES

The high barrier of the Sierra Nevada on the windward side of the Great Basin so reduces the rainfall on the lower Basin Ranges as to make the forest growth thin and scattered or wholly absent. No large and dense forests occur in the province, Fig. 61. The existence of a

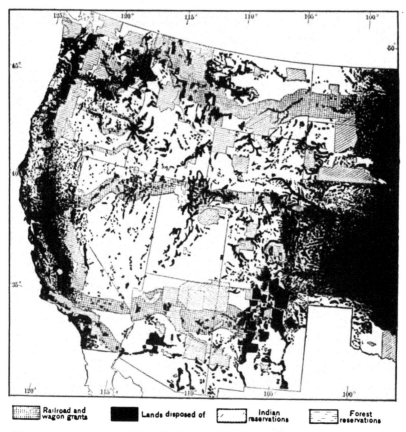

| Railroad and wagon grants | Lands disposed of | Indian reservations | Forest reservations |

Fig. 61. — Location of vacant public land. Note the disproportionately large amount of vacant public land in the Great Basin. (Newell, Irrigation in the U. S.)

forest is conditioned by the amount of rain that falls on each range. As a whole the province is not one whose forest growth is of great importance to lumbermen; but the very scarcity of forests in so arid a region makes doubly important the study of the physical conditions surrounding the isolated forests that do exist.

In the arid Great Basin a lower timber line and an upper timber line are clearly defined on most of the mountain ranges. The position of the upper or cold timber line is determined mainly by the annual temperature of 32° F., but has some variation depending upon differences in snowfall, soil conditions, severity of winter storms, and exposure to the sun. The vegetation above the cold timber line is commonly alpine

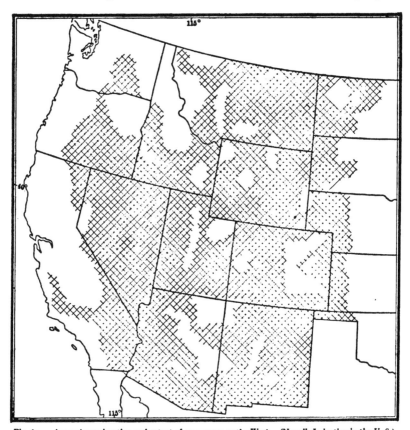

Fig. 62. — Approximate location and extent of open range in the West. (Newell, Irrigation in the U. S.)

flowers and grasses of various sorts up to the lower limit of snow.[1] The lower limit of tree growth may be called the dry timber line, although to drought is added the influence of cultivation, soil conditions, alkali, hot winds, exposure to the sun, etc. Unlike the cold timber line,

[1] This line is at sea level in Alaska and northern Canada, where it defines the polar limit of the subarctic forest and may be called the "continental timber line." North of it is a zone of tundra and barren grounds corresponding to the zone of alpine flowers above the cold timber line on the mountain slopes and summits of temperate latitudes.

the dry timber line may be independent of latitude and altitude, its position depending almost entirely upon the amount of rainfall. Below the dry timber line of the Great Basin are, in some cases, treeless, grassy to arid plains and valleys, while in other cases the dry tree line merges into the zone of juniper. In central Idaho the cold timber line is at 10,000 feet, the dry at 7000 feet. Its position in the Prairie Plains province in eastern Kansas and Nebraska is dependent upon the gradual increase of rainfall eastwardly from the Rockies and High Plains. The dry timber line disappears in humid regions, as in

Fig. 63. — Typical view of desert vegetation, southern Great Basin, near Goldfield, Nev.
(Ransome, U. S. Geol. Surv.)

the Adirondacks and the White Mountains. The meeting of the dry and the cold timber lines on those ranges in the Great Basin that are both cold above and excessively dry below, as the White Mountains of western Nevada, Plate IV, results in the complete absence of forest growth.[1]

While the upper timber line is determined usually by temperature, other factors may have a determining local influence and in nearly all cases have an important influence. Among these factors are the slopes of the surface, the degree of exposure to the sun, the depth of the snow, the

[1] I. C. Russell, Timber Lines, (Abstract) Bull. Geol. Soc. Am., vol. 14, 1903, pp. 556–557.

severity of the winter storms, etc., though the dominant cause is the low temperature.[1]

The most abundant and most generally distributed trees from the western foothills of the Wasatch Mountains in eastern Utah to southeastern California, northern Arizona, western Colorado, and southern Wyoming are the juniper and the nut pine (*Pinus monophylla*). In central Nevada the juniper often descends into the valleys and forms open stunted forests at elevations of about 5000 feet. It is more abundant and of larger size on arid slopes at elevations of 8000 feet above the sea, where it occurs in dense and nearly pure forests.[2] The nut pine, or piñon, occurs on dry gravelly slopes and mesas throughout the same territory, often forming extensive open forests at elevations between 5000 and 7000 feet.[3]

From the commercial standpoint the most important trees in the Great Basin are the yellow pine and Douglas spruce, but their growth is limited to the higher ranges of the provinces where they form open woodland, never a true forest. A typical growth is found on the Snake Range which lies just west of the Nevada-Utah line, trends north and south, and is about 135 miles long. It contains the highest peak between the Wasatch and Sierra Nevada, Jeff Davis or Wheeler Peak, more than 12,000 feet high, besides being one of the most rugged ranges in the Great Basin.[4] The intermediate slopes, Fig. 64, are covered with a tree growth. Alpine fir, white fir, Douglas fir, and Engelmann spruce are the principal species, yellow pine being relatively scarce and limited to the lower elevations. The higher portions of the range are almost devoid of vegetation owing to the low temperature. The cold timber line lies at 10,500 or 11,000 feet. It should not be considered a definite line, however, since the forest disappears on the dry spurs at much lower elevations than in the wet canyons. The dotted line near the summit in Fig. 64 represents the upper limit of growth in the canyons. The foothill belt below the dry timber line is covered with sagebrush and bunch-grass. The valleys have a better growth of grasses, while the springs and streams are lined with shrubs and aspen and a few cottonwoods.

By far the greater number of the Basin Ranges are desert or support

[1] For a discussion of this matter see I. C. Russell, Timber Lines, Nat. Geog. Mag., vol. 15, 1904, pp. 47–49; for a criticism see C. H. Merriam, Nat. Geog. Mag., vol. 14, 1903. A "wet timber line" may also be identified about the borders of lakes, swamps, etc.
[2] C. S. Sargent, Manual of Trees of North America, 1905, p. 89.
[3] Idem, p. 12.
[4] J. E. Spurr, Descriptive Geology of Nevada South of the 40th Parallel, Bull. U. S. Geol. Surv. No. 208, 2d ed., 1905, p. 25.

only a scanty growth of sage-brush and juniper and a few stunted pines near the summit. They have only a very scanty supply of water and a few widely separated springs.[1] The Cedar Mountains west of Salt Lake Valley afford typical conditions.

The Montezuma Range in western Nevada also illustrates this type of range. Only a few stunted pines and junipers are found and these grow only on the upper slopes and in the more sheltered canyons. The

Fig. 64. — East side of Snake Range, Nevada. Jeff Davis or Wheeler Peak from Robinson's Ranch. Yellow pine comes in on the lower edge of the timbered belt and Alpine fir, white fir, and Engelmann spruce are the principal species at higher levels. (U. S. Geol. Surv.)

range lies not far from the great Sierra Nevada on the west and in spite of its bold appearance provokes but little rainfall from the winds that pass the higher topographic barrier on the west. The range is one of the driest in Nevada.[2] Excellent grass is abundant, however, and has high value for grazing, but in order to supply stock precautions must be taken to prevent spring waters from running to waste.

Many of the Basin Ranges with intermediate elevations have important forest tracts even though a continuous forest cover is wanting.

[1] S. F. Emmons, Desert Region, Descriptive Geology, vol. 1, 1877, p. 462 (King Surveys).
[2] Arnold Hague, Descriptive Geology, vol. 2, 1877, p. 752 (Hayden Surveys).

The Schell Creek Range, for example, has closely restricted patches of yellow pine and fir in the moister canyons and small upper basins. While these are not important in a large way, they at least supply a most important need on the part of the ranchmen, farmers, and miners engaged in developing resources near by.

The most prominent range between the Sierra Nevada and the Wasatch is the East Humboldt Range in central Nevada. It is a bold, single range about 80 miles long with many summits reaching over 10,000 feet. Because of its relatively high altitude and its greatly dissected condition it has a more alpine aspect than the other Basin Ranges and, as compared with the lower ranges about it, receives more rainfall and snowfall. In response to the heavier precipitation it supports an open tree growth. Its higher canyons and upper slopes are covered with scattered forests including several varieties of pines and firs, among which the limber pine (*Pinus flexilis*) is the prevailing species. The trees do not, however, supply much valuable timber since they are knotty and rarely over 50 feet high.[1]

[1] Arnold Hague, East Humboldt Range, Descriptive Geology, vol. 2, 1877, p. 528 (King Surveys).

CHAPTER XV

LOWER COLORADO BASIN

On account of the scarcity of arboreal vegetation in the lower Colorado Basin and the close genetic relationship of its forms with those of the Great Basin already somewhat fully discussed we shall devote but a few paragraphs to its physiography. The province includes (1) an eastern section of low, residual mountains, piedmont slopes, and intermontane basins forming the southwestern portion of Arizona, and (2) a western section of interior basins west of the Colorado and north of the mountains of dry southern California.

The mountains of the eastern section are regarded as of the basin-range type — fault-block mountains originally like those of southeastern Oregon. They are, however, much older than the latter and are so thoroughly dissected that their original asymmetry has been lost. The broad structural valleys between the ranges have been in part floored by piedmont and basin deposits in part extended by rock planation, a result attributed to sheet-flood erosion.[1] The basin floors in the Sonora district of Mexico and Arizona, where the half-buried mountains rise above broad plains, appear at first to be wholly alluvial. More intimate examination shows that only half of their surface is covered with alluvium; the other half is in reality planed rock. Two-fifths of the entire area including both plains and mountains is smoothly-beveled rock floor, the rock being granite, schist, and other types, planed off in a belt from 3 to 5 miles wide which merges with the alluvial portion of the basin on the one hand and from which the mountains rise sharply without any intervening foothills on the other. The graded character of the floor is no doubt to be ascribed to water action, as was insisted by McGee; but the fact that the surface of the rock floor on the basin margins is kept relatively free from alluvium should probably be ascribed in large part to wind action which is universal and almost constant.

A peculiar feature of many of the basin floors of the arid region is the gravelly appearance of the surface. Gravels and small bowlders

---

[1] W J McGee, Sheet-flood Erosion, Bull. Geol. Soc. Am., vol. 8, 1897, pp. 87–112.

are found scattered over the higher slopes of nearly all the intermont valley plains. The general appearance is that of a vast gravel bed. It is not uncommon to find an area several acres in extent covered with small angular stones as closely and evenly set as mosaics. The pebbles constituting the gravel are, however, but a thin surface veneer. The wind constantly blows away the fine material and is unable to remove the coarse, which accumulates as a protective cover. The pebbles are sometimes only one deep, and below them there is often a fine porous loam which may be of great fertility.[1]

It is estimated that about 85% of the entire surface of the eastern section of this province (east of the Colorado River) is plain and about 15% is mountains. Such an excess of low over high country in an arid region means that the mountain-born streams will quickly wither on the plains and that trunk streams will be either rare or wanting altogether. Where the mountains are low, the streams are insignificant in size and disappear almost at the mountain bases. Except the Gila and the Colorado, which have their sources in high and well-watered country, the Lower Colorado Basin has no through-flowing streams; the majority of its streams are of the type of "lost rivers" which disappear by absorption and evaporation before reaching the sea.

The western section of the province beyond the Colorado is exceptionally arid and hot and includes the Mohave desert. The rivers terminate on the piedmont slopes of the desert ranges or feed permanent salt lakes or the temporary lakes of salinas and playas (see Fig. 27).

TYPES OF LOWLANDS

The portion of the Lower Colorado Basin that lies in Arizona contains three distinct kinds of lowlands: (1) valleys and canyon floors now containing running water such as the Colorado and Williams valleys and Santa Maria Canyon; (2) old, deeply filled, alluvial valleys such as the Sacramento and the Big Sandy; and (3) plains of erosion such as those that in many places border many of the desert ranges and are due to sheet-flood erosion. Among the intermontane plains are Cactus, Posas, and Ranegras plains, etc. They are in part old, deeply filled valleys that have a general altitude of about 2000 feet toward the plateau region on the east and gradually descend westward to about 400 feet at the Colorado River. All of them have been somewhat modi-

---

[1] C. R. Keyes, Rock Floor of Intermont Plains of the Arid Regions, Bull. Geol. Soc. Am., vol. 19, 1908, pp. 63–92. See also C. F. Tolman, Erosion and Deposition in the Arizona Bolson Region, Jour. Geol., vol. 17, 1909, p. 14.

fied by crustal disturbances and basaltic extrusions from many local centers of igneous activity.[1]

One of the largest of these valleys is the Detrital-Sacramento Valley which extends north and south parallel to the Colorado River for more than 100 miles. It is interrupted here and there by lava masses, but its material consists chiefly of gravel filling of great depth. It is in a region of profound faulting and warping, and may have originated as a succession of structurally depressed areas. Whatever its origin it has been greatly modified by a stream of considerable size which has almost filled the entire bottom of the valley with an enormous amount of detrital material. This work may have been accomplished by the Colorado or by some stream now extinct, a fact which has not yet been safely determined.

SPECIAL DRAINAGE FEATURES

The lower valley of the Colorado River, that portion which crosses the Lower Colorado Basin, is remarkable for extreme irregularity of topography within short distances. It consists of a series of narrow, steep-walled gorges, and broad alluvial basins through which the river winds in an exceedingly irregular course. It is concluded[2] that at one time the Colorado River ran through the present Detrital-Sacramento Valley, that it filled this valley and adjacent depressions with a prodigious quantity of alluvial material, and when, on account of changed climatic or geologic conditions or both, the river began again to degrade its channel it occupied its old valley throughout the greater part of its course; but at certain places, as in Pyramid Canyon, Eagle Rocks (Fig. 65), and other localities farther south, its course at the moment of change from aggradation to degradation was directed across alluvium-buried mountain spurs and knobs. With the progress of down-cutting these spurs and knobs were uncovered and the course of the river across or through them is now marked by narrowness, bank declivity, hard rock, and steepened gradient, instead of the flat gradients of the alluvial bed and the wide flat-floored valleys that elsewhere characterize its course.

The Colorado River is to-day carrying immense quantities of silt which it is spreading over its rapidly aggrading flood plain. The river carries more suspended matter per unit volume than any other stream in North America — 2000 parts of sediment per 100,000 parts of water, or enough in one year to cover 164 square miles 1 foot deep with mud.

[1] W. T. Lee, Geologic Reconnaissance of a Part of Western Arizona, Bull. U. S. Geol. Surv. No. 352, 1908.
[2] Idem.

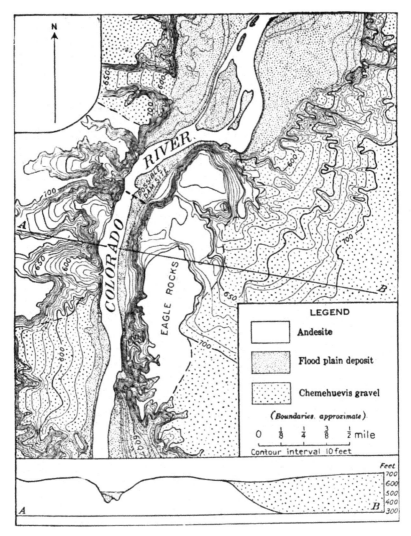

Fig. 65. — Map of part of Colorado Valley, showing old gravel-filled channel on right and rock channel on left. (Lee, U. S. Geol. Surv.)

The rapidity of filling of the Colorado River on its flood plain is indicated in many places where lateral cutting exposes the roots of trees and shrubs now buried to a greater or less degree. In many places living arrow weeds may be seen standing in five feet or more of silt. The material is well stratified and exceedingly fine. When deposited in thick beds it dries and cracks in great columns two or three feet in diameter, the cracks themselves being several inches wide and two feet or more deep.

The river frequently changes its course over the wide bottom lands which it drains, sometimes by normal lateral cutting and sometimes by more intense erosive action during a period of high water. These changes result in the formation of new channels, the abandonment of old ones, the formation of cut-off meanders, sloughs, lagoons, ox-bow lakes, and dry channel courses.

But little of the land along the Colorado is irrigated because the river and its tributaries are in general so far below the bordering lands as to render diversion extremely difficult or impracticable. At Yuma two pumping plants lift water from the river for irrigation and several other lifting plants are located below this point. The Imperial Canal diverts water from the river at a point about 10 miles below Yuma.[1]

The Colorado is subject to annual overflow from April to June and may spread out over the Salton region of the Colorado desert forming lakes. Mearns noted (1892–1894) that these lakes eventually dried up, but for a long time after the water had disappeared the region was green and vegetation throve in the rich surface deposits. Cattle were driven in and the owners endeavored to make a breach in the Colorado River bank at each annual overflow, so that the region became flooded through the channels of New and Salton rivers, causing a fresh crop of forage plants to spring into life.[2]

SALTON SINK REGION

That portion of the Lower Colorado Basin known as Salton Sink is of special interest because it contains one of the two tracts of land in the United States below sea level.

The Salton Sink region contains two fertile valleys, the Coachella Valley in Riverside County northwest of Salton Sink, and the Imperial Valley in Imperial County southeast of Salton Sink. Lying partly in each of these two counties is Salton Sea, the bottom of which is 273.5 feet below mean sea level.[3]

In recent geologic time Salton Sea was a part of the Gulf of California which then extended about 200 miles farther northwest than at present. At that time the mouth of the Colorado was near Yuma, 60 miles from its present location, and was gradually building a delta

[1] Freeman and Bolster, Surface Water Supply of the United States, 1907–08, Colorado River Basin, Water-Supply Paper U. S. Geol. Surv. No. 249, pt. 9, 1910, p. 34.

[2] E. A. Mearns, Mammals of the Mexican Boundary of the United States: A descriptive catalogue of the species of mammals occurring in that region, with a general summary of the natural history and a list of trees, Bull. U. S. Nat. Mus. No. 56, pt. 1, 1907, p. 28.

[3] Freeman and Bolster, Surface Water Supply of the United States, 1907–08, pt. 9, Colorado River Basin, Water-Supply Paper U. S. Geol. Surv. No. 249, 1910, pp. 46–51.

that extended southwest toward the Cocopa Mountains. Deposition continued until the upbuilding of this delta had completely separated the head from the rest of the gulf and converted its floor into an inland sea. Delta growth continued until the inland lake became not only entirely independent of the Gulf but also actually raised to a higher level than the sea. Consequently one may see to-day faint terraces in favorable places on the margin of the depression, and on rocky points is a thin deposit of calcium carbonate and slightly cut sea cliffs about 40 feet above sea level. Even some of the alluvial cones formed on the shore line had beaches which, although easily eroded, are even now well preserved, an indication of their recent formation; and over the floor of the desert and along the sandy beaches are thousands of shells of fresh- or brackish-water mollusks.[1] The water of the lake was not perfectly fresh, for it is estimated that the evaporation from its surface nearly equaled the average annual inflow from the Colorado, and even if the flow of the river exceeded this evaporation it could not have done so by a large amount; in either case the waters of the lake would be markedly alkaline. This is also shown by the fact that wherever the lake waters broke in spray and evaporated more rapidly than usual, carbonate of lime was deposited. It is known from the extent of the delta that the river broke out of its channel many times while building it and alternately discharged into the Gulf of California and the Salton Sea. During those periods in which the river discharged into the Gulf of California, Salton Sea must have contracted and become more and more alkaline. The last natural discharge of the Colorado into Salton Sink was of very recent occurrence. It is probable that the lake which it supplied existed but little more than a thousand years ago. In recent years we have had well-known instances of changes.

During the summer of 1891 the Colorado overflowed into Salton Sink at the time of high water to such an extent as to endanger the Southern Pacific Railway line; and in the summer of 1905, after a number of winter and spring floods in the Gila River and a heavy summer flow in the Colorado, the floods were repeated on a much larger scale. The gravity of the situation was increased by the existence of diverting canals which conveyed water to the Imperial Valley from the Colorado. The canals were not provided with protective headworks, and had a gradient much greater than that of the river, so that after the flood of 1905 and in July the main canal was carrying 87% of the total flow of the river, and the water was deepening and widening the Alamo River, along which the canal extended, to a great gorge. Strong efforts by the Southern Pacific Railway Company resulted in the control of the Colorado in the early fall of 1906, but it broke out again on December 7, and was only closed finally in February, 1907. On December 31, 1908, the surface of the Salton Sea was still far above its normal level, being only

[1] R. E. C. Stearns, Remarks on Fossil Shells from the Colorado Desert, Am. Nat., vol. 13, 1879, pp. 141-154.

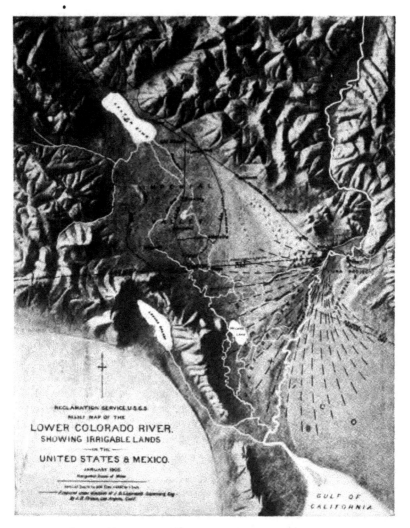

Fig. 66. — Map of the Salton Sink region, California.　(U. S. Geol. Surv.)

206 feet below mean sea level.　It then had a maximum depth of 67.5 feet and an areal extent of about 443 square miles.　Its temporary enlargement necessitated shifting the Southern Pacific tracks over a stretch about forty miles long, a stretch that was originally constructed on the 200-foot contour below sea level.　Another line has since been surveyed and graded on the 150-foot contour below sea level for possible use in the future.　The Sea has also completely submerged the plant of the New Liverpool Salt Company and a few ranches near Mecca.[1]

---

[1] Freeman and Bolster, Water-Supply Paper U. S. Geol. Surv. No. 249, 1910, pp. 50–51.

So completely cut off from the rest of the Gulf of California was the Salton Sink region before 1905 that the waters of the beheaded section of the ancient gulf were evaporated to the point where only a fragment of the original water body remained. Its waters were until recently exceedingly salty and the shores of the lake or sink were fringed with wide, white belts of salt-incrusted land. Only a few trifling streams descend from the higher portions of the San Bernardino and San Jacinto mountains to feed the dwindling remnant of what was once a great water body. In Fig. 66 is represented the maximum size of the ancient water body as recorded in a well-preserved strand line that contours evenly about the margin of the basin.

CLIMATE, SOIL, AND VEGETATION

The following data on the relief, climate, and vegetation of the southern border of the Lower Colorado Basin are compiled mainly from the excellent work of Mearns.[1] They are typical of the relations of topography, rainfall, and vegetation over a much wider area and deserve close examination by the student of forestry.

The mountains of the southeastern portion of the Lower Colorado Basin form continuous ranges rising very sharply from the plains. They are seldom well forested because of the steepness of their slopes, from which the soil is blown or washed away almost as fast as formed, leaving the bare rocks without vegetation except in crevices, benches, and hollows.

The soils of the region are mainly of colluvial, alluvial, and lacustrine origin, modified by the addition of recent stream sediments; they are without important amounts of humus owing to the aridity. They occupy mountain foot slopes, alluvial fans, débris aprons, or sloping plains of filled valleys, sloping or nearly level plains, and bottoms of stream valleys or sinks and drainage basins. Since the climate of the arid Southwest is characterized by semi-tropical desert conditions, the soils have little or no value save as they can be irrigated or as they occur in limited amounts in rock crevices, etc., and as a thin veneer on higher slopes where the rainfall makes a thin forest growth possible.[2]

There is a very thin population along the entire boundary, the only towns near it being Bisbee, Santa Cruz, Nogales, Yuma, and San Diego. Except for these towns and a score of small settlements in the principal valleys the zone of 24,000 square miles along the boundary, 20 miles on either side, contains less than 100 permanent inhabitants.

[1] E. A. Mearns, Mammals of the Mexican Boundary of the United States: A descriptive catalogue of the species of mammals occurring in that region, with a general summary of the natural history and a list of trees, Bull. U. S. Nat. Mus. No. 56, pt. 1, 1907.

[2] Soil Survey Field Book, U. S. Bureau of Soils, 1906.

The average precipitation along the entire boundary is about 8 inches and on the Yuma and Colorado deserts it is but 2 or 3 inches. For 700 miles between the Rio Grande and the Pacific, the boundary line is crossed by only five permanent running streams although it crosses the mountain ranges nearly at right angles, a direction most favorable for encountering existing streams. There are two periods of rainfall, one in midwinter and one in midsummer, the midsummer rainy period being known as the rainy season. The summer rains generally begin about the first of July and last until the middle of September. Soon after the first rain falls the vegetation assumes a spring-like aspect, leaves burst forth, hills and valleys are covered with grass, and a bewildering profusion of wild flowers covers the surface. The plants grow with great rapidity, their seeds mature before the rains cease, and in a month or so after the rains have stopped they have the somber colors typical of fall and winter.

On the whole the Mexican boundary district of the Lower Colorado Basin is treeless; the forests are confined almost entirely to the mountain ranges and the stream courses, but those in the latter situation are few in number and of insignificant size. On some of the desert spaces arboreal cacti and yuccas form open groves. The streams are lined with Fremont cottonwood, black willow, box elder, walnut, sycamore, oak, mulberry, ash, etc. Among these the cottonwood and willow are found on every permanent stream, and are usually flanked by a broader zone of mesquite. The desert willow, hackberry, and indigo tree are found in arroyos in which there is a slight amount of moisture.

There are a few large alkali flats perfectly bare of vegetation and a number of spots in the desert are without plants, but over the great stretch of desert country between the Gila Mountains and the Colorado River there are found almost everywhere four species of plants — the creosote bush, the sage, an ephedra, and a grass. The prickly, thorny shrubs and bushes together with the cacti and yuccas are usually disposed in groups or thickets surrounded by more or less open spaces. In the sheltered situations are found more or less tender shrubs, grasses, and other herbaceous plants.

Shrubs and grasses increase in number and variety on the foothills and there is often an abundance of shrubbery in the ravines near timber line. On the whole the rocky soils are much richer in plant food than the sandy soils because they retain moisture better. The desert vegetation with the exception of a few green-bark trees and shrubs is dull and dusty and in general the plants have pulpy leaves and exude gums and resins for retarding evaporation. The leaves are usually small and many are covered with waterproof dermal structures.[1]

Under 4000 to 6000 feet the rainfall is so low and the evaporation so high that true desert conditions prevail. Upon the higher mountain slopes are limited areas where much more mesophytic conditions are found — less evaporation and greater precipitation; hence islands of vegetation occur on the mountains surrounded by great desert plains. The greater portion of the area is occupied by true desert species equipped for life under arid conditions — structures for preventing evaporation and other structures for rapid absorption, great storage, and long retention of a scanty water supply.[2] The highest portion of the province lies in south-central Arizona, where a few mountain ranges — Baboquivari, Carobabi, and Cobota ranges — break the continuity of the plains. The Gila, Mohawk, and Growler mountains are important ranges farther west. None of them has a sufficient summit extent to provoke large quantities of rainfall, hence even the highest portions are very scantily covered with tree growth.

[1] Mearns, loc. cit., pp. 32–34.
[2] D. T. Macdougal, Across Papagueria, Bull. Am. Geog. Soc., vol. 40, pp. 724–725.

CHAPTER XVI

ARIZONA HIGHLANDS

TOPOGRAPHY AND DRAINAGE

THE Arizona Highlands cross Arizona from northeast to southwest as a broad zone of short and nearly parallel mountain ranges separated by valleys deeply filled with river and lake deposits. The width of the zone is from 70 to 150 miles and the lengths of individual mountain ranges such as Santa Catalina, Pinal, Dragoon, and Ancha rarely exceed 50 miles, while the elevations are rarely above 8000 feet. The northeastern portion of the Arizona Highlands is continuous with the ranges of the Great Basin in Nevada and Utah. On the east the common line of division of the Arizona Highlands and the Trans-Pecos Highlands is roughly the Mimbres River just west of the Rio Grande. The ranges east of this line trend north, those west of the line trend northwest. Between these two divergent lines, and on the north, is the lava-fringed southern border of the Colorado Plateaus; on the south the ranges of the two provinces have no well-defined dividing line.[1]

The mountain structures are very similar to those of the Great Basin. They are usually monoclinal, and in the Chiricahua and Pinal ranges the monoclinal structure is demonstrably due to faulting as shown by Gilbert and by Ransome.[2] The greater number of ranges consist mainly of sandstone, quartzite, and limestone (Paleozoic) that rest upon schists and granites (pre-Cambrian). Both types of rocks are partly covered by volcanic flows, Fig. 67.

As a result of the monoclinal structure of the mountain ranges and the prevailing northwest strike of the beds, the southwestern slopes are longer and somewhat less steep than the northeastern slopes, the latter consisting of a series of steep scarps and benches that give the individual ranges notably bold mountain fronts.

The larger creeks of the region have broad, sandy or gravelly beds of distinctly even gradient, and the tributaries of the main creeks exhibit similar features on a smaller scale. The regular gradients of the stream

---

[1] G. K. Gilbert, The Geology of Portions of New Mexico and Arizona, U. S. Geol. Surv. West of the 100th Meridian (Wheeler Surveys), vol. 3, 1875, p. 508.

[2] F. L. Ransome, Globe Folio U. S. Geol. Surv. No. 111, 1904.

channels and the fact that the channels are dry for the greater part of the year result in their use by man for purposes ·of travel and transportation; they are the natural roads of the region. Throughout the Arizona Highlands a considerable part of the small annual precipitation (15 to 20 inches and less) falls in sudden rainstorms or "cloud-bursts," which are common in July or August. The stream channels are rapidly filled with turbulent waters that wash along great masses of loose detritus swept down from the hill slopes above. The cloud-bursts are incredibly violent and do a remarkably large amount of work. It is through their energetic action that the mountains are dissected and the basins filled with alluvium. For this reason the erosive work due to water action is very important in the aggregate in spite of the semi-arid character of the climate.

SOILS AND VEGETATION

With the exception of the timbered slopes of the mountains and of the alluvial areas along the main arroyos, the surface of the region is almost without soil. The grass and shrubbery occur in such small amounts as to exercise but little retaining influence upon the land waste during short periods of heavy rainfall. Furthermore, the deficient vegetation results in the formation of very small quantities of humic acid, and the rock is therefore not affected by such acid to anything like the degree to which it is affected in humid regions. The granitic masses crumble into particles of quartz, fragments of mica, and angular fragments of crystals of rather fresh feldspar. The quartz and mica are washed down the larger streams by the sudden rains; but the larger fragments of feldspar often accumulate upon the alluvial fans and give them a very distinctive appearance.

The occurrence of arboreal vegetation in response to greater rainfall and its zonal distribution in response to temperature are here as everywhere in the Southwest interesting subjects of study. The general geographic distribution of the many types of vegetation has been worked out as follows:

Fig. 67. — Geologic section from Colorado River to Colorado Plateaus (right). The basement rock is crystalline; the Yampai Cliffs are of limestone; Truxton Plateau and other areas in black represent lava flows; the valleys are deeply filled with alluvium; the Cerbat Mountains are composed of sedimentary rocks. (Lee, U. S. Geol. Surv.)

(1) Zone of cactus, yucca, agave, scanty grass, 3000 to 3500 feet. More luxuriant vegetation in the vicinity of water.

(2) Zone of Obione and Artemisia (greasewood and sage-brush), poor grass, diminished growth of cactus, altitude 3500–4900 feet.

(3) Zone of cedar (*Juniperus occidentalis*), few cactus, 4900 to 6800 feet.

(4) Zone of pine and fir, 6800 to 10,800 feet.

Fig. 68. — Waste-bordered mountains of the Arizona Highlands, Camelsback quadrangle, U. S. Geol. Surv. Note the lack of a permanent stream, the large number of intermittent streams as shown by the dotted lines, and the ragged character of the mountain slopes. Contour interval, 250 feet.

These zones are lower on eastern and northern than on southern and western slopes and the amount of canting increases with increasing elevation (p. 293). The quaking aspen is seen below 7500 feet, likewise the fern (*Pteris aquilina*). Above 7000 feet the white oak accompanies the pine but is never found in great quantities, principally in small patches or groves. In some instances the pines occur in splendid forests.[1] The character of the tree growth in the extreme southern part of the Arizona Highlands has been analysed by Mearns.[2] He found (1893) the Mexican white pine (*Pinus strobiformis*) at the summits of the main peaks of the San Luis Mountains south of the boundary line (7874 feet); also on the Animas peaks in New Mexico (8783 feet); and in the San José Mountains (8337 feet), where a few trees grow close to the summit of the main peak. It is a common tree on the highest peaks of the Huachuca Mountains where it occupies a considerable area and descends as low as 6550 feet. It is an interesting fact in the distribution of this tree that it belongs to the Canadian life zone and is usually associated with the Douglas spruce, aspen, etc. The yellow pine has a vertical range (on the Mexican line) from 6200 feet in the San José Mountains of Sonora to 8500 feet in the Huachuca Mountains of Arizona. The Douglas spruce is found on the San Luis, Animas, and Huachuca Mountains, on all of which it reaches the summits and extends as low as 6000 feet, in cold wet ravines, and is not found on any other mountains of the boundary strip.[3]

On the Dog Mountains in New Mexico the regular juniper zone extends from 6000 feet up to the summits, but in most canyons it descends to the base of the mountains. There is a well-marked juniper zone at 7500 feet on the west side of the San Luis Mountains also. The desert yucca in the fifty-mile desert west of El Paso occurs in open forests spread over large areas where the largest trees grow to a height of 16 feet. The aspen or quaking asp has a vertical range on the Mexican line from 7690 feet in the San José Mountains of Sonora, Mexico, to 9472 feet in the Huachuca Mountains in Arizona. The Fremont cottonwood (*Populus fremontii*) is the most common, beautiful, and valuable shade tree in the whole Mexican boundary region. It grows naturally on almost every stream along the boundary and is found planted around the houses and along the irrigation ditches of almost every ranch. In deep narrow canyons its stem is tall and slender, but in open spaces the isolated trees have full round tops with spreading and often drooping branches. The vertical range of the cottonwood is from sea level to 6100 feet in the Huachuca Mountains near the boundary.[4]

[1] Report upon Geographical and Geological Explorations and Surveys West of the 100th Meridian (Wheeler Surveys), Geology, vol. 3, 1875, pp. 603–604.

[2] E. A. Mearns, Mammals of the Mexican Boundary of the United States, etc., Bull. U. S. Nat. Mus. No. 56, pt. 1, 1907.

[3] Idem, pp. 38, 39, 40.

[4] Idem, pp. 43–48.

A common tree or shrub through the desert Southwest is the mesquite (*Prosopis glandulosa*). The vertical range is from sea level and even below sea level in the Colorado desert up to about 5500 feet. In the deserts of New Mexico, Arizona, and California it is a shrub which obstructs drifting sand, thus forming mounds of sand and lines of sand hills; in the most fertile places along the Colorado River and its tributaries it is a tree of considerable size. Along the Santa Cruz River in Sonora are forests of unusually large mesquite, with some individuals 2½ feet in diameter and 50 feet high.[1]

The San Luis Mountains are composed largely of calcareous rock and are steep and rough. Where a soil covering has been formed they are wooded from a well-marked dry timber line at 5250 feet to the summit at 7870 feet. Below the lower timber line the country is covered with grass and in places with patches of mesquite and chaparral. The forest trees at the lower timber line of the San Luis Mountains are mostly evergreen-oak (*Quercus emoryi*), though in the low canyons grow cypress, walnut, cherry, sycamore, and gray oak (*Quercus gresea*).[2]

The trees of the Animas Mountains (northward continuation of the San Luis Range) in New Mexico are the same as those of the San Luis, with the addition of a zone of quaking aspen. The lower timber line in the Animas mountains is at 5250 feet except where springs occur that support a belt of fine oak timber in the moist canyons far below the main timber line. In one instance a straggling line of oaks is actually continuous across the valley between the Animas and San Luis Mountains, joining the two timber lines of these mountains down two long canyons.[3]

REGIONAL ILLUSTRATIONS

CLIFTON DISTRICT

In portions of the Arizona Highlands, as for example in the vicinity of Clifton, near the eastern border of the state, the topography becomes far less regular than is generally the case. Looking north of Clifton it is impossible to discern any well-defined mountain system. The whole region north of Gila Valley at this point appears as a maze of short ridges, small plateaus, and insignificant peaks.

The topographic complexity of the highlands in the Clifton district is explained by the geologic structure. A core of older rocks (granites, limestones, and sandstones) was deeply and irregularly eroded, and at

---

[1] E. A. Mearns, Mammals of the Mexican Boundary of the United States, etc., Bull. U. S. Nat. Mus. No. 56, pt. 1, 1907, pp. 59–60.

[2] Idem, pp. 89–90.

[3] Idem, p. 92.

a later time (Tertiary) was partially covered by great masses of volcanic flows (rhyolites and basalts), with great variations in thickness and character. The drainage developed upon the lavas after their extrusion was consequent upon the lava flows; those portions of the region not affected by lava flows preserved their original drainage. The consequence was that extremely irregular drainage courses were made still more irregular by the differences in hardness between flat-lying basalt and deformed rock of older age.[1] The only regular features of the Clifton district are the small plateaus due to volcanic accumulations or to the regular and broad uplift of sedimentary rock; but even these plateaus are but dimly discerned and all of them are deeply dissected by canyons and furrowed by a maze of shallow and wide-spreading ravines.

The arboreal vegetation of the Clifton district is found at elevations above 6000 feet, though below this elevation the ridges generally support a certain amount of agave, yucca, and low cactus. Above the 6000-foot level stunted juniper and cedar are quite common and are used as firewood; on the higher slopes a growth of manzanita bushes and stunted oak is also found. The heaviest timber grows in the sheltered mountain basins at altitudes of 5000 to 6000 feet and consists largely of yellow pine. Along some of the river bottoms are large groves of cottonwood. Many of the dry mountain spurs are covered with piñon and juniper.[2]

BRADSHAW MOUNTAINS

The higher peaks of the Bradshaw Mountains district are composed of gneiss, granite, and schist, the granite having been intruded into the schists. Differential erosion has resulted in a high relief, due to the more resistant character of the intrusive granite. In some places quartzite combs in the granite stand prominently above the general level of the wide valleys formed upon the less resistant schist and may be traced for miles by their distinctive and bold relief.

Toward the northwest the Arizona Highlands are also marked by a large number of rather extensive lava flows which have covered over the older topography and simplified the contour of the surface. Among them is Black Mesa, 10 miles east of the Bradshaw Mountains, Bigbug Mesa, 15 miles northwest of Black Mesa, and many others of lesser extent. They are striking forms, for they lie in a region of generally rugged topography whose irregularities have been devel-

[1] Waldemar and Lindgren, Clifton Folio U. S. Geol. Surv. No. 129, 1905, p. 1.
[2] Idem, p. 2.

oped upon rocks of complicated structure and variable hardness.   They
are almost without soil, since they consist of durable basalt recently
formed.   The basalt weathers into spheroidal fragments of variable size
which cover the surface and make the so-called "malpais" of the region.
Between the fragments finer waste accumulates in small quantities and
supports a thin growth of grass in the rainy season.

The schists of the district weather very slowly and their soils are thin,
so that the outcrops of the steeply inclined or vertical strata are visible

Fig. 69. — Bradshaw Mountains, looking northwest near Goddard. The mountains are composed of
granite; the plain in foreground is underlain by basaltic agglomerate.   (U. S. Geol. Surv.)

for miles.   The quartz-diorite and granite weather more rapidly and are
covered with a sandy soil which supports a good forest growth in the
higher mountains of the region toward the north.   The quartz-diorite
weathers most easily of all and is noted for the characteristic basin or
park-like forms developed upon its outcrops.

The semi-aridity of the plains and valleys of this district has resulted
in the development of a characteristic arid vegetation.   The common
vegetation of the lower slopes consists of cactus, yucca, maguey,
paloverde, "cat claw," etc.   In favored localities there are stunted
growths of oak and manzanita, and in the arroyos one may find larger
oaks and sycamores in some quantities.   On the mountains where
greater rainfall is precipitated on account of the greater elevation pine
and fir find a congenial habitat.

In the Bradshaw Mountains the precipitation is greater than on the surrounding plains. Heavy thunder showers occur almost daily in July and August, and during the winter months the mountains are frequently covered with snow. The heaviest timber grows in the mountain basins at 5000 and 6000 feet and is largely yellow pine and its varieties. Some of the river bottoms have by contrast thickets of willow, mesquite, and alder, and groves of cottonwood, while the mountain spurs are frequently covered with thick mats of shrubs or dense stands of pin oak, nut pine, greasewood, and juniper.[1]

SANTA CATALINA MOUNTAINS

The Santa Catalina Mountains ten or twelve miles north of Tucson are among the most impressive ranges of the province and indeed of the Southwest. Their deeply dissected, bold, picturesque slopes rise to a series of exceptionally ragged peaks. Perpendicular cliffs and sharp ridges are common, among them "The Needles," a series of long slender precipitous points which crown the summit of a sharp granite ridge that rises 3000 feet above the plain at the mountain base. The highest point in the range is Mount Lemmon, which rises to 10,000 feet, or 7000 feet above the plains. The ranges composing these mountains are subparallel, extend nearly east-west, and the whole belt is about 50 miles long and almost as wide. The southern slope is especially rugged, so that even the cacti, hardy yuccas, and Spanish daggers have a hard struggle to maintain themselves on the barren rocks. Oaks and juniper are occasionally found in some sheltered alcove, and the summits of the higher mountains and large portions of the northern slopes are covered with pine and fir.[2]

EASTERN BORDER FEATURES

The northeastern edge of the Arizona Highlands is formed by the Grand Wash cliffs and their continuation — the Yampai Cliffs — which rise in a precipitous manner 4000 feet or more above the plains that on the north stretch westward from their base, Fig. 67. The continuity of the westward-facing escarpment is broken by several canyons. Here in a few places the confluent alluvial fans constitute a piedmont foreland of slight development. The lower portion of the cliffs is steep but not even approximately perpendicular. It is composed of granite, while the upper portion of the slope is of limestone and decidedly precipitous.

In the northern portion of the Arizona Highlands and in the Sacra-

[1] Jaggar and Palache, Bradshaw Mountains Folio U. S. Geol. Surv. No. 126, 1905, p. 1.

[2] J. W. Toumey, La Ventana, Appalachia, vol. 8, 1897, pp. 225–232.

mento Valley region in northwestern Arizona the mountain escarpments face the plateau and are due to faulting. Specific localities are Globe, Arizona, and the mouth of the Grand Canyon. Farther from the edge of the plateau sedimentary rocks do not occur in many cases, and it is difficult to tell whether the mountains, composed of crystalline rock, are the products of local uplift or of circumdenudation. In general the mountains parallel the bordering plateau and become smaller and more isolated the greater their distance from the plateau. As the mountains grow less conspicuous the valleys broaden to greater and greater width and finally blend into each other in such a manner as to form a plain completely surrounding isolated mountain groups. The topographic character of this portion of Arizona lies between two extremes of well-defined fault-block mountains and narrow gorge-like valleys near the Colorado Plateaus and broad alluvial plains surrounding single mountain groups at a distance from the plateau.

Truxton Plateau serves as a type of the eastern border features. It is comparatively level and extends from the Yampai cliffs on the east to the Cottonwood and Aquarius cliffs on the west. It lies about 5000 feet above the sea and consists of denuded granite whose depressions have been almost filled with eruptive rock but whose higher portions project above the lava. Truxton Plateau is described as a lava-covered peneplain which has been slightly dissected by a few streams that have cut narrow canyons.[1] The recent uplift of Truxton Plateau is indicated by the rapid deepening of the stream valleys as they approach the cliffs to the west, while their courses within the plateau are distinctly shallow.[2]

For general remarks on Soil Investigations and the Soils of New Mexico and Arizona see G. K. Gilbert, U. S. Geol. Surv. West of the 100th Meridian (Wheeler Surveys), vol. 3, 1875, pp. 594-597. For climatic conditions, idem, pp. 598 et seq. Also idem, pp. 603 et seq., for the geographical distribution of plants in this region.

RAINFALL AND RINGS OF GROWTH

The conditions of growth of the forests of yellow pine in the northern Arizona region have suggested that they might form climatic registers of great importance and that a study of the rings of growth may indicate the character of the climate during the life of the individual tree. This matter has recently been studied with some very interesting results.[3]

[1] W. T. Lee, A Geologic Reconnaissance of a Part of Western Arizona, Bull. U. S. Geol. Surv. No. 352, 1909, p. 21.

[2] For a brief discussion of the physiography of Arizona and the topography of the Globe quadrangle with some excellent cross sections see F. L. Ransome, The Geology of the Globe Copper District, Arizona, Prof. Paper U. S. Geol. Surv. No. 12, 1903.

[3] A. E. Douglass, Weather Cycles in the Growth of Big Trees, Weather Rev., June, 1909, pp. 225-237.

It has been shown that at the 7000-foot elevation at which these trees grow the seasons are very sharply defined; the mean temperature for January is 29°, that for July 65°. Consequently there is a sharply seasonal character to the tree growth. A narrow red ring is formed during the autumn and winter and a broad, soft, white ring during the summer. Under the microscope the winter cells look lean and emaciated; the summer cells are round and well fed. The winter ring is thin, hard, and pitchy; the summer ring is wide, white, and pulpy. It appears probable that the red winter rings are governed directly by the low temperature and the white summer rings by the abundance of moisture.

About twenty sections have been measured by micrometer scale and the result is thousands of readings covering a period of from two to five centuries. On theoretical grounds it would seem that these rings of growth ought to register the rainfall, for they are a measure of the food supply, which depends entirely upon moisture, especially where the supply is limited and the life struggle of the tree is against drought and not against its fellows or members of other species. The measurements show that the rainfall curves and curves of growth have a really remarkable resemblance. An analysis of the curves of growth for longer periods during the life of the tree indicates that there is a direct connection between curves of growth and curves of rainfall in 21.2 and 32.8 year periods[1] with suggestions of shorter periods, especially in the case of the 11-year period already determined by Bigelow.[2]

[1] E. Brückner (Vienna), Klimaschwankungen seit 1700, nebst Bemerkungen über die Klimaschwankungen der Diluvialzeit, 1890. In this paper the author assembles the data for the changing level of the Caspian Sea and its tributaries to show a strikingly regular rise and fall in cycles of about 35 years. Further investigations by Brückner show that these oscillations hold for much larger areas than the Caspian region, probably for the whole world. Still later analyses of rainfall curves in numerous localities have brought out the generality of this fact and the wide application of the law of climatic oscillations.

[2] F. H. Bigelow, Studies of the Diurnal Periods in the Lower Strata of the Atmosphere, Weather Rev., July, 1905.

CHAPTER XVII

COLORADO PLATEAUS

THE physiographic province known as the Colorado Plateaus is roughly circular in shape and embraces portions of the states of Utah, Arizona, Colorado, and New Mexico. The Grand Wash Cliffs on the west, the Uinta Mountains on the north, the Colorado ranges and the Trans-Pecos Highlands on the east, and the Arizona Highlands on the south are the most conspicuous border features. The boundaries are definite on the west, north, and south, but the eastern boundary is in many places indefinite, partly because of its topographic character, partly because it is the least-known portion. The outline displayed on the general physiographic map, Plate II, follows rather closely that assigned by Dutton.[1]

Its most prominent and general characteristics are (1) the flat and but partially dissected upper surfaces of the various members of the province; (2) the distinct boundaries of the individual members as determined by (a) strongly developed, northward-trending fault scarps and (b) eastward-trending lines of cliffs developed most strikingly upon the out-cropping strata of the High Plateaus of Utah; and (3) the great canyons that ramify through almost every large section but are thoroughly developed in the central portion. The most striking physiographic features of the province are the Colorado River and its world-famous canyon. An early view of the course of the river ascribed its origin to a great fault which the river followed and modified. The detailed studies of later years show that certain portions of it are demonstrably located upon faults but that the course as a whole is independent of faulting; and the facts as at present known point to a consequent origin upon either (1) the deformed surface of a peneplain or (2) the present structural surface when it was in a different attitude.[2]

While the surfaces of the individual members of the Colorado Plateaus are smooth or gently undulating, level or slightly tilted, and so undis-

[1] C. E. Dutton, Mount Taylor and the Zuñi Plateau, 6th Ann. Rept. U. S. Geol. Surv., 1884-1885, pp. 114-117, and Plates 11 and 12.

[2] H. H. Robinson, The Tertiary Peneplain of the Plateau District and the Adjacent Country in Arizona and New Mexico, Am. Jour. Sci., vol. 24, 1907, p. 129.

CO

0

B

NEVADA

A R I

H I G

POATES

sected as a whole that the level sky line, the broad expanse, t h e great extent of structural surface are among the most important elements of the relief, yet it must not be thought that the region can be described even in general phrase as level. The great depth to which the canyons have been cut and the great breadth to which they have been widened by the action of secondary erosional processes have resulted in strong relief so that the province is one of the ruggedest in the West. Unlike t h e mountain sections of the West its relief is not to any important degree the product of upward departures from the general level of the region but of downward departures. Instead of mountain flanks we have here prodigious canyon walls diversified by innumerable cliffs; instead of lofty ridges and peaks we have here profound chasms. Exceptions to this rule are the volcanic and laccolithic

Fig. 70. — Ute Mountains, from Cortez, Colorado. Also called El Late Mountains (Hayden Surveys). A group of laccolithic mountains breaking the evenness of the plateau surface. Looking from the same point as in Fig. 71, but in a southwest direction. (Photograph by H. Brigham, Jr.)

masses that rise above the general level to a truly mountainous height, such as Mount Taylor (southeast), the San Francisco Mountains (San Francisco Plateau), Mount Dellenbaugh (Shiwits Plateau), Mounts Trumbull and Emma (Uinkaret Plateau), and the Ute Mountains of southwestern Colorado, Fig. 70. They are not in the aggregate of great extent, but they are important not only through their relief but also through their association with those great lava sheets that although greatly eroded still cover large portions of the Colorado Plateaus.

It is important to see at the outset that the degree of dissection of the different portions of the province is not everywhere the same. The San Francisco Plateau is so little dissected that within its margins there is little hint of the great erosion forms to be found elsewhere in the region, and the flat plateau quality comes out strikingly in every general view; on the other hand the northeastern, north-central, and central portions of the province are dissected by a maze of ramifying canyons. The great breadth of the canyons in this portion of the province might be made the basis for a separate division of the plateau country — the Canyon Lands, as Powell termed the large district included in the valleys of the upper Colorado and the lower Green and Grand.[1] So thoroughly dissected is this portion that the walls of the labyrinthine canyons are the dominating elements of the relief, and movement across the country is by tortuous and extremely toilsome routes, now paralleling some small stream in the depths of a profound abyss, now crossing a plateau spur of mountainous proportions. The degree of topographic development which each district has attained will be the subject of further discussion; here it is sufficient to note that although the strata in their horizontal development are decidedly uniform over the whole area, the topographic aspects of the various sections are as decidedly variable.

A special feature of the plateau country is the manner in which erosion takes place. It is directed against the almost vertical edges of the strata much more than upon the almost flat upper surfaces. Plateau erosion is by stripping through cliff recession. Wash is almost ineffective on the flat gradients of the plateau summits; on the steep and alternating slopes and cliffs of canyons and the cliffed edges of outcropping strata everywhere it is vigorous, as is suggested by the name "Colorado," whose colored, turbid current carries the waste of its

[1] J. W. Powell, Lands of the Arid Region of the United States, U. S. Geog. and Geol. Surv. of the Rocky Mountain Region, 1879, p. 105.

great tributary system of cliffs and slopes with their streams of land waste.[1]

With this special erosion feature in mind it will be easy to understand how extensive the canyon systems of the plateau and the associated terraces are, and yet how within the borders of each plateau there is in many cases so little dissection; each plateau is attacked practically on its margins only and so preserves almost until extinction a marked summit flatness. Conversely we should not understand because such large expanses of flat plateau exist, and each is so little dissected within its borders, that erosion of the general surface is not active, for in the province as a whole, and between the flat-topped plateaus, erosion is taking place at a most rapid rate.

The Colorado Plateaus consist of a large number of individual members separated from each other in various ways but in most cases by strongly defined lines, Plate II. The most striking line of division is the Colorado Canyon itself, which separates a northern

[1] For a discussion of erosion by cliff recession see J. W. Powell's various reports on the Colorado Plateaus.

Fig. 71. — Panorama of Mesa Verde, southwestern Colorado; an example of mesa and scarp topography in the Colorado Plateaus. Looking southeast from Cortez. The border of the mesa is a ragged cliff formed on Cretaceous beds. Formerly these beds extended farther toward the foreground; in time they will be pushed still farther back. Sandstone caps the cliff and underlies the surface in the foreground. Between them are layers of shale. (Photograph by H. Brigham, Jr.)

from a southern series.　Crossing the canyon almost at right angles are a number of northward-trending faults and associated fault scarps which block out an east-west series of plateaus.　A third type of plateau boundary is exhibited in central Utah where the worn edges of outcropping, northward-dipping plateau strata constitute a line of remarkable cliffs of great physiographic interest and unusual scenic beauty.　Besides these well-marked divisions are others of lesser definition.　The northeastern and southeastern sections, although of great extent, have never been studied in the same detail as the rest of the plateau province, and their mutual boundaries are therefore less definitely established.　They are roughly separated from the other divisions by the Green-Colorado Valley on the northwest and the Little Colorado on the southwest, and from each other by the valley of the San Juan. The various lines of division block out four large and important districts or sub-provinces, a fact that needs emphasis, for the elementary student usually regards the great plateaus north of the canyon and in the Kaibab district as the only important parts of the province.　The separate districts may be called for convenience:

I High Plateaus of Utah: Awapa, Aquarius, Paria, Kaiparowits, Markágunt, Paunságunt, Tushar, Wasatch, Sevier, Fish Lake, etc.

II Grand Canyon District: Shiwits, Uinkaret, Kanab, Kaibab, San Francisco, Coconino, etc.

III Southern District: Zuñi, Natanes, Taylor, etc.

IV Grand River District: White River, Roan or Book, Uncompahgre, Dolores, etc.

HIGH PLATEAUS OF UTAH

In any general view of the plateau country, as shown in Plate II, two distinctly different scarp systems may be seen.　The north-south system as just described is caused by great faults with downthrow on the west.　Almost equal to the north-south system of escarpments in magnitude is the east-west system that crosses the plateau country several hundred miles north of the Grand Canyon, a system due to plateau stripping, each cliff marking the outcrop of a resistant stratum. The individual scarps of the latter system break the northward continuity of the Colorado Plateaus in a most decided manner and block out the country into a series of north-south plateaus, among which are the Colob, Markágunt, and others, the whole series known as the High Plateaus of Utah.[1]　They are all characterized by great ruggedness of

---

[1] C. E. Dutton, The Tertiary History of the Grand Canyon District and Geology of the High Plateaus of Utah, Mon. U. S. Geol. Surv., vol. 2, 1882.

outline, pronounced declivity, and a rude parallelism, although they are most irregular in detail. Named in order from south to north the principal ones are the Shinarump Cliffs, Vermilion Cliffs, White Cliffs, and Pink Cliffs, while many lesser cliffs block out plateaus of smaller extent.

The cliffs and intervening plateaus constitute a group of great terraces from 30 to 40 miles in extent north and south and 100 miles east and west. Among the great cliffs perhaps the most remarkable in form are the White Cliffs, while among the most remarkable in color are the Vermilion Cliffs. The latter are from 1000 to 2000 feet high, more than 100 miles long, and consist of evenly stratified layers of sandstone and shale with gypsiferous partings. In color they are brick red, which at twilight takes a strong vermilion hue; in form they are very ornate and architectural, with many vertical ledges rising tier above tier with intervening talus slopes through which the fretted edges of the cliffs project.[1] Though the profile is complex it never loses its typical character and is always extremely picturesque because of the numberless alcoves and alternating promontories where streams cut into the edges of the plateaus which these cliffs terminate.

The escarpments of the northern plateau country are all of a different type from those to the south. They consist of the outcropping edges of resistant northward-dipping strata that act as cliff makers and are being slowly stripped off the plateau surface by wind and water erosion. The former greater extent of the strata is inferred from the remnants left out upon the surface of the plateau south of the main cliffs in the form of isolated buttes and mesas. The map abounds in illustrations, of which perhaps the most conspicuous are the mesas that occur south of the Virgin River and west of Canyon Spring.

In general the drainage of the region in which the east-west line of great cliffs occurs is from the north southward, and it is therefore in this direction that the main canyons run in contrast to the east-west course of the Grand Canyon farther south. There are three principal streams in this district: the Virgin flows west to the Great Basin and finally to the Colorado; Kanab Creek flows southward through a deep narrow gorge to enter the Colorado midway of the Kanab Plateau; and Paria River enters the Colorado at the head of the Marble Canyon. The beds of the plateau streams retain pools of water in the depressions provided they are flooded with material that is not too coarse. These pools are called "water pockets," "lakes," "pools," or "tanks." They are scattered and few in number though always important features of the

[1] C. E. Dutton. Tertiary History of the Grand Canyon District, Mon. U. S. Geol. Surv. vol. 2, 1882, pp. 17, 52, 53.

whole region. They are much more numerous in the higher plateaus of Utah than in the lower country. They are an important source of supply for travelers, settlers, and stockmen, and are the resort of bands of wild horses that roam the uninhabited desert tracts.

The High Plateaus consist of three principal members, a western, a central, and an eastern, Fig. 72. The western member is composed of the

Fig. 72. — Principal relief features of the High Plateaus of Utah. For Colb, lower left-hand corner, read Colob. Scale, 30 miles to the inch. (Dutton, U. S. Geol. Surv.)

Pávant, Tushar, and Markágunt plateaus, named in order from north to south; the central member is composed of the Sevier and Paunságunt plateaus, named in the same order; and the eastern member consists of the Wasatch, Fish Lake, Awapa, and Aquarius plateaus.

We shall describe only a few of the great plateaus of this northern region, selecting those which from our standpoint appear most important as types in the series.

AQUARIUS PLATEAU

The Aquarius, the grandest of all the High Plateaus, is about 35 miles long, of variable width, and 11,600 feet high. Its summit is clad with dense spruce forests sprinkled with grassy parks and exceptionally beautiful lakes.[1] These are not small pools but broad sheets of water from 1 to 2 miles long. Their existence is due to differential erosion by local glaciers that originated in the higher portions of the plateau; 8500 to 9000 feet was the lower elevation of the glaciers of the region and it is at this level that the terminal moraines are usually found. The high elevation of the Aquarius Plateau, 10,500 to 11,600 feet, favored the exceptional development of glacial forms in the Pleistocene, just as to-day it favors a greater rainfall and better forest cover than occur on the neighboring lower plateaus.

The upper surface of the Aquarius is developed upon a 1000- to 2000-foot layer of basalt, which gives rise to marginal cliffs of exceptional height and steepness, as on the northwestern flank. Again, on the eastern border, a great wall from 5500 to 6000 feet high overlooks the lower country and owes its boldness largely to the hard lava cap at the summit.[2]

AWAPA PLATEAU

The Awapa Plateau is 35 miles long and about 18 miles broad; its elevation is about 9000 feet. The western border of the plateau is a wall from 1800 to 3000 feet high, but the other boundaries are less distinct. The slopes of the surface everywhere converge toward a central depression, Rabbit Valley. Although its altitude is that at which moisture and vegetation usually occur, only sage-brush and grasses grow and not a spring or a stream is found upon its entire surface. It is an endless succession of hills and valleys and shallow canyons (400 to 500 feet). It consists entirely of a great variety of volcanic material which has been poured out upon a sedimentary base or platform. Some of the grandest and most massive trachytic beds of the plateau region are found here. The irregular distribution and varied dissection of the many kinds of lavas that occur upon it are in large part the cause of the irregular relief of the surface.[3]

[1] C. E. Dutton, Geology of the High Plateaus of Utah, U. S. Geog. and Geol. Surv. of the Rocky Mountain Region, 1880, p. 5.
[2] Idem, pp. 292–293.
[3] Idem, pp. 272–276.

PARIA PLATEAU

The Paria Plateau terminates on the south in a semicircular line of cliffs which are really a great southward prolongation of the Vermilion Cliffs and include the same strata (Triassic). It lies almost in line with the great Kaibab series of plateaus, yet topographically it is a part of the High Plateaus of Utah and like them is a great structural terrace. It is scored by a labyrinth of sharp narrow canyons which cut deeply into the platform-like surface. The course of the main drainage feature, the Paria River, is independent of the dip of the strata; the courses of the smaller streams are all dependent upon the structural dips.[1] Only the channel of the Paria, however, carries water. The rest are all dry channels which appear to have been formed during the moister glacial period and to have become functionless with the advent of the drier postglacial climate. Growing aridity has extinguished the smaller streams and increased the area drained by the living streams.[2]

KAIPAROWITS PLATEAU

Between the Henry Mountains and the Paria Plateau, Plate II, is a broad area of plateau and canyon country known as the Kaiparowits Plateau and the Escalante Canyon. The canyons of both the Escalante and its numerous tributaries are a network of deep narrow chasms hemmed in by great unscalable cliffs. The depression carved by these streams is bordered on the north and west by a line of cliffs which terminate the Aquarius and Kaiparowits plateaus respectively. The Aquarius Plateau is forest clothed because high and relatively well watered; the Kaiparowits Plateau has only a scattered tree growth of very limited development; the depression below them on the southeast is waterless and treeless — a desert country, swept bare of soil. The cliffs which border the Kaiparowits Plateau on the northeast are 60 miles long and almost 2000 feet high. Their summit constitutes a divide from which almost no streams descend to the barren country below them on the east and only a few traverse the gentler western slope.

MARKÁGUNT PLATEAU

The Markágunt Plateau, Plate II, is a broad plateau expanse south of the Tushar Plateau. It is limited on the west by the Hurricane fault scarp; the eastern base lies at the foot of the great Sevier fault on whose

[1] C. E. Dutton, Tertiary History of the Grand Canyon District, Mon. U. S. Geol. Surv., vol. 2, 1882, p. 201.

[2] Idem, p. 202.

eastern side the Paunságunt Plateau is found. The greater part of the area is covered with eruptive rock (trachyte) resting on sedimen-taries (Tertiary). The southern margin is a line of cliffs not made by faulting but by erosion of hard cliff makers in the northward-dipping series that constitutes the High Plateaus of Utah. The surface of the Markágunt consists of rolling hills and ridges and grassy slopes, the greater part covered with scattered groves of pine. Few canyons are sunk below its general level.[1]

PAUNSÁGUNT PLATEAU

The Paunságunt Plateau is the southernmost extension of the High Plateaus of Utah and fronts south, with a marginal line of cliffs like its neighbors both east and west. These are the Pink Cliffs, often resembling in a most striking way well-known architectural forms such as ruined colonnades, buttresses, and panels. Its strata are sensibly flat and are wholly of sedimentary origin, not volcanic flows capping sedimentaries as is the case with so many of the plateaus of this region.[2]

TUSHAR PLATEAU

The Tushar Plateau is highly inclined and is a transition type between the flat plateaus general to the Colorado Plateaus province and the fault-block mountain type of the Great Basin. It is transitional in position as well as topography and structure, and, strictly speaking, lies within the geographic limits of the Great Basin. Its eastern front is steep and mountainous, for it has been developed largely across the edges of the strata; the western slope is down the dip of the strata and though considerably dissected is far less bold. Its summit is crowned by a cluster of peaks, true erosion remnants, which reach above timber line.[3]

WASATCH MOUNTAINS

Although the Wasatch Mountains are not a part of the High Plateaus they are in such close relation to them as to demand a word of explanation at this point. They stand as a great wall upon the northwestern margin of the plateau country overlooking the Great Basin, and consist of a number of abrupt ranges crowned with sharp peaks that attain altitudes of 10,000 to 12,000 feet. Their boldness gives rise to a moderately heavy rainfall, and the mountain slopes bear forests of spruce, pine, and fir, while the broken and drier foothills support a growth of

[1] C. E. Dutton, Geology of the High Plateaus of Utah, U. S. Geol. Surv., 1880, pp. 195 et seq.

[2] Idem, pp. 251 et al.

[3] Idem, p. 173.

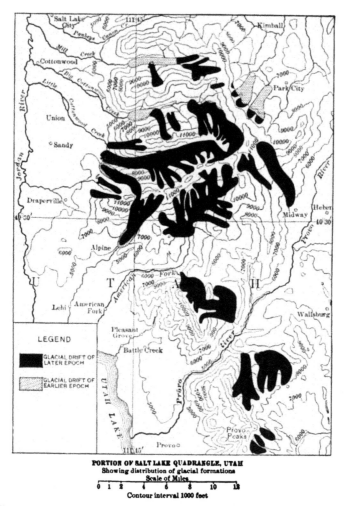

PORTION OF SALT LAKE QUADRANGLE, UTAH
Showing distribution of glacial formations
Scale of Miles
0 1 2   4   6   8   10   12
Contour interval 1000 feet

Fig. 73.— Former glacier systems of the Wasatch Mountains. (After Atwood, U. S. Geol. Surv.)

piñon pine and cedar. In the valleys there are many natural meadows but no forest growth, merely groves of aspen about the springs and lines of willow, box elder, and cottonwood on the borders of the streams.[1]

The Wasatch Mountains extend about 100 miles south of the meeting point with the Uintas, or about to Mount Nebo, 75 miles south of

[1] J. W. Powell, Lands of the Arid Region of the United States, U. S. Geog. and Geol. Surv., 1879, p. 96.

Great Salt Lake. Beyond this point the western margin of the plateau country is the western edge of the High Plateaus. These in turn give way to the broad platform of Carboniferous rock that constitutes the surface of the Grand Canyon District, whose western margin is the Grand Wash Cliffs. The eastern slope of the Wasatch Mountains falls off gradually as a 15 to 20 mile belt of broad ridges and mountain valleys whose waters reach Great Salt Lake through gorges that cut across the main western range of the mountains. The gentler eastern slopes are generally well clothed with vegetation. On the west the mountains present a bold abrupt escarpment which rises suddenly out of the broad flat plains of the Utah basin. The degree of abruptness may be appreciated from the fact that the mountains attain elevations of 10,000 feet within 1 or 2 miles of the western base.[1]

The main crest of the Wasatch Mountains is near the eastern border of the range and the western valleys are therefore much longer than the eastern valleys, — generally from two to three times as long. The loftier peaks — 11,000 to 12,000 feet — that are developed upon crystalline or highly metamorphic rock have rugged, pinnacle-like forms, those developed upon horizontal sedimentary beds have pyramidal outlines with alternating cliffs and talus slopes. In both cases the sharpness of form is due to glaciation. Summits that do not reach above 9000 feet, and hence were never glaciated, are rounded and softened and bear a heavy cover of land waste.

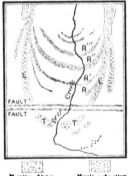

The elevation necessary for the development of Pleistocene glacier in the Wasatch Mountains was 8000 to 9000 feet. Over 50 glaciers were formed exceeding a mile in length. Of these, 46 were west of the crest, and but 4 east of it. Of the 10 exceeding 5 miles in length, 9 lay on the western slope, 1 on the eastern. The greater number and size of the western glaciers were determined by the larger catchment areas and by the heavier snowfall, the west slope being the windward or exposed slope. The western valleys were therefore more completely cleared of loose material, the exposed surfaces more rounded, the main canyons deepened by a greater amount, and more massive moraines developed than in the eastern valleys. The typical relations of the moraines to each other and to

Fig. 74. — Sketch map of morainic ridges near the mouth of Bell Canyon, Wasatch Mountains. (Atwood, U. S. Geol. Surv.)

[1] S. F. Emmons, U. S. Geol. Expl. of the 40th Parallel (King Surveys), vol. 2, 1877, p. 340.

the drainage features in a single valley are shown in Fig. 74. The extent of the glacial systems and their relation to the topography and drainage are shown in Fig. 73.[1]

GRAND CANYON DISTRICT

The most celebrated and best-known portion of the Colorado Plateaus is the Grand Canyon district. The great north-south crustal fractures of this part of the plateau region are lines of faulting which block out the separate members of the district. Named in order from west to east the plateaus are the Shiwits, Uinkaret, Kanab, and Kaibab. Their elevations increase in the same order: the Shiwits is

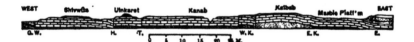

Fig. 75. — East-west (top) and north-south (bottom) sections of the Colorado Plateaus, Grand Canyon district, showing both structure and topography. G.W., Grand Wash Cliffs; H., Hurricane Ledge; T., Toroweap fault scarp; W.K., West Kaibab fault; E.K., East Kaibab monocline; E., Echo Cliffs and monocline. (Davis, Bull. Mus. Comp. Zoöl., modified from Dutton.)

5000 feet and the Kaibab, the highest of all, is 8000 to 9000 feet above the sea. These several plateaus lie on the northern margin of the Grand and Marble canyons and constitute in the aggregate a great platform of nearly horizontal strata (Carboniferous) which is bounded on the east and north by the cliffs of the High Plateaus developed upon younger strata (Mesozoic) and on the west and south by its own terminal escarpment that descends to the older rock (Silurian and Archæan) of the Great Basin and Arizona Highlands.[2]

The elevations of all the plateaus except the Kaibab are not great enough to cause a precipitation adequate for a forest growth. In contrast to the hot, dreary, and barren plateaus about it the Kaibab plateau is moist, and bears meadows and parks and forests of spruce and pine. The descent from one flat-topped plateau to another is over a

[1] W. W. Atwood, Glaciation of the Uinta and Wasatch Mountains, Prof. Paper U. S. Geol. Surv. No. 61, 1909, pp. 73–93.

[2] C. E. Dutton, Tertiary History of the Grand Canyon District, Mon. U. S. Geol. Surv., vol. 2, 1882, p. 19.

high, ragged, westward-facing cliff whose front is constantly battered back by wind and rain. The cliffs are great fault scarps which are strikingly linear and continuous. They are of such prominence as to cause a local rainfall which concentrates the erosive work of the running water; detrital material, due to cliff recession principally, is washed forward from the foot of each cliff in the form of long, sloping, waste fans. The cliffs retreat without being obliterated, in part because of the horizontally bedded rocks, in part because of their recent origin and the small rainfall of the region.

SHIWITS PLATEAU

The Grand Wash Cliffs are the westernmost of the series in the Grand Canyon district and form the western border of the Shiwits Plateau. They are from 1000 to 2000 feet high and overlook the Grand Wash and Hualpi valleys and the rugged sierras and stern deserts of the Great Basin. They form a continuous and bold line of cliffs from the upper Virgin Valley southward across the Colorado River almost 150 miles, to Music Mountain, where they are replaced by the Cottonwood and Yampai Cliffs. There is no river along this western wall of the plateau, "only occasional deluges of mud, whenever the storms from the southwest are flung against the lofty battlements and break in torrents of winter rain."[1] The Grand Wash Cliffs mark the western end of the Colorado Canyon, 5000 feet deep; toward the west the Colorado suddenly changes its character and continues seaward without extraordinary features. Eastward of the Grand Wash Cliffs but still on the western border of the Shiwits Plateau is a second line of cliffs similar to the Grand Wash Cliffs in form but not in origin. It represents merely the eroded upper layers of the geologic series exposed by faulting and pushed back by ordinary erosion to their present position.

The surface of the Shiwits Plateau is diversified by a number of volcanic masses and a few large mesas and buttes, outliers of higher strata capped and so preserved by basalt. The most conspicuous height is Mount Dellenbaugh, 6750 feet high, a central mass of basalt in a large basaltic field.[2]

UINKARET PLATEAU

The Uinkaret Plateau, next east of the Shiwits, is separated from the latter by Hurricane Ledge, a fault scarp 1000 feet high which is con-

[1] C. E. Dutton, Tertiary History of the Grand Canyon District, Mon. U. S. Geol. Surv., vol. 2, 1882, p. 12.
[2] Idem.

tinued south of the Canyon by the Aubrey Cliffs, of similar height and origin.[1]

The Uinkaret Plateau is the narrowest of the four and its southern portion is the most strongly diversified by basaltic eruptions. Indeed this portion (20 miles from north to south) is so broken as to be known as the Uinkaret Mountains in contrast to the Uinkaret Plateau farther north.[2] In sharp contrast to the rectilinear outlines and vivid colors of the greater part of the plateau region are the irregular profiles and gloomy aspect of the basaltic plateau of Mount Trumbull, a great mesa consisting of horizontal sedimentary strata capped by 500–600 feet of basalt. Its summit (2000–3000 feet relative altitude) reaches to such a height that the climate is cool and moist and the plateau sustains a forest growth of yellow pine. About the mesa or mountain and covering the whole southern end of the Uinkaret Plateau are great lava flows — jagged masses of black basalt, desolate except for an occasional grove of cedar and piñon.[3]

From the summit of Mount Trumbull 120 to 130 cinder cones are visible and upon the whole Uinkaret are 160 to 170 vents in all. None of them in the main field is of great size; the highest are only 700 or 800 feet in altitude, with a diameter of a mile.

KANAB PLATEAU

The Kanab Plateau is separated from the Uinkaret by an inconspicuous boundary. For about 20 miles north of the canyon it is clearly marked by the line of cliffs along the Toroweap fault which gradually dies out northward, and beyond this point no prominent topographic feature forms a dividing line.[4]

The Uinkaret and Kanab plateaus are separated from each other by the twenty-mile long Toroweap Valley, the locus of the Toroweap fault that causes the eastern wall of the valley to be several hundred feet higher than the western. The greatest displacement along the line of the fault is only 700 feet. It extends but 18 or 20 miles north of the canyon, hence is a much weaker boundary than the neighboring faults.

[1] The heights of the various fault scarps should not be taken as an indication of the amount of faulting. While the Grand Wash Cliffs are from 2000 to 3000 feet high, the fault which gave rise to them exhibits 6000 feet maximum vertical displacement. Likewise Hurricane Ledge is from 500 to 2000 feet high, but it is associated with a maximum displacement of about 12,000 feet 10 miles north of the Virgin River Valley. (Dutton, idem, p. 20.)

[2] See the Mount Trumbull quadrangle, U. S. Geol. Surv.

[3] C. E. Dutton, Tertiary History of the Grand Canyon District, Mon. U. S. Geol. Surv., vol. 2, 1882, pp. 81–84, 103.

[4] Idem, p. 13.

KAIBAB PLATEAU

The Kaibab and Kanab plateaus are separated by the West Kaibab fault. The eastern border of the Kaibab Plateau is not formed mainly by faulting, hence it is an exception to the general rule. Here occurs not a fault but a great monoclinal flexure that is partially and locally faulted on a small scale and with a total descent of about 4000 feet. About mid-length it becomes a double monocline and finally a double fault before reaching the Colorado. The whole displacement dies out south of the Colorado. The Kaibab Plateau is thus a great uplifted block between two parallel lines of displacement, the eastern a compound flexure, the western a fault. The western border fault continues southward for but a short distance before splitting into three secondary faults with associated scarps and these extend almost to the brink of the Grand Canyon.[1]

The northern end of the Kaibab is a great cusp which terminates a little south of the town of Paria. Its southern border is the mile-deep Grand Canyon of the Colorado. Structurally and topographically the Kaibab is continued across the canyon into the southern district, but the strong break of the canyon has caused the adoption of a different name for the southern section, which is called the Coconino Plateau.

It would be reducing to common terms that which is in the highest degree uncommon if no mention were here made of the extraordinary scenic features displayed in this greatest wonderland of North America. Even superlative terms are feeble, and ordinary language is wholly inadequate for the expression of forms and colors that have been the theme of every enthusiastic scientist and every poet who has beheld the great cliffs, profound canyons, and vast expanses of the plateau region. Dutton's rich vocabulary enabled him to write a description which is probably the most satisfactory expression of the ideas and feelings which these great scenic features inspire. We therefore restrict ourselves to two or three quotations from his classic memoirs, *Tertiary History of the Grand Canyon District* and *Geology of the High Plateaus of Utah*. They are not the most enthusiastic passages that may be found; they are rather the most restrained and careful and may therefore be accepted not only as scientific but also as imaginative and interesting. The first description presents the region of terraces and canyons east and south of the High Plateaus of Utah as seen from one of the latter at an altitude of more than 11,000 feet.

[1] C. E. Dutton, Tertiary History of the Grand Canyon District, Mon. U. S. Geol. Surv., vol. 2, 1882, p. 13.

". . . the eye ranges over a vast expanse of nearly level terraces, bounded by cliffs of strange aspect, which are truly marvelous, whether we consider their magnitude, their seemingly interminable length, their great number, or their singular sculpture. They wind about in all directions, here throwing out a great promontory, there receding in a deep bay, but continuing on and on until they sink below the horizon, or swing behind some loftier mass, or fade out in the distant haze. Each cliff marks the boundary of a geographical terrace sloping gently backward from its crest line to the foot of the next terrace behind it, and 'each [in northward succession] marks a higher and higher horizon in the geological scale as we approach its face. Very wonderful at times is the sculpture of these majestic walls. Panels, pilasters, niches, alcoves, and buttresses, needing not the slightest assistance from the imagination to point the resemblance; grotesque forms, neatly carved out of solid rock, which pique the imagination to find analogies; endless repetitions of meaningless shapes fretting the entablatures are presented to us on every side, and fill us with wonder as we pass."[1] "Isolated masses cut off from the main formation, and often at considerable distances from it, lie with a majestic repose upon the broad expanse of the terrace. These sometimes become very striking in their forms. They remind us of great forts with bastions and scarps nearly a thousand feet high. The smaller masses become regular truncated cones with bare slopes. Some of them take the form of great domes where the eagles may build their nests in perfect safety. But noblest of all are the white summits of the great temples of the Virgin gleaming through the haze. Here Nature has changed her mood from levity to religious solemnity, and revealed her fervor in forms and structures more beautiful than anything in human art."[2]

"But of all the characters of this unparalleled scenery that which appeals most strongly to the eye is the color. The gentle tints of an eastern landscape, the rich blue of distant mountains, the green of vernal and summer vegetation, the subdued colors of hillside and meadows all are wanting here, and in their place we behold belts of fierce staring red, yellow, and toned white, which are intensified rather than alleviated by alternating belts of dark iron gray."[3] "As the sun nears the horizon the desert scenery becomes exquisitely beautiful. The deep rich hues of the Permian, the intense red of the Vermilion Cliffs, the lustrous white of the distant Jurassic headlands are greatly heightened in tone and seem self-luminous. But more than all, the flood of purple and blue which is in the very air, bathing not only the naked rock faces but even the obscurely tinted fronts of the Kaibab and the pale brown of the desert surface, clothes the landscape with its greatest charm. It is seen in its climax only in the dying hours of daylight."[4]

SAN FRANCISCO PLATEAU

The term is here employed to include the whole region between the Grand Wash Cliffs, the Arizona Highlands, the Grand Canyon District, and the Little Colorado River. It includes the Coconino Plateau on the north, and on the south the broad platform of much greater extent upon which the San Francisco Mountains stand.[5] Its general

[1] Geology of the High Plateaus of Utah, p. 8.

[2] Description of the scenery of the White Cliffs, developed upon the white Jurassic sandstone. From Tertiary History of the Grand Canyon District, p. 37.

[3] From description of the same region noted in the first quotation from Geology of the High Plateaus of Utah, p. 8.

[4] Impressions of a sunset on the Kanab Plateau in Tertiary History of the Grand Canyon District, pp. 124–125.

[5] It is somewhat confusing to use the term Colorado Plateau for this portion of the district as in Dutton's standard description. The term Colorado had already been established for the entire province. Both the Land Office maps and the maps of the Topographic Branch of the U. S. Geological Survey follow Dutton's usage however. In using the term San Francisco Plateau I am following the excellent suggestion of Robinson (Am. Jour. Sci., vol. 24, 1907).

altitude is from 7000 to 7500 feet, which is distinctly higher than the plateaus immediately north of the canyon, except the Kaibab. The strata of the district are nearly horizontal though with a broad regional slope toward the southwest, and, with few exceptions, the surface is not deeply or extensively scored by canyons. Unlike the Kaibab it has no strongly marked divisional boundaries, since the great fault scarps which form such prominent features of the one are either absent or diminish in height or disappear in the other. Its surface is diversified in some localities by (1) low, lava-capped, and usually forest-clad mesas, as Black Mesa and Mogollon Mesa, and (2) volcanic cones and peaks such as San Francisco Mountains (13,000) and the associated lava flows about their bases.[1] Some of the volcanoes are very recent, as in the San Francisco Mountain region, an example 16 miles north of San Francisco Mountain being probably not more than 100 or 200 years old.[2]

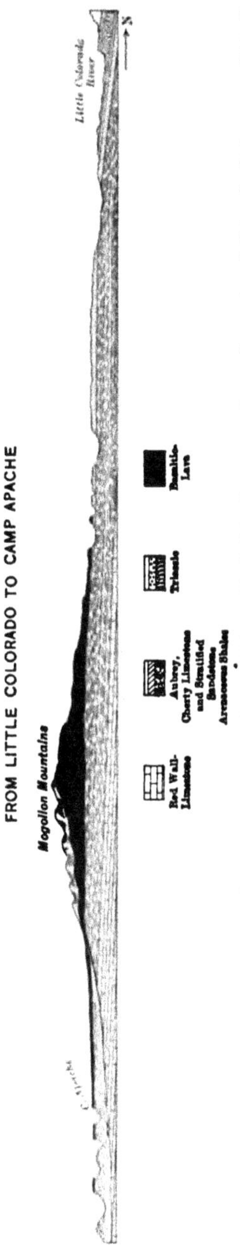

SOUTHERN PLATEAU DISTRICT

ZUÑI PLATEAU

The Zuñi Plateau (locally called the Zuñi Mountains) is an illustration of a type of structure seen in a number of places in the plateau country, for example in the San Rafael Swell on the eastern margin of the High Plateaus, and is of special physiographic interest. The otherwise flat plateau strata have been somewhat locally

[1] C. E. Dutton, Tertiary History of the Grand Canyon District, Mon. U. S. Geol. Surv., vol. 1, 1882, pp. 14–15.
[2] D. W. Johnson, A Recent Volcano in the San Francisco Mountain Region, Arizona, Bull. Geog. Soc. Phil., vol. 5, 1907, pp. 2–6.

Fig. 76. — Topographic and structural section of the Mogollon Mountains and Mesa. Typical border topography southern edge of Colorado Plateaus, see Plate II. (Modified from Marvine, Wheeler Surveys.)

uplifted, raising the surface well above the general level and exposing it to pronounced erosion. The cross section, Fig. 77, exhibits the main structural features. It will be readily seen that the margins of the uplifted tract have been structurally bent and in places faulted, but that the summit of the uplift is almost flat. The erosion of the superior part of the broad uplift has removed portions of the capping layers and exposed the geologic section ranging from rather young (Cretaceous) to very old rocks (Archæan). Each stratum ends in a cliff facing the central axis of the uplift, and each cliff may be traced in the form of a great oval entirely around the tract. Each stratum also dips outward from the cliff summit and extends below the next cliff of the younger formation overlying it. The heart of the uplift is composed of Archæan gneisses and mica schists, and these constitute the main summit platform or plateau over a considerable portion. Since the summit has this constitution the oldest (stratigraphically the lowest but topographically the highest) sedimentary formation, the Carboniferous, ends at the shoulder or margin of the main platform. Many portions of the summit have strongly marked topographic features and are very rugged and diversified, being deeply scored with canyons. The principal drainage lines cross portions of the uplift regardless of the structure and show that they were determined before the uplift, a feature of common occurrence throughout the plateau country, while it is equally common to find all the lesser stream-ways in close sympathy with the broad structural surfaces of the main plateaus.[1]

MOKI-NAVAJO COUNTRY [2]

That part of Arizona bounded by the San Juan, the Colorado, Little Colorado, and Puerco rivers reveals in slightly modified form the typical characteristics of the plateau province. The river valleys are canyons or wide, open washes, the rocks are chiefly sedimentaries of Mesozoic age, the climate is arid, the vegetation sparse, and there is little prospect that the area can be made profitable to civilized man. The only permanent streams are the San Juan, the Colorado, and the headwaters of

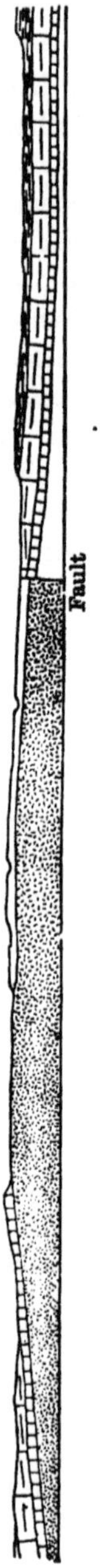

[1] C. E. Dutton, Mount Taylor and the Zuñi Plateau, 6th Ann. Rept. U. S. Geol. Surv., 1884-1885, pp. 162-163.

[2] In the absence of published data on this section of the plateau country Prof. H. E. Gregory has kindly prepared this brief description from his field notes.

Fig. 77. — Section from northeast to southwest across the Zuñi Plateau. The strata are Archæan, Carboniferous, Jura-Trias, and Cretaceous, in upward succession. (Dutton, U. S. Geol. Surv.)

a few rivers which come from the highlands; for example, Chinlee, Moencopie, Navajo, and Little Colorado, Plate II.

During the rainy season (July and August) the run-off from the highlands is so rapid and uncontrolled and the flat valley washes are so thoroughly drowned that travel is exceedingly difficult and often dangerous; and the utilization of the extensive alluvial bottoms for agriculture on a large scale is rendered impossible. None of the important streams is reduced to grade throughout its course. All of them are carrying prodigious quantities of waste to be added to the burden of the Colorado. In the absence of running streams the Navajos and Hopis depend for themselves and their stock upon springs and ephemeral lakes, and the supply at best is meager and commonly unpalatable. It is remarkable how little water is used by man and beast.

The broad structural features of the region are easily comprehended. The highlands along the Arizona-New Mexico line include Carrisos Mountains at the north, the Lukachukai, Tunichai, and Choiskai mountains, and the hogback ridges between Fort Defiance and Gallup. The Carrisos (elevation 9420 feet) rise 5000 feet above the San Juan at their base, and constitute one of the most prominent landmarks of the plateau country. The mountains are laccolithic in origin and are caused by irregular intrusions of quartz-monzonite which deformed Triassic, Jurassic, and Cretaceous sandstone into a group of domes. Subsequent erosion has removed the cover here and there, exposing the igneous core, and deeply trenched the heavy sedimentary beds which form the flanks of the uplift. Lukachukai, Tunichai, and Choiskai mountains are composed of sedimentary strata lying approximately horizontal; their elevation is due to a fold of great amplitude, the western limb of which forms the eastern side of the Chinlee Valley. Caps of lava rest upon the Tunichais as well as upon a number of detached buttes at the south, while the covering beds of the Choiskais are Tertiary sediments. The ridges south and east of Fort Defiance are formed by the resistant members of Mesozoic strata which here dip to the east at a high angle.

The area between the lower Chinlee and Colorado rivers is part of a great dome of Triassic and Jurassic sediments intersected by the San Juan River. At Marsh Pass the strata plunge beneath the Cretaceous of the great Zilh-le-jini (Navajo for Black Mountain) mesa, and the hogback rim of the dome is prominently exposed from Marsh Pass, Arizona, to Elk Ridge, Utah. The elevation at Skeleton Mesa and the region about the head of Pahute Canyon is attained by a series of monoclinal folds with steep eastern limbs. As one travels from the Chinlee westward he ascends a series of giant steps until an elevation of nearly 8000 feet is reached. Standing on the plateau at this elevation his view is unobstructed except for Navajo Mountain (10,400 feet), a laccolith with its sedimentary cover still intact. It stands as an island covered with vegetation in the midst of a sea of barren red rock. From the top of Navajo Mountain a more comprehensive view of the plateau province may be obtained than from any other point. From Lee's Ferry the Leupp Echo Cliffs and their southern continuation mark the line to which the Jurassic and Upper Triassic have been stripped by the tributaries of the Colorado and the Little Colorado.

In the center of the Navajo Reservation and including most of the Hopi Reservation is the Zilh-le-jini mesa, attaining an elevation of 6000 to 8000 feet, and bordered on all sides by steep escarpments. In structure this highland is a shallow syncline of Cretaceous strata in which are included valuable beds of coal. On the narrow finger-like mesas forming the southern border of Zilh-le-jini are located the six villages of the Hopis which constitute the ancient province of Tusayan. Between Tusayan and the Little Colorado is an extensive area of lava-capped mesas and volcanic necks and dikes, the remnants of early Tertiary lava flows.

The vegetation is characteristic of the plateau province. Above 7000 feet on the highlands along the Arizona-New Mexico boundary line there grows an open stand of yellow pine which is used for the manufacture of rough lumber for various government projects. Pine in limited amounts grows on the northern rim of Zilh-le-jini mesa and covers the upper slopes of Navajo Mountain. Between 6000 and 7000 feet, piñon and cedar are commonly found, but over wide areas it averages not more than two or three trees per acre. Sage-brush and greasewood grow in limited amounts at elevations below 6000 feet, and grass, which is fairly abundant at higher elevations, is also found in limited amounts below 6000 feet, and in a few favorable localities forms a continuous sod. The extent of bare rock and sand floor is very great, and probably not more than 10% of the Navajo-Moki region is actually covered with vegetation. In fact it is possible to walk from Gallup, New Mexico, to Tanner's Crossing on the Little Colorado, or from the Carrisos Mountains to Lee's Ferry, without stepping on a twig or a spear of grass.

GRAND RIVER DISTRICT

West of the Rocky Mountains of central Colorado, Plate II, and on meridian 107° 30' is the northeastern edge of the Colorado Plateaus. This portion is called the Grand River district after the main drainage

line. It consists of the White River Plateau, Grand Mesa, the Book or
Roan Plateau, the Uncompahgre Plateau, and the Dolores Plateau, named
in order from north to south. The strata upon which these plateaus are
developed as a rule dip toward the west and away from the Rocky
Mountains. In the southwestern part of Colorado, along the San Juan
Valley, six well-marked groups of hard rock (sandstone) occur and a
corresponding number of soft shales alternate with them and form lines
of weakness for the attack of the weather. Each group of strata slopes
and dips gently westward until a break occurs and a precipitous descent
is made across the edges of the rock layers to the next lower group of
strata,[1] a succession that is typical of the district, although the number
of alternations of strata and to some extent the topographic expression
varies from section to section.

Like the other districts the Grand River district exhibits a number of
mountain groups of igneous origin scattered upon its plateau surfaces.
These include the San Miguel, La Plata, Carriso, Abajo, and other
groups, which are as a rule of laccolithic origin. Erosion has removed
the once overlying sedimentary beds and exposed the intruded rock.

WHITE RIVER PLATEAU

The White River Plateau is the northernmost of the series in the northeastern part of the plateau province. The general level of its surface
is interrupted in many places by higher summits (500–1000 feet above
the general level) and ranges of mountains on the one hand and by
deep valleys and canyons on the other. While plateau surfaces are
still conspicuous features of the region, the upland has been so largely
cut away by erosion, so largely modified by uplift or faulting of the once
flat-lying strata, so complicated by intruded masses of igneous rocks,
that the plateau feature can scarcely be said to be the dominating one,
and by some the plateau would be classified as a part of the Rocky
Mountains. The best-preserved sections of the plateau are those whose
flat upper surfaces have been unbroken and but slightly tilted.

The White River Plateau has been formed largely by lava flows,
the borders of which have been cut into deep gorges and ravines in
which some of the headwaters of the White and Yampa rivers take
their rise.[2] On the west especially this plateau falls off in abrupt and
high cliffs to the long slopes of the bordering spurs; the eastern border
is marked by high detached masses which give it a mountainous aspect

[1] C. E. Dutton, Mount Taylor and the Zuni Plateau, 6th Ann. Rept. U. S. Geol. Surv.,
1884-1885, p. 242.
[2] F. V. Hayden, U. S. Geog. and Geol. Surv. of the Terr., 1876, p. 7.

when viewed from the east. North of the White River Plateau the country is mountainous and so broken that no distinct, well-defined topographic system has been determined.[1] On the south the plateau forms the divide between the White and the Grand rivers and loses its distinctive mesa-like quality, being cut by profound canyons.[2]

SPECIAL BORDER FEATURES

In portions of northwestern Colorado the plateau character is largely replaced by hogback or monoclinal ridges, strike valleys, and other features related to moderate complexities of structure developed in a series of alternating hard and soft strata.[3]

One of the most remarkable physiographic features of northwestern Colorado is Grand Hogback, which extends southward from the Danforth Hills to a point beyond Grand River. Indeed under the name of Colorado Ridge it continues far beyond Grand River to the point where its identity is lost in the more rugged West Elk Range. The Grand Hogback is bordered on both sides by long continuous valleys; the ridge itself is at some places a single-, at others a double-crested hogback ridge formed by the outcrop of massive sandstone ridges inclined eastward at steep angles. Its steep inclination has afforded opportunity for the development of an exceedingly rugged topography whose principal elements are precipitous marginal slopes. The most striking features of the Grand Hogback and the Colorado Ridge are their topographic continuity and structural uniformity. The stream valleys in the Hogback Ridge contain a small amount of red fir, while the upper slopes and crest, 2000 feet above the marginal valley bottoms, are covered with a dense growth of oak brush, chokecherry, and juniper.[4]

ROAN OR BOOK PLATEAU

One of the most striking topographic features of the Grand River district is the southern margin of the Roan or Book Plateau, the Roan or Book Cliffs.[5] In places the cliffs rise almost vertically from the edge of the Grand River Valley to their full height; elsewhere the ascent is by a succession of broken steps. Their mean height is about 8600 feet, or

---

[1] S. B. Ladd, U. S. Geog. and Geol. Surv. of Colorado and Adj. Terr. (Hayden Surveys), 1874, pp. 437-438.

[2] Idem, p. 439.

[3] H S. Gale, Gold Fields of Northwestern Colorado and Northeastern Utah, Bull. U. S. Geol. Surv No. 415, 1910, p. 24.

[4] Idem, p. 31.

[5] Roan, from their prevailing color; Book, from the resemblance of their characteristic form to a bound book.

3500 feet above the Grand River. They are the southern escarpment of the plateau of the same name which slopes rather gently to the north and northeast and is drained by tributaries of the White River. Nearly half the original surface of the plateau is still intact, so that a rather flat aspect is more common than in the more dissected White River Plateau on the east or the Uncompahgre Plateau on the south. The western is the most dissected portion of the plateau, and here the divide is in places only 30 or 40 feet wide, bordered on the cliff side (south) by a precipice and on the plateau side by a strong slope. The crest of the cliffs has very little water on account of the low elevation (8600 feet); it is covered in the main with grass and sage. Almost the only arboreal vegetation is quaking aspen, which occurs but sparingly; there are but a few occurrences of spruce and pine.[1]

UNCOMPAHGRE PLATEAU

The Uncompahgre Plateau lies between the Uncompahgre Mountains and the Grand River. It is 90 miles long and from 15 to 25 miles wide; its elevation is from 8600 to 10,200 feet (northwest). It is in the form of a broad arch of sedimentary beds about 1000 feet thick over a central core of granite, the latter being well exposed in the steep and beautiful canyon of the Unaweep, where it forms two-thirds of the height of the walls. The tributaries of the Unaweep have all cut down to the granite, and in the season of floods due to the melting of the snows their waters drop from 300 to 2000 feet to the bed of the main canyon below. On the west the plateau is bordered by steep cliffs which descend to the valleys or canyons of the Grand and other streams.

One of the most profound canyons of the whole mountainous western half of Colorado is that of the Gunnison on the northeastern border of the Uncompahgre Plateau. It is 56 miles long and 3000 feet deep in the deepest part. The plateau in which it is incised consists of granite-gneiss capped by 1000 to 1200 feet of sedimentary strata. The canyon is cut through the overlying sedimentary rock and deep into the gneiss, the contact being marked by a sloping bench below which are rough, ragged, and nearly vertical walls which extend to the river margin; above the contact are steeply sloping talus and vertical cliff in alternating series. The tributaries of the Gunnison have incised their courses but little into the gneiss and therefore have very steep descents where they join the master stream.[2]

---

[1] F. V. Hayden, U. S. Geog. and Geol. Surv. of the Terr., 1875, p. 346, and 1876, p. 69.

[2] Henry Gannett, U. S. Geol. and Geog. Surv. of Col. and Adj. Terr. (Hayden Surveys), 1874, p. 425.

The influence of elevation upon rainfall and thus upon the character and amount of the vegetation is admirably illustrated in the Uncompahgre Plateau. In the interior of the plateau and down to an elevation of 7000 feet or more are streams and springs in some number, good pasturage, a sprinkling of aspen groves, and game in some quantity; below 7000 feet aspen gives way to piñon, grass to sage, cacti, and bare rock, the streams become muddy torrents or dry up altogether, and in place of the rolling plateau surface are deep narrow canyons and steep precipices.[1]

The influence of elevation upon the character of the vegetation is also well shown on the plateau between the Gunnison and the North Fork of the Gunnison. This plateau has a smooth unbroken summit sloping in a direction slightly west of north. It ranges in elevation from 9000 to 5400 feet, and owing in part to its marked slope, which drains the surface water rapidly away, and in greater part to its low elevation, the plateau is without timber in contrast to the bordering mountains of slightly greater elevation. Its vegetation is piñon pine, cactus, sagebrush, and scrub oak in contrast to the timbered plateau on the east.[2]

DOLORES PLATEAU

The Dolores Plateau lies between the Dolores and San Miguel rivers about halfway between the San Juan and Grand River canyons. The highest portion is known as Lone Mesa and is at 10,000 feet elevation, with an area of about 40 square miles. About it are a number of high flat-topped buttes which like it are erosion remnants with steep bordering scarps. The superior elevation of Lone Mesa gives it a greater rainfall than falls on adjacent tracts and meadows, and forests of pine abound.[3]

The canyon of the Dolores is very narrow and precipitous, with almost no alluvial bottoms except in its shallowest portion at the great bend, Fig. 78. Here are a rich growth of grass, some cottonwood groves and bushes, and vines. On the borders of the canyon below this point is a rather heavy growth of piñon pine and cedar, and on the various headwater branches are forest and meadow, aspen groves, and rich grassy parks in contrast to the desert canyon below.[4]

One of the most remarkable cases of stream direction not in accord with the present slope of the plateau surface (p. 284) is that of the

[1] F. V. Hayden, U. S. Geog. and Geol. Surv. of the Terr., 1875, pp. 340, 341, 349.
[2] Henry Gannett, U. S. Geol. and Geog. Surv. of Col. and Adj. Terr. (Hayden Surveys), 1874, p. 426.
[3] Idem, p. 266.
[4] Idem, pp. 266–277.

Dolores River, a tributary of the Grand. The Dolores rises in the San Juan Mountains, runs south and west for more than 30 miles, then flows northwest against the inclination of the surface for about 60 miles to the Grand River in western Colorado, in a gradually deepening canyon, Fig. 78. At the turning point it is in a canyon only 100 feet deep; above and below this point the canyon deepens to 2000 feet. Evidently the river gained its course sometime before the present surface conditions were established. The surface is structural in origin, being developed on a great sandstone layer (Dakota). Either the surface was peneplaned and the river now pursues an antecedent course with re-

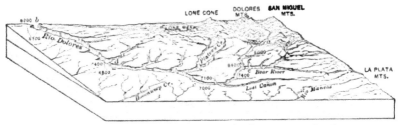

Fig. 78. — Canyon of the Dolores. The canyon is 2000 feet deep at *a*, 2100 at *b*, and 100 at *c*.
(Hayden Surveys.)

spect to later uplifts that deformed the peneplain, or its course was developed in response to a structural slope that has since been reversed. While the main stream has thus maintained its earlier course the smaller streams of the region show marked responses to the present attitude of the surface. The tributaries of the San Juan (south) rise almost on the brink of the Dolores canyon, for they have been extended in response to the structural surface now existing and have encroached on the weaker Dolores tributaries until the latter are all but extinguished south of the canyon.[1]

The Yellow Jacket tributary of the San Juan has encroached so far upon the Dolores at the great bend that a tunnel several hundred yards long has been constructed through the intervening barrier and a large part of the water of the Dolores turned into the Yellow Jacket for the purpose of irrigating the valley flats at Cortez, Plate II.

PHYSIOGRAPHIC DEVELOPMENT; EROSION CYCLES

An understanding of the physiography of the Colorado Plateaus requires at least some knowledge of its recent history as expressed in a number of topographic cycles through which the region has passed. With the long periods of deposition in the plateau region when it stood

[1] F. V. Hayden, 9th Ann. Rept. U. S. Geog. and Geol. Surv. of the Territories, 1875, pub. 1877, pp. 263–264, and Plate 42, p. 264.

at or below sea level we have little to do; nor is it necessary for an understanding of existing topography to take account of the ancient and now buried surfaces of erosion that are so well exposed in the walls of the Grand Canyon. It is sufficient for our purpose to begin with rather late movements in the evenly bedded mass of strata accumulated by long-continued erosion of the Great Basin and adjacent mountains.

In the Tertiary (latter part of the Eocene or possibly in early Miocene) the plateau region was uplifted and the uplift was accompanied by monoclinal folding. The result of these first deformations was an elevation of the western plateau country above the eastern, and the descent from plateau to plateau was at that time from west to east and not as at present from east to west. The folding probably gave rise to a number of closed basins in which lakes were formed and sediments laid down, although the drainage of the region was probably on the whole through basins draining to the sea. Since this first deformation of the plateau country (in Tertiary time) the region has suffered continuous erosion down to the present, although the character of the erosion, the geologic structure, and the topography, have been modified repeatedly, as outlined in the following paragraphs.

In response to internal forces of the earth a period of faulting followed the first period of monoclinal folding and by it the plateau district was marked out and the eastern and western borders clearly defined through the lowering of the country on either side. The date of this period of faulting has not been closely ascertained, but it may be regarded with a certain degree of accuracy as having occurred at the close of the Miocene.

FIRST EROSION CYCLE

The two geologic events of monoclinal folding and of faulting lent to the relief of the plateau region a pronounced character. But the topographic expression of these structural features was gradually and

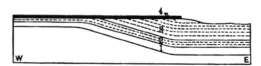

Fig. 79. — Black Point Monocline, Colorado Plateaus, showing remnant of base-leveled surface capped by lava. 1. Kaibab; 2. Moencopie; 3. Shinarump; 4. Basalt. (Robinson).

at last completely obliterated at the close of a cycle of erosion known as "The Period of the Great Denudation." The surface of the entire southern plateau country, and probably also of the northern plateau country, was reduced to a peneplain. At the close of the great denudation cycle there occurred widespread eruptions of basalt. The lava flows capped portions of the surface of the peneplain and thus protected them from the effects of later denudation.

An instructive locality for the study of those features upon which the foregoing description rests is at Black Point in the Little Colorado Valley, where the Black Point monocline dips eastward as shown in the cross section, Fig. 79. The black basalt which caps the surface of the country and preserves a portion of the ancient peneplain rests upon an exceedingly smooth, almost flat surface, a degree of smoothness that could not have been developed across strata of such variable hardness except at a base level of erosion. The Kaibab cherty limestone is very resistant, while the Moencopie soft sandstone and shales and the Shinarump marls are very soft and easily weathered; above the marls at the eastern end of the section is a compact sandstone of most resistant quality. A similar occurrence is at Anderson Mesa, eight miles southeast of Flagstaff, where a smooth surface was once developed across rocks of very different hardness such as the Kaibab cherty limestone and the soft Moencopie shales. The peneplain is also well shown in the Mount Taylor region on the walls of both the Mount Taylor and the Prieta mesas where it bevels across beds of slight dip. Its character as inferred from the once buried remnants now reëxposed was that of a surface of rather faint relief.[1] Huntington and Goldthwaite[2] have described it in the Toquerville district, Utah, and Davis has interpreted phenomena of the same significance in a number of other localities.

The boundaries of the known portions of the base-leveled surface include about 25,000 square miles of country, and it seems at present as if these peneplain remnants in various portions of the plateau country were at one time united so as to form a surface of very slight relief, out of which the existing plateaus were blocked by faults. Recently the work of Gregory[3] in the Navajo country and the valley of the San Juan has shown the extension of the great peneplain surface far to the north. An excellent locality is between Tuba City and Oraibi, where a remarkably flat surface only partially dissected bevels regularly across strata of pronounced dip and structural variation. The entire extent of the peneplain will only be known when the now little explored portions of the province (which aggregate the greater part of it) are examined and the physiography interpreted.

SECOND EROSION CYCLE

The next important geologic event in the plateau region was the inauguration of a second period of faulting near the close of the Pliocene, a period of faulting in which the plateau region was again strongly blocked out and given those features that are most prominent at the present time. The second period of faulting increased the relief of the plateau region and resulted in the pronounced step-like relation of the different members of the plateau. While the folding of the first period of deformation operated in such a manner as to cause a descending series of plateaus from west to east, the major faults of later dates reversed the order of descent, an order that has been maintained down to the present.

The second period of faulting introduced the second or post-peneplain cycle of erosion, in which there was developed a widespread system of

[1] D. W. Johnson, Volcanic Necks of the Mount Taylor Region, New Mexico Bull. Geol. Soc. Am., vol. 18, 1907, pp. 307–308. See also idem, cross section, Fig. 2, p. 309.

[2] Huntington and Goldthwait, The Hurricane Fault in the Toquerville District, Utah, Bull. Mus. Comp. Zoöl., vol. 42, Geol. Series, vol. 6, 1903.

[3] Personal communication.

shallow but mature valleys, one of the most persistent features of the better-known portions of the province. The drainage characteristics of this partial cycle have been noted by several writers, among whom Robinson was the first to demonstrate their persistence and their meaning in terms of erosion cycles.[1] The same feature has been described by Noble,[2] who says:

"The drainage system of the plateau surface (Coconino) consists of a series of mature open-floored valleys with gently sloping sides, which contain no living streams. . . . In tracing one of these mature valleys toward its head . . . [it is] common . . . to find it truncated as a hanging valley by the wall of the Grand Canyon. . . . The same system of mature valleys covers [the] surface [of the Kaibab plateau] which slopes southwesterly to the rim of the canyon."

The surface drainage of the Kaibab now runs through this system of mature valleys into the Grand Canyon; the surface drainage of the Coconino runs through a similar valley system away from the Grand Canyon, and both plateaus have only temporary streams. These mature drainage systems offer the most striking contrast on the one hand to the youthful topography of the deeply trenched canyon developed in the third or next erosion cycle, and on the other hand to the base-leveled surface now preserved in fragments beneath the basalt caps of various mesas.

It should not be thought that the second cycle of erosion brought about any great vertical reduction of the surface. On the contrary the depth to which the streams cut in response to the new base level was only a few hundred feet as against the few thousand feet of the third and last, or canyon cycle of erosion. The chief result of denudation in the second cycle was the broad horizontal stripping back of the strata outcropping on the surfaces of the plateaus and the development of mature valley systems. The stripping of the gently inclined strata proceeded on structural planes, so that the present flat plateau surfaces are structural surfaces developed upon the upper surface of resistant strata. Generally the structure dips uniformly toward the southwest at the rate of about 200 feet to the mile, so that the surfaces of the Coconino and San Francisco plateaus south of the Grand Canyon slope to the southwest from the canyon rim at the rate of about 200 feet per mile. The Kaibab plateau north of the Colorado Canyon slopes southwesterly in the same degree to the rim of the canyon. It is for this reason that the southern plateau drains away from the canyon and that the northern plateau drains into it. The plateau surfaces are

    [1] H. H. Robinson, A New Erosion Cycle in the Grand Canyon District, Arizona, Jour. Geol., vol. 18, 1910, pp. 742–763.
    [2] L. F. Noble, Contributions to the Geology of the Grand Canyon, Arizona, The Geology of the Shinumo Area, Am. Jour. Sci., vol. 29, 1910, pp. 374–380.

everywhere accordant with the rock structure.[1] These facts must be thoroughly appreciated because, while the plateau as a whole has been base-leveled, the remnants of the ancient base-leveled surface total a very small area, and in general are preserved only beneath lava caps near the summits of mesas.

THIRD EROSION CYCLE

The third period of faulting and of regional uplift inaugurated a third cycle of erosion, and, because the most striking effects of erosion in this cycle are the great canyons of the region, it has been called the "Canyon Cycle of Erosion." During this cycle a certain amount of plateau stripping has taken place and the cliff profiles have been refreshed, although the chief result has been the development of the profound canyons of the great Colorado and its principal tributaries.

During the canyon cycle of erosion there occurred a third period of volcanic activity characterized by eruptions of basalt from many small volcanic cones. The later eruptions took place, as a rule, before or during the period of glaciation, since they have been glacially modified, although a few cones and flows are of very recent geologic age. The later basaltic flows may be distinguished from the earlier flows capping the surface of the peneplain by their freshness, their undissected condition, and the absence of those displacements which affected the earlier volcanic flows.[2]

With these historical events in mind we may now turn to some features of the existing topography of the Colorado Canyon that require

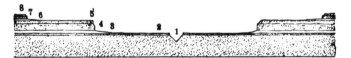

Fig. 80. — Cross-profile of the Grand Canyon of the Colorado River. Scale, one inch = 15,000 feet. 1, inner gorge; 2, 5, 7, sandstone; 4, 8, limestone; 3, 6, shale. (Gilbert and Brigham.)

more detailed study. The main section of the canyon is known as the Grand Canyon. Above the Grand Canyon is the Marble Canyon (cut into Carboniferous limestone) a feature of great magnitude but dwarfed by the adjacent highly diversified Kaibab division of the Grand Canyon, more than a mile deep (6000 feet). The length of the Marble Canyon is 65 miles, that of the Grand Canyon about 125 miles.

The two main divisions of the Grand Canyon are the Kaibab and the Kanab, which have certain topographic contrasts of importance. In

[1] L. F. Noble, Contributions to the Geology of the Grand Canyon, Arizona, The Geology of the Shinumo Area, Am. Jour. Sci., vol. 29, 1910, pp. 374–380.

[2] H. H. Robinson, Tertiary Peneplain of the Plateau District and Adjacent Country in Arizona and New Mexico, Am. Jour. Sci., vol. 24, 1907, pp. 110–112.

the Kaibab division of the canyon the descent of the wall is unusually abrupt over the entire series of sedimentary rock. The only well-defined terrace is that developed upon the summit of the basal sandstone of the Tonto group near the bottom of the canyon — a terrace known as the Tonto platform. Below the platform is an inner gorge formed upon the basement schists (Algonkian). The platform averages a mile wide on either side of the canyon and is well enough defined to make travel upon it possible. In the Kanab division to the west the lower terrace has disappeared and a different terrace has been developed upon the summit of the Red Wall limestone about a thousand feet below the level of the canyon rim. Through it as through the Tonto platform is cut a deep narrow inner gorge. This platform or terrace is known as the Esplanade; it averages two miles in width on either side of the canyon.

Still further differences may be seen in the aspect of these two sections of the canyon. The topography of the Kaibab division exhibits much greater dissection than that of the Kanab. The former is diversified by great amphitheaters thronged with buttes and outliers and trenched by a multitude of tributary gorges. In the Kanab division a very simple topography is found. There is a fairly regular broad outer canyon in which is cut an inner gorge, and the fantastic scenery of the Kaibab division is here wholly absent.[1]

CLIMATIC FEATURES AND VEGETATION

The distribution of temperature and rainfall according to relief or elevation is brought out strikingly in the plateau province. An inch

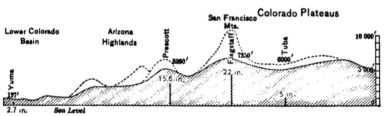

Fig. 81. — Topographic profile in relation to rainfall distribution from southwest to northeast across the three physiographic provinces of Arizona.

of rainfall is the normal increase for every 500 feet of altitude on the border of the plateaus; about half this increase in altitude is required for an increase of an inch of rainfall within the border of the plateaus. Yet most of the plateaus are dry, almost as dry as the plains of southwestern Arizona at half the altitude above sea level. The dryness is explained chiefly by the effect of the abrupt southwestern and western border in draining the passing winds of water vapor.

[1] L. F. Noble, loc. cit., p. 375.

Below an altitude of 7000 feet the rainfall of the Colorado Plateaus is probably not more than 8 inches a year. With increase of altitude above that figure there is increasing rainfall and the highest of the High Plateaus receive the equivalent of about 24 inches of rain, though a large part of it occurs in the winter months and is mostly in the form of snow.[1] They are favored however with a summer maximum in July and August. Indeed Dutton estimates their summit rainfall at 30 inches.[2] The High Plateaus are therefore not arid; they are the first prominent topographic barrier which the winds strike east of the Sierra Nevada.

The natural vegetation of the Rio Grande Valley on the southeastern border of the plateau country is limited to scanty grass, cottonwoods, and willows on the river bottoms, and to cacti and other desert forms on the higher slopes. At still greater elevations the junipers begin to come in, at first as gnarled stunted growths, then in better form, and on the slopes of the volcanic mountains, as Mount Taylor and the San Francisco Mountains, and on the summits and to some extent on the slopes of the higher mesas such as Mogollon Mesa, Black Mesa, and the Zuñi Plateau are thousands of square miles of magnificent forests of yellow pine and spruce.[3]

This is the usual succession of vegetation in the plateau country. The altitudinal range of timber species is sometimes not clearly defined on account of diversity of climatic conditions and exposure. At the headwaters of the East Clearwater, a tributary of the Little Colorado, the shady northward facing canyon slopes and walls are timbered with red fir and white fir, while the drier southward facing slopes are covered entirely with yellow pine.[4] The most valuable tree of the region is the yellow pine; the sugar pine is found only on the southern plateaus and is rarely of commercial size; the piñon pine and the pitch pine are common but not valuable growths and usually occur on talus slopes. Engelmann spruce occupies the highest elevations and is the only timber above 11,000 feet, attaining its best growth at 10,000 feet and disappearing at 11,500 feet. The great height at which it grows makes it difficult to secure; and to the altitude is added the difficulty of the

[1] A. H. Thompson, in Powell's Lands of the Arid Region, U. S. Geog. and Geol. Surv. of the Rocky Mountain Region, 1879, p. 151.

[2] C. E. Dutton, Geology of the High Plateaus of Utah, U. S. Geog. and Geol. Surv., 1889, p. 41.

[3] C. E. Dutton, Mount Taylor and the Zuñi Plateau, 6th Ann. Rept. U. S. Geol. Surv., 1884–1885, p. 125.

[4] F. G. Plummer, Forest Conditions in the Black Mesa Forest Reserve, Arizona, Prof. Paper U. S. Geol. Surv. No. 23, 1904, p. 16.

broken canyoned country between its plateau and mountain home and the valleys where the people live.[1]

The growth of grass is in most places scant, and at lower elevations diminishes in quantity and disappears in the lower and more desert country except where springs occur. The grasses grow characteristically in bunches and are thus in part protected among themselves from wind and blown sand. They have large strong stems and are not easily broken down by the infrequent rains and snows. They cure on the stalk and are highly nutritious, furnishing winter pasturage of great value to stockmen.[2]

The range of climate between the summits of the plateaus and the lower desert country of the canyon terraces and canyon bottoms is great indeed. The high precipitation of the Aquarius Plateau, for example, 25 to 30 inches, is chiefly in the form of snow which accumulates in the forest to a great depth. Settlers find it very difficult to live at an altitude over 7000 feet; their farms are usually found below that level where the climate is hot, arid, or semi-arid, and where irrigation is a necessity. From the cool, lake-besprinkled forests and meadows of the higher plateaus one looks down upon a country of extreme heat and dryness. The range of climate between the two situations is as great as one may find in most regions only by traveling through a considerable number of degrees of latitude, — a common range, however, in the great intermontane country between the Rockies and the Pacific Mountains.

The range in climate between the Kaibab and the bottom of the canyon, for example, is as great as the climatic range between the mountains of Colorado and the Mohave desert. The winters on the Kaibab Plateau are extremely severe and from November until April the snow lies deep in the woods, often accumulating to a depth of 10 feet. Even in midsummer the nights are chilly and the days are cool. The winters in the depth of the canyon are mild, freezing temperatures are rare, and snow rarely falls below the level of the Esplanade (4500 feet), while snow never falls on the Tonto platform (2100 feet).[3]

In contrast to the cool summer days of the plateau is the intense heat of the entire canyon below the Red Wall. The bare rocks become so hot as to burn the hand, and by nightfall a wind like a furnace blast

[1] J. W. Powell, Lands of the Arid Region of the United States, U. S. Geog. and Geol. Surv., 1879, pp. 98–103. See these pages for a more extended description of the tree species of the plateau province, their habitats and relative commercial importance at the time of Powell's surveys.

[2] J. W. Powell, Lands of the Arid Region of the United States, U. S. Geog. and Geol. Surv. of the Rocky Mountain Region, 1879, p. 110.

[3] Nokle, loc. cit.

escapes through the canyon. But the heat is not enervating, for the relative humidity is very low and moisture is rapidly evaporated from the body. The bottom of the canyon is decidedly arid, for much rainfall evaporates before it reaches the great depths. The Coconino Plateau on the south side of the canyon has a lesser altitude than the Kaibab Plateau, and while the latter is decidedly moist the former is semi-arid. Often no rain falls for a month at a time. In both cases there are distinct summer and winter maxima. Powell Plateau on the southwestern border of the Kaibab has a high eastern portion, and a low western portion, and from end to end there is a transition in climatic conditions similar to the transition that is experienced in passing from the Kaibab to the Coconino Plateau. In winter it is a resort for game and wild horses driven out of the Kaibab by the snow. The higher eastern end has an abundant rainfall; the lower western end is semi-arid.

One of the finest forests of the whole region is found on the high and better-watered Kaibab. The trees of the Kaibab forest are mostly yellow pine, but at the higher elevations spruce also is common. Pines are found on the sunny slopes of the ravines, spruce on the shady slopes, and both grow only scatteringly upon the valley bottoms. These and the aspens are the three principal genera, but about the borders of the plateau are patches and scattered individuals of cedar (*Juniperus occidentalis*), mountain mahogany (*Cercocaspus ledifolius*), and piñon.[1]

The influence of elevation upon vegetation through temperature and rainfall is admirably shown in the Grand Canyon, where it has been observed by Noble, from whose excellent descriptions the following paragraphs are taken[2]:

"The surface of the Kaibab Plateau is covered with a magnificent open forest of yellow pine; the trees grow large and far apart and the ground is free from undergrowth, giving its surface the aspect of a great park; Engelmann spruces grow on the north slopes of the washes, and cottonwoods, aspens, and scrub oaks in their bottoms; a minor flora of flowering plants, exceedingly rich in species, covers the floor of the forest. The flora of the plateau surface or the south rim of the canyon differs completely from that of the Kaibab; it is covered with a forest of gnarled and stunted trees of juniper and piñon, with here and there a buckbrush bush; the trees never form thickets, but grow wide apart; while the open stretches are covered with sagebrush and mormon tea, with occasional cactus, mescal, and plants of the century family. This difference between the floras of the north and south rim is due to the differences in precipitation and temperature, which vary directly with the altitude. For this reason the floras of the plateaus furnish an almost unfailing index of the elevation. This is beautifully shown on the southwestward-sloping surface of Powell Plateau, the whole eastern half of which lies at an elevation of from 7000 to 7500 feet and is covered with the open pine forest characteristic of the Kaibab. At about 7000 feet the character of the flora changes, and passes into the

[1] C. E. Dutton, Tertiary History of the Grand Canyon District, Mon. U. S. Geol. Surv., vol. 2, 1882, p. 132.

[2] L. F. Noble, Contributions to the Geology of the Grand Canyon, Arizona (The Geology of the Shinumo Area), Amer. Jour. Sci., vol. 29, 1910, pp. 374-380.

gnarled and stunted forests of juniper and piñon characteristic of the southern plateau across the canyon.

"Within the canyon itself the variation in the flora is just as great, and is again an index of the elevation.

"The flora of the Esplanade platform, a thousand feet below the south rim, consists of stunted bushes of juniper and piñon, with greasewood as the ground bush in place of the sagebrush of the Coconino Plateau. The cactus, mescal, and plants of the century family are present in greater abundance than on the plateau, but in less abundance and in more stunted development than in the bottom of the canyon. This is due to the fact that the Esplanade level is within reach of the winter snows and frosts.

"The flora of the Tonto platform, three thousand feet below the south rim, and of all the interior of the canyon below the Red Wall, is the flora of a hot and arid desert in its most characteristic form. The dominant plants are the greasewood bush, the mormon tea, and the cactus. The mescal and the plants of the century family here attain their greatest development and size. The cacti are particularly rich in species. Every plant in the flora is either prickly or aromatic; leaf surfaces are reduced to a minimum; devices for storing water attain the greatest perfection; and the dominant color is a somber gray. The somber colors and the reduction of leaf surface are apt to deceive the observer, both in regard to the richness of the flora in species and the abundance of plant life, which is far greater than one would suspect. The only tree is the screw-mesquite, which grows in the beds of those washes that contain living or intermittent streams.

"The vegetation in the bottoms of those canyons of the north side in the Shinumo Amphitheater which contain living streams is beautiful beyond description, and in refreshing contrast to the desert flora of the Tonto platform. Tall cottonwoods grow in the lower canyons; the walls are hung with maidenhair fern in the shady places; and willow thickets border the stream. Grass grows on the banks where there is soil. Higher up in the canyons, oaks, maples, and other deciduous trees come in, and often beds of tall rushes. The most characteristic bush of these upper north-side canyons is the manzanita, which does not grow on the south side of the Grand Canyon."

In the northeastern or Grand River district there is a characteristic change in vegetation on ascending from the valley or canyon floors to the plateau summits. In the low valleys and on the dry ridges sagebrush occurs; in the moist valleys are found willow, buffalo berry, service berry, and along the larger streams cottonwood. The quaking aspen requires more water and is found only above 7500 feet, on the plateau summits or in the vicinity of springs or on cool and moist slopes. All the low bluffs and ridges support some piñon and juniper of low height growth, yellow pine occurs infrequently, and what is locally called white pine (*Abies engelmanni*) is found only in some of the gulches and ravines leading down from the summit of the Book Cliffs.[1]

MOUNTAINS OF THE PLATEAU PROVINCE

HENRY MOUNTAINS

The mountains of igneous origin that occur locally throughout the plateau country have been referred to in a number of preceding paragraphs in this chapter. Some of the most important features of the

[1] F. V. Hayden, U. S. Geog. and Geol. Surv. of the Terr., 1876, pp. 68–69.

Fig. 82. — Relief map of the Henry Mountains, a group of eroded laccolithic domes in the plateau country, southeastern Utah. (Gilbert, U. S. Geol. Surv.)

larger groups deserve a word of detailed explanation. Our first consideration will be the Henry Mountains, which, with the San Francisco and Mount Taylor groups, have been studied in more detail than others of their kind.

The Henry Mountains are in southern Utah and are on the right bank of the Colorado River between the Dirty Devil and Escalante tributaries. They are a group of five individual mountains separated by low passes. Although they are in an arid region their height (7000 to 11,000 feet) is such as to cause them to have a rainfall sufficient to support forests. Mount Ellen is the highest of the five mountains and has a continuous crest line 2 miles long, with radiating spurs and bordering foothills.

The Henry Mountains were formed by the intrusion of great masses of molten rock into surface beds in the form of a laccolith. The overarching strata that were lifted up as a consequence of the intrusion have been so extensively removed by erosion that there is now revealed the heart of the laccolith, and the displaced beds are exposed on the flanks and borders of the mountains for the most part. There is considerable diversity in the degree of erosion; some of the mountains are still largely covered on one or more sides by overlapping sedimentary rocks.

The Henry Mountains in southern Utah stand upon a desert plain having a mean altitude of 5500 feet. A large part of the rain that falls in the region is distributed over the mountains, a fact that is reflected in the distribution of the springs and the vegetation. On the surrounding plain the vegetation is extremely meager, grasses and shrubs and in favored localities the dwarf cedar (*Juniperus occidentalis*). The highest peak of the group is Mount Ellen, 11,250 feet, which bears cedar about its base, piñon a little higher up, then yellow pine, spruce, fir, and aspen. The summits are naked; the upper untimbered slopes are covered with a luxuriant growth of grasses and herbs. Of the forest growths noted the pines grow in a scattering manner, but the cedars and the firs are in dense groves.[1] Among the other four mountains of the five comprising the group, Mount Pennell (11,150 feet) reaches above timber line and has a forest growth similar to that on Mount Ellen except that the timber reaches almost to the summit; on Mount Hillers (10,500) the timber reaches to the principal summit, but is less dense than on the other higher mountains; Mount Ellsworth is so low (8000) that it bears none of the forest trees that grow on the others, its vegetation is cedar and piñon right to the summit, and the grasses are less rank and grow in scattered bunches; Mount Holmes (7775) has a similar growth except that a few spruce trees grow high up on the northern flank.

Among the extinct volcanoes and associated lava flows Mount Taylor, San Francisco Mountain, Sierra Mogollon or the Mogollon Mesa, the Panguitch Lake Buttes of Utah and the Marcou Buttes of New Mexico are important members. These have all been formed by the extrusion

[1] G. K. Gilbert, Report of the Geology of the Henry Mountains, U. S. Geog. and Geol. Survey of the Rocky Mountain Region, 1877, p. 118.

of lava in one or another form, either as volcanic mountains or as lava flows. Like the laccolithic mountains they have no range forms or structures. Their features are all grouped about centers of structural disturbance, and radiating slopes and spurs and drainage lines are the rule. They have the utmost differences of position with respect to timber lines. Many do not reach above the dry timber line, some do not reach the cold timber line, a few extend above the cold timber line and have unforested summits. Those having a forest cover at all are encircled by a band of forest vegetation from the 7000 or 8000 foot level to the 11,000 foot level, which represents the upper limit of tree growth. In all of them there is a notable canting of the timber lines on passing from the warmer and drier southern slopes to the colder and moister northern slopes. The difference of level amounts in some instances to a thousand feet and is always several hundred feet, being lower on the northern slopes. This condition has been especially well described by Merriam.[1]

More recently Lowell has noted that the degree of canting increases with increasing altitude and appears to be a function of radiation and insolation as controlled by area. With increasing altitude there is decreasing area. The two sides of a cone will therefore show increasingly greater differences in the limiting elevations of the tree zones and a more marked canting of the vegetation belts and timber lines.[2]

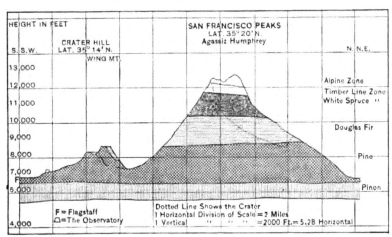

Fig. 83.—Timber zones on San Francisco Peaks, Ariz., showing increase in the degree of canting with increasing elevation. (Lowell.)

[1] C. H. Merriam, Results of a Biological Survey of the San Francisco Mountain Region in Arizona, North Am. Fauna, No. 3, 1890.
[2] Percival Lowell, The Plateau of the San Francisco Peaks in its Effects on Tree Life, Bull. Am. Geog. Soc., vol. 43, 1911, p. 380.

SAN FRANCISCO MOUNTAINS AND MOUNT TAYLOR

SAN FRANCISCO MOUNTAINS

The San Francisco Mountains are the center of a volcanic field in northern Arizona that includes seven large peaks and several hundred small peaks. Cones and lava flows together cover about 2200 square miles of country. San Francisco Mountain, the highest peak in the group, rises to 12,700 feet above the sea and has 5000 feet of relative altitude. Other prominent peaks are Kendrick Mountain, 10,500 feet, and Bill Williams, Sitgreaves, Elden, and O'Leary mountains, which do not exceed 9500 feet. The smaller volcanic cones are scattered about irregularly and have great variety of topographic detail. The plateau about the base of these mountains and throughout the volcanic field maintains a greater altitude than elsewhere, a fact due to (1) greater initial height of the locality before volcanic activity began and (2) the protective influence of the hard lavas which have preserved the surface from that denudation which later brought the surrounding plateau to a lower level.

San Francisco Mountain is a composite cone built up by the lavas and breccias of five distinct periods of eruption. The chief vents of these periods are near each other, so that the total effect has been to give the mountain a symmetrical outline. The principal lavas are dacite, rhyolite, and andesite.

Twelve principal watercourses drain radially outward from the higher parts of the area. They are irregular in the lava field, but become more direct beyond the borders of the field, where deep canyons generally occur. After a cloud-burst or heavy shower water runs for a few hours in the canyons or washes within the limits of the storm area and then the channels become dry again. But one stream, Oak Creek, runs throughout the year; its more even plow is due to several large forest springs at its headwaters. In an area of 1000 square miles about San Francisco Mountain only about 25 springs and water pockets or "tanks" occur, and at least two-thirds of these are situated near the mountain, so that the surrounding country is extremely dry and difficult to traverse. The "tanks" and lakes which usually occur at the heads of the washes have a most variable supply as regards both quantity and quality. The lakes are due to the damming of the watercourses by lava flows. Some of them have become dry by the cutting down of their outlets and the accumulation of sediments; their floors are now grass-covered and picturesque glades.

The soils of the San Francisco volcanic field vary from "adobe" soils of the floors of temporary lakes and ponds to the pervious cinder and scoriaceous soils of the slopes of volcanic cones and ridges. Between these two extremes are gravelly loams of several varieties moderately pervious to water and best adapted of all the soils to the growth of forests. The coarse cinder soils lose water so rapidly that they are extremely sterile, while the adobe soils crack open when dry and swell when wet and are not adapted to forest requirements.[1]

The San Francisco Mountains are encircled by barren plateau country, but they are themselves covered with a beautiful forest of juniper, pine,

Fig. 84. —'Mesa forest of western yellow pine in'the Mogollon Mountains near Iron Creek, Gila National Forest, New Mexico. (Photograph by DeForest.)

and spruce. It surrounds the base of the mountains and stretches westward to the borders of the escarpment of the Colorado Plateaus and down it nearly to its foot. The forest accompanies the escarpment southwestward to the boundary of Arizona and extends in a northwest-southeast direction for over 200 miles. Its greatest breadth is opposite the San Francisco Mountains, where it is nearly 50 miles wide. Elsewhere its breadth is from 12 to 25 miles, thus giving it an area of from 12,000,000 to 23,000,000 acres. It is from all points of view the finest forest in the Southwest and is composed of an almost pure growth of yellow pine except in the higher altitudes of the San Francisco Moun-

[1] Leiberg, Rixon, Dodwell, and Plummer, Forest Conditions in the San Francisco Mountains Forest Reserve, Arizona, Prof. Paper U. S. Geol. Surv. No. 22, 1904, p. 15.

tains. It is an open forest with little undergrowth and the trees are of good size for lumber purposes.[1]

The yellow-pine zone extends from 7000 to 8200 feet. Above it and between elevations of 8500 and 9800 feet and occasionally to 10,000 feet is the transition forest type, composed principally of red fir, white fir, aspen, and a few Engelmann spruce. The subalpine forest type extends from 9800 to 12,400 feet and is found only on Kendrick Mountain and San Francisco Peaks. It is composed of aspen, Engelmann spruce, etc. Above the subalpine forest is a treeless belt which is represented only on and near the summit of San Francisco Peaks, Fig. 83.[2]

Below the yellow-pine zone and between 5700 and 6200 feet is a woodland belt where both climate and soil are very dry. Juniper and piñon are the chief species. They stand as a transition type of vegetation between the desert sagebrush and the true forest of the next higher belt.

MOUNT TAYLOR

Northeast of the Zuñi Plateau is Mount Taylor one of the most prominent volcanic elevations of the southeastern section of the province. The central mass and associated lava flows constitute the summit of a mesa whose base consists of sedimentary rock. The mesa is 47 miles long and has an extreme breadth of 23 miles, with an average altitude of about 8200 feet. Mount Taylor itself is 11,390 feet high. It is a lava cone principally and was a central pipe or vent from which the surrounding flows were derived. It is now very much eroded, though it still maintains a roughly amphitheatral form. It is heavily timbered and deeply covered with soil and talus.

In the Mount Taylor district and east of the main volcano are many lesser volcanic elevations now greatly denuded so that only the central core remains. They stand from 800 to 1500 feet above the general level of the plain as steep-sided sharp-crested buttes of great interest as the roots of once active volcanoes now all but swept away.[3]

Among the other areas of higher relief in the Colorado Plateaus is Mogollon Mesa, a timbered plateau 7000 feet above the sea and extending southward from the San Francisco Mountains (Plate II). Limestones and shales form the basement rock upon which the basaltic lavas that

---

[1] Henry Gannett, 19th Ann. Rept. U. S. Geol. Surv., 1897-98, p. 47.

[2] Leiberg, Rixon, Dodwell, and Plummer, Forest Conditions in the San Francisco Mountains Forest Reserve, Arizona, Prof. Paper U. S. Geol. Surv. No. 22, 1904, pp. 18-19.

[3] C. E. Dutton, Mount Taylor and the Zuñi Plateau, 6th Ann. Rept. U. S. Geol. Surv., 1884-85, especially Panorama in the Valley of the Puerco, Plate 21, p. 171; also D. W. Johnson, Volcanic Necks of the Mount Taylor Region, New Mexico, Bull. Geol. Soc. Am., vol. 18, 1907, pp. 303-324.

constitute the mass of the mountains have been distributed. These strata dip northward and their edges are exposed on the southern border of the district where the lava cap has been most extensively frayed by dissection, Fig. 83. Here the main margin of the elevation consists of long lava-capped promontories of the plateau fronted by a series of detached remnants in the form of mesas and buttes. The mountains have been extensively dissected and present a bold and varied relief in contrast to the more regular margin of the mesa above which the mountains rise as from a pedestal. Thrifty forests of yellow pine cover both the mesa and the higher slopes and owe their existence to the favorable temperature and rainfall induced by the elevated position of the mountains on the windward border of the plateau country.[1]

[1] Wheeler Surveys, Geology, vol. 3, 1875, pp. 217-587.

CHAPTER XVIII

ROCKY MOUNTAINS. I

NORTHERN ROCKIES

BOUNDARIES AND SUBDIVISIONS

BETWEEN the sensibly flat and only slightly dissected basalt plains of the Columbia River and Snake River regions and the almost equally flat Great Plains of Montana is a broad belt of wild, rugged, mountainous country composed of many northward- and northwestward-trending mountain ranges. Though several transcontinental railway lines cross the region and occasional valleys, as the Bitterroot, are rather thickly settled, the forested mountain slopes and the higher ranges have few trails and fewer roads, and are almost unknown except to the hunter and the explorer.[1]

The chief members of the northern part of the Rocky Mountain System[2] in the United States are the Lewis and Livingston ranges, which form the eastern and front members of the system in northern Montana; the Galton, Flathead, Purcell, Cabinet, and other ranges east of the continental divide; the central Bitterroot Mountains on the boundary between Montana and Idaho; and the Clearwater, Salmon River, Priest River, and Cœur d'Alene mountains on the west, Fig. 86. The close proximity of these ranges to each other, in general their similar alignment and their common participation in certain important events in the physiographic history of the region, are conditions which form an adequate basis for the common association of the separate units under the group name of the northern Rockies of the United States; but there are wide differences in local detail, in the quality of the mountain slopes, the length and disposition of transverse valleys, and the degree of dissection which the individual ranges have suffered. It therefore becomes necessary to distinguish the essential elements of form and the limits of each unit in the system.

[1] F. C. Calkins, A Geological Reconnaissance in Northern Idaho and Northwestern Montana, Bull. U. S. Geol. Surv. No. 384, 1909, p. 12.

[2] The United States Geographic Board makes the term Rocky Mountain System embrace the whole of the mountainous region between the Rio Grande and the 49th parallel, specifically the ranges of western Texas, New Mexico, Colorado, Wyoming, Idaho, and Montana. In this book the Trans-Pecos Highlands are set off as a separate province. (See p. 387.)

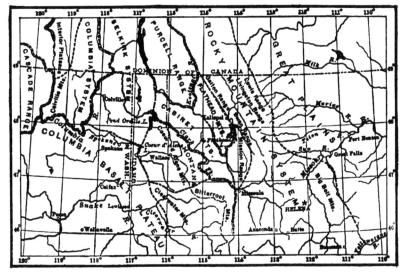

Fig. 85. — Mountain systems and ranges and intermontane trenches, northern Rockies.
(Ransome, U. S. Geol. Surv.)

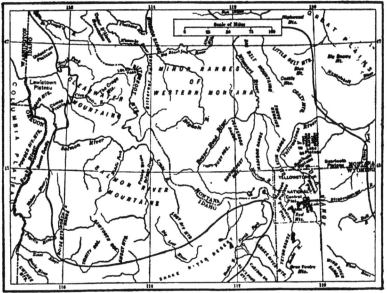

Fig. 86. — Location map of a part of the Northern Rockies.

In naming and describing the different mountain ranges of the northern Rockies that cross the international boundary line it has been proposed that a double and purely topographic principle be employed, the principle of the continuity of crest lines and the positions of the major erosion valleys.[1] Erosion troughs or intermontane trenches are here the natural lines of demarcation between the different ranges, for they have a highly exceptional and remarkable development on a great scale.

INTERMONTANE TRENCHES

In respect of size the major valleys of the region are often quite out of proportion to the streams that now drain them; the longest depression in the whole system is occupied by relatively small streams, the headwaters of the Kootenai, Columbia, Frazer, and others. They have steep though seldom precipitous walls, and rather flat, lake-dotted floors. They are called intermontane or valley trenches. It is certain that they had their direction determined in many cases primarily by fault lines or zones, so that some of them, for example Kootenai Valley, mark the boundary between different rock formations. They have probably been brought to their present form chiefly by the long-continued erosion of valley glaciers powerful enough to override the divides between the heads of the larger streams and to degrade them to a common level.[2]

In this connection it is in point to indicate that valley glaciers of great size supplied by many headwater tributaries are much more powerful agents of erosion in effecting topographic discordances such as appear in the hanging valleys and the mountain slopes adjacent to steep-sided U trenches, and in opening up valleys such as these trenches are to-day, than are continental ice sheets. A continental ice sheet is spread over the entire surface, and though it may erode to some extent differentially there is no such concentration of ice action as is the case in a region of heavy alpine glaciation. The valleys of

[1] R. A. Daly, The Nomenclature of the North American Cordillera between the 47th and 53d Parallels of Latitude, Geog. Jour., vol. 27, 1906, pp. 586–606. This excellent paper outlines the difficulty of understanding the mountain nomenclature of the Rockies, and on pages 589 and 590 summarizes the various names applied by prominent authorities to the Rocky Mountains. In all at least 26 different names have been employed. It is suggested that greater differentiation be aimed at in designating the various members of the Rocky Mountain system. It is also noted that the authorities variously designate the terminal points of the different mountain ranges (pp. 592–593). The nomenclature is still further confused owing to the fact that some atlases and map sheets give different titles to the same range.

[2] F. C. Calkins, A Geological Reconnaissance in Northern Idaho and Northwestern Montana, Bull. U. S. Geol. Surv. No. 384, 1909, p. 32.

Lake Chelan, the Okanogan, and the Cœur d'Alene are illustrations. In all three instances there are extensive catchment basins which fed prodigious glaciers.

Among the eleven longitudinal valleys or intermontane trenches which serve as convenient lateral boundaries for the members of the Pacific Cordillera are four of first rank; three of these occur in the northern Rockies, and lie west of the front ranges, the Purcell range and the Selkirks respectively. Each is designated a "trench," and as so used the

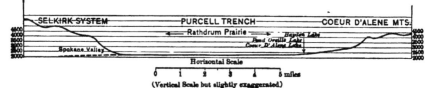

Fig. 87. — East-west section, showing flat floor and steep bordering slopes of a typical intermontane trench. (Data from Cœur d'Alene quadrangle, U. S. Geol. Surv.)

term means "a long, narrow, intermontane depression occupied by two or more streams (whether expanded into lakes or not) alternately draining the depression in opposite directions."[1] The easternmost and much the longest of the series of trenches is the wide valley occupied at the international boundary by the Kootenai River.[2] It extends from the southern end of Cœur d'Alene Lake northward to Liard River in British Columbia, about 900 miles, and throughout it has the form of a narrow and remarkably straight depression lying between the easternmost member of the Rocky Mountains and the rest of the system. This depression is unique among all the mountain valleys of the earth for its remarkable persistence. It is not drained by a single great river, but is occupied in turn by the headwaters of the Flathead, Kootenai, Columbia, Canoe, Frazer, Parsnip, Finlay, and Kachika rivers. The name Rocky Mountain trench is now applied to it.

The major valley next in order to the west is occupied at the boundary by that portion of the Kootenai River that returns into Canada after rounding the great bend at Jennings, Montana. This trench begins on the south near Bonners Ferry, Idaho, and is occupied north of Kootenai Lake by the Duncan River. In line with the Duncan River Valley is the 50-mile trough occupied by Beaver River which enters the Columbia at the Canadian Pacific Railroad crossing. The length of this topographic unit, called the Purcell trench, is about 200 miles.

The third major trench is known as the Selkirk trench or the Selkirk

[1] Daly, loc. cit., p. 596.
[2] F. L. Ransome, Geology and Ore Deposits of Cœur d'Alene District, Idaho, Prof. Paper U. S. Geol. Surv. No. 62, 1909, p. 13.

Valley, and is occupied southward by the Columbia River. Its northern extremity is near the 52d parallel, where it is confluent with the Rocky Mountain trench. The southern end of the trench is about 60 miles south of the 49th parallel, where the Columbia River enters the great basalt plateau of Washington.

Besides these three first-rank valleys or trenches in the northern Rockies are a number of second-rank longitudinal trenches. Waterton River, Flathead River, and Zigzag River occupy parallel valleys that still further divide the northern Rockies into the Lewis, Livingston, MacDonald, and Galton ranges. Cross trenches or valleys are employed in indicating the minor subdivisions of the principal mountain systems and ranges, as the Slocan Mountains are separated from the rest of the Selkirk System by a depression occupied by Slocan River, Slocan Lake, and the valley of a creek whose mouth is at Nakusp or Arrow Lake.[1] In a similar way cross valleys divide the Colville Mountains, Pend Oreille Mountains, etc., from the larger ranges.

GEOLOGIC FEATURES

The detailed topographic qualities of the northern Rockies as discussed in succeeding paragraphs may be better understood by a brief consideration of certain fundamental geologic conditions. From Kootenai River or Purcell trench westward to the Columbia there is a much disturbed, highly folded, rock complex in which the metamorphism of the sediments generally diminishes in intensity westwardly.[2] To this metamorphosed complex the rocks east of the Purcell trench are in striking contrast. They consist of a thick series of arenaceous sediments (pre-Cambrian) which show no pronounced regional metamorphism. The essential structural features are open folding and extensive faulting. They are known to extend from the Belt Mountains westward to the Purcell trench. Some of the ranges (Purcell, Cabinet, Cœur d'Alene, Livingston, and Lewis) are composed almost entirely of this group of rocks. In central Idaho the same strata are invaded, displaced, and probably for the most part cut off by a great granite batholith.[3] The most striking structural feature of the western part of the sediments east of the Purcell trench is of course the existence of great faults of northwest trend that determine the courses of the trenches and rivers and block out the individual mountain ranges.

[1] Daly, loc. cit., p. 599.

[2] Ransome and Calkins, Geology and Ore Deposits of the Cœur d'Alene District, Idaho, Prof. Paper U. S. Geol. Surv. No. 62, 1908, p. 16.

[3] W. Lindgren, A Geological Reconnaissance across the Bitterroot Range and Clearwater Mountains in Montana and Idaho, Prof. Paper U. S. Geol. Surv. No. 27, 1904, p. 16.

Fig. 88. — Cœur d'Alene Mountains, looking from above Wardner. Shows the mature dissection of a plateau-like uplift. (Ransome and Calkins, U. S. Geol. Surv.)

Fig. 89. — Cœur d'Alene Mountains, looking toward the crest of the range, from valley of South Fork of Cœur d'Alene River. Shows characteristic equality of ridge lines. (U. S. Geol. Surv.)

CŒUR D'ALENE RANGE

The Cœur d'Alene range, rudely triangular in outline, extends from Lake Pend Oreille on the northwest, southeastward to Lolo Pass (lat. 46° 36′ on the continental divide between Idaho and Montana).[1] Its southern boundary is vague but should probably be considered as corresponding to the divide on the northern border of the Clearwater drainage basin, Fig. 86. Its western margin adjoins the Columbia Plateaus; its eastern boundary is constituted by the valleys of Clark Fork, Flathead River, and Jocko Creek. The Cœur d'Alene range, Fig. 89, appears as a rather monotonous expanse of ridges nearly equal in height and with somewhat level crests that do not bear prominent summits.[2] Its broad aspect is that of a maturely dissected plateau, whose general level in its central and highest part is a little above 6000 feet. The rough equality of summit levels becomes less marked toward the west, dissection having there progressed so far that the original regularity of level has now been almost lost. It is tentatively concluded that this topographic uniformity depends upon former base-leveling, but much further work is required for a complete demonstration.

The Cœur d'Alene region had been folded and faulted into essentially its present structure and its topography had been developed by probable base-leveling, uplift, and extensive and deep dissection to essentially the present conditions by the time the great Miocene lava flood occurred, for great quantities of basalt flowed up the existing valleys. This fact for example explains Cœur d'Alene Lake, which occupies a large valley that was partly filled with basalt and afterward recut by the original river, thus leaving terraces of basalt on the older slopes. Later a deposit of Pleistocene gravels dammed the valley on the north, originated the lake, and gave it approximately its present outline, backing up the waters of Cœur d'Alene and St. Joseph rivers.

PRIEST RIVER RANGE

The Priest River range south of the international boundary and east of the Pend Oreille range has been deeply sculptured by glaciers and streams, so that canyons, cliff-bordered cirques, and narrow ridge crests are common. The rocks of the range consist principally of highly fissured granite and syenite; a certain amount of slate and gneiss occurs at the southern end. The average altitude of the range is 5000 to 6000 feet, with a number of elevations attaining 8000 feet. The irregularly fissured condition of the granites results in the better retention of the precipitation than is the case in the schistose rocks of the Pend Oreille range. The latter are either water-tight or else afford too rapid drainage, depending on the dip of the planes of schistosity. About 91 % of the total forest in the Priest River National Forest is white pine and

[1] W. Lindgren, Prof. Paper U. S. Geol. Surv. No. 27, 1904, p. 23.
[2] F. C. Calkins, A Geological Reconnaissance in Northern Idaho and Northwestern Montana, Bull. U. S. Geol. Surv. No. 384, 1909, p. 14.

tamarack; yellow pine is found below 3500 feet, white pine between 2400 and 4800 feet (best development between 2800 and 3500 feet), and the subalpine fir, of little economic importance, grows above 4800 feet:[1]

CABINET RANGE

This range extends from Bonners Ferry southeastward to the junction of Jocko Creek and Flathead River. Its western border is definitely marked by the Purcell trench and its southwestern border by the nearly straight valley occupied chiefly by Clark Fork, Columbia River, and

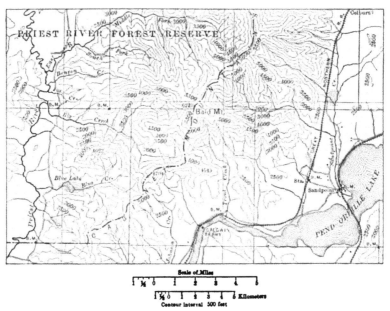

Fig. 90. — Southern end of the Cabinet Mountains, Idaho. The depression on the right is part of a typical intermontane trench. (Sandpoint quadrangle, U. S. Geol. Surv.)

Flathead River. The eastern border is constituted by the valley of Little Bitterroot River, a short section of Flathead River, and Fishers Creek.

As a whole the Cabinet range is somewhat loftier than the Cœur d'Alene range and far more diversified in character. The eastern part is a dissected plateau, but the quality of the dissection is markedly different from that in the Cœur d'Alene Mountains, for the differences in the hardness of the rock are so pronounced and the region has been so deeply scored by rivers and so intensely glaciated that its

[1] J. B. Leiberg, Priest River Forest Reserve, 19th Ann. Rept. U. S. Geol. Surv., pt. 5, 1897–98, pp. 218–224.

details of sculpture are highly picturesque. The eastern portion of the Cabinet range has a number of prominent mountain peaks and a deeply serrate sky line in striking contrast to the lower, even-crested ridges toward the west. Some of the peaks are composed of resistant quartzite and overlook many great steep-walled amphitheaters or glacial cirques developed upon their northern, eastern, and western slopes. In the northern cirque of Bear Peak (8500 to 9000 feet) a small glacier may be found to-day, a remnant of those larger Pleistocene glaciers that once flanked all the higher peaks of the region and gave rise to cirques and other glacial features.[1]

PURCELL RANGE

The Purcell range trends north-northwest and is for the most part bounded on the east by the Rocky Mountain trench and on the west by the Purcell trench; the rest of its boundary is defined by the valley of Kootenai River. The greater part of the range is in Canada; only the southern end projects into the United States and is embraced in the great bend of the Kootenai. The section south of the boundary is divided into three subdivisions by Mooyia (west) and Yaak (east) rivers. Five miles south of the international boundary the westernmost range is crossed by a remarkably flat-bottomed, low-grade valley or trench, one of the many of its kind in the region. The crests of the ridges are comparatively even in height and have no conspicuous peaks. The mountains occupying the interstream area between the Mooyia and Yaak rivers are for the most part of gentle profile and are heavily wooded except where they have been swept by forest fires. The main divide is a rocky ridge bearing Mount Ewing just south of the boundary, besides a group of jagged summits (7500) at the southern end of the chain.[2] The easternmost mountain ridge has the same general character as the main central ridge west of it.

Among the other mountain ranges of the boundary section of the northern Rockies are the Pend Oreille on the west and the Flathead and the rugged Mission ranges on the east. These mountains have not been explored to any extent and generalizations concerning them would be too broad to be of any practical value. The easternmost ranges of the region are the Lewis and Livingston ranges, which form the sharp western boundary of the Great Plains. These have been studied with more care than the ranges west of them and may be discussed in greater detail.

[1] Calkins and McDonald, A Geological Reconnaissance in Northern Idaho and Northwestern Montana, Bull. U. S. Geol. Surv. No. 384, 1909, p. 15.
[2] Idem, p. 16.

LEWIS, LIVINGSTON, AND GALTON MOUNTAINS

The Lewis and Livingston mountains in western Montana constitute the front ranges of the Rocky Mountains in that state and a part of the adjacent province of Alberta. They consist in the main of stratified rocks of Algonkian age; igneous rocks occur but sparingly. The

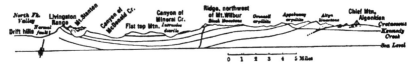

. Fig. 91. — Topographic and structural section across the front ranges in Montana. (Willis.)

strata are disposed in the form of a broad, northward or northwestward trending, well-defined syncline between two marginal and poorly defined anticlines. The northeastern margin — the Lewis Mountains — fronts the Great Plains; the southwestern margin — the Livingston Mountains — is in part eroded, and in part downfaulted by a normal fault (North Fork Valley). The valleys and ridges of both ranges are closely related to the syncline and anticlines thus defined.

The eastern border of the Lewis range is an overthrust fault; Algonkian strata have been moved northeastward over Cretaceous strata on the plane of the fault; the displacement on the thrust surface is not less than 7 miles, and the vertical movement is estimated at 3400 feet or more. It is to the upward movement on the fault plane that the Lewis Range owes its present elevation above the Great Plains, and, to a large degree, the abruptness of its eastern border.[1]

The eastern margin of the Lewis range is deeply sinuous and is marked off from the Great Plains by cliffs of prodigious size. The crest of the range is everywhere narrow and in many places is "a knife-edge of jagged rocks." The precipices are frequently more than 1000 feet high and in some instances have an altitude of 4500 feet, with a slope that is nowhere below 50°, Fig. 92. These cliffs are the walls of profound amphitheaters which enclose small mountain basins commonly occupied by lakes. The elevations of the highest summits range from 8500 to

---

[1] Bailey Willis, Stratigraphy and Structure, Lewis and Livingston Ranges, Montana, Bull. Geol. Soc. Am., vol. 13, 1902, pp. 307–308.

The eastern border of the Rocky Mountains near the common border of British Columbia and Alberta is also marked by the presence of a great overthrust fault, the continuation northward of the fault at the eastern border of the Lewis Mountains. It runs north by west as indicated on the map of part of Alberta and British Columbia by Dowling. (D. B. Dowling, Cretaceous Section in the Moose Mountain District of Southern Alberta, Bull. Geol. Soc. Am., vol. 17, 1906, pp. 295–302.)

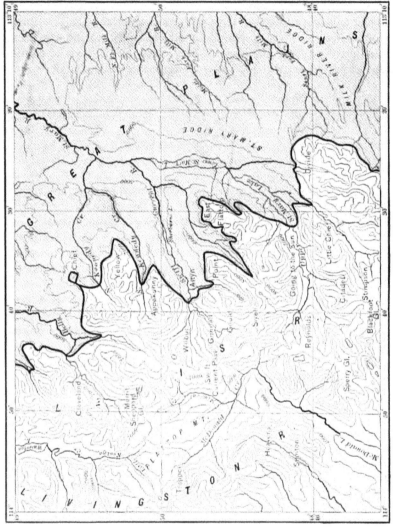

Fig. 92. — Map of Great Plains and front ranges, western Montana. Heavy black line indicates the outcrop of the Lewis thrust plane.

10,400 feet; the elevations of the wind gaps are from 5400 to 6000 feet.

The Livingston range extends from Mount Heavens north of Mc-Donald Lake northwestward to Mount Head in British Columbia. It lies west of the Lewis range in the United States, but the latter range ends near the international boundary and in Canada the front range is

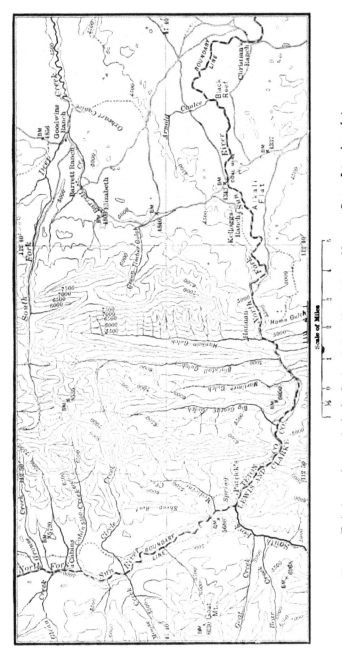

Fig. 93. — Hogback type of mountains border, Lewis and Clarke National Forest, Montana. Contour Interval, 500 feet. (Saypo quadrangle, U. S. Geol. Surv.)

the Livingston. The main continental divide follows the Livingston range for some distance southward of the boundary, descending to Flat Top Mountain and finally to the Lewis range, which it follows southward to latitude 46° 45′. Like the Lewis range the Livingston is often narrow and presents massive mountain peaks of pyramidal outline. Deep valleys diversify its western slope and contain long narrow lakes which vary in length from two to ten miles. The lakes are bordered by slopes of gravel or talus, except at their heads, where cliffs rise precipitously to the range summits. The western limit of the Livingston Mountains is definite, but it has the aspect of a bold mountain face rising from foothills rather than the almost sheer face of the eastern margin of the Lewis Mountains rising abruptly from the plains.

The Lewis and Livingston ranges are characterized by the dominant influence of structure on altitude. In the northern part of the Lewis range and in the Livingston range the greatest altitudes are in general related to the two anticlines; the master valleys are in the intervening syncline, and Flat Top Mountain is the former floor of a broad synclinal valley. A peneplain was formerly developed over the Great Plains and over the Galton range. On the soft rock of the plains it was well developed, but on the harder rocks of the Galton Mountains it was probably imperfect. In the Lewis range it is notable that each peak approaches in height that of its neighbors which stand along the strike in a similar structural position. However in a broad view and taking the Lewis and Livingston ranges as a unit, no general upper limit of heights common to widely distributed peaks may be discerned. If the base-leveled surface was ever in existence in the range, the extreme localized deformation of the mountains has so warped the ancient surface, and intense erosion by both water and ice has so completely dissected it, as to make its determination very difficult if not impossible.

It is truly remarkable how abruptly the well-developed and but little dissected peneplain of the Great Plains of Montana terminates at the foot of the front ranges. The line is almost as definite as a shore. The long-continued erosion which the peneplain represents must have affected the present mountainous area west of it, but was probably offset by repeated deformation terminating or culminating in the great overthrust to which the present mountain height and the deep dissection are largely due. In the Galton range the old surface may be safely inferred; the older features have not been so completely destroyed owing to the relatively small amount of local deformation and the less intense action of water and ice. For the Galton range, although bounded by

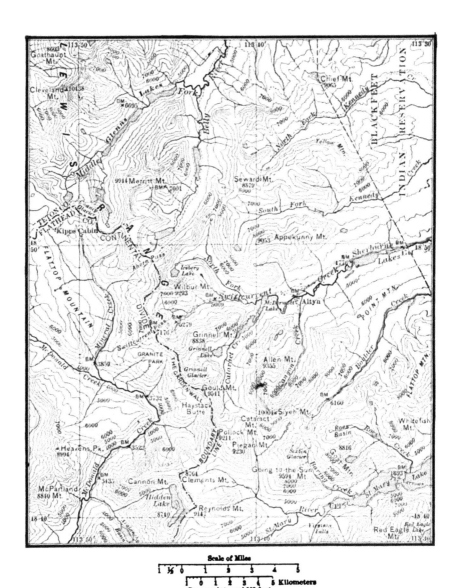

Scale of Miles

Kilometers

Contour interval 500 feet

Fig. 94. — Map of a part of the **Lewis Mountains**, western Montana, representing typical glacial features of the range.   (Chief Mountain quadrangle, U. S. Geol. Surv.)

structural limits, is internally but a simple uplifted block whose minor flexures or faults are not sufficiently pronounced to interrupt the ancient surface.

Some of the most remarkable erosion features produced by alpine glaciers are to be found in the Lewis range on the continental divide in western Montana. They are represented on the Chief Mountain quadrangle, Montana, Fig. 94, one of the most interesting topographic sheets yet published, not only because of the exceptional nature of the region but also because of the unusual faithfulness with which the topographer has represented the landscape. The strata of the Lewis Mountains are strongly jointed, a quality which is highly favorable to the plucking action of glacier ice; furthermore they lie nearly flat and boldly overlook the adjacent plains: a set of conditions exceedingly favorable for the development of glacial forms on an unusual scale.

Three major topographic features are apparent in the region: (1) numerous reversed slopes occupied by lakes and lakelets; (2) strikingly deep wall-like valley sides; and (3) huge amphitheatral valley heads in which a number of living glaciers are to be found. All these features are normally associated with the work of alpine glaciers whose former existence in large numbers has also been determined by the familiar phenomena of eroded and plucked rock surfaces, terminal and lateral moraines, and hanging tributary valleys. Former glaciers plowed down all the main valleys and built up morainic accumulations which at Blackfoot and Browning, p. 412, merge into the terminal deposits of one of the great continental ice sheets.

An examination of the valley systems in plan shows a remarkable degree of headward cutting on the part of the glacier ice. The amphitheatral valley heads or alcoves on opposite sides of a divide have in many instances cut back to the point where a knife-like ridge has been formed, as south of Gould Mountain and on the continental divide two miles south of Chaney glacier. The process has resulted in the formation of pyramidal mountains such as Going-to-the-Sun Mountain, Cataract Mountain, Little Chief Mountain, Heavens Peak, and others, or in the formation of skeleton mountains of irregular outline such as Almost-a-dog Mountain, Appekunny Mountain, and Merritt Mountain.

The manner in which such headward cutting has taken place has been well set forth by Johnson,[1] who has observed that in glaciated mountains the great curving bergschrund of the snow-field penetrates to the foot

[1] D. W. Johnson, The Profile of Maturity in Alpine Glacial Erosion, Jour. Geol., vol. 12, 1904, pp. 569–578.

Mount Gould, Lewis Range, Montana, looking southwest from South Fork of Swift Current. See Fig. 86 for
. Characteristic cliff of limestone overlooking argillite. The dark band is intrusive diorite. The valley head is
amphitheater developed along joint planes. It is 4670 feet from lake to summit. (Willis.)

of the precipitous rock slope constituting the wall of the amphitheater
enclosing the snow-filled cirque. He concludes that a causal relation determines the coincidence in the position of the bergschrund and
the foot of the cliff wall. The opening allows air to come into contact
with both ice and rock at the bottom of the crevasse. By day there is
thawing, by night freezing, and blocks of rock are wedged off and the
cirque wall riven. The bottom of the crevasse is therefore a "narrow
zone of relatively vigorous frost-weathering." The result is a sapping
of the foot of the cirque wall and its gradual steepening and retreat.

The continuance of this action steepens the slopes of the glacial
amphitheaters and pushes them back until the slopes on the open sides
of the divide meet. Thus a more or less rounded divide such as may
be found a few miles southwest of Point Mountain, at Flat Top Mountain, and other places, was destroyed, the mountain top was reduced to a
pinnacle or needle and the divide to a sharp crested arête. It is this
action which appears to have taken place in the Lewis range, and to
the varying degrees to which the basal sapping was carried may be
attributed the variation of mountain forms ranging between pinnacles
on the one hand and flat-topped mountains with small bordering
cirques on the other. An extreme case of such extension of headwater
amphitheaters may be seen in the Hayden Peak quadrangle representing
a portion of the Uinta Mountains of Utah. In this case the amphitheatral walls have been pushed back to the point where only skeleton
ridges remain, and in one instance at least, as west of Hayden Peak, the
divide has been almost completely obliterated, so that the snow-fields
on opposite sides must have been confluent during the glacial period.

It may readily be inferred from these considerations that cirques
may be (a) but slightly developed and the present expression of the
mountains closely resemble the preglacial expression; or (b) they may
be so extensively developed that the upland or mountain region in which
they were formed has a fretted appearance; or (c) the cirques may be
developed headward to a still greater extent and the dividing ridges
trimmed to a row of serrate peaks or skeleton ridges. Further contrasts in the present characters of the cirques will depend upon the
amount of postglacial cutting or filling that has taken place. Most
cirques as left by the ice contain but little loose material. Whether or
not loose material is now present will depend upon the friability of the
rock, the steepness and height of the cirque walls, the precipitation, etc.
In the Lewis Mountains the cirques are especially well preserved and
still exhibit, Fig. 94, great expanses of little-modified cirque wall, lakes
of considerable size and number, and expanses of bare, glaciated rock.

Concentrated sapping at the foot of the cirque wall results not only in headward retrogression but also in downward excavation, and with the melting of the glacier on account of a warmer period the reversed slope at the foot of the cirque wall is occupied by a lake. It will be noted on the Chief Mountain sheet, Fig. 94, that lakes commonly occur in this position. Lakes are also commonly found some distance down valley where tributary glaciers have caused local overdeepening or the bottoms of the glacial channels and the production of reversed slopes back of which the present drainage is impounded. A third group of lakes is frequently found some distance down valley where the drainage has been blocked by morainal accumulations that mark the former limit of glacial ice.

The valley sides and bottoms in this region usually bear one, two, or more terraces. The lower ones represent a valley filling which the streams are at present cutting away. The higher ones represent interrupted valley widening in horizontally bedded rock.[1]

The mountain valleys of these ranges are all noted for their steep walls and lake-dotted and rather flat floors. Their forms are chiefly the result of earlier glacial action. The main glaciers were wide and thick and eroded their channel floors below the channel floors of their tributaries. Upon the disappearance of a glacier the former glacial channel became a large part or all of the present valley. Hence the characteristic features of the glaciated valleys and the hanging condition of their tributaries. (See Figs. 96, 97, and 98.)

GALLATIN, MADISON, JEFFERSON, AND BRIDGER RANGES

In southwestern Montana is a group of mountains whose most important members are the Gallatin, Madison, Jefferson, and Bridger ranges. A small part of the southern end of the Madison range and the greater part of the northern end are composed of gneiss. The central part of the range is an immense laccolith, the uplift due to intrusion having carried portions of the sedimentary rocks almost to the summit of the range and overturned, broken, and changed the strata along the western and northwestern edges of the range, causing a varied and highly irregular topography. The Jefferson range consists in the north largely of granitic rocks. The southern section is divided into two parts; the eastern is plateau-like in character, the western is a monocline in which the sedimentary rocks are overturned to greater or lesser degrees from place to place. The Bridger range also consists of three sections, the southern extension being an overturned monocline of stratified beds that rest upon gneisses which form the western spurs and foothills and also a large part of the main moun-

[1] Bailey Willis, Stratigraphy and Structure, Lewis and Livingston Ranges, Montana, Bull. Geol. Soc. Am., vol. 13, 1902, p. 310.

Fig. 96. — A normally eroded mountain mass not affected by glacial erosion. (Davis.)

Fig. 97. — The same mountain mass as in Fig. 96, strongly affected by glaciers which still occupy its valleys. (Davis.)

Fig. 98. — The same mountain mass as in Fig. 97, shortly after the glaciers have melted from its valleys. (Davis.)

tain mass.   The overlying stratified beds form a sharp crest with peaks. The central portion of the range is composed chiefly of sedimentary beds and gneisses are absent.   The northern section is also composed of stratified beds of sandstones, shales, and limestones which curve around the ends of the range.

The Gallatin range is plateau-like at its summit and is composed largely of volcanic breccias which dip eastward and have their greatest elevations, about 10,000 feet, along the western border.   Mount Blackmore, one of the most prominent peaks of this range, rises to a height of 10,196 feet.

During the general elevation of the region in which these mountains occur the strata were folded and eroded (Cretaceous) and lakes, some of them of great extent, were formed in enclosed fresh-water basins. The lake period lasted for a long time (Neocene to Pleistocene), and during the earlier part of the period there was tremendous volcanic activity.   Great quantities of volcanic dust were carried hither and thither by the winds and at length deposited in part on the lake floors as white dust beds; deposition of vast amounts of dust took place upon the adjacent land surfaces, whence the deposits were later washed in large part into the lakes.   Later cutting down of the lake outlets allowed these water bodies to become drained and the lake beds themselves to be dissected.   The last phases of volcanic activity in the region were flows of basalt and rhyolite, and these now form the summits of mesas, as in the southern part of the Three Forks district.[1]

The last geologic episode which has had an influence on the topography and drainage of the district has been glaciation, but the glaciers were local and the drift deposits in the valleys are of local origin.   The low elevation of the ranges, 7000 to 10,000 feet, did not allow vigorous glaciation, and glacial forms have weak expression except in a few favorable localities.

MINOR RANGES OF WESTERN MONTANA

In western and southwestern Montana is an extensive area hemmed in by the Madison, Jefferson, and other ranges on the south and the Lewis, Livingston, and other ranges on the north.   On the east it extends to the Big Belt Mountains, on the west to the Bitterroots, Fig. 86.   The tract includes no prominent mountain chains, only short ranges which reach up to a more or less common level.   The geologic conditions are somewhat complex, the rocks consisting of greatly deformed sedimen-

[1] A. C. Peale, Three Forks Folio U. S. Geol. Surv. No. 24, 1896, p. 1, col. 4.

Fig. 99. — Effects of slope exposure on forest distribution, western Montana, looking east from Mt. Belmont; see topographic map below. A and C are cool, moist, forest-clad northern exposures, B and D are warm, dry, unforested southern exposures. The trees in the foreground grow on a north-eastern exposure. For the position of these slopes on the map see corresponding letters in Fig. 100. (Barrell, U. S. Geol. Surv.)

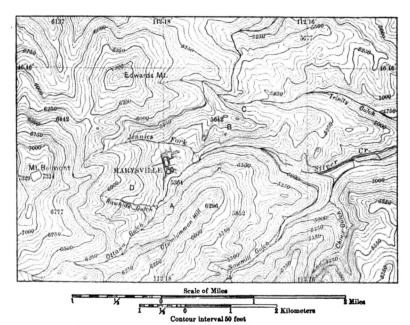

Scale of Miles

Contour interval 50 feet

Fig. 100. — The positions of the letters A, B, C, D, on the map correspond to the positions of the same letters on the photograph. Culture omitted except in case of Marysville.

taries intruded by granite and other igneous rocks. After the deformation of the rocks of the region erosion swept away great quantities of the surface material and reduced the topographic profiles to maturity, Fig. 99.

From the standpoint of forestry this great district in western Montana is of special interest because of the strong topographic control of forest distribution, not by control of rainfall distribution but by control of the water supply through variations in slope exposure. Its special significance may be appreciated by contrast with the physical conditions in the Cascades. The western slopes of the Cascades are wet, the eastern slopes are dry, a rainfall distribution that is directly dependent upon topography and that has a marked effect upon the distribution of the forest trees (p. 164). Among the minor ranges of Montana, on the other hand, there are no great topographic features to obstruct the rainfall and to occasion the climatic contrasts so marked in the Cascades. Slope exposure is here the factor of primary importance in forest distribution. The rather evenly distributed rainfall is more quickly evaporated on the sunny southern slopes than on the shady northern slopes, hence the latter are moist and forest covered, the former dry and almost treeless. Since the daily maximum temperature of soil and air is generally attained about two or three o'clock in the afternoon, the driest slope is the one facing southwest. Hence eastern and northeastern slopes are also forested, while western slopes are forestless. The effect of these conditions is heightened by the action of the prevailing southwest winds which not only dry the southwest slopes but also sweep them clear of snow in winter. The snows accumulate on the northeastern or leeward slopes, where they linger until midsummer and supply the ground and the vegetation it supports with the necessary moisture. These features are exceptionally well developed about Marysville and the district therefore merits a somewhat detailed description.

In the Marysville district the Rockies are developed in the form of a broad tract extending westward nearly to the Bitterroot Mountains. The eastern portion is a rather flat-topped granite batholith. The entire district has the features characteristic of topographic maturity — "fairly steep slopes, rounded hill crests, and few cliff exposures. On the lower elevations the topography is . . . softened, the view showing successive tiers of well-dissected .foothills with slopes of 10° to 20°."[1] The surface is in general covered with a residual soil so thin on the upper slopes as to show a large number of rock outcrops especially on

[1] J. Barrell, Geology of the Marysville Mining District, Montana, Prof. Paper U. S. Geol. Surv. No. 57, 1907, p. 4.

the more resistant formations. The lower elevations are covered with a slightly deeper soil, while the main valley floors are deeply filled with alluvium.

The Marysville region is in the transition belt between the lower, drier Great Plains on the east and the higher, abundantly watered and forested mountains on the north and west. While the soil covering, though thin, is suitable for a forest growth, the rainfall and snowfall are so light as to support a forest only in favored situations. Marked contrasts in climate occur and are due to marked variations in elevation and exposure. The result is a striking contrast in vegetation on different slopes and at different elevations, Fig. 99.

"The trees are confined largely to the northern and eastern slopes, since these suffer least from the drying action of the summer sun on the thin soil and also in spring hold the snow longest around the roots. The prevailing winds are from the southwest, with the result that the southwestern slopes are swept more or less bare of snow, which accumulates on the leeward side of the hills. Here, protected by the evergreen trees, stray banks linger until about the first of July.

"The bottoms of the deeper gorges are especially picturesque, offering the contrast of dark, forested southern [1] or western walls and grassy northern slopes, while cottonwoods and willows grow in clumps and lines along the courses of the streams.

"On the northern hill slopes the trees continue down to elevations of about 5000 feet, the limit varying considerably with the nature of the soil. Below this level the low hills of the northern half of the district are bare of trees and almost without grass, but in the gulches which trench them scattered pines have found enough moisture to give them a foothold. On the lowest levels the prickly pear and bunch grass hold sway. The most desolate portion of the district is north of Little Prickly Pear Creek, for here the sandy, porous nature of the surface renders it doubly difficult for vegetation to maintain a foothold, and large areas are covered with nothing but shaly shingle or ancient river cobbles." [2]

MOUNTAINS OF NORTH–CENTRAL IDAHO

The mountain ranges of the northern Rockies that we have examined thus far lie chiefly east of the Bitterroot axis. An examination of Fig. 86 will show a vast mountain region north of the Snake River and west of the Bitterroots, a labyrinth of sharp peaks and ridges whose steep slopes descend to deep steep-sided canyons. As a rule the mountains rise suddenly from the bordering plains and ultimately to sharp ridges that attain 11,000 and even 12,000 feet.

The origin and nature of a large portion of this wild mountain region may be appreciated best from a view that embraces the contrasting features of the plains country formed upon the flat basalt sheets that rim about the margins of the mountains and extend far westward into Washington, as far as the eastern wall of the Cascades. The summit of Bald

---

[1] It should be remembered that the southern wall of a valley has the same slope exposure as the northern slope of a hill.

[2] Idem, p. 6.

Mountain near the common border of the two regions (lat. 46° 25′) affords an extensive view eastward over the level crests and maze

Fig. 101. — View south across Salmon River Canyon, from south slope of Caseknife Mountain, showing plateau character of Salmon River Mountains. (Lindgren, U. S. Geol. Surv.)

Fig. 102. — Profile across Bitterroot Mountains, Clearwater Mountains, and Columbia Plateau, showing the plateau quality and the rock character. (Lindgren, U. S. Geol. Surv.)

of ridges and canyons that constitute the principal features of the Clearwater Mountains (lat. 46°). So thoroughly dissected are these mountains that little of their original flat outline may now be seen. For the first 80 miles eastward toward the Bitterroot Valley the lonely trail does not disclose a settlement or even a miner's cabin.[1] The irregular canyon courses are the chief routes for transportation by pack mule. It is the worst part of the Kentucky and West Virginia "mountains" set upon the western fringe of the Rockies.

Four thousand feet below the level of Bald Mountain is a scene of far different quality. The lava plain of the Columbia, only gently undulating, stretches out apparently without limit westward. The undulating Camas and Kamiah prairies are checkered with waving wheat fields or wild grass, and cultivation and prosperity are brought into close contact with mountain wilderness.

Here we have two regions of great difference in relief but also with many features that indi-

[1] W. Lindgren, A Geological Reconnaissance across the Bitterroot Range and Clearwater Mountains in Montana and Idaho, Prof. Paper U. S. Geol. Surv. No. 27, 1904.

cate peculiar similarities. The basalt plain is comparatively flat to-day; once it was still flatter; as time goes on it will become more and more irregular, will become indeed very much like the Clearwater Mountains are to-day. The even accordance of hill and ridge top levels in the Clearwater Mountains and the manner in which the plane of the ridge tops cuts across rock of diverse hardness and structure are clear indications that the region was once base-leveled and has since that time been uplifted and maturely dissected so that flat land is nowhere visible. Dissection has progressed so far that the once flat tabular summits have been transformed into sharp ridges, but it has not yet progressed far enough for the rivers to have begun to form valley flats. The result is a typical hill-and-valley country, whose resources of forest and mine are difficult of access, and where agriculture and related industries are practically unknown.

The sudden descent of the Clearwater Mountains to the level of the lava plain is due not to differential erosion but to pronounced crustal warping or faulting, for there are no structural features that would enable erosion to work out a mountain border of this character. The sudden and notable descent from the level of the plateau to the level of the plain was brought about probably at the time of uplift, for deep canyons were cut following the uplift and before the extrusion of the lava (Miocene) that runs up the old valleys now extending westward under the lavas. A similar abrupt border characterizes the plateau of the Boise Mountains where these descend to the lower valley of the Snake River. About the western base of the Clearwater Mountains the basalt flooded the foothills to a height of 3000 feet and greatly reduced the relief of the region, for the foothills in many cases had a sharply accentuated topography. Above the level to which the lava rose the courses of the streams draining the Clearwater Mountains have been steadily deepening.[1]

Both the Clearwater and the Salmon River mountains are portions of a once gently undulating plateau that has been so deeply eroded by powerful streams that they stand out with great distinctness. The plateau which their crests outline is from 1000 to 3000 feet below the summit of the Bitterroot Mountains, causing the latter to appear as a boldly raised block.[2]

Throughout the mountain region of central Idaho the canyons are so deep and steep-sided that many of them are quite impassable.[3] The

[1] W. Lindgren, A Geological Reconnaissance across the Bitterroot Range and Clearwater Mountains in Montana and Idaho, Prof. Paper U. S. Geol. Surv. No. 27, 1904, p. 78.

[2] Idem, p. 13.

[3] G. H. Eldridge, A Geological Reconnaissance across Idaho, 16th Ann. Rept. U. S. Geol. Surv., pt. 2, 1894-95, p. 220.

canyon walls are precipitous, sheer drops of 1000 feet being quite common, and the adjacent mountain slopes rise by steep ascents 2000 to 3000 feet higher. The most rugged canyons are those of the South Fork of the Boise River, numerous branches of the Clearwater, and the Middle Fork of the Salmon. A stream with typical characteristics is the Salmon River, which heads at the foot of the Bitterroot range, then flows westward through a wide open valley to a point near Shoup, where it enters a profound canyon through which it flows without interruption for about 250 miles to its junction with the Snake. The canyon extends through one of the wildest and least-known parts of the state, and is itself so narrow and abrupt and so deep (3000 to 5000 feet) as to be almost untraversable.[1]

The mountainous country north of the Snake River and west of the Bitterroot divide may be divided on the basis of structure into three parts:

(1) A great central granite area 100 miles wide and 300 miles long, extending from the Snake River plains northward to an unknown distance but at least as far as 45° 30′; it probably ends near the northern border of the Clearwater drainage.[2]

As thus outlined it forms one of the largest granite batholiths on the continent. Near the lower Salmon River and also near the Seven Devils in the Snake River Valley it is margined by sedimentary rocks. A similar contact occurs on the eastern border of the granite, and from the nature of the intruded beds it is known that the granite mass is probably of Cretaceous age and is an intrusive body similar to the great granitic batholiths of the Sierra Nevada. The granite is remarkably uniform in character except in places upon its margin as on the eastern border of the Bitterroot Mountains where it has a gneissoid structure owing to metamorphism at the time of the formation of the range.[3]

(2) Partly metamorphosed rocks of sedimentary origin — slates and limestones accompanied by schists — occur on the western border of the granite area and form the western border of the mountains.

(3) Partly metamorphosed rocks — quartzites, conglomerates, slates, shales, and limestones — occur on the eastern margin of the great granite batholith.[4]

Two types of mountains have been developed upon these rocks and structures: (1) those consisting of masses of rock without any definite range trend and developed upon the almost structureless granite of

[1] W. Lindgren, The Gold and Silver Veins of Silver City, De Lamar, and Other Mining Districts of Idaho, 20th Ann. Rept. U. S. Geol. Surv., pt. 3, 1898–99, p. 78.

[2] W. Lindgren, A Geological Reconnaissance across the Bitterroot Range and the Clearwater Mountains in Montana and Idaho, Prof. Paper U. S. Geol. Surv. No. 27, 1904, p. 17.

[3] Lindgren and Drake, Silver City Folio U. S. Geol. Surv. No. 104, 1904, p. 1, col. 4.

[4] W. Lindgren, The Gold and Silver Veins of Silver City, De Lamar, and Other Mining Districts in Idaho, 20th Ann. Rept. U. S. Geol. Surv., pt. 3, 1898–99, pp. 79, 86–89.

homogeneous texture. The Clearwater Mountains north of the Salmon River and the Salmon River Mountains south of the Salmon River are the best representatives of this type. The second type of mountain occurs in ranges and is the result of erosion of sedimentary rock and foliated schists, or a rock complex, whose secondary structures on erosion give a rough trend to the elevated portions.

In some of the granites of the western half of Idaho east-northeast-trending structures show chiefly in lines of jointing, in the strike of foliation planes, or in the trend of the fissures. West of the Sawtooth range the divides between the drainage basins are with few exceptions due to early structural features — folds, faults, joints — but over most of the tract the granite is homogeneous, the structural features are only slightly pronounced, the disposition of the topographic elements is a response to the disposition of the early drainage systems, and no well-defined outlines of a range system can be identified.[1]

The irregular ranges that rise above the general level as in the Sawtooth Mountains seem to owe their existence to greater resistance to erosion rather than to folding and faulting. A few exceptions to this rule are (1) Boise Ridge, which seems to have been partly outlined by orographic disturbances, and (2) smaller ranges which have been developed where the granite is strongly foliated.

The mountains of the range type occur chiefly in the eastern part of Idaho near the continental divide in a region of pronounced uplift. They trend in two different directions, east-northeast and west-northwest, depending upon the strike of the beds upon which they are developed. They are composed of altered or unaltered quartzites, schists, and limestones. The Smoky Mountains are an illustration of this type; the Bitterroots are a larger and better known unit and receive more extended description in the succeeding paragraphs.

BITTERROOT MOUNTAINS .

From Hamilton or Missoula in the Bitterroot Valley and in the heart of the northern Rockies one may look west at the bold front of the Bitterroot Mountains. For many miles it maintains a quite remarkable regularity of form and straightness of trend. The slope descends at angles between 18° and 26°. The rectilinear quality is owing to a fault whose locus is the foot of the scarp and whose throw approximates the present difference in elevation between range top and valley bottom.

---

[1] W. Lindgren, The Gold and Silver Veins of Silver City, De Lamar, and Other Mining Districts in Idaho, 20th Ann. Rept. U. S. Geol. Surv., pt. 3, 1898–99, p. 77

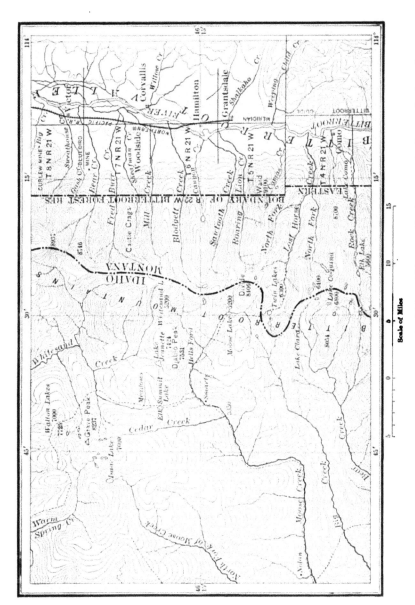

Fig. 103.— Topographic map of a part of Bitterroot Mountains, Idaho and Montana. Sketch contour only, about 300 foot interval.
(Lindgren, U. S. Geol. Surv.)

At least two groups of facts strongly support this explanation. (1) Faulting occurred in the region so recently as 1898, when for 1500 feet along the base of the mountains a displacement of from 1 to 2 feet was effected and may be observed in favorable localities to-day. (2) All the streams flowing eastward down the regular mountain front have remarkably straight courses, and all have steepened gradients in the last mile or more before debouching upon the valley flat bordering the Bitterroot River. The steepened gradients are an expression of progressive faulting which has prevented the attainment of a profile of equilibrium, approach to such a profile being counteracted by a repetition or continuance of faulting.[1]

The remarkable regularity of the eastern slope of the Bitterroot Mountains is the more interesting because of the preservation of the fault plane throughout the slope up to the eastern summit, a fact which appears to be explained by the flat angle at which the slope was formed, the uniform character of the crystalline rock (gneiss and schist) composing the eastern front of the range, and the original regularity of the fault.

The main divide of the Bitterroots is a succession of sharp craggy peaks alternating with deep saddles at the heads of the large canyons where glacial cirques have opened up the valley heads. The western slope is more rugged than the eastern, although even the latter is almost unknown except to prospectors. The immense steep-sided gorges, sheer precipices, and extensive rock slides make the western slope entirely impassable toward the crest.[2] The trails, originally located by the Indians, follow the divides or primary ridges as closely as possible, tortuous as they are; only grading would make the canyons passable.

CLIMATIC FEATURES; VEGETATION

The character of the primeval forest of northern Idaho and northwestern Montana varies according to latitude, altitude, and conditions of exposure. The size of the trees decreases toward the north; on the slopes of the Cœur d'Alene Mountains it is not uncommon to find trees 2 or 3 feet in diameter, while tamarack, spruce, and lodgepole pine along the international boundary are seldom more than 1 foot in diameter. The valley terraces along the principal rivers support large groves of yellow pine and tamarack, almost without underbrush. The moun-

[1] W. Lindgren, A Geological Reconnaissance across the Bitterroot Range and Clearwater Mountains in Montana and Idaho, Prof. Paper U. S. Geol. Surv. No. 27, 1904, p. 49.
[2] J. B. Leiberg, Bitterroot Forest Reserve, 20th Ann. Rept. U. S. Geol. Surv., pt. 5, 1900, p. 319.

tain slopes have a denser cover of fir, spruce, and tamarack, which with increasing elevation becomes an open growth of spruce, with a luxuriant cover of grass. Snow does not persist throughout the year upon any of the slopes much exposed to the direct rays of the sun, though perennial banks form on the steep northward-facing sides of the higher peaks and one or two small glaciers occur in the Cabinet range. There is a wet and a dry season, the latter beginning about October 1, the former about June 15, and the summers are very mild and agreeable. The mountainous part of northern Idaho and north-western Montana has a thin population engaged in lumbering, agriculture, and mining.[1]

The vegetation of the mountainous area immediately north of the Snake River plains consists of forests of fir and pine, above an elevation of 5000 feet and gradually increasing in luxuriance northward. The southern foothills of the main mountain area below 5000 feet are barren. The agricultural population of the Snake River Valley is concentrated chiefly where the tributaries issue from the mountains and on the limited flats along the main river, where irrigation is possible. Besides these plains settlements are others in the intermontane valleys on the Weiser, Payette, and upper Salmon. Scattered mining settlements are also found from south of the mountains to Florence, and from the Seven Devils to Challis. They are often in places very difficult of access and at elevations ranging from 4000 to 8000 feet.

In the Bitterroots the soils of the mountain slopes are composed of granite débris below and loam above; in the canyon bottoms similar conditions obtain, but the top layer is usually heavier; in the subalpine meadows the subsoil is a pure granite gravel and the surface layer a loam varying from 6 inches to 6 feet in depth.[2]

Forests of economic value are found in the Bitterroot Mountains, in the upper valleys of the branches of the Bitterroot River, in the canyons of its tributaries farther north, and on the lower slopes of the mountains. At greater altitudes and upon the sides and summits of the mountain spurs the forests are thin and of little value. Two zones of forest distribution may be distinguished, the dividing line lying about 5800 feet above sea level. The lower is the yellow-pine zone, the upper the alpine-fir zone. The timber of the lower zone consists

[1] F. C. Calkins, A Geological Reconnaissance in Northern Idaho and Northwestern Montana, Bull. U. S. Geol. Surv. No. 384, 1909, pp. 20–21.

[2] J. B. Leiberg, Bitterroot Forest Reserve, 19th Ann. Rept. U. S. Geol. Surv., pt. 5, 1897–98, pp. 262–267.

mainly of red fir and yellow pine in the proportions of 60% and 30%; in the subalpine zone nine-tenths of the timber consists of lodgepole pine of little commercial value.[1] The former type of growth prevails on the lower slopes and in the canyons; the latter occurs on the summits and ridges and on the steep upper slopes.

[1] Henry Gannett, 19th Ann. Rept. U. S. Geol. Surv., 1897–98, p. 57.

CHAPTER XIX

ROCKY MOUNTAINS. II

CENTRAL ROCKIES

BETWEEN the northern and southern Rockies is a group of ranges whose principal members are the Absaroka, Wind River, Gros Ventre, Teton, Laramie, and Medicine Bow mountains. Part of them form a belt of broken, rugged country with alpine characteristics on the western border of Wyoming, such as the Teton and Gros Ventre ranges; another part, including the Wind River and Laramie ranges, extends eastward and southward, making a great curve through central and southeastern Wyoming in continuation of the Colorado Range of central Colorado.

All these ranges (and many others of lesser extent not described here) are sufficiently alike in certain general characteristics and in geographic position to form a family of ranges with prominent traits. Many of them are anticlinal in structure, the anticlinal uplift being directly related to the mountainous relief, so that uplift along axial lines is responsible for the trend of the range heights. The asymmetry of the folds is another group quality and in the anticlinal ranges always causes one mountain flank to be relatively less steep than the other. Erosion is of course variable in degree, because the degree of folding and the height of the folded strata are not everywhere the same nor are the resistances of the different strata uniform.

It is sometimes ventured that these structures and forms resemble those of the long, narrow, Appalachian ridges of Pennsylvania. The comparison is not good as regards structure, for the Wyoming ranges have a far less regular structure and trend than the Pennsylvania ridges. In regard to form the comparison is wholly at fault. No central core or axis of crystalline rock is found in the latter case, and no such heights are reached. The Pennsylvania mountains are ridges; the Wyoming mountains are ranges. In the one case the even sky line is the result of base-leveling; base-leveling has not been generally determined in the other case and the sky line is serrate, the crests and peaks rugged and in some places lofty. Some of the foothill ridges developed upon the

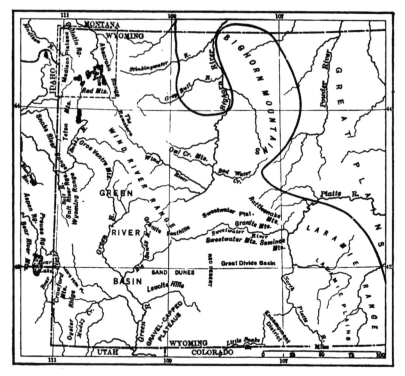

Fig. 104. — Outline map of the central Rockies, showing location of principal ranges.

edges of flanking strata are nearly level for miles, have subsequent valleys
developed upon soft strata alternating with hard, and in general show
a rough resemblance to Appalachian mountain topography; but on the
whole there are far more contrasts than resemblances.

LARAMIE MOUNTAINS

The great Colorado or so-called Front Range terminates near the Colo-
rado-Wyoming line and the mountain axis of which it is but a part
continues northward in a double line of uplifts known as the Laramie
Mountains on the east and the Medicine Bow Mountains on the west.
The southern end of the Laramie Mountains is near the North Fork
of the Cache la Poudre River and is surmounted by a confused mass of
peaks of considerable height. The front range is also elevated north of
the Laramie River, where it rises in rugged granite hills. But between
the North Fork of the Cache la Poudre on the south and the Laramie

River on the north is an 80-mile division of the front range[1] of the Rockies whose topographic features are in strong contrast to the rugged groups both north and south. The average elevation of this section is only 7800 to 8300 feet and all but one of the few isolated peaks that rise above the general level are under 9000 feet above the sea. Sanders Peak, 9077 feet, Central Peak, 8744 feet, and Arrow Peak, 8683 feet, are the principal eminences.

The so-called Laramie Mountains are really an elevated plateau some 1500 feet above the Laramie Plains on the west and somewhat more above the Great Plains that lie beyond their eastern foothill ridges. The plateau is cut by many canyons and roughened somewhat by numerous knobs and short ridges, but these features do not wholly obscure the expression of the uplifted peneplain (early Tertiary) formed alike upon granite and schist. The interstream areas are smooth or gently rolling and even the ridges have rounded summits. The latter together with the knobs are residual with respect to the peneplain. The rocks are deeply decayed and thick residual soil mantles the remnants of the ancient surface.

The main divide of the Laramie Mountains is a distinct ridge of sandstone and limestone bordered on the east by a scarp from 50 to 200 feet high; on the west this ridge descends regularly to the Laramie Plains as a dip slope. East of the main divide is a broad area of gneiss and schist, the core of the anticlinal axis. Farther east are the hogback ridges of sandstone and limestone that form the eastern foothills. The Laramie Mountains are structurally a great anticline whose summit has been worn away. To the fact that it is an asymmetrical anticline is due the gentle western and the steep eastern slopes of the mountains. Locally, faulting has occurred on the eastern border and heightened these topographic contrasts. At other places the contrasts are lessened by an extensive overlap of Tertiary deposits which on the east extend far up the mountain slopes concealing the underlying harder rocks and affording easy ascents to the mountain or plateau summit.[2]

There is a noticeable lack of timber over this entire section. Small groves of pine, aspen, and other types of tree growth are found in sheltered localities, as in the larger basins, but on the whole the timber is of little practical importance.[3]

[1] The name applied to the easternmost great range of the Rocky Mountain System. In Colorado the front range is the Colorado Range; in Wyoming it is the Laramie and other mountains; in Montana it is the Lewis Mountains; etc.

[2] Darton, Blackwelder, and Siebenthal, Laramie-Sherman Folio, U. S. Geol. Surv. No. 173, 1910, pp. 1, 2, 13.

[3] Arnold Hague, U. S. Geol. Expl. of the 40th Parallel (King Surveys), vol. 2, 1877, pp. 5-7.

WIND RIVER RANGE

The Wind River Range trends southeast and is one of the eastern border ranges of the central Rockies, Fig. 104. It continues toward the north in line with the Laramie Range, and like it is one of the front ranges of the system. The general structure of the Wind River Range is that of a complex anticlinal whose short and steep dips and corresponding steep slopes are toward the southwest; the northeast descents are less steep. The latter aspect is varied by two parallel mountain ridges formed upon hard strata, the intervening soft strata having been cut away by strike or subsequent streams. The main summit of the mountains consists of a broad plateau-like tract of granite and gneiss whose northeastern and southwestern faces are precipitous and deeply scored by large canyons.[1]

In the Wind River Range spruce and fir constitute the main tree species, with occasional groves of yellow pine and quaking aspen. Below the foothill elevations the willow and the quaking aspen grow along stream courses; at still lower elevations and in dry interstream situations sage-brush and cacti are the characteristic plants, as in the low country east of the Wind River Range and north of the Sweetwater Plateau. On dry ridges and steep dry bluffs piñon and cedar, or more properly, juniper, are found.[2]

BEARTOOTH AND NEIGHBORING PLATEAUS

The Yellowstone River after leaving Yellowstone Park makes a great northerly bend between which and the Bighorn basin are the Beartooth Mountains or Plateaus, East and West Boulder plateaus, Lake Plateau, etc., and the northern end of the Absaroka Mountains. In fact the three last-named tracts, together with the Buffalo Plateau and other minor subdivisions, constitute the northern part of the Absarokas as shown in Fig. 105. They are drained northeasterly by the Stillwater, Clarke Fork, and other tributaries of the Yellowstone, through deep canyons bordered by wall-like cliffs from 1500 to 3000 feet high. The region is everywhere extremely rough, and scored by deep, narrow, rocky canyons between which are (1) narrow ridges often only a few feet wide at the top or (2) broad, massive, rolling, bowlder-strewn, plateau-like surfaces like the East and West Boulder plateaus. The

---

[1] Orestes St. John, U. S. Geol. and Geog. Surv. of the Terr. (Hayden Surveys), 1877, pp. 228 et seq.

[2] F. M Endlich, U. S. Geol. and Geog. Surv. of the Terr. (Hayden Surveys), 1877, pp. 59–60.

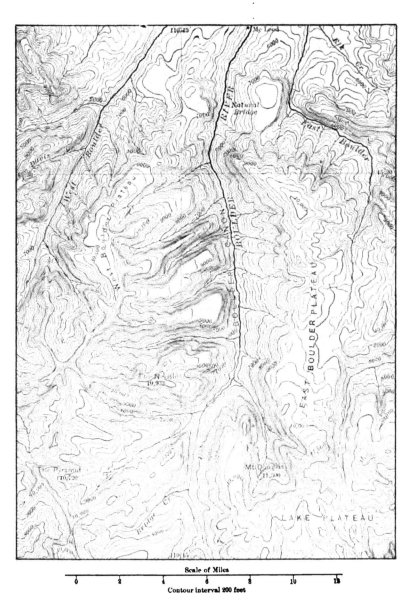

Fig. 105. — Topography of East and West Boulder plateaus, northern end of the Absaroka Range, south-western Montana.   (Part of Livingston quadrangle, U. S. Geol. Surv.)

average elevation is about 8000 feet; the canyon floors are at 4000 and the highest peaks at 11,000 feet.

The Beartooth Plateau, the main mass of high, rolling, inter-canyon country, and the associated ranges or plateaus, lie east of the Stillwater and at higher levels, 9000 feet, with peaks reaching to 13,000 feet. The canyons are here also steep and deep and end in rock-bound glacial cirques whose development about the bases of the mountains has sharpened the latter from a more rounded form into peaks and pinnacles of the matterhorn type that give an aspect of profound relief. The main plateau is pitted with glacial depressions, many of which have lakes and marshes bordered by alpine meadows. On the western border of Beartooth Plateau is the deep canyon of the Stillwater; on the east the plateau ends in a great frontal scarp in some places more than 3000 feet high. The steepness of most of the forms is owing to the late uplift of the region, the profound dissection by both water and ice, and the resistant gneisses and other crystallines out of which they have been carved. Some of the canyons are mere rifts between almost perpendicular walls; most of the headwater cirques are sheer cliffs; the peaks are generally glaciated and their bases carved by ice and water to such an extent as to give them great steepness and sharpness of form both in general outline and in detail.

The plateaus and canyons in the entire region as well as in the Absaroka and Gallatin ranges on the west were entirely covered with snow and ice during the glacial period; in addition all were profoundly dissected even before glaciation set in. Each range therefore repeats the general features of its neighbor. All are characterized by broader or narrower plateau-like tracts between deep canyons; all have some surmounting peaks; all have glacial lakes (1) on the cirque floors, (2) on the plateau summits, and (3) in the valleys, where either unequal cutting or morainic damming originated basins of variable size.

The forests of the region are almost wholly coniferous. Limber pine grows in belts from 5500 to 6000 feet high; lodgepole pine grows at elevations ranging from 5000 to 8000 feet; red fir is most commonly found on dry rocky slopes; and Engelmann spruce grows along the canyon bottoms or along seepage lines. Above the lodgepole pine forest are subalpine fir, white-bark pine, and Engelmann spruce. Timber line is at 9300 feet on northern and western slopes and at 9800 feet on southern slopes. Toward the east it rises and on the eastern edge of the Beartooth Plateau is at 11,000 feet. At the upper limit of tree growth the spruce and the white-bark pine decrease in height and become mere crooked shrubs. Grass grows in the alpine zone in such abundance as to form a thick turf capable of storing water and adding to the effect of lakes and tarns in preventing too rapid run-off and gullying. Overgrazing immediately starts gully erosion and results in effects as important as the overcutting of the forests at lower elevations.[1]

ABSAROKA RANGE

The Absaroka Range extends in a north-south direction for over 80 miles, with an average width of nearly 50 miles, Fig. 105. The southern end of the range is closely related to the Wind River Plateau and the Owl Mountains and is made up of enormous volcanic flows. The Absaroka mountains are to be considered as a broad deeply-eroded plateau rather than a sharply outlined range, Fig. 105. Profound erosion has carved the former more extensive summit into a topographic complex of rugged peaks and jagged pinnacled mountains bordered by bold escarpments which rise to heights of hundreds and even thousands of feet above the surrounding country. The walls encircling the mountain groups are usually abrupt and owe much of their steepness to ice action.

After the volcanic materials of which the Absaroka mountains are largely composed had accumulated the streams cut the range into isolated mountains and broadly-spreading spurs and its deep broad valleys were occupied by glaciers. All the upper valleys on the eastern slopes of the mountains have been sculptured by ice action, and lateral and terminal moraines are abundant. In addition, many narrow canyons are broadly benched a thousand feet or more above the valley floors, and rounded and polished forms are characteristic features of the scenery wherever gneisses and granites occur.

The chief effects of glaciation are, however, to be found upon the western slopes of the Absaroka Range, where a heavy ice cap developed, — one of the largest local glacial centers developed in the Rocky Mountains south of the great continental ice sheet. This local ice cap was confluent with that in the Yellowstone National Park, where a broad area of elevated country supplied favorable conditions for extensive glacial accumulations which fed marginal glaciers deploying down the valleys.[2] In the deep basins of some of the higher valleys facing northeast, a few snow-fields are still to be found and even small glaciers. The glaciers lie in broad,

---

[1] J. B. Leiberg, Forest Conditions in the Absaroka Division of the Yellowstone Forest Reserve, Montana, and the Livingston and Big Timber Quadrangles, Prof. Paper U. S. Geol. Surv. No. 29, 1904, pp. 10–21.

[2] W. H. Weed, The Glaciation of the Yellowstone Valley North of the Park, Bull. U. S. Geol. Surv. No. 104, 1893, p. 13.

Fig. 106.— Topography of the Absaroka Range 10 miles east of Yellowstone National Park. Scale, 2¼ miles to the inch; contour interval, 100 feet. (Ishawooa quadrangle, U. S. Geol. Surv.)

rock-bound amphitheaters in great measure protected from the direct rays of the sun between high walls, and in localities where the prevailing southwest winds deliver vast quantities of snow during the winter season.[1]

The greater part of the Absaroka Range in the Yellowstone National Forest is clothed with coniferous forests broken by open glades. The isolated peaks and irregular crests of the main ridges are above timber line, and bear only scattered and stunted growths of weather-beaten trees. The western side of the range has a more continuous forest cover than the eastern. Lodgepole pine is the prevailing tree, limber pine is found at higher altitudes, and balsam and spruce are scattered widely though they nowhere attain great height or size. The most stately and vigorous tree of the Absarokas is the Douglas spruce, but it has a scattered growth. None of the timber of the Absaroka region is of superior quality, though sufficient for local requirements may be obtained by judicious use of the forest.[2]

MEDICINE BOW RANGE

This range is about 100 miles long and diverges slightly with respect to the front range. Included between it and the front range is a high intermont basin, so-called Laramie Plains. Its broadest part, in the region of Medicine Peak, is from 30 to 35 miles across, but the southern end is only 10 or 12 miles wide. The highest peaks named from the southern end northward are Mount Richthofen, 13,000 feet; Clark's Peak, 13,100 feet; Medicine Peak, 12,200 feet; and Elk Mountain, 11,500 feet. North of the 41st parallel the range becomes double crested in response to the double anticlinal structure, but toward the south its unity is preserved. Between the double crests of the northern section is a high, gently undulating, intermont plateau 10,000 feet above the sea, whose surface is covered with timber and dotted with lakes. All the higher portions of the range have been glaciated, and some of the glacially carved amphitheaters are very striking, with steep walls 1500 feet high cut in extremely hard quartzite.

The greater part of the range is covered with coniferous forest which is in places quite dense. Douglas spruce, Engelmann spruce, and yellow pine are the chief species. The cold timber line is about 11,000 feet above sea·level.[3]

[1] Arnold Hague, Absaroka Folio U. S. Geol. Surv. No. 52, 1899, p. 6, col. 3.
[2] Idem.
[3] Arnold Hague, U. S. Geol. Expl. of the 40th Par. (King Surveys), vol. 2, 1877, pp. 94–97.

BASIN PLAINS OF SOUTHERN WYOMING AND MINOR RANGES ON THEM

An accurate relief map of the Rocky Mountains represents a broad and flat tableland between the Laramie Mountains and the ranges west of the Green River. Through the eastern portion of this tract runs the Medicine Bow Range and through the central portion run the Leucite Hills, etc., so that the tract is partially divided into three sections. The portion east of Medicine Bow Range is called the Laramie Plains;

Fig. 107. — Laramie Plains, looking southwest from near Mandel, Wyo. (U. S. Geol. Surv.)

the central section is known as the Red Desert, etc.; and the westernmost section is known as the Green River Basin. These three main divisions are, however, continuous about the northern end of their mountainous boundaries.

The Laramie Plains are 7000 feet above sea level, have a relatively smooth and gently undulating surface, Fig. 107, with gentle slopes and rounded outlines, so that they appear practically level over broad expanses except where a few bench-like ridges or very low buttes of circumdenudation occur. Shallow lakes, none more than a few square miles in extent, are scattered over the surface of the plain, whose waters range from fresh to brackish or strongly alkaline; some of them disappear during the dry season and leave saline incrustations. The whole tract is a great natural pasture; trees grow only along the broad stream valleys.[1]

The Laramie Plains apparently owe their flatness to the general horizontal attitude of the underlying strata (Cretaceous). The highest observed dips in the central portions of the basin are 12°, the average 5° to 8°. On the borders of the basin the dips increase somewhat and the eroded strata present steep bluffs toward the older rocks that form the cores of the bordering ranges. One of the most typical portions of the high intermont plains of southern Wyoming is the central division, which extends west of Rawlings. There are but few exposures of the flat-lying beds, and these occur only in escarpments a few feet high. The barren surface is but slightly dissected by the dry, shallow watercourses that traverse it. It is a flat, monotonous region as far as the Leucite Hills and is broken only at long intervals (1) by hills of eruptive material, or (2) by lines of moving sand dunes and drifts, as west of the Red Desert and north of the Leucite Hills, or (3) by local uplifts of small extent as at Rawlings. No marked drainage features are found in this section, but the section west of the 109th meridian is drained by the upper tributaries of the Green and along their courses an important amount of dissection has taken place.[2]

The westernmost section of the Wyoming Plateau is also underlaid by nearly horizontal strata (Tertiary and Cretaceous). The general nature of the relief is similar in most respects to that of the eastern divisions. The central depression of the tract is occupied by the valley of Green River, which on the north flows through the basin in a wide alluvial bottom; farther south it flows through a 1000-foot canyon cut in nearly horizontal strata (Tertiary). It then enters the Uinta range at the Flaming Gorge, where it cuts through the hardest quartzite, turns out of the mountains, flows eastward, and finally cuts straight across the Uintas in a superb 3000-foot quartzite canyon known as the canyon or gate of Ladore.[3]

The Green River drains a great basin whose eastern portion is somewhat diversified by low, strike ridges of irregular outline and whose western portion bears ridges of similar structural nature but of more regular topographic development, the horizon being always bounded

[1] Arnold Hague, Rocky Mountains, U. S. Geol. Expl. of the 40th Par. (King Surveys), vol. 2, 1877, pp. 73–75.

[2] S. F. Emmons, The Green River Basin, U. S. Geol. Expl. of the 40th Par. (King Surveys), vol. 2, 1877, p. 164; and Geol. Map by Peale, St. John, and Endlich, accompanying Report.

[3] Idem, pp. 192–193 and 287, with Plate 8.

by an almost perfectly horizontal line.[1] The southern portion of the region north of the Park Range, known as the Savory Plateau, likewise presents a horizontal sky line, but for a different reason. The surface strata (Tertiary) are horizontal here and overlie the upturned and eroded edges of Cretaceous and older beds and present smooth plain surfaces.[2] From these stratigraphic relations it is reasonable to conclude that a mountain-making period at the close of the Cretaceous or

Fig. 108. — Terrace and escarpment topography of nearly horizontal beds (Eocene), Green River Basin, southwestern Wyoming. (Veatch, U. S. Geol. Surv.)

in early Tertiary deformed the underlying beds and erosion planed them off, some perfectly, others imperfectly, and that upon the subdued, partially reduced surface thus developed, Tertiary beds of great thickness (up to 1000 feet) were deposited. Erosion dissected (1) the Cretaceous strata that were never buried, causing the development of a ridge and valley type of topography, and (2) the Tertiary cover, in some

[1] S. F. Emmons, The Green River Basin, U. S. Geol. Expl. of the 40th Par. (King Surveys), vol. 2, 1877, p. 192.

[2] Idem, p. 164.

instances, laying bare the Cretaceous beneath, in others exposing only Tertiary material in the valley sections.

Similar relations have been identified in southwestern Wyoming and on the southwestern border of the Green River Basin. The basin floor is here developed chiefly upon Tertiary strata dipping gently eastward, and in general has been eroded so that the strata present their outcropping edges as westward-facing escarpments, a typical terrace topography. In places the mantle of nearly horizontal Tertiary deposits has been worn away and the underlying older beds with steep inclination are now exposed. Since there is considerable difference in the exact degree of dissection of the capping strata and of the exposed portions of the underlying, deformed rocks, the topography also varies

Fig. 109. — Hogback topography of inclined beds (pre-Eocene), southwestern Wyoming. The crests of the ridges are developed upon sandstone underlain by shale. (Veatch, U. S. Geol. Surv.)

from place to place. Two main types of topography have been developed: (1) flat table-like forms in places with bordering escarpments of considerable length between benches that rise in regular succession, a topographic type developed upon the uppermost and horizontal beds; (2) long ridges, often sharp, separated by equally long valleys, the whole in a markedly parallel arrangement and developed upon the steeply inclined older rock composed of alternating hard and soft layers. Hard sandstones are the ridge makers; soft shales underlie the valleys and the lowlands.[1]

The greater portion of the Green River Basin presents a gravel-strewn surface almost everywhere in process of dissection. The gravel cap was formed (1) by the weathering in place of the partly consolidated or wholly unconsolidated deposits filling the basin, or (2) by detrital accumulations washed into the basin during an earlier period of aggradation. The latter material is abundant about the mountainous border of the basin and originated in the form of desert-fan deposits. Its origin may be related to the renewed growth of the bordering ranges. Its deposition was preceded by a long erosion interval in which a local peneplain was formed, a fact which partly explains the extent of the gravel cap. It was followed by erosion now in progress.[2]

The Leucite Hills, which partially break the continuity of the plains of southern Wyoming, consist of a number of small conical peaks of volcanic origin. Some of them are capped by lava flows, others have crater-like forms and appear to have been centers of eruption.[3] They have no great topographic prominence and are of little interest in the present connection.

SIERRA MADRE RANGE

Continuing with the mountain ranges of the central Rockies it should be noted that in south-central Wyoming (Encampment District, south of Rawlings and west of the Medicine Bow Range) the mountain topography is on a whole of a subdued type. Steep slopes are confined to the middle courses of a few of the main streams and to a few basins or amphitheaters near the main divide formerly occupied by glaciers. The Sierra Madre mountains here form the main or continental divide. Their summits are generally broad and form a flat surface whose elevation ranges from 9200 feet in the lowest pass to 11,000 feet on the highest summit. The steep headwater declivities on opposite sides of the divide are often a mile or more apart, and the soft, broader-spaced contours of the mountain summits are frequently continued out upon the marginal spurs for several miles. The high portion of the range, Fig. 110, thus appears as an elevated plateau submaturely dissected by existing streams. The topography is subdued rather than rugged, and wagon

[1] A. C. Veatch, Geography and Geology of a Portion of Southwestern Wyoming, Prof. Paper U. S. Geol. Surv. No. 56, 1907, pp. 34–35, and Plate 1.

[2] J. L. Rich, The Physiography of the Bishop Conglomerate, Southwestern Wyoming, Jour. Geol., vol. 18, 1910, pp. 601–632.

[3] S. F. Emmons, U. S. Geol. Expl. of the 40th Par. (King Surveys), vol. 2, 1877, p. 236.

roads have been constructed on almost every portion of the area without that excessive expense that is usually required in opening transportation routes in mountainous regions.[1]

The general structure of the Sierra Madre Mountains is a low arch or anticline the axis of which is parallel with the mountain crest. This arch was gradually uplifted and eroded and the Mesozoic rocks removed from the axial portion, revealing older pre-Cambrian rock which forms the main mass of the mountains. The Mesozoic formations outcrop on the foothills on either side and dip away beneath the surrounding plains.

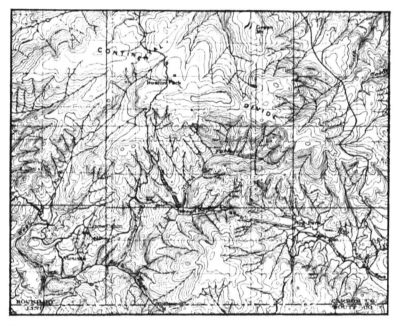

Fig. 110. — Mature profiles, long gentle spurs, and rounded summits of the mountains of the Encampment district, south-central Wyoming. (Encampment Special quadrangle, U. S. Geol. Surv.)

TETON RANGE

The Teton Range extends almost due south from the southwestern corner of Yellowstone Park and just east of the Idaho-Wyoming line. It is about 40 miles long and from 10 to 15 miles wide, with a central cluster of exceptionally high peaks, the highest of which, Mount·Hayden and Grand Teton, attain elevations of 13,700 feet and 13,800 feet respectively. The only practicable pass is Teton Pass at 8400 feet. There are still in existence several small glaciers occupying cirques high

[1] A. C. Spenser, The Copper Deposits of the Encampment District, Wyoming, Prof. Paper U. S. Geol. Surv. No. 26, 1904, pp. 12-15.

up in the range. The Tetons consist of a great longitudinal axis of uplift and folding; the steep and short descent is on the east, the longer and gentler is on the west. The steep eastern front is cut into a number of great buttress-like spurs faced by bold precipitous cliffs. On the west are a number of long narrow canyons separated by broad westward-descending slopes. These features are very uniformly developed along the entire eastern and western mountain fronts and are the most important in the range. The main mountain forms have a very intimate relation to a single great anticlinal uplift greatly eroded, exposing a central core or nucleus of crystalline rock, chiefly gneiss and granite, the latter forming the sharp peaks and aiguilles as well as the highest peaks as in Mount Hayden; the gneiss tends to erode into the form of sharp ridges. The crest of the range does not everywhere follow the granite however. At the south it is developed chiefly upon sedimentary rocks that have not yet been stripped from the crystalline foundation.[1]

In general the climatic conditions in the Tetons favor forest growth, yet but little forest exists owing altogether to the repeated and destructive fires which have swept over the tract. The lodgepole pine has been temporarily driven out in many places and former timbered areas have become grass-covered parks or aspen groves. Besides lodgepole pine and aspen the principal species are Engelmann spruce and red fir, the former in damp, the latter in dry situations.[2] Timber line is at 10,000 feet in this locality, and as the average altitude of the range is about 12,000 feet, large portions of the Tetons rise well above the upper limit of the forest.

GROS VENTRE RANGE

The Gros Ventre Range consists of two parallel mountain folds or anticlines about 5 miles apart, the axes trending southeast. Erosion has unroofed both anticlines and given them corresponding topographic features. Had the anticlines been symmetric, long, gentle, dip slopes would, in the present state of erosion, be found upon the outer flanks of the ridges, and steep, short slopes would be found cutting across the inner edges of the strata. The folds are, however, not symmetric, for the axes lie near the southwestern margins, hence the southwestern· slopes are steep, the northeastern slopes relatively gentle and uniform in character. The contrast is identical in general kind, though

---

[1] Orestes St. John, U. S. Geol. and Geog. Surv. of the Terr. (Hayden Surveys), 1877, pp. 411–416.

[2] T. S. Brandegee, Teton Forest Reserve, 19th Ann. Rept. U. S. Geol. Surv., pt, 5, 1897–98 pp. 195–197.

somewhat dissimilar in detail to that afforded by the eastern steep and western less steep slopes of the Tetons. As in the latter case also, erosion has denuded the sedimentary cover and exposed portions of the underlying crystallines. These do not, however, form any notable portion of the mountain crest, for denudation is here far less advanced and the mountain summits are still largely developed upon sedimentary rock.[1]

EXTRA-MARGINAL RANGES

On the eastern and the western borders of the central Rockies are two mountain ranges which stand out prominently at a little distance forward from the main mountain front. The eastern range is the Bighorn Mountains; the western range is the Uinta Mountains. Although they are here classified as a portion of the central section of the Rockies they should be regarded as having no very close geologic or geographic affinities with the Wyoming ranges described above.

UINTA MOUNTAINS

The Uinta range is peculiar in trending in an east-west direction in a mountain region where the prevailing trends are north-south. It is not altogether unique in this respect, however, for the Owl Creek range is a fold having a similar trend, and others in Wyoming and elsewhere having the same trend have been discovered.[2]

The Uintas form a rather flat, elliptical dome or elongated arch 150 miles long from east to west and with an average width of 20 to 25 miles. The interior of this elongated dome is a deeply dissected, plateau-like region, in general about 10,000 feet high, from which rise narrow ridges and peaks 12,000 to 13,000 feet high developed upon horizontally bedded quartzites.[3]

The strata of the interior plateau nowhere depart more than 5° or 6° from a horizontal position except in a small number of local instances. On the margins of the mountain belt the beds dip more steeply, in some cases over 45° and in extreme instances nearly 90°. The marginal zones of highly inclined strata are also affected by minor faults and folds, the principal displacement being along the northern side of the arch, Fig. 111.[4]

[1] Orestes St. John, U. S. Geol. and Geog. Surv. of the Terr. (Hayden Surveys), 1877, pp. 208 et seq.

[2] N. H. Darton, Senate Document No. 219, 1906.

[3] S. F. Emmons, Uinta Mountains, Bull. Geol. Soc. Am., vol. 18, 1907, pp. 287–302.

[4] W. W. Atwood, Glaciation of the Uinta and Wasatch Mountains, Prof. Paper U. S. Geol. Surv. No. 61, 1909, p. 9. See also J. W. Powell, Geology of the Uinta Mountains, 1876.

The physiography of the Uinta Mountains is closely related to the geologic structure. The broad open valleys of gentle contour depend upon sculpturing of soft horizontal strata beneath a capping of harder beds. The steeply inclined strata on the flanks of the range have been eroded to the point where they stand out as a series of hogback ridges separated by parallel trough-like valleys. The streams run in a series of rapidly deepening canyons with nearly vertical walls 3000 to 4000 feet high, and are deepest where they cross the hogback ridges on the

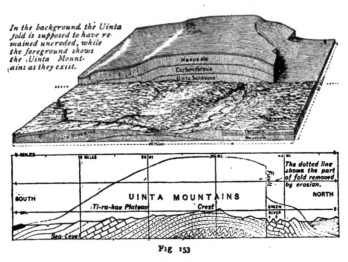

Fig. 111. — Stereogram and cross-section of the Uinta Mountains arch. (Dryer, adapted from Powell.)

margins of the anticline. The axis of the Uinta uplift is near the northern margin and the canyons of the northern slope are therefore shorter than those of the southern slope. The inter-canyon spurs are broad and flat and are overlapped by gently sloping Tertiary beds in places up to elevations of 10,000 feet.

The Green River enters the northern flank of the Uinta Mountains at the Flaming Gorge, then swings out again, only to return and cross the main axis of the range at the eastern end through the Ladore Gate or Canyon. Powell described the Green River as an excellent illustration of an antecedent stream that had maintained its course across the Uinta Mountains during the period of their uplift.[1] The later studies of Emmons show that Powell's hypothesis must be modified, for it involves physical impossibilities.

[1] J. W. Powell, Geology of the Uinta Mountains, 1876.

The upturned and truncated edges of the various formations of the Uinta arch are partly covered by overlapping Tertiary beds in a nearly horizontal position. These beds reach altitudes of 9000 to 10.000 feet at various points on either flank of the higher western portion of the range. The Green River Canyon, at the Flaming Gorge, has an elevation in the eastern portion of the range of 7500 to 9000 feet. From these conditions it seems clear that (1) the mountain-making movements which gave rise to the Uintas were accomplished principally at the close of the Cretaceous, though a small subsequent movement is allowable on stratigraphic grounds (each of the three series of Tertiary beds is marked by an erosion interval between successive beds and by a slight upturning on the flanks of the Uinta Mountains); (2) the Green River had its course directed as a consequent stream upon the surface of the overlying Tertiary deposits; (3) the dissection of the Tertiary beds allowed the river to become superposed upon the buried portion of the Uinta arch; (4) further erosion both within and on the borders of the range has intensified the topographic expression of the Uinta arch and to some extent obscured the origin of the course of the Green River.[1]

FORMS DUE TO GLACIATION [2]

In addition to the influence of alternating hard and soft beds on the valley outlines, glaciation has operated to widen and deepen the canyons and in many cases to give them characteristic U-shaped profiles. The floors of the basins in the western part of the range are about 9000 feet in elevation and every large canyon that heads near this part of the crest has been glaciated. The glaciers in the central portion of the range were 20 to 27 miles long; elsewhere they were but 4 or 5 miles long. The ice slopes about Bald Mountain and Reed's Peak near the western end of the range coalesced to form a great ice cap, Fig. 112. The greater portion of the divide was, however, not covered by ice and the loftier peaks rose above the snow-fields. The basins at the heads of the canyons on the northern slope vary in area from 1 to 12 square miles, while many of those on the southern slope are 20 to 30 square miles in extent, a difference which is due to the fact that the southward-flowing streams have developed valleys in the plateau-like summit where the beds are nearly horizontal and where the conditions are therefore most favorable for glacial plucking and sapping and for the development of broad flat-bottomed cirques. The northward-flowing glaciers worked upon steeply inclined strata and were resisted by every hard stratum in the section instead of a single hard stratum; hence they were but little assisted by sapping, for the soft beds whose removal gave rise to the sapping of the harder beds above them soon dipped down beneath the plane of effective action. The result was not only a greater amount of work to be performed by the northward-flowing streams but also a restriction of the fields of nourishment which correspondingly decreased the intensity of glaciation.

[1] S. F. Emmons, Science, n. s., vol. 6, 1897, p. 131 et al.

[2] W. W. Atwood, Glaciation of the Uinta and Wasatch Mountains, Prof. Paper U. S. Geol. Surv. No. 61, 1910.

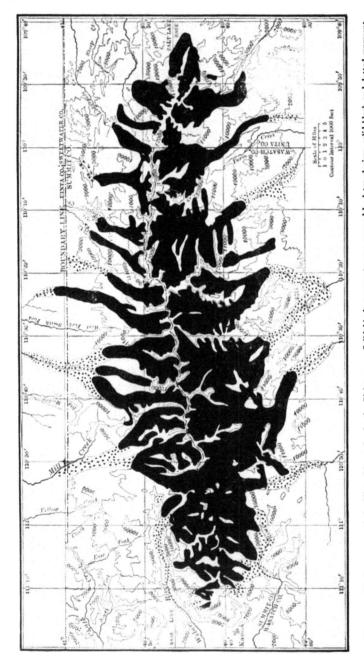

Fig. 112. — Glacier systems of the Uinta Mountains in the Pleistocene. Solid black represents area covered by later glaciers, solid black and dotted represent area covered by earlier glaciers. (After Atwood, U. S. Geol. Surv.)

In the higher portions of the range the forest growth is extremely scanty, but in the lower valley basins there is a heavy growth of coniferous forest. The upper limit of forest growth is about 11,000 feet, the lower limit on the southern slopes is beyond the base of the range in the Uinta Valley and is less than 7000 feet, while on the northern slopes it is 9000 feet owing to differences of relief and to corresponding differences of exposure. The prevailing species are white pine, yellow pine, Engelmann spruce and Douglas spruce, besides species of lesser importance.

"The view from one of the mountain lakes, with its deep-green water and fringe of meadowland, set in the somber frame of pine forests, the whole enclosed by high walls of reddishpurple rock, whose horizontal bedding gives almost the appearance of a pile of cyclopean masonry, forms a picture of rare beauty." [1]

BIGHORN MOUNTAINS

The Bighorn Mountains rise from 4000 to 5000 feet above the Great Plains to from 10,000 to 13,000 feet absolute altitudes. The range trends north-northwest in its northern portion, due north and south

Fig. 113. — East-west section across highest part of Bighorn Mountains, showing extent of erosion and the contrast in dip on the two slopes of the range.

in its central portion, and east and west where it joins the main Rocky Mountain System whose border ranges are known here as the Bridger Range and the Owl Creek Mountains.[2] It is in effect an extension or prodigious spur of the Rockies jutting far out beyond their border.

The Bighorn Mountains are a great anticline lifted many thousands of feet and composed of a thick series of sedimentary rocks (Paleozoic and Mesozoic) and a central core of granite and schist (pre-Cambrian).[3] The sedimentary rocks arch over the northern and southern portions of the uplift, and give rise to elevated plateaus about 9000 feet high at the north and 8000 feet high at the south. On the divides these plateaus often present broad tabular surfaces though they are trenched by numerous large canyons. The central plateau terminates in high cliffs at certain places in the border region, but in others and especially

[1] S. F. Emmons, Geol. Expl. of the 40th Par. (King Surveys), vol. 2, 1877, pp. 194–195.
[2] N. H. Darton, Geology of the Bighorn Mountains, Prof. Paper U. S. Geol. Surv. No. 51, 1906, pp. 10–11.
[3] Darton and Salisbury, Cloud Peak-Fort McKinney Folio U. S. Geol. Surv. No. 142, 1906, p. 1, cols. 2, 3.

Fig. 114. — Sharp upturn of strata on the east side of limestone front ridge of Bighorn Mountains, Wyo. The topographic expression of the upturned beds are clearly shown in the sharp ridges of the photograph. (Darton, U. S. Geol. Surv.)

along the eastern side of the mountains it is flanked by a distinct ridge of limestone that rises slightly above an inner valley and slopes steeply toward the plains. These plateaus are covered by forests with many park-like openings and are extensively used as grazing grounds for cattle and sheep during the short summer season.

The subordinate topographic features of the Bighorns depend upon local flexures on the flanks of the main uplift and a number of faults, the latter being especially marked northwest of Buffalo and on the eastern border of the range, where one of the faults has a throw of about 9000 feet. The effect of the sharper flexing and of the pronounced faulting has been to increase the development of precipitous slopes, and to cause a greater development of hogback topography especially on the eastern side. The hogback ridges rise from 100 to 200 feet above the adjoining valleys, and are due to the outcrop of sandstone of moderate hardness, while the valleys follow the outcrop of the "Red Beds" composed largely of soft shales.

The general configuration of the central cluster of high mountains in the Bighorns is very rugged and presents some of the boldest alpine scenery in the United States. There are many precipices over 1000 feet high, particularly in the glacial cirques along the higher summits. Some of the cirques still hold glaciers; one on the eastern side of Cloud Peak is nearly one-half mile long. Extensive snowbanks remain the entire summer in many higher portions of the range.[1]

GLACIAL FORMS

At the time of maximum glaciation in the last glacial epoch the snow and ice accumulations of the Bighorns were two or three times as extensive as those of Switzerland to-day and the largest glacier was considerably larger than the largest existing Swiss glacier.[2] For thirty miles the crest of the range has been glaciated and signs of glaciation are everywhere abundant. The topographic and climatic conditions governing the distribution and growth of former glaciers were unusually favorable. Glaciers started from approximately one hundred sources, but in descending they combined in many cases, so that the number of separate glaciers or glacier systems was but nineteen. A specific instance is the Pear Rock Glacier, which started from not less than twenty sources.

---

[1] Darton and Salisbury, Cloud Peak-Fort McKinney Folio U. S. Geol. Surv. No. 142, 1906, p. 1.

[2] Idem, p. 9, col. 4.

Fig. 115. — Wall at head of cirque, upper end of a tributary of West Tensleep Creek, Bighorn Mountains. A. View from above rim, showing old rounded surface. B. View from below rim, showing granite walls nearly 1000 feet high. (Blackwelder, U. S. Geol. Surv.)

During both the advance and the retreat of the ice the number of separate glaciers was greater than during the time óf maximum glaciation, for the ice was melted out of the lower glacial troughs, while it still lingered in the separate tributary valleys. The glaciers on the

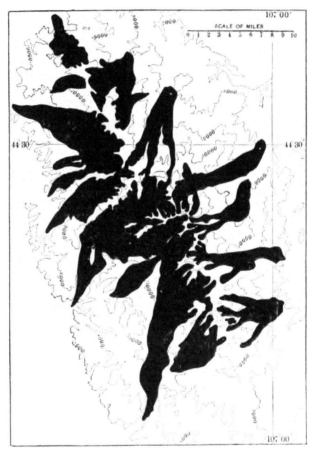

Fig. 116. — Glacier systems of the Bighorns at the time of their maximum development. (After Blackwelder, U. S. Geol. Surv.)

west side covered almost twice as great an area as those on the east and they were also somewhat longer, a difference due to the heavier precipitation on the west side (prevailing westerly winds) and the wider catchment basins. The surviving glaciers are now found wholly on the east side, a condition probably due to the better protection of the cirques on that side from the sun and in part to the westerly winds which drift snow over the crest of the range and into the cirques and alcoves on the lower side. The lower limits of the glaciers were from 6500 feet to 10,500 feet, the necessary elevation for the generation of glaciers in the Bighorn Mountains during the last glacial epoch ranging from 9500 to 11,500 feet.[1]

Glacial cirques and hanging valleys are of frequent occurrence, though arêtes and needles are absent because glacial sapping did not continue long enough to allow the meeting of glaciers on the divides.[2] The cirque walls have not been notably denuded in postglacial time because of the compact and relatively structureless granite in which they were carved and in part because of the recency of formation. The cirques have been formed on a huge scale, for the original valleys were often far apart and neighboring cirques were thus developed without mutual interference on the lateral divides.[3]

Below the cirques are steep valleys whose gradients range from 600 to 800 feet per mile. Where the valleys cross the great central granite plateau they are shallower and more open; but where they cross the sedimentary rim they become narrow gorges. Some of the longest glaciers entered these gorge-like marginal valleys, but none of them reached out upon the plains. Among the most important features of the present drainage are the large numbers of lakes due to glacial over-deepening or to the presence of morainic dams. Some of the terminal moraines of the Bighorns are of great size, the undulations of their surface reaching as high a value as 800 feet. The terminal moraine of North Fork is 300 feet above the level of the valley floor and its topography is very irregular.[4]

FORESTS

Nearly all the trees of the Bighorn National Forest (which includes most of the forested area of the mountains) are lodgepole pine (*Pinus murrayana*, locally called white pine), yellow pine, and jack pine. Another species is limber pine of scattered growth. Engelmann spruce occurs in some of the moister areas along the mountain slopes and the higher portions of some of the canyons. The dry or lower timber line

[1] Darton and Salisbury, Cloud Peak-Fort McKinney Folio U. S. Geol. Surv. No. 142, 1906, p. 10, col. 1.

[2] F. E. Matthes, Glacial Sculpture of the Bighorn Mountains, Wyoming, 21st Ann. Rept. U. S. Geol. Surv., pt. 2, 1899–1900, p. 175.

[3] Idem, p. 176.

[4] Darton and Salisbury, Cloud Peak-Fort McKinney Folio U. S. Geol. Surv. No. 142, 1906, p. 11, col. 3.

occurs at 6000 feet, and the cold timber line at 10,000 feet, so that the upper portions of the range do not generally support a timber covering of any importance, but on the whole expose bare rock-strewn summits. The wood of the white pine, the predominating species, is coarse-grained, knotty, and small. The lumber warps and cracks considerably, and it is therefore not regarded as valuable.[1]

[1] For detailed distribution of the timber of the Bighorns, etc., see 19th Ann. Rept. U. S. Geol. Surv., pt. 5.

CHAPTER XX

ROCKY MOUNTAINS. III

SOUTHERN ROCKIES

IN contrast to the irregularly arranged mountains of the Central Rockies with variable trend, numerous offsets, and plain and plateau interruptions are the northward-trending, somewhat regular, and continuous mountain ranges that constitute the southern Rockies. There

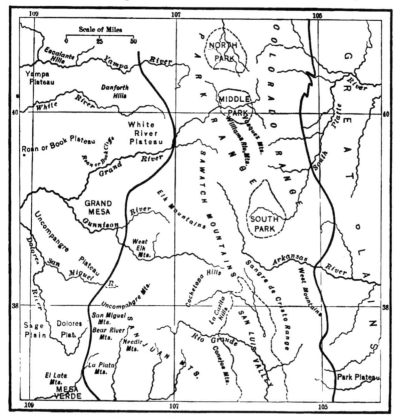

Fig. 117. — Location map of the southern Rockies.

appears to be a rather well-marked and consistently developed mountain plan in the eastern and central ranges of the district; the westernmost ranges are of more diverse structure, topographic texture and origin, and occur in groups of longer and shorter lengths arranged on the whole with less regularity.

Fig. 117 represents the locations of the main topographic features of the region, which fall readily into five categories:

(1) The eastern foothills, of varied origin and commonly exhibiting the hogback type of topography.

(2) The Colorado or Front Range and the Wet Mountains, consisting in a broad way of a great anticlinal from whose central granite axis the flanking sedimentary beds dip east and west.

(3) The Park, Sawatch (Saguache of some maps), and Sangre de Cristo ranges, all of which lie nearly on the same meridian, trend northward, have roughly equivalent structures and topographic forms, and are parallel to the Colorado Range, from which they are separated by

(4) a chain of "Parks," North, Middle, South, and Huerfano, four intermont basins of exceptional character, primarily of structural and secondarily of alluvial origin (San Luis Park, the largest of all the intermont parks of Colorado, lies between the Sangre de Cristo range and the southern continuation of the Sawatch Mountains — the Conejos, etc.).

(5) Irregular mountain knots or groups of igneous origin which lie beyond the Park-Sawatch axes, such as the Elk, La Plata, San Juan, and Uncompahgre mountains.

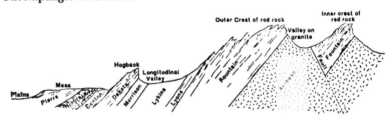

Fig. 118. — Generalized east-west section near Boulder, Colorado, showing the structure and topography of the South Boulder Peaks and the hogback ridge of Dakota Sandstone. (Fenneman, U. S. Geol. Surv.

EASTERN FOOTHILLS

HOGBACK TOPOGRAPHY

The eastern foothills of the southern Rockies are generally formed upon a belt of sedimentary beds upturned at steep angles along the mountain slopes. The most characteristic feature is the hogback ridge, Fig. 118, developed in the form of long narrow ridges of harder beds that stand like a fringing and often a serrated wall at a little distance from the base of the mountains. They occupy a belt about two miles wide involving portions of all valleys intervening between them and the main range. The easterly dip of the sedimentary beds of the foothills is quite general, and ordinarily varies from 35° to 50°, though locally it is from 80° to 90°

Besides the general flexing of the strata, which is due in part at least to the uplift of the Colorado Range, there has been a considerable amount of minor crumpling along lines parallel with the general trend of the range or diverging from it to some degree. The latter process resulted in a series of secondary folds reflected in the topography by successive offsets. In such localities the ridges are arranged *en échelon.* Their general appearance is shown in Fig. 119, which represents a series occurring between St. Vrains and Cache la Poudre creeks, north of Denver.

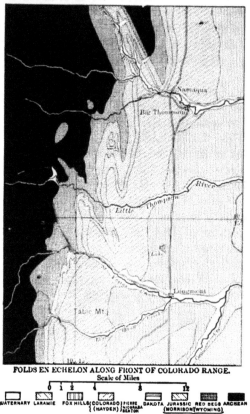

FOLDS EN ECHELON ALONG FRONT OF COLORADO RANGE.
Scale of Miles

QUATERNARY LARAMIE FOX HILLS(COLORADO | PIERRE DAKOTA JURASSIC RED BEDS ARCHEAN
                     { (HAYDEN) | NIOBRARA   (MORRISON(WYOMING)
                                  BENTON   JURA-TRIAS

Fig. 119. — Cross folds on eastern border of Colorado Range.
(U. S. Geol. Surv.)

VEGETATION

The vegetation of the Pueblo region is closely related to climate and soil, and in this respect and in its general character it is typical of the vegetation of the hogback ridges of Colorado. The uplands above 6500 feet bear a forest of yellow pine, and straggling pine growths occur on sheltered hillsides down to 5300 feet; in the same zone rocky slopes are sometimes covered with aspens. Hackberries and other hard-wooded trees grow to moderate size in the moister valleys, while cottonwoods fringe the lower streams; juniper and piñon pine extend down to 5000 feet and are associated chiefly with dry rock-strewn ledges of resistant limestone and sandstone.

OTHER TYPES OF FOOTHILL TOPOGRAPHY

While the hogback type of mountain border or foothill topography is the predominating one there are a number of other types that require examination. They are not all confined to the southern Rockies, but are assembled here for comparison with the well-developed foothill types of Colorado. The hogback ridges are absent (1) where the harder beds rest directly upon the crystalline rocks of the mountain slopes, without the presence of a soft valley stratum between the crystallines and the hard sedimentary formations overlapping them, of which some of the foothills on the eastern border of the Bighorns are perhaps the best illustrations; or (2) where the upturned edges of other sedimentary (Mesozoic) beds are still covered by a capping of younger (Tertiary) beds in a horizontal position; or (3) where faulting has brought the sedimentary rock in a nearly horizontal position against the deformed crystallines as at Pueblo, Colorado. (4) The eastern border of the Rocky Mountains in Montana is developed upon strata that have been deformed in such a manner that they dip westward and present a reversed border, (a) a single, steep, precipitous, eastern scarp overlooking the Great Plains, such as marks the eastern front of the Lewis Mountains, Fig. 92, or (b) a series of scarps and flats formed upon alternate soft and hard layers — short, steep, eastern slopes and long, gentle, mountainward or western slopes, Fig. 93. In the Bighorn Mountains and in the Colorado Range, on the other hand, the marginal strata in general dip steeply eastward, with the consequence that typical hogback ridges are characteristic. (5) Lava flows have occurred locally and have caused the formation of a mesa type of topography or (6) volcanic cones surmount the eastern border. Besides the hogback type six types of foothill topography are thus distinguishable along the eastern border of the Rockies. Instead of being everywhere formed simply upon the basset edges of the sedimentary beds exposed along the mountain flanks, the border has been formed in diverse ways and exhibits diverse topographic features. While each of the types indicated has distinctive features the first five types are so closely related as not to require detailed explanation. The sixth type — the igneous border — is well developed in the southern Rockies and has rather complex features that merit further consideration.

IGNEOUS BORDER

SPANISH PEAKS

In southern Colorado and locally elsewhere in that state, as at Golden, the strata have been deformed or to some extent concealed and the border given an unusual character by the occurrence of igneous masses which in the case of the Spanish Peaks consist of volcanoes and associated dikes and lava flows. The type of mountain border exhibited is wholly unlike that found anywhere else along the eastern face of the Rockies from Alaska to New Mexico.

The most prominent feature of this kind is the volcanic nucleus known as the Spanish Peaks. The two culminating points, West Peak

Fig. 120. — West Spanish Peak, from the northwest. One of the large dikes of the region is seen in the foreground. (U. S. Geol. Surv.)

and East Peak, are about 2 miles apart and have elevations of 13,623 feet and 12,708 feet respectively. They rest upon a small platform that descends eastward gradually and terminates in an irregular and deeply dissected line of steep bluffs that descend abruptly 500 feet to the gently rolling plains. The average elevation of this platform is about 7500 feet. Its surface is a succession of mesa-like ridges and narrow valleys of extremely rugged development. The peaks themselves consist of stock-like masses occupying nearly vertical fissures, and from them the flanking strata dip steeply away with various structural complexities. The general ruggedness of the country about them is emphasized by many dikes which have a rude radial arrangement with respect to the peaks which were the centers of eruption. They were formed in crustal fractures related to the deformation of the surface strata, deformations produced by the intrusion of great masses of igneous rock.[1] They are from 2 to 100 feet in thickness. Some of them are practically vertical and project above the surface as smooth wall-like masses from 50 to 100 feet high. Their crests are sometimes straight for long distances, although curved crests are most common. They range in length from a few hundred yards to 10 and 15 miles, and intersections are common.[2]

The upland portion of the region adjacent to the Spanish Peaks is rather heavily timbered and the eastern borders of the plateau support a dense growth of piñon and juniper, scattered pines, and scrub oak, with occasional park-like openings. The more elevated central, southern, and western borders support forests of pine; a certain amount of spruce, fir, and aspen grows at the base and on the lower slopes. The summits of the peaks are from 1000 to 1500 feet above the cold timber line. The plains eastward of the plateau upon which the Spanish Peaks stand are practically destitute of timber except for fringes of cottonwood along the running streams. The whole plateau portion is subject to frequent summer showers. The summits of the Spanish Peaks are rarely free from snow for more than two months in the year, while along the narrow valleys cut below the level of the plateau irrigation is necessary, as also upon the surface of the plains country toward the east.[3]

NORTH AND SOUTH TABLE MOUNTAIN

Among the various types of foothill topography one of the most important is the flat-topped mesa type, which is well shown in the vicinity of Golden. Here the hogbacks completely disappear and in their place are two basalt-capped, sedimentary masses known as North and South Table Mountain. They were originally continuous but have since been cut through and isolated (Clear Creek). The basalt was extruded after the strata had been flexed into steep attitudes and somewhat eroded. It consists of two fissure flows with no sedimentary rock between them.[4] They protect the underlying beds from erosion; their present elevation indicates the degree of dissection that has taken place in the vicinity since their formation.

[1] R. C. Hills, Spanish Peaks Folio U. S. Geol. Surv. No. 71, 1901, p. 3, col. 1.
[2] Idem, p. 3, col. 4.
[3] Idem, p. 1, col. 1.
[4] Emmons, Cross, and Eldridge, Geology of the Denver Basin in Colorado, Mon. U. S. Geol. Surv. No. 27, 1896, p. 285.

The foothill terraces are among the important features on the eastern border of the Colorado Range. They are mesalike remnants of stream terraces carved in rock, Fig. 118, standing out with special prominence along the base of the foothills south of Boulder where they rise from 100 to 300 feet above the lower plains. They vary in width from a fraction of a mile to three miles, and descend eastward with slopes varying from 1° to 10°. They appear to be due to a leveling of the rock margin of the plains by meandering streams and the deposition upon the plains surface of a thin sheet of unassorted gravel.[1] The terraces always slope toward the streams, with which they once had flood-plain relations.

COLORADO OR FRONT RANGE

The Colorado or Front Range is the easternmost of the imposing chains that form the southern Rockies. Its elevation above the surrounding country is represented in Fig. 121.

On the north the Colorado Range is a high, heavily timbered, plateau-like range separated from the Medicine Bow Mountains by the Big Laramie River, and from the Laramie Mountains by the North Fork of the Cache la Poudre; but throughout most of its extent it is alpine in character, with sharp serrated summits bordered by great glacial amphitheaters.[2] The broad anticlinal of the range has a very flat summit arch, with increasing dip toward the flanks of the anticline on the borders of the range.[3] The rocks of the Colorado Range are almost exclusively crystalline granites, gneisses, and schists.[4] The topographic features are not uniform throughout and we shall therefore describe them by groups.

West and northwest of Denver and north of Arapahoe Peak, the serrated crest of the Colorado Range is exceedingly formidable, with many peaks rising to 13,000 and 14,000 feet. Even the passes are high and difficult, and on either side are steep-sided and deep mountain gorges. The irregularity of the mountains and their steep descents are suggested by the northeast face of Long's Peak (14,270), which consists of an almost perpendicular cliff over 3000 feet high which

---

[1] N. M. Fenneman, Geology of the Boulder District, Colorado, Bull. U. S. Geol. Surv. No. 265, pp. 13-15.

[2] S. B. Ladd, U. S. Geol. and Geog. Surv. of Colorado and Adj. Terr. (Hayden Surveys), 1873, p. 436.

[3] Clarence King, U. S. Geol. Expl. of the 40th Par. (King Surveys), vol. 1, 1878, p. 21.

[4] Arnold Hague, Rocky Mountains, U. S. Geol. Expl. of the 40th Par. (King Surveys), vol. 2, 1877, p. 22.

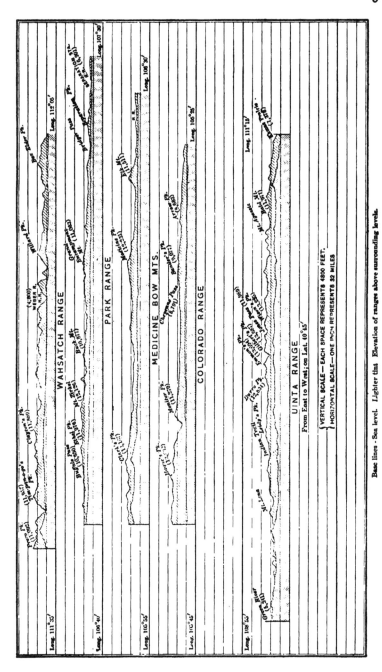

Base lines - Sea level. Lighter thui Elevation of ranges above surrounding levels.

Fig. 121. — Longitudinal profiles of five prominent ranges in the Rocky Mountain province. (King Surveys.)

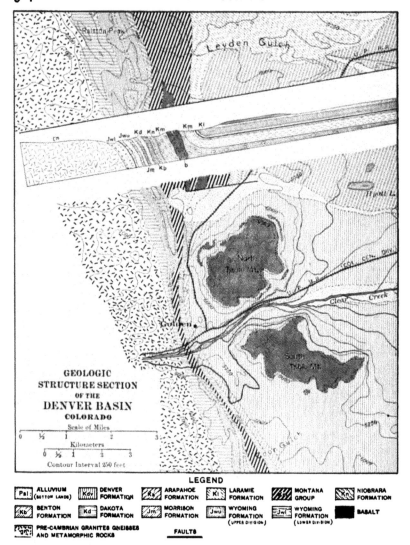

GEOLOGIC
STRUCTURE SECTION
OF THE
DENVER BASIN
COLORADO

Scale of Miles

0   ½   1   2   3

Kilometers

0   ½   1   2   3

Contour Interval 250 feet

LEGEND

Pal ALLUVIUM (BOTTOM LANDS)	**Kdv** DENVER FORMATION	**Ka** ARAPAHOE FORMATION
Kb BENTON FORMATION	**Kd** DAKOTA FORMATION	**Jm** MORRISON FORMATION
Kl LARAMIE FORMATION	**MONTANA GROUP**	**NIOBRARA FORMATION**
Jwu WYOMING FORMATION (UPPER DIVISION)	**Jwl** WYOMING FORMATION (LOWER DIVISION)	**BASALT**

PRE-CAMBRIAN GRANITES GNEISSES AND METAMORPHIC ROCKS   **FAULTS**

Fig. 122. — Section on the common border of Great Plains and Southern Rockies, showing structure and topography and the lava-capped mesas near Golden, North Table, and South Table mountains. (U. S. Geol. Surv.)

extends from timber line (11,000 feet) to the mountain summit. Its eastern face presents an almost continuous line of steep-walled amphitheaters; the western side of the range is a zone of high mountains 5 to 10 miles across which have steep but not precipitous slopes that descend to the valley of the Grand and are cut by profound canyons.

Southward for 12 miles from Arapahoe Peak the topography changes. The crest presents a very uniform ridge but little above timber line. Like the section north of Arapahoe Peak the eastern face of this section is diversified by amphitheatral valley heads between which are broad rounded spurs of such regular descent that wagon roads ascend some of them. The western face of this portion of the range is marked by massive spurs with rounded forms not separated by canyons.[1]

The whole eastern border of the Colorado Range is usually well defined and even sharply defined, and all along it the mountain spurs end abruptly. They rise from 5000 or 6000 feet, the level of the border of the plains, to 8000 feet at the foot of the main range, and involve a belt about 8 miles wide. The streams draining the border have cut deep parallel valleys and the intervening ridges therefore have an east-west alignment. They are not sharp but massive and have rounded or level summits and steep margins. Their crests fall into a general level in a quite remarkable manner. On the western border of the Colorado Range the descent is also rather abrupt to a general plateau between 8500 and 10,000 feet high which stretches westward for many miles.[2]

These border plateaus have been studied in detail in the Georgetown region west of Denver and their topographic relations clearly defined. It has been found that the Colorado Range here exhibits three sets of topographic forms: (1) an old, mountainous upland, (2) young V-shaped valleys incised in the upland, (3) glacial cirques developed in the valley heads.

The old upland surface, but little modified by recent erosion, occurs in remnants preserved on the ridge crests, which in part accounts for their regularity of elevation, form, and alignment. This mountainous upland was an ancient land surface whose relief in the preceding topographic cycle would be adequately shown by filling the deep valleys cut into it.[3] Dome-shaped mountains and broad soft-contoured ridges stood where sharp peaks and more rugged ridges now are, and the valleys between were broader and less steep than those of the present streams. The surface was well adjusted to that of the underlying rocks and to their varying resistances to erosion. Precipitous slopes were rare and occurred under special conditions, as on the southeastern border of the Alps Mountains of Colorado. Shallow valley heads led

[1] F. V. Hayden, U. S. Geol. and Geog. Surv. of Col. and Adj. Terr., 1873, pp. 86–87.

[2] Idem, pp. 88–89.

[3] Spurr and Garey, Geology of the Georgetown Quadrangle, Prof. Paper U. S. Geol. Surv. No 63, 1908, pp. 31–36. N. M. Fenneman some time ago called attention to the fact that these features appear to indicate an imperfectly base-leveled surface that has been trenched by streams invigorated by recent uplift: Geology of the Boulder District, Col., Bull. U. S. Geol. Surv. No. 265, 1905.

down gently from the dome-shaped mountains of the old upland and
joined wide gentle valleys that were broad and basin-like. All these
members of the ancient stream systems had normal profiles and no
reversed slopes existed; consequently lakes did not then exist. Although
both normal and glacial erosion have modified this old upland, rem-

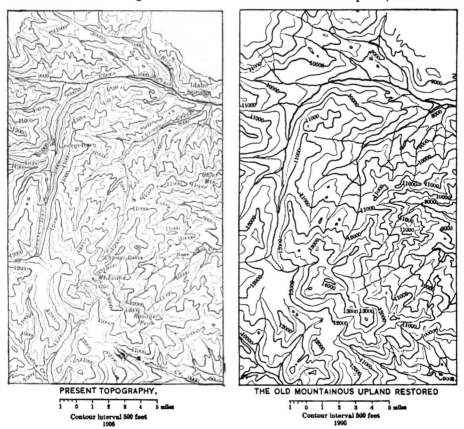

PRESENT TOPOGRAPHY.   THE OLD MOUNTAINOUS UPLAND RESTORED

Contour interval 500 feet   Contour interval 500 feet
          1906   1906

Fig. 123. — Old mountainous upland of the Georgetown district, Colorado, and present topography.
(Spurr and Garey, U. S. Geol. Surv.)

nants of it may still be seen and such remnants bear residual elevations
and in many places are covered by deep soil, conditions that were
probably general over the larger portion of the area before uplift and
glaciation took place, so that rock exposures were comparatively uncommon except on the steepest slopes.[1]

[1] Spurr and Garey, Geology of the Georgetown Quadrangle, Prof. Paper U. S. Geol. Surv.
No. 63, 1908, p. 52.

The mountainous upland of the Colorado Range, formed in late Tertiary time, was then uplifted and tilted eastward so that valley carving was begun. Canyons of considerable size were opened up before the glacial period and these the streams are still enlarging. Clear Creek is in places 2½ miles wide, while the gorges along the Deer, Elk, and Bear creeks, all in the eastern slope of the range, are in general not more than one-half mile wide. The valley walls approach each other upstream and the valleys finally become narrow V-shaped notches in the old upland surface. Rock projects from the valley sides and huge bowlders lie along the courses of the swift streams. The old upland surface is comparable with similar uplands (a) in the vicinity of Pikes Peak,[1] (b) in the Sierra Madre Mountains of south-central Wyoming,[2] (c) remnants of an ancient surface widely distributed over other portions of Colorado,[3] and (d) mature profiles widely developed in other portions of the Rockies as in western Montana, p. 318. The character of the old mountainous upland is brought out clearly in the maps of the Hayden Surveys of 1877, where the summits of the long table-topped spurs of various plateau and mountain groups appear to indicate a rather general development of moderate slopes. All of these surfaces are approximately contemporaneous and of late Tertiary age. The mountainous surface is now exposed only in remnants on the spur tops and in certain of the unglaciated valleys.

The present-day expression of the detailed topography of the Colorado Range west of Denver is affected rather intimately by the character of the rock wherever dissection has partly destroyed the older graded surface. The country underlain by schists has very distinctive topographic qualities. Where the dip of the schistosity is low, smooth-contoured slopes are developed upon the exposures; where it is approximately vertical the topographic forms are craggy and serrate as on the north side of Chief Mountain. The gneisses weather into exposures which at a distance have the bedded aspect of sedimentary rocks except where there is a prominent development of joints. The rugged crags found almost continuously along stream valleys carved in gneiss are controlled in their minor details by rectangular joint systems and have very characteristic features. Turrets and buttresses and castellated forms are seen throughout the area covered by the granite-pegmatite, which is one of the most resistant massive rocks of the region.

[1] W. Cross, Mon. U. S. Geol. Surv., vol. 27, p. 202.
[2] A. C. Spenser, Prof. Paper U. S. Geol. Surv. No. 26, 1904, p. 12.
[3] F. V. Hayden, U. S. Geol. and Geog. Surv. of the Territories, Atlas of Colorado, 1877, Sheets 5, 6, 7.

Below timber line the granite weathers into typical dome-shaped forms, the domes being from 200 to 300 feet high and from 300 to 400 feet in diameter; above timber line the domes are absent owing to the strong temperature changes and their shattering effect.[1]

In the Pikes Peak district of the Colorado Range the chief topographic feature is a great plateau developed upon granite and gneiss and modified by large extrusions of volcanic rock that form mountains upon it. Sedimentary beds, formerly of greater extent, are now confined chiefly to the borders of the tract and to small local basins on the plateau.[2]

GLACIAL FEATURES

The higher parts of the main ranges of Colorado were for the most part covered with ice during the last glacial epoch, only the narrow crests of the inter-valley ridges projecting above the glacial cover.

The ice was deepest and most powerful in the larger mountain valleys down which it extended as valley glaciers for great distances, in some cases out over the piedmont plains or plateaus below.[3] The number of glaciers which existed in the Colorado Range during the glacial epoch may be appreciated by the fact that within the Leadville quadrangle alone, p. 371, there were 26 large glacial systems and 11 smaller ones, or 37 in all, in an area of 945 square miles.[4] The glaciers ranged in elevation from a minimum height of 8800 feet to a maximum height of 13,700 feet. Since many peaks of the Park Range reach elevations approximating 13,000 feet it follows that all the larger valleys were occupied by glaciers. The larger and more vigorous glaciers developed on the eastern slopes, a condition due probably to the greater size of the preglacial valleys rather than to any advantages of exposure or precipitation. As a result of glaciation in the Colorado Range cirques and glacial lake basins were formed. The cross sections of the valleys were developed into a pronounced U shape, so that a shoulder now occurs between a lower, younger, and slightly modified glaciated portion and an upper, older, unglaciated portion. The floors of both cirques and glaciated valleys are very uneven and

---

[1] Spurr and Garey, Geology of the Georgetown Quadrangle, Col., Prof. Paper U. S. Geol. Surv. No. 63, 1908, pp. 35–36.

[2] W. Cross, Pikes Peak Folio U. S. Geol. Surv. No. 7, 1894, p. 3.

[3] The extent of any glacier depends upon the following features: (a) the height of the surrounding mountains; (b) the shape and size of the catchment basins; (c) the number and size of the tributary valleys; (d) the slope and shape of the valley floor; (e) the exposure; and (f) the precipitation.

[4] S. R. Capps, Jr., Pleistocene Geology of the Leadville Quadrangle, Col., Bull. U. S. Geol. Surv. No. 386, 1909, p. 9.

smooth rock ledges protrude above the till. Lateral moraines, sometimes double, form prominent narrow ridges of till on either side of the valley walls and are here the most striking types of glacial deposits. The terminal moraines have an irregular surface marked by hillocks, ridges, and depressions and many of them have very steep fronts over which the streams descend in a succession of waterfalls. The cirques are the most striking feature of glacial erosion in the region; the deepest are in the vicinity of Mount Evans, where one cirque wall rises 1700 feet in less than a mile. The steeper cirques have bare, rocky, unscalable walls and frequently have postglacial talus accumulations along their bases. Some of them are compound, as in the case of the three well-defined cirques at different elevations in the valley of the stream that heads in Summit and Chicago lakes, a mile west of Mount Evans. In a number of cases a knife-edge ridge or arête was formed by the headward erosion of two cirques on opposite sides of a divide in the mountainous upland.

TREE GROWTH

The main portion of the Colorado Range was heavily timbered below the cold timber line (11,250 to 12,000 feet) when the valleys of the

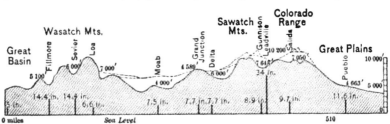

Fig. 124. — Topographic profile and distribution of precipitation across the Wasatch and the southern Rockies.

region were first settled. At altitudes over 8500 feet western yellow pine and Douglas spruce grow on dry areas. On higher and moister ground lodgepole pine and Engelmann spruce are found; these are superseded in turn by dwarf spruce and finally near timber line by mountain white pine. Large areas have been burned over and are now covered with second-growth lodgepole pine and aspen. Above timber line there is an alpine flora of many species.[1]

PARK RANGE

North of the Grand River Valley the Park Range is not high and rugged but a great, rolling, heavily timbered, even-contoured plateau of

[1] Spurr and Garey, Economic Geology of the Georgetown Quadrangle, Col., Prof. Paper U. S. Geol. Surv. No. 63, 1908, pp 30–31.

crystalline rock with massive marginal slopes. For 50 miles south of the Grand River Valley its character is entirely different. The summit ridges are rugged and sharp and bordered by great amphitheaters with steep, cliffed sides. Deep glacier-scored canyons alternate with the secondary ridges; at their mouths are belts of morainic drift formed by ancient glaciers whose courses were the present valleys or canyons and whose névé fields occupied the existing amphitheaters of the valley heads.[1] Glacial action was carried so far that parts of the range were thoroughly skeletonized. Some of the eastern spurs are flat-faced and have gentle descents. At the higher elevations the western slopes are precipitous being formed upon hard crystalline rock, but at lower elevations where sedimentary rock occurs they become gentler. On both flanks of the range the sedimentary formations seldom rise more than a few hundred feet above the bordering plains and plateaus. The culminating summits of the range are of granite.[2] In structure the Park Range is a great overthrust extensively eroded. The movement was directed westward, hence the western border is abrupt, for it has been developed upon the edges of the overthrust beds in a manner in some respects similar to the eastern border of the Lewis range of Montana (p. 311).[3] The highest peaks are much lower than those of the Colorado Range; Mount Zirkel alone reaches above 12,000 feet, and this in spite of the equivalent elevations of the main summits of the two ranges.[4]

TREE GROWTH

The Park Range is covered with large timber similar to that found on the Colorado Range. The main types are pines above and aspens and small low trees along the lower or dry timber line. In the local alluvium-floored basins timber is wanting, as in Egeria Park and the parks along the Yampah, except for a fringe of cottonwoods bordering the streams.[5]

SAWATCH RANGE

The Sawatch Range extends for over 80 miles from the Mountain of the Holy Cross (lat. 39° 28') southward to the San Luis Valley. "For this entire distance [it] literally bristles with lofty points about 10 of which rise above 14,000 feet and many more are 13,000 feet above sea

---

[1] Hayden Reports, 1873, pp. 178, 188, 189, 436, 437.
[2] Idem, pp. 65–66.
[3] Idem, 1874, p. 71.
[4] Arnold Hague, U. S. Geol. Expl. of the 40th Par. (King Surveys), vol. 2, 1877, pp. 130–132.
[5] S. B. Ladd, U. S. Geol. and Geog. Surv. of Col. and Adj. Terr. (Hayden Surveys), 1874, p. 441.

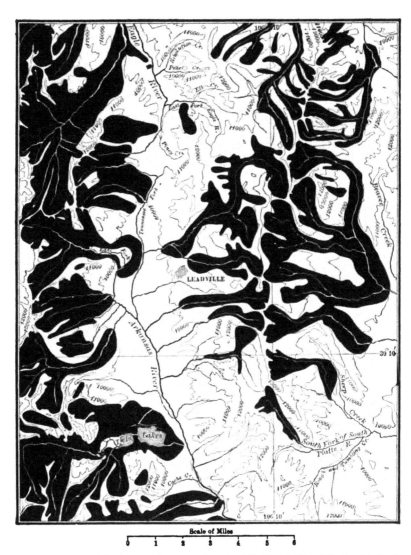

Fig. 125. — Extent of the former glacier systems in parts of the Park (east) and the Sawatch (west) ranges of Colorado, Leadville quadrangle. (After Capps, U. S. Geol. Surv.)

level." It has a rather symmetrical outline, and its pointed summits vary but little in either form or height.[1] On the east is the valley of the upper Arkansas; on the west is the valley of the Gunnison. Like the neighboring ranges on the north and east it is a great anticlinal uplift whose central axis is composed of crystalline rocks from which sedimentary formations dip away at various angles. This type of structure is, however, complicated on the southwestern flank of the range by volcanic flows which wrap about the Sawatch crystallines, as the Cachetopa Hills — a portion of the great igneous field of the San Juan region. Because of this and other structural irregularities, the range is not so simple as the Park and Colorado ranges, and is still more irregular in the southeastern extension of this mountain axis known as the Sangre de Cristo Mountains.

The higher portions of the Sawatch Range have been glaciated in common with the other high ranges of Colorado. Glacial cirques or headwater amphitheaters were developed and peaks trimmed to sharp, sometimes pyramidal outlines.

La Plata peak in the Sawatch Range has been glaciated to such an extent that it now stands as a sharp peak between the encroaching heads of three glacial channels. Its steep slopes merge into the steep slopes of the glacial cirques that occur northwest, southwest, and east of the peak. The forms of the mountain are all sharp in contrast to the rounded forms of the unglaciated lower portions of the adjacent mountains. Mount Elbert may be cited as a type of peak in which the glacial cirques have not encroached so far and where the summit of the mountain still retains a massive dome-like outline such as La Plata Mountain had before the erosion on its flanks had been carried so far.[2]

SANGRE DE CRISTO RANGE

The Sangre de Cristo Mountains, Fig. 117, rise abruptly on the east from Wet Mountain Valley and Huerfano Park, among the larger parks of Colorado, and on the west from the San Luis Valley. They are from 12,000 to 14,000 feet high with from 3000 to 5000 feet relative altitude. The range follows an almost straight line for 40 miles and presents a very imposing sight, the dark color of the rock bringing out its relief to advantage. Its average width is not much over 10 or 12 miles, which is small compared with its length and altitude. The range extends southward to Sangre de Cristo Pass, and beyond this point the axis of elevation extends to and beyond the New Mexico boundary to Santa Fe, where it ends in a number of minor ranges.

[1] Hayden Surveys, 1874, p. 54.
[2] W. M. Davis, Glacial Erosion in the Sawatch Range, Col., Appalachia, vol. 10, 1899, p. 403.

North of Poncha Pass the Sangre de Cristo Range is a true sierra. The Sierra Blanca group of peaks, for example, have a serrate and jagged crest line; their precipitous front rises abruptly from the flat San Luis plain to heights of more than a mile, thus giving the range a bold and majestic appearance.[1]

The Sangre de Cristo Range is a great and somewhat complex anticline. In general the axis of the range consists of intrusive granite flanked on both sides by conglomerate chiefly and also by sandstone, shale, and limestone. Faulting has locally caused the sedimentary formations to abut squarely against the basal granite.[2] In places the faults trend at an angle with the main axis of the anticlinal uplift, a structure that causes the topographic condition of offsets that diversify the mountain border.[3] Local variety is given the border relief by dikes of igneous rock which traverse the flanking sedimentaries. The streams have cut deep gorges in the upturned sandstones and limestones; steep bluffs occur along the valley sides and only an occasional patch of flat land occurs on the valley floors.[4]

WESTERN BORDER RANGES

The eastern and central divisions of the southern Rockies are characterized by the presence of granitic masses and extensive areas of metamorphic schists typically represented in the Colorado and Park ranges. But in the westernmost ranges of the district vast quantities of volcanic material have been extruded, which have in many notable instances profoundly modified the topography. The San Juan, La Plata, Needle, and Elk mountains may be taken as typical illustrations. They are so rugged and access to them is so difficult that permanent settlements have been formed in very few places. There are few trails, and at great distances from each other are scattered prospectors' cabins, now commonly deserted.[5]

---

[1] C. E. Siebenthal, Geology and Water Resources of the San Luis Valley, Col., Water-Supply Paper U. S. Geol. Surv. No. 240, 1910, pp. 34-35.

[2] Idem.

[3] For structural features see J. J. Stevenson, The Geology of a Portion of Colorado, U. S. Geog. Surveys West of the 100th Meridian (Wheeler Surveys), vol. 3, 1875, pp. 490 et seq.

[4] F. M. Endlich, U. S. Geol. and Geog. Surv. of Col., and Adj. Terr. (Hayden Surveys), 1873, pp. 323 ff.

[5] So little known is the region that during the survey of the Needle Mountains quadrangle, in 1900, a large number of peaks and other prominent features were given names for the first time.

SAN JUAN MOUNTAINS

The term " San Juan region," Fig. 117, is generally applied to a large tract of mountainous country in southwestern Colorado and an adjacent zone of undefined lower country bordering it on the west and south. The greater part of the district is a deeply scored or dissected plateau drained on the north by branches of the Gunnison, on the west by branches of the Dolores and San Miguel, on the south by the San Juan, and on the east by the Rio Grande; the eastern part consists of high tablelands that represent old plateau surfaces; the San Juan Mountains proper extend from the great plateau region of Colorado and Utah on the west to San Luis Park on the east and from the canyon of the Gunnison on the north to the rolling plateaus of New Mexico on the south. They have a length of 80 miles east and west and a north-south width of nearly 40 miles, their summits forming a great group rather than a range. The elevations of the highest peaks exceed 14,000 feet, and hundreds of peaks exceed 13,000 feet. Some of the valleys in the heart of the mountains have been cut down 9000 feet or more, and the configuration, especially on the west, is extremely rugged and presents an almost infinite variety of slopes. Several of the small groups of high peaks in the border zone have special names, such as Needle Mountains on the south, La Plata and Rico mountains on the southwest, and San Miguel Mountains on the west. The geologic structure and to a certain extent the topographic character of the bordering groups are sufficiently unlike the San Juan Mountains to constitute them subordinate topographic divisions.

The southwestern border of the San Juan Mountains is marked by bluffs with vertical faces where the great upland crowned by mountains descends to the Navajo, San Juan, Piedra, and adjacent stream courses, Fig. 117. When viewed from this direction the upland appears as a rugged range, but at the general summit level its true character as a great plateau crowned by isolated summits becomes apparent. The relatively unbroken character of the surface over a considerable number of summit areas prevents thorough drainage, and swamps abound above timber line.

A special feature of the San Juan region remarkably developed in the San Juan Mountains and in many places in the Needle Mountains is the large number of landslides determined by the alternation of soft shale and beds of limestone or sandstone. They consist of great blocks of rock tilted at various angles, with fine material collected in the intervening spaces, and have very uneven surfaces without regular drainage systems. Stagnant pools and morasses alternate with deep trenches. The slides occur at the heads of the ravines and on the sides of the steep ridges separating the smaller valleys.[1]

The landslides of the Telluride and Rico districts of the San Juan region are supposed to have been caused in part by earthquake shocks on such a prodigious scale that the sliding and fracturing of solid rocks occurred. Possibly also the melting of supporting ice masses after they had deepened the valleys and steepened the valley slopes may have

Fig. 126. — Landslide surface below Red Mountain, near Silverton, Colorado.

added to the effect of the structural conditions.[2] Earthquake shocks have occurred in recent years in the Telluride and Red Mountain districts within the landslide area, and fresh fractures have also been observed in the Red Mountain mines.[3] In the Telluride region the rocks involved in the landslides have been those of the volcanic series which rest upon soft Cretaceous shales.[4]

[1] Ernest Howe, Landslides in the San Juan Mountains, Colorado, Prof. Paper U. S. Geol. Surv. No. 67, 1909.

[2] Cross and Howe, Silverton Folio, Col. U. S. Geol. Surv. No. 120, 1905, p. 25, col. 1.

[3] H. C. Lay, Trans. Amer. Inst. Mining Engineers, vol. 31, 1902, pp. 558–567.

[4] F. L. Ransome, A Report on the Economic Geology of the Silverton Quadrangle, Col., Bull. U. S. Geol. Surv. No. 182, 1901, pp. 27–28, et al.

RAINFALL AND VEGETATION

The San Juan Mountains are surrounded by arid country, but they themselves have an abundant precipitation, the higher peaks and basins being seldom entirely free from snow. The abundant water supply supports a heavy forest growth in many localities upon the western and northern slopes. The upper timber line lies between 11,500 and 12,000 feet, and as a consequence large areas in the lofty interior of the San Juan Mountains are without tree growth, although a low alpine flora is found in favored situations. Spruce and aspen cover the higher slopes, and below them white pine, scrub oak, piñon pine, and cedar are found. On the great volcanic plateau of the San Juan region the precipitation is less and supports only an herbaceous vegetation. Prominent yet low cone-shaped peaks present a more desolate appearance and are sometimes so strewn with rock fragments as not to support plants of any sort. High plateaus of volcanic rock appear in places with a grassy vegetation so different from the forests on the older sedimentary and igneous rock to the east and locally in the deep canyons of the volcanic area, as to form in a certain sense a guide to the geologist.[1]

LA PLATA MOUNTAINS

The La Plata Mountains are located between the well-watered, well-forested San Juan Mountains and the arid mesa, plateau, and canyon country that extends into Utah, Arizona, and New Mexico. The La Plata Mountains are a dissected dome and their drainage systems have a radial arrangement. Many of the peaks rise a thousand feet or more above timber line and are characteristically rugged.[2] On the west the slopes descend rather steeply 4000 or 5000 feet to the rolling plain known as the Dolores Plateau, deeply dissected by a number of deep canyons, Fig. 78. The simple character of the La Plata Mountains is complicated on the western and northwestern sides by intruded igneous rocks which occur in soft shales overlying sandstone, the intrusive rock having prevented the uniform erosion of the shales. On the southwest the dip of the slope of the sandstone is uninterrupted between 8000 and 10,000 feet, and the descent in this quarter is regular and broad. The intrusive rock is for the most part porphyry, and the masses are large enough to cause peaks or ridges which appear as irregularities in the

[1] F. M. Endlich, U. S. Geol. and Geog. Surv. of Colorado (Hayden Surveys), 1873, p. 306.
[2] Cross and Spencer, La Plata Folio U. S. Geol. Surv. No. 60, 1899, p. 1.

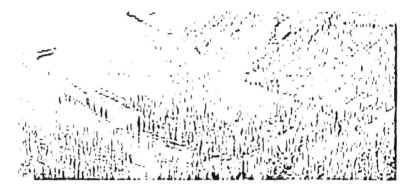

Fig. 127. — Looking down La Plata Valley from the divide at the head of the valley. The steep slopes are characteristic. Landslide in left foreground. (La Plata Folio U. S. Geol. Surv.)

Fig. 128. — Western summits of La Plata Mountains from the divide at head of La Plata Valley. This view is panoramic with the view above. It shows the rugged summits within the eroded volcanic stocks of the mountains. Diorite Peak on left. (La Plata Folio U. S. Geol. Surv.)

377

general slope of the dome. The porphyries occur as sheets or small laccoliths;[1] later intrusions took place which cut the sedimentary beds and the sheets of porphyry forming stocks which constitute several of the highest peaks of the La Plata Mountains.

In connection with the formation of the dome of the La Plata Mountains a number of faults were developed but these are neither numerous nor important. The upthrust is commonly on the side near the center of the dome and has resulted in the formation of a number of steep slopes facing southward.[2]

The arid plains or plateaus rising west of the La Plata Mountains support a growth of white pine, piñon, and cedar, transitional between the barren expanses of the plateau and the forested slopes of the La Plata Mountains. In general the heaviest timber grows on the western and southern slopes since it is from those directions that the winds commonly blow. The forests extend up to 11,500 and 12,000 feet; the growth consists of spruce, fir, and aspen; above the timber line there is a scanty vegetation of alpine character.[3]

NEEDLE MOUNTAINS

The rocks composing the Needle Mountains are chiefly granite, schist, and some quartzite. Four of the summits exceed 14,000 feet in elevation and many exceed 13,000. Some of the mountains are deeply dissected and extremely bold. North of the Needle Mountains is the great curve of the Grenadier Range, which is chiefly of quartzite; on the north and east the surface is almost entirely covered by volcanic or other rocks; on the south sedimentary rocks were laid down upon a pre-Paleozoic land surface of moderate relief now exposed (by the removal of the sedimentary cover) in the form of a southward-sloping tableland consisting of isolated mesas between deep-cut valleys. There is thus afforded a very strong contrast between the sharp peaks and needles of the Needle Mountains and the comparatively low relief of the reëxposed and ancient land surface extending southward from them.[4]

Of great topographic importance was the formation (Tertiary) of three series of volcanic accumulations: (1) a tuff which attained a thickness of 3000 feet east of Ouray (San Juan), (2) lava flows, tuffs, and agglomerates (Silverton Series), also reaching a thickness of about half

---

[1] Cross and Spencer, La Plata Folio U. S. Geol. Surv. No. 60, 1899, p. 8, col. 1.
[2] Idem, p. 10, col. 3.
[3] Idem, p. 2, col. 1.
[4] Cross and Howe, Needle Mountains Folio U. S. Geol. Surv. No. 131, 1905, p. 1, cols. 2-3.

a mile in places; and (3) later thin lava flows and tuff of a third period of volcanic activity (Potosi Series). In connection with the formation of these materials there were intruded into the sedimentary rocks various dikes, sills, laccoliths, and irregular masses which are the deeper-seated equivalents of the surface volcanics.

On the flanks of the Needle Mountains dome a simple consequent drainage was developed during the doming period. The southern flanks

Fig. 129.—Mount Wilson group in background. Looking down Howard Creek from south of Ophir Pass.

of the dome were not covered with volcanic material and the consequent drainage developed on these slopes was not affected by volcanic action. The northern slopes of the domes, on the other hand, were affected by volcanic outpourings and the drainage greatly modified. After the last great eruptions the streams proceeded to dissect the surface deeply.

The consequence of further uplift of the region and resulting deep dissection has been the formation of a confusing variety of slopes in intimate relation to the detailed geologic structure. Within short distances of each other one may find (1) ancient granites, (2) sedimentaries of more recent origin, (3) Tertiary conglomerates, and (4) more recent volcanic accumulations. The development of slopes upon these varied rocks has naturally been very complex and few generalizations may be applied to individual mountains or to the group as a whole.

Fig. 130. — Part of Needle Mountains, Colorado. Grenadier Range on west, White Dome in center; about the head of Vallecito Creek. These two views are panoramic with each other.

GLACIAL FEATURES

A phase of the recent dissection of the region is associated with glaciation; the conditions prevailing in the Needle Mountains during the portion of the glacial period of which the best records are left were nearly the same as those which exist in the Alps at the present time. From the positions of well-defined terminal and lateral moraines it is inferred that during the last recognized period of glaciation alpine glaciers descended the more favorably located valleys some 25 miles. The rock débris in the terminal moraines consists largely of material derived from the Needle Mountains. The higher and sharper peaks and ridges were not buried beneath ice and snow, but at lower elevations the bare rock surfaces are grooved and polished and indicate vigorous glacial action.

ELK MOUNTAINS

The principal range of the Elk Mountains lies north of the San Juan region and west of the northern end of the Sawatch Range. The mountains have great diversity of color, due to the presence of light-gray trachyte, red, maroon, and brown sandstone, etc. They are from 13,000 to 14,000 feet high, and are characterized by sharp, conical peaks and ragged, serrated ridges, pinnacles, and spires. The main range is bordered throughout a large part of its extent by broad, high, flat-topped spurs or secondary ranges.[1] It is about 40 miles long and dissected by gorges and canyons whose extreme ruggedness and picturesqueness are owing to the complicated structure, the variegated rocks, and the youthfulness of the forms themselves. Enormous amphitheaters at the heads of the bordering valleys have been cut back so far as to give the main crest a zigzag course and to make it so narrow as to be almost impassable.[2]

PARKS OF THE SOUTHERN ROCKIES

Between the two main ranges of the Rockies in Colorado is a line of basins of exceptional interest. The three northernmost, North, Middle, and South parks, are geologically a unit. Their southern continuation is not San Luis Park, as generally supposed, but Wet Mountain Valley and Huerfano Park between the front range axis on the one hand and the Sangre de Cristo axis on the other. The San Luis basin lies between the latter axis and the Sawatch Mountains. The former depression began to take shape at least as far back as the Triassic; the San Luis Valley is occupied chiefly by late Tertiary and Pleistocene sediments.[3]

[1] H. Gannett, U. S. Geol. and Geog. Surv. of Colorado and Adj. Terr. (Hayden Surveys), 1874, p. 417.

[2] Hayden Surveys, 1874, pp. 58, 71.

[3] C. E. Siebenthal, The San Luis Valley, Colorado, Science, n. s., vol. 31, 1910, p. 745.

San Luis Park is the largest of the five main parks of Colorado. It is in many respects like the Sage Plains of the Green River Valley or the valley of the San Joaquin in California, consisting of a depression floored by young sedimentary strata and still younger alluvium; the lost rivers on the piedmont alluvial plain flanking the mountainous margin of the basin are also characteristic. Each of these parks drains to the sea by way of a stream which cuts through a deep mountain gorge on the margin of the basin. North Park is drained by the Platte, South Park by the South Platte, Middle Park by the Grand, and in the mountains enclosing San Luis Park the southward-flowing Rio Grande takes its rise. These parks are all the more striking because they lie in the midst of a very rugged mountain region. Their unique character is due primarily to broad intermont structural depressions whose low relative elevation and high outlets have made them the seat of aggradation, not dissection, while the neighboring uplifts are among the loftiest ranges in the country and have been profoundly dissected, giving rise to a rugged and, locally, even an alpine topography. With these general relations in mind we shall now note briefly the special features of each park.

NORTH PARK [1]

North Park, 7500 feet above the sea, is nearly quadrangular in shape; it is about 50 miles in extent from east to west and 30 miles across from north to south. Its surface, although somewhat rugged, is marked by broad bottoms along the streams, especially the North Platte and its tributaries. The lofty and, for a part of the year, snow-covered mountains enclosing it like a gigantic wall, are covered with a dense growth of pine, but scarcely a single tree may be found over the whole 1500 square miles of the park itself save along the stream courses. Grass is found in abundance, and the park was originally the feeding ground of thousands of antelopes. The bordering strata dip under the park, becoming nearly or quite horizontal toward its center. In this respect North Park resembles the three other major parks of Colorado in being primarily a great structural depression sufficiently free from mountainous elevations and hence sufficiently drier to form a striking contrast to the high forest-clad country about it.

MIDDLE PARK

Although Middle Park has on the whole a basin-like aspect, yet in detail this feature is often lost in the prominence of many of the ridges separating its secondary drainage lines. In this respect it is in sharp

---

[1] F. V. Hayden, U. S. Geol. Surv. of the Terrs., 1867, 1868, and 1869, pp. 87–88.

contrast to both North and South parks on either hand. The basin is unique in being the easternmost region in the United States where Pacific waters take their rise; it drains westward through the Grand River to the Pacific. The borders of the basin are composed of crystalline schists and granites on the eastern, southern, and western sides, while the northern border as well as much of the floor of the basin is composed of younger sedimentaries.[1]

SOUTH PARK

South Park is about 45 miles long and about 40 miles wide at its widest portion near the southern end. Its elevation at the north is about 9500 feet, at the south about 8000 feet. Its surface is regular only by contrast with the more rugged country surrounding it. Numerous parallel ridges with northerly trend cross the park and make its surface irregular. At the southern end are many isolated buttes, most of them of volcanic origin. Portions of the park are underlain by rather flat-lying sandstone, as is the case with San Luis Park, and such portions have as a rule a more uniform surface than the rest.[2]

SAN LUIS PARK

The San Luis Valley basin owes its origin primarily to the broad, gentle, trough-like depression of its underlying sandstones. In late geologic time it probably contained a lake of considerable extent, a

Fig. 131. — Cross section of San Luis Valley from foot of Blanca Peak to foot of Conejos Range. (Adapted from Siebenthal, U. S. Geol. Surv.)

condition inferred from the character of the fine sediments accumulated in it.[3] The draining of the lake by the cutting down of the outlet stream, the building up of great alluvial fans, among which that of the Rio Grande is the largest, and the deposition of glacial deposits locally about its border are still later occurrences of physiographic importance.

[1] Emmons, Cross, and Eldridge, Geology of the Denver Basin in Colorado, Mon. U. S. Geol. Surv. No. 27, 1896, p. 4.

[2] A. C. Peale, U. S. Geol. and Geog. Surv. of Col. and Adj. Terr. (Hayden Surveys), 1873, p. 212.

[3] F. M. Endlich, U. S. Geol. and Geog. Surv. of Col. and Adj. Terr. (Hayden Surveys), 1873, pp. 334, 339.

San Luis Valley appears nearly level over great areas but in fact it departs from the horizontal by important amounts. The marginal slopes descend by regular gradients to an axial depression located well toward the eastern margin of the valley; they are developed upon a series of coalescing alluvial fans and the longest and flattest slopes are those built by the largest streams; the Rio Grande has built an alluvial fan so large as to throw the axis of the valley to one side (east) of the center of the depression.[1] The eastern side of the basin is bordered by

Fig. 132. — Looking eastward from Hunt Springs across the north end of San Luis Valley; San Luis Creek in the middle distance, Sangre de Cristo Range in the background. (Siebenthal, U. S. Geol. Surv.)

the Sangre de Cristo Range, on whose flanks a great alluvial plain has been formed whose deposits above an elevation of 9000 to 9500 feet consist of fluvio-glacial débris formed in connection with Pleistocene glaciers. Concentric terminal moraines surmount the crests of the alluvial fans and cones, Fig. 131.[2]

[1] C. E. Siebenthal, Geology and Water Resources of the San Luis Valley, Col., Water-Supply Paper U. S. Geol. Surv. No. 240, 1910, p. 10.
[2] C. E. Siebenthal, Notes on Glaciation in the Sangre de Cristo Range, Colorado, Jour. Geol., vol. 15, 1907, p. 15. Idem, The San Luis Valley, Colorado, Science, n. s., vol. 31, 1910, p. 746.

Northeastward from Antonito there stretches a line of flat-topped to rounded basaltic hills which represent late flows upon the alluvial floor of the valley. They form the chief topographic break in the continuity of the basin floor. The smooth valley surface is also interrupted in a number of localities by sand dunes. The highest occur between Medano and Sand Creek and these also cover the most extensive single tract, 40 square miles. They consist of rather coarse, white, quartzite sand blown into place by the heavy southwest winds that occasionally blow for a two or three day period across the valley. Elsewhere the sand is commonly blown out of place between patches or clumps of bush and grass, leaving small hollows or basins that contain lakelets in the wet season.[1]

OTHER TYPES OF PARKS

Besides these large intermont parks there are scores of smaller ones. They have been formed in at least four main ways. (1) San Luis Park, as we have seen, is a broad structural basin drained by the Rio Grande River, and the same is true of North, South, and Middle parks. (2) In the Pikes Peak region and specifically in the south-central portion of the Pikes Peak quadrangle, is a depression known as Shaw's Park; it is formed upon sandstones, marls, and limestones in the form of a syncline whose margins have been in a number of cases down-faulted in a pronounced manner. The weathering of the softer sedimentaries has accentuated the flatness of the depression normal to the synclinal structure and thus developed a lowland 5 or 6 miles in width. A similar park has been developed under similar structural conditions immediately west of Shaw's Park and is known as Twelve Mile Park. (3) A third type of park may be seen very commonly throughout the region, a typical occurrence being Low Park, shown in the Needle Mountains topographic sheet, in the valley of the Florida River, where a broad valley lowland is formed by combined glacial erosion and the present drainage.[2] Similar valley lowlands of small extent are not uncommon at the junction of tributary and master streams and are commonly known as "parks." (4) A fourth type of park is described by Powell,[3] who calls attention to the fact that the great anticlinal folds in the Park Mountains have a north-south trend and that on the flanks of each fold there is developed a zone of maximum dip so that the rocks at the flanks of the ranges are turned abruptly down. In such cases small parks are formed by the wearing out of the softer beds, leaving

[1] C. E. Siebenthal, Geology and Water Resources of the San Luis Valley, Col., Water-Supply Paper U. S. Geol. Surv. No. 240, 1910, pp. 47-48. The detailed features of the San Luis basin are represented in Sheet 10, Atlas of Colorado, Hayden Surveys, where the course of the Rio Grande is shown as well as the course of San Luis, San Isabel, and other creeks which empty into San Luis Lake instead of having a surface channel to the Rio Grande.

[2] Pikes Peak Folio U. S. Geol. Surv. No. 7, 1894.

[3] Physiography of the United States, Nat. Geog. Soc. Monographs, 1896, pp. 88-89.

the harder strata standing as great steep walls parallel to the axes of the principal ranges. The famous Garden of the Gods near Pikes Peak is explained in this way.

TREE ZONES AND TYPES

Among the smaller parks Estes Park shows a typical zonal arrangement of the tree species of northern Colorado. There are three well-defined forest types: (1) the yellow pine type, developed typically as an open woodland which reaches from the upper foothills to an altitude of 9000 feet; it occupies the ridges and slopes of the lower part of the park, but at higher elevations occurs only in the open valleys. In the open it occurs on small rocky knolls and ridges, whence it invades the grasslands of the park floor. (2) The lodgepole pine-Douglas spruce type occurs above the yellow pine type between 8000 feet and 10,000 feet, though below the lesser altitude the growth is mixed to some extent with the yellow pine, while above the greater altitude it is mixed with Engelmann spruce. (3) The Engelmann spruce-alpine fir type constitutes the forest at timber line and usually some distance below it. Timber line is here at 10,500 feet. From 8500 to 9000 feet the grass land of Estes Park is in the form of typical mountain meadows.[1]

Among the larger parks San Luis exhibits a rather typical distribution of vegetation. On the high mountain sides flanking San Luis Valley are pine, aspen, and spruce. Lower down are piñon and cedar; in the valleys and along the streams are cottonwood and willow. Away from the streams and on the arid basin floor greasewood is the principal growth.[2]

[1] F. E. Clements, The Life History of Lodgepole Pine Forests, Bull. U. S. Forest Service No. 79, 1910, pp. 7–8.

[2] C. E. Siebenthal, Geology and Water Resources of the San Luis Valley, Col., Water-Supply Paper U. S. Geol. Surv. No. 240, 1910, p. 26.

CHAPTER XXI

TRANS–PECOS HIGHLANDS

MOUNTAINS AND BASINS

THE Trans-Pecos region embraces an assemblage of topographic forms which individually resemble the features of adjacent provinces, the Rocky Mountain province on the north, the Arizona Highlands province on the west, and the Mexican plateau on the south, and it may therefore be regarded as a transition province. The Trans-Pecos ranges do not have that continuity which marks the main mountain ranges of the Pacific Cordillera. They exhibit a great variety of slopes produced at different geologic times and in many different ways; the province includes a mixed group of topographic forms not conveniently classified with the forms of adjoining provinces. In this respect it is unlike the other physiographic regions of the United States, each of which has a certain essential unity of structure or topography or both.

Mountains and intermontane plains of several types and of variable extent are the predominating features of the Trans-Pecos country. The latter are in part plains of degradation, in part broad constructional plains built up by detritus from the bordering mountains and called bolsons (Spanish *bolsón*, for "large purse"). The trend of the mountains and intervening basins is north-south, except near and south of the New Mexico-Texas boundary where they trend northwest-southeast. In general the elevations are distinctly lower than those of any other portion of the Pacific Cordillera in this longitude. Only two peaks rise above 8000 feet and the general elevation of the lowlands is from 3500 to 4500 feet.

While the greater number of the Trans-Pecos mountains have a fault-block origin, as a whole they are of diverse origin, structure, and topography. They exhibit many irregular forms of relief and all have sharp and rugged outlines. Few of them rise to the height of the dry timber line (6000 feet). On the summits and in the higher canyons of the Sacramento, Chisos, and Davis mountains, and a few others, this elevation is exceeded and a certain amount of timber is found. On the lower and drier mountains the vegetation consists of edible pines, a variety of junipers, and several species of maguey. The lower eleva-

Fig. 133. — Principal mountain ranges of the Trans-Pecos province, chief drainage features, and the bordering Llano Estacado on the east. (U. S. Geol. Surv.)

tion of the group as a whole is expressed in the lesser rainfall and general barrenness of the mountains in contrast to the partly timbered Rocky Mountains to the north. The highest summit is Sierra Blanca (11,880 feet) in the Sacramento range in southern New Mexico, but in general the peaks fall short of this altitude.

MOUNTAIN TYPES

Four types of mountains may be identified: (1) mountains of deformation, consisting of structural folds or tilted fault blocks in which the outline of the mountain is in sympathy with the structural breaks or

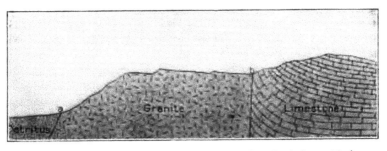

Fig. 134. — Western face of the Fra Cristobal Mountains, New Mexico. Two faults, *a* and *b*, give prominence to the mountain front. They are represented in the section below the photograph. (Lee and Girty, U. S. Geol. Surv.)

deformations, for example the Sandia and Manzano mountains east of Albuquerque; (2) plateau mountains, consisting of nearly horizontal plateaus without important deformations. This type of mountain occurs in the form of summits or shoulders and in close proximity to the higher relief features of the province.[1] (3) Mountains due to the upthrust of a granitic core through sedimentary rock that now forms the borders of the range, as the Black and Mimbres ranges. (4) Mountains consisting chiefly of volcanic material like San Mateo and Valles

Fig. 135.— East-west section across the Trans-Pecos Highlands north of El Paso, Texas. pC, pre-Cambrian quartzite and porphyry; COS, Ch, gs, etc., paleozoic limestone, sandstone and gypsum; K, cretaceous limestone and sandstone; gr, intrusive granite; Qb, Quaternary basin deposits. Scale, 1 inch represents nearly 50 miles. (Richardson, U. S. Geol. Surv.)

mountains and others. The members of the last-named group fall into three subtypes: (a) old igneous vents such as dikes and necks, (b) volcanic craters, and (c) mesas capped by sheets of lava that have protected the underlying rocks from the denudation that carried away the surrounding material.

The Trans-Pecos province is one of the least-known physiographic provinces in the United States and few generalizations may be made concerning details of mountain form beyond those in the preceding paragraph. It is known that the province as a whole is separated from the Rocky Mountains province at Bernal, New Mexico, by a broad level plain or plateau, and it is not until about 100 miles south of Bernal that the mountains of the Trans-Pecos begin in the Jicarillas, the northern outliers of a chain of mountains, the loftiest ranges of the province, that extends with marked continuity through New Mexico and Texas as far as the 32d parallel. The individual members of the chain are the Jicarillas on the north, and next in order from north to south are the Sierra Blanca, Sacramento, Sierra Guadalupe, Comanches, Caballos, and the Sierra de Santiago. These constitute the front ranges of the province and they give a distinctive character to the eastern edge of the highland belt in which they lie. The rocks composing these mountains are chiefly limestone and sandstone. The general structure of the eastern border ranges is an eastward-dipping monocline ending at the Rio Pecos, but there are many lateral ridges and separate peaks. On the west the mountains of the eastern border terminate in a steep scarp 1000 to 2000 feet high. The long eastern slopes correspond with the dip of the underlying strata.

West of the front ranges of the Trans-Pecos country are other ranges of the fault-block type. They occur as short sierras or groups of sierras in the form of isolated mountains or elongated chains surrounded by extensive and nearly level desert plains. They have lower altitudes

[1] R. T. Hill, Physical Geography of the Texas Region, Folio U. S. Geol. Surv. No. 3, 1900, p. 3.

than the front ranges, and so isolated are the various members that no group name has been given them. The principal ranges are, however, arranged in two lines on either side of the Rio Grande Valley. On the east side are the Sandia, Manzano, Oscura, San Andreas, and Organ ranges; on the west are the Nacimiento, Limitar, Magdalena, Cristobal,

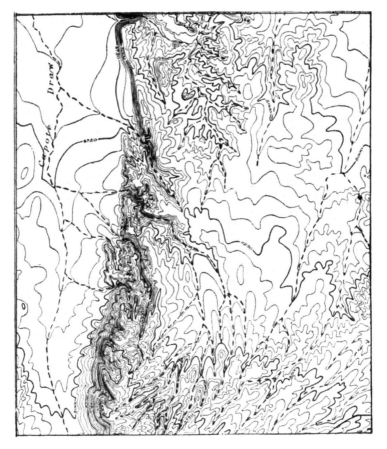

Fig. 136. — Fault-block mountain in the Trans-Pecos Highlands, Texas. (Marfa quadrangle, U. S. Geol. Surv.)

Caballos, and Cuchillo Negro ranges. The fault scarps of these ranges are always steep and in general face inward toward the Rio Grande Valley, which is thus a great structural depression or series of elongate basins, though there are some important exceptions to this rule. They are all high and in process of vigorous dissection, and hence display a varied relief. The trend of the group is, however, definite, due to their fault-block origin, and of similar definiteness are the valleys that occur among them. Westward of these block mountains are many short independent blocks surrounded by bolson deserts and not capable of union into related chains. They rise from desert plains and in general

Fig. 137. — Western escarpment of the Caballos Mountains, New Mexico, and the dissected alluvial slope at their base. (Lee, U. S. Geol. Surv.)

have steeper descents on their western than on their eastern sides. They consist for the most part of fault blocks with monoclinal dips, although in places extensive folded structures also occur.

FRANKLIN RANGE

The most detailed study of mountain form and structure in the Trans-Pecos region is by Richardson, whose structural sections of the Franklin range are shown in Fig. 138. They will serve as illustrations of a type of mountain found in the province. The Franklin range is a long, narrow, crust block resembling the mountains of the basin-range type in its general features but differing from many though not from all of them in having a more complex system of internal faults. The strata strike parallel to the trend of the range and dip westward at steep angles. On the steeper eastern face of the Franklin range are exposed the eroded edges of the strata composing the fault block, while the gentler westward slope coincides to an important degree with the dip of the underlying strata. The range is broken into several secondary blocks by normal parallel faults some of which strike with the trend of the range, but there are also several transverse fractures, and the strike of a few of the faults is a curve. Both the eastern and the western margins of the range are marked by faults; that on the western border consists of two parallel faults at the base of the range between the foothills and the main mountain mass. The greatest displacement is in the central part of the range, where a throw of 2500 feet has been determined.

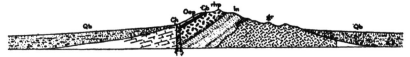

Fig. 138. — Section across Franklin Mountains, showing relation of structure to relief. In, quartzite; rhp, rhyolite porphyry; Cb, sandstone; Oep, Om, Sf, Ch, and Kcm, limestone; gr. intrusive granite; Tap, andesite porphyry; Qb, basin deposits. Scale 1 inch represents 2 miles. (Richardson, U. S. Geol. Surv.)

Fig. 139. — East-west section across the Franklin Mountains north of El Paso, Texas. To supplement Fig. 138. Legend and scale as in Fig. 138.

The fault at the eastern foot of the range is completely concealed by wash and its position is hypothetical. That the mountain front is the locus of a fault is suggested by the evidences of geologically late (probably Quaternary) faulting at El Paso, where displacements in alluvial deposits may be seen, and by the benches west and northwest of Fort Bliss at elevations ranging about 4000 feet. The benches are the upper portions of broken alluvial slopes which in places fringe the base of the range in an uneven eastward-facing scarp varying from 16 to 50 feet in height. In many places the bench has been destroyed by erosion along the many arroyos that descend from the adjacent mountains. The original fault along the base of the range is probably of ancient origin, for the drainage has diversified the eastern border and to a large extent obliterated that inequality of slopes that obtains in the case of a young block mountain.[1]

VOLCANIC MOUNTAINS AND PLATEAUS [2]

In addition to the mountains formed out of tilted crust blocks of sedimentary material there are many eccentric and irregular mountains of volcanic origin that rise from broad plains. Of this type are the Chisos (ghosts), Corazones (hearts), the Sierra Blanca, and the Davis mountains, among which the last-named are the most extensive. They consist of a group of volcanic necks and dissected volcanic plateaus that end

[1] G. B. Richardson, El Paso Folio U. S. Geol. Surv. No. 166, 1909.
[2] R. T. Hill, Physical Geography of Texas, Folio U. S. Geol. Surv. No. 3, 1900.

Fig. 140. —Fisher's Peak and Raton Mesa. Shows vertical cliffs of basalt, and the flat-topped mesa rising 3000 feet above the plain in the foreground. The prominent terrace is developed on a hard flat-lying stratum (Laramie).

at the south in a great escarpment over 1000 feet in height that overlooks the continuation of a broad intermontane plain toward the south.

The mesa and cuesta types of mountains are capped in some instances with old igneous material in places 500 feet thick. The volcanic
caps are remnants of former sheets of similar material that have been
all but removed by denudation. Elephant Cuesta and Nine Point
Mesa are illustrations of this kind of mountain. Besides these are
other volcanic mountains of recent origin, cinder cones or true volcanic
craters accompanied by sheets of lava which flowed from them. The
craters are of interest because they are the most easterly known in the
United States and the only ones lying east of the front of the Cordillera.
The cinder cones have been formed since the degradation of the Mesa
de Maya and Ocate plateaus, for they rise out of the newer and lower
plain below the latter. The most conspicuous crater is Mount Capu-
lin, a very symmetrical cinder cone with a vast crater at the top of it.

Within the Trans-Pecos Highlands are extensive areas of uplands
with sublevel summits bordered more or less completely by pronounced
escarpments. They occur as large benches that border the mountain
ranges or that rise from the bolson deserts. They are either completely isolated from the mountains or project bench-like from the
bases of the mountains. They are lava-capped mesas whose summit
layers of hard volcanic rock have protected the more friable material
underneath them; they are distinguished from the flat-topped mountains
noted above only by their greater extent and lesser elevation. They
are cut by deep canyons, bordered by steep escarpments, and owe their
present outlines to marginal erosion. Northeastern New Mexico, Fig. 133,
is therefore not a part of the Great Plains but an eroded plateau of
Cretaceous rock surmounted by basaltic flows.[1]

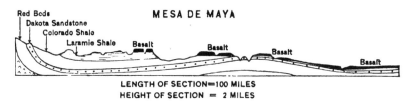

Fig. 141.— Upturned strata on the eastern border of the Rockies, the anticlinal structure of the bordering
strata underneath the Mesa de Maya, and the protective influence of the basalt cover. (After Keyes.)

The three most conspicuous plateau plains of northern New Mexico
are known as the Mesa de Maya, the Ocate Mesa, and the Las Vegas

    [1] R. T. Hill, Notes on the Texas-New Mexico Region, Bull. Geol. Soc. Am., vol. 13, 1892,
pp. 85–100.

Mesa. The western portion of Mesa de Maya is also known as Raton Mesa or Raton Mountains. The mesa is a long dissected plateau almost due east of Trinidad, Colorado, which extends southward into New Mexico and Texas. Its border rises nearly 4800 feet above the adjacent valleys and the massive summit is composed of thick beds of old lava now dissected into remnants of a once more extensive plateau. Underneath the lava are less resistant sandstones and shales (Upper Cretaceous).

Raton Mesa is 8 miles long, 4 miles wide, and includes about 20 square miles. It has been entirely separated by erosion from a similar area

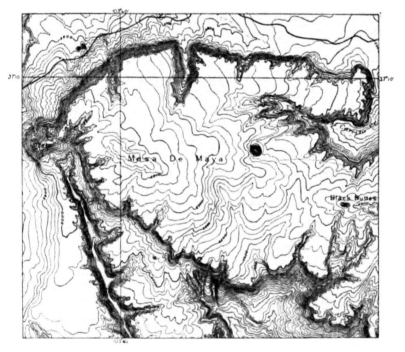

Fig. 142. — Topography of Mesa de Maya, south-central Colorado. Typical lava-capped mesa with steep bordering scarps and flattish summit. (U. S. Geol. Surv.)

to the south and from the main lava field to the east. The aggregate thickness of the basalt flows which constitute the cap rock of Raton Mesa and of the Mesa de Maya is from 250 to 300 feet about their borders, and increases to 500 feet toward the central part of the western mesa. At least 8 distinct beds of lava have been identified, which probably represent the same number of independent eruptions. The different beds are 30 or more feet in thickness, though they vary greatly from place to place.[1] Raton Mesa has a mean elevation of about 1800 feet and is bordered by an almost vertical escarpment from 200 to 300 feet high, an escarpment so sheer as to render it almost inaccessible, Fig. 141.

There is a dense growth of piñon and juniper along the base of the Raton Mesa; the small streams that head in the plateau are fringed with cottonwoods; on the steep border of the plateau, pine and spruce trees are scattered through a dense undergrowth of scrub oak, with aspen appearing locally near the base of the escarpment. The entire district up to 7500 feet affords a scanty growth of nutritious grasses well suited for sheep farming, which is one of the chief industries. The tableland on the summit of the mesa supports a growth of bunch grass.[2]

The Ocate Mesa has a cap rock of thick sandstone and it is the remnant of a lower-lying mesa that extends westward to the foothills of the Rocky Mountains between the Arkansas River in Colorado and the Cimarron River in New Mexico. The largest development of the Ocate Mesa is between the Cimarron Valley and the Mora, where from a broad platform rises Ocate crater (8900 feet). The eastern border of Ocate Mesa is an escarpment nearly 500 feet above the plain to the east, known as the Las Vegas Mesa, which extends south of Mesa de Maya to the great cliffs of the Canadian valley and east toward the Great Plains near the Texas line. The surface of Las Vegas Mesa slopes gently to the east; the western border has an average altitude of about 8600 feet. It is a vast stratum plain underlain by a chalky sandstone (Colorado formation). Many volcanic craters and dikes rise out of it to diversify and complicate its relief.

The cliffs which terminate the eastern border of Las Vegas Mesa form one of the longest and most remarkable escarpments in America. They are continuous with the cliffs of Galistes Mesa and extend in an indirect manner from the 104th meridian southwest to the 106th meridian, or nearly 300 miles. The cliffs are developed upon sandstone and overlook the deep Pecos Valley.

DRAINAGE FEATURES

A detailed map of New Mexico and western Texas, Fig. 142, displays many features characteristic of surface drainage in arid lands. A large number of enclosed basins of structural origin occur, and these are in the main floored with loose sediments derived from the surrounding mountains. About the margins of the basins are long talus slopes, huge bowlder and alluvial fans, and dry arroyos. The basins are without

[1] R. C. Hills, Elmoro Folio, U. S. Geol. Surv. No. 58, 1899, p. 3, col. 2.
[2] Idem, p. 1, col. 1.

surface streams of any extent, a few watercourses entering them from surrounding highlands, but the porous nature of the soil, the high rate of evaporation, and the low rainfall, cause the streams to disappear before they have traveled far from the margins of the basins. The annual rainfall is in some places less than 10 inches, over the greater part of the region it does not exceed 15 inches, and everywhere it is chiefly in the form of summer thunder showers of short duration and limited extent.

All of the basins or bolsons, as they are sometimes called, are characterized by "lost rivers,"— streams that disappear about the borders of the basins. The floors of some of the basins are occupied by salt marshes or temporary lakes. In the largest basins, those between the Sierra Madre and Guadalupe mountains, the salt deposits are of great extent and have been used for hundreds of years by the Mexican population as one source of a salt supply. Around the margins of some of the bolsons are benches consisting of fan-shaped heaps of land waste derived from the mountains and deposited by torrential streams.

The word "bolson" is of Spanish origin and according to Hill it designates structural valleys between mountains or plateau plains which have been partially filled with débris from adjacent eminences.[1] Bolson originally meant a constructional plain bordered by mountains or plateau escarpments that supplied the material with which the bolson is floored; the term did not cover the bordering mountains, though the definition implied that the bordering mountains are features normally associated with the plain, and that a bolson plain was part of an interior basin.[2] In short, a bolson plain is the floor of an interior basin with such variations in its expression as are naturally associated with variations in climate, disposition of waste, size of basin, length of existence, etc.

The basin floors are characteristically flat and in broad view are somewhat similar in appearance to the plateau plains just described. They are, however, to be distinguished from the plateau plains by the fact that they are, at least in part, surfaces of aggradation, while the plateau plains are on the whole either destructional surfaces or surfaces that have been at least partly preserved from destruction by a cap of resistant lava.

LONGITUDINAL BASINS

The basins of the Trans-Pecos region occur in four longitudinal belts in sympathy with the belted arrangement of the intervening mountains, Fig. 133. The easternmost lies along the eastern fronts of the Guadalupe and Santiago ranges. The second belt lies between these ranges and the Sierra Diablo and Cornudas mountains. The third line of basins lies between the Sacramento and Sierra Blanca mountains on the east and the Sierra Oscura and San Andreas ranges on the west.

[1] R. T. Hill, The Physical Geography of the Texas Region, Folio U. S. Geol. Surv. No. 3, 1900.

[2] W. G. Tight, Am. Geol., vol. 36, 1905, pp. 271–284.

The fourth line lies between the last-named mountains and the Sierra de los Caballos on the borders of the Rio Grande.

Among these four longitudinal basins the Hueco basin, Fig. 133, is the largest. It lies partly in New Mexico, partly in Texas, is about 200 miles long, about 25 miles wide, and stands about 4000 feet above the sea. The mountains that border and enclose it are 2000 to 5000 feet higher. A few miles north of the New Mexico-Texas boundary the basin is crossed from west to east by a low divide. The southern half is drained by the Rio Grande; the northern half is a closed basin with salt marshes and unusually interesting dunes of white gypsiferous sands, a district known as the Tularosa Desert. In the vicinity of El Paso the Hueco basin is a structural basin or trough deeply filled with detritus. The streams wither at the foot of the mountains, and the mouths of the mountain arroyos are marked by huge detrital cones and fans that spread radially outward and finally coalesce with the wash from the intervening slopes.

JORNADO DEL MUERTO

The most noted basin of the Rio Grande Valley is the Jornado del Muerto Bolson. Its name means "the journey of death" and was given to it in pioneer days because of the great difficulty then involved in crossing its dry, barren, and inhospitable wastes.

The Jornado basin lies in south-central New Mexico. It extends north and south a distance of more than 200 miles, and is from 30 to 40 miles wide. It is a flat-bottomed basin except along the Rio Grande where the river has cut a valley 400 feet below the general level. Its margins are upturned steeply on all sides to heights of 2000 and 3000 feet above the general level. In the central part of the basin are a number of shallow depressions which hold storm waters that sometimes linger as lakes for several months, seldom through the year. At various points the even surface of the basin floor is broken by low hills of volcanic origin, among which may be mentioned the Doña Ana Hills, San Diego Mountains, and Cerro Roblero in the southern part of the basin. All of the volcanic cones are of recent origin and some of them have perfectly preserved craters. From some of the cones basalt flows have extended out over the surrounding plain, the one south of San Marcial covering more than 100 square miles.

It has been shown recently[1] that certain interior basins of the Trans-Pecos country, among them the Jornado del Muerto, instead of being structural valleys deeply floored by mountain waste, have a rock surface developed on the beveled edges of the strata, representing a total

[1] C. R. Keyes, Rock Floor of Intermont Plains of the Arid Regions, Bull. Geol. Soc. Am., vol. 19, 1908, pp. 63-92.

thickness of thousands of feet, and that for the most part the rock floors of these plains are
covered by a thin veneer of detritus only instead of the thick alluvium usually ascribed to
them. Rarely does the thickness of the detrital mantle exceed 100 feet. The general slope
of the Jornado del Muerto plain for example is only 2° or 3°, while the dips of the strata are
in many cases as high as 30° in the same direction, and even vertical. On the beveled edges
of these steeply inclined beds alluvium and also broad sheets of basaltic lavas have been
laid down.

BASINS OF THE RIO GRANDE VALLEY

Among the most important basins of the Trans-Pecos country are
those drained by the Rio Grande. This river from the San Luis
Valley in southern Colorado to the point where it cuts the easternmost
mountains of the province (long. 103° W.) flows almost continuously

Fig. 143. — Valley of Rio Grande, El Paso, Texas, showing passage of the river from the Mesilla Valley
to the Hueco basin across the southern end of the Franklin range. Note the terraced margins of
the dissected basins. (Hill, U. S. Geol. Surv.)

through a series of old structural basins connected by canyons that
increase in length and depth toward the southeast. The present course
of the Rio Grande is owing in part to the basin feature and in part
to volcanic action, and the activities of the river itself.

Long after the partial filling of the broad structural troughs of the region there was an early
period of valley cutting which was followed by a second period of valley filling, probably on
account of changed topographic and climatic conditions. Near the close of the second period of
aggradation great sheets of basalt were extruded. During this time the stream courses were in
many instances violently changed and the Rio Grande, which formerly ran southward through
the Jornado del Muerto basin into the interior basins of northern Mexico, was diverted into
Engle and Las Palomas valleys, which are much narrower than the first-named basins. At
the same time topographic changes due to volcanic action farther south probably forced the
Rio Grande eastward to its present course south of the Franklin range at Hill Pass and southeastward to the Gulf. The volcanic eruptions and changes of river courses were followed by

a second period of erosion in which were formed the present narrow elongated valleys that stand at an elevation several hundred feet lower than the basins in which they were cut. A last phase of river activity is expressed in the form of silt accumulations on the floors of the flood plains, a filling with a maximum depth of 85 feet in the El Paso canyon. The river is carrying an immense quantity of silt to-day; on the average it transports about 14,580 acre feet of mud a year.[1]

The flood plain of the Rio Grande is cut well below the level of the Jornado del Muerto and is known in part as the Mesilla Valley. This division is about 45 miles long and 5 miles wide. Along the valley margins terraces of notable extent and height have been formed. Some of them are of structural origin; their upper surfaces correspond with the dip of the rock in many instances, and rock outcrops along their valley-ward margin. At a lower level are alluvial terraces representing complexities in the down-cutting of the Rio Grande since gaining its present course across the basins.

The Rio Grande is a storm-water stream subject to great and sudden floods. The rainfall occurs principally in the form of violent showers or cloud-bursts which fill the dry stream beds with turbulent floods of short duration. When such floods occur simultaneously at many points they are likely to cause destructive floods on the valley floor where the fertile irrigable lands are located and where most of the population and the principal towns are to be found.

CLIMATE, SOIL, AND VEGETATION

The rainfall of New Mexico varies from 20 to 25 inches in the mountains, as above Las Vegas and Cloudcroft, to about 15 inches on the Great Plains to the east in the vicinity of Roswell and Carlsbad, so that the Pecos River, forming the eastern boundary of the Trans-Pecos Highlands, receives practically no tributaries from the east. The effects of the small rainfall of lower elevations are reënforced by the porous nature of the soil, upon which there is no extensive surface drainage; the water also drains rapidly away underground in many places through fissured limestone rock.[2] At El Paso the mean annual precipitation is 9.85 inches, ranging between a maximum of 18.30 inches in 1884 and a minimum of 2.22 inches in 1891. The average humidity is 38.8 %, ranging between 23.2 % in May and 47.3 % in January.[3] The mean annual precipitation throughout western Texas is about 12 inches. It falls mainly in the summer months in the form of brief showers and is variable and uncertain.

1 W. T. Lee, Water-Supply Paper U. S. Geol. Surv. No. 188, 1907, pp. 20–24.

2 Freeman, Lamb, and Bolster, Surface Water Supply of the United States, 1907–08, pt. 8, Western Gulf of Mexico, Water-Supply Paper U. S. Geol. Surv. No. 248, 1910, p. 114.

3 G. B. Richardson, El Paso Folio U. S. Geol. Surv. No. 166, 1909, p. 2.

In the lower Pecos Valley irrigation has reached a very high stage of development.[1] Thousands of acres are under cultivation, beginning a short distance above Roswell and continuing into Texas; below this fertile belt but little irrigation is carried on because the seepage water contains a large amount of alkali, which renders it unfit for irrigation purposes.

The floors of the arid basins in the Trans-Pecos region generally consist of fine detrital material and support a growth of stunted shrubs and grasses, such as mesquite, greasewood, and cactus. Along the river bottoms are cottonwoods and an undergrowth of shrubs. The coarser talus slopes are covered with a growth of yucca, cactus, and other desert flora and a scattered scrubby growth of oak, cedar, and juniper. These forms are gnarled and stunted at lower elevations, but in higher

Fig. 144. — North end of San Mateo Mountains, Trans-Pecos Highlands, New Mexico. Shows characteristic tree-covered slopes. Mountains composed of rhyolite flows and tuffs. (Gordon, U. S. Geol. Surv.)

situations they become increasingly more exuberant in growth.[2] Upon the slopes and summits of the highest mountains, the Chisos, Davis, Capitan, and Sacramento ranges, a thrifty tree growth of pine and fir is found; upon the lower mountains the tree growth becomes more scattered and upon the lowest ranges no true forests appear. The Black range bears a good growth of pine upon its upper slopes; the Magdalena range and San Mateo Mountains have poorer growths of

[1] G. B. Richardson, El Paso Folio U. S. Geol. Surv. No. 166. 1909, p. 115.
[2] C. E. Dutton, 6th Ann. Rept. U. S. Geol. Surv., 1885, p. 125.

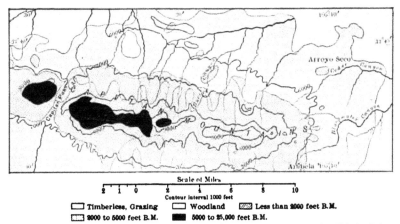

Scale of Miles

Contour interval 1000 feet

☐ Timberless, Grazing ☐ Woodland ▨ Less than 2000 feet B.M.
☐ 2000 to 5000 feet B.M. ■ 5000 to 25,000 feet B.M.

Fig. 145A. — Timber belts, Capitan Mountains, New Mexico. Note (1) the manner in which the timber belts follow the waterways, (2) the increase in growth with increase of elevation, and (3) the island-like outlines of the densest growths. (Adapted from U. S. Geol. Surv.)

pine interspersed with cedar and juniper, though in the better-favored situations good stands are found.[1]

The ranges of the species of trees occurring in the Trans-Pecos region vary as to rainfall, soil, and slope exposure. The lowest and best growth is found on the northern cooler and moister slopes and wherever the best soils have been developed. Except on the broader summits the soil has little humus for the undergrowth is scattered and light, oxidation of decaying vegetation is rapid and fairly complete, and erosion is in general active. There is a notable banding of the various species of forest trees, though the ranges of the different species overlap, as shown in Fig. 145B, which illustrates distributions in the Lincoln Forest Reserve of New Mexico (Capitan and Sierra Blanca ranges). The highest, or subalpine zone is found in this district between 9000 and 11,000 feet. Engelmann spruce is the principal tree of this zone with subordinate amounts of white fir, red fir, Mexican white pine, and aspen, generally in groves. The yellow pine zone between 6400 and 9000 feet includes, beside the dominant yellow pine, red fir, white fir, Mexican white pine, and oak. Between 5000 and 6400 feet is the woodland zone in which the principal species are piñon, juniper, cedar, scrub oak, and, along the streams, ash, box elder, and walnut.

The Trans-Pecos forests seldom form dense stands; open scattered forest growth is the prevailing type. Since the forests grow only at the higher elevations, they are in relatively inaccessible situations, for the settlements are found in the valleys. The principal value of the forests

[1] Lindgren, Groton, and Gordon, The Ore Deposits of New Mexico, Prof. Paper U. S. Geol. Surv. No. 68, 1910, p. 217.

is in relation to water conservation. Many small perpetual streams are limited to the forested zone; only a few advance even two or three miles into the bordering deserts. Forage grasses form a ground cover of great extent both in the parks of the forested zone and below the

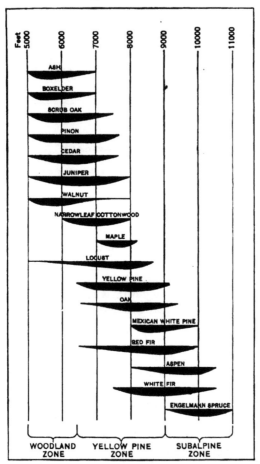

Fig. 145B. — Range and development of tree species in Lincoln National Forest, Trans-Pecos province. (U. S. Geol. Surv.)

7000 foot level where the forest ends. In the more accessible situations these have been irreparably damaged by overgrazing, for with the destruction of the grasses deep and extensive erosion has taken place and rich pastures have been laid waste.[1]

[1] Plummer and Gowsell, Forest Conditions in the Lincoln Forest Reserve, New Mexico, Prof. Paper U. S. Geol. Surv. No. 33, 1904, pp. 10, 11, 18.

CHAPTER XXII

GREAT PLAINS

TOPOGRAPHY AND STRUCTURE

THE Great Plains province slopes eastward from the foot of the Rocky Mountains to the valley of the Mississippi, where it merges into the Prairie Plains on the north and the Coastal Plains on the south. The Great Plains appear as wide areas of relatively flat tabular surfaces crossed by broad shallow valleys drained by scores of eastward-flowing streams tributary to the Mississippi. Certain portions, as in central Montana, are drained by streams sunk in narrow canyons several hundred feet deep. The general expanse of smooth surface is also broken in some places by buttes, mesas, domal uplifts, extended escarpments, and local areas of "badlands." In some districts extensive areas are covered with sand hills, as in northwestern Nebraska, where is found a typical area of this character several thousand square miles in extent.

As a whole, the province descends toward the east about 10 feet per mile and from altitudes of about 6000 feet on the west to about 1000 feet on the east, though the elevations and the general regional slope have considerable variation from place to place. An important illustration of variation in altitude is Pine Ridge, an irregular escarpment which extends from the northern end of the Laramie Mountains in Wyoming eastward to the northwestern corner of Nebraska and the southern part of South Dakota. It marks the northern boundary of the higher portions of the Great Plains, and from it cliffs and steep slopes descend northward 1000 feet into the basin of the Cheyenne River. North of the Cheyenne River the divides are much lower and the surface as a whole does not attain the level of the High Plains to the south.

The plains topography is developed on a great mass of (1) rather soft deposits — sands, clays, and loams — spread out in the form of thin and extensive beds that slope gently eastward and (2) gently inclined and more indurated stratified rock varying from limestone and gypsum through shale and sandstone to conglomerate. The material of the first class is found chiefly in that subdivision of the Great Plains called the High Plains which extend from the southern margin of the bad-

lands to the parallel of 32° south in central-western Texas, and east
and west extends between the 100th and 104th meridians. The hard
rock elsewhere underlying the Great Plains is for the most part thinly
cloaked with rock waste, and its topography is in large part responsive
to structure save (*a*) where the effects of base-leveling are still topographically expressed or (*b*) where glaciation has modified the surface
or (*c*) where local alluviation has concealed an earlier structural surface.
The two classes of material forming the Great Plains were derived

Fig. 146. — Geologic map of the Texas regions, showing the relations of the Great Plains formations to
those of the surrounding provinces. 1, Older granite; 2, Paleozic and Mesozoic; 3, Cambrian-Silurian;
4, Carboniferous; 5, Permian; 6, Jurassic; 7, Lower Cretaceous; 8, Upper Cretaceous; 9, Nonmarine
Tertiary; 10, Marine Eocene; 11, Coast Neocene; 12, Later igneous. (Hill, U. S. Geol. Surv.)

mainly from the west and deposited in stratified condition upon flood
plains and lake floors and in part on the sea floor. The region has
suffered broad uplift and depression a number of times in its geologic
history, but these deformations have been regional in character and
have not deformed the sedimentary series in a complex manner except
in local instances.[1]

[1] Darton and Salisbury, Cloud Peak-Fort McKinney Folio U. S. Geol. Surv. No. 142,
1906, p. 2, col. 1.

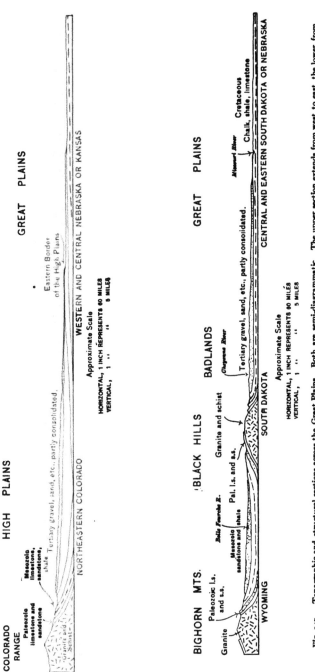

COLORADO
RANGE

HIGH PLAINS

GREAT PLAINS

Paleozoic
limestone and
sandstone;

Mesozoic
limestone,
sandstone,
shale

Tertiary gravel, sand, etc., partly consolidated.

Granite and
Schist

Eastern Border
of the High Plains

NORTHEASTERN COLORADO

WESTERN AND CENTRAL NEBRASKA OR KANSAS

Approximate Scale

HORIZONTAL, 1 INCH REPRESENTS 60 MILES
VERTICAL, 1 '' '' 5 MILES

BIGHORN MTS.

BLACK HILLS

BADLANDS

GREAT PLAINS

Paleozoic l.s.
and s.s.

Granite

Belle Fourche R.

Mesozoic
sandstone and shale

Pal. l.s. and s.s.

Granite and schist

Cheyenne River

Tertiary gravel, sand, etc., partly consolidated.

Missouri River

Cretaceous
Chalk, shale, limestone

WYOMING

SOUTH DAKOTA

CENTRAL AND EASTERN SOUTH DAKOTA OR NEBRASKA

Approximate Scale

HORIZONTAL, 1 INCH REPRESENTS 60 MILES
VERTICAL, 1 '' '' 5 MILES

Fig. 147.—Topographic and structural sections across the Great Plains. Both are semi-diagrammatic. The upper section extends from west to east, the lower from northwest to southeast.

Although the strata of the plains lie sensibly flat within short distances, the structure of the hard rock, as shown in Fig. 147, has the general character of a great geosyncline which rises on approach to the eastern border of the Rocky Mountains, where the steeply inclined, eastward-dipping strata appear in the form of hogback ridges, fringing the front ranges of the Rockies. This regional westward dip of the Great Plains strata is distinctly characteristic; but the general structure is modified in places by structural deformations of a certain degree of importance. In southeastern Colorado, for example, there is an arch in the form of an anticline which extends southeastward from the Greenhorn and Wet Mountain ranges and passes under the Mesa de Maya, Fig. 141. Again, the continuity of the monocline of the foothills on the western border of the Great Plains, formed upon the upturned edge of the Plains strata, is broken by a number of small offsets, a series of three occurring at Greeley, Colorado, Fig. 117.[1]

Over the larger portion of the Great Plains the westward-dipping strata end in long, low, ragged, eastward-facing escarpments; the topography thus has a "stair step" composition, the escarpments stand out as conspicuous risers and the broad, flat, inter-escarpment areas represent the treads. Of such origin are the Flint Hills, the Dakota Sandstone Hills, and the gypsum hills of Kansas and Nebraska; and similar examples occur in many other places. These features are most prominently developed where the hard strata (sandstones) are thick and the soft strata (shales) thin. Where the reverse is true the escarpments may be so low as scarcely to be identifiable and the inter-escarpment tracts dissected by a maze of branching systemless streams. For example, the Pierre shale occupies many thousand square miles of country adjoining the Black Hills and gives rise to a monotonous landscape of rounded hills thickly covered with grass instead of the more common tabular relief bordered by escarpments.[2] The Great Plains of Texas are developed largely upon an extensive limestone stratum (Edwards) which is completely surrounded by escarpments of erosion. In the Llano Estacado, which is the southern end of the High Plains subdivision, the surface is for the most part capped by alluvium from the mountains.[3]

[1] N. H. Darton, Geology and Underground Water Resources, Central Great Plains, Prof. Paper U. S. Geol. Surv. No. 32, 1905, pp. 74–75.

[2] Idem, pp. 22–23.

[3] R. T. Hill, Physical Geography of the Texas Region, Folio U. S. Geol. Surv. No. 3, 1900, pp. 6–7.

STREAM TYPES

Two types of streams cross the Great Plains, (1) streams whose headwaters are in the mountains and (2) streams whose headwaters are on the plains. Those streams whose headwaters are in the mountains are supplied with water more or less regularly either through rain or snow or both. In addition, their headwaters often drain glacial lakes and these tend to regulate the flow. The result of the somewhat regular water supply is that although the streams may become very low and even dry up occasionally, they are for the most part through-flowing the entire year. This result appears the more striking when it is known that their plains tributaries are intermittent and feeble and that the master streams suffer heavy losses by evaporation in crossing the semi-arid plains country.

A stream of this type is the Arkansas River, whose headwaters are in the Sangre de Cristo, Culebra, and Sawatch mountains, in each of which there are summits exceeding 14,000 feet in altitude. The precipitation along the crests of these high ranges is mainly in the form of snow and amounts to 20 or 30 inches of rain each year. From the foothills of the mountains to Arkansas City the rainfall ranges from 12 to 35 inches; it is 25 to 35 inches in the last 100 miles below Hutchison. Natural storage in the Arkansas basin is limited to a few mountain lakes of glacial origin. The streams are subject to two floods per year — the annual spring floods caused by the melting of the mountain snows, and the summer floods due to cloud-bursts in the foothills and plains regions.[1]

The Missouri River resembles the Arkansas in that its upper tributaries drain a forested region, while the main stream flows through a country almost wholly devoid of forests. The precipitation in the mountainous portion of the basin is mainly in the form of snow, but a great part of the area lies within the arid and the semi-arid regions, and it is probable that the mean annual precipitation throughout the entire basin is less than 20 inches. The river notably decreases in volume by evaporation in crossing the dry plains, though it never disappears, as is the case with many neighboring streams whose headwaters do not reach into the mountains.[2]

In contrast to the Arkansas and the Missouri is the Red River, whose sources are on the plains of northern Texas. The flow of this river is

[1] Freeman, Lamb, and Bolster, Surface Water Supply of the United States, 1907–08, pt. 7, Lower Mississippi Basin, Water-Supply Paper U. S. Geol. Surv. No. 247, 1910, pp. 29–31.
[2] Follansbee and Stewart, Surface Water Supply of the United States, 1907–08, pt. 4, Missouri River Basin, Water-Supply Paper U. S. Geol. Surv. No. 246, 1910, p. 30.

very uncertain, the run-off consisting chiefly of flood water from heavy rains. The flow ceases entirely in the late summer and fall of ordinary dry years. The drainage area consists of semi-arid plains varied by small areas of sand hills. If the summer is dry, the flow of the river ceases altogether, although water sufficient for stock always remains standing in pools. During long-continued or unusually heavy rains, or directly after such rains, the river becomes wide and deep, its bottom lands are flooded, and considerable damage is often done to livestock, railroads, and plantations.[1]

Those streams that descend the eastern slopes of the uplands south of the Canadian River and near the New Mexico-Texas boundary line are of an extreme type and die out just eastward of the boundary line. Their waters are absorbed in the surface material or gathered in the form of temporary lakes. It is not until the central portion of the Panhandle region of Texas is reached that the drainage is through-flowing and reaches the Gulf of Mexico by way of the Red River and other streams.

REGIONAL ILLUSTRATIONS

NORTHERN GREAT PLAINS

Many features of the northern part of the Great Plains in the United States owe their origin and character either to base-leveling or to glacial accumulation or both.[2] Westward from the valley of Red River of the North and the prairies of the Lake Winnipeg region is a plains country which rises gradually northwestward to the foot of the Rockies. The chief ascent of this plain west of the basin of Lakes Manitoba, Winnipeg, and Winnipegosis is over an abrupt escarpment developed upon Cretaceous strata that form the eastern border of the continuation of the Great Plains in Canada. The escarpment is from 200 to 1000 feet high, and its sections are named in order from south to north, Coteau des Prairies, the Pembina, Riding and Duck mountains and the Pasquia Hills or Mountains. West of these hills and so-called mountains the broad expanse of plains is broken here and there by valleys and irregularly dissected tracts, but in general the surface appears to be very moderately rolling, rising 4 or 5 feet to the mile to a height of 4000 feet at the foot of the Rocky Mountains in Montana and Alberta.

At the beginning of Tertiary time this region became a land surface

---

[1] Follansbee and Stewart, Surface Water Supply of the United States, 1907-08, pt. 4, Missouri River Basin, Water-Supply Paper U. S. Geol. Surv. No. 246, 1910, p. 85.

[2] W. Upham, Tertiary and Early Quaternary Base-leveling in Minnesota, Manitoba, and Northwestward, (Abstract) Geol. Soc. Am., vol. 6, pp. 17-20. See also the American Geologist, vol. 14, 1894, pp. 235-246.

and has remained a land surface down to the present. During this long period (Tertiary) the great tract of country was base-leveled except for isolated residuals, here and there, consisting of remnants of horizontal Cretaceous strata elsewhere eroded. Turtle Mountain on the northern edge of North Dakota and the southern edge of Manitoba is an illustration of such a residual. It is about 40 miles long, about two-thirds as wide, and has an altitude of 200 to 800 feet above the surrounding country. It consists of nearly horizontally bedded shales and lignites capped by 50 to 75 feet of glacial drift.[1] West of Turtle Mountain the depth of the Tertiary base-leveling was greater and attained a tremendous value on the plains of Montana; in the Highwood and the Crazy Mountains districts the country was planed down 3000 to 5000 feet. These mountains owe their superior elevation to the great resistance of their rocks as compared with the strata wrapping about them; they now stand as embossed forms rising conspicuously above the general expanse of the monotonous plains.[2] The greater denudation of the western portion of the Great Plains was due to greater initial uplift; it appears to have been raised 1000 to 5000 feet.

The Tertiary topographic cycle of erosion was closed by a great uplift, after which vigorous stream erosion set in. In the eastern part of the Great Plains region erosion was carried to the point of partial base-leveling and the development of wide flat valleys. The Red River developed a broad plain more than 200 miles in extent from south to north and from 200 to 500 feet below the general level of the country. In Manitoba the Cretaceous beds were eroded down to the underlying Archæan and Paleozoic rocks over a large area and the eastern marginal escarpment formed. The depth of post-Tertiary erosion is indicated by the White Hills west of Lakes Winnipegosis and Manitoba, a long line of escarpments. During the period in which this escarpment was forming in Manitoba and in which its tributaries were deeply incised the plains of Montana were partly dissected by the deeply intrenched streams.

These plains now present one of the best illustrations of an uplifted base-leveled surface to be found in the country. The surface bevels the strata of the region regardless of changing dip and hardness, consequently neither the structure of the rock nor alluviation can be appealed to in explanation of the general topographic uniformity. The

[1] W. Upham, Tertiary and Early Quaternary Base-leveling in Minnesota, Manitoba, and Northwestward, (Abstract) Geol. Soc. Am., vol. 6, pp. 17-20. See also the American Geologist, vol. 14, 1894, pp. 235-246.

[2] W. M. Davis, The United States in Mill's International Geography, 1900, p. 756.

interstream spaces are rather flat and give no indication of the marked depth and steep walls of the intrenched streams which are practically invisible until one is almost at the canyon brink. The narrow valley of the Missouri is from 300 to 600 feet below the general level of the plains, and that of the Yellowstone and many others have been cut to comparable levels.

GLACIAL FEATURES

The map, Fig. 148, shows how small a portion of the Great Plains has been glaciated. Besides glacial features similar to those of the northern

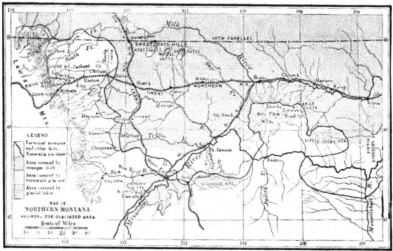

Fig. 148. — Note the large proglacial lake west of the Highwood Mountains, caused by the glacial damming of the Missouri River. Its discharge across the northern border of the Highwood Mountains formed Shonkin Sag, a temporary outlet. (After Calhoun, U. S. Geol. Surv.)

part of the Prairie Plains may be noted the two terminal moraines which cover the country north of the Little Rocky Mountains. The surface is a rolling plain with rounded flat-topped ridges and broad and low intervening hollows. The larger drift is chiefly Laurentian, but the bulk of the material is made up of quartzite drift from the Rocky Mountains, and consists of well-rounded pebbles and small quartzite bowlders of different colors.[1]

The Keewatin ice sheet extends southward into the northern Great Plains as far as the Highwood, Bear Paw, and Little Rocky Mountains. These elevations stopped its southward progress locally, while on the

[1] Weed and Pirsson, Geology of the Little Rocky Mountains, Jour. Geol., vol. 4, 1896, p. 402.

plains between it moved somewhat farther south in great lobes, Fig. 148. The Sweetgrass Hills, almost on the international boundary line (long. 111° W.), were completely surrounded by the ice, above which they projected 2000 feet as great nunataks. Terminal moraines therefore completely encircle the hills, while the higher mountains farther south and east are but partly encircled by moraines. The position of these moraines along mountain slopes and on north-south lines makes it possible to establish the fact that the slope of the ice surface was from 30 to 50 feet per mile. A similar basis is afforded by the disposition of the moraines formed on the margins of lobes that extended up many of the valleys of the region, the divides between (150 feet high) remaining uncovered — with a similar result.

Terminal moraines fringe the border of the drift-covered country in places only. Elsewhere there is no well-defined terminal ridge, only a broad low rise usually too slight to be noticeable. The terminal moraines are monotonously rough and their hollows contain large numbers of smaller lakes and ponds. In places the moraines are bowldery ridges, in others they are broad belts of till of variable composition. Near the edge of the glaciated tract the drift is thin, though locally it is from 50 to 200 feet thick.

The drainage of the glaciated northern portion of the Great Plains is toward the northeast and east, and ice invasion was therefore bound to block the streams either temporarily by the ice or more permanently by moraines or both. The lakes are of two classes. Those of the first class were in a few instances of great extent, and their floors are noted for the scattered bowlders rafted into position by detached ice blocks. The name Great Falls Lake has been given to one such temporary water body, formed by the ponding of the upper Missouri between the southward-facing ice sheet and the northward-facing slopes of the Highwood Mountains. Its outlet along the northern mountain flanks at 3900 feet cut a broad channel nearly a mile wide and 500 feet deep, known as Shonkin Sag. It is one of the most striking topographic features in the region, since it is entirely independent of rock structure and crosses the present drainage lines at right angles.

After the disappearance of the ice the terminal moraines continued for some time to act as barriers to the drainage, but outlets were soon cut down and the nearly bowlderless clays of the lake floors exposed. Most of the existing lakes of the region occur in the hollows of the morainic belts and are individually of small extent.

Glacial features of considerable topographic prominence are also found upon the northwestern border of the Great Plains of the United

States west of the limit of continental glaciation where valley glaciers from the front ranges of the Rockies deployed, building terminal and lateral moraines and supplying material for local alluviation (p. 312). The glaciers extended as individual tongues of ice down the main valleys, but 14 of them united in groups along the base of the mountains to form piedmont glaciers of great size which extended out over the plains for from 30 to 35 miles. In other cases the ice extended only part way down the valleys, deposited valley moraines, and caused the formation of related features such as lakes, outwash plains, and valley trains.

The piedmont glaciers and the continental ice sheet overrode common ground in two localities — the headwaters of the Marias and St. Mary's rivers — though these ice bodies were not contemporaneous. The valley glaciers had retreated before the continental ice sheet reached its greatest extension, since the deposits of the latter ice body overlie those of the former. The terminal moraines of the piedmont glaciers are of great size, some of them attaining heights of 200 feet and breadths of a mile or more.[1]

BADLANDS OF THE BLACK HILLS REGION

Among the badland areas of the Great Plains the largest and most important is in South Dakota and Nebraska on the southern, southwestern, and southeastern borders of the Black Hills. The portion of the badlands showing the greatest topographic complexity lies near the southeastern border of the Black Hills between the White and Cheyenne rivers and is known as the Big Badlands. It is continuous with a second badlands areas which extends eastward, southward, and westward along the upper White River forming the high Pine Ridge escarpment which extends through western Nebraska and into Wyoming. Outside the limits of the main badlands area are many remnants of the strata that formerly extended over the region. They now occur as mesas or tables which stand at various heights up to three hundred feet or more above the adjacent basin or valleys. Some of them are of large size and those east of the Cheyenne River have been given individual names by the settlers who occupy them, as Sheep Mountain Table, White River Table, Kuba Table, etc. The badlands were at first thought to be quite uninhabitable, the name itself being derived from the French name "Mauvaises Terres" applied by the early hunters and trappers. The phrase is meant to signify a country difficult to cross, chiefly because

---

[1] F. H. H. Calhoun, The Montana Lake of the Keewatin Ice Sheet, Prof. Paper U. S. Geol. Surv. No. 50, 1906.

of the rugged surface and the general lack of water.   Later exploration and development have shown that much the greater portion of the area within the badlands is level and fertile and covered with abundant grass; and that water may be obtained, especially on the higher tables, by sinking shallow wells in the surface mantle of gravel.   As a whole the region has considerable agricultural and grazing importance.

The chief factors controlling the development of the badlands topography have been the great extent of slightly consolidated, fine-grained strata lying at a considerable altitude above the sea in a region of low

Fig. 149. — Details of Badlands in Brule clay at foot of Scotts Bluff, Nebraska.   (U. S. Geol. Surv.)

rainfall and sparse vegetation.   The scarcity of deep-rooted vegetation enables the soft material to be rather easily eroded.   While short grasses are abundant over large areas they do not have deep root penetration and do not form a sufficiently continuous cover to prevent cutting. There is a small amount of vegetation of a higher order, but it is even less effective than the grass in preventing the formation of gullies.   A few gnarled cedars occur on the highest points and bushes of various kinds occur in the valley floors in favorable places, but they offer little obstruction to the development of the gulches and canyons which diversify the scarped margins of the area.

It is also noteworthy that the rainfall is more or less concentrated into heavy showers of short duration and these act most vigorously in denuding the surface, forming channels, and transporting accumulated rock waste. The clays which compose a large part of the badlands expand greatly when wet and contract when dry (p. 415), so that their surfaces tend constantly to break up and form an easily eroded mass of material. Frost action tends toward the same effect, with the result that the clay is constantly being loosened in the dry as well as in the wet season, and is thus prepared for rapid erosion when rain falls.

The topographic complexity of the badlands is explained chiefly by the unequal resistance to erosion on the part of alternating beds. Individual layers vary horizontally as well, and the softer portions are worn back into the form of gulches and alcoves while the harder portions remain projecting as spurs or headlands. A hard layer tends to persist longer than a soft layer above or below it, thus forming typical scarp and terrace topography. In many cases the hard layer is undercut to such an extent as to develop a precipice. Joints are developed here and there in the clays and these tend to accelerate erosion along vertical planes with the result that cave-like excavations are formed at the heads of vertical walled gulches. Many beds contain large numbers of concretional masses whose resistance is greater than that of the surrounding material, thus tending to the production of a surface whose irregularities reflect the irregularities of the concretions.

Among the important topographic elements of the badlands are the alluvial fans which cling to the base of every pillar, mound, or table, and have a large total extent. Their surfaces have low gradients and they represent the lag of transportation behind disintegration. Most of the streams of the badlands area are intermittent. They have a continuous flow for a short time after the rainy season but their sources rapidly fail, and soon nothing is left in their channels but bars and banks of sand and silt, and a few trifling pools of water, either so strongly alkaline as to be of little value, or so turbid as to have the consistency of mud. Only two streams have a continuous flow, Cheyenne River and White River; but their flow varies greatly in volume, being high in the rainy season, when their tributaries are full, and very low in the dry season, when their tributaries fail. Many streams, especially the streams of medium size, flow in box-like trenches, the material of their banks standing up in bluff-like form as about the border of the area.[1]

[1] For a summary description of the badlands of South Dakota, illustrated by well-selected photographs, see The Badland Formations of the Black Hills Region, C. C. O'Harra, Bull. S. D. School of Mines No. 9, 1910.

HIGH PLAINS

The map, Plate IV, indicates the extent of the High Plains, the most important subdivision of the Great Plains province. The interior topographic and drainage features of the High Plains are typically represented in the Llano Estacado (Staked Plains) of Texas, the surface of which appears extremely flat to the eye, though it actually slopes eastward at the rate of 8 or 10 feet to the mile. Only gentle depressions occur

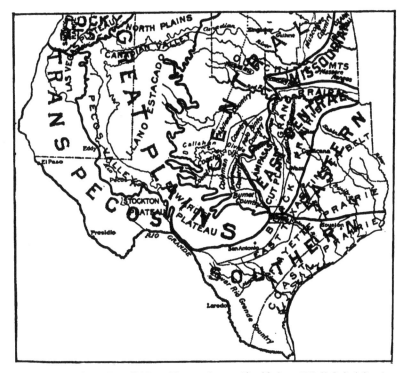

Fig. 150. — Physiographic subdivisions of Texas and eastern New Mexico. (Hill, U. S. Geol. Surv.)

here and there — the products of underground solution and of unequal wind erosion. Local storm floods tend to level these irregularities by filling the hollows and eroding the surrounding surface.[1]

The High Plains strata were laid down as a series of great compound alluvial fans or a piedmont alluvial plain at the eastern base of the Rocky Mountains. The eroded bedrock floor upon which the materials were deposited and the eroded condition of the materials themselves

[1] R. T. Hill, Physical Geography of the Texas Region, Folio U. S. Geol. Surv. No. 3, 1900, p. 6, col. 3.

imply a succession of changes in the character of the stream action from degradation to aggradation and back again to degradation. These changes harmonize with the conception of climatic changes of known

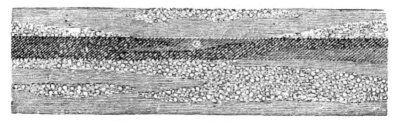

Fig. 151. — Ideal structure of the Tertiary deposits of the High Plains. The dark band indicates the position of a partly consolidated portion of the section known as a mortar bed. (Gould, U. S. Geol. Surv.)

occurrence, the dry Tertiary (aggradation) being succeeded by the wet Pleistocene (degradation), which was in turn succeeded by the dry

Fig. 152. — Typical view of the High Plains of western Kansas. (Gilbert, U. S. Geol. Surv.)

Present in which the streams are aggrading their valley floors.[1] It is also conceivable that cutting power may have been gained or at least increased (1) by broad uplift on the west which would increase the

[1] D. W. Johnson, The High Plains, 20th and 21st Ann. Repts. U. S. Geol. Surv., 1898–1899, 1899–1900.

stream gradients and (2) by a cooling of the climate so as to reduce the evaporation.

The greater part of the Tertiary deposits of the High Plains consists of clay, sandstone, and conglomerate, with clay largely predominating. The materials are usually arranged in a heterogeneous manner, as shown in Fig. 151, a characteristic dependent upon their origin as coarse channel or finer flood-plain deposits.[1] A large part of the deposits is composed of smooth rounded white or yellow quartz grains locally cemented by lime into coarse rough sandstones, but more commonly unconsolidated and in places blown by the winds into dunes. In many instances the sand is mingled with small pebbles and clay cemented with lime, forming the so-called "mortar beds." The pebbles vary greatly in size, shape, and color, and generally occur in more or less lenticular cross-bedded layers, some of which are at least 50 feet thick. In many places they are mixed with fine sand and locally are scattered through the clay formations.

The clays are generally white, but locally buff and pink in color. In places the lime cements the clay into irregular and grotesque forms, which, being harder than the surrounding formations, are left standing in relief by erosion and form picturesque elements in the local scenery.[2]

The average thickness of the Tertiary deposits in eastern Colorado and western Kansas is 200 to 300 feet, but increases to nearly 1000 feet in portions of western Nebraska and southwestern Wyoming. Originally the entire region was probably covered with late Cretaceous deposits that extended far up the flanks of the Rocky Mountains, the Bighorn Mountains, and the Black Hills, a conclusion indicated by the occurrence of outliers of these deposits at high altitudes, the intervening portions having been extensively stripped off.[3]

PHYSICAL DEVELOPMENT

After the great Rocky Mountain System on the west had been outlined and during late Tertiary time there was a long period in which streams of moderate gradient drained across the central portion of the Great Plains. Locally, extensive lakes were formed, but the larger part of the deposits (Oligocene) of that time consists of coarse sandstones, sands, etc., deposited on river flood plains and aggrading stream channels. Aggradation took place chiefly in Oligocene time, and the area of Oligocene deposits, Plate V, extends across the eastern part of

[1] C. N. Gould, Geology and Water Resources of the Panhandle, Texas, Water-Supply Paper U. S. Geol. Surv. No. 191, p. 33.

[2] Idem, p. 32.

[3] N. H. Darton, Geology and Underground Water Resources of the Central Great Plains, Prof. Paper U. S. Geol. Surv. No. 32, 1905, pp. 169-170.

Colorado and Wyoming, western Nebraska and South Dakota, and probably northward into Canada. The streams of this time shifted their courses from side to side and laid down a sheet of débris in the form of low-grade and confluent alluvial fans of vast extent and with a thickness in places of 1000 feet. The period of deposition ceased with the Miocene, and uplift and erosion followed which removed large portions of the accumulated material. During this period of deposition (Oligocene) in the northern part of the High Plains the southern part was probably a region of erosion, but with the uplift and erosion of the northern High Plains (at the close of the Miocene) the southern portion became a region of deposition and the streams began depositing thin mantles of late Pliocene sands in southern Colorado, southern Nebraska, Kansas, and still more southerly localities.[1]

It is important to recognize that these changing conditions of late Tertiary deposition and erosion first in the north and then in the south were determined by differential uplift. The uplifted region suffered erosion, the depressed region suffered deposition. It is equally important to recognize the fact that these occurrences took place in the Tertiary under more or less stable climatic conditions and before the advent of the glacial ice. We have in these conditions clear evidence of broad differential uplifts and depressions which produce alternating conditions of aggradation and degradation over large areas during the Tertiary period, and which were, so far as we may judge, wholly unrelated to climatic influence except the general influence of aridity to leeward of the high Rocky Mountain Cordillera. The broad changes of later date are, however, undoubtedly to be associated with climatic change, though the exact locality in which erosion or deposition was strongest was probably determined by crustal movement. During the early Pleistocene uplift of the land there also occurred an increased precipitation which resulted in widespread degradation of the preceding deposits; the Tertiary deposits were entirely removed in the eastern portion of the area and widely and deeply trenched in the western portion.[2] There was deep erosion about the Black Hills dome, and the High Plains, whatever their extent in that direction, were largely removed and their northern edge left as at present in the great escarpment of Pine Ridge facing the Black Hills uplift. The streams of Pleistocene time also cut deeply and removed widely the high plains of Nebraska, Colorado, Kansas, and Texas, though wide areas of tabular surfaces are still exposed.[3]

Descriptions of various portions of the High Plains vary greatly with reference to the present character of the stream work. It is frequently asserted that the High Plains are a region of denudation, although as frequently one finds the rivers described as building up the bottoms of their valleys. This contradiction is only apparent. The greater portion of the High Plains and many portions of the Great Plains are being denuded by the ramifying headwater streams, and the

[1] N. H. Darton, Geology and Underground Water Resources of the Central Great Plains, Prof. Paper U. S. Geol. Surv. No. 32, 1905, pp. 185–186.

[2] W. D. Johnson, The High Plains and their Utilization, 21st and 22nd Ann. Rept. U. S. Geol. Surv., pt. 4, 1899–1900, 1900–1901.

[3] N. H. Darton, Geology and Underground Water Resources of the Central Great Plains, Prof. Paper U. S. Geol. Surv. No. 32, 1905, pp. 186–188.

detritus is being carried down to the valley bottoms, where it accumulates in the form of broad valley flats or flood plains which to a large extent control the curves of the rivers, as along the valley of the Platte. Such aggradation is a normal result of the return to the drier conditions prevailing both now and in the Tertiary. The process will continue until a gradient is established which will allow the streams to carry all their waste to the sea. Dissection by the streams of Pleistocene time caused such a degree of valley cutting that the present aggradation on the floors of the valleys is small compared with the degradation that preceded it. Consequently the tributaries of all the master streams of the High Plains region join valleys far below the general level. The tributary streams are therefore eroding the surface actively, and they drain by far the larger portion of the High Plains. Active erosion has stopped completely and active aggradation taken its place on almost all the larger valley floors.

BORDER TOPOGRAPHY

In some localities the erosion of the border of the High Plains has given rise to the formation of "holes" with local badland topography. A typical instance is Goshen Hole on the border of Wyoming and Nebraska. The formations which outcrop in this area consist largely of clays and sandstones. The clay (Brule) is easily eroded in such manner as to keep the sandstone (Arikaree) covering it in constant retreat as a nearly vertical wall about the edges of the Hole, a process greatly hastened by the issuance of ground water at the line of contact of the two formations. The escarpment separates the dissected lowland of the floor of the hole from the upland or general undissected surface of the High Plains. It is best defined upon the western side of Goshen Hole, owing to the absence of streams upon this portion, though the escarpment is here in constant retreat as the result of the sapping action of the ground water at the heads of the minor streams to which it gives rise. Upon the lowland constituting the floor of the hole are small mesas and tablelands, remnants of the upland that once occupied the region and that has since been worn away. Wherever the relations of drainage and geologic formations are such as not to bring the streams below the soft strata or where the upper strata are soft and the lower strata hard, either a lowland has not been formed or, if formed, its borders are not so well defined as in those cases where the softer members underlie the harder and the drainage has been developed distinctly below the level of the soft formations.

The eastern border of the High Plains in Texas and Oklahoma is in

most places a distinct scarp 200 to 500 feet above the eroded plains on
the east, though ordinarily the descent is more gradual and takes place
in a belt 5 or 6 miles wide — a belt with distinct topographic features
in high contrast to the flatter plains to east and west.[1] This escarp-

Fig. 153. — Jail Rock with the valley of North Platte in the distance, looking east. The capping stratum
is sandstone (Gering) while the slopes are of clay (Brule). Typical border topography of a portion
of the High Plains. (Darton, U. S. Geol. Surv.)

ment is locally known as "The Breaks," Fig. 155. It occurs along the
valley margins, passing in broad eastward-looping curves from one
drainage system to another, and is characterized by badland erosion

Fig. 154. — Eastern erosion escarpment of the High Plains of Texas, Oklahoma, and Kansas. The eroded
bed-rock surface has been covered with Tertiary deposits and later reëxposed in part. (Johnson,
U. S. Geol. Surv.)

forms, short ridges, steep talus slopes, isolated buttes and peaks, and
an intricate system of narrow V-shaped valleys that sometimes develop
into impassable canyons, Fig. 154. It is a region very difficult to trav-

[1] C. N. Gould, Geology and Water Resources of the Eastern Portion of the Panhandle of
Texas, Water-Supply Paper U. S. Geol. Surv. No. 154, 1906, p. 9.

erse, and can be crossed with a wagon only at infrequent intervals and by specially selected routes. The marginal escarpment is most typical along the Canadian River and in Palo Dura Canyon in Armstrong County, Texas.

Fig. 155. — Details of form, eastern border of the High Plains, Texas. The scarp is known as "The Breaks of the Plains." (Hill, U. S. Geol. Surv.)

The eroded plains east of "The Breaks" have suffered such extensive dissection that the Tertiary and Pleistocene deposits have been entirely removed from their surface and the streams are actively dissecting the underlying Red Beds. The region is a rolling plain into which the streams are cutting rather deep steep-sided valleys, and outlining hills generally of conical shape but often elongated and attaining heights of 200 to 800 feet or more. These hills are capped by resistant ledges of sandstone, gypsum, or dolomite that have resisted the erosion which swept away the softer clays and shales above them.[1]

SAND HILLS AND LAKES

The sand hills of the Panhandle region of Texas range in size from small mounds to ridges 30 or 40 feet high. They are oval, crescentic, or elongated in shape and extend in various directions. They are interspersed with broad, shallow, basin-like depressions from 1 to 10 acres in extent, probably representing great "blow-outs" or wind-eroded hollows. In a few localities there are migratory dunes, as on the west

[1] C. N. Gould, Geology and Water Resources of the Eastern Portion of the Panhandle of Texas, Water-Supply Paper U. S. Geol. Surv. No. 154, 1906, pp. 9–10.

side of the Canadian River in Roberts County. The dune sands are derived from sandstone ridges chiefly, but also in part from river sand.

The rainfall on an important part of the southern High Plains is collected into shallow saucer-shaped depressions known as "lakes" or playa lakes, scattered irregularly about. The depressions vary in size from a diameter of a few feet and a depth of a foot or two to lakes several hundred rods in diameter and from 20 to 40 feet below the general level. Some of these depressions are more than a square mile in area. Many lakes in the depressions are perennial and afford an

Fig. 156. — Precipitation in the Texas Region. I, over 50 inches; II, over 45; III, over 40; IV, over 35; V, over 30; VI, over 25; VII, over 20; VIII, over 15; IX, over 10. (Hill, U. S. Geol. Surv.)

abundant supply of water for stock. Others are ephemeral, that is, they are filled during a period of rain, but soon dry up; and still others contain water only at long intervals. Before wells were constructed and in the early settlement of the country the lakes were often in intimate relation to the location and welfare of settlers, for they constituted practically the only water supply. Wagon trains in crossing the Llano Estacado camped beside them, and they were centers to which cattle were driven until comparatively recent times.[1]

[1] C. N. Gould, Geology and Water Resources of the Panhandle, Texas, Water-Supply Paper U. S. Geol. Surv. No. 191, 1907, p. 50.

The sand-hill country is not limited to the High Plains but is found (1) in many cases with limited development along the valley floors where sands deposited in the stream channels during high water become available to the winds during the low-water stages, as along the Arkansas, and (2) over the outcrop of unconsolidated material. The largest tract lies in Nebraska[1] and was developed upon the outcrop of a sandy unconsolidated formation which the present sand cover largely conceals. The thickness of the sand cover is rarely over 100 feet and generally much less. The total area of the tract is about 18,000 square miles. A peculiar feature of this great area is the relative absence of streams, except such as are supplied from outside sources. In spite of the small surface drainage there is considerable spring flow and seepage along the main river valleys. Most of the rainfall is absorbed by the loose dry sands and percolates downward, to be added to the ground water, which is surprisingly large in amount and is in general of good quality. The large amount of ground water is reflected in the numerous lakes scattered through the sand-hill country. Since the lake surfaces represent the surface of the ground water the lake levels rise and fall in response to the variable rainfall just as does the ground water. In times of light rainfall they are smaller and shallower; in times of heavy rainfall they are deeper and broader. Some of them disappear in extremely dry seasons; others are so permanent that their waters are stocked with fish. They are of great importance as a source of stock water, since the bunch-grass of the sand hills gives the area high value in respect of grazing, the chief industry of the region

VEGETATION

Bunch-grass is the characteristic vegetal growth of the High Plains. Its growth is extended by roots and more rarely by seeds. The dryness and coarseness of the surface soil and its wind-swept condition render reproduction by seed very difficult. The result is not a turf so dense that the ground is not visible but a series of tufts or bunches separated by large and small bare spaces.[2]

It is the general conclusion of students of the plains streams that floods are growing in volume and frequency, a state of things that is thought to be related largely to the breaking up of the protective grass cover by over-pasturage in the headwater regions. A grass cover has almost as great influence in checking run-off and favoring absorption

---

[1] G. E. Condra, Geography of Nebraska, 1906, pp. 85–94.

[2] Report upon Geographical and Geological Explorations and Surveys West of the 100th Meridian (Wheeler Surveys), Geology, vol. 3, 1875, p. 606.

by the soil as a timber cover, and this relation is of such importance throughout the Great Plains region as to make it a point of great interest to the forester interested in general conservation, for the combination of grass-land and timber is thoroughly practical.

In the middle courses of the Great Plains streams and on numerous tributaries in the mountains there is a protective timber covering. In a similar way a timber growth is found on the borders of the escarpments. and its maintenance is a matter of the liveliest concern to every

Fig. 157. — Vegetation of the Texas regions. 1, Atlantic forest belt; 2, Rocky Mountain forest; 3, Chaparral; 4, Black Prairie; 5, bolson desert flora; 6a, Grand Prairie; 6b. Great Plains; 7, transitional, with plains, prairie, and Atlantic flora; 8, coast prairies; XXX, yucca belts. (Hill, U. S. Geol. Surv.)

one in any way related to the régime of the streams. The amount of washed soil running off in the floods that follow cloud-bursts in the Great Plains is enormous, especially where trees are absent and the grass covering thin; the maximum extent of grass and forest cover ought therefore to be maintained. This is seen especially in the lignitic belt of Texas, which is a rough broken country with sandy soils. But for the shortleaf and post-oak forests which cover them the soils would be washed in quantities from the steeper slopes. Annual plants with a superficial root system can not flourish in such soils without abundant rainfall. More than that, soils in these positions have little capacity for retaining moisture, and forest growths are slow in beginning. The hills bordering the Edwards Plateau would become arid and worthless if they were stripped of their forest cover. The inch-deep soil débris would be rapidly washed away and the restoration of the forest become impossible.

The Great Plains contain but little timber, though the supply is in many cases sufficient for local and limited use. In the Camp Clarke district of Nebraska the ridge extending west from Redington Gap bears scattered pine trees of moderate size, and there are also a few Rocky Mountain pines (*Pinus ponderosa*) on the slopes ascending to the high table at the southern margin of the district. The largest are from 1 to 2 feet in diameter. A moderate number of young pines begin growth in favorable situations on the ridges, though few of them attain maturity. The zone of cottonwoods is characteristic of most western streams, though it is absent along North Platte River, where only a few small trees and bushes are found. The principal deciduous growths are found in ravines; they include cottonwood, box elder, wild plum, and a few other varieties.[1]

The Great Plains include much of the central treeless region of the United States where forest planting has been a part of the progress in agriculture, so that the areas of most extensive agricultural development are also those where greatest tree planting has taken place. Approximately 840,000 acres have been planted in the central treeless region, but the best conditions can be obtained only by planting about 14,000,000 acres in all. Although forest planting had been in a period of decline in this region, it has recently revived. It has been demonstrated that about 5% of the prairie region should be forest covered, and farther west, in the Great Plains proper, 3% could profitably be devoted to tree growing. South Dakota has planted about 122,000 acres; Nebraska, about 192,000 acres; Kansas, 175,000 acres; Oklahoma, 21,000 acres; Texas, 13,000 acres. The chief purpose of such tree planting has been for shelter, for post production, for protection from the hot winds of summer and the blizzards of winter, and to a limited extent for fuel.[2]

A point of great interest in connection with the limited tree growth of the central plains is the possibility of an earlier and more extensive timber cover which may be restored by proper effort. If the rainfall was more abundant for a time during and after the retreat of the ice, it follows that timber may have grown upon areas now too dry to support it. The

[1] N. H. Darton, Camp Clarke Folio U. S. Geol. Surv. No. 87, 1903, p. 1.
[2] Yearbook Dept. Agri., 1909, pp. 340–342.

thought is the more plausible when we recognize the great length of time involved in the disappearance of the ice sheet, long enough, we may suppose, for considerable progress to have been made in natural reforestation. The conclusion appears to be supported by the finding of remnants of old pine forests in geologically recent deposits in the valley of the Niobrara, 50 miles southeast of existing forests of pine. It has been concluded that these forests formerly extended down near the mouth of the Niobrara.[1] Likewise in the sand-hill country of Nebraska the stumps of cedars and pines — trees now confined almost wholly to the river bluffs — are found some distance away from the rivers, thus suggesting former more extensive tree growth.[2]

The fundamental condition governing the treelessness of the Plains and the Prairies is the deficiency of rainfall in spite of the fact that the larger portion of the rainfall of the central plains region occurs in the summer or growing months, Fig. 22. Beyond the Mississippi the rainfall shades off rapidly; in sympathy with this increasing dryness is the gradual thinning out of the forest and its final disappearance. In the transition belt it is obvious that secondary forces will hold the balance of power. Among these undoubtedly the most important are the fineness of the soil, forest fires started by the Indians and by lightning, the prolonged droughts of exceptionally dry years, the high rate of evaporation of soil water, and the low humidity.

The importance of the fineness of the soil has been strongly urged,[3] since almost all forms of vegetation require an aerated soil, and this is true especially of all but a small number of forest trees. Prairie soils are notably fine and compact, especially in their natural state, and this tends to keep rainfall near the surface for a longer time than in the case of coarse soils, hence a larger part of the rainfall is reëvaporated. The deeper-lying tree roots are thus deprived of a large part of the rainfall, besides being deprived by grass roots of such rainfall as is left available to plants. Prairie and forest fires are of frequent occurrence, and before the coming of the whites were often started by Indians either in sheer ignorance, carelessness, and wantonness, or for purposes of war and the chase. The number of thunderstorms that develop on the plains during the summer months is excessive. Lightning starts a fire in the dry half-withered grass and this spreads with great rapidity and is often not extinguished by the ensuing rain, which may be very light. Indeed

[1] S. Aughey, U. S. Geol. and Geog. Surv. of Col. and Adj. Terr. (Hayden Surveys), 1874, p. 266.

[2] G. E. Condra, Geography of Nebraska, 1906, p. 93.

[3] J. D. Whitney, Plain, Prairie, and Forest, Am. Nat., vol. 10, 1876, pp. 577-588 and 656-667.

many summer thunderstorms are not accompanied by rain and fires burn unchecked. That this was formerly an important natural agent is shown by the extension of the natural forest upon prairie land in the Edwards Plateau of Texas.[1] Settlement has stopped the periodic burning of the grasses, which were harmed by fires much less than were the shrubs, the vanguard of the timber vegetation. The over-pasturing of the ranges of this region has prevented the grass from forming a continuous sod, thus throwing further influence in the direction of the timber growth. In recent years the results have been marked. The shrubs have come in rapidly, trees have grown in their wake, and there is now in progress a gradual but extensive transition from grass to woody growth.

The prolonged droughts of exceptionally dry years are effective in preventing the germination of the seeds of forest trees and in harming tender seedlings. At least three-fourths of the rain that sinks into the ground never gets beyond the surface layer of two feet,[2] while the increasing distance to the ground water with increasing aridity is well known. When these naturally hard conditions are enforced in exceptionally dry years the roots of seedlings and mature trees are confronted with grave difficulties in their search for a water supply. The water in the surface layer is soon evaporated in large part or left in the form of a film surrounding the soil grains where it ceases to be supplied by capillarity and gravitational flow to the root zone of forest trees. Evaporation is hastened by the high summer temperatures of the plains and the continued sunshine and the rather steady and often high winds.

It should not be forgotten in the analysis of the problem that the underlying condition is regional dryness. Were the rainfall heavy, or even moderately heavy, a forest cover would undoubtedly exist in the plains country. Besides this factor the others appear secondary, yet the sum total of the secondary forces definitely restricts the forest. (1) There are far windier situations on forested mountain slopes that one may find on the plains, (2) forest fires do not eliminate trees from many relatively dry regions, (3) heavy clay lands bear luxuriant forests where rainfall and temperature conditions are normal. But when the conditions of rainfall are critical the balance of power is held by otherwise feeble influences.

That the cause is not wholly climatic is reasonably inferred from the success with which tree planting has been conducted in the treeless

---

[1] W. L. Bray, The Timber of the Edwards Plateau of Texas, Bull. Bur. For. No. 49, 1904, pp. 14-15.

[2] F. H. King, Productivity of Soils, Science, n. s., vol. 33, 1911, p. 616.

settled portions of the western country where windbreaks, snowbreaks, thrifty groves in city parks, etc., testify to the possibilities of at least a restricted reforestation. Even the sand dunes of Nebraska appear to be reclaimable, as shown by the at least temporary success that has attended the reforestation efforts of the government. The whole plains country can not be reforested since the greater part of it is too dry, but a large part will support a limited forest cover and the Prairie Plains can be made to grow trees almost everywhere, where now the growth is limited, at least toward the western border, to the valley floors and borders, and to tracts of porous soil with a good water supply.

CHAPTER XXIII

MINOR PLATEAUS, MOUNTAIN GROUPS, AND RANGES OF THE PLAINS COUNTRY

EDWARDS PLATEAU

THE combined Edwards Plateau and Llano Estacado (the portion of the High Plains that lies in western Texas) is the most extensive relief feature of the non-mountainous part of Texas, with an area of about

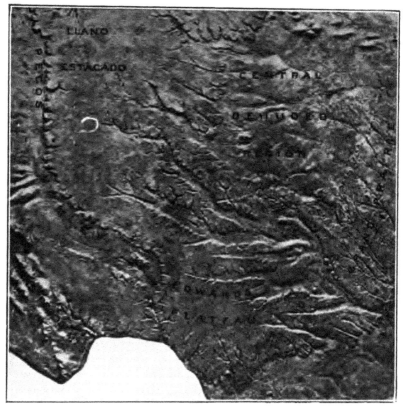

Fig. 158. — Llano Estacado, Edwards Plateau, and adjacent territory. The central denuded region was once covered with the Edwards limestone now exposed about the border as a frayed escarpment of erosion. (Hill, U. S. Geol. Surv.)

60,000 square miles. The two merge into each other and no sharp line can be drawn between them. They are surrounded on all sides by pronounced escarpments. The eastern margin is an escarpment of headwater erosion, the southern margin is the dissected Balcones fault scarp, and the western and northern margins descend steeply to the drainage grooves of the Pecos and the Canadian respectively.

While the Edwards Plateau continues the High Plains of western Texas or the Llano Estacado southward, it presents structural conditions wholly different from those found in the High Plains and, except for the general plain-like character of much of the surface, is quite different in its topographic character. It should be regarded as a subdivision of the

Fig. 159. — Summits of the Lampasas Plain, Texas.   (Hill, U. S. Geol. Surv.)

Great Plains province and not as a portion of the High Plains. The surface of the Edwards Plateau is capped by the Edwards limestone, the most conspicuous and extensive sedimentary formation in Texas (and a large part of Mexico) inland from the coastal plain. The Edwards limestone is also important topographically, for its harder strata resist erosion to a greater degree than associated formations, and it is on this account the chief structural factor in the formation of the scarps, mesas, and buttes of the Grand Prairie, Edwards Plateau, and large portions of the limestone mountains of Mexico.[1]

The wide, flat, upper surface of the Edwards Plateau is terminated on the borders of the province by a pronounced descent, a ragged scarp due to the dissection of the draining streams whose irregular interlocking headwaters have cut up the border into innumerable circular mesas

[1] Hill and Vaughan, Nueces Folio U. S. Geol. Surv. No. 424, 1898, p. 3.

and buttes.   The descent from the level plateau to the canyons is over a cornice layer of hard rock that weathers into a nearly horizontal vertical bluff.   At intermediate levels on the walls of canyons and valleys cliff makers occur; the intervening spaces between cliff makers consist of more easily weathered beds that yield a sloping talus.   Fig. 155 represents typical features in the eastern border of the plateau.

The drainage of the plateau summit except in the case of the largest streams is of the intermittent variety, and in places even some of the larger streams disappear for a short distance.   This is due partly to

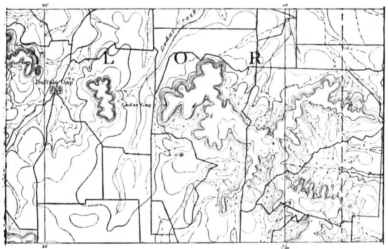

Fig. 160. — Summits of the Callahan Divide on the Great Plains of Texas (Central Province of Fig. 158).
(U. S. Geol. Surv.)

evaporation, which is especially strong under the cloudless skies of western Texas, but in larger degree to absorption in the waste-filled valley floors or in the fissured limestone.   The smaller streams have been called streams of gravel rather than streams of water, so clogged do their valley floors appear.   The streams draining the border of the plateau have a very much more regular flow than might be expected because of the large number of springs which are fed by fissures and underground channels into which the absorbed water is directed.

A subdivision of the Edwards Plateau is the Stockton Plateau west of the Pecos and east of the front ranges of the Trans-Pecos region. The southern border of the Stockton Plateau is tilted and faulted into low monoclinal blocks, while the scarps of the northern border overlook the Toyah basin.

Formerly the Edwards Plateau extended farther eastward, a condition

shown by the erosion remnants now occurring as outliers beyond the
eastern border of the plateau.   These remnants consist of circular flat-
topped hills and groups of hills on the interfluves.   The summits of the
outliers are all of the same geologic formation as the adjacent plateau
and are in vertical alignment with the normal coastward slope.[1]   The
altitude of these outlying mesas is about 500 feet above the principal
stream courses and about 250 feet above the surrounding plains.   They
form less than 10% of the total area, although they are widely distributed.   The principal group occurs on the 31st parallel on the divide
between the Brazos and the Colorado and is known under the collective
name of the Callahan Divide.   As a group these remnants represent a
former broad topographic level that once extended from the mountains
of the west to the eastern border of the Grand Prairie of Texas, p. 493.
During and since the Tertiary this old level was largely destroyed in the
central region of Texas and two opposing escarpments formed — the
eastern border of the Edwards Plateau and the western border of
the Grand Prairie — which are retreating from each other at the present
time.   Although the Balcones escarpment is commonly referred to as a
fault scarp it should be noted that a portion of the descent is to be
attributed to the increased steepness of dip along the front of a monoclinal fold which has been still further accentuated by erosion, Fig. 162.

The northeastern extension of the Edwards Plateau forms a third subdivision of the province.   Its surface is structural in origin, as indicated
by the large number of flat-topped remnantal summits which dominate
the tract.   They are developed most extensively on the divides between
the drainage lines and in places have sufficient height to be called
mountains.   In general the plateau remnants are bordered by bare
white limestone cliffs above and by gentler waste slopes below.   The
flat summits consist of weathered limestone in places without soil, in
other places bearing a thin but rich soil cover.   The largest continuous
area of these remnants is called the Lampasas Plain which extends for
115 miles from the Colorado in Travis County to the northeast corner of
Comanche County, Fig. 160.   As shown in Fig. 159, there is an intimate
relation between the topography and the geology on the one hand and
the vegetation on the other.   From the scrub oak and post oak growths
of the summit with its thin soils one passes in downward succession over
the nearly soilless cliffs with a growth of shin oak to the deeper soils of the
sandy formations with their good forests of black jack and post oak.[2]

[1] R. T. Hill, The Physical Geography of the Texas Region, Folio U. S. Geol. Surv. No. 3,
1900, p. 6, col. 4.
[2] R. T. Hill, Geography and Geology of the Black and Grand Prairies, Texas, 21st Ann.
Rept. U. S. Geol. Surv., p. 7, 1899–1900, p. 77 et seq.

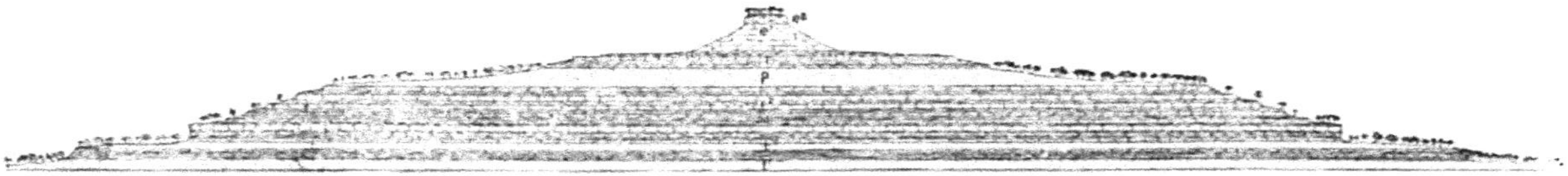

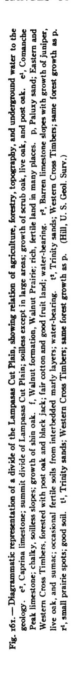

Fig. 161. — Diagrammatic representation of a divide of the Lampasas Cut Plain, showing relation of agriculture, forestry, topography, and underground water to the geology. e², Caprina limestone; summit divide of Lampasas Cut Plain; soilless except in large areas; growth of scrub oak, live oak, and post oak. e¹, Comanche Peak limestone; chalky, soilless slopes; growth of shin oak. f, Walnut formation, Walnut Prairie; rich, fertile land in many places. p, Paluxy sand; Eastern and Western Cross Timbers, forested with post oak and black jack; fair cotton and good fruit land; water-bearing. r², Barren limestone slopes with growth of juniper, live oak, and sumac; occasional fertile soils from interbedded marly layers; water-bearing. t², Trinity sands; Western Cross Timbers; same forest growth as p. r¹, small prairie spots; good soil. t¹, Trinity sands; Western Cross Timbers; same forest growth as p. (Hill, U. S. Geol. Surv.)

Fig. 162. — Edwards Plateau on left, Balcones Fault Zone in center, Black Prairie on extreme right. Note details of fault zone and increase of dip toward the east (right). Horizontal and vertical scales, 5000 feet to the inch. Edwards limestone on extreme left, Austin chalk on extreme right. (Hill and Vaughan, U. S. Geol. Surv.)

The main mass of the Edwards Plateau was elevated into permanent dry land at the close of the Cretaceous period. During the Eocene period there was extensive erosion and the upper Cretaceous formations (3000 feet) were swept away. Near the close of the Eocene, folding and faulting took place and the Balcones fault scarp that limits the plateau on the south was outlined. During the Miocene the plateau was still further stripped and the Edwards limestone that now forms the surface of the plateau was exposed. Active erosion has continued from the early Pleistocene down to the present day. At the time of its appearance above sea level at the close of the Cretaceous the Edwards Plateau region probably had topographic features similar to those exhibited to-day. The strong differences between the alternating strata resulted in wide stripping wherever the upper surface of a harder stratum of wide extent was exposed.

SOIL COVER

The soils of the Edwards Plateau are residual and consist in large part of chert nodules that have resisted solution and ordinary wear much better than the soluble limestone. With these are found impurities common to limestone everywhere and even a certain proportion of the calcareous element. Upon the higher levels where flattish summits prevail the soils are good, but along the stream courses they are usually rough and stony.

VEGETATION

The most important topographic elements of the Edwards Plateau from the standpoint of vegetation are (1) the flat-topped summits of the plateau on the divides, (2) the "breaks" or scarps and related slopes of its ragged canyoned border, known locally as "the mountains," and (3) the streamways and their tributaries. The ragged escarpment which borders the plateau rises from 400 to 1000 feet above the coastal plain, and it is chiefly on this border that rainfall occurs.[1]

The Edwards limestone has a high absorptive capacity because of its low dip and the extensive system of fissures and caverns developed in it. These structures operate also to convey a large part of the water to the deeper strata, from which it discharges as spring water on the valley

---

[1] W. L. Bray, The Timber of the Edwards Plateau of Texas: its Relation to Climate, Water Supply, and Soil, Bull. U. S. Bur. For. No. 49, 1904, p. 9.

margins.   The border region of the plateau is so deeply dissected that were the water not detained by a vegetal covering it would flow off after a heavy rainfall before it had time to enter the limestone, and the streams would have such volume and velocity as to cause swift and destructive floods.

The Edwards Plateau is not covered with continuous forests even in the most favorable situations.   The timber is much interrupted by open

Fig. 163. — Escarpment timber of the Edwards Plateau. Conral River near its source.
(U. S. Bur. of For.)

grassy uplands.   Tongues of luxuriant forest follow the stream-ways into the center of the limestone region and in the deeper, well-watered, sheltered canyons the trees attain large dimensions.   The forest is in the form of a thick-canopied, shady cover protecting many shade-loving shrubs.   This floral community is altogether unlike that found in the adjacent country.   It is distinctly like the Atlantic type, from which it is separated by miles of treeless country.   The chief representatives are the American elm, sycamore, pecan, cottonwood, walnut, black cherry, etc., and in some places cypress with a diameter of 5 feet or more and a corresponding height. The timbered belt is confined rather strictly to the deeply eroded portions of the plateau, gradually giving way to prairie on the level uplands toward the west and to grassy plains toward the east. There is a gradual dwarfing and thinning out of the heavy timber as one passes from the generally heavy growth of the watered canyons to the stunted forests of the hills and bluffs and the still scantier tree growth of the loose stony slopes. Finally there remain only scattered chaparral and the vegetation of the low Sotol Country, Fig. 157, whose principal representatives are sotol, cactus, yucca, and agave.

The hill and bluff forest occurs on the "breaks" of the Colorado along the escarpment front from Austin westward and on the Guadalupe, the Pedronalles, and the Freio. It also extends northward upon the breaks of the Grand Prairie and the jagged hills of the granite country. The timber of this type varies in density with local conditions. On lower flats where deep black soil occurs there is a heavy mixed growth of cedar, live oak, elm, hackberry, mountain oak, shin oak, etc., and the type is extended to the side gorges and draws leading from the main stream-ways. Ten miles northwest of Austin on the Colorado are some timber-capped buttes, and similar occurrences are found in a few other localities. On the unstable talus slopes, where the natural shortage of water is emphasized by the excessive porosity of the loose débris, no timber covering is found except a scattered growth of mountain cedar.[1] Juniper and laurel occur along the vertical slopes of the scarps and sometimes encircle the hills with bands of evergreen. On the floors of the drier valleys are tongue-like extensions of the chaparral flora of the Rio Grande Valley. They are characterized by thorny deciduous trees, mostly acacias, with an undergrowth of Mexican nopal. Thick-skinned yucca, ixtle, and cacti are also found as true desert flora on the bare limestone of the numerous buttes about the borders of the plateau.

On the broken limestone areas are the "oak shinneries," marked by a predominating growth of oak or dwarf shin oak. These are simply dense thickets, scarcely more than tall shrubbery, as on the divides between the Colorado and the San Gabriel drainage country in Burnett County, where much of the growth is so dense as to make it impossible to ride through on horseback. Though it has no value as timber it is a soil retainer of great value. Mixed with this growth is a certain amount of live oak, mountain oak, plum, sumach, holly, and a number of climbers.[2]

---

[1] W. L. Bray, The Timber of the Edwards Plateau of Texas: its Relation to Climate, Water Supply, and Soil, Bull. U. S. Bur. For. No. 49, 1904, pp. 15-17.

[2] Idem, pp. 17, 18.

Among the most valuable assets of the Edwards Plateau region are the cedar brakes conspicuous on the white arid hills of crumbly limestone where the cedar is the dominant and practically the only species. It also grows in mixture with other species, attaining its largest growth in the mixed forests of the lower uplands, where the water supply and soil conditions are better. The most extensive bodies of cedar are those of the Colorado River brakes from Austin to the San Saba country.[1] The cedar brakes are dry and as likely to burn as a prairie of tall grass. Evidences of ancient or recent fires are found almost everywhere.

Many of the trees of the Edwards Plateau region are peculiar products of their environment.

"The eastern red cedar here becomes the mountain cedar; black walnut is represented by the Mexican walnut, whose nuts are tiny balls scarcely half an inch in diameter. Texas oak becomes mountain oak. The common live oak becomes a new form in its mountain habitat. The common persimmon is represented by the Mexican persimmon, whose fruit is a dark blue-black; Canadian redbud is here also a characteristic 'Judas-tree,' but of a different species. The same is true in the case of the buckeye, mulberry, hackberry, and still others."[2]

The type of vegetation in the Edwards Plateau is undergoing a transition from grass to woody growth. The mesquite is capturing the open pastures, and the scrub oak occupying uplands that were formerly open prairies. The transition is due to a number of causes. The ridges have been over-pastured and the balance of power thus thrown to the shrubs which are the vanguard of a timber covering. Twenty-five years ago the prairie held sway over large areas where now one finds scrub oak on every side. A great deal of the "shinnery country" undoubtedly represents a recent gain in timber on the prairie divides; from the edge of the brush each year new sprouts or seedlings are pushed out a few feet farther, and the new growths soon offer shelter for others. These scattered vanguards were formerly killed by the prairie fires and the timber growth held in check; but with the spread of settlements prairie fires have been reduced and a means of holding its position withdrawn from the grass covering.

One of the most striking aspects of this encroachment of the timber upon the prairie land has been the spread of mesquite over the cattle country. Pastures have often been covered with a thicket as close as that of scrub oak, in which, however, the mesquite forms an open orchard-like growth with which are finally associated various species of chaparral and very commonly the prickly pear and the Opuntia. The final result of the spreading of mesquite is a heavy covering of vege-

[1] W. L. Bray, The Timber of the Edwards Plateau of Texas: its Relation to Climate, Water Supply, and Soil, Bull. U. S. Bur. For. No. 49, 1904, p. 19.
[2] Idem, p. 15.

tation which serves as a protector of water supply and soil, although the original grassy growth would be of far greater value to man.[1]

Some of the interrelations of forests, water supply, temperature, and soils are peculiarly interesting. The forest covering furnishes shelter to the ground beneath from the sun's rays and prevents intense heating of the rocks and soils. A reduction of temperature is also afforded by the constant transpiration of water vapor from the leafage of the forest. A far more important effect, however, is the influence of the forest upon the soils. The foliage breaks the force of the rain and compels it to run harmlessly down the trunk or to drip slowly through the leaves. The direct impact of the falling raindrops upon the soil is thus prevented, erosion greatly diminished, and the absorption of the rain water promoted. Tending to the same result is the organic material of the forest floor, which holds back the fallen water until it has had time to soak into the soil. Thus gullying is prevented and frequent and destructive floods reduced in number and violence. These results are beneficial not only to the inhabitant of the region but also to those whose welfare is in any degree bound up with the régime of the streams. The rice planter on the coast is just as eager as the ranchman of the plains to have a constant and large flow of water, and the ranchman of the hills wishes his soils preserved and the soil moisture retained. For these purposes a forest cover, especially in a semi-arid region like the Edwards Plateau, is a vital necessity.[2]

BLACK HILLS

The Black Hills of southwestern South Dakota and adjacent portions of Nebraska and Wyoming rise several thousand feet above the

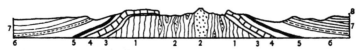

Fig. 164. — Ideal east-west section across the Black Hills. Vertical scale six times the horizontal. 1, slates and schists; 2, granite; 3, sandstone (Potsdam); 4, limestone (Carb.); 5, sandstone (Trias.); 6, shales (Jura); 7, shales (Cret.); 8, shales (Tertiary). (After Newton.)

surface of the surrounding plains. They have, by reason of their elevation, an abundant rainfall and as a result are well wooded, have many streams, and are in effect an oasis in a semi-arid region. They are

---

[1] W. L. Bray, The Timber of the Edwards Plateau of Texas: its Relation to Climate, Water Supply, and Soil, Bull. U. S. Bur. For. No. 49, 1904, pp. 23, 24.

[2] Idem, pp. 26, 27, 29.

carved from a dome-shaped uplift of the earth's crust; the uplift has been sufficient to cause vigorous erosion, so that the upper layers once forming the top of the dome, and corresponding in age and character with the strata on the surface of the plains, have been removed by erosion, and the sedimentary rocks now exposed on some of the mountain summits of the Black Hills are older than those forming the surface of the Great Plains.   The length of the Black Hills dome is about 100 miles, the width 50 miles.   The chief features are (1) a central area of

Fig. 165. — Western slope of Black Hills southeast of Newcastle, Wyoming looking southeast. Steep-dipping beds are limestone, which spread out in a plateau at foot of slope. (Darton, U. S. Geol. Surv.)

high ridges culminating in Harney Peak (7216 feet).   This central area is composed of intrusive crystalline schists and granite of several varieties. (2) About the central crystalline area occur various concentric rings of sedimentary rock separated by well-developed valleys; the innermost mass of sedimentary rock occurs in the form of a limestone plateau with an infacing escarpment, Fig. 164.

The limestone plateau slopes outward and extends around the Black

Hills. Near its base there is a low ridge of limestone with a steep infacing escarpment from 40 to 50 feet high. The escarpments and slopes are sharply notched here and there by canyons which form characteristic "gates." Between this limestone plateau and the hogback ridge which constitutes the outer rim of the Hills is a depression, the Red Valley, which extends continuously about the uplift. The Red Valley is in many places two miles wide, though it is much narrower where the strata dip more steeply. It is one of the most conspicuous features of the region and is called the Red Valley on account of the red color of its soil. The outer hogback rim presents a steep face toward the Red Valley, above which it rises several hundred feet; on the outer side it slopes more or less steeply down to the plains that encircle the hills. It is crossed by numerous gaps and canyons which divide it into subordinate ridges of various lengths.

The Black Hills dome was first developed either in early Tertiary or in late Cretaceous time, but the first uplift was to a moderate height. After this first uplift the larger topographic outlines of the region were established, the dome truncated, and its largest encircling valley excavated in part to its present depth. Later deposits (Oligocene) were laid down by streams and in local lakes, and these deposits extend far up the flanks of the Black Hills, as at Lead and Bear Lodge mountains, respectively. After the deposition of these beds the Black Hills dome was raised several hundred feet higher and more extensively eroded, an erosion that has continued through Quaternary time and is in progress to-day.[1]

SOILS AND FORESTS

The soils of the Black Hills region are closely related to the underlying rocks and are of residual origin except on the valley floors and in a few cases where eolian action has taken place. On the limestone plateaus calcareous material forms the greater part of the rock. The soluble portions have been removed and the insoluble portions of clay and sand have collected as a soil mantle which varies in thickness with the character of the limestone, being thin where the limestone is pure, and thick where it contains many impurities. The amount of soil in a given region depends also on the rate of erosion. On many slopes erosion removes the soil as soon as it is formed, leaving bare rock

---

[1] Darton notes the occurrence of outliers of Tertiary (Oligocene and Miocene) deposits high up on the slopes of the Black Hills (N. H. Darton, Geology and Underground Water of South Dakota, Water-Supply Paper U. S. Geol. Surv. No. 227, 1909, pp. 26–27).

surfaces; on others the surface is flat, erosion is not active, and the soil has accumulated to considerable depth.

In the central core of crystalline schists and granites the rocks have been decomposed most by the dehydration of portions of their feldspar, with the result that the derived soil is usually a mixture of clay, quartz grains, mica, and other materials. The shales yield a sandy soil where they are sandy and a clayey soil where they themselves have been formed of relatively pure clay. In many cases the geologic formations alternate in short distances; there are corresponding abrupt transitions in the character of the soil in narrow parallel zones. Lack of sympathy between soils and underlying rock may be seen in the river bottoms, in sand dunes, in areas of high river gravels, and on slopes where soil derived by slope wash are mingled with or covered by soils derived in place.

The Black Hills forest in general terminates abruptly at the broad valley known as the "Race Track," which lies between the main mass of the uplift and the lines of ridges which encircle the hills. The higher portions of the encircling ridges are also clothed with forest. Sometimes the growth in the latter case is good, sometimes it is in the form of a narrow summit fringe of trees. The main portion of the Black Hills forest includes about 2000 square miles of densely timbered territory. In many places the continuity of the forest is broken by parks and mountain prairies, and large tracts have been destroyed by forest fires which have swept the Black Hills periodically for years if not for centuries.

The yellow pine is the only species of commercial importance; the others are either too small or have too specialized a use to have any great value. A few small bodies of spruce occur, and aspen has come in on some of the burned tracts. The best-quality pine is found in the side ravines and canyon bottoms, where soil and water supply are most favorable and where protection is afforded by the topography. On the steep slopes the soil is stony and thin, the drainage excessive, and the trees shorter and smaller. The forests on the north slopes are in general better than those on the south slopes because of (1) better protection from fire and (2) greater water supply on account of lessened insolation.[1] In general the limestone soils are more fertile than those derived from other kinds of rock and bear the most vigorous growths of trees.

[1] H. S. Graves, Black Hills Forest Reserve, 19th Ann. Rept. U. S. Geol. Surv., pt. 5, 1897–98, pp. 72–75.

Outlying Domes

On the northern border of the Black Hills the Great Plains are diversified by a number of picturesque elevations that are unique topographic features besides affording illustrations of a very peculiar type

Fig. 166. — Devil's Tower from the north. The steep laccolithic mass is phonolite; the gentler slopes about it are developed on shales and sandstones. (Darton, U. S. Geol. Surv.)

of structure. Each hill owes its existence to the injection from below of a column of molten rock into stratified beds. They are not to be considered as volcanic necks, which they in some respects resemble, for the injected rock did not reach the surface so as to form either coulees or cinder cones. Named in general order from east to west the principal hills are Bear Butte, Custer Peak, Terry Peak, Black Butte, Crow Peak, in South Dakota, and the Inyan Kara, the Sun Dance Hills, Warren Peaks, Mato Tepee or the Devil's Tower, and the Little Missouri Buttes in Wyoming.

As a group these hills display variety in the degree of erosion and hence in the degree of preservation of the original structures. In some cases the dome of stratified beds covering the injected rock is still unbroken and the plutonic rock concealed; in others erosion has exposed the plutonic core, while in some erosion has progressed to the point where the central core stands forth as a tower of columnar rock (phonolite) several hundred feet in height. The first type is illustrated by Little Sun Dance dome, which has very regular outlines, is about a mile in diameter, and deeply scored by erosion but not sufficiently eroded to expose the core of igneous rock that presumably lies underneath.[1]

The eroded type is illustrated by Mato Tepee, in which the arch of sedimentary rock has been entirely removed, exposing a column of injected rock. The tower stands on the west bank of the Belle Fourche River, quite by itself, and rises to a height of 626 feet above the platform on which it stands. The shaft of this magnificent natural column is composed of cluster prisms which extend from base to summit without cross divisions, each prism being a uniform unbroken column more than 500 feet high. Since the plateau on which Mato Tepee stands is itself about 500 feet above the river, it is clear that the minimum amount of erosion that must have taken place to expose the column is somewhat over 1000 feet. It was probably much over this amount, for none of the material of the Butte was extruded at the surface. There is good reason for believing that the amount of erosion in the region is but little under three-fourths of a mile, a conclusion which means that this great tower must have been buried under at least a half mile of sedimentary rock.

LITTLE BELT, HIGHWOOD, AND LITTLE ROCKY MOUNTAINS

In central Montana, just east of the front ranges of the Rocky Mountains, are a number of isolated mountain groups with structural and topographic qualities totally different from those of the northern Rockies. On the east they break the continuity of the Great Plains of Montana and are prominent topographic features long before the traveler sights the Rocky Mountains. Lying east of or on the eastern border of the main Cordillera and in a region of deficient rainfall, their own sum-

---

[1] I. C. Russell, Igneous Intrusions in the Neighborhood of the Black Hills of Dakota, Jour. Geol., vol. 4, 1896, pp. 23–43.

mits rise to such heights, 5000 to 9000 feet, as to induce a greater rainfall than occurs on the surrounding plains, and they are therefore clothed with forests. Among these mountain groups are the Little Belt, Highwood, Little Rocky Mountains, and others.

Little Belt Mountains

The Little Belt Mountains are an elevated and eroded plateau, as shown on the accompanying topographic map, Fig 167. They are about 60 miles wide from east to west and 40 miles from north to south on the west side, and taper to a sharp point at Judith Gap on the east, so that the group is roughly triangular in shape. Individual peaks along the northeastern border of the group rise above the general level and form an uneven crest line visible from the open plains. As compared with the well-defined ranges of the Rockies the Little Belt group is relatively low and wide and composed of many spurs radiating from a central point.[1] The border of the mountains on the west is defined by the deep canyon of Smith River, beyond which the country has gentler relief.

On the summit of the Little Belt arch the rocks are gently inclined or horizontal, but on the flanks or shoulders of the arch the rock dips steeply away from the uplift. The intrusions which arched the beds are laccolithic in character. Since the development of its structural features the range has suffered extensive denudation. Erosion has laid bare the larger laccoliths and worn down the general level differentially; the harder igneous rocks have been left in relief to form the higher summits, while the softer sedimentary rocks have been eroded, although the members of each group have been eroded at very different rates owing to differences in degree of resistance. On account of deficient height the Little Belt Mountains did not support local ice sheets during the glacial period, nor were they covered by the continental ice sheet during the time of its maximum southward extension.[2]

Throughout the greater part of the Little Belt region a plateau-like topography prevails; broad flat summits are characteristic. The average elevation is 7600 feet, though the summit level from which the spurs radiate is 8000 feet high. The highest summit of the group is not at the center but along the northeastern border, where Big Baldy reaches an altitude of 9000 feet. The Little Belt Mountains are

---

[1] W. H. Weed, The Geology of the Little Belt Mountains, Mont., 20th Ann. Rept. U. S. Geol. Surv., pt. 3, 1898–1899, p. 273.

[2] Idem, p. 277.

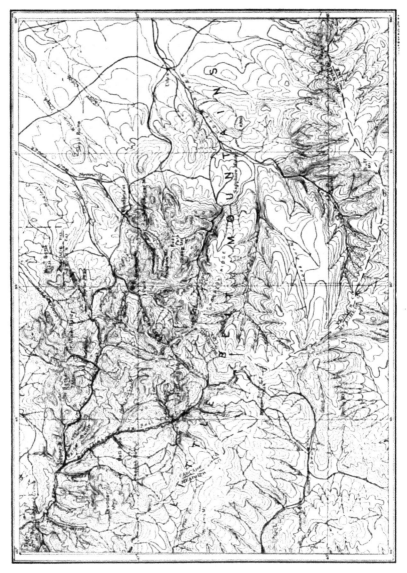

Fig. 167. — Topographic map of the Little Belt Mountains.   (U. S. Geol. Surv.)

bounded by relatively soft rocks to whose deep erosion the prominence of the mountains is chiefly due.

The relation of the detailed topographic features to the rock character is so intimate that it is impossible to distinguish the one feature without distinguishing the other. Summit plateaus are bordered by steep escarpments with towering limestone cliffs along the stream gorges. Only the highest peaks along the northeastern border and the steep limestone gorges lend picturesqueness to the scenery. In the center of the mountain area where the beds are horizontal or only gently inclined, secondary structural plateaus are commonly found. Upon some of these the rock is so resistant as to determine broad ridges which in some cases are emphasized by differences in soil and vegetation, as is the case in Belt Park and other timberless parks near Neihart, which have been developed on quartzite and which are in contrast with the wooded and soil-covered slopes above and below. There are no broad valleys within the mountains; for the most part the streams flow in deep trenches or narrow canyons in the harder limestone and in less narrow valleys in the shale belts.[1]

CLIMATE, SOIL, AND VEGETATION

Both rainfall and snowfall are relatively abundant in the Little Belt Mountains. Intermittent streams which flow only in wet weather or at times of melting snow are common. These are especially characteristic in limestone areas where the waters are absorbed by the porous and fissured rock and in those regions where the catchment areas are small. As a consequence of the greater precipitation due to the greater height of the Little Belt Mountains over the surrounding plains they are in general forest-clad, their dark slopes being in strong contrast to the arid treeless plains about them. Lodgepole pine is the prevailing species; in some places it forms forests with individual trees 10 to 40 inches in diameter, but usually it is much smaller. On the plateau summits the white pine is found, and spruce and fir grow along the wet stream bottoms and on moist and cold northern exposures.

There is a very intimate relation between the character of the tree growth and the nature of the slope exposure. On southward-facing slopes, which are relatively dry because of the greater insolation, the growth is sparse and open, and interrupted by grassy parks; on northward-facing slopes thick and dark forests occur. The growth varies somewhat also with the character of the rock and the physical nature

[1] W. H. Weed, Geology of the Little Belt Mountains, Mont., 20th Ann. Rept. U. S. Geol. Surv., pt. 3, 1898-99, p. 275.

of the derived soil. The shales produce but little soil and support scanty vegetation; they usually underlie the park regions. The sandstones and the igneous rocks are covered with land waste and are generally densely wooded.

HIGHWOOD MOUNTAINS

The Highwood Mountains are a prominent member of the group of elevations that break the continuity of the northern Great Plains. They lie in the great bend of the Missouri in central Montana about 20 miles north of the Belt Mountains, and about 50 miles southwest of the Bearpaw Mountains, Fig. 148. About them stretch the smooth monotonous plains above which they rise about 4000 feet to summit elevations of 7600 feet. The Highwoods are a group of old and now much eroded volcanoes which broke through the once flat-lying Cretaceous strata of the plains. The mountains are composed chiefly of volcanic flows and breccias and a number of stocks or central cores, the remnants of former more lofty cones. The outer foothills are low and rounded; toward the center the country becomes more rugged. The descent to the plains is abrupt on the south; the chief feature of the northern slopes is an old glacial spillway known as Shonkin Sag whose origin is explained on p. 416. The drainage is in the main radial and consequent upon the original slopes of the volcanoes. The streams are rather constant in flow in the mountains but become sluggish and alkaline on the plains, some of them drying up in summer to such an extent that water remains only in pools in the deepest portions of the stream channels. The outer foothills and valley openings have been occupied by ranchmen who also utilize the water for limited irrigation. Extensive pastures are found on the higher slopes, though these give way to true forests of small pines on the northern exposures. In many places there are dense thickets of lodgepole pine. The name "Highwoods" undoubtedly had its origin in the forest growth of the northern slopes.[1]

LITTLE ROCKY MOUNTAINS

The Little Rocky Mountains lie about 200 miles east of the Rocky Mountain Cordillera and between the Missouri and the Milk about 60 miles south of the 49th parallel. They rise from 2000 to 3000 feet above the treeless plains of central Montana, forming a conspicuous topographic feature in a plains region that is generally without

[1] L. V. Pirsson, Petrography and Geology of the Igneous Rocks of the Highwood Mountains, Montana, Bull. U. S. Geol. Surv. No. 237, 1905, pp. 1–22.

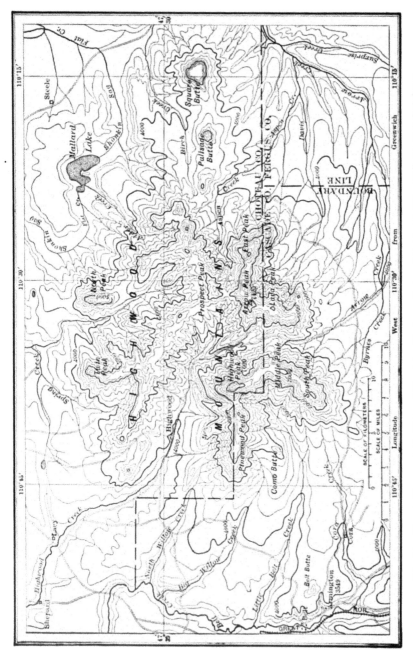

Fig. 168. — Topographic map of the Highwood Mountains, Montana. Contour interval, 200 feet. (Bull. 237, U. S. Geol. Surv.)

prominent landmarks. Their topographic prominence led the Indians to call them "Eah hea Wwetan," or the "Island Mountains."

The Little Rocky Mountains have an undulating crest without sharp peaks. The highest summit is more than 6500 feet above the sea, though only about half that height above the surrounding plains. The scenery is attractive but not grand, for the mountain summits are generally rounded, without that boldness that is the striking feature of most alpine scenery; and to this softness of outline is added the softening effect of a thick growth of small pines, which covers almost all of the summits. Perhaps the most picturesque aspects of the mountains are developed about their borders, where heavily bedded limestones (Carboniferous) are cut by deep narrow canyons variegated by a vegetation that stands in pleasing contrast to the dry plains. The limestones form a white wall encircling the mountains, and stream and cliff erosion has cut the thoroughly jointed beds into huge white scarps visible from points 50 miles away. Within the mountains the streams flow in deep V-shaped gorges.

These mountains are formed upon a single dome-shaped uplift having a central core of crystalline schists overlain and marginally wrapped about by limestones (Paleozoic) and softer beds (Mesozoic).[1] Hard and soft layers encircle and overlap the mountains and are crossed one after the other by streams consequent upon the original slopes of the uplift. Subsequent streams have developed along the strike of the softer beds and usually join the consequents where the valleys of the latter are broadest. The central core of crystalline schist is exposed in the headwater gorges of all the larger streams and in the deep-cut side slopes of the main crest. The gorges have heavily timbered slopes; the prevailing type of timber is the lodgepole pine, from 3 to 20 feet high. It is characteristic of the granite to have a covering of young pines; pines are found also upon the débris slopes, while the limestones and schist areas are covered by the big-leaf pine, which forms groves alternating with open parks.[2]

OZARK PROVINCE

The approximate limits of the Ozark region are the Missouri and Osage rivers on the north, the Arkansas on the south, the Neosho on the west, and the Black River on the east; the Shawnee Hills of southern

---

[1] Weed and Pirsson, Geology of the Little Rocky Mountains, Jour. Geol., vol. 4, 1896, pp. 399–428.

[2] Idem, p. 410.

Illinois, a continuation of the Ozark plateau, extend eastward to Shaw-
neetown on the Ohio River.

The Ozark region may be described as a broad, relatively flat-topped
dome somewhat extensively dissected and consisting of three subdivisions, the Salem Platform, the Springfield Structural Plain, and the
Boston Mountains. The Springfield Plain inclines at a low angle
toward the west in Missouri and toward the southwest in Arkansas, a
slope which corresponds with the dip of the underlying formation. It

Fig. 169. — Subdivisions of the Ozark region and relations to surrounding provinces.
(Taff, U. S. Geol. Surv.)

is deeply dissected by the large streams which flow through it in
narrow valleys; the interfluves are large tracts of broad flat structural
surfaces from which younger formations have been eroded. The
Boston Mountains are capped by thin layers of sandstone and shale,
the more resistant sandstone governing the physical features of the
mountains. The rocks dip in the main to the south at an angle
slightly greater than the general southward slope of the surface. From
hilltops on the Springfield Plain the Boston Mountains appear as a
bold, even escarpment with a level crest, but on closer examination the
escarpment is seen to have many finger-like extensions to the north

in the form of ridges and foothills. Toward the west the Boston Mountains decline in elevation and in ruggedness as the sandstone beds become thinner and more shaly in that direction.

The Boston Mountains on the southern border of the Ozark province owe their dominating height in part to the excessive erosion of the region north of them and in part to differential uplift.[1] The result of erosion

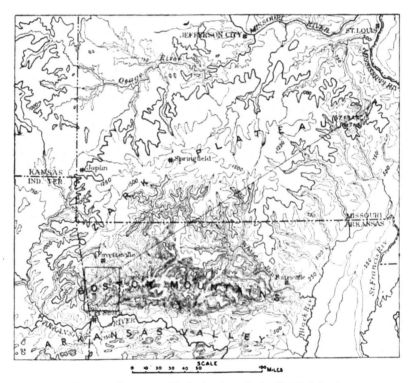

Fig. 170. — Topography of the Ozark region.   (Purdue, U. S. Geol. Surv.)

was to reduce a large part of the Ozark region in Missouri and Arkansas to a comparatively low altitude, leaving the Boston Mountains as re_siduals and their front as a rather bold escarpment.   The former greater extent of the Boston Mountains is indicated by remnantal outliers standing several hundred feet above the general level of the area about them.   In contrast to the northern part of the Ozark region, which is a low flat dome with only local faulting and minor undulations, the

[1] A. H. Purdue, Winslow Folio U. S. Geol. Surv. No. 154, p. 5, col. 4.

Boston Mountains have a monoclinal structure and a correspondingly steeper border topography.

The topographic forms of the Ozark region are those characteristic of early maturity in a region of nearly flat-bedded rocks of varying hardness in which the dip of the rock and the slope of the surface are in many places coincident. The entire northern slope of the uplift is a succession of broad flat plateaus separated by more or less ragged escarpments which mark the margins of the harder formations. Forward from the escarpments there are scattered outliers, the fragments of more extensive layers separated from the escarpments by circumerosion. The greatest dissection of the region is on the south and east, the least on the north. On the south and especially at the eastern end of the Boston Mountains the topography is rougher and the plateau character is all but lost. On the north the streams run commonly through broad and rather shallow valleys between which lie extensive areas of relatively undissected plateau. The drainage is in the main radial and consequent, the stream directions corresponding with the

BOSTON MOUNTAINS   OZARK PLATEAU   ST FRANCIS MTNS.

Fig. 171. — Topographic and structural section across the Ozark region along line A–A in Fig. 170. Shows the crystalline rocks of the St. Francis Mountains, the limestones of the Ozark Plateau, and the shales, sandstones, and limestones of the Boston Mountains.

restored slopes of the structural dome, but a certain amount of subsequent stream development has also taken place where tributaries have excavated valleys along the outcrop of the softer formations.

Seen from commanding points the surface of the Ozark region presents the appearance of an almost unbroken plateau due in part to the flatness of the structure but also and in larger part to peneplanation that once brought diverse structures to a common level which the vigorous erosion of the region since uplift has not yet wholly destroyed.

The two main subdivisions of the Ozark Plateau, the Salem upland and the Springfield Plain, are separated by the Burlington escarpment, which runs in a general north-south direction.[1] The escarpment is formed upon the border of a limestone layer (Mississippian) which constitutes the surface rock of the western part of the plateau. East of the escarpment, outliers of the limestone occur in the form of small residual areas between which are dissected plains bearing chert accumulations derived from the weathering and erosion of the limestone

[1] C. F. Marbut, Physical Features of Missouri, Missouri Geol. Surv., vol. 10, 1896, pp. 11–110.

that once extended over the entire region.   The streams of the Salem upland flow in narrow valleys some of which are 250 feet or more deep. The degree of dissection is in some places so great that the former extensive plain is not easily recognizable, though large interstream tracts still preserve their plain-like character.

The Springfield upland on the other hand is largely a structural plain developed on the surface of the Mississippian limestone.   Near the eastern border of the upland the streams flow in shallow trough-like valleys which deepen westward as the streams cross low anticlines and domes. Those that drain the upland eastward have deep valleys where they cross the edge of the Mississippian limestone, which they have dissected to a ragged fringe.

SOILS AND TREE GROWTH

The soil of the Salem upland is cherty to a high degree, the chert having been derived from the overlying limestone.   On weathering, the chert breaks into angular blocks and, because of its greater durability, forms a surface layer of débris.   The finer soil particles produced by the weathering of the limestone are carried downward to the base of the weathered zone and are therefore at too great a depth to be accessible to agriculture, though they are of first importance to the forests, which seem to thrive in spite of the surface accumulations of loose stones.   The most luxuriant forest is formed upon soils derived from the light-blue limestone and blue shales of the Morrow formation (Carboniferous).   Walnut, locust, and other types found naturally only on fertile soils occur on this formation in spite of the fact that it outcrops usually on steep slopes.   In general it may be said that in spite of the rather poor soils of the entire region, due to the cherty residual products resulting from the erosion of the limestone and also to the considerable dissection which the region is undergoing to-day, the surface is rather well occupied by forests, since it is too stony and steep to serve for other purposes and since the finer soil is washed down to too great a depth below the surface to be available for the ordinary plants of agriculture.

ARKANSAS VALLEY

South of the Boston Mountains is the Arkansas Valley district, which is structurally much more complex than the Ozark region, although standing at a lower elevation.   The underlying strata have been thrown into overlapping folds which have been beveled off by erosion so completely as to form a local peneplain approximately 800 feet above the sea. It is surmounted, however, by residual peaks and ridges of small area and from 1700 to 2500 feet above sea level. Since the formation of the ·Arkansas Valley peneplain, uplift has occurred and the soft shaly beds have been worn away, leaving the inclined sandstones as low, narrow, and sharp-crested ridges whose summits are generally horizontal, a condition indicating the former elevation and topographic character of the area.

OUACHITA MOUNTAINS

The Ouachita Mountains lie south of the Arkansas Valley and extend from Little Rock, Arkansas, westward into eastern Oklahoma. The range is 200 miles long and nowhere rises more than a few thousand feet above the surrounding valleys and plains. The topographic forms of the range are developed upon an Appalachian type of structure, p. 585. Near the center of the range, long wide folds are developed upon massive sandstone; on the borders of the uplift are shorter and more complex folds developed upon shales and limestones as well as sandstones.[1] The regular development of hills or ridges and valleys is a striking feature of the group as contrasted with the irregularities of the igneous knobs and ridges of the Wichita Mountains of western Oklahoma. The regularity of the topographic details is absent in many places along the border, however, where faults with vertical displacements of several thousand feet have combined with the overturning of the folds to make the relief more irregular both in general plan and in detail.

ARBUCKLE MOUNTAINS

The Arbuckle Mountains form a triangular area approximately 30 miles on each side with a westward extension. The western part of the Arbuckle Mountains has an elevation of about 1300 feet or 1400 feet above the plains on either side. The mountains were uplifted (late Carboniferous) and base-leveled in common with the great Appalachian province east of the Mississippi. Upon the base-leveled surface, then an almost flat plain, Cretaceous deposits were laid down. Later uplift caused the removal of the Cretaceous strata over large areas, but the superior hardness of the underlying rock has resulted in its preservation in the form of a low plateau in the central part of the mountain group, while the bordering strata have been eroded to a lower level. The minor topographic details of the Arbuckle Mountains are due chiefly to the varying resistances of the rock formations. The lime-

---

[1] J. A. Taff, Structural Features of the Ouachita Mountain Range in Indian Territory, Science, n. s., vol. 11, 1900, pp. 187–188.

stones in general are hard and form level-topped and narrow ridges with crests approximating the general plane of the regional uplift.   Intervening soft cherts and shales are found in the wooded valleys that separate the ridges.   In the more elevated part of the uplift erosion by swift streams has etched the border of the plateau into a frill of deep gulches.

Fig. 172. — Mixed hardwoods, etc., in typical relation to topography, Arbuckle Mountains, Oklahoma. On the dry limestone formations forests are absent, except along the lower and moister valley slopes and the valley floors. (Reeds, Oklahoma Geol. Surv.)

A second and lower plain (probably Tertiary) occurs in the Wichita-Arbuckle region and descends approximately with the grade of the rivers.   This plain merges into the Tertiary deposits of the coast on the one hand, and on the other hand stretches westward into Oklahoma and northward across Indian Territory.   It is preserved through eastern and southeastern Oklahoma in innumerable ridges and hills, occurring at elevations approximating 1800 feet.   A third erosion surface has been identified in the plains surrounding the Arbuckle Mountains; the larger streams have cut wide and flat valleys 200 feet below the general level of the Tertiary peneplain.   This erosion surface is, however, much more fragmentary than either of the two surfaces just described. It is found bordering the Ouachita Mountains in the form of wide flat valleys developed upon the softer rocks.[1]

Scrub Oak, Red Cedar, and Red Bud are the common trees on the mountain slopes, while red and white elm, Bois d'Arc or Osage orange,

[1] J. A. Taff, Preliminary Report on the Geology of the Arbuckle and Wichita Mountains in Indian Territory and Oklahoma, Prof. Paper U. S. Geol. Surv. No. 31, 1904, pp. 15-17.

hickory, pecan, hackberry, honey locust, river plum, papaw, mulberry, sycamore, willow, and cottonwood grow in the creek valleys and on the flood plains of the rivers like the Washita. There is also a striking increase in the timber growth (1) on the weaker formations (Sylvan and Simpson) which weather to lower levels, and (2) on those formations (Woodford and Reagan) which contain appreciable amounts of phosphate and iron. The chert, shale, and sandstone strata, together with the granite and porphyry of the East and West Timbered Hills, are all covered with timber. In the narrow creek valleys and canyons, which have been developed on limestones (Arbuckle, Viola, and Hunton), trees appear on the alluvial deposits and near slopes, but rarely on the upland surface, Fig. 172. The rainfall is notably heavier in the mountains than on the adjacent plains. Where the slopes are cleared for cultivation the soil washes so badly that the clearings are soon abandoned. While the clearings are not abundant, yet in the aggregate they include an important and always a conspicuous part of the total area.[1]

WICHITA MOUNTAINS

The Wichita Mountains are the westernmost of the three mountain groups of the southern Great Plains. They are composed of igneous rock,

Fig. 173. — Border topography, Wichita Mountains, Oklahoma. (Gould, Oklahoma Geol. Surv.)

chiefly granite, and have been eroded into peaks varying in height from a few hundred feet to 1500 feet above the surrounding plains.

[1] The data for this paragraph have been supplied by Prof. C. A. Reed of Bryn Mawr College and Prof. C. A. Gould of the University of Oklahoma.

The main range is about 30 miles long and 12 miles wide.   West of it are a number of scattered smaller ranges and peaks, such as Mount Tepee, Quartz Mountain, and Headwater Mountain.   North and east is a 30-mile parallel range of hills composed chiefly of hard massive limestone, and the same limestone on the south and east occurs in the form of small rounded knobs.   The limestone hills represent remnants of a series of rocks which once extended as a dome over the igneous rocks but which have since been deeply eroded.   Between the granite and limestone ridges on the borders of the uplift are softer "Red Beds," shale and sandstone with local deposits of conglomerate, and these have been eroded to form valley lowlands whose extent corresponds to the outcrop of the less resistant strata.[1]

Mounts Scott and Baker are the highest mountains in the group and, like the numerous other peaks in the region, they have steep, rugged, bowlder-strewn slopes.   Although the relative altitudes of peaks, ridges, and valleys are small, the mountains have a distinctly rugged appearance, with sharp outlines and narrow passes.   There are more than 250 detached areas of igneous rocks in the Wichita Mountains group, and these are expressed topographically in the form of a main mountain mass 150 square miles in extent and a large number of neighboring isolated sharp knobs which rise like islands above the smooth plains about them.   The whole group forms an archipelago of granite mountains and peaks rising rather abruptly from the sea-like plains developed upon softer rock.[2]

[1] C. N. Gould, Geology and Water Resources of Oklahoma, Water-Supply Paper U. S. Geol. Surv. No. 148, 1905, p. 15; J. A. Taff, Prof. Paper U. S. Geol. Surv. No. 31, 1904.

[2] J. A. Taff, Preliminary Report on the Geology of the Arbuckle and Wichita Mountains in Indian Territory and Oklahoma, Prof. Paper U. S. Geol. Surv. No. 31, 1904, pp. 54, 77.

CHAPTER XXIV

PRAIRIE PLAINS

Extent and Characteristics

BETWEEN the great Laurentian area of Canada on the north and the Ozark-Appalachian provinces on the south and extending east and west from the Great Plains to the Appalachian Plateaus is a broad expanse of moderately rolling country known as the Prairies or Prairie Plains. The province includes the greater part of the Middle West, but it is not confined to that section of the country. Its western portions are thinly timbered, the forest growth shading off to scattered groves of timber in lower and wetter localities or to the valley floors and the stream margins. Its eastern, better-watered portions are covered with a denser arboreal growth, though clearings are so extensive as to leave but insignificant patches of the primeval forest, and everywhere the timber is confined chiefly to wood-lots of limited and generally decreasing extent. The timber of the southern two-thirds of the Prairie Plains is prevailingly hardwood; on the north is a belt of coniferous forest which fills the gap between the northern edge of the hardwoods and the spruce forest belt of Canada.

Although the Prairie Plains support a dense agricultural population and supply a disproportionately large part of the corn, wheat, oats, etc., of the country, they do not have an exceptionally favorable rainfall. It is certain that the agricultural products are far below the possibilities of the soil were a heavier and better distributed rainfall to occur. There is good reason for believing that were the surface less flat, the absorption of rain water less pronounced, the original forest would have been much more restricted.

The Prairie Plains belong to the vast central plains region between the Atlantic and the Pacific Cordillera and present from commanding points a number of characteristic views. In many places the province has the appearance of a limitless expanse of grove-dotted, gently undulating country, here and there trenched by rivers and surmounted by low hills; or it stretches away as far as the eye can see without either of these relief features—a grass-covered, farm-dotted, smoothly contoured prairie. North of the Missouri and Ohio rivers it is glaciated. Morainic belts cross it in looped pattern, and between them are notably flat till plains.

In that portion covered by glacial ice in the last (Wisconsin) glacial invasion, lakes, ponds, and undrained hollows in great numbers are scattered freely about, stream courses are disorganized, falls and rapids abound and alternate with swampy depressions often of considerable extent. The southern portion of the glaciated tract and a part of the unglaciated area beyond are covered to a variable degree by loess deposits, — fine,

Fig. 174. — Typical view of the Prairie Plains in the Great Lake region. Woodland tract in left foreground.

light, wind- and stream-deposited detritus of great importance in relation to both run-off and soil fertility. East of the Mississippi the Prairie Plains grade southward into the Appalachian Plateaus; west of that river and south of the limit of glaciation they have a distinctive quality unlike any other portion of the province. In the Osage Prairie of eastern Kansas, and specifically about Independence, they have the appearance of an ancient surface of erosion. Here broad valleys separated by low divides are characteristic features; none of the irregularities of glacial topography or drainage is found; in fact there is so little relief that water storage by dam construction is impracticable on account of the breadth of the valleys and their gentle gradients.

THE PENEPLAIN OF THE PRAIRIES

The most general topographic feature of the Prairie Plains is the Tertiary peneplain now standing at various elevations above its former level. It may be traced from central Texas northward through Oklahoma and Kansas into Wisconsin, Indiana, and southern Michigan. While it is now a dissected and therefore a discontinuous surface its former character and wide extent may reasonably be inferred because large tracts are still in an excellent state of preservation. It is fair to assume that the Tertiary peneplain of the region was far more extensive than its visible remnants, now for the most part in process of dissection. The Tertiary peneplain appears to have been much more extensive in the upper Mississippi basin than was the corresponding plain in the Appalachian tract; the latter was limited largely to the belts of softer rock or to the larger streams.

The uplifted peneplain was dissected to various stages of maturity before the Pleistocene period, so that by the beginning of that period pronounced valleys had been formed: in some places the valleys were narrow and the intervening divides wide, while in other places the valleys were widely opened and the divides reduced to narrow ridges. The slopes of many valleys were steep and covered with a thin veneer of decayed rock, and almost the whole of the now glaciated country had a somewhat more rugged appearance than at present.

The surface of the peneplain was not everywhere affected in the same way after its development. In some cases it was uplifted and broad valleys opened at lower levels, in other cases there appears to have been no uplift, in still others uplift appears to have been progressive and no opportunity given for the formation of a peneplain. Even without peneplanation the surface would nowhere have great relief, for this is distinctly a region of low-lying and nearly flat strata. The depth of erosion conditioned by the rainfall, the elevation, and the character of the rock would be a measure of the relief, unlike many mountain regions where the relief in the earliest stages of an erosion cycle is commonly related directly to original structural irregularities.

A description of a few typical occurrences will serve to show the general development of the peneplain of the prairies and its present condition. An ancient topographic level has been clearly determined in southwestern Wisconsin (Driftless Area). Fragments of the plain slope gently southward across the outcropping edges of shales, dolomites, and sandstones. In spite of the strong variations in resistance to erosion which these beveled strata display the plain truncates them with striking uniformity. It is therefore not a structural plain but a plain of erosion. It is typically developed about Lancaster, Wisconsin, where it stands about 1100 feet above the sea. Since its formation it has been broadly uplifted and extensively dissected. The main valley bottoms lie several hundred feet below the general level of the uplifted peneplain and well-developed flood plains are common. Dissection has progressed to the point where the original plain surface now remains only along the stream divides, and the cubic contents of the valleys approximately equal the cubic contents of the ridges between the valleys.[1]

South of the Great Lake region the uplifted and dissected peneplain may be clearly identified. In eastern Missouri and in Illinois it has been described as a plain of wide extent now much dissected, its remnants being represented by the accordant hill and ridgetops. In a general view this regularity of level of the summit areas makes the sky line nearly flat, a dominating topographic feature. Such striking accordance of summit levels is the more significant when it is realized that the plane thus denoted truncates inclined strata offering very unequal resistances to erosion.[2]

In Indiana the surface appears to display two erosion levels forming an upper and a lower plain. The level summits of the higher hills and ridges mark the position of the now greatly dissected older topographic level about 600 feet above the sea. One hundred feet lower are rolling uplands which exhibit smooth, gently rounded hills and ridges or flat divides separated by broad and relatively shallow valleys.[3]

In the prairie plains of eastern Kansas the uplift appears to have been insufficient to occasion dissection of the peneplain formed during the Tertiary. The beveling of the strata is here still in progress. The divides are smooth and broad, truncation is now in progress; if the land retains its present level there will be even more complete leveling of the surface, and the region will finally exhibit an ultimate base-leveled plain. All relation between structure and topography has, however, not yet vanished. In the Osage Prairie, for example, are broad structural terraces surmounted by a few outlying and isolated hills, though the relief is far less gentle and faint as compared with the relief of the region during the earlier periods of the present erosion cycle. The terraces face eastward and have frontal scarps from 100 to 200 feet

---

[1] Grant and Burchard, Lancaster-Mineral Point Folio U. S. Geol. Surv. No. 145, 1907, p. 2, cols. 1–2.

[2] N. M. Fenneman, Physiography of the St. Louis Area, Bull. Ill. Geol. Surv. No. 12, 1909, pp. 17, 52.

[3] Fuller and Clapp, Ditney Folio U. S. Geol. Surv. No. 84, 1902, p. 1, col. 3.

high. Each terrace has a long and relatively gentle westward slope whose total descent is less than the descent of the frontal scarp. Each terrace therefore stands above its neighbor on the east and below the next terrace on the west, so that the series rises in westward succession and the rate is about 10 feet per mile. The north-south drainage lines show a marked tendency to migrate down the dip of the individual terrace, i.e., westward; hence the eastern valley slopes are commonly gentle and low, the western, steep and high. Stream courses at right angles to the terrace fronts show alternations of broad open stretches on the terrace tops and narrow steep-sided stretches on the terrace fronts.

Both the topographic and the drainage features are related to the structure in still other respects. The westward-dipping shales, sand-

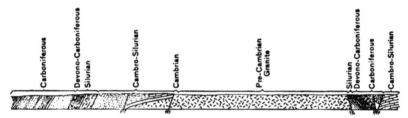

Fig. 175. — North-south section near Tishomingo, Oklahoma, showing the peneplaned surface of the Prairie Plains on the south. (Taff, U. S. Geol. Surv.)

stones, and limestones are of variable thickness. Where the resistant sandstones are thick the topography is irregular and has marked relief; where the soft shales are thick and the resistant strata thin or absent the surface has very gentle gradients, broad valleys, and but little relief. The average elevation of the district is about 900 feet, and the difference in elevation between adjacent hills and valleys is from 50 to 200 feet.[1]

Toward the south, as in Oklahoma and northern Texas, the Prairie Plains once more take on the character of an uplifted and dissected peneplain, Fig. 175. The uplift was not great, however, probably not more than 100 feet, and large remnants of the former featureless plain occur on the interfluves. The valley topography is a feature of the present cycle of erosion; the inter-valley topography exhibits many characters inherited from the preceding cycle of erosion in which a peneplain was formed.[2]

[1] F. C. Schrader, Independence Folio U. S. Geol. Surv. No. 159, 1908, pp. 1–5.

[2] J. A. Taff, Coalgate Folio U. S. Geol. Surv. No. 74, 1901, p. 4, col. 4., and C. N. Gould, Geology and Water Resources of Oklahoma, Water-Supply Paper U. S. Geol. Surv. No. 148, 1905, p. 12.

CENTERS OF GLACIATION [1]

The ice sheets which overrode the northern Prairie Plains originated in two centers of glaciation known as the Keewatin, west of Hudson Bay, and the Labrador, in the Labrador peninsula. A third, the Cordilleran center in northern British Columbia, is of little interest here, for it lay too remotely to the west to affect to an important degree

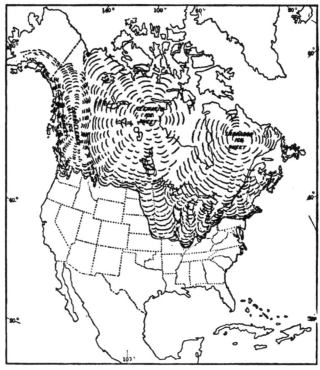

Fig. 176. — The centers of ice accumulation, the position of the Driftless Area, and the southern limit of glaciation. (Alden, U. S. Geol. Surv.)

the topography of the Plains, Fig. 176. The Keewatin glacier formed in a great gathering ground close to the western coast of Hudson Bay, from which the ice radiated in all directions — eastward into Hudson Bay, northward to the Arctic Ocean, westward to the Mackenzie Valley, and southward toward Manitoba and the Great Plains. At or near the center of glaciation the striæ are very indefinite and appear to have changed in direction as the center slightly shifted its position, but no evidence of more than one glaciation has been found, nor has it been

[1] J. B. Tyrrell, The Genesis of Lake Agassiz, Jour. Geol., vol. 4, 1896, pp. 811–815.

determined that the ice left the country uncovered at any time during the glacial epoch.

The Keewatin glacier advanced southward and southwestward until it came in contact with the high escarpment of Cretaceous shales in eastern Manitoba, followed the course of Lake Winnipeg and Red River and in this direction advanced far into Minnesota, Dakota, and Iowa. The eastern margin of this lobe did not extend very far east of the present eastern shore of Lake Winnipeg, and it is probable that throughout its advance there was a free drainage eastward, probably into Hudson Bay.

The evidences of the time of advance of the three continental ice caps which formed over northern North America go to show that these three seem to have reached their widest extent and to have retired in succession from west to east, and that a fourth ice cap, probably similar in character to those that have disappeared from the American continent, covers Greenland at the present time.

Glacial and Interglacial Periods

The different glacial and interglacial stages of the glacial period are numbered in the order of their occurrence below.[1] Their deposits are represented in part on the accompanying maps showing relations of the drift sheets in the Middle West.

XIII. The Champlain substage (marine).
 XII. The glacio-lacustrine substage.
 XI. The Later Wisconsin, the sixth advance.
  X. The fifth interval of deglaciation, as yet unnamed.
 IX. The Earlier Wisconsin, the fifth invasion.
VIII. The Peorian, the fourth interglacial interval.
 VII. The Iowan, the fourth invasion.
 VI. The Sangamon, the third interglacial interval.
  V. The Illinoian, the third invasion.
 IV. The Yarmouth or Buchanan, the second interglacial interval.
 III. The Kansan, or second invasion now recognized.
  II. The Aftonian, the first known interglacial interval.
   I. The sub-Aftonian, Jerseyan, or Nebraskan, the earliest known invasion.

I. The sub-Aftonian or Nebraskan [2] glacial stage is represented in Iowa as a very old drift sheet whose surface bears sand and gravel, peat,

---

[1] Chamberlin and Salisbury, Geology, vol. 3, 1906, p. 382 ff.
[2] B. Shimek, Bull. Geol. Soc. Am., vol. 20, 1909, p. 408.

old soil, and other products of prolonged weathering. The Nebraskan drift consists largely of compact blue-black bowlder clay. In many places it is thickly set with woody material gathered from forests that were overwhelmed by the ice. The wood is largely spruce, cedar, or coniferous species that indicate a cool-temperature flora in advance of glaciation. The drift filled the deep preglacial valleys to depths of nearly 350 feet, but it formed only a thin cover on the uplands, where it was nearly removed by erosion before the next glacial invasion.[1]

The interglacial stages are of little interest in the present connection. Rather typical conditions are represented by the first epoch in the series.

II. The Aftonian interglacial stage is represented by sand and gravel deposits and by beds of peat and muck, with stumps and branches of trees, and shows prolonged erosion and weathering. Several bones of the camel have been found in the deposits of this epoch; also the antler and fragments of bones of a large stag, three species of elephant, the American mastodon, and a large variety of molluscan remains of both aquatic and terrestrial species. The fauna and flora indicate a climate not greatly different from the present, though the fossils merely represent the conditions at the time of their burial, and still leave largely to conjecture the full range of temperature of the period.[2]

III. The Kansan glacial deposits, p. 473, occupy a large area in Kansas, Missouri, Iowa, and Nebraska and probably extend under the later glacial formations far to northward. The Kansan till is pronouncedly clayey and has very little water-laid material either within the body of the till or on its margin, where the usual fluvio-glacial deposits of sand and gravel are very meager. The surface of the Kansan till although originally quite flat has been considerably eroded and its topographic features notably altered from their original expression.

It is estimated that the Kansan drift occupies only from 10% to 30% of the original plain. The streams flow in valleys with broad and low (3° to 5°) slopes and bottoms. The average drift removed by erosion in the exposed portion of the Kansas drift is estimated to be not less than 50 feet. The material is deeply weathered and it has been deprived of its finer and more readily dissolved calcareous material and even limestone pebbles down to 5 or 8 feet.[3]

V. The Illinoian till is chiefly clay and is similar to the Kansan till in showing a marked absence of assorted drift in most places. Drainage modification brought about through the Illinoian glacial invasion in the Mississippi Valley is of uncommon interest. The western edge of the Illinoian ice lobe crossed the Mississippi Valley between Rock Island and Fort Madison, and pushed the Mississippi River out of its course a score of miles (p. 475).

[1] Frank Leverett, Comparison of North American and European Glacial Deposits, Zeitsch. f. Gletscherkunde, Berlin, vol. 4, 1910, p. 250.

[2] Idem, p. 253.

[3] Idem, p. 258.

The bowlder clay of the Illinoian drift appears originally to have been very calcareous but to have been leached since deposition to a depth of from 5 to 8 feet and nearly all the limestone pebbles completely removed. Much of the deposit has become partially cemented because of the large calcareous element in the drift. Where erosion has been most rapid it has destroyed scarcely more than half the original plain.[1]

VII. The Iowan till is thin and is noteworthy for the exceptional number of large granitoid bowlders which lie chiefly on the surface. This drift was formed by a lobe of the Keewatin ice sheet. The moraines formed about the borders of the lobe were very feeble and the outwash scant.

IX. The two Wisconsin ice sheets are the most important of all, not because of their greater original extent or their thicker drift deposits but because their drift deposits are the last of the series and have therefore been more extensively preserved. The interval since the occurrence of these last glacial stages has been so brief, furthermore, that but little modification of topography, drainage, or soils, has been possible. The outermost limits of the Wisconsin ice sheets are marked in most places by pronounced terminal moraines, and in addition the surfaces of the till sheets are diversified by terminal moraines due to the periodic recession of the ice. Their surfaces are also diversified by various glacial and fluvio-glacial forms not developed, as a rule, on the deposits of the earlier glacial stages. The chief varieties of these deposits are the outwash plains formed by streams discharging from the ice front, also kames, eskers, and drumlins. In a number of noteworthy cases these forms have a marked tendency toward aggregation. The Sun Prairie topographic sheet of Wisconsin represents the remarkable drumlin accumulations of that and adjacent portions of the state; and a similar swarm of drumlins may be seen in central New York between the Finger Lake district and Lake Ontario. The drumlins are oriented in the direction of ice movement and are composed of unstratified and very compact till.[2]

The two glacial advances of Wisconsin time were separated by a rather short interval of deglaciation, and the readvance in the second of the two stages was along slightly different lines, so that the relative size and relations of the ice lobes appear to have undergone consider-

[1] Frank Leverett, Comparison of North American and European Glacial Deposits, Zeitsch. f. Gletscherkunde, Berlin, vol. 4, 1910, p. 279.

[2] For discussion of drumlins or, more properly, rocdrumlins, formed out of shale and earlier till deposits in Menominee County, Michigan, and that drumlins are forms of aggradation in some cases and of degradation in others, see I. C. Russell, The Surface Geology of Portions of Menominee, Dickinson, and Iron Counties, Michigan, Rept. Geol. Surv. of Mich. for 1906, and H. L. Fairchild, Drumlins of Central Western New York, Bull. New York State Mus. No. 111, 1907, pp. 393 et al.

able change, and the moraines of the later stage cross those of the earlier at distinct angles, as at some points on Long Island (p. 470). The drift sheet of the later of the two stages is characterized by enormous terminal moraines, great bowlder belts, and an unusual development of kames, eskers, drumlins, outwash plains, valley trains, etc., all formed in such a manner as to show the disposal of the ice in the form of great lobes which in turn reflect the influence of the basins in which they were formed. The late Wisconsin drift is in many notable instances thin and overlies the interglacial fluvial deposits as a thin veneer of till.

TOPOGRAPHIC, DRAINAGE, AND SOIL EFFECTS

TOPOGRAPHY

In the successive periods of glaciation the ice advanced over the up-lifted and dissected peneplain of the northern Prairie Plains described in the preceding pages. In places it increased the relief by irregular deposition on a flat plain, in other places it decreased it by deposition in the valleys of a dissected plain. Had the country been mountainous an overriding ice sheet would not have produced till plains of such topographic uniformity and continuity. The present relief is therefore a function not only of glacial but also of preglacial topographic form.

The topographic and drainage effects of the successive ice invasions are of the first magnitude in the region in which they occur, while the distribution of the soils is almost everywhere related to the material which the ice laid down or which the draining streams swept forward in the region immediately beyond that covered by the ice. The southern limit of ice advance is indicated upon the map, Plate , but it must not be thought that a single vast sheet of ice at any one time occupied the entire region north of the line indicating the limit of glaciation. There were in all at least six ice sheets which swept from the north at as many different times, from two of the three recognized centers of ice accumulation. The deposits produced by the successive ice invasions occur as distinct drift sheets between which, in intervals of deglaciation, soils and beds of peat and marl were formed (p. 467). In addition to these lines of demarcation between successive till sheets there are differences among the sheets themselves; they vary in extent and physical constitution in a marked way. The lines of flow in successive invasions of ice were in many cases far from coincident, so that ice fields in some places fell short of earlier ones and in other places exceeded them.

Another important feature is the existence of pronounced morainic ridges along the fronts of some of the till sheets, although this is not a constant feature, since a wide stretch of country in Illinois, Kansas, and Iowa is marked by the absence of a terminal moraine. Where the moraines are present it is noteworthy that the earlier ones have been very extensively eroded, so that many irregularities found upon the later moraines are wanting in the earlier. A typical occurrence is found in southern Illinois east of St. Louis, where practically all the kettle holes and minor depressions of the Illinoian moraine have been either filled up, or drained by the cutting down of their outlets, or both, while the detailed irregularities of the slopes of the hills and ridges have been so completely smoothed and rounded as to appear to a certain extent artificial. This group of qualities is in marked contrast to the numerous sags, depressions, kettle holes with lakes or swamps on their floors, and the extreme rugosity of moraines formed during the last (Wisconsin) ice advance.

TERMINAL MORAINES AND TILL SHEETS

The terminal moraines marking the margin of the last Wisconsin invasion are pronounced features of the landscape and often rise from

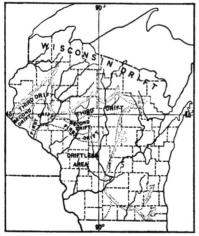

Fig. 177. — The four drift sheets of Wisconsin. (Weidman, Wisc. Geol. Surv.)

100 to 200 feet above the surrounding surface. They may be traced almost continuously westward from Long Island to and beyond the Great Lakes. Not all the ice invasions of previous glacial epochs resulted in the development of pronounced terminal moraines. Some of the till sheets have distinct marginal ridges, while others are marked by mere swells of thicker drift at the margins, and still others have very thin edges.

The marginal deposits of the Wisconsin drift are in the form of bold terminal and interlobate moraines of such topographic importance that they have served to modify or to form entirely the divides between the smaller rivers. They fill a number of preëxisting river channels and deflect many of the small streams and even many larger ones from their former courses.

The concentric trend of the morainic ridges is to be ascribed to the lobate character of the ice margin, a feature imposed upon the continental ice sheet by the irregularities of the topography which it covered. The depressions appear to have been lines of maximum flow, and in the

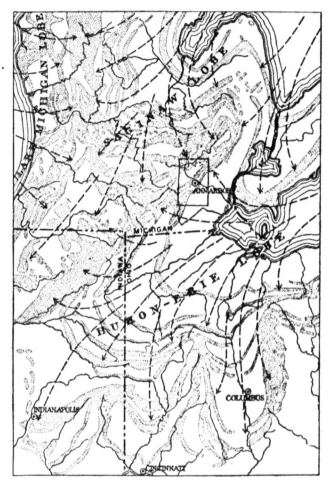

Fig. 178. — Distribution of glacial moraines and direction of ice movement in southern Michigan and northern Ohio and Indiana. (Russell and Leverett, U. S. Geol. Surv.)

Great Lake region this resulted in the strongly looped quality of the morainic systems as shown in Fig. 178.

The extent to which drift deposits dominate the glacial topography is further indicated by the fact that the present divide between the Great Lakes or St. Lawrence drainage and the Mississippi drainage is determined very largely by moraines and thick drift deposits. The drift is heaped up in greatest amounts at the ends and along the sides of the Great Lake basins where it was accumulated on the margins of tongues or lobes of ice that once occupied the basins. Such modification of not only the drainage but also the topography and the soils is clearly indicated by the nature of the drift covering in the south-

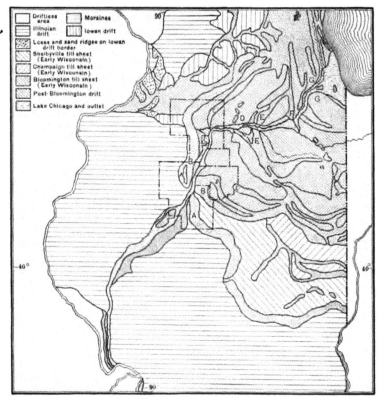

Fig. 179. — Generalized glacial map of northern Illinois. (Barrows, Bull. 15, Ill. Geol. Surv.)

ern peninsula of Michigan, northwestern Ohio, northern Indiana, and northeastern Illinois. The till deposits here attain their greatest thickness. The southern peninsula of Michigan is quite remarkable in this respect since the geographic position and relations of the Great Lake basins determined a maximum convergence of the ice lobes of this district. The northern half of the peninsula has a drift cover from 700 to 800 feet thick; the surface rises in places to 1000 or

1100 feet above the surface of the lakes; there seems to be no rock more than 250 feet above the lakes.[1] It has been estimated that the entire southern pen-insula of Michigan has an aver-age of about 300 feet of drift. The drift cover in the adjoining states on the south is about 100 feet thick. These heavy glacial deposits do not consist of a single sheet, but embrace three, and in places four, distinct drift sheets which are the products of repeated glaciations at widely separated in-tervals. The upper surfaces of the older sheets are marked by beds of peat and river and lake deposits containing remains of life forms similar to those now existing in the region, an indica-

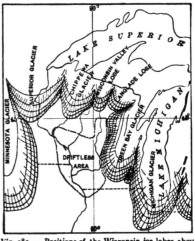

Fig. 180. — Positions of the Wisconsin ice lobes about the Driftless Area. (Weidman, Wisc. Geol. Surv.)

tion of a temperate interglacial climate not markedly unlike the climate of to-day.

Farther west, as on the plains of eastern North Dakota, are many

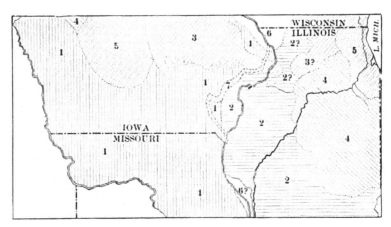

Fig. 181. — Relations of the drift sheets of Iowa and northern Illinois. 1, Kansan; 2, Illinoian; 3, Iowan; 4, Early Wisconsin; 5, Late Wisconsin; 6, Driftless area; 7, course of the Mississippi River during the Illinoian glacial epoch. (After Leverett, U. S. Geol. Surv., and Calvin, Iowa Geol. Surv.)

[1] W. F. Cooper, Water-Supply Paper U. S. Geol. Surv. No. 182, 1908, Plate II.

surface features which also show the characteristic effects of glaciation. The country is here for the most part level, but it also presents long rolling slopes rising from 300 to 800 feet above broad valleys. The most important topographic elements are massive ridges or mesas due to preglacial erosion. The mesas are in many instances bordered or crowned by long morainic ridges representing halts in the glacial advance or retreat. The morainic material is derived chiefly from the underlying shales (Cretaceous), and is a compact clay which also contains erratic fragments and bowlders of crystalline rock derived from northerly localities. The upper 5 to 10 feet of till has been oxidized to a light yellow color. The intermorainic tracts consist of rolling plains of till from a fraction of an inch to a hundred feet thick or of more level plains due to alluviation on the floors of glacial lakes. A typical instance of lake-bed topography is found in the upper James River and the valley of the Red River.[1] Postglacial stream dissection has further diversified the topography,

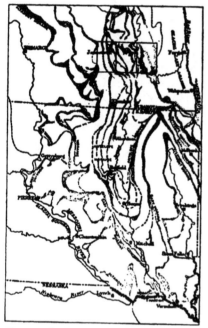

Fig. 182. — Southern limit of the Pleistocene ice sheet and distribution of moraines of the Dakota glacial lobe, North and South Dakota. (Willard, U. S. Geol. Surv.)

the Missouri River having cut a trench several hundred feet deep whose sides are for the most part steep.

DRAINAGE MODIFICATIONS

One of the interesting effects of glaciation was the diversion of streams from northward to southward courses; an instance of this is the Mississippi in the southwestern corner of the Driftless Area below Dubuque. The valley maintains an even width of 1¼ miles for about 12 miles, then widens gradually, and in a stretch of about 60 miles its width increases, except for a slight local contraction near the mouth of the Wisconsin River, and finally becomes 3¾ miles. This decided widening of a low-gradient valley upstream suggests that in preglacial time the valley was occupied by a northward-flowing river.

[1] J. E. Todd, Aberdeen-Redfield Folio U. S. Geol. Surv. No. 165, 1909, pp. 7 et al.

Other cases of stream diversions are discussed in connection with the Appalachian Plateaus, page 685, the Great Plains, page 405, and the Connecticut Valley, page 638.

More general and quite as con-spicuous are the influences of glaciation upon the drainage sys-tems laid out upon the surfaces of the various till sheets or dis-posed along their margins. The Ohio and the Missouri have courses in marked sympathy with the glaciated tract, as if they had been pushed bodily out of pre-glacial courses, and such indeed is the case at many points on both streams. The Ohio in particular exhibits a valley whose most strik-ing characteristics are open-broad stretches alternating with rock-walled gorges which closely con-fine the stream; and abandoned channel stretches are exhibited at many points. Many less con-

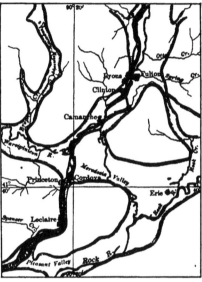

Fig. 183. — Old and new channels of the Mississippi at the upper rapids, Fulton, Ill.

spicuous illustrations of stream diversions along the border of the glaciated country have been discovered.[1]

Within the borders of the till sheets influences no less striking are found. The moraine crests are the chief minor divides of the glaciated country and in many cases they are also the major divides. Number-less tributaries starting on the slopes of moraines are gathered into trunk streams which flow for long distances along the margin of a moraine or between parallel moraines before finding an outlet across one or the other of the moraines into a transverse stream (pp. 471 and 479). Stream courses within the morainic belts are characteristically irregular and marked by many lake-like expansions alternating with steep stretches and often by falls and rapids of great economic value. These conditions are in sharp contrast to the regular drainage lines and the general absence of lakes and waterfalls in the streams draining the smooth outwash plains that in many cases front the moraines.

[1] Frank Leverett, The Illinois Glacial Lobe, Mon. U. S. Geol. Surv., vol. 38, 1899, pp. 102–103 et al.

The preglacial drainage of the Great Lake region was probably very unlike the drainage of to-day, for borings along the line of the buried channel that runs across Michigan from Saginaw Bay to Lake Michigan indicate a fall toward the southwest and also a widening of the channel in that direction. Similar deep borings in western Indiana between the head of Lake Michigan and the Wabash show a lower rock floor than any yet discovered across Illinois, and suggest a preglacial drainage to the southward toward the Wabash and Ohio.[1] It seems probable that there was a divide on the line of the present St. Lawrence below Lake Ontario where the river now flows among the Thousand Islands from which the drainage was southwestward.

The outlines of all the lake basins have been notably affected by the deposition of drift in part on their margins, in part on their floors. Near the mouth of the Cuyahoga River at Cleveland the drift extends more than 470 feet below lake level and to within 100 feet of sea level, while the greatest depth of water of Lake Erie is but 210 feet and the average depth only about 70 feet. These figures indicate a very notable modification of the western end of Lake Erie by the deposition of till, and such modification when brought about by a large ice mass must inevitably be associated with very notable changes in the preglacial drainage.

Fig. 184 is a profile across Lake Michigan, showing the extent to which drift forms a barrier to the lake waters at the present time.

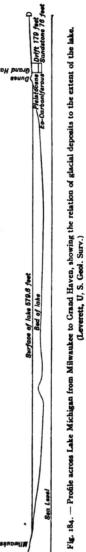

Fig. 184. — Profile across Lake Michigan from Milwaukee to Grand Haven, showing the relation of glacial deposits to the extent of the lake. (Leverett, U. S. Geol. Surv.)

LAKE PLAINS

The so-called Lake Plains of the Great Lake region are special features of the Prairie Plains province. They have an aggregate extent of several thousand square miles and form long narrow strips of smooth country bordering the

[1] Frank Leverett, Outline of History of the Great Lakes, 12th Rept. Mich. Acad. Sci., 1910, p. 22.

Great Lakes in sharp contrast to the rough morainic country about them. Many portions of them have low gradients, smooth surfaces, regular consequent drainage, fine soils, and extensive marshy tracts that remind one strongly of portions of the Atlantic and Gulf Coastal Plain. They have an aggregate extent of several thousand square miles, but this is far less than the area usually assigned to them. There is no doubt that their extent as represented on Powell's physiographic map of the United States [1] is greatly exaggerated. For the most part they are closely confined to the borders of the Great Lakes. In places they extend inland for 50 or 60 miles, but such breadths are exceptional. Commonly their width is under 20 or 30 miles.

The Lake Plains lie between morainic ridges and the lake shores. The morainic ridges are concentric with respect to the existing lakes since the lake basins were lines of movement of the glacial lobes along whose margins the moraines were developed. The lakeward slope of the innermost moraine formed the border of the lake and against this slope beach ridges and wave-cut platforms were developed in many cases. Where the innermost moraine was feebly developed or lower than its outer neighbors it did not determine the lake border. Furthermore, the lakes once lying over the lake plains were constantly changing in level and hence in outline. Shore lines are found between the outermost abandoned shore line and the present lake shores. These were developed not upon moraines but upon smooth lake floors uncovered by the retreat of the lake waters as they fell to lower levels.

The shallowness of the glacial-marginal lakes and the gradual movements of the shore zone of maximum wave action over the entire lake floor exercised a smoothing effect upon the lake bottom elevations. These lake bottoms, now exposed, in many cases exhibit a topography varied only by recent stream channels opened in them and by shallow depressions filled with small swamps, ponds, and lakes.

PROGLACIAL LAKES [2]

It was known even to the first settlers that lakes formerly occupied not only the present lake basins but also the bordering lands thus overlying the present lake plains and forming the prominent beaches developed

[1] J. W. Powell, Physiographic Regions of the United States, in Physiography of the United States, Nat. Geog. Mon., 1896, pp. 98-99.
[2] Whittlesey and Warren, The Great Ice Dams of Lakes Maumee, Am. Geol., vol. 24, 1899, pp. 6-38. See Frank Leverett, The Glacial Formations and Drainage Features of the Erie and Ohio Basins, Mon. U. S. Geol. Surv., vol. 41, 1902, for an excellent brief historical review of the problems of the abandoned shore lines of the Great Lake region as well as for a compact summary of the physical changes and a good working bibliography. See also J. W.

in numbers about the existing lakes at both high and low levels. Many of the beaches once served as trails for the Indians. Their level, dry, and moderately straight courses are also ideal for modern purposes. They are generally known as "ridge roads" and are fairly common in the southern part of the Great Lake region.

It was early demonstrated[1] that the lakes owed their origin to ice dams now extinct, and that the water impounded in front of the ice dams rose to the level of the lowest point on the rims of the enclosing basins and then overflowed into valleys draining to the sea. The ice dams were nothing more or less than the frontal lobes of the retreating continental ice cap of late glacial time.

To understand these relations one should conceive of a great continental ice sheet whose front was being melted back from a southern limit, Plate IV, so that it finally lay on the northern side of the great divide which separates the St. Lawrence drainage system from the Mississippi and Atlantic drainage. From the pattern of the terminal moraines shown in Figs. 178 and 182, the border of the continental ice sheet is known to have been lobate. Each lake basin as well as each subsidiary bay, such as Green Bay or Saginaw Bay, was occupied by an individual lobe of ice connected at the north and east with the main ice sheet.

In front of each ice lobe and in the southwestern and western portions of the St. Lawrence basin there eventually formed a glacial-marginal or proglacial lake which was at first crescent shaped, but with continued retreat of the ice its shape would be modified by the general outlines of that part of the basin that was not occupied by the ice lobe. It is also clear that the lake water could not discharge over the thick ice sheet or under it, but must have discharged along its front or across the divides on the distal margins of the basin. Lake Chicago is the name applied to the crescent-shaped lake that formed in front of the Lake Michigan lobe and discharged through the Desplaines-Illinois rivers. In a similar way Lake Maumee was formed in front of the Erie lobe and dis-

Goldthwait, The Abandoned Shore Lines of Eastern Wisconsin, Bull. Wisconsin Geol. and Nat. Hist. Surv. No. 17, 1907, for historical data of interest and for unusually accurate and detailed studies of abandoned shore lines in Wisconsin. More recently observation has been extended into Canada by Taylor and Goldthwait and the conclusions based on studies in the United States correlated with the observations of Coleman and others on the Canadian Geological Survey. The results of the refined studies of recent years point conclusively to the truth of Spencer's early contentions that the deforming movement of the region was differential and greater toward the north, a conclusion long neglected outright or regarded with undeserved skepticism chiefly because of the author's retention of the overthrown hypothesis of marine submergence.

[1] G. K. Gilbert, On Certain Glacial and Postglacial Phenomena of the Maumee Valley, Am. Jour. Sci., vol. 1, 3d ser., 1871, pp. 339-345.

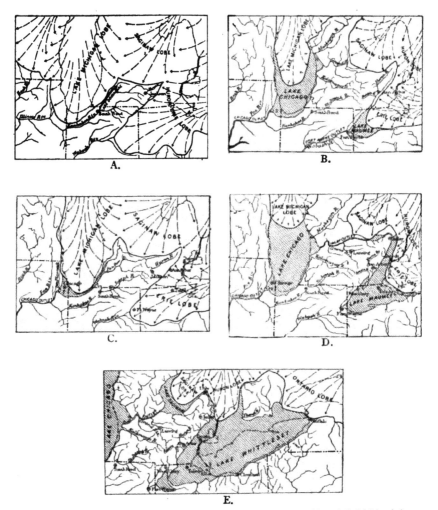

Fig. 185. — Drainage history of the southern Great Lake district. A, position of glacial lakes before lakes were formed; B, beginning of Lake Chicago; C, D, E, F, later stages in retreat of ice and expansion of proglacial lakes; G, retreat of ice from Mohawk valley and discharge of proglacial lakes down the Hudson; H, present drainage.

charged through what is now the valley of the Maumee into the Wabash past Fort Wayne. Lake Duluth was formed in front of the Superior lobe and discharged down the Chippewa to the foot of Lake Pepin in the Mississippi Valley.

On the borders of the glacial-marginal lakes, deltas, shore lines, and offshore deposits were formed, and on the bottoms, muds, sand, and silts were deposited. One side of each lake had a temporary shore

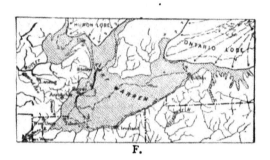

F.

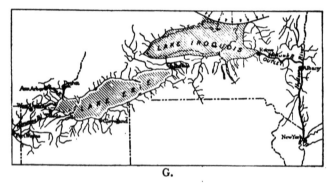

G.

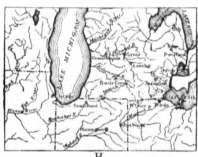

H.

Fig. 185. Continued.

line — the unstable ice wall which marked the front of each glacial lobe. It must also be noted that this ice wall was not fixed. Sometimes it advanced slightly, overriding the beaches that had been formed by the lake water in front of it; sometimes it remained stationary long enough to accumulate a water-laid moraine on the lake floor and a land-

Laid moraine on the adjacent land surface; but the net result of all the changes of ice front was a gradual retreat northeastward. Associated with this retreat was the gradual extension of the glacial-marginal lake waters and to a certain degree their coalescence along the ice front, as in the case of the water bodies at the heads of Saginaw Bay and Lake Erie, Fig. 185-E. Numerous changes in outlet also took place, among which only the principal changes are noted below. The principle which controlled these changes is, that with each marked retreat of the ice lower surfaces were exposed which would have different relations to each other with respect to elevation.

Thus Lake Maumee for a time discharged into Lake Chicago by way of the Saginaw basin and the Grand River Valley. Lakes Michigan and Huron for a time discharged through Georgian Bay and the Trent Valley into Lake Iroquois (Ontario), and a still later and lower outlet was afforded through the Ottawa River Valley on the uncovering of this drainage way through the further retreat of the ice. This stage is known as the Nipissing stage.

At this time Lake Erie drained across the Niagara escarpment and Lakes Huron, Michigan, and Superior drained eastward through the Ottawa Valley to the St. Lawrence. The small amount of water discharging over Niagara Falls during this period resulted in the production of a very narrow gorge, but the gradual tilting of the land toward the southwest caused the Ottawa outlet to be abandoned and the low col south of Port Huron to be occupied by a new drainage channel marking the outlet of these three great lakes into Lake St. Clair and Lake Erie. Associated with this change in outlet and the greater discharge of water over Niagara Falls was the widening and still more marked deepening of the gorge between the whirlpool and the Suspension Bridge, changes which are conceived to have occurred in the last 5000 years, judging by the present rate of retreat of the falls. The St. Clair outlet of Lake Huron is thus seen to be a very recent geologic event, as may be seen clearly also by the youthful character of the valley of that river and the recent formation of a feature which is anomalous in physical geography — the delta at the head of Lake St. Clair, formed by the short St. Clair River. A river draining a lake is singularly free from sediment at the intake, and it might be expected that a very small delta or none at all would be formed at the mouth of so short a stream as the St. Clair. Instead, the recent occupation of the valley of the St. Clair by the St. Clair River has caused very rapid channel cutting that is still in progress, an action that is reflected in the large delta at the head of Lake St. Clair. Tending toward the same result are the shore currents of Lake Huron, which deliver a large amount of beach sand to the intake of the St. Clair.[1]

The former small size of the lake in the Erie-basin and the fact that it did not then cover the western part of the basin are established by the deeply submerged lower courses of the streams in the western part of the basin, especially that of Sandusky River, which has been traced entirely across the bed of Sandusky Bay. More recent changes of water level at this end of the basin are shown by the submerged level of the valleys discharging into Lake Erie and the submerged stalactites in the caves of Put-in Bay Island at the western end of the lake.

During still later stages of their history the lakes discharged through the Mohawk and down the Hudson. The great sand plain at Albany and Troy was the delta of this large-volume, temporary stream. Through the later uncovering of the St. Lawrence Valley by continued ice melting a lower point of discharge was offered to the lake waters and the Mohawk channel abandoned. The present Mohawk River occupies but a small portion of the ancient glacial channelway.

The earliest formation of glacial-marginal lakes in the Lake Superior drainage basin appears to have been in the area now drained by the St. Louis River, where the waters were ponded at

---

[1] L. J. Cole, The St. Clair Delta, Mich. Geol. Surv., 1905.

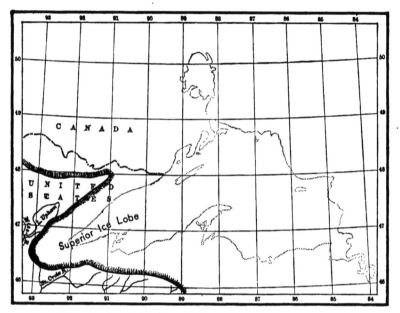

Fig. 186. — Superior ice lake and glacial marginal Lake Duluth. (Leverett.)

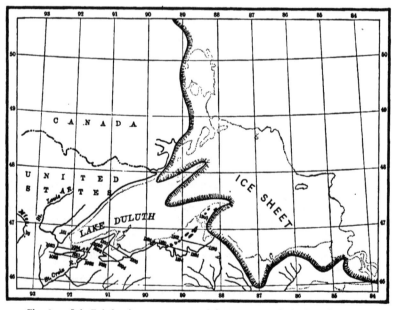

Fig. 187. — Lake Duluth at its greatest extent and the contemporary ice border. (Leverett.)

an altitude of about 700 feet above Lake Superior, Fig. 186. To this small transient body of water the name of Lake Upham has been given. The fringing ice later afforded a lower passage about 530 feet above Lake Superior, and the resulting lake with outlines quite different from those of Lake Upham and with a different line of discharge has been called Lake St. Louis. Still further retreat of the ice border resulted in the formation of a lower lake about 500 feet above Lake Superior, known as Lake Nemadji, which discharged a large volume of water (from Kettle River Valley) into the St. Croix. These lakes lie quite beyond the border of Lake Superior, and well above it. The name Lake Duluth has been given to the extremely large body of water draining southward from the head of Brule River into the St. Croix River and occupying a large portion of the area now occupied by Lake Superior. The highest shore line of Lake Duluth at the western end of the basin shows a marked rise toward the northeast. At Duluth it is 465 feet above Lake Superior, while at Calumet on the Keeweenaw peninsula it is 702 feet above it. The rise of 237 feet seems to be largely due to differential uplift, though, as in the case of all the other deformed beaches of the region, a small portion is referable to ice attraction, though this will account for but 37 of the 237 feet and was effective only within a few miles of the ice border. The difference of 200 feet seems explainable only on the ground of differential movements since the formation of the earlier shore lines.

During and since the retreat of the ice, tilting of the Great Lake region has been in progress. Elevated and now tilted and abandoned shore lines occur in numbers about the lakes of the Great Lake region. In places they are very numerous, as about the head of Lake Superior, where may be counted more than 30 distinct strand lines, representing as many successive lowerings of water level through the vicissitudes associated with the melting of the ice and the uncovering of lower and lower outlets. The beaches all rise in altitude toward the northeast, with a slightly increasing divergence in that direction, showing that the uplift of the land which gave the shore lines their present attitude was differential and greater toward the north; also that the tilting was not that of a simple plain but of a warped surface.[1]

With a view to determining the degree of stability of the Great Lake region, Gilbert compared the elevations of certain bench marks as determined in 1876 and again in 1896. After a careful elimination of all irregularities of lake levels due to wind, tide, etc., it was found that there was a small but definite increase in height in a northerly direction, and that the tilting is greatest in a direction south 27° west, and is at the rate of 0.42 foot per 100 miles per century.

The prolonged tilting of the Great Lake region is also shown by the drowned lake shores on the west side of the lower peninsula of Michigan, where many of the rivers discharge into pouch-shaped lakes separated from the main lake by long slender sand bars, as the Black River, the Muskegon, the Pentwater, and in a similar way at the western end of Lake Erie, the Maumee, the Rouge, the Raisin, and the Huron. In

[1] J. W. Spencer, Deformation of the Algonkian Beach and Birth of Lake Huron, Am. Jour. Sci., vol. 141, 1891, pp. 12–21.

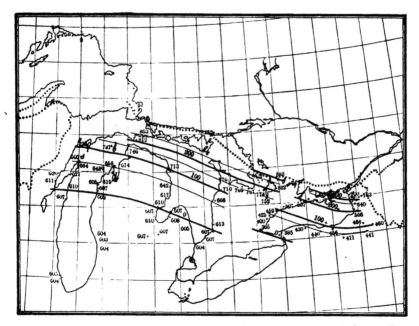

Fig. 188. — Isobasic map of the Algonkian and Iroquois beaches, showing the broader features of warping. Numbers indicate elevations of raised beaches above the sea. (Goldthwait.)

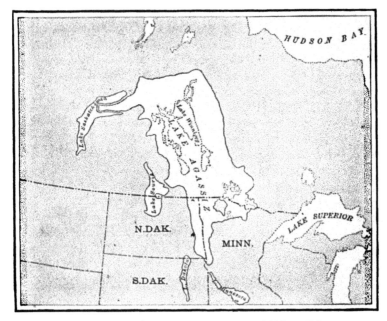

Fig. 189. — Map of extinct Lake Agassiz and other glacial lakes. (Upham U. S. Geol. Surv.)

contrast to these features of shore lines of subsidence are the features indicating elevation on the shore lines of the eastern side of Lake Huron at Kincardine and Goderich, Ontario, where the rising land keeps the lake always at work on new levels, and till, bowlders, cobblestones, and pebbles are everywhere in evidence. The near-shore bottom is stony and the stream discharging into Lake Huron at Kincardine has a stony bed, caving banks, and impetuous flow, indicating uplift and constantly renewed energies.[1] If the tilting continues at the present rate, the water of Lake Michigan will ultimately discharge across the low divide that separates the lake from the Desplaines Valley. Its channel will then be reoccupied and Niagara will become dry. It has been calculated that this change would take place naturally in from 5000 to 10,000 years; but man has anticipated the change by cutting a drainage canal for diverting the sewage of Chicago from Lake Michigan to the Illinois River — a project enforced by the urgent need of a pure supply of lake water for a concentrated population. The old outlet channels of the glacial-marginal lakes are among the most distinctive topographic and drainage features of the entire region. At the present time the Desplaines has a wide steep-walled valley with a flat marshy floor thronged with lakes, bayous, irregular depressions, and stream channels. The present *valley* was the *channel* of the outlet of Lake Chicago, and the irregularities of the present valley are•nothing more than the irregularities normal to the channel bottom of a large stream.

In the Red River Valley similar drainage changes of equal importance took place. Upon retreating gradually from the Lake Winnipeg-Red River Valley the Keewatin glacier was met by the gradually advancing Laurentide glacier, and the union of the two caused a ponding of the waters between their fronts, which produced glacial-marginal Lake Agassiz. The lake waters rose until they overflowed southward into the valley of the Mississippi and then gradually declined. The Laurentide glacier advanced to the western shore of Lake Winnipeg; this extreme advance was some time after the advance and retreat of the Keewatin glacier, and between the two till sheets stratified deposits occur near the mouth and on the banks of the Saskatchewan River. The large areas of flat fertile soil in the Red River Valley are largely lacustrine deposits — silts and clays formed on the floor of the ancient glacial-marginal lake and now exposed by the extinction of the lake, Fig. 189.

[1] Mark Jefferson, The Geography of Lake Huron at Kincardine, Ontario, Jour. Geog., vol. 2, 1903, pp. 144–155.

Soils of the Glaciated Country

Each type of glacial topography in a glaciated country has not only its own distinctive slopes but also its distinctive soils. A map of the glacial forms of a region is therefore to a large extent a soil map of the region. Terminal moraines are noted for stony till, outwash plains for their porous, gravelly, stratified, alluvial nature, eskers for their cobbly material, and kettle and kame topography is marked by sandy hill-slopes and undrained hollows with soils largely of organic origin. These are not hard and fast criteria but they are in general true. Exceptions to the rule are the sandy moraines near the Great Lakes, the stony out-wash where a late glacial readvance of the ice veneered the surface with a thin ground moraine, etc. Between the concentric terminal moraines , formed during the retreatal stages of glaciation one finds in many cases extensive tracts of sub-level land formed of till, the so-called ground moraine. These tracts are in many cases poorly drained and their soils therefore have a high content of organic material. They form the flattest and richest parts of Illinois and Indiana, and lie in the so-called "corn belt" of these and other states.

The unweathered condition of the soil minerals throughout the most recently glaciated area insures prolonged fertility of the soil. The older glacial deposits have been weathered to an important degree. Both the Kansan and Illinoian drift sheets, Fig. 181, have been greatly leached and large portions of their original calcareous content removed. The Wisconsin drift is fresh and unaltered, one of its distinguishing features. It is also noteworthy that the older drift sheets have a smaller proportion of organic matter in their soils than is the case with the youngest or Wisconsin sheet. Since the deposition of the Wisconsin drift there has accumulated large quantities of organic material along the shores of the small lakes and ponds, indeed some of the water bodies have disappeared through the filling of the basins by sediments, plant remains, and the cutting down of the outlets. The older till sheets have been so extensively eroded, on the other hand, that the accumulated organic remains in the small basins have been largely removed and the soils to that extent impoverished. Corresponding contrasts are exhibited in the value of the soil products, the density of population, and the natural vegetation.[1]

Over the old lake bottoms the soil varies considerably from place to place according to (1) the nature of the underlying glacial deposits, (2) proximity to a well-defined strand line, (3) character of the post-

[1] Frank Leverett, The Illinois Glacial Lobe, Mon. U. S. Geol. Surv., vol. 38, 1899, pp. 734–736 et al; also An Instance of Geographical Control upon Human Affairs, Geog. Jour. (London), vol. 24, 1904.

glacial drainage, etc. In some places the soil is a fine silt, loam, or clay, and is free from stones and bowlders and very fertile. In other localities it is marked by the presence of stones and bowlders, as where a beach was formed against the slopes of a rocky moraine, although for the most part the beaches are composed of sand and gravel. The poorly drained portions of the lake plains are in the aggregate of considerable extent and are marked by post-lake accumulations of vegetation in sufficient amounts to form a surface layer of peat or muck of great fertility when properly aerated by drainage. Many of the most productive of the market gardens that supply the cities of Port Huron, De roit, and Chicago are formed upon these extensive, poorly drained muck swamps.

Some of the beach levels are so warm on account of their porous soils and excellent natural drainage as to be ideal sites for vineyards and orchards, a feature well marked in western New York south of Lake Ontario. The lake plains on the eastern side of Lake Michigan and Lake Huron are noted for their large production of fruit, but this development is dependent to a greater degree upon climate than upon soil, for the Great Lakes are exceptionally large bodies of water and mitigate both the heat of summer and the cold of winter. The eastward or leeward side is favored by sufficiently moderate winter temperatures to permit the development of peach, plum, cherry, and apple orchards in great numbers.

In the northern part of the southern peninsula of Michigan the glacial deposits are exceptionally sandy and infertile, and large tracts, aggregating nearly one-sixth of the state, are known as the "pine barrens." Six million acres of these lands have been thrown on the delinquent tax list and are a burden to the people. Formerly they were covered with magnificent forests of white pine. These were so thoroughly cut down, the surface so extensively and repeatedly burned, and young growth and humus so devastated, that great areas have been rendered all but worthless.

Sandy tracts are also found along the lake margins and in some cases, as at the southern end of Lake Michigan, appear to be due to the fact that beach and dune material, formed on the borders of the interglacial Great Lakes, was reworked and redeposited during the last (Wisconsin) glaciation. Further modification has resulted from wind action, the light sands having been blown into dunes some of which attain an enormous size. "Creeping Joe" on the eastern shore of Lake Michigan, near Muskegon, is one of the largest dunes in the world, with a relative altitude of several hundred feet. The dunes are most numerous on the eastern shores of the lakes, as on the eastern shores of Lake Michigan and Lake Huron, where sandy deposits are exposed to the full force of the wind sweeping in from the lakes.

LOESS DEPOSITS

The student of soils must take account of an important deposit in the Mississippi Valley related to glaciation and known as *loess*. The loess deposit of the Prairie Plains extends westward from a point east of the Mississippi and has its greatest development in Illinois, Iowa, Nebraska, and the states to the southeast. The distribution of loess is one of its most important features. It is thick about the borders of the area occupied by the Iowan ice sheet, but thins out on interstream areas when traced away from this border tract, though it retains its thickness along the larger valleys.[1] It follows the Mississippi nearly to the Gulf, and is particularly well developed along this stream and the Missouri. It is habitually found on bluffs immediately overlooking the valleys, and in this position has more than average thickness and coarseness of grain, becoming thinner and finer at a distance from the valley margin.[2] The northernmost limit of the loess in the central part of the United States is about 35 miles below St. Paul, Minnesota. In central and eastern Illinois and south-central Ohio it passes under the Wisconsin drift. In the Mississippi basin the loess is rarely over a score or two of feet in thickness, though exceptional thicknesses of nearly 100 feet have been noted. It is composed of angular undecomposed particles of calcite, dolomite, feldspar, mica, and a certain number of rarer minerals. It is generally light brown in color and is a variety of silt intermediate in size of grain between the finest sand and clay. Stones are generally absent except at its base. On erosion it often exposes vertical faces for long periods. It is markedly porous, owing in part to vertical tubes usually found in it and supposedly due to root action.[3]

The loess is preponderatingly on the east sides of the main rivers, and this fact suggests the hypothesis of an eolian origin, which has perhaps more to commend it than the rival aqueous hypothesis. It appears that the loess is chiefly wind derived and that favorable opportunities for its formation were afforded by the periodic flooding and drying up of the alluvial deposits along the streams of the region either in the in-

[1] Chamberlin and Salisbury, Geology, vol. 3, 1906, p. 407.

[2] Frank Leverett, Comparison of North American and European Glacial Deposits, Zeitsch. f. Gletscherkunde, Berlin, vol. 4, 1910, p. 297 ff.

[3] For an excellent discussion of the various phases of the loess problem, and for recent data of great consequence, see B. Shimek, Jour. Geol., vol. 4, pp. 929-937, and various loess papers, Bulls. Lab. Nat. Hist., Univ. Iowa, 1904.

terglacial period preceding the Iowan glacial invasion or during the Iowan invasion itself. The absence of fluvio-glacial deposits in important amounts on the margin of the Iowan till sheet would appear to favor the idea of a dry glacial stage, strange as this may seem.

It is found that the loess is thicker on eastern or lee sides of ridges and prominent hills than on the windward sides, a feature that harmonizes with the thicker deposits on the east sides of the stream valleys in pointing to wind action as the chief factor in loess deposition. The diminishing amount of loess in an eastward direction suggests that a considerable part of the material was derived from the Great Plains east of the Rocky Mountains, where dust storms are frequent even to-day.

That the loess was deposited in an interglacial period is supported by the fact that buried soil occurs between the loess and the glacial deposits, and appears to indicate a long period between the melting of the ice and the deposition of the loess. This view is strengthened by the finding of a molluscan fauna in the loess strikingly similar to the existing fauna, showing that the conditions which prevailed when the loess was forming were not greatly different from those now existing in the region.[1]

The shells found in the loess are almost exclusively those of land species or such as frequent isolated ponds. There is a practical absence of forms that frequent rivers and lakes. Fossils of land mammals, chiefly in the forms of bones and teeth, have also been found.

Tree Growth of the Prairies

By definition a prairie is an extensive tract of land, level to gently rolling, without a timber cover. The definition includes the idea of a meadow or at least a grassy vegetation. The distinction between *plain* and *prairie* as the terms are actually employed in the United States is not sharply drawn, but in general they are considered to differ in two main respects. The prairie country is dotted with groves of trees and has a continuous grass cover or sod; the plains country is without timber except along the streams, on prominent elevations, and in places about springs; the grassy vegetation of the plains is disposed in clumps separated by bare spaces. A view in any direction on the typical prairies includes groves and bands of trees and even tracts of true forest. The phrase "Prairie Plains" as here used means the whole great region in which patches of prairie occur and not merely the true prairie land exclusive of wooded tracts (see Plate I and Fig. 23).

Since the whole prairie country lies in the belt of moderate to light rainfall between the semi-arid West and the humid East, Fig. 19, it follows that there are great differences in the relative amounts of woodland and true prairie. On the east, north, and south the province borders

[1] See especially Proc. Iowan Academy of Science, vol. 15, 1909, pp.57-75.

heavily forested tracts; on the west it borders an all but treeless region. In general the largest tracts of true prairie within the province are found toward the west in the direction of diminishing rainfall. Locally, however, there are extensive tracts well within the western border.

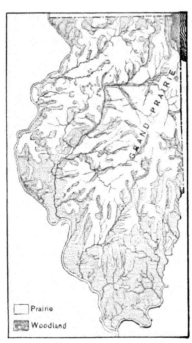

Fig. 190. — Original distribution of prairie and woodland in Illinois. (Barrows, Ill. Geol. Surv., modified from Gerhard.)

South of Chicago, for example (Fig. 190), is a large tract known as the Grand Prairie with a rainfall of 35 inches a year. Extensive prairie tracts occur also in northern Indiana, southern Michigan, etc.

On the north the Prairie Plains province is covered with the southern edge of the great pine and hemlock forests of the Lake Region. The individual trees attain great size, especially on the sandy glacial material south of the Straits of Mackinac in Michigan; the virgin growth of this district consisted of extensive stands of superb forest. Farther south is the zone of hardwoods with many outliers of the northern pine forests, especially where the land is either too wet or too dry. The sandy, dry situations have a growth of pines; in the wettest situations are growths of cedar and hemlock. In the intermediate situations are woodland and a true forest growth of mixed hardwoods. The occurrence of the hardwoods on the richest land is notable and in part explains why they were so extensively destroyed in clearing of land for agriculture. It is a common saying that the soil supporting a good growth of beech and maple forest is better than open prairie soil or land so wet that it must be drained before being cultivated. While the saying has a great deal of truth in it, it must also be noted that the peaty soils of low situations have a high fertility when they are properly drained and sweetened and are better than the upland soils for the production of many vegetable products. What measure of truth the saying contained led the settlers rapidly to clear away the hardwood forests until now the supply of hardwood is greatly diminished and in urgent need of conservation.

Many owners are beginning to realize the value of their wood lots and are no longer cutting them for fuel but for timber.

The strongly marked topographic and drainage features of the heavily glaciated country about the Great Lakes have had an important control over the forests not only in the distribution of the forest flora but also in the development of the forest products. It is noteworthy that the steeper slopes of the terminal moraines are wooded because too steep for cultivation, just as bottom land and upland are in many places cleared of their original timber cover while the steep undercut river bluffs are allowed to remain tree-covered. The terminal moraines are not only rough, they are also stony and clayey; many miles of terminal moraine are composed of tough stony clay too difficult of cultivation to compete with better favored tracts of outwash and ground moraine in the intermorainic belt. Such tracts are commonly left wooded or used for pasture-land or for special crops such as flax and rye. It is also noteworthy that the steep slopes of drumlins in the glaciated country are commonly left in their original wooded condition. Some drumlins have narrow summits, others have broad rounded tops. In the latter case a farm may be located on the summit, in the former case the timber cover may be left undisturbed. In the same way the borders of marshy tracts and the floors of undrained depressions are covered with a growth of swamp vegetation including the swamp oaks and maples, buttonwood, elm, cedar, and other types. The aggregate of such land is large and with suitable improvement might be made to grow a steady supply of valuable timber.

On the east the forests of the Prairie Plains grade off into the more or less continuous stands of the Appalachian region in the belt of heavier rainfall. On the south they merge into the pine forests of the Coastal Plain and on the west into the scattered hill and escarpment timber of the Edwards Plateau and its outliers in central Texas. Farther north, as shown in Fig. 23, long finger-like extensions of the Prairie Plains timber follow the stream valleys. For some distance this growth covers both bottom lands and valley slopes and bluffs; farther west it is confined to the bottom lands wholly and still farther west it is found only along the margins of the streams.

The tree growth of the Prairies of Texas has such a close relation to the geology, water supply, and soils as to warrant a more detailed description of its occurrence. The Prairie Plains are here but 100 miles wide and consist of two main subdivisions, — the Black Prairie on the east and the Grand Prairie on the west. The Black Prairie is developed upon a series of marls, sands, and limestones that dip gently eastward. The marls have weathered into a uniform and very gently undulating type of topography terminated on the west by a low inward-facing escarpment 'that never exceeds a relative altitude of 200 feet. It is known as the Black Prairie. West of the Black Prairie is the Grand Prairie, developed upon nearly horizontal limestones of great extent as a series of flat structural plains terminated in succession on the coastal side by low inward-facing escarpments. The western border of the Grand Prairie is a pronounced escarpment with a lobate or crenulated outline that extends from Arkansas, west and south through Oklahoma and Texas, to the Rio Colorado, whence it curves about to the west and north and becomes the eastern escarpment of the Great Plains.

The most important forest tracts of this region are the Western and Eastern Cross Timbers, two narrow forest belts developed upon the inner edge of the Grand Prairie and Black Prairie respectively. They are of peculiar interest since they form long narrow ribbons of forest in what is otherwise a prairie country. Both belts have been developed upon the outcrop of two sandy formations (Cretaceous). The open sandy soils not only permit thorough aeration, but also favor the rapid absorption of water during periods of rainfall, whereas the close-textured soils of the intervening prairies shed water to an exceptional degree and are too dry for tree growth. The limited rainfall of the district thus makes the porosity of the soil a determining factor in forest distribution. The Eastern Cross Timbers consist of a belt of timber whose extent is shown in Fig. 191. It has many outliers towards both the east and the west and like the Western Cross Timbers sends finger-like extensions up and down the stream-ways. Within the Cross Timbers are many local prairie tracts. The Western Cross Timbers are about 10 miles long and have about the same extent north and south as the Eastern Cross Timbers. The various irregularities of the belt in response to geologic changes and to relief are clearly indicated in Fig. 191. The western border of the Western Cross Timbers is the more indistinct since it merges into local forest tracts developed upon other sandy formations. Wherever compact marls and clays outcrop there is a marked absence of forest growth except in cases where the mesquite grows. In general the surface is without tree growth between and on either side of the Cross Timbers; it is here that the prairies of Texas have their most typical development.[1]

Among the effects of glaciation none is perhaps more directly interesting to the forester than the overwhelming effects of the continental

[1] R. T. Hill, Geography and Geology of the Black and Grand Prairies, Texas, 21st Ann. Rept. U. S. Geol. Surv., pt. 7, 1899–1900, pp. 65–85.

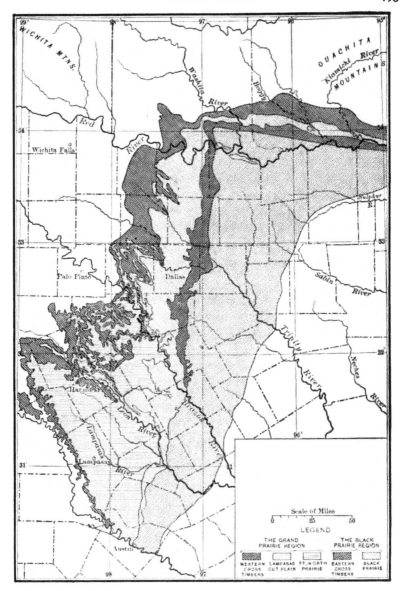

Fig. 191. — Cross Timbers of Texas.

ice caps upon the forests whose regions they invaded.[1] During the period of maximum development the ice had a most disastrous effect

---

[1] C. H. Merriam, The Geological Distribution of Life in North America with Special Reference to the Mammalia, Proc. Biol. Soc. Washington, vol. 7, 1892, pp. 42 and ff.

upon animals and plants, not only in the great area occupied by it, but far south of its actual border. As the ice advanced from the north, northern species were driven southward and more southerly species correspondingly displaced or exterminated. It appears that the gradual southward movement of the cold zone greatly restricted the range of the plants, compressing them from north to south within very narrow limits. Perhaps the most interesting evidence of the strength of this hypothesis is the presence of colonies of Arctic species on isolated mountain summits in southerly latitudes where at a high altitude abnormally low temperatures exist similar to those which exist in their northern homes. Such colonies could not in many cases have reached their present position during existing climatic conditions; following the retreat of the ice at the close of the glacial period many boreal species were stranded on mountains where their survival was conditioned by their ability to migrate with a congenial climate.

DRIFTLESS AREA

Lying well within the southern border of the general region once covered with glacial ice is a small district, Fig. 192, which has never suffered glaciation. It is called the Driftless Area. Its extent is about 10,000 square miles and it lies in the states of Wisconsin, Illinois, Minnesota, and Iowa, the greater part of it occurring in southwestern Wisconsin. The soil of the Driftless Area is residual, derived directly from the decay of the underlying rock. Like residual soils everywhere the surface detritus grades gradually into solid rock and there is not that sharp dividing line that is found between glacial or other overplaced soils and the rock surface.

The surface of the Driftless Area is topographically mature and the drainage is well organized, without falls and rapids and lakes or swamps except within narrow limits and along the river bottoms. The aimless flow of streams developed upon the irregular topography characteristic of till sheets or terminal moraines, as well as the abundant swamps, lakes, and ungraded stream courses of such localities, are here practically absent. At numberless points within the area overridden by glacial ice the bedrock is scratched, grooved, and smoothed, while within the Driftless Area glacial markings are entirely absent.[1] The thoroughly dissected upland plain of the Driftless Area is in striking contrast to the more even country with drift-filled valleys about it. The valleys and ravines tributary to the main streams have been thoroughly organized and the divides reduced to narrow ridges. The

---

[1] Grant and Burchard, Lancaster-Mineral Point Folio U. S. Geol. Surv. No. 145, 1907, p. 1.

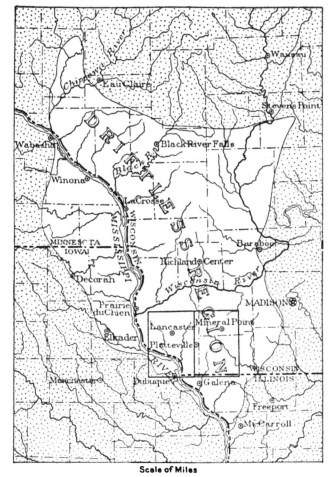

Scale of Miles

Fig. 192. — Driftless Area of Wisconsin.

valley bottoms toward the south are from 600 to 700 feet above sea level, and the relief of the area is from 200 to 300 feet.[1]

In spite of the extreme irregularity of the drainage of the glaciated region its relief is less than that of the Driftless Area, a condition due to the manner of distribution of the drift, which in general was lodged in the valleys more freely than on the heights. Such plain-like qualities as the drift-covered region exhibits are due less to the glacial denudation

[1] J. E. Carman, The Mississippi Valley between Savanna and Davenport, Bull. Ill. State Geol. Surv. No. 13, 1909, p. 29.

of prominences than to the grading up of depressions. In a rough manner the driftless region therefore presents the essential features of the preglacial topography of this portion of the United States.[1]

The purely residual products in the Driftless Area have their greatest expression upon the flat-topped upland areas. On the valley slopes the residual products tend to be removed by erosion, and on the valley bottoms there is alluvial filling derived from the uplands in part during an earlier period of alluviation (probably glacial) and in part by wash under existing conditions.[2]

The Driftless Area has created widespread interest since early in the eighteenth century, largely because it has not that exceptional height which on the border of the glaciated country is the general cause of immunity from continental glaciation. Its average summit level falls a

Fig. 193. — Diagrammatic section in the Driftless Area, showing relation of the mantle rock to the solid rock beneath. (Alden, U. S. Geol. Surv.)

little short of 1200 feet, while the effective height of the highlands lying between it and Lake Superior is from 1700 to 1800 feet. The summit level between the Driftless Area and the plains of Iowa and southern Minnesota is about 1300 feet, so that as a whole the Driftless Area is somewhat lower than the surrounding tracts. Furthermore, the immunity from glaciation enjoyed by the area was due not to some accidental condition but to a geographically fixed cause, for none of the repeated invasions of the ice affected it.[3]

The most important elements of the explanation of the Driftless Area are that it is in the lee of the elevated territory of northern Wisconsin and Missouri, which acted as a wedge, forcing the ice away on either side and protecting the region south of it,[4] and that the great valleys of Lake Superior and Lake Michigan lie in such position with reference to the Driftless Area as to tend to divert the glacial streams

[1] Chamberlin and Salisbury, The Driftless Area of the Upper Mississippi, 6th Ann. Rept. U. S. Geol. Surv., 1884–85, p. 307.

[2] S. Weidman, The Geology of North-Central Wisconsin, Bull. Wisc. Geol. and Hist. Surv. No. 16, 1907, p. 554.

[3] Chamberlin and Salisbury, The Driftless Area of the Upper Mississippi, 6th Ann. Rept. U. S. Geol. Surv., 1884–85, p. 315.

[4] N. H. Winchell, 5th Ann. Rept. Geol. and Nat. Hist. Surv. of Minn., 1877, p. 36.

to right and left and thus to increase the effect of the highlands lying between the lakes in turning the ice from the driftless region.[1]   A third cause is found in the retarded feeble flow and relatively increased wastage caused by the thinning ice through the diversion of the main lobes down the lake troughs.   An important factor was the broad zone of glacial wastage which probably includes approximately 150 miles of the ice border; on this assumption the meager ice streams that crept down across the highlands south of Lake Superior and the enfeebled tongues which crept down the adjacent depressions were profoundly controlled by the topography and easily directed this way and that according to the relations of the major topographic depressions.

It is also maintained with much reason that the outlines of the ice front would be modified to an important degree by the law of self-perpetuation of climatic conditions; by which is meant that the ice-clad regions increased the snowfall upon themselves and tended toward self-perpetuation, just as the ice-free and therefore relatively warmer area resisted the advance of the ice by excessive melting.   While this last agency was not an originating one it would undoubtedly produce a modifying effect and tend to coöperate with the topography in holding the surrounding ice to the depressions down which the invasion first took place.

The Driftless Area is interesting ecologically because of the occurrence therein of many plants peculiar to it alone.   Among these mosses and other low forms predominate.

---

[1] R. D. Irving, Geology of Wisconsin, vol. 2, 1877, pp. 608, 611, 632, 634.

CHAPTER XXV

ATLANTIC AND GULF COASTAL PLAIN (INCLUDING THE LOWER ALLUVIAL VALLEY OF THE MISSISSIPPI)

GENERAL FEATURES AND BOUNDARIES

THE greater part of the eastern and southern coasts of the United States is bordered by a lowland whose physiographic features are developed upon a mass of soft sands, silts, and clays, chiefly of Cretaceous and Tertiary age, and disposed in strata that incline (dip) gently seaward. The structural features are in the main very simple, though in places there are important departures from the average condition. On the north shore of Long Island the clays (Cretaceous) were disturbed by ice action and quite generally concealed by a cover of glacial and fluvioglacial material; in Louisiana important faults complicate the structure, topography, and drainage; in Oklahoma and Texas some of the coastal plain deposits are indurated, not soft and friable as generally; and in many localities the dip of the strata, almost never great, varies from a mean sufficiently to cause important differences in the width of the coastal tract, the character of the shore line, and the general topographic expression.

The Coastal Plain terminates on the north at Cape Cod, on the south in Mexico. Both ends are narrow and the northern is glaciated and partly submerged. The physiographic features of the province are varied at two places by special features of the first order — the peninsula of Florida and the lower alluvial valley of the Mississippi. Its width varies from a fraction of a mile to 500 miles measured normal to the coast; its greatest inland development is in the Mississippi Valley, where it swings northward into western Kentucky, southeastern Missouri, and eastern and southern Arkansas. The inner edge of the northern portion of the Coastal Plain is bordered by water bodies — Cape Cod Bay, Long Island Sound, etc.; by an inner valley lowland in New Jersey, Alabama, and Arkansas; and these features in turn are bordered by various topographic provinces — the Piedmont Plateau developed on crystalline rocks at the north and east and in succession westward and southward, the Appalachian Mountains and Plateaus, the Ozark Highlands, the Ouachita Mountains, and the Prairie Plains of central Texas.

FALL LINE

The Atlantic division of the Coastal Plain borders the Piedmont Plateau along the fall line, as it is commonly termed. This phrase imperfectly describes the boundary, which is not really a line but a zone of appreciable width; hence "fall belt" would be a more accurate designation. As the streams cross from the old land or Piedmont Plateau to the new land or Coastal Plain they descend over rapids and falls of moderate height. Between the Raritan and the Roanoke the rivers discharge directly from rock-walled channels into tidal estuaries, since here the depressed land has brought the sea to the very borders of the Piedmont Plateau; but farther south the fall line is above sea level; on the James it is at tide level; on the Neuse 100 feet above the sea; on the Wateree, near Camden, it is 125 feet; on the Savannah, near Augusta, 125 feet; on the Chatahoochee, near Columbus, 210 feet; and on the Tuscaloosa at Tuscaloosa, 150 feet.

On the interfluves, Coastal Plain and Piedmont Plateau merge into each other in an intermediate zone from a fraction of a mile to 10 or 12 miles wide. The boundary is never a cliff and seldom even a well-defined scarp, and the lowland hills are nearly as rugged and more than half as high as the piedmont hills; but chiefly on account of differences of soil the two provinces present some of the most strongly marked contrasts in the United States, if not in the world. In the Piedmont Plateau the rocks are crystalline, the soils residual, the stream courses flow in narrow gorges with cataracts and rapids; while in the Coastal Plain the soils are derived from clay, sand, and gravel deposits, and the streams flow in shallow valleys and discharge into broad tidal estuaries or coastal swamps. In the piedmont region the topographic details, the soils, and the watercourses reflect the character of the rock, while in the coastal region these features represent the work of streams born upon plains newly uplifted from the sea, or reflect the general attitude of the lowland rather than its rock character. Among the important cities located upon the common boundary of these two great provinces may be mentioned Raleigh, Camden, Columbia, Augusta, Macon, Columbus, Montgomery, Tuscaloosa, Little Rock, Austin, San Antonio, etc.

RELATION TO THE CONTINENTAL SHELF

The seaward limit of the coastal lowlands on the southeastern border of the United States is a steep scarp which marks the transition from the outer edge of the shallow continental shelf to the abysmal depths of the main ocean floor. This scarp forms the true continental margin and is

comparable in height and extent though not in topographic variety
with some of the most majestic mountain ranges. Were sea level
lowered to its foot, present sea level in this latitude would be a region
of ice and snow. The seaward slope of the Coastal Plain is continued
as the continental shelf beneath the level of the sea to distances varying
from 100 to 200 or more miles; the Coastal Plain is therefore to be
considered as the landward extension of a greater plain whose outer
border is submerged. The present position of the shore line upon this
plain is a purely accidental one and subject to change, indeed is now
changing at what from even a human standpoint may be considered
a rapid rate. The steady depression of the larger part of the coast is
so pronounced that if continued the sea will in a very short time again
cover the coastal lowlands. In more recent periods of the earth's history such transgressions of the sea over the land have been frequent
and important, and no less important have been the various retreats of
the sea which have brought the coastal border of the continent out
near the border of the continental shelf and turned what is now continental shelf into dry land.

The repeated invasions of the sea are clearly indicated by the marine fossils found in the
flat-lying sediments which constitute the Coastal Plain; and the surface of the continental
shelf is deeply scored opposite the mouths of many of the coastal rivers by trenches or submarine canyons whose position and character have led to the conclusion that they represent
old river valleys formed at a time when the land stood higher than now.

MATERIALS OF THE COASTAL PLAIN

ATLANTIC SECTION

The lowest strata of the Atlantic Coastal Plain, as illustrated in the
Maryland section, are composed of arkosic sands and clays derived
from a deep mantle of disintegrated gneiss and phyllite such as now
form part of the Piedmont Plateau. The sands and clays have detached outliers west of the main border of the Coastal Plain which
show by their position, altitude, and composition that the plains strata
formerly extended farther west over the adjoining portion of the Piedmont Plateau. Higher in the section the strata consist of variegated
clays and coarse, irregularly bedded sands. Then follow sands and clays
in alternation with but slight variations in character and a gradual
transition toward sandy and marly deposits.[1]

[1] C. Abbe, Jr., A General Report on the Physiography of Maryland, Including the Development of the Streams of the Piedmont Plateau, Md. Weather Serv., vol. 1, pt. 2, pp. 74-75.

GULF SECTION

The Gulf Coastal Plain section is rather typically represented in the Arkansas-Louisiana district where the lowermost beds are sandy and contain vegetable remains and brackish-water shells. Above these basal beds are limestones and marls; higher in the section are sands, lignitic clays, marls, and chalks, a succession indicating a gradual deepening of the water in which the sediments accumulated. Succeeding the deep-water deposits are limestones, marls, and lignitic sands, showing a return to shallow-water conditions.[1] Then came complete and final uplift of the region and the formation of a coastal plain out of what had long been a continental shelf. The migration of the shore zone, characterized by sand deposits, over the surface of the plain from its inner to its outer edge was the last important geologic event and is the cause of the prevailingly sandy nature of the surface deposits.

SUBDIVISIONS

(1) Including a broken and glaciated northern section the Coastal Plain may be said to fall into seven well-defined districts. The first includes Cape Cod, Martha's Vineyard and Nantucket islands, Long Island, and a number of lesser islands.

(2) The second extends from the Hudson to the Potomac, and is characterized by a cuesta, an inner lowland, and a gentle outward slope or lowland merging into great estuaries that nearly isolate it from the mainland.

(3) The third extends from the Potomac to the Neuse. It is a low eastward-sloping plain characterized by long arms of the sea reaching far into it, by broad terraced plains of loam, and broad coastal sounds defended on the seaward border by long tenuous reefs partly of sand, partly the narrow and wave- and current-modified outcrops of clayey formations (Cretaceous) that dip seaward.[2]

(4) The fourth section extends from the Neuse to the Suwanee as a gently sloping plain characterized by great stretches of pine-covered sands with swampy marginal development also fringed by coastal reefs.

(5) The fifth section extends from the Suwanee River to the bluffs at the eastern border of the Mississippi flood plain. This division has more important topographic irregularities than the Georgia-South Carolina section of the Coastal Plain, its inner border having been dissected in

[1] A. C. Veatch, Geology and Underground Water Resources of Northern Louisiana and Southern Arkansas, Prof. Paper U. S. Geol. Surv. No. 46, 1906, pp. 20-28.

[2] Collier Cobb, Notes on the Geology of Currituck Banks, Jour. Mitchell Soc., vol. 22, No. 1, p. 17.

such manner as to develop a rather typical inner lowland and cuesta. Along the coast are long narrow keys, some of which are scarcely able to maintain themselves owing to the depression of the land and to the encroachment of the waves, while some are completely submerged and have the form of offshore shoals. The seaward border of the Coastal Plain of this section is marked by broad lagoons, low coastal swamps, and extensive savannas.

(6) The sixth section is the low, poorly drained, swampy terminus of the flood plain of the Mississippi, an area whose southern end is scarcely yet reclaimed from the sea.

(7) The seventh section extends from Atchafalaya Bayou to the Rio Grande. Its interior development is topographically weak except in Arkansas and Louisiana, where a somewhat regular series of cuestas and inner and outer lowlands has been developed. The seaward margin of this district is bordered by swamps, extensive sounds, and long narrow lagoons.[1]

CAPE COD-LONG ISLAND SECTION

CAPE COD

The northernmost section of the Coastal Plain is broken by dissection and drowning into a number of islands, shoals, and capes of which the most important members are Cape Cod and Long Island. Lesser members are Staten Island, Block Island, and Martha's Vineyard and Nantucket islands.

That these are all members of a formerly more extensive and less dissected coastal plain fringe is inferred from (1) the evidences that stream courses on the present mainland border are superposed from a former cover of gently and regularly sloping marine deposits; (2) the actual finding of fragments of such deposits at Marshfield,[2] Boston,[3] and Third Cliff,[4] at the first two localities below sea level and at the third above it; (3) the occurrence of known Tertiary deposits[5] on Georges Bank east of Cape Cod in line with the general trend of the strike of the unsubmerged portions of the plain to-day; and (4) from the general likeness of the shoal and channel outlines of the sea floor north of Cape Cod to drowned valleys.[6]

[1] For a characterization of these various districts and a clear description of the coastal plain of the southeastern portion of the United States, see W J McGee, The Lafayette Formation, 12th Ann. Rept. U. S. Geol. Surv., 1890–91, pt. 1, pp. 353–521, with four excellent maps, photographs, and cross sections.

[2] C. H. Hitchcock, Final Report on the Geology of Massachusetts, vol. 1, 1841.

[3] F. G. Clapp, Clay of Probable Cretaceous Age at Boston, Massachusetts, Am. Jour. Sci., vol. 23, 1907, pp. 183–186.

[4] I. Bowman, Northward Extension of the Atlantic Preglacial Deposits, Am. Jour. Sci., vol. 22, 1906, pp. 313–325.

[5] A. E. Verrill, Occurrence of Fossiliferous Tertiary Rocks on the Grand Bank and Georges Bank, Am. Jour. Sci., 3d Series, vol. 16, 1878, pp. 323–332.

[6] N. S. Shaler, The Geology of the Cape Cod District, 18th Ann. Rept. U. S. Geol. Surv., pt. 2, 1896–97, pp. 516, 578, 580.

Cape Cod is the most extreme projection on the eastern coast of the United States north of the peninsula of Florida. In general outline it roughly resembles a great flexed arm, the southern portion being the humerus, the eastern portion the forearm, and the northernmost extremity the clenched fist. It is so narrow at the base where it adjoins the mainland that a canal is building to connect Cape Cod Bay and Buzzards Bay; the land section will be only 8 miles long. Its exposed position, the lack of good harbors, and its featureless shores combine to give the Cape notoriety as one of the two graveyards of the Atlantic. The projection of the Cape eastward and northward would not seem so remarkable if the material composing it were solid rock of great resistance to wave erosion; as a matter of fact no hard rock occurs upon it, only soft, easily washed gravel, sand, and clay.

The extreme northern end of Cape Cod, the wrist and fist of the Cape Cod arm, is composed of wave- and current-derived material entirely. It is low, covered with sand dunes of irregular outline, bears a scanty growth of shrubs and grasses or is entirely bare of vegetation, and has well-rounded coastal outlines where the wear of currents and waves is constantly reshaping the beach. The material added to the Cape at this point is derived from the crumbling eastern side of the arm. The attack of the waves on the steep eastern shore is vigorous and sustained, and at the present rate of retreat, 3.2 feet per year, the Cape will be entirely destroyed 8000 to 10,000 years hence. On the western side of the Cape the shore is low and flat and bordered by wave-built sand reefs which enclose extensive lagoons. Cape Cod is relatively quiet, a great natural harbor, yet the shore currents are here attacking the border of the Cape and assisting in its demolition.[1] The same action is exhibited at Monomoy Point, the southeastern extremity of the Cape, where a portion of the material torn from the bold eastern margin of the Cape is drifted southward and built into a long flying sand reef.

The foundation of a large part of Cape Cod is a mass of preglacial sands and clays presumably of Cretaceous or Tertiary age, revealed in well sections, clay pits, cuts, and bluffs. These deposits form the substructure of the east-west section of the Cape where they occur above sea level. The north-south section nowhere reveals similar deposits above sea level, but they have been encountered in well borings. The morainal ridge that forms so prominent a feature of the first-named section rests upon these preglacial deposits and its height is enhanced thereby.[2]

The surface material of the Cape is, however, not derived from the basement deposits; it is of glacial derivation. It occurs in the form

[1] W. M. Davis, The Outline of Cape Cod, Proc. Am. Acad. Arts and Sci., vol. 31, 1896, pp. 331–332.

[2] N. S. Shaler, The Geology of the Cape Cod District, 18th Ann. Rept. U. S. Geol. Surv., pt. 2, 1896–97, p. 534.

of a thin veneer of till except (1) where actual morainic ridges of exceptional thickness occur, or (2) where outwash deposits were laid down in front of the moraine. Cape Cod is therefore a mass of glacially derived material resting upon a preglacial basement of sands and clays. Its detailed surface features are due chiefly, and in some cases wholly, to ice action. Lakes are relatively abundant; the surface is characteristically pitted with kettle holes; unsorted material, till, is common; and there are many erratic ice-rafted bowlders. The glacial material is prevailingly sandy and this also is the character of the outwash plains that front the moraine. Lying well forward of the main line of the coast where no topographic prominences afford shelter from the wind the Cape is exposed to the fury of every tempest and its porous, light, dry material has been largely blown into dunes and drifts of sand. This is especially true on the immediate shore, where dunes of considerable height occur and where serious effort has been made to stop the progress of the wind-drifted material.

CONTROL OF SAND DUNES

The problem of dune control is a paramount one in the utilization of large portions of the surface of Cape Cod. Similar efforts are required in many other localities in the United States, and since these are more or less alike in natural features and involve the same problems, the following general discussion is included here.

The chief areas of shifting dunes to be found along the Atlantic coast are on Cape Cod, near Provincetown; southern New Jersey, near Avalon and Stone Harbor; Cape Henlopen, near Lewes, Delaware; Cape Henry, Georgia; Currituck Banks, North Carolina; Isle of Palms, near Charleston, South Carolina; and Tybee Island, near Savannah, Georgia. Sand dunes are also found on the Pacific coast in Ventura, Monterey, and Mendocino counties, California, and along the coast of Oregon.[1]

In California drifting sand has played an important part in obstructing the channels in certain harbors on the coast; for example it has been blown into the channels every year near Eureka from a narrow sand spit several miles long between Humboldt Bay and the ocean.[2] Very

[1] A. S. Hitchcock, Controlling Sand Dunes in the United States and Europe, Nat. Geog. Mag., vol. 15, 1904, pp. 43-47.

[2] S. B. Kellogg, Problem of the Dunes, Cal. Jour. Tech., vol. 3, 1904, pp. 156-159. See this paper for a discussion of various kinds of schemes for controlling sand dunes by sand-binding grasses, such as beach grass (*Ammophila arenaria*) and "rancheria grass" (*Elymus arenarius*); and for a photograph of a dune covering a forest above Fort Bragg, Mendocino County, Cal.

troublesome dunes are developed on a large scale along the Columbia River in Oregon and Washington from Dallas to Riparia, from sand blown about over the flood plain of the Columbia, which is widely exposed during the dry season. Dunes occur in large numbers also about the southern end of Lake Michigan, on the eastern side of Lake Michigan, and east of Lake Huron.

The most important principle of control of sand dunes is the establishment of a cover of material that will prevent drifting, the type of cover usually depending upon climatic conditions, cost of material, etc. Where possible it is best to produce a forest cover, for this is permanent and, if properly managed, yields an income. For this purpose the land must be temporarily fixed in some other manner than by seedlings, and a temporary binding of inert material such as brush, sand sedges, etc., may suffice. Beach grass has given most satisfactory results in the Great Lake region, North Carolina, and Europe. The grass is transplanted in the spring or fall and set 2 or 3 feet apart in the sand. It has the power of continuing to grow up through a cover of drifted sand by establishing root systems at constantly higher levels. If the sand is temporarily fixed by these or other means, young trees, usually conifers, may be planted and a forest established. In southwestern France a forest has been established by sowing the seed of *Pinus maritima* upon the sand and covering with brush. In France and on the coast of Prussia it has been found necessary in places to construct long artificial barrier dunes between the ocean and the forest and to fix the barrier dunes by beach grass and maintain them by constant oversight. In northern Europe the trees employed for reclamation of sand dune tracts are *Pinus montana, Pinus laricio, Pinus austriaca,* and *Pinus sylvestris*.[1] Extensive pitch-pine plantations are now being experimented with on the sandy areas of Cape Cod, Martha's Vineyard, and Nantucket.[2]

NANTUCKET AND MARTHA'S VINEYARD

The relief and drainage of Nantucket Island afford many parallels to the conditions on both Cape Cod and Long Island. There is a terminal moraine on the northern shore of the island and a very irregular primary or inner shore; the same kind of outwash plain is built forward of the moraine, pitted with kettle holes and creased by shallow drainage ways dry for the greater part of the year; and long tenuous sand reefs and flying spits occur like those on the coasts of the adjacent islands. Underneath the thin mantle of glacial materials lie preglacial deposits

[1] S. B. Kellogg, Problem of the Dunes, Cal. Jour. Tech. vol. 3, 1904, p. 47.
[2] Yearbook Dept. Agri., 1909, p. 336.

which in places project above the level of the sea, so that were the glacial accumulations removed a number of small islands would still exist where Nantucket stands. For the most part the older surface is masked by glacial till and outwash gravels. The original outline of the island has been much altered by wave and current action; portions of the shore have been cut away, while other portions have been fringed with recently formed marine deposits.[1]

Fig. 194. — Diagrammatic section of Martha's Vineyard. (Shaler, U. S. Geol. Surv.)

Martha's Vineyard lies west of Nantucket and has closely allied physiographic features. Not only are the deposits of the island of the same general nature but they are also disposed in roughly the same way, so that the northeastern and northwestern sides of the triangular island are hilly and morainic, with irregular shores, while the southern side is an almost straight eastward-trending line. It presents almost wholly those features typical of a morainal island. The surface drift is almost everywhere thin (about 10 feet) and as on adjacent islands it to some extent masks the older basement material, Fig. 194. The straight southern side of the island is a sand reef enclosing a large number of water bodies with a branching pattern closely resembling that of a stream system and tributaries, a condition signifying a slight submergence of the land after the formation of the outwash plain bordering the shore.[2]

LONG ISLAND

GENERAL GEOGRAPHY

To a forester Long Island is of peculiar interest. It supports a considerable variety of forest trees under special conditions, and the relations of these to the physical conditions are the more easily determinable because of the detailed studies made by the United States Geological Survey and the Soil Survey. The value of the natural resources of Long Island is unusually great because of their proximity to a great consuming center; and detailed studies of its climate, soil, water supply, and physiography have therefore been made.

The greatest length of Long Island (the "Lange Eyelandt" of the early Dutch) is 118 miles and its greatest width 23 miles. Its outline suggests the shape of a huge fish; the flukes

[1] Curtis and Woodworth, Nantucket, a Morainal Island, Jour. Geol., vol. 7, 1899, pp. 226–236; N. S. Shaler, The Geology of Nantucket, Bull. U. S. Geol. Surv. No. 53, 1899.

[2] For geologic description see N. S. Shaler, Report on the Geology of Martha's Vineyard, 7th Ann. Rept. U. S. Geol. Surv., 1885–86, pp. 303–363.

of the tail are represented by Orient and Montauk points on the east; the head is represented by the blunt western end with its mouth-like extension at Coney Island. The south shore is double: the inner or primary shore line is the border of a broad lagoon — Great South Bay and its extensions; the outer shore line consists of narrow sand reefs of remarkably regular outline. The eastern half of the northern shore is without notable indentations, but the western half is deeply embayed by a dozen or more well-developed fiords with steep sides and noteworthy depth of water.

GLACIAL TOPOGRAPHY

TERMINAL MORAINES

The principal relief features of Long Island are associated with a double line of eastward-trending glacial moraines fronted by extensive outwash plains of loose alluvium, Fig. 195. The southernmost is known as

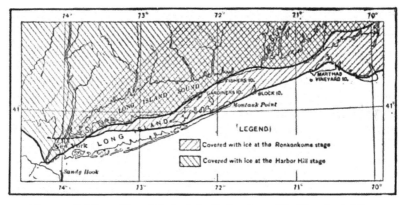

Fig. 195. — Relative positions of the ice during the two stages of the Wisconsin glaciation.
(Veatch, U. S. Geol. Surv.)

the Ronkonkama moraine. Its eastern extension gives character to that end of Long Island, and with the eastern extension of the Harbor Hill, or northern, moraine encloses a large body of water known as Peconic Bay. The Ronkonkama moraine is remarkable for the large body of water it encloses and after which it was named, Lake Ronkonkama. This is by far the largest lake on Long Island and extends about 25 feet below sea level. The two moraines are often referred to as the backbone of Long Island, though it should be noted that it is a compound backbone, for east of Huntington the two moraines diverge more and more and enclose broad sub-level tracts between them.

The seaward slope of a large part of the eastern half of the northern moraine has been largely removed by wave and current erosion, and in a few localities this has been carried so far that in the wave-cut cliffs are exposed sections of the outwash plain through the complete removal of the moraine once fronting and guarding it on the north. Everywhere along this shore the projecting headlands are being modified by vigorous wave action. Cliffs are kept perpetually freshened and so steep that landslides are a common occurrence, bars are being built across the bay mouths, and in the mill of the surf the bowlders and stones of the crumbling moraine are fast being ground to pieces.

The Montauk moraine lies chiefly in the interior of the island and is not subjected to such general attack by shore processes though its eastern extension, exposed to all the fury of Atlantic storms, is being battered rapidly to pieces. That the extreme eastern end must once have extended farther east is self-evident. Erosion has now progressed so far that portions of the once more extensive outwash plain have been destroyed.

Both moraines bear the usual marks of terminal accumulations formed by a continental ice sheet. Their surfaces are deeply pitted by large and small kettle holes, and enclosed, formless pits and depressions, and a maze of mounds, knobs, and ridges of till and other glacial débris still further diversifies their surfaces. Upon the floors of some of the enclosed depressions lakelets or swamps occur, and all others are dis-

Fig. 196. — Section showing the relation of outwash to terminal moraine. (U. S. Geol. Surv.)

tinctly moist, but by far the larger number are without standing water. Large portions of both moraines are composed of till of rather typical composition, but there are also large portions which are composed largely and in some localities almost exclusively of sand. The sandy phases are developed chiefly at the eastern end, the clayey phases at the western end of the island. Bowlders, large and small, are scattered freely upon the surfaces and throughout the mass of both moraines. They are of variable composition; gneisses and schists, sandstones and quartzites predominate.

OUTWASH PLAINS

Long gently sloping outwash plains extend southward from both moraines. Their northern margins are in many places pitted by kettle holes, and a considerable number of old broad drainage grooves cross them from north to south. The grooves are not now effectively occupied by streams, for the present drainage consists chiefly of small wet-weather streams extremely diminutive as compared with the ancestral streams whose channels they occupy. The old broad channels have a markedly asymmetric development, the steepened slope being on the west, a response, it is thought, to the right-handed deflective influence of the earth's rotation in the northern hemisphere.[1]

The porous nature of the outwash material greatly reduces the run-

[1] G. K. Gilbert, The Sufficiency of Terrestrial Rotation for the Deflection of Streams, Am. Jour. Sci., 3d Series, vol. 27, 1884, pp. 427-432.

off, and instead of a normal run-off of about 40% the run-off is about 60% of the total precipitation. The reduced run-off has retarded the dissection of the frontal plains, and the small elevation has favored the

Fig. 197. — Outwash plain in foreground, the hills in the background are preglacial and have but a slight morainal covering. (Veatch, U. S. Geol. Surv.)

same result. They extend mile after mile as almost flat plains of alluvium in striking contrast to the ruggedness and picturesqueness of the terminal moraines.

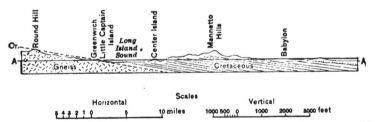

Fig. 198. — Cross section of Long Island, showing glacial deposits (dotted) and pre-Cretaceous peneplain (Cr.). (Veatch, U. S. Geol. Surv.)

TOPOGRAPHIC FEATURES RELATED TO STRUCTURE

The present topography of Long Island can not be fully understood without some knowledge of the substructure, Fig. 198. Cretaceous and Tertiary sands of great thickness occur up to heights of several hundred feet above sea level; their outcrops are exposed principally along the north shore, where the long, deep, and narrow fiords or reëntrants reveal good sections of the white and red sands and white clays of the Cretaceous formations. In the Half Hollow and Mannetto hills and distinctly beyond the limit of the ancient ice sheet preglacial sections also are exposed, and the strata (Tertiary) consist of fine fluffy sands without glacial material of any sort. From such exposures of preglacial material above sea level it follows that an island would now exist even if the glacial material were removed; and in fact such an island did exist before the glacial period, but it was an island of far gentler topography and much more regular outline.

Upon a base-leveled and down-warped rock floor Cretaceous and Tertiary materials were laid down and afterward eroded as the result of uplift of the sea floor. The erosion was at first normal and resulted in the formation of an inner lowland where now Long Island Sound occurs. Following this there were a number of emergences and submergences of complex character, during which the drainage of the inner lowland was first developed to the west and later, through the tilting of the land, to the east.

The essential features of the present topography were developed after the Lafayette submergence. The erosion of the inner lowland was continued to the point where a well-developed cuesta was formed, and it is to the further accumulation of glacial material upon the summit of this cuesta that Long Island owes its present marked contrast between rugged moraine and abrupt high northern shores on the one hand and smooth southern plain with low, reedy, flat southern shores on the other. The scouring action of the ice sheet further deepened the inner lowland, and with the final disappearance of the ice from the region a slight sinking of the land brought out even more strongly the features owing to depression. At a higher elevation the inner lowland of Long Island would have similar relations to the cuesta and the old-land as the inner lowland of New Jersey now has to the cuesta of the New Jersey coastal plain on the east and the old-land on the west.

SOILS AND VEGETATION

SOIL TYPES

A soil map of Long Island, Fig. 199, shows four broadly defined types of soils: (1) stony loams and gravels which occupy the terminal moraines and the narrow plateau between the northernmost moraine and the north shore, (2) coarse sandy loams that constitute the greater part of the outwash plain, (3) fine sandy loams which form the outer fringe of the outwash plain and those portions of it that are adjacent to the old drainage ways, and (4) clay loams that form a transition type between the upland and the salt marsh and the beach sands. While the clay loams have greater natural fertility than the sandy loams they are often found on lands too rough for cultivation, and the great market gardens of the island are found on the outwash principally, where many of the conditions of cultivation are ideal. Agriculture has become so highly specialized in the western part of the island on account of proximity to New York City that the natural fertility of the soil is a far smaller factor than position with reference to the market.

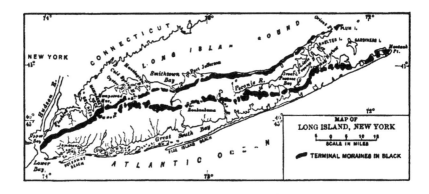

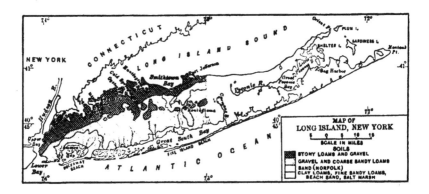

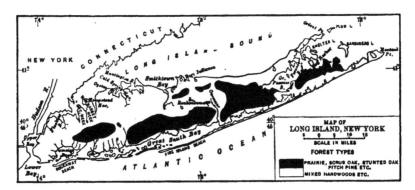

Fig. 199. — Terminal moraines, soils, and vegetation of Long Island.

Fig. 200. — Characteristic growth of pitch pine (background) and scrub oak (foreground) on the outwash plains of eastern Long Island.

Fig. 201. — The effects of repeated fires on soil and vegetation, Long Island. The raw soil with the humus almost entirely burnt out has the appearance of snow in the photograph.

Fig. 202. —Typical growth of hardwood on the clayey portions of the Harbor Hill moraine, Long Island.

In the disposition of the natural vegetation we have a far truer index of soil fertility and water-supply conditions than in the artificial disposition of farms. On the poor, dry, sandy, porous soil of the outwash plains the characteristic growth is pitch pine and scrub oak (*Quercus nana* and *Quercus prinoides*), as shown on the accompanying map, Fig. 200. The natural prairie for which Long Island is famous is also located on the southern outwash plain in the vicinity of Hempstead and Garden City. It bears a growth of prairie grass and has never been covered with forests. In Dutch colonial days it was a famous pasture ground for sheep, horses, and cattle. On the better soils of the outwash plains a better growth of pitch pine is found, and even a small stand of white pine has been reported southeast of Sag Harbor. Where a slight admixture of clay or loam appears, as in a number of scattered localities, a few species of hardwoods grow with pine and oak.

Upon the moister and more clayey moraines excellent growths of hardwoods occur, Fig. 202. Chestnut, oak, elm, beech, and locust constitute the principal types. In a few localities, as several miles southwest of Port Jefferson, the moraine and accompanying outwash are occupied by oak, but it is a stunted growth and reflects the infertile character of the Norfolk sand on which it grows, the distribution of the stunted species corresponding almost precisely with that of the Norfolk sand. The sandy phase of the Ronkonkama moraine south of Riverhead is occupied by pitch pine, which occurs up to the summit of the moraine, Fig. 203. A similar exception on the outwash is the occurrence of hardwoods along the southern shore, where the nearness of the water table to the surface has resulted in hardwood growths and the exclusion of pitch pine and scrub oak.

NEW JERSEY-MARYLAND SECTION

The coasts of New Jersey, Delaware, and Maryland are great peninsulas formed by the drowning of the major valleys, and here the lesser waterways discharge through reedy swamps or shoal inlets into landlocked bays. In Delaware and Chesapeake Bays these secondary reentrants are fronted by small banks and bars similar to those that fringe the outer shore except where low cliffs have been formed at the ends of finger-like extensions of land between bays. Formerly the whole section of the coast between Cape Hatteras and New Jersey stood at a higher level and was faintly sculptured by the draining streams, but later depression has submerged the lower ends of the valleys, where bays now exist, and the bays still preserve the characteristic dendritic plan of the Coastal-Plain drainage. While the submergence affects a large extent of the Coastal Plain the actual amount of depression has been astonishingly slight. The waters of Chesapeake Bay are so shallow that it is sometimes more miles to the shore from a given point in the bay than it is feet to the bottom of the bay. The depth of the water is seldom more than 18 feet and averages only 10 feet. Twenty-five feet of elevation would cause it to become a low coastal terrace.[1]

NORTHERN PORTION

In the northern portion (New Jersey) of this section of the Coastal Plain there are extensive flats both along the coast and in the interior, but these range in elevation from about 40 feet on the coast to 130 and 150 feet farther inland, and 200 feet in the highest part of the Coastal Plain. The tidal marsh of New Jersey lies principally between the beach of the Atlantic coast and the mainland; but there is also a tidal marsh bordering Delaware Bay which is not fronted by a beach. The width of the marsh varies greatly from place to place and is from less than a mile in its narrowest portion to 5 or 6 miles between Great Bay and Atlantic City. It has its greatest development at the mouths of the larger streams, and along Delaware Bay attains a width of 5 miles.

Fig. 204. — Sand reef, salt marsh, and coastal plain upland, coast of New Jersey.

The sand reefs owe their height chiefly to dunes. During times of storm the sandy barrier reefs are piled up above normal water level and on becoming dry are blown into hills and ridges by the wind. As in the Maryland section, the shallow lagoon between the beach and the mainland is gradually being filled up with wind-blown material, vegetation, and sediments. The most striking feature of the Coastal Plain of New Jersey is the line of

[1] W J McGee, The Geology of the Head of Chesapeake Bay, 7th Ann. Rept. U. S. Geol. Surv., 1885–1886, p. 552.

elevations extending in a northeast-southwest direction from the Navesink highlands on the northeast to Mount Holly on the southwest, elevations which include heights that approach 400 feet.[1] These are due, curiously enough, to widely contrasted conditions. (1) The clays, sands, marls, etc., do not have great inequalities in hardness, but locally the material is cemented into more or less solid rock. This occurrence is most common at the junction of beds of different texture, and in some cases has reached the point where the rock is quarried. Many of the most prominent elevations are capped by such cemented beds of gravel, sand, or marl, and owe their prominence to a protecting cap. (2) Many other prominences in the highest belt of the district by contrast owe their height to the extremely loose and porous condition of the material. The rainfall sinks into the material as into a sponge and does not run over the surface and erode it; consequently the hills formed on the outcrop of the most porous beds are at elevations comparable to the hills formed upon the hardest material. Few elevations of note in the highest part of the New Jersey Coastal Plain are without such a protecting cap of rock or loose gravel.[2]

The strata on the inner edge of this part of the Coastal Plain have been stripped from the crystalline rocks beneath in such manner that a valley lowland has been developed parallel to the highlands that cross the state diagonally. The lowland extends from Raritan Bay to Trenton, and marks the outcrop of a series of less resistant formations whose removal goes forward more rapidly than that of the crystallines and hard sedimentaries on the west or the higher coastal plain formations on the east.

SOUTHERN PORTION

The outer edge of the coastal plain of Maryland is bordered by long narrow sand reefs caused by shore drift and wave action. Behind them are shallow lagoons of variable width ranging from a fraction of a mile to 4 or 5 miles in Maryland. The eastern portion of the lagoon is formed by shallow marshes along the western edge of the sand reef; the western shore is formed by the low, half-submerged topography of the mainland, somewhat modified by salt marshes. The floors of the lagoons are very shallow and flat and are composed (a) of sand blown over from the beach dunes, (b) of mud deposited by the rivers and tides, and (c) of matted roots of marine vegetation.[3]

The surface of the coastal plain of eastern Maryland is broad and

[1] R. D. Salisbury, The Physical Geography of New Jersey, Final Rept. of the Geol. Surv. of New Jersey, vol. 4, 1895, p. 54.

[2] Idem, p. 64.

[3] C. Abbe, Jr., A General Report on the Physiography of Maryland, Including the Development of the Piedmont Plateau, Md. Weather Service, vol. 1, pt. 2, p. 82.

even and resembles a smooth or gently undulating sea floor.   Many portions are characterized by long interstream stretches of plane surface of considerable breadth.[1]   The inequalities of the outer border of the Maryland plain were produced during a submergence which took place in very recent geologic time (Pleistocene); the plain has been so recently raised above the sea and to so small a height that time enough has not elapsed for the streams of gentle gradients to drain the swamps and lakes located upon them.   It is characteristic of the swamps that

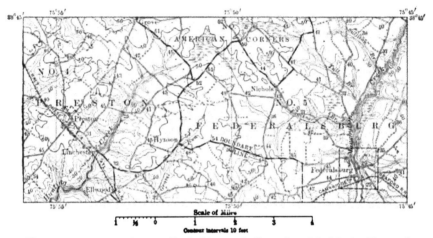

Fig. 205. — Swampy divides in eastern Maryland between the Chesapeake and the Atlantic. The coastal plain is here so young and so little dissected that many of the original irregularities have not yet been destroyed. (Hurlock quadrangle, U. S. Geol. Surv.)

they are disposed chiefly along the main divides, as though dissection had not yet progressed to the headward sections of the streams, Fig. 205. The surface has a gentle seaward slope upon which has been developed a characteristic drainage system; the pattern is irregularly branching or dendritic, with the small streams commonly making almost a right angle with the general trend of the larger streams at the junction.   Small lakes and swamps dot the surface and are due to inequalities on the sea floor produced by wave and current action.

Except in their expanded lower courses the streams are small and unnavigable and are characterized by broad, shallow valleys with very gentle side slopes and smooth contours.   In the interstream areas of the southern counties of Maryland one may travel for miles and never cross a well-marked valley.   Where forests grow they have still further re-

[1] C. Abbe, Jr., A General Report on the Physiography of Maryland, Including the Development of the Piedmont Plateau, Md. Weather Service, vol. 1, pt. 2, p. 84.

tarded the run-off, so that wet swamps occur in the original inequalities of the surface. Signs of stream sculpture are rare; the surface seems to preserve the outlines originally imposed by waves and currents. In contrast to these streams are the tidal estuaries and the slow meandering creeks that cross the salt marshes and have low though steep banks deeply fringed with reeds.

VIRGINIA-NORTH CAROLINA SECTION

The physiography of the third district of the Coastal Plain is characterized by a gentle seaward slope, smooth and monotonous, interrupted by long tidal inlets. The rivers expand towards their mouths in reedy

Fig. 206. — Cypress trees of the Dismal Swamp. (Norfolk Folio, U. S. Geol. Surv.)

marshes or in broad shallow estuaries barred from the open ocean by wave-built reefs that stretch almost continuously from Cape Lookout to Cape Henry. The uplands and lowlands of this district have so little difference in altitude that they could be scarcely distinguished from each other if the lowlands were not almost at tide level. The lowlands are the bay bottoms or the tidal marshes or the broad savannas which the highest tides barely fail to reach; the uplands rise in irregular scarps from a few feet to 15 feet in height to form stretches of excessively flat plain. Farther from the coast there is again a zone of undulating surface; the depressions containing the waterways between undulations are less conspicuous than in New Jersey and lower Maryland. Toward the fall line the surface is characterized by broad terraced plains with irregular margins, and smooth monotonous interiors whose borders are diversified by labyrinthine ravines.

Fig. 207. — Albemarle and Pamlico Sounds and bordering sand reefs, east coast of North Carolina.

Fig. 208. — Chesapeake Bay and Delaware Bay and the principal bays tributary to them.

SOUTH CAROLINA-GEORGIA SECTION

That portion of the Coastal Plain between the Neuse River of North Carolina and the Suwanee of southern Georgia and Florida is a land of gentle slopes which incline from the fall line to the coast. It is a pine-clad, sandy plain with dendritic drainage. The seaward margin of this section is a wave- and current-built sand reef at the north and a line of sea islands at the south, that merge at the extreme south into long low islands locally known as "keys."

Along the Altamaha River the overflowed river bottoms have been reclaimed to some extent by diking and ditching and are cultivated. Their position is excellent for rice culture, since irrigation water is easily applied. Agricultural operations have been confined principally to the diked river lands since the early part of the nineteenth century, and relatively small areas have been cropped on the uplands since the early abandonment of indigo growing and the decline and cessation of cotton production.[1] Near Savannah these diked rice fields have been badly neglected in recent years, but the fertility of the tide-marsh soils is beyond question, and were drainage reëstablished they would constitute a valuable addition to the farming lands of the coast.[2]

In portions of eastern North Carolina the seaward margin of the Coastal Plain bears drainage ways which are merely natural depressions marked by a quick growth of water-loving shrubbery and in some areas no well-developed drainage courses occur. These are the savannas, or open flat lands with badly drained soils that support a poor growth of pine and an undergrowth of berry bushes and bay and pitcher plants.[3] Inland from the low coastal zone is the broad belt of pine-clad sands which more than anything else characterizes the section. It is a vast plain with slight undulations, the depressions slightly accented by streams occurring in the form of old terraced scarps much dissected owing chiefly to the friable character of the material, and now having rounded bottoms and softly contoured sides. With increasing distance from the sea the land stands higher and higher, the streams are more active, the valleys deeper, and the surface more undulating, often rising into rounded hills. Again, as in the north, the hills may be isolated and flanked by terraces or may constitute parts of broad interstream areas, and sometimes the terraces extend a short distance into the Piedmont Plateau. The common boundary of plateau and plain is ill-defined as to topography but rather sharply marked as to drainage and soil characters.

ALABAMA-MISSISSIPPI SECTION

The great section of the Coastal Plain between the Suwanee and the Mississippi is topographically diversified by two pronounced scarps; the first one is the steep river bluffs along the eastern border of the Mississippi flood plain, bluffs due to stream planation; the second is the cuesta of the Alabama-Mississippi section of the plain. The cuesta begins on the river bluffs overlooking the Mississippi in extreme western Kentucky, crosses the Tennessee-Mississippi boundary 50 miles east of the river bluffs, curves southeastward to within 50 miles of the head of Mobile Bay and dies out eastward. This feature represents the inward-facing slope of the dissected Coastal Plain of Alabama and Mississippi. It

---

[1] Milton Whitney, Soils in the Vicinity of Brunswick, Georgia, Cir. U. S. Bur. Soils No. 20, p. 3.

[2] Milton Whitney, Soils in the Vicinity of Savannah, Georgia, Cir. U. S. Bur. Soils No. 19, p. 2.

[3] Milton Whitney, Soils of Pender County, North Carolina, Cir. U. S. Bur. Soils No. 20, pp. 1–2.

stands 600 or 700 feet above sea level. The outer lowland averages 200 feet lower, while the inner lowland is a broad trough often 100 and sometimes 200 feet lower than the border of the cuesta. The inner lowland is from 20 to 25 miles on the average and extends eastward a short distance beyond Montgomery. The densest population of the state of Alabama outside the cities is found in the inner lowland where the Selma chalk has weathered into a tract known as the "black belt" because of its prevailingly black soils. These are highly calcareous residual clays and have a black color where they contain much organic matter. They are among the most fertile lands in the South.

In the eastern counties of Alabama a limestone (Gladden) of the coastal-plain series, 200 feet or more thick, is extensively developed into caves and lime sinks with which are associated big springs. This formation is marked by the occurrence of strong limy black soils similar to the black prairie soils of the Selma chalk, but the topography is so broken and deeply eroded as to be in sharp contrast to the smooth-floored Selma lowland.

After the Tertiary period a blanket formation was deposited upon the Coastal Plain. This is known as the Lafayette formation and is a mantle of reddish and light-colored loams and sands with frequent beds of water-worn pebbles in the lower parts. It is from 25 to 30 feet thick on the average and formerly covered the entire Coastal Plain, resting unconformably on the older formations. In general it is sympathetic to the topography, though in many large areas it has been in great part removed by erosion. Because of its widespread development it constitutes about four-fifths of the cultivated soil of the entire Coastal Plain of Alabama, and is the chief factor in affecting the character of the soils.[1]

The Cretaceous and Tertiary beds of the Coastal Plain of Alabama have an average dip toward the Mississippi and the Gulf of Mexico of 30 to 40 feet a mile. The surface of the Coastal Plain descends in the same direction at a much less rapid rate, about a foot per mile, so that in going toward the south from the Appalachian Plateaus one passes in succession over the beveled edges of these deposits from the oldest to the newest. Each formation with few exceptions occupies the surface in a belt proportional in width to its thickness and running nearly east and west.

The sandy formations of the outer lowland are commonly characterized by short steep slopes and frequent ravines, the shales by long

[1] E. A. Smith, The Underground Water Resources of Alabama, Geol. Surv. of Alabama, 1907, p. 12.

slopes and few ravines, and the calcareous formations by smooth-contoured ill-drained valleys known as "black prairies."

The inland margin of this district is even less definite than the neighboring region to the east. In western Georgia and eastern Alabama the rivers cascade over hard rocks to form sluggish stretches in the lowlands, but commonly an arm of sedimentary material extends miles into the adjacent plateau in an ancient estuary and the river transition is seldom sharp. In central Alabama, where the Coastal Plain overlaps the southern end of the Appalachians, long fingers of lowlands stretch into the valleys between the ridges. In northwestern Alabama the line of demarcation between the Cumberland Plateau and the Coastal Plain is so poorly defined that it can not be drawn except as a zone from 10 to 20 miles wide. Still farther north the boundary between the Coastal Plain and the older formations coincides approximately with the course of the Tennessee River, though here and there occur outcrops of the older harder rock west of the river and outcrops of softer coastal-plain deposits east of the river.

The trunk streams of the Alabama section of the Coastal Plain flow across the Cretaceous and Tertiary strata, while the tributaries flow in general parallel to the strike of the outcrops. The infacing or northward-facing slopes of the hills are precipitous, while the southward-facing slopes are gentle. The tributary streams generally flow at the base of the steep infacing slopes. The major streams of the region thus run transversely to the cuesta and preserve their ancient consequent courses gained after the last emergence of the coastal lowlands from the sea. The tributary valleys on the other hand respond to a large degree to the detailed geologic structure and have excavated abnormally large subsequent valleys along belts of weaker strata. They thus join the master streams almost at right angles, and their direction of flow conforms to a notable degree to the outcrop of the strata upon which they have been developed.

The outer edge of the Coastal Plain of Alabama and Mississippi is rather sharply defined by a line of more or less dissected bluffs and hence appears as an upland when viewed from the south. The undissected interfluves are flat everywhere except for shallow depressions (of uncertain origin) sometimes containing ponds and bordered by a shrubby growth of gums. The border depressions are without standing water and are usually grassy savannas or pine meadows without undergrowth.[1]

The southern border of this section of the Coastal Plain is in part not a coastal plain but a flood plain under the dominance of the Mississippi River. It is commonly skirted by keys separated from the mainland by narrow sounds, but the keys are narrow and low and the sounds commonly broad, shallow, and irregular in outline, and pass here

[1] E. A. Smith, The Underground Water Resources of Alabama, Geol. Surv. of Alabama, 1907, pp. 250–251.

and there into grass-grown marshes, landlocked bays, and tidal flats. The sand reefs commonly lie 10, 15, or 20 miles off shore. The bays and marshes are partly bounded on the Gulf side by low sand banks, while between the bays friable sands and loams stand in vertical cliffs 5, 10, or 15 feet high. Instead of the high grounds and the low grounds of the Carolinas the entire surface of this portion of the alluvial valley of the Mississippi is low ground. The savannas are the most prominent features — broad tracts bounded by low scarps sloping steeply down and overlooking swamps, bays, and sounds. About the margins of the savannas are a certain amount of shrubbery and forests of pine or magnolia, but their interiors are often broad and imperfectly drained tracts of flat grass-land. The swamps are covered with reeds, sedges, and coarse swamp grass on their coastal sides, with live-oak groves on the coast rivers, and with canes and tangled shrubbery toward the interior.[1]

The border of the higher portion of the plain on the inner margin of the coastal swamps is often a confused belt of knobs, crests, divides, spurs, peaks, and buttresses smoothly rounded, divided by flat-bottomed valleys with innumerable ramifications. On the floors of the valleys streams wander through broad plains of sandy alluvium. The summits of the hills and knobs reach elevations of 200 feet above tide and the valley flats to half that height. The upper plain represents the older surface, the lower the younger, and it is evident that after the excavation of the valleys and gorges depression ensued and the ravines were clogged with débris washed down from the hills, the final episode being an uplift of the land sufficient to drain but not deeply to erode the savannas or grass lands of the valley flats.

The 600-mile western border of this district is a line of bluffs of complex character which marks the border of a broad terrace, the bluffs consisting of a series of truncated spurs separated by ravines and broader valleys. At Memphis the bluffs are rounded and about 100 feet in height. At Yazoo and at Vicksburg they are 200 feet above the Mississippi flood plain. The slope of the Coastal Plain of Tennessee is eastward from the Mississippi bluffs, so that the highest portion of the Coastal Plain is immediately on the bluff, a physical feature which in the early settlement of the region determined the position of the roads, for it was the highest and driest portion of the country. The outer border of the district, the edge of the Mississippi flood plain, undulates very gently in long low sweeps sculptured into a labyrinth of rounded hills

---

[1] For a description of the shell hammocks and the coastal marshes, the character of their soil, etc., along the coast of Mississippi and Alabama, see E. W. Hilgard, Agriculture and Geology of Mississippi, 1860, p. 373 ff.

and long low valleys with a local relief not exceeding 200 feet and sometimes as low as 50 feet.

One of the most important physiographic conditions of this section of the Coastal Plain is related to the abandoning of the old fields and the removal of the forest cover. It appears that in a state of nature the drainage was accomplished without unduly rapid dissection of the fertile surface soil — a yellow loam from 3 to 7 feet thick. But when the oak forests were removed in the settlement of the country and the plantations abandoned during the Civil War, the hills, no longer protected by a forest foliage, and the soil, no longer bound by the forest roots, were vigorously attacked by the streams and gullied and channeled in all directions. Year by year the formerly fertile fields were invaded by gullies of ever-increasing width, the soil of long geologic growth disappeared down the stream-ways, and the land was gashed and harmed beyond belief.

"The washing away of the surface soil . . . diminished the production of the higher lands, which were then commonly 'turned out' and left without cultivation or care of any kind. The crusted surface shed the rain water into the old furrows, and the latter were quickly deepened and widened into gullies — 'red washes.' . . ."

"As the evil progressed, large areas of uplands were denuded completely of their loam or culture stratum, leaving nothing but bare, arid sand, wholly useless for cultivation; while the valleys were little better, the native vegetation having been destroyed and only hardy weeds finding nourishment on the sandy surface.

"In this manner whole sections, and in some portions of the state [Mississippi] whole townships of the best class of uplands have been transformed into sandy wastes, hardly reclaimable by any ordinary means, and wholly changing the industrial conditions of entire counties, whose county seats even in some instances had to be changed, the old town and site having, by the same destructive agencies, literally 'gone down hill.' " [1]

Specific names have been given to the erosional features of this district: a "break" is the head of a small retrogressive ravine; a "gulf" is a large break with precipitous walls of great depth and breadth, commonly being one hundred or one hundred and fifty feet deep; a "gut" is merely a road-cut deepened by storm wash and the effects of passing travel.

MISSISSIPPI VALLEY SECTION

The sixth division of the coastal lowland is the great flood plain, or delta and flood plain combined, of the lower Mississippi. It is bounded on the east by a continuous line of bluffs and on the west by a conspicuous ridge known as Crowley's Ridge (north) and by an alternating line of irregular bluffs and valleys (south). This alluvial district is one of the most extensive of the really low areas of the continent, lying practically at base level. In the southern part of the district the surface approximates tide level and indeed the outer border consists of permanent tidal marshes. The surface is very ill-drained, and bayous, lakes, and abandoned channels constitute an irregular maze of water upon a

[1] E. W. Hilgard, Soils, 1906, pp. 218-219.

plain with scarcely perceptible slope. Between the waterways are ridges of slightly higher land, some of which are the natural levees of channels long since abandoned. In the western part of the district the interstream areas lie so high that they are invaded only by the highest floods, but the surface material is fine and compact and the surface itself ill-drained; consequently the trees are either drowned by the floods or withered by the sun in the droughts and the surface is without a forest cover. It supports a patchy growth of coarse grass, the patches being known as the "black prairies" of southern Arkansas and Louisiana.

In the northern portion of the lower alluvial valley of the Mississippi extensive land tracts were converted into lakes, flowing rivers transformed into stagnant bayous with uplifted areas between, and some

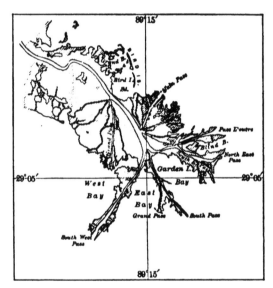

Fig. 209. — Finger-like extensions of the Mississippi delta. The entire area shown as land is swamp-land.

stream courses actually diverted during the series of earthquakes between 1811–1813. This region includes the "sunk country" of Missouri and Arkansas and the "Reelfoot Lake district" of Kentucky and Tennessee, and forms the uplifted land of Lake County, Tennessee, one of the few sections of the Mississippi flood plain that escape the inundations of the highest floods.

Between Lake Borgne and Mobile Bay the Gulf is advancing upon the land so rapidly that the coastal rivers are nearly submerged and occur as narrow mud banks like the Chandeleur Islands, or as completely submerged bars and shoals that parallel the coast. A further effect of submergence is shown in the excessive breadth of the lagoons.

On either side of the immediate delta of the Mississippi the alluvium deposited by the great river has been modified by waves and currents into bars and reefs and the shore has a smoothly trimmed effect in contrast to the finger-like extensions of the actively growing portion of the delta. In the latter case the rate of deposition exceeds the rate of current wear and the land is being built seaward.[1]

Meander development and sidewise bodily movement of the river as a whole have opened up a river lowland so flat as to make the coastal plains appear as uplands when viewed from the river. This lowland, the lower alluvial valley of the Mississippi, is from 30 to 60 miles wide and about 600 miles long. Over its flat surface the great Mississippi meanders in an extremely indirect course. Receiving the contributions of a vast network of tributaries themselves major streams, it is little wonder that the river is subject to numerous floods transmitted from the tributaries. Formerly they inundated a vast expanse of lowlands and were of yearly occurrence; now a system of restraining dikes or levees and expensive revetments partially restrain the great river, and floods are in a measure under control. At present the levee system comprises about 1500 miles of structure and is about 71% completed. The number of square miles of overflowed land in 1903 was half the mileage for 1897. In 1882 there were 284 crevasses recorded; in 1903 there were only 9 of importance.[2]

The extent of alluvial bottom land that escapes inundation is extremely limited. The higher tracts are confined chiefly to the river border, hence this is where the principal plantations are located. From the low natural levees the land slopes gradually away on either side to swampy back country covered with heavy forests of cypress and gum. The first problem in the utilization of these back swamps is drainage, and until the present canal system is perfected and extended but little development of the forest can be expected. The possibilities are suggested by the large shipments of logs, staves, headings, etc., from the Yazoo basin, where there was practically no lumber industry

[1] The Mississippi is estimated to carry to the Gulf of Mexico enough land waste to cover a square mile 268 feet deep, an amount of material that would require a train of 44 loaded cars arriving at the Gulf every minute, the specific gravity of the sediment being taken at 2.5 and the capacity of a railroad car being 50,000 pounds. (J. E. Carman, The Mississippi Valley between Savanna and Davenport, Bull. Ill. State Geol. Surv. No. 13, 1909, p. 23.)

[2] R. M. Brown, The Protection of the Alluvial Basin of the Mississippi, Pop. Sci. Mo., Sept., 1906, pp. 248-256. See also W. S. Tower, The Mississippi River Problem, Bull. Geog. Soc. Phil., vol. 6, 1908, pp. 83-100.

Fig. 210.—The lower alluvial valley of the Mississippi.

until the present levee system was constructed, the floods reduced, and swamp drainage begun.

At the northern end of the lower alluvial valley of the Mississippi are two main belts of lowland separated by a long ridge of varying width. The two lowlands are joined across the ridge by a broad belt of lowland and by several narrow stream valleys. The lowlands are called the Advance (west) and the Cairo (east) lowlands, Fig. 210. The Advance lowland is essentially a unit, while the Cairo lowland is divided into two subordinate belts by a long ridge known as the Sikeston Ridge. The Advance lowland is from 2 to 8 miles wide and is bordered in many places by relatively steep bluffs. The surface deposits of the lowland are sand and clay. A few long, low, flat-topped, sandy "islands" lie upon it at various places and furnish the only bits of land favorable for agriculture.

The Cairo lowland on the east is nearly level and is bordered by prominent bluffs. The surface deposits consist of sand and silt, and on both this and the Advance lowland are numerous sloughs containing varying amounts of water. The small streams draining both lowlands have banks so low that they overflow during ordinary floods and the land is often submerged. The ridge between the two main lowland belts varies in height, form, and material, and is broken into a number of isolated sections. The main ridge extends southwestward, and to it the general name of Crowley's Ridge has been given. At its broadest part Crowley's Ridge is about 20 miles wide. Its upper surface is generally in the form of a rolling plain sloping gently westward. All the creeks flowing upon it have broad flood plains, with high land only in the form of narrow ridges on the main and secondary divides. On the main divide the valleys are deep and narrow and the surface maturely dissected. The eastern side of Crowley's Ridge is a definite and steep bluff throughout; the western edge is not so definite, only a part of its course being marked by bluffs.

The western lowland (Advance) was once occupied by the Mississippi River. The eastern lowland (Cairo), once occupied by the Ohio River, is now occupied by the Mississippi. The change of course in the Mississippi was brought about through stream capture by tributaries of the Ohio working westward into Crowley's Ridge. A second capture of the Mississippi took place at a later time and brought about a further adjustment. The length of the valley abandoned by the first change of the Mississippi was more than 200 miles; the second change effected an abandonment of about 50 miles. The second change took place so recently that the river has not yet had time to clear its channel of rocks, much less to form a flood plain. The first change in the Mississippi brought its point of junction with the Ohio to a short distance south of New Madrid, Missouri; the second change brought it into its present relations with that river.[1]

[1] C. F. Marbut, The Evolution of the Northern Part of the Lowlands of Southeastern Missouri, Univ. Mo. Studies, vol. 1, 1902.

LOUISIANA-TEXAS SECTION

The next section of the Coastal Plain extends from the Mississippi flood plain to the Rio Grande; the coastal margin of this section consists of wave-built ridges or reefs of exceptional continuity, Padre Island, south of Corpus Christi Pass, being 100 miles long. The lagoons back of these reefs are equally continuous; light-draft vessels may sail from the mouth of the Rio Grande through the Laguna de la Madre and through shorter sounds to Matagorda, 250 miles away. The depression of the coast is so rapid that the sand reefs are drowned and the lagoons are increasing in breadth. The combination of lagoons, fringing reefs, drowned river courses, complex estuaries, and clean sea cliffs is an expression of a sea advancing on a low-lying land, and the width of the belt in which these features are developed is in a general way proportional to the rapidity of depression.

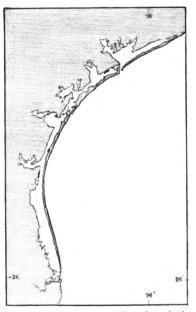

The border of the plain is made up of alternating savannas and swamps or shoal bays. The low grounds are abandoned to reeds and sedges. The savannas are so low as to be clothed only with coarse grass and dotted with scrub pine or palmetto. Broad, low, natural levees like those of the Mississippi occur throughout the savannas, and these are commonly wooded, while the interstream tracts are prairie lands. Narrow belts of forest thus occur along the waterways except where the surface has been cleared

Fig. 211. — Coastal features of Texas, long, simple sand reefs enclosing narrow lagoons.

for agriculture. Along Red River are well-wooded tracts with oak and hickory on the uplands, poplar and liquid amber on the lowlands, cypress and tupelo in the swamps.

In the coastal plain of Louisiana the rivers are steep and sluggish, such as Vermilion Bayou, Calcasieu River, etc. Each of these streams is narrow, deep, and clear, has scarcely any current, expands into a broad shallow lake, and enters the Gulf through a shallow bay. All of them have features characteristic of drowned streams. In southeastern Texas the streams of the Coastal Plain are similar to those of Louisiana; but west of the Nueces the coast drainage fails almost absolutely, the whole stretch of the coast to the Rio Grande containing only two small creeks. Though the greater part of the Louisiana-Texas section of the Coastal Plain is crossed by large rivers, a considerable part of the surface is poorly drained. Water stands in a multitude of small lakes or ponds throughout the year and there are large tracts covered with water during the wet season.

The innermost belt of country belonging to the Coastal Plain rises rather rapidly from the adjacent seaward belt and has a more broken surface with numerous, small, rounded hills. The general elevation of this belt does not exceed 175 or 200 feet above sea level. The surface is in general timbered.[1]

SOILS AND VEGETATION

In Texas the Coastal Plain consists of a western sandy subdivision and an eastern clayey subdivision; the line of separation is the Guadalupe River in Victoria County. Soil distinctions while of great importance to vegetation in the interior of Texas are of little importance on the low Coastal Plain, where the chief consideration in tree growth is the water supply. The outer margin of the eastern and clayey subdivision is swampy and flat and but little above sea level. Patches of sandy land occur on the borders of the clay area. Forests of pine, oak, and magnolia fringe the northern border of the coastal plain on the higher grounds, various species of gum occur on the benches, and heavy forests of black and red cypress are found on the low river flood plains. The greater part of the Texas section of the Coastal Plain consists of groups of prairies separated by forest tracts.

One of the most interesting forest trees of the low coastal margin is the loblolly pine, which grows on slightly elevated mounds of the Texas lowlands where it forms forest islands. The surrounding prairie is covered with water several months each year and is wet nearly the entire year, so that loblolly pine develops almost undisturbed by fires. The gradual filling in of the prairie by the loblolly groves causes the land to become drier, and strips of young growth bridge the space between groves and finally develop into large bodies of forest. The seeds of the loblolly are scattered partly by the wind in a southeasterly direction if they ripen at the end of September or early October when northwest winds prevail, and partly by the southeastward drainage toward the Gulf during periods of high water. The drying up of the prairie and

[1] Hayes and Kennedy, Oil Fields of the Texas-Louisiana Gulf Coastal Plain, Bull. U. S. Geol. Surv. No. 212, 1903, pp. 10 ff.

the reclamation of the land is a natural process common to the whole coast of Texas and Louisiana. The swamp vegetation adds by its decay to the surface material and gradually the surface is built upward to the point where shrubs and trees come in. The process is rapid enough to cause marked changes in the distribution of timber in 40 or 50 years, so that the marshes within the loblolly pine forest of eastern Texas, which a few years ago were impassable, are now accessible to man and to cattle.[1]

Southward from San Diego the Coastal Plain is composed of a belt of brown sand probably 25 miles in width. It is a rolling country more or less covered with mesquite and chaparral. Still farther south is a gray sand belt having a width of 50 or 60 miles and, except for a few live oaks, practically without trees. It has been called the Great Texas Desert, and across the face of it stretch two belts of moving sand hills in an east-west direction. Each belt consists of a double row of dunes from a half mile to a mile apart, moving westward under the prevailing easterly winds. Some of them (northern Star County) are from 90 to 100 feet high, but the size of the dunes decreases eastward and near the coast they appear only as white spots lying but slightly above the plain. Some of them are almost circular with a central depression, others are oval, and still others are in the form of great crescents, the typical symmetrical dune shape. The dunes are composed of extremely fine sand of snow-white color, and the lightest wind sets the fine grains in motion. In several instances tall live oaks have been buried so deeply that only the dead tops of the highest branches indicate the fate of the groves invaded by the dunes.[2]

From one to two hundred miles west of the coastal tract the forests disappear or occur chiefly in the form of scraggly growths of blackjack and Chickasaw plum. Still farther west the mesquite is found in low orchard-like groves on the interfluves, with hackberry and acacia along the streams. Eventually these forms give place to the sage and the cactus of the deserts, except where the fertilizing streams have led to reclamation or support a more prosperous native growth.

PHYSIOGRAPHIC DEVELOPMENT

The structure and physiographic history of the Coastal Plain of Louisiana and Arkansas bear such an intimate relation to the major topographic features that a word in regard to them is necessary at this

[1] R. Zon, Loblolly Pine in Eastern Texas, Bull. Forest Service, U. S. Dept. Agri., No. 64, 1905, pp. 9-10.
[2] Idem, p. 16.

place. In the Arkansas-Louisiana district the relatively soft formations of Cretaceous and Tertiary age which form the Coastal Plain overlie a peneplain developed upon older and greatly deformed strata.

In Miocene and early Pliocene times erosion was active at a level distinctly lower than that of the old peneplain and resulted ultimately in the formation (late Tertiary) of local base-leveled surfaces, essentially coincident with the Coastal Plain and continuous with an uneroded outer area lifted so slightly above sea level as to suffer no important modification during this erosion period. This Tertiary base-leveling while extensive was not complete, and many remnants of the older and higher surface still project above the common level.

After the partial development of a Tertiary peneplain there came a depression of the entire region of sufficient amount to allow the formation partly by river aggradation, partly by marine deposition, of a great blanket of silts, sands, and gravels (the Lafayette formation), which still occurs widely distributed throughout the region, its materials forming the surface soil to a large extent.[1]

The process of excavation of the material constituting the Coastal Plain has gone on with but one interruption since the deposition of

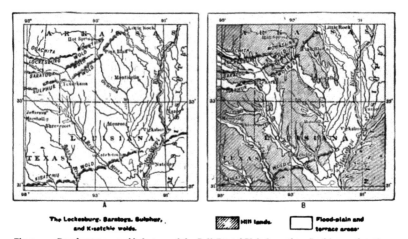

The Lockesburg, Saratoga, Sulphur, and Kisatchie wolds.   ▨ Hill lands.   ☐ Flood-plain and terrace areas·

Fig. 212. — Prominent topographic features of the Gulf Coastal Plain in northern Louisiana and southern Arkansas. (Veatch, U. S. Geol. Surv.)

the Lafayette formation. Through either climatic change or crustal deformation or both the streams of the region after cutting out valleys began to aggrade their valley floors, but after the partial filling of the valleys reëxcavation was begun. The result is that the streams are

[1] A. C. Veatch, Geology and Underground Water Resources of Northern Louisiana and Southern Arkansas, Prof. Paper U. S. Geol. Surv. No. 46, 1906, p. 46. .

generally intrenched below the level of the upper surface of the alluvium which now occurs in terrace form fringing the borders of the valleys and constituting an intermediate level of bench land of considerable extent between the general surface of the Coastal Plain and the flood plains of the streams.

Besides the terrace and flood-plain areas are extensive areas of upland that are interesting chiefly because of the development of alternating belts of higher and lower lands in the form of cuestas and inner lowlands. The best development of these features is in Arkansas and Louisiana where there are four more or less persistent cuestas that follow the general strike of the formations. The three northerly ones are the Lockesburg, Saratoga, and Sulphur cuestas; the southernmost is distant from the other three and is called the Kisatchie cuesta. It is formed on the outcrop of the hard sandstone formations known as the Catahoula (Oligocene), which have been brought to the surface and exposed by a fault, so that the cuesta is not a typical one but should be regarded rather as a modified fault scarp very similar, however, in its topography to a cuesta formed by ordinary differential erosion, Fig. 213. The Sulphur cuesta is formed on the harder members of the lower Eocene, and the Saratoga and Lockesburg cuestas on the Cretaceous.

SPECIAL FEATURES

MOUNDS

Two important local features merit description at this point, for they constitute notable departures from the normal types of topography and drainage with which we have so far had chiefly to deal. These are (1) the two types of mounds which exist in the district and (2) the lakes at the lower ends of the tributaries of the Red

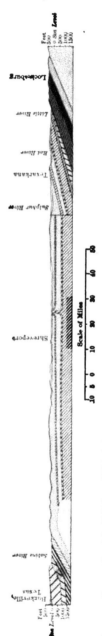

Fig. 213. — Cross section of the Gulf Coastal Plain in Louisiana and southern Arkansas. The formations range in age from Paleozoic and Lower Cretaceous on the right to Tertiary (Oligocene) on the left. Note the structural variations and the topographic complexities to which they give rise. (Veatch, U. S. Geol. Surv.)

River. The first type of mound occurs in valleys where erosion has revealed the presence of domes with steep marginal dips attributed to igneous intrusions which did not reach sufficiently far toward the surface to be exposed by erosion. Only the top of the dome has been removed, and the topographic expression is commonly not a mound but a depression rimmed about by harder limestone, from beneath which the softer formations have been removed by erosion.

The second type of mound is not only a structural uplift but also a topographic elevation and has an extremely wide distribution, being well developed on the prairies and pine flats along the coast of Louisiana and Texas, where are found the "pimple prairies" popularly but erroneously associated with the oil deposits. They occur irregularly throughout the Coastal Plain and are best developed along the river terraces, though they are frequently found on the upland as well. No satisfactory theory has yet been formulated to account for them. On account of the wide distribution and unsettled character of the problem they represent, a partial bibliography of references is given at this point for the convenience of the reader. The different theories are discussed at some length by Veatch,[1] from whose paper the following list of references is taken.[2]

RED RIVER RAFTS

Peculiar interest attaches to the great rafts of the Red River Valley on account of the unique combination of conditions which they represent and the important changes they have effected in the hydrography of the Red River and its tributaries. The name *raft* is applied to the natural accumulations of timber along the river caused by the caving of the banks on the outside of the river bends. The main raft of the river began or at least was first known at Natchitoches, a town located below the raft and at the head of navigation. This was known as the Great Raft, and it grew steadily upstream with the constant addition of

---

[1] A. C. Veatch, Geology and Underground Water Resources of Northern Louisiana and Southern Arkansas, Prof. Paper U. S. Geol. Surv. No. 46, 1906.

[2] John C. Branner, Science, n. s., vol. 21, 1905, pp. 514–515; E. W. Hillgard, idem, pp. 551–552; W. J. Spillman, idem, p. 632; A. H. Purdue, idem, pp. 823–824; and C. V. Piper, idem, pp. 824–825. Branner and Purdue suggest that these mounds may represent immense concretionary formations. Spillman refers certain mounds in southwest Missouri to unequal weathering of limestone containing large chert masses. Branner gives many references to the mounds of the Pacific coast, for which he states the following theories have been advanced: (1) surface erosion, (2) glacial origin, (3) æolian origin, (4) human origin, (5) burrowing animals, including ants, and (6) fish nests exposed by elevation. D. I. Bushnell, Jr., Science, n. s., vol. 22, 1905, pp. 712–714, has suggested the human origin theory, and this phase of the matter has been discussed by Veatch, Science, n. s., vol. 23, 1906, pp. 34–36.

Fig. 214. — One of the timber jams composing the great Red River raft. In such a jam silt accumulates very rapidly and effectually fills the channel. (Veatch, U. S. Geol. Surv.)

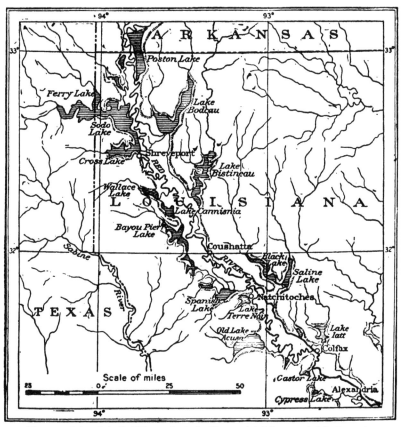

Fig. 215. — Lakes of the Red River valley in Louisiana at their fullest recorded development. (Veatch, U. S. Geol. Surv.)

Fig. 216. — Timber deadened in temporary raft lake which was drained by the removal of the raft. (Veatch, U. S. Geol. Surv.)

Fig. 217. — One of the Red River rafts after partial recutting of filled channel, 1873. (Veatch, U. S. Geol. Surv.)

material in that direction until in the latter part of the fifteenth century it had reached a point near Alexandria. The effect of the natural dam which the raft created was to raise the level of the river on the upstream side of it and cause a ponding not only of the main river but

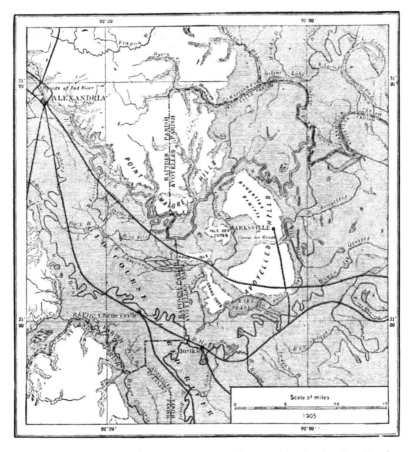

Fig. 218. — Showing diversion of Red River below Alexandria, La., and location of rapids. Map also shows typical drainage features in the Red River and Mississippi River flood plains. (Veatch, U. S. Geol. Surv.)

also of the tributaries within the reach of the ponded portion of the main stream. In some cases the rise of water in the main stream was sufficient to cause it to discharge about the natural dam and through the timbered bottom lands on one side or the other. Driftwood would be quickly accumulated about the new point of discharge and again the channel would be shifted. At the end of about 200 years (estimated) the lower part of the raft began to decay and the front of it to move upstream as a great irregular accumulation of log jams and open water about 160 miles in length. Its average rate of advance was about four-fifths of a mile a year during the period between 1820 and 1872, though the rate was intermittent and to a large extent dependent upon the discharge of the stream from year to year.

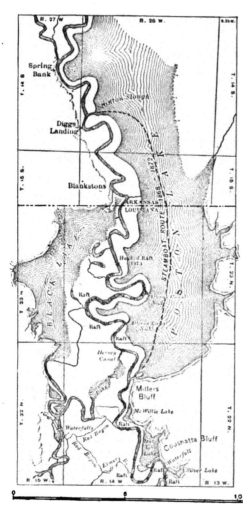

As the head of the raft moved upstream it blocked all the tributary streams in succession and caused the formation of lakes at the points of junction with the main stream. In similar fashion the tributaries that were freed by the decay and retreat of the raft material along the front of the raft again discharged in the normal way and proceeded to dissect the deposits that had been accumulated on the floor of the temporary lake at their mouths.

Fig. 219. — Growth and drainage of the raft lakes at the Arkansas-Louisiana state line. (Veatch, U. S. Geol. Surv.)

Since the removal of the rafts from the Red River the water has gradually resumed its old level. From 1873 to 1892 the river had lowered its bed about 15 feet at a point above Shreveport; the lake areas are rapidly draining and the extensive lake system along the course of the Red River shown on most maps is practically nonexistent. The continuance of some of the lakes and the occurrence of rapids and small falls in some of the tributaries are due to the drainage deflections that resulted from the silting up of the floors of the temporary lakes and the assumption of a new channel by the recreated stream. Superposition was inevitable under these circumstances, the stream in certain instances flowing over projecting and now covered Tertiary spurs that have been revealed by erosion. The most notable instance of such diversion is in the case of the Red River itself, which has rapids immediately above Alexandria. The main features are shown on Fig. 218.

SOILS

Although the soils of the various districts of the Coastal Plain have been incidentally mentioned in the discussion of topographic and drainage features, a connected description of their qualities is essential in understanding their geographic distribution and origin. The soils of the Atlantic and Gulf Coastal Plain are for the most part composed of sands and light sandy loams, with occasional deposits of silts and heavy clays. The heavy clays are found principally near the inner margin of the Coastal Plain. The silts, silty clays, and black calcareous soils upon which the rice and sugar-cane industries of southern Louisiana and Texas are developed have no equivalents in the Atlantic division. Differences in the method of deposition, subsequent erosion, and drainage conditions are responsible for a great diversity of soil types with complicated relationships.

The following are the most important series that have so far been recognized: (1) Light-colored sandy soils underlain by yellow or orange sand or sandy clay subsoils. Where the drainage is insufficient, the subsoil is often mottled. With few exceptions these are special-purpose rather than general farming soils, and constitute the most important truck soils of the coastal plain. (2) Dark-gray to black surface soils, underlain by yellow, gray, or mottled yellow and gray subsoils. The dark color of the soils is due to an accumulation of organic matter during an earlier or existing swampy condition. This series is intermediate between the light-colored soils on the one hand and the peat and swamp areas on the other, and occupies depressed areas, or areas so flat that the water table is at or near the surface, except where the country is artificially drained. In this series the fine sandy-loam type supports a heavy growth of cypress, gum, magnolia, and other water-loving trees and undergrowth. It is characterized by level or slightly depressed surface features. Lack of drainage is responsible for the existence and peculiar characteristics of the type. In most cases artificial drainage is impracticable, owing to the low gradient.

(3) A third series is derived largely, but not entirely, from the Lafayette mantle of gravels, sands, and sandy clays. The surface soils are usually gray to brown in color and are invariably underlain at a depth of 3 feet or less by a red or yellowish-red sandy clay. The prevailing red color of the subsoil is the characteristic feature. (4) A fourth series includes the barrier islands or bars, shore-line deposits, and low-lying marshes of the immediate coast line. The barrier bars consist of sand accumulated by wave action and further modified by winds. The soils of the marshes, consisting of sandy loams, loams and clays, have been built up by the deposition of silt and clay carried in by streams, by wind-blown sand from the adjoining sand areas, and by the decay of coarse salt grasses and other native vegetation. The agricultural value of these lands is very low, depending mainly upon pasturage and the coarse hay, and they are a distinct menace to health, as they form the breeding places of disease-carrying insects. Efforts to drain and reclaim these marshes have been attended with some success. The possibilities of successful reclamation depend upon the keeping out of the tides and the subsequent efficient drainage of the land.

(5) The soils of the black, calcareous prairie regions of Alabama, Mississippi, and Texas are characterized by a large percentage of lime, especially in the subsoil, which in some of the

types consists of white, chalky limestone. They have been derived from the weathering of calcareous clays, chalk beds, and "rotten" limestones (Cretaceous). In some localities remnants of later sandy and gravelly deposits have been mingled with the calcareous material, giving rise to gravelly and loamy members. The soils are very productive and are at present devoted chiefly to the growing of cotton and corn.

(6) The next series includes dark-gray soils found upon gentle slopes or undulations adjacent to streams and upon level or depressed areas in the uplands. Their formation is due largely to the peculiar topographic conditions resulting from the sinking of the limestones which underlie, in some of the areas, the materials from which other soils have been derived. They may be considered as colluvial soils formed by the creeping or washing of material from higher-lying areas. The sandy type has a considerable admixture of organic matter and lies on gentle slopes or undulations adjacent to streams. It is mainly hammock land, supporting a growth of hardwood forest, and is very productive.

(7) A series consisting of gray and brown surface soils underlain by heavy, plastic, red, mottled subsoils. Where the basal clays are exposed by erosion they show brilliant colorings, often arranged in large patches of alternating liver-color, red, and white. These clays are remarkably plastic and constitute the oldest marine deposits along the inland margin of the Coastal Plain. The soils are usually of low crop-producing capacity and are covered chiefly with pitch pine, scrub oak, and other trees of little commercial importance.[1]

Tree Growth of the Coastal Plain

The special vegetal features of the various sections of the Coastal Plain have been described in the foregoing pages. It remains to note certain general features more or less common to the whole province. Two points are of chief interest in this connection: (1) the effects of water supply and (2) the effects of texture and chemical properties on the native vegetation. Speaking in general terms one may say that the Coastal Plain exhibits (1) a number of inner belts more or less clayey in character and heavy, (2) a broad expanse of sandy land forming the long outer slope of the plain and almost all of the so-called upland, (3) river bottom lands fringed in many places with terraces or higher lands called "second bottoms" or hammocks, (4) a border of marshy land consisting of fresh-water marshes on the landward side and of salt-water marshes on the seaward side, and (5) a line of long, narrow reefs but little above high tide except where blown by the wind into higher sand dunes.

The reefs are for the most part sandy, though coral reefs fringe a part of the coast of Florida and some of the reefs of the Gulf region are covered with shells accumulated by the Indians. The sand reefs generally bear a growth of pine which may be of good quality as originally on the reefs of North Carolina, or it may be stunted and mixed with low, gnarled cedar, etc., as on many of the reefs of New Jersey and Texas. The shell hammocks of Alabama and Mississippi are not restricted to the reefs but are found also on inlets and bayous easily

[1] Soil Survey Field Book, U. S. Bur. of Soils, 1906.

accessible to the water. Pitch pine, live oak, red cedar, sweet gum, and prickly ash are the most common shell-hammock trees.

On the marshes and wet pine barrens the deciduous cypress (*Taxodium imbricarium*) and the common swamp cypress are the ordinary growths along a large part of the coast line. In addition many of the smaller marshes have a growth of stunted pine, maple, and black gum. The bottom lands along the rivers, the "first bottoms," generally have a stiff heavy soil with an excess of water and are subject to overflow. Their timber is generally luxuriant and consists of chestnut, white oak, sweet gum, black gum, magnolia, bottom white pine, cypress, black walnut, tulip, hickory, and ash. Among these the black and the sweet gums are most numerous and characteristic, hence the name "gum swamps" commonly applied to the bottom lands. The second bottoms or hammock lands are slightly higher than the first bottoms, are never overflowed, have a silty not a clayey soil, and have a poorer tree growth, among which the most common types are white oak, water oak, bottom white pine, magnolia, ironwood, post and black-jack oaks, willow oaks, etc., all more or less stunted and scattered in their growth.

The upland trees are economically of greatest importance and are at the same time most interesting from the ecologic standpoint. The great sandy expanses of the outer slope of the Coastal Plain are covered mainly with longleaf pine. The seaward margin of the sandy belt is, however, level prairie in most cases, on which the trees are disposed in clumps or groves. In Texas the islands of higher lands are covered with loblolly pine. In Louisiana groves of honey locust, red haw, and live oak dot the outer prairie region. Inland from the great longleaf pine belt the calcareous layers of the Coastal Plain outcrop and are covered with a distinctive and for the most part a lime-loving vegetation. Oak and shortleaf pine are the most important types; mixed with them are red cedar, red haw, crab-apple, and honey locust. It is noteworthy that the heaviest clay soils on the inner margin of the Coastal Plain of Texas, Louisiana, Mississippi, and Alabama are marked by large numbers of prairie tracts alternating with groves of stunted trees, mostly red cedar, Chickasaw plum, and scrubby post and black-jack oaks.

The most unproductive soils occur on the so-called ridge lands toward the inner margin of the Coastal Plain where some of the ridges are developed on the outcrop of sandy formations. Scarlet and post oak, black-jack oak, and stunted pines are the principal trees. They are characteristic growths on all the more infertile phases of the coastal-plain soils. The calcareous soils, by contrast, are notably fertile and

have a good growth of hickory, ash, sweet gum, and honey locust.   So
faithfully do the distributions of these types follow the outcrops of the
respective formations that a vegetation map on the one hand, and a
soil map or a geologic map on the other, show striking correspondences
in the positions of the boundary lines.[1]

[1] For a description of the tree types of the Coastal Plain in Texas see R. M. Harper, Contr.
Dept. Bot., Colum. Univ., Nos. 192, 215, 216, 1902–1905; for the timber belts of the Coastal
Plain in Texas see W. L. Bray, Forest Resources of Texas, Bull. U. S. Dept. Agri., Bur. For.
No. 47, 1904; for the trees of Mississippi see E. W. Hilgard, Geology and Agriculture of Miss.
1860, and Soils, 1906, Chaps. 24 and 25; for Alabama see Charles Mohr, Plant Life of Ala-
bama, Ala. Geol. Surv., 1901.

CHAPTER XXVI

PENINSULA OF FLORIDA

GENERAL GEOGRAPHY

THE peninsula of Florida is remarkable for its projection southeastward into the Atlantic Ocean for 350 miles, its smooth, well-developed eastern shore line, its great keys and swamps, and its slight relief. It ranges in altitude from sea level to 200 feet above the sea at various points on the broad flat-topped tract which forms the center of the peninsula and to about 300 feet in the northwestern counties of Florida. A depression of 50 feet would cover all of southern Florida except the tops of sand hills and ridges, while an elevation of 50 feet would extend the shore line westward 20 miles from Cape Romano and make dry land of Biscayne Bay and Bay of Florida.

The northern and western parts of Florida consist of a narrow, deeply eroded limestone (Vicksburg) upland which descends southward rather abruptly to a low coastal region and northward by more gentle descents to the adjacent Coastal Plain. The rivers of the northern region are consequent upon the initial slopes of the land as it emerged from beneath the sea except where they have removed the thin mantle of surface sand and superimposed themselves upon the older strata beneath. The southern and central portions of Florida form a great lake and swamp district, the lakes occupying sinks or depressions in the underlying limestones or shallow and broad depressions in the surface of the deposits overlying the limestones. Extreme irregularity of outline characterizes the shores of the lakes and extreme irregularity of direction the courses of the rivers. Many streams of the peninsula have their sources in springs that occupy underground courses in the limestone. These bring to the surface large quantities of mineral matter, and it has been estimated by Sellards[1] that the rate of solution is sufficient to remove about 400 tons per square mile per year, an amount which would lower the surface of the limestone about 1 foot in 5000 to 6000 years. The underground passages are sometimes several hundred feet in diameter and several miles in length. The level char-

[1] A Preliminary Report on the Underground Water Supply of Central Florida, Bull. Florida State Geol. Surv., No. 1, 1908, p. 16.

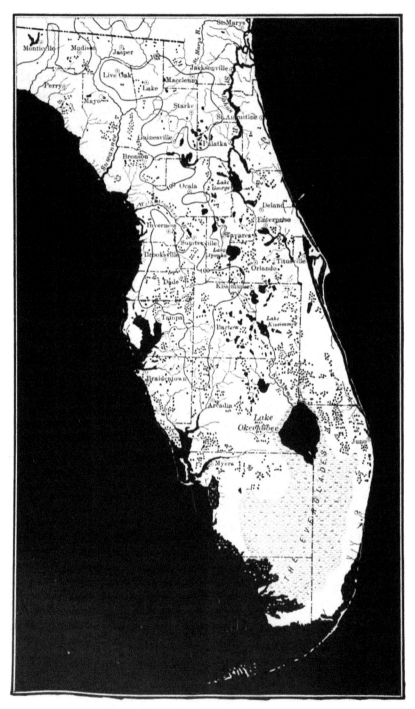

Fig. 220. — Principal lakes and coastal features of Florida.

acter of the surface and the porous soil which mantles the bedrock afford an excellent opportunity for the formation of caverns, since they result in a large absorption of the heavy rainfall. The sink holes are in process of formation to-day and instances are known where sinks have been formed by the collapse of cavern roofs in different parts of the lake region. The same process has resulted in the formation of many natural bridges, as those of northern Walton County, at St. Marks River, and Santa Fé River.

GEOLOGIC STRUCTURE

In respect of structure Florida is an elevated crust block modified by a number of minor folds.[1] The present peninsula is regarded as resting on a much more extensive foundation of Eocene limestone, forming a plateau which formerly extended from the southeastern margin of the continent to Cuba and the Bahamas, and possibly to Yucatan. The isolation of the peninsula may possibly be due to faulting, and in part to current scour. In a number of places are gentle folds whose axes are generally parallel with the trend of the peninsula; these succeed one another in series between Lake Okechobee and Gulf of Mexico.[2] One such fold is near the Atlantic coast, another near the Gulf coast, and a third in the vicinity of Brooksville and Plant City. The eastern ridge includes the well-known Trail ridge and forms the eastern boundary of the central lake basin. The western ridge forms the western boundary of this basin and passes through Lakeland.[3]

PHYSIOGRAPHIC DEVELOPMENT

Recent stratigraphic studies have determined the fact that the Floridian region was outlined in somewhat its present form in pre-Oligocene, probably Eocene time, but that it then existed in the form of a shallow submarine platform swept by ocean currents and blanketed by both organic and terrigenous deposits.[4] During the time between this early period and the final emergence of the crust-block as a peninsula the region was subjected to many changes of level — a series of four emergences and a corresponding number of submergences. During the progress of these changes the surface of the platform was never carried far below and never far above sea level. The submergences were usually about 100 feet and never more than 200 feet, while the latter figure expresses the

[1] W. H. Dall, Neocene of North America, Bull. U. S. Geol. Surv. No. 84, 1892, pp. 85-87.
[2] Idem, p. 88.
[3] Matson and Clapp, 2d Ann. Rept. Fla. Geol. Surv., 1909, p. 48.
[4] T. W. Vaughan, Sketch of the Geologic History of the Floridian Plateau, Science, n. s., vol. 32, 1910, pp. 24-27.

maximum elevation above sea level, a maximum that was attained during the Pliocene emergence. It was the fourth Pliocene emergence that gave the peninsula the outline that it has to-day. In the time since then the living coral reefs have developed, the Everglades have been formed, and the shores given their detailed characteristics. The net result of all the changes of level has been to leave the eastern side of the peninsula higher' than the western and to bring into existence a number of small low folds whose axes run north-south in sympathy with the main axis of the peninsula. The Kissimmee River for example occupies an anticline flanked on both sides by low anticlines.

TOPOGRAPHY AND DRAINAGE

Although the mainland of Florida has slight relief it exhibits a considerable variety of topographic types, and since even small differences of elevation have a marked effect on the vegetation, it is possible roughly to divide the mainland of the peninsula into districts whose topographic features are intimately related to the vegetal growth. On this basis Florida may be divided into pineland and swamp. The pineland includes "hammocks" — isolated elevated patches supporting hard- wood trees of several genera — and many prairie or grassy tracts. The swamps include the marshy borders of the inland lakes and the coastal swamps with their sedges and black or red mangroves. It must be re- membered, however, that this line of demarcation between swamp and pineland is extremely irregular and, on account of the low relief, varies with the seasons to a certain extent.

The total area covered by pine forests is probably in round numbers 300 square miles; on the east coast they extend in a narrow belt be- tween the Everglades and the coastal swamp from northern Palm Beach County to near Homestead. This belt is about 20 miles wide at the north, 6 miles wide near Jupiter Inlet, and from 2 to 8 miles wide towards the south. In northern Monroe County the pines grow in disconnected areas alternating with stretches of cypress.[1] The pine- lands may be divided according to the relief into dunes, rolling sand plains, flat land, and rock ridges.

DUNES

The dunes of southern Florida lie near the coast and occur as a dis- continuous series of irregular mounds and ridges separated by inter- vals of flat or gently rolling country or by stretches of shallow water.

---

[1] The best brief description of the topographic features of southern Florida and the one on which the following paragraphs are chiefly based is by S. Sanford, 2d Ann. Rept. Fla. Geol. Surv., pp. 177–231, where the results of many other investigators are assembled.

They seldom reach more than a few miles inland and but rarely face the ocean. An important group of dunes are those that occur near Jupiter Inlet at West Palm Beach, and on the east side of Lake Osborn. On the west coast of Florida there are few dunes south of the Caloosahatchee River. The dunes are not in movement to-day, and there is no leeward march which overwhelms trees and threatens dwellings as at Cape Henry, Virginia, and on Hatteras and Currituck Banks in North Carolina. When cleared of pine timber and palmetto scrub they may be utilized in the growing of tropical fruits.

ROLLING SAND PLAINS

The rolling sand plains include sandy stretches of the mainland with broad swells and low ridges, the swells being occupied by shallow lakes or lagoons and wet prairies or cypress swamps. On the east these sand plains form a belt with a maximum width of 6 miles and extend from the north side of Palm Beach County nearly to the Miami River and merge inland into the monotonous level of the flat lands and prairies bordering the Everglades. On the seaward side they are bounded by swamps; on the east coast the higher ground and the ridges are frequently covered with a straggling growth of spruce pine. In the hollows are many fresh-water lakes, some of which are several miles long. Most of them are less than 10 feet deep and some are so shallow that they entirely disappear during seasons of deficient rainfall. The rolling sand plains are believed to be the result of wind action and to constitute but a broader development of the present dune topography near the coast.

FLAT LANDS

The name "flat lands" is used to designate the imperfectly drained pinelands lying between the narrow belt of rolling sand plains and the Everglades and their bordering prairies. The soil is a white sand which bears a thin growth of pine trees, with many prairie expanses a mile or more wide. In the rainy season these prairies are shallow lakes. There are also a number of sloughs and shallow ponds that in places support good growths of cypress, the pine and cypress often intermingling in very irregular fashion. On the west side of the lower end of the peninsula the surface between the Everglades and the Gulf is even more monotonously level than on the eastern coast, and the relations of swamp and dry land are more irregular. Much pine occurs in patches and strips separated by cypress swamps, a combination often designated as "pine islands and cypress straits." In places prairies are scattered through or fringe the pinelands, and some of them make excellent cattle ranges.

This term is applied to those outcrops of solid rock that rise above the general flat expanse of country, though they may not rise more than 2 feet above the general level and probably in no case does their elevation exceed 35 feet. Their extent is estimated at 200 square miles. On the east coast the rock ridges are of oölitic limestone and separate the great saw-grass swamp of the Everglades from the fringe of mangrove swamps and salt prairie along the western shore of Biscayne Bay. Even in the Everglades some of the keys have a rocky foundation, but the only ones which expose bare rock are Long Key and those related to it. On the west coast of Florida hard rock outcrops are more scattered than on the east coast, but cover a wider area, running through the pinelands in strips varying in length up to several miles.

SWAMPS

THE EVERGLADES

The swamp land of southern Florida includes the great saw-grass morass of the Everglades, the cypress swamps about it, and the salt marshes and mangrove swamps of the coast. The most important swamp is of course the Everglades with its wide expanse of sedge, its broad strips of shallow water, its scattered clumps of bushes, its many islands, and its underlying limestone floor. The Everglades reach from Lake Okechobee on the north to Whitewater Bay on the south, and are 50 miles wide in their widest part. Their area is estimated at 5000 square miles.[1] The border of the Everglades is well known, but the vast interior of water- and sedge-covered muck has been visited by few geologists and crossed by none.

The Everglades tract lies in a huge shallow sink, or a series of more or less connected sinks.[2] Countless shallow ponds of clear water are found in which grow bulrushes, lilies, and other water plants, saw grass, flags, and cane. The Seminole name for the Everglades is "Grassy Water." Scattered here and there in the sea of grass are islands of bushes and trees, called keys, which owe their origin to accumulations of vegetable matter; and the slight relief of the region is brought out strikingly by the fact that such slight accumulations of material should produce islands of drier land. The whole region appears to be very young; it is almost without soil or definite surface drainage. As a result of the slight relief there is no sharp dividing line

[1] L. S. Griswold, Notes on the Geology of Southern Florida, Bull. Mus. Comp. Zoöl., vol. 28, 1896.

[2] A. Agassiz, The Elevated Reefs of Florida, Bull. Mus. Comp. Zoöl., vol. 28, 1896, p. 30.

between the Everglades and the surrounding country, and a difference of two feet in water level means the difference between shallow lake and dry land for hundreds of square miles. No two of the maps of Florida agree either in the number, position, or outline of the lakes of the Everglades because of the variation in these features with the height of the water. Much of the eastern and northern shore of Lake Okechobee is bordered by cypress swamps, some of these containing the tallest cypresses to be found in Florida. On the east the Everglades are bordered by

Fig. 221.—Swamp lands of the United States with degrees of swampiness shown in two shadings. (Gannett, U. S. Geol. Surv.)

prairie and cypress swamps; in a few places patches of hardwood grow on slight elevations on the western border of the Everglades. Here are low irregular elevations, measured by inches, that diversify the distribution of water and sedge; here, too, are narrow winding sloughs some of which extend for miles.

Lake Okechobee itself is drained by a canal through the saw grass to Lake Hicpochee and the Caloosahatchee River. It is also drained by a few short streams that flow southward from the southern edge of the lake but are closed up after a few miles by thick growths of saw grass. The flatness of the Everglades may be appreciated by the elevations as determined at different times, which run from 6 to 23 feet in various portions, and it has been stated that the damming up of the main canal on the west of Lake Okechobee by raising the water three feet inundated the marginal prairies on many of the east coast ridges and seriously hindered the growing of vegetables. Bedrock lies at or near the surface toward the edge of the Everglades, and near Miami the rock forms bare ridges with a maximum height of 15 feet above the sea. The depth to bedrock in the central portion of the Everglades is 3 or 4 feet, and the flatness is so great that the word "basin" is inappropriate in describing it.

COASTAL SWAMPS

The coastal swamps include the wet lands along the lagoons or the so-called rivers back of the barrier beaches of the east coast. Many of the swamps on the east side support a scrubby growth of mangrove, but on the west side, especially in the Shark River archipelago and the southern part of the maze of land and water known as the "Ten Thousand Islands," the mangrove forms a notable forest, the trees reaching to a height of 60 feet or more, with green smooth trunks 2 feet or more in diameter at the base and without limbs for 30 feet above the ground. The mangrove forest rises from the Gulf like a green wall and is one of the most striking features of the shore line of southern Florida.

FLORIDA IS NOT A CORAL REEF

The work of Agassiz and others has shown that the older and popular view that the peninsula of Florida is an elevated reef is incorrect; the thickness of the coral reef formed since Pliocene times is probably about 15 feet, as determined by borings at Key West in 1895. Beneath the coral reef are Pliocene rocks, and beneath these in turn at a distance of 700 feet are rocks of Eocene age. The coral reefs now visible on the seacoast have but a moderate development inland. An elevated coral reef of notable width constitutes the outer edge of the peninsula of Florida. It has been traced from 20 to 30 miles inland at some points, although in other localities it has a very much more limited occurrence. This great development of reef rock should not, however, obscure the fact that the interior of the peninsula is of totally different origin, consisting of limestone, chiefly of marine origin, although with certain æolian facies. In the shore zone in which the reefs occur there are also found great patches of æolian rocks filling intermediate sinks and alternating with patches of reef rock of variable extent.[1]

DRAINAGE FEATURES DUE TO KARSTING

The drainage of Florida is characterized by the presence of large numbers of sloughs, shallow ponds, and lakes. The interior is chiefly swamp, Fig. 220, with no well-defined river systems or stream valleys, and in spite of the low elevation of the larger part of the peninsula the streams that flow from the Everglades into the Atlantic have rapids in their upper courses wherever bedrock occurs. The drainage irregularities

---

[1] The most comprehensive account of the geology and topography of the keys of southern Florida is by A. Agassiz, The Elevated Reef of Florida, Bull. Mus. Comp. Zoöl., vol. 28, 1896, pp. 29–62.

are due in part to the irregularities of the beds that form the land surface and in part to their general features and slight elevation above sea
level. Additional factors controlling the stream systems have been the
extensive karsting[1] of the limestone areas that underlie so large a portion
of the state, with the production of many sink holes, underground channels, caverns, etc. The higher portion of the state is honeycombed with
underground passages, but these decrease in number on approach to
the coast. No small part of the irregularities is probably due to the
irregular manner in which the grasses grow, which in turn affects the
distribution of land waste.

The extreme irregularity which characterizes the drainage features of the peninsula of
Florida must therefore be ascribed to three causes; first, the youthfulness of the entire surface
and lack of time in which the drainage might become organized and the lakes drained; second,
to the extremely slight relief of the surface and the absence of any dominating slope, as in the
Coastal Plain of the Atlantic region or the Great Plains of the central part of the United States;
third, to the influence of the dissolving action of the ground water which proceeds in detail
with almost a total disregard of the surface of the slopes of the land. The result is that underground water passages are opened from sink to sink and from lake to lake, oftentimes in
opposition to the surface slopes, as in the celebrated case of the Danube and the Rhine in
Europe, in which the upper part of the Danube is to a large extent absorbed by porous limestones, which in turn deliver the water to underground passages that lead to the Aach, a
tributary of the Rhine system. The result is that the upper Danube belongs almost wholly
to the Rhine system.[2] To add to the complexity of such regions the covers of the underground channels are dissolved by percolating water to the point of collapse, passageways are
blocked, and the waters rise to the upper level of the obstruction. In brief, it may be stated
that the irregularities are in large measure due to the combination of surface and underground
drainage systems which originate in different ways and which to a large extent obey different laws.

COASTAL ISLANDS

The islands that fringe southern Florida are of several types. Some
are long, narrow, barrier beaches crowned with coconut palms and
bordered by mangrove swamps; some are true mangrove islands that
resist wave action; some are low-lying sand banks supporting a scanty
growth of beach grass and weeds; and some are of rock that reaches
above sea level, these being covered with scrubby hardwoods, palms,
and pines. Within the outer chain of islands that fringe the mainland
or dot the Bay of Florida are other keys in all stages of development,
from banks below sea level to banks bare at low tide and covered with
mangroves that arrest the movement of sand, seaweed, and driftwood,
and contribute to the reclamation of the sea floor.

The existing keys and islands south of the peninsula of Florida
are regarded as once having been continuous or practically so. It appears that a process of gradual disconnection has taken place between

[1] The name applied to the process that results in a *karst* topography, i.e., a topography
marked by sink holes, underground channels, vertical shafts, etc., as developed on many limestone tracts and typically in the Karst Mountains of Austria.

[2] Petermann's Mitteilungen, 1907, and Natur Wissenschaftliche Wochenschrift, 1908, No. 7.

the Florida keys and the mainland proper. It is ascribed to the erosive and solvent action of the sea. The sounds may have originally been sinks similar to those of the Bahamas and the Bermudas, the action of the sea having broken through the barriers separating the sounds from the ocean. Once channels were formed the action of the sea would increase their width and depth, and the sea would thus gradually encroach upon the floors of the sinks, forming more and more distinct sounds out of them, or even huge open bays like Key Biscayne Bay.

The keys forming the small group immediately east of Key West, Fig. 220, have a northwest-southeast alignment. They are not coral reefs but are underlain by oölitic limestone which has an irregular surface more or less covered with marl and calcareous sand. The breaches between the islands are due to tidal currents caused by differences in time and height of the tides of the Gulf and the Strait of Florida. The great curve of the sand reefs on the eastern coast is caused by longshore current action, the currents moving from north to south as an eddy between the Gulf Stream and the mainland.

There is a bedrock floor of limestone in the Biscayne pineland on Long Key and on adjoining keys in the Everglades. The coast of this part of Florida is either stationary or sinking, but if sinking, the rate of depression is extremely slow. At the present time there is in progress an extensive reclamation of the sea by the accumulation of large quantities of shells of marine organisms and by the abundant organic life associated with the mangrove swamps of the coastal border.[1]

There is the utmost difference in the vegetal covering of the islands, due to slight differences in elevation and to differences in the character of the surface. In some places it is bare rock, in others rock with a thin veneer of sand, marl, and leaf mold, and in still others a cover of calcareous sand. On low land near the water's edge mangroves are found, on the beaches coconut palms, on the low marl flats grasses and sedges, while the higher ground, called "hammock," supports a dense growth of scrubby hardwood trees, buttonwood, ironwood, madeira, etc., while on three keys patches of pine are found. The trees seldom reach a height greater than 20 feet.

SOILS AND VEGETATION [2]

The widespread occurrence of Pleistocene sand as a surface deposit a few feet thick results in an intimate relation between the soils of the

[1] For an excellent description of these mangrove swamps, their great importance in the winning of lands from the sea, and their peculiar forms of vegetation, see N. S. Shaler, 10th Ann. Rept. U. S. Geol. Surv., pt. 1, 1888–9, pp. 291–295.

[2] The common and much more abundant types of vegetation are described in detail in connection with the topographic types with which they are so intimately related, pp. 546 to 550.

peninsula and this formation. Where erosion has been exceptionally active both the Pliocene and the Pleistocene deposits have been removed, leaving older geologic formations to form the soils, but such occurrences are relatively rare and unimportant.[1] In the southern part of Florida and in isolated patches elsewhere in the state peat and muck soils occur. These have their greatest development in the Everglades. They consist of organic matter mixed with a certain amount of sand and clay and are of recent origin. Their occurrence is confined to low upland areas with imperfect drainage. In the upland portion of the state the Lafayette formation is found as isolated areas but these are of little importance except as tobacco soils in the northwest. At various places along the east coast Pleistocene marls and coquina in a more or less decomposed state form the subsoil, and the same is true on the west coast south of Bradentown. The clay and loam soils cover a very limited portion of Florida and are not of much importance. Clay soils are confined chiefly to small areas along the stream courses and are not tilled. The greater part of Florida has a sandy or sandy loam soil of low natural productivity but quickly responding to proper treatment. It is naturally deficient in moisture in spite of the rather abundant rainfall, and irrigation is practiced in a few places.[2] Extensive drainage operations are in progress in the Everglades region which will ultimately mean the reclamation of large tracts of peat soil whose natural productivity is high and which is suitable for the production of a large variety of tropical fruits.

Florida extends so far southward as to exhibit subtropical or antillean forms of vegetation. The subtropical belt encircles the southern half of the peninsula from Cape Malabar on the east to Tampa Bay on the west.[3] The royal palm, Jamaica dogwood, manchineel, mahogany, and mangrove are among the uncultivated tropical plants, and the banana, coconut, date palm, pineapple, grapefruit, and cherimoya among the cultivated plants of this district.[4] It is of interest to note here that the subtropical region enters the United States at two other points besides Florida — the lower Rio Grande region in Texas and the lower Colorado River Valley in western Arizona and southeastern California.

[1] Matson and Clapp, 2d Ann. Rept. Fla. State Geol. Surv., 1909, pp. 37–43.
[2] Idem, p. 45.
[3] C. H. Merriam, The Geographical Distribution of Life in North America with Special Reference to the Mammalia, Proc. Biol. Soc. Wash., vol. 7, 1892, p. 33.
[4] C. H. Merriam, The Geological Distribution of Animals and Plants in North America, Yearbook Dept. Agri., 1894, p. 211; idem, Life Zones and Crop Zones of the United States.

CHAPTER XXVII

LAURENTIAN PLATEAU AND ITS OUTLIERS IN THE UNITED STATES

LAURENTIAN PLATEAU

TOPOGRAPHIC FEATURES

THE Laurentian Plateau includes all the country north of Lake Superior, Lake Ontario, and the St. Lawrence line, as far as Hudson Bay and the Arctic shore of Canada; east and west it extends from Lakes

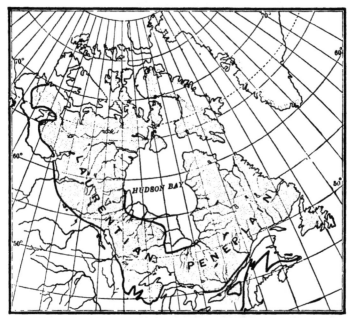

Fig. 222. — Chiefly metamorphic and igneous rocks within the heavy lines, chiefly sedimentary rocks without; the Laurentian peneplain is dotted. (Wilson.)

Athabaska, Manitoba, etc., to the Labrador coast. Its boundaries are indicated on Fig. 222 and include about one million square miles of land. It embraces practically the whole northeastern section of North America. Two outliers of the province lie in the United States:

554

(1) the Superior Highlands of northern Wisconsin, northwestern Michigan, and northern Minnesota, and (2) the Adirondacks of northeastern New York, while it is practically continuous with the New England Plateau of related origin and character. Indeed the history of this great topographic province is intimately associated with, and its forms are in many respects similar to, those of a considerable portion of the Appalachian region, and it is therefore necessary briefly to sketch its salient features.

The Laurentian Plateau may be considered as a topographic unit, its chief features being those characteristic of a peneplaned region of crystalline rock — granites, gneisses, schists, etc. — bordered by overlapping Paleozoic sedimentary layers which have been eroded into the form of an ancient belted coastal plain, then uplifted, extensively dissected by ordinary erosional processes in preglacial time, and, more recently, heavily glaciated. Two of the three great centers of glacial dispersion lay in the Laurentian province, the Keewatin and the Labradorian, and the effect of the great ice sheets that radiated therefrom was practically to denude the tract of its residual soils. With this set of physiographic conditions and changes we shall find associated practically all the topographic forms found in the region to-day.[1]

The dominating feature of the Laurentian region is the remarkably even sky line whose character is maintained in spite of the great diversity of rock structures and hardnesses which occurs from place to place. Almost anywhere in the interior of the tract the horizon, as seen from an elevation, is almost as level as that of the sea and, also like that of the sea, is almost circular. The surface from point to point and in detail is decidedly broken, but in a general broad view it is strikingly flat and plateau-like. The degree of evenness may be appreciated from the fact that residual elevations of only 50 or 100 feet relative altitude stand out as prominent landmarks visible for many miles. The Laurentian area has been traversed by the officers of the Geological Survey of Canada in many different directions, and all the descriptions and photographs are alike in representing it as once having been one of the most typical peneplains in North America. Even in those parts of the now uplifted peneplain which have been most thoroughly dissected and glaciated, as in the region south of James Bay, the dominant topographic feature is still the even sky line intercepted only here and there by an occasional

[1] A. W. G. Wilson, The Laurentian Peneplain, Jour. Geol., vol. 11, 1903.
The true character of the great Laurentian area of eastern Canada was first set forth by G. N. Dawson, then Director of the Geological Survey of Canada, at the Toronto meeting of the geological section of the British Association in 1897, and in 1893 a systematic description of the region was prepared by Wilson.

isolated residual or monadnock which was able during the period of peneplanation to maintain a slightly greater elevation because of favorable position on stream divides or superior hardness or both. The southeastern margin of the peneplain, where it borders the St. Lawrence River, is more rugged and uneven than the rest and is known as the Laurentide Mountains. Toward the extreme northeast and at the base of the Ungava peninsula there is a narrow range of mountains which extend to heights of 6000 to 8000 feet and stand prominently above the surface of the peneplain. The range includes the highest mountains

Fig. 223. — The Laurentian Plateau as an uplifted, dissected, and glaciated peneplain in Labrador, east of Hudson Bay. Note the even sky line, the network of lakes, and the absence of a soil cover. (Low, Can. Geol. Surv.)

in eastern North America. Broader and gentler undulations which give but slight diversity to the surface of the now uplifted peneplain also occur widely scattered in the form of low rounded domes and ridges roughly parallel to themselves and with their longer axes conforming in general to the strike of the rocks.[1]

Differences in elevation between different portions of the peneplain are roughly indicated by the numerous lakes, large and small, which occur at levels but little below that of the general surface. From the elevations of these lakes it is found that the average gradient of the valleys is from 1 to 4 feet per mile. East of Cree Lake as far as the junction

[1] A. W. G. Wilson, The Laurentian Peneplain, Jour. Geol., vol. 11, 1903, p. 628.

of Churchill and Little Churchill rivers the surface has an average gradient of 1.8 feet per mile for 450 miles; on the Hamilton River the gradient for about 300 miles, partly above and partly below the Grand Falls, has a mean value of approximately 1 foot per mile. Considering the large number of falls and rapids in the courses of the streams draining the Laurentian Plateau these values appear astonishingly small, especially when compared with such gently sloping surfaces as those developed in piedmont areas. The average gradient of the plains of Alberta is from 2 to 3 feet per mile, a value which distinctly exceeds that of the general slope of the surface in the Laurentian province, although we are accustomed to thinking of the latter as rugged and the former as smooth. The Laurentian peneplain has been dissected to such an extent as to be rough but not rugged, though even its roughness is not expressed in the figures representing the gradient of the hilltop plane.

Although the derivation of the great Laurentian area from a peneplain is clear, it is worth while to see that it now exhibits but few of the characteristics of a peneplain. A peneplain is an old erosion surface of slight relief which has been worn down by normal physiographic agencies almost to the level of the sea. The typical peneplain is one which also bears upon its surface an occasional residual elevation, is mantled by deep residual soils, and is absolutely without lakes, falls, rapids, or any other features indicating a youthful condition of the drainage courses. It is noteworthy that scarcely any of these qualities are present in the Laurentian area to-day. Instead of residual soils there are here almost no soils at all, only locally developed patches of glacial till or of water-laid material glacially derived. Instead of well-organized and low-gradient drainage courses, the Laurentian area exhibits the most irregular drainage features and is, *par excellence*, the lake region of North America; instead of standing near sea level the region is now at elevations ranging from sea level to more than 2000 feet. Although it is one of the oldest portions of the continent, a large number of its features are due to a very recent geologic event — glaciation. In short, the Laurentian region is one which through unequal uplift and differential stream and ice erosion has been so modified from its condition at the end of the preceding topographic cycle that its former peneplain character is recognizable only by the pronounced and uniform discordances that prevail throughout its entire area between general surface and geologic structure — a discordance that can be explained only by the assumption that it was once a region of prolonged erosion and that a pronounced, possibly a mountainous relief, was reduced practically to the level of the sea.

With these general features in mind we may now turn to the more detailed features of representative portions of the Laurentian area, beginning with the Laurentide Mountains that border the St. Lawrence Valley on the northwest. Perhaps the most important point in this district is that relating to the deformation which the Laurentian peneplain suffered in the uplifting process that raised it to its present level. The uplift was differential and is still in progress; the general nature of the first uplift was a tilting of the whole northeastern corner of the continent upward at the north; the present movement is in har-

Fig. 224. — The Laurentide Mountains north of St. Lawrence River at St. Anne de Beaupré. The lowland in the foreground is formed on limestone which terminates at the foot of the mountain slope.

mony with and in continuation of the earlier movement. The Laurentide Mountains are due only in small part to their residual quality. They are due chiefly to the greater uplift of the peneplain along the border of the St. Lawrence Valley, so that its southeastern margin as viewed from the valley does not appear as the margin of a plateau but as a range of hills of sufficient height to be known as mountains. This marginal swell roughly parallel to the margin of the peneplain is succeeded toward the interior by a number of circular depressions occupied by lakes, of which the St. John at the head of the Saguenay River is the most important. The streams of this portion of the plateau cross the marginal swell or the Laurentides through deep, steep-sided canyons of which the broadest and deepest is the fiord of the Saguenay.

From the Laurentide Mountains northwest to Hudson Bay the Laurentian area is a plateau formed upon ancient crystalline rock with an average elevation of about 1500 feet above sea level. It rises slowly from 1000 feet near the margin to about 2000 feet in the interior. The surface is dotted with low rounded hills arranged in a series of ridges parallel to themselves and to the general strike of the rocks. These hills are the stubs of extensive and elevated mountain chains subjected to prolonged erosion which brought them down to their present state.[1]

To the south of James Bay in the Nipissing and the Temiskaming regions the surface of the plateau consists of a succession of more or less parallel rock ridges with intervening valleys occupied by swamps and lakes and with a general elevation rarely falling outside the 900 to 1200 foot limits. There are no very pronounced hills, the highest seldom attaining more than 300 feet above the adjacent country, while hills 100 to 150 feet high are rather conspicuous topographic features. The areas of Laurentian granite and gneiss in the southern and southeastern portions of the district have weathered into a monotonous succession of hills and intervening rocky valleys; in the northwestern part of the Nipissing area there are more important elevations owing to the presence of resistant quartzite in country rock consisting of granite, diabase, and slate, the last-named member of the series weathering into a characteristically flat surface.[2]

East of Hudson Bay is the Labrador peninsula, a rough tableland having an elevation of about 700 feet above sea level near its margin and about 2000 feet in its higher interior portions.[3] The interior of the Labrador peninsula is so nearly flat that in an area of 200,000 square miles there is a difference in level of not more than 200 to 400 feet, and the highest general level of the interior is less than 2500 feet above the sea. The divides are often ill defined and never prominent.

[1] A. P. Low, The Mistassinni Region, The Ottawa Naturalist, vol. 4, 1890, pp. 11–28.

[2] A. E. Barlow, Report on the Geology and Natural Resources of the Area included between the Nipissing and Temiskaming. Map Sheet, comprising portions of the district of Nipissing and of the county of Pontiac, Can. Geol. Surv., vol. 10, n. s., Rept. I, 1899, p. 21.

[3] A. P. Low, Report on Explorations in James Bay and the country east of Hudson Bay drained by Big, Great Whale, and Clear Water rivers, Can. Geol. Surv., vol. 3, n. s., pt. 2, Rept. J, 1888, p. 16.

Likewise the "Barren Lands" in the far north of Keewatin, p. 565, although developed upon rock of most complex structure, are extraordinarily uniform in general elevation and in a broad view closely resemble the great plains of undulating grass-covered country of western Manitoba, Alberta, and Saskatchewan. Only here and there does a hard granite knoll projecting above the general surface indicate the character of the underlying rock and suggest the past history of the region.

BORDER TOPOGRAPHY AND DRAINAGE

Along the outer or southern and western border of the Laurentian peneplain is a group of land forms which represent on a great scale the general features of an ancient belted coastal plain. These features are best preserved in the region of the Great Lakes. The most continuous development of the cuesta feature is that associated with the outcrop of Niagara limestone which appears in the form of a pronounced but ragged escarpment extending from near Rochester on the southern shore of Lake Ontario diagonally northwestward through Ontario, forming the peninsula between Georgian Bay and Lake Huron. Thence it extends through Manitoulin Island and along the southern shore of the northern peninsula of Michigan. Farther west and southwest it constitutes the two peninsulas between Green Bay and Lake Michigan, and its final expression in the United States is in the southwestern corner of the Driftless Area, where it forms the boundary of an upland approximately 100 feet above the general level of the country adjacent to it on the north.[1]

West of Lake Winnipeg the Niagara escarpment again appears and fades out gradually to the north. Lakes Erie, Huron, Michigan, Manitoba, and Winnipegosis are on the outer lowland of the ancient coastal plain; Lake Ontario, Georgian Bay, Green Bay, and Lake Winnipeg lie upon the inner lowland between the cuesta[2] and the old-land;[2] Lake Superior occupies a depression not associated with either lowland or cuesta.

Upon the Hudson Bay side of the Laurentian Plateau a similar margin of Paleozoic sediments may be seen, although they are more extensively concealed by the deposits of the present coastal plain than is the case

---

[1] U. S. Grant, Lancaster Mineral Point Folio U. S. Geol. Surv. No. 145, 1909. See geologic map and descriptive text.

[2] The term *cuesta* is applied to the culminating ridge on the inner side of a dissected coastal plain. Its outer slope is long and gentle, its inner slope short and relatively steep. The term *old-land* is applied to any extensive tract of older and higher land that supplied land waste out of which the young land, or bordering coastal plain, was formed.

on the southern margin. Over the peninsula southwest of Cape Henrietta Maria in the angle between Hudson Bay and James Bay these Paleozoic sediments are developed to their greatest extent, but the cuesta feature is topographically very indistinct. On the eastern shore of Hudson Bay and James Bay the Paleozoic sediments have been dissected and submerged to the point where they now appear as chains of islands in some places continuous with anvil-shaped peninsulas. At one time the land stood higher than it now does, and dissection progressed so far that with a later submergence of the region and the invasion of the eroded lowlands by the waters of Hudson Bay the cuesta was almost submerged; the degree of submergence was so great that the cuestas still appear as islands in spite of the great uplifts which the region has suffered in postglacial time.

In a large number of instances the streams of the Laurentian Plateau discharge across its borders in gorges and canyons of noteworthy size and beauty. In places the gorges and valleys are steepsided, narrow, and long; in other places they are short. They have their best development along the southeastern border, where the numerous streams that enter the Gulf of St. Lawrence from the northwest have deep canyons (sometimes with unscalable walls) cut to a depth of over 1000 feet below the general level of the plateau. A specific case is the Saguenay, through which Lake St. John discharges into the St. Lawrence. Lake Temiskaming and Ottawa River, as far down as Mattawa, likewise occupy long narrow depressions cut beneath the surface of the plain. All of these canyons and deep, steep-sided valleys are marked by topographic unconformities, that is, well-defined shoulders where the steep slope of the valley wall intersects the rather flat slope of the upland surface, a clear indication that the canyon is of later origin than the upland surface below which it is incised. Where the deep preglacial canyons and gorges trend in the direction in which the ice flowed in the last glacial invasion the sides of the gorges have been steepened and the bottoms deepened. This is especially true of the Saguenay, which may probably be regarded as a typical glacial fiord, although it must be remembered that preglacial erosion had to a large extent prepared the valley for the maximum effects of glaciation as expressed in hanging valleys and deep channels below sea level.

The fact that some of the long, narrow, and steep-sided gorges or depressions are at various angles with the direction of ice movement suggests that they owe their origin to the downfaulting of long narrow crust-blocks. This view would seem to be supported by the fact that Paleozoic sediments are preserved in the bottoms of many of the valleys.

The sediments in the Lake Temiskaming and Lake Nipissing basins are in valleys lying below the level of the plateau. The margin is well defined, often cliffed, and the tributary streams spill over the edge of the basin in a series of waterfalls and cascades where they pass from the more resistant to the less resistant rock.

The long narrow depressions may be explained also on the assumption of an early Paleozoic period of valley cutting followed by a period of valley filling. In this view later valley excavation would proceed along the outcrop of the relatively softer filling in the older valleys. It seems quite probable that further work will show the validity of both explanations each being applicable to a number of valleys to the exclusion of the other.

SPECIAL FEATURES OF THE NIPIGON DISTRICT

The topography of the Nipigon area north of Lake Superior is of special interest because the normal erosion features have been either partly masked or greatly complicated by lava flows. To the northeast of Lake Nipigon is the Archæan old-land of crystalline rock containing numerous outliers of sedimentary strata. The southwestern portion of the present basin of Lake Nipigon represents a maturely dissected portion of the inner lowland of the ancient coastal plain.

At a time when the relief of the coastal plain was at a maximum there occurred extensive intrusions of diabase and extrusive flows of basalt which to a large degree filled the inner lowland and even in some places poured out over the adjacent uplands. This period of lava flows followed the erosion that in turn followed peneplanation. Since its formation the diabase has been very extensively eroded so that only from 5% to 10% of it remains, but its occurrence dominates the region to such an extent that practically every prominent feature of the topography for one hundred miles along the southwest shore of Lake Nipigon and the adjacent country is associated with the trap sheets. Their upper surface is usually a tableland or mesa, and gorges have been cut to a large extent through the sheets by the draining streams. Beneath the basalt may be seen from place to place portions of the old peneplain which were still in existence when the flows occurred and which were covered over and so preserved by the resistant rock. The scenery is varied, bold, and picturesque, as determined by the sharp outlines of the igneous rocks whose cliffs were still further cleaned and freshened by glacial action.

CHANGES OF LEVEL

The change of level which the Laurentian Plateau is now undergoing and which it has undergone in postglacial time is shown in a variety of most convincing ways. At Cape Nachvak, one hundred and forty miles south of Hudson Straits, are old strand lines reaching up to 1500 feet above sea level. Also on the east coast of Hudson Bay are well-preserved and numerous terraces cut in till and other deposits reaching to an elevation of 300 feet above the sea. Deep-water species of shells are washed out of the present beach and are found in stratified clays up to elevations of 500 feet; and long, continuous lines of driftwood, chiefly spruce and cedar, occur 30 feet above tide. These data combine to prove that the region has suffered important changes of level so recently as still to preserve in the clearest manner the features indicating uplift. Besides the physical features there are a number of important and convincing pieces of human evidence. The Eskimos of the region catch fish by constructing weirs of stone which at low tide impound fish that have come up close inshore during high tide for feeding purposes. These fish weirs and traps are found at all elevations up to 70 feet above present sea level, and as their use is invariably associated with the shore line of the time when they are employed, it is clear that the level of the land has changed not less than 70 feet since the Eskimos have inhabited the region. Hudson mentions wintering in a bay full of islands where now only a canoe may pass, and the salt marshes about the border of Hudson Bay are drying up to such an extent that the geese and ducks which formerly made their home among them have extensively abandoned them, and the residents of the district find it increasingly difficult to secure the eggs of these wild fowl. Finally, the Indians often remark the growing distances between old buildings, forts, trees, and the shore line, although this piece of evidence is of course far less convincing than the others.[1]

On the east coast of Labrador along the 1100 miles from St. John to Cape Chidley no careful measurements of postglacial uplift have been made, although Daly has approximated the amount by observations upon the distribution of bowlder clay and bowlders.[2] He concludes that the higher bowlder-covered zone has never been submerged since the ice sheet retreated from the country, and that the lower bowlderless zone is a wave-swept zone. He finds that the smooth unbroken surfaces

[1] R. M. Bell, Proofs of the Rising of the Land around Hudson Bay, Am. Jour. Sci., vol. 1, 4th ser., 1896, pp. 219-228.

[2] R. A. Daly, The Geology of the Northeast Coast of Labrador, Bull. Mus. Comp. Zoöl., vol. 38, 1902, p. 254.

of the roche moutonnées in the bowlder-covered zone (about 75 feet above sea level) contrast strongly with the jagged and riven wave-swept ledges below that level. Low, steep, and rugged cliffs are sometimes found associated with pebble and bowlder beaches of limited development. The elevatory movements continue in both Labrador and Newfoundland, as determined by the testimony of the inhabitants in regard to the decreasing depth of water on the beach and the gradual shoaling of water over ledges and bowlders off shore. The recency of the shore uplift is also shown in the numerous fresh-water ponds lying back of the barrier beaches — ponds that at one time were true coastal lagoons on the landward side of submarine bars. The forms are strikingly fresh, the glaciated, bordering slopes having contributed but little land waste to shore filling.

THE LAKE REGION

The Laurentian area of Canada constitutes the greater part of the lake region of North America whose boundaries are shown on p. 565. It has been stated that there are areas within it 25% of whose surface is occupied by lakes.[1] In order to get a closer value for this interesting condition the actual areas of the lakes represented upon a detailed map by Collins,[2] Fig. 225, were carefully measured. Some thirty independent measurements of the lake areas afford a mean that is close to accuracy. The result: about 16% of the total area of the district is covered with water. When it is considered that this value is for a representative district the figure is very striking indeed. However, even on this map only the more prominent lakes were represented on the original sheets. Were even the smallest lakes included, at least 20%, and possibly 25%, of the surface would be found covered with water. To secure a comparable figure for total lengths of drainage systems composed of lake, about 2000 miles of river course were measured, with the result that the rivers examined showed lake in 57% of their total length.

To a large extent the abundance of lakes in the prevailingly rocky Laurentian Plateau is due not alone to glaciation but to glaciation of a peneplain of such perfect development that the slightest degree of roughening by differential erosion would produce enclosed basins in large numbers. If the region were mountainous the effect of glacial erosion would be in the main to deepen the valleys and basins already formed and accentuate the preglacial differences of level and topographic form. We have only to recall the appearance of the glaciated mountain region of

[1] A. W. G. Wilson, The Laurentian Peneplain, Jour. of Geol., vol. 11, 1903, pp. 645–650.

[2] W. H. Collins, Report on a Portion of Northwestern Ontario traversed by the National Transcontinental Railway between Lake Nipigon and Sturgeon Lake, Ann. Rept. Can. Geol. Surv., 1908.

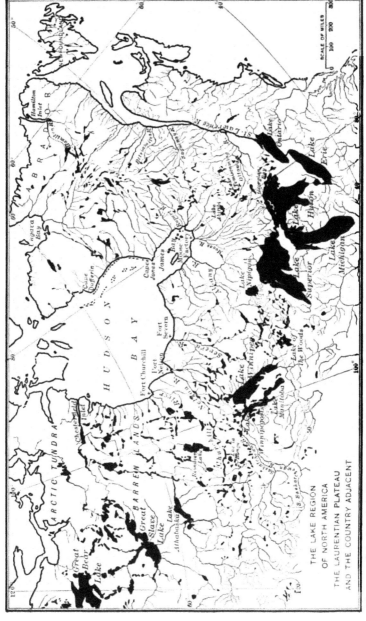

Fig. 225.—Only the principal lakes are indicated. See Fig. 229 for drainage details on a large scale of an area shown here as without lakes. There are at least 20,000 small lakes south of the international boundary that can not be represented on a map of this scale.

the northern Cascades and the northern Rockies, with their strong topographic discordances, to appreciate how different the Laurentian area
would appear if its local relief had not been so slight as to be almost
negligible. Even rocks of the same geologic age and general character
vary sufficiently from place to place to cause important differences in

Fig. 226. — Details of drainage in a limited portion of the Laurentian Plateau. (Based on Can. Geol.
Surv. Maps and Railway Surveys by Collins.)

topography; and when it is considered that so large a variety of rocks
as granites, gneisses, schists, slates, and trap sheets that occur here extend
these differences in resistance from place to place, it is clear that some
roughening of the old peneplain must follow upon erosion of any sort and
that in the case of ice the roughening would result in the formation of
a very large number of lake basins. The preglacial slopes were almost
never so strong as to overcome the general irregular scooping action of
the ice, and every local overdeepening either as the result of decreased
rock resistance or of glacial convergence produced reversed slopes that
were occupied by lakes as soon as the glacial cover withdrew.

It is clear that the glaciation of the Laurentian Plateau would result in the formation of a large number of lakes, and that the lakes would be extremely irregular in contrast to the regularity of the lakes that occupy overdeepened portions of glaciated mountain valleys. Labyrinthine shores and shapeless islands are the rule in the Laurentian of Canada. The maze of waterways displayed by this region is the marvel of every traveler in it.

The readiness with which some of the lake basins were formed may be judged from the fact that but a slight change of level may effect a change of outlet, as has been suggested many times in connection with the two-outlet phenomena commonly displayed by many of the Laurentian lakes. The feature was regarded as a novel one when first described, but the repeated finding of lakes with two outlets shows it to be a very much more general occurrence than was at first supposed. It has already been noted that the whole northeastern corner of the continent is suffering a change of level and that the movement is upward on the north. The result, in the case of lakes lying in depressions with low rims, is to cause a new outlet to come into existence as the old outlet ceases to function through uplift. The abandonment of an old outlet would of course be gradual and for a time two outlets would function. Even without such a change of level it has been thought that two-outlet lakes may have been caused by the formation of exceptionally shallow basins on a peneplain surface. The two-outlet feature is indeed thought to be on such a scale as to be independent evidence of the tilting of the land. In some cases, however, where two and even three outlets are in existence, they are found on the same side of a lake. Lake Chibougamou (50 miles south of Lake Mistassini), for example, is drained by two outlets both of which are on the east side.[1] In such cases the two-outlet feature is unrelated to regional tilting and appears to be due instead to coincidence in the level of low cols on the rim of the lake basin. The condition is perpetuated by the hard rock which is but little affected by the clear streams.

Besides the island-dotted lakes with ragged shores there is a distinct though comparatively small class of lakes that has more regular features. These lakes are usually longitudinal and are caused in a number of different ways. In some cases they owe their existence to the erosion of soft dikes, as, for example, the peninsulated northern shore of Lake Superior with its long narrow bays and points. The dikes are commonly of greenstone, though sometimes they are of pegmatite, and in both cases are usually soft and their outcrop covered with lakes or streams. Sometimes the correspondence of position of former dike and later river or lake is so close that where the dike narrowed the lake or river was narrowed and where it widened the drainage features widen. It must not be forgotten, however, that the dike material is not always softer than the country rock; where it is harder it stands out as a prominent ridge instead of a lake basin or a river valley. The straightness of trend of the dikes within short distances is directly related to the straight-line topography and drainage so characteristic in many portions of the area as to give a pronounced appearance of artificiality to it. Lake Temiskaming and the "Deep River" of the Ottawa drainage sys-

[1] Data from Bateman, Can. Geol. Surv.

tem are examples of longitudinal lakes, but their origin may have been due to crustal warping or to block-faulting of the *graben* type. In still other cases the longitudinal lakes are due to blocking of valleys by glacial material.

In addition to the lake types of the Laurentian that we have noted there is another class dependent upon broader and less local conditions. These lakes lie on the border of the region and their occurrence is coincident with that of large isolated areas of sedimentary rock within or on the border of the crystalline area. Whether these sedimentary masses represent downfaulted blocks of rock formerly more extensive

Fig. 227. — View on the shore of Lake St. John, Quebec. The level sky line in the background represents the Laurentian Plateau; in the foreground is a lowland developed on limestone; in the middle distance are alluvial terraces.

than now, and preserved by downfaulting because carried below the general level of the peneplain surface, is not certainly known, but the explanation has a high degree of plausibility, for faulted zones have been found about the borders of some of them and everywhere the sediments are sharply localized and extend to great depths below the level of the ancient peneplain. Whatever their origin they were areas of weakness during the glacial period and indeed ever since the elevation of the peneplain, with the result that not only do lake basins lie in them but the outcrop of the softer sedimentaries is always marked by the existence of a lowland whose border is sharply determined by the crystalline rocks that rim about it. A typical occurrence is at Lake St. John, which occupies a portion of an outlier of the Paleozoic sediments that border the St. Lawrence Valley on the north. The outlet of Lake St. John is over a harder rock rim to the

Saguenay, and falls and rapids mark the transition from one rock belt to the other. Hamilton Inlet, Great Bear and Great Slave lakes, and many other depressions have a similar origin, and in all cases small portions of the older sedimentary rock still cling to the border of the depressions in the crystalline tract, or, as at Lake St. John, occur in larger masses forming lowlands.

Lakes are known to be ephemeral features of the earth's surface and to be the easy prey of geologic processes. Commonly the streams that feed lakes bear such quantities of material into them and the draining streams cut down the rim at the outlet so fast and vegetation accumulates so rapidly that the life of a lake is in a geologic sense short. It is therefore interesting to find in this great lake region developed on the Laurentian of Canada a set of conditions that insure the longevity of the lakes within it, so that at least a whole geologic period would seem to be a conservative estimate of the length of time these lakes will persist. The principal conditions that are the basis for this conclusion are as follows. The fresh-rock rims have an exceptionally high degree of resistance. Instead of rotten rock the rock enclosing these lakes is firm. The products of decay were swept off by the continental ice sheet, leaving the undecomposed rock of the subsurface. Although falls and rapids abound the average gradient of the draining streams is only one to two feet per mile. In addition to the practical absence of land waste over vast portions of the Laurentian area, lakes are so abundant that the water is thoroughly filtered and the sediments are thoroughly entrapped. The streams are thus deprived of their cutting tools and waterfalls and rapids retreat with amazing slowness in the extremely hard rock. More powerful than these factors in extinguishing some of the Laurentian lakes is the vegetation that grows in such abundance about the borders of the shallower lakes as sometimes quickly to effect their reclamation. Lakes are turned into swamp or muskeg and then finally reclaimed, but the process is largely limited to the smaller and shallower lakes and does not affect to an important degree the conclusions we have just stated concerning the life of the lakes as a group.

VEGETATION

There is a limited forest growth in the hollows and along the borders of the Labrador region, but large portions of the interior are barren tracts of rock and muskeg. Exceptionally good forest tracts are found in a number of places. On the 51st parallel and east of Lake Mistassini, whence the Porcupine Range extends toward the north, is a well-wooded tract. The black spruce is here the predominating species. Other trees are the white birch, poplar, willow, alder, and Banksian pine. The growth is surprisingly large; the diameters of the largest trees exceed 2 feet and the height attains 70 or 80 feet.[1] Quebec and Ontario south of James Bay are densely forested, the forests extending westward to within a short distance of Great Slave Lake. Beyond this point the region may be described as a treeless moss-covered tundra whose subsoil is perpetually frozen. The northern boundary of the great transcontinental spruce forest closely follows the western shore of Hudson Bay from the mouth of Churchill River for a few miles, then curves gently inland; thence it extends northwesterly, crossing Island Lake, Ennadai Lake on Kazan River, and Boyd Lake on the Dubawnt. The next dividing point is just north of 60° on Artillery Lake. From this point the line curves south-westerly, crossing Lake Mackay south of latitude 64°. The banks of the Coppermine are the boundary to 67°. Tongues of timber follow the northward-flowing streams, with their warmer water, well into the Barren Grounds. The most remarkable stream of this kind is the Ark-i-linik, a tributary of Hudson Bay. From a point near latitude $62\frac{1}{2}°$ north, within the main area of the Barren Grounds, a more or less continuous belt of spruce borders the river to latitude $64\frac{1}{2}°$, a distance of over 200 miles by river. A few species of woodland-breeding birds follow these extensions of the forest to their limits. Alders occur in more or less dwarfed condition in favorable places well within the treeless areas, and several species of willows, some of which here attain a height of 5 or 6 feet, border some of the streams as far north as Wollaston Land. These are the only trees which occur even in a dwarfed state in the Barren Grounds proper.

The principal trees of the spruce forest, whose northern limit is thus defined, are the white and black spruce, whose range is coextensive with the forest limits, the canoe birch, tamarack, aspen and balsam poplars, Banksian pine and balsam fir common in the southern part of the belt and terminating from south to north about in the order given. With these are associated, generally in the form of undergrowth, a number of shrubs. The tree limit on the western mountains in latitude 56° is about 4000 feet. The head of the Mackenzie delta is marked by islands well wooded with spruce and balsam poplar. Lower down these trees give way to willows, which continue to the Arctic shore.[2]

South of the line noted above is a belt of conifers stretching across

---

[1] R. McFarland, Beyond the Height-of-Land, Bull. Phil. Geog. Soc., vol. 9, 1911, p. 31.

[2] E. A. Preble, A Biological Investigation of the Athabaska-Mackenzie Region, North American Fauna, Bureau of Biol. Surv. No. 27, 1908.

northern North America from Hamilton Inlet to Great Slave Lake, the trees increasing in size and variety toward the south. This is the great spruce forest belt of Canada, whose characteristic growths are

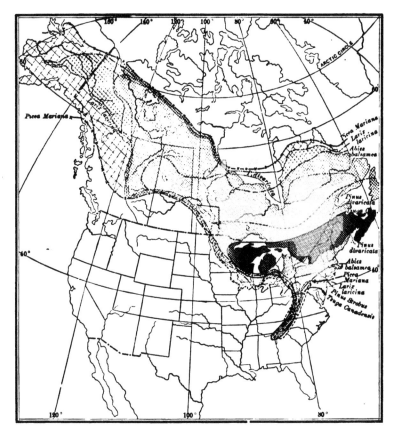

Fig. 228. — Map showing distribution of the dominant conifers in Canada and eastern United States. (Transeau.)

spruce, fir, poplar, willow, alder, tamarack, etc., and in the Great Lake region its borders are covered with forests of deciduous hardwood.[1]

The uniformity of level, the prevailing absence of soils, and the widespread occurrence of lakes, favor uniform life conditions through-

[1] R. M. Bell, The Geographical Distribution of Forest Trees in Canada, Scottish Geog. Mag., vol. 13, 1897, pp. 281–296.

out the region, a fact well shown by the widespread distribution of the fur industry, which is also favored by the equally widespread forest growth, while the network of navigable streams and lakes makes communication relatively easy as unsettled countries go.

SUPERIOR HIGHLANDS

After consideration of the topographic, drainage, and soil conditions of the Laurentian area of Canada the features of the Superior Highlands may be easily appreciated, for this physiographic province is but a part, strictly speaking, of the great Laurentian area of Canada. Like the latter, its sky line is distinctly even and gives little hint of the moun-

Fig. 229. — Typical drainage irregularities in the Lake Superior Highlands.
(U. S. Geol. Surv.)

tainous structures that prevail almost everywhere within it. Its principal topographic feature is an uplifted and dissected plain of erosion which has been glaciated and hence has many secondary features due to ice erosion. It lies neither in the region of pronounced glacial aggradation nor in that of intense glacial denudation; hence those forms that are of glacial origin are due in some cases to ice scour, in others to ice accumulation. A certain amount of glacial detritus occurs here and there; in other localities the surface is swept practically clean by gla-

cial erosion. The glacial material is irregularly disposed in characteristic fashion and blocks the drainage to such an extent that lakes and
ponds occur in large numbers. The most common type of depression

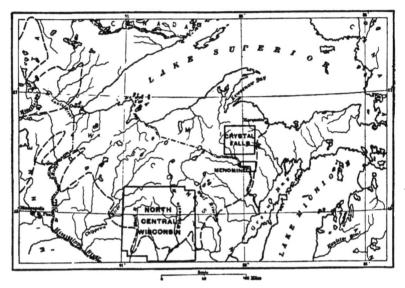

Fig. 230. — The heavy broken line is the boundary between the Superior Highlands on the north and the
Prairie Plains on the south. It is, however, more prominent as a geologic boundary than as a topographic boundary. Note the connection at the western end of Lake Superior between the Superior
Highlands and the Laurentian Plateau of Canada. The special districts indicated are portions of
the region described in the text.

is in the form of a small basin partly grown up to bushes and grasses, a
swampy tract called a "muskeg."

In describing the detailed features of the area we shall select three
tracts that have been well explored and that are representative of the
whole area. In examining them it is well to remember a number of
general features. The region consists of crystalline rock, — gneiss,
granite, and schist as well as slate, dolomite, sandstone, limestone, and
other types. The crystalline rock predominates throughout the greater
part of the area; the sedimentary rock occurs chiefly upon the borders.
The crystalline rock dips at so high an angle that it is often vertical,
thus exposing rock types of different hardnesses side by side in narrow
belts. The result is that a valley carved in the soft rock may not be
very deep and yet may have nearly vertical sides, so that the country

often has a very rugged appearance in detail that is the result of recent erosion and yet not lose on the interstream areas that plateau quality that is the result of prolonged erosion in early (pre-Cambrian) and later (Jurassic-Cretaceous) geologic time. The structure of the sedimentary rocks (Paleozoic) is in sharp contrast to that of the crystallines. The former lie in comparatively flat attitudes, dipping only slightly outward (southward), Fig. 232. In many places outliers of these flat to gently inclined strata occur beyond the main border of the sedimentaries and in such instances cap the vertical crystallines and accentuate the plateau quality of the region. Their nearly vertical cliffed borders add also to the steepness of the valley sides.

Almost the entire zone of decomposed rocks formed in preglacial time in the Superior Highlands was removed by glacial erosion; indeed glacial truncation was so pronounced as seriously to reduce the amount of available iron ore in the Lake Superior region. In many places the ice cut deeper into the soft ore bodies than into the adjacent harder rocks and thus produced subordinate valleys in sympathy with the ore deposits, a feature that is well illustrated in the Mesabi district. The soft ores were so comminuted that they are not easily distinguishable in the drift, but the hard ores occur in the glacial drift farther southeast and indeed in the whole region so plentifully that it is clear that a great portion was swept away by the ice.[1]

REPRESENTATIVE DISTRICTS

CRYSTAL FALLS DISTRICT [2]

The Crystal Falls district in the Superior Highlands is a somewhat rolling plain sloping gently downward to the southeast at an average rate of a little less than 20 feet per mile. Its average elevation is 1200 to 1300 feet above the sea. This plain is formed in part upon soft and gently inclined sandstones (Upper Cambrian) and in part upon crystalline and much harder and more highly tilted rocks (pre-Cambrian) of varying characters, though it everywhere maintains a very uniform slope regardless of the underlying rock formations. The minor topographic features of the plain have a large variety of form and detail yet very small relief. There are no commanding eminences and the sky line is generally even. Portions of the border of the mountain

---

[1] C. R. Van Hise, The Iron Ore Deposits of the Lake Superior Region, 21st Ann. Rept. U. S. Geol. Surv., 1899–1900, pt. 3, pp. 333–336.

[2] H. L. Smythe, The Crystal Falls Iron-bearing District of Michigan, Mon. U. S.-Geol. Surv. No. 36, 1899, pp. 331–333.

area are composed of crystalline rock, and here the surface is dotted with rocky knobs generally elongated east and west, which is the direction both of the gneissic foliation and of the ice movement. The elevations rise 5, 10, 20, and in some cases 60 feet above the intervening depressions and have steep smooth walls. The slight relief of the district is indicated by the fact that the more prominent elevations are deposits of modified drift. Only occasionally are they due to small rock masses of harder material, as Michigamme Mountain, which reaches an elevation of 200 to 300 feet above the general level, an elevation sufficiently great in a region of very slight relief to result in the application of the name "mountain" to what would generally be regarded as a very insignificant topographic feature. The details of the topography are primarily glacial in origin and only secondarily a response to geologic structure.

The almost innumerable muskegs or basin-like depressions are the work of either irregular glacial cutting or filling. Where softer rock occurs the ice gouged out a depression; where the rock is exceptionally resistant a knob or ridge is the result. Hills of glacial drift are scattered about the surface in large numbers. A glacial soil is also developed over a large part of the tract but has been entirely washed away and the rock surface laid bare on many steep slopes where it was originally thick and where fires and lumbering operations have removed the forest cover and allowed soil erosion to take place. The structural domes of crystalline rock have a characteristic topographic expression, their margins being frequently abrupt and in places marked for considerable distances by scarp-like slopes in the granites where vertical contacts with softer formations occur. These broad zones of harder rock are separated by broad and slightly lower lying plains, in many of which a valley character is still distinguishable in spite of extensive glacial modification. The existing drainage follows in a general way the older (preglacial) valleys, although the relation is oftentimes much confused in its details since glacial drift now lies irregularly disposed on the floors of the preglacial valleys.

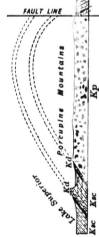

Fig. 232. — Structure and topography of the southern border of the Superior Highlands. (Weidman, Wisc. Geol. Surv.)

The southern portion of the Superior Highlands (central-northern Wisconsin) consists of a main upland, an uplifted and dissected peneplain with a very even summit below which the Wisconsin River and its tributaries have eroded their valleys and above which project a few isolated hills and ridges. The plain gradually descends southward and finally diverges; one plane runs under the Potsdam sandstone which forms the northward margin of the Paleozoic deposits of the region and another extends across the Paleozoic formations. In the vicinity of Grand Rapids, Wisconsin, and approximately at the border of the sandstone district, 1000 feet above the sea, the valley bottom of the Wisconsin River practically coincides with the level of the lower or pre-Cambrian plain. The pre-Cambrian and the main crystalline rocks are very much folded and crumpled and the dips are at various angles and frequently almost vertical, Fig. 233.[1]

The structures are such as are associated, in the early stages of a first topographic cycle, with topographic features of mountainous proportions and in their present planed condition indicate that widespread peneplanation has taken place. The stratigraphic conditions show that this peneplain was of pre-Cambrian age and was buried by Paleozoic sediments to be reëxposed by the long interval of erosion between the Cretaceous and the Quaternary. During the Jurassic-Cretaceous cycle of erosion the marginal deposits were in part stripped from the pre-Cambrian peneplain and in part

[1] S. Weidman, The Geology of North-Central Wisconsin, Bull. Wisc. Geol. and Hist. Surv. No. 16, 1907, pp. 592 ff.

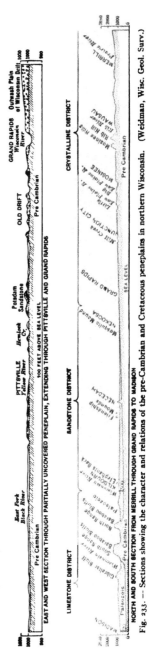

Fig. 233. — Sections showing the character and relations of the pre-Cambrian and Cretaceous peneplains in northern Wisconsin. (Weidman, Wisc. Geol. Surv.)

beveled to the form of a lowland plain. The Cretaceous peneplain made but a slight angle with the pre-Cambrian peneplain, and the widespread occurrence of the later peneplain in this region is probably to be ascribed in large part to the earlier peneplanation which had largely obliterated the relief.

ADIRONDACK MOUNTAINS

The Adirondacks are a small group of mountains in northeastern New York whose northern and southern slopes descend to tidal valleys, the Hudson and the St. Lawrence. The relative altitudes are therefore great and the absolute altitudes are such that the dominating peaks rank with the highest in the eastern half of North America. Mt. Marcy (the "Tahawus" or "Cloud-Splitter" of the Indians) rises to 5344 feet above the sea, and Whiteface is not much lower with an altitude of 4900 feet. For a long time after the fertile and easily traversed valley of central New York had been inhabited the Adirondacks were still an untamed wilderness. Dense forests clothed slopes and summits alike; deep valleys with steep descents and tillable land of small extent did not tempt the farmer.

GEOLOGIC STRUCTURE

The whole of the Adirondack region consists of a great series of metamorphic rocks, — sediments, gneisses, quartzites, and coarsely crystalline limestones on the one hand and igneous rocks, such as syenites, granites, and gabbros on the other. Roughly speaking, the region is composed essentially of gneiss, with numerous and rather generally scattered limestone belts, the whole cut through on the east by immense intrusions of gabbro. The rocks and their relations are very similar to those of the Laurentian area of Canada. Excepting the limestones all of the rocks are hard and resistant to erosion, so that their weak points are structural features — to a small degree their planes of schistosity and to a greater degree their joints and faults. Wrapping all about the Adirondack nucleus are sedimentary strata, chiefly of marine origin, whose material was largely derived (in so far as it is land-derived) from the Adirondack massif during the prolonged periods of erosion in which the Adirondacks existed as land masses. In some cases, as on the west and northwest, the sedimentary rocks (Paleozoic) overlap the crystalline rocks (pre-Cambrian) of the mountain mass; in other cases, as on the northeast and east, the transition from the rocks of the mountains to those of the bordering plateaus and plains is more abrupt — the result of block-faulting somewhat modified in detail through the influence of post-faulting erosion and glacial action.   In general the sedimentary rocks—sandstones, limestones, and shales—have little to do with the physiography of the mountains, for with few exceptions they lie on their border.

TOPOGRAPHY AND DRAINAGE

A striking feature of Adirondack relief is the strong contrast between the eastern and western halves of the mountain tract.   The western half of the Adirondack area is really an extensive upland; the eastern half alone is truly mountainous.   Broad and moderately deep valleys are characteristic of the western area, and the accordance of summits makes the plateau feature almost as prominent as in the Appalachian Plateaus and decidedly more so than in the Catskill Mountains.   The prevailing flatness is diversified solely by the broad valleys and certain low hills which are probably residuals upon the Cretaceous peneplain whose level the even upland seems to represent.   The eastern area was never base-leveled, or, if so, all traces of a former erosion surface appear to be lost in the changes that have since occurred.   That the area should have escaped base-leveling is so remarkable, considering the perfection of development of the base-leveled surface in other localities around it, that a great deal of attention has been paid by geographers to this fact. Burial and reëlevation have been suggested, with the inference that, during burial, erosion ceased here while it continued elsewhere; later erosion, the stripping off of the sedimentary cover, and the reëxposure of a once buried peneplain or old surface are other assumptions required by this suggestion.   Perhaps the most plausible suggestion is that the Adirondacks represent the locus of repeated movement and that the forces of erosion and uplift have been nearly balanced, so that while erosion has been tending toward the production of a base-leveled surface this tendency has been effectively offset by repeated uplift.   It is known that the Adirondacks have been free from extensive folding since early geologic time (Ordovician), though they have probably been subjected to broad uplift, for sediments younger than the Ordovician have been uplifted by considerable amounts and the mountains have been uplifted by equal and possibly greater amounts.

The rugged eastern half of the Adirondacks is composed of mountains that have certain very striking features.   Mounts Marcy and Whiteface, the two principal mountains of the Adirondack tract, are composed of granite and stand well above the general level.   The central cluster of high peaks to which they belong is surrounded by a plateau-like tract interrupted by valleys with rather gentle slopes developed upon softer beds and now largely drift-filled.   The granites are, however,

not abrupt in form; they have a gentle, rather flowing outline, due to
the fact that denudation has passed the stage of greatest activity.
Though the Adirondacks were not obliterated during the long erosion
cycles in which the surrounding tracts were peneplaned they were greatly
reduced in height and softened in form.   Those valleys that were developed upon the softer rocks were broadened and reduced to base level,
while even those valleys that were developed in the crystallines acquired

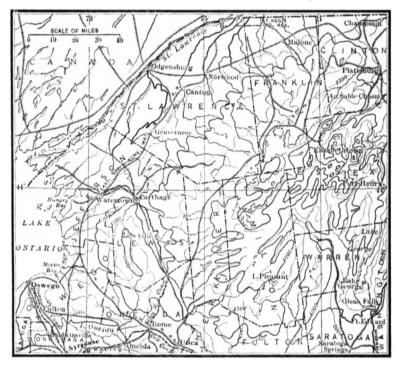

Fig. 234. — The plateau-like western portion of the Adirondack Mountains in contrast to the
mountainous eastern portion.   Contour Interval, 500 feet.   *

moderate slopes.[1]   This marked roundness of the Adirondack mountain
forms has no doubt been slightly increased by the action of the continental
ice sheet which overrode the highest peaks of the mountains, although
the glacial effects should be regarded rather as finishing touches than as
important modifications.[2]

   [1] I. H. Ogilvie, Glacial Phenomena in the Adirondacks and Champlain Valley, Jour. Geol.,
vol. 10, 1902, p. 410.
   [2] R. S. Tarr, Physical Geography of New York State, 1902, pp. 46–47.

The main outlines of the eastern Adirondacks were determined by faulting, while the detailed features are due to minor structural conditions such as fissures, joints, etc. The eastern section of the mountains, as shown in Fig. 234, is most affected by faults which have given the topography much sharper outlines than are as a rule found elsewhere in the Adirondacks. The faults trend northeast-southwest and northwest-southeast in two major fault systems and north and south in a third system of lesser importance. These faults control the trends of both the mountains and the intervening valleys and give them their most distinctive character. The mountain profiles are typically serrate, with a gradual slope on one side cut off by a steep slope on the other. The slopes that look to the southeast or northwest are particularly steep if not precipitous, with less pronounced steep slopes at right angles to them. So deep are some of the intermontane valleys and so steep their margins, it has been suggested that they look like undrowned fiords. Lower Ausable Lake and the central part of Lake George are illustrations of valleys which occupy depressed blocks or *graben* between the relatively elevated crust blocks that constitute the mountains. The valley trends are curiously related in the case of both tributaries and master streams and even in different sections of a single valley. There is generally a pronounced angularity or trellis pattern to the drainage, which is commonly associated with mountains of the Appalachian type that have been base-leveled and rejuvenated. In the Adirondacks, however, the pattern is unrelated to rock belts; it is due entirely to block-faulting, and had the pattern of the faulting been more irregular the drainage courses would now be more irregular.

The faults of the crystalline portion of the Adirondacks can not be worked out either in detail or with much accuracy, but on the borders of the mountains where sediments overlap the crystallines a means for determining the amount of faulting is supplied. One such fault in which the surface of a down-thrown block forms the bottom of a topographic depression or trough occurs near Northampton. The bordering parallel faults run north-northeast and have well-defined abrupt scarps about 6 or 7 miles apart. They involve up to 1500 feet of structural displacement, and the topographic depression thus created is about 1000 feet deep.[1] The occurrence is of particular interest because it is so clear-cut as to strengthen the previous explanations of the fault origin of scarps and troughs of similar character in adjacent tracts where a stratigraphic means for the determination of the amount of faulting is not at hand.[2]

On the western shores of Lake Champlain and on the shores of many other larger lakes of the Adirondacks, block-faulting has given a characteristic outline to the coastal topography. Sharp headlands with definite

---

[1] See the Broadalbin quadrangle, U. S. Geol. Surv.
[2] W. J. Miller, Trough Faulting in the Southern Adirondacks, Science, n. s., vol. 32, 1910, pp. 95-96.

trend project from the general course of the coast and in many cases form the extensions of block-faulted mountains whose intervening valleys, formed on the downthrown blocks, are submerged.   In general the faults

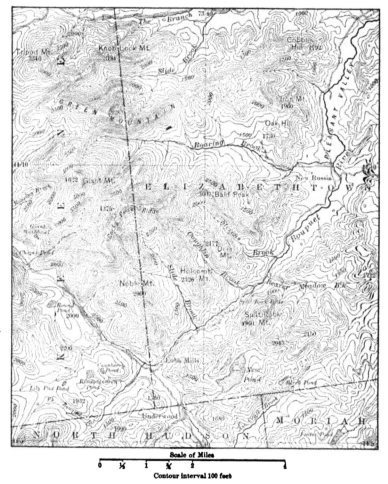

Fig. 235.—Rectangular pattern of relief and drainage lines in fault-block mountains of the eastern Adirondacks.   (Part of Elizabethtown quadrangle, U. S. Geol. Surv.)

are compound, giving a terraced quality to the mountain slopes.   On the slopes bordering Lake George the trees grow in pronounced rows or bands on these structural terraces, with thinner lines or bands between where the steeper slopes, thinly covered with soil, supply but little available plant food or moisture. The ridge from Black to Elephant Mountain is likewise of this character.

In addition to the fault valleys just described are a second set of valleys of quite different characteristics and origin. These exhibit broad expanses with gentle slopes and a mature topography. They depart from the courses common to the fault type of valley and are generally arranged in two lines, north-south and east-west. The valley of Schroon Lake, parts of the valley of Lake George, and the valley of the Hudson are of this type. The mature characteristics of these valleys suggest far greater age than the fault valleys of more recent origin. They may date back to early geologic time and represent ancient valleys which have not suffered that wholesale transformation of character and direction that the fault valleys appear to have experienced through the agency of faulting and associated forces.

GLACIAL EFFECTS

The faulted and jointed character of the eastern half of the Adirondacks enabled the ice of the glacial period to produce marked topographic effects. The faults and associated escarpments were freshened by plucking and the relief was heightened, while the valleys, which in the main represent the loci of the faults, were in some instances deepened and steepened, in others deeply filled with drift. Certain drainage changes were also effected by the glacial ice. Some streams, as the Sacondaga, a tributary of the Hudson, and the upper Hudson itself, were turned aside from their preglacial valleys and caused to flow across low preglacial divides. Glacial drift was accumulated almost everywhere along the valley floors and spread as a thin mantle over the hill slopes, thinly on the steeper slopes and sometimes to considerable thickness on the marginal terraces.[1]

The greater part of the drift in the Adirondacks is stratified and is in the form of deltas and sand-plains; morainic material is scarce. This is due to vigorous ice movement in the glacial period and to the fact that the late glacial events were associated with submergence in the Mohawk, St. Lawrence, and Champlain valleys. A small number of local valley glaciers existed after the withdrawal of the main ice sheet from the region and formed (1) local moraines which overlie the stratified drift, and (2) cirques at the valley heads.

---

[1] For a general description of the Adirondacks see J. F. Kemp, The Physiography of the Adirondacks, Pop. Sci. Mo., March, 1906. See also for rock character and structural relation, Van Hise and Leith, Adirondack Mountains, in Pre-Cambrian Geology of North America, Bull. U. S. Geol. Surv. No. 360, 1910, pp. 597–621.

CLIMATE AND FORESTS

The greater height of the Adirondacks over the surrounding regions causes the precipitation to be decidedly greater. It is almost twice as great as in the lowland district in the western part of the state. In contrast to the temperature maxima of 100° and over, in the Hudson and Mohawk valleys, the summer maxima of the Adirondacks are so low that there is seldom a day that is uncomfortably warm. The winter minima are lower, of course, than elsewhere in the state. In January, 1904, when the minimum in New York City was − 1°, the temperature was − 26° at Binghamton in the plateau of southern New York, elevation about 850 feet, while in the Adirondacks, at Saranac Lake, about 1500 feet above the sea, it was 46° and probably much lower in the higher points.[1]

The influence of the greater height and lower temperature of the central portion of the Adirondacks is strikingly brought out in any detailed map of the forest regions, on which is represented an island of northern spruce forest surrounded by hardwoods which occupy the St. Lawrence, Mohawk, and Hudson-Champlain depressions that rim completely about the tract. To the temperature factor is due the primary grouping of the forests of the region, while secondary influences are the irregular distribution of the drift and the steepness of slope as determined (a) by glacial erosion along the main valleys and (b) by faulting along the margins of the great crust blocks that form the principal mountain units in the more rugged portions of the Adirondacks, notably along the southeastern border.

The predominating trees of the Adirondack forest[2] are spruce, hemlock, and balsam. There are also scattered individuals and groves of pine on the mountain slopes and cedars on the lake shores and in the swamps. Tamarack grows on the beaver meadows, dense tangled thickets of alder border the rivers, and birch and poplar are scattered here and there on mountain slopes and valley bottoms. Patches of maple, beech, and birch grow on tracts that have been bared by fire, wind, or lumbermen, and are the only important hardwood types that the region supports.

[1] A. J. Henry, Climatology of the United States, Bull. Q. U. S. Dept. of Agr., Weather Bureau, 1906, pp. 176–177.

[2] C. H. Merriam, The Mammals of the Adirondack Region, 1884, p. 21.

CHAPTER XXVIII

APPALACHIAN SYSTEM [1]

(Introductory to the four succeeding chapters)

GENERAL FEATURES, SUBDIVISIONS, AND CATEGORIES OF FORM

THE Appalachian System, Plate IV, includes the highest, roughest, best watered, most densely forested, and some of the most thinly populated sections of the eastern half of the country. Except the Adirondacks no other part of the eastern half of the United States has slopes of so great declivity and streams of such steep gradients. For a long time the western movement of the American colonists was hindered by the rough barrier of the Appalachians, and communication across their roughest sections is still difficult. In their remoter coves and valleys, as in the mountains of western North Carolina and those of eastern Kentucky, rude mountaineers still follow customs and employ a speech as unlike those of the valley peoples about them as the topography and soil of the one situation are unlike those of the other.

The earliest settlers found the land covered with a great forest mantle; it was an almost illimitable wilderness of forest-covered hills and mountains.

"Up to the doorsills of the log huts stretched the solemn and mysterious forest. There were no openings to break its continuity; nothing but endless leagues on leagues of shadowy, wolf-haunted woodland. . . . On the higher peaks and ridge crests of the mountains there were straggling birches and pines, hemlocks and balsam firs; elsewhere oaks, chestnuts, hickories, maples, beeches, walnuts, and tulip trees grew side by side with many other kinds. The sunlight could not penetrate the roofed archway of . . . leaves; through the gray aisles of the

---

[1] The term "system" is here employed to denote the whole territory affected by related mountain-making movements through successive geologic periods, whether these are in the nature of plateau uplifts or acute mashing and folding in restricted belts. Since crustal movements of sufficient magnitude to produce mountains in many cases result in the uplift of bordering areas of some extent many belts of acute deformation are bordered by transitional upland belts. Though included in a system they are set apart as separate provinces from the mountain ranges formed upon rock belts of great structural complexity. Thus the Appalachian Plateaus and the Piedmont Plateau are included in the Appalachian System. The former was only slightly and locally deformed, then peneplaned, and later uplifted unevenly. Dissection of the more highly uplifted portions has been so deep as to give a relief of mountainous proportions — West Virginia and eastern Kentucky. Since peneplanation in the Tertiary the Piedmont Plateau has suffered only moderate uplift and dissection. It is an extensive upland whose inclusion in the Appalachian System is based not only on its former participation in older mountain-making movements but also on its participation in later and broader uplifts common to the whole territory between the Prairie Plains and the Coastal Plain.

forest men walked always in a kind of midday gloaming. . . . Save on the border of a lake, from a clifftop, or on a bald knob . . . they could not anywhere look out for any distance. All the land was shrouded in one vast forest. It covered the mountains from crest to river bed, filled the plains, and stretched in somber and melancholy wastes towards the Mississippi." [1]

To-day the forester finds great interest in the possibilities which the region suggests as to the scientific treatment of soils and trees, nowhere more important than in a rough country where the soil is thin and its maintenance bound up with the care of the forest. Some of the most baneful effects of excessive forest cutting are illustrated in the Appalachians. This great forest has been, and even now is, a great source of wealth, but its exploitation is attended by such a disregard of the influence of deforestation upon the soil as to affect not only its productivity but also its very existence. The sociological and political aspects of the forestry problems which await solution here are no less interesting than the scientific aspects. The increasing demands for greater food supplies on the part of a rapidly growing population are now extending the areas of cultivated land at the expense of the forest just as in the past. The state may say how far this shall go; whether clearings shall be made or not; prescribe the conditions of forest cutting; control fires; and guide the population along lines of greatest practical economy. These activities are involved in the question of the Appalachian forest reserves. It is implied in the argument for their maintenance that the state should exercise proper political means for securing the best scientific results; on the other hand is the contention that local experience will guide developmental projects along proper lines. With the political aspects we have here nothing to do; the groundwork for an understanding of the scientific aspects lies in a comprehension of the conditions of drainage, relief, and soils, which are the chief themes of the succeeding sections.

The territory east of the lower Mississippi River consists of six different physiographic provinces — the low and poorly drained Mississippi Valley, the forest-clad Appalachian Plateaus, the parallel and regular mountains of the Great Appalachian Valley, the irregular, forested, and narrow Appalachian Mountains, the gently undulating Piedmont Plateau, and the low, flat-lying Atlantic and Gulf Coastal Plain. The rocks of the Mississippi Valley are nearly horizontal limestones, shales, and sandstones deeply covered with river alluvium, Fig. 210; those of the Appalachian Plateaus are similar except that they are more sandy and have been lifted to a higher altitude, with the consequence that they are more deeply dissected. In the Great Appalachian Valley the

[1] Theodore Roosevelt, The Spread of English-speaking Peoples (in The Winning of the West), ed. of 1905, vol. 1, pp. 146–147.

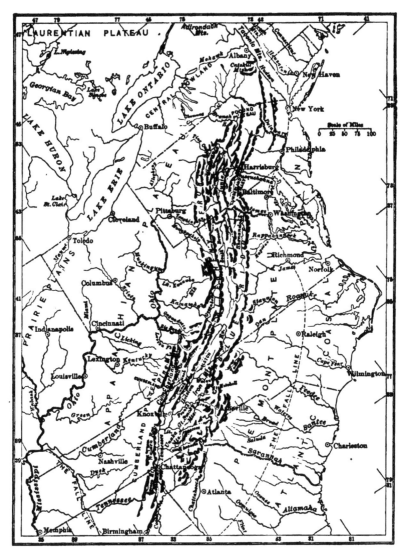

Fig. 235a. — Drainage map of the Appalachian region.

rocks have been folded into long, symmetrical, parallel folds which are among the most notable physiographic features in the world. In the Appalachian Mountains and the Piedmont Plateau the rocks are chiefly schists, gneisses, and granites of extremely complex structure, — ancient crystalline rock deeply decayed and mantled with residual soil. Seaward from the Piedmont Plateau and on both the Atlantic and Gulf slopes of the Appalachian and Ozark districts is a vast coastal lowland, the Atlantic and Gulf Coastal Plain.

The lines of division between these physiographic provinces are everywhere topographically distinct except in the case of the common border of Piedmont Plateau and Coastal Plain. Although the latter is everywhere lower than the former, yet on the common border it is but little lower; and more distinct than the differences in topography are the differences in soils and the steeper descents of the eastward-flowing streams on passing from the Piedmont to the Coastal Plain.

The great Appalachian System includes the four central members of the series of six between the Mississippi and the Atlantic coast. It extends roughly from the Ohio Valley on the northwest to the Atlantic Coastal Plain on the southeast and from the St. Lawrence Valley on the north to central Alabama and Georgia on the south. Its subdivisions and border relations on the north are somewhat different from those on the south. The Coastal Plain terminates at Cape Cod, north of which the Older Appalachians extend to the coast. A narrow upland plain, corresponding to the Piedmont Plateau of the south, borders the southern and eastern margins of the Green Mountains and extends from western Connecticut through Massachusetts and Vermont. The Green Mountains correspond to the Great Smokies, Unakas, etc., and the valleys of the Hudson and the Champlain are the counterparts of the Tennessee and the Coosa respectively at the south. The Catskills and the Cumberland Plateau are also corresponding elements. The White Mountains and bordering uplands are exceptional features in that they have no southern representatives. At the south there is a single mountain and a single plateau element in the Older Appalachians; at the north there are two mountain axes and two bordering upland plains with a great valley — the Connecticut — between them.

In general the Appalachian System includes two great belts composed of broadly different rock types — a southeastern belt composed chiefly of crystalline rock and a northwestern belt composed chiefly of sedimentary rock, Plate V. Both trend northeast and both have their broadest development toward the southwest. The crystalline belt, called the Older Appalachians, includes the Piedmont Plateau and the Appalachian Mountains; the sedimentary belt embraces the Newer Appalachians, including the Great Appalachian Valley (in places so filled with even-topped ridges as to be a mountain and not a lowland belt) and the Appalachian Plateaus, a group name for a number of separate plateaus such as the Cumberland Plateau, Allegheny Plateau, etc., and interrupted by local basins such as the Nashville Basin and the Blue Grass Country. A knowledge of the geographic positions and relations of these members is indispensable for the further discussion of their physiography, Fig. 235a.

The various provinces of the Appalachian System stand in a very interesting geologic relation to each other, a relation that is of physiographic importance chiefly through its effect upon the drainage. While the details of this relation are complex the essentials may be roughly outlined as follows. The southeastern crystalline belt has very irregular structures which were gained during several pre-Paleozoic and Paleozoic mountain-making periods that deformed the strata and elevated the mass of the province. This whole area has been called the Older Appalachians in contradistinction to the Newer Appalachians formed upon the regularly folded strata.[1]

The Older Appalachians existed as a narrow land area of unknown limits eastward but bordered on the western side by an arm of the sea that spread over a large portion of the cen-

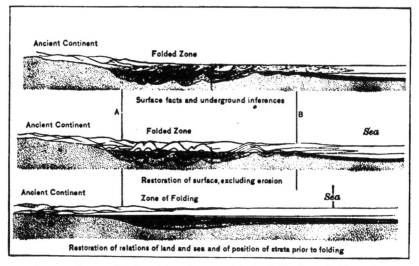

Fig. 236. — Structural Relations of the various parts of the Appalachian System, looking south.
(Willis, U. S. Geol. Surv.)

tral lowland of the continent. The erosion of the Older Appalachian land mass gave rise to land waste that was swept into the neighboring seas. On the east the sediments then accumulated have since been buried and are now concealed by Cretaceous and Tertiary strata of later origin; on the west the heavy marine and fresh-water sediments accumulated in great beds which were (1) somewhat regularly folded at the end of the Paleozoic era to form the Newer Appalachians or (2) broadly uplifted to form the Appalachian Plateaus. The folding force was directed westward, so that the folds are in general more compressed, more notably faulted, and the dips steeper on the eastern than on the western margin of the Newer Appalachians. On the western border in fact the folds die out gradually. Cumberland Plateau, Walden Ridge, and other forms are flat synclines, while the Sequatchie Valley occupies an eroded anticline, and low folds of similar nature occur elsewhere within the eastern border of the Appalachian Plateaus, as in the Tioga and Elkland districts of north-central Pennsylvania. On the whole, however, the change from the strongly folded to the gently folded rocks of the

[1] W. M. Davis, The United States, Mill's International Geography, 1900, pp. 727-729.

northwestern sedimentary belt is rather sharply localized along a line now represented by the eastward-facing escarpments known on the south as the Cumberland Escarpment and on the north as the Allegheny Front.

It must not be supposed that the strata of the Appalachian Plateaus are flat merely because the strongly developed features of the Newer Appalachian belt east of them practically cease along the definite line indicated. They are often sensibly flat, but never absolutely level over considerable areas. Low folds occur here and there and a number of important faults have also been identified.

These distinctions between the various belts of the Appalachian System serve a useful purpose in making physiographically simple what is geologically very complex. While the student may require for very detailed geographic and geologic work a minute knowledge of the geologic history of, for example, the Older Appalachians, the general features of the belt become very simple if the net results of all the complex history and not the details of the history itself are kept in mind. The broad features of the region have been developed with striking uniformity upon a mass of greatly deformed strata; the detailed topography and drainage are minutely irregular and reflect the detailed structural features. The broad features of structure and origin are of chief importance in the present connection, but we shall nevertheless keep in the background of our thought the fact of geologic complexity in order that the more detailed descriptions that follow may be adequately explained.

It is possible to assign all the topographic forms of the Appalachian System to a few simple categories. (1) The uplands and the even crest lines of the ridges are remnants of a Cretaceous peneplain above which are (2) hills and mountains — Unakas and Great Smokies on the south, White and Green mountains on the north, etc. — that have survived as residuals. (3) The third category includes the slopes, valleys, and open lowlands sunk into the Cretaceous lowland after its elevation in early Tertiary time. If the valleys and lowlands were filled up to the level of these remnants and the whole surface depressed to a lower level we should have a representation of the geography of the tra near the end of the Cretaceous period. (4) This category includes those locally base-leveled areas of small extent developed in late Tertiary time. They are much more limited in extent than the early Tertiary lowlands and were developed only upon the softest rocks near the largest streams. (5) The fifth category includes the narrow and young trenches and valleys with their associated terraces below the level of the late Tertiary peneplain.[1]

PHYSIOGRAPHIC DEVELOPMENT

The physiographic development of the great Appalachian System has been complicated by great variations in structure, by pronounced

[1] W. M. Davis, The Geologic Dates of Origin of Certain Topographic Forms on the Atlantic Slope of the United States, Bull. Geol. Soc. Am., vol. 2, 1891, pp. 578-579. The assignment of the relief forms to categories was first clearly brought out in the paper cited; but the number of the categories is increased here to include the fact of a third local peneplanation later than the early Tertiary, — a local peneplanation resulting in the development of a topographic level identified by Hayes, Campbell, and others in the Chattanooga district as the Coosa peneplain, in the Nashville basin as the Nashville peneplain, in western Pennsylvania as the Worthington peneplain, in northern New Jersey as the Somerville peneplain, etc.

differences in elevation, and by the varying distances of different districts from the sea. Many features of the topography are unified, however, through their association with the most important single fact related to the system, the fact of former peneplanation. This makes it necessary always to keep in mind two groups of facts, those which make for diversity and those which make for uniformity of topographic expression.

On the whole the province has been a land area since the close of the Carboniferous period and in all that time has been a region of erosion. The uplifts which gave rise to erosion were not simple but complex and were probably halted a number of times by depression, so that the present altitude of the region represents the algebraic sum of a number of uplifts and depressions. The uplifts do not appear to have been regular but intermittent, for, as we have just seen, the surface was stable during several periods long enough to allow the formation of either extensive or local peneplains.

A remarkable feature of the province is the unity of expression and origin of a large number of the most prominent topographic forms. Related features are continued over a wide area throughout Alabama, Tennessee, Kentucky, West Virginia, Georgia, North Carolina, South Carolina, Virginia, and Maryland, and indicate that the conditions that produced them must also have been uniform over wide areas.[1]

The most widely distributed feature is the Cretaceous peneplain that constitutes the highest surface of reference in the province. It is also the oldest topographic feature that can be identified with certainty and is a plane of reference for the forms of later origin. The remnants of the Cretaceous peneplain are sufficiently numerous so that a number of important generalizations can be determined. It is the most perfectly base-leveled surface ever developed in the tract and was exceptional for its extent and regularity. It must not be supposed, however, that its surface was perfectly horizontal. It was most nearly level where erosion progressed under highly favorable conditions, as near the sea margin, or along the largest streams, or where the rocks had little power of resistance to weathering and erosion. In favorable localities hard and soft rocks were reduced to a common level and the

[1] For details see (1) Hayes and Campbell, Geomorphology of the Southern Appalachians, Nat. Geog. Mag., vol. 6, 1894, pp. 63-126; (2) C. W. Hayes, Physiography of the Chattanooga District in Tennessee, Georgia, and Alabama, 19th Ann. Rept. U. S. Geol. Surv., pt. 2, pp. 9-58; (3) Bailey Willis, The Northern Appalachians, Nat. Geog. Mon., 1896, pp. 169-202; (4) W. M. Davis, The Geologic Dates of Origin of Certain Topographic Forms on the Atlantic Slope of the United States, Bull. Geol. Soc. Am., vol. 2, 1891, pp. 545-586.

rivers flowed in meandering courses and with slow currents across the underlying deeply weathered and soil-covered strata.

The outcrops of the harder strata or the originally higher masses were generally developed in relief upon the Cretaceous peneplain, especially in the case of those located on broad divides distant from the sea. In western North Carolina and northern New England, for example, subdued mountains stood at altitudes varying from 3000 to 3600 feet above sea level and somewhat less than this amount above the level of the adjacent portions of the Cretaceous peneplain. The map, Fig. 237, brings out clearly the unreduced portions of the land surface that projected above the general level of the Cretaceous peneplain. Such areas were on the whole relatively small, and although they were not reduced to the general level they were eroded to such a degree as to stand at only moderate elevations. Their present height is due to later uplift and the development of local or partial peneplains as well as deep valleys below and around them. They constitute, for example, the highest portions of the Blue Ridge on the western border of the Piedmont Plateau and the groups of higher mountains, principally in western North Carolina, the Great Smoky Mountains, the Unakas, the Iron Mountains, the Pisgah range, etc.

Among the mountains the forces of erosion were able to reduce local tracts only, thus giving rise to prairie-like country among the mountains, adjacent river basins being separated in many instances by low divides. Such a locally base-leveled surface is found all the way from Roanoke to Cartersville, Virginia, in the heart of the Smoky Mountains. The upper basins of the Coosawattee and Etowah, Georgia, consist of broad undulating plains partly enclosed by mountains; island-like residuals with gentle slopes rise above the level of the plains.

Later erosion has in many instances completely destroyed the Cretaceous peneplain over wide areas, and this is true in general around the margins of the province, but especially in central Tennessee and Kentucky, where the uplift of the peneplain brought about its destruction over wide areas, so that there are only a few widely separated outliers of the Cumberland Plateau whose summits still indicate the surface of the peneplain. A typical outlier of this kind is Short Mountain in central Tennessee, which rises a thousand feet above the surrounding plain. It is 20 miles from the Cumberland Plateau, has the same altitude, and is capped by the same hard sandstone. The low plain intervening is formed upon limestone which was easily eroded upon the removal of the sandstone cap. In eastern Pennsylvania, portions of southern New England, and over the greater portion of the Piedmont

Plateau the Cretaceous peneplain has also been dissected in the formation of local peneplains at lower levels.

That the uplifted Cretaceous peneplain is with certainty recognizable so long after its uplift is due to the fact that interstream surfaces which are in the aggregate of great areal extent waste very slowly. The rivers of a region denude the land surface by relatively small amounts, for most of the land is not occupied by streams. Wasting takes place chiefly through rain-wash, infinite gullying, and slumping, etc., so that the interstream surfaces, less affected by these forces, may carry clear records of denudation cycles long closed.

The uplift of the oldest topographic level, the Cretaceous peneplain, has been accomplished in large part by a series of deformations of true

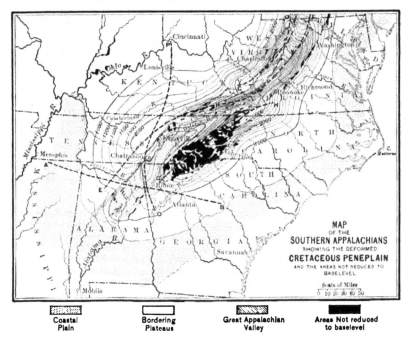

Coastal Plain | Bordering Plateaus | Great Appalachian Valley | Areas Not reduced to baselevel

Fig. 237. — Axes of deformation represented by broken lines, AB, CD, EF, GH, and OP. Contours represent elevations of restored surface. (Hayes and Campbell.)

orogenic character affecting comparatively narrow areas along certain well-defined axes. One of the most important of these is that which extends from Cincinnati to Cape Hatteras, a transverse uplift which is believed to have a prominent part in the great projection of Cape Hatteras, on the eastern side of the United States. A second and more closely defined axis of elevation extends from Chattanooga to Cincinnati. A third prominent axis passes near Atlanta and forms a tangent to the great northwestward bend of the Tennessee. In the northern part of the Appalachian System the restoration of the remnants of the Cretaceous and Tertiary peneplains shows an axis of deformation in western Pennsylvania, as shown in Fig. 238. This deformation was in the nature of a broad bowing up of the earth's crust along a southwest-northeast axis which appears to be parallel to or continuous with the axis that passes through the summit of the Cumberland Plateau. A similar pronounced axis is recognizable in western Massachusetts and Connecticut and corresponds to the structural axis of the Green Mountain range.

The orogenic movements which deformed the surface of the peneplain determined the concentration of erosional energy along certain axial lines indicated on pp. 593 and 688. Where the elevation was slow erosion was moderate; where the elevation was rapid erosion was rapid and the peneplain was here quickly dissected.

The movements which terminated the Cretaceous cycle of erosion inaugurated the succeeding or early Tertiary cycle of erosion. The crust of the earth was maintained at a fairly constant elevation so long that the surface was again reduced to a peneplain in those regions where conditions of erosion were exceptionally favorable. Broad valleys developed upon soft rock belts of the interior portions of the southern Appalachians were reduced to base-level lowlands. The Harrisburg peneplain, standing about 500 feet above the sea, east of Harrisburg, Pennsylvania, is the northern representative of this erosion level. Like the Cretaceous peneplain, the early Tertiary peneplain has been greatly modified by erosion and by uplift, which carried both the Cretaceous and early Tertiary peneplains above their former level. The second uplift introduced a third or late Tertiary cycle of erosion, which had progressed to an even less advanced stage than its immediate predecessor when uplift again intervened and brought the land approximately to its present level. After this level was attained erosion partially destroyed the latest peneplain and continued the destruction of the two higher peneplains.

In a sense it is incorrect to speak of the two Tertiary lowlands as peneplains, for a relatively small portion of the whole region was reduced to base level. They are, strictly, local peneplains, and the durations of the erosion cycles they represent were very short as compared with the duration of the Jurassic-Cretaceous cycle.

The Cretaceous peneplain is called the Cumberland peneplain at the south and the Kittatinny (sometimes Schooley) peneplain at the north

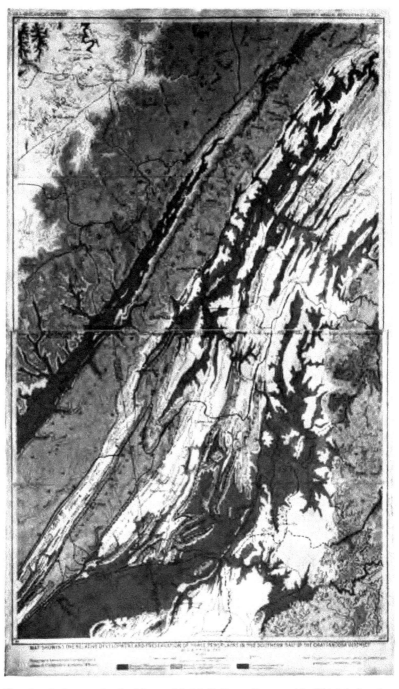

Fig. 238. — Darkest areas, base-leveled areas of the late Tertiary or Coosa peneplain; lightest shade, early

(Pennsylvania, New Jersey, etc.), in the former case because the summit of the Cumberland Plateau is the best preserved remnant of it, in the latter case because the even crest of the Kittatinny Mountains of northern New Jersey is due to its former development in that district. The early Tertiary peneplain is called the Highland Rim peneplain at the south because well preserved beyond (west of) the edge of the Cumberland Plateau on a surface called the Highland Rim, with respect to the Nashville Basin below it; the same .level is called the Harrisburg at the north because well developed east of that city. For comparable reasons the late Tertiary peneplain is called the Coosa at the south and the Somerville and Worthington at the north. We shall generally refer to these three peneplains as Cretaceous, early Tertiary, and late Tertiary, implying their correlation over the whole Appalachian System. The student will find reference to them facilitated by the use of these terms instead of the local names ordinarily employed.

RELATION OF TOPOGRAPHY TO ROCK TYPES

Since lithology and rock structure control topographic form to a large degree during the youthful and mature stages of landscape development, it is of fundamental importance in the study of the Appalachian region, where vigorous erosion is now going on, to determine the principal rock types and their degrees of erodibility. In the southern part of the Appalachian System the beds of conglomerate, quartzite, and siliceous shale are composed of nearly insoluble materials that powerfully resist erosion. The outcrops of these rocks are marked by high ridges, as in Beans Mountain, an outlier of the Unakas, and Indian and Weisner mountains south of the Coosa River.

The limestones alternating with more or less calcareous shales are in general easily eroded, a large part of the erosion being by solution. Portions of them have sufficient resistance to erosion to stand as somewhat higher ridges, but both the height and the number of such ridges are nowhere great. One of the members of the limestone group is the Knox dolomite, which contains a large proportion of relatively insoluble chert which, on the removal of the more soluble calcium carbonate, remains behind as a heavy residual mantle that has a protective influence on the remaining portion of the formation, thus giving rise to moderately high hills and irregular ridges.

A third group of rocks (Silurian, Devonian, and Carboniferous) consists of sandstone and chert of great resistance to erosion. These strata therefore stand out in some localities as residuals of superior height, for example, Oak and Chattanooga mountains. Chert beds have also been instrumental in preserving the Highland Rim.

The Coal Measures conglomerates, sandstones, and sandy shales form a group with distinct topographic characteristics. The conglomerate is the most resistant member and constitutes the cap rock of much of the Cumberland Plateau and the most important factor in the long preservation of its base-leveled surface. These various rock groups as a whole show a tendency to grow thicker and less calcareous toward the southeast, so that in a few cases strata of the same age weather into entirely different topographic forms on the two sides of the district.

The igneous and metamorphic rocks of the region contain a large proportion of feldspar and may be designated the feldspathic group. The feldspar which they contain is an element of weakness, for on exposure to the weather it decays quite rapidly, and the rock of which it is a part disintegrates. The Piedmont Plateau is composed largely of such easily weathered feldspathic rock, chiefly igneous, and a number of large valleys in the mountainous portion of the crystalline rock belt, of which Mountaintown and Talking Rock valleys are illustrations, owe their existence to the occurrence of large areas of highly feldspathic rock. It is not uncommon to find granite and diorite with a high percentage of feldspar weathered to depths of 50, 70, or 100 feet from the surface.

On the other hand the slates, graywackes, and conglomerates of the metamorphic terranes are nonfeldspathic and have a high degree of resistance to weathering and erosion. They have as a rule been greatly deformed, standing on edge in much of the mountainous section of the southern Appalachians. The result of combined hardness and vertical or nearly vertical position is shown in long, narrow, steep-sided ridges, or in rather sharp and high peaks with many radiating finger-like spurs separated by narrow, V-shaped valleys, a type of irregular topography seen to good advantage at the southern end of the Unakas on the boundary between North Carolina and Tennessee. The accompanying illustration, Fig. 239, brings out the relation between the topography on the one hand and the composition and erodibility of the rocks on the other. The relative thicknesses of the different groups is indicated by the vertical distances, relative erodibility by the lengths of the horizontal lines.

A similar relation between rock types and topography is exhibited in the northern and central portions of the Appalachian System. In the zigzag ridges of Pennsylvania the hard ridge makers are thick, hence the mountain feature is strongly developed; in the lowlands of the Hudson east of the Catskills the soft formations are thick, the hard forma-

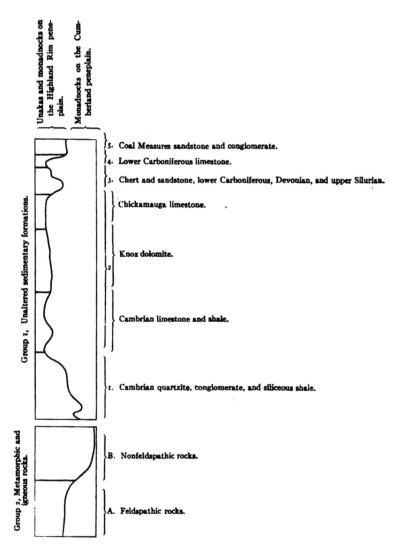

Fig. 239. — Curve illustrating the relation of topographic relief to lithologic composition in the southern part of the Appalachian System. The term *unaka* is applied to a massive residual, *monadnock* to an isolated residual surmounting a peneplain. Each curve toward the left denotes a less resistant rock, each curve toward the right a more resistant rock. (Hayes, U. S. Geol. Surv.)

tions thin, hence the lowland feature is strongly developed. The principal hard formations are the Pocono sandstone, the Oneida-Medina sandstones, and the Pottsville conglomerate; the soft formations are the Hudson River shales, the Mauch Chunk shale, and the Coal Measures. The Pocono sandstone forms Second, Peters, Mahantanago, Line, and Little mountains northeast of Harrisburg in the splendid series of zigzag ridges developed there; the Oneida-Medina sandstone forms Blue Mountain in the same locality; and the Pottsville conglomerate forms Third Mountain and Big Lick Mountain. The intervening valleys are formed upon the soft formations, principally upon the Mauch Chunk shales and the Coal Measures, as in the case of the upper valleys of Mahonay and Shamokin creeks. These soft formations are narrow as compared with their development toward the north, and the valleys formed upon them are likewise narrow for the most part; in eastern New York they are thick and have weathered into broad valley lowlands.

GLACIAL EFFECTS

Reference to the map, Plate IV, will show that the northern part of the great Appalachian System was glaciated but that the southern and central portions were not covered with ice. It follows that the interpretation of the topography and drainage of the latter districts is much easier than in those regions where glaciation has interfered with the normal development of the landscape or has partially buried the surface underneath a cover of glacial till.

One of the most important effects of glaciation was the changing of river courses either by the bodily diversion of

Fig. 240. — Probable preglacial drainage of western Pennsylvania, the limit of glaciation is shown by the broken crossed line. (Modified from Leverett.)

streams, the ice occupying the valleys long enough to enable a new course to be cut by the river in a different situation, or by the blocking of the preglacial channels with glacial drift, causing streams to be deflected into adjacent valleys. One of the most striking instances of glacial diversion is that of the Allegheny, which formerly flowed north-ward into Lake Erie northwest of Pittsburg, Fig. 240. The continental ice sheet displaced the river and kept it to a southern course so long and so modified the ancient valleys by drift deposits that with the re-treat of the ice the streams had established themselves in new channels draining in the opposite direction and forming a system tributary to the Mississippi.[1]

Many other tributaries of the Ohio suffered similar changes of posi-tion and direction of flow through glacial action. The Ohio itself was in large part diverted from its preglacial channel. The variation in width of the present channel of the Ohio is the result of the formation of a single channel out of a number of sections of different stream channels. It is significant that the general course of the stream corresponds roughly with the southern limit of glaciation, a condition similar to that found in the course of the Missouri and due to similar causes. Under these conditions it is not surprising that the fall of the Ohio River is not uniform; it varies from 0.2 foot to at least 5 feet to the mile. The greater fall commonly occurs where the river crosses the old rock divides which are not yet reduced to a graded profile. The original courses of the streams of the till-covered country north of the Ohio have now been determined in some detail by borings for oil and gas. Well records are so numerous in the region as to enable faithful restorations of the bedrock surface and the character of the drainage. Those streams which discharged northward toward or against the ice margin also suffered extensive modifications too intricate to examine in detail in this connection.[2]

In at least one locality the ice had an important effect in impounding the drainage of a district and causing the formation of lake clays and marginal deltas now at some height above the drained lake floor. Lake Passaic, an extinct glacial-marginal lake of northern New Jersey, Lake Neponset in eastern Massachusetts, and many other similar lakes in central New York are illustrations in point. Lake Passaic was formed within (west of) the curved Watchung trap ridges and its extinction followed upon the disappearance of the ice and the cutting down of the outlets. The chief importance of such phenomena to the student of forest physiography lies in their relation to the character of the soil;

---

[1] Topographic and Geologic Survey of Penn., 1906–1908, pp. 123–124.

[2] For a discussion of both typical and detailed features see W. G. Tight, Drainage Modi-fications in Southeastern Ohio and Adjacent Parts of West Virginia and Kentucky, Prof. Paper U. S. Geol. Surv. No. 13, 1903, who discusses the changes in the courses of the rivers of the region, reconstructs the old courses, and analyzes the causes that have led to the drain-age changes.

an understanding of the distribution of the bottom clays and the gravelly and sandy beach and delta deposits rests upon a knowledge of the existence and extent of a former lake.

While the northern ends of both the Older and the Newer Appalachians were glaciated, Plate IV, the topographic effects of glaciation are of minor importance. The preglacial relief is still prominent in valley and upland while the glacial relief is in the nature of minor irregularities —

Fig. 241. — Maximum stage of Lake Passaic. All outlets except that at Moggy Hollow were either blocked by ice or filled with drift. (U. S. Geol. Surv.)

the detailed features of valley slopes and floors and the drumlins, eskers, sand plains, and low morainic ridges scattered irregularly about.

The amount of till deposited by the ice sheet in the Appalachian Plateaus is very slight. This is due to the brief period in which the ice overlay the region, to the elevated nature of the country, and to the fact that hard rock yielding little waste constitutes the cap rock of a large part of the province. The valleys received the chief contributions of drift, and many of them are also partially filled with fluvio-glacial deposits now terraced by the streams. The terraces occur without as well as within the southern border of the glaciated country and are among the most conspicuous and persistent elements of the valley forms.

Detailed features of relief and drainage that owe their origin to glaciation will be discussed in connection with the general physiography of the various regions of the Appalachian System.

CHAPTER XXIX

OLDER APPALACHIANS

SOUTHERN APPALACHIANS AND PIEDMONT PLATEAU

APPALACHIAN MOUNTAINS

THE southeastern belt of crystallines in the Appalachian System consists of two very unlike portions, a western portion of strong relief, the Appalachian Mountains, and an eastern of low relief, the Piedmont Plateau. The mountain belt is broad at the south and narrow at the north. In western North Carolina, northwestern South Carolina, and eastern Tennessee the Appalachian Mountains are 50 miles wide and are bordered by declivities of the first order; in western Virginia, central Maryland, and south-central Pennsylvania the belt narrows to a single ridge known in its various parts as the Blue Ridge (Virginia), Catoctin Mountain (Maryland), and South Mountain (Pennsylvania). Farther north the Appalachian Mountains are represented by the highlands of New Jersey, the highlands of the Hudson, the Green Mountains, etc. These northern representatives of the system are narrow and their relief is not great as compared with the mountains of North Carolina, yet they have notable relief, for both the crystalline rocks and the limestones, shales, and sandstones of varying structure and age that border them on either hand have been worn to lowlands, valleys, or upland plains of relatively slight relief.

Attention will here be given chiefly to the mountains of the broad southern portion in western North Carolina and eastern Tennessee, whose most notable qualities as compared with other portions of the Appalachian Mountains are (1) their great height and ruggedness, (2) their great areal extent, and (3) their dense and valuable forests. This is by far the most important member of the southern Appalachian district and is here designated the southern Appalachian Mountains.

The southern Appalachian Mountains include the highest point in the eastern half of the United States (Mount Mitchell, N. C., 6711 feet) and are more truly mountainous than any other portion of the country east of the Rockies. So unsettled are they that certain sections have only the merest sprinkling of population; and in general one finds only scattered cabins and small clearings, except in a few localities where areas of less resistant feldspathic rock have enabled the forces of erosion to form rolling plains of limited extent or valley flats that permit more extensive clearings.

The ridges and ranges of this mountain group nowhere display exceptional declivities, and in few places may one find any approach to the wild mountain forms that usually attract the lover of mountain scenery. The mountains have never been glaciated, and cirques and broad, flat-bottomed, steep-walled valleys on a scale comparable to the higher valleys of the Sierra Nevada and Lewis mountains of the Pacific Cordillera are here wanting. The slopes have a smoothed and softened appearance enhanced by the dense forest cover, while the blue haze that hangs over the mountains perpetually gives the effect of great distance and height to ranges but a few miles away.

The mountains of this group have suffered great erosion and have passed the point where they exhibit the profounder aspects of mountain scenery. On the whole they are topographically subdued and are well cloaked with waste except here and there where bare rock slopes may be seen. Furthermore, the drainage has become well adjusted to rock structure and hardness, the whole group being drained by an intricate and thoroughly ramifying system of streams. For example, the sides of the mountains in the Pisgah range are steep but are composed of smooth, flowing slopes and rounded crests as a rule free from cliffs. Lines of small cliffs and ridges occur only upon the harder and limited outcrops of an extremely resistant mica-gneiss. The granites make cliffs near drainage lines or along the slopes of the Blue Ridge, as at Cæsar's Head. In general the even slopes of weathered rock of the mountain group are seldom broken and the cover of heavy forest is continuous on high and low ground alike.[1]

The height, irregularity, swift drainage, and general ruggedness of the southern Appalachian Mountains are features which distinguish this mountain system only because of the contrast they afford to the flatness, fertility, easy cultivation, and good means of communication of the low, rolling, piedmont and coastal plains on the east and the broad flats of the great valley of eastern Tennessee on the west. As mountains go, the southern Appalachians are neither notably rugged nor lofty nor are they remote from the centers of consumption as are large tracts of equal size in the Pacific Cordillera, but they possess these qualities in sufficient degree to form a strong contrast to the other mountains of the eastern half of the United States.

[1] A. Keith, Pisgah Folio U. S. Geol. Surv. No. 147, 1907, p. 1.

Interest in the southern Appalachians is heightened by the fact that they contain one of the largest bodies of fine timber to be found on the Atlantic slope of the country and are of almost supreme interest to the eastern forester, who sees in them one of the country's greatest forestry possibilities.   Nowhere else in the United States, not even excepting the arid and untimbered West, do the correlated problems of drainage, soil maintenance, forestry, and agriculture have a larger immediate importance or their solution a more far-reaching effect.   In a region made up chiefly of steep mountain slopes each man's acts must be in a

Fig. 242. — Pisgah Mountains from Eagles Nest near Waynesville, N. C., looking S. 70° E.   Cold Mountain (6000 feet) and Big Pisgah Mountain (5749 feet) are on the sky line.   Beatty Knob in the middle distance shows characteristic details of spurs and ridges.   (U. S. Geol. Surv.)

high degree adjusted to the welfare of his neighbor if both have interest in the productions of the soil; each man must become his brother's keeper as a matter of economic as well as of moral necessity.

The distinctly greater elevation of the southern Appalachians above the surrounding country gives them a notably cooler climate.   The temperature is that of south-central Pennsylvania or New Jersey. Above 5000 feet it ranges usually from 45° to 75° in summer and from −10° to 45° in winter.   The low temperature is expressed among other ways in the presence of red and black spruce, northern birch, white pine, and fir, while the magnolia, mulberry, papaw, and persimmon trees are representative southern species which occur at lower elevations and have reproductive associations with large areas of such species in the surrounding plains and valleys.

Because of greater elevation, the degree of cloudiness and the rainfall are both much greater than on the surrounding plains and plateaus. This being the case we should expect also to find differences in the distribution of these climatic conditions within the mountain area, on account of the rather wide topographic differences. The most important generalization of this kind is in the distribution of heat and rainfall. The southeastern slopes, with a declivity of 10° to 30°, are almost exactly normal to the sun's rays for several summer months and receive almost as much heat from the sun as do equatorial lands during the same period. The northwestern slopes depart from the normal by an equal amount but in the other direction, and may be said to receive an amount of direct insolation no greater than that received by Newfoundland or southern Norway, though in each case the net result is nowhere near such extreme temperatures as these comparisons would seem to show on account of the mitigating effects of the wind, which tends constantly to equalize the distribution of the heat. Nevertheless the southeastern slopes are notably warmer than the northwestern slopes, and being warmer are also drier on account of the greater evaporation.

A second and more important cause for the comparative dryness of the southeastern slopes is the effect of the prevailing westerly winds which precipitate about 60 or 70 inches of rain on the windward (northwest) slopes and but 40 or 50 inches on the leeward (eastern) slopes. This climatic contrast is sufficiently marked to affect the distribution of forest fires, which are found to be much more frequent, or at least do the greatest amount of damage, on the drier southeastern slopes than on the moister northwestern slopes. About 80% of the total area, or 4,500,000 acres, have been burned over, the greater number of the fires being on the ridges, where they are often set to improve the pastures, or to effect partial clearings.

Besides these differences in rainfall and temperature on the two slopes of the southern Appalachians there are important differences dependent upon altitude. These are roughly expressed on the accompanying map showing the rainfall of the United States, but a larger number of observations would undoubtedly show even greater differences, as have been shown by the refined observations of late years in the Alps of Europe and the mountains of Wales.

PHYSIOGRAPHIC DEVELOPMENT

The southern Appalachians have a very complicated structure. The rocks are chiefly crystalline and are in part igneous, in part metamorphic. The igneous rocks occur in the form of ancient dikes, sheets, etc. The metamorphic rocks consist of altered sedimentary strata which were deformed in several periods of mountain making but chiefly in the Ordovician period when there was formed a very complex structure, the folds due to compression being arranged in all directions, though the northeast-southwest direction predominates. So complicated is the structure, and so diverse is the rock character within short distances, that great irregularity of drainage direction and of topographic form is the consequence.

While irregularity of structure thus prevails there is a marked predominance of trend toward the northeast and it is in a northeast-southwest direction that the principal stream courses, valleys, and ridges lie.

Fig. 243. — Geologic structure of the Appalachian region where extreme faulting has occurred. Length of section, 8½ miles. (Keith, U. S. Geol. Surv.)

Structural irregularities are expressed in the many cross ranges, the highly irregular trend of mountain and range spurs, and the minute irregularities in the slopes and crests of individual mountain ranges. The character of the rock also has been an important factor in making the slopes irregular in detail. The areas of softer rock such as the feldspathic Cranberry granite, extensively exposed about Asheville and in smaller areas in many other localities, weather down much more rapidly than less feldspathic rock, and as the outcrops of the former are extremely irregular, the dependent erosion forms are correspondingly irregular. It is therefore only in a broad view that the mountains appear to possess any system whatever, and even in such a view there are large tracts in which no general arrangement can be discovered. The larger members of the group, such as the Unakas, the Great Smoky Mountains, the Black Mountains, the Iron Mountains, the Pisgah range, and many others, all run in a northeasterly direction, Plate III, but this feature is less clearly seen in the details of the mountain slopes than in a general view of the ranges such as a group map affords.

The smoothest contours in the southern Appalachians are in the southeastern portion, occupied almost exclusively by igneous rocks which have given rise to broad and massive domes; the northwestern portion is carved out of metamorphic rock of greater resistance to erosion, and in contrast to the subdued mountain forms west of the Blue Ridge are the Unakas and Great Smoky mountains, whose peaks are prevailingly rather sharp and rise to greater heights. The summits of the Unakas are capped for the most part by hard quartzite; the mountains close to the Blue Ridge are carved out of massive granite. This adjustment of topography to the hardness of the rock is illustrated everywhere and is one of the features indicative of the great geologic age of the region, though, had the topographic cycle been carried farther, these adjustments would in their turn have given way to peneplanation in which

Fig. 244. — Roan Mountain, Tenn. Hills in foreground are composed of granite; the broad rounded summits of the mountains are characteristic of the Roan gneiss. (U. S. Geol. Surv.)

rocks of varying hardness all but cease to have topographic expression, as is the case in the Piedmont Plateau on the east or the Cumberland Plateau on the west.

The absence of adjustments to structure is exhibited in the courses of the westward-flowing streams which in some instances cross the main mountains of to-day, the Unakas and Great Smokies, in deep gorges and canyons that divide these ranges into a number of more or less separate units. A certain degree of this transection is, however, to be attributed to upheaval of the mountains in the paths of the streams, an upheaval that continued during the two Tertiary uplifts that followed the development of the Cretaceous peneplain in the Appalachian region. As already described, these uplifts were sharply localized along the present mountain axes and were orogenic in nature, thus giving a certain antecedent quality to the westward-flowing streams.

In the southern Appalachians the Cretaceous peneplain was developed only upon the highly feldspathic rocks and the less altered slates, while the nonfeldspathic conglomerates and the harder slates formed considerable elevations above the peneplain. These elevations were sometimes of mountainous proportions, and were of such size and such degree of resistance to erosion and so favorably placed at the headwaters of the dissecting streams as to have suffered relatively little erosion since the uplift of the Cretaceous peneplain. The best-preserved residuals are the Great Smoky Mountains, the Unakas, and a dozen or more neighboring groups.[1]

The largest and best-preserved remnants of the Cretaceous plain formed within the southern Appalachians are those about Asheville, where a local peneplain was developed upon the feldspathic Cranberry granite, and east and south of James Mountain and south of the Unakas in Mountaintown and Talking Rock basins. Over a large part of the southern Appalachians the Cretaceous level is expressed only in dissected remnants of a former plateau surface at the heads of the main streams as in the plateau immediately west of the Blue Ridge. Many smaller remnants of such a leveled surface are to be found here and there among the mountains, for example in the plateau of the French Broad River from 2200 to 2300 feet above the sea, and similar cases are to be found in a number of other river systems at comparable but slightly different altitudes.

The plateau bordering Pisgah River near Waynesville and Sonoma is between 2700 and 2800 feet above the sea; that of French Broad River is about 2200 feet. Remnants of the plateau west of the Blue Ridge pass entirely around the head of French Broad Valley and along the heads of the minor streams and continue southwestward across the headwaters of Toxaway and Horse Pasture rivers, which are tributary to the Atlantic.

The valleys intervening between the mountain ranges of the Pisgah region are sharp, narrow, and V-shaped at their heads but widen out at lower levels where they are bordered by rounded, plateau-like tracts only slightly varied by shallow valleys. They are alike in form and origin but vary considerably in altitude, rising gradually toward the heads of the rivers, so that each main stream has its own particular set of altitudes. All of these basins, but the larger ones more conspicuously than the smaller ones, are now being dissected by the streams which drain them. In the case of the largest and most interesting of them all, the Asheville basin, the main stream has intrenched itself well below its former level and the tributaries are intrenched by smaller amounts. The former plain has been carved into hills and valleys with thoroughly organized and graded waste slopes. But the dissection of the former plain has not yet reached the point where the earlier flatness can not be safely inferred, for some areas of flat land can still

[1] The term "monadnock" is applied to a single isolated residual such as Mount Monadnock in New Hampshire, which stands alone. More massive residuals of greater size and height would seem to require a different name, and the term "unaka" has been proposed for such large residuals as the Unaka Mountains which illustrate the type.

be made out here and there and the hilltops still reach accordant elevations, though the structure and to a lesser degree the rock character vary from point to point. East of the Black
Mountains, in the valley of the South Toe River, is a basin smaller than that of the French
Broad at Asheville but made in a similar manner and now in process of dissection similar
to that exhibited about Asheville, the river being intrenched about 200 or 300 feet below
the general level of a comparatively even upland A third basin of this kind is that of the
Caney River west of Mount Mitchell, but it is much smaller than the other two.[1]

A topographic and drainage feature of the southern Appalachians
that has a dominating influence in the distribution of the population
and a direct relation to the growth and development of the forests is
the distribution and character of the basins, gorges, and coves. The
basins have already been defined in terms of topographic cycles and
rock character as the result of local peneplanation of areas of more
feldspathic rock. They vary in size from the many small, even tiny,
flat or gently rolling areas such as those found along the courses of the
minor rivers to such large tracts as those on the Nolichucky and the
Holston and the Asheville basin in the valley of the French Broad.

Less important than the basins just described are the small plains
that frequently occur at the headwaters of the streams. They are
perched well up on the mountain slopes and reëntrants, where the brooks
or "branches" unite to form creeks. They appear upon irregular areas
of softer rock, at stream junctions, and where land waste has been washed
into the hollows of the mountain slopes. These basins are a common
feature of the whole southern Appalachian mountain region. Between
the high coves and the basins at an intermediate level and also between
the basins and the great valleys and plains that border the southern
Appalachians the streams descend through gorges or steep valleys. Thus
between the Asheville basin and the Great Appalachian Valley the French
Broad leaves the wide valley in which it flows above Asheville and
enters a gorge at Hot Springs, Tennessee; the Linnville flows through a
broad valley above the falls at the Blue Ridge, then plunges down in a ·
reversed curve to the quieter stretch of the Piedmont; to the same class
belong the gorges of the New River above Ivanhoe, the Doe River above
Elizabethtown, Tennessee, the Nolichucky above Unaka Springs, Tennessee, the Tallulah at Tallulah Falls, Georgia, and the Nantahala,
Little Tennessee, and Hiwassee at the points where they make their
steepest descents. In like manner the streams descend from the high-
level coves to the level of the larger intermontane basins through gorges
and canyons often rather steep-walled though seldom precipitous.

[1] For an interesting discussion of the Stream Contest along the Blue Ridge see a paper
with that title by W. M. Davis, Bull. Geog. Soc. Phil., vol. 3, 1903, pp. 213-244. The paper
also contains a short discussion of the general character of the southern Appalachians and
the Piedmont Plateau.

In these two features of plains and gorges one has a large part of the physiography of the southern Appalachians that enters directly into relationship with man, for it is in the coves, basins, and plains that the 350,000 people of the region are chiefly found and it is through these to a large extent that the development of the forest must proceed. Since the total area of the basin and valley lands is small, the population is small and as scattered as the relatively flat lands they occupy. Only one-fourth of the tract is under cultivation of any kind, the wagon roads are chiefly ruts, and but one railway crosses the entire mountain

Fig. 245. — The uplifted and dissected local peneplain known as the Asheville Basin.
(Keith, U. S. Geol. Surv.)

region from east to west, though there are a score of feeders that enter it for considerable distances. It is largely to these features that the backward condition of the region is to be ascribed. Its life is but an eddy of the life of the surrounding plains; the mountaineer of the southern Appalachians is almost as crude as his somewhat more isolated brother in eastern Kentucky in what are locally known as the Kentucky Mountains. In both cases life is largely a struggle against space, whose vertical and horizontal elements are both discouragingly large.

The succession of basins and gorges with falls and rapids is not ideal for the agricultural development of the southern Appalachians, but it has positive advantages for the mineral and forestry interests, since it affords an unrivaled source of energy. More than 6400 acres of land along the Blue Ridge, Unakas, and intermediate highlands have an elevation over 6000 feet, and about 54,000 acres are more than 5000 feet above the sea. The Piedmont Plateau on the east is from 1000 to 2000 feet high and the Great Appalachian Valley on the west from 1500 to 2000 feet high. Under these conditions the gradients of the streams are necessarily very steep; they fall from 2000 to 4000 feet within the mountains. It is estimated that about 1,000,000 horse power could be developed on the principal streams — the New, Kanawha, Holston, French Broad, Nolichucky, Little Tennessee, Coosa, Chattahoochee, Catawba, and a dozen others. At present water power is used in almost every settlement for grinding and sawing, and the larger towns are becoming to an increasing degree dependent upon it, but as a whole the greater part of this splendid resource is unused.

In the development of the timber resources of the region two advantages are required that will not be long delayed — a denser population for labor supply and for markets. Already the manufacturing South is a reality. North Carolina cotton mills require more cotton than is grown in the state, those of South Carolina consume nearly three-fourths of the home-grown cotton,[1] and the concentration of population to which these industries lead will greatly enlarge the demand for cheap lumber. At present some of the finest wood in the country is being sold for astonishingly low prices. For example, oak, cherry, walnut, and hickory commonly sell for $5 to $10 per thousand feet, and in scores of localities the difficulties of marketing the forest products limit the development of the timber to that required for local use.[2]

SOILS AND VEGETATION

The soils of the southern Appalachians reflect the variations in rock character quite as faithfully as do the slopes of the mountains, the trends of the ranges, or the courses of the streams. Not only are the siliceous gneisses and the quartzites prominent topographically but they also have the thinnest soils; the feldspathic granites have the deepest soils and underlie the largest basins, as the Asheville basin in the valley of the French Broad.[3] The soils of the upper part of the Little Tennessee River basin are sandy, being derived from granite. On Little Tennessee River around and above Franklin most of the good farms

[1] E. R. Johnson, Sources of American Railway Freight Traffic, Bull. Am. Geog. Soc., vol. 42, 1910, p. 246.

[2] Ayres and Ashe, The Southern Appalachian Forests, Prof. Paper U. S. Geol. Surv. No. 37, 1905.

[3] Hall and Bolster, Surface Water Supply of the United States, Water-Supply Paper U. S. Geol. Surv. No. 243, 1910, p. 165.

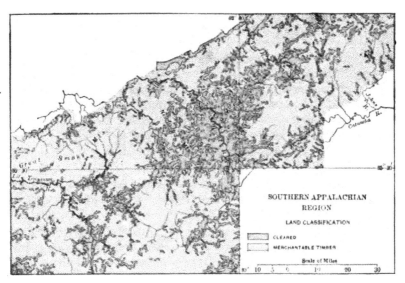

Fig. 246. — The large cleared areas lie chiefly in local basins such as the Asheville basin and the valley basins of the Little Tennessee. The smaller isolated areas are largely upon the hillslopes and in the upper and smaller basins or coves.

Fig. 247. — Grassy "bald" and border of spruce forest, White Top Mountain, Virginia. (U. S. Geol. Surv.)

are located on deep fertile red loams derived from schists. In the narrow valleys among the high mountains, where sandstones, quartzite, and conglomerate prevail, the soils are generally thin and sandy and have little agricultural value; but on the north slopes and hollows they bear well-developed forests. In the Hiwassee River basin deep valleys extend from the rivers far into the mountains between spurs five to twenty miles long. The mountain sides are steep and in many places rocky, while the creek valleys have considerable areas of alluvial flats and rolling foothills. The foothill soils are almost entirely clay and the alluvial flats along the river and creeks have a large percentage of clay. The soils of the mountain slopes are loamy and moderately fertile; the ridges are covered with a light and stony soil that precludes agriculture.[1]

The distribution of the timber of the area is in general sympathetic with the major natural features. The principal timber is the oak of several varieties, found chiefly on the ridges, and the pines, found chiefly on the plains. The hemlock is found in strips or bands along the shaded ravines and on the better-watered northern or northwestern slopes between 3000 and 5000 feet. The northerly exposures also support beech, maple, birch, etc. Shortleaf and pitch pine and hickory are found chiefly along the lower slopes of the Blue Ridge. Individual oaks reach their best development in the coves, though as a whole the stands are there too dense for the best reproduction. The soils of the ridge crests are generally too stony and thin for the best stands, though the conditions are very favorable for reproduction. Northerly exposures at the higher altitudes have the added difficulty of being too cold, a condition that limits the productive power of the soils. Finally, on the higher summits of the Great Smoky, Pisgah, and Balsam mountains are a few thousand acres of black spruce occupying a habitat similar to that of the spruce forests of New England but very strictly limited in range because of the small proportion of land located at the requisite elevation for favorable conditions of temperature and humidity.

The higher coves of the region have singular value and interest. In them the forest litter and leaf mold are generally thick, being washed down from the surrounding slopes. The soil of such localities is rarely ever wanting in humus, unless it has been recently reburned, as is sometimes the case with those coves located on the southern slopes, where there are less rainfall and a higher temperature and consequently more frequent and more destructive forest fires.

Great havoc is being wrought in the region by the pursuit of cultural

---

[1] Hall and Bolster, Surface Water Supply of the United States, Water-Supply Paper U. S. Geol. Surv. No. 243, 1910, p. 190.

and forestry methods which permit the too rapid wastage of the soils. At first the clearings were all located on the basin and valley floors, but as the population gradually increased and the old fields became depleted the clearings extended farther and farther up the valley sides beyond the point where natural fertility is long maintained in the soil and where the land yields so poor a return that it ought always to remain in forest. As the cleared farms became unproductive the clearings were extended farther into the forest and the old fields allowed to revert to natural growth. Widespread erosion of hill and mountain slopes resulted, and great areas have in this way been rendered worthless, either through gullying or through the growing of useless brush. This is the rule within the mountain area, but along the western border of the Unakas and the eastern border of the Blue Ridge young quick-growing pines are the pioneers and cover the mountain slopes before gullying becomes too far advanced. Only the highest state of cultivation and the maintenance of brush dams and stone and earth terraces will preserve the soil upon the mountains and valley slopes once the natural forest cover is removed. So vigorous is erosion on the unprotected areas that the streams from them are often nearly half earth. Tending to the same destructive results is the overgrazing of considerable areas of the forest. As a result the young growth is checked or prevented altogether, the humus is depleted, the roots broken and bruised, and a rapid run-off ensues, which accumulates in effect and soon gains on the forest beyond the point of natural control.

About 74% of the total area is still forested. Even this amount, however, is considered too small. Under the given conditions of slope gradient, soil texture, rainfall, etc., it is considered unsafe to have more than 18% to 20% of the surface cleared. Recent studies[1] have supplied clear-cut illustrations of soil erosion in a variety of situations: (1) on sodded "balds" where overgrazing and trampling by cattle have broken the turf and started landslides that quickly developed into gullies; (2) on all slopes where lumbering has removed the original protective covering and hastened the action of rain-wash; (3) on cleared and abandoned slopes once used for agriculture. In the last-named case the harm has been underestimated in the past. Undercutting and caving, once started in a cleared area, often extend upward into forested tracts and the débris derived in this manner is washed downward into the forest below. Some porous soils are erosion-resisting but their aggregate area is not relatively great. The allowable limit of steepness for cleared lands,

[1] L. C. Glenn, Denudation and Erosion in the Southern Appalachian Region, Prof. Paper U. S. Geol. Surv. No. 62, 1911, p. 137. Five plates, 1 fig.

15°, is almost everywhere exceeded, and in many places greatly exceeded. Terracing is practiced on a wholly inadequate scale. There is increased silting on the flood plains and in the stream channels, a conclusion applied to all streams draining catchment areas that have been extensively cleared.

A great deal has been written concerning the evil effects of deforestation upon soils and stream flow upon the assumption that the removal of a forest cover will inevitably and under all circumstances cause vigorous soil washing and floods. That some damage results upon the removal of a forest is axiomatic, but superlative terms can not be applied to all regions. Upon large areas of the sandy plains of the northern part of the southern peninsula of Michigan the removal of the white-pine forests has indeed affected the régime of the streams to a large though not a disastrous degree; the sandy soil imbibes the rainfall as readily as forest litter and retains it, without washing, almost as effectively as the roots and ground litter of a forest. In general it may be said that the deforestation of a plains area with a sandy soil produces minimum effects upon stream flow.

The vital feature of the forest physiography of a mountain region is the balance of power which the forest holds in the contest between soils and run-off. If in the run-off of a mountain region we grant any retarding effect at all, it follows that in those mountain regions in which the waste slopes are delicately organized and supply and demand sensitively balanced, the presence of the forest throws the advantage to the side of the soils and moderate and normal soil removal takes place at a rate compensated by the decay of the rock and soil formation. If on the other hand the forests be removed, their influence is withdrawn from the contest and run-off gains upon soil formation and disastrous soil wastage results.

The primeval forests have a peculiarly sensitive relation to these processes. In such regions the forest influence is expressed in the fashioning of the slopes, in the depth of decay of the rock, the rate of run-off, the amount of rain beating to which the soil is exposed, etc. Grant any degree of influence at all, however small, and the conclusion must follow that upon the removal of this influence — the forest — readjustment of relations must take place. The rain beats directly upon the soil, the retarding influence of the ground litter and tree roots and trunks is withdrawn, and more rapid soil removal occurs.

When once these evil effects have been allowed to take place mankind is deprived practically for thousands and even millions of years of the favorable conditions that preceded the epoch of destruction. In

Fig. 248. — Protection against erosion by parallel ditches. (Ayres and Ashe, U. S. Geol. Surv.)

Fig. 249. — Erosion checked by covering gulleys with brush, Longcreek, Va. (Ayres and Ashe, U. S. Geol. Surv.)

a hundred years man may achieve such baneful results as nature will compensate only during a geologic period of hundreds of thousands of years. Soil is a resource of priceless value. On resistant rocks its formation is excessively slow. The mills of the gods grind nowhere with more exceeding slowness than here. Many glacial striæ formed on resistant rock during the last glacial epoch, roughly 60,000 to 75,000 years ago, are still preserved as fresh as if they were made but yesterday. In that time man has come up from the cave and the stone hammer. Seventy

Fig. 250.— Erosion checked by brush dams, Walnut Run, N. C. (Ayres and Ashe, U. S. Geol. Surv.)

thousand years is a very short time for the development of a soil cover; for man it means a period so great that his mind can hardly appreciate it. The earth as we find it in the geologic to-day must be treated with care if the human race is to have a fair distribution of its wealth in time. To the geologic mind there is something shocking in the thought that a single lumber merchant may in 50 years deprive the human race of soil that required 10,000 years to form.

These considerations apply with peculiar force to the southern Appalachians, which are in a subdued state of topographic development.

Bare rock ledges are the exception; waste slopes are the rule. The eye may roam over hundreds and thousands of square miles of country and see only well-graded waste slopes, delicately organized waste removal. This means that rock decay has gained on soil removal, that the interior forces of the earth are relatively feeble, and that weathering has gained on uplift. In this interplay of forces the forests have had a prominent part and have contributed to the formation and holding in place of the soil cover, an effect reciprocally helpful to the forest. The forest cover removed, the waste slopes are dissected, soil removal gains on soil formation, and the effect is of such magnitude as to deserve the name "geologic." Nor is the effect mitigated by systems of lakes such as occur in the glaciated mountains of the northern Appalachian region, the White and the Green mountains of New England.

Lakes act as strainers and deprive the waters that flow from them of their cutting tools, so that erosion by lake-fed streams is generally far less vigorous than in the case of a lakeless land. A region of lakes is also commonly a glaciated region, and if the topography is mountainous the soil is often largely removed on the mountain slopes. It is patchy in distribution and alternates with areas of rock outcrop and ledge. Such a composition — lakes, clear streams, rock outcrops abundant, and soil patchy in development — means that the removal of the forest cover is not expressed in widespread denudation such as takes place under other conditions.

BLUE RIDGE

The Blue Ridge, so-called, forms the eastern escarpment of the southern Appalachians and extends northward from their northern extremity as far as Pennsylvania. At the north it is a true ridge; at the south it is a great scarp that descends steeply from the upper level of the mountains to the lower level of the Piedmont Plateau. The gradients of the streams that flow eastward from it are excessively steep; descents of 2000 feet are in many places made in 3 miles. So vigorous is the assault of these streams upon the divide between them and the long roundabout streams that flow westward to the Tennessee and the Mississippi that the Blue Ridge divide is rapidly retreating westward.

The Linnville River of western North Carolina has pushed its headwaters so far westward as to invade the valley of a stream flowing southwest to the Nolichucky. The valley side has been broken down and the feebler westward-flowing stream diverted to the Linnville and the Atlantic by a course about 2000 miles shorter than its former one. The profile of the Linnville shows a reversed curve at the point where it crosses the Blue Ridge, an indication that the capture has taken place so recently in a geologic sense that the gradient has not yet been worn down to normal form. In a good many other instances, as near Rutherford, North Carolina, where the South Fork of the New River is being endangered by Elk Creek, a headwater tributary of the Yadkin, capture of a similar sort is imminent, and, as the earth counts time, to-morrow will see similar diversions take place at these localities. So strong is the contrast between the gentler descents of the westward-flowing streams and the torrential descents of the eastward-flowing streams that steep headwater alcoves are often separated from well-developed meanders by only 2 or 3 miles, or less, of low divide. This great wall-like scarp has been an effective barrier to the works of man, and only one railroad, that through the valley of the French Broad, via Asheville, crosses it.   Elsewhere and. in a large number of places the railroads run to the foot of the scarp, where they end abruptly,

Eastward for several miles from the Blue Ridge escarpment, Fig. 251, the country is exceedingly broken and forms a wilderness of hills and long trailing spurs that enclose headwater coves where many rural population groups are found.   Villages and individual farms are numerous, while the "mountains" between the coves or the streams that head in the coves are wholly without inhabitants.   To the people in the coves the Blue Ridge and the mountains beyond them are the "Land of the Sky."   The escarpment is not a straight line nor a straight wall but consists of a labyrinth of coves, hills, and spurs that represent strong differential erosion of a mass of highly irregular rock.   Formerly the

Fig. 251. — Plateau and escarpment of the Blue Ridge, looking southwest from Cæsar's Head, S. C. (U. S. Geol. Surv.)

highland west of the ridge extended much farther east, but the shorter eastward-flowing streams have been rapidly extending their headwaters westward since the uplift that followed or marked the close of the Cretaceous cycle, and the spur and hill crests are but the reduced remnants of the eastern border of the upland.

Typical spurs of this sort are Haines' Eyebrow and Singecat Ridge, and among the isolated hills a typical occurrence is Pilot Mountain, whose top is covered or protected by a remnant of quartzite from which fragments are constantly breaking off and cluttering the slopes and aiding in their resistance to erosion.

North of the southern Appalachians the Blue Ridge changes its character completely.   In western Virginia and in its northern extensions in Maryland and south-central Pennsylvania it is a true ridge, but its topographic aspects are gentle and it is characterized by rounded spurs and knobs.   It is soil covered throughout and bears forests, or cultivated fields, or pastures, up to the summit.

The highest portion of the Blue Ridge north of North Carolina is about 65 miles south of the Potomac, where Stony Man and Hawk's Bill opposite Luray are 4031 and 4066 feet respectively above sea level. This is also the widest portion of the Blue Ridge, 10 to 16 miles. Northward the ridge crests are lower, and from Mount Marshall to the Potomac (50 miles) there are three deep gaps cut down to about 1000 feet — Snickers, Ashby, and Manassas gaps. Southward from Mount Marshall (3150 feet) 100 miles to the James there are numerous smaller gaps, but they are all at higher elevations, about 2300 feet above the sea.

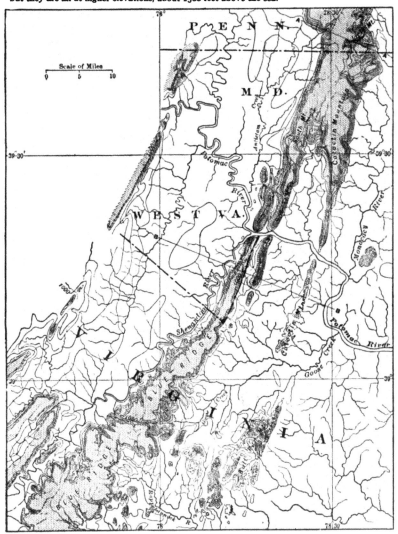

Fig. 252. — Blue Ridge, Catoctin Mountain and Bull Run Mountain in Virginia. Blank areas represent Tertiary peneplain; shaded areas represent remnants of Cretaceous peneplain now appearing as residuals above the Tertiary level. (Keith, U. S. Geol. Surv.)

The Blue Ridge extends northward
from the Potomac still as a quartzite ridge of rounded contour known
as Catoctin Mountain in Maryland
and finally into south-central Pennsylvania near Carlisle, where it is
known as South Mountain, reaching
an elevation of 2000 feet, or about
1000 to 1200 feet above the Cumberland Valley which here borders it
on the west. In New Jersey the
Blue Ridge is represented by the
highlands above Morristown, and
the Highlands of the Hudson are
really the northward continuation
and expansion of the Blue Ridge.
The latter form a narrow upland
belt 1200 feet high and 12 miles wide
that crosses the Hudson between
Fishkill and Peekskill and continues
east, merging finally with the upland
of western New England.

The Blue Ridge owes its ridge-like
quality in Virginia, Maryland, and
south-central Pennsylvania to the
harder rocks of which it is composed and to their superior resistance
to erosion. These rocks are not
everywhere uniform in character,
and variations in altitude, breadth,
and shape, of the Blue Ridge, occur
in response to variations in rock
character. Across Maryland and
into Virginia as far as Berryville the
Blue Ridge is formed upon resistant
sandstones (Lower Cambrian), and
over this portion its summit is rough
and rocky and usually sharp-
crested. South of Berryville the
main crest is formed of an epidotic
schist (Catoctin) and to some ex-

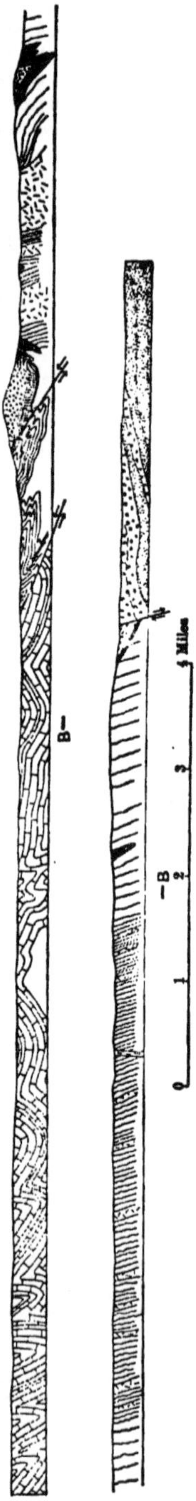

Fig. 252 a. — Cross section of the Catoctin Belt, western border of Virginia. (Keith, U. S. Geol. Surv.)

Fig. 252. — Cross sections of the Catoctin Belt, western part of Virginia. (Keith, U. S. Geol. Surv.) For location see Fig. 252.

tent of granite. The portion formed by the Catoctin schist is broad and has no very definite linear arrangement, as shown on the map, Fig. 252, so that the Blue Ridge through an expanse in breadth loses its simplicity and becomes markedly irregular in form.

East of the Blue Ridge in western Virginia and central Maryland and parallel with it is a similar ridge of resistant material known as Catoctin Mountain and Bull Run Mountain, formed for the most part of hard sandstones (Lower Cambrian), with the exception of a small part of the southern end of Catoctin Mountain, where a schist (Catoctin) makes the summit of the range. Bull Run Mountain has crests developed upon resistant sandstone.[1]

The structure of both Blue Ridge and Catoctin Mountain is synclinal, complicated by many faults; while the valley between them is anticlinal, and a similar structure is found in the Shenandoah Valley west of the Blue Ridge; so that while the Shenandoah Valley is at a lower elevation than the adjacent ridges, it is composed of younger rocks (Siluro-Cambrian limestone with occasional troughs of Lower Silurian shale). East of Catoctin Mountain the Piedmont Plateau is developed in the main on Newark strata composed of red sandstone and conglomerate.

PIEDMONT PLATEAU

The Piedmont Plateau extends southwestward from northern New Jersey and eastern Pennsylvania, its width gradually increasing from 50 miles in Maryland to a maximum of about 125 miles in North Carolina. Beyond this point it narrows and in central Alabama finally passes beneath the sediments of the Coastal Plain. Its surface dips gently eastward at the rate of about 20 feet per mile from an altitude of 1000 to 1200 feet on the west to 400 or 500 on the east.

The Piedmont Plateau is named from its position at the foot of the mountains that lie upon and beyond its western border. The name does not signify, however, that it has any of the characteristics of a piedmont alluvial plain, for its soils, topography, drainage, and physiographic development are in contrast to these features as developed on a foreland plain subject to alluviation.

The Piedmont province is a plateau only by reference to the low and flat Coastal Plain on the east, from which it rises by low bluffs or more strongly marked slopes than those developed upon the Coastal Plain. This border is marked by steepened stream descents and low falls and rapids. It is commonly designated the fall line, or fall zone. It probably marks a simple monoclinal flexure or a series of slight faults whose down-throws are toward the east.[2] The fall line is at variable distances from the coast; it is several hundred miles inland in Georgia and at the head of tidewater in the Chesapeake Bay region. Were the province otherwise situated it might never have received the name of plateau,

[1] A. Keith, Geology of the Catoctin Belt, 14th Ann. Rept. U. S. Geol. Surv., pt. 2, 1892–93, pp. 285–395.

[2] C. Abbe, Jr., A General Report on the Physiography of Maryland, Including the Development of the Streams of the Piedmont Plateau, Maryland Weather Service, vol. 1, pt. 2, pp. 115 et seq.

for its absolute elevations are but little above those of the great plains of the earth. Indeed, as seen from the crest of the Blue Ridge (looking east) it appears as a low rolling plain. From the coastal side it appears essentially as an upland tract between a lower plains region and a region of mountainous relief.

The western border of the Piedmont Plateau is more varied than its eastern border. On the south it extends to the outliers of, or to the foot of, the strongly developed eastward-facing escarpment of the Blue Ridge. The ascent to the higher province is here from 1000 to 2000 feet in short distances. Northward the Blue Ridge becomes a true ridge and not only the eastern border of a mountainous country, a character which it retains as far as South Mountain, Pennsylvania. Farther north the whole Piedmont province changes in character as described in later paragraphs (p. 629).

From any commanding point within the Piedmont Plateau the province appears not as a smooth plain but as a broadly undulating surface extending in every direction as far as the eye can reach; upon this general surface are low knobs and ridges rising above the general level of the plateau, while below the general level are numerous rather deep and narrow stream valleys and channels. The most striking feature of the piedmont topography is the even sky line formed by the rounded hilltops which fall almost into a common plane that is not dependent upon or the result of the structural features of the region. The highly inclined and folded crystalline rocks of which it is chiefly composed and the softer Triassic sediments that occur in its central portion are both beveled off or truncated by the surface of the upland. It is inferred that the region has been subjected to long-continued and active erosion involving the reduction of lofty mountain ranges that once existed here. In this respect the Piedmont Plateau is but the seaward portion of a broad, gently rolling surface that once extended westward across the Blue Ridge, north along the Appalachians into New York and New England, and south across the Cumberland Plateau of Tennessee. Since its development the surface of this lowland has been elevated and much dissected; the present higher mountain crests are the residuals which were never reduced to a lowland because of (a) their greater initial height, or (b) their greater hardness, or (c) their favorable position on stream divides, or (d) a combination of two or all of these conditions.

The even contour of the surface of the Piedmont Plateau is not due to the underlying rock formations, for their structures are so diverse and complex as to produce a complex topography. The plain surface is due to peneplanation. An interesting survival of drainage conditions since peneplanation is shown in the courses of the larger streams, which are quite independent of the structure and character of the rock floor, while many of the tributary streams show adjustment to the character of the rock floor. The heterogeneity of the rock and the complexity of its structure are reflected in the indirect courses of the tributary streams and in the complex character of their valleys; but these structures are without expression in the case of the master streams.

The Tertiary cycle of erosion in which the peneplain of the piedmont region was formed was closed by uplift and in consequence the gradients of all the streams were steepened and their erosive powers increased. Down-cutting was the immediate result, and since this has been accomplished a certain amount of lateral swinging also has been accomplished, so that many of the larger streams have limited flood plains and the smaller ones have graded waste slopes. The piedmont has been well dissected by the erosion of the present cycle, but dissection has not yet gone far enough to destroy the accordance of the hilltops, whose approach to a common altitude in a region of disordered rocks is the strongest evidence of the former existence of a base-leveled plain throughout the region. In harmony with this condition is the deeply weathered character of the rock which mantles all the hill slopes and even the hilltops, and under the influence of gravity and rain wash is slowly creeping down to the streams.

So deeply decomposed was the rock of the Piedmont Plateau during the last stages of the long erosion cycle which resulted in its peneplanation that in spite of its later uplift and dissection many of the streams have not yet cut down to fresh rock.[1]

The middle and lower courses of the piedmont streams are gorge-like; the headwater tributaries commonly flow in broad shallow valleys separated by low rounded divides.[2] With the gradual deepening and widening of the lower courses more marked dissection will take place along the headwater streams and their drainage basins will be roughened accordingly.

The rocks of the Piedmont Plateau consist of a number of types each of which has had an important influence on the topography. They are chiefly crystalline and are derived in part from original sediments and in part from original igneous masses. They include crystalline gneisses and schists associated with crystalline limestones, quartzites, and phyllites intruded by granite, and a large variety of other types.

[1] L. C. Grafton, Reconnaissance of Some Gold and Tin Deposits of the Southern Appalachians, Bull. U. S. Geol. Surv. No. 293, 1906, p. 13.

[2] N. H. Darton, Washington Folio U. S. Geol. Surv. No. 70, 1901, p. 1.

The gneisses, granites, and gabbros all offer about the same resistance and form by far the greater part of the general surface of the plateau. They are developed in the form of rounded hills and gentle slopes on the upland surfaces. Bands of varying resistance are distributed through these rocks, but they are very irregular in size and position and are therefore expressed in the topography by alternate softening and intensification of the contours of the valley walls and in expansions and contractions of the gorge floors. The phyllites, somewhat softer than the gneisses, form more rounded hills, gentler slopes, and valleys of broader proportions. Lenses of limestone and marble in the phyllite, with a high degree of solubility, have resulted in the development of broad flat-bottomed valleys. These limestone and marble bands are the weakest topographic factors among the piedmont rocks.

The more resistant rocks and those most prominent topographically are serpentines, slates, quartz-schists, and quartzites, which stand out as ridges or rounded knolls above the surrounding gneiss. The serpentine is most striking topographically where it is crossed by streams; at such points, steep, bowlder-strewn, rocky gorges and rough channels have been developed. While the form and size of the piedmont valleys have a relation to the geologic structure, the courses of the streams on the whole have not been strongly influenced by the arrangement of the bed-rock, and this is true particularly among the larger streams, which flow markedly independent of the structure.

TRIASSIC OF THE ATLANTIC SLOPE

From the Minas Basin in the Bay of Fundy to the northern boundary of South Carolina there occurs at intervals a geologic formation of great physiographic importance, for its topography, soils, and vegetation are to a large degree exceptionally developed and are unlike the topography, soils, and vegetation of the bordering crystalline rock. The formation is of Triassic age and is commonly known as the Triassic formation of the Newark system of rocks, and occupies about 10,000 square miles of territory. The general distribution of the formation is shown in Fig. 254, from which it will be observed that it does not occur as a continuous body of rock but as a series of elongated and detached areas occupying local basins of sedimentation. The longer axis of each area is roughly parallel to the main trend of the formation as a whole. There are in all thirteen major areas besides a group of smaller areas. As a whole the belt is about 1200 miles long and always less than 100 miles wide.

The principal rock members of the Newark system are sandstones, shales, and conglomerates, with local beds of slate and limestone and a certain marginal development of arkose and breccia. Almost all the areas of Newark rocks have associated sheets and dikes of basic rock, chiefly basalt and diabase known under the collective name of trap. The dikes trend as a rule from northeast to southwest, cross indifferently

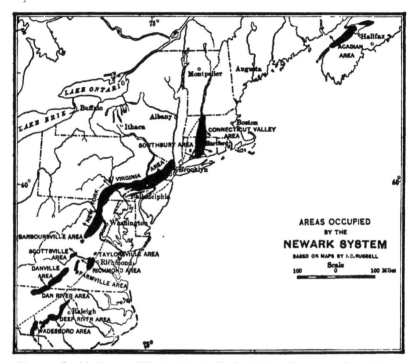

Fig. 254. — Local development of Triassic rock in the older Appalachians. The arrows indicate the direction in which the strata dip.

from sedimentaries to crystallines, and are narrower in the crystalline rocks bordering the area than they are in the sedimentary rock.

In part the trap sheets are extrusive and in part intrusive. The trap of Nova Scotia is extrusive in origin; most of that in the Connecticut Valley is extrusive, with important intrusives along West Rock Ridge and the Barndoor Hills north of the Ridge; along the eastern edge of the New Jersey Triassic are the Palisades, also of intrusive origin. In the Richmond and Catoctin areas the igneous rocks are of intrusive origin and usually occur in the form of sheets or sills, and dikes.

The widespread occurrence of faults of all degrees of displacement from a few inches to several hundred feet has caused the repeated outcrop of individual strata. The faulting has almost everywhere resulted in either westward or eastward dips, roughly at right angles to the trend of the basins in which the formation lies. The larger faults are believed to affect the underlying crystalline rock as well as the sandstone and the trap.

In general the Triassic rocks have a monoclinal structure throughout and all have marginal faults; in some cases this is pronounced along one

Fig. 255. — Relations of the igneous rocks to the sedimentary strata (Newark) in New Jersey, Paterson quadrangle. Vertical scale three times the horizontal. True profile in lower section. (U. S. Geol. Surv.)

border and in all cases it is developed to some degree. Some of the faults are large enough to bring to the surface the crystalline rock floor on which the sediments were deposited; others expose the basal conglomerate; and still others expose only the lower trap sheets or the

Fig. 256. — Palisades of the Hudson from the Jersey side, looking south. The vertical cliff of diabase and the sloping talus are characteristic. (U. S. Geol. Surv.)

intermediate finer shales and sandstones. In addition to faults, broad undulations occur, but no true folds have been observed. The faults in many notable instances pass directly into the crystallines, where they produce marginal features of importance, although within the crystal-

lines the structural and textural differences are not sufficiently marked
to give rise to a type of topography as distinctive as that formed in
the areas of Triassic rock where uniform structures prevail over wide
areas and where strong and sudden alternations in hardness are the rule.

The topographic expression of these belts of Triassic rock is not
everywhere the same, but depends upon the nature of the surrounding
country rock and the elevation.   In the Connecticut Valley the Triassic rock rests upon and is bordered by crystalline rock which is very
much harder than the sandstone.   The elevation of the region above
the sea is in the main from several hundred (near the sea) to 2000 feet
and less (in the interior), and erosion has therefore been sufficiently
active to degrade the soft rocks to a lowland, while making so little
impression upon the hard rocks as to leave them in the form of uplands.
On the other hand the Richmond basin of Virginia is a plain continuous with the plain developed upon the surrounding rock.   In th:s case
it is not possible, except in a few localities, to infer any geologic change
from a change in the topographic level,[1] a condition due not to the
character of the surrounding rock, which is crystalline as in New England,
but to the absence of pronounced elevation.   The Richmond Triassic is
on the eastern edge of the Piedmont Plateau and only a few hundred
feet above sea level.

The characteristic features of the Piedmont Plateau as developed in
Georgia, South Carolina, etc., are modified to an important degree or
wanting altogether over large portions of northern New Jersey, eastern
Pennsylvania, and central Maryland, Virginia, and North Carolina because of the occurrence of Triassic sandstone, the largest body of which
occurs in New Jersey and Pennsylvania, Fig. 254.

In New Jersey and Pennsylvania the Triassic is a gently rolling plain
from 100 to 400 feet above the sea.   It is lowest along its southeastern
margin.   The hills for the most part show a distinct northeast-southwest
trend, which coincides with the strike of the underlying strata, and in
general reflects the slight variations in texture in the different sandstone
layers.   The relief is uniformly slight.   Practically all the slopes are
long and gentle and covered in most places with a thick, fertile soil.
The general continuity of the plain is interrupted by valleys cut below
the general level and by hills, ridges, and plateaus of harder rock that
surmount the plain.   The region was reduced to a peneplain during the
Tertiary, and it is this that accounts for the general plain-like character
of the region, but the reduction was only partial and the hard outcrops

---

[1] Shaler and Woodworth, Geology of the Richmond Basin, 19th Ann. Rept. U. S. Geol.
Surv., pt. 2, 1897-98, p. 393.

of the included trap were in this cycle but little reduced below the level to which they were brought during the Jurassic-Cretaceous cycle of erosion. The crystalline rocks of the Piedmont are overlain and concealed by the Triassic sandstones and shales except, so to speak, on the four corners of the depression in which the Triassic rocks were deposited. Here the crystallines run down into long tapering bodies of

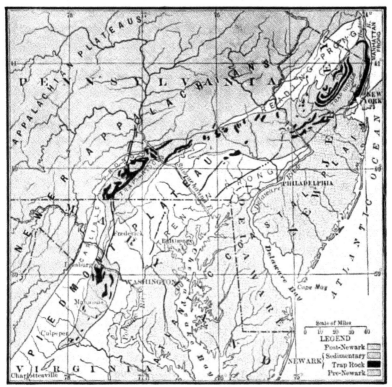

Fig. 257. — The four crystalline prongs of the older Appalachians which enclose the New York-Virginia area of Newark rocks. (Adapted from Russell, U. S. Geol. Surv.)

rock upon which is developed a topography of a type far different from that developed upon the Triassic strata.

To these four narrow, tapering tracts the name *prong* has been applied by Professor Davis and each has been given a distinguishing name after the name of the town near its terminus. The prong that extends across southeastern Pennsylvania to Trenton is called the Trenton prong; the prong whose local name is South Mountain and which terminates near Carlisle, Pennsylvania, is the Carlisle prong; the narrow upland of crystalline rock which crosses northwestern New Jersey and terminates near Reading on the Schuylkill is the Reading prong; and the prong that includes and terminates Manhattan Island is called the Manhattan prong. Each prong has certain distinctive features which will now be described.

The Highlands of New Jersey, the Reading prong, are made up of long, parallel, continuous ridges, separated by limestone valleys.

"The side slopes are often steep and the soil on them is generally thin, rocky, and poor. Between the ridge hills lie secluded vales of rich farming lands, opened out on soft limestones, more or less hilly on a small scale, and directly comparable to the larger valleys of New Jersey. Similar valleys, though less numerous, are also found in other parts of the range, as Saucon Valley, occupied by the creek of the same name, and Oley Valley, near Reading." [1]

Toward the west the Reading prong consists of short, rounded, semi-detached hills, often stony and rugged, ranging roughly northeast and southwest. Although the hills over the greater part of the area are generally rounded, the southern slopes are in most cases somewhat steeper, due to the inclination of hard strata at a high angle toward the south. The rather level summits of the highest hills reach a fairly uniform elevation between 900 and 1000 feet; occasional crests rising 100 or more feet higher represent residuals on the old peneplain surface of which all the summits were a part.[2]

The Carlisle prong (South Mountain, Penn.) extends south from opposite Carlisle, in Cumberland County, has a greater altitude and a more marked development of the ridge and valley type of surface than the Reading prong, and merges southward into the distinct single crest of the Blue Ridge. The general elevation is not less than 1000 feet throughout the range, though it increases toward the south, until along the Maryland line it reaches 2100 feet. As a result of greater elevation the hills toward the south appear less subdued than in the Reading prong, though there are no rocky peaks and few bare cliffs. The uniformity of upland level is also less marked over the area in general than in the case of the Reading prong, though from a distance it presents a smooth, even sky line.[3] The prong rises steeply from the surrounding valley plains and exposes a core of ancient volcanic rock and quartzite. The general structure is that of a broad uplift with minor folds on its surface which produce offsets of considerable magnitude.

[1] W. S. Tower, Regional and Economic Geography of Pennsylvania (Physiography), Bull. Geog. Soc. Phil., vol. 4, 1906, p. 21.
[2] Idem.
[3] Idem.

From its narrow northeastern extremity at the rapids of the Delaware just above Trenton, where the hills are very low, the Trenton prong extends southwestward, widening gradually. The general elevation of the upland level slowly rises from about 400 feet to 600 feet on the south, and the surface has an eastward slope to the Delaware River; the streams have cut many tortuous valleys and shallow ravines below the general level to depths of 100 or 200 feet. The area may be described as a rolling country of rather even-topped hills and shallow dales. The hills are generally low. They are flat, to rounded or dome-shaped, with summits at a generally uniform level, expressive of the former base-leveling, and have gentle slopes coated with a thick layer of soil.[1]

The Manhattan prong descends from the level of the rugged uplands east and south of the Highlands of the Hudson to sea level at New York. The submergence of its outer border is marked and has made the lower Hudson a tidal estuary, broadened and deepened the East River, and given the coast an embayed character similar to that exhibited along the whole coast of New England.

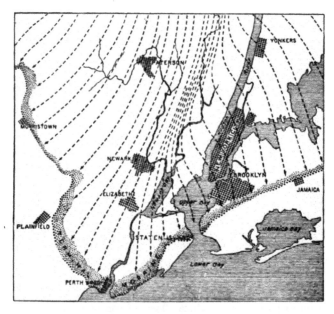

Fig. 258. — Terminal moraine and direction of ice movement in the vicinity of New York.
(U. S. Geol. Surv.)

[1] W. S. Tower, Regional and Economic Geography of Pennsylvania (Physiography), Bull. Geog. Soc. Phil., vol. 4, 1906, p. 19.

In northern New Jersey lakes and ponds were formed in the glaciated belt by the irregular disposition of the glacial drift in such a manner as to make basins; while glaciation also gave rise to minor topographic features in marked contrast to those produced by erosion. The drift in northern New Jersey is thin, its average thickness being probably not over 15 feet, while the general relief is measured in terms of several hundred feet. Had the drift been disposed uniformly over the surface it would have had small effect on the topography. But in some

Fig. 259. — Characteristic terminal-moraine topography. (U. S. Geol. Surv.)

places the rock was left bare, while in others the drift was accumulated to a depth of scores of feet. In general the valleys received a heavier drift covering than the uplands, a covering which consists not only of glacial deposits themselves but of fluvio-glacial deposits related to them. The result of the filling of the valleys and the erosion of the intervalley ridges and uplands was to diminish the relief, though such diminution is nowhere marked.[1]

SOILS OF THE PIEDMONT PLATEAU [2]

The soils of the extreme northern part of the Piedmont region, in New Jersey, are glacial, but elsewhere they are purely residual in origin, and have been derived almost exclusively from the weathering of igneous and metamorphic rocks. The chief exceptions are the detached areas of Triassic sandstones and shales. Marked differences in the character of the rock and in the methods of formation have given rise to a number of soil types, those derived from crystalline rocks being the most numerous and widely distributed.

The most important and widely distributed soils of the Piedmont Plateau are known as the "red-clay lands" and are characterized by red clay subsoils, with gray to red soils ranging in texture from sand to

[1] R. D. Salisbury, The Physical Geography of New Jersey, Final Report of the Geol. Surv. of New Jersey, 1895, pp. 155–156.

[2] Data chiefly from Soil Survey Field Book, U. S. Bur. Soils, 1906.

clay, the lighter colors prevailing with the sandy members. The red-clay soils are of residual origin, derived from the degradation of igneous and metamorphic rocks which have been weathered generally to great depths, so that outcrops are rare. Fragments and bowlders of the parent rocks are, however, found on the surface to a variable extent.

An important type of red clay soils is a gravelly loam which has been derived from the breaking down of granites chiefly of a coarse-grained variety; it represents a less complete weathering of the rocks than some of the other types. The soil is a brown sandy loam about 7 inches deep, carrying variable quantities of feldspathic or quartz gravel. The subsoil is a heavy, micaceous, red loam or clay loam containing considerable gravel. Outcrops of granite appear in many places. A prominent feature of this type is a lack of tenacity in both soil and subsoil, as a result of which the land erodes and gullies in a serious manner. As a rule it occupies high broken uplands and the drainage is good. The characteristic timber growth is hickory, shortleaf pine, and some cedar.

The fine sandy loam type of red sand soils is formed chiefly from the weathering of talcose schists and slates. It is light gray to pale yellow in color. Hickory, oak, and pine are the growths found on the better-drained areas, while gums are characteristic of its poorly drained phases. The Indian or purplish red soils derived from the weathering of red sandstones and shales (Triassic) are developed in detached areas in shallow basins in the Piedmont from New England to South Carolina.

The occurrence of a large number of fragments of shale (from 10% to 40%) in those soils derived from Triassic shales is characteristic, whence the local name "red gravel land" for the shale loams. The Triassic sandstones weather into gravelly and sandy loam types where they are coarse and into pure loams and silt loams where they are fine. A stony loam type consists of very stony land, hilly to mountainous in character, generally covered with a natural forest of chestnut and oak. The soil consists of a rather heavy red loam, 8 to 10 inches deep, containing from 30% to 60% of red or brown sandstone fragments. The subsoil is of much the same character to a great depth. This type is derived from the more siliceous or hardened phase of the Triassic sandstone. It is well adapted to forestry and orcharding, and the more level areas, when the stones are removed, to general farm crops.

The soils of residual origin, derived principally from mica schists, have yellow or only slightly reddish subsoils and gray or brown surface soils. They are also micaceous and subject to erosion. Locally they are known as "gray lands," to distinguish them from the "red lands" derived from the Triassic sandstones.   The topography is in general not so rough, being rolling to moderately hilly.   They are not deeply weathered, and the underlying rock is often encountered within 2 feet of the surface on slopes where erosion is pronounced and rarely more than 10 to 15 feet below the surface.

The gneisses and schists of Maryland, Pennsylvania, and Virginia weather into a stony loam containing from 30% to 60% of stony material.   The type occurs on the summits of hills and ridges and on steep slopes where the drainage is good.

CHAPTER XXX

OLDER APPALACHIANS (*Continued*)

NORTHERN OR NEW ENGLAND DIVISION

THE New England physiographic province is continuous with New Brunswick and Nova Scotia and that part of Quebec lying south of the St. Lawrence River. To a large degree it has shared in the physiographic history of these districts, the Laurentian Plateau of Canada, and the great Appalachian province of the United States of which it forms a large division. Like the adjacent regions a large part of the New England province was peneplaned (Jurassic and Cretaceous time), then uplifted, extensively dissected, and glaciated. In general terms it may be described as an upland, for it possesses a somewhat uniform upper surface which extends from an elevation of several hundred to two thousand feet above the sea. But this term can scarcely be applied to its outer margin. Recent depression has submerged the border of the region, and the upland character is not therefore conspicuous on the immediate shore. The word upland or plateau, often applied to the New England province, also requires important modification when applied to certain interior portions of the New England region; for while most of the southern half of New England is accurately described as a dissected upland plain or plateau, the northern portion of New England bears important mountain groups, the White Mountains of New Hampshire, the Green Mountains of Vermont, and other groups of lesser height and size, besides a number of isolated mountain peaks such as Mount Katahdin in central Maine and Mount Monadnock in southern New Hampshire. Indeed, so mountainous are large parts of New Hampshire and Vermont that the mountain feature and not the plateau feature is conspicuous, and over large areas the plateau feature is not expressed at all.

The western portions of Massachusetts and Connecticut are also more rugged than the central and eastern portions. The principal elevations are the Hoosac Mountains, the southern extension of the Green Mountains. Even in southern New England almost every general view embraces one or more residuals of former lofty mountains. As elevations they are scarcely ever of notable scenic interest, though in a moderate way they are attractive with their forest-clad summits and slopes, drained by a large number of cool, spring-fed, torrential brooks. They are but the unimpressive remnants of a once more mountainous country. In place of these trifling elevations were mountain peaks and ranges of alpine height bearing upon their flanks snow fields and perhaps glaciers.[1]

UPLAND PLAIN OF NEW ENGLAND

The detailed topographic forms of the *southern* New England region have been carved in the Cretaceous peneplain since its elevation; for this reason a knowledge of the peneplain feature is of great importance in the study of the present topography. The argument for peneplanation is sustained by the striking accordance of level between different portions of the old peneplain developed alike on the crystalline rocks of the eastern and western uplands and the tilted lava sheets of the Triassic area of Massachusetts and Connecticut. Furthermore, no other process but peneplanation would give so uniform a slope to the surface southward toward Long Island Sound.[2] An imaginary plane passed through the hill summits of the region would slope southeastward gently and would rest on nearly all the summits of both the eastern and western uplands.

CRETACEOUS AND TERTIARY PENEPLAINS

During the Jurassic and early Cretaceous periods the New England region, of rugged aspect because of displacements at the end of the

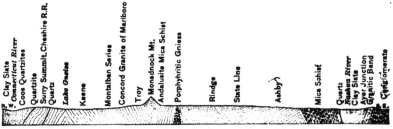

Fig. 260. — Profile across central New England, showing uplifted peneplain and the Mt. Monadnock residual. (Hitchcock.)

[1] Among the isolated residual elevations of New England Mount Monadnock in southern New Hampshire is typical. The name of the mountain is therefore applied to any isolated residual. Mount Monadnock rises 3866 feet above mean sea level and about 2000 feet above the surrounding plateau of southern New England. The peak does not stand alone, but is one of a family of monadnocks which includes Wachusett, Watatic, Asnebumskit, and others. All of them are believed to owe their survival to their position on the divides between the major streams rather than to the greater resistance of the rocks composing them. See J. H. Perry, Geology of Monadnock Mountain, New Hampshire, Jour. Geol., vol. 12, 1904, pp. 1-14.

[2] W. M. Davis, The Geological Dates of Origin of Certain Topographic Forms on the Atlantic Slope of the United States, Bull. Geol. Soc. Am., vol. 2, 1891, p. 557.

Triassic, was eroded practically to sea level. Topographic irregularities disappeared, hills melted under the forces of erosion until only their roots remained. A slight depression then occurred which submerged the outer border upon which were deposited Cretaceous sediments of considerable depth and horizontal extent. At the beginning of the Tertiary period New England was again uplifted, but the uplift was of a broad regional variety. During uplift a slight tilt was given the land surface in the direction of the sea, so that the interior valleys were cut far below the general level. Dissection followed upon the uplift of early Tertiary time, the overlapping Cretaceous sediments of the southern border of the region were entirely removed, and the limestones of the Housatonic Valley, less resistant than the crystalline rocks in which they were enclosed, were reduced to a narrow lowland. The soft Triassic shales and sandstones of the Connecticut Valley were also worn down, forming a local peneplain bordered by crystalline uplands and surmounted by resistant trap ridges and hills.

Again uplift occurred, which invigorated the streams and led not only to the deeper dissection of the remnants of the uplifted Cretaceous peneplain but also to the dissection of the lower (Triassic) and more local lowland. The dissection of both the Cretaceous and Tertiary peneplains has everywhere brought into relief areas of softer and harder rock, even among the prevailingly hard crystallines where a relatively small degree of variability in resistance is the rule. Since the period of maximum uplift and dissection of southern New England there appears to have been a depression, for the outer border of the region is now drowned, as shown by the estuarine bays at the mouths of all the coastal streams and by Long Island Sound, which now occupies the inner lowland of the Coastal Plain, whose remnants are Long Island, Martha's Vineyard, Nantucket, and Cape Cod.

GEOLOGIC FEATURES

The chief structural features of New England are (1) a western mountain axis extending through Vermont and western Massachusetts into Connecticut, (2) an eastern mountain axis extending through Maine and New Hampshire southward to the Sound, (3) a long, narrow, structural depression between these axes — the Connecticut Valley, and (4) two ancient basins on the eastern border of the province, the Boston and Narragansett basins. Each mountain axis is a line of topographic elevation on the north but has been worn to a lowland plain on the south, and, more recently, raised to form an upland. Since dissection accompanied and followed uplift the plain has become a dissected upland.

Thus the Green Mountains extend through Vermont as a prominent mountain range but change from a range-like elevation to a plateau near the northern boundary of western Massachusetts. In a similar way the White Mountains terminate near the northern border of Massachusetts. The Connecticut Valley is everywhere a structural as well as a topographic depression, but it too has a northern and a southern section of unlike characteristics. On the common border of Vermont and New Hampshire it is a great synclinorium[1] of crystalline Paleozoic rocks whose valley character has been emphasized by the predominance of rather easily denuded mica schists[2]; its southern continuation across Massachusetts and Connecticut is developed upon a block-faulted and much dissected mass of sandstones, shales, and conglomerates (Triassic) and intercalated trap sheets.

The mountains and uplands east and west of the Connecticut Valley are composed mainly of crystalline rock — schists, gneisses, and granites of many varieties. In the western part of the province a considerable body of limestone occurs, a fact of physiographic importance since the limestones are characterized by topographic depressions and deep soils. The schists, gneisses, and granites are of variable composition and have been variously affected by the weather. The granite gneiss of Lighthouse Point, New Haven, is a very resistant rock and is but little weathered; the chlorite schist west of that city and on the eastern edge of the western upland is more deeply decayed, the planes of schistosity offering relatively easy means for the penetration of weathering agencies. In general the north-south trend of the structure is brought out rather clearly by the differences in resistance among the various rock types. The main valleys have been carved on north-south lines and the intervening ridges trend in the same direction. Railway construction on north-south lines is comparatively easy; construction on east-west lines is almost as difficult as in a mountainous country.

Differences in rock character are largely responsible for differences in the valley widths in both the western and eastern uplands. The upper Housatonic and the Deerfield illustrate this relation admirably. The one flows southward in western Connecticut through a limestone belt and has developed a broad, deep valley of exceptional size; the other traverses a belt of resistant gneiss and its valley has a canyon-like aspect. The limestone is so soft relatively that good exposures of it are rare in individual stretches of a mile or more; many of the schist

---

[1] Pumpelly, Wolff, and Dale, Geology of the Green Mountains in Massachusetts, Mon. U. S. Geol. Surv., vol. 23, 1894.

[2] C. H. Hitchcock, Geology of the Hanover Quadrangle, Rept. State Geol. Vt., 1905.

and gneiss outcrops are so resistant that the striæ of the glacial period are still clearly defined upon them.

EFFECTS OF GLACIATION

TOPOGRAPHIC AND DRAINAGE MODIFICATIONS

The last geologic event of topographic importance was the invasion of New England by the continental ice sheet which covered the northern portion of America. The ice removed the deep soils of the peneplain from the interstream areas where they probably persisted even after preglacial uplift and dissection had removed large portions of them. The bed-rock was scoured and its weathered upper portion largely removed and fresh rock exposed. Rock ledges were commonly grooved, scratched, polished, and rounded by the land waste which clogged the lower portions of the ice, and minute irregularities of form and drainage were thus imposed upon the landscape. Glacial action produced a characteristic roughening of the surface, areas of soft rock were excavated to a greater depth than surrounding areas of hard rock, and reversed slopes were formed not only in the localities of glacial scour but also in those of glacial accumulation. A rearrangement of stream channels resulted in the filling of many depressions, and the formation of lakes and ponds and numberless falls and rapids.

TILL DEPOSITS

Of great interest in any glacial region is the distribution and extent of the glacial débris. New England was neither covered with heavy sheets of till as was the case with Illinois, Indiana, and other central-western states, nor so denuded of soils as Labrador and portions of Newfoundland. Although rock outcrops are common and even abundant in southern New England, the surface consists in the main of a thin sheet of glacial material largely of local origin and of very complex composition. A part of this material was brought into place by the glacial ice as subglacial drift, but the larger part is probably englacial material that was dropped in a confused and irregular manner upon the country as the ice sheet gradually melted away and disposed in large part regardless of the underlying topography. The lower few hundred feet of ice must have been heavily clogged with drift after the manner of the Greenland ice sheet to-day; and during the last stages of melting the stagnant ice must have dropped its enclosed material upon the surface in the most irregular manner and largely without regard to the form of the surface upon which the ice rested. The relative absence of a drift cover on the higher elevations, as on the trap ridges of the Connecticut Valley, is probably owing (1) to the greater freedom of the ice at this elevation from drift, and (2) to postglacial erosion. The lower hilltops and many of the hill slopes bear quantities of glacial material. Hillside and hilltop farms are common, and their soils, when cleared of bowlders, are not sterile and easily impoverished, as is so often stated, but include some of the most naturally productive lands of the region. Their chief defects are an excessive amount of bowlders and a tendency to erosion when kept cleared of the native vegetation. Locally there exist extensive areas of glacial till, as in Aroostook County, Maine, and in all the valley lowlands, especially the Connecticut Valley lowland. Such areas are as a rule intensively farmed, whether they are of large or of small extent. They and the drained swamps and valley floors of the coastal regions represent the chief basis of the agricultural interests of New England.

The more rugged northern portion of New England bears a smaller quantity of glacially derived land waste. Its more northerly position subjected it to more intense erosion because the southward-flowing ice was there thicker and hence more powerful, while the greater relief of the section gave abundant opportunity for the thorough scouring of the more exposed and higher masses against which the ice impinged. Its hilltops and sides are bare or the rock is thinly veiled with a bowldery till. The most notable accumulations of material are in the valleys and especially at the point of convergence of several valleys where englacial material, often in large amounts, was dropped at the time of the disappearance of the continental ice sheet.

ESKERS, DELTA PLAINS, AND SAND PLAINS

Eskers occur in greatest abundance in southern Maine, where they were formed during the waning stages of the glacial invasion. They represent channel deposits of subglacial and englacial streams that discharged across the zone of wastage of the continental ice sheet. They form a not inconsiderable part of the total area and are conspicuous relief features, often extending for a score of miles without a break of any sort and with even summits, again crossing hills and valleys, the level undulating in sympathy with the general topography. Their composition is rather uniformly of sand, gravel, and stones of medium size, the whole with various degrees of stratification but always somewhat sorted. Their slopes are commonly so steep as to be uncultivated and wooded. The tops are so narrow that although flat they are not of sufficient area to make cultivation worth while. In the other New

England states they are found here and there, but they do not generally form important elements of the relief nor have that great length that marks the eskers of Maine.

In a flat region glacial sand plains are often inconspicuous topographic features, but in so rough a country as New England every flat tract, be it sand plain, salt marsh, or valley floor, introduces into the landscape an unusual element that catches the eye and enhances the view. The general roughness of the surface of New England is responsible for the formation of a large number of sand plains, glacial deltas, and allied forms that are important topographic features. During the period of deglaciation a large number of small water bodies developed in front of the ice as the ice impounded the drainage of some local basin draining toward it, and into these water bodies streams were discharged and deltas were accumulated. Later drainage of such lakes by the disappearance of the ice exposed the deltas and allowed their dissection.

Where alluvial material was accumulated in front of the ice and not in a body of standing water it commonly formed a sand plain or valley train, the one if deposition took place upon a rather flat plain, the other if deposition took place in a narrow valley. Locally sand plains are sometimes found along the borders of valleys where deposition took place at the mouth of some tributary while the main valley was still filled with ice. Such plains are commonly associated with rude terraces of which they themselves form a part. The terraced effect was gained by the disappearance of the ice, which left the outer edge of the accumulated material unsupported.

STREAM TERRACES

The influence of glaciation is also expressed in stream terraces built during a period of stream aggradation associated with the retreat of the ice. It now seems probable that aggradation was enforced as much by the lower gradient of the valleys during the waning of the ice sheet as by the abundance of land waste contributed to them. In this view the change from aggradation to that degradation which resulted in the terracing of the aggraded material was invoked by both decreased waste supply and tilting of the land. The tilting is common in a very large region including at least the Great Lake district, New England, and the Laurentian Plateau.

The Connecticut and Merrimac valleys on the New Hampshire boundary have been to some extent mapped and leveled and the terraces identified. The highest terraces and deltas of tributaries represent the remnants of an ancient flood plain; they stand quite uniformly about 200 feet above the Connecticut River, except where a tributary enters.[1]

Farther south the attitude of the Connecticut Valley was such as to favor the formation of a succession of lakes. There were three main bodies of water — Montague, Hadley, and Springfield lakes — separated from each other by the central trap ridges. Tributaries accumulated sand and gravel deltas on the lake margin, fine laminated clays (in which arctic leaves have been found) were spread over the lake floors, and shore lines were formed. On the northern border of Massachusetts these accumulations are now 380 feet above the sea, at Northampton they are at 300 feet, on the southern border of the state they are at 180 feet, and at New Haven correlative deposits are near sea level. The southward tilting of the land which these southward-diminishing elevations imply, drained the lakes, gave the Connecticut greater erosive power, and resulted in the formation of the great terraces that now border that stream and enhance its scenery as well as the productivity of the valley.[2]

While the New England terraces are everywhere the product of either the tilting of the land or decreased waste supply or both, their disposition, size, and field relations offer a great variety of conditions. In the Connecticut Valley they seem in the main to have been controlled in their size and shape by natural unrestrained meander growth and stream deflection. In the Westfield Valley, and presumably in a large number of others not yet examined, their development appears to be guided largely by the rock sides of the preglacial valley.[3]

LAKES

Lakes are among the most characteristic features of a glaciated surface and are of special interest in connection with forests and stream flow. They are very abundant in New England, Connecticut alone including more than 1000, and together with swamps and marshes they occupy 145 square miles, or about 3% of the surface of that state. In Maine the proportion of surface occupied by lake is greater than in any other state in New England; 5% to 10% of the total area of the larger catchment basins is the rule. The lakes have a very beneficial effect upon both the climate and the run-off. They act as great reservoirs, and though they rise and fall with the seasons, they do so over larger areas

[1] C. H. Hitchcock, The Geology of New Hampshire, Jour. Geol., vol. 4, 1896, p. 61.

[2] B. K. Emerson, Holyoke Folio U. S. Geol. Surv. No. 50, 1898, p. 3, col. 4.

[3] W. M. Davis, River Terraces in New England, Bull. Mus. Comp. Zoöl., vol. 38, 1902, pp. 281-346.

than the rivers and therefore by smaller vertical amounts. They steady the river discharges by holding large volumes of water but little above the level of the outlet both in time of flood and in time of drought.

It is perhaps the general conception that of two groups of streams occurring in regions of equal rainfall and comparable relief, that group that has the larger number of lakes will have the more even flow through the year. In this view the rivers of New England should have fluctuations of lesser value than the lakeless streams of the broken portions of the Middle Atlantic States. To test this generalization the author made some computations of the average discharge of 30 New England streams from the St. John to the Connecticut during the months of May and November, 1908, the months which appear by rough estimation to represent the greatest departures in opposite directions from the normal discharge for 1908.[1] It was found that the ratio of discharge in the month (November) showing least discharge to the month (May) showing the greatest discharge is 1 : 11.6. Similar computations for the 14 streams between the Susquehanna and the Rappahannock, almost wholly outside the glaciated area and the region of the lakes, is 1 : 7. The influence of the lakes would seem theoretically to throw these ratios in the other direction. The cause for this curious condition is probably found in the facts (1) that the northern streams are fed by the rapidly melting snows of spring, the discharge lagging behind the time of maximum melting, and (2) that the thinner soil of New England enforces a more rapid run-off.

A partial test of this explanation would be the examination of two stream systems in the glaciated area, one in the region of no snow or of early melting snows and the other in the region of late snows.

Were it not for the lakes the fluctuations of stream discharge would be still greater in the lake region than they now are, for the effect of the rapid melting of the snows would not then be mitigated. A thorough test of this conclusion would require an examination of streams throughout whose catchment areas the forest cover is in the same state of preservation and the topography is of the same quality but whose drainage systems contained widely different percentages of lake basin. A partial test is the contrast afforded by the St. Croix and the Penobscot. The St. Croix basin is well covered with forest, the Penobscot is about two-thirds covered with forest; the former has about 10% of lake and pond surface, the latter about 6%. In harmony with these figures are the ratios of discharge for May and November, 1908, which are 1 : 5 for the St. Croix and 1 : 8 for the Penobscot. The Merrimac basin has about 3.6% of lake and pond surface, and the ratios of discharge for May and November are also 1 : 8 respectively for this river.

These figures clearly show the influence of the lakes in retarding the flow and also how disastrous the floods of New England would be if

---

[1] The basis of these computations is the summary table by Barrows and Bolster, The Surface Water Supply of the United States, 1907–8, pt. I, North Atlantic Coast, Water-Supply Paper No. 241, 1910, pp. 342–344.

the lakes did not exist. Though New England is a lake country, its forests are needed to help equalize stream flow to an even greater degree than more rugged southern districts where snows are absent or light or melt at more regular intervals and where the soils are deep.

SUBREGIONS OF THE NEW ENGLAND PROVINCE

Having examined the general physiographic features of the New England province we shall now consider the special features of the topography, drainage, and soil of a number of exceptional subregions.

The largest of these subregions are:

(1) White Mountains and bordering uplands.

(2) Green Mountains and bordering uplands.

(3) Connecticut Valley lowland and associated trap ridges.

WHITE MOUNTAINS AND BORDERING UPLANDS

The most interesting and important mountains in New England are the White Mountains of New Hampshire. Their highest summit, Mount Washington, is 6290 feet above the sea and is the nearest New England rival of Mount Mitchell, N. C. (6711 feet), the highest peak in the eastern part of the United States. The most important single feature of the White Mountains is the Presidential Range, which besides Mount Washington includes Mount Jefferson, 5725 feet, Mount Adams, 5805 feet, Mount Madison, 5380 feet, Mount Monroe, 5390 feet, and a number of other peaks of comparable altitude. This range extends roughly north and south, which is the prevailing trend of the other ranges that constitute the group, although there are many departures in detail from the general condition, such as the northeast trend of the Dartmouth Range, the northwest trend of the Belknap Range, the great curve from east-west to north-south in the Randolph Range, etc. The valleys about and among the White Mountains lie at elevations ranging from 500 to 1000 feet, and the plateau remnants that lie about their bases are from 1000 to 2000 feet above the sea. Their relative elevations are therefore but little less than their absolute elevations and their influence as cloud gatherers and rain condensers is conspicuous. The average precipitation of the mountains is perhaps from 10 to 15 inches more than that of the surrounding valleys and plateaus, although no accurate observations of this condition for a period of years have yet been made.[1]

[1] For incomplete data as to the rainfall and the snowfall of the White Mountains consult A. J. Henry, Climatology of the United States, U. S. Weather Bureau, Bull. Q, 1906, p. 120; and S. A. Nelson, The Meteorology of Mount Washington, Geological Survey of New Hampshire, 1871; Water-Supply Paper U. S. Geol. Surv. No. 234, 1909, pp. 7-76 et al, map, Plate 1.

The White Mountains constitute the most important part of the eastern mountain axis in New England, which has two points of difference when compared with the western or Green Mountains axis: (1) it is composed largely of igneous rocks while the western range consists largely of metamorphic rocks, and (2) it has no southern representative in the Older Appalachians. Like the western range the eastern is bordered on one side by sedimentary rocks; the Carboniferous conglomerates, sandstones, and shales of the Boston and Narragansett basins lie on the eastern flanks of the New Hampshire-Maine axis, just as limestones, sandstones, and grits lie on the western flanks of the Green Mountains axis. The geosynclinal structure combined with the lesser resistance of the basin sediments as compared with the bordering igneous rocks have resulted in the development of lowlands, and the drowning of portions of the lowlands thus developed has given rise to Boston Bay and Narragansett Bay.

The eastern mountain axis of New England extends northward and northeastward through New Hampshire and northern Maine where it forms a belt of rough country, Plate V. Mount Katahdin, Maine, is a part of the White Mountain district. The White Mountains are the only part of this mountain axis that have not been reduced to a lowland, then uplifted, dissected, and glaciated. Elsewhere the surface has been so completely base-leveled that only isolated residuals occur, such as Mount Katahdin in Maine, Mount Monadnock in New Hampshire, Mount Wachusett in Massachusetts, and other monadnocks of less importance.

The whole of Maine southeastward of the White Mountains axis is a gently inclined upland sloping with marked regularity toward the sea. The degree of base-leveling which it reached at the end of a previous

Fig. 261. — Section south of Blue Hill, Maine, showing the base-leveled surface of the uplands bordering the White Mountains on the east. 2, granite; 3, diorite, diabase, and gabbro; 4, schist. (Emmons, U. S. Geol. Surv.)

erosion cycle is indicated in Fig. 261. It must of course be remembered that isolated unreduced masses occur even here where the general reduction of the rough topography appropriate to a disordered mass of rock such as the figure indicates is so complete. Mount Desert Island is perhaps the best-known residual of this sort. Its highest peaks are but little over 1000 feet above the sea.

In so far as the geology of the White Mountains has been deciphered there is warrant for the conclusion that the general structure of the main or Presidential Range is that of a great overturned anticline.[1] The subsidiary ranges are in general also anticlinal, while the main valleys or depressions are considered as synclinal in structure. If these interpretations are correct, the White Mountains as a whole may be described as a geanticline or an anticlinorium.[2]

The general mountainous condition in so far as it depends upon elevation may then be assigned to uplift of the rocks of the White Mountains in the form of an anticlinorium; this is to be regarded as the net result of a long series of very complex mountain-making movements. During the long erosion interval which in southern New England produced a peneplain, the White Mountains were worn lower than previously but not completely reduced. The White Mountains of to-day should therefore be regarded as but the remnants of far higher mountains worn down to moderate elevation. With the uplift of the peneplain in late Cretaceous and early and late Tertiary times the White Mountains were also broadly uplifted, and from the gradual rise of the peneplain in their direction we may reasonably infer that the uplift was more pronounced in the mountain belt than elsewhere. It may have amounted to 2000 feet.

Following uplift, vigorous erosion set in, with the result that the profiles were steepened and the relief developed on bolder lines than before. Still further variety was given the relief by glacial action. It is noteworthy that in general the relief is steeper along the main lines of glacial flow, as if from more pronounced scour of the preglacial valley borders. The combined effects of preglacial and glacial scour and erosion were the fashioning of the slopes on bolder lines, so that the relief at times becomes almost alpine. The degree of steepness is suggested by the destructive landslides that have occurred in the past and may occur in places again.[3]

Details of slopes in the White Mountains are as complicated as the structure upon which in most cases they have a certain dependence. General statements are of little value in this connection. The wedge-shaped mass of granite in the White Mountain Notch is more easily weathered than the flinty rock of anticlinal structure and high dip in Mount Willey and Mount Webster on either side, so that it has been excavated as a deep valley bordered by excessively steep walls, and is one of the most notable scenic features of the region. Granite seems

[1] C. H. Hitchcock, Geological Survey of New Hampshire, 1871, p. 8; also the same author's Geology of the White Mountains, Appalachia, vol. 1, 1879, pp. 72–74.

[2] For definition see Chamberlain and Salisbury, Geology, vol. 1, p. 485.

[3] C. H. Hitchcock, The Recent Landslide in the White Mountains, Science, vol. 6, 1885, pp. 84–87.

everywhere to be the least resistant rock and to underlie the valleys.[1] Where it occurs as a narrow band the valleys are steep-sided, being bordered by more resistant schists, etc.; where it occurs in broad bands the valleys are wide.

During the closing stages (Champlain) of glaciation a climate sufficiently Arctic existed, to produce local glaciers upon the Laurentides and the Green and the White mountains.[2] From each central *mer de glace* alpine glaciers moved radially outward, into the intermediate tracts which may have been submerged.[3] The detailed occurrence of transported bowlders and lateral and terminal moraines upon which these conclusions rest is indicated in the two last-named references. In one stretch of about two miles north of Bethlehem village sixteen terminal moraines have been identified.

Agassiz notes that the number of the moraines is here greater than in the case of a spot long regarded as particularly favorable for such phenomena, the valley of the Rhone below the Rhone glacier; and he believes that no one who had studied similar phenomena in connection with living glaciers could doubt the glacial explanation of these forms. Besides these evidences there have been found by numerous observers glacial cirques of subperfect development which must be assigned to the action of local glaciers such as those that have occurred in the Mount Toby district, Massachusetts, and on Mount Katahdin, Maine. The chief valleys affected by the local glaciers, which in one instance attained a length of 12 miles, are the Peabody, Ellis, Saco, Ammonoosuc, Pemigewasset, and others. With the removal of the mountain forests and the more detailed observations of the topography which this makes possible, many more evidences of a similar nature will probably be found.

GREEN MOUNTAINS AND BORDERING UPLANDS

Vermont, the Green Mountain State, contains the largest portion of the Green Mountains, which extend across it from north to south as a rugged chain of ridges and hills. Save for the alluvial terraces and patches of till on the western border of the Connecticut Valley, and for the so-called valley of Vermont on the western border of the state, there is little arable land. The total length of the Green Mountains is about 250 miles; their width varies from 25 to 50 miles. The range extends from central-western Connecticut through western Massachusetts, northward across Vermont, then turns northeastward, and terminates in eastern Quebec and northern New Brunswick on the southern border of the St. Lawrence

[1] C. H. Hitchcock, The Geology and Topography of the White Mountains, Am. Nat., vol. 4, 1871, p. 568.

[2] A. S. Packard, Evidences of the Existence of Ancient Local Glaciers in the White Mountain Valleys, Am. Jour. Sci., 2d ser., vol. 43, 1867, pp. 42–43; Louis Agassiz, The Former Existence of Local Glaciers in the White Mountains, Am. Nat., vol. 4, 1871, pp. 550–558; C. H. Hitchcock, Glacial Markings among the White Mountains, Appalachia, vol. 1, 1879, pp. 243–246.

[3] C. H. Hitchcock, The Geology of New Hampshire, Jour. Geol., vol. 4, 1896, pp. 60–62.

Valley in Quebec near the meridian of 70° west longitude. The northern termination has a strong northeastward trend parallel with the bordering St. Lawrence Valley.

The Green Mountains are the northernmost representatives of the mountainous western border of the older Appalachians and correspond in position with the Great Smokies of western North Carolina, South Mountain, Pennsylvania, and the Highlands of New Jersey. The axis as developed in the United States consists of three distinctly unlike parts: (1) a northern, discontinuous section, (2) a continuous range section extending through the southern two-thirds of Vermont, (3) a plateau section in western Connecticut and Massachusetts.

The northern section of the Green Mountains in the United States contains the highest peaks of the range and might therefore be thought the most rugged and difficult to cross. On the contrary, four valleys cross it, and dissection along them has been so pronounced that roads are obliged to ascend to an elevation of only 500 feet.

The four valleys, named in order from north to south, are: the St. Francis, in Canada; the Missisico, near the international boundary; the Lamoille, near Mount Mansfield; and the Winooski at Bolton; and in them all the drainage is from the eastern border of the mountains toward the west and northwest, unlike the drainage of the southern end of the Green Mountains axis, which is from the western border eastward. The highest peak in the Green Mountains is Mount Mansfield, 4430 feet, and Mount Killington is but a little less at 4380 feet.[1]

The second section of the Green Mountains extends from Hoosac Mountain northward for about 100 miles with an absence of passes and considerable uniformity of summit level, so that roads crossing this section not infrequently rise to an altitude of 2000 feet above the sea. The highest peaks of this section are under 4000 feet high and in general range between 2500 and 3500, with the larger number reaching to a little over 3000 feet.

In Vermont the main axis of the Green Mountains consists of a series of sharply compressed folds striking approximately north and south in sympathy with the general trend of the range. The series is overturned to the west in most localities. Steep westerly slopes are characteristic and occur in the form of terrace scarps on a large scale or as mountain brows of precipitous quality. The eastern slopes are characteristically formed on the dip of the planes of the schistosity or the dip of the planes of stratification and are steep or gentle as the dips vary between these extremes, although in general they may be called

[1] C. H. Hitchcock, Glaciation of the Green Mountain Range, Rept. of the State Geologist of Vermont, 1904, pp. 69–71; A. D. Hager, Physical Geography and Scenery, Geology of Vermont, pt. 2, 1862, pp. 144–145.

relatively gentle. West of the Green Mountain range is a great lime-stone valley with island-like ridges of folded schist with synclinal structure, the remnants of former more extensive formations largely removed by erosion.[1]

That portion of the Green Mountains in the northwestern corner of Massachusetts and near the Vermont boundary is called the Hoosac Mountains and forms the divide between the Hoosac and Deerfield rivers, branches of the Hudson and the Connecticut respectively.[2]

The Green Mountains extend into Massachusetts and Connecticut, but their form undergoes a radical change south of Hoosac Mountain, Massachusetts, where they are developed as a broad plateau extending from the Connecticut Valley on the east to the upper valleys of the Hoosic and Housatonic rivers on the western border of Massachusetts and Connecticut. Its western border is a bold continuous scarp about 2000 feet high, which forms the eastern border of the Berkshire Hills country and is a divide between the tributaries of the Connecticut and the Housatonic; eastward the plateau descends by a more gradual and undulating slope as far as the western border of the Connecticut Valley lowland, where it descends from the 1000-foot level to the 400-foot level of the lowland floor in distances of a half mile to a mile.

The Green Mountains of western Massachusetts and northwestern Connecticut (the Green Mountain Plateau) are mountains of the second generation. Their original mountain forms were obliterated by the peneplanation which affected the region in Jurassic and Cretaceous times. Whatever of mountain form they possess to-day has been derived by uplift and erosion since peneplanation. Occasional residuals rising above the ancient peneplain as low eminences are now the highest points of the region, affording fine panoramas over a broad expanse of upland.

The Tertiary lowlands developed upon the softer limestones of western Connecticut and Massachusetts and the sandstones of the Connecticut Valley establish a surface which is of flatter gradient than the surface of the Green Mountain Plateau (a portion of the old Cretaceous peneplain) and appears to indicate that the uplift of the region after peneplanation was not uniform throughout but that it was somewhat sharply localized along the old mountain axis, after the manner of the uplifts that have been determined in the southern Appalachians (p. 593). When this is combined with the fact that most of the residual elevations on the divide between the Connecticut and the Housatonic lie upon the Green Mountains axis it will be readily understood that dissection since uplift has acted with unusual vigor here and given the region a rugged and mountainous appearance.

[1] Pumpelly, Wolff, and Dale, Geology of the Green Mountains in Massachusetts, Mon. U. S. Geol. Surv., vol. 23, 1894.
[2] Idem.

Between the Green Mountains axis in western Connecticut, Massachusetts, and Vermont on the one hand, and the Connecticut Valley on the other, is a plateau or upland fringe bearing the same relation to the bordering mountains as the Piedmont Plateau of the southern Appalachians bears to the mountainous western border in Georgia and North and South Carolina. It is an upland composed in part of somewhat metamorphosed granite and diorite, in larger part of slates, schists, etc., derived by metamorphism from sedimentary rocks. Although much narrower than its southern representative it maintains continuously its identity as a border feature and possesses a well-marked physiographic character. It forms a transition upland between the eastern flanks of the Green Mountains and the lowland of the Connecticut Valley in Connecticut, Massachusetts, and Vermont, and throughout its whole extent is much dissected by transverse southeastward-trending valleys such as the Farmington in Connecticut, the Westfield in Massachusetts, and the White and the Williams rivers in Vermont. The uplands rise from a low southern margin on the shore of Long Island Sound to an elevation of 1000 feet at the northern boundary of Connecticut and to 1400 or 1600 feet at the northern boundary of Massachusetts. For the first score of miles the average rate is 20 feet per mile and is maintained with tolerable regularity. The ascent of the upland in Massachusetts is somewhat more gentle, from 10 to 12 feet per mile.

Lying well within the border of the glaciated country of North America and constituting one of the two principal lines of elevations in New England it is natural that the Green Mountains should bear the marks of heavy glaciation. Striæ and rounded knobs and bosses of rock, erratic bowlders and sheets and patches of till are the common features, while many of the lakes of Vermont within and without the Green Mountains are due either to glacial overdeepening of preglacial valleys or irregular deposition of drift or to both.

The mountain summits are often bare rock ledges; the mountain flanks, especially on the west, are steep slopes with abundant rock outcrops and naked precipitous cliffs. Rarely are the mountain flanks soil-covered up to their summits, although a few such cases occur. Mount Stratton, one of the two highest peaks in southern Vermont, is completely covered with glacial débris even to its summit.[1]

The eastern slopes of the Green Mountains are the gentler and it is natural that we should find them more heavily cloaked with glacial

[1] C. H. Hitchcock, Glaciation of the Green Mountain Range, Rept. State Geologist of Vermont, 1904, p. 75.

material.   Especially do they have a covering on the lower slopes, which are cleared and farmed and form a pleasing contrast to the western slopes of the White Mountains on the opposite side of the Connecticut Valley.   The narrow fringe of upland between the Green Mountains and the Connecticut Valley, the northern representative of the Piedmont Plateau, is also blanketed with glacial waste and both hills and valleys are dotted with homesteads.

The abundant rainfall of the Green Mountains, a condition which they share with the rest of New England, offsets to some degree the dis-

Fig. 262. — Forest growth on a steep and rocky New England hillside, Jamaica Plain, Mass.

advantages of a thin soil, and though the forest growth is never luxuriant, it is, or was, in the main continuous.   The lower valley slopes offer the deepest soil and the most sheltered positions and hence are most heavily timbered.   The soil of the higher exposed situations supports a relatively thin forest of spruce and birch.   The most favorable situations for tree growth are the coves developed upon the older granites of the range, where the decayed rock is heavily cloaked with glacial till.

Both the White and the Green mountains, as well as many other portions of New England, have great tracts of ultimate forest land, that is to say, land which will bear trees more profitably than anything else and hence will be kept timbered when man's purposes become well adjusted to the soil. Indeed much of their surface will yield nothing but timber. It is surprising how healthy a growth will sometimes be developed upon a steep and almost barren hillside, Fig. 262. The clearing of such a situation rarely does great damage to the soil because the soil is so thin and stony (see p. 6), though if kept cleared it is subject to harmful erosion. Vigorous soil erosion is not, however, common on the unforested mountain slopes of New England. On the other hand, the effects of the forests on streams flow is clearly apparent. The slopes of the surface are sufficiently steep to enforce a dangerously heavy and rapid run-off were the timber cover and lake systems less extensive.

CONNECTICUT VALLEY LOWLAND

GENERAL FEATURES

The Connecticut Valley, or valley lowland, as it is sometimes called, is about 95 miles long. Its width varies from 5 miles at each end to 15 or 18 miles in the middle. Its area is about 1000 square miles; a third of this is in Massachusetts, the rest in Connecticut. A slight uplift since Cretaceous peneplanation and the general occurrence of soft sandstones and shales have led to the development of a lowland (Tertiary) between an eastern and a western upland of crystalline rock of sufficient resistance to preserve traces of the peneplain. The greater part of the lowland is drained by the Connecticut River, although large portions are drained also by the Quinnipiac and the Farmington.

GEOLOGIC STRUCTURE

The lowermost rocks which constitute the Triassic formation of the Connecticut Valley consist of a series of 5000 to 6500 feet of coarse sandstone and conglomerate and a limited amount of shale. The material consists of the waste of granite and other crystalline rocks similar to those upon which the deposits lie. Intruded in these and generally not more than 200 to 300 feet above the base of the formation are the trap sheets of West Rock Ridge in the south and of the Barn-door Hills in the north. The sheets are from 500 to 600 feet thick in

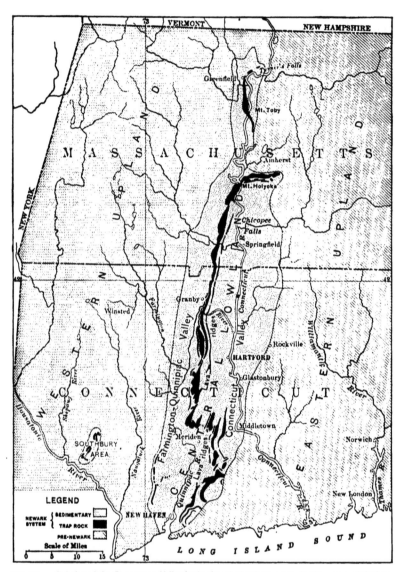

Fig. 263. — Relation of the Connecticut Valley lowland to the bordering uplands, the extent of the low-land, principal relief features, and drainage systems. (Russell, Davis, and others, U. S. Geol. Surv.)

places, but generally somewhat less. Succeeding the lower members with their intruded trap sheet is a series of sandstones, shales, impure limestones, intercalated with a series of three trap sheets. The lower trap sheet is called the anterior sheet and is generally about 250 feet thick, but thins out and disappears before reaching the northern boundary of Connecticut. Above the anterior sheet are shales and sandstones

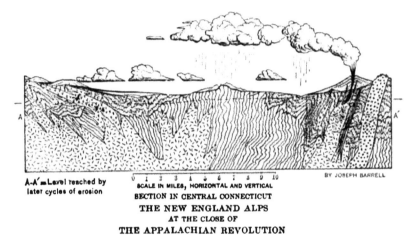

A-A′=Level reached by later cycles of erosion

0 1 2 3 4 5 6 7 8 9 10
SCALE IN MILES, HORIZONTAL AND VERTICAL

BY JOSEPH BARRELL

SECTION IN CENTRAL CONNECTICUT
THE NEW ENGLAND ALPS
AT THE CLOSE OF
THE APPALACHIAN REVOLUTION

Fig. 264. — The geologic and physiographic history of the Connecticut Valley lowland and adjacent portions of the bordering uplands. Vertical and horizontal distances to same scale.

A. A time of great regional metamorphism and inferred volcanic activity.

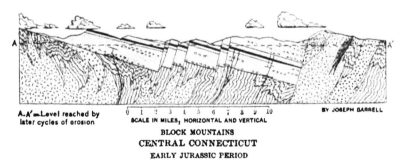

A-A′=Level reached by later cycles of erosion

0 1 2 3 4 5 6 7 8 9 10
SCALE IN MILES, HORIZONTAL AND VERTICAL

BY JOSEPH BARRELL

BLOCK MOUNTAINS
CENTRAL CONNECTICUT
EARLY JURASSIC PERIOD

B. After sedimentation had filled a structural depression and the strata had been block-faulted. Contemporaneous lavas are shown as black bands conformable with the strata.

SCALE IN MILES, HORIZONTAL AND VERTICAL

BY JOSEPH BARRELL

CROSS SECTION

CENTRAL CONNECTICUT

IN THE CRETACEOUS PERIOD

C. Appearance of the surface at the end of the Jurassic-Cretaceous cycle of erosion.

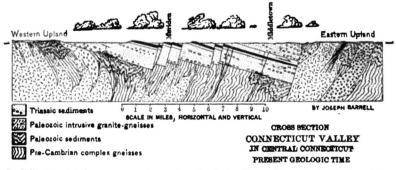

Triassic sediments

Paleozoic intrusive granite-gneisses

Paleozoic sediments

Pre-Cambrian complex gneisses

SCALE IN MILES, HORIZONTAL AND VERTICAL

BY JOSEPH BARRELL

CROSS SECTION

CONNECTICUT VALLEY

IN CENTRAL CONNECTICUT

PRESENT GEOLOGIC TIME

D. Following uplift there was more vigorous dissection of the softer sandstones, leaving the trap ridges and crystalline uplands in relief.

from 300 to 1000 feet thick. Then comes the thickest and main trap sheet, 400 to 500 feet thick, another series of shales 1200 feet thick, and an uppermost or posterior trap sheet 100 to 150 feet thick.

Originally the sediments were deposited in a nearly horizontal position, and had this position been maintained up to the present, the topography and drainage of the region would be far more simple than is the case to-day. It was while the sandstones were being deposited that the lava flows of the eastern trap sheets took place, and after they had been deposited that the igneous intrusions of the western trap sheets took place. There followed a period of deformation in which both sandstones and trap sheets were extensively faulted and tilted, thus exposing the basset edges of both.

FAULTS AND ASSOCIATED OVERLAPS

The western range of trap ridges including West Rock Ridge and the Barndoor Hills of northern Connecticut have their successive members arranged *en échelon*. The southern ridges are arranged in advancing order, each northern ridge standing to the west of the next southern ridge, the ends overlapping by a moderate amount. In the northern group the arrangement is reversed; the order is receding and each northern ridge in succession stands farther east than the next southern ridge; and instead of the overlapping feature of the southern ridges there are gaps between the different members of the northern ridges, Fig. 265.

The eastern trap ridges are greatly diversified as to form, size, and arrangement on account of differences in the thickness of the trap sheets and their greater number. Two features predominate. In the southern section two ridges are curved into the form of an irregular crescent convex to the northwest, the horns of the crescent extending almost to the eastern crystalline area. The other ridges, about 20 in number, are arranged in the same advancing and receding order that is exhibited in the western range, the advancing order being displayed exceptionally well near Meriden, the retreating order about Tariffville, and the change from one to the other may be noted west of Hartford.

Along the eastern ridges the main or middle trap ridge is of greater topographic prominence because of its greater thickness, 400 to 500 feet.

The advancing and retreating order of the trap ridges is explained by displacement along the faults that cross the Triassic formation diagonally from northeast to southwest. The ridge overlaps on the south and the gaps on the north are both the expression of a movement on the fault planes. In the one case (overlapping ridges) displacement resulted in *offset with overlap*, in the other case (gaps between ridges) displacement resulted in *offset with gap*. The movements in the two cases were in parallel lines but in opposite directions. The application of this simple principle explains all the more prominent outlines of the topography of the district. The transverse diagonal faults just described are not confined to the Connecticut Valley. Just as

Fig. 265. — Displacement of trap ridges near northern end of West Rock Ridge and corresponding displacement in the crystallines of the bordering uplands. (Davis.)

they cross sandstones and traps, lowland and ridges, so they pass from the valley to the upland, into which they may in a few cases be traced. Upon the border of the upland they produce important indentations, the chief ones being from South Glastonbury to South Manchester and from Vernon to Rockville, where the border of the eastern crystallines strikingly corresponds in northeast trend with the direction of the lines of major faulting in the Triassic. On the western border the faults of the lowland pass into the crystalline upland in a much less distinct manner, causing slight offsets and overlaps in the western boundary, Fig. 265.

CRETACEOUS PENEPLAIN

Except on the trap ridges the Cretaceous peneplain is now completely destroyed in the Connecticut Valley. Upon the peneplain at the time of its full development the streams must have flowed in courses that were within certain wide limits independent of rock structure and therefore of belts of hard and soft rock. The whole region was blanketed with a cover of residual soil upon which the streams were free to meander, perhaps only slightly controlled by the harder trap which may have persisted in the form of low flat-topped divides even during the time of most complete denudation.

SUPERPOSED COURSE OF THE LOWER CONNECTICUT

The Connecticut River, after flowing in a rather direct manner from north to south through the Connecticut Valley lowland, turns almost at right angles at Middletown and cuts through the crystalline rocks

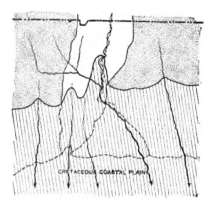

Fig. 266. — Inferred Cretaceous overlap on the southern shore of Connecticut. (Davis.)

of the eastern upland as far as the point of discharge near Saybrook. Such a course could not have been gained during the period of adjustment of streams to structures before the formation of a penepflaned surface.

The most reasonable explanation applied to the lower Connecticut is as follows. The Cretaceous clays and sands outcropping upon the northern shore of Long Island formerly extended farther north, overlapping the southern border of the Cretaceous peneplain, Fig. 266. It is probable that they formerly overlapped the southern border of New England as far as Hartford.

The retreat of the shore line after the deposition of the Cretaceous overlap would allow an extension of all the streams down the slope of the coastal plain thus formed. The general slope of the plain would determine the general direction of the extended streams and the more detailed courses would be determined by the local irregularities of the surface. The stripping off of the overlapping Cretaceous cover would result in the gradual superposition of the lower courses of all the streams upon the underlying floor. It is clear that the lower courses might be directed across points of hard rock, across ridges, and from crystallines to sandstones and vice versa, in a manner only visible when the cover causing these complexities was extensively removed. After uplift and while the lower Connecticut was developing a narrow valley in the resistant crystallines, its tributaries developed a broad valley in the soft sandstones.

It need not be supposed that the Cretaceous overlap was of great thickness to cause a complete turning aside of the streams from a course coincident with that which they formerly had upon the peneplain beneath the Cretaceous cover. The base-leveling was so nearly completed that a cover of a hundred feet thickness, perhaps fifty feet, would conceal all but a very few of the low rounded hills that appeared as residuals above the general level of the peneplain. It is even conceivable that the master stream of the region, the Connecticut, might have accumulated an exceptional amount of material upon the sea floor at its mouth and that when uplift occurred it would flow down the steepest slope of a broad fan, a course which might be on any radius. Conceiving the radius most favored as that directed most nearly to the east, the river would be turned from its former direction to one that would cause it later to be incised in the crystallines.

TERTIARY PENEPLAIN

The uplift of the Cretaceous peneplain and of the overlapping border of Cretaceous sediments enabled erosive agencies completely to remove the overlap and everywhere to dissect the uplifted peneplain. The uplift was accomplished not in one period of deformation but in two periods; for between the old Cretaceous level of the trap ridges and the present level of erosion one finds an intermediate level, a local Tertiary peneplain developed upon Triassic sandstones. It stands out conspicuously in certain views, being well developed north and northeast of Hartford, south of Mount Carmel, west of Mount Tom, and in many other localities. During the time in which the Tertiary lowland was being formed the resistant trap rocks of the valley lowland and the only slightly less resistant crystallines of the uplands maintained approximately their old positions.

The short Tertiary cycle was terminated by a second uplift which
brought the land approximately to its present level and enabled the
dissection of the uplands to be invigorated and that of the Tertiary
plain to be begun.   The last geologic event of importance in the region
was glaciation, which, however, did not markedly affect the principal
topographic and drainage outlines.   The slopes of the hills and valleys
were more or less thinly cloaked with rock waste, the débris of the continental ice sheet.   The tops of the ridges were in many cases notably
rounded, smoothed, and striated, although they were presumably but
little reduced in height.   The lowland plain and the river valleys were
everywhere made more irregular, and in many cases reversed slopes were
produced back of which lake waters now lie.

The most conspicuous drainage change was in the case of the Farmington River, which formerly ran south from Round Hill and Farmington through the present valley of the Pequa-
buck, through Plainville and Southington, and thence through the present Quinnipiac Valley
to the Sound.   But a low dam of glacial material at Plainville diverted the water of the river
northward, with the result that from Farmington the river runs almost due north for about
12 to 15 miles to Tariffville, where it makes a sharp turn to the east and southeast, crosses
the three trap sheets of the valley and pursues a more or less irregular course toward the southeast to the Connecticut at Windsor.[1]

SOILS AND VEGETATION

The soils of the Connecticut Valley are of many kinds, depending in
part upon the many differences in rock character from place to place and
in part upon different modes of origin.   To understand the first cause of
difference it is necessary to recall that though we commonly speak of
a glacial soil as composed of foreign material, this is true only within
certain rather narrow limits.   Analyses made by Leverett and by Alden
show that about 85 % of the till of the Great Lake region is derived
from the underlying sedimentary rock.   An even higher proportion of
locally derived material is found in the Connecticut Valley.

Such glacial forms as drumlins, eskers, sand plains, moraines, etc.,
have distinctive soil characters that are easy of identification.   On
account of their flatness and areal extent the sand plains of the Connecticut Valley deserve special consideration.   The material of the sand
plains is commonly loose and porous, varies in texture from fine to
coarse but is always prevailingly sandy, may have moderate natural
fertility but is generally decidedly infertile as compared with the more

[1] Rice and Gregory, Manual of Connecticut Geol. Bull. Conn. Geol. and Nat. Hist. Surv.
No. 6, 1906, pp. 251–253.

clayey soils of the till plains, and has a high absorptive capacity because of its flatness and porosity. Its loose nature, however, prevents it from retaining the absorbed water in large enough quantities and for long

Fig. 267. — The North Haven sand plain, or "desert," five miles south of Wallingford, Conn. See the New Haven quadrangle, U. S. Geol. Surv. Part of Mt. Carmel in the background. Note the tufted grass (*Andropogon scoparius*) and the extent of bare surface. (Photograph by Beede.)

enough periods as a rule to allow a maximum or even a favorable plant growth, and sand plains are commonly local semi-arid tracts in the midst of more fertile areas.

One of the best illustrations of these features found in New England is the North Haven sand plain which stretches from Montowese (New Haven topographic sheet) to Wallingford, Connecticut, and beyond. It is about 15 miles long, with an average width of 1 to 2 or more miles. Its surface is in general flat or gently sloping; its soils vary in texture from a fine to a coarse sandy loam, with large areas of pure sand without a loamy admixture. The yellow sand shows distinctly through the thin cover of vegetation and gives such tracts a strikingly desert-like appearance. The water table stands from about 10 feet to 20 feet below the surface in spite of proximity to the river (Quinnipiac) and the fact that it receives the drainage from the adjacent upland portions of the drainage basin in which it occurs. Even after a heavy rain one can find little water in the soil except in the most favored portions. In places the scanty vegetation has a prosperous appearance, but in general distinctly xerophilous characteristics are displayed. Certain tracts support only grasses (chiefly Andropogon) which grow in scattered bunches. An elaborate botanical study of the vegetation of this area has been made with some interesting results.[1] It has been found that while the lack of water is pronounced it is not this lack but the burning heat of the sun on the bare sand that enables only xerophytes to exist. Other plants perish soon after their seeds germinate. Among the perennial grasses, *Andropogon furcatus* has thickened root nodes in which food and moisture are stored up and carried through the winter. *Andropogon scoparius* is present in tufts, is the most abundant, and has a leaf whose upper surface is composed of an epidermis made up of water cells that constitute about one-third the total thickness. Similar or comparable adaptations have been found on nearly a dozen other annuals found within the area.

On certain areas there is a regular order of occupation of the bare sand. The reindeer moss (*Cladonia rangiferina*) covers the bare sand, and where this lichen becomes established other plants spring up because the moss prevents the sand from shifting and entangles the seeds blown across it. Upon this undisturbed surface there accumulates in time a layer of leaf and vegetable mold that retards evaporation and enriches the soil, enabling, through these more fortunate conditions, the germination and development of other more delicate plants. Sweet fern (*Comptonia peregrina*) may possibly follow the reindeer moss as the next stage in the development of a vegetal cover. The black cherry, with a tendency to form colonies by root sprouting; the common milkweed, also grows in colonies; and scrubby black oaks widely scattered throughout the area are common forms of larger growth. The prosperous condition of the larger trees and especially the oaks is noteworthy and appears to be due in the case of the oaks to the length of the characteristic taproot.

Each tree forms a sort of anchorage ground about which and under which acorns, grasses, mosses, and lichens grow or accumulate in some numbers. In time a mold is formed even at some distance from the parent tree, and in it acorns may sprout and thus gradually extend the vegetal covering. Once started the long taproot quickly reaches the ground water, and once in touch with this source of supply the life of the tree is assured barring accident. The normal development of vegetation thus outlined is interfered with by fires, which burn the leaves and grasses and even burn out the mold from the surface soil, the element most needed for the reclamation of the area. One sample of the sand-plain soil at Montowese was found to contain but .09 of 1% of nitrogen, and it is probably to the nitrogen deficiency to which this points as much as to the dryness of the area that the absence of plant growth is due at any one time, though the more fundamental cause in the long run must be the relative dryness of the area and the shifting character of the surface.

The soils of the Connecticut Valley (between Springfield and Hartford) are of many varieties and are most irregularly disposed. The so-called Suffield clay (glacial and interglacial), the Triassic stony loam, and the Holyoke stony loam (strictly glacial) are distributed in the most hit-and-miss manner imaginable. The fine sandy loams are com-

[1] W. E. Britton, Vegetation of the North Haven Sand Plain, Bull. Torrey Bot. Club, vol. 30, 1903, pp. 571–620

monly found along the stream courses and in terraces and valley flats, while the larger areas of undrained or poorly drained meadow land are found along the valley floors. Among the strictly glacial soils the Triassic stony loam is the most important. It is generally a fine sandy loam mixed with gravel and bowlders, the whole derived chiefly from the underlying Triassic sandstones. The amount of gravel and unde-composed rock in it exceeds 5% in all cases and in some cases exceeds 50%. The stoniest loams of the Connecticut Valley sometimes contain from 10% to 50% of bowlders ranging in size from 1 inch to 15 inches in diameter. They are relatively infertile and are but little farmed, being given up mainly to stony pastures, wood lots, and orchards, with occasional patches of corn, oats, and rye.

The exposed floor of the old glacial lake and river terraces in the Connecticut Valley are composed of yellowish-red or brown sand that contains less than 5% of clay. About 5% of the soil consists of coarse gravel and is inclined to be leachy and dry, though it is valuable for truck farming. The surface of the area is level or gently rolling in Connecticut; in Massachusetts it is much more rolling. The type is not very extensively cultivated, and in Connecticut there are but few houses upon it, the roads are deep and sandy, and along them are many old and unsuccessful fruit farms. Many areas of Windsor sand which were formerly cultivated are now grown up to a characteristic forest growth of pine. The soil is open and porous, offers little resistance to rains, and is so flat and so little washed that there are old well-preserved corn rows running through a forest in which the trees must be at least 50 to 80 years old.[1]

On either side of the Connecticut River from Holyoke south to Long-meadow, Massachusetts, and from Warehouse Point to South Glaston-bury, Connecticut, are the Connecticut meadows or the present flood plain of the Connecticut River and its tributaries. The lower portions are frequently wet and swampy and subject to overflow, but in spite of this condition there is considerable farming on them at some risk. The character of the material is very uniform; it is a fine sand and silt 16 to 18 inches deep, containing a large amount of organic matter. Below Merrick the meadows are diked to keep out the high water and to insure against overflow. The Connecticut meadows are among the most extensive and most important soils in the valley, with marked differences in texture, water-holding capacity, and warmth, and an equally marked difference in the quality of the products.

[1] Dorsey and Bonsteel, Soil Survey in the Connecticut Valley, Rept. U. S. Bur. Soils No. 64, 1900, p. 133.

Scattered over the entire Connecticut Valley are considerable areas of swamp land and wet meadow. They generally occur along the scarps between the valley flats and upland where the ground water appears. The degree of swampiness precludes cultivation except where special drainage conditions are maintained.

CHAPTER XXXI

NEWER APPALACHIANS

INTRODUCTORY

THE Newer Appalachians are the most striking member of the Appalachian group of physiographic provinces. The subdivision includes rather regularly folded strata and long narrow valleys separated by nearly parallel ridges. It is marked by the presence of a great valley

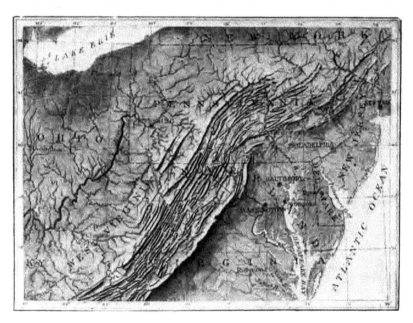

Fig. 268. — Relief map of the central part of the Appalachian System. (U. S. Geol. Surv.)

which extends with but local and minor interruptions from end to end of the long province. This is not a single river valley but a composite of many valleys to which the name Great Appalachian Valley has been applied. The Coosa, Tennessee, Shenandoah, Cumberland, Middle Hudson, and Champlain valleys are its chief members. The Newer Appalachians province is bordered on the east by the Unaka Mountains, the Great Smoky Mountains and the Blue Ridge at the south, and by the Highlands of New Jersey and the Green Mountains at the north; on the west it is bordered by the Cumberland Escarpment, the Allegheny Front, the Catskills, and the Adirondacks.

The southern portion of the Great Appalachian Valley is limited on the west by the eastern edge of the Appalachian Plateaus, whose various sections are here designated Walden Ridge, Lookout Mountain, and Sands Mountain. These plateau remnants have a heavy sandstone cap and their margins are defended by heavy and extremely resistant beds of sandstone and conglomerate. They are commonly from 800 to 1000 feet high and on the east overlook the Great Appalachian Valley as a bold scarp — the Cumberland Escarpment — which forms as definite a border to the Great Valley on the west as the southern Appalachians do on the east.

In a broad view the Newer Appalachians consist more largely of valleys and valley lowlands than of ridges in the Chattanooga region on the south and the Champlain-Hudson region on the north. In the central portion of the province the mountain and not the valley feature is on the whole the more prominent, Fig. 268; only the eastern side of the central district is marked by broad valleys, as the Shenandoah Valley of Virginia, developed on less resistant limestones, the Cumberland Valley of Maryland and Pennsylvania, and the Lebanon Valley of eastern Pennsylvania. On the basis of these topographic differences it will be convenient to subdivide the Newer Appalachians into three districts, a southern, a central, and a northern district.

The rock formations change from point to point somewhat, but their most striking characteristics are their continuity and lack of variation through long distances. In general they are more limey at the south and sandy and conglomeratic at the north. There are also important differences of structure such as the presence of great overthrust faults in the Chattanooga district and the general absence of these structural features and related topographic forms in the Pennsylvania ridges. In the former district the presence of great thicknesses of easily eroded strata such as limestone and shale, structurally deformed so as to expose on erosion the edges of the strata, has resulted in profound denudation and the carrying away of at least 10,000 feet of rock.

SOUTHERN DISTRICT

A prominent feature of the southern district of the Newer Appalachians is the number of ridges that follow exactly in line with the topographic level maintained by Walden Ridge and Cumberland Plateau,

Fig. 238. These ridge summits constitute, however, but a small portion of the entire area, for since the uplift of the Cretaceous (Cumberland) peneplain the greater part of it has been removed by erosion. This is due both to the relatively soft rock of the district and to the compressed nature of the folds, which reveal the beds in nearly vertical attitudes and so permit greater erosion. Such ridges as occur are, however, very even-crested in spite of considerable diversity in rock character, and are unquestionably the remnants of a former more extensive plain.

In places the ridges depart somewhat from the general type and seem to rise above the level of the peneplain. In such cases the wind gaps probably represent the old base level, for they have a constant altitude, whereas the intervening portions of the ridges rise irregularly from 100 to 300 feet above them, and were probably a series of knobs projecting above the peneplain level. In contrast to these exceptional features are the ridges composed of less resistant rock or occupying more exposed positions. These have been so reduced by erosion following upon the uplift of the Cretaceous peneplain that no point along their crests attains the altitude of the peneplain. On the whole, however, the ridges are surprisingly accordant in altitude and their level is nearly always harmonious with that of adjacent, better preserved portions of the plain.

The early Tertiary (Highland Rim) peneplain was not developed so extensively as the Cretaceous peneplain within the borders of this district. Its chief development was along the larger valleys where narrow belts of rock were planed to a more or less level expression. Remnants of it may still be seen at altitudes above 1000 feet, where the great majority of hills and ridges may be seen to reach nearly to a common level. Standing above this topographic level are a number of residuals among which White Oak Mountain, Tenn., is the most prominent. Like the Cretaceous cycle the early Tertiary cycle was closed by irregular uplift, the first result of which was to invigorate the streams and cause them to incise valleys below the general level. Gradually the valleys were widened to form local lowlands upon areas of softer rock. The result was a partial planation of a region distinctly smaller than that peneplaned during the early Tertiary cycle and only a fraction as large as the great area peneplaned during the Cretaceous cycle. This local lowland has been called the Coosa peneplain because of its excellent development along Coosa River (see Fig. 235).

The Coosa "Flat Woods" are the largest unit of this peneplain. They form a belt 10 to 12 miles wide and but little above the narrow flood plain of the river though never reached by the present floods. The peneplain is developed upon soft or soluble rocks, limestones, and limey shales; where the Coosa flows upon more resistant formations the valley is comparatively narrow. The altitude of the Coosa peneplain is more than 700 feet at the southern margin of the Chattanooga district and about 800 feet at the northern margin. The slope is little more than the normal grade of a base-leveled surface, which shows that but slight local deformation has occurred, although when examined in detail considerable variation in altitude is expressed in the various portions of the peneplain.

STREAM TYPES

The topographic changes outlined above took place during immensely long intervals of time, and if were call the physiographic principles of stream adjustment, the constant shifting of stream courses through the more rapid development of those which flow upon belts of softer rock, we shall be prepared to appreciate the fact that very extensive stream changes have occurred in the Appalachian region. These have been conditioned by (1) the exposure of softer beds by the erosion of harder beds overlying them and (2) crustal warping which terminated each cycle and deformed the successive peneplains.

A number of types of streams may be identified; the characteristics of these we shall sketch here only in the briefest manner. The first type is that represented by the Hiwassee, the New-Kanawha, and other streams which flow westward from the southern Appalachian mountains. The most striking member of this group is the New-Kanawha, which drains portions of all three physiographic provinces in succession from the Appalachian Mountains to the Appalachian Plateaus, Fig. 235. It appears to be an antecedent stream in that it has maintained its course westward against all the deformations which have been produced in its path. Its age must be measured in millions of years, for it probably dates from the great Appalachian revolution of Permian time, and though it has suffered many slight vicissitudes and its course has been opposed by mountain-making movements, the river has been victorious in all its contests and still pursues its ancient northwestward course.

The present course of Tennessee River has topographic relations of equal interest. After flowing southwest for several hundred miles, as far as Chattanooga, it turns sharply west, crosses the high plateau known as Walden Ridge, then turns sharply southwestward again, and flows in another straight stretch for over 200 miles before turning north to join the Ohio. Its course across Walden Ridge corresponds to the course of the river upon the Cretaceous peneplain. It is very strikingly meandering, and under other circumstances the meanders would be accepted without question as inheritances from a low-gradient strongly meandering stream later incised in a formation so hard as to preserve them in somewhat their original form or a derived form not departing much from the original pattern. When taken with the other evidence these meanders may be regarded as safely referable to this mode of origin. That they have been preserved for such a long period as all of Cretaceous time does not seem unreasonable when it is recalled that only the lower 100 to 200 feet of the 1000-foot gorge across Walden Ridge is composed of soft limestone, while all the rest of this great section exposes extremely resistant sandstone.[1] The great thickness of the resistant strata has prevented the river from widening its valley in precisely the manner in which the hard strata of the Kittatinny anticline in New Jersey have restrained valley development along the Delaware and made possible the formation of the Delaware water gap. During the time that the Delaware River was cutting the gorge at the water gap the tributaries and their auxiliary streams were widening the soft formations on either side and developing them into the form of a broad valley lowland. The strong differences in rock character between the Sequatchie anticline and the Appalachian Valley on the one hand, and the Walden and Cumberland plateaus on the other, are not less than the differences among the valley widths of the streams or portions of streams flowing through these formations.

A second type of drainage is that which represents the effects of wandering during the later stages of an erosion cycle. This is considered to be an explanation of the course of the Ocoee where it crosses or is superimposed upon the point of Bean Mountain.

A third class of stream is that found in the longitudinal valleys of the Newer Appalachians. It requires little explanation, for it is obvious that it belongs to the type known as subsequent rivers. When the Cretaceous peneplain was uplifted erosion went forward rapidly along belts of weak rock, and tributary streams originating in such belts rapidly extended their valleys headward. Those streams which crossed harder strata were prevented from lowering their channels at the same rate and eventually numbers of them were beheaded and made tributaries of their more powerful rivals. The abandoned channels across the hard ridge makers stood at higher and higher levels as the softer rocks were lowered on both sides; they now appear as wind gaps. Many streams are made up of sections which belong to different types. Their headwaters descend the western slopes of the southern Appalachians as originally, their middle courses occupy subsequent valleys arranged longitudinally, and their lower courses are superposed upon structures across which the streams flowed while working at the level of the Cretaceous peneplain.

[1] The data for this paragraph are chiefly from the admirable discussion of the Tennessee problem in a paper by D. W. Johnson, The Tertiary History of the Tennessee River, Jour. Geol., vol. 13, 1905, pp. 194–231. The paper also contains a bibliography of the literature bearing on the Tennessee problem.

CENTRAL DISTRICT

The eastern boundary of the central district of the Newer Appalachians is the Blue Ridge; the Allegheny Front forms the western boundary. The latter is rugged and bold and faces southeast. It is the eastern edge of the Appalachian Plateaus and extends from the northern end of the Cumberland Plateau through West Virginia, Maryland, and Pennsylvania, and is continued northeastward more irregularly as far as the Catskill Mountains.

Prominent points upon it are Dans Mountain, Maryland, 2100 feet above the sea; the Pinnacle, West Virginia, 3400 feet; Roaring Plains, 4400 feet; the Big Black Mountains of Virginia and Kentucky, 4000 feet; southward the summit becomes lower and at Cumberland Gap is but 1600 feet high.

It is crossed by many streams such as the New, the Potomac, etc., but not enough dissection has taken place to destroy its wall-like character throughout most of its extent. Narrow thousand-foot canyons are cut in it here and there, and between them are straight stretches of precipice often crowned by a resistant sandstone that stands out under weathering and erosion as a steep brow.[1]

From the Maryland line northward a distance of about 50 miles the crest of the Front is rather broad and flat and nearly straight, at a fairly uniform elevation of between 2500 and 2700 feet. Above this level occasional knobs rise to heights of 3000 or more feet, as Blue Knob, 15 miles south of Gallitzin, 3136 feet, the highest point in the state.

From above or from a distance the lower part of the Allegheny Front in Pennsylvania appears as a grand terrace, made by the even tops of a series of projecting spurs that break the descent of the scarp. They are due to the presence of a second resistant formation, and as the crest of the Front becomes lower toward the northeast the terrace is entirely absent. This feature is nowhere better shown than near Altoona. The lower plain in that locality stands at 1200 feet, from which there is a moderately steep ascent to 1600 or 1700 feet, over a terrace belt from 1 mile to 1½ miles wide. From the terrace there is an abrupt rise to the crest of the escarpment 2400 to 2700 feet high.

The Allegheny Front crosses Pennsylvania in a sweeping curve toward the northeast for a distance of 230 miles, until it merges obscurely with the zigzags of the anthracite region in the northeastern corner of the state. The escarpment is broken only by narrow ravines through which flow the northern branches of the Susquehanna, the North Branch, Muncy, Loyalsock, Lycoming and Pine creeks, the West

[1] Bailey Willis, The Northern Appalachians, Nat. Geog. Mon., 1896, pp. 172-173.

Branch, and Beech Creek. South of the gorge of Beech Creek the Front is unbroken by any large stream, though many steep ravines notch its crest and offer practicable routes of communication with the regions beyond.

The larger main streams have cut narrow and steep-sided valleys 500 or more feet deep, as well shown by the transverse course of the Youghiogheny. The smaller tributaries, especially in their upper portions west and north of the escarpment, have wider valley floors, and flow in relatively shallow valleys, often not more than 100 feet deep.[1]

The western part of the central district of the Newer Appalachians consists of closely folded strata that outcrop in the form of ridges with alternating narrow valleys, and the mountain, not the valley, feature is most prominent. Furthermore, the ridge crests almost everywhere reach to the level of the plateau on the west and exceed the elevation of the Blue Ridge on the east.

The most persistent characteristic of the Appalachian ridges of the central district is the even character of the sky line as determined by the level-topped ridges and the accordant altitudes of the hilltops. Only here and there are the ridges interrupted by gaps where superposed streams have maintained their courses across the ridges as they were gradually brought out to a strong topographic expression by differential erosion. A commanding point such as the summit of a residual surmounting the general level of the country affords a fine view out over a broad landscape. If the valleys and lowlands, now sunk below the general level, were filled up, the country would appear as a vast gently rolling plain of slight relief — an approximation to the condition that existed at the end of the Jurassic-Cretaceous cycle of erosion when the mountains formed during the great Appalachian Revolution were reduced to the condition of an almost featureless plain. This fact forms the starting point in all the physiographic considerations that follow concerning the Appalachian region as a whole, for it is in the erosion cycles that have followed and out of the peneplain that existed at the beginning of the Tertiary periods of degradation that all the later forms have been carved and the drainage features developed.

The distinctive features of Appalachian topography in Pennsylvania, Maryland, and Virginia are long, parallel, sharp-crested ridges, often with zigzag pattern, Fig. 268, and with narrow valleys intervening. Adjoining ridges are often markedly parallel. They are in places developed upon anticlines that expose resistant strata, in places upon

[1] W. S. Tower, Regional and Economic Geography of Pennsylvania (Physiography), Bull. Geog. Soc. Phil., vol. 4, 1906, pp. 205-206.

synclines, and in others upon strata with a more complex internal structure. All the larger valleys have their positions and directions determined by the more yielding rocks and have been developed subsequent to both the deformation of the strata and the base-leveling that is responsible for the even-topped character of the ridges.[1]

Eleven principal mountain folds occur between the Blue Ridge on the southeast and the Allegheny Front on the northwest, in a distance of 49 miles. Claypole suggested that the amount of crustal shortening involved in the flexing of these folds meant a reduction to 65 miles of a surface that originally measured 153 miles.[2] Chamberlin's later and more accurate measurements show a compression into 66 miles of an original surface of 81 miles in a section west of Harrisburg.[3]

Single folds more than 300 miles long are known in the Appalachian region, but the lengths of individual folds are more commonly from 25 to 50 miles. The intensity of the folding increases from east to west throughout the length of the province. In the Appalachian Plateaus the folds are very gentle, with dips generally less than 10°, and there is a close approach to horizontality on the west. The rocks are unaltered, even the shales being free from cleavage planes. In the Newer Appalachians the folding was more intense, the dips are generally 30° or more, and in many areas the rocks are nearly vertical. Most of the folds are symmetrical, with shorter, steeper northwest sides, and longer, gentler southeast sides, and many of them are overturned. The result is that the northwestern is the shorter and steeper and the southeastern the longer and gentler of the mountain aspects. The structural folds are of considerable magnitude, reaching 5 miles or more in vertical dimension between the larger folds, and are not simply a unit, being composed of numerous minor folds and these in turn of still smaller folds down to minute wrinkles.[4]

A number of facts are essential to the understanding of the physiography of the zigzag ridges of the Newer Appalachians: (a) the Appalachian type of structure prevails throughout the region, that is to say, a series of rather regularly folded strata, the folds being in the form of more or less regular anticlines and synclines; (b) these folds have been base-leveled or peneplaned, so that by the end of the Cretaceous cycle of erosion the surface of the country had been worn down nearly to a plane surface; (c) the fact of base-leveling of these folds means, further, that hard and soft rocks were at one time exposed in belts but with only the faintest topographic expression; and (d) naturally all the rock strata would be exposed almost in the same plane, for the topographic cycle was long enough not only quickly to bring down the soft rocks to base level but also finally to reduce even the most stubborn members almost to the general level.

(e) Uplift then occurred in the region and opportunity was afforded for the rejuvenation of the streams, the belts of soft rock were worn

[1] Topographic and Geological Survey of Pennsylvania, 1906–1908, p. 111.

[2] E. W. Claypole, Pennsylvania before and after the Elevation of the Appalachian Mountains, Amer. Nat., vol. 19, 1885, pp. 257–265.

[3] R. T. Chamberlin, The Appalachian Folds of Central Pennsylvania, Jour. Geol., vol. 18, 1910, pp. 228–251.

[4] G. W. Stose, Mercersburg-Chambersburg Folio U. S. Geol. Surv. No. 170, 1910, p. 13.

quickly down approximately to the new base level, while the harder rock belts stood out as ridges whose summits now present to our belated sight the ancient level of the Cretaceous peneplain. The ridges are even-topped because they were all worn even by the end of the earliest erosion cycle, and time enough has not elapsed since the uplift of the region and the development of extensive lowlands by differential erosion for the ridges to be very much affected by erosion. The material composing them is most resistant conglomerate (Pottsville) and a stubborn sandstone (Medina and Pocono), and when compared with the soft Coal Measures and the slates (Hudson River) and shales (Mauch Chunk) these offer incomparably greater resistance.

A sixth and last fact must be observed: (f) the axes of the folds are not horizontal for any distance, but pitch below the level of the peneplain, now at steep angles, now at gentle angles. Upon this feature depends the degree of divergence of the ridges. If the axes of the folds pitch at a steep angle the more strongly divergent will the ridges be formed; and conversely, the gentler the pitch the more narrow the angle between the ridges, the limit being parallelism, which would appear only when the folds became actually horizontal. According as the original folds were broad and gently pitching, or narrow and steeply pitching, the zigzags are long and wide or short and narrow.

The best example of zigzag ridges is the double series at the western end of the anthracite coal region, where the ridge crests loop back and forth as Catawissa and Line, Manhantango and Berry, and Peter's and Second mountains. Following the same course, essentially parallel to the first, and contained within them, is the second set, Big and Mahanoy, Coal and Lock, and Stony and Sharp mountains. Another example is in the Buffalo or Seven Mountains (Center and Clinton counties, Pennsylvania), where a group of narrow folds, steeply pitching, has given a series of short zigzag ridges with 7 loops to the northeast and 7 to the southwest.[1]

The student who will keep these groups of facts before him will be able to understand practically all the problems that the Appalachian ridges afford. A great variety of relief features can be assigned at once to their proper categories and order maintained in the examination of a group of data that at first sight may seem very complex. They furnish the key to the solid geometry of the region, for it is solid geometry, or a conception of three space dimensions, that is necessary in understanding the zigzag ridges of Pennsylvania.

[1] W. S. Tower, Regional and Economic Geography of Pennsylvania (Physiography), Bull. Geog. Soc. Phil., vol. 4, 1906, p. 14.

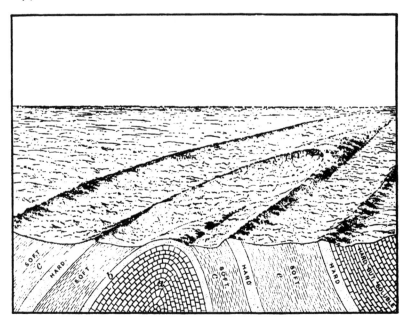

Fig. 269. — The half-cigar-shaped mountains developed on the hard rocks and the arches formed by the beds of an anticline. (Willis, U. S. Geol. Surv.)

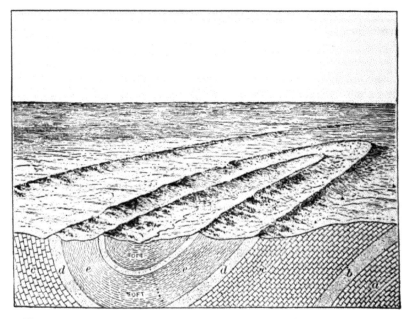

Fig. 270.— The canoe-shaped ridges of hard rocks and the arches formed by the beds of a syncline. (Willis, U. S. Geol. Surv.)

Because of the gentle inner slopes and steep outer slopes the topographic forms of the zigzag ridges have been likened to a canoe, and the resemblance is the more striking when it is recognized that the end of a synclinal mountain where it often meets in a rather sharp V is doubly resistant and usually stands out as a not quite reduced portion of the mountain region, a terminal knob suggesting the high prow of a canoe.

In the case of the anticlinal mountains as in Fig. 269 the strata dip outward as represented, and the unroofing of the anticline by peneplanation and subsequent valley excavation in the belts of softer rock has produced a group of mountain forms in sharp contrast to the forms of synclinal mountains. The steep slopes are here on the inside of the fold and the gentle ones on the outside. Fundamentally the law is the same in both synclines and anticlines, for in both cases the gentle slopes are down the dip of the strata and the steep slopes are those formed across the strata. It is the difference of direction and dip in the two cases that has produced the slope contrasts. Were the strata arched across the gap in the heart of the fold the resulting form would roughly resemble a cigar tapering down at one or both ends, so this type of mountain is described as cigar-shaped. Variations of form and degree of contrast of opposite slopes depend, as in the case of the synclinal mountains, upon the degree of dip of the strata. And like the synclinal mountain the meeting of the two ridge makers at the terminal point of the mountain doubles the resistance at that point and produces a terminal knob which in many cases exists as a residual or monadnock upon the surface of the peneplain.

It is important to see how a region may exhibit either or both synclinal and anticlinal mountains. The actual condition at a given place will depend upon the relation of the plane of base-leveling to the hard and soft strata. In A, Fig. 272, the plane of base-leveling is in such relation to the hard layer 1 that during the cycle of erosion terminating in the complete reduction of the land surface, the hard layer 1 is completely removed and hard layer 3 exposed but not removed from the underlying soft layers. When uplift opens the next cycle of erosion all the mountains will be for a time anticlinal and all the valleys will be synclinal. In B, by the same process of reasoning, half the mountains would be synclinal and half anticlinal and the valleys would be correspondingly disposed. In C all the mountains would be synclinal. In a region never base-leveled but for the first time passing through a period of subaerial denudation the changes of form and their relations to structure are brought out in Fig. 271, provided the region is still above base level.

Since a large number of the mountain systems of the earth have passed through one or more partial or complete cycles of topographic development, it is clear that a knowledge of the position of the plane

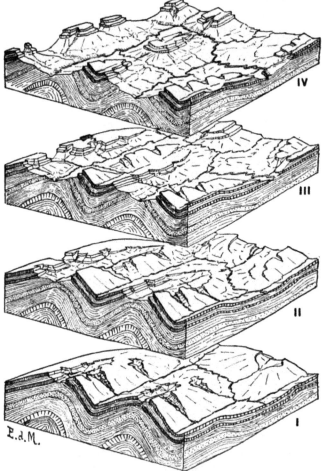

Fig. 271. — The development of anticlinal valleys and synclinal mountains from an original consequent drainage has been established in a region with Appalachian structure. (Martonne, Traité de Géographie Physique, Armand Colin.)

of base-leveling to the resistant rock strata, or ridge makers, is a matter of fundamental importance. In the Appalachian region the plane of base-leveling appears to have cut through the strata in such a manner as to form a larger number of anticlinal than synclinal mountains, though the latter type are numerous. The diagram Fig. 272–C roughly represents the actual conditions in the central Appalachians. Horse Valley opposite Chambersburg, and Bear Meadows north of Huntingdon, Pennsylvania, are good examples of synclinal valleys.[1]

The most striking features of the valleys of the zigzag ridges, whether of one structure or another, are their linear extent and shut-in or cove-like character. Bald Eagle and Black Log Valleys, Penn., are typical. Kishicoquillis Valley, between Stone and Jack's Mountains in Mifflin

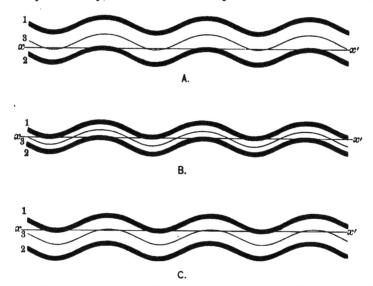

Fig. 272. — Varying positions of the plane of base-leveling to hard and soft strata and their relation to anti-clinal and synclinal mountains. 1 and 2 represent hard layers; 3 represents a soft layer; $x - x'$ represents the plane of base-leveling. In A the plane of base-leveling lies below 1 and intersects 2; in B it intersects both 1 and 2; in C it lies above 2 and intersects 1. After uplift of the base-leveled surface the early stages of the erosion cycle will be marked by anticlinal mountains in A, anticlinal and synclinal mountains in B, and synclinal mountains in C.

County, 53 miles long and 4 miles wide, is completely isolated except for the single outlet of Logan's Gap, near Lewiston. Tuscarora Valley is 50 miles long and 5 miles wide. Path and Nittany valleys are both over 30 miles long and from 2 to 5 miles wide, with no easy outlets except through an occasional water gap or over a higher wind gap. The sharp contrast between the linear extent and the width of the valleys is a direct result of the attitude of the strata, giving broader valleys where the strata are gently inclined and narrower valleys where steeply inclined.[2]

[1] W. S. Tower, Regional and Economic Geography of Pennsylvania (Physiography), Bull. Geog. Soc. Phil., vol. 4, 1906, p. 128.
[2] Idem, p. 128.

Variations upon the simple scheme outlined above are not difficult to understand. If for example each anticline and each syncline has several hard ridge makers, instead of one, several parallel ridges will come into existence as in Figs. 269 and 270, which represent the actual conditions in the northern Appalachians in the region north of Harrisburg. The number of ridges that will occur in a given place will depend upon the size of the folds and the number of alternations of hard and soft strata. This rule may be stated in another way as follows: As many hard layers as are truncated by the plane of base-leveling will stand forth as ridges in the following cycle, and theoretically this number is limited only by the ability of the rock to be compressed into folds. Obviously a mass of strata may be so thick that it can not be compressed in such a manner as to form folds exposing the entire section in the form of regular anticlines and synclines. The limit may be placed somewhere around 50,000 feet. A section across such a repeated series of ridge makers in a single fold appears as in Fig. 269. If the hard layers are sufficiently far apart, then each fold will repeat the ideal features shown in the figures above. If they are separated by a very thin soft layer the valleys between the ridge makers will not be deep unless the dip of the strata is unusually great. The two sides of a given ridge maker under these circumstances are also apt not to be so sharply contrasted as in the case where sufficient space occurs between the ridges for the formation of a large valley.

No less striking than the topographic features of the zigzag ridges of the central district are the drainage features of the region. The master streams flow roughly at right angles to the trends of the ridges and cut across them through water gaps of notable depth and often of pronounced beauty. The course of the Delaware through the Delaware water gap, the Susquehanna through the gaps of the central Pennsylvania ridges above Harrisburg, and the prominent gaps of the Potomac through the same or similar ridges farther south are illustrations of this feature. Where they cross the ridges in the gorge-like water gaps the main streams are swift, often descending short rapids, while across the intervening valleys they often flow lazily and in regularly meandering courses.[1]

Perhaps the chief cause for the wholesale modification of the drainage of a given region such as will bring the streams into courses directly across the grain of the country, as in the case in hand, is the warping or bowing of the surface that commonly takes place after or during the late stages of peneplanation and inaugurates a new cycle of erosion or forms one of the late substages of the first cycle. The effect of such warping or bowing is to cause a migration of the divides toward the main axis of uplift as determined by Campbell[2] and away from the area of subsidence. The antecedent drainage tends to become adjusted to the warped condition; streams come to occupy axes of depression, and divides finally become located on the axes of elevation. We have seen in Fig. 237 and accompanying text that the Appalachian

[1] W. S. Tower, Regional and Economic Geography of Pennsylvania (Physiography), Bull. Geog. Soc. Phil., vol. 4, 1906, p. 133.

[2] M. R. Campbell, Drainage Modifications and their Interpretation, Jour. Geol., vol. 4, 1896, pp. 567–581, 657–678.

region was bowed or warped up along a southwestward-trending axis located in western Pennsylvania and probably continuous with the axis in the southern Appalachians located by Hayes and Campbell in the vicinity of the Cumberland Plateau. While all the master streams do not have divides on this axis — the exceptions are the Tennessee, the New-Kanawha, and others — most of the streams conform to the general law applicable to the case.

As intrenchment progresses after uplift and other erosive agencies are set into operation differential erosion will follow, hard rock will be exposed in patterns sympathetic with respect to structure, and there will be brought about a most unsympathetic relation between the topography and the drainage, precisely the sort of drainage that now exists between the main streams and the main lines of relief within the central district. The weaker tributary streams will for a time flow across the harder ridges like the master streams, but the steady development of subsequent streams along belts of weak rock that occupy the inter-ridge spaces will in time effect an almost complete adjustment of tributary streams to structure. This wholesale readjustment means stream capture on a most extensive scale. The gradual headward growth will be accompanied by progressive capture of streams at the disadvantage of crossing the harder ridges to reach the main streams. The old water gaps will become wind gaps and a diminished river will flow in the channel of the beheaded stream. Here and there a larger tributary or one with exceptional advantages will persist like its master stream. From the map, Fig. 235, one may see all degrees of adjustment as outlined above.[1]

NORTHERN DISTRICT

The northern district of the Newer Appalachians consists of a number of well-defined valleys and ridges whose structural features are more complex than those of the central district. The topography is not capable of an analysis as simple as in the case of either the Pennsylvania zigzags or the ridges and valleys of the Chattanooga district; the ridges and valleys are here less regular both in general and in detail. In respect of border features the northern district is unlike either of the other districts of the Newer Appalachians. On the west the province is terminated not by a single plateau but by an outlier of the great Lauren-

[1] For an excellent description of the features of northern Pennsylvania and New Jersey with respect to drainage, see Davis and Wood, The Geographic Development of Northern New Jersey, Boston Soc. Nat. Hist. Proc., vol. 24, 1890, pp. 365–423, and W. M. Davis, The Rivers and Valleys of Pennsylvania, Nat. Geog. Mag., vol. 1, 1889, pp. 183–253.

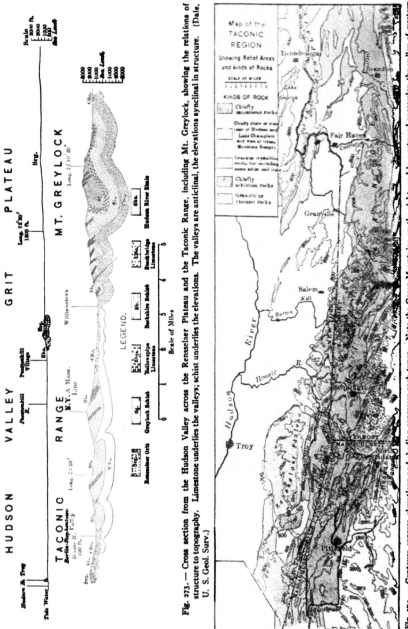

Fig. 273. — Cross section from the Hudson Valley across the Rensselaer Plateau and the Taconic Range, including Mt. Greylock, showing the relations of structure to topography. Limestone underlies the valleys; schist underlies the elevations. The valleys are anticlinal, the elevations synclinal in structure. (Dale. U. S. Geol. Surv.)

Fig. 274. — Contours express elevations and shading expresses rock types. Note the Green Mountains on the right, and in westward succession the valley of western Vermont, the Taconic Range, the Rensselaer Plateau, and the Hudson Valley. See structure section, Fig. 273. (Modified from Dale. U. S. Geol. Surv.)

tian area of Canada, the Adirondacks, and by an exceptionally high and
rugged portion of the Appalachian Plateaus, the Catskill Mountains.
The northern district of the Newer Appalachians is also unlike the
other districts in that the mountains upon its eastern border consist
partly of metamorphosed rock, the schists of the Taconic and Mount
Greylock ranges.

The Newer Appalachians have a very restricted development in
northwestern New Jersey, but shortly after entering New York state
and specifically in the Walkill and Middle Hudson valleys the Hudson
River shales thicken greatly and the whole belt has a notably broader
development. Between the Highlands of the Hudson and the Catskills
a broad valley lowland has been formed which continues northward
along the eastern border of the state to a point between Hudson and
Rensselaer, where the southern outliers of the Taconic mountains begin.
From this point northward the relatively soft Hudson River shales are
restricted to a narrower belt and the mountain feature of the Newer
Appalachians becomes more prominent. The principal topographic feature of the district is the Taconic Range. Its geographic position and
subdivisions are shown on the map, Figs. 273 and 274. The broadest
development of the mountain group of which it forms a part is between
the Hudson and the Hoosic valleys, roughly on the parallel of Troy.
At this point the Newer Appalachians consist of the following members
named in order from west to east: (1) the Hudson Valley, (2) the
Rensselaer Plateau, (3) the Little Hoosic Valley, (4) the Taconic Range,
including the Mount Greylock spur of the main Taconic Range, and (5)
the upper portions of both the Hoosic and Housatonic valleys. From
this point northward the Taconic mountains extend as a narrower
range as far as northern Vermont, where they terminate. The total
length of the range is about 200 miles, its width is from 5 to 10 miles.
Its course is somewhat serpentine, with a north-northwest trend
near Great Barrington, north-northeast to Dorset, and similar turns
farther north.[1]

Many of the forms of the Taconic region are directly or indirectly

[1] The name "Berkshire Hills" is so commonly employed to designate portions of the region
here discussed that a word as to its usage is in point. Dale (The Rensselaer Grit Plateau in
New York, 13th Ann. Rept. U. S. Geol. Surv., pt. 2, 1891-92, p. 297) implies a restriction of
the term to the Taconic Range and the Greylock offshoot or the rugged country between the
headwaters of the Hoosic and Housatonic rivers on the east and the western foot of the Taconic Range. Emerson's usage (Holyoke Folio U. S. Geol. Surv. No. 50, 1898, p. 1) corresponds to that of the inhabitants of the region and includes all the hills and mountains of
Berkshire County, Massachusetts. In this usage the Hoosac Mountains, the southern continuation of the Green Mountains of Vermont, or what is called in Massachusetts the western edge
of the Green Mountains Plateau, are included with the Taconics in the term "Berkshire Hills."

related to deformations acquired in three periods of folding: (1) at the close of the Cambrian, affecting the central portion of the area; (2) at the close of the Ordovician, with more far-reaching effects; and (3) in post-Silurian time (Devonian or Carboniferous).[1] The general structural features of the Taconic region are a succession of major and minor folds, as shown in the accompanying illustration, Fig. 273. These correspond approximately to the trend of the ranges. The general broad aspect of the structure of the Taconic Range is that of a synclinorium, in contrast to the geanticlinal character of the Green Mountains. The valleys are generally developed upon the softer limestone, the hills upon the harder schist.[2]

The easterly dipping cleavage of the schist determines the character of the eastern slope of the hills. There is a marked arrangement of the drainage along north-south or longitudinal lines both within the range and in the bordering valleys. There are 5 main transverse valleys — the valley of the Hoosic, the Mettawwe, the wide valley of the Walloomsac, the Battenkill, and the valley of the Castleton. Of these the Hoosic and the Walloomsac are deeply intrenched in wide valleys and the Taconic Range is greatly dissected in their vicinity.[3]

The main forms of the Taconic mountains are narrow ridges of harder rock separated by valleys developed upon softer rock. There is considerable variation in the lengths and forms of the ridges. Some are long, others short. The longer ones in many cases sag gently toward the center in response to variations in the axial pitch of the anticlines, or to the exposure of softer limestones in the heart of the anticlines. The shorter ones are in some cases roughly pyramidal in outline, irregular spurs with amphitheater-like hollows between them. There is also a small number of plateau-like masses. Few cliffs occur though some of these attain heights of 1000 feet.

The rugged hilltops and ridge tops of schist thinly veneered with soil have few agricultural resources and constitute areas of ultimate forest land. In general the valleys are deeply covered with drift and alluvium derived in large part from the underlying limestone. They are fertile and have an important agricultural population.

The Rensselaer Plateau lies between the Taconic mountains and the Hudson River in Rensselaer County, New York, and extends northward in scattered remnants toward Castleton, Vermont. Its structural fea-

[1] T. N. Dale, Taconic Physiography, Bull. U. S. Geol. Surv. No. 272, 1905, p. 48.
[2] Idem, p. 29.
[3] Idem, pp. 20-91.

tures and geographic position ally it with the Taconic mountains on the east. It rises from 700 to 1200 feet above the adjacent valleys and from 1400 to 2000 feet above sea level. Its structure is broadly synclinal, its rocks massive, and its relatively flat surface is the product of base-leveling. The lakes that dot its surface are chiefly due to irregular glacial erosion. It bears little good soil and in this respect is in marked contrast to the fertile Hudson Valley on the west, the Berlin and Berkshire valleys on the east, and even the Berkshire Hills. It was once thickly timbered, but has since been deforested, and is to-day an unattractive region with relatively steep slopes on the east, north, and south, and with a general westerly inclination.

The extreme northern end of the Newer Appalachians is the St. Lawrence Valley, which drains out to the northeast longitudinally like the Coosa on the southwest. In this respect the extremities are in contrast to the greater part of the province, which is drained only by tributaries, the master streams crossing the mountain ridges and valleys alike at right angles. The limestones on the northern side of the St. Lawrence Valley have been weathered down into a lowland extensively covered with glacial detritus and estuarine deposits formed when the land stood lower than now and the sea washed the foot of strand lines now elevated 1500 feet above sea level. The lowland strips become narrower toward the northeast and finally disappear before reaching the mouth of the Saguenay. Beyond this point the bay of the St. Lawrence occupies the entire breadth of the lowland.

TREE GROWTH

The various sections of the Newer Appalachians display quite different types of tree growth. The rich limestone valleys which form so large a portion of the southern district were originally covered with an excellent growth of hardwoods interspersed with open parks or natural prairies where the limestone is most fissured and porous and therefore most dry. At present the limestone valleys are extensively cleared owing to the exceptional fertility of their soils and the abundant supply of timber available on the uncultivated ridges of the province. Likewise the lowland portions of the northern district are cleared and farmed. The zigzag ridges of the central district, on the other hand, are still for the most part tree covered. Their bordering slopes are quite too steep and their summit areas are quite too small to tempt the farmer. They constitute ultimate forest land of great value when kept in trees, and of little value, even as sheep or goat pastures, when cleared of their timber. The drier, sandier ridges formed on resistant sandstone, such as the Pocono, are marked by stunted growths of scrub pine and red and black oak; the more fertile valleys underlain by limestone have a heavy native growth of walnut, blue ash, etc. The best growth is found on the highest ridges of eastern West Virginia where the rainfall is notably greater than elsewhere in the province.

CHAPTER XXXII

APPALACHIAN PLATEAUS

THE Appalachian Plateaus consist of a number of subdivisions of the first rank: the southern or Cumberland district; the central district including the so-called "mountains" of eastern Kentucky and West Virginia; the northern district, the Allegheny Plateau of western Pennsylvania and southern New York; and the extreme northeastern portion — a special category— including the Catskill Mountains. The Catskills and the plateau of West Virginia and Kentucky are the loftiest members of the series and have truly mountainous relief and proportions. The Allegheny Plateau has been dissected into a systemless maze of spurs by the almost infinite branches of the deeply intrenched streams. The high and rugged eastern border region of the Allegheny Plateau is also known as the Allegheny Mountains, a term which includes not only the Allegheny Front but also the rough uplands immediately west of it, the northern representatives of the Cumberland Plateau.[1] Like the Cumberland Plateau the upper altitudes of the region represent an old erosion level, but dissection has progressed to the point where practically no flat land exists on the divides as an inheritance from an erosion cycle long closed. The tremendous resistance of the thick border rock of the Cumberland Plateau and the lower altitude have practically preserved it from destruction and flat uplands still persist.

NORTHERN DISTRICT

The northern district of the Appalachian Plateaus consists of the plateaus of western and northern Pennsylvania and southern New York and include such specialized tracts as the Pocono Plateau in northeastern Pennsylvania and the Catskill Mountains in east-central New York.

The western border of the district is a well-developed scarp that swings southwestward along the southern shore of Lake Erie, finally disappearing in central Ohio. At Cleveland it stands out as a ragged scarp several hundred feet high, between which and the lake shore there occurs but a narrow strip of lowland. Farther southwest the

---

[1] See the Bedford quadrangle, Penn., U. S. Geol. Surv.

province has a less definite border than elsewhere except at the extreme south, where it descends gradually and dips beneath the Gulf Coastal Plain.

Its northern border is a low ragged northward-facing escarpment (Fig. 275) stretching eastward from Lake Erie to the Hudson across central New York. The eastern end of this escarpment is formed by the Helderberg Mountains, where the extremely hard Helderberg limestone outcrops in a stratum about 500 feet thick. Farther west, in the Finger Lake district, the escarpment makes a bend southward, a change in trend due to the great thickness (1000 feet) of the soft and

Fig. 275. — North-south section across the northern edge of the Appalachian Plateaus, Chemung River to Glenwood, N. Y. Vertical scale 5 times the horizontal. (Tarr, U. S. Geol. Surv.)

easily eroded Salina shales. At this point too the resistant Helderberg limestone is only 90 feet thick. Farther west the shales thin out again, the sandstones and limestones become thicker, and the escarpment again swings back to a more northerly position. The present position of the escarpment is in a geologic sense merely temporary, for the sapping of the hard layers which are responsible for its prominence is going on now just as in the past. By the same reasoning it was once farther north than now, and has been steadily pushed southward, so that this part of the Appalachian Plateaus province, like so many other portions with scarped margins, is suffering a reduction in area by the extension of the lowlands about it.

An interesting consequence of this process of escarpment retreat is progressive stream capture exhibited in many forms. The drainage of the plateau is southward almost from the very edge. The shorter, steeper, and more powerful northward-flowing streams are cutting back into the drainage systems of the Susquehanna and the Allegheny. In places they have diverted one tributary after another until almost the entire headwater systems of individual tributaries have been deflected to northerly courses. The junction of the deflected and the deflecting streams is marked by a sharp turn, so that the two stand in a curious relation designated as barbed drainage. From the extent to which this feature is developed conclusions may be drawn as to the former position of the plateau margin in recent time. An example is West River above the head of Canandaigua Lake. The southward-flowing plateau streams whose headwaters have been captured are diminished in volume and in many cases their headwaters flow as tiny brooks in broad valleys. Indeed the upper portions of some valleys on the plateau margin are without a living stream.

The eastern margin of the plateau of southern New York is also being pushed back rapidly by the short precipitous tributaries of the Hudson that descend the great scarp on the eastern aspect of the Catskills. Stream capture is here both vigorous and general. In places it means merely the westward retreat of the escarpment without sudden and great changes in the courses of the streams; at other places the valley sides are broken down by lateral attack, as in the escarpment of central New York, and barbed drainage relations developed. The best-known case is that of the eastward-flowing Kaaterskill, which has captured the headwaters of westward- and northward-flowing Scoharie Creek and turned down a 5-mile course waters which formerly flowed 50 miles to the Mohawk and down the Hudson to reach the same point.

The plateau of northern Pennsylvania and southern New York consists of a broad elevated region so extensively and deeply dissected that only small remnants exist here and there of what was once a fairly even surface.[1] The Cretaceous peneplain which by uplift became a plateau has by that erosion which is dependent upon uplift become a dissected plateau. Above its general level stand distinctly higher ridges with comparatively flat tops.

The degree of dissection of this portion of the Appalachian Plateaus is so great that the features of the central district are repeated in kind though not in degree. The lesser elevation of the northern district has resulted in shallower valleys and a less mountainous aspect than occurs in eastern Kentucky, but the relief is decidedly rugged. There is the same kind of dependence upon the valley ways in both the newer and the older systems of transportation.

"Wherever the surface is underlain by the same set of strata, it is cut into hills and valleys of the same general style. One valley can not be called the counterpart of another, nor are the hills and uplands always alike, yet the type of topography is the same. A view from the top of any well-exposed upland gives a good idea of what the country is like. Below the upland lies a valley as variable in the nature of its slopes as it is irregular in its course. Here steep walls rise from the stream on both sides. There a sharp descent on one side is faced by a long gentle slope on the other. Numerous ravines, some short, some long, some deep, some shallow, are occupied by the smaller streams which flow in from either side. As far as the eye can see in all directions the uplands stretch away in a broad, undulating tableland, unbroken by ridges, but on every hand bearing the deep scars of a multitude of valleys and ravines."[2]

"Over the entire area the streams branch again and again, until there is hardly a square mile into which one or more has not worked its way . . . the surface is that of a well-dissected plateau, varying from place to place, both in elevation and in surface detail, yet everywhere preserving the general feature of more or less rugged relief produced by the trenching valleys of innumerable streams."[3]

The topographic studies thus far made in Pennsylvania seem to show that three erosion levels can be identified. The first is the level of the ridge and hilltops in northern Pennsylvania, a feature equally well shown on the summits of the zigzag ridges of central Pennsylvania and northern New Jersey. This is the Cretaceous peneplain, and was developed widely upon rocks of diverse structure and resistance. The

[1] M. R. Campbell, Geological Development of Northern Pennsylvania and Southern New York, Bull. Geol. Soc. Am., vol. 14, 1903, pp. 277-296.

[2] W. S. Tower, Regional and Economic Geography of Pennsylvania, pt. 1, Physiography, The Central Province, The Plateau Province, Bull. Geog. Soc. Phil., vol. 4, 1906, p. 30.

[3] Idem, p. 36.

second peneplain was early Tertiary and is known as the Harrisburg peneplain. It was developed upon the Chemung rocks of the northern part of Pennsylvania, and is also well shown in the Monongahela Valley on the Brownsville, Missiontown, Connellsville, and Union Town topographic sheets, where the surface is a very gently undulating plain which in a distant view has an almost horizontal sky line at 1250 feet. The Harrisburg peneplain is best developed east of Harrisburg, where it

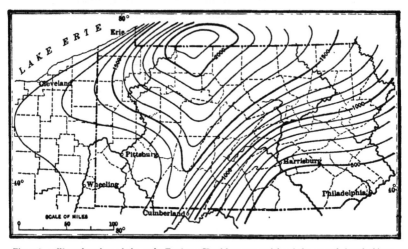

Fig. 276. — Warped surface of the early Tertiary (Harrisburg) peneplain of the central Appalachians. (Campbell.)

is at an altitude of 500 feet. It rises steadily upstream to about 800 feet in the vicinity of Sunbury, and to 1200 and 1300 feet at Pittston. Its present warped attitude is shown in Fig. 276.

In the Schuylkill the rocks are considerably disturbed and consist of a heterogeneous mass of shales; yet the hilltops are very regular indeed, with occasional monadnocks rising as high as 800 feet above the general 500-foot level. On the northwestern side of the Shenandoah Valley is a region of low flat-topped hills which appears like a great plain trenched by many small valleys.

In the Potomac Valley (Hancock quadrangle) the Harrisburg peneplain is from 600 to 700 feet above the sea in the southeastern corner and about 800 feet in the northwestern corner. The geologic structure consists of broad open folds, with many minor wrinkles, and the planes of stratification are generally inclined. In spite of these structural complications the peneplain represented by the hilltops cuts across the beds whatever their angles of inclination.

The third topographic level is that formed in late Tertiary time; it is known as the Somerville or the Worthington plain. In New Jersey it was developed on the rocks of the Kittatinny Valley and on the wide outcrop of Triassic rocks which form the lowland belt across New

Jersey and Pennsylvania, Fig. 276. It is, however, of exceedingly local development, and along the Susquehanna Valley as at Harrisburg it stands at an altitude of 400 feet, at Lancaster at 350 feet, and on the Potomac River near Harpers Ferry at 500 feet. It represents a partial cycle only, and nowhere was developed extensively across rocks of different hardnesses, but was etched out upon the softer formations only.

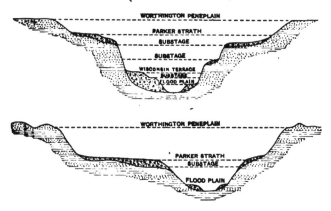

Fig. 277. — The upper section illustrates the terraces of the Ohio Valley, the lower the terraces of the Allegheny Valley. (Top. and Geol. Surv. of Penn.)

In the uplift of the third and lowest plain to the present level there was one main halt which permitted the development of broad valleys. Fragments of these valleys now occur as benches along the valley margins. All these features as well as those related to the glaciation of the region on the north are shown in Fig. 277.

Among the more prominent effects of glaciation was the development of an extensive system of abandoned channels whose origin was long in doubt. They are of widespread occurrence in western Pennsylvania, West Virginia, and eastern Kentucky. Detailed surveys and studies in recent years have at last supplied a basis for an acceptable explanation.[1] The abandoned channels appear to have been upbuilt by the streams during a period of stream aggradation associated with the waning stages of glaciation. All the southward-flowing streams heading in the glaciated country were so abundantly supplied with material that they aggraded their valley floors. The northward-flowing streams joining the southward-flowing aggrading streams were therefore compelled to aggrade their courses to the same level. In many cases they developed

[1] E. W. Shaw, High Terraces and Abandoned Valleys in Western Pennsylvania, Jour. Geol., vol. 19, 1911, pp. 140–156.

courses to one side or the other of the older courses, crossed low points in former upland spurs, and now exhibit most striking anomalies with respect to earlier channels. Typical conditions are represented along the Monongahela and its tributaries, as shown in Fig. 278.

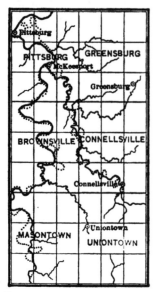

Fig. 278. — Present and pre-Pleistocene courses of Monongahela (left) and Youghiogheny rivers. (Top. and Geol. Surv. of Penn.)

POCONO PLATEAU

The Pocono Plateau, a separate division of the northern district, lies in the northeastern corner of Pennsylvania, almost completely separated from the rest of the plateau areas of that state by the deep synclinal trough of the Wyoming Valley. The Pocono Plateau covers Monroe, Pike, Wayne, and eastern Carbon counties, merges southwestward into the ridge area of the anthracite coal region, and extends east and north to become the Catskill Plateau of New York state. Almost the whole of it is underlain by the broad strata of nearly horizontal hard sandstone and conglomerate at the bottom of the Coal Measures series.

The southern edge of the plateau is often known as Pocono Mountain and presents a close analogy to the Allegheny front, both in origin and in character. It is an erosional escarpment, 1000 feet high, with a step-like ascent resulting from the horizontal position of the strata. The plateau back of the boundary escarpment (Pocono Mountain) is a nearly level upland wilderness, known to the early settlers as the "Great Beech Woods" and the "Shades of Death." Over most of its extent it stands from 1400 to 1800 feet above the sea. Below this level the streams have cut valleys from 100 to 200 feet deep, and an occasional knob rises 200 or more feet above the upland. It is often described as one of the wildest parts of the state, "a wilderness of forest and swamp," but it is made picturesque by the numerous lakes and cascading streams that are the result of the glacial action to which the region has been subjected.[1]

[1] W. S. Tower, Regional and Economic Geography of Pennsylvania (Physiography), Bull. Geog. Soc. Phil., vol. 4, 1906, pp. 216–217.

CATSKILL MOUNTAINS

The Catskill Mountains stand upon the northeastern border of the Appalachian Plateaus as a conspicuous group of ridges and peaks of mountainous proportions. Their summits reach to heights almost double those of the adjacent plateau. Some of the principal peaks are Slide Mountain (4220 feet), Hunter Mountain (4052), Black Dome (4000), Windham High Peak (3809), etc. From the upper portions of the Hudson Valley their whole elevation may be seen in a single view, from which point they appear to have imposing form and height. The blue haze that generally hangs over them — a feature of rare beauty in autumn weather — lends to their height a majesty and to their outlines a softness which in a distant view blend to form one of the most charming sights of the Atlantic slope.

The structure of the Catskills is as simple as that of the neighboring plateau on the west. The strata lie almost flat, with slight dips to the west, northwest, and southwest in various places. Anticlinal and synclinal structures are practically absent, and even when present they trend not with the ranges of the Catskills but at right angles to them, showing no relation to the present topographic forms. Shale commonly outcrops on the lower slopes of the valleys, but sandstones occur higher in the section, and on the summits of the principal peaks the rock is generally a conglomerate, very durable and thick. The flatness of the strata is expressed in the flat summits of the mountains, a characteristic feature and one that often interferes with the view, since these mountains are all heavily wooded. The tops are often of considerable extent and are never sharp-pointed peaks as in a region of alpine forms. While the valleys among the mountains are broad and open their sides are often cliffed to a notable extent for some distance. This is due to the system of almost vertical joints, which are the principal lines of weakness along which secondary erosion and valley widening take place. Abrupt ledges are frequent and are often a source of great difficulty in ascending a peak by unusual paths. In a few places these ledges are of great height and afford splendid panoramas over the surrounding country, as at the Catskill Mountain House and Overlook Mountain. To the vertical jointing and erosion along the joints is also to be attributed the successive steps which are common features of the valley floors and give rise to numerous picturesque cascades.

There are two main ranges in the Catskills, the line of division being marked by Esopus Creek. The southern Catskills consist of a massive central chain which bears the highest mountain in the Catskills, Slide

Mountain. The roughness of the topography and the unbroken forest that covers the mountain makes the penetration of this part of the Catskills very difficult. The northern ranges of the Catskills trend northwest and the principal range is about 35 miles in length. It makes a sharp curve, bending back upon itself in a sort of secondary range. A number of lateral spurs or secondary ranges trend at right angles to the main ones. Toward the south and southwest the main range falls off in long slopes and heavy spurs. It is divided into four sections by three deep gorges or " cloves " which give access to the interior valleys.

The drainage of the Catskills is chiefly to the west by tributaries of the Schoarie system. This appears to be a consequent drainage developed upon the original surface at the time of uplift of the region. Base-leveling, which is so common a feature of the Appalachian Plateaus, appears to be absent here; the Catskills appear to have survived as residuals because of their original superior height and the protection of the heavy cap of horizontal conglomerate. Erosion produced important effects, however, and opened up the broad valleys that are characteristic of the region. A common feature of the valleys is the presence of a pronounced shoulder halfway up the valley slopes, a feature which appears to be related to later uplift and dissection in the Tertiary. Were the valleys filled almost to the level of the shoulder we should probably have a picture of the Catskills as they appeared at the close of the Cretaceous cycle of denudation. They would then present smooth flowing outlines without any important number of steep ledges; the latter are commonly found below the level of the shoulder, where vigorous erosion is now taking place. The effects of Tertiary erosion are also shown in the form of local lowlands of limited extent opened up here and there within the mountain borders.

Although the Catskills were overridden by ice, signs of which are everywhere abundant, the ice appears not to have had any important effect upon the topography; rather it conformed to the broad slopes, only slightly molding them here and there by the deposition of small quantities of glacial till or by the erosion of the sharper forms.[1]

SOILS AND VEGETATION

The most uniform type of soil in the glaciated northern portion of the Allegheny Plateau is the till sheet which veneers the hills and uplands. It is a smooth, thin, and locally stony sheet of bowlder clay.

[1] For an excellent topographic description from which the above is largely derived see Arnold Guyot, On the Physical Structure and Hypsometry of the Catskill Mountain Region, Am. Jour. Sci., 3d Series, vol. 19, 1880, pp. 429-451.

The soil is much deeper in the valleys and of more variable form and composition, especially where morainic accumulations occur side by side with fluvio-glacial material—kames, eskers, outwash plains, and the like. In both the uplands and the valleys a large proportion of the glacially derived material is from underlying shales which weather into clay and increase the fertility of the soil by increasing its water-holding capacity. The varying proportion of this soil element causes great variation in the rapidity and thoroughness of soil drainage, so that in a dry spell one part of a field may be covered with a fresh green growth while another part near by may be parched and brown.[1] Swampy tracts

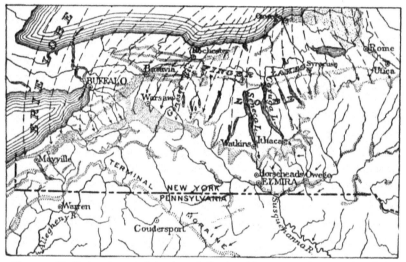

Fig. 279.—Distribution of morainal deposits and direction of ice movement in western New York. (Tarr, U. S. Geol. Surv.)

are characterized by a fertile black muck of great value when drained. Besides these soil types are narrow strips of flood-plain deposits and fan-shaped alluvial accumulations where the hill streams descend abruptly to the main valley floors.

While practically none of the primitive forest can now be found in the region, yet the uplands and valley slopes, too steep for cultivation, are in the main covered with extensive forests. The forest cover is extending naturally and encroaching on the cleared lands. The thin, stony upland soils are relatively infertile, and the extension of the forest

[1] R. S. Tarr (Williams, Tarr, and Kindle), Watkins Glen-Catatonk Folio U. S. Geol. Surv. No. 169, 1909, p. 33.

over them would benefit the region not only from the standpoint of forest products but also from that of the agricultural interests of the valleys, which suffer from increasingly destructive floods.[1]

CENTRAL DISTRICT

The central district of the Appalachian Plateaus lies in eastern Kentucky and West Virginia and is the ruggedest portion of the entire province. The highest portions are from 3000 to 4000 feet above sea level and are known locally as mountains, a name they fully deserve. The eastern rim of the district in Kentucky is Pine Mountain, whose steep eastern escarpment rises from 800 to 1500 feet above the Great Valley

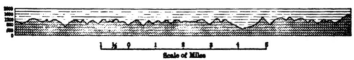

Fig. 280.— Maturely dissected Allegheny Plateau in West Virginia. The large depression on the right is the valley of the Kanawha at Charleston. Note the prevalence of slopes and the absence of flat land.

on the east and has but one water gap in 150 miles. The Kentucky "Mountains" are structurally a part of the Cumberland Plateau and formerly had a flattish summit; the original nearly flat surface has been so greatly dissected however that but few remnants remain to indicate the former level. These, however, show a remarkable uniformity of elevation on northeast-southwest lines, so that the eastern escarpment of the district has an almost perfectly straight sky line.

The western slopes of the district are in sharp contrast to the eastern. There is a bordering escarpment, but it is highly irregular, the streams having carved valleys far back into the elevated portions of the upland, leaving long narrow spurs running out toward the west. The valleys are steep and gorge-like; the main streams such as the Kanawha occupy canyons; flat land is seldom found to any extent either upon the hill summits or on the valley floors. It is a hill-and-valley country, Fig. 280, where so little flat land occurs that the water everywhere falls upon a slope and the run-off is heavy; floods due to spring rains and the melting of late winter snows are common and rise to great heights in the confined valleys, destroying roads and bridges, washing away the valley soils or covering the soils of the narrow flats or flood plains with heavy deposits of coarse waste. Almost all the roads and trails follow the valleys, to which the railroad lines themselves are strictly confined.

[1] R. S. Tarr (Williams, Tarr, and Kindle), Watkins Glen-Catatonk Folio U. S. Geol. Surv No. 169, 1909, p. 33.

Travel by any means is often suspended during time of high water. The whole section is a great forested wilderness whose resources of coal and timber constitute its chief wealth, resources slow in development because of the almost insuperable topographic obstacles to transportation. Even the railways on the margin of the country have been built since 1880.

Magnificent forests once covered the central district of the plateau region. They are still untouched in the remoter localities. Oak, walnut, poplar, chestnut, maple, ash, and tulip trees grow to great size. The making of staves of white oak is a considerable industry among an isolated and backward mountaineer folk. Since the roughness of the country limits the railways to the principal streams, the lesser waterways are almost everywhere utilized for rafting both timber and lumber to the railroads and the lowlands. In many sections remoteness from transportation lines forbids any attempt at forest exploitation. As in the southern Appalachians a wasteful system of agriculture is practiced. Hillside farms are cleared of the finest timber by combined girdling and later burning, then cultivated a few years and abandoned for a new site. The abandoned clearing grows up to useless brush or is deeply gullied and the thin soil washed away.[1]

SOUTHERN DISTRICT

CUMBERLAND PLATEAU, WALDEN RIDGE, LOOKOUT MOUNTAINS, ETC., AND THE HIGHLAND RIM

A first inspection of the flat-topped plateaus of the southern district leads one to the conclusion that the flatness is a function of the structure, for the strata appear to have a roughly horizontal attitude. A closer examination, however, reveals the fact that the plateau surface does not generally coincide with a particular stratum however resistant it may be; the surface is found to be composed of very soft shale as

[1] For extremely interesting and accurate descriptions of both the physical geography of the region and the Kentucky mountaineers see the writings of John Fox, Jr., as for example Hell-fer-Sartin, Blue Grass and Rhododendron, and The Trail of the Lonesome Pine. On Horseback to Kingdom Come, Scribner's Mag., vol. 48, No. 2, 1910, p. 175, discusses later industrial development. The best scientific description of the country and the people is by E. C. Semple, The Anglo-Saxons of the Kentucky Mountains, A Study in Anthropogeography, Geog. Jour., vol. 17, 1901, pp. 588–623. In Theodore Roosevelt's The Winning of the West (The Spread of English-speaking Peoples), vol. 1, ed. of 1905, pp. 146–147, is a fascinating description of the forest of pioneer days. For a good topographic description of portions of eastern Kentucky see Ky. Geol. Surv., vol. 5, n. s., 1880. For a discussion of the natural water routes of the Kentucky mountains see N. S. Shaler, The Transportation Routes of Kentucky and their Relation to the Economic Resources of the Commonwealth, Ky. Geol Surv., vol. 3, pt. 5, 2d series, 1877.

well as of very hard sandstone. In short, the surfaces of the various plateaus truncate hard and soft beds and are much more nearly horizontal than the strata upon which they have been developed.

Following the principles we have already applied so frequently in the study of the physiography of the United States, we shall conclude that the surfaces of Walden Plateau, Cumberland Plateau, Lookout Mountain, Sand Mountain, etc., are portions of an uplifted peneplain in process of more or less rapid dissection. Along certain lines narrow anticlinal folds had developed during the period of structural deformation, with broad synclines between. The projection of the anticlines above the level of the peneplain resulted in their erosion and the exposure of softer underlying beds lying in the heart of the anticlines. When later uplift occurred, opportunity was afforded for the dissection of the softer exposed beds. The result has been that the synclinal basins of an earlier period have been converted into the mountains and plateaus of the present period.

One of the most important departures from the plateau topography is known as the Sequatchie Valley, which separates Walden Plateau from Cumberland Plateau. It lies parallel with the Great Appalachian Valley, has remarkable continuity and regularity of expression for over 100 miles, and appears to be an outlying anticlinal fold of the Appalachian system of folds.

The Sequatchie anticline, like the anticlines of the Newer Appalachians, has a typically unsymmetrical form, the beds dipping much more steeply on one side of the axis than on the other, and the gentler dips are upon the eastern side. Near the upper end of the Sequatchie Valley the strata have been broken by a thrust fault developed along the steep side of the arch. Walden Plateau on the east shows a distinct synclinal structure. Of similar structure is Lookout Mountain, but it is much narrower than Walden Plateau. Wills Valley, developed upon an anticline, separates the syncline of Lookout Mountain from the syncline of Sand Mountain, and Sand Mountain is in turn separated from Cumberland Plateau by the valley of Tennessee River, which has been developed in the southwestward extension of the Sequatchie anticline.

Along the eastern edge of Cumberland Plateau the strata dip westward at a steep angle, but these dips are maintained for very short distances, usually not more than a few rods, where they change to dips that are sensibly flat, a condition maintained across the Cumberland Plateau and the Highland Rim. Though apparently horizontal the beds dip toward the southeast from 20 to 30 feet per mile.

It must not be supposed that Cumberland Plateau and Walden Plateau were perfectly peneplaned. Along the western edge of Walden Plateau, the northern end of Lookout Mountain, the eastern edge of Cumberland Plateau, and at a large number of isolated points elsewhere, residuals in the form of isolated knobs or mesas rise from 100 to 300 feet above the general level of the plateau. In places, the residuals are composed of more resistant beds of massive conglomerate, but the residuals within the borders of the plateaus are composed of horizontal strata in some cases capped by a bed of conglomerate but more often composed entirely of rather soft sandstones and shales.

Following the uplift of the Cretaceous peneplain there ensued a period of crustal stability sufficiently prolonged to enable the forces of erosion to develop a partial peneplain at a level from a few hundred to a thousand feet lower than the Cretaceous. The peneplanation accomplished during this period was chiefly upon the softer rock of the region and was so incomplete as to leave large portions of the Cretaceous peneplain standing above the early Tertiary peneplain in the form of massive unakas, as shown in Fig. 238. The early Tertiary peneplain is called the "Highland Rim" peneplain because the Highland Rim, so-called, between the Nashville basin and the Cumberland Plateau, is the best-preserved portion.

As in the case of the well-preserved remnants of the Cretaceous peneplain, the remnants of the early Tertiary or Highland Rim peneplain have been preserved largely because of the presence of resistant beds along the outer margin, but the peneplain as a whole truncates beds of widely differing degrees of resistance to erosion. The peneplain is preserved upon rocks of intermediate resistance, chiefly siliceous limestones and sandy shales, for rocks of greater resistance than these were never base-leveled, and rocks of less resistance were dissected in the uplift which closed the early Tertiary cycle of erosion.

The elevation of the Highland Rim peneplain is about 1000 feet, west of the Cumberland Plateau in the Appalachian Valley it is about 1150 feet, toward the northern edge of the Chattanooga district 950 feet. Elevations above the topographic level developed on the Highland Rim are in the form of isolated residuals similar to those we have noted upon the Cumberland Plateau, as long irregular projecting spurs along the ragged west border of the Cumberland Plateau, or as massive unakas such as Cumberland and Walden plateaus themselves.

On the western and southern borders of Cumberland Plateau many long spurs and isolated knobs, products of circumdenudation, project irregularly westward over the surface of the Highland Rim, breaking the continuity of the eastern portion. Between the spurs are great gorges from 800 to 1000 feet deep, which, on account of their depth and narrowness, are known as "gulfs." At their heads coves are found, headwater alcoves which usually contain a pocket of limestone soil that supports a better timber growth than the flat upper surfaces of the upper and lower plateaus with their thin soils formed upon sandstones

and shales.   Both the distribution of the soils and the character of the
topography on the border of the Cumberland Plateau are due to the
structure.

Between the hard sandstone and conglomerate capping most of the Cumberland Plateau
and the rock of the lower country about it are soft limestones and shales which are so easily
eroded as to result in the sapping and undermining of the hard formations above them.   It is
this process which maintains the steepness of the border scarps (especially their upper portions,
which are frequently vertical cliffs) and results in the sharp line of delineation between the
Appalachian Plateaus on the one hand and the Appalachian Valley and Highland Rim on the
other.   These features are persistent, being found on the projecting portions of the plateau as
well as at the heads of the coves.[1]

The "barrens" of Tennessee are developed chiefly upon shales
(Waverly) which yield a white, siliceous, and unproductive soil.   Scarcely
more productive are the soils derived from sandstones which are thinly
inhabited and have thin native forests of pine and oak.   The surface
of Cumberland Plateau, consisting of sandstones and shales, is covered
with a thin poor soil; the Highland Rim is also far from having a productive soil, though on its western margin, where a limestone (New-
man) outcrops, a soil of greater but not of high fertility occurs.   In
general the timber covering of the flat plateau summits and remnants is
thin, but in the coves, hollows, and gorges, where a richer soil and more
abundant water supply are found, hickory, chestnut, and oak reach a
good size and grow in first-class stands.   Near the watercourses, pine,
hemlock, and spruce find the necessary elements of their environment,
but their growth is everywhere second in importance to the members
of the first-named series.[2]

LIMESTONE SOILS OF THE APPALACHIAN VALLEYS

"The limestone soils are among the most extensively developed of any in the United States
and occur in both broad upland and enclosed narrow valley areas.   The greatest upland development is seen upon the Cumberland Plateau in eastern Tennessee and Kentucky and
upon the Carboniferous formation in central Tennessee and Kentucky, northern Alabama and
Georgia, and in Missouri.   The valley soils are found principally in Pennsylvania, Maryland,
and Virginia, and in the mountain section of eastern Tennessee and Kentucky and northern
Alabama and Georgia."[3]

The limestone soils are residual in origin, being derived from the
weathering in place of limestone of several ages and variable composition.
This is accomplished by the removal through solution of the calcium carbonate of the limestone.   Limestone soils are remarkable for the fact

[1] C. W. Hayes, Seuanee Folio U. S. Geol. Surv. No. 8, 1894, p. 1.

[2] A. Keith, Wartburg Folio U. S. Geol. Surv. No. 40, 1897, p. 4, col. 3, and M. R. Campbell,
Standingstone Folio U. S. Geol. Surv. No. 53, 1899, pp. 4-5.

[3] Soil Survey Field Book, U. S. Bur. of Soils, 1906.

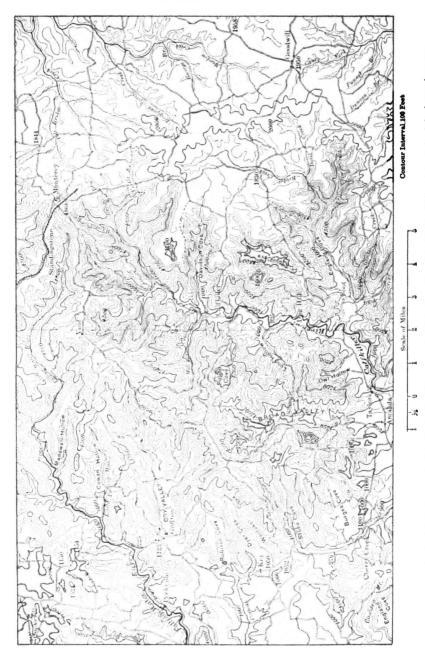

Fig. 281. — On the right is the summit of the Cumberland Plateau; on the left is the surface of the Highland River; between is a belt of rugged country which marks the descent from the higher to the lower level. (U. S. Geol. Surv.)

that they constitute but a small percentage of the original limestone rock, the larger part having gone into solution, leaving behind the more resistant siliceous minerals. It has thus required the solution of many feet of rock to form a foot of soil. They have a naturally heavy character. Solution and subsequent filtration of pure massive limestone of Cambro-Silurian age have given rise to a soil which as a rule occurs in valleys bordered by areas of the more resistant sandstones and shales. The series is typically developed in the limestone valleys of the Great Appalachian Valley and in the central basins of Kentucky and Tennessee, but smaller areas are found as marginal deposits in the Piedmont section and in the deep valleys of the Appalachian Plateaus, where the underlying limestones have been exposed to weathering by deep erosion. The most productive valley phase occurs in the Great Appalachian Valley.

The cherty and fossiliferous limestones (St. Louis) of the region have given rise to a second soil type which occurs on both the level and the undulating uplands and in rough, hilly country with steep valleys. Where the latter features predominate the soils are generally unproductive and very stony, but in some sections are adapted to fruit, especially apples. The soils formed from beds of purer limestone occupying level and gently rolling areas are as a rule very productive.

LOCAL LOWLANDS

BLUE GRASS AND NASHVILLE BASINS

The upland quality of the general regional slope of the Appalachian Plateaus is interrupted in Kentucky and Tennessee by lowlands of unusual size and importance. The Kentucky lowland is the Blue Grass country, famous for its rich limestone soils and its nutritious blue grass; the Nashville lowland or Central Basin is of equal importance, though it has for some reason never gained such general renown. In respect of soil fertility and easy cultivation both are in happy contrast to the uplands about them, which are as a whole either too broken to permit easy tillage or too thinly covered with soil of inferior quality to tempt men in large numbers. These two lowlands are essentially alike in structure and origin in spite of many detailed differences. Both are great structural domes which extended above the general level of the surface of erosion once developed here. The Blue Grass lowland was developed in the early Tertiary cycle of erosion and was so denuded by the base-leveling of that period as to have its cap rock either partially or wholly removed. With later uplift, erosion quickly cut below the hard and into the underlying soft strata, Plate V and Fig. 282. The result was a lowland, while the surrounding areas, underlain by the hard strata once arching over the dome, remained as uplands. The central basin of Tennessee or Nashville Basin was developed in the late Tertiary cycle of erosion under similar conditions.

BLUE GRASS COUNTRY

The general altitude of the Blue Grass country of north-central Kentucky is from 800 to 1000 feet above sea level. The region may be described as a broad plain on whose southern margin the hills rise abruptly; in the hill country the large streams have cut deep narrow gorges which only become slightly less deep and narrow throughout the Blue Grass region itself. Although in short distances the surface of this plain appears to be structural and to correspond with the bedding of the limestone rocks (Ordovician) which compose the greater part of the surface, a large view discloses the fact that the surface cuts across rocks of different ages and varying degrees of hardness, and is an uplifted peneplain, called the Lexington peneplain.[1]

The hills which rise above the Lexington peneplain have a fairly constant altitude of about 1500 feet. They generally have round or sharp tops and a regular altitude despite the variation of the underlying rock, hence it is inferred that they too represent a former peneplain, the Cretaceous peneplain of the Appalachian region.

It is noteworthy that the valleys of the Lexington peneplain of the Blue Grass region exhibit a topographic unconformity showing two episodes of erosion. Long gentle slopes lead down from the surface of the Lexington peneplain to an inner valley with steep walls. The gentle slopes evidently constitute the borders of an older broad valley in the bottom of which the modern narrow gorge has been cut. The floors of the older valleys bear deposits of sand, while the sides of the modern valleys are in many cases rock cliffs.[2]

The Blue Grass country of Kentucky is developed upon a broad structural arch known as the Cincinnati anticline, which extends from Nashville through Lexington, nearly to Cincinnati. Its occurrence north of the latter point will not be described here, for it is of lesser topographic importance than glaciation which has largely concealed it. The Cincinnati arch south of Cincinnati may be divided into two broad domes, one of which culminates near Nashville, Tennessee, and the other in Jessamine County, central Kentucky.[3]

[1] M. R. Campbell, Richmond Folio U. S. Geol. Surv. No. 46, 1899.
[2] Idem.
[3] G. C. Matson, Water Resources of the Blue Grass Region of Kentucky, Water-Supply Paper U. S. Geol. Surv. No. 233, 1909, pp. 26–27.

A large part of the Blue Grass region, and particularly that part of it in Woodford, Franklin, and Fayette counties, Kentucky, has flat-topped divides with practically no surface drainage. This condition furnished exceptionally favorable opportunities for the formation of caverns and the development of underground drainage systems for which Kentucky is noted. The topography is marked by a series of sink holes which

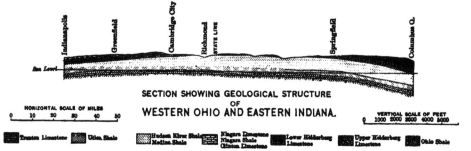

SECTION SHOWING GEOLOGICAL STRUCTURE
OF
WESTERN OHIO AND EASTERN INDIANA.

HORIZONTAL SCALE OF MILES
0   10   20   30   40   50

VERTICAL SCALE OF FEET
0   1000 2000 3000 4000 5000

Trenton Limestone | Utica Shale | Hudson River Shale, Medina Shale | Niagara Limestone, Niagara Shale, Clinton Limestone | Lower Helderberg Limestone | Upper Helderberg Limestone | Ohio Shale

Fig. 282.—Northern portion of the Cincinnati arch, showing exposure of the lower and softer beds of shale and limestone.

receive a large part of the rainfall. It is inferred that before the uplift of the Lexington peneplain so large a part of the surface was drained by underground streams that surface drainage was probably limited to the larger streams and their principal tributaries.[1] The sink-hole topography of the Blue Grass region favors the quick removal of surface water and its rapid absorption by the rock.[2]

The residual soils of the Blue Grass region range in thickness from 3 to 5 feet on the average, but with extreme ranges from a few inches up to about 30 feet. The residual material absorbs and stores the rainfall and delivers it gradually to the underground channels. Were the rock surface exposed, springs would flow only a short time after each rain, and the stream flow would be exceedingly irregular.[3]

The Blue Grass region has been famous for its excellent soils ever since the first settlers under Daniel Boone penetrated the region. The finest types of soils occur where the Lexington limestone is from 140 to 160 feet or more thick. It is composed of bluish finely crystalline limestone in thin irregular beds frequently separated by intervals of calcareous shale.[4] The soils of the Lexington limestone are loamy.

[1] G. C. Matson, Water Resources of the Blue Grass Region of Kentucky, Water-Supply Paper U. S. Geol. Surv. No. 233, 1909, p. 30.
[2] Idem, p. 60.
[3] Idem, p. 62.
[4] M. R. Campbell, Richmond Folio U. S. Geol. Surv. No. 46, 1899, p. 2.

The upland soil of the Blue Grass region is adapted to a large variety of crops such as blue grass, corn, wheat, tobacco, and hemp. In the Ohio Valley occur small areas of loam which are well adapted to general farming, but they do not equal the upland areas either in extent or in value. The purest soils of the region are those derived from the Eden shale, which usually occupies the hilly areas. It is subject to rapid erosion because of its softness, and in many places attempts to farm this soil have been abandoned and it has been converted into pasture and timber.

It is not always the case in the Blue Grass region that each rock formation yields a distinctive type of soil. It has been found that a soil derived from a calcareous shale may be little different from one derived from a clayey limestone. The large amount of limestone dissolved to form the soils of the Blue Grass region naturally means that a certain amount of carbonate of lime is found in the soils to-day, which gives it a slight tendency toward aggregation into soil crumbs. The analyses show that the sizes of the openings between the crumbs of the Blue Grass soils are very small and that the soils have great capacity for retaining moisture. The small size of the pores also brings the ground water into contact with a large amount of soluble material and favors the solution of the various elements of plant food. Chemical analyses of the Blue Grass soils show that the plant elements are so abundant in them that it is generally unnecessary to add fertilizers. The average of 32 analyses of soils from the Lexington limestone is represented in the following table.

AVERAGE COMPOSITION OF SOILS FROM LEXINGTON LIMESTONE

	Per cent
Organic and volatile matters	6.211
Alumina, iron, and maganese oxides	11.200
Lime carbonate	.749
Magnesia	.644
Phosphoric acid (P_2O_5)	.328
Potash extracted by acids	.404
Sand and insoluble silicates	73.380

Another feature of these soils that gives them a high degree of fertility is the large quantity of phosphorus contained in them in the form of phosphate of lime derived from Ordovician limestones. They are richer than[1] the average in those mineral elements which support plants.[1]

The low productivity of the soils developed upon the Ohio shales is

[1] G. C. Matson, Water Resources of the Blue Grass Region, Kentucky, Water-Supply Paper U. S. Geol. Surv. No. 233, pp. 33-35.

to be attributed not to a deficiency in the elements of plant food, but to an excess of moisture. They are so clayey that they retain too large quantities of rainfall and tend to remain wet and sour.[1]

NASHVILLE BASIN (CENTRAL BASIN OF TENNESSEE)

From the steep and ragged western escarpment of the Cumberland Plateau the country extends westward as a more or less deeply dissected plateau about 1000 feet above sea level, known as the Highland Rim. This in turn terminates on the west in an escarpment which practically surrounds a lowland known as the Central Basin of Ten-

Fig. 283. — Section across the Nashville Basin of Tennessee and the country adjacent, from the Cumberland Plateau (right) to the Harpeth River. C, Cambrian; Ot, Trenton limestone; On, Nashville shale; Sn, Niagara limestone; Db, Berea shale; Ms, sandstone; Mlm, Mountain limestone; C, Coal Measures. Length of section about 120 miles. (Safford, Geol. of Tenn.)

nessee. It is about 70 miles across, extends north and south about 60 miles, and stands about 600 feet above sea level. It is drained by the Tennessee and its tributaries through a narrow gorge-like valley that cuts through the surrounding uplands. It has a gently undulating surface save on the border, where spurs from the surrounding plateau and isolated hills, which represent detached and greatly eroded spurs, occur in numbers. The deeper and richer soils of the basin, its lowland character, and its relation to the surrounding uplands that stand about 400 feet above it, give it a distinctive character. These features all rest back upon the structure and late physiographic development of the region.

The present lowland was originally an upland with respect to surrounding tracts. This relation was owing to its domed structure, the rocks dipping outward in all directions though at varying rates. During the erosion cycle in which the Highland Rim was formed (early Tertiary) the topographic effects of the doming were obliterated and the cap rock, a cherty limestone, reduced in thickness because of its domed attitude. The base of the formation was, however, below the plane of base-leveling, hence the formation, though thinned, was not broken through at any point.

[1] For the soils of this region see R. Peter, Comparative Views of the Composition of the Soils, Limestones, Clays, Marls, etc., of the Several Geological Formations of Kentucky, etc., Geol. Surv. of Ky., 1883.

The early Tertiary erosion cycle was interrupted by uplift which enforced dissection. Soon the intrenched streams cut through the thin cherty limestone capping the dome, but were not able elsewhere to breach the formation because of its greater thickness. The underlying limestones are free from chert, are soft, shaly, and easily eroded, and were worn down to a lowland while the surrounding country capped by the hard cherty formation maintained its level. Though the whole Highland Rim is more or less dissected, considerable areas of dissected remnants occur which preserve the ancient level; in the lowland no trace of the ancient level remains except on the margins. After the development of the lowland there occurred two later uplifts of the land. After the first uplift the valleys were broadly opened several hundred feet below the general level. The second uplift caused the streams to cut narrow valleys within the broad ones, leaving the remnants of the broad valleys as terraces along the upper levels of the lower valleys. The descent from the lowland to any valley floor is over a terrace which marks a topographic unconformity or break between a younger and lower and an older and higher set of slopes.

In the Nashville Basin the soil characters are very intimately related to the underlying geologic formations. The conditions in the Columbia district in the southwestern part of the basin are typical. The former great red-cedar glades of middle Tennessee were here formed upon the Lebanon limestone, which has a shallow and rocky but fertile soil. The purest soil of the region is derived from a cherty shale and limestone known as the Tullahoma formation, which is flinty and nearly always occurs on steep slopes. When thoroughly leached and light in color it constitutes the "barrens" of the Highland Rim. If the formation is underlain by clay so that the calcareous matter can not be readily leached out, the soil is very good and capable of producing abundant crops. A number of good soil makers have a limited outcrop or outcrop on steep slopes where the soil is readily washed away, and are of little importance to vegetation. Among the best of natural blue-grass soils in the Nashville Basin is the Bigby limestone, which is a crystalline phosphatic limestone from 30 to 100 feet thick in the Columbia region.[1]

The surrounding uplands, or the Highland Rim, have a thinner and less fertile soil. Toward the north the upland is underlain by sandstone, and here occur the famous " barrens " of Kentucky and Tennessee, long without more than a mere sprinkling of agricultural population. The first settlers in Kentucky also regarded the untimbered limestone lands of the western part of that state as infertile, and gave them the

[1] Hayes and Ulrich, The Columbia Folio U. S. Geol. Surv. No. 95, 1903, p. 6.

name of "barrens" from their previous experience of the relative infertility of untimbered lands. Several years passed before the true character of these "barren lands" was ascertained and their fertility recognized.[1] So, too, the sandstone barrens to the east of the lowlands have had their day of neglect, but are now cultivated to an increasing degree, and when properly fertilized yield profitable though not as a rule abundant härvests. On steep slopes they are too thin and cultivation is mechanically too difficult, and such belts along stream valleys are commonly left timbered; the interfluves are formed on gentler slopes and have a deeper though rarely a fertile soil.

[1] N. S. Shaler, The Origin and Nature of Soils, 12th Ann. Rept. U. S. Geol. Surv., pt. 1, 1890-91, p. 325.

CHAPTER XXXIII

LOWLAND OF CENTRAL NEW YORK

THE most noteworthy topographic depression across the Appalachian tract is the lowland of central New York. It is remarkable for its continuity from Lake Erie to the Hudson, its low gradients and elevation, its thoroughgoing transection of the upland in which it occurs, besides its swarms of drumlins and its narrow, deep, picturesque lakes and bordering glens. Among its exceptional features are to be noted also its dense population, numerous and busy railways, its canals, and its long and romantic historical development. In many of these respects it is a happy contrast to the adjacent provinces. The Adirondacks on the north are more grand, but they are thinly peopled, have few railways, and limited resources. On the south the uplands of the Appalachian Plateaus are also thinly populated. In recent years they have suffered a relapse into the ways of a back country and exhibit an increasing number of abandoned farms.[1] Between these two rugged uplands extends the relatively narrow central valley with its deep, fertile soil, pleasing countrysides, and prosperous homes.

The geologic map of New York, Fig. 285, largely explains the central valley. The strata outcrop in westward-trending belts and the thickness and interrelations of the hard and soft members from place to place determine both the local topographic expression and the width of the central valley (Plate IV). The southern border of the lowland is the frayed northern border of the Appalachian Plateaus. Spurs from 10 to 20 miles long, from 2 to 5 or more miles wide, and from 100 to 500 feet high extend north on the interfluves between the deep narrow valleys whose floors are occupied by lakes. The border is maintained definitely because of a capping layer of hard sandstone. The central lowland is developed upon the Salina and Hudson River shales and other soft formations. Formerly the strata extended farther north, overlapping and wrapping about the Adirondack old-land. When first uplifted above the sea the sediments formed a simple coastal plain. In time this plain was dissected, and since the soft formations weathered more rapidly,

[1] R. S. Tarr, Decline of Farming in Southern-Central New York, Bull. Am. Geog. Soc., vol. 41, 1909, pp. 270-278.

they were worn to a lowland, while the hard formations stood in relief. Similar features extend westward across Ontario and through northern

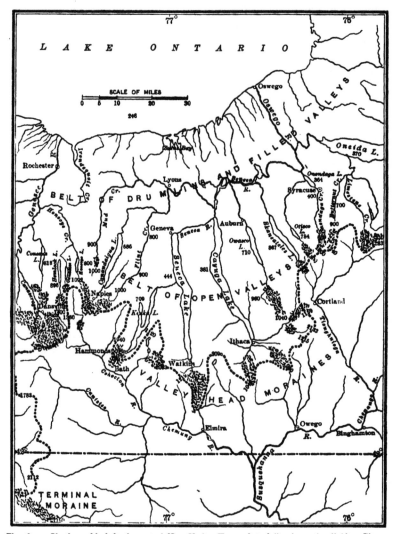

Fig. 284. — Physiographic belts in central New York. Heavy dotted line is on the divide. Figures represent elevations above sea level. (Modified from Fairchild.)

Michigan and eastern Wisconsin, as shown in Plate V. They represent an old coastal plain so eroded as to present both its structure and its relief in rudely parallel belts, hence an ancient belted coastal plain.

The upper Susquehanna still pursues a consequent course; the Mohawk is a subsequent stream, while the normal courses of most of the streams farther west have been modified by glaciation. In such a description Lake Ontario and Georgian Bay lie upon the inner lowland, Lakes Erie and Michigan upon the outer lowland of the plain. In western New York the plain is a double depression separated by the Niagara escarpment which dies out east of Rochester.

Many of the detailed topographic features of the depression of central New York, such as drumlins, lakes, glens, old abandoned channels, etc., depend chiefly upon glaciation. The ice advanced southward as far as northern Pennsylvania, overriding and somewhat modifying the divide between the Ontario and the Susquehanna-Allegheny drainage.

Not everywhere in southern New York and northern Pennsylvania were moraines accumulated at the margin of the ice. In few localities does one find moraines developed on the uplands between valleys, though they are well developed in the valleys themselves.[1] As shown in Fig. 284 the valley heads draining north everywhere have well-defined morainic accumulations which were formed in front of local ice tongues that extended along each valley after the margin of the great continental ice sheet had been melted from the inter-valley tracts. In addition there are three major features, more or less directly due to glaciation, whose character and mode of formation deserve attention. These are (1) the Finger Lakes, (2) the glacial-marginal drainage channels, and (3) the drumlins of the central lowland.

FINGER LAKES

The basins of the Finger Lakes were first generally explained as the result of ice erosion, a kind of glacial overdeepening, but with the action confined to a single glacial period. Later and more detailed studies have shown that the forces which produced these basins are complex in their nature and relations. Glacial erosion is indeed the most important element of the explanation, as the steepened sides of the basins, the hanging tributary valleys, and the presence of lateral and terminal moraines at the valley heads testify. The last-named feature is clearly shown in Fig. 284, and proves that the valleys were highways of more active glacial motion.

The most important facts pointing to complexity of origin relate to the hanging valleys tributary to the Finger Lakes. On the steepened slopes of the main valley sides a series of buried gorges has been found.

---

[1] Williams, Tarr, and Kindle, Watkins Glen-Catatonk Folio U. S. Geol. Surv. No. 169, field ed., 1909, p. 124.

The gorges are occupied by drift deposits of the last (Wisconsin) ice invasion, which indicates that they were formed before that invasion.[1]

From the known facts it is concluded that before the glacial period there was a system of mature drainage with main valleys along the axes of Cayuga and Seneca lakes and with tributaries entering them at grade. With the overspreading of the region by glacial ice there was begun a process of exceptional deepening in the main valleys because they served as lines of most rapid glacial flow. At the end of the first ice

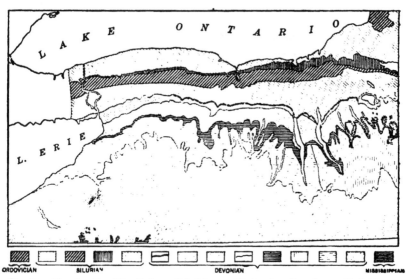

Fig. 285. — Map of portion of New York. 1, Ordovician; 2–5, Silurian; 6–13, Devonian; 14, Mississipian.

invasion the valleys had been broadened and deepened, the amount of the deepening being about 500 feet. Lakes may have formed in these overdeepened valleys after the ice had been melted away. At any rate the discordance of level between tributary and master valleys was pronounced and the tributaries began to cut down their valleys to the level of the main valleys, making gorges of notable breadth and depth before the second glacial invasion filled them with drift, reëxcavated and deepened the main valleys, increased the discordance between tributaries and master streams, and so altered the topography in detail that the postglacial streams do not flow everywhere in the interglacial courses. Wherever a postglacial stream enters one of these buried gorges with its easily eroded drift in contrast to the hard rocks of the rest of the valley

[1] R. S. Tarr, Watkins Glen and Other Gorges of the Finger Lake Region of Central New York, Pop. Sci. Mo., vol. 48, 1906, pp. 387–397.

there the valley broadens suddenly; where the stream leaves the buried valley the valley contracts. It is these features, together with the irregular as well as regular jointing of the bedded rock — shales and sandstones — that give to the gorges so much of their surprising variation from place to place and to the beautiful waterfalls and cascades of the many glens their wild, picturesque quality.

The lower portions of the valleys of the tributaries to each of the Finger Lake basins are for the most part more or less completely drift-filled gorges, so that the rock floor is in many cases far below the drift floor of the valley, a condition brought about not by overdeepening of the main trough but by the clogging of the tributary valley with drift. There are, however, perfect examples of hanging valleys with rock floors in the Cayuga and Seneca troughs and in the Tioughnioga and Cayuta valleys.[1]

The abrupt descents of the valley slopes in the Finger Lake region is illustrated by the change at the 900-foot contour just west of Watkins in the Cayuga trough. The slope from the 900-foot contour to the valley bottom, which has an elevation of 477 feet, takes place in a distance of a little over a half mile, while west of the 900-foot contour the valley side rises only 700 feet in a distance of 5 miles.[2]

ABANDONED CHANNELS

Among forms indirectly due to glacial action an important part is taken by the features relating to marginal drainage of the ice, such as have been described for the Great Lake region and will now be noted for the central New York region. In central New York the northward slope of the margin of the Appalachian Plateaus furnished unusual conditions for the impounding of glacial waters in front of the retreating ice cap, when that front had receded northward so as to lie beyond the divide between the Lake Ontario and the Susquehanna drainage.

The earliest accumulations of ice dammed the waters at the heads of the great valleys on the south. Upon the gradual retreat of the ice lower outlets were offered past the ice margin and across the ridges or plateau spurs between the lakes, so that high valley lakes drained into lower valley lakes; as the ice front receded still farther the lakes were successively lowered and shifted northward and rivers formed in the inter-valley or spur tracts. In central New York all the glacial waters escaped westward to the glacial lake in the Erie basin and ultimately to the Mississippi River, from a point west of Batavia; and the same is

[1] R. S. Tarr, Watkins Glen and Other Gorges of the Finger Lake Region of Central New York, Pop. Sci. Mo., vol. 48, 1906, p. 233.

[2] R. S. Tarr, Drainage Features of Central New York, Bull. Geol. Soc. Am., vol. 16, 1905, pp. 220-242.

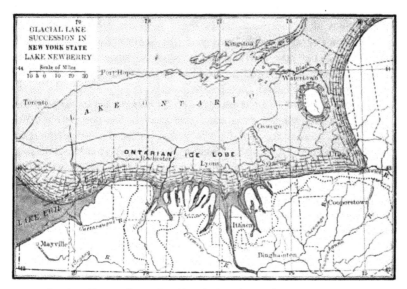

Fig. 286. — Note overflow southward into the Susquehanna. Elevation over 1000 feet.

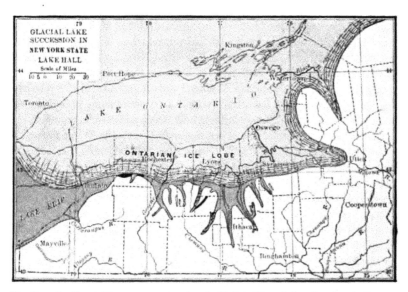

Fig. 287. — Overflow at this stage was westward to the Mississippi instead of southward to the Susquehanna. Elevations from 1000 feet on east to 900 feet on west.

true of all the waters held in the Genesee region under about 1200 feet, as well as in several other lake regions. But all the drainage under 900 feet at a later period was eastward past Syracuse to the Mohawk Valley. In the general region of the Finger Lakes between Batavia and Syracuse at least 13 separate valleys sloped to the north and these lie from Oatka Valley on the west to the Onondaga at the east, and include such important valleys as the Genesee, Canandaigua, and Cayuga. The higher and more local glacial waters in these valleys

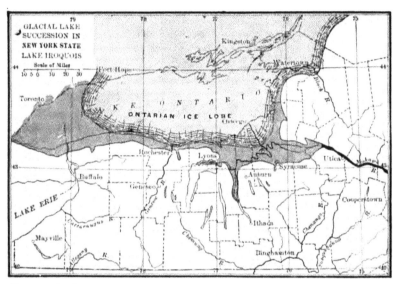

Fig. 288. — Overflow eastward to the Mohawk. Elevation under 360 feet. A number of intermediate stages are omitted from this series. The three stages shown here are of chief importance. (After Fairchild, Bull. N. Y. State Mus.)

escaped southward, but at later stages the water of the broad area collected here mainly into two large lakes, one of which occupied the large low central valleys of Seneca, Cayuga, and Keuka, with an outlet to the Susquehanna, and another in the Genesee Valley region which escaped by a different route to the Susquehanna, Allegheny, and Mississippi. In a later stage of development and before the beginning of the extensive spur channeling so prominent in central New York, these two bodies of water were united into one extensive lake overflowing westward into the Mississippi drainage above 900 feet and eastward to the Hudson below that level.

A study of the channels which drained the valleys of the separate lake waters through a single outlet shows that they lie in series falling

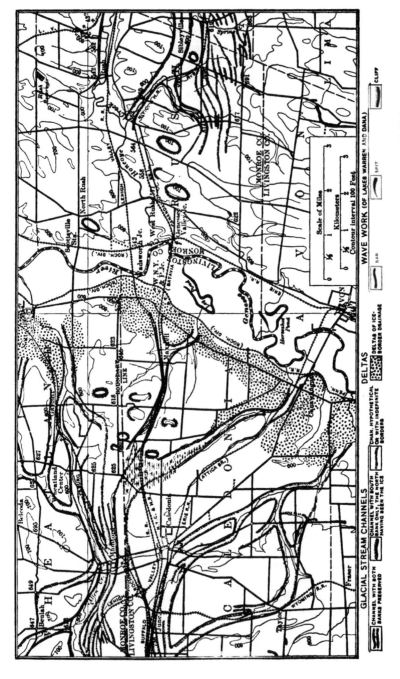

Fig. 289. — Channels and deltas of a part of the ice-border drainage between Leroy and Fishers, New York. (After Fairchild, N. Y. State Mus.)

northward on the theory of a receding barrier, though not as controlled by a steady continuous single recession of the ice front but by an oscillation of the ice front and a certain amount of seesawing between Batavia and Syracuse.[1]

Typical features of the channel series may be seen at Spread Rock and Jamesville and on the meridians of Mumford and Rush, etc. They were carved directly in front of the ice, and in such a position that the

Fig. 290. — Gulf channel, looking southeast (downstream) near mouth of channel. Four miles north of Skaneateles, New York. The depth of the gorge is 100 to 150 feet, the width from an eighth to a quarter of a mile. The walls are of shale. The gorge ends in a huge fan delta. (Gilbert.)

streams that occupied them must in many cases have laved the ice front, in which case only the southern banks are now in existence, since the northern banks were formed by the glacier ice. In some instances, as in the case of the Fairport-Lyons channel, the channel that was initiated on the ice front remained effective long after the ice had retreated from the region.

The most compact and remarkable set of cross-ridge channels is north of the parallel of Jamesville; the lowest is the finest glacial lake outlet

[1] H. L. Fairchild, Glacial Waters in Central New York, Bull. New York State Mus. No. 127, pp. 7-10.

channel in the state. It is 2½ miles long, 800 to 1000 feet wide at the bottom, and 125 to 155 feet deep in rock. The channel sides are composed of nearly vertical limestone. The highest channel of the Jamesville group is associated with a cataract, a semicircular amphitheater about 100 feet in diameter, with steep limestone walls 160 feet high; Jamesville Lake, 60 feet deep and 400 to 500 feet across, occupies the plunge basin at the foot of the cataract. Below the cataract is a gorge cut in limestone, and above it the limestone is worn and terraced in a manner characteristic of rapids.

NIAGARA FALLS

Niagara Falls came into existence during the retreatal stages of the continental ice cap of the Wisconsin epoch. For a time the glacial marginal waters were confluent over both the Erie and Ontario basins and extended northward as far as the ice barrier, but with the lowering of the water level the escarpment of Lockport limestone gradually emerged. This escarpment held up the Erie waters to its upper level, while the level of the Ontario water was controlled by the relations of the ice and the topography at the northern escarpment in lower country. Thus the waters of Lake Erie came to cascade over the cliff of Niagara limestone and drop into the water of the Ontario basin. As the Ontario waters receded the river cascaded over the scarp and formed a gorge which by upstream retreat had gradually been brought into its present position and character. The first spilling of the Erie waters over the escarpment took place at at least two points of overflow, one at Lockport and one at Lewiston; but by the more rapid development of the Lewiston channel the Lockport channel was abandoned.[1]

DRUMLIN TYPES AND BELTS

Among topographic forms due to glaciation drumlins are scarcely less important in areal extent and importance to soils than terminal moraines. They deserve a word of detailed description not only for this reason but also because of the remarkable development of drumlins in certain areas in New York, Wisconsin, etc., where their slopes constitute the most important elements of the relief. That the forms of drumlins are of glacial origin is evident from the location of drumlins only within glaciated areas; that their material is of glacial origin is shown also by their composition, which is for the most part compact till or ground moraine, and their position in the zone in which the transporting power of the ice was incompetent to carry along all the material

[1] H. L. Fairchild, Glacial Waters in Central New York, Bull. New York State Mus. No. 127, p. 30.

within it. In form they vary from mounds to long slender ridges, their most general form being a smooth oval; in size they vary from massive conspicuous hills from 100 to 200 feet high to indefinite swells of drift surface. A less common type of drumlin than the one described above has been known as the drumloid or, as has been lately proposed, roc drumlin.[1] The term is employed to designate ice-made rock masses whose form is so closely allied to that of the drumlin as to deserve correlation with it.

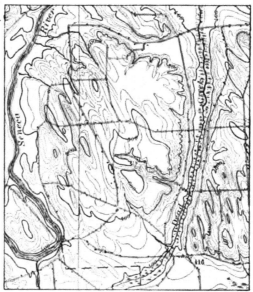

Fig. 291. — Roc drumlins or drumloids. The drumlin-shaped forms in the upper part of map are developed on shale (Salina). (Baldwinsville quadrangle, U. S. Geol. Surv.)

They are distinct from the more common type of drumlin in that they are due to erosion, while the ordinary drumlin is a product of upbuilding and of shaping. Roc drumlins occur in much smaller numbers than ordinary drumlins, but are sometimes found in the same general field, as in central New York and northern Wisconsin and Michigan. The last-named occurrence has been described by Russell.[2]

Three general regions of great drumlin development have been identified in the United States: (a) the New England area, including southern New Hampshire, where about 700 drumlins have been mapped, Massachusetts with about 1800, and Connecticut with an unnumbered amount; (b) the Michigan area, which includes eastern Wisconsin and adjacent

[1] H. L. Fairchild, Drumlins of Central-Western New York, Bull. New York State Mus. No. 111, 1907, p. 393.
[2] I. C. Russell, Rept. Mich. State Geologist, 1906.

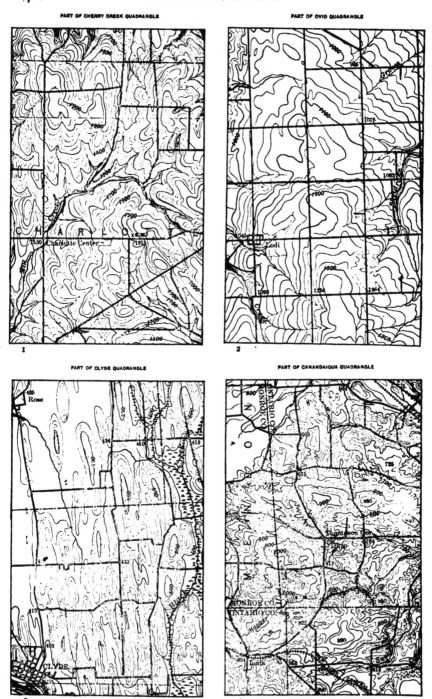

PART OF CHERRY CREEK QUADRANGLE

PART OF OVID QUADRANGLE

PART OF CLYDE QUADRANGLE

PART OF CANANDAIGUA QUADRANGLE

1

2

3

4

portions of Michigan, where the estimated number is 5000; and (c) the drumlin area of central New York. The last-named is a belt about 35 miles wide bordering the southern side of Lake Ontario and about 140 miles long, from the Niagara River to Syracuse; this area probably includes not less than 10,000 drumlin crests; 15 drumlins to the square mile is common, though the average is about 3 to the square mile.

On the south the drumlin area of central New York reaches up the north-facing slope of the Allegheny plateau, where it fades off into smooth drift or is lost in the bolder relief of the rock hills. The most abrupt ending of the drumlin topography is along the crests of ancient drainage levels, as between Victory and Geneva, New York. The most massive development is on the low ground north of Finger Lakes and south of Lake Ontario, chiefly under 500 feet altitude. The greatest development lies over the greatest thickness of the soft Salina shales, where the drift is most clayey and adhesive. Those on the southern border of the drumlin belt are attenuated, while those toward the north and under the deeper ice are broad. The greatest steepness and greatest regularity of form seem to occur in the middle of the drumlin belt.

The drumlins have been but little modified in postglacial time by ordinary stream erosion, but in the zone of wave erosion on the southern shore of Lake Ontario, as at Sodus Bay, the drumlins have been extensively cut or entirely removed, and similar shore cutting was accomplished, though on a smaller scale, during the existence of temporary glacial-marginal lakes that occurred at a higher level than Lake Ontario during the later stages of the glacial period. The general relations of the drumlin area of central New York to the surrounding topographic features are shown in Fig. 292. The various categories of form are indicated in the following table:

" 1. Domes or mammillary hills and low broad mounds.

2. Broad oval drumlins.

3. Oval drumlins of high relief.

4. Long oval drumlins, commonly bolder on the north or struck end; the dolphinback or whaleback hills.

. Short ridge drumlins.

6. Long ridge drumlins. This includes two extreme varieties of form: (a) the long broad ridges or rolls or gentle swells which are not generally recognized as belonging in the drumlin class, and commonly fail of representation on the contoured maps; (b) the small, close-set, parallel ridges which lie as minor moldings between the larger and conspicuous ridge drumlins, or those which form the attenuated edge of a drumlin belt.

7. Abrupt struck slopes.

8. Low or gentle struck slopes.

9. Sharp-crested hills with steep, or even concave, side slopes. Many occasional or peculiar forms and characters might be noted, but they are not regarded as genetically important."[1]

[1] H. L. Fairchild, Drumlins of Central-Western New York, Bull. New York State Mus. No. 111, 1907, pp. 422-423.

In regard to the relation of drumlins to moraines it may be said that
the moraines are weak where the drift was left in drumlin form and
strong where the drift was not drumlinized.[1] The drumlins are believed to be shaped by the sliding movement of the lowest ice, a movement that was produced by thrust on the marginal ice, caused by the
pressure rearward applied in such a manner that the margin of the ice
was pushed bodily forward. It is believed that this condition of ice movement is fundamental to drumlin formation.[2] In this manner the drumlins were constructed by a plastering-on process on the obstructions, a
process favored by the plastic and adhesive drift. As masses of drift
the drumlins were produced by the accretion of drift, but their peculiar
form is due to erosion. The whole process has been aptly compared
to clay modeling, a process of plastering on and rubbing away.[3]

[1] H. L. Fairchild, Drumlins of Central-Western New York, Bull. New York State Mus.
No. 111, 1907, p. 425 et al.

[2] Idem, p. 430.

[3] Idem, p. 432.

APPENDIX A

Soil Class; Soil Type; Soil Series [1]

SOIL CLASS

Because soils are made up of particles of different sizes, they may be grouped according to texture, that is, according to the relative proportions of the particles of different sizes which they contain. This grouping is known as the "soil class." By means of mechanical analyses the particles less than 2 millimeters in diameter are separated into 7 grades and the various percentage relationships of the different grades determine the class of soil — sand, sandy loam, loam, clay, etc., as in the table below. In addition to the fine earth, of which a mechanical analysis is made, many soils contain larger particles, which if of small size are called "gravel," and if of larger size are called "stones," so that in the soil classification it is possible to have a gravelly sand, loam, or clay, and likewise stony members of the various classes.

In outlining soil boundaries, it is necessary to (1) make preliminary borings in sufficient number to show the location of a considerable body of soil material of uniform character, (2) record the general description of one or more borings, (3) select a color to represent this description and color in so much of the map as undoubtedly corresponds with the description, (4) work away from this identified area until soil materials are found which manifestly do not fit the former description, (5) select a second color for this new set of soil characteristics and color in on the map only where the new material undoubtedly occurs, (6) work in between the areas of the two classes thus established until a zone or line is found where all material on one side becomes increasingly characteristic of the one class and on the other side of the other class, (7) draw a line on the map to represent this line or to represent the center of the zone of gradation of soil characteristics. This line will constitute a soil boundary. Usually the distinctions between adjacent soil classes are sharp and clear in a moderately broken country and grade into each other in a flat country, although there are many exceptions to this rule. In a glaciated region the mixing action of the ice may thoroughly confuse the rock waste, so that the most minute examination is required for the separation of many soil classes distributed almost at haphazard. Sometimes this is true to such an extent that a new designation is required, such as the geologic term "till," although this should be avoided if possible. Drainage differences and attendant differences in the content of organic matter, depth of soil, etc., may constitute a basis of distinction. Postglacial wash is sometimes responsible for considerable differentiation of material, which may be made the basis of distinctions between soil classes. When minor differences of texture, structure, organic-matter content, or succession of materials occur in the soil sections representing single areas of 10 acres or more, such variations may be described in the report as phases.

The soils of different classes grade into each other and therefore the line of separation between the different classes is necessarily an arbitrary one. The particles also may be very irregularly distributed between the different grades, so that it is not possible to make a rigid classification according to the mechanical analyses. The following table is the result of an examination of thousands of soil samples from all over the United States by the U. S. Bureau of Soils and may be accepted as a standard for the classification and description of the soils of a given area, large or small. Uniformity and close adherence to the standard are the chief considerations which it is desired to secure. The table therefore constitutes a codification and

[1] Based on the Soil Survey Field Book, 1906.

arrangement of facts reported up to this time to the Soil Survey. It has been found convenient to number the different grades into which the soil is separated by mechanical analysis. The name of the grade to which these numbers refer is given in the table. Care should be taken in interpreting the table not to confuse these grades for either the soil class or the soil type.

SCHEME OF SOIL CLASSIFICATION BASED UPON THE MECHANICAL COMPOSITION OF SOILS.

Class.	1. Fine gravel. 2–1 mm.	2. Coarse sand. 1–.5 mm.	3. Medium sand. .5–.25 mm.	4. Fine sand. .25–.1 mm.	5. Very fine sand. .1–.05 mm.	6. Silt. .05–.005 mm.	7. Clay. .005–0 mm.
Coarse sand.	More than 25 per cent of 1+2.					0–15	0–10
	More than 50 per cent of 1+2+3.					Less than 20 per cent of 6+7.	
Medium sand.	Less than 25 per cent of 1+2.					0–15	0–10
	More than 20 per cent of 1+2+3.					Less than 20 per cent of 6+7.	
Fine sand.	Less than 20 per cent of 1+2+3.					0–15	0–10
						Less than 20 per cent of 6+7.	
Sandy loam.	More than 20 per cent of 1+2+3.					10–35	5–15
						More than 20 per cent and less than 50 per cent of 6+7.	
Fine sandy loam.	Less than 20 per cent of 1+2+3.					10–35	5–15
						More than 20 per cent and less than 50 per cent of 6+7.	
Loam.							15–25
						Less than 55 per cent of 6.	
						More than 50 per cent of 6+7.	
Silt loam.						More than 55 per cent of 6.	Less than 25 per cent of 7.
Clay loam.						25–55	25–35
						More than 60 per cent of 6+7.	
Sandy clay.						Less than 25 per cent of 6.	More than 20 per cent of 7.
						Less than 60 per cent of 6+7.	
Silt clay.						More than 55 per cent of 6.	25–35 per cent of 7.
Clay.							More than 35 per cent of 7.
						More than 60 per cent of 6+7.	

SOIL TYPE

In mapping a small area the soil class is the matter of chief and possibly even of ultimate interest. The determination of the class tells about all the facts of texture that a forester requires. When, on the other hand, large areas are under consideration it may be necessary to distinguish soil types. A soil type is conceived to embrace all soil material in any region which is marked to corresponding depths by identity or close similarity in texture, structure, organic-matter content, and color, and by similarity of origin and of topography. A type comprises all soil material which may properly be included in one general description covering these points. In the humid regions description covers the material to an average depth of 3 feet; in the arid regions to a depth of 6 feet. In the determination of a type of soil there are many factors to be considered in addition to texture, such as the structure, which deals with the arrangement of the particles and their chemical composition, the organic-matter content, origin, color, depth, drainage, topography, native vegetation, and natural productiveness. The value of the type idea lies in the possibility of distinguishing between two soils of, let us say, practically identical texture, whose chemical composition, depth, humus content and drainage conditions, etc., are markedly unlike. Both may be, for example, sandy loams, but the one may possess other characteristics than textural which make it infertile, while the other may possess a high degree of fertility.

SOIL SERIES

In working in a very large territory such as the whole United States it has even been necessary to recognize still larger groups or soil series. It has been found that in many regions the members of a given set of soil types are so evidently related through source of material, method of formation, topographic position, and coloration, that the different types constitute merely a gradation in the texture of an otherwise uniform material. Different types that are thus related constitute a series. A complete soil series consists of material similar in many other characteristics but grading in texture from stones and gravel on the one hand through sands and loams to a heavy clay on the other.

In arranging the soils in series the same factors should be considered that are used in separating soils of the same class into different types. For example, the Marshall silt loam and the Miami silt loam have been separated because of the difference in the amount and condition of the organic matter in the surface soil and the essential differences in coloration. The former is dark brown to black, while the latter is light brown to almost white. This relation has been found to exist between soils of other classes in the glacial regions, and has been used as a basis for separating the glacial soils into the Marshall and Miami series.

On account of the very different processes of their formation, residual and recent alluvial soils should not be included in the same series. Soils may, however, be very similar in origin and texture but occupy such entirely different topographic positions that their relation to crops is entirely changed, and this fact should be recognized by the use of another serial name. An example of this is found in the separation of the soils of the Piedmont Plateau and the Appalachian Mountains into the Cecil and Porters series.

The color of the soil is one of its most noticeable physical features, and is often a factor in separating the soils into different series. The soils of the Orangeburg series, for example, have been formed in a manner very similar to the Norfolk series, but are distinguished from the latter by the red color of the subsoil and by associated differences in agricultural value. Soil series may grade into each other in a manner similar to the intergradation of the types within the series. Thus the Marshall series may grade into the Miami series and the Norfolk series into the Orangeburg or Portsmouth series.

UNCLASSIFIED MATERIALS; SPECIAL DESIGNATIONS

There are certain conditions of soil, or in many areas even local absences of true soil, which do not readily fall into any general classification. They may be due to excessive erosion, to overflow, to insufficient drainage, or to wind action, or the soils may be infertile on account of their texture or their present topographic position. Areas of this kind are as follows:

ROCK OUTCROP

Areas consisting of exposed rock or fresh accumulations of stone, entirely unfit for cultivation. The most extensive areas of this type of surface are found in the strongly glaciated portions of northeastern Canada. The type is also common in mountain regions undergoing vigorous dissection and in wind-swept arid regions, etc.

ROUGH STONY LAND

Areas so stony and broken as to be nonarable, although permitting timber growth and pasturage. They frequently consist of steep mountain ridges, bluffs, or narrow strips extendng through definite soil types. These areas differ from rock outcrop by supporting vegetation of economic value, and from the stony loams in being nonarable.

GYPSUM

The surface consists of a light-brown or reddish-brown sandy loam or loam underlain by soft saccharoidal gypsum at a depth of from a few inches to 6 feet. Gypsum is often present at the surface. The type occupies level bench land. It is derived from disintegration of gypsum deposits and possesses remarkable power of transmitting seepage waters by capillarity and gravitational flow. Where the irrigation water possesses a high salt content this is not a desirable land for agricultural purposes. It often contains large quantities of alkali.

PEAT

Vegetable matter consisting of roots and fibers, moss, etc., in various stages of decomposition, occurring as turf or bog, usually in low situations, always more or less saturated with water, and representing an advanced stage of swamp with drainage partially established.

MUCK

This type consists of black, more or less thoroughly decomposed, vegetable mold from 1 to 3 feet or more in depth and occupying low, damp situations, with little or no natural drainage. Muck may be considered an advanced stage of peat brought about by the more complete decomposition of the vegetable fiber and the addition of mineral matter through deposition from water or from æolian sources, resulting in a finer texture and a closer structure. When drained, muck is very productive and is adapted to a large variety of agricultural crops. Extensive areas of it occur in the glaciated Middle West, where it grades into peat or swamp on the one hand and into conventional soil classes on the other.

MADELAND

Areas are occasionally encountered where filling has taken place over considerable tracts. The arrangement of the materials and even the materials themselves may be artificial and not in harmony with any soil classification. In many instances such areas are extensive and should be represented by a color on the map.

DUNESAND

Dunesand consists of loose, incoherent sand forming hillocks, rounded hills, or ridges of various heights. Dunes are found along the shores of lakes, rivers, or oceans, and in desert areas. They are usually of little value in their natural condition on account of their irregular surface, the loose, open nature of the material, and its low water-holding capacity. Dunes are frequently unstable and drift from place to place. The control of these sands by the use of windbreaks and binding grasses is frequently necessary, as at Cape Cod and on the coast of California, for the protection of adjoining agricultural lands. In certain regions they have been improved for agricultural purposes or employed as catchment areas in city water supplies or planted to pine forest for the protection of agricultural land and for revenue.

SANDHILL

This term is used to describe ridges and uneven areas of sand not in motion, either on account of partial consolidation or because the sands are fixed by a natural growth of grasses such as the Sand Hills tract of western Nebraska. Such areas sometimes occur in the vicinity of old shore lines of lakes and seas; again they may be related to river action, as where flood plains are alternately flooded and drained and river sands exposed to the winds.

RIVERWASH

Sand, gravel, and bowlders, generally in long narrow bodies, but occasionally spread out in fan-shaped areas. These areas occupy river bottoms or flood channels, and occur where the streams are intermittent or liable to torrential overflow. They are of such recent origin as not to be covered by vegetation and are subject to modification during the next season of high water. The flood plains of the Missouri and the Platte supply examples.

MEADOW

Low-lying, flat, usually poorly drained land, such as may occur in any soil type. Frequently used for grass, pasturage, or forestry, and can be changed to arable land if cleared and drained. The present character of meadowland is due to lack of drainage, and the term represents a condition rather than a classification according to texture. The soils vary frequently in texture, even within small areas, and on account of occasional overflow the character of the soil at any one point is subject to change. Wherever it is possible to separate such areas into distinct soil types the term *meadow* should not be used.

MARSH

This term is used to designate low, wet, treeless areas, usually covered by standing water and supporting a growth of coarse grasses and rushes. Marsh areas occur around the borders of fresh-water lakes and the lower courses of streams. They can seldom be drained without diking and pumping. When this is done the soil is usually productive.

SWAMP

Areas too wet for any crop and covered with standing water for much or all of the time. Variations in texture and in organic-matter content may occur. Swamp frequently occupies areas which are inaccessible, so that detailed mapping is impossible. The native vegetal growth consists of water-loving grasses, shrubs, and trees. Many areas of swamp are capable of drainage, and when this is properly accomplished they not infrequently constitute lands of high agricultural value. Drainage may be employed also to improve the soil by aerating the organic matter and permitting humification, or it may make possible the introduction of new forest types. The salt-water swamps of the Coastal Plain, the swamps of the glaciated region of northern United States and Canada, and the river swamps are the three chief occurrences. Wherever small areas of swamp occur within a definite soil type and the texture of the soil is known to be the same as that of the surrounding type, they should be mapped with the type and the swampy condition shown by symbol.

APPENDIX B

The various factors of importance in a study of soils are summarized in the accompanying outline. It is intended to present only a suggestive outline. A categorical adherence to this outline is very undesirable, since a condition or a brief list of conditions may be of such predominating influence in determining the value of a soil as quite to obscure other conditions. Not all the suggestions are applicable to a given small area; as many as possible should be identified and others not in the table should be found. The relative value of the soil qualities should be strongly brought out in every soil survey and greatest attention given to those of most prominence. Under special conditions this principle may properly be carried so far as to ignore even the standard mechanical classification of soils.

OUTLINE FOR A SOIL SURVEY IN FOREST PHYSIOGRAPHY

Location and Boundaries of Area —
> Present condition as to settlement.
> Chief towns.
> Transportation facilities.
> Markets, etc.

Topography and Drainage —
> Geologic structure and rock types.
> Topographic forms and stage of physiographic development.
> Brief description of the regional surface drainage in relation to form and structure; stream gradients and sizes, characteristics of valley sides and floors in relation to run-off, seepage, floods, etc.

Climate —
> Direction and strength of winds.
> Relation to plant distribution.
> Rainfall: amount, frequency, and distribution in the year.
> Disposal as controlled by geologic and physiographic conditions; the regional run-off, absorption, evaporation.
> Daily and yearly temperature variations, extremes and means; relief controls of temperature and rainfall; proximity to the sea, latitude, etc. The growing season, temperatures necessary for germination and growth, degree to which climatic conditions meet the needs of forest types.

Soils —
> (1) A general study of the soils of the area, showing their broad relation to the geologic formations and to each other, to drainage, erosion, and other agencies, their classification and distribution. (2) A detailed and full description of the classes of soil and subsoil, noting texture, structure, color, depth, and ease of cultivation; follow this with a statement as to the location of various soils in the area, topographic and drainage features, origin and process of formation, peculiar mineral or chemical features — as alkali, its chemical composition and vertical distribution, and approximate area; native vegetation, its value in determining soil classes and its responses to them, etc. Unclassified materials that require special designation: rock outcrop, talus, marsh, dunesand, etc.

Underground water conditions.

    Depth of water table.

    Fluctuations of level.

    Occurrence of springs and seepage lines.

    Forms and effects of subirrigation.

    Character of ground water and reclamation possibilities.

Soil temperature as related to:

    Air temperature.

    Slope exposure.

    Underground water conditions.

Soil fertility; cultural methods and conditions that have effected or will effect improvements.

The soil humus, forest litter, drainage effects upon, maintenance by shading, by protection from the wind.

APPENDIX C

(From Merrill's *Rocks, Rock-weathering, and Soils.*)

Constituents	I [1]	II	III
Ignition	1.22%	3.27%	4.70%
Silica (SiO₂)	69.33	66.82	65.69
Titanium (TiO₂)	not det.	not det.	0.31
Alumina (Al₂O₃)	14.33	15.62	15.23
Iron protoxide (FeO)	3.60	1.69	......
Iron sesquioxide (Fe₂O₃)	......	1.88	4.39
Lime (CaO)	3.21	3.13	2.63
Magnesia (MgO)	2.44	2.76	2.64
Soda (Na₂O)	2.70	2.58	2.12
Potash (K₂O)	2.67	2.04	2.00
Phosphoric acid (P₂O₅)	0.10	not det.	0.06
	99.60%	99.79%	99.77%

[1] (I) fresh gray granite, (II) brown but still moderately firm and intact rock, and (III) the residual sand.

ANALYSES OF FRESH AND OF DECOMPOSED GNEISS, ALBEMARLE COUNTY, VIRGINIA

Constituents	Fresh Gneiss	Decomposed Gneiss	Calculated Amounts Saved and Lost		
	I	II	III	IV	V
	Bulk Analysis	Bulk Analysis	Loss	Percentage of Each Constituent Saved	Percentage of Each Constituent Lost
Silica (SiO₂) { in HCl } { in Na₂CO₃ }	60.69%	45.31%	31.90%	47.55%	52.45%
Alumina (Al₂O₃)	16.89	26.55	0.00	100.00	0.00
Iron sesquioxide (Fe₂O₃)	9.06	12.18	1.30	85.65	14.35
Lime (CaO)	4.44	Trace	4.44	0.00	100.00
Magnesia (MgO)	1.06	0.40	0.80	25.30	74.70
Potash (K₂O)	4.25	1.10	3.55	16.48	83.52
Soda (Na₂O)	2.82	0.22	2.68	4.97	95.03
Phosphoric acid (P₂O₅)	0.25	0.47	0.00 [1]	100.00	0.00 [1]
Ignition	0.62	13.75	0.00 [1]	100.00	0.00 [1]
	100.08%	99.98%	44.67%	......	......

[1] Gain.

. 728

ANALYSES OF FRESH AND OF DECOMPOSED DIORITE FROM ALBEMARLE COUNTY, VIRGINIA

Constituents	Fresh	Decomposed	Calculated Loss for Entire Rock	Percentage of Each Constituent Saved	Percentage of Each Constituent Lost
Silica (SiO₂)................	46.75%	42.44%	17.43% loss	62.69%	37.31%
Alumina (Al₂O₃)............	17.61	25.51	0.00 "	100.00	0.00
Iron sesquioxide (Fe₂O₃) ¹....	16.79	19.20	3.53 "	78.97	21.03
Lime (CaO)................	9.46	0.37	9.20 "	2.70	97.30
Magnesia (MgO)............	5.12	0.21	4.97 "	2.83	97.17
Potash (K₂O)..............	0.55	0.49	0.21 "	61.25	38.75
Soda (Na₂O)..............	2.56	0.56	2.17 "	15.13	84.87
Phosphoric acid (P₂O₅).......	0.25	0.29	0.00	80.11	19.87
Ignition...................	0.92	10.92	0.00	100.00	0.00
	100.01%	99.99%	37.51% loss	......	......

ANALYSES OF FRESH AND OF DECOMPOSED ARGILLITE, HARFORD COUNTY. MARYLAND

Constituents	Fresh Argillite	Residual Clay	Percentage of Loss for Entire Rock	Percentage of Each Constituent Saved	Percentage of Each Constituent Lost
Silica (SiO₂).................	44.15%	24.17%	25.34%	42.4	57.57%
Alumina (Al₂O₃).............	30.84	39.90	0.00	100.00	.00
Iron oxide (FeO and Fe₂O₃)......	14.87	17.61	1.23	91.22	..18
Lime (CaO).................	0.48	None	0.48	0.00	100.00
Magnesia (MgO).............	0.27	0.25	0.08	71.84	28.16
Potash (K₂O)...............	4.36	1.24	3.39	22.04	77.95
Soda (Na₂O)...............	0.51	0.23	0.33	0.36	99.64
Ignition (C and H₂O)..........	4.49	16.62	0.00	287.37	None
	99.97%	100.02%	40.83%	.....	......

ANALYSES OF FRESH LIMESTONE AND ITS RESIDUAL CLAY

Constituents	Fresh Limestone	Residual Clay	Percentage of Loss for Entire Rock	Percentage of Each Constituent Saved	Percentage of Each Constituent Lost
Silica (SiO₂).....................	4.13%	33.69%	0.00%	100.00%	0.00%
Alumina (Al₂O₃).................	4.19	30.30	0.35	88.65	11.35
Ferric iron (Fe₂O₃).............	2.35	1.99	2.13	10.44	89.56
Manganic oxide (MnO).........	4.33	14.98	2.49	42.41	57.59
Lime (CaO)...................	44.79	3.91	44.32	1.07	98.93
Magnesia (MgO)...............	0.30	0.26	6.25	10.62	89.38
Potash (K₂O).................	0.35	0.96	0.23	33.63	66.37
Soda (Na₂O).................	0.16	0.61	0.085	46.74	53.26
Water (H₂O)...................	2.26	10.76	0.95	58.37	41.63
Carbonic acid (CO₂)............	34.10	0.00	34.10	0.00	100.00
Phosphoric acid (P₂O₅).........	3.04	2.54	2.73	10.24	89.76
	100.00%	100.00%	97.635%	......	......

APPENDIX D

THE GEOLOGIC TIME TABLE[1]

Old classification	New classification	
Cambrian	Georgic / Acadic / Ozarkic or Cambric	} Paleozoic
Ordovician or Lower Silurian	Canadic / Ordovicic / Cincinnatic	
Siluric / Devonic	Siluric / Devonic	} Neopaleozoic
Mississippian or Sub-Carboniferous	Mississippic / Tennesseic	
Pennsylvanic- / Permic	Pennsylvanic- / Permic	
Triassic / Jurassic	Triassic-Jurassic	} Mesozoic
Cretaceous	Comanchic / Cretacic	
Eocene / Oligocene	Eogenic	} Neozoic
Miocene / Pliocene / Pleistocene	Neogenic	

Old classification groupings:
- **Paleozoic** — Cambrian; Ordovician or Lower Silurian; Siluric, Devonic; Mississippian or Sub-Carboniferous; Pennsylvanic-Permic
- **Mesozoic** — Triassic, Jurassic; Cretaceous
- **Tertiary or Cenozoic** — Eocene, Oligocene; Miocene, Pliocene, Pleistocene

New classification groupings:
- **Paleozoic** — Georgic, Acadic, Ozarkic or Cambric; Canadic, Ordovicic, Cincinnatic
- **Neopaleozoic** — Siluric, Devonic; Mississippic, Tennesseic; Pennsylvanic-Permic
- **Mesozoic** — Triassic-Jurassic; Comanchic, Cretacic
- **Neozoic** — Eogenic; Neogenic

[1] The new classification is suggested by Charles Schuchert, Paleogeography of North America, Bull. Geol. Soc. Am., vol. 20, 1910.

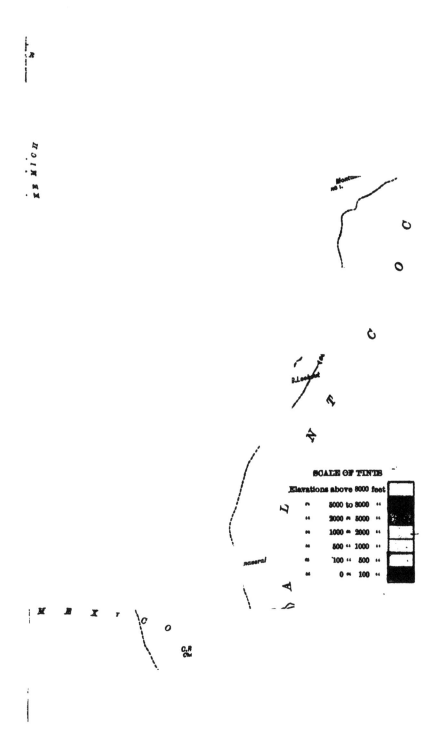

SCALE OF TINTS

Elevations above 8000 feet
"   5000 to 8000 "
"   3000 " 5000 "
"   1000 " 2000 "
"   500 " 1000   "
"   100 " 500   "
"   0   " 100   "

INDEX

BABOQUIVARI RANGE, 245.

BACTERIA, in soil formation, 17; action of in relation to nitrogen, 86; size of, 86; functions and value of, 86; fed upon by protozoa, 88; nitrogen changes in soil produced by, Fig. 9, p. 89; and direct fixation of nitrogen from soil air, 90; in symbiosis with algæ, 90; in symbiosis with legumes, 90; anaerobic or denitrifying, 90; in root nodules, 92.

BADLANDS OF THE BLACK HILLS REGION, factors controlling the development of, 415; scarcity of deep-rooted vegetation in, 415; details of, Fig. 149, p. 415; rainfall of, 416; topographic complexity of, 416; important topographic element of, 416; stream flow in, 416.

BALCONES FAULT ZONE, section of, Fig. 162, p. 435.

BALD EAGLE VALLEY, 677.

BALD MOUNTAIN, 321.

BALDWIN LAKE, 136.

BALLY CHOOP MOUNTAINS, 141.

BALSAM MOUNTAINS, spruce forests of, 614.

BARLOW, A. E., *Report on Geology and Natural Resources of Area between the Nipissing and Temiskaming*, 559.

BARRELL, J., *Geology of Marysville Mining District, Montana*, 319.

BARREN GROUNDS, relation to transcontinental spruce forest, 570.

BARREN LANDS, elevation of, 560.

BARROWS AND BOLSTER, *The Surface Water Supply of the U. S.*, 644.

BARTLETT, W. H., *Experiments on the Expansion and Contraction of Building Stones*, 15.

BASIN RANGES, longitudinal profiles of, Fig. 55, p. 219; structure of, 220; geologic history of, 221; fault-block mountains of, 221; variation in topographic development of, 222; of central portion, 222; evidences of fault-block origin of, 223; mountain border and internal structure of, 223; continuity of range crest of, 224; evidences of progressive and recent faulting, 225; broken waste slopes of, 225; stream profiles and recent faulting in, 226; terminal facets of mountain spurs of, 227; springs and fault lines of, 228; Death Valley region of, 228; and Arizona Highlands, 246.

BATEMAN, *Can. Geol. Surv.*, 567.

BATTENKILL VALLEY, 682.

BAY OF FUNDY, 626.

BEAR BUTTE, 445.

BEAR LAKE, and San Bernardino Range, 136; water of, 212.

BEAR PAW MOUNTAINS, 412.

BEAR RIVER MOUNTAINS, 198, 202.

BEAR RIVER, volume of, 217.

BEARTOOTH AND NEIGHBORING PLATEAUS, canyons of, 332; topography of East and West Boulder Plateau, Fig. 105, p. 333; glacial features of, 334.

BEAR VALLEY, and the adjacent country, Fig. 27, p. 137.

BEECH CREEK, 671.

BELKNAP RANGE, trend of, 645.

BELL, R. M., *Proofs of Rising of the Land around Hudson Bay*, 563; *Geographical Distribution of Forest Trees in Canada*, 571.

BELT MOUNTAINS, 302.

BERLIN VALLEY, 683.

BERKSHIRE HILLS, usage of the name, 681; fertility of, 683.

BERKSHIRE VALLEY, 683.

BERRY MOUNTAIN, 673.

BERTHELOT AND ANDRE, *Comptes Rendus Academie de Paris*, 88.

BETTS AND SMITH, *Utilization of California Eucalyptus*, 145.

BIGBUG MESA, and lava flows, 251.

BIGELOW, F. H., *Studies of Diurnal Periods in Lower Strata of Atmosphere*, 255.

BIGHORN, height of, 157.

BIGHORN MOUNTAINS, section across highest part of, Fig. 113, p. 349; structure and topography of, 349; east side of limestone front ridge of, Fig. 114, p. 350; subordinate topographic features of, 351; high mountains in, 351; glacial forms of, 351; wall at head of cirque, Fig. 115, p. 352; former glacier systems of, Fig. 116, p. 353; surviving glaciers of, 353; lakes of, 354; forests on, 354; cretaceous deposits on, 419.

BIG MOUNTAIN, 673.

BIRCH, red, temperature requirements of, 55; and aspen in relation to soil erosion, 78.

BISCAGNE BAY, 543.

BITTERROOT MOUNTAINS, and Snake River Valley, 202; and Blue Mountains, 207; in Northern Rockies, 298; profile across, Fig. 102, p. 321; map of part of, Fig. 103, p. 325; eastern border features of, 326; main divide of, 326; soils of, 327; forests of, 327.

BITTERROOT VALLEY, 321.

BLACK BUTTE, 445.

BLACK DOME MOUNTAIN, 691.

BLACK HILLS, ideal east-west section across, Fig. 164, p. 440; structure of, 441; western slope of, Fig. 165, p. 441; soils of, 442; forest of, 443; outlying domes of, 444.

Henry H. Howorth, H. Howorth Henry

History Of The Mongols

Part: III

Henry H. Howorth, H. Howorth Henry

History Of The Mongols
Part: III

ISBN/EAN: 9783348014151

Printed in Europe, USA, Canada, Australia, Japan

Cover: Foto ©ninafisch / pixelio.de

More available books at **www.hansebooks.com**

HISTORY *of the* MONGOLS

FROM THE 9th TO THE 19th CENTURY.

PART III.

THE MONGOLS OF PERSIA.

BY

HENRY H. HOWORTH, M.P.

CORR. MEM. ROY. ACAD. LISBON, M.R.A.S., F.S.A.
AUTHOR OF 'CHINGHIZ KHAN AND HIS ANCESTORS,' 'THE MAMMOTH
AND THE FLOOD,' ETC. ETC.

LONDON:

LONGMANS, GREEN, AND CO.

AND NEW YORK: 15 EAST 16th STREET.

1888.

MAJOR-GENERAL SIR A. CUNNINGHAM, K.C.I.E., C.S.I,

AND

M. CHARLES SCHEFER,

MEMBER OF THE INSTITUTE OF FRANCE, ETC., ETC

I·T is very grateful to me to be able to associate this volume of a long and laborious work with the names of two friends whom I hold in high esteem, and whose reputation is world-wide—Sir Alexandei Cunningham, who for forty yeais has studied so well, and so much to oui profit, the archæology and history of India, and who is not less known for his urbanity and high character than for his deep and scientific knowledge of the East ; and M. Charles Schefer, the *doyen* of French Orientalists, whose encyclopædic knowledge of the history, literature, and art of the Muhammedan world are unrivalled, and who has always been ready and willing to put his knowledge and his skill at the service of others.

My two friends will, I know, somewhat qualify their fastidious standards in judging of a work so multiform and so full of perpetual difficulty as the present, and they will not be displeased that a student who has learnt so much from their labours should wish to associate their names with his own.

May the golden autumn of their lives be prolonged, and may we continue for many years to profit by their ripe and matured judgment, and by the harvest which they have sown and reaped.

PREFACE.

FIVE years have elapsed since the appearance of the previous volume of this work, a longer period than I had intended, and justified only on the ground of my indifferent health and very engrossing occupations in other fields. Those who have been over the ground, or some part of it, will perhaps admit that this new instalment contains a good deal of honest work. Its author knows too well that it is full of imperfections. His original purpose was not to produce a final work; finality is not the fate of any human undertaking. We cannot control the languid attention, the dull eye, and the hasty thought. They are the continual companions of the best of us at times, and more especially do they stand by us when our head is aching, and our health is fragile. To travel away from the beaten track of scholars into the great tangled jungle of Eastern history is in itself a toilsome task. To bring together and to reconcile the various versions of the same tale, told by imaginative Eastern story-tellers, is too often a despairing duty. Having done so, to make a chronicle of the whole so that it can be understood even, is by no means the easiest of labours. Beyond this it is nearly hopeless to venture. When the matter has to be packed so closely that there is no room of any kind for ornament or rhetoric, and every sentence is the statement of a new fact, it is impossible to indulge in the luxury of style. All this binds down the weary author, and robs his work of its philosophy. Nay, more, it not only clips his wings and makes him cling to the humble ground, but it too often leads to obscurity, and nearly always to dulness. All this I know and feel better than my critics can. Beyond the ordinary blemishes which disfigure all histories when carefully tested, this one has the further weakness which it shares with much Eastern story, in that the spelling of the proper names is uncertain, and in many cases irregular and inconsistent. When myriads of names, recorded by various writers with various modes of transcription, have to be reduced to a common orthography, it is almost impossible to avoid many slips; and unfortunately, when the reader looks over the pages he cannot realise or guess the tremendous labour involved in the compilation of a few phrases, even where the matter has had to be sifted from the reading of many books, nor can he recognise the very large number of cases where a mistake has been corrected and a blot removed. His careful eye naturally sees only "the flies in the ointment," and these motes in his brother's eye are too often made the excuse for ignoring the beam in his own. I have no fear that real students will be too exacting. They will accept the imperfect as inevitable, and try to improve it. If their path has been lightened by my labours, I cannot wish for a better justification of my own.

Jingis Khan was succeeded as Imperator of the Mongols by his second son, Ogotai, and he by his son Kuyuk. By a curious fortune the supreme rule in the Mongol world, on the death of Kuyuk, fell into the hands of the descendants of Jingis Khan's youngest son, Tului. For a while the families of Ogotai and Tului struggled, but the former were presently overwhelmed, and the Mongols, properly so called, still obeyed princes descended from Tului when they were conquered by the Manchus.

While one branch of this family controlled the furthest east, another became supreme in Persia. When Tului's son Mangu Khakan succeeded Kuyuk, as Imperator of the Mongol world, he dispatched his brother Khulagu to conquer Persia. The empire thus founded by Khulagu, which lasted for a hundred years, is the subject matter of the following pages. The story is an interesting one in many ways. Khulagu was a mere nomade chief, with the antipathy to civilisation and to town life which was shared by his grand-father Jingis. His path was marked by ravage and destruction, and his army was a plundering horde—ruthless, cruel, disciplined, brave, and, indeed, endowed with the usual virtues of the desert. He swept away the pernicious power of the Ismaelites or Assassins, who, under the mysterious chieftain known in Europe as the Old Man of the Mountain, had planted a number of strong fortresses in the hilly country of Demavend. He compelled the various petty princes of Kerman, Luristan, Yezd, and Fars to do him homage. He did the same with the Christian rulers of Georgia and Little Armenia, and the Mussulman rulers of Iconium. His greatest feat, however, and one which greatly altered the course of Eastern history, was his destruction of Baghdad, and of its famous line of pontiff rulers, the Abbassidan Khalifs. The Khalifate was presently revived as a mere shadow by the Mamluk rulers of Egypt, with its seat at Cairo, but the institution as a potent fulcrum and focus of Muham-medan power was extinguished by Khulagu. His merciless troops also laid waste Mesopotamia and Syria, and reduced the princes of those flourishing cradles of the arts to the condition of the desolated province of Khorasan. The followers of Khulagu were really a nomade army, moving each season into winter and summer quarters, and not a settled body of colonists. Their trade was that of herdsmen, qualified by that of soldiers. They were Shamanists by religion, their aristocracy largely patronising Buddhism.

Nomades the Mongols remained until their conversion to Muhammedanism, a conversion which had a very important effect upon Eastern history. What might have been the course of Asiatic annals if they had remained Shamanists is not easy to picture. Muhammedanism, in the first place, brought in its wake culture. Never, probably, did literature flourish so marvellously in Persia as during the reign of Ghazan Khan and his successors. No nobler specimens of Eastern architecture exist than the magnificent ruins of Sultania, which was built by the Mongols; while the finest brass work of Mosul dates from the same era. In the second place it induced the invaders to adopt a sedentary life instead of a nomadic one. They planted cities, and largely ceased to move hither and thither with their herds. In the next place, it broke the ties which the Mongols had with the Christian principalities in the East, the Crusaders, the kings of Cilicia or Little Armenia, the rulers of Georgia, the emperors of

Constantinople and Trebizond. It also interfered very largely with the permeating influence of the crowds of Franciscan and Dominican friars, who planted convents in almost all the great towns of the East, under the tolerant shield of the early Mongol khans, and thus interwove for a while threads of European culture with the web of Eastern life. Lastly, it tied together once more, if somewhat loosely, the various states of the East which accepted Islam, into a virtual confederation of allies, and eventually, no doubt, broke down and enervated the power of the conquering caste, as was the case in China, and led to the rapid emancipation of the country from their yoke, a result which proved to be by no means an unmixed blessing. The Mongol supremacy in Persia was also marked by a remarkable succession of able administrators. Whether this was due to the central authority being a strong one, and affording opportunity for skill in this respect, I will not profess to decide; but it would be difficult to find in Eastern history a more remarkable example of good government, and of its best theories put into practice, than that presented by the reign of Ghazan Khan, whose laws and regulations remind us of the far-seeing prudence and wisdom of Akbar. Of course the lives of even the best of these men were continually in peril, and few viziers of the Ilkhans died peaceably. Their very ability and uprightness made the best of them the eventual victims of jealous and envious masters. As has been well said: " In the East the death of an official is not too often the result of his ill deeds, but only a means of appeasing the cupidity of an avaricious tyrant."

While Muhammedanism went through these vicissitudes in Persia, it reached its highest point of culture and prosperity in Egypt and in India, where the most active spirits of that faith naturally took refuge, and under their patronage were built the magnificent tombs of the Mamluk Sultans at Cairo, and the equally splendid palaces and mosques erected contemporaneously by the Muhammedan rulers of Delhi.

Abel Remusat has summed up in some graphic paragraphs the general effects of the Mongol invasion of the West, which I shall not hesitate to appropriate. He refers to the great moral revolution in the affairs of the world caused by bringing together the civilisations of the East and West, which had hitherto grown apart, without communication, and without mutual influence. It was not only by means of the many stately embassies which passed to and fro, but also by the more humble journeys of merchants, missionaries, and of those who naturally follow in the wake of armies. The invasion of the Mongols opened the various roads which had hitherto been closed, and brought men of all races together, while one of the chief consequences of their invasion was the transportation of whole peoples to and fro. Among the royal princes who made their way to the heart of Asia, to press their interests there, were Sempad, the Orpelian; Haithon, King of Armenia; the two Davids, Kings of Georgia; Yaroslaf, Grand Prince of Russia, and many others. Italians, Frenchmen, Flemings, &c., went on the same errand as envoys to the Great Khan. Mongols of distinction came to Rome, Barcelona, Valencia, Lyons, London, and Northampton. A Franciscan from Naples became Archbishop of Pekin, and was succeeded by a Professor in the Faculty of Theology from Paris. These were all famous people, whose names were likely to be preserved; but what crowds of obscure folk must have followed the same way, drawn by the double temptation of gain or curiosity to visit the unknown and romantic East. Accident has preserved the names of some of them. The first envoy to the King of Hungary on the part of the Tartars, was an Englishman who, having been banished from his country for various crimes, became a vagabond in Asia and eventually joined the service of the Mongols. A Flemish Franciscan met in the depths of Tartary with a woman from Metz, named Paquette, who had been captured in Hungary, a Parisian goldsmith, whose brother had a shop on the great bridge at Paris, and a young man from Rouen, who had been present at the capture of Belgrade. He also met with Russians, Hungarians, and Flemings there. A chanter, named Robert, after traversing Eastern Asia, returned to die at the cathedral of Chartres. A Tartar made helmets for the army of Philip le Bel, as we learn from the receipts of the Treasury between 1296 and 1301, preserved in a manuscript in a French library. Plano Carpini found a Russian at the Court of Kuyuk, acting as interpreter there; and the Franciscan friar himself tells us how he was accompanied on his journey by merchants from Breslaf, Poland, and Austria. Others accompanied him on his return by way of Russia; among them Genoese, Pisans, and Venetians. Two Venetian merchants, whom accident took to Bukhara, joined an embassy which Khulagu sent to his suzerain, Khubilai. After spending some time in China and Tartary they returned with letters from the Great Khan for the Pope, returning again to the Great Khan, taking with them one of their sons, the famous Marco Polo, whose narrative is such a mine of materials for Eastern history and geography. Both uncles and nephew returned again to Venice. Similar journeys were not less frequent in the next century, as we know from the fantastic story of Sir John Mandeville, Odoric of Friuli, Pegoletti, William de Bouldeselle, and others. Many such adventurers doubtless remained and died in the far East. Many others returned home as obscure as they went, and no doubt told famous stories in the monasteries and the lordly castles, where such visitors were always welcome. Such travellers would take with them a knowledge of many handicrafts, as well as precious objects. Silks and porcelain from China and from India thus probably became familiar objects in the West of Europe, which had been shut off from intercourse with the East since Roman times. Curiosity was everywhere stirred, and curiosity is the great mother of progress. It was proposed to found a chair at the University of Paris for the study of Tartar; and how far reaching the effects may have been we can perhaps gather from the fact that it was in search of the "Zepangri" of Marco Polo that Christopher Columbus set out to discover a new world. Nor was the influence only on one side. The Mongols probably introduced Indian figures into China, as they introduced the Mussulman methods of astronomy. The New Testament and the Psalms were translated into Mongol by the Latin Archbishop of Khanbaligh. It was the Mongols who founded the hierarchy of Lamaism in Tibet, in which they seem to have combined the dogmas of Buddhism with the ritual of the Nestorians. It was the Mongols who probably introduced the knowledge of the mariner's compass, which had long been known in China, into Europe Gunpowder had been used by the Hindus and Chinese from early times. It apparently first became known in Europe after the Mongol invasions, and doubtless through their influence. Paper money was another early Chinese invention, the introduction of which by the Mongols into Persia forms an interesting incident in the following pages; and it is curious to find the traveller Josaphat Barbaro telling us how he learnt from an intelligent Tartar whom he met at Azof, in 1450, and who had been as envoy to China, how this kind of money was annually printed there, as he says, "con nuova stampa." Lastly, playing cards, whose origin is so interesting, because they would appear to have been among the first efforts of engraving on wood, were known in China in 1120, and were very likely introduced through the Mongols. It is at least curious that the earliest playing cards used in Europe, in the so-called "Jeu de tarots," were in their shape, their designs, their size and number, similar to those used by the Chinese. It may be that printing from wooden blocks also came to us from the far East by the same channel.* Thus, again, the Shan-pan, or arithmetical machine, of the Chinese was introduced into Russia and Poland, where it is still universally used by the women who cannot read, in their calculations. In speaking of this interchange of most fertile ideas and inventions through the agency of the Tartars, Remusat says it was by the mortal struggle of nations that the dark clouds of the Middle Ages were dissipated. Catastrophes which seemed to bring only suffering to the human race, in fact awoke it from the lethargy into which it had fallen for centuries, and the destruction of twenty empires was the price which Providence exacted from Europe for the civilisation which it now enjoys.†

One of the main difficulties I have had in writing the following volume, a difficulty which has prevented the story from flowing evenly, has been the necessity of incorporating the history of the various subordinate principalities under the suzerainty of the Mongols in Persia, with the main story. It must be remembered that while the invaders in many cases conquered large districts, and exacted allegiance from the conquered, they allowed them to continue subject to their own rulers; and I have felt bound, therefore, to collect and interweave in my narrative the events which occurred in the dependent principalities. So that what follows is not merely the history of the Ilkhans in Persia, but also that of Herat, Kerman, Fars, Luristan, Mardin and Hosn Kaifa, Rum, Little Armenia, and Georgia, all of which had rulers of their own during the Mongol domination.

I have also endeavoured to arrange and tell in detail the confused and yet important history of the various fragments into which the Ilkhans' empire was broken up, a story which has been hitherto almost entirely neglected; so that this volume may be accepted as a fairly complete history of Iran and its borders, from the invasion of Khulagu to the conquests of Timur.

The next volume will comprise a more obscure, but perhaps more interesting, section of the work, namely, the history of the descendants of Jagatai, who ruled at Almaligh and Kashgar, and also the history of Timur and his descendants. It is partially written, and I hope I may have health and strength

---

* This is the view of Paul Jovius, who has the following very remarkable sentence on the subject : "Cujus generis volumen à rege Lusitaniæ cum elephante dono missum Leo P. humaniter nobis ostendit, ut huic facilé credamus hujus artis exempla, antequam Lusitani in Indiam penetrarant, per Scythas et Muscovitas ad incomparabile litterarum præsidium ad nos pervenisse."
† Memoirs French Academy, vii. 409-419.

to finish it. The main authority for the history of the Ilkhans is of course the great history of Rashid ud din, the vizier and historiographer of Ghazan Khan, and the continuation of it, devoted to Ghazan's successors. This work was partially printed, translated, and edited, with very elaborate notes, by Quatremere. It seems a duty incumbent on French Orientalists to complete what he so well began. The same work has been digested by the two Western historians of the Mongols; D'Ohsson, who has devoted the two last volumes of his work to this dynasty, and Von Hammer, who wrote a monograph upon it, in two volumes. They have also incorporated much material from other sources, Eastern and European. I need not say that these three works have been continually before me, as has also Major Raverty's edition of the " Tabakat-i-Nasiri," the notes to which are so much disfigured by rancour and bitterness, and so wanting in references. I have carefully used both the Syriac and Arabic chronicles of Abulfaraj Bar Hebræus, and the annals of Abulfeda, which are accessible in translations. My friend Mr. Guy Lestrange has generously placed at my disposal a MS. translation of the " Tarikeh Guzideh" of Hamdullah. In addition to these authorities, I have continually used the very valuable " Georgian Chronicle " which has been published by Brosset since the works of D'Ohsson and Von Hammer were written ; and have incorporated the material preserved by the Armenian historians, especially the Chronicle of Haithon, and those of Makaria, Guiragos, &c., only recently made accessible. For the later story I have chiefly used Dorn's memoir on the Serbedarians, Quatremere's paper on the Muzaffarians, the Chronicle of Herat, published in the " Journal Asiatique," &c., &c. It will be seen that I have made considerable use of coins in fixing chronology, &c., and have searched the writings of Fræhn, Dorn, Tiesenhausen, and Stanley Poole. The last of these, a close friend of mine, has also supplied me with notes from his own examination of the Russian collections. Another generous friend of mine, Dr. Rieu, has continued his kind services. To him I owe the notices of the later history of Great and Little Luristan, of Hosn Kaifa and of Mardin, and other help. I hope I have duly acknowledged all particular obligations whenever I have used an authority.

I must ask my readers, before they pass judgment upon any statement in this volume, as in the previous ones, to see that it is not corrected or modified in the too long list of notes and errata. The fact is, my scanty means, upon which this publication is a heavy burden, will not allow of repeated corrections of proofs. The consequence is that, between my indifferent sight and writing and the human frailty of my friend the printer, the list of errata has grown considerably. As previously, no doubt my critics will find fault with the absence of an index ; for this I must ask them to wait until the conclusion of the work is reached.

In again sending out a volume dealing with an unattractive and seldom traversed field of human inquiry, I hope I may have eased some student's burden, and done somewhat to build up, or at least to supply materials for, that vast palace of history which it will take many generations of patient workers to complete. Meanwhile, I will conclude with the words of a much greater man than myself: " Nescio benevoli auditores, au vestram patientiam his nugis fatigaverim, meam certe eas scribendo fatigavi."

CHAPTER I.

THE PREDECESSORS OF KHULAGU.

WHEN Jingis Khan drew off his forces beyond the Oxus, he left a terrible waste behind in Khorasan and Afghanistan. His campaign south of that river had been one of revenge against the Khuarezm Shah rather than one of conquest. He had chased him with the pertinacity of a blood-hound, till he brought him down and had driven one of his sons into the recesses of India and the other into those of Southern Persia, but he does not seem to have treated the intervening country as a permanent addition to his Empire. Khorasan and the country east of it, as far as the Indus, was virtually reduced to a wilderness, and as we shall see, was shortly after, at least nominally, re-occupied by the Khuarezmians. The Mongol possessions towards Persia, on the withdrawal of Jingis, may be roughly drawn at the southern limits of the modern Khanates of Bukhara and Khiva. South of this, the land was reduced to desolation.

Well might the Mussulman and Christian world shrink down upon its knees in the presence of such a terrible visitation. "We pray to God," says Ibn al Athir, "that he will send to Islam and to the Mussulmans some one who can protect them, for they are the victims of the most terrible calamity, the men killed, their goods pillaged, their children carried off, their wives reduced to slavery or put to death, the country, in fact, laid waste."* Juveni says that in the country traversed by the Mongols, only a thousandth part of the population remained, and where there were previously 100,000 inhabitants there remained but a hundred. "If nothing interferes with the growth of population in Khorasan and Irak Ajem from now to the day of resurrection," he adds, "it will not be one tenth of what it was before the conquest."† Pachymeres also reports how the terror of the Mongol arms reached the Court of Byzantium, where the Emperor John Ducas put his fortresses in order, and where popular rumour painted the invaders as having dogs' heads, and eating human flesh.‡

In order to understand the subsequent movements of the Mongols in Persia, we must enter in some detail into the history of the sons of the Khuarezm Shah Muhammed. Of these, Rokn ud din had been killed by

---

* D'Ohsson, i. 350. Note. † *Id.*, 351. Note. ‡ Stritter, iii. 1028.

A

the Mongols in the fortress of Sutun-avend. Jelal ud din was a fugitive
in India, while Ghiath ud din had taken refuge in the strong fortress of
Kharendar, in Mazanderan.   After the retreat of the Mongols, Irak
again became the scene of a struggle between two Turks, the Atabeg Togan
Tayissi* and Edek Khan, who divided the province, the latter taking
Ispahan.   Edek was speedily defeated and killed by Togan, whereupon
Ghiath ud din, who repaired to Ispahan, gave him his sister in marriage
and received his submission.   In a short time he found himself master of
Irak, Khorasan, and Mazanderan.†

Meanwhile, Jelal ud din, when he heard of the retreat of Jingis, having
received an invitation from some officers in Irak, who were discontented
with Ghiath ud din, determined to return to his paternal dominions. Leaving
his General Uzbeg in charge of his conquests in India and Hasan Karak,
styled Vefa Malik of the countries of Ghur and Ghazni, he traversed
Makran, leaving a portion of his followers in its unhealthy climate, and with
but 4,000 men arrived in Kerman, where Shuja ud din Abul Kasim, who
nominally commanded there on behalf of Ghiath ud din, was having a
struggle with a rebel named Borak, styled the Hajib or Chamberlain. Borak
was a native of Kara Khitai, and a near relative, probably the brother of Jai
Timur i Baniko or Taniko, the son of Kalduz, who commanded the forces
of the Gurkhan of Kara Khitai, and was defeated and made prisoner in
1210 A.D., by Muhammed Khuarezm Shah.   Borak and his brother,
Husam ud din Hamid-i-Bur, had been previously sent by the Gurkhan
to Khuarezm during the reign of Sultan Takish to collect tribute. They
had settled there, become Muhammedans, and Borak himself rose in the
Sultan's service to the position of a Hajib or Chamberlain.‡   After the
retreat of the Mongols, Borak had joined Ghiath ud din, and, according
to D'Ohsson, had been appointed Governor of Ispahan.   With him
Ghiath ud din marched into Fars, where he defeated the Atabeg Said, and
committed great ravages and afterwards withdrew.   This was in 620 HEJ.
He seems now to have quarrelled with Borak, who left him and set out
intending to go to India to the Sultan Shams ud din Altamsh, who, like
himself, was a native of Kara Khitai.   D'Ohsson says he quarrelled with
Ghiath ud din, and asked permission to go and join Jelal ud din Khuarezm
Shah in India.   As he traversed Kerman he was attacked, near Giruft, by
Abul Kasim, who held Kuwashir, otherwise called Kerman, the capital of
that province, on behalf of Ghiath ud din.

Borak defeated and slew his assailant, who was captured and put to
death, and he was about to attack Kuwashir, whose citadel still held out
under Abul Kasim's son Shuja ud din, when he heard of the arrival
of Jelal ud din, to whom he offered presents and the hand of his

---

' *Tayi*, says D'Ohsson, means maternal uncle in Turkish, and *Tayissi* means the uncle. Togan
was Ghiath ud din's maternal uncle. (D'Ohsson, iii. 13. Note.)
† D'Ohsson, iii. 2-3.   ‡ Tabakat-i-Nasiri, 283, 1118. Notes.

daughter. Kuwashir opened its gates to the Sultan, who appointed Borak his deputy in Kerman. Borak presently began to show signs of treachery; but, being advised that it would not be prudent to punish the first chief who had submitted to him, Jelal ud din determined to move on to Fars, and confirmed Borak in his authority as Governor of Kerman, which he seems to have held as a dependent of Jelal ud din till the latter's death, and then on his own behalf. His descendants ruled for 86 years, the dynasty being known as that of the Kara Khitaians of Kerman.*

At this time the Atabeg Abubekr Said, son of Zengui, ruled at Fars. He was descended from a Turk named Salgar, whence the dynasty was known as that of the Salgarids. Said sent his son with 500 horsemen to welcome Jelal ud din to Shiraz, but he excused himself from going in person, on the ground that he had made a vow never to present himself there. He was, in fact, much irritated against the Khuarezmians, on account of the raid Ghiath ud din had recently made; but he sent his son, Salghur Shah, with 500 horsemen to do the Sultan honour. With him he sent splendid presents, among which Habashi, Hindi, and Turkish slaves are especially mentioned. Said was given the title of Farzandkhan, and confirmed in his authority. Jelal ud din also cemented this friendship by marrying his daughter.

Thus did he prudently make his harem a bond of union between himself and his greater dependents. He also took Said's son with him. Quitting Shiraz, Jelal ud din marched on towards Irak to oust his brother. The latter, a feeble and voluptuous prince, was incapable of repressing the anarchy which had followed the invasion of Jingis. Each district had its petty tyrant, who had the *khutbeh* said in the name of Ghiath ud din, but paid him no tribute. While he, having no money to pay his Turkish mercenaries with, was constrained to let them plunder. Their officers when discontented were rewarded with higher titles—an Amir became a Malik and a Malik a Khan.† On his way, Jelal ud din was joined by the Atabeg Ala ud daulah, who had ruled over Yezd for 60 years, and was a lineal descendant of the last of the Dilem rulers of the Buwiah dynasty.‡ Jelal ud din having reached Ispahan, advanced upon Rai, where Ghiath ud din was collecting his forces. The former had given his men white standards, like those of the Mongols. The latter had mustered a force of 30,000 cavalry, with which he, however, withdrew. Jelal ud din sent his brother a friendly message, to say he had merely come to visit him, but seeing he was hostile to him he proposed to retire again. Taken in by this message, Ghiath ud din returned to Rai and disbanded his troops. Meanwhile, Jelal ud din had corrupted his generals and sent them rings as pledges of his goodwill. News of this having reached his brother, he had his agent arrested, but Jelal ud din, who felt sure the

* D'Ohsson, iii. 5-6. Tabakat-i-Nasiri, 283, 295. † D'Ohsson, iii. 7-8.
‡ Tabakat-i-Nasiri, 296. Note.

troops were with him, marched on, although with but 3,000 men. His brother fled, but presently went to his camp and offered his submission.

Jelal ud din was now generally recognised, and the various petty chiefs of Khorasan, Mazanderan, and Irak deemed it prudent to offer their allegiance, and were treated with generosity. He then marched to punish the Khalif Nasir, who had been most unfriendly towards his father, and was accused of having invited the Mongols to invade his dominions. He ravaged Khuzistan, laid siege to its chief town, Shuster, and advanced as far as Ya'kuba, or Bakuba, seven parasangs from Baghdad. The army of the Khalif, commanded by Kush Timur, consisted of 20,000 men. A battle ensued, in which, although his troops were much weaker, Jelal ud din planted an ambuscade, and the result was that Kush Timur was defeated and killed. This defeat was followed by the capture of Dakuka, and of the Prince of Erbil, who had marched to the assistance of his suzerain, the Khalif.* This is Mirkhond's story ; Ibn al Athir says nothing of his capture, but tells us he made peace with Jelal ud din.†

Jelal ud din, for some reason, now abandoned his enterprise against Baghdad, and turned towards Azerbaijan, then subject to the Atabeg Uzbeg. Having reached Meragha, he proposed to rebuild it, but set out again to encounter Togan Tayissi, already named, who was maternal uncle to Ghiath ud din, and who, having been invested with the districts around Hamadan by the Khalif, had made a profitable raid upon Arran and Azerbaijan, and had a great collection of cattle, &c.; the result of his foray around his camp. Jelal ud din made a night march, and at dawn Togan, who had married his sister, disconcerted by the unexpected appearance of the Imperial umbrella, which marked his presence and that of his troops, deemed it best to submit, and returned with him to Meragha. Meanwhile, Uzbeg Ibn Alpehluvan, the Prince of Azerbaijan, who, as we have seen, had been very accommodating to the Mongols, left Tebriz and went towards Gandza, or Kantzag, the capital of Arran, leaving his wife, a Seljuki princess, in charge of his capital. Jelal ud din attacked it. In five days the citizens surrendered. The Sultan reproached them for having put to death the Khuarezmian fugitives the year before, when they sent their heads to the Mongols. They laid the blame on Uzbeg. Having occupied Tebriz, he made over the town of Khoi and some other possessions in Azerbaijan to Uzbeg's wife, and then set out for Georgia.‡

When the Mongols invaded the steppes north of the Caucasus, the Kipchaks who lived there dispersed. One section of them retired through Derbend, and lived for awhile in the country of the Shirvan Shah, much to the discomfort of the latter.§ They eventually took possession of Derbend, and then marched to Kabala, a town of Georgia, situated on

* D'Ohsson, iii. 11-12.   † Weil, iii. 390. Note 3.   ‡ D'Ohsson, iii. 14-15.
§ Ibn al Athir, Journ. Asiat., 4th ser., xiv. 463-466.

the left bank of the Kur, near Berdaa. Its chief made overtures to them to join him in conquering the neighbouring districts. They thereupon refrained from molesting his people for a few days, when their predatory instincts overcame them, and having plundered after their wont, they passed on to Arran, and settled near Kantzag. Kushkareh, a freed slave of Uzbeg, Prince of Azerbaijan and Arran, who then ruled there, treated them well by order of his master, and they were assigned a camping ground on the mountain of Kielgun (*i.e.*, "like a navel"). The Georgians, who were then at constant feud with the Mussulmans, and doubtless feared the proximity of such marauders, attacked them, but were defeated with terrible losses.*

The following year (1224) they were amply revenged, however. The Kipchaks were defeated and dispersed. Many of them were waylaid by the inhabitants of the country, the Georgians and the Lesghs joining in the work, so that Kipchak slaves were sold at Derbend for very small prices.† The Georgians, apparently animated by this success, invaded Arran, and attacked the town of Bailekan, whose inhabitants were busy restoring it after its devastation by the Mongols. Having captured it, they slaughtered the inhabitants, and behaved even worse than the Mongols.‡ The Georgians now attacked Surmari, a dependency of Ashraf, Prince of Khelat, where they suffered severely. In 1225 they advanced against Kantzag, but were obliged to raise the siege after a short time. They were not more fortunate in a campaign against Shirvan, whose ruler had appealed to them to assist him against his revolted son. They were defeated here, as they also were in a raid they made upon Azerbaijan.§ At this time they seem to have been a scourge in fact to their Mussulman neighbours all round. They were preparing a fresh expedition to revenge their recent defeat in Azerbaijan, when Jelal ud din arrived at Meragha, as we have mentioned. They thereupon made overtures to Uzbeg for a common alliance against the Khuarezm Shah.‖ The latter was burning to revenge the wrongs suffered by the Mussulmans. He sent a messenger to declare war against the Georgians, and they bravely replied that the Tartars who had destroyed his father had been forced to withdraw before them. They mustered a force of 70,000 men. Having captured Tovin, he sacked the country round. The Constable Ivaneh informed his mistress, Queen Rusudan, of his approach, and was ordered to go and meet the enemy. Jelal ud din encamped his army at Karni, or Garhni, one of the most ancient towns of Armenia, situated in the district of Kegh'arkunik and the province of Siunik.¶

The two brave brothers, Ivaneh and Shalwa of Akhal Tzikhé, were put in the advance guard. The Constable was apparently jealous

---

* *Id.*, 468. Journ. Asiat., 5th ser., xi. 201-202. † Ibn al Athir, Journ. Asiat., 4th ser., xiv. 468-470.
‡ *Id.*, 472.   § *Id.*, 472-481.   ‖ *Id.*, 481.   ¶ Vartan, Journ. Asiat., 5th ser., xv. 280.

of these two chiefs, and refused to march the main army to help them when hard pressed. The Georgians were accordingly defeated. Shalwa was captured and treated well for some time by the Sultan, but was put to death a few months later for refusing to apostatize. Nissavi says for carrying on treasonable correspondence with the Abkhazians. His brother was killed in the fight by a stone rolling down upon him.* Another account says that Shalwa had smeared his face with blood, and lay down among the corpses to escape detection, when he was captured. The Georgians are said to have lost 20,000 out of their army of 70,000.† Vartan accounts for the defeat of Ivaneh as a punishment for a gross act of sacrilege which he had committed in disinterring and burning the remains of a saint named Parcecht, and immolating a dog on his tomb in derision of the crowd of pilgrims who had gone there attracted by the saint's relics. Ivaneh was attached to the Georgian Church, which was in union with that of Constantinople, and had a strong antipathy to the Armenians, who were tainted with the Eutychian heresy, and were not deemed orthodox.‡ After his defeat Ivaneh retired to the fort of Kheghi (the *Georgian Chronicle* says to Bejni). The Khuarezmians now overran Georgia as far as the frontier of the Abkhazians, and would have marched to Tiflis, but Jelal ud din was recalled to Tebriz by an impending revolt in favour of Uzbeg.§ Leaving his army in Georgia, under his brother, Ghiath ud din (Ibn al Athir says Tiflis was left in charge of Ak Sonkor, a mamluk of Uzbeg), he went there, put to death the *reis* or mayor of the town, and arrested the conspirators. Having married Malika, the wife of Uzbeg, who was divorced from the latter by a legal fiction, he captured Kantzak, the capital of Arran (whence Uzbeg fled to Alenjik, near Nakhchivan), after which he returned to Georgia.|| Uzbeg shortly after died.¶ The Georgians, after their defeat, had mustered a force, comprising Alans, Lesghs, Kipchaks, &c., which was speedily crushed, and the district of Somkheth was devastated. Jelal ud din then marched upon Tiflis, whence Rusudan had retired to Kuthathis, leaving a garrison in the place in charge of two chiefs, named Memna and Botzo. Some Persians, who guarded one of the gates of the city, proved treacherous, and opened a way for his men. The citizens were mercilessly slaughtered, except those who would accept Islam. The men were circumcised in large numbers, and the women ravished. The *Georgian Chronicle* gives some ghastly details, and compares the catastrophe to the capture of Jerusalem by Titus. The churches were ruined, and the sacred images torn down. This was a terrible blow to the cause of Christianity north of the Caucasus, of which the Georgians were a famous bulwark, and Ibn al Athir speaks of the event with corresponding elation.

---

* Brosset, Hist. de la Géorgie, i. 497-500.   † Journ. Asiat., 4th ser., xiv. 482-483.
‡ Journ. Asiat., 5th ser., xvi. 280-281.   § Brosset, Hist. de la Géorgie, 308. || D'Ohsson, iii. 17.
¶ Brosset, Add., 312.

The *Georgian Chronicle* describes how Jelal ud din proceeded to ravage the surrounding districts, Somkheth, Kambejian, the banks of the Yor, Karthli, Trialeth, Javakheth, Artan, Samtzkhe, Tao, Carniphora (Kars), and Ani.* The wretched Georgians were pursued into the country of the Abkhazians, into which numerous raids were made, and Georgian slaves were sold for two or three gold pieces each.†

The Ayubit prince Ashraf, lord of Harran and Roha, was brother to Moazzam, Sultan of Damascus, and Khamil, Sultan of Egypt, all three being nephews of the great Saladin. Moazzam had a very high opinion of Jelal ud din, and used to wear a robe and to ride a horse which he sent him, and used also to swear at his banquets by the head of Jelal ud din. Being at issue with his brothers, he sent to urge the Khuarezmian prince to attack Khelat, also called Akhlat (situated on the northern shores of Lake Van), which was subject to Ashraf.‡ This was a sufficient temptation, and the Sultan accordingly marched thither, but hardly had he begun the siege when he raised it, on hearing that Borak was meditating revolt in Kerman, and had informed the Mongols of his (Jelal's) increasing power. He thereupon marched against him, reaching Kerman on the eighteenth day from leaving Tiflis, only 300 horsemen having kept up with him. Borak retired to a strong fortress, and sent envoys with his submission to Ispahan. These Jelal received affably, and confirmed him in his government.§ Meanwhile, some of Jelal ud din's troops having made a foraging expedition towards Erzerum, were attacked on their return by the people of Khelat, who secured the booty they were carrying off.

Nissavi tells us that during the Sultan's absence, Sherif ul Mulk, his Vizier, remained in charge of Tiflis, and devastated the country by numerous raids. He was prodigal in the largess he distributed, but was not favourably looked upon by the generals of Jelal ud din (who are referred to as "the Khans" by our author), except Ur Khan. News having arrived that the Vizier was being pressed, at Tiflis, by the Georgians, the latter went to his assistance with 5,000 men, but the tidings proved to be false. Presently the Sultan returned in person, and the prodigal Vizier gave 4,000 gold pieces to the messenger who brought the news, and a fresh devastation of Georgia was the consequence.‖ Jelal ud din now marched to attack Ani, where the Constable Ivaneh had sought refuge with the *débris* of his army. He invested it, as well as Kars, but found them too strong, and again returned to Tiflis, whence, by way of a ruse, to persuade the people of Khelat that he was a long way off, he made a ten days' raid into Abkhazia, and then speedily advanced upon Khelat, which he would have captured but for some traitors in his camp, who duly informed its governor. He arrived there on the 5th of November, 1226, and attacked

---

* Op. cit., i. 507.   † Ibn al Athir, Journ. Asiat., 4th ser., xiv. 494-495.   Brosset, Add., 313.
‡ Novairi, in D'Ohsson, iii. 18.   § D'Ohsson, iii., 18-19.   ‖ Brosset, Add., 315-316.

it vigorously; but the citizens, knowing what they might expect at the hands of the Khuarezmians, resisted bravely. Meanwhile Ashraf went to Damascus, and persuaded his brother Moazzam to ask Jelal ud din to withdraw. This he did not do, however, until compelled by the severity of the weather, and by a raid which some Turkomans, called Ivanians, had made into Azerbaijan. He thereupon hastily left, cut off the retreat of the freebooters, put them to the sword, captured their wives and the booty they had made, and then returned to Tebriz. This was in December, 1226.*

Jelal ud din now went again to superintend the siege of Khelat (called Akhlat by Nissavi). In the autumn, Sherif ul Mulk went with his troops into winter quarters at Kantzak. Presently, profiting by the absence of the Sultan and the weakness of the garrison, the Georgians at Kars, Ani, &c., assembled a force, with which they attacked the capital. It was abandoned by Kar Mulk, who was then in charge. Knowing that they were not strong enough to hold it, they set fire to the town.†

Jelal ud din now invaded the territory of the Ismaelites or Assassins, to punish them for having killed one of his officers who had been given the fief of Kantzag. He next attacked a body of Mongols which had traversed the desolated districts of Khorasan and appeared at Dameghan. This, we are told, he pursued for several days. It was doubtless a small reconnoitring body merely.

Jelal ud din's temper, and the asperity of his troops, having caused discontent, which was fanned by the intrigues of his wife, recently the wife of Uzbeg, who regretted her new position, induced Hussam ud din Ali, who commanded at Khelat for Ashraf, to enter Azerbaijan, where he captured the towns of Khoi, Merend, and Nakhchivan, with other fortresses, while he carried off the discontented lady with him to Khelat.‡ Jelal ud din had to postpone his revenge, on account of the approach of a more dangerous body of Mongols.

It would seem that it was an army sent by the Mongol Governor of Transoxiana, or of Khuarezm. Rashid ud din says the invaders marched with five divisions, under the Generals Taji, Baku, Assatogan, Taimaz, and Tainal, and drove before them a detachment of 4,000 men, whom Jelal ud din had posted towards Rai and Dameghan. He himself made his head quarters at Ispahan, which was approached by the Mongols. He was very self-possessed, and when his generals reported the enemy's approach, created much confidence by his *sang froid*. Having pressed his officers to prove themselves men, he had the armed citizens mustered by the kadhi and reis of the town. The Mongols detached 2,000 horsemen to the mountains of Lur for foraging purposes. They were waylaid in this difficult country, and 400 of them captured. Jelal ud din handed a portion

---

* D'Ohsson, iii. 20-22.   † Nissavi, in Hist. de la Géorgie, Add., &c., 316.
‡ D'Ohsson, iii. 22-23.

of them to the rabble of Ispahan, who killed them, while others he slew with his own hand in the palace court. Their bodies were left to the vultures and dogs.* The astrologers having fixed the 26th of August, 1227, as a fortunate day for the fight, Jelal ud din ranged his men in order of battle, when he was treacherously abandoned by his brother Ghiath ud din, and by the General Jihan Pehluvan Ilchi (*i.e.*, by the Uzbeg already named, as left by him in command of the troops when he left India), with their troops. He nevertheless determined to fight. The battle was fought in the evening. His right wing, under Otuz or Uz Khan, drove back the left of the enemy as far as Kashan. Jelal ud din was reposing on his laurels, when he was urged to pursue his enemies by one of his officers. Advancing confidently, his left and centre were attacked by a body of Mongols placed in ambush in a ravine, a favourite stratagem of theirs. His officers died at their posts like men, and he himself fought desperately, and with his own hand slew his standard-bearer who was attempting to fly; but it was of no avail—there was a general flight. Some went towards Fars, others to Kerman, others again to Azerbaijan,† while those who had lost their horses remained at Ispahan. The successful division, which had advanced towards Kashan, having turned and learnt what had taken place, also disbanded. The Mongols had suffered too severely, however, to renew the fight, and withdrew by way of Rai and Nishapur. They lost a great many men in the retreat, and recrossed the Oxus much weakened.‡ For eight days Jelal ud din lay *perdu*, and it was proposed at Ispahan to elect a fresh ruler, and to plunder the harem and goods of the Khuarezmians. The Kadhi persuaded the citizens to wait till the feast of Bairam, when at the hour of prayer, if the Sultan had not returned, he proposed they should put the Atabeg Togan on the throne. Jelal ud din, who had fled to Luristan (according to Ibn al Athir to Khuzistan, whence, not being well received by the Khalif's deputy, he went to the Ismaelites), returned on the day of the feast, and was received with great joy. He delayed a few days at Ispahan, to await the return of the fugitives, and rewarded his generals of the right wing, conferring the title of Khan on those who were merely Maliks. Those who had misbehaved were promenaded through the town with women's veils about their heads. Meanwhile, Ghiath ud din had retired to Khuzistan, where he sought an alliance with the Khalif. The ill-will between the two brothers had come to a head a few days before the recent battle. Muhammed, son of Kharmil, of an illustrious family, and a favourite of Jelal ud din, had taken into his service some troops who had detached themselves from Ghiath ud din, on account of arrears of pay. The latter, annoyed at this, had, after an altercation at a banquet given by his brother, run a poignard into Muhammed.

---

* D'Ohsson, iii. 24-25. † *Id.*, 26. ‡ *Id.*, 27.

Jelal ud din, who was naturally enraged, declared that he no longer
felt any obligations towards him, and should not protect him if the
relatives of the murdered man claimed the blood-penalty, while he ordered
the latter's funeral *cortège* to pass twice in front of the door of his assassin.*
Ghiath ud din, as I have said, sent to the Khalif, offering to serve him
faithfully, and asking for his aid. His envoy was well received and given
a subsidy of 30,000 dinars.

Meanwhile, Jelal ud din, having sent a body of troops in pursuit of the
Mongols as far as the Oxus, repaired to Tebriz. He was playing at his
favourite game of polo in the great square of the town, when he heard
that his brother was marching against Ispahan. He threw down his
mallet and hastened to the rescue. Learning *en route* that Ghiath ud din
had retired to the fortress of Alamut, in the country of the Assassins, he
demanded his surrender from the famous chief of the Ismaelites. The
latter replied that Ghiath ud din was a Sultan, and the son of a Sultan,
and he could not think of surrendering him. He would, however, guarantee his good behaviour, and he gave Jelal ud din leave to ravage his
territory if his guest behaved badly while he harboured him. Jelal ud din
professed to be satisfied, and was ready to overlook the past, but Ghiath
ud din was apparently not reassured, and preferred to retire to Kerman.
There the ambitious Borak insisted upon marrying his mother, who
accompanied him. She refused for a long time, but as he was all-
powerful she had to give way. Presently, two of Ghiath ud din's
dependents having plotted to kill Borak, the latter heard of it and had
them cut in pieces before his eyes. Ghiath ud din himself was then
strangled with a bowstring, and his mother suffered the same fate, while
the 500 companions who had gone with him were also put to death.†
Thus perished another son of the Khuarezm Shah Muhammed. Borak
sent the head of the murdered prince to Ogotai Khan as a peace-offering,
which secured the friendship of the Mongols, who confirmed him in the
possession of Kerman.‡

The Kankalis and Kipchaks were closely connected with the Khuarezm
Shahs, who intermarried frequently with their chiefs, whence the pertinacity of Jingis Khan in attacking them. Jelal ud din, after his defeat at
Ispahan, had sent to ask assistance from them. They assented ; and we
are told that Kurkhan, one of their leaders, embarked on the Caspian
with 300 picked men, and went to join him at Mughan, where he
passed the winter. It was arranged that Jelal ud din should secure the
Pass of Derbend in the Eastern Caucasus, by which alone a substantial
force could reach him from Desht Kipchak. A body of 50,000
Kipchak families marched to aid in its capture, while the Sultan tried
to negotiate with the young prince who ruled at Derbend, and with his

* D'Ohsson, iii. 30.   † *Id.*, 33.   ‡ Tabakat-i-Nasiri, 284 Note.

Atabeg, Al Asad (*i.e.*, the lion), for the surrender of the place in lieu of certain fiefs, &c., but the plan failed.* We now read of Jelal ud din securing the district of Gushtasfi, situated between the Araxes and the Kur. This belonged to the Shirvan Shah, and was made over to the latter's son, Jelal ud din Sultan Shah, whose father had sent him apparently under constraint into Georgia, with the intention that he should marry the daughter of the famous Queen Rusudan. When the Khuarezm Shah overran Georgia he was released.† Jelal ud din claimed tribute from the Shirvan Shah, as the successor of the Seljuki who, when master of Arran, had exacted tribute from him.‡

Having spent some time at Mughan, Jelal ud din sent an army under Ilek Khan, which captured Lôré, in the district of Tashir, the principal town of the Orpelians, and then advanced along the Lake of Erivan. The Georgians fell on him at night and defeated him, whereupon the Sultan withdrew his army.§ Meanwhile, the Georgian Queen Rusudan and her Constable Ivaneh had assembled a force of 40,000 men, consisting of Georgians, Armenians, Alans, Serirs, Lesghs, Kipchaks, Suans, Abkhazes, and Janits (the Jiks of the *Georgian Chronicle*). Jelal ud din, although his army was very inferior in numbers, marched against them, and pitched his camp at Mendur. His Vizier, Sherif ul Mulk, counselled a delay, but received a blow on the head with a writing-case for his pains, and was told that a lion should not fear a flock of sheep like that. He was also mulcted in a fine of 50,000 dinars. When the two armies drew near together, the Sultan made overtures to the Kipchaks, who to the number of 20,000 were in the right wing of the opposite army, and recalled the services he had done their people, whereupon they drew off. He then proposed to the Georgians that they should enter into a truce, and that each side should send a champion and let them fight in view of both armies. He himself went out to encounter the Georgian hero, and pierced him through with his lance. He also killed three of his sons, as well as a fifth champion, a man of gigantic size, after which he gave the signal for the struggle, and notwithstanding the treaty, charged the Georgians, who were defeated.‖ The *Georgian Chronicle*, in describing these events, says that when Rusudan heard of Jelal ud din's approach, she summoned all the troops from both sides of Mount Likh. Shahanshah, the chief of the Mandators ; the Generalissimo Avak (son of Ivaneh) ; Varam-Gagel, chief of the Makhurs ; those of Hereth, Kakheth, Somkheth, Jawaketh, Meskhia and Tao, the Dadian Tzotné, the Abkhazians, and the Jiks. She opened the Gates of Dariel, and summoned the Osses, the Durdzuks, and all the mountaineers. Having gathered them together at Najarmagef, they traversed Tiflis, and encountered the invaders in the Valley of Bolnis. The Georgians were defeated and fled, and Jelal ud din re-entered Tiflis, and re-enacted the massacres and ravage of his previous visit.*

He now proceeded to waste the territory of Vahram, the Armenian Prince of Shamkor (a town situated in the Province of Udia, west of the Kur), the Varam-Gagel of the *Georgian Chronicle*, who had recently plundered the environs of Kantzag, close by. The Sultan captured Sékan, or Sagam, and Ali Abad. He then besieged Kak (Gaga) and another fort, which were constrained to sue for peace, and to pay a ransom. Sending his baggage through Kakezvan, or Gaghzvan, a town situated in the district of Gapéghean, north of the Araxes, he himself went by way of Nakhchivan, and again defeated the Georgians near Pdchni, or Bejni, and having delayed for a few days to arrange the affairs of Khorasan and Irak, went on to renew the siege of Khelat.† This he pressed during the winter of 1228, during which he received a visit from the Seljuk Prince of Erzerum, Rokn ud din, Jihan Shah, who presented him with 10,000 dinars, and a more valuable gift in the shape of a great catapult and some shields and weapons. The Princes of Amid and Mardin also submitted, and consented to have the khutbeh said in his name. The Khalif Nasir had died in 1225, soon after the defeat of his General Kush Timur, and was succeeded by his son Zahir, who in nine months gave place to his son Mostansir. The latter sent an envoy, requesting that the Sultan would not exercise any rights of suzerainty over the Princes of Mosul, Erbil, Abuych, and Jebal, who were his feudatories, and that he would re-insert the Khalif's name in the public prayers in Persia, whence it had been excluded by his father Muhammed. Jelal ud din consented willingly to these requests, and sent an envoy in turn to the Khalif, who soon returned with some officials bearing the robe of investiture of Persia for Jelal, together with some rich presents. The Khalif styled him Khakan, and also Shahin Shah, but would not consent to give him the title of Sultan. Thenceforward he called himself servant of the Khalif in his letters, and styled the latter his lord and master.† He now ordered a splendid tomb to be prepared for his father's remains at Ispahan, and pending its building, had them removed to Erdehan, near Demavend, and ordered his aunt, with a grand *cortège*, to escort them from their resting-place in the island of the Caspian, where he had died. Muhammed of Nissa, the biographer of Jelal who wrote her the order, did it unwillingly. He was afraid the Mongols might return, and desecrate the tomb, for, deeming the graves of all kings they met with connected with the Khuarezm Shahs, they treated them accordingly. Thus they tore Mahmud, the great Ghaznevid chief, who had been dead for more than two centuries, from his sepulchre. Muhammed's fears proved to be justified, for eventually the Mongols captured Erdehan, and the ashes

* Brosset, Hist. de la Géorgie, i. 510.
† Journ. Asiat., 4th ser., xiv. 510-511.   ‡ D'Ohsson, op. cit., 35-37.

of the great Khuarezm Shah were sent to the Khakan in Mongolia, who had them burnt.

At this time, Jelal ud din had a correspondence with the Seljuk ruler of Rum, Alai ud din Kai Kobad, and the latter asked that his son Kai Khosru might marry his daughter, and thus unite more closely the buttresses of Islam, in the east and west; and that Jelal would surrender to him his cousin and vassal, the Prince of Erzerum, who had behaved badly to him. Jelal ud din refused either to give his daughter to the Seljuk chief, or to surrender his guest; while his Vizier, Sherif ul Mulk, who was annoyed at the paucity of their master's presents, treated the envoys with marked incivility, and boasted that if the Sultan would permit him, he would enter their country with his own troops only, and conquer it. When they returned home, Kai Kobad, disgusted with this treatment, resolved to ally himself with his rival Ashraf.[*]

Meanwhile, Jelal ud din continued his feud with the latter prince, and especially pressed the attack against Khelat. The siege continued for a long time, and at length, after an obstinate resistance, the town fell, on the 2nd April, 1230,[†] one of its Amirs having surrendered it by treachery. Contrary to the wishes of the Sultan, and under pressure from his generals, the place was given up to be sacked for three days, and a great number of the inhabitants perished. The garrison had suffered severely, and the fare of the besieged citizens had gradually deteriorated. Ibn al Athir thus enumerates the descending scale: Sheep, cows, buffaloes, horses, asses, mules, dogs, cats, and even mice; and he goes on to declare that God, the Most High, to punish Jelal ud din for his conduct at Khelat, did not permit him to survive its capture long.[‡] Abulfaraj says a Damascus pound of bread cost an Egyptian gold piece. Thamtha, daughter of the Constable Ivaneh, the Georgian wife of Ashraf, who was living at Khelat, was captured there, and was married by the conqueror, who also took prisoners Yakub and Abbas, two young brothers of Ashraf. He distributed the lands of the district of Khelat among his generals.[§] Ashraf was the brother of the Sultan of Egypt, who had appointed him to the Principality of Damascus, and received in exchange Harran, Roha, Suruj, Reesain, Rakka, and Jemelein. On the capture of Khelat, he accepted the overtures of the Seljuk Sultan of Rum, and also demanded help from the Princes of Aleppo, Mosul, and Mesopotamia. Kai Kobad joined him at Sivas, and together they marched towards Khelat. On his side, Jelal ud din sent out Chaushes and Pehluvans to summon his own dependents, and on the advice of the Prince of Erzerum, marched to meet the advancing enemy to Khartpert, hoping to attack each army separately; but he fell dangerously ill there, and his enemies succeeded accordingly in uniting. Kai Kobad had 20,000 horsemen, and Ashraf 5,000 picked men. He, on

the other hand, had not recalled the contingents from Arran, Azerbaijan,
Irak, and Mazanderan, whom he had dismissed to their homes: while one
division of his army was delayed at Manazguerd under his Vizier, and a
second body was besieging Berkeri. Nevertheless he resolved upon a
fight, and met his opponents at Erzenjan. He was very badly beaten,
and lost most of his men. Among the prisoners was the Prince of
Erzerum, who had promised to hand over to Jelal ud din a portion of the
territory of his cousin Kai Kobad, and instead, lost his capital, his fortresses,
and treasures. The Khuarezmian officers captured were put to death,
while the fugitives fled to the mountains of Trebizond and to Georgia.
Alai ud din was received with an ovation by his people, Christians and
Mussulmans alike. Jelal ud din fled to Manazguerd, and drawing off the
troops who were laying siege to it, retired upon Khelat, where he pillaged
what could be carried off, and burnt the rest. He also took with him
Ashraf's brothers and his Georgian wife, and departed for Azerbaijan,
leaving his Vizier at Sekman Abad, to watch the enemy. He at length
halted near the town of Khoi, and found himself deserted by his generals.
Meanwhile, however, the two allies, who apparently deemed him their
best bulwark against the Mongols, did not press their advantage. Ashraf,
in fact, made overtures for peace, which were at first rudely spurned, and
Jelal for some time also refused to entertain a friendly disposition towards
Kai Kobad, whom he deemed a traitor to himself in having joined the
Prince of Syria, and only consented to do so in view of another formidable
Mongol invasion.[*]

On the death of Jingis Khan, and in the spring of 1229, Ogotai was
nominated his successor at a great kuriltai held on the banks of the
Kerulon, as I have described.[†] At this kuriltai, it was determined to
send two armies towards the west—one against Kipchak and Southern
Russia, whose doings I have chronicled,[‡] the other against the family of the
Khuarezm Shah. The latter was commanded by the Noyan Churmagun,
or Charmaghan.[§] Von Hammer says he was a Jelair, and Major Raverty
a Mangkut—I know not on what authority, for Rashid ud din distinctly
tells us, that like several other great Amirs, he was a Sunid. He had
belonged to Jingis Khan's body guard.[||]

As he was nominated to such a responsible post, he was doubtless a
person of great reputation. The Armenian historian, Chamchean, gives
a list of the Mongol chiefs who accompanied him (I give it in his corrupt
orthography, which I have no means of correcting): "Benal Noyan and
Mular Noyan, Ghataghan, Chaghata, Tughata, Sonitha, Jola brother of
Charmaghun, Asutu, Bachu (Baichu), Tutu, Khuththu, Asar or Aslan,

---

* D'Ohsson, iii. 46-47. Abulfaraj, Chron. Arab., 307.   † Ante, i. 116.
‡ Ante, i. 137-155. ii. 38, &c.
§ He is called Charman, Charma, Chorma, Chormakhan, and Charmaghan, by various
Armenian authors. Rashid ud dun always calls him Charmaghun. Abulfaraj, in his "Syriac
Chronicle," Sharmagun ; and in his Arabic one, Jurmaghun. St. Martin Memoires, ii. 272.
Note 31. || Erdmann, 178.

Ogota, Khola or Khoga, Khurunji, Khunan, and Ghatapughaor Karabugha."
Stephen Orpelian mentions Charman, Chagatai, Arslan, Asavur (*i.e.*, Yassaur)
and Ghadaghan.[*] The *Georgian Chronicle* mentions but four commanders,
Charmaghan, Chaghata, Yosur (*i.e.*, Yassaur), and Bechui (*i.e.*, Baichu),
each at the head of 10,000 men.[†] Others tell us the army was 30,000
strong, and comprised contingents from the various Mongol appanages.
Chin Timur, who governed Khuarezm for the family of Juchi (*i.e.*, the
princes of the Golden Horde), was also ordered to join him with his troops.
The latter accompanied him to Khorasan, where he remained as governor,
with four colleagues representing the four branches of the family of Jingis,
namely, Kelilat, Keulbilat, or Kalbad, nominated by the Khakan, Nussal
by Batu, Kul Tuga by Jagatai or Chagatai, and Tunga by the widow and
children of Tului.[‡] The author of the "Tabakat i Nasiri" says the force
under Charmaghan numbered 50,000 Mongols, together with those of other
races of Turkestan and captives of Khorasan, in all about 100,000 men.[§]

Charmaghan speedily traversed Khorasan, and advanced by Esferain
and Rai. The *Georgian Chronicle* says his men were much molested by
the Mulahids or Assassins.[||] Meanwhile Jelal ud din, under the impression
that the Mongols would winter in Irak, went from Khoi to Tebriz, but
withdrew on learning from one of his pickets that they had reached the
district between Zanjan and Ebher. Leaving his harem at Tebriz, he
thereupon repaired to the plain of Mughan, where he proposed to muster
his men, and awaiting their arrival he, with but a thousand followers, spent
his days in hunting and his evenings in dissipation. Meanwhile he sent
the governors of Khorasan and Mazanderan to watch the enemy, with
orders to plant post-horses at Erbil and Firuzabad. He was suddenly
attacked by a body of them, near the fort of Shirkebut, situated
on a height in the Mughan plain. He barely escaped, and fled
towards the Araxes, whence he turned towards Azerbaijan, and on
arriving at Mahan, which was well stocked with game, he sent his
prisoner, Yakub, to his brother Ashraf, to bid him march to the rescue.
Jelal ud din's Vizier, Sherif ul Mulk, who, as we have seen, had a grudge
against Ashraf, and was not faithful to his own master, being ordered to
send an envoy to accompany Yakub, gave him perverse instructions, at
issue with those of Jelal ud din. The Vizier had conveyed his master's
harem into Arran, and lodged it in the fortress of Sind-Surakh, and
deposited his treasures in several forts belonging to the chief of the
Turkomans of Arran. He then repaired to Khizan, where he raised the
standard of revolt, his grievance being that the Sultan had interfered with
his management of the revenue. When the latter was surprised at
Mughan he had written to the Sultan of Rum and the Prince of Syria,
offering, if they would make over Azerbaijan and Arran to him, to do

---

[*] St. Martin Memoires, ii. 123 and 272. Note 31.
[†] Brosset, Hist. de la Geor., i. 511. [‡] D'Ohsson, ii 103-104. [§] Op. cit., 1116. [||] Op. cit., 511.

homage for those provinces, and to have the khutbeh there said in their names. In his letters he had referred to Jelal ud din as "the fallen tyrant." He now endeavoured to tamper with the various Khuarezmian officers, and to induce the Turkoman chief already named to keep a firm hold on the harem and treasures in his care. Jelal ud din, convinced of his treachery, issued orders that he was no longer to be obeyed. This is the account given by Nissavi. Novairi reports that the Vizier's discontent was due to the extreme prodigality and extravagance of his master, which also alienated from him the goodwill of some of his generals.* Having passed the winter of 1231 in the plain of Mahan, he heard the Mongols had left Aujan to search him out, and set out for Azerbaijan. On passing the fortress where the Vizier was living, he summoned him to his presence, professing to be ignorant of his treachery. He appeared with a cord about his neck, and Jelal ud din did him the unusual honour of offering him a cup of wine, it not being usual for the Khuarezmian sovereigns to feed with their viziers. But this was only an apparent civility, and he was really deprived of all authority. Meanwhile revolts broke out in various parts of Azerbaijan and Arran, where the people presented the heads of the Khuarezmian officers as a peace-offering to the Mongols. Nissavi succeeded in collecting a considerable contingent in Arran, whereupon the Mongols again retired to Aujan. They were speedily busy again, however, and a Mussulman officer in the service of Taimaz, one of their generals, was sent to summon Bailekan. Taken before Jelal ud din, who promised him his life if he reported truly the strength of the enemy, he said that when reviewed at Bukhara, Charmaghan's army numbered 20,000 fighting men. The Sultan ordered the man to be killed, for fear his troops should be disheartened by the statement.† Meanwhile he repaired to Jarapert, near Kantzag, in the mountains of Artsakh, where he issued orders for the arrest and execution of the Vizier. On seeing the guards who were commissioned to put him to death, he asked for a few minutes' respite. Then, having performed his ablutions, said his namaz, and read a piece of the Koran, he remarked on the fate of those who relied on the word of an ungrateful person. Being asked if he preferred to die by the sword or the rope, he chose the sword. "It is not usual to decapitate grandees," was the reply, and he was strangled.‡

Meanwhile, a revolt broke out at Kantzag, where the Khuarezmians were killed. Jelal ud din marched to the town, which, after a show of resistance, surrendered, and thirty of the principal malcontents were beheaded. The Sultan spent fifteen days at Kantzag, and, much against his inclination, determined to ask help from his recent foe, Ashraf, the Prince of Syria, who, hearing that his envoys were on the way on this errand, withdrew to Egypt, and sent them courteous but insincere letters

---

* D'Ohsson, iii. 50-51.   † Id., 52-54.   ‡ Id., 54-55.

to Damascus, offering to send help, but really meaning to stand aloof.
Jelal ud din then sent to Ashraf's brother, Mozaffer Gazi, who had been
appointed the Chief of Khelat by Ashraf, to go to him with the Princes
of Amid and Mardin.   Nissavi was chosen as his envoy, and he was
to promise to reward Mozaffer with a large accession of territory ;
but he did not expect much from these Turkish princes, whose policy
was generally limited to their own advancement.   Mozaffer said he
could do nothing without the consent of his brothers, the King of
Egypt and the Prince of Syria ; that his contingent would be so small
that it would be of little assistance to the Sultan ; that he could not
do homage to Jelal ud din without also doing it to Kai Kobad, the Seljuk
ruler of Rum ; and that the Princes of Amid and Mardin were not
subject to him.   Nissavi warned him that by standing neutral, he would
fail to share in the division of the spoil if Jelal ud din succeeded,
while if defeated, he would be at the mercy of the Mongols.   He merely
replied he was not his own master.   It seems they had written to the
Khalif and other princes, counselling them not to assist Jelal ud din.*

Meanwhile the Mongols continued their advance.   A letter, borne by a
pigeon from Perkri, announced that they had passed that town; and
Nissavi, on returning from his embassy, found only the harems and
baggage of the army at Hany, the Sultan himself having withdrawn to
Jebel Jor.   He had been joined by a Mongol officer, who had deserted on
account of some punishment he had undergone.   By his advice, Jelal ud
din abandoned his baggage and posted his men in ambush, so that he
could fall on the Mongols while they were pillaging.   Otuz Khan was
commanded to make a feint with 4,000 men, and to draw them on
into the ambush; but he was afraid, and returned with the misleading
message that they had abandoned the district of Manazguerd.   Jelal ud
din thereupon left his retreat and went to Hany, where, after an interview
with Nissavi, who reported the result of his fruitless mission, it was
determined to go to Ispahan.   While *en route* thither a messenger came
from Masud, Prince of Amid, who tried to persuade him to conquer
Rum, which he urged would be easy ; master of this, and secure of an
alliance with the Kipchaks, he might then make head against the Mongols.
Masud himself promised to join him with 4,000 horsemen.   This
suggestion was made out of revenge, Kai Kobad having conquered several
fortresses from him.   Jelal ud din approved of the plan; and went towards
Amid.   On the way he had been spending an evening in drinking, when
a Turkoman arrived and reported that he had seen some strange troops
at the place where the Sultan had passed the previous night.   Jelal ud din
declared this to be a lie, and a trick of the Prince of Amid, but he was
undeceived in the morning, when a body of Mongols surrounded his

---

* *Id.*, 57-58.

tent while he was in a drunken sleep. Their commander is called Baimas Noyan by Abulfaraj.* His general, Orkhan, charged them with a body of his men, while some of his officers rushed into his tent, put a small white tunic upon him, and seated him on horseback. He only thought of one of his wives, the daughter of the Prince of Fars, whom he ordered two of his officers to escort. He himself fled towards Amid with only 100 followers. Its gates he found closed against him, so he sped on to Mesopotamia. The Mongols were in pursuit, and by the advice of Otuz (called Uz Khan by Raverty) he determined to double upon them. He arrived at a village of Mayafarkin, and dismounted at a farm, intending to spend the night. Otuz Khan left him there, and at dawn he was again surrounded by them. He had barely time to mount, and most of his people were killed. The Mongols having heard from their prisoners that the Sultan was there sped after him to the number of fifteen. Two overtook him, but he killed them both, and the rest could not reach him. He then escaped to the mountains (one of the mountains of Sophane, says Abulfaraj), and was captured by some predatory Kurds. They stripped him, as was their wont, and we are told his saddle, girdle, and quiver were more than usually loaded with precious stones. They were going to kill him, when he disclosed himself to their chief, asked him to conduct him to Mozaffer, Prince of Erbil, who would reward him, or else to escort him to some part of his dominions, and promised to grant him the title of malik if he saw him safe. He therefore took him home with him, and left him with his wife while he went to look for his horses. Meanwhile a Kurd who came up asked who this Khuarezmian was, and why he was not put to death. She told him who he was, and said he was under the protection of her husband. The Kurd thereupon said, "Jelal ud din, at Khelat, killed my brother, who was a better man than himself," and he struck him dead with his javelin. This was on the 15th of August, 1231.

Thus perished the last of the Khuarezm-Shahs. His biographer describes him as brave to excess, calm, grave, and silent, laughing only at the tips of his lips. He spoke both Turkish and Persian.† He was of middle stature, with a Turkish face and a dark complexion, his mother having been a Hindoo. As D'Ohsson says, he was rather a brave and reckless Turkoman chief, than a skilled general or sovereign. Pillage, drinking, and music were not put aside, even in the presence of the Mongols. He did not know how to conciliate his troops, who being paid irregularly, had to eke out their income by rapine, which again increased his unpopularity. While at Tebriz there died a young eunuch slave to whom he was much attached; he had a magnificent funeral prepared for him, followed the corpse himself on foot, and ordered his troops

* Chron. Arab., 308 (? a corruption of Taimaz). † D'Ohsson, iii. 63.

to do the same. He was angry with the people of Tebriz because they did not show sufficient concern for the corpse, and ordered that when his meals were brought to him, some meat should also be taken to the body, while he had a slave put to death who ventured to tell him his favourite was dead.* He was, in short, a fickle, reckless, eastern Sybarite, with a great deal of courage and energy.

Some time after his death, Mozaffer, Prince of Erbil, sent for his bones, which were buried in a mausoleum, but the rumour arose (his death having been so obscure) that he was still alive, and it was reported that he had been seen in several places, especially in Persia. A person at Ispahan professed to be him, and the Mongols had him seized and examined by people who knew the Sultan, and then they put him to death. Twenty-two years after his disappearance, a poor man dressed as a fakir, in crossing the Jihun, told the boatmen : " I am the Sultan Jelal ud din, Khuarezm Shah, who it is said was killed by the Kurds in the Mountains of Amid. It was my squire who was thus killed, and I have travelled for many years without letting it be known." The Mongols seized him and put him to the torture, but to his last breath he continued to affirm the truth of his story.† Major Raverty reports a more circum-stantial tale. He says, Sheikh Ala ud Daulah al Byabanki of Simnan relates as follows :—"When at Baghdad, I used daily, at noon, to wait upon the pious and venerable Sheikh Nur ul Hak wa ud din, Abd ur Rahman i Isferaini. May his tomb be sanctified. I happened to go upon one occasion, at the usual hour, and found him absent from his abode, a rather unusual occurrence at that time of the day. I went again on the following morning, and inquired the cause of his absence on the previous day. He replied, ' My absence was caused through Sultan Jelal ud din Mangbarni having been received into the Almighty's mercy.' I inquired, ' What ! has he been living all this time ?' He answered, ' You may have noticed a certain aged man, with a mole upon his nose (mangbarni means with a mole on the nose), who was wont to stay at a certain place,' which he named. I had often remarked the venerable devotee in question. ' And that was the heroic but unfortunate Sultan Jelal ud din.'" According to this account, he could not have died till 688, i.e., about 60 years after the date above mentioned.† These stories are of course mere stories, and doubtless largely arose from the fact of his having a mole on his nose, a feature which would draw attention to others similarly endowed, and easily give rise to imposition.

Abulfaraj says that after attacking Erbil, the Mongols went to Nineveh, and laid siege to Khamalic (?), the citizens fleeing. Thereupon they burnt the churches. They placed two of their leaders at two of the city gates, one of whom gave life and liberty to those who passed him, while the other put

---

D'Ohsson, iii. 63-64.   † Abulfaraj, Chron. Arab., 309.   ‡ Tabakat-i-Nasiri, 299. Note.

the fugitives who endeavoured to escape by his gate to the sword. Thence they went to Shigra, and plundered and killed a great number of merchants on their way to Syria.[*]

Orkhan, after leaving the Sultan as I have mentioned, was joined by some troops, and reached Erbil with 4,000 men. Thence he went to Ispahan, which he captured, and which was, shortly after, again taken by the Mongols.[†] A large portion of Jelal ud din's men took service after his death with the Seljuks of Rum and the Syrian princes. Many others were waylaid and killed by the Kurds, Bedouins, &c. We have seen how, when Jelal ud din captured Khelat, he secured Thamtha, the daughter of Ivaneh, the Georgian Constable, whom he married. On his death she fell into the hands of the Mongols, who sent her, according to Guiragos, to Ogotai Khan, in Mongolia. Brosset suggests that she was really sent to Batu Khan. She lived several years in Tartary.[‡]

On the death of Jelal ud din the Mongols proceeded to ravage the districts of Amid, Erzerum, and Mayafarkin.[§] After a siege of five days they captured Sared, two days' journey east of Mardin, and put its inhabitants, to the number of 15,000, to the sword. Tanza, and Mardin itself, except the citadel, suffered a similar fate. The district of Nisibin, save its capital, was ravaged. The Mongols then entered Sinjar, and laid waste Al Khabur and A'raban. Another division of them went towards Mosul, and pillaged the town of Al Munassa, situated between it and Nisibin. Its citizens, as well as the peasants from the country round, had taken refuge in a *khan* in the middle of the town, where they were all slaughtered. A native of the place, who secreted himself, told Ibn al Athir, the historian, that when they killed anyone they shouted "La illahi," and their cruelties were accompanied with laughter and merrymaking.[||]

Another division marched upon Bidlis, whose people escaped, partly to their citadel and partly to the mountains. The town was burnt. The strong fortress of Balri, in the district of Khelat, was now captured, and all who were found in it were killed. The same thing happened at the large town of Argish.

A third body attacked Meragha, which submitted on condition of its people being spared, but a great number perished. Azerbaijan was laid waste, and then Erbil, where the Ivanian Turkomans, the Kurds, and Cheburkans were trampled upon, and where terrible atrocities were committed. Mozaffer ud din, Prince of Erbil, collected his men, and received aid from the Prince of Mosul, whereupon the prudent invaders drew off and went towards Dakuka. Within two months after the disappearance of Jelal ud din, Diarbekr, Mesopotamia, Erbil, and Khelat

---

* Op. cit., Chron. Syr., 513.
† D'Ohsson, iii. 65-66.   ‡ Brosset, Hist. de la Géorgie, 505-506. Notes.
§ D'Ohsson, iii. 67. || Id., 68.

were desolated, without encountering any resistance. The rulers of these small districts hid away in their fastnesses, while the people were stupefied. I have related some anecdotes reported by Ibn al Athir, showing the fatuous conduct of the inhabitants.*

It was now three months since Jelal ud din had been seen, and it was unknown whether he was dead or merely hiding away. The Mongols meanwhile were in the heart of Azerbaijan. Tebriz was summoned, and offered a ransom of silver, of rich stuffs, &c., and of wine. The kadhi and mayor went to their camp, and the town agreed also to send a number of artisans. Persian artisans were a most welcome present to the Great Khakan at Karakorum, who was a patron of the arts. They also sent him a splendid tent, and agreed to pay an annual tribute.†
Meanwhile the Khalif mustered his supporters to the rescue, while Khamil, the Egyptian sultan, marched from Cairo with a considerable army into Syria. He passed Damascus, and went towards the Euphrates, losing many men between Salamiyat, north-east of Hims, and that river. Having learnt at Harran that the Mongols had evacuated Khelat, he went towards Amid, then ruled by Masud, of the Ortokid stock, the capture of which, and not the defeat of the Mongols, was apparently the main object of his march. He was accompanied by his brother Ashraf, by the various Ayubit princes, and by the Sultan of Rum. The siege lasted but five days, when the voluptuous Masud surrendered the place, which was made over to Khamil's son, Salih, while Masud received an appanage in Egypt. Khamil also attacked Hosn-Keifa, which was the term of his expedition. These events took place in 1232.‡ Meanwhile the Mongols proceeded systematically to ravage Azerbaijan, Dilem, and the other western provinces which had been subject to the Khuarezm Shah. They made the fertile plain of Mughan their winter quarters, and thence sent out expeditions in various directions.§

In the year 1233 they laid siege to Kantzag, called Gandja or Guenja by the Persians, the Jelizavetpol of the Russians, the capital of Arran. Guiragos tells us the greater part of its inhabitants were Persians, but that there were a few Christians there, who were subjected to constant insult and contumely, and quotes as an example that crosses were put on the ground at the gates so that they might be trodden under. Its destruction was presaged by some unusual phenomena. The earth opened and vomited out a torrent of black water. A very tall cypress outside the town was seen to stoop down and then become erect again. This happened three or four times, after which the tree fell down altogether. The Mongols assailed the place with their battering engines, destroyed the vines in the environs, and eventually breached the walls. As they delayed the assault,

---

* Ante, vol. i. 131-132.
† D'Ohsson, iii. 70-71.   ‡ D'Ohsson, iii., 72.   § Journ. Asiat., 5th ser., xi., 213.

the inhabitants set fire to their houses and property. This greatly exasperated the invaders, who rushed in, sword in hand, and made a general
massacre of men, women, and children. Only a body of troops, which
cut its way through, and some who were reduced to servitude, escaped.
The Mongols spent some days in digging among the ruins for treasure,
and then withdrew. Fugitives afterwards returned to look for hidden
furniture, &c., and many objects in gold and silver, bronze and iron, were
thus recovered.

Kantzag remained in ruins for four years, when the Mongols ordered it to
be rebuilt.* Meanwhile they made another attack upon Erbil, which they
captured, with a great booty. The citizens withdrew to the citadel, where,
although many perished from want of water, they successfully resisted the
attack, and the Mongols at length withdrew, after receiving a sum of
money. They overran the northern part of Irak Arab, as far as Zenk
Abad and Surmenrai. This district belonged to the Khalif who put
Baghdad in a state of defence. He also put it to the Ulemas
which was more meritorious, a pilgrimage to Mekka, or war against
the infidels. They unanimously replied the latter, whereupon a holy
war was preached. The grandees and expounders of the law joined
in the exercises of the troops. They marched out and inflicted a defeat
on the Mongols at Jebel Hamrin (i.e., the Red Mountain), on the Tigris,
near Takrit, and released the prisoners who had been carried off from
Erbil and Dakuka.† Another body of 15,000 invaders, who had advanced
as far as Jaferiya, now withdrew. A similar division had a more
fortunate engagement at Khanekin. Near Holvan they encountered
7,000 troops of the Khalif, under the orders of Jemal ud din Beilik,
drew them into an ambush, and killed them nearly all, including their
commander. To revert to their operations further north. We find Charmaghan now setting out from Mughan, and methodically overwhelming
Arran and Great Armenia, which were distributed among his chiefs or
noyans, who, we are told, proceeded to take possession of the portions
thus assigned them, accompanied by their wives, children, and baggage,
and consumed all the herbage in the fields with their camels and flocks.

When the Mongols invaded Armenia, that province was assigned as an
appanage to Arslan Noyan. Elikum, the chief of the once powerful family of
the Orpelians, fortified himself in the impregnable fortress of Hrashkaperd.
Seeing he could not capture it by force, Arslan sent a messenger to Elikum,
to tell him he was irrevocably settled in Armenia, and that it would be
better for him to come down from his fastness, where he would starve,
and make friends with him. Elikum received these overtures favourably,
and having exacted an oath from Arslan, went to visit him, with great
presents. The latter treated him well, and numbered him among his

* Guiragos, ed. Brosset, 116-117.   Journ. Asiat., 5th ser., xi. 213-216; xvi. 282,
† D'Ohsson, iii. 73-74. Ilkhans, i. 110.

generals. He then advanced to Ani, which he conquered, as well as the country of Vaio Tzor and Eghégik, as far as the town of Eréron, opposite Garhni, all of which he gave to Elikum. He told him what he conquered by the sword was as much his property as what he bought with money, and he freely gave it to him, on condition that he should be faithful to him and the Grand Khan. Thenceforward Elikum was a good friend to the Mongols. He took part in the siege of Mayafarkin, where he fell ill and died. It was reported he had been poisoned, by order of the Georgian Prince Avak. He was succeeded by his brother Sempad, of whom we shall have more to say.* Let us now turn to Georgia.

At the time of the Mongol invasion, Georgia was in every way the most powerful kingdom subject to the Christians. Defended by its mountains, says Remusat, its line of rulers had never been interrupted. The generals of the Khalifs had only made momentary raids, or gained a very precarious footing there. The Seljuki Turks had laid their hands more heavily upon it, but at the end of the eleventh and beginning of the twelfth century, David, surnamed the Restorer, took advantage of the disunion among the Turkish princes, recaptured Tiflis, and drove the Turks beyond the Araxes. His successors followed in his steps, and numbered among their vassals all the Armenian princes north of the Araxes, whom they rescued from the Mussulman yoke. The family of the famous Ivaneh, Constable of Georgia, which ruled in the greater part of the country from the Araxes to the Kur, the Princes of Shamkor and Khachen, &c., recognised the suzerainty of the Georgian kings, who at the beginning of the thirteenth century dominated from the Black Sea between Trebizond and the possessions of the Krim Tartars as far as Derbend and the junction of the Kur and the Araxes, i.e., over Colchis, Mingrelia, the land of the Abkhazes, Georgia, properly so called, and Northern Armenia, with many small adjacent districts.† George Lasha, King of Georgia, died January 18th, 1223.‡ He was succeeded by his sister Rusudan, famous for her beauty and her peccadilloes. Her subjects became noted for their debaucheries, and she gave herself up to pleasure.§ The country was virtually ruled by the Constable Ivaneh. Rusudan married the Mussulman Prince Mogit ud din Tughril Shah, son of Kilij Arslan, the Seljuk Prince of Erzerum, who was a handsome person, and by whom she had a daughter, Thamar, and a son, David. This marriage took place, according to Wakhoucht, in 1228 A.D.‖ She was very unfaithful to her husband, who on one occasion surprised her in bed in the arms of a Mamluk, and duly imprisoned her. Having later heard of the beauty of two Alans, she sent for and eventually married one of them. She also fell in love with a Mussulman of Kantzag, whom she could not, however, persuade to abjure his faith.¶ Rusudan's

---

* Hist. de la Siounie, 227-228. St. Martin, ii. 123-127. † A. Remusat, Mems. French Acad., vi. 400
‡ Brosset, Hist. de la Géorgie, i. 496. Note. § Id., 496.
‖ Brosset, Hist. de la Géorgie, i. 501. Note 3. ¶ Abulfeda, sub. ann. 620, i.e., 1223.

daughter, who was also a great beauty, was named Thamar. She attracted the attention of Ghiath ud din Kai Khosru, the second son of the Sultan of Rum, who, although a Mussulman, was readily accepted by the diplomatic Queen as a suitable partner for her daughter. Ghiath ud din promised not to interfere with her religion. This marriage probably took place about 634 HEJ., *i.e.*, 1236 or 1237. She received Atskur as an appanage, and was accompanied by her cousin David, son of Lasha, who acted as her *paranymph!!!* He was a dangerous aspirant for the Georgian throne, and at the instance of his aunt Rusudan was imprisoned by Thamar. She shortly after became a Mussulman, and, according to Abulfaraj, became the mother of Alai ud din Kaikobad, who had a separate appanage, and whose name appears on the coins with those of his half-brothers, Iz ud din and Rokn ud din.

David, son of Lasha, was the next heir to the throne. Rusudan was exceedingly jealous of the young prince, and according to the *Georgian Chronicle* she sent more than one message to Thamar and her husband urging that they should put the young man away. As this was not carried out she became very irritated, and even had the wickedness to write to Ghiath ud din to suggest to him that his wife, her own daughter, was carrying on an intrigue with her nephew.* Ghiath ud din, on hearing this calumny, began to treat his wife very badly, dragged her by the hair, kicked her till she was blue, broke the sacred images, &c., before which she said her prayers, and threatened her with death unless she abjured her faith, which she was constrained to do.† This statement of the *Georgian Chronicle* is confirmed by Abulfaraj. The former goes on to say that Ghiath ud din, having ill-used the young Prince David, ordered the captain of a ship to take him out to sea, and when he had got him fairly away from the land to throw him into the water. They accordingly set out for Pelagon (*i.e.*, for the Ægean). He was duly thrown out, but was given a plank by a benevolent sailor. With the assistance of this he made his way towards the land, whence he was seen by a traveller, who sent a good swimmer to his rescue. He then took him home, provided for him, and kept him for six months. All this having come to the ears of Ghiath ud din, he was greatly enraged. He ordered the young prince to be thrown into a dark pit, tenanted by reptiles and vermin, whose mouth was closed by a stone. One of his father's dependents, Sosna, "a Rowth" (Brosset suggests a Russian) by nation, dug a hole secretly at night, by which he passed victuals into the pit, and thus fed him for five or, according to another paragraph, seven years. He used to pass down two bags by cords to him, one containing bread and the other water. Our author, in reporting the saga, makes out that the serpents in the hole did him no harm, he being preserved like Daniel in the lions' den. One

* Hist. de la Georgie, 524. † *Id.*

of them having bitten him, in consequence of his having leaned on it heavily, was consumed by the rest.* We shall revert to the distinguished prisoner, and meanwhile turn again to Georgia.

Its beautiful and amorous queen was dominated by a crowd of courtiers. Her most trusty counsellors were the Generalissimo Ivaneh and his son Avak, Shahanshah son of Zak'aria, Vahram, and others. Georgia was not in a position to resist the Mongols, having been so terribly ill-used by Jelal ud din, Khuarezm Shah, as we have shown. When she heard of their approach, therefore, she quitted Tiflis and went to Kuthathis, leaving Goj, son of Mukha, in charge of the capital, with orders, if the enemy should appear, to set fire to Tiflis, except the palace and the quarter called Isanni, so that they could find no shelter there. When Goj heard of their approach he fired the place, not even sparing the palace and the Isanni.† Chamitch tells us Rusudan took refuge at the fortress of Usaneth, but Brosset suggests that this was too dangerous a locality for a place of refuge, and argues that she retired to the district of Suaneth.

Meanwhile the various chieftains withdrew, and each one sought safety in some retired place. Guiragos compares the swarms of Mongols who overran the country to flights of locusts and drops of rain. Fear and decrepitude overcame the people. " He who had a sword hid it, for fear that if found upon him he might be pitilessly killed ; children were broken to death upon the stones, and young maidens cruelly ravished. The Tartars had a hideous aspect, and bowels without pity ; they were insensible to mothers' tears, or to the white hairs of age, and they sped to carnage as to a wedding or an orgy. Everywhere were unburied corpses, the services of the church ceased, while the people preferred the night to the day. The avarice of the invaders was insatiable, and what they could not carry away they destroyed. Having wasted the open country they attacked the towns. As their campaign was undertaken in the summer, and without warning, the latter were speedily reduced by want of water, and their inhabitants were duly slaughtered or reduced to slavery." ‡ The district of Shamkor belonged to Vahram (i.e., Vahram Gagel) and his son, Akbuka, who had captured it from the Persians. It now fell to Molar Noyan. Setting out from Mughan, he sent on an advanced guard of 100 men, and forbade the inhabitants to pass in or out of its gates. They sent for aid to Vahram, and informed him of the small number of the invaders, but he would not move. When Molar himself and the main army arrived, he had the ditch filled with fascines ; these were, however, burnt by the citizens. He then ordered each man to carry a load of earth in his robes and to throw it into the ditch, which was speedily filled up. The Mongols stormed the

---

' *Id.*, 526-527.   † Hist. de la Georgie, i. 514.   ‡ Op. cit., Journ. Asiat., 5th ser., xi. 216-218.
‡ *Id.*, 219-221.

place, massacred the inhabitants, and burnt the houses. They then invested the remaining fortresses belonging to Vahram, Derunagan or Terunakan, Erkevank, and Madznaperd, all situated near Shamkor, in the district of Kartman, in Armenian Albania. The last town belonged to Kyrikeh the Fourth, of the dynasty of the Bagratids of Dashir. They also captured Kartman, in the district of Udi. Meanwhile, another Mongol chief, named Ghataghan Noyan, conquered Charek and Kedapag, or Getabac.* Vartan says he conquered the four cantons of Kedabag and Vartanashad.† Vahram, who was at Kartman, fled and escaped. Having imposed a tribute upon them, the Mongols withdrew. The army which had taken Shamkor also subdued Tavush, Kadzareth, Norpert, Kak or Gag, &c.‡ At this time, the great Vartabied, or doctor, Vanakan, had made himself a retreat with his own hands on the summit of a high rock, opposite the village of Olorut, south of Tavush, where he had sought refuge when Jelal ud din destroyed his monastery at Erkevank. There he lived with a crowd of disciples and a fine library, and there he had built a church and some cells. When Molar Noyan arrived, a crowd of men, women, and children sought refuge in his cavern, where they were blockaded by the Mongols, and presently food and water ran short, while the terrible heat made the place most unhealthy. The Mongols cried out to them to come down from their vantage and surrender, and that they would be well treated. They begged the Vartabied to go and conciliate the terrible invaders. He accordingly went down with his two disciples Mark and Sosthenes, and found the Mongol chief on a height opposite the cave, with an umbrella held over his head, as it was fiercely hot. They were ordered by the guards to bend the knee three times, "in the fashion which camels do," and when they were admitted, they were bidden to prostrate themselves towards the East, that is, towards the great Khakan. Molar addressed the white-bearded doctor, and asked him why he had not gone to offer his submission, as he had ordered that he and his people were to be well treated. He replied that they were unaware of his good intentions, for they did not understand his language, and that no one had in fact gone to acquaint them with his wishes. When they knew them they had complied. "We are neither soldiers nor rich people," he said, "but strangers and pilgrims collected from various places to study religion together. Do with us as you will." Molar bade them be seated and at ease. He inquired about Vahram's whereabouts, and about his various fortresses. He then ordered the rest of the refugees to come down, and promised them safety under chiefs he would appoint over them. Guiragos, who tells the story, was among those who now went down. They felt, he says, like sheep going among wolves. Each one, expecting to be killed, repeated his

---

* Guiragos, ed Brosset, 120.   Journ. Asiat., 5th ser., xi. 221.
† Journ. Asiat., ser. v., xvi. 287.   ‡ Id., xi. 222.   Guiragos, ed Brosset, 120.

profession of faith in the Holy Trinity, and before leaving the cave they all partook of the Sacrament. The Mongols, however, treated them fairly. They first gave them water to assuage their thirst, and then put them in custody of some guards. In the morning they stripped them of all that they could, and proceeded also to plunder the grotto and the church of its ornaments—copes, cups, silver candlesticks, and two gospels encrusted with silver. Having selected such of the men as they wished to transport elsewhere, they sent the rest to live in the neighbouring village and monastery, and set a person over them to protect them. Among those who had to go away was Guiragos, the Vartabied, and a young priest named Paul, the nephew of the latter. They were dragged over a rugged country without roads, on foot, and were escorted by Persians whose hands had been dipped in Christian blood, and who treated them insolently. They were hurried on, and any who lagged were beaten with rods. "There was no time to draw thorns out of the feet, or to drink by the wayside." When they halted, they were shut up in small houses, whence they were not allowed to go out, even to satisfy nature, and were closely guarded. Guiragos and some of his companions were employed as secretaries, to write letters for the invaders. He enlarges on the miseries of the way. At the approach of autumn, and as they neared the frontier of Armenia, individuals began, at all risks, to escape. Those who thus ventured all got away except two priests, who were re-caught and executed before Guiragos and his companions. The chronicler tells us his master offered them horseflesh to eat; for the Mongols ate all kinds of animals, pure and impure—even rats and serpents. The Vartabied replied they wanted no such food, but if he wished to do them a kindness he might let them return, as he had promised, for he was old and ill, and could be of no service to them either as a soldier or a herdsman. He said he would consult his major-domo Chuchughan, who was then absent on a plundering expedition. This man of the world insisted upon a ransom being found, and urged that the alms which went to buy repose for the dead might be reasonably used to ransom the living. The Vartabied declared they had been stripped of all their goods, and had nothing left, but if they were conducted to one of the neighbouring fortresses, the Christians there would ransom them. They were accordingly taken to Kak, or Gag. There the Vartabied was ransomed, but they refused to let Guiragos go, as they said they needed him to write their letters. Guiragos says, there was at Kak a famous cross, which performed miracles, especially in favour of captives, and that those who invoked it faithfully saw the martyr, St. Sargis, himself open the prison doors. The Vartabied promised to go and invoke the saint in his behalf. He was ransomed for eighty dahegkans, fifty more, says Vartan, than what Judas sold the Saviour for. Molar Noyan, who evidently, like the other Mongol chiefs, valued a clever writer, consoled Guiragos for the loss of his old master. He promised to promote him over his own chiefs; if he already had a wife he would send for her—if not, he bade him choose one from among them; and he gave him a tent and two boys to wait on him, and promised to give him a horse on the following day. But as they passed the Monastery of Kedig, or Getic, in Eastern Armenia, where he had been brought up, and which had been sacked by the Mongols, he managed to escape.*

Turning elsewhere, we find that the district of Khachen was also ravaged at this time. Its strongholds fell by force or stratagem. A great many of the people who had sought safety in difficult retreats were duly followed there and put to the sword, thrown down precipices, and their bones whitened the ground for a long time after. The Mongols also marched against Hassan, styled Jelal, son of Vakhtang, Prince of Khachen, and of the sister of the Constables Zakaria and Ivaneh. He was pious and charitable, had the virtues of an anchorite, and was a faithful attendant at the church's services, and a scholar. After the death of his wife, Vakhtang's mother brought up his three sons, Jelal, Zakaria, and Ivaneh. She eventually went as a pilgrim to Jerusalem, and died there. When the Mongols drew near, Jelal assembled his people in the fortress of Khoiakhan, or Khokan (called Khokhanaberd in Persian), in the province of Artsakh. When summoned, he went to their camp, with rich presents. The Mongol chief to whom he submitted, was Chola,, or Jola, brother of Charmaghan. He was well treated, and not only restored to his principality, but it was increased in size. He was ordered to join the Mongols every year in their campaigns, and to be faithful and obedient. By his prudence and conciliation, and by adapting himself to the insatiable habits of the invaders, and meeting their greed with continual presents, he secured an immunity from their attacks, which was most exceptional.† His daughter Rusudan was married by Chola to Bugha, son of his brother Charmaghan.‡

In another direction, another subordinate of Charmaghan, called Jagatai, marched upon Lôrhé, capital of the district of Tashir, in the province of Kukark, the treasure city of Shahan Shah. The latter, on the approach of the Mongols, withdrew with his family, and took shelter in the caverns in the neighbouring valley, and committed the defence of Lôrhé to his father-in-law. His people were effeminate persons, and gave themselves up to dissipation, and, in the words of Guiragos, trusted to the strength of their walls rather than in God. The Mongols undermined the ramparts, which fell down, and they then entered the place, and as usual with them, commenced an indiscriminate slaughter. They discovered Shahan Shah's treasures, which he had amassed by the oppression of his people, and which he had concealed in a chamber with a very

* Guiragos, ed. Brosset, 120-124. Journ. Asiat., 5th ser., xi. 222-231.
† Id., 245-248. Guiragos, ed. Brosset, 131-133.
‡ Hist. de la Géorgie, i. 514, Note 4 and Additions, 346.

small entrance, which, like a child's money-box, enabled things to be put in, but not easily taken out again. His father-in-law was put to death. The remaining fortresses of the district were then captured either by force or craft, and were similarly devastated. These included Dmanis and Shamshuildé, in the province of Kukark, and Tiflis, the metropolis.* Avak, son of the Constable Ivaneh, seeing the country overwhelmed by this flood of enemies, sought shelter in the very strong fort of Gaian, or Kaian, in the district of Tzorophor, in the province of Kukark, to which the inhabitants of the surrounding district also fled. One of the Mongol chiefs, named Tughata Noyan, with a force of Mongols, beleaguered him there, built a wall of circumvallation round the foot of the fortress, and sent several messages to Avak, offering him terms if he would acknowledge his supremacy. He offered him his daughter Khochak and some of his riches in the hope of thus buying him off. The Mongols accepted these, but insisted more strongly upon his going to them in person. Water began to run short, and the crafty besiegers allowed many of the people who had sought refuge to pass through their lines in safety to water their horses at the river (i.e., the Débéda, the Kamenka of the Russians, on whose left bank Lôrhé was situated); they would not, however, let them return, but told them to summon their families out. They thus planted their foot upon them and despoiled them, taking such of their women as suited them and killing such of them as they disliked. At length Avak, finding that their attack continued, and also their cruelties, determined to surrender, and thus buy a respite for his people. He accordingly sent Gregory, familiarly called Tgha, or the infant, his major-domo (according to A. Remusat, his nephew; and to Brosset, his cousin), to Charmaghan, who was then encamped on the Keghakuni, otherwise called by the Armenians the Lake of Kegh'am and the Lake of Sévan, and now known to the Turks as the Blue Sea, and to the Persians as the Beautiful Sea. The *Georgian Chronicle* says the Mongol leaders were at this time in their winter quarters at Berdaa, their summer ones being in the mountains of Gélakun, and near the Araxes.† Charmaghan was immensely pleased at this embassy. Avak's envoy promised, on his behalf, that he would faithfully serve the Mongols, and pay them the kharaj, or feudal dues, for his domains. He also asked them to swear solemnly that he should be safe if he went to them. This they agreed to do. Their religion, says our chronicler, was to adore the only God, and to make three genuflexions to him daily, at sunrise, towards the east. In swearing an oath, they dipped a piece of gold in water, which they afterwards drank. This kind of oath, we are told, was never broken, and they told no lies. They gave Avak's messenger a golden tablet or paizah, guaranteeing him a safe conduct. On Avak's arrival Charmaghan

rebuked him for not having at once submitted, and quoted the proverb,
"I went to the window, but you did not come. I then went to the door,
and you hastened to me." He caused him to be seated below his
grandees, and gave a grand feast in his honour, in which the flesh of pure
and impure animals, quartered and roasted, was served up, while kumis
was liberally served out of skins. Avak would not eat or drink, saying that
Christians only ate clean animals, which had been properly killed, and
drank only wine. These were furnished him. On succeeding days, his
seat at table was promoted, until he was seated among the principal
Mongol officers, while, out of consideration for him, a number of his
people who had been made captive were released, and his former
appanage was restored to him, and even enlarged.* The *Georgian
Chronicle* says that Mongol commissaries were placed in his towns.

Charmaghan, accompanied by Vahram and Avak, now marched
against Ani, the ancient capital of Armenia, which was fortified, had a
strong garrison, and was well provisioned. It was so full of churches
that it was usual to swear by the thousand-and-one churches of Ani. It
was subject to Shahan Shah. The envoys sent by the Mongols, calling
upon it to surrender, were murdered by the citizens. This was speedily
revenged. The town was attacked with vigour, and numerous war engines
were planted around it. It was soon captured. Some of the principal
citizens, who had probably been traitors, were spared: the rest of the
people were ordered to go out of the town in the method practised by
Jingis. They were then divided among the troops in squads and massacred.
Only a few women, children, and artisans were spared, and reduced to
slavery. The town was now sacked, its churches pillaged, and its monuments defaced. Guiragos describes in lurid colours the horrible sight,
the ravishing of chaste nuns, the slaughter of helpless priests, &c. One
of his phrases is grim. "Delicate bodies," he says, "accustomed to be
washed with soap, were lying about damp and livid."† The devastation
must have been dreadful. In a work published at Venice in 1830, entitled
"Patmutiun Anuoi," and written by the Father Minas Bjechkian, we
are told that some of those who escaped on this occasion, found shelter at
Kaffa and Trebizond, where their posterity still remain ; a larger number
went to Astrakhan and Ak Serai. These, in 1299, being hard pressed by
the Tartar Khan, sent to the Genoese, at Kaffa, to ask for an asylum. They
then traversed the country of the Tartars with arms in their hands, and
settled in the Krim. They multiplied so much, that they eventually had
100,000 houses and 1,001 churches about Kaffa, as they had had about Ani.‡
When the people of Kars saw what had befallen Ani, they hastened to
give up the keys of their town. But the Mongols, whose appetite for

 * Guiragos, ed. Brosset, 126-127.   Journ. Asiat., 5th ser., xi. 233-236.   Hist. de la Géorgie, i. 516.
       † Guiragos, ed. Brosset, 127-128.   Journ. Asiat., 5th ser., xi. 237-238.
       ‡ Brosset, in Lebeau Histoire du Bas Empire, xvii. 456.   Note.

plunder had been whetted, did not in consequence spare them, but pillaged Kars as they had done Ani, appropriated its riches, and carried off its population into captivity. They withdrew from it, leaving a few humble people in possession, who were afterwards exterminated or carried off by the Turks of Asia Minor. The Mongols who captured Kars, also took the town of Surp Mari, or Surmari, situated on the Araxes, south of Echmiadzin. It had only a few years before been captured from the Muhamedans by Shahan Shah and Avak. The contingent which now took it was commanded by Kara Baghatur.*

When they had completed their conquest of the country, they issued orders for the fugitive inhabitants to return to their villages and homes, and to rebuild and re-occupy them under their new masters. Their campaign in these parts was undertaken in the summer, when the crops were not all gathered, and they trod a great deal under foot with their horses and cattle. The subsequent winter proved to be mild, and although there was no possibility of sowing fresh crops, or of tilling the ground, it produced a scanty crop nevertheless, while succour was afforded by the Georgians, whose general conduct, however, towards the Armenian fugitives who sought refuge among them may be gathered from the epithet of "the pitiless nation of the Georgians," applied to them by Guiragos.

Shortly after this, Avak was dispatched to visit the great Khakan Ogotai. He was accompanied by the prayers of his people, who hoped he would obtain a surcease of their terrible sufferings. He duly arrived at the Court and presented the letters of the Mongol chiefs, disclosing the object of his journey, which was to offer his submission. Ogotai received him well, gave him a Mongol wife, and sent him home again. He also ordered his generals to reinstate Avak in his dominions, and with his aid to reduce those who continued to resist. These orders they carried out, and secured the submission of Shahan Shah, son of Zakaria, of the Prince Vahram and his son, Ak Buka, of Hasan, surnamed Jelal, Prince of Khachen, and of many others.

In the *Georgian Chronicle* we read how, when Shahan Shah saw the security which Avak had brought his people by submission, he sent to tell Avak that if he counselled it, he would also submit. The Mongols were very pleased with this, and conferred a golden tablet on him, and also made over Ani and all its dependencies to him. The Georgians who submitted were well treated, while those who were obstinate were trodden under. Meanwhile, however, says our author, Hereth, Kakheth, Somkheth, Karthli, and all the country towards Karnukalak, was cruelly devastated, and the inhabitants slaughtered or reduced to slavery. Tiflis was also captured. In winter, the Mongols encamped at Berdaa, on the banks of the Mtsuar, towards Gag. They pillaged all Karthli, Samtzkhé, and

* Guiragos, op. cit., 128-129. Journ. Asiat., 5th ser., xi. 238-239.

Jawakheth, and as far as Greece (*i.e.*, Rum), Hereth and Kakheth as far as Derbend. Overwhelmed by these disasters, the Georgian mthawars submitted. Among these were those of Hereth, Kakheth, Karthli, Gamresel of Thor, and Sargis of Thmogvi, a wise philosopher, endowed with great gifts. The Georgian Queen, Rusudan, had taken refuge in the mountains. To bring her to her knees, we are told that Jagatai Noyan made a cruel raid upon the province of Samtzkhé. The Meskhes in terror fled to their fortresses, and a great number of the people were captured or killed. Ivaneh, son of the commander of Tzikhis-Juar, also named Kuarkuareh, asked the Queen's permission to be allowed to submit, so as to save Samtzkhé from utter ruin. He had the title of the Chief of the Armourers, and was the mthawar of the province. The Queen having consented, he went to Chaghatai, or Jagatai, who received him well and placed overseers in the province, which was thus spared.* At first, the Mongols allowed the princes who were submissive, as above described, to retain their authority in peace, but presently began to harass them by perquisitions, demands for military service, &c. Nevertheless, they did not put any of them to death. In the course of a few years, Avak also became the victim of their exactions, for they were most avaricious, and demanded not only meat and drink, but also horses and rich garments ; horses especially were their delight, and no one could keep one, or a mule, except secretly, for wherever they met with one they appropriated it. Each horse thus captured was marked with a hot iron with the tamgha, or private mark of the owner. Thus, if it strayed it was returned to its owner ; anyone keeping such a marked horse being punished as a thief. These exactions became more frequent after the death of the Mongol General Jagatai, who was assassinated, as we shall presently show. He was the friend of Avak, and when he died many of the other Mongols declared against the latter. One day, one of these chiefs of inferior rank, named Joj-Buka, having entered the room where Avak was seated, and the latter not having risen to greet him, he struck him on the head with his whip. The attendants would have fallen upon the intruder, but were restrained by Avak. After this outrage, he collected his men, with the intention of assassinating Avak in the night ; but the latter fled, and sought refuge with the Queen of Georgia, who he thought was at issue with the invaders. When Avak fled to Rusudan, the Mongols affected to be distressed, sent to ask him to return, and blamed those who had caused his withdrawal. His principality they made over to Shahan Shah as to a brother. Meanwhile, Avak wrote to Ogotai, to tell him he had only fled to escape ill-usage, and was always at his service While he awaited the Khakan's reply, the Mongols made a search for his treasures, which they found hidden in his fortresses. Afraid of the anger

* Brosset, Hist. de la Géorgie, 517.

of the Khakan, they sent message after message bidding him return. When he at length reached their camp, he was met by a messenger from Ogotai bearing letters and presents for him, and also orders that he was not to be molested, and that he might go wherever he pleased. He was then sent, with an officer named Tonghuz Aka, who had been specially deputed as commissary of taxes in Georgia, to invite the Georgian Queen Rusudan to submit. He acquitted himself well in his mission, and a treaty was agreed upon, by which the Queen and her infant son David, whom she caused to be crowned, were to be subject to the Mongols, while the latter were not to molest her.* The famous beauty was not inclined to be so submissive as her various nominal dependents. She wrote to the Pope, asking for the aid of a Christian army, with which to repel the Mongols, and professed a complete submission to the Roman Church. Gregory the Ninth, in his reply, congratulated her on the latter decision, but held out small consolation otherwise. He perhaps doubted her sincerity, and we are in fact assured by Bar Hebræus that she renounced Christianity and became a Moslem.†

Malakia has a curious story, which is not reported, so far as I know, elsewhere. He tells us that the three leaders of the Tartars at this time were Chorman (i.e., Charmaghan), Benal, and Molar Noyan. One evening, at a kuriltai, where it was resolved to make a fresh invasion of the west and a fresh massacre, the three were not of one mind. Charmaghan, who was of a more humane disposition than the other two, urged that by the order and with the help of God, they had ravaged the land sufficiently, and that it was better that the population which remained should take one-half of the produce of its labour for its sustenance, and pay over the other half to them. Night coming on, the kuriltai came to an end, and each retired to rest. When morning broke, two of the chiefs were found dead, and Charmaghan alone remained. He set out with witnesses for the Court of Chankzghan (i.e., Jingis Khan, but really of Ogotai), to whom he related what had passed. The Khan was astonished, and declared that the death of the chiefs was good proof that their course was not grateful to God, while his was, and that the will of God was, that in conquering the earth they should cherish and protect it—people it—and impose their laws upon it; and also the four taxes, tghghu, mal, thaghar, and ghphtchur. Those who would not obey or pay these taxes ought to be killed and to have their lands devastated, while the others should be spared. Chorman was sent back, and the Khakan, we are told, gave him one of his own wives, named Ailthana Khatun, in marriage. He accordingly returned, and settled on the plain of Mughan.‡

---

* Guiragos, ed. Brosset, 130-131. Journ. Asiat., 5th ser., xi. 240-245.
† A. Remusat, Mems. de l' Institute, vi. 406-407.
‡ Malakia, Brosset, Hist. de la Géorgie, Add., &c., 444-445.

Guiragos tells us that at this time a Syrian doctor, named Simeon, who was styled Rabban Athor (a mixed title, rabban meaning doctor in Syrian, and athor meaning father in Mongol), gained great influence over Ogotai. He asked the great Khakan to issue an order exempting the innocent people who did not resist the Mongol arms from massacre. Ogotai assented to this, and sent him westwards, amidst great pomp, and bearing a note for the Mongol commander, ordering him in these matters to conform to the wishes of the Syrian doctor. On his return, he greatly eased the condition of the Christians. He built Christian churches in the *Mussulman towns, where hitherto no one dared pronounce the name of Christ*, notably at Tebriz, and at Nakhchivan. In these two towns their condition had been particularly humiliating, and they dared not show themselves even. He built churches and raised crosses there, while the jamahar (*i.e.*, the substitute for a bell, consisting of a sonorous piece of wood, which was struck by another), was heard by night as well as by day. Christian funerals, accompanied by the cross and gospel, and the surroundings of the liturgy, openly paraded the streets. All who opposed were liable to be put to death. The Mongol troops treated him with great deference, while his tamgha, or seal, attached to a document, was a free passport for his compatriots. No one dared touch those who invoked his name, and the Mongol generals gave him a portion of the booty they captured. He was modest and temperate, and only took a little food in the evening. He baptised numbers of the Mongols.*

Guiragos condenses in a few graphic phrases some of the chief characteristics of the invaders, whom he knew so intimately. He describes them as having horrible and repulsive countenances, and as being (except in the case of a few who had a little) without beards. On the upper lip and chin were a few hairs, which might be counted. They had small, piercing eyes, and a shrill, piercing voice. They were long lived. So long as they had abundant food, they ate and drank gluttonously, and when this was scarce, they as easily supported hunger. They fed on the flesh of all kinds of animals, pure and impure, but preferred that of the horse. They cut the animals into quarters, and then boiled or roasted them without salt. They then cut them into small pieces, and having dipped them in salt water, ate them. Some knelt while eating, like camels, while others sat down. Masters and servants had equal shares at their feasts. In drinking kumis or wine, a large vessel was produced, out of which a man took a portion in a cup, and threw some of it towards the sky and towards the four points of the compass. After the libation, having tasted, the cupbearer handed some of it to the principal chiefs, who, to prevent being poisoned, made the person who carried it taste any meat or drink he offered. They had as many wives as they pleased, and punished adultery

---

* Guiragos, ed. Brosset, 137-138. Journ. Asiat., 5th ser., xi. 253-254.

mercilessly with death. They punished theft in the same way. Guiragos says they had no religion and no religious ceremonies, although they had the name of God on their lips on all occasions. They often declared that their ruler was the equal of God, who had taken heaven himself, while he had given the earth to the Khakan, and to prove it declared that Jingis Khan had not been produced in the ordinary way, but that a ray of light, coming from some invisible place, had entered by the roof into the house of his mother, and had said, "Conceive, and thou shalt have a son who will be ruler of the world." This story Guiragos says had been told him by Gregory, son of Marzban and brother of Arslanbeg of Sargis and Amira, of the family of the Mamigonians, who had heard it from the lips of Khuthu Noyan, one of the principal Tartars, while he was teaching the young people. When a Tartar died, or was put to death, they carried his corpse about with them for several days, since they believed that a demon entered the body, and made a number of statements; they then burnt it Sometimes also they buried it in a deep grave, with its arms and apparel, and the gold and silver belonging to the deceased. If he was a chief, they also buried some of his male and female slaves, that they might wait on him, and also some horses, since they believed there were great fights in the other world. In order to perpetuate the memory of the deceased, they slit open the belly of his horse and took out all the flesh through the opening. They then burnt the bones and entrails, and afterwards sewed up the skin as if its body was whole, and thrust a pole through it, which came out of its mouth. This memorial they hung on a tree or in an elevated situation. Their women, he says, were magicians, and cast their incantations everywhere. It was only after a decision by their magicians that they undertook a march.*

We have now reached the term of Charmaghan's career, but before describing his end it will be well to sum up the result of his administration, and also to relate what took place in Khorasan and elsewhere during his term of office in Persia. The main results of Charmaghan's campaigns, were the thorough subjection and parcelling out among his fellows of Azerbaijan, Armenia, Irak-Ajem, and Arran, the last of which provinces, with its beautiful grassy plains, became the real head-quarters of the Mongols for a long time. Georgia, as we have seen, was severely punished, but retained, although in a dependent position, its own line of princes, whose history continued closely entwined with that of the conquerors. Kerman and Fars were spared devastation by timely submission. We have seen how the Hajib Borak obtained possession of the former. We are told by Juveni that he carried on a long struggle with Ghiath ud din, the Atabeg of Yezd.† He agreed to pay the Mongols an annual tribute, and received from the Khalif the title of

---

* Guiragos, ed. Brosset, 134-135. Journ. Asiat., 5th ser., xi. 248-250.   † Ilkhans, i. 66.

Kutlugh Sultan. When Tair Behadur, as we shall see, attacked Seistan, he sent orders to Borak to send him some troops, and to go and acknowledge the Khakan's supremacy. He replied that he would undertake to capture the place himself, and that the Mongols need not trouble themselves about it. As to visiting the Khakan, he was too old, but he would send his son, Rokn ud din, in his place. That young prince set out, and *en route* heard of his father's death, and that the throne of Kerman was now filled by his cousin.[*] Borak died in the year 632 HEJ. (*i.e.*, 1235), and was succeeded by his nephew, who was also his step-son and son-in-law, Kutb ud din Abul Fath, the son of his elder brother Taniko or Baniko, of Taraz, to whom he left the succession by his will.[†]

The same year, some Khuarezmian chiefs who had sought shelter at Shiraz, went to Jiraft in Kerman, a town described by Tavernier as one of the largest cities of Kerman, having a trade in horses and wheat. They were named Aor Khan, Sunj Khan, and Timur Malik, the famous defender of Khojend. Having attacked Kutb ud din, many of them were killed, and the rest captured or dispersed. Kutb ud din gave his prisoners state robes, and sent them back to Shiraz, whose Atabeg made apologies for the raid, which he said had been made without his knowledge. In 1236, Kutb ud din went to Ogotai's court to receive investiture. He was well received there, but was deprived of his sovereignty in favour of Borak's son Rokn ud din, while he himself was sent to China, to serve under Mahmud Yelvaj. Rokn ud din retained the sovereignty of Kerman till the year 650 HEJ., *i.e.*, 1252 A.D., when he was deposed by order of Mangu Khan, and his cousin Kutb ud din was reinstated.[‡]

We must now say a few words about Fars. We have seen how it was ruled by the Atabeg Said of the Salgarid family.[§] He died in the year 625 HEJ., *i.e.*, 1228, and was succeeded by his son Abubekr, who, we are told, "annexed the greater part of the tracts lying on the side of the Gulf of Persia, such as Hormuz, Katif, Bahrain, Oman, and Lahsa, perhaps the Al Hasa of Ibn Batuta, which he says was previously called Hajar."[||] We are further told that he sent his brother Tahamtan with rich presents to Ogotai, and received investiture from him. The author of the "Tabakat-i-Nasiri" says, "He engaged to pay tribute to them (the infidel Mongols), and brought reproach and dishonour upon himself by becoming a tributary of the infidels of Chin, and became hostile to the Dar ul Khilafat."[¶] Abubekr is famous as the prince to whom Saadi dedicated his famous "Gulistan." He retained the sovereignty of Fars for thirty-three years.[**]

We must turn aside for an instant to see what had taken place in Khorasan during Charmaghan's control of the army. I have mentioned

---

[*] D'Ohsson, iii. 131-132.   [†] Tabakat-i-Nasiri, 1118. Raverty's note.
[‡] Tabakat-i-Nasiri, 1118-1119. Notes. D'Ohsson, iii. 132.   [§] Ante.
[||] Tabakat-i-Nasiri, 179. Note.   [¶] Op. cit., 180.   [**] Id.

how Chin Timur was nominated governor of that province. He proceeded to treat it from the point of view of a farmer of taxes, and to grind out of the remaining inhabitants of the unhappy country the little remaining property they had. The Mongols, we are told, did not value money or precious stones ; but he did, and extracted what he could, by torture and otherwise, and then slew the victims of his tyranny. The few who escaped him had to pay a ransom for their houses.* While this was the character of the civil administration of Khorasan, it was also the scene of some military exploits. Two of the Sultan Jelal ud din's officers, named Karaja and Tughan-i-Sunkar (called Togan Sangur by D'Ohsson), at the head of 10,000 Kankalis, made their way to the mountains in the neighbourhood of Nishapur and Tus, whence they made attacks on the country round, and killed the governors appointed by Charmaghan. The latter ordered Chin Timur, with his deputy, Kelilat, or Kalbad, to march against them.† Chin Timur attacked them three times without result, when Kelilat defeated them near Sebzevar, after a three days' struggle, which cost him 2,000 men. Karaja thereupon fled towards Sejistan, or Seistan, and Tughan towards Kuhistan. Three thousand Kankalis found shelter at Herat, in the great mosque. Kelilat sent 4,000 men after them, who forced their way in and killed them all.‡

Meanwhile, Tair Baghatur, who commanded the Mongol troops about Herat and its dependency Badghis, had been ordered to march against Karaja, and to lay waste the country where he had sheltered. In regard to this last part of the order, Juveni quotes the Persian proverb, "Wolves know well enough how to tear ; it is necessary to teach them how to sew."§ He was already on the march, when he heard that Karaja had been beaten by Kelilat, and had taken refuge in the fortress of Arak, or Uk of Seistan, which we are told lies north-east from the Shahristan of Seistan. There Tair beleagured him for nineteen months, when, a pestilence having broken out, it succumbed. Major Raverty says Uk is situated between Farah and Zaranj, and that it has been in ruins for many years.‖ The author of the "Tabakat-i-Nasiri" tells a curious story of the siege. How, on a certain night, the defenders of the place had determined to plant an ambuscade in some kilns outside the northern gate of the town while a sortie was made from the eastern gate. When the Mongols attacked the latter body, the kettle-drums were to be sounded at the citadel, whereupon those in ambush should emerge, and take the Mongols in rear. During the night 700 men, natives of Tulak, accordingly planted themselves, fully equipped, in the appointed place, while at daybreak the other contingent, after performing its religious exercises, made the appointed sortie. When they had engaged the enemy, the kettle-drums made the appointed signal, which was repeated, but no one issued from

---

† Tabakat-i-Nasiri, 1116. Note. ‡ D'Ohss
Note 2. ‖ Tabakat-i-Nasiri, 1122. Notes.

the ambush. The Malik Taj ud din Binal Tigin, who was then ruler of
Seistan and Nimroz, sent trusty men to inquire the reason for this, who
reported that the whole 700 were dead. Our author says "they had
surrendered their lives to God, and there was no sign of life in any one of
them." And he explains in the context that they had died from the pestilence which then raged at Uk. He says it began by a pain in the mouth,
which on the second day was followed by the teeth dropping out, and on
the third day the patient died. A woman having been seized, feeling her
teeth loose, and knowing that her end was near, summoned her little
daughter, and applied henna to her feet and hands. It was usual, he
says, for women in doing this, to wet their fingers with their tongue, and
then to rub the henna. Having done this, the woman resigned herself to
death, but in the morning her teeth became fast and the aching passed
away. It was thus discovered that henna was a specific for the pestilence,
and in consequence a menn of the drug was sold for 250 golden dinars.
After some time, Malik Taj ud din Binal Tigin was struck in the eye with
an arrow, and presently, while directing the defence of the fort from the
top of one of the towers, he lost his footing, fell down, and was captured.
The fortress then fell. "The inhabitants were martyred after a great
number of the infidels had gone to hell."* Taj ud din Binal Tigin
was taken from Seistan to the fortress of Safhedkoh, where he was put
to death underneath the walls.†

Thus was suppressed this dangerous outbreak of one part of the disbanded soldiery of the Khuarezm Shah, consisting mainly of Turkomans.
Another portion found its way to Syria and Egypt. Meanwhile, after the
fall of Uk, Tair Behadur wrote to Chin Timur to say he had been
intrusted by the Khakan with the government of Khorasan, which he
called upon him to surrender. The latter reproached him with his
cruelties in destroying the innocent people, with the misdeeds of Karaja,
and added that he had sent to report his conduct to head-quarters.
Meanwhile, Chin Timur and his officers received a summons from Charmaghan to go to him (Raverty says to return to Khuarezm with the agents
of the princes who were with him), and to give up the government of
Khorasan and Mazanderan to Tair Behadur. A council was held, and it
was determined that Kelilat, or Kalbad, who represented Ogotai's special
interests, should repair to the Imperial Court, to solicit his master's decision in favour of Chin Timur. Some princes of the country accompanied
him. Among these were Malik Baha ud din Saluk, one of the principal
chiefs of Mazanderan, who submitted at this time, and the Asfahed Ala
ud din (or Nusrat ud din), of the Kabud Jamah.‡ It was the first time
that any of the Maliks of Iran had gone to do homage, and Ogotai,
who was much pleased, contrasted Charmaghan's conduct in this respect

---

* Op. cit., 1123-1125.   † Id., 202.
‡ (?) Tabaristan and Rustamdar. Vide Tab. Nas., 263. Note,

with Chin Timur's. Ogotai thereupon rewarded the latter, and appointed him supreme governor of Khorasan, with Kalbad (or Kelilat as he is called both by D'Ohsson and Von Hammer) as his associate, making them both independent of Charmaghan. He conferred the tract extending from the Kabud Jameh territory to Asterabad on the Asfahed Ala ud din; and the districts of Isterain, Juven, Baihak, Jajurm, Khurand, and Arghaian upon Baha ud din, and gave each of them a golden paizah.* Chin Timur appointed Sherif ud din, of Yezd (Von Hammer says of Khuarezm), his Ulugh Bitikji, or Chief Secretary, or Master of the Seals; and Baha ud din Muhammed Juveni, the father of the famous author of the "Jihan Kushai," his finance minister. In the latter's office was a representative of each of the three other princes who had furnished contingents for the Persian war as I have mentioned, and who had a joint interest in the revenues of Khorasan.†

Chin Timur died in 1235, and was succeeded by a very old Mongol, named Nussal, Tusal or Usal, who, we are told, had been appointed by Jingis as joint guardian of the Ulus of Juchi.‡ He was soon after displaced by a Buddhist Uighur, named Kurguz (i.e., blind-eye), who had risen successively from being tutor and writing-master to the children of Juchi Khan to be the secretary of Chin Timur (like himself a Uighur) when the latter was Governor of Khuarezm. He had been sent with Muhammed of Juveni to Ogotai's Court, to report to him the condition of Khorasan and Mazanderan, which he described in inflated Persian figures, inter alia, saying that where winter formerly reigned, there was now spring, and that the country was as full of flowers and perfumes as paradise. These phrases, mixed with flattering speeches, won him the favour of Ogotai, whose minister, Chinkai, also a Uighur, favoured him. During the rule of Nussal, Kurguz was summoned to the Court, where he had enemies as well as friends, to give an account of the affairs of Khorasan. While Chinkai supported him, and argued that the principal people of Khorasan also wished to have him, Danishmend Hajib, another official at the Court, urged the claims of Ungu Timur, son of Chin Timur. Kurguz at length obtained a temporary authority in Khorasan and Mazanderan, with orders to make a census, and receive the taxes in the two provinces. The order appointing him deposed Nussal, who had been a mere puppet, the real authority having been controlled by Kelilat, who now found himself put into the shade. Kurguz proceeded to repress a good deal of exaction, &c., in his government. Meanwhile, Kelilat and Sherif ud din, the vizier, secretly supported Ungu Timur, and incited him to send complaints of the doings of Kurguz to the Khakan. Their attacks were parried there by Chinkai, and Ogotai at length sent Arghun, with two officers, to report on the state of things.§ They were met at Fenakat by Kurguz, who had set out to report in person,

---

* Tabakat-i-Nasiri, 1120-1121. Notes. D'Ohsson, iii. 106-107.
† D'Ohsson, iii. 107-108.   ‡ Ilkhans, i. 113.   § D'Ohsson, iii. 110-111.

and had left Bahu ud din in charge of his administration. The Imperial commissaries asked him to return with them, and as he refused, a disturbance took place, in which he had a tooth broken. Although compelled to accompany them, he dispatched a messenger to Ogotai, who carried his coat marked with blood. On the arrival of the commissaries in Khorasan, Kelilat, Ungu Timur, and Nussal drove out the secretaries and other officials of Kurguz from his palace with sticks, and carried them off. The latter's messenger soon after returned. Ogotai, who was irritated at the sight of the bloody garment, summoned the disputants to his presence. Kurguz at once set out and was shortly followed by Kelilat and Ungu Timur. Kelilat was assassinated as he passed through Bukhara. I have described how the Khakan was entertained by the two rivals.* Chinkai, who was Kurguz's patron, was appointed to report on the matter. The latter was himself a shrewd man of business, while Ungu Timur was young and inexperienced and had lost his most sagacious adviser in Kelilat. Ogotai tried to reconcile the two parties, and ordered the rivals to be deprived of their arms, and to live in the same tent and drink out of the same cup. This mode of reconciliation failed. Chinkai at length made his report, and Ogotai decided in favour of Kurguz. As Ungu Timur was a subject of Batu's, his father having been Governor of Khuarezm, as we have seen, he was ordered to be handed over to him for punishment. I have described how he asked to be punished by Ogotai himself.† Some of his supporters were bastinadoed, others were handed over to Kurguz to be punished with the *cangue*,‡ and to return with him, their lives being spared for the sake of their wives and children. Kurguz was given authority over all the country south of the Oxus which had been conquered by Charmaghan. He took back with him, so as to have him under his eye, the vizier, Sherif ud din, whose secret intrigues on behalf of Ungu Timur had been disclosed to him.

Kurguz returned to Khorasan in 1239-40, and fixed his residence at Tus, where he assembled the grandees of Khorasan and Irak, and the Mongol generals, and celebrated his installation with grand fêtes, at which the new Imperial edicts were published.§ He sent his son to deprive the creatures of Charmaghan (who were ruining Irak and Azerbaijan by their exactions) of their posts. He protected the Persians from the ruthless Mongol soldiery, and was everywhere respected. Tus had but fifty houses left in it. He proceeded to restore it, and the various Persian grandees built new houses there, and we are told the price of furniture increased a hundred-fold in one week.‖ Herat, too, began to revive. Since its destruction, in 1222 to 1236, it had remained practically a waste. In the latter year, Ogotai having ordered the restoration of

---

* Ante, i. 134.   † Ante, 134.
‡ This was a Chinese instrument of punishment, consisting of a heavy wooden collar, through which the head and hands were thrust and then locked.
§ D'Ohsson, iii. 115-116.   ‖ *Id.*, 117.

Khorasan, a native Amir of the place whom Tului had transported with
1,000 families to Bishbaligh, where they exercised the craft of weavers, and
supplied the Court with robes, and who was named Iz ud din, was ordered
to return there with 100 families, where for want of provisions they
suffered greatly. As they had no oxen, the men dragged the ploughs in
pairs, while the canals being choked, the land had to be irrigated by
hand. After the first harvest twenty strong men were sent, each with
twenty menns of cotton, into Afghanistan to buy ploughs and long-tailed
sheep. In 1239, 200 more families were sent from Bishbaligh to
settle at Herat. Fugitives and others who had escaped the general
massacre in the campaigns of Jingis Khan, collected round them from
various parts of Khorasan, and the following year, a census having been
taken, it was found that its inhabitants had increased to 6,900, after which
it continually grew.*

Iz ud din had died at Farab, while conducting the second batch of
emigrants from Bishbaligh, and was succeeded as superintendent of
Herat by his son, Shems ud din Muhammed. He went to Ogotai's
Court, and asked that a Shahnah or Intendant, and a darugha or
Mongol commissary, should be appointed for Herat. A Karluk Turk,
whose name is not recorded, was appointed to the former post, and a
Mongol named Mangasai to the latter, while Shems ud din himself
retained the chief control of civil matters. They proceeded to open the
Jui Injil canal, and to take it into Herat, and they built the Burj-i-Karluk,
named after the Karluk Shahnah. In 1241 Shems ud din was displaced
as governor, in favour of Malik Majd ud din, the Kalyuni, who, in concert
with the Karluk, opened the Alanjan canal.† These events took place
while Kurguz was governor of Khorasan, and we are told that after
Ogotai's death he had Majd ud din put to death, and his head taken to
him at Tus. He was succeeded as governor of Herat by his son,
Shems ud din Kalyuni, who a year later died from poison.‡

Kurguz had put his enemy the vizier, Sherif ud din, in the *cangue*,
and extracted from him confessions, which he sent on to the Court. His
messenger heard *en route* of the death of Ogotai. He had himself set
out to make a report to his master, and in passing through Mavera un
nehr had quarrelled with an official there, and aroused the anger of the
princes of the house of Jagatai, who were further incited by a messenger
sent by the wife of the imprisoned Sherif ud din. They accordingly
dispatched Kurbuka and Arghun, who was, as we have seen, no friend
of his, with orders to carry him off by force. The latter, on hearing
this, respited the life of Sherif ud din, who had already been handed over
to the governor of Sebzevar for execution. Kurguz himself, after a show
of resistance, was arrested in his house at Tus, with his vizier, Usseil ud

---

* *Id.*, iii. 117-118.   † Tabakat-i-Nasiri, 1127-1128. Raverty's notes.   ‡ *Id.*, 1128.

din Rogdi. The removal of his strong hand was the signal for renewed
anarchy in Khorasan and Mazanderan. Kurguz was taken to the ulus of
Jagatai, and thence to Mongolia, to the Court of the Regent, Turakina.
His patron, Chinkai, was now in disgrace there, and, as we read, having
no money he could get no justice. He was remitted back to the Jagatai
princes for trial, and was at length put to death by order of Kara Khulagu,
by having his mouth filled with earth. He had recently abjured Buddhism,
and become a Muhammedan.* Turakina appointed Arghun in his place.
We shall revert to him presently. Among the stories recorded of the
Khakan Ogotai we read that he was very fond of wrestling, and entertained at his Court a large number of Mongol, Kipchak, and Chinese
athletes. Having heard of the renown of the Persian wrestlers, he
ordered Charmaghan to send him some. The latter forwarded him
thirty, under two famous leaders, Pileh and Muhammed Shah. Ogotai
was much struck with the size and physique of Pileh. Ilchikadai, who it
would seem had charge of the Mongol wrestlers, ventured to question
if the cost of bringing them so far would be repaid. Ogotai replied
that he would back him against Ilchikadai's men for 500 balishes against
500 horses. The following day the latter produced a champion to
struggle with Pileh. The Mongol succeeded in throwing his adversary
down and in falling on him. "Hold me fast," said Pileh playfully, "and
take care I don't escape." At the same time he raised him up, and
threw him to the ground with such force that his bones were heard to
crack. The emperor then rose and told him to hold his opponent fast,
and turning to Ilchikadai claimed his bet. Pileh was rewarded with many
gifts, together with 500 balishes. Ogotai presented him shortly after with
a young damsel, and asked her some time after, laughingly, how she had
found the Tajik (i.e., the Persian). She replied that they did not live
together, and when Ogotai wanted an explanation, Pileh told him that
having acquired a reputation at the Khakan's Court, and never having
been beaten, he wished to preserve his powers so as to merit the emperor's
favour. The latter replied that he wanted his like perpetuating, and
excused him from further combats.†

We must now shortly consider the doings of the Mongols in the districts
east of Khorasan, bordering on India, during Charmaghan's campaigns in
the west. According to Vassaf, when Jingis Khan withdrew northwards
he ordered each of his four sons to furnish 1,000 men, who were to plant
themselves in the districts of Shiburghan, Talikan, Ali-Abad, Gaunk,
Bamian, and Ghazni.‡ The author of the "Tabakat-i-Nasiri" tells us
how, on the accession of Ogotai, when Charmaghan was intrusted with
the army which overran Western Persia, other Mongol armies were sent
into the districts of Kabul, Ghazni, and Zabulistan, and how the Malik

* D'Ohsson, iii. 120-121. Ilkhans, i., 115. Tabakat-i-Nasiri, 1149. Note.
† D'Ohsson, ii. 96-97.　‡ Id., 280.

Saif ud din Hasan, the Karluk, who it would seem held the fief of Bamian, together with the Maliks of Ghur and Khorasan, submitted, and consented to receive Mongol Shahnahs or commissaries.* Notwithstanding this they attacked Saif ud din and drove him from Karman† (this was about 636 HEJ.), Ghazni, and Bamian. He thereupon went towards Multan and Sind. His son, Nasir ud din Muhammed, went on to Delhi, and was granted the fief of Baran, but presently joined his father, and seems to have fallen into the hands of the Mongol commissaries, with whom he remained some time.‡ When the Malik Saif ud din Hasan withdrew across the Indus the districts of Ghazni and Karman fell under the complete control of the Mongol Shahzadahs or Shahnahs, and we may take it that Afghanistan was incorporated with the Mongol Empire.

Let us now turn once more to Charmaghan. He had some time before this been attacked by an illness which caused him to become dumb, and which was probably some form of paralysis. He left two sons, Shiramun, who became a famous general and was called the Golden Column by his countrymen because of his successes, and Baurai, who was put to death by Khulagu because of his evil character.§ According to Guiragos, in the beginning of the year 691 of the Armenian era (i.e., Jan. 20th, 1241, to Jan. 19th, 1242), an Imperial edict of the Khakan superseded Charmaghan, and appointed Baigu or Baichu in his place. We are further told that Baichu was chosen by some magical process, as was customary with the Mongols.‖ I believe rather that his appointment, which took place in 1241, on the death of Ogotai, was due to the policy of his widow, Turakina, who, on her accession, placed her creatures in various places of trust. It was probably as the protégé of Turakina and her son Kuyuk that Baichu aroused the jealousy of Batu and Khulagu, as we shall see further on.

Baichu (called Baichu Kurchi by Guiragos) belonged to the tribe Baisut (called Yissut by D'Ohsson), and was a relative of Chepe or Jebe Noyan, who made the famous campaign in the west with Subutai. He commanded a hazarah under Charmaghan and, as we have seen, was promoted to command his tuman.¶ His first efforts after his appointment were directed against the Seljuki rulers of Asia Minor or Rum. This dynasty had been founded about the year 1080, by Suliman Shah, who had been sent into Asia Minor with 80,000 Ghuz or Turkomans, and had conquered the central part of the peninsula from the Byzantine emperors. He fixed his capital at Iconium, and his dominion was known as that of the Seljuks of Rum. Kai Kobad, the seventh successor of Suliman, was on the throne in 1235-6, when a

* Tabakat-i-Nasiri, i. 119. Hasan had been a feudatory of Jelal ud din Khuarezm Shah. On p. 2, following Abulfeda, I called him Hasan Karak, but his real name was doubtless Hasan the Karluk.
† Major Raverty explains this as meaning a darah or long valley, watered by a tributary of the Shahezan. It has, of course, nothing to do with the Persian province of Kerman. (Op cit., 499. Note 7.) ‡ Tabakat-i-Nasiri, 1119-1129. Notes. § Malakia, op. cit., 449.
‖ Guiragos, ed. Brosset, 138. Journ. Asiat., 5th ser., xi. 426. ¶ Erdmann's Temudjin, 229.

Mongol envoy, named Shems ud din, went to his Court, bearing a yarligh or Imperial order summoning him to submit, which he accordingly did. Notwithstanding this, a body of 10,000 Mongols invaded his dominions.* When Baichu received the command of the Mongol armies in the west he prepared to strike a heavy blow against the Seljuk monarchy. At this time (i.e., 1243) Ghiath ud din Kai Khosru, son of Kai Kobad, had been its ruler for some years. As we have seen, he had married Thamar, the daughter of the Georgian Queen, Rusudan. Baichu first marched into that part of Armenia which was subject to the Seljuki, and attacked Karin, the ancient Theodosiopolis, called Karno Kaghak by the Armenians, and better known as Erzen-er-Rum or Erzerum, which W. de Nangis identifies with Uz, the land of Job. Its commander was Sinan ud din Yakut. Having invested it, they summoned the citizens to surrender. They refused, drove out their envoys, and jeered at them from the walls. The Mongols thereupon battered the ramparts with twelve catapults. They speedily destroyed its churches and monasteries, made a general massacre of its inhabitants, and then pillaged and fired it. It had a numerous population of Christians and Mussulmans, and many peasants from the country round had also sheltered there. *Inter alia*, the Mongols captured a great number of bibles, martyrologies, and liturgical books, delicately written in letters of gold, which they sold at a small price to their Armenian and Georgian allies, who sent them as presents to the churches and monasteries in their own country. These Christian auxiliaries also redeemed many men, women, and children, bishops, priests, and deacons, and we read that Prince Avak, Shahan Shah, and Akbuka, son of Vahram, Gregory of Khachen, son of Tuph, who was sister to the great Atabegs Ivaneh and Zakaria, as well as their troops, gave their freedom to their captives, and allowed them to go where they pleased. The Mongols not only sacked the town, but also a number of the surrounding districts. The Sultan of Rum did nothing to help them, but hid away in fear, and it was even said he was dead. The Mongols withdrew with their booty to spend the winter in their rendezvous on the plain of Mughan.†

While they were encamped there Kai Khosru sent their commander a boastful message. " Do you think," he said, " because you have ruined one of our towns that you have vanquished the Sultan and laid low his power ? My cities are innumerable, and my soldiers cannot be counted. Remain where you are and await my arrival. I will come in person to see you, sword in hand." The Mongols were not disturbed at this message, and Baichu merely said, " You have spoken bravely. God will accord the victory as he pleases." After having got his horses and other cattle in good condition, he set out by easy marches towards where the Sultan was

* D'Ohsson, iii. 79. Note.
† Guiragos, ed. Brosset, 138-139. Journ. Asiat., 5th ser., xi. 426-428. D'Ohsson, iii. 79-80.

encamped, not far from Erzenjan.* There he was encamped with his wives and concubines, and great store of gold, silver, and other treasures. He also had with him a menagerie of wild animals to be used in hunting, and including rats, cats, and even reptiles. He wished to show his troops that he had plenty of confidence.† The King of Little Armenia and the Princes of Hims and Mayafarkin,‡ who had promised him assistance, failed to send it ; but he had 2,000 Frank auxiliaries under the orders of John Liminata, from Cyprus, and Boniface de Castro, a Genoese. Sanuto calls the latter Boniface de Molinis, a Venetian. Abulfeda tells us he was also joined by a contingent from Aleppo, under Naseh ud din Persa.§ Baichu divided his army into various sections, which he intrusted to his most valiant subordinates, and distributed his auxiliaries among them so as to avoid treason.‖ In regard to the date of this famous battle (namely, the Armenian year 692), Vartan tells us that the letters forming this number, make up the word Oghb (meaning woe or lamentation), which, he adds, was well borne out by the terrible sufferings of Armenia, not only those of its inhabitants, but also of its plains and mountains, which were deluged with tears and blood.¶ Abulfaraj tells us the fight took place in June and July, 1243.** Abulfeda says in 641 HEJ., which began June 20th, 1243. Rubruquis tells us that he was informed by an eye-witness that Baichu had only 10,000 men with him. Haithon says 30,000. Malakia tells us the Sultan, on the other hand, had 160,000.†† Before the fight, according to Chamchean, Baichu sent home many of the Georgian and Armenian auxiliaries, retaining only those princes on whom he could depend, such as Avak, Shahan Shah, Elikum the Orpelian, and Akbuka, son of Vahram.‡‡

According to the *Georgian Chronicle*, the advance guard of the Sultan's army was commanded by Dardan Sharwashidzé Apkhaz, promoted on account of his great valour. He was a Christian. With him was Pharadaula, son of Shalwa, lord of Thor and Akhal Tzikhé, who, according to Malakia, had been a refugee with the Sultan for many years. A large contingent of Georgians fought willingly enough in the Mongol ranks, in the hope of exacting vengeance from their bitter foes, the Mussulmans. The Sultan's army was very numerous, but this did not cow their opponents, who were accustomed to fight against great odds. "What

---

* The place was called Tchman Katuk, or Asechman gadug. Guiragos calls it a town, but there was no town of this name, but north-east of Erzenjan is a mountain called Chimenkedik (*le defile herbu*) by the Turks. Bar Hebræus (Chron. Syr., 519, Chron. Arab., 314) calls the place Kusa tagh (*i.e.*, "Mons sordidus"), while Novairi approximates more nearly to the Armenian historian in calling it Aksheher in the plain of Erzenjan. (D'Ohsson, iii. 81. Guiragos, ed. Brosset, 140. Note 1. Journ. Asiat., 5th ser., xi., 429.)
† Guiragos, ed. Brosset, 139-140.
‡ In 1241 Shihab ud din, Prince of Mayafarkin, had received a summons, commencing "The Lieutenant of the Lord of Heaven upon Earth, the Khakan," and which offered him the title of *Selahdar*, or cupbearer, and bade him raze the walls of his fortresses. He pleaded that he was a very small person, and asked the Mongols to address themselves to the rulers of Rum, Syria and Egypt, whose example he would follow. (Makrizi, in D'Ohsson, iii. 85-86.)
§ Abulfeda, iv. 473. Haithon Chron., 34. D'Ohsson, iii. 80-81. Ante, i. 166.
‖ Hist. de la Géorgie, i. 518-519. ¶ Brosset, Hist. de la Géorgie, Add., 508.
** Chron. Syr., 520. Chron. Arab., 314. †† Op. cit., 446. ‡‡ Hist. de la Géorgie, i. 519. Note.

shall be my reward," said Baichu to Sargis, a brave and renowned warrior, the grandson of Kuarkuareh-Jakel, "for the news that I bring you? The Sultan has learnt that we were coming, and has set out himself. His camp is not far off, he has an innumerable host, and proposes to attack us to-morrow." Sargis replied, "I know your warlike ardour and your successes, oh Noyan, but this vast host does not seem to presage any good." "You know not," said Baichu, smiling, " our Mongol people. God has given us the victory, and we count as nothing the number of our enemies. The more they are, the more glorious it is to win ; the more plunder we shall secure. Meanwhile make ready, for in to-morrow's fight we shall see what will become of them." It is thus, adds the chronicler, that they dared all nations. Malakia tells us the son of Shalwa (*i.e.*, Pharadaula) defeated the Tartars opposed to him, and killed many of them, but on the other side Akbuka, son of Vahram, and grandson of Blu Zakaria, fought valiantly with the battalion of noble Georgians and Armenians, his companions. They defeated the right wing of the Sultan's army, and killed several of his Amirs. Night soon after intervened, and the two armies encamped close together on the plains between Erzerum and Erzenjan. The following morning the Tartars, Armenians, and Georgians made a rush upon the enemy's camp. They found it abandoned, and secured a great booty. The Sultan's tent was splendidly decorated inside and out, and they found, *inter alia*, a panther, a lion, and a leopard chained at its entrance. The Sultan, we are told, had fled during the night, afraid of his Amirs, who wished to submit. Leaving a guard to watch over the camp, the Mongols went in pursuit.* The *Georgian Chronicle* says that Dardan Sharwashidzé Aphkhaz having been killed in the battle, the Sultan's people fled, when there was a terrible carnage, while a great many prisoners were made. The Sultan was much exasperated, and put to death Pharadaula out of hatred for the Georgians. The conduct of the latter won the hearts of their allies, who liberally divided the booty with them.† When the Sultan fled he sent his harem to Iconium, abandoned his baggage, and himself went to Ancyra.‡ The Turks were pursued for some distance mercilessly, and the victors then returned to plunder the dead. They ravaged the country round, and collected a great quantity of gold and silver, of rich vestures, of camels, horses, mules, and cattle.

The authorities differ as to the order of the next proceeding of the invaders, but it is natural to suppose they attacked Erzenjan, which resisted bravely. The citizens were, however, inveigled into a surrender, when they were mercilessly slaughtered, except the young people, who were reduced to slavery. W. de Nangis says that two Franks were made

* Brosset, Hist. de la Géorgie, i. 518-519. Adds., &c. 446-447.
Hist. de la Géorgie, i. 519. ‡ Abulfaraj, Chron. Syr., 511. Chron. Arab., 314.

prisoners in the town who were famed for their valour. The Mongols determined to pit one against the other, and having armed and horsed them, stood round to watch the fight. The two champions, however, turned upon them, and before they were killed had destroyed fifteen and wounded thirty Tartars.[*] Tephriké, the modern Divirigi, paid heavy black mail, and was spared.[†] Sivas or Sebaste was also submissive, and purchased at least a respite by surrendering a portion of its wealth. The Mongols put shahnahs there, imposed the taxes of thal and talar, burnt the war engines they found, and destroyed the walls. They then apparently advanced upon Cæsarea, the citizens of which resisted for some days ; but the town being at length captured, the grandees and rich people were put to death after having been tortured, while the women and children were carried off as slaves. Meanwhile the Sultan's mother took refuge with her daughter and dependents in Cilicia.[‡] Seeing that resistance was useless, one of Kai Khosru's generals and the Kadhi of Amasia went at their own instance to the Mongol camp, which was then at Sivas, and undertook to pay an annual tribute of 400,000 pieces of money, and a certain number of rich cloths, horses, and slaves. According to the missionary friar Simon, as reported by Vincent of Beauvais, the Seljuki undertook to pay 12,000,000 hyperperes, 500 pieces of silk, 500 camels, and 5,000 sheep annually, which were to be transported free of cost to the Khakan's Court. Besides this tribute, a sum equal in value was to be disbursed in presents, while the various Tartar envoys who visited Rum were to be supplied with what they needed, free of cost. The Sultan, who was meanwhile at Iconium, gladly accepted these terms.[§] The Sultan's notary computed that the cost of entertaining the Tartar envoys (perhaps shahnahs or commissaries is meant) during two years at Iconium, independent of the meat and wine they used, was 60,000 hyperperes. The treaty was made at Sivas, in the presence of Constantine, Lord of Lampron.[||] In this campaign the Mongols became the virtual masters of Rum as far as Angora, Gangra, a town of Paphlagonia, and Smyrna, while, as we know from Rubruquis, the ruler of Trebizond became their vassal.[¶]

The Mongols after these successes once more returned to winter in the plain of Mughan, and their Christian auxiliaries and allies again ransomed numbers of their co-religionists. At this time the Greeks and Latins were struggling for the Empire of the East, John Ducas, Vataces being the Greek Emperor, and Baldwin the Latin one. Both of them entered into negotiations with the beaten Seljuki sovereign for an alliance. The latter naturally preferred the stronger rival, Vataces, whose greater proximity to the Mongols made him a more certain ally. A meeting was

---

* Dom. Bouquet, xx. 342.
† Guiragos, ed. Brosset, 140. D'Ohsson and Von Hammer call the place Tokat.
‡ Abulfaraj, Chron. Syr., 520. Chron. Arab., 314-315. § D'Ohsson, iii. 82-83. Note.
|| Vincent of Beauvais, Spec. Hist., xxxi. 18. ¶ St. Martin Memoires, ii. 121, and Notes.

arranged at Tripoli on the Mæander, where the Sultan built a wooden bridge as a means of communication between the two camps. An offensive and defensive league was entered into between them, after which the Sultan returned to Iconium, and the Emperor to Philadelphia.*

The campaign of the Mongols against Rum naturally took them close to the famous town of Malatia, then governed by Rashid ud din Al Juveni, who, collecting such treasures as he could, withdrew with a number of the principal people towards Aleppo. Abulfaraj tells us his own father was wishful to accompany them, and had brought together some sumpter cattle to carry his treasures. He adds that a mule belonging to him, having bolted when being strapped to its burden, was caught and pillaged by the town boys, which is assuredly a naive story to occur in such a grim narrative. His father eventually stayed behind, and arranged with the Metropolitan for the defence of the place, Mussulmans and Christians meeting together to consult in the great church, and agreeing to man the walls, &c. The party which fled from the town were attacked ten parasangs off, at a place called Beth Goza in the Syrian, and Bajuza in the Arabic chronicle of Abulfaraj. Many of them were slaughtered, and the young people made prisoners, only a few regaining the town.† The following year (i.e., 1244) a detachment of Mongols under Yassaur Noyan made an attack upon Syria, and by way of Mayafarkin, Mardin, and Edessa or Urfa, crossing the Euphrates, they advanced as far as Hailan (?), near Aleppo. They did not actually reach the latter city, as they were obliged to withdraw on account of the dryness and heat, which injured their horses' feet. Yassaur demanded black mail from the governor, which having been paid, he approached Malatia, where he laid waste the vineyards and orchards, and put to death everybody met with outside the town. Its governor, Rashid ud din, collected together gold and silver ornaments, &c., to the value of 40,000 gold pieces, together with sacred vessels from the churches, reliquaries, thuribles, candlesticks, crosses, covers of sacred books, &c., which he gave the invaders, and they returned home again. Abulfaraj tells us how, about this time, his father took his family, including himself, to Antioch, where they continued to live for some time.‡ The campaign just mentioned is named by Guiragos, who tells us the Mongols made a raid upon Mesopotamia, Amid, Urha (i.e., Urfa or Edessa), Nisibin, Syria, &c. Although unopposed, they lost many men from the heats. On their return they ordered Erzerum to be restored, intrusting the work to Sargis, bishop of that town, and to Shahan Shah, son of Zakaria.§ In the autumn of 1244, as Matthew Paris tells us, Bohemund the Fifth, Prince of Antioch, received a summons from the Mongol commander

* Nicephorus Gregorias, and Akropolita. Stritter, iii. 1031-1033, and Notes. Lebeau, xvii. 411-412, passim.
† Abulfaraj, Chron. Syr., 521. Chron. Arab., 315.
‡ Abulfaraj, Chron. Syr., 522-523. Chron. Arab., 318-319. § Op. cit., ed. Brosset, 145.

ordering him (1) to level the walls of his fortresses, (2) to send him all the revenue of his kingdom, and (3) to send him 3,000 young damsels. Bohemund refused, and Yassaur had too many men prostrate by the heats to enable him to compel him, and retired to Asia Minor. The following year Bohemund and the other Christian princes, his dependents, were constrained to submit and to pay tribute. Thenceforward they continued subject to the Mongols.

In the year 1245 the Mongols invaded the districts north of Lake Van. •Having captured Khelat they made it over, with the surrounding districts, to Thamtha, the sister of Avak and widow of Malik Ashraf, to whom it had formerly belonged. After she had been captured by Jelal ud din, the Khuarezm Shah, she had fallen into the hands of the Mongols, and had visited the Court of the Khakan, where she had lived some years. When Hamad-ud-daula, the envoy of Rusudan, visited Ogotai, he was allowed to take her back with him. The Khakan then ordered that the possessions she held while her husband, the Malik Ashraf, was living, should be restored to her.* Haithon, King of Little Armenia, seeing how matters were going on, and probably not sorry to break the yoke of the Seljuki, now sent envoys with magnificent presents to the invaders. These envoys, we are told, were presented to Baichu, to Charmaghan's widow, Ailthina Khatun, and to the other officials. They demanded the surrender of the Seljuki Sultan's mother, wife, and daughter, who had sought shelter in Cilicia. As I have mentioned, Haithon professed to be greatly distressed at this demand, and said he would rather they had asked him to give up his son Leon, but he was constrained to obey. The Mongols were much pleased at his conduct, and sent him a tamgha, or official seal, constituting him a vassal of their empire.† He shortly after had to make head against Constantine, the Lord of Lampron (now called Nimrun Kalesi, situated two days' journey west of Tarsus, in one of the gorges of Mount Taurus). He had rebelled, and allied himself with the Sultan of Rum, who was naturally aggrieved at his harem having been surrendered. Together they invaded and ravaged Cilicia, but they were badly beaten, and their army almost destroyed.‡ Abulfaraj tells us they attacked Tarsus, where they were assailed by terrible rains, which converted the country round into mud, and made it very harassing for their horses. They were in this plight when news arrived of the death of their master, the Sultan Ghiath ud din Kai Khosru. This happened in 1246. Thereupon the grandees put his eldest son, Iz ud din Kai Kavus, on the throne, associating with him the latter's two younger brothers. Messengers now came from the Mongols demanding that Iz ud din should go to the Khakan's Court to do homage. He excused himself on the ground that he was afraid of the Greeks and Armenians, who were his enemies,

---

* *Id.*, 145.   † Guiragos, ed. Brosset, 141.   ‡ *Id.*, 142.

promised to go later, and offered to send his younger brother, Rokn ud
din.* It seems a number of partisans of the latter wished to raise him to
the throne. When the Grand Vizier, Shems ud din, of Ispahan, learnt
this he had them seized and put to death. He then presumed to take the
mother of Iz ud din into his harem, by whom he had a son, and finally
dispatched Rokn ud din with rich presents for the Khakan.† Rokn ud din
is called the Sultan of Khelat by the Georgian chronicler.

Meanwhile the Mongols gradually enlarged their borders. Bedr ud
din Lulu, Prince of Mosul, on behalf of the Prince of Damascus, made a
treaty with them, by which the people of Syria were taxed, the
richest at ten dirhems per head, the middle class at five, and the poor at
one. This tax was duly levied in 1245. The same year a detachment of
them entered Sheherzur, eight days north of Baghdad, and sacked the
town. News of this reached Baghdad by pigeon post. The following
year they advanced as far as Yakuba, but were defeated by the troops of
Baghdad, under the so-called Little Devatdar, who took some prisoners.‡

Let us now turn again to Georgia. According to the *Georgian
Chronicle*, the Mongols, after their campaign against the Seljuki of Rum,
went to their summer quarters of Gelakun and Mount Ararat, whence they
sent messengers to Rusudan offering her their alliance, and bidding her
send her son David to their camp, as they wished to confer the
sovereignty of Tiflis and of all Georgia upon him. This authority tells us
the Queen was charmed to comply, inasmuch as the Mongols never broke
their promises, and always treated those well who submitted. She
accordingly came down (*i.e.*, from her mountain retreat), with Shahan Shah
and Avak, who were much esteemed by the Tartars, Shotha Kupri
Vahram, chief of Thor, Grigol Suramel, eristhaf of Karthli, Kuarkuar,
commander of Samtzkhe and of Tzikhis-Juar, and chief of the armourers,
and Sargis, commander of Thmogvi, with the people of Shawkheth, of
Klarjeth and of Tao, who all went to meet the young Prince David. The
latter was accompanied by Tzotné Dadian, a virtuous man and illustrious
warrior ; by the Bédian, the Eristhaf of Radsha, the Guriel, and the most
distinguished people. They all went to Tiflis, and thence to Berdaa,
where the Mongol Noyans were encamped. They received him well, and
conferred on him all Georgia and Samshwildé (which had been previously
conquered by Yassaur Noyan), and Angurga (also written Agurnaga),
assisted by Avak. So great was the honour paid to the young prince that
he was called Narin David (*i.e.*, David with the august countenance).
Wakhucht says Narin means "arrived" (venu).§ The Mongols now sent
news of their victories to the Khakan at Karakorum, and forwarded to
him a richly ornamented head-dress, a suit of armour, &c. They
reported also how the Georgians, king and people, had submitted ; that

* Abulfaraj, Chron. Syr., 524.   † Abulfaraj, Chron. Syr., 526. Chron. Arab, 320.
‡ D'Ohsson, iii. 88-89.   § Hist. de la Géorgie, i. 520-521, 528. Note 3.

they professed a good religion, were truthful, and did not practice sorcery or magic, while the Persians were false, traitors, and breakers of their word, and much given to magic and sodomy. The Grand Khan sent word back that they must employ the Georgians, who were trustworthy warriors, to exterminate the Persians, and ordered their chiefs to be sent to him. Jaghatai Noyan therefore sent on Avak, who had been created Atabeg and commander-in-chief by Rusudan.* He travelled in company with the Seljuk Prince of Rum, Rokn ud din. We are told they traversed unknown kingdoms, where no Georgian had hitherto put his foot. They eventually reached the camp of Batu, who is described as singularly handsome. Avak had with him his chamberlain, David, son of Ivaneh of Akhal Tzikhé, who said to him: "As we are going into strange lands, and there is no knowing what may happen, it would be perhaps prudent that I should act the part of your master, and you that of my slave, and if they intend to kill you I shall be taken and executed. They will not heed a servant." After some entreaties, Avak consented, and on arriving before Batu, David passed himself off as his master. Batu treated them very well, and, seeing they had nothing to fear, on a further interview Avak himself passed in front. His host was astonished, and on having the matter explained to him, greatly praised David, saying: "If this be the quality of the Georgian race, I order it to have pre-eminence over all the races subject to our khanate;" and he issued a special order in this sense, and gave him an introduction to the Khakan.† Shortly after, the Mongol Noyans determined that the young King of Georgia should also visit Karakorum. He was accordingly sent there, and was accompanied by Bega, son of Grigol Suramel, eristhaf of Karthli; and the senior chamberlain, Beshken, son of Makhunjag Gurcelel, to whom were confided two pearls of great price. The party followed in the footsteps of Avak and Rokn ud din, and first went to the camp of Batu and then to the capital, where David was well received, and where he stayed with Avak.‡ Meanwhile, his mother, Rusudan, was living in the mountain district of Suaneth and Abkhazia. We are told she was pressed by Batu§ and by Baichu to go to their Courts. Having sent her submission to the former, he gave orders that she was to go and live at Tiflis. She was, however, much chagrined at the course of events and the absence of her son, and is said to have taken poison in her embarrassment how at the same time to conciliate Batu and Baichu, who were very jealous of one another. She was buried in the tomb of her family in the monastery of Gelath, and is still to be seen represented in rich costume on the walls of the church there.‖ The date of her death is not quite certain. The Georgian annals give the impossible year 1231. Wakhucht gives

---

* Hist. de la Géorgie, 522. † *Id.*, 522-523. ‡ *Id.*, 528.
§ He is called the chief of the army which occupied the country of the Russians, of Ossethi and Derbend, by Guiragos.
‖ Hist. de la Géorgie, 528-529. Note.

1237, but there is a letter extant sent by Pope Gregory the Ninth to her, showing she was living at that date. Chamchean says she poisoned herself in 1247, and it is not improbable that it really occurred in 1245-1246.

Georgia was now without a sovereign, Rusudan's son being away in Mongolia, while her nephew, according to the Georgian annals, was still a prisoner with the Sultan of Rum. The country was accordingly partitioned by the Noyans, who nominated chiefs of ten thousand, or *Thumnis-mthawars*. The first of these, we are told, was Egarslan Bakurtzikhel, a great orator and warrior, but *without any worldly goods*. To him they confided the forces of Hereth, Kakheth, of Kambejian, and the country from Tiflis as far as the mountains of Shamakha. Shahan Shah was given the appanage of Avak, in addition to his own. Vahram Gagel was given all Somkheth. Gregol got Suramel and Karthli, Gamrecel, of Thor, the rival in bravery of Egarslan, commanded in Jawakheth, in Samtzkhe, and as far as Karnukalak. And, lastly, Tzotné Dadian and the Eristhaf of Radsha, in all the kingdom beyond the mountain of Likh.* Their various troubles, and the harsh rule of the Mongols, drove the Georgians to despair, and we are told the Mthawars of Imier and of Amier held a meeting at Kokhta. There were present Shahan Shah, Egarslan, Tzotné Dadian, Vahram Gagel, Kupro Shotha, and the chiefs of Hereth, Kakheth and Karthli, with Gamrecel of Thor, Sargis of Thmogwi, the Meskhians, and those of Tao, and they decided to band themselves against the Tartars. Karthli was fixed upon as the place of meeting, and all withdrew to make preparations. When news of this plot reached the Noyans, Baichu and Angurg (? Arghun), they hastened to the borders of Kokhta, where they found the Georgian leaders, who had not yet collected their people. They were captured and taken to Shirakawan, in the district of Ani. On being brought before Charmaghan† they declared that they had no intention of rebelling, but had merely met to settle their own affairs, and to arrange the levying of the kharaj, or tax. The Noyans did not credit this, ordered them to be stripped, to be bound together, made them sit down naked and in chains, notwithstanding the heat, and threatened them with death if they did not confess. These punishments were repeated on several days.

Meanwhile, Tzotné Dadian, who lived a considerable distance away, and had gone to bring his people to the general rendezvous appointed by the conspirators, reached Rcinis-Juar, between Samtzkhe and Ghado, where he heard how the princes had been carried off to Ani. He dismissed his people, and traversing Samtzkhe and Jawakheth, went himself to that town determined to share their fate. The Noyans had reached Ani, and their prisoners, the Georgian mthawars, were seated in the hippodrome there, naked and with their arms bound. Seeing them in

---

* Hist. de la Georgie, 529.
† This is an anachronism, for, as we have seen, he was dead.

this miserable condition and condemned to death, Tzotné Dadian dismounted, took off his clothes, had himself bound, and seated himself among them. The Tartars, who were astonished at this, and knew him well, asked for an explanation. He replied that they had merely assembled to regulate the kharaj, but had been treated as malefactors, and he thought it right to come as a witness. If they had done anything worthy of death he wished to die with the rest, while if they were innocent he wished to share in their justification. The Tartars, we are told, were astounded at so much virtue. "As the Georgians," they said, "are so good that they do not betray each other, and this prince has come from Abkhazia to sacrifice himself for his friends, and to devote himself to death, they are innocent of the crime, and we remit their punishment." The various chiefs were accordingly allowed once more to return to their homes.*

We now read that the didébuls of the kingdom met and blushed to have Egarslan as their head, who was of no better blood than their own. Thereupon Shahan Shah, Vahram Gagel, Kuarkuaré-Jakel, Sargis Thmogwel, Grigol Suramel, eristhaf of Karthli, Gamrecel of Thor, the Orpelians, and several mthawars, met together and concerted about a ruler, and especially about a strange rumour which had reached them that David, son of Lasha, was still living and a prisoner in Rum. They reported what they had heard to the Mongol commanders, and begged that they would restore to them the imprisoned prince. They consented, and Angurag was accordingly deputed to fetch him home. With him went Vahram Gagel and Sargis Thmogwel. When they reached the Court of the Sultan of Rum he told them that he had put the young prince in the pit seven years before, and that he must have died long ago. They then assured him how they had learnt he was still living. A man was accordingly sent to see. David was drawn up out of his retreat. He was half dead and demented—stiff and cold as one dead; his skin was yellow, his hair reached to his heels, while his nails were grown of an immense length. Vahram and his companions were moved to tears by the piteous sight. He was duly bathed, and dressed in suitable clothes and ornaments. Ghiath ud din professed to be greatly distressed at what had occurred, asked him to pardon him, and sent him back.† The story, which as told in the *Georgian Chronicle* contains several anachronisms, is also referred to in the history of the Orpelians, where we read that Rusudan made two attempts on her nephew's life, in one of which he was put into a chest and thrown into the sea, and in the second the people who had orders to kill him threw him into a deep pit. She afterwards shipped him to a distant country, and he eventually reached Mangu Khan.‡ Guiragos tells us that Rusudan having refused

* Hist. de la Géorgie, 533-535. † *Id.*, 536-537.
‡ St. Martin Memoires, ii. 155. Brosset, Hist. de la Siounie, 235.

to go to the Mongol camp, or to send her son, and Baichu being jealous of her intercourse with Batu, determined to set up her nephew, the son of Lasha, who was living with the Seljuk ruler, Ghiath ud din, and who had imprisoned him so that he might not plot against her.* Malakia tells us that Vahram, Lord of Gag, together with a Tartar chief and an escort of 100 men, were sent in all haste to Cæsarea for him. They duly found the young prince in a deep well, where he had been preserved by the divine will. He was tall and fair to look at, with a brown beard, and full of wisdom and divine grace. Having dressed him appropriately, and seated • him on horseback, they took him with them to Tiflis, whence, by order of Baichu Noyan and Ailthina Khatun, he was sent on to the Court of the Grand Khan.†

The *Georgian Chronicle* says that although the Georgian grandees, Shahan Shah, his son Zakaria, Kuarkuarch-Jakel, Grigol Suramel, eristhaf of Karthli, the Orpelians, Gamrecel, Shotha-Kupri, and all the mthawars, except Egarslan, went to meet David Lasha, and received him with joy, they did not recognise him as King, but sent him to Batu, in company with Shahan Shah and Zakaria, Akbuka, son of Vahram, and Sargis Thmogwel. The author of this work evidently treats Batu as the supreme ruler of the western possessions of the Mongols to the south as well as north of the mountains. When David and his companions reached Batu's camp, he detained Zakaria, son of Shahan Shah, and Akbuka, son of Vahram, and sent David on to the Imperial Court, escorted by Sargis Thmogwel and other Georgians. There were thus two Davids, aspirants to the Georgian throne, both at Karakorum. When David Lasha arrived he was met by Avak, the Suramel Gamrecel, and the first chamberlain, Beshken, and they made a long stay at Karakorum.†

Meanwhile the Mongols had begun a campaign against the Ismaelites or Assassins, which proved a very protracted one. They advanced against their chief fortress, Alamut, taking a body of Georgians with them. During the siege the citizens sent one of their number, who, evading the guards, made his way to the tent of Chaghatai, or Jagatai, one of the principal Mongol leaders, and assassinated him. In the morning the guards, having discovered the dead body of their master, began to weep for him. His troops also collected about his tent. It was not known who had done the deed, and it was declared that the Georgians, who had been much ill-used by the Mongols, had done it. Charmaghan (?) opposed this view, and declared that the Georgians were not a race of homicides. The exasperated soldiers, however, made their way to the Georgian camp, some of whose occupants prepared to defend themselves, while others, feeling too weak, awaited the turn of affairs. Thereupon Grigol Suramel, eristhaf of Karthli, spoke out,

* Op. cit., ed. Brosset, 157. Journ. Asiat., 5th ser., xi. 437-438.
† Brosset, Add., 449. ‡ Op. cit., 537-538.

and said they were too weak to resist, and that resistance would assuredly lead to their being exterminated, while if they refrained the Mongols would merely revenge themselves on a few thawads like himself, and spare the rest. He advised them all, meanwhile, to go down on their knees, and in sets of three to implore the aid of the Virgin. Our naive chronicler says that when they had done so, and the Tartars were advancing to overwhelm them, a man came out of the reeds holding a poised lance soiled with blood. Raising his arm on high, he cried out, "Man kuchem Chaghatai! man kuchem Chaghatai," which in Persian means, "It is I who killed Chaghatai." Thereupon the Tartars rushed upon him. He fled again among the reeds. These were fired, and he was driven out and captured. Brought before Charmaghan (?), Yassaur, and Baichu, and being interrogated by them, he said he was a distinguished Mulahid (i.e., an Assassin), that his chiefs had given him plenty of gold, and bidden him go and kill one of the four Noyans. On being asked why, having hidden among the reeds, he had come forward and confessed his crime in the face of all, he replied that while he was in the thickest part of the reeds a beautiful woman had met him, saying, "What have you done? You have killed a man, and many innocent people will suffer death for it." "What should I do, Queen?" I replied. "Go forth and say you did it, and thus save a crowd of people." "I thereupon rose and followed her, and she led me towards you. When I had made my confession she disappeared. I know not whence she came." The Tartars thereupon clave him in two. The *Chronicle* compares the beneficent act of the Virgin on this occasion with her intervention to save Constantinople when attacked by the Khakan of the Avars in 626.*

Having described the various troubles brought upon Georgia by the Mongols, it is well to recall them in a more humane capacity. Guiragos tells us how, in 1247, the Vartabied Hoseph, who went about repairing the damage done by the Turks and the Georgians, visited a Tartar chief named Angurag, who had his summer quarters near the tomb of the Apostle Thaddeus, who gave him permission to clean the church and re-dedicate it. He also restored the monastery, and assembled a crowd of worshippers. The Tartar, we are further told, caused roads to be prepared in various directions to it, and issued orders that the monks were not to be molested by his people, many of whom had their children baptised.† This is not the only instance we have of the very considerate treatment of the Christians by the Mongols. The Syrian doctor previously named having mentioned Ter Nerses, the Catholicos of the Aghuans, to Ailthina Khatun, widow of Charmaghan, he was summoned to her camp. He was then living in the monastery of Khamshi, in the district of Miaphor, and was subject to Avak. He duly

* Op. cit., 530-531.
† Op. cit., ed. Brosset, 154-155. Journ. Asiat., 5th ser., xi. 446-447.

went to Mughan, carrying with him suitable presents. The Syrian doctor was then absent at Tebriz. He was nevertheless well received by Ailthina, who gave him a seat above her principal officers. They were assembled to celebrate the wedding of her son, Bora or Basra Noyan, with the daughter of a chief of high rank, Khutan Noyan, and of her daughter with another chief named Usur Noyan. She gave the Catholicos an introduction to her brothers Sadik Agha and Gorgoz, who were Christians and had lately arrived from their country, and who treated him with great consideration. She also gave him presents, and a tamgha protecting him from imposts, and assigned him a Mongol as an escort, who conducted him back to the country of the Aghuans (*i.e.*, to Arran), and went with him about his diocese, where for a long time he and his predecessors had hardly dared to show their faces on account of the Mussulmans.*

The inauguration of Kuyuk Khan, in 1246, was attended by a very remarkable body of persons of rank and consequence, from many latitudes; an assemblage which, better than aught else, proves the far-reaching power and influence of the Mongols at this period. Abulfaraj tells us that in addition to his relatives, the descendants of Jingis Khan, it was attended by the Amir Masud Beg, from Mavera un Nehr and Turkestan; by Arghun, from Khorasan, who was accompanied by the grandees of Irak, Lur, Azerbaijan, and Shirvan; by Sultan Rokn ud din, of Rum; by "The Constable," *i.e.*, Sempad, brother of Haithon, King of "Cilicia;" by the two Davids, the Greater and Lesser, from Georgia; by the brother of the Malik of Syria, Al Nasir; by the chief judge, Fakhr ud din, from Baghdad (representing the Khalif); and some chiefs of Kuhistan, representing Alai ud din, Lord of Alamut.† From other sources we learn that the famous kuriltai was also attended by Yaroslaf, Prince of Russia; by the son of Bedr ud din Lulu, Lord of Mosul; by Kutb ud din, cousin of the ruler of Kerman; and by a Prince of Fars. A notice of the visit of Sempad is contained in a letter he wrote to Henry, King of Cyprus (*i.e.*, Henry of Lusignan), his sister Emelin, and his brother John de Hibelin. He tells us that, journeying to further the cause of Christianity, he arrived at Sautequant (otherwise read Saussequant, *i.e.*, no doubt, Samarkand). He saw many large and opulent cities which had been laid waste by the Tartars, some three miles in circuit, and more than 100,000 mounds of bones of those whom the Tartars had killed. He says he crossed one of the rivers of Paradise, called Geon (*i.e.*, the Jihun). After journeying for eight months, he had barely traversed one-half of the dominions of the Tartars, whom he describes as excellent archers, terrible to look at, and very numerous. Five years, he says, had elapsed since their Great Khan had died, and a general

---

* Guiragos, ed. Brosset, 144. Journ. Asiat., 5th ser., xi. 438-441.
† Op. cit., Chron. Arab., 320.

assembly took place of all their notables to elect a successor. They came to this meeting from various directions—some from India, others from Cathay, others from Russia, others from Cascat (*i.e.*, Kashgar) and Tangath (*i.e.*, Tangut). This is the land, he says, whence came the three kings to worship Christ at Bethlehem, and the people of that land were Christians. He had been in their churches, and seen Christ painted with the three kings making their offerings, one of gold, one of incense, and the third of myrrh. He says further, the people of Tangath had been converted by the three kings, and their Khan had thus become a Christian. At the doors of the Tartar tents were churches, where bells were rung after the fashion of the Latins, and paintings after the manner of the Greeks, and it was customary to attend service in the early morning, and afterwards to pay respect to the Khan in his palace. He found many Christians scattered throughout the east, and saw many churches which had been devastated by the Tartars before they became Christians. He tells us the Tartars had made an invasion of India, and carried off 500,000 Indians, so that the East was full of Indian slaves. He also heard that the Pope had sent to the Khan to inquire if he was a Christian, and why he had sent his people to destroy the Christians and others. To this he had replied that God had ordered him to send his people to destroy the bad, and as to whether he was a Christian or no, if the Pope wished to know he had better go and inquire for himself. This last paragraph doubtless refers to the mission of Carpini and Benedict of Poland.* Malakia, speaking of these events, tells us that Haithon, having determined to submit to the Tartars, and to pay them tribute and the *khalan*, so that they should not enter his country, entered into an arrangement with Baichu Noyan, after which he sent his brother, the Baron Sempad, Generalissimo of Armenia, to Sair. Khan (*i.e.*, the Good Khan, meaning Batu), who then ruled over the dominions of Jingis. He set out and had an interview with Sain Khan, who greatly loved the Christians. He received Sempad very graciously, and gave him the title of Sgamish (?) and a Mongol Khatun for a wife, named Bkhtakhavor. He was furnished also with a great yarligh and a golden paizah.† Sempad was very well received, and returned with letters patent for his brother, and an order for the restitution of various districts which had once belonged to King Leon, and of which he had been deprived by the Sultan of Rum, after the death of that prince.‡ Sempad was accompanied on his way home by Rokn ud din, the Seljuk Prince of Rum. On the latter's arrival at Kuyuk's Court, one of his officers, named Baha ud din, the interpreter, had accused the Vizier of Rum of having set up Iz ud din without the Khakan's consent, and abused him for his other recent acts. Kuyuk thereupon ordered the deposition of Iz ud din, and his replacement by Rokn ud din, and also that Beha ud

* William de Nangis, Gesta Sancti Ludov. Dom. Bouq., xx. 360-362. † Malakia, 448. ‡ Guiragos, ed. Brosset, 157-158. Journ. Asiat., 5th ser., xi. 452-453.

din should have the post of Vizier. The latter, on his return, proceeded with 2,000 Mongols to proclaim Rokn ud din at Erzenjan, Sebaste, Cæsarea, Malatia, and in the fortresses of Saida and Amid.* The Vizier, Shems ud din, is, perhaps, the same person as the brother of Ghiath ud din, who, we are told by Guiragos, had married a daughter of Leshkar, Sultan of Greece, who reigned at Ephesus (*i.e.*, Lascaris, the Emperor of Nicæa), and who had usurped power at Iconium (?), thanks to the assistance of his father-in-law, while his young brother had done so at Halaia, a town of Western Karamania.†

When Shems ud din heard of the decision of the Khakan, he sent Rashid ud din, the Prefect of Malatia with a quantity of treasure to the Khakan, to obtain a revocation of the order; but having heard of the rapid approach of Baha ud din, he deposited the treasure at Kamah, and fled to Aleppo. Shems ud din now tried to escape with his protégé, Iz ud din, from Iconium, so as to set him up in the maritime district ; but he was seized and imprisoned, and presently Baha ud din sent a body of Mongols, who tortured him until he disclosed where his treasures were, and then put him to death. Abulfaraj tells us he was a learned man, and wrote some elegiac verses on his own fate, which were elegant and steeped in pathos. It was now arranged, by the influence of an ascetic named Jelal ud din Keratai, who had great influence at Iconium, and who had been instrumental in arresting Shems ud din, that the empire should be divided between the two brothers : the western parts, with Iconium, Akserai, Ancyra, Anatolia, &c., being assigned to Iz ud din; and the eastern districts, including Cæsarea, Sivas, Malatia, Erzenjan, Erzerum, &c., being given to Rokn ud din ; while large private domains were made over to Alai ud din. The partisans of Rokn ud din wished to insist, however, on the Khakan's decision being carried out to the letter. An interview between the brothers to settle matters was arranged at Axara or Cæsarea, where Rokn ud din and his chief supporter, Baha ud din, were treacherously surprised by some partisans of Iz ud din, who carried them off to Iconium. He did not treat them badly, however, and eventually the empire was jointly ruled, and the coin was struck in the names of all three brothers, with the inscription : "The very great Kings, Iz, and Rokn, and Alai."‡ Brosset says the names do, in fact, occur together in the year 647 HEJ. (*i.e.*, 1249).

To return again to Kuyuk's inauguration as Khakan. It was there decided that the two Davids should occupy the throne of Georgia after one another, the older of the two, David, son of Lasha, reigning first.

---

* Abulfaraj, Chron. Syr., 526-527.   Chron. Arab., 321.

† Guiragos, ed. Brosset, 158.   Brosset says that he could not find any confirmation of this match in the "Familiæ Augustæ" of Ducange, nor in the articles on Lascaris and Vataces, nor in that devoted to the Sultans of Iconium, nor, lastly, in the two chronicles of Abulfaraj.   At this time John the Third Vataces, and not Theodore Lascaris, was Emperor of Nicæa. He reigned 1222-1255.

‡ Abulfaraj, Chron. Syr., 527-528. Chron. Arab., 321-322.

Kuyuk ordered a splendid throne, belonging to the Georgian kings, and a marvellous crown, which had belonged to Khosru the Great, the father of Tiridates II., King of Armenia, and had been taken to Georgia for safety with other things, to be sent to him. The remaining objects in the treasury were to be divided between the two princes.* On the return of David, the principal chiefs in the Mongol service, Avak, who had the rank of generalissimo, Shahan Shah, son of Zakaria, Vahram, and his son, Akbuka, took him to Medzkhitha, where they summoned the Georgian Catholicos, and had him consecrated. In gratitude to Vahram, he styled himself Vahramul (*i.e.*, enthroned by Vahram).† David, son of Lasha, lived at Tiflis, and the other David in Suaneth.

At the kuriltai above named, the envoys of the Assassins were ignominiously expelled, while those of the Georgians, Franks, and of the Khalif were sharply upbraided.‡ Kuyuk superseded Baichu as generalissimo of the forces in the west, and appointed in his place Ilchikidai, called Elchi Gaga by Guiragos. He was the son of Khadjiun, Jingis Khan's brother, and had distinguished himself at Herat in Jingis Khan's invasion. Abulfaraj, who calls him Iljiktai, says he was given charge of Rum, Mosul, Syria, and Georgia.§ He was authorised to receive the taxes there, and each of the princes of the blood was ordered to furnish two men out of every ten to form his army, and he was, on arriving in Persia, to make a similar levy there. Kuyuk announced his intention of himself marching to the west, and the army of Ilchikidai was to act as his advance guard.‖ We have seen how, on the death of Ogotai, Arghun was nominated Governor of Khorasan, &c.¶ Having left several commissaries in Khorasan to receive the tribute, he hastened on to Irak and Azerbaijan to relieve those provinces from the exactions of the Mongol commanders, who treated them as if they were their private property. At Tebriz he was met by envoys from the rulers of Rum and Syria, who tendered their masters' homage, and he sent deputies to collect the taxes there. Meanwhile, the general control of the finances was left in the hands of Sherif ud din, the Ulugh Bitikji, who obtained his post through the influence of Fatima, a favourite of the regent, Turakina.** He behaved in a very cruel and arbitrary manner, put spies in the houses of the people, kept them without food, or put them to the torture, in order to extract more from them. The ministers of the Muhammedan faith, the widows and orphans who had been treated

---

* Guiragos, ed. Brosset, 157.   Journ. Asiat., 5th ser., xi. 451-452.
† Malakia, 449.   Lebeau, xvii. 460.   ‡ Abulfaraj, Chron. Syr., 525.   Chron. Arab., 321.
§ Chron. Arab., 320.   ‖ D'Ohsson, ii. 205.
¶ Arghun and his family filled a famous rôle in the history of the Ilkhans. He was a Uirad by birth, and had been sold by his father during a famine to a Jelair officer, named Iluki, from whom he passed into Ogotai's service. As he could write the Uighur character he entered the Chancellery, and was appointed jointly with Koban on an important mission in China. It was apparently his address on this occasion which caused him to be selected as arbitrator between Ungu Timur and Kurguz, as we have described. (D'Ohsson, iii. 121-122. Ilkhans, i. 89.)
** He was the son of a porter at Khuarezm, and entered the service of Chin Timur, whom he accompanied to Khorasan as secretary, his knowledge of Mongol making his services invaluable. (D'Ohsson, iii. 122-123.)

with tenderness by Jingis Khan, were now trampled upon. At Tebriz, people pledged and sold their children, and a teacher even had to sell the shroud of one of his tenants who was dead. At Rai the proceeds of the various exactions were piled up in the mosques, into which the sumpter beasts were driven, while the sacred carpets were used to cover up the goods. Happily Sherif ud din died in 1244, and Arghun tried to alleviate the misery he had caused by remitting some of the taxes and releasing some of the victims. He now set out for Tartary with a great crowd of functionaries and many presents to attend the inauguration of the Khakan Kuyuk, to whom he handed over, much to the Khakan's satisfaction, a great quantity of illegal assignations of revenue, &c., which had been issued during the regency. He was retained in his government, and was nominated as civil governor and head of the finances of Khorasan, Irak-Ajem, Azerbaijan, Shirvan, Kerman, Georgia, and that part of Hindustan then subject to the Mongols (i.e., the Punjaub, as far as the Biah). The post of Ulugh Bitikji was conferred on Fakhr ud din Bihishti. Arghun was met on his return at Merv by the various grandees of the country, who welcomed him at a great feast.*

Guiragos says that on his accession, Kuyuk sent commissaries to collect a tithe of the property secured by the Mongol armies in Persia, as well as to levy taxes on the various conquered countries. He says that Arghun, who had attended the kuriltai, where Kuyuk was elected, was at their head, and under him was a very tyrannical official named Bugha. Surrounded by a crowd of Persians and other Mussulmans, he made heavy exactions from the grandees then in camp without anyone daring to oppose him. He seized the Armenian Prince Hasan, surnamed Jelal, and put him to the torture; seized and demolished his strong fortresses of Khoiakhan, or Khokhanaberd (now ruined, and situated near Kantza Sar), Degh, or Tet, Dzirana-Kar (the two latter near Khokhanaberd), &c., and so completely destroyed them that their traces were not to be seen when he wrote. Hasan barely saved his life by the payment of a large sum of money. Bugha tried to seize Avak also and to put him to the torture, but having been warned, the latter showed such a bold front with his people that he was cowed. Bugha shortly after died of an ulcer in the throat.† Arghun had enemies at head-quarters whom it was necessary to appease, and had reached Taraz, on his way thither, when he heard of Kuyuk's death. He received orders from Ilchikidai to return to make provision for the campaign, which he proposed making. In 1249-50, the Mongols made another raid upon the territory of Baghdad, and advanced as far as Dakuka, where they killed the Prefect Bilban. The next year Nasir, Prince of Damascus, received letters of safety from the Khakan, which he carried in his girdle, and for which he showed his gratitude by sending handsome presents.‡

---

* D'Ohsson, iii. 126. Tabakat-i-Nasiri, 1152. Note.
† Guiragos, ed. Brosset, 155-156. Journ. Asiat., 5th ser., xi. 447-449. ‡ Novairi, D'Ohsson, iii. 91.

Meanwhile, Georgia continued to suffer from the Mongol depredations. Malakia tells us that the pious and good King David and his Court passed their time in enjoying themselves, and in drinking. One day there was a grand feast, and as the Georgians were great boasters, and fond of using big phrases, a Georgian prince began to sum up the number of the various princes subject to the King, and boasted that there were a thousand grandees, and that one of them had 700 soldiers ready to defend their master. These words were re-echoed in the crowded feasts, and they began to count the forces which the Armenians and Georgians could bring together against the invaders. The Tartars were, meanwhile, very exacting, and demanded much from the Georgian princes and generals : from some, gold or cloth ; from others, gerfalcons, a good dog, or a horse, &c., all of which were demands in excess of the regular imposts, the *mal*, the *thaghar*, and the *khalan*. These exactions were the cause of the murmurs that arose at the feast, which were duly reported by some traitor, and led to a fresh invasion of the country and fresh pillage. The King and principal grandees, including Avak (who being ill and not able to ride, was dragged off in a coffin), were taken to the tent of their chief.*

Guiragos says that their intention of putting them to death was prevented by Jagatai, one of their principal commanders, who was a friend of Prince Avak, and who adjured them at their peril not to kill those who were peaceable subjects of the Khakan without the latter's authority. Khochak, Avak's mother, who had gone to their camp, offered to guarantee his fidelity. The Mongols proceeded to tie them together with cords, and kept them thus for three days, jeering at them meanwhile, to show their contempt. Having then made them give up their horses and pay a ransom, they let them go ; but they nevertheless invaded Georgia and plundered a number of districts, indifferent whether they were rebellious or not. They made a great many prisoners, and, we are told, threw a crowd of children into the rivers. This took place in 1249, and the next year Avak died, and was buried with his father, Ivaneh, at P'ghntzahank.† He left only an illegitimate son, and a daughter Khochak, who was very young. His principality was given by the Mongols to his cousin Zakaria, the son of Shahan Shah. They soon after deprived him of it, and made it over to Avak's widow, Vartoish Kontsa. Sempad had been nominated guardian to Avak's children. He soon quarrelled with her, and by order of Khulagu had her drowned. Khochak, Avak's daughter, was eventually married to Shems ud din Muhammed, Khulagu's Vizier, and brother of the historian, Juveni.‡

Mangu Khan's inauguration took place on the 1st of July, 1251. Guiragos has a curious story to tell about this election, which is interesting as that of a contemporary, and which I had overlooked in

---

* Malakia, op. cit., 450.　† Guiragos, ed. Brosset, 158-159.　Journ. Asiat., 5th ser., xi. 456.
‡ Brosset, Hist. de la Siounie, 234.　St. Martin, ii. 151 and 288.

writing the previous volumes.   He describes Batu as occupying the vast
plains of Kipchak with an immense army, and as living under tents,
which, during the migrations of his people, were transported on waggons,
drawn by great teams of cattle and horses.   The princes of his family
recognised his supremacy, and he who became khan had need of his
countenance.   On the death of Kuyuk, he goes on to say, they offered the
post to Batu himself or to the one he should nominate for it.   He set out
for the purpose of fulfilling this duty, leaving his son Sertak in command
at home.   When he nominated Mangu some members of his family were
displeased that he did not either mount the throne himself or place Khoja
Khan (*i.e.*, the Khoja Ogul of the Persian writers), the son of Kuyuk, on it.
They did not dare to openly oppose him, but revolted against Mangu,
whereupon he ordered several of them to be put to death, including Elchi
Gaga (*i.e.*, Ilchikidai), who had been nominated generalissimo in Persia
in the place of Baichu.   He was denounced to Batu by the chiefs of
the army, who were afraid of his haughty temper, was accused of
refusing to support Mangu, taken before Batu in chains, and by him
was put to death.*   Ilchikidai was arrested at Badghiz, in Khorasan.
His two sons, who were at the Imperial Court, were put to death, by
having stones thrust into their mouths.†   After this, we are told by Guiragos,
kings, princes, and great merchants, together with those who had been
molested or plundered of their goods, sought out Batu, who decided
impartially among them, and committed his decision to writing sealed
with his tamgha, and no one dared disobey his orders.   Guiragos says
positively that Batu's son, Sertak, was brought up by Christian governors,
and that when he grew up he embraced Christianity, and was baptised by
the Syrians, who had brought him up.   He was very good to the
Christians, and with the consent of his father he freed the Christian priests
from the payment of dues, and extended the same privilege to the
mosques and those who served them.   His camp was constantly
visited by Christian prelates, and attached to it was a tent where the
sacred mysteries were constantly performed.   Among those who visited
his camp was the Armenian Prince, Hasan, familiarly styled Jelal, who
was courteously treated by Sertak.   There also went the princes Gregory,
habitually styled Tgh'á, *i.e.*, infant, although he was an elderly man ; the
Prince Desum, the Vartabied Mark, and the Bishop Gregory.   Sertak
conducted Jelal to his father, Batu, who restored him the fortresses of
Charapert, Agana, and Gargar, in the district of Khachen, and the
province of Artsakh, of which he had been deprived by the Turks and
Georgians.   He also received a diploma in favour of the Catholicos of
Aghovania, or Albania, Nerses, granting him exemption from taxes,
and a free right to traverse the various dioceses of his patriarchate.   Jelal
returned, well satisfied, but presently, harassed by Arghun and his people,

---

* Guiragos, ed. Brosset, 172. Journ. Asiat., 5th ser., xi. 458.   † D'Ohsson, ii. 259.

he repaired to the Court of Mangu Khan.* We described the doings of Arghun until the death of Kuyuk. During the interregnum which followed, fresh and illegal assignations were issued to the various princes of the blood, who again settled like gad-flies upon the unfortunate country. With his subordinates, Arghun received a summons to attend the kuriltai, where Mangu was elected Khakan in 1251. He reported how the country was being ruined by the issue of indiscriminate taxing orders, and Mangu ordered the various intendants of Persia to present each a separate report on the evils which affected their districts, and the remedies they proposed. They were all agreed that the best plan was to introduce a capitation tax proportioned to the means of those who had to pay, similar to the one Mahmud Yelvaj had established in Transoxiana. This was decided upon, and a poll-tax varying from one dinar to ten per head was appointed, the proceeds of which were to be devoted to paying the soldiery and keeping up the postal communications, and on no pretence was more to be exacted. Arghun's skill and prudence secured his re-appointment, and he received a paizah or official tablet marked with a lion's head. Baha ud din Juveni, the famous historian, was nominated finance minister, while a second finance minister, named Sarraj ud din, was nominated as his coadjutor by Nikbey, who ruled over the dominions of Jagatai. The Khakan nominated two commissaries in addition, while each of his brothers, Khubilai, Khulagu, Arikbuka, and Moga had his agent at his Court. Persia was divided into four governments, the governor of each being styled Malik, and having a paizah marked with a lion's head. Their subordinates had tablets of gold or silver according to their rank. The Khakan, in sending them to their appointments, presented them with robes of Chinese brocade.†

We saw how Sempad succeeded his brother Elikum as head of the Orpelians, and ruler of a large district in Armenia. He was a very accomplished person, and we are told could speak five languages, namely, Armenian, Georgian, Uighur, Persian, and Mongol.‡ The Orpelians were at feud with the family of Avak, who secretly intrigued against them. The Mongol general Baichu, we are told by Stephen the Orpelian, was at this time encamped at the entrance of Tzagé Tzor, in the province of Haband. He says he took by force David, the Little King of Siunia (i.e., David, son of Rusudan), and detained him prisoner in his camp, but he some time after succeeded in escaping at night with three companions. David had with him a beautiful precious stone, of great size and brilliancy, and of a red colour, probably a ruby. He also had a piece of the true cross, which was valued more than all his kingdom. He passed through Kudeni, which belonged to one of the nobles of Sempad, named Tankreghul (i.e., servant of God), who tried to arrest

---

* Guiragos, ed. Brosset, 173.   Journ. Asiat., 5th ser., xi. 459-460.
† D'Ohsson, iii. 126-128.   ‡ Ante, 22-23.   St. Martin Memoires, ii. 127

him, whereupon he drew out a little leathern bag, which was suspended
about his neck, containing the precious objects already mentioned.
This he gave to his captor, and told him to give it to his master Sempad,
for it was worth more than his kingdom.   He was to tell him to keep it,
and that if he once more regained his kingdom, he would reward him
with any town or district he might ask for.   If he should not succeed
in this venture he might keep it for himself.   Sempad, when he received
this present, thanked God for it ; but fearing he might not be able to hide
it, he thought it better to make a present of it to Mangu Khan, and at
the same time secure his pity for his countrymen.   He accordingly went
to Baichu, and asked him to take the jewel for himself if he wanted it, and if
not to let him go and offer it to the Khakan.   He bade him do the latter,
and provided him with an escort.   This was in 1251.   On the way he
visited the monastery of Noravank'h, where he offered prayers for a safe
journey and a happy termination of his mission.   He then went on, and
after a long journey reached Karakorum.   Mangu, Stephen says, was
pious, and had at the gate of his great palace a church, where services
went on continuously without molestation.   The Mongols, he says, loved
the Christians, whom they called Ark'haiun, and all the country professed
Christianity ! ! !

When he arrived, he visited the various grandees, and communicated to them the object of his journey.   They presented him to
the Khakan, to whom he gave the precious stone.   He was much pleased
with it, and inquired whence he had come.   Sempad then enlarged
on the desolation of Armenia, the loyalty of his brother, who
was in the service of Arslan Noyan, and the possessions he had
lost.   Mangu listened attentively to him, and then handed him over
to his mother, Siurkukteni (called Surakhthembek by our author),
who was the daughter of the Kerait Prince Jakembo, and gave him the
title of enchu (i.e., lord). *   He asked him to stay awhile at the Court, and
ordered his officials to supply his needs.   He lived there three years,
during which he was very diligent in his religious services.   He had with
him a small consecrated wafer, before which he said his prayers.
He was thus saying them on one occasion when, as Stephen says, a
luminous cross appeared, which shed its light over the place.   Mangu, we
are told, was informed of this, and himself went to see it.   Sempad was
unconscious of it.   When Mangu summoned him to explain he could only
produce the small host, whereupon Mangu descended from his throne,
bent his knees, uncovered his head, and declared that the cross upon it
was like the luminous cross he had seen.   After this he paid great
deference to Sempad.   He gave him a golden paizah, or official tablet,
and also a yarligh, or diploma, and conferred on him all the district which

---

* St. Martin says this title still subsists among the Mongols and Manchus in a slightly altered
form, namely, as edshan among the former, and edshen among the latter.

had been conquered by Arslan Noyan, together with Orodn, the fort of Borodn, and its revenues. He also obtained privileges for the clergy of Armenia. He now returned home again. With the help of Baichu he once more occupied his heritage of Orodn (as far as the frontiers of Borodn and Bghen), in which was situated Tathev, the episcopal see of Siunia, then in ruins. He also took Eghékis, and all the district of Vaio-Tzor, Phogha-Hank, Urdz, and Védi, with its dependent valley, as far as Eréron, and many places in the country of Kotaikh and Geghak'huni, and emancipated the clergy of his province and of all Armenia. He founded monasteries and restored ruined churches. For a long time the residence of the bishops of Siunia had been in ruins. The bishop, John, and his nephew, Hairapied, had begun to build a monastery with the permission of Baichu's wife, but could not continue it on account of their poverty, there only remaining one house out of all the property of the Church. Sempad now devoted all his efforts to this work.*

Stephen, the Orpelian, tells us that Arghun, the administrator of Persia, was summoned to Mangu's Court to answer a charge of treason, and that when Sempad arrived he found him in chains. He says that the charges against him were preferred by Sevinjbeg and Sharaphadin (i.e., Sherif ud din Khuarezmi, his naib or lieutenant, but the latter apparently died in 1244, vide ante). Sevinjbeg was also an enemy of Sempad's, and had some intention of poisoning him. On his arrival Mangu inquired from him about Arghun's proceedings, if he had ruined the country, put to death the priests, and been an assassin, as was reported. Sempad justified Arghun completely, and charged his enemies with being the real offenders. Thereupon Mangu summoned a council, and Sevinjbeg and Sherif ud din were put to death. Arghun was released from prison and promoted. He recommended Sempad to him, and they returned together.† Our author dates these events in 1256, during Sempad's second visit to the Court, but, as St. Martin argues,‡ they clearly refer to the first one, in 1251-4.

It was in 1254 when Arghun, who had been reinstated, as we have seen, arrived once more in the west, accompanied, according to Guiragos, by an official attached to Batu's Court, named K'ura Agh'a, or Thora Agha, charged with making a census. They inscribed all males above the age of ten on their registers, and insisted upon all paying taxes. The people again began to be ground down, torments were applied freely, and those who could not pay had to part with their children. The tax collectors were escorted by Muhammedan Persians, and they were assisted in their miserable work by those grandees whose property had been spent. These exactions did not suffice. They made all artisans pay a licence tax, they taxed the lakes and ponds where fish were caught, iron

---

* Hist. de la Siounie, ed. Brosset, 229-232. St. Martin Memoires, ii. 129-139.
† Hist. de la Siounie, ed. Brosset, 232-233. St. Martin Memoires, ii. 141-115.
‡ Op. cit., ii. 282. Note 4.

E

mines, smiths and masons—Brosset adds perfumers. They destroyed the canals which belonged to the native chiefs, and seized the salt mines of Koghb, situated at the foot of Mount Takhaltu, in the district of Surmalinski, south of the Araxes. They also extorted gold, silver, and precious stones from the merchants. Thus they reduced the country to great distress. One man alone remained rich. This was a merchant named Umeg, called Asil by the Mongols, who had been spared at the sack of Karin. At Tiflis, where he lived, he was styled "Father of King David." Having presented Arghun with some valuable gifts, he was treated with great consideration, as were also the clergy, about whom the Mongols had no orders from the Khan, also the sons of Saravan, Shnorhavor, and of Mkrtich.* The *Georgian Chronicle* tells us Arghun caused an inventory of everything to be taken, men and animals, ploughed lands and vineyards, gardens and orchards, while one peasant in every nine was inscribed on the rolls for military service ; the number of Georgians thus enrolled amounted to 90,000, which gives a male population of about a million, and as the clergy, both Christian and Muhammedan, were exempted, this would give a population of about 5,000,000 for the provinces of Karthli and Kakheth, in which David alone ruled—a number which seems impossible, for in the census of 1836 the whole population of these provinces was only 225,395. Our author says that each village furnished a lamb and a piece of gold for every chiliarch, and two sheep and a gold piece for each myriarch, as well as three whites per day for the keep of a horse. M. Brosset says the white is a mere money of account, and in modern times is of the value of the hundredth part of an abaz, an abaz being worth eighty kopeks.† According to the *Georgian Chronicle*, Arghun was a protégé of Batu Khan, of Kipchak, and it makes him employ him in all parts of his empire—in Russia, Khazaria, Ossethi, Kipchak, as far as the Land of Clouds (*i.e.*, the Arctic country), in the east and in the north, and as far as Khatai. It calls Arghun a friend of equity, very truthful in his language, a deep thinker, and profound in counsel, and says he was employed by Batu to make the census, to fix the military conscription, and to pay to each, great and small, according to his position, the dues for the horses, &c., furnished for the posts on the great roads. It also says that Batu sent him to Karakorum, to Khubilai Khakan, who employed him in a similar way in his dominions. Thence he went to the capital of Jagatai, in Turan, where Ushan (?) reigned, and having regulated matters there, crossed the Jihun, and did the same in Khorasan, Irak, and Romgor (? Rum), whence he passed, under the patronage of Khulagu, into Georgia and Greece (*i.e.*, the Seljuki territory).†

---

* Guiragos, ed. Brosset, 175. Journ. Asiat., 5th ser., xi. 462-463.
† The date of this census is not easy to settle. Vartan, Malakia, and Guiragos all date it in 1254. The *Georgian Chronicle* puts it after Khulagu's arrival in Persia; while St. Martin, basing his conclusion on the authority of Abulfaraj and Rashid ud din, dates it in 1250. (See Guiragos, ed. Brosset, 175-176. Note 6.) ‡ Op. cit., 550-551.

Mangu was visited by Haithon, the King of Armenia, whose journey has previously occupied us.* We shall have more to say about it presently, and will now turn to that of another Armenian prince. We have seen how the Georgian prince Avak and his family were at feud with Sempad, the head of the Orpelians. They constantly incited Arghun against him, offered him presents if they might be allowed to destroy him, promising at the same time not to appropriate to themselves any of his territory. He would not consent, but nevertheless they captured several of his towns and ravaged the remainder. He thereupon determined once more to visit the Court of the Khakan Mangu, and having obtained the permission of Arslan Noyan, he duly set out. This was in 1256.† He was well received by Mangu. On his return the favours he received from the Mongols disconcerted his enemies, and he continued to prosper under the patronage of Khulagu. The latter sent him to the country of Pasen, to cut wood for the palace he was building at Alatagh.‡ Haithon and Sempad the Orpelian were close allies of the Mongols. The former had his capital at Sis, in Cilicia, and the latter at Ani, situated at the junction of two streams which fall into the Araxes. It is said in the eleventh century to have had 100,000 inhabitants and 1,000 churches.§ Haithon's eldest daughter was married to Bohemund IV., Prince of Antioch, others married the Sieur de Saiete, the Sieur de la Roche, and Guy d'Ibelin, son of Baldwin, seneschal of Cyprus, respectively, which allied him closely with the Crusaders. His younger son, Toros, fell in Syria, in the Mongol campaign against the Mamluks in 1266, to be described presently. Purthel, nephew of Sempad, similarly perished on the Terek, in the struggle with Bereke. |

One of the complaints made against Baichu by Khulagu was that he had done little to push forward the fortunes of the Mongols, and it must be said that not much was certainly done during the later years of his authority, when he was, however, subordinate to Ilchikidai. In 1252-3 a Mongol division entered Mesopotamia, pillaged the districts of Diarbekr and Mayafarkin, advanced as far as Rees-Ain, and Suruj, and killed more than 10,000 people. They waylaid and plundered a caravan which was on its way from Harran to Baghdad, and thus secured *inter alia* 600 loads of sugar and of Egyptian cotton, besides 600,000 dinars. They then returned to Khelat.¶ The same year Yassaur, who had eight years before devastated Malatia, went once more there. He laid waste the country with fire and sword. Some of the Mongols passed by the town of Guba, assailed the monastery of Makrona, and demanded money and food from the monks. These miserable people in their simplicity refused to give any, thinking the invaders would withdraw.

* Ante, ii. 88-89.   † Brosset, Hist. de la Siounie, 232.   St. Martin Memoires, ii. 141.
‡ Brosset, Hist. de la Siounie, 233.   St. Martin, op. cit., ii. 145.   § Ilkhans, i. 165
‖ Journ. Asiat., 5th ser., xviii. table 2.   ¶ Novairi, in D'Ohsson, iii. 92-93.

They did withdraw for a while, but soon returned again, and again asked for something. As they were again refused, they attacked the monastery and set fire to the tower. All the monks, old and young, with 300 refugees from the neighbourhood, perished, but a large quantity of cotton, of wax, and oil, which was stored there was saved. Abulfaraj tells us he was at this time bishop of Guba.* We are told that, by the Khakan's express orders, Hindujak, a Mongol general commanding a tuman or 10,000 men, who had unjustly put the governor of Kum to death, was executed outside the gates of Tus, while his family, slaves, and other property were confiscated to the treasury, and partitioned among the four branches of the Imperial family. His father, Malik Shah, who belonged to the Sunid tribe, had entered Persia at the head of a tuman, consisting of Uighurs, Karluks, Turkomans, Kashgarians, and Kuchayens (*i.e.*, natives of Kucha, east of Kashgar).†

We have seen how the Seljuk kingdom of Rum was partitioned between Iz ud din and his brothers.‡ In 1254 the former was summoned to Mangu's presence. Afraid that his brother, Rokn ud din, would take advantage of his absence, he determined to send another brother, Alai ud din Kai Kobad, who set out, bearing many presents, by way of the Black Sea and the steppe of Kipchak, accompanied by Seif ud din Tarentai, one of his principal generals, and Shuja ud din, the governor of the maritime districts. Iz ud din excused himself on account of his fear that the Armenians and Greeks would attack his country if he were absent. Meanwhile the partisans of Rokn ud din forged a letter from Iz ud din to Tarentai and his colleague, ordering them to hand over Alai ud din and the presents he had with him to the chancellor, Shems ud din, and the amir Seif ud din Jalish, who bore the letter, and who would accompany the young prince to the Imperial Court. The two messengers overtook the travellers at the Court of Batu, whom they informed that Tarentai having been struck by lightning, could not present himself before the Grand Khan, while Shuja ud din was a doctor, skilled in necromancy, and meant to poison Mangu, and that consequently the Sultan had recalled them. Batu ordered the baggage of the two suspected officials to be searched. Some medicinal roots, *inter alia*, scamony, were found, and Shuja ud din was ordered to taste them, which he did, except the scamony, which aroused Batu's suspicion. This was allayed, however, by his doctors. He decided that all four should go on to the Court, the newly arrived messengers escorting Alai ud din, and those originally appointed bearing the presents. They set out separately. Alai ud din died *en route*. His mother was the daughter of the beautiful Queen Rusudan. When the rival officers arrived at Mangu's Court they pleaded the cause of their respective patrons. It was decided that Iz ud din

* Op. cit., Chron. Syr., 536-537.
† Rashid ud din, in D'Ohsson, iii. 128-129.   ‡ Ante. p. 58.

should retain that part of Rum west of the river of Sivas (Kizil Ermak), and Rokn ud din should hold the country thence to Erzerum; the tribute they were to pay was also duly fixed.   While the officials just named were absent, Rokn ud din's supporters raised some troops, and tried to surprise Conia, or Iconium.   They were beaten, and he was captured and imprisoned in the fortress of Davalu.   The following year, 1255, Baichu Noyan, annoyed at Iz ud din's tribute not being regularly paid, sent him a message demanding the surrendering to him of some fresh winter quarters, as Khulagu had appropriated those he had formerly used in the plain of Mughan.   The Sultan refused to do so, and treated Baichu cavalierly.   The latter, with the Armenian king Haithon, marched upon Conia, and defeated the Sultan's army between that town and Ak Serai.   Iz ud din took refuge with his family in the citadel of Anthalia.   Baichu thereupon took Rokn ud din from prison, and seated him on the throne.   Iz ud din now fled to the Emperor Theodore Lascaris, who was living at Sardis, and who, afraid of attracting the revenge of the Mongols, advised him to return home.   He accordingly did so, and sent in his submission to Khulagu, who maintained the division of the Seljuki kingdom fixed by his brother Mangu.   Iz ud din thereupon returned once more to Conia, while Rokn ud din went with Baichu into winter quarters in Bythinia.*

We will now continue the notice of the Mongol doings east of Khorasan.   We have seen how they became masters of Afghanistan.   In 639 (*i.e.*, 1242) Tair Baghatur, who was commander-in-chief of the forces about Herat and Badghiz, and other Noyans from Ghur, Ghazni, Garmsir, and Tukharistan, marched towards the Indus.   At this time the Malik Kabir Khan of Ayaz was the feudal chief of Multan.   On hearing of the bold front he had assumed, they advanced towards Lahore, where the Malik Ikhtiyar ud din Karakush was the feudal chief.   We are told that he was unprepared with either stores, provisions, or war materials, while the citizens were disunited.   Most of them were traders, and had been in Khorasan and Turkestan, where they had obtained safe conducts, and were careless about the fate of the Malik Kara Kush.   Meanwhile, the latter's feudal chief, Sultan Muiz ud din, Bahram Shah of Delhi, was at issue with his Turk and Ghuz troops, and there was, therefore, some delay in sending assistance from Delhi.   The Mongols proceeded to invest Lahore, and bombarded and destroyed its walls with a number of mangonels.   The Malik Kara Kush, feeling that from the disaffection and disunion inside it would not be possible to defend the city, made a sortie at night with his men, and cut his way through.   Some of the harem and of his retinue got separated from him in the darkness, and in the tumult dismounted and hid away among the ruins and graves.   The

* Abulfaraj, Chron. Syr., 542-543. Chron. Arab., 329-331. D'Ohsson, iii. 95-99.

following day the Mongols captured the place. Conflicts arose in all directions. "Two bands of Mussulmans in that disaster girded up their lives like their waists, and firmly grasped the sword, and up to the latest moment that a single pulsation remained in their dear bodies, and they could move, they continued to wield the sword and to send Mongols to hell, until the time when both bodies, after fighting gallantly for a long period against the infidels, attained the felicity of martyrdom," while among the latter a vast number perished, and we are told there was not a person among them who did not bear the wound of arrow, sword, or nawak (some projectile is here meant).* Two of the principal of these heroes were named Ak Sunkar, the Seneschal of Lahore, and Din dar Muhammed, the Amir-i-Akhur of Lahore.† The former is said to have had a single combat with the Mongol commander, Tair, in which both were killed, "one company to heaven; one to the flaming fire."‡ In regard to Tair, the statement that he then died is probably a mistake. The capture of Lahore was followed by the usual massacre of the old and useless, and the making captive of the young. Kutb ud din, Hasan the Ghuri, who had been sent with an army from Delhi to the relief of the place, arrived too late, and after the Mongols, who had suffered great losses, had retired.§ When he learnt of their retreat, Kara Kush retraced his steps towards the River Biah, where in his flight he had hidden some treasure, gold, &c. This he recovered, and then went on to Lahore, where he put to death the Hindu Khokhars and the Gabrs, who were committing destruction there.||

Minhaj-i-Saraj, the author of the "Tabakat-i-Nasiri," reports in regard to this campaign, that when he himself was about seven years old he used to go to the Imam Ali, the Ghaznivi, in order to learn the Koran, and from him he heard the tradition how the Imam, Jemal ud din, while he preached at Bukhara, during Ogotai's reign, used often to say, "Oh, God, speedily transport a Mongol army to Lahore that they may reach it," the explanation of which became evident when the Mongols captured Lahore in the month Jamadi ul Awwal, in the year 639 HEJ. A number of the merchants and traders of Khorasan and Mavera un Nehr afterwards declared that Ogotai died on the second day after the capture of Lahore.¶ Meanwhile, the Sultan of Delhi, Bahram Shah, was killed by some of his generals This was in May, 1242. His nephew, Alai ud din Masud Shah, mounted the throne in his place.**

The next incident in the Mongol dealings with India is wrapt in some obscurity. Minhaj-i-Saraj speaks of an invasion of Sind in 643-644 by a leader named Mangutah, whom he describes as an old man with dog-like eyes (i.e., with eyes aslant in the Mongol fashion), who was one of Jingis Khan's favourites. At the beginning of Kuyuk Khan's reign he

* Tabakat-i-Nasiri, 1133-1134.   † Id.   ‡ Id., 1135-1136.   § Id., 1135. Note
|| Id., 1136.   ¶ Id., 1140-1142.   * Id., 660.

held command at Ghazni, Tokharistan, and Khatlan.* He is not mentioned *eo nomine* so far as I know by Rashid ud din. It may be that he is to be identified with Ilchikidai, Mangutah being merely an appellative meaning flat-nosed. As we have seen, Ilchikidai was at this very time nominated supreme commander in the west, and he was a famous general of Jingis Khan. On the other hand Mankadhu, or Mankadah, is named as a Noyan who was sent by Ogotai into Seistan during Jingis Khan's campaign,† while a Mongol named Mangatai, a favourite attendant of Tuli, was by him nominated Shahnah of Herat. We, however, read of his having been killed shortly after.‡ Minhaj-i-Saraj says that in 643 HEJ. (*i.e.*, 1246) Mangutah marched an army against Uch and Multan. The former town was at this time governed by Hindu Khan Mihtar-i-Mubarak, the Khazin or Treasurer, as a feudatory of the ruler of Delhi, Sultan Alai ud din Masud Shah. Hindu Khan was not then in the town, which was under the control of his deputy, the Khoja Salih, the Kotwal or Seneschal.§ When the Mongols reached the Indus, Malik Saif ud din, Hasan the Karluk, whom we have mentioned before, abandoned Multan, and having embarked on the Indus, which then flowed east of the town,|| set sail for Diwal and Sindustan, or Sewastan. The Mongols attacked Uch and destroyed its environs. The place was bravely defended by the inhabitants. The breach was at length forced by a famous Baghatur, who led a storming party in the third watch, when the men on guard were reposing, and appeared on the top of the breach. The people inside, however, had prepared a great pit, into which they had poured much clay and water, so that it was in fact a quagmire more than a spear's length in depth. Into this the storming party stumbled, whereupon the defenders raised a shout, brought out torches, and armed themselves, and the attacking party withdrew.¶ Their leader, however, had been suffocated in the slough. The Mongols outside thought he had been captured, and offered to retire if he was surrendered. They eventually retreated without taking the place, a very unusual circumstance with them. This was on hearing that an army was advancing from Delhi to the rescue. Minhaj-i-Saraj tells us he was himself at this time in the service of the Sultan of Delhi.** The Mongols, on hearing of the concentration of the army of Delhi, withdrew in three divisions, and many of their captives, both Hindus and Mussulmans, escaped.†† When he found the Mongols had retired, the Sultan of Delhi turned aside into the hills to punish the Ranah of the Jud country, near the river Jhilam, who had acted as guide to the invaders. He ravaged the country between the Jhilam and the Sind, or Upper Indus, "so that all women, families, and dependents of the infidels, who were in those parts, took to flight." A body of Mongols came to the rescue, and advanced as far as the ferries of

---

* Tabakat-i-Nasiri, 1152-1153. † *Id.,* 1047. ‡ *Id.,* 1037-1049. Notes.
§ *Id.,* 1153. || *Id.* ¶ *Id.,* 1154-1155. ** Op. cit., 1155-1156. †† *Id.,* 817.

the Jhilam, but on seeing the Sultan's well-appointed army they withdrew again.* The Mongols virtually remained masters of the country west of the River Biah, whence they seem to have made periodical raids into India. We read that in 648 HEJ. Delhi was decorated on account of the capture of a large number of Mongol prisoners by Ikhtiar ud din from Multan.†

At the Council held at Lyons in 1245 it was determined to send two missions to the Tartars to try and convert them to the Christian faith, one of Franciscans and the other of Dominicans. They were sent to induce them to be less cruel to the Christians. One of these, under Carpini, has' already occupied us ; ‡ the other was headed by the four Dominican friars -- Anselm (or Ascelm) of Lombardy, Simon de St. Quentin, Alberic, and Alexander. They received a special commission from Pope Innocent IV., with orders to repair to the nearest Mongol camp in Persia. Vincent, the author of the famous "Speculum Historiale, or Historical Looking-glass," knew Simon de St. Quentin, and received from his lips an account of his journey.§ They were joined by Andrew de Longiumello, who had already visited the East as an evangelist, and Guiscard of Cremona, who joined them at Tiflis, and left them again at the same town on their return five months later, remaining in the Dominican convent there for seven years.|| The travellers arrived at the Mongol camp, situated at an unknown place called Sitiens, fifty-nine days' journey from Acre, on the day of the translation of St. Dominic, 1247. Baichu was seated in his tent dressed in rich brocade, ornamented with gold, as were his principal councillors. He sent some of his people to summon the travellers. They asked them whose envoys they were. Anselm replied they were the envoys of the Pope, who was esteemed among Christians as the first among men, and to whom they paid the reverence due to a father and a lord. The Mongols professed great indignation at this, saying that their Khakan, who was the son of God, was much higher, as were his princes, Baiothnoi (i.e., Baichu) and Batu, whose names were familiar everywhere. Anselm professed that the Pope had never heard the Khakan's name nor that of his lieutenants, but had merely heard that a barbarous nation called Tartars had come from the furthest east, conquered a great many countries, and destroyed an infinite multitude of people. If the Pope had known their names they would certainly have appeared on his letters. He added that they had come to exhort the Tartar chief in their name to cease his carnage, and to expiate by penance their evil deeds, and they wished to know if Baichu had any answer for their master.¶ These officers having returned to Baichu, changed their dress, and returning again, asked what presents the friars had brought. They replied that the Pope was not in the habit of making presents, but of

* Op. cit., 815.   † Tabakat-i-Nasiri,   ‡ Ante, ii. 68-75.
§ Vincent, op. cit., lib. xxxi., ch. 11. || Mosheim, Hist. Tart. Eccl., 45.
¶ Vincent, op. cit., ii. 2, and 40.

receiving them, both from the faithful and infidels. They thereupon again withdrew and again returned, and were told that no one ever appeared before Baichu with empty hands, upon which Anselm said if they could not have any audience without presents they must be content without one, and simply hand over the Pope's letters to them to be passed on. The officers made numerous inquiries about the Franks, of whom, as they had heard from their merchants, a large army was being transported into Syria, and with whom they professed to wish to be on friendly terms.*

After a short delay they again visited the brothers, having meanwhile again changed their costume. They reported that if the friars wished for an audience they must consent to make three genuflexions before Baichu, as if they were before the Khakan himself, the son of God and master of the earth. The friars debated together what Baichu meant by this ceremony, and Guischard of Cremona, who, we are told, knew the Tartar customs well, having learnt them from the Georgians, assured them he meant by it to signify that the Pope and all the Roman Church was to be subject to the Khakan. Thereupon the friars agreed that they would rather be decapitated than go through the ceremony, and cause exultation among the enemies of their church—Georgians, Armenians and Greeks, Persians and Turks.† Anselm said that they were not moved in this by arrogance, and were prepared to do whatever was seemly in envoys of the Pope; that they would pay Baichu the same respect they paid their own princes; nay, if he would become a Christian, then, for the sake of the faith, they would not only prostrate themselves before him, but before them all, and kiss the soles of their feet and their poorest garments. This stirred the indignation of the Mongols greatly. "Are we dogs like you?—the Pope is a dog, so are all you," they said, and then withdrew and went to report to their master.‡ Baichu would have killed them all when he heard what their reply was. Some, however, recommended him to kill two of them, and to send the others back to the Pope; others, again, that he should flay the senior envoy, and forward his skin to his master. Others suggested that they should be put in front of the battering engines, so that they should be killed by the latter, and not soil Mongol hands with the blood of ambassadors. Baichu's counsel at length prevailed that they should be put to death. He was eventually dissuaded from this, however, by his principal wife and by the chamberlain, who managed the introduction of envoys, and who threatened to report the whole proceedings before the Khakan if the envoys were killed.§ He now became more reasonable, and sent his people again to inquire from the friars how they were wont to honour their own princes. Anselm thereupon drew back his hood and inclined

* Vincent, op. cit., ii. 41.　† Id., 42.　‡ Id., 43.　§ Id., 44.

his head somewhat. They then asked how Christians reverenced their God. "Some by prostration, some in other ways," was the answer. They finished up by scoffingly inquiring how they, who stooped to wood and stone, refused to thus honour the representative of the Khakan, the son of God. Anselm replied that they did not worship the cross, but only reverenced the symbol of that on which their Saviour had poured out his blood. They presently withdrew, and told the friars that it would be better they should go in person to the Khakan's Court, and there deliver their letters, and see what was now veiled from their eyes, namely, how' great his power and glory were. Anselm, who suspected Baichu's motives, replied that the Pope had never heard of the Khakan, and had merely ordered him to visit the first Tartar chief he met with, and that he did not care to go further. If Baichu wished he would present him and his people with the Pope's letters, if not he preferred to return with them. The Mongols once more jeered at him, saying, "How can you claim that the Pope is so much higher than other men? Who has ever heard of his having conquered so many kingdoms as the Khakan, the son of God? Who has heard that the Pope's name, like that of the Khakan, is diffused from the limits of the East to the Black Sea? Surely he is greater than your Pope?" When Anselm proceeded to explain that the Pope's greatness depended on his being the representative of St. Peter, the unsophisticated Mongols laughed and jeered so loudly that he could not continue his conversation.* At the request of Baichu's messengers, Anselm sent the Pope's letters to him. They were remitted to them to be translated into Persian, whence they were retranslated into Mongol, and were then again presented. Baichu now informed the friars that two of their number must be prepared to accompany a secretary of the Grand Khan, who was about to return to Mongolia, so that they might present their letters in person, and see for themselves what a great person their master was. Anselm replied that they had received no commission to go on thither, and should not go on unless forced, and that they did not want to separate.† The friars, whose tent was pitched a mile away from the Mongol camp, now solicited leave to return, and asked,if Baichu had any letters for them, but they could get no answer. They were spurned and treated with contumely by the Mongols, who looked upon them as viler than dogs. Three times Baichu, we are told, gave orders for their execution. Day after day they went and stood in the broiling heat of June and July from sunrise to sunset without shelter, awaiting a reply. Thus matters continued for nine weeks.‡ At length Baichu granted his permission for them to leave, but revoked it again on the ground that he had heard an envoy from the Khakan named Angutha, who had been given authority over all Georgia (this was probably Arghun), was coming.

Vincent, op. cit., ii. 46.　　† *Id.*, 47.　　‡ *Id.*, .

For three weeks they patiently awaited his arrival, living on bread and water and occasionally a little goat's milk.* As the winter was coming on, when it was dangerous to navigate the Mediterranean, Anselm once more pleaded through a friendly official for Baichu's permission to depart. This was at length granted, and they were about to leave when their departure was again postponed by the sudden arrival of Angutha, with the uncle of the Sultan of Aleppo, and the brother of the Sultan of Mosul, who had been to the Khakan's Court bearing the homage of their relatives and many rich presents. They now performed the triple genuflexion before Baichu, and offered him gifts. Then followed a feast of seven days, in which drink and dissipation prevailed, and the business of the departure was once more postponed. This over, they were at length allowed to depart with a letter addressed by the Khakan to Baichu, which they called the letter of God, and a separate one from Baichu to the Pope. They arrived safely at home after an absence of more than three years.

Vincent has preserved us copies in Latin of the two letters. The letter of the Khakan to Baichu, which was called the letter of God, and was apparently a copy of Jingis Khan's general instructions to his officers, has an incoherent sound, due probably, as D'Ohsson suggests, to the ignorance of the interpreters : "By the order of the living God, Jingis Khan, the son of God, the gentle and venerable. The Great God is Lord over everything, and on earth Jingis Khan is alone master. We would have this known in all our provinces, obedient or otherwise. It behoves thee, therefore, O Baiothnoi, to let it be known by them that this is the will of the Living God. And let this be known everywhere where an envoy can go, that whoever disobeys you shall be driven out, and his land shall be laid waste. And I declare to you that whoever does not hear this my command must be deaf, and whoever sees it and does not obey must be blind, and he who knowing it does not carry it out must be halt. This, my order, will reach everyone, wise and ignorant. Whosoever, therefore, hears and neglects to obey it shall be destroyed, lost, and killed. Make this known, therefore, O Baiothnoi, and whoever obeys, wishing to save his house, and undertakes to serve us, shall be saved and treated honourably ; and whoever shall oppose it, do according to your will and destroy him.† The other letter ran as follows : "By order of the Divine Khan, Baichu Noyan sends these words. Pope, do you know that your envoys have been to. us and have brought us your letters ? Your envoys have spoken big words. We know not whether this was by your orders or at their instance. In your letters it is written, 'You have slaughtered and destroyed many men.' But this is the command of God, who rules the earth, to us,

---

* Vincent, op. cit., ii. 49.   † Id., 51-52. D'Ohsson, ii. 232-233.

'Whoever hears my words shall retain his land, water, and patrimony; but those who disobey are destroyed and lost.' We accordingly send you this message. If you, Pope, wish to retain your patrimony, you must come to us in person, and present yourself before the master of the whole world. If you disobey, we know not what will happen. God knows. Before you come it will be well to send messengers to say whether you mean to come or no, and whether you mean to be friendly or otherwise. This order, which we send you by the hands of Ibeg and Sargis, we write the 20th of July, in the district of the Castle of Sitiens."*

In reference to the conversations of the friars with the Mongols, above reported, and especially in regard to the delicate question of the ignorance of the Pope of their chiefs' names, we have a story preserved showing that the friars were ready and witty diplomatists. This story is reported in the "Peregrinatio de Fr. Bieult," who tells us how one of them was at an audience with the Khakan, when the latter asked what presents he had with him. The reply was he had none, as he was not aware of his great power. "How is that? Have not the birds which visit your country told you anything of our power?" "Sire, it may well be that they have," said the traveller, "but I understood not what they said;" an answer which appeased the Khakan.†

The result of the mission to the Pope is not told us by Vincent, whose account ends abruptly, but Matthew Paris tells us how, in 1248, two Tartar envoys, doubtless the same as those mentioned by Vincent, had an audience of the Pope. Their letters were thrice translated from less known into better known languages. The Pope gave them precious garments called "robas," made of scarlet cloth and furred, and also presents of gold and silver. The interviews were formal, interpreters only being present, and neither clerics, notaries, nor others, and Matthew Paris suggests that their object was to obtain help against Vataces, the schismatic ruler of Nicæa, but, as Remusat says, he was much too obscure a person for the Mongols to want aid in opposing him, and their message was much more probably a peremptory order to submit.‡

The friendly intercourse of the Christians with the Mongols was naturally very distasteful to the Mussulman princes, who put obstacles in its way. The Governor of Erzenjan gave express orders that provisions were not to be supplied to those who came from among the Franks, nor to the envoys of Haithon of Armenia, or of Vastak. Similarly we read how the missionaries who went to the Court of Malik el Mansur Ibrahim, Prince of Edessa, were refused permission to go on to the Mongols, among other reasons because he was satisfied they meant to incite the Mongols against the Mussulmans.§

It was in 1247, when Saint Louis had summoned his notables

preparatory to starting on his crusade, that a letter arrived from the
Mongols summoning him to submit, and stating that they were the people
of whom it had been stated that God had given the earth to the children
of men.* There have not been wanting speculations as to what might
have been the fate of the world if St. Louis, instead of attacking the
strong power of Egypt, had turned his arms against the Seljuki Turks, at
this time much weakened and broken by their conflict with the Mongols.
He would, no doubt, have crushed them and been then brought face to
face with the terrible Tartars. The Mussulmans who intervened between
the latter and the Christians at this time probably saved the world from
disaster.

According to Haithon and W. de Nangis, it was Ilchikidai (called
Erchalchai and Ercaltai in contemporary writings) who, when St. Louis
reached Cyprus, sent some envoys to that last of the Crusaders. Mangu
afterwards repudiated them, and De Guignes has treated them as
impostors. Joinville distinctly says they went to assure Louis that the
Mongols were ready to assist him in the conquest of Jerusalem and the
Holy Land. Louis received them well and sent some of his people back
with them. Odo, or Hugh, bishop of Tusculum, another contemporary,
tells us the envoys landed at Cyprus on the 19th of December, 1248 ; that
they duly reached Nicosia, and presented the king with letters written in
the Persian tongue and in Arabic characters. After the translation of
these letters, Hugh himself reported their contents to the king. Vincent
of Beauvais and William of Nangis call the chief envoy David, and tell
us he was recognised by the brother Andrew de Longiumel, already
named, who had met him among the Tartars. Another chronicler tells
us that the king had the letter, when translated, sent on to France, to his
mother Blanche.† Another copy was sent to Pope Innocent by his legate,
Cardinal Hugh, of Chateau Royal. Vincent of Beauvais has preserved a
Latin translation of this letter :—

"By the power of the High God, the letters of the King of the Earth,
the Khan, the words of Erchaltai, the great king of many provinces, the
vigorous defender of the world, the sword of Christian victory, the
defender of the Apostolical faith, the son of the Evangelical law, to the
King of the Franks. May God increase his kingdom and preserve it to
him for many years. May his wishes be gratified now and in the future
by the truth of the divine power, the director of mankind and of all the
prophets and apostles. Amen. A hundred thousand salutations and
blessings. I hope he will accept these greetings, and that they may be
welcome to him. God grant that I may see this magnificent king who is
coming near. The exalted Creator can well bring about our friendly
meeting together.

* A. Remusat, Mems. Fr. In., vi. 435.
† W. de Nangis, Bouquet, xx. 358, &c. A. Remusat, Mems. Fr. Acad., vi. 439-440.

"Let it be understood that in this our greeting we mean nothing more than the benefit of Christianity and the strengthening of the king's hands, God being willing; and I pray that God will grant victory to the armies of the Christian king, and will give him victory over his enemies who contemn the cross. On behalf of the exalted king, may God exalt him, namely, of Kiukai (*i.e.*, Kuyuk). May God increase his splendour. We have come (*i.e.* into Persia) with authority and power to announce that all Christians are to be free from servitude and taxes, dues, and tolls (*a servitute et tributo et angaria et pedagiis*), &c., and are to be treated with honour and reverence. No one is to molest their goods, and those of their churches which have been destroyed are to be rebuilt, and to be allowed freely to sound their plates (*pulsenter tabulæ--i.e.*, the substitutes for bells, already named). No one must dare to prevent them freely and with a quiet mind praying for our kingdom. So far we have provided for the advantage and protection of the Christians. In addition, we beg to send our faithful envoys, the venerable Sab ed din, Mufat David, and Mark, that they may announce these good tidings. My son, hear their words and believe them. In his letters the King of the Earth (*i.e.*, Kuyuk)—may his splendour increase—orders that there may no difference be made between the different classes of believers—the Latins, Greeks, and Armenians, the Nestorians and the Jacobites, and all who reverence the cross. All are one with us, and thus we pray the Magnificent King to make no difference between them, and to extend his beneficence over all Christians. May his piety and beneficence endure. Given in the end of Maharram, with the approval of the exalted Lord."*

According to Bishop Hugh, above quoted, Louis asked the envoys how their master knew of his arrival. They said the Prince of Mosul had sent to Ilchikidai some letters he had received from the Sultan of Egypt, and at the same time falsely pretending he had captured sixty of the Frank ships. Ilchikidai reported his intention of marching the following summer against the Khalif, and asked St. Louis to make a diversion against Egypt so as to keep its ruler employed. The envoys mentioned that Kuyuk Khan's mother was a Christian. She was called Kuiotai, and was a daughter of Prester John, and that he had, at the instance of a pious bishop, named Malassias, been baptised, with eighteen kings' sons and many grandees of the Court. They added that Ilchikidai had been a convert for many years, although many of the Tartars were not so, and, that although not of the royal blood, he had much power; that Baichu was a Pagan, and surrounded by Mussulman councillors, hence his harsh treatment of the Pope's messengers, but that his power was now much curtailed, and he was subordinate to Ilchikidai. There are some misstatements in this report, and many suspicious

---

* Vincent de Beauvais, xxxi. cap. 91. D'Ohsson, ii. 238-239. Note.

circumstances about the letter, such as its unusually civil tone, its ignoring questions likely to interest the Mongols, and entering into the rival policies of the various Christian sects in the East, and its reference to the request that Louis would make no difference between the Latin Christians and their Eastern brothers. This led De Guignes and others to suspect that the whole embassy was an imposture made up by some of these Eastern Christians to further their own aims. Remusat concludes that the embassy was a genuine one, but that the envoys, for diplomatic or other purposes, either concocted a letter of their own or interpreted it after their own fashion.* Louis determined to reply to the message, and organised an embassy in return, of which Brother Andrew de Longiumel, "who," says Joinville, "knew the *Sarrazinois*," was at the head, and with him joined a French friar named John of Carcassonne.† The presents Louis sent to the Khakan comprised a chapel made out of good scarlet (*i.e.*, of embroidered scarlet cloth), ornaments for the service, a piece of the true cross for the Khakan and another for Ilchikidai. With these, Joinville says, there were also sent pictures of the chief events in the life of Christ—the annunciation, nativity, baptism, passion, ascension, &c.† With these things were sent letters, according to one account, exhorting the Khakan to imitate the example of his mother, and to become a Christian, and to another bidding him, as well as Ilchikidai, persevere in the faith. The legate Odo also sent letters to the Khakan, to his stepmother, to Ilchikidai and the bishops who were with him, saying that the Roman Church received them gladly, and had learned with joy of their conversion, that they should cling to the orthodox faith, recognise Rome as the mother of all churches, and the Pope as its head.§ These letters, as Remusat says, must have been a surprise to the Court at Karakorum. The envoys set out from Nicosia on the 27th of January, 1248 (D'Ohsson says on the 10th of February, 1249). They apparently made their way to Antioch, and thence to the camp of Ilchikidai, whence Andrew despatched a letter, together with one from the Mongol general, which were translated into Latin, and sent to France to Queen Blanche. These letters seem to be no longer extant.‖ The envoys then went on to the Imperial Court, travelling at the rate of ten leagues a day. There they arrived at the end of 1248 or the beginning of 1249. Kuyuk was dead, and it was the Regent Ogul Gamish who received them. She received the presents of Louis affably, and gave the friars some in return, including, in Chinese fashion, a piece of silk brocade. The Regent also intrusted a letter to them. According to Joinville the presents sent by Louis were treated, much to his chagrin, as tribute, while the letter in reply demanded an annual tribute in gold and silver, menacing the French king with destruction if he refused to pay it.*

The notice given by Joinville of the reception of this embassy is so quaint, and so exactly represents the Mongol mode of dealing in such cases, that it is worth while printing it in full. "Le roi des Tartarins," he says, "fit tendre la chapelle, et dit aux rois qui se trouvaient à sa cour. Seigneurs, le roy de France est venu en nostre sujestion, et vezci le treu que il nous envoie. Avec les messagers le Roy vindrent . . . . . si apportèrent lettres de leur grant roy au roy de France qui disaient ainsi. 'Bone chose est de pèz, quar en terre de pèz manguent cil qui vont à quatre pied l'erbe pèsiblement ; cil qui vont à deus labourent la terre, dont les biens viennent pèsiblement ; et ceste chose te mandons nous pour toy aviser ; car tu ne peus avoir pèz si tu ne l'as à nous, et tel roy et tel (et moult en nommoient), et tous les avons mis à l'espée. Si te mandons que tu nous envoies tant de ton or et de ton argent chascun an, que tu nous retieignes à amis ; et se tu ne le fais, nous destruirons toy et ta gent aussi comme nous avons fait ceulx que nous avons devant nommez.' Et sachez qu'il (i.e., the king) se repenti fort quant il y envoia."†

The envoys returned two years after setting out and found the king at Acre, and notwithstanding the ill success of his previous venture, he determined to send another embassy. This was headed by William of Ruysbrock, or Rubruquis, of whom we have written much in the earlier volumes. Joinville declares that Rubruquis repudiated the character of an envoy, and that in preaching in the church of Saint Sophia, while on his journey, he declared he had been sent neither by Louis nor any other sovereign, but went in accordance with the statutes of his order to preach the gospel to the infidels ; and it would seem that he took up this position at the instance of St. Louis himself, who, no doubt, wished to guard himself against his acts being misinterpreted as acts of submission. Rubruquis reached Karakorum on the 27th of December, 1252, having traversed the Steppes of Kipchak, as I have mentioned.‡ Rubruquis tells us the year before he was at Karakorum there was a cleric there from Acre, who called himself Rammud, but whose real name was Theodolus. He travelled from Cyprus to Persia with Andrew, taking with him an organ from Amoric (?) (ab Ammorico ?). When Andrew went home again he remained behind and repaired to Mangu, who asked him what his business was, and he replied he had come from a bishop named Odo, in the kingdom of King Louis (the text has Moles, but this is clearly a clerical error), who, if the ways had been open, and if the Saracens had not been posted between them, would have sent envoys to make peace

* W. de Nangis, Bouquet, op. cit., 448.   † D'Ohsson, ii. 244. Note.   ‡ Ante, i. 4.

with him. Mangu asked if he was willing to conduct some envoys to that king and bishop. He said he was, and also to the Pope. He then caused a very strong bow to be made which two men could barely pull, and two whistling arrows called bozunes or bousiones with silver heads full of holes, which when thrown whistled like a flute, and he bade a Mongol, whom he had chosen as his envoy, go to the King of the Franks, and tell him that if he made peace with him he would if he acquired the country now held by the Saracens as far as his borders, make over to him the remainder as far as the west. He also told him ominously to point out to the King the bow and arrows, and to tell him such a bow shot a long way and such arrows pierced very deeply.* He also bade his Mongol conductor explore well the roads, districts, and castles, and the men he should pass, and also their arms. The interpreter, who was a European, suggested that Theodolus should drop his inconvenient companions into the sea *en route*, so that no one would know what became of them, for they were merely spies. Mangu gave the Mongol a gold tablet, a palm in width and half a cubit in length (*i.e.*, a paizah). He says that anyone who bore it could order and obtain anything he pleased. Theodolus duly arrived at the Court of Vastaces, or Vataces, the Emperor of Nicæa, wishing to go to the Pope to deceive him as he had deceived Mangu Khan. Vataces asked him where his letters were, which was the envoy and which the conductor. As he would not produce his letters he imprisoned him. The Mongol fell ill and died there, and Vataces thereupon sent the golden tablet back to Mangu Khan. Rubruquis says he met these messengers at Erzerum, who told him what had happened to Theodolus.†

On taking leave, Rubruquis was intrusted with a letter for St. Louis, of which he gives the purport, "so far," he says, "as he could understand it through the interpreter." This letter was phrased in the usual peremptory fashion of the Mongols. *Inter alia*, it denounced David, already mentioned, as an impostor, and characterised the late regent, the mother of Kuyuk, to whom Louis' envoys had gone, as viler than dogs. Rubruquis reports that Mangu Khan had declared to him that she was given to necromancy, and had destroyed all her relatives by her sorceries.‖ The letters stated that it was not convenient and safe for him to send envoys, but that he expected Louis to send him some to state whether he wished for peace or war, and threatening him accordingly.§ I have already described Rubruquis' journey back to Serai, the capital of Batu Khan.‖ The Mongols furnished him with an escort of twenty men, to protect him from the Lesghs and other robbers in traversing the Iron Gates. In regard to the arms of these Mongols Rubruquis has a very interesting sentence. Two had haubergions (*i.e.*, coats of mail). These, they told him, they had obtained among the Alani, who were accounted good makers of such suits, and splendid smiths.* Rubruquis concluded that the only arms indigenous with the Mongols were quivers, bows and arrows, and pelliciæ (? felted armour, or armour made of skins). Among the presents he saw offered to them were iron plates, or scales, and iron helmets, from Persia, and he saw two Alans present themselves to Mangu in tunics of fish skin (de peccaisiis), made from stiff hides, which were very inconvenient.†

He describes Derbend as hanging between the sea and the mountains. No road passed below or above the town. The only road traversed the city itself, and was closed by an iron gate, whence its name. It was well fortified, and dominated by a fortress which the Tartars had captured. Two days further on he reached the town of Samaron (? Shirvan), where there were many Jews, as there were in many of the mountain recesses on this coast, and also in the towns of Persia. Presently he reached a great town called Samag (i.e., Shamakhi), and then on the following day entered the plain of Moan (i.e., Mughan), through which flowed the Kur, from which he says were named the Kurgi, called Georgians in the West. In this plain he again met with Tartars. Travelling along the Araxes, he passed the camp of Baichu, in whose house he was entertained and given wine. His host, however, drank kumiz, which he says naively he would have freely drunk if it had been offered to him. He followed the Araxes to its sources near Erzerum. On leaving Baichu, Rubruquis' guide and his interpreter went to Tebris to see Arghun. Baichu caused the friar to be taken to Naxua (? Nakhchivan),‡ the former capital of a great kingdom, once a great and beautiful city, which the Mongols had converted into a waste. There were once eighty Armenian churches there, but at this time they had been reduced to two small ones. The Armenians professed to recount to him some prophecies of one of their saints, named Acacron, who foretold the advent of the archers (i.e., the Mongols).§ He tells us how he passed near Mount Ararat, which, although it seemed so accessible, none had been able to climb ; and that a monk, who was very anxious to do so, had a piece of the ark brought him by an angel, which the Armenians professed to keep in one of their churches. An old man had told him that the reason why the mountain ought not to be climbed was that its name was Massis in their tongue, and it was of the feminine gender, and no one should ascend it since it was the mother of the world ! ! ! ‖

Four days after leaving this town Rubruquis reached the territory of Sahensa (i.e., Shahan Shah), formerly the most powerful of the Georgians, but then tributary to the Tartars, who had destroyed all the fortresses his father Zakaria had conquered from the Mussulmans. Shahan Shah,

---

* The Kubechi, in the Caucasus, were a famous tribe of armour makers.
† Rubruquis, 381.   ‡ Id., 384.   § Id., 385-386.   ‖ Id., 387.

with his wife and his son Zakaria, received Rubruquis with honour. The last, an amiable boy, asked him if he went to St. Louis whether he would receive him, for although he had plenty of all he needed he preferred to travel to a foreign land than to wear the yoke of the Tartars.* They claimed to be faithful to the Roman Church, and if the Pope would send them some help they would subject all the surrounding districts to the Church. In fifteen days thence he reached Erzenjan, all whose inhabitants were Christians, Armenians, Georgians, and Greeks, but the Muhammedans were masters of it, and its governor, as we have seen, had been forbidden to supply food to any Frank and to any envoy from Armenia or Vataces, so that Rubruquis had now, till he reached Cyprus, to buy his food. He passed through Ani, also subject to Shahan Shah, a very strong fortress containing 1,000 Armenian churches and two mosques. The Tartars had a bailiff there. There he met five Dominicans, who had no interpreter except a feeble servant who knew Turkish and a little French. They had letters from the Pope for Sertak, Mangu Khan, and Buri, but on hearing Rubruquis' story, instead of going on, went to consult their companions at Tiflis. "What they afterwards did," says Rubruquis, "I know not." Leaving the valley of the Araxes, he crossed into that of the Euphrates, and mentions a terrible earthquake which had destroyed 1,000 people at Erzenjan. He crossed a valley, where he tells us the Sultan of the Turks was defeated by the Tartars, the former having 200,000 horsemen and the latter but 10,000, and in regard to the earthquake says quaintly and grimly : "Dicebat michi cor meum quod tota terra illa apperuerat os suum ad recipiendum adhuc sanguinem Sarracenorum."† He passed through Sebaste, and visited the tombs of the eighty martyrs. Thence he went on by Cæsarea and Iconium. There he met, inter alios, with a Genoese merchant from Acre, Nicholas de Sancto Siro, who, with a companion, a Venetian, named Benefatio de Molendino, had the monopoly of exporting all the aluinun (?alum) from "Turkia," and had raised its value in the proportion of 15 to 50. Rubruquis was presented to the Seljuki Sultan, and received permission to go on through Cilicia, or Little Armenia. He made his way to Kurta (?), the port of that kingdom, and having deposited his goods on board ship, went to pay a visit to Haithon's father, who he heard had had letters from his son. He found him at Asium with all his family except a son named Barunusin, who had been appointed governor of a fortress. The Court was delighted at the news that King Haithon was on his way home, having received a remission of part of the heavy tribute they had to pay, and other privileges.‡ The old man had Rubruquis conducted to a port named Auax (?Ayas), whence he passed into Cyprus to Nicosia, where he had an interview with one of King Louis' officials, who conducted him

---

* Id., 388.   † Id., 391.   ‡ Op. cit., 393.

to Antioch and Tripolis, and Acre. He complains that he was not allowed to visit the King in person, and that it was not possible to report the results of his journey *viva voce*.* He ends up by a survey of the various Muhammedan powers which he had encountered, and explaining how easy it would be for the Christians to overwhelm them. That a large proportion of the population of Turkia (*i.e.*, of the Seljuk Empire) were Greeks or Armenians. The Sultan had three sons, one by a Georgian wife, a second by a Greek, and a third by a Turk. The first of these he wished to succeed him, but the Turks and Turkomans wished for the success of the third. They had twice risen in his support, but he had been beaten, and was then imprisoned. The son of the Greek mother also had partisans, who declared the son of the Georgian mother, who had been sent to the Tartars, was a feeble person. This rivalry created great confusion. There was no money in the treasury, few soldiers, and many enemies. The son' of Vataces also was feeble, and had a war with the son of Assan (? Jelal-Hasan), who was also ground down by the Tartar yoke. So that if it was thought well that the army of the Church should march to the rescue of the Holy Land, it could easily subdue or traverse that district. From Cologne to Constantinople was only a forty days' journey by chariot. Thence to Little Armenia not so much. It was more safe, and quite as cheap, to go thus by land as by sea, and, adds our traveller, "I speak faithfully; if your peasants (I speak not of kings and knights) would travel as do the kings of the Tartars, and be content with the same food, they could conquer the whole world."† Rubruquis must have been a delightful companion; so full of genuine hatred for the Saracens and the Tartars, and so full of confidence in himself.

Note 1.—The coinage of the district comprised in the old Empire of Khuarezm during the interregnum between Jingis Khan's campaign and that of Khulagu is an interesting but obscure subject. There are certain coins published by Thomas in his "Coinage of the Pathan Sultans, 91 and 97," which bear the name of the great conqueror himself, and strangely enough have the Khalif's name and titles on the other side. As Major Raverty suggests, these were probably issued by some of the Muhammedan princes on the borders of India, who acknowledged the supremacy of Jingis Khan. One of them has the mint city Kurman. It need hardly be said that the Mongols themselves had no stamped money until a later date, and merely used bullion in the form of ingots, called balishes. These coins are very like in fabric to those issued by Jelal ud din, the Khuarezm Shah, when in the east, and by Nasir ud din Muhammed ibn Hasan Karluk.‡

* Op. cit., 393-394. † *Id.*, 394-395. ‡ Op. cit., ç

I know of no coins with the name of Ogotai. "The earliest coin of the Mongols with Arabic inscriptions," says my friend, Mr. Stanley Poole, "and probably their earliest with any inscription, is that struck at Tiflis in 642."[*] The year 642 falls within the regency of Turakina, Ogotai's widow, and Mr. Poole says this coin may have been struck by Arghun after his appointment to the Governorship of Persia, or it may have been struck by some pretender to the throne, who considered the interregnum, and the dissatisfaction caused by Turakina's rule, a favourable opportunity for striking a blow for sovereignty. A second coin in the same collection of the same date has apparently the mint place Kenjeh (? Kantzag). "The second of these coins," says Mr. Poole, "has the familiar Anatolian and Georgian device of a mounted bowman, with dog, and presents no indication of a striker's name except an obscure inscription which has been doubtfully read Alush Beg by M. Bartholomaei, while M. Gregorief, omitting the points, reads it (coin) of the great Mongol Ulus," which seems to me to be an exceedingly probable reading. Three specimens in the Jena collection were apparently minted at Nakhchivan. Of Kuyuk there are in the British Museum only coins struck in his name by his vassal David V., of Georgia.[†]

Of Mangu, written Möngkó on the coins, we have specimens struck both in silver and copper. His name occurs alone on five coins in the British Museum struck in 652 and 653, in all cases where the mint mark is legible, at Tiflis.[‡] There are also coins extant of Bedr ud din Lulu, the ruler of Mosul, with the name and titles of Mangu upon them.[§]

Note 2.—Western Armenia at the time of the Mongol invasion was so much broken up into feudal principalities that it is not easy to follow their history, and it will be convenient to give a short conspectus of the most important family, that of the so-called Mkargrdzels. Guiragos and Vartan agree in giving them a Kurdish origin. They consisted of two branches. One of these, to which the famous Constables of Georgia belonged, we are told by Guiragos, conquered from the Persians and Turks several districts of Armenia, of which they remained masters; that is to say, the district surrounding Gelarkuni, Tashir, Ararat, Bejni, Tovin, Anberd, Ani, Kars, Vaiotz-Tzor, the country of Siunia, and the fortresses, towns, &c., in its neighbourhood. They also made tributary the Sultan of Karin, or Erzerum.

The second or collateral branch, which is deduced from the same ancestor by M. Brosset, captured from the Persians the fortresses of Kartman, Karhertz, Ergevank, Tavush, Kadzareth, Terunakan, Gag, and eventually Shamkor.[‖]

M. Brosset has criticised the pedigree of this family as given by Guiragos, Vartan, and in inscriptions, and the following is the result :—

* It is in the British Museum. Catalogue Oriental Coins, vol. 6, liii.
† *Id.*, liii. and liv. ‡ *Id.*
§ Yule's Marco Polo, i. 62. ‖ Hist. de la Géorgie, Add., &c., 415.

THE MKARGRDZELS.

Khosrov.

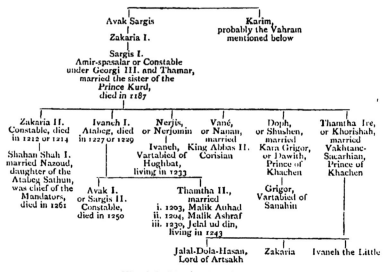

Hist de la Géorgie, 362 and 417.

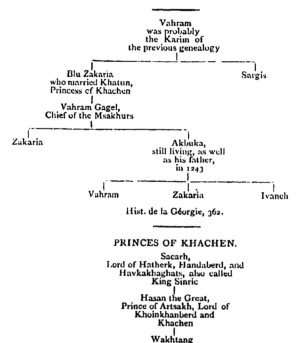

Hist. de la Géorgie, 362.

PRINCES OF KHACHEN.

Sacarh,
Lord of Hatherk, Handaberd, and
Havkakhaghats, also called
King Sinric

Hasan the Great,
Prince of Artsakh, Lord of
Khoinkhanberd and
Khachen

Wakhtang

Jelal, styled Hasan　　　Zakaria　　　Ivaneh the Little

Hist. de la Géorgie, 339-349.

THE ORPELIANS.

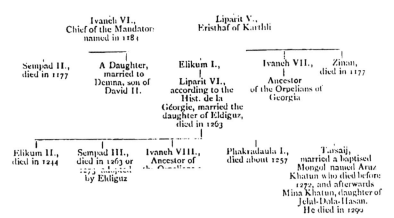

Sempad I.,
died before 1177

Ivaneh VI., Chief of the Mandator; named in 1184

Liparit V., Eristhaf of Karthli

Sempad II., died in 1177

A Daughter, married to Demna, son of David II.

Elikum I., Liparit VI., according to the Hist. de la Géorgie, married the daughter of Eldiguz, died in 1263

Ivaneh VII., Ancestor of the Orpelians of Georgia

Zinan, died in 1177

Elikum II., died in 1244

Sempad III., died in 1263 or 1273 .... by Eldiguz

Ivaneh VIII., Ancestor of the Orpelians

Phakradaula I., died about 1257

Tarsaij, married a baptised Mongol named Aruz Khatun who died before 1272, and afterwards Mina Khatun, daughter of Jelal-Dola-Hasan. He died in 1290

Hist. de la Géorgie, 350, &c.

Note 3.—In describing the struggle of the Seljuki ruler Iz ud din with the Mongols, I overlooked the fact that at this time Michael Palæologus, who governed Nicæa and Bythinia on behalf of the Emperor Baldwin II., having aroused the suspicions of his master, fled to Iconium to the Turkish Sultan, who gave him command of a contingent of Christians. With them he fought against the Tartars, and Michael wounded their commander with his own lance, and drove back the enemy. Meanwhile, one of the Sultan's officers deserted with his men, which turned the tide in favour of the enemy. The allies were beaten, and Michael, with a Turkish general, fled for several days, and was closely pursued as far as Castamonia, in Paphlagonia, where the Turkish general lived.*

Note 4.— Guiragos has preserved for us a short vocabulary of the Mongol language as spoken when he wrote (*i.e.*, about 1241), which is very interesting as a proof of the conservative character of the language, and the little alteration it has undergone in the six centuries which have since intervened. I have here printed it, together with the corresponding words in Buriat, Kalmuk, and Mongol, as given by Brosset and Schiefner in Brosset's edition of Guiragos.†

Akropolita, in Stritter, iii. 1038-1039. Lebeau, xviii. 18-22.
† Vide op. cit., 135-137.

	Guiragos.	Buriat.	Kalmuk.	Mongol.
God	Thanghri	Tengere, Tengeri (the sky)	Tenggri (the sky)	tengri, tegri
Man (Vir)	Harérian (var Ere)	ere	ere	ere
Woman	Apdji (var Eme Apdji)	eme	eme	eme (1)
Father	Etcheka	etsege, esege	etchige	etchige
Mother	Ak'a	eke, ekhe	eke	eke
Brother (older)	Agha	akha, aka	akha	akha
Sister (sister)	Ak'adji	egetche, egeche	egetchi	egetchi
Head	Thiuron (var Thiru)	tologoi, turnun (beginning)	tologhoi	(tologhai) terigun
Eye	Nitun (var Nitu)	(pl) nyideng, nyndeng	nūdūn	nidun
Ear	Tchikin (Dchih)	(pl) tchike chikeng	tchikin	tchikin
Beard	Sakhal	sakhal, hakal	sakhal	sakhal
Face	Hioq (var Hiugh Niur)	nyur nyur	niur	nighur
Mouth	Aman	aman, ana	aman	aman
Tooth	Skhur (var Skhursitun)		shidun	shidun, sidun
Bread	Othmak		otmok (2)	
Ox (Bœuf)	Akar (var Ok'är)		tsar	shar shir, uker (beast with horns) (3)
Cow	Unen	unye, unyen	ukur	uniyen
Sheep (le Mouton)	Qoina	khutsa, kusa	khutsu	khucha (4)
Sheep (la Brebis)	Qurqan	khonye, khonyen	khoin	khonin (5)
	Iman	yamang	yaman	imaghan (roebuck)
Horse	Mori	moring, morye	morin	morin
Mule	Losa (var Lusa)		lus	
Camel	Thaman		temen	temegen
Dog	Nokha	nokhoi nokoi	nokhoi	nokhai
Wolf	Tchina	tchono, chono	tchino	tchino
Bear	Ait'ku	otokoi (ursa)	ayu	ötege
Fox	Hok'an (var honk'an)	unege, unegen	ünegen	ünegen
Hare	Thaplqa (var Thoblqa-Thula)	tulai, tulei	taulai	taolai (6)
Bird	Thakia		takiya	takya (a fowl)
Pigeon	Kokatcha (v K'ok'uchin)		kokoltchirgene	kegündjighene (7)
Eagle	Qush (var Burkui Qush)	burgut, khanardyi		burghut, shcha-ghun (8)
Water	Usun (var Sun)	oso, uhun	usun	usun
Wine	Tarasu			darasu, darasun
Sea	Tangez (v Naur-Tangez)	dalai	tenggis	dalai
River	Ulan Su (var Moran Ulan-Su)	gol	ghol	ghol, muren
Sword	Eoltu (var Ialtu)		ildu	ildu, toltu (knob of a sword-guard)
Bow	Nmu (var nmo)	nomo, nomon	numun	numun
Arrow	Ornu	somo, homon	sumun	sumun
King	Malik			
Baron	Nuin (var E'ka Nuin)			noyan (9)
Earth	El (var El-Ercan) (10)	gadzer, gaser		
Sky	Gogai (var Gog)	tenggere, oktorgoi	tenggere, oktorghoi (11)	
Sun	Narhan or Naran	nara, naran	naran	naran
Moon	Sara	sara, hara	sara	saran
Light	Otur		geren, gerel	
Star	Saqra (pl.) (var Hutut-Sarqa)	odo, odon	odon	odon
Night	Soini	sönyi hönyi	söni	suni
Scribe	Bitikchi			bitchiktchi
Satan	Barhauri (var Barha-hurh-Elep)			

(1) Eptyi among the Koibals and epchi among the Sagayans=woman. (2) Turkish, etmek. (3) Tartar, oquz. (4) Tartar, qoi. (5) Khuraghan among the Mongols and kuragan among the Koibals=a lamb. (6) Tartar, taushan. (7) Tartar, gogurtchin (8) Karakush means black eagle among the Koibals. (9) Yeke noyan=the great noyan. (10) El=people, subjects among the Koibals. (11) Tartar, kök.

A few Mongol words have also been preserved by Malakia, which have been examined by M. Schiefner. The four names of taxes imposed by the Mongols, as given by him, are Tghghu, Mal, Thaghar, and Ghphtchur. The first of these has not been traced in Mongol. Mal in Mongol means cattle; in Persian riches, and more especially cattle, and it is clear this was a tax on cattle. Thaghar in Mongol is a sack for straining whey, a woven purse, a sack or measure of grain. This, then, was a tax on grain. Ghubtchighur in Mongol is a net. It was a tax on the revenues of the land, or, more generally, on the products of the soil. Khalan has apparently some analogy with the Mongol Khalkhu, to attack, or with Khulusu, hire, interest, or rent, and probably means a war subsidy.

Kesikth, Mongol kia, Turkish késhik; a body-guard.

Bauka or buka, Mongol buke; an athlete, a wrestler.

Kunah has perhaps some connection with the Mongol khonok, " a degree or section of a circle."

Bichikchi in Mongol means a scribe or copyer.

Dzarghuchi, or yarguchi, means in Mongol a judge.

The words Thagia (name of an idol), Sghamish, Yam, and Themachi (meaning a myriarkh) cannot at present be explained in Mongol.

CHAPTER II.

KHULAGU KHAN.

KHULAGU was the fifth son of Tului, the youngest son of Jingis Khan. His mother was Suirkukteni, the niece of the Kerait chief, Wang Khan, and daughter of his brother Jakembo. He was thus own brother of the two great Khakans, Mangu and Khubilai, and of Arikbuka, who contested the claims of Khubilai to the empire of the Mongol world. He was born about the year 1216. He is first mentioned in the winter of 1224 and 1225, when he was nine years old and his brother Khubilai was eleven. Jingis Khan was returning home after his great campaign when he was met near the river Imil by the two boys just named, his grandsons. Khubilai had killed a hare on the way, and Khulagu had captured a deer, and as it was customary for the Mongols to draw blood from the middle finger of boys when they first engaged in hunting, and to mingle it with some food and fat, the operation we are told on this occasion was performed by Jingis Khan in person.* He was thirty-five years old when, at his brother Mangu's bidding, he undertook his famous campaign in the West, to which we shall now turn. This was one of the expeditions decided upon at the great kuriltai held at the accession of Mangu, in 1251 ; the other one being directed against China, under the orders of Khubilai.

As a preparatory measure a Naiman, called Kitubuka, styled baverji or the butler, was sent in July, 1252, with an advance column of 12,000 men. The Georgian annals also speak of Elgan the Jelair (probably the Kuka Ilka of other writers, the Kuok'an of the Yuan shi,† the Kulkhan of Chamchean) as a commander of 10,000 men under Khulagu.‡ The first object of the Mongol attack was the famous community of Muhammedan schismatics known as the Ismaelites or Assassins, subject to the so-called Old Man of the Mountain, a translation of Sheikh ul Jibal, the name by which the Arabs knew him. They were called Ismaelites from Ismael, a son of the fifth Imam, to whom they were devoted.§ They were called Assassins from their use of Hashish, an intoxicating preparation of hemp, and were styled Mulahids or heretics by the orthodox Mussulmans.

---

* D'Ohsson, i. 323. Ilkhans, i. 79-80. † See Bretschneider, Notes, &c., 60.
‡ Hist. de la Géorgie, 540. D'Ohsson, iii. 135-136. § Ante, i. 15.

Their country has a curious synonymy. Kuhistan, by which it is some-times described, merely means the mountain land, and connotes the country of the Shavarkoh range, south of Asterabad and Ghilan. In a more limited view the focus of the Ismaelites was the district called Rudbar by some writers, which was watered by the Shahrud. This district was situated north of Kazvin, and contained some fifty fortresses, the chief one being Alamut, the Ismaelite capital. Its name was corrupted into Alah Amut (*i.e.*, the eagle's nest). Ibn al Athir tells us the district about Alamut was called Talikan, others called it Dilem. In the narrative of Chang ti's journey, we are told that in the country of the Assassins all the oxen were black and had a hump on their necks; the low country was destitute of water, wells were accordingly dug on the summits of the mountains, whence water was conducted for a great distance in order to irrigate the plains.*

Marco Polo has a curious account of the Ismaelite chief, which Colonel Yule says is virtually the same as that current all over the East, and of which other versions are preserved by Odoric, in the narrative of Chang ti, and in an Arabic version translated by Von Hammer. Marco Polo tells us their chief had caused a certain valley between two mountains to be enclosed, and had converted it into a garden, the largest and most beautiful that ever was seen, "filled with every variety of fruit. In it were erected pavilions and palaces, the most elegant that can be imagined, all covered with gilding and exquisite painting. And there were runnels, too, flowing freely with wine and milk, and honey and water, and numbers of ladies and of the most exquisite damsels in the world, who could play on all manner of instruments, and sang most sweetly, and danced in a manner that it was charming to behold." He wished the people to believe that this was actually Paradise, as described by Muhammed, and his people really believed it. The entrance to this garden was protected by a strong fortress. The Old Man kept about him a number of warlike youths from twelve to twenty, who believed in him as the Saracens believed in Muhammed. These he would first make drunk with a certain potion, and then have them conveyed, six or ten at a time, so that when they awoke they found themselves inside, where they deemed themselves in Paradise. When he wanted to send one of these devotees on a certain mission, he again administered the potion, and had him carried from the garden to the palace, where he was brought before the Prince, and when asked whence he came he would reply that he came from Paradise, which was just as Muhammed described it, which gave the others who stood by, and had not yet entered, a great desire to do so. When the Old Man wanted a prince slaying, he would say to such a youth, " Go thou and slay so and so, and when thou returnest my angels

---

shall bear thee into Paradise, and shouldst thou die, natheless even so
will I send my angels to carry thee back to Paradise." So he caused
them to believe, and there was no order of his they were not willing to
obey, and thus he murdered anyone he wanted to be rid of, and thus he
inspired the neighbouring princes with great dread."* It is curious that
one of the Ismaelite fortresses destroyed by the Mongols was called
Firdus, or Paradise.†

The Ismaelites were close neighbours of the citizens of Kazvin, who
were good Sunnis, and between them there had been a long feud. We are
told that the Imam Kazi, Shems ud din, of Kazvin, made several
journeys between Kazvin and China. Although an ecclesiastic, he wore
a coat of mail under his clothes as a precaution against assassination.
This having attracted the attention of Mangu when he was at his court,
gave him an opportunity of denouncing the heretics, whom he also
described as a danger to the Mongols themselves.‡

As we have seen, the Georgian chronicles assure us that the Mongols
had already for some time been attacking the Ismaelites, or.Assassins,
and had lost one of their chiefs named Jagatai, who had been assassinated
by them. I have in a previous volume traced out the origin of these
famous schismatics and of their chiefs the old men of the mountain, § and
shall here take up the story at a later point. When Jingis Khan invaded
the West, the first Mussulman sovereign to send him his submission was
Jelal ud din Hassan, the chief of the Ismaelites. Jelal ud din died in
1221, and was succeeded by his son Alai ud din Muhammed, who was
only nine years old. He received no education, for as Imam he was
infallible ; whatever he did was right, and no one could give him advice.
In his youth he had some struggles with the Khuarezm Shah, Jelal ud din.
That prince, on his return from India, gave the district of Khorasan in
charge to his general Orkhan. The latter's lieutenant made a raid upon
the Ismaelite districts of Nun and Kain, or Kuhistan. Alai ud din
thereupon sent an envoy to Khui to complain. The Khuarezm Shah
summoned Orkhan and the envoy to his presence to explain. Orkhan
drew from his boots and girdle several daggers in view of the envoy, who
had used some threats, saying, " See our daggers ; besides these we have
swords which are sharper and more pointed, which you have not seen."
As he could get no satisfaction, the envoy returned ; but shortly after
Orkhan was assailed near Kantzag by three Ismaelites, who killed him.
They then went through the town with their bloody daggers, exclaiming,
" Long live Alai ud din." They penetrated into the Divan, intending to
assassinate the Vizier, Sherif ul Mulk, but he happened to be in the Sultan's
palace at the time, and thus escaped. They wounded one of the guards
and then sallied out brandishing their daggers, and were eventually killed

---

* Yule's Marco Polo, i. 145-149.   † *Id.,* 154.
‡ Tabakat-i-Nasiri, 1189-1196.   § Ante, i. 15-16.

by stones thrown from the roofs of the houses, and died crying, "We are the victims of our lord Alai ud din." Presently another envoy named Bedr ud din Ahmed went from the Ismaelites to Jelal ud din. He said that his people merely wanted guaranteeing from attack. Jelal ud din, in reply, demanded the return of Dameghan, which had been seized by the Ismaelites during the Mongol troubles. It was agreed that it should be ceded to them on a payment of 30,000 dinars annually. The envoy, after this arrangement, accompanied Jelal ud din into Azerbaijan, and one day in his cups in the Vizier's presence boasted that there were fidayis (i.e., devotees) of Alai ud din among the Khuarezmians, among their servants and their generals, even in the Vizier's own office, and among those in the service of the heads of the chaushes or ushers. Sherif ul Mulk begged him to summon them, and gave him his handkerchief as a pledge of their safety. Five of them were accordingly brought. One of them, an Indian, strong and determined, said to the Vizier, "I should have killed you on a certain occasion, but that I waited for further orders." "And why?" said the Vizier, throwing off his tunic and seating himself in his shirt. "What does Alai ud din want with me? What have I done that he would have my blood? I am his slave as I am the slave of the Sultan. I am at your service. Do what you will with me." The Sultan on hearing of this was very angry that his Vizier should thus have humiliated himself, and sent him orders to burn five of the fidayis before his tent. The Vizier made excuses. Thereupon the Sultan had an immense brazier set up in front of his tent, and had five of them put into it, who cried out as they were dying, "We are the victims of our lord, Alai ud din." The Sultan then had the head of the chaushes put to death for having such people in his service.

When Jelal ud din afterwards went to Irak, the Vizier remained at Berdaa, when there came a fresh envoy from the Ismaelites demanding a payment of 2,000 dinars for each of the fidayis who had been burnt to death. The Vizier, who was delighted to be let off so easily, ordered the Chancellor Muhammed of Nissa, to whom we owe the account, to draw up a rescript reducing the tribute which Alai ud din had agreed to pay by 10,000 dinars. After the battle of Ispahan, while Jelal ud din was at Rai, and his troops were pursuing the Mongols towards Khorasan, he received an envoy from Alai ud din, who was accompanied by nine fidayis. To prove their goodwill, they asked him to point out those whom he wished to destroy. Some of his councillors were for accepting this offer, but Sherif ud din, the Vizier's substitute in Irak, urged that Alai ud din only wanted to know who his enemies were so that he might intrigue with them, and he accordingly replied, "You must know who are our friends and who are our enemies. If you wish to do what you propose there is no need of instructions, and if it so pleases God, our sabres will enable us to dispense with your daggers." Soon after this Ghiath ud din, Jelal ud din's brother, sought refuge with the Ismaelite chief, as we described, and was supplied by him with horses and arms, which greatly displeased his brother ; and as, instead of sending the tribute he had promised, he only sent 20,000 dinars in two years, Muhammed of Nissa was sent to expostulate, and to demand that Alai ud din should have the khutbeh said in the Sultan's name. If he failed to pay the arrears, Nissavi was authorised to ravage his borders with fire and sword. The Sultan's letter was couched in rather peremptory language, and Nissavi was ordered not to enter Alamut unless Alai ud din came out to meet him, not to kiss his hand, and to omit all the usual marks of respect or politeness. Nissavi set out. The Ismaelite chief did not come out to meet him, but he was met by the Vizier, Amad ud din El Meuhteshem, who asked that the message might be communicated to him. This he refused, and it was four days before he was eventually admitted to an audience at midnight on the top of the mountain. The Vizier was seated on the Prince's right, while Nissavi was offered the seat on his left. He asked that the Sultan's name might be inserted in the khutbeh, as it was in the days of his father. With this demand Nissavi handed in a written declaration from the Kadhi Mojir ud din, who was still living, and who had been employed by the late Sultan to secure this right. At first they pretended it was a forgery, but they did not persist. " The thing," says Nissavi, " was too patent and too recent. Everyone knew that they formerly paid an annual tribute of 100,000 dinars to the Sultan." The subject of the arrears was then raised, and they pleaded that the commandant of Firuzkuh had seized a sum of 15,000 dinars, which was being transported from Kuhistan to Alamut. When Nissavi urged that this was before the late treaty, they said: " When have we been the enemies of the Khuarezmians, or, rather, when have we not been their friends ? The Sultan has proved it both in ill fortune and good fortune. Did not our companions help him in India after passing the Indus, when he was reduced to the lowest state ? " The fact was afterwards admitted by Jelal ud din. When Nissavi said this was no reason for reducing the tribute, they produced the Vizier's attested agreement for its reduction, as we have mentioned. Nissavi said this did not bind the Sultan. " The Vizier disposes," they replied, " of all the Sultan's revenue. He spends it as he likes without any restriction, and according to his whim. Are his hands only tied in regard to us." It was eventually agreed they should pay 20,000 dinars, the rest being left over for further consideration.*

As Alai ud din grew up he showed signs of mental aberration, but his physicians dared not acknowledge it nor prescribe for it, for fear of being massacred by the fanatics, who would not credit that the Imam could

---

* Nissavi, quoted by D'Ohsson, iii. 174-185.

suffer thus. He grew more imbecile, and as he was not contradicted or corrected, his passion was unbounded. Meanwhile his senilities were accepted as divine inspirations, while brigandage flourished, and his subjects were greatly oppressed. When he was eighteen years old he had a son named Rokn ud din Khurshah, whom he instituted as his successor, and who, when he had passed the age of infancy, was treated with the same honours as his father. The latter presently grew jealous, and wished to supersede him by another son, but his followers declared this impossible, the first nomination being irrevocable. He therefore began to torment his son. He in turn intrigued against his father among those who were growing weary of the latter's absurdities. He declared that his father's conduct would bring down the Mongols upon them, and proposed to separate from him and to send his submission to the Grand Khan. The greater part of the grandees agreed to support him to the last drop of blood against his father's adherents, but with the reservation that if his father marched in person they could not raise a hand against him. Soon after this, Alai ud din being one day drunk, was sleeping in a hut made of wood and reeds, adjoining a sheep pen, in a place called Shirkuh, where he used to go to enjoy his favourite relaxation of a shepherd; about him were lying his cameleers and servants. There he was found dead in the middle of the night, his head being separated from his body. An Indian and a Turkoman who slept beside him were both wounded. A few days later, when several innocent people had suffered, it was discovered that the deed had been done by his confidante and constant companion, Hasan Mazanderani. Rokn ud din did not have him tried, but had him assassinated, which confirmed the suspicions that rested upon him; and he had the cruelty to throw his three children, two sons and a daughter, into the brazier in which the assassin's body was burnt. Shems ud din, Ayub of Tus, wrote a poem on his death. On his accession, Rokn ud din enjoined a strict adherence to the Muhammedan law, and took measures to secure the safety of the roads.

Meanwhile, as we have seen, Kitubuka had been sent on with the advance guard of Khulagu's army, to deal him some hard blows. He crossed the Oxus early in March, 1253, and penetrated into Kuhistan, where he captured several strong places. Thence advancing with 5,000 horsemen and 5,000 foot soldiers, the former probably Mongols and the latter Tajiks or Persians, he assailed Girdkuh. It was also called Derikunbed (*i.e.*, the vaulted gate),* and was situated three parasangs west of Dameghan, to which town it was in fact a kind of fortress, where its inhabitants could take refuge.† It is called Tigado by Haithon, and Ki du bu gu in the Yuan shi, where we read it was situated on a very steep rock, which could not be reached either by arrows or by stones

---

* Ilkhans, i. 93. Quatremere reads it Diz-gunbedan.   † Quatremere, 278. Note.

from catapults. It was so steep that in looking up one's cap fell off.[*]
Having put double lines of circumvallation about it, so that his army had
a rampart and ditch both before and behind it, Kitubuka left an army to
blockade it, under an Amir named Buri, and proceeded to attack Mehrin
and Shah. Meanwhile Hirkutai, one of his subordinates, devastated the
districts of Tarem and Rudbar. The Mongols afterwards assailed
Mansuriah and Alabeshin, or Alah beshin, and continued the slaughter
for eighteen days. The garrison of Girdkuh now made a sally and killed
100 Mongols, including Buri. Kitubuka meanwhile harried all the herds
in the districts of Tun, Tershiz and Zirkuh, while Mehrin and Kemali
both fell.[†] Having heard that his famous arsenal, Girdkuh, was afflicted
by pestilence and likely to surrender, Alai ud din, the King of the Assassins,
sent a body of 100 men, under Mubariz ud din Turan and Shuja ed din
Hasan Sarabani, each bearing three menns of salt and one of henna (the
latter, well known as a dye to dye the nails, was made of the powdered
leaves of the *Lawsonia inermis*.[‡]) On this occasion [§] it was welcome on
account of its medicinal qualities.

Shortly after, namely, on the 2nd of December, 1255, Alai ud din was
murdered, as I have mentioned. In the spring following Kitubuka and
Kuka Ilka received orders from Khulagu, who was rapidly advancing,
to attack the remaining fortresses of Kuhistan. This they did in the
course of a month, during which they committed great ravages there;
inter alia, they captured Tun after an attack of twelve days, and killed all
the inhabitants except the artisans, after which they joined Khulagu, who
had advanced to Tus, near Meshed.

Meanwhile, let us turn to Khulagu himself. We are told he was
accompanied by two of his ten sons, namely, Abaka and Yushmut; a
third, Jumkur, he left at his brother the Khakan's Court in charge of his
interests there; while another son, Temkian, was left at home in charge of
his yurt. With him also went his brother Suntai, or Sitai Oghul, the
ninth son of Tului. Nigudar[||] represented the Ulus of Jagatai. The
Golden Horde was represented by Khuli, son of Orda, eldest son of
Juchi; by Balakhen, or Balakan, also called Bulgha, Bulga, and Bulga
Kabli, son of Sheiban, son of Juchi; and by Tutar (called also Tumar,
Kotur, or Kotar, and by St. Martin, Bukan), son of Mankadi, son of Tual,
son of Juchi. These princes apparently joined him when he arrived in
Iran. He was accompanied also by Buka Timur, son of Jijeghan
(called Jehakan Begi by Abulfaraj), the daughter of Jingis Khan, whom
she bore to the Uirad chief Turalji, and who was step-brother to Kubak
Khatun, and Oljai Khatun, two of Khulagu's wives. Buka Timur took

---

[*] Bretschneider, Notes on Chin. Trav., 78. Notices of Med. Geog., 203. Note 341.
[†] Quatremere, 171, 3. [‡] *Id.*, 172. Note. [§] Ante, i. 194.
[||] Called Thagudar by Malakia (op. cit., 451), and Tacudar by Abulfaraj, who calls him the son
of Buchi Ogul. (Op. cit., Chron. Arab., 329.) Buchi Ogul is called Juchi by Von Hammer.
(Ilkhans, i. 86.)

with him a contingent of Uirads.* Khulagu also had with him his wives, Yisut and Oljai, and his stepmother, Tokuz, whom he eventually married. It will be noted that the Ulus of Ogotai, which was at feud with Mangu, was not represented at all.

The princes above mentioned commanded contingents supplied by their several hordes, which commands were hereditary, and the general notion seems to have been that the enterprise was a joint one, in which the fruits of victory were in fact to be shared among all the Mongol ulusses. Each one was accordingly called upon to furnish two men out of every ten for the campaign, while, as we have seen, 1,000 skilled Chinese arbalisters and men accustomed to hurl fire arrows (the ho pao of the Chinese), in which naphtha was a main ingredient, were also supplied. In regard to this section of the army Major Raverty has translated an interesting notice, in which we are told it consisted of a thousand families of *Chinese Manjanik chis (manganel workers), naft andaz (naphtha throwers), and charkh andaz (shooters of fiery arrows worked by a wheel), and they took with them a vast quantity of ammunition. They had with them also charkhi kamans, i.e., arbalists worked by a wheel, so that one bowstring would pull three bows, each of which discharged an arrow three or four ells long. The arrows or bolts, from the notch of the bowstring to near the head, were covered with feathers of the vulture and the eagle, and the bolts were short and strong. These machines would also throw naphtha. The manganels were made of ash, very tough and strong, and covered with the hides of bullocks and horses (to prevent them being burnt), being thus enclosed like a dagger in its sheath, and each manganel was so constructed as to be capable of being separated into five or six pieces, and easily put together again. The machines were brought from China into Turkestan on carts, and were under the direction of skilled engineers.† A thousand pounds of meal and a skin of kumiz were also ordered to be provided for each man.‡ Orders were issued to reserve the pastures west of the Tungat Mountains (identified by Bretschneider with the range now called Tangnu), and lying between Bishbaligh and Karakorum. Roads were repaired and bridges made, and to prepare Khulagu's way more effectively, the troops of Baichu Noyan were told to draw near to Rum (i.e., Asia Minor), so that the pastures in the Mughan plain might be fresh. Before leaving, Khulagu gave a feast in Mangu's honour, and was feasted in return by his young brother Arikbuka and others at Karakorum. The Khakan Mangu bade him obey the counsels of Jingis Khan, to treat those who submitted kindly, but to exterminate those who resisted, and he commissioned him to conquer the land from the Oxus to the borders of Egypt, to subdue

---

* Abulfaraj, Chron. Arab., 329. Ilkhans, i. 86.
† Tabakat-i-Nasiri, 1191.   ‡ Quatremere, 137.

Kuhistan, the Kurds, and the Turks, and to compel the Khalif to be submissive. When he had accomplished his mission he told him to return again. He poured rich presents upon him and his amirs, and bade him take council with his stepmother Tokuz. Mangu sent splendid presents in gold, robes, and horses to Khulagu, his wives, and children, and to the principal noyans and amirs ; and he also arranged that his younger brother, Suntai or Sitai Oghul, was to accompany him, probably to look after his immediate interests. Khulagu having repaired to his own *ordu*, at length set out in October, 1253, leaving a portion of his harem behind. The amirs in charge of the different districts had duly provided provisions at the various stations. Stones and other impediments were removed from the roads, while the different princes and generals who were to take part in the expedition employed themselves in exercising their troops. He set out in February, 1254, and marched from station to station till he approached Almaligh.* There he was probably met by Nigudar with the Jagatai contingent. He was feasted by Orghana, or Irghana, the widow of Kara Khulagu, ruler of the Ulus of Jagatai (who was a granddaughter of Jingis Khan, her mother having been Jigeghan, already named, and she was consequently stepsister of Oljai, Khulagu's wife). We are told he left a large portion of his family "in Turkestan, near Almaligh."† He was again fêted further on by Masud Bey, the Governor of Mavera un Nehr and Turkestan, and arrived at Samarkand in September, 1255. Khulagu's mission was merely that of a general who commanded an army, and Mangu's purpose in dispatching him westwards was not to make over to him any *independent* authority over the western countries or their peoples, but only to head a great campaign against the enemies of the Mongols. We must understand this when we read that the various contingents of troops from the Indus to the borders of Syria were placed under his control, while the different feudatory princes and the civil governors of Mavera un Nehr and Khorasan were put at his service. Mavera un Nehr and Khorasan were treated as imperial appanages, and remained so at least until the days of Khubilai Khakan. East of Khorasan the country was controlled by maliks or princes, who paid tribute to the Mongols, and were largely controlled by Mongol commissaries at their Courts. Much the most important of these maliks was the Chief of Herat, and to him we must devote a longer notice.

The best authority available for the history of the family of Kert or Kurt,‡ is the "Chronicle of Herat," composed by Muyin ud din Muhammed, surnamed es Zemji, who was a native of Esfizar, near Herat. His work, entitled "The Celestial Garden," is a description of the town

---

* Quatremere, 145-147.   † *Id.*, 99. Ilkhans, i. 88.
‡ A soubriquet said to mean greatness or magnificence. Raverty, 1198. Note 8.

of Herat, and was finished in 897 HEJ. (*i.e.*, 1491-2). The portion which interests us at present was translated by M. Barbier de Meynard in the sixteenth and seventeenth volumes of the "Journal Asiatique." Iz ud din Omar Meraghani was the favourite Minister of the Ghurian Sultan, Ghiath ud din. He is styled Malik of Khorasan by Minhaj-i-Saraj.[*] He conferred on his brother, Taj ud din Osman, the fortress of Khaisar, and he occupied the post of chief armour-bearer at the Court of Mahmud, Iz ud din's son. Osman, on his death, transmitted his fief to his son, Rokn ud din Abubekr, who married a daughter of Sultan Ghiath ud din,[†] who was reigning when Jingis Khan invaded the west. He seems to have conciliated that conqueror. His heritage was unmolested when the rest of Ghur was overrun, and he apparently was confirmed in its possession by Jingis Khan. According to Muyin ud din Esfizari, Jingis Khan was about to attack the fortress, and had a plan of it first made, when, afraid of being beaten, he left it in the hands of Rokn ud din Kert.[‡] It was the strongest fortress of Ghur, and its citadel still remains north-east of Teiverch, at the foot of the peak of Chap dalan, on an inaccessible rock.[§]

We are told that when Malik Rokn ud din used to attend the camp of Jingis Khan, of Ogotai and the Mongol Noyans, he used to take his son Shems ud din with him, so that he became acquainted with the Mongol usages and regulations.[||] Rokn ud din died in 1245, and was succeeded by Shems ud din, who is found the next year accompanying Sali Noyan (perhaps the Mangutah previously mentioned)[¶] in his invasion of Sind, and treating with the governors of Multan and Lahore. The former was, at his instance, ransomed for 100,000 gold pieces, and the latter for 30,000 dinars, 30 loads of fine cloth, and 100 slaves. We are told that in consequence of his success on this occasion he was made military governor of Lahore, but presently the Mongol chiefs grew jealous of him, and accused him of having secret negotiations with the infidels of the country. They said he had accepted 50,000 dinars from the governors of Multan and Lahore, and had promised he would join the troops of the Sultan of Delhi if they should approach. Shems ud din, on hearing of these accusations, determined to escape to Tair Baghatur, Sali's superior officer, and accordingly fled with a few soldiers, and took shelter in a pagoda near Guejuran. From the people there he begged some arms, &c., to present to Tair Baghatur, but Fakhr ud din, the chief of Guejuran, having been told that he meant to possess himself of that district, sent Emad ud din with some troops to seize him and lodge him in the fortress of Guejuran. Emad ud din first consulted Tair Baghatur, who, remembering his friendship for Rokn ud din, the father of the

* Tabakat-i-Nasiri, 193. † Journ. Asiat., 5th ser., xvii. 440-441.
‡ Journ. Asiat., 5th ser., xvii. 455. Note.
¡ See Ferrier's Travels, ii. 9. || Tabakat-i-Nasiri, 1200. Note. ¶ Ante, p. 71.

fugitive, ordered him to be taken before him to be tried. Tair's tent was pitched on the crest of a hill, and when Shems ud din was admitted he asked him, "These towns and villages on the right, to whom do they belong?" "To you, prince," said the culprit. "And these fields and orchards in front of us?" "To you also." And he made the same reply to a number of similar questions. He then turned to Emad ud din and asked him to point out his property. He diplomatically said he owned only one poor house there, and had few connections with the country. "Know then," said Tair, laughing, "that it belongs very largely to Shems ud din, and that he is at liberty to levy requisitions without being treated as a rebel." Emad ud din thereupon withdrew, and left the Mongol camp the same night, while Shems ud din remained with *éclat* with his protector till the return of Sali Noyan from India with a large booty.*

Tair Behadur died in 645 (*i.e.*, 1247), whereupon his son Halkatu Noyan (Arkatu), in concert with Kara Noyan, who, it seems, had a grievance against Shems ud din, reported him to Jagatai Khan. He set out to justify himself, but on his arrival Jagatai was dead.† He seems to have been driven away by Jagatai's son, Yissu Mangu, and fled to the Court of Batu, whence he made his way to the kuriltai where Mangu Khakan was inaugurated.‡ The officers who introduced him exalted his virtues and the services of his ancestors, and Mangu received him with special honour, and conferred on him as a fief the whole province of Herat, with Jam, Bakherz, Kusuyeh, Fushenj, Tulek, Ghur and Khaisar, Firuz Kuh, Gharjistan, Murghab, Meruchak, Fariab, as far as the Oxus ; Esfizar, Ferrah, Sigistan, Kabul, Tirah, and Afghanistan, as far as the Indus and the borders of the Hindus.§ Besides granting him the great fief I have mentioned, Mangu issued an order to Arghun Aka, the civil governor of Khorasan, to make over fifty tomans of money to the intendants of Shems ud din. The next day, in a private audience, the Khakan presented the Malik with one of his own robes, gave him a paizah, or official tablet, 10,000 dinars, and arms, including an Indian sabre, a lance of Alkhatt, a mace with a bull's head on the top, an axe, and a dagger. Shems ud din left for Herat, accompanied by an officer of the Khakan. He turned aside to pay a visit to Arghun Aka, to whom he presented the Khakan's order, who duly handed him the fifty tomans.‖ After occupying Herat he put Sherif ud din, the Bitikji, whose tyranny had ruined the country, to death, and severely reprimanded Korlogh, the military governor of the place. He also obtained possession of Bakar, a fortress of Sijistan, which no one had been able to capture by force since the days of Nushirvan ; and in 647 HEJ. (*i.e.*, 1249) he slew Saif ud din,

---

* Journ. Asiat., 5th ser., xvii. 442-445.　　† *Id.*, 445.　　‡ Ilkhans, i. 276-277.
§ Journ. Asiat., 5th ser., xvii. 445.
‖ *Id.*, 445-446. D'Ohsson, iii. 129-131. M. de Sacy says Alkhatt is a plain in the district of Yemama, or Bahrein, where the handles of lances are made that come from India. (Chrest Arab., ii. 79. Note 12.)

the Malik of Gharjistan, who had apparently refused to acknowledge his authority. He sent 400 men against him, whereupon Saif ud din fled to Arghun, who would not listen to him, but sent him bound in chains to the Malik of Herat, and he was put to death by being trampled under foot by horses near the gate Khosh, and his corpse remained exposed for three days in the great bazaar.* The date of this event is clearly wrong, since Mangu did not mount the throne till 1252.

To revert to Khulagu. While he was encamped in the meadows of Kan Ghul, near Samarkand, he was visited by the Malik Shems ud din Kert, and the subordinate chiefs of the district, who duly did homage. He was also feasted there for forty days in a tent of golden tissue furnished by Masud. At this time he lost his brother and companion, Sitai Oghul, and, according to Abulfaraj, he received news of the death of another brother named Balador, who is not otherwise known to me.† At Kesh, the birthplace of the Great Timur, he was met by Arghun, the Mongol Governor of Khorasan. There also went at his bidding the two joint Sultans of Rum, Iz ud din and Rokn ud din ; from Fars there went Said, son of the Atabeg Muzaffar ud din, while other chieftains greeted him from Irak, Khorasan, Azerbaijan, Arran, Shirvan, and Georgia.‡ While at Kesh, where he stayed a month, Khulagu issued a firman, or order, to the various princes of Western Asia to march and aid him against the Mulahids, or Assassins, or take the consequences. The boats and boatmen on the Oxus having been impounded, the army safely traversed the river on the 2nd of January, 1256. The boatmen were rewarded for their zeal on this occasion by being relieved of the dues they had previously paid.§ Guiragos says Khulagu's army was so large that it took a month to pass over the Oxus.‖ Having crossed the river he, by way of amusement, held a review on the banks. Suddenly several lions came out of a forest. Khulagu ordered his horsemen to form a ring and surround the animals, and, as the horses were afraid of the lions, they mounted on camels, and succeeded in killing, according to the Jihan-Kushai, ten—Rashid says two. Quatremere argues that the lion was unknown to the Manchus and Mongols, who borrowed a name for it (arslan) from the Turks.¶ The next important halt was at Shiburghan, corrupted into Shibrghan, a town situated about ninety miles west of Balkh, and now containing about 12,000 houses. It is a very old place. Its earliest recorded name is Asapuragan, while the Arabs called it Saburkan or Shaburkan. Its famous dried melons are mentioned by Marco Polo.** A fall of snow and a frost, lasting seven days, caused so many horses to die that it was determined to delay there during the winter. There, in the spring, Arghun Aka entertained Khulagu in a tent

---

* Journ. Asiat., 5th ser., xvii. 446.　　† Abulfaraj, Chron. Arab., 330.
‡ Quatremere, op. cit., 153.　§ Quatremere, 153.　‖ Journ. Asiat., 5th ser., xi. 482.
¶ Op. cit., 153-155.　Notes.　** Op. cit., ed. Yule, i 156-157.

of golden tissue, pinned down by 1,000 golden pegs. It had a rich
pavilion as an ante-chamber, while the hall of audience was furnished
with gold and silver vessels decked with precious stones. A grand feast
was given on a day fixed as auspicious, during which Khulagu was seated
on a throne, while the various princes and grandees who surrounded him
did him honour. After the feast Arghun, by Khulagu's order, set off for
Mangu's Court. He left his son, Kirai Malik Ahmed, the Bitikji, and
Alai ud din Ata Malik, in charge of the affairs of Iran in his absence.*

Khulagu now dispatched the Malik Shems ud din Kert, the Lord of·
Herat, to summon the Mohtesshim (Preceptor) of Kuhistan, Nasir ud din,
who was then at Sartakht. The latter set out, and was duly submissive,
whereupon he gave him a paizah or official tablet, and a yarligh or
diploma, with the command of the town of Tun, but he shortly
after died. Tun was situated near Kain or Ghain, whence the two
towns were joined together by Marco Polo, under the name of
Tunocain.† It is described as a fine town, with a moated castle in the
centre, surrounded by houses and a market-place, outside which were
cornfields and melon gardens. Khulagu now advanced to Zawah, the
modern Turbat-i-Hadari, and Khavuz or Khaus, where he was taken
slightly ill. Thence he went on to Tus, where he was rejoined by Kitubuka
and Kuka Ilka. Tus was the head-quarters of the civil governor of the
Western Mongol possessions. There he was feasted, and then went
on to Mansuriah, which had been restored by Arghun, and where the
latter's wives and the Amir Khoja Iz ud din Tahir entertained their
powerful guest. At Radekan, between Tus and Khabushan, he feasted
on the rich products of Merv, Baverd† (or Abiverd, situated between
Sarrakhs and Nissa), and Dahistan. At Khabushan he restored the
ruins caused by the previous Mongol invasion, the cost of which he
defrayed out of the public purse. Canals and workshops were made, and
a garden laid out near the principal mosque. Saif ud din, the Vizier,
superintended these works. On the order of Khulagu the amirs and
principal courtiers also built themselves houses there.§ Raverty says
they were not canals which were made, as here stated, but that kahrezes,
or subterranean aqueducts, were repaired.‖ On the 2nd of September,
1256, Khulagu reached the environs of Kharakan, or Kharkan, and
Bostam. The latter, situated in the valley of the upper Attrek, in the
east of the district of Kumus, was the birthplace of several famous men;
among them of the mystic Sheikh Bostami, the founder of the order of
dervishes named after him, Bostami.¶ From Bostam Khulagu sent two
envoys, named Merketai and Menklemish, to menace Rokn ud din, the
chief of the Ismaelites, with his vengeance. At this time the famous poet

---

* Quatremere, 165.   † Quatremere, 157-159. Ilkhans, i. 95. Note 3. Marco Polo, i. 84.
              ‡ Called Yesrud by Von Hammer. Ilkhans, i. 97.
    § Quatremere, 181-183.   ‖ Tabakat-i-Nasiri, 1196. Note.   ¶ Ilkhans, i. 98.

and astronomer, Khoja Nasir ud din Tusi, with several doctors, were living against their inclination among the Ismaelites, and having determined to put an end to the oppression of the chief of the latter, they, in concert with some other Mussulmans, persuaded him to be submissive.* He accordingly sent from Maimundiz, where he was living, an officer to Yassaur, the Mongol Noyan, who was then at Hamadan, to assure him of his submission. He advised him to repair to Khulagu. Rokn ud din said he would send his brother, Shahin Shah, to him. The latter, in fact, set out, and Yassaur commissioned his son to accompany him on his return. Notwithstanding this, he entered a few days later, viz., in June, 1256, the district of Alamut, with an army composed of Turks and Persians, and attacked that fortress; but after a sharp conflict his troops were obliged to retreat, and wreaked their vengeance in destroying the crops and ravaging all the country round.† Meanwhile Shahin Shah repaired to and was well received by Khulagu, who in turn sent four envoys to the Ismaelite ruler, among whom was one called Bakhshi by Rashid ud din. The Buddhist Lamas were so styled, and he was perhaps one of them.‡ They were to tell him to dismantle his fortresses, and to go to him in person, and meanwhile Yassaur, with his Mongols, was to withdraw from his territory. He partially complied, and began to overthrow the ramparts of Maimundiz, Lembeser, and Alamut, and offered to accept a Mongol baskak or commissary at his Court ; but in regard to going in person to his Court, he pleaded that he would do so in the course of a year. Khulagu had determined to destroy him, however, and on this pretext he ordered his troops to come together from Irak and the adjacent provinces.

While Khulagu was advancing towards Kuhistan, the three princes of the house of Juchi had apparently traversed the pass of Derbend from Kipchak with their contingents. They advanced, says Guiragos, with their chariots, having levelled and made passable all the roads.§ Khuli, one of the three, styled himself "Son of God." Malakia tells us that he was a merciless persecutor of the Christians ; that he caused all the crosses on the roadsides and mountains he met with to be burnt, and treated with especial brutality the inmates of the various monasteries they passed. One of his chiefs went to the Monastery of Gereth, whose abbot was called Stephanos, and was a very old man with grey hair, distinguished for his sanctity. On the approach of the Tartar chief he took a glass of wine, and offered him the tghghu (i.e., the usual tax or offering insisted upon by the Mongols on such occasions), and conducted him to the monastery, where he killed a sheep and distributed it with the wine to the leaders of the band. They went on drinking till night, when they returned to their quarters, which were close by. On rising the

---

* Quatremere, 183-185.   † D'Ohsson, iii. 186-189. Ilkhans, i. 99.
‡ Quatremere, 184.   Note 51.   § Guiragos, ed. Brosset, 182.

next morning their chief was very ill, and charged the monks with having poisoned him, his illness really being the result of his gluttony. They nevertheless seized and chained Stephanos. They tortured him to extract a confession, and this not being forthcoming, they fixed four stakes, to which they fastened him, spread some earth over him, and then lit a fire over all until they roasted his flesh and he gave up the ghost. Malakia goes on to report what is usually told of Armenian martyrs, that a light hovered over his remains, while the cruel chief was driven by the demon which possessed him to tear his own flesh with his teeth, and several of his companions perished from the complaint which had seized them. This epidemic spread to Khuli himself. Malakia then tells a very grim story, viz., that Khuli summoned a doctor, who is elsewhere said to have been a Jew, and who declared that there was no other remedy for this disease than to thrust his feet into the warm entrails of a child who was to be cut open for the purpose. They accordingly seized some thirty Christian children in the streets. They killed them with arrows and cut them open. Khuli's pain was not, however, assuaged, and in a rage he had the doctor himself cut open and his entrails thrown to the dogs. Khuli presently died, and was succeeded by his son Migan, also called Mizan, or Mishan.*

Let us now return to Khulagu, who, as we have seen, had ordered a general muster of his troops. The right wing, commanded by Buka Timur and Kuka Ilka, marched by way of Mazanderan ; the left, under Nigudar and Kitubuka, went by Khowar and Semnan ; while Khulagu commanded the centre, called kul by the Mongols, in person.† Meanwhile, he dispatched the doomed prince another warning. The Khurshah in reply sent his Vizier, Kaikobad, and other envoys, who met the invading army at Firuzkuh. This place was visited by Morier, to whom the ruins of the castle of the Ismaelites were pointed out as a windmill and baths of the time of Alexander the Great.‡ Quatremere has a long note on the place, which was situated under the famous mountain of Demavend, and near Rai. Clavigo describes it as situated on a high rock rising precipitously from a plain, and as, in reality, comprising three fortresses girdled by walls and bastions.§ It was, as we have seen, perhaps "The Paradise" of Marco Polo. The envoys offered to surrender all the towns in the country except their ancestral strongholds of Alamut and Lembeser, and again pleaded for a year's delay, after which they promised that their master, who meanwhile gave orders for the surrender of Girdkuh and the fortresses of Kuhistan, would visit Khulagu in person.

The Mongols continued their advance, and reached Lar and Demavend. The latter is one of the oldest cities of Iran, and is situated at the foot of

* Malakia, 451-453.   † Quatremere, 191-193.   ‡ Ilkhans, i. 99.   § Quatremere, 274-276.

a famous volcanic peak in the Elburz chain, which is 20,000 feet high, and which bears the same name. It was the residence of the tyrant Sohak, the carbuncle on whose shoulder, which appeared when Satan kissed him, could only be eased by the brains of two men, killed daily; and from whose tyranny the people were delivered by the smith Giawe, whose leathern apron, fixed on a lance, was the gathering point of those who opposed him. The 31st of August is kept as a festival in the Mussulman world in memory of the deliverance from Sohak's tyranny.* The mountain of Demavend was the scene of much early romance, and Quatremere has devoted a long note to it.† From Demavend Khulagu advanced to Shahdiz, which he captured in two days. Fresh envoys were thence sent to the recalcitrant chief, who now consented to send his son with a contingent of 300 soldiers, and to demolish his fortresses. The Mongols delayed at Abbasabad, on the main route from Demavend to Sari, awaiting the performance of these promises. Rokn ud din sent a young son he had had by a Kurdish concubine, who was then eight years old, and who in consideration of his youth was allowed to return. Khulagu now asked him to send an older prince, namely, his second brother, Shahin Shah. The latter accordingly went, and reached the Mongols at Rai. He was deluded by a fair-sounding yarligh or diploma, setting out their goodwill, and stating that if Rokn ud din duly demolished his fortresses he would have nothing to fear. The Mongol troops, however, kept advancing. When Buka Timur and Kuka Ilka neared Aspendan—called Ispidar by Von Hammer, and, doubtless rightly, Astadar by Major Raverty—the Khurshah sent to ask what motive they had for going there, since he had submitted to their master and was occupied in demolishing his fortresses. Their enigmatical answer was, "As we are at peace with one another, we have come to search for pasture."‡

On the 21st of October Khulagu left the district of Rudbar by the pass of Baskal or Yaskal, and took the route of Talikan,§ situated between Kazvin and Abher.‖ He advanced with his troops, and planted them round Maimundiz. As the walls were very strong, a council was held as to whether they should press the siege or withdraw. Most of those present urged that it was mid-winter, that their horses were thin, and that it would be necessary to get provender for them from Armenia and Kurdistan, and urged a retreat. Buka Timur, Kitubuka, and Saif ud din, the bitikji, on the other hand, advised the siege to be pressed.¶ Meanwhile the contingent of 300 men sent by Rokn ud din were put to death near Kazvin. A summons was sent into the town bidding it surrender in five days. The reply was that Rokn ud din was then absent. Trees

* Ilkhans, i. 100.   † Op. cit., 200-204.   ‡ Quatremere, 207.
Von Hammer has confused this place with the more famous Talikan in Tokharistan.   Ilkhan
i. 101.
‖ Quatremere, 278.   ¶ Quatremere, 211.

were now cut down to make catapults with, which were dragged to the top of the neighbouring heights. Meanwhile the besieged returned a heavy fire. The following day the duel was renewed, but the chief of the Ismaelites proposed a cessation of hostilities. Khulagu insisted on immediate surrender, and Atha ul Mulk of Juveni was ordered to draw up the form. Meanwhile a tumult occurred among the citizens (to which their chief was probably privy), who did not wish to surrender. Rokn ud din sent word to the Mongols of what had occurred, and stated that his life was in danger, and the bombardment recommenced. The vigour of the attack and the uncommon mildness of the season made the besieged at length lose heart.* Rokn ud din accordingly sent his brother, Shah Kiya, with the astronomer Khoja Nasir ud din of Tus, two sons of Rais ud daulat, from Hamadan, who were famous as doctors, with many grandees, bearing rich presents, and a few days after, on the 19th of November, he went in person and, in the words of Rashid ud din, "kissed the ground before his August Majesty." Khulagu treated him kindly, and this induced others to submit. Sadr ud din, or according to Abulfaraj, Shems ud din, prefect of the fortresses of Kuhistan, was sent to demolish the various Ismaelite fortresses in Kuhistan, Kumus, and Rudbar. There were a hundred, well provisioned and armed. The governors of the fortresses of Dilem also agreed to demolish their walls, and all were thus razed except Girdkuh and Lembeser. The latter held out for a year, when a disease broke out there and it had to surrender. Girdkuh held out longer; the Yuan shi says it was captured by Kitubuka in 1257.

In the biography of Kuo Khan (*i.e.*, of Kuka Ilka) he is made to capture it. We there read that it was only accessible by suspended ladders, which were guarded by the most valiant troops. It was battered with catapults, when its commander, Bu-jo na-shi-r, elsewhere called Da-dje na-shi-r, surrendered.† Other writers make out that Girdkuh held out for many years. To reconcile these notices we must suppose that after its surrender it again rebelled. The author of the "Tabakat-i-Nasiri" tells us that when he wrote it had been besieged for ten years, but still held out.‡ It apparently finally surrendered in December, 1270.§ The treasures collected by the Ismaelite princes at Maimundiz, which were less valuable than was expected, were distributed among Khulagu's soldiers, who then advanced to Alamut by way of Sheherek, the ancient capital of the Princes of Dilem, where he celebrated his success in a feast of nine days.‖ Alamut was ordered to surrender by the Khur Shah himself, but its governor, Sipah Salar, sent an uncivil reply, and refused.¶ Bulghai was left with a considerable contingent to attack it, and after three days it surrendered. The Mongols entered and broke the war engines on

* D'Ohsson, iii. 196.   † Bretschneider, Notes on Med. Travellers, &c., 78 and 79.
     ‡ Op. cit., 1207-1210.   § See Yule's Marco Polo, i. 153.
   ‖ D'Ohsson, iii. 197.   ¶ Tabakat-i-Nasiri, 1209-1210. Notes.

the walls, removed the gates, and pillaged the place. Khulagu himself entered the fortress, and was astonished at the extent of the mountain, which was compared in shape, by Eastern writers, to a camel kneeling with its neck stretched out, the fortress being built on the summit, and approachable only by one narrow path.* He sent his Vizier, Athamulk Juveni, to inspect the archives and library there. The astronomical instruments, Korans, and some other valuable works were put aside, including one with the title, "Serguseshti Sidina ; or Adventures of our Lord and Master," giving an account of the founder of the sect, Hasan Sabbah, from which Juveni drew the main portion of his account of the Ismaelites. All the works dealing with the tenets of the sect were given to the flames. The solid vaults of the fortress were found stored with great quantities of provisions; *inter alia*, were wine, vinegar, and honey, which, it was said, had been there since the time of Hasan Sabbah, and were still wholesome after 160 years. A Mongol officer was assigned the tedious duty of destroying the strong walls of the fortress.† Khulagu now went to Lembeser, or Lemser, where his winter quarters were, and where he left Tairbuka to prosecute the siege while he went to pass the New Year's feast at the Grand Ordu, seven parasangs from Kazvin. A whole week was spent in festivities, and the grandees were rewarded with robes of honour. The Khurshah was given a yarligh and a paizah, and a Mongol damsel for his wife, and Kazvin was assigned as a depôt for his treasures and wealth. Thence he dispatched two or three confidential men in company with the Mongols to order the governors of the Ismaelite fortresses in Syria to surrender. Khulagu apparently held his hand until these various fortresses, which might have taken years to capture, were in his power, when he was disembarrassed from a promise he had made to spare his life by his request to be allowed to visit the Khakan Mangu. He set out with some messengers of Khulagu, with whom he had some sharp words at Bukhara. When they reached Karakorum, Mangu would not see him, and said he ought not to have been sent on to him, as it unnecessarily fatigued the horses. Abulfaraj says he ordered him to return and surrender the fortresses of Girdkuh and Lembeser, which still held out. He set out on his return, but when near the mountains Tungat (*i.e.*, Tangnu) he was put to death with his suite. Orders were sent to Khulagu to exterminate the Mulahids. Thereupon he sent word to Kazvin, and the two sons of the Khurshah, with his daughters, brothers, and sisters, and their attendants, who had been moved to a place between Abher and Kazvin, were put to death. Orders were given to eradicate the rest, even to children in their cradles ; and we are told the Mongol Governor of Khorasan assembled the Ismaelites of Kuhistan under pretence of taking a census for a military levy, and put

* *Id.*, 1210. Note. Quatremere, 215.
† Ilkhans, i. 102-103. D'Ohsson, iii. 198-199.

them to death to the number of 12,000. They were similarly slaughtered elsewhere.* Alfi says that a number of the Khurshah's offspring and relatives were made over to Salghan Khatun, the daughter of Jagatai, in order to take blood revenge upon them for the murder of Jagatai, who had been killed by Mulahid assassins.† All the Khurshah's people were apparently not destroyed, for in 674 HEJ. (*i.e.*, 1275) a body of the Mulahids, combined with one of Khurshah's sons, seized the fortress of Alamut. Abaka sent an army against them, which defeated them, and the fortress was razed.‡ Muhammed of Esfizar, in his history of Herat says that at the beginning of the 16th century some of the people of the district were still attached to the errors of the sect. They levied among themselves a tax called the money of Hassan Sabbah, which was devoted to the decoration of his sepulchre, and the old women put aside one out of every ten spindles of yarn which they had spun, and which they called the tenth of the Imam (*i.e.*, of Hassan Sabbah).§ The author of the *Georgian Chronicle* says that many of the Mulahids took refuge in Egypt, where their descendants remained when he wrote.‖ A very interesting and graphic account of the descendants of the Ismaelites as they exist now in India has been given by Colonel Yule.¶

Having overwhelmed the Ismaelites, Khulagu set out in March, 1257, for Hamadan (the Ecbatana of the Greeks), the famous capital of the ancient Medes; famous also in Muhammedan times as an opulent and beautiful city. Among its noted monuments were the tomb of the Gazelle of Bahram-gur, and a colossal stone lion which stood over the pillar of one of its gates, and has been described in detail by Masudi. It was reported to have been put there by Alexander, and was looked upon as a kind of palladium, like the famous stone of Scone. It was broken to pieces by Merdavij in the year 319 of the HEJ., when he captured the town at the head of the troops of Ghilan and Dilem, and perpetrated a terrible massacre, so terrible that, according to the author of the "Mujmal Altawarikh," fifty asses were laden with the drawers of the dead. It again revived, and is reported to have been 12,000 paces in circumference, and to have contained 1,600 fountains, and several shrines which were objects of pilgrimage, and it abounded in fruits, flocks, and merchandise.**

The astronomer, Nasir ud din, of Tus, with the two doctors above mentioned as captured at Maimundiz, were now taken into Khulagu's service. This famous astronomer had formerly been in the service of Nasir ud din Abdur Rahim, governor of Kuhistan on behalf of the Khurshah, to whom he had dedicated a work entitled "Akhlak Nasiry,"

---

* Quatremere, 215-221. Abulfaraj, Chron. Arab, 332. D'Ohsson, iii. 201-202.
† Tabakat-i-Nasiri, 1211. Note. This was the Noyan Jagatai, whose death we have previously described, and not, of course, the Khan Jagatai.
‡ *Id.*, 1212. Note.   § D'Ohsson, iii. 202. Note.   ‖ Op. cit., 544.
¶ Yule's Marco Polo, i. 153-155.   ** Quatremere, op. cit., 220-223. Note.

or the Ethics of Nasir.   It was divided into three parts, the substance of one, treating of moral perfection, had been written by an Arab named Abu Ali Meskuyah, and was much esteemed by Muhammedans. The other two books, on economics and political society, the author declares were chiefly taken from Greek sources. The author of this work having sent an ode in praise of the Khalif to Baghdad, the latter's vizier, Ibn Alkamiyi, a zealous Shia, wrote a verse on the back of it, in which he advised the governor of Kuhistan to keep his eye on him, as he was corresponding with the Khalif. He had accordingly put him under arrest, and sent him to Maimundiz, where he was when it was captured.* Khulagu now treated him and the sons of Rais ud daulat and Muvaffek ud daulat, famous physicians, with especial favour, and having learnt that they were natives of Hamadan, gave them horses on which to transport their families, servants, and slaves.   They and their descendants retained positions of trust for some time in the household of the Ilkhans.†

Let us now go on with our story.   At Hamadan Khulagu was met by Baichu, who in answer to his reproaches that he had done so little with his army, replied on his knees that he had conquered all the country from Rai to the borders of Rum and Sham (*i.e.*, Asia Minor and Syria).   As to Baghdad, he enlarged upon the power of the Khalif and the difficulty of approaching his dominions.   " Nevertheless," he said, "it is for the prince to command, and his slave will punctually obey his orders." Appeased by this reply, Khulagu bade him return and conquer the country as far as the sea, and to take it from the Franks (*i.e.*, the Crusaders) and the infidels.   He set out on this errand, defeated Ghiath ud din Kai Khosru at Kusch tagh, and gave up the Seljuki dominions in Rum to pillage.   Meanwhile Khulagu, with the Princes Khuli, Bulghai, and Tutar, and the great Amirs, Buka Timur, Kadsun, Katar Sunjan (called Sunjak by Von Hammer), and Kuka Ilka, encamped in the meadows of Khaneh-abad, in Kurdestan, near Hamadan, and proceeded to organise and equip his troops afresh.‡

The *Georgian Chronicle* make the Georgian chiefs, with Egarslan at their head, and the Mongol Amirs who had preceded Khulagu, meet the latter at Tebriz.   Khulagu mounted them on horseback, and gave them commands in the army.   One was named uldachi (*i.e.*, sword-bearer) ; another was girded with a scimitar and ordered to stand guard at the door, with the title of evdachi (Schmidt, who has explained these words, says it means porter) ; another was named sukurchi (*i.e.*, umbrella-bearer).   The Georgian writer says this umbrella, which was apparently new to him, was held over the Khakan, was round, and attached to a large pole; only the Khan's relatives were privileged to have the sukur over them.   Others were called qapchak (*i.e.*, those charged with the

---

* D'Ohsson, iii. 205-206   † Quatremere, 216-217.   ‡ *Id.*, 223-225.

clothes and boots. Schmidt says qapchaki means keeper of the clothes) ; others were doorkeepers or evchis ; others were quiver and bow-bearers (*i.e.*, korchi). It was by such patronising favours the Khan rewarded the great mthawars of Georgia.*

We read how at this time, the revenues of the churches, of Mtzkhetha, and the other monasteries, as well as of those dependent villages and land, were unprotected, as each of the grandees contented himself with looking after his own interests. In consequence the Catholicos Nicoloz repaired to Khulagu, who, we are told, was struck with his character, for he had hitherto, of the Christians, only known the Arkauns (*i.e.*, probably the Nestorians). Khulagu gave him a yarligh, and assigned him a a shahnah, or overseer. He had two gold bejewelled crosses made, of which he gave one to the Catholicos and the other to the Superior of Wardzia, his companion. He also gave the former a gilt baton, surmounted by a cross. He then bade them good-bye, and gave them charge of the churches and monasteries.†

Guiragos tells us the condition of the Georgians now became worse. The invaders "ate and drank without ceasing, and brought the people within two fingers' breadth of death." Among other things, whereas Arghun had imposed the two taxes of mal and khaphchuri on the people, Khulagu added that called thaghar. All the people entered on the royal registers had to pay one hundred litras of wheat, fifty of wine, two of rice, two sacks of dzgndjat (?), three topraks (?), two cords (probably bow-strings), one white (*i.e.*, a piece of money), one arrow, and one horse-shoe, besides a twentieth of cattle, and money and other presents. Those who could not pay were robbed of their sons and daughters.‡

There is some confusion in the authorities in reference to the doings of the two Georgian kings at this time. It would seem, however, from the narrative of the author of the *Georgian Chronicle*, that they lived on good terms with one another. He tells us that he had himself seen numbers of charters headed "David and David, Bagratids, kings by the will of God," with their double signature.§ Vartan tells us Khulagu was visited by the two kings, who were well treated. The *Georgian Chronicle*, on the other hand, assures us it was only David, son of Lasha, who was his favourite. He was a big man and stout, and could draw the strongest bow, was simple, frank, and credulous; while his cousin, the son of Rusudan, was small, puny, and fair to look upon. He had beautiful hair, and was a skilled hunter, was eloquent of speech, generous, and modest, a good horseman and brave warrior, just, and stirred by an active ambition. The son of Rusudan, we are told, was much disliked by Khulagu, who, when at Alatagh, had him arrested and sent to the winter camp of the Tartars, at Berdaa. When they reached Nakhchivan he escaped, with

---

the Gurcelel amir Ejib and Bega-Suramel, whom the Mongols called Salin (Sain) Bega, or the Good Bega, and fled to Aphkhazeth. When he reached the district ruled by Avak, dressed in humble costume, he was seen by Sempad the Orpelian, who was then hunting. He begged him not to betray him, and gave him a precious stone which had belonged to his mother. Sempad accordingly sent him disguised to Thor, where the Liparit prince, Thorel, surnamed Dewis Kur (*i.e.*, the camel's ear), gave him horses and clothes, and conducted him to Kuthathis. Thereupon the Aphkhaz, the Suans, the Dadian Bedian, the Eristhaf of Radsha, and all those beyond Mount Likh, assembled together and saluted David as King of Aphkhaz, as far as that chain of mountains which thenceforward separated the two principalities.* The story of the feast is no doubt the same which has been already told, and it is a proof of the impossibility of reconciling the various accounts of these transactions. The *Georgian Chronicle* reports how, shortly before the campaign against Baghdad, Batu Khan, of Kipchak, of Osseth, Khazaria, Russia, Bulgaria, and all the country between Servia, Derbend, and China, sent an express to summon David, who left with rich presents, leaving behind him as Regent the Queen Djigda Khatun, and the mestumrét Jikur, during whose rule brigandage and robbery ceased. He built a magnificent palace at Isanni, imposed a tribute on the Phkhoels, and used this savage people as muleteers.

David went, according to the Georgian annalist, to Batu : if it was really to Batu it must have been before 1256, when the latter died, but the dates of our author are so crooked that they are not to be relied upon, and it was more probably Bereke to whom he went. He was well received, and remained at the Golden Horde for some time. David, in setting out, had appointed deputies in his various provinces, and among others, gave Kakheth to Thorgua Pancel (*i.e.*, chief of Pancis), with orders to obey the Queen. Imagining that David would never return, he retired to the citadel of Pancis, and usurped authority in Kakheth for himself, and ceased to obey the Queen and Jikur, the mestumré. Batu, we are told, conferred the sukur, or umbrella, which the Khan and his family had alone the right to use, upon David. He also asked Khulagu to give him precedence over every one but himself. We are told that among the Tartars no one could sit in the Khan's presence, not even at meals. David now returned again to Karthli. He was received with rejoicings at Hereth, and thence went to Tiflis. Thorgua was summoned to his presence. He demanded a safe conduct, and it was granted him ; but he was, nevertheless, taken to Cldé-Karni, and put to death.† David now repaired to Khulagu, who granted him the privileges of the other Noyans

---

* Op. cit., 545-546.
† The person so called issued invitations in the king's name, and welcomed the guests.
‡ Hist. de la Géorgie, 547-548.

in regard to standing and sitting, with the title of Yaraguchi (*i.e.*, in Mongol, a judge), the right to try cases and give judgment.*

The Georgians were not the only Christians who were very considerately treated by the Mongols at this time. I have already described how Haithon, King of Little Armenia, visited Batu.† The account given of his journey by his relative, the Armenian Prince, Haithon, in his chronicle, verges on the improbable. He says he asked Mangu Khan to become a Christian, and cause his people to be converted ; and goes on to say that this demand, with six others, having been laid before Mangu, he assembled his council, and the King of Armenia being present, addressed him in these terms : "Since the King of Armenia has come a very long distance without being compelled, it is reasonable to satisfy his wishes, at all events, in what is just. We tell you, then, O King of Armenia, that your requests are agreeable to us, and by the help of God they shall be carried out. In the first place I, the Emperor of the Tartars, will be baptised. I hold the Christian faith, and will urge my people to conform to it also, although I will use no force to compel them to do so." Haithon goes on to say that Mangu, in fact, had himself baptised by a certain bishop who was chancellor to the Armenian King, together with his household and many grandees of the Empire. D'Ohsson remarks, in regard to this, that "it is quite possible he was baptised, for he supported without favour the various religions practised at his Court, without professing any positive faith, and the Mongols doubtless looked upon baptism as a form of purification." Haithon reports that the King secured the exemption of the Christian priests from taxes ; but the exemption had already been specially provided for by Jingis Khan. The towns captured from his people by the Mussulmans, and re-captured by the Mongols, were to be restored to him. The Mongol generals in the west were ordered to help him when in need. They were, lastly, to attack the Khalif, and to unite themselves with the Christians for the emancipation of the Holy Land from the Muhammedan yoke.‡ I shall reserve an account of Haithon's itinerary from Karakorum to the Oxus for the next volume, and will here merely say that after crossing that river he went on by way of Mrmn (Merv), Sarakhs, and Tus ; then, entering Mazanderan, he passed by Bostam, and thence to Irak, on the borders of the Mulahids, or Assassins. He then passed successively the towns of Dameghan, Rai, Kazvin, Abher, or Ahr, Zenguian, Miana, Tebriz, and eventually reached the Araxes. At Sisian (?) he met Baichu Noyan, who conducted him to Khoja Noyan, to whom he had deputed his command, while with the bulk of the army he had set out to meet Khulagu. On arriving at the village of Vartenis, where lived Prince K'urth, and where he had left his suite and baggage, he awaited the return of the priest

---

* Hist. de la Géorgie, 548.   † Ante, ii. 88-89.   ‡ Haithon Chron., 23. D'Ohsson, ii. 312-313.

Basil, whom he had sent to Batu to take him the letters with which he had been intrusted by Mangu. There he was met by various ecclesiastics, to whom he presented some rich vestments and other presents. The ingenuous king reported to his friends some of the marvels which he had seen or heard tell of on his journey. How beyond Khatai was a race among whom the women were as they were elsewhere, while the men were shaped like dogs, were big and hairy, and had no reason or were dumb. These dog-men allowed no one to enter their country; they hunted, and lived on the game they caught, which they shared with their wives. Of the offspring of these people, the males followed the appearance of their fathers, and the females that of their mothers. He also spoke of a sandy island where grew a bone of great value in the form of a tree, which they called a fish's tooth. When this was cut down another grew in its place like a stag's horn. The former story may be compared to the tales about Burtechino, the wolf ancestor of the Mongols, and of Tsena, the ancestor of the Turks, and their intercourse with women, while the latter, as Brosset says, seems a distinct reference to mammoth ivory.* Haithon also brought home stories about people who worshipped clay statues, which were very large, and called Sakia munim (*i.e.*, they were statues of Buddha). They reported to him that this god had lived 3,040 years, and still had 35 tumans of years (*i.e.*, 350,000) to live, when he would lose his divinity in favour of another god named Madri (*i.e.*, Maitreya), to whom they raised enormous clay statues in a magnificent temple. All this people, men, women, and children, were clerics, and were called tuins (this is the Mongol name for the Buddhist clergy). They had their chins and heads shaved. They wore a yellow mantle like the Christians, with this difference: that it hung from the neck, but not the shoulders. They were temperate and chaste. Haithon reached his house in Armenia eight months after leaving Mangu. This was in the year 1255.†

Let us now turn more directly to Khulagu's own doings. Of the principal commissions he had received from his brother he had amply fulfilled one, viz., the crushing of the Ismaelites, and he now turned to accomplish the other—the destruction of the Khalif. Matters were going on badly at Baghdad. In the autumn of 1256 a terrible downfall of rain had flooded the town and submerged many of the houses, while one-half of Irak remained untilled. The Khalif Mostassim was a weak prince, and passed his life in debauchery—musicians, dancers, tumblers, &c., being his chief companions. His arrogance was a match for his imbecility. The princes who went to Baghdad to do homage were not admitted to his presence. They had to be content with holding to their lips a piece of black silk, representing the lappet of the Khalif's gown, which was

---

* Guiragos, ed. Brosset, 180. Note.
† Brosset, op. cit., 176-181. Journ. Asiat., 5th ser., xi. *Id.*, 470-473.

H

suspended at the palace gate, and to kiss a stone placed on the threshold, like the pilgrims to Mekka, who similarly kissed the black stone and the veil of the Kaaba.   When he sallied forth on horseback on solemn occasions his face was covered with a black veil.*   The great vassals who formerly received investiture at his hands were the Sultans of Egypt and Rum, the Atabegs of Fars and Kerman, the Princes of Erbil, Mosul, &c.; but the chiefs of Rum, Fars, and Kerman were at this time feudatories of the Mongols.   The Khalif's principal officers were Suliman Shah, the generalissimo of his army, which was said to consist of 60,000 cavalry; the Great Devatdar, or chancellor, the Devatdar i Kuchuk, or Little Devatdar, i.e., the vice-chancellor, the Sharabi, or cupbearer, and the Vizier, Muayad ud din Muhammed, son of Abdul Malik el Alkamiyi. The Khalif's most trusted officer was the Little Devatdar, Eibeg, who, notwithstanding, plotted with some of the principal people to dethrone him in favour of some other prince of the house of Abbas. The Vizier having heard of this reported it to his master, who was infatuated by Eibeg and told him what he had heard, and said he should not credit the accusations.   Although the Devatdar Eibeg continued his intrigues, he wrote a memoir in his own hand, declaring all the accusations against him to be calumnies.   This was publicly proclaimed in the streets, and the Devatdar's name was inserted in the khutbeh, or Friday prayer, directly after the Khalif's.†   Eibeg, in his turn, charged the Vizier with having secret negotiations with the Mongols.   This charge had some truth in it, and Wassaf distinctly states that he sent his submission to Khulagu, and invited him to invade the country.‡

Abulfeda, Wassaf, and others tell us why he was dissatisfied.   They say that the village of Karkh, near Baghdad, was occupied almost entirely by Muhammedans, of the sect Ranefi (i.e., Shias), between whom and the Sunnis there arose a dissension, whereupon the Baghdad troops, under the command of Abubekr, the Khalif's son, and Rokn ud din, the Devatdar, proceeded to ill-use the Ranefitis shamefully, to drag their women out of their harems, and to carry them on their horses' cruppers with their faces and feet bare in the public streets.   The Vizier, who belonged to this sect, was outraged, and sent a letter to the Seyid Taj ud din Muhammed, Ibn Nasir el Hoseini, the rais of Hillah, a famous seat of Shia influence, complaining, inter alia, that Karkh had been plundered, that the sons of the house of Ali had been robbed, the people of the stock of Hashim made prisoners, and the dishonour which had formerly been put upon Hussain, the grandson of the Prophet in the plundering of his harem, and the accompanying bloodshedding, had been renewed. The Seyid replied in the names of all the relatives of the Prophet : " The heretics must be put to death and destroyed, and their race be uprooted.

---

* D'Ohsson, iii. 208.   † Quatremere, 229.   ‡ Op. cit., 57-58.

If you will not side with us you will be lost. You will be despised in Baghdad, as henna, which delights women, is despised by rough men, and as a ring is despised by him who has had his hand cut off."* Khulagu at this time had captured the Ismaelite fortress of Alamut, and the Vizier wrote to him pointing out the weakness of Baghdad, and inviting him to march thither. Khulagu was naturally a little anxious about a struggle with a power so formidable as that of the Khalif, whose troops had already twice defeated the Mongols, and he consulted Husam ud din, an astrologer, who had accompanied him at the instance of the Khakan. He was apparently a Mussulman (friendly to the Abbassi dynasty), and foretold that an expedition against Baghdad and the House of Abbas would be followed by six grave events : (1) all the horses would die, and the soldiers be attacked with pestilence ; (2) the sun would not rise ; (3) rain would not fall ; (4) there would be violent hurricanes and earthquakes ; (5) plants would cease to grow; (6) the Emperor would die during the year. Khulagu insisted on the astrologer putting these lugubrious prophecies down in writing. On the other hand, the Mongol bakshis and the amirs declared that the expedition would have a fortunate issue, an opinion also propounded by the famous astronomer, the Khoja Nasir ud din, of Tus, who was a Shia. He had a personal grievance against the Khalif and also against the Vizier. It seems that on one occasion he sent the Khalif one of his poems, on the back of which the Vizier wrote a note addressed to Nasir ud din the Mohtesshim, in which he sneeringly said that the composer had the knack of putting into his letters and writings the thoughts of other people, a jibe which was highly resented by Nasir ud din, who was the most learned man of his time.† Elsewhere Von Hammer gives a different version of this, and says that while Mostassim was one day sitting by the Tigris, Nasir ud din took him a poem, in which he expressed his devotion. Instead of rewarding him the Khalif, in consequence of a sharp criticism of the Vizier's, had it thrown into the Tigris. He thereupon left Baghdad in a rage and went to Sertakht, to the Ismaelites.‡

Meanwhile things were going badly at Baghdad. The Vizier, probably in preparation for his intended treason, persuaded the Khalif to reduce his army, urging that with so many powerful princes as his vassals, he had no need of such a large force, which continually drained his resources. He urged also that with the money thus saved he might buy off the invaders for a while, and persuaded him to reduce his army from 100,000 to 20,000.§ Meanwhile earthquakes and some terrible fires desolated the country. These were apparently caused by lightning. One of them laid waste the district of Hara, near Medina, over a district of four parasangs. Medina itself was burnt, and afterwards plundered by the Arabs. In this last fire its famous library perished. "Thus," says Von Hammer, "there were

---

* Abulfeda, iv. 551. Wassaf, 53-54.   † Wassaf, 56.   ‡ Ilkhans, i. 140.
§ Abulfeda, iv. 551. D'Ohsson, iii. 212.

destroyed in one year two of the most famous libraries in the East, that
at Alamut and that at Medina."* Khulagu having determined to crush
the Khalif, now sent him a summons from Hamadan couched in haughty
phrases. He began by denouncing him for not having assisted the
Mongols in their campaign against the Ismaelites ; reminded him of the
success which had attended the armies of the Mongols from the time of
Jingis Khan, and how the Khuarezm Shahs, the Seljuki, the rulers of
Dilem, the Atabegs, and others had all succumbed, all of whom had been
masters of Baghdad. Why should its gates be closed to him? He
warned him not to strike with his fist against an iron spike, nor to mistake
the sun for a taper, and bade him dismantle the fortifications of Baghdad,
to leave his son in charge there, and go to him in person, or, at least,
send the Vizier, Suliman Shah, and the Devatdar to confer with him. In
that case he should preserve his dominions ; if not, the Mongols would
march on Baghdad ; and where would he hide—in the heavens or the
depths of the earth?† The Khalif received the envoys with courtesy, and
sent back Sherif ud din ibn Duzy, or Juzy, an eloquent person, Bedr ud
din Muhammed, and Zanghi Nakhjivani, who was probably an Armenian,
with his reply, which was by no means a cringing one:—"O, young man
only just commencing your career, who show such small regard for life,
who, drunk with the prosperity and good fortune of ten days, deem yourself
superior to the whole world, and think your orders equivalent to those
of destiny, and irresistible. Why do you address me a demand which you
cannot secure? Do you think by your skill, the strength of your army, and
your courage, that you can make captive even one of the stars? You are
probably unaware that from the east to the west, the worshippers of God,
religious men, kings and beggars, old men and young ones, are all slaves
of this Court, and form my armies ; that after I have ordered these isolated
defenders to gather together, I shall first settle the affairs of Iran, and will
then march upon Turan and put each man in his proper place. No doubt
the earth will be the scene of trouble and confusion in consequence, but I
am not greedy for vengeance nor eager to win the applause of men. I am
not anxious that through the tramp of armies men shall have occasion
either to bless or curse. I, the Khakan, and Khulagu all have the same heart
and the same language. If, like me, you would sow the seed of friendship,
what have you to do with meddling with the intrenchments and ramparts
of my servants? Follow the road of goodness and return to Khorasan.
If, however, you desire war, I have thousands of troops who, when the
moment of vengeance arrives, will dry up the waves of the sea."‡ This is
apparently the message reported by Guiragos in somewhat different
terms. He says the Khalif was very arrogant, styled himself Jehangir,
master of the sea and land ; boasted that he possessed the standard of

---

* Ilkhans, i. 142-143.   † Quatremere, 231-233.   D'Ohsson, iii. 215-217.
‡ Quatremere, 235.

Muhammed, and if he set it in motion he and all the universe would perish. "You are only a dog and a Turk, why should I pay you tribute or obey you?"* Hardly were the envoys outside the walls of Baghdad when they were attacked by the fanatical mob, who tore their clothes, spat in their faces, and would have killed them if the Vizier had not sent some people to rescue them.† Khulagu, who was at Panj Angusht (the five fingers), on hearing of this declared that the Khalif was as crooked as a bow, but he would make him as straight as an arrow ; and sent back his envoys with the message that God had given the empire of the world to the descendants of Jingis Khan, and as their master refused to obey there was nothing for it but that he must prepare for war.‡

Meanwhile the Khalif was perplexed by the varying counsel of his Ministers. While the Vizier advised him to propitiate the Mongols by rich presents, including 1,000 Arab horses, 1,000 camels, and 1,000 *asses, laden with treasure and richly caparisoned, and by offering to have the khutbeh said, and money coined in Khulagu's name, his rival, the Devatdar, bade him rely on his army, and on the assistance of the faithful. The latter at length prevailed. He and his supporters professed great contempt for Mostassim, whom they accused of being fond of musicians and buffoons, and of being unfriendly to the army. The amirs complained that they had lost everything in his reign which they had acquired in his father's, and their chief, Suliman Shah, spoke out bravely that if troops were only summoned from the various provinces and he was put at the head of them, he thought he could break the Mongol army, and even, if beaten, it was well for a brave man to perish with glory and honour in the midst of the fight.§ The Khalif approved of these words, ordered largess to be distributed to the soldiers, and told the Vizier to give the command over them to Suliman Shah. The Vizier prepared to carry out these orders, but only in a languid fashion, which strengthened the suspicion that he was in league with the Mongols, a view which the Devatdar widely proclaimed. The Khalif's avarice prevented sufficient money being spent, and it was five months before the troops were ready. He now dispatched Bedr ud din Diriki and the Kadhi of Bindinjan, a town of Kurdistan, with a fresh mission to Khulagu to remind him of the fate of many who had formerly attacked the sacred Abbassidan House. "How Yakub ibn Leith, of the family of Saffar, had died while on his way to attack Baghdad. How his brother, Amru, who had the same intention, was captured by Ismail ibn Ahmed, the Samanid, who sent him in chains to Baghdad. How Besasiri had marched from Egypt with a large army and had captured the Khalif and kept him prisoner at Hadithah, and for two years the khutbeh had been said and the money struck at Baghdad in the name of Mostansir, the

---

* Op. cit., ed. Brosset, 185.   † Quatremere, 257.
‡ D'Ohsson, iii. 218-219. Tabakat-i-Nasiri, 1231. Note.   § Quatremere, 237-247.

Ismaelite Khalif of Egypt; and how then Besasiri was attacked and put to death by Tughrul Bek, the Seljuk. How the latter's successor, Muhammed, had to retreat after his venture on Baghdad, and died on the way ; and, lastly, how the Khuarezm Shah Muhammed, who had determined to uproot the family of Abbas, had been almost overwhelmed in the defile of Asad abad by a storm, in which he lost most of his troops, and was forced to retire, and how he had ended his days miserably in the Isle of Abisgum, chased thither by the Mongols." The envoys concluded by reminding Khulagu that he had no cause of quarrel with the Khalif, and bidding him take warning.* This portentous retrospect only aroused the anger of Khulagu, who is said to have quoted in reply some lines from the great Persian epos, the Shah Nameh :

> Build about yourself a town and a rampart of iron ;
> Erect a bastion and a curtain-wall of steel ;
> Assemble an army of Peris and of Jins ;
> Then march against me, inspired by vengeance,
> If you were in heaven I would bring you down,
> And spite of yourself I will reach you in the lion's den.†

Khulagu knew it was a serious matter to assail a town so renowned as Baghdad, and he took precautions accordingly. Hearing that Husam ud din Akah—who on behalf of the Khalif commanded at Daritang (i.e., the narrow defile), a fortress commanding the main route from Hamadan to Baghdad, and the key to Irak Arabi—was dissatisfied, he summoned him to his presence. Leaving his son Said in the town, which was famous for its beauty and strength,‡ he obeyed. Khulagu received him well, and gave him as an appanage the castles of Warudah, Merj, &c.§ He proceeded to occupy these fortresses. Having collected a considerable force about him, he seems to have repented of his treachery, and communicated with the Khalif through Taj ud din Ibn Salayeh, of the family of Ali, who governed the town of Erbil, offering to raise an army of 100,000 Kurds and Turkomans, with which to overwhelm the invaders. His proposition was not accepted by the Khalif. Meanwhile the intrigue had reached the ears of Khulagu. He was naturally greatly enraged, and ordered Kitubuka to march with 30,000 cavalry to forestal the traitor. This officer sent him word he wanted to concert common measures against Baghdad. He unwittingly went to his camp, whereupon Kitubuka arrested him, and told him if he wanted to save his life he must order his wife and son, and all his adherents and soldiers, to march out of the fortress, that a census of them might be taken for the poll tax. Husam ud din had to issue an order to this effect, and also to demolish his fortresses, after which he was put to death with all his adherents.

    * Quatremere, 249-251.   † Rashid ud din, by Quatremere, 253.
‡ The defile in which it lay was watered by the Diala, which some miles higher up flowed past Kasr Shirin, the ancient Artemita. Ilkhans, 145.
    § Quatremere, 255. Von Hammer calls these two places Disser (i.e., the Golden Castle) and Dis Merjh (i.e., the Meadow Castle). Ilkhans, i. 145.

Only one of his towns escaped, viz., that governed by his son Said, who refused to surrender, and afterwards made his way to Baghdad, where he fell fighting.*

Khulagu now summoned the various contingents of his army to converge on the doomed city. Baichu was sent for from the borders of Rum, while Bulghai and the other princes, who then commanded contingents belonging to the other ulusses, with Sunjak and Buka Timur, took the road from Shehrsor to Dakuka.† Kitubuka Noyan, Kadsun, called Kurusun by Von Hammer,‡ and Nerkilka, arrived from Luristan, Beiat, Takrit, and Khuzistan. Khulagu himself, leaving his family and greater baggage in the meadows of Zek, not far from Hamadan, in charge of Kaiak Noyan, advanced with the centre towards Kerman-shahan and Holwan. He had with him the great amirs Kúka Ilka, Arkatu, Arghun-aka, the bitikchis Karatai and Seif ud din, his favourites the astronomer Khoja Nasir ud din, Alai ud din Ata Mulk (i.e., the historian, Juveni), as well as all the sultans, kings, and secretaries of Iran.§ He passed by way of Asad abad, a small town seven parasangs from Hamadan, which still exists, and is mentioned by Kerr Porter.|| Thence he sent a fresh message to the Khalif, who only replied by evasions. When the army reached Dinawar, twenty parasangs north-west of Hamadan, Ibn Juzi came with fresh threats from the Khalif in case Khulagu did not retire; but he replied that, having come so far, he could not go back without having an audience of the Khalif, and that after conferring with him and receiving his orders, he could then retire. Khulagu marched through the Kurdish mountains (Kuh-Girdaa), captured Kerman shahan and pillaged other places on the route. At Tak Kesra he was joined by Sunjak, Baichu, and Suntai, with whom he held a consultation; and we read how, after leaving him, the Mongol officers consulted the burnt shoulder-blades of sheep which were used by them in divination.

We must here make a short digression, to bring up the story of the Mongol doings in Rum to this point. We have seen how Rum was divided between the two brothers, Iz ud din and Rokn ud din. Iz ud din was very suspicious of Baichu, and, we are told, began to collect some forces, and sent a messenger to Malatia and Khartabert, or Saida, to bring together a contingent of Kurds, Turkomans, and Arabs. Two Kurdish chiefs, named Sherif ud din Ahmed ibn Bilas, from Al Hakkar, and Sherif ud din Muhammed ibn Al Sheikh Adi, from Mosul, came to him, and he appointed the former governor of Malatia and the latter of Khartabert. The Malatians having sworn allegiance to Rokn ud din, refused to receive the Kurdish chief, and as he besieged the place, until great want prevailed there, they attacked him and killed 300 of his followers. He himself withdrew through the district of Klaudia, and burnt the monasteries of Madhik and Mar Asia, and plundered that district and Guba. He then went on towards Amid, where he was attacked and killed by the governor of Mayafarkin. The other Kurdish chief was on his way to join Iz ud din when he was attacked and killed by the Noyan Angurg. Iz ud din now nominated Ali Behadur as governor of Malatia. He was small of stature but of great vigour, and speedily reduced the neighbourhood to order, and severely punished the Turkomans who infested the neighbouring mountains and continually harried the country round. Malatia had, however, been assigned to Rokn ud din in the partition already named, and Baichu marched from Bithynia, with his Mongols who were scattered in Cappadocia and Galatia, to secure it for him. He first attacked Abulestin, which he captured, killed 7,000 people, and carried off the boys and girls into captivity. When he approached Malatia, Ali Behadur, its governor, fled. The citizens then surrendered the place. He made them swear allegiance to Rokn ud din and pay a fine. Fakhr ud din Ayaz was appointed its governor.* It would seem from Guiragos that Haithon, the King of Little Armenia, took part in this campaign of Baichu's. The latter afterwards sent him with an escort to Sis, his capital.

On the departure of Baichu, Ali Behadur again obtained possession of Malatia, after a siege in which the inhabitants were reduced to great want. He put to death Rokn ud din's deputy and some of his supporters, and presently, fearing the return of the Mongols, again abandoned the place. Baichu meanwhile advanced upon Mosul, where he arrived in the beginning of 1258. Malik Salih, son of Bedr ud din Lulu, Prince of Mosul, who was an ally of the Mongols, had recently returned from visiting Khulagu, and had married a daughter of the Khuarezm Shah, Jelal ud din. According to Minhaj-i-Saraj, both he and the ruler of Fars had furnished a contingent to the Mongols for the campaign. The people of the country round sought refuge in the town at Baichu's approach, but he left again without doing them any harm.† He crossed the Tigris and joined Khulagu as I have mentioned. The advance guard of the Khalif's troops which was stationed at Yakuba, or Bakuba, was commanded by a Turk from Kipchak, called Kara Sonkor (i.e., black falcon), while in the Mongol army there was a Khuarezmian Turk named Sultan Juk. The latter now wrote to his compatriot, counselling him if he wished to save his family, to do as he had done, viz., to submit to the Mongols, who had treated him well. Kara Sonkor, in reply, vaunted the long history and prosperity of the Abbassidan House, and having denounced the threatened advance of Khulagu, offered complacently to ask the Devatdar to obtain the Khalif's pardon for him if he would

---

retrace his steps and be penitent.* Khulagu laughed when this letter was read to him, and, according to Rashid, replied in poetry :

> In my eyes the ant, the fly, and the elephant are alike indifferent ;
> So are the springs, the rivers, the seas, the Nile.
> If our measures contravene the orders of God,
> Who can tell but Himself what the end may be.†

Abulfaraj apparently refers this incident to Eibeg al Halebi, an envoy of the Khalif himself.‡ Khulagu sent a fresh demand for the Khalif's submission, and orders for him to send the Vizier, Suliman Shah, and the Devatdar to him to arrange terms. If he was determined to resist, however, he bade him prepare for war, and the next day he pitched his camp on the River Holwan, where he remained for thirteen days, while the Amir Kitubuka conquered the greater part of Luristan.

Meanwhile, Baichu Noyan, Buka Timur, and Sunjak crossed the Tigris. Bedr ud din, Prince of Mosul, had supplied Baichu with a bridge of boats, which he put on that river at Takrit. The people of Takrit sallied out and burnt it, and killed some of the invaders. The next day, however, they repaired the bridge, and crossed over to the west bank of the Tigris, and pushed on towards Kufah, Hillah, and Karkh, and martyred the people.§ Elsewhere we read that Baichu, with Buka Timur and Sunjak, went to encamp on the Nahr Isa, or the canal of Isa. Sunjak took command of the advance guard of this division, and speedily arrived at Harbieh. The inhabitants of the district of the Little Tigris (Dojeil), of El Ishaki, and the canals of Malik and Isa fled precipitately, and freely gave the boatmen bracelets, brocaded robes, or large sums of money to transport them in safety to Baghdad. When the Devatdar and the general Fath ud din Ibn Korer (Minhaj-i-Saraj says Fath ud din's son, Iz ud din), who were posted between Yakuba and Besheriyeh, on the way to Holwan, learned that the Mongols had thus approached Baghdad on the western bank of the Tigris, they also crossed that river. Minhaj-i-Saraj says they summoned the men of Karkh and other towns to assist them. The forces of the Khalif were chiefly infantry, and sustained the attack bravely, and killed many Mongols.‖

Elsewhere we read that the Khalif's officers fought the Mongols under Sunjak, near Anbar, before the Koshk Mansur, above Madrikah or Mezzrikah, on the east bank of the Euphrates, about nine parasangs from Baghdad. Wassaf merely says the fight took place near the Dojeil, or Little Tigris. Abulfaraj says the struggle took place at the tomb of Ahmed.¶ It was fought on the 9th Muharrem, 656 (i.e., the 16th January, 1258). The Mongols were defeated, or perhaps merely pursued

---

* Quatremere, 269-277.   † Id., 279.   ‡ Chron. Arab., 337-338.
§ Tabakat-i-Nasiri, 1237-1238.
‖ Quatremere, 279-281.   D'Ohsson, iii., 229-230.   Tabakat-i-Nasiri, 1237-1238.
¶ Ilkhans, i. 148.   Chron. Syr., 549.

their usual Fabian tactics, and having made a detour joined their main army under Baichu at Besheriyeh.* The Devatdar wrote to his master to tell him he would complete the victory next day, and exterminate the enemy. Meanwhile a discussion arose between the Khalif's two principal officers. Fath ud din, who was a skilful soldier and feared some stratagem, counselled delay ; while his civilian companion, the Devatdar, urged an immediate pursuit, while the enemy was distracted.† Fath ud din allowed his judgment to be overborne by his imprudent friend. The Mongols having reached the Dojeil turned about, and a second and more terrible struggle followed, to which an end was put by the darkness, when each army bivouacked on its own ground. In this struggle Fath ud din had ordered the feet of the mule on which he rode to be shackled with iron splints, so that he could not well escape.‡ Minhaj-i-Siraj says that "near the battle-field was a piece of water, called the Nahr i Sher, which was connected with the Euphrates, and the land through which it flowed was' elevated, while the Mussulmans were encamped on the low ground. During that night the accursed *rafizi* Vizier dispatched a body of men and turned the water of the canal on the Mussulmans, and the whole was flooded with water, and their arms and armour were spoiled, and they became quite powerless. Next morning at dawn the infidels returned, and another battle ensued." The Khalif's people were defeated and driven across the Little Tigris, and posted themselves where the great Sanjari mosque and kazr (castle) was situated.§ Wassaf, Rashid ud din, and Abulfaraj, who wrote under the shadow of the Mongol rulers, do not suggest the breaking of the dykes as the work of the Vizier, which is indeed most improbable. With these authors it was the Mongols them- selves who cut the dykes, so that the plain behind the Khalif's army was flooded. They then attacked and routed the latter. Fath ud din and Kara Sonkor, with 12,000 men, were killed, without counting those who were drowned and smothered in the mud.|| The Devatdar reached Baghdad again with only a few—one account says three—persons. Others found refuge at Hillah and Kufah. Meanwhile Khulagu, leaving his baggage at Khanekin, pitched his tent to the east of the city. This was on the 11th Muharrem (*i.e.*, the 18th of January, 1258).¶ He planted himself opposite the gate Ajami. The Noyan Kuka Ilka, with the two princes Tutar and Kuli, of the Golden Horde, faced the Kalwaza gate, while the princes Bulghai, Tutar, Aroktu, and Shiramun posted themselves opposite the gateway of the Suk-i-Sultan (*i.e.*, the Sultan's market-place). Meanwhile, on the western bank of the river, Buka Timur was on the side of the citadel, near Dulabi-Bakul (Abulfaraj says

* D'Ohsson, iii. 230.
† Minhaj-i-Saraj reverses the positions of the two men, but here, as so frequently elsewhere, cannot be trusted.
‡ Wassaf, 63-64.   § Op. cit., 1241-1244.
|| Quatremere, 281. Wassaf, 64. Abulfaraj, Chron. Syr., 549.   ¶ Quatremere, 281.

near the kitchen garden), and Baichu and Sunjak were on the west, where the Uzdi hospital (called Adad by Quatremere) was situated (Abulfeda says in Karia, near the Sultan's palace).*

Meanwhile the Khalif continued in a state of mental imbecility. When the Little Devatdar returned to him after the slaughter of his army, accompanied by only three men, he merely said, "God be praised that Mushahid-ud din is safe," as when the Mongols made a previous invasion of Irak Arab, and had advanced to Jebel Hamrin, he had said, "How can they ever pass it?"† The walls were ordered to be repaired and barricades made, and the citizens were told off to man the defences, and the two Devatdars, the Munjenk, Suliman Shah, and other leaders of the army and the Mamluks encouraged them. The attack was pressed. The bricks that lay about outside the city were collected and built into great mounds, upon which were planted battering engines and machines for shooting burning naphtha.‡ The Khalif now sent the Vizier with one of his favourites, named Ibn Darnus, and Makiko, the Nestorian patriarch, with presents. Khulagu told them that the conditions which would have satisfied him at Hamadan were no longer enough, and he must insist on the Devatdar and Suliman Shah, the latter of whom had won more than one victory over the Mongols, being surrendered. The next day the Vizier, the Sahib Divan, or Minister of the Interior, and a deputation, consisting of the principal inhabitants of the city, went to Khulagu's camp. He would not, however, receive them. The attack was closely pressed, and the bombardment continued for six days. As there were no stones near Baghdad to ply the machines with, they were sent for from Jebel Hamrin and Jelula, and palm trees were also cut down to furnish projectiles, while letters were shot into the place offering their lives to the kadhis, doctors of the law, sheikhs, Alévis, and other non-combatants.§ At length, on the 28th Muharrem (i.e., the 4th of February‖), the Burj-i-Ajami, or so-called Persian Tower, was battered down, and presently the Mongols stormed this part of the wall. Khulagu having reproached his relatives who were posted before the gate Suk Sultan with being dilatory, they also stormed the wall in front of them, and during the night the whole of the defences of the eastern part of the city were in the Mongol hands. The invaders had taken care to close the Tigris with bridges of boats, on which were planted war engines.¶ Buka Timur was dispatched with a tuman (i.e., 10,000 men) towards Modain and Basrah, to cut off the retreat of any who might try to escape by the river. Minhaj-i-Saraj says the Devatdar tried to persuade the Khalif to embark on a boat with his treasure, and to make his way down the Little Tigris to Basrah, and to take shelter in the islands in the delta of the Euphrates and Tigris till the

* Tabakat-i-Nasiri, 1243. Raverty's notes.   † Wassaf, 64.   ‡ Id., 66. D'Ohsson, iii. 233.
§ D'Ohsson, iii. 233-234.   ‖ See Weil, iii. 477.   ¶ D'Ohsson, iii. 234-235.

danger had passed. The Vizier argued against this counsel, and persuaded the Khalif that he was himself arranging terms with the Mongols.* Rashid ud din says nothing about the proposed escape of the Khalif, but that the Devatdar himself made an attempt to get through and to reach the town of Sib, but when he reached Karia ul Ukab (*i.e.*, the eagle village), also called Karia ul Ghaffar, a shower of arrows, stones, and stink pots drove him back, after losing three of his boats, the men on which were all killed, and the Devatdar had to make his way back to Baghdad.†

The Khalif now began to lose heart. He sent Fakr ud din, of Dameghan, and Ibn Darwish to the Mongol camp, to try and appease Khulagu, sending only a few presents with them, as he feared to excite his cupidity. These not having been received, it was determined that the Khalif's second son, Abu fazl Abd ur Rahman, should go to Khulagu's camp. He is called Abubekr by Minhaj-i-Saraj, who says he was sent* at the instance of "the accursed Vizier," who at the same time sent a messenger to Khulagu to tell him to pay the young prince special attention, so as to secure his object with the Khalif. He goes on to say that the prince was met by a crowd of Mussulmans and Mongols as he neared Khulagu's camp, who paid him the usual deference. When he came to the place of audience Khulagu advanced four steps to meet him, took him to a seat, and said that his uncle (relative) Bereke had become a Mussulman at the hands of the Sheikh Saif ud din, the Bakhurzi. He then went down on his knees before him, affirming he had gone to Baghdad in order to accept Islam under the Khalif himself. All this is a most unlikely proceeding, as anyone who has any acquaintance with Mongol ways will allow. The prince, we are told, returned to his father thoroughly deceived by these promises.‡ Rashid ud din says, on the contrary, that Khulagu would not receive him nor his elder brother, who went out with the principal citizens to beg for clemency. Khulagu sent them back, and ordered the Khoja Nasir ud din to go with Itimur and open negotiations directly with the head of the Faithful. They returned on the 7th of February, and were followed by Fakhr ud din, of Dameghan, and Ibn Darwish, who were armed with a yarligh and paizah and were told to summon Suliman Shah, the Khalif's commander-in-chief, and the Devatdar. Abulfeda says Khulagu wanted to treat the Khalif generously, and wished to marry his own daughter to his son Abubekr.§ Having received safe conducts, Suliman and the Vice-Chancellor went at length to the Mongol camp. They were ordered to go back into the city and fetch their relatives and retainers, as Khulagu intended to send them with some of his own people against Syria and Egypt. They accordingly went in to bring them out. On their return they were distributed among

* Op. cit., 1245.   † Quatremere, 291-293   Tabakat-i-Nasiri, 1245.   Notes.
‡ Tabakat-i-Nasiri, 1247-1248.   § Op. cit., iv. 553.

the Mongol soldiery. At this stage, an arrow having struck Hindu,[*] who was a bitikchi or secretary, and a favourite of Khulagu's, in the eye, he ordered the siege to be pressed, and told the Khoja Nasir ud din, of Tus, to station himself at the gate Halbeh, and to receive those who came out of the city to surrender.[†]

On the 8th of February, Eibeg, the Little Devatdar, was put to death. Suliman was summoned to his presence by Khulagu, who said to him, "You are an astrologer, who understand the portents, good and evil, of the stars. How is it you did not foresee these events, and forewarn your master?" "The Khalif," replied the warrior who had already twice defeated the Mongols, "was led by his destiny, and would not heed the counsel of his faithful servants." Khulagu had him put to death, with all the people of his household, to the number of 700. The Amir Haj ud din, son of the Great Devatdar, suffered the same fate. The heads of the three chiefs were sent to Salih, who commanded the Mosul contingent, in which were enrolled the Shias from Karkh, with orders to send them to his father, Bedr ud din, who had been an old friend of Suliman Shah, and now with tears in his eyes had to give orders for the three heads to be exposed.[‡]

On the 10th of February the Khalif left the town with his three sons, Abd ur Rahman, Ahmed, and Mobarek, with 3,000 other people—Seyids, Imams, Kadhis, and grandees. Khulagu, on his arrival in his presence, asked after his health. He was told to order the citizens to lay down their arms, an order which was proclaimed in the streets. A special tent was set up for him before the gate Kalvaza, in the quarters of Kitubuka, where he was guarded by some Mongols, and on the 15th of February, Khulagu having entered the city to visit his palace, had him summoned, and said to him: "You are the master of this house and I am your guest. Let us see what you can give us." The trembling Khalif broke some locks, and offered Khulagu 2,000 complete sets of robes, 10,000 gold dinars, and a quantity of precious stones. He would not take them, but said, "It is unnecessary to point out what is patent; disclose your hidden treasures." The Khalif then bade them dig in a certain place, where they found a cistern filled with gold pieces, each weighing 100 miskals.[§] Sunjak was ordered to make an inventory of the treasures. These were taken to the Mongol camp, and piled up like mountains about Khulagu's tent. The Mongols, says Wassaf, treated the gold and silver vessels which they had carried off from the Khalif's kitchens as if they had been lead. Many of these treasures in this way reached Shiraz, and those who had been wretchedly poor became very rich. The soldiers secured so much money, rich stuffs, and products of Greece, Egypt, and China, Arab horses, mules, Greek, Alan, and

* Tabakat-i-Nasiri, 1246. Note. † Quatremere, 297. ‡ D'Ohsson, iii. 237-238. § Quatremere, 299 and 301. D'Ohsson, iii. 239-240.

Kipchak boys; Turkish, Chinese, and Berber slave girls, that it was impossible to count them. *Inter alia*, Wassaf mentions that they secured a beautiful bowl, decorated with gold, and engraved by Mostansir and Nassir. It was a curious circumstance that the Khalif En Nassir lidin illah left behind him two basins or cisterns filled with gold. His grandson, Mostansir, was one day with one of his most trusted followers, and expressed the wish that he might not live until it was necessary for him to spend this money. His companion laughed. The Khalif was angry, and asked him the cause. "One day," he replied, "I came into your grandfather's presence here when one of those two basins was not full, when he said, 'I wish to live only until I have filled up these two basins.' I was contrasting the two wishes." Mostansir spent all this money in good works, and, *inter alia*, built the famous college, Mostansiriyeh. "The point of this story," adds Wassaf, "is that when Mostassim came to the throne he once more filled up these basins, or rather reservoirs, by his avarice, and finally emptied them as well."* Khulagu now gave orders that the Khalif's harem should be numbered, and it was found to contain 700 wives and concubines, and 1,200 servants. He thereupon implored that 100 of the females, on whom the sun and moon had never shone, should be handed back to him, and this being granted he selected his relatives.†

The Georgians especially distinguished themselves in the capture of Baghdad, where Guiragos tells us Zakaria, son of Shahan Shah, was present. It was a grand opportunity for them to repay on their Mussulman enemies the terrible sufferings they had long borne at their hands. We are told in the *Georgian Chronicle* that it was they who breached the walls, and having entered the place commenced a great slaughter, the troops of Baghdad having great dread of the Georgians. The latter are made to open the gates through which the Tartars entered. The booty captured, we are told, was so great that Georgians and Tartars succumbed under the load of gold and silver, precious stones and pearls, rich stuffs, gold and silver vessels, &c., while as to the vases from China and Kashan (*i.e.*, porcelain), and those made in the country of iron and copper, they were deemed of scarcely any value, and were broken and thrown away. The soldiers were so rich that the saddles of their horses and mules and their most ordinary utensils were inlaid with stones, pearls, and gold. Some of them broke off their swords at the hilt and filled up the scabbards with gold, others emptied the body of a Baghdadian, refilled it with gold, precious stones, and pearls, and carried it off from the city.‡ The place was cruelly ravaged ; the only people to whom consideration was shown were the Christians, who were sheltered in one of the churches by the Nestorian

---

patriarch. This fact seems to give some foundation to the remark of Minhaj-i-Saraj, that they had been in communication with Khulagu. Abulfaraj says that many rich Muhammedans handed over to the patriarch their treasures in the hope of securing their lives, but all perished.* The place was now gutted, and the Khalif's palace was reduced to ashes, together with the Great Mosque. The tombs of Musa-Jewad and of the Khalifs were burnt. Nearly all the inhabitants, to the number, according to Rashid ud din, of 800,000 (Makrizi says 2,000,000) perished, and thus passed away one of the noblest cities that had ever graced the East—the cynosure of the Muhammedan world, where the luxury, wealth, and culture of five centuries had concentrated. Presently the wretched remnant of the population sent Sherif ud din Meraghi, Shehab ud din Zengani, and Malik dil rast to beg that the carnage might cease. Khulagu gave orders accordingly, and, we are told, he had to withdraw to the villages of Wakhf and Jelabieh to avoid the tainted air.† As a proof of the horrors that took place at this time, a story told by Hamdullah may be cited, viz., that a Mongol, named Mianju, found, during the massacre, in a small street of the city, upwards of forty motherless sucking babes, and thinking to himself that without mothers' milk they would perish, put them to death to deliver them from their sufferings.‡

It is probable that Khulagu would have spared the Khalif's life, impressed by the lugubrious prognostications of the faithful Mussulmans about him, if he had not been dissuaded from this course by the Shias who were with him, and who had a bitter resentment against the Abbassidan dynasty. Minhaj-i-Saraj tells us that the Malik Bedr ud din, Lulu of Mosul, and other infidels (thereby probably meaning Shias) represented to Khulagu saying, " If the Khalif continues alive, the whole of the Mussulmans among the troops, and the other Mussulman peoples who are in other countries, will rise and bring about his liberation, and will not leave thee alive."§ Wassaf says that Khulagu was afraid of releasing him, since the Mussulmans looked upon him as the successor of the Prophet, and the true Imam, and the absolute master of all life and property, and would have gathered round him a very powerful army.‖ On turning to the Vizier for counsel the latter replied, "The Vizier has a long beard." This was a joke which had been used against that official by the Devatdar, and is derived from the Arab proverb, "Long in beard, short in wit."¶ Some of the orthodox Mussulmans affirmed that if the Khalif's blood was shed upon the ground there would be an earthquake.** Another account attributes the warning about the portents that would happen if the Khalif were executed to the astrologer, Husam ud din, and tells us that these predictions were answered by his brother astrologer, Nasir ud din, of Tus, who was a Shia, and who said that no such portents had occurred

* Op. cit., Chron. Syr., 550. † Quatremere, 303. ‡ Tabakat-i-Nasiri, 1250. Raverty's Note.
§ Op. cit., 1252. ‖ Op. cit., 76. ¶ Ilkhans, i, 154. ** Tabakat-i-Nasiri, 1252.

when John the Baptist, the Prophet Muhammed, and the Imam Hussain
were killed, and that they were not likely to happen then.* It was
determined, therefore, to put him to death, and we are told that Husam
ud din was himself executed on the 23rd of November, 1262, his prophecies
having proved false.† The mode of the Khalif's execution is wrapped in
some obscurity. Rashid ud din says that, having lost all hope of saving his
life, he asked permission to make his ablutions. Khulagu ordered five
Mongols to attend him, a *cortége* " of infernal guards " to which he
objected.‡ He recited two or three verses of a poem beginning thus : *

> In the morning we dwelt in a house like paradise or heaven,
> In the evening we had no longer a dwelling as if we had been homeless.

On the 20th of February, we are told, he was put to death in the village
of Wakf, with his eldest son and five eunuchs who remained faithfully
with him. The mode of his execution is not stated by Rashid ud din, and,
Quatremere suggests, it was in fact probably kept secret.

The *Georgian Chronicle* tells us that on being brought into the presence
of Khulagu, the Khalif was ordered to bend the knee. This he refused,
and remained standing, saying: " I am an independent sovereign, who am
dependent on no one. If you choose to set me free I will submit to you ;
if not I will die before becoming any man's slave." To make him
stoop they tripped him up by the foot, so that he fell on his face. As he
remained obstinate, Khulagu told Ilka Noyan to take him out and kill
him and his sons. " The Khan pities you," that officer said to him.
" Does he propose then to restore me Baghdad ?" " No," said Ilka,
" but he will kill you with his own hand, while his son Abaka will perform
the same office for your relatives." " If I am to die," he replied, " it
matters little whether it be a man or a dog who kills me."§ Wassaf and
Novairi say he was rolled up in carpets and then trodden under by horses
so that his blood should not be spilt. This was in accordance with
the yasa of Jingis Khan, which forbade the shedding of the blood of
royal persons. Guiragos, whose account, as he tells us, was derived from
the lips of Prince Hasan, son of Vasag, surnamed Brosh, who was an
eye-witness, and employed by Khulagu as one of his envoys to the Khalif
tells us that when Khulagu had summoned him to his presence, he asked
the Khalif, " Are you God or man ?" The latter replied, " I am a man,
the servant of God." " Did God order you," said the Mongol chief, " to
treat me with contumely, to call me a dog, and to refuse me, the dog of
God, something to eat and drink? Verily, I am the dog of God, and I am
very hungry, and will devour you." He then killed him with his own
hand, telling him it was as a special honour he did so, instead of remitting
the work to another. He ordered his son to similarly kill a son of the
Khalif's, and to throw a second one into the Tigris. He afterwards put

---

* Tabakat-i-Nasiri, 1253. Raverty's note.   † D'Ohsson, iii. 225.
‡ Koran, ch. xcvi., verse 18. Tabakat-i-Nasiri, 1253. Note.   § Op. cit., 550.

to death many of the grandees, while his men for forty days continued a horrible butchery of men, women, and children. Tokuz Khatun, Khulagu's Christian wife, redeemed the lives of the Nestorian and other Christians.*

Another and much more romantic story is told by Nikby and Mirkhond. They tell us that when the Khalif presented his treasures to Khulagu the former put him before a trencher covered with gold pieces and bade him eat. "I cannot eat gold," was the reply. "Why then have you hoarded it instead of giving it to your troops? Why have you not converted these iron gates into arrow points and advanced to the Jihun to prevent my crossing it?" The Khalif replied that it was the will of God. "What will happen to you is also the will of God," was the grim answer.† A similar story is told in his inimitable language by Joinville, who calls Khulagu the Lord of the Tartarins, and speaks of the Khalif as the apostle of the Saracens. He says the former insisted on the Khalif entering into matrimonial relations with him, that when he consented he urged him to send forty of his principal people to attest the marriage, and afterwards forty of his richest men, and that, having thus secured the leading people in Baghdad, he made sure of overwhelming the place. He goes on to say : " Pour couvrir sa desloiauté et pour geter le blasme, sur le Calife de la prise de la ville que il avait fète, il fist prenre le Calife et le fist mettre en une cage de fer, et le fist jeuner tant comme l'en peust faire home sanz mourir, et puis li manda, se il avait faim. Et le Calife dit que oyl ; car ce n'estoit pas merveille. Lors le fist aporter le roy des Tartarins, un grand taillouer d'or, chargé de joyaus à pierres precieuses et li dit. 'Cognois tu ces joiaus? Et le Calife respondi que oyl,' il furent miens. 'Et il li demanda si les amait bien, et il respondi que oyl.' Puisque tu les amoies tant, fist le roy des Tartarins, or pren de celle part que tu vourras et manju. 'Le Calife li respondi que il ne pourrait; car ce n'estoit, pas viande que l'en peust manger. Lors li dit le roy des Tartarins. Or peus veoir, ô Calife, ta défaute ; car se tu eusses donne ton tresor d'or, tu te feusses bien deffendu à nous par ton trésor se tu l'eusse despendu, qui au plus grant besoing te faut que tu eusses onques."‡ This is much like the report of the Armenian historian, Malakia, who says that Khulagu ordered him to be imprisoned for three days without food or drink. He then summoned him and asked him what he needed· The Khalif denounced his inhumanity, and said he had lived three days at the bottom of a pit. He had boasted to his people before the siege how he would put Muhammed on his standard and disperse the enemy. Khulagu then sent for a salver with some gold coin on it and bade him eat it, and thus satisfy his hunger. The Khalif replied that 'one cannot support life on gold, but needs bread and meat and wine." "Why, then, did you not send me a lordly present of gold so that I might

---

* Journ. Asiat., 5th ser., xi. 490-491. † D'Ohsson, iii. 243. Note. ‡ D'Ohsson, iii. 245. Note.

I

have spared your city and not captured you, instead of spending your time
in eating and drinking," and he had him trodden under foot.* The
Khalif's death took place on the 21st of February, 1258. His tragical end
forms one of those grim episodes which Longfellow delighted to put into
verse. He makes Khulagu address the avaricious Khalif thus :

> I said to the Khalif, "Thou art old,
> Thou hast no need of so much gold ;
> Thou shouldst not have heaped and hidden it here,
> Till the breath of battle was hot and near,
> But have sown through the land these useless hoards,
> To spring into shining blades of swords,
> And keep thine honour sweet and clear."
>
> *　　*　　*　　*　　*　　*
>
> Then into his dungeon I locked the drone,
> And left him there to feed all alone,
> In the honey cells of his golden hive ;
> Never a prayer, nor a cry, nor a groan,
> Was heard from those massive walls of stone,
> Nor again was the Khalif seen alive.

On the morrow after the Khalif's death all his attendants were killed, as
well as nearly all of the family of the Abbassides, except some obscure
individuals, and Mobarek Shah, the Khalif's youngest son, who was
spared at the request of Khulagu's wife, Oljai Khatun. She sent him to
Meragha, to the Khoja Nasir ud din. He afterwards married a Mongol
woman by whom he had two sons.† Minhaj-i-Saraj reports that a
daughter of the Khalif was also spared, who, with some females from his
harem and some rarities from his treasure, were set aside to be presented
to the Khakan Mangu, and were dispatched towards Turkestan. Other
things were sent to Bereke, the Khan of the Golden Horde, who refused
to accept them, and, according to this author, put the messengers who
took them to death, thus causing enmity between him and Khulagu.
When the booty meant for Mangu Khan reached Samarkand, the daughter
of the Khalif asked leave to visit the tomb of Kusam, son of Abbas, in
that city. He had accompanied Said, the son of the Khalif Osman, who
had been sent to Mavera un Nehr with an army, and had died and been
buried at Samarkand. There she performed the customary rites, made a
prayer of two genuflexions, and said, "O God, if this Kusam, son of
Abbas, my ancestor, hath honour in Thy presence, take this Thy servant to
Thyself, and deliver her out of the hands of these strange men," whereupon
she died.‡

It is curious to contrast these accounts of the famous campaign against
Baghdad with the accounts given by the Chinese. In the "Si shi ki" we
are told how the city, which is there called Bao da, a name like M. Polo's
Baudas, was divided into an eastern and a western part, separated by the
Tigris, the eastern city having walls of large bricks, the upper part of

---

* Malakia, ed. Brosset, 453.　　† Quatremere, 306-307.　　‡ Op. cit., 125.

splendid construction, and the western having none. A great victory was won against 400,000 men (!!!) beneath the walls. The western city first fell, and its population was slaughtered ; then the eastern city was assailed, and after an attack of six days it was captured, and a terrible slaughter ensued. The Ha li fa (*i.e.*, the Khalif) tried to escape in a boat, but was captured.* In the biography of Kouo Khan (*i.e.*, of Kuka Ilka) we read that this chief during the siege built floating bridges, to prevent the retreat of the enemy down the river. When the place was taken the Khalif tried to escape in a boat, but finding the way thus barred went to the Mongol camp and surrendered. Kouo Khan then went in pursuit of a general of the Khalif's, named Judar (*i.e.*, the Devatdar), captured and put him to death.† In the " Si shi ki " we are told the Khalif's palace was made of fragant and precious woods, viz., of aloë-wood *(aloëxylon agallochum)*, sandal-wood *(santalum album)*, ebony *(diospyrus ébenum)*, and a red fragrant wood called hiang chen hiang by the Chinese, and whose botanical name is not apparently known.† The biography of Kouo Khan states that when the palace was burnt the fragrance impregnated the air for a distance of 100 li.§ The walls of the palace were built, according to the " Si shi ki," of black and white jade (*sic.*, but surely porcelain tiles are meant). Great stores of gold and immense pearls, precious stones, and jewelled girdles, worth a thousand liang, were found there. The people of Baghdad were famous for their goods, and the horses there were called tolicha. The Khalif, we are told, did not drink wine, but sherbet, made of orange juice and sugar. His people used guitars with thirty-six strings. On one occasion when the Khalif had a bad headache, a man was sent for who played on a guitar of seventy-two strings, when the headache immediately left him.‖

Muayid ud din Alkamiyi retained his post as Vizier, the reward doubtless of his dubious loyalty. Fakhr ud din Dameghani was made Sahib-divan, or chief of the administration. Ali Behadur, who was the first to enter the city when assailed, was given control of the merchants and artisans, with the title of Shahnah (*i.e.*, governor), Imad ud din Omar Kazvini, deputy of the Amir Karatai, caused the mosque of the Khalif and the Meshed of Muza Jewad to be rebuilt. Nejm ud din Ibn-Abu Jafar Ahmed Amran, who was entitled Vizier-rast-dil (the sincere Vizier), was given command of the districts east of Baghdad, including the country towards Khorasan, Khales, and Bendinjein.¶ Nizam ud din Abd

---

* Bretschneider, Notes on Med. Travellers, &c., 82.   † *Id*, 83.
‡ *Id.*, 83. Note 110.   § *Id.*   ‖ *Id.*, 84.
¶ Quatremere, 307-309. His is a good example of the rapid fortunes that often attend men in the East. Mirkhond reports how when Khulagu arrived he was in the service of the Governor of Yakuba, employed to tickle the soles of his master's feet to lull him to sleep. One day he told his master how he had dreamt that he had become Governor of Baghdad, and received a kick for his pains. When the siege of Baghdad took place, Ibn Amran shot a letter attached to an arrow informing the Mongols, who were then in some stress for provisions, that if they would send for him he would tell them something useful. Khulagu having accordingly asked the Khalif to let him go to his camp, he conducted the Mongols to some hidden granaries at Yakuba, where they found corn to keep them going for fifteen days. His reward was the command already mentioned, which was an approximate realisation of his dream. Wassaf, 79. D'Ohsson, iii. 247-248. Note.

ul Mumin Bendinjein was made Kadhi of the Kadhis, or chief judge. Ilka Noyan and Kara buka, with 3,000 Mongol horsemen, were sent into the city to restore order, and rebuild the houses. The bazaars were rebuilt, and the corpses of men and animals removed.* The devastation must have been dreadful, and when Wassaf visited the place sixty years later not a tenth part of the old city remained.† Master of the city, Khulagu proposed this question to the Doctors of the Law there : "Who is to be preferred, a just, unbelieving ruler, or a Mussulman ruler, who is unjust?" The Ulemas, who had assembled in the college of Mostansir to deliver their fathva, or decision, on this question, hesitated to reply, when a famous doctor, called Razi ud din Ali Ibn Tavus, took the paper and wrote the words, "The infidel who is just is preferable to an unjust Mussulman," and his example was followed by the rest.‡

Khulagu having left Baghdad, encamped near the tomb of the Sheikh Makarem, and afterwards marched by easy stages to rejoin his ordu in the town of Khanekin.§

During the siege of Baghdad, some of the chief people of Hillah, where the Seyids or descendants of Ali were influential, sent an embassy to Khulagu with their submission, and stating that it was a tradition among them derived from their ancestors, Ali and the twelve Imams, that he (Khulagu) would become the master of that district (i.e., of Irak Arab). Khulagu thereupon dispatched Buklah, or Tuklah, and the Amir Bijel-Nakhchivani (called Alai ud din by Wassaf and Ali by Von Hammer), and eventually Buka Timur, brother of his wife Oljai Khatun, to secure the towns of Hillah, Kufa, and Vassit. The people of Hillah put a bridge on the Euphrates, and went to meet him gladly. He therefore passed on towards Vassit, where he arrived seven days later, and where he was resisted. He speedily captured the place and slaughtered its male inhabitants.‖ Buka Timur now advanced towards Khuzistan, taking with him Sherif ud din ibn Juzi. He captured Shuster, where the soldiery were put to death, while Basrah and other places submitted willingly. Meanwhile Seif ud din, the bitkichi, with the approval of Khulagu, sent a body of one hundred Mongols to protect the tomb of Ali at Nejef. Buka Timur rejoined his master on the 12th of Rabi the First (i.e., the 19th of March).¶ When Khulagu marched against Baghdad he dispatched Arkatu (called Oroktu by Von Hammer and D'Ohsson) to attack the fortress of Arbil or Arbela,

---

* Quatremere, 309.   † Ilkhans, 155.   ‡ D'Ohsson, iii. 254-255,   § Quatremere, 311.
‖ Quatremere says 40,000 were thus killed.   D'Ohsson says, more probably, 4,000.   Quatremere, 309-311.   D'Ohsson, iii. 255.   Ilkhans, i. 155-156.
¶ Quatremere, 311.   Minhaj-i-Saraj here tells a story which is not confirmed by the other authorities.   He says that some of the Khalif's people who had retired into the Wâdi (i.e., the low marshy ground near the river) to the number of 10,000, suddenly crossed the Dijlah, and attacked Baghdad, cut the Vizier and the Mongol Shahnah to pieces, together with all the supporters of the Mongols and Christians they could lay hands upon.   When news of this reached the Mongol camp, a body of cavalry was sent to reinstate matters, but the assailants had withdrawn, and not one among those holy warriors of Islam was taken.   Tabakat-i-Nasiri, 1260-1261.   I look upon this story as a fable.

whose fame dates back at least to the time of Alexander the Great. Rashid says it had not its equal in the world. It was situated between the Greater and Lesser Zab, two days' journey from Mosul. It owed its chief importance to the Turkoman chief, Kukebusi ibn Abul Hasan Ali, entitled Malik Moaasem Mozaffer ud din, who had died about twenty-eight years before. He was famous for his beneficence, and made Arbil one of the finest towns of the Persian Irak. He founded several institutions there, such as had not been patronised by Islam before—a foundling hospital, an institution for wet nurses and for suckling babes, a house for widows, a common hospital, a special hospital for the blind, a karavanserai, in which travellers were not only provided during their stay, but had the expenses of their further journey defrayed; a sort of monastery (probably for dervishes), a medressah, or school, in which both the Hanifi and Shafi rites were taught; and, lastly, a mosque, where the birth of the Prophet was annually celebrated with great pomp. During this feast visitors, preachers, orators, poets, koran readers, and sofis flocked thither from the surrounding towns. A month before, twenty dome-shaped buildings of three storeys high, and made of boards, were erected between the monastery and mosque. From their galleries poets and orators addressed the crowd, while others exhibited magic lanterns. Mozaffer ud din himself repaired to one of these buildings for the mid-day prayer, spent the night in the monastery with the dervishes, and in the morning went out hunting. At the birth-feast itself, a great number of camels, cattle, and sheep were taken to the square, and there killed and cooked amid music. At night the town was illuminated, and in the morning the guests sat down at two tables—one for the more distinguished, the other for the crowd. The dervishes danced, and prayers were sung from the minarets, while dancers and singers were rewarded with alms. Such was Arbil; the town itself was situated on a plain, and its castle on an isolated hill close by.* The Vizier, Taj ud din ibn Salaia, who apparently governed it, went to his camp. Arkatu said he would believe in his sincerity when the town had submitted, but the Kurdish garrison refused to surrender it. He was thereupon sent on to Khulagu, and was put to death. The garrison meanwhile resisted the Mongol attack bravely. They made a sortie, and destroyed their siege apparatus and killed many of their men. Bedr ud din Lulu, the Prince of Mosul, who had sent a contingent of troops to help the Mongols, was asked his advice as to what should be done. He counselled the abandonment of the siege till the summer, when the Kurds would seek shelter from the heat in the mountains. The siege was confided to him. He captured it in the summer, and it was made over to him.† Bar Hebraeus says Lulu bought the town and its contents from the Mongols for 70,000 dinars, but his people were not long in possession of it, and the unruly Kurds there speedily gained

---

* Ilkhans, i. 158-159.   † Quatremere, 316-317.

the upper hand, and a Kurdish amir, named Sherif ud din Jelali, drove out the garrison, and secured the place, but having shortly after marched with a Tartar army against some rebels at Gulmeragi, Bedr ud din sent some Kurds, who assassinated him while sleeping in his tent. A Christian called Muktez, the brother of a famous doctor of Arbil, named Saphi Solimani, now secured the place, and on his death was succeeded by his son, Taj ud din Isa, a good and faithful person.*

Meanwhile the Malik of Herat, Shems ud din Kért, who was perhaps the most powerful of the Mongol vassals, and who had not taken part in the campaign against Baghdad, was having some adventures of his own in the far east. We read that he attacked Mustebij, a town of Guermsir.† The chiefs of Guermsir, Shahin Shah, Bahram Shah, and Miranshah, shut themselves up in the fortress of Khasek with 5,000 men. The fortress was blockaded for ten days, when it was reduced to extremity. Miranshah escaped with some men in the night. The next day the place was captured, and the two other chiefs, with ninety of their adherents and relatives, were put to death. Shems ud din then attacked Hissar Tiri, another fortress of the Afghans, which was taken after an attack of two months, and its Afghan governor Almar was cut in two, and his principal officers were either executed or bastinadoed. Three other fortresses, named Kehberar, Duki, and Saji, the last of which was razed, fell into the hands of the Malik of Herat. A great number of Afghans perished in this campaign. In 1258 Shems ud din had another adventure, which shows that Khulagu's authority in these regions was an administrative one merely, and shows also what a powerful person Batu Khan of Kipchak, who was the Aka, or senior prince of the Mongol world. We are told that in that year the Malik returned from Badghiz from visiting Bulghai and Tumar (both princes of the Golden Horde), when some officers of his army informed Batu that he disregarded the yarlighs or Imperial orders, and despised the envoys of the Mongol, Shah zadehs. Batu sent one of his officers named Guerai-beg to Bulghai to tell him to arrest the Malik. Bulghai, who was then in Mazanderan, forwarded the order to Kebtuka (*i.e.*, Kitubuka) to apprehend and take him to Mazanderan. A little before this Shems ud din had set out for Sijistan to strengthen his authority there. *En route*, he met his deputy in that province, the Malik Ali Masud, who was not friendly to him, and who was now on his way to Kitubuka's camp, where he professed to have a pressing engagement. When he reached it he urged upon Kitubuka that if the Malik of Herat were left at liberty he would presently displace the Mongols from all Khorasan, that already his power extended as far as India, and that he was master of the principal fortresses of Khorasan. This intrigue was reported to Shems ud din by a secret agent, who

---

* Abulfaraj, Chron. Syr., 551.
† Yakut says Mustebij is a town of Sind, a day's journey from Khandabil, seven days' journey east of Bost.

hurried to let him know, and was speedily followed by Masud and Dendai, who, with 10,000 men, was charged by Kitubuka to arrest Shems ud din. The latter shut himself up in his castle, and decided to defend it to the last extremity. Dendai secured an audience with him, and urged him to come out and receive the letter and robe of honour which Kitubuka had sent him, but he was not to be taken in with such a trick, nor would he leave the limits of his castle, asking that the yarligh and the robe of honour might be given to him there. Various expedients were tried, but without avail. Eventually Masud, having secured an interview, determined to try and assassinate the Malik, and said to his men : " When you see his head roll down from the walls rush into the castle." Shems ud din, who evidently suspected something, ordered ten men to be posted at each one of the gates, and that all Masud's men were to be detained, and when he reached the fourth gate he found himself with but three companions. Shems ud din, who was concealed behind a veil, sprang upon him, killed him, and threw his head over the wall of the audience chamber. The soldiers of Kitubuka, and the Sinjars, or people of Sijistan, mistaking the head for that of Shems ud din, pressed their attack, but on seeing the Malik himself still alive they retired in disorder. Shems ud din then came out and proclaimed Mangu Khan as his suzerain. The next day he put to death the three chief kelaunters (calendars) of Sijistan, disarmed the Sinjarians, and then distributed a large number of khilats or robes, and 30,000 pieces of gold to the learned men and poor. He then went to the camp of Khulagu, and met Tumar and Bulghai, who were on their way to arrest him. Although he told them he must hasten on to Khulagu, and could not stay with them, " yet the Turks," says the chronicler of Herat, " with their natural brutality, tried to detain him." He struck the Mongol who seized his bridle over the face with his whip. Things were becoming critical, when Khulagu's envoys arrived and conducted him to the Imperial camp.*

According to other authorities, the events just related are told very differently, and we are assured that Shems ud din, having incurred Khulagu's resentment, the latter ordered Tegur to march against him and to bring him his skin stuffed with straw. Shems ud din defeated Tegur, and also a second army sent against him, at Shelaun, on the borders of Herat, but afterwards sent an embassay with his submission and with presents. He had put to death the Governor of Nimroz (i.e., Sijistan), by whom Masud is no doubt meant. Khulagu demanded why he had done so, when he answered, " I slew him so that the Khan might inquire of me, ' Wherefore didst thou kill him,' and not inquire of him why he had killed me." Major Raverty makes him give this answer to Mangu.† Of the two accounts the former seems to be the most reliable. We are told that Khulagu sent Shems ud din back again covered with honours.

---

* Journ. Asiat., 5th ser., xvii. 447-451.   † Ilkhans, i. 278. Tabakat-i-Nasiri, 1200. Note.

In 657 (*i.e.*, 1258) the Malik laid siege to Bikr, a fortress of the Afghans, built on a rock in the midst of *the sea*.* It was deemed impregnable, and was thence known as Bikr (*i.e.*, the Virgin). In eighteen days the Malik built thirty large vessels and one hundred boats. He then attacked the place on two sides, and after a struggle of twelve days, during which most of the officers in the army of Herat were killed, the inhabitants submitted, and agreed to pay the capitation tax. Its governor also offered a sum of 10,000 dinars, ten loads of silken goods, five Arab horses, and fifty slaves loaded with precious objects. After this Shems ud din entered Zamin Daver, and pursued Miranshah (previously named),† who, on his approach, left Khasek. He was captured and put to death.‡

We must now return again to Khulagu. On the 17th of April he once more reached Hamadan and Siah Kuh, where he rested from the fatigues of the Baghdad campaign. He ordered the Malik Majd ud din Tebrizi to build a treasure-house, where the various treasures he had captured in the fortresses of the Ismaelites, of Rum, of the Georgians, Armenia, Luristan, and Kurdistan were to be guarded. We can hardly picture the number of valuable objects thus despoiled. Wassaf, as usual, is rhetorical on the subject, but here, at least, his rhetoric seems justifiable. He speaks of the gold, of the rich stuffs and cloths from Greece, Egypt, and China, Arab horses, rare mules, Greek, Alanic, and Kipchak boys, Turkish, Chinese, and Berber slave girls, &c.§ The site of Khulagu's treasure-city is discussed by Quatremere in a long note. The place itself was called Tela, and was, situated in the great lake of Azerbaijan, the Lake of Urmia, called Spauta by Strabo, and Gabodan by the Armenians, whence the name of Kabudan given it by some of the Arabic and Persian writers. Its waters are said to be very salty, and to contain no fish, but its banks were strewn with towns and were well cultivated. Large numbers of boats traversed it to and fro. From its saltness it was also called Deriai Shur (*i.e.*, the salt lake), while the districts of Urmia, Oshmiah, Dehwarkan, Tasuj, and Silmas, which bordered it, also gave it their name. It is now called indifferently the Lake of Shahi and the Lake of Urmia.|| The fortress itself of Tela, according to Von Hammer, is the modern Gurchinkalaa, a great rock inaccessible on three sides. It is compared by Porter, who visited it, to Konigstein, in Saxony.¶ Abulfeda says that Khulagu garrisoned his fortress with a thousand men, and that its commander was changed every year.** Hulaju was sent to the Khakan Mangu with a share of the spoil. He also took him word that Khulagu, having conquered Iran, now proposed to attack Syria and Egypt, news which was very grateful to the Khakan.††

---

* *i.e.*, the Lake of Abistandeh, the only one in Afghanistan. Its waters are salt like those of the streams Paltsi and Jilga, which feed it, and it is situated three or four miles S.S.E. of Ghazni. Journ. Asiat., 5th ser., xvii. 451. Note.
† Ante.   ‡ Journ. Asiat., 5th ser., xvii. 451-452   § Wassaf, 73-74.
|| Quatremere, 316-321. Note. Ilkhans, i. 160.   ¶ Porter, ii. 592-593.
** D'Ohsson, iii. 257. Note.   †† Quatremere, 321.

A few weeks after the capture of Baghdad, and on the 2nd of Jumada (*i.e.*, May), the Vizier Alkami died, and was succeeded by his son, Sherif ud din.* Eastern opinion has been divided in regard to the merits of the Vizier, but the majority of his critics denounce him as a traitor. For a long time it was customary to inscribe in books used in colleges, &c. : " May he be cursed by God who will not curse Ibn ul Alkamiyi."† The historian Ibn Tagriberdi says expressly that the Vizier, belonging to the sect of the Rafizis, designed the ruin of the Abbassidan house and the transference of the Khalifate to that of Ali.‡ Wassaf speaks more diplomatically, but refers to the chagrin the Vizier felt when he was made to play a second part to Ibn Amram, the Governor of Yakuba, after his own ignoble behaviour and disobedience of the Khalif.§ Minhaj-i-Saraj continually refers to him as the accursed Vizier, and invokes God's curse upon him, charging him with continual treachery to his master.|| On the other hand, a contemporary Arab writer, the author of the " Fi adab is Soltaniyet," or " Qualities of the Sovereign," defends him warmly. He tells us how he studied the *belles lettres* in his youth, wrote well, and had an excellent memory. He describes him as accomplished, generous, able as a governor, equitable and honest. He was a patron of men of letters, and had acquired a library of 10,000 volumes, of which several had been dedicated to him. The household of the Khalif were jealous of him, and he was accused of treachery; " but his best character," says our author, " was the confidence Khulagu reposed in him. He would never have trusted him if he had betrayed his master."¶ His title of Alkamiyi was derived from the fact that he had made the canal Alkami in Egypt, of which country he was a native, and which was afterwards known as Kazani.** About this time there also died the Khoja Fakhr ud din, who held the post of Ulugh bitkichi. This was given to Hosam ud din, although he was the youngest of his sons, but he could speak Mongol and write Uighur, " which," says Juveni, " was considered a paramount accomplishment." ††

We are told how the astronomer Nasir ud din ventured at this time to suggest to his peremptory master, who was at Meragha, that he should do something else than destroy, and told him that once when the Khuarezm Shah was being pursued by the Mongols, and his troops were pillaging Tebriz, he answered the protests of the people with the words, " I came as a world conqueror, and not as a world preserver." Khulagu replied, " Thank God I am both a world conqueror and a world preserver, and no weakling like Jelal ud din Khuarezm Shah."‡‡ Nasir ud din was therefore commissioned to build an observatory. Rashid ud din tells us that Mangu Khan was distinguished among all the Mongol sovereigns by his prudence, tact, sagacity, and wit, and was sufficiently intelligent to have

* Quatremere, 313.   † D'Ohsson, iii. 249.   ‡ *Id.*, 253. Note.   § *Id.*, 252. Note.
|| Tabakat-i-Nasiri, 1234, &c.   ¶ D'Ohsson, iii. 250-251. Note.   ** *Id.*, 251-252. Notes.
†† D'Ohsson, iii. 268.   ‡‡ Quatremere, 324-325.

mastered several problems of Euclid, and he desired that during his reign an observatory should be built, and charged Jemal ud din Muhammed ibn Tahir ibn Muhammed Zeidi, of Bukhara, to build one, but the difficulties proved insurmountable. The reputation of Nasir ud din having reached Mangu, the latter, in saying goodbye to his brother, asked him, when he had destroyed the Ismaelite fortresses, to send Nasir to him, but as Mangu was then occupied with his campaign in Southern China, Khulagu ordered him to build an observatory in Persia, which was completed in his seventh year. With him were associated four learned men—Muayid-ud-din-Aradi (or Urzy), Fakr ud din Meraghi, Fakr ud din Akhlati, and Nejm ud din Denran Kazvini.* Nasir ud din had convinced his master of the desirability of such a work, since it was necessary to calculate some new tables and to make some new observations, if the daily position of the sun, stars, and planets was to be duly calculated, for the purpose of drawing horoscopes, &c. ; and as the stars had a certain motion of precession it was necessary to continue these observations for at least thirty years, to cover the revolution of Saturn. Khulagu wished him to complete the work if he could in a dozen years, and he said he would try to do so with the help of the earlier tables, including those of Enerjes drawn up fourteen centuries before, those of Ptolemy 275 years later, others made at Baghdad in the reign of the Khalif Meimun, others again by Tebani, in Syria, and lastly those of Hakemi and Ibn ul A'lem in Egypt, made 250 years before, which were the latest. These facts are recorded by Nasir ud din himself in the preface to his own tables. Abd ulla Beidavi tells us that Khulagu took with him many learned men from China, astronomers, &c., and it was from one of these, Fao mun ji, better known as Sing-Sing, or the learned, that Nasir learnt about the Chinese era and their mode of calculating tables.† The observatory was built on a hill north of Meragha, and was duly furnished with armillary spheres, astrolabes, &c., including a terrestrial globe, in which the earth was divided into seven climates, while a slit in the cupola allowed the sun's rays to record on the pavement the height of the meridian, &c. Some of the learned works captured at Baghdad were sent there. The tables which were now made were published during the reign of Khulagu's successor, Abaka, under the title of Zij ul Ilkhani and they showed an error of forty minutes in the position of the sun at the beginning of the year as calculated by previous tables.‡ The building of the observatory, the instruments for which alone cost 20,000 dinars, was a costly matter. Nasir ud din, to further convince his master that the money was well spent, rolled a metal bowl down the hill. At the noise made by this the soldiers, who did not know its cause, rushed from their tents, while the astronomer and his patron, who both knew it, remained

---

* Quatremere, 325-327. D'Ohsson, iii. 266-267. Note.   † D'Ohsson, iii. 263-265.
‡ Id., 265.

tranquil; the moral being that events do not cause panic when they can be foretold.*

While Khulagu was at Meragha he was visited by Bedr ud din Lulu (*i.e.*, the full-moon pearl), Prince of Mosul, who was then ninety years of age.† He had been the slave of Nur ud din Arslan Shah, of the dynasty of the Sunkars, rulers of Diarbekr, who on his death appointed him tutor to his son Masud, and he ruled the principality during that prince's life, which ended in 1218. Nur ud din left two infant sons, who died within two years of his own death, whereupon Bedr ud din was acknowledged as ruler, and had many struggles with the Ayubit princes. He had now exercised authority for 39 years.‡ He had apparently taken care not to declare himself too openly for the Mongols until their success was quite assured; the contingent of 1,000 men under his son Salih only arrived after the fall of Baghdad. Having aroused their suspicion in consequence he was constrained to pay Khulagu a visit in person.§ Some writers say that his family wished to dissuade him from going, as he had incurred Khulagu's anger, but he replied that he hoped to conciliate this redoubtable warrior, and to rub his ears.|| When he arrived, he presented Khulagu with rich presents, and went on to say, " I have reserved a gift which I mean for you personally." He then produced two splendid ear-rings, each containing a large pearl, and asked permission to put them on Khulagu's ears himself, which would bring him great credit among the other princes. Khulagu having consented, he turned to the people of Mosul, who had accompanied him, and pointed out how he had kept his word.¶ Bedr ud din thereupon returned to Mosul, where he shortly after died. Abulfeda says he died in the year 657 HEJ. (*i.e.*, 1258).** Makrizi says the same. He left three sons. Malik al Salih al Ismael, the eldest, succeeded him at Mosul, Alai-ud-din at Sinjar, and Saif ud din at Al Jezireh.†† Salih was confirmed in his authority by Khulagu, who gave him a daughter of Jelal ud din Khuarezm Shah, named Turkan Khatun, in marriage.‡‡ We shall hear of him again.

Bedr ud din Lulu's visit to Khulagu was followed by that of Said, son of Abubekr, atabeg of Fars, who congratulated him on the capture of Baghdad.§§ Khulagu having moved his quarters to Munik, in the district of Tebriz, was also visited by the brothers, and now reconciled rivals, Iz ud din and Rokn ud din, the joint Seljuk rulers of Rum. The former had not only defied Baichu Noyan, as we have seen, but had ventured to attack him, and it was necessary he should now be humble. He accordingly had a splendid boot, worthy of a sovereign, made, upon the sole of which was painted his own portrait, and presenting it to Khulagu, he begged him to put

---

* D'Ohssou, iii. 266. † Quatremere, 321. ‡ D'Ohsson, iii. 258.
§ Abulfaraj, Chron. Arab., 344-345. Chron. Syr., 553.
|| This is an eastern expression, meaning to vanquish, and was used by Molière in " Tartuffe "—
" Jour de Dieu ! je saurai vous frotter les oreilles."
¶ Novairi, &c., quoted by Quatremere, 236-237. Notes. ** Op. cit., iv., 567.
†† Abulfaraj, Chron.:Arab., 347. ‡‡ Ilkhans, i. 194. D'Ohsson, iii. 306. §§ Quatremere, 323.

his foot on the head of his slave. Khulagu was moved at this, and Tokuz Khatun, his Christian wife, having asked for his pardon, it was granted.[*] Rokn ud din ruled over the district between Cæsarea and Great Armenia, with his capital at Sebaste, while Iz ud din held the country thence to the sea, with his seat of power at Iconium. The two princes accompanied Khulagu on his march towards Syria as far as Mesopotamia, and then returned home again.[†]

The Greater Lur was now ruled by a line of Atabegs, founded about a hundred years before. Its ruler at this time was the third Atabeg, Tekele, the son of Hasarsif or Hasarasp. When Khulagu advanced against Baghdad, he went to do homage, and was ordered to join the advance guard under Kitubuka. Unable to restrain his indignation at the capture of Baghdad and the death of the Khalif, he incurred the displeasure of Khulagu, and hearing that he was suspected, escaped from the Mongol camp. Khulagu reprimanded Kitubuka for allowing this, and sent him with the Noyan Sidak and some troops in pursuit. His brother, Shems ud din Alp Arslan, counselled him to allow him to go to the Mongol camp and intercede for him. He set out, but was waylaid by the Mongols, and, notwithstanding his errand, was put in irons and his men were killed. Tekele sought refuge at the fort of Manjasht. He presently offered to surrender if Khulagu would send him his ring as a guage of safety. This was sent, but notwithstanding he was put to death on his arrival at Tebriz, and his brother, Shems ud din Alp Arslan, was given the throne of the Greater Luristan.

The Lesser Lur was ruled by another dynasty of Atabegs, founded seventy years before. The fourth of these Atabegs, Bedr ud din Masud, had been driven out by his cousin and rival, Suliman Shah, who was supported by the Khalif's troops. He went to implore the help of Mangu, and accompanied Khulagu in his westward march. At the siege of Baghdad his rival, Suliman Shah, commanded the Khalif's army. He asked to have him surrendered to him if captured. Khulagu replied, " That is a great promise to make to you by God." When Baghdad fell, however, and Suliman Shah was killed, the latter's family were handed over to Masud. He behaved so well to them that when he presently gave them the choice of remaining in Luristan or going to Baghdad, few of them went. He was renowned for his judgment, and knew by heart 4,000 juridical maxims of the Shafi Rite. He died two years after the taking of Baghdad. On his death his sons, Jemal ud din Bedr and Nasir ud din Omar, struggled with Taj ud din Shah, son of Hosam ud din Khalil, for authority in the Lesser Luristan.[‡]

The generally favourable treatment meted out to the Christians by the Mongols had an exception in the case of Takrit, whose Christian

* Quatremere, 323.   † Abulfaraj, Chron. Syr., 554.
‡ Tarikhi Guzideh, cited by D'Ohsson, iii. 259-261. Ilkhans, i. 161-163.

inhabitants had applied to the Catholicos to secure that they should have a prefect sent to protect them. When the Mongols proceeded to slaughter the Arabs, the Christians, who remained for six weeks concealed in one of their churches, were accused by an Arab, called Ibn Duri, of having killed many of his co-religionists, and appropriated their property. When the Prefect brought this charge before them they would not deny it, and sent him what things they had secreted. The facts having been reported to Khulagu, he ordered them, in accordance with the Draconic Mongol code, to be put to death, and an officer was sent with a contingent of troops, who took the Christians in parties of twenty to the citadel under pretence of making them assist at its demolition, and then put them to death. Only the old people were spared, and the boys and girls whom the Mongols carried off captive. The Muhammedans once more occupied the cathedral of Takrit, while the few Christians who had escaped were ministered to by two Carmonian priests (presbyteri Carmonenses), in the other churches. Presently Ibn Duri was in turn put to death by a Christian named Bahram, who had been nominated Prefect of Takrit.* Altogether, however, the condition of the Christians was greatly improved all over the East by the Mongol invasion. They were relieved from many indignities which the Moslems had heaped upon them, and they looked upon Khulagu as a deliverer. After the capture of Baghdad the Georgians—David and his people—having been a long time under arms, asked permission to return home. Khulagu gave his permission, and they set out loaded with presents and booty, and returned to Tiflis by way of Azerbaijan. This was in 1259. At this time the grandees of the kingdom were greatly distressed that David's wife, Jigda Khatun, had no son, and in order to secure a successor he married a pretty Ossetian damsel called Althun, engaging not to see her again if she should have a male child. She had a son called Georgi, who was adopted by the Queen. Althun presently had a daughter named Thamar. Soon after the Queen was buried in the Royal Sepulchre of Mtzkhétha.†

The Mongols were now masters of all the country from the Oxus to the Tigris, but Khulagu's commission was that he should lay his hands upon all Asia, as far as the farthest west, and the next objects of attack therefore were Mesopotamia and Syria. These two countries, with Egypt, had been dominated over by the famous Ayubit dynasty, of which Salah ud din was the greatest name. Egypt had been lost to the family, however, and was now controlled by the Mamluks, but six branches of the Ayubit stock ruled over Arabian Irak and Syria. These were the princes of Mayafarkin, Hossnkief, Karak, Aleppo, Hamath, and Hims. If they had been united in a common policy they might have offered a reasonable resistance to the Mongols. But who ever heard of union

---

* Abulfaraj, Chron. Syr., 552-553. D'Ohsson, iii. 270-271.   † Hist. de la Géorgie, 553-554.

among the Kurds? The most powerful of these princes was Malik Nasir
Yusuf, Prince of Aleppo and Syria, and great grandson of Salah ud din.
Mayafarkin, Hossnkief, and Karak were ruled by descendants of Malikol
Aadil, Salah ud din's brother ; Hamath by a descendant of Shahin Shah,
another of his brothers, and Hims by a descendant of Shirkuh, Salah ud
din's uncle.* Nasir, Prince of Aleppo, had succeeded his father in 1236,
when only six years old, and in 1250 conquered the principality of
Damascus from the Mamluk Eibeg, who had usurped authority in Egypt.
He intended driving Eibeg completely out of Syria, but having been
defeated by him in 1251 he ceded to him, on the mediation of the Khalif,
Jerusalem, Gaza, and the coast as far as Nablus.† Nasir had sent his
Vizier, the Sahib Zain ud din Hafedi, to the Court of Mangu, with
precious gifts worthy of a sovereign, and had been granted a yarligh and
paizah.‡ On Khulagu's arrival in the West, in 1258, he sent his son Aziz,
with the same Vizier, to conciliate him.§ Khulagu asked him why his
father had not gone in person, and was conciliated by the reply that he
feared to leave his dominions lest they should be attacked by
the Franks.|| Makrizi says the young prince offered Khulagu the
presents which he bore, and also asked for his aid to help
him to drive the Mamluks out of Egypt. Khulagu ordered that the
prince on his return should be escorted by 20,000 horsemen.¶ Novairi
tells us that Nasir had also sent a letter to the Prince of Mosul to intercede for him.** The young prince took back with him for his father a
letter which was composed by Nasir ud din of Tus, and which Wassaf
praises as a *chef d'œuvre* of this kind of Arabic composition, in which
brevity and condensation, sonorous phrases, short periods, cadences and
rhymes, alliterations and puns, and apt citations from the Koran, are
greatly admired.†† We have this letter in several copies differing from
one another greatly. It has been preserved by Rashid ud din, Wassaf,
Abulfaraj, Makrizi, and by Ibn Arabshah in his history of Timur.‡‡ In
the first volume of this work I gave the letter as reported by Wassaf,
Here I will transcribe the version preserved by Makrizi :—

"We inform Malik Nasir, Prince of Aleppo, that by the strength of the
sword of the most high God, we have conquered Baghdad, exterminated
the warriors of that town, destroyed its buildings, and made its inhabitants
prisoners, according to the maxim which God has enshrined in the sacred
volume. 'When the Kings enter a town, they cause ravage there and
reduce to the direst humiliation the most distinguished of the inhabitants.'
We summoned the Khalif to our presence, and addressed him questions
which he answered deceitfully. But he presently repented of his conduct,
and has well deserved the death which we inflicted on him. The perverse
man devoted himself only to amassing riches, and hoarding up precious

---

* Ilkhans, i. 168-169.   † D'Ohsson, iii. 290-291.   ‡ Quatremere, 327.
§ Abulfeda, iv. 565. || D'Ohsson, iii. 294.   ¶ Makrizi, i. 78-79.   ** D'Ohsson, iii. 294.
†† *Id.*, iii. 302-303, &c.   ‡‡ *Id.*, iii. 302-305   Note.

objects, without caring at all for his subjects. His reputation had spread very widely, and he occupied the highest rank. May God defend us from perfection, and the fate attending grandeur.

> When a thing has reached its highest point it begins to descend.
> When you hear a man say there is perfection beware of a catastrophe.
> If you are prosperous be on your guard,
> For crimes continually undo goodness.
> How many men have spent the night happily,
> Without suspecting that death would suddenly overtake them?

"When you have considered my letter hasten to submit to the King of Kings, Lord of the World, and to subject to him your person, your people, your warriors, and your riches. Thus you will avoid his anger and deserve his benevolence. As God the most high has declared in his august work, 'Yes, man shall only reap the reward of his works, and God who knows his works, will not fail to recompense him with the greatest zeal.'[*] Mind you do not, as you have done before, imprison our envoys, but treat them according to the laws of justice, and send them back with proofs of goodwill. We have heard that some Syrian merchants and others have taken refuge in a caravan-serai with their wives and riches. If they were to retire to the mountains we would tear them down ; if they hid beneath the ground we would root them up.

> Who shall escape, for no one shall find a hiding place.
> The two elements—the land and the water—belong to us.
> Our redoubtable strength has overcome lions,
> Amirs and viziers are subject to us." [†]

The young prince, with this somewhat truculent letter, started homewards about a month after the capture of Baghdad, and we are not surprised that it should have aroused some panic in Syria. Nasir sent his wife to seek refuge at Karak, many of his people fled towards Egypt, and many were robbed and plundered *en route* thither. Among other fugitives was the Prince of Hims, and Wassaf has preserved the text of a bitter letter which he says he sent to Khulagu in reply to the one the latter had sent. This I have already abstracted.[†] This no doubt inflamed the wrath of the Mongols, which would not be made calmer when they heard how Nasir had lately given a welcome to 5,000 deserters from Khulagu's army, who are called Sheherzurs, whence it is probable they were Kurds from Sheherzur. Presently they deserted him in turn and went over to one of his rivals, Moguith, the Prince of Karak, with whom he was at this time in strife. Having at length made peace with Moguith, on condition that the latter surrendered all the Bahri Mamluks in his service, and dismissed the Sheherzurs, he returned to Damascus. Some of the Sheherzurs went on to Egypt, and as far as Maghreb (*i.e.*, Africa).[§]

Khulagu set out on his new campaign on the 12th of September, 1259, and was accompanied, *inter alios*, by Salih, the Malik of Mosul. Kitubuka

---

* Koran, ch. liv. verses 40-41.
† Op. cit, 83-85. ‡ Ante, i. 206-208. § Makrizi, i. 79-80.

with some Amirs.*   Nasir sent to ask help from the Egyptian
authorities. When his envoy reached Cairo, the throne was filled by
Mansur, a son of Eibeg, who was a mere child.   His council was
convened, and the grand judge, Bedr ud din Hasan, and the Sheikh Iz ud
din Ibn Abd us Salam, were asked for their opinion, whether it was lawful
under the circumstances to levy a war tax on the nation. They replied
that when the enemy had entered the territory of Islam, it was the duty of
every Mussulman to arm, and that a levy might be made.   This decision
was accepted. Meanwhile the times were not favourable for a boy to
fill the throne. The threatening aspect of affairs, and the appearance of
Khulagu, gave a pretentious excuse to Kuttuz, Mansur's atabeg, or tutor, to
seize the throne, at least until the Tartars were driven back. He professed
to be a nephew of the Khuarezm Shah, Jelal ud din, and that he had been
formerly captured by the Mongols, and sold as a slave at Damascus,
whence he was taken to Cairo. Having imprisoned the discontented, and
received the allegiance of the army, he wrote to Nasir a humble letter
offering him the throne of Egypt, and treating himself as his lieutenant
there. He said he would march into Syria to his help if he wished it, but
if this would embarrass him offered to send an army under any general he
might name.†

Khulagu having secured his rear by the conquest of Mesopotamia, now
continued his advance, accompanied by his wife, Dokuz. He captured
the fort of Bire or El Biret, on the Euphrates, where the Ayubit prince
Said had been imprisoned for nine years. Having set him at liberty,
and given him the command of Subaibah and Banias,‡ Khulagu crossed
the river by four bridges of boats, viz., at Malatia, Kalat ur Rum (i.e.,
the Roman castle), El Biret, and Kirkesia, respectively on the sites of
the ancient Melitene, Zeugma, Birtha, and Kirkesion. The guards
stationed at these fords were killed. He captured Menbedsh, the ancient
Hierapolis, also called Bombyce, and once famous for its temples (especially
that to Astarte) and cotton products.§ Various places on the Euphrates
were taken, and their inhabitants slaughtered, such as Mabug Nejm
(i.e., the star castle), Rakka (the ancient Kallinike or Nicephorium),
Jaaber (famous in the history of the Osmanli as the place where Suliman,
one of their early sultans, was drowned, and where we are told his grave,
called the Turkish grave, still remains), and Lash (?).|| When a division
of the Mongols reached Salamiyet, near Aleppo, some of the garrison,
with a rabble of the citizens, went out to oppose them, but seeing that the
enemy offered a firm front they withdrew. Presently they made a similar
effort, and posted themselves at the mountain Bancussa, under the Ayubit
prince Moazzam Turan Shah. The Mongols drew them into an ambuscade,

Makrizi, i. 87-88.   † D'Ohsson, iii. 315-316.   ‡ Makrizi, i
§ Abulfaraj, Chron. Syr., 555. Chron. Arab., 347. Ilkhans, 181-182.
|| Abulfaraj, Chron. Syr., 555. Ilkhans. 182. D'Ohsson, iii. 316.

killed a considerable number of them, and then marched to Azaz, north of Aleppo, which capitulated. Other divisions secured the towns of Maaret Naaman, Hamath, and Hims, and also the town of Bab Ali or Babela, near Antioch, so called from St. Babylas.*

The Mongols now approached Aleppo. Aleppo is famous both for its ancient prosperity and trade, and for its products—cucumbers, water melons, figs, apricots, and especially pistachio nuts, which are called by the Arabs the daughters of memory, since eating them is thought to strengthen that faculty. As the *entrepôt* for Indian goods, it was known as the Little India. Before the Jewish Gate, also called the Gate of Delight, is a great and ancient stone on which Jews and Christians used to swear. Mussulmans reverence Aleppo as the abode of Khizr, the guardian of the sources of life (the legends about whom are mixed up with those of Saint George), and also because Abraham is said to have milked his herds here. The latter legend has possibly arisen out of the corruption of the ancient name of the city, Khalybon, into Alep, which in Arabic means milk.† The Mongols now approached the famous city. Khulagu sent the Prince of Erzerum to Moazzam Turan Shah, its governor, to say that they did not wish to do it or its inhabitants any harm, their quarrel being merely with Nasir, and requesting only that two Mongol Shahnahs might be allowed—one in the town, the other in the citadel—to await the impending battle which was to decide to whom the place should belong. If the Mongols won it was to be theirs, if the Sultan won then they might put the Shahnahs to death. Moazzam replied that there was only a drawn sword between them—a rash reply, which brought a sharp Nemesis.‡

The place was now beleaguered. Arkatu Noyan was posted at the gate of the Jews, Kitubuka at that of Rum (*i.e.*, of the Greeks), Sunjak before that of Damascus, and Khulagu himself before that of Antioch.§ The town was surrounded by lines of circumvallation, on which were planted the battering engines, consisting of twenty catapults, and the attack was sustained for seven days, being chiefly pressed against the so-called gate of Irak. The place fell on the 25th of January, 1260,‖ and a general massacre ensued, which lasted for five days, and was at length put an end to by a proclamation of Khulagu. He had issued a firman, in virtue of which the houses of Shihab ud din Ibn Amru, of Najm ud din, the brother of Mazdekin, of Bazdiad, and of Alm ud din Kiasari, of Mosul, were to be spared. In these, in the khanoka or monastery of the Sufis, where Zain ud din Sufi lived, and the synagogue of the Jews, upwards of 50,000 people found shelter.¶ The citadel, whither Moazzam Turan Shah had fled, held out for two months longer,

* Ibn Tagri Birdi, in D'Ohsson, iii. 317-318.   Ilkhans, i. 182.   Abulfeda, iv. 573-575, and Note 413.
† Ilkhans, i. 183.   ‡ Abulfeda, iv. 577-579.   D'Ohsson, iii. 318-319.   § Quatremere, 334-335.
‖ Weil, iv. 13.   ¶ Abulfeda, iv. 579.

during which some people who were suspected of carrying on a correspondence with the Mongols were put to death.   At length, however, the
garrison deemed it better to surrender.   Haithon says it was captured by
sapping, and their lives, including that of Moazzam, who was a very
old man, and who died a few days later, were spared.   The Mongols
released some Mamluks who had been imprisoned there, *inter alios*,
Sonkor Ashkar, Seif ud din Tenkez, Seif ud din Beramak, Bedr ud
din Bekmesh Masudi, Lajin jamdar Salchi, and Kijadi the Little.
They were handed over to a Kipchak in the service of the Mongols,
named Sultan Jak, of whom we have already spoken in describing the
campaign against Baghdad.*   The prisoners were sold to the Armenians
of Cilicia and to the Europeans.†   Makrizi says the streets were so
encumbered with corpses that the Mongols marched over them.   The
number of women reduced to slavery he calculates at 100,000.   The
citadel was razed, and the walls of the city, the jamis, mosques, and
gardens were destroyed.‡   Rashid ud din tells us that during the siege
several chiefs, such as the Amirs Kurjan, Uju Sokurji and Sadek Kurchi,
were wounded in the face and elsewhere.   Khulagu congratulated them,
saying that as a rose colour is the prettiest *parure* of a woman, a man
can have no nobler ornament than some crimson blood strewn over his
face and beard.§

Among the prisoners were several of Nasir's children with their
mothers.‖   Haithon says that Khulagu presented the King of Armenia
with the spoils he had captured at Aleppo, and also made over to him
some of the lands he had conquered, so that the King secured several castles,
which he fortified.   Khulagu afterwards sent him some presents by the
Prince of Antioch, and restored to him the districts he had captured from
the Muhammedans, and which they had retaken,¶ a heritage which
brought its Nemesis when the Mussulmans were again dominant.
Vartan tells us that Haithon accompanied Khulagu on this campaign, and
redeemed many Christians, both lay and cleric, who had been made
prisoners.**   Abulfaraj, the historian, was at this time the Jacobite
metropolitan of Aleppo.   He tells us the upper part of his church had been
destroyed by the Balbecenses (*i.e.*, the people of Baalbek), and in fear of
these events he had gone to visit Khulagu.   He had been detained by him
at the fortress of Nedjm, on the Euphrates, and deplores that in
consequence he could not protect his flock as he would have wished.
The Mongols apparently found their way into a Greek church and killed
a crowd of Christians who had sought refuge there, only a few being saved
by the exertions of an Armenian priest named Turus, the brother of the
Catholicos Mar Constantine, and by the monk Khurakh.††

Khulagu now issued a proclamation, in which he appointed Amad ud

---

* *Id.*, 583.   Makrizi, i. 90.   † D'Ohsson, iii. 320.   ‡ Makrizi, 90.   § Quatremere, 357.
‖ D'Ohsson, iii. 320.   ¶ *Op. cit.*, ch. 29.   D'Ohsson, iii. 321.   Note.
** Journ. Asiat., 5th ser., xvi. 293.   †† Abulfaraj, Chron. Syr., 555-556.

din, of Kazvin, his *locum tenens* at Aleppo, and intrusted the citadel of the same place to Fakhr ud din; Tukal Bakhshi being appointed Shahnah or Mongol commissary.* On the arrival of the Mongols in Syria, Mansur, son of Mozaffer, Prince of Hamath, left that town in charge of the Tavashi Mureshid, and went to Damascus. Mureshid, on hearing of the fall of Aleppo, rejoined his master, whereupon the notables of Hamath, taking the keys of the city with them, went and submitted to Khulagu, asked him to spare their lives and property, and to appoint a shâhnah. Khulagu appointed a Persian named Khosru Shah, said to have been descended from Khaled, the son of Valid, as governor of the town, and Mojayed ud din Kaimaz as commander of the citadel. † When the news of the fall of Aleppo reached Damascus, the Malik Nasir was still there. He had collected an army of 100,000 Arabs and Persians, who now disbanded, each one sold his furniture, and prepared to fly in hot haste. Nasir left the camp at Berzah, near Damascus, on the 29th of January, and taking with him the Prince of Hamath, and the few retainers who stood by him, went towards Gaza. The citizens were thus defenceless, and so great was the anxiety to get away that the hire of a camel rose to 700 pieces of silver. There was a general stampede, and it was thought the Day of Resurrection had arrived.‡ On the departure of Nasir the Vizier Zain Hafidi, already named, took possession of the city, and closed its gates, and having summoned the citizens, it was agreed to surrender it to the Mongols, and it was duly made over to Fakhr ud din Merdegai, to the son of the commander of Erzerum, and the Sherif Ali, who had been sent as his envoys to Nasir by Khulagu. They informed Khulagu, who sent a Mongol corps to take possession of it, forbidding his soldiers to touch anything belonging to the citizens. Khulagu meanwhile appointed the Kadhi, Muhi ud din ibn Zaki, Kadhi of all Syria, and gave him a robe of honour, made of golden tissue. The Kadhi thereupon returned to the city, and having assembled the chief inhabitants in the great mosque, on the Sunday, the 3rd of February, dressed in his khilat, read out his diploma of investiture, with Khulagu's order granting a general amnesty.§ He was shortly followed by two commanders, one Mongol and the other Persian, who had received orders to treat the people well, and to obey him, and they were followed on the 1st of March by Kitubuka Noyan with a body of Mongols. The act of amnesty was again read, and also a diploma conferring on Kamil ud din Omar Tiflisi the office of Kadhi alkodat in all the towns of Syria, at Mosul, Mardin, and Mayafarkin, and also the superintendence of the mosques and pious foundations. This order was publicly read in the Maidan-akhdar, or green square.‖

· Abulfeda, iv. 585. Quatremere, 339. † Abulfeda, iv. 581. D'Ohsson, iii. 322.
‡ Makrizi, i. 91-95. D'Ohsson, iii. 322. § Abulfeda, iv. 585. Makrizi, i. 96-97.
‖ Makrizi, i. 97-98. D'Ohsson, iii. 323-324.

Makrizi tells us the Christians at Damascus now began to be in the ascendant. They produced a diploma of Khulagu guaranteeing them express protection and the free exercise of their religion. They drank wine freely in the month of Ramazan, and spilt it in the open streets, on the clothes of the Mussulmans and the doors of the mosques. When they traversed the streets, bearing the cross, they compelled the merchants to rise, and illtreated those who refused. They carried the cross in the streets and went to the church of St. Mary, where they preached sermons in praise of their faith, and said openly, " The true faith, the faith of 'the Messiah, is to-day triumphant." When the Mussulmans complained they were treated with indignity by the governor appointed by Khulagu, and several of them were by his orders bastinadoed. He visited the Christian churches, and paid deference to their clergy.* The governor here meant was no doubt Kitubuka, who was a Kerait and a Christian. " Guiboga," says Haithon, "loved the Christians, for he belonged to the race of the three kings who went to worship Our Saviour at his nativity."†

Meanwhile Zain Hafidi levied immense sums on the inhabitants, with which he bought costly stuffs, and gave them to Kitubuka, to Baidara, and to the Mongol amirs and generals, and sent them daily gifts of various kinds.‡ The citadel of Damascus still held out under Bedr ud din Muhammed ibn Karmjah and the Amir Jelal ud din ibn Sairafi. Kitubuka laid siege to it. Meanwhile, however, there came on a terrible storm of rain and hail, with a hurricane of wind, and an earth-quake which shook the district, and the siege was accordingly protracted. Twenty catapults battered the walls without ceasing, until the wearied garrison agreed to capitulate. The Mongols then entered the place, secured all the valuables, demolished a large number of the towers, and set fire to the siege machinery.§ Zain Hafidi sent to ask Khulagu what was to be done with the commander of the fortress and his deputy, and having received orders to put them to death, he did so with his own hand, at the Mongol camp of Merj Bargut.||

At this time the Ayubit prince Ashraf, who had been deprived of Hims by Nasir twelve years before, and had been given Telbashir in exchange, having presented himself before Khulagu was by him reinstated in his principality. He was also given a diploma constituting him viceroy of Damascus and of all Syria, and Kitubuka and the other amirs were constrained to obey him.¶

While the siege of Aleppo was proceeding a summons was sent to Harim, a town situated between Aleppo and Antioch, often mentioned in the history of the Crusades, and famous for its pomegranates ; and as it did not surrender it was attacked. The citizens offered to submit if a Mussulman in whom they could trust was sent to swear on the Koran

---

' Makrizi, i. 98. † Op. cit., 30. D'Ohsson, iii. 325. ‡ Makrizi, i. 98-99.
§ Makrizi, 99. || D'Ohsson, iii. 326. ¶ Makrizi, i. 99.

that they would be spared. Khulagu inquired whom they wished, and they replied Fakhr ud din, who was the commander of the citadel of Aleppo. Khulagu accordingly sent him, but piqued at their refusal to trust his word, he had them all slaughtered—old and young, women and children—except an Armenian goldsmith.* Fakhr ud din was also put to death, having been charged with tyranny by the people of Damascus.† Abulfaraj says he was upbraided by Vali ud din, son of Safi ud din, Prince of Aleppo, who said, "He killed my father and brothers, to whom he had said, 'surrender the town, lest you be put to death.'"‡ Zain Hafidi, from Damascus, was given his command, and we are told that Mogul, with three Persian assessors, Alai ud din Jashi, Jemal ud din Karkai Kazvini, and the Kadhi Shems ud din Komi, were given charge of Damascus.§

While Khulagu was encamped at Aleppo, Sinktur Noyan arrived from the Imperial head-quarters in Mongolia with the news that Mangu Khan was dead. Khulagu was greatly distressed, and determined to return home.‖ Haithon says that it was his intention to have returned to Mongolia to put in a claim for the Imperial throne, but that when he reached Tebriz, hearing that his second brother Khubilai had been raised to that dignity, he did not go on.¶

Note 1.—The spelling of Mongol names is a subject of great difficulty. For the most part Western writers have followed the spelling used by the Persian authors, to whom we owe so much of our information about the Mongols, in which the names are presented in a decayed fashion. The name of the founder of the power of the Ilkhans is generally written Hulagu or Hulaku. I have spelt it Khulagu, and have followed in so doing the excellent example of Schmidt,** who was one of the first Mongol scholars of our time. Fraehn wrote a paper which he entitled, "De Ilchanorum seu *Chu*lagidarum numis commentatio." M. Renaud wrote the name, Kulagu.†† Remusat has the interesting note: "Houlagou (mieux Khoulakou) est nommé par les Grecs χυλαυ par nos historiens Olaon, par les Armenians Houlav."‡‡ Von Hammer writes: "Hulagu oderwie die Mongolen den Namen schreiben und sprechen, Chulagu."§§ Schiefner also writes the name in the same way. This is assuredly ample authority, and in view of it I cannot resist quoting the following characteristic phrase of Major Raverty, *apropos* of the way I spelt the name in a former volume: "Anyone who understands a single letter of Oriental tongues knows that Khubilai is as impossible as Khulagu for Hulaku, and is incorrect, whatever the 'Mongol' professors may say."‖‖ While the Armenians retain the initial

* Abulfaraj, Chron. Arab., 348. Chron. Syr., 556. † Quatremere, 349. ‡ Chron. Syr., 556. § Quatremere, 339-340. ‖ Quatremere, 341. ¶ Op. cit., ch. 29. D'Ohsson, iii. 328. Note 2. ** Journ. Asiat., 1st ser., iii. 108. †† Journ. Asiat., 3rd ser., xiii. 114, &c. ‡‡ Recherches sur les Langues Tartares, 168. Note. §§ Ilkhans, i. 79. ‖‖ Tabakat-i-Nasiri, 1215. Note. See also same work, pp. 1225-1226. Note.

aspirate in the forms Houlav, Haulaou, Hulaou, the Georgians have dropped
it in that of Ulo, which is like the form given by Marco Polo, Alau. The
Chinese write the name Hiu lie wu. The name in Mongol means simply
thief.*

Note 2.—Some Mongol words which habitually occur in these pages deserve
a little notice here. 1. *Noyan.* Noyan or Noin is a Mongol word, and means
the leader of a tuman or division of 10,000 men.† The Jihan-kushai says one
of Jingis Khan's sons was called Ulugh Noyan, that is to say, Great Amir. In
the Zafer Nameh we read in one place, "The amirs and noyans of Jagatai."‡
At the present day, noin or noyan among the Mongols and Kalmuks means a
prince or any member of the Royal Family.§ 2. *Taiji* or *Taishi.* The over chief
of a ulus or tribe was formerly styled Taishi among the Mongols and Kalmuks.‖
Now the title is used nearly co-extensively with the word Noyan, and is
applied to those of royal blood.¶ Among the Tumeds and Karchins the
Taijis are styled Tabunans, which was possibly the primitive title in use in
Mongolia. Among the Mongols generally by tabunan is now understood the
sons-in-law and brothers-in-law of emperors and princes, answering to the
Oefu of the Manchus.** The title is of Chinese origin,†† and corresponds to
the very primitive title of Tai si or Taishi, which means Grand Master, a title
which was borne by the senior official of the empire.‡‡ Rashid ud din says
that Taishi was a Chinese word, meaning a great teacher, or a skilful writer.§§
Abulghazi also says it was a Chinese word, meaning the same as Hafiz in
Arabic, *i.e.,* one who knows something by heart.‖‖ 3. *Baghatur.* The word
Behadur or Bahadur, says Quatremere, is not of Persian origin, but is derived
from the Mongol word Baghatur, meaning brave, warlike. Clavigo speaks of
"los valientes ebahadures," and a little lower down he says that the man who
drinks the most wine is styled Bahadur. The word is written bagator in
Jehosaphat Barbaro's Travels, and later, in those of Hanway, a derivative of
the word is behaduri, meaning an act of courage. The Akbar Nameh mention
a corps of behadurs. In later times the name was applied to a class of functionaries.¶¶ The Russians have adopted the word in the form bogatur. 4. *Shahnah.*
This word, according to Quatremere, meant at different times a governor, one
who had charge of the police in a town, a head man. In the Kamus we read
the word Shahnah, in speaking of a town, indicates the person who governs it
on behalf of the Sultan. According to the supplement to the Borhani Kati,
p. 1060, the term among the Persians was chiefly used to connote the person
charged with superintending the night patrols, and otherwise called kutual.
Ibn Khalidun, speaking of the Mongols, says: "They established one of their
amirs in each town, who, with a body of troops, was charged with the protection
of the country, and they gave this officer the name Shahnah;" and in another
passage he says: "A Shahnah who represented the supreme chief of the

---

* See Guiragos, ed. Brosset, 179. Note 12.
† Masalak Alabsir, quoted by Quatremere, 76. Note. ‡ *Id.*
§ Pallas, Saml. Hist. Nach, i. 186. Rytschkof Orenburgishe Topographie, i. 32. Bergm:
Nomadische Streifereien, ii. 30.
‖ Pallas, loc. cit. Bergmann, *id.* ¶ See Hyacinth's Mongolia, by Borg, 320.
** *Id.*, 321. Note. †† Pallas, Saml. Hist. Nach, i. 186. Note.
‡‡ De Mailla, i. 179-181. Notes. §§ Erdmann, Temudschin, 196. ‖‖ Op. cit., 67.
¶¶ Quatremere, 307-308. Note.

Mongols, resided constantly at Baghdad. When Gazan Khan mounted the throne, he drove the Shahnahs out, and caused his own name alone to be graven on the coin." In the history of Bedr ud din Aintabi we read that Timurlenk established a Shahnah at Damascus to command there in his name. Firishta applies the name to the head of the council and also to one intrusted with superintending the police of the markets, and speaks of a Shahnah of the bazaar.* Von Hammer seems to imply that the word, which he writes Shohné, is Arabic and not Mongol, and he makes it the equivalent of the Mongol baskak, which means a governor, and occurs frequently in this sense, as "Arghaman, the Great Baskak of Volodomir," &c.† Darugha was a similar title, which originally meant one in charge of the police. The Byzantines knew the title in the form Daregas; the Kalmuks use it in the form Darghui. The Tunguses also employ it in the sense of head of a tribe.‡ 5. *Bitkichi.* The word bitkichi, or bitikchi as it is sometimes written, is a Mongol word meaning a scribe or secretary. In the "Jihan-kushai" we read, "the bitkichis, the maliks, and the other amirs," and elsewhere, "among the Mussulman bitkichis was the Amir Imad ul Mulk." There is also mention made in the same work of the chief secretary, Ulugh bitkichi. Rashid, who has frequent references to the title, says : " Formerly the key of the Grand Seal was in the hands of the bitkichi." In the " Zafer Nameh " we read of "the bitkichis of the Chancellary." The word is still in use among the Mongols, for we read in Hyacinth's work, "Denkwurdigkeiten ueber die Mongolie," 306, "bitschet-schi," that is to say, scribe ; and in " Timkofski's Journal," " bitkhechi," that is to say, secretary of the seventh class.§

Note 3.—The Eastern Christians were at this time broken into so many sects that it is not easy to exactly realise or follow the details of their organisation or to understand the nature of the dignities filled by their higher clergy. A few words on this subject may not be uninteresting. There were three great schisms in the Eastern Church, arising out of different ways of viewing the Incarnation. First and most important was that of the Nestorians, dating from the fifth century, who held that Christ had two distinct natures, one divine and one human, and that the Incarnation was not a natural and complete union of the Divine Word with human nature, but a simple residence of the Word in a man, as in a temple. Almost contemporaneous in origin with the Nestorians were the Monophysites, also called Jacobites, who held that there were not two natures mingled together in Christ, but one nature only. The Monophysites prevailed chiefly in Egypt and Syria, while the Nestorians had their chief seats further east. Lastly, the Armenian Church separated itself at the Council of Chalcedon, on the question of the single nature of Christ, and other matters.

The orthodox were styled simply Greeks or Melkites (*i.e.*, Imperialists), since they recognised the civil supremacy of the Emperor of Byzantium, and the ecclesiastical superintendence of the Patriarch of Antioch, who was once the

---

* Makrizi, by Quatremere, ii. 195-197. Note.
† Von Hammer, Golden Horde, 238. Note 10. ‡ *Id.*, 238-239. Notes. Georgi Besch, &c., 307.
§ Quatremere, Rashid ud din, 113-115.

supreme head of all the Asiatic Christians. The Jacobites had a patriarch of their own, with his seat either at Amid, or in the monastery of Barsuma, near Malatiya. They also had a Maphrian or Primate, whose dignity was intermediate between that of the Patriarch and the Archbishop, who lived at Takrit, and ruled over the more eastern dioceses. In Syria, Asia Minor, and the countries watered by the Euphrates and the Tigris, there were as many as 121 Jacobite bishoprics.

The Patriarch of the Armenians had his seat at Kalaat ur Rum, on the Euphrates, and ruled over sixty-four dioceses. The Nestorian patriarchs, in the time of the Seleucidan Empire, had their seat at Koché, near Seleucia, a town separated by the Tigris from Ctesiphon. When the Abbasidan Khalifs made Baghdad their capital the Nestorian patriarchs moved their seat thither Before the schism these patriarchs had been suffragans of the patriarchs of Antioch, with the title of Archbishops of Seleucia. After the separation from the orthodox church, about the year 498, it became the practice for them to be elected by a synod, composed of a number of metropolitans and the bishops of the dioceses nearest to Baghdad, and after having received the confirmation of the Khalif, they were duly consecrated according to ancient custom in the church of Koché. The Nestorian patriarchs had secured the exclusion from Baghdad of the suffragan of the orthodox Patriarch of Antioch, and of the Maphrian of the Jacobites. The Jacobites had only a bishop there, and the orthodox Greeks at Baghdad were only occasionally visited by one of their own bishops. The Nestorian Church in Asia comprised twenty-five provinces and seventy dioceses, and included Irak Arab, Mesopotamia, Diarbekr, Azerbaijan, Syria, Persia, India, Transoxiana, Turkestan, China, and Tangut.* The Nestorian patriarch was not only the spiritual head of his co-religionists, but had civil authority over them also, and also held from the Khalif the right to judge disputes among the Jacobites and Melkites; and even the clergy of the two latter rites were subjected to him by two diplomas, written in Arabic, of which the terms have come down to us. They ran as follows: " The ruler of the faithful has thought proper to constitute you Catholicos of the Nestorian Christians living in the City of Salvation (i.e., Baghdad) and of all other countries, and to set you over the Jacobites and Greeks living in the country of the Mussulmans, or who happen to be travelling there. Cause your orders to be respected by all Christians."† Many Christians during the Khalifate followed the profession of doctors, and although it was contrary to law, were also employed as scribes in the administration. These officials acquired considerable influence over their co-religionists, and thus influenced the elections of the patriarchs. During the first two lines of Khalifs, Christians even secured sometimes the government of provinces. They prospered greatly for some time in Egypt and Syria, but presently aroused the envy and cupidity of the Mussulmans, and as they became richer they became also the more certain victims of the orthodox officials and of outbreaks of fanaticism. The Christians therefore looked upon the Mongols as saviours and friends, and for a while the latter, no doubt finding them useful allies against a common foe,

---

D'Ohsson, iii. 278-281.   † Id., 281-282.

treated them with consideration, and until they themselves became Muhammedans the condition of the Christians was no doubt greatly improved.*

Note 4.—The unfamiliar titles and dignities in use among the Georgians, several of which occur in these pages, make it useful to devote a few words to them. Immediately below the King were the thawads or princes, a word derived from thawi, head, and equivalent to the Latin princeps (*i.e.*, primum caput). They were of three classes : the first class, styled didebuls (*i.e.*, " the great " or the glorious), comprised in Karthli the heads of the six aristocratic families (*i.e.*, the Eristhavs of Aragvi and of Ksan, the heads of the Amilakhors, Orpelians, and Tzitzishivili, and the Maliks of Somkheth), but in its widest sense the term included all the princes having large domains, or an influential position either in Karthli or the other Georgian districts. 2. The mthawars, an administrative rank, including the Dadian, the Guriel, the Atabeg, the Sharwashidzé of Abkhazia, and the principal grandee of Suaneth. 3. The eristhavs (*i.e.*, literally, chiefs of the people), also rather administrative titles than titles of nobility, included, besides those already named, those of Radsha, of Bar in Imereth, and certain great vassals of the Prince Guriel.

Next to the thawads or princes were the aznaurs or nobles. These were sometimes in the feudal service of the thawads, and could pass into that rank. A class somewhat apart was that of the mokalakós or *bourgeois*, who were found chiefly at Tiflis, and were almost entirely of Armenian origin, that town being essentially an Armenian town. The traders of Gori, another Armenian town, formed a similar body. Nearly all the larger merchants in Georgia were Armenians, except in Western Georgia, where there were many Jewish traders. Few Georgians followed the vocation of trade in the towns. Next to the nobles were the msakhurs, a class standing above the serfs or slaves, and employed in the service of the ruler or the State either as soldiers or with civil functions. The chief of the msakhurs was the third official in rank in the State. "It would seem," says Brosset, "that, as in Russia, the fact of a man's being in the public service gave him a special status."

The title " son of an aznaur " had the same significance in Georgia that that of " son of a boyard " had in Russia. The head of a family represented its nobility; the other members were known as sons of aznaurs; so it was also with the thawads.

Lastly, were the mchas or slaves, and qmas or serfs, who were attached to the soil, and worked it, performing in return certain services, paying certain imposts, &c. They were royal serfs (*i.e.*, belonging to the Treasury) or private. Some could have property of their own, others not.†

---

* *Id.*, 283-285. † Hist. de la Géorgie, Introduction, lxxix., lxxxi.

CHAPTER III.

B EFORE describing the subsequent doings of Khulagu and of his
lieutenant in Syria, Kitubuka, it will be well to bring the story
of Mongol aggression in other directions up to this point. And
first in regard to the Arabian Irak, and the old frontier fortresses of the
Byzantines and Persians.

Mayafarkin was a famous old town situated to the north-east of
Diarbekr, on the site of the Karcatiocerta (the capital of Sophiene) of the
ancients. It was situated on the Nymphius, now called the Golden
River (the boundary between the Roman and Persian Empires),
and north of it flowed the stream Bekr. Malakia and Guiragos call it
Mufarghin; Stephen the Orpelian, Nephergherd; while it was called
Maifkerkat by the Syrians. The Greeks called it Martyropolis, after its
bishop, Marutha, who had assembled there the relics of many martyrs he
had found in Persia, Armenia, and Syria, and had fortified it. It filled a
very important *rôle* in the wars of the Romans and Sassanians, and in
early Muhammedan times, and was the capital of the province of
Diarbekr. Quatremere has a very valuable note on the town.* At this
time it was ruled by the Ayubit prince, Malik el Kamil Nasir ud din
Muhammed, the son of Muzaffar Shihab ud din Gazi, who was the son
of El Adil Abubekr, son of Ayub, who gave his name to the dynasty.
Kamil had succeeded his father in 642 HEJ. (*i.e.*, 1244). When the
Mongols under Charmaghan and Baichu were harassing the west, we are
told that Kamil determined to visit Mangu Khan, by whom he was
presented with an honorary dress. This was because he had refused to
drink kumiz at a feast, urging that it was forbidden by his religion, and
he therefore would not act contrary to his faith. When Khulagu set out
westward Kamil accompanied him to Irak. When he advanced to
Baghdad he ordered him to furnish 7,000 horse and 20,000 foot for the
campaign. Kamil pleaded that he could not furnish more than about
2,000 horse and 5,000 or 6,000 foot. Khulagu was annoyed at this, and
told the Vizier he must be put to death. He was then apparently at the
Mongol Court, and the Vizier, who was friendly to him, informed him of

---

* Op. cit., 360-365. Note. See also Ilkhans, i. 186-187.

Khulagu's resentment. He therefore made an excuse that he wanted to go on a hunting excursion. He set out with eighty followers, and hastened away to his own country, which he reached in seven days. There he gave orders to put to death the various shahnahs who had been placed in the towns he ruled over during the control of western affairs by Charmaghan and Baichu. These shahnahs were executed, we are told, by having five spikes driven into them—one into their foreheads, and four others into their feet and hands. Abulfaraj says that Kamil also put to death a Syrian priest from Badlis, who had been sent to him by the Khakan with a yarligh or diploma. When Khulagu heard of his flight he sent in pursuit, but in vain. Kamil now asked Nasir, the Prince of Damascus, to make common cause with him against the Mongols, to aid the Khalif, and to prevent the invaders entering Syria, but he was put off with empty promises. He set out for Baghdad himself, but heard *en route* of its fall. He accordingly returned home, put his strong places in order, and warned the nomads who lived in his land to seek shelter in places of strength, and himself took up his quarters at Mayafarkin.*

After the capture of Baghdad, Khulagu ordered his son Yashmut to march upon Mayafarkin. He was accompanied by the Amirs Ilka Noyan and Suntai. Stephen, the Orpelian, says they were also joined by the Armenian princes, Avak, Shahan Shah, and Elikum, the last of whom fell ill there, and was eventually poisoned, it was said, by a doctor at the instigation of Avak.† The Malik Kamil having been in due course summoned, sententiously replied that it was useless for the Mongol chief to hammer cold iron, that he had no faith in his promises, and was not to be taken in by his smooth phrases. He did not fear the Mongol army, and should fight, sword in hand, as long as he lived. He charged Yashmut with being the son of one who had broken his promise to the Khurshah, the Khalif, and others, and said he did not mean to court their fate by imitating their ingenuousness. Having dispatched these brave words, he busied himself with putting the place in defence. Addressing the citizens, he said : "You shall have the gold and silver in the treasury and the corn in the granary. It shall all be distributed among those who have need. Thanks be to God, I am not like Mostassim, a worshipper of gold and silver." His spirit animated that of his people. They made several successful sorties, in which two famous champions, named Saif ud din Arkali, or Lukhili, and Kamr i Habash or Anbare Habashi, killed many Mongols and greatly distinguished themselves. *Inter alios*, they killed a Georgian hero, called Aznawari, who fought on the side of the Mongols. A famous and large piece of siege artillery inside the city was also admirably worked, and the Mongol amirs began to despair. At length Bedr ud din Lulu, Prince of Mosul, ordered a famous engineer in the besiegers' army to build a great

* Tabakat-i-Nasiri, 1262-1268. Abulfaraj, Chron. Syr., 553 ; Chron. Arab., 345.
† Hist. de la Siounie, 228.

catapult to oppose to that inside, and we are told that both machines being discharged at the same moment the stones cast by either met in mid-air, and were broken to pieces.	All were astonished at the skill of the engineers, but presently the machine outside was set on fire with naphtha. As the siege was very protracted, Khulagu sent Arkatu with a division of troops to help his people, and ordered that the attack should not be relaxed till the money and food began to run short.	Arkatu or Oroktu* went with this view, but when he reached the camp the two champions from the garrison, who had already performed prodigies of valour, again made a sortie, and created some confusion.	Arkatu, who happened to be drunk, ordered the attack to be pressed, but the sortie was more or less successful: the two heroes killed a number of the enemy, and, we are told, the Mongol general, Ilka Noyan, was himself unhorsed.	Every day the two warriors, who were doubtless protected by heavy armour, made their attack.	Thus a year passed by.	Food began at length to run short, and live provisions came to an end; the people began to live on carrion; dogs, cats, and rats, and even human flesh were put under requisition. Guiragos says a pound of human flesh sold for 78 dahekans.	Like fish, says Rashid ud din, they devoured one another.	The two champions having no barley or straw left, killed their horses and ate them.	Malakia says an ass's head was sold for thirty pieces of silver.	Vartan says they first ate various animals, pure and impure, then the poor, then their own children, and lastly each other.	The dean and chief of the clergy, in his terrible hunger and rage, ate some of the flesh of his relatives.	"He wrote the confession," says our author, "on a slip of paper, hoping that it would fall under my eyes, and that he would obtain pardon from the merciful Being who created us.	Then giving himself up to lamentation and grief, he took it so to heart that he died.	We have seen, as he hoped we would, his written confession, and we have confidence that he will obtain grace from Him who is goodness itself.	All of you into whose hands this book may come, implore God with all your hearts, saying Amen for him and for the vartabed Thomas, the transcriber."†	To revert to the siege, treachery presently began to appear.	A letter was written to the Mongols describing the condition of things, and urging an immediate attack.	We are told by Rashid that, having entered, they found the place almost full of corpses piled on one another; but seventy remained alive.	Malik Kamil and his brother were both captured, and the place was given up to pillage. The two champions we have named meanwhile climbed to the roof of a house and shot with their arrows those who came near.	Arkatu ordered a number of brave men to dislodge them.	They now came down, and holding their bucklers in front of them, fought desperately until both were killed.

---

* He is called Jagatai by Guiragos, who says he was accompanied by Prince Hasan, surnamed Prho h. Op. cit., ed. Brosset, 188.
† Journ. Asiat., 5th ser., xvi. 292-293.

Kamil himself was taken to Telbashir, the Turbeysel of the Crusaders, not far from Aleppo, and on the other side of the Euphrates, and brought before Khulagu, who had reached that town on his return from Syria, and who reproached him bitterly, and recalled the various favours he had received at the hands of the Khakan. He was then cut into small pieces, which they thrust into his mouth till he died. Rashid says he was a pious and austere man, who, although a prince, followed for humility the craft of a tailor.* The head of the hero, for he surely deserved the name, was paraded round the Syrian towns of Aleppo, Hamath, and Damascus, with music and singing, and hung out at last from one of the windows of the gate known as the Gate of Paradise at Damascus. On the victorious entry of Kuttuz into that city (*vide infra*), it was buried in the tomb of the martyr Hussein. His memory was celebrated in some stirring verses by the poet Sheikh Shihab ud din Ibn Abu Shamah, in which he apostrophised the two heads which met together thus gloriously in one tomb.† Of the mameluks who were captured with Kamil, seven were put to death, while the eighth, who had been Master of the Hunt, was taken into his service by Khulagu.‡ Vartan has a curiously worded paragraph about the siege, showing what curious alliances the religious animosities of the time brought about. He says "there perished there a fine young man, Sevata of Khachen, son of the Grand Prince Gregory. After prodigies of valour, he won an immortal crown, always faithful to God and the Ilkhan. He will be associated in the triumph with those who have shed their blood for Christ, and who have preserved their faith and the fear of our Saviour. Amen."§ On the sack of the town, the churches as usual were spared, as well as the innumerable relics which the holy Marutha had collected there. We are told that the Christians who fought in alliance with the Tartars had let them know the veneration in which these relics were held, and recounted to them the numerous apparitions of the saints which had shown themselves on the ramparts with luminous bodies.

Minhaj i Saraj, who was then writing in India, reports similar stories. He says that while the siege progressed several horsemen, clothed in white and wearing turbans, used to sally out and attack the enemy. "They used to dispatch about a hundred or two hundred infidels to hell, while no arrow, sword, or lance of the infidels used to injure them, until about 10,000 Mongols had been destroyed by them." Khulagu sent Ilka Noyan to complain to his son that while he himself had captured Baghdad in less than a week, he had not been able to capture the small fortress after such a prolonged attack. Yashmut replied that his father had taken Baghdad by treachery, while he had to fight hard, and that it

---

* Quatremere, 361-375. Tabakat-i-Nasiri, 1270-1274.
† Abulfeda, iv. 589-591. Ilkhans, i. 188-189. ‡ D'Ohsson, iii. 357.
§ Journ. Asiat., 5th ser., xvi. 294.

was not fair to compare the place with Baghdad. Khulagu, we are told, was much displeased at the answer, and sent him word to keep out of his sight or he would kill him, and swore he would take Mayafarkin in three days. He hastened there. The attacks of the supernatural champions continued, and many Mongols continued to be killed, whereupon Khulagu remarked that the fortress belonged evidently to the Tengri (*i.e.*, to heaven, or to the gods), and he would spare it, but he asked the citizens to tell him who the white-robed champions were. They swore they knew not any more than himself. He thereupon offered a propitiatory offering of 1,000 horses, 1,000 camels, 1,000 cattle, and 1,000 sheep, but the citizens would not accept his offer. He thereupon raised the siege, leaving behind the offering he had made, and went towards a verdant plain called the Sahra i Mush (*i.e.*, the rats' plain, which is marked on the maps, according to Raverty, on the eastern branch of the Euphrates, fifty miles west of Lake Van). There was soft mud and stagnant water, and he sank in it.* This story doubtless grew out of the exploits of the champions already named, and out of the prolonged resistance offered by the place, and is interesting only as showing the kind of tales that found their way to India about the Mongol doings. The same author reports that the son of Bedr ud din Lulu, of Mosul, who was with Yashmut, saw in a dream, several times, Muhammed appear on the ramparts of Maya-farkin and draw the hem of his garment about the fortress, declaring it was under the protection of God and of himself, and that eventually he became so frightened that he left the Mongol camp and went to his father, who reproached him for the danger and trouble he would bring upon his kingdom. "I cannot war with Muhammed, the Apostle of God, the Almighty bless him and guard him," was the reply. And he wrote out an account of what he had seen, and departed.†

Mayafarkin fell in the early spring of 1260. Malakia tells us how some of the Christians were enriched by the capture. Thus, we are told, the Armenian Grand Prince Thaghiatin, who was one of the Bagratid princes of Loré, secured the cross of St. Bartholomew, which a Syrian prince was carrying off, and which he was afterwards obliged to cede to the Grand Prince Sadun the Ardzrunian, who deposited it in the monastery of Haghpat, which belonged to him, and with it the right arm of St. Bartholomew.‡ The Mongols nominated one of Kamil's amirs, named Abdulla, to govern the city.§ At the same time as the capture of Maya-farkin the country of Sanasun, or Sasun, a mountainous district in the province of Aghetznik, north of Armenian Mesopotamia, south-east of Mayafarkin, also submitted. This was through the influence of Prince Sadun, son of Sherparok and grandson of Sadun, a strong and brave

* Op. cit., 1271-1275.   † Tabakat-i-Nasiri, 1279-1280.   ‡ Op. cit., 454.
§ Abulfaraj, Chron. Arab., 349.

warrior, who was a Christian, and had gained the goodwill of Khulagu by his skill as a wrestler. The district of Sasun was made over to him.*

About the same time that Kamil, Prince of Mayafarkin, came to his end, there also perished his cousin, Mowahid, son of Turan Shah, the last Ayubit Sultan of Egypt, who ruled over the strong fort of Hosnkeif (the castle of Kiphas of the Byzantines), which derived its name from the Syrian word kifo, a rock.† According to the Arab legends the name was derived from a certain brave man named Hasan, who, confined there as a prisoner, asked the Amir's permission one day to exercise one of his mares in the castle garth. He galloped round and round, and eventually leaped desperately over the wall into the Tigris below, over which he safely swam. The name of the place, it was said, was derived from the exclamations of the bystanders, "Hasan Keifa." (Go on, Hasan.)‡ The geographer Seif ud daulah ibn Hamdan speaks of it as very strong, and protected by defiles on all sides save the east. It was situated on the western bank of the Tigris, and opposite to it was a high bridge of stately architecture, re-erected in 510 HEJ. It contained beautiful streets, shops and houses, markets and baths, all built of stone and with lime ; but its climate, especially in summer, was unhealthy. The author of the "Nozhat alkolub" tells us "that the town, although largely ruined, had a customs receipt of 82,000 pieces of gold." The Portuguese traveller Teixeiro calls it Arcengifa ; Josaphat Barbaro, Hassan Chifh. He mentions the caverns cut in the mountain close by, which are also named by the modern traveller, Kinneird. A merchant, who travelled in Persia in the sixteenth century, and whose voyage was published by Ramusio, calls it Asanchif. He places it four days' journey from Mardin, and mentions a magnificent bridge of five arches over the Tigris, no doubt the one above-named.§ Before it was fortified by the Ayubic Amir Merd Mahmar, it was called Rasol Ghul (i.e., the Demon's Head) by the Arabs, probably from the position of its citadel. Hosnkeif was now captured by the Mongols, and its prince killed.||

When Khulagu recrossed the Euphrates on his return from Syria, he sent to command the ruler of Mardin to go to him in person. This he refused. His son Mozaffer, who was at Khulagu's Court, was then specially commissioned to urge his father to go, and to point out to him the danger of obstinacy. The old man, instead of listening to him, put him in prison.¶ The town was built in terrace above terrace on the mountain called Izale, Judi, or Masius, the last name derived from its oak woods (from the Persian, masu). In Moslem tradition Noah's ark rested on its highest summit, and thence Noah and his sons went forth to repeople Mesopotamia. There Sunni and Shia,

---

* Guiragos, ed. Brosset, 188.   Journ. Asiat., 5th ser., xi. 495-496.
† Quatremere, 333.   Note.   ‡ Ilkhans, i. 189-190.   § Quatremere, 333-334.   Note.
|| Ilkhans, 189-190.   On p. 193, Von Hammer says that the Ayubits continued to rule there.
¶ Abulfaraj, Chron. Arab., 349.   Chron. Syr., 557.

L

Catholic and Schismatic, Armenians, Jacobites, Nestorians, Chaldæans, and Jews, sun, fire, calf, and devil worshippers still live over each others' heads. The most numerous are the devil-worshipping Kurds, called Yesidi, perhaps descended, says Von Hammer, from the Mardi, who gave their name to the town.* Arkatu summoned the place, and bade Malik Said remember that if his head reached the sky it would, when trodden under by the Mongol army, be as dust, and that if he refused to listen the Eternal God knew what would happen. The old chief replied that it had been his intention to submit, but the fate of several of his friends who had done so deterred him, and that, thank God, the town was well provided with arms and provisions, and defended by a good garrison of Turks and Kurds.† Arkatu therefore planted his siege apparatus, and commenced a bombardment. For eight months the place held out bravely, while the Mongols plundered the neighbouring towns of Duniasar and Arzan. At length pestilence, preceded by famine, began to devastate the place, and Malik Said himself fell ill, and as he still refused to submit, his son Mozaffer gave him a bowl of poison and he died. Mozaffer now surrendered. Khulagu reproached him for the base crime, which he defended on the ground that his father's obstinacy was causing a terrible calamity to the town and its inhabitants. Khulagu pardoned him and gave him command of Mardin, where he ruled till the year 695 H.J., and was succeeded by his son and grandson, the latter of whom was a favourite of Gazan Khan, who also made him governor of Diarbekr and Diar rabiah.‡ Thus the family of the Ortokids subsisted here as vassals of the Mongols, as the Ayubits did at Hosnkeif and Hims.§ Wassaf calls the Mongol commander who attacked the town Shamaghar. Soyuk Kotoghtai and Tenghur were the names of the envoys he sent to summon the place. He says Mozaffer had been imprisoned by his father, and adds that Said was submissive, but notwithstanding was put to death with his seven viziers, and Mozaffer having been put on the throne in his place, the three envoys above-named were appointed baskaks of the town.||

Malik Ashraf Musa, the former Prince of Hims, or Emessa, had been deprived of that fief by Nasir, the Prince of Syria, and been given Tel bashir in exchange. On the fall of Aleppo he retired towards Egypt, but changing his mind determined to submit to Khulagu, who reinstated him at Hims, and presently appointed him Viceroy of Syria. Shortly after he received orders from his new master to dismantle the walls of Hims and of Hamath. He accordingly went to the latter town (whose prince, Malik Mansur, son of Mozaffer, had fled to Egypt with his family), destroyed the walls of the citadel, burnt the arsenal, and sold the library. He would also have destroyed the city walls but for the warning of Ibrahim, styled

* Ilkhans, i. 190.   † Quatremere, 377.   ‡ Id., 377-379.
§ Ilkhans, 192-193. || Op. cit., 91-92.

Ibn el Afrangia from the fact of his mother having been a Frank, who was in the Mongol service as a tax collector, and who reminded Khosru Shah, the Mongol prefect at Hamath, that the presence of the Crusaders at Hesn el Akrad (? Acre) made it imprudent to do so.[*] The activity he showed at Hamath covered an excusable lack of that quality at his own city of Hims, where he only very nominally carried out his master's orders. A few days after Ashraf was nominated Viceroy of Syria an outbreak took place at Damascus, in which the governor of the citadel was the leader. Kitubuka proceeded at once to attack it, amidst a terrible hurricane of rain and hail. The place held out obstinately for forty-five days, but was at length battered by twenty war engines, and the garrison sued for capitulation. The place was pillaged, and many of the towers, with the war machinery and arms of the garrison, were destroyed. The Mongols then marched upon Baalbek. Kitubuka's camp was in the beautiful valley of Ghuta, deemed by Orientalists one of the four paradises of the East.[†] D'Ohsson calls the place Merj Bargut.[‡] There he received deputies from the Franks (i.e., the Crusaders), who were accompanied by Dahir, the brother of Prince Nasir, who was confirmed in the possession of Sarkhad.

Nasir himself continued his retreat towards Egypt. He halted a few days at Nablus, the ancient Neapolis, and having left a garrison there under Mojir ud din ibn Abu Zakr and the Amir Ali ibn Shogga ud din, went on to Gaza, where he was joined by his brother Dahir, and by the Mamluks who had recently deserted him. A few days after he left Nablus the Mongols arrived there under Kushluk Khan, and the garrison having made a sortie were put to death with their two commanders.[§] They continued their advance as far as Gaza, Beit-Jebrail, Khalil (Hebron), the lake of Zira, and the town of Salt, killing or making captive the people, and carrying off a great booty. They then returned to Damascus.[||] Malakia says Jerusalem also fell into their hands, but he somewhat mars the credibility of this statement by telling us that Khulagu went there in person, and having entered the Church of the Resurrection, prostrated himself before the tomb.[¶] Nasir meanwhile arrived at Katia. The Egyptian Sultan, Kuttuz, was not well pleased that an Ayubit prince with such prestige as Nasir should come so near Egypt. He had, it appears, some reason for suspecting that he had some designs on that goal of many fugitives,[**] and when he reached Katia he went with his troops and encamped at Salahiyet, where Nasir was deserted by his Kurdish and Turkoman followers. Some of them ranged themselves under the banners of Kuttuz. Others went to Belbeis, and there only remained with him his brother Dahir, the Malik Salih Nur ud

---

[*] Abulfeda, iv. 587.   [†] Ilkhans, i. 196.   [‡] Op. cit., iii. 329.
[§] Abulfeda, iv. 581. D'Ohsson, iii. 329-330.   [||] Makrizi, i. 91
[¶] Op. cit., 458.   [**] Weil, iv. 13.

din Ismael, son of the ruler of Hims, and three amirs of the tribe Kaimeris. The deserters had been seduced by offers of rewards, &c., but they were not well treated. The Amir Jemal ud din Musa ibn Yagmur was imprisoned, while Nasir's pages and secretaries were plundered.* Nasir himself, not daring to advance further towards Egypt, crossed the desert to Shubek, being robbed of his baggage *en route*. Thence he went towards Karak, whose prince, like himself an Ayubit, sent him horses, tents, and clothes, and offered him an asylum either at Shubek or Karak. He did not accept this, but went on to Balka. His place of retreat was disclosed to Kitubuka by two Kurds in his service.† He was captured on Lake Ziza, and taken before Kitubuka, who was then engaged in besieging Ajalon. He bade him order the governor of the fortress to surrender, which it did after some resistance, and its walls, which had been built by Iz ud din, an amir of Saladin's, were razed.‡ The Mongols had a short time before secured the possession of Baalbek, which they ruined, as well as its citadel. Malik es Said, son of Azis, son of Aadil, son of Ayub, the ruler of Subaiba, or Sabib, and Banias, who had been incarcerated at Biret for nine years and been released by the Mongols, was invested with that district. He supplied his patrons with suggestions for punishing the Mussulmans.§ Nasir, with his brother Dahir, and the Malik Salih above named, were sent on to Khulagu at Tebriz. The Sultan of Karak also sent his infant son Azis with them. They passed through Damascus, Hamath, and Aleppo. When he saw the ruins of the last of these towns, Nasir wept. Khulagu treated him well, and promised to restore Syria to him when he had conquered Egypt.‖

The Mongols were disposed to be friendly towards the Crusaders of Sidon and Beaufort, who were the enemies of their enemies, the Mussulmans; but the Christians brought vengeance upon themselves by plundering some of them, and then killing a nephew of Kitubuka, whom he had sent to get the plunder restored. He revenged himself by harrying Sidon, and destroying a portion of its walls. This *contretemps* impaired the confidence which previously existed between the Mongols and Christians, and which was due to Dokuz Khatun's influence, and to the friendship which existed between Khulagu and Haithon, the King of Little Armenia. The latter, however, obtained for his son-in-law, the Prince of Antioch, the restitution of all the places of which the Moslems had deprived him.¶ The successes of the Mongols in Syria were not altogether reassuring to the Christians. We are told how the people of Acre cut down all the gardens about their town, while urgent letters were written to the Sovereigns of Western Europe to come to the rescue. A rumour spread that Antioch and Tripolis had been taken by the Tartars, and an envoy

---

* Makrizi, i. 100.　† Makrizi says he was betrayed by the halbardier, Hosain Kurdi.
‡ Abulfeda, iv. 591.　D'Ohsson, iii. 330-331.　§ Abulfeda, *id.* Makrizi, i. 88.
Abulfeda, 591-593.　D'Ohsson, iii. 330-332.　Ilkhans, i. 196-197.　Quatremere's Rashid 341.
¶ Ilkhans, i. 197-198.

reached England, where a council was held, and prayers and fasting enjoined. St. Louis held a similar council at Paris, where a like discipline was enjoined, and orders given that no games were to be played except archery and shooting with the cross-bow. The next year (1261) the Pope tried to arouse the Christians to make some opposition to the Tartars, both in Persia and Hungary.* Egypt, which had been a refuge and retreat for the various victims of the Mongols, now found itself threatened, and the greater part of the Africans (we are told) who lived there withdrew. Khulagu, according to Rashid ud din, as he was leaving Syria sent an envoy named Ilchi Mogul with forty subordinates to summon the Egyptian ruler to submit.†

Makrizi has preserved a copy of this minatory message, which is couched in the usually arrogant language of the Mongols. It was addressed from the King of Kings of the East and West, the Supreme Khan, to Malik Mozaffer Kuttuz, of the race of the Mamluks, who had fled to escape their sword. It bade him and his people remember that the Mongols were the soldiers of God on earth, who had created them in his anger and delivered into their hands all the objects of his wrath. It bade him take warning from what had occurred in other countries, and not to oppose them, but to submit before the veil was torn, for they were insensible to tears or entreaties. "You have heard," says the letter, "how we have conquered a vast empire, how we have purified the earth of the disorders which tainted it, and have slaughtered the greater part of its inhabitants. It is for you to fly and for us to pursue, and whither will you fly, and by what road shall you escape us? Our horses are very swift, our arrows sharp, our swords like thunderbolts, our hearts are hard as the mountains, our soldiers numerous as the sand. Fortresses will not detain us, nor arms stop us. Your prayers to heaven against us will not avail. You enrich yourself by vile means and break the most solemn promises. Revolt and disobedience are in your midst. And you are about to receive a terrible punishment for your pride. Those who have been unjust are going to learn their fate. Those who dare to make war upon us are about to repent. Those who seek our protection will alone be safe. If you will submit to our orders and the conditions we impose you shall share our fortune. If you resist you will perish. Do not commit suicide. He who has been warned ought to be on his guard. You are persuaded we are infidels while we look upon you as criminals, and God, whose orders are irrevocable and whose decrees are perfectly just, has caused us to triumph over you. Your strongest forces are in our eyes mere small bands of men, and your most distinguished people we contemn. Your kings we despise. Do not delay long. Hasten to reply to us before war lights its fires and throws their sparks upon you, or you will find no refuge from the terrible

---

* Remusat, Mems. Ins., vi. 467. † Quatremere, 341.

catastrophe that will overwhelm you, and you will make a desert of your country. We mean well by our warning. It is to arouse you from your slumber. At present you are the only enemy against whom we have to march. May safety be with us and you and all those who follow the divine commands, who fear the issue of death and submit to the orders of the Supreme King. Say to Egypt: 'Holaun is about to come, escorted by naked swords and sharp blades. He is going to humiliate the great ones of this land, and will send the children to join the old.'"* According to Novairi it was Kitubuka who sent the message.†

When the envoy arrived with this insolent letter Kuttuz summoned a council of his officers. He told them how Khulagu had been everywhere successful and was already master of Damascus, and asked them to consider whether they should resist or obey him. Thereupon Nasir ud din Kaimeri, one of the six Khuarezmian leaders who had abandoned Nasir, spoke out, and said that in the presence of such a power it would be no disgrace to give in, but they must remember how faithless Khulagu was, and he recounted the names of the various princes who had trusted him and been undone. Kuttuz then replied that all the country from Baghdad to Rum was laid waste; that unless they took time by the forelock and attacked the Mongols, that Egypt would share the same fate. There were only three courses open to them—to submit, to fight, or to abandon their country, and the last was impracticable, for "the Maghreb," (i.e., North-Western Africa), their only resource, was too far off; while peace with those who never kept treaties was also undesirable. Some amirs urged that they had not resources with which to oppose the enemy, but asked him to do as he pleased. Kuttuz then summed up his resolve. "I am of opinion," he said, "that we should march together to the combat. If we win, we shall gain our end; if we lose, men cannot reproach us." The same night the Mongol envoy and three of his companions were executed; one in the horse-market, at the foot of the famous Castle of the Mountain; the second outside the gate of Zavila; the third beyond that of Nasr; and the fourth in the place called Ridania. Their heads were hung at the gate of Zavila, and, we are told grimly, were the first Tartar heads which were suspended in that place. Only one of the envoys, who was a young man, was spared, and was enlisted among the Mamluks. The next morning the Egyptian army set out.‡ To pay the expenses of the expedition, Kuttuz had recourse to sources of revenue forbidden by Muhammedanism. He levied an income and a capitation tax, but these only produced 600,000 dinars. He confiscated the property of all the adherents of Nasir, who had abandoned the latter to join him (assuredly a curious kind of gratitude). The wife of Nasir was obliged to produce her jewels, of which a portion were taken. The

* Makrizi, 101-102.   † D'Ohsson, iii. 333. Note.
Makrizi, 102-103. Ilkhans, i. 202-203. Quatremere, 345-347. D'Ohsson, iii. 335.

wives of other amirs had to make similar sacrifices, and some of them were badly treated, and even put to death. Kuttuz set out from his fortress, called the fortress of the mountain, on the 26th of July, 1260. His army of 120,000 men consisted (independently of the Egyptian troops) of the Syrians who had joined him, of Arabs and Turkomans, and of the *débris* of the Khuarezm Shah's troops who had sought shelter in Syria and Egypt. A general levy for the defence of Islam was made, and those who hid away were bastinadoed. A summons was sent to Ashraf, Prince of Hims, Khulagu's deputy in Syria, and to Said, Prince of Sabib, to ask them to aid him in the enterprise. Said illused the envoy, and received him with insulting phrases. The messenger then went on to Ashraf, who gave him a private audience, and then prostrated himself before him, offered him a seat, and told him to do obeisance in his name to his master, to tell him that he was at his service, that he thanked God that He had raised him to aid their common faith, and to go on and fight the Tartars, for the victory would be his.[*]

At Salahiyet Kuttuz held a council of war. Most of his generals were for halting there. "Oh, chiefs of the Mussulmans," he said, "you who have lived for so long out of the public purse, do you now shrink from a holy war? I mean to advance. Those can follow who please, while those who remain behind, God will not forget them ; on their heads rest the dishonour of the Mussulman women." He then took an oath from the generals he knew to be faithful, to follow him to the war, and the next day the cymbals sounded the advance, and none presumed to stay behind. The advance guard was commanded by the Mamluk Rokn ud din Bibars Bondukdari, a dependent of Nasir's.[†] Baidar, who commanded at Gaza for the Mongols, informed Kitubuka, who was at Baalbek, of the advance of the Egyptians. He was ordered to stand firm, but was beaten before Kitubuka could arrive, and pursued to the River Asi. Gaza was occupied by Bibars. The army rested there awhile, and Kuttuz received a deputation from the Knights of St. John, offering him a contingent of troops and also presents. He distributed robes of honour among them, and made them promise that the people of Akka, or Acre, would remain neutral.[‡] When he approached the enemy, Kuttuz roused the enthusiasm of his troops by appeals to their faith and patriotism, calling upon them to rescue Syria, and to deal a great blow for the faith. His officers shed tears, and promised to use every effort to drive out the hated Tartars. Bibars having gone on ahead with a body of troops, was the first to encounter the Mongols, and began a skirmish with them. This was at Ain-i-Jalut (*i.e.*, the Springs of Goliath), between Nablus and Baissan. Kitubuka and Baidar, on hearing of the march of the Egyptians, had brought together all the Mongol forces in Syria, and

---

* D'Ohsson, iii. 335-336.   † Makrizi, 103.   D'Ohsson, iii. 336-337.
‡ Makrizi, 103-104.   Quatremere's Rashid, 347.

had marched against them. The two armies were in presence of one another on the 3rd of September, 1260. Makrizi tells us the Egyptians went into the battle with little confidence ; that it was sunrise, and that the cries of the labourers in the villages were mingled with the martial sound of drums.* The Mongols poured in a shower of weapons, and one wing of the Egyptian army gave way.† Thereupon Kuttuz pulled off his helmet and threw it to the ground, shouting out, " O, Islam ! " and threw himself, with those about him, upon the enemy, who were in turn broken.‡ Rashid says that Kuttuz had planted a section of his men in ambush, and that when his main army was beaten, and was being hotly pursued and losing many men, those in ambush sprang out and restored the battle, which lasted till mid-day, when the Mongols broke and fled.§

Wassaf has an improbable story, in which he makes out that the Mongols were taken by surprise in their camp by the Egyptians, who displayed white standards, such as were used by themselves,‖ and dressed themselves in white overcoats called burkas, such as are still used and so called by the Circassians, and which, he says, were also used by them-selves.¶ The Egyptian historians and Haithon declare that Kitubuka was killed in the battle, and Makrizi adds that Malik Said, who fought in the Mongol ranks, also perished.** Rashid, who was naturally a flatterer of the Mongols, reports matters differently. He tells us that during the fight Kitubuka fought desperately. He refused to surrender. " Go and tell Khulagu I refused to retreat disgracefully, and sacrificed my life in consequence. As for the rest, the loss of a Mongol army ought not to distress the King. What does it signify? If the wives of his soldiers or the horses in his stables have young ones during one season, it will replace this loss. The monarch himself is safe, and this is a sufficient balance to all the rest. The life or death of us his slaves matters nothing." Although abandoned he fought on alone desperately. At length, his horse having fallen, he was captured. His hands were tied, and he was led before Kuttuz, who jeered at him, saying, " Perfidious man, after having shed so much innocent blood, after having undone a host of warriors by your vile double-dealing, and overturned so many ancient houses with your lies, you have at last fallen into a trap yourself." Kitubuka, like a true Mongol to whom fear of death was unknown, replied with dignity, " Do not be too much elated with your momentary victory. If I perish it is by the hand of God, and not by yours. As soon as the news of my death shall reach the ears of Khulagu Khan, his wrath will boil over like an angry sea. From Azerbaijan to the gates of Egypt the whole land shall be trodden under by the hoofs of Mongol horses, and our soldiers will carry off in the sacks of their horses the sands of Egypt. Khulagu Khan has among his followers 300,000 warriors equal to Kitubuka.

---

* Op. cit., 104.   † Quatremere, 349.   ‡ Makrizi, 104-105.   § Quatremere, 349.
‖ Op. cit., 89.   ¶ Id., 89-90.   ** Op. cit., 105.

My death will only make them one less." Kuttuz replied, " Do not boast of the valour of the horsemen of Turan, for they only succeed by treachery and chicane. None of them have the courage of Rustem, son of Destan." Kitubuka answered again, " From my birth I have been the slave of the King. I am not like you, a traitor and murderer of my master. Make haste and put an end to me that I may no longer hear your reproaches." Kuttuz then ordered him to be decapitated.* Abulfeda makes Kitubuka be killed in the fight, and his son be made prisoner.

Makrizi tells us how, during the battle, the young envoy of the Mongols who had been enlisted among the Mamluks, put an arrow to his bow and aimed it at the Sultan, but before it was shot he was cut down : others reported that the arrow in fact struck the horse of Kuttuz, and that he was dismounted.† The Mongols were sharply pursued, and many were killed and others captured ; one body of them took refuge in a thicket of reeds, which Kuttuz ordered to be fired, and they perished. The main body was pursued as far as Baisan, where they turned round, and a second fierce struggle followed, more animated than the previous one, during which the Sultan is reported to have cried out three times : "O, Islam, O God protect thy servant Kuttuz, and make him triumph over the Tartars." The Mongols were again defeated, whereupon Kuttuz dismounted, laid his head in the dust, and offered a prayer of thanksgiving, accompanied by two rikahs.‡ The Mongols were everywhere driven out of Syria, and as far as the Euphrates. The camp of Kitubuka was pillaged. His wife, children, and dependents were captured. The various deputies and governors, except those at Damascus, were put to death.§ Zain Hafidi and the other authorities at the latter town fled hastily when they heard the news, and thus escaped slaughter, but their goods were pillaged by the villagers. The Mongols had been at Damascus altogether seven months and ten days. Kitubuka's head was sent to Cairo.‖

In reference to this campaign, Malakia tells us that " Kitubuka had the presumption to advance ten days' journey beyond Jerusalem, but the doglike and impure Egyptians, knowing that the Tartar troops were not on their guard, marched against and massacred many of them, making some prisoners, and causing others to fly." They recaptured Jerusalem, Aleppo, and Damascus, and were aided by the Frankish knights, who, he says, had not as yet allied themselves with the Tartars.¶

The victory of the Egyptians was a turning point in the world's history. It was the first time for a long while that the Mongols had been fairly beaten, and although the defeat was probably largely due to the smallness of their numbers, Kitubuka having apparently only 10,000 men with him, it was none the less decisive. It stopped the tide of Mongol aggression and probably saved Egypt, and in saving Egypt

---

* Quatremere, 349-353.   † Op. cit., 105.   ‡ Makrizi, 106.
  § Quatremere, 353.   ‖ Makrizi, i. 106.   ¶ Op. cit., 458.

saved the last refuge where the arts and culture of the Mussulman world had taken shelter ; where, under the famous Mamluk dynasties, and under the new line of Khalifs, it blossomed over in wonderful luxuriance, and not only made Cairo the cynosure of eastern cities, but was eventually the means of distributing culture to the Golden Horde, and very largely also to the Empire of the Ilkhans itself.

The march of the Egyptian army had greatly elated the citizens of Damascus, and the Mongols, who had imprisoned the Naib and Vali of the town, apparently for encouraging this feeling, had then hanged them. The Christians had during the domination of Kitubuka, who was himself a Kerait and a Christian, behaved themselves with great arrogance towards the Moslems, and had openly beaten in the streets the wooden clappers, called nakus, used instead of bells for summoning people to church, and even taken wine into the great mosque. Their day of humiliation was now at hand, and the infuriated Mussulmans, on the victory at Ain-Julat becoming known, destroyed the church of the Jacobites, and also the famous great church dedicated to the Virgin. This was the church which the Khalif Omar II., Ibn Abd el Aziz, had surrendered to the Christians to compensate them for the loss of that of St. John, which on the capitulation of the city to Omar I., Ibn Khattab, had been made over in perpetuity to them, but had been taken from them again by Valid, the son of Abdul Malik, and converted into the Great Mosque, the master-piece of Saracenic art.* The Mussulmans also put to death a great many Christians, and reduced the rest to slavery, and thus revenged themselves upon those who had lately pulled down the mosques and minarets near their own churches, and otherwise aggrieved them. The Jews were the next victims, and their houses, shops, and synagogues were plundered or destroyed. Lastly came the turn of those Mussulmans who had supported the invaders. Among others was Hussain, the Kurd, who had betrayed his master, the Prince Nasir. Thirty of the Christians were put to death, and a contribution of 150,000 drachmas was levied on their community. Makrizi, who was a Mussulman, says the town offered a terrible spectacle.†

It was not only the Christians who suffered severely by the Egyptian victory. The Ayubit princes of Syria also had cause to regret it. We have seen how Said, the son of Azis and grandson of Malik el Aadil, who had been granted a fief at Sabib and Banias by Khulagu, received the overtures of Kuttuz for an alliance, with contumely. On the defeat of the Mongols he surrendered himself, and offered to kiss the hand of Kuttuz. The latter, however, struck him in the mouth with his foot, and thereupon one of the Egyptians decapitated him.‡ A similar fate overtook Nasir, the Prince of Aleppo. He had taken refuge, as we saw, with the Mongols,

* Abulfeda, iv. 595. Ilkhans, i. 207.   † Makrizi, 107. D'Ohsson, iii. 342-343.
‡ Novairi, in D'Ohsson, iii. 340-341.

and had gained the good opinion of Khulagu, who had restored him to the government of Damascus, and had dispatched him thither with 300 horsemen, the very evening when he heard of Kitubuka's defeat. A Syrian who was present suggested that Nasir only wished to join Kuttuz, who owed his victory to his machinations, whereupon a party of horsemen was dispatched in pursuit.* Bar Hebræus reports what followed, on the authority of one of Nasir's companions. He reported that while Nasir was sitting in his tent with himself, whom he had ordered to draw his horoscope, there arrived about noon a Mongol chief, with fifty followers. He spoke to Nasir, who had gone out to meet him, and told him that Khulagu was giving a grand feast that day, and had sent to ask him, with his brother, sons, and grandees to attend it. Thereupon he set out with twenty followers. Shortly after a body of twenty horsemen came up to the tent and summoned the rest of the party, except the servants, cooks, and herdsmen. They mounted accordingly, and rode on to a deep valley where the Mongol chiefs were assembled. The latter approached them and took them severally into custody. The individual who reported the matter to Abulfaraj, and whom he calls Mohar ud din, is called Mej ud din by Rashid, who says he came from Maghreb, or Africa. He let them know that he was an astrologer, and could interpret the stars, whereupon they spared him. All the rest, including Nasir, except two of Nasir's sons, who were taken into his harem by Khulagu, were put to death.† Rashid ud din confirms this.‡ The astrologer was sent to join the staff at the observatory of Meragha. Makrizi reports that Khulagu, in addition to Syria, had invested his *protégé* with the government of Egypt, had loaded him with presents and honours, and given him a seat by himself. He says the party was overtaken in the mountains of Selmas, and that, besides Nasir, there perished his brother Malik Dahir Gazi, Malik Salih, son of Malik Ashraf, Lord of Hims, and many others. Malik Azis, a son of Nasir, who was very young, was spared on the intercession of Dokuz Khatun.§ This slaughter took place on the 29th of October, 1260.‖ Abulfeda reports that Nasir implored Khulagu to spare him, and was rebuked by his brother, Dahir, who bade him meet his fate, which was inevitable, in a manly fashion. Nasir, he adds, fell by Khulagu's hand, who shot him with an arrow.¶

He had been a very powerful prince. Not only had he ruled over all Syria, but also over a large portion of Mesopotamia, including Harran, Roha, Rakka, Ras Ain and later over Edessa, Damascus, Baalbek, Cœle-Syria, and Palestine, as far as Gaza. He lived very luxuriously, and 400 sheep were daily killed for his kitchens. His clemency was so great that the country was overrun with robbers, and men needed a military escort

---

* Quatremere, Rashid, 353-355.
† Abulfaraj, Chron. Arab., 350. Chron. Syr., 558-559. ‡ Quatremere, 355-359.
§ Makrizi, 108-109. ‖ Ilkhans, 208. Note 4. ¶ Op. cit., iv. 621.

in going from Damascus to Hamath ; while the Arabs and Turkomans in his service greatly plundered and illused the people, and the miscreants who were brought before him for punishment were lightly treated, his policy being, we are told, to preserve the living, and not to increase the dead. He was a poet, and Abulfeda has preserved some of his verses. He also built a school at Damascus, which was called after him, and prepared himself a grand tomb at Salahiyet, in which he was not buried himself, but it became the tomb of the Mongol amir Karmun.*

Meanwhile let us turn again to Kuttuz. Abulfeda tells us he had been accompanied from Egypt by Malik Mansur, the Prince of Hamath, and by the latter's brother, Malik el Afdal, who was Abulfeda's father. After his victory he proceeded to distribute fiefs among his followers. Malik Ashraf, Prince of Hims, who had conciliated the Mongols, and had also sent him a friendly message, was pardoned and restored to his principality, to which were added Palmyra and Rahbah. Mansur was re-appointed Prince of Hamath, and was also given Barin and Maarah, the latter of which the Prince of Aleppo had appropriated twenty-four years before. Salamiah was taken from him, however, and given to the Arab Amir Sherif ud din Isa ibn Mohanna. Mansur, with his brother Afdal, now returned to Hamath, and imprisoned some of those who had sided with the enemy, their advent being celebrated at Maarah in some verses by the Sheik Sherif ud din Sheik es Shoiush, who congratulated them on their victory.† Alem ud din Sanjar, of Aleppo, was appointed governor of Damascus, and the Prince of Sanjar, Mozaffer Alai ud din Ali (called Malik es Said by Abulfeda), son of Bedr ud din Lulu, Prince of Mosul, was made governor of the district of Aleppo. Shems ud din Albarli (Von Hammer says Berlas), a Turkish dependent of Nasir's who had abandoned him and joined Kuttuz in Egypt, and fought at Ain-Julat, was appointed governor of the Sahél and Gaza.‡ Hussain Kurdi, the tabardar who had betrayed Nasir, was strangled. Thirty Christians were put to death, and a fine of 150,000 dirhems was imposed on their co-religionists.

We have seen how the Prince of Karak, Moguith, sent his son Azis to Khulagu with his submission. He was then but six years old. Novairi tells us how he heard from his lips the adventures he went through. He was presented to Khulagu at Tebriz, and, although so young, was given a seat by the great conqueror. The Empress (i.e., Dokuz Khatun) then spoke to him through an interpreter, asked him if his mother was still living, and whether he preferred to stay or to return to his parents. The boy answered that his mother was alive and with his father, and that as to his return it did not depend on himself, who had merely gone on behalf of his father to secure his safety, and that he was at her orders. On her

---

* Abulfeda, iv. 621-625.　　† *Id.*, 597-603.　　‡ *Id.*, 597-603. Makrizi, i. 107-108.

intercession, Khulagu granted his prayer on behalf of his father, and he then knelt down and withdrew. He set off homewards with a Mongol who had been nominated Prefect of Karak. He was at Damascus when Kitubuka was defeated, was captured there by the Egyptians, and taken to Egypt, where he was detained for two years, when he was sent back to his father, with whom the Sultan formed an alliance. This did not prevent him from afterwards inviting him to Egypt under pretence of friendship, executing him on a charge of holding communications with the Mongols, and then appropriating the principality of Karak.* After his victory, Kuttuz entered Damascus in state. It had been in the Mongols' hands for seven months and ten days. He dispatched the Mamluk chief Bibars towards Hims in pursuit of the Mongols, of whom he killed a great many, and then rejoined his master. Rashid says the Mongol Noyan Ilka, with a number of his followers, found refuge in Rum.† Vartan says the fugitives went to Haithon, King of Little Armenia, who supplied them with horses, clothes, and victuals, and they then returned to their master, both Tartars and Christians pouring blessings on his head. "Thus was the name of Christ glorified in the person of the King, both by strangers and our own people."‡ From the Euphrates to the borders of Egypt Syria was now free from the Mongols, and Kuttuz turned his steps homewards to meet with a singular fate. He had been a traitor to his master, as Kitubuka charged him, but had assuredly done the Mussulman world such a service as might condone many crimes. The Mamluk, Bibars, who had fought so well in the late battle, having been refused the government of Aleppo, was much irritated, and formed a plot with some of his friends to murder the Sultan. The latter was hunting near Kosseir, a day's journey from Salahiyet, and had just returned to his tent when Bibars entered it and asked for the hand of a female captive who had been taken from the Tartars. The Sultan assented, whereupon Bibars kissed his hand and took the opportunity to fall upon him with his companions and kill him.§ This took place on the 25th of October, 1260. The body of Kuttuz was removed to Cairo.

Meanwhile Bibars, with the other conspirators, returned to the camp at Salahiyet, and entered the royal tent. They were about to proclaim their senior Bilban Rashidi as sultan, when the Atabeg Fars ud din Ogotai, called Aktai Mostareb by Makrizi, who had been left in charge of Egypt, arrived and asked what they were doing. "We are about to proclaim Bilban." "What is the fashion among the Turks in such a case? That the murderer should succeed. Which is he?" They thereupon pointed out Bibars, whom he accordingly conducted to the throne. The latter said, "I sit here by the will of God; kneel down and swear

* D'Ohsson, iii. 351-352.   † Quatremere, 359.   ‡ Journ. Asiat., 5th ser., xvi. 294.
§ Makrizi, 110-113.   D'Ohsson, iii. 345-346.

allegiance." "It is you," said Ogotai, "who must swear first that you will treat them loyally and as your equals, and will promote them."* After this grim comedy, Bibars set out for Cairo, which was *en fête*, preparing to welcome the victorious Kuttuz, and the people there were naturally startled to hear the criers in the streets shout out, "O people, pray for divine pity on the soul of the Sultan El Mozaffer (*i.e.*, Kuttuz), and pray for your Sultan, Ez Zahir Bibars.† Bibars (*i.e.*, the panther beg) was a Turk of the Kipchak tribe of Berlas, called Albarli by Abulfeda. He had been sold at Damascus for 800 drachmas. The purchaser, noticing that he had a white spot on his eye, repudiated the purchase. He was then bought by the Amir Idekin el Bundokdari, whence, after the custom of the Mamluks, he was styled Bibars al Bundokdari. His master having been disgraced in 1246, he entered the service of the Ayubit Sultan Salih, by whom he was successively promoted to several posts, and ended by becoming one of the chiefs of the Bahri Mamluks.‡ His full name was Rokn ud din Bibars, and he first took the title of Sultan Kahir (*i.e.*, the vanquisher), and afterwards that of Sultan Zahir (*i.e.*, the glorious). A curious legend was apparently current among the Armenians about the origin of Bibars, for Malakia tells us that when the Tartars captured Baghdad there were two slaves of the Egyptian Sultan there, named Phentukhtar (*i.e.*, Bondukdar) and Sghur. They managed to secure horses and escaped. They were pursued by the Tartars. The former, who was grown up, rode a miserable horse, while Sghur, who was younger, rode an excellent one. As they were being overtaken Sghur exchanged horses with his companion and bade him flee, saying that if captured, as he was young, the Tartars would not harm him, but reduce him to slavery, and that Phentukhtar could redeem him. Sghur was, in fact, captured, while his companion arrived safely in Egypt. The Sultan being then dead, they made Phentukhtar sultan in his place.§ Bibars was acknowledged by the several chiefs who obeyed Kuttuz, except Alem ud din Sanjar, the governor of Damascus, who set up authority on his own account, adopted the title of Malik Mujahid, and had his name inserted on the coin and in the khutbeh jointly with that of Bibars. Presently he went still further, had the *gashia* borne before him, and took the title of Sultan.||

When the Mongol general Baidar heard of the assassination of Kuttuz he marched at the head of 6,000 troops, consisting of the *débris* of the army defeated at Ain Jalut, and some other troops from Mesopotamia, to try and restore his master's fortune. When he reached the fortress of El Biret, on the Euphrates, Prince Said, the governor of Aleppo, already

---

* Makrizi, i. 116. Shafi, in his Life of Bibars, quoted by D'Ohsson, iii. 345-346.
† D'Ohsson, iii. 347.
‡ Shafi's Life of Bibars, D'Ohsson, iii. 347-348. Another account is followed by Wolff, Geschichte der Mongolen, 403, and I have also followed it in an earlier page, ante ii. 115.
§ Op. cit., 459-460. || Makrizi, i. 120-121.

named, sent a small body of troops under Sabuktigin against him, contrary
to the advice of the Mamluk chiefs of Aleppo, who thought that a disaster
was invited by sending such a small contingent. The Mongols were
victorious, and Said's people had to seek shelter at El Biret.* This
defeat exasperated the Mamluks, who were further estranged, according
to Abulfeda, by the cruelties and ill-conduct of Said himself. They
accordingly seized him, plundered him of all his money, and pillaged his
tents, and after supplanting him by the chief amir, Husam ud din, they
sent him in chains to Shogr and Baka. Husam ud din received a
diploma from Bibars constituting him ruler of Aleppo ; meanwhile the
Mongols had marched upon that town, which they re-entered in
November, 1260,† and he sought refuge with Malik Mansur at
Hamath.‡ Thither the Mongols now marched, whereupon the prince
withdrew towards Hims, and in conjunction with Malik Ashraf, ruler of
that town, and his own brothers, Afdal and Mobarez ud din, set out at
the head of 1,400 horsemen, who were joined by a large body of Arabs
under the Amir Zamil ibn Ali, and attacked the Mongols near Restin on
the 10th of December, 1260. Although the latter numbered 6,000, the
confederates defeated and destroyed many of them- -according to Ez
Zehebi, with the loss of only one man, proving, if true, that the fight was a
surprise and massacre rather than a battle. The heads of the slain were
taken to Damascus.§ This victory was won on the 4th Muharrem, 659.

Baidar now retired by way of Famia, and was attacked and punished
by the governor of the citadel. Damascus being relieved by this victory,
Mansur, Prince of Hamath, and Ashraf, Prince of Hims, put up at their
own palaces there. We have seen how Mujahid had usurped authority
at Damascus. Makrizi tells us that Bibars dispatched Jemal ud din
Muhammed with 100,000 pieces of money and an array of robes to gain
over the principal people of Damascus. This he succeeded in doing, and
they proclaimed Bibars as sultan, whereupon Mujahid with his supporters
marched against him, but they were defeated, and Mujahid himself was
wounded and sought refuge in the citadel. Meanwhile an army marched
towards Damascus under the Amir Idekin Bundokdari, Bibar's former
master, and now his major domo, and who had been nominated Governor
of Egypt by the Sultan, and Mujahid fled towards Baalbek. He was
pursued and captured, and sent to Egypt, where he was confined for
a while, and eventually released.‖ All Syria was now completely subject
to Bibars, who proceeded to rebuild the various fortresses there which had
been ruined by the Mongols—viz., the citadels of Damascus, Salt,
Ajelun, Sarkhad, Bosra, Baalbek, Shaizer, Subaibah, or Sabib, Shemaimis,
and Hims. Their towers were restored and their ditches cleared, and they

---

* Abulfeda, iv. 609-611.   † Makrizi, i. 121-122. Abulfeda, iv. 611. D'Ohsson, iii. 359.
‡ Abulfeda, iv. 613.   § Makrizi, i. 132-133. Abulfeda, iv. 613. D'Ohsson, iii. 360. Note.
‖ Makrizi, i. 138-139. Abulfeda, iv. 615.

were supplied with garrisons and provisions, and *inter alia* there was built near Ain Julat, as a memento of the recent fight, a monument named "The Meshed of Victory."[*]

Let us return to Baidar and his Mongols. After their defeat they withdrew by way of Famia to Aleppo. According to Abulfaraj, in his "Syrian Chronicle," their leader was called Khukhalaga Noyan, and in his Arabic one Gugalki, both being probably corruptions of Kuka Ilka. A crowd of fugitives from the country round had collected at Aleppo. The Mongols ordered the people to leave the place, and that those of each district and village should collect apart. They assembled, according to Abulfeda, at a place called Makar al Anbiia (*i.e.*, the seat of the prophets, which was corrupted into Karnabia). D'Ohsson says at Babili. The country people who had sought refuge in Aleppo were mercilessly slaughtered, on the plea that they had not trusted the Mongols, and among them many Aleppins who had joined them, including some of Nasir's relatives. The Aleppins themselves who had not fled were spared. The invaders now withdrew towards the Euphrates. The town was given up to various excesses, and the ill-fortune of the citizens was completed by the arrival of an Egyptian army, which levied a contribution of 1,600,000 drachmas.[†]

We must now devote a few words to a very important event, viz., the revival of the Khalifate. On the capture of Baghdad, Abul Kassim Ahmed, son of the Khalif Dahir Abu Nasir Muhammed, and uncle of Mostassim, who was killed by Khulagu, fled, escorted by some Arabs. After living for some years among the Arabs of Irak he determined to go to Egypt, to the Court of Bibars. The latter gave orders that he was to be received *en route* with the honours due to a relation of the Prophet, and he himself went out to meet him, accompanied by the grandees and the principal people of Cairo and Fostat. The Jews bearing the Pentateuch, and the Christians the Gospels, also went out to greet him, a piece of timely diplomacy. Ahmed entered Cairo on the 19th of June, 1261, dressed in the costume of the Abbasides, and rode through the streets, accompanied by Bibars, to the Castle of the Mountain, where splendid apartments were prepared for him, and where the Sultan sat beside him without any symbols of his dignity—neither throne, nor dais, nor cushion. Proofs of his identity were formally examined and attested, and then the various dignitaries, headed by the Sultan, did homage. He in turn invested Bibars with the government, not only of all the lands subject to Islam, but of all such as he should by the aid of God conquer from the infidels. All classes in turn swore allegiance to the new Imam. Messages were sent out to the

---

* Makrizi, i. 141-142.
† Abulfaraj, Chron. Syr., 560 ; Chron. Arab., 351-352.　Abulfeda, iv. 611-613.　D'Ohsson, iii. 361-362.

different provinces, calling upon them to follow the example of Egypt, and it was ordered that the new Khalif's name should be inserted in the khutbeh, or Friday prayer, and on the coin. He adopted the same surname as his brother, the predecessor of the late Khalif, viz., El Mostansir Billahi, which was an innovation on the previous practice of the Abbasides.* He himself read the khutbeh in the great mosque in the Castle, which he ended by imploring the blessings of heaven upon the Sultan. The latter then strewed gold and silver pieces over him and, amidst tears, repeated the prayer with his people. The Sunday following, the two made a progress in state on the Nile, where a sham fight took place between the galleys. The next day the Sultan was dressed in the robes of the Abbasides, which were given him by the Khalif, consisting of a black turban embroidered with gold, a violet robe, a golden collar, a golden chain, which was fastened about his legs, and a sword, two pennons, two long arrows, and a buckler. He mounted a white horse, with a black scarf about its neck and a horsecloth of the same colour on its back. Ibn Lokman, chief secretary of the chancellery, then got into a pulpit and read out the formal diploma containing the investiture granted by the new Khalif to the Sultan.† It is given at length by Makrizi. It begins with the usual language of praise to God (who had once more displayed the glory of his pearls, hidden for a while in a rough shell), and to the Prophet. It then goes on to describe the virtues of Bibars, and especially his beneficence in restoring the family of the Abbasides to prosperity. It then duly makes over to him the sovereignty of Egypt, Syria, Diar-Bekr, Hejaz, Yemen, the borders of the Euphrates, and all the lands he might conquer ; bids him cherish his people, and beware to-day of ambition, for to-morrow he could demand nothing, but it would be from him that demands would be made ; tells him to cloak himself with piety as with the provisions for a journey, and to devote himself to virtue and justice. The preacher reminded him that the various provinces needed governors, both civil and military, and as he would be responsible for them he ought to have confidential people to report to him their doings ; that he was to choose virtuous men as his subordinates, who would follow the precepts of clemency and moderation, and not let private affection interfere with justice ; who would listen to the complaints of the poor with a bright face ; who would treat those subject to them with kindness, for every Mussulman, whatever his rank, ought to deem himself the brother of another Mussulman. Let them try and win legitimate praise, which, at whatever sacrifice it is secured, is always underpaid, and to remember that riches extracted by crime are a load which presses heavily on a prince, and that no one is more unfortunate than he who at the day of Resurrection shall have the crowd

---

* D'Ohsson, iii. 366.   † Makrizi, i. 148-150.

M

for his enemies. The sermon afterwards went on to enjoin the duty of fighting the infidel, which was indispensable to all Mussulmans, and which God had promised to reward magnificently. It reminded him how he had already distinguished himself in this way, and how his sword had dealt incurable blows to the heart of the unbelievers, and that it was his duty to restore the throne of the Khalifs. It bade him look well to the fortresses on the frontier, especially those on the borders of Egypt, and also to cherish his fleet. God would not fail to reward him, for reward is the outcome of good deeds.* When this address was finished, Bibars made a grand tour of Cairo, accompanied by a cavalcade, the streets being carpeted with rich rugs. The Sultan then devoted himself to providing his august *protégé* with an army and a suitable Court, the various officers of which are enumerated by Makrizi. He supplied the necessary arms, and, we are told, also bought him a hundred Mamluks, great and small, and gave them each three horses, and camels to carry their baggage; besides these there were pages, doctors, surgeons, secretaries, horses, palfreys, mules, camels, &c.; while he gave to the various persons who had come from Irak in the Khalif's suite, diplomas granting them fiefs.

At length the Khalif and Sultan set out together for Syria, accompanied by all the army. This was on the 4th September. They made a solemn entry into Damascus, whence Bibars returned home while the Khalif went on.† Bibars intended to give his *protégé* a body of 10,000 troops to see him seated safely on the throne at Baghdad, but was dissuaded by one of his followers, who urged that he would then be too strong and would try and deprive him of Egypt. He accordingly only gave him an escort of 300 horsemen. He set out accompanied by the three sons of Bedr ud din, the Prince of Mosul, who had been to pay Bibars a visit, but they all left him *en route* to go to their several appanages. At Rahbah he was joined by the Amir Ali ibn Hodhaifah with 400 Arabs, by 60 Mamluks from Mosul, and some 30 horsemen from Hamath.‡ They went on towards Baghdad, along the western bank of the Euphrates, and at Meshed Ali met El Hakim, who belonged to the stock of the Abbasides, and claimed to be a rival Khalif. Mostansir invited him to make common cause with him to restore the fortunes of the family. This he agreed to do after he had been abandoned by 700 Turcomans who escorted him, and they went on together. Mostansir was well received at Anah and Haditsé, but at Hit the gates were closed against him, and the place had to be stormed. This was on the 24th of November. The Jews and Christians were duly plundered.§

Meanwhile Karabuka, the Mongol general whom Khulagu had put in command of the troops of Irak Arab, hearing of his approach, set out to attack him with 5,000 troops. He fell suddenly upon Anbar and slaughtered all its inhabitants. He was then joined by Behadur Ali, the Governor of Baghdad, and met the Khalif close to Anbar on the 29th of November, 1262. The latter placed the Turcomans and Arabs on either flank, and reserved a corps of picked troops for the centre. He fell in person upon the Mongols and broke their advance guard, but he was betrayed by the nomades already named, who fled. The troops about him were thereupon surrounded and killed ; only a small body escaped. The fate of the Khalif is unknown. According to some he was killed during the struggle ; according to others he was only wounded, and took shelter with some Arabs, among whom he died.* Well may the biographer of Bibars, Shafi, declaim against the absurdity of spending a million and sixty thousand dinars in inaugurating a new Khalif with becoming honours, and then sending him home with such an insignificant escort, that it could not make head against 1,000 Mongols, a race which had made so many conquests.† It would almost seem as if Bibars was chiefly aiming at establishing his own magnificence and power on a better basis, and that he cared little for the *protégé* whom he had so patronised. It gives point in fact to the doubts of Abulfeda about the Khalif's origin, for he speaks of Ahmed as "a certain Egyptian of a black colour, called Ahmed, who was said to be a son of the Imam Dahir."‡ Among those who escaped in the struggle with Karabuka was Hakim, whom we have described as having rival pretensions to Ahmed. He claimed to be fourth in descent from the Khalif Mostereshed, who was assassinated in 1135 by the Ismaelites. He now fled to Egypt, where Bibars gave him welcome. He was pleased to entertain and be the patron of one so reverenced in the Moslem world as a scion of the house of Abbas. He gave him a lordly home in the palace called Menasirolkebesh. His duties were those of giving legitimacy and a good title to those in authority, otherwise his power was a mere shadow. He was styled "Shadow of God upon earth, Ruler by the command of God, Hakim biemr-illahi." He lived thus for forty years, and was the first of a line of Egyptian Khalifs who were mere puppets of the Egyptian Sovereigns, and were only displaced when Egypt was conquered by the Ottoman Sultan, Selim I. Shortly after the accession of Bibars, Said, brother of Salih, Prince of Mosul, who had been driven away from Aleppo by the Mamluks, as we have seen, and had gone to Egypt, wrote a letter to his brother, the Malik of Mosul, advising him to repair to Bibars, who, when he had conquered the Tartars, would constitute him ruler not only of Assyria, but of all the East. This letter was surreptitiously acquired by one of his father's magnates, Shems ud din Muhammed Ibn Yunus al Baashiki, who put his hand under the coverlet and abstracted it. He then set out for Baashika, in the province of Nineveh. When Salih missed the letter he sent two slaves after him; but afraid of punishment if caught, he fled towards Irbil, and at Bakteli or Bartella, advised Abad ullah, son of Kushu, to escape at once with his people, as Salih meant to destroy the Christians there, and then escape to Egypt. They accordingly fled towards Irbil.[*] Meanwhile, Salih, afraid that Ibn Yunus might inform the Mongols, set out from Mosul with his son Alai ul Mulk, and withdrew towards Syria. His wife, Turkhan Khatun, refused to go with him. She remained behind with the Mongol prefect, Yasan. They shut the gates and prepared to defend themselves. One of Salih's officers, named Alam ud din Sanjar, left him as he retired through Syria with a troop of soldiers, and returned to occupy the town. He found the gates barred, and attacked it for some days, when Mohai, son of Zebellak, and a number of the citizens arose within, and opened the gates. Sanjar thereupon entered, and the Mongol prefect, with the princess, were obliged to seek shelter in the citadel. Sanjar commenced a cruel persecution of the Christians, killing those who would not become Mussulmans; and we are told that many priests, deacons, grandees, and others, except those belonging to the families of Said, Dekhukh, and Naphis the goldsmith, renounced their faith. At the same time the Kurds made an attack on the surrounding district, and slaughtered many Christians. They stormed a nunnery at Khudida, and put to death many who had sought refuge there, and for four months they attacked the monastery of Mar Matthew with 1,000 horsemen and foot soldiers, and attempted to storm it, but the monks burnt the scaling ladders with naphtha. The Kurds now rolled two great rocks against the walls from a height above. One of these stuck fast in the wall, we are told, like a seal in a ring; the other pierced the wall. When they tried to force their way through the opening the monks and others opposed them bravely with stones and darts, and repaired the breach with stones and lime. In these struggles the Abbot Abunser lost an eye. The weapons and strength of the defenders began presently to fail, and eventually the Kurds, who were afraid of a Mongol attack, were bought off by the sacrifice of the gold and silver ornaments in the churches, &c. Their booty weighed 1,000 golden denarii. The Amir Kutlughbeg perpetrated another slaughter of men and women at Irbil.[†] Alam Sanjar, who had secured possession of Mosul, as we have seen, having heard that the Mongols were advancing upon that town, marched against them, but was defeated and killed.[‡]

Let us now return to Salih, the Malik of Mosul. He made his way, with his son Alai ud din, to the Court of Bibars, who was then apparently at Damascus with the Khalif. He was there received with great

---

honours, as were his brothers Malik Mujahid Saif ud din Ishak, Prince of Jezirah, and Malik Mozaffer Alai ud din Ali, Prince of Sanjar. They were presented with robes of honour, banners, horses, &c., and were granted diplomas confirming them in possession of their states, which were further confirmed by the new Khalif. Salih was appointed Prince of Mosul, Nisibin, Akr and Shush (both near Mosul), of Dara, and the fortresses of the district of Amadiah. Mujahid was styled Prince of Jezirah, and Mozaffer Prince of Sanjar.* The three brothers set out from Egypt, as I mentioned, in company with the new Khalif, but they all left him *en route* to go to their several principalities. Salih repaired to Mosul. He was speedily followed by a Mongol chief named Samdaghu called Sadagun by Makrizi, and Shidaghu by Raverty), who, we are told, was a Christian, and therefore probably a Kerait. He was also young and amiable. He attacked the place with a tuman of troops and twenty-five battering engines, while Malik Sadr ud din, of Tebriz, assisted with a tuman of Tajiks.† The siege began in December and lasted till summer. The garrison consisted of Kurds, Turcomans, and Shuls (a tribe of Luristan, on whom Quatremere has a long note‡), and Salih distributed largess freely among them, and promised that Bibars would speedily send to their assistance. The place was bravely defended. One day eighty Mongols succeeded in scaling the walls, but they were all killed, and their heads shot among their companions. Sadr ud din, of Tebriz, was himself wounded, and was allowed to return invalided. He went to Alatagh and reported what was going on, whereupon Khulagu sent a second army to relieve Samdaghu. Meanwhile Bibars ordered Agush Arbarlu (called Barlu in his "Arabic Chronicle" by Abulfaraj) to succour the place. He sent a pigeon, with a note fastened to its wing, to inform the garrison that help was at hand, but by a singular fatality the bird alighted on one of the Mongol battering engines. Samdaghu having had the letter read, dispatched a tuman of troops to surprise the Egyptians. They planted themselves in ambush in three sections, near Sanjar, almost destroyed their army, and then took vengeance upon the people of Sanjar, many of whom they killed, carrying off the women and children prisoners. They then dressed themselves in the uniforms of the Syrians, and let their hair hang down after the fashion of the Kurds. When they neared Mosul many of the citizens saw them and went out to meet them, fancying they were friends, whereupon they were surrounded and all killed, The siege had now lasted six months, and the terrible heat of summer had made each party desist a while from attacking the other. Famine and pestilence raged inside. Salih sent out a letter, offering to surrender if Samdaghu would send him on to Khulagu and intercede for him. §
Abulfaraj suggests that it was the Mongols who made overtures and

---

* Makrizi, 164-167.   † Tabakat-i-Nasiri, 1280-1281. Notes.   ‡ Op. cit , 380-381.
§ Quatremere, 383-389.   Abulfaraj, Chron. Syr., 565.

fair promises. Salih was a dissipated person, and came out of the city accompanied by dancers and tumblers, and amidst the playing of cymbals and sistras.* Samdaghu would not see him. The Mongols entered the town on the 25th of June, 1262, when the whole population, except the artizans, were put to death. The latter were carried away captive. The place was completely depopulated, and it was only after the withdrawal of the enemy that 1,000 fugitives, who had sheltered in the mountains and caverns, returned.† In reading these accounts we can realise why Mesopotamia ceased to be a civilised land, and how it came about that a country once so thriving and prosperous became the home of bitterns and pelicans.

Salih was sent to Khulagu, who treated him with great cruelty. He was wrapped in a fresh sheep's skin, which was fastened tightly round him, and in this condition he was exposed to the sun. In a week's time the foul skin produced horrible vermin, which attacked his flesh, and he perished after a month's sufferings. His son, Alai ud din, only three years old, was sent to Mosul.‡ Having made him drunk, they fastened a cord so tightly round his body that they squeezed his entrails towards his face. They then clove him asunder into two pieces, which they hung on two gibbets on each side of the Tigris; Bar Hebræus says, on each side of the city gates. Rashid ud din says dolefully :

> He rotted and fell down from that place.
> O, Heaven, thou art not satisfied then with this act of vengeance.
> Thou hast delicately nurtured this lovable man,
> And hast then given him over to the tooth of the worm. §

Mohai, son of Zeblak, was also decapitated. The traitor, Shems ud din Ibn Yunus, called Bar Yunus by Bar Hebræus, was appointed Governor of Mosul. Samdaghu, after his success at Mosul, went on to Jezirah, called Gazarta by Abulfaraj, which he beleagured during all the winter and into the summer. At that time Hananieshua (meaning the grace of Jesus) was Nestorian bishop there. He knew Khulagu personally, who patronised him on account of his knowledge of alchemy. He went to the Mongol ruler, from whom he obtained a yarligh securing their lives to the inhabitants. The town gates were opened, and Samdaghu entered and ordered the walls to be razed. They then withdrew to Shemam, in the district of Irbil. They set over the place Jemal ud din Gulbeg, an officer of Seif ud din, the Ayubit Prince of Jezirah, who, as we have seen, had been a refugee in Egypt. But some time after, the latter having sent to Gulbeg to forward him a quantity of gold which he had hid when he retired, the news coming to Samdaghu's ears, he arrested Gulbeg and let him know that he had been appointed ruler of Jezirah by the Mongols and not by the Egyptian exile. He then had him executed, and replaced him by Mar Hasia.‖ This took place in 1263. The next year Bar Yunus,

* Abulfaraj, Chron. Syr., 565.　† Quatremere, 389.　‡ Id., 389.
§ Id., 389. Abulfaraj, Chron. Syr., 565.　‖ Abulfaraj, Chron. Syr., 565-566.

the ruler of Mosul, whose treacheries we have described, was charged before Khulagu by a spy named Al Zaki, or Zacchæus, of Irbil, with secreting a quantity of treasure with the intention of sending it to Egypt. Knowing that Al Zaki was aware of his design, he had tried to poison him, but he had been saved by a Christian doctor named Muphek, who gave him an antidote. Khulagu, who was enraged, ordered him to be bastinadoed. When he was stretched out a document fell from the folds of his cloak, which was written over with the following sentence from the Koran : "If their tongues clave to their mouths, their feet were shackled, and their necks in chains, we should be delighted and greatly pleased." Khulagu having asked the meaning of these words was told it was an incantation directed against himself, upon which he had him killed. Al Zaki was appointed in his place.* About this time (i.e., June, 1262) Salar, Prince of Vassit, Kufat, and Hillet, a feudatory of the Khalif's, who, after the capture of Baghdad, had sought refuge in the desert of Hejaz, and had remained there six months, received a message from Khulagu reinstating him in his former dominions. When Bibars mounted the throne of Egypt he had summoned him more than once to his presence, and he had professed that he would go when he had collected his wealth. This having reached the ears of Khulagu, he was summoned in turn by him. Afraid to obey, he left his family and goods and retired to Egypt, where Bibars gave him a military command and a fief.†

We must now turn aside to consider what was taking place in Rum. We have seen how the two brothers Iz ud din and Rokn ud din made friends, submitted to the Mongols, and divided Rum between them.‡ They remained good friends till the death of their common vizier, Shems ud din Mahmud, when each prince got a vizier of his own. Rokn ud din's vizier, Moyin ud din Suliman, better known by his Persian title of Sahib Pervana, or keeper of the seals, and called the Peishwa of Rum by Wassaf, determined to make his master ruler of the joint kingdom, and endeavoured to win over Khulagu's lieutenant in those parts, the Noyan Alijak, who, under his instructions, informed his master that Iz ud din was conspiring with the Egyptian sultan, and meditated a revolt ; and, in fact, Iz ud din had sent a missive to Bibars offering, for his assistance, to surrender half his kingdom to him, and sending him a number of blank patents, which Bibars might fill up and confer on whom he pleased some fiefs in Rum. The latter ordered his men to march from Damascus and Aleppo to the aid of Iz ud din, and prepared several diplomas conferring fiefs in Rum on his friends. But he presently heard from his ally that in view of his alliance his enemies had withdrawn, and that his own people were attacking Conia (or Iconium). This was during the year 1262. Khulagu issued orders that Iz ud din was to be suppressed. At first

---

* Abulfaraj, Chron. Syr., 567 ; Chron. Arab., 354-355. † D'Ohsson, iii. 375-377.
‡ Ante, 139-140.

he thought of going to Khulagu's Court in person. He was setting out, when a messenger came to announce that his brother, Rokn ud din, with his minister, the pervana, and the Mongol Noyan Alijak were marching against him, and intended to capture him and to take him to the Mongol Court. He accordingly abandoned his camp and fled. Alijak Noyan entered Iconium, and Rokn ud din was constituted sole ruler of Rum, with the pervana as his minister.* The Alijak Noyan of this notice is no doubt the Aljakta Noyan of Minhaj i Siraj, who says Rokn ud din married his daughter.† Meanwhile, Iz ud din sought shelter with the Greek Emperor Michael Palæologos, who had recently regained that city from the Latins, and who had already offered an asylum to several of his chiefs and found them employment in his service. Having collected his treasures, he accordingly set out from Iconium for Nymphæa, where the Emperor was staying with his treasures. The latter behaved with marked duplicity. While he professed to welcome his guest and to treat him with hospitality, assigning him a guard equal to his own, and also the use of the imperial insignia, he meanwhile negotiated with the Mongols, who wished him to keep Iz ud din under durance, so that he might not disturb their control of Iconium. To secure this end still more, he had the Sultan's wives and children sent to Nicæa. Iz ud din, to whom Michael was under obligations, was deluded by one excuse after another. He accompanied Michael in his various journeys, and enjoyed a kind of imperial servitude, the wily emperor deeming anything better than the imprudence of drawing upon himself the Tartar arms, even to the extent of disarming the natural buckler of the Greek Empire on its eastern flank, namely, the Seljukian kingdom.‡ Abulfeda says that presently Iz ud din, having conspired against the life of his host, was imprisoned, and the eyes of his supporters were seared with a hot iron.§

Let us now turn to the Mongol doings at the other end of Iran. Khulagu had been nominated generalissimo of all his forces in the west by Mangu. He had a commission to destroy the enemies of the Mongols, but he was not apparently endowed with any territory or special jurisdiction. This came afterwards. He and his descendants nominally exercised jurisdiction in the country east of Khorasan. The troops cantoned in Afghanistan and its borders were, nominally at all events, part of their command, and when Khulagu marched westwards we are expressly told that Sali, or Sari, who had previously commanded the troops on the borders of India, was ordered to put himself at his disposal. Sali was a *true* Tartar, and belonged to the Tartar tribe Tutukaliut. When his race was virtually exterminated by order of Jingis, he owed his life to the intercession of Yessulun and Yessugat, the two Tartar wives of Jingis. Sali, we are told by Rashid, had effected the conquest of Kashmir, whence he

* Abulfaraj, Chron. Syr., 564.   † Tabakat-i-Nasiri, 164.
‡ Stritter, iii. 1042-1044. Lebeau, xviii. 79-81.   § Op. cit., v. 12.

had carried off many thousand captives. He adds that all the troops which were under his orders, wherever they were stationed, became by right of inheritance the special property of the King of Islam (*i.e.*, of the Ilkhan).* The conquest of Kashmir by the Mongols is also mentioned in the biography of Kuo Khan (*i.e.*, the Kuka Ilka of Rashid ud din) in the "Yuan shi," where we read that Ki shi mi and a sultan named Hu li, or Khu li, surrendered to the Mongols.† In the "Yuan shi" we are told that in 1263 the generals Sa-li-tu-lu-hua and Ta-ta-r-dai (*i.e.*, Sali and Tair) were sent to Hin-du-sze and Kie-shi-mi-r.‡

In the year 654 Sali and his Mongols invaded India. We are told that after crossing the Indus they sent Shems ud din Muhammed Kert, who was then the ruler of Khaisar in Ghur, to Multan, on a mission to the Sheikh of Sheikhs, Bahai ud din Zakaria, also known as Bahai ul Hakk, whose tomb Major Raverty says was much battered by the English in the siege of Multan, in 1848-9. He agreed to pay 100,000 dinars, and to accept a mamluk of Shems ud din as hakim of the town.§ The invaders dismantled the fortifications of Multan.‖ After this they marched upon Lahore, or what remained of it after its sack by Tair Baghatur in 639 HEJ. After that event it is said to have been occupied by the Khokars and also by Kurt Khan, who now agreed to pay a ransom of 30,000 dinars, thirty ass-loads of soft fabrics, and 100 captives.¶ After this Sali plotted against Shems ud din Muhammed, who retired from India and went towards Ghur, but was apprehended *en route* by Malik Imad ud din, the Ghuri, and we are told he thereupon sent a messenger to inform Tair Baghatur (who, according to this account, still governed those parts). He ordered his release, and afterwards had him living near his person.**

At this time Iz ud din Balban, originally a Turkish slave from Kipchak, was the governor of Uch and Multan.†† He was not loyal to his suzerain, the Sultan of Delhi, but entered into negotiations with Khulagu through the medium of the Malik Shems ud din Kert; virtually threw off his allegiance, and asked the Mongols to send a shahnah or intendant to superintend his country. He also sent one of his grandsons to Khulagu as a hostage. Soon after this Balban marched against Delhi, but after making a demonstration before its walls withdrew again, and was abandoned by many of his men. Having reached Uch in safety he went to pay Khulagu a visit in Irak, whence he returned again to his fief.‡‡ In consequence of this he felt constrained to go to Sali who about this time arrived, and proceeded to dismantle the walls of Multan. News of the Mongol invasion having reached Delhi, the Sultan Ulugh Khan i A'zam summoned his troops. On hearing of this concentration they

* Quatremere, 130-131. D'Ohsson, ii. 280-282.
† Bretschneider, Notes on Chinese Travellers, &c., 82. ‖ *Id.* Note 103.
§ Journ. Asiat., 5th ser., xvii. 442. Tabakat-i-Nasiri, 1201. Note. ‖ *Id.*, 844.
¶ *Id.*, 1201. Note. ** *Id.*, 1202. Note. †† *Id.*, 784. ‡‡ *Id.*, 786.

did not advance further, but harried the frontiers of Sinde and Lahore, as far as the River Biah.*

About this time a curious diplomatic intercourse took place between Khulagu and the ruler of Delhi. According to Minhaj i Saraj the Malik Nasir ud din Muhammed, son of Hasan, the Karluk, who held authority about Banian, had sent secret overtures to Ulugh Khan i A'zam, the Sultan of Delhi's most valiant feudatory, for a marriage between their families. The latter sent a Khalj Turk named the Hajib i Ajall (the most worthy chamberlain) Jamal ud din Ali with his answer. *En route*, he was detained at Multan, and closely questioned about his journey by the Malik Iz ud din Balban and the Mongol shahnah or intendants. He was allowed to go on, and reached Banian safely, but the news of his arrival having reached the Mongol shahnah there, the Malik Nasir ud din was forced to send him on to Khulagu to Irak and Azerbaijan, and in addition wrote letters and sent presents by him in the name of Ulugh Khan i A'zam. They reached Khulagu's presence at Tebriz ; the Hajib Ali was well received, and his letters were translated from Persian into Mongol. It was customary in writing letters to the inferior Mongol and other dependent chiefs to alter the usual designation of Khan, borne by many of the grandees of India and Sinde, to Malik, since Khan was among the Mongols a title of supreme dignity. We are told by our courtly author that Khulagu having noticed this alteration in the case of Ulugh Khan i A'zam, enjoined that in his case the title Khan should be used—a very improbable story. When the Hajib returned, Khulagu ordered the Shahnah of Banian, who was a Mussulman, to accompany him, and we are again told a questionable story, viz., that Khulagu sent orders to the Mongol troops under Sali Noyan, saying : "If the hoof of a horse of your troops shall have entered the dominions of the Sultan of Sultans, Nasir ud din Mahmud Shah (God perpetuate his reign), the command unto you is that all four feet of such horse be lopped off."† When Khulagu's envoys arrived near the capital orders were given that they should be detained at Barutah.‡ After a while they were conducted to the capital to be presented to the Sultan, and a magnificent review of troops was held, when, according to Minhaj i Saraj, 200,000 foot soldiers and 50,000 horse, fully equipped, were present, and the imposing armament "was paraded in twenty lines of men one behind another, like the avenue of a pleasure garden with the branches entwined, placed shoulder to shoulder, row after row."§ This spectacle was doubtless arranged to create a feeling of respect on the part of the Mongols when they heard of it. There is a curious bit of local colour in the remark that some of the emissaries were thrown from their horses when the trumpeting elephants charged.‖ The

---

* Tabakat-i-Nasiri, 846-850.   † *Id.*, 859-863.   ‡ Perhaps Merut. *Id.*, 851.
§ *Id.*, 856.   ‖ *Id.*, 857.

envoys were conducted to the capital and received with the honours due to distinguished guests. They were conducted to the Kasr i Sebz, or Green Castle. The castle was decorated with various kinds of carpets and cushions, and a great number of rare articles of gold and silver, with two canopies, one red and the other black, adorned with costly jewels over the throne. The distinguished maliks, amirs, and sadrs, &c., and the handsome young Turk slaves, with golden girdles, stood round about. On the throne sat the Sultan, "as a sun from the fourth heaven, with Ulugh Khan i A'zam in attendance as a shining moon, kneeling upon the knees of veneration and reverence, the maliks in rows like unto revolving planets, and the Turks in their gold and gem-studded girdles like unto stars innumerable."[*] Minhaj i Saraj unfortunately closes his narrative at this point, and we do not know what was the issue of the embassy.

Let us now turn to Khulagu's doings nearer home, especially in •Georgia. According to Vartan, on his return from Syria he went to winter in the plain of Mughan.[†] There, according to the Georgian history, he was visited by King David of Georgia, who afterwards went to his summer camp, and was thence sent to Karthli to prepare to assist in the campaign against Egypt. He traversed the territory of Avak, the son of the Atabeg Ivaneh. Avak was then dead. He had left no son, and only a daughter named Khoshak. David visited Bejni to mourn for him, and having seen his widow, Gontza, who was of the family of Kakhaber, eristhaf of Radsha, and very beautiful, he shortly after married her and gave her the title of queen. Khoshak was left behind in charge of Sadun Mankaberdel, a prudent and sagacious counsellor, fortunate in his undertakings, and famous for his bodily strength, and his skill as an archer and wrestler. Chamchean says he belonged to the princely family of the Ardzrunians, and was the grandson of Kurd of Sasun. Malakia calls the latter the Amir Kurd. Sadun visited Khulagu, and challenged any man in the Mongol army to wrestle or draw the bow with him, and no one was found who could compete with him.[‡] Malakia reports a curious story of him, viz., that Mangu was visited by an adventurous character, who had a repulsive appearance, was very high, and had great shoulders, a neck like a buffalo, hands like a bear, and who devoured a sheep daily. He was a famous wrestler. He committed to him a letter and a robe of honour of great value. The letter was addressed to Khulagu, and stated that if any wrestler overcame him he was to have the robe, but if his champion proved unconquerable he was to have it, and to be sent back to Mongolia. Khulagu, on his arrival, summoned his chiefs, and asked if they knew anyone who could cope with him. The Armenians and Georgians said they knew such an one, upon which he sent for him. This was Sadun. He was of great stature and well skilled,

---

but was disconcerted by the invitation, as he had never wrestled before the Khan, and had heard of the prowess of his adversary. He repaired to some hermits to pray for him, went to Kak, to the Church of St. Sargis, the dispenser of justice, and received the blessing of the Vartabied Mesrop, and having made a vow and an offering at the Church of the Holy Cross then went on to Khulagu, who was delighted with his appearance. He ordered the two athletes to live together for nine days, and furnished them daily with a sheep and a skin of wine. They were at length matched, and struggled for three hours without either getting the advantage, when Sadun, in the name of God, by a sudden throw overturned his opponent. Khulagu was delighted, and gave him a yarligh freeing him and nine of his descendants from taxes.* He is mentioned in several Armenian inscriptions in the monastery of Haghbat.† One of these is on a cross, set up in 1279, and dedicated to St. Sargis, the general, to whom Sadun chiefly addressed his prayers before encountering the Mongol champion.

At this time the kingdom suffered greatly from the exactions of Arghun and his tax collectors. We read that three whites (*i.e.*, silver pieces) were levied on every 100 sathers (a derivative of the Greek ϛατηρον, a talent) of anything sold at Tiflis. A certain Khoja Aziz, who was a Persian by race and religion, was the tax superintendent at Tiflis, and was very exacting, and even levied the tax on every sheep and lamb for the royal kitchen. David, the Georgian king, was much annoyed at this, and determined to revolt. When he summoned his supporters and told them his views, some of the grandees joined him, but the greater part, such as Ivanch, son of Shahinshah ; Grigol Suramel, the Orpelian ; Kakha Thorel, eristhaf of Akhal-Kalah, and the majority of the chiefs of Héreth and Kakheth remained faithful to the Mongols, and went and joined Khulagu. He nevertheless determined to prosecute his plans, and sent an invitation to Sargis, the commandant of Jak and Tzikhis Juar, who had the title of General of Samtzkhé, to join him. He agreed, and the king went to Samtzkhé, where he was royally entertained, and where he spent the summer with a few followers, while his queen, Gontza, and his son Dimitri stayed at Bejni, in the house of Avak.‡ Khulagu, on hearing of this revolt, summoned Arghun and 200 other captains, whom he placed under his command, as well as 20,000 horsemen, and also ordered the Georgians friendly to him to assist. Arghun traversed Kantzag (*i.e.*, Arran) and Somketh, and came to Tiflis, where he was joined by the chieftains above named as siding with the Tartars, and then marched towards Samtzkhé. Meanwhile the king assembled the Meskhes and the people of Khawkheth (*i.e.*, the Caucasus) and of the Clarjeth, who had remained faithful to him, and managed to

---

* Op. cit., 457-458.　　† Hist. de la Géorgie, 555. Note.　　‡ *Id.*, 556-557.

collect a small army of 8,000 men, of which Sargis Jakel, whose courage, military reputation, and physique were exceptional, was in command. This army set out, and reached the valleys of the Mtsuar. Arghun crossed Karthli, and having halted at Suram, detached a body of 6,000 men as an advance guard, which proceeded to the valley of the Shola, and himself remained at Shindara. Sargis had also detached an advance guard. This was 1,500 strong, and went boldly across the bridge of Akhal Daba, unaware of the proximity of the Tartars. It was then midwinter, and there was a hard frost. Scarcely had they emerged from the defile when the Tartar advanced guard was seen and very bravely charged by the Meskhes, who broke it and pursued it for a long distance, returning to Sargis with a number of Tartar heads as trophies. Sargis determined to at once march against Arghun with the main body of the army. The latter would have retired, but his Georgian allies would not allow him. "We know how to fight these people," said Cakha Thorel, "we will defeat them for you."* A hard struggle ensued, in which the Parthian tactics of the Mongols were put in force. They professed to retreat, but presently turned round on their pursuers, who were scattered, and pitilessly slaughtered them. Few escaped, and they were pursued to the bridge of Akhal Daba, or even further. Among the prisoners were Murwan Gurcelel and others. Arghun now returned to Khulagu, and Sargis to his unfortunate patron, King David. David having spent part of the winter in Samtzkhé, went afterwards into Khawkheth (i.e., the Caucasus) and Clarjeth (? Abkhazia), and entered into the valley of Nigal (probably situated near Artanuj). When the spring grass was ready for the horses, Khulagu again dispatched Arghun with his Georgian allies, who laid waste Samtzkhé and besieged Tzikhis Juar, otherwise called Juaris Tzikhé, which he did not, however, succeed in taking. He was a Mussulman, and no doubt enjoyed the task of harrying the Christians. He was hastily summoned away by a message from Khulagu, saying that the Khan of Turan (i.e., the chief of the Jagatai horde) was meditating an attack upon him in Khorasan.† Juveni, who describes these events very shortly, places them in the autumn and winter of 1259-60.‡

According to Guiragos, in the campaign just described, Arghun pursued the king, but could not overtake him, and proceeded to cruelly ravage Georgia. The famous monastery of Gelath in Imeretia (the burial-place of the Georgian kings) was razed to the ground, as was Atzghur, the residence of the Catholicos. Arghun having returned to his master incited him to imprison Gontza, the wife, and Khoshak, the daughter of King David, together with the Grand Prince Shahin Shah, Jelal ud din Hasan, lord of Khachen, and many others, under pretence that they were

---

behind in paying their tribute. A large sum of money was extorted from them as the price of their lives. Jelal ud din was treated with especial cruelty. He was called upon to pay a much larger sum than he could afford, and as he could not meet the demand, he had a wooden collar, or cangue, fastened about his neck, while his feet were chained. This treatment was suggested by the fanatical Mussulmans, who knew that Jelal was a very fervid Christian, and whom they described as the greatest enemy of their faith. He was removed to Kazvin. Meanwhile his daughter Ruzukan, who had married Pora Noyan, the son of the Mongol general Charmaghan, went to entreat the good offices of Dokuz Khatun, but Arghun, having heard of this, sent some executioners, who, having in vain asked him to apostatise, put him to death during the night and dismembered his body. This took place in 1261. His son Atabeg sent furtively to collect his father's remains, which had been thrown into an empty cistern. They were removed to the monastery of Kantza Sar, the burial place of the princes of Khachen, situated on a mountain near the town of Kantzag, in the province of Artsakh. A bright effulgence is said to have surrounded the body of the prince. With the consent of Khulagu and Arghun, he was succeeded in his principality by Atabeg, who was much given to religious exercises, and harmless as an anchorite. Zakaria, son of Shahin Shah, the Lord of Ain, having been accused before Khulagu, was also put to death. He was in the Mongol service, and had won the favour of its chiefs by his bravery. He was with the army which ravaged Georgia, on which occasion, without informing Arghun, he went to pay a visit to his wife, who was living with her father, Sargis, here called Prince of Ukhthik (a town and district in the province of Daik), who had taken part in the rebellion of King David. This having been reported to Khulagu he was put to death, and his dismembered remains were thrown to the dogs. When his father, Shahin Shah, heard the news, he became so depressed that he also died. He was buried in the monastery of K'opair, some distance to the south-east of Sanahin and Otzun. Its Arabic name means a tomb.* The *Georgian Chronicle* makes out that Zakaria, understanding that he was charged with being privy to Bereke's invasion, to be mentioned presently, fled to King David, son of Rusudan, in Kuthathis, where he was well treated. Presently the Khan having sent for him, and sworn not to kill him, he trusted himself at his Court, where he was nevertheless put to death.†

Let us now return to the fugitive king, David. We are told that on the retreat of the Tartars, he returned to Samtzkhé, and summoned his friends to ask their advice. Sargis said the province was too small for him to live in, and he advised him to cross the mountains of Likh, and to go to his cousin David, son of Rusudan, for each part of the kingdom really

---

* Guiragos, ed. Brosset, 190-192. Journ. Asiat., 5th ser., xi. 499-503.   † Op. cit., 568.

belonged jointly to the two princes, and added that he would devote himself and all his wealth to his service, and if his cousin refused to have anything to do with him he (Sargis) would at all events cling to him. He accordingly sent a messenger to his cousin, who promised to receive him. He went and was treated hospitably. There he remained a year. Feeling, however, that he was treated as a stranger, Sargis plotted with Kakhaber, Eristhaf of Raja, with the son of Kakhaber, and the Phardjainans, sons of Kwabul, to nominate him as King, and he having consented there arose a strife in Likht Imereth (i.e., the country beyond the mountains of Likh). Some espoused the cause of David Lasha, and others that of his cousin. The Dadian Bedian, son of Juansher, attached himself to the latter. The Suans were also divided into parties. Meanwhile, however, the two kings, notwithstanding that their partisans were at issue, distributed no arms to them. They eventually decided to divide the kingdom and the arsenal in two, as well as Tiflis and Kuthathis. They also divided the thawads and eristhafs of Nicophsia at Derbend. The famous necklace of diamonds, the precious stone cut in the shape of an anvil, and the large pearl, which it would seem were celebrated Georgian state jewels, fell to David, son of Rusudan. This partition applied only to the part of Georgia beyond the mountains of Likh. The rest was too closely controlled by the Tartars to be the subject of such arrangements. Meanwhile Khulagu, who wished to make terms with David, ordered Arghun to send him a messenger with a guarantee for his safety. Another account reports that David's wife Gontza and his son Dimitri, having been removed as prisoners to the Ordu, and the Khan having determined to do the latter harm, the wife of a noyan who had no children sent an express to David suggesting that peace might be made. A treaty was therefore entered into. David was to be restored to his kingdom, his wife, and younger son, while his elder son Giorgi was to be detained as a hostage. Khoja Aziz, the author of his troubles, was to be surrendered to him, to pardon or kill him as he wished, while David was to go in person to Khulagu's Court. Enuk Arkun (i.e., the Christian) acted as Khulagu's agent in this negotiation, and answered with his head for the safety of the king and his son. He was allowed to take the young prince with him. He took him to Tiflis, where all the mthawars and eristhafs went to meet him. Among others was an Armenian, named Badin, who had charge of Tiflis and of the throne. Arghun made Giorgi some rich presents. He then went on to Khulagu's Court. The latter sent his wife, Tonghul Khatun (the Kerait Dokuz Khatun), who was a Christian, to meet him. When she saw him, we are told, she loved him, for he was beautiful, she herself being the same. The Khan was also pleased with him. He remained at the Mongol Court for a year. David himself had not yet been to the Ordu, and excused himself through Arghun on the ground that Khoja Aziz had not yet been surrendered to him. At this the Khan was greatly irritated, and was counselled by Arghun to put the young prince to death, and to intrust him with an army; and promised to bring him back in chains. Khulagu thereupon ordered Giorgi and his attendants to be killed. At this news Enuk Arkun at once repaired to Dokuz Khatun, who reproached her husband with his cruelty in ordering the death of a descendant of such a long race of kings, who had gone to him on the strength of his oath. She added that he and Enuk Arkun were prepared to offer their lives for the prince. She also urged that the ruler of the Khanate of Kipchak was habitually intriguing to persuade the Georgian king to open for him the passes of Darialman and of the west, which were in his power. "What was the life of a Persian merchant to the danger of having the Ulus of Batu in alliance with the King of Georgia?" We see here how it was that the Georgian kingdom survived so long. The fact is the rival policies of the Khans of Kipchak and of the Ilkhans made it easy to secure an ally against either power when it assumed a threatening attitude. Khulagu was disconcerted by these home truths. He at once handed over Giorgi to Dokuz Khatun and Enuk Arkun to do what they pleased with him, and they in turn offered to secure the attendance of David if Khoja Aziz was handed over to them to take to him. This was done. Enuk Arkun set out for Tiflis with his charges, and the king went to the extremity of Kwishkheth, and stopped between that place and Suram. As soon as Khoja Aziz was handed over to him he had him beheaded, and his head was sent to Tiflis, where it was placed on a stake. Rashid ud din mentions the death of Khoja Aziz. He calls him one of the governors of Gurgistan (*i.e.*, Georgia). This event happened, according to the Persian historian, in November, 1262.*

David now went to the Ordu, and was accompanied by his friend Sargis, lord of Tzikhis-Juar and Jak, who insisted upon going, although the safe conduct did not include him. The King was much pleased at his devotion, and conferred on him the district of Kwabulian, in Samtzkhé, and the church of Tbeth in the Khawkheth.† They presented themselves to Khulagu at Berdaa, his winter quarters, without knowing what their fate would be. He, a few days later, summoned David and the mthawars to an interview. Everyone thought they would be put to death. Khulagu, however, offered the king wine out of a gold cup with his own hand, after the fashion of the Khans. The King and the mthawars having seated themselves with their legs crossed, Khulagu then asked him why he had rebelled and disobeyed his orders, and fought against Arghun, and reminded him how, when an exile and condemned to death, he had been drawn out of the pit filled with serpents and placed on the throne. The King had nothing to say, and turned to Sargis, who had

---

* Quatremere, 395.   † Hist. de la Géorgie, 563-564.

been the chief abettor of the attack on Arghun. Seeing that the blame was going to be put upon him, Sargis rose and advanced towards Khulagu, and said boldly, " It was I, great king, who attacked Arghun, but the only blameworthy person was Khoja Aziz. He took the king's domains, towns, and villages, and took possession of everything. He ruined the churches and fortresses, most blessed Khan, and in order that no one should suspect him he closed by corrupt means the avenues to your Court. This is why I carried off the King, in order that the Khan might cause inquiries to be made, and that his eyes might be opened as they are now. Know too, oh Khan, that from time immemorial the Persians have been the enemies of the Georgians. If I therefore fought against Arghun it was because I could no longer bear the injustice of Khoja Aziz towards the King. The King is innocent. It was I who prevented him coming to your Court." The conversation was prolonged, and diverged into various matters, Sadun Mancaberdel, who was an excellent orator, acting as interpreter. Everyone expected that the King would be punished and that Sargis would be put to death; but happily, as the interview was still in progress, one of the Khan's sentinels on the route of Derbend arrived, saying : " This is no time for disputation. The Grand Ulus of the Khan Batu is in motion, and the Khan and his son Barkai are advancing towards Derbend with an innumerable army."* The mention of Batu is an anachronism. He was then dead, and it was his brother Bereke whose march was thus announced.

I have described the struggle between Khulagu and Bereke in a previous volume.† Since writing that account I have met with fresh materials. As we have seen, Bereke had become a convert to Muhammedanism. The very orthodox and inaccurate author of the "Tabakat-i-Nasiri" gives us some characteristic details about him. He tells us he was born about the time when his father captured Khuarezm (i.e., 618 HEJ., A.D. 1221), and adds that the latter from the first was determined that he should be brought up as a Mussulman, and that accordingly his nurse severed his navel string in the Mussulman fashion, that he was suckled by a Mussulman nurse, and taught the Koran by a Mussulman doctor. Some reported that he had studied the Koran at Khojend with a pious ulema of that city. On arriving at the proper age he was duly circumcised, and when he arrived at manhood was set over the Mussulmans in his father's ulus. Batu retained him in the same position, and confirmed him in his command, fiefs, vassals, and dependents.‡ The " Shajrat ul Atrak" has a similar story. It says Bereke refused to take the breast of any female except that of the Mussulman woman who brought him up. It also says his mother was a Mussulman. When he grew up, his brother ordered him to go to various parts of the

---

* Id., 564-565   † Ante, ii. 113, &c.   ‡ Tabakat-i-Nasiri, 1283-1284.

N

empire. On one occasion he went to Bukhara, where he fell in with the Sheikh Hazrat Seif ud din Bakhurzin, a disciple of the Sheikh Nejm ud din Kobria. He remained for some time under his tuition, when he was ordered to return home, which he did by way of Chaji Turkhan (? Shash or Tashkend).* Minhaj i Saraj tells us how in the year 631 HEJ. (*i.e.*, 1234) a number of envoys went from Bereke to the Indian Sultan Iyal tamsh, taking various rarities with them. The Sultan always refused, however, to have communication with the Mongol chiefs, and these envoys were sent to the fortress of Gwalior. They were Mussulmans, and used every Friday to be present in the mosque there, and, our author tells us, used to say their prayers behind his own *nawab*. Eventually the envoys were removed from Gwalior to Kinauj, where they were restricted to the limits of the city, and there died. Minhaj i Saraj tells us further that Bereke made a pilgrimage to visit the illustrious ulemas at Bukhara, and also sent envoys to the Khalif, who was said on two occasions during his brother Batu's reign to have sent him robes of honour. All his army consisted of orthodox Mussulmans, and trustworthy persons reported that every one of his horsemen had a prayer carpet, and that they refrained from intoxicating drinks. Bereke made companions of the great ulemas, consisting of commentators, traditionists, theological jurists, and disputants, and had many religious books. Most of his receptions and debates were with ulemas, and in his place of audience debates on moral science and ecclesiastical law constantly took place.† These exaggerations are very pardonable when we consider what a notable event in Mussulman history the conversion of such a potent Mongol chief as Bereke must have been. We can well believe, too, that to a recent convert the slaughter of the Khalif and his family by Khulagu must have been a great outrage. In addition to this, Bereke, as we have pointed out, had other grievances.‡ In regard to the death of his relatives the Armenian monk, Malakia, reports matters somewhat differently. He says that after the capture of Baghdad the Khan's seven sons, who were gorged with riches, gold, and pearls, would not obey each other, but each one followed his own way, and pillaged and laid waste the country. Khulagu, their senior, thereupon wrote to his brother, the Khakan Mangu, in these terms: "We seven chiefs of tumans, thanks to the grace of God and to yours, have arrived here, and have taken with us the former temashis (*i.e.*, chiefs of tumans, meaning Baichu, &c). We have advanced and captured Baghdad, the city of the Tajiks, and have returned thence laden with treasures, thanks to the grace of God and yours. Meanwhile, what are your wishes? These people are lawless, and living in anarchy; the country is devastated, and the ordinances of Chanks Khan are not carried out, for he ordered us to cherish the lands

---

* Op. cit., 229.   † Tabakat-i-Nasiri, 1283-1286.   ‡ Ante, ii. 114.

subject to us or conquered by us, and not to lay them waste. If you have any other commands, give them, and we will obey." The bearers of this letter were questioned by Mangu as to what had occurred. He now ordered his arghuchis or judges to proclaim Khulagu as Khan in the countries where he was, and that anyone who disobeyed him was to be summoned to answer for it in Mangu's name. The arghuchis thereupon summoned a kuriltai, to which all the various chiefs, as also the King of Georgia with his suite, were summoned. The Khan's sons, such as Balakhain, Bora, Tegudar or Nigudar, and Mighan, son of Khuli, were summoned by special messengers. It having been announced that Khulagu was to be supreme, four of the chiefs, viz., Balakhain, Tutar, Ghataghan, and Mighan, became rebellious. Nigudar and Bora were more submissive. The arghuchis ordered Balakhain, Tegudar,* and Tutar to be strangled with a bowstring (such was the Mongol method of putting Khans to death), while Mighan, son of Khuli, on account of his youth was arrested and imprisoned on an island in the Lake of Urmia, called the White Sea by the Mongols. They also ordered the troops of Khulagu, together with the Armenians and Georgians, to march against the contingents of the guilty princes and exterminate them, which was accordingly carried out. So many were killed that the mountains and plains were infected with the Tartar corpses. Two of the chiefs, however, named Nukhakuun and Aradamur (the Ala Timur of the *Georgian Chronicle*) fled, taking with them twelve horsemen and considerable treasure. They crossed the Kur, and went with all speed to the country whence they had come (*i.e.*, the Kipchak), where they were protected by Bereke Khan, and for ten years committed great depredations. Malakia goes on to say that Mangu's arghuchis then proceeded to duly instal Khulagu (*i.e.*, to instal him as ruler of the western countries he had conquered).†

The *Georgian Chronicle* calls the Kipchak princes who were put to death Tutaré, or Khutaré, Balgha, and Kuli.‡ In regard to the families of the slaughtered princes it adds some graphic details. We there read of the wives of Tutar, Kuli, and Balgha, who were living in Greece (*i.e.*, among the Seljuki), escaping with their baggage under charge of a certain distinguished person named Ala Timur, in the direction of Samtzkhé. They were pursued by Khulagu's people, and in a struggle which followed many of the latter were killed. Ala Timur fought several successful engagements with his pursuers before he reached the mountains of Kola. He at length arrived at the village of Glinaf, in Lower Artan, where he met Murvan Gurcelel, son of Makhujag. They would have killed him, but he promised to conduct them to Imereth, whence they might escape to Bereke Khan. But instead of this he treacherously led them to the

---

* This is a mistake ; perhaps Ghataghan is meant.
† Op. cit., 454-455.   ‡ Op. cit., i. 572, and note.

forest of Gurcel, in Samtzkhé, whence he sent on couriers to Sargis
Thmogwel, to Shalwa, son of Botzo, and all the Meskhes and people of
Sargis Jakel, to come and seize them. Ala Timur having heard of this,
took his charges across the Mtsuar (Kur) and went towards Jawakheth,
passed a place called Eladi, and reached Lerdzavni, below Oshora, where
Sargis barred his way ; but as soon as he saw the Tartars he and the
army retired. Murvan Gurcelel encountered them with a small force, but
was beaten and lost many men. The enemy then traversed Jawakheth
and Trialeth, and crossed the Kur at Rusthaf. A succession of fights
followed, in which Ala Timur was continuously successful. He traversed
Kambejian, Kakheth, and Hereth, and took the route to Belakan, entered
Ghundzeth (*i.e.*, Kunzag, the country of the Avars), whose king gave him
battle, but he was again victorious, and eventually reached the Court of
the Khan (*i.e.*, of Bereke) covered with glory. It was certainly a marvellous march, and we are told the survivors who accompanied Ala Timur
were given the style of aghnarghoms (superiors or elders).* Tutar, who
is also called Kutar, was put to death, according to the authority followed
by Major Raverty, for having caused Balgha's death by sorcery, on the
17th Safar, 658. The Sadr, Sauchi, he adds, was also put to death,
as he was charged with having prepared a charm for Tutar.†

In addition to the various causes enumerated in a former volume for
the strife between the two cousins, Guiragos tells us there was this—that
Bereke supported Arikbuka, while Khulagu was a champion of Khubilai
in the struggle for the Khakanship, thus confirming my conjecture.‡
Guiragos names the princes of the ulus of Juchi who were killed by
Khulagu, Kuli, Balakha, Tutar, Méghan, son of Kuli, Ghatakan, and
many others, who he says were exterminated with many of their followers ;
old and young all securing the same fate. Some escaped to Bereke.§ He
also adds the interesting statement that Alghui, the son of Jagatai, had a
feud with Bereke, because the latter had incited Mangu to destroy his
family,‖ and that he accordingly wrote to Khulagu offering him his
alliance against the common enemy.¶ On the other hand, Makrizi tells
us that Bibars, the Sultan of Egypt, having heard that Bereke had become
a Mussulman, sent to ask him to march against Khulagu.**

The envoys of Bibars and Bereke were welcomed at Constantinople,
where the Greeks had recently driven out Baldwin and his supporters,
who were allied with the Crusaders, the friends of Khulagu. The Emperor
asked Bibars to send a patriarch to take charge of the Melkits,†† while he
received the Sultan's envoy, who presently escorted the patriarch, very
graciously, and allowed an old mosque which had formerly been

---

* Op. cit., 570-571.   † Tabakat-i-Nasiri, 1286.   Note.
‡ Guiragos, ed. Brosset, 192. Journ. Asiat., 5th ser., xi. 504.   Ante, ii. 113.
§ Journ. Asiat., 5th ser., xi. 504.   ‖ Brosset has misread the passage.
¶ *Id.*   ** Op. cit., i. 170.
†† *i.e.*, the Greek Christians who acknowledged the Patriarch of Antioch as their head.

occupied by the Mussulmans, and which was said to have been built in the year 96 of the HEJ., to be restored. Bibars also allied himself with Manfred, the famous King of Naples, who was styled Imperator by the Arabs, and who was at deadly issue with the Pope.*

To revert to Bereke. We are told he assembled an army of 30,000 men to revenge his various wrongs. On his part Khulagu also collected his forces, and divided them into three sections. He confided the first to his son Abaka, associating with him the experienced Arghun, and sent them to Khorasan, to join hands with Alghui, the ruler of the Khanate of Jagatai. He posted a second division at the gate of the Alans (i.e., the pass of Dariel); while he himself, with the third division, went towards Derbend.† Bereke's army, under Nogai, who was a near relative of Tutar, or Kutar, and therefore appropriately helped to avenge him, had already crossed the mountains and was in the neighbourhood of Shirvan.‡ It attacked the advanced guard of Khulagu, commanded by Shiramun, the son of Charmaghan Noyan, and Semagher, or Shamaghu, and defeated it with great carnage near Shamakhi, a chief named Sultan-juk being drowned. This was on the 11th of November, 1262,§ and the reverse was apparently attributed to the flight of the troops of the Kipchak princes Khuli and Bukan (? Mighan).|| This defeat was avenged by Abatai, who arrived shortly after, and Nogai was in turn badly defeated. Thereupon Khulagu set out in person from Shamakhi for Derbend Some of the Inaks, we are told, denounced the vizier, Seif ud din bitkichi, as well as Khoja Azis, one of the governors of Georgia, and Khoja Mejd ud din Kerman. They were all put to death, as well as Hosam ud din, the astrologer, who was also denounced. Malik Sadr ud din, of Tebriz, and Ali Malik, governor of Irak Ajem, escaped the last penalty and were bastinadoed. We are not told what the offence of these officials was, but it was probably some conspiracy on behalf of the Mussulman princes who were at this time opposing Khulagu. The latter approached Derbend, where some of the enemy showed themselves, but were driven away by a flight of arrows, and the place was stormed. A fight took place with a Kipchak army outside the walls, which lasted till nightfall. It ended in the defeat of the latter.¶ It would seem that when Shiramun was beaten Khulagu sent his son Abaka with a force to the rescue. I have already described what followed.** I would only add here that among the victims in the battle on the Terek was Biurthel, nephew of the Orpelian prince, Sempad.††

The *Georgian Chronicle* in describing this campaign, in which the Georgians took part, makes them as usual fight in the advance guard. Sargis, after his recent revolt, was evidently put upon his mettle to prove

---

* Weil, iv. 43-44.   † Guiragos, ed. Brosset, 193.   Journ. Asiat., 5th ser., xi. 505.
‡ Ante, ii. 116.   § Quatremere, 393. || St. Martin, ii. 285.
¶ Quatremere, 395.   ** Ante, ii. 116.   †† Hist. de la Siounie, 233.

himself a hero. When he had ranged his men, a so-called shweli, or wild goat of the Caucasus, ran along the lines, and was killed by Sargis, who was armed *cap à pie.* Presently he killed a fox, and last of all a hare. Khulagu, having seen this, complimented him greatly. When the two armies were close to one another a tall archer left the ranks of Bereke and approached King David. The King put an arrow to his bow, and struck this man's horse in the chest. The Georgians thereupon raised a cry of joy, and charged. When Bereke's men were defeated and being pursued, Khulagu was left with but four men on a small eminence. The fugitives noticing this, seven of them rushed upon him. Sargis, who was passing at the time with three other warriors, went to the rescue and killed four of the assailants, the rest escaping. On his return to Bardaa, Khulagu, we are told, covered the Georgian king with honours, and gave Sargis Sarnukalak, with all the surrounding district, and also Erak (? Irak).* This addition to his importance aroused the jealousy of some of the Georgians, who urged on the king that Sargis would now be so powerful that he would not obey him. He listened to them, went to consult the Noyan Elgon (called Engin by Wakhusht), and told him that if the Khan gave Sarnukalak to Sargis he might as well make him king. Elgon was surprised, and said the Khan had thus rewarded him because he had defended him, but if it displeased the King he would no doubt withdraw his gift. "You Georgians," he added, "do not know how to reward those who behave bravely in battle. Don't you know that Sargis saved the Khan's life, and fought most gloriously?" On Elgon's report to the Khan, Sarnukalak was taken away from Sargis, who was much irritated. The King was detained at Bardaa during the winter by the Khan, while Sargis, who was disaffected, went to Samtzkhé.†

Meanwhile, it would seem that Guantsa, or Gontsa, the widow of Avak, and wife of the Georgian King David, remained among the Tartars, among whom she was killed. According to the *Georgian Chronicle,* her death was instigated by her daughter, Khoshak, who was married to the Seviphaddar Khoja Shems ud din. Chamitch says it was Stephen the Orpelian who ordered her death, seized her goods, and gave her daughter in marriage to the Sahib divan of Khulagu. Stephen the Orpelian himself tells us Sempad, by order of Khulagu, caused her to be drowned, and appropriated the inheritance of Avak, of which he had been trustee. It is probable that her death really occurred in 1262-3.‡ The *Georgian Chronicle* says that the King, having thus lost his wife, married Esukan, daughter of the great noyan, Charmaghan, and sister of Shiramun, and left for Tiflis, where he celebrated his wedding with great rejoicings.

At this time there arrived at the Georgian Court, as fugitives from the country of Bereke, two wonderful women named Limachav. They had some young children with them of the race of Akhasarphasaian, the elder called Pharejn and the younger Bakathar ;* there also went several chiefs, who were sent on to Khulagu. He remitted them again to David, who assigned them lands at Tiflis, Dmanis, and Jinwan.† Khulagu himself, on the approach of October, went to Shirwan, to a place called Chalan Ussuri, or White Water, where he formed the entrenchment of a camp, which was called Siba (an Arabic word, meaning an entrenchment). He went there as he expected an attack on the part of Bereke Khan. From this time, we are told, the Tartars and Georgians began to live at Siba from October till the spring.‡ Guiragos says that the war lingered on from the year 1261 to 1265, the two sides coming to blows every winter, but remaining quiet in the summer on account of the great heats and the swollen rivers.§

Rashid ud din says that Khulagu, having heard that Nogai was meditating another attack, ordered the Sheikh Sherif Tebrizi to cross the mountains of Lesghistan and to spy out what he was doing. The Sheikh having ventured into Nogai's camp, was made prisoner. Nogai asked him if it was true that Khulagu in his fury had slaughtered his sherifs, his grandees, his holy men, anchorites, and merchants. "It is true," said the Sheikh, "that he was much irritated, and has burnt the green with the dry ; but now," he added, diverting into poetry, "by his justice the fire no longer burns the silk, and the kid sucks the lioness. Quite recently envoys have arrived from Khitai with the news that Khubilai has mounted the throne, that Arikbuka has recognised his authority, and that Alghui is dead. A yarligh, addressed to Khulagu, gives him authority over all the lands from the Oxus to the borders of Syria and Egypt, and 30,000 young Mongols, picked men, are marching to his assistance." On hearing this Nogai was much disturbed, and the Sheikh returned again to his master, who rewarded him handsomely.‖

At this time Jelal ud din, the son of the Little Devatdar, of whom we heard much in the account of the overthrow of the Khalifate, who had been much patronised by Khulagu, was nevertheless treacherous. He urged that there dwelt in the territory of the Khalif many Kipchak Turks, who knew perfectly the laws and customs of their country, and requested permission to collect them together, so that they might form the advance guard in the contemplated campaign against Bereke. Khulagu approved of the notion, and sent a yarligh and a paizah ordering the governors of Baghdad to make over to Jelal ud din what he should require in the shape of arms and war engines, and that he was to be free

---

* Like Brosset, I am at a loss to explain this sentence.

† Hist. de la Géorgie, 569.   ‡ Op. cit., 569.   § Op. cit., ed. Brosset, 193.

‖ Quatremere, 399-401.

to do as he pleased.   He accordingly went to Baghdad, and having assembled those whom he deemed suited for military service, told them that Khulagu was enrolling them so that they might form a buckler and shield from the blows of the enemy.   He said that death was their probable portion in the campaign, and if they survived it, it would only be to be dragged to another elsewhere. "You know," he said, "my origin, my family, and the ties which bind me to you.   Although Khulagu has shown me very great favour, I cannot permit you to be slaughtered. I mean, with your help, to break the Mongol yoke.   We must act together."   They agreed to follow him.   Thereupon, crossing the bridge of Baghdad, he fell on some Arabs of the tribe Khafajah, and captured a number of buffaloes and camels, and took from the treasury at Baghdad the horses, arms, and money necessary for the equipment of his men. Soon after, having told them to hold themselves in readiness, with their wives, children, slaves, servants, and goods, he again beat the drum for departure, and crossing the bridge of Baghdad, said to them, " Let us take our wives and families to visit the sacred places, for otherwise they will have no other dwelling-places than Derbend, Shirvan, and Shamakhi. As for the rest of us, let us provision ourselves from the Arabs of Khafajah, who are our enemies."   After crossing the Euphrates, he said to them, " I mean to go to Syria and Egypt ; those who care may follow me, while the rest may return."   They were all afraid to speak, and went on together by way of Anah and Hadithah towards Syria and Egypt.   Khulagu was naturally very much irritated when he heard of this treacherous act.*

Let us turn once more to Egypt, where a new Khalif was at this time inaugurated.   This was the Amir Abdul Abbas Ahmed, who had escaped from the combat at A'nbar, as I have described.   His inauguration took place on the 8th Moharrem, 661, and he took the title of the Imam Hakim bi Amr Allah, and when he had stated his genealogy, which was attested by the Kadhi Mohai ud din, the Sultan swore allegiance to him, and to his duty as a faithful Mussulman, whereupon the Khalif in turn invested him with the empire "over kingdoms and men" (*i.e.*, with universal empire).   His example was followed by the various grandees, all in turn doing homage to the new head of the faith.   At the grand audience when this ceremony took place the subjects of Bereke who had fled to Egypt, as I have described,† were present.   After dispatching the envoys whom he sent to Bereke, and nominating Jemal ud din Akush as his viceroy at Damascus, Bibars set out for Gaza, in Syria, where he regulated the affairs of the Turkomans, and wrote to the ruler of Shiraz (*i.e.*, of Fars) and the Arabs of Khafajah, urging them to make war upon Khulagu, and encouraged them by telling them how he had heard of Bereke's recent victories over him.‡   From Gaza Bibars went on to Tur (*i.e.*, Mount Tabor),

---

* Quatremere, 405-415.   † Ante, ii. 114-115.   ‡ Makrizi, i. 189-190.

where he received a visit from Malik Ashraf, Prince of Hims, whom he treated with great courtesy. Not so Moguith, Prince of Karak, who was charged before the great officials, the judges, the ambassadors of the Franks, &c., with having carried on a correspondence with the Mongols, inciting them to invade Syria. From an intercepted letter of Khulagu it would seem the latter had offered him the government of Gaza. He was sent off prisoner to Egypt, where he was afterwards put to death.* Karak was soon after conquered, while on another side Bibars laid a heavy hand on the Crusaders, the allies of the Mongols, and returned home again after what was really a triumphal progress. I have described the embassies that passed at this time between Egypt and Kipchak at some length, and how, in the autumn of 1263, a large body of Bereke's people arrived in Egypt.† In the following spring there also went thither for shelter several officers of the army of Fars, some chiefs of the Arab tribe Khafajah, and the Amir of Irak Arab. They were rewarded with fiefs.

While Bereke and Bibars were united in their alliance, Khulagu was befriended by Haithon, the King of Little Armenia, who made an incursion upon the Egyptian territory, and advanced upon Aintab (Makrizi says upon Sarfand). He had previously formed an alliance with Rokn ud din, the Sultan of Rum. Bibars, who was kept well informed of the doings of his neighbours, ordered the troops belonging to the principalities of Hamath and Hims to advance upon Aleppo. The Egyptian troops followed them. This was in 1262-3. The Armenians were surprised and defeated, whereupon Haithon summoned to his help 700 Mongols who were encamped in Rum. With them he advanced into Syria, and was joined by 150 horsemen from Antioch. This little army encamped near Harim, but was obliged to withdraw by the severity of the weather. Haithon tried to deceive the Egyptians into believing that he had received a reinforcement, by dressing 1,000 of his men in Mongol capes and caps, but it availed him nothing, and the Egyptians revenged themselves by ravaging his borders and those of Antioch.‡

The rivalry of Khulagu and Bibars extended to the realms of diplomacy, and each one sought diligently for allies against the other. While Bibars drew the ties with Kipchak closer, and offered a ready asylum in Egypt to fugitives from the Ilkhan's dominions, the latter tried to win over some of Bibars' dependents, and allied himself with the various Christian communities, including the Crusaders, and with the rulers of Asia Minor. About the same time Bibars heard from his secret emissaries in Irak that Khulagu had dispatched two agents to try and tamper with his officers, and that they had set off by way of Sis, the capital of Little Armenia. He afterwards heard of their departure from Acre for Damascus, and ordered them to be arrested there. They were sent on to Cairo, where they

* Makrizi, 191-193, 242. D'Ohsson, iii. 395. Abulfeda, v. 3-9.
† Ante, ii. 117-122. ‡ Makrizi, i. 234-236. D'Ohsson, iii. 393-394.

were interrogated, and not being able to clear themselves were duly hanged.*

Khulagu had, as we have seen, put his vizier to death in the spring of 1263, when he marched to Shamakhi, and appointed Shems ud din Muhammed, of Juveni, in his place. He was given entire charge of the affairs of the empire, while his brother, Alai ud din Atta Mulk, the historian, was appointed governor of Baghdad.† The same year, Zain ud din Abul Muayid Suliman, son of the Amir El A'sarbani, better known as El Hafidi, was charged with embezzling some of the revenues of Damascus when he was governor there. Khulagu also accused him of an attempt to betray him, as he had betrayed his former masters, the Princes Nasir and Hafiz, and still earlier the Prince of Baalbek. He was executed, with all his family—his brothers, sons, relatives, and dependents, to the number of fifty; only one of his sons and one of his nephews escaped.†

Khulagu's attention was now turned to Fars. We have seen how its ruler, Muzir ud din Abubekr, sent his brother Tahamtan with rich presents to the Khakan Ogotai,§ with his submission. Ogotai granted him a diploma of investiture, with the title of Kutlugh Khan. Fars, by this submission, was saved from Mongol attack. Its ruler paid an annual tribute of 30,000 gold dinars, which was not much, considering the revenues of the province. In addition, the prince generally sent a member of his family every year with presents to the Court of the Grand Khan. When Khulagu marched west he was met on the Oxus by Seljuk Shah, the nephew of Abubekr, who was well received by him. Abubekr died in 1260, and was succeeded by his son Said, who died twelve days later, leaving a young son in the care of his widow, Turkhan Khatun, sister of Alai ud daulat, Atabeg of Yezd. This infant, named Muhammed, died two years later, whereupon one of his uncles, named Muhammed Shah, who had commanded the contingent of Fars in Khulagu's campaign against Baghdad, succeeded. He was brave, but cruel, and his tyranny caused discontent. He had married Turkhan Khatun (Von Hammer says he married her daughter, Selgham). The Khatun, who disliked him, had him arrested as he was passing her harem, and conducted to Khulagu, with a message to the effect that he was not fit to reign. She then, with Khulagu's consent, released Seljuk Shah, brother of Muhammed (so named because he was descended from the Seljuki on the mother's side), from his imprisonment in the citadel of Istakhr, married him, and put him on the throne. He had a vile temper, and one day, when drunk, having been taunted with what he owed to Turkhan Khatun, he ordered a eunuch to go and decapitate her. Presently, the negro returned with the head of the beautiful princess in a golden basin, whereupon her brutal husband tore two pearls from her ears, and threw them to the musicians.

---

* D'Ohsson, iii. 394.   † Quatremere, 403-405.   ‡ Novairi, in D'Ohsson, iii. 397.
§ Ante, 36.

He then fell upon the two Mongol commissaries at his Court—Ogul Beg, or Oghlubeg, and Kutlugh Bitikji—killed one with his own hand, and had the other put to death, as well as all their people. Thereupon Khulagu ordered Muhammed Shah, whom he was about to release, to be put to death, and sent his generals, Altaju and Timur, together with the contingents of Ispahan, Luristan, Yezd, Kerman, and Ij,* to march upon Fars. They sent a messenger to call upon Seljuk to submit, and offering him pardon. The latter was cruelly maltreated. The Mongols accordingly entered Fars, with the ruler of Kerman, the Atabeg of Yezd, who was brother to Turkhan Khatun, and the Prince Ilk Nizam ud din Hasneviyeh, who ruled a small mountain district of Fars. Seljuk Shah retired with his troops to the borders of the Persian Gulf. The magistrates of Shiraz went out with banners, korans, and provisions to meet Altaju, who, having promised them safety, forbade his people to plunder and marched on. They met Seljuk Shah at Kazerun, or Kiarsun. He fought desperately, but had to give way, and took shelter in the mausoleum of a holy sheikh, named Morshed, where he was duly beleaguered. Going up to the saint's tomb, he struck its cover with his mace and broke it, saying, "O, sheikh, come to my help," for it was known there that the sheikh had bidden those who were menaced by any danger to let him know it at his tomb. The Mongols soon forced their way in and captured the Atabeg, who was put to death at the foot of the castle of Sifid. This was in 1264.† There only remained of the Salgar dynasty two daughters of Said, son of Abubekr. One of these, named Abish Khatun,‡ who was the daughter of Turkhan Khatun, was placed by Khulagu on the throne of Fars. Timur, one of the Mongol generals, wished to exterminate the people of Shiraz, as an example, but was restrained by his colleague, who declared its citizens were innocent, and that the army could not do this without an order from their master. He contented himself with carrying off some of the notables to Khulagu's Court.

Meanwhile, the grand judge of Fars, Sherif ud din, who was one of the chief Seyids or descendants of the Prophet in Fars, and therefore a person of much consequence, having become ambitious, called upon the people of the province to do him homage. In the various towns and villages he passed through many attached themselves to him, believing him to be the Madhi expected at the end of the world by the Shias, and that he performed miracles. Having adopted the insignia of royalty, he went from Shebankiareh to Shiraz with a crowd of followers. The Mongol prefect and Abish Khatun's chief minister sent an army of Mongols and Mussulmans against him. The rival forces met each other at Guvar. It was thought that spirits fought for the Sheikh, and that whoever struggled

---

* The capital of the Shebankiar princes.
† Mirkhond, quoted by D'Ohsson, iii. 397-402. Weil, iv. 131-132. Ilkhans, i. 241-243.
‡ See Ilkhans, i. 243. Note.

against him would be paralysed. At first the people of Shiraz were in consequence afraid to fight, but two soldiers having ventured to shoot their arrows others followed their example. The Mongols now charged the insurgents, who fled, and the Seyid with the greater part of his people were killed. This was in May, 1265. When Khulagu heard of the revolt he ordered Altaju to be bastinadoed for having interfered with his colleague's wish to destroy the people of Shiraz, and he sent a tuman of soldiers to wreak his vengeance on the place ; but hearing that the Seyid had been killed, and that the Shiraz people had not taken his part, he revoked his order.*

After reigning for a year, Abish Khatun, who is called Uns by D'Ohsson, was summoned to the Ordu to marry Mangu Timur, son of Khulagu. From this time Fars was governed by the Mongol Divan, in the name of Abish, who brought her husband a handsome dowry, comprising a sixth of the domains of Shiraz, with an annual charge of 8,000 ducats upon them. She reigned nominally for twenty years, but the authority was really in the hands of the Mongol baskaks and maliks. On her death, in 1287, at Tebriz, the Salgarid dynasty came to an end.†

We will now turn to an obscure corner of our subject. A turbulent tribe of Kurds, named Shebankiars, occupied one of the five districts of Fars called Darabgherd. They succeeded in forming a separate principality, under Nizam ud din Mahmud, son of Yahia, grandson of Hasuieh, or Hasnuieh. In 624 the principality was ruled by Muzaffer ud din Muhammed, son of Almarz, son of Hasuich, who increased his territory by the conquest of several towns and districts bordering on Hormuz. The district of Shebankiareh was bounded towards Fars by Hasuieh, Rabir, and Khireh. In another direction were the towns of Mishkanat, Lar, Babek or Sanek, and Guristan, seven parasangs from Hormuz.‡ The turbulent Shebankiars made raids on Fars, burnt the palm and other fruit trees, destroyed the crops, &c., and the troops of Fars sent against them made really no impression. In the year 658 HEJ. (i.e., 1260), Khulagu sent Tekucheneh, of the Jelair tribe, with orders to capture the fortress of Ij, the capital of the Shebankiars. He advanced with 17,000 men to attack it. Malik Muzaffer ud din Shebankiari and the garrison bravely defended it, when at length he was struck in the eye with an arrow, and died. His children and the principal inhabitants of the place thereupon determined to submit. Tekucheneh, having thus secured the capital, marced upon Esfid (the White Castle), situated to the south. This he captured, broke down its walls, and destroyed its cisterns. Ij itself, we are told, contained 17,000 houses, squeezed close together round the citadel. It was much favoured by nature, and produced the fruits both of warm and cold countries, and its purple

---

* Mirkhoud, in D'Ohsson, iii. 402-404.   † Id., 404. Ilkhans, i. 243.
‡ Quatremere, 445-446.

oranges, figs, and apricots were especially noted. Having secured the treasures in the place, Tekucheneh conferred the principality of Sheban-kiareh on Kutb ud din Mubarez, son of Malik Mozaffer ud din, and appointed Mongol darughas to be with him. Eleven months later, on the 10th Zulhijah, 659, Kotb ud din was assassinated by his brothers. He was succeeded by Nizam ud din Hasuieh, or Hasnuich, son of Ghiath ud din Muhammed, and grandson of Muzaffer ud din. He fell on Zebr the 2nd, 662, in a fight with Seljuk Shah, near Kazerun, and was succeeded by his brother, Nusret ud din Ibrahim, by a special edict of Khulagu, and as his brother had been killed in fighting for the Khan he received orders, according to Mongol custom, to marry his widow. She eventually married in succession two of her husband's brothers. Nusret ud din died on Zebr the 2nd, 664, and was succeeded by Jelal ud din Taib Shah, who was on the throne for seventeen years.*

• In the latter part of 1264 the Mongols laid siege to El Biret, where the Amir Jemal ud din Akush commanded on behalf of Bibars. It was deemed the key of Syria, and they proceeded to fill up its ditch with wood. The besieged mined underneath and set fire to the wood, and the battery of seventeen catapults which was brought against the walls was met by a vigorous resistance on the other side, in which even the women took part. It seems that the Franks had written to the Mongols advising them to invade Syria in the spring, when the Syrian troops were dispersed in their several fiefs and their horses were out at grass. When Bibars heard that El Biret was being assailed he sent a contingent under the Amir Iz ud din Aigan, and four days later a second body, under Jemal ud din Aidogdi, to its assistance. Bibars set out in person on the 27th of January, 1265. He arrived six days later at Gaza, and there heard of the precipitate retreat of the enemy, the fact being that on the approach of the Amir Aigan, in alliance with Mansur, Prince of Hamath, they had raised the siege and hastily retired. Bibars ordered El Biret to be supplied with arms, provisions, and everything that would enable it to sustain a siege for ten years. He sent 200,000 drachmas and 300 robes of honour to be distributed among its defenders.† Meanwhile, he proceeded to press the Crusaders once more. Abulfeda tells us that this year (*i.e.*, 1264) he captured Carkesia (*i.e.*, Circesium, one of the towns on the Euphrates) from the Tartars, which was governed in his name by Rahaba.‡

Let us return once more to Khulagu. He spent the year 1264 in reforming the administration of his dominions. He charged his eldest son, Abaka, with the government of Irak, Mazanderan, and Khorasan, as far as the Oxus. To Yashmut, his third son, he intrusted Arran and Azerbaijan, as far as the Araxes. The provinces of Diarbekr and Diar-

---

* Quatremere, Rashid ud din, 447-448.   † D'Ohsson, iv. 404-406.
‡ Op. cit., v. 17.

rabia, from the Tigris as far as the Euphrates, were made over to the Amir Tudan. Rum was assigned to Mo'yin ud din Pervaneh, Tebriz to the Malik Sadr ud din, Kerman to Turkan Khatun, and Fars to the Amir Ankianu.* This is Rashid's statement, but in reference to Kerman it is certainly a mistake, for Kerman was subject to the family of Borak, the Kara Khitaian.

Vartan tells us how, during the year 1264, the great Ilkhan, Khulagu, summoned him by a man named Shnorhavor (*i.e.*, the gracious), who had acquired considerable influence at the Courts of Khulagu and Baju. Shnorhavor transported him and his companions, viz., the Vartabieds Sargis and Gregory and the married priest Avak, from Tiflis. They arrived at the solemn season of the Mongol new year, that is, in July, when the Tartars spent a month in feasting, and held their kuriltai or grand assembly, which was attended by the various chiefs and by the subject princes. Each day, says our author, those who attended the meeting wore a costume of a different colour.† Vartan noticed at the Court, Haithon, King of Little Armenia, David, King of Georgia, the Prince of Antioch, and a number of Sultans from Persia. When he was admitted the ceremony of prostration was excused. Khulagu caused his visitor to bless the wine which he received from his hands. "I have sent for you," said the Mongol chief, "that you might see me and make my acquaintance, and pray for me with all your heart." He caused him to be seated, and offered him wine, while the monks, his companions, chanted hymns. The Georgians, the Syrians, and the Greeks all celebrated their offices. Khulagu, noticing what a crowd of clergy had assembled, said he had summoned only him, and wanted to know why they had gone at this particular time to see and bless him—a most unusual phenomenon—and concluded that it was by the special favour of God, a view in which Vartan diplomatically concurred. On one occasion he sought a more private interview with our chronicler, making his people stand at a distance, and related to him, in the presence of two others only, the various events of his life from his childhood. He told him that his mother was a Christian, and that he felt much attached to the Christians, and, taking his hand, bade him speak frankly if he wanted him to do anything. Vartan says he answered as he was inspired : "Just as you are raised above other men, so are you more like a god. The throne of God reposes on justice. He gives to each nation the empire of the world, and puts it to the proof. Hitherto these nations have ravaged the earth and been pitiless towards the unfortunate. They now have to bear a heavy servitude themselves, and their plaints are laid before God, who has taken the power away from them and given it to others. If you are benefactors to the people and pitiful to the weak, He will not take it from

---

* Quatremere, Rashid ud din, 403.   † Compare the statements of Rubruquis, &c.

you, but will let you keep what He gave you, for He takes away from one and gives to another as He pleases. Place about your gates men who fear God and are faithful to yourself. When the unfortunate come to you with tears in their eyes, and with nothing to offer, send them home again satisfied. Cause your realm to be inspected by honest men, who will not take bribes, and will report the truth to you." Khulagu replied that it was singular these views should have been already impressed on his heart, and asked if God had appeared to his visitor or spoken to him. Vartan said no, he was but a poor fisherman, and had merely read the books of men who had spoken on God's behalf, while kings were in the hands of God, who it was clear had spoken to him (Khulagu) personally. He went on to tell him that all the Christians who lived on land or sea were devoted to him at heart, and did not cease to pray for him. " I believe it is so," said Khulagu, "but the Christians are not in God's presence. What good will it do if they pray to Him for me? Can they secure a favourable hearing? Can the Christian priest cause God to come down on to the earth. Those alone pray to God who follow His ways. On these questions we and our brothers are at issue, for we love the Christians and their faith is favourably looked upon by us, while they favour the Mussulmans. Why don't you wear a robe of golden tissue, instead of one of sheep's skin?" Vartan replied, he was not a grandee, but a simple monk; that gold and dust were to him of equal value, and what he would prize much more would be to secure his goodwill for the people. The conversation ended by Khulagu offering him money to buy incense for his church. At a subsequent interview, where he bade him adieu, Khulagu gave him a balish and two dresses. He also gave him a yarligh or diploma, which he bade him read to see if it contained what he wished, and if not to alter it, and told him he had confided the care of his country and person to Sakhaltu and Shampandin, with orders to obey his behests.*

Abulfeda thus enumerates the provinces ruled by Khulagu : Khorasan, whose capital was Nishapur ; Irak Ajem, commonly called Belad al Jibal (i.e., the mountainous region), whose capital was Ispahan ; Irak Arab, whose capital was Baghdad ; Azerbaijan, with its capital, Tebriz ; Khurestan, with its capital, Tostar (i.e., Shuster) ; Fars, with its capital, Shiraz ; Diar Bekr, with its capital, Mosul ; and Rum, whose capital was Conia (i.e., Iconium).† Rashid says he had been told by the Sheikh Shems ud din, of Ispahan, that Khulagu never enjoyed absolute authority. He ruled as the viceroy of his brother Mangu, and could not strike money in his own name, but the dinars and dirhems were struck in the name of Mangu Khan, a practice which was followed out by Khulagu's son, Abaka. Arghun, the son of Abaka, was the first to add his name to the coins, and

Gazan eventually excluded the Khakan's name altogether. During the reign of Khulagu and his successors the Khakan had an amir who lived at their Courts as his representative, and who was treated with great consideration and honour,* but after the step taken by Gazan, as above mentioned, this office virtually ceased, and became of small repute. The author of the "Mesalik Alabsar" confirms the statement that Khulagu never had absolute authority. During Khulagu's reign and those of his successors the Khakan used to send an amir into Iran, who lived there as his representative, and who was treated with every honour and consideration by the reigning Ilkhan.†

The actual power of Khulagu was no doubt greatly strengthened by the struggle which took place for the supreme Khanship of the Mongols in Mongolia. I described this struggle in a former volume.‡ We are told that among those who supported Arikbuka against Khubilai was Jumkur, the son of Khulagu, who had his camp in Mongolia, in Mangu's old country, and was therefore constrained to side with Arikbuka, who represented the Nomadic Mongols against Khubilai, whose settlement in China, and adoption of Chinese habits, had probably irritated his less sedentary subjects. When, subsequently, Arikbuka declared war against Alghui, the chief of the Jagatai Horde, Jumkur accompanied him ; but, feigning illness, he left him when near Samarkand and went to join his father, who all through took the side of Khubilai, as his rival, Bereke, did that of Arikbuka. Jumkur had been previously ordered by Khulagu to remain neutral.§ Khubilai rewarded Khulagu's constancy by appointing him, as we have seen, ruler of all the country from the Oxus to the borders of Egypt,‖ and he adopted the title of Ilkhan, on which see the note at the end of the chapter.

Khulagu died on the night of the 19th Rebi, 663 (i.e., the 8th of February, 1265), at the age of 48. "A comet," says Rashid ud din, "appeared in the sky, in the shape of a pointed column, and showed itself for several nights. When it at length disappeared there happened the great catastrophe."¶ Makrizi also describes this comet in some detail, and from his account it must have been a very imposing object.** Malakia and Guiragos also describe it at some length. The former says that Khulagu, as soon as he saw it, knew that it referred to him. He touched the ground with his head and adored God, and his fears increased greatly as its light diminished. Khulagu survived the appearance of this comet only for a year, when he died, leaving thirty sons.†† Vartan, in referring to the great chief's death, uses the turgid phrase, "Death, with his great foot, overthrew this lofty mountain, and levelled it with the

---

* Quatremere, 12. Note.   † Quatremere, Rashid ud din, 12-13. Notes.
    ‡ Ante, i. 217, &c.   § Quatremere, 101. Ilkhans, i. 222.
‖ Rashid ud din, quoted by Quatremere, 12-13. Notes.   ¶ Quatremere, 417.
                        ** Op. cit., i. 241-242.
†† Malakia, 458-459. Guiragos, ed. Brosset, 194. Journ. Asiat., 5th ser., xi. 508.

plain." This contrasts sharply with the terse words of Abulfeda, " In the year 683, 4th month and 9th day, there died the cursed chief of the Tartars, Hulaku, son of Tului, son of Jingis Khan, near Meragha."* It happened in his winter quarters, on the banks of the River Chagatu Nagatu, that is, says Rashid, the Zerineh rud, or golden river.† (See Note 3.) According to the Egyptian historian, Ibn Tagri berdi, Khulagu was subject to fits of epilepsy, which perhaps account for his strange cruelties at times, and the attacks became more and more frequent, until he had two and three fits in one day. He at last became worse, and after lingering for two months, died.‡ He was buried on the summit of the mountainous island of Shahu, opposite Dihkhawarkan (the Deschawakan of Von Hammer). D'Ohsson identifies it with Yala, on Lake Urmia, but Von Hammer questions this, and quotes Rashid to show it was near Mount Sehend (i.e., near Kazvin).§ According to Mongol custom, they placed gold and precious stones in the tomb, while some young and beautiful damsels, in rich garments, were buried with him, and funeral meats were offered at his grave for several days.‖

Vartan tells us that on one occasion Khulagu said to him, " I have not sent for you to obtain exemption from death for me, for I know it to be inevitable, but that you will pray God that I may not die by the treason of my enemies." " God alone knows," says this chronicler, " whether this wish was gratified, for the news spread abroad that he was in fact poisoned."¶ This is not altogether unlikely, when we consider the animosity he and his wife showed to the Mussulmans. His death was quickly followed by those of two of his wives—one of them the mother of his eighth son, Ajai, eight days after his own death ; and four months and eleven days later the famous Princess Tokuz, or Dokuz Khatun, who had been his father's wife before she joined his harem.** She was a Christian, and through her influence, as Rashid ud din says, the Christians were much favoured by Khulagu, who, profiting by this patronage, built many churches in various provinces.†† Vartan says the Tartars carried about with them on their journeys a cloth tent in the form of a church, where the jamahar or rattle called the faithful to prayers, and where priests and deacons performed the services daily. There were also schools for the children. The ecclesiastics were well treated at their Court, and consequently crowded thither from all parts.‡‡ Malakia says much the same.§§

The deaths of Khulagu and his wife were naturally much regretted by the Christians. Rashid tells us how, to please her, Khulagu had loaded them with favours, that churches daily arose in various parts of the empire, and that one was always stationed at the gate of her ordu,

---

* Op. cit., v. 15.   † Quatremere, 401. D'Ohsson, iii. 406-407.   ‡ Id., 406. Note 3.
§ Quatremere, 417. Ilkhans, 229.   ‖ D'Ohsson, iii. 406-407.   Wassaf, 97.
¶ Journ. Asiat., 5th ser., xvi. 308.   ** Quatremere, Rashid ud din, 419-423.
†† D'Ohsson, iii. 407.   ‡‡ Journ. Asiat., 5th ser., xvi. 290.   §§ Op. cit., 455.

O

where bells were sounded.* Bar Hebræus writes, in 1265, "At the beginning of Easter died Khulagu, whose wisdom, magnanimity, and great deeds are not to be matched. The following year died the very faithful queen, Dokuz Khatun. The grief of Christians in all the world was very great at the departure of these great lights and protectors of the Christian religion."† Stephen Orpelian speaks in even stronger terms. "The great and pious king, the master of the world, the hope and stay of the Christians, Khulagu Khan, died in the year 1264. He was soon followed by his respected wife, Dokuz Khatun. They were both," he adds, "poisoned by the crafty Sahib Khoja. The Lord knows they were not inferior in well-doing to Constantine and his mother Helena !!!"‡ Haithon, speaking of Dokuz Khatun, says she was devoted to the Christians and very zealous in destroying the temples of the Saracens, and so illused the latter that they dared not show themselves.§

This sympathy for the Christians seems to have reached the ears of the Roman Pontiff, and Odoric Raynald has published a letter without name or date, but which he assigns to Alexander IV., and the year 1261. The Pope expresses the pleasure with which he had heard from a certain Hungarian, named John, how he had been commissioned by Khulagu to report to him his willingness to become a Catholic, and his wish that some one would go and baptise him. "O what joy," says the Pope, "fills our heart when we consider how your presence will delight your Maker and Redeemer, who gave Himself up for the salvation of mankind to the punishment of the Cross, if you present yourself on the day of judgment with the mark of baptism and the other emblems of Christianity ; not only you, but all your subjects, who will no doubt follow your example, a fact which will increase your merit and your eternal recompense. Surrounded by this crowd, rescued from the very throat of the enemy, with what safety will you await this terrible judgment. Consider, my son, consider how transitory is this life—how quickly and easily the body decays. If you have any such intention, therefore, it would be well to lose no time. See how it would enlarge your power in your contests with the Saracens if the Christian soldiery were to assist you openly and strongly, as it could, with the grace of God. You would thus increase your temporal power, and inevitably also secure eternal glory. For the rest, as the aforesaid John has not produced clear proofs of his commission, we have addressed our letters to our venerable brother, the Patriarch of Jerusalem, telling him to inform himself of your Serenity's intentions, and then to write to us. This is why we request you to confide your wishes to the patriarch, so that we may make all the necessary arrangements."||

Guiragos tells us how Khulagu was much imposed upon by the Tartar

---

* Quatremere, 95.   † Op. cit., Chron. Syr., 567.   Chron. Arab., 355.
‡ Hist. de la Siounie, 234-235.   St. Martin Memoires, ii. 151-153.   § Op. cit., 44.
‖ Mosheim, Hist. Eccl. Tart., Appendix, xvii.   D'Ohsson, iii. 410-412.

magicians, whom they called Tuins, who he says could make horses, camels, and the idols made of felt* talk. The priests had their heads shaved, and wore yellow mantles fastened about their necks. They adored everything, but especially Sakyamuni and Madri (i.e., Maitreya). The former, says Vartan, was a god, and had already lived 3,040 years ; he still had 37 tumans of years (i.e., 370,000), when Madri would evict him. They persuaded Khulagu that he himself would live to a great age in his present body, and would then pass into a new one.† He regulated his conduct by their decisions, and halted, marched, or mounted on horseback when they pronounced it propitious. He daily prostrated himself several times before their chief. He ate meats consecrated in the idol temples, and treated the priests with greater consideration than anyone else, and was very lavish in gifts to their temples. His wife, Dokuz Khatun, reproached him frequently, but she could not turn him aside from these magicians.‡

Malakia tells us that many kings and princes having offered him rich presents, Khulagu became so powerful and rich that his horsemen and troops were innumerable. His riches, precious stones, and pearls were like the sands of the sea, without counting all kinds of precious things, a great quantity of gold and silver, and horses and cattle without number. Personally, he says, he was a man of great intelligence and justice, that he was very cultivated, and although he shed a great quantity of blood, he only put the wicked and his enemies to death, and not good men ! ! ! He loved the Christians more than any others. He reports that when he once levied a tax of 200,000 heads of sows§ upon Armenia, he sent 200 swine into each of the towns belonging to the Tajiks (i.e., the Persians), with orders that they were to eat them, and ordered a special report to be furnished him of those who ate the pork. If any Tajik, great or small, refused to eat, he had him decapitated. He did this (i.e., this outrage upon Muhammedan feelings) to please the Georgians and Armenians in his service, who, from their bravery, he named bahadurs. His body guard, who took charge of the entrance of his tent, was made up of the sons of the great Armenian and Georgian princes, whom he named kesiktoi. They were armed with bows and arrows. He began the restoration of the places which had been devastated, and selected certain artisans from each town, whom he called yam, one from each of the small ones and two from the great, and sent them in various directions to repair the ruins, exempting them from all taxes except to supply bread and soup to the Tartars who should pass that way.||

Novairi reports some singular judgments given by Khulagu. Several people went one day to ask justice from him against a manufacturer of

---

* The idola or imagines de feltro of Rubruquis and Plano Carpini, and les dieux de feutre of Marco Polo. † Journ. Asiat., 5th ser., xvi. 306.
‡ Guiragos, ed. Brosset, 193-194. Journ. Asiat., 5th ser., xi. 507.
§ Brosset says the translation of the word is doubtful. | Op. cit., 456-457.

files, who had killed one of their relatives, and who they demanded should be given up to them for punishment. Khulagu having inquired if there were many manufacturers of files in the country, and learning there were only few, ordered the aggrieved to avenge their relative's blood upon a manufacturer of pack saddles, for they were numerous. As they insisted that this would not do, he made over a cow to them in satisfaction. On another occasion, a gold embroiderer having thrust out the eye of a man in a quarrel, Khulagu ordered an arrow maker to be deprived of an eye, and when asked the reason for this curious decision, he said the embroiderer had need of both eyes, while the other had need of but one, since he closed the other to see if his arrow was straight.* He was very fond of architecture.† The year before his death he divided between his architectural works and the administration of the empire. He also pushed on the completion of the observatory at Meragha, which we have already named. Rashid ud din says he was also very fond of philosophy. 'He encouraged learned men to discuss science and history, and granted them pensions and gifts, and was especially fond of alchemy. In the pursuit of this hobby, says the matter-of-fact historian, his assistants burnt a great number of different substances and, "without any real gain, caused many large and small volumes of smoke, and made some large earthen crucibles. But all this produced nothing, and merely earned them their morning and evening meals. Nor could they produce a single piece of gold or silver which they had made, from their laboratories. The amount of money wasted in this search was so enormous that the unfortunate Karun, during all his life, and with the aid of the philosopher's stone, could not have replaced it."‡

Khulagu had six wives and twelve concubines. The first wife was Dokuz Khatun, of whom we have said so much. She had been betrothed to his father, Tului, shortly before the latter's death, but Tului had not consummated the marriage. According to Mongol custom, Khulagu married his stepmother. Her niece Tukuri, or Tukiti, became one of his concubines, and inherited her ordu.§ Dokuz was the daughter of Ittiko, or Iku, second son of Wang Khan, and was therefore niece of Khulagu's mother. They neither of them had children. He also married two wives from the Uirat tribe, viz., Kubak, Koyuk, or Kuik Khatun, and her half-sister Oljai, the former the mother of his second son, Jumkur, and the latter of his eleventh son, Mangu Timur. They were both daughters of the Uirat chief, Turalji. Kuik's mother was Jijegan, daughter of Jingis Khan. Kuik was the first wife married by Khulagu. She died in Mongolistan. He also had two wives from the Kunkurat tribe— Kutui Khatun, the mother of Tekshin (called Bikin and also Tekshi by Quatremere), his fourth, and of Ahmed Takudar, his seventh son ; and

Mertai Khatun, who was childless. The former had the ordu of Kuik given her when she died. Yisut, or Yisunchin (Quatremere calls her Sunjin), of the Suldus tribe, was the mother of Khulagu's eldest son and successor, Abaka, who was only a month older than Jumkur.* Yashmut and Tuzin, Khulagu's third and sixth sons, were by a Chinese concubine, named Tukaji, or Bukajin Ikaji, who was a slave in the household of his wife Kutui. Tarakai, who was killed by lightning in Persia, was the son of Burkajin, also a slave in Kutui's household; he was Khulagu's fifth son. The mother of Ajai, his eighth son, was Irtikan Ikaji (called Arikak by Quatremere), also in the household of Kutui. He remained at the head of her establishment when Khulagu marched westwards. Ajuji or Juji Ikaji, the mother of his ninth son, Kuikurtai, or Kunkurtai,† was a slave in the household of Dokuz. She was afterwards decked with the boktak or pyramidal cap, which was the symbol and privilege of a wife, as distinguished from a concubine.‡ Yesudar, the tenth son, had Uwishjin (Quatremere calls her Hesijin, sister of Akrabeighi), the Kurlas, for his mother; Hulaju, his twelfth son, had Il Kaji, a slave in the household of Dokuz, for his; while the mother of Sherbaweji, or Siauji, the thirteenth, and Taghai Timur, the fourteenth son, was also a slave in the household of Kutui. Her name is illegible in the MS. followed by Quatremere. Of Khulagu's seven daughters, Bulughan Aka was the daughter of Kobak, or Kuik. She married her uncle, Jume Kurkan, the son of the Tartar Juji, brother of Bokdan Khatun, chief wife of Abaka Khan; on her death he married Jemi, Khulagu's second daughter by his wife Oljai (Quatremere calls her Hami). His third daughter, Mengelugan, or Manglukan, also by Oljai, was married to Jaku, or Jakir, Kurkan, the son of the Uirat Buka, or Tuka, Timur, and the brother of Oljai; she therefore married her uncle. His fourth daughter, Tutukaj, or Budakaj, by a slave in the household of Dokuz, was first married to the Uirat Tengkir, or Tenker, Kurkan; she afterwards married his son Sulamish, and lastly his grandson Jijak Kurkan, so that she was married to the father, son, and grandson. The fifth, Tarakai, whose mother was the concubine Irtikan Ikaji (Quatremere here calls her Baganikaji) already named, was married to the Kunkurat Musa Kurkan, the grandson of Jingis Khan, by his daughter Tumalun. The sixth daughter, Kutlukan, or Kotlkan, whose mother was Menklikaj ikaji, was first married to Yisubuka Kurkan, of the tribe Durban, and afterwards to his son Takel. Baba, Khulagu's seventh daughter, whose mother was Oljai, was married to Legsi, or Lekzi, Kurkan, the son of the famous amir, Arghun Aka, of the Uirat tribe.§

During Khulagu's reign coins were struck both in silver and copper. They may be divided into two series—those struck during the supremacy

---

* Ilkhans, i. 82-83. † Quatremere, 107. ‡ Id.
§ Ilkhans, i. 83-86. Quatremere, 107-113.

of Mangu (spelt Mönghe on the coins) as Supreme-Khan, on which Mangu's name and title occur in full ; and those struck during the reign of Khubilai, in which only the Grand Khan's title occurs, and on which the inscription reads : "The very great Kaan ; the great Hulagu Ilkhan." During the former period Khulagu styles himself Khan on his coins, and during the latter Ilkhan. On his coins in the British Museum, Mosul, El Basrah, Mardin, El Mubarakiyeh, El Jezireh, Irbil, and Jorjan occur as mint places. As in the case of some of the Khans of the Golden Horde, previously referred to,* coins occur with Khulagu's name on them struck in the years 665-669, that is, after his death.† These posthumous coins are probably to be explained in the same way as those of Janibeg.‡

Note 1.—On his coins struck during the reign of Mangu, as we have seen, Khulagu styled himself Khan, but after the accession of Khubilai he called himself, as did his successors, Ilkhan. This title has received more than one explanation. Quatremere has devoted a learned note to it. *Il*, in the dialect of Jagatai-Turkish, and in Persian, means nation, tribe, or people. The word is used also as an adjective, and then means subject, dependent. As used by the Mongols, Ilkhan doubtless meant Khan of the people, or of the nation. It must be remarked, however, that in a marginal note to a passage in Wassaf *Il* is glossed as meaning "Great," Ilkhan thus meaning Great Khan. The former meaning is, however, much more probable.§ Fraehn, in a memoir on the coins of the dynasty, supposes the title conveys a notion of dependence on the Grand Khan. He also explains it as meaning the strong Khan, the energetic Khan. It has also been explained as meaning Prince of Peace. Vartan gives the title as Elghan, whence Brosset suggests the Georgians derived the forms Eldjin and Ehdjin.‖

Note 2.—The Mongols were essentially nomades, and they never ceased to be more or less nomades during their occupation of Persia. Their armies had wintering quarters and summering quarters, between which they moved to and fro. Many stories are told to show their attachment to their nomadic life. *Inter alia*, we read how the powerful amir Nuruz, wishing to seek shelter at Herat, was warned by some of his people, who said to him : " The Mongols have the immensity of the desert in lieu of fortresses and citadels. It behoves the amir, therefore, to renounce his project, and that he takes care not to confine himself between four walls."¶ The author of the " Mesalek Alabsar" says: "The town of Aujan is situated near Tebriz. It is surrounded by splendid meadows, and has capital supplies of water. There is a palace there, built by the later Mongol chiefs, and all around it the principal amirs have built themselves houses. The princes, amirs, and chiefs also form

* Ante, ii. 204.   † Poole's Catalogue of Oriental Coins, vi. 8-16.   ‡ *Vide* Ante, ii. 204.
§ Quatremere, Rashid ud din, 14-15. Note.   ‖ Hist. de la Géorgie, i. 539. Note.
¶ Quatremere, 69-70. Notes.

pounds or parks, inclosed with reeds and wattles, in which they keep their cattle, and in which they live during the winter season, which they pass at Aujan. They also make tents of felt and horsehair. During the winter the place looks like a vast town, with great streets and markets in it, but when they leave Aujan for their summer quarters they set fire to all the huts, for otherwise a vast number of serpents would accumulate there. The princes of Iran (*i.e.*, the Ilkhans) passed the winter either at Aujan or at Baghdad. The summer camp is at Karabagh, meaning in Turkish the black garden, and so called from the colour of the soil. In that district are many settlements. The air is pure, the water excellent, and the pasture abundant. When the ordu, or Ilkhan's camp, is fixed there, and the princesses and amirs have built their houses, jamis or mosques are also built, and bazaars, where objects of all kinds are sold, are constructed. There are also houses for courtesans. Although food, utensils, clothes, &c., are there in any quantity, they are very dear, in consequence of the cost of transport, which doubles their value. The Sultan always has with him in his journeys some of the principal wise men and doctors, who receive pensions from the treasury. Each one is accompanied by several fakihs (jurisconsults) and disciples, and they are known as itinerant doctors. The chief officers of state, commanders of troops, tax officials, scribes, and workmen of all kinds follow the ordu, so that it resembles a large town. Tents, ready made and furnished, and of various sizes, can be bought by those who need them."[*]

To revert to Aujan, the wintering quarters of the Mongols. Ghazan Khan gave a grand fête there, which is described in detail by Rashid ud din. Wassaf also names a general kuriltai or diet as having been held there. It was at Aujan that Uljaitu was proclaimed Ilkhan, and it was near there that Adil Shah Khatun, wife of Abu Said, died in 732 HEJ. Timur, after his expedition against Baghdad, spent eight days at Aujan, in Arghun's palace. Lastly, in 823 the famous Turkoman chief Kara Yusuf died there.[†] Rashid ud din refers to the Zerineh rud, or Golden River, called Chagatu Nagatu by the Mongols, as the wintering quarters of Khulagu.[‡] According to the author of the "Nozhat Alkolub" this river springs in the mountains of Kurdistan, not far from a town called Siahkuh. It traverses the district of Meragha, and unites with the rivers of Safi, and Bagatu or Nagatu, and eventually falls into the salt lake of Tasuj (*i.e.*, the Lake of Urmia).[§] The summer quarters of the Mongols in the time of Khulagu and his successors, according to Rashid, were in the neighbour-hood of Alatagh. Khulagu halted there in his campaign against Syria, and, we are told, was so pleased with the pastures in the neighbourhood that he gave it the name of Lebnasagut.[‖] According to the "Jihan Numa" the Alatagh range is that in which the Murad chai, or Euphrates, springs. Alatagh, in Turkish, means the spotted mountain.[¶] Rashid ud din tells us that Khulagu built a palace there, and idol temples (*i.e.*, Buddhist temples) at Khoi.[**] Malakia, in referring to this palace, says it was built in the plain of Darhin Dasht, adding the odd comment, which is no

---

[*] Quatremere, 21-23. Note. [†] *Id.*, 23. Note. [‡] *Id.*, 400.
[§] *Id.*, 103-105. Note. [‖] *Id.*, 329. [¶] D'Ohsson, iii. 580. [**] Quatremere, op. cit., 411.

doubt a mis-translation, that he called it after his own name, Alatagh. He
adds further that this place was formerly the residence of the kings of
Armenia, of the Arkhakunian dynasty.* Chamitch identifies this plain with
the famous plain of Mughan, whose pastures are so famous, and which is
situated partly in Azerbaijan and partly in Arran and Shirvan.† Stephen the
Orpelian, speaking of the same palace, calls the plain where it was built Darhan
Dasht, and says the Tartars called it Aladagh. Sempad, we are further told,
went to Basen, by order of Khulagu, to get red cedar-wood for this palace.‡
St. Martin also identifies this plain with the plain of Mughan, south of the
Kur and the Araxes.§

Note 3.—The fact of Khulagu marrying his stepmother, Dokuz or Tokuz
Khatun, sounds strangely in our ears, but it was quite in accordance with the
Mongol law. Quatremere says that when a man died among the Mongols, and
above all a prince, his wives became the property of his eldest son, who could
marry which he liked, except his mother, and dispose of the rest as he pleased.
Juveni says this was the custom among the Mongols and Uighurs, and he is
confirmed by Abulfaraj. According to Haidar Razi, the Mongol law prescribed
that the senior wife had the disposal of the others, but Quatremere suggests
that the senior wife is here a mistake for the eldest son. Rubruquis says that
on the death of a Mongol chief his son married his various wives, except his
own mother. Similar statements are made by Carpini and Marco Polo. Among
the Mongols a man's sons took rank according to the rank of their mother.‖
Khulagu only married Dokuz Khatun after he had crossed the Oxus, when he
already had several other wives, but as his father's widow she retained her
pre-eminent position.

Note 4.—In 1259 Khulagu was visited by an envoy from his brother Mangu,
named Chang ti. The narrative of his journey has been published by Remusat
and Pauthier, whence I abstracted it in the first volume of this history.¶ Since
then there has appeared Dr. Bretschneider's admirable edition of the same
story, which has corrected many mistakes, and is illustrated by his usual wealth
of notes, and I feel bound to abstract the story again from his pages, in so far
as it deals with the Ilkhan's dominions, leaving the earlier part of the journey
to be illustrated in the next volume. Two days after leaving Talas, he says, he
reached Bie-shi-lan (? Tashkend), where a fair was held by the Muhammedans
like the fairs in China. Next day Chang ti crossed the Ilukien, or Sir Daria,
in a boat resembling a Chinese lady's shoe. Jade, we are told, was produced
in the mountains at the sources of the Sir Daria. In the district our traveller
now entered there were post stations, and inns like bathing-houses (*i.e.*,
caravanserais), whose windows and doors were glazed. Ten golden dinars
was the maximum poll tax paid by each individual there. Sün sze kan (*i.e.*,
Samarkand) our diarist describes as large and populous, and the country
round as very fertile, roses and other flowers, vines, rice, winter wheat, and
many medicinal plants growing abundantly. Chang ti crossed the Anbu,
(*i.e.*, the Amu Daria or Oxus) on the 14th of the 3rd month. In this district,

---

* Op. cit., 456.   † *Id.* Note 1.   ‡ Hist. de la Siounie, 233.
§ Memoires, ii. 283. Note 10.   ‖ Quatremere, Rashid ud din, 92-93. Notes.   ¶ i. 280.

he tells us, it did not rain in summer, only in autumn, when the ground became very moist. There were large swarms of locusts there, and of birds which ate them. Bretschneider identifies the latter with the *pastor roseus*.[*] Five days later he passed Lich'u (?), where mulberry trees and jujubes abounded. In this place he tells us the Mongols halted for some days in their march westward. Our traveller successively passed Ma lan (? Merv),[†] Na shang (? Nishapur),[‡] where lucerne was the chief grass and where cypresses were used for hedges, and T'i-sao-r.[§] Thence he went by Gi-li-r (? some place in eastern Mazanderan).[||] Here large lizards, five feet long, and with black and yellow bodies, were found;[¶] and passing A-la-ding(?) and Ma-tze-t'sang-r (?), where the people wore dishevelled hair, red turbans, and black clothes, thus resembling devils, he apparently terminated his journey and reached the Court of Khulagu (who was then at Tebriz) in the latter part of April, 1259. We have no account of his further progress, or of his dealings with Khulagu.[**] I ought to add that the whole of the narrative of his journey, with nearly all the illustrations, I owe to Dr. Bretschneider.

* Op. cit., 77. Note 78.   † *Id.* Note 81.   ‡ *Id.*
§ Bretschneider suggests Sebzevar. He remarks that Conolly mentions mines of rock salt near Nishapur. *Id.*, 78. Notes 83-84. || *Id.* Note 96.
¶ Dr. Bretschneider suggests they were a species of *stellio*. *Id.*, 80. Note 97.   ** *Id.*, 81.

CHAPTER IV.

ABAKA KHAN.

VARTAN, who was in the confidence of Dokuz Khatun, Khulagu's Christian wife, tells us that before the Khan's death she consulted him as to whether they should say masses for his soul. He replied that this would not be proper, but that they should distribute charity and remit taxes. The Syrians on the contrary argued that such a mass was allowable. Dokuz Khatun also consulted him as to whether Abaka should be put on the throne in accordance with his father's will, or no, and he advised that he should be so.*

When Khulagu died a courier was dispatched to summon Abaka, who held the post of Governor of Khorasan, and who was then in winter quarters in Arran,† with his vizier, Arghun, and the princess Oljai Khatun. All the routes leading to the imperial residence were meanwhile closed, and travellers were stopped. Yashmut, his younger brother, who was encamped in the neighbourhood of Derbend, arrived seven days after his father's death, and hastened to sound the amirs as to his own prospects, but meeting with no encouragement he left again, two days later, for his government. Abaka arrived at the ordu of Chagatu on the 9th of March, 1265,‡ and was welcomed by the amirs; the marshal of the ordu, Ilkai, offering him the funeral meat and wine usual on such occasions. Having presented their devotions to the spirit of the dead chieftain, the khatuns, princes of the blood, and generals met to elect a successor. The principal chiefs who thus came together were : Ilga or Ilkai, Sugunjak, Suntai, Abatai, Temagu (called Semaghar by Von Hammer), Singtur or Shiktur, and Arghun Aka.§ Singtur, who had received the last wishes of Khulagu, attested that he had nominated Abaka as his successor. As was usual, he professed to decline the honour, and offered it to each of his brothers, but they as regularly on their knees pressed it upon him. He then said he could not mount the throne without the concurrence of his uncle, the Khakan Khubilai. The whole assembly replied that no one had a better right to the throne than himself, who was the eldest brother and had been

---

* Journ. Asiat., 5th ser., xvi. 308.   † D'Ohsson says in Mazanderan.
‡ Weil, on the authority of Novairi, says the 8th of February.
§ Wassaf says Arghun Aka, Ilkan, Peder i Shiktur, Borghan Oghul, Amir Sansis, Shiramun, son of Charmaghan, &c. Op. cit., 102.

nominated by his father, and that no one knew better what the yasa prescribed. Thereupon, on the 19th of June, which the kams and astrologers declared to be a lucky day, Abaka, whose name means "maternal uncle" in Mongol, was duly inaugurated at Chaghan Nur (*i.e.*, the White Lake), in the district of Berahan.*

It is curious to read that among those who attended the obsequies of Khulagu and the inauguration of Abaka, was Mar Ignatius, the Jacobite Patriarch of Antioch, who obtained a diploma confirming him in his post.† Vartan says a prince of the blood named Ilkhan Takudar (? Abaka's brother, so called) placed him on the throne, and all the army ratified the choice and did homage,‡ The princes of the blood each went with his girdle over the back of his neck and prostrated himself seven times before the sun. The fêtes lasted for several days, during which dissipation of various kinds was rife. Abaka did not wish to adopt his full style until the authorisation came from Khubilai, and till then abjured a throne, and would only consent to be seated on a stool, whence he dispensed justice. Having distributed gifts among the various officials, messengers were sent out in different directions to announce his accession, and that the yasa of Jingis Khan would be rigorously carried out. He then distributed the various great appointments. He was born in March, 1234, and was therefore 31 years of age. His brother Yashmut was given command of the troops on the northern frontier towards Derbend, Shirvan, the plain of Mughan and Alatagh. Tekshin, another of his brothers, had charge of the eastern frontier of Khorasan and Mazanderan. Tughu, or Tuguz, the bitikji or secretary, son of Ilkai Noyan, and Tudan, brother of the Noyan Sunjak, or Sugunjak, commanded the troops stationed in Rum, where they were afterwards relieved by the Amirs Semaghar and Kehurkai. Durtai, or Dutai Noyan, led those at Diar Bekr and Diar Rabia, on the Syrian frontier. Shiramun, son of Charmaghan, commanded those in Georgia, while those in Baghdad and Fars were confided to Sugunjak Noyan, whose deputy at the former town was Alai ud din, brother of the vizier. The management of the Crown demesnes was made over to Baltaju Aga, and that of the Imperial dues (makatir) to Arghun Aka. Shems ud din Muhammed, of Juveni, was nominated vizier and head of the divan at Tebriz, and his son Khoja Bahai ud din was put at the head of affairs at Ispahan. The administration of Khorasan was confided to the Khoja Iz ud din Tahir, and after him to his son the Khoja Weji ud din. Fars was governed in the name of the Atabeg Abish, while Tasiku was sent thither to superintend the dues claimed by the Imperial treasury. Kerman was subject to the Princess Turkan Khatun ; Nimruz

---

* Berahan, or Ferahan, was a town of moderate size situated near a lake 16 parasangs square, where, according to Persian tradition, Tahmuras the Div tamer built himself a palace. D'Ohsson, iii. 413-415. Ilkhans, i. 245-247. Shajrat-ul-Atrak, 248.
    † Bar Hebraeus, Chron. Eccl., i. 760-762.   ‡ Journ. Asiat., 5th ser., xvi. 308-309.

to the Malik Shems ud din Muhammed Kert ; Georgia to Abd (doubtless a corruption of David) and his son Sadren ; Armenia (*i.e.*, Little Armenia) to Haithon. Diar Bekr was governed by Jelal ud din Tarsi, Diar Rabia by Mozaffer Fakhr ud din Kara Arslan, Kazvin and a portion of Irak by Iftihshar ud din Kazvini, and Tebriz by Sadr ud din.

As Von Hammer says, this enumeration, which is that given by Rashid ud din, proves the administration to have been like that of the Ottomans in later times, which was in many points imitated from that of the Mongols. The military and civil administration were in different hands, as was the direction of the finances from the other functions of state. There were, as we have seen, six frontier commanders, stationed respectively in Shirvan, Khorasan, Georgia, Rum or Asia Minor, Fars, and the Arabian Irak ; three viziers, heads of the divan, at Tebriz, Baghdad, and Ispahan ; three superintendents of taxes and revenue, and five overseers of internal affairs. Rashid enumerates, as we see, five princes who still retained their sovereignty under Mongol suzerainty. These were the rulers of Kerman, Nimruz (*i.e.*, Herat), Georgia, Little Armenia, and Fars. To these should be added the Malik of Herat, the Atabegs of Great and Little Luristan and of Yesd, the Princes of Mardin of the Ortokid stock, and those of Hosnkief of the Ayubit family.[*]

Soon after his appointment as head of affairs at Baghdad Alai ud din, of Juveni, the historian, was the victim of an intrigue. The prefect of the town, Karabuka, and his deputy, the Armenian Isaac, having a grudge against him, plotted to undo him, and suborned a certain Bedouin,[†] who falsely accused Alai ud din of intending to escape with his family and property into Syria, and declared that he himself was to conduct them. Karabuka thereupon had him confined in his house. But the matter having been examined at the ordu, the Arab confessed under torture that he had been incited to say what he did by the Armenian Isaac, and he and Isaac were put to death. Alai ud din was restored to his honours.[‡]

Under the fostering care of the vizier, Shems ud din Muhammed, the empire began to revive. "The sheep," says the inflated Wassaf, "recovered the blood-tax which the wolves had so long taken, and the partridge exchanged loving looks with the falcon and hawk. Through him the good name of the Padishah was inscribed in fortunate characters on the white and black pages of the day and the night." Under his patronage Baghdad, which was immediately governed by his brother, Alai ud din, began once more to flourish. He spent a hundred thousand gold pieces in digging a canal leading from the Euphrates to Meshed, near Kufa, and the neighbourhood of Nejef. Upon this canal Taj ud din Ali, the son of the Amir Dolfendi, who was intrusted by the vizier with its

---

[*] Ilkhans, i. 248-249. D'Ohsson, iii. 416. Vassaf, 98-102.
[†] In the Chron. Syr. he is called a Mede.
[‡] Abulfaraj, Chron. Syr., 568. Chron. Arab., 355-356.

construction and with the cultivation of the desolate land, wrote a special treatise.* While the vizier, Shems ud din, devoted himself to restoring the country to prosperity, his eldest son, the Khoja Bahai ud din, who was over the divan at Ispahan, conducted himself very differently. He was a person of considerable attainments, especially cultivated philosophy and the *belles lettres*, and studied music under Safr ud din Abdul Mumin. As we have seen, he was given charge of the Persian and Arabian Irak, and of Yezd, and had his seat at Ispahan. He governed it with the greatest rigour ; a word spoken contrary to his wishes was followed by the overwhelming of the household root and branch, many thousands were tortured or put to death, great and small all trembled for their lives, and the people of Ispahan when they went to bed at night were in mortal dread of what might befal them in the morning. Nevertheless, he put down all kinds of ruffianism. Bloodshed and outrage had been common in the city, open robberies in the bazaars, and the workpeople had had neither rest nor safety. The ill-doers now became so cowed and humble that the peasants used to leave their agricultural implements in harvest time in the fields at night. If anyone dared to remove them "the harvest of life of the delinquent was speedily cut off with the sickle of destruction."† The overseers and leaders were so carefully checked that the market folk used at night to leave the booths strewn with goods and food without anyone in charge and no one took the smallest thing. As a proof of this it is said that on one occasion at night as the watchmen were going their rounds, one of them entered the booth of a sugar baker or confectioner, took a sweet cake, and left two silver dirhems, which was double the price, in the corner of the booth. On the following morning, when the owner of the booth found a dirhem more than his due he dared not conceal it, could not rest, showed the silver piece to the treasurer, and gave information of what had happened. It was immediately ordered that the watchman who had transgressed the rigid law should be hung to a hook like meat at the butcher's.†

Bahai ud din had a slave named Nikpei, whom he employed as a spy upon the watchmen and police. Wassaf tells a story how he reported of three men : that one was vigilant and dutiful, that another was found sleeping at his post, and the third, instead of being on duty, had wandered away. The governor ordered all three to have 71 strokes of the cudgel. The Sheikh of Islam, Jemal ud din, protested that he who had done his duty had not deserved this, and should not be treated like the two delinquents. The governor replied, "The reason for their punishment is negligence ; his fault was that when Nikpei came to him furtively in the night he did not punish him as an evil doer, and did not make

---

* Wassaf, 112. Ilkhans, i. 250. † Wassaf, 116. ‡ *Id.*, 116-117.

inquiries into his reason for being abroad at such an hour."* On one occasion when the Khoja was riding out with his Court amidst great pomp, he was annoyed that one of the common people should stare at him, summoned him and asked him what he was looking at. " The poor man's tongue was bound in a knot," says Wassaf, meaning he was silent. The tyrant thereupon gouged his eyes out with his knife, and tore out his eyelashes. The following quatrain was written upon this lugubrious act :—

> An eye in wrath was torn out since it gazed on you,
> And why, since many thousands do the same ?
> The angel of death has removed you from office.
> How many souls does death not overwhelm ?

To show his passion, the same author mentions that one of his boys, a favourite child, having in play touched his beard, he swore a terrible oath, and ordered him to be put to death. As none of the grandees, the imaums, or the queens interceded for the child, he was seized by the executioner and put to death. This terrible act stirs the rhetoric of Wassaf into unwonted vigour, and seems to have made a great impression. The rigorous measures of Bahai ud din certainly produced order at Ispahan, and it is reported that after his death disorders again broke out, and Wassaf was told that after an outbreak there more dead bodies were found about the streets than. all the victims of his severity put together. He was a great worker, and distributed his time in the active duties of his position and in the patronage of learning, devoting little to sleep or to his harem. He built many palaces and other buildings, and laid out pleasure grounds. His weakness seems to have been wine, in which he indulged with his brother Khoja Hasun and his intimates. At these parties the great musician, Safr ud din Abdul Mumin, already named, was generally present. On one occasion Hasun having had too much to drink, addressed the musician, and called him familiarly by his proper name, Safr ud din. This familiarity greatly displeased Bahai ud din. Hasun replied, " I am a son of the head of the Divan, and have wedded a pearl from the mussel of the Khalifate. My name is Hasun, and that of my son Mamun, and I am now Governor of Baghdad, where the khalifs ruled, and where there are innumerable excellencies. Is it strange, therefore, that I should adopt the mode of the khalifs and address him as Safr ud din ?" This reply was unanswerable. Soon after this Bahai ud din died. He was only 30 years old. His death took place on the 23rd of December, 1279,† and he was much regretted by his father. His was a strange type of a Draconic nature, in which hardness and cruelty were prominent factors. He nevertheless secured a short respite of peace in a very turbulent community.‡

Abaka chose Tebriz for his capital, appointed Alatagh and Siah kuh

---

* Wassaf, 117-118.   † Id., 118-125.   ‡ Wassaf, loc. cit. D'Ohsson, iv. 11-13.   Note.

(*i.e.*, the Black Mountains) as his summer quarters, and Arran, Baghdad, and Chagatu for his quarters in winter. We have seen how Khulagu asked the hand of one of the daughters of Michael Palæologus in marriage, and how he accordingly sent one of his natural daughters, named Maria. Her mother belonged to the family Diplovatatze. She was escorted by Theodosius, of Ville Hardouin, Archimandrite of the convent of Pantocrator, and brother of the Prince of Achaia and the Peloponnesus.* Her father gave her some splendid presents, including a tent of silken hangings, which was destined for a church, and containing golden figures of the saints, crosses, sacred vessels, &c.† On arriving at Cæsarea the princess heard of Khulagu's death, but she continued her journey and married Abaka.‡ Vartan says that before marrying she wished him to be baptised, and the rumour went abroad that he was so baptised.§ Guiragos expressly says that the Patriarch of Antioch and other bishops, Sargis, bishop of Ezenga, and the Vartabed Pener, baptised Abaka, and then married him to the princess.‖ The Mongols called her Despina, from her Greek title of princess, and she is so called by Rashid ud din, who makes her a daughter of the ruler of Trebizond.

The first important event in the reign of Abaka was the war he sustained on his northern frontier against Nogai, the general of Bereke Khan of the Golden Horde. This was in the spring of 1266. The Ilkhan had passed the previous winter in Mazanderan, whence he moved to Tebriz. It was while there he heard of Nogai's invasion. I have described what followed elsewhere.¶ According to Vartan, Bereke defeated Abaka and his son, and afterwards crossed the Kur and repaired to pay his respects at the tomb of a Mussulman saint. The troops who were stationed there had built a solid rampart with a ditch, which they called Shipar (*i.e.*, the Siba previously named), and employed the winter in making all kinds of preparations. Bereke, losing confidence, retired. Vartan says he was reputed to be of a pacific nature, and averse to shedding blood.** The " History of Herat " tells us how, on the invasion of the Tartars of the Golden Horde, Shems ud din, the Malik of Herat, was at Abaka's Court. The latter promised him a handsome reward if he would march with him and take command of a picked body of 200 cavalry, each man having a coat of mail, cuirass, sword, and javelin. The Malik swore to sacrifice his life if necessary to secure victory, and we are told he took off his helmet and rushed bareheaded into the fight. He was badly wounded on this occasion. Abaka was much struck by his bravery, and sent his private surgeons to attend him, and after he had beaten the enemy he conferred on Shems ud din a special diploma, and the drums and banners

* Abulfaraj and Guiragos call her conductor Euthymius, Patriarch of Antioch.
† Pachymeres, Stritter, iii. 1045.   ‡ Abulfaraj, Chron. Syr., 567-568. Chron. Arab., 355.
§ Journ. Asiat., 5th ser., xvi. 309.
‖ Op. cit., ed. Brosset, 194. Journ. Asiat., 5th ser., xi. 508.
¶ Ante, ii. 123-124.   ** Journ. Asiat., 5th ser., xvi. 311.

which were the insignia of royalty, and returned to Herat with a rich booty.*

At this time the Georgian king, David, went to Abaka's Court, and was well received by him. When Bereke made his invasion David was summoned to attend Abaka with his troops. The *Georgian Chronicle* says that the Ilkhan, on discovering the strength of Bereke's army, instead of crossing the River Mtsuar (? Kur) contented himself with an inspection of all the fords, and planted garrisons at the confluence of the Mtsuar and the Araxes, and thence as far as Mtzkhetha. Bereke, when he , had ravaged Shirvan, Héreth, Kakheth, and the borders of the Yor, advanced as far as Tiflis, and many Christians were killed, but he died while in the mountains of Garesja, and his people withdrew beyond Derbend with their booty. Fearing a repetition of the invasion the people of the Ilkhan repaired annually, in October, to Sibat (*i.e.*, the Shipar of Vartan). The rampart there is said by eastern writers to have been bordered by a wide ditch, and to have reached from Dalan or Valan, or Dalai Nur (? the Caspian), to the Kurdish waste. Wassaf calls it Assia. It was garrisoned by Mongol and Mussulman troops.‡ At this time David, the Georgian king, who had grown jealous of Sargis-Jakel-Tzikhis-Juarel, summoned him to his palace, and imprisoned him in the arsenal. The arsnaurs in Sargis' service repaired to the Khan, who demanded from Abathai Noyan that the king should be punished, that Abaka should be informed of what had happened, and that the captive should be released. Abaka consented to this. Sargis was sent for to Tiflis, and thenceforward, until the reign of George the Brilliant, who mounted the throne in 1318, the princes of Jak were immediately subject to the Ilkhans, and not to the Georgian kings.§

Abaka spent the winter of 1266 in Mazanderan and Jorjan, the ancient Hyrcania, and the next year went to Kebud Jameh (*i.e.*, "Blue cloth"), in Taberistan, to meet his mother Yisunchin Khatun, who arrived from Mongolia with Kutui Khatun, another widow of Khulagu, with Tekshin and Takudar, two sons of Khulagu, with Jushkaf, son of Jumkur, and Badu, son of Tarakai. Jumkur was Abaka's younger brother, and had been left in charge of his ordus as we have mentioned. He afterwards set out to join Abaka, and died *en route*, leaving two sons, Jushkaf and Kinkshu. Abaka gave the revenues of the district of Mayafarkin as an appanage to Kutui Khatun for pin money (called tonlik by the Mongols); Diarbekr and Jezireh to Oljai Khatun, another widow of Khulagu, and other domains to the sons Khulagu had left by his various concubines.‖ He spent the winter at Changanlu, near Meragha, and the summer in the meadows of Alatagh at the sources of the

---

* Journ. Asiat., 5th ser., xvii. 452-453.   D'Ohsson iv. 180.
† Hist. de la Géorgie, i. 572-573.   ‡ D'Ohsson, iii. 419.   Von Hammer, Ilkhans, i. 255.
§ Op. cit., 572-573.   ‖ D'Ohsson, iii. 419-420.

Euphrates and Siahkuh and the following winter (*i.e.*, that of 1266-7) in Arran.

We will now revert to the progress of the Egyptian arms, whose recovery of Syria we have previously traced. After the death of Khulagu, Bibars, the Egyptian Sultan, attacked the Crusaders in Palestine vigorously, and during the years 1265-6 captured the towns of Cæsarea, Arssuf, Safad, Yafa, and Shakif, and the fortresses of Meluhat, Haifa, Jeliba, Arka, and Kaliat from them.* Bibars now turned upon Haithon, the King of Cilicia, or Little Armenia, who under the ægis of the Mongols had, as we have seen, considerably enlarged his borders at the expense of the Mussulmans. He demanded the surrender of these conquests, the payment of tribute, the opening of commercial communications with Syria, and the exportation of horses, mules, grain, and iron, from his country. Not having received a satisfactory answer he sent an army against Cilicia, commanded by Al Mansur, the Prince of Hamath, under whom were the two generals, Iz ud din Ighan and Saif ud din Kalavun.† Malakia tells us that Haithon left his army under his two sons, Leon and Toros, and himself went with a body of troops to secure the assistance of the Tartars between Ablastan and Cocosn.‡ Abulfaraj says he went to make his appeal to the Mongol commander in Rum, called Naphshi, who replied he could do nothing for him without the orders of Abaka. He thereupon dispatched a messenger to the Khan himself ; but, meanwhile, his army had been attacked by the Egyptians. The same author tells us that Haithon's brother, Gondu Setbal, as well as his two sons, were with the Armenians. Leon had posted his men in the pass of Iskanderun, near the sea, which Abulfeda says he protected with catapults. The Egyptians however, forced the heights which commanded it, and which were thought safe, the forts which the king had planted there being very strong. They attacked Leon near the fortress of Serund, or Hajar Surwand. The Armenians were defeated, Toros and one of his uncles were killed, and his other uncle, the Constable, fled, leaving his sons in the hands of the Mussulmans. Leon himself was made prisoner, and the Armenian army, which comprised twelve princes, was completely dispersed.§ Malakia charges the Armenian princes with treachery, saying they gave the heirs to the throne into the hands of the infidel wolves, and themselves fled to their mountain fortresses.‖ The following day the victors reached Tel Hamdun, devastating the country on the road. They crossed the river Jihan and captured the fortress of A'mudin (called Arsaf by Weil), which belonged to the Knight Templars. Two thousand two hundred people were in the fortress at the time ; the men were killed, the women and

---

* D'Ohsson, iii. 420. Von Hammer, Ilkhans, i. 257-258.
† D'Ohsson, iii. 421. Vartan, in Journ. Asiat., 5th ser., xvi. 311. Lebeau, xviii. 472. Abulfaraj, Chron. Syr., 568 ; Chron. Arab., 356.
‡ Malakia, op. cit., 460.
§ Abulfaraj, Chron. Syr., 568-569 ; Chron. Arab., 356. D'Ohsson, iii. 421-422. Ilkhans, i. 257.
‖ Op. cit., 460.

P

children were carried off, and the fort was then burnt. Vartan says that the Sultan captured Sis, the Armenian capital, and discovered the royal treasure, which was contained in a storeroom, and it is said that out of one vase or cistern 6,000,000 gold tahegans were taken. He advanced as far as Adana, and eventually retired with 40,000 captives, but our chronicler chiefly laments the death of Toros, whom he greatly praises, and who, he says, on being captured, refused to give his father's name, in order that they might not spare him and use him against his country.* Abulfaraj tells us the Great church and all the others there were burnt, except the Jacobite churches of Deipara and Barsuma, and this because they were made of stone.† While the Prince of Hamath superintended the ruin of Sis, the General Ighan moved towards the frontiers of Rum, and Kalavun destroyed Ayas, Massissa (Mopsuetia), and Adana, and burnt many ships. They carried off a great number of captives. They burnt the monastery of Paximatus, but they did not molest Guiechat, since there was a monk there who could speak Arabic and parleyed with them, nor did they go to Tarsus. After wasting the country for twenty days, and advancing as far as Adana, they withdrew. Vartan says they carried off 40,000 captives, ‡ and another author that the number of cattle secured was so great, that although oxen were offered at two drachmas each they found no purchaser.§ King Haithon, who was getting an old man, did not shine in these transactions. Malakia says he repaired to the hermitage of Acants, where he remained with the monks till the withdrawal of the enemy, and afterwards gave way to exaggerated laments. Meanwhile Leon, his son, was carried off to Egypt, where Bibars, who had been jeered at as a slave by Haithon, said to him, " Thy father called me a slave, and refused to be at peace with me. Now it is thou who art my slave." || Abulfaraj says that a few days after the retreat of the Egyptians, Haithon returned with a body of Turks from Rum, and of Mongols, who finally destroyed what the invaders had spared.¶ Malakia says Haithon summoned his grandees, and when they were assembled asked if all were present. When they replied yes, he asked pathetically where Leon and Thoros were, whereupon the assembly broke out into lamentations in the spirit and language of Jeremiah. He afterwards consulted with them as to the best way of securing the young prince's release. He told them that he was informed by Armenians at the Court of Abaka that the bitikchis, or secretaries, there, who were doubtless chiefly Mussulmans, were secret partisans of the Egyptians, and were writing to Bibars to say they would urge Abaka to overrun Armenia and trample it under. It would seem that their method of operations was to suggest to the Khan that Haithon was carrying on a secret correspondence with the

* Journ. Asiat., 5th ser., xvi. 311-312.   † Chron. Syr., 569. Chron. Arab., 556.
‡ Op. cit., 311.   § D'Ohsson, iii. 423. || Op. cit., 460-461.
¶ Chron. Syr., 569 ; Chron Arab., 356.

Egyptian ruler, a charge which the death of his two sons, one would suppose, would have saved him from.[*] This reported communication from the Armenian princes at Abaka's Court may have been genuine, but I confess it looks like a clever tactical move on the part of the King to assist his proposed negotiations with Bibars.

Meanwhile the Pervana, who administered Rum, desired to ally himself with the Armenian king by marriage, and suggested the matter to the monk Persig, who was Haithon's envoy at the Ilkhan's Court, who suggested that when the King passed through his territory he should approach him, pay him great honour, and make his request in person, which he thought would not be refused. When the King, therefore, was returning from the ordu, and arrived at Kertai, the Pervana went to him with his grandees and many presents, and preferred his request. The king, who was afraid he would be waylaid *en route* if he refused, promised to give him his second daughter. When he reached home and the Pervana pressed for her to be sent, he replied it was not seemly that her marriage festivities should be in progress while her brother was still a prisoner in Egypt. Meanwhile she died, and the Pervana wreaked his vengeance on Persig, the monk.[†] Haithon now appealed directly to Bibars as to the terms upon which he would surrender his son. He replied that if Haithon would procure the liberty of a friend of his, Shems ud din Sonkor el Ashkar, of Samarkand, called Sangolascar by Haithon, to whose good offices he owed his lucky escape from Baghdad, and who had been captured at Aleppo by Khulagu, Leon should be exchanged for him. Bibars' friend is called Sghur by Malakia, who says the Armenian King collected a great number of valuable presents, with which he repaired to Abaka at Mosul, and laid his difficulties before him, and secured the release of Sghur, who had been in confinement at Samarkand. In addition to this Bibars also insisted on the surrender of the fortresses of Bahasna, Darabsak (or Darbasak), Marzaban, Roban (called Ra'nan by D'Ohsson), Er Rub, and Sikh ul Hadid, and undertook to release Haithon's son and nephew and their dependents. A treaty to this effect was signed at Antioch, and was duly carried out. This took place in 1267.[‡] The country ceded by this treaty was the district included between the River Jihan and Syria. The Jihan is a well-known river falling into the Gulf of Iskanderun. Haithon says that the King surrendered the citadel of Aleppo which the Armenians had held since the days of Khulagu. He also gave up Tempesack and dismantled two other fortresses.[§] On the return of Leon from captivity, Haithon set out for Abaka's Court at Baghdad to thank his patron for the assistance he had rendered in securing his son's release, and to ask that

* Op. cit., 463-464.   † Abulfaraj, Chron. Syr., 570 ; Chron. Arab., 357.
‡ Abulfaraj, Chron. Syr., 569-572.   Malakia, 464.   Abulfeda, v. 23.   D'Ohsson, iii. 423-425.
§ Op. cit., 50.

on account of his great age and infirmities he might be allowed to resign
the Royal dignity in favour of Leon.   To this Abaka consented, and Leon
went to the Ilkhan's Court and was duly invested with the kingdom;
Haithon himself becoming a monk under the name of Macarius.   He
shortly after died and was buried in the monastery of Drazark.   Leon
was duly consecrated at Tarsus, and devoted himself to restoring some
prosperity to his country, which had been so terribly shattered by the
Egyptian inroads.   Haithon, according to Brosset, died on the 28th of
October, 1270.*

We may well believe that the Egyptian attack on Cilicia was very
distasteful to the Mongols, and we find Abaka in 1269 sending envoys to
Bibars, who received them at Damascus, as well as those sent by the
Greek Emperor and by Mangu Timur, Khan of the Golden Horde.   In his
letters Abaka reproached him with the murder of his master Kuttuz, and
demanded how he, a mere slave, who had been sold at Sivas, dared to
resist kings and sons of kings.   He menaced him with his vengeance, and
told him if he mounted to the clouds or descended into the ground he
should not escape him.   Bibars acknowledged that he had killed Kuttuz,
but he added that he had been elected by the people.   As to his threats,
he was ready to receive him, and hoped to recover what the Mussulmans
had lost.   Abaka's envoy was sent back with this answer.†

Abaka's intentions in regard to Egypt were frustrated, for a while at
least, by an invasion of his eastern borders by Borak, the ruler of the
Jagatai Horde.   Borak had sent Masud, the famous governor of Transoxiana, on a special mission to the Ilkhan.   The professed motive of his
journey was to look after the special domains belonging to Kaidu, Ogotai's
grandson, and to himself within the jurisdiction of Abaka, and as the bearer
of a friendly message for the latter ; but he was secretly instructed to make
inquiries about the armies of Irak and Azerbaijan, and also about the
roads traversing those provinces.   Masud crossed the Oxus, and posted
onwards, leaving two horses with a trusty man at each post-station on his
route.   When he drew near Abaka's residence the latter's vizier,
Shems ud din, went out to meet him with presents, and notwithstanding
his elevated rank he dismounted and kissed Masud's stirrup.   The
latter asked him if he was the chief of the divan, and then rudely
said that his reputation was in excess of his worth.   Shems ud din,
who was a proud man, dissembled his rage, for, as Khuandemir
says, the place was not a suitable one for explanations, and he remained
silent.   At his interview with Abaka, Masud was treated with special
honour.   He was given a seat above the other amirs, except the
Noyan Ilka, and was dressed in the tunic of Jingis Khan.   He acquitted
himself gracefully, using diplomatic and courtly phrases, and gained the

    * Abulfaraj, Chron. Syr., 572.   Malakia, 464-466.   Lebeau, xviii. 474.   D'Ohsson, iii. 425.
Abulfeda, v. 29.   Haithon, 51.   † D'Ohsson, iii. 426.

confidence of Abaka; but feeling presently that he was an object of
suspicion, he asked for his *congé*. Abaka ordered that the information
asked for by Masud was to be ready in eight days, whereupon he hastily
set out on his return. The day after he left, news came from Khorasan
that Borak was preparing for war, and that Masud was only a spy.
Abaka dispatched a messenger to arrest him, but, as we have seen, he
had arranged relays of fresh horses at the post stations, and retired so
rapidly that, according to Wassaf, he reached the Oxus in four days and
nights. Having crossed that river, he reported to Borak what he had
learnt.* This journey of Masud's was apparently made in the winter of
666 (*i.e.*, 1267-8).† Von Hammer dates it a year earlier ; Weil a year
later. Before he set out Borak tried to secure the alliance of Nigudar, one
of the princes in Abaka's service. It does not clearly appear whether he
sent a special envoy for the purpose or intrusted Masud with the
commission. Nigudar was the eldest son of Juchi, the eldest son of
Jagatai, and had accompanied Khulagu in command of the contingent
furnished by the Ulus of Jagatai. The *Georgian Chronicle* says he
commanded two myriads (*i.e.*, 20,000 men) ; that he had his summer camp
in the mountains of Ararat, and his winter one on the banks of the
Araxes and at Nakhchivan. It calls him Thaguthar Khan, and makes
him Borak's brother.† Malakia, the Armenian historian, also calls him
Thagudar, and says he was very rich in men and treasures ; that it
required 300 camels and 160 carts to carry his riches, while his flocks
and herds were innumerable. He had 40,000 horsemen under him, brave
and intrepid warriors, who were accustomed to plunder the caravans.
They also attacked and ravaged the villages, plundering their contents,
and killing their inhabitants. They assailed the monasteries, hung the
monks up by their heels, and having mixed salt and soot thrust it into
their nostrils, saying, " Bring us a sea of wine and a mountain of meat."
In many places they forced the monks in the monasteries who said they
had no wine to hold a dog's tail in their mouth while making the statement,
this being a mode of swearing with them. In consequence of these
indignities the Armenian and Georgian princes went to Abaka, and putting
their swords down before him, demanded either that he would deliver
Nigudar and his people over to them or make them put him to death in his
presence. Other Tartars also presented their complaints that the people
of Nigudar plundered them and carried off their horses. Malakia makes
Abaka declare to them that Nigudar was too strong for him to punish.§
This author knows nothing of the negotiations with Borak, to which we
must now revert. Among the presents taken by the latter's envoy was
one of the arrows, called tugané by the Mongols, which concealed a

* Khuandemir, Journ. Asiat., 4th ser., xix. 252-255.   Wassaf, 132-134.   D'Ohsson, iii. 432-433.
Von Hammer, Ilkhans, i. 255.   † Khuandemir, op. cit., 252.
‡ Hist. de la Géorgie, i. 575-576.   § Malakia, 465-466.

letter. On presenting it the envoy made a certain sign which Nigudar understood, and on breaking the arrow he found in it a letter from Borak announcing his intended invasion of Persia, and expressing a wish that he who, like himself, was descended from Jagatai, would not fight against his relative. Nigudar, who was at the Court, accordingly asked permission to return to his quarters in Georgia. Presently, on more alarming news arriving from Khorasan, Nigudar was summoned to Abaka's presence to take part in a council of war. He made various excuses, and presently set out for Derbend, in order to reach Borak by the north of the Caspian.*

The *Georgian Chronicle* makes the negotiations start with Nigudar, who suggested to Borak that by attacking the dominions of Abaka on either side they might secure them, and makes him send the arrow. Borak is made to answer his overtures by a similar missive, and to suggest that in the course of two months he should be ready to rise. The time was very short, but Nigudar managed to assemble his women and baggage and over 10,000 of his men, and afraid of being discovered he set out for the mountains of Ghado, Kartzkhalni, and Kars. When he reached Phijutha he urged Sargis to let him pass into Abkhazia, as he wanted to have an interview with the Georgian king, and offered to reward him handsomely if he let him pass. Sargis summoned his troops, together with the great Shahin Shah, son of Ivaneh, chief of the mandators, and Shiramun, son of Charmaghan (styled, says Malakia, the Golden Column), who was very friendly to the Christians, and who, with other Mongol chiefs, was encamped in the mountains of Artan, whom he sent in pursuit. The *Georgian Chronicle* here has one of its marvellous tales. It says that Nigudar wished to plunder a rich hermitage dedicated to John the Baptist called Opiza, and situated in the mountains of Ghado, where was preserved as a relic the saint's windpipe. It was stored with rich images, lamps, &c., and Nigudar sent 1,000 of his men to pillage it, but there came on a great storm, induced by the saint to protect his shrine, and the would-be plunderers all perished except one individual who, like Job's herald, went to announce what had happened to his master.† As Nigudar was near the mountain Arsian he was informed that his pursuers were on Mount Artan, and would arrive the following day. He accordingly posted his women and baggage on the mountain of Kars, and himself crossed the mountain Arsian. At the mountain Kuel he found himself in the presence of Shiramun and his men. The principal chiefs on the side of Nigudar were Segzi, Jolaki, Abib Khanui (the name is also given as Abibkhanokhi and Abib Akha), and Thelka Démur. After a fierce struggle Nigudar was beaten, and fled to his women at Jinal, in the mountains of Ghado. Shiramun pursued

---

* D'Ohsson, iii. 434.   † Op. cit., 577-578.

him, and another struggle, two days long, ensued, after which Nigudar escaped secretly. Some of his people retired towards the Adshara, and others towards the valley of Nigal, which was considered almost impassable for men—much more, therefore, for horses—so rugged was it, and so incumbered with thick woods and prickly shrubs. In crossing one portion of the wood which was planted on loose soil, the whole gave way, slipping over the rocks like an avalanche, and overwhelmed a thousand men and women, who were precipitated into the valley of Adshara, "where," says the chronicler, "the people still dig for and find women's ornaments in gold and silver." Crossing the valleys of Adshara and Nigal, they reached Guria and came to Kuthathis, to King David, who, we are told, prepared a great feast for his guest, at which 500 oxen were boiled, in addition to pigs and sheep : 600 horses, 1,500 oxen, 2,000 sheep, and as many pigs were devoted to feeding the army, while wine was given without measure. We are told that the gift greatly touched Nigudar, especially as the King adopted the humble tone of a slave, while his wife, who was the natural daughter of Michael Palæologus, showed the same consideration to the wife of Nigudar, the two ladies treating each other familiarly and on equal terms, while David paid his guest several visits. Wassaf says Nigudar gave the King one of his daughters in marriage. Meanwhile Shiramun had returned to Abaka, just before a messenger arrived to say that Borak, with all the army of Turan, had crossed the Jihun. On hearing the news, Abaka summoned all his vassals, including the other King of Georgia (David, son of Lasha), who, notwithstanding the recent death of his son, set out for Khorasan with his troops, to join his suzerain.*

Let us now revert to Borak. Before setting out he asked assistance from his nominal suzerain, Kaidu, the grandson of Ogotai, who set up rival claims to the Empire of the Mongol world against Khubilai. Kaidu gladly assented, and ordered Ahmed ibn Buri, son of Moatugan, son of Jagatai ; Nikbei Oghul, son of Sarban, son of Jagatai ; and Balighu, or Yalgu, the son of Kaidu, son of Jagatai, to cross the Oxus by the ford at Termed ; Chabad, son of Hukur, or Huku, son of Kuyuk Khakan ; Mobarek Shah, the son of Kara Khulagu, the predecessor of Borak on the throne of Jagatai ; and his own son Kipchak, to cross the river with Borak at the town of Amuye : that Kokaju Buzurg (called Gueuk Achui the Great by D'Ohsson) and Bainal, or Banial, were to cross the river at Khiva ; and Kokaju Kuchuk (called Gueuk Achui the Little by D'Ohsson) was to cross it at Ming Kishlak, which was the most frequented fording place in Khuarezm. They were to unite together beyond the river and join Borak.† Khuandemir says that when he gave orders to these princes to march he also gave them secret instructions that they were

---

* *Id.*, 579-580. D'Ohsson, iii. 434-435.
† Wassaf, 134. D'Ohsson, iii. 435-436. Von Hammer, Ilkhans, i. 262.

to return before Abaka and Borak actually came to blows.* Besides the chiefs ordered by Kaidu to join him, Borak was joined also by the two Yasaurs, the Great and Little (the former, called Besmar by Wassaf, was Borak's brother, and was also called Yesas by Rashid ud din; the latter was the son of Juchi, son of Kaidu), and by Merghaul and Jelairtai, who was the son of Hindu, son of Jagatai, son of Juchi, of the Golden Horde.† Borak forbade his soldiers to ride on horseback, the horses being needed for other purposes. Each horse was supplied with seven ménns of barley and corn per day. The cattle were all killed, and shields were made from their hides, and Borak wished to make special requisitions upon Bokhara and Samarkand, but was prevented by the entreaties of Masud.‡ Some time before Borak sent a message to Tekshin, or Tushin, called Tebshin Oghul by Khuandemir (Weil reads the name Buchin), the brother of Abaka, who had been granted the government of Badghiz, east of Herat, by his father, Khulagu, and been confirmed in that post by Abaka, to tell him that the district between Badghiz, Ghazni, and the Indus having belonged to his ancestors, he (Tekshin) must evacuate it. Tekshin said he had received it as a patrimony from his aka, or elder brother, Abaka, to whom he must first appeal. Abaka, on being appealed to by his brother, said that Badghiz belonged to the dominion of Khulagu, and that he would defend it.

Borak now crossed the Oxus, leaving his son Bey Timur, or Beg Timur, with 10,000 men, to defend his dominions during his absence. He crossed the river on a bridge of boats, and encamped near Merv.§ Malik Shems ud din Kert, of Herat, was summoned to do homage to him, in order to save his district from being ravaged. Orders were given to lay waste all the country subject to Khubilai Khakan or his nephew Abaka. Abaka's army was commanded by his eldest son, Arghun, who was intrusted with the government of Khorasan. Among his officers was a leader of 1,000 men, named Sijektu, who had been formerly a dependent of Kipchak Oghul. When he heard that the latter was in Borak's army he deserted, and sent him a present of some beautiful horses, with some others for Borak. The next day Kipchak being at the latter's quarters, was addressed by the general Jelairtai, who remarked sarcastically that it would seem the expedition had been made for his (Kipchak's) special profit. " What do you mean?" the latter replied. " Why," said Jelairtai, " if Borak had not come hither you would not have received a present of these from Sijektu." He went on to suggest that he had taken advantage of his position and received a number of horses, which ought to have been Borak's, for himself, while the inferior horses which had been passed on to Borak ought in reality to have been his. Kipchak, getting enraged at this, asked him how he, a karaju (i.e., a subject), dared to use such language to a descendant of Jingis Khan. He also went on to compare him to a dog. "If I am a dog," said Jelairtai, "I am Borak's dog, and not thine." "I would hew thee in twain," said Kipchak, "only that my aka (*i.e.*, Borak) would blame me." "If thou comest near me," said Jelairtai, laying his hand on his dagger, "I will rip thy belly open." As Borak did not speak, Kipchak fancied he approved of his adversary's conduct, took offence, returned to his quarters, two leagues off, and having consulted with his officers, withdrew during the night, and retired rapidly towards the Oxus with 2,000 horsemen. He left his family, however, behind, persuaded that Borak would do them no harm, and his wife was the first to inform Borak of his flight. The latter, fearing a surprise, collected his people, and at daybreak sent his three brothers after the runaway, to persuade him to return, or at all events to detain him till Jelairtai, whom he dispatched with 3,000 men, could overtake him. The three princes overtook Kipchak and rushed to embrace him. "Borak is troubled at your departure," they said, "and does not know how he has offended you. Justly irritated against Jelairtai, you left without hearing what he had to say. He intended punishing this insolent officer the following day. He begs that you will return, and will punish him as you may direct." "I am not a child," said Kipchak, "to be led away by your fair words. I set out originally by order of Kaidu; I return home because you do not care for me. I have left my family behind; send it on to me, or I will seize yours." The three brothers, seeing they could not persuade him, asked him to drink a glass of their wine before separating. "People drink wine," said Kipchak, "when they are going to make merry. Now is not such a time; but I see plainly that some troops are coming after me, and that you wish to detain me. Leave quickly, or I will take you with me." The three princes accordingly left, and Kipchak entered the desert of Amu. Jelairtai, who was short of provisions, was obliged to return, while Borak presently sent Kipchak's family back again. Kaidu was apparently irritated at the treatment his son received, and made friends with Abaka, the two princes styling themselves Ortak (*i.e.*, companions).* Soon after, Chabat, grandson of Kuyuk Khan, taking advantage of a journey Borak made towards Herat, also fled. Borak did not send in pursuit of him, but complained to Kaidu, and demanded the punishment of the two princes. Chabat remained for a while near Bukhara, and his presence there was made known to Beg Timur Oghul, whom Borak had left in command of Transoxiana, by an amir of the Tajiks or Persians. He asked the latter if he could not arrest him with 500 men. The Tajik replied he was a karaju (*i.e.*, a subject), and could not attack an urugh (*i.e.*, one of the royal house). Therepon Beg Timur himself went after and defeated

---

him. He barely escaped with ten men, after destroying the bridge of Chiramegan. After being pursued thirty leagues, he at length reached Kaidu's camp, and eventually died from the results of the terror he had suffered.[*]

Borak now entered Khorasan, and we are told he ravaged the whole land from Badakhshan, Kishim, Shaburghan, Talikan of Benda, Mervjuk (i.e., Meruchak), and Merv Shajan, as far as Nishapur.[†] Rashid says that after some fights with Prince Tekshin he occupied the greater part of Khorasan. His cavalry horses fed in the best pastures of the province, and he forbade his soldiers to mount them, so that they might grow fat, and they accordingly went to and fro riding on bullocks and asses. The army was living in clover. Borak took up his quarters at Talikan. His troops entered and reached Nishapur, which they abandoned the following day. He would have done the same thing at Herat, but Kutlugh Timur assured him he would thereby alienate Shems ud din Kert and all the grandees of Persia. Shems ud din, who had been invested, as we have seen, by Khulagu with the districts of Herat, Sebzevar, Ghur, and Garja, had also occupied Seistan, and his dominion extended to the Indus. He lived at the fortress of Khaisar, east of Herat, whither Kutlugh Timur, with 500 men, went to him. He told him that Borak was marching into Irak, and if he would embrace his cause with zeal he should be invested with authority over all Khorasan. He consented, accompanied Kutlugh Timur on his return, and was well received by Borak, who gave him Khorasan as a fief, and promised to add to this the provinces which he should conquer. Borak's people boasted largely of what they would do, and talked of advancing to Baghdad and Tebriz. After these fair promises Borak demanded from Shems ud din the names of the richest men in Khorasan, and then dismissed him, but he sent him several Mongol commissaries, who were ordered to raise a contribution of money, arms, and cattle from the district of Herat. Shems ud din had Borak's orders carried out, and having heard of Abaka's advance he withdrew to the fort of Khaisar, to await the turn of events.[‡]

Borak had now secured the greater part of Khorasan. A few days after the plundering of Nishapur, viz., on the 28th of April, 1269, Abaka set out from Azerbaijan to oppose him. He ordered his brother Yashmut to leave 40,000 men, Mongols and Mussulmans, for the defence of Derbend, and to join him with 10,000 picked horsemen. The Sultan Mozaffer ud din Hajaj received orders to march with the troops of Kerman. Tekshin, who had withdrawn, waited with 10,000 men in Mazanderan for his father's arrival. The vizier, Shems ud din, supplied a body of 1,000 horsemen, and 10,000 horses in addition.[§] Abaka strictly forbade his troops to touch the growing corn. When he reached Sherubaz (called Kungkur-ulang by

* D'Ohsson, iii. 440. Von Hammer, Ilkhans, i. 266.   † Wassaf, 135.
‡ D'Ohsson, iii. 440-442.   § D'Ohsson, iii. 442 and adds. and corr., p. 624.

the Mongols), a district of Irak Ajem, between Zengan and Ebher, famous for its pastures, and where was afterwards built the town of Sultania,* he met Meka Bey (called Tekajhek by Von Hammer), the envoy of Khubilai, who had been waylaid by Borak, but had escaped, and who furnished information about the condition of his army. On reaching Kumis he was joined by Tekshin, who, after being beaten by Borak's advance guard, near Herat, had retired to Mazanderan. With him were his son Arghun, Arghun Aka, and Hajaj, the Sultan of Kerman. On the way to Tus (Von Hammer says in the district of Radegan) Abaka distributed largess among his soldiers. Thence he passed through Bakhers, the district lying between Nishapur and Herat, and famous as the country of the cele- brated author Bakhersi. Near Faryab, he sent out flying parties and distributed his army in various sections. Yashmut was appointed to command the right wing, Abatai Noyan remained with himself and the centre, while Tekshin was sent to Beljaghran, where the yurt of Merghaul, one of Borak's commanders, was stationed, and who informed Borak of the approach of the enemy. · Borak ordered him to go and stop Abaka's advance until his people were got ready. From Badghiz Abaka sent envoys to Borak with offers of peace. He offered to give up the country of Ghazni, as far as the Indus. If this offer were accepted, he might return in peace ; if not, he must get ready for a struggle. The Prince Yassaur advised that they should accept these terms, rather than measure themselves against such a powerful ruler as Abaka, while Kipchak and Chabat had both fled, and their horses were weak. The astrologer Jelal also urged a delay of a month, as the stars were not propitious. Merghaul argued, on the contrary, that they must not allow themselves to be thus overcome by fear. "Where is Abaka?" he said. "Is he not occupied in Syria? It is Tekshin and Arghun Aka who have spread the false rumour of his arrival." "We came here to fight," said Jelairtai. "If we wished for peace we should have remained in Transoxiana." These speeches decided Borak, who, boiling with rage, said : "What does it matter whether the stars are or are not propitious? We must remember that the enemy is coming to destroy us in our camps." It was determined therefore, to give battle, and to send spies to Abaka's camp.†

In regard to these spies Khuandemir has a good deal to say. He tells us that Abaka, having set out for Herat, against which he was irritated because of the assistance it had given his enemies, and which he had given orders to sack (an order which he recalled), the news of his march was brought to Borak, whose men were further disconcerted by the defection of the princes who had been told by Kaidu to join him. He accordingly sent three spies to inquire if he was really with the army, or whether he had intrusted it to one of his princes. The spies found Abaka's people encamped on a vast plain bordered by the mountains called Karasui by the Mongols, and which his general, Burgur, had chosen as a battle-field. Having been captured by some of Abaka's men, they were conducted before him, were fastened to the pillar of his yurt, and under terrible menaces one of them confessed. Abaka then caused a false rumour to be spread abroad that Azerbaijan was in a state of confusion in consequence of an attack by an army from Kipchak, or the Golden Horde,* and himself repeated publicly that the safety of the empire demanded the withdrawal of the army. He then ordered the troops to retreat, and calculated he would reach Tebriz in ten days. The camp and baggage were abandoned, and the army set out for Mazanderan. Abaka also shouted out loudly that the spies were to be put to death. He, however, gave secret orders that the one who had confessed was to be allowed to escape, and the other two alone were to be killed. The spy who thus escaped fled to Borak as quickly as he could, and reported how the plain of Hazar Jérib was dotted with tents, pavilions, stuffs, and carpets, while not a soldier belonging to the army of Azerbaijan remained there. Thereupon Jelairtai and Merghaul both entered the audience chamber in high glee. Before dawn Borak and his amirs mounted their horses, and set out for the plain of Hazar Jérib. Having found that district crowded with abandoned tents and booths they passed the day in feasting. "In the morning, when the sun, the King of the East, puts his chariots in the sky, and chases the army of the stars, Borak Khan, like an impetuous torrent, again broke forth in pursuit of Abaka," and when he reached the village of Shekendian he was surprised to find encamped there the army of Irak and Azerbaijan.† Abaka, we are told, had encamped on the plain of Jiné, five or six parasangs from Herat, and he sent to the Kadhi of Herat ordering him not to open the gates of the city to Borak. When Borak's army neared Herat, Masud Bey went ahead, and, surprised to find the gates closed, he summoned the Kadhi, Shems ud din, who cried out from the walls that Abaka had intrusted him with the defence of the place, and that he had sworn not to surrender it. Masud Bey returned after having menaced him, and Borak did not deem it prudent to delay, but, having crossed the river of Herat and pillaged the valuables abandoned by Abaka, speedily came upon the latter's forces set out in battle array. He was naturally taken aback. His courtiers, especially Merghaul and Jelairtai, offered him consolation, and devoted the night to preparations for the struggle on the following day. Abaka exhorted his men to fight bravely. He told them how he had deceived Borak, and that it was now their turn to show themselves, and that they were about to fight for their families and their sovereign, whose ancestors had conferred so many

---

* D'Ohsson says that he arranged that a messenger should arrive hastily with this misleading news.
† Khuandemir, Journ. Asiat., 4th ser., xix. 258-260. D'Ohsson, iii. 442-445.

benefits upon them. His generals replied with a cheer, and repaired to their posts.* Ibn Farat tells us that the two armies were in presence of one another when an astrologer, skilled in foretelling events, deserted Borak, and announced to Abaka a certain victory, which he foretold after examining the fissures on the shoulder bones of sheep in the approved fashion. Abaka treated him with every honour, and promised to make a village over to him if he was successful—a promise which he afterwards carried out.†

The *Georgian Chronicle*, of course, enlarges on the great doings of its special *protégés*, the Georgians, in this campaign. We are told how their king, David, was sent ahead with the advance guard as the two armies approached each other, in the plain of Amos, near Her (*i.e.*, Herat). Other noyans also marched four or five miles ahead to report on the measures of Borak. These advanced patrols were called karauls by the Tartars. Having gone ahead in this way, the King and the Mongol karauls noticed a great dust, and were certain it proceeded from Borak's army. The King and Sikadur (Samaghar) made their preparations. The latter wished to retreat his soldiery, instinct telling him that advanced patrols have no business to fight, save when compelled. The King replied it was not the custom with the Georgians to turn their backs without fighting when they saw the enemy coming. "Ought we to fight?" At these words the Tartars, who were rigid disciplinarians, replied that they had received orders from Abaka not to fight without him against the Grand Khan. "You Georgians," they added, "are mere ignorant people, and do not know how to behave;" and they threatened in the name of Abaka to ill-use the King and his people, but without effect. They thereupon sent an express to the Khan, to tell him Borak was approaching, and that the dust raised by his army was to be seen in the plain of Amos; that they wished to withdraw in accordance with his orders, but that the Georgians, who understood nothing, would not retire, saying it was not their wont to turn their backs on the enemy, and entreating him to come to their aid or they would be lost. Abaka ordered his people to mount, and hastened forward to find the advanced patrols set out in battle array. He summoned the King, and said, "I know the bravery of the Georgians. You are unruly, like real demons. If one of my noyans had behaved thus I should have had him killed, but you do not understand our methods. Meanwhile, take the advance-post with your men." Descending from his horse, the King bent the knee, and repeated that it was not customary for the Georgians to turn their backs when they had seen the enemy, and that the Khan should see how death could be faced. He then left, and took his post with the advance guard.‡

Let us now revert shortly to the Mussulman authorities. They tell us

---

* Khuandemir, 257-260. D'Ohsson, iii. 444-447. Ilkhans, i. 267-268.
† Quatremere, 268. Note. ‡ Op. cit., 580-581.

that Abaka gave the command of the right wing to his brother Tekshin, or Buchin, with the Noyan Semghur or Samaghar; the left wing to Prince Yashmut, who had under him the generals Sunatai, Mingtur Noyan, Burultai Abdulla Aka, and Arghun Aka. Arghun Aka had in his division the troops of Kerman and Fars, which were led by the Sultan Hajaj and the Atabeg Yusuf Shah; Abatai commanded the centre. In the beginning of the battle Merghaul was killed while fighting bravely, being shot with an arrow. Meanwhile, Jelairtai asked permission to charge the left wing of Abaka's army, which he routed, and drove back with great slaughter as far as Pushenk, or Fushenj, four leagues from Herat. The centre and right of Abaka's army held their ground bravely, and he ordered Yashmut to go and rally the broken left wing. Jelairtai's men having got into disorder in the ardour of the pursuit, he could not hold them in hand, and finding his retreat cut off he was obliged to flee. His success at the beginning of the fight stirred the zeal of the aged Sunatai, who was over 90 years old, and who, seating himself on a stool in the middle of the battle-field, cried out to the officers who surrounded him : "To-day we must show what we owe to Abaka, victory or death." Abaka himself charged at the head of some of his men, and his troops gathered themselves together and made a desperate effort. At the third charge Borak's line was broken and he himself was dismounted. He cried out to his officers, who sped past him in their flight : "I am Borak, your sovereign ; give me a horse." They were too frightened to stay. At length one of them offered him his horse in exchange for some arrows, which Borak threw to him, and he thereupon hastily fled. Abaka pursued the defeated army, giving no quarter. They would nearly all have perished but for the courage and presence of mind of Jelairtai, who rallied them and led them to the desert of Amu, protecting their retreat with a body of troops like Ney so often did in the famous retreat from Moscow. He thus saved the *débris* of the Jagatai army, which recrossed the Oxus. Some men had sheltered in a kiosk. Abaka ordered this to be fired, and all perished in the flames.*

The *Georgian Chronicle* describes the battle with some detail. It says the Grand Noyan Abathan, whom it also calls Abathai, who was generalissimo, commanded the left. With him was Sirmon (*i.e.*, Shiramun), Sikadur, Tougha-Bugha-Jinilis (the same person otherwise known as Bugha Chingsang ; Chingsang being a Chinese title). Arghun Aka and Yasbugha were posted on the right, while the other noyans were distributed between the two wings. When the men were ranged in their ranks a centurion, named Alinak, of great size, courage, and of comely appearance, asked permission from the Generalissimo Abathan to be

<hr />

' Khuandemir, op. cit., 260-262. D'Ohsson, iii. 446-449. Von Hammer, Ilkhans, i. 268-270.

aHowed to fight in the front rank where he should please. This was granted him. Twice with his companions he cut his way through the enemy's ranks, crying out, " Allah ! Allah !" and compelled them to retire. Abathan also fought very bravely, and we are told " he dragged one of the enemy in full armour from his saddle and held him on the pommel of his own during the rest of the fight like an eagle holds a partridge." Shiramun also behaved well. Meanwhile Borak attacked Abaka's right wing, where Sikadur, Tougha-Bugha-Jinilis, and Arghun Aka found it impossible to hold their ground, and were pursued till the following day. Abaka similarly pursued the wing opposed to him for two days, and during this long interval it was not known what had happened. Presently both Borak and Abaka retraced their steps, and a fresh struggle ensued. Abathan Noyan at the head of his men charged those of Borak and made a dreadful slaughter, and captured many prisoners.*

Let us now shortly revert again to Nigudar, whom we left as the guest of David, son of Rusudan, in Guria. Wassaf says that the latter gave him his daughter in marriage,† but this is not mentioned in the *Georgian Chronicle*, which has so many details about his adventures. There we read that while Borak was invading Khorasan, Nigudar contrived to send some of his officers, viz., Segzi-Badur, Abib Akha, Tholak-Demur, and Jolak, with their wives and baggage, apparently to make a diversion. They set out, and reached the mountains of Likh (? Lesghistan), and crossed its western portion called Ghado. They stopped at a place called Lomis Thaf, and ravaged Jawakheth as far as Phanawar. The corps of Tartars which was posted in the district descended into the valley of Eser, crossed the ford of the Mtsuar (Kur) above Atskur, and having penetrated into Jawakheth, carried off a stud of horses belonging to Sakha Thorel, entitled chief of the armourers (? of the Kubechi or Sirgherans, a famous tribe of Lesghistan, whose name means armourers), and another belonging to Kurumchi, a commander of 1,000 men, and his son Aralikan, and returned with them to Lomis Thaf. Kurumchi and Sakha set off in pursuit. The two bodies met by the Kur, near the outfall of the river of Gursel. The people of Nigudar were greatly outnumbered, but Tholak Demur having crossed the Kur without being seen, and with but thirty men, mounted a hillock and unfurled another standard, and advanced with loud cries. Kurumchi, fancying this was a fresh army, and that he was being attacked in front and rear, fled hastily, and a number of his followers perished, among others the two chiefs of Sokhta. Samdzimar lost his horse, and swam over the river. Kurumchi fell by the hand of the Tartar chief, and many of his followers lost their way, and went towards the mountain of Rugeth,

---

* Op. cit., 581-582. † Op. cit., 137.

which is nearly impassable. The success of this raid made Nigudar more audacious, and he descended into Karthli, where he committed great excesses.

Meanwhile Abaka sent word to David, son of Rusudan, that he was no longer to protect Nigudar, making him at the same time generous promises, whereupon the King placed guards so as to prevent his sudden escape. Abaka also sent Shiramun and some other noyans to secure him. They entered Thrialeth and summoned the other Georgian King, David, son of Lasha, to go to him, but he was then ill, and sent his officers, who accompanied Shiramun to Karthli. A struggle now took place, which was prolonged into the following day, and in which Nigudar's people were utterly beaten and dispersed, and Nigudar himself, his wife, and his son were captured, stripped of their belongings, and taken before Abaka, who pardoned him and sent him to live in Irak, gave him abundance of food and raiment, falcons, &c., and a lordly establishment, with a guard to prevent him moving elsewhere. He terminated his days in peace, while King David was rewarded with numerous presents, and was given Aténi, with its appurtenances, and other villages in Karthli.* Wassaf says that Nigudar excused his conduct to Abaka on the ground that he had been invited to do what he did by Borak, that he was pardoned, but that six of his chief supporters were put to death, and his troops were incorporated with those of the Ilkhan. Nigudar himself was put under the surveillance of the Noyan Kurumchi.† Von Hammer says he was guarded by fifty Mongols and imprisoned at Deriar Kebudan, whence he was released on the defeat of Borak.‡

After his defeat Borak retired to Transoxiana. Thereupon Abaka, leaving Tekshin in command of the army of Khorasan, went homewards. On the way a party of people from Dilem tried to assassinate him. Yusufshah, the son of Shems ud din Alp Arghun, the Atabeg of Luristan, who lived at Abaka's Court and had taken part in the recent campaign (his own country being governed meanwhile by prefects appointed by himself), sprang from his horse and saved him. For this, and his brave conduct in the war with Borak. for which he had supplied a contingent, he was rewarded with the district of Khuzistan, the mountain Kiluya, and the two towns of Firuzan and Jerbadakan. The former took its name from the Sassanian monarch, Firuz, who built it ; it was famous for its cotton, corn, and fruit, and was situated in the Persian Irak. The latter was also called Derbayekan, or Güljadkian, and was situated between Kerj and Hamadan. Yusufshah repaired to the mountain Kiluya, where he defeated the Shuls, a Kurdish tribe, the victory costing the life of his brother, Nejm ud din.§ Abaka reached Meragha on the 18th of October, 1270, and twenty days later

---

* Op. cit., 582-583.   † Wassaf, 137-138. D'Ohsson, iii. 433-435.
‡ Ilkhans, i. 261.   § Id., i. 274. D'Ohsson, iii. 454-455.

arrived at his ordu at Chagatu, where he received from the envoys of his uncle, the Khakan Khubilai, a crown, a mantle of investiture, and letters-patent conferring on him the government of Iran, which his father had held. The ceremony of investiture was accompanied by the usual rejoicings. He also received envoys from Mangu Timur, of the Golden Horde, congratulating him on his victory, and taking him presents of falcons.

When Abaka was hunting on one occasion outside Meragha, he was wounded in the neck by the horn of a wild ox. To stop the flow of blood one of the aidajis or cooks made a light ligature about the wound with a bowstring, which caused it to become cicatrised. A tumour having supervened, which caused him much pain, none of his doctors dared to open it until the famous astronomer, Nasir ud din, offered to answer with his head if any harm came of the operation, whereupon it was cut open and he was relieved.* This was shortly followed by the death of the two princes, Yashmut and Tekshin, who had so distinguished themselves in the war against Borak ; and six months later Yisunchin, the mother of Abaka, also died. Her household was made over to his wife, Padishah Khatun, the daughter of Kutb ud din, Sultan of Kerman. It was in this year also that Girdkuh, the fortress of the Assassins, identified by Von Hammer with the Gilgerd of the Byzantines (a veritable castle of Lethe or oblivion), is said to have been finally captured.† The next event of any moment in Abaka's reign was the expedition he sent to ravage Trans-oxiana, which did its work very effectually. Its issue I shall relate in the next volume. The Turcoman, Akbeg, who took an active part in this campaign, would have gone with the booty he had secured to Kaidu, but one of his brothers disclosed his intentions to Arghun. He was arrested and sent on to Abaka. He was executed, *en route*, at Kökje denis, or the Blue Sea (*i.e.*, the Sea of Aral); the intendant, Malik Sadr ud din, was also put to death at Rai. Jenglaun Bakhshi, Khulagu's and Abaka's secretary, and the Amir Arghun Aka, son of Charmaghan, died a natural death at this time. There was also an earthquake at Tebriz which caused much destruction there. The same year there died at Iconium the famous mystic sheikh, Sadr ud din of Konia.‡

The *Georgian Chronicle* tells us how, after his return from the campaign against Borak, Abaka went to Siba with King David (the son of Lasha), who spent the winter there, where the king fell ill and died. He was succeeded by his young son Dimitri. The date of his death has been discussed by Brosset.§ He fixes it in 1269, but it would seem that Malakia's authority must be right, and he tells us David died the same year and the same month as Haithon, King of Armenia (*i.e.*, October, 1270). The youth of the young king, Dimitri, caused many of the eristhafs to join the Mongol ranks. About this time also Aghalar and Sakhaoer, eristhaf of Raja, determined to break the yoke of David, son of Rusudan, and to go over to Abaka. They discussed matters with Alikan Behadur, who lived in the mountains of Jawakheth. He informed the monarch, who made them large promises. The Khan complained of the way in which David had given shelter to his enemies, such as Nigudar and Yalkhur, and it was necessary to punish him. Thereupon Sakhaber said he and his companions knew the country well, and would willingly act as his guides. Abaka thereupon ordered Shiramun, Alikan, Taicho, and Abchi to march against the King. They collected a force of 30,000 men, traversed Trialeth, crossed the mountain Likh, and fell on David, who was then bathing at Kuthathis. He had barely time to mount, covered with a single garment. Meanwhile the Tartars pillaged the churches, killed or made prisoners a large number of Christians, and returned without any loss to Abaka. Two years later, Shiramun and Alikan, called Alinakh by Stephen the Orpelian, and no doubt the Alinak already named, were again ordered to invade Georgia. David at once withdrew, and allowed them to plunder at their will. Presently, having heard that he was collecting a force to attack them, they withdrew hastily with their prisoners and booty. At this time Sadun Mankaberdel had become the first of the mthawars. Abaka attached him to his person, and gave him the surveillance of Georgia, and also confided to him the daughter of the atabeg Avak, who appointed him her chamberlain. Meanwhile the Georgian thawads conveyed the young king Dimitri, son of David Lasha, to the ordu, and were accompanied by Ivané, son of Shahin-shah, the chief of the Mandators. Abaka invested Dimitri with the whole kingdom, except the territory belonging to Sargis Jakel, and caused him to be escorted home by Sadun, whom he appointed atabeg, and on reaching Tiflis he was duly consecrated there.*

We will now turn to Abaka's intercourse with Egypt. The Mamluk Sultan Bibars, who, as we have seen, now ruled there, was a terrible foe to the Crusaders, whom he had determined to drive out of Syria. In 1268 he captured Antioch, which belonged to Bohemond, Prince of Tripoli. After the capture the citizens were killed or reduced to slavery, while other districts subject to the Crusaders were devastated, and Bibars' intentions were strengthened by the result of the ill-starred expedition of Louis IX. to Tunis. Driven to desperation, the Crusaders appealed to Abaka to go to their rescue.† By his orders a division of 10,000 men under Samaghar, the commander of the Mongols in Asia Minor, with a body of Turks from Rum, under the Pervana, first minister of the Sultan of Rum, advanced into Syria. Their advance guard, consisting of 1,500 Mongols, commanded by Amal, son of Baichu, which went

---

* Op. cit., 585-586.   † D'Ohsson, iii., 458-459.

by way of Amk, in the district of Aintab, surprised and cut to pieces a tribe of Turkomans encamped between Harem and Antioch, ravaged the districts of Harem and Al Muruj, and advanced as far as Apamia. Abulfeda says they attacked Aintab, Rug, and Camit, near Famia. The garrison of Aleppo retired upon Hamath. A thousand pieces of silver was asked for a camel, while to hire one for a journey to Egypt cost 200. Bibars, who was at Damascus, sent orders to Baisari to march at once with 3,000 men from Cairo. That officer duly arrived at Damascus on the 12th of November, 1271. The Sultan, with his troops, set out for Aleppo, but the Mongols had already retired. He, however, dispatched the Amir Ak Sonkor Farekani, with a large number of Arabs for Merash (the Germanicia of the ancients), while another division was sent to Harran and Roha (i.e., Edessa), which opened its gates to them. At Harran they put to death the Mongols who were in the town, and caused the rest to take to flight.* The Egyptians did not permanently occupy Harran, and on their withdrawal the principal inhabitants, afraid of the vengeance of the Mongols for having surrendered the place, left it, and scattered themselves in various parts of Syria. A Mongol division re-entered it on the 26th of April, 1272, razed the walls, destroyed the buildings, carried off the greater part of the citizens, and the town was thus ruined.†

Meanwhile, supported by the Crusaders, the Mongols attacked the fortress of Kakun, not far from Cæsarea, which its governor, Bejka Alai, abandoned. Bibars secretly left Aleppo, and arrived at Damascus with a large number of Mongols captured at Harran. He dispatched Akush Shemsi with the troops of Ain Jalut to relieve Kakun, whereupon the Crusaders fled. They were pursued and severely punished. While Bibars was at Damascus envoys came from the Mongol general Samaghar and the Pervana to negotiate for peace, and asking that some one should be sent to treat. He accordingly sent Mobariz ud din Tuzi, the amir tabardar, and Fakhr ud din Mukri, the hajib. They found Samaghar encamped in the province of Sivas, and presented him with nine bows and nine maces, excusing the poverty of their present on the ground that they had had to ride post haste. On the following day they were received by the Pervana, and gave him the costly stuffs with which they had been intrusted by Bibars. With him they went on to Abaka, to whom they presented a cuirass, a helmet made of hedgehog's skin, a sword, a bow, and nine arrows, and reported that their master had received several envoys from Mangu Timur, of the Golden Horde, asking him to make a joint attack upon himself. This news naturally distressed the Mongol ruler, and two days later he sent back the envoys. The fact is that the bitter feud which separated the Khans of the Golden Horde and the

---

* Makrizi, i. (part ii.) 100-101. D'Ohsson, iii. 459-460. Weil, iv. 73-74.
† D'Ohsson, iii. 461.

Ilkhans caused the former to seek the assistance of their co-religionist, the Sultan of Egypt, while the Ilkhans as naturally clung to the Crusaders, and the Emperors of Byzantium.

In September, 1272, some fresh envoys went to Damascus from Abaka, and others from Rum. Makrizi says the former were told to go through the Juk or Kow-tow (*i.e.*, the well-known Mongol form of prostration) before the two naibs of Aleppo and Hamath. They had been charged to ask that Sonkor Ashkar should be sent to the Imperial Court, but they now changed this message into a summons to Bibars either to go himself or to send his first subject. The Sultan replied that as it was Abaka who desired peace he had better go in person, or send one of his brothers to go to him. He also ordered his troops, in complete equipment, to perform their evolutions in the Meidan, outside Damascus, before the envoys. Soon after, news arrived that the Mongols were attacking the fortresses of El Biret and Er Rahbet, and had seized the fords of the Euphrates. The people at Biret sent messengers to Hamath and Hims by pigeon post, asking for help. The Sultan dispatched Fakhr ud din, of Hims, from Harim, with one division, while Alai ud din Alhaj Taibars Waziri went in another direction with a second one. He himself set out from Damascus and Hims, taking with him some boats which were mounted on carts. On reaching the Euphrates, he was deceived by the Mongols, who had moved from the ford where they had previously been posted, and had intrenched themselves opposite a deep part of the river. Bibars launched his boats, filled with soldiers, and a sharp hail of arrows followed from each side. Presently, Kelavun crossed the river by a ford, and defeated the Mongols. The rest of the troops then swam over the river, the horsemen being close to each other, holding their horses' bridles with one hand, and using their lances as oars. The Sultan was one of the first over. The enemy's camp was captured, and he thanked heaven for his victory in a prayer, accompanied by two rik'ats. The enemy were 3,000 or, according to Novairi, 5,000 strong, and lost their commander, Haifar (read Chabakar by D'Ohsson), and many of their number. The Euphrates was crossed at Menbej, after which troops were dispatched up and down the river, who captured and killed many others. Meanwhile Derbai, with the Mongol army that was besieging El Biret, hastily withdrew, abandoning their catapults, baggage, and provisions. Having waited to see if they would return, Bibars once more recrossed the Euphrates and repaired to El Biret, which he entered by a bridge of boats that had been prepared by the Mongols. He rewarded the governor with a robe of honour and a thousand gold pieces, while he distributed 100,000 dirhems and other marks of favour on the inhabitants; and having strengthened the garrison returned to Damascus, which he entered in triumph, preceded by his amirs.* D'Ohsson

* Makrizi, i. (part ii.) 110-111. D'Ohsson, iii. 463-464. Weil, iv. 76.

says the Egyptian army numbered 12,000, and that to enable him to cross the Euphrates Bibars threw 35,000 camels into the river, whose bodies formed a bridge over which his men advanced ;* but this is corrected by Von Hammer. What Wassaf says is that the camels were linked together by their bridles.

The Egyptians now once more assailed Little Armenia, of which Leon III. was ruler. They complained that the citizens of Kinuk molested the Mussulman merchants and travellers. They accordingly crossed the frontier and suddenly appeared before the town. The inhabitants fled to the citadel, which was taken in July, 1273. The men were killed, and the women reduced to slavery.† Tarsus was also sacked. Leon himself, who suspected the fidelity of his vassals, withdrew to the mountains, whence he, according to Chamitch, inflicted a defeat on a second Syrian army which invaded the district. These events took place in the latter part of 1273.‡

While his troops were ravaging Cilicia, Bibars learnt that Abaka was making preparations to attack him, and prepared in turn to repel him. He set out from Cairo on August 12th, 1273, and heard at Askalon that the Ilkhan had left Baghdad on a hunting excursion towards the Zab, and he sent to Egypt to summon his troops. A division of 4,000 accordingly set out under his general, Taibars, and as the news from Persia became daily more alarming, the Sultan ordered all the Egyptian forces, including the Arabs, to march, and whoever had a horse was instructed to obey the sacred call and set out. Bibars reached Damascus on the 2nd of September, but no enemy appeared.§ A few months later Sherif ud din Issa, son of Mohanna, the chief of the Bedouins of Syria, made a raid by his orders into Irak Arab, and advanced as far as Anbar. The Mongols, who fancied it was the Sultan in person, retired fighting, and rejoined Abaka.||

The Malik Shems ud din Behadur, Prince of Semsat, son of the Malik Ferej, chief cupbearer of the last Khuarezm Shah, after the death of that prince, had occupied the strong fortress of Kirat and six others in the district of Nakhchivan, and had then gone to Rum, where he received the town of Akserai as an appanage. He had begun a secret correspondence with Bibars, to inform the latter of what was going on among the Mongols. He had also joined in a plot with the Sultan for the destruction of the Catholicos of the Christians (i.e., of the Nestorians) at Baghdad. The latter lived in the palace of the Khalif, and had treated the Mussulmans with contumely. The Sultan wrote him a letter, saying he had heard how much he had at heart the wellbeing of the Christians in his states, and that it was in consideration for him that he (Bibars) treated him so kindly, and went on to say, " Thanks to you, we are well

---

* D'Ohsson, iii. 464. Wassaf, 167.   † D'Ohsson, iii. 465.   ‡ Id., 466-467.
§ Id., 467. || Id., 466-467. Makrizi, i. (part ii.) 117.

acquainted with the most secret affairs of the Mongols." The letter then went on to make some imaginary statement, as : "We grant what you have asked for such an one. We promise to promote the person you name. We shall know how to treat the person you have in view. You ask us for some balm, and some relics of the Messiah. We send them to you, as well as a portion of the Cross. We have sent these things to Rahbet, and we have communicated to the naib there the secret sign between us ; send a confidential person who knows the sign to fetch them." The Sultan gave this letter to the naib of El Biret, with orders to hand it to an Armenian, who was to pass it on to the Catholicos. He then informed Shems ud din Behadur of what he had done. The latter had the messenger arrested and sent to Abaka, who ordered the Catholicos to be put to death. Shems ud din did several other services for the Sultan, but the Mongols having discovered his intrigues, arrested and conveyed him to the ordu. His mamluks and attendants, to the number of 200, fled to Egypt, where they were well received by the Sultan. Shems ud din himself managed to escape, and went to El Biret, where the people went out to meet him. Thence he passed into Egypt, where he and his followers had appanages granted to them.* This story is a good example of the diabolical ingenuity of the eastern princes in intrigue. It also shows a curious phase of relic-dealing, when ingenuous bishops were ready to buy Christian relics and pieces of the true Cross from the Sultan of Egypt, the great antagonist of the Crusaders, and who would no doubt have gladly supplied such articles *ad libitum*. The story above told is preserved for us by Novairi, but, as D'Ohsson says, there was at this time no Catholicos living at Baghdad, nor did any Catholicos perish by order of Abaka. Abulfaraj tells us that in 1268 Henan Yishua, bishop of Jeziret, was put to death by his orders, his skull being broken with a stone when he was asleep, and his head then exposed at the gate of Jeziret. He was accused of unnatural crimes, and of interfering in matters of state. Perhaps he is the person mistaken by Novairi for the Catholicos.† Although no Catholicos was actually killed, the Christians seem to have had rather a bad time of it at this period, in consequence of the intrigues of the Muhammedans. The Catholicos of the Nestorians left Baghdad in 1268, after an outbreak there. He was called Mar Denha, and was the successor of Makisa. He had seized a Nestorian from Takrit, who some years before had turned Mussulman, and had threatened to drown him in the Tigris. The people appealed to Alai ud din, the civil governor of the town, who demanded the release of the apostate, and on the refusal of the Catholicos they attacked his house, burnt the entrance, and tried to get in and kill him. Mar Denha escaped by some tortuous streets to the house of Alai ud din. He laid his complaints

---

* Makrizi, i. (part ii.) 116-117, and note 143. D'Ohsson, iii. 467-469.
† Abulfaraj, Chron. Syr., 571-572.

before the Mongol Court, but no one there supporting him, he retired to Irbil. In 1271 some Ismaelites (*i.e.*, Bedouins) tried to assassinate Alai ud din. They failed, and were cut in pieces. The Muhammedans declared the attempt had been made by some Christians, emissaries of Mar Denha. This sufficed to cause a general imprisonment of the bishops and the heads of the regular and secular clergy at Baghdad. At the same time Kutbuka, the governor of Irbil, imprisoned the Catholicos and his bishops, and they were only released after some weeks, and by order of the court. Thereupon the Nestorian patriarchs fixed their residence at Ashnu, in Azerbaijan.*

Sempad, whose journeys to the Mongol Court we have described, apparently died about 1273.† We are told that he built many churches, monasteries, &c., and that having repaired to Tebriz, to visit Arghun Aka and the Sahib Divan, he fell ill and died. He was succeeded by his brother Tarsaij, with the permission of Arghun and the Sahib Divan.‡ In 1273 Abaka sent an army to invade Khuarezm, which was then subject to the Khans of the Golden Horde. It laid waste the towns of Khiva, Urgenj, and Karakush.§

Abulfaraj describes how, in 1274, a Nestorian monk from St. Michael's monastery, near Mosul, having had an intrigue with an Arab woman, became a Mussulman, much to the distress of the Christians. His fellow monks, including his uncle, repaired to Tepash, the leader of the Mongol soldiery, who at their instance went to Mosul, where they seized the renegade. Thereupon a tumult arose, and a crowd of Mussulmans went to the palace armed with clubs and torches, and threatened to kill the Mongol leaders unless they released their co-religionist. He was accordingly released, and was taken in triumph on horseback round the city, which was a greater grief to the Christians than his apostacy.‖ We are further told that at this time the Christians of Irbil, intending to celebrate the festival of palms, and afraid they would be molested by the Arabs (*i.e.*, the Mussulmans), asked a number of Christian Mongols who were encamped not far off to escort them. They did so, and they marched out, the Mongols carrying crosses on the points of their spears. The crowd marched headed by the Nestorian metropolitan, and accompanied by Mongol horsemen. They were assailed with stones and dispersed, however, by the Mussulmans, and for some time after dared not come out of their houses.¶ During the same year (*i.e.*, 1274) some of Bibars' officers having been found corresponding with the Mongols were arrested, and twelve of them were put to death.**

In 1273 we read in Abulfaraj that Intab and Birha, from Syria, made a raid into the district of Claudia. They marched continuously, without

* Abulfaraj, Chron. Syr., 571-573.
† Hist. de la Siounie, 235. Note 4. St. Martin, ii. 291. Note 38.
‡ Hist. de la Siounie, 235. St. Martin, ii. 153. § Weil, iv. 135.
‖ Abulfaraj, Chron. Syr., 574-575. ¶ *Id.*, 575-576. ** D'Ohsson, iii. 471.

bivouacking, for fear the Mongols might attack them, and carried off many prisoners.* Two years later some fakirs went to visit the tomb of the Khalif Mamun, at Tarsus. It was suspected that the Egyptian Sultan, in person, was among them, having gone to explore the country, and they were accordingly arrested and imprisoned by order of the Armenian king. Several people were sent from Egypt to inquire why they were imprisoned, which only increased Leon's suspicions. This was not the only grievance which Bibars had. He plausibly charged the Armenian king with several offences—that he had not sent him the presents agreed upon ; that he had built some new fortresses and added to the old ones; that, contrary to his promise, he had not furnished him with useful intelligence ; and, lastly, that Armenians, invested with Mongol sarakuchis, had assailed his caravans, falsely pretending to be Mongols—acts which had led to the destruction of Kinuk, as we have mentioned. Bibars left Cairo on the 1st of February, 1275, and Damascus on the 6th of March. At Hamath he was joined by Mansur, prince of that country, and later by the Arab Sherif ud din Issa, son of Mohanna. He had kept his purpose secret. When his army reached the country between Derbessak and Bagras he divided it into bodies of 1,000 men, which forced the mountains at various points. The soldiers carried torches, and thirty boats followed the army, with which to cross the rivers. The Sultan encamped beyond the defile of Iskanderun, behind a wall which Haithon, father of Leon, had built, whence he advanced to Merkes (not Mankab, as D'Ohsson has it), which takes its name from the river, the ancient Kersos. His people captured and burnt Massissa, the ancient Mopsuetia, on the Jihan (the Pyramus of the classics). There they secured a large booty, and caused much slaughter. We are told that some clans of Arabs and Turkomans, who were owners of great herds here, submitted to Bibars, who moved them into Syria. Once more advancing, he crossed the defile separating Cilicia from Rum, where he captured the families of some Mongols. Thence on to Sis, which was burnt, the palace of the king, with its pavilions and gardens, being destroyed ; the citizens had abandoned the city and sought shelter in the citadel. He dispatched the prisoners and cattle he had captured home-wards, and sent detachments to the maritime towns of Tarsus (where they recited the Mussulman prayers on the Friday), Adana, Barin, and Ayas. Ayas, situated two days' journey from Baghras and one from Tel Hamdun, was in the hands of the Crusaders. They transported their wealth on to ships which were anchored in the harbour. The Egyptians burnt the town and killed many of the citizens. Two thousand Franks and Armenians who tried to escape by sea were drowned. Some of its citizens fled to the Lesser Jeziret, which was not far off ; others who

---

* Chron. Syr., 574.

escaped were pillaged even of their coats by the Franks; "they did not kill any of them, however," says the considerate Abulfaraj. At Adana the men were killed and the women carried off. Abulfaraj says the invaders advanced, plundering, burning, and murdering, to Cyric. They killed twenty-one monks in the monastery of Paximiatus, with an illustrious old monk named Salomon, and the prætor of the patriarch Mar Ignatius. They burnt the monastery, as well as the one at Gujekhat, and the rest of the monasteries of the Armenians and Greeks, except the small one belonging to the Jacobite patriarch at Sis. That dignitary had fled to Behgâ, where he hid till the invaders withdrew.* On returning to Massissa the Sultan set fire to the two parts of that town on the banks of the Jihan, and when all his men had assembled again, and the Turkomans and Arabs who were subject to him had passed the defiles on the Syrian frontier, he continued his retreat. He stayed awhile on the plains of Antioch, which were covered with immense numbers of cattle, and proceeded to distribute the booty, every functionary, both of sword and pen, participating; he did not retain any for himself. He there learnt that the division which he had sent to Biret had advanced as far as Rees Ain, driving before it the Mongols it met *en route*, and had returned loaded with booty. This terrible raid is said to have cost the lives of 60,000 people, while a larger number were made captive.†

The next year the Mongols made an attack upon El Biret. They invested the fortress on the 29th of November, 1275, and bombarded it with catapults. They were led by the noyan Abatai (called Antai by Abulfeda). They had to withdraw, however, on account of the scarcity of provisions and the severity of the weather, which caused the death of many horses; and Bibars, who had already set out from Damascus and distributed largess to the troops, again returned.‡ Abulfaraj tells us the invaders on this occasion numbered 70,000, and that after his failure Abatai returned to Assyria, where he became very ill. The withdrawal of the Mongols, as here described, was followed by a raid upon Cilicia by a number of Turkomans, in which Sempad, Leon's uncle, and several grandees were killed.§ The Pervana, who governed Rum, had a critical part to play. His Mussulman inclinations inclined him to be friendly towards the Sultan of Egypt, whereas the way in which he had been virtually cheated out of his wife by her father, King Haithon, made him ill disposed towards the Armenians, although Haithon's son, Leon, had tried to appease him by a marriage between his natural daughter and the Pervana's son.‖ To keep himself in good odour with the Mongols, he repaired to their ordu with the daughter of Rokn ud din, Sultan of Rum, to save her, as he alleged, from being taken off to Egypt.

* Abulfaraj, Chron. Syr., 578.
† Makrizi, i. (part ii.) 123-124. D'Ohsson, iii 471-474. Ilkhans, i. 290-293.
‡ D'Ohsson, iii. 474-475. § Chron. Syr., 580. ‖ Abulfaraj, Chron. Syr., 574.

He also warned the Mongols that a certain amir, called Mar Khetir, had got possession of the young Sultan Ghiath ud din, and was conveying him to Egypt. He asked for some troops with whom to rescue him. These were given him, under the command of Kongurtai, Abaka's brother. They overtook Mar Khetir, with the young Sultan, at Ablestin, killed the former, and rescued the latter, whom they made over to the Pervana, who thus acquired great credit among them.* The young Seljukian Sultan, Ghiath ud din, had been taken to Nakidah (now called Nigdeh), between Marash and Konia, for safety, by the Amir Sherif ud din Masud Ibn Alkhatir, a strong opponent of the Pervana. Bibars, who had gone to Aleppo, sent a division under Seif ud din Bilban Azzeini, with orders to march upon Nakidah; but when he reached the Koksu, or Blue River (probably an affluent of the Jihan, falling into it south-east of Marash), he was attacked by a Mongol contingent and forced to retire upon Aintab. The result of this expedition was that the Pervana once more secured the custody of Ghiath ud din.†

About the year 1276, Alem ud din Yakub, a great merchant, who was a Christian, and a native of Berkut, in the district of Irbil, and had been on a visit to Khubilai Khan's Court, died on his way home in Khorasan. Yashmut, an envoy of the Khakan's, who was his companion, and also a man of great consequence and illustrious birth, of Uighur origin, and who had been a monk, took charge of his sons, and went with them to Abaka's Court. Abaka received them well, and appointed the eldest of them, Masud, governor of Mosul and Irbil, while Yashmut became his chief minister.‡ According to St. Martin, Arghun Aka, the famous Mongol official, died in 1275. In the "Shajrat ul Atrak" we are told he died in 673 HEJ., at Tus.§ Another important personage died at this time, viz., Shems ud din Kert, the Malik of Herat. His enemies at Abaka's Court had intrigued against him, and the latter, grown suspicious, determined to secure his person. He had to set about it diplomatically, since the malik's fortress of Khaisar was impregnable. Accordingly, in 1275, he sent him a khilat or state robe, a paizah or official tablet, and a yarligh or diploma. This last was thus phrased ; "The Malik Shems ud din Muhammed Kert knows that we are very fond of him ; that his words and acts have always won our approbation and praise ; that all which he has sent to the foot of our throne has been very welcome ; that the statements of his detractors and those envious of him have not been listened to ; and that we have several times told our brother, Tekshin Oghul,‖ to send him some of his most distinguished officers to invite him to quit his inaccessible home, this abode of lions and tigers, this nest of eagles and vultures (i.e., Khaisar), and to take up his residence at Herat.

* Abulfaraj, Chron. Syr., 582.　　　† Weil, iv. 81.
‡ Abulfaraj, Chron. Syr., 582-583. D'Ohsson, iv. 24. § Op. cit. 254. See Ante, p. 241.
His name is otherwise read Tebsin and Bishin.—See Journ. Asiat., 5th ser , xvii. 454. Note.

He must, on the receipt of this order, repair at once to Herat and rule there firmly, and make the frontier provinces, as far as Afghanistan on the one hand, and of Shiburghan and the Amu (*i.e.*, the Oxus) on the other, flourish ; he must reside in the flourishing city of Herat, and there punish those who have been oppressive and tyrannical." Abaka finished his letter with numerous expressions of his goodwill, and swore never to injure him. Shems ud din thereupon assured him of his obedience, and sent rich presents to him, as well as to Tekshin Oghul and the great amirs and chiefs of the administration, and he went to Herat, where the maliks and grandees of the surrounding district went to meet him. Shortly after, there arrived letters from his namesake, the vizier, and from the vizier's son, the Khoja Bahai ud din, the governor of Ispahan, inviting him to go to Irak. He accordingly set out. Bahai ud din, with a crowd of grandees, went to the borders of his province to meet him, and conducted him to the ordu, but Abaka gave him a very cold welcome, and his suspicions having been aroused, determined that he should not return again to Herat. He was detained at the ordu, while his son, Rokn ud din, was sent to join the army at Derbend.* Von Hammer has translated a number of epigrams and poems which he and his namesake, the vizier, wrote in answer to one another at this time.† Abaka himself refused him an audience. He was detained at Ispahan, and his two sons were enrolled among the troops stationed at Baku. The amir Bahai ud din, supported by Tikneh, one of the chief dignitaries of the Court, in vain recalled the services of the family of Kert. Abaka refused to see him, and the following year (*i.e.*, 1277), while he was in a bath at Tebriz, he ate a water melon which Abaka had sent him, and which was poisoned, and died. This was in January, 1278. The Ilkhan's suspicion of him was so great that, fearing a trick and that he was not really dead, he ordered one of his courtiers to superintend the laying out of the corpse, and to fasten the coffin with iron chains. He was buried at Jani, in a turbeh, or funereal chapel.‡ The death of the malik caused matters at Herat to fall into disorder, and we are told that the following year (*i.e.*, 1278) Prince Tekshin, on his return from Ghazni, reported the state of things to his brother Abaka, and persuaded him to nominate Shems ud din's son, Rokn ud din, who as we have seen had been sent to Derbend, in his father's place, and ordered him to bear, like his father, the title of Shems ud din, adding to it that of "the Little," to distinguish him from his predecessor. He reigned without dispute for three years, and received the homage of the chiefs of Khorasan, except of the governor of Kandahar, against whom he marched in 680 (1281), Khuandemir says in 677 (*i.e.*, 1278). The inhabitants defended themselves bravely, but after thirteen days' attack, finding that the gates

---

of their fortress had been fired, they agreed to submit and to pay a money fine.*

Let us now turn our attention to the other end of the empire. We have seen how Iz ud din, the joint ruler of Rum, fled to Constantinople, leaving his brother Rokn ud din in complete possession of that empire.† We have also traced his subsequent fortunes until his death.‡ Rokn ud din was only nominally ruler. The real ruler was the Pervana, who, fearing that his nominal master meant to assert himself, charged him at the Mongol Court with intending to revolt, and having received due authority, he had him strangled with a bowstring, after he had corrupted the Mongol generals with large presents. This took place in 666 HEJ. (1267-8), at a banquet to which he had invited the Sultan and the Mongol generals. Rokn ud din's son, Ghiath ud din, who was only four years old, was put on the throne. Nine years later, troubles broke out in Rum. Some of the amirs there, who had combined with the Pervana to secure an Egyptian alliance, having been betrayed by that treacherous person, fled to Egypt, and incited the Sultan to invade Rum ; among them Seif ud din Jenderbek (the Haidarbek of D'Ohsson), Prince of Ablestin, an old amir called Bishar, and others. Bibars remitted the matter for discussion to some of his own amirs, and ordered two of them, Baisari and Anes (D'Ohsson calls him Akush), to report to him the result. Bibars himself repaired to Egypt, where we are told great exertions were made, and the artisans were fully employed in preparing arms, &c. Bibars presented splendid equipments to his mamluks, and held a grand review, a sham fight, and a feast, which are described in picturesque detail by Makrizi.§ The "Shajrat ul Atrak" reports how, before he set out against the Mongols, Bibars went as a spy to Rum. On his return to Egypt he sent a message to Abaka to tell him he had been to Rum for his amusement, and had left a ring in pledge with a certain cook or confectioner for provisions supplied to him, and he coolly asked Abaka to return it. The latter, who was astonished at his rashness, ordered the Pervana to send Bibars a friendly answer.‖ When the Armenian king heard of the Egyptian preparations he warned the Mongols, but the effect of his messages was discounted by the Pervana, who disliked him, and had a crooked policy of his own to carry out, and who suggested that Leon had some corrupt motive in what he was doing, whence they distrusted him. Bibars, on setting out, left the amir Ak Sonkor Farekani as his deputy in Egypt, with the title of naib algaibah, and left in his charge his son Said Bereke Khan, whom he had nominated as his successor. When he reached Aleppo he ordered its governor to march to Sajur, on the Euphrates, to guard the fords and prevent the Mongols from invading

* Journ. Asiat., 5th ser., xvii. 454-455.   † Ante, 184.
‡ Ante, part ii. 122-123. I ought to say here that Abulfeda attributes his rescue rightly to Mangu Timur, and not to Bereke Khan. Op. cit., v. 27.
§ Op. cit., i. (part ii.) 135-138. D'Ohsson, iii. 480-481.   ‖ Op. cit., 253-254. Wassaf, 164.

Syria. In conjunction with the Arab amir, Sherif ud din Issa Ibn
Mohanna, this general defeated a body of Khafajah Arabs sent against
him by the Mongols, and captured 1,200 camels.* Bibars himself went
by way of Heilan, Aintab, Dulek, Merj Dibaj, and Kinuk (*i.e.*, by the
pass in the Taurus mountains, still followed by the caravans of pilgrims).
He passed the defile of Akcha, where he stationed guards, and thence
sent on an advance guard, under the general Soukor, which met
and defeated a contingent of 3,000 Mongols, and captured many
prisoners.† Bibars advanced to the Jihan, where the Mongols and
Seljuki were assembled, and when he had crossed a mountain range he
found the enemy ranged in the plain of Ablestin. The Mongols were
divided into eleven divisions, each having more than 1,000 men in it.
The Seljuki cavalry formed a division apart ; the Mongols probably
deemed them uncertain allies in a struggle with their co-religionists.
They were under the command of Tukuz (called Tanaun by Abulfeda),
son of the noyan Ilka, his brother Urugtu, and Tudun (the Behadur
Thudan of Abulfaraj), brother of Sugunjak, or Sughurjak (called Thonda,
of the family Saldukh Bahadur, governor of Gartha, in the *Georgian
Chronicle*), who were encamped on the frontiers of Rum ; Wassaf says
of Temghur Noyan and Tudaun Behadur.‡ The battle took place on
Friday, the 16th of April, 1277, on a very cold day. The impetuous
charge of the Mongols gained an advantage at first, but Bibars in person
urged his men that this was a holy war. He thrust himself into the thick
of the fight. The fierce struggle ended in the victory of the Egyptians.
Tukuz and Tudun, two of the Mongol commanders, were killed, as well as
6,770 Mongols. According to Abulfaraj, out of 3,000 Georgians who
fought with them, 2,000 perished, and the rest were dispersed. One of
their champions, named Morghul, is specially named as having thrown
himself alone on the enemy's ranks and cut his way through. He escaped,
his horse carrying him for three days, after being mutilated by being cut
above the pastern of the hind leg.§ Bibars occupied the enemy's camp,
where the Mongol prisoners were taken and put to death, except some of
the superior officers. He also spared the Seljukian officers, whom he,
however, reproached with having fought on the side of the infidels.
Among the prisoners were a son and nephew, and the mother of the
Pervana. In regard to the latter, we are told that before the fight
he supplied the Mongols freely with meat and drink, especially with
drink, so that when the Egyptians arrived they were so drunk they
could hardly guide their horses ; but inasmuch as their laws forbade
them to fly until they had attacked the enemy, they rushed against
the Egyptians and were defeated.|| The *Georgian Chronicle* tells

---

* D'Ohsson, iii. 481.   † Makrizi, i. (part ii.) 140.   D'Ohsson. iii. 481-482.
      ‡ Op. cit., 169.   § Georgian Chronicle, 588.
|| Abulfaraj, Chron. Syr., 583-584.   Chron. Arab., 358-359.

us how in the struggle the Georgians surprised the Tartars by their valour.

Abulfaraj tells us that after the great battle in Rum, Bibars sent his younger brother, whom he calls Bar Khetir, to examine the dead bodies of the Mongols to see which of them of any distinction had been killed. A Mongol who lay concealed among the dead, and hoped to escape at night, fearing he would be discovered, shot the prince in the back. Those who were with him thought the weapon had come from heaven. He was taken to his brother, and shortly afterwards died.* The next winter (*i.e.*, that of 1278) the weather was again terribly severe, and many cattle, &c., perished. This was followed by great scarcity. The Pervana himself, with his *protégé*, the young Sultan, Ghiath ud din, fled first to Cæsarea, and then towards Tokat. Bibars dispatched Sonkor in pursuit of the fugitives, and intrusted him with a missive to the inhabitants of Cæsarea bidding them submit, and ordering them to hold markets outside the town. Sonkor met and dispersed a body of Mongols who were travelling with their kibitkas. He was followed by Bibars in person, who marched along a route which had been much devastated. He received the submission of the fortresses of Semendu, Darenda, and Devalua. The people of Cæsarea came out to meet him and *fêted* him in some royal tents belonging to the ruler of Rum, and when he entered the Seljukian capital in triumph on the 23rd of April, a canopy was held over his head like the one used by the Seljuks. He sat himself on the royal throne in the palace with a crown on his head, and sent a respectful message to the princesses who were in the harem. He then again seated himself on the throne, when the great religious and civil functionaries were admitted, and ranged according to their rank by an officer of the Seljuki, who wore the largest robe and the biggest turban. The royal air, only played in the residence of the ruler of the Seljuks, was played ; pieces from the Koran were read, and verses in Arabic and Persian in praise of Bibars were recited. After a feast he repaired to the mosques. His name was inserted in the Friday prayer in the various mosques of the town, and his name was stamped on the coin. The vast wealth which the Pervana and his wife, Gurji Khatun (who was the daughter of the Prince of Erzerum, Ghiath ud din Kai Khosru, and his Georgian wife, Thamar),† and the other fugitives had left at Cæsarea was divided among the victors. The Pervana himself, with his usual versatile loyalty, sent to congratulate Bibars on mounting the throne of the Seljuki. When he withdrew to Tokat, his wife, Gurji Khatun, who had also set out for Cæsarea with a following of 400 female slaves, died at four days' journey from that town. In his reply to the Pervana, Bibars invited him to go to Cæsarea, intending to give him control of Rum during his

---

* Chron. Syr., 586.　　　† Georgian Chron., 587. Note.

absence. He asked for a delay of fifteen days, while he urged the Ilkhan to march at once to the rescue, hoping that in the meantime Bibars, informed of the approach of Abaka, would withdraw. He did, in fact, retire, setting out from Cæsarea on the 28th of April. He had expected that the grandees of the country would have hastened to him to escape from the Mongol yoke, but they held aloof, fearing the vengeance of Abaka, and he therefore deemed it prudent to return. He put to death many Christians before doing so, but his troops did not otherwise maltreat the inhabitants, paying for everything they took, even straw for their horses, as he said he had gone to Rum not to devastate the country, but to rescue it from the Mongol yoke. On leaving Kai Kobad, he sent Taibars to punish the Armenians who lived in the town of Roman, which had sheltered a body of Mongols. It was burnt, and its inhabitants killed or made captive. At this time one of his officers, Iz ud din Eibeg, of whose prowess he was jealous, and whom he had struck a blow, deserted to the Mongols. On passing the battle-field of Ablestin, which was still encumbered with the slain, he asked how many there were. The Mongol dead numbered 6,770. He ordered most of his own people to be buried, to make it appear his loss had been much less than that of the enemy.

At Cæsarea he was visited by Shems ud din Muhammed, the chief of Karamania (the founder of the Turkoman dynasty of Karamania), at the head of 3,000 Turkomans. He received standards and letters of investiture for himself and his brothers. He had recently rebelled against the Seljuki and the Mongols.* The district ruled by him was in the south of Asia Minor, and it is now known as Ich il. On leaving Bibars he marched upon Conia, or Iconium, which he obtained possession of, as well as the citadel and the town, where he set up a Pretender, who he made out was Iz ud din Kai Kobad, who, as we have seen, had gone to the Krim, where he had died. Shems ud din only remained in the place, however, for thirty-seven days, for having heard that Abaka was advancing against him he withdrew to the mountains with his Turkomans. Bibars, to avoid pursuit, gave it out on leaving Cæsarea that he was going towards Sivas, and went across the river Kizil Irmak (i.e., the Red River), which Weil identifies with the upper Sihun.† When he reached Harim, he received a letter from the Turkoman chief just named, saying he was going to him with 20,000 cavalry and 30,000 foot soldiers, but he was too late. Bibars had already set out for Damascus (taking with him the Pervana's aged mother and his eldest son), where he died on the 30th of June at the age of fifty-five. Haithon says he was poisoned ; Abulfaraj that he was struck in the leg with a weapon, the head of which remained in the wound, and was extracted by the doctor when he died. He adds another report, that the treasurer mixed some poison with mare's milk, and gave it

* D'Ohsson, iii. 486-489. Makrizi i. (part ii.) 144-145. Weil, iv. 83-84.
† Op. cit., 83. Note.

him to drink. When he felt a pain he ordered that the treasurer should drink also, and they both accordingly died. He was a brave soldier of fortune, tall, with blue eyes. He was full of energy, and frequently passed from Egypt to Syria, and *vice versa*, on post horses or swift dromedaries, to see for himself what his subordinates were doing. He had 12,000 Mamluks divided into three bodies, stationed respectively in Egypt and in the districts of Damascus and Aleppo. The Egyptian Mamluks were his own private property, whom he had himself bought. They formed his bodyguard, and their officers occupied the chief positions in the State. His entire army was about forty thousand strong. His death was concealed till the army reached Cairo, the news being given out that he was sick, and had therefore to be carried in a litter. Meanwhile, his body had been buried at Damascus. He was succeeded by his son Said who was nineteen years old.*

Bibars had few scruples when anyone stood in his way. Thus Almalik Almuvahid, son of a former ruler of Egypt, Almalik Assalih Nedjm ud din Ayub, having some claims to the succession, was an object of great jealousy to him. Inasmuch, however, as he had submitted to the Mongols, and been appointed by them governor of Hosn Keif, he could only ruin him by creating a dislike against him on the part of the Mongols. He accordingly incited some of the amirs to commence a correspondence with some of his friends, suggesting that it was worth his while to make an attempt upon Egypt, where the Ayubits still had many supporters. One of the answers being somewhat compromising, was dexterously conveyed to Abaka, who ordered the writer of the letter to be put to death. Malik Muvahid himself was imprisoned, and remained in confinement for seven years, when he was restored to his principality, which he retained till his death in 682 HEJ. Another of his victims was the amir Shems ud din Sallar, a Turkish mamluk of the former Khalif Zahir, who had been nominated by him governor of Kufa, Vasith, and Hillah, which post he filled during the reigns of his two successors. When Khulagu attacked Baghdad he fought against him, in alliance with the princes of Khuzistan, and fled eventually to the Arabian desert. Presently Khulagu invited him to return, and reinstated him in his former position. When Bibars became the ruler of Egypt and Syria he invited him to go to him. He accepted the invitation, but delayed setting out, as he wished to put his treasures in safety. He now dispatched a messenger with a letter, bidding him go, and followed this up with a second one, ordering the bearer of the second letter to overtake the first, to kill him, and leave his body, with all that was upon him, near the frontier guards of the Mongols. As he expected, the body was duly found, and the letter upon it. Khulagu ordered him to be sent for, but he heard of this, and fled to Egypt, where

---

* D'Ohsson, iii. 489-495.

Bibars treated him in a very friendly manner. Again, Zain ud din Alhafiz, who had been sent by Nasir, the Prince of Syria, to Khulagu, with his son, Almalik Alaziz, and who afterwards induced the surrender of Damascus, had gone after the battle of Ain Jalut to Khulagu, who treated him in a very friendly manner. In order to undo him, Bibars ordered his brother Imad ud din, who was in his service, to write and ask him to go to Egypt, and to betray Khulagu. He suspected the letter, and showed it to Khulagu. Bibars, not to be beaten, wrote a second one in person, in which he commended him for showing the former one to Khulagu, as a good means of gaining his confidence, and thus more easily betraying him. He ordered the bearer to drop his coat off, with this letter in it, as he crossed the Euphrates. The coat was duly found by a Mongol, and the letter having reached Khulagu, Zain ud din and all his family were put to death.*

We can hardly realise the misery and destruction caused by the internecine struggle of the two great rivals, the Ilkhan and the Egyptian Sultan. Thus, in the year 660 HEJ., Bibars, dreading a fresh invasion, caused all the women and children to be removed from Northern Syria, while the country was laid waste from Aleppo as far as Mesopotamia and Asia Minor, and the bushes and trees were burnt, so that the Mongols should find no food for themselves or forage for their cattle. The results of such policy are quite obvious, and we read that in 1271 the famine was so pressing that the people of Hamath had to seek shelter at Damascus, and among others the parents of the prince-historian, Abulfeda, who was born that very year at Damascus.

Let us revert again to Abaka. The slaughter of his men by the Egyptians was a heavy blow for him, and he set out from Tebriz to revenge it. At Ablestin, the site of the battle, he was met by the Seljuk chief, Ghiath ud din, who went to him with his vizier, Fakr ud din, of Ispahan. He shed tears on the battle-field, and was surprised at the discrepancy between the number of his own people who were dead and those of the army of Rum, not knowing the trick which had been played on him. In his rage he put to death several Seljuki officers, to whom he ascribed the disaster. He caused the Egyptian camp to be measured with the handle of a mace to test what numbers the enemy had mustered, and he reproached the Pervana for not having given him due warning of the force of the Egyptians. The latter tried to excuse himself on the ground that they had arrived suddenly, and he had not, therefore, had means of informing himself. The amir, Iz ud din Eibeg, who had deserted Bibars and gone over to Abaka, pointed out to him, by thrusting his lance into the ground, where the centre and wings of the Egyptian army had been posted. Abaka remarked that it must have outnumbered

---

* Weil, iv. 37-40.

the army which he then had with him, which was, nevertheless, 30,000 strong. Abaka's troops now, by his orders, spread themselves over the country between Cæsarea and Erzerum, which they wasted with fire and sword over a distance of seven days' journey, so that there perished more than 200,000 souls, or, according to some accounts, 500,000. The kadhis, fakirs, &c., perished in the common slaughter, which lasted for seventeen days. This punishment was limited to the Mussulmans, however, and we are expressly told by Makrizi that no Christians were killed.* Many captives were also redeemed by the vizier, Shems ud din Juveni. Half the town of Sivas was destroyed, the other half being spared on his representation that it was wrong to punish a whole people for the faults of a few.† Nur ud din Khasneyi and Sahir ud din Ibn Hush were executed.‡ Bar Hebræus contradicts the statement made by Makrizi. He tells us that although Abaka had issued orders that the Christians having sheltered and otherwise served the Mongols, were not to be molested, yet the latter through their cupidity killed some and reduced others to slavery. He accordingly supplied a priest and a monk with a yarligh or order, and told them to traverse his camp and release the Christians who were of Rumean origin (i.e., from Asia Minor).§ Haithon says that the Mongols pursuing the Egyptians overtook a body of them at a place called Pasblank, and captured 2,000 prisoners and much booty, and also secured 5,000 Kurdish families who encamped in that district.‖

Abaka wished to follow Bibars into Syria, but his amirs urged him that this was imprudent in the heat of the summer, and that it was better to postpone the campaign till the spring. He contented himself, therefore, with sending him a menacing message, " You pounced like a robber on the advance guard of our army, and have defeated it, and when we drew near you fled like a thief." He also bade him come and meet him like a man, and not slink away like a fox. This letter reached Bibars at Damascus shortly before his death.¶ Having advanced as far as Aksha Derbend, in the mountains of Cappadocia, he left his brother Konghuratai in command of the troops in Rum, ordered Tokat and the castle of the Pervana to be destroyed, and then returned to his head-quarters at Alatagh.** As he passed the fortress of Baiburt, in Armenia, famed for the beauty of its women, it was reported that a sheikh asked permission to speak frankly to him, a permission which was freely granted. " Sire, your enemy entered your borders, but did no harm to your people, nor did he shed any of their blood. You marched against this enemy, and because he has escaped you you have slaughtered your own subjects and ravaged this land. Which Khan among your ancestors behaved

* Abulfaraj, Chron. Syr., 584. Makrizi, i. (part ii.) 145. Weil, iv. 84. D'Ohsson, iii. 495-496.
† Ilkhans, i. 298. D'Ohsson, loc. cit. ‡ Ilkhans, i. 298.
§ Abulfaraj, Chron. Syr., 584. ‖ Op. cit., 52. ¶ Ilkhans, i. 298. ** Id., i. 298.

thus?" These words, we are told, had a great effect upon Abaka, who ordered 40,000 Mussulman captives to be released.*

On reaching Alatagh the Pervana was tried by a council of generals, and was found guilty upon three charges: (1) of having fled before the enemy; (2) of not having warned Abaka of the invasion of the Egyptians; and (3) of not having, immediately after the battle of Ablestin, repaired to the Ilkhan. Meanwhile, the messengers whom the latter had sent with the above-mentioned menace to Bibars, returned and reported that they had been told at Cairo that the expedition of Bibars had been instigated by the Pervana, who had nevertheless treated him falsely, and fled on his approach, instead of handing over the kingdom to him. He was accordingly strangled with a bowstring. This was on the 23rd of July, 1278.† Novairi says that Abaka intended sending the Pervana to Egypt, but that the widows of the Mongols who had perished in the recent fight went to the royal palace in tears, imploring that the manes of their relatives might be avenged upon him, whereupon he determined upon his death. He ordered one of his officers, named Gunkji Behadur, to carry out his wish. The latter told his victim that the Khan wished to go out riding, and wanted him and his companions to accompany him. The Pervana, with thirty-six of his people, were accordingly escorted by Gunkji and a body of 200 horsemen, and on arriving at the place appointed were surrounded. The Pervana, conscious of his fate, asked for a momentary respite, which he employed in praying a namaz of two rek'ats, after which he was put to death, and with him all his companions.‡ Abulfaraj says he was invited to a feast by the Mongols, and liberally supplied with mare's milk, and that while he went out for an interval some Mongols, on a signal from the Khan, dismembered him. He did not give way to lamentations when he knew his fate, but poured imprecations on his murderers.§ The Armenian historian, Haithon, makes out that his body was cut in pieces, and a portion of his flesh was mingled with the food which Abaka and his chief officers ate.||

He was a native of Dilem, whence his father, Mohazzab ud din Ali, had gone to seek his fortune in Rum, where he was patronised by Said ud din, finance minister of the Seljuk Sultan, Alai ud din Kai Kobad, who gave him his daughter in marriage. He afterwards became vizier, and left the post to his son Suliman, known as the Pervana. Having conquered Sinope, it was granted him as an hereditary fief by the Sultan Rokn ud din Kilij Arslan. After his execution this fief, in fact, passed to his son, Moyin ud din Muhammed, who, dying in 1297, left it to his son, Mohazzab ud din Musud, who also took possession of Janik and Samsun. In 1299 two European ships arrived at Sinope with

* Novairi, quoted by D'Ohsson, iii. 497-498.   † D'Ohsson, iii. 498.   Ilkhans, i. 299.
‡ D'Ohsson, iii. 498-499. Note.   § Chron. Syr., 585. || Op. cit., 52.

merchandise. One day those who manned them fell suddenly on the Bey's palace, and captured and carried him off. He redeemed himself for 900,000 aspres, and returned home, where he died in 1300. The territory of Sinope then passed into the possession of the Beys of Castiamuni.*

Haithon tells us that Abaka now summoned the Armenian King and offered to make him ruler of Rum as a reward for the faithful services of himself and his father to the Mongols. The King was very grateful for the offer, but prudently declined it, urging that he could not easily govern two kingdoms, and that in view of the ill-will of the Egyptian Sultan he must devote himself to protecting the Armenian frontiers. He urged Abaka, however, that before he withdrew from Rum he should pacify that province, and not hand it over to a Muhammedan administrator. He also urged him to rescue the Holy Land from the Mussulman yoke, an appeal which was not ungrateful to Abaka, who told him to write to the Pope and the Christian princes to ask for their aid in such a campaign.†

The *Georgian Chronicle* tells us that the post filled by Pervana was conferred on Erinj, a distinguished person descended from Onk Khan (*i.e.*, Wang Khan, of the Keraits), and related to the earlier Khans. From this charge there was excepted Atskur, in Samtzkhe, which the Pervana had held in right of his wife, it having been made over as a dowry to her mother Thamar. This was given to Sargis Jakel and his son Beka.‡

Six weeks after the execution of the Pervana, Abaka sent his vizier, Shems ud din, to Rum, to restore prosperity to that desolated land. He rebuilt the ruined towns, and also introduced a stamp duty which was previously not known there. The predatory prince of Karaman having concealed himself in pathless woods, they were fired, and he was burnt in them. Iz ud din Ibek, the Syrian, was nominated governor of Malatiya. Having settled matters in Rum, Shems ud din went home by the Caucasus, Derbend, Elburz, and the country of the Lesghians, whom he subjected to the Mongol yoke.§ These good offices of the vizier, which marked his character as a statesman, were coincident with the beginning of his collapse. This was brought about by Majd ul Mulk, son of Safr ul Mulk, the former vizier of the Atabegs of Yezd. At first in the service of the Khoja Bahai ud din, Governor of Ispahan, he afterwards was employed by the Khoja's father, the vizier, Shems ud din, in various important commissions. He superintended a census of the inhabitants of Georgia, and afterwards, by the influence of Bahai ud din, he was employed in Rum. The vizier was not, however, very fond of him, and he therefore determined to ruin him, and began accordingly to tamper with the Mongol amirs. He

* D'Ohsson, iii. 499-500. Notes.   † Op. cit., 52-53.   ‡ Hist. de la Géorgie, 588-589.
§ Ilkhans, i. 300.   D'Ohsson, iii. 500-501.

suggested to Yisubuka Kurkan, the husband of Abaka's sister, Kutlukan, that Majd ud din Athir, the deputy of Alai ud din Juveni (the vizier's brother), was carrying on a treacherous correspondence on behalf of the two brothers with the Egyptians, and meditated handing Baghdad over to them. Informed by his brother-in-law, Abaka had the deputy arrested, but, although he received 500 strokes with a stick, no confession fell from him. Hoping to disarm his enemy, the vizier nominated Majd ul Mulk Governor of Sivas, and gave him a golden balish, and a charge of 10,000 dinars on the revenues of Rum ; but this did not appease him, and he intrigued still more against the two brothers.

Abaka having left Tebriz for Khorasan, in March, 1279, his son Arghun went to meet him at Kazvin, where Majd ul Mulk got an introduction to him. He assured him that he had a secret which he had wished to convey to the Khan for more than twelve months, but the mouths of the grandees of the Court had been closed by the gold which the vizier had liberally given them out of the treasury. " If they sell their sovereign's rights you won't sell yours," he added. He then went on to accuse him of appropriating immense sums from the treasury ; of being in correspondence with Bibars ; of having incited the Pervana to invite him to his recent raid on Rum ; and of being the real cause of the death of so many of his people. He also accused the vizier's brother, Alai ud din, of seizing absolute power at Baghdad, and that he had had made for himself a crown, garnished with precious stones fit for a royal head. He accused the vizier of having appropriated 400 tumans (*i.e.*, 4,000,000) worth of the public domains, and of being possessed in addition of 2,000 tumans in money, jewels, and cattle ; while the whole treasures belonging to the Ilkhan, including the booty from Baghdad and the Ismaelites, only amounted to 1,000 tumans. " It is for this reason," he added craftily, "that the vizier wishes to close my mouth, by offering me a sum of money and the government of Sivas."[*] Arghun reported this to his father, who recommended secrecy, so that effective measures might be taken. While Abaka was at Sheruyaz, a fertile district in the north of Irak Ajem, between Zenjan and Ebher, where Soltania was afterwards built, Majd ul Mulk, through the intervention of the general Togachar and of Sadr ud din of Zenjan, secret enemies of the vizier, obtained an audience with him while he was having his bath. He repeated his accusations with the insinuating diplomacy of an eastern courtier, charged the vizier with never having furnished an account of the revenue, and of treating the State as his private property ; while he accused his son, Bahai ud din, the governor of Irak, with appropriating 600 tumans, without devoting a dinar to the public service. Abaka listened to these complaints, and rewarded their author with the present of a cup and a state robe. He also replied

---

* Wassaf, 173-176. D'Ohsson, iii. 502-505. Ilkhans, i. 300-301.

so ably to his questions about the administration that he was made superintendent of the finances, and ordered to make an examination into the receipts and expenditure of the previous few years. He was given a patent marked with a lion's head, a favour never before conferred on a Mussulman, not even a sovereign prince. In it everybody, including military commanders, khatuns, and princes of the blood, were forbidden to put any obstacle in the way of the complete accomplishment of the commission. The naibs, or lieutenants of the vizier, were summoned to Tebriz with their registers. Meanwhile the informer basked in the sunshine of royal favour. He surrounded himself with pages with golden girdles, mounted on Arab horses, and built himself a tent of satin of Shuster, supported by forty pillars.* The vizier, growing alarmed, repaired to his patron and protector, the Khatun Oljai, who spoke in his behalf, and he sought an audience with Abaka. The latter said to him : "You served my father for a long time. I retained you in your old position. Majd ul Mulk has made these accusations against you. How could you be so ungrateful?" He saw that Abaka's prejudices were too much aroused to make it prudent to accuse his enemy of calumniating him, and he was most submissive. "My life and goods," he said, "are my master's. Without doubt, with my brother and sons, I have shared his munificence, and have dispensed it to others. Part of it has been expended upon the royal princes, the khatuns, and grandees ; another part has been spent in alms, to secure a long reign for your majesty. What I possess now in land, in goods, in slaves and herds, I owe to your favour. At your command I will surrender it all, and only ask that I may be allowed to serve my master to the end of my days." These words appeased Abaka, who took him again into favour, and ordered the release of his naibs. But Majd ul Mulk was not satisfied. Professing himself in danger, he asked for Abaka's protection from the vizier, who was now in power again, and requested to be put under the protection of one of the great amirs. Abaka accordingly ordered him to live with the Amir Togachar. He continued his intrigues, and at length, through the support of Sadr ud din of Zenjan, he was, in the spring of 1280, appointed joint vizier with Shems ud din. Abaka ordered the ordinance appointing him to be read aloud in the idol temple at Meragha (not Mekka, as Von Hammer says), in presence of the various princes and princesses, and the grandees, and it was remarked that never had a Mongol prince treated a Persian thus. He was given control of the finances of the treasury and the administration, and was told to remain near his sovereign, and that if anyone attempted his life, Abaka said he would have to answer for it to himself.†

Majd ul Mulk was now the object of general respect, and had his agents everywhere, while the various decrees issued from the royal divan

---

* Wassaf, 178-179. D'Ohsson, iii. 505-507. Ilkhans, i. 301.
† D'Ohsson, iii. 507-509. Ilkhans, i. 302-303.

bore the vizier's name on the right and his own on the left. It was at this time he sent the vizier a waspish verse, which may be thus translated :—

> I wish to dive into the ocean of thy spleen,
> And there to drown, or bring with me a pearl ;
> It is not safe to strike thee, but I will dare,
> And my face or breast shall crimson o'er.

A blushing face means a happy issue, while a red breast means a violent death.* To this the vizier replied :—

> As we may not lay our grief before the Shah,
> We must endure the kicks of fortune ;
> But mark that in the task you are engaged in
> Both neck and face will redden o'er.†

The vizier saw his influence gradually decline, and some anecdotes are preserved showing the indignities which, under such circumstances, the Mongol officials had to submit to. Abaka having summoned him one day to answer a charge made by Majd ul Mulk, the two appeared as usual before the throne, and knelt down opposite one another. Abaka ordered the vizier to kneel further away. On another occasion, at a feast, Shems ud din three times offered his master the cup without his deigning to notice it, and without his losing his composure. On his offering it a fourth time, Abaka presented him in return with a morsel of swine's flesh at the tip of his knife. The vizier, after kissing the ground, ate this (to a Muhammedan) most unclean food, upon which the Khan took the cup and pointed him out to his courtiers. "He was not offended when I refused the cup, and if he had in turn refused the meat I should have thrust his eye out with my knife." Such was the subservience demanded by these autocratic masters, and such the incense his subjects were willing to offer him, rather than lose their positions.

In July, 1281, the vizier's brother, Alai ud din, arrived from Baghdad to pay his court to Abaka. He offered a large sum of gold as the regular proceeds of the taxes for the year, and a second sum representing the increase these taxes had made during the year. Majd ul Mulk accused him of having annually received twenty gold tumans more than he had accounted for during the twelve years he had been governor of Irak Arab and Khuzistan, and several naibs who were under deep obligations to the vizier for various favours nevertheless ungratefully joined in the denunciation of his brother. In vain he protested that it was impossible to save such a sum, considering the expenses of his government, and that the revenue was invaded by the assignations made to the royal princes, the khatuns, &c., and by the profuseness of the sovereign. That notwithstanding he had had an actual deficit the year before, yet he had presented the full amount of royalty due from him as the farmer of the tax. This year he had done even more, for the excess which he presented was imaginary, and had to be paid for by himself, while the revenue of

---

* Ilkhans, i. 303. Note.   † Wassaf, 182. D'Ohsson, iii. 510. Note. Ilkhans, i. 303-304.

the last two years had been curtailed by extraordinary expenses. His enemies, afraid that he would prove his case, changed their mode of attack. They declared that in the year 669 (1270-1) the officials charged with receiving the revenue of the different provinces had found a deficit of 250 tumans in that of Alai ud din, which was still unpaid ; but they forgot to remind Abaka that this matter had already been before him, when he had been convinced that the deficit could not be made up without ruining the inhabitants, that he had remitted it, and had sent Alai ud din back to his government with honour.*

The immediate cause of this pressure put upon Alai ud din was the need of money for a campaign against Egypt. Abaka sent a division under his brother Mangu Timur in that direction. He also reinforced the command of his son Arghun, in Khorasan, and sent assistance to the garrisons about Derbend. In the middle of September he set out by way of Irbil and Mosul for Baghdad, where he intended to winter, and sent on Alai ud din to prepare relays of horses and provisions. The very day the latter set out, and when Abaka was having a great hunt at Deviasir, in the district of Rahbet, Majd ul Mulk laid before his master the charge about the deficit above named. Officers were sent after him to inquire into the matter, who sequestered all his property. The vizier obtained permission to set out at once for Baghdad, and to appease Abaka's wrath he brought together all the precious stones, the gold and silver vases which he had in his house and those of his children, precious carpets and rich hangings, slave girls and palfreys, horses, mules, and camels, oxen and sheep, drums and trumpets, and obtained from his naibs all the money and precious objects they could furnish. He repaired to Abaka, who was at Dojeil, with them as a present. The Ilkhan was expecting a much larger sum, and was by no means satisfied, and it was suggested that it was because the vizier had been in collusion with his brother that he now sacrificed his private fortune to save him. Abaka was still more embittered, and Togachar, the grand judge, was sent to Baghdad, where the trial commenced. All conditions of people were interrogated about the supposed secret hoards of the governor. A visit was then paid to his pious foundations, and to the tomb where his family was buried. The most minute search was made, but in vain. He was nevertheless carried off from his house. Instead of being manacled, as was ordered, his enemies had him fastened with a cangue, or wooden collar, a Chinese punishment, introduced into the West, no doubt, under Mongol patronage. His life was spared on his acknowledging himself debtor to the treasury to the extent of 300 golden tumans. His brother had advised him, in order to avoid being put to torture, not to dispute any of the sums claimed from him.†

* Wassaf, 182-186. D'Ohsson, iii. 510-514.
† Wassaf, 186-194. D'Ohsson, iii. 510-516. Ilkhans, i. 304-305.

This was not a solitary instance of the same kind. Abulfaraj tells us how Masud, the governor of Mosul, having contemplated the death of a certain Persian, named Papa, he accused him in turn of wasting the province of Mosul and maladministering its affairs. Abaka ordered the charges against Masud and his minister, the Uighur Yashmut, to be investigated. Papa suborned some false witnesses and corrupted the judges. The two Christian governors were accordingly displaced, and Papa put in their post. Masud is also called the son of Kota by Bar Hebræus.* This went on till 1280, when the two displaced grandees succeeded in obtaining a new inquiry. Abaka sent two of his relatives to investigate matters, and the result was that the former judges confessed they had been bribed, Papa was beheaded, and Yashmut and Masud were reinstated at Mosul and Irbil.† We are told that at this time the Christians not daring to go out at Epiphany-tide to bless the waters, on account of the Mussulmans, the Khatun Kotai went to the town of Meragha, and ordered that the Christians were as usual to bear crosses at the ends of their spears. When they went out, our chronicler tells us that the wintry cold subsided and the weather became genial again, much to the delight of the Mongols, on account of feeding their horses, while the Christians rejoiced in the victory of their faith.‡

At this time troubles broke out in Fars, caused by an incursion of the Nigudars. These were the followers of Nigudar, grandson of Jagatai, who had gone with Khulagu, as we have described, and on the disgrace of their leader had escaped and settled in Seistan. They now invaded Fars, and defeated a combined army of Mongols, Shuls, Turkomans, and Kurds, at Tenk Shikem, on the frontiers of Kerman, and caused them a loss of 700 men, and having pillaged the town of Kerbal, retired again loaded with booty. Three months later they again invaded Fars, and advanced as far as the Persian Gulf, whence they retreated loaded with booty.§ Abulfaraj, speaking of this campaign, says that 5,000 fugitive Tartars, who had gone to the borders of India, invaded the district of Shiraz and made a great slaughter there, but could not enter the town. The garrison made a sortie, but was overwhelmed and almost destroyed. They also assailed the lion hunters of Shiraz, who were men of wealth, and despoiled them. Abaka's son (i.e., no doubt, Arghun) went against and killed many of them.|| In the spring of 1279 a certain informer or spy, who was in the employment of the prefect of Baghdad (? of Alai ud din) committed great excesses, corrupted the women, and mocked at the prætor (i.e., the mayor of the town). The latter took the opportunity of the prefect's absence on a hunting expedition, had him seized and bound hand and foot, and paraded in a waggon round the streets of Baghdad, with a jeering crowd as an escort, two large pins being driven through

---

* Chron. Syr., 586-587.   † Id., 590-591.   ‡ Id., 587.   § D'Ohsson, iii. 516-517.
|| Chron. Syr., 587.

his tongue. Behind him in the waggon stood a boy, who with a shoe brushed away the flies from his face. He buffeted him, saying : " Thus are punished those who make sport of the grandees." They then took him to the Tigris, where they decapitated him, put his head on the bridge, and burnt his body.*

We now read of the Egyptians making an attack upon Kelat ur Rum (*i.e.*, the Roman fortress), which occupied the site of the ancient Zeugma, and guarded one of the fords of the Euphrates. They had 9,000 cavalry and 500 foot soldiers with them, and were led by Basar and Hosm ud din, the latter of whom commanded the Syrian contingent. They sent two messengers, an Armenian and an Arab, to the Catholicos, and said to him : " The Sultan orders you to surrender the castle, and to take your monks and remove them to Jerusalem. He will give you lands where you may live. If you prefer to go to Cilicia, he will provide you with horses and mules. If you refuse, the blood of all these Christians God will demand at your hands." The Catholicos replied : " I will die then, and will fight, for I will not be faithless to God and the King." They there-upon occupied the surrounding gardens, and cut down the trees, from which they made scaling ladders. They attacked the city on the Sabbath, and drove away the Armenians from the lower walls. They then stormed the place, which they pillaged and burnt. The citizens had meanwhile taken shelter in the citadel, which they failed to take. They retired after a delay of five days, having destroyed what they could, laid waste the the vineyards and orchards, and carried off the baths to Beroea.†

Meanwhile, affairs in Egypt were very unsettled. Bibars, as we have seen, had been succeeded by his son Said, over whom the chiefs of the Mamluks acquired considerable influence ; and the feudal military chiefs, who filled such an important part in Egyptian polity, growing jealous, determined to depose the young prince, who received the fortress of Karak as an appanage, where he died in April, 1280. He was replaced by another son of Bibars, who was only seven years old, and over whom Kelavun exercised complete control as the young prince's atabeg. He was given command of the army, and his name was read out in the public prayer after that of Selamish. The latter only reigned 100 days, until Kelavun had put his own creatures in the various places of trust, and put under arrest the partisans of the family of Bibars. The young prince was then deposed and sent to Karak. This was on the 27th of November, 1279. Kelavun was a Kipchak Turk of the tribe Burj Oghli. He had been sold to a Mamluk when a child for 1,000 dinars, whence his name of Elfi (*i.e.*, the millenarian). The Ayubit Sultan, Salih, placed him among the Bahrit Mamluks, and in honour of that patron he added to his name, on mounting the throne, that of Es Salihi. Kelavun had appointed

---

* Abulfaraj. Chron. Syr., 587-588.   † *Id.*, 588-589.

Sonkor Ashkar (or the Red) governor of Syria. When the latter heard of his patron's mounting the throne, he himself rebelled in Syria, but he was beaten in two fights, in which his troops abandoned him, and Damascus opened its gates to the new Sultan. The amir Bektut was appointed its governor, and Sinjar governor of Aleppo.* Sonkor fled to the fortress of Rahbet, on the Euphrates (to the Bedouin amir, Issa, son of Mohanna), which he seized, together with Borzaia, Blatanusa, Shogr, Bakas, Akkar, Shaizar, and Famia, and then wrote to Abaka, offering his submission,† and asking him to seize Syria. After his revolt, Alai ud din, the civil governor of Baghdad, had, in concert with the military authorities there, sent him and his fellow rebel, Issa, an invitation to submit to the Mongols, but before the messenger arrived Sonkor had repented and withdrawn. Issa sent the envoy back with his brother, who was given a robe of honour, with a charge on the revenues of Baghdad, by Abaka. Sonkor meanwhile shut himself up with his family and treasures at Sihiun, which was one of the strong fortresses belonging to the Assassins.

Deeming this a favourable opportunity for attacking Syria, and recovering the former Mongol possessions there, and hoping much from the partisans of Sonkor, Abaka ordered his people to march. This was in the autumn of 1280. One body of them, under the generals Sagaruni and Turunji, went towards Rum ; another came from the east, under Baidu, son of Targai, son of Khulagu, accompanied also by the Prince of Mardin ; while the main body was commanded by Abaka's brother (Abulfaraj calls him Konghuratai ; other authorities say it was Mangu Timur). Meanwhile the Egyptian troops in Syria, together with a contingent from Cairo, united together at Hamath, and sent an invitation to Sonkor to co-operate. He sent a division of troops, but remained himself near Sihiun. The Mongols were accompanied by Dimitri, the Georgian king, with a contingent of his people. Mangu Timur also sent a summons to Beka, who complained of Arghun Aka's inroad into his country, but who offered to join him if his safety was guaranteed. Mangu Timur swore to protect him in the usual method, viz., by drinking water in which gold was mingled, and gave him the ring on his finger, which was deemed the most solemn engagement. Beka then assembled his Meskhes, and set out. Mangu Timur covered him with honours, and he was warmly received by the Khan. The Mongols entered the province of Aleppo on the 18th of October, 1280. They speedily captured Aintab and Derbessak, where Beka and his Meskhes greatly distinguished themselves, being the first to enter the place, and being duly rewarded with robes, &c., by Mangu Timur. They then went on to Bagras, and advanced as far as the town of Aleppo itself, which was defenceless, and which they entered, burning mosques and colleges, the Sultan's palace, and the

---

D'Ohsson, 519-521. † Abulfeda, v. 53.

houses of his generals. They killed many men there, and reduced the women and children to slavery. The sack lasted for two days. Many of the inhabitants of the province had previously, however, escaped to Damascus, and thence to Egypt.* After this exploit the Mongols withdrew again from the city. Abul Mahasin tells us that the cause of their sudden withdrawal was that a native of Aleppo, who had remained in the place, mounted a minaret in despair, and shouted out, "God is great, and has sent us aid." Thereupon he began to wave a cloth as a signal of approaching succour, and entered the houses as if he were a woman. The Mongols now took their departure.† Kelavun set out from Egypt to meet the invaders, leaving his son Salih as his deputy there, and after distributing a gratuity of 1,000 dinars to each of his officers, and 500 drachmas to each of his soldiers ; but having heard at Gaza that the Mongols had withdrawn, he returned to Cairo. The next spring he set out to deal with his vassals who had rebelled. Issa, the son of Mohanna, who had gone from Irak to Egypt to implore his pardon, was treated generously, while Sonkor, who demanded that Shogr and Bakas, Famiat (Apamia), Kaffartab, Antioch, Sahiun, Blattanus (the ancient Banias), Berziyet, and Ladakiya (Laodicea) should be made over to him, and that he should also be given command of 600 cavalry, whose officers he was to choose himself, was granted the conditions he asked for. Abulfeda says he received Shogr and Bakas.‡

Let us now turn again shortly to Georgia. Its young king, Dimitri, ruled there under the superintendence of Sadun, the famous wrestler, of whom we have already spoken, and who had the confidence of the Tartars, by whom he was invested with Thélaf, Belakan, and other districts ; and during his rule the Tartars abstained from doing violence in Georgia, which began once more to become prosperous. Dimitri himself visited the Ordu, where he was constrained to pay some large sums to the ever-craving Tartars. Sadun himself became constantly more powerful. He asked for the district of Dmanis, which Dimitri was constrained to make over to him, and he was surrounded with wealth. He won the favour of the monks by his benevolence to the monasteries, and we are told that during his rule he paid the two Mongol taxes of kalan and mal for the twelve monasteries of Garesja. He married Khoshak, the daughter of the atabeg Avak, and also apparently a daughter of Sargis Jakel, and bore the title of grand Sahib divan of Avak. He resided at Kars, which had been made over to him, and had control of all the Georgians, except the grandees of the Court, of Karthli, Somkheth, Hereth, and Kakheth, who were subject to the King.§ Dimitri was married about the year 1277, to a daughter of the Emperor of Trebizond.|| We now find

---

* Makrizi, ii. 25-26. D'Ohsson, iii. 522-523.   † Weil, iv. 122. Note.
‡ D'Ohsson, iii. 523-524. Ilkhans, i. 311. Abulfeda, v. 55.
§ Hist. de la Géorgie, 589-590. || *Id.*, 590-591.

Arghun Aka, who had made the former census of Georgia, as we saw, sent to repeat his work, and he found that the previous calculations had been greatly marred by the desolation caused by Bereke's invasion, especially in Hereth, Kakheth, and the plains of Kambej. While he was at Tiflis he asked for the hand of Dimitri's sister, Thamar, for his son, a union which had been consented to by his father, David. Dimitri strongly objected to this union of a Christian princess to an infidel, but he was too weak to resist, and the wedding having been celebrated with due rejoicings, Arghun returned to the ordu, leaving his son in Georgia. *

Ghilan was still independent of Abaka, and did not pay him the kharaj. Shiramun was accordingly sent with a force of Mongols and Georgians into this difficult country, protected on one side by the sea, and on the other by difficult mountains. A fierce fight ensued, and we are told when the Ghilanians fired their arrows, Shiramun dismounted and sat on the ground with his back to the enemy. His men also dismounted. When the arrows were exhausted they remounted, and he charged "like a tiger." He lost two fingers of his right hand in the struggle, but otherwise the Tartars and Georgians suffered no loss. Seeing the country was too strong and difficult to be conquered, he returned again to Abaka.†

We now read of Arghun Aka going with an army of 20,000 men to visit Sargis Jakel, in the country of Samtzkhé. The latter was old and very decrepit. The Mongols traversed Somkheth, Tiflis, and Karthli, committing excesses on the way. This, we are told, was not by Arghun's wish, but was caused by the necessity of providing for such a large force. Having reached Atskur, Sargis and his son Beka visited him. The former was taken to the ordu, while Beka was left in Samtzkhé.‡ This rueful visit of Arghun was followed by a terrible series of earthquakes in those parts. Arghun himself, having returned to Abaka, fell ill and died.§ His death, according to St. Martin, took place at Radekan, near Tus, on the 21st of June, 1275.‖ Arghun's son was now deserted by his wife, Thamar, who professed to detest him as an infidel, and more probably as a Turanian of not very gainly appearance. She fled to Mthiuleth. As she did not wish to return to her husband, Sadun negotiated for her purchase (sic) from Abaka, who approved of the negotiations, and the King thereupon gave over his sister to Sadun, who, notwithstanding that he was a Christian, and in spite of being anathematised by the Catholicos Nicholas, thus became the husband of three wives, the other two being the daughters of the atabeg Avak and of Sargis. About this time we are told that Sargis and his son, for some unknown reason, revolted against the Mongols. Sargis was then old, and his feet were bad (? with gout). Buka Noyan (called "The Eye") sent his brother Arukha, with 20,000 men, to ravage Samtzkhé. Beka withdrew to the mountains between

* Id., 591.   † Id., 592-593.   ‡ Id., 593.   § Id., 594.
‖ Memoires sur l'Armenie, ii. 282.

Ajara and Guria, and the Meskhes took refuge in the fortresses, caverns, and woods. The enemy traversed the country, and remained there for twenty days, doing no harm.*

In the year 1280, Masud, who had been reinstated at Mosul, as I have described, was accused by Jelal ud din Turan, a native of Khoten, who was connected with the treasury there, with having appropriated a large quantity of gold and precious stones. Being put to the torture, he promised to refund 500,000 darics. His cousin Suidat was condemned as an accomplice, and put to death, while a Kurdish leader, named Abubekr, who had for some years been rebellious in the mountains of Assyria, and had been pacified by Masud, was also put to death, as was his son, Sheikh Ali, who had fled to Syria, and then returned to the ordu to excuse his flight. Masud himself was taken to Mosul, so that he might pay over the money which he had promised to do. When he had been there a few days, however, he escaped at night.† Scarcely was this matter settled, when news arrived that the Mongols were again advancing on Syria in two bodies. One, 30,000 strong, under Abaka himself and the Prince of Mardin, was marching on Rahbet, while the other, commanded by his brother, Mangu Timur, was advancing by another route, and had encamped between Cæsarea and Ablestin. Abaka was joined by Leon, the King of Little Armenia. Wassaf says that Abaka and Mangu Timur had with them the amirs Ayaji, Arghasun, and Alinak, and three tumans of troops.‡ An Egyptian detachment, sent in advance from Aintab, captured an equerry of Abaka's, who had been sent on to report on the state of the pastures. He was taken before the Sultan, and reported that the Mongols intended to invade Syria, 50,000 strong, towards the middle of October. Thereupon the people of Aleppo emigrated in large numbers towards Hims and Hamath, so that the place was deserted.§ Mangu Timur advanced leisurely and by short stages, contrary to the usual Mongol tactics, by way of Aintab, Bagras, and Harim, and reached the environs of Hamath, where he plundered the palace and gardens of Malik Mansur. Kelavun, who was at Hims, was there joined by Sonkor, who had recently rebelled, and who had consented to join him on condition• that after the fight he might be allowed to go back to his fortress of Sahiun. He arrived with seven amirs who followed his fortunes, each of whom headed a contingent of troops, and whose arrival greatly raised the spirits of the Egyptians. The two armies faced one another on the 30th of October, 1280, in a plain situated between Hamath and Hims, near the tomb of Khalid, son of Valid, known as "the Sword of God," who ravaged Syria in the time of the Khalif Omar, and who died at Hims in the year 642 A.D. The army of Mangu Timur comprised 25,000 Mongols, 5,000 Georgians, a

---

* Hist. de la Géorgie, 594. † Abulfaraj, Chron. Syr., 591-592. ‡ Op. cit., 170.
§ D'Ohsson, iii. 524-525. Abulfaraj, Chron. Syr., 592. Haithon, 54.

contingent under the King of Armenia, and another of Turkomans from
Rum. Makrizi also mentions Franks among his allies. Abulfeda says
the Mongol army was 80,000 strong, 50,000 being Mongols, and the rest
Armenians, Georgians, people from Irak Ajem, &c.,* and says that a
Mamluk deserter pointed out to Mangu Timur the most vulnerable points
in the array of the Egyptians. The latter, whose numbers were about the
same as their opponents', passed the night on horseback, and the following
day were reviewed by Kelavun. He put the Prince of Hamath, with the
generals Baisari, Alai ud din Taibars, Iz ud din Ibak al Afram, and
Keshtagdi, with their troops, and also the governor of Damascus, in the
right wing ; while in front of them were those admirable skirmishers,
the Syrian Bedouins, of the tribes Al Fazel and Al Mora, under the orders
of Issa, son of Mohanna. To the left wing were attached Sonkor, Bedr
ud din Bilik, Bedr ud din Bektash, Salah, Sinjar, Bekjha, Bektuk, and
Cherek or Khabrek ; while its front was protected by the Turkomans and·
the troops of the Castle of the Kurds. In the jalish, or advance guard of
the centre, were placed Tarantai, viceroy of Egypt, the generals Ayaji and
Bektash, son of Keremun, with the Sultan's mamluks to the number of 800.
Kelavun himself remained with the royal standards, surrounded by his
guards, the officers of his household, and the civil functionaries. This
body, the *élite* of the army, consisted of 4,000 troops. There were with
him many Kurdish and Turkoman chiefs not belonging to the army of
Egypt and Syria. His entire force was also estimated at 50,000 men.
Kelavun, we are further told, wanted to await the enemy's attack near
Damascus, but his amirs insisted on advancing to Hims, and threatened,
if he did not go, to kill him on their return. He therefore went to that
town, and planted himself on a hill, whence he could survey and somewhat control the battle.† The battle began by the Mongol left wing
making a furious charge which was well met by the Egyptian right, which
charged in turn, and forced it back upon the centre. Meanwhile the left
of the Egyptians and the left of the centre were utterly broken by the
onslaught of the Mongol right, which Abulfaraj says was composed of
avarithei (*i.e.*, Uirads), in which the Armenian king and his army were
incorporated, and also of 5,000 Georgians. They pursued them to the
gates of Hims. These were closed, and the wretched camp followers and
other non-combatants who crowded there were mercilessly slaughtered.
This Mongol wing, Rashid ud din says, was commanded by Mazuk Aka,
Hindukur, and Alinak, and secured a vast booty in darics, mules, &c.
Haithon says a division of the Egyptian army was routed by the Mongol
chief Almack (? Alinak), and fled to a town called Tara.‡ The victories of
the Mongols at these points caused a panic in various parts of Syria, as

---

* Op. cit., v. 59.
† Makrizi, ii. 34-36. D'Ohsson, iii. 525-527. Abulfeda, v. 57-59. Weil, iv. 126.
‡ Op. cit., 54.

some of the fugitives made their way to Safad, others to Gaza, and others again to Damascus. The victors dismounted under the walls of Hims, and proceeded to pillage the baggage of the Egyptians, and then to refresh themselves, awaiting the arrival of Mangu Timur. As he did not arrive, they sent to inquire, and found to their natural surprise that he had fled. They accordingly remounted, and retired precipitately.[*]

The cause of Mangu Timur's rout is variously assigned. The courtly Rashid ud din, who naturally glosses over the ill fortune of his patrons, says merely that the centre of the Mongol army, commanded by Mangu Timur, a young prince who had no experience of war, was broken, that the Mongols fled disgracefully, and many were killed. Ibn Tagri Berdi tells us that Iz ud din Aitimur Alhaj, one of the first of the Egyptian generals, made his way into the midst of the Mongol army, pretending he was a deserter, and asking to see Mangu Timur, he rushed at, wounded, and unhorsed him. The Mongols, seeing their prince fall, dismounted. The Egyptians took advantage of this position and charged, whereupon Mangu Timur fled and his people followed his example. The Arab amir, Issa, son of Mohanna, contributed to the final defeat by falling upon the Mongols suddenly with his 300 Bedouins, and proceeding, more suo, to pillage the baggage.[†] Wassaf says that Prince Bakurmishi and Kumishi having fled were followed by Mangu Timur. Some one shot an arrow after the latter which killed him.[‡] Abulfaraj assigns to the Bedouin attack the panic and rout of the Mongol centre.[§] Haithon makes the Armenian king command the Mongol right wing, which had been victorious. He adds that Mangu Timur, who was inexperienced in war, seeing the Bedouins, became frightened, recalled the Armenian king and Almack, and fled. These two chiefs returned; the former, finding Mangu Timur had retired, followed his example, Almack did the same after a delay of two days.[||] Meanwhile Kelavun remained where he had planted himself, as we have seen, on a hill, one half of his army being dispersed and the other in pursuit of the Mongols. There only remained with him 1,000 Egyptians (Haithon says but four armed men). When the victorious Mongols returned from Hims he ordered the tymbals to be struck and the royal standards to be raised, but they were in no mood to stay. They hastened past him, and were in turn pursued by the Egyptians. The Mongols lost a considerable number of men, among them the noted general Samaghar, who had made several attacks upon Syria. The Egyptians also lost twelve noted officers, among others Ai Timur, who had wounded and unhorsed Mangu Timur.[¶] The victory was a very complete and a very fortunate one. For the second time, the Egyptians gave a heavy blow to the Mongols, and again prevented their desolating influence from overwhelming the only refuge left for Mussulman

* D'Ohsson, iii. 527-528.　　† Id., 528-529.　Weil, iv. 127.
‡ Op. cit., 172.　§ Op. cit., 592.　|| Op. cit., 53-54.　¶ D'Ohsson, iii. 530.

art and culture. But it was pure good fortune. No doubt that but for the panic of which Mangu Timur was the victim the victory would have been on the other side. As Makrizi frankly says, " It was a wonderful proof of the divine protection afforded to the Mussulmans, for if it had pleased Him that the enemy should return, the troops of Islam were not in a position to resist."* Kelavun himself expected their return, and ranged his men in order to meet them, until the troops which had pursued the Mongols returned.

The Mongols had retired in two bodies, one towards Salamiyet and the Syrian desert, the other towards Aleppo and the Euphrates. Of the former, who were 4,000 in number, we are told their retreat was cut off by the commandant of Rahbet, and taking to the desert they perished of thirst and want, except 600 horsemen, who were attacked by the garrison of Rahbet, and partly killed and partly made prisoners. The prisoners were decapitated at Rahbet. That town had been besieged, as we said, by a division under Abaka himself, but the day after the battle a pigeon with the news reached the place, when the commandant caused victorious strains to be played, and the besiegers withdrew.† Of those who fled towards the Euphrates, we are told many sought shelter in the caverns bordering that river, and were burnt out, as the Kabyles were in Algeria by Pelissier.‡ A body of Mongols that was laying siege to Biret was also attacked by the garrison there ; 500 of them were killed, and all the rest were made prisoners. Mangu Timur himself crossed the Euphrates, and went to Jeziret, his mother's appanage.§ These incidents prove how very disastrous one defeat is to armies constituted like those of the Mongols, even when possessed of long prestige and discipline.

Notwithstanding their victory, the Egyptians had lost the greater part of their baggage, which had been pillaged after the rout of their left wing by its custodians, but none of the coin Kelavun had taken with him was lost, as he had taken the precaution of distributing it among his mamluks, who carried it in their girdles.

The *Georgian Chronicle* limits its account of the battle mainly to the doings of the Georgians, who, there can be small doubt, distinguished themselves by their usual bravery. Their young king, Dimitri, fought in the advance guard, and was attacked by the *élite* of the Egyptian army, 12,000 strong, under Kara Songhul (*i.e.*, Sonkor) and Yakub Aphrash. A terrible struggle and slaughter followed. Dimitri's body guard of 200 men was cut in pieces, his own horse being killed by a lance thrust from Kara Songhul, but he was speedily remounted by one of his followers, and the Georgians about him fought so desperately that the 12,000 Egyptians who had charged them were thrown back. Meanwhile Mangu Timur, with his Mongols, had retreated, and the Georgians had to do the same.

' Op. cit., ii. 38.   † D'Ohsson, iii. 531.   ‡ Makrizi, ii. 39.
§ *Id.*, 41-42. D'Ohsson, iii. 531-532.

Their king escaped almost miraculously, although most of their number were killed.* The Armenians also suffered greatly in the retreat, their horses having been quite worn out by the bad roads and scant forage. Many of their soldiers were thus separately overtaken and killed by their pursuers; and thus the greater part of the Armenian army, and especially of its chiefs, perished.

During the period of uncertainty before the victory was known, the people of Damascus had passed an anxious time. Prayers were fervently offered up in the Great Mosque, and in the oratory outside the town, for the victory of the Mussulmans, while the Koran of the Khalif Osman was borne on the head of one of the clergy. In the midst of the excitement a pigeon alighted, after the Friday prayer, on the day after the fight, bearing news of the victory. Music at once resounded in the citadel, and both it and the town were decorated, and the crowd loudly expressed their joy. Presently there arrived some fugitives, who related not a victory, but the defeat of which they had been witness. The wave of excitement now collapsed, and a rush was made for the open gates, to try and escape elsewhere, but a few hours later a special courier brought the true account. It was publicly read in the principal mosque. A similarly fitful mood had passed over Egypt. There also people read the Koran diligently, recited the Salih of Bokhari, &c., when a pigeon from Kakun, a town situated between Lejun and Ramla, brought news of the arrival of the fugitives there. The agitation was very great. Prince Salih sent some Turkish and Arab troops to Kattiya to drive back the runaways and prevent any of them from entering Cairo. A second pigeon post speedily relieved the public mind. Great rejoicings took place in Egypt, and Salih wrote to his father asking him to pardon the fugitives, and urging Baisari to intercede for them.

Among the captives made by Tarentai, the viceroy of Egypt, in his pursuit of the Mongols, was a man who had charge of Mangu Timur's portfolio, or valise. In this were found letters from Sonkor and some of his amirs, urging the Mongols to invade Syria, and promising them their help. Kelavun magnanimously ordered the writing to be erased, renewed his pact with Sonkor at Hims, and sent him back to his fief of Sihiun. He then set out for Damascus, into which he made a triumphal entry, headed by the prisoners, of whom several carried the tymbals and standards which had been captured, while the poets poured out a deluge of compliments.†

When Mangu Timur invaded Syria, Abaka, as we have seen, advanced hunting towards Rahbet. He did not, however, cross the Euphrates, but after destroying some forts returned to Sinjar on the 25th of September, and in the beginning of November rejoined his ordus at

---

* Hist. de la Géorgie, 595-596.   † Makrizi, ii. 39-41.   D'Ohsson, iii. 532-535.

Mahlibiya, near Mosul. There he learnt of the defeat of his army. He was greatly irritated, and announced that at the next kuriltai, to be held in the spring, those who were found blameworthy for the recent disaster would be punished. He also announced his intention of marching in person upon Egypt.* The *Georgian Chronicle* tells us the news of the disaster to his army was disclosed to Abaka by a Tartar, who addressed him in verse in his own tongue, describing how each of the chiefs had behaved. He said Alikan had attacked like a falcon pouncing from the clouds, and compared Mangu Timur to a ram ; Abagan, son of Shiramun, to a tiger, which bounds ; Yasbugha to a young bull ; Buka to a buffalo, while as to the Georgian king he expressed himself thus in the Tartar tongue : "Thengari methu kaurkurbai, bughar methu buirlaji" (*i.e.*, "They growled like God ; they bit like the camel"). Abaka received the Georgian king with honour and sent him home.

When the defeated Mongols retired from Syria, a body of Mussulmans Turkomans, Kurds, &c., made a raid upon Cilicia. They advanced as far as Aias, which they plundered and burnt. It had been deserted by its inhabitants, who had taken refuge in a fortress they had built on a neighbouring island. The Muhammedans withdrew with their booty, but returned again three times, the last time advancing as far as Tel Hamdun. On this occasion they were attacked, while retiring with their gains, by the Armenians, who had occupied the defiles. They captured their arms and stripped them of their scalps, and sent Abaka several loads of arms, lances, swords, and scalps. A few days later the governor of Biret, named Haidar, having collected 2,000 horsemen, captured the castle of Saida. Many Christians who had sought refuge in a large mosque were released. Others had fortified themselves in a place called Alastona, which the invaders had not been able to capture, as the approach to it was like a cavern. They nevertheless carried off 4,000 women and children, with whom they safely crossed the Euphrates, and went towards Malatia, where they laid waste the country, and carried off many Christians from the town of Erka, with whom they again withdrew to Syria.†

We described the machinations of Majd ul Mulk against the vizier and his brother, the governor of Baghdad. The latter, to save his position, had promised to restore 300 golden tumans to the treasury, which his enemy went to Baghdad to receive. He sold his wives and children, and undertook to pay with his head for the least prevarication proved against him. Abaka pitied him and released him from prison, but soon after his relentless enemy again went to Baghdad, with the generals Togachar and Ordukaya, to drag from him, by torment, if necessary, the hundred tumans he charged him with appropriating. Alai ud din could not pay, and was tortured and promenaded naked about the

* D'Ohsson, iii. 535.   † Abulfaraj, Chron. Syr., 594.

town.* His enemies were relentless in their attacks. His correspondence with Sonkor and Issa ibn Mohanna, the Egyptian refugees, whom he had tried to persuade to submit to the Mongols, was turned into a charge of treasonable correspondence with the Sultan, and employed as an instrument of their plots, the Arab messengers whom he had employed being corrupted into making false charges against him. An unknown Jew having written several times with saffron and cinnabar on a piece of paper, as if it were a talisman, this was hidden away among his clothes while his house was being searched. During all his troubles the famous historian and administrator consoled himself by composing verses, satirical and elegiac, many of which became famous. Several are contained in "The Consolations of the Brothers," a kind of Arabic Boethius. One of his poems was glossed by many poets. As Alai ud din was being led off by his enemies from Baghdad to Hamadan, accompanied by his faithful friends, his nephew, Khoja Bahai ud din; Ali ibn Issa of Irbil, and Nuruddin Abdur Rahman of Shuster, and when he had reached the heights of Asadabad, he learnt of the death of Abaka, which brought him and his brother considerable respite, and proved, for a while at least, a new turn in his fortunes.†

Abaka set out for Baghdad on the 13th of February, 1282. He arrived at Hamadan on the 18th, where he put up at the palace of Fakr ud din Minochéer. There he fell ill. According to the Persian historians he was habitually given to drinking to excess, and having one night got very drunk went out, and believed he saw a raven sitting on the branch of a tree. He ordered one of his guards to shoot it with an arrow. The bystanders looked attentively, but could see no such bird. Suddenly Abaka closed his eyes and died. This was, Wassaf says, on the 20 Selhije (i.e., the first of April). Abulfaraj reports that Abaka had passed the previous Easter Sunday with the Christians, and taken part in the service in the church at Hamadan ; that on the Monday he dined with a Persian gentleman named Behna, and the night following he saw visions in the air, and died on Tuesday morning, the 1st of April. The "Shajrat ul Atrak" says he died of excessive drinking, after a feast at the house of the vizier, Shems ud din.‡ Novairi says that according to some, Abaka, after his defeat, fell into a state of melancholy, which was increased when news arrived that the famous treasure house which his father had built on Lake Urmia had collapsed, and that the various treasures it contained had sunk into the lake. When he heard this he was going to his bath. As he came out he heard a raven croak, and declared that it presaged his death, while his favourite hunting dog barked at him, which he accepted as another ill omen, and he died shortly after.§ Abulfeda says it was reported he had been poisoned.* Other accounts say that Mangu Timur died fifteen days after his brother, at Jeziret.† Haithon says both brothers were poisoned, that Abaka, having determined to avenge his recent disasters in Syria, was about to set out, when a Mussulman arrived at his Court with rich presents, and in conjunction with some of Abaka's courtiers arranged his death.‡ Abulfaraj says that a certain informer, named Al Saphi Karbuki, accused some of the grandees of Jeziret to Mangu Timur. He accordingly punished them, and they plotted with his butler, who, when he one day came out of his bath, mixed poison with his cup. He set out ill from Nisibin for Jeziret, and died on the way. The informer was duly put to death.§ Rashid ud din and Wassaf say nothing of this, but it must be remembered they were good Muhammedans, and that Abaka was an infidel. We, at all events, learn from Novairi that Mumin Aga, the commander of Jeziret, was accused of poisoning Mangu Timur, and fled to Egypt with his two sons, where he was rewarded with some fiefs. His wife and children were put to death by the prince's relatives. It was said that the poisoning was arranged by Alai ud din Juveni, who had ample reasons for wishing Abaka's death.‖ Abaka and his brother were buried with their father in the fortress of Tela, or Teke, on the Lake of Meragha.¶ Soon after this the Georgian chief, Sadun, died. The *Georgian Chronicle* says that his son, Kutluk Shah, succeeded to his father's domains, and was raised to the rank of a generalissimo.**

The Christian writers speak of Abaka in terms of considerable praise. The *Georgian Chronicle* calls him good, generous, and clement, soft and modest, a lover of justice, charitable to the poor, and so forgiving that whatever a man's faults he would not sacrifice his life. " God has given me the empire of the world," he said, " I must not take away that which I cannot give." He urged that theft was an effect of poverty, and several times refused to punish with death those who had stolen from him. As he was surrounded by people who plundered the treasury by securing immense sums, he chose a man who was charitable, just, attentive to his religious duties, and a patron of pious people, named Aghubagha, and charged him to protect the weak and the poor.†† Haithon speaks of Abaka as a prudent and prosperous ruler, fortunate in all things save two: first, that he failed to become a Christian as his father had been, but was devoted to idols and their priests, and secondly that he was continually at strife with his neighbours, and did not in consequence molest the Egyptian Sultan as he might have done, whose power consequently greatly increased. His people were so weighed down with exactions that many of them fled to the Sultan, who showed his sagacity by

---

* Op. cit., v. 63. † Abulfaraj, Chron. Syr., 595. ‡ Op. cit., 55.
§ Chron. Syr., 594. Chron. Arab., 361. ‖ D'Ohsson, iii. 536. ¶ Id. Ilkhans, i. 313.
** Op. cit., 597. †† Op. cit., 572.

his close alliance with the Tartars in Cumania and Russia (*i.e.*, in the Kipchak), which prevented Abaka from attacking the Egyptians as he might have done, and thus the Christians lost Antioch and several other fortresses which they had possessed in Syria.*

Abaka had eleven wives and three concubines. These were (1) Oljai, who came out of his father's harem; (2) Durji Khatun; (3) Tokini, the cousin of Khulagu's wife, Dokuz, who on the death of Durji was given the baghtak, or wife's head-dress, and made his head wife; (4) the Tartar Nukdan, the mother of Gaikhatu; (5) Iltirmish, the daughter of Timur Kurkan, and belonging to the Konkurat tribe; (6) Padishah Khatun, daughter of Kutb ud din Muhammed, the Khan of Kerman; (7) Mertai and (8) Kuti, both Konkurats and both widows of Khulagu : they were sisters of Musa Kurkan, and their mother was a daughter of Jingis Khan ; (9) Tudai, also a Konkurat; (10) Bulaghan, a relative of the chief judge, Nokar (? a mistake for Buka) ; and (11) Maria, styled Despina, the daughter of Michael Palæologus.† Abaka left two sons, Arghun, by a concubine named Katmish Ikaji,‡ and Gaikhatu, by Nukdan.

Abaka, like his father, had considerable intercourse with the Christians. In fact, it was necessary the Mongols should begin to look out for allies. The world of Islam was naturally incited against them, and it had received great encouragement by the Mongol defeat at Ain Jalut. Although nominally their subjects, it was hardly likely that the Seljuks of Rum, that the Prince of Mosul, the chiefs of the Kurdish mountains, or the Ortokids of Mardin and Hosnkeif, should have felt any great loyalty for the Mongols, who were infidels and strangers. The latter naturally leaned more and more on the Christians. The princes of Georgia and Great and Little Armenia clung to them as their natural allies, in the face of the hereditary enemies of their faith. So did the Crusaders, who probably, as Remusat says, expected to become their deputies in Syria, where the climate was so ungrateful to them, and expected also to be relieved from taxation, as the Christians of Armenia and Georgia were.§ We must also remember that Abaka was married to Maria, the daughter of Michael Palæologus, who doubtless used her influence to draw the Christians of the West and the Mongols nearer together. We find the Ilkhan engaged in a correspondence with the supreme Pontiff. In a letter dated 1267, at Viterbo, Clement IV. says he had received his letter, but as it was written in Mongol no one at his Court could read it, and he expresses his regret that it was not written, as previous letters had been, in Latin, and that he had therefore been constrained to employ his messenger as an interpreter, who apparently somewhat sophisticated in his report his master's religious views, for the Pope begins his reply by thanking God that Abaka recognised the Eternal God, and humbly

---

* Op. cit., 49.   † Ilkhans, i. 252-253.
‡ *Id.*, 360. Erdmann Temudjin, 243.   § Acad. des Inscript. vii. 335-337.

adored His crucified Son. He continues : "You rejoice, you say, in the victory we have gained in Sicily, where the presumptuous usurper, Manfred, natural son of Frederick, Emperor of Rome, fell on the field of battle, with a great number of perfidious Christians and of Saracens, deprived of his life and throne by the same blow, by our very dear son in Jesus Christ, Charles, to whom we have given the kingdom. The kings of France and Navarre, followed by a great number of counts and barons, a multitude of soldiers, and others, taking to heart the condition of the Holy Land, are preparing valiantly and powerfully to attack the enemies of the faith. Many others, lords and commons, in other countries are wishful to follow his example, to exalt with all their power the name of Christ, and to destroy the power and sect and even the name of the Saracens. You have written to say you intend to join your father-in-law to help the Latins. We shall do everything to help you, but we cannot say, until we have made inquiry from these princes, what route our people propose to take. We will communicate to them your intentions and those of your father-in-law, so that they may develop their plans, and we will instruct your magnificence by a trusty messenger. Persevere, therefore, in your admirable plans. If you trust in God, he will strengthen your throne. His is the power and the dominion. He holds in His hands the hearts of kings, and humiliates and raises whom He wishes ; no one can resist Him."* D'Ohsson doubts, very naturally, whether a letter was ever written by Abaka himself to which this was a reply. The clause about the fate of Manfred and the statement about his conversion are hardly consonant with Mongol ways of thought, and it is more likely that the letter was either composed or sophisticated by the hand of some Eastern Christian, not improbably by some dependent of the Greek Emperor of Byzantium.

In 1269 the envoys of Michael Palæologus and of Abaka visited James, King of Arragon, at Valencia. Surita says that James had not previously heard of either of the two princes, and that it was suspected that the real object of the mission was far from being a pious one, and was merely to rid themselves of some domestic foe. Mariana, on the other hand, says James had already received some Tartar envoys, and had sent in return a certain John Alaric, a native of Perpignan, with whom the new envoys came. They promised, on behalf of their master, his help if he would join his forces to those of the other princes. The envoys stayed at Barcelona, but Alaric went to Toledo, where he laid before the Junta a full account of his doings. James, although so old, determined to go to the war, and would not be dissuaded by the prayers of his relative, Don Alphonso, and the Queen of Castile, who pointed out the treachery of the Greeks and the ferocity of the Tartars. Alphonso promised

---

* D'Ohsson, iii. 539-542. Remusat, Acad. des Ins., vii. 339-340.

to send subsidies. Michael Palæologus had offered, by his envoys, ships and provisions. The expedition was wrecked, however, by a storm at "Aigues mortes" (?), and obliged to return. The fatal expedition to Tunis, in 1270, postponed any active alliance between Western Europe and Abaka. After his return from his campaign against Borak, Abaka seems to have again made overtures to the Western princes. To this, it is said he was urged by the King of Little Armenia, who wished him to relieve the Holy Land. His envoys, according to Remusat sixteen in number (one having died *en route*), arrived at Lyons in 1274, where Gregory X. had called a council. They were admitted to the council at its fourth session, on the 6th of July, 1274. The Pope made them sit opposite to himself, and at the feet of the patriarchs, and their letters, or rather the version they chose to give, were read out at the succeeding session on the 16th of July. The chief envoy, with two distinguished Tartars, was baptised by Peter of Tarantaise, cardinal of Ostia, afterwards Innocent V., and they were presented with precious garments. This was the sole result of the embassy, for the continued advance of the Mussulmans and the decay of the crusading spirit made a great effort at this time impossible.* Abaka's letter was sent to Edward I. by David, chaplain and familiar of Thomas, patriarch of Jerusalem and legate of the Holy See. Edward's answer, dated the 26th of January, 1274 (? 1275), at Bellus locus (? Beauchamp), is given by Rymer. It runs as follows : " Brother David, of the order of preachers, chaplain and familiar of brother Thomas, patriarch of Jerusalem, legate of the Apostolic See, has arrived at our Court, and presented letters sent through your envoys to the Holy Father and other Christian kings. We note in them the love you bear to the Christian faith, and the resolution you have taken to relieve the Christians and the Holy Land from the enemies of Christianity. This is most grateful to us, and we thank you. We pray your magnificence to carry out this holy project. But we cannot at present send you any certain news about the time of our arrival in the Holy Land, and of the march of the Christians, since at this moment nothing has been settled by the Sovereign Pontiff, but we will inform your excellence as soon as we learn. We commend to your puissance both this matter of the Holy Land and of all the Eastern Christians."†

Two years later, under the pontificate of John XXI., two fresh envoys, named John and James Vasalli (? Vasili), went to Rome from Abaka, and were admitted to audience in an assembly of the cardinals. These envoys invited the Christians to invade Palestine, and promised them aid if they went. They were sent on to the Courts of France and England. To Philip III., king of France, they promised that if he would go to Acre, with a view of invading Palestine, their master would help him. William of Nangis, speaking of them, says: "Were they really envoys or spies God knows. At least they were not Tartars, either by birth or manners, but Christians of the sect of the Georgians."* They were taken to St. Denis for Easter, and then passed on to the Court of Edward, the English king. What befel them in England we have apparently no record of. The two envoys reported that their master, as well as his uncle the Khakan Khubilai, wished to be instructed in the Christian faith. This persistent report was partially due, perhaps, to the welcome sound it naturally had at Rome, but more, as Remusat suggests, to the open patronage which the Mongol Khans, as we have seen, extended to the Christians, according to their cosmopolitan notion that there is only one religion, the forms of which have been varied according to time and place by the wise men of each country.† The Pope determined to verify this report, and five Franciscans, viz., Gerhard of Prato, Anthony of Parma, John of St. Agatha, Andrew of Florence, and Matthew of Arezzo, were selected to go and preach the faith in the East. John XXI. having, however, died during the year 1277, the mission was delayed, and only set out the following year with letters from Nicholas III. to Abaka and Khubilai. The former expressed the joy felt by the Roman see at the news brought by the two brothers Vasalli, and acknowledged gratefully the Ilkhan's offers of assistance to any Christian army that might land in Syria. The Pope went on to say that to secure the salvation of Abaka and of the Khakan, of his sons and of his people, he had sent the friars mentioned to administer baptism to those who wished for it, or who had not had it duly administered before, and had ordered them, if he thought right, to go on to the Khakan's Court and do the same there. By letters patent special powers were conferred on the friars. They were authorised to preach the Word of God in all the land of the Tartars; to baptise Abaka and those of his sons and his people who should wish to become Christians; to absolve those excommunicated who wished to return to obedience; to confess and exact penance; to absolve the murderers of clerics and monks, if they gave due satisfaction to the churches, monasteries, or persons injured by their crimes; to found new churches in extra-diocesan places, and to confide them to meritorious men; to allow converts already married to people within the prohibited degrees to continue to cohabit; to decide matrimonial cases brought before them; to perform mass and other divine offices, where there was neither church nor oratory; to consecrate cemeteries, grant indulgences, dissolve vows, bless the sacred vestments, altars, &c., where there was no Catholic bishop, and, in fact, to do singly or collectively everything that could contribute to the glory of God's name and the furtherance of the faith.‡

The Chronicle of St. Denis says Grégeois, or Greeks. † Remusat, op. cit. 350.
‡ D'Ohsson, iii. 545-549.

Abaka's reign was coincident with a very flourishing epoch in Eastern literature.   The most famous among his *protégés* was the great astronomer, Nasir ud din, of Tus, of whom we have already spoken. Nasir ud din died at Baghdad, on the 25th of June, 1274. As Bar Hebræus says, he excelled in all sciences, especially in mathematics.   He refers to the famous astronomical instruments which he constructed, and which have already occupied us, and Meragha, the seat of his observatory, became the goal of a great number of learned men.   As he had been assigned the revenues from the temples, *i.e.*, mosques, and schools (the so-called vakfs) of all Baghdad and Assyria, he distributed with a free hand assistance to indigent scholars. He wrote works on many subjects—on logic and natural science—as well as commentaries on Euclid and the Almagest of Ptolemy, and on the ethics of Plato and Aristotle, besides his famous ephemerides, dedicated to the Mongol Khan, and entitled "Zij Ilkhani." It was reported that he died by poison.*   Next to him were the two brothers Shems ud din and Alai ud din Juveni, whose ill fortune during the latter years of Abaka I have described. Wassaf has preserved a number of poetical compositions which were exchanged between the former of these, the Vizier Shems ud din, and Shems ud din Kert, the ruler of Herat.† In Rum there lived the poets and philosophers, Sadr ud din, of Konia, and Jelal ud din, of Rum.   At Shiraz there still flourished the famous and now very aged Persian poet Saadi, in close friendship with Mejd ud din Semeki, known as the king of the poets, with Imami, of Herat, and with Khoja Hemam ud din, who was clerk to Nasir ud din, of Tus, and was well known for his prodigality, having on one occasion given a splendid feast to the son of the vizier, Shems ud din, at Tebriz, in which 400 porcelain bowls were used.   There also flourished the following poets : Purbeha Jami, whose verses were composed in a mixture of Persian and Mongol ; Abulmadhi Raigani, so called from the village of Raigan, near Kazvin ; Jemal ud din, of Kashan ; Jemal ud din Rastak ol kotu, who lived to the age of ninety, in the reign of Abaka, and was so called from Rastak, one of the quarters of Kazvin ; the judge, Bahai ud din Senjani, the panegyrist of the vizier, Shems ud din, who also, like Purbeha, mingled Mongol and Turkish words with his Persian ; Rasig ud din Bela, who had charge of the revenues of Diarbekr, of whom Von Hammer quotes a verse, complaining of Abaka having deprived him of his post in favour of the amir, Jelal ud din ; Nejm ud din Serkub (*i.e.*, the goldsmith), who flourished also in the reign of Arghun ; and, lastly, Nisam ud din, of Ispahan.‡ The chief poetry then fashionable is marked largely by puns and play upon words, and by adulation, clothed in inflated imagery, which Wassaf also introduced into his prose, and of which his work is an exaggerated

---

* Abulfaraj, 576. D'Ohsson, iii. 538-539.   † Op. cit., 154-158. Ilkhans, i. 278-283.
‡ Ilkhans, i. 317-319.

example. Besides these poets there also flourished during Abaka's reign the famous geographer, Jemal ud din Yakut, whose work is still so deservedly esteemed ; and the musician, Safr ud din Abd ul Mumin Al Urmeir (*i.e.*, from Urmia).*

In the year 1278, the Metropolitan of the Nestorians having died, the patriarch, John Denha, ordained Simeon, called Bar Kalig, formerly bishop of Tus, to this post. Before he left, however, he had a feud with the Catholicos (*i.e.*, with Denha), whom he treated badly, and was accordingly imprisoned. Having tried to escape, he presently, with some other bishops and monks, came to a violent end.† Meanwhile, two Uighur monks passed through Mesopotamia, *en route* for Jerusalem They had come from China, and had gone by order of Khubilai Khan, for the sake of visiting the Holy City. Mar Denha ordained one of them as Metropolitan of China, and gave him the name of Yaballaha. Before he set out for his post, however, Mar Denha died. Thereupon Abaka having been informed by Yashmut, who, like the two travellers, was a Uighur, of the death of the Catholicos, asked the Christians living there (*i.e.*, at Baghdad) to accept Yaballaha as their Catholicos, and issued an edict to this effect, whereupon some twenty-four bishops assembled together from Seleucia and Ctesiphon, and ordained him as Catholicos. Bar Hebraeus says naïvely that this Mar Yaballaha, "although he was too little versed in doctrine and in the Syriac tongue, was nevertheless of a naturally good disposition, endowed with the fear of God, and showed much charity to us and our people."† Yaballaha now consecrated his late companion as Bishop of Uighuria. He was called Bar Suma. This very friendliness and patronage of the Christians was no doubt a great cause of offence to his Muhammedan subjects and employés, who doubtless looked with much more favourable eyes upon his rival, the Sultan of Egypt, and made it easy to suspect that his end was hastened by some sinister act on the part of those who treated him as a heretic.

In the British Museum there are silver coins of Abaka struck at Mosul and Tebriz, and copper coins struck at Mosul, Irbil, and Baghdad. They occur of various dates, from 666 HEJ. to 680. Most of them bear Arabic legends, but some have an inscription on the obverse side in Mongol characters, and in the Mongol language, which was read by Schmidt, " Khaganu darugha Abagha Khan deled keguluksen" (*i.e.*, " The Great Khan's Viceroy Abagha-Khan, his coinage ").§ Mr. Poole questions the reading " darugha," as did De Saulcy, and says the letters read " arab " or " arun."‖ Although anything but a Mussulman, Abaka's coins often contain a formula from 'the Koran, notably the sentence, " There is no God but the one God who has no equal."¶

* Wassaf, 103. D'Ohsson, iii. 539.   † Mosheim, 69-70.
‡ Abulfaraj, Chron. Eccles., ii. 453-454.   Mosheim, 70.
Frahn Resentio, 637.   ‖ Catalogue, Oriental Coins of the British Museum, vi. 46-49.
¶ Frähn, op. cit., 636.

Note 1.—In a MS. translation of the " Tarikhi Guzideh," for which I am indebted to my friend Mr. G. L'Estrange, I find it stated that in the year 674 a body of Mulahids, or Assassins, joined the son of Khurshah, gave him the name of No Daulat, and seized the fortress of Alamut, and their insurrection having spread, Abaka sent an army, which captured and ruined the fortress.

Note 2.—I find I overlooked an interesting passage in the *Georgian Chronicle* referring to Mangu Timur's campaign in Syria. We read there that before setting out he sent a summons to the mthwar of Samtzkhé (*i.e.*, to Beka) to accompany him. " My enemies," he replied, " have aroused the anger of your brother, Abaka Khan, who sent Arghun to devastate my country. As I was innocent I withdrew, but now I dread the Khan. If, therefore, you will promise to forget the past, to cause my lands in future to be respected, I will go and join you with my troops." Mangu Timur gave an undertaking accordingly, and ratified it by drinking water in which gold had been mingled, and also sent Beka the ring he had on his finger as a further pledge. Beka then assembled his Meskhes, and went and joined Mangu Timur, who received him heartily and introduced him to Abaka. They then set out for Egypt. At Darabsak Beka and his Meskhes distinguished themselves in repelling a sortie from the town, and were the first to enter the place. Mangu Timur rewarded him, his didebuls, and arsnaurs with presents of horses and garments.[*]

* Hist de la Géorgie, i. 935

CHAPTER V.

SULTAN AHMED KHAN.

BY his will Abaka had nominated his son Arghun, whom he had previously appointed governor of Khorasan, as his successor. But this was clearly contrary to the yasa or law of Jingis Khan, which, in regulating the succession to the throne, appointed that the eldest living prince of the house should succeed. This position was now occupied by Abaka's brother, Tagudar (called Tongudar by Haithon and Nigudar by Wassaf), the son of Khulagu by Kutui Khatun. He had remained behind in China when his father set out for the West, as we have seen, and was sent to Persia by Khubilai Khakan during the reign of Abaka. According to Haithon, he had been baptised in his youth under the name of Nicholas.* Later in life he greatly favoured Muhammedanism, and this, together with his patronage of the two viziers, Shems ud din and Alai ud din, who controlled Tebriz and Baghdad, and against whom Arghun had great animosity, created a powerful party in his favour among the Muhammedans. He was supported by Khulagu's three sons, Ajai,† Konghuratai, and Hulaju ; by Juskab, or Chuskab, and Kunkju, sons of Chumkur, the second son of Khulagu ; and by the amirs Singtur and Sughunjak, Arab, and Karabukai. Arghun was supported by the two brothers Buka and Aruk, and by Akbuka, amirs attached to his father's household, and by the great amirs, Shishi Bakhshi, Doladai Aidaji, Jushi, and Ordukia. A third party gathered about Oljai Khatun, who tried to create a diversion in favour of her son, Mangu Timur, but as he died twenty-five days after his father, she and Kutui Khatun joined the party of Arghun.‡ The "Shajrat ul Atrak" says that Arghun was of opinion that Mangu Timur should succeed his father, but Tagudar having assumed the office he was reluctantly compelled to submit.§

On Abaka's death, Tagudar set out for Tebriz from Kurdistan, and Arghun, who was already on his way from Khorasan, halted at Meragha, and heard from Singtur of his father's death. There the funeral ceremonies, including the offering of the bowl of kumis, were gone through, while Buka ordered the officers of Abaka's household to do homage to him. After the funeral the assembly adjourned to Chagatu, where Shishi Bakhshi, who saw that the majority of the princes favoured

---

* Op. cit., 56. ‡ Rashid, however, says Ajai was already dead. See Quatremere, 107.
‡ D'Ohsson, iii. 551. Ilkhans, i. 323. § Op. cit. 257.

Tagudar, persuaded Arghun to submit gracefully, and Tagudar was unanimously elected on the 6th of May, 1282. Three days later Arghun set out for the Siah Kuh, or Black Mountain, to secure his father's treasures.* Wassaf calls the place Alatagh.† Abulfaraj says that on his accession to the throne he showed great generosity and clemency. He distributed his treasures among his brothers, the great amirs, and the army, and treated his people kindly, especially the heads of the Christian religion, and issued edicts in their favour, exempting their churches and monasteries, their priests and monks, everywhere from tribute.‡ The vizier, Shems ud din, who was in the hands of Arghun, was summoned to Tagudar's presence, and on the 21st of June, the various princes having sworn the oath of allegiance, he was duly enthroned, his brother Konghuratai and the amir Singtur Noyan taking him by the right and left hand respectively. This was on the 21st of June, 1282.§ As he was a Muhammedan, he adopted the name Ahmed (the Acomat of Marco Polo) and the title of Sultan. He then sent to the castle of Shahu-tela for the treasures, and distributed largess liberally, each soldier receiving 120 dinars (Von Hammer says twenty). Arghun now returned, and complained that they had not waited his arrival before going on with the ceremony of inauguration. Tagudar received him graciously, and presented him with twenty golden balishes, which he said he had specially reserved for him. It was during this visit that he formed a close friendship with Konghuratai, and swore mutual oaths of attachment in the ordu of Tuktai Khatun (called Tuktini by Von Hammer), one of Abaka's widows.||

Ahmed, soon after his accession, made a public profession of his faith, and addressed a brief to the authorities of Baghdad, which has been preserved to us. It runs as follows: " In the name of the clement and merciful God. There is no other God but God, and Muhammed is the Prophet of God. We who are seated on the throne of sovereignty are Mussulmans. Make it known to the inhabitants of Baghdad. Let them patronise the medressehs (colleges), the wakfs (religious foundations), and their other religious duties as they were accustomed to do in the time of the Abbassidan khalifs, and let everyone who has claims upon the various charities attached to the mosques and colleges be reinstated. Do not transgress the laws of Islam, O people of Baghdad. We know that the Prophet (may God grant him peace and pity) has said: 'This faith of Islam shall not cease to be triumphant till the day of resurrection.' We know that this prophecy is true, that it emanated from a true prophet, that there is only one God unique and eternal. Rejoice, all of you, and make this known throughout the province."¶ Haithon says his conversion

---

* D'Ohsson, iii. 552.   † Op. cit., 201.   ‡ Chron. Syr., 596.
§ Wassaf, 202.   ‖ D'Ohsson, iii. 553.
¶ Life of Kelavun, in Makrizi, ii. 185-186. D'Ohsson, iii. 553-554. Ilkhans, i. 325.
Weil, iv. 137-138.

to Islam was followed by that of many of his people, and also by a persecution of the Christians.* Ahmed appointed the Noyan Sughunjak his lieutenant-general, and Shems ud din Muhammed his finance minister. He stayed a few days at Siah Kuh, and sent to Hamadan to summon Majd ul Mulk Yezdi and his deadly rival, Alai ud din Juveni, the latter of whom was still in prison.†

We have seen how Alai ud din learnt of the death of Abaka at Asadabad, while on his way from Baghdad to Hamadan. His enemies instructed the commissary who was in charge of him not to release him, and he remained in chains till the accession of Ahmed, who ordered his release. When he reached the Court with Majd ul Mulk, the latter, supported by a Mongol grandee, renewed his intrigues, and was on the point of again obtaining the farming of the taxes; but Shems ud din having secured the protection of Ermeni Khatun, Ahmed's second wife, also obtained the favour of her husband, and poured a series of accusations, true and false, upon his bitter rival. He, in turn, wrote to Arghun, saying : "The vizier poisoned your father, and now wishes to take away my life because I am aware of his crime. If I die you will know the reason." Shems ud din, in turn, employed a nephew of Majd ul Mulk, named Said ud din, who had been deposed by his uncle from his post of mestufi, or president of finances, in Irak and Persia, on account of his dishonesty, and who now accused him of having had secret correspondence with Arghun. Ahmed ordered Alai ud din's confiscated property to be restored to him, but he would not take it, and gave it up to the crowd for pillage, a very diplomatic movement. Meanwhile, Majd ul Mulk was hoist with his own petard. Sughunjak Aka and Aruk (called Sunjak and Arukaka by Von Hammer), who were sent to apprehend him, found among his effects a piece of lion's skin with certain unknown figures upon it in cinnabar and saffron (i.e., red and yellow), which had been secreted there by Abdur Rahman, a friend of Shems ud din. The Mongols were very timid in the presence of such necromantic charms. Their bakhshis and shamans declared that the skin was to be dipped in water which the accused was to drink, whereupon he acquired magical powers. This he stoutly denied, but he was found guilty.‡ Sughunjak would have spared him, but, unfortunately, he was laid up with a bad foot, and Abdur Rahman urged his execution strongly upon him. Ahmed ordered him to be handed over to his enemies. We are told by Rashid that Shems ud din wished to spare him, but that this was opposed by Alai ud din and his son Harun. Wassaf, on the other hand, says that Alai ud din would have pardoned his enemy, but was opposed by the Treasury officials. He was taken to Alai ud din's tent, where from the mid-day till the evening prayer he was called to account for all his extortions, and especially for the 300 tumans he had appropriated at

---

* Op. cit., 56.   † Ilkhans, i. 325.   D'Ohsson, iii. 554.
‡ Quatremere, Rashid ud din, 359-360.   Note.

Baghdad, and also for the various warrants; diplomas, &c., he had unlaw-
fully issued. As Alai ud din went to the evening prayer, he was brought
out of the tent, and was torn in pieces by the crowd. This was on the
14th of August, 1282.* He was dismembered, and his head was sent to
Baghdad, upon which Rashid ud din writes—

> The head with so much proud ambition girt,
> That it aspired even to the viziership ;
> I saw it the toy and sport of the hangman,
> A portion held in either hand.

Some one paid a hundred gold pieces for his tongue, and took it to Tebriz.
His feet were sent to Shiraz, where he had made his entry so proudly,
and his hands to Ispahan. Whereupon the famous poet Pur-beha Jami
said :—

> He would have raised his hand to heaven ;
> It did not reach thither, but it has come here.

Alai ud din was now restored to the government of Baghdad. Ahmed
gave him one of his own robes, and a paizah or tablet of office,
hoping thus to secure a continuance of his services, he having expressed
a wish to retire into private life. Shems ud din recovered his authority
as vizier, while the superintendence of the affairs of Islam was made over
to the Sheikh Kemal ud din Abdurrahman er Rafii. The latter, according
to Bar Hebræus, was the son of a slave of the Khalif Mostassem, and a
Turk of Rum by origin,† who escaped from the massacre at Baghdad and
went to Mosul, where he carried on the trade of a joiner. Thence he
went to A'madiyah, and informed its ruler, Iz ud din, that the spirits had
taught him magic. Being taken by him to Abaka, he told him that if he
were taken to the castle of Tala, where the imperial treasures were stored,
he would show him something wonderful, and having measured the
ground from one side to the other, he ordered them to dig in a place he
pointed out, while he stood some distance away. A valuable precious
stone was found and taken to Abaka. He acquired great repute in
consequence, and became head of the administration during the short
reign of Ahmed, and had control of the wakfs, or pious foundations, in all
his dominions, from the Oxus to the borders of Egypt, with orders to
restore them to their original purpose, and to detach from their registers
of pensions the Christian and Jewish astrologers and doctors whose
names had been inscribed on them during the previous reigns, and whose
salaries were ordered to be paid out of the Treasury. Arrangements were
made for the comfort of the caravans of pilgrims to Mekka, and the
sending of provisions to the Kaaba, while Ahmed ordered the Buddhist
temples and the Christian churches to be converted into mosques.‡
Haithon says he caused the Christian churches at Tebriz to be destroyed.
He threatened to decapitate the Christians who refused to adopt Islam,

---

* Ilkhans, i. 326-327. D'Ohsson, iii. 557-558.
† According to the Persian writers he was a native of Mosul, of humble rank.
‡ Abulfaraj, Chron. Syr., 614-615. D'Ohsson, iii. 561.

and summoned the kings of Armenia and Georgia to his Court. He adds that they preferred to risk a struggle rather than go, which can only apply to the Armenian king.* Besides Abdur Rahman, he had another favourite, named Mingueli, a Muhammedan saint, whom he called his son. He spent a part of the day with these two doctors, listening to their lessons, and occupied himself very little with affairs of government. His mother, Kutui Khatun, who according to the biography of Kelavun was a Christian, used to go to him there and discuss state affairs with him. Ahmed neglected Sughunjak Aka and Singtur Noyan, to whom he owed his elevation to the throne.†

Shems ud din began his new career by economising the expenditure of the imperial household. The cost of the cooks' department, which was superintended by the head cook, Fakr ud din, had hitherto been eighty tumans annually. This was reduced to one half, and was dictated by his jealousy of Fakr ud din, who at the accession of Ahmed had applied for the post of vizier, and thus threatened to displace him. Wassaf, the historian, tells us how he had also committed himself in making accusations against Shems ud din, and tried to make amends by an outflow of his very profuse rhetoric in apologies. Towards the end of Abaka's reign, viz., in 1279, the government of Shiraz had been entrusted to the amir Sughunjak, whose sagacity and equity Wassaf enlarges upon. Discontented with the farmers of the taxes, he selected the Khoja Nisam ud din, the one who had embezzled the least public money, and set him over the rest. He nominated Ibn Muhammed Yahya Imad ud din to be chief judge, although the citizens desired Seyid Abdallah, the author of many works on exegesis and hermeneutics, on tradition and jurisprudence, on dogmatics and philosophy. Sughunjak went to the Court with some tax collectors who were short in their accounts. During his absence his vizier and chief judge quarrelled. He sent orders that the former should be confined in the house of the latter. The judge Seyid thereupon repaired to Buka, one of the secret tax appraisers of Abaka, who was then at Shiraz, who sent him with the tax farmers, who were deputies of Shems ud din, to the Court. There they laid before Abaka their complaints against the administration of Sughunjak, and of his vizier, Nisam ud din. Abaka offered them a beaker of wine, and ordered that Nisam ud din should make good 200 tumans that were wanting of the public money. The amir Toghachar was sent to execute this order. Nisam ud din was at this time under arrest in the house of Imad ud din, but the tax farmers, becoming alarmed, made common cause with him. They did not rest until they had released Nisam ud din, and caused Toghachar great embarrassment. Toghachar, on the accession of Ahmed, repaired to the Court, taking with him the malik Shems ud din and

---

* Haithon, 57.   † Wassaf, 209-211.   D'Ohsson, iii. 559-561   Ilkhans, i. 328-329.

T

Imad ud din. Ahmed appointed the latter to the viziership of Shiraz, and promoted the governor, Bulghuvan, who had openly taken sides against Toghajar. The latter accordingly went to join Arghun in Khorasan.*

After his conversion, Ahmed naturally desired to draw nearer to the Egyptian Sultan, who controlled the head-quarters of Islam at this time. As his envoys he selected the chief judge, Kutb ud din Mahmud of Shiraz, the Kadhi of Sivas, and the Amir Bahai ud din (called Seba by Abulfaraj), Atabeg of Sultan Masud of Rum, while the Prince of Mardin also sent his vizier, Shems ud din ibn Sharf ud din ibn Tenesi. The envoys set out with a magnificent *cortège*, leaving Alatagh on the 25th of August, 1282. When the Sultan heard of this embassy he became suspicious, and sent two of his hajibs or chamberlains to meet it at Biret, with orders to exercise the greatest vigilance, not to permit its members to communicate with any-one, and to cause them to travel by night. The remembrance of Mongol treacheries was so recent and so keen that the Egyptians were naturally timid. The envoys entered Aleppo by night, and their arrival there was kept secret. Passing by Damascus, they arrived at Misr, opposite Cairo, at night, and in the month of October. Admitted to an audience by Sultan Kelavun, they submitted their master's letter, as well as a verbal message. The former is such an interesting document that I will transcribe it at length from the copy printed by Quatremere in his appendix to Makrizi. It contained neither subscription nor seal, but was marked with thirteen tamghas in vermillion, and was as follows :—

" In the name of God, the most clement and pitiful. By the power of God, and under the auspices of the Khakan. The Firman of Ahmed to the Sultan of Egypt. The Supreme Being worthy of all praise has, by his grace and the light of his supreme direction for a long time, and since our youth, caused us to know his divinity, to confess his unity, to proclaim Muhammed. May God be propitious to him and grant him peace, to venerate the saints whom he has chosen as his disciples and placed among his creatures. *God opens and purifies the heart of him whom he intends to direct, so that he may adopt Islam.*† For a long time, and until the death of our august father and brother opened the succession to us, we have ever exalted religion and wished well to Islam and the Mussul-mans. God has deigned to confer on us all the favours and benefits which we would hope for from his munificence. He has opened this empire to our eyes, and made it over to us as a bride. We have assembled a kuriltai (a meeting where the friction of opinion produces light), where we have collected all our brothers, sons, the princes of the blood, the great amirs and generals, and the governors of the towns. At this meeting it was resolved to carry out the work of our elder brother, and to send against you such a multitude of men that the earth could scarcely hold

---

* Ilkhans, i. 329-331.   † Koran.

them, with a zeal that would level the highest mountains and soften the hardest rocks, and whose courage and fury would fill men's hearts with fear. We have sifted the cream which rose from this discussion, and have found it contrary to the wish of our heart for the general good. In order to strengthen the foundations of Islam, and being determined that no orders should come from our hand, save those which would tend to prevent the shedding of blood, to calm the troubles of men, and to cause the soft zephyrs of peace and security to blow over every country, so that the Mussulmans should recline in peace on the couch of our favour and beneficence, wishing thereby to show our respect for the Most High, and our love for his people, God has inspired us therefore to quench this fire, to re-establish calm, and to make known to those who have advised us that we should devote ourselves to furthering man's wellbeing, and postpone indefinitely an appeal to the last resort. We have no wish to draw the spear till we have gauged the end for which we are doing it, nor to throw it till we have satisfied ourselves that our cause is right. We have fortified our resolution to make peace, and to do what is necessary to secure it by the advice of the Sheikh ul Islam, the model of wise men, Kemal ud din Abdur Rahman, who is our good helper in matters of religion. We have published this in the hope that God will pity those who call upon him, and will punish the disobedient. We have sent the chief judge, Kutb ud din, the pole of law and religion, and the Atabeg of Bahai ud din, both of whom have our confidence, in order that they may convey to you our good faith and good intentions towards all Mussulmans, that you may know that God has opened our eyes, that Islam may blot out what has gone before, and that God has inspired our heart to follow the ways of truth, and to accept as guides those who know it. You will recognise in the intentions he has inspired us with, a great favour of God to men, and you will not thrust aside our peaceful message because of what happened in the past, for every day has its special character.

"If you wish for a proof of what we say, contemplate our acts, which are well known, and whose effect has been universal. By the grace of God we have unfurled the banner of the faith, and proved our belief in different ways—in recommending the observance of the Muhammedan law and in pardoning those who have incurred penalties. We have given orders for the re-organisation of the wakfs or pious foundations attached to the mosques, tombs, and colleges ; to rebuild the hospitals and the ruined ribats ; to restore their incomes to those who have title to them according to the wishes of the founders ; and have given orders that nothing is to be taken from the recent charities and nothing altered from the old ones. We have ordered the pilgrims to be well treated, and their needs to be supplied, to guard the routes by which they travel, and to find escorts for their caravans. We have given complete liberty to the merchants who visit your country, and have expressly forbidden the soldiery, the karaguls (*i.e.*, the guardians of the roads), and the governors of provinces to molest them, either on setting out or in returning. Our karaguls having seized a spy, disguised as a fakir, we have remitted the punishment of death which was his due, and out of respect for the divine commands we have released him and sent him back again. Nevertheless, you will not forget how prejudicial to the cause of Mussulmanism the sending of spies is. For a long time our soldiers having found that these spies disguise themselves as fakirs, anchorites, &c., have formed a very bad opinion of these religious, and have killed those they have laid their hands upon. Thank God, since a free'passage to merchants has been granted through our dominions, there is no longer any need for such disguises. If you consider all these acts, you will see that they are innocent and natural, and quite inconsistent with artifice and chicanery. This being so, there are no longer any causes of ill-will between us. Our anxiety has had as its source a zeal for religion and for the defence of the land of the Mussulmans, but by divine grace our reign has been lit up with a true light. We declare that whoever follows the way of reason will find in us a friend and a defender. We have raised the veil and speak freely. We have made known to you our sincere views, which have the Almighty God for their object, and have forbidden our soldiers to act contrary to these views, so that we may gain the favour of God and His Apostle, so that the Mussulmans may be spared from the consequence of our discords ; that the mists of enmity may be dissipated by the light of good harmony, and that townsfolk and rustics may equally repose under its tutelary shade ; that the hearts which have been forced by fear to the gorge may be tranquillised, and that old grievances may be forgotten. If, by the grace of God, the Sultan of Egypt is inspired to do that which shall secure peace to the world and the well-being of men, he will follow the right path, and open the way to union and good friendship. Thus shall his country prosper, troubles will be appeased, swords will return to their scabbards, the earth will become calm again. The necks of Mussulmans will be relieved from the chains of ignorance, but if evil thoughts prevail over the designs of the God of pity, and you refuse to appreciate our benevolent offers, God will recompense our efforts, and take note of our excuses. *We shall not inflict punishment before sending an Apostle.* But may God point out the right way, and grant success. He is the protector of countries and people, and alone suffices for us.—Given in the middle of jumada the first, 681 (21st August, 1282), at our camp of Alatagh."*

To this rhetorical epistle the Egyptian Sultan replied in these terms :

"In the name of God the clement and pitiful. By the power of the Most High, by the fortune of the reign of the Sultan Malik Mansur. The

---

* D'Ohsson, iii. 563-570. Quatremere, Makrizi, ii. 187-191. Abulfaraj, Chron. Arab., 361-365. Wassaf, 215-218. Von Hammer, Ilkhans, i. 331-335.

reply of Kelavun to Sultan Ahmed.   Praise be to God who has opened
for us and by us the path of truth, who in bringing us hither has made
divine help and victory follow our steps, so that men have in crowds
joined the religion of God.   May his blessing rest on our Lord, our
Prophet Muhammed, whom God has made greater than all the prophets,
by which means he has saved the people.   May this blessing lighten
those who are in darkness, and overwhelm the hypocrites.   By the favour
and devotion of the Imam, Hakim biemrillah, the amir ul muminin, the
offspring of the khalifs, who has trodden the right path, the cousin of the
Lords of the Prophets, the Khalifs, who were the protectors of religion.
We have received your noble letter, in which you report your conversion
to the faith, and your separation from those of your family and your nation
who are its enemies.   On the opening of this letter certifying your
Islamism to the Mussulmans, thanks have been offered to the Eternal,
and prayers have gone up to him praying that he will make you persevere
in your resolution, and cause to grow in your heart a love for religion,
as he makes the tenderest plants grow in the most arid soil.   We have
attentively read the first part of your letter, in which you announce that
from your early youth you have confessed the unity of God and the truth
of Islam in thoughts, words, and acts.   May God be praised that he has
thus opened your heart to the faith, and favoured you with his holy
inspirations.   We thank God that he drew us to this sacred goal even
earlier, and has strengthened our steps where we have acted and fought
for his glory, for without him our steps totter.   If you have taken
possession by right of inheritance on the death of your father and elder
brother, if God has conferred on you his surprising favours, if you have
mounted the throne which your faith has purified, and to which your
power has given additional lustre, it is God who has given it to whom he
has chosen among his servants, and has realised in him what he has
promised, the graces belonging to the saints of God and holy men.

" You say that at the kuriltai where your brothers and the other princes
of the blood, the grandees of the empire, the chiefs of the army, and the
governors of the provinces unanimously determined to send an army
against us, having reflected on this decision you found it contrary to
your convictions, which were solely devoted to the public good and the
general peace, that you, therefore, sought to calm the troubles, and to
quench this fire.   This is the conduct of a pious sovereign, who looks
tenderly to the safety of his subjects, and prudently calculates the results
of things.   If, in fact, your people had followed out their intentions and
abandoned themselves to their illusions, their exploit would assuredly
have brought them a terrible reverse ; but you have acted like a man
fearing God, who is not misled by passions nor shares the ways of evil
men, nor of those blinded by illusions.   You say that you do not wish to
rush into war before you have traced out your path and justified it by

argument, but now that you have joined the faithful your efforts and ours should be directed against those whose idolatry prevents them following this route, and God and man know that we have only armed ourselves in order to protect the Muhammedans, and have only acted for the glory of God and man. You have embraced the faith. All animosity has disappeared. The past is forgotten. Mutual aid has succeeded to aversion, for the faith is like a building, each part of which supports another. Wherever it unfurls its banner there should be one family.

"You say you have taken these steps by the counsel of the Sheikh ul Islam, model of doctors, Kemal ud din Abdur Rahman. We hope that by his benign influence, and by the merits of the past, all countries may be won over to Islam, and that the scattered fragments of the faith may be re-united, and we do not doubt that one who has begun so well will complete his noble work. As to the mission of the grand judge, Kutb ud din, and the atabeg Bahai ud din, they have delivered your messages, and they have reported a thousand interesting things about your situation, your ideas, and projects.

"You call our attention to the proofs of your justice and equity, especially in the good administration of the wakfs of the mosques, &c. These are acts worthy of reward, and of a great prince who desires the stability of his empire, but such matters are too little for a great prince to glorify himself about them. They are but elementary duties. The glory of great sovereigns is to restore empires to their rulers. See what your father did. The Seljukian sultans and other princes were not of his religion, yet he confirmed them in their sovereignty; he did not expel them from their kingdoms. If you find a right violated it is your duty to correct it, and not permit the oppressor to continue his oppression, so that your empire may become consolidated, and your reign be embellished by acts of piety.

"The order you have given to your soldiers, your karaguls, and the governors of the different provinces, to protect travellers from all vexations, has been reciprocated by ourselves, who have issued similar orders to the governors of Rahbet, Aleppo, Biret, and Aiptab, as well as to the commanders of the provincial troops.

"As to the spy disguised as a fakir, whom you have released, and as to the suspicion attached to his profession from such disguises, and the number of dervishes and others who have consequently perished from suspicion, it is from your side that this kind of thing began. How many people disguised as fakirs have come to spy out our land? We have arrested many and spared their lives, and have not tried to learn things hidden under their mendicant robes.

"You say that our union will bring peace to the world and well-being to man. One ought certainly not to turn aside when the door of reconciliation is opened, and he who turns aside to avoid an encounter is as worthy as he who offers the hand of friendship ; peace is assuredly the first of the commandments.

"The general matters you enter into are indeed necessary as a basis on which the social edifice may be built, and by which we can learn whether peace exists or no ; but we need other matters more specifically settling ; and in regard to these we have charged our envoy to treat with you about them *viva voce*, for what is contained in the breast of a messenger is better than what is written on a scroll. You cite this passage from the word of God : '*We have never punished anyone without first warning him by an envoy.*' This sentence does not savour of friendly intercourse like ours, and is not to be commended. The man who has the advantage of priority on the road and in the defence of the faith, has rights which should be respected, and prerogatives which belong to him. However many follow him, the first will retain his pre-eminence.* We have heard the message which has been delivered to us by the grand judge, Kutb ud din. It accords with your letter, and confirms the news that you have embraced the true faith, and have taken rank among the true believers, and are everywhere the patron of justice and right, qualities which deserve the praise of men. May God have the glory. Did not he in revealing to the prophet what concerns those who accept the faith, say : 'Do not think you do me service by being converted to Islamism, it is the grace of God that draws you thither.'

"According to the message you have sent us, God has given you so much that you do not covet other territories, and you say that if we are ready to treat on this basis you, too, are willing. We reply that when things are settled on the basis of a common accord they become stable, and the foundations of friendship. God and man know how we exalt our friends, and abase our enemies. How many allies have we, when we have neither father, nor brother, nor relative? In the early days of Islam the faith was founded by the co-operation of the companions of the prophet. If you wish to be on friendly terms, make an alliance with us against our common foes, and lean on those who can offer you at all times a strong succour. You remind us that if we covet any part of your territory we shall merely injure the cause of Islam, by sending hostile forces into your territory. We reply that if you close the hand of hostility against travellers, and leave the Mussulman princes in peaceable possession of their own, calamities will cease, and so will bloodshed. Nothing is more just than to abstain from doing ourselves what we forbid another to do, and more unjust than to prescribe a good action to others and to forget to do it ourself. At this moment Konghuratai is in the land of Rum, which is subject to you and pays you taxes ; nevertheless he has shed blood there, has dishonoured women and reduced children to slavery, sold free men, and continues his devastating course.

---

* This is doubtless a reference to Ahmed having been a recent convert, while Kelavun was a Mussulman of older standing.

" You send us word that if strife is not to cease between us, that we had better choose a battle-field, and that God will give victory to whom he will. Here is our answer : Those of your troops who survived their last defeat are not anxious to revisit the former battle-field. They fear to go there again to renew their misfortunes. As to the day of battle, God can alone fix that, and the victory will be to him whom God chooses, and not to him who feels himself secure. We are not of those who can be deceived, nor are we anxious about the result. As to the hour of victory, it is like that of the last judgment : it arrives unexpectedly. God does what is best for his people, and he is strong enough to do right.

" Written in Ramazan (December)."

When the envoys had concluded their mission, and received robes of honour and magnificent presents from the Sultan, they again set out. They were subjected to the same surveillance as on their arrival. No one was allowed to see them. They arrived at Aleppo on the 6th of Sheval, 681.*

During this year there arrived in Egypt the Sheikh Ali, of the tribe of the Uirads or Kalmuks, who had become a Mussulman, and had adopted the profession of a fakir, and, according to Makrizi, miracles had been performed by his hand. Finding himself followed by a number of Mongol children, he passed at their head first into Syria, then into Egypt. He was there presented to the Sultan, together with his brothers, Akush, Timur, Tukhi, Juman, and others. They were well treated, and some of them, including the three brothers, Akush, Timur, and Omar, were enrolled among the irregular troops, or khasséki (*i.e.*, kazaks), and promoted to the rank of amirs ; but presently Sheikh Ali, having misconducted himself, was put in prison with Akush. Timur and Omar died in the exercise of their functions.†

Let us now return to the more intimate affairs of the Mongols. We have seen how Arghun aspired to the throne on the death of his father, Abaka. He continued to nurse his resentment against his uncle and the vizier, Shems ud din. After repeated requests of Sultan Ahmed, he sent the amir Buka, who married Kutui Khatun, Abaka's widow, to him.‡ He was residing at Sughurlak, in the district of Baghdad, where there encamped a tuman of predatory Karaunas who had belonged to the military household of Abaka, and had their winter quarters at Siah kuh. Wassaf, who calls them a kind of demons, and the most fearless of the Mongols, says that Arghun made the General Toghachar (on whom he conferred the insignia of drums and standard) their commander. Under Toghachar were Gaikhatu and Baidu, the brother and cousin of Arghun, as well as the generals Chaukur, Chongutur, Doladai, Idaji, Iji, Tetkaul, Juchi,

---

* Makrizi, op. cit., 199-200. Wassaf, 218-223. Abulfaraj, Chron. Arab., 365-367. D'Ohsson iii. 570-580. Ilkhans, i. 335-342. Weil, iv. 139-142.
 † Makrizi, ii. (part i.) 53.   ‡ Ilkhans, i. 342. Wassaf, 223.

and Kimshkabal. The principal supporters of Abaka were thus devoted
to the cause of his son. This was doubtless as champions of the Mongol
and Shamanist elements of the community as against the Muhammedans,
who, under the new *régime*, were again holding up their heads ; and
Haithon says Arghun sent to inform their suzerain, the Khakan Khubilai,
how Ahmed had deserted the ways of his fathers, and how he and his
followers were becoming Mussulmans, which greatly displeased Khubilai,
who sent Ahmed his reproof.* We are told that at this time two faithless
dependents of Alai ud din Juveni, named Ali Chinsang and Kutluk
Shah, went to Arghun to tell him he had sent word to Wejih ud din, the
Vizier of Khorasan, to poison Arghun. He had him at once arrested and
imprisoned at Kerker, appropriated his property, and only spared his life
on the solicitation of Bulughan Khatun, Abaka's favourite widow, who
had joined his harem. Arghun then went to winter at Baghdad. As he
passed through Rai he treated the Malik Fakhr ud din with honour, and
appointed him governor of the district. When Sultan Ahmed heard of
this he had him arrested and taken to Shirvan, where he paid for his
recent honours by being tortured. Arghun, greatly enraged at this, sent
word to the Vizier Shems ud din and the amirs that Abaka had made
Fakhr ud din over to him, and that he would revenge the injuries done to
his *protégé*. When he reached Baghdad he demanded from Alai ud din's
deputy, Nejm ud din Asfer, that he should pay over the sums which had
been declared owing to the treasury in the reign of Abaka, and which had
not yet been paid. What there was in the treasury was seized, and
torture was applied to extract more from the officials, Arghun's efforts
being seconded by those of Sishi Bakshi, Pulatamur, and Toghachar,
whose names prove them to have been of Mongol extraction. Alai ud din,
on hearing of these persecutions, had an attack of apoplexy and died.
This was, according to Abulfaraj, at Mughan, in Arran. He was buried
at Tebriz on the 5th of March, 1283. Thus passed away the famous
historian, who has preserved us such graphic accounts of the earlier
Mongol doings, and who, under the name of Juveni, has a world-wide
reputation. He was succeeded in his office by his nephew Harun.†

The author of the Georgian annals, in referring to Ahmed, tells us he
was wanting in all the qualities of a sovereign, which, in ordinary English,
means he favoured the Muhammedans and not the Christians. At the
commencement of his reign the Georgian king visited the ordu, where he
gave his daughter Rusudan in marriage to the son of the Great Buka, an
alliance which greatly distressed the Catholicos Nicholas, and brought on
the king a severe reprimand.‡ After his accession to the throne, and in

* D'Ohsson, iii. 581-582. Ilkhans, i. 344. Note 2. Haithon, 57.
† Ilkhans, i. 343-344 ; and 344, Note 1. D'Ohsson, iii. 582. Abulfaraj says : " He wrote an
admirable chronography in Persian on the kingdoms of the Seljuks, Khuarezmians, Ismaelites,
and Mongols, from which we have taken what we have hitherto related in this work." Chron.
Syr., 604.
‡ Hist. de la Géorgie, i. 598.

April, 1282, Ahmed sent his brother Konghuratai with an army to the
borders of Rum to guard that unruly frontier, after having married him to
Tuktai Khatun, the widow of Khulagu, and niece of the famous Princess
Dokuz Khatun.* As he heard that Konghuratai was corresponding with
Arghun, he posted a force at Diar Bekr to prevent them from uniting
their forces. It would seem that Prince Jushkab, the son of Jumkur, the
second son of Khulagu, a partisan of Arghun's, also wintered in that
town. Ahmed now sent Alinak, the Governor of Georgia, to summon
Arghun to a kuriltai. He was won over by the latter, and did him
homage, and on his return to the Court he tried to make excuses for
Arghun's non-appearance there; but the vizier saw through it all, and
Alinak was again won over by being given the hand of Kuchuk, Ahmed's
eldest daughter, and being raised by an edict to a higher rank. When
in the spring Arghun left his winter quarters at Baghdad for his summer ones
in Khorasan, he took the Prince Jushkab with him.† Having reached
Rai he had Ahmed's deputy there bastinadoed, put a cangue on him, and
sent him thus to his master to remind him that Shems ud din had not yet
paid the sum which was found to be owing by him to the treasury. So
Von Hammer reads the story. Wassaf says he sent Juchi to ask Ahmed
to state that the vizier, although he had controlled the finances so long,
had not given any account of them, and he seems to have raked up the
rumour about Abaka's and Mangu Timur's death having been caused by
poison administered by the vizier, and asked that the latter might be sent
back with Juchi, but the Khan replied that Shems ud din could not be
spared from the divan, as there was no one to take his place, and Juchi
returned with this answer.‡ Meanwhile, Arghun was strengthening his
position in Khorasan. When he reached the borders of Mazanderan, on
his way thither, he met Yankaji Noyan, who commanded a tuman there,
and Hindu Noyan, who commanded two tumans on the frontier of the Oxus,
and told them how, on his father's death, being without an army, he could
not seize the throne, but now, if they would help him, he could carve his
way to it with the sword. Hindu replied that Ahmed, as the aka, or
eldest prince of the house, was entitled to the throne, while he (Arghun),
God be praised, was ruler there. He bade him be content with this
position, and follow the advice of those who had grown grey in his father's
service. If Ahmed ventured to attack him, however, they promised to
side with him.§ Arghun also needed money, and it was always possible,
under these circumstances, to make charges of embezzlement against the
officials of his treasury. Some of the amirs now incited Ali Chekbin and
others to accuse the Sahib Weji ud din Sengi el Furumdi, son of the
Sahib Iz ud din Tahir, of having misappropriated money.‖ When thus

---

* Wassaf, 236. Ilkhans, i. 342. D'Ohsson, iii. 583-584.
† Ilkhans, i. 344. D'Ohsson, iii. 584.   ‡ Wassaf, 224-225. D'Ohsson, iii. 585-586.
        § Ilkhans, i. 345.   ‖ Wassaf, 230.

charged, the vizier wrote a pathetic letter to Tughan, the ruler of Kuhistan, which, together with the answer he received, is reported in inflated verse by Wassaf.* The vizier did not fly, we are told, to the women in the Serai for succour, but he boldly faced the charges, and offered, if the slightest misappropriation was proved, to replace every piece by a thousand.† Arghun was not to be moved by any answer save the production of some coin, nor, says Wassaf, would he accept golden words in lieu of gold. He had him arrested, and insisted on his finding 500 tumans, 300 tumans in gold bars and 200 in kind (*i.e.*, in cattle, fruit, clothing, materials, &c.) "Eye-witnesses report," says Wassaf, "that on a single day as much as 3,000 menns of gold were paid down, while the treasuries of Firuzkuh, Herat, Merv, and other places were stripped of jewels and rich robes to meet the demand."‡ Arghun then gave the vizier a robe of honour, and left him in charge of Khorasan. He now demanded from the Ilkhan, in addition to that province, the cession of the royal domains in Irak and Fars. "As you hold in virtue of your right and the general suffrage the throne of my father, it is necessary I should have a province sufficient to support the troops which I command. If you make over to me the provinces which pertain to the private domain,§ the best feeling will exist between us ; if not, the contrary." The Khan replied: "We have from our affection and solicitude given him Khorasan, his appanage. If he wishes for another province, let him come to the kuriltai. After having consulted about matters, we will not refuse him our favours, but if he persists in his disobedience we shall march against him."‖

Ahmed had summoned Konghuratai to a kuriltai. The latter accordingly went to Alatagh. We have seen how he was on intimate terms with Arghun, and we read how he sent the latter some rarities from Rum, and received from his nephew a present of two pairs of hunting panthers in return. The Ilkhan further heard that Konghuratai had made a conspiracy with the two amirs, Kuchuk Anukji and Shadi (or Shashi) Akhtaji, to seize upon him during the feast of the New Year, when, according to Von Hammer, the Khan and all his Court went through the emblematic process of tempering iron in memory of the march out from Irgene Kun.¶ The plot was revealed by one of the conspirators. On the morning of the 18th of January, 1284, the day fixed for the carrying out of the conspiracy, Konghuratai was arrested by the General Alinak, and was put to death by having his backbone broken. Being a prince of the blood, it was unlawful to shed his blood, according to the yasa of Jingis Khan. This was in January, 1284. His two accomplices were taken to Karatagh, in Arran, and after six days' trial were condemned and executed.**

---

* Op. cit., 232.   † *Id.*   ‡ *Id.*, 233-234.
§ The greater part of Irak and Fars did so.   ‖ D'Ohsson, iii. 587.
¶ See Vol. I. of this work, p. 35.   ** D'Ohsson, iii. 588.

The *Georgian Chronicle*, in referring to these events, says Ahmed committed an execrable deed. He summoned his brother Kongharda from Greece, and put him to death. Two brothers, sons of Abuleth, who had escaped from the hands of Sadun, met the same fate by order of Kutluk Shah, son of the latter.* Stephen Orpelian says that, in addition to Konghuratai, there were put to death the two sons of Tsaghan, who, Brosset suggests, belonged to the family of the mthawars of Dzaganis Dzé, mentioned by Wakhusht.† He also says that Ghiath ud din, Sultan of Rum, was killed by him, but the other authorities assign the deed to Arghun, Ahmed's successor.

The chiefs of the troops encamped at Diar Bekhr received orders to arrest Arghun's officers in the district of Baghdad, and accordingly the generals Toghachar, Chaukur (called Jaughir by Wassaf), Jinkutur, Doladai (or Tuladai), Iji Ilchi (or Anji), Tetkaul, Juchi, and Kunjukbal, Abai (the son of Sunatai), and Jenghatu (the son of Juchi) were taken to Tebriz, and put in irons. Chunghur was sent to Yusuf Shah, the Atabeg of Luristan, to tell him to prepare an army to co-operate with his suzerain's and to guard his frontiers. The Prince Gaikhatu (called Kenjatu by Von Hammer), with the Amir Batmaji (called Temaji Aktaji by Wassaf) and others, managed to escape, when they reached Sawa, and made their way to Arghun in Khorasan. The latter was informed of what had happened by the judge of Kazvin Rasiuddin. Meanwhile Ahmed was married to his niece, Tudui Khatun, the daughter of Musa Kurkan, the husband of Tarakai, Khulagu's fifth daughter.‡ Ahmed was well seconded in his efforts to oppose his rival by his vizier, Shems ud din, who knew well that his own life and fortune depended on the issue. Yusuf Shah, the Atabeg of Lur, received orders to guard his frontiers carefully, while a large force of all races and religions, Mongols, Mussulmans, Armenians, Georgians, and Turkomans, was got ready, the army consisting of 80 tumans (*i.e.*, 80,000 men). Ahmed was also accompanied by the Georgian King Dimitri, who took with him Yoané, the chief of the Mandators, son of the great Shahin Shah, and the Generalissimo Kutluk Shah, son of Sadun. A heavy fall of snow at the end of January (1284) delayed the march of the army, which at length started from Mughan, the advance guard of 15,000 men being commanded by the Ilkhan's son-in-law, Alinak, by Baisar Oghul, and Tagai Kokoltash (Wassaf, instead of the last, mentions the Prince Hulaju). Other chiefs in it were Arghasun Tekta, Narin, Ahmed, and Ashghan Asan. The advanced posts reached Talikan on the 27th of April, and advanced against Kazvin, where 300 families of Uzes (*i.e.*, of Turkomans) belonging to Arghun were seized. As soon as he heard of Alinak's advance Arghun went to his treasury in Gurgan (Hyrkania), and to the cities of Nishapur,

---

* Hist. de la Géorgie, i. 598. † Hist. de la Siounie, 238.
‡ Ilkhans, i. 345-346. D'Ohsson, iii. 588-589. Wassaf, 240.

Tus, and Isferain, for money, clothes, and weapons, which he distributed among his amirs. Fakhr ud din, of Rai, who kept the register of these things, wrote above it, "Account of the sums distributed among the victorious army." Arghun, who was accidentally present, took up the pen and wrote the word "victorious" in a beautiful Persian hand, of which he was master. Kawam ud din, the Persian vizier, thereupon prognosticated a happy issue to their venture.* On the following day Arghun heard of the capture of Kazvin, that Ahmed had ravaged the province of Rai, and ruined the Serai of Lur, which was his private domain, whence he had carried off his people into Azerbaijan. He swore to be revenged,† and dividing his army into three bodies, set out. His advanced guard was sent on under Yula Timur, Jorghodai, and Bulughan, while he himself set out on the day fixed by the astrologers at the end of May, 1284.‡ With him were the amirs Amakaji, Nokai Yarghuji, Tawtai, Kasan (the son of Kutluk Buka), Baitmish Kushji, Sertak, Alghu, Oladai, Kadughan, Aghman, and 4,000 horsemen,§ and he left Sishi Bakhshi in charge of his baggage and impedimenta, while he summoned Nuruz to join him with his tuman of Karaunas. From Irbil, Ahmed dispatched Kurimshi, the son of Alinak, to his father to tell him not to engage the enemy unless he was superior to him in numbers, otherwise to await his arrival. He now advanced again, leaving his baggage in charge of Abukian. The advanced guards of the two armies met at Khiel büzürg (called Khail buzurk by D'Ohsson), situated about half-way between Rai and Kazvin. One of Arghun's spies, who was caught and made drunk, disclosed the strength and position of his master's people. Alinak marched against him, and a battle was fought on the plain of Ak Khoja, near Kazvin, on the 4th of May, 1284. The "Shajrat ul Atrak" calls the plain Fuhwacheh. Yula Timur and Amakaji commanded Arghun's right wing, Bulughan the left, and Tawtai the centre ; while on the side of the Ilkhan the centre was commanded by Prince Hulaju, the twelfth son of Khulagu, the right by Alinak, and the left by Basaraghul.‖ Notwithstanding the disparity of numbers, Arghun's people fought well, and as is often the case in Eastern battles, where the picked troops are put in the right wing, the left wing of each army was defeated, and the division of Basaraghul was pursued as far as the walls of Kazvin. His wife and son were captured, and the village of Gurgan was plundered.¶ The fight lasted from mid-day to sunset. At length, seeing that his people were overmatched, he withdrew with 300 horses towards Firuzkuh, where he hoped to meet the body of Karaunas whom he had summoned to his aid. Meanwhile his men, on hearing of his flight, disbanded. The Karaunas arrived at Ak Khoja after the

---

* Ilkhans, 346-347. Wassaf, 259. D'Ohsson, iii. 589. Shajrat ul Atrak, 258. Hist. de la
Géorgie, 600.
† Wassaf, 242. ‡ Id., 241. § Id.
Id., 241. Ilkhans, i. 344. D'Ohsson, iii. 590. ¶ Wassaf, 245.

battle, and proceeded, after their truculent fashion, to plunder and
burn Damaghan, and to waste the country round.* On his precipitous
retreat, Arghun was joined by an officer who had been dispatched by
Ahmed before the fight to tell him that he had not instructed Alinak to
attack him, but merely to secure his presenting himself at the Court, and
asking him to lay down his arms and submit. Arghun sent Kutluk Shah
Noyan and Legsi Kurkan (Rashid calls them Legsi and Ordubuka) with
his answer, which was submissive, but reminded the Ilkhan that if he
drove Arghun to extremity, and he was joined by the Karaunas, things
might be very awkward, and complained of the ravage committed by the
Ilkhan's troops, especially near Damaghan. The matter was remitted to
the vizier, Shems ud din, who said it was impossible to stop the ravage by
the army, which was necessary to keep up its spirit ; "predatory birds
preferred to seize their prey rather than to live on regular rations," a
sentiment which, Wassaf says, brought the vizier no good, while the
State speedily suffered.†

It is reported that Arghun on his retreat towards Bostam made a
pilgrimage to the tomb of the Sheikh Abu Yezid, while Ahmed similarly
went to the grave of Babi to ask their aid, which, as Von Hammer says,
is certainly remarkable in the case of the former, who was not a
Mussulman. Ahmed now ordered his brother Hulaju to go to Rai with a
tuman of soldiers. He also ordered all the leading officers of the army to
subscribe a document stating that they would not obey any other
commander but Buka, whom he appointed generalissimo. All signed
this, including Alinak.‡ Ahmed's officers tried to persuade him to pardon
Arghun's conduct as due to youthful indiscretion, and to end the campaign,
for the heat was terrible, and many horses had perished. He would not,
however, listen to them, and when Sadr ud din and Arsil ud din, sons of
the famous astronomer, Nasir ud din of Tus, declared that the stars were
unfavourable, he was angry with them. When the Ilkhan reached Surkh,
near Semnan, the Surikkala of Fraser, there went to him from Arghun the
latter's famous son Ghazan, with Omar Oghul, son of Nigudar, of the
family of Jagatai, together with Nokai, the yarghuji (i.e., the superior
judge), and Sishi Bakhshi (i.e., the Secretary of State). Ahmed, in reply
to this embassy, dispatched his brother Togha Timur, and his nephew
Suke, the third son of Yashmut, son of Khulagu, with the amirs Buka
(called Buka Gizbara by Abulfaraj) and Doladai to tell Arghun that if he
was sincere he should go to him in person. Buka suggested, in setting
out on this mission, that it would be better if the army were meanwhile to
halt. Ahmed said he would go as far as Kharkan (called Khojan in the
"Shajrat ul Atrak"), where there was good pasture, and there await his
return. His men ravaged the country as they marched, especially the

---

* Wassaf, 245. Ilkhans, i. 348. D'Ohsson, 590-591.
† Ilkhans, i. 351. Wassaf, 246-247.   ‡ Ilkhans, i. 351.

Georgians, and plundered in the town and district of Damaghan what the
Karaunas had spared. Kharkan was famous as the birth and burial place
of the Sheikh Abul Hasan Kharkani. The amir Jirkudai, his brother
Yesuder, and Bulughan, the Governor of Shiraz, went there to do homage
to Ahmed. The latter did not stop at Kharkan, as he promised, but
dispatched Alinak ahead with an advance guard, while he himself went to
Kalpush and Kebud Jameh (i.e., Bluecloak), a district of Dahistan, rich in
corn, grapes, and silk. He was now joined by the envoys he had sent to
Arghun, viz., the Princes Togha Timur and Suke, and the amirs Buka
and Poladai, who took Gaikhatu, Arghun's brother, with them. Buka
complained that Ahmed had not halted at Kharkan, as he promised.
Nuruz and Buraligh, Arghun's envoys, who accompanied them, returned
without securing anything. Shortly after the amir Yula Timur and
Imkajin, son of Suntai, went to Ahmed with their submission. He grew
impatient at the delay in Arghun's submission, displaced Buka by Akbuka,
and naturally made the former even more a secret partisan of Arghun.*

Let us now turn to Arghun. With only a hundred followers he repaired
to the strong fortress of Kelatkuh, famous in early Persian history and in
later times as a treasure place of Nadir Shah. It is situated in a beautiful
valley rich in horses and game, between Kazermian, Serkhaz, Abiverd,
and Tus. The Georgian history calls it Kala. Here Arghun sought
shelter with his favourite wife Bulughan.† Nuruz, a faithful dependent,
urged him to cross the Oxus and take shelter with Kubinji, who had his
yurt there. On the other hand, another of his officers, Legsi, went over to
Ahmed and asked him for an army with which to attack him, and
with which he in fact harried the yurt of Arghun's wife, Kutlukh Khatun.
He also tried, but in vain, to persuade Nuruz to imitate his example. On
reaching Ahmed's camp with his booty he was richly rewarded.‡
Meanwhile Alinak with his troops reached Kelatkuh. Arghun thereupon
came out of the fortress alone, and shouted with a loud voice to him.
He prostrated himself, and said his uncle wanted to see him. Arghun
replied that he also wished to see his uncle. Alinak made him a present
of a white horse. They then entered the fortress together and had a
long interview, at which the young prince was much pressed to submit.
Seeing that there was no other course open, he set out with Alinak, and
joined Ahmed at Guchan, or Ghujan, on the 29th of June.§ He was
treated with scant civility by his uncle : entered the camp by the entrance
on the left side, and was deprived of his girdle, nor was he admitted to
an audience for some time, but was exposed to the sun till the perspiration
covered his face, until his sister Taghan, who loved him dearly, came
out from the royal tent to shade him. Presently his wife Bulughan
Khatun was allowed to enter, and Ahmed presented her with a bowl of

* Ilkhans, i. 353.   † Wassaf, 247-248. Ilkhans, i. 353-354.   ‡ Ilkhans, i. 354.
§ Wassaf, 248. D'Ohsson, iii. 594.

kumiss. He then went out hawking for two hours. When at last Arghun was admitted, he knelt down and did homage in the usual Mongol fashion. Ahmed embraced him, and promised that he should retain the government of Khorasan, as in his father's reign. Nevertheless he had him closely watched by a guard of 4,000 men, commanded by Aruk, the brother of Buka, who surrounded him, says Wassaf, like the circling planets.*

Ahmed now set out to join his new wife, Tudai Khatun, to whom he was much attached, and, according to Wassaf, left orders with Alinak to put Arghun to death when the royal banners had withdrawn.† Rashid says that it was Alinak who urged the Ilkhan to put an end to him at once. He asked what harm he could do with neither army nor treasure, and said he would first ask the advice of his mother, Kutui. Meanwhile he ordered Arghun's amirs, Sishi Bakhshi, Kadan, and Buraligh to be executed.‡ The *Georgian Chronicle* tells us how, on the march home, some bands of Mussulmans attacked and killed several Georgians, whereupon Rat, son of Bega Suramel, pursued and dispersed them ; many of them were killed, and others taken to the Georgian king. The latter was much thanked by Ahmed for his services in Khorasan. Ahmed granted him many honours, and gave him all the Georgian aznaurs, after which, we are told, he went to his kingdom to show his victorious self to his wives. He was accompanied by the Georgian king, while Alinak was left to look after Arghun, with orders to presently put him to death.§ Buka obtained permission to stay behind under pretence of assisting at the marriage of his intimate friend, Kipchak Oghul, a descendant of Juchi Khasar. He had been attached to Arghun's household during the reign of Abaka, and had only left him with regret after considerable pressure from Ahmed, who had treated him with distinction and given him one of Khulagu's robes. He had been displaced, as we have seen, by Karabuka, and had a corresponding grievance. He now persuaded several officers, some of whom were his relatives, that Ahmed, with his confidantes, Hugai, Kara Buka, Alinak, and Abugan, intended to make an end of them near Isferain. " It is necessary to protect ourselves. Ahmed has determined to exterminate the descendants of Jingis Khan. Through the influence of the vizier he favours the Mussulmans. It is in order to destroy the Mongols that he has placed the Georgians under the orders of Alinak, and that he has raised him above the other generals and courtiers." Seduced by these words, the officers in question, as well as the Princes Jushkab and Hulaju, determined to carry out their purpose that very night. Buka invited Karabuka, Biak, and Alinak to a feast. The last of these excused himself from drinking on the ground that the following night his regiment

(Kázik) was to act as Arghun's guard. Jushkab offered to take his place. He accordingly went, and at midnight was dead drunk. Buka, followed by three horsemen, now entered Arghun's quarters, and sent one of them to go quietly and awake him, and announce to him that he had formed a party in his favour, and was there to save him. Arghun was frightened, and thought it part of a plot, but being reassured on his taking the most solemn oaths, he came out of his tent, and Buka bade him mount. On leaving the camp the Mongol sentry asked how it was they were only four when they entered, and were now five. They assured him he was mistaken, and reached Buka's camp safely.* Arghun having put on his armour and mounted, they repaired to Alinak's tent, in which they cut him to pieces, together with the mosquito net with which he was covered. Some of his men seized their bows. Buka cried out to them, " Hitherto we have obeyed Ahmed. We have killed Alinak by order of Hulaju." Whereupon the guards threw down their arms and prostrated themselves.† Messengers were now sent to Hulaju and Bektu, who were at Firuzkuh, bidding them do to Basar Oghul (the Yessar of D'Ohsson) and to Abukian what they had done to Alinak. Basar, or Yessar, was thereupon killed while drunk in his tent with some of his companions. Karabuka, Biak, and Tabui were also arrested on the following morning. Some of these were put to death, while others were released.‡

A horseman, named Mama, meanwhile escaped from the camp and went to warn Ahmed, who was then four parasangs from Isferain. With him were Kinshu, the son of Jumkur, son of Khulagu, and the amirs Ak Buka and Legsi. He had already turned back to punish the rebels when he heard of the fate which had overtaken his officers. There was nothing for it now but a rapid flight. He passed the night at Kalpush with his wife Tudai, and then went by way of Kumus and Irak in order to reach the ordu of his mother, Kutui, near Serab, a town of Azerbaijan, between Tebriz and Irbil. As he retired, his officers and the petty princes who were with him broke away. The " Shajrat ul Atrak " says that at Kazvin he put to death Tilai Timur, one of Arghun's chief amirs, and his sons. The Vizier himself arrived at Jajerem with only one attendant, and went on to Ispahan. " It was a veritable rout," says Wassaf, " in which the confusion and fear were such that gold and silver balishes, vessels ornamented with precious stones, rolls of golden tissue, and Chinese silk were strewn along the roads like stones or leaves, without anyone staying to collect them. The fugitives threw away the pearls and jewels round their necks and in their ears, and went to hide themselves in caverns, &c."§ Sughunjak, with Agharuk Sultan, escorted Ahmed's treasures on camels and other beasts towards Mosellemi, intending to join his master at Serab, but he was attacked *en route* by Taiju Kushji and Kitbuka

---

* D'Ohsson, iii. 597-599.   Ilkhans, i. 356.   † D'Ohsson, iii. 599.   ‡ Id. Ilkhans, i. 356.
§ Wassaf, 255-257.   D'Ohsson. iii. 599-601.

(called Kituga Kuruji by D'Ohsson). They secured the treasures, which were detained at Mosellemi.* Buri was sent to order the Karaunas who were at Siah kuh to seize Ahmed, while Jerik, the amir of the ordu of the murdered Prince Konghuratai, was also dispatched with 4,000 men to avenge his late master.†

The princes of the blood, amirs, &c., were met together at Kharkan with Arghun to decide upon a successor to Ahmed. There they were joined by Hulaju and Kinshu. There were three parties. Buka, and, according to the *Georgian Chronicle*, Yas Buka, chief of the Uirads, supported Arghun. His brother Aruk and Kurmishi were in favour of Jushkab, the brother of Kinshu, who, they said, held the great yurt ; and Bekta, or Tekia, of Hulaju, who was a son of Khulagu, and who, therefore, according to right, had superior claims to the younger men. Buka said that the Khakan, who was the master of the earth and also the aka of the house of Jingis, had given the sovereignty of Iran to Abaka on his father's death, and that it now ought to belong to his son ; and when Bekta interfered he drew his sword, and swore that so long as he bore it no one should be ruler but Arghun. He then turned to Tenguiz, or Tengir Kurkan, the husband of the Princess Tudukash, the fourth daughter of Khulagu, and the father of Arghun's wife, Kutluk Shah, and asked him what Abaka's own intentions had been. "I and Singtur," he replied, "heard him say that he left the throne to his brother Mangu Timur, and afterwards to Arghun." "You have invented these words," said Bekta ; "where did you hear them spoken?" Arghun said he did not want the throne, and would be content with the government of Khorasan. Buka then interposed with the sage remark that their enemy was yet at large, and that when he had been captured they should meet in the yurt of Oljai Khatun and the other princesses to elect a new khan. All agreed to this, and Arghun and Buka set out on the 11th of July with the advance guard, while the other princes followed with three divisions.‡

Meanwhile Ahmed, having reached Sheruyaz, called Kunkur Olang by the Mongols, pillaged the ordu of Buka, and would have put his family to death but for the interference of Sughunjak. He reached his camp on the 18th of July, and having embraced his mother and told her what had happened, proposed escaping towards Derbend ; Kutui advised him to stay in her ordu, and to try and secure the support of the generals there. But the news of his ill luck had already spread. Karabuka and Singtur having gone to do him homage, asked him why he had returned in this haste, without escort. He said that, having secured Arghun, he had returned to arrange about provisioning the army. Naitan, or Natian, who was seated outside the tent, having heard this, said in a loud voice, "It is not so ; ten princes of the blood and sixty superior officers have

---

* D'Ohsson, iii. 601.   † Ilkhans, i. 357.   ‡ D'Ohsson, iii. 601-603.   Ilkhans, i. 357.

leagued themselves with Arghun, while Ahmed has come here as a fugitive. The good of the State and the public peace demand that he should be arrested." The two generals thereupon placed a guard over his tent. Singtur committed his custody to the Princess Kutui, who appointed 300 men to guard him.* We have seen that the Karaunas had been ordered by Buka to fall upon Kutui's ordu, where Ahmed now was. They arrived under Bureh, entered the tents of the princesses, and robbed them of their clothes and jewels. Everything in the royal camp—clothes and furniture, gold, silver, &c.— became their prey. They took the jewels from the neck and ears of Kutui herself, and dragged the boots from her feet, and she was left naked with the princesses Tudai and Ermeni Khatun. It was against the yasa of the Mongols to thus illuse women and children, but the Karaunas (Mongolian demons, as Wassaf calls them) were not subject to such scruples. They ended by seizing Ahmed, stripped him of his robes, and guarded him in his tent.

Meanwhile, Arghun himself, fearing that his victim might escape if he waited till his horses were in condition and his men all with him, set out with but 300 horsemen. When he arrived near Mosellemin he was met by Karabuka and Singtur at the head of the Karaunas, who had Ahmed with them. It was the fashion with the Mongols, when engaged in archery, that the one who won stretched out or clapped his hands, and shouted " Morio." When Arghun saw Ahmed bound he cried out " Morio," and his officers followed his example, and they celebrated their good fortune beaker in hand, and congratulated Arghun.† Arghun having crossed the River Moor on the 26th of July, reached Abshur, near Yuz Agach, on the Sunday following. His adherents, the generals Toghachar, Kunjukan, and Doladai, who had been imprisoned at Tebriz, were now released, and some of the officers of the late Prince Konghuratai, with Bektu, were told off to act as Ahmed's judges. They charged him with ingratitude towards those who had raised him to the throne, and with ill-treating Konghuratai and Arghun. He confessed his guilt. Arghun and the amirs wished to spare his life on account of his mother, Kutui Khatun, who was much respected ; but as the mother and six sons of Konghuratai demanded the blood penalty, and as Yesubuka Kurkan, the husband of Kutulun, Khulagu's sixth daughter, reported that the two princes Hulaju and Jushkab were collecting an army at Hamadan, Arghun gave orders for his execution, and he was put to death in the same way that Konghuratai had been, viz., by having his back broken. This, according to Rashid, was on the 10th of August, 1284. Abulfaraj says the 16th.‡ The " Shajrat ul Atrak " says they broke his back, and at the same time the hearts of the people of Islam. Marco Polo, in relating the

* Wassaf, 258-259. D'Ohsson, iii. 603-604.
† Wassaf, 260. D'Ohsson, iii. 605-606. Ilkhans, i. 358.   ‡ Weil, iv. 145. Note.

downfall of Ahmed, speaks of Arghun's chief supporters as Boga (*i.e.*, Buka), Elchidai, Togan, Tegana, Tagachar, Ulatai, and Samagar. He also says that Acomat (*i.e.*, Ahmed), when he fled, was trying to escape, to take shelter with the Sultan of Babylon (*i.e.*, of Egypt), attended by a small escort, and that he was arrested and taken to the Court by an officer in charge of a pass by which he had to go.* He was buried at Kara Kapchilghai. Wassaf says that it was Arghun who incited Timur and Ilduz, or Ildir, to demand the death of Ahmed.† There can be small doubt that Ahmed's death, which was a remarkable event when we consider the constancy and loyalty of the Mongols to their princes, was mainly due to his patronage of Muhammedanism, which set against him the conservative feeling, both political and religious, of the Mongol chieftains, and also to his injudicious behaviour.

We must now revert somewhat, to relate some other events that occurred during his reign. The year before his death he sent a second embassy to Egypt, which was presided over by the Sheikh Abd ur Rahman, his tutor, who had persuaded him to become a Mussulman. The envoy took with him as presents, precious stones and pearls, rich stuffs, and gold tissues. Leaving Alatagh, he went to Tebriz, where he stayed a month, and enlisted a number of skilled artisans (goldsmiths, shoemakers, &c.), and prepared right royal surroundings for himself. Thence he reached Mosul, and having sent for a large quantity of gold from Baghdad, went to Mardin. There he was met by a messenger from the Egyptian Sultan, bidding him hasten on to Damascus, where he had been awaiting him for some time, and whence he must speedily set out homewards, as the district could not maintain such a large army. Abd ur Rahman, in reply, said he was ready to go, but he asked that he might be treated in a becoming manner, and not have to travel in the night, as the preceding envoys had had to do. The Sultan sent him an assurance that this would be so, and he accordingly set out in January, 1284, with the Mongol general Samdagu, and with Shems ud din Muhammed, vizier of the Prince of Mardin.‡ He marched with a suite of about fifty persons, including secretaries, doctors of the law, &c., lawyers, guards, servants, and slaves, and was escorted by a detachment of Mongols, while the Prince of Mardin also joined him with his troops, to see him safely to the Euphrates.§ When he reached Harran, he was met by an Egyptian amir, named Jemal ud din Akush Faresi. He expected that the latter would have dismounted and kissed his hand, but he contented himself with saluting him from a distance. He also demanded that the Sheikh should send back his Mongol escort, which he accordingly did. He had also to lay aside the State umbrella which he

* Yule's Marco Polo, ii. 470-471.

† Wassaf, 260-261. Abulfaraj, Chron. Syr., 600-602. D'Ohsson, iii. 606-607. Ilkhans, i. 358-359.
‡ Abulfaraj, Chron. Syr., 596-597. D'Ohsson, iii. 608-609. § D'Ohsson, iii. 609.

carried, and also his arms, nor was he permitted to advance by the ordinary route. When he reached the Euphrates the troops of Mardin wished to return home, but were told that the Sultan wished to see them at Aleppo, and they accordingly crossed the river. When they had encamped on the other bank the Sheikh retired to rest after his evening meal, but was awakened in an hour. The Amir was already on horseback, and said they must set out at once, and when the envoy said he should not start till morning he was told that the orders were that he was not to be allowed to travel except by night. "You may kill me," said the Sheikh, "but you will not make me travel by night." "We shall not kill you," was the unmistakable reply, "but we shall compel you to do our way." The Sheikh, indisposed to making a scene, consented to go. The Amir had ordered his men not to exchange words with the strangers. These indignities prove the intolerant arrogance of the Egyptian authorities at this period, and also show the dread which the Mongols everywhere inspired. The envoy entered Aleppo on the night of the 7th January, 1284, so secretly that no one knew of his arrival. The Mardinians were there rewarded with 200 zuzæ each, and sent home. The travellers, again travelling by night and by unfrequented routes, reached Damascus on the 2nd of March, and were lodged in the Hall of Ridwan. Orders were again given not to speak to the strangers ; they were to be listened to, but not answered. One thousand silver pieces were assigned daily for Abdur Rahman's needs, and a similar sum to buy meat, sweets, and fruits for his table.* At Damascus he had to await the return of the Sultan Kelavun, who left Cairo on the 17th of July, and learnt at Gaza of the death of Ahmed. On reaching Damascus, on the 20th of August, he at once gave him an audience. This was also at night, in the presence of 1,500 Mamluks, dressed in red embroidered atlas (i.e., satin), with turbans made of golden tissue, and with golden girdles, each bearing a torch. The Sheikh presented himself, with the Mongol amir Samdagu, and with Shems ud din Muhammed, son of Sherif ud din Beiti, surnamed Ibn-alsahib, vizier of Mardin.† The Sheikh was dressed as a fakir. He was ordered to prostrate himself, and on his refusal was rudely thrust down. He made several prostrations, while the Sultan took no notice of him. He nevertheless received Ahmed's letter from his hands, and ordered the three envoys to be given kaftans. Ahmed's letter was dated from Tebriz, in June, 1283. It was written in Arabic, and merely con-tained a number of friendly phrases. Among the presents offered by the Sheikh were sixty strings of large pearls, a piece of yellow yakut (i.e., topaz) weighing more than 200 mithkals, a red yakut (a ruby), and a piece of balkhash (i.e., the precious ruby of Badakhshan) weighing 22 dirhems. The envoys having delivered their message were sent to their quarters.‡

' Abulfaraj, Chron. Syr., 596-598.   Makrizi, ii. 64-65.   D'Ohsson, iii. 609-61
† Makrizi, ii. 69-71.   ‡ L., ii. 69-72.   D'Ohsson, iii. 612-614.

While Ahmed and Kelavun were carrying on this diplomatic intercourse, the latter did not scruple to commit acts of aggression on his neighbour. He sent a body of men from the fortress of Karkar to lay siege to Katiba, one of the fortresses of the province of Amid. It was blockaded until the garrison surrendered it, and was then garrisoned with troops from Biret, Aintab, and Revandant, and became, says Makrizi, one of the strongest bulwarks of Islam. The Egyptians also this year secured the fortress of Kakhta. Its governor was put to death by the citizens, who then gave the place up freely. It formed a capital base in the Egyptian operations against the Armenians in Cilicia, who, when at Aleppo, two years before, had burnt the great mosque there. To avenge this, Kelavun now ordered his troops to invade Cilicia. They advanced as far as Ayas, plundering, and having defeated a body of the enemy near the defile of Izkanderun, arrived at home again safely with their booty.*

The Armenian historian, Haithon, in describing the struggle between Arghun and his uncle, tells us that the conversion of Ahmed to Muhammedanism, and his efforts to convert the Mongols, were reported to the Khakan Khubilai, who was much irritated by the news, and sent to reprove him; that he in turn was annoyed at the rebuke, and although he dared not oppose himself to the Khakan, he proceeded against his own brother (i.e., Konghuratai) and nephew (i.e., Arghun).†

Ahmed had several wives. The first of these was Dokuz Khatun, the Konkurat. The second one was also of the same tribe, and was called Ermeni. The third one was Tudakun Khatun, the daughter of Musa Kurkan. The fourth, Baitegin, the daughter of Huseinaga. The fifth, Ilkotlogh, the daughter of Shadi Noyan; she was the mother of Tughanjik, who, being suspected of magic, was drowned in the River Kur. The sixth was Tudai Khatun. By these wives Ahmed had three sons—Kaplanshi, Arslanshi, and Nukajiyeh—and six daughters.‡

Mr. S. Poole only describes one coin of Ahmed's as being in the British Museum. This was struck at Mosul. The date is obliterated. It bears a curious type, which is also represented on one of Abaka's coins from the same mint, viz., a figure seated cross-legged, holding the crescent moon in its uplifted hands.§

Note 1.—Sultan Ahmed's name before he became a Mussulman is written in several ways by the authorities, the confusion caused by misplacing the diacritic points being the main cause. Hence several Persian writers have called him Nikudar. His name, as we learn from Rashid ud din and Abulfaraj, was really Takudar, and it is possible that the great-grandson of Jagatai, who accompanied Khulagu to Persia, was also called Takudar, and not, as he is

* Makrizi, ii. 61-63.　D'Ohsson. iii. 615-616.　† Op. cit., 57.　‡ Ilkhans, i. 325-326.
§ Catalogue Oriental Coins, Brit. Mus., vol. vi. 24.

generally styled, Nikudar, or Nigudar. The Georgian annalist calls the Ilkhan whom we have been discussing, Thaguthar. Wakhusht calls him Thenguthar.

Note 2.—The Egyptian historians have a curious story about the pilgrimage of a Georgian king to Jerusalem at this time, which is not referred to in the native annals. Novairi, Abulmahasen and the *soi disant* Hasan ibn Ibrahim report the fact under the year 1272, in the reign of Bibars. Makrizi puts it in the reign of Kelavun, as does the biographer of the latter Sultan. He is variously called Buba suta, Tauta sutena, and Tuma suta, and is said to have been the son of Gulbaz, or, as it is otherwise written, of Kiliari. We are told he was one of the most faithful allies of the Tartars. He had an old wound, caused by an arrow, on his neck, wore a golden ring on his right hand, and was about forty years old ; was pale in colour, with black eyes and a narrow forehead. Quatremere reads the name of his kingdom as Chavaketi. Brosset reads the name of his companion Thamgha, son of Abgar ; Quatremere, as Tibaga son of Ankavar ; and we are told he had a round face, a cicatrice over his right and left eyes, a long beard of a russet colour, and a tall and stout body. His interpreter was a prince of Abkhazia. Some writers say he went overland to Sis, and thence set sail for Acre or Ptolemais. Another story is that he sailed from Poti. When the Sultan heard of his travelling *incognito*, he gave orders that his steps were to be watched. He was arrested while *en route* by Bedr ud din, Governor of Jerusalem, who handed him over to the amir Rokn ud din Mankuris, who took him to Bibars at Damascus, and he was eventually imprisoned in the Castle of the Mountain.*

' Makrizi, ii. 56. Notes. Hist. de la Géorgie, 596, and notes.

CHAPTER VI.

ARGHUN KHAN.

ARGHUN was the eldest son of Abaka by one of his concubines named Kaimish Igaji (the latter being the title given by the Mongols to the concubines of their princes, and meaning elder sister). Immediately on Ahmed's death, on the 22nd of June, 1284, the khatuns Oljai and Takteni, the amirs Buka, Singtur, and Toghachar, &c., having met together at Abshur, near Yuz Agach (called Kamsiun, between Hesht er Rud* and Kurban Shira, by Wassaf) unanimously elected Arghun as his successor. The festivities which took place on this occasion are described by Wassaf in more than his usually inflated sentences, and with less than his usual modicum of tangible fact. He tells us that news of the event was dispatched in various directions, from the sources of the Oxus to the borders of Egypt. The princes Hulaju, Jushkab, Kinshu, Baidu Oghul, and Gaikhatu had not arrived when this election took place, nor was Shems ud din, the vizier, whom Arghun cordially hated, present. A gracious message was, however, sent to him, and he sent a reply by Yusufshah, of Lur, and Malik Imad ud din Kasvini. Hulaju was the son of Kumukur, second daughter of Khulagu, and was the agha or senior prince of the royal family, and as such entitled to succeed, and a party was favourable to him. Arghun sent him a present of a rich tent, with a message telling him how he had been chosen by the princesses, &c., to mount his father's throne, and offering to share that throne with him. Hulaju had no wish to do this, and joined Arghun, who, with the other members of the kuriltai, adjourned to Kurban Shira (Rashid ud din says to the yurt Suktu), where Gaikhatu had also arrived. The final ceremony of installation was completed on the 11th of August, 1284, Hulaju taking Arghun by the right hand, and Anbarji by the left, and seating him on the throne, while the various grandees prostrated themselves before him, putting their girdles about their necks, like slaves ready to be strangled for their lord, holding their caps in the air, and drinking his health.† Three days later there arrived the princes Kinshu and Jushkab, sons of Jumkur, son of Khulagu, who had supported Hulaju for the khanship, and also proffered their allegiance. Abukian, son of Shiramun son of the famous general Charmaghan who was yarghuji, or chief judge, was put to death, as one of Ahmed's most intimate advisers. The rest of Ahmed's supporters were each granted a yarligh of indemnity. Wassaf makes no exception, and says that Bukia, Tinai, Abkian, son of Shiramun, and Hulaju, the baskak of Tebriz, all received this favour.* Baidu, son of Targai, son of Khulagu, was made governor of Baghdad ; Jushkab, of Diarbekr ; Hulaju and Gaikhatu, of Rum ; Ajai, eighth son of Khulagu, of Georgia ; while Arghun's own son Ghazan was given Khorasan, Mazanderan, Rai, and Kumus, the Prince Kinshu and Nuruz, son of the famous Arghun Aka, being nominated his assistants. On the 18th September, Arghun appointed his faithful friend Buka vizier, and ordered as much gold to be poured over his head as would entirely cover him.†

Arghun's accession was naturally very grateful to the Christians, and we are told how the Georgian king, Dimitri, was especially favoured by him ; and by the influence of Buka, who was his friend, he was given authority over all the land of the Armenians, including the principality of Avak, that of Shahan Shah, with those of the families of Vahram and of Sadun, the last of whom had died in 1282. Arghun also favoured Dimitri's right-hand man, the atabeg of Armenia, Darsaij.‡ The " History of Georgia " adds that Dimitri sent his young son David to take possession of the country of Avak, and to hold it as an appanage.§ Ghiath ud din, the Sultan of Rum, had been deposed by Ahmed, who had appointed Masud, son of Iz ud din Kai Kavus, in his place, and sent Ghiath ud din to Erzenjan. There he was strangled by order of Arghun, for having been privy to the death of Konghuratai.‖ Hamdullah says he was put to death by Ahmed, for being implicated in Konghuratai's rebellion. Abulfaraj says he was poisoned there by his grandees on account of his prodigality and ill-government.¶ Abulfeda and Makrizi agree that this was in the year 1283. Hamdullah says that the affairs of Rum at this time were much disturbed, and that the children of Muhammed and Tughrul Shah seized on the coast of Anlakiah (?), Alasah, and Ladakiah by force, and took Basara saraf, a neighbouring province to Sis, from the Mongols. To put this down, the princes Gaikhatu and Hulaju were sent to Rum, and the viziership was conferred upon Fakhr ud din Muhammed, otherwise called the Khoja Fakhr ud din Mestofi, who by his goodness and wise measures compelled some of these people to submit, and destroyed others, and made the kingdom of Rum the cynosure of realms. Presently Fakhr ud din visited the ordu, and showed Arghun an elaborate calculation, in which the various expenses of the empire were tabulated under several heads, as the treasury, the camp, &c. This aroused the jealousy of Arghun's Jewish vizier, Said ud Daulat, who, when

* Wassaf, 267. Ilkhans, i. 359.　　† D'Ohsson, iv. 4. Ilkhans, i. 359.
‡ Hist. de la Siounie, 238. St. Martin Memoires, &c., ii. 163-165.
§ Op. cit., i. 602.　　‖ D'Ohsson, iv. 203-204.　　¶ Chron Syr., 604.

Arghun was drunk, got his permission to execute him, and he was accordingly executed on the first of Ramazan, 689.

To revert, however. Abulfaraj tells us that on the accession of Arghun fear fell upon everybody, and there was a general rush from Syria towards Egypt, and the price of a camel for the transport rose accordingly. Arghun re-appointed the Uighur Masud, whose former administration we have described,* governor of Mosul.† This was greatly to the delight of the Christians. His friend Yashmut had been assassinated a short time before by the sons of Jelal ud din Turan, whose death he had brought about. Abulfaraj tells us that when the Egyptian Sultan heard of Ahmed Khan's death, and the accession of his successor, he released Abd ur Rahman, the former's envoy, who had been imprisoned at Damascus, gave him a pension, and assigned him a residence.†

Makrizi says the envoys were thrice summoned to his presence by the Sultan, who, having abstracted from them the information he needed, told them their master Ahmed was dead. They were conducted to less stately rooms in the citadel, and their rations were reduced to mere necessities. They were then ordered to disgorge. The Sheikh did not long enjoy his liberty, for he was presently sent to the fortress of Saphda, at Damascus, and detained there. It is curiously like modern journalism, and its great autumn gooseberry, to find Abulfaraj digressing at this point to tell us how a pigeon at Bertellus laid an egg as big as that of a goose, and another which was long and crooked like a cucumber.§ Makrizi goes on to say that Abd ur Rahman was forced to surrender the treasures he had with him belonging to the Ilkhan, including a great quantity of gold, pearls, &c., *inter alia*, being a neck-lace of pearls belonging to himself, worth 100,000 dirhems. He and his companions were put in prison, where he died on the 18th Ramazan. His companions were presently released, except the Amir Shems ud din Muhammed, who was sent to Egypt, and imprisoned in the so-called Castle of the Mountain.‖

The new *régime* at the Ilkhan's Court was naturally fraught with danger for the late vizier, Shems ud din, who had called down upon himself Arghun's resentment in many ways, and he at once took flight from Jajerem, where he was living. Mounted on a dromedary, and with but two companions, he hastened across the desert to Ispahan. News of what had happened had not yet reached there, and the maliks (the governors of provinces were so styled under the Ilkhans), amirs, kadhis, and a crowd of people of all ranks went out of the town with presents to meet the vizier. So says Wassaf. Von Hammer, apparently quoting Rashid ud din, says that having heard of the revolution which had taken place, they consulted with the Atabeg of Yezd, who had been arrested by the governor of Ispahan during Ahmed's reign, as a supporter of Arghun, as to what should be done. Shems ud din, informed of this, under pretence of going to pay a visit to a burying place outside the town, fled on a swift horse towards Kum, to the famous sacred tomb of the sister of the Imaum Riza, which for a thousand years has been a place of asylum for those who entered its walls. Its sacred character was respected by the Mongols, no less than by the princes of the Seljuki and Buyid families. Chardin described its magnificence in detail, and it is still famous for its silver lattice-work and gold-plated doors, and its treasury, whose riches chiefly date from the time of the Safevi dynasty, but were much augmented by the gifts of Feth Ali Shah, who, *inter alia*, dedicated there a head-dress of his mother's, as Croesus did his wife's necklace and girdle at Delphi. Morier says the town is now famous for three things : the gilt cupola over the tomb, the market-place, and also for its ruins. Its old walls had a circuit of 40,000 ells, *i.e.*, 40 ells more than those of Kazvin. Kum gave its name to rich silk stuffs, called kumash, which still bear that name. It is famous for its high cypresses and its blue drinking mugs.

At Kum Shems ud din was joined by his friends, who advised him to go to Hormuz, on the Persian Gulf, and thence take ship for India,* but he thought it would be wrong to abandon his family and supporters to Mongol vengeance, and preferred to go to Arghun's Court, hoping to secure the good offices of his old friend Buka.† He therefore delayed a few days, and was then apparently joined by Imad ud din, of Kazvin, and Yusuf Shah, the Atabeg of the Greater Luristan, who had been sent by the Ilkhan. Yusuf Shah had been forced to join Ahmed in his campaign against Arghun. Now that the latter was on the throne the Lurs had left Khorasan, and turned their faces homewards by way of Tabs, but the greater part of them perished of thirst on the way.‡ Yusuf had married a daughter of Shems ud din, and he took his father-in-law with him. When near Sava they were met by the amir Khumar, who brought the Vizier word that the past was forgotten, and that he was restored to favour. He at once sent word round to inform the various chiefs of Irak of the welcome news. He reached Kurban Shira, where Arghun was, on the 21st of September, and repaired to Buka, with whom he was formerly on friendly terms. The meeting was full of superficial good feeling, but this was feigned on both sides. Buka presented him to Arghun, who received him coldly, but restored him to his former post as Vizier. He now held it, however, jointly with Buka. Shems ud din confessed that he only wished to be the latter's substitute ; but inasmuch as presents and gratitude poured continuously upon the elder occupant of the office, Buka's jealousy was aroused.§ This was further fanned by some of the Court officials whose enmity he had secured, viz., Ali Tamghaji, Fakhr ud din

* Wassaf, 265-266. Ilkhans, i. 362. D'Ohsson, iv. 4-5.   † Wassaf, 267-268.
‡ D'Ohsson, iv. 5   Ilkhans, i. 363.   § Ilkhans, i. 364. D'Ohsson, iv. 5-6.

Mestofi, and Hosam ud din Hajib, who urged that he would be speedily
eclipsed by Shems ud din, who would not be quiet until he had thrust
him into the shade, as he had Arghun Aka. Buka now urged upon his
master that the Vizier had been unfaithful to his father and would be the
same to him, and being already irritated against him for various reasons,
he gave orders that he was to be tried by the two amirs Kadagai and
Ogotai. He had already been ordered to find the 200,000 gold pieces
which he had been declared to be deficient, and had replied that he had
no ready money, since he had not been accustomed to bury it like some
people, but had bought properties with it which brought him in 1,000
dinars a day.*

Abulfaraj tells us that when the Vizier said he could not find the sum
demanded, unless he were restored to his former income, he was bidden
to borrow it. He thereupon borrowed what he could among his friends
and servants, but could not get together more than 40,000 gold pieces,
and said if they were not satisfied, they must put him to death.† When
he had been manacled he was subjected to the jeers of the Turks and
Persians, but stoutly affirmed his innocence. He was bastinadoed, but
without effect, and was therefore conducted, on the 10th of October,
to the place appointed for his execution, viz., Munia, near Ebher, north of
Kazvin. Having obtained a few minutes' surcease, he performed the
prescribed ablutions, and opened the Koran he had with him at haphazard,
looking for an omen, and then sat down and wrote to the heads of his
faith at Tebriz : " Having consulted the Koran, I have found this passage,
'Truly those who have said, "God is our master," and who have afterwards been constant to the faith, will see angels descend upon them.
Therefore have no fear. Do not regret, but, on the contrary, rejoice in
the paradise into which you are about to enter.' God, who has greatly
favoured his servant and granted him all his wishes, has come to introduce
him to life eternal. I deem it right, therefore, to inform the mulanas
Mohai ud din, Afdhal ud din, Shems ud din, and Humam ud din, and the
other great sheikhs whom this is neither the place nor time to mention
more particularly, that I am about to quit this world, and wish them to
aid me with their prayers. Let them look to my sons, whom I make over
to God as a pledge ; for God does not forget his pledges. I had hoped
to see them again, and help them with my counsel." As this was not to
be, he commended them to their care (i.e., of the mulanas), bidding them
protect them, and see that they needed for nothing ; that they led good
lives, and did not forget what God had done for them. If his son
the Atabeg, and his mother Khoshek wished to return home, they
were to be allowed to do so. His two sons Nuruz and Masud, with
their mother, were to remain with the Princess Bulughan, and were to

---

* D'Ohsson, iv. 7-8.　　† Chron. Syr., 603.

stand at either end of his tomb. **They were to support the eating-house
and cloister of the Sheikh Fakhr ud din, and to repair thither. Ferrukh
and his mother were to attend on Atabeg. Sekeria was to work in the
Ilkhan's service. In regard to the rest of his property, he left it to the
amir Buka. If he chose to return some of it, well and good ; if not, he
would still be satisfied. "May God grant his pity and his blessing. I
now resign my life into the hands of God, who will not forget me. If the
Almighty gives anything to my sons, may they take it and be content.
Whatever happens to the great harem at Tebriz, this is my wish : May
he be happy who seeks the right path."* When he had finished writing
the Vizier said, "What comes from thee, O Lord, is right, be it weal
or woe." He was seized by the hands and feet by his executioners,
raised up and struck against the ground three times. He was then
trodden under foot till he was dead, after which he was castrated, and his
head was decapitated. This was on the 16th of October, 1284.**

• "Such was the end," says Abulfaraj, "of this most powerful man, who
supported on his finger the whole Mongol world. He was a prudent
man, and endowed with natural capacity, and well cultivated and
polished."† This is remarkable testimony from a Christian, and it is
not strange, therefore, that the Muhammedan writers should speak in
hyperbolic terms of his memory. He had been a very powerful factor in
the Mongol polity for thirty years, and with his brother and other
relatives had done much to restore prosperity after the desolating wars of
Khulagu. He was, nevertheless, a vindictive, crafty, and not very
scrupulous person, and the misfortune which eventually overtook him
he had not hesitated to bring upon his rivals. One of the contemporary
poets said of his death—

> On the departure of Shems (the sun) from the heavens it rained blood ;
> The moon scarred her sweet face, and Venus tore her hair.
> Night clothed herself in mourning for him,
> And the morning sighed deeply and tore her veil.‡

His death was followed by the ruin of his family. Buka sent the amir
Ali Tamghaji to Tebriz to seize his son Yahia, and he was put to death
there. His other sons Ferrukhshah, Masud, and Atabeg were some time
after also put to death. Yusuf Shah, Prince of Luristan, the Vizier's
son-in-law, died on his way back to Luristan. He left two sons,
Afrasiab and Ahmed, of whom the former was given his father's state of
the Greater Luristan, while the latter remained with Arghun as a hostage.
Shems ud din and his sons were buried at the burying-place of Cherendab
at Tebriz, where his brother, Alai ud din, already reposed.§ With the
death of the Vizier, Abulfaraj closes his Arabic chronicle.

During the year 1285 we read how a body of Syrian, Kurdish, Turkish,

---

* Ilkhans, i. 366-367. Wassaf, 269-271.
† Op. cit., Chron. Syr., 603-604. D'Ohsson, iv. 9-10.   ‡ Shajrat ul Atrak, 261.
§ Wassaf, 273. Ilkhans, i. 367.

and Bedouin vagabonds, 600 in number, made a raid upon the district of Irbil, and killed many Christians in the towns of Emkabad, Surhegan, &c. Behai ud din, a Kurdish amir at Irbil, went out against them and was defeated, and they carried off a great booty and many women. Other plunderers made another attack on the province of Turabden, and killed many people at Kalesht, Beth Mana, and Beth Sebrina, and carried off many captives from Beth Resha. In 1286 another band of Kurds, Turkomans, and Arabs, 10,000 strong, with whom were 300 Mamluks, attacked Mosul and its neighbourhood. Masud went out against them, but finding himself quite outnumbered he withdrew again, crossed the Tigris, and sought refuge in the monastery of Mar Matthew. They entered the town, and were regaled there by the Arabs (*i.e.*, the Mussulmans) with rich meats and cooling drinks, in the hope that they would wreak their vengeance on the Christians. They proceeded to 'plunder and ravish at their will, and carried off many prisoners, both male and female, and including both Mussulmans and Christians.* This' shows the confused condition of affairs while the Mongol sceptre was in uncertain hands.

After Shems ud din's death, Arghun went to his palace of Mansuria in Arran, where, on the 23rd of September, 1284, he was joined by Pulad Ching Sang, whom he had sent to the Khakan Khubilai. A kuriltai was held between Serah, Irbil, and Sain, and nine days later (*i.e.*, on the 2nd of October) he returned to Tebriz, and thence again went to his winter quarters in Arran, where a solemn assembly was held for the trial of the Princess Abish, the widow of Mangu Timur, in whose name, as we have seen,† as the heiress of the Salghurid family, Fars had been governed by the Mongols for many years. After her marriage with Mangu Timur, she had lived at the Ilkhan's Court, and the country was really under the control of the Mongol baskaks, or governors. During Ahmed's reign this post had been filled by the Noyan Toghachar, who had command of 10,000 men, and who sided with Arghun in the latter's quarrel with his uncle.‡ He was succeeded by Bulughan, or Bulghuran, whose rule was very unsatisfactory, and he was accordingly displaced in favour of Tashmenku. Bulughan was not disposed to give up his post without a struggle. He put to death Hosam ud din, son of Muhammed Ali of Lur, an employé in the office of Crown demesnes, whom Tashmenku had sent as his forerunner, but when the latter called in the aid of the Atabeg of Luristan Bulughan seized what there was in the treasury and fled to Khorasan with his agents, Kawam ud din and Seif ud din. Tashmenku busied himself in administering the province, but was deposed in the course of a year for heading the orders issued in Ahmed's name with the formula, "Ahmedaga," which was contrary to all the rules of the Mongol

* Abulfaraj, Chron. Syr., 606-608.   † Ante, 204.   ‡ Ilkhans, i. 344.

chancellary. On his deposition the Princess Abish was herself appointed governor, a position she owed to the influence of Oljai, the mother of Mangu Timur, who had much influence at the Court. There was great rejoicing in Fars at the news of her return. As her substitute, or deputy, she nominated her relative, Jelal ud din Arkan, the son of Malik Khan, son of Muhammed, son of Zengi, while she appointed the Khoja Nisam ud din Abubekr, who had an old feud with the chief judge, Imad ud din, to be vizier.* Nisam ud din, who was a clever financier, proposed to the princess to acquire, through a diploma of the Ilkan, power to redeem such of the property of her family as was in strangers' hands. Ahmed granted this diploma thoughtlessly. Nisam ud din, however, speedily confused the princess's property with that of the Crown demesnes and that belonging to private people, and treated the people of Shiraz, both the crowd and aristocracy, as if they were slaves. The beginning of the administration of Abish and her finance minister fell in the latter part of Ahmed's reign. On the accession of Arghun, Buka's *protégé*, the Seyid Imad ud din, repaired to the Court to lay before it the true condition of affairs in regard to the treasury at Fars, and through the influence of Buka he obtained a diploma constituting him sole administrator of Shiraz, both sea and land (*i.e.*, including the islands in the Persian Gulf), with very large powers, his diploma being sealed with a lion's and a cat's head.

Meanwhile the two employés of Bulughan who had fled with him to Khorasan, as I have described, viz., Kawam ud din of Bukhara and Seif ud din Yusuf, had returned to Fars and been intrusted by the Princess with the management of the finances. They aroused her hatred against the Seyid before his arrival, and as he began his work by hanging one of her bailiffs to a tree, and summoning her to the Ilkhan's presence, her anger was still further inflamed. As soon as the Seyid reached Shiraz he erected himself a royal throne. Eight days later there commenced the cast of Bairam, at which the Princess did not appear as usual. Presently there came news that the Nigudars were threatening an invasion, and that it would be prudent for her to take shelter in the Castle Istakhr. This she declined, as she was afraid he wished to imprison her. One evening after this the Seyid repaired to her house with a great following. On the road he was met by a party of Mamluks belonging to her household. A struggle ensued, in which the leader of his own men, on whom he had heaped his favours, and who was called Zeraj ud din Fazli of Lur, and was doubtless a Kurd, struck him the first blow. He was killed, his head was cut off, and his house was plundered.† His death was followed by that of his cousin, the Seyid Jemal ud din Muhammed, whom the Mamluks murdered in the night, and then spread the report that he had fled to Kerker. The pestilence and famine which visited

---

* Ilkhans, i. 368-369.   † *Id.*, 371-372.

Shiraz shortly after, and in which over 100,000 people are said to have died, was deemed a punishment from heaven for this double crime, for the Seyids were a sacred family.

The Seyid's son, who was a minor, now repaired with his complaint to his father's protector, Buka.* Buka called the Ilkhan's attention to these occurrences, and he accordingly summoned the Princess, with the opponents of the Seyid, to the Court, and also sent word to Oljai, by whose influence Abish had been made governor of Fars. They loaded the messengers with presents, but did not obey the summons. Doladai Yarghuji, Jiyurghutai, and Hosam ud din were commissioned to inquire into the whole affair. Kotan Ataji was sent to bring her by force, while her finance officers were imprisoned. She arrived at Arghun's Court at night, and was conducted by Buka's major-domo to one of his master's tents. It was contrary to Mongol custom for a princess thus to enter the tent of a karaju, or subject, and the unfortunate major-domo was ordered to be bastinadoed for his pains. Oljai made excuses for her *protégé*, laying the blame for what had happened on her relative, Jelal ud din Arkan. The three head men of her treasury, Kawam ud din, Seif ud din, and Shems ud din, each received seventy-two strokes on the soles of his feet. Imad ud din's Mamluks, who had been treacherous to her, were unsparingly punished. Jelal ud din justified himself at the cost of the Princess. She and her relatives were ordered to pay fifteen gold tumans as a fine, together with twenty tumans to the orphans of the murdered Seyids. The Princess outlived these events but two years. On her death, prayers, readings of the Koran, and distributions of alms took place in the mosques at Shiraz. According to her will, her private estate was divided into four portions, of which two fell to her daughters, the Princesses Gurdujan and Algharji, another to her Mamluks and freedmen, and the fourth to Taiju, the son of Mangu Timur, to whom she also left 10,000 gold pieces. With her the famous dynasty of the Salghurs came to an end. Fars lost even the semblance of independence, and was incorporated with the Mongol empire.†

We will now return to the more immediate affairs of the Ilkhan. On the 24th of February, 1286 (Quatremere says on the 25th of June), the general Ordu Kia, who had been sent by Arghun to the Khakan Khubilai with the news of his elevation, returned with the title of Khan for Arghun, and that of Chingsang for Buka. The *fêtes* and rejoicings of the installation were accordingly renewed. Ten days later a body of 16,000 men were sent against the Kurdish tribe Hakari, under the Amir Masuk Kushji (*i.e.*, the fowler) and Nurinaga the Jelair. A month later the Princess Bulughan died on the banks of the Kur, and her remains were taken to the mountain Sejas. On the 20th of April

---

* Ilkhans, i. 372-373.   † *Id.*, 372-374.

Arghun went to Tebriz, where he was handsomely entertained by Buka, and afterwards went to Sughurlak by way of Meragha. There he was met by Aruk, Buka's brother, with the bitekjis, or Mongol secretaries of Baghdad, with whom was Harun, the son of the vizier, Shems ud din, whose strict rule at Ispahan we have previously described. Aruk, on the strength of the support given him by his brother Buka, had put to death Mestofi Said ud din, the brother of Fakhr ud din, and Majd ud din, son of Asir, without the Khan's permission. The latter was a *protégé* of Gaikhatu, Arghun's brother, who poisoned the Ilkhan's mind against Aruk. Aruk had another enemy in Yesu Kurkan, the husband of the Princess Tudukash, Khulagu's fourth daughter. Buka, on the other hand, supported his brother, and the death of the son of Asir was attributed to the instigation of Harun, who was accordingly put to death.* Wassaf says that Aruk, who commanded in Irak Arab under the Prince Baidu, put to death Khoja Harun, whom he accused of being in league with Majd ud din Asir, one of the richest and most influential men of his time, to charge him with peculation.† But this seems at issue with a previous statement of the same author that Harun, whose remarkable career at Ispahan we described under the reign of Abaka, died in his father's lifetime.‡ Soon after this Yesu Kurkan, Aruk's enemy, died, and for a while he and his brother basked in prosperity.

On the 27th of September, 1286, Arghun went to Tebriz. Two months later, when he was one day in Arran, combing himself, an unusual quantity of hair came out in the comb. This, according to Mongol notions, was caused by his having taken poison, and Wejih, the son of Iz ud din, was executed on suspicion of having administered it. On the 7th of January, 1287, Tudai Khatun, the Konkurat princess, who had passed from Abaka's harem to that of his son, was crowned with the head-dress of the royal wives (baghtak). Arghun, during the year 1287, lived respectively at Pilsuvar, at Tebriz, in the summer camp of Alatagh, and the winter one of Arran. In March, 1288, he lost his wife, Kutluk, the mother of his youngest son, Khatai Oghul ; while in April Buka's envoys brought back with them to Persia one of the relics so much esteemed among the Buddhists, called sharil. These are hard pieces of a substance which is said to be found in the ashes of some saintly persons when cremated. Von Hammer says that Buddha's heart was supposed to be made of bone and not of flesh, similarly with the hearts of great men, and that the sharil is really held to be the ossified heart of the cremated person. Arghun, we are told, treated this relic with the greatest honour, gold was strewn over it, while a feast was duly celebrated.

Later, viz., in May, 1288, news arrived that Nogai, the famous leader of the armies of Kipchak, was making an invasion, by way

---

* Ilkhans, i. 374-375.   † Wassaf, 272.   ‡ Id., 125.

of Derbend, at the head of 5,000 men, and was putting to death such of the merchants as he could meet with. Arghun at once marched to the rescue, crossed the Kur, and halted at Shamakhi. Buka and Kunjukbal were sent on with an advance guard, and returned in a few days with the news that the enemy had retired.* A few months later, viz., in the spring of 1290, the Khan of Kipchak made a fresh invasion by way of Derbend. The amirs Siktur Noyan, Kunjukbal, and Toghachar, were ordered to march, and Arghun followed to Pilsuvar, and thence pushed on to Shaburan with the heavy baggage, &c. The two armies met at the Karasu. The army of Arghun was commanded by the amirs Toghachar, Kunjukbal, Toghrulji, and Taiju, the son of Bukuwa ; and that of Kipchak by Nogai, and by the two sons of Mangu Timur, Abaji and Mengli. The enemy was defeated, 300 of them were slain, and many were captured. The " Shajrat ul Atrak " says that Choban, the Sulduz, of whom we shall hear much presently, greatly distinguished himself in the fight, which, according to Hamdullah, was the first in which he took part. The victory was celebrated by a feast at Pilsuvar, and was announced by special letters throughout the empire by the Vizier.†

The " History of Georgia " has a notice of a campaign which seems to be the same as this. It says the people of Derbend having revolted, Arghun marched against them in company with Dimitri, whereupon the rebels retired to the strong fortress of Anik, which, having resisted obstinately, Dimitri was told to attack it. The Georgians were not long in storming it, Rat Begashwili leading the assault. The principal people in the fort were put to death, and a large number of prisoners, riches, and women were captured, while the citadel was burnt. The Khan, who was a spectator of the whole affair, conceived a violent jealousy against the king, who surrendered to him a famous suit of armour which he wore.‡

We must now describe the collapse of the great Mongol chieftain, Buka, called Buka Gizbara by Bar Hebræus. We have seen how the Khakan conferred on him the title of Chingsang. In addition to this he had secured special privileges. He was not to be punished until he had committed nine offences, and was only to be called to account by the Khan himself. The ordinances of the Khan were not to be valid unless his red seal was attached, while his own orders did not need the royal sanction. These privileges no doubt aroused the animosity and envy of the other amirs, for he was a haughty and impetuous person. Wassaf describes him as a terrible Turk, whose severity and prudence were remarkable. As an instance of his severity, it is reported that he put one of his own ostlers to death for stealing an apple from a fruit stall. Abulfaraj says that the various princes and princesses had to stand at his gate while he distributed their salaries to them. Among his principal

---

* Ante, ii. 139. Von Hammer, Ilkhans, i 376. Golden Horde, 265.
† Ilkhans, i. 388-389. Shajrat ul Atrak, 263.   ‡ Op. cit., 602-603 ; and 603, note 2.

enemies were Sultan Aidaji and Tughan, son of Tarakjai, Governor of Kuhistan, Arghun's intimate companions, who had been twice cudgelled, and had had many indignities put upon them by Buka's orders, and they lost no opportunity of denouncing him to Arghun. The arbitrary doings of his brother Aruk, who governed the provinces of Babylon, Azerbaijan, and Mesopotamia, also reflected on himself. Aruk treated the Ilkhan's envoys with scant courtesy. Three officials having brought an accusation against him were ordered to be put in custody till his arrival. He had them executed mercilessly before any trial, and appropriated the revenues which should have gone into the State treasury. The officers of Ordu Kia, Sherif ud din, and the Jew, Said, handed over 500 tumans to Aruk, who did not account for a farthing of it to Arghun.* Abulfaraj calls this Jew Said ud daulet, and says he was the father-in-law of the Prefect of Baghdad, who had recently died, and that he told the Mongol amirs if they would prevent Aruk from going to Baghdad he would undertake ·to double the income of the treasury, whereupon Aruk was ordered not to go there, and the Jew was nominated Procurator of Baghdad. " Behold," says the Syrian chronicler, " there sits a Jew there to-day as prefect in the capital of the Abbassidæ. How humbled is the position of the Arabs (*i.e.*, of the Muhammedans)." From this time various accusations began to pour in upon Aruk—*inter alios*, we are told a Persian named Abd ul Mumin declared that he and his creatures had so plundered the country, that if he were arrested a million golden dinars might be extracted from him.†

Turning to Fars, we read that Sadr ud din Zenjani, the financial secretary of Toghachar, complained to his master of Buka's continual demands for money, and that he was virtually the ruler, the Khan having quite a secondary authority.† The financial affairs of Fars had been a source of irritation for a long time. Fakhr ud din Hassan, one of the illustrious Seyids of Shiraz, who during Abaka's reign was attached to the Court of Arghun, had often told him that much property in the province of Shiraz, which had belonged to his ancestor, the grand judge, Seyid Sherif ud din, and which he had inherited from the daughter of the Sultan Azd ud Devlet, of the Salghur dynasty, had been unjustly confiscated by the atabeg, Abubekr, Prince of Fars, and its revenues appropriated to the public treasury. He produced abundant documentary proofs of this, and urged Arghun to press his father to have these lands transferred from the registers of the general treasury, and assigned to his special domain (inju). Abaka consented, and sent one of his officers with the Seyid to carry out the transfer ; but the treasury officials put obstacles in the way, and were supported by the Mongol commanders, and the affair was not carried out. Fakhr ud din

---

* Ilkhans, i. 377. † Abulfaraj, Chron. Syr., 610-611. ‡ Ilkhans, i. 377-378.

returned to Arghun. When the young prince mounted the throne, he issued an order withdrawing this property from the public registers, and assigning it to his own domain, according to the title he had proved, and the Seyid summoned the finance officials of Fars, who were then at the Court, and peremptorily demanded the restoration of the property according to the Ilkhan's orders. Buka urged reasonably that inasmuch as the province of Shiraz now belonged to him (Arghun), what was the necessity of separating these lands, and making a special department of them. Arghun would not listen, and ordered that Buka should not meddle with Fakhr ud din's affairs, nor with those of his own private domain. At the same time, the latter property was confided to the Noyan Togachar as administrator. Buka was thus deprived of the greater part of his authority. Fakhr ud din was ordered to repair to Shiraz with Yul Kutlugh, son of Arghun Aka, to superintend the transfer of the disputed property from the register of the public treasury to that of the private one of Arghun, and as no one was in a position to answer their assertions on the subject, they succeeded in thus transferring one-fourth of the villages, fields, gardens, corvées, irrigating canals, and windmills in the province of Fars, so that in a short time these private domains were farmed for the large sum of 600,000 dinars, and many families who had been in possession of their property for a century were compelled to surrender it. Fakhr ud din himself died eighteen days after his arrival at Shiraz, which looks suspicious. His son, Seyid Kutb ud din, was invested with his father's authority in the Ilkhan's name by Yul Kutlugh, already named.*

This matter, thus carried out in spite of Buka's views, was no doubt a large invasion of his prestige. Another attack on it was made by the appointment of the amir Kunjukbal to superintend the affairs of the army. Meanwhile, Tughan continued to poison Arghun's mind against him, urging that, although Ahmed had completely confided in him, he had, nevertheless, betrayed him, and he was now appropriating all the power in the kingdom. One day Buka and Bekta quarrelled over their cups in the presence of Arghun, and the latter not having reprimanded Bekta, Buka was further irritated, while a more real grievance was forthcoming when his principal dependents, including especially the Amir Ali, collector of the customs at Tebriz, were deprived of their posts. It is not to be wondered at that under these circumstances he should have begun to conspire with a number of discontented princes and amirs. Among these were the Princes Hulaju, Toka Timur, Karabuka, Kingshu, and Gabarchin (Von Hammer calls the last three Karankai, Konghshir, and Anberjin), with the Amirs Aruk, Kurmishi (the son of Haidu Noyan), Machu, Tamdui, and Tughluk Karauna. Rashid ud din adds to this

---

list, Aujan, the armour-bearer ; Kadan, the envoy ; Zengi, the son of
Babu Noyan, amir of the camp of the Khatun Oljai ; Ghazan Behadur,
Ishik Togli, and Ashak Togli.* Buka also wrote to Prince Jushkab, who
was encamped by the Euphrates, to urge that although Arghun owed his
throne to him, he had shown himself very ungrateful. Jushkab, who felt
sure Buka meant to make a tool of him, refused to join in the conspiracy
unless he saw the written agreement (called muchalga in Mongol) between
the conspirators. This Buka incautiously sent to him. Jushkab now
repaired to Arghun, in Arran, to acquaint him with the plot. He would
not believe in it until the written proofs were produced, and then broke
out into bitter complaints against Buka, whom he had raised to such
honour, and who had, nevertheless, been so treacherous. He ordered
his three officers, Sultan Aidaji, Doladai, and Tughan, to go and arrest
him. He had had timely warning, however, and had crossed the Kur and .
sheltered in the camp of the Khatun Oljai, one of Arghun's wives. She
refused to receive him, but the amir of her camp, Zengui (Von Hammer
says the son of Zengui), allowed him to shelter in his tent. Doladai and
Tughan speedily crossed the Kur in pursuit of him, and having secured,
conveyed him to the Court. There he was brought before Siktur,
who sarcastically asked him if he wanted to have a fresh master every
day, and charged him with being the cause of much mischief. He denied
having done anything against the Ilkhan, and said that he had only been
plotting against his own personal enemies, Aidaji and Tughan, who had
so persistently maligned him, but being confronted with the incriminating
papers he trembled and fainted. Arghun ordered him to be at once put
to death, and Jushkab asked the favour of being his executioner. When
they reached the place of execution Tughan kicked him in the chest,
saying, " This is the reward for thy ambition to mount the throne."
Jushkab himself cut off his head, and slit the skin of his back into strips.
Bar Hebræus says he was dismembered. The head was stuffed with
straw, and then displayed on the bridge Jaghan, while the troops were
ordered to pillage his quarters. This execution took place on the 17th of
January, 1289, and was followed by the deaths of the various amirs who
had joined in the conspiracy, Maju, Tughluk, Ashak, Toghli, Zerwana,
Nokhshi, Tushkina, Hosam ud din of Kazvin, and the Malik Ali, the
temghaji of Tebriz. Kadan, the Khakan's envoy; the Bitekji Naghai,
who had spoken out the truth; and another, for whose life the amirs
interceded, were spared. Among those who perished were also the
astronomer, Imad ud din ; the Christian, Simon of Rumkalaa ; Bahai ud
daulet Abul Kirem, and the King of Georgia, Dimitri, to whom we shall
revert presently.† Abulfaraj says there perished Simeon the priest, and
a doctor and scribe from Irbil called Abu Alkherem, and many Mongols.

Buka's wives and daughters were distributed among the army. The corpses of the slain were put together in piles, and left for the wolves and dogs to devour their flesh, after which their bones were buried.

Betmish Kushji, Tamudai Aktaji, and Shadi, son of Buka, were sent to Diar Bekhr to fetch Buka's sons and brothers. In six days they reached Irbil, and killed Ghazan, Buka's eldest son, who was living with his uncle Aruk. Aruk himself, who did not know what had happened to his brother, seeing the Mongol garrisons of the district of Amid approaching, fled to the fortress of Keshaf. Betmish summoned him to surrender. He said he had no intention of resisting, but wanted to know why they had thus come. Betmish then told him what had happened to his brother, and that he had been ordered to take him to the Court. He then left the fortress, and was carried off in chains. When he saw his brother's head on the bridge Jaghan, he merely asked where that of his armour-bearer, Aujan, was. He was then put to death. This was on the 3rd of February, 1289. His head, with that of his relative Kurmishi, was exposed near that of Buka. Zengui, who had sheltered Buka in his tent, was handed over to Oljai, as one of her dependents, for punishment. She ordered him to be decapitated, saying she would have done so if he had been her eldest son, Anbarji. Four sons of Buka, named Abaji, Malik, Teikhan Timur, and Kutlugh Timur, fled to Tughan. He gave them shelter until he thought Arghun's anger was appeased; but the latter, when he heard of it, ordered them to be put to death, and thus exterminated the family. A proclamation was now issued announcing far and wide how Buka, having been guilty of the basest ingratitude, had been destroyed, with his wives and children, his friends and relatives; while his wealth, which he had acquired through the Ilkhan's munificence, had been pillaged, and all who were suspected of having abetted him, Mongols or Mussulmans, paid the last penalty.*

The sudden downfall of such a powerful chief as Buka, whose dependents filled places of trust in various directions, naturally entailed a reaction, when envy, cupidity, and revenge had a free field. Under the orders of Betmish, Abdul Mumin, whom we have previously named, paid the sum of money which each of Aruk's officers at Mosul was charged with having robbed. Torture and death were freely applied to extort this money and to punish the wrong-doers. Among those who suffered from the revolution was Masud, the Governor of Mosul. Devoted to Buka, and probably fancying that that minister's position was impregnable, he had neglected to be civil and attentive to Arghun's own creatures. Accordingly the very day of Aruk's arrest he was put under surveillance until the arrival of Abdul Mumin, which was the signal for a cruel persecution against the Christians. Bethag Alden, son of Mohtesi, who was Prefect of Irbil, was

---

* D'Ohsson, iv. 18-23. Ilkhans, i. 380-381. Abulfaraj, Chron. Syr., 610-612.

among those put to the torture. They put him on the rack and sat down upon him, and thus tortured him till they had extorted 50,000 dinars from him. He at length escaped and plunged into the Tigris. Masud was himself ill, so they did not torture him, as they feared that if he died they would not be able to secure the wealth he was supposed to have concealed. The Mongol commissaries promised that he should retain his position and be released if he paid ten tumans of gold. But fancying they were afraid of him, he dealt hard words to them. Thereupon he was cudgelled and threatened until they had extracted what they could, and he was then taken to Irbil, where he was put to death on the 4th of April, 1289. His son was branded and imprisoned, while his brother, Shihab úd din, fled, and a villager named Dobis, from Beth Sehar, who was suspected of sheltering him, although he swore he had never seen him, was killed, and his corpse was stoned by the crowd. This series of cruelties was no doubt instigated by Muhammedan hatred of the Christians. A young Christian accused of an illicit intercourse with a Muhammedan was also put to death. His body was dragged through the streets and burnt, while his head was carried round in procession past the various church doors to humiliate the Christians "The cruel persecutions," says Abulfaraj, "which the people of Mosul suffered during these two months, tongue cannot describe nor pen indite Awake, O Lord, and do not sleep! Look at the blood of thy servants shed without mercy. Have pity on thy Church and flock, which are being torn by persecution." Abdul Mumin, the prime actor in the movement did not long survive. Denounced by an Egyptian scribe employed at Mosul, named Faraj Allah, he was tried and executed.*

These were not the only troubles of the wretched inhabitants of Mesopotamia and its borders at this time. We read how 2,000 Syrian cavalry crossed the borders of Shogr and Arabia, and went as far as the town of Pishabur, near the Tigris. Crossing that river, they approached Vassit, where the Nestorians were very strong. The town was captured, 500 men were killed, and 1,000 boys and girls carried off, as well as great herds of oxen and flocks of sheep. When the invaders reached the River Habora, which was traversed by a very narrow bridge, they had great difficulty in passing over. News of this having reached the Mongol amir at Mosul, he set out, and caught the intruders while still embarrassed at the bridge. They killed all who had not yet crossed, and rescued 300 boys and girls. The same summer 1,000 Syrian horsemen approached Malatia. Karbenda, the Mongol governor there, attacked them, but was defeated, and many of his people were killed, while several of his friends and relatives were captured. He himself, with forty followers, retired to the fortress of Hesnun.† At this time, Alaalmish, the governor of Mayafarkin, greatly persecuted the Christians, and especially laid heavy hands on the monks of the monastery of Mar Koma. One of the monks repaired to Arghun, who listened courteously to him. When the Ilkhan was one day crossing a bridge over the River Khorer, this monk seized his bridle, and swore he would not let him pass until he had ordered Alaalmish to be put to death, which he accordingly did.*

Among those who were compromised when Buka fell, perhaps the most important was the King of Georgia, who had been his close friend. On his execution, we are told, a messenger was sent to summon Dimitri to the ordu. Greatly distressed, he called together the Catholicos Abraham, the bishops, the monks from the monasteries and from the hermitages of Garesja, and all the mthawars of the kingdom. Seating himself on his throne, while they were seated also, he bade them listen, and told them that when his father David died he was left very young, and at the mercy of the Tartars; that God, the all-powerful, our Saviour, Christ, the most holy mother of God, and the adorable cross which had been presented to him by the emperor had protected him (so that he had arrived at the age of manhood), had granted him the kingdom, the sceptre, and the purple, while with their help his reign had been a prosperous one, and there had been general peace. Now the Khan was irritated, and had exterminated the mthawars, and summoned him to his presence, no doubt with evil intentions against him. If, instead of going to the ordu, he withdrew to the strong positions of Mthiuleth, he would save his life, but his kingdom would be at the mercy of the Tartars. " How many Christians would be enslaved, how many churches profaned or pillaged, how many images and crosses broken? If I go," he said, " they are certain to kill me. Decide, therefore, according to your wisdom. As for me, I look upon the world as a troubled sea, and life as a dream, a shadow, and in spite of ourselves we must leave it. What advantage is it to me to live if many have to die in consequence? If I must some day leave life bearing the burden of my sins, I would prefer to go to the Khan; the divine will will then be accomplished. If I am put to death my country will not suffer." The assembly, touched by the proof of his devotion, declared that there was no one who could replace him. " God preserve us from seeing you massacred by the Tartars. The country will be desolate, your sons dispersed. No one can replace you. Our advice is that you seek the fastnesses of Mthiuleth and Aphkhazeth, as your father did. We must not despair of your safety. We will remain faithful to you." He still insisted that he could not leave his people to become the victims of the Mongols, and declared he should go to the ordu; and, notwithstanding their arguments, he determined to go in great pomp, and to take the Catholicos Abraham with him. He

---

* Abulfaraj, Chron. Syr., 617.

assigned fitting portions to each of his sons, and confided them to the care of the mthawars, who remained behind, and sent some of them to Mthiuleth and others to Kakheth, while the young George was sent to Asparakhen, to the citadel of Ishkhan in Tao.

When the King arrived in the territory of Khochak, son of Avak, he met his son David, whom he took with him, so as to disarm Arghun's suspicions by another proof of loyalty. As he drew near, Arghun, who had not expected that he would go, sent the Noyan Siukol, son of Yas Buka, who deprived the King of his baggage and wealth, and led him captive to the Khan, and he was put under arrest. Arghun, we are told, was troubled, because he did not know who to put in his place, while he could hardly spare one who had been an accomplice in the treacheries of Buka. Kutlugh Buka thereupon suggested that he should put Wakhtang, son of Narin David, who, as we have seen, was the ruler of Abkhazia, on the throne. Arghun thereupon sent Buka to David, bidding him send his son, whom he intended to put on the throne, and to give him his sister Oljath in marriage. Dimitri was ordered to draw up a list of his possessions, his arms, cattle, sheep, and all that he had. He was compelled to obey, and a Mongol was sent to bring the enormous wealth, the sight of which seemed to appease Arghun, who released him, and the mthawars advised him to take advantage of his more conciliatory mood to escape ; but he refused to put his people in jeopardy by such a step. Presently news arrived that Kutlugh Buka was returning with the young Wakhtang, whereupon Dimitri and his son David were again arrested and put in separate prisons. The King was now bastinadoed, which people thought meant he was not going to be executed, for we are told it was not the Mongol custom to put those to death who had been thus punished. Nevertheless, the Khan's anger was not appeased. He was taken to the Hall of Justice (called the divan khana), where he was asked if he had taken part in Buka's conspiracy, and, we are told, he was found innocent, which seems like a diplomatic phrase on the part of the chronicler. His death was decided upon. Twelve horsemen came to conduct him to the place of execution. He spoke cheerfully to his supporters, and asked Kutlugh Buka to plead with the Ilkhan for his young son David. Having said his prayers and taken the sacrament, he offered his neck to the executioner, and was decapitated on the plain of Mughan. The day of the execution was marked by a solar eclipse, which seems to fix it as in March, 1289. Dimitri's son David was made over to Tachar (i.e., Togachar) for execution. He remonstrated with his master on thus putting a boy to death, and asked him to make him a present of him. This the Ilkhan consented to do, and he accordingly confided him to his khoja, who afterwards became sahib divan, or vizier. Dimitri's body was conveyed to Mtzkhetha, and buried with those of his fathers.

By his principal wife, the daughter of the Emperor of Trebizond, Dimitri left four sons, David, Wakhtang, Lasha, and Manoel, and a daughter named Rusudan. By Sorghala, the Tartar, he had two sons, Badur and Yadgar, and a daughter, Jigda Khatun; while by Nathila, daughter of Beka, he had a son, Georgi. On Dimitri's death David, we are told, was well treated by Tachar Noyan. The queen and other widows of the king hid themselves. Sorghala went to her father's home in Tartary; the daughter of Beka also to her father in Samtzkhé. Wakhtang was in Mthiuleth. He assigned the valley of Scoreth for the support of his mother, who kept by her her two young sons, Manoel and Lasha. Badur and Yadgar followed their mother home, and "our country," says the annalist, "was without a king."

Arghun now summoned Kutlugh Buka, and said to him, "I have destroyed my enemies, and Dimitri, who was your enemy. There is now no king; go and fetch the son of the King of Abkhazia, that he may mount the throne." He accordingly went and brought his *protégé*, Wakhtang, to Tiflis. His father, David, collected his troops and escorted him thither. Having crossed the mountains, they stopped in the plain of Kwishkheth, at Tasis-Sar. Konchiba, son of Shiramun Noyan; Kurumchi, son of Alikan, who lived in the mountains of Jawakheth, between Artan and Samtzkhé; and all the didebuls of Georgia, assembled together and swore an oath of unalterable fidelity to the young prince. Thereupon the King returned to Kuthathis. Arghun was much pleased with Wakhtang, and gave him the kingdom of Georgia, and after having married him to his sister Oljath, sent him to Tiflis.* The death of Dimitri reminds one of those of the various Russian princes, at Serai, as related in a former volume. It is not unlikely that he may have been compromised in Buka's conspiracy, Buka having been a great friend and patron of his, but it seems more likely that he was also the object of personal animosity, for we are expressly told that the son of Khoja Azis, whose father had been killed by Dimitri's father, David,† plotted with Kutlugh Buka for his destruction.‡ Wakhtang now ruled over Georgia, from Nicophsia to Derbend, except the domains of Beka Jakel and Tzikhisjuarel. He appointed Kutlugh Buka atabeg and generalissimo of the united kingdom. Meanwhile David, son of Dimitri, remained with the Mongols, suffering great troubles there. His mother was at Sacureth, or Scoreth, while his brothers were dispersed abroad. Georgi, the youngest, whose mother was daughter to Beka, was living with his grandfather.§

The next important victim of Arghun's jealousy was the Prince Jushkab, who had divulged Buka's conspiracy, and had killed the traitor with his own hand. Arghun, suspecting that he had plans of his

---

own, sent some troops after him under the Amirs Yatmish Kushji, Gharbetai Kurkan, Burju (the son of Duriai), Boghdai, and Arkasun Noyan, who overtook him on the River Karaman (D'Ohsson calls it Kumar), between Erzen and Mayafarkin. He resisted and escaped, but was pursued and captured three days later, and was taken before Arghun, who had him put to death. This was on the 6th of June, 1289. He had been Governor of Diar Bekr. This year, when Arghun had gone to his summer quarters at Kongorolang, Ordu Kia and the Jew Said went to him a second time, taking with them a good store of money, the proceeds of the taxes at Baghdad. Said reported that the amount would have been double, but for the interference of the Mongol officials. Their execution was ordered, and their heads were sent to Baghdad. Mansur, the son of Khoja Alai ud din of Juveni, was also taken from Hillah, and put to death. Jelal ud din, of Semnan, who held the post of vizier, and who had aroused the suspicions of Tughan, would also have been executed but for the intercession of the Bakhshi Berendeh. He was replaced by Said, and presently being accused of complaining that he had been displaced in favour of a Jew, he was condemned to die, and was put to death on the 7th of August.*

The elevation of a Jew to such a high position as vizier, in which he controlled the lives and fortunes of so many true believers, was indeed a proof of the terrible degradation of Islam during the Mongol supremacy. Said, styled ed daulat (*i.e.*, felicity of the empire), was the son of Hebetollah Ben Mohesib of Ebher. He had five years before been appointed to an important post in the exchequer at Baghdad, and by his activity had greatly increased its receipts. Kutlugh Shah, the Governor of Baghdad, and others had become jealous, and complained to the vizier about him. They also praised Said's unusual skill in medicine, and suggested that as a doctor he would be very useful at the Court. Thither he went, and attached himself to Ordu Kia, by whose influence he was sent back to Baghdad to collect the arrears of revenue, which now amounted to 1,600 tumans.† He spoke both Mongol and Turk, while his long residence at Baghdad made him familiar with the condition of that province. According to Wassaf, Arghun's attachment to him was due to his having cured him by administering on one occasion a strong dose of medicine, and having gained his confidence he inveighed against Buka and Aruk for their depredations on the exchequer, declared their houses to be full of treasures, and complained especially against Aruk for his arbitrary acts—among other things, that he had demolished a mosque, and several colleges and hostels, in order to use their materials in building his own palace and the houses of his friends. Under the protection of Ordu Kia, as we have seen, he worked at

Baghdad in spite of Aruk, and having got together a considerable sum by collecting arrears, by new taxes, &c., he took it to Arghun, who was greatly pleased, presented him with the cup with his own hand, and granted him a robe of honour, and he was now appointed controller of the finances of Baghdad, under the authority of Ordu Kia. He worked so well that presently his patron and superior was able to lay a larger sum at his master's feet. Ordu Kia praised the zeal and other qualities of Said, and suggested that as he had done so well at Baghdad he might be intrusted with the finances of the kingdom, which he accordingly was, displacing, as we have seen, Jelal ud din of Semnan.*

The Jewish minister now became very powerful. He appointed his own relatives as farmers of the taxes. Irak Arab he confided to his brother, Fakhr ud daulat ; Diar-bekr and Diar-rabia to Shems ud daulat ; Tebriz to his cousin or nephew, Abu Mansur Mohesibed daulat (*i.e.*, the realm's purifier), the doctor ; and Azerbaijan to Lebid, son of Abi-rabi. In Khorasan and Rum alone did he fail to put his dependents, for those provinces were the appanages of Arghun's son and brother, Ghazan and Gaikhatu.† Bar Hebræus says that Arghun, wearied by the continual intrigues, perversity, and audacity of the Mussulman officials, ordered that only Christians and Jews should be employed in controlling the finances. The new minister therefore conferred the prefecture of Baghdad, which he was himself resigning, upon one of his brothers. To another he confided Mosul, Mardin, and Diar-bekr, associating with him Taj ud din, son of Moktadh. When these two arrived at their post, a Kurdish amir named Mobariz, who had been displaced from the prefecture of Irbil, feeling himself insecure, went to the ordu to seek protection. They sent such accusations after him, however, that Arghun resolved upon his death ; but he lulled him for a while with false hopes, until he had secured his sons, parents, clients, and servants, wishing to put them all to death as well. He managed to escape, sped on in advance of the messengers, and carried his family off to the mountains. Troops were sent in pursuit, but the heavy snow prevented the Mongols securing their prey. They returned to the plain, and revenged themselves by pillaging and killing the Kurdish peasants they met there. Their chief allies in this work, we are told, were the Christian mountaineers called Kiashis, who hated the Muhammedans cordially. The country was terribly wasted, men being killed, and women and children carried off. This aroused the animosity of the Arabs, who felt sure that the Mongols, many of whom were now Mussulmans, would not have committed these ravages but for the Kiashis, or unless they had been urged on by their chiefs. On the return of summer they left the country of Mosul and Irbil. Thereupon the Kurds descended in great numbers from their

---

* D'Ohsson, iv. 31-33. Ilkhans, i. 382-383.   † D'Ohsson, iv. 33-34. Ilkhans, i. 348.

mountains. The people rushed from the plains to take refuge in the towns and fortresses. Those of Irbil itself fled to the citadel, where they were beleaguered by the Kurds for seventeen days, when there arrived at Mosul, by way of the Tigris, 200 Franks, who had gone by orders of Arghun to Babylon, where they were to obtain boats, and thus to descend by Basrah and on to the Persian Gulf, to attack the Egyptians. Seven hundred others, who had gone overland, had stopped to winter at Baghdad. The Kurds, fancying these Franks had come to attack them, withdrew. This passage, as D'Ohsson says, shows the state of anarchy to which the Ilkhan's dominions were reduced by the contests and bitter animosities of the Christians and Muhammedans.*

Inasmuch as Said feared the influence of the Mongol generals Siktur, Toghachar, Samaghar, Kunjukbal, &c., he was crafty enough to solicit from Arghun the services of some colleagues of influence. He chose for this purpose Ordu Kia, to whom he allotted the military government of Baghdad. Karajar (called Kujan by D'Ohsson) was made governor of Arran, while Juchi was similarly given that of Fars. Said reduced the other generals to impotence by concentrating the general administration in his own hands. He himself was virtually supreme and uncontrolled. He tried to reform abuses ; urged that civil affairs should be judged by the Muhammedan code ; forbade the military commanders interfering with the decisions of the courts, and enjoined them to support the cause of justice, and to protect the weak and innocent. Requisitions of food and of post-horses for the grandees were put a stop to, as were the periodical visits of special commissaries to collect taxes, who used to tyrannise over the people, and make great exactions. The taxes were in future to be sent to the Court at stated times by the civil and military governors of the provinces. Said largely increased the pious foundations, and gave many charitable gifts. He surrounded himself with men of letters, and many pieces in his praise were composed in prose and verse. A number of these productions were collected in a work to which his name was attached. In imitation of the princes of the Buyid dynasty, he and his brothers took a surname compounded with Dévlet, or Daulat, meaning "realm" or "court." Thus his own name meant "The future of the realm."† Well might the Jews fancy that their Messiah had arrived, when they saw one of their long-despised race treated as an equal by the princes and noyans, and even by the Ilkhan himself. A notable instance of this is recorded. One day, when he had been playing at backgammon with the Ilkhan for a long time, and had reclined himself on the sofa, he extended his foot with great unconcern. One of Arghun's wives, who came in, asked him how he dared to stretch out his foot in the presence of a Khan, whose very slaves played with the stubborn sky as if it were

* Abulfaraj, Chron. Syr., 618-620. D'Ohsson, iv. 34-36. Note.
† D'Ohsson, iv. 34-38. Ilkhans, i. 384-385.

a bowl of paste. He excused himself on the ground that he had the gout, and his excuse was allowed by Arghun.*

When Said was placed at the head of the finances, Mahmud and Ali, grandsons of the late vizier Shems ud din, and sons of his son, Bahai ud din, were reduced to want. Arghun, having heard of their distressed condition, ordered that a portion of their father's property in Irak Arab should be restored to them. Ali had gone with his mother, the daughter of Iz ud din Tahir, to Ispahan. At this time Mejd ud din Muminan, of Kazvin, having represented that the revenues were considerably curtailed in consequence of the income which had to be paid to Shems ud din's family, an order was issued that the latter's sons were to be put to death. Two of them, Masud and Farajulla, were accordingly executed at Tebriz. His grandson Mahmud was spared, because the edict only mentioned sons and not grandsons, but this did not save Ali, who was put to death at Ispahan. But one son of the late famous vizier escaped execution.†

The Egyptian, Faraj Allah, who had exposed the ill-doings of Abd ul Mumin at Mosul, and brought about his death, encouraged by his success, again repaired to the Court, and made complaints there about Taj ud din, son of Moktadh, the substitute of the vizier of Diar-bekr, brother of the vizier, pretending that he had misappropriated 400,000 pieces of gold. Said ud daulat, alarmed at these accusations, flattered Faraj Allah, and pointed out to him that the accusations were as fatal to his (the vizier's) brother as to his deputy. Faraj Allah was now embarrassed, afraid on the one hand of braving the vizier, and on the other of being condemned to death by the Draconic judges of the ordu if he should retract his charges. He was persuaded to say that he had made them when drunk, and that Taj ud din and his patron were both innocent, being promised that if he made this statement the vizier would see him harmless. The latter, however, no sooner had obtained the retractation in writing, than he laid it, with the other circumstances, before Arghun, who ordered the unfortunate Egyptian to be put to death at Mosul, as well as two of his companions. Some days later, says Bar Hebræus, Matthew, a generous person, and one of the principal buttresses of the Christians, was sent from the ordu to Mosul to collect the tribute of that town. The citizens, who hated the Christians, deemed this an indignity. They attacked him in his house and killed him. This was in July, 1290. His sons fled to the Court. They were authorised to put their father's murderers to death, and to exact a fine of 10,000 gold tumans from the town. They accordingly returned, and put seven or eight of the principal inhabitants to death.‡
We have seen how Juchi was appointed governor of Fars under the ægis

· Ilkhans, i. 385.   † D'Ohsson, iv. 39.   Ilkhans, 383-384.
‡ Abulfaraj, 622-623.   D'Ohsson, iv. 39-41.

of the vizier. He was assisted there by two functionaries, and the son of Sunjak, as Zerwans or Chaushis (*i.e.*, as executive officers). Having gone there to collect the taxes, Jelal ud din of Sirustan undertook to pay them a sum exceeding that fixed as the levy of that province by 400 tumans. The farmers of the taxes undertook, on the other hand, to furnish 500 tumans if Jelal ud din was made over to them bound, but as they failed to keep their engagement the two Mongol officials ordered them to be put to death, and Jelal ud din was released, and a monstrous sum was squeezed out of the community by confiscations and other pressure.*

In the autumn of 1289, Arghun, who was fond of astronomy and alchemy, went to Meragha to consult the stars. In the following May he presented his wife Bulughan, the daughter of Otaman, son of Obotai Noyan, with the household of another Bulughan, who had been the wife of his grandfather and father, and had eventually passed into his own harem. Four days after this, viz., on the 28th of May, 1290, news arrived that the people of Kipchak had made a raid by way of Derbend as previously described.†

Let us now turn our attention further east. The famous administrator of Khorasan, Arghun Aka, had died peaceably near Tus in 1278. His son Nuruz had been intrusted with the superintendence of Khorasan and Mazanderan, for the young Prince Ghazan, son of Arghun, whose appanage they constituted. Although he had been long faithful to Arghun, he seems to have become afraid of his future on the downfall of Buka, and had perhaps taken part in the latter's conspiracy. He accordingly determined to trust only to his own strong arm. Under pretence of going to review the troops under his command, and of making preparations against a threatened raid from Transoxiana, he left Ghazan at Merv, leaving behind him his wife Tughan, the fourth daughter of Abaka ; his mother, Surmish ; and his two brothers, Ordai Ghazan and Narin Haji. Von Hammer says his three brothers—Uiratai, Ghazan Haji, and Narin Haji. Ghazan passed the winter of 1288 in his usual quarters at Jesir, and Prince Kinshu, son of Jumkur, at Herat. In the spring of 1289 Ghazan moved towards Merv, and encamped in the neighbourhood of Sarakhs, which was famous as a pasture ground. He sent several times to summon Prince Kinshu, who had married Nuruz's sister, and Nuruz himself. The latter excused himself on the ground that his foot was bad. Fearing some movement against himself, he called together the millenarians and centurions belonging to his army, the officers of his household, and his guards (khassékis), and told them that he knew positively Ghazan meant to put them all to death as accomplices of Buka. He told the same story to the Prince Kinshu, and inspired him with

---

* D'Ohsson, iv. 41·42. Ilkhans, 386. † Ante, 322.

similar fears to his own.   Meanwhile the family of Nuruz asked his permission to go to the wedding of one of his daughters with Nikpei, son of Sarban, the eighth son of Jagatai, and left him under that pretext.

Ghazan set out from Karatepe towards Tus and Radikan, in the end of March, and sent word to Nuruz to meet him at the River Keshf. Von Hammer says the River of Ferghana (i.e., the Heshtrud).   Knowing that the messenger had recently been on a mission to Arghun, Nuruz put him to the torture, to extract some information about the intentions of the Court towards himself, and as he learnt nothing in this way he would have put the unfortunate man, who was called Sadak Terkhan, to death, but for the intercession of his wife and mother.   He detained him prisoner, however, raised the standard of revolt, and marched to surprise Ghazan on the Keshf.   He surrounded the quarters of three of his generals, who were encamped on the river, hoping to secure the young prince himself.   The three generals were captured and their tents pillaged.   Ghazan was not there, however, having escaped towards Nishapur, where he was joined by the amirs Satilmish and Mulai, and with them went to Mazanderan to try and secure the person of Prince Hulaju, who was said to be leagued with Nuruz, the latter having in his letters spoken of Kinshu and Hulaju as allies.   Putting himself at the head of the troops of Mazanderan, he marched to Hulaju's residence.   The latter fled to Jorjan, near the mountain Kortaghu, where he was captured by Mulai, and taken before Ghazan.   He firmly denied having any part in the plot.   Nevertheless he was taken before Arghun, and put to death at Damaghan, on the 7th October, as well as Karabuka, son of Yashmut, who was accused of the same offence.*   Von Hammer says that Hulaju and Karabuka were captured by Muktil, brother of Ordu Kia, who was in the latter's service.   They were lodged for some time at Girdkuh, the famous fortress of the Assassins, and were put to death four months later.†

On the 19th of June, 1290, Arghun lost his son Yesutum, and about this time the Amir Sunjak and his son Shadi died at Meragha.   Two months later Majd ud din Muminan was put to death at Tebriz, while Arghun went to his summer quarters at Alatagh.   Thence he returned to Van and Vastan, where he was met by the Shah Kotb ud din of Shiraz, who took with him his atlas of the Western seas, with a description of their coasts and islands, and of the empire of Rum and the Mediterranean borderlands. Arghun had it all explained to him, and, we are told, he was specially interested in the town of Amuria, the birthplace of the Emperor Theophilus, and famous for its capture by the Khalif Moteassim. Amirshah, Fakr ud din, and his son, Haji Leila, three administrators of the province of Rum, who had apparently misbehaved themselves, were

---

* D'Ohsson, iv. 42-45. Ilkhans, ii. 10-12.   † Ilkhans, i. 381-382.

now apprehended. The first owed his life to the intercession of Kotb ud din and the Vizier. The other two were put to death. This year the Ramazan fast was observed with great solemnity at Tebriz, much to the satisfaction of the Muhammedans. Four minarets were erected, and the kadhi and imams, the khatibs and sheikhs, were all assembled together there.*

After the retreat of the army of Kipchak, Arghun sent the general Toghachar to help Ghazan. This was in the beginning of May, 1289. Eight days later two commissioners were sent to Khorasan to distribute the revenues of that province among the troops who were defending it. Meanwhile Ghazan, with the amir Kutlugh, marched against the rebels Kinshu and Nuruz, and having made a forced march of eighty fersenkhs in seven days, attacked him in the fertile plain of Raigan, or Radkian, in the neighbourhood of Tus. The troops behaved badly and fled, and Ghazan retired to Kalpush with the *débris* of his army, to await his father's orders.† Von Hammer says he retired to Juven, the birthplace of the two famous brothers Shems ud din and Alai ud din. There he was not welcomed by anyone, and accordingly felt grateful when the mehter, or tent maker, Nejb ud din, treated him very hospitably at Lirabad—a fact which he did not forget, and, when he mounted the throne, he made the village over to him as a fief, gave him a yarligh as Terkhan, nominated his family as managers of the richly-endowed monastery of Busijerd, near Hamadan, and appointed him besides to an office in his household. At Jajerm, near Izferain, which had been spared in various struggles, the country for two days' journey round being remarkable for its unwholesome herbage, a council of war was called, when it was decided to go to Kalpush, and await the return of the messengers who had been sent to Arghun. At Samatkam and Jermakan the advance guard met Nisam ud din Yahya, of Beihak, with a welcome supply of camp furniture, weapons, sumpter cattle, &c.‡

A few days before the above struggle a body of Karaunas, under the command of Alaju, or Aladu, had pillaged the baggage of Nuruz, which was posted near Kelat. He started in pursuit. The unruly robbers, after their attack, had refused to obey their leader, and had divided into three bodies, the largest one of which joined Nuruz, while the other two withdrew to their quarters. Aladu went to join Ghazan, who, after halting at Kalpush forty days, had been joined by a contingent of troops from Irak and Azerbaijan, under Prince Baidu, the son of Tarakai, and Nurin Aka, and had marched to Kabushan. Nuruz went to meet him as far as Charmagan, but feeling himself too weak, withdrew again, and was pursued. The people of Ghazan found all the country from Jam to Herat strewn with cattle which had been abandoned by the enemy, and part of

---

* Ilkhans, i. 389.   † D'Ohsson, iv. 45-46.   ‡ Ilkhans, ii. 12-13.

Y

which they had harried from the Arabs and Turkomans of the province. So much booty was captured that a sheep was sold for a penny. Nuruz fled with a few followers towards Sebzevar, crossing in the heat of summer a dry desert, which prevented Ghazan's pursuit. The latter went to Hera', encamped near the bridge Malan, and sent some amirs to Badghiz, to summon Prince Kinshu, who had prudently retired to the mountains of Ghur and Gharjestan. His chief officers, with Tekin and his Karaunas, went to Ghazan, at Herat. He now took up his quarters at Shutur kiuh ("the camel hill"), where he passed the summer of 1290.

Prince Baidu, the amir Nurin, and the other amirs were profusely feasted, and Ghazan himself having partaken too freely of drink on the occasion, was laid up with illness for forty days at Anjujan. The next summer and spring were passed between Radkian and Khaujan (Khabushan), and the spring near Nishapur. Ghazan had his own quarters at Moeyidi ; Baidu at Shankyan, between Nishapur and Baihak. The severity of the winter was very fatal to the horses, and the following spring many of the troops were unmounted. As this severe winter was followed by a summer of corresponding drought, Baidu and his troops, on the orders of Arghun, returned to Irak and Azerbaijan. A body of plundering Karaunas from Juven were driven away by the Amir Mulai. During the winter Ghazan dammed the River Kialteshen for the purpose of irrigation. Khuarezmi, the Terkhan, went from Arghun to superintend the taxes of Khorasan, and all the secretaries and tax-collectors were arrested. In the spring the Karaunas rebelled at Merv (D'Ohsson says at Sarakhs). Aladu Noyan was sent against them, while Ghazan marched towards Derei and Murgha. Ghazan encamped for a while in the district of Darjah and Shevkyan, whence he went to Sarakhs and Karatepe (i.e., the black hill, also called Eshirsil).*

Meanwhile Nuruz had gone towards Badakhshan, and had eventually joined Kaidu, who was then a rival Khakan, representing the claims of the house of Ogotai against Khubilai.† Kaidu asked him why he had fled. "Because," he said, "an honest man is obliged to imitate the fox in the fable, who one day running as hard as he could, was seen by a jackal, which asked him why he ran. 'Because the king is hunting wild asses.' 'But you are not a wild ass,' said the jackal. 'My friend,' said the fox, 'before it was discovered that I was not a wild ass I should have received many wounds.'" Kaidu was much amused at the answer, kept him at his Court, and treated him with honour ; but Nuruz had been too important a person while his father had ruled Khorasan for thirty years, and he had inherited too much wealth from the latter to submit easily to the position of a pensioner, while his haughty attitude brought him into trouble with Kaidu's officers. Meanwhile Kaidu furnished him with an

---

* Ilkhans, ii. 14-15.　　† Vide Ante, i. 173-181.

army of 30,000 men, under the command of the Princes Abughan and Uzbeg Timur. Von Hammer says under Sarban, Abukaan, and Oreg Timur. They were also allowed to make use of the garrisons encamped on the banks of the Oxus and at Shaburgan.*

Ghazan dispatched some troops towards Jebekis to get information, and on the news that the enemy was marching, inasmuch as his men were scattered, he withdrew from Karatepe to Merghana, on the River Keshfrud, to await the arrival of the Amir Kutluk Shah and the troops of Herat. Kunjuk was sent to Mazanderan, and Mulai to Kuhistan, to collect troops there. Ghazan was presently joined at Meshed by Kutluk and his division, whom he received with great honour. He sent his harem for safety to Isferain. Meanwhile, news arrived that the enemy had advanced towards Nishapur by way of Habak and Ishakabad. The amir Aladu at the council of war quoted the Mongol proverb, "It is easy to quarrel with an enemy, but difficult to vanquish him." It was determined to withdraw first to Tus, where a struggle took place. Ghazan now began to feel one of the inconveniences of a fading fortune. Some of his amirs found specious excuses for withdrawing from him. Aladu asked to be allowed to go and collect some troops at Juven ; Shirin Ikaji to remove his family from Nishapur ; Arghurtai Ghazan, who had a close connection with the Jagatai princes, also withdrew. Ghazan encamped for a while at the village of Kisragh, near Izferain. A body of Karaunas having revolted, the Amir Kutluk was left behind to suppress them, while Ghazan moved to Jorbed. He ordered the Karauna families who lived there to be moved to Jajerm and Bostam. Nuruz and his troops committed ravages in Khorasan. They blockaded Nishapur, and carried off many prisoners from the district round. Near Nishapur is a strong position called Barubaka, where a crowd of people with their herds had retired. The invaders cut off their retreat both at the head and foot of the valley, and killed more than 1,000 folk who were not Mussulmans. They then plundered the sacred tombs at Tus, and carried off four golden ornaments which adorned the fountain. At Badghiz the invaders mustered their men, and as the numbers were 5,000 horsemen short, Nuruz was bastinadoed at the instance of his allies, the princes of the house of Ogotai. Ghazan withdrew again from Jajerm towards Bostam and Damaghan, at which latter place he heard of his father's death.† In regard to the ravages of the Jagatai troops under Nuruz, Wassaf says he was so feared in the country that when cattle were seen rushing to water people said it was because they had seen the spirit of Nuruz.‡

Let us now return to Arghun. He abandoned the cares of State more and more to his Jewish vizier, who governed the country prudently for two years, repressing disorders, and accumulating 1,000 gold tumans in

---

* D'Ohsson, iv. 46-49.   Ilkhans, ii. 14-15.   † Ilkhans, ii. 15-16.   ‡ D'Ohsson, iv. 46-49.

the treasury, and his authority grew daily. But he was not without enemies, the most dangerous of whom was Arghun's favourite, Tughan. Tughan was the son of Tarakai, formerly Governor of Kuhistan. He was very accomplished. In judgment and sagacity, in wit and eloquence, he had no equal among the Mongols, and was a good letter writer, book-keeper, poet, and astronomer. He had been sent with a body of troops in October, 1289, to Khorasan, to oppose Nuruz. When he arrived there Nuruz had withdrawn, so Tughan returned. He was thereupon charged by a bakhshi, at the Vizier's instance, with having impressed more post-horses than he was authorised to do by his warrant (kara tamgha). This having been proved, he was, according to the inexorable yasa, or law, ordered to receive seventeen strokes with the bastinado. Tughan, who was always self-possessed, and had a ready answer at command, looking round the room, and seeing more than seventeen amirs present, asked what harm it would do if the seventeen strokes awarded to him were distributed among them. Arghun laughed, and quoted a verse of Motenebi—

> If the lion winks with a lion's strength,
> Do not imagine that the lion laughs.

By his happy remark on this occasion, and his general good-humour, he escaped punishment; but he did not forget that the Jew was the foundation of the affront which he had received, and determined upon his ruin, allying himself for the purpose with the amir Kunjukbal and others, who had grievances against him. Their plans were not easy to carry out, since Arghun was much attached to his vizier, who had not only kept his coffers well filled, but had flattered him by a much more potent offer. He had tried to persuade Arghun to become, like Muhammed, the founder of a new religion. Wassaf reports that he had been told by the vizier Sadr jihan, that he had one day met Said ud daulat on the main road, and had asked for a short consultation. Said alighted, and they talked together confidentially. Having prepared him by some preliminary remarks, he showed him a memoir, in which it was set out that there should always be on the earth, as was assured by the conjunction of the stars, a man who dominated his contemporaries, whose existence was necessary to maintain order among men, and who introduced, according to the exigencies of the day, and the needs of the people, new religious laws, employing either force or persuasion, as needs be, to secure their obedience. The qualities, he added, of this heavenly apostle were to be found united at that moment in the person of the Ilkhan the Just. To the foot of this document were appended the signatures of the chief imaums, or doctors of the law, who confirmed the position therein maintained by several quotations. One of them had written, "The people follow the faith of their sovereigns." Said had asked Wassaf's friend, Sadr Jihan, to add his signature also, but he had excused himself. It

was said that Said proposed to convert the sacred Ka'aba into an idol
temple to force the Muhammedans to become Pagans, and that he had
already begun to prepare for a campaign against Mecca. Wassaf adds
(but we must remember that he was a Mussulman, and speaking of a
despised Jew, who held authority over true believers) that Said sent
another Jew, named Khoja Nejb ud din, the oculist, with a list of 200
proscribed persons, including the most noble and best in the land, whom
he was to put to death; while Shems ud daulat, who commanded at
Shiraz, was ordered to put to death seventeen prelates and grandees of
that city. (Von Hammer says to pay the penalty inflicted by the yasa
which he interprets as the bastinado.) To him also were attributed the
various sanguinary acts of Arghun, who, when a youth, had been very
gentle, and had on one feast day expressed his pain on seeing a number
of dead sheep in the shambles, saying it was dreadful to have to kill so
many innocent animals merely for our food.*

Arghun was a great believer in alchemy, and in its professors, the
bakhshis or lamas, who claimed to have found the object of quest of
mediæval philosophy, viz., a potion for securing long life, and which we
are told in their case was a salivating decoction compounded of sulphur
and mercury. Mirkhond tells us that a lama who came from India in
698 HEJ., prepared such a decoction for Arghun, who took it for eight
months, and was then advised to retire for forty days to Tebriz, where he
saw no one but the general Ordu Kia, the vizier Said ud daulat, an
officer of his household named Kujan, and the bakhshis, who surrounded
him day and night, and disputed on the mysteries of their faith. He then
went to his winter quarters in Arran, where he became ill, and as the
medicine given him by the doctor, Armu ud daulat, did not cure him,
one of the bakhshis gave him a large dose of some potion. This
brought on a partial paralysis, and as he got no better, inquiries were
diligently made as to the real cause of his disease. The Shamans having
consulted the burnt shoulder-blades of sheep, as was their custom,
declared that his illness was due to sorcery, and one of his wives, named
Tukchak, or Tughanjak, the daughter of a sister of Jushkab, was
accused of being the cause of the mischief, and was interrogated and
tortured before the other Khatuns. She replied that she had done
nothing except employ a charm written on a piece of paper, such as was
commonly used to secure the Ilkhan's affection. She was thereupon
drowned, with all her maids. The Ilkhan's continued illness was
naturally very distressing to Said ud daulat, who knew that his own
fortunes depended upon it, and after consulting with other courtiers it was
determined to dispense liberal alms, and to open the prisons. A month
passed by, and Arghun gradually sank, while the Vizier multiplied

* D'Ohsson, iv. 49-53. Ilkhans, i. 386-388.

various acts of benevolence. He dispatched seventy letters in one day, recommending the authorities to protect the oppressed, relieve the poor, and release prisoners. He made a present of 30,000 dinars to the people of Baghdad, while 10,000 were distributed among the poor and the fakirs, &c., at Shiraz. Similar alms were distributed in other places, while the khatuns, princes, and princesses were forbidden to appropriate anything which had been devoted to good works.

When inquiries were instituted in view of releasing various prisoners, it was found that Kara Buka and Hulaju had been put to death at Girdkuh and Damaghan respectively, while thirteen princes of the blood had been put to death by order of Sultan Aidaji. Meanwhile, the principal military chiefs, Toghachar, Kunjukbal, Tugal, and Ilchidai, banded themselves together to exact vengeance from their personal foes. *Inter alios*, Aidaji had put to death the young children of Hulaju and Kara Buka, as well as Tukchak Khatun. He said he had merely carried out his master's orders. Ordu Kia was sent to inquire from Arghun, and brought back word that the latter knew nothing of the matter. " How could he have said so, since he has not spoken for some days?" " If so," was the reply, "he could not have given you orders. You have committed these cruelties without his knowledge, and now you lay them to his charge." They accordingly put him to death. This was on the 4th of March. Only Juchi and Said ud daulat, both of whom were in the greatest jeopardy, had access to the Ilkhan's tent. The latter had sent a messenger to Ghazan, bidding him hasten home, hoping he would arrive before his father's · death, or at least before it should have become generally known, and thus escape the fate he felt to be impending over him ; but the generals who had leagued themselves together, suspecting from the prohibition to enter Arghun's tent that the latter's case was getting desperate, determined to complete their work. Juchi and Ordu Kia were killed at a feast given by Toghachar. Tughan killed Kujan, brother of Ordu Kia, with a sabre cut, in the ordu of Uruk Khatun, one of Arghun's wives. Said ud daulat himself was arrested and conducted to Toghachar's quarters by Tugal and Kurumishi, son of Alinak, where he was decapitated. His palace at Tebriz was ransacked, and a large quantity of treasure was found stored in jars, &c., in the walls. Arghun not seeing his usual courtiers about him, asked the reason. They made an excuse, but Wassaf says he guessed their fate.* Said's death took place on the 29th of February, 1291.

Thus perished a very remarkable man. " From the beginning of the Arab domination in Asia," says Abulfaraj, " Jews had never been intrusted with responsible posts. The humbler among them kept shops, dye works, or rope walks ; those who were better off became doctors or

---

* D'Ohsson, iv. 53-57. Ilkhans, i. 391-393. Quatremere, 194. Note.

scribes in places where Mussulmans contemned these occupations. The Mongols," he says, "on the other hand never honoured those who were worthy, nor confided the government of their towns or districts to those of high birth. They made no difference between a slave and a free man ; between a Muhammedan, a Jew, or a Christian. They treated men of all nations with the same asperity. Whoever went to them with presents in his hands obtained the office he asked for, irrespective altogether of his fitness. They only exacted perpetual deference. Said ud daulat directed in person all the affairs of the State. He neglected the grandees of the Ordu. He prohibited them from giving or receiving. He treated the amirs, generals, and great dignitaries disdainfully. He himself was the only avenue to every suppliant. There gathered about him from all parts of the earth numbers of Jews, who declared it was for their special safety and his own glory that God had given this man to the Jews. They became greatly inflated in consequence."* It is certainly singular that a Jew should in these times have been able to climb to the vizierate of the Mongol empire of Persia, and apart from the natural capacity which this success proves, there is something remarkable in the Vizier's suggestion that Arghun should found a new religion, and the reasons urged for the step, which makes us wish to know something more about him. Arghun himself died, after a five months' illness, on the 10th of March, 1291, at his palace of Bagchi Arran.

The "History of Georgia," speaking of his illness, says that his limbs shrivelled up, his flesh and bones fell, and he became like a corpse, and eventually, after an illness of four months, his body began to putrify from his head to his feet, whereupon his noyans fell on him and choked him to death under his tent. We are further told that he died on the 12th of March, the same day on which King Dimitri had been put to death.† Abulfeda agrees in the date here given, and says Arghun died in the third month of 690 (i.e., in March, 1291).‡ Wassaf's notice of his death is characteristic : "The parroquet of the soul of the Ilkhan left the cage of his body to go and dwell among the peacocks of the palace of the sublime garden."§ Rashid says that the magicians or bakhshis having been consulted on the cause of his death, reported, after examining the burnt shoulder-blade of a sheep, that his illness was caused by sorceries.|| Stephen Orpelian, in the 70th chapter of his "History of Siunia," reports that Arghun was poisoned by one of his concubines whom he greatly loved. This happened, he says, on the plain of Mughan, and on the feast of St. Theodore. This story is doubtless founded on that already related about one of his wives having used a charm to gain his love. Orpelian adds that his death was marked by the massacre of several chiefs—Khojan, the grand treasurer ; Sultan, superintendent of meats and

---

* Chron. Syr., 624-625. D'Ohsson, iv. 60-61. Note. † Op. cit., i. 608-609.
‡ Id., 609. Note. § Quatremere, 9. Note. || Id., 268.

drink ; Chishu and Ordu Khan, chiefs of tribunals ; the virtuous Jit ; Sada daula, of the grand divan, director of taxes in Georgia and elsewhere.*

The " Shajrat ul Atrak " says of Arghun : " The religion of Muhammed by his death became flourishing as a garden by the breezes and flowers of spring, and the hearts of its enemies were broiled on the fire of grief and despair." It says that he was buried on the mountain of Zobir, near the burying-place of Kyde, one of the prophets, in a tomb made for the purpose, and the environs were made *kuruk*, or sacred. The place, says the author of that work, is still known by the name of the Tuffuruk of Arghun.† Others say he was buried on the mountain of Sijas, which the Mongols called Avizeh, and for three days the commanders of ten men in his regiment sent funeral meats to his grave, which, after the usual Mongol fashion, was unmarked and undisclosed until his daughter placed a domed tomb and a chapel over it. The place itself was a famous pasture ground, called Endjrud, and Abaka had built a palace there. The mountains Sijas took their name from a village which was destroyed by the Mongols. They ran through beautiful meadows, which they called Kungkur ulong.†

Rashid ud din, who had to temper his Muhammedanism with considerable servility, says that Arghun had an agreeable temper, but was irascible. Ez Zehebi, who is more candid, says he was unjust, cruel, ignorant, hardy, courageous, strong and active. If three horses were placed together, he would vault over two of them and seat himself in the saddle on the third.§ Haithon says he was very handsome, and a powerful and prudent ruler.|| Arghun was fond of building. He erected a suburb to the west of Tebriz, which was called Shem, or Shenb, and where at a later day his son Ghazan built his tomb. D'Ohsson says he built two palaces there, between which was a town, to which he gave the name of Arghuniya. He built another town near Sheruyaz, north of Kazvin, in the meadows of Kungkur ulong (meaning in Mongol the pastures of the sorrel-coloured horse), which was afterwards known as Sultania. Abaka had built a palace a day's journey from this spot, where there was a natural lake whose water did not grow or lessen. At Alatagh, Arghun built another serai, or palace, while he put up a summer palace at Lar, or Larjan, the chief town of the district of Laristan, at the foot of the mountains of Demavend. This was called Arghun's Kiosk, and retained the name for some time, being so called in the " Zafer Nameh." The bazaar of the town is still reputed the most beautiful in Persia, while the existing ruins of the palace are remarkable.¶

Arghun's great passion, however, was for alchemy, and his Court was the rendezvous of the various professors of the art in the East. Not-

* Hist. de la Siounie, 259.　Hist. de la Géorgie, i. 609.　Note.　† Op. cit., 264-265.
† D'Ohsson, iv. 5-8.　Ilkhans, i. 293.　§ D'Ohsson, iv. 58.　Note 1.　|| Op. cit., 59.
¶ D'Ohsson, iv. 53-59.　Ilkhans, i. 389-391.　Quatremere, 203.　Note.

withstanding the cost of their experiments, he did not complain, but readily paid what they needed.   One day, when the alchemists had been discussing the mysteries of nature in his presence, he turned to the mollah Kutb ud din, of Shiraz, a learned man of repute, and said : " You, who are a learned man, believe, because I am a Turk, that these people are deceiving me ; but it is quite certain that there is a science of alchemy, and there is *some one* somewhere who knows its secret.   If I ill-treat and put to death these ignorant people, *that one man* will be afraid to come and see me."   Nevertheless, after seeing the result of many futile experiments, his faith in them began to falter.*

As soon as his death became known the troops proceeded to plunder the houses of the Muhammedans and Jews who were about, and dug holes in their tents to find hidden treasure.   The catastrophe which overwhelmed the vizier Said ud daulat was the subject of great rejoicing to the Muhammedans, and was followed by a terrible reaction against the Jews, who were everywhere treated with renewed cruelty.   At Baghdad their dwellings were pillaged, and more than a hundred of their principal people were reduced to poverty.   Abulfaraj tells us how, when they had killed the Vizier, they sent messengers in various directions, who seized and imprisoned his brothers and relatives, and robbed them of their property.   Such of the Jews as were not killed resumed their humble mode of living again.   A general attack was made on the Jewish quarter at Baghdad, but the assailants were beaten back, and many on both sides were killed.†

We are told that when Kutlugh Buka and his partisans learnt of the serious illness of Arghun they determined to put David, son of Dimitri, on the throne of Georgia, and to dethrone Wakhtang.   In this they were supported by Tachar (*i.e.*, Toghachar) and by Pharejan, son of the King of Ossethi, and they met to deliberate as to which of the two should be king. The other mthawars were not partial to David, and remained attached to Wakhtang.   David, instead of the kingdom, secured a grant of some villages and districts.‡

Armenia proper was at this time a mere geographical expression, its independence having been absorbed by the Mongols and Georgians. The annals of the country dwindle down to notices of ecclesiastical jealousies and disputes.   Darsaiji had been appointed atabeg of all the country between Tiflis, Ani, and Kars (*i.e.*, of Armenia), with the guardianship of his sons, by King Dimitri of Georgia.   His son Stephen was in 1287 made bishop and metropolitan of Siunia.   Some of the bishops were jealous of him, and caused considerable trouble.   Stephen thereupon repaired to Arghun, and showed him a letter from the patriarch, so that he could see for himself how matters stood.   Arghun received him

---

* D'Ohsson, iv. 59-60.   † *Id.*, 60.   Abulfaraj, Chron. Syr., 625.
‡ Hist. de la Géorgie, i. 608.

kindly, and sent him a yarligh which gave him authority over the church, the country, and the bishops. He also assigned him a person from his Court, and gave him a golden paizah or tablet, and sent him back to his see. Darsaij died in his palace of Arpha, in 1290, and was buried at Noravankh by his brother Sempad. A feud arose among his sons about the succession. They brought the matter before Arghun, who selected the eldest, Elikum, put him in his father's place, and made him prince over his brothers. He divided what he had with the bishops, vartabeds, and nobles, gave a portion to his brother Jalal, and another to his cousin Liparit. Peace reigned everywhere, and the ruined monasteries and devastated land began to revive. "The family of the Orpelians," says the historian of the house, "was at this time, by the grace of God, like Noah's ark amidst the waves that were desolating the world, and it is to be hoped the Lord will continue it so until the end of time."*

Arghun had nine wives. 1. Kutlugh, daughter of the Uirat Tengir, who was a Kurkan (i.e., son-in-law of the ruler, inasmuch as he married Tudukash, a daughter of Khulagu). 2. Oljatai, daughter of Tengir's son Sulamish by the same princess, Sulamish having married his father's widow in the recognised Mongol fashion. 3. Uruk, daughter of Saruji, and sister of the Kerait amir, Irinjin. 4. Seljuk, daughter of Rokn ud din, Sultan of Rum. 5. Bulughan, the relative of Buka Yarghuji, who had been one of his father's wives. 6. On her death he married a second Bulughan, the daughter of Utaman, son of Obotai Noyan of the Konguruts. 7. Mertai, the Kongurut, who had been the wife of Khulagu and Abaka. 8. Tudai, the daughter of Musa Kurkan and of Tarakai, daughter of Khulagu; she had been Ahmed's widow. 9. Kultak Ikaji. By Kultak, Arghun had his eldest son, Ghazan. Uruk was the mother of his two sons, Yesu Timur and Uljaitu, and of his three daughters, Oljaitai, Oljai Timur, and Kutlugh Timur. Kutlugh was the mother of Khatai Oghul, while the second Bulughan, the favourite among his wives, was the mother of his daughter Dilenji.†

Marco Polo has a story about the elder Bulughan, whom he calls Bolgana, a name quaintly and accurately translated Zibellina by Colonel Yule. He says that when she died she desired in her will that no lady should take her place, or succeed her as Arghun's wife, except one of her own family who still lived in Cathay (i.e., one of the tribe of Bayaut). Arghun therefore dispatched three of his barons—Ulatair, Apuska, and Koja, as envoys to the Great Khan, with an escort, to bring home some lady of Bulughan's family. The Great Khan received them well, and sent for a lady named Cocachin (Kukachin), who belonged to the same tribe as Bulughan. She was seventeen years old and very beautiful, and when she reached the Court she was presented to the

* Hist. de la Siounie, 238-239. St. Martin Memoires, ii. 169.
† Von Hammer, Ilkhans, i. 360-361.

three barons, who declared that the lady pleased them well. At this time Marco Polo himself, who had been sent by the Great Khan as envoy, returned from India, and reported the various things he had seen in his travels, and described the seas which he had crossed. The three barons determined to return to Persia in the company of Marco, his father Nicolo, and his uncle Maffeo, who were all then at the Great Khan's Court, and they determined to return by sea, to avoid the great fatigue which the land journey would be for the lady they were escorting. The Great Khan gave the Poli two golden tablets or paizahs, to secure them liberty of passage through all his dominions, and securing them all necessaries, &c., for their journey. He also sent messages by them to the Kings of France, England, Spain, and the other rulers of Christendom. He then had thirteen ships equipped, each of which had four masts and often spread twelve sails, among which were four or five which carried 250 or 260 men. They were provided for a journey of two years. After sailing for some three months, they reached the island Polo calls Java, but which was really Sumatra, and after a voyage of eighteen months more they reached their destination, to find that Arghun was dead. Ramusio's version says that 600 of the mariners and others on board had died on the way, while of the envoys only Khoja survived. The envoys, besides the Princess Kukachin, also had in charge a daughter of the King of Manzi (i.e., a princess of the Sung dynasty) who had also been sent to Arghun.* They probably learnt of Arghun's death at Hormuz. They thereupon sent word to Gaikhatu, Arghun's brother, who directed them to convey her to Ghazan, Arghun's son, who was then in the province of the Arbre Sec (i.e., Khorasan), guarding the passes with 60,000 men. Having done so they returned to Gaikhatu, with whom they stayed nine months. We learn from other sources that it was at Abher, near Kazvin, that Ghazan met the princess, whom he married. She died twelve months later, in June, 1296.†

During Arghun's reign we read of no intercourse with the Egyptians of a peaceful kind. The fact was that Arghun's patronage of the Jews and Christians created a breach with the Egyptian Sultan, which probably would have ended in war, but that each had other troubles on his hands. Arghun's we have described. Kelavun's were caused by fear of the supporters of Bibars, the so-called Zahirits, and who made Karak, which a son of Bibars held, their centre. The Sultan contented himself therefore with strengthening his fortresses, the securing of Karak and of Sehyun, in Syria, where Sonkor headed another body of the discontented, and with assailing the remaining fortresses of the Crusaders,‡ and it was not long before the famous stronghold of Tripoli was in his hands. He also bitterly pressed the King of Little Armenia, whose land as far as Ayas was devastated, and who in 1285 sought a ten years' truce, by agreeing to pay an annual tribute of a million dirhems, undertaking not to build any new fortresses in his country, and to release any Mussulmans he held prisoners.*

On the other hand, Arghun tried to draw closer his ties with the Christians. Haithon says he undertook to rescue the Holy Land from the infidels, while he rebuilt the churches which had been destroyed by Ahmed.† A letter is extant, in very canine and broken Latin, which professes to have been sent by Arghun to the Pope Honorius IV. It is probably a translation by one who did not know his original very well. It is dated the 18th of May in the year of the Hen (*i.e.*, 1285), " in ʿCoriis,'' which Remusat suggests is a mistake for Tebriz. It is written in the name of Christ, and is courteously phrased. The Ilkhan recalls the good feeling which the Mongols, since the time of Jingis Khan, the first father of all the Tartars, had had for the Pope, the serene King of the Franks, and the serene King Charles (*i.e.*, Charles of Anjou), and the protection they had always extended to Christians, whom they had exempted from paying dues, &c. He speaks gratefully of the reception accorded to his envoys, Ise turciman (*i.e.*, Ise the interpreter), Boyagok, Mengilik, Thomas Banchrui, and Ugeto turciman, who had, if we understand the very clumsy Latin, been rewarded with presents of precious robes and perfumes (*roba et tus*). Arghun goes on to say that his mother was a Christian, while his grandfather and father, Khulagu and Abaka, had greatly cherished the Christians, as did at that time Khubilai Khakan. He had determined himself to send the holy father a present of precious robes and incense, and to guarantee the Christians the advantages they had previously enjoyed. The long interval which had elapsed since the previous embassy was due to the apostacy of Ahmed. He recalled the fact that the people of Scam (*i.e.*, Shem, or Syria) intervened between his folk and those of the Pope, and suggested a joint campaign against Egypt.‡ " It is singular," says Remusat, "and certainly a matter of regret, that we cannot illustrate this singular letter from other sources." A few years later, viz., in 1288, other envoys from Arghun arrived at Rome. They were Bar Sauma, a Uighur monk, who had been nominated bishop of Uighuria by Yahaballaha, patriarch of the Nestorians ; Sabadin Arkhaun (Arkhaun is the term by which the Christians were known to the Muhammedans) ; Thomas de Anfusis, and the interpreter Ugeto, the two latter being apparently the same persons as are mentioned in the previous letter, and there called Thomas Banchrui and Uguet. These envoys were received at Rome by the newly-elected Pope, Nicholas IV. They reported how the faith was flourishing in Tartary; that Arghun was

---

* Weil, 159.　　† Haithon, 59.
‡ Remusat, Mems. Acad. Inscr., vii. 356-357 and 426-427.　Mosheim, Hist. Eccl. App. N xxv.

determined to prosecute his campaign against the Mussulmans, and to rescue the Holy Land ; that the Grand Khan was not averse to the faith, and that missionaries might be sent to him ; many of Arghun's officials were Christians, many again were Muhammedans, and it was the latter's influence which prevented him from becoming a Christian ; two of his queens, viz., Elegag and Tuktan, had been baptised, while Arghun himself was determined to be so at Jerusalem, after he had taken it from the Saracens ; finally, the Nestorians and Franciscans were engaged in spreading the faith everywhere in Tartary.   At Arghun's Court were several Christian interpreters, such as Johannes de Bonachia, Hugo Gantelini, Petrus de Molina, Girardus Kasmuri, from Constantinople ; Balaba de Yanua (? Genoa), Girardus de Caturco, Georgius Cusi, Johannes de Barlara, and Johannes de Casaria.*

These matters having been reported to the Pope, he replied in a letter, dated the 10th of April, 1288, in which he expresses his satisfaction at having received the Khan's letter, and also at its contents, and the assurances of the envoys, which had been delivered in the presence of his brothers (*i.e.*, the cardinals).   He was especially thankful to the Saviour, who held in his hand the hearts of worldly sovereigns, that he had inspired Arghun, not only to treat the Christians kindly, but also with a desire for the spread of Christianity.   He sent him his thanks, and also an exposition of the principal articles of the Catholic faith—the sacrifice, resurrection, and ascension of Christ, and the remission of the keys and the power of binding and loosing to St. Peter and his successors.   As the vicar of Christ on earth the Pope exhorted the Khan to enter the only path which led to eternal safety.   In a second letter the Pope mentioned that he had heard from his envoys that it was his purpose to deliver Jerusalem from the infidels, and that he proposed to be baptised there.   He praised this intention, but wisely, if not craftily, suggested that his previous baptism would facilitate the conquest, while considerations of his own safety would not brook delay.   He urged him to be baptised at once ; it would lead to many following his example.   The Pope also wrote to Tuktan,† and to another princess named Elegag (*i.e.*, Ilkutlugh), saying he had heard of their conversion and their zeal for the faith, which he urged them to redouble.   He also wrote to Denis, bishop of Tebriz, a letter from whom had accompanied that of the Ilkhan.‡

The object of the Mongols was no doubt more political than religious. They wanted help against the Egyptians more than fraternal advice.   It seems that the Pope sent on the Khan's message to Philippe le Bel, king of France, who is in turn found sending envoys in 1288 to Persia.   We learn about them, chiefly through Arghun's subsequent complaints, that they

---

* Mosheim, op. cit., 75-76.
† D'Ohsson identifies her with the widow of Abaka of that name, but Von Hammer says, on the authority of Rashid ud din, that she died in Abaka's lifetime.
‡ Remusat, op. cit., 359-360.   D'Ohsson, iv. 66-69.

would not do him homage after the usual fashion, on the excuse that he was not a Christian, and it would therefore be an indignity to their master if they prostrated themselves as had been demanded. Eventually they were received in their own fashion, and were treated with courtesy. In the year 1289 some Franciscans returned to Italy after an evangelising tour in the East. John of Monte Corvino was among them; he had been away ten years, and reported how a certain Isola, a noble of Pisa, had acquired great authority among the Tartars. He presently returned again, and the Pope sent with him a letter, dated the 15th of July, 1289, in which he repeated his former congratulations, and recommended John of Monte Corvino himself to his favour.* Other letters were sent for Prince Caidon (? Baidu) and for Khubilai Khan. Shortly after there arrived another envoy from Arghun at Rome, in the person of a Genoese, named Buscarel de Gisulf, reporting that he (the Ilkhan) was ready to march to the rescue of the Holy Land at the time fixed for the crusade. Buscarel was sent on to Edward I. of England, with a letter of recommendation from the Pope, dated the 30th of September, 1289. On his way he presented Philippe le Bel with a letter from Arghun which is still preserved in the French archives. Remusat says it is nearly six feet and a-half long and ten inches wide, written on paper made of cotton, and is in the form of a roll. On one side it has thirty-four lines of black writing and a thrice-repeated mark of a red seal, five and a-half inches square. The letter is written in the Mongol language, and in Uighur characters, in vertical lines. The seal bears six ancient Chinese characters. This Mongol roll has attached to it a document in French, to explain the nature and contents of Arghun's own communication.† Arghun's letter has naturally attracted much attention. It has been published and translated both by Remusat and Schmidt. The former says it contains the earliest known specimen of Mongol. This was so when he wrote, but we have since recovered the vocabulary preserved by Malakia, which is considerably earlier.‡ The letter is written in a dialect differing somewhat from either standard Mongol, or the Kalmuk dialect, with less complicated phrases, and in more simple grammar.§ The following is a translation of it as given by Schmidt :—

"By the power of the Almighty God. Under the auspices of the Khakan Arghun, our word.

The King of France,
by the envoy,
Mar Bar Sauma‖
Sakhora.

You have told us you will set out to join us when the Ilkhan's troops

---

* D'Ohsson, iv. 69-70.   † Remusat, op. cit., 363-364.   ‡ Ante, Notes to chapter i.
§ Remusat, 365.
‖ *Mar* is a Syriac title of honour, meaning Mr. or Esq., and is given to all respectable people. *Bar Sauma* has been already mentioned. (Remusat, op. cit., 369.)

begin their march towards Egypt. Pleased with this communication, I say that we propose, confiding in God, to set out in the last month of winter, in the year of the panther (*i.e.*, January, 1291), and to pitch our camp before Damascus on the 15th of the first month of spring (*i.e.*, towards the 20th of February). If you keep your word and send your troops at the time fixed, and God favours our undertaking, when we have taken Jerusalem from these people we will make it over to you. But unless you join us at the rendezvous it is useless to send your people at all, and if in consequence we do not know what to do, what good will it do? We have sent Muskeril, the Kuruji, to say that if you will send us envoys knowing several languages, and bringing us presents, rarities and images of different colours from the land of the Franks, we will be very friendly with you, by the power of God and the fortune of the Khakan. Our letter is written at Kundulen, the 6th of the first month of summer, of the year of the cow."

* It will be noticed that Buscarel's name is written Muskeril. This is due to the usual substitution of M for B by the Mongols and Turks. Kuruji means one who has charge of the arms (*i.e.*, of the prince) and answers therefore to our western armiger and the Persian selahdar. The sovereign's body guards are also called kurujis.* Kundulen is an unknown site, according to D'Ohsson. Remusat suggests that it may be a corruption of Kongorlan, the earlier name of Sultania, but prefers to identify it with some place north of Lake Urmia, and says there is a river Kundalan which falls into the Araxes. St. Martin says that Soyuthi, in his geographical dictionary, mentions a place called Kondelan, near Ispahan.† In regard to the mark of the seal attached to the document, Remusat urges that the seal which made it was probably sent to Arghun with his investiture as Ilkhan. The inscription is in the early Chinese character called *chuan*, which is otherwise known as the seal character. The inscription does not impress one that the position of Ilkhan was held in high estimation at the Grand Khan's Court. It reads, "Seal of the Minister of State, pacificator of peoples." Remusat calls attention to the singular result of the Crusades on the one hand and the Mongol conquests on the other, in bringing together the affairs of the furthest East and those of Jerusalem and Egypt, and to the curious fact that part of the story should be preserved in ideographic characters which, so long extinct in the valley of the Nile, still survive in China.

The letter in French accompanying the Mongol original is more diffusely written, and more courteously phrased, a sophistication doubtless due to the courtly and diplomatic interpreter. In this note Arghun adds that in his expedition against Jerusalem he intended taking with him the two Christian kings of Georgia, who were his vassals, and could

---

* D'Ohsson, iv. 71-73. Remusat, op. cit., 369-372.   † Remusat, op, cit., 372-373.

muster 20,000 horsemen. As it would be very inconvenient for the King of France to cross the sea with a great number of horses, he (Arghun) was prepared to furnish him with 20,000 or 30,000, either by way of gift, or at a reasonable price. Arghun could also, if the French king pleased, order in Turkey (*i.e.*, Rum) for him, cattle, cows, and camels, grain and flour, and such other provisions as he would need. As a proof of his good intentions he mentioned how, having heard of it, he had greatly rejoiced at the news of the recent disaster of the Christians at Tripoli, and that he had had four of the principal people among the Muhammedans put to death, and refused permission for their bodies to be buried, but ordered them to be given to the dogs and birds of prey. He mentions how his sister had recently married the King of Georgia, and that she had become a Christian. That during Lent just past Arghun had caused the mass to be chanted in a chapel where the Nestorian bishop Rabanata had officiated, and had also caused several of his grandees to take the sacrament. He then went on to express his surprise that the envoys of the French king had refused to salute him in the prescribed fashion, and had declared that they could not kneel before him, since he was not a Christian ; that he had summoned them to do so by his great barons three times, and finding them obdurate he had admitted them and treated them with great honour. Arghun said if this was by the king's wish he had nothing to say, and that everything that pleased the king would please him also, but he desired that if he sent him envoys in future they would salute him in the fashion customary at his Court, without crossing fire (*i.e.*, without being purified by being made to walk between two fires, as in the Mongol ceremonial).*

Buscarel, having delivered his message, apparently went on to England, where he had been preceded by a brief of Pope Nicholas, addressed to King Edward, stating how one Zagan, who had recently been baptised by the Archbishop of Ostia under the name of Andrew, with his nephew Dominic, formerly called Gorgi (? Kurji), and Buscarel de Gisulf, a Genoese citizen, as well as Moracius, envoys from Arghun, the illustrious King of the Tartars, bearers of presents, would presently arrive at his Court. He begged him to receive them well, to listen attentively to what they had to say, and to send them back as quickly as possible, for on their return he proposed to send them home again, accompanied by his legate at that time in England. Remusat says there are two briefs extant, one dated the 30th of September, 1289, and the other the 2nd of December. D'Ohsson speaks of one only, which is dated the 10th of December, 1290, and of which the above is the abstract.† From the wardrobe accounts of Edward I. we learn that Buscarel arrived in London on the eve of the Epiphany, January 5th, 1290, accompanied by three esquires, a cook,

---

* Remusat, *id.*, 430-432. D'Ohsson, iv. 74-76.   † D'Ohsson, iv. 76-77. Remusat, 381.

eight horses, and six *garçons*. He remained thirteen days at the English Court, and in all twenty days in England. His expenses were defrayed by Edward. On Buscarel's departure Edward gave him a letter for Arghun, in which mention is made of the attachment which his father had always shown towards the Christians. The English king compliments his correspondent on his intention of arming against the Soldan of Babylon (by which title the ruler of Egypt is meant) in aid of the Holy Land and the Christian faith, and thanks him for the offer of horses, &c., when he shall reach Palestine, and assures him that as soon as he had obtained the assent of the Roman pontiff to the passage of himself and his army beyond the sea he would inform him by envoys, and also send him some gerfalcons and other treasures of his land (*de nostris Girofalcis et aliis jocalibus nostre terre*), as Arghun had requested.* Edward's wars in Scotland prevented him from carrying out the crusade, to which he was persistently urged by the Pope.

In 1291 Arghun renewed his message, and sent fresh letters to Nicholas IV. and the King of England. His envoy on this occasion was called Chaghan, or Zagan.† The faith was spreading, it seems, among influential people in Tartary, for we read how two queens, whose names are given as Dathanicatum and Anichohamini, had become Christians, as had also Arghun's son "Karbaganda." In his letters to the Pope the Ilkhan chiefly refers to a joint expedition against Syria. The English king had already taken the cross, but the fall of Ptolemais, the only remaining vantage held by the Christians on the coasts of Syria, the news of which reached Western Europe about the time of Chaghan's embassy, damped the spirits of the Christians, and, as Remusat says, led to the strange result that the most eager crusading spirit was actually found among the Mongols, who desired the alliance of the Franks against their powerful enemies, the Egyptian sultans.‡

In his reply, dated the 21st of August, 1291, the Pope acknowledged the receipt of the letter which Chaghan had taken, and said he had forwarded the one addressed to Edward. He seems to have said little about Palestine, but repeated his entreaties to Arghun to be baptised, excusing himself from sending the presents which the Ilkhan had asked for on the ground that he was an ecclesiastic, and concluded by recommending to his favour William de Sherio, the penitentiary, and Matthew "de civitate Theatina," a Franciscan, who carried letters with them for the Ilkhan. In a second letter, dated two days later, on the occasion of the capture of Tyre and Acre by the Mussulmans, Nicholas told the Ilkhan how he had urged the kings and princes who were Catholic to unite in recovering the Holy Land ; that Edward, king of England, had crossed the sea with an imposing force ; that he had ordered a crusade

---

chæological Journal, vii. 48-49. Article by T. Hudson Turner.
† Mosheim, op. cit., 79. ‡ Remusat, op, cit., 382.

to be preached throughout Christendom against the Mussulmans, and that he was convinced if Arghun would assist that the venture would come to a fortunate issue. *He again urged him to be baptised*, and to exert himself for the recovery of Palestine as his own royal prudence judged best. The Pope also sent by the two friars a letter for Uruk Khatun, who was great grand-daughter of Wang Khan, of the Keraits, who, according to Haithon, was a Christian, and had a chaplain and a chapel. Her son Karbaganda (*i.e.*, Kharbanda) was baptised under the name of Nicholas, but on his mother's death became a Mussulman, and afterwards mounted the throne as Sultan Uljaitu. The Pope expressed his gratification at hearing she was a Christian, and urged her to induce Arghun's sons Saron* and Kassian (*i.e.*, Ghazan), to whom he wrote directly, to embrace the faith. He also commended the two friars to her. In his letter to Ghazan, which is dated the 23rd of August, 1291, he also exhorted him to become a Christian, expounded to him the principal articles of the faith, thanked him for his kindness to the Christians, of which he prayed the continuance, and introduced the friars to his notice. A similar letter was addressed to Toghachar, the general, and one to Isolus, already named. In writing to Nicholas, the son of Arghun, who was already a Christian, he bade him fulfil zealously the demands of the faith, to change nothing of his mode of living, costume, or food, for fear his people should be estranged from him, but to keep up the same customs he had before his baptism. The Pope also expounded to him at length the dogmas of the faith, and recommended the bearers of his letter to his regard.†

Remusat has some apt remarks on this correspondence of the Pope with the Christians. The supreme Pontiff was no doubt very anxious to obtain a set-off against the losses Christianity had recently sustained in the East by the victories of the Mussulmans, and such a set-off would certainly have been the conversion of the Mongols. But these nomades were for the most part indifferent both to Islam and the religion of Christ. If they had openly embraced either faith earlier, they might probably have prolonged their dominion in Persia considerably, instead of being evicted after a short supremacy there, without leaving scarcely any traces of their nationality. Far different was the fate of the Turks, who became zealous Mussulmans, and retained, with varying vicissitudes, the real control of the affairs of Asia for many generations.

Arghun was on terms of friendship with Christians nearer home. Stephen the Orpelian, the historian of Siounia, whose work we have often quoted, was one of the sons of Darsaij, the atabeg of Armenia.‡ Stephen became a priest, and in 1285 was consecrated metropolitan of Siunia by Ter Hakob, at the Court of Leon III., King of Cilicia.§ He returned

---

* No such name occurs in Rashid ud din.   † D'Ohsson, iv. 77-81.
‡ Ante, 345-346.   § Hist. de la Siounie, 238 and 264.

home in 1287, and tells us how he repaired to Arghun, who treated him with great kindness and honour. He showed him his encyclical, which the Ilkhan caused to be translated and read, and also the documents by which Darsaij had made over to him for his sustenance the monasteries of Tathev, Noravank, Tsaghats-Kar, Aratés, &c. Arghun confirmed him in his authority, both spiritual and temporal, and ordered him to remain with him, and to consecrate in his palace the church (probably a movable tent) which had been sent to him by the Pope. He tells us he consecrated this church with great pomp, with the assistance of the Catholicos Nestor and twelve bishops. Arghun with his own hands dressed him in his pontifical ornaments. Holding the wooden rattle which did office for a bell, the Ilkhan himself made the round of the camp, and compelled everyone to receive benediction. Another bishop, sent by the same pope some time after, baptised Arghun's young son, whom he christened Theodosios, in the Mongol language Kharbanda (the name is written Khudabandah on his coins), and put him under the care of a Frankish prince named Sirchol, whom father Chanazarien has identified with Buscarel de Gisulf, already named.*

Arghun's coins occur both in silver and copper. Those in the British Museum were struck at Baghdad, Tebriz, Mardin, Mosul, Arrajan (?), Kashar, and Jeziret. Two in that collection are posthumous coins, and were struck in the year 691. Such posthumous coins have already been mentioned in the history of the Golden Horde, &c.† It is curious to find Arghun's coins struck at Baghdad with the orthodox Sunni symbol upon them, and then to find on a coin which Frachn with great probability suggests was struck in Georgia, the sentence: "In the name of the Father, of the Son, and of the Holy Ghost, the one God."‡ The Mongol inscription on some of Arghun's coins is like that on the coins of Abaka already mentioned.§ Since I wrote the latter description I have met with an interesting passage by Schiefner, showing that the reading there criticised of *darugha* is not in fact maintainable as Burnouf and Poole have both argued, the word so read by Schmidt really reading *arebri*, *arebari*, or *arebchi*.|| Schiefner views in this word, which he reads *arebchi*, a corruption and contraction of the Tibetan *rab-mdzes-rin-tch'en-rdordge*, pronounced *rab dze rin-tch'en dordge*, being the translation of the Sanscrit *sundararadaratnavag'ra* (very beautiful, precious diamond).¶

Note 1.—I have made more than one reference in the previous chapter to Abulfaraj which ought to have been to the continuator of that chronicler. Abulfaraj, whose full name was Almufrian Mar Gregorius Abulfaraj Binalhakim Harun Almalaty, was the son of a doctor named Aaron, or Harun, and was born at Malatia, in 1226. At the age of 20 he became bishop of Gobos, next year bishop of Lacabene, and later of Aleppo, and was finally, in 1264, elected maphrian or primate of the Jacobites, a dignitary intermediate between that of patriarch and metropolitan. He wrote an abridgment of universal history, in Syriac, which was translated into Latin by Bruns and Kirsch, in 1789. The part relating to the Mongols is partially based on Juveni, and partly on his own observations. He also translated his history into Arabic, or rather re-wrote it in Arabic, under the title, " Tarikh oktasir ud Dual," which was translated by Pocock in 1663. This history closed in 1285 or 1286, but was continued in the Syriac version by an anonymous author, who carried it down to 1297, and whose notices of Arghun's two successors are very valuable. It will be understood that all references to this author dated after 1286, ought to be to the continuator.

Note 2.—-The word *yarligh*, which I have used more than once in this work, is so written in Arabic, Persian, and Turkish, while in Syriac it is written *yarlik*. It is derived from the Mongol word *yarlıkh*, a law, decree, or ordinance, from *yar*, meaning law.* It came to be used definitely, however, for an order or patent emanating directly from the sovereign. The missionary, Ricold de Mont Croix, speaking of these orders, describes one peculiarity which Quatremere says is quite exact. " The Tartars," he says, "honour their rulers so much that they do not insert their names with the other words, but leave a blank, and insert the name in the margin."† The yarlighs or patents were generally accompanied by metal tablets, called paizahs, which I have described in vol. i. pp. 271 and 530.

CHAPTER VII.

GAIKHATU KHAN AND BAIDU KHAN.

GAIKHATU KHAN.

THE name of this Ilkhan has been corrupted by most authors. The best reading is that on his coins, which is Gaikhatu, which in Mongol means "the surprising, the admirable."[*] The "Shajrat ul Atrak," quoting the "Ulusi Arba," or history of the Four Ulusses (*i.e.*, Ulugh Beg's history), calls him Kenjatu, and says that he was originally called Unkatu, which in Mongol meant "astonishing," or "wonderful."[†] Abulfeda calls him Kanachtu[‡]—Marco Polo, Kiacatu.[§] Haithon calls him Regayto, which is clearly a corruption of Kaikhatu.[||] Stephen Orpelian names him Kégathoi, and the Georgian history Kéghato and Keghtu.[¶] St. Martin calls him Kantchatu, while Klaproth calls him Khultho Khan.[**] At his accession, we are told, the bakhshis conferred on him the name Irinchin Durji, or Arinjin Turji, which occurs on his coins and also on his paper money. Burnouf explains this name by two Tibetan words, *rintchhen rdô-rdjé*, meaning "precious diamond."[††] This is doubtless the correct explanation of it. Wassaf says it is Chinese, while Von Hammer explains it as old Turkish.[‡‡] Khuandemir says the name was conferred, not by the bakhshis, but by the Emperor of China.[§§]

Wakhtang, son of Narin David, king of Karthli, died soon after Gaikhatu's accession, viz., in 1292, and was buried at Gélath, in the royal sepulchre, whither he was conveyed at the desire of David, son of Dimitri, the other ruler of Georgia. His father took his death much to heart, and himself died the following year. Narin David left three sons, Constantine, Michael and Alexander, whose mother was a daughter of the Emperor of Byzantium. We are told that Constantine and Michael had a lifelong feud, and that during their lives Imeuthi was in a state of confusion.[||||] On the death of Arghun, the generals Singtur, Toghachar, and Bekta nominated governors for the various provinces, in order to secure order during the interregnum, but anarchy was not averted.[¶¶]

We must now make a short digression to bring up the history of "Great Luristan" to this point. We have seen how Shems ud din Alp Arghun

---

* D'Ohsson, iv. 82. † Op. cit., ed. Miles, 270. Ilkhans, i. 394, &c.
‡ Op. cit., v. 101. § Op. cit., i. 35. | Op. cit., ch. xxxix. ¶ Op. cit., 610-611.
** See St. Martin, Memoires, &c., ii. 298. Note 8.
†† Journ. Asiat., 3rd ser., xiii. 131-132. ‡‡ Ilkhans, i. 405. Note 2.
§§ Journ. Asiat., 3rd ser., xiii. 124. || Hist. de la Géorgie, i. 610-611. ¶¶ D'Ohsson, iv. 61-62.

secured the throne there.* Hamdullah tells us when he arrived in Luristan he found the country ruined and the peasants scattered. By his good arrangements he soon restored it to prosperity. After the manner of the Arabs and Mongols, he shifted to and fro between his summer quarters on the skirts of the mountains, and his winter ones near Shuster. After reigning for fifteen years, Shems ud din died, and left two sons, Yusuf Shah and Imal ud din Pehluvan, apparently also called Nejm ud din.† The former succeeded his father, but lived at Abaka's Court, governing his own country by deputy. He took part in the campaign against Borak, and for this, and for saving Abaka's life when he was attacked by a body of marauders from Dilem, he was rewarded by the Ilkhan.‡ On Abaka's death, and during Ahmed's struggle with Arghun, Yusuf Shah was summoned to his aid by the latter, and took with him 2,000 horsemen and 10,000 foot. On Ahmed's defeat the Lurs set out by the desert road of Tabas, to try and reach Luristan as speedily as possible, but a great number of them perished there of thirst. Arghun sent to summon Yusuf Shah, who duly went to the Court and married the daughter of the vizier, Shems ud din. He died in the year 680 HEJ, and left two sons, Afrasiab and Ahmed. Afrasiab succeeded him as ruler of Luristan, and sent his brother to live at the Ilkhan's Court. Hamdullah says he greatly persecuted the viziers (of Lur), the khojas Nisam ud din, Jelal ud din, and Sadr ud din, in whose family the viziership had frequently been since the days of Hazarasp. He brought down their beneficent houses, and when some of their relatives fled to Ispahan he sent Kizil after them, who was his father's uncle.§ Afrasiab had, during Arghun's reign, in spite of the remonstrance of the governor of Shiraz, attacked the mountain district of Kiluyeh, which was the frontier between Fars and Lur, had captured the fortress of Manjasht, and put his nephew Kizil over the newly-conquered district, while Kizil's eleven brothers were put at the head of as many bodies of troops. Presently, a dispute arose between the uncle and nephews in regard to the administration of the district. Kizil was defeated and fled to Shiraz, but presently returned and made peace with his, uncle, who conciliated him by executing his vizier, Jelal ud din, and paying a handsome sum to him. At Kizil's instigation Afrasiab, during the troublous time which preceded Arghun's death, made a raid into Irak. He put to death at Ispahan the deputy of Jelal ud din, and also the Mongol governor, Baidu, the brother-in-law of Toghachar; and Kizil's brother Salghurshah, with a following of turbulent Lurs, took possession of the city. Salghurshah mounted the throne in the house of the Khoja Bahia ud din, and money was struck at Shiraz in the name of Afrasiab of Lur. His authority at this time extended from Hamadan to the Sea of Fars, over which he

---

* Ante, 140, where I have inadvertently given his name as Alp Arslan.
† Hamdullah *passim.* ‡ Ante, 240. § Hamdullah *passim.*

distributed his countrymen as governors. He now sent a body of 2,000 cavalry, under Jelal ud din, the son of the atabeg Tekele, of the Lesser Luristan, and Malik Nasret, against the tuman or division of the Mongol general Arghasun. In the struggle which ensued they were at first successful, and captured considerable booty, but the Mongols turned upon them and defeated them badly, practising, doubtless, the Fabian policy for which they were famous. In this struggle we are told that a Mongol heroine herself killed ten Lurs.[*] The council of regency now sent Doladai Aidaji against him with a tuman of soldiers, and ordered the Mongol and Mussulman troops in the governments of Ispahan and Shiraz to reinforce him. On his approach, Afrasiab's deputy at Ispahan fled. The town of Yezd, which had revolted under the atabeg Yusuf Shah, was sacked by the Mongols. Afrasiab himself sought shelter in the fortress of Manjasht, and Luristan was ravaged. Hamdullah says he was like a gnat tossed by a strong wind. He presently sent in his submission and was taken before Gaikhatu, who, on the intervention of Uruk Khatun and Padishah Khatun, pardoned him and restored him to Luristan, whither he went with his brother Ahmed, and put to death Kizil Salghurshah, and a number of his relatives and of the great nobles, and his licence was unrestrained.[†]

To revert to the Mongols. Five days after Arghun's death the generals sent the news to Ghazan in Khorasan. The next day Baitan was sent to Baidu, at Baghdad, and Lékézi, or Legsi, to Gaikhatu, in Rum. Lékézi was told to offer the throne to Gaikhatu, and to press him to go and occupy it, but after his departure several of the generals repented what they had done. Toghachar urged that Gaikhatu would prefer to surround himself with his present officers ; nor did he favour Ghazan, whose vigorous and severe disposition he dreaded, but he declared for Baidu, the son of Tarakai and grandson of Khulagu. The various commanders of the left wing, viz., Singtur, Shamagar or Samaghar, Tuladai or Doladai, Bekta or Tekne, Ilchidai, Kunjukbal, Tughan, Timur buka, Toghdai, and Tugal, sided with him, and they sent an officer named Balizad to inform Gaikhatu of Baidu's threatened elevation to the throne. Gaikhatu put this messenger to the torture, forced him to disclose the names of the authors of the scheme, and sent a division under the orders of Baitmish Kushji towards the royal residence. Meanwhile, the generals had sent to summon Baidu, and to urge that the throne belonged to him by virtue of seniority. He was timid and prudent, and said that according to the yasa of Jingis Khan the throne belonged to the son or brother of the last sovereign, and he preferred to decline the honour. What could he say, was his reply, to his good ancestors, who had made of the State a bridge of gold, and had decided which member of the family should pass over

---

[*] Hamdullah. Ilkhans, i. 401-402.   [†] Hamdullah *passim*.

first. He sent Gaikhatu the act of submission which the generals had forwarded to him, and advanced slowly towards Kurban Shira, whence he took the road to Kuit bulak (*i.e.*, cold spring). The various generals who went to do homage to him there, having learned his determination, withdrew disconcerted, having earned the resentment of Gaikhatu. Tugan, the most zealous partisan of Baidu, fled to Ghilan, but he was arrested and taken to the Court. Toghachar was arrested by order of the noyan Siktur. Kunjukbal withdrew towards Alatagh, and Tugal towards the Georgian frontier.

When news arrived that Gaikhatu was *en route*, the Prince Sugai, or Suka, son of Yashmut, son of Khulagu, the generals Choban and Kurumishi, the son of Alinak, and the ordus of the khatuns set out on the 23rd of May, by way of Alatagh, to meet him. They were followed by a chief of 4,000 men, and furtively by other leaders. Presently, the eyu oghlans (body guards) and the other generals went the same way. This movement in his favour was the work of Uruk Khatun, the widow of Arghun, and niece of Dokuz Khatun, the wife of Khulagu. She was therefore a Kerait, and doubtless a Christian.* As Gaikhatu neared Alatagh he was met by Khatai Oghul, the youngest son of Arghun, and by other princes. He was enthroned on the 22nd of July, 1291, in a place near Akhlat, where the various princes, khatuns, and generals had assembled. Soon after, the officers who had exercised authority during the regency were arrested. They were sharply questioned about the circumstances of Arghun's death, and that of his Jewish minister and other supporters. Gaikhatu was present at the first sitting of the court, and ordered Singtur, who was the highest in rank of the conspirators, to give an explanation of their conduct. "The generals are present," said Singtur, "for the Ilkhan to interrogate them. He knows my faults and those of the rest." The generals all agreed that Toghachar and Kunjukbal were the real authors of the plot, that they had induced Shamagar and Bekta to join them, and that these four had communicated their plan to Singtur, who had then offered to make common cause with them. "What could I do," said Singtur, "against so many powerful chiefs? If I had resisted, I should have met the same fate as Juchi and Ordu Kia." Gaikhatu admitted this excuse, and set him at liberty. The others having confessed their fault were also liberated, Toghachar and Kunjukbal receiving three strokes with a cudgel, according to the yasa. They eventually had their full of revenge, as we shall see. Toghachar's tuman, or division, was made over to Baiju Tetkaul (called Bighaul by D'Ohsson), that of Kunjukbal to the noyan Singtur, and Tugal's to Narin Ahmed. Tughan remained in prison, as the relatives of Juchi and Ordu Kia demanded his blood. Gaikhatu wished to pardon him also,

---

* D'Ohsson, iv. 61-66. Ilkhans, i. 396-398.

but Uruk Khatun urged strongly that such clemency would only encourage the guilty, and the Ilkhan said that Tughan in fact deserved to die, whereupon Akbuka and the Khatun Uruk went out and informed the sons of Ordu Kia, who speedily put him to death.* Gaikhatu returned to Alatagh on the 7th of August, and having received the homage of the chiefs he had pardoned, he assigned Khorasan to Anbarji, the son of Mangu Timur, and named the noyan Singtur as his lieutenant-general in the kingdom, with control both of civil and military affairs.

On the 1st of September he set out to repress an outbreak in Rum, where the Karamanian Turkomans had attacked the Mongol garrison. The continuation of Abulfaraj calls them the Augæian Turkomans.† The "History of Georgia" says he set out to suppress the revolt of the *Greek town* of Thonguzalo (called Tangezlu by the author just quoted). He was joined by David, the son of King Dimitri, with all the mthawars of Karthli, the latter having left his wife and home in charge of the noyan Elgoz. Gaikhatu posted Kutlugh Buka and the other Karthlian mthawars at Mughan, as he expected an attack from Kipchak, and having set out with David, as just mentioned, attacked Tonguzalo for four months. In the fifth month it was stormed, by David and his Georgians forcing their way in. In one corner of the place were the Christians, who implored the King to spare them and to plead for them with the Khan. The latter acceded and withdrew, after securing a kharaj and immense riches.‡ The continuation of Abulfaraj says he afterwards pursued and punished the Turkomans, who had fled into the country.§

The withdrawal from his capital so soon after his accession led to the forming of ambitious projects. Toghachar spread the rumour that Gaikhatu's army had been destroyed in Rum by the Karamanian Turkomans. This news was sent to Kutb ud din, brother of Sadr ud din Ahmed Khaledi, the latter of whom was in league with Toghachar. Kutb ud din was then in the employ of Anbarji, the new ruler of Khorasan, whom it was wished to put in Gaikhatu's place. He told the false news to Jemal, Sheikh of Shiraz, who was a confidante of Anbarji. The latter was too prudent and circumspect to enter upon such a dangerous venture without due inquiry. He sent the Sheikh on a friendly embassy to Singtur. On the way the Sheikh met Toghachar and Sadr ud din. They urged him to turn back with them, and to persuade Anbarji to advance at once. He expressed himself as willing to do so, but said he wished first to pay his home a visit. On this excuse he made his way to Singtur, to whom he disclosed what he had heard. Singtur sent him back with a friendly message and presents for Anbarji, and proceeded to arrest Toghachar and Sadr ud din, keeping them in durance until he heard of Gaikhatu's return in triumph, when he sent Toghachar, guarded by an escort of 500

---

* D'Ohsson, iv. 84-85. Von Hammer, Ilkhans, i. 398.   † Chron. Syr., 626.
‡ Op. cit., 611.   § Chron. Syr., 627.

men (D'Ohsson says 2,000) to Erzerum. Gaikhatu once more pardoned him and his accomplices.*

In reward for his services at the capture of Tonguzalo, Gaikhatu conferred on David the throne of Karthli, which had been filled by his father Dimitri, and had him conducted to Tiflis by Shahinshah, Kutlugh Buka, and all the didebuls. Beka was summoned from Samtzkhé. He refused to go, but sent his son Sargis, the general of Samtzkhé, with all the wealth which Dimitri had confided to him, including a very precious girdle. David was duly consecrated by the Catholicos Abraham. The Khan married him to his sister Oljaitai, the widow of Wakhtang. This is so said in the Georgian history, but Rashid apparently makes Oljaitai marry the noyan Tukal, and afterwards Kutlugh.†

Soon after his return to Alatagh, Gaikhatu was taken seriously ill. Ulemas and imaums, bishops and monks, as well as the Jewish rabbi, were summoned and ordered to pour out prayers for the Ilkhan's recovery. Meanwhile, abundant alms were also distributed.‡ On his restoration to health he was solemnly enthroned with the usual ceremonies. The installation had been postponed, because the astrologers declared the stars to be unpropitious. The festivities lasted a month. The treasury chests, which had been filled at the cost of much blood during Arghun's reign, were lavishly emptied. Gaikhatu distributed to the khatuns and princesses the precious jewels brought together by the Khans his predecessors, saying such things were only fit for women to wear. He ordered the prisons to be opened, alms to be dispensed, and exempted from all taxes the ulemas, the descendants of the prophet, and learned men.§

In the midst of these festivities news arrived that the fortress of Kalat ur Rum, situated on the right bank of the Euphrates, a little distance north of Biret, had been captured on the 29th of June by the Egyptians, commanded by Sultan Ashraf Khalil in person. Kelavun, his predecessor, had pursued with ardour the project of Bibars for expelling the Franks from Syria, and had captured Markab, the chief stronghold of the Hospitallers, and also Laodicea, and had demolished the flourishing town of Tripoli. He died at Cairo on the 10th of November, 1290, at the age of 68, when he was setting out to attack Acre, the strongest of the Christian settlements. His eldest son Salih was dead, and he was succeeded by his second son, Malik al Ashraf Salah ud din Khalil, who, in pursuit of his father's intentions, captured and sacked Acre in 1291, and followed this up by occupying Tyre, Sidon, Tortosa, Haifu, Athlith, and Beyrout, the last remaining possessions of the Crusaders in the East. The next year a holy war against the Mongols was preached three times by the Khalif at Cairo, and also in Syria, while at the head of the armies

* Ilkhans, i. 399-400. D'Ohsson, iv. 85-86. Abulfaraj, Chron. Syr., 627.
† Ilkhans, Genealogical Table.   ‡ D'Ohsson, iv. 86.   § D'Ohsson, iv. 86-87.

of Egypt and Syria the Sultan attacked Kalat ur Rum. The walls were bombarded with twenty catapults, and mines were also sprung. The amir Sanjar, naib of Damascus, fastened one end of a chain to the battlements of the citadel, while the other was firmly fixed in the ground, and his men clambered on to the walls by its means. The place was captured, after an attack of thirty-three days, on the 29th of June, 1292. The garrison, consisting of Mongols and Armenians, was put to the sword, and 200 men, with the women and children, were reduced to slavery. From the year 1268 Kalat ur Rum had been the seat of the Armenian patriarch. The patriarch's palace and church were burnt, and he himself, named Stephen IV., was taken with his monks to Jerusalem. According to some, says the continuation of Abulfaraj, he was crucified, and his companions were taken off in chains to Egypt. When the Armenians heard of this they elected a fresh Catholicos, whose seat was at Sis, in Cilicia, or Little Armenia. The name of Kalat ur Rum (*i.e.*, Castle of the Romans) was changed to Kalat ul Muslimin (*i.e.*, Castle of the Mussulmans). The news of its capture was received with great rejoicings at Damascus. The amir Sanjar was appointed naib of Syria, with the duty of restoring the fortress, one-fourth of which was in ruins. Novairi tells us that in the bulletin of victory published at Damascus there occurs the following sentence : "The way is open to us after this victory, if God pleases, to conquer the East (*i.e.*, Persia), Rum, and Irak, and to take possession of the land from the furthest east to the furthest west." Gaikhatu had sent some troops to the relief of Kalat ur Rum, but they arrived after its fall.[*]

Some months later an envoy arrived in Egypt from Gaikhatu, to say that he proposed to fix his residence at Aleppo, which had been conquered by his ancestor Khulagu, and threatened that if he were not allowed to do so he would occupy all Syria. The Sultan replied : "The Khan has similar intentions to my own. I intend to take Baghdad, and to put the garrison to the sword, for I hope to restore to its former eminence the capital of Islam. We will see which of us will first invade his enemy's land." Orders were sent to Syria to review the troops, and to prepare provisions.[†]

Let us now turn to Kerman. We saw how, in 1252, Rokn ud din, the ruler of that province, was displaced by his cousin Kutb ud din, by order of Mangu Khan.[‡] After four months, according to Hamdullah, Kutb ud din married Kutlugh Turkhan, who had been one of the women of Borak the Hajib, the founder of the dynasty, by whom he had many daughters. Rokn ud din set out to win the Khalif over to his side, but hearing that the latter's people, fearing the resentment of the Mongols, were unwilling to give him an asylum, he determined to set out for Karakorum. His

* D'Ohsson, iv. 87-89. Weil, iv. 184. Makrizi. ii. 141. Chron. Syr., 626-628.
† Makrizi, ii. 150. ‡ Ante, 36.

cousin followed him there, and the two competitors pleaded their cause before Mangu, who decided in favour of Kutb ud din. He ordered that his cousin should be given up to him. Kutb ud din put Rokn ud din to death with his own hand in the year 651 HEJ, and returned to Kerman. At this time a man who resembled the last Khuarezm Shah, Jelal ud din, in figure, and was well informed in his affairs, gathered a number of people round him and rebelled. Kutb ud din marched against him and dispersed his people, after which he set out to punish the robbers of " Luch and Baluch," whose bands boldly paraded the highways with drum and flag. Kutb ud din surprised them at night, and slaughtered them. He reigned for six years, and spent his time, says Hamdullah, in justice and equity, and building lofty buildings. He died in Ramazan, 655 (*i.e.*, 1257). His son, Sultan Hajaj, received the investiture of the country from Mangu. As he was only an infant, his father's widow, Kutlugh Turkhan, was appointed regent, and exercised this authority for fifteen years. When Sultan Hajaj took the reins of power there arose a violent strife between him and Kutlugh Turkhan, who accused him, *inter alia*, of so forgetting the reverence due to her as on one occasion, at a feast, to have led her out in a dance. She went to implore Abaka's protection. Abaka had married her daughter, Padishah Khatun, and he gave her absolute authority in Kerman. Sultan Hajaj during her absence, we are told, had sought aid from the sons of Ogotai Khakan, of which she heard on her return. He now, in 669 HEJ., fled to Delhi, whence he returned ten years later with an army furnished him by the Sultan Jalal ud din Khalj, but died before reaching Kerman. This was in 1270. Kutlugh Turkhan reigned in peace for twelve years longer, when Jelal ud din Siyurghatmish, the second son of Kutb ud din, having done homage to the Ilkhan Ahmed, obtained from him, through the influence of his mother, Kutui Khatun, and of the noyan Sughunjak, the investiture of Kerman. On his arrival there, in 1282, Kutlugh Turkhan left and set out for the ordu. There she was supported by the khatuns, the amirs, and by the vizier Shems ud din Muhammed, who wished that she should be associated with Siyurghatmish. The partisans of the latter suggested that their patron, discontented with this arrangement, would join Arghun in Khorasan, and urged that the decision should be postponed until Siyurghatmish himself came to the Court to settle his affairs, when the matter could be arranged. Meanwhile Kutlugh Turkhan died at Tebriz, after having reigned twenty-five years in Kerman with great prudence and sagacity, and her daughter, Bibi Turkhan, conveyed her body back to Kerman. When Arghun mounted the throne he summoned Siyurghatmish to the Court, no doubt to be tried as a partisan of Ahmed's. He owed his safety to the good offices of Buka, who confirmed him in his government, while the revenues of Kerman were farmed to him for the sum of 600,000 dinars; 19,000 were devoted to the expenses of the administration, the rest being assigned to Siyurghatmish for the expenses of his Court. He married Kurdujin, the daughter of the atabegin Abish, wife of Mangu Timur, son of Khulagu, who was herself the last of the house of the Salghurids, rulers of Fars. His sister, Padishah Khatun, was married to Abaka, as we have seen, so that he was very closely connected with the Mongol Imperial house. On Abaka's death Padishah Khatun had joined the harem of Gaikhatu, who apparently gave her some authority in Kerman, but her brother's vizier, Fakhr ul Malik Muhammed, who was the real ruler there, refused to admit her claims, and said, " If the Sultanate come to thee, thou mayest cut me in twain with a butcher's knife." This aroused her rancour. Gaikhatu now (*i.e.*, in 691 HEJ.) gave his wife the Sultanate of Kerman, while the late Vizier fled to India. She inveigled him back by fair promises, and then put him to death. She was witty, and wrote verses, among others the following :—

> Was ever a fleck of musk seen on a ruby (lip)?
> Or did ever civet taint the golden honey?
> My soul, that black mole on thy lip
> Is like the meeting of darkness with life's fountain.

She made her brother her vicegerent, but when she discovered that he was plotting to recover the throne she imprisoned him. His wife, Khudavand Zadeh Kurdujin and his daughter Shah Alam Khatun, sent him a coil of rope in a water skin into the castle, by means of which he escaped and went to the camp; but he was speedily seized and put to death, by his sister's orders, in 694 HEJ.*

We must now turn again to the affairs of Khorasan. We described the raid made by Nuruz, and the retreat of Ghazan to Damaghan, where he heard of his father's death. The citizens had left the town on his approach. Some fled to Girdkuh and others to the strong fort of Dih Muyan. He could therefore obtain neither provisions nor shelter there. He sent word to those who had fled to Dih Muyan to return, and as they refused he sent to attack them. They submitted after three days, and sent him plenty of provisions. The fort was razed, and only rebuilt again after Ghazan's accession. On Arghun's death the customary Mongol mourning was ordered by Ghazan, a feature of which was that all ornaments and feathers were discarded from their caps. Ghazan was now joined by the amir Mulai, who had come from Kuhistan across the desert to meet him. He was gladly welcomed, and given the sister of the amir Satilmish in marriage.

We have seen how, when Ghazan reached Tus on his retreat, several of his amirs asked for furlough and left him. One of these, Aighurtai, went over to the enemy's cause, and we are told that he set out from Sultan Meidan, traversed Kabushane and Jorjan, and went as far as Devin and Asterabad, proclaiming Kaidu as the supreme ruler of the land. As

---

* Hamdullah *passim.*

none of the great amirs were there, some of lower rank, named Saighan, Abaji, Mamluk, &c., undertook to drive him away. Ghazan now moved from Semnan to Firuzkiuh, and to the borders of Demavend, as far as Menshan. There the Princess Bulughan was confined, and died in childbirth.

When Gaikhatu mounted the throne, Ghazan sent the amir Kutlugh with his submission, and apparently also to inform him of the deplorable condition of Khorasan and to ask for help. The summer of 1291 he passed at Ezran, also called Nekatuilak, which was situated between Firuzkiuh and Semnan, where he passed the time in hunting and feasting; and in the autumn he went by way of Damaghan and Bostam to Kialpush, where Nisam ud din Yahya Kutlugh Khoja was called to account for various malpractices, and was imprisoned at Andemed. Ghazan then went on to Sultan Devin, near Asterabad, where he was joined by the Prince Anbarji, or Enbarji (a grandson of Khulagu, by his son Mangu Timur), and by the amirs Doladai, Kunjukbal, and Iltimur, who had been sent to his aid by Gaikhatu. They fixed their winter quarters at Karatughan.* The winter was not over when news arrived that Nuruz was approaching with the intention of releasing Nisam ud din Yahya. He contented himself, however, with plundering Juven. In the spring, Prince Anbarji was ordered to march by way of Dahistan, Yasu, Nissa, and Abjurd. There was a famine in the land, corn became exceedingly dear, and the army had to have recourse to hunting. Ghazan pitched his camp for a while at Jukjuran, on the river of Herat, but retired thence to Badghiz on hearing that the enemy was about. The pressure of famine became so great that the troops began to kill their horses. He therefore returned to Herat, and encamped near the bridge of Pulmalan. Shems ud din Kert, who had imprisoned his elder son, Fakhr ud din, in the strong fort of Hissar, sent the younger one, Alai ud din, to express his submission. Ghazan was very considerate with the people of Herat, who had suffered much during the recent famine, but he ordered the town of Fushenj, which had not furnished the provisions he had demanded for the troops of Irak, to be attacked. On its capture there was an overflow of booty and prisoners, but the women and children were presently released. Gaikhatu left the army of Khorasan without pay. He wished to go there in person, but afterwards changed his mind. When Prince Anbarji returned home with the troops of Irak and Azerbaijan, Ghazan went to the summer quarters of Shutur-kiuh, and built the kiosk Murad. A mob having killed the sons of Malik Susen and several other grandees, at the village of Jizerd, near Khavas, the amirs Sutai and Mulai were sent against them. They were met by Shah Ali, the son of the ruler of Seistan, who had surrounded

---

* Ilkhans, ii. 17-18.

Khavas. Ghazan's troops fell upon him, and cut his army in pieces. We next read of the execution of Amad ud din, a preacher, of Nishapur, who had declared for Nuruz in the recent troubles. The winter of 1292 was passed at Devin and Asterabad, and in the spring the army moved towards Jorjan and the towns of Shereknev and Murjabad, and it was further decided to march through Azerbaijan. At Temish the amir Kutlugh was married to the daughter of Jinghutai. From this place to Shuril, a distance of thirty parasangs, was traversed in one night. Ghazan dispatched the amirs Satilmish and Khoja Said ud din to bring the revenues of Khorasan, Mazanderan, Kumis, and Rai, from Demavend. He also ordered provisions to be collected, and then set out for Tebriz, but at Abher he received orders from Gaikhatu, who was probably suspicious of his intentions, to return at once to his government. He took no notice of this, or of a second message, and it was only on receipt of a third, which reached him at Tebriz, that he withdrew, sending Gaikhatu word that he wished to see his uncle a hundred times more eagerly than his uncle wished to see him. He went to Yusagaj, where he married Eshel, or Ishil Khatun, the daughter of the amir Toka Timur, and a month later set out for Khorasan.*

We have seen how the Poli were sent from China in charge of a princess for Arghun, and how when they arrived in Persia they found him dead. This news they apparently learnt on their arrival at Hormuz. Having sent news of their arrival to Gaikhatu, he ordered them to conduct the lady to Ghazan, who was then living in the region of the Arbre sec (*i.e.*, of Khorasan), guarding the frontier with 60,000 troops. They did so, and then went to Tebriz, to Gaikhatu, and stayed at his Court for nine months.† Von Hammer, apparently quoting Rashid ud din, tells us that it was while at Abher that Ghazan received the lady Kukachin from the envoys he had sent to the Great Khan. The marriage festivities were celebrated with great splendour. She did not long survive, and died in June, 1296.‡ When the Poli left Gaikhatu's Court on their journey westward, the Khan gave them four golden tablets or paizahs, two of them bearing gerfalcons, one bearing a lion, and another plain, with inscriptions stating that the envoys should receive the same honour and service as was rendered to the prince in person, and that horses, provisions, &c., should be supplied them. And so it came about, as Marco Polo says, they frequently had as many as 200 horsemen to escort them. "And this was all the more needful," says our author, "for Gaikhatu was not the legitimate lord, and therefore the people had less scruple to do mischief than if they had had a lawful prince."§ The Poli now hastened on to Trebizond, Constantinople, Negroponti, and Venice.||

‘ Ilkhans, ii. 19-20.   † Yule's Marco Polo, i. 38 Note 6.   ‡ *Id.* Ilkhans, ii. 20.
§ Op. cit., i. 35. || *Id.*, i. 36.

Gaikhatu appointed Akbuka to be generalissimo of the forces, and nominated Singtur and Toghachar as his lieutenants. He confided his private domain to the charge of two of his favourite officers, Hassan and Taichu, while the post of vizier was conferred on Sadr ud din, of Zenjan, deputy of Toghachar, and a co-conspirator with him, as we have seen. Enriched by the spoils of the Mongol grandees who had perished at the close of the previous reign, he distributed large presents to those who could assist him, and had gained the patronage of Akbuka. A list of persons well fitted to fill the post of vizier had been prepared and presented to Gaikhatu. In this list the name of Sadr ud din Ahmed el Khalidi did not appear. Gaikhatu noticed this, and said that he knew no one more worthy than he. Thereupon the khatuns and grandees sang his praises. Von Hammer says he owed his promotion to the influence of Gaikhatu's favourite, Buzaljin Ikaji, and to Sherif ud din, of Semnan, who had great influence with Akbuka. He was duly installed as vizier, or sahib divan, on the 19th of November, 1292. He changed his name to Sadr Jihan, or president of the world. He was given a golden seal (*al tamgha*), with a lion's head on it, a tuk or horsetail standard, a drum, and command of a tuman of soldiers ; while it was forbidden to the amirs, khatuns, or princes of the blood to interfere in any way with financial matters. The dignity of Kadhi ul Kudhat, or chief judge, was conferred on his brother, Kutb ud din Ahmed. He had control over everything relating to Muhammedanism—the administration of the wakhfs and the charitable institutions. He took the title of Kutbi Jihan, or pole of the world. His uncle, Kavam ud din, was appointed governor of Tebriz, with the style of Kavam ul Mulk (*i.e.*, the strength of the kingdom). Fakhr ud din Aidaji was nominated to look after the supply of food and the provisioning of the army. He begged to be excused, urging that he had already had this duty for thirty years, and had got into debt in consequence of the demands of the princes and princesses. Gaikhatu ordered three tumans to be devoted to paying off his debts, and afterwards treated him with the greatest liberality. Gaikhatu's aim was to emulate the mild administration of Ogotai Khan.*

He was extremely lavish in his expenditure." We are told he sometimes made presents to the khatuns to the amount of thirty tumans at a time. When he received presents from other princes, or from his great vassals, he made them over at once to his khatuns or the young princesses without looking at them, or he distributed them to his officers. He devoted himself to debauchery and lasciviousness ; wine, women, and boys were his main delight. He abused without shame or restraint the sons and daughters of his grandees. Many women fled away to avoid his lust, while others sent their boys and girls to distant places in order

---

* D'Ohsson, iv. 95-97. Ilkhans, i. 400-401.

to escape him.* He abandoned the direction of affairs to the vizier, Sadr Jihan, whose authority became supreme, and who decided as he pleased, displacing the prefects and others according to his whim. He deprived Hassan and Taiju of the control of the private domain of the Ilkhan (*inju*), which he united with the public domain (*délai*). These two officers, in concert with Daulat Shah and some of the notables of Tebriz, made an attempt to destroy Gaikhatu's good opinion of his Vizier. In November, 1293, when he was in one of his hunting boxes, they urged upon him that Sadr Jihan devoted the public revenue to his own private expenditure, and neglected the troops and the proper provision for the khatuns, adding that the treasury was empty; and as an example they cited the case of Tebriz, where they declared the Vizier had appropriated to his own use thirty tumans of the eighty which the taxes of the province produced. Gaikhatu was unmoved by these charges, told Sadr Jihan what they were, ordered his accusers, with their wives and children, to be made over to him for punishment, and issued an order that anyone in future accusing the Vizier of malpractices should be put to death without trial. The accusers of the Vizier, having confessed their fault and demanded pardon, were forgiven. A new edict was issued, re-affirming that the administration of the kingdom, from the Jihun to the borders of Egypt, was made over to Sadr Jihan, that he was authorised to appoint whom he pleased to public offices, and that all the deputies of the khatuns and generals were subject to him. It was also forbidden the princes of the blood and the military chiefs to take anything from the revenue for their sustenance, or the pay or keep of their people.†
After the death of Arghun a disease called *yut* in Mongol had carried off the greater part of the cattle in the kingdom, especially in the districts of Baghdad, Mosul, Diarbekr, and Khorasan; meanwhile, the treasury was exhausted by the extravagance of Gaikhatu and his minister, the latter of whom, with the wish of conciliating everyone, distributed largess widely, especially to those devoted to religion. In the course of two years Sadr Jihan had to borrow about 500 tumans, the whole revenue of the kingdom being 1,600 tumans. Seven were required for the ordinary expenses of the government, and what remained was not sufficient to meet the prodigalities of the Ilkhan.‡ We are told by Wassaf that whereas the whole cost of the kitchen in the reigns of Abaka and Sultan Ahmed only amounted to 40 tumans, it now rose to 165 tumans. The continuation of Abulfaraj tells us that the needs of the treasury were so great that money was not forthcoming even to buy a sheep for Gaikhatu's table. A Jew, named Rashid ud daulat, was accordingly ordered to get what was needed for the Khan's table in whatever way he could. He applied himself diligently to this duty, and

---

* Chron. Syr., 628-629.　　　† D'Ohsson, iv. 97-99.
‡ *Id.*, 99-100. Extract from Wassaf, Ilkhans, i. 423-424.

A*

spent in it the greater part of his fortune, buying a large number of oxen and sheep, engaging cooks, &c. He had been promised that the money he advanced should be repaid him at the end of each month, but the treasury was empty, and the orders given him by the Vizier on the provinces were dishonoured, for there was no money there. The Jew having spent all his fortune, and unable to perform his commission any longer, fled.[*]

Amidst these difficulties a man named Iz ud din Mozaffer Ibn Muhammed Amid, who is referred to in terms of reproach and indignity by Wassaf, suggested to the Vizier that he had a plan which would restore the public prosperity and be without reproach. This was to put in circulation, in lieu of metallic money, a paper currency like the chao in China. The Vizier, pleased with this advice, repeated the proposition to Gaikhatu, who took counsel with Pulad Chingsang, the representative of the Khakan at his Court, and a plan was accordingly arranged for its issue. In vain the noyan Singtur, the most intelligent of the Mongol grandees, warned him of the dangers of the plan. Sadr ud din only suggested that Singtur had mercenary motives for his opposition, and in May, 1294, an order was issued for the creation of the chao. The 3rd of July following, the generals Akbuka, Toghachar, and Tamaji, and the Vizier, left for Tebriz, where the new money was to be made. On the sides of an oblong piece of paper were written several words in Chinese characters. At the top on both sides was the profession of the Muhammedan faith : " La illahi ill' Allahi Muhammedun rassul Ullahi " (There is no God but God, and Muhammed is his prophet) ; and below, " Irenchin Turji," or " Irichi Turichi," *i.e.*, the name which the bakhshis had given Gaikhatu on his accession.[†] In a circle in the midst of the paper was marked its value, the varieties being from half a dirhem to ten dinars, and then came the sentence : " The sovereign of the world has issued, in the year 693, this propitious chao. Whoever defaces it will be punished with death, with his wives and children, and his goods will be confiscated." A mint for the issue of the chao was founded in each province, which had its governor, scribes, cashiers, &c. The use of a metal currency was prohibited in all the kingdom, as was that of gold and silver for bowls, &c., and also for the manufacture of golden tissue, except what was needed for the Ilkhan's own wardrobe and those of his chief officers. The goldsmiths and others who were thus deprived of a living were given orders on the chao banks. These banks were ordered to exchange the old notes for new ones on payment of a discount of 10 per cent. The merchants of the Persian Sea who trafficked with foreign lands were alone to be allowed to exchange their notes for gold at the treasury, but orders were given to look after them closely.[‡]

---

* Chron. Syr., 631-632. D'Ohsson, 100. Note.　　† Vide supra, 283 and 355.
‡ D'Ohsson, iv. 102-103. Wassaf, Ilkhans, i. 428-429.

Gaikhatu had been told that when gold had been displaced by the chao there would be no more poor in the land, while provisions would be very cheap, and poets sang the praises of the paper money. The first issue of chao took place at Tebriz on the 12th of September, 1294, and it was accompanied by an edict declaring that whoever refused to accept it, whoever bought and sold for other money than chao, and whoever did not take his coin to the mint to be exchanged for paper money, was to be punished with death. This was shouted in the streets by criers.* The fear of punishment caused the order to be obeyed for eight days, but afterwards the shops and markets were deserted. Nothing was to be bought in the city, and people began to leave. The famished citizens rushed to the neighbouring gardens to get fruit. Gaikhatu one day traversing the bazaar, and noticing that the shops were empty, inquired the reason why. The Vizier said that a great magistrate was dead, and that it was customary with the citizens to leave the bazaars on such occasions. The authorities and troops had great difficulty in restraining the crowd. The Mussulmans met in their mosque on the Friday, and broke out into lamentations. Presently open murmurs were heard, and imprecations were flung at Iz ud din Mozaffer and the other authors of the innovation, and eventually attempts were made on the lives of the Vizier and his people. As in the famous panic at the time of the issuing of the French assignats, prices became quite arbitrary, and Wassaf tells us how the sellers of a horse not worth more than 7½ gold pieces asked 750 in paper. In the panic that ensued the Vizier's brother, Kutb ud din, was compelled to sanction the purchase of provisions with coin, and several were put to death for having taken part in the disturbances. The Vizier presently saw the ill effects of his experiment. An ordinance permitting the use of coin in buying provisions was issued, and coin appeared again in other commercial affairs. Finally, the chao itself was suppressed, amidst universal joy. For two months commercial dealings had virtually ceased, the shops were empty, the roads were deserted by traders, while wits and poets emulated each other in constructing lampoons and gibes at the expense of the paper money and its authors. Much money was wasted also in building the various mints, that at Shiraz having cost five golden tumans. In that town no one could sell a sheet of paper without the permission of the bank.† Prince Ghazan did not wish the chao to be introduced into his appanage, and when the official arrived with the paper and materials for its fabrication he sent word to the Ilkhan that the air in that part of the country, especially in Mazanderan, was so damp that arms and armour could not resist it for twelve months, while a piece of paper when used became as fragile as a

---

* Cont. of Abulfaraj, 632. D'Ohsson, iv. 104.
† D'Ohsson, iv. 106. Cont. of Abulfaraj, Chron. Syr., 632-633. Ilkhans, i. 432-433.

spider's web, and he ordered it all to be burnt. Von Hammer assigns this story to Oghul, and not to Ghazan.*

The dissipation and extravagance of Gaikhatu were wearying the whole nation, and opened a way for the ambition of Baidu, who had been a competitor for the throne with himself. The continuator of Abulfaraj tells us that in July of the year 1294 Baidu was at a banquet with Gaikhatu. They were eating, drinking, and laughing, when the latter, who was apparently drunk, said something insulting to his relative, who retorted, calling him a product of adultery. Gaikhatu, much enraged, ordered the attendants to take Baidu outside the ordu, and there to put him to death. Wassaf says he ordered one of them to hit him with his fist. They fell on him accordingly, and put him in a small tent with the intention of killing him, but Gaikhatu having awoke after a short sleep, relented, and sent his people to ask him how he had dared to say such a thing to the king of kings. Baidu professed to be still drunk, feigned not to understand what they said, and asked where Gaikhatu was. "Bring some wine," he said, "and let us drink. What has happened, and why am I here?" Gaikhatu was taken in by the artifice, and repented of the indignities he had put upon his relative, and determined to appease him. After he had slept for a while he sent some attendants to ask him if he knew what he had said when drunk. He said he did not, and that if anyone had struck him he was unaware of it, and pressed them to say whether they were joking or speaking seriously. They thereupon related what had happened. He appeared to be stupefied, and said, "Truly, Gaikhatu is most good to me, or he would have ordered me to be cut in pieces on the spot." When this was reported to the Ilkhan he was much touched, went to Baidu in person, took him to the ordu, and dressed him in royal robes. Wassaf says he took off his kulah or cap, and put it on Baidu's head. The latter declared himself a guilty criminal, cursed drunkenness, and said, " I am not conscious of having committed the offence. I did it unwittingly. I beg you to give my flesh to the dogs without pity." Gaikhatu was now more effusive than ever, and gave him in three or four days things of the value of forty tumans, in gold and silver, in dresses of golden tissue, precious stones, horses, and mules. For this he was reprimanded by his courtiers, who told him he ought not to have wounded Baidu's honour, nor illtreated him, nor handed him over to brutal people, who had pulled him by the hair, cuffed and wounded him ; but that having done this, neither his caresses nor his gifts were of any avail, and that he must be on his guard against him. Some suggested that he should put him to death, or he would do him harm ; others that he should be retained as a prisoner for the rest of his life in the ordu. Gaikhatu determined to

' D'Ohsson, iv. 106. Ilkhans, i. 405.

insist on Baidu leaving his son with him as a hostage. Having sent the boy, he withdrew to the mountains of Hamadan, as if on a hunting excursion, and sent to inform Ghazan of the indignities he had suffered.* Rashid ud din says Baidu's life was spared at the entreaty of Gaikhatu's nurse, Borakchin Igaji, who had considerable influence. In the spring following, Baidu returned to his winter quarters at Dakuka, nursing his revenge. He opened his grievances to several generals who had their quarters in the neighbourhood of Baghdad, such as Tudaju, the superior judge; Jijek Gurkan Legsi, the son of the famous administrator, Arghun; and Iltimur, the son of Hindukur Noyan, and made a conspiracy with them. Jemal ud din of Destajerd, the chief secretary of the treasury at Baghdad, went over to him, and thus secured him provision for his men. Having collected some troops, he marched upon Mosul, where he had the governor seized and put to death. His people also went to Baghdad, to put to death Muhammed Sikurji, who governed the town in the name of Gaikhatu, and the standard of revolt was openly unfurled.† The continuator of Bar Hebræus tells us emissaries were sent to inform the Khakan himself,‡ to tell him that Gaikhatu had abandoned the customs of the Mongols and contemned Jingis Khan, that he gave himself up to debauchery and extravagance, and that the grandees had therefore determined to put him away, and to raise him (i.e., Ghazan) to the throne. The latter replied, according to the continuation of Abulfaraj, in biblical phrases, telling Baidu that he was greatest in Israel, and that he was ready to submit to him, and to follow his counsel. He bade him do what seemed best to him to rescue the State from its condition of decrepitude. It behoved their future king to eschew a lascivious life, gluttony, drinking, &c., extravagance and inordinate gifts, and to protect the kingdom.§

Let us revert shortly to what had passed recently in Khorasan. We have seen how, after his unfortunate campaign against the rebellious general, Nuruz, Ghazan set out to pay Gaikhatu a visit, and how when he refused to see him he returned. When he reached Firuzkuh he learnt that his general and *locum tenens* during his absence, Kutlugh, had defeated Nuruz, and compelled him to seek shelter in the mountains of Nishapur. Kutlugh met him at Bostam with a considerable booty, and they went along the road from Huirmabehrud towards Jorjan, and encamped at Sultan Devin, near Asterabad. It is quaint to read that at this place the amir Kutlugh, having become ill through drinking to excess at a feast, took a pledge not to drink wine any more, and kept it. Kia Salah ud din, who had once before rebelled, submitted, and been forgiven, was again rebellious. He was defeated, and a considerable booty was divided among the army. Ghazan passed the spring and summer

* Cont. of Abulfaraj, Chron. Syr., 629-631. D'Ohsson, iv. 107-109. Note.
† Cont. of Abulfaraj, Chron. Syr., 633. Ilkhans, i. 405-406. D'Ohsson, iv. 107-110.
‡ Clearly a mistake for Ghazan, as D'Ohsson has read it.   § Op. cit., 634.

near Damaghan, Sultan Maidan, and Firuzkuh. The amirs Mulai and Hirkudak having reported that the people of Nishapur behaved haughtily, and took no heed of his messengers, he set out in October, 1294, for that town, and encamped at Moeyedi, close by. The Khoja Said ud din negotiated with the deputies of the town for the surrender of the peace-breakers, but as the townsfolk refused to give them up it was determined to attack the place. A large portion of the citizens sheltered in the mosque Migh, whose walls were undermined. They now implored mercy, which was granted them. Having put down the rebellion, Ghazan returned to Jorjan, and wintered at Sultan Devin.*

Meanwhile Nuruz had quarrelled with his patron, the Khakah Kaidu. After his defeat by Kutlugh Shah he had retired to Seistan, where he was supported by the so-called Nigudarians, or troops who had followed the rebellious Prince Nigudar. Thence he sent several expeditions into Khorasan. At the end of 1294, discontented with Kaidu, he agreed with Prince Uzbeg Timur (called Uruk Timur by Von Hammer), who was his brother-in-law, to drive Kaidu's troops beyond the Oxus. They marched together against Yassavur, but, obliged to give way before superior forces, they went towards Herat. Notwithstanding this check, Nuruz continued to make himself feared in Khorasan, and sent orders in various directions in the name of Uzbeg Timur, which he had countersigned. The two proceeded to assail Nishapur, which was about to surrender when they quarrelled. Uzbeg Timur, fancying that Nuruz meant to supersede him, left with his men. Nuruz, who was thus in a difficulty, was persuaded by his wife Tuganjuk to send his submission to Ghazan. He accordingly sent his relative Satilmish, and presently a number of officers. He asked for pardon and oblivion as to the past, and promised to be faithful in future. Ghazan was pleased to secure such a redoubtable person as a friend, and offered to pardon him. The messenger then begged him to advance with his troops as far as Mervchak, so as to protect Nuruz if his opponent should attack him. Ghazan said he would go in person, making pretence of a hunting excursion, and sent them away with tokens of his goodwill. The Mongol new year was spent at Sarakhs, and afterwards Ghazan, with the generals Nurin and Kutlugh Shah and a large body of troops, went towards Merv, hunting. At Bagshur (called Mori Shiburghan by the Mongols, Von Hammer calls it Yaghshu), Nuruz, with his wife Tuganjuk (Von Hammer says with Prince Tughan), went to meet him, and offered him nine beautiful horses. Ghazan promised him every favour if he were faithful to him, and they swore an eternal friendship. The feast lasted for three days, but as there was little wine the healths were drunk in water. In token of the reconciliation one of the memorial cairns called *obo* by the

Mongols was set up.* Nuruz took the road of Shahrevan and Mervchak, while Ghazan marched to Fariab by Anjui. There the troops of Trans-oxiana had established their magazines. A large quantity of cattle was captured and taken off towards Shiburghan. Ghazan went himself by way of San and Harik, and encamped on the banks of the river of Shiburghan. The amirs who went in pursuit of the enemy inflicted a severe defeat upon him in the neighbourhood of the mountains of San and Harik. This was on the 19th of January, 1295. The enemy lost many men and much booty, and the magazines were filled with grain. Having waited there for twenty days, Ghazan went to Firamarsan, and thence by way of Sarakhs to Karatepe. There he was met by Boghdai Aidaji, Gaikhatu's messenger, who brought news of the revolt of Baidu, already described.†

Meanwhile Gaikhatu's son-in-law, Ghurantai, or Gartebai Gurkan, informed his father-in-law of the conspiracy, and warned him that the generals Doladai Aidaji (i.e., the butler), Kunjukbal (who had married Arghun's eldest daughter, Oljatai), Tukal (the husband of Arghun's second daughter, Oljai Timur), Ilchidai (chief falconer and huntsman, according to Stephen Orpelian), and Ildar, a grandson of Khulagu (Von Hammer omits Ildar, and mentions Bukdai), who were at his Court, were in league with the rebels. Gaikhatu had them arrested at Kiawabari, or Gaupari, and sent to Tebriz. His close friends, Hassan and Taiju, counselled him to nip matters in the bud by putting the rebel amirs to death. Toghachar obtained a reprieve for them on the plea that it would be well first to see what Baidu intended doing, and that if the latter were summoned to the Court, and refused to attend, it would then be time to push matters to extremity with the generals. Gaikhatu agreeing to this, they were handed over in chains to Toghachar, who conveyed them to Tebriz. Khurumchi, we are told, was confined for a while in the Armenian monastery of Tathef, and regaining his liberty, was restored to his post by a miracle of the Church. Tukal went to Georgia. Meanwhile Baidu was summoned to the Court, while Toghachar sent him word that if he hastened thither his partisans would seize Gaikhatu, whereupon he set out with his troops. Orders were now sent to Baibuka, who commanded at Diarbekr, to seize Baidu. When the messengers reached Irbil, they learnt that Baibuka had already been seized by Baidu's people. They thereupon returned hastily to inform their master.‡ Fancying that Baidu would make for Khorasan to join Ghazan, troops were sent to seize the roads leading thither. Tughachar (doubtless the Tagrus of the continuator of Bar Hebræus) was sent to prevent Baidu crossing the mountains of Sheharzur, and shortly after Akbuka, who was Gaikhatu's father-in-law, was sent with another army in the same direction. Other writers tell us that Taitak was sent ahead

* D'Ohsson, iv. 118-120. Ilkhans, ii. 21.   † Ilkhans, ii. 22.
‡ Ilkhans, i. 406. D'Ohsson, iv. 111. Hist. de la Siounie, 259-260.

with an advance guard of 5,000 men, while Akbuka and Toghachar each commanded a tuman. Near Hamadan there was a struggle of outposts. Toghachar meanwhile marched ahead of his colleague, and when the latter sent to inquire his reason, he replied that he was obliged to hasten, because of the want of pasture. As he continued to draw away, Akbuka reminded him that he was acting against his master's orders. Toghachar now threw off the mask. "Hitherto," he said, "Akbuka has ruled the kingdom in the name of Gaikhatu ; henceforward I shall rule it in the name of Baidu." The continuator of Abulfaraj tells us that this took place at the foot of the mountain Sheharzur. Akbuka now found himself deserted by most of his men, and returned with barely 300 men to Gaikhatu, who was encamped at Abher. That prince wished to fly to Rum, but his courtiers urged him not to run away while he still had so many resources. He summoned Berim, one of his officers, and entreated him not to desert him, and presented him with a robe. He then withdrew to his ordu in Arran, accompanied by a few horsemen, leaving his permanent quarters to be pillaged by Berim. He was soon abandoned by Hassan and Taiju, the former of whom went over to Baidu, whose example was followed by other Mongols. At Mughan he alighted in the quarter of his equerries. He fancied that Kunjukbal and Doladai, who had been unfaithful to him, were at this time in prison at Tebriz, but Toghachar had already released them, and they had made their way to the royal ordu with the intention of plundering it, while the amirs Irinjin and Taijik having banded themselves with others, released Baidu's son Kipchak (who, as we have seen, had been left at Gaikhatu's Court as a hostage), and sent him to his father.

Three days after the skirmish outside Hamadan, above-mentioned, Taitak and Toghrulji had another struggle with Baidu's people, commanded by Bashmak Oghul and Karaju, in which the victory rested with Taitak. But at this time Tukal, who had collected an army in Georgia, drew near. He sent word to the various amirs that he had declared for Baidu, and bade them join him on the Kur. A body of one thousand men, who were at Pilsuvar, apparently under the Berim already named, now seized the unfortunate Khan, and handed him over to the amirs who had revolted.* Abulfeda calls the place Salasalar, in the district of Mughan. He begged for his life, saying he had never been ambitious to occupy the throne, and had only done so at the instance of his generals, and that if they wished to dethrone him his duty was to obey. This was not the kind of language to inspire much respect among the Mongol soldiery, whose chiefs had been made of very different stuff. They replied with gross insults. They dragged him into his tent, where he was strangled with a bowstring, on (according to Rashid) the 23rd of

---

* Cont. of Abulfaraj, Chron. Syr., 635-636. Ilkhans, i. 407-408.

April, 1295.* Marco Polo says he was poisoned.† He was buried at Karabagh by order of his successor Baidu.‡ Wassaf, drawing his simile from one of Gaikhatu's vile habits, in describing the catastrophe by which his life and reign came to an end, has an extraordinary rhetorical outburst. "A la fin l'empire montra a Gaikhatu ce qu'il aimait, c'est a dire *le derriere*." His singular weakness and overstrained clemency, which probably lost him his throne, are attributed by Wassaf to an opinion of the Shamans, who being asked by him how it was that Arghun had enjoyed such a short life, replied because he put to death so many princes, officers, and soldiers.§

All the authorities agree in describing Gaikhatu as a dissolute and lecherous person. This is the character Marco Polo gives him. Haithon's graphic denunciation has been well translated by Colonel Yule. "A man without law or faith, of no valour or experience in arms, but altogether given up to lechery and vice, living like a brute beast, glutting all his disordered appetites ; for his dissolute life hated by his own people and lightly regarded by foreigners "|| This opinion of the Christians is for once fully shared by the Muhammedans, Wassaf, Abulfeda, &c. In the "Shajrat ul Atrak," after reciting his dissolute habits, the author adds : "It may be stated in his favour that he was the most generous of all the descendants of Khulagu Khan, and that during his reign he never caused the execution of an innocent man, or permitted the punishment of the guilty otherwise than was prescribed by the law.¶

Gaikhatu had six wives, viz., Aisha, daughter of Tughu, son of Ilkai Noyan ; Dundi, daughter of Akbuka, son of Ilkai ; Ilturmish, daughter of Kutlugh Timur Gurkan, the Konkurat ; Padishah Khatun, daughter of Kutb ud din, the ruler of Kerman ; Bulughan, and Uruk. He left three sons, Alafreng and Iranshah by Dundi, and Jiuk pulad by Bulughan. He also left four daughters.**

De Saulcy published two gold coins of Gaikhatu, and Mr. S. Poole has published three silver ones. On these coins his name is replaced by the formula, Arinchin Turji (very precious diamond), previously discussed.†† These coins bear the Sunni symbol, and were struck at Baghdad and Tebriz, between the years 691 and 693.||

BAIDU KHAN.

On the murder of Gaikhatu, Akbuka, Tamaji, Sertak, and several others of his favourites were put to death. Taiju was conducted to the ordu and there examined. Baidu accused him of having received many favours from Gaikhatu, notwithstanding which he had not

* D'Ohsson, iv. 113. Hist. de la Siounie, 260. † Yule's Marco Polo, 474.
‡ Shajrat ul Atrak, 270. § D'Ohsson, iv. 113-114.
|| Yule's Marco Polo, ii. 474. ¶ Miles, op. cit., 270. ** Ilkhans, i. 400.
†† Ante, 283 and 356. ‡‡ Journ. Asiat., 3rd ser., xiii. 129-130. Poole Cat., vi. 32.

helped him in his distress, and he suggested that he himself might expect the same treatment, and he had him put to death. Hassan was condemned to death for the same reason, but was pardoned. Baidu summoned Ayet-Kali, who had struck him at the banquet by order of Gaikhatu, and apostrophised him about his audacity. He replied that Gaikhatu being his sovereign, if he had ordered him to kill his brother or his son he should have felt bound to obey him, and that now he was the servant of Baidu he should feel obliged to obey him in the same way. His excuse was accepted, and he was allowed to retain his post.* The throne on which Abaka and Arghun had been enthroned was brought from Tebriz and set up at Aujan, where Baidu was duly proclaimed in April, 1295. Stephen Orpelian says the installation took place in the plain of Srav. After the usual festivities Baidu set out for Siahkuh, and issued an edict, and had it sent to various parts of the empire, in these words : " As Gaikhatu had a distaste for the affairs of government, and acted contrary to the yasa of Jingis Khan, we have dethroned him, in concert with our akas and degus,† with the khatuns and the amirs. May the pensions and payments fixed by our father be paid punctually to those who are entitled to them."

The noyan Toghachar was appointed generalissimo and head of the administration, while Kunjukbal, Tugan (the Tukal of Von Hammer), Chichak, Legsi Gurkan, and Tudaju were appointed his deputies. Jemal ud din of Destarjirdan was put over the finances, and adopted the title of vizier in lieu of that of sahib divan, deeming it would bring him good fortune. Considering that the amirs were discontented with Gaikhatu because they were not employed in the administration, he gave them the government of the provinces, recalling the fact that during Abaka's reign, when each province was controlled by one of the ruler's familiars, order and tranquillity generally prevailed, and the troops were quiet. Baghdad was confided to Tudaju, Rum and Diarbekr to the noyan Toghachar, Irak Ajem and Lur to Doladai Aidaji, and Shiraz and Shebankiareh to Kunjukbal. Each was absolute in his own province. ‡

Let us now revert shortly to the doings of Ghazan in Khorasan. When he heard that Baidu had revolted he took no notice of it, but having permitted Nuruz to leave for his camp, he himself went to Badghiz, and after a halt of some time repaired again by way of Eshja, Shevkian and the valley of Kharir to Radkian, to hunt. Envoys now arrived from Baidu with the news that the amirs and other grandees had adjudged him the throne. Ghazan consulted with his amirs, and sent for Nuruz, whom he appointed governor of Khorasan, with powers equal to those held by his famous father, but took no other immediate steps. After

---

* D'Ohsson, iv. 116.
† *Akas*, a term which in Mongol and Turk means elders ; *degus*, the Turkish *ini*, means the younger members of the royal stock ; together, all the members of his family.
‡ D'Ohsson, iv. 115-118. Ilkhans, i. 408-410.

spending some time at Radkian and the springs of Kebseb, he went to Duber, where he was met by Nuruz. Meanwhile, Prince Suka, with the greater part of Gaikhatu's troops in Mazanderan, went by way of Shehreknev and Jorjan to Sultan Devin, where they joined Ghazan, and feasted for several days.* Before leaving Sultan Devin, Ghazan sent word to Baidu that he was coming to see him, and then advanced by way of Jehardih to Damaghan. He would have stayed to rest at Girdkuh, but was urged to go on by his amirs, who pronounced it to be of unhappy augury if he stayed to sup there. The sons of Taj ud din Ilduz, the former commander, were brought out of the castle, and the astronomer Said ud din Habesh was told to look after them. At Semnan they were met by Ardubuka, who had been sent by Gaikhatu with a sack of paper money, and machinery for making it. Ghazan ordered it to be burnt. At Hetran (D'Ohsson says at Khailbuzurg, between Rai and Kazvin) he was met by Timur Aidaji, whom he had sent to Baidu, and who brought word that he had mounted the throne, and that the amirs Toghachar, Kinduskul (? Kunjukbal), Doladai, &c., had declared for him. At the council which was now summoned Nuruz said it was not wonderful that the amirs who had usurped power should have declared against him (Ghazan), as they were afraid he would avenge the death of his uncle Gaikhatu, of Ordu Kia, and of Juchi ; but they feared still more his great qualities, and wished to retain the power in the hands of a feeble and timid prince, whom they could control as they pleased. He advised Ghazan to send successive envoys to Baidu, and to act according to their report of what took place. As he had never dreamt there might be a struggle, Ghazan had only brought a few troops and very little baggage with him. He determined to send the two amirs Mulai and Yaghmish to say he was coming to have a personal interview with him, assuring him of his goodwill, but also saying that according to the yasa of Jingis Khan it was forbidden for karajus (*i.e.*, those not of royal blood) to shed the blood of princes, and demanding the surrender of the guilty beys, so that they might be tried and punished. When he reached Akkhoja, near Kazvin, he was met by Shadi Gurkan, the son of Bukua, and the Princess Yelturmish, who came on behalf of Baidu, to say that he had not desired the throne, but as on the death of Gaikhatu he (Ghazan) was so far off, the khatuns, generals, and noyans, to put an end to the confusion which was desolating the country, had united in electing him, that all would be well. Ghazan was further urged not to fatigue his troops, but to turn back home again.†

Ghazan, who had a very small force with him, was for returning, but Nuruz urged him to go on. "Man must end by dying," he said. "As death is certain, let it be honourable." Ghazan decided to follow the

---

* Ilkhans, ii. 22. D'Ohsson, iv. 121.   † D'Ohsson, iv. 121-122. Ilkhans, ii. 23-24.

counsel of Nuruz, and to advance with his little army of 6,000 men.[*]
He assigned to each officer his part in the coming struggle, and promised
to reward them each with the command of a district or a province if they
should prove victorious, while he urged on his men that it was better to go
on than to retrace the long journey they had already made. On arriving
at the station of Robat Moslim, the place was summoned, and Shadi
Gurkan was sharply questioned about the usurpation of Baidu. Prince
Ildar, the second son of Kiukurtai, son of Khulagu, who was with Shadi
Gurkan, boasted when drunk that if Ghazan would not submit to Baidu,
there would be open war between them. The envoys were now dismissed.
The amirs Nurin and Kutlugh Shah were told to have the army ready,
and Isen Buka, who was a subordinate judicial officer, was sent to Baidu
with the laconic message, "We are coming." Ghazan advanced by way
of Tuka, Turuvan, and Sipidrud.[†] Baidu learnt at Heshtrud, near
Meragha, that his rival was approaching. He sent his equerry, Bughdai,
to report, and as he did not bring encouraging news, a council was called,
including the amirs Toghachar, Doladai, Kunjukbal, Ilchidai, and Tugal,
and it was determined to offer battle. The two armies were before one
another on the 19th of May, near Kurban shira (which was a few
leagues west of the Sipid rud, i.e., the White River), and Karieh
Shirguiram. Nuruz advised an immediate attack. Prince Ildar and the
amirs Ilchidai and Chichek commanded Baidu's right wing, and Kutlugh
Shah and Nurin Aka the right wing of Ghazan. The Princes Suka and
Buralighi, with Nuruz, were with Ghazan in the centre. The battle began
by Kutlugh Shah charging Ildar before the great trumpet sounded. Ildar
and eight hundred of his men were killed, while Arslan Oghul (Von
Hammer calls him Arslan Arghun) was captured, and led with a cord
about his neck before Ghazan. Prince Burultai was for cutting him down,
but Ghazan would not permit it, and ordered a kaftan, a cap, girdle, and
trousers to be given him out of his wardrobe. He also ordered medicine
to be distributed to the wounded. All the army of Khorasan was about to
charge, when the equerry Bugdai rushed in between the two forces, and
prostrating himself before Ghazan, said, "Prince Baidu sends to tell you
that relatives should not fight. If you will divide the kingdom you may
have Khorasan and Mazanderan, Irak, Kerman, and Fars."[‡]

The continuator of Abulfaraj says nothing about any battle, but that
Ghazan sent to inquire why he had been summoned, and why, further,
having been so, Baidu had not awaited his arrival, so that he might mount
the throne with his consent. He adds that the latter consented to make over
to him Khorasan, Shiraz, Behrin, and Shirvan, and also gave him the camp
and effects of his father Arghun, reserving only one tent for himself.
Nurin and Kutlugh Shah pressed Ghazan to accept the offer, but Nuruz

---

advised him to take advantage of his rival's weakness to attack him suddenly at night. When his men were duly mounted, however, a terrible hurricane came on, and they lost themselves and each other, and when daylight broke they were very weary, and much disturbed at finding themselves close to Baidu's people, who marched out to attack them. Ghazan seeing the position of affairs agreed to a truce.* Thereupon the two princes advanced to meet one another between the armies, accompanied by ten men each. With Baidu were Toghachar, Doladai, Kunjukbal, and Ilchidai; while Ghazan was accompanied by Nuruz, Nurin, Kutlugh Shah, and Sutai. They all dismounted on the battle-field, and the two princes embraced one another. Thereupon Baidu repeated what he had already sent as a message to Ghazan several times, and they promised in future not to disturb the public peace by their quarrels, and in the usual Mongol fashion wine, in which gold was mingled, was drunk from golden cups. Nuruz refused to drink wine in swearing, pleading that he was a Mussulman. The generals on each side followed the example of their masters in swearing friendship. It was decided that Baidu should be duly enthroned the following day. He promised to do all that Ghazan should require, and in the evening the two princes returned to their camps. The two armies marched together as far as Kurban Shira, but there was continual watchfulness and jealousy on either side, and each man kept his hand on his bridle. Meanwhile the Sheikh Hurkasan, who was one of the pages of the Princess Bulughan, had gone over with a considerable following and joined the party of Ghazan. As the latter's troops were passing through a narrow defile, Kukatu and Kunjukbai urged Baidu to close the entrance, and thus catch them in a trap, but he refused to do so. At length, by the intervention of the officers on either side, it was arranged that the two princes should have another interview in a royal tent pitched between the two armies. To this they repaired with the same grandees who had accompanied them before. This interview was brought about by the invitation of Tudai Khatun, the widow of both Abaka and Arghun, whose yurt was in the neighbourhood, and who wished the reconciliation to be effected in her own presence. A long debate took place, lasting from morning till sunset. It was arranged that all the property of Arghun, including that of the Princesses Bulughan and Uruk, and of Prince Khudabendeh, Ghazan's brother, should belong to the latter, and that he should control the country beyond the Sipid rud, viz., Irak Ajem, Khorasan, Kumuz, Mazanderan, and one-half of Fars. The "Shajrat ul Atrak" adds Khoristan (?) and Luristan. Baidu was to keep Irak Arab, Diarbekr, Azerbaijan, Iran, Armenia, Georgia, and Rum. Ghazan was to retain the royal domains there. He asked in addition for the tuman of

---

* Cont. of Abulfaraj, Chron. Syr., 637-638.

Karaunas commanded by Toghachar, which had been attached to Arghun's appanage, but Baidu urged that Arghun, who had treated him as one of his own sons, had assigned him the tuman of the Karaunas of Baghdad for his retinue. The amirs urged that the matter should rest as it had bee fixed. While the negotiations were in progress the army of Baidu was being continually reinforced. There arrived troops from Baghdad and from Mughan, and his generals urged him to take advantage of his opportunity to put an end to Ghazan. Tugal was so discontented with his refusal to entertain this that he hurriedly returned to his quarters in Georgia. Nor would Baidu listen to those who demanded the blood penalty for the relatives who had been killed by Ghazan's people. Kunjukbal, we are told, urged that, as in case of Baidu's discomfiture the amirs would turn to Akbuka, who was in chains, it would be well to put him to death. Baidu was apparently quite loyal to the arrangements he had made. He sent Doladai with meat and drink to attend upon Ghazan, and to offer him the cup.

Ghazan, alarmed at the increasing forces of his rival, determined to withdraw by way of Siah kuh, Sipid rud, and Sukurluk, but Baidu feared that the Karaunas who were encamped in that direction would desert to him, and also that he would lose a considerable treasure which was there, and he sent the famous amir Pulad Chingsang to urge him to return by the same route he had come. The next day, Baidu's son Kinjak (called Kipchak by D'Ohsson) went to Ghazan's camp at Kurban Shira to offer him the cup again, and to invite him to a parting feast, after which he might return to Khorasan. Ghazan was naturally suspicious. We are told his generals swore mutual oaths, Nurin, Kutlugh Shah, and others doing so by drinking out of gold cups, while Nuruz, Buralkai, and Mulai did so with their hands on the Koran. Nuruz pressed his patron to become a Mussulman, saying he would thus attract all the followers of Islam to his side. He presented him with a beautiful ruby ring, of the weight of ten miskals, on his knees, and said that although it was not seemly for a karaju, or subject, to make such presents to a prince of the blood, he had been made bold by his goodness to him, and begged him to retain the ring as a memento of the promise he had made. Baidu pressed for another interview, but the generals of Nuruz, fearing some treachery (the " Shajrat ul Atrak " says Baidu's amirs had determined to put Ghazan to death next day at a display of fireworks), urged him to refuse, and he sent word that the astrologers had declared the day fixed for the interview to be unpropitious, but that he would be at the rendezvous the day after. During the night, however, he hastily left with the advance guard, and marched so quickly that he had crossed the Sipid rud before morning, and taking the road Dih Minar, reached Zenjan. He left Nuruz and Tuktimur behind to receive the act of investiture of the provinces of Irak and Fars, and to superintend the removal of the ulusses of the wives of his father and uncle, as well as of the troops belonging to the military establishments of Arghun and Abaka. They were also to try and learn Baidu's plans, and to prevent him from pursuing. From Mosellim (called Robat Moslim, on the Kur, by Von Hammer), Ghazan sent word to Baidu, begging him to execute the treaty they had made, and to make over to his generals what he had promised to give him. Meanwhile Kunjukbal, Ilchidai, and Doladai went in pursuit of him with 5,000 men. When the latter reached the river Kiuh (not Kiere, as D'Ohsson reads it, says Von Hammer), called Turkan Muran, or the Turk's river, by the Mongols, east of Kazvin, there came Kesher Bakhshi with a message from Baidu, expressing his concern that they had not had another interview. Ghazan sent him back his spear bearer, Ibrahim, to again urge the fulfilment of the treaty. It seems Baidu had ordered Jemal ud din to deliver up a part of Fars to Ghazan, but when the latter's officers went to take possession a counter order was produced. He now went on to Demavend to await Baidu's answer. Nuruz and the other officers of Ghazan who had been left behind went with Baidu to Sheruyas (Sultania). They were arrested, and menaced with all kinds of threats, but Nuruz remained unmoved. Thereupon Baidu's officers changed their tactics, and charged his brother Legsi to try and win him over. But promises were no more effective than threats; nor was he to be induced to break his oath to Ghazan when the amirs went to him in troops to threaten him. He, on the other hand, tried counter intrigue. He succeeded in gaining over Toghachar, and even persuaded Baidu that he was friendly to himself, and promised solemnly to hand Ghazan over to him bound hand and foot. The continuation of Abulfaraj says he promised to send him his head on a charger. He was released, while his son was given the amirship of Yezd, with an assignation of 10,000 dinars on the revenues of that town. Ghazan's amirs were also allowed to leave, and Baidu himself encamped in his yurt of Sughurlik. Nuruz now hastened, with Tuktimur, to rejoin Ghazan, and traversed the distance from Meragha to Firuzkuh, near Demavend, in four days. He told him that to save his own life he had feigned an intention to betray him. In order to keep his word to Baidu he sent him a cauldron, bound in cords, in a sack. (Ghazan means, a cauldron.) On receiving this, Baidu and his people were naturally outraged, and regretted having let such a dangerous enemy escape them.

At a council which Ghazan now called, Nuruz again urged him to become a Mussulman. "The astrologers, doctors of the law, and other holy people have foretold," he said, "that about the year 690 HEJ. (i.e., 1291) there will appear a ruler who will protect the Muhammedan faith, will restore to it its ancient glory, will bring prosperity to his subjects, and will reign for many years. I have always thought that this meant you. If you will embrace Islam you will become the ruler of Iran. The Mussulmans raised by you from the degradation they have suffered at the hands of the pagan Tartars will be devoted to your cause, while God, recognising that you have saved the true faith from extinction, will bless your arms." Ghazan was moved by this address. He sent for the ruby which Nuruz had given him, and which he had kept as a gauge of his promise. He gave a great feast on the 19th of June in the meadow of Lar-Demavend, near a country house where his father had often lived. Having purified himself by a bath, and dressed himself in new clothes, he entered the house, where, standing at the foot of the throne, he repeated several times the profession of faith which was recited by the Amir Sadr ud din Ibrahim, the son of the great Sheikh Said ud din, descended from Hamuyah, the disciple of Mohiyeddin Arabi, the pole star, says Von Hammer, of Arabian mystics, who was born in 1164, and died in 1240. Sadr ud din Ibrahim was the contemporary of Sadr ud din of Konia, and the author of some famous mystical works, as the "Kitabi Mahbub" (*i.e.*, the book of lovers), the "Sijilol-ervah" (*i.e.*, the records of spirits), and of many mystical verses. The example of Ghazan was followed by his officers and soldiers. He distributed largess to the imaums, sheikhs, and seyids, or descendants of Muhammed, and gave rich alms to the poor. He visited the mosques and tombs of the saints, asking God to grant him victory over his enemies. Couriers were sent with the news of his conversion to Irak and Khorasan, whence many imaums and sheikhs hastened to his camp. He observed the fast of Ramazan, and every evening had many Turks and Persians to sup with him at his table. The auspicious event took place on the 16th of June, 1295.* According to the "Shajrat ul Atrak" the adoption of Islam by Ghazan took place at Firuzkuh. It adds that in commemoration of the event, Nuruz raised a white marble pillar on the spot, which, it says, is between Ubeh (?) and Gurgistan, or Georgia, which stone was known as Mir Nuruz's pillar when the author wrote.† Ghazan now adopted the name of Mahmud. He had been up to this time a devoted Buddhist, and had built a large Buddhist temple at Khabushan. His conversion on the present occasion was doubtless in a large measure due to political rather than to religious considerations, but with whatever motive the ,result was exceedingly important and fraught with the greatest consequence for the history of the East. Not only did it re-arrange entirely the forces of politics in Persia, but it drew into the service of the State in much closer fashion all the best elements and the greatest political insight in the East at this time. Its immediate effect on Ghazan's own fortunes were naturally very marked.

Baidu made Toghachar governor of Rum, and appointed as his assistant the vizier Sadr ud din, of Zenjan, surnamed the Chaoyian, or introducer of paper money. This was a virtual deposition of the latter officer, whose

---

* Ilkhans, ii. 24-29. D'Ohsson, iv. 122-133. Shajrat ul Atrak, 273, &c.
† Op. cit., 273.

post as vizier was given to Jemal ud din, of Destarjerd, who owed the
greater part of his fortune to him, but nevertheless became his open
enemy. The deposed vizier saw that his best chance of revenge was to
join the party of Ghazan, and he incited Toghachar, who was at this time
irritated in consequence of a quarrel with Todaju, one of Baidu's generals
to join him. They determined to act in concert. At this time there was
at the Court an envoy of Ghazan's, named Ariktimur, who had gone to
urge the departure of the Princess Bulughan, whom Baidu had detained
on the ground that she ought not to make the journey in that unfavourable
season. Sadr ud din persuaded the princess to send the sheikh Mahmud,
of Deinavar, to Ghazan, to make her excuses. He was also instructed to
assure him secretly of the goodwill of the amirs Toghachar, Choban,
Kurumishi, and Bogdai, as well as of the ayu-oghlans (meaning
good fellows in Turkish, and probably connoting pages.*) The sheikh set
out with an officer named Kutlugh Shah.† On his arrival he delivered the
message from the princess openly and before his colleague, and then had
a private interview with Ghazan, and told him how all the military chiefs,
except Kunjukbal, Tugal, Doladai, and Ilchidai, who were too guilty to
expect pardon, were ready on the news of his march to join him. Ghazan
was delighted with this unexpected news, and asked for information
about what was passing at Baidu's Court. At this time the forces
of Transoxiana made an invasion of Khorasan, but withdrew again on
the approach of Nuruz, who marched against them. On the return of
the sheikh to the Court he reported this invasion, which had necessitated
the departure of Nuruz. The news was welcome to Baidu, who disbanded
his troops. The sheikh also reported to the generals the result of his
secret interview with Ghazan. He had been one of Arghun's companions,
and disliked Baidu because he favoured the Christians, and supported
Ghazan in the interests of Muhammedanism.‡ Sadr ud din, pretending
that he was on his way to Rum, stayed at Tebriz. He sent to fetch his
brother, Kutb ud din, or Kutb Jihan, from Kazvin, and fled with him and
his cousin, Kavam ul Mulk, with all the gold and silver which he could
carry. He was pursued by the troops of Kunjukbal, and his baggage was
pillaged, but he arrived safely at Firuzkuh, where he was received with
great honour by Ghazan. He promised that Toghachar, over whom he
had great influence, would faithfully keep his promise of joining him with
his troops when he should advance, and obtained a promise from Ghazan
that he himself should be made vizier if their plans did not miscarry.

Kutlugh Shah was again sent by Baidu to Ghazan's camp. He was
arrested, and under torture confessed that he had gone to report
whether it was Ghazan's intention to advance. He was imprisoned in

---

* D'Ohsson, iv. 155. Note.
† Not the general of that name, who was one of Ghazan's amirs.
‡ D'Ohsson, iv. 133-135. Ilkhans, ii. 29.

the fort of Hebl rud, and Ghazan set out on the 26th (Von Hammer says the 28th) of August for Rai. Nuruz, with Kutlugh Shah, or, according to the "Shajrat ul Atrak," Sadr ud din, and Khoja Ahmed Khabdi, went on with the burunghai, or advanced guard, of 4,000 horsemen, and with them marched Sadr ud din, of Zenjan. After the flight of the latter, the generals Choban and Kurumishi Kurkan, son of Alinak, who were attached to the tuman of Tudaju, asked permission from that chief to go and exercise their horses, so as to have them ready, since news was continually arriving of Ghazan's march. They accordingly took 500 horses, and setting out at nightfall went and joined Ghazan at Hebl rud, and were rewarded by him with kaftans and jewelled girdles. At their request they were sent to join the advance guard. A few days later Ghazan reached the river Kiuheh (called Kuma by D'Ohsson). Nuruz was intrusted with the organisation of the army. By his advice news was spread abroad that he (Ghazan) was marching at the head of twelve tumans to secure his father's throne ; that the sword having been drawn, those who opposed him would be put to death with their families. This proclamation had the desired effect. The advance guard stopped all whom they met, so that news of the march might not reach Baidu. Nuruz hastened on to Sijas and Sohravrad, on the banks of the Sipid rud. This was reported to Baidu by Doladai, who commanded his advance guard. After the promises which had been made to him, he was very indignant, and especially resented the conduct of Nuruz, who had so completely deceived him. He at once sent to inform Toghachar, his generalissimo, and to ask his advice. He reassured him, and promised to disperse the enemy if he marched against him. But during the night he furtively escaped with some of his officers and joined Nuruz. This defection was followed by that of numerous soldiers. When Baidu learnt of this the following day he lost heart and fled. Leaving the Sipid rud, he reached Maidan Sulimanshah the same day with some officers. The next day Eltimur, with his tuman, and Khodabandeh, brother of Ghazan, went to join Nuruz. Learning this new defection, Baidu continued his flight towards Aujan and Merend with the amirs Kunjukbal, Chichak, and Ilchidai, and several officers, with the intention of reaching Georgia, where he hoped for support from Tugal.

Ghazan meanwhile reached Sijas, where he received the submission of Khodabendeh and Ildar. Further on, on the Sipid rud, he was joined by the generals Doladai, Iltimur, and others. He then went to Merend, to await the return of Nuruz and Kutlugh Shah, who had gone towards the Aras in pursuit of Baidu. Nuruz advanced very rapidly, and his horses being worn out, he dispatched Kurumishi and Shadi with 4,000 men, who overtook the fugitive near Nakhchivan, in Armenia. When he was brought before Nuruz, that unscrupulous believer in blood and iron jeered at him, saying : "Did we not agree that I would bring you to Ghazan?

You see I meant what I said. I have kept my word, but why have you not kept yours? Why did you turn your back upon us when we came?" Baidu begged to be conducted before Ghazan, who had halted at Aujan. News was sent on to the latter that his rival had been captured, and would arrive the following day, escorted by 200 men. Ghazan, who was not anxious for an interview, sent one of his equerries with some troopers to put him to death. They met him and his escort near Tebriz, and having in the Mongol fashion given him a feast, which lasted until night, and during which he was treated with the honours due to one of his rank, put him to death on the 5th of October, 1295. His son, Kipchak Oghul, was put to death at Keshur, near Meragha. Ildar fled towards Rum, and Tugal to Georgia.*

Baidu had barely reigned eight months. The continuation of Abulfaraj praises his good qualities, and tells us he was free from the debauched habits and unnatural offences of his predecessor. He was prudent, gentle, and modest, and patronised the distinguished men of every country, giving them splendid presents and robes. He had passed much time in the society of the Greek wife of Abaka, the Princess Despina, from whom he had acquired a good opinion of the Christians. He had allowed a Christian church to be set up in his camp some years before, and also allowed the bells to be rung. At this time many of the Mongols had become Mussulmans, were circumcised, and practised the ablutions and said the prayers enjoined upon the faithful; and, we are told, to please them Baidu also became a Mussulman, but according to the continuator of Abulfaraj, he was really a Christian, and wore a cross suspended about his neck. Although a nominal Muhammedan, he did not join in their services, but to please the faithful he sent his son to pray with them.†

Haithon has a notice confirming these statements of Bar Hebraeus. He says he was a good Christian, rebuilt the Christian churches, and forbade the preaching of Muhammedanism among the Tartars, and as there were many of them who had become converts to that faith, they were much opposed to him. This was why they made secret overtures to Ghazan, and why when Baidu marched against him those who were Muhammedans went over to him.‡

Abulfeda's notice of Baidu§ adds nothing to the accounts already given, which were derived by D'Ohsson and Von Hammer from Rashid ud din and Wassaf.

In the Georgian history a very prominent rôle is assigned to Tukal, or Tugal, who, as we have seen, had his camp in Georgia, and whose doings it confuses with those of Baidu. It calls him khan, and says he lived in the mountains of Ararat, where he summoned the

* D'Ohsson, iv. 133-140. Ilkhans, ii. 29-30.   † Op. cit., 642-643. D'Ohsson, iv. 141.
‡ Op. cit., xl. D'Ohsson, iv. 141-142.   § Op. cit., v. 121-126.

Georgian king to go and see him. He treated him courteously, and gave him Dmanis, the widow of Kutlugh Buka, brother of Mangasar, in marriage. This authority makes Nuruz and Kurumishi, the son of Alinak or Alikan, flee from Tukal, who had killed the latter's brother Buka, and go over to Ghazan in Khorasan.* They united their forces with his, while Tukal, we read, joined Baidu, and the rivals came to an issue near the little town of Zangan. The result was indecisive. Ghazan was not present at this fight, but was represented by Nuruz and Kutlugh Buka. Nuruz was captured after a severe struggle by Jalirman, and Ghazan withdrew to Khorasan.

Stephen the Orpelian says Baidu was a Christian, but was persuaded by his generals to become a Mussulman. He says he was a feeble administrator. He sent Jelal, son of Darsaij, with some Mongol troops to Amaras, the former residence of the Catholicos of Arran, situated south-east of Gantzasar, in the district of Haband and the province of Artsakh, and carried off " the cross" of St. Gregory, grandson of St. Gregory the Illuminator, and another richly jewelled cross. The Greek princess, Despina, who had been foster mother to Baidu, begged it, and it was sent to Constantinople.† The same authority tells us that when Ghazan marched against Baidu he did so at the head of the troops of Khorasan called Gharavunas (i.e., the Karaunas), under the command of Kutlugh Shah, who was of Armenian origin.‡

Baidu left a son named Ali, whose son Musa at a later date became Ilkhan. In the British Museum there is a silver coin of Baidu's, dated in the year 694, and struck at Tebriz, and a copper one without mint or date.

Note 1.—The Karaunas. Under this name we have constant reference in the historians of the Ilkhans to a body of freebooters whose exact nationality is not easy to fix. Their name has been connected by Von Hammer § with Karaun Jidun, by which the Mongols knew the mountains separating the Gobi desert from China and Manchuria. In this district were encamped the Turkish tribe of the Kunkurats, one of whose branches is called Karanut by Rashid ud din. The "t" in this name is merely the Mongol plural, and if we remove it we have a name singularly like that of the Karaunas. In support of this view it may be mentioned that Niki Behadur, a grandson of the Kunkurat chief, Tukujar, whom Jingis Khan left in command of his ordus when he marched against China, is expressly said to have commanded a hezareh of the Karaunas which was encamped at Badghiz.‖ Von Hammer says they were employed in working the machines for throwing naphtha, in which they were much skilled.¶ In 1278 we find Abaka visiting Herat, and receiving the submission of the amirs of the Karaunas.** Arghun passed the winter of 1283 at Baghdad,

* Op. cit., 613-614.	† Hist. de la Siounie, 260.	‡ Id., 261.	§ Ilkhans, i. 309.
‖ Erdmann Temudschin, 200.	¶ Ilkhans, i. 17 (note) and 309.	** Id., 309.

where 10,000 Karaunas, described by Wassaf "as a kind of demons, the fiercest of the Mongols," were posted.* The next year, when Arghun was at war with Ahmed Khan, the former sent word to Nuruz to send him a tuman of Karaunas. They arrived too late to take part in the battle of Ak Khoja, but ravaged that district and burnt Damaghan.† Later in the year the rebels against Ahmed sent Buri to Ispahan to the Karaunas to tell them to seize Ahmed.‡ When Ahmed fell, we are told these robbers plundered his harem at Sughurluk, and drove out his wives and mother naked.§ Under a leader named Aladu they plundered the camp of Nuruz in 1289, and later in the year we read of their going to Herat with their leader Tekne to pay their respects to Ghazan. Their summer quarters at this time were at Shuturkuh (*i.e.*, the Camel mountain).‖ In 1290 a party of these robbers were driven away from Juven, and later in the year they seized on Merv.¶ Somewhat later Ghazan removed a body of them from Jorbed to Jajerm and Bostam.** Baidu claimed that the tuman of the Karaunas at Baghdad had been made over to him by Arghun.††

D'Ohsson treats the Karaunas as identical with the Nigudarians,‡‡ the latter name being taken from their original leader. Although this view is disputed by Von Hammer there is some probability in it. They were both robbers, and both apparently infested the same district, and it is certainly curious that Marco Polo should identify them in his not very satisfactory account of them. He says, speaking of the plain between Kerman and Hormuz, that the villages there were protected by lofty walls of mud, against the banditti, who are very numerous, and called Caraonas. "This name," he says, "is given them because they are the sons of Indian mothers by Tartar fathers." He describes them as producing darkness when they pleased, by their enchantments. They knew the country well, and rode abreast, keeping near one another, sometimes to the number of 10,000, and thus extended across the whole plain they were harrying, and secured everything. They killed the old men, and sold the young ones and the maidens into slavery. "The king of these scoundrels," he says, "is called Nogodar, who went to Jagatai's Court with 10,000 of his men, and abode with him, Jagatai being his uncle."§§ The details he gives of this Nogodar's adventures I shall return to on another occasion.

To revert to the Karaunas. I am not at all certain that they are not to be identified with the Kara Tartars, who are mentioned in Khorasan at the break up of the power of the Ilkhans, and whom I shall mention presently. Von Hammer suggests that the word carbine, for which no satisfactory etymology is forthcoming, was derived from the naphtha-using Karaunas, whose name he writes Karavinas. Colonel Yule says that a link in such etymology is perhaps furnished by the fact that in the 16th century the word carbine was used for some kind of irregular horseman.‖‖ Marco Polo's own etymology of Karauna is possibly a confusion founded on the existence of an Indian term, Karani, for children of mixed parents.¶¶

Ilkhans, i. 344. † *Id.*, 348. ‡ *Id.*, 357. § *Id.*, 358. ‖ *Id.*, 13-14.
¶ *Id.*, 14-15. ** *Id.*, 16 †† *Id.*, 26. ‡‡ Op. cit., ii. 516.
§§ Op. cit., ed. Yule, 99-100. ‖‖ Yule's Marco Polo, 103. ¶¶ *Id.*

Note 2.—Paper Money. In describing the experiment in paper money made by Gaikhatu, I mentioned that it was a close imitation of that used in China, and that it was known by the same name, *i.e.*, chao. My friend M. Terrien de la Couperie says the Chinese traditions date the first introduction of paper money in China as early as the beginning of the Empire, in 2697 B.C. The statement in the annals is that Pöh ling, minister of Hien yüen (*i.e.*, Huang ti), began to make use of fabrics (pa pöh) as substituted money (ch'u pi).* But, as M. de la Couperie says, this refers rather to the use of various forms of material used in lieu of coin, and not to a paper currency as we understand it. The first paper money, properly so called, was issued in the reign of the Emperor Hian tsung, of the Tang dynasty, about the year 806 A.D., and a specimen of it from the Tamba collection is preserved in the British Museum, and consists of an oblong piece of paper, less than two inches long, with the representation of an ordinary Chinese coin (cash) on it, and the inscription " current value of the Tang's counting-houses."†

This paper money of the Tang was called *fei thsian*, or flying money. It was introduced to remedy the scarcity of copper coin, which had also led to an order forbidding the use of copper for making vases, bowls, &c. People were allowed to deposit what copper money they had in the Government banks, and received in exchange these notes, which were cashed on presentation elsewhere, thus obviating the necessity of merchants and others having to carry large quantities of coin about the country. For some reason they were withdrawn from circulation in the capital within three years of their issue, and only allowed to circulate in the provinces.‡ Tai Tsu, the founder of the Sung dynasty, who mounted the throne in 960, issued similar notes on the same terms. They were called *piau thsian*, or convenient money. They were much used. In 997 A.D., the issue of such notes amounted in value to 1,700,000 in silver, and in 1021 to 1,130,000 ounces.§ Meanwhile the plan was abused, and we read how in the district of Shu in Sechuan, one called Chang yung introduced notes answering to assignats, having no deposits to represent their value. These were introduced to replace the iron coinage, which was deemed too clumsy and inconvenient. These assignats were called *chi tsi*. During the reign of Chin Tsung, of the Sung dynasty (997-1022), this example was followed, and notes called *kiao tsu* were issued, which were redeemable in three years. Each kiao tsu was worth a string of a thousand cash, and represented in value an ounce of pure silver. Sixteen wealthy houses undertook the operation, but speedily became bankrupt, which led to much litigation, and an order forbidding private individuals from issuing notes was promulgated, and a public bank for this money was founded at I chau. In 1032 there had been issued 1,256,340 ounces in value of kiao tsu. False ones having appeared, the makers were punished in the same way as those forging Government orders. Presently banks for assignats were founded in different provinces, those of each province not circulating outside it, and the terms of payment and mode of circulation were frequently changed.¶ In 1131, during the reign of Kao tsung,

---

* Numismatic Chronicle, 3rd ser., ii. 337.   † *Id.*, 334-336.
‡ Klaproth, Mems. Relatifs a l'Asie, 378-379. La Couperie, op. cit., 334-335.
§ Klaproth, op. cit., 379.   ¶ *Id.*, 380-381.

certain notes called *kuan tsu* were issued at U Chau for the payment of those supplying the army with provisions, which were cashed at a special bureau; but they were liable to abuse, and caused discontent. They were nevertheless issued in other provinces. In 1160 a new kind of assignats was issued in Chekiang and its neighbourhood, called *hoei tsu*. Eventually these also were issued generally. The paper for them was first made at Hoei Chau and Chi Chau in Kiang nan and afterwards at Ching tu fu in Sechuan and Lin gan fu in Che kiang. At first each hoei tsu was worth 1,000 cash, but presently others were issued of the value of 200, 300, and 500. Between 1160 and July, 1166, as much as 28,000,000 ounces in value of these assignats had been issued, while in November of the same year they had been augmented by 15,600,000 ounces. During the continuance of the Sung dynasty the number continually increased, as did that of the kiao tsu and other provincial issues, and the value constantly fell, while that of provisions rose, notwithstanding the efforts of the Government to prevent it. In 1264 the minister, Kia szu tao, tried to substitute a new form of assignat, apparently redeemable in silver, and called *in kuan*, offering one of these for three of the discredited notes; but the depreciation continued.*

Meanwhile the Sung dynasty lost hold on Northern China, which became subject to the Kin Tartars. There, also, copper money became very scarce, in consequence of the disturbed state of the country, and assignats were introduced. These were of the value of 100, 300, 700, and 900 cash respectively and others of two, four, eight, and ten strings of 1,000 cash each. They were to be in circulation for seven years, when they were to be exchangeable for new ones. There were banks in all the provinces, and the Government, to cover their cost, retained 15 cash out of every 1,000 presented for exchange.

The Mongols having conquered a large part of China, soon adopted this ready expedient of raising money. Khubilai Khan issued in the period Chung tung (1260-1263) notes, which were called *pao chao*, or precious paper money. Fresh notes, issued between 1264-1294, were exchanged for the previous issue at one-fifth of the value, those of 1,000 cash replacing the notes of 5,000 cash. This process was again repeated during the years Chi ta (1308-1311), showing how fast they depreciated. The total issue during the 34 years of Khubilai's reign was equivalent to 249,634,290 ounces, or 124,827,144 pounds sterling The issue was continued during the Ming dynasty, but was gradually replaced by metal coinage again. So far had the depreciation gone that in 1448 the chao of 1,000 cash was worth only three.† There is no further mention of paper money in China after 1455.‡ No paper money of the Yuen period survives, but some of those of the Ming do, and one has been figured by Colonel Yule.

Marco Polo has a graphic account of these notes which I shall transcribe He says: "The Emperor's mint is in the city of Cambaluc (*i.e.*, Khanbaligh or Peking), and the way it is wrought is such that you might say he hath the secret of alchemy in perfection, and you would be right, for he makes his money after this fashion: He makes them take the bark of a certain tree, in fact, of the mulberry tree, the leaves of which are the food of the silkworm

---

* *Id.*, 381-383.   † Yule's Marco Polo, 412-414.   ‡ Klaproth, op. cit., 387.

Then what they take is a certain fine white bast or skin, which lies between the wood of the tree and the thick outer bark, and this they make into something resembling sheets of paper, but black. When these sheets have been prepared they are cut up into pieces of different sizes. The smallest of these sizes is worth a half tornesel; the next, a little larger, one tornesel; one, a little larger still, is worth half a silver groat of Venice, another a whole groat, others yet two, five, and ten groats. There is also a kind worth one bezant of gold, and others of three bezants, and so to ten. All these pieces of paper are issued with as much solemnity and authority as if they were of pure gold or silver, and on every piece a variety of officers, whose duty it is, have to write their names and to put their seals, and when all is prepared duly, the chief officer deputed by the Kaan smeareth the seal entrusted to him with vermillion, and impresses it on the paper, so that the form of the seal remains printed upon it in red. The money is then authentic. Anyone forging it would be punished with death, and the Kaan causes every year to be made such a vast quantity of this money, which costs him nothing, that it must equal in amount all the treasure in the world. With these pieces of paper, made as I have described, he causes all payments on his own account to be made, and he makes them to pass current universally over all his kingdoms and provinces, and territories, and whithersoever his power and sovereignty extends; and nobody, however important he may think himself, dares to refuse them on pain of death, and indeed everybody takes them readily, for wheresoever a person may go throughout the Great Khan's dominions he shall find these pieces of paper current, and shall be able to transact all sales and purchases of goods by means of them, just as well as if they were coins of pure gold, and all the while they are so light that ten bezants worth does not weigh one golden bezant."

Polo adds that merchants who brought gold, gems, &c., from India and elsewhere were prohibited selling to anyone but the Emperor, who paid a good price for their wares in paper money, with which they could trade in any part of the Empire. "And it is a truth that the merchants will, several times in the year, bring wares to the amount of 400,000 bezants, and the Grand Sire pays for all in that paper, so he buys such a quantity of these precious things every year that his treasure is endless, whilst all the time the money he pays away costs him nothing at all. Moreover, several times in the year proclamation is made through the city that anyone who may have gold or silver, or gems, or pearls, by taking them to the mint shall get a handsome price for them, and the owners are glad to do this, because they would find no other purchaser give so large a price. Thus the quantity they bring in is marvellous. Those who do not choose to do so may let it alone; still, in this way nearly all the valuables in the country come into the Kaan's possession. When any of these pieces of paper are spoilt—not that they are so very flimsy neither—the owners carry them to the mint, and by paying three per cent. on the value they get new pieces in exchange; and if any baron or anyone else soever have need of gold or silver, or gems, or pearls, in order to make plate, girdles, or the like, he goes to the mint and buys as much as he lists, paying in this paper money."[*]

* Klaproth, op. cit., 409-411.

CHAPTER VIII.

GHAZAN KHAN.

KHULAGU and Abaka were two important figures in Asiatic history. They conquered and controlled a vast empire with vigour and prudence. Their successors, until we reach the reign of Ghazan, were for the most part weak and decrepit rulers, whose authority was gradually disintegrating. Had it not, in fact, been for the utter desolation and prostration caused by the campaigns of Jingis and Khulagu in Persia, they would undoubtedly have been driven out and displaced ; and, as it was, a very little more aggressive vigour on the part of the Egyptian rulers who controlled the various forces of Islam would no doubt have led to the collapse of the empire of the Ilkhans. With the accession of Ghazan that empire entered upon a new lease of power. He was not only the greatest of the Ilkhans, but also one of the most important figures in Eastern history, while he was fortunate in having as his vizier the famous historian, Rashid ud din, to whom we are indebted for so many details of the Mongol polity, and we consequently have abundant materials for illustrating his reign. He was the son of Arghun, by Kultak, the daughter of Bitekji Kehin, of the tribe Durban (D'Ohsson calls her Kutluk Igaji), who was very beautiful, and thus matched her husband, of whom Haithon says : "Iste vero Argonus fuit aspectu pulcherrimus."* He married her when but twelve years old, and was very fond of her. We are told that he would have gone out to meet the bridal procession on its way to the camp, contrary to Mongol etiquette, but for the advice of his amirs, Sertak and Jujegian. Von Hammer's eastern phrase, "the oyster of purity bore the pearl of royalty," means that she was the mother of Ghazan, who was born at Abisgun (D'Ohsson says at Sultan devin), on the borders of Mazanderan, on the 4th of November, 1271 (D'Ohsson says the 30th), when the horoscope showed the conjunction of two lucky stars. His nurse was Moghaljin, the wife of the Chinese Ishik, who came to Persia with Kultak. She was an *usenin* (*i.e.*, one well versed in old songs and tales), and had a pleasant expression. It was a rule among the Mongols that one

---

* Op. cit., lib. 38.

who was suckling a prince was to keep away from her husband, but as, notwithstanding this prohibition, she was found to be pregnant, the child was removed from her and given in charge to the mother of Hasan, who became the chief of the Tukjis or standard bearers. When the boy was three years old he mounted a horse for the first time. When his grandfather, Abaka, heard of this he sent for him. Arghun took him himself to Konghuralank,* where Abaka was, and the latter met him, took the boy off the horse, and seated him beside himself on his own saddle. He took such a fancy to him that he asked Arghun to allow him to undertake his education, and he was given in charge of the famous Princess Bulughan, who had no son, and only a daughter named Malik, and who was delighted, saying it was a present from God, and that she would take care of the child as if he were her own son. Arghun took him accordingly to Sughurluk, where Bulughan lived, and leaving his two dependants Kukamachar and Kalkai Altun Buka with him, he returned to Khorasan. Abaka had the boy constantly by him when hunting, and at his meals when he was feasting or drinking. The boy did not care for boys' games, but loved weapons and horses, and we are told that, like Napoleon, he engaged in mimic fight with his companions. D'Ohsson says they made puppet soldiers of felt, and ranged them against one another. They also used to imitate courts of justice. When he was five years old he was handed over to the Chinese Barik Bakhshi, to learn the Mongolian and Uighurian writing, and the other studies in the Lamaist curriculum. At ten he changed the pen and the ruler for the bow and sword, in using which, and playing at mall or polo, he afterwards greatly excelled.†

When Abaka marched against the Karaunas in Khorasan he was met at Semnan by Arghun, who was much pleased to again see his boy Ghazan. In the mountain of Akhid, between Semnan and Damaghan, Abaka engaged in hunting, and Ghazan, who was but eight years old, killed a stag, and was duly smeared with its fat in the approved Mongol fashion by the famous hunter, Kurji Buka. This was celebrated by a three days' feast. As it was still winter, Abaka went off towards Bostam, while the Princess Bulughan, with Ghazan, took the route of Mazanderan towards Nervbirun, and joined Abaka again near Radkian. The Ilkhan now went towards Kutujan and Herat, and having sent Arghun towards Ghur and Garja to oppose the Karaunas, Ghazan asked to be allowed to act as his father's cup-bearer. Abaka was pleased with the boy's request, and we are told the ceremony of installation as cup-bearer, equivalent among the Mongols to the western conferring of knighthood, was gone through in the garden of Husein, near Tus.

Abaka commended his grandson to Seljuk, the wife of Arghun, and to

---

* D'Ohsson calls it Kunkur ulong.   † Ilkhans, ii, 1-5. D'Ohsson, iv. 153-154.

the Chinese Okbakhshi, so that he might under their teaching improve
his writing, while he was in summer quarters at Demavend. In the
autumn he rejoined his grandfather at Rai, who used sometimes to put on
an old cap and creep away furtively to his tent to romp with him. On
the advice of Isht Ikaji he was not allowed, as was usual with Mongol
princes, to have a cushion on his saddle. He became, as the Mongol
proverb has it, " like a tooth in tender flesh," and was named in consequence, Ghazan, or the tooth.* When his grandfather died he was ten
years old. He spent the next winter with Bulughan at Baghdad, and in
the spring went to his father in Khorasan. The latter now married
Bulughan, as we have seen, and Ghazan lived with them. When Ahmed
Tekudar marched against Khorasan, Arghun sent Ghazan to meet him at
Semnan, and got permission to return to Bostam. When on one occasion
Ildar made some false accusations against Arghun in Ahmed's presence,
Ghazan answered him so eloquently that they were all surprised. When
Arghun mounted the throne he was accompanied to Azerbaijan by
Bulughan, leaving Ghazan as his deputy in Khorasan, with the ogruks,
his personal effects, and also the ayu-oghlans (i.e., good boys, or pages).
On Bulughan's death Arghun took to himself some gold and silver vessels
which belonged to her, but her pearls and jewellery were made over to
Ghazan, as Abaka had arranged. These were very valuable, since Abaka
was in the habit of presenting her with the most valuable objects he
secured in his campaigns. D'Ohsson says these objects were made over
by Arghun to another of his wives, named Bulughan, who was the
daughter of Utuman, a Konkurat, and not to Ghazan, as Von Hammer
says.† On Arghun's accession to the throne Ghazan was made governor
of Khorasan. The history of his administration there, and of his
struggle and eventual reconciliation with Nuruz, has been related in the
account of the reigns of Gaikhatu and Baidu.

We will now resume our story. We have seen how Ghazan sent his
generals in pursuit of Baidu. He himself set out from Aujan, and made
his solemn entry into Tebriz on the 5th of October, 1295. Sadr ud din
of Zenjan, the vizier, who had monopolised a great deal of power, set out
to meet him, surrounded with great pomp. The amir Doladai, or Tuladai,
told him to put this aside, and as he took no heed, that truculent general
struck him over the head with a whip and compelled him to return.‡
Prince Sukai and the chief judge, at the head of a number of imaums,
sheikhs, ulemas, and seyids, went out to meet Ghazan. He alighted at the
palace built by his father in the meadows of Shems. Von Hammer calls
the palace Shenb, and tells us that a suburb there still bears this name.
His first edict commanded his people to live in peace with one another,
forbade the grandees to oppress those below them, and enjoined upon all

---

* So says Von Hammer, but Ghazan, as we have seen, meant a kettle in Mongol.
† Ilkhans, ii. 5-8. D'Ohsson, iv. 154-5.   ‡ Ilkhans, ii. 30.

the observance of the precepts of religion.   Orders were issued to destroy
all the idol temples, the churches, synagogues, and fire towers, in fact, all
buildings prohibited by the Mussulman law.   The idols were broken and
tied to pieces of wood, and promenaded through the streets of Tebriz.*
Nuruz had already, before this, when pursuing Baidu, given orders to
destroy these buildings, to kill the Buddhist priests, to treat the clergy with
contumely, and to insist on their paying taxes like other people.   Christians
were not to appear in public without having a zonar, or peculiar girdle,
about their waist, nor Jews unless they wore a special head-dress.   The
people of Tebriz destroyed all the churches in that town, and it is impossible,
says the continuator of Abulfaraj, to enumerate the indignities the Christians
had to endure, especially in Baghdad, where none of them dared appear
in the streets.   Their women used, in consequence, to do all the buying
and selling, as they could not be distinguished from the Muhammedan
women, but if they were recognised they were insulted and beaten.   The
Christians were asked jeeringly : "Where is your God?   We will see if,
you have a protector or a liberator."   The same persecution extended to
the Jews and to idolatrous priests, says the same author, and it was
particularly hard for the latter, who had been so tenderly treated by the
Mongol sovereigns, who used to spend immense sums in making gold and
silver idols.   Many of these priests now became, at least outwardly,
Mussulmans, though secretly favouring the old faith.   Presently, Ghazan
issued a yarligh, and sent commissaries to destroy the churches and
monasteries.   His emissaries were easily bribed not to do their duty.·
Thus, at Arbela, or Irbil, they waited for twenty days to see if anyone
would offer them some money, and as no one came, not even the
metropolitan, they allowed the rabble to begin their work of destruction,
and on one day, the 28th of November, two beautiful churches, one
belonging to the Jacobites and the other the Nestorians, were destroyed.
At this news the people of Mosul were much disturbed.   When the
commissaries reached the latter town they were met by those who
offered them a large sum of money.   To make this up they took the
sacred vessels, the censers and crosses, and stripped the volumes of the
Gospels of their rich coverings, and also took money from the Christians
in the neighbourhood, and thus got together a sum of 15,000 dinars, with
which they redeemed the churches, and none of them were in consequence
injured.†   During these troubles the Mussulmans seized the church
which the Nestorian patriarch Makika had had built in the palace of the
Devatdar, at Baghdad, and which had been made over to him by
Khulagu.   Haithon says these harsh measures against the Christians
only characterised the beginning of his reign, and were meant to conciliate
the Mussulman faction, to which he owed his elevation to the throne, and

---

* D'Ohsson, iv. 144-145.   † Chron. Syr., 644-645.   D'Ohsson, iv. 145-147.

he was afterwards much more tolerant.* Stephen the Orpelian gives a grim account of their immediate effect. He says many churches were destroyed and priests killed, and those who escaped were plundered, while their wives and children were made slaves of. At Baghdad, Mosul, Hamian, Thavrej, and Meragha the persecution had most dire results. In Armenia the churches and monasteries of Nakhchivan were plundered, their doors broken, and the altars overturned, but out of respect for their Georgian allies the churches were not destroyed. The metropolitan church of Siunia, of which the chronicler was then bishop, paid a ransom for its safety. The Nestorian Catholicos of Meragha was captured and put to great indignities. Ter Tiratsu, bishop of the Church of the Apostles (probably that of Dadi Vank or Kutha Vank, dedicated to St. Thaddeus,) was tortured and robbed, while his monastery, in which reposed the remains of St. Thaddeus, was ruined.†

As we have seen, Ghazan had built several Buddhist temples at Kabushan, or Kaushan, where he spent much of his time talking, eating, and drinking with the Bakhshis, and worshipping the idols. After the Mongols settled in Persia there went thither a great number of Lamas from Kashmir, India, Uighuria, and China, who had built Buddhist temples in many places, and Buddhism for a while threatened to resume the status which it possessed six centuries before, when it was driven out by Muhammedanism. Rashid argues that Ghazan's conversion was not one of mere policy, and quotes a conversation he himself had had with the Khan, in the course of which he said : "There are sins which God never forgives. The greatest of these is the worship of idols. I have been guilty of this myself through ignorance, but God has enlightened me. Those who first made an idol only wished to commemorate the memory of some man more perfect than the rest. Full of confidence in his merits, they accepted him as their intercessor, and supplicated him to secure their prayers being answered, forgetting that this man when alive had never asked for such deference, nor would he in fact have permitted that anyone should prostrate himself at his feet. Living in humility, he would have thrust down into the nether regions those flatterers who give birth to pride in the heart. They address their prayers to him, but how can he tolerate men who adore the image of his body? It will not hear them. The body is nothing without the soul which animates it ; one is the image of hell, the other of paradise. An idol is only fit to be used as a threshold upon which travellers may tread. The soul will be delighted to see the image of the body in that position of humility by which it attained its perfection when they were united. Men would thereupon say, 'Since the body belonging to a soul so perfect is reduced to dust, and its very image to a doorstep, what should our body

---

* Op. cit., xli.   † Hist. de la Siounie, 261-262.

be to us, who are so far from perfection?' Such reflections will cause
them to neglect their perishable bodies. They will think only of the
soul—of life eternal, and where the blessed dwell. It is thus they
may improve the present life, for man was only made to pass from this
land of shadows to the land of light." This prince, says Rashid, frequently
spoke thus, in a more lofty strain than any of the philosophers.[*]

Ghazan having learnt that many of the bakhshis, as was natural, still
clung 'o their old faith, gave permission to such of them as wished to go
back to Kashmir, Tibet, &c., to return, while those that remained he
insisted should honestly conform to Islam. Anyone who made a fire
altar or an idol temple was to be put to death. As many still remained
unchanged he said that his father, who had lived and died an idolater,
had built himself an idol temple, which he (Ghazan) had reduced
to ruins. They were at liberty to go and live there upon the income
which had been devoted to its service. The amirs and princesses
then addressed him, saying, "Your father built a monastery, on whose
walls was painted his portrait. Now that the building is decayed the rain
and snow are ruining these pictures. Send some Lamas to live there, and
thus secure repose for your father's soul." This advice having been
rejected, it was suggested that it should be converted into a palace. He
said, " I intended building a palace, but I would not build one in a place
where idolatrous monks have lived." Eventually he forbade the Lamas,
on pain of death, practising their religion openly. Otherwise they might
continue to live in the country, and many of them accordingly did so.[†]

When Ghazan was safely seated on the throne he proceeded to punish
several chiefs who had compromised themselves in the troubles of the
previous years. Among these was Ilchidai Kushji, who went to swear
allegiance with Prince Alafreng, the eldest son of the Khan Gaikhatu,
and was put to death without trial. Stephen the Orpelian calls him an
excellent man in all respects, and commander of 10,000 men. Nuruz and
Kutlugh Shah arrived on the 12th of October from their expeditions
towards the Aras, in pursuit of Baidu's officers, several of whom they took
back with them in chains. Bulughan, Ghazan's wife, interceded vigorously
for Kunjukbal, but Nuruz, who had to avenge his father-in-law Akbuka's
death, as well as a private grievance, in that the amir had urged Baidu to
put him (Nuruz) to death, pressed for his head so eagerly that he was
executed on the 15th of October.[‡] Tukal, who as we have seen had filled
an important position in Georgia, and had been closely allied with its
king, David, also suffered the same fate. We are told he had fled to
Beka, a Georgian prince who was practically independent of David, and
sent his son to the latter. Ghazan sent to David and Beka, demanding
the surrender of Tukal. Twice they refused to give him up, and pleaded

---

* D'Ohsson, iv. 147-150.   † Quatremere, 195. Note.   ‡ D'Ohsson, iv. 150.

for him. At length, Ghazan having given his word and sent his ring as a proof of good faith, he was sent towards Tebriz, but he was waylaid *en route* at Nakhidur, by Kurumchi (*i.e.*, Kurmishi), who killed him. David also surrendered Tukal's son, with all the wealth he had deposited in the fortress of Ateni, in charge of Barejan, mthawar of Ossethi. Prince Ildar, who fled towards Karin, hid away in the house of a Sheikh, but was also put to death.* Doladai, Chichak, and Idaju were released after being bastinadoed.†

Ghazan now went for a short time to Kara tepe, whence he returned to Tebriz. The amir Mulai was appointed governor of Diarbekr and Diar Rabia, while Nuruz was given charge of all the uluses from the Oxus to the Euphrates, as the lieutenant-general of Ghazan, who set off for Mughan, in Arran, where, on the 17th October, 1295, he married his father's widow, Bulughan, with Muhammedan rites. A halt was made at Puli Khosrau (D'Ohsson says at Karabagh), where Nuruz with the various princes and princesses arrived, and a formal installation took place on the 3rd of November, 1295 (the 23rd day of the 9th month of the year of the sheep), which had been fixed by the astrologers. Ghazan took the title of sultan, and the name Mahmud. The festivities lasted eight days. When Nuruz was promoted as I mentioned above, Ghazan told him to ask some favour. He knelt down and asked that in future the name of God and the Prophet should appear at the head of all the royal ordinances, while the altamgha, or royal seal, should in future be round, and not square, the circle being the most perfect figure. He also asked that their various ranks should be assigned to the different members of the divan. This was granted, and in future the coinage also bore the profession of the Mussulman faith, and according to the author of the "Shajrat ul Atrak" the names of the Rashidi Khalifs, who during the rule of the Abbasidan Khalifs were included in the kutbeh after the two professions of faith were restored.‡ Ghazan appointed Sadr ud din of Zenjan vizier, and Sherif ud din of Semnan ulugh bitkiji, *i.e.*, secretary or keeper of the seals.

After the installation, Toghachar, whose activity was considered somewhat dangerous at head-quarters, was sent to command the troops in Asia Minor. Meanwhile the Mongols of Transoxiana, taking advantage of the defenceless state of Khorasan, invaded that province. News of the invasion reached the Ilkhan's Court on the 8th of December. Prince Sukai, son of Yashmut, the third son of Khulagu, with the amir Nuruz, were ordered to oppose them. Sukai had meanwhile withdrawn to his yurt, and did not come when sent for, nor until Ghazan dispatched the amir Hirkudak to fetch him. He had used some treasonable language when drunk, but Ghazan took no notice of it, and treated him very

---

graciously. The recent change of rulers and the lavish expenditure of the late Court had denuded the treasury. A special requisition in advance had therefore to be made in the various provinces, and especially Fars, and two out of every ten of the cattle there were appropriated, while the officials at Tebriz had to furnish several tumans of gold in advance. Nuruz now set out. With him went Prince Sukai (to whose person were attached two chiefs of tumans, viz., Barula the Olkonut, and Arslan Oghul, the grandson of Juchi Khasar, Jingis Khan's brother), and the general Hirkudak, Nuruz's brother Hajfu Narin, Satilmish, and others of his old and faithful supporters, were appointed their substitutes.* Sukai and Barula, who marched with the advance guard, having halted on the river Kiuhe (called Kéré by D'Ohsson), also known as Turkan muran, or the Turkish river, made a plot to kill Nuruz and to dethrone Ghazan, whom they reproached for having become a Muhammedan. Sukai was to be put on the throne. They sent a message to Prince Taiju, son of Mangu Timur, to join them. He pretended to join them, but in reality informed Nuruz, who put his troops in ambush and left his camp. The conspirators assailed his tent, whereupon his men came on from their ambush. Barula was killed, and Sukai fled towards Kharkan and Sava. Hirkudak was sent in pursuit, and captured him near Kharkan, and ordered Sati, who had had a hand in the conspiracy, to put him to death in the mode prescribed for Mongol princes, but Sukai ran a knife into his stomach. This weapon was taken from him by Beitimur, a follower of Hirkudak, who dispatched him in turn. On the 7th of March, 1296, the Prince Ildar, son of Kunkuratai, ninth son of Khulagu, fled with 300 men. Shadi Gurkan was sent in pursuit of him with 3,000 men. He overtook and put him to death at Erzenrum. Yedutai, the son of Tashiminku, who had created a disturbance at Diarbekr, and Buralghi Katai, the lance bearer, were also put to death. These executions were doubtless the result of the conversion of the Court to Islam.

Ghazan had set out from Abubekrabad on a hunting expedition to Aktagh, when he heard of Sukai's outbreak. He at once returned, stopped at the bridge of Mangu Timur, and sent the amirs Kutlugh Shah, Satilmish and Sutai, who were with him, to the assistance of Nuruz, and in order to capture Issen Timur, the brother of Ildar, and Kurmishi, the son of Barula, who had taken up arms to avenge the deaths of their relatives. They were brought to trial and executed, as were also Chichek and Doladai, who had been implicated in the revolt. Of the various rebels, Arslan Oghul, who is described by Stephen the Orpelian as of royal blood, alone remained at large. He marched and encamped with his people near the palace of Mansuriah. (Stephen says at the Ilkhan's palace at Mughan.) Ghazan did not feel himself strong enough to oppose him,

---

* D'Ohsson, Iv. 156-157. Ilkhans, ii. 32.

but as he knew that his only safety was in showing a bold front, he acted as if he was unaware of the enemy's proximity, ordered his men to engage in hunting, and himself followed leisurely, performing his usual occupations. His dissimulation and firmness saved him, for before his men were aware of their danger he was joined by a large contingent under the amir Choban. The latter, with Sulamish, Toghrilji, Taitak, and Kurmishi, son of Alinak, attacked Arslan near Bailekan. They were beaten, but having received a reinforcement of 2,000 men under Hirkudak, were prepared to renew the fight, when the rebels submitted. Many chiefs had fallen on both sides, but victory leaned to Ghazan's men. Arslan Oghul and Prince Tulek, the son of Aujan, were captured and put to death. This was on the 28th of March. In the short space of one month five princes of the blood and thirty-eight amirs were executed, thus clearing away a large number of competitors for power, and also the most energetic and restless spirits in the army.*

While these proceedings were going on in one part of Ghazan's dominions, a horde of Uirats, or Kalmuks, encamped in the province of Baghdad (Rashid ud din says in Diarbekr), passed into Syria. Their leader, Targai Gurkan (called Tukai by Von Hammer, Tongai by Quatremere, and Taragaih by Abulfeda), who had the position of chief of a tuman, and had married a grand-daughter of Khulagu and daughter of Mangu Timur, whence his title of Gurkan, had been threatened with death for having assisted Baidu against Gaikhatu, and Mulai, the new governor of Diarbekr, had received orders to arrest him and the other leaders of the Uirats. He thereupon crossed the Euphrates, and when pursued by Mulai defeated and killed many of his men.† The continuator of Abulfaraj tells a different story. He calls the Uirats Avirathei, and says they wintered near the monastery of Mar Matthew, and that during the reign of Baidu they had taken from the Turkomans a great number of cattle, horses, sheep, swine, mules, and camels. Ghazan issued orders that this plunder should be restored on pain of death, but inasmuch as a large proportion of it no longer existed, the Uirats, who were hard pressed by the commissaries of Ghazan and of the Turkomans, put them to death, and then migrated, to the number of 10,000 warriors, with their families and property.‡ The Egyptian historians tell us that the fugitives numbered about 18,000 tents. The Sultan Ketboga, when he heard of their arrival, sent orders to the naib or viceroy of Syria to dispatch the amir Alem ud din Sanjar, the devatdari, to the town of Rahbeth to meet them, and sent two of his officers, viz., Kara Sonkor Mansuri and Seif ud din Alhaj Behadur Halebi, the hajib, from Cairo, to meet them at Damascus. There the leaders of the Uirats, to the number of 113, arrived on the 30th of January, with Tugai, Alus, and Kakbai at their

* D'Ohsson, iv. 155-159. Ilkhans, ii. 32-34. Hist. de la Géorgie, 620. Note.
† D'Ohsson, iv. 159-160. Abulfeda, v. 127. ‡ Chron. Syr., 646. D'Ohsson, iv. 160. Note.
C*

head. The naib and amirs went out to meet them amidst considerable pomp. They then set out for Cairo, conducted by Kara Sonkor. The Sultan received them very graciously, and conferred on several of them the rank of grand amir ; but they remained idolaters, nor did they take heed of the feast of Ramazan, and the orthodox Egyptians shrank from taking their posts beside them when on guard at the entrance of the castle. They fed on the flesh of horses which had not had their throats cut, and which had merely had their limbs tied and were then killed by being struck over the head. The people grumbled at the honour paid to these strangers, and the prince was blamed for showing them so much consideration. Orders were given to settle the horde in the province of Sahel. When they reached Damascus they were encamped outside, and provision dealers were ordered to go to them and plant their booths in the meadows near the village of Damin, and at Kisueh. The Uirats were forbidden to enter Damascus itself. The amir Sanjar remained with them. Many of them died. We are told that their children were beautiful, and that the officers and troopers adopted the boys and married the girls, while the men were distributed among the various troops, and having become Mussulmans were speedily absorbed.* Abulfeda tells a different story, and says these Uirats, urged by their leader, became Mussulmans, and that they were settled at Cacun, which was a place near Cæsarea, in Palestine, called Caco by William of Tyre.† In regard to the invasion from Transoxiana, we are told that the enemy withdrew with a great booty on the approach of Nuruz.

Ghazan was now visited by the King of Cilicia, or Little Armenia. Leon III., the king of that country, had died on the 7th of January, 1290, leaving five sons—Haithon, Thoros, Sambat (or Sempad), Constantine (called Condin by Abulfeda), and Ushin.‡ Haithon succeeded him, and is known as Hethum or Haithon II. When the Egyptians captured Tyre, in 1291, he had sent to the Pope and the other principal rulers of Europe to ask for aid. The Pope sent a special request to King Philip of France, and also dispatched a general summons to the Western Christians, to go to the assistance of their brothers in the East, but it was in vain. In 1293 the Egyptian Sultan Ashraf marched upon Armenia. The King sent envoys to implore his pity. The Sultan insisted upon the surrender of Behesna, Merash, and Tel Hamdun, as the price of peace. The demand was immediately granted. Behesna had, before the invasion of Khulagu, belonged to the rulers of Syria, but at that time the officer who commanded in these parts for Nasir sold it to the Armenian king for 100,000 dirhems. After reigning for four years Haithon surrendered the crown to his brother Thoros, and went into a monastery of Franciscans, and took the name of John. There Thoros and the great chiefs used to

* Makrizi, ii. (part ii.) 29-30. D'Ohsson, iv. 161-162.
† Abulfeda, v. 129, and Note 84.   ‡ Id., v. 139. Journ. Asiat., 5th series, xvii. 385.

go, however, to consult him on matters of State policy, and at length, under pressure from Thoros and the other grandees, who had assembled at Sis to celebrate the marriage of his sister Isabella with Amauric, Count of Tyre, brother of the King of Cyprus, Haithon in 1295 resumed the reins of government. Having heard of Baidu's accession, and of the favour he showed the Christians, he set out to pay his respects to him. After a journey of about two months he reached Siah kuh (*i.e.*, Sughurluk). At this time Baidu was being hard pressed by Nuruz, and sent word to Haithon to return to Meragha and wait there till things were more pacified. Haithon had only been there a few days, where he was subjected to some insult. Baidu was finally defeated. Ghazan having pitched his camp at Tel Ukhama, or Okma (*i.e.*, the black hill), near Dihburkan, Haithon went to him, bearing magnificent presents. Ghazan reproachfully said that he had gone to see Baidu, and not himself. Haithon diplomatically replied that it was his duty to pay homage to the whole family of Jingis Khan, and to pay his respects to the person who occupied the throne. Ghazan, who laid the blame of much of the recent action against the Christians upon Nuruz, was pleased at this answer, and presented him with some royal robes, gave him a diploma of investiture, and ordered that his wishes were to be carried out. Although orders had gone out for the general destruction of churches, Haithon urged that these buildings should be spared, since they were used in the worship of God. Ghazan consented to their being spared, issuing commands at the same time that the idol temples were to be converted into mosques and colleges. Haithon left the ordu again on the 9th of October, 1296, well satisfied with the services he had rendered to the Church.*

Nuruz was now all powerful. He proceeded to depose the vizier, Sadr Jihan, on the pretext that during the recent troubles he had taken upon himself to issue orders in his own name, although he had acted faithfully in his office, and devoted the taxes obtained to the payment of the troops. He was replaced by Jemal ud din of Destarjerd. His brother, Haji Bey, was made general controller of finances and head of the royal chancellery, while another brother, Nasir ud din Satilmish, was appointed to counter-sign his private dispatches with a red seal. Sadr ud din soon felt another reverse of fortune. He himself told the historian Wassaf that he had dreamt on the night of his arrest, and when he expected to be put to death, that a bright figure with a halo of light about it alighted in the hand of death! At all events, being accused of complicity with the rebels, and this charge being supported by the employés of the Divan, who feared that he knew of their incapacity and dishonesty, he was arrested, put to the torture, and orders were given to put him to death without further trial. Two men accordingly seized him, and

---

* Hist de la Siounie, 262. Cont. of Abulfaraj, 643-644. D'Ohsson, iv. 162-164.

led him, naked and bound, on a pack-horse to the place of execution, which was in a thick forest. Fortunately, his two conductors were under obligations to him for some good service he had rendered them in the reign of Gaikhatu. They delayed his execution till night, when he saw the same apparition which had appeared to him in his dream, and fell from his horse to the ground. At this fortunate moment the amir Hirkudak, who was returning from his expedition against Sukai, came by, and inquired what had taken place. He ordered the execution to be stayed. The next day a list of the rebels was presented to Ghazan. Sadr Jihan's name was not among them, and he was accordingly pardoned, at the instance of the Princess Bulughan, and assigned a residence near the ordu.*

Hirkudak, who had been appointed governor of Fars, set out on the 13th of April, 1296, to arrange the affairs of that province. Its finances he specially confided to Jemal ud din, the mufti of Shiraz, while he himself went to Hormuz, to quell the disorders which had arisen there.† At this time the struggle between Toktu Khan and Nogai, which I have described elsewhere,‡ was rife, and we are told that in consequence of it Chini, the wife of Bukai (? Nogai), with his younger son Turi, fled to Ghazan, who received them well and treated them with honour. Ghazan himself, on the 31st of May, 1296, married his sister Oljaitu (called Oljai Timur by D'Ohsson), the widow of the amir Tukal, to the amir Kutlugh Shah.§ This statement is, however, contrary to the Georgian history, which makes her marry David, King of Georgia.

We have seen how the energetic Toghachar was appointed governor of Asia Minor. Ghazan was not satisfied of his loyalty. The author of the " Shajrat ul Atrak" says he suspected him of having fomented the revolt of Sukai and Arslan. He determined therefore to put an end to him. Khurmenji (called the amir Jirmchi by the author just cited) was charged with this delicate duty. He was first to gain over the military chiefs in Rum, and to lull Toghachar into security by presenting a friendly letter from himself, and then to destroy him. Ghazan's policy was as cold and selfishly calculating as Napoleon's, and on this occasion, according to Rashid ud din, he related to his courtiers a story how, in former times, two princes contended for the throne of China. One of them being defeated fled before his enemy. An officer having found him, took pity on him, and hid him away in a dry well. When the soldiers who were in pursuit arrived at the place, uncertain which way to take, since a violent wind had covered the tracks with sand, the officer urged that further pursuit was needless, and they all accordingly left. Presently the concealed prince came out of his hiding place, collected a body of followers, and defeated and killed his rival. He thereupon promoted the

---

* D'Ohsson, iv. 165-166.　Ilkhans, ii. 34.　　† Ilkhans, ii. 35.
‡ Ante, ii. 143, &c.　　§ Ilkhans. ii. 45.

officer to whom he owed his life to high honour. One of the courtiers
having expressed his surprise that he should have rewarded, and not
punished, a man who had betrayed his prince and caused his overthrow,
he, after thinking a moment, ordered his favourite to be put to death.
" It is I who saved your life," said the officer. " I feel you have reason
to complain," said the emperor, in tears, " I am greatly distressed,
but justice and the interests of the empire demand your death," and
the order was carried out. " It costs me much," said Ghazan, " to put
anyone to death, but if a ruler is not severe when the good of the State
demands, he cannot reign." It must be said that he did not scruple to
carry out his philosophy with rigour. The execution of Toghachar did
not prevent the outbreak of a rebellion in Rum. Baltu, son of Tebsin, son
of Khulagu, had held a command there since the days of Arghun, and
had been very powerful, especially after the death of his colleague, the
amir Shamaghar, who, according to the Syrian chronicle, was a Christian.
He had been summoned several times to the Court, but had always
excused himself, and on the death of Toghachar he rebelled. Kutlugh
Shah was sent in 1297 against him, at the head of three tumans. He
defeated Baltu in the plain of Amasia (Von Hammer says at Maliyeh),
and then returned to Arran, leaving Sulamish behind to pursue him.*
On the 1st of June, 1296, Ghazan set out from Pil Suwar for Tebriz, and
Ainabeg, who had been brought from Khorasan, was put to death. A
report on the state of that province was made to him by the amir Oladu
, and he afterwards went to the place called Sain (*i.e.*, the good), between
Sireh and Irbil, where he held a kuriltai.†

Ghazan's theory about breaking his subordinates was now going to be
tested in the person of no less a grandee than Nuruz, to whose help he
owed so much, and whose arrogance had been aroused with his success.
He naturally created enemies. Among these was Nurin Aka, the military
governor of Khorasan and Mazanderan, who was a great favourite of the
Ilkhan, owing to his reputation and his descent, for he belonged to the
famous stock of the Kiats. Nuruz complained that he had slighted his
brother Uiratai. He attributed the success of the recent invasions of
Khorasan to the negligence of Nurin, and in careless conversation let fall
some bitter phrases about him, and hinted that Prince Taiju, the son of
Mangu Timur, who at this time arrived at Temisheh, in Mazanderan, had
gone there to report about himself. This caused them both to have an
ill feeling towards him. On the 6th of June he sent Taiju and Nurin to
Radkian, where an outbreak had taken place, and then suddenly left his
post in Khorasan, and set out for Azerbaijan, with the excuse that he
wanted to see his sick wife (Tughan, Abaka's daughter and Ghazan's
aunt), and also the Ilkhan. A number of troops whom Ghazan had sent

to reinforce him thereupon dispersed, or went with him, and, *inter alios*, the amir Sum returned to his yurt at Irbil with 400 horsemen. Ghazan was much displeased at this, and gave orders that he was to return. He, however, asked to see his wife, and arrived at the Court on the 23rd of June, 1296, where he was received by his master with at least outward courtesy. Kutlugh Shah and other amirs urged that he was too dangerous and ambitious a person to be allowed to retain the government of Khorasan, while the sons of the judge Nogai, whom Nuruz had put to death, clamoured for the blood penalty. Ghazan resisted these importunities, saying he could not violate his word on simple suspicion. He was ordered to return to Khorasan, and soon after his wife, the Princess Tughan or Tughanjuk, died, and Nurin returned to the Court, and was well received by Ghazan, who appointed his own brother, Khodabendeh, commander of the troops of Khorasan.* On the 11th of July Ghazan left Sain, and went by way of Naurdul to Tebriz, where he began the building of a country house and garden. Executions meanwhile continued to be frequent. On the 2nd of September, 1296, Dundi, the widow of Gaikhatu and mother of his son Alafreng, and Baighut, son of Shiramun, son of Charmaghan, the famous pioneer of Mongol conquest in Western Asia, were put to death at Sehkünbed (*i.e.*, "the three domes"), in Mazanderan.†

We must here shortly digress to consider the affairs of the Lesser Luristan. We have seen ‡ how this principality, on the death of Bedr ud din Masud, in the year 658 HEJ., was divided between his two sons Jemal ud din Bedr and Nasir ud din Omar, who struggled with Taj ud din Shah, the son of Hosam ud din Khalil, the brother and predecessor of Masud. The two former were put to death by order of Abaka, who nominated Taj ud din as ruler of the country. He was a famous caligraphist and a good ruler, and was also put to death in the year 677 HEJ., by order of Abaka Khan, who divided the Lesser Lur between Falak ud din Hasan and Iz ud din Hussein, sons of Bedr ud din Masud. They ruled the country for fifteen years, and were very successful in struggling with their neighbours, and we read that the country from the province of Hamadan to Shuster, and from Ispahan to the frontiers of Arabia, was subject to them. They had an army of 17,000 men. Hasan, we are told, was wise, witty, and devout, while his brother was revengeful, and had no pity for criminals. Both died in the year 692, during the reign of Gaikhatu Khan. Each of them left a son, but by orders of the Ilkhan the province was intrusted to Jemal ud din Khizi, son of Taj ud din Shah. He was opposed by a considerable party of his relatives. They surprised him at a hunting party, and slew him and his family, so that the stock of Hosam ud din Khalil was exterminated. The leader of the rebels, Hosam ud din Omar, the son of

---

* D'Ohsson, iv. 172-174. Ilkhans, ii. 36-37.　　† Ilkhans, ii. 38.　　‡ Ante, 140.

Shems ud din, succeeded to the throne, but the usurper was in turn opposed by several princes, who declared he had no right to it, not being of royal blood, but that the kingdom really belonged to Samsan ud din Mahmud, son of Bedr ud din Muhammed, son of Iz ud din Gurshasp, the predecessor of Khalil, above named. He was a soldierly man, and brought an army from Khuzistan to Khoramabad. It seems that Hosam ud din only controlled part of the country, Shihab ud din Elias, one of his co-conspirators, and apparently of the royal stock, having also a large authority. Hosam ud din was persuaded to resign. Shihab ud din fought for his position, but was defeated, and fled to a snowy mountain, whence he was brought down after some trouble and was slain. Samsan ud din was now acknowledged as ruler of the Lesser Luristan. Meanwhile, a grandson of the Sheikh Kamuyah Buzurg reported to Ghazan the recent revolutions. Samsan ud din and Hosam ud din were summoned to the Court to answer for the deaths of Khizr and Elias. They were both executed, and Iz ud din Mahmud, son of the amir Iz ud din Husein, was appointed atabeg of the Lesser Lur. This little dynasty survived till the sixteenth century.*

On the 18th of September Ghazan set out from the neighbourhood of Meragha by way of Hamadan, to his winter quarters at Baghdad. He halted at Rek, near Hamadan, for a month, and displaced Jemal ud din, of Destajerd (who had been appointed vizier under the auspices of Nuruz), by Sherif ud din, of Semnan. At this time Afrasiab, the Prince of the Greater Lur, went to him to obtain confirmation of his position as atabeg of Great Luristan. Unfortunately for him, as he was returning home after a successful visit, he was met by the General Hirkudak, who was on his way from Fars, and who compelled him to return with him to the ordu. He complained to Ghazan that when he passed through Afrasiab's territory, while on his way to Fars, he had not supplied him with a measure of barley or a bundle of hay, that his officers had treated the tax collector whom he had sent into the mountains of Kiluyeh with incivility, and refused to allow him to collect the taxes there, saying they belonged to their master. He also recalled the troubles the atabeg had formerly caused, and expressed his astonishment that he should be allowed to return home. Ghazan ordered his execution, which took place before the Royal tent. This was in the year 669 HEJ. The principality was given to his brother, Nasir ud din Ahmed, who restored prosperity to the province, and, we are told, made it the envy of paradise. Ahmed reigned for 38 years.†

Two days later, Jemal ud din of Destajerd, the former vizier, was put to death. His fall made the fortune of Jemal ud din of Shiraz, whose financial skill is described by Wassaf in hyperbolic terms, and who at this

---

* Hamdullah *passim*. Ilkhans, ii. 37-38. D'Ohsson, iv. 171.
† D'Ohsson, iv. 169-170. Ilkhans, ii. 38-39.

time went to the Court to vindicate his conduct of the finances and to silence his accusers. The latter charged that he had levied 200 gold tumans, or more, in taxes on land and sea, and had similarly levied a sum of 1,500 tumans from the pearl fishery. During the six years he had had the farming of the taxes he had received by order of Arghun forty-two tumans, and in addition had arbitrarily taken thirty-three tumans, making altogether seventy-five tumans. Jemal ud din replied that in the best years the pearl fishery only supplied 750 pearls, in ordinary years 450, and in the worst ones only 250. Ghazan and the judges of the court, the amirs Nurin and Satilmish, as well as the inaks Sutai and Taremtas, were convinced of his innocence, and the diploma he had received from Arghun was renewed. One hundred tumans of the revenue were to be forthcoming at the new year, and Jemal ud din was made controller of all the taxes in Irak for three years, as well as of those of Shebankyareh and Shiraz. He was given a robe of golden tissue, five round golden paizahs of the first order, marked with the lion's head, three of the finest sonkors or falcons, together with a golden tent, and his reward was proclaimed three times at Baghdad, Shiraz, Kaiz, and Bahrein. While Ghazan was at Hamadan he sent his brother Khodabendeh as his deputy to Khorasan. At Baghdad he alighted at the palace Mobni, and hunted for four-and-twenty days in the neighbourhood of Naamaniyeh and Hilleh, whence he returned again to Maasaniyeh. He visited the tomb of the Imaum Abu Hanifeh, and was present at the Friday prayer in the mosque. At this time, viz., on the 14th of March, 1297, his son Oljai Kutlugh was born at Shehraban.*

The enemies of Nuruz continued their intrigues against him. When Jemal ud din of Destajerd was tried, it came out that at the time Nuruz was very zealously working for the conversion of Ghazan, and to secure him the throne, he had sent as his agent a Baghdad trader, called Alem ud din Kaissar, who often went to and fro between Syria and Egypt, to invite the co-operation of the Egyptian Sultan, as a good Mussulman, on behalf of his *protégé*. When the messenger returned, Baidu had been deposed, and Ghazan was on the throne. The answer of the Sultan was apparently not quite what Nuruz expected, and he determined not to show it to his master, but persuaded Jemal ud din to compose another answer in the Sultan's name. This having been transcribed by a clever writer, was duly presented to Ghazan. The enemies of Nuruz having heard of the deception, and of the journey of Kaissar, accused Nuruz of having had secret intrigues with the Sultan. Nuruz had noticed when last at the ordu that he was not in favour, and when he got to Khorasan he sent one of his confidantés, named Sadr ud din, son of the mufti of Herat, to be his agent at the Court. Ghazan determined to employ him to counteract the supposed plot of Nuruz. He first sent him to Baghdad,

---

* Ilkhans, ii. 39-40.

where he invited Kaissar to a banquet, and having mixed some soporific with his food, he fell into a profound sleep, and he and his suite were bound with cords.

Sadr ud din, of Zenjan, the vizier who had replaced Jemal ud din, and who was among Nuruz's bitter enemies, now in concert with his brother, Kutb ud din, forged letters addressed to the Egyptian amirs, stating that although Ghazan was a Mussulman, his officers were not, and thwarted the faith in every way, and inviting the Egyptians to march and put down the infidels, and to seize upon Iran. He promised his assistance and that of his brothers Haji and Legsi, and said he sent some robes as a present. These letters, with seventeen complete sets of robes, were furtively concealed among Kaissar's effects, and were found among them when he was arrested. Sadr ud din also forged a letter from Nuruz to his brother, Haji Narin, and concealed it among the latter's things when he paid him a visit. When the presumed plot was divulged to Ghazan, he at once returned from near Hamadan to Shehraban, covering a distance of thirteen parasangs in one day. He arrived there on the 17th of March, 1297. There Kaissar arrived, guarded by Sheikh Mahmud and Kutb ud din. His effects were examined, and the letters and robes were duly found; the former it was declared were in the handwriting of Haji, the brother and secretary of Nuruz. Ghazan, in a great rage, ordered Kaissar and his three companions to be pounded to death with maces, while Nurin and Tainchar were ordered to put to death all the relatives of Nuruz and their dependants. As Von Hammer says, this was a very strong step, for the family of Nuruz comprised the most powerful amirs in Persia. His father, Arghun Aka, the famous governor of Khorasan, left nine sons, who were naturally very influential people. Two of them had married royal princesses, viz., Nuruz, whose wife, as we have seen, was Ghazan's aunt, and Lekesi (or Legsi), who married Baba, the seventh daughter of Khulagu. The other seven were Satilmish, Haji Narin, Barghun Haji, Bulduk, Kerrai, Erdhaighan, and Mengb Baba. The execution began with those about the Court. Satilmish, who represented Nuruz there, was put to death, with his son Kutlugh Timur and Ordubuka, the son of Nuruz. Haji Narin was arrested at Khanikin, and was at once compromised by the forged letter found in his portfolio; he was paraded naked about the royal ordu, and then put to death, his goods being pillaged. His young son, Taghai, fled to the amir of the camp of Bulughan, Ghazan's wife, who was a granddaughter of Arghun Aka. There he was concealed until Ghazan's revenge had been satiated by the slaughter of his relatives. Legsi was executed in the square of Mardesht; Keshlik and Yul Kutlugh, nephews of Nuruz, were killed in the square which was afterwards called after Kutlugh, and the amirs Shidun and Ilbuka were executed near Kasr Shirin.*

---

* Ilkhans, ii. 41-42. D'Ohsson, iv. 174-177.

Ghazan now summoned several of the amirs. His brother Khodabendeh went to Bisutun from Khorasan. Kutlugh Shah joined him at Esedabad from Mughan, and the amirs Choban and Pulad Kia from Rai. Kutlugh Shah was dispatched from Esedabad with an army to arrest Nuruz ; his subordinates, Sunatai and Hirkudak, each with a tuman, marching in advance. When he reached Damaghan he learnt that the generals just named had already put to death the commanders who had been left at Rai, Veramin, Khawar, Semnan, and Bostam, by Nuruz. At Izferain he was met by the sons of Buka Timur and Alghui, who demanded the vendetta for their father's blood, whom Nuruz had slain. His troops were conducted by Danishmend, a deserter from Nuruz's service. The latter having been defeated in a struggle near Ispahan, in which his two sons, Ahmed and Ali, were taken and killed, his camp and treasures fell into the hands of his pursuers. Pursued by Hirkudak, he arrived at night at Jam, where his horse-herds were. Having put his people in ambuscade behind some high walls, he awaited the approach of his pursuers, and when they rushed at the horses he fell upon them with his people and killed many of them, after which he again took flight towards Herat, where the Malik Fakhr ud din Kert offered him an asylum.*

It will be well here to make a short retrospect, and to see what had taken place recently at Herat. We have seen how Abaka having got rid of the famous Malik Shems ud din, nominated the latter's son, Rokn ud din, in his place. After Abaka's death Rokn ud din withdrew to his, stronghold of Khaisar, appointing his son Alai ud din (D'Ohsson says Ghiath ud din) as his deputy at Herat, with the title of naib. The following year the young prince was visited by Arghun Khan. An important occurrence presently again created ill blood. Hindu Noyan, one of Arghun's principal generals, having revolted, sought shelter at Khaisar. Ordered to surrender him, Rokn ud din did so, and was rewarded with a khilat, a drum, and standards, the insignia of royalty. The relatives of Hindu Noyan did not forgive him, and knowing they were intriguing against him he determined not to leave his fortress, and also withdrew his son from Herat. Abandoned in this way, the town was given up to disorder, the richer inhabitants migrated, while a leader of the Nigudars, named Amaji, at the head of a number of his truculent people, fell upon the place, pillaged and virtually ruined it. Quarters where previously there had been 100 ketkhôdas remained with only two or three.

When Ghazan received Khorasan as an appanage, in 1291, he sent Nuruz with 5,000 men to restore order there. He took a quantity of cattle which he had captured in the neighbourhood of Dereguez thither,

---

and sent orders to Esfizar, Ferrah, and Seistan, to recall those who had migrated, and exempted the place from the payment of taxes for two years. Nuruz having thus somewhat restored prosperity to the place, sent to Rokn ud din to come down from Khaisar and take charge once more of the city. He excused himself on the ground that he had quite determined to withdraw from the world.* Rokn ud din's eldest son, Fakhr ud din, had been a turbulent person, and had in consequence been imprisoned at Khaisar for seven years. He had then broken his chains, killed his guards, and fled to the mountain which dominated Khaisar, saying he meant no longer to obey his father. Nuruz having heard of this, and of the vigorous qualities of the young man, sent his brother, the amir Haji, with a letter to the malik, to ask him to forgive his son. Rokn ud din replied that his son was very foolish, and that if he was set at liberty he would cause great trouble. Haji, notwithstanding, went to the young prince and informed him of the good offices of Nuruz. He would not leave his retreat, however, until he was assured that he would not be seized by his father. The malik only consented to give him permission to go to Nuruz on receiving an undertaking that he would not be held responsible for anything the young prince might afterwards do. Nuruz replied that he himself undertook to be responsible for Fakhr ud din's behaviour. Fakhr ud din now went to Herat, where Nuruz received him with open arms, and gave him proofs of his goodwill in the presence of the military chiefs and the grandees of Khorasan. Rokn ud din continued to live at Khaisar until September, 1305. The young prince soon gave proof of his capacity. Muhammed Jeji having revolted in the district of Khaf and ravaged it, Fakhr ud din joined the amir Dendai, brother of Nuruz, who marched against the rebel with 5,000 men. Muhammed thereupon withdrew to his fort of Jejd, where he was attacked for four months. Fakhr ud din proved himself very energetic in this campaign, and restored the district of Khaf to obedience. In reward for his services he was nominated ruler of Herat, and Nuruz gave him his niece, the daughter of his brother Tergan Haji (the Barghun Haji of Von Hammer), in marriage.† At this time malik Inaltekin, wishing to revenge his brother, Jemal ud din, who was kept prisoner by Nuruz in a fortress of Gharjistan, left Irak with a large army, and appropriated several of his towns and killed a great number of his partisans. He then fortified himself at Ferrah, but dreading a struggle in which he would be opposed by Fakhr ud din, he undertook to submit if the latter would in turn undertake to obtain the release of Jemal ud din, which was secured without difficulty.‡ We now hear of an inroad made upon Khorasan by Dua, the chief of the ulus of

---

* Journ. Asiat., 5th ser., xvii. 455-456. D'Ohsson, iv. 183-184.
† D'Ohsson, iv. 184-186. Journ. Asiat., 5th ser., xvii. 473-474.
‡ Journ. Asiat, 5th ser., xvii. 473-474.

Jagatai, who sent his relative, Bérĕket, to get the submission of Fakhr ud din in Gharjistan. Bérĕket, with a number of his officers, was seized and taken to Nuruz, who was at Tus, and who conducted them to Ghazan's camp in Irak. Fakhr ud din was also presented to the Ilkhan. He was rewarded with a splendid royal robe, a diploma confirming him as Lord of Herat, a drum and banner, a tent, a special pavilion for his harem, a Mongol regiment of 1,000 men, and ten tumans in money, while Nuruz was ordered to show him every attention.[*]

This digression will show that Fakhr ud din was under deep obligations to Nuruz, and it is not strange that he should have sided with the latter when he revolted. He was captured in a struggle by Ghazan's general Sunatai (a name otherwise given as Sutai and Sukai), but managed to escape, and made his way to Herat, where he offered Nuruz shelter, as we have seen. The latter's officers tried to dissuade him from trusting himself with the malik, but he replied that it was three days since he had said the "namaz," and that he could not put it off any longer. Thereupon several of his adherents left him, and he entered Herat with 400 men. The malik took him to the citadel, which was called Ikhtiar ud din.[†] Four days after his arrival at Herat there appeared before the town the amir Kutlugh Shah, with 70,000 horsemen. On his way thither he had visited the famous tomb of Ali Riza, the eighth imaum of the family of Ali, near Tus, and said a namaz of two rek'ats,[‡] praying God to let his enemy fall into his hands. He invested Herat, but the town was so well fortified and presented such a bold front, while the heat was so excessive, that some of his chiefs advised him to withdraw. But he was not made of yielding material. He caused a letter to be written to Fakhr ud din by the latter's father-in-law, the mufti of Jam, and sealed with his seal, summoning the malik to surrender Nuruz if he wished to prevent the destruction of Herat, and sent it by a spy. Fakhr ud din at once showed the letter to his guest, who was thus convinced of his good faith. Haji, the secretary of Nuruz, nevertheless urged his master to seize the malik and to make sure of him, and although Nuruz refused, their conversation was overheard and reported to Fakhr ud din, who took alarm and advised with his ministers and the chief people in the town. However this was, feeling that sooner or later the place must fall and their wives and children be carried off captive, and that Nuruz had broken his solemn promise never to take up arms against his master, the malik determined to surrender him to his enemies. He told Nuruz that the garrison was discouraged, and that he had better distribute his men in different parts of the town, so as to encourage the citizens. He fell into the trap, and distributed two of them among each ten men of Fakhr ud din, remaining

---

* Journ. Asiat., 5th ser., 475. D'Ohsson, iv. 186-187.   † D'Ohsson, iv. 178-179.
‡ D'Ohsson explains this. He says the five namaz, or daily prayers, consist of from four to eight rek'ats or attitudes, but travellers or soldiers on the march need only say a namaz of two rek'ats.

' almost alone in the citadel. His guards were suddenly arrested, and he himself was bound with cords. "What harm have I done you," said Nuruz to his host, "that you desire my life? At least leave me my horse and my sword. I will rush naked among my enemies and perish sword in hand." "Henceforth," said the treacherous and ungrateful malik, "you will only see a sword in the hands of other people." He then sent the head of Haji to Kutlugh Shah, with the news of the arrest of Nuruz and his people, and demanded in return, by the oath which Ghazan had sworn to him, that Herat should be spared. This assurance was sent to him by the amir Fulad Kia, the khoja Alai ud din, and the mufti of Jam. Thereupon Nuruz was sent bound, with an escort of Ghurian troops, to the camp of Kutlugh Shah. The latter put some questions to his prisoner, who replied haughtily, "It is Ghazan, and not you, who have the right to question me." He was thrown on the ground and cut in two. Fulad Kia set out at once to take his head to Ghazan, whom he found at Baghdad. The head of the brave and turbulent chief, to whom Ghazan was so much indebted, was suspended for some years in front of the prison in that town. His brothers Arghun Haji and Kulduk were also executed. Three days after the death of Nuruz Kutlugh Shah raised his camp and returned to Irak. This is the account given by Rashid ud din.*

Mirkhond has a somewhat different version of the concluding episodes of the life of Nuruz. According to this author, Fakhr ud din having determined to surrender Nuruz, told him that the garrison was composed of such a mixed body of troops that any one of them might treacherously give up one of the gates of the place, and to be secure in that respect he urged Nuruz that his people should be employed in the work. Accepting this as a fresh proof of the malik's friendship, Nuruz sent his people as desired, and remained almost alone. Thereupon the malik charged four of his officers, viz., Taj ud din Ilduz, Jemal ud din Muhammed Sain, Saraj ud din Omar Harun, and Muhammed Na'man, to take a body of Ghurians and arrest him. They approached the citadel by a covered way, each one with a cord in his hand. The amir, having only four companions with him, was engaged in firing some arrows at the besiegers, but his bowstring broke, and he threw it down angrily. Seeing his four assailants, he addressed them, saying, "Pehluvans, what are you come to do, and why these cords?" Ilduz replied that the malik had sent them to build a shelter, behind which he might be safer from the enemy's weapons. Nuruz pointed out to them where he would like it to be, and proceeded to mend his bow. Ilduz now advanced, and struck him on the head with his mace. Muhammed and Omar then bound him and conveyed him to a house. The malik, who was at this time hiding close

---

* D'Ohsson, iv. 187-192. Ilkhans, ii. 42-44.

by with 200 horsemen, sent word to Nuruz's men that their master wanted them. They went accordingly to the citadel, and were forthwith killed by the Ghurians.*

Fakhr ud din was liberally rewarded by Ghazan for his recent services. He sent him one of his own robes, and again confirmed him as malik of Herat and its dependencies, and granted him, at his request, the privilege of omitting to pay the usual visits to the ordu, on his promising to remain a faithful vassal and to furnish his quota of troops when the Ilkhan should be at war. But presently he conceived the notion of breaking the Mongol yoke. Trusting in the fortifications of Herat, which he had strengthened, and in his forces, which reached the number of 60,000 men, he began by evading the payment of his annual tribute, and excused himself on various pleas from furnishing the requisitions demanded by the troops of Kutlugh Shah. A more daring act of rebellion at length brought him under the scourge of the Mongols. Ghazan had assigned the famous Nigudarian bands, whom we have described, summer quarters in Seistan and winter quarters in Irak Ajem, and exacted from these robbers an undertaking to cease their raids, and declaring themselves worthy of death if they violated it. They were nevertheless accused of the various outrages that took place in Irak, and harassed by the charges and fear of punishment, they moved without leave into Kuhistan, and then asked Fakhr ud din to give them shelter within his principality. He acceded to this, gave them horses, arms, and clothes, and employed them in his service, and sent them to make raids into the districts he meant to appropriate, where they killed many people. Deputations from the sufferers went to Ghazan, who ordered his brother Khodabendeh to take with him the troops of Mazanderan and Khorasan, and to demand the extradition of the Nigudars. If the malik refused to comply he was to attack Herat, put the turbulent Nigudars and Ghurians to the sword, but to spare the innocent inhabitants. Khodabendeh having reached Nishapur duly demanded the surrender of the Nigudarian chiefs. Fakhr ud din had exacted a promise from them—solemnly made, according to their custom, upon their swords—that they would not leave him without permission He then presented Khodabendeh's messenger with one of his robes and thirty prisoners, and told him to tell his master that Buka and the other leaders of the Nigudars had set out on an expedition, but that on their return they would be surrendered. Khodabendeh having received this answer from a prince whose resolute character he well knew, resolved to march at once upon Herat. This was in the summer of 1299. He pitched his camp on the river, close by the walls, and had begun the siege, when he learnt that the malik had left for the fortress of Amankuh, now called Ishkéléji, and he proceeded to invest it. Having postponed

---

* D'Ohsson, iv. 191-192. Notes.

his attack for four days, in the hope that the malik·would submit, he ordered an assault to be made, which cost him many men. The following night Fakhr ud din made a sortie and traversed the Mongol lines, re-entering Herat. Having confided its defence to the amirs of the Ghurians and Khallajes, he himself set out for Ghur with 100 horsemen. The next day Khodabendeh renewed his assault, but again unsuccessfully. Hearing that the malik had left the fort, he returned to Herat, and began to besiege that town. Under him were the generals Reis Kutlugh, Hulaju, Hirkudak, Mulai, and Danishmend Behadur. The malik's chief officers, Iftikhar ud din, Muhammed Haruni, Jemal ud din Muhammed Sam, Ilchi Khoja, Omar Shah Khuarezmi, Pehluvan Yar Ahmed, and others, left the place with a large force, and fought a terrible battle outside. The struggle was more or less continuous for seventeen days, and several thousand men perished on each side. Eventually, the Sheikh ul Islam Shihab ud din Jam went to Khodabendeh's camp and told him, after much peaceful entreaty, that there were 50,000 men capable of bearing arms in the place, who had decided to resist to the death; that it was probable the town could not be captured, which would be a disgrace to the royal forces, and that it would be better to make some arrangement. His proposal having been accepted, he sent to tell the chiefs that the prince was ready to pardon them, and they must send him 100,000 dinars. The town duly sent 30,000, and undertook that the remainder should be forwarded when Khodabendeh had raised the siege. Fakhr ud din returned to Herat, and proceeded to repair and increase its fortifications.[*] In the year 1301 Fakhr ud din put down the revolt of Hosam ud din, the governor of Esfizar. That chief died just as he was going to be attacked, and his son Rokn ud din concentrated his forces in the citadel of Dubah. The malik summoned some Mongol mercenaries and 3,000 Ghurians, and captured Esfizar, and those of its inhabitants who escaped death were removed to Herat.[†]

Fakhr ud din, *inter alia*, built two great bastions at Herat of the height of fourteen guez, separated by a slope of six guez wide, and also laid out a vast open space, surrounded by a circular wall, outside the city walls, where popular fêtes, &c., could be held. He also built a monastery in the royal square, and repaired the mosque of Abd Allah, son of Amir, as well as that called Tereh Farushi, on the Amankuh, and also founded the famous market-place, afterwards known as the royal market. Every month 1,000 dinars were distributed to the poor and the dervishes, and 1,000 blankets were given them at the beginning of winter; 10,000 menn of bread and ten sheep were also divided daily among the indigent. These measures made the malik very popular. In the year 700 HEJ. (*i.e.*, 1300) he forbade women to appear in public under the penalty of being promenaded about

---

* D'Ohsson, iv. 193-196. Ilkhans, ii. 73-75.   † Journ. Asiat., 5th ser., xvii. 480, &c.

the streets and markets in black veils. He also forbade the use of
wailers and chaunters at funerals, as prescribed by the Koran. Those
who engaged in gambling were led through the streets with their beards
and hair shaven. Drunkards, besides suffering the penalty ordered by
the religious law, were to have their feet manacled, and to be made to
work in the tile works at Herat. Poets and learned men were patronised,
and Seifi Heravi, who composed 80 odes and 150 rubais or quatrains in
his honour, says in his chronicle that forty poets lived at his Court.*
Among all these poets the only name which has reached us is that of
Mola Sadr ud din, of Fushenj, surnamed Rebii, whose career is a
curious instance of the Bohemian lives which men lived at the Courts of
the great Mongol vassals in the 13th century. He filled the post of pulpit
orator or preacher at Fushenj, when he was summoned to Herat by Fakhr
ud din, and commissioned to write the Kert nameh, on the plan and in
the metre of the Shah nameh of Firdusi, in which he sang of the doings
of the malik and his grandfather. He received a salary of 1,000 pieces
of gold (Von Hammer says of silver) as Court poet, but, like many other
poets, he was given to intemperate habits, and his orgies scandalised
Fakhr ud din. To avoid punishment, he sought refuge in Kuhistan with
Shah Ali, son of Nasr ud din Seistani. There he one day allowed
himself to abuse the malik so violently that Shah Ali gave him 200 dinars
and, telling him he was not his man, ordered him to take his departure at
once, and when the courtiers expressed surprise at this treatment of the
first poet of Khorasan, he replied, " Rebii has all the gifts of intelligence,
but he lacks a heart. If he now abuses one who has for ten years covered
him with favours, do you think he will in the future spare my name?" The
poet now went to Nishapur, where he led a precarious life for a while,
and then went to Irak. Afraid that he might use his talents to discredit
him with the Ilkhan, Fakhr ud din sent him an invitation to return, and
having obtained a safe conduct, signed with the malik's own hand, he did
so. The latter was not given to forgiving injuries, and having heard of
the poet's biting sarcasms, which were in everybody's mouth, sought an
opportunity to destroy him. One day the poet, who gave himself up
more than ever to debauchery, invited some of his young friends to a
feast. Here many indiscreet things were said. Among others, Rebii
declared that if they would support him he would in a few days become
the ruler of Herat. This statement was applauded, a mock coronation
was gone through, and the hero of the evening distributed dignities and
ranks among his companions. This performance having been reported
to Fakhr ud din, the culprits were summoned to his presence. They all
pleaded guilty, but urged that they had been drunk. The malik, however,
was inexorable. Some of them were flayed alive. Others had their

---

* Journ. Asiat., 5th ser., xvii. 478-479.

tongues and ears cut off. Rebii himself was thrown into a dungeon. During his long captivity he composed many odes and *mesnevis* in honour of the king, but without moving him. We do not read what became of him, but it is probable that he died in prison. Von Hammer has translated a number of his verses. They depend, like most Persian poetry, upon assonance and other graces, almost impossible to translate, and sound very vapid, stilted, and invertebrate in another language.[*]

We have seen[†] how the Padishah Khatun, the wife of Abaka and Gaikhatu successively, was made ruler of Kerman by the latter, and how she put to death her brother Siyurghatmish. He left but one daughter, Ismet ud din Aalemshah, who married Baidu Khan, and who, with her mother, Kurdujin, determined upon the overthrow of their strong-minded and truculent relative. On Baidu's accession they accordingly marched an army to Fars, and put Padishah Khatun to death. This was in 694 HEJ. (*i.e.*, 1295). Hamdullah philosophises in verse on this event—

> If thy cross be thorns, verily thou hast sown them ;
> If it be China silk, verily thou thyself hast spun it.

Next year Ghazan nominated Muhammed Shah, the son of Hijaj, brother of Siyurghatmish, who was then fifteen years old, as her successor. The young prince lived at Ghazan's Court, while the country was administered by the judge Fakhr ud din of Herat. He treated the young prince's brothers, Mahmud and Hasan Shah, with scant courtesy, and reduced their allowances tenfold, whereupon they rose against and killed him. Ghazan then sent Timur Buka as governor of Kerman, who summoned troops from Irak and Fars. The siege lasted for a year and a half, and famine began to press the garrison. But as it held out bravely the amirs sent to Ghazan to beg him to send Muhammed Shah, to whom they thought the citizens would give it up. Meanwhile the town surrendered. The Khoja Sadr ud din of Abher was now given charge of Kerman. He sent one of the rebel princes, Mahmud Shah, to the Court. As he was going there he met Muhammed Shah at Ispahan, and as the latter refused to see him, he took poison. When Muhammed reached the city he put to death all those he could lay his hands upon who had fomented the outbreak. He also seized his brother, Siyuk Shah, whom he sent to Tebriz, where he was put to death with torture. Sadr ud din of Abher, Muhammed Shah's atabeg or guardian, began to grow suspicious of him, and taking advantage of his absence on a hunting excursion he fled to Shirkhan, whose governor assisted him, and he thus reached Fars. Muhammed sent the vizier Bahai ul Mulk to persuade him to return, and he did so and was well treated, but he still felt uncomfortable, and presently secured an appointment as Muhammed's deputy at Ghazan's Court, where he closely watched his master's interests. Muhammed Shah

---

[*] *Id.*, 479 and 480. Notes. Ilkhans, ii. 75-78.　　[†] Ante, 365.

P *

died of excessive drinking, in the year 703 HEJ. (*i.e.*, 1304, the same year as Ghazan), after a reign of eight years.*

We will now turn to the affairs of Fars, which are omitted by D'Ohsson, while Von Hammer's account in the "Ilkhans" is fitly described by Colonel Yule as frightfully confused. During the supremacy of the Salghurid atabegs of Fars, Hormuz formed one of their subordinate governments. When that dynasty came to an end, Mahmud Kalhati, who had been governor of the island, established himself as prince of Hormuz. He is called Rukn ud din Mahmud in Teixeira's extracts from the chronicle of Turan Shah. He is said to have reigned from 1246 to 1277, and to have been succeeded by his son Saif ud din Nazrat, who reigned apparently till the year 1290, when he was put to death, with his wife, by his brother Rokn ud din Masaud (called Amir Masaud by Teixeira). Thereupon Bahai ud din Ayaz, a Mamluk in Nazrat's service, who had been appointed vizier of Kalhat on the Arabian coast by his master, took up arms to avenge him, assisted by Siyurghatmish, the ruler of Kerman ; while the Mufti, who had had to seek temporary shelter in the island of Kish, was similarly helped by Jemal ud din, the governor of Fars, who assigned him 12,000 gold pieces annually out of the revenues of the province for the furnishing of his army. Having driven out Masaud, and secured a treasure of 200 tumans, in gold, silver, and rich stuffs, he had Fakhr ud din Ahmed ibn Ibrahim al Thaibi (apparently a relative of Jemal ud din of Fars, already named) proclaimed as ruler of Hormuz. A MS. history, quoted by Ouseley, assigns to this Fakhr ud din the foundation of New Hormuz, on the island of Jerun, the old one having been deserted in consequence of the attacks of predatory neighbours.† In regard to the Arab family of Al Thaibi, of which Fakhr ud din was a scion, Colonel Yule remarks that they seem to have been powerful at this time on the shores of the Indian Ocean. One of them, Jemal ud din Ibrahim, was farmer-general of the taxes of Fars, and quasi-independent prince of Kais and other islands in the Persian Gulf. We shall have more to say about him. Fakhr ud din above named, as he is called ibn Ibrahim, was no doubt his son. His brother Taki ud din Abdur Rahman was vizier at Marzban, in Maabar.‡ Von Hammer says that Jemal ud din went to Hormuz, and that a quarrel arose between him and his son, Fakhr ud din. A struggle ensued, in which Bahai ud din Ayaz took the part of his former patron, Jemal ud din, and Fakhr ud din was driven from the island. He was sent in 1297 as his envoy to the Great Khan at Khanbaligh by the Ilkhan, and died on his return voyage, on the Coromandel coast, in 1305.§ Bahai ud din now seems to have occupied Hormuz, whence Jemal ud din tried to drive him out, but the latter was defeated, and his opponent, who

* Hamdullah. Ilkhans, ii. 49.
† Yule's Marco Polo, i. 124-125. Pauthier, *id.*, i. 86. Note. Ilkhans, ii. 50.
‡ Yule's Marco Polo, ii. 316. Note. Elliott, Hist. of India, iii. 32.
§ Yule, op. cit., i. 125. Ilkhans, ii. 50.

seems to have been more or less of a piratical chief, landed upon and plundered the island of Kais. This defeat, and the fact that the season was approaching when the trading vessels would return from Maabar, induced Jemal ud din to make advances to Bahai ud din, and an accommodation was the result. We are told that Jemal ud din, the tax farmer of Fars, carried on an active trade with the district of Maabar. Fourteen hundred horses (Von Hammer says 400) were by agreement sent annually from the island of Kish or Kais. Horses were similarly exported from the other islands, viz., Hormuz, Bahrein, Khatif, and Lahsa. The price of each horse was fixed from old times at 220 dinars of red gold, and those which died *en route* had their value paid out of the treasury. In the reign of the Salghur atabeg, Abu Bekr, 10,000 horses were so sent to Maabar, Kambayat, and other ports, and their value, which amounted to 2,200,000 dinars, was paid for out of the funds belonging to the Hindu temples and the tax upon courtesans attached to them Wassaf says, instead of being fed on raw barley, these horses were given roasted barley and grain dressed with butter, and boiled cow's milk to drink, and he adds his appropriate commentary in verse—

> Who gives sugar to an owl or a crow?
> Or who feeds a parrot with a carcase?
> A crow should be fed with a dead body,
> And a parrot with candy and sugar.
> Who loads jewels on the back of an ass?
> Or who would approve of giving dressed almonds to a cow?

The Indian climate was very fatal to these horses, which became rapidly weak and worn out from the heat, &c. "Their loss," says the chronicler, "is not without its attendant advantage, for it is a providential ordinance of God that the West should continue in want of Eastern products and the Eastern world of Western products, and that the North should with labour procure the goods of the South and the South be furnished in like manner with commodities brought in ships from the North. Consequently, the means of easy communication are kept up between these different quarters, as the social nature of human beings necessarily requires and profits by :—

> "Thou wert called a man because thou wert endowed with love."[*]

Jemal ud din made a treaty with the Diwar of Maabar, which secured him trading rights there. When he was intrusted with the government of Irak and Fars, the messenger who took the news was sent to the Diwar with a present of a falcon. Wassaf has preserved the letter written on this occasion, which has been translated by Von Hammer, and is a good specimen of the hyperbole then fashionable in diplomatic correspondence. The Diwar sent a still more elaborate and pompously phrased reply, and with it some singers and musicians. This reply was written in Shaaban, in the year 700 HEJ. (*i.e.*, 1300), and was

---

[*] Elliot, Historians of India, iii. 33-34. Ilkhans, ii. 51-52.

duly acknowledged by an answer written in October of the same year.'
The letters are a mere farrago of stilted phrases, and absolutely wanting
in historical details.* " Notwithstanding the immense wealth he (*i.e.*, Jemal
ud din) acquired by trade," says Wassaf, "he gave orders that when
commodities and goods were imported from the remotest parts of China
and Hind into Maabar, his agents and factors should be allowed the
first selection, until which no one else was allowed to purchase. When
he had selected his goods he dispatched them on his own ships,
or delivered them to merchants and shipowners to carry to the island of
Kais. Even then it was not permitted to any merchant to contract a
bargain until the factors of Malik ul Islam had selected what they
required. Then the merchants were allowed to buy whatever was
suited to the wants of Maabar. The remnants were exported on ships
and beasts of burden to the isles of the sea and the countries of the East
and West, and with the money obtained by their sale such goods were
purchased as were suitable for the home market. The trade was so
managed that the produce of the remotest China was consumed in the
furthest West. No one has seen the like of it in the world."†

Jemal ud din, the Mufti of Shiraz, to whom we have so frequently
referred, was styled Sheikh ul Islam (*i.e.*, the Sheikh of Islam), and also
Melikol Islam (*i.e.*, the farmer-general of Islam), because he had the
farming of the various taxes in Irak and Fars. Through the favour of the
Vizier, Sadr ud din, he was two years behind in his payment, but in
November, 1298, when the Court was at Jukin, he was summoned there
and ordered to pay up the arrears immediately. This he did to the
amount of 45 tumans, each consisting of 10,000 dinars. The taxes of the
province were now arranged on a more equitable basis. The whole
province was divided into sixteen parts, and was assigned a tribute of
1,000 tumans, of which 10,800,000 dinars were set aside for the costs of
collection, &c. Orders were sent to Fars that 20,000 portions of land
(*fedan*), as Crown lands, had been exempted from taxes. Of these 3,000
were to be assigned to the governor and 3,000 more to the tax farmer, who
were to pay to the treasury 61 gold pieces and 4 daniks (?) for each estate
or portion. Whenever the taxes were put up for sale, these lands, with
their appurtenances of seed and agricultural implements, were to be
surrendered by the previous farmers, who were to be paid 85 dinars for
each plough ox, and the seed which was reckoned with it. The rate at
which exchange was to be valued in collecting taxes was that the gold
miskal was to be counted at four dinars and the silver miskal at one
dinar. The gold and silver money in the provinces was to be accepted
at the same value as in the capital, and the payment of earnest money
for difference of exchange was forbidden. The farmers-general were to
produce the wealthiest and most considerable amirs as bail. The

---

* Ilkhans, ii. 52-63.   † Elliot, iii. 35.

amirship and government of the land was confirmed to the Terkhan
Sadak. Sherif ud din, of Semnan, was made supreme director of the
taxes. He was to hear the complaints of the people, and to see that the
tax collectors did their duty honestly. These reforms were due to the
vizier, Sadr ud din.*

Let us now return again to Ghazan. His reign was marked by a
terrible roll of executions, and, as D'Ohsson says, there is hardly a page
of Rashid ud din at this time without a notice of the execution of
some public functionary. In August, 1297, the rebel governor of Rum,
Baltu, who had fled, was captured by the generals Sulamish and A'rab,
who took him to Tebriz, where he was executed on the public square, on
the 14th of September, the day after Ghazan's return to his capital, after
spending the spring and summer at Alatagh. Having been told that
some person had predicted that Prince Taichu, son of Mangu Timur,
would be on the throne before forty days were over, he had him, together
with the prophet and all those who were present when he made the
prophecy, put to death.† On the 5th of October, 1297, Ghazan laid the
foundations of the famous tomb in the garden Aadiliyeh, in the suburb of
Shem, whose ruins still form the most precious monument at Tebriz. He
superintended the workmen himself, and as the base rose above the ground,
the architect questioned him about where he should put the windows so
as to light up the cool room or cellar (serdat), which was under the
ground.

On the 2nd of November, 1297, Ghazan and his Court formally adopted
the Muhammedan head-dress (i.e., the turban), which, according to
Mussulman tradition, was the head covering of Abraham, who is accounted
a Hanefi (i.e., one acquainted with the true faith). This ceremony was
followed by a feast. On the 7th of the same month the Ilkhan set out for
his winter quarters in Arran. En route he heard that an outbreak had
taken place in Georgia.‡ We have seen how David, the sixth son of
Dimitri II., obtained the throne of Karthli from Gaikhatu Khan, and how
he was patronised by the officers of Baidu. Although he had sent back
Tukal and his son at the bidding of Ghazan, he did not like to venture
himself in person to his Court. He had been restrained by his fear of
Nuruz, who was such an aggressive enemy of the Christians. The
"History of Georgia" tells us that Nuruz seized the aged Nestorian bishop
of Meragha and had him beaten and cruelly tormented, in order to make
him abjure his faith, and when he found this was unavailing he exiled
him, notwithstanding his great age. He also sent one of his relatives to
lay waste all the churches of Georgia, including the church of the Virgin
at Vardzia, and to carry off the rich plate, &c., belonging to them. He
went to Nakhchivan, but was miraculously struck by lightning and
consumed, even to his bones, as he was about to lay his hands on the

---

* Ilkhans, ii. 65-68.   † D'Ohsson, iv. 196-197.   ‡ Id., 197.

cross in the church of Vardzia, just named.* Stephen Orpelian says that through the efforts of Nuruz many churches were ruined, priests killed, and Christians put to death, and those who escaped death were robbed of their goods, while their wives and children were sold into slavery. At Baghdad, Mosul, Hamian (? Hims), Thavrej (? Tebriz), and Meragha, and in the countries of Rum and Mesopotamia, the persecution caused great troubles. The churches of Nakhchivan were sacked and the priests sold into slavery, the doors of the sanctuaries were demolished and the holy tables overturned, but the buildings themselves were spared out of consideration for the Georgian troops, who were held in high honour by the Mongol chiefs. At Noravank, the capital of Siunia, they would have overturned the churches, but were appeased by presents, &c., while that part of Armenia beyond the Araxes was entirely spared. The Syrian Catholicos at Meragha (i.e., the Nestorian dignitary before named) was captured and reduced to great stress by the cruelty of the Mongols, while Ter Tiratsu, bishop of the Church of the Apostles, was put to outrageous tortures and pillaged of his goods, and his monastery (i.e., that of Dadi Vank), where the apostle Thaddeus was buried, had its walls completely overturned and destroyed. Haithon, King of Little Armenia, having gone to complain to Ghazan, the latter said all this had been done without his orders, and he took measures to stop the devastations. This author also states that Ghazan married the daughter of the King, but according to Langlois he really married the niece of Haithon, daughter of his brother Sempad.† The "History of Georgia" tells us that the King of Georgia, having received a fresh summons to go to the ordu, set out as far as Hereth and Kakheth, where the didebuls and aznaurs met him, and among others the famous warrior, the cristhaf Samadaula. He consulted with them as to whether he should go on or no. They advised him to do so, but he was afraid, and went to Mthiuleth, and established an entrenched camp at Jinwan, whence he sent his young brother Wakhtang as his envoy to the Khan, Othakha, grandson of Batu (i.e., Toktogu, Khan of the Golden Horde), and offered to open him a road by way of Derbend if he would attack Ghazan. He was well received, and promised troops, lands, and riches. Having heard of David's defection, the Ilkhan sent his general or beglarbeg, Kutlugh Shah, with a considerable army against him. He went to Tiflis, whence he dispatched messengers to David, to urge him to send some responsible people with his submission and a promise to be faithful. The king accordingly sent the Catholicos, Abraham, Joané Bursel, and the Kadhi of Tiflis, who in turn asked for hostages. The Mongols, we are told, not only gave an oath according to their faith, but also sent the son of Kutlugh Shah, named Sibuchi ; Arpa, brother of Kurumshi, and some other sons of noyans, and also offered Ghazan's ring as a gage for the King's safety. Thereupon David went to

---

* Hist. de la Géorgie, 616-617.   † Hist. de la Siounie, 261-262.

them and promised to repair to the ordu. Kutlugh Shah received him honourably, and on his return he released the hostages, to whom he had given great presents.* Stephen Orpelian, in relating these events, says that David, when summoned by the Ilkhan, fled with all his nobles to Mthiuleth, and entered the impregnable fortress of Maslénakhé (Modanakhé, *i.e.*, "come and see me.") He also seized the gate of the Alans, formerly called Dariala, and now, says our author, Jasanin Cap, whence he expelled the Tartar garrison. It then belonged to the successors of Bereke. Stephen says their chiefs at this time were Thutha Mangu and Nukha (*i.e.*, Tuda Mangu and Nogai), grandsons of Batu and Sartakh. They sent him several envoys, and made protests, but he refused to leave. Thereupon Kutlugh Shah entered the country with a numerous army, and posted himself in the plain of Mukhran, near the impregnable citadel, where the King had retired to. It was only when he sent his own son Shipauchi (the Sibuchi above named) and three other considerable personages as hostages, that the King, when he had put them under a strong guard, ventured to go to his camp with rich presents, which were reciprocated. The Georgian patriarch acted throughout as mediator. After the King's return home, and the release of the hostages, Kutlugh Shah went to his winter quarters in the plain of Ran (*i.e.*, Arran).†

On the approach of spring, Ghazan again summoned David to the ordu, but being still afraid, he re-opened negotiations with the Khan Othaka (*i.e.*, Toktogu), through his brother Baadur. When Ghazan heard of this he once more sent Kutlugh Shah, who went to Somketh and again urged the King to go, sending him rich presents. He sent the Catholicos Abraham, the Kadhi of Tiflis, and Joané Bursel, and promised to go himself. Seeing he made these constant excuses, the Tartars put Bursel to death, letting his two companions go free. They then marched to Mthiuleth, and ravaged Somkheth, Karthli, Thrialeth, and Ertso, and planted themselves successively at Mukhnar, Kherk, Bazaleth, Ertso, and Thrialeth. They made so many prisoners that their number was unknown, and they committed great slaughter and devastation. This year (*i.e.*, 1298) there appeared a great comet, in the shape of a lance, in the north, where it was visible for four months—a presage, says the chronicler, that the Georgians would perish by a weapon of this shape. The Tartars now offered terms on condition that the King would not offer a passage through his territory to the ruler of the Golden Horde. After solemn reciprocal promises, David sent the Queen (his mother), his young brother Manuel, the Catholicos Abraham, and his wife Oljath,‡ who were received with honour by Kutlugh Shah ; and having made over Georgia and its capital, Tiflis, to the King, he set out for the ordu with Oljath. In the spring of 1299 a messenger or spy was sent by the Mongols to see if

---

* Hist. de la Géorgie, 618-619.   † Hist. de la Siounie, 263-264.
‡ We have seen that the Persian authors make her the wife of Kutlugh Shah.

David was keeping his word. He reported the overtures the King had made to the Khan of the Golden Horde, through his brother Baadur, offering him a passage through Georgia. Ghazan was naturally enraged, and dispatched Kurumshi and Alinji, one of his favourites, together with Shahinshah Mkhragrdzel. Their mission was to depose the King, and to set up his young brother Giorgi, who we are told was like the unicorn, the only son of his mother. She was the daughter of the Beka already named. Beka had become a very powerful person, and ruled over the country from Tasis-Sar to Sper, and as far as the sea, including Samtzkhé, Adshara, Khavkheth, Klarjeth, and the valley of Nigal. Alexis II., Comnenus Emperor of Trebizond, who mounted the throne in 1297, had apparently married his daughter,* and had made over to him all Janeth. He thus possessed the greater part of Tao, Artan, Kola, Karnéphora, Kars, and all the surrounding land and fortresses, including Artanuj and the hermitages of Klarjeth. All the didébuls, arznaurs, and monasteries were dependent on him. He paid tribute to the Mongols, and furnished them with a contingent of auxiliary troops. His grandson Giorgi, David's half brother, was living with him. The Khan's messengers asked that he might be sent with them to be installed as King of Karthli. He was accordingly sent, escorted by a large army, and notwithstanding his youth was duly crowned at Tiflis.†

These events took place in 1299. In the following spring Ghazan again sent Kutlugh Shah with several noyans and a large army, who once more ravaged Karthli. David had taken refuge at Khada. His dependant, Chalwa Kwéniphnéwéli,‡ eristhaf of Ksan, submitted to the invaders, and in fact acted as their guide through the valleys of Tzkhrazma or Lomisa, and the mountains of Tzkhavat. Informed of his treason, David withdrew to the impregnable fort of Tzikaré. Kutlugh Shah went successively to Khada, Khévi, and Gwéleth (or Gélath), where he expected to find David, and laid siege to the town of Stéphan Tsmida, where David was supposed to be. Having failed to capture this place, they agreed to descend again to Khada on being supplied with some provisions. Meanwhile David, in his safe retreat, was supported by Ahmada Suramel (the first chamberlain), Jila Abazas-Dze, and the aznaurs of Karthli and Somkheth. Kutlugh Shah determined to overwhelm the rebel King. He sent two divisions to attack Tzikaré. Five hundred Tartars were waylaid in a defile, and either killed or captured. A struggle now ensued close to Khada. The Tartars again suffered severely, since it was not possible in this hilly ground to fight on horseback. They were assisted by Shahin Shah, son of Ivané, by two troops of Meskhes sent by Beka and other Georgians. The difficulty of the ground prevented either side from properly assailing

---

* The Georgian Chronicle wrongly calls the emperor Kir Michael.
† Hist. de la Géorgie, i. 620-622. Lebeau, xix. 86. Note.
‡ .e., The master of Lower Iphnev, a place on the left bank of the Upper Ksan.

the other, and the Tartars withdrew to Khada. Seeing it was not possible to penetrate to Tzikaré, Kutlugh Shah returned once more to Karthli, and killed or made prisoners all whom he found there, after which he rejoined his master, Ghazan. The spring following (*i.e.*, in 1301) he again invaded Georgia. He went to Tiflis, intending to march upon Mthiuleth. The didébuls of the country wishing to preserve what still remained undevastated, urged David to repair to the ordu after receiving proper hostages. Kutlugh Shah undertook that he should return safe and sound if he would only go and visit his master; but although the people of Karthli, of Hereth, and Kakheth urged him strongly that he would be well treated, he dared not trust himself there, but remained at Gwéleth. His chief didébuls, however, went with his consent. They were well received by Kutlugh Shah, who, having again harried the land, except Mthiuleth, returned once more to the ordu.

David now determined to revenge himself upon the traitor Chalwa, marched into the valley of Tzkhrazma, and devastated it. Chalwa, driven to straits, asked Ahmada Suramel, the head of the eristhavs of Karthli, to intercede with the King for him, who pardoned him, but made over his appanage to Ahmada. At this time David's brother, Wakhtang, became troublesome. He had married the daughter of Shabur, and chafed at having no heritage of his own. He accordingly went to Ghazan's ordu, who conferred the Georgian kingdom upon him, and had him escorted to Tiflis by Kutlugh Shah with a considerable force. On arriving there he was supported by all the Tartars and Georgians, the General Sargis, son of Beka, Prince of Samtzkhé; and the people of Tao, Thor, Thmogwi, and Somkheth, as well as by Shahin Shah, so that there were now three kings of Georgia: David VI., son of Dimitri II.; George V., whose name seems to disappear after his accession till the year 1318; and Wakhtang III., who now (*i.e.*, in 1301) mounted the throne.* On learning of the patronage the Tartars had extended to his brother, David again offered, through Joané Bursel, to go in person to the ordu if he received proper assurances as to his safety and hostages. These were promised by Kutlugh Shah, and he thereupon sent his queen, Oljath, who was met at Bazaleth by Kutlugh Shah, and treated with special honour, she being a Mongol princess. She was given the most solemn assurances for the king's safety, as well as the ring and the napkin, the latter being, as we are told, a gage of pardon, while Sibuchi, son of Kutlugh Shah, was offered as a hostage. The Queen, however, was detained, and the Catholicos, the Kadhi, and Bursel were sent back to bring the King. He refused to go with them, and on their return without him Kutlugh Shah was so enraged that he put Bursel to death.† Shahin Shah buried him in his own sepulchre. Wakhtang was duly confirmed, and the Queen was sent to the ordu.

---

* Hist. de la Géorgie, 622-625.
† The same authority, as we have seen, makes him be put to death some time before.

The Tartars now made an invasion into the difficult country of Gwéleth, but were constrained to retire, after suffering considerable losses by its impregnable position, and the bad weather having proved too much for them. They withdrew again to Karthli. As we have seen, the Queen Oljath had been sent to the ordu. The Mongols determined that she should not again return to her husband. When he learnt this, he, in 1302, married the daughter of Ahmada, eristhaf of Karthli, who was very beautiful. The recent devastations caused great suffering in Georgia, and a terrible famine ensued. The Georgian chronicler attributes the ills which overwhelmed his country to the vices of the people, which he compares, apparently literally, with those of Sodom. A crowd of people from Karthli passed into Samtzkhé, where Beka ruled, and the latter's wife distributed charity widely among the fugitives. Meanwhile Wakhtang continued to rule at Tiflis over Somkheth, Dmanis, and Samshwildé.* Beka had a considerable struggle at this time with the Turks of Asia Minor, which is detailed in the Georgian history, but which forms no part of our subject.

We must now revert again to Ghazan himself. In January, 1298, as the Court was on its way towards Tebriz, and had reached Dalan Naur, his youngest brother, Khatui Oghul, died, while a month later his second son Alju was born. Another of his great officials was now going to be displaced, viz., Sadr ud din of Zenjan, who held the post of vizier, and against whom charges of treason had been laid by Kutb ud din of Shiraz and Moin ud din, who had returned from Khorasan. Rashid ud din, the historian, who was one of his subordinates, reports that he and the Vizier had been on good terms, but some officers of the Divan tried to sow disunion between them. Rashid paid no heed to them, but the Vizier was disposed to believe them, and complained to Ghazan, who told him Rashid had not spoken a word against him, and he further urged the latter not to defile his tongue by answering the slanders, but to go on his way as before. Presently, while the Court was at Dalan Naur, beyond the Kur, fresh charges against the Vizier were made by Kutlugh Shah, who was returning from Georgia, where he commanded, and complained of the financial position of the province. Sadr ud din, in reply, declared that the amir's troops had ruined it. Kutlugh Shah having accused the Vizier of speaking ill of him to the Khan, the latter said it was Rashid. "Kutlugh Shah," says the historian, "coming out from an audience with Ghazan, asked me how, after we had been friends so long, I could thus slander him. I denied the charge, and asked him who had made it, and as he refused to tell me, I said that I would speak to the Khan. I spoke to Ghazan about it during a hunt. Ghazan thereupon sent for the amir, and said to him, 'My son, tell me who spoke to you thus.' He thereupon named Sadr ud din. 'I can never cure this man's untruthfulness, for it is

---

* Hist. de la Géorgie, 625-626.

ingrained in him.'" He accordingly ordered his arrest, together with that of his brother, Kutb ud din. This was on the 30th of April. When brought before the Court he answered his interrogators with skill, and if time had been allowed him would probably have escaped his fate, but he was handed over to Kutlugh Shah for punishment, who cut him in two while two officers held his hands. This was on the 4th of May, 1298. " Such was the end of this vizier, who, after employing so many means and causing so much trouble to secure his fortune, enjoyed it only for an instant."* On the 4th of June the Kadhi Kutb ud din, brother of Sadr ud din, and their nephew Kavam ul Mulk, were executed at Tebriz. Their relative, Zain ud din, escaped to Ghilan, and only suffered the same fate two years later, when he ventured to return home.

Ghazan now appointed the Khoja Said ud din of Saveh as vizier,† and Rashid was apparently nominated as his colleague. This took place in the year 697 HEJ., according to Hamdullah, but Wassaf and Mirkhond * date it in 698. Ghazan, as we said, had at his accession ordered many churches, synagogues, and Buddhist temples to be converted into mosques. His acceptance of the turban was the signal for another outbreak of fanaticism, and the Christians who, during the previous reigns, had often found protection under the patronage of the Christian wives of the khans, suffered a great decline of fortune.‡ The Court duly set out in the autumn of 1298 for its winter quarters in Irak Arab. It halted, en route, at Murghsarek, on the borders of Hamadan, and at Burejerd, where Abubekr, of Dadukabad, was put to death for embezzlement ; then crossed the Kurdish mountains, near Wasith, to the plain of Jukin, where Jemal ud din, the farmer-general of taxes, was summoned to account for the taxes overdue. Here news first arrived of the outbreak of Sulamish in Rum. After the execution of Baltu, Ghazan had made over the army in that province to Buchkur, Kiur Timur, and Tainjar, under the general supervision of Sulamish, who was the son of Afak, son of the amir Baichu. At the same time he deposed the puppet Seljukian Sultan, Masud, the second son of Iz ud din Kaikavus II., who had been placed on the throne, as we have seen, by the Ilkhan Ahmed, and who was now accused of complicity with Baltu. This deposition took place in 1295. Masud was confined in a fortress, and Rum was divided into four prefectures, respectively controlled by the Pervanéji, Muhammed Bey ; the grand vizier, Jemal ud din ; his lieutenant or ketkhuda, Kemal ud din, of Tiflis ; and the Defterdar, or Finance Minister, Sherif ud din, who were jointly to account to the Treasury for a revenue of sixty tumans. They committed considerable exactions, and two years later (i.e., in 1297) Ghazan raised Masud's nephew, Alai ud din III. Kaikobad Firamurz, to the throne.§ He set out for his

---

* Rashid ud din, in D'Ohsson, iv. 198-199. Ilkhans, ii. 69-70.
† D'Ohsson, iv. 199-200. ‡ Ilkhans, ii. 70-71. § Ilkhans, ii. 72-73. D'Ohsson iv. 204.

appanage, accompanied by the generals above-named. The next winter
was very severe, and a heavy snow fell. Sulamish took advantage of the
difficulty caused by this bad weather, spread a report that Ghazan had
been deposed by a party of rebels, and himself revolted. He surprised Kiur
Timur and Tainjar, raised a body of 10,000 men, was joined by Mahmud
Bey, prince of Karamania, with 10,000 Turkomans ; by the troops cantoned
in the plain of Ak Sheher, in the district occupied by the Danishmendi
Turkomans, and by the various vagabonds in the country. He thus collected
an army of 50,000 men, among whom he liberally divided the revenues
of Rum, appointed officers, and distributed standards and drums, while
the Egyptian Sultan, to whom he wrote, also promised to help him.
The town of Sivas having declared against him, he laid siege to it.
In March, 1299, Ghazan sent Kutlugh Shah against him. His army
marched in three bodies : the advance guard commanded by Choban, the
centre by Kutlugh Shah himself, and the rear guard by Sutai Aktaji
(i.e., the equerry). Makrizi says the Mongols numbered 35,000, and were
commanded by Bulai. They met the enemy in the plain of Aksheher,
in Erzenjan, on the 25th of April. The rebel was deserted by his
Mongol troops, as well as by the contingent from Rum and the
Karamanian Turkomans, who returned to their mountains, so that he
was left with only about 500 followers, and accordingly fled to
Behesna, on the frontier of Syria, then in the hands of the Egyptians.
He had sent one of his officers, named Mokhlis ud din Rumi, to press the
Egyptian Sultan to send him help, and the latter accordingly sent orders
to Damascus that 5,000 men from Hims, 5,000 from Hamath, and as
many from Aleppo should at once march ; but as they were starting
news reached Damascus of the flight of Sulamish, who shortly after
appeared there in person, accompanied by the amir Iz ud din El
Zurdkash, governor of Behesna, and a suite of twenty persons. The
garrison of Damascus and the principal citizens, with the naib or governor
at their head, went out to meet them. They were treated with unusual
pomp, and Sulamish was assigned quarters on the Meidan, and he was
invited to go on the night of the mid-month to see a grand illumination
in the great mosque of the Ommiades. He presently set out on post-
horses, with Katkatu and Mokhlis ud din Rumi, for Cairo. Katkatu was
granted an ikta (i.e., appanage), while a salary was assigned to Mokhlis ud
din. Sulamish determined to return home, and asked that the amir
Bek Timur might accompany him. He marched by way of Aleppo, whence
he took a contingent of troops with him, and went towards Sis.
En route, he was attacked by the Tartars. The amir Bek Timur was
killed, and he himself escaped to a fortress, where he was shortly after
captured, and sent to Ghazan by order of the King of Armenia. He was
thereupon put to death.*

---

* Makrizi, vol. ii., part ii. 130-133. Ilkhans, ii. 72-73. D'Ohsson, iv. 200-203. Abulfeda, v. 143-144.

Alai ud din, the sovereign of Rum above named, was deposed in 1300, and Masud was reinstated. He died four years later, and with him passed away the last scion of the famous stock of the Seljukians of Rum, and thus was another famous monarchy put an end to by the ceaseless appetite for aggrandisement of the Mongols. The influence of the conquerors was as pernicious in Asia Minor, where culture had greatly flourished under the Seljukian princes, as it was elsewhere. Anarchy and ruin seemed to follow their footsteps. Each year the subsidies from the province demanded by the Imperial treasury increased, while it was also charged with payments and assignations to the princesses, the grand officers, &c. Whoever wanted a fief in Rum paid liberally for it, and if a fresh bidder came forward it was made over to him, with the consequence of intro-ducing terrible feuds and quarrels. Offices were sold to the highest bidder, the wretched people having to furnish the price in the taxes squeezed out of them by the fortunate purchaser. Rich people were* falsely accused of crimes to extort money from them, and commissaries were sent to press for arrears which might be overdue, and which were ground out of the poor cultivators without mercy.* In consequence of the turbulent condition which was thus induced, and of the frequent revolutions at the Mongol head-quarters, the leaders of the various Turkoman hordes of Asia Minor seized first upon the remoter districts, and presently of the districts in the interior of the peninsula, and founded each a separate petty dynasty. One of these, which swallowed up the rest, gradually grew into the famous Ottoman Empire.

Ghazan now began to have intercourse of a more hostile character with Egypt. He meditated an invasion of Syria, in Moharrem, 698 HEJ., and the naib of Damascus and other amirs made preparations to resist, but according to Novairi the Mongol troops which were collected for the purpose of the invasion were struck by lightning, many of them being killed and the others dispersed.†

Before we describe Ghazan's campaign it will be well to take a short retrospect of the last few years of Egyptian history. We saw how Kelavun died in 1290, and was succeeded by Malik Ashraf, who had a considerable struggle with the Ilkhan Gaikhatu.‡ Ashraf himself was assassinated on the 13th of December, 1293, while hunting, by a party of thirteen amirs, at whose head were Lachin (i.e., the hawk), Kara Sonkor, who had been deprived of the government of Damascus, the deposed governor of Aleppo, and Baidara, who called the Sultan uncle and was his viceroy. Baidara was put to death by Ketboga, who, in concert with the amir Sinjar Es Shujayi, now placed Ashraf's brother, Malik Nasir ud din Muhammed, who was only nine years old, on the throne. Ketboga reigned in his name as viceroy, and presently put Sinjar, who meditated deposing him, to death. The principal chiefs, resenting being ruled by a child, raised

---

* D'Ohsson iv. 204-205.　　† Makrizi, ii. (part ii.) 83.　　‡ Ante, 362-363.

Ketboga himself to the throne, and Nasir was remitted to a residence in the castle. Ketboga's elevation is a good instance of the extremely democratic tendencies of the Egyptian polity during the supremacy of the Mamluks. Born a Mongol, he had been captured by the Egyptians in the first battle of Hims, in 1261. He was then an infant. Sultan Kelavun took charge of his education, gave him his freedom, and enlisted him among his Mamluks. He reigned for two years (during which Egypt was visited by a famine), and was then deposed. His viceroy, the amir Lachin, with some others, tried to assassinate him. He escaped and reached Damascus, but lost his throne, which was now (*i.e.*, on the 13th of November, 1296), occupied by Malik Mansur Lachin, who had formerly been a slave, or Mamluk, of the Sultan Al Mansur, son of Eibeg. Ketboga presently swore allegiance to him, and retired to the castle of Sarkhad, which was appointed his residence. Lachin appointed his slave and favourite, Mangu Timur, his viceroy. In the early spring of 1298 an expedition was sent against Cilicia, which was commanded by Bedr ud din Bektash, chief of the Sultan's Silahdars. The army comprised troops from Egypt, Damascus, Aleppo, Tripoli, and Hamath, the last under their prince Mozaffer.

Haithon II. had ceased to reign in Cilicia. On his return from visiting Ghazan in 1295, he went to Constantinople with his brother Thoros to see his sister Maria, who had married the Emperor Michael. He left another brother, Sempad, behind him as regent, who gained over the grandees, seized the throne, and was duly consecrated at Sis by the Patriarch Gregory in 1297. Ghazan not only confirmed him in the possession of Cilicia, but also gave him one of his relatives in marriage. Sempad sent a notification to the Pope of the change of ruler, and, with the concurrence of the Patriarch, placed himself and his kingdom under the protection of the Roman See. When Haithon and Thoros returned home in 1298 he drove them out again. They returned to Constantinople for help, but only received some money. They then proposed to go to Ghazan's Court, but were arrested at Cæsarea, and imprisoned at Barzberd, where a few days after, by Ghazan's orders, Thoras was put to death and Haithon blinded. This last operation was only partially effective, and Haithon recovered the sight of one eye. These acts were supposed to have been incited by Sempad. In 1298 Sempad fled to Constantinople.* It was against him that, according to Abulfeda, the Egyptians marched. D'Ohsson, following Chamich, makes them march against his successor. They invaded Cilicia in two divisions, one marching through the pass of Izkanderun, and the other through the pass of Meriı. The two divisions united again on the Jihan, and laid waste the open country. The towns were not attacked, a policy which created a feud between Alai ud din Sinjar (the devatdar) and Bektash, the two Egyptian

---

* D'Ohsson, iv. 212-213.

commanders, the latter of whom was in favour of attacking them, and reported in this sense to the Sultan, who supported him, and while the army, having retreated for some distance, was encamped at Ruj, orders arrived that it was not to return until Tel Hamdun had been captured. The troops accordingly retraced their steps, and again crossed the Bagras mountains. Tel Hamdun had been deserted by the Armenians, and was duly occupied on the 18th of June. Another body of troops took possession of Merash. The Egyptians now forced their way into a valley defended by the strong fortresses of Nejimet and Hamus, and where a large number of Armenians had sought refuge. Many of these fled for protection to the two fortresses ; of the rest the men were killed and the women made prisoners. Nejimet was now attacked, and after a siege of forty-one days, during which many assaults were repulsed, the garrison was compelled by want of water to drive out the fugitives who had sought shelter there. The men who thus went out were mercilessly killed to the number of several hundreds by the Egyptians, who divided the women and children among them. Water became so scarce in the fort that the garrison fought for it sword in hand. This scarcity at length forced it to capitulate, the defenders securing their lives. Serfenkiar, Hims (? Hamus), Shoghlaa, a Tesiru, and Anawer (Ainsarba), according to Wassaf, also succumbed to the Egyptians in this campaign. Abulfeda says the Armenians attributed their disasters to Sempad, who was of a proud disposition, and would not conciliate the Egyptians. They accordingly appealed to his brother Constantine, who marched upon Sis, and drove him out. He fled to Constantinople. Constantine now mounted the throne. These events took place in 1298.* Constantine, we are told by Abulfeda, made overtures to the Egyptians, offering to consider himself as their vassal, and to cede to them all the country south of the river Jihan, or Pyramus. Among the towns thus surrendered were Hamus, Tel Hamdun, Kubar, Nakir, Hagr, Shoglaa, Serfendkiar, and Merash. D'Ohsson says the Egyptian army received the keys of eleven strong fortresses. Bektash appointed the Georgian, Seif ud din Asan Timur, their commander. He made Tel Hamdun his head-quarters. Seif ud din Asan Timur retained his post as commandant till the Mongol invasion, when he sold the stores in the towns, and abandoned them, whereupon they were re-occupied by the Armenians.†

The real cause of this war was the anxiety of the Sultan Lachin to find employment for some of his amirs, who were jealous of the rising fortunes of his favourite mamluk, Mangu Timur. That ambitious person persuaded him to arrest the amirs who commanded the troops in Egypt. There remained those in Syria. On the 9th of October, 1298, news was published at Cairo that the Mongols were preparing to enter Syria. Orders were at

---

* *Id.* Weil, iv. 211-212. Journ. Asiat., 5th ser., xvii. 385. Abulfeda, v. 139.
† Abulfeda, v. 138-141. D'Ohsson, iv. 213-217. Ilkhans, ii. 82-83.

once dispatched to Kipchak, the governor of Damascus, to march on Aleppo. This was merely an artifice to get him out of the Syrian capital. He speedily discovered that no Mongol invasion was threatened, and the amir Chaghan, who had meanwhile occupied Damascus, having been so ordered, refused to allow him to re-enter. Orders were also sent to the governor of Aleppo to arrest the amirs Begtimur, Elbegui, Bezlar, and A'zaz, and to poison those whom he could not seize. They were duly warned, however, and proved too vigilant to be captured. *Inter alia*, we are told how, it being customary when a royal order was read for the amirs to assemble at the foot of the citadel, and to dismount and prostrate themselves, it had been arranged to secure them on such an occasion ; but they held their bridles, and went through the process so quickly, surrounded by their Mamluks, that it was not deemed prudent to attack them. The governor then summoned them to an extraordinary council, to deliberate on the news which had come by pigeon post from Biret, that the Mongols were ravaging the environs of that town. They promised to go, but instead, mounted their horses, and fled to Hamath, to join Kipchak. The fugitives were the amir Seif ud din Beg Timur, one of the Egyptian generals ; the amir Faris ud din El Begui, governor of the province of Safad, and the amir Saif ud din A'zaz. Kipchak sent a messenger to Cairo to secure pardon for the fugitives, and also sent to Chaghan at Damascus for money and clothes for them. This was refused. Chaghan, on the contrary, received repeated orders to arrest them, and was threatened with punishment if he failed to do so. He was not furnished with money to pay the troops, who began to desert him and return to Damascus, and he was presently left with but few men and little money, and determined to seek shelter with the other amirs among the Mongols in Persia. He accordingly left his camp at Hims on the night of the 14th January, 1299, with Beg Timur, El Begui, and A'zaz, and went towards the Euphrates, by way of Salamiyet, followed by more than 300 horsemen, and taking with him as far as Kariétéin the governor of Hims. On reaching that place he released him, after depriving him of his horse.* When it was ascertained at Aleppo that the amirs had fled, it was felt to involve very serious consequences. Two armies were sent in pursuit, one towards the Euphrates and the other towards Hamath, but they did not overtake the fugitives. Their property was duly plundered, however, and Kipchak's house at Damascus was sequestered. Meanwhile, Mangu Timur's enemies, feeling that his position was secure so long as the Sultan lived, formed a plot against the latter, who was assassinated on the 15th of January, 1299, by Gurji, the captain of his guards. Mangu Timur shared his fate. Gurji insisted on the amir Tugji, one of the chief conspirators, being raised to the throne, he himself retaining the post of lieutenant-general. His reign was very short, for having gone out to meet the

---

* D'Ohsson, iv. 217-221.

general Bektash, who was returning with his victorious army from the Cilician campaign, he was put to death with Gurji. This was on the 19th of January, but four days after the death of Lachin.

The Mamluks were divided into several bodies. One of these, constituted by Sultan Kelavun, consisting of 3,700 Circassians and Ossethi, had its quarters in the famous "Castle of the Mountain," and was called Burjiyan, from the Arab word "burj," meaning a tower. Another body was styled Ashrafian, from Ashraf, Kelavun's son; a third was called Mansurian, and was Kelavun's special body, he having styled himself Malik Mansur; a fourth body was called Salihiyan, probably from the Ayubit Sultan Salih. On the murder of Tugji, the Burjiyan Mamluks wished to put the amir Bibars, the cupbearer, on the throne, while the Salihiyans and Mansurians wished to appoint the amir Salar. Both parties eventually united in choosing the young prince Nasir, son of Kelavun, who had already, as we have seen, had one ephemeral reign. Couriers on swift dromedaries were sent to fetch him, and meanwhile a council of regency was appointed, composed of eight amirs, who signed all official documents. Nasir soon after reached Cairo, and once more mounted the throne. He was now fourteen years old, and Salar was appointed viceroy.* Orders were sent to Syria to arrest Aidogdi, Chaghan, Hamdan, and the other amirs who had supported Lachin, and also to inform Kipchak of the recent revolutions. The messengers reached the latter and his companions at Reesain, or Reisolain (i.e., the fountain head), which is mentioned by Ammianus Marcellinus under the name of Resania. It was subsequently called Theodosiopolis. Its site is still marked by the ruins of a fine temple.†

The fugitives repented their flight, but had gone too far to turn back. They were received with honour on the Mongol frontier by Jenkli, son of Albaba, governor of Diarbekr, and by the governor of Mardin. They were taken by way of Mosul to Baghdad, whose garrison went out to meet them. One of the Ilkhan's officers was sent to conduct them to his camp at Es Sib, in the district of Vasith. Ghazan went in person to meet them with a great *cortège*. They were supplied with tents and all they needed. Ghazan invited them to a feast, and he sent Kipchak and Begtimur each a present of 10,000 dinars, each dinar of the value of 210 dirhems. A'zaz and El Begui received 6,000 dinars, while every Mamluk, down even to the palfreymen, received 100. Ghazan also sent them horses and other presents. He ordered that his great officers should each give them a feast, so that a round of rejoicing should go through the Ordu. Kipchak was of course delighted with his entertainment, and was joined by some of his family. Begtimur, on the contrary, was not very comfortable among his new friends. Ghazan offered him the town and district of Hamadan as an

---

appanage, but this he refused, saying he had merely gone to pay his
respects to the Khan. Wassaf says the fugitives promised to assist
Ghazan if he made an attack upon Egypt and Syria, and that the ruler of
Mardin, who also had grievances to revenge, would do the same, and he
pointed out how the campaign was to be fought.* On the 24th of March,
while Ghazan was at Baghdad, Bular joined him with a number of
Egyptian dependants. Here news also reached him of the defeat of the
rebel Sulamish, in Rum, by the amirs Choban and Bashgerd, as we
mentioned. The Ilkhan's head-quarters were now moved to Kujin-
büsürg, near Sughurluk, and thence moved on again to the neighbourhood of Aujan, where he was met by his brother Khodabendeh, who
came to see him from Khorasan. A kuriltai was held, and there was also
feasting. On the 25th of June (Von Hammer says the 31st of May)
Gürseh, Cherkes, and Isen, supporters of Sulamish, were put to
death. A few weeks later, viz., on the 17th of July, Ghazan married
Keramun, daughter of Kutlugh Timur, son of Abatai Noyan, who received
a dowry, or marriage portion (mahr), of sixty tumans from her father ; she
succeeded to the establishment which had successively belonged to Dokuz
Khatun, wife of Khulagu ; to her aunt, Tukini ; and to Kukaji, who, like
Keramun, was a relative of the powerful Princess Bulughan. Khodabendeh having set out again for Khorasan, Ghazan went to Tebriz, and
while he was there the rebel, Sulamish, was brought from Rum and
executed. This was on the 27th of September, 1299. Ghazan here had
an affection of the eyes, to cure which rosemary was burnt, and public
prayers were ordered. News now arrived that a body of 4,000 Syrians
had invaded the province of Diarbekr. They captured Mardin, where
they are said to have violated young girls and drunk wine in the mosques
in the holy fast of Ramazan. They also made an attack on Reesain, or
Reesolain, but failed to take it. They, however, carried off many
captives which they made during the raid. The invasion is not
mentioned by the Egyptian historians, but Abulfeda denounces it bitterly
as an infamous act to have been committed by a Mussulman, and as the
cause of Ghazan's terrible invasion. The leader of the freebooters he
calls Bilban Attabbashi, governor of Aleppo.

Ghazan had other wrongs to avenge—the attacks made by the Egyptians
on Cilicia, their capture of Kalat er Rum, and the welcome they gave to the
fugitive Uirads and to Sulamish. His ambition was further supplemented
by the advice of the Egyptian fugitives, and by the distracted state of
affairs in Egypt, where the usual anarchy incident to a government by a
military aristocracy prevailed, and also by the zeal of a recent convert to
Islam at the outrage above mentioned.† Wassaf tells us that when he was
converted Ghazan sent word to the Egyptians as follows : "If my good

---

* Makrizi, ii. 124-125. D'Ohsson, iv. 224-226.
† D'Ohsson, iv. 206-208. Ilkhans, ii. 84-85. Abulfeda, v. 161.

fathers were the enemies of your country, it was because of the difference of faith. Do not entertain for the future any fear that you will be attacked by our victorious troops. May the merchants of both countries freely traverse each. Contrary to what has been hitherto, consider peace with us as the principle of your eternal prosperity. Be assured that all countries now owe us obedience, and particularly Egypt, where the throne has passed from kings to slaves, and where there is no longer any difference between masters and servants."* The last sentence of this message contains, in fact, the main justification of the new war. To put down the turbulent Mamluk dynasty might well be deemed worthy of a Mussulman prince, and having called together the imaums and ulemas, all of them declared in their fethvas that it was the duty of such a prince to repress violence exercised against the Faithful by truculent oppressors. Ghazan issued orders for an army to assemble in Diarbekr, and his generals dispersed in various directions to make preparations. Five of every ten men in the army were to march. Each man was to have five horses, to be fully equipped, and to take six months' provisions with him. Five thousand camels were set aside for carrying provisions. Ghazan fixed a safe place where his wives were to plant their ordus. The noyan Nurin was appointed to guard the frontier of Derbend and Mughan. Sadak Terkhan, the bravest officer in the army, was given charge of the frontiers of Fars and Kerman, as far as Ghazni and Seistan, while Rum and the troops there were committed to Apishka.† Wakhtang, King of Georgia, was called upon to march with his troops, while Beka furnished a contingent of cavalry.‡ Kutlugh Shah commanded the advance guard, and Ghazan himself marched with the main body called Sieg, consisting of three tumans, or 30,000 men. The Georgian history says of thirteen myriads. He set out from Tebriz in the autumn. Rashid says on the 16th of October, 1299, and Wassaf on the 22nd of November.§ He marched by the fort of Alinji to Meragha, and thence through Kurdistan to Irbil and Kieshaf, where the Arran-rud was crossed. At Diarbekr he was joined by the contingent from Rum under Bashgerd Behadur and Kierbuka, or Kertua Behadur. His wives accompanied him as far as Mosul. He reviewed his troops near Nisibin, and was there fêted by Nejm ud din, Sultan of Mardin, who did homage to him and supplied the army with provisions on the route to Reesain and the fort of Jaahuzad, where he crossed the Euphrates. Rashid says this took place on the 7th December. There he left a body of 10,000 men under Prince Balarghu, Mamai, and the Sultan of Mardin. He foresaw that the army

---

* D'Ohsson, iv. 227. Note. † Id., 228-229. Ilkhans, ii. 85-86.
‡ Hist. de la Géorgie, i. 630.
§ Wassaf thus enumerates his chief officers : Kur Timur, Teremtar, Nakuldar, Habak, Doladai, Aktaji, Fenjaghatai, Taghar (son of Sutai), Badrangirai, Kutlughia, Belarghi, Yemish, Yusuf Buka, and the amirs Satilmish, Mirsadeh, Melai, Walid Dostai, Sultan Wejilar, Toka Timur, Kurmishi (son of Alinak), Tebbak, Ilbasmish, Chichak, Toghrulji, Beitas, and Yeman. Ilkhans, ii. 86. Note.

on its return would have some difficulty in recrossing the river, which
would be swollen with the rains, and a great raft, supported by inflated
skins and fastened to each bank by chains, was ordered to be made, the
Sultan of Mardin taking charge of its construction. Having crossed the
Euphrates, Ghazan reviewed his army, which was 90,000 strong. Kutlugh
Shah and the amir Mulai set out with the advance guard, and three days
later Ghazan himself arrived at Jil, near Aleppo, where a halt of two days
took place. The people of Ram, Aintab, and Aleppo had largely
deserted those towns on the news of the Mongol approach. It was
determined not to stop to attack the citadel of Aleppo, but to pass it by
and march on. Another parade of the troops was held. Ghazan
inspected the ranks on foot, and noticing that he especially regarded the
horses, the amir Choban suggested to his companions that each of them
should present him with his best horse, which they accordingly did. The
march now continued through a very fertile district, and the soldiers
would have allowed their horses to feed on the corn, but Ghazan forbade
it on pain of death, saying it was not right for horses to eat man's food.
On reaching Jebles Sumak, a fortress of the Ismaelites situated near
Harim, they captured a spy, who informed them that Bilban Tabbakhi,
governor of Aleppo, with Kara Sonkor, governor of Hamath, had retreated
and gone to join the Egyptian army at Hims. On reaching Zermin, a
day's journey south of Aleppo, and famous for its olives, Mogoltai Ajaji,
with a body of Kipchak's dependants, deserted. Maaret Naaman was
found abandoned by its people. The citadel at Hamath, like that 'at
Aleppo, was passed by, and they reached Salamiyet, east of the
Orontes, a day's journey from Hamath. Here they met the first patrols
of the enemy, who were posted in force at Hims (the ancient Emessa).
The Sultan had sent Kertai, Kadlubek and other amirs to Syria, to make
head against the invaders, and had himself set out from Cairo on the
22nd of September.* His turbulent amirs were at issue with one another,
and when he reached Tel-el-Aajul, north of Gaza, disturbances broke out.
The Uirats who had sought shelter in Egypt, as I described, aggrieved
by the execution of some of their chiefs in the reign of Lachin, by the
deposition and exile of Ketboga, their protector, and jealous of the
supremacy of the Burji Mamluks, made a plot to murder the Sultan and
his chief amirs, Bibars and Salar, who were leaders of the Burjis, and to
restore Ketboga. The plot failed, but not till the conspirators had actually
penetrated into the Sultan's tent. The Uirats were arrested, and fifty of
them were strangled. They were led to execution in their turbans and
dresses, a crier proclaiming in front of them: " See the just punishment
of those who cause troubles among the Mussulmans, and who dare to
attack their ruler." Four days later their corpses were taken down from

‚the gibbets where they were exposed.* The Sultan now set out, by way of Ascalon and Karita, towards Damascus. While the army was encamped at the former place, a terrible flood of water destroyed the baggage, and reduced many of the troops to destitution. This calamity was followed by the appearance of great flights of locusts, and both were deemed evil portents for the coming struggle. The Sultan entered Damascus on the 3rd of December, where soon after the fugitives arrived from Aleppo, and the news also came that Ghazan was encamped on the Euphrates with a large army. Largess was distributed to the troops, each soldier receiving from thirty to forty dinars, but presage of defeat caused a general discouragement. The Sultan was joined by Asen Timur Karji, who governed the districts recently conquered from the Armenians, and who took the Armenian king and the taxes he had collected at Tel Hamdun with him. The army now set out for Hims, whence a number of Bedouin scouts were sent. The soldiers remained under arms for three days, and provisions began to fail.†

Let us now return again to Ghazan. We are told he made a special prayer of two rek'ats with his army to implore divine aid. Many of the horses had been disabled by their long march, a fact which, according to Haithon, had been communicated to the Sultan by the fugitive Kipchak.‡ The dismounted troopers called the Ilkhan's attention to their forlorn condition on the very eve of battle. He thereupon gave orders that they were all to fight on foot, urging that they could in this way by flights of arrows best meet the furious charges of the Mamluks, whose policy was to break their enemy's ranks mace and scimiter in hand. He set out from Salamiyet, and advanced to within a day's journey of the Egyptian army. Orders were issued to the troops to make ready, but as the Ilkhan did not wish to engage on a Thursday, which is an unlucky day, especially when, as on this occasion, the last Thursday in the month, it was determined to halt for a day near a stream that flowed by, and the troops dismounted and lay under their armour. Suddenly the scouts reported the approach of the enemy. Ghazan had only the centre with him, consisting of 9,000 men. He ranged it in order of battle, and called up the two wings which had remained behind, as the battle was not expected to take place so soon.§ The Egyptians were posted near the mausoleum of Khalid, son of Velid, at the foot of a hill which was known as the Hill of Victory, and where the Mongols had, on a former occasion, suffered a severe reverse at the hands of the Egyptians. Ghazan's commander, Kutlugh Shah, on the present occasion, boasted that they intended to convert the hill of victory for the enemy into one of defeat.|| The centre of the Mongol army was commanded by Ghazan himself, with whom were the amirs Choban and Zerban; the

---

* Makrizi, ii. (part ii.) 141-144.　　† Id., 144-146.　　‡ Op. cit., ch. xli.
§ D'Ohsson, iv. 233-234.　　|| Ilkhans, ii. 91.

right wing consisted of the tumans commanded by the amirs Mulai,
Satilmish, and Kutlugh Shah; and the left those of the amirs Toghrilji,
Ilbasmish, Chichek, Kurmishi (son of Alinak), and Kurbuka Behadur.
Each of their divisions was posted one behind the other. Ghazan
detached Sultan Yassaul with a tuman to turn the enemy's right wing,
and make him abandon the Hill of Victory. The struggle then began.
The battle-field was called Mojma ul Muruj (*i.e.*, junction of the meadows),
now known as Wadi al Khazinadar (*i.e.*, valley of the treasures), and it
lay between the Hill of Victory and a stream called Ab Barikh by the
Persians and Narin su by the Mongols. Abulfeda says it was half
a day's journey from Hims.* Wassaf says the Egyptians numbered
about 40,000. Makrizi says twenty odd thousands. Novairi says 25,000,
while he gives the Mongols 100,000. The amir Isa ibn Muhanna,
with his Arabs, and Bilban Tabbakhi, naib of Aleppo, with the troops
of that town and of Hamath, composed the right wing. The amir
Bedr ud din Bektash, with a number of amirs, was on the right; while
the centre was commanded by Bibars, Salar, and other chiefs, the
Sultan's own Mamluks forming one of its wings. The young Sultan
himself, with Hosam ud din Lachin, the ostadar, were posted some
distance off in a place of safety. Five hundred picked Mamluks formed
the advance guard. On the eve of the struggle the amir Bibars had a
violent attack of diarrhœa, and left the battle-field in a litter. The amir
Salar, with the hajib, the amirs, and the fakih inspected the men, the
hajibs exhorting them to be firm, and causing many of them to weep.
Meanwhile Ghazan's men stood still, he having ordered them not to move
until he himself advanced, and then to charge altogether. The attack
was begun by a body of Mamluks armed with javelins, whose heads
contained naphtha (? stink pots). This was doubtless to throw the enemy's
horse into confusion, but as he retained his ranks, and the intervening
distance was considerable, the naphtha was extinguished. The Mamluks
charged before Ghazan's men were fairly in order. As we have seen, he
had ordered them to dismount, and forming a rampart with their horses,
they threw a great flight of arrows into the advancing Egyptians.
Suddenly there was heard the noise of the great Mongol war cymbals,
which came from the right wing under Kutlugh Shah. Fancying this
was the advance of the main army under Ghazan himself, the Egyptians
turned thither, and charged with all their force. Kutlugh Shah's men,
who were caught while re-mounting their horses, were overthrown, and
5,000 of them were put *hors de combat*. Kutlugh Shah himself, with some
of his people, now joined Ghazan. The latter had caused the centre and
left wing to advance, 10,000 archers on foot being in front. Their
arrows caused great havoc among the Arab auxiliaries, who fled, and were

followed by the men of Aleppo and Hamath. Eventually the whole of the Egyptian right wing was routed. The left wing, as we have seen, had been more fortunate. Ghazan was about to flee, but his courage was revived by the naib Kipchak, and having gathered his men around him, he charged the Egyptian centre, which broke. Salar, Bek Timur, the jukendar, Borloghi, and the other Burji amirs fled, and the Mongols pursued so closely that their arrows struck fire on the helmets of the fugitives. The young Sultan, who, as we have seen, had taken up his quarters a little distance off, wept, and addressed prayers to heaven. He had only a handful of Mamluks with him. Meanwhile, the Egyptian left wing, which had been successful, returned from its pursuit, and found the rest of the army broken. According to Novairi, the Egyptians had then only lost 1,000, while the Mongols had lost 14,000, which made Ghazan fear that the flight was a ruse, and he halted his people. Makrizi says this was a divine mercy, for if he had continued his pursuit the Egyptians would have perished to the last man. The battle had lasted from nine in the morning to two in the afternoon.*

Ghazan greatly distinguished himself in this battle, charging the enemy lance in rest, until his officers had to restrain him. The *Georgian Chronicle* tells us the struggle was so vigorous that Kutlugh Shah, seeing a great body of men attacking the Khan, planted himself before him with 1,000 horsemen, of whom 600 perished, and there only escaped 400, who beat off the Egyptians. He was supported by Choban (called Chophon Suldur by the chronicler) with 500 cavalry. Wakhtang and his Georgians also fought very bravely.† Abulfeda says the Sultan took refuge at Hims, but towards night his followers fled towards Egypt. They were pursued as far as Gaza, Kodsa, and Karak, and the Mongols captured great spoil.‡ Wassaf says Ghazan used his victory with moderation. He advanced slowly as far as a fersenkh from Hims, and at nightfall ordered the carnage to cease. The ground was covered with weapons, armour, &c. The Egyptians retired by way of Baalbek to Damascus. The Sultan himself went to Cairo. During the height of the struggle a body of 5,000 Bedouins appeared from the direction of the desert, and tried to take the Mongols in rear, but Ghazan, afraid that the same manœuvre might be practised upon him that overwhelmed Mangu Timur, had ordered Kur Buka with the rear guard to keep watch in that direction. When he saw the Arabs approaching, he charged and broke them.

After the battle Apishka arrived from Rum with the King of Cilicia, who brought 5,000 troops. This king was Haithon II. I have described how he had been blinded by order of Ghazan, and how his brother Constantine mounted the throne. The blinding was apparently very clumsily done,

---

* Makrizi, ii. (part ii.) 147-149.   Haithon, 62-63.   D'Ohsson, 236-237.   Ilkhans, ii. 92-93. Weil, iv. 228-229.   † Op. cit., 631.   ‡ Op. cit., v. 165.

for we are told Haithon recovered his sight, and was replaced on the throne for the third time by the Armenian barons. Constantine, whom this movement dispossessed, proposed releasing his other brother and predecessor, Sempad, from prison, and creating troubles, but they were both seized and sent to Constantinople, where they remained till their death.* Abulfeda says it was Constantine who released Haithon from prison, and describes the latter's conduct as ungrateful.†

The Egyptian fugitives fled to Damascus, where the citizens were more or less panic-stricken and joined them in their flight, and many of them were plundered by the predatory Ashir and the Arabs. Makrizi tells us among those who perished were the amir Kert, naib of Tarabolos, or Tripoli ; the amir Nasir ud din Muhammed, son of the amir Ai-Timur Halebi ; Malian Takwi, one of the amirs of Tripoli ; Bibars Gatmi, naib of the citadel of Markab ; Uzbek, naib of Balatonos ; Bilik Taiar, one of the amirs of Damascus, and about 1,000 Mamluks. The chief judge, Kadhi Alkodat, Hosam ud din Hasan, Kadhi of the Hanefis of Damascus ; and the secretary (muwakhi), Imad ud din Ismael. According to the Egyptian authorities, the Tartars lost 14,000 men in the battle.‡ The Georgian Chronicle says the fugitives were pursued as far as the holy city of Jerusalem, where many Christians and Persians were killed.§

Ghazan having pitched his camp near Hims ("in quodam loco, qui vocatur Cametum," says Haithon‖), received the congratulations of his generals, and distributed rewards amongst those who had distinguished themselves. Wassaf himself wrote the letter in which the victory was announced to the principal places in the empire. His own inflated statement is that Wassaf " decorated the robe of this royal letter with the border of good deeds and acts which form the subject matter of this fortunate news."¶ The governor of Hims presented Ghazan with the keys of that town, where the Sultan had deposited his treasure, and where the baggage of his army was stored. Ghazan distributed these among his officers, dressing several of them in the robes of the Sultan Nasir. The chronicler Haithon, who praises Ghazan, hyperbolically tells us he was an eye witness of his generosity on this occasion, and that the Ilkhan only reserved to himself a sword, and a sack containing all the title deeds of the kingdom of Egypt, and the muster roll of its armies.**

The news of the Sultan's defeat caused great excitement at Damascus. Women went about the streets unveiled, with children in their arms ; men abandoned their shops and goods to escape from the place, and the crush was so great that many were smothered at the gates. Some fled to the mountains and villages, but the greater number went to Egypt. People became more reassured when it was known that the Khan was a

---

* D'Ohsson, iv. 238-239. † Abulfeda, v. 173.
‡ Makrizi, ii. (part ii.) 149-150. D'Ohsson, iv. 239. § Op. cit., 631. ‖ Op. cit., 63.
¶ D'Ohsson, iv. 240. Note. ** Haithon, 6, D'Ohsson, iv. 241. Note.

Mussulman, and that the greater part of his troops professed the same faith; that they had treated the fugitives kindly and had not put them to death, but had been content to deprive them of their horses and weapons, and then let them go free. Presently the fugitive soldiers began to arrive, and changed their costume to avoid being insulted by the people. Some of them went to the length of cutting off their hair, and only remained long enough at Damascus to secure their wives and such property as they could hastily move, and then went on towards Egypt, many of them being pillaged *en route* by the Bedouins. There were no longer any police in the place, and one night the prisoners set fire to a door of the prison, and escaped, to the number of 150; and bands of robbers pillaged the houses. The people who remained in the place collected before the great mosque, and agreed to send a deputation to Ghazan. The chief judges of the sects, Shafiyi and Maliki, the prefect of the town and the district, the prefect of police, and a great number of notables, doctors of the law, and readers of the Koran accordingly set out as a deputation to Ghazan to a place named En Nebek (Makrizi says Nebl). They encountered him while on the march, and dismounted, and many of them kissed the ground. Ghazan halted, his Mongols dismounted, and meanwhile the interpreter, Fakhr ud din ibn Es Shirji, arrived. They asked his clemency for the people of Damascus. He replied, "What you have come to ask has already been granted," but he would not accept the food which they offered. Ghazan had already granted protection to the citizens on the request of one called El Sherif el Gatémi, who had gone to him with three other inhabitants of Damascus before the grand deputation. El Gatémi arrived there on the 31st of December armed with the Khan's orders, and accompanied by four Mongols. The grand deputation did not return till later. They arrived on the Friday afternoon after the hour of prayer, and, we are told, on this Friday there was no prayer made for the Sovereign. On the 2nd of January, 1300, an officer named Ismail entered the city with a detachment of Mongols, went to the mosque of the Ommiades, where the people had been convoked, and had the following edict read from the pulpit by one of his companions:—

"By the power of God. Let it be known to the chiefs of tumans, of thousands, of hundreds, and to all our victorious troops, Mongols, Tajiks, Armenians, Georgians, and others subject to our rule, that God when he enlightened our hearts with the light of Islam, and guided us towards the religion of the Prophet (on whom rest the most excellent blessings of God and of peace), it was in these celestial words: *He whose heart God has opened to the light of Islamism follows the light of the Lord. Cursed are those whose hearts are hardened against the divine will. They are in manifest error.** When we learnt that those who ruled over Egypt

---

* Koran.

and Syria had abandoned the path of religion, that they did not obey the precepts of Islam, that they broke their promises, were perjured, without faith and law. That there was no order among them, and that each of them on attaining power *sought only how to satisfy his evil desires, destroying the seed* (of plants) *and the fertility of animals. God loves not disorder.** That their acts spread terror among the people, that they laid hostile hands on the women and property of their subjects, that they abandoned the path of justice and equity to follow that of violence and oppression. Therefore we were compelled by our zeal for the Faith to march against these countries with a numerous army in order to put a stop to these evils. And we have vowed if God enables us to conquer these lands to stop such misdeeds, and to spread over all the blessings of justice and benevolence, conformably to the divine commands which declare that *God enjoins justice, kindness, and liberality towards one's relatives. He forbids crimes and ill-deeds. He exhorts you in order that you may take good heed.†* In accordance with the announcement of the Prophet (may God be propitious to him and grant him peace), *The just shall sit on thrones of light on the right hand of God, and his two hands will aid those who follow justice in their dealings with their relatives and subjects.‡*

"As it is our intention to attain this laudable aim and to accomplish our vow, God has granted us a signal victory, and has heaped his benevolence upon us. We have defeated an audacious enemy and unjust legions. We have dispersed them as error vanishes before the truth, for error must needs perish. Our heart has consequently been still more opened to receive the Faith, fortified by the truth of the precepts given *to those to whom God has granted the love of the Faith, whose hearts he has inspired with its charms, in whom he has inculcated a horror for unbelief, wickedness, and rebellion. Such march in the right way by the grace of God.§*

"Under an obligation to observe these firm resolves and this sacred vow, we have forbidden everyone in our army, to whatever class he may belong, to do any harm to the city or territory of Damascus, or to the towns of Syria subject to Islam; to refrain from touching the persons, harems, or goods of the inhabitants, and from prowling about their houses, so that they may in perfect security follow their employments of merchants, agriculturists, &c. Among the crowds of our warriors, some having been found who, notwithstanding our prohibition, have dared to pillage and to reduce the people to slavery, we have had them put to death as an example to the rest of what they may expect; and as a proof that we mean this to be carried out rigidly, we also forbid them to molest those of other faiths—Jews, Christian, or Sabæans, *For they pay tribute, so that their goods may be as our goods, their blood like our blood.‖*

---

* Koran, c. ii. † Koran, c. xvi. v. 92. ‡ Koran. § Koran, c. xliv. v. 7. ‖ Koran.

"Rulers owe protection to their tributaries as they do to their Mussulman subjects, as the Prophet says. The Imam placed over men is in the position of a shepherd; and a shepherd has to give account for his sheep. The kadhis, khatibs (preachers), sheikhs, doctors of the law, sherifs, lords, notables, and all our subjects in general are invited to rejoice in our glorious victory, and to address their prayers to God with cheerfulness for the strengthening of our power."

The document is dated the 30th of December, 1299.* The proclamation somewhat calmed the people in the town. Meanwhile, however, Alem ud din Sinjar Arjevash, who commanded the citadel, closed its gates and determined to defend it. On the 5th of January the amir Ismail who had taken command of the place, ordered the ulemas, sheikhs, and magistrates to press him to surrender it, as the Mongols, would otherwise sack the town. His reply was to cover them with imprecations, and to tell them that he had received news by pigeon-post that the Sultan, had defeated his pursuers, had rallied his men at Gaza, and would speedily arrive at the head of the army. The amirs Kipchak, Beg Timur, Elbegui, and Azis, who had accompanied Ghazan and assisted at the battle of Hims, entered the city the following day, and also pressed their old companion to surrender, but it was without effect; as was a summons signed by the chief of the sheikhs, by a Mongol general, who styled himself Ghazan's foster brother, and by Kipchak.† The same day Ghazan himself encamped at Merj Rakith, the eastern part of the district of Guttat, often cited by Orientalists as a second paradise, from its wonderful trees, gardens, vineyards, &c. The principal people of Damascus went there to do him homage. Ghazan was never weary of expressing his contempt for the parvenu ruler of Egypt. He now asked his visitors who he himself was. They replied, "Shah Ghazan Khan, son of Arghun Khan, son of Abaka Khan, son of Khulagu Khan, son of Tului Khan, son of Jingis Khan." He asked who then was the father of Nasir, the Egyptian Sultan. They replied, "Kelavun, son of Elfi." "And who was Elfi?" They confessed he had been a slave, bought with a thousand ducats, whose father was unknown. "Your living ones are good for nothing," was the reply, "but your dead (i.e., the famous and holy men buried at Damascus) are indeed worthy, and for their sakes I granted you pardon." Ghazan now entered the city, and went to visit the Meidan ul hassa. He was charmed with the place, and to preserve it from harm he closed seven of its gates and placed a guard at the eighth, viz., the Gate of Baghdad. Jemchi and Tulka Bakhshi were given charge of it, but having failed in their duty they were bastinadoed, and replaced by Jihurgutai. Ghazan forbade the troops to enter the gardens in the environs. On Friday, the 8th of January, the khutbeh was said in Ghazan's name. In it he was styled

---

* Novairi, &c., Quatremere's notes to Makrizi, ii. (part ii.) 151-154. D'Ohsson, iv. 245-249.
† Makrizi, ii. (part ii.) 155. D'Ohsson, iv. 249-250.

"Our Lord, the Great Sultan, the Sultan of Islam and of the Mussulmans, the victorious Mahmud Ghazan." After the divine service, Kipchak and Ismail mounted the gallery, whence the muezzin was sounded, and an ordinance was publicly read, constituting Kipchak governor of Syria and authorising him to appoint prefects, judicial and ecclesiastical officers. The text of this proclamation is extant, and after reciting the usual pious and other phrases, goes on to say that Ghazan, after mature consideration, had appointed, as most fit for the post, the naib Saif ud din Kipchak, distinguished by his noble qualities, and uniting in his person every excellent title, who had been driven into exile and had sought shelter at his stirrup. He had conferred on him the august rank of Naib Assallanah (viceroy), and put under his authority the provinces of Damascus, Aleppo, Hamath, Shaizar, Antioch, and Bagras, with all the fortresses, the province adjoining the Euphrates, Kalaat ur Rum, Behesna, as well as all their dependencies, where he was to have absolute authority, taking counsel in his appointments and acts with the king of amirs and of viziers, Nasir ud din. He had conferred on the Naib a sword, an august standard, a drum, and a paizah marked with a lion's head ; and all the amirs, commandants, the amirs of the Arabs, Turkomans, and Kurds, the officers of the chancellery, the sadrs, and all others were to take heed of the power thus conferred on him, and to be subordinate to him, their fortune depending on gaining his goodwill and approval, and paying him all due respect and honour. He, on the other hand, was to have the fear of God before him in his decisions; to hold a strong hand over his subordinates, so that they might duly fulfil their duties ; was to see the decisions of the judges carried out, to strengthen the hand of justice, and punish the evil-doers, &c.* This appointment was a very popular one. The proclamation was accompanied by the usual shower of dinars and dirhems over the crowd. The finance officers and other civil officials were for the most part retained in their posts. Ghazan's generals demanded to be allowed to pillage the place, on the ground that the citadel had not surrendered. He peremptorily refused, and forbade any soldier to enter the town unless armed with a proper passport issued by the Divan ; but the inhabitants were called upon to pay a sum of 100 gold tumans, or a million of dinars. One MS. says 3,600,000 dinars, besides all the sumpter beasts in the city, weapons, and many costly stuffs, &c.†

Notwithstanding Ghazan's orders, his troops could not be restrained, especially the Armenians and Georgians, and we are told how they pillaged Salihiyat, at the foot of Mount Kassiun, an hour's journey from Damascus, whose environs were famous for their gardens and country houses. They plundered even the carpets and the lamps in the

---

* Quatremere's notes to Makrizi, ii. (part ii.) 156-159. Notes. Ilkhans, ii. 94.
† Weil, iv. 233. Note.

mosques, the sepulchral chapels, and colleges. They burnt some of the buildings, and dug up corpses in search of treasure, and many of the inhabitants were killed or reduced to slavery, numbering, it was said, as many as 9,900. The town was thus ruined. This devastation was attributed to the King of Cilicia, who thus sought to avenge the cruel raids of the Egyptians upon his country. He intended laying waste Damascus, but Kipchak opposed, and gave up to him instead Salihiyat. The towns of Mezzat and Daria were also laid waste. The Sheikh Taki ud din, son of Timiah, went to complain to Ghazan, who was then at Tel Rahet, but he could not get an audience, as he was drunk. Another account says he was dissuaded from speaking to the Ilkhan himself, since the latter would certainly order some of the evil-doers to be put to death, and thus arouse the animosity of the chiefs against the people of Damascus. He accordingly appealed to the Vizier, Said ud din, and to his colleague, the historian, Rashid ud din, who replied that as some of the Mongol generals had not shared in the prize money it was necessary they should be rewarded in some way, and ordered a further contribution and the release of the prisoners. This fresh tax was distributed among the various corporations or guilds of citizens, each class being placed under a Mongol officer, who, to exact payment, did not scruple to torture and maltreat the people. The price of everything rose very greatly ; wheat was sold at 360 dirhems the ghirárah, and barley at 180. A *rotl* of bread cost 2 dirhems, of meat 12, of cheese 12, of oil 9 ; four eggs cost a dirhem. The girdle makers' bazaar had to pay 130,000 dirhems, the lace makers 100,000, and the copper smiths 60,000. The principal inhabitants had to pay 400,000 dirhems. Massacre and pillage prevailed throughout the city, and the number of soldiers and civilians who perished is said to have reached 100,000. Kemal ud din ibn Kemal ud din ibn Kadhi Shohbah wrote on this occasion : " The vicissitudes of fortune have let loose seven scourges upon us, and no one can protect us from their assaults—famine, Ghazan, war, pillage, perfidy, apathy, and a continual grief."*

Such was the terrible penalty of being defeated by the Mongols, even after they had become Mussulmans. We are told that the result was that 3,600,000 dirhems were poured into Ghazan's treasury, exclusive of the arms, rich stuffs, sumpter beasts, and grain, and of all which the Tartars had pillaged. Every day 400 ghirárahs were sent to them by the eastern gate. Ghazan had ordered that the horses and camels were to be impounded. More than 20,000 were accordingly seized. Istabal, son of the famous astronomer, Nasir ud din of Tus, who was inspector of the wakfs, or charitable foundations, appropriated 200,000 dirhems. Kipchak and the other amirs were also liberally rewarded, while a daily contribution was exacted for Ghazan's own expenses.† Ghazan had

---

Makrizi, ii. (part ii.) 160-161.   † *Id.*, 162.

vowed before the battle to present golden lamps, turbans, and carpets to the mausoleum of Seif ud din Khalid, son of Velid, the famous Arab general who gained a signal victory over the Emperor Heraclius. This vow he fulfilled. He assigned the revenues of several villages near Damascus to the support of Abraham's tomb at Hebron. The Egyptian Sultans had employed the revenues of the wakfs, consecrated to the two holy cities of Mecca and Medina, in paying the escort of the pilgrims' caravan to Mecca, authorising this mode of spending the money by a special fethva. Ghazan deemed this assignation illegal, and devoted it to its primitive purpose. This statement of Rashid's probably explains what Makrizi says about Istabal appropriating 200,000 of the revenues of the wakfs at Damascus.* After the battle of Hims the King of Cilicia and the amir Molai were sent in pursuit of the Egyptian Sultan, who fled along the coast road. The former was recalled to Damascus by Ghazan, but Molai continued his pursuit with 15,000 horsemen, or, as Haithon says, with 40,000. He had advanced as far as Gaza, killing the Egyptian soldiers he met *en route*, and ravaging the country. Many of the fugitives, we are told by Haithon, who found their way to Tripoli in Mount Lebanon, which was inhabited by Christians, were cruelly put to death there. Having heard that he had entered the desert with 2,000 or 3,000 horsemen (Haithon says he reached Babylon escorted by Bedouins), he retraced his steps.†

Ghazan now determined to retire. According to the more unprejudiced testimony of Haithon, his retreat was caused by the invasion of his eastern borders by the Jagatai Mongols, to which we shall presently refer. His own historians attribute his withdrawal to the approaching heats of summer, which is rather ridiculous, as it was only February.‡ Before setting out he nominated Kutlugh Shah commander-in-chief in Syria. Molai, with a tuman, was appointed governor of Gaza, and subordinated to Kutlugh Shah. Kipchak, as we have seen, was appointed naib of Damascus, a province which extended as far as Hims ; Bektimur Shami, naib of northern Syria, *i.e.*, of Aleppo, Hamath, Aintab, and the mountains of Sumak, Biret, and Rahbet. Albegui was set over middle Syria, which included Tripoli and Akka, Salamiyat and Maaret naaman. Each of these officers had a contingent of Mongol troops with him. Malik Nasir ud din Yahya, son of Jelal ud din Toridi, was appointed head of the finances.§ Haithon says that Ghazan, before leaving Syria, summoned the King of Armenia and told him he would make over the places he had conquered to the Christians if they would come and occupy them (*i.e.*, to the Crusaders), and he ordered Kutlugh Shah to supply them with funds to rebuild them.|| Ghazan raised his camp on the 4th of February, and

---

* D'Ohsson, iv. 255. † Haithon, ch. xlii. D'Ohsson, iv. 257-258.
‡ Haithon, ch. xliii. D'Ohsson, iv. 256.
§ Makrizi, ii. (part ii.) 162. Ilkhans, ii. 95. D'Ohsson, iv. 256. Note 1. || Op. cit., ch. xliii.

recrossed the Euphrates on the 16th, at the fortress of Jaaber, by a floating raft made of bark, fastened with ropes, which was his own invention.* Before we follow him we will conclude the adventures of his officers in Syria, which ended by no means so fortunately as they had begun.

Soon after he retired, the Mongols ordered the inmates to leave the college Adeliah, at Damascus. They stripped each one as he came out, and then entered the place, broke open the doors, pillaged it, and proceeded with their robberies in other parts of the city. Many houses and colleges were burnt, among which were the Dar alhadith (house of traditions) Ashrafiah, and all its appurtenances : the Dar alhadith Nuriah, the little college of Adeliah and all about it, the college Kameriah and its environs, as far as the Dar assaat (house of happiness) and the Maristan (hospital) Nuri, and from Dimaghâat to the gate of Ferej.† We have seen how the citadel at Damascus held out under Arjevash. (Abulfeda* calls the governor Saif ud din Argovani Mansuria.‡) The Mongols ordered the quarters of the city nearest to the citadel to be evacuated, and used the roofs of the houses as a vantage whence to shoot their arrows. Thereupon Arjevash set fire to them, and the conflagration consumed many public buildings, colleges, &c., including, according to Abulfeda, the Dar as Sadaat, where the sultans of Egypt were accustomed to lodge when in Syria. A skilled engineer, named Hazrawi, was employed to make catapults, a large one being erected on the roof of the Great Mosque of the Ommiades. Arjevash saw that he would have to converge his weapons upon this structure, and that the famous mosque might in consequence be destroyed. He accordingly sent some of his people, whose religious zeal made them face any danger, and who went and sawed in pieces the beams which had been got together for the catapult. The engineer re-commenced his work protected by a guard of Mongol soldiers, who were posted in the mosque and committed all kinds of debauchery there, so that the evening prayer was omitted on several days. They also.pillaged the bazaar close by. Arjevash having set a price of 1,000 ducats on the engineer's head, a Shia penetrated into the mosque, struck him with his sword, and killed him. He had some companions with him who were prepared to fall upon the Mongol guard, but it dispersed, and the bold author of the deed succeeded in escaping with his victim's head to the citadel.§

When Ghazan left Damascus, Kipchak was left behind as nominal ruler, but the real one was Kutlugh Shah, who was in command of the Mongol contingent, and we are told how the very day the Ilkhan left the place the Mongols plundered the citizens of a larger sum than the ransom they had recently paid. Ten days later Kutlugh Shah withdrew

* D'Ohsson, iv. 257. † Makrizi, ii. (part ii.) 163. ‡ Op. cit., v. 165.
§ Makrizi, ii. (part ii.) 160. Ilkhans, ii. 95-96. D'Ohsson, iv. 258-259.

from Damascus, and Kipchak drew from the inhabitants a considerable
sum as a parting gift. Haithon says he was recalled by Ghazan;
Makrizi that he was persuaded by Kipchak and Bektimur to take up
his residence at Aleppo. He left Molai in charge of the Mongols in
Syria. Kipchak now took up his residence in the Kasr Ablak (i.e., the
White Castle), and sent criers round inviting people to return to their
homes. The bazaars and gates were again thrown open, and to inspire
confidence he ordered a number of soldiers to promenade the town with
an itinerant wine shop, and the consequence was that drunkenness and
indecency greatly prevailed. Meanwhile the Mongols ravaged the
districts of Gaur, and penetrated as far as Jerusalem in one direction and
beyond Gaza in another. They killed fifteen people in the great mosque
of the latter town. They also visited Baalbek and Al Bakaa, whence,
on hearing that the Sultan Nasir was about to march from Egypt at the
head of a fresh army, they withdrew again towards Persia,* and Syria was
once more free from the Mongols.

After his defeat, the Sultan had retired, as we have seen, with but two
amirs and a few followers. He re-entered Cairo on the 12th of January.
Fugitives from the battle came thither in small parties, and in great
distress, some of them having not only lost their weapons and uniforms,
but also shaved their hair and. beards. They were insulted by the people,
who jeered at them for having fled before the Tartars. A funeral service
for those who had perished was performed, and preparations were
immediately begun to avenge them. Artisans were collected from all
sides to manufacture arms. The Vizier made a requisition of money to
pay the expenses of a new campaign. Orders were sent to the various
provinces to send horses, dromedaries, lances, and swords. Horses and
mules which were working the mills were impounded, a handsome price
being paid for them. A horse worth 300 dirhems in ordinary times now
rose to a thousand. The price of arms rose seventy or even a hundred
fold. Soldiers were ordered to rejoin their ranks, and the pay of those
who had perished was duly paid. Each commander of 1,000 men was
assigned ten licensed soldiers (soldats licenciés?), for whom he was
to provide; each amir of tabl khanah was to furnish five, and each
amir of ten two. Many of the amirs raised bodies of volunteers. Mejd
ud din Isa ibn Alhabbab, naib (substitute) of the Mohtesib, was told
to obtain from the fakihs a fethva, or order, authorising an extraordinary
levy of money. A canonical decision of the Sheikh Iz ud din abd
Esselam was produced, imposing a tax of a dinar on each individual. The
Sheikh Taki ud din refused to concur in the fethva, and declared that the
decision of Iz ud din had only been given after the military chiefs had
surrendered all they possessed in gold and silver, the jewels of their wives

---

and children, and when each one had sworn that he had nothing more, and that the sum collected was still insufficient. "But we know," said the exacting sheikh, "that now the amirs have plenty of money, that they can give their daughters strings of pearls and jewels, that the shoes of their wives are garnished with gems, while their wash-hand basins are made of silver." As the obdurate authority refused to concur, a requisition was made on the merchants and traders, and by these exertions a fresh army was on foot in the beginning of February. The various towns of Egypt were crowded with soldiers who came from different parts of Syria. The houses were too small to hold them. They encamped in the Karafah, about the mosque of Ibn Tulun, and at the extremity of the quarter Hosainiah. Meanwhile the price of provisions remained very moderate.* The Sultan dispatched express posts to the governors of the various fortresses of Syria announcing to them his speedy march, and bidding them hold on bravely. Ghazan had summoned them to surrender by special firmans, headed "By the divine power, and under the auspices of the Muhammedan faith," but his acts were contrary to his words, says Novairi, for he permitted the excesses of the Armenians and others.†

News now reached Egypt that Ghazan had retired, and that Kipchak had been left in charge there. This news was very welcome. The Sultan wrote to Kipchak, to Bek Timur, to El Begui, and other officers who had joined the enemy, asking them to return to their allegiance. They replied that they were prepared to do so, and Kipchak, with all the people of his suite, accordingly set out for Cairo. This was in the middle of April. The Tartars were naturally greatly alarmed, for the hot season was close at hand when they could neither fight nor expect succour. We have seen how Molai withdrew with his men beyond the Euphrates. Arjevash, who had held the citadel of Damascus so bravely, came out and occupied the town itself on the 8th of April, the name of the Sultan was again inserted in the public prayer, after having been erased from it for 100 days, and the wine shops, &c., were again repressed. An order had been issued at Cairo that the army was about to march, and that whoever remained behind would be strangled. The price of the dinar, which had formerly been 25½ dirhems, and in the recent cheapness of money had fallen to 17, was fixed at 20.

The Sultan set out on the 31st of March, and halted at Salihiyet, whence his lieutenant Salar and some other grandees set out for Damascus. Between Gaza and Askalon they met Kipchak and his cortège, a meeting which must have been singularly embarrassing. Makrizi says both parties dismounted and shed tears, and that the visitors were received with every attention, and sent on to the Sultan. Ibn Tagri Berdi, on the other hand, says they were received with bitter reproaches,

---

* Makrizi, ii. (part ii.) 165-167. D'Ohsson, iv. 260-262.
† D'Ohsson, iv. 259-260 and 262-263.

F*

for the ills they had brought on their country. They pleaded that they were obliged to fly from the destruction which awaited them at the hands of Lachin and his Mamluk, Mangu Timur, and that when they heard of Lachin's death they would have returned, but were unable to do so. They went on to Salihiyet, where the Sultan heaped fresh reproaches upon them and then pardoned them. The army went to Damascus, where Arjevash had taken charge of the treasury on behalf of the Sultan, and closed the wine shops which Kipchak had opened and farmed out for 1,000 dirhems daily. Akhush al Afram, who was appointed governor of the city, laid a heavy hand on those who had acted as Mongol agents. Some were crucified, others hanged, others again had their hands, feet, or tongues cut off, or were deprived of their eyes. A special civic guard was organised for the protection of the city, from which no one could claim exemption. The sheikhs of the Druse tribes from the mountains Kesrovan, who had plundered the fugitive Egyptians, were sent for, and compelled to make restitution. A division was sent to occupy Aleppo, and put to the sword the Mongols who were found there. Kara Sonkor was nominated governor of that town, and replaced at Hamath by the ci devant Sultan Ketboga. Tripoli was made over to Kadlubeg Almansuri, Safed to Saif ud din, Kerai and Karak to Jemal ud din Akush Alashrafi. Kipchak was made commandant of El Shaubek, Arjevash was rewarded with a robe of honour and a present of 6,000 dirhems, Bek Timur was given 100 Mamluks and set over a regiment of 1,000 men, and El Begui was given 100 Mamluks. The Mongols had been altogether masters of Syria for 100 days, and during that time the khutbeh, or Friday prayer, had been said in the name of Ghazan there. Damascus was ruined. A large part of its wealth passed into Egypt, but that country was already so prosperous that it was hardly affected by this accession.* Abulfeda tells us that one effect of the campaign was that the Egyptians lost a number of towns which they had conquered some years before from the Armenians. Among these were Hamus, Tel Hamdun, Kubara, Saifandkar, and the various towns south of the Jihan, except Hagr Shaglan.†

While Ghazan was invading Syria, Fars was devastated by an incursion made by the Mongols of Jagatai from beyond the Oxus, which I shall describe in a later chapter. We are told that the year 698 of the hejira (i.e., 1299) was a particularly unlucky year, and had been foretold to be so by the astrologers, since the two stars of ill omen, Mars and Saturn, were in very unfortunate conjunction. Wassaf, who explains this conjunction at some length, points out, first, the dearth of water, and then the famine which prevailed widely over the country, the Oxus, Tigris, Euphrates, and Nile all running short in their supply of water. This was

---

* D'Ohsson, iv. 263-267. Weil, iv. 236.   † V. 173.

followed by the black and the red death, from which more than 50,000 died at Shiraz alone. Many learned men fell victims to the pestilence, the most celebrated among them being Ahmed ibn Abi Ghazan, and the father of Wassaf. The following year was just as fortunate, so that the corn which the year before cost thirty gold pieces fell to six. This was counterbalanced at Shiraz, however, by the exactions of the officials, which now passed all bounds. First came commissaries to inquire into and punish the conduct of the controllers of the granaries; next, overseers of the coin, to fix the correct standard; then the publishers of the tax papers in the mosques and baths; then those who were sent to call in the weapons of all kinds from those who were not Mongols, so that the craft of armourer, for which Fars was anciently and still is famous, was almost suppressed; and, lastly, the exactors of the arrears of taxes. Some of the grandees of Shiraz went to the Court. The general overseer of the taxes of Islam, Jemal ud din, had guaranteed taxes to the amount of 283 tumans during the years 697 and 698. They offered to increase it by 22 tumans, and in addition to pay 17 tumans of the arrears of the year 696. Said ud din, the Negro or the Moor, received the commission of the tax collector in Fars, and was given a golden paizah, while Jelal ud din Kurdestani was ordered to make a new survey of the land and a new distribution of the taxes. The grossest tyranny and oppression prevailed, and Wassaf has piled up the rhetoric of which he is a master in denouncing the evils of the times.*

. On his return from Syria, Ghazan devoted himself for a while to the arts of peace. Having arrived at Meragha early in June he paid a visit to the famous observatory there, where he examined the instruments and had their use explained to him. He expressed his intention of building a fresh observatory near Tebriz, and described the instruments he wished to put in it, several of which we are told were of his own design. From Meragha he went to Aujan, where he summoned a kuriltai. It lasted for five weeks, on the conclusion of which his son Alju died, and was taken for burial to Tebriz.† Ghazan himself repaired to Tebriz, where he superintended the erection of the various buildings which he had ordered, among which the principal one was his own tomb. These will be described in a subsequent note. He made over considerable charitable foundations (wakfs) to supply these institutions with furniture, such as carpets, perfumes, lights, wood, &c., and also the various expenses necessary for conducting them. There were also special endowments, one for an entertainment to be given on the anniversary of the founder's death, to the employés, as well as to the Imaums and other considerable people who went there from Tebriz, and who in return undertook to read the Koran there. Other sums of money were dis-

---

* Ilkhans, ii. 101-104. † D'Ohsson, iv. 271. Ilkhans, ii. 98.

tributed in alms. Moneys were assigned for buying sweets for the officials
of the mosque, the monastery, and the colleges ; for the keep of 100
young boys, who were to be circumcised, and were to learn the Koran by
heart ; they were committed to five tutors, five guardians, and five women ;
others again for the care of foundlings, and the expense of bringing them
up to manhood ; others for the burial of strangers who died at Tebriz
without leaving anything to pay for their funerals ; to buy corn and millet
to place on the roofs during the six months of winter for the birds to eat,
and the founder cursed anyone who should hurt these birds or take them.
Money was also left to supply 500 poor widows with cotton to spin, and
to replace the vessels which the slaves of either sex should have the
misfortune to break when carrying water ; a trusty man was appointed at
Tebriz to superintend this last curious charity. Money was also left for
clearing the roads of stones and for making bridges over the brooks for
eight fersenkhs round the capital. Ghazan set aside a portion of the royal
domain for the endowment of these charities, after it had been declared
legitimate to do so by a judicial decision of the muftis, the kadhis, and the
chief ulemas. Seven copies of the deed of gift, all attested by the judge,
were made. One was intrusted to the general administrator of the
establishments, another was deposited in the Kaaba (*i.e.*, the famous holy
temple at Mekka), a third in the archives at Tebriz, a fourth in the
tribunal at Baghdad, &c. Every kadhi, on entering office, was to confirm
this document, and to attach his seal to it. Ghazan appointed the Khoja
Rashid ud din, the famous historian, whom he raised to the rank of vizier,
to superintend these charities, which, according to Wassaf, had a revenue
of more than 100 tumans (*i.e.*, of a million gold pieces). About these
pious foundations, which were surrounded by gardens, there presently
arose a town even larger than Tebriz, which was called Ghazania. Near
each of the gates of this town there was built, by order of the Ilkhan, a
caravanserai, a market, and baths. Thus merchants, from whatever
quarter they came, found their conveniences ready before entering the
town, and the Customs officers inspected their goods there. Ghazan had
brought to Tebriz various kinds of fruit trees which had not hitherto been
seen there. The old wall of the town was small, and was girt about
outside with houses and gardens. It was now inclosed with a new wall,
four fersenkhs and a-half in circuit and ten guez (cubits) in thickness. This
was paid for out of his own privy purse, so that the people's attention
should not be diverted from the benefits that would accrue to them by
contemplating merely the cost. Ghazan's views in the matter were
enlightened, and his motive in the change was largely to make the place
more healthy, for, as he said, nothing is more unhealthy than narrow
streets and high houses. Rashid tells us the town increased, by taking in
the environs, from 6,000 to 25,000 kulajis in circuit, and the new wall
enclosed the two hills known as Valian and Sinjan. The quarter of the

Mount Valian (Valian kuh) was built by the Vizier Rashid ud din, and was called Raba Rashidi.* Ghazan also built markets and baths at Aujan, which was eight fersenkhs from Tebriz, where he passed the spring. There also arose summer-houses, gardens, &c. He surrounded Shiraz with a high wall and a deep ditch. In the district of Halla he constructed a canal to convey water from the Euphrates to the tomb of Hussein, and to irrigate the dry and desert plain of Kerbela. The environs of the tomb now became very fertile, and produced more than 100,000 tughars of grain, better than that of the province of Baghdad. Orders were given to distribute annually a certain quantity of corn to the poor Seyids who lived about the tomb. The canal there was known as the higher canal of Ghazan, to distinguish it from another which irrigated the neighbourhood of the tomb of the Seyid Abul Véfa. Hunting one day near the latter tomb, he could find no water for his horses, and remarked that the wild asses and deer were thin in consequence of the want of water and pasture. He ordered a canal to be made, therefore, which was known as the lower canal of Ghazan. A third was made on the eastern verge of the desert, and was called simply the canal of Ghazan. The product of the land fertilised by these canals was devoted partly to keeping up the tomb of Abul Véfa, and partly to the endowment of the foundations at Shenb. To protect the environs of the tomb of Abul Véfa from pillage, he had them surrounded by a wall. Baths and other buildings were also erected there, and there arose in the midst of the desert a town surrounded by gardens and cultivated fields. The example of Ghazan was followed by his Mongol subjects—by those, says Rashid, who formerly were wont to destroy and not to build, so that the price of houses and gardens increased tenfold. There were many villages in the empire in which there were neither mosques nor baths, so that the people could neither worship properly nor perform the prescribed ablutions. An edict was issued ordering this need to be supplied, which was carried out in the course of two years. The receipts of the baths were devoted to the service of the mosques and the relief of the local rates. Mirkhavend reports that the commissioners who were intrusted with supplying them caused fresh exactions, especially in the province of Fars.

Ghazan wished that in all the large towns, such as Tebriz, Ispahan, Shiraz, Baghdad, &c., houses should be built for the accommodation of the descendants of Ali, under the name of Dar us Siyadet, and he also assigned endowments for their entertainment. It was reported that Muhammed had twice appeared to him in dreams, accompanied by Ali and his two sons Hasan and Husein, whom he presented to him, telling him to treat them as his brothers, and ordered him to embrace them.

---

* D'Ohsson, iv. 272-277. Quatremere, xvii.

This caused him to have a special liking for the descendants of Ali, (*i.e.*, the Seyids). It thus appears he was a Shia, and D'Ohsson suggests he invented these dreams to justify himself with the Sunnis.*

When Ghazan, at his accession, destroyed the idol temples which had been patronised by his predecessors, he wished the Bakhshis to follow his example and become Muhammedans. This they pretended to do. Seeing it was only a pretence, however, he gave permission to those who wished to retain their old faith to return to Mongolia, and insisted that those who remained should conform to Islam, threatening to put to death those who built funeral pyres or idol temples. "My father," he said, "built a temple and endowed it richly. I have destroyed it like the rest, but you may go and live on the land attached to it." 'On this occasion the Khatuns and Amirs said to him, "Your father had himself painted on the walls of this temple. Now that it is in ruins his portraits are exposed to the rain and snow. Since he was an idolater you ought, for the repose of his soul, to restore what he founded." To this he would not consent. They suggested that he might convert it into a palace, but he would not have a palace built on such a site. "There still remain some of these Bakhshis in Persia," says Rashid, "but they dare no longer openly avow themselves, but, like the Mulahids (*i.e.*, the Ismaelites or Assassins), they are obliged to practise their religion secretly."†

In the autumn of the year 1300, Ghazan determined upon another campaign in Syria. He doubtless chose the winter season as less trying to his own men as well as to his Armenian and Georgian allies, and sent Kutlugh Shah forward with the advance guard. The latter set out on the 16th of September. Ghazan was at Tebriz. He first went to Mosul, where his wife Tughanshah, the daughter of Mubarek Shah, died. This was on the 17th of December. On the 3rd of January, 1301, he crossed the Euphrates at Jaaber, and encamped at Ziffein, renowned for the famous battle fought there between Ali and Moawiyeh. When he reached Jebul, or Habul, near Aleppo, he was joined by Kabartu Behadur with some Syrian prisoners. Kara Sonkor, the governor of Aleppo, fled to Hamath, where Ketboga held his ground, and where some contingents of troops gathered. We are told the population of the province of Aleppo abandoned it at the enemy's approach. From the 17th to the 19th of January Ghazan encamped at Kinesrin, situated a day's journey south of Aleppo, on the lake into which the Koweik, or River of Hamath, flows, and which was a famous place in the struggles of the Crusaders. Thence he sent troops into the mountains of Antioch and Sumak. As these districts had escaped attack the year before, a crowd of people fancying this would again be the case had sought refuge there. The troops of

* D'Ohsson, iv. 277-281.   † *Id.*, 281-282.

Ghazan secured a great number of prisoners, men, women, and children, and also carried off many horses, cattle, and sheep. The captives were so numerous that a man or woman only sold for ten dirhems. We are told the Armenians bought many of them and transported them to the neighbouring islands, which were held by the Franks. Abulfeda says the invaders laid waste Saruima, Maarah, Taizma, Amka, and other places, and for three months did their worst upon the country round Aleppo.

When the Sultan of Egypt heard of the Mongol attack he ordered a contribution of money to be levied, and 100,000 dinars were raised, causing much discontent among the people, who denounced the soldiery. "Yesterday you were in full flight; to-day you rob us of our goods," was the cry, and if answer was made, a jeering reply was sent back in regard to the want of courage that had been shown in the face of the enemy; to put a stop to these insults an order was issued decreeing confiscation of goods against anyone who spoke to a soldier. A similar contribution was levied at Damascus, and there also it proved a terrible hardship to the poor, who had even to cut down their finest trees and sell the wood to raise the money. The valley of Gutah was depopulated, and a large part of the inhabitants fled to Egypt. The money was devoted to enrolling troops. Eight hundred Kurd muleteers were enrolled, each receiving 600 dirhems, but they for the most part deserted. At Fostat a large number of artizans and others were enrolled. A general levy of the citizens of Damascus took place, and the amirs pitched their tents in the Maidan Alkabak, where an inspection of the troops, their horses and lances, was held. This was on October 28th, 1300. Novairi says they went to Bedrish, near Gaza; Abulfeda and Makrizi to Al-Auja. A great number of fugitives crowded into Egypt. At length the Sultan set out from Cairo. The troops suffered terribly on the march; the rains were so excessive that it was most difficult to provision the army. Straw and barley rose excessively in price, and we are told three round loaves cost a dirhem, while meat was three dirhems the rotl. The rain was followed by a great inundation, which swept away the baggage, while the roads were covered with thick mud.

This bad weather was equally fatal to the Mongols. A section of them, under the amir Sutai, and a part of the troops of the amir Sheibawaji, who had come from Rum and were encamped on low ground, would have been swept away by the flood if the amir Mulai had not rescued them. A great mortality began among their horses and camels. Ghazan's special horses, which numbered 12,000, were reduced to 2,000. A large part of his army marched on foot, and a general retreat was ordered, many of the troopers having to sit behind their companions on the saddles. Ghazan recrossed the Euphrates at Rakka on the 3rd of February, and having visited the martyrs' graves at Siffein reached his harem at Chehartak, near Sinjar, on the 23rd of the same month, when he was greatly troubled to hear of the death of Satilmish, the son of Buralighi, a relative of Altaju aka.* Rashid ud din diplomatically covers his master's forced retreat by the excuse that he did not wish to shed the blood of true believers, and we should not have known the real cause but for the Egyptian chroniclers, who do not fail to point out what a terrible visitation such raids as the present were, not only on account of the exactions made upon the people, but of the panic which caused a general flight and shaking of the community, disturbing trade, the price of commodities, &c.   Haithon, the Armenian, in describing this war, tells us that Ghazan in this campaign sent Cotulossa (Kutlugh Shah) with 30,000 Tartar cavalry, and told him when he arrived in the district of Antioch he was to inform the King of Armenia and the other Christians of the East and of Cyprus that they were to go and join him. He accordingly did so, and the King of Armenia joined him, while the Christians from Cyprus went as far as the island of Antarados.   There was the King of Tyre, brother of the King of Cyprus, who was generalissimo of the army, as well as the commanders of the Templars and the Hospitallers. These were all ready to join, when news arrived that Ghazan was ill, and that the doctors despaired of his life, whereupon Kutlugh Shah withdrew, and the King of Armenia and the other Christian princes returned home again, and thus the expedition to the Holy Land (which was looked upon doubtless as a new Crusade by the Christians) was abandoned.†

On his return home, Ghazan spent some days in the favourite Mongol amusement of hunting. On this occasion he gave a signal proof of his skill as an archer. We are told he shot at a doe with a three-pointed arrow, called *tuin* by the Mongols. Although he seemed to have missed his aim, the quarry fell to the ground in a heap. He had shot it while springing in the air, with its four feet together, all of which were wounded. Wounds had also been made in its flank, belly, breast, neck, and throat, making altogether nine wounds. This wonderful shot was witnessed by more than 2,000 people, and was deemed to have excelled the famous shot of Bahramgur, who was reported to have hit a wild ass when in the air with his arrow, and to have pinned its hoof to its ear.‡ On the 29th of April there arrived at Ghazan's Court an embassy from Toktu, the Khan of Kipchak, which was well received, and returned again.§ Ghazan now crossed the Tigris, and having punished some bands of Kurds who infested the roads, returned again to Aujan on the 1st of June. There, on the 13th of July, the Khoja Said ud din, who had hitherto shared with Rashid ud din the supreme control of affairs, was appointed vizier. Three weeks later the general

* D'Ohsson, iv. 282-285. Ilkhans, ii. 106-108. Makrizi, ii. (part ii.) 174-176. Abulfeda, v. 175.
          † Op. cit., xliii.   ‡ Ilkhans, ii. 108.   § Id., 108.

Kutlugh Shah married Ilkotluk, the daughter of the Ilkhan Gaikhatu. During the summer a conspiracy was formed against the Vizier and his colleague, Rashid, by Sain of Semnan, and the Sheikh of Sheikhs, Mahmud. With them were also joined Said Kutb ud din, of Shiraz, who was chief secretary of state, and Moin ud din, of Khorasan, the head of the Exchequer. Rumours which reached him aroused Ghazan's suspicion that the plot was meant against himself as well as his ministers. He summoned Said Kutb ud din, who was the administrator of the revenues of Shiraz, and who was in league with the conspirators, plied him with wine, and presently got some disclosures from him. The conspirators were arrested, and some of them were put to death. Mahmud obtained his life at the intercession of the Princess Bulughan, but only on condition of not again appearing at the Court.* The tale of victims continued to grow, and meanwhile Ghazan professed to be so humane that he would not kill a fly, but would take it out of his food and let it rest on his hand till it had strength to escape. To kill a fly cost him more pain, he said, than to put to death a guilty man.† This morbid mixture of sentiment and cruelty reminds one of Robespierre and others of the same kidney. In September Ghazan went to Alatagh, and Kutlugh Shah went with the army to Diarbekr. In November he went into winter quarters in Arran.‡

Three months after his return from Syria, Ghazan sent Kemal ud din Musa, the chief judge of Mosul, and Nasir ud din Ali Khoja, of Tebriz, on an embassy to Egypt. When the Egyptians heard that they had crossed the Euphrates, they sent the amir Seif ud din Kerai to meet them. When they arrived at Damascus they were lodged in the citadel, and the two envoys, with a Turkish attendant, went on to Cairo, where they arrived on the 22nd of August, and were received in audience on the following day, after the last prayer. The amirs and the troops assembled in the "Castle of the Mountain." The Mamluks were dressed in their richest clothes; their caps and the borders of their coats were made of gold brocade. The Sultan was seated on his throne, while a thousand lighted torches were held in front of him. The Mamluks were ranged in a double row, from the gate of the fortress to that of the Iran. The envoys were now presented, the Kadhi of Mosul wearing one of the turbans worn by men of the law, called *tarhah*, which differed from the others in that one of the ends of the muslin scarf fell over the shoulders. He spoke tersely and eloquently, expressed a desire for peace, and addressed prayers to heaven for the Sultan, for Ghazan, and the amirs. He also presented a letter in the name of Ghazan, which was sealed. The envoys now returned to their lodgings, where they remained till Monday. The letter was then opened. It was written in Mongol characters, on a half-sheet of Baghdad paper. When it had been translated into Arabic

---

* Ilkhans, ii. 109. Quatremere, xi. and xii.   † Ilkhans, ii. 108-110.   ‡ *Id.*, 110.

it was read aloud in the presence of the government officials.* The text of the letter has been preserved in two versions, which differ considerably, by Novairi and Ibn Tagri berdi. The former version has been translated by D'Ohsson. Beginning with the usual pious phrases, it goes on to complain that the year before the troops of the Sultan had invaded his (Ghazan's) borders, and had ill-used his subjects, the people of Mardin. Irritated by these evil acts, out of his zeal for Islam he had set out to punish this conduct, but in accordance with the divine precepts he had sent Yakub the Sikurji,† with kadhis and imaums worthy of trust. To this the Sultan had replied by obstinacy, and had drawn upon himself and the Mussulmans all kinds of evils. He had treated his envoys with ignominy and put them in prison, and had forgotten the true path of kings, which was to follow the ways of rectitude. God had presently given him (Ghazan) the victory, and he had hoped that when the Sultan came to consider how matters stood, he would have tried to undo the past, and that on his return to Egypt he would have sent envoys to him to treat for peace. He had consequently delayed at Damascus, like a man who is master of the situation and all-powerful, but the Sultan was steeped in indolence and did nothing. On returning home Ghazan said he had been informed that he had been boasting that he would march to meet him with his troops at Aleppo, or on the Euphrates. He had accordingly set out again and gone as far as Aleppo, and was surprised at his delay. He then learnt that he (the Sultan) had retired, and it was clear that he had evaded an encounter with him. Having considered that if he continued his advance further the country would be devastated, and the people would suffer, he had withdrawn, in order to prevent such an end. Now he was busy assembling his men, and preparing his catapults and other war engines, and he would march after sending warning. " For we do not punish until we have sent an apostle." He had accordingly sent the two envoys, Nasir ud din and Kemal ud din, and charged them to deliver an oral message. He bade the Sultan heed their words, prepare presents, and come to terms, or the blood of Mussulmans would be shed and their property be destroyed by his fault, and he would be responsible before God. The letter was dotted with quotations from the Koran, and was dated between the 10th and 20th of Ramazan, 700 (i.e., between the 19th and 28th of May, 1301), from the Kurdish mountains.‡

Some days after the reading of the letter the amirs who were at the head of the administration summoned the Kadhi of Mosul and said to him : " You are one of the chiefs of the ulemas, the *élite* of the Mussul-

---

* Makrizi, ii. (part ii.) 182.
† *Shikur* means umbrella in Mongol. and it would appear that *Sikurji* means umbrella-bearer. The Egyptian author, Soyuti, tells us that at the end of the 15th century, when he wrote, the Sultan of Egypt, when he rode out on horseback in state, had an umbrella shaped like a vault, covered with yellow silk, embroidered with gold, and on the summit of which was a bird made of silver gilt. It was borne by one of the amirs commanding a hundred men, who rode beside the Sultan. D'Ohsson, iv. 289. Note.
‡ Makrizi, ii. (part ii.) 295-298. D'Ohsson, iv. 288-293.

mans, and you know the duties which the Faith imposes upon you. Tell
us plainly if this be only a ruse, and we will swear not to betray your
answer." He declared that so far as he knew Ghazan desired peace, and
that commerce should again flourish between the two countries. " You
have only to keep on your guard, and march to the frontier with a considerable force, as you do every year. You will then see if it be a ruse,
or if the proposition is honestly made, and be able in either case to act
prudently".*

On the 24th of September the Sultan went hunting with all the amirs,
and after some time went to Salihiyet, whither Ghazan's envoys had been
conducted to receive their *congé*. They were admitted into the Sultan's
tent at night. It was brilliantly lighted, and there were assembled there
420 officers of all ranks, dressed in splendid state robes which the Sultan
had given them. The envoys were given robes of honour and a present
of 10,000 dirhems, and other gifts. They were also intrusted with the ,
Sultan's answer.† This letter has also been preserved, and is phrased in
the pompous language, much imbued with religious aphorisms drawn
from the Koran, that then prevailed. I shall condense and paraphrase it.

The Sultan said he had received Ghazan's letter with the reverence due
to him, and had read it with attention, and found that it thrust the blame
for his own acts on the shoulders of others, and that it tried to excuse his
misdeeds by imputing them to others. He protested that it was not his
fault if marauders on either side made incursions across the frontier in
spite of their treaty. That the chief of Mardin had protected some of
these plunderers. If he (Ghazan) was moved to punish these doings, he
ought to have attacked their real authors, and not to have marched a
mixed body of various faiths into a Muhammedan country, nor allowed
the worshippers of the Cross to enter and profane the holy temple of
Jerusalem, which was second in sacredness only to the Mosque of the
Prophet. "If you charge us with being the authors of these ills, we can
easily reply that it is the want of a treaty of peace between us that
compels us to follow such a way. You claim to follow the example of the
Prophet in sending envoys before fighting. We would answer by
remarking that these messengers were not sent until tent was pitched
against tent, until arrows were flying against arrows, and only a few hours
intervened before the struggle. We are not of those who avoid the
fight, nor again of those who reply to friendly overtures by violent hatred.
The high God has said, ' If they incline towards peace, do thou incline
also, like a rolled manuscript inclines towards its title.' If your envoys
had arrived before the swords were unsheathed, while the lances were not
in rest, the arrows were not already on the string, and the bridles loose,
they would have been listened to, and their message would have been

* D'Ohsson, iv. 293-294.   † *Id.*, 294.

answered." In regard to Ghazan's insulting phrase, "We have tolerated patiently your attachment to obstinacy and rebellion," he asked what kind of patience was that which invaded an enemy's country before sending messengers of peace. As to the vaunts about his victory, he would remind him that if the Egyptians could have assembled all their forces not a vestige of his army would have remained. He reminded him also of the exploits of the beginning of his reign, and bade him ask his own soldiers what had happened. Ghazan had no doubt been successful now, but in the various struggles of kings defeat, by divine decree, had been followed by victory; the victor ought not, therefore, to be too exultant, nor the defeated too depressed. As to your complaints that we sent you no envoys when you awaited us at Damascus, we reply that we devoted our time to collecting our forces, and distributing largess among them. "Those who give their goods for the service of Islam, are like a grain which produces sevenfold." When we marched we found that you had withdrawn against your will, and we halted as a man halts who finds that fear of him dispenses with a hasty march. We sent on a division to fight with those of your troops who still remained in the country. Our people captured those who had incautiously delayed, but they found no trace of your army, although they advanced as far as the Euphrates. As to the boast that you again advanced as far as Aleppo, and there awaited our arrival, we, when we heard of your coming, set out, accompanied by the Prince of the Faithful, Hakim bi amir Allah. We arrived in Syria, crossed plain and mountain, and went as far as Hamath, where we awaited your coming, but you came not, and when you withdrew we returned to get our forces ready. You say you advanced no further because you did not wish the country to be ravaged and its people injured. When have you by your acts shown such solicitude? The Prophet says : "The Mussulman is he whose hand and tongue men need not fear." Mussulmans are kept in chains by you, and have been handed over to the Armenians and to the Takfur,* facts which contradict your humane sentiments. When the Mussulmans defeated Abaka and killed a great number of Tartars, and conquered the kingdom of the Seljuks, neither in advancing nor retreating did they ill-use the inhabitants of the country. They paid for what they needed. Such is the way of good Mussulmans, and of him who desires to perpetuate his power. As to your menaces and your boasts of preparation, we reply, "God is enough for us ; he is the best of protectors." You say that if you had not acted thus the blood of Mussulmans would have been shed. How can you, without invoking divine vengeance, use such words? "The intention of a man," says the Prophet, " is better than his acts ; " and again, " He who wilfully kills a single Mussulman will go to hell, which will be his eternal home ;

---

* Takfur is Armenian for king, and the Armenian kings are frequently so styled by Eastern writers.

the anger of God and his most terrible punishments will fall upon him."
Things being so, we have announced to our people that we are using all
our efforts to make ready and to collect our forces, who will, if it pleases
God, have the august angels as allies. Has not the Apostle of God said :
" My people shall never cease to triumph over its enemies until the Day
of Resurrection." As to your ambassadors, we have received them with
honour, and have replied to them. We know that they are humble
people, and that they would not have been chosen for such a risky
mission if they had not committed grave crimes. They are not the kind
of people who ought to have been sent by one like yourself to a sovereign
like us. Such matters of importance ought to have been confided to those
who combine eloquent speech with eminent merit. As to the presents
you demand, if you had sent us presents we should have returned you
presents of greater value. When your uncle Ahmed sent an embassy to
our father, the martyred Sultan, he sent him presents from a distant
country and a message couched in gracious phrases, and they were
accordingly received with the greatest regard. In conclusion, we say if
the King inclines towards peace we incline also. If he has really adopted
the Mussulman faith, is obedient to God's command, abstains from what
he forbids, is ready to fulfil the obligations of religion not as a merit, but
because he feels honoured by doing so. (" Do not attribute your
conversion to Islam to me. It is God who has given you grace to lead
you towards the true faith.") * If the King's acts are consistent with his
professions, if he separates himself from the infidels with whom he ought
no longer to consort, if he will send us an envoy stating the conditions of
peace, and to say plainly what he wishes, then we will act in concert
against those who oppose us, and our alliance shall crush the polytheists of
all countries, and those who see our friendship will contemplate these
words : " Remember the blessings which God has granted you. You
were enemies. He has reconciled your hearts, and now you are
brothers."† And if it pleases God the most stable peace shall be created
between us. The 28th Moharrem, 701 (i.e., 3rd October, 1301). This
letter was written by the hand of the Mollah and Kadhi, Ala ud din Ali,
son of the deceased Mollah Mohai ud din Abdallah ibn Abd Eldâher.‡

The envoys duly arrived in Arran, where Ghazan was passing the
winter, on the 19th of December, 1301. The Ilkhan was engaged in
hunting in the mountains of Shirvan and Lesghistan. He chased wild
swans at Kiavbari, and then went to Khalise, whose name by his orders
was changed to Kush Koyun. There, on the borders of the lake, he
shot cranes and other wild fowl. At this time, Toktu, the ruler of
Kipchak, defeated his rivals, who retired towards the Eastern Caucasus,
as they expected that Ghazan was going to Derbend. As he did not do so

---

* Koran, ch. xlix., verse 17. † Id., ch. iii., verse 98.
‡ Makrizi, op. cit, 298-306. D'Ohsson, iv. 295-309.

the commercial intercourse, which had been for some time suspended, was again resumed between the peoples of the two khanates. Ghazan took the opportunity while in their quarters to punish the Lesghian chiefs who had been rebellious, and also to seize and put to death a number of brigands from Azerbaijan, who had taken up their quarters there, and infested the neighbouring districts. He now returned to Pilsuvar, and then went towards Talishan and Ispehbed. Here he held a grand battue. The game was driven into a triangular inclosure made of palisades. The entrance to the park so formed was a day's journey across, while the other sides of the triangle converged to an apex. Into this were driven the various wild animals, aurochs, buffaloes, bears, wild asses, wolves, foxes, shakals, deer, &c. Ghazan and his wife, Bulughan, sat in the midst of the inclosure, in a lofty kiosk or summer-house, whence they watched the hunt for some time, and then gave orders for the release of the animals that remained. He returned by various stages, hunting *en route*, to Tebriz, whence the citizens came out to welcome him with banners flying, and he in return promised various remissions of taxes. In July, 1302, he went to Aujan, where in the midst of well-watered meadows, at the intersection of two avenues of willows and alders, was built a beautiful summer palace, with baths, kiosks, towers, and other buildings. The whole were inclosed by a square boundary, containing various entrances for the various classes of people who had access to it. The summer palace itself was a movable tent, made of golden tissue, at the making of which the best artists had worked for three years. It took more than a month to erect, with its hall of audience and appendages, so large was it. In it was placed the royal throne, which glittered with precious stones. To its inauguration the imaums, sheikhs, &c., of the Muhammedan faith, and also the ministers of other religions, were invited. Ghazan addressed them, saying : " I do not wish to enter here with feelings of pride. Ask God to pardon our sins, yours and mine, and to incline our hearts to humility. We will commence by reading the Koran, we will pray to God, and afterwards give ourselves up to pleasure." He pronounced the name of God and the Prophet as he alighted, and then seated himself on the throne and addressed them as follows: "I am a feeble servant of God, who confess my many sins, and declare myself unworthy of his goodness. The pity, grace, and favours with which the Almighty covers his servants are beyond all their gratitude. I cannot too highly recognise his favour towards me in committing to me the sacred trust of governing all the peoples of Iran. I must not let pride overtake me in regard to a power which has come to me after so many princes. Of all God's favours to me there is one which he did not grant to any other sovereign, and which was longed for by my predecessors. He has permitted me to see my subjects happy, content with my rule, and friendly towards me. I cannot be sufficiently grateful." After having given the assembly a feast, Ghazan distributed robes and gold to those present. The Muhammedan clergy spent three days and nights in reading the Koran. The ministers of other religions also devoted themselves to religious exercises. After these services to God began the general festivities. On the fête day Ghazan appeared with a jewelled crown (turban) on his head, and dressed in a robe of gold tissue, bound by a magnificent girdle. In imitation of him, the khatuns, princes, and grandees dressed themselves in their best robes, and, mounted on splendidly caparisoned horses, followed their master. Ghazan now held a kuriltai, where it was decided that his brother Khodabendeh should retain command of the eastern provinces, with his summer quarters at Tus, Abiverd, Merv, Sarakhs, and Badghiz, and his winter quarters in Mazanderan. The amir Nurin Aka was to control the frontier of Derbend. The amir Kutlugh was to go to Georgia, and having been joined by the Georgian army was to march to Diarbekr, on his way to Syria. Hulaju, with his tuman, was to post himself on the borders of Fars and Kerman, and when necessary to unite with the amir Sadik and the Sultan of Kerman; and Mulai was to hold himself in readiness to march again into Syria. A fresh envoy was sent to Cairo.*

On the 26th of August, 1303, Ghazan set out from Aujan, by way of Hamadan, towards Syria. At Heshtrud the amir Nurin went and paid his respects to him before setting out for his command in Arran. Here also he was visited by the sons of Sherif ud din Abderrahman, the former commander of Tebriz, who had also been employed as an envoy in Egypt. They approached the Ilkhan, dressed in black, and complained to him that Nisam ud din, the son of Wejih Khoja, had put their father to death. He was accordingly executed on the feast of Aashura, between Yusagaj and Heshtrud, and with him Devlet Shah, the son of Abubekr, of Dakukaba, and Arabshah, the grandson of Hijaj, Sultan of Kerman. This was on the 4th of September. At Hamadan Ghazan stayed in the monastery of Businjerd, which he had built and richly endowed. Thence he went to Navur Ferhan and Bisutun, where three Syrian amirs did homage to him. Here he paid a visit to the spot, where, seven years before, during the revolt of Nuruz, and when the latter's brother Legsi was still at large, he had rested under a tree in considerable distress. He visited this place with his wives and amirs, and shed tears when he remembered the miserable night he had spent there; and in thankfulness to heaven for having listened to his prayers he said a namaz of two rek'ats, and then prostrating himself with his head to the ground, prayed God to help him always, and addressed a suitable homily to his followers, bidding them obey God and never despair of his goodness, nor to trust in their own strength. He also prayed the Almighty to

---

* D'Ohsson, iv. 310-313. Ilkhans, ii. 114-115.

grant that he himself might always be just. Those who were present fastened various ribbons to the tree, around which the amirs danced to music, a curious Shaman rite that it is certainly interesting to find practised by good Mussulmans. Pulad Chingsang told Ghazan that on one occasion Jingis Khan's predecessor, Khubilai Khakan, whose bravery had become proverbial, on his way to attack the Merkits dismounted before a tree on his route, and prayed God fervently, undertaking if he were victorious to return and deck the tree with beautiful pieces of cloth. He did so return, and decked out the tree, while he and his troops danced around it, after returning thanks to the Eternal. This story delighted Ghazan, who said that if his ancestors had not been pious people they would not have become great kings, and he himself joined in the dance.[*]

When Ghazan arrived at Bendlejin, or Bendinjin, a small town near Beiat, there came to him envoys from the Greek Emperor Andronius, soliciting his help against the Turks, who were pressing the Empire very hard, and offered him in marriage the hand of a young princess who passed at Constantinople as his natural daughter.[†] Ghazan received them well, and promised to repress the Turks. This alliance seems to have had a restraining influence on the latter.[‡] Rashid ud din reports the same event, but hardly in courtly language. He says: "The ambassadors of Fasilius, Emperor of Istambul, sent presents to Ghazan on the part of their Sovereign, who offered to send him his own daughter as a concubine."[§] Ghazan now spent some days hunting in the districts of Sab, Vasith, Meshed, and Sidi Abulwefa. At Meshed he ordered a canal to be dug to convey water from the Euphrates. He then went on to Hillah, where he was joined by Nasir ud din, of Tebriz, and Kemal ud din, the judge of Mosul, whom he had sent to Egypt as his envoys. This Hillah, situated between Baghdad and Kufa, was known as Hillah of the Beni Masid, to distinguish it from others of the same name. It was once famous for its stuffs and for its porcelain, which was like that of China.[‖] Mirkhavend tells us that Ghazan had insisted by his envoys that the only condition of peace was that the Egyptian Sultan should acknowledge his suzerainty by the payment of an annual tribute, the insertion of his name in the khutbeh, or Friday prayer, and that the coinage should on one side bear the name of Sultan Mahmud Ghazan below that of the Khalif, and on the other that of the Egyptian ruler under the profession of the Mussulman faith. His envoys on their return were accompanied by Hussam ud din Azdémir, the Kadhi A'mad ud din, and Shems ud din Muhammed on behalf of Nasir, the Egyptian Sultan. According to Mirkhavend, they said their master could not accept these terms. That as to tribute, the revenues of Egypt were already completely assigned, and devoted either

* D'Ohsson, iv. 313-314. Ilkhans, ii. 116-117.   † Pachymeres, Stritter, iii. 1086.
‡ Id., 1087.   § D'Ohsson, iv. 315. Note 1.   ‖ Ilkhans, ii. 117.

to the purpose of the holy wars, the protection of the frontiers of Islam, or the payment of troops employed for the defence of the faith, to whom assignations of fiefs were made; that there was no spare money, therefore, in the treasury, and if demands were made upon it these sacred needs would suffer. When they had delivered their message, the envoys presented Ghazan with a box locked and sealed. He asked them what it contained. They replied on their knees that they did not know. On opening it, it was found to be filled with all kinds of arms. Ghazan was greatly enraged on seeing this, but repressed his anger, which had been previously aroused by some informality in the Egyptian letter, and by the fact that Nasir's name was written in it in golden letters. After the New Year's festivities, which followed closely upon the arrival of the Egyptian envoys, Ghazan had them conducted to Hamadan, where they were to remain until his return from the campaign he meditated making in Syria. Rashid says they were sent to Tebriz, where they were detained prisoners on parole.* About this time an embassy also arrived at the Court from Toktu, the Khan of the Golden Horde, which I have already described.† The festival of the New Year was celebrated this year with special pomp, probably in honour of the Kipchak and Egyptian envoys, and lasted from the 17th to the 30th of January, 1303.

Ghazan determined to have another campaign in Syria. When his preparations were made, he crossed the Euphrates at Hillah on the 30th January, 1303, and on the 8th of February went to visit the tomb of the Imaum Hussein, son of Ali. It was a famous place of pilgrimage for the Shias, situated on the plain of Kerbela, a day's journey west of Hillah, where Hussein was killed by the partisans of the Ommiade Khalif Yézid. There he hung up the carpets, or veils, which he had promised to present, and distributed alms, and assigned 3,000 menns of bread daily of the product of the lands fertilised by the water of the Upper Canal of Ghazan, which flowed from the Euphrates to Hillah, for the sustenance of the Imaums and Seyids who lived there.‡ Ermenibuka arrived from Khorasan with the news that 3,000 rebels had been defeated in Khorasan, news which was followed by that of the death of Nurin Aka in the winter quarters in Arran.§ Ghazan now followed the river northwards to Haditsé, situated at the distance of a fersenkh from Anbar, where he issued orders for his harems and oghruks to go to Sinjar, and there await his return. He then went on with his army to Aana, as far as which he was accompanied by his favourite wife, Bulughan, and some others of his wives. Aana, the ancient Anelhot, the reputed birthplace of Jeremiah, is situated on a headland jutting out into the Euphrates in the midst of woods and olive-yards; thence a space of 90 fersenkhs from Feluje (near Anbar), and as far as Saruj (Von Hammer says Harran), was strewn with marble fountains, summer-

---

* D'Ohsson, iv. 314-317. † Ante, ii. 145-146. ‡ D'Ohsson, iv. 324-325. Ilkhans, ii. 119.
§ Ilkhans, ii. 119-120.

houses, and villages, palm groves, and cornfields. While at Aana the historian Wassaf presented Ghazan with the first three books of his history of the Mongols from the death of Jingis Khan to the accession of Ghazan himself. His talents had been noticed by the two viziers, Said ud din and Rashid ud din, and at the instance of the Ilkhan some tumans of gold pieces had been given him to pay the cost of his pilgrimage. Wassaf was then forty years old. He now had the honour of presenting the first three books of his work, as I have said, to Ghazan, who greatly encouraged him, and deputed him also to write an account of the two Syrian campaigns, of which he was an eye-witness, and also to add an account of Jingis Khan.*

After a stay of ten days at Aana, Ghazan's wife, Bulughan, left him and went to Sinjar, and we are told that, in accordance with Mongol etiquette, the drums were for three days after her departure beaten with especial loudness. The head-quarters of the army now advanced to Rahbat, a fortress situated on a mountain on the right bank of the Euphrates, between Aana and Rakka (Nicephorium). It was a halting place of the Syrian caravans. It had been put in a state of defence, and catapults, &c., had been planted on its walls. The amirs Sutai and Sultan, the vizier Said ud din, and the famous doctor and historian, Rashid, were sent to summon the place to surrender.† The commandant, Alem ud din Sinjar Al Gatmi, had withdrawn with the inhabitants into the citadel. The letter summoning him recalled the provocation which the Mongols had suffered in their envoys being sent back with unfriendly answers, how they had borne with these things patiently for a while, attributing them to ignorance and want of experience, but that matters having passed all bounds, they had been forced to march to exact vengeance. "We have been compelled to traverse Syria," concluded the note, "but we wish to do you Syrians no harm. Consult your own interests therefore, submit speedily, and as it is clear that justice is on our side, do not by a vain resistance throw yourselves into the abyss of death." This letter, with the royal tamgha attached, was taken into the citadel by a herald. Those to whom it was addressed remarked that the style was very lofty, and asked for a night's consideration in order to understand it, promising to reply on the following day. After deliberation they sent two deputies with their submission. Presently, the commandant himself, with his son and the leading citizens, went in person to Ghazan, who distributed letters patent freeing them from payment of taxes, among the military and civil officers in the place, and took the territory under his protection. Novairi, who has to diplomatically cover a reverse, says the

* Ilkhans, ii. 122-123.
† Rashid on this occasion writes : " The author of this work accompanied the Khan as secretary, to write his orders in the Arab tongue. Everything he needed was supplied from the treasury, by order of the sovereign, who also deigned to present him with a mule from his stables, and ceased not to cover him with proofs of his goodwill in a way to excite the envy of all the world." D'Ohsson, iv. 326. Note.

Egyptian commander went with presents to Ghazan's camp, and under-
took to surrender the place as soon as his army had conquered Syria.
Ghazan consented, took his son as a hostage, and repassed the
Euphrates.* While Ghazan was at Rahbat, news reached him that
Kaidu, the famous ruler of the Khanate of Ogotai, and the rival of the
Khakan Khubilai, had died, and that Dua, the chief of the Ulus of
Jagatai, had been wounded. The amirs Kutlugh Shah, Choban, and
Mulai, who had crossed the Euphrates at Rakka, were encamped at
Deirgesir (the ancient Thapsacus). The Mongols now advanced towards
Aleppo, the governor of which, Kara Sonkor, submitted, and thus
avoided the occupation of the town by the enemy. The summer heats
being now at hand, Ghazan, having held a grand feast, ordered the amirs
Sutai, Alghui, and Naghuldar to join Kutlugh Shah with their troops, and
himself recrossed the Euphrates, and having rewarded the messengers
who brought him news about the death of Kaidu, he advanced hunting
towards Sinjar, whence his wives went to meet him. He made over
Diarbekr and Diar rabia to the Sultan Nejm ud din, of Mardin, with the
title of Malik al Mansur, and ordered him to go to Mosul and inquire
into the complaints made by the Mussulmans there against their Christian
governor, Fakhr Isa El Ghiath. Having crossed the Tigris, he awaited
at Keshf the issue of the Syrian campaign, in which his generals were
engaged.† Keshf is a strong fortress two days' journey west of Ardebil,
near the confluence of the Zab and the Tigris.‡ Haithon explains Ghazan's
long delay on the Euphrates on account of the Syrians having burnt up
the fodder, &c., on the route he would have to march, which made him,
he says, postpone his attack till the spring, when the young grass had
grown. " For," he adds, " the Tartars take much more care of their
horses than they do of themselves, and are content themselves to eat the
vilest food." While he was waiting there Ghazan summoned the King of
Armenia. We are further told his army was so large that it stretched
over a journey of three days, from a certain fortress called Kakkabe, to
another named Labire,§ which De Guignes interprets as meaning from
Rakka to Biret.‖

When news reached Egypt that the Tartars meditated another invasion
a small contingent of troops was dispatched under the amir Bibars, the
jashenkir, to strengthen the Syrian garrisons. Kutlugh Shah, according
to the Egyptian historians, commanded 80,000 men. Haithon says
40,000. He sent to summon Iz ud din Aibek Alafram, the naib of
Damascus, to submit. Meanwhile Bibars, in April, 1303, arrived there,
and wrote to press the Sultan to march in person. Fugitives from Aleppo
and Hamath, who were in dread of the Mongols, also reached Damascus,
where preparations were made for a vigorous defence, and where it was

---

* D'Ohsson, iv. 325-327. Ilkhans, ii. 124-125. † Ilkhans, ii. 126. D'Ohsson, iv. 327-328.
‡ D'Ohsson, iv. 338. Note. § Haithon, 69. ‖ Op. cit., iii. 274.

proclaimed that if anyone left the place his life and goods were to be at the, mercy of those who chose to take them.   The amirs Behaduras and Katlubek, and the jemdar Anes, at the head of a division, went towards Hamath, and were joined by the troops of Tarabolas (or Tripoli) and Hims.   Together they reached Hamath, where the malik Adel Ketboga was encamped.   Meanwhile the Mongols detached about 4,000 men, which waylaid a body of Turkomans, near Kariatain.   Asandemur, the Georgian naib of Tarabolas, with some of the chiefs, and about 1,500 men, surprised the Mongols in the camp of Ord (or Aradh), pressed them hard as far as Asr (or Arz), cut them to pieces, and released the Turkomans, with their wives and children, who numbered 6,000.   The Egyptian loss was only fifty-six men, with the amir Anes, the jemdar Mansuri, and Muhammed ibn Bashkird Nasiri.   The Mongols left 180 prisoners in the hands of the enemy.   News of this skirmish, which took place on the 30th of March, 1303, was sent to the Sultan, and the drums at Damascus were beaten to announce the good news.*

Abulfeda tells us how the Mongols advanced with about 10,000 men as far as Kariatain, intending to plunder its neighbourhood.   At this time the Egyptians, who were encamped at Hamath under Ketboga, who was not well, sent Asandemur, the Georgian, the Prefect of Tarabolos or Tripoli, with a portion of the troops of Aleppo and Hamath, among whom was Abulfeda himself, who was a prince of Hamath.   They attacked the invaders at Koron, near Arud.   This was in April, 1303.   A sharp struggle ensued, in which the Mongols were defeated.   A section of them who were dismounted were offered their lives if they surrendered, but this they refused, and poured a volley of weapons into the Egyptians, who outnumbered them, and used the saddles of their horses as shields.   The struggle continued until the Mongols were all killed.   After this successful skirmish, which was the precursor of a more important victory, the Egyptians retired upon Hamath.   Thither they were followed by the invaders under Kutlugh Shah, who was eager to wipe out the disgrace of the recent defeat.   The governor deemed it prudent to withdraw towards Damascus.   Abulfeda himself was left at Hamath in the hopes that he would restrain the fury of the Tartars, but, as he tells us frankly, when they drew near and pitched their camp by the city, at the sight of their enormous numbers it was not deemed prudent that he should remain, and therefore, abandoning his camp at Aliliat, he joined Ketboga at Kotaif, to whom he reported their movements.   They then retired together to Damascus.

The Sultan left Cairo on the 23rd of March, accompanied by the Khalif Mustakfi-billah-Abu'rebi Suliman, with a large army, leaving as his deputy in Egypt Iz ud din Aibek Baghdadi.   Meanwhile his deputy,

Makrizi, ii. (part ii.) 197-199.   D'Ohsson, iv. 328-329.   Ilkhans, ii 127.

Adel Ketboga, who was ill, was carried in a litter. A discussion arose at Damascus whether they should make a sortie or await the Sultan's arrival, and it was eventually decided to retire further. The citizens were naturally much disconcerted, and the price of conveyances went up accordingly. An ass sold for 600 dirhems, and a horse for a thousand. Many of the men abandoned their wives and children, and withdrew to the citadel. Hardly had night arrived when the Tartar vedettes appeared in the environs of the town. The Great Mosque was crowded with people praying for victory to the Egyptian arms. The Mongols made no direct attack upon the place, but passed it by and encamped in the valley of Gutah. D'Ohsson says they marched on Kesvet. Meanwhile the streets and mosques of Damascus resounded with prayers, and we are told women mounted the flat roofs of their houses with their children, and with their heads uncovered implored the divine aid.* The Mongols were posted at the foot of a mountain called Kénef ul Mizri (i.e., the side of Egypt). Abulfeda says at Shakhab, near Marj us Safar. Quatremere says at Shakhab, at the foot of the mountain Ghabaghib. Their army was 50,000 strong, comprising two contingents of Georgians and Armenians, and was under the supreme command of Kutlugh Shah, under whom were the chiefs of tumans Mulai, Choban, Tittak, Kurmishi, Sunatai, Tugan, Apisha, Ajai, and many others. The Sultan ranged his men in order of battle in the green plain of Marj us Safar (or the yellow plain), famous for the terrible defeat inflicted by the Arabs on the Greeks in the year 634 A.D. According to Makrizi, he planted himself in the centre, having by him the Khalif, the khazindar or treasurer, Seif ud din-Bektimur, the silahdar, Jemal ud din-Akush-Alafram, who was the naib or viceroy of Syria, Bulurghi, Aibek-Hamavi, Bektimur-Bubekri-Katlubek-Nugai, the silahdar; and Aghirlu Zein. In the right wing were Hosam ud din Lajin, the *ostadar;* Mubariz ud din Siwar, the amir-shikar (*i.e.*, grand huntsman), Yakuba Shehrizuri, Mubariz ud din Ulea ibn Karaman, and also the amir Kipchak with the troops of Hamath and with the Arabs. In the left wing were the amir Bedr ud din Bektash Fakhri, amir-silah, the amir Kara Sonkor, with the troops of Aleppo; the amir Bedkhas, naib of Safad, Togril-Igani, Bektimur, the *silahdar,* and Bibars, the devatdar.† D'Ohsson, apparently following Novairi, distributes these chiefs somewhat differently. The Sultan advanced on foot, having the Khalif beside him, who bade them take no heed of their ruler, but fight for their wives, for the defence of religion, and for the Prophet. They were accompanied by readers, who recited the Koran, and incited the soldiery to fight bravely for the rewards of Paradise. Many of the soldiers broke into tears at these exhortations. Bibars, the Grand Marshal, and Salar, the viceroy of Egypt, made mutual promises to stand firm; while Nasir, addressing

---

* Makrizi, ii. (part ii.) 199. D'Ohsson, iv. 328. † Makrizi, ii. (part ii.) 199-200.

the Mamluks behind him, told them to kill anyone who fled, and to appropriate his arms and goods. The camels with the baggage formed a barrier behind the army.

The battle began when the Egyptian lines were hardly in order, with a charge by Kutlugh Shah on the enemy's right, in which eight Egyptian officers and 1,000 of their men perished, but several bodies from the centre and left went to support them. Salar cried out, "Great God! Islam is going to perish." He summoned Bibars and the borji Mamluks, who collected round him. Kutlugh Shah now turned upon them. Salar and Bibars behaved with great heroism, and inspired their companions, and presently Kutlugh Shah was hurled back. Choban and Karmeji, two commanders of tumans among the Mongols, had marched to help Mulai, who had pierced the Egyptian ranks, and was now posted behind them. Noticing the repulse of Kutlugh Shah, they rushed to his help, and faced Salar and Bibars, who were in turn reinforced by Asandemur, Katlubek, and Kipchak, with the Sultan's Mamluks, and the enemy was driven back, and threw himself upon Burlughi. Meanwhile the Egyptian right wing had been broken, and was being hotly pursued, and a panic began to spread among the non-combatants behind. The Sultan's treasure-chest was broken open and robbed. The women and children who had come from Damascus after the departure of the troops, added to the confusion; the former unveiled themselves, dishevelled their hair, and cried to heaven for succour. Meanwhile the struggle had been suspended. Kutlugh Shah, whose division had been severely handled, withdrew to re-form to a neighbouring mountain. There he was joined by the other divisions. He fancied that victory was in his hands, but on surveying the plain found it still occupied by the Egyptians, whose left wing had stood firm, and whose standards were unfurled. He waited until his people had joined him, including those of Mulai, who had been in pursuit, and who returned with a crowd of prisoners, among whom was Iz ud din Aidemur, nakib of the Sultan's Mamluks, who informed his captors of the presence of the Sultan himself. A council of war was summoned, but was interrupted by the drums and cymbals of the Egyptians sounding the advance. Mulai, called Bulai by Abulfeda, deeming the issue too uncertain, or perhaps afraid of being caught in a trap, waited till sunset, and then withdrew. The Sultan and his people passed the night on horseback, while the drums were beaten and the cymbals sounded to direct the fugitives to the rallying place, and the mountain on which the Mongols had taken refuge was speedily blockaded. Salar, Kipchak, and the other amirs spent the night in going round the ranks encouraging the men. At sunrise the Egyptian army was seen ranged in order, the baggage being some distance away. The whole presented an imposing spectacle. Presently the Mongols descended to meet them, and a vigorous struggle recommenced, several of the Sultan's Mamluks having three horses shot under them. This combat lasted till noon, when Kutlugh Shah withdrew again to the mountain, after losing eighty killed and many wounded. His people suffered greatly from thirst. The Sultan now determined to open a passage through which they might march, and then to fall upon them. The Georgian Asandemur, with his division, therefore stood aside. The Mongols marched through; first a division under Choban, then the centre under Kutlugh Shah, and lastly a third division under Taitak, and went down towards the river, into which they plunged. Many of the horses were bogged in the morasses, and they lost a great number of men in the pursuit.* Wassaf says that in the previous struggle the amirs Choban, Taitak, Ebrenjin, Kineshin, and Tersa were separated from Kutlugh Shah, and, unaware of his withdrawal to the mountain, continued to struggle desperately, and that Taitak, covered with wounds, together with Irinjin (? Ebrenjin), Kineshin, and Tersa, were all made prisoners.†

The account of the struggle here given is that followed by D'Ohsson and Von Hammer, and is based chiefly on the reports of the Egyptian historians ; but we have an independent notice by Haithon, who claims to have been present, and to have been, therefore, an eye-witness. According to him, Kutlugh Shah, after entering Syria, marched with the King of Armenia upon Hims, where he believed the Egyptian army to be posted. There he learnt that it had not, in fact, left the neighbourhood of Gaza. Having captured Hims, and put all the Mussulmans there to death, and secured much treasure and a store of arms, the Mongols advanced towards Damascus, where the people sent out to ask for a three days' truce, which was granted them. The Tartar advance posts, which had proceeded a day's journey beyond Damascus, having captured some of the enemy, learnt that a small body of them, 1,200 strong, were posted not far off (? the Turkomans of the other notice). Determined to surprise them, Kutlugh Shah and the Armenian king advanced, and arrived near there at nightfall, but they found the Sultan with the main army had also come up. A council was called, when it was advised to wait till the morrow, but Kutlugh Shah, who despised the enemy, determined to attack at once. The Egyptians were posted in a position where they were protected on one side by a lake and on the other by a mountain. In front of them was a river, only fordable in certain places. This delayed the Mongol advance, and as the Egyptians would not leave their vantage, the fight was, in fact, postponed until the next day, the Mongols encamping on a mountain close by. The next day they again tried to draw the Sultan from his position, but failed, and as they were greatly in need of water and much wearied, they began to withdraw in small parties towards the plain of Damascus, where pasture and water abounded, and where they determined to recruit before again attacking

---

* Makrizi, ii. (part ii.) 200-202. D'Ohsson, iv. 332-334.   † Ilkhans, ii. 129.

the Sultan. Thereupon, the citizens of Damascus cut the dykes of the river and flooded the country, thus compelling the enemy to withdraw, but a great number of men and cattle were lost, while the quivers and bows of the Mongols, with which they chiefly fought, and their other arms, were rendered useless. They were naturally much distressed, inasmuch as if they had been pursued they must in this condition have been overwhelmed. They withdrew gradually towards the Euphrates, which was much flooded, and as they crossed it on horseback, numbers of them and their horses perished, including many men belonging to the King of Armenia, Georgians, and others; "and thus it happened," says our diplomatic witness, "that not by the strength of the enemy, but by accident and bad judgment, the ill-chance befell them, for Kutlugh Shah would not take anyone's advice," a conclusion which enables the chronicler to moralise on his own report of these events, whose length he excuses on the ground that the experience might not be lost. His notice, it will be seen, glosses over the real defeat sustained by the Mongols, which he, in fact, does not mention; while he actually goes on to say that the enemy dared not pursue them.

The *Georgian Chronicle* says the Tartars, after the first day's fight, which was indecisive, retired to a hill, under Kutlugh Shah and Hussein Sevinj, while Sibuchi, son of Kutlugh Shah, planted himself in the plain with the people of Beka. Each one held his horse's bridle. The Sultan determined to make a stream flow behind the Tartars (? by opening some dykes). During the night a fog arose, so that men and beasts were completely enveloped. At daybreak, seeing this fog behind them and the Egyptians in front, the Tartars were much embarrassed. They, however, found a small space uncovered by the fog, which was speedily crowded with corpses, but through which they managed to escape, although after a terrible slaughter. Wakhtang and the Noyan (*i.e.*, Kutlugh) both escaped.* Abulfeda tells us a division was sent in pursuit of the Mongols, under Salar, which followed the fugitives as far as Kariatain, and that many of those who survived the dangers of this flight were overwhelmed by the Euphrates in trying to cross it on horseback when it was flooded. Many others, he says, were waylaid and put to death by the Arabs, "and thus," he adds, "did God avenge the wrongs we had suffered, and repaid them the injury they did us in the year 699, at Mojma ul Muruj, in the district of Hims."†

The Egyptians are said by Wassaf to have captured 10,000 prisoners, and 20,000 cattle, among whom were Taitak, Sunatai, and Kinju, and many superior officers. The Turkish prisoners were incorporated in the frontier garrisons and the Arab cavalry. The strict discipline of their enemies very much surprised the Egyptians. Having heard of the bravery of Taitak, who had received four wounds, Nasir summoned him,

---

* Op. cit., 634.   † Op. cit., 187.

and asked him how much Ghazan gave him a-year as a reward for his devotion. "The Mongol," was the reply, " is the slave of his sovereign. He is never free. His sovereign is his benefactor ; he does not serve him for money. Although I was the last of Ghazan's servants, I never needed anything." On being asked how much each soldier received annually, he replied, from two to five tugars. "But," adds Mirkhavend, " what is most to be admired is that although 5,000 horsemen in the late war had lost their horses, they cheerfully shouldered arms and articles for a two months' march ;" and he further relates that after such a march, if, when they arrived at home, and even before they had unloosened their mantles, they were ordered upon a distant campaign, they at once set out without demur."* No wonder the historian, whose experience of discipline had been gathered among Arab mercenaries, &c., was much surprised. Notwithstanding this discipline the retreat was very disastrous. The amir Salar was sent in pursuit, and followed the retreating Mongols as far as Kariatain. The Mongol horses were worn out with fatigue. The soldiers lost heart, and hardly offered resistance, and many of them were killed by the camp followers, who secured a large booty, and in some cases we are told as many as twenty Tartars fell to one of these creatures. The Arabs who offered themselves as guides led them astray into desert places, &c., where they perished of thirst. Others were conducted to Gutah, near Damascus, where the people set upon and killed them. Wassaf says many of them, to avoid dying of hunger, sold themselves as slaves. Others exchanged their own slave girls for mares. Nasir ordered the bodies of all Mussulmans who had perished in the struggle to be buried together, without being washed, and without shrouds, and a circular tomb was erected over them.†

The victory was naturally the subject of great rejoicing. Pigeons conveyed the news to Gaza, and also an order that the runaways from the army were not to be allowed to enter Egypt, while those who had plundered the Sultan's treasure were to be sought out and detained. The amir Bedr ud din Bektut-Fattah was selected to take details of the good news of the success to Egypt, and left immediately. News was also sent to Damascus and other fortresses. The Sultan entered Damascus on the 23rd of April, and was received with the greatest rejoicings as a deliverer, the whole place being splendidly decorated. He rewarded his generals with robes of honour and other gifts. He refused at first to see the amir Burlughi, one of those who had fled, and only did so under the solicitation of his other officers, when he was admitted and pardoned. One of the amirs of Aleppo, who had been corrupted by the Mongols, and had acted as their guide, was nailed to a camel and perambulated about the town. The naib of Gaza arrested and searched the runaways, and discovered on

---

* D'Ohsson, iv. 334-335. Note.
† Makrizi, op. cit., ii, (part ii.) 203. D'Ohsson, iv. 335-336.

them many bags of gold and silver which had been appropriated, and which were still sealed. The amir Alem ud din Sanjar Jauli went towards Damascus with the officials of the treasury, and recovered from the camp followers a large quantity of purloined treasure. The guilty were imprisoned, nor did the search cease till a large part of what had been lost was recovered. When Bektut Fattah reached Cairo, he gave orders that the city was to be decorated from the gate of Nasr to that of the Chain, and he summoned Arab musicians from the various towns of Egypt. The news of the victory had already reached them by pigeon post from Katia, but as Bektut was delayed *en route* by an affection of his hand the people began to be uneasy. The markets were closed, and bread sold at a dirhem for four *roll*, while a skin of water cost four dirhems. On his arrival everybody went out to meet him. It was a regular fête, and each one rivalled his neighbour in decorating the place. Balconies with seats were raised, the ostadars of the amirs divided the chief street of Cairo among them, each taking a certain length to decorate, the crier went round the town to proclaim that anyone employing an artisan for another purpose than that of erecting balconies was to be deemed to have committed an offence against the Sultan ; and the price of wood, reeds, and joiners' tools rose very greatly. The country people came into the town, and jewels, precious stones, pearls, and silks were used to ornament the houses. Nasir ud din Muhammed Ibn Alshaikhi, the vali, made a balcony at the gate of Nasr, which contained all kinds of rare objects. By his orders basins were filled with sugar and citrons, and about these were ranged Mamluks, holding glasses of lemonade in their hands to give drink to the soldiers. The Sultan presently arrived, and we are told that fifty or even a hundred dirhems were paid for the hire of a house whence the triumphant procession could be seen. The various amirs met him at the gate. The aged Bektash relieved him of his arms, and notwithstanding his great age and the entreaties of the Sultan, insisted on carrying them on foot ; the amir-shikar, or chief huntsman, carried the umbrella and the falcon, the amir-jandar the sceptre, and the amir-bajmakdar the mace. Carpets were spread from each house to the next one, over which the Sultan marched his horse, and the procession stopped at each to examine the beautiful things. The Mongol prisoners marched in front in chains, bearing about their necks the heads of their companions who had fallen in the battle. In addition, 1,000 heads were held aloft on lances. The prisoners were 1,600 in number, who had a similar number of heads tied about their necks. Before them were carried their drums, which were broken. The Sultan rode over silk carpets all the way to the citadel, where he presented the amir Burlughi with 30,000 dirhems, and nominated him amir of the caravans.*

---

* Makrizi, op. cit., 203-213. D'Ohsson, iv. 336-337.

Kutlugh Shah recrossed the Euphrates with only a small portion of his army, and there was great grief in the cities of Hamadan, Tebriz, &c., when the sad news arrived. Makrizi says Ghazan was so affected by it that blood flowed from his nose. He was on the point of dying, and secluded himself. Only one man in ten had returned. Abulfeda attributes his death, which occurred shortly after, to chagrin at his defeat. The Persian historians tell us Kutlugh Shah arrived at Keshf, where Ghazan awaited his return, on the 7th of May, and the day following he left for Ardebil, where on the 4th of June he received Choban, and praised him greatly, inasmuch as he had remained behind to rally the dismounted troopers, and to conduct the débris of the army to Baghdad, sustaining their courage during their long march. After spending a month in the mountains of Schend, he arrived at Aujan on the 26th of June, having been preceded thither by his wives and oghruks. The following day began the trial of the officers who had commanded in the Syrian campaign. Choban accused Kutlugh Shah of not having supported him when he broke the Egyptian right, and was charged by him in turn with having been too rash. The trial of the officers lasted till the 17th of July, and according to Rashid, two of them, Agathai Takhan and Tughan Timur, were put to death.* The Egyptian historians say that Ghazan was so irritated at the disaster, that he had Kutlugh Shah, Choban, Sunatai, and other officers arrested. He condemned Kutlugh Shah to death, but spared his life on the entreaty of the other officers. He, however, subjected him to great indignity. He was held by some ushers at a distance from the throne, while all those present came forward and spat in his face. He was then exiled to Ghilan. Mulai was bastinadoed, and treated most ignominiously. Mirkhavend says all Kutlugh Shah's officers received a certain number of blows with a stick, and they were forbidden access to the Ordu for several days, and we are told that Choban, although he had been so signally praised by Ghazan, and had been presented with one of his own robes, was beaten like the rest, so that there might be no exception.†

The King of Armenia, before returning home to his own country, also went to see Ghazan, who received him well and consoled him for the loss of so many of his men, and even made over to him a guard of 1,000 Mongols, who were to be maintained at his own expense, and assigned him a portion of the revenue of the kingdom of Turkey to keep another body of 1,000 horsemen. He then returned again to Armenia.‡ Chamitch says that Pilarghu (i.e., the Barlogi of Abulfeda and the Barlagu of Sanutius) was the commander of the contingent assigned to the Armenian king. He died in 1307.§

---

* D'Ohsson, iv. 338. Ilkhans, ii. 131.
† D'Ohsson, iv. 337-340. Note. Makrizi, op. cit., 204-205. ‡ Haithon, 72-73.
§ Georgian Chron., 631. Note.

The bitterest pang of all to the vanquished Mongol chief was doubtless the scornful letter which the Sultan wrote him :—

"Praised be God, who has renewed his goodness towards us, and has given us his grace in full measure ; who has again illumined the moon with his light, and granted us unmeasured joy ; who by his rich gifts has stilled the longing of the soul and given repose to the heart ; who has lighted up brightly the sun of wisdom, and has caused its moon to rise with renewed fortune ; who has wiped out the mishaps of Time and scared away hard Fate, so that our eyes are radiant with delight at the sight of all our hopes and wishes fulfilled. May the Lord be praised, so long as the lightning flashes and the stars wander through the night. The exalted king, the assembler of armies, knows that he paraded as the patron of Islam, yet he enveloped himself in fraud and forgot the right. He accordingly accomplished what God's will determined, and has courted the fate which no mortal can escape. Only the other day he sent messages of peace, inviting concord in the name of Islam, and declaring he would cause no harm to the land, nay, that he wished to fill up the gap separating us, as the holy law requires. We knew his real object, yet at the very prospect of peace, which shone upon us like the full moon, our suspicion was dissipated, we received his envoys with honour, as it became us, listened to their message, and gave our reply to the demands which we could not accede to. We then sent them back again, and left it to him to develop the evil intentions which he concealed in his breast. We also sent him an embassy, as he demanded, although we knew his designs, and thus gave him an opportunity to commit his shameful outrage. For hardly had our envoys reached him when he threw off the mask, marched his troops, and commenced the war which became his ruin. They crossed the Euphrates, while he, trusting to his heels for safety, prudently returned home again. His army invaded our land, and did there what he had ordered it—cut down the trees and destroyed the seed, planted itself by Aleppo, and thence made raids.

"Meanwhile our troops were ambushed in Syria. They advanced with God in their hearts, as to a holy war, and our brave divisions marched against the enemy, who through our weakness had entered the land. Two thousand of them attacked two divisions of the foe near Alurdh, where they had plundered the Turkomans. After a short struggle their corpses were strewn about ; God sent their souls to hell, and none of them remained, while Armenians, Georgians, and others were made prisoners. The rest of the hostile army went towards Damascus, not anticipating that they would soon be girdled about with lances, that our horses would be planted around, while our troops would be so near them as to be able to watch them every hour. When they came in view of Damascus they contemplated entering it victoriously, and did not know that they would find a road to hell near that city. They went over the hill of Mani, but when they saw our army fear possessed their hearts, and they then realised that destruction is the sequel of faithlessness, and ruin of violence. They then sought safety in flight, but before sunset we had spread them on the ground like carpets. Many withdrew to the mountain and sought shelter there from death, and spent the night there till Sunday. They feared none of them would escape, and they had to seek comfort in repentance. They were hopeless unless they could secure our pardon, and it was in despair they appealed : ' Spare us, merciful king, and pardon us at this holy season; we are true believers.' We thereupon ordered our troops to let them pass through. They fled amazed, like a sheep before wolves. The father did not look back for his sons. If you, O King, had seen this day of battle, you would not have slept peaceably for a long time, for your friends were killed or enslaved. It was a bad day for the unbelievers—a day on which the wolves and the vultures went out together for rapine. If you had seen how your comrades were eaten by the wolves you would have exclaimed, ' O, that I were changed into dust.' Thank God you did not see these things with your eyes, and only learnt them through your ears. If you had been witness of the evil that overwhelmed your friends you would not have survived the horrors, for it was a doom of which the angels were witness.

"I gave you candid advice, but you paid no heed. You would not refrain from your violent purpose, and the result is the destruction of your army. We told you, 'Whoever uses the sword of wrong commits suicide.' But you imprudently despised my words, and the reward of your ill-doing is that God has made yourself and your army a by-word till the day of resurrection. Turn again to the true faith, and do not let Satan deceive you any more into vile courses that bear ill fruit. If you are a Mussulman, act as such. Let us both, conformably to the holy script, which deems violence a sin, renounce all pretensions to Baghdad and to Irak, restore them again to the Khalif, and obey his commands. Thus shall we bend the bow of our faith. If this be not done, you will secure God's curse till the day of doom. If you act otherwise, then is your ruin inevitable, and Persia as well as Irak will be lost to you, when you sink into nothing. We have now pointed out to you what is right ; deviate not from it. Receive our envoy kindly, and see him escorted safely back to Syria. Then take the rest of your troops to Khorasan, and heed not the suggestions of Satan. A man once told us that your horsemen and your foot soldiers would come to Egypt. This prophecy has now been fulfilled, but in another sense. The riders are sitting sideways, and the foot soldiers are assembled with drums about their necks, and revolving banners in their hands. Make haste, therefore ; retire from Asia Minor, and from Irak, and go to Khorasan. We shall soon follow you with our troops, who will speedily scatter your people in terror, and before us you will tremble and quake. He who gives warning must not afterwards be blamed. Hail to him who follows advice."[*]

After the guilty chiefs had been punished, Ghazan held a kuriltai, where he distributed liberal largess. At his accession to the throne he had found the treasury empty and the country ruined, the revenues were absorbed by his officers, and the taxes were obtained with difficulty. The treasures which Khulagu captured at Baghdad, in Syria, and in the country of the Ismaelites, and which by his orders had been deposited in the strong fort of Téla, had been plundered by its successive guardians, who sold the golden balishes and precious stones to merchants. As they were all guilty the secret was well kept. One tower of the castle situated on the borders of the Lake of Urmia having fallen, the guards seized the opportunity to appropriate still more treasures, which they declared had fallen into the water with the tower. What remained, and it was valued at only 150 tumans, was distributed among the troops by Ahmed when he wished to march against Arghun. The treasure amassed by Arghun disappeared after his death. When the military chiefs, who rebelled during his illness, put his ministers and courtiers to death, they appropriated a portion of the treasure, and divided a portion of it among the soldiers. Gaikhatu neither saved anything nor left anything. Accordingly, on Ghazan's accession, he had nothing to give the troops whom he had brought with him from Khorasan, and whose tents and cattle had been appropriated by the invaders from Transoxiana. Hardly anything reached the treasury. Nuruz, Sherif ud din of Simnan, and Sadr ud din all tried in vain to restore the finances. Not only was there not money enough to pay the troops, but not even sufficient for a present to an envoy. No one could understand how this was, and Ghazan was accused of avarice and carelessness. One day he said to his officers, " You fancy that these mules which follow the camp are laden with gold. You are mistaken ; they only carry wooden instruments and various tools which I like to use, as you know, and you can test it for yourselves. When I have nothing I can give nothing. My predecessors have left me nothing. I inherited a ruined country, and have not received the revenue."

But in the course of two years, after he had organised the army, protected the frontiers, and cleared the kingdom of the robbers who infested it, he devoted himself to regulating the finances and to reforming the administration, as we shall presently describe. Only trustworthy people were given charge of the provinces, and they held their posts for at least three years if they behaved well. Order being thus restored, the treasury began once more to fill. He distributed the first two or three hundred tumans which were received among the officers, fixing how much

---

[*] Weil, iv. 260-263.

each corps was to receive, but at the kuriltai held at Aujan, which we mentioned just now, he distributed largess with his own hand. Seated in a large tent, where were collected the revenues of various provinces, and surrounded by his principal officers, he distributed robes, rolls of gold and silver of different sizes, inscribed with their contents and with the corps for which they were destined, calling out each corps himself by its name. This distribution lasted for fifteen days, during which time there were distributed 300 tumans of gold in coin, 20,000 sets of robes, 50 girdles decorated with precious stones, and 300 golden girdles. Rashid tells us Ghazan had the art of rewarding his followers according to their merit without arousing jealousies. There never passed a day on which he did not give away from ten to a hundred thousand pieces of money, and from a hundred to three hundred robes. Notwithstanding this generosity, such was the order in the administration that the treasury was not emptied, and none of his predecessors distributed so much as he.*

From Aujan, he set out for Tebriz, where he arrived on the 8th of September, and began preparations for another campaign in Syria, but he was suddenly attacked with ophthalmia, which lasted some time. Meanwhile, on the 11th, there arrived at his ordu the dowager Princess Ilturmish, daughter of Kutlugh Gurkan and widow of Gaikhatu, who had afterwards married Ghazan's brother, Khodabendah. She took her two sons, Bestam and Abuyezid, with her. Ghazan betrothed his daughter Oljai Kutlugh, who was only six years old, to Bestam. In order to cure his eyes, the Chinese doctors had made scarifications or wounds in two parts of his body. In the end of October he left Tebriz riding on an Indian elephant, which had been sent him as a present by the Sultan of Delhi. The sight was a new one for the people of Tebriz, who came from all sides to see it, and he amused himself on the animal during the whole day in the public square. On the 31st of October he set out again for Aujan. The pain which his recent operation caused him, prevented him riding on horseback, and he travelled by short stages in a palanquin. From the station of Yusagaj he sent Kutlugh Shah to take command of the troops in Arran, a command which had become vacant the year before by the death of Nurin Aka. On the 13th of November he arrived at the palace of Jome Gurkan. Jome Gurkan, who was by origin a Tartar, had married two daughters of Khulagu in succession, namely, Bulughan and Jome. Abaka married Jome's aunt Nukdan, while his mother was Chichegan, the daughter of Ochigin, Jingis Khan's youngest brother. This shows how curiously involved the marriage relations of the Mongol prince were. From the palace of Jome Gurkan Ghazan went to Sughurluk. He intended going to Baghdad, but the heavy snow made the roads impracticable, and decided him to pass the winter on the banks of the Hulan muran, where he spent the

---

* D'Ohsson, iv. 339-344.

time in feasting and dispensing charity. He allowed the poor to feed in his presence and distributed clothes among them. On one occasion there came ten dervishes. Ghazan ordered ten sets of robes to be given them. Nejm ud din, the mehter or upholsterer, who had the distribution of these things, declared that two of the dervishes were not Mussulmans at all, but were Christians, and it turned out that in the hope of getting food and clothing they had feigned to be so. Ghazan ordered that they were, nevertheless, to have the robes, since the Padishah of Islam could not break his word.* While in winter quarters, Ghazan went into seclusion for forty days, during which he took little food, gave himself up to meditation, and was waited upon by the dervishes. His residence was inclosed by a palisade, and no one except the khoja of the serai or palace had access to him. While he was going through these penitential exercises the dervishes of Tebriz, headed by the Pir (i.e., the master of contemplative life), Yakub Baghban, formed a conspiracy to depose him and put on the throne his cousin Alafrenk, the eldest son of Gaikhatu. One of them, named Mahmud, went to the camp and reported that a gigantic figure, forty ells high and with a breast five ells broad, had come down as a messenger from heaven, and had alighted on the mountain Merend, and declared it to be the will of heaven that Alafrenk should mount the throne. When the vizier, Said ud din, heard of this he had him seized, and sent the akhtaji (equerry) Khani to Tebriz to seize the envoy of the Khakan Khubilai, Nasir ud din, together with the heads of the conspiracy, Pir Yakub, Seyid Kemal ud din, and the sheikhs, Rashid and Sadr ud din. When they were brought before Ghazan he remembered that they were partisans of the deposed vizier, Sadr ud din, of Zenjan, whom they doubtless wished to avenge by his own deposition. They were questioned by him in person, and it was proved they were adherents of the doctrines of Mezdek, who lived during the reign of Kobad, father of Nushirvan, king of Persia, and who were attached to a form of Manicheism. Pir Yakub was thrown down from the summit of a neighbouring mountain, and his accomplices were executed. Alafrenk was pardoned, and ordered to go to Khorasan to join Khodabendah. Touched by this clemency, he confessed that he had been induced to go to Tebriz on pretence of a hunting expedition, and having joined with Yakub and his companions in their pious exercises, was persuaded by them that he should mount the throne. Yetmish, the deputy of Taitak, who had been captured in the Egyptian war, was implicated in the conspiracy, and was put to death. Taitak's son, Akbuka, who was also a party to it, was pardoned on account of his youth and of his father's services.†

The new year's feast was held on January the 10th, 1304, in the ordu or

---

* Ilkhans, ii. 131-133. D'Ohsson, iv. 346-347.   † Ilkhans, ii. 133-135. D'Ohsson, iv. 347-349.

the Princess Ilturmish. This was spent with especial festivities, in consequence of the recent escape from the conspiracy, and we are told Ghazan specially honoured the vizier Said ud din, and the historian Rashid, on the occasion. To the former he appointed a command of 1,000 life-guards, with horsetail standards and kettledrums, and ordered the various amirs to wait upon him with their congratulations. Eight days after the new year Keramun, Ghazan's wife, died. She was buried at Tebriz. Her death greatly affected him, since she was in the bloom of youth, and he often wept for her. One day he asked his amirs what was the most grievous thing in life. Some said a defeat, others imprisonment, others poverty, or sickness, or death. Ghazan replied : "The most grievous is to be born at all, for life is but a string of misfortunes, closed by death." " When two are on a journey," he asked, " one on foot, the other sitting down, which is the more tranquil ? " " He who sits," they replied. " And when one is sitting and the other lying down ? " " The one who is lying down." " If one is awake and the other asleep, which is the more tranquil ? " " The one who sleeps," they said. " Quite true," he said, " and the only real peace is in death. True is the Prophet's word : ' The world is the prison of the faithful, and the paradise of unbelievers.'"*

Before we conclude the personal history of Ghazan we will devote a few words to the two remote provinces of Fars and Kerman, at this time about which Von Hammer collected some interesting details. The province of Fars was divided in very early times into five districts, Ardeshir, with Shiraz as its capital ; Istakhr, with Persepolis as its chief town ; Shabur, with its chief town of the same name ; Kobad, with Aujan as its chief town ; and Darabsherd. The last of these was situated in its eastern parts, and was in later times called Shebankiareh. Its capital Darab, so named, says Von Hammer, from its founder Darius, was situated on a number of hills, surrounded by a wall a parasang in circuit. The district round was famous for gold, silver, markesite, fluospath, iron, salt of seven colours, quicksilver, and the famous resin, *mumia*, used in bone-setting. The district contained some other famous towns, which are enumerated and described by Von Hammer. It was governed from early times by the family of Fasluych, whose importance dated back to the time of the early struggles of the Arabs and the Sassanians. At the time of Khulagu's campaign in the West, the head of the house was Mozaffer ud din, son of Nizam ud din. He was the contemporary and rival in power of the Salghurid chief, Mozaffer ud din Abubekr, and his authority extended from Khasuyeh Runis and Khair as far as the villages of Mishkusabad, Lur, Sanik, and Guristan, the last of which was only seven parasangs distant from Hormuz. Khulagu in 1259 sent Tekyujin with a tuman of Jelairs to attack Darabsherd.

---

* Ilkhans, ii. 135-136.

H *

Mozaffer ud din was killed by an arrow during the attack, and the town was taken and destroyed; 17,000 houses which surrounded the citadel like a girdle were levelled with the ground. Tekyujin then appointed Mozaffer ud din's son, Kutb ud din, to govern the district of Darabsherd, under the control of a Mongol commissary. Eleven months later, viz., on the 5th of November, 1261, Kutb ud din was murdered by his brothers. Two of them, Nisam ud din and Nasir ud din, succeeded him quickly. The former was killed at Kiarsun by Seljuk Shah in February, 1264. The latter died on his wedding-day, when he married Seljuk Shah's daughter. He was succeeded by Jelal ud din Taibshah, who reigned for seventeen years, when he was put to death by order of the Ilkhan Ahmed. His brother Behai ud din reigned for seven years, and then sickened and died. These two brothers were succeeded by their sons, Malik Ghayas ud din and Nisam ud din III. They were succeeded by Ardeshir, under whom the dynasty of the Shebankyarehs was swallowed up and succeeded by that of the Beni Mozaffer (*i.e.*, by the dynasty so called in Fars). The long lease of virtual independence from Mongol control enjoyed by this district now came to an end. On the return of Ghazan from his Syrian campaign, an impost of twenty tumans was laid upon the province of Fars. The amir Mingkutluk and the great sahib Iz ud din Al Kohedi were sent to Shiraz to regulate the taxes there, and each ten dinars of the land tax had an additional dinar added to it. The additional sum was called *tebghur*. This was found so onerous that the vizier ordered it to be cut down to one-half, and we are told that, taking advantage of an ambiguity in the phrase in which the order was embodied, the lucky inhabitants managed to retain the tax at the old amount.[*]

Let us now turn shortly to the affairs of Kerman. We have seen how, on the murder of Padishah Khatun, Ghazan nominated her nephew Mozaffer ud din Muhammed Shah, son of Hijaj, as her successor; after doing homage, he remained at Ghazan's Court. The disorders which had arisen in Kerman during the commotions following the death of Padishah Khatun were set right by the judge Fakhr ud din, of Herat, who went to Kerman as the young Sultan's representative. He treated the royal princes there with haughtiness, and cut down their allowances, whereupon Mahmud and Hasan Shah rose in rebellion and killed him. Ghazan now sent Timur Buka as governor of Kerman, with an army which beleaguered the town for a year and a-half. The princes submitted, and Sadr ud din Khoja, the successor of Fakhr ud din as vizier, sent Mahmud to the Court. He arrived at Ispahan when his brother Muhammed was on his way to his appanage, but as the latter did not send for him as he expected, he poisoned himself. When Muhammed arrived at Kerman he had his other brother, Sujukshah, sent to

---

Ghazan, who had him put to death.* This is one account. Another is so entirely different that it is difficult to find any common features. According to this other account, as Muhammed spent most of his time at the Court, a capable Sahib and Pishaver was sent to take charge of the province. The most learned man of his time in Kerman, Abdallah ibn Muhammed el Beyari, was chosen for the post. Mahmud, Muhammed's brother, and several grandees, with a body of Turkomans, fell upon and killed him and his two sons, and plundered the archives and library. Mahmud Shah did not stop here, but proceeded to raise an army, seize fortresses, establish magazines, and levy taxes. Seyid Moaasem Jelal ud din, Beyari's son, called in the Mongol general Sadakbeg to avenge his father and brothers. Ghazan, who was meditating his Syrian invasion ordered the amir Juyurghatai and Khirmenju, the governor of Ispahan, to march with the forces bordering on Kerman, and also ordered the malik Nisam ud din Hasan, brother of malik Ghayas ud din Shebankyareh, and the great atabeg Nusret ud din Pir Ahmed, with the amirs of Luristan and Fars, to march. A struggle followed, in which Mahmud, at the head of the Turkomans and Kermanians, resisted bravely the attacks of the Lurs and Afghans. The next day the town was blockaded. The siege lasted for many months, the garrison holding out bravely, notwithstanding the want of food and firewood, and to supply the latter the most beautiful palaces were dismantled, and the wainscoting burnt. Sujuk Shah, the instigator of the revolt, thought to save himself when things were at the worst by deserting to the Mongols, but Mahmud, having heard of this, had him arrested and sent off to them as the author of the troubles. Sadak sent him to Tebriz, where he was put to death. The siege dragged on ten months, during which five-and-thirty combats took place, which lasted from dawn till sunset, and famous builders of siege machinery had to be summoned from Mosul and Shiraz. At length the town was stormed from all four sides. Mahmud Shah and his Turkomans were taken bound into Sadak's camp. A yarligh was issued by the Ilkhan ordering that all who had taken part in the murder of Beyari and his sons should be put to death. Ten great amirs and eight-and-twenty of their associates were executed, otherwise the town was spared, and the terrible ravage of a ten months' siege was not augmented by a general sack. Sadak was granted the amirship, with command of the armies of Irak, Great and Little Luristan, and Kerman, with a special diploma sealed with a lion's head, with one of the Ilkhan's robes and a sword, and with the hand of Tulan Khatun, the widow of Prince Yesen Timur. These events took place in the year 1300.† Which of these accounts is correct, who can say? The latter, which is that given by Wassaf, from its circumstantial detail, would seem to be the more trustworthy. Muhammed was

---

* Ilkhans, ii. 49. † *Id.*, ii. 140-142.

now duly installed at Kerman. Differences arose between him and his
vizier, Sadr ud din, which ended in the latter taking up his residence at
Tebriz, where he looked after his interests faithfully. The Sultan died of
hard drinking, according to Hamdullah, in the year 703 HEJ., after a
reign of eight years. We are told, his brother Hasan Shah was installed
in his place, but he died a month later at Hadisi. There only remained
one prince of the royal family, viz., Kutb ud din Shah Jihan, the son of
Siyurghatmish, who was now appointed ruler of Kerman, and duly given
a diploma, state robe, &c.*

In March or April, 1305, Ghazan set out from the yurt of Hulan Muran,
meaning the Red River in Mongol, a name which the Turks write Kizil
Ussun. This river rises in Mount Elvend, north of Hamadan, and falls
into the Caspian. In Persian it is called Sipid Rud, or the White River.
On it was a yurt or camp of the Mongols, which Uljaitu named
Boinuk or Yamuk. Leaving his harem and baggage near the castle of
Juk, which was only one stage distant from the palace of Jome, Ghazan
set out with the jeride, or light baggage, and his head-quarters, for Merak.
He hunted for some days in the mountains of Kharkan and Masdekan,
and eventually reached Sava. Here the Vizier gave a grand feast, and
offered the cup to all the princes and princesses, and was treated by the
Ilkhan with marked honour. The Home Secretary, Khoja Shihab ud
din Mubarek Shah, who had a house at Sava, and whose father, Khoja
Sherif ud din Saadan, one of the chief people of Irak, was still living, also
gave a similar feast, and distributed presents among the princes,
princesses, and high officials. Ghazan had largely recovered from his
illness, and could again mount on horseback and eat in public. On
leaving Sava, and on the way to Rai, he had a relapse. His sight and
appetite were again affected, notwithstanding which he continued to ride
on horseback. He stayed some days at Rai. When he arrived at Khial-
buzurk his illness became more critical, and he went on by short stages
to Pishkaleh, near Kazvin (called Yeskele rud by D'Ohsson), whence he
sent messengers to summon his favourite wife, Bulughan. After her
arrival he summoned all the grandees of the empire, among whom are
specially named Kutlugh Shah noyan, who had held the post of first of
the amirs and commander-in-chief since his accession; Choban, Baidu,
Bilar, Ostai, Molai, Ramazan, Ilghu, Kur Timur, and Taremtan ; the two
viziers, Said ud din and Rashid ud din, and various officers whose names
are given by Von Hammer, and included the pages, the master of the
horse, the commanders of the life-guards Biludai, the quartermaster-
general Masuk, the amirs of the life-guards, the secretaries, silentiars,
head cooks, falconers, treasurers, treasury clerks, &c. He addressed an
exhortation to them, bidding them do justice and right. He appointed

his brother Khodabendeh, whom he had already, four years before, nominated for that post, to succeed him, and conjured his friends to see to
the carrying out of his will in its integrity. After having performed this
duty, he passed the greater part of his time in retirement. Although so
much weakened by disease, he preserved his faculties and natural
eloquence until the moment of death, which happened on Sunday, the
17th of May, 1304. He was but thirty-three years old, and had reigned
nine years.*

An author, quoted by De Guignes, calls the place where he died Sham
Ghazan (*i.e.*, the Damascus of Ghazan). It was a town he had himself
built, and to which he had united his name, as he did to others, which he
called Cairo and Aleppo.† His corpse was taken on horseback to Tebriz,
and was followed by his khatuns and amirs. The people of the towns
and villages *en route* came out of their houses with bare heads and feet,
clad in sackcloth, and with dust on their heads. The minarets throughout
the kingdom were covered with sackcloth, and straw was strewn in the
streets, bazaars, and public squares. The people of all classes for seven
days dressed themselves in tattered clothes or in sackcloth. They
of Tebriz wore mourning of deep blue, and went out to meet the *cortège;*
soldiers and citizens walked beside the bier uttering groans. The body
was at length deposited in the mausoleum which he had built for
himself near Tebriz.‡

Ghazan is described by Haithon, who must have frequently seen him,
as of small stature, to which he refers, and also to his ungainly looks, in
the quaint sentence : " Et hoc precipue erat admirandum qualiter in tantillo
corpusculo tanta virtutum copia invenire poterat. Nam inter xx mille
milites vix potuisset staturæ minoris aliquis reperiri, neque turpioris
aspectus ; omnes tamen alias in probitate & virtutibus excedebat."§ The
Georgian Chronicle says Ghazan died deplored by all on account of his
justice, which was such that his kingdom was free from robbery, rapine,
and injustice, and he was generally regretted.‖ Pachymeres says he
enjoined his brother not to alter anything which he had settled for at
least three years, after which he was free to do as he pleased.¶

Ghazan had seven wives : 1. Kurtika, daughter of Mangu Timur
Gurkan the Suldus. 2. Bulughan, the daughter of the amir Tesu. 3.
Eshel, daughter of Toka Timur, son of the chief judge Buka. 4. Kokaji,
who was a dependant of the great Princess Bulughan. 5. The latter
princess married in turn three of the Ilkhans—first, Ghazan's father,
Arghun, by whom she had a daughter, Dilanji ; secondly, Arghun's
brother, Gaikhatu, by whom she had a son, Jing pulad ; and, thirdly,
Ghazan himself, by whom she had a son, Alju. Alju therefore stood in
the double relationship of brother and nephew to Dilanji. Besides Alju,

---

* Ilkhans, ii. 145-147.   † De Guignes, iii. 276.   ‡ D'Ohsson, iv. 349-350.
§ Haithon, 65.   ‖ Op. cit., 635.   ¶ Stritter, iii. 1094.

Ghazan also had by her a daughter, Oljai Kutlugh. 6. Tundi, daughter of, Akbuka, the Jelair. 7. Keramun, daughter of Kutlugh Timur, son of Abatai Noyan.*

Stritter quotes a curious story about Ghazan, which was derived from the report of a certain Florentine, who lived at this time in Persia, viz., that he married a daughter of the King of Armenia, by whom he bore a son, who was so deformed that there was scarcely anything human about him, and as it was said she must have had some unclean connection, it was decided that she with the child should be thrown into the fire. She begged that before this was done she might make confession and take the sacrament from some priest, and that the child should be baptised ; and it happened that directly he was touched by the sacred water his looks changed, and he became very fair to look at, upon which Ghazan revoked his order, and the Christian faith was greatly benefited.†

When Ghazan became a Muhammedan he definitely broke off his allegiance to the Supreme Khan in the furthest East. Hitherto the Ilkhans had been merely feudatories of the Khakan of Mongolia and China. They were now to become quite independent, and it is natural that the formulæ on the coins should accordingly be changed. On Ghazan's coins we find in Mongol characters the words, "Tegrini Kuchundur Ghasanu deledkeguluksen." (By God's power Ghazan's coinage.) In addition to this inscription there occur three characters, which M. Terrien de la Couperie has shown to be in the Bashpa character, the first native. form of character introduced among the Mongols, which was the invention of the Lama Bashpa, or Pakba, and was first introduced by Khubilai Khan in 1269.‡ The three characters represent the words, "Ma Kha san," the first being a contraction for Mahmud, and the second and third forming the nearest approximation to Ghazan which the characters were capable of representing.§ When the name occurs in Arabic letters it is written "Ghazan Mahmud." Pachymeres extols the beauty and purity of his coins. In the collection of the British Museum are coins of Ghazan struck at Tebriz, Shiraz, Baghdad, Kemnaz'ar, B'ar'an, Sinj'ar, Mosul, Hamadan, Armiiah, Kashan, Tiflis, Berdaa, Jezireh, Sultania, Lulueh, and others doubtfully attributed to Kerm'an, Arzenjan, and Erzerum.‖ Fræhn, in his "Recensio," published a coin of Ghazan's struck at Basrah and another at Damaghan ;¶ and De Saulcy one struck at Nakhchivan.** Another coin, published by De Saulcy, is dated, not in the year of the hejira, but in the second year of the Ilkhanian era (i.e., the new era founded by Ghazan Khan). This year would answer to 602 HEJ.††

---

* Ilkhans, ii. 8-10.   † Stritter, ii. 1093. Note 2.   ‡ Ante, i. 223.
§ De la Couperie, in Poole's Catalogue Oriental Coins, vi. l.-lii.   ‖ Poole, op. cit., 34-43.
¶ Op. cit., 639.   ** Journ. Asiat, 3rd ser., xiii. 136.   †† Id., 134.

CHAPTER IX.

GHAZAN KHAN.—*Continued.*

GHAZAN, considering his years and the short time that he reigned, was certainly one of the most remarkable sovereigns that the East has produced. He was no doubt fortunate in having as his special biographer the great historian, Rashid ud din, but he was well deserving of having his deeds described by such a pen. Pachymeres, the Byzantine historian, speaks of him with no stinted praise. " With him," says he, " departed the flattering hope of seeing tranquillity restored to the Roman Empire, and when he was taken—(he who alone, not only had the power, but also the will, as he had begun to prove by his acts, to prevent the barbarians, his subjects, from devastating our provinces)—evils greater than had ever befallen fell upon us, and particularly upon Philadelphia, which the Karamanians proceeded to assail."* In recounting his many virtues this historian becomes quite hyperbolical. He tells us that he made Cyrus and Darius, and especially Alexander, his patterns, and delighted to read their lives, &c., &c. His reign was famous as the acme of the literary culture of the Persian Mongols. He himself, we are told, beside his mother tongue, Mongol, understood a little Arabic and Persian, and something also of the language of Kashmir, Tibet, and China, and even, as it seems, of Latin also, which is apparently what Rashid ud din means by the language of the Franks ; and the best proof of the attainments of some of his people is to be found in the pages of Rashid ud din and his epitomiser, Binaketi, in which a certain familiarity is shown, not only with the history of Byzantium, but also with that of Germany, France, and Sicily. This knowledge was doubtless derived from the envoys from the various sovereigns who visited his Court. These embassies prove to us what a centre for the meeting of all the nations Tebriz was. We will now describe some of them.

In the year 1298 Ghazan sent the malik Moazzam Fakhr ud din Ahmed and Bokai Ilchi with magnificent presents—great pearls, precious stones, and cheetahs, or hunting leopards—as envoys to his suzerain, Timur Khakan, the ruler of China. The malik took from his own stores rich jewels to give away, and we are told he was intrusted with ten tumans of gold pieces to purchase Chinese products for Ghazan. The expenses of

---

* Stritter, iii. 1087-1092. Marco Polo, ii. 477.

this embassy, from the Chinese frontier (*i.e.*, the mountains of Kanghai),
as far as the Imperial residence, were entirely defrayed by the Imperial
exchequer.   When he was admitted to an audience at Taidu (*i.e.*, the
capital), the malik Fakhr ud din, after presenting his sovereign's presents,
offered his own.   The Khakan offered him a cup of wine with his own
hand, and orders were issued to supply the envoys during their stay with
provisions, robes, servants, and forty-five horses.   The Khakan's answer
to Ghazan's letter was full of friendly expressions.   It seems that when
Khulagu went to the West he had a joint share in certain Imperial
manufactories in China, and we are told that the produce of this share
had been carefully set aside and was now sent to his great-grandson.   It
consisted chiefly of beautiful silk brocades, and was taken by a Chinese
officer who accompanied the embassy.   Fakhr ud din, the chief ambassador, died on the way home again.*   The "Yuan shi" refers to an
embassy as having gone from Ghazan to Timur Khakan with tribute, but
dates it in 1304.   In this notice he is called the Chu wang (prince)
Hadsan.†   We have already mentioned the embassy from the Sultan of
Delhi, which brought Ghazan a present of elephants.   Envoys also went
to him from the rulers of Kipchak and of Egypt, from the Greek Emperor,
and the King of the Armenians.‡

His conquest of Syria brought him a letter of congratulation from
James II., King of Aragon, which was taken by a citizen of Barcelona,
named Peter Solivero.   The letter was dated in May, 1300, from Lerida.
It was addressed to "the very great and powerful King of the
Mongols, Ghazan, King of Kings of all the East."   In it the King informs
Ghazan how he had heard with pleasure of his successes against the
enemies of God, and offered him aid in the shape of ships, galleys, men-
at-arms, horses, and all kinds of provisions which could be of use to the
Tartar army, and bade the Mongol prince inform him by his messengers
what he wished in this respect.   "We have ordered," he said, "that all
our subjects who wish to go to these countries in honour of God, and to
reinforce your army, shall do so without hindrance."   The King went on
to ask that Ghazan would make over to him one-fifth of the Holy Land
recently conquered by him, as well as of the lands he should presently conquer, and also asked that his Aragonese subjects might have the privilege
of visiting the Holy Sepulchre and other places without paying tribute.§

The "Chronicle of St. Denis" expressly says that the Mongols occupied
Jerusalem, which must have been a very welcome fact for the Christians,
inasmuch as Ghazan, according to Haithon, had promised to make over
to them such lands as he conquered in Palestine.|| He also sent envoys to
the sovereigns of Western Europe, and renewed his ancient offers of alliance

---

* D'Ohsson, iv. 320-321.   † Bretschneider, Notices, &c., 104.   ‡ Ilkhans, ii. 148.
§ Remusat, Memoires de l'Académie des Inscriptions, &c., vii. 386-387.   D'Ohsson, iv. 321-323.
|| Haithon, xliii.

and also stated his willingness to embrace Christianity.* These envoys also passed over into England, as we learn from the letters written in reply by Edward I.—one in answer to Ghazan's letter and the other addressed to the Patriarch of the Eastern Christians. Ghazan's principal ambassador was the same Buscarel who had been sent by Arghun, as we have described, and who on this occasion is called Buscarellus de Guissurfo. In his letter Edward (probably in answer to the Ilkhan's complaints of delay) justifies himself for not having engaged in a crusade by the wars which were then raging in Western Europe. The envoy of the English king, who was named Geoffrey de Langley, was attended by two esquires, one of whom was Nicholas de Chartres. They joined Buscarel at Genoa, and travelled from thence to the Persian Court with him, his nephew Conrad, and Percival de Gisolfi. The original roll of their itinerary is extant, and has been abstracted by Mr. T. Hudson Turner. The Mongol Court was constantly on the move, so that the route of the envoys had to be, frequently changed. When the embassy started the Court was supposed to be at Cassaria (the ancient Cæsarea, in Armenia). The envoys are found successively at Sebaste, Tebriz, Mardin, Erzerum, Coya (?), Papertum (i.e., Baiburt in Armenia), and Sarakhana. It is curious to read of the outfit they provided themselves with at Genoa, which included furs, cloths, armour, carpets, silver plate, and fur pelisses. The silver plate they bought cost, we are told, the large sum of £193. 12s. 7d. currency (English) of that time. Their carpets, fifteen in number, which would have to serve as beds, cost £15. 15s. 6d. The armour, including seven iron plates, eleven basinets, &c., cost £44. 5s. On their journey through Asia Minor, the Saracens (i.e., Muhammedans) acted as their porters and servants. At Trebizond, Buscarel provided himself with a *parasole (sic)*. As the weather grew warmer a second one was bought at Tebriz. These, says Mr. Turner, with two shillings worth of paper, were their most remarkable purchases. On returning home to England they brought with them a leopard in a gabea, or cage, which was fed on sheep throughout the journey, several being put on board for its use at Constantinople.†

Ghazan's will was worded as follows : " Khatuns, princes of the blood, generals, leaders of thousands, and soldiers of all ranks ; sultans, courtiers, (inakan), governors (muluks), kadhis, imaums, sheikhs, prefects, receivers of taxes, and in general the inhabitants of our hereditary dominions from the river Amuyé as far as the western frontier, know that God, who, by his grace, has illumined our heart with the light of Islam, and has by his favour seconded the efforts which we have not ceased to make during our reign to maintain the precepts of the faith, to exalt the divine word, to govern the people well, and to cause justice to reign everywhere,

---

* Remusat, op. cit., 388.　　　† Archæological Journal, viii. 49-50.

having decreed that we should be struck with sickness, and the time approaching when we must pass from this perishable world to the eternal home, it has been our wish in our love for our subjects, among whom the weak are particularly committed to our charge, to remove entirely the wounds of oppression, to strengthen the bases of justice, and to cause the revival of the neglected precepts of Islam, but it has not been permitted us to complete this noble purpose to which we have devoted all our efforts. We entreat all to take care that after our departure troubles do not arise ; that our dear brother (whom we nominated four years ago as our heir and successor, convinced that he was worthy, an appointment which we have on several occasions confirmed) be promptly installed on the throne, and that all our subjects do him obeisance. Our successor will not act contrary to the decrees which we have issued after mature deliberation. He will exert himself to make Islam prosper, and to protect the Mussulmans. He will cause whatever is commanded or forbidden by the religion of Muhammed to be scrupulously observed, and will not allow any enemy of the Faith to do it injury. He will follow our example and maintain peace and security among the people, a trust which God has for a while confided to his feeble servant. He will not exact from them taxes in excess of those we have fixed, levy fresh taxes, or re-impose those we have remitted. He will not divert from their objects the funds we have devoted to alms nor the revenues of our pious foundations. He will continue to pay the pensions which we have awarded, and which the officers of the treasury must not appropriate. He will not neglect benevolence, for our destiny is to do good. He will punish the evil doers, and when our decease shall be announced in all the provinces he will cause a funeral namaz to be said, and that the faithful shall aid us with their prayers.' *

The character of Ghazan has been described for us in detail by his great minister and historian, Rashid ud din, whose narrative has been translated by D'Ohsson. "He observed," says Rashid ud din, "during all his reign the precepts of Islam with fervour, whence it is clear he had adopted the faith from conviction. This explains why the descendant of a line of Pagan sovereigns, who had conquered the world, and who had no other motive, should have changed his faith. He caused his army, composed partly of Mongols, who," says Rashid, "were Unitarians, and partly of other races of Northern Asia, who were polytheists, idolators, Buddhists, &c., to do the same. Whenever the kadhis, sheikhs, devotees, &c., appeared before him he gave them some advice. On one occasion, when he had summoned the chiefs of the Muhammedan clergy, he addressed them thus : ' You are robed in the religious dress, and seek no doubt to appear perfect in the eyes of God rather than in the eyes of men.

---

* D'Ohsson, iv. 351-353.

They may be deceived by appearances, but God sees the core of the heart. He hates what is false, and will punish it in this world and the next. He unveils the hypocrites, tears off their cloak and reputation, giving them up to the contempt and amusement of the world. Your costume shows that while you are otherwise like other men, you have, in virtue of your dress, a reputation for virtue not shared by them. You have also established this reputation by your speech and your austerity. Search yourselves well to see if you strictly carry out the duties this dress imposes upon you. If you do so carry them out, you will be, in the eyes of God and men, superior beings. If otherwise, you will deservedly incur shame. Be sure that God has elevated me to be a ruler, and has confided his people to me in order that I may rule them with equity. He has imposed on me the duty of doing justice, of punishing the guilty according to their crimes. He would have me be most severe with those who hold the highest rank. A ruler ought especially to punish the faults of those most, highly placed, in order to strike the multitude by example. I ought, therefore, most closely to watch your faults. Do not think I will pay attention to your dress. Let all your actions conform to the law and to the precepts of the Prophet. Mind that everyone does his duty and leads others in the path of safety. You ought not to expect support from each other by a mere appeal to *esprit de corps*, nor should you exact from other men what God does not command, for it is not seemly that you should torment your neighbour in order to gain repute for yourself, nor yet that you should be more zealous than God himself and his Prophet in the causes of humanity. Warn me if I do anything contrary to the law and to religion, and be sure that when your heart is the true reflex of your profession, your words will persuade me, for they will be dictated by sincerity, true zeal, and courage, and will have corresponding weight; if otherwise, they will only arouse my anger. I might add much to what I have said, but I limit myself to these general rules. If you approve of my words, they will be profitable to us both; but if they wound you, hatred will fill your hearts, and I shall conceive an aversion for you, much to the harm of religion as well as of the things of this world.' "*

Rashid describes his hero as very courageous and firm, and as having been trained well in his youth from his having command of Khorasan, which was incessantly exposed to attack from the Mongols of Transoxiana, and against whom there was a campaign or two every year. He often exercised his troops, and used to address them in stirring words. "Death," he said, "attends us at home and while travelling, in the hunt as in the battle. Why then fear the enemy? Fear will only deprive us of resources. It may be useful to lose some blood, lest it should fester and produce fever. Blood, again, is the paint of man. No one

---

* D'Ohsson, iv. 354-357.

may remain in this world. He who dies in his bed grieves by his sickness his wife and children, and inspires his friends only with pity; but the warrior who dies fighting is honoured, while the prince looks after his family." He often, like Jingis Khan, his great ancestor, gave special instructions to his officers. "When you make an incursion," he said, "take care the enemy is not informed beforehand. Advance with all speed, night and day, so that you may arrive unexpectedly, and withdraw again before the enemy has time to assemble his troops. If you make incursions annually, make them at different seasons, so that your opponents may not be expecting you and be on their guard. It is well also to use different routes, and above all things to secure good guides. If, instead of a mere incursion, you advance with a great army, the wider you spread the news of your approach and the better, for it is not possible in such a case to keep the enemy ignorant, while such news often discourages him and gives rise to disunion. It is important that the source of supplies should not be cut off, and consequently you ought to know before advancing where game, water, and forage are to be found, and to halt in such places, so as to save your provisions for more desolate spots. Always learn the situation of the enemy well by means of spies, for to act without knowledge is to strike with your fist against a shadow. But the chief importance of such knowledge is that you are always free to accept or refuse a combat, as you deem best. Exact rigid discipline; do not allow an individual to leave his standard to take the smallest thing. You can do nothing with an army habituated to licence, for at the moment of battle it amuses itself with plundering, and nothing will prevent it. This is the cause of the greater part of defeats. When the enemy is destroyed, then appropriate his wealth. Do no hurt to your own country, for the prayers of the people will bring you good fortune. Do not despise the enemy; do not exaggerate your strength, and above all things avoid boasting."[*]

He knew the history, character, and habits of other sovereigns, both ancient and modern, but especially of his contemporaries. He spoke to individuals of other nations in a way which surprised them, but what he was chiefly attached to, like his countrymen, was the history of the Mongols. He knew by heart better than any Mongol, except Pulad Aka, the names of his ancestors of both sexes, the names of Mongol leaders, ancient and recent, and their genealogies. "It was from this ruler," adds Rashid, "that I learnt in great part what I have related in this work, but he knew of many Mongol amirs and historical facts unnamed here. There was no art or handicraft, such as those of the smith, carpenter, painter, founder, turner, and others, at which he did not work with his own hands better than the artisans themselves, and he often directed

---

them himself." This somewhat hyperbolic sentence of the great vizier is to some extent confirmed by Pachymeres, who says : " I have omitted to add in praise of this great man the attention he paid to the least details of various handicrafts, at which, all-powerful monarch as he was, he liked to work with his own hands. His philosophical spirit judged it praiseworthy to know how to do everything, and to excel in every calling, so no one could manufacture more elegantly than himself saddles, stirrups, bridles, boots, sabres, and helmets ; he could hammer, stitch, and polish, and he was accustomed to devote his leisure when he was not occupied with his military duties to these mechanical works."[*]

To return to Rashid ud din. He says : " He acquired in a short time a knowledge of chemistry, of all arts the most difficult, and brought about him those who were proficient in it, but instead of wasting huge sums like his predecessors in the search for the philosopher's stone, &c., he rather devoted himself to the more practical part of making enamel, dissolving talc, melting crystal, making condensations and sublimations, and producing substances like gold and silver, and said his object was not to learn how to make these latter metals, which was a most difficult art, but to learn how to make, and to make himself, various chemical experiments. He was also acquainted with medicine, knew plants and their properties, and instead of studying them like the doctors did in the shops of the herbalists, he found them for himself in the fields. He found many in Persia which were thought peculiar to Turkestan, India, and China, and which merchants brought from those countries and sold at great prices. He summoned to his Court several famous botanists, Turks and Tajiks. He took them with him on his walks and hunting excursions, and it was with their aid he acquired his knowledge of plants. He added to the Pharmacopœia twenty-four simple drugs, each one of which was a capital remedy. He was also well versed in the natural history of animals. He knew enough of mineralogy to distinguish minerals, and knew the operations of mining and of extracting metal from the ore, and himself worked at these occupations. He knew the charms and magical formulæ proper for all ills. He often foretold the future, announcing that an envoy would arrive from such a place with such an appearance, or that he would receive good news ; and although most of the princes styled Sahib Kuran (*i.e.*, favoured by heaven) have this faculty, one never saw one endowed to the same degree. Greedy of knowledge, he had learnt the art of taking auguries by means of geomancy *(ramel)*, by the shoulder blades of sheep, by the teeth of horses, and other methods employed by different nations to foretell the future. He knew the planets and constellations, the times of their rising and setting, as well as their various astrological properties. As he often visited the observatory at Meragha, and had the

---

* Stritter, ii. 1091.

instruments there explained to him, he had learnt a good deal of astronomy, and gave orders for the foundation of another observatory near Tebriz, and designed a cupola for observing the sun's disc, which he described to the astronomers. They told him they had never seen such an instrument, and that it was very ingenious. In building the observatory of Shenb, near Tebriz, which was built in the form of a cupola, this idea was carried out."*

"Before the reign of Ghazan," says Rashid, "it was the amirs and viziers who ruled. The sovereign passed most of his time in hunting and other amusements. We can understand what measures were in consequence passed by a divided ministry. It was they who negotiated with foreign envoys. With Ghazan things were very different. He ruled himself, and established order in the administration, which for years had been in confusion. He listened to the advice of no one, but gave his own orders. Young and old, superior and inferior officers, all obeyed him cheerfully. Recognising his superiority and wisdom, they felt humbled before his genius. When a foreign envoy arrived, he received him himself, instead of remitting him to one of his ministers. This was an advantage, for only men of merit, experience, and knowledge were then selected as ambassadors. All the envoys and learned men who went to his Court were astonished at his eloquence and the grace of his conversation. He quoted incidents from their own history; spoke to them of the character of their masters and the customs of each nation. He greatly delighted in the society of learned men and philosophers, and was a good judge of them; nor did he see them again if their behaviour belied their conversation. In the great assemblies he astonished people of all classes by the questions he put to the different philosophers and learned men. At parties of pleasure, when he had drunk a little too much, instead of saying and doing foolish things like other young people, he dived into subtle and profound questions. He knew the dogmas of other religions, and discussed them with their professors, but of ten questions which he put they could not answer more than one. He completely nonplussed them. He had great discernment of character, and used to note those who made perfidious and malignant insinuations, and watch until they had declared themselves more openly, when he had them punished. When he discovered good men, he gave them his confidence, and was then deaf to the calumnies of hate and envy. He augmented their power and supported them, so that they might carry out their plans. On the other hand, he was very severe with wrong doers." Rashid adds that those engaged in affairs declared on oath that he had never put anyone to death who did not deserve it, and whose existence was not hurtful to society. He recommended his ministers and yargujis to beware of listening to the

---

* D'Ohsson, iv. 359-563.

accusations brought against any governor or public officer. "For it is possible," he said, "that those complaining have been made to pay taxes which were not exacted previously, that they have lost some post, or have some other private grudge. In such cases it is better to consult public opinion, which is the best judge. Do the people like him or no? There are few functionaries who know how to conciliate the affections of the people, and who are inclined to justice. If an official has many good qualities and only a few faults—if, above all, he be not avaricious—if he be firm and loyal, it is not necessary to displace him." When hunting, if he needed provisions he ordered that twice their value should be paid for what were obtained. If he heard that the troops had committed disorders in some district, he summoned the subordinate officers and had them bastinadoed, and severely reprimanded the generals. "You wish," he one day said to his officers, "that I should let you pillage the Tajiks, but what will you do after you have destroyed the cattle and . seed of the labourer? If in such a case you should come to me asking for food, I would punish you. Remember, when you would strike or maltreat their women and children, how dear our own are to us, and that they are men like ourselves." He was generous in rewarding his servants, and in distributing to the poor. Finally, his morals were pure. No one accused him of adultery, and if by chance he looked at a woman lecherously, he did not go beyond looking at her. In time of war, when he was far from his harem, his amirs used to offer him the most beautiful damsels they had secured, but he would never accept them. "He never committed," says Rashid ud din, "any of the sins which in our law are styled fornication, adultery, unnatural crimes, &c., and was as severe in this matter towards others as he was himself strict. He issued stringent regulations about these offences, and punished those who were guilty of them with death."*

We will now turn to the various reforms, &c., which were introduced by Ghazan, and which are described for us by the great Vizier, who took such an active part in their introduction. Before the reforms introduced by him, the hakims (intendants or governors) were charged with the levying and collection of the taxes. The amount each had to account for was fixed for him, as were the charges upon this sum. But these hakims were most extortionate. They made as many as ten and in some cases twenty levies (coichur) during the year. They began by putting aside the amount which was needed to make up what they had to account for. Then every time a commissary (ilchi) arrived to demand money, or for some other reason, they levied a fresh coichur; and the more of these officials (whose exigencies were very great) arrived the better pleased was the hakim, for while he levied a contribution for their lodgings and enter-

---

* Id., 363-368.

tainment, for their food, or for the presents they demanded, he took care
that a large portion of the levy was retained for himself, while he gave
the rest to the *shahneh* or commandant, and to the *bitikchis* or treasury
clerks, to secure that these malpractices should not be entered in the
official accounts. The revenues of the provinces were largely absorbed
by the current expenses, and by a quantity of assignations. In Khorasan
four-fifths of these assignations were unpaid. Those to whom they had
been made, together with the commissaries, returned to the Divan with
their *bérats* or treasury orders in their hands. They were told that the
province still owed a part of its contribution, which must be paid. A
severe ordinance (*al-tamgha*) was issued, ordering it to be paid at once.
They accordingly returned to the province, and the hakim, on their report,
immediately made a fresh levy, telling those who had to pay : " You see
that many commissaries (*ilchis*) are waiting here. We must supply their
needs and expenses until we have time to arrange the matters they have
come about." And no one dare tell him that he ought to pay these out
of the levies he had already received during the year, which greatly
exceeded the proper quota. Two-thirds of the new contribution were
appropriated by the hakim and his creatures, and those whose assignations
were unpaid, and for whom the levy had been made, had to be content
with a third portion, which barely went to cover the expenses of their
continual journeyings to and fro. The officials of the Divan did not
make inquiries as to the receipts of each province, but granted assignations
of revenue indiscriminately, which were not paid, but which temporarily
satisfied the importunate. The Vizier and the hakims of the provinces
had a secret understanding, and if some secret marks were not attached
to the *bérats* or orders of the treasury, money was not forthcoming when
they were presented. The public treasury did not, in fact, receive a
farthing from the provinces, while they were only charged with one-
fifth of the usual payments. Neither pensions nor salaries, nor the
expenses of the local administration, were paid by them, although these
last were made the first charge on their revenues by the special instructions
of the hakims. In the spring they put off those who asked to be paid
with the excuse that the treasury must first be satisfied, and promised to
pay them at harvest time, and then they were told they must stand aside
until the crowd of ilchis and other similar claimants had been paid. Those
whose salaries were due waited, hungry and naked, from year's end to
year's end for them. The more adroit among them managed to get half
of what was due to them from the hakim's officers, by giving receipts for
the full amount, and then they had to accept in payment articles at double
their real value. Those who in this way secured a fourth of what was
due to them deemed themselves fortunate, since so many got nothing at
all. If one of these unfortunates went to the Ordu, after a harassing
journey, and laid his case before the Divan, and if in consequence inquiry

was made from the governor how such a claim was not discharged out of the first receipts, as it ought to be, the latter laid the blame on his not having been able to get in the whole of the levy, and then gave an order of payment upon the arrears, which in fact were merely arrears of illegal exactions, for he had already received many times his real due.

The unfortunate people who would not meet the repeated demands for dues abandoned their villages and families. If in consequence the last levy or one-half of it was remitted, the bitikchis, or secretaries, nevertheless entered in their books that the money was still owing. The heads of the treasury knew well what took place, but they were bribed by the hakims with a large proportion of the ill-gotten money to say nothing. "All the viziers who have held authority to this time," says Rashid, "have shared in this guilt; but it was chiefly Sadr ud din Chaoyi who carried this iniquitous system to its further limits. During his sway assignations of revenue, pensions, &c., were mere mockeries." Often, we are told, a dervish, a sheikh, or some other deserving poor person would be given an assignation of 500 dinars by him. Overjoyed with his fortune, he would devote 100 dinars to securing a good mount and the expenses of his journey, and hasten to secure the charge which had been assigned him upon the revenues of some province. In his eagerness he would forget his office of sheikh and adopt the *rôle* of a courier or a tax gatherer. But his energy was of no avail, and he ended by taking flight to escape his creditors. The overtaxed people were obliged to emigrate; towns and villages were deserted. Commissaries were sent after the fugitives, but neither menaces nor ill-usage, nor yet further exactions, would prevail upon the runaways to return. Those who remained in the towns blocked their doors with stones and entered their houses by the roof. When the tax collectors visited a certain quarter they employed as guide some vagabond who knew the district, and enabled them to ferret out the unfortunates who had sought refuge in the sewers, where they were hidden; and if they could not find them they took their wives and drove them from quarter to quarter like flocks of sheep. They were hung up by the feet with cords and beaten, and the air resounded with their lamentations. "We have more than once seen a person," says Rashid, "who, on seeing a tax gatherer on his roof, would run and jump into the street, breaking his legs." In the province of Yezd things were so bad that you could go through villages without meeting anyone. The few people who remained employed vedettes, and on a signal from these that some one was coming they hid themselves underground. It is said that in the year 691 (*i.e.*, 1292) a proprietor from the province of Yezd went to Firuz Abad, one of the large villages of that province, to see if he could obtain any rent for the lands he owned there. Notwithstanding all his efforts, he could not for two or three days meet with any of the husbandmen, but he saw a tax collector seated in the midst of the village,

I *

with a *bérat* or order in his hand, and before him were three peasants. They were hung up by their feet, and were being bastinadoed to compel them to supply some food, for these exacting officials insisted upon being furnished with food, forage, wine, and boys and girls. In some districts they were so numerous that they were said to outnumber their victims two to one. While the hakims and their subordinates plundered the provinces in this fashion, nothing found its way to the treasury, which was empty, and when means were urgently needed for the army, the defence of the frontiers, or the necessities of the State, there was no other resource than another set of violent exactions.*

Ghazan felt how difficult it was to alter practices which had taken root so firmly, and which had outlived the various edicts of his predecessors. "How are we," he said, "to bring into the path of honesty these governors and tax collectors, who have become habituated to exact more than is due, and to pay nothing into the treasury; who are habitually tried and habitually secure exemption from punishment by a distribution of money; and when one of their colleagues is put to death, attribute it rather to his evil star, or to some one's malice, for if it were not so many others would suffer the same fate? We must devise some plan by which the provincial governors shall be prevented from handling the public moneys, and the best way will be to take away their power of levying a single farthing." He accordingly ordered a bitikchi to be sent to each province, who was charged to draw up a report on all its communes, to re-adjust the taxes according to the last census, taking care that the imposts should be moderate. He was to draw up an account of the *ingu*, or private domain, and of the *vakfs* or charitable funds, as well as to make out a list of the persons who had enjoyed for thirty years uncontested these funds.

The bitikchis went to the several provinces, to make a survey of the communes and to fix the taxes. As it was impossible in such a great work to have perfect exactitude, when their reports had been sent in to Ghazan permission was given in cases where considerable errors were alleged to appeal to the Divan. From the various registers they prepared a general summary (*kanun*), which was deposited in the Divan, and from which at the beginning of the year the amount of the taxation of each commune was drawn up, which was sealed with the seal of the high officials of the Divan and the golden tamgha. The taxes were to be paid in two ways, first to those who had assignations upon them, certified with golden tamghas, and the balance to the treasury. Everything was to be discharged in cash, and not in kind. Each person learnt from the official tax-paper issued by the Divan how much he had to pay, and knew accordingly to a farthing how much could be exacted from him. The

* D'Ohsson, iv. 370-379.

maliks or prefects, bashkaks or commandants, and bitikchis or secretaries were forbidden, on pain of death, to grant any assignations, and the person who should draw up such a document of assignation was to have his hand cut off. Hitherto, whenever the ruler granted a village or an estate, either as a reward, a fief, or otherwise, or whenever he made the same over to a charitable purpose; each time, again, that a district obtained exemption from paying taxes from a khatun, a prince of the blood, or a Mongol grandee; or when, again, some village became deserted, it had been customary for the receiver of taxes to deduct from his account a much larger sum than the quota paid by the place, and as the treasury officials had no means of checking his statement they were obliged to accept what he and his friends said. After the reforms, the amount to be received from each province was as well known as the amount of money issued by the mint. Henceforth there was no demand made by the central authority upon any province for corn, straw, wine, animals, &c., upon the treasury, in the shape of taxes, and all that went to furnish the army, or was given by the State to private individuals, was paid in specie by the treasury. Nevertheless, the magazines were continually full, and by a comparison of former registers it was seen that in no former reign had as many presents of gold and robes been made in any five years as were made by Ghazan in one year; and while formerly the ungathered harvest was always hypothecated, "now," says Rashid, "there is always a year's supply of grain in the magazines in advance"

Looking into the future, Ghazan wished to secure that his reforms should outlive his own days, and survive the disturbances that might follow on his death. To secure matters against a collapse or laxity at head quarters, a special edict was issued, which was to be enrolled in each commune, forbidding anyone to be taxed beyond the amount fixed by the Divan, and forbidding also the local authorities levying the smallest sum at their own instance. The authorities were ordered to give the widest publicity to the new regulations. The various local returns were deposited in the State paper office which Ghazan built near Tebriz. They were given in charge to several archivists or official guardians, and a sum was set apart for their care. An edict cursing anyone who destroyed these returns was issued, while counterparts of the more important of them were preserved in the offices of the Grand Divan. Ghazan ordered further that the list of expenses of the local treasury of each commune, as contained in the *kanun* or general official summary, and in the orders signed with the golden tamgha, was to be handed by the governor of the province, in the presence of the kadhis, seyids, imaums, and notables of the town, to the various communal authorities, who were within twenty days to have it inscribed on wood, stone, copper, iron, or plaster, as they preferred, and to see that it did not get obliterated. These inscriptions were to be put up at the entrance of the villages, before the mosques, &c.

This publicity, this letting in of the bright sunlight of public criticism, upon financial corruptions, was assuredly a most enlightened and brilliant idea to have been formulated in Persia in the fourteenth century. Those places which had hitherto paid their taxes in coin were to continue to do so, and those which had paid them in kind were to continue as they had been accustomed. On the inscribed tables were also to be entered the customs-charges, and alongside of each list of charges was to be written a copy of the order empowering it, so that each one should know what was taxed, and when and how it was to be paid. The payments in kind were to be paid in either by the mayor or by those who contributed them, in a tent set out in the public square, and the receiver was to forward the amount received five times a day to the treasurer of the province. The receivers of taxes were forbidden to accept presents under any pretence. Those who had not paid what was due at the end of the specified time were to be fined a certain sum, and those who did not pay at all were to receive seventy strokes of the bastinado. The country people were to pay the *koijur* and other taxes during two seasons of twenty days each, the first after the *nevruz jelali* (*i.e.*, the vernal equinox), and the other after the autumnal equinox. The nomades were to pay theirs all at once, at the new year. The *kharaj*, or capitation tax, was generally to be paid once a-year, within twenty days after the spring equinox. At some places, as at Baghdad, it was to be paid at harvest time. Twenty days were also allowed there. The taxes paid in grain were to be taken to the magazines which were placed in the various cantons; forty days were allowed during which to pay them. The customs-dues and the tolls were classed under the title *tamga*; the cattle tax was called *koijur*. All four taxes, according to Von Hammer, still survive under different names among the Osmanli Turks. In every canton given as an appanage to the khatuns, princes of the blood, or Mongol grandees, or given as a military fief, or made over to some private individual as a reward or otherwise, or classed as a vakf (*i.e.*, as a pious trust), a table inscribed with the dues which according to the kanun it had to pay was to be drawn up, so that in such cases also there should be no illegal exactions. The edict by which Ghazan regulated these various fiscal matters is given by Rashid ud din *in extenso*, and dated the 22nd of February, 1304, at his palace of Uljaitu-imuk, near the Hulan muran or Red River.*

Rashid tells us that during the reign of Abaka Khan, when he says peace, order, and security prevailed, who was a just prince, and who was essentially a *sahib kuran*, some merchants (*ortaks*) went to his Court with arms, armour, and good horses, which they had bought with their own moneys and sold again to the *kurujis* and *atajis* (knights and squires) at a price which brought them considerable profit. This induced others

---

to follow their example, and those who had not money borrowed it, intending to repay it out of the profit which they made, securing the capital for themselves. They went to the Divan with the receipts of the kurujis and aktajis, and duly obtained upon them assignations upon the revenue. This brought large gains to many who hitherto had nothing, and they were suddenly to be seen mounted on Arab horses or beautiful mules, dressed like princes, and surrounded with handsome slaves and a great number of muleteers (serhenks), &c., who led the mules and laden camels. When it was discovered how this surprising change of fortune had come about others were eager to imitate them. Thousands of Mussulmans and Jews who followed the occupation of vagabonds, or carried sacks of vegetables suspended from their necks, or wretched weavers and others who never possessed a dank of gold and had scarcely bread to live upon, began to borrow money, but instead of buying horses and arms they spent what they borrowed on rich clothes, &c.; and although the officers of the Divan knew what was taking place, they gave them assignations on the revenue on receiving hush money. Those who had already made money began to lend it, and so the custom of lending, not only money, but also furniture, clothes, &c., at interest began to prevail, and was accompanied by a free forging of assignations. They even bribed the bitikchis to grant them real assignations on agreeing to pay them in return so much per cent. on the amount so granted, with which they boldly faced the officers of the Divan, and we are told that in this way receipts were current for larger sums of gold and silver than existed anywhere, even counting what was in the bowels of the earth. In effect to translate the naive language of Rashid into that of the nineteenth century, a vast system of most dishonest credit had suddenly sprung up. The treasury officials were no doubt embarrassed when receipts were presented to them for many times the number of articles which had really been supplied. Abaka might well ask where these things had been delivered, in what arsenal these arms were preserved, and in what fields all these horses were being pastured. The officers who should have checked such public robbery had their mouths closed, while each of the ill-doers obtained the protection of an amir or a khatun by the gift of some petty present—a sheep or a skin of wine. On the other hand the governors of provinces, eager to pay the assignations presented to them, passed off upon their holders things worth one-third of the sum asked, which were gladly taken in payment, for if they had not taken in payment these precious stones and pottery they would probably have obtained nothing. Having obtained these things, they sold them again at a low price, or raised money upon them, thus reducing the value of fine stones. But this state of things had its natural Nemesis. Presently, the profits on such transactions became nothing at all, and the merchants found themselves without either food or clothes. The money-lenders were

chiefly Mongols and Uighurs, and their debtors who were unable to pay
their debts ended by becoming, with their wives and children, the slaves
of the usurers.   The worst result of all (says Rashid) was that men of
birth and consideration, who were the proper people to be superintendents
of finances (muluks) and farmers of the taxes (motassarifan), avoided such
employments, while those who had spent their lives in poverty, and hoped
to become rich in a few days, undertook to farm the finances under
onerous conditions, and in order to supply themselves with slaves, horses,
clothes, and also to meet their expenses at the ordu, they were obliged to
misappropriate money.   Increased risks naturally led to enhanced interest;
300 and 400 per cent. was offered for money by these new taxing-officers,
and it was natural that all the revenues of the provinces upon which they
had a lien did not suffice to meet their debts.   They were obliged,
therefore, to have recourse to extortion of various kinds.   The officials of
the Divan were kept quiet by a liberal flow of presents, and the treasury
hardly ever received a fourth of what ought to have gone there.   The
army was in need of necessaries.   Honourable people among the higher
classes shunned these public employments, and referring to this Rashid
says, "Wise men have declared that a State falls into decrepitude when
those most worthy of fulfilling public duties avoid them."   Surely we
could point this moral in the nineteenth century, by some conspicuous
examples in more western and self-sufficient latitudes.

Under Sadr ud din Chaoyi, the employés of the treasury were the
vilest of men, says Rashid (who, it must be remembered, however, was a
Muhammedan vizier criticising a *Jewish* predecessor), and as they knew
his habit of selling a cow for its ear, when they had borrowed a large
sum of money they made him a present of it.   They took what was
worth 10 dinars and gave 20 for it, and then presented it to him again at
the value of 30.   Sadr accepted the present, and then said to the tax-
farmer, "The treasury needs money."   He replied he had had the
greatest difficulty to raise enough to pay the present which he had made
to him.   "Very well," he replied, "in order that you may not be a loser,
you must set off the value of what you have given me at the price which
it cost you."   They accordingly, in paying the amount they owed to the
treasury, the price in fact they had paid for their office, set off the value
of the presents they made the Vizier at the valuation they themselves put
upon them, and thus things which were really worth but 10 dinars were
palmed off upon the treasury at 40.   On the other hand, when Sadr ud
din was pressed for money, his own subordinates cheated the public
purse in another way by declaring that such objects, although valued at
10 dinars, would not really fetch more than 6, and thus pocketed 4, so
that of 40 dinars which were professed to be paid into the treasury, it
really received 4.   Rashid quotes an actual example of what happened.
He had ordered some thousands of sheep to be bought at 5 dinars each,

payable in two months. When the money became due, there was none forthcoming. Many of the sheep had meanwhile become very thin, while others were dead. He had them all sold at a low price to pay the interest which had accrued due during the two months, and the obligation was renewed for another two months.

Things grew so bad that during Gaikhatu's reign neither assignations on the revenue, salaries, nor expenses, &c., were paid, which was the chief cause of the discontent in the army. Sadr ud din never had any money, and the worst feature of the case was that the worthless people who had got the taxes into their own hands, as farmers, &c., were protected and patronised by the princes of the blood, khatuns, amirs, viziers, bitikchis, &c. Some of them owed money to these people, others were partners in their gains, others again had received presents to keep them quiet, and it naturally required great firmness and prudence to combat such a deeply-rooted dishonesty.

Ghazan, by an edict dated in May, 1299, prohibited the charging of interest. Those who had been in the habit of practising usury, and their patrons, complained loudly, saying that all commercial transactions would be stopped. Ghazan replied that his only object was to put a stop to illicit transactions; others said that the treasury was constantly in need of money, and that if the farmers of the revenue were not to be allowed to borrow they could not supply it with funds. Ghazan and his viziers replied that they did not expect the farmers to pay in advance, and an ordinance forbade the farmers of the tax making up either capital or interest due by borrowing. He more than once advised the khatuns, princes of the blood, and amirs not to lend these people any money, and had the following notice proclaimed by criers, "As we do not allow one who has lent money to a tax-farmer to recover it, neither during the life nor after the death of the debtor, so we do not demand payment in advance from the tax-farmers, and if they squander the revenue, their goods, movable and immovable, will be answerable for it." As these patrons of the old state of things raised various other excuses, Ghazan asked them if God and his Prophet knew more of worldly affairs than we ourselves do. On their saying "Yes," he said, "Then God and his apostle have so ordered it, and we will listen to nothing against their commandment." From that time the taking of interest was forbidden. Some tried to evade this by charging an excessive rent for furniture, pretending that this was only a bargain. Ghazan in a rage declared that if these artifices were not given up he should forbid the repayment of capital as well as interest. He asked why those who had money did not buy articles with it, or employ it in cultivating the soil, or in legitimate commerce. "Now," says Rashid, "Good faith and justice have returned. The greater part of men's wealth is honestly acquired, abundance reigns, and

industrious people engage in agriculture, commerce, and other useful
occupations."*

We will now turn to another abuse which had grown up, and to
whose remedy Ghazan applied himself. This was the great number of
messengers and couriers who were constantly going about, and who
were a great source of harass to the people. They traversed the country
with large suites, and lived at the expense of the much-enduring
inhabitants. It had become customary (says Rashid) for the khatuns, the
princes of the blood, the commanders of tumans (or 10,000 men),
commanders of 1,000 and of 100 men, the shahnahs, or governors of
districts, officers of the Court, and even equerries and hunters to send
couriers on the smallest pretext. If a dispute arose between two
individuals, the party won it who could secure the services of some
powerful man's courier to go to headquarters on his behalf, and it came
to be a custom for people in such cases to give their sons up to slavery to
the khatuns, princes of the blood, or amirs for the smallest gift in order
to get their patronage. The person who lost the case, if he had to save
himself from ruin, had to secure the services of a courier, who speedily
returned from headquarters with the means of revenge. This move had
to be met by the sending of another courier. Thus a number of agents
were kept going to and fro on these trivial errands, for the protectors of
each side made it a point of honour to secure the triumph of their clients.
People who wished a favour from the mayor of a village, dispatched
couriers, others were sent merely to receive presents. The idajis, or
officers of the table, sent so many couriers for provisions, that the halls
where the divans in the large towns met could not always hold them, and
things became so bad that more couriers were met on the great roads
than other travellers, and a thousand horses at each posting station did
not suffice for their needs. They even began to employ the horses of the
Mongol soldiery encamped in the district. If they met caravans coming
from abroad, or public officers or private people travelling on their own
business to the capital, they took their horses, and left them unmounted
with their baggage, sometimes in dangerous places. Not content with
being fed, they even forced quarrels on their hosts in order to extort
money, while their muleteers appropriated what they could lay hands
upon—dresses, turbans, &c. They took more horses than they had need
of. In the villages and the cantonments of troops they extorted more
provisions than the law permitted, and more than they could consume,
making merchandise of the rest. It is indeed wonderful, as our author
says, that the miserable inhabitants could support existence under the
constant attacks of these myriads of gadflies.

The couriers generally styled themselves brothers or sons of this or that

---

great amir, and declared that they were sent on most urgent business, but
no one believed them. Their calling had become contemptible, and when
an officer of mark, really commissioned with some important duty, happened
to come, the chances were that the people on the route who should
furnish the horses had withdrawn for safety to the mountains, or else he
was supplied only with poor horses, so that his journey took two or three
times the proper time to accomplish. Of 500 horses at each post station
not more than two were really efficient. The money set aside for meeting
the expenses of posting and the couriers was largely appropriated by the
prefects to their own use, and although all the customs receipts, which
were the most satisfactory part of the revenue, were appropriated to the
expenses of posting stations, they did not suffice. The customs officers,
afraid of the unruly couriers, often fled, leaving them to fight for the
money. Thus it became the custom for them to have a party of their
friends, relatives, or clients with them. It seems hardly credible to read
that the less important couriers had a suite of two or three hundred men,
while those of greater distinction had as many as 500 or 1,000. Often
when several of these couriers met in a town, the prefect asked them in a
loud voice, "Whose business is the most pressing? I will begin with
him." And they would then fight with one another for precedence, and
he who won was first served. Meanwhile the highwaymen profited largely,
as it became the custom for one courier to stop and take the horses of
another, whom he declared to be of lower rank, so the highwaymen,
pretending to be couriers, did the same, and not only took horses, but
stripped those whom they met of everything, including yarlighs or
diplomas, and paizahs or tablets, which were equivalent to passports.
Armed with these official documents, they proceeded to plunder the
caravans, being mistaken for official persons.

Ghazan felt he could not stop these abuses all at once. He began by
founding a special post service for his own private couriers. Stations for
these were placed at intervals of three parasangs apart. Those on the
most frequented routes were to have fifteen horses, and on the other roads
fewer. These horses were only to be forthcoming when the person
asking for them could produce an order with his own golden tamgha
attached. He appointed officers to supervise specially these relays, who
were themselves closely watched, and he devoted large sums to making
the service efficient. In order that the frontier commanders might send
immediate news of importance, Ghazan confided to them a number of
these special orders for two, three, or four horses, and forbade the
use of more than four horses to a courier, even when he was the son
of a noyan. In pressing cases the same courier was not to take the
message all the way himself, but the dispatch was to be sent from one
station to another, and in this way sixty parasangs could be covered
within the twenty-four hours, and a message sent from Khorasan

to Tebriz in three or four days, which it would take a courier six to compass. Some time after this he ordered that his officers and the public functionaries should travel with their own horses, and at their own cost. Then the public posts (*Yamha-i-Tuman*) were suppressed. Lastly, he forbade anyone except the sovereign sending couriers, and ordered the governors of provinces to arrest and imprison all whom they met with. This put an end to the custom of private people sending couriers. Another order assigned definite pay to the Government couriers, and forbade them exacting anything *en route*. In this way, in the course of two years the people were delivered from the scourge of having to find horses and provisions for a host of men. "At present," says Rashid, "they are not more than a thirtieth of what they were, and as they cannot make exactions on the people as they travel, they cannot be distinguished from ordinary travellers."* Wassaf says that the mounted couriers were expected to travel continuously for twenty-four hours and to cover sixty parasangs at a stretch ; unmounted ones forty. They were to carry passports, describing themselves and stating the time of their starting, which was to be checked by the recorders of the post stations.†

There were always in each town (says Rashid) one or two hundred messengers (*ilchis*), who were billeted on the citizens. There were also many others, friends of the prefect, &c., who were similarly privileged. "Valets de place," in numbers, gained their livelihood in finding lodgings for these people, taking them from house to house and receiving black mail to persuade them to go on a little further, and eventually dropping their charge where most convenient to themselves. They appropriated beds and other furniture for these messengers in various houses, which seldom returned to their owners, the excuse being that the travellers had taken them with them ; or if they were returned at all it was generally in very bad condition. Every commandant (*bashkak*) going to his post generally took with him more than a hundred families, who were billeted as above described. Rashid says that to his own knowledge, when Toghai, son of Yasudar, was deprived of the government of Yezd his suite occupied 700 houses. It was of course natural that both ilchis and bashkaks should choose the best houses for their lodgings. Building ceased, or if it went on a person would call his house a college or a hospice. Nor did this even avail ; the doors were blockaded and narrow underground entrances made to the houses. The ingenious travellers thereupon did not scruple to break holes in the walls, while they forced the inclosing boundaries of the gardens and pastured their horses there, and a garden which had taken ten years to beautify was desolated in a single day. If one of the horses so pastured fell into a hole or a ditch, or escaped, the host was

---

* D'Ohsson, iv. 397-404.   † Ilkhans, ii. 168-169.

made to pay its value. In winter the trees and even the doors of the houses were burnt for fire. We have heard it said (again reports Rashid) that one of the Imaums of Yezd having, in 1296, received at his house Sultan Shah, son of Nuruz, with his mother, they stayed four months, and on leaving left behind none of the furniture they found there. He had the damage which he had sustained valued by experts, and it was found that, besides other mischief, they had burnt the outer door of the house, which might have cost 5,000 dinars, as well as many of the internal doors of exquisite workmanship. When such things could happen to the mufti of the town, it may be imagined what befell ordinary individuals. In addition to this, many thousands of coverlets and mattresses, with furniture and house gear, were annually taken from the houses, on the plea that the newly-arrived ilchis needed them. The day one messenger left a house another one entered it, and his arrival caused desolation to all the neighbourhood, for his people made their way by the door or the roof into the neighbouring houses, killed the fowls and pigeons with their arrows, and often wounded the children. They also carried off all the eatables and forage they could find, and it was no use complaining, for the complaints were not listened to.

One day a good old man, father of a family, went to the Divan, and said, " My lords, can you approve what has happened to me. I am old, and have a young wife. My sons, who are away, have left their wives with me. I also have daughters. Several young and handsome ilchis have come to my house. They have been for some time with me, and have been seen by these women, who are no longer content with me and my sons, who are absent, and I cannot guard them night and day. The greater part of the fathers of families are in the same difficulty. If this continues there will not presently be any more legitimate children. They will all be bastards, half-breeds, and sons of Turks." The old man then told Ghazan the following story : " In the days of the Seljuki, in the royal capital of Nishapur, the Turkish officers were billeted on the inhabitants as is the case now. A Turk was billeted in a house whose mistress, young and beautiful, had just been married. He wished to compel the husband to go out of the house, which the latter, knowing his motive, refused. The Turk struck him, and ordered him to go and water his horse. Thereupon his wife seized the bridle, and herself went with the horse to the brook. At that moment the Sultan was passing, and seeing such an unusual thing as a bride in her bridal costume watering a horse, asked her for an explanation. ' It is because of your tyranny,' she replied, and on her telling her story at length, the Sultan was so touched that he forbade the Turks in future billeting in Nishapur, and ordered that they should build houses for themselves outside the town, which was the foundation of Shadiak." The old man ended his story in tears, but they did not touch the amirs or viziers.

Ghazan had already limited this abuse by reducing the number of couriers, who were only now dispatched on urgent business, and were obliged to use diligence on their journey, only stopping to change horses or to hastily take some food. Nevertheless, as commissaries had still to be dispatched occasionally to look after the finances, he ordered that in each town a special residence should be built for these officers (Ilchi Khaneh), furnished with proper furniture. Von Hammer says that the Elchikhan at Constantinople, where envoys were formerly entertained, was an imitation of this.* The citizens, freed from their tormentors, proceeded to decorate their houses and plant gardens, and many who had left their native towns and become vagabonds, returned home again.†

Ghazan forbade people to speak to him on business matters when his reason was in any way obscured by wine, nor would he allow them at such times to press insidious and malicious insinuations against others, nor would he do things in a hurry. In order that he might not be thus induced to sign orders whose contents had escaped his memory, and to prevent plausible people from obtaining orders on the treasury by surprise, he himself kept the key of the casket where the great seal (tamgha) was preserved. Formerly it had been in the custody of the bitikchis. When he had a certain number of assignations (bérats) and orders to seal, the secretaries, viziers, and other officers of the Divan asked for the key, and returned it again after affixing the tamgha in his presence to the various documents. These documents were then countersigned on the back with the black tamgha of the four heads of the four kiziks (i.e., body guards). They were then looked over by the officials of the Divan, who appended the tamgha of the Divan. No royal ordinance was valid which had not passed through these formalities. A special secretary registered every yarligh, word for word, with its date, the name of the writer and the reporter, so as to guard on the one hand against any alteration, and on the other against its being disavowed by the officials who had drawn it up. There was a register for each year, which was kept in the casket with the great tamgha. The aljis, or officers of the seal, were not permitted to receive anything from individuals. Formerly they had been most exacting. There were different seals. The grand tamgha was made of jade (yeshim). It was used for the letters of investiture of sovereigns, the diplomas of general officers and of civil governors (muluks). Another seal, also of jade, but a little smaller, was used by the kadhis, the imaums, and sheikhs. A great gold tamgha was used in matters of smaller importance. Another gold tamgha, with the same inscription as the preceding ones, but around which were engraved a bow, a mace, and a sword, was used by the generals, and the troops were not to obey anyone who was not possessed of such a tamgha. A

<hr>

Ilkhans, ii. 168. † D'Ohsson, iv. 404-409.

smaller tamgha was employed for sealing assignations upon the treasury and other treasury documents issued in conformity with the orders (pervaneh) of the sovereign.* As it was not possible for Ghazan to read all the documents issued in his name, he had forms drawn up very carefully upon each subject, and also forms of answers to the various questions that might be sent to the authorities. By this means documents were simplified, and as they were all more or less of one type, it reduced litigation and disputes. When these various forms had been drafted, Ghazan summoned the grand officers and said to them, " We are about to read them through very carefully, and each one who has corrections to suggest must point them out. When they are entirely approved by you and me, they will serve as models for all my decrees, and there will be greater uniformity in business matters." All these forms were transcribed into a register, which was entitled " The Canon of Business." If a new case occurred a new formula was prepared, which was submitted for the prince's approval. It only needed a few alterations to adapt it to various persons, places, or circumstances.†

In regard to paizahs, or tablets of office, whose use we described in a former volume, Ghazan ordered that the large ones, with a lion's head upon them, were to be given to sultans, military governors, and prefects. These were to bear their names, and were to be cancelled when their commission ceased. There was one assigned to each province, which might not be taken to another. Formerly, if twenty governors succeeded each other during twenty years, each one received a paizah, which he retained after he lost his office, and continued to use. The paizahs issued to commandants and prefects of the second order were smaller, and bore particular marks. It was forbidden to make these paizahs in the provinces, and when one was conferred the goldsmith who made it engraved on it a special mark, difficult to counterfeit, with a steel punch. There were special tablets used by couriers and messengers. Five in copper were sent to governors of the first class, and three to those of the second, for the couriers whom they might have occasion to employ. The princes of the blood and military chiefs no longer employed their paizahs, which were formerly employed by them in covering the land with their couriers.‡ It was the custom for each sovereign on mounting the throne to send round a commissary to collect the yarlighs and paizahs issued by his predecessor, and those who had them were bound to give them up on pain of punishment. But these commissaries were generally bribed, as we have seen, and did not return with more than one in a hundred of the various yarlighs which had been issued. It often happened that fresh yarlighs were issued contradicting previous ones, which caused endless litigation, especially as these yarlighs were passed on from father to son.

---

* D'Ohsson, iv. 409-411. † Id., 411-412. ‡ D'Ohsson, iv. 412-413.

Often, too, the secretary whose duty it was to draw up the yarligh was bribed to insert some words extending its authority, or especially favouring the person to whom it was given. So many contradictory yarlighs and paizahs were current that the Mongol judges (*yargujis*) and administrative authorities (*hakims*) found matters too complicated for them to decide. Thus people took the law into their own hands, and murders increased greatly. Ghazan, by a special order, annulled all yarlighs and paizahs which were current, ordered the governors of the provinces to disregard them when presented, and cancelled all those issued in the first three years of his reign; for at first, in order to conciliate a certain number of friends, he was obliged to confirm the former yarlighs, while afterwards, Nuruz, the vizier Sadr ud din, and other ministers, abused their power and issued a multitude of these documents. None were in future to be regarded which were issued before the order above named. Thus only those who had a good title to these documents, and who therefore had no difficulty in getting them renewed, retained them.*

Persia had been terribly devastated by the invasions of Jingis Khan and those who came after him. The borders of the Euphrates and Tigris were similarly laid waste. From the Oxus to the Syrian borders were almost continuous ruins, and lands lying fallow. If rulers like Khulagu, Abaka, Arghun, and Gaikhatu (says Rashid) had the desire to build, and did in fact proceed to build, palaces at Alatagh, Arminiya, Sukurluk, Nejas, Khojan, Zindan, and Mansuriah (in Arran), or if they wished to build a bazaar, to found a town, or open a canal, they merely ruined the country in doing so. They spent immense sums without carrying out their object, and thus we are told that barely one house in ten was inhabited in a large number of towns. The land was largely uncultivated, whether belonging to the public domain or in private hands; no one dared cultivate it, for fear he should be dispossessed of the product of his labour after devoting his money and energies to it. Ghazan felt the necessity of encouraging this kind of enterprise, and issued an edict assuring to those who should cultivate the land the fruits of their toil. The public domains which had been waste so long were given to anyone who would undertake to till them, without charging any dues for the first year, and in the second only a third, one-half, or two-thirds of the ordinary charges, according as the state of canals which irrigated them rendered them more or less easy to cultivate, and in view of this the domains were divided into three classes. A special divan was appointed to lease out these crown lands and look after the taxes which they were to pay. In regard to lands in private hands, it was decreed that those which had been deserted for a certain number of years might be cultivated by anyone, and if the real

---

* D'Ohsson, iv. 414-416.

owner came and proved his ownership, the treasury was to surrender to him one-half of the taxes and to retain only the other half, the cultivator taking the produce of the land for himself. The treasury officials made repeated reports that the land was untilled and the people reduced to misery because they had neither the necessary bullocks nor seed. These reports had hitherto been neglected, but Ghazan ordered that a certain proportion of the money paid by each farmer of taxes, &c., was to be devoted to buying bullocks, seed, and other necessities for agriculture, and each one was to give a written assurance that the bullocks so bought should be used in his own province, and to encourage agriculture. At the same time it was forbidden to use the people's donkeys for posting. Before this they had been arbitrarily seized and used, and they were seldom recovered. Ghazan forbade his falconers to take the pigeons and fowls of his subjects, saying it was necessary to exact compliance in small matters if bigger ones were to be carried out. If it was not forbidden to take pigeons, why forbid the taking of sheep and oxen? In accordance with this, sportsmen were forbidden to spread their nets in any place where there was a dovecote. Many of the lands thus restored to cultivation were portioned out as military fiefs, thus enabling the army to be rewarded, while a large margin remained over, with the income of which the treasury was replenished.* In regard to these military fiefs, Ghazan introduced important reforms. Before his reign the Mongol soldier had received neither pay nor clothes, nor land nor food. On the contrary (says Rashid), as among all the nomadic nations, a contribution (*coïchur*) of horses, sheep, cattle, pieces of felt, skins, &c., had been made from the whole army, the proceeds of which were made over to the ordus, and the tribes reduced to indigence. Ghazan began by giving a small quantity of wheat to the troops who were habitually cantoned about his residence, which he afterwards increased ; but the State farmers, who were to furnish this grain, were loth to comply, and frequent messengers had to be sent to urge the matter forward, causing great harass to the cultivators. All this cost much money, without benefit to the troops. The various officials, farmers, bukauls, high stewards (or commissaries of food), bitikchis, and idajis (or commissariat officers) were all more or less corrupt, and although the soldiers had the bérats or orders for food in their hand, they could not obtain it, and there were constant disputes with the idajis. At length, after four or five years, weary of the complaints he received from all sides, Ghazan issued a special order on the subject, which also extended the privilege to the whole army. The order issued on this occasion was as follows :—

" Let it be known to the Seyids, khatuns, princes and princesses of the blood, relatives, chiefs of tumans, millenarians, centurions and decurions,

---

* *Id.*, 417-420.

sultans, maliks, bitikchis, and all others our subjects, from the banks of
the Jihun to the frontiers of Egypt,* that our great ancestor, Jingis Khan,
assisted and guided by the Divine power, caused his smallest wish to be
carried out, without allowing a step to be taken away from the path of
obedience. This was why, at the head of his Mongol armies, he
conquered the whole earth, from the East to the West, and filled the page
of history with his great deeds. He left his empire to his sons, of whom
those who governed wisely acquired a great renown, and as this is the sole
advantage we can secure from our transitory voyage of life, we deem it
well to try and secure it in the few years that we shall occupy the throne,
by working for the happiness of the ulusses which have passed under our
rule, so that we may engrave on the leaves of time merits of which an
eternal memory shall be the reward, and to perpetuate the renown of our
justice. It is well kown that in the days of our ancestors the ulusses were
weighed down with taxes and imposts which we have totally abolished,
while the troops used to receive for the most part no gifts of food.
Nevertheless, satisfied with their condition, they served faithfully and
supported patiently the fatigues of distant campaigns. Since God gave
us the kingdom of our fathers we have paid every attention to improving
the condition of the army, and assuring its comfort now and in the future.
It is well known that only a section of the troops have hitherto received
small grants of food, while the greater part of the army has received
nothing from the State. We wish it all to share equally in our munificence, so that it may all show equal zeal and valour in the defence of the
State, of which it is the support. We therefore order that the land
belonging both to our private and to the public domain, both cultivated
and uncultivated, shall be distributed to each soldier in the form of fiefs
(*akta*), under the following conditions :—

"1. The peasants on the lands belonging to the private or the public
domain shall continue to cultivate it, and shall pay to the soldiers all the
contributions in coin (*mal*), in cattle (*coichur*), and otherwise, which they
have hitherto paid to the treasury.

"2. The soldiers are not to appropriate either the land or water belonging
to private people, nor to the vakfs or pious foundations.

"3. In regard to the ruined villages and untilled lands belonging to the
public domain which are comprised within their territory or yurt, and the
soil of which has become grass-grown, they are to cultivate a portion of
it themselves, and to cause the rest to be cultivated by their slaves and
servants. They are to employ their own bullocks and seed, and to claim
what produce the land may give.

"4. Those people who have been absent from ruined villages, or villages given to the soldiery, for a less time than thirty years, and who have not been enrolled in some other district, are to return home, no matter with whom they are living, and if the soldiers find men on their fiefs belonging to some other district they are to send them home. They are not to give asylum on any pretext to those of other districts ; they are not to transport the people from one village to another, on the plea that all are their vassals, but, on the contrary, the people of each district are to cultivate that district. The soldiers must not consider that the peasants have been made over to them with their fiefs as if they were serfs. They have no other rights over the peasants than to see that they cultivate their lands, and to receive the quit-rent and other dues which they have to pay. Nothing more is to be exacted from them, while as to others who are not cultivators, if they pay the soldiers the imposition fixed by the Divan, they are not to be made to work on the land, nor to be treated harshly.

"5. The soldiers are not to usurp the villages neighbouring upon their fiefs on pretence of obtaining water there, and they are to leave them enough pasture for their cattle, donkeys, and sheep.

"6. Having granted these favours to our troops, with the intention of securing the well-being of our subjects, of making them love our memory, and establishing order and justice, and the commanders of tumans, millenarians, centurions, and decurions having in writing (*mojelga*) given us their assurance that they will use all their power to cause justice and good faith to prevail, that they will no longer commit acts of oppression and iniquity as they have done hitherto, they ought to keep their promise, and to extort nothing on any pretence.

"7. The treasury is never to grant assignations on military fiefs, but they are to pay into the treasury a tax of fifty menns, according to the standard of Tebriz, for each military colonist.

"8. When the fiefs in gross, comprising cultivated land, land lying fallow, and running streams, are assigned to the various regiments of 1,000 men, the notables of each canton are to meet the official bitikchi. Thereupon lots are to be drawn, and the whole distributed in this way among the companies of 100 men each, and these again among sets of ten men. The bitikchi, or registrar, is to inscribe separately on the registers the portions assigned to each decuriate and company. A copy of this register is to be deposited at the treasury, and another handed over to each commander of a regiment. Similar registers of smaller portions are to be assigned for custody to the centurions. The registrar is to make an annual inspection, and those soldiers who have neglected to cultivate their lands are to be punished. The fiefs are not to be sold or given away, nor assigned to a sworn friend (*anda kuda*), to an older or younger brother (*aka or ini*), or other relative, nor granted by way of jointure or otherwise, on pain of death.

K *

"9. On the death of a military man, his fief is to pass to one of his sons ; if he have none, to one of his old slaves (*gulam*), and, in default of slaves, to a man chosen from his company. A fief forfeited for misconduct is to be made over by the officers to a person competent to serve, and who is to be inscribed on the register in his place. The registers are to be examined annually. The inspector is not to allow a military man to exact anything beyond the quit-rent, the *coïchur*, and the other payments which are assigned him in the register, and he is to report the cases of extortion he meets with."*

It would seem from clause 8 of these regulations that the military fiefs were granted out to small colonies of soldiers, who tilled and worked them in common, and were not individual holdings.

Ghazan felt the necessity of increasing the army, so that the frontiers when attacked might be more easily defended, and in order that the troops of one frontier should not be marched away to another, thus leaving it unprotected, he formed a kind of special active division, recruited from the territorial army, each hut or tent of several men furnishing a quota of one or two soldiers. He also ordered that when a frontier was menaced the force nearest to it should at once march to the rescue. In regard to the frontiers where there were mountains and defiles, as these could be guarded by infantry he placed there the Tajiks or Persians, who received pay and also fiefs. Hitherto a grant had been made regularly for a body of Persian soldiers, but as they had until now been only men in buckram this had merely profited the officers, who bravely drew the money. Ghazan ordered this to be remedied, and the Persian soldiers in future were organised in bodies of a thousand and a hundred, and were inspected every three months.† Ghazan also increased his own guards (*kul*). Hitherto his *kul*, or regiment of body guards, had numbered only a thousand men, but in future it was raised to two or three thousand. They were reviewed by the Khan annually. He gave them presents, pay, and fiefs. During the wars with the princes of the houses of Juchi and Jagatai many Mongol children were captured, and were sold as slaves by the victors. Ghazan was outraged that the descendants of officers who had served under Jingis Khan in his conquests should be sold as slaves to Persians, and determined to upset this practice, which destroyed the respect formerly entertained for the name Mongol, and humiliated the race in the eyes of strangers. He ordered that all Mongol children offered for sale within his kingdom should be purchased on his own account, and in the course of two or three years a body of nearly 10,000 of these young people was formed, for whose sustenance the district of Meragha was assigned. Ghazan nominated officers over them from the officials of his Court, and made Pulad Ching

---

* D'Ohsson, iv. 420-429. † *Id.*, 429-430.

•Sang general superintendent of them, with the rank of commander of a tuman. This body, which was daily augmented, was attached to the royal guard.*

There were in each province and town many armourers (*ozan*), both Persians and Mongols, who made bows, arrows, quivers, swords, &c., who received annually a salary from the State, and were in return to furnish a certain number of arms. In addition, there were in certain towns regular manufacturers of arms, under the control of officers appointed by the kurujis, and to whom were assigned large grants from the revenues of the province ; but these moneys had, as in so many other cases, been grossly misappropriated and embezzled by the various officials. Ghazan ordered that in future the manufacturers of each article should form a separate company or guild in every town; and that they were to receive no fixed salary, but should be bound to supply a certain number of articles at a fixed price. He said that, although they were his slaves, he wished them to be paid the value of their productions, just as private and independent artisans were who sold theirs in the bazaar. He placed a governor over each of these guilds. The money necessary to pay them all was made a charge on the revenue of a single province, so that it should no longer be needful to send messengers at great cost in all directions, to secure payment for what was supplied. Every year 10,000 complete suits of armour were to be supplied. Hitherto not more than 2,000 had been annually furnished. Ghazan ordered that fifty were to be supplied for his own special use, while he stored his arsenals with many thousands of bows, arrows, and coats of mail. The same system was adopted in regard to saddle and bridle makers, as well as those who made the necessary equipments of the sekurjis (halbardiers or lance-bearers) and idajis. Previously, instead of having them in store, if an article of 50 or 100 dinars value was needed a messenger had to be sent for it.†

Hitherto there had been no regular account kept of the receipts and outgoings of the royal treasury of the Mongol sovereigns. There were a number of treasurers whose duty it was to receive what was paid into the treasury, and to pay it out collectively. The articles in the treasury were so badly taken care of that they were not even put under tents, but exposed to the air, or merely covered with felt. From this mode of guarding them the general method of administration may be guessed. Every time money was paid into the treasury it was customary to give the employés of the treasury who asked for them presents according to their rank. When the cupbearers or butlers, the housekeepers or grooms, brought any food or something to drink, the treasurers, after a consultation, gave them what they asked. Whoever, in fact, went and begged the custodians of the

---

* *Id.*, 430-431.　　† *Id.*, 431-433.

magazines to give him something was pretty sure to get it. These
custodians were also in the habit of making mutual presents, and thus
every year four-fifths of the contents of the treasury were dissipated.
Knowing of this dishonesty, the governors of the provinces, when they
sent their contributions to the treasury, used to obtain by the payment of
a small bribe a receipt for twice the amount they really contributed ; and
although the ushers (*tangauls*) had strict orders to arrest all plunderers,
they were so lax that a culprit was hardly ever seized, and if he was it
was some one against whom they had a spite.

Ghazan began by ordering the various classes of objects in the treasury
to be separately classed.   He caused the Vizier to draw up an inventory
of his precious stones, which he locked up himself in a special case, sealed
it with his own seal, and gave it in charge to the double custody of a
treasurer and of a major domo (*khoja serai*).   Gold coin and the precious
stuffs which had been made in the royal manufactories, or sent as presents
from distant countries, were also confided to the same individuals, after a
list had been made of them by the Vizier, and it was forbidden to dispose
of any of them except under authority of a special order (*pervanch*) of the
Ilkhan himself.   Another treasurer, with a major domo, was intrusted
with the silver coin, and with the robes destined for ordinary gifts.   These
were not to be disposed of except under a pervanch of the Vizier, countersigned with the signature (*nishan*) of the Prince, written with his own
hand, which was to be registered by the Vizier's substitute, and they were
forbidden to take notice of a pervanch which was not countersigned in this
way.   The former section of the treasury was styled *narin* (*i.e.*, small, fine),
the latter *bidun* (*i.e.*, great, thick).   The former distinguished what was
reserved for the Prince, the latter what was destined for the public.   An
inspection was to be made by the Vizier every six months, or every year,
to see that the treasury contained what it ought to do.   Formerly, the
treasurers used to lend money to the grandees, or their own friends.   It
was now forbidden to lend anything without a royal ordinance (*pervanch*).
All the fabrics deposited in the treasury were marked with a special
mark, so that they could not be changed for others.   It was ordered that
the various housekeepers employed in the treasury should in future
limit themselves to their proper duties, and should not meddle with
the disposition of the objects, which duty was to be performed by the four
officials just named, and by them alone.   They were to be paid 2 per
cent. on the revenues sent in from the provinces, and were to exact
nothing more from anyone.   Ghazan also instituted a third department,
under a major domo (*khoja serai*), who was to receive a tithe of all the
gold and the fabrics sent in to the treasury, which he was to distribute in
charity.   When the Prince changed his residence, at the approach of
summer or winter, he went himself to the treasury to order what things
were to be transported with him.   The Vizier drew up a list of the trunks

which remained at Tebriz, and the magazine was then sealed. Rashid adds that no prince made so many presents of gold and robes as Ghazan.*

Formerly, the expenditure of the moneys devoted to providing the Ilkhan's table was a perpetual subject of contest among the officials charged with this duty (*Bitikjian Idaji*), who were always making mutual accusations. As in other cases, the moneys allotted for this purpose were appropriated by various officers, and otherwise dissipated, and consequently those whose duty it was to supply the table had to borrow money at heavy interest, and to pay exorbitantly for what they purchased; so that, for instance, wine whose normal price was fixed at ten dinars for every hundred menns, and which under a better administration ought not to have cost more than five, was bought sometimes at twenty and even forty dinars, and often the royal table was largely in debt to the various dealers, who begged in vain for the payment of their accounts. Ghazan ordered that in future the moneys necessary to supply his table were to be furnished by the treasury six months in advance, so that things might be bought with ready money, and in this way order was restored in the kitchens and other domestic offices.†

The camels and sheep which belonged to the royal domain were formerly under the inspection of certain officers called *kanjis*, who were uncontrolled, and although the pastures were abundant, and the herdsmen numerous and exempted from the payment of all public taxes in view of their occupation, instead of the flocks having increased a hundred-fold, it was found that they had diminished almost to nothing. Ghazan fixed the number of sheep which the kanjis were bound to provide annually. The baggage camels were confided to special officials, and this part of the public service was so well organised (says Rashid) that no sovereign either Mongol or Mussulman, ever had so many beautiful camels to transport his baggage, while their pack saddles and equipment were extremely elegant.‡ The Ilkhan's herds of sheep had a particular name.§

Formerly, orders were issued annually to the falconers (*kushjis*, *i.e.*, birdcatchers), and those who had to procure hunting leopards (*barsjis* or *parsjis*), when they were to capture the falcons and leopards, which they were to hand over to the officials of the hunt. They were paid by assignations from the revenue, which they exacted from the most fruitful sources, and by liberal blows if necessary, making also outrageous levies for food and forage; and when *en route* for the capital with the animals they had in charge, they requisitioned post horses liberally for themselves and their baggage from the various towns, post stations, and cantonments of troops on the way.

---

* D'Ohsson, iv. 437.　† *Id.*, 437-439.　‡ *Id.*, 439-440.　§ Ilkhans, ii. 157.

They freely gave away many of the animals they had charge of, and the few leopards and falcons they eventually accounted for cost the public revenue a very large sum, besides the private rapine we have mentioned. A person who succeeded in securing an animal useful for the chase, either by trapping it or otherwise, demanded a patent as *terkhan* (*i.e.*, one exempting him from all taxes), armed with which he proceeded to persecute the much-enduring people. There were many such patents issued, and each holder had an entourage of a crowd of dependants. The falconers and officials of the hunt, who lived at the Court, had in their wake a number of palfreymen (*kuteljis*), muleteers, cameleers, &c., all of whom had their girdles ornamented with feathers, bunches of feathers on their heads, and carried poles, professedly for the birds of prey to perch upon, with which they belaboured those whom they met before addressing them. If they met anyone who wore owls' feathers in his turban or cap they took them from him, on the pretence that everybody was not allowed to wear them, and those who had to pass the tents or huts where the huntsmen lived were generally plundered. They appropriated sheep and fowls for their own food and that of their charges from the various villages, and straw and barley for their horses, &c. They seized on post horses, which they sold, and if they saw good asses, kept them for themselves, and to create terror they sometimes seized a proprietor and cut off his beard, while on every side vagabonds and ne'er-do-weels attached themselves to them. They were, in fact, organised brigands.

Ghazan set to work to reform these abuses. He ordered that no more than 1,000 falcons and 300 leopards should be annually supplied to the Court, that a list of the authorised falconers and hunting-leopard men should be prepared, with the name of the province where they were to live, and that no one else was to be considered as of their craft. The price to be paid for hunting animals, trained or untrained, was fixed, and also the pay of the huntsmen, which was to be according to the number of animals of which they had charge. They were each given yarlighs, with a golden seal (*altun tamgha*) attached, and were given proper salaries on condition that in travelling they made no exactions of post horses, provisions, &c. This regulation was ordered to be published by proclamation in the various provinces. When a calculation was made of the cost of this part of the public service it was found that although three times the number of animals were now supplied, it was at half the cost, not counting the exactions which had been made from individuals, which were of course incalculable. In regard to the falconers who lived at the Court, it was ordered that their wages, and what was needed for the food, &c., of the animals in their charge, was to be paid in advance, so that there was no pretext for their making exactions. It was ordered that whenever they were sent to a distance to exercise their animals they were to be supplied with post horses (*ulagh*), and were to be granted assignations, sealed with a golden seal, upon the revenues of the province to which they were sent, greater or smaller, according to the distance or the season. To prevent all wrongdoing, fowls and pigeons, kept in cages, were provided for the food of the royal falcons. It having come to his ears that some of the huntsmen had made exactions in excess of the sum permitted by their bérats, Ghazan sent a special commissioner to make inquiries, and those who had been guilty received seventy-seven strokes with the bastinado. "Frightened by this example," says Rashid, "their companions ceased their ill deeds. It is rare now that a falconer or a huntsman gives rise to just complaints, and although the wolf never becomes a sheep, abuses have considerably diminished."*

The Mussulman kadhis were all subject to the chief judge, and the Mongol judges to the great yarghuji.† Ghazan issued instructions to his judges in the following edict :—

"In the name of the most clement and merciful God.

"By the divine power and under the auspices of the Muhammedan religion.

"Ordinance of Mahmud Ghazan.

"Let it be known to the commandant (bashkak), the prefect (malik), and the other functionaries who exercise authority in our name in such a province, that we have conferred upon X the office of judge in such a place, with its dependencies, so that he may judge and decide causes in his jurisdiction according to the law of religion, and that he may carefully protect the property of orphans and those who are absent. With these duties no one else has any right to interfere, nor yet to release from prison whoever has been legally condemned. The grand yarligh having laid it down that the kadhis, doctors of the law, and the descendants of Ali should pay neither *kolan* nor *koichur* (*i.e.*, taxes and tithes), we order that they be exempt, and that in future they are not to be called upon to furnish post horses (*ulagh*) or food (*sussun*). That neither Turks (couriers) nor *ilchis* (envoys) are to be billeted upon them, and that the full amount of their salaries is to be punctually paid them every year. Anyone failing either in word or action to pay proper respect to the kadhi is to be punished according to his fault ; while, on the other hand, we enjoin the kadhi, according to his written engagement (*mojelga*), to refrain from receiving anything from anybody for the exercise of his functions. When a new contract is made the old one must be produced in court and put into the *thas* (basin) of justice, so that the water may efface the writing. Pretensions which have not been urged for thirty years, and contracts thirty years old, are not to be admitted, but to be remitted to the basin (*i.e.*, washed out).‡ Anyone convicted of having used violence to another in order to compel him to do an illegal act, is to

---

* D'Ohsson, iv. 440-445.   † Ilkhans, ii. 163.
‡ This is surely a very extraordinary anticipation of our statutes of limitation.

have his beard cut, to be promenaded about the town on a cow, and to be severely chastised. Attestations in common, *i.e*, made collectively (*mahzars*), are abolished, and if any exist they must be washed out. If parties to a suit appear in court with people whose aid and protection they have invoked, their case is not to be heard until these patrons have withdrawn—the cause is in no case to be commenced in their presence.

" In causes between two Mongols, between a Mongol and a Mussulman, and others difficult to decide, we order that a council of justice (*divan ul mozalim* = the divan of grievances) comprising the commandants, prefects, treasury officers, kadhis, doctors of laws, and descendants of Ali shall meet two days in each month in the great mosque, decide the matters together, judging offences according to the Muhammedan law, and signing the sentence jointly, so that no one in future shall be able to disturb the decision.

"When a piece of land is in dispute, neither the mother nor grand-mother (of the Ilkhan) nor his sons—nor the khatuns, the princes, princesses, the relatives of the ruler, the chiefs of tumans, millenarians, centurions or decurions, nor the ordinary Mongols, nor the officers of the Great Divan, nor the kadhis, descendants of Ali, doctors of the law, sheikhs, nor mayors are to purchase it. The kadhis are to take care to prepare no contracts transferring to any of these persons the land in dispute, and if they learn that others have done so they are to exert themselves to get it annulled.

"The seal of the kadhi is to be charged at nineteen dinars and a half and no more.

" If the districts belonging to the jurisdiction of X are too far away from his head-quarters, and he deems it well to place deputies there, he is to take care to select magistrates worthy of confidence, and to impose upon them the prescribed obligation. He must look over all they do monthly to secure that justice is being done according to our wishes. He may authorise them to draw up contracts and to decide causes, and each month they are to send him a report of what they have done."

The edict seems then to say that the deputy kadhis placed in the rural communes *were not* to try causes, give judgment, or prepare contracts of sale, but were to confine themselves to reciting the Friday prayer or khutbeh, preparing deeds of partition on succession to property, and deeds securing dowry (*sadak namch*) ; while if difficult cases came before them they were to remit them for trial to the kadhi of the principal town. Perhaps this means that unless specially authorised, the deputies were only to try the limited cases, and do the less important work just mentioned. The kadhi was to have a person worthy of confidence to date contracts and to keep an exact register, so that if anyone sold or mortgaged a property, and wished to sell or mortgage it again, the first transaction might be easily verified. Any person guilty of fraud in these transactions was to have his beard cut off, and to be promenaded about the town, while a registrar found guilty of abetting him was to be punished with death.

In this edict Ghazan warns the judges against the practice which had grown up of setting up false titles to property. These false titles were thus obtained : The original person who first tilled the land, or who inherited it after it had been cultivated, obtained from the proper authorities a number of documents of title. He then made over his property to some one else, who sold it to a third, and so on ; but the original documents were meanwhile retained by the first owner ; and when presently, in many cases through carelessness, or the troubled times, or other causes, it was thought that the person in possession had no documents to show his title, these original deeds were produced, and with them a number of suborned witnesses, and the kadhi having nothing brought before him on the other side to weigh against these official • documents, declared them genuine, and, thus armed, their possessor proceeded to oust the real proprietor. Similar practices had long before attracted the notice of the famous Seljuk sultan, Malik Shah, and his vizier, Nizam ul Mulk, and he had in consequence issued an order by which documents of title which had been in abeyance for more than thirty years were not to be received in evidence. This order had been sent to all the muftis in Khorasan, Irak, and the district of Baghdad, in order that they might give their *fethva*, or judicial decision, about it, which was sent to the Khalif for his imprimatur. If such things took place (says Rashid) in the days when magistrates were just and enlightened men, how much more during the domination of the grandsons of Jingis Khan? In Persia the Mongols only recognised those who followed the profession of the law by their turbans and costume, and they had no knowledge of jurisprudence or equity. It was natural, therefore, that ignorant men should usurp the distinctive insignia of the magistrature, and by presents, &c., secure the patronage of some Mongol grandee, and thus obtain judicial offices, which led to the greater part of the more worthy magistrates resigning their posts, as they could not bear to have these men their equals. Presently, the evil deeds and injustices of these parvenus brought the whole profession of the law into contempt, and with it the Muhammedan faith also suffered. The Mongol commanders became the patrons of these judges, and presented or deprived them at their pleasure, and in some cases received money for them. It was chiefly during the reign of Gaikhatu and his minister, Sadr ud din, that these things happened.* Sadr ud din nominated his own brother to the office of chief judge (*Kadhi ul Kodhat*). Sheikh Mahmud was at that time head of the sheikhs (*Sheikh ul Moshaikh*). They sold the judicial

---

* We must always remember that Sadr ud din was a Jew, and therefore especially hateful to our authority, Rashid ud din, who had been accused of being a Jew himself.

offices. In consequence of fraudulent documents, those who had estates were in greater danger than if they had a hundred enemies, for dishonest men, dying of hunger, would not scruple, when armed with obsolete titles and false witnesses, to attack the most respectable people. When judicial offices were objects of traffic these same unworthy processes were much encouraged by the judges, who stimulated the aggressors, and delayed judgment for months and even years. Meanwhile they appropriated the income of the estate, and made both sides pay black mail. Many estates were thus for a long time the subject matter of litigation, and the unfortunate proprietor spent in law more annually than his income from his property ; while those who attacked the judges received presents and made themselves feared, and a regular system of levying black mail on the proprietors, with the alternative of being dragged into the courts, became the custom. These pestilent people sought out others who happened to hold any old title deeds, or in other cases fabricated them. Ordinances of former sovereigns were produced—titles dating 150 years back—and although these required to be attested by witnesses, the claimant nevertheless proceeded, having first secured a powerful Mongol as a patron. The threatened proprietor was obliged to find a protector too, and thus the area of litigation and of dispute widened. Every proprietor, we are told, was threatened either in his estate or his honour.

The Tajik viziers of Khulagu had brought the order of Malik Shah before his attention, and he had issued a corresponding edict forbidding the production of deeds more than thirty years old in evidence, which edict had been renewed by Abaka, Arghun, Ahmed, and Gaikhatu ; but these edicts were not obeyed, the interest of the officials, whose object was to obtain the properties of the proprietors at a low price for themselves, being to throw obstacles in the way of carrying it out. When Ghazan issued his edict, he had it specially drawn up with the help of the most learned kadhis, with full explanations of the principles upon which it professed to proceed, and also the various regulations and laws which had been passed on the subject.

This edict, prohibiting the reception in evidence of documents dated more than thirty years before, is dated from Keshaf, near Mosul, the 29th of March, 1300. It concludes with these words : " If a powerful man urges a kadhi to act in opposition to this edict, and if the kadhi sends me his name, I will see that he is punished in such a manner as will serve as an example." This edict also contains the formula which each kadhi was to sign. It enjoins the magistrates to decide cases according to the law, and with strict equity. They were to examine most critically old contracts and deeds, and to verify the witnesses, so that false documents might not be palmed upon them, and the kadhi bound himself to perform his duties strictly, on pain of being punished or deposed.

Another edict provided that before the sale of immovable property, &c., the right of the vendor should be properly made out. It ran thus :—

" In the name of God, &c. By the divine power, &c.

"Order of the Sultan Mahmud Khan.

"Let it be known to the commanders, prefects, judges, officers of the treasury, governors, notables, proprietors, and to all our subjects in general that according to the word of David 'we have appointed you our lieutenant upon earth. Judge between men with equity,' and according to the sentence of the Prophet 'An hour of justice is worth more than seventy years of prayer.' We exert all our care to secure the wellbeing of our people, and desire that the shadow of our justice may be generally spread everywhere, so that the powerful may not be able to oppress the feeble, that right may not be undone by wrong, and that the greater part of the differences among private individuals may disappear. Among the evils which afflict our kingdom are the numerous disputes between our subjects about old titles and based upon old contracts. In some cases the vendor who has more than one title deed retains a portion, and after selling his property has the dishonesty to produce these documents, and to lay claim to what he has already been paid for, and succeeds in getting witnesses who have been won over by his solemn declaration, or are prepared to suborn themselves. If the vendor does not do this his descendants sometimes do it, and produce this really false evidence either knowingly or innocently. It is natural that when these old documents are brought before the kadhi duly attested, that he should pronounce them genuine. How is he to know that the rights claimed under them have been already legally parted with. In order therefore to prevent these miscarriages, it is decreed that on a sale of land the vendor and the purchaser are to appear before the kadhi. The vendor is then to produce his title, and witnesses to swear that they have never known anyone raise a legal claim to the estate. The titles thus produced are then to be washed in water and cancelled. If the vendor has no documents, the witnesses must attest that he has been since such a date in possession of the property, while he is to declare that he has no title deeds, and that if any are afterwards found they will be considered as null and void. A document is then to be drawn up affirming his right to the property. The witnesses are to attest it, and the kadhi is also to add his attestation. Below this is to be written the contract of sale. If in future any other title deeds relating to this property are produced by the vendor or his heir, or anyone else, the kadhi is not to admit them, but he is to take possession of them, and immerse them in water. If the owner of the document refuses to give it up to the kadhi, he is to report the matter to the commandant of the town, who is to take it and wash it in the court. The contracts for the sale of immovable property are to be prepared by the registrar of the court. The kadhi is to have beside him a basin of water on a stool. This basin is to be styled the basin of justice (*tass-i-adl*). When a dispute of the nature above-mentioned has been decided, the old titles produced and rendered obsolete by the decision are to be washed.*

" If a witness to a contract of sale or mortgage eventually raises pretensions to the thing sold or hypothecated, or if a person tries to sell a *part of* what he has already sold or mortgaged to another, he is to have his beard cut, and to be promenaded about the town on a donkey ; while a person who tries to sell a property which he has already sold or mortgaged, is to be put to death."

The kadhis were not to accept anything for settling contracts or deciding causes, but were to be content with the salaries they received. The registrar was to receive a dirhem for each document (*heujet*) which he prepared, if the matter it referred to was of the value of 100 dinars, and one dinar if it exceeded 100 dinars, and was to take nothing more. Every police officer (*vakil*) who took money from both sides was to be scourged, deprived of his post, and to have his beard cut, and for a second offence was to be ridden through the town on a donkey, with his beard cut off. " We desire," says Ghazan, " that the commandant, or the prefect, in each town is to inform the kadhis that they are to enter into a written undertaking in regard to performing their duties." Ghazan appointed special commissaries, who were to search out and report to him the names of those who had made a trade of attacking the titles of proprietors by means of false documents, and to take care that they did not shield themselves from public justice. All those who were thus pointed out to him were sent to the capital, and were put to death.

Ghazan introduced fresh regulations in regard to the coinage. Up to this time several types of coins had been current in Persia. " It is not long ago," says Rashid ud din, " that the princes of Rum, Fars, Kerman, Georgia, and Mardin still enjoyed the right of coinage, and coined money of various standards." Again, although according to the edicts of Arghun and Gaikhatu the silver coins were to contain nine-tenths of silver, there was really only a proportion of four-fifths. Outside the Mongol world things were very bad. The coins of Rum, which were of higher value than those of other countries, were so reduced that where there ought to have been ten dinars worth of silver there was only two, that is, one-fifth ; all the rest was copper, besides which they were clipped. · This difference of intrinsic value in the coins of the two countries had to be taken into account in various trade transactions. He who bought merchandise lost 10 or even 20 per cent. upon the transaction, while there was some difficulty in passing off the money in the villages and the military cantonments, where the very ignorance of the people as to the amount of alloy in the coins made them more suspicious.

---

* D'Ohsson explains this reference to washing by saying that the soluble ink used by Orientals is removed from the polished cotton paper used by Orientals when a document is dipped in water.

Ghazan had a new die (*sikkeh*) made, with a mark (*nishan*) which could not be easily counterfeited. The gold and silver coin was re-struck, so that in future they might bear the name of God and of the Prophet. They also bore the name of Ghazan. In Georgia the names of God and the Prophet had never been placed on the coins; they were now to appear there also, according to Rashid ud din, since no other coins than those so stamped had currency in Persia. In regard to the standard, Ghazan said that if he permitted any alloy to be mixed with the gold and silver in the coins, as it was in those of the Khalifs of Egypt and of Africa, it would speedily happen that more and more would be so mixed with it, and his inspectors would either not notice the bad practice, or would be bribed to keep their eyes shut. It was much better that the coin should be struck without alloy, so that when the gold coin was tested with borax, and the silver was either tried with mercury or in the fire, the sophistication might be easily detected. He ordered that the gold pieces of Hormuz, which were not of greater value than those of Africa, and other baser gold coins, were to pass current at a lower value, so that the bankers might be induced to buy and melt them down into ingots, and it followed that in the course of a twelvemonth there was not a *miskal* of base gold coin to be found in the country. Formerly gold and silver were rare in the market place, and when a little was forthcoming there was a rush for it. The quantity had diminished in consequence of the increase in the use of gold tissues, &c., during the Mongol domination, as well as in consequence of the exportation of gold to India; but now (says Rashid) it circulates abundantly in the bazaars, even in the hands of the peasants. Ghazan wished the silver dinar to weigh three miskals. Pieces of gold (*durusthai thila*) were struck of the weight of a hundred miskals, upon which his name was engraved in the letters of different countries, so that it might be known everywhere that he had struck these coins. They bore, in addition, verses from the Koran and the names of the twelve imaums. The coins were so beautiful that those who obtained one had not the heart to part with it. Ghazan liked to give these medallions—for so they really were—as presents, so that they might be known to foreigners. Pachymeres speaks of the gold coins of Ghazan as having been of exquisite purity. This kind of money, he adds, was called *kazaneos*.*

There were formerly a great number of weights and measures current; they differed even in the various districts of the same province. Commerce was neglected in favour of the mere transporting of objects from one district to another, the difference of weight in the two districts affording a profit in itself. From this cause some materials became very scarce, and even unprocurable in certain countries. There were two or

---

* Stritter, iii. 1092.

three kinds of measure (*kufiz*) current in each village. The villagers employed the longest one among themselves, but in trading with strangers they used the shortest, and whether the latter knew the fact or no they had small redress, since the natives swore boldly together that this was the legal standard. The provisions, which ought to have been supplied to the troops by the *kuban* of 100 menns, were really weighed to them by the *kuban* of 70, or 60, and even of fewer menns, while powerful people exacted the full measure by the liberal use of the bastinado. This variety of weights was a constant subject of disputes.

Ghazan issued an order on the subject, which, after the usual initiatory formula, went on to say that in the markets of the Ordu and of the towns each one used at his pleasure weights made of stone, bone, iron, &c., and increased or diminished them arbitrarily. "We order that in all our kingdom, from the River Amu (*i.e.*, the Oxus) to the frontier of Egypt, the weights and measures shall be tested and verified ; that they be made of iron, and be properly marked."

1. The weights of gold and silver coins were to be according to the standard of Tebriz, so that money should not in future be transferred from place to place to make a profit out of the mere difference of weight. In future it was to be everywhere of the same weight, as it was of the same value. Fakhr ud din and Behai ud din, of Khorasan, were charged to make special weights for the gold and silver coins, which were to be of an octagon shape. They were to appoint two of their people for each province to revise the weights there, who were to be assisted by an expert (*amin*), and to carry out their work in the presence of the inspector of the markets (*mohtesib*). In consequence of this order various private individuals were to have iron weights, made similar to the patterns furnished by Fakhr ud din and Behai ud din, which were to be taken to the experts in each province, just mentioned, so that they might be verified, and that they might attach their marks to them. The counterfeiting of these marks was punishable with death.

2. All who were provided with these stamped weights were to be registered, and the weights were to be verified monthly. Anyone who had unjust weights, or counterfeited the official mark, or who in buying or selling used weights which were not marked, were to be brought before the commandant (*shahneh*), who was to have him punished as ordered by the Khan's ordinance.

3. The weights used for selling merchandise were also to be fashioned after a type of an octagon shape, in iron, and were to be duly marked and verified by the same experts. These weights were to consist of a series of eleven, from 10 menns down to a drachma, in this order: 10 menns, 5, 2, 1, ½, ¼, ⅛ ; 10 drachmas, 5, 2, 1. For objects weighing more than 10 menns the tamgajis were to make kubans, or hundredweights, weighing 100 menns.

4. As there was in each province a variety of measures under the names of *kil, kofiz, jerib, tugar,* &c., and each one in fact used his own measure, and as the Mongol soldiers, merchants, and strangers who went to receive the provisions which had been apportioned them by the treasury, or to buy things, had constant quarrels with the inhabitants, the consequence being that might generally prevailed, it was accordingly decreed that there should be a uniform standard of weights (*kilch*), viz., that of Tebriz, of which the unit weighed ten menns, each menn being of 260 drachmas. Ten of these kilehs were to go to a tugar. Different kinds of grain, such as wheat, barley, rice, sesamé, and millet were each to be measured by a separate measure, which was in every case, however, to be based on a standard quantity of grain weighing ten menns of Tebriz. Each of these measures was to be marked as the kileh of such a grain, and the experts already mentioned were to attach their marks to them in the presence of the inspector of markets (*mohtesib*). They were to be inspected every month in the town and country districts; anyone found with an unmarked measure was to be taken before the commandant, and, if found guilty, was to have his hand cut off or to pay a penalty.

The measure of liquids used for the ordus, and for distributing in presents, was to contain five péimanés, weighing fifty menns, while that used for feasts (*thui*) was to be of the capacity of four péimanés, weighing forty menns. All the ell measures (*guez*) used in measuring cloth for sale were to be of the length of the guez of Tebriz, excepting, however, the measure used in Rum, which was very different. The two sides of the ell measures were to be marked with a mark composed by the two officials, Fakhr ud din and Behai ud din. These measures were also to be periodically inspected by the experts already mentioned.

The kingdom, through the recent ill-government, was overrun with highway robbers, Mongols, Tajiks (Persians), Kurds, and Shuls, who were joined by runaway slaves and other Bohemians. The peasants, as in Southern Italy and in Ireland, acting as their guides and their spies, informed them of the route taken by travellers. If a brigand fell into the hands of justice he easily found powerful friends, as he does still in Calabria, and these friends urged that it was not right to put to death *a man who was so brave.* By an old decree those who travelled together were to make a joint defence against any robber band which attacked them; but the brigands, who knew their victims well, used to shout, like the chivalrous Jack Sheppard, that they did not want to molest those who had nothing, who consequently withdrew, leaving their richer brethren in difficulties. The robbers became so bold that they attacked travellers close to the military cantonments, the towns, and villages. They had friends among the peasants and the nomades, and those who denounced them were visited with vengeance. The peasants, and even the village mayors, were so closely bound up with the robbers that they supplied them with everything they needed, and gave them asylum in their houses. Their friends in the towns disposed of the stolen booty for them, and they sometimes spent a month or two in the towns with their accomplices, thus enjoying the fruits of their rapine. The Tangauls (Tetkauls of Von Hammer), whose duty it was to protect the great routes, only aggravated the evils. They stopped caravans on pretence of searching for thieves, and thus gave the latter timely warning to escape or plant ambuscades. Instead of pursuing the robbers they levied black mail themselves on travellers, who feared their protectors more than the thieves, since the latter were only encountered occasionally and at haphazard ; while the Tangauls were sure to be met with at every station, and many caravans, rather than run the risk of being plundered by them, adopted unfrequented and difficult roads.

To remedy these matters Ghazan ordered, first, that anyone abandoning his companions when attacked by robbers was to be pursued, and to forfeit his goods and life. The cantonment or village near which a robbery occurred was to be made answerable, especially if it had been warned. Anyone convicted of connivance with the robbers was to be put to death. Ghazan ordered Inkuli, one of the officers of his household, known for his integrity, to carry out this scheme. Many robbers were captured and punished, and the few who escaped, forsook their occupation. Those who denounced them were created *terkhans*, and Inkuli was rewarded for his zeal with all that was recovered from the robbers. Ghazan ordered that the Tangauls should only be posted in dangerous places, or where the caravans needed instructions as to the route they ought to take. They were to be paid a stated fee of half an akchi for every four mules or pair of laden camels, and no more. Nothing was to be paid for animals without burdens, or carrying only provisions. The places where Tangauls were to be stationed were marked with stone columns, on which were inscribed the number of the guards, the duties of their chief (*bashkak*), and the dues to be paid by travellers. These small monuments were called tables of justice. They prevented the exercise of the office of Tangaul by unauthorised persons, which had hitherto prevailed. The guards who failed to arrest the robbers after a robbery committed near their post, had to make good the property stolen. They entered into written engagements to fulfil these conditions. If a caravan wished to halt near a village or military cantonment, it was to make inquiry if there were robbers in the neighbourhood. If the answer was yes, the caravan was to be permitted to enter the village for safety, and no one was to prevent it. If the answer was no, and in consequence the caravan stopped in the open country, and robbery ensued, they were to answer for it. This order was not made applicable to towns, because of the great difficulties in such cases. The amir Buralghi was appointed head of the Tangauls. When he inspected the rolls he found that nearly

10,000 men with their officers were engaged in protecting the lives and goods of travellers.*

As drunkenness and consequent quarrels and murders prevailed largely, Ghazan issued an order, in which he said : "Our legislator and other heavenly messengers have forbidden the use of wine, nevertheless the prohibition does not prevent its use. An absolute interdiction on our own part would have no better result. We merely order, therefore, that anyone found drunk on a public road is to be stripped naked, and tied to a tree in the midst of the public square, as an example." But he forbade private houses being entered for the purpose of looking for drunkards, on account of the abuses which his officials might be tempted to make.† In the great towns houses for prostitution had grown up close to the mosques, monasteries, and private dwellings, and as those who kept them paid a higher price for young slave girls than private individuals, and therefore commanded the market, much to the horror of such of these girls as had any feelings of chastity and self-respect, Ghazan said that all such establishments ought to be closed, as contrary to religion and morals ; but as they had been tolerated for a long time it was not possible to abolish them suddenly. He, however, forbade any girls who objected to this mode of life being sold to the keepers of such houses. Those who had objection and were nevertheless in houses of ill-fame were to be liberated, the State paying their ransom, and they were then to be married to whoever wished to marry them.‡

Formerly, a well-dressed man (says Rashid) could not traverse the bazaars without a crowd of muleteers, cameleers, or servants pressing him for money for their needs or pleasures, and if this were refused he was insulted, and sometimes even struck. These people collected in groups in the streets and markets, and scarcely was a traveller free from one gang than he encountered another. They were generally in the service of the khatuns, princes of the blood, and amirs. On solemn days, as those of Bairam and Nevruz, they decked out their mules and camels, and took them to the doors of people of mark. If the person in charge were at home they extorted money from him by repeated and energetic solicitation. If he were away they seized what they could and pawned it, and when the proprietor tried to recover his own he had to redeem it at a large price, and to listen to much insulting language. These scenes were renewed annually during the five days preceding and succeeding the festivals, and it became the fashion for officials and others to imitate the custom, and to pass for the nonce as muleteers, cameleers, &c., and to go about with animals dressed out, levying their contributions, until decent people avoided the streets and bazaars, while shopkeepers were made

---

* D'Ohsson, iv. 470-474.   † Id., 474.   ‡ Id., 474-475.

victims of the same artful system of compulsory almsgiving. Ghazan
forbade such exactions on pain of death, and the guards were ordered to
strike with their maces any camels and mules on the head and legs, when
on feast days they heard the sound of their bells, and also to punish their
conductors. This order caused the nuisance to cease.*

Ghazan's financial reforms enabled him to distribute alms generously.
Twenty tumans (i.e., 200,000 dinars) were distributed in this way annually,
of which Shiraz furnished 40,000. Funds were set apart for the pilgrims
to Mecca, and by a yarligh issued in 1300 the Seyids, sheikhs, imaums,
and sacristans at Mecca were secured all their privileges. The young
amir Kotlokia was appointed general conductor of the pilgrim caravans,
and for their safety a guard of a thousand horsemen, with officers, drums,
kettle-drums, and standards, was set apart. Ghazan sent thither himself
a cover for the Kaaba, embroidered with all the titles of the Ilkhan,
and an imperial camel-sedan, in describing which Wassaf has an opportunity for pouring out his hyperbolic phrases with gusto. Twelve gold
tumans were set apart for the pensions of the Arab sheikhs at Mecca
and Medina. The camel-sedan still sent annually by the Turkish Sultan
seems, like a good many other Turkish institutions, to be derived from
Ghazan's example, which itself was doubtless an imitation of older
Mussulman sovereigns, like the Sultans of Egypt.†

Von Hammer remarks that one of the great causes of corruption and
of confusion in the finances was the double year, founded respectively on
a lunar and a solar basis. The difference between the two became so
great that at the end of the seventh century of the hejira there was a gap
of nine years between the two reckonings, causing constant strife between
the tax collectors and their victims. Ghazan therefore determined to
rectify the calendar and to fix a new era, which commenced on the 13th
Rejeb 701 (i.e., 14th March, 1302),‡ and was known as the Ilkhanian or
Ghazanian era. We shall have more to say about it in the accompanying
note. Ghazan left no son, but had three daughters —Otoghloghshah,
Aljui, and Oljai, who married Bestam.

Note 1.—Ghazan was a great builder, and a large number of his foundations
are recorded. Hitherto the Mongol sovereigns had been buried in the manner
prescribed by old custom, in a retired place which was kept secret from the
people, was planted over with a wood and access forbidden. Ghazan,
who had become a faithful Mussulman, followed the practice of Mussulman
sovereigns. He had visited many tombs of saints in various parts of Persia,
and is said to have declared that a man who died in a state of grace, and
whose tomb was thus revered, was happier, though dead, than living men.
' That I may have some part with these holy men I will plant some pious

---

* D'Ohsson, iv. 475-477.   † Ilkhans, ii. 172-173.   ‡ Id., 174-175.

foundations about my own tomb, which will perhaps secure me the divine pity."
He accordingly laid the foundations of some of these establishments at Shenb,
two miles to the south of Tebriz, which was thenceforward known as Sham
Ghazan or Shenb Ghazan. They were completed in the course of a few
years. These buildings, which were on a greater scale than the famous
cupola of Sultan Sanjar, the Seljuk, at Merv, which passed among the
Mussulmans as the largest of their buildings, consisted of (1) a mausoleum.
The foundations of this were laid in the third year of his reign. Its walls were
of the thickness of thirty-three bricks, each brick weighing ten menns.
Fourteen thousand four hundred workmen were engaged upon its construction,
of whom 13,000 were continually employed, and 1,400 assisted. The height of
the building, as far as the vault of the cupola, was 130 ells. Within three years it
rose to a height of 80 ells. The pediment was 10 ells high, and the cornice
also 10 ells. The perpendicular measure of the cupola was 40 ells, its
circumference 1,530 ells. It formed a dodecagon, and on it was represented
the signs of the zodiac.* The mausoleum was girdled about with a golden
inscription, composed by Wassaf in his usual exaggerated language. It
terminated with these phrases: "Thanks be to God that he has permitted him
(*i.e.*, Ghazan) to build these beneficent buildings, and to found these all-
embracing charities, and to cover over the traces of decay with these high-
reaching and deep-sinking gardens, stretching widely like heaven's blue
vault, before which the pyramids must hide away, before which the two
eagles of heaven, the rising and setting one (*i.e.*, the zodiacal signs of the
swan and of lyra) must writhe like the crab and scorpion. Whose battlements, like those of heaven, rise stage upon stage, whose lamps shed their
light like the sun and moon, whose summit is high as Arcturos, whose diameter
is like that of the Pole of Heaven, which runs like an Eden balcony from the
Garden of Paradise, where dwell pity and kindness and Rizwan, the guardian
of Paradise." Among the many objects of gold which it contained was a
great lamp, which weighed 1,000 miskals. For the decorating of its walls with
lapiz lazuli, not less than 300 menns of that substance were used.

In addition to his tomb, a number of Ghazan's other foundations are enumerated, as (2) a monastery for dervishes, (3) a college where the doctrines of the
Shafi sect were taught, (4) a similar one devoted to the Hanefi sect, (5) a
hospital, (6) a palace for the administrator of the various establishments, (7) a
library, (8) an observatory, (9) a philosophical academy, (10) a noble fountain,
(11) a house of residence for the Seyids, or descendants of the Prophet. The
majority of the carpets for the rooms in these buildings were made at Shiraz. (12)
The garden and kiosk or summer-house of Aardiliya (*i.e.*, of Justice), in comparison with which, says Wassaf, "the palace of Khosroes was only rubbish, the
Khavrnak a mere bagatelle, and the Ledir a wilderness." The cost of keeping
up the mausoleum and its dependent buildings was over a hundred tumans
(*i.e.*, a million) annually, and a special divan was appointed to superintend
them, presided over by the great amirs Kur Timur and Terentai. In addition
to the buildings here named the whole town of Aujan was rebuilt, and

---

* *Id.*, 153.

supplied with markets and baths. Tebriz was surrounded with a wall, which inclosed within it the suburbs of Jerendab, Surkhab, and Beilankuh. This wall was 10 ells wide and 54,000 paces in circuit, with five gates, compared by Wassaf with the five senses of the body. Shiraz was also surrounded with a wall, while a monastery was built at Hamadan.*

Chardin, in describing Tebriz, tells us that when he wrote, the mosque built there by Ghazan's first vizier, Khoja Ali Shah, had been almost entirely destroyed. The base had been repaired, and thither people went to pray. The minaret, also, which was very lofty, had been similarly repaired. Of Ghazan's tomb there only remained a large ruined tower, which was still called Minar Khan Ghazan.† Ker Porter thus refers to these ruins: "To the south-west of the new walls of the city, but far within the remnants of the old boundaries, stand the magnificent remains of the sepulchre of Sultan Kazan. It is situated about two miles from the town, the whole way being marked with shapeless ruins, even stretching beyond the sepulchre to a great extent; but the tomb itself is an object too pre-eminent in desolated grandeur to descry without approaching. Its appearance now is that of a huge mound of mingled lime-dust, tiles, and bricks, but surrounded with spacious arches of stone and other vestiges of departed majesty.‡

Note 2.—The era of Ghazan. Among Eastern peoples there were two ways of calculating the year, one based on the sun's motion, the other on the moon's. The former commenced with the entrance of the sun into Aries, when day and night are exactly equal, and terminated when it reached the same point again, and was reckoned by Wassaf 365¼ days. The lunar year, based on the reciprocal positions of the sun and moon, was calculated on the basis of twelve lunar revolutions. This year numbered only 354 days and a fraction It is very clear that these two years speedily fell asunder, and left a considerable gap between them. To rectify the lunar year, and make it consonant with the solar one, various devices were at different times adopted. The Persians, whose era, says Wassaf, commenced with the reign of Yezdejird, the son of Sherujar, calculated their year on a basis of twelve months of thirty days, adding five days to the end of February (Izfendiarmed), which they named the five added. In the course of 120 years this reckoning was a month wrong. The Romans, who were well skilled in science, says the same writer, and made observations in their observatories, included thirty days in November, April, June, and September; twenty-eight in February, and thirty-one in the other months, but every fourth year counted twenty-nine days in February. That is, their calendar was the same as ours, and they thus made their lunar year commence with their solar one. The sixteenth Abbasidan Khalif El Motedhadbillah Ahmed founded an era based on a calculation of thirteen lunar months to the year. This era began on the 11th of June in the year 1207 of the Alexandrian era, the 218th year of the hejira, the sun being then in the zenith. The solar and lunar years had since the commencement of the hejira accumulated a gap of nine years between them. This was now rectified, and they were brought together again. Seventy years later the secretary of state, Iz ed daulet Bakhtiar, of the Buweyid dynasty, wrote a treatise on this era. From its commencement to the year 702 of the hejira, when Ghazan's era commenced, the differences between the solar and lunar years had accumulated to 13 years. Ghazan's new era, which began on the 14th of March, 1302, rectified this, and again brought the solar and lunar years into unison. It was known as the Ilkhanian era.*

Note 3.—John of Montecorvino in a letter dated at Cambalec, on Quinquagesima Sunday, in the month of February, 1306, written to the Vicar-General of the Franciscans, to the Master of the Order of Preachers, and to the Friars of either order abiding in the province of the Persians, mentions how, in the January of the previous year, he sent a letter to the father vicars and friars of Gazaria, by a friend of his attached to the Court of the Lord Kathan Khan, and he had since learnt from some messengers of the same Kathan Khan to the "Cham" (*i.e.*, to the Emperor of China) that his letter had reached its destination, its bearer having gone on to Tebriz from Serai This Kathan Khan, as Colonel Yule urges, can be no other than Ghazan Khan, who, however, died in 1304, which makes the chronology hard to reconcile.†

---

* Ilkhans, ii. 175-177 and 357-365.   † Cathay and the Way Thither, 204.

CHAPTER X.

ULJAITU KHAN.

WITH the death of Ghazan we lose the assistance of the great historiographer, Rashid ud din, whose work concludes at that point. It was continued by Masud, son of Abdullah (at the instance of Shah Rukh, the successor of Timur), who wrote an account of the reigns of his two successors.

On Ghazan's death his brother Khodabendeh, whom he had nominated to succeed him, was in his appanage of Khorasan. General Mulai advised him to keep the news secret, and to take measures for preventing troubles, especially on the part of Alafrenk, son of Gaikhatu, who raised claims to the throne, and was supported by Hirkudak, whose wife was the daughter of Kutlugh Shah and of a sister of Alafrenk, and who had recently been appointed commander-in-chief in Khorasan. It was determined at a council that before publishing the news of Ghazan's death these dangerous people should be removed. Three distinguished amirs, Isenbuka, Gurji, and Kartokabuka, were told off for the work. When they arrived at the ordu of Alafrenk that prince was not aware of Ghazan's death. They sought a secret interview with him, during which Gurji killed him with his sword. Hirkudak apparently fled, but was pursued. A vigorous struggle ensued, during which Gurji fell, but at length Hirkudak and his people were captured and taken before Khodabendeh. Ismail Tarkan was ordered to put him to death. His head fell off at the second blow. This was on the great military road, and the troops were marched over his body. His two brothers, Toghuz and Ala Timur, and his three sons, Malik, Arab, and Ramzan, were likewise put to death. Hirkudak's widow, Shahaalem, the daughter of Siurghutmish, sent her son Jihangir, who was a minor, to Khodabendeh, to ask his mercy. He granted him his life, and spared her house from being plundered, as was customary in the case of executed criminals. Alafrenk's troops swore allegiance to Khodabendeh, who appointed Musasadeh Beiktuk, the son of Aladai Noyan, as his deputy, and also as commander-in-chief, and made over the viziership of Khorasan to Alai ud din, the son of Said Weji ud din. He then set out for Tebriz with his most faithful noyans, Husinbeg, Sevinj, Isen Kutlugh (called Uveis Kutluk by D'Ohsson), and by the amirs Mulai (the Melaidu of Von Hammer), Jelair, Yas the son

of Juchi, Alikushji, and a large contingent of troops.* At each station
he was met by amirs and troops who went to do homage, and in six
weeks, on the 10th of July, 1304, he arrived at Aujan, where he performed the ceremonies usual after the death of a Muhammedan ruler, and
distributed funeral meats to the officers, soldiers, and public. Ten days
after his arrival was fixed upon as a fortunate day by the astrologers for
his enthroning, which was carried out with the usual ceremonies. The
princes Kutlugh Shah, Choban, Pulad, Sevinj, and Isen Kutlugh stood on
the right of the throne, with their girdles bound round them, while the
princesses stood on the left, and in front were the amirs. Outside the
tent the troops were ranged in ranks, the military music played, while the
various grandees in order offered the cup and did homage.†

Khodabendeh adopted the style of Sultan Uljaitu (*i.e.*, the fortunate or
rich Sultan).‡ He was the third son of the Ilkhan Arghun, and was now
twenty-three years old, having been born in 680 HEJ. His mother, Uruk
Khatun, the daughter of Prince Sarijé, brother of Dokuz Khatun,
gave birth to him in the midst of the desert separating Merv and
Sarakhs. There had been a great scarcity of water, and her suite were
troubled how to procure it, but directly the prince was born an abundant
rain fell, whence they called the newly-born infant Uljai-buka. Presently
the name Tamudar was substituted for this, according to a Mongol
custom. They believe that a change in the name of their children protects
them from the evil eye. When, in more modern times, Shah Sefi fell
into a state of languor, and it was suggested that it was due to the
witchcraft of the Jews, he was recommended to change his name to
undo the effects of this sorcery.§ Uljaitu was a Shia, and was surnamed
Khodabendeh (*i.e.*, servant of God) by his co-religionists, while the Sunnis
played on the word and called him Kharbendeh (*i.e.*, servant of the ass),
a name corrupted by Pachymeres into Karmpantes. On most of the State
documents he was called Uljaitu Muhammed Khodabendeh. Makrizi
says he was called Ghiath ud din Muhammed.‖ In his infancy he had
been married to Kunjuskat Khatun, daughter of Shadi Gurkan, with
whom he was brought up. She persuaded him to become a Mussulman
after the death of his mother, the Kerait Uruk Khatun, who, according
to Haithon, had taught him the Christian faith, and had had him baptised
with the name Nicholas.

Three days of feasting followed the enthronement, after which an order
was issued strictly enjoining the observance of the commandments of
religion and the precepts of Muhammedanism, while an adherence to the
yassa of Ghazan was also required. Robes of honour were distributed
among the principal officers. Kutlugh Shah was appointed commander-
in-chief of the army and given the first position at Court as Beglerbeg,

* D'Ohsson, iv. 478-480. Ilkhans, ii. 178-180. † Ilkhans, ii. 180.
‡ Remusat, Mems. Acad. Inst., vii. 392. § Quatremere, 31. Note. ‖ Op. cit., ii. part 2, 233.

and his tamgha or mark in red was attached to all orders. The noyans Choban, Pulad, Husein, Sevinj, and Inskutluk were put under him. Khoja Said Rashid ud din, the historian, was appointed first vizier, and Khoja Said ud din of Saveh his coadjutor. They had control of the finances, and authority over the Tajiks (i.e., native Persians). The administration of the vakhfs was assigned to Kutlugh Kaya and Behai ud din Yakub. They were enjoined to distribute these funds according to the will of the founders, and not to take a tenth for themselves, as their predecessors had done, contrary to the religious law. A considerable largess was distributed, by which the treasury was emptied.* Directly after the festivities were over Khodabendeh released the Egyptian envoy Hosam ud din Mojiri and the judge Imad ud din Ali, who had been placed in confinement in the latter years of Ghazan's reign.

On the 8th of August he set out from Aujan for Tebriz, where he visited his brother's grave, and prayed, shed tears, and distributed alms. On the 19th of September he received in audience the envoys of Timur Khakan, the Emperor of China, and nominally, at least, his suzerain, as he was of all the Mongol world. They took presents and congratulations. He also received envoys from Chapar and Dua, the heads of the ulusses of Ogotai and Jagatai, who went to report the treaty by which the long feuds between the rival princes of the house of Jingis had been terminated,† and offering their friendship. The Khakan's envoy presented a yarligh, in which he expressed the hope that thenceforth there might be concord and good understanding among the princes of the house of Jingis, &c.. A great feast was given. In front of the throne were placed the golden sideboards, shaped like lions, and decked with cups and beakers. On the right sat the princes ; on the left the royal princesses and their daughters. The tent was encompassed by a tuman of troops. Immediately before it were the Sikurjis or lancebearers, the masters of the horse, and riding masters (akhtajis), with led horses in rich bearings ; the fowlers (kushjis) with their falcons ; the four amirs commanding the four sections of the lifeguards. Sentinels with cudgels were employed in keeping back the pressing crowd, and in maintaining order. Kumiz and kara kumiz flowed freely, amidst singing and music. The reunion of the Mongol world was made the subject of congratulations. The Khakan's envoys were treated with the greatest honour, and presented with rich gifts, and Yaghmish, who had been governor of Ispahan during Ghazan's reign, was commissioned to take the Ilkhan's homage, or, perhaps, rather congratulations, to China.‡ Uljaitu now set out for Meragha to inspect Khulagu's observatory. There he installed the Khoja Assil ud din, son of the great astronomer, Nasir ud din, as director. Returning once more to Tebriz, he set out for Mughan to pass the

---

* D'Ohsson, 481-483. Ilkhans, ii. 180-182.　　† Ante, i. 288-289.　　‡ Ilkhans, ii. 183-184.

winter. There he, on the 9th of December, received envoys from Toktu, the ruler of Kipchak ; and on the 7th of January dispatched an embassy to Egypt, with which were sent the two envoys recently released from prison, with the request that Irinchin, brother of Sevinj, whom the Egyptians had captured in the recent campaign and detained, might be released. A friendly message was also sent by the Ilkhan. Uljaitu had asked for the hand of Irinchin's daughter, Kutlugh Shah, in marriage. The contract was signed on the 23rd of March, Pulad Ching Sang representing the Ilkhan and Rashid ud din the princess, and the marriage was celebrated on the 29th of the same month. On the 20th June the bakhtak (the word means a helmet in Persian) was placed on Kutlugh Shah's head, and she was installed in the grand ordu of Dokuz Khatun. At this marriage Rashid acted as vakil to the princess. Three days later Uljaitu married Bulughan Khatun Khorasani, widow of Ghazan, to whom he gave a marriage gift of 90 menns of silk.*

We have seen how Shah Jihan was nominated ruler of Kerman by Ghazan. He was summoned to his presence by Uljaitu, was reproached with his want of courtesy to the representatives of his suzerain, his care-lessness in paying tribute, and his cruelty towards the grandees of the country. The Ilkhan, touched by his youth and beauty, spared his life, but he would not allow him to return to Kerman, which was thenceforth governed in the name of the Mongol divan. Shah Jihan reconciled himself to his fate, and withdrew to Shiraz, where he amassed great wealth and acquired great influence. With him ended the dynasty of the Karakhitais of Kerman, which had reigned there since the year 1223.†

Uljaitu allowed the great sheikh Safi ud din of Ardebil to sit on his right at table, and on his left the great sheikh Alai ud daulet of Semnan. One of these ate while the other touched nothing. When the tables were removed, Uljaitu said, "The piety of the two great sheikhs is pre-eminent, but why did one eat of all meats while the other ate not at all ? If it was lawful, why did they differ ? If unlawful, why did one of them not refrain like the other ?" Safi ud din replied, " His eminence the Sheikh Alai ud daulet is a sea, which nothing can soil," and Alai ud daulet replied, " His eminence the Sheikh Safi ud din is a royal falcon, which soars over everything." The Ilkhan was much pleased with these reciprocal compliments of the holy men, and commended and rewarded them accordingly.† At this time, we are told, several Jewish physicians embraced Muhammedanism. In order to test the sincerity of their con-version the Vizier, Rashid ud din, urged that they should be asked to eat some mule's flesh (Quatremere says camel's flesh), boiled in sour milk, the Mosaic law having prohibited both the eating of mule's flesh and the cooking of flesh in sour milk. This test was accordingly applied to them.

* D'Ohsson, iv. 484-485. Ilkhans, ii. 184. Quatremere (Rashid), xiii.
† D'Ohsson, iv. 485. Ilkhans, ii. 142. ‡ Ilkhans, ii. 185-186.

As Von Hammer says, it may have been suggested by Rashid ud din to prove his own faithfulness to Islam, some having accused him of being a Jew by origin.* Wassaf reports an intrigue which now arose against Rashid ud din and his colleague, Said ud din, who were accused by Kursirh, the former secretary of Hirkudak, and others, of some malpractices. Kutlugh Noyan was deputed to inquire into the matter, and the accusations having proved groundless, some of the accusers were put to death and others bastinadoed.† On the 8th of March, 1306, fresh envoys came from the Khakan Timur with a present of gerfalcons.‡ This was apparently followed some time after by another embassy, headed by Uljaitu Buktimur, who escorted a princess of the imperial house as a wife for the Ilkhan, and also took a present of 1,500 horses. This envoy, on his return home, was detained by Isenbuka, the Khan of Jagatai.§ Rashid says when envoys arrived from the Khakan's Court they became Mussulmans, and submitted to circumcision, though they would have to bear severe reproaches on their return.||

In the second year of Uljaitu (i.e., 1306) a struggle broke out in the further East between the houses of Ogotai and Jagatai, which will be described in detail in the next volume. One of the consequences was that Sarban, son of Chapar, who was at the head of the ulus of Ogotai, with a number of princes and other supporters, fled to Persia, and sought shelter at the Court of Uljaitu. Among the fugitives were two whose relatives had filled a notable *rôle* in the history of the earlier Ilkhans, named Behadur Kazan, son of Kurkuz, and Ordu Kazan, the brother of Nuruz. They sent word to the frontier to explain the reason of their flight. The noyan Yasaul, commanding the frontier, to guard against surprise first sent forward a regiment of karauls, or frontier guards, and followed himself with his amirs, fully armed. When they drew near Konduzbaghlan, in the neighbourhood of the Oxus, there was some misunderstanding between their outposts, which led to a momentary struggle, but the real state of things having been disclosed, there was a speedy peacemaking and feasting. When Uljaitu heard of their arrival, he summoned them to the Court. Sarban, who was the most important among them, died in passing through Khorasan ; but Timur Oghul, a descendant of Juchi Khasar, with his son Minkian, and Sarban's two sons, Buruntai and Bojir, with the amirs Behadur Kazan and Ordu Kazan, leaving a portion of their forces in their yurts, repaired with the rest, and the amirs of the *hazarehs* (i.e., the commanders of regiments) to the Court, where they were royally entertained. Prince Buruntai was, with the concurrence of the Grand Prince Suntai, who was master of the horse, appointed governor of Georgia. His brothers, Bojir and Hirkodak, were given a body of 500 Karaunas, at the request of the vizier Rashid ud din. Timur Oghul

* D'Ohsson, iv. 487-488. Ilkhans, ii. 186. Quatremere, xiv.   † Ilkhans, ii. 186-187.
‡ D'Ohsson, 487-488.   § Quatremere, 13-14. Note. || Id.

died about this time. Lastly, the amirs Behadur Kazan and Ordu
Kazan, were given suitable yurts, forces, and pensions.*

We described in a previous chapter the earlier doings of the famous
farmer-general of the taxes of Fars, the Seyid Jemal ud din, who lived in
regal splendour in the island of Kish in the Persian Gulf.† On Uljaitu's
accession he went to do homage to him, and was offered the government
of Fars; but he declined it, and returned to his principality of Kish. His
son, Fakhr ud din, had been sent, as we have seen, by Ghazan to China
to look after his share of the imperial revenues there, and also to buy
silk, and died, on his return home, on the coast of Coromandel, where a
tomb was erected over him near that of his uncle. This uncle, who
died early in the year 1303, was Taki ud din Abdur Rahman Al Thaibi,
who was the chief minister of the Diro, or ruler of that coast—a man of
sound judgment, to whom Fitan, Mali Fitan, and Kabil were made over.
On his death his master would have confiscated his property, but his son,
the Malik Muazzain Siraj ud din, having secured his goodwill by the
payment of 200,000 dinars, obtained his father's post as Vizier, and also
his property.‡ In 1305 Jemal ud din was summoned to Shiraz, the
government of which was conferred on him in order that he might restore
order to its finances ; but he was ill at the time, and died the following
year (*i.e.*, 1306), to the great regret of the people of Shiraz, who raised a
handsome tomb over his remains, and composed an elegy on his death.
Wassaf himself, who had known him, composed another.§ On the 14th of
April, 1306, Rashid ud din presented Uljaitu with his famous work,
entitled "Jami ut Tevarikh" (or Collection of Annals), and was duly
praised.||

Uljaitu now turned his attention to the strip of mountain country lying
north of his favourite residence, Sultania, called Ghilan, which, thanks to
the difficult passes and forests which it contained, had hitherto maintained
its independence against the Mongols. This district, lying between the
Caspian and the high mountains of Dilem, although only thirty parasangs
in length, was divided into twelve districts, subject to as many princes.
Ardai Kazan, son of Arghun Aka, Khulagu's famous minister, who
arrived at the Ilkhan's Court on the 25th of December, 1306, with the
news of the death of Dua Khan, the chief of the ulus of Jagatai, told
Uljaitu that Dua and his officers used to speak scornfully of the fact
that the ruler of Persia could not conquer such a small district as Ghilan,
inclosed as it was in his dominions. Piqued at this remark, he determined to subdue it, and consulted with Kérai, the governor of Tharem,
who knew the country. Four divisions were accordingly ordered to
march. Choban commanded one which went by way of Ardebil,
Kutlugh Shah by Khalkhal, Tugan and Mumin by Kazvin, and Uljaitu

* Ilkhans, ii. 189-190.   † Ante, 418-421.   ‡ Elliot, iii. 45.
§ Elliot, iii. 47.   Ilkhans, ii. 197-198. || D'Ohsson, iv. 488.

in person by way of Lahejan.   At Sitaré Choban was met by Rokn ud din
Ahmed, prince of the district, who had prepared provisions for his troops,
and took him presents.   He was ordered to guide the Mongol forces by
the easiest route, and was promised that he would be confirmed in his
principality when the country was conquered.   The Mongols then
marched on Kesker, or Kierkier, ravaging, making prisoners, and killing
those they found in arms.   When Choban arrived near Kesker, the amir
Sherif ud daulet went to him with presents.   This part of Ghilan having
thus been conquered without resistance, Choban took the two princes
with him, and joined Uljaitu on the road to Lahejan.   Kutlugh Shah had
received the submission of the prince of Khalkhal, named Sherif ud din.
Having asked him for information on the country he was going to march
through, and which he proclaimed he intended desolating with fire and
sword and conquering, Sherif ud din recommended him to be prudent,
as the country was difficult and the people not easily frightened ; but
Kutlugh Shah, who was of a reckless disposition, took no heed, and sent
Pulad Kaya in advance.   The Ghilanians defended their passes bravely,
and after three encounters, in all of which they were beaten, they sent to
offer their submission to Pulad Kaya, who sent on the news to Kutlugh
Shah.   He would have assented, but his son Sipauji dissuaded him,
saying that having entered the country it was better now to conquer it,
and to exterminate the inhabitants, and that if Pulad Kaya's advice were
followed they would lose the glory of the campaign.   The latter was
therefore recalled, and Sipauji given command of his detachment.   He
continued the advance, putting everyone he met to the sword, and
arrived at Tumin after slaughtering a great number of people.   Reduced
to despair, the Ghilanians concentrated between Tulem and Resht, and
marched against him.   Sipauji was encamped on marshy ground.   After
a stubborn fight the Mongols were defeated, and their horses being bogged,
most of them perished.   Kutlugh Shah wished to advance with his division
to avenge this disaster, but his men refused to obey, and retreated in
disorder, nor could he stop them, although he put several of them to
death.   Although he had but forty companions left he would not retire,
but continued to fight on bravely.   He at length fell from his horse,
struck by an arrow, whereupon a Ghilanian officer ran up to him, and
said, " Nuruz did not find an avenger in Persia ; God has sent you here
to receive punishment at my hands."   He thereupon killed him.   The
immense booty which the Mongols had made in the district of Dibaj fell
into the hands of the Ghilanians.

Meanwhile, Togan and Mumin, who had advanced by way of Kazvin,
received the submission of Prince Hindushah, whom they took with them
to Uljaitu's camp, promising to secure his confirmation on the throne.
Uljaitu himself left Sultania in May, 1307, leaving Pulad Ching Sang in
command of his oghruks, or domestic establishment.   He traversed

Tarem, and entered Dilem on the 21st of May by the route of Kurandesht and the village of Lussan, and pitched his camp on the Sipid rud. His troops pillaged Khashjan, although that town had submitted. On the 29th Talish had the same fate. Many of the people of Dilem who had withdrawn to the woods were killed, and their wives and children carried off. On the 2nd of June the royal head-quarters were moved to the banks of the river Deiléman. On the 6th Uljaitu broke through the pass of Ghilan, and entered the country of Nu-Padishah by Russita, which is on the main road to Kazvin. The troops could only traverse the defiles in small parties, and suffered considerably. Near Shiruyei-Talish, which was situated amidst forests and mountains, the inhabitants fell on the baggage train, and captured many arms. When he arrived near Lahejan, Uljaitu summoned the Padishah to submit, so as to spare the blood of his people. "Trust not in the height of your mountains nor the denseness of your forests, for my army, can fill up the sea and overturn the mountains." The Ghilanian prince went with his sword and a sheet to implore the clemency of Uljaitu, who, at the instance of the general Isen Kutluk and the vizier Rashid ud din, received him well. When he entered Lahejan the Padishah gave him a feast. He stayed four days there, and left again on the 13th of June. Having crossed the Sipid rud, he encamped at Kerjian, whose environs were laid waste. The following day he received the submission of Prince Soluk, and after sending a body of troops into the district of Témijan, he withdrew again by way of Kutem and the Sipid rud.

When he heard of the fate of Kutlugh Shah he was much distressed, and sent a picked body of 3,000 men under Sondavé Behadur, Behlul, and Abubekr to revenge him. They found the people of Tumen, Resht, and Tulem had assembled to resist them. A battle ensued on the 18th, in which both sides fought desperately. Sondarvé and Abubekr, the latter of whom commanded the men of Khorasan, both fell. Sheikh Behlul was wounded. The Mongols, weakened by their losses, withdrew from the battle-field, and entrenched themselves to await reinforcements, which duly arrived under Hussein and Sevinj. A second struggle, more bloody than the previous one, now ensued. This time the Ghilanians, having lost half their men, fled to the woods and mountains. Tumen, Resht, and Tulem were sacked, the men there killed, the women and children carried off into slavery, and the whole district ravaged. Another body of Mongols marched on Temijan, whose prince, Amir Muhammed, offered tribute. The Mongol chiefs would have granted him peace, but were dissuaded by a certain Mamshaki, who urged them to plunder the town, which was believed to be very rich. Muhammed's envoy was sent back with a negative answer. Driven to bay, he met the invaders on favourable ground, and won such a complete victory that few of them were able to regain Kazvin. Muhammed now sent to explain the circumstances to

Uljaitu, and offered to go to the Court. Uljaitu accepted his statement, blaming those who had refused to accept his advances.

Having conquered Ghilan, he set out again on his return on the 29th of June, taking with him the Ghilanian princes who had submitted, among whom the Padishah was the most powerful, and also Soluk, who was renowned for his bravery, and Jelal ud din, brother of Dibaj. They offered to pay an annual tribute of silk, and also redeemed the prisoners who had been carried off, and who were sent back to Ghilan. The Padishah was given a robe of honour and special diploma, and also a royal princess in marriage, and with Soluk and the other chiefs was allowed to return home. Dibaj, who was descended from the Sassanian princes, on whose territory Kutlugh Shah had perished, feared to meet Uljaitu, but sent his submission. He afterwards went, however, and subsequently paid several visits to his suzerain, who always treated him with distinction. Uljaitu, on reaching Sultania, issued a commission to inquire as to who had been to blame for the deaths of so many brave officers. After strict inquiries the yargujis pointed out Sipauji, Mamishki, and several other amirs. They were condemned to death, but the Sultan remitted the punishment in the case of Sipauji, on account of his father, in favour of 120 strokes of the bastinado, while his father's tuman, or division, was made over to Choban. Other officers received 120 strokes of the baton on their back and chest.*

The " History of Georgia " describes the events of this campaign in some detail. It tells us the Georgian king Wakhtang was attached to Uljaitu's army corps. That Kutlugh Shah was accompanied by the people of Beka ; that Usen, the Jalair (i.e., Hussein), marched with the Osses or Ossetes, living at Gori ; and that Choban, the Sulduz, went with a fourth army. Among the Ghilanians there were four princes—Erkabazn, Ubash, Rostan, and Asan, the last of whom occupied the most inaccessible defiles. Erkabazn opposed the Khan's army. A sharp fight took place, in which Wakhtang commanded the advance guard. The Mongol advance was much impeded by woods and marshes. A hand-to-hand fight ensued under the Khan's eyes, and the carnage was great, only one-fifth of the Mongols escaping. All those of Somkheth perished close to the king, many aznaurs leaving no heirs, since fathers, sons, and brothers were all killed. The king himself, who fought very valiantly, was wounded in the thigh. The Khan, seeing how desperately the country was defended, determined to retire. He was pursued, and again lost many men. Ushish, prince of the Ghilanians, closed one of the defiles with a gate, and thus cut off Kutlugh Shah's retreat. He thereupon dismounted, and ordered his people to do the same. Presently he was shot by an arrow while sitting on a bench. His son Sibuchi (i.e., Sipauchi)

---

* D'Ohsson, iv. 488-497.

thereupon fled, but the greater part of his people were killed, many of them being bogged in an inundated rice field, and slaughtered. Meanwhile Choban defeated the Ghilan chief, Rostan Malik, who retired to a secure vantage, where Choban did not pursue him, but retired slowly. Gamresel Javakhis Shiwili fought bravely in this struggle. The Tartars succeeded in withdrawing without any inconvenience. Usen, with the Osses, fought an indecisive battle, after which both parties withdrew. When the troops reunited again at Aujan, Uljaitu appointed Choban generalissimo in the place of Kutlugh Shah. Wakhtang, the Georgian king, and all his didebuls, were handsomely rewarded for their brave conduct in the campaign, and were then sent home again.*

When Wakhtang returned, the people of Khodris, who were Mussulmans, incited the Khan against the Christians. Uljaitu, seduced by their advice, sent a noyan to Georgia to cause the King and his people to apostatize, and to destroy the churches. He arrived at Nakhchivan, as the King, was leaving there. Instead of ceding to his menace, he appealed to his companions to remember the faith of their ancestors and the glorious end of the Holy Martyrs, and bade them accompany him to the Khan, where they might sacrifice their heads for him who had laid down his life for them. His soldiers gladly obeyed, and Uljaitu was astonished to see the King's return. Kneeling down, Wakhtang said, "High and powerful Khan, you know that all the Christian chiefs have been ordered to abandon their faith. Listen, now, to me. My father and grandfather served thy grandfather and father while they professed this faith, yet no one said to them, 'The Georgians hold a wicked creed,' and we Georgians are distinguished among all. The religion of the Persians, on the other hand, was regarded as abominable by preceding Khans, who exterminated those professing it, because they were poisoners, péderasts, and homicides. If you listen to them, O, Khan, I, the King of the Kharthlis, and all the Georgian mthawars are ready to be executed." He then presented his neck. At these words the Khan treated the King honourably, raised him up, and addressed him some sweet words, and repenting of what he had done, blamed his counsellors. The Georgian history says that while the Christian churches were being dismantled four of Uljaitu's sons died on four successive days. The King now returned to Nakhchivan, where he fell ill and died. This is the account in the Georgian history. Wakhusht, on the other hand, says he was actually put to death by the Khan with cruel tortures.† Wakhtang was taken and buried at Dmanis, and left two sons, Dimitri and Giorgi, the former of whom possessed Dmanis, and the latter Samshwildé.

When Choban succeeded Kutlugh Shah (also called Kutlugh Buka) he marched with a large army to Kola, thence to Artan, and as far as Arsian.

He summoned Béka to go and see him or the Khan. The latter, instead of going, sent his young son Shalwa, whom the Mongol commander took with him when he traversed Somkheth, not molesting Béka's country.* Béka died in 1308,† and his son David VI., brother of Wakhtang III., in 1310. The latter left a young son Giorgi, known as Mtziré, or the Little, and who is referred to as Giorgi VI. He was only two years old on his father's death. The boy was summoned to the Ordu, and was nominated ruler of Tiflis by the Khan. He was placed under the tutelage of his uncle Giorgi, called Brtsqinwaléh, or the Brilliant, son of Dimitri the Devoted, who was brother of David VI. and Wakhtang III. When Giorgi VI. was sent to Tiflis, Uljaitu sent with him a certain Zaal Malik, a Persian, of Khorasan, and also Akhrunchi, brother of Choban's father, who had orders to bring all the Georgians under his rule.‡

When Uljaitu set out for his campaign in Ghilan he left his harem at Konkurolang (i.e., the place where Sultania was afterwards built). His favourite wife, Iltermish, was left in charge of Rashid ud din, who had orders to remain with her till her recovery. Meanwhile Taifur, a son of Uljaitu, who was a child, had a narrow escape of being burnt to death. Some one having planted a torch near his tent, it was upset, and the tent got on fire, and had it not been for the devotion of two attendants, who threw themselves on the burning tent cover, which they clasped, and at great risk to themselves, put out the flames, he would have perished. Iltermish thereupon distributed alms among the poor.

Let us now turn our view further East. We have seen how Fakhr ud din, the Malik of Herat, had proved rebellious, and had supported the turbulent Nigudarians during the reign of Ghazan, and how Uljaitu, who was then governor of Khorasan, was sent against him. When the latter mounted the throne, he summoned the malik, like the other great vassals, to go and do homage, an invitation which he evaded. Presently the Ilkhan dispatched Danishmend Behadur at the head of 10,000 men to bring him to his senses. On arriving at Herat he sent Tutak bela and Hindujak into the place with the Sultan's orders, which were that he must surrender the Nigudarians and send home again the people of Merv, Abiverd, Sarakhs, Jam, and Khavaf, who had rebelled at Herat, and remit to Danishmend's agents the produce of the three previous years' customs and the profits of the mints. The "Chronicle of Herat" says the demand also included that the Ilkhan's name should appear on his seal and coins. In case of refusal he threatened to besiege the place. "Tell Danishmend," said the malik, "that if he wishes for presents due to my generosity I am willing to satisfy him, but if he has come here sword in hand to acquire renown, and to subject me to his authority, he has come with a vain hope." Danishmend thereupon summoned the great feudal chiefs of Khorasan with their troops, and there speedily arrived the lords of Ferah, Déréi, Izfézar, Tulek, and Azab, each with some cavalry and infantry. The mollah Vejih ud din Nesséfi, grand judge of Herat, having left that town with Fakhr ud din's consent, met Danishmend at Nishapur, attached himself to his person, and urged him strongly to attack Herat, which he told him he would capture if he cut off its supplies. A close blockade was accordingly instituted. Fakhr ud din prepared for a vigorous resistance. He opened the magazines which his predecessor had stored, and distributed largess among his troops, who, thus encouraged, made some vigorous sorties and killed many of the besiegers. In the course of ten days Danishmend, who was somewhat discouraged by these reverses, sent the Sheikh ul Islam, or mufti, Kutb ud din, of Jesht, or Chesht (now called Khajeh Chisht), a place near Herat, to say that he had no personal animosity against him, nor did he wish to ruin the country or shed Mussulman blood; but that he had received orders, and that if he did not execute them he would be held responsible. "I suggest to you, therefore," he continued, "whom I look upon as my son, to defer to the royal wishes, and to retire for a few days to one of your fortresses, while one of my sons occupies Herat; and rest assured this proposal has no other motive than to secure peace between us." Fakhr ud din said he was willing to do anything which the Sheikh ul Islam suggested. The latter therefore advised him to drive out the pestilent Nigudars, and to withdraw himself to the fortress of Amankuh, whence he could return when the Mongol troops had withdrawn. When the malik suggested that Danishmend meant to deceive him, and to waylay him *en route*, the sheikh urged that he should ask for hostages. He therefore asked that Danishmend's son Togai should remain in the city, while his other son Laghiri accompanied him to Amankuh, saying he would send him back on his return. Danishmend, on the return of Kutb ud din, called a council of the maliks and amirs, and in spite of their advice he ordered Vejih ud din to draw up the formula of his oath. It ran thus in Persian: "By the God, before whose majesty the powerful and the weak, the king and the pauper, are prostrate. By the Almighty Being and the All-powerful. By the God of heaven and earth. By the God who knows what is hidden, and knows one secret from another, and by his Apostle. When the Malik Fakhr ud din shall have departed for Amankuh, I, Danishmend Bahadur, will not do, nor attempt, nor order to be done, any harm to his ministers nor his officers. I will, on the contrary, be kind to the people of Herat; nor will I attempt to capture its citadel." The sons and relations of Danishmend, the maliks, and amirs became guarantees of this promise, and signed it with their signatures. Kutb ud din took the document to the malik, who sent him a corresponding engagement in these terms: "By the substance of God and the soul of Muhammed. By the high veneration due to Islam and the

M *

glory of the Holy Faith. By the purity of every precept of the divine law,
By the interpretation of every letter of the Koran. I, Malik Fakhr ud din,
will do no hurt to the amir Danishmend, and when I arrive safe and
sound at Amankuh I will send back the amir Laghiri, and so long as the
amir Danishmend remains faithful to his promises, and treats me as a
father, I will not oppose him, and if I violate this engagement may I be
abandoned by God and become the object of the severest punishment
of the Most High " Danishmend now sent Laghiri, with ten of his chief
officers, to accompany the Malik to Amankuh, and Togai with others to
enter the town, and ordered the latter to act prudently, to win over the
maliks and officers, and to secure the goodwill of the inhabitants by his
affability, and to hold out hopes until the town was fairly in his possession,
when he might proceed to punish the guilty.

On the arrival of Togai the Malik confided the command of the town
and citadel to Jemal ud din Muhammed Sam, one of his old officers,
whom he instructed to carefully guard the citadel, known as the Castle of
Ikhtiar ud din, and beware of any stratagem on the part of Danishmend ;
that he was not to allow any of the garrison to go, nor to go himself to him,
but to plead that the Malik had made him swear not to leave the citadel
without his permission. If Danishmend asked for presents he was to
send him 10,000 dinars, fifty sets of robes, some loads of provisions, an
Arab horse, and a Turkish slave, whom he pointed out. Having given
these orders, he summoned the Heratian, Seistanian, and Ghurian
officers in his service, distributed robes among them, and made them
promise to be united, and to strictly obey the orders of Muhammed Sam,
to whom he gave his sword, saying, " If anyone disobeys you, cut off his
head." He also handed over to him the arsenal of the citadel, which was
well supplied with swords, cuirasses, coats of mail, bucklers, and bows and
arrows, and left at nightfall, with his coat of mail and helmet on, and
escorted by two hundred horsemen and three hundred foot soldiers. He
arrived at Amankuh at night, and the following day sent Laghiri back to
his father, begging him to keep his word, and to treat the people of Herat
well.

The following day Danishmend entered Herat at the head of his troops,
with trumpets and cymbals playing, and with a standard representing a
dragon. He was astonished when he saw the fortifications there,
which Fakhr ud din had recently augmented. The Molla Nesséfi urged
upon him that it was these fortifications which made the people of Herat
so hard to manage, and that it would be well to destroy them. Danishmend caused the gate of Khosh, by which he had entered, to be dismantled, drove away with blows of maces those who were guarding
it, and posted a detachment of his troops there. A proclamation was
issued, stating that the city belonged to the Sultan Uljaitu, that Danishmend governed it in his name, and that the inhabitants could count on

his goodwill and should peacefully continue their occupations. The following day he sent Tutak Bela to summon Muhammed Sam to his presence, and receiving an insolent answer he swore in his rage that he would make a terrible example of him, and told his officers and the Persian grandees that the citadel must be assailed that very day. "General," replied the Molla Nesséfi, "it would be better to take it without a struggle." "Without doubt," said Danishmend, "if it be possible." "Send word, then, to the Malik, by Kutb ud din Cheshti, that you are going to send your son Laghiri to the Sultan to announce to him that the Prince of Herat has submitted to his orders and surrendered the town, and to beg that he will deign to confer on the Malik the grand diploma and robes from the royal wardrobe, and that the country of Herat cannot be better governed than by this Ghurian prince ; but do not fail to add to Fakhr ud din, 'The citadel of Herat has become so famous since the catastrophe that overwhelmed Nuruz, that the Sultan will not fail to ask if it has been surrendered. I beg of you to order Muhammed Sam to allow my son Laghiri to enter it with twenty men, so that he may truly say that he has been received within the citadel.'" Danishmend readily adopted this advice, and ordered Kutb ud din Cheshti, Tutak Bela, and one of his relatives to set out. When Fakhr ud din heard the proposal he at once replied, "I always said that this cursed Turk was not to be trusted, and would not keep his word." Tutak Bela assured him that his master meant no deception, and swore that it was in the interest of both parties. "That may be," he said, "but the demon of pride may overcome him, and he may try to secure the garrison, which will cause great trouble, for they are all determined men, especially Muhammed Sam, the bravest of the Ghurians." He ended, however, by giving his consent, and wrote a letter to Muhammed Sam, which he intrusted to them, stating that his father, Danishmend Behadur, was coming to see the citadel, and that he must show him the greatest honour. It is said that by another letter, which he dispatched secretly to him, he told him to be on his guard against any artifice that Danishmend might contemplate.

Muhammed having received these orders, planted three hundred men in ambuscade in various parts of the citadel, and made preparations for a feast he was going to give to Danishmend. Presently Kutb ud din Cheshti arrived with the announcement that the general was coming. The commander replied that he was ready to obey the malik's order, and would open the gates of the citadel whenever Danishmend wished. The latter's officers were greatly elated, and fancied themselves already masters of the place. Danishmend asked Kutb ud din how many men there were inside. He replied there might be 250 Seistanis and 50 Ghurians, who in a struggle would fly before ten men. Vejih ud din said that there were not so many as this, and that his spies informed him there were not more than thirty armed men, the others being merely servants or

custodians of the magazines.   Thereupon Kutb ud din said to the
general, " If the amir is going to the citadel with sinister views, he may
repent of it. I know Muhammed Sam and his companions Ilduz, Lokman,
Ferrukhsad, and Abul Feth. They are resolute men. God grant that
no harm comes to you, and that you lose not in one moment the whole
fruit of your negotiations." " Rest assured," said Danishmend, laughing ;
and taking his sons apart, he said to them, " Keep your eyes always on
me. When I ask for my bow from my squire seize Muhammed Sam and
his men." He then went to the bath of Chehar Su, whither he summoned
a Hindu geomancer, and asked him whether he should or should not go
to the citadel. The Hindu having gone through his calculations, said
neither the aspect of the stars nor the lines of the *reml* were favourable,
and tried to dissuade him. Danishmend, affected by his reply, was about
to return to his quarters when he met Vejih ud din, who said to him, " Pay
no heed to the words of the geomancer. He professes to know the
future, which is known to God alone. Remember this divine sentence :
'Whoever believes in the influence of the stars is an infidel.' Very
often there happens the exact reverse of what these people predict."
Danishmend being reassured by these words, first sent his son Laghiri
with twenty picked men to the citadel. They were followed by Kajui
with ten other men, while Minkui, one of his relatives, conducted a third
body. Muhammed Sam received the amir's son with marks of respect
and conducted him to the malik's tent. In the course of an hour, the
other two detachments having arrived, there were collected some eighty
of Danishmend's men inside. Dinner was served to them, and
Muhammed Sam did the honours, presented the cup, and received the
congratulations of his guests. Meanwhile Kajui, who was half drunk,
having left the room, noticed four armed Ghurians hiding in ambush
among the fortifications. Entering the room again, he said to
Muhammed Sam, "My brave friend, I have seen some armed men
behind a wall. Have you put them there to arrest us ? " " God preserve
me," he said, " from having any ill designs against you," and seizing a
mace he drove the men outside the citadel with great blows. Danishmend, having heard of this incident, had his confidence increased, and
arrived at the citadel with 180 picked horsemen. The governor went out
to meet him with every token of respect, holding his stirrup while he
dismounted. But Danishmend, who had his previous uncivil answer still
on his mind, addressed him, saying, " Insolent Tajik, how did you dare to
refuse to come to me ? Your prince did not disobey my orders, while a
dog like you, who have sheltered yourself behind four walls, proud of a
few Tajiks whom you command, dare to place yourself among the
enemies of the ruler of the world. I will have you cut in pieces, and
your castle razed to the ground." " Servants ought to faithfully obey
their masters," said Muhammed, " and I did not come because Fakhr ud

din ordered me to stay in the citadel." Professing to be satisfied with this reply, Danishmend embraced him, said he pardoned his temerity, and received in return assurances of his devotion. Danishmend alighted in the square in front of the citadel, and advanced towards the door of the grand saloon, having beside him the Molla Vejih ud din and the amir Kirai, who had arrived the same day from the Court with the appointment of governor of Herat. Danishmend's officers wore coats of mail under their cloaks, and as they could not openly wear arms they had secreted daggers in their girdles, and other weapons in their boots, and followed their commander. The road leading up to the citadel was carpeted with costly stuffs. Muhammed Sam had ordered his people to kill Danishmend when he arrived at the foot of the staircase leading up to the grand saloon. When he reached the fatal place, Taj ud din Ilduz came to meet him, and having kissed his hand, let him pass in front of him. He then seized him by the collar with one hand, and struck him on the head with his mace. At the same moment Abubekr Sédid, another of the malik's officers, came out from behind a parapet and struck off Danishmend's head with his sabre. Thereupon the Molla Vejih ud din, Hindujak, Kirai, the Indian astrologer, and the rest rushed for the gate, but it had been closed. Men who lay in ambush rushed out from all sides. Terrible cries filled the air, and they were all speedily overwhelmed by the Ghurians. Meanwhile Kutb ud din Cheshti, who had remained between two gates, cried out "Fear the anger of God. Do not act against the malik's orders, and bring misfortune on the town." Laghiri and his companions, who were still in the room where they had been feasting, barricaded the doors, but the windows were speedily broken, and they were killed with volleys of stones and other missiles, Laghiri having rushed out, sword in hand, was slaughtered ; others tried to escape by the walls, and were broken to pieces Shirin Khatun, Danishmend's wife, with his daughters, and the wives of his sons and brothers, who had gone to the *fête*, and were witnesses of the tragedy, raised terrible cries. Muhammed Sam abandoned them to his officers. Outside, people were ignorant of what had taken place. Inaltikin, Prince of Férah, and Tutak Bela, with a group of officers, were assembled before the gate of the citadel. The "Herat Chronicle" says at the reservoir called Filbend. A Seistanian, who was a friend of Inaltikin, having passed out, as if with a message from Muhammed Sam, was asked by the Prince of Férah if Danishmend had finished his meal. He replied in the language of his country that they had given Danishmend the same feast as Nuruz. At these words Inaltikin and Tutak, who were naturally greatly agitated, withdrew hastily to regain their quarters. They found the gate of Firuz Abad closed, but they broke the chain and lock with hatchets, and left the town with one hundred horsemen.*

* D'Ohsson, iv. 497-513. Journ. Asiat., 5th ser., xvii. 481-485.

Hardly had they escaped when the Ghurians cried out from the walls of the citadel to close the city gates and announce to the inhabitants the death of Danishmend and his companions. They lit a great fire as a signal to Fakhr ud din at Amankuh. Muhammed Sam descended at the head of his warriors, and put to the sword all the Mongols he could meet with, and a terrible massacre was the result.

Delighted as Fakhr ud din was with what had occurred, he was too prudent to show any exuberant feeling; but while he professed in the presence of his followers to be displeased with Muhammed Sam, he wrote to him, saying, " It would have been better not to have done this, but now it is done you will defend the city at all hazards, and say that Danishmend entered the citadel with the intention of killing you, and that you only acted in self-defence." At the same time he sent 100 well-armed men from Amankuh to reinforce the garrison of Herat. These events took place in September, 1306. When Uljaitu heard of them he dispatched Yassaul to take the command of the troops of Khorasan, with orders to establish his head-quarters on the Jihun or Oxus, and ordered Bujai, son of Danishmend, who was then in Rum on the Greek frontier, to go and avenge the slaughter of his relatives. Another of Danishmend's sons, Togai, who was encamped at Tus, marched with his men upon Herat, took command of the troops in its neighbourhood, and blockaded it, pending the arrival of reinforcements. By the influence of the Sheikh Kutb ud din Cheshti, and after long negotiations, an order was issued by the Malik Fakhr ud din for the release of Shirin Khatun, Danishmend's widow. On leaving the citadel she had 200 people of Herat killed, and further ordered everybody within forty parasangs of the city to be put to death. While the Mongols were investing it they were attacked several times from Amankuh, and many of them killed. Bujai arrived before Herat in the beginning of February, 1307, five months after his father's death. In conjunction with his brother Togai, the funeral ceremonies were renewed, and according to the Mongol custom nine days were spent in lamentations and mourning. On the 10th Bujai wrote to Fakhr ud din at Amankuh, saying : " Jemal ud din Muhammed Sam has killed my father and three hundred of his people. Let me know if it was done by your orders. If not, write to the officials at Herat that in order to avoid great troubles they must surrender Sam and the other enemies of my blood, and restore the money, goods, horses, and arms which they have taken from us. If not, all this country shall be wasted with fire and sword." " I declare," said Fakhr ud din in reply, " that I never ordered Jemal ud din or any other person to kill your father, and that I have never approved this act. Sam did it at his own instance to protect his own life, but he is at the head of 2,000 well-armed men. How can the people of Herat execute any order I may give them for his surrender. It is a matter for arrangement between you and me." Irritated by this answer, Bujai sent out couriers in various directions to summon the great vassals. He had brought with him men from the country of the Franks, skilled in using the balista. In the course of forty days he assembled an army of 30,000 men, who were brought to him by the princes and lords of Esfizar, Aazab, Herat rud, Kussuyé, Bakharz, Jam, Khavaf, Sarakhs, and other districts of Khorasan. The attack on the town began in the early days of March. Muhammed Sam commanded its garrison of 2,000 men, who were clad in mail, and to whom he had distributed gifts according to their rank. For three days a fight took place under the walls of the bastion Khak ber ser (afterwards called Khakister), when Bujai, seeing he had lost many men, withdrew to a distance and converted the siege into a blockade, determining to take the place by starving it out, for it was so well fortified that he despaired of its capture. The various approaches were closely guarded, but we are told Muhammed Sam made nightly sorties, and carried off many hundred horses. During these events Fakhr ud din died at Amankuh. Muhammed Sam tried to keep the news secret, and even produced a fabricated letter from the Malik, stating that he had been ill, but thanks to God he had recovered, and hoped the people of Herat would second all the efforts of Muhammed Sam. This letter was read to the magistrates and people, but the very night of the Malik's death one of his squires, Mozaffer of Esfizar, left Amankuh and repaired to inform Bujai, who was so pleased that he took off his robe and cap and gave them to his guest. He was also invited to a feast, attended by the maliks and other officers, who in their delight at the news spent the night drinking. The following day Bujai's camp resounded with the noise of musical instruments calling the troops to the assault. The defence was obstinate. During the fight Mozaffer of Esfizar cried out to the townsfolk, " Do not sacrifice yourselves ; Fakhr ud din died yesterday morning. I have come from Amankuh." Seeing the impression this news produced upon his people, Lakman the Ghurian shouted out from the top of a tower, " Miserable impostor, we received last night a letter written in Fakhr ud din's own hand." He then poured out invectives upon Bujai and the officers of his army, so that he began to doubt of the Malik's death, and Mozaffer had to confirm his statement on oath.* After an attack lasting for five days Bujai, again despairing of capturing the city by force, determined to sow discord among its defenders. He wrote a letter in his own hand to a Seistanian chieftain, named Shah Ismail, in these terms : " You promised some days since that you would arrest Muhammed Sam. If you were serious in making this promise, it should be fulfilled this week. Promise the inhabitants that I will show them pity, so that they will unite with you." Bujai then sent for one of his Herat prisoners, and said, " I intended

---

* D'Ohsson, iv. 513-518. Journ. Asiat., 5th ser., xvii. 486-487.

putting you to death, but I grant you life for the prolongation of that of our sovereign. You must go this very day into Herat. Say you have escaped, and let this letter drop on the threshold of the door of Shah Ismail." Bujai wrote another letter in the name of the people of Herat who were in his camp, informing Muhammed Sam that Shah Ismail was in correspondence with himself, and that he had better be on his guard. This letter was fastened to an arrow and shot into the town, and was taken to Muhammed. The next day Bujai's own messenger appeared at the gate of Herat. He was taken before the commander, and said he had escaped; but on being pressed about the reason why Bujai had let him go he produced the letter. Muhammed Sam at once deemed the whole thing an artifice, summoned Shah Ismail, showed him the letter, and said, "He wishes to create a difference between us by means of this." The two chiefs renewed their mutual promises to remain united, and following Bujai's example wrote a letter to a Heratee in his camp, named Fakhr ud din Zengui, saying, "It is a long time since you left us with the intention of killing Bujai. Why do you delay?" They wrote in a similar way to other people of Herat who had withdrawn to the Mongol camp. These letters showed Bujai that his ruse had been seen through.

There was in Herat an officer named Yar Ahmed, whose great bravery had gained him the favour of the Malik Fakhr ud din. He commanded 200 warriors. Envious of Muhammed Sam, he formed a plot with two other officers, Mahmud Féhad and Nikpei, to kill him and seize the citadel. To them he said that Bujai had offered, if he would betray Muhammed, to make him governor of Herat, to get the appointment confirmed by the Sultan, and to make over to him 10,000 dinars for distribution among his companions. The plot was to have been carried out on the morrow, but it was disclosed by Nikpei. Muhammed Sam informed his other officers of what he had heard, and they said it was necessary to arrest the conspirators. On the following day, as appointed, Yar Ahmed, with his sword by his side and his khanjar in his girdle, went up to salute Muhammed Sam, who sat in his hall of audience, with his guards ranged in ranks on either side. He had five companions with him. Muhammed received him in a friendly way, and told him laughingly to sit down. Yar Ahmed gave his sword to one of his followers, and seated himself on the sofa beside him. Directly after, Muhammed said to him, "Pehluvan, take out your khanjar." Ahmed did so, and gave it to his follower. The former then continued, "Pehluvan, is it thus that brave men act?" "What have I done to displease you?" said Yar Ahmed. "Have you not formed such a plot?" He denied it, but was arrested with Mahmud Féhad, and the next day they were executed on the public square. The same day 200 of his people left the town and went over to Bujai. Khuandemir tells us, on the contrary, that these 200 men were decapitated, and that their heads thrown over the ramparts informed Bujai that the plot had been discovered.*

Meanwhile Yassaul, the commander-in-chief, had arrived in Khorasan, and had sent several thousand men, under the orders of Muhammed Duldai, to reinforce Bujai. Duldai sent to Muhammed Sam a message from his leader, saying he would place him and the people of Herat under his own protection if they submitted. Muhammed Sam declared he had merely acted in self-defence in his attack on Danishmend, and was ready to submit to the general. Hearing of these negotiations, Bujai feared he might lose the fruit of his exertions if the town submitted to Muhammed Duldai, and after a consultation with his officers and the Malik's he sent word to Muhammed Sam that if he would send him Kutb ud din of Tulek, who had been captured by the garrison, and would promise to surrender the town and citadel to him, and not to Duldai, he would pardon the shedding of the blood of his father and brothers, and would bind himself , by a most solemn oath to do him no hurt, nor to any of the people of Herat. Muhammed destroyed Bujai's letter publicly, heaped insults on his messenger, and declared only that he would fight. Bujai's answer to this truculence was to press the siege. Famine now began to be felt, and was so severe that we are told a load of wheat cost eighty dinars, and 6,000 people perished of hunger. A crowd of starving people went to the gate of the citadel to implore the pity of Muhammed Sam. Another account says they entered the Great Mosque and cursed him, and asked in loud voices that the gates of the city might be thrown open. The commander allowed those to march out who had no food, and we are told that 5,000 thus left ; but they were driven back again with swords and sticks by Bujai's people, and the greater part of them perished on the banks of the Kartébar, on the roads, and along the walls. The next day Muhammed Sam removed the manacles from the feet of the Malik Kutb ud din Tulek, dressed him in a robe of honour, and sent him to Bujai with an offer of capitulation. Bujai gave his promise in writing to spare Muhammed's life, and the maliks, as well as his principal officers, also signed the document. On the following day Bujai's brother Togan had an interview with Muhammed on the banks of the Kartébar, and assured him that the engagements entered into with him would be faithfully carried out. The next day the city gates were thrown open, and the Mongols marched in. The inhabitants were told to march out, and orders were given to demolish the ramparts, towers, and walls. On Sunday, the 23rd of June, the people left the town and scattered themselves on the banks of the Kartébar. Muhammed Sam still held the citadel with 200 men. He now went in person to Bujai, who embraced him, seated him beside him, called him his son, and said to him, " I

---

forgive you the blood of my father and all your crimes. Rest assured, and open the gates of the citadel, so that I may send some of my men in." "All that you command shall be done," was the reply. Bujai gave him his own robe, and ordered his officers to offer him the cup, and afterwards gave him a feast in his tent. When Muhammed saw that Bujai was overcome with wine he withdrew under some pretext, and said to his people, " Bujai is drunk ; he has but a handful of men with him in the tent ; let us go in and kill them." They dissuaded him, on the ground that vengeance would be exacted from the people of Herat, and that they were too few to escape easily from the camp. Towards evening Muhammed Sam returned to the citadel. The next day Shah Ismail was well received by Bujai. Each day a fresh officer paid him a visit, and returned well satisfied with presents of horses and robes of honour. Bujai wished all the garrison to come out together and surrender, but Muhammed objected, and his small following daily diminished by desertion until he only had 100 men with him. Muhammed had sent word to Yassaul that if he would come the citadel and city should both be given up to him, and that he would place himself at his orders. Yassaul reached Herat with 5,000 men, four or five days after Bujai had taken possession of it. He sent word to Muhammed to go to his camp, promising to protect him from the resentment of Bujai. Trusting in this promise, he went with all his garrison. Yassaul had him and his people arrested, and handed them over to Bujai, in spite of his promise, telling him to execute them according to the Sultan's orders, and then to leave Herat, since he was merely to punish the assassins of his father and brothers, but not to exercise authority in the place. He put Taj ud din Yilduz, Lokman the Ghurian, and twenty other brave warriors, to death, near the Puleh Malan, and then raised his camp. Yassaul issued a proclamation to the inhabitants, bidding them return to the town and resume their ordinary occupations. They found it much dilapidated. Bujai took Muhammed Sam with him in chains, and ordered one of his relatives to conduct him to Uljaitu's Court, certain that Muhammed would lay the blame of the death of Danishmend upon Fakhr ud din, and thus obtain for him (Bujai) the Malik's inheritance. But Yassaul, afraid that Muhammed might charge him with betraying him after receiving splendid presents from him, was determined to prevent him reaching the Ordu. He accordingly sent a body of 100 horsemen after him, who overtook him near Tus, and conducted him to Béshuran, where Yassaul was. The same day Bujai arrived there from the Murghab, and demanded that he would surrender his prisoner, but Yassaul, feigning that he had an order from the Court, put him to death.*

Fakhr ud din was hardly dead when his brother Alai ud din seized his treasure, and began to behave in a sinister way towards his other brother,

---

* D'Ohsson, iv. 522-527. Journ. Asiat., 5th ser., xvii. 487-488. Note.

Ghiath ud din, who repaired to Uljaitu, by whom he was well received, and invested with the territory from Khorasan as far as India. This was a safe gift, much of it being altogether beyond the power of the Ilkhan to give away, being now in the hands of the Jagatai princes. But perhaps we must merely understand by the phrase the principality of Herat. Presently, Uljaitu, moved by some intrigues, put him in prison. After Bujai's capture of Herat, and revenge on the murderers of Danishmend, the Ilkhan's irritation was appeased, and Ghiath ud din was authorised to return to Herat as Malik. This was in the year 717 (*i.e.*, 1308).*

Let us now return to Uljaitu. In the beginning of 1307 an envoy came from the ruler of Kipchak, or the Golden Horde, to ask for the return of a beautiful maiden who had been captured in the war between Toktu and Nogai, and been sold into slavery in Persia. She was sent back again, and at the same time an order was issued that in future no Mongol maidens were to be sold as slaves within the Ilkhan's dominions, and that loose houses and wine shops were to be put down, one only being allowed in each district for the use of strangers.† On his return from Ghilan, Uljaitu spent some days feasting at his new capital of Sultania, and left there on the 7th of September for Hamadan, for hunting. On the 30th was celebrated the wedding of his daughter Dulendi with the amir Choban. The succeeding month he passed at Gaubari.‡ The festivities this autumn were enlivened by the dancing and singing of a famous performer named Rebiiol-Kolub (*i.e.*, the Spring of Life), who belonged to the tribe Ikdish, and was born at Baghdad. The Sahib Taj ud din Ali Shah Tarkhani, who afterwards became vizier, married this singer. He also presented Uljaitu with a splendidly equipped ship to navigate the Tigris, whose sails were decorated with stars, with white curtains and a canopy. This ship was fitted up with various luxuries, including fountains and a dancing-stage.§ A Kurd named Musa at this time gave himself out to be the Mahdi, the Messiah of the Shias. He drew a large number of Kurds after him, and the matter might have been serious but for the activity of the Mongol officials stationed in the country, who captured the false prophet and some of his followers, and sent their heads to the Royal Ordu.

We must now revert a little to bring up the history of the Armenian kingdom of Cilicia to this date. We have seen how the partially blinded Haithon II. was replaced on the throne by the Armenian barons. He almost immediately, however, probably owing to his infirmity of vision, abdicated in favour of his nephew, Leo IV., the son of his murdered brother Thoros, retaining the position of atabeg or governor of the young prince.|| The little principality of Armenia had been, as we have seen, greatly cherished by the Mongols, to whom it formed a kind of advanced post towards their mortal enemies the Egyptians. This alliance naturally

---

* Journ. Asiat., 5th ser., xvii. 487-489.   † Ilkhans, ii. 198.   ‡ D'Ohsson, iv. 529-530.
§ Ilkhans, ii. 198.   ¶ Abulfeda, v. 173. Journ. Asiat., 5th ser., xvii. 385.

brought upon it the bitter hatred of the latter, and whenever an opportunity arose it was cruelly punished. Two months before Ghazan's death the Egyptian Sultan sent an expedition to Cilicia, under pretence of avenging a body of troops which had attacked that district from Aleppo, and had been in turn assailed by the Mongols. The amir Bedr ud din Bektash was appointed commander of this expedition. He left Egypt in March, 1304, and was joined *en route* by contingents from Damascus, Hims, Hamath, Tripoli, and Aleppo. Bektash was delayed in the last of these cities by illness, but his troops marched on under his son. They were divided into two bodies. One advanced by way of Kalat ur Rum and Malattiya, and the other by Derbend. After laying waste the country, and killing and making prisoners many of the inhabitants, these two divisions united under the walls of Tel Hamdun, which capitulated on the 17th of June. After this the Egyptian army retired.* The following year (*i.e.*, in July, 1305), Shems ud din Kara Sonkor, the governor of Aleppo, under pretence that the Cilician king had failed to send the usual tribute, sent a body of 2,000 men under Kush Timur, one of his Mamluks, again to invade Cilicia. The King offered him a considerable sum of money if he would retire.† The Egyptians continued their course of rapine and destruction, burning a number of villages, until threatened by a body of 6,000 Armenians, Mongols, and Franks. A struggle ensued at Arasa, in which they were defeated, and were pursued by the Mongols and Armenians. Only a few returned home again with Kush Timur, and Marino Sanuto says that of 7,000 who reached Tarsus only 300 escaped.‡ The regent or atabeg, Haithon, now wrote to Kara Sonkor to say that it was the Mongols, and not his people, who had attacked the Egyptians. He promised to ask Uljaitu to release four Egyptian officers who had been captured and taken off to the Ordu. He also sent rich presents, and promised to pay his tribute regularly. Kara Sonkor sent word on to the Sultan, who accepted the presents and excuses.§ The same year the historian Haithon, who was a grandee of Corhicos, retired to Cyprus, with the permission of his namesake, the King of Armenia, and joined the order of the Premonstantensians. Having gone to Rome in 1306, he thence went on to Poictiers, in France, where he dictated in French his Oriental history, which we have so much quoted. This was translated into Latin by Nicholas Salcon, by order of Clement V.||

Bilarghu, a relative of the noyan Togachar, was the commander of a small body of troops under Irenchin, who in 1306 was appointed governor of Rum. He was encamped in Cilicia, and being a fanatical Mussulman persecuted the Armenian king bitterly. In the spring of 1308 he went

* D'Ohsson, iv. 530-531.　　† Makrizi, ii. (part ii.) 254.
‡ Abulfeda, v. 197, and note 121.　D'Ohsson, iv. 532.　　§ D'Ohsson, iv. 532.
|| Hist. de la Géorgie, 631. Note 1.

with his patron Irenchin to the Ilkhan's Court, where their conduct was approved, and whence they returned with their authority more firmly established. Bilarghu had heard that Leo had complained of him at Court, and resolved to ruin him. He entered the district of Sis with 500 men, and asked that twenty-five of his men might be allowed access to the strong fortress of Anazarba. This the King was obliged to grant. Leo, who paid tribute both to the Mongol ruler and also to the Sultan of Egypt, sent word secretly to the latter that Bilarghu, who disposed of the revenues of the kingdom as he pleased, prevented him from paying his tribute. The Sultan sent an officer to inquire into this, who informed Bilarghu of Leo's message to his master. The Mongol commander sent for him to have a conference with them. The King accordingly went, accompanied by his uncle, the late king, Haithon II. (who acted, as we have seen, as his atabeg or tutor), by the generalissimo Oshin, Haithon's brother, and some forty barons. He was introduced to Bilarghu alone. The latter, after a few minutes, rose as if he was going to make his *namaz*, when, drawing his sabre, he decapitated the King while he was himself saying the *tekbir*. When his people heard him invoking God they proceeded to slaughter Leo's attendants. The atabeg Haithon, Leo's uncle and predecessor, was also put to death. This was on November the 17th, 1307. At the news of this treachery the governor of Anazarba put Bilarghu's soldiers to death, while he ordered watch fires to be lighted as a signal to the other strong fortresses. Bilarghu now went to Anazarba, hoping his soldiers would open the gates to him, but he was received with a shower of stones and other missiles. He was allowed to withdraw, we are told, because the Armenian troops did not wish to harm the troops of their suzerain. Meanwhile Oshin, Leo's uncle, set out for the Ordu, but was arrested at Sivas by order of Bilarghu. Irenchin, who was on his way from the Court, set him at liberty, and reported to his master how matters stood. Both parties were cited to appear. They pleaded their cause before Uljaitu, who pardoned Bilarghu, but shortly after was prejudiced against him, and had him put to death. Oshin, the youngest of the five brothers of Haithon II., was now raised to the throne of Cilicia, and was duly consecrated at Tarsus.* Abulfeda says he was nominated king by Bilarghu, and that it was his brother Alinak who went to complain to Uljaitu of the murder of his relatives.†

Uljaitu, as we have seen, was in his young days a Christian, and had been baptised with the name Nicholas. On his conversion to Muhammedanism by his wife he attached himself to the Hanéfi sect, by the Imaums of which he had been surrounded in Khorasan. He naturally favoured this sect, and had the names of the first four Khalifs inscribed

---

* D'Ohsson, iv. 532-536. † Op. cit., v. 205.

on his coins. Under this patronage the sect grew insolent, and attracted the hatred of several powerful people, notably of the vizier, Rashid ud din, who was attached to the Shafiyi rite, and patronised and protected its professors. He, however, concealed his resentment in consideration for his master. Notwithstanding his preference for the Hanéfis, Uljaitu appointed Nizam ud din Abd ul malik, of Meragha, who was a learned Shafiyi doctor, to be chief judge of Iran, thus raising him above all the Hanéfi magistracy. Nizam ud din seems to have owed this appointment to the skill he showed in his polemics with the doctors of the rival sect. In 1309 the son of Sadr Jihan of Bokhara, who was a devoted Hanéfi, having visited the Ordu, determined to undermine the position of Nizam ud din, and one day, in the presence of Uljaitu, pressed him, with a question relative to marriage with a woman who had been born in adultery. The judge repudiated the interpretation put upon the opinions of his sect, and in turn assailed his opponents with the charge that they permitted a man to marry his mother or his uterine sister, and further cited against them the axiom of the " Manzuméh," a Hanéfi treatise : " Pederastry is not forbidden ; if you marry your sister, do not consummate it." This discussion was much to the distaste of Uljaitu and his officers, who said : " What have we been doing to abandon the religion of our ancestors, the faith of Jingis Khan, for this Arab religion, which is divided into so many sects, and which permits a man to cohabit with his mother, sister, or daughter ? Let us return to our ancient religion." Although such marriages were really forbidden by the Muhammedan law, an opinion spread among the Mongols that it was not so, and the princesses were especially outraged. They began to despise those bearing a doctor's turban, and the Mongols in general took a dislike to Muhammedanism. While this feeling was still warm, Uljaitu, on his return from Arran, stopped at Gulistan, in a summer-house built by his predecessor, and witnessed a terrible storm, during which several people who sat about him drinking were killed by lightning. Frightened at this, he set out hastily for Sultania. After this he always carried eagles' feathers, jade, and other stones, which were supposed to protect people from lightning. The Mongol grandees declared that, according to the national practice and the yasa of Jingis Khan, the Sultan should pass between two fires, and they summoned some Bakhshis or Shamans to preside at the ceremony. The Lamaist priests declared that misfortune had overtaken him because he had become a Mussulman, and entreated him to abjure it. Uljaitu spent three months in hesitation, saying to those who urged him to recant his new opinions, " How can I abandon the Muhammedan faith, which I have professed with so much zeal ?" The amir Taremtaz replied that Ghazan, the ablest man of his time, had embraced the religion of the Shias, and that the Sultan would do well to imitate him. " How, wretched creature," said the prince, " do you wish me to be a *rafizi* (heretic) ?" Taremtaz, who was a ready person, made a judicious reply, and contrived to exalt the Shias at the expense of the Sunnis. "See," he concluded finally, "in what they differ. It is as if the Shias were to maintain that the succession belonged of right to the descendants of Jingis Khan, while the Sunnis pretended that it also belonged to his generals, the karajus (*i.e.*, to his subjects)." This neat and judicious speech made an impression on the Sultan. He was also moved by the declamation of several Alévi Imaums at the Ordu, who continually attacked the Sunnis, but they were powerfully answered by the Mollah Nizam ud din. During the latter's absence on a journey to regulate the vakhfs of Azerbaijan, in 1310, Uljaitu visited the tomb of Ali. There he had a dream, after which he determined to embrace the doctrine of the Shias, and desired his generals and courtiers to follow his example. They obeyed, except Choban and Issen Kutlugh, zealous Sunnis. They resisted all the importunities of the Shia Imaums and Seyids, who feared these powerful grandees, to convert them.

The conversion of Uljaitu was followed by a change in the formula of the khutbeh, or Friday prayer, in which the names of the three first khalifs were suppressed, and those of Ali and his two sons, Hassan and Hussein, and that of Ali Muhammed, the Mahdi, alone retained. The type of the coins was also changed. The principal doctors of the Alévi sect were summoned to the Court, where the Ilkhan liked to discuss the dogmas of the faith. Wishing to encourage learning, he founded a kind of nomadic college, which was held under tents in his ordu. He attached to it as professors five of the most learned men in Persia. Their disciples, to the number of one hundred, were entertained at the cost of the Ilkhan, who was followed about the country by this movable college, for which special horses were supplied. He had already founded near his tomb at Sultania a college with sixteen professors and assistants, which could accommodate two hundred students.* In regard to his change of religion, the author of the "Shajrat ul Atrak" tells us that Uljaitu was reputed to be the most just of the descendants of Jingis Khan, and a great protector of seyids and learned men ; but in the year 709, at the instigation of some persons of the Imaumia sect, as Sheikh Jemal ud din Hussein, the son of Seyid Bedr ud din Mizhuhur Hulubi, &c., who was making proselytes for the Imaumia religion, he was induced to add to the profession of the faith inscribed on the coinage from the time of Sultan Ghazan, the words, "*Ali wuli Allah*"—"Ali the friend of God." He also expunged the names of the three khalifs from the khutbeh, and after the name of Ali, the fourth khalif, he inserted those of Hassan, Hussein, and Ali Muhammed, the Mahdi. He also inserted in his sekka the names of the leaders of the twelve sects, and for this the Sunnis called him in disgust Sultan Muhammed Kharbundeh, or "the slave of an ass," but the Shias Khodabendeh, or "the slave of God."

---

* D'Ohsson, iv 536-542. Ilkhans ii. 216-219. Quatremere, 437.

We now read of Uljaitu quarrelling with his principal vizier, Said ud din Sauji, or Savaji (*i.e.*, a native of Sava). He had forbidden him giving assignations on the revenue, or granting pensions and salaries, and wished that the revenue should be handed over to him intact, so that he himself might dispose of it. The Vizier had also drawn upon himself the hatred of several powerful chiefs, notably of Tokmak, who was a favourite of the Sultan; Ali Shah, who was also a *protégé* of the Ilkhan; and of Rashid ud din, formerly a friend of Said ud din. The Vizier had, in fact, courted this feeling by his arrogant behaviour. His employés were so numerous that anyone wishing to see him had to see successively some thirty-five officials, to whom it was necessary to make presents by way of blackmail. This bureaucracy absorbed the revenue, which amounted to 30,000,000 drachmas. The Sultan's anger at getting so little of the revenue received an impulse when two employés of the Vizier, having quarrelled at Sultania, accused each other of having taken a considerable sum from the treasury. Said ud din, alarmed at the consequences that such imprudence might bring, ordered the Seyid Taj ud din Uj to summon them before him in order that they might be reconciled, and to inform them they must not in future let a word drop about the revenue. Shortly after the Vizier met two others doing the same thing, and ordered them to go to Taj ud din, and to hear his admonition as if it had been made by himself. They accordingly went, and had the same homily preached to them; but on going out they went and reported the whole matter to Rashid ud din, who informed the Ilkhan that several of the employés of the Divan had sworn to make no disclosures about the affairs of the treasury. Uljaitu sent from Baghdad an order for the apprehension of Said ud din, and to try his dependants. Five of the latter were condemned to death, and Said ud din himself was executed the same day (*i.e.*, the 19th of February, 1312). Their goods were all confiscated, and large sums were extorted from their subordinates by torture.[*] Wassaf says that with the Vizier, the Great Khoja Bedr ud din Luli and seven great amirs named Kairbuka, Urba Kerim, Khuljin, Daujah, Sain ud din Museri, Shiab ud din Mubarek, and Nasir ud din Yahyah were put to death, their houses plundered, and their estates confiscated.[†]

About this time Uljaitu laid the foundations of a new town, in the *Jemhal* (*i.e.*, the defile), on the way from Baghdad to Sultania.[‡] He called it Sultan Abad. Choban, the greatest of the amirs, was sent to Sultania, armed with a golden bull, to superintend the dedication of the high school, called Seyar, there. Four thousand workmen were employed in making a huge bazaar, 300 ells in diameter and 100 ells wide, which was supplied with sweet water, and became the residence and rendezvous of the dealers in cloth. At this time also an order was

---

* D'Ohsson, iv. 542-544.   † Ilkhans, ii. 219-220.   ‡ *Id.*, 220. D'Ohsson, iv. 544-545.

given that the various works of Rashid ud din, including not only his history, but also his exegetical and juridical works, the description of the seven climates, genealogical tables, &c., were to be transcribed in ten volumes, consisting of 3,000 leaves. The writing out, gilding, and binding of the work cost more than 600,000 dinars. Quatremere says 60,000 dinars (*i.e.*, 900,000 francs). These famous works were ordered to be deposited in the newly-founded mosque.* Rashid ud din's income at this time must have been enormous, and Uljaitu lavished great sums upon him. The author of the "Mesalek Alabsar" cites the Sheikh Mahmud, a native of Ispahan, for the following story: One day Rashid presented his master with one of his works, and said to him, "Aristotle having offered Alexander a work he had written, received from him a present of 1,000 gold pieces; a prince as magnificent as yourself should deem it unworthy not to equal the liberality of Alexander." The Sultan answered this challenge by making over to him a domain worth three times the amount of Alexander's gift.†

Ali Shah had been a jewel merchant and dealer in precious stuffs and other things, and he had been brought into commercial relations with the amir Hussein Gurkan and the Prince Oljitai, who introduced him to the Ilkhan. He was adroit, insinuating, and witty. The vizier, Said ud din, who was jealous of him, sent him to take charge of the royal manufactory at Baghdad, where he introduced many improvements. When the Ilkhan went there Ali Shah gave him a splendid feast, and, among other presents, he offered his master fourteen *rotl* of the richest jewels and pearls, a richly embroidered cap, decked with stones, and having on its summit a ruby of four-and-twenty miskals in weight; nine beautiful boys, with girdles made of chrysoliths, in robes of moghribin; and gold-harnessed horses. Quatremere also mentions a splendid ship.

We are told that the Baghdad singer previously named, who was probably a gipsy, whom he married, contributed to his rise by the admiration she inspired the Sultan with. Ali Shah erected a splendid bazaar and other buildings at Sultania, which pleased Uljaitu very much. The late vizier, Said ud din, had treated the favourite with scant courtesy, refusing to rise and meet him when he entered the room, &c., while his colleague, Rashid ud din, on the other hand, had showed him marked attention. On one occasion, Ali Shah, having given the Sultan and his grandees a splendid feast, offered Rashid ud din three pieces of rich stuff, and afterwards a similar present to Said ud din. The latter, who had taken too much wine, professed to consider it a slight that his colleague should have been first considered, and abused him in an unbecoming manner. Rashid ud din kept a discreet silence, which won him the good opinion of his master, who was indignant at the Vizier's

---

* Ilkhans, ii. 220-221. Quatremere, xix.   † Quatremere, xviii.-xix.

conduct. The latter, as we have seen, presently fell a victim to his two rivals.* When he was beheaded, Ali was, at the instance of Rashid, given his post.†

Soon after this, a Jew named Nejib el devlah, who was in the service of the Amir Togmak, concerted with another Jew a plot against the Vizier. A letter was written in Jewish characters, in Rashid's name, and addressed to a jeweller, who was a confidante of one of the chief amirs. In this letter Rashid urged his correspondent to poison the Sultan. The letter was allowed to fall into the hands of the amir Lulu, who passed it on to Uljaitu. The latter, in a violent rage, summoned Rashid, whom he sharply attacked. He asked for a delay of three days in which to discover the authors of the plot. This was disclosed by the amir Muhammed, who had been devatdar of Said ud din, and who confessed that the note had been written by a Jew at the instance of Said to undo Rashid. The Jew was summoned to the Royal presence, and having been duly examined, confessed. He was executed. Some time after Muhammed Zerker, nephew of Said ud din, Taj ud din Uj, the leader of the Shias, the amir Muhammed Seyar, and several other amirs, who, according to the Jew, had been accomplices in the plot, were also put to death. This was on the 10th of April, 1312.‡ The Seyid Ainad ud din had his eyes seared, but did not lose his sight.

Another tragical event which occurred at this time, if it took place as reported by one historian, a descendant of Ali, reflects considerable discredit upon Rashid. There lived at Baghdad a Seyid or descendant of Ali, named Taj ud din Abu fazl Muhammed, who had formerly been a *vaiz*, or preacher, and having secured the good opinion of Uljaitu was by him promoted to the post of Nakib, *i.e.*, head of all the descendants of Ali, who were scattered in Irak, Rai, Khorasan, Persia, and generally in the Ilkhan's empire. The historian above referred to, Omdat Altalib, reports that not far from the Euphrates, between Hillah and Kufah, is a town where, according to tradition, was buried the body of the prophet Ezekiel. The Jews had a great reverence for this place, often went there on a pilgrimage, and gave large alms there. The Nakib forbade them access to this town, and having put a pulpit within the sacred structure had the Friday prayer said there. Rashid, according to this author, was much grieved at this prohibition, and was jealous of the Seyid and the favour Uljaitu showed him. He determined therefore to ruin him and his son Shems ud din Hussain, who was head of the Seyids of Irak. He had by his tyranny and ill-conduct disgusted many of them, and through them Rashid succeeded in poisoning the Sultan against Taj ud din and his children. Uljaitu having appealed to the Vizier as to what should be done, he, to avoid all controversy, suggested that Taj ud din should be

D'Ohsson, iv. 545-546.　　　† Quatremere, xxiii.　　　‡ *Id.*, xxiii-xxiv.

tried by the Seyids themselves. Meanwhile, he incited a violent and
sanguinary person to kill his rival and his two sons, promising him the
posts of nakib, kadhi, and sadr of Irak. He recoiled, however, from such
a crime, and said he would never assassinate a descendant of Ali, and the
same night fled towards Hillah. Rashid then tried another Seyid, but
equally in vain. Eventually a third, also of the sacred caste, descended
from Ali, Taj ud din bin Mokhtar, who was in Rashid's service, and
completely devoted to him, having secured Taj ud din and his two sons,
Shems ud din Hussein and Sharf ud din Ali, conducted them to the borders
of the Tigris, and there had them put to death by his satellites. By the
Vizier's suggestion the two sons were killed before their father's eyes. The
people of Baghdad and the Hanbalis wreaked their vengeance on the body
of the Seyid, which was cut in pieces, and the pieces themselves were
eaten. His hair was pulled out, and each hair in his beard was sold for a
golden piece. To appease Uljaitu, who was much grieved at these events,
Rashid declared that it had come about with the assent of all the Seyids
of Irak. The Sultan wished to suspend the Kadhi of the Hanbalis from
a gibbet, but spared his life on the representation of several eminent
persons. He nevertheless had him mounted on a blind ass, with his head
turned towards its tail, and promenaded about the town, and ordered
that in future the Hanbalis should have no kadhi. Quatremere urges
strongly that this story has many incredible features, and argues that it is
incredible that Rashid, who was of a tender disposition, should have
outraged all decency. It is more probable that Taj ud din had committed
some grave offences and been duly found guilty, and that the details of the
story as it affects Rashid were made up by the descendants of Ali, who
were enraged at the violent end of their chief. Wassaf certainly supports
this view. He tells us how in the year 711 (1311) the principal amirs and
viziers, in the presence of the kadhi of the kadhis and of a large number
of Imaums and Seyids, formed a court to try Taj ud din, who was accused
of many crimes : *inter alia* of having appropriated a sum of 300,000 gold
pieces belonging to the Seyids and others, of having continually tried to
seduce the wives of the descendants of Ali, of having committed several
murders, &c. Having been duly convicted, he was handed over with his
two sons to the inhabitants of Meshed Ali, who were charged with his
punishment. These Seyids accordingly took him to the borders of the
Tigris, and beat him till he died. His sons shared his fate. The crowd
pressed so much to take part in the execution that two or three people
were seriously injured. Wassaf says that all present—Mussulman men
and women, Jews, Christians, and men of all classes—filled the air with
acclamation, and regarded the death of Taj ud din as assuring the triumph
and the stability of religion and the empire. As Quatremere says, the
fact that he was a descendant of the Prophet, and yet convicted by a
court comprising the heads of religion, and that there was general joy on

his execution, certainly seems to point to Rashid having been guiltless of mere personal vengeance in this matter. Soon after these events, Jelal ud din, son of Rashid, was appointed governor of Ispahan.* On Friday, the 24th of Moharrem, 712, Wassaf was presented to Uljaitu by Rashid, who greatly praised him and his work, of which Uljaitu consented to hear him read a chapter. This was done in the presence of the two viziers and the various grandees of the Court, who duly applauded him.† The obligations Wassaf was under to the great Vizier must always, however, be remembered when he is quoted in his praise.

In August, 1312, there arrived at Sultania several Egyptian officers, who sought asylum there from Sultan Nasir. Nasir had grown weary of the condition of tutelage in which he was held by the amirs Salar and Bibars, who had acquired great renown in the Mongol wars, and who exercised authority in his name, and he had abdicated in 1309 and retired to his principality of Al Karak. Thereupon Bibars, styled the cupbearer, a Circassian by birth, became Sultan of Egypt, with the title of Malik Mozaffer. A year later Nasir, recalled by a party opposed to Bibars, mounted the throne for the third time, and had Bibars strangled in his presence. Salar died the next year.‡ Among Nasir's supporters the most powerful was probably Kara Sonkor, who had had a strangely turbulent life. A slave of Sultan Kelavun, he was a party to the murder of his son Ashraf, and afterwards, with Lachin, made an attempt to kill Ketboga. During Lachin's reign he was appointed viceroy of Egypt. He then fell into disgrace. He was governor of Aleppo during the reign of Kelavun, and of Hamath during Nasir's second reign, after which he again became governor of Aleppo, and soon, as a reward for his services, was appointed governor of Damascus. Nasir, notwithstanding this, did not trust him. At his own request he transferred him again to Aleppo, and sent an officer named Arghun el Dévatdar to invest him with the post, but bearing secret letters to the various chiefs at Damascus to arrest him. He suspected that something was wrong, and kept a close watch upon Arghun, and cleverly succeeded in getting a promise from him, in the presence of his chief supporters, that he had no sinister intentions, and had only come to invest him. Thereupon, having distributed whatever money and precious things he had among his Mamluks, so that they might carry them in their girdles, he set out, accompanied by Arghun and surrounded by 600 Mamluks. He entered Aleppo on the 8th of June, and thence sent Arghun home again after giving him 1,000 dinars, a state robe, a horse, and other gifts. Kara Sonkor was by no means reassured, and determined to secure some friends. He made overtures to Hosam ud din Mohanna, prince of the Bedouins of Syria, and his son Musa, having first persuaded the former that he had received orders to arrest him

---

* Quatremere, xxiv.-xxxi.　† Id., xxxi.　‡ D'Ohsson, iv. 547.

which he did not mean to carry out. Having got Nasir's permission, he set out on a pilgrimage to Mekka; but hearing that the Sultan had posted some Mamluks on the road to waylay him, he determined to return. Meanwhile the gates of Aleppo had been closed against him, and it was only when Mohanna threatened to attack the place that he was allowed to take away his property. He then went towards the desert. Kara Sonkor, after a while, went to Rahbet with Mohanna, Mogoltai, three amirs from Damascus, and a party of Mamluks, whither they were pursued by the Sultan's troops. From Rahbet he sent his wives, his son Ferej, and horses to Egypt. Akush el Afrem, the governor of Tripoli, who had joined him, also sent his son Musa. These people were ordered to prostrate themselves before the Sultan, and to assure him that it was only the fear of his displeasure that had made their masters pass into the enemy's country. Having obtained Uljaitu's permission they set out for Sultania, and were treated with great honour *en route* and supplied with every necessary, the amirs Kutlugh, Kiaf, and Uji receiving orders to go to Diarbekr to welcome them. They arrived at Sultania in August, 1312, with an escort of 1,000 horsemen, the heads of the administration, and the Muhammedan clergy having gone by the Sultan's orders to meet them. Uljaitu presented them with splendid khilats, comprising a robe, cap, and girdle, ornamented with precious stones. He also distributed appanages among them. Kara Sonkor was given Meragha; Akush el Afrem, Hamadan; Serdkesh, Nehavend; and Sonkor Afrem, Asadabad. The name of Kara Sonkor (*i.e.*, black falcon), who was an old man, was changed to Ak Sonkor (*i.e.*, white falcon), a change aptly compared by Von Hammer with the change of the name Thunichtsgut into Thugut by Maria Theresa. The Bedouin chief Hosam ud din Mohanna was presented with Arab horses and falcons, and Uljaitu gave him one of his own robes, and an assignation of 3,000 loads of corn, which were to be supplied by Irak Arab and Diarbekr. Later, when Mohanna's son, the amir Suliman, submitted to the Ilkhan, at Mosul, he was rewarded with robes, &c., and a tuman of gold. Hilleh, Kufa, Shifateh, and their revenues, which amounted to forty tumans, were also made over to Mohanna.* In 1315 Suliman, with a body of Tartars and Arabs, made an attack on the Arabs and Turkomans near Tadmor, and retired again behind the Euphrates with an immense booty.†

Uljaitu now determined to invade Syria, and preparations were made on a great scale. Timber and inflated skins, for making bridges or rafts, were brought together at Sinjar and Mardin, at a cost of 50,000 dinars; siege machinery was sent from Baghdad, 1,500 suits of armour and helmets were bought from European merchants; 260 of the most beautiful Arab horses, from Nejd, were secured as led horses for the Ilkhan's

---

* D'Ohsson, 547-555. Ilkhans, ii. 223-227.   † Abulfeda, v. 299.

riding, and 2,400 camels as sumpter beasts; 90 catapults or mangonels,
70 grappling machines, filled with iron hooks, for attacking the towers,
100 bottles of naphtha, 100 kettle-drums, 100 banners, 360 sappers,
and 15,000 hides as covers for the baggage. Uljaitu set out on the 15th
of December, 1312, accompanied, *inter alios*, by the viziers and the three
noyans Choban, Sevinj, and Isenkutlugh, and ten tumans (*i.e.*, 100,000 men),
and also by the Georgian king with his men. At Mosul, where they
arrived in the middle of December, the armour was distributed. At the
end of the month they reached the Euphrates, which was crossed at
Kirkesia (the ancient Kirkessium), where the Khabur (the Khaboras, or
Mygdonius) falls into the main river. They then proceeded to attack
Rahbet or Diar Rahbeh, a fortress situated on a hill. The place
was summoned by the amirs Ali Kushdji, Jemal ud din Afrem, and
Haji Dilkandi; but its commander, Bedr ud din Musa ibn Uskeshi,
refused to open its gates, and trusted to the strength of its twelve
bastions, the height of its walls, and its ditch, 30 ells deep and 15 wide,
with its scarp lined with stone. The army was ordered to cut
down wood with which to fill this ditch. The wood was carried
on a kind of carts called *tewarch*, and in three days the ditch was
filled up to the height of the walls. Nejm ud din, the judge of the place,
now came out, with three companions, to treat. He was presented with
a kaftan, and negotiations were begun for the withdrawal of the garrison
and the surrender of the town; but meanwhile weapons continued to be
fired from the walls, which the sappers were mining. Uljaitu having
ridden round the town, and deeming it hopeless to capture the place,
fearing also a want of provisions, offered to allow the garrison to go out,
with its property, on the condition that seventy-one Egyptian amirs in the
fortress would give a written undertaking not to take up arms against the
Padishah in future, but would be friends of his friends and enemies of his
enemies. He also offered to give 5,000 dinars towards the completion of
the mosque there. According to Wassaf, who was a poet and had
a poet's licence, the negotiations were on the point of succeeding when a
Mongol applied fire to the piled-up wood, which burnt up as high as the
top of the towers, whereupon the army withdrew. Von Hammer says it
is more probable that Uljaitu, seeing that it was inevitable the siege must
be raised, had the wood fired before his withdrawal.[*] Novairi, who cites
the report of the commandant of Rahbet to the Sultan, says the siege was
begun on the 23rd of December, and continued till the 25th of January,
when the Mongols retired (as they were just going to capture the
place), leaving all their siege machinery, baggage, and horses behind
them,[†] which is no doubt an exaggeration. Abulfeda says a pestilence
broke out among the besiegers, and that the machines they left behind

---

* Ilkhans, ii. 229-230.   † D'Ohsson, iv. 555.

were dragged into the town by the citizens.* The writer of the continuation of Rashid ud din says the siege was commenced on the 4th of January, 1313. Its commander, Bedr ud din Kurd, at first wished to surrender, but the next day changed his mind and made a vigorous defence. The Mongols proceeded to batter the walls with their catapults, and at last the garrison offered to give in, and in fact capitulated on the 13th of February. Uljaitu allowed it to remain in the place.† Nasir, with his Egyptians, was already *en route* for Syria when he heard, on the 6th of January, of the retreat of the Mongols. He disbanded his men, but went on himself to Damascus, whence he sent orders to Cairo to sequester Kara Sonkor's house, and to take anything precious there to the treasury. This comprised a considerable sum in gold and silver, and other objects of great value.‡ He then went on to Mekka on a pilgrimage.§ The retreat from Rahbet was evidently not treated by the Mongols as a serious reverse. Six days after crossing the Euphrates a feast was held, at which Wassaf was openly congratulated by the Ilkhan on the sensation which his history had created, and was presented with a cup of unfermented wine by him in person.||

In 1312, Kurmishi, son of Konghuratai, revolted in Rum. He was captured by the general Taremtaz, who was sent against him, and was put to death with his four sons. Uljaitu having heard in June, 1314, that the Karamanian prince, Mahmud bey, had captured the city of Conia, or Iconium, sent the amir Choban to Rum at the head of three tumans. The revolt of these Turkomans was attributed to the tyrannical conduct of Uljaitu's uncle, Irenchin, who governed that province, and who now repaired to the Ilkhan's Court. Choban went to invest Conia, where the rebel prince had shut himself up, and was accompanied by Georgi, the Regent of Georgia, and his troops. Rum was at this time devastated by famine, caused by a swarm of locusts, and as his provisions began to fail, Choban had recourse to negotiation. Mahmud bey offered to capitulate. He asked only for a delay of a few days, and to prepare his presents, but on the night of the last day of the truce he escaped by way of Larenda. Pursued, and unable to escape, he went with a shroud about his neck and a sword in his hand to ask Choban's pardon. He granted him his life, and after occupying Conia, returned to Persia by order of the Ilkhan.¶ The "History of Georgia" calls the rebels the sons of Phariman, or Pharwana (*i.e.*, makes them sons of the Pervaneh, whose career occupied us in the chapter on Abaka Khan), and says Choban was accompanied by the Georgians, and by Georgi (*i.e.*, probably, by Georgi V., son of Dimitri). The partisans of Pharman only resisted for a year. All the Georgians who were found there were made over to King Giorgi, who, with his people, had behaved very valiantly in the struggle.**

' Op. cit., v. 269.   † *Id.*, Note 2.   ‡ *Id.*, 556.   § Weil, iv. 309
| Ilkhans, ii. 230.   ¶ D'Ohsson, iv. 576-577.   ** Op. cit., 642.

In 1313 Uljaitu nominated his son Abusaid, who was only nine years old, governor of Khorasan, which was a kind of Dauphiny of the empire. The young prince was accompanied by the amir Sevinj, who had brought him up, and the amir Algu, two experienced and skilful generals, who were to act as his atabegs or governors. The grandees of the Court, on Uljaitu's order, attached one of their relatives to the suite of the young prince, and Abd ul Latif, son of Rashid ud din, was appointed his vizier. The two Grand Viziers were ordered to supply him with money, precious stones, and stuffs. He was given drums and cymbals, the Além and Sanjak (i.e., banners), arms, and armour, Arab horses, with harness decorated with stones, and everything necessary for his equipment. This is D'Ohsson's account, based apparently on that of the continuator of Rashid.* Wassaf reports, according to Von Hammer, that Abusaid was accompanied by Sevinj Noyan as beglerbeg, Selasun as atabeg, and Abdulla as administrator of the Divan. Sitai Kutlugh, son of Kutlugh Shah, who was killed in Ghilan, was commander of the right wing, and Nuvinsadeh Hassan of the left wing. Tokal, son of Istan Kutlugh, and Rustem, son of Melai, had charge of his administration. There were also falconers (kushji), cooks (andaji), taxing masters (khasanji), standard bearers (aalemdar), chief marshals (yesaul), waiting men (odabashi), masters of the horse (akhtaji), and confidential servants (inak). The several names are given by Wassaf. The vizier, Khoja Taj ud din Ali Shah, had charge of the general administration and the equipment of the army.† Uljaitu accompanied his son as far as Abher, where he gave him a parting feast and addressed some complimentary phrases to the amirs, especially Sevinj : " It is because I know your ancient services and I count entirely on your fidelity that I confide my son to you and the relatives of my amirs and ministers. You will treat them with paternal care, and they will be duly submissive, but beware lest you be overcome with pride at having brought up a sovereign and at being the guardian of his son, and that this thought, which is enough to excite ambition, do not lead you into enterprises which may trouble the State, and that I have to punish you severely." Sevinj bent his knees and protested his zeal and goodwill.‡ In the very beginning of his rule in Khorasan there was an invasion on the part of the Jagatai princes, which will be described under that Khanate. The Ilkhan's generals, Yesaul and Ali Kushji, who were defeated, joined Abusaid at Yulak Koshuk Murad.§ After the withdrawal of the Jagatai troops one of the princes of Jagatai, named Yassaur, having quarrelled with the Khan, appealed to Uljaitu for help. The latter sent an army to his assistance, under the command of Yassaul, which overran Mavera un Nehr. Yassaul was accompanied on this expedition by the Prince of Herat, Ghiath ud din.

We will now bring up the history of Herat to this date. We are told that in 1310 Ghiath ud din received at Herat several of the Ilkhan's officers—Yassaul, Tukal, Khajeh Alai ud din Hindu, Jemal ud din, Shah Muhammed, &c.—whose friendship he tried to secure; but Alai ud din Hindu, who was of a jealous disposition, profiting by the absence of his colleagues, wrote a letter in concert with Dildai and Bujai, sons of Danishmend, denouncing Ghiath ud din as a traitor who meditated revolt, for which he was preparing, and intended eventually to shut himself up in the fortress of Khaisar. This letter, with others written by the Hindu to his friends, was laid before Uljaitu when under the influence of drink. The latter ordered one of his chamberlains, called Eutek, to go and summon him. The Prince of Herat at once obeyed, leaving the place in charge of his uncle, Shems ud din Omar Shah Khonduri. The King left Herat in August, 1311. Uljaitu would not receive him, and deputed several of his officers to try him. He answered frankly that he had always been faithful, that in increasing his army it was only to strengthen the frontiers, and that the best proof of his innocence was his coming in person. Uljaitu was pacified, but he had Ghiath ud din put under surveillance, pending his being confronted with his accusers, and he was thus detained for a long time. Meanwhile, the sons of Danishmend, who remained at Herat, committed great disorders there. At this time the Malik's brother, Alai ud din, who was one of his greatest foes, died, and the princes of Jagatai having crossed the Oxus with 50,000 men, marched upon Fariab, and encamped at Murghab. Yassaul, with the two sons of Danishmend, marched against him with 80,000 men; but when within five parasangs of their enemy they fled. This released Ghiath ud din from his most dangerous enemies. Meanwhile, the Sheikh Nur ud din Abd ur Rahman Ezferain urged upon the two viziers, Rashid ud din and Taj ud din Ali Shah, the necessity of a reconciliation between the Sultan and Ghiath ud din, and an exposure of the misrepresentations of the latter's enemies. This led to an ordinance by which all Khorasan, from the Oxus to the borders of Afghanistan, was made over to him.* This appanage comprised the districts of Fushenj, Jezeh, Kussuyat, Aazab, Tulek, Herat Rud, Firuz kuh, Gharchestan, Ferah, Ghur, and Guermsir, most of which had their own princes, who were feudally subject to the Malik of Herat. Uljaitu presented him with one of his own robes, some Arab horses, precious tunics, mantles of gold tissue, a cap decked with precious stones, golden girdles, Egyptian arms, tents from Rum, five golden paizahs, seven banners with figures of a dragon on them, seven pairs of cymbals, and three great drums (*kurga*), with other instruments completing the royal orchestra; also a royal seal (*tamgha*) of white onyx (*i.e.*, of jade), a present which no feudatory prince of Persia had received. The Malik returned to Herat

---

* Journ. Asiat., 5th ser., xvii. 489-402.

in October, 1315, where all the maliks and prefects of Khorasan went to greet him.* Before leaving he obtained the deposition of the amir Ali Mustapha, grand judge of Herat, whose incapacity was notorious, and nominated Nasr ud din of Khaisar in his place. Next year (*i.e.*, in 1316) the Malik marched against a Nigudarian chief named Avjibéla, who invaded Guermsir from Kuhistan, and easily defeated him.† We read at this time of an invasion of Khuarezm, which was subject to Uzbeg Khan of the Golden Horde, by Baba, a prince of that horde, who had sought refuge in Persia. This invasion we have already described, as also the messages that passed between Uzbeg and Uljaitu.‡

In April, 1315, the Egyptians under Seif ud din Tenker (or Tenkiz), Governor of Damascus, entered Cilicia by way of Aintab, and marched on Malattiya, whose governor surrendered it. Nevertheless a terrible massacre ensued. The town was fired, and many captives carried off; among others were a large number of woollen weavers, who were transported to Aleppo. Three days after the Egyptians left the place the Armenians and others who had escaped slavery and massacre came out of their hiding places, when the neighbouring garrisons of Kakhta and Karkar attacked them, killed 300 Armenians, captured 100, and made a considerable booty. Soon after this Choban, who had received the town of Malattiya as an appanage, arrived there, put a garrison of 1,000 men in the place, and ordered it to be rebuilt. This expedition of the Egyptians against Malattiya was followed by others, in one of which, in February, 1316, they captured the fort of Dérendeh, near Malattiya, where they slaughtered about 1,000 Armenians, who formed its garrison, and carried off the women and children.§ They also captured Arfekin, in the province of Amid. Berzali says Malattiya was captured in consequence of an understanding between the Egyptians and the Mussulmans who were inside. Makrizi has a more definite story. He says that the Sultan having dispatched some assassins to kill Kara Sonkor, a collector of taxes, a Kurd named Mendu discovered and had them arrested. Nasir was so annoyed at this that he aroused the jealousy of Bedr ud din, the Governor of Malattiya, against Mendu, whom he charged with wanting to supplant him. This began a correspondence, which ended in the surrender of the town and Mendu.‖

Prince Abusaid was continually sending to the Court for means with which to meet the expenditure of his army. Uljaitu remitted the requests to his two Viziers. Rashid ud din declared he had never had the management of the finances, and had never affixed his seal to the assignations made upon the revenue, and was not therefore responsible. "I only possess the robe that covers me, and have not a single coin, and inasmuch as we govern the empire together," replied Ali Shah, "why should we separate from one another when it is a question of paying ?" "Because you have undertaken that responsibility yourself," said Rashid. "You are the guardian of the Great Seal, and are charged with the carrying out of the Sultan's orders." "Why then not affix your seal after mine?" was the reply. "I do not want to join myself with you who profess poverty when asked for money, while each of your employés has made a hundred tumans and become a Carun (i.e., a Koran, the type of wealth among the Arabs, as Crœsus is with us)." When Uljaitu had heard them dispute in this fashion for some time he ordered the empire to be divided into two separate administrative districts. Rashid took Irak Ajem, Khuzistan, the two Lurs, Fars, and Kerman, and Ali Shah Azerbaijan, Irak Arab, Diarbekr, Arran, and Rum. Each of them had a deputy assigned him. Alai ud din was Rashid's and Iz ud din Kuhedi Ali Shah's. Ali Shah again insisted that they should jointly seal all assignations, but Rashid refused, saying he should constantly have to be answering for his colleague, who, whenever money was demanded from him, pretended to have none. The fact was that Ali Shah, who was a good-natured and well-meaning man, was dominated by others who were self-seeking and avaricious. He speedily quarrelled with the amir Togmak, who accused him and his creatures of having plundered the treasury even more than Said ud din and his clients. His complaints were seconded by those of Jevheri, Ali Shah's deputy, who coveted the post of Vizier for himself. Meanwhile couriers kept coming from Khorasan, asking for money, to whom Ali Shah replied that he had not a dirhem in the treasury. "What has become of the money?" said Uljaitu. "Rashid has appropriated it," was the reply. The latter was confined at home by an attack of gout, which prevented him from going out for four months. The Ilkhan ordered Choban and the two deputies above-named to examine into Ali Shah's accounts. They demanded 300 tumans (i.e., three million gold pieces) from his subordinates, Dahir ud din Savaji, Fakhr ud din Ahmed, and Imad ud din Foleki, who during the three previous years had managed the finances, and who now implored Ali Shah to save them from utter destruction. He accordingly repaired to the Sultan that very night, and laid his case before him amidst tears, and with the most touching appeals, which won his master's heart as he was about to carry out the finding of the court. "This poor Ali Shah," he said to the amir Irenchin the next day, "knows neither how to write nor keep accounts. He has employed this money in the service of the State, but has forgotten how. He has promised to explain, and it will be all right." Irenchin reported this conversation to Choban, and said to him, "In the days of Khulagu and Abaka a Tajik could not speak to the sovereign except through the intervention of the amirs, and now it has come about that a Tajik has a midnight interview with the Sultan and puts an end to all our projects." Choban was in a great rage, but was appeased by a large present which

Ali Shah gave him. The latter was no longer molested with inquiries. He therefore turned his animosity against Rashid ud din. He first assailed Jelal ud din, Rashid's son, whom he accused of appropriating the revenues of the young prince, his *protégé*, Uljai Kutlugh, and this charge having been disproved, he charged Rashid himself with feigning sickness and taking a fourth of the revenue, and with monopolising the income of the princesses, and even the money devoted to pious uses. These repeated and insidious attacks began to have effect, and Ali Shah gradually superseded his rival in the Sultan's confidence. Rashid had to secure the patronage of Togmak, Uljaitu's favourite, by a large bribe. The two Viziers now received orders to make friends, which, we are told, they obeyed.*

In 1316 Homaiza, prince of Mekka, sought refuge with Uljaitu. Since the year 1202 the district of Mekka had been governed by princes of the dynasty Kattada, who were styled Sherifs. They were descended from Hassan, son of Ali. Iz ud din Homaiza and Asd ud din Rimaitha (called Remisha by Von Hammer), sons of the Sherif Abu Noma, were ruling jointly, when the complaints of the people of Mekka induced Nasir, the Egyptian Sultan, who was their suzerain, to send their brother Abulgaith with some troops to displace them. This was in January, 1314. On his approach Homaiza left Mekka. When he was established Abulgaith sent home the Egyptian troops, and was soon after driven out by Homaiza. The latter sent an envoy to Nasir, who, however, imprisoned him. Two years later (*i.e.*, in September, 1315), Rimaitha went to Cairo, submitted, and asked for aid against his brother Homaiza. The Sultan gave him some troops, with whose assistance he drove out the latter, who thereupon fled to Irak, to Uljaitu, as we have mentioned. He offered to consider himself his vassal, and was given a body of 1,000 cavalry, under Haji Dilkandi (called Darkandi by Abulfeda), who had orders to reinstate him. On passing Bassora, whence he took 100 tumans, according to orders, for the expenses of the expedition, he was attacked at night by Muhammed, son of Issa, brother of the Prince of the Bedouins of Syria (Mohanna), who had 4,000 men with him. The greater part of Homaiza's escort was killed. That prince and Haji Dilkandi managed to escape on swift horses, leaving their treasure and baggage in the hands of the Bedouins. It had been reported that the Ilkhan had given orders to Dilkandi for the exhumation of the bodies of Abubekr and Omar, who had been buried near Muhammed, and who were deemed unworthy of such a sepulture by a zealous Shia like Uljaitu.†

Towards the end of 1316 Uljaitu was attacked near Sultania, when out hunting, with violent pains in his joints, the result of sexual excesses. He

---

* D'Ohsson, iv. 579-583. Ilkhans, ii. 337-339. Quatremere, xxxii.-xxxiv.
† D'Ohsson, iv. 583-585.

was already convalescent, after having practised a restraint on his diet, when one day, after visiting his harem, he remained a long time in his bath, and afterwards ate some indigestible food. His doctors disagreed as to the remedy. The majority recommended a slight purgative, but an old one, who was very obstinate, insisted on plying him with numerous tonics, and the illness proved mortal. During his last illness he issued two ordinances. In one of these the names of the four Khalifs were replaced in the khutbeh or Friday prayer, whence they had been excluded in favour of that of Ali alone when he became a Shia. By the second ordinance one-half of the confiscated property of the late vizier, Said ud din, was returned to his sons.* By his will he appointed his son Abusaid his successor, and nominated the vizier, Rashid ud din, to superintend the empire. He then detached his two ear-rings, lustrous with diamonds, and his signet ring, and rolling them with the title deeds of his estates in his turban, gave them, with his will and the insignia of office, to the amir Issen Kutluk, whom he made his executor. To the amir Choban he said : "As you have supported so faithfully the empire of my father and brother, you have earned thanks and gratitude. Continue to protect the realm and the army during the reign of my son and successor." He then said the profession of faith, and died on the 16th of December, 1316. Wassaf says the 15th. The "Shajrat ul Atrak" says on the 1st Shevval (*i.e.*, the 17th of September). He was in the thirty-sixth year of his age. The Georgian history and Chamich both say Uljaitu was blind of one eye. The continuer of Rashid's history tells us he was good, liberal, and seldom accessible to calumny ; but, like all the Mongol princes, he drank spirits to excess, and spent his time chiefly in pleasure. His coffin, made of pure gold and silver, garnished with stones, was placed on the throne, and there received the last homage of the officers of the palace and of the army. His subjects wore mourning for eight days. They sat on the ground, dressed in deep blue, giving forth loud lamentations. The minarets and pulpits of the mosques were robed in frieze, or crape.† This blue mourning and these lamentations have parallels elsewhere. In the Shah-nameh, blue is described as that of mourning among the ancient Persians. Violet was the colour used for mourning at the Court of Byzantium. The same colour was that appointed by the regulations of Napoleon for the mourning of the emperor and princes of the blood. The Jews still sit on the ground wailing for the dead.‡

Uljaitu, like more than one of his predecessors, had intercourse with the sovereigns of Western Europe, and an interesting proof of it is preserved in a letter contained in the French archives. This document consists of a roll of cotton paper, nine feet long and eighteen inches wide, containing forty-two lines of Mongol writing in Uighur characters, like

---

* Ilkhans, ii. 240.   † D'Ohsson, iv. 583-587. Ilkhans, ii.   ‡ Ilkhans, ii. 240-241.

those in which Arghun's letter, previously cited, were written. On it are impressed five square seals in red ink. On the reverse and on one of its sides there is an Italian translation in a very small hand. Remusat, in describing the document, says Orientals attribute importance in these letters to the size of the paper, the length of the lines, and the width of the margins and intervals. Arghun's letter contained hardly any margins or blank spaces, and was only six and a-half feet long, and was only marked with a seal three times. The more important names in these letters, including the Khan's own, are raised above the general level of the lines, while that of the prince to whom it is addressed is somewhat below the line. The tamgha or seal on Uljaitu's letter, as on that of Arghun, shows he acknowledged the supremacy of the Khakan at Peking. This seal was engraved with Chinese characters of the ancient form known as shang fang tu chuan, and the inscription may be translated, " By a supreme decree, seal of the descendant of the emperor, charged with reducing to obedience the 10,000 barbarians."* From this it would seem that the suzerainty of the Khakan, which had been repudiated, at least as far as his coins went, by Ghazan, was re-asserted by Uljaitu. The letter we are now describing is headed : " My words, Uljaitu Sultan. ' To Iridfarans (i.e., Re de France) Sultan, and to the other Sultans of the Firankut peoples.'† In former times all of you Sultans of the Firankut were in alliance with our good great-grandfather, our good grandfather, our good father, and our good elder brother, and notwith-standing the distance which separated you, you regarded each other as neighbours, and interchanged messages, embassies, and presents. You cannot have forgotten it. Now that, by the power of God, we are seated on the Great Throne, we shall not depart from the commands of our predecessors—our good grandfather, our good father, and our good elder brother. We shall follow their precepts, and what these good ancestors have promised you we will keep it as if it were our own oaths. We will unite in friendship with you even more than in the past. We will send you envoys. We, older and younger brothers, have been at issue, disunited by the intrigues of wicked vassals (karajus). Now, however, Timur Khakan, Tuktukha, Chapar, Tugha (i.e., Dua), and ourselves, who have been at issue for forty-five years, are reconciled by the inspiration and with the aid of God, so that from the land of the Nankiyas or Angkias (China) in the east, as far as the setting sun and the ulus of Kundalan on the lake of Talu (i.e., the camping ground of the ulus of Khulagu on the plain of Mughan), our people are united and our roads are open. We have agreed to fall jointly upon any of us who proves aggressive.‡ Unable to forget the bonds which tied you to our good grandfather, our good father, and our good elder brother, we have sent you two envoys, Mamlakh and

---

* Mems. Acad. des Inscrip., &c., vii. 390-392.   † Ut is the Mongol plural only.
‡ Id., 395-396.   D'Ohsson, iv. 588.

Tuman, who will explain our intentions *viva voce.*" Uljaitu went on to
say he had heard with pleasure of the termination of the strife among the
sultans of the Firankut or Franks (*i.e.*, the war of Guienne). " Peace is a
good thing," he said, adding that, thanks to their proposed alliance,
they might, with the power of heaven, unite against those who would not
join them, and would thus fulfil the will of heaven (in which expression
the Khan probably hinted at the war with the Egyptian Sultan, upon
which the envoys were, no doubt, to enlarge). " My letter (is written) in
the year 704, the year of the serpent, the 8th of the first month summer
(*i.e.*, the 13th or 14th of May, 1305), in our residence of Alijan."* In the
Italian translation the envoy named Tuman is styled Tomaso, and is
designated as an ilduchi (*i.e.*, a swordbearer). These are not the only
points in which the Italian translation differs from the Mongol text.

We don't know (says Remusat) what kind of reception Uljaitu's envoys
had, and this letter is the only evidence of their presence in France. No
historian mentions them, nor is any answer of the French king recorded.
The envoys, however, passed into England, where they arrived after the
death of Edward I. (*i.e.*, after the 7th July, 1307), nearly two years after
the letter was written. The answer Edward II. made to the ambassadors
is still extant, and shows they must have brought a similar letter with
them to the one that they took to Philippe-le-Bel. This letter of the
English king is dated at Northampton, October 16th, 1307. It is headed,
" To the most excellent Lord Prince Dolgieto, illustrious King of the
Tartars," and addressed to the " Pacificator of the ten thousand
barbarians," as he is styled on the seal. It recites that he (Edward) had
admitted to audience the messengers which the Khan had sent with
letters " to the lord Edward, of glorious memory, late king of England,
our father, who before their arrival had ended his days. We have noted
your letters, and also the message which your envoys have brought us.
We thank your royal Magnificence for the goodness and friendship which
you and your ancestors have shown towards our father, and that you now
show towards us, for the sending of your envoys, for the hope you express
of seeing union and affection grow between us, and especially for the
memory which you preserve of the friendship which existed between your
noble predecessors and our father, as well as other facts which you name.
We rejoice in the peace wrought between you by the grace of God from
the bounds of the East as far as the sea. In regard to the wish you
express that on this side of the sea peace and concord should reign
among us who have been disunited, we wish it to be known to your royal
Excellency that we believe and hope firmly that concord and peace will
shortly, by the help of God, succeed all the quarrels and divisions which
have arisen among us."†

---

* Remusat, op. cit., 396-397. D'Ohsson, iv. 588-589.
† Rymer, Foedera, ed. 1818, vol 2, part i. 8. Vide Remusat, op. cit., 400 and Note. D'Ohsson,
iv. 591-593.

In a second letter, dated the last day of November, 1307, and written at Langley, Edward writes to the King of the Tartars to say that he would employ all his efforts to extirpate the abominable sect of Mahomet if the distance and other difficulties did not prevent him, for the time was favourable for such an enterprise. "If we are well informed," he says, "the books of this abominable sect predict its approaching destruction. Pursue then your laudable design, and may you succeed in your intention of exterminating this villainous sect. Some religious, good, and learned men are on their way to your Court with the intention of converting, by the help of God, your people to the Catholic faith, outside which no one may be saved ; to instruct them in this religion, and to exhort them to make war on the detestable sect of Mahomet. These are the venerable brother William, of the order of preachers, bishop of Lidd, with his venerable suite. We commend them to you, and pray you to receive them well."* Rymer also prints two letters of the same date, commending the same missionaries to the Pope and the King of Armenia.† The letter of the English king, as D'Ohsson says, shows the Khan's envoy had not only concealed the fact that Uljaitu was a Mussulman, but had even imposed on him so far as to persuade him he would aid in extirpating the abominable sect of Mahomet.

Pope Clement V. was similarly deceived by the same envoy, as we gather from his letter addressed to Uljaitu, and dated from Poictiers, the 1st of March, 1308. "We have received in audience," he said, "with the habitual kindness of the apostolical see, your envoy, Thomas Ilduchi, and the letters he has brought from you, and we have carefully noted their contents, and have also taken note of what this your messenger has said on your behalf. We have noticed with pleasure, from these letters and communications, that appealing to our solicitude on behalf of the Holy Land, you have offered us 200,000 horses and 200,000 loads of corn, which will be in Armenia when the army of the Christians arrives there, and in addition to march in person with 100,000 horsemen to support the efforts of the Christians to expel the Saracens from that Holy Land. We have received this offer with satisfaction. It has strengthened us like spiritual food. We are assured this angel has come from the same whose angel sent Abakuk to furnish Daniel with strengthening nourishment when in the lions' den. You have indeed given us sweet nourishment in offering us the hope of such a magnificient aid. Having duly deliberated with our brothers on this offer of succour, and about the desired recovery of the Holy Land; among our most cherished wishes this is the one which exercises us continually, to see, by the grace of God, the land once trodden by the feet of the Saviour restored to the Christian faith, and we have always felt it most important to examine by what means such a pious design could be

---

* Rymer, id., 18.   † Id., 17-18.

accomplished. This is why I and my brothers have deliberated
seriously about the matter, and hoping in the justice of the Most High,
who strengthens his servants, we will execute so far as we can what God
has inspired us to do, and when a favourable season for crossing the sea
shall come we will advise you by letters and messengers, so that you may
accomplish what your magnificence has promised. But do you in faith and
works turn to Christ, who is the way, the truth, and the life. To serve
him is to rule. Persevere in your laudable intention in regard to this
Holy Land, and secure on earth by this and other means the approbation
of Christ the Redeemer, so that you may deserve to obtain from him an
ample share of his delights in heaven and of glory in this present world.
We and the Apostolical See shall rejoice in your success and glory.*
D'Ohsson suggests that Thomas Ilduchi, who misled the Pope and the
Christian princes in this fashion as to the real sympathies of Uljaitu, was
probably inspired by the Eastern Christians, and especially the Armenians,
who no doubt desired at all costs that another Crusade should be
organised to aid them against their mortal enemy, the Egyptian Sultan.
The same motive probably moved the King of Aragon, James II., to send
his envoy, Peter Desportes, to Uljaitu. His letter was addressed:
" Illustri et magnifico Olvecacu, Dei gratia regi dels Mogols, &c." He
asked that all his subjects might have the privilege of buying and selling
arms, horses, provisions, and other things which were necessary for the
Aragon army when it should have passed the sea, and also that any
Armenian, Greek, or other Christian who wished to visit the army might
do so without hindrance. The letter, in the published copy as given by
Martin Fernandez de Navaretto, is dated in 1293 ; but this is clearly a
mistake, for Uljaitu had not then mounted the throne, and Remusat is
disposed to assign it to the year 1307.†

One can see (says Remusat), in comparing the evidence, that Uljaitu
had long before the defection of the Egyptian amirs, in 1312, made up
his mind to attack Egypt, and had sought allies in the West. This
Western assistance was no doubt especially pressed by the Christians
of Cyprus and Armenia, who were closely interested in the revival
of the Crusades and of the European settlements in Palestine and its
borders. Haithon, the Armenian author whom we have often quoted,
was directed by the Pope to draw up a memoir on the subject, and from
his account we can gather what the basis of Eastern politics then was.
Having defended the policy of the Crusades, and given a rapid survey of
the revolutions which had occurred in Egypt and Syria, he goes on to
show how opportune the time was for giving the Sultan a final blow,
especially as Karbanda (*i.e.*, Uljaitu), the ruler of the Tartars, was ready
to help them, and had sent envoys to say he would employ all his forces

---

* D'Ohsson. iv. 594-597.   † Remusat, Op. cit., 402-403. Note.

O *

to overwhelm the enemies of the Christian faith.   It was necessary to
meet this offer quickly, and it would be well to send him envoys to ask
two things especially.   First, that he would publish a prohibition to the
supplying of food or other merchandise to the enemy, and, secondly, that
he should be asked to send a force to devastate the district of Aleppo;
while the Christians, in conjunction with their allies from Armenia and
Cyprus, assailed the land of the Saracens by sea and land.   Haithon
enters at some length into the strategy of the campaign, and advises, *inter
alia*, that an advanced expedition should first set out and try and seize
on the district of Tripoli, in which they might count on the assistance of
the Christians of Mount Lebanon, who could raise a contingent of 40,000
archers.   Having seized on this district, they would do well to rebuild
the city and to make it their base, and thus be ready when the Tartars
had completed their conquest of the Holy Land to take over from them
the towns there, which he was confident they would make over to the
Christians for custody, because they could not endure the heat of the
summer in those parts ; nor did they fight with the Sultan to conquer
more lands, for they were masters of all Asia, but because the Sultan was
very unfriendly to them and always doing them some injury, especially
when they were at war with the neighbouring Tartars (*i.e.*, those of
Kipchak and Jagatai).   He urged that letters should be sent to the
Christian kings of Georgia and Armenia.

Haithon says there were three routes to the Holy Land.   One by way
of Barbary, in regard to which he left others to advise who knew it ; the
second by way of Constantinople, the same followed by Godfrey of
Boulogne, which was safe and feasible as far as Constantinople, but
beyond this, through Turkey (*i.e.*, Asia Minor) to Armenia (*i.e.*, Little
Armenia, or Cilicia), it was not free from danger because of the Turks ;
but the Tartars would secure a safe transit across it, and also
abundance of provisions, horses, and other necessaries.   The third
route was by sea, through Cyprus.   In regard to their proceedings,
if they adopted the last of these routes, he enters into considerable details.
He suggests that an auxiliary army of 10,000 Tartars would be as large
as would be required, since the very fear of the Tartars would prevent the
Bedouins and Turks from attacking them.   They would supply them with
provisions and other necessaries, and act as spies for them, for which they
were specially adapted from the agility of their horses.   They were also
well skilled in attacking fortresses.   Haithon then supplies his patron
with some touches of worldly wisdom.   In urging that too great a body
of Tartars should not be employed, he says, " If Karbanda, or some one
sent by him, should invade Egypt with a very large army, it would be well
to avoid him, for the Lord of the Tartars would deem it derogatory to
follow the counsel of the Christians, and would insist on their following
his commands.   Besides which, the Tartars were all mounted and marched

rapidly, and a Christian army, much of which marched on foot, could not keep up with them. The Tartars, again, when in small numbers were humble and obsequious, but when in large numbers were overbearing and arrogant, insulting to their allies who were weaker than themselves, and would be found unbearable by the Christians." As Remusat says, it is clear Haithon understood well the advantages and drawbacks of an alliance with the Tartars, and he suggested that the campaign should be a double one—the Tartars should advance, according to their usual plan, by way of Damascus, and occupy the country there, while the Christians advanced on Jerusalem. He adds a singular phrase, in which he describes one advantage which the Egyptians had over the Tartars, in that they kept their own counsel, while their foes decided at a public gathering, in the first month of the year, what their plans for the ensuing season were, and thus gave their enemies warning.[*] Sanuto also speaks of the advantages of an alliance between the Christians and the Tartars. "The King of Armenia," he says, "is under the fangs of four ferocious beasts—the lion, or the Tartars, to whom he pays a heavy tribute ; the leopard, or the Sultan, who daily ravages his frontiers ; the wolf, or the Turks, who destroy his power; and the serpent, or the pirates of our seas, who worry the very bones of the Christians of Armenia." He attributes the war which the Tartars waged upon Egypt at the request of the most Christian King of Armenia, to the vengeance of God for the cruelties exercised upon the Christians of Acre and of Syria.[†]

The Western princes of Europe were not the only Christians who found it convenient to cultivate a close friendship with the Tartars. Pachymeres tells us that the Emperor Andronicus, to protect himself from the assaults of the Turkomans of Asia Minor, made an alliance with Uljaitu, whom he calls Kharmpanta, and gave him his sister Maria in marriage. She was known as "the Lady of the Muguls," and was conducted to Mecca by a suitable escort, where she seems to have taken up her residence for a while, and sent a complaint to Uljaitu of the doings of Othman (called Atman by Pachymeres). The Ilkhan thereupon dispatched an army of 30,000 men from the interior of Persia towards the eastern borders of Rum, to repress him ; but it only gained the empire a temporary respite, and Othman gradually extended his borders.[‡]

Maria was known as Despina Khatun to the Mongols, and she was given the yurt or appanage which had belonged to the Great Despina, the wife of Abaka.[§] Another foreign wife of Uljaitu was Dunya, the daughter of Malik Al Mansur Gazi, son of Kara Arslan, Lord of Mardin, whom he married at the Ordu in the year 1309. Her brother, Malik es Salih, succeeded his father as ruler of Mardin.[||] When Uljaitu was quite a boy

---

* Haithon, op. cit., 99-107.   Remusat, Mems. de l'Academie, vii. 403-406.
† Remusat, op. cit., 406-407.   ‡ Stritter, ii. 1101-1102.   § D'Ohsson, iv. 536.
|| Abulfeda, v. 227.

he was married to Kunjuskab, the daughter of Shadi Gurkan. It was through her influence that he became a Mussulman, having first, as we saw, been baptised a Christian. A year after his enthronement he married Kutlugh Shah, the daughter of Irenchin, who was given the yurt of the great Dokuz Khatun. A month later he married Bulughan, his father's widow, who, to distinguish her from the two other princesses of the same name, was called Bulughan the Khorasani.* In addition to these, Uljaitu had several other wives, as Terjughan, the daughter of Legsi Gurkan, whose mother was a daughter of Khulagu ; Iltirmish, daughter of Kutlugh Timur Gurkan ; Haji Khatun, daughter of Sulamish, who was the mother of his successor, Abusaid ; Oljai, sister of the previous princess ; Siyurghatmish, daughter of the amir Hussein ; and Kutuktai.

Uljaitu had seven sons—Abusaid, Bestam, Bayezid (Abu Yezid), Taifur, Sulimanshah, Aadilshah, and a second Abusaid, who, with four of his brothers, died when young. He also had four daughters—Satibeg, who herself occupied the throne, Dulendi or Devletmendi, Fatima, and Mihrkutlugh.

The coins of Uljaitu occur in considerable variety. In the British Museum are gold coins struck by him at Basrah and Sultania ; silver ones struck at Vasith, Samsun, Irbil, Mardin, Mosul, Shiraz, Tebriz, Hillah, Kashan, Kara Ghaj, and Damaghan ; and copper ones struck at Baghdad, Hillah, Sultania, Saweh, Erzenjan, and Shiraz; and they occur of various dates, from the year 704 HEJ. to 714 HEJ.† The only *Mongol* words on his coins reads Uljitu Sultan. The Arabic inscriptions on his coins present considerable variety. As he became a very devoted Shia, we accordingly find the formulæ of this sect, and find them also amplified in the most unusual fashion, but this is only on coins struck after the year 707. Before that date the names of the four khalifs, treated as usurpers by the Shias, are found on his coins, together with the Sunni symbol.‡ One of these earlier coins in the British Museum Poole calls a most remarkable piece. It bears (he says) two Koran inscriptions (ch. xlviii. 29 and ch. xxiv. 54) which are not found on any other coins, together with the formula of the four orthodox khalifs.§ On the later coins we find not only the phrase about Ali, but also the names of the twelve imaums. One of these coins is described by Fraehn. On the obverse we read : "Struck under the Lord Supreme Sultan, defender of the Faith, Ghiath ud dunya w'eddin Uljaitu Sultan Muhammed, whose reign may God perpetuate," with the date ; and on the reverse : "There is no God but God. Muhammed the Apostle of God. Ali is the Vicar of God. O God, bless Muhammed and Ali and Hasan and Hussein and Ali and Muhammed and Jafar and Musa and Ali and Muhammed and Ali and Hasan and Muhammed."‖

---

* Ilkhans, ii. 182-183.   † Poole's Catalogue, vi. 44-61.
‡ De Saulcy, Journ. Asiat., 3rd ser., xiii. 139.   § Poole, op. cit., lvi.
‖ Fraehn, Resensio, 181.

Note 1.—The following notice in Ibn Batuta I overlooked when writing the previous chapter. He tells us Uljaitu had with him a jurisconsult of the sect of the Rafidhites (*i.e.*, the Shias), named Jemal ud din, son of Mothahher. He it was who persuaded Uljaitu to attach himself to the Shias, and thereupon the Ilkhan sent messengers to various towns to press the inhabitants to follow his example. The first places where this order arrived were Baghdad, Shiraz, and Ispahan. At Baghdad, the people of the Gate of the Dome (*Bab Alazaj*), who were the most notable among its garrison, who were Sunnis, and for the most part followed the doctrines of the Imaum Ahmed, son of Hanbal, refused to obey. They took up arms, and went on the Friday to the great mosque, where the Ilkhan's messenger was, and when the Khatib mounted the pulpit they, to the number of 12,000, pressed round him, all armed, and swore to the Khatib that if he altered the khutbeh in any way from what he had been wont to say it they would kill him, as well as the Khan's deputy, and would then submit to the will of God. Although Uljaitu had ordered the suppression of the names of the four khalifs, and that only Ali and his people should be named, the preacher was afraid, and the khutbeh was said as usual. The people of Shiraz and Ispahan behaved in the same way, and when the matter was reported to Uljaitu he ordered the Kadhis of the three towns to be summoned. The first to arrive was Medjd ud din, of Shiraz. The Ilkhan was staying at the time at Karabagh, where he generally spent the summer, and he ordered them to throw the Kadhi to the dogs which were about the palace. These animals were very big, and had chains about their necks, and were trained to devour men. When some one was brought in and given to the dogs, the unfortunate person was put on a wide plain, where he was set free. The dogs were then let loose upon him, and soon tore him in pieces. In this case they merely fondled him and wagged their tails. When Uljaitu learnt this he set out barefoot from the palace and prostrated himself before the Kadhi, kissed his hand, and gave him the robes which he was wearing, which was the greatest honour he could do him. He then took him by the hand and conducted him into the palace, and ordered his women to treat him with respect, and to look upon his presence as a blessing. The Sultan, we are further told, renounced the doctrines of the Rafidhites, and joined the Sunnis. He gave the Kadhi splendid presents, and sent him home again with great honours.*

Note 2.—On the 24th of July, 1305, Uljaitu began to build the famous town of Sultania, whose foundations had been already laid by his father Arghun. It was situated in the beautiful meadows of Kongoralank (or Kunkur ulong, as D'Ohsson calls them). It contained several mosques, the principal one of which, built at the Sultan's expense, was richly ornamented with marble and porcelain. He also founded a hospital, duly supplied with a dispensary and all other necessaries, and to which he attached several doctors. Also a college, on the plan of that of Mostansir at Baghdad. The grandees rivalled each other in building beautiful houses at this place. One whole quarter, containing 1,000 houses, was built at the cost of the vizier, Rashid ud din, who, in addition, put up a great building flanked by two minarets, comprising a college,

---

* Ibn Batuta, ii. 57-61.

a hospital, and a monastery, all richly endowed.　The citadel was surrounded by a square wall, flanked with towers.　Each side was 500 guez in length.　It was made of dressed stone, and the wall was so thick that four horsemen could easily ride abreast upon it.　Uljaitu also built himself a mausoleum in the fortress.　It was of an octagonal shape, each side being 60 guez in length, and was covered with a cupola 120 guez in height.　It was pierced with a large number of windows, with rich iron grilles in front of them.　One of them was 30 arishes in height and 15 in breadth.*　According to De Guignes, the mausoleum had three splendid doors, made of polished and damascened steel.　The grille or trellis surrounding the actual tomb was made of the same metal, and so skilfully that, although the network was made of bars as thick as a man's arm, no joints could be found in it, and the Persians said it was made in one piece, which took seven years to complete, and was brought whole from India.†　Near this tomb was built a hospice and a house for the Seyids, which he endowed handsomely.　The royal palace consisted of a high pavilion or kiosk, surrounded at a certain distance by twelve smaller ones, having each a window looking into the court, which was paved with marble; a chancellary sufficiently large to accommodate 2,000 people, and many other buildings.　During all his reign Uljaitu devoted fifty tumans every year towards the building of Sultania, which, if he had lived long enough, would have become one of the most beautiful cities in Asia.　Novairi says in regard to it: " We learnt in the year 713 (*i.e.*, 1313) that it was finished and inhabited.　The Ilkhan transferred thither from Tebriz a great number of merchants, weavers, and other artisans, who were forced to go there whether they liked it or no.　Eventually the greater part of these artisans returned to Tebriz."‡　To the consecration of the mausoleum Seyids, sheikhs, and imaums were summoned from all parts of Persia.§　Hamdulla tells us that the walls of the city as originally planned by Arghun were twelve miles in circuit.　This was enlarged by 3,000 paces by Uljaitu.　Within these walls was a citadel made of dressed stones, which was two miles round, and where Uljaitu lived.　There were also other buildings.　The district was cold, and well watered by canals, &c., with abundant pasture and plenty of game, and people of all nations and religions were to be found there.　The town contained finer buildings than any other, except Tebriz.　The contribution to the taxes of the place was not more than twenty or thirty tumans.　It was ravaged and virtually destroyed by Timur.　Besides Sultania, Uljaitu also built the town of Jemjalabad, also called Sultania Jarmejan, at the foot of the mountain Bisutum, referred to by the Persian geographer Mestufi, and also by Rich in his first journey.‖　Speaking of Sultania, Ker Porter says in his travels: " An hour's ride brought us into the midst of the ruins, amongst whose broken arches and mouldering remains of all sorts of superb Asiatic architecture I discovered the most wretched hovels of any I had seen in Persia."¶　He says that Uljaitu built it as a sacred shrine for the remains of the Khalif Ali and the no less holy martyr Hussein, intending when it was finished to translate the bodies, with

---

* D'Ohsson says that the *guez* and *arish* were Persian measures, equivalent to a cubit or ell. *Id.*, 487. Note 1.　　　　† De Guignes, iii. 279.
‡ D'Ohsson, iv. 485-487.　§ Ilkhans, ii. 185.　‖ *Id.*, 187.　¶ Travels, i. 276.

all religious pomp, from Meshed Ali and Kerbela to their new tomb. But the pious Sultan did not live to complete his object, and instead of the venerated relics his own ashes occupied the place, which he had hoped, when tenanted according to his intentions, would cause Sultania to be a point of future pilgrimage as revered by the faithful of his own empire as Medina had heretofore been to Mussulmans in general. Porter says of the building: "The height of the dome certainly exceeds 130 feet; the diameter of the circle below is 33 paces. The whole interior of the building presents one uninterrupted space; but to the south is a large distinct chamber, choked up with rubbish, under the floor of which, I was told, are three immense vaulted rooms. The entrance to them is now lost under the ruins above, but in one stands the tomb of the Sultan Mahomed Khodabund, raised from the earth. The inside of the whole mosque which covers these royal remains is beautifully painted, and tiled with varied porcelain. Much gilding is yet to be seen upon the upright and transverse lines of decoration, amongst which it is said the whole Koran is written in ornamented characters. It required a Mussulman's eye to find them in the varied labyrinth of arabesque patterns with which they were surrounded. Formerly, the whole building was inclosed within a square of 300 yards. Its ditch is still visible to a great depth, and at the north-west angle stands part of a large tower, and a wall 40 feet in height, built of fine large square masses of hewn stone, excellently cemented together, the thickness of the wall being 12 feet. On the top still remain a number of the pedestals, belonging to the machicolated parapet. Two Arabic inscriptions are yet distinct on the wall and the tower, but I could not find any person to translate them. The late mollah of the place, who might have been the interpreter, had recently paid the debt of nature; but one of the natives told me, from memory, that the purport of the inscription was merely to say that the mosque had been built 575 years ago. All the proportions and decorations of this vast structure are in the most splendid Asiatic taste; but the blue, green, and golden tiles with which it has been coated are rapidly disappearing; yet enough remains to give an idea of the original beauty of the whole. The ruins of other superb mosques are still conspicuous in many parts of the city, and all seem to be on so extensive a scale that we can only stand in amazement at the former magnitude of a place which at present scarcely numbers 300 families. When the Holstein ambassadors were in Persia (A.D. 1637), even then the waning city contained 6,000 people. How has it been reduced since, in little more than a century and a half! The walls of its ancient houses, and spacious gardens, cover a great stretch of the plain, and in some places we find large black mounds of earth, where I imagine the public baths stood."[*]

Fergusson says of Uljaitu's tomb at Sultania: "Its general plan is an octagon, with a small chapel added opposite the entrance, in which the body lies. The front has also been brought out to a square, not only to admit of two staircases in the angles, but also to serve as a backing to the porch which once adorned this side, but which has now entirely disappeared. Internally the dome is 81 feet in diameter by 150 feet in height, the octagon being worked into a circle

* i. 277-280.   † Fergusson's Hist. of Architecture, ii. 438-439.

by as elegant a series of brackets as were perhaps ever employed for this purpose. The form of the dome, too, is singularly graceful and elegant, and much preferable to the bulb-shaped domes subsequently common in Persian architecture. The whole is covered with glazed tiles, rivalling in richness those of the mosque at Tebriz, and with its general beauty of outline this building affords one of the best specimens of this style to be found either in Persia or any other country."† Fergusson gives pictures of the exterior and interior of the building, which quite bear out his eulogium. I may add here what I overlooked before, viz., the same author's notice of the remains of Ghazan Khan's mosque at Tebriz. " In plan it is not large," he says, " being only about 150 feet by 120, exclusive of the tomb in the rear, which as a Tartar it was impossible he could dispense with. In plan it also differs considerably from those previously illustrated, being in reality a copy of a Byzantine church, carried out with the details of the thirteenth century—a fact which confirms the belief that the Persians before this age were not a mosque-building people. In this mosque the mode of decoration is what principally deserves attention, the whole building, both externally and internally, being covered with a perfect mosaic of glazed bricks of very brilliant colours, and wrought into the most intricate patterns, with all the elegance for which the Persians were in all ages remarkable. Europe possesses no specimen of any style of architecture at all comparable with this. The painted plaster of the Alhambra is infinitely inferior, and even the mosaic painted glass of our cathedrals is a very partial and incomplete ornament, compared with the brilliancy of a design pervading the whole building, and entirely carrried out in the same style. . . . The entrance door is small, but covered by a semi-dome of considerable magnitude, giving it all the grandeur of a portal as large as the main aisle of the building. After the date of this building this mode of building portals became nearly universal in the East. The mosque was destroyed by an earthquake at the beginning of the present century, but it seems to have been deserted long before that, owing to its having belonged to the Sunnis, while the Persians for five centuries have been Shias."*

CHAPTER XI.

ABUSAID KHAN.

ABUSAID was the seventh of the Ilkhans, and seven being a fortunate number in the East, a prosperous reign was foretold for him. The event proved clearly enough the vanity of such predictions, for his reign virtually saw the collapse of the power of the Ilkhans in Persia. He was but twelve years old when he succeeded his father, and among the turbulent communities of the East minorities are generally periods of special trial. Troubles began early in the jealousies of the amirs Sevinj (or Sunej) and Choban, the former of whom was his governor, while the latter had acquired great influence through his fifty years' service, and through marrying Abusaid's sister Dulendi. Directly the amirs and viziers were satisfied that Uljaitu's illness was mortal, they sent a message to Abusaid to go in all haste to the Court, hoping he would go without his governor, who, they were afraid, would try to monopolise power, and use it to favour his private ends Summons after summons was sent to the young prince, who was in Mazanderan; but the officers of his household, who were devoted to Sevinj, would not set out without his permission, and sent him word to Radégan of Tus that the Sultan had sent for his son. Suspecting the motives of the amirs, he professed not to believe the bad news, and joined his charge in Mazanderan, and when he heard the Sultan had actually died he did not hasten his steps, hoping to secure a more stable authority for himself, so that he might have the amirs under control. The amirs having duly performed the funeral ceremonies for the dead Khan, each sent a relative to meet the young prince. They encountered him near Bostam. The amirs relied on Choban to counterbalance the influence of Sevinj. That general, on hearing of Uljaitu's death, set out from Bailekan in Arran with his troops, and duly reached Sultania, where the amirs gave him the title of Amir ul umera, or generalissimo. Meanwhile Sevinj dispatched an officer named Zanburi from the neighbourhood of Rai to inquire what was going on, and to sound Choban. He was well received by the latter, by the amirs and khatuns. Having received a reassuring message from his agent, Sevinj went on to the capital with the young prince. Those who were about Abusaid wished Sevinj to be Amir ul umera, but he declined the honour, saying he must devote himself to assure the tranquillity of the empire until the prince was grown up, and that if he began to dispute about this dignity with Choban, their quarrel would end in civil war. He said he wished rather to remain near the person of the ruler, while if he became commander of the troops he must be away, and lastly that his health was too feeble to enable him to support the fatigue of war. He then went to Abusaid, and told him that though his long services might command that he should be appointed over all the other amirs, he wished to sacrifice his right for the good of the State and the young prince, and he therefore urged the confirmation of Choban in the post. This was before they reached Sultania. Choban met the *cortège* at the head of the amirs, and having dismounted, saluted Abusaid, who came out of his tent to meet him, with several genuflexions, and kissed his hand. They then all mounted on horseback and rode, into the city, where the young prince went to see his father's corpse, and gave the funeral feast. After the usual days of mourning, the ceremony of enthronisation was proceeded with.* This took place on the 4th of ·May, 1317. The ceremony was the same as that gone through at the installation of the Grand Khan. The seven senior princes of the blood seated him on the throne, a form which, according to Von Hammer, was borrowed by the Emperor Charles IV. when in his golden bull he prescribed the ceremonial for the election of an emperor, in which the same *rôle* was assigned to the seven electors, and he suggests that the ceremony may in the latter case have been directly borrowed from the Mongols through the missionaries. Of the seven princes, four held the corners of the felt on which the Khan was lifted up, two held his arms, and the seventh offered him the cup.† Such ceremonials are generally very conservative in their surroundings. We accordingly read that the two noyans, Sevinj and Choban, conducted the Khan to his throne, while gold and precious stones were strewn over his head. This was a very early custom, and still survives in the East in marriage ceremonies, where money is sprinkled over the head of the bride. The various grandees kissed the ground in front of the throne nine times, with their girdles loosened and put about their necks, and their caps in the air to signify that their lives were at the Khan's service. Abusaid adopted the title of Alai ud dunia ved din (*i.e.*, the sublimity of the world and of the faith) Abusaid Sultan.‡

Abusaid was born at Berkui on the night of the 2nd of June, 1304, and was confided to the tutelage of the amir Sevinj and his wife Ogul Kandi. When he was five his father ordered that he should learn to ride, so the khatuns, amirs, &c., assembled before the khaneh or quarter of Sevinj to assist at the ceremony, and to pay honour to the young prince. He mounted at the time fixed by the astrologers, the horse's head was turned towards the East, and a bowl of kumiz was duly poured over its

---

* D'Ohsson, iv. 599-603. Ilkhans, ii. 252-255. † Ilkhans, ii. 255.
‡ *Id.*, 256. D'Ohsson, iv. 603.

neck and croup. When he was nine, as we have seen, his father nominated him governor of Khorasan. There he continued his studies. He used
to go on foot from his ordu to the school, and forbade his masters to rise
on his approach. After six months' lessons in writing he wrote, in the
year 1314, a letter to his father in his own handwriting, which was taken
round to the ordus and the houses of the various amirs and ministers.
As we have seen, on his accession he confirmed Choban as generalissimo,
and gave the command of the troops in Rum to his son Timurtash, who
took with him to superintend the finances the Khoja Jelal ud din, eldest
son of the vizier Rashid. Diarbekr he confided to the amir Irenchin,
Armenia to Sunatai, while the amir Isenkutlugh was sent to Khorasan to
replace, Yassaul, who had been killed before Abusaid's accession, as I
shall describe in giving an account of the khanate of Jagatai. Ali Shah
was nominated as first Vizier, and he was given charge, not only of the
grand treasury, but also of the public buildings, the palaces, royal stables,
and arsenals. This meant the virtual displacement of Rashid ud din
from the supreme control which he had hitherto held.*

The "History of Georgia" tells us that as soon as Giorgi, the King of
Georgia, heard of Abusaid's installation he went to the Ordu, where his
arrival greatly pleased Choban, who treated him with all the affection of
a father for a son. He made all Georgia over to him, with all its
mthawars, the sons of King David (*i.e.*, David VI.), the Meskhes, and the
sons of Beka. This took place in the year 1318. The Georgi here named
was George, the son of Dimitri, known as George V., and he succeeded
George the son of David VI., called the Little, whose death is nowhere
mentioned, and whose disappearance we only learn from this installation
of his successor, George V., styled Brtsqinwalé, or the Brilliant.† The
Georgian history describes how he repressed the Ossetes and attacked
the tribes of the Caucasus.‡

One of the first acts of the new reign was the arrest of the amir
Togmak, who was accused of an intrigue with Kutlugh Shah Khatun
Uljaitu's favourite wife. He was presently liberated, however, and
Choban made him his lieutenant.§ We have described the rivalry that
existed between Rashid ud din and his colleague, Ali Shah. The latter,
afraid of Rashid's influence with the mighty amir Choban, tried various
plans to discredit him. Their quarrels made the lives of their subordinates embarrassing, since they could not serve one master without
offending the other. Three of their leaders, Daial Mulk, Iz ud din
Kuhedi, and Alai ud din, asked Rashid's permission to formulate an
accusation against his rival, who they offered to show had been guilty
of great malversation. As he would not consent, and afraid that he would
inform Ali Shah against them, they went to the latter and sided with

* D'Ohsson, iv. 603-605. Ilkhans, ii. 256.   † Hist. de la Géorgie, 643, and Note 3.
‡ *Id.*, 645-646.   § Hamdullah, notice of Abusaid Khan.

him against Rashid. Ali Shah distributed money freely among the
followers of the greater chiefs, in order that they might poison their
master's minds against his rival, and one of these, Abubekr Aka, the
deputy of Choban, served him so well that Rashid was actually deposed
in October, 1317, and left Sultania and went to Tebriz. Sevinj, who
disapproved of this, was ill, and he gave it out that when he recovered
Rashid should be reinstated ; but having followed the Sultan in a litter to
Baghdad, he died near that town in January, 1318, lamented by his young
charge Abusaid. He was taken to Sultania for burial. In the spring the
Court set out for Sultania, and when it arrived near Tebriz Choban
summoned Rashid, who had retired to that town, and wished to replace
him at the head of affairs, saying that his talents were as indispensable to
the State as salt was to meat. Rashid excused himself. " My career has
been a long one," he said, " and I have been Vizier longer than anyone. I
have thirteen sons in the public service. I wish to devote the few days that
remain to me to my eternal salvation." Choban would not be persuaded by
these excuses. He said he intended to speak to the Sultan, and that he
was going to get his diploma as Vizier prepared. Pressed in this way,
Rashid ud din consented to be again reinstated. Ali Shah and his
creatures, the members of the Divan, no sooner heard of this than they
recommenced their intrigues, and began to bribe the officers of the great
amirs, and notably Abubekr aka, the confidante of Choban, who promised
to work his ruin with his master. The conspirators now accused Rashid
ud din of having administered a draught to Uljaitu which had caused his
death, and it was further said that the cup with the potion had been
administered by Rashid's son, Sultan Ibrahim, who was cupbearer. This
is the story told by Mirkhavend. Sakkai gives us some details which
show what foundation there was for it. He says that being summoned
for trial before Choban, Jelal ud din, son of Harran, a doctor who had
attended Uljaitu, was accused of poisoning the late Sultan. He said that
Uljaitu suffered from a violent indigestion, accompanied by great diarrhœa
and frequent vomiting. In concert with the other doctors he had prescribed an astringent medicine for this complaint. Rashid was of a
different opinion, and urged that the complaint was really due to gluttony,
and he prescribed a laxative, which increased the diarrhœa, and thus the
Sultan died.* At this trial the accusers were Togmak and the amir
Iz ud din Talib, surnamed Dilkandi, who accused both father and son.
They were condemned to death by the Sultan. Juskeder, a small town
near Tebriz, was fixed as the place of execution. Khoja Ibrahim, a youth
of beautiful figure and an amiable disposition, was executed before his
father's eyes. Rashid himself was then put to death, saying before he
died : " Tell Ali Shah that he sheds the blood of a man who has never

done him any harm, but that I shall be avenged." Hardly had he spoken
these words when Dilkandi cut him in two. Dilkandi, who was a relative
of Ali, gladly accepted, we are told, the office of executioner, in order to
avenge the death of Taj ud din. The quarter of Tebriz built by Rashid,
named Raba Rashidi, including his house, was given up to pillage, and his
family were seized as slaves by the first comers. All his lands and goods,
and those of his sons, were confiscated, even those he had devoted to
pious uses. His head was taken to Tebriz, and carried about the town
for several days amidst cries of "This is the head of the Jew who has
dishonoured the word of God ; may God's curse be upon him." One
account says that he was dismembered, and that his limbs were exposed
in various places, while his body was burnt ; but it would seem that he
was buried at Tebriz, near the mosque which he had built there, for we
are told how, a century later, Miran Shah Mirza, Timur's grandson, when
governor of Tebriz, and after he had become partially demented, had his
bones exhumed and buried in the Jewish cemetery.* Quatremere has
examined the story of his having been a Jew, which is reported by Sakkai
and others, and was evidently widely believed, and has shown that it is in
all probability baseless, and due to his having, among his other researches,
interested himself in Jewish history, &c. Rashid tells us himself that his
father was a zealous Mussulman, while his grandfather was a friend of
Nasir ud din, of Tus, which would be incredible if he had been a Jew.†

Thus, on the 17th of Jamada the First, 718, terminated, at the age
of 73, the career of Fazlallah, son of Abulkhair, son of Ali, the
physician, philosopher, historian, the author of Ghazan's famous code
of laws (kanun), the greatest vizier of the Ilkhan dynasty, and
one of the greatest men the East has produced, to whom the
students of Eastern history are under immeasurable obligations. The
famous college which he built, and which was known as the Rashidia,
bore this enigmatical inscription : "It is more difficult to destroy
this beautiful building than to build another." Besides his famous history,
he was also the author of exegetical, geographical, and mathematical
works. These he wrote, as he tells us, during the two earliest hours,
from the morning prayer till sunrise—the only two hours he could
spare in the morning from his exacting duties to the State.‡ A few days
after Rashid's execution, the amir Isen Kutlugh arrived from Khorasan,
and reproached the amirs warmly, and asked them how they hoped to
benefit by putting to death an old man so near his grave. Isen Kutlugh
died almost directly after, which, like the death of Sevinj, seems
suspicious. The chief instruments in Rashid's downfall speedily met
with retribution. Dilkandi having plotted with Zenburi and some others
to assassinate the amir Choban at the gate of the palace, was seized and

---

* Quatremere's Rashid, xliv.   † Id., v.-viii.
‡ Ilkhans, ii. 259-261. D'Ohsson, iv. 608-612.

executed, and Zenburi was sent to Asia Minor to Timurtash. The vizier Ali Shah was naturally exultant, and we are told by an historian of Mekka, Taki ud din Fasi, that among the treasures of the Kaaba were two gold rings, each weighing 100 miskals, and enriched with six great pearls and six rubies. These were offered by Haji Bulavaj in the name of Ali Shah in the year 718 of the HEJ. He had vowed to suspend them at the door of the Kaaba in case he overthrew Rashid.*

News arrived at this time that Uzbeg, Khan of Kipchak, was marching towards Derbend at the head of a numerous army, and that an Egyptian army had entered Diarbekr. This action, which was doubtless concerted, was probably induced by Abusaid's youth. It was decided in a grand council that the amir Irenchin should defend the last province, that the Sultan in person should oppose Uzbeg, while the amir Hussein was sent against Yassaur, who had revolted in Khorasan. Hussein halted at Rai ,to await reinforcements, since Yassaur had twice as many men with him in Mazanderan. On the arrival of reinforcements, he continued his advance in spite of the snow and severe cold, but learnt at Damaghan that Yassaur had retired. The latter's forces had been exaggerated, and had caused some alarm at the Court, and Choban himself had determined to advance into Khorasan. Having reviewed his army near Bailekan, he was about to move on when he heard that Uzbeg Khan had already arrived at Derbend, and that the amir Taremtaz, who had gone to oppose him, had returned again to the Ordu. The Sultan had only a thousand men-at-arms with him, and an equal number of servants, muleteers, and cameleers. He advanced, notwithstanding, to the Kur, where he ranged all his tents in a line to make believe to the enemy encamped on the other bank that his forces were much larger. As soon as Choban heard of his master's danger he hastened with 20,000 men from Khorasan. Before his arrival Uzbeg's army had withdrawn. He pursued it and captured many prisoners. On his return he was much praised by the Sultan, but the officers who had withdrawn before the enemy was in sight were severely punished. Some were bastinadoed and others degraded by order of Choban, who afterwards, as we shall see, suffered from the effects of their revenge.†

Soon after the accession of Abusaid there was a succession of serious troubles in Khorasan, caused by the ambition of the Jagatai prince, Yassaur, above named, who had been granted an important appanage by Uljaitu, and who now aspired to a much more important position. I have postponed a notice of these matters till the next volume, when the history of the Jagatai princes at this time is related. Here it will suffice to say that Yassaur, after a protracted struggle, in which Ghiath ud din, the ruler of Herat, bore a prominent part as an ally of the Ilkhan, was compelled to

---

* Quatremere, op. cit., xlv.-xlvi. † D'Ohsson, iv. 613-614.

withdraw to his appanage, and eventually killed there by his relative, Kepek Khan, of Jagatai, who had an old grudge against him. On the retreat of Yassaur from the neighbourhood of Herat, Ghiath ud din sent a mollah to the Court, which was then in Arran, to report what had happened. Choban ordered 50,000 dinars to be sent to the malik as a mark of the royal munificence, and issued an edict excepting Herat from the payment of all taxes for three years, and made over to Ghiath ud din all the domains of the maliks of Khavaf, Esfizar, Tulek, and other chiefs who had sided with Yassaur ; made over to him also all the free people and slaves he had captured from the son of Bujai, placed the amirs of the Nigudarians under his authority, and attached to his Court the archers (*charkh°endazan*) of Khorasan.* We read in the history of Herat that the malik Kutb ud din Esfizari was on very bad terms with Ghiath ud din. He had tried to bribe the officers of Abusaid to invest him with the district of Esfizar, and not having succeeded, had obtained from. Ghiath ud din the title of commissary (*mubashir*) in that country. When Yassaur occupied that district with 5,000 men he broke out into open rebellion against the ruler of Herat, shut himself, his family, and partisans in the citadel, and tried to surprise the fortress of Abkal. In this he failed. Meanwhile the amir Ali Khototai, who governed Esfizar on behalf of Ghiath ud din, besieged him in the citadel. He now appealed to Inaltekin, chief of Ferrah, who promised to send 10,000 men to his assistance. Thereupon the ruler of Herat arrived within a parasang of Esfizar with a large army, and proceeded to invest it. Inaltekin arrived about the same time, but deemed it prudent to withdraw again, leaving only 4,000 men behind. Without leaders, they were attacked by Ghiath ud din. They were dispersed ; many of them fled to the mountains, and 2,000 of them, perishing with hunger and fatigue, were made prisoners. Ghiath ud din ordered them to be exposed in chains under the walls of the citadel in order to intimidate the rebels. It did in fact discourage them, disunion broke out among them, and they agreed to surrender at discretion. Kutb ud din was duly tried and condemned to the bastinado, which he underwent in the middle of the Cheharsu. Many of his accomplices were put to death with torments. The rest were amnestied at the prayer of some venerable sheikhs, and some of them were sent to the tile works in Herat. Ghiath ud din appointed his nephew Muhammed, son of Alai ud din, governor of Esfizar, making Khototai his naib, or lieutenant.† Notwithstanding his ruinous wars and other occupations, Ghiath ud din greatly beautified Herat, which had been much neglected since the reign of Muhammed, son of Sam, and had become in parts ruinous. Able architects were employed to restore it, and it soon became finer than before. A medresseh (or college) was built to the north of the

---

* D'Ohsson, iv. 629-630.   † Journ. Asiat., 5th ser., xvii. 499-501.

great mosque, which was called Ghiathiah. A hall of reception (*bargulah*) was also built north of the citadel, and some skilful painters were employed in decorating it. On its western wall they represented the victorious army of Abusaid, and on the other Yassaur fallen and in the midst of a crowd of corpses, a scene which was not long in being realised. Ghiath ud din also erected beautiful baths near the ditch of the citadel. Two caravanserais and a bazaar, which extended from the citadel, reached to the Chehar Su ; a cistern and a caravanserai not far from the mosque of Tereh Furushi. He also caused many towns and fortresses to be renovated.*

Let us now turn again to Abusaid. His youth encouraged insubordination among his officers, who showed their temper during the campaign against Uzbeg. The Ilkhan complained to his old friend and supporter, Choban, who had a grudge against one of the most important of them, viz., Kurmishi. He ordered him and others to be bastinadoed. It would seem that the amirs who had charge of this punishment contented themselves with inflicting a few light blows only. Choban, having heard of this, had him stripped, and the punishment repeated more seriously. At a feast which took place some time after, when the amirs drank to one another, Kurmishi offered the cup kneeling to Nejm ud din Abubekr, one of Choban's chief nawabs, who excused himself from kneeling in return on account of his age, which added to Kurmishi's irritation.† The latter now formed a plot with Ghazan and Buka Ilduzdi, who had also suffered corporal punishment. "He wishes to dominate over us. Our fathers never marched under the banners of his ; they were, on the contrary, their superiors in rank. Rather death than have him for our master." They accordingly determined to assassinate him. When Abusaid returned to Sultania, Choban, having dismissed his troops, went to his summer quarters in Georgia. Leaving his oghruks in command of his son Hussein, he set out for Gukché Tenguiz. (Wassaf says for his summer quarters of Akseran.) The conspirators, deeming the opportunity favourable, pursued him with a troop of picked cavalry. One of them, Kara Toghai, however, informed him of what was taking place. He was taken aback, and sent some officers to make inquiry. They were seized and put to death. He was for waiting quietly until their return, but was persuaded to leave at nightfall with two amirs. He intrusted his camp to Togmak, but scarcely had he left when his troops began to plunder it, and were joined by Togmak, who had been offended because the daughter of the amir Irenchin, whom he had courted, had been by Uljaitu's orders married to Dimashk Khoja, the son of Choban. Several of Choban's officers were killed in the *melée*. Hasan, another son of Choban, had gone to his father's rescue with 500 horsemen. They were pursued by

Kurmishi, Mama Khoja, Sati son of Timur, Kia, &c., with 20,000 men.
A terrible struggle ensued on the Blue Lake, near Nakhchivan, in which
Mama Khoja was killed, and Aghrukji and Kurmishi were wounded and
had to retire. On the other hand, Choban himself was wounded, and fled,
accompanied by only fifty horsemen. He was pursued by Hali Uruz (called
Aras, Togmak's brother, by Hamdullah), with 300 men, but he was not overtaken.* This is Wassaf's account. The continuator of Rashid says that
Choban and his small escort passed by a party of people roasting a sheep
in the midst of a meadow, and were asked to partake, but they sped on.
Togmak, an officer of Kurmishi, who was following the fugitives with
fifty men, having arrived at the same place, accepted the invitation. This
gave time to Choban to escape. When he reached Nakhchivan Choban
asked help from the malik Ziai ul Mulk, who excused himself and merely
gave him some provisions, conduct for which he was afterwards severely
reprimanded, and he would have been put to death if he had not bought
off the penalty by a payment of 100,000 dinars. The vizier, Ali Shah,
who was at Tebriz superintending the building of a mosque, behaved
very differently. No sooner did he hear of Choban's danger than he
went to rescue him with a body of well-armed horsemen. He met him at
Merend, and escorted him to Tebriz, where the people declared for him;
but he did not delay there more than a day, and hastened on to Sultania.
In passing Aujan he ordered Siyurghatmish and Kublai to defend this
post, and to send daily intelligence of the movements of the enemy.†
We read how at this time the Princess Kordojin, who held the government of Fars, sent a present of standards, drums, led horses, tents, and
armour to the Sultan, who rewarded her by giving her the title of
Terkhan, which exempted her from the payment of all taxes.‡

Meanwhile Kurmishi and his companions persuaded the amir Irenchin,
who was a Kerait, like Kurmishi himself, to join them. He had become an
enemy of Choban's since the latter had deprived him of the government
of Diarbekr and given it to Sunatai. Wanting to have it believed that
they were acting in the name of Abusaid, they showed everywhere a false
order of that prince commanding Irenchin and Kurmishi to kill Choban
and those whom they found with him. Many distinguished people,
deceived by this artifice, joined them. From their camp between
Nakhchivan and Tebriz, they sent word to Abusaid that Choban had
revolted, and that they were obliged to march against him. Their
messenger arrived at Sultania before Choban, and put up with Sheikh
Ali, son of Irenchin, who was a great favourite of Abusaid. Sheikh Ali
at the first impulse was for killing Dimashk Khoja, son of Choban, but
postponed his purpose until he could make fresh inquiries. Ali Shah
arrived the following morning, and had an interview with Abusaid, whom

---

* D'Ohsson, iv. 630-632. Ilkhans, ii. 274-275.   † Id., D'Ohsson, iv. 632-633.
‡ Ilkhans, ii. 275-276.

P *

he informed of the falseness of the report, and praised the great services of Choban, who himself speedily arrived at Sultania and exposed the designs of his enemies.   The rebel amirs had arrived near Tebriz, intending to plunder that city to satisfy their troops ; but after second thoughts deemed it better not to begin by irritating the people, so they passed it by. On approaching Aujan, Siyurghatmish, feeling unable to resist them, retired upon Sultania, where he gave the alarm. Abusaid marched against them.   He commanded the centre of his little army, Choban commanded the right wing, and Yassaur, with the grand amirs, Kerenj, Alghui, Satai, Sheikh Ali Behadur, and Siyurghatmish, the left, where was also Kara Sonkor with a body of Egyptians.   On the side of the enemy were Kunjuk, the daughter of the Ilkhan Sultan Ahmed, with her husband Irenchin, who commanded the centre ; on the right were Kurmishi and his sons, and on the left Togmak, Uruz, and other leaders. When the Sultan was a day's march from the rebel army, Kutlugh Shah Khatun, daughter of Irenchin, begged him to halt while she sent some officers to persuade her father to submit.   The Sultan accordingly halted at Zenjan ; but on the return of the officers with the report that they had not been able to persuade Irenchin, he continued his march. The next day the enemy was seen in the distance, near the village of Minare'dar. Wassaf says in the place called Miane, near the village of Minareh.   The troops on each side passed the night on horseback. Kutlugh Shah Khatun now made another attempt to persuade her father to submit, promising that he would obtain mercy.   " In that case," said Irenchin, " the Sultan has only to unfurl his white standards ; this shall be the signal for peace and pardon."   The princess told Abusaid, who the next day ordered the white banners to be unfurled.   On seeing this, Irenchin, inflated with arrogance, thought Abusaid was afraid.   Summoning Kurmishi, he told him the imperial troops could not resist their first charge, and they advanced, deeming themselves certain of victory and of empire. Thereupon Abusaid caused Sheikh Ali, Irenchin's son, a young man of very prepossessing appearance, to be beheaded, and ordered his head to be exposed on the end of a lance, crying out, " Thus perishes whoever is the Sultan's enemy."   Irenchin, on seeing this head, rushed upon the foe, regardless of his great age. Several warriors fell under his blows and those of his wife Kichik (or Kunjuk, as Von Hammer calls her), who followed him, sword in hand, excelling the bravest warriors in valour. The two armies joined issue—father fought against son, brother against brother.   Abusaid's men began to give way, when he charged in person, and his officers followed his example, and the enemy was routed.   The amirs Uruz, Hasan, Hussein, Tamuk, Sighaneh, Seli, Begtimur, and Karauna were killed in the fight. Irenchin himself was captured in the village of Kiaghidkunan, and was put to death by the guards ; so were Kurmishi and Togmak. One of Kurmishi's sons fell with his father, the

other fled to Uzbeg, the Khan of Kipchak. Wusdadar, son of Irenchin,
• a giant in size and courage, was put to death with the sons of Togmak
and other amirs. Some were cleft in four, others placed in pans of
burning naphtha. The princess Kunjuk, in accordance with the yasa of
Jingis Khan, was stoned to death. Her naked body was then thrown
out on the road, and was trodden under by horses and cattle.* This is
apparently the account given by Wassaf. In the continuation of Rashid
we are told that only four amirs escaped from the fight—Kurmishi, his son
Abd ur Rahman, Buka Ilduzji, and Choban Karaunas. All the rest were
either captured or killed. A tent was pitched on the field of battle, and
when the Sultan had returned thanks for his victory the prisoners were
brought in. Aras and his relatives were put to death at once, but
Irenchin, Togmak, and Issenbuka were reserved for a more ignominious
end. They were conducted to Sultania, and hung up to hooks, and wood
fires were lighted below them. Their relatives, friends, and clients were
also put to death, as were other people who lived far away, and were
accused of being their partisans. The four amirs above-named were
arrested by Sunatai, who had set out, on hearing of the outbreak, from
Diarbekr for the royal ordu. He killed three of them at once, and sent
Kurmishi to Sultania, where he shared the same fate.†

The Egyptian historian Novairi has another story. Under the year
719 HEJ. he reports how the vizier Taj ud din Ali Shah, having preceded
Choban to the Court, vaunted his excellences, and said that his enemies
• were moved only by envy and ambition, and that they wanted to destroy
him merely to reign in his stead, and that the amir Irenchin thought he
had a claim to the throne, as he belonged to the Imperial stock. Abusaid
thereupon allowed Choban to return, and he went to him in sackcloth and
tears, saying, "They have killed my principal officers, men whom I have
selected for your service; they have appropriated the moneys I have
amassed by your bounty. I have lost the consideration which your favour
had surrounded me with. If you wish for my death, I am here; I am only
one of your slaves." Abusaid said he was far from wishing him any
harm, and knew the motives which had impelled the rebels to rise both
against himself and him. Choban then asked for troops with which to
fight the rebels, and 10,000 men, commanded by the amir Taz, son of
the noyan Kitubuka, who had perished at Ain Jalut, were confided to him,
while the Egyptian Kara Sonkor presented himself with 300 horsemen,
armed cap-à-pie, in the Egyptian fashion. Abusaid marched in person
at the head of his military household, so that Choban might be assured
that he was fighting for him and not for his enemies. Meanwhile
Kurmishi, Irenchin, and Togmak had pursued Choban to Tebriz, whose
gates were closed by the citizens; but its commander, El Haji, went out

---

* Ilkhans, ii. 276-277. † D'Ohsson, iv. 636-637.

to meet them with an offering of meat and drink, and provisions for their
men.   They demanded 70,000 dinars from him on account of the gates
having been closed, and also of its people having, at the instance of the
vizier, Ali Shah, gone out to welcome Choban.   They left the same day
and went towards Mianét, passed Zenjan, and met the Imperial forces
near a village called Minaré.   Irenchin, at the sight of the Imperial
standard, was taken aback ; but Kurmishi urged that they must fight, as
Abusaid was really their friend.   The two armies being ranged opposite
one another, Kurmishi shouted out to Choban to show himself, as he
wished to go and make his submission to him.   Having unfurled his
standard, Choban took the precaution to withdraw.   Kurmishi made a
terrible charge, expecting to find the general near those defending the
standard.   The struggle was very fierce.   The amir Taz and Kara
Sonkor both showed great bravery.   At length Irenchin and his people
took to flight, and the greater part of the rebels passed over to the other
side.   Irenchin, Kurmishi, Togmak, his brother, and other officers were
captured and taken to Sultania.   Taken before a military council (*yargu*)
and interrogated about their revolt, they replied unanimously that they
had acted by order of Abusaid.   Kurmishi said to Choban, " Yusuf Beka
and Muhammed Herzeh came to me from Abusaid to engage me to attack
and kill you."   Choban summoned the two men just cited, who confirmed
the statement.   Abusaid denied it and said that they lied, and ordered that
they were to be treated as men who had calumniated their sovereign.   All
were condemned to death according to the yasa of Jingis Khan.   Thereupon Irenchin drew a paper from his portfolio, and said to Abusaid, "See
the order you gave me to kill Choban."   He then cast reproaches upon
him, presuming on the fact that he was uncle to the Sultan's mother.   The
Sultan denied the accusation, and said to Choban, "Act towards them in
accordance with the yasa ; they have rebelled against me and you."
Choban had them accordingly executed.   He began with Irenchin,
whom he hung by his sides to iron hooks.   While in this position he
cursed Abusaid with terrible imprecations.   They tried to cut out his
tongue, but as they could not seize it they thrust an iron spear under his
chin, which traversed his palate.   His body was exposed naked for two
days.   Eventually his head was cut off and promenaded about Khorasan,
Azerbaijan, the two Iraks, Rum, and Diarbekr.   Kurmishi and Togmak
were hung from hooks.

Novairi reports these facts from the history of the Sheikh Alem ud din
el Berzali, who heard them from the Sheikh Muhammed, son of Abubekr
el Kattan, of Irbil, who came from Damascus.   He adds others from the
statements of a merchant named Alai ud din Ali, who was at Sultania
when the tragedy occurred.   The latter said that about thirty-six amirs were
put to death, and their goods appropriated by Choban, who thus recouped
his losses.   The body of Irenchin was, according to him, exposed for

three days, and with him were hooked Togmak and his brother Ersemeh, as well as the amir Bektut. The next day Yusuf Buka, Bektut's brother, and the amir Yumai, suffered the same fate. Two sons of Togmak, only seven years old, were put to death on the third day. On the fourth, Vefadar, son of Irenchin, who was only fifteen. Ali, another of his sons, had perished in the battle. His head was thrown to his mother, Kichik, daughter of Sultan Ahmed, son of Abaka, who took part in the fight, and was captured and trodden under horses' feet, by order of Abusaid. Kurmishi was put to death on the seventh day. His chin was first shaved, and a painted cap, called *turtu*, was put on his head. He was promenaded with nails inserted in his body through the town of Sultania, and conducted before Choban, who had him killed with volleys of arrows. Kutlugh Shah, son of Irenchin, and one of the widows of Uljaitu were condemned to death by Abusaid, on the charge of having poisoned his father. Her life was spared at the intercession of the vizier Ali Shah, who married her to Khoja Damashk, son of Choban. The amir Taz married the widow of Togmak, and succeeded Kurmishi in command of Taberistan. The bodies of those put to death were all burnt.* The report of the Egyptian historian that Abusaid had something to do with the revolt was perhaps not altogether an invention, and it may well be that, wanting to rid himself of the patronage of Choban, he incited what he was afterwards afraid to support.

Von Hammer, apparently following Wassaf, tells us how the vizier, Taj ud din Ali Shah, having been the first to break through the enemy's ranks in the late fight, was given the juldu, which in the Mongol hunters' phrase was the best portion of the quarry, viz., the back, head, fore and hind feet, and the skin, and which was given to the one who first wounded the game. In his case this was the harem of Irenchin. Kutlugh Shah, the latter's daughter, the widow of Uljaitu and stepmother of Abusaid, he made over to the amir Pulad Kia. Ilghi, the widow of Togmak, was first given to a man of middle-class station. She redeemed herself from this disgrace by a payment of 5,000 gold pieces, and then married the amir Siyurghatmish. In consequence of Abusaid's conspicuous bravery in the battle, the title of Behadur was added to his name in public documents. Wassaf, the historian, wrote a kassidet on the victory.† It would seem that the rebels had sympathisers elsewhere, and especially in Rum, where several amirs, viz., Kurbuka, the brother of Issen Kutlugh; Berbai, brother of Sai Noyan; Muhammed, son of Budin Noyan, and others, broke out in rebellion, and committed various depredations for eighteen days, until the news of the Sultan's victory arrived, when they quarrelled with one another. Those who were not killed in this struggle were put to the sword by Timurtash, son of Choban, who then governed Rum.‡

---

* D'Ohsson, iv. 637-642. Note.　† Ilkhans, ii. 277-278.　‡ *Id.*, 277.

After the events above described Abusaid set out for his country quarters at Karabagh, in Arran, and Choban received a new proof of his favour. His wife Dulendi, sister to Abusaid, whom he had married in 1304, had died, and he begged Abusaid to give him the hand of his other sister, Sati beg, which was accorded to him, and the marriage took place on the 6th of September, 1306.* The following year Abusaid was delivered from a turbulent neighbour, the Prince Yassaur, at the hands of the Khan of Jagatai, as I shall describe in a later volume. At this period there was great suffering in the southern provinces of the empire. Diarbekr, Mesopotamia, and Kurdistan were, during the year 1318, ravaged by famine. Many of their inhabitants emigrated, and others perished of hunger and disease. Dead bodies were eaten ; people sold their children, the price of boys varying from five to fifty dirhems. They were for the most part purchased by the Mongols, and as it was forbidden for Mussulmans to sell their children, the mothers declared that they were Christians. The towns of Mardin, Jeziret ul Omar, Mayafarikin, Mosul, and Irbil, were largely depopulated. Irak Arab also suffered. The first cause of the misfortune, which began in Diarbekr and Sinjar, was the passage of a flight of locusts. The drought in the spring of 1318 aggravated the evil, which went on increasing till the spring of 1319, when it was at its height. In August, 1320, there was a hailstorm at Sultania, in which the hail was of exceptional size, one piece weighing eighteen dirhems. This hailstorm killed many cattle. It was followed by a great inundation in the town. The panic was general, and the divine pity was invoked. The doctors of the law, on being consulted as to the cause of the visitation, attributed it to the oppression and tyranny that prevailed, and especially to the existence of wineshops and houses of ill fame near the mosques, colleges, and monasteries. Abusaid accordingly ordered taverns to be suppressed in the kingdom, and the tavern keepers had to take all their wine in hogsheads to the foot of the castle, and we are told more than 10,000 of these casks were collected there. Thereupon the vizier, Ali Shah, came with his employés, and the barrels were duly emptied into the ditch, which was certainly a large experiment in enforced total abstinence. The casks were burnt, and the fire lasted two days. Novairi says he took the story from the history of Alem ud din el Berzali, surnamed El Moctafi, who related it on the authority of a merchant of Mosul, who was at Sultania, and passed on to Tebriz, where he tells us the streets ran with wine, as they did at Mosul, to which he afterwards repaired. At this time Abusaid also abolished the tax on grain.† About the year 1320 a prince of the Golden Horde, named the amir Asghir Oghlan, who was brother of Uzbeg Khan, seized and fortified himself in a strong position in Karabagh, a northern province of Persia, bordering

---

* D'Ohsson, iv. 641-642. † Id., iv. 644-646.

on the Caspian. Abusaid sent the amir Pulad Kia against him with 10,000 men, who speedily killed the intruder.*

On his accession, Abusaid informed the Egyptian Sultan of his desire to see a firm peace established with that country, and received assurances of a similar kind in reply ; but the welcome continually offered in Persia to Egyptian refugees, greatly irritated the Egyptian Court. The amir Hosam ud din Mohanna, who, on the death of his father, Sherif ud dia Yasa, in 1284, had been invested by Kelavun with the government of the Syrian Bedouins, had not dared to pay his respects to Nasir after having assisted the rebel amir, Kara Sonkor ; and, although the Sultan had shown him every consideration, and sent him letters and presents, given him new fiefs, and distributed largess to his sons and relatives, he was not more submissive. Nasir accordingly, in May, 1316, conferred the command of the Syrian Bedouins on Shuja ud din Fazl, brother of Hosam ud din, and decked him, as was customary, with a robe made of satin of Mardin, embroidered with fur. Hosam ud din then went to Irak Arab, to the ordu of Uljaitu at Baghdad, who rewarded him handsomely, granted him an appanage, and gave him permission either to settle in his dominions or to leave. He preferred to return to Syria to reconcile himself with the Sultan, who, in May, 1317, re-established him by letters patent. Nasir asked him to pay him a visit, but he feared to go. He also suspected the largess which Nasir had caused to be distributed among the Bedouins, his subjects. Abulfeda tells us that in the year 1319, Fazl, son of Mohanna, son of Isa, went with a present of Arab horses to Abusaid and his patron Choban, by whom he was invested with the fief of Bassora in addition to the fiefs he held in Syria.† We saw how Homaiza, Prince of Mekka, was frustrated in trying to recover his patrimony, and was obliged to escape to the Bedouins.‡ He remained for some time in the Hijaz. Presently the Egyptian Sultan sent two officers to summon him to his presence. He asked for money to defray the journey, which they gave him. Thereupon, however, he fled. A year later (i.e., in the spring of 1318) he surprised Mekka, whence he drove his brother Abulgaith, and substituted the name of Abusaid for that of Nasir in the Friday Prayer. Homaiza was assassinated by one of his own slaves in 1320.§

When Nasir, in 1320, ordered an invasion of Cilicia, Hosam ud din thought the expedition was directed against himself, and migrated a second time. He again went to Irak Arab with his relations and their clients of the tribe Al Fazl. The Sultan thereupon ordered the sequestration of the fiefs belonging to the fugitives, and nominated the amir Shems ud din Muhammed, son of Abubekr, chief of the Bedouins, and ordered all the family of Mohanna to be driven out of Syria.|| Nasir

---

* Ilkhans, ii. 284. † Op. cit., v. 323. ‡ Ante, 572. § D'Ohsson, iv. 585-586.
|| Id., iv. 646-648.

retained a violent hatred for Kara Sonkor, and tried to have him
assassinated. He sent thirty Ismaelites, or "Assassins," to Tebriz in
1320 upon this errand. They were put at his service by the head of the
Ismaelites of Syria, who lived at Massiat. One of them having betrayed
the rest, several of them were seized and put to death. This did not prevent
one of the survivors, however, from attempting Kara Sonkor's life when
he was one day riding in the neighbourhood of his camp. At the news
of this attempt there was a general alarm, and it was feared the Ismaelites
intended killing, not only Kara Sonkor, but also the Sultan, Choban,
Ali Shah, and the principal Mongol officers, and Abusaid secluded
himself for eleven days. Choban summoned El Majd Ismael es Sélami,
agent of the Egyptian Sultan, and said to him: "At one time and another
you have brought us presents, and you hope that we shall live at peace
with the ruler of Egypt. This is that he may more easily put an end to
us by means of the Ismaelites." He had him seized and threatened to
put him to death, but he was released by the vizier, Ali Shah. News
now came that an Ismaelite had tried to poniard the governor of
Baghdad, but had failed, and having been pursued had committed suicide.
This had such an effect upon Choban that he determined to propose a
treaty of peace with Egypt. An envoy was nominated to go and treat,
and Majd es Sélami was sent in advance to announce the fact. Orders
were issued to the governors of Damascus and Aleppo to go and meet
this envoy, and to treat him with the honours due to his position. The
conditions which he offered for a treaty of peace were that the Egyptian
Government should employ no more assassins (*fidaviyet*) in the Mongol
territory ; that it should not demand the extradition of fugitives who had
gone to settle in the dominions of Abusaid, who in return was not to ask
for the extradition of his own subjects similarly placed ; that the Arabs
and Turkomans should not be encouraged to invade the Mongol territory ;
that there should be free intercourse between the two kingdoms, so that
traders might pass freely from one to the other ; that the pilgrim caravan
should be allowed to go freely every year from Irak to Mekka, bearing a
standard with the name of the Egyptian Sultan upon it, and another with
that of the Sultan Abusaid ; and that the extradition of Kara Sonkor
was not to be asked for. Nasir having summoned his council, it was
decided to agree to a peace on the terms proposed, and presents were
ordered to be prepared for Abusaid. The fact of his having prohibited
the use of strong drink, and ordering the drink to be poured out on the
ground, and that those who evaded the order were to be put to death ;
that houses of ill fame were closed, and singers and dancers expelled,
that the taxes levied on foreign merchants were suppressed, the churches
near Tebriz destroyed, and the mosques restored, all made the Egyptians
more favourable to this pact. Nasir was not to be outdone in rigour by the
Mongol chief. He also forbade the sale of fermented liquors, and ordered

them to be poured away, closed taverns, and redeemed those whose lives had been devoted to prostitution.*

Abulfeda has preserved some special details of these negotiations, and names several embassies as having passed between Abusaid and the Sultan. The first was led by Maj ud din Ismal Salameti. It took splendid presents, mamluks, girls, &c.† In 1321 a second embassy went from Timurtash, son of Choban, who was then governor of Rum.‡ The next year the Sultan sent a return embassy under Atamish Nasir.§ Fresh envoys went on behalf of the Mongols in 1323.|| In December of the same year, Abulfeda himself was in Egypt when the more famous embassy above referred to arrived there. He tells us it had two leaders—Tughan on behalf of Abusaid, and Hamza on behalf of Choban—and Tavashi Rihan, the kharendar (i.e., the treasurer), was charged with the imperial gifts. They were all present at a grand feast, at which the amirs and mamluks were dressed in magnificently decorated robes and caps, among whom the Sultan was conspicuous by his simple costume. Abulfeda says he saw the presents they brought, which consisted of three noble chargers of the cross between Arab mares and Persian stallions called ikdish, whose saddles were decorated with Egyptian gold plate, inlaid with precious stones; three saddle girths made of gold tissue, inlaid with gems, and a sword with a sheath made of the same costly materials; also striped cloth, with golden threads interwoven; a splendid girdle similarly embroidered; and eleven sumpter camels bearing chests covered with Persian cloth. There were also 700 pieces of cloth with the name and titles of the Sultan interwoven. The Sultan accepted the gifts, and conferred robes and other marks of his favour on the envoys, who two days later were invited to take part in the famous feast held on the day of the great Mekka sacrifice, and which was celebrated with great pomp. The envoys were then decorated with golden girdles, and we are assured that altogether the Sultan spent upon their entertainment more than 100,000 dirhems. The envoys were then sent home again.¶

In 1322, Timurtash, the son of Choban, who was governor of Rum, revolted, coined money, had the Khutbeh recited in his own name, and gave himself out as the Mahdi, who was to appear at the end of the world. He sent envoys to inform the Egyptian Sultan that he intended to conquer Persia, and asking for his help. Choban, when he heard the news, was naturally taken aback, and asked to be allowed to march against his son, whom he promised to bring to the foot of the throne if he should freely submit, and if not, undertook to bring his head. Although it was winter, and Choban was suffering from gout, he set out. Timurtash prepared to resist his father, but was urged by the magistrates and clergy to submit. He accordingly did so. Choban had him bound, and took

* D'Ohsson, 648-652. † Op. cit., v. 345. ‡ Id., 347. § Id., 351.
|| Id., 355. ¶ Id., 357-359. Ilkhans, ii. 285-286.

him with him to the Sultan, who pardoned him for the sake of his father,
and reinstated him in his command.   Several of the counsellors of
Timurtash, however, were put to death.*   About this time we read that
Kutlugh, the daughter of Abaka Khan, made a pilgrimage to Mekka, and
was treated with great honour *en route* by the Egyptian officials.†   The
negotiations with Egypt, above-mentioned, led to a peace, which was
duly ratified in 1323 by Abusaid, and published from the pulpit at Tebriz.
The Egyptian amir Itmish returned to Cairo with the treaty, which was
ratified by the oath of the Sultan and of the vizier Ali Shah, while an
envoy of Abusaid went to Cairo to receive Nasir's oath, and from this
time frequent embassies passed between the two rulers.   In making peace
with Egypt, Abusaid interposed his good offices on behalf of the
Armenian kingdom of Cilicia, which had been terribly ravaged by the
Egyptians.‡

We have seen how, on the murder of Leo IV. and his uncle Haithon,
'in 1308, Oshin, Haithon's brother, was appointed king of Little Armenia.
He was duly consecrated at Tarsus, and reigned till the year 1320, and
was succeeded by his son Leo V.§   As he was only ten years old he was
under the tutelage of the Bailiff Oshin, who married his mother, Joan of
Sicily, and who caused him to marry his daughter.   The treaty with
Egypt having nearly expired, the Armenian King asked for its renewal
on the same terms; but Nasir insisted on the restitution by the Armenians
of several places they had appropriated under their strong ruler, Lachin,
and as they would only surrender one of these fortresses the Egyptians
entered Cilicia under the amir Shihab ud din Karttai, governor of the
province of Tripoli.   After crossing the Jihan, in which about 1,000
horsemen were drowned, these troops divided themselves into several
bodies, which ravaged the country for some days, and retired with
their booty.   This was in June, 1320.   The King and the Bailiff Oshin
wrote to the Pope to implore the aid of Christendom.   John XXII., who
was then Pope, replied that the sovereigns of Europe were at war
with one another, and could not therefore help him, but that he would
himself send some troops to their aid.   But before their arrival Nasir,
who had heard that Leo proposed to arm Europe against Egypt, took a
cruel vengeance.   At his instigation Timurtash entered Cilicia, when an
attack from the Mongols was unexpected, took Sis, the capital, and
advanced as far as the maritime stronghold of Aras, ravaged the country,
killed many people, and carried off a number of prisoners.   After which a
Turkish amir named Omar pillaged the unfortunate land, opening the
tombs to find treasure, and throwing the bones found in them into the fire.
He burnt the standing corn and that in the magazines, carried off the
cattle, and retired, after a raid lasting for twenty-five days, with the

---

* D'Ohsson, iv. 658-659.   Ilkhans, ii. 288-289.   Abulfeda, 355-357.
† Abulfeda, v. 355.   ‡ D'Ohsson, iv. 659.   § Journ. Asiat., 5th ser., xvii. 385.

spoils of Little Armenia, which was then again invaded by the Egyptians. They captured several towns, burnt Adana, razed its citadel, captured a rich booty there, and carried off 20,000 prisoners. Meanwhile the various Armenian grandees, instead of uniting together, were engaged in mutual quarrels.

On hearing of these terrible events, the Pope addressed a general brief to the Faithful, exhorting them to aid the Armenians with men and money, ordered solemn prayers in all the churches, and collections for those who would take the cross. He sent Leo a sum of money with which to raise troops, and sent a letter to Abusaid, dated from Avignon on the 13th July, 1322, in which he told him how his ancestors, who had been friendly to the Christians, had always been the allies of the Armenian kings, and had supported them against the Turks and other enemies, and begged him to do the same and to aid the king of Armenia, whose enemies had invaded a portion of his kingdom and appropriated the rest. The Pope went on to exhort the Ilkhan to embrace Christianity, on which subject he wrote him a special letter, dated on July the 12th. In this he recalled the ancient friendly relations which had existed between Abusaid's predecessors and other Christian rulers. "We remember having been often told that your ancestors, illustrious by their magnanimity, have sent envoys at various times to show their veneration for the Roman Pontiffs and the Apostolic See, and that they were on friendly terms also with the kings of the Franks and other Christian rulers, who had received these envoys in a friendly fashion. These kings and your ancestors mutually honoured each other with gifts and presents. You will complete your meritorious qualities if you also will send envoys to us and the Holy See, and if you will renew the friendly intercourse with our dear son in Jesus Christ, the illustrious king of the Franks. We rejoice in God, and desire you may always do what shall be pleasing to the Lord." These letters are both subscribed with the name Boyssethan, a corruption of Bu Said Khan.* The Pope, by a Bull dated the 1st of May, 1318, had founded an archbishopric at Sultania, to which he appointed Francis of Perusia, a Dominican who had already preached the faith in the East.* He confided to the new archbishop the charge of all the subjects of Abusaid who were Catholics, as well as those inhabiting the country ruled by Kaidu, and the Kings or Princes of *Ethiopia and India*. He appointed six Dominicans to be his suffragans, and they received orders to proceed at once to Persia. Francis of Perusia resigned his archbishopric in 1323, and was succeeded by William of Ada; but the Christians in Persia had latterly, and since the conversion of the Mongols to Muhammedanism, greatly fallen from the state of prosperity in which they were during the reigns of Khulagu, Abaka, and Arghun.†

* D'Ohsson, iv. 660-663. Abulfeda, v. 349.   † D'Ohsson, iv. 663-664.

Leo, the Armenian king, had himself asked the aid of Abusaid, who sent 20,000 men to his help, and also begged Nasir to make peace with him; but before the arrival of the Mongol troops a band of plunderers entered Cilicia from Asia Minor, pillaged and burnt the town of Ayas, and committed great ravages in the country. Shortly after, Constantine, the patriarch of the Armenians, went to Egypt, and obtained from Nasir a treaty of peace for fifteen years. It had been concluded when the Mongol troops reached Cilicia.* In the year 1328-29, Leo rid himself of his patron, Oshin, who is called the Lord of Kurk by Abulfeda, and was duly installed as King of Armenia. It was certainly a sad spectacle to the eyes of Christendom when the young king was constrained to accept investiture from the Egyptian Sultan, who sent him a royal robe, a sword, and a charger. Having kissed the ground in honour of his Muhammedan protector, who sent the amir Shahib ud din Ahmed to invest him, he sent presents back by him to Egypt.†

We will now revert more immediately to the Ilkhan and his people. Choban, his famous general, who had occupied such a prominent position for so long, was now at the height of his prosperity. We read how at this time his wealth was greatly increased by a curious means. It seems that when the Mongols overran Kurdistan under Khulagu, Choban's father secured as a prize Naz Khatun, the daughter of the ruler of that country. The judge of Hamadan, in order to please Choban, now produced a number of grants of land, &c., which professed to have been made by this princess, and urged that as she had become his father's slave, this property, or the reversion in it, belonged to him. He thus secured a large property in the neighbourhood of Kazvin, Kharkan, and Hamadan; and it also came about that people who were displeased with their landlords, professed that they had formerly occupied under the Khatun Naz. This was deemed a sufficient title, and thus properties worth thousands of golden pieces were disposed of for quite a few, for fear that some claim might be set up to them on behalf of the Khatun Naz. The vizier Ali Shah tried to stop these iniquities, and claimed on behalf of the Treasury to have the real title to these reversions. To settle matters, he offered Choban 200,000 gold pieces for all his claims.† Meanwhile, however, the Vizier died. This was in the beginning of 1224. He was the first Vizier of the Ilkhans who had ended his life by a natural death. He was buried in the Great Mosque at Tebriz, which he had founded, and which is still the largest and finest in the city. Abulfeda says he bound together the Tartars and Mussulmans (i.e., the Mongols and Egyptians) by numerous embassies. Abusaid was greatly attached to him, and went to see him when he was ill, and the ablest doctors were summoned to attend him. His authority was divided between his two

---

* D'Ohsson, iv. 664-665.   † Abulfeda, v. 381-383.   ‡ Ilkhans, ii. 289-290.

sons, Ghiath ud din Muhammed and Khalifeh, who speedily quarrelled.
The officers of the Divan took sides with one or the other, and their
quarrels became so violent that they were both dismissed, and only saved
their heads by the sacrifice of the great fortune which their father and
they had accumulated through many years. The post of Vizier was then
conferred on Rokn ud din Sain (Hamdullah calls him Malik Nuzrat ud
din Adil Nessavi, and says he took the title of Sain Vizier), whose grandfather had been inspector-general of the forces of Muhammed Khuarezm
Shah. Rokn ud din himself was a *protégé* of the amir Choban. Having
gained over the latter's officers, he persuaded them to urge on Choban
that it had always been the custom for the most powerful amirs to secure
the viziership for one of their clients. Thus, in the reign of Arghun,
Ordukia had secured the post for Said ud devlet ; under Gaikhatu,
Toghachar for his *protégé*, Sadr ud din ; under Ghazan, Nurin Aka for
Said ud din Sanji ; and under Uljaitu, Hussain Gurkan for Ali Shah.
"As our master is not inferior to these amirs, it is seemly that he appoint
Rokn ud din Sain." Choban accordingly nominated him, but he was not
long in proving his incapacity.*

Choban had long determined to revenge the invasion which Uzbeg, the
Khan of the Golden Horde, had made some time before. He accordingly
in 1325 broke through the pass of Derbend, and advanced as far as the
Terek, whence he returned with many captives and much booty.† He
was now going to feel the usual reverse of fortune that overtakes rapid
promotion in the East. His position to the outside world must have
seemed almost unassailable. He had successively married two of
Abusaid's sisters, was everywhere acknowledged as the head of the amirs,
while his own *protégé* was Vizier ; but "the night," as the Arab proverb
says, "is pregnant, and brings forth many things ere dawn." Abusaid,
who now was twenty years old, had conceived a great passion for his
beautiful niece Baghdad Khatun, the daughter of Choban, who was the
wife of one of the greatest amirs in the country, viz., Sheikh Hassan.
According to the yasa of Jingis Khan, the sovereign could always claim
the wife of any of his subjects if she took his fancy, and Abusaid would
have been well within the law in claiming Baghdad Khatun for himself ;
but Choban, relying on his great power and authority, and forgetting how
entirely beyond control and direction such passions are, fancied that
absence and time would cure Abusaid's fancy. He professed to take no
notice of what he had heard, and persuaded Abusaid to go and pass the
winter at Baghdad, while Sheikh Hassan and his wife were sent to
Karabagh. When the young prince reached Baghdad his passion seemed
to grow upon him. He seldom left his tent, and saw hardly anyone.
Choban in vain tried to distract his attention by hunting parties. Abusaid

took no pleasure in them. One day Choban took the liberty of asking him the cause of his depression. Abusaid turned his complaints upon his son Dimashk Khoja, who was very extravagant and dissolute. Choban summoned and reprimanded him for his ill behaviour. Dimashk replied that he was well aware the prince's sentiments towards himself were changed, and he attributed it to the intrigues of the Vizier. "I have learnt from an undoubted authority," he said, "that this ungrateful man, who owes his fortune to us, has urged upon the Sultan that the amir Choban and his family dispose of the kingdom as they please, and that no one else has any power." As a matter of fact, the Vizier, who had been given the title of Nuzret ud din, and who was incapable of managing his department, was jealous of the authority of the family of Choban, and seized every opportunity to vilify them to the Sultan. He told him they absorbed all the revenue, and that even what little money came to the Treasury could not be spent without their interference. That he (the Vizier) had not power to dispose of a single dinar, and that the Sultan had better put a stop to these usurpations before it was too late. These repeated attacks aroused Abusaid's jealousy, and as, whenever he went out on horseback, he was followed by a crowd of suppliants demanding justice, he imputed the public distress to the ill-conduct of the Chobanians.*

It was feared that in the spring the Jagatai Mongols would renew their invasions of Khorasan, which was bare of troops. Towards the end of winter Choban set out for that province, accompanied by the Vizier and by the generals Ekrenj (or Okrunj), Issen Kutlugh (Von Hammer says Mahmud, son of Issen Kutlugh), Muhammed (brother of Ali Padishah, uncle of Abusaid), and others, leaving the control of the Court in his absence to his son Dimashk Khoja. Choban was met *en route* by the kielenters (or calenders) and the kiotvals (*i.e.*, the governors of the towns and fortresses) bearing presents, and went on to Herat, where he was met by an officer sent by the Great Khakan in China bearing a robe of honour for him and the title of Amir ul Omera of Iran and Turan, and a letter full of praise of his administration. Choban sent this envoy back again with magnificent presents for the Khakan. Termésherîn, the Khan of Jagatai, now crossed the Oxus, but was defeated near Ghazni by the amir Hussein, son of Choban, as I shall describe in a later chapter. After plundering Ghazni, Hussein rejoined his father at Herat. This campaign took place in the autumn of 1326.†

Meanwhile, as we have said, Dimashk Khoja remained in charge of affairs at the Court. He allowed himself to be controlled by four dissolute favourites, and during the winter of 1326, while the Court was at Baghdad, committed all kinds of excesses, appropriating the wives and children

---

* D'Ohsson, 667-669. Ilkhans, ii. 292-293.   † D'Ohsson, iv. 669-670. Ilkhans, ii. 293-294.

as well as the property of the citizens. In the spring of 1327 the Court
returned to Sultania, where his excesses continued. The famous traveller,
Ibn Batuta, who was then travelling in Persia, has preserved us an
interesting account of what followed. He tells us that the Sultan was
much galled at the tutelage in which he was held by Choban and his
family. One day one of his father's widows, Dunia Khatun, said to him :
" If we were men we would not allow Choban and his sons to occupy the
positions they do." Abusaid asked what she meant to convey by these
words. Upon which she replied that the assurance of Dimashk had
reached the point when he even dared to have connection with the
Sultan's stepmothers. The very night before, he had passed with Thaghy
or Taki Khatun, and had sent word to herself, " I will spend to-morrow
night with you." " Collect the amirs and troops, and when he goes up to
the citadel secretly to pass the night you can have him arrested." Abusaid
was much enraged, and followed this advice. The citadel was duly
surrounded, and the next morning Dimashk was leaving it, accompanied
by a soldier named Alhaj al Misri (*i.e.*, the Egyptian pilgrim), when he
found that the door was blocked by a chain, which stopped his horse.
His attendant struck the chain with his sword and broke it. The two
then went out together. One of the Sultan's amirs, named Misr Khoja,
and a eunuch named Aka Lulu, killed him, and took his head to the Sultan,
who trampled it under the hoofs of his horse, and orders were given to
pillage his house.*

Other authorities give more details. Thus we read that the intrigue
which Dimashk carried on was in the harem of Taki Khatun, with
Kunkutai, one of the concubines of Uljaitu, Abusaid's father. Abusaid
took counsel with the amir Gunjuskat (his maternal uncle), Narin Togai,
and Tash Timur. Dimashk was duly watched by spies and was seen to go
to visit Kunkutai, and orders were given to kill him. Knowing this, he
remained in the palace where she lived, and tried to persuade the military
chiefs to side with him, but none of them would do so, and the next day
siege was laid to the citadel. Meanwhile the heads of a number of
highwaymen who had been caught and decapitated were brought into
Sultania. The Sultan had it declared that those were the heads of
Choban and his clients, who had been killed at Herat. (Von Hammer
says of the murderers of Choban.) This rumour having reached Dimashk,
he left the citadel with ten followers, cut his way through the troops, and
fled on a swift horse he had reserved for such an occasion; but Lulu Aka,
who went in pursuit of him, overtook him at the village of Virdgan.
Misr Khoja, who followed him, arrived with orders to put him to death.
Lulu Aka was loth to do this, as Dimashk was a person of such
importance, whereupon Misr made a journey to the Sultan and speedily

* Ibn Batuta, ed. Defremery, &c., ii. 117-119.   Ilkhans, ii. 295.

returned with his ring, which was accepted as ample authority. Dimashk
Khoja was now decapitated, and his head exposed on one of the gates of
Sultania. This was on the 25th of August, 1327. His four favourites
were treated in the same fashion, and his goods were given up to pillage.
Dimashk left four daughters : 1. Dilshad Khatun, who married Abusaid
and afterwards Sheikh Hassan, by whom she had two sons, Kematdin
Todan and Sultan Uveis. 2. Sultan bakht, who married the amir Ilkan,
son of Sheikh Hassan, and afterwards Masud Shah Inju. 3. Dendi Shah,
who married the Sheikh Ali Kushji, and became the mother of the amir
Missir Malik. The fourth was called Alem Shah.*

Abusaid now felt the necessity of crushing Choban before he had time
to take revenge. It was decided at a council that letters should be
written to the amirs Ekrenj, Issenkutlugh, Nuruz, and others, on whose
fidelity he could rely, to announce the punishment of Dimashk, and
ordering them to put Choban to death before he heard of it, and to inform
them that an army was on the march against Choban's sons, Timurtash
and Sheikh Mahmud, who commanded the troops in Rum and Georgia
respectively, and that orders had been issued to put the Chobanians to
death wherever found. This letter was sent by a trustworthy person.
On his arrival at the head-quarters in Khorasan the generals held a
council. Accustomed to look upon Choban as their absolute master, and
intimidated by his authority, they determined to go to Badghiz, and to
inform him of what had occurred. They showed him the Sultan's order,
with which they professed to be indignant, and promised to be faithful to
him and to assist him in revenging himself. When they had withdrawn,
Choban summoned his son Hassan and others to a consultation.
"Inasmuch as Abusaid has become our enemy," said Hassan, "we have
no other appeal than that to arms. But let us be on our guard against
these amirs who now promise us their devotion. It would be better, in
order to save our lives, that they should be executed. Khorasan is in
our hands, and we dispose of the revenues of Kerman and Fars. Let us
march against Abusaid. Timurtash is master of Rum, Mahmud of
Georgia ; we will surround him with our armies." Choban disapproved
of this advice. He thought he had nothing to fear from the amirs or
anyone else in the kingdom. He had taken the Vizier with him, so that
he should no longer poison the mind of the Sultan against Dimashk
Khoja, and remembering all at once the conduct of that minister, he had
him summoned and put him to death, thus avenging his son.† When
his forces were united, he found himself at the head of 70,000 men, with
whom he marched upon Irak. Abusaid also assembled an army.
Directly they heard of the death of Dimashk, Sunatai, governor of
Diarbekr, Devlet Shah, Ali Padishah, and other frontier commanders,

---

* D'Ohsson, iv. 671-672. Ilkhans, ii. 294-295.   † D'Ohsson, iv. 672-675.

made for the Royal Ordu with their forces. Abusaid set out from Sultania, and encamped on the plain of Kazvin. When Choban reached the holy city of Meshed he made the generals renew their oath of fidelity to himself. He advanced thence as far as Simnan, his troops committing terrible ravages on the way without his interfering. At Simnan he went to pay a visit to the Great Sheikh Alai ud devlet (the " Shajrat ul Atrak " calls him the Sheikh Rokn ud din Simnani), and again assembled his officers in the presence of that holy prelate, and bound them by a fresh oath. He took the Sheikh with him, and charged him to negotiate peace for him with the Sultan. He was to tell the Sultan that he had grown old in his service, and could not remember having committed any crime which should bring upon him the Sultan's anger. If Dimashk deserved death, the Sultan was too just to include his innocent father and brothers in his punishment; but he had been told that Dimashk had been put to death by some officers without the Sultan's consent. If it were so, he begged him to send him these officers that he might have them tried, and then treated as the Sultan should himself desire. He told the Sheikh, in addition, to see each one of the amirs and viziers, except two or three whom he regarded as the authors of his son's death, and try and gain them over. The Sheikh, having received these instructions, set out. Abusaid received him with great honour, rose at his approach, and gave him a seat beside himself. After a pious exhortation, he opened his commission, and pressed the Sultan, for the sake of peace, to hand over to Choban the authors of the recent troubles, and to pardon him. The Sultan replied, in the presence of all the officers summoned to the audience : " The arrogance of Dimashk Khoja, the ambition and power of Choban and his sons, exceed all measure. I have for a long time borne the authority which they arrogate to themselves, in the hope that they would behave differently and remember what my ancestors did for them ; but the more patience I have shown, the greater has been their audacity. They have illtreated my principal servants ; they have dealt with the revenue as they liked. These are the reasons that have moved me, but if Choban wishes sincerely to regain my goodwill, if he will come to me alone I will assign him a retreat where he may pass the rest of his life in works of piety. If not, the sword shall decide between us." The Sheikh renewed his exhortations in an eloquent speech, filled with citations from the Koran and Mongol aphorisms. He answered the amirs who argued against him, and tried to pacify them with the charm of his words. At length these officers said to him, " If the Sultan forgives Choban he must send us all, with our heads and beards shaven, into a country where he cannot follow us, for if he comes back here he will make an end of us." The Sheikh, seeing that argument was useless, now left.

Choban continued his march, and his troops behaved worse than an enemy's army would have done, and reached Kuhar, or Kurha. Abulfeda

Q *

calls the place Sari Kemash. Hamdullah says it was at Ibrahim Abad, on the confines of Rai. The place was only a day's march from the Sultan's camp. Several of the amirs, such as Chichek (the maternal uncle of Abusaid), Muhammed beg, Niknuruz, and others, seeing they were going to fight on behalf of a karaju, or subject, against a sovereign prince—a serious breach of all good citizenship, according to Mongol morals—abandoned Choban in the middle of the night with 30,000 men, and went over to Abusaid. In the morning, when he discovered this desertion, feeling he could not rely on his other officers, he regretted bitterly not having followed his son's counsel, and determined to retire at once towards Khorasan with his wives and relatives. He accordingly set out by way of the desert, abandoning his rich baggage, which was pillaged by his people. The amirs Ekrenj and Mahmud Issenkutlugh, who had been so much compromised, remained faithful to him, but the greater part of their troops went over to the Sultan. Truly, as De Guignes says, a man in disgrace has few friends. After a flight of three days he reached Saveh, and seeing that his wife Satibeg, the sister of Abusaid, with another of his distinguished wives, Kordojin, could not go with him, he told them to go to Abusaid. Satibeg took with her her son Shiburgan Shireh, who was named after the Sulduz who had saved the life of Jingis on one occasion, and from whom Choban claimed to be descended. Choban took with him his other son, Jelaukhan, whom he had by Dulendi Khatun, also a sister of Abusaid, and went towards Tabs with some horses and dromedaries laden with boxes filled with jewels. At each station he lost some of his following, and presently had only seventeen persons with him. He intended passing into Turkestan, but when he reached the Murghab he changed his mind, and determined to go to Herat to the Malik Ghiath ud din, with whom he had long been on friendly terms. In vain it was pointed out to him that Herat had already proved fatal to Nuruz and Danishmend Behadur, who were both similarly circumstanced to himself, and that it would be better to go to China, India, or Rum. He persisted in his resolution, and sent one of his officers, named Dulkandi, to have an interview with the malik, and followed himself. Ghiath ud din received him kindly, but presently received a letter from Abusaid, ordering him to put his guest to death, and promising to reward him with the hand of the Princess Kordjin and with the domains of the atabegs of Fars. On learning Choban's plight, Abusaid had dispatched Togai, the son of Sunatai, in pursuit of him with 2,000 men. He went as far as Saveh, when he turned back on hearing that Choban had fled into the desert. He returned with the two princesses above-named. Meanwhile the two generals, Ekrenj and Mahmud Issenkutlugh, had deemed it wise to send in their submission to the Sultan, who deprived them of their commands, but restored them to them some time later. His letter to the malik of Herat had put that prince in a great difficulty. He, on the one hand, was unwilling to break the laws of hospitality, and, on the other, was afraid of disobeying his suzerain. After long deliberations with his counsellors, his personal interests carried the day. He sent Choban the order he had received, and had him arrested. Choban reproached him with ingratitude, and begged him to postpone his death for a while, but to send word to the Ilkhan that his order had been executed. Fearing, however, he had already done too much to escape Choban's revenge if the latter escaped, he remained firm. He also refused his request to give him an interview so that he might commit his last wishes to him, and ordered the executioners to proceed with their work. Choban now summoned his son Jelaukhan, whom he embraced closely, amid tears. He then sent his last wishes to the malik : First, that he would not have his head sundered from his body, since he had been guilty of no crime ; but had, on the contrary, rendered the State great service. If proof of his death was needed, he asked that one of his fingers, the nail of which was very long, should be sent. Secondly, he asked that Jelaukhan should be conducted to the Sultan, who would, no doubt, have compassion on his youth and innocence. Lastly, he desired that his body might be laid in the tomb which he had built for himself at Medina. He then said a namaz of two rek'ats, and having repeated the profession of the Faith, gave himself up to his executioners, who strangled him. The officers of his suite were also put to death.

During these events, which had caused Abusaid great anxiety, he retained his passion for Baghdad Khatun, and now deputed the Grand Judge, Mobarek Shah, to demand her from her husband, Sheikh Hassan, who was constrained to surrender her, and having waited the time prescribed by law, which was three months, under the direction of the Kadhi, he gave a magnificent feast to his new bride. The finger of Choban was taken to him in November, 1327, when he was at Karabagh, and was suspended by his order in the market-place of the Royal Ordu. Shortly after, Ghiath ud din, the malik of Herat, set out for the Court. He learnt at Rai that Baghdad Khatun, Choban's daughter, had become the Sultan's wife, which news fell upon him like a thunderbolt. He sent word back to Herat to have Jelaukhan, who was a youth of rare beauty, put to death.* When he reached Karabagh he was admitted to prostrate himself before the Sultan. Baghdad Khatun—who was surnamed Khudavendigar in Persian (i.e. sovereign)—had already gained great ascendancy over Abusaid, and she would not permit him to carry out his promises to the Malik until the latter had brought the corpses of her father and of Jelaukhan from Herat. After having again washed and laid out the bodies, and said the funeral prayer for the repose of their souls, she had them sent by the caravan

---

* Both Ibn Batuta and the author of the Shajrat ul Atrak say his mother was Satibeg.

which went to the Hijaz. The Sultan set aside 40,000 dinars for their transport. The two bodies made the usual circuit about Mekka with the pilgrims, and accompanied them in the other prescribed processions. On the great day of sacrifice, after the divine service, all who had repaired to the Kaaba from the various countries of Islam, made a common prayer for the soul of the amir Choban, imploring God's pardon for his sins, and especially since he had made an aqueduct to convey water to Mekka; while they heaped curses on his murderer. The bodies were then conveyed to Medina, and buried near the tombs of the khalifs Osman and Hassan. This is one account. Ibn Batuta says he was not buried in the mausoleum he had built himself near the mosque of the Prophet, but in the Baki cemetery at Medina.

Choban was a pious and courageous man, and had shown great fidelity to his sovereigns, the Ilkhans. Among other monuments of his goodness are quoted the inns he had built on the route to Syria, which, we are told, surpassed everything of the kind which had ever been made by the Cæsars and the Chosroes (i.e., the rulers of old Rome and Persia). Raynald quotes a brief addressed by Pope John XXII. to the noble Zoban Begilay, by which name, as D'Ohsson says, it is probable that Choban Beïlerbeg is meant. It is dated from Avignon, the 22nd of November, 1321. In it he says that he had learnt from the Minorite friars, James and Peter, who had taken him presents, that Choban had treated the Christians living in Persia well, which gave him hopes that his eyes would be opened to the light, and that he would be delivered from the error of his former life to the knowledge of God. He begged "His Prudence" to continue to protect the Christians, and recommended the two friars, who were on their way to the Khan's dominions, to work for Choban and the people subject to the Khan.*

Choban had nine sons. The eldest, Hassan, was governor of Khorasan and Mazanderan. Hassan's eldest son, Talish, had charge of the provinces of Ispahan, Kerman, and Fars. When Choban fled from Rai, Hassan and Talish went to Mazanderan, where they would have been captured by the troops but for the generosity of a friendly grandee, who supplied them with horses and provisions, a generosity which cost him his life. The two fugitives reached Khuarezm by way of Dahistan. Closely pursued, they had lost a portion of their escort every day, and were reduced to five individuals when they were overtaken by seven horsemen, of whom they wounded three with their arrows. Eventually they reached Khuarezm, and were well received there by Kutlugh Timur, who governed the province on behalf of Uzbeg, Khan of the Golden Horde. Uzbeg himself invited them to his Court, and treated them with honour, and they marched some time after with an army supplied them by that prince against Majar and the Circassians, where they gained much renown. Hassan, however, was wounded in the campaign, and died shortly after his return to Uzbeg's Court, leaving, besides Talish, two other sons, Haji Bey and Kuch Hussein. Talish died a natural death ; the other two came to violent ends. Haji Bey was poisoned by his cousin, the Little Hassan ; while Kuch Hussein was killed by Suliman Khan.*

Timurtash, the second son of Choban, was governor of Rum, where he attacked the Karamanians, and pushed his conquests as far as the Mediterranean, which the Mongol arms had never previously reached, and fought by turns with the Greeks and the rebellious Turks. He had set out from Egerdür, or Egridur, on the 22nd of August, 1327, leaving his family and baggage there, while he had dispatched a body of 5,000 men under Eritai, to attack Kara Hissar. Meanwhile, a courier arrived at Egridur from Diarbekr with the news of Dimashk's death. He was sent on to Eritai, who left at once to rejoin Timurtash and to convey the news to him. He found him attacking the town of Bugurlu. Timurtash kept the news secret, but left in the course of three days, and arrived on the 13th of October at Egridur. There he disbanded his army, keeping only 5,000 men with him, with whom he went to Cæsarea, where he waited some days impatiently expecting news of his father (Von Hammer says five days, D'Ohsson says fifty) ; but all the routes to Persia were closed, and strange rumours were abroad. He presently left for Sivas. At Nigdur or Nigdui, situated seven parasangs from Sivas, where he spent the night, he received a message from Timurbuka, his agent, who was at the Court on business, telling him of his father's flight, which frightened him. He returned at once to Cæsarea, uncertain what he ought to do. Several of his officers suggested he should shut himself up in a strong fortress until he could appease the Sultan's anger, and regain his good opinion by promises of fidelity. Timurtash was for following this advice, and selected some rich articles to present to Abusaid. Then, reflecting that he had small hopes of reconciling a prince who had put his brother to death and driven away his father, he sent a messenger to the Egyptian Sultan to ask him if he would give him shelter. Meanwhile, he distributed his officers among the fortresses of Rum, and selected that of Kuh Larenda for himself. The Sultan Nasir received his messenger courteously, and told him his armies, treasure, and country were at the service of Timurtash. The latter, after some indecision, selected some rich presents, levied a heavy contribution from the people of Rum, and left Cæsarea on the 22nd of December, 1327, with his treasure and 700 picked young horsemen. In seven days he reached the fortress of Larenda on the Syrian frontier, where he would have delayed for further

---

* D'Ohsson, iv. 685-686. Ilkhans, ii. 300-301.

news, but was obliged to move on by lack of provisions and forage.
When he reached Behesna, the first town in Syria, the commandant and
magistrates went out to meet him, and pigeons were sent to Cairo to
announce his arrival, and thenceforward he received a maintenance of
1,500 dinars daily. The governor of Aleppo went out a league from that
town to welcome him, and supplied him with twenty post-horses, and told
him the Sultan was so anxious to see him that he wished him to hasten
on. At Damascus, the generalissimo of Syria went out to the Grand
Square to meet him, and they embraced each other on horseback. The
Sultan sent his cupbearer as far as Ghazzat with a mihmandar, some
tents, &c., and when he neared Cairo the amirs went out to meet him.
He reached that city on the 21st of January, and an amir duly conducted
him to the Sultan, who was at Jizet, on the other side of the Nile.
Timurtash kissed the ground three times before the Sultan, who made
him seat himself beside him, addressed him some friendly words, and
asked him about matters. He clothed him in a splendid state robe
embroidered with gold, and gave him also five Bedouin horses, with
saddles and bridles decorated with gold and silver. He then took him
out hunting. They crossed the Nile together, and Timurtash was lodged
in the palace of Chaoli in the fortress of Cairo. The next day he was
presented with fresh robes, a turban, a golden girdle, and a sword. His
establishment was maintained on a splendid footing at the expense of the
Sultan. Timurtash's presents arrived three days after himself, and
consisted of one hundred saddled horses, eighty Bactrian camels, five
mamluks, and five parcels of beautiful robes, one of which inclosed a
splendid satin dress, decorated with superb stones. The Sultan would
only accept this dress, a horse, and a string of camels. He ordered his
chamberlain to seat Timurtash on the right of the throne below the amir
Seif ud din el Malik, and, hearing that he was discontented with this
position, he told one of his officers to make his excuses, and to say that the
Sultan did not know his exact rank, but he had seated him amongst his
father's old friends, to whom he owed his own elevation, which satisfied
him. Some days later the Sultan reviewed the suite of Timurtash,
consisting of 700 horsemen, whom he distributed among his amirs,
permitting ninety of them, at the request of his guest, to return again to
their own country. Timurtash himself was given the command of the
Mamluks, who had formerly been led by Sinjar el Chomakdar. His
family had been left behind in a strong fort in Rum. Nasir wrote to his
vassal, the prince of Karamania, to have them sent on to Egypt. On this
message being sent to them, they replied they had no wish to go to
Egypt, and the Karamanian chief's son declared that this reply had been
secretly dictated by Timurtash himself, who, he added, had shed much
innocent blood and put many Mussulmans to death ; and that, judging
from his audacious character, he could have no other motive in going to

Egypt than to seize its throne. This letter was sent to Egypt by Nejm ud din Ishak, the Rumian Lord of Antakka, a fortress which Timurtash had captured. Nejm ud din accused that amir of having killed his father, and demanded vengeance upon him. This message and these reports aroused the Sultan's suspicions. He informed Timurtash of them, and ordered both parties to clear themselves in the presence of the amirs. At this trial it was decided that the father of Nejm ud din had been killed in a fight. Nejm ud din himself was sent back with Nasir's answer to the Prince of Karamania, but the Egyptian ruler retained a suspicion that his guest had some sinister design against him.

A month after Timurtash reached Cairo, a letter arrived there from Abusaid, which contained friendly sentiments, explained how Choban had been put to death on account of the design he had formed against himself, and of his unbridled assumption of authority. The Sultan questioned the messengers who brought this note about Timurtash. They said they had not heard anything about him until they arrived at Damascus. Nasir sent them to see him, but he did not wish to see them. Nasir himself had meanwhile sent a friendly note to Abusaid, informing him of the arrival of Timurtash, and that he had given him asylum, since he thought it better for the interests of the Ilkhan that he should be a guest at a friendly Court ; to which Abusaid replied that he hoped he would no longer give him shelter, that it was very inconvenient for criminals to find protection from one country in another, and that it encouraged desertion, and he begged him to surrender him. Meanwhile, Nasir had sent him two embassies, one openly, through which he asked Abusaid to send the family of Timurtash to Egypt, while the other asked for a secret interview, and explained that the other embassy was merely formal, that he had no other wish than to satisfy Abusaid, even against his own interests, and that Timurtash was at his disposal. Abusaid replied that he had already sent Abaji as his envoy to demand the extradition of the fugitive, which would alone satisfy him. Nasir accordingly summoned Timurtash to his presence on the 5th of July. On his arrival at the palace he was deprived of his sword, which had not been done before. The Sultan said to him, " You have asked us to send for your family, while you have secretly instructed them not to leave the territory of Abusaid. This proves you are not sincerely attached to us." Timurtash, seeing the Sultan meant him no good, remained silent. He was taken to prison and put in chains. His chief officers were arrested, and his mamluks distributed among the amirs. Abaji, or Ayaji, Abusaid's envoy, offered to see him conducted to Persia if he were supplied with an escort ; but Nasir, thinking that Timurtash would perhaps secure pardon, through the intervention of Baghdad Khatun and the vizier, Ghiath ud din Muhammed, who was an old friend of his, and would not fail to try and revenge himself upon him for the affront he had put upon

him, said that as the Kurds infested the roads on the way to Persia
it was better he should put him to death, and that Abaji should take
his head with him. "The Sultan ordered me to bring him alive," was
the reply. "What I wish to do is in the interest of Abusaid as it is of
my own, for I have taken a careful gauge of my prisoner." He accordingly
sent some of his people at night with Abaji to put Timurtash to death.
Abaji presented himself respectfully before the son of Choban, and
seeing he remained silent he said to him, "Amir, you are a man, and
liable to human vicissitudes ; resign yourself to the will of God." " I am
a man," said Timurtash, "and I believe I have proved it. I know the
decrees of the Eternal are irrevocable, a good reason why we should
submit without grief. But why have you come ? Is it to conduct me
alive before my sovereign, or to see me put to death here ?" Abaji
replied that he had been ordered to take him to Persia. " I suffer terribly
from the weight of my chains," said Timurtash, "cannot you relieve me
a little ? O, Abaji, I committed a grave fault in coming here. I ought
to have gone to my sovereign, and if he had condemned me to death I
should at least have died at the feet of his charger." Abaji then withdrew,
and Nasir's men completed their work. This was on the 22nd of August,
1328. His head, stuffed with straw, was put into a box and confided to
the son of Abaji, who reached Aujan on the 13th of September, and
who conveyed a letter from Nasir in these terms : " Having observed the
conduct of Timurtash, and having penetrated the secrets of his heart, I
was convinced that his existence would prove pernicious to you and me.
I have exposed myself to universal blame to prove my attachment and
friendship, for they will not fail to accuse me of having neither generosity
nor humanity ; but when the evil doers in both countries see the perfect
agreement between us our Courts will not serve as asylums for them, and
they will be less disposed to foment troubles between us." Timurtash
left four sons, Sheikh Hassan Kuchuk (*i.e.*, the Little), Malik Asraf, Malik
Ashar, and Misser Malik.* In reference to this tragedy Hamdullah
quotes the apposite lines :—

> He who seeks shelter with Amr in his distress
> Is like one who seeks shelter from the fire in the burning sands.

Nasir, in promising to surrender Timurtash, had asked in return for
the extradition of Kara Sonkor, which Abusaid refused, notwithstanding
the counsel of his advisers ; but he was relieved from embarrassment by
the death of that turbulent Egyptian at Meragha a few days before the
arrival of the head of Timurtash, whereupon he thanked heaven that
he had abstained from the ungenerous action which had been pressed
upon him. On hearing of the death of Kara Sonkor, Nasir said, " I had
hoped that he would have fallen by my own sword." During the rest of
his reign, Abusaid remained on good terms with Nasir. They styled each

---

* D'Ohsson, iv. 686-697. Abulfeda, v. 379-380.

other brother, and frequent embassies passed between them. Those of Nasir's traversed the country of Abusaid with their escorts, and with music playing and banners flying. One of these envoys named Timurbuka went to Cairo in 1328 to ask for one of the Sultan's daughters for his master, Abusaid. The envoy took with him a large sum of money to pay for the usual feasts given at betrothals, so sure was he of the success of his plans; but the Sultan excused himself on the ground of his daughter's youth, and said when she was grown up the request should be granted. The envoy did not return home by the way he went, but passed through Hamath. Abulfeda was then prince of Hamath, and this account forms the penultimate paragraph of his famous annals.*

As a proof that this friendship was genuine on the part of the Sultan, may be quoted a truculent act committed by him in the year 734 HEJ. Yasaur, a Mongol chief, of whom Abusaid was jealous, went on a pilgrimage to Mekka, and the Ilkhan urged Nasir to put him to death, as he might drive him from the throne, and thus put an end to the good ' understanding between Persia and Egypt. Nasir hoped, in the first instance, to persuade the sherif Rumeitha to put him away ; but, as he would not, he urged Bursbogha, the leader of the Egyptian caravans, to carry out this purpose. He bribed a Bedouin to do the deed, which was duly carried out on the second feast day, outside Mekka, during the ceremony of the casting of the stone. To remove suspicion from himself when the murderer fled to the mountains, he had him pursued by his mamluks and cut in pieces.†

When Dimashk Khoja, Choban's third son, was killed, Abusaid sent an army against another of his sons, Sheikh Mahmud, who was the governor of Armenia and Georgia. He was seized and put to death at Tebriz, and left four sons—Pir Hussein and Shirun, who were poisoned by Sheikh Hassan Kuchuk ; and Chamargan and Dua Khan, who were killed by the amir Ilkan, son of Sheikh Hassan Buzurg. The amir Hassan, Timurtash, Dimashk Khoja, and Sheikh Mahmud, were all own brothers of Baghdad Khatun, the favourite wife of Abusaid, by the same mother. Jelaukhan, killed at Herat, and Shiburgan were sons of Choban by Dulendi Khatun and Satibeg respectively, who were both daughters of Uljaitu. Besides these, Choban had three sons, whose mother is unknown, viz., Siukshah, Yaghibasty, and Nuruz.‡

After the death of Dimashk Khoja, who had been intrusted with the vizierate in the absence of Nusret ud din, who had gone to Khorasan with Choban, as we have mentioned, his post as Vizier was jointly held by the Khoja Ghiath ud din Muhammed, son of the great vizier Rashid ud din, and a grandee of Khorasan, named Alai ud din Muhammed ; but the latter was displaced in the course of eight months, in May, 1328, and

---

* Vide op. cit., v. 383-385.   † Weil, iv. 379-380.   ‡ D'Ohsson, iv. 699-700.

appointed controller-general of the finances. Ghiath ud din, who was a man of great ability and a zealous Mussulman, devoted himself to his duties with zeal, put the finances, which since his father's death had been in great disorder, straight again, and greatly encouraged agriculture. We are told he made himself loved even by the enemies of Rashid ud din, who, instead of being punished as they expected, received only kindnesses from him ; and several learned men dedicated their works to him.* Hamdullah, who praises him in hyperbolic terms, says he refrains from giving his titles—

For of what use is moonshine on the night of the Epiphany.

Among the works dedicated to the Vizier just named in the reign of Abusaid, which was marked as a flourishing literary period, was the "Mevakif," of Adhad ud din Iji, which is one of the most important metaphysical works that has been reprinted at Constantinople in recent times. Ibnol Hajib similarly dedicated his compendium, and Kutb ud din Razi his commentary on the logical treatise, "Shemsiyet." The "Fevaidi Ghiathideh" embodied his name in its title. Ewhadi, one of the greatest mystical poets of the East, dedicated to the same Vizier his poem, "Jems," while the Khoja Kermani Muhammed Ibn Ali Morshidi, surnamed the Palm Branch, similarly dedicated to him his romantic poem, "Humayi and Humayun" (*i.e.*, Augustus and Augusta), which in later days was translated or imitated by the Ottoman poets Shemali and Moeiyed.† To these we must add the "Tarikhi Guzideh" of Hamdullah, and the "Kusideh" of Suliman.

News kept coming to the Court that the Khan of Jagatai was meditating an attack upon Khorasan, and towards the end of May it was said that he was on the march, and prompt reinforcements were asked for ; but while measures were being taken in regard to this invasion, Narin Togai, who commanded in Khorasan, quarrelled with Ghiath ud din, the malik of Herat, over whose province he endeavoured to extend his jurisdiction, which had hitherto been independent of the governor of Khorasan. Ghiath ud din, who happened to be at the Court, reporting the services he had recently rendered the Sultan, obtained an order forbidding Narin Togai to exercise any jurisdiction in the principality of Herat ; but Narin paid no heed to this order, which only exasperated him against the malik. The latter, who was afraid of him, dared not return to his own country ; and as Abusaid and his council desired to do him a favour, as they considered him the bulwark of their Eastern dominions, who had always been zealous and faithful, it was decided to send an officer to Khorasan with a military rank superior to that of Narin Togai, and that the malik should return with him. The choice fell on Ali Padishah, the maternal uncle of the Sultan, who left the environs of Irbil on the 10th of June

---

* D'Ohsson, iv. 699-701. Ilkhans, ii. 295-296.　　† Ilkhans, ii. 296.

with Muhammed Bey and Tash Timur, each one at the head of some
troops. On hearing of their approach, Narin Togai, who felt he was
about to be superseded, sent courier after courier to say that he had been
misled, and that there was no real danger of an invasion of Khorasan.
At this news the generals halted at Sultania, intending to return ; but
Abusaid sent Turjan to order them to continue their march. Ali Padishah
evidently did not care to go, and sent some excuses, urging that after
Narin Togai's message their presence in Khorasan would probably only
lead to trouble. Abusaid told Turjan to return with peremptory orders,
and reproached the generals for their disobedience ; but Ali Shah,
fancying that it was the policy of the courtiers to get him and others as
far away as possible from the Court in order to push themselves, conspired
with the other generals, and marched with them towards Aujan, whither
Abusaid had transferred his residence. The Ilkhan, irritated at this, sent
Sheikh Ali to order them to retrace their steps, while Haji Khatun, mother
of the Sultan, sent to tell Ali Padishah he had better not advance if he
wished to avoid the Sultan's anger Sheikh Ali met the generals at
Heshtrud, and showed them the royal command. They said that as they
were so near the Court, they desired once more to kiss the Sultan's feet,
and to explain their purpose to him, after which they were ready to go
wherever he ordered them. Khoja Lulu was now sent with 5,000 men to
bar the route to the rebel generals ; but meanwhile several of their
subordinates, afraid of being implicated with them, went over to the
Sultan, and disclosed their plans to him. He said, " Inasmuch as they
are unwilling to go to Khorasan and to supersede its governor, and are
oblivious of the fate of Choban, who was the most powerful amir there
has been since the foundation of our dynasty, I deprive them of the
command of all their tumans, and they shall go and serve in Khorasan
under the orders of Narin Togai." Haji Khatun interceded for her
brother, who, she urged, had been led astray by evil counsels, and begged
the Sultan to content himself with exiling him to his own yurt.
Abusaid conceded this. "Ali Padishah," he said, "is still young and
without experience—I told him, when I first appointed him to the
command of a tuman, ' I raise you to a higher rank, but do not presume
on our relationship, for it counts for nothing when the interests of the
State are at stake.' If he had listened to this advice he would not have
incurred this disgrace. Muhammed Bey is also a frank, hasty-tempered
man. I expected more from Tashtimur, who has a sounder judgment,
and has experience, and lived a long time at my father's Court. I told
him when he left, 'Although several of you are charged with this commission, it is you to whom I look ; I rely on your zeal and fidelity.' And
it is he who has led the others astray. I consent, out of regard for my
mother, that Ali Padishah shall go and pass the winter at his yurt, near
Baghdad. Muhammed Bey must go to Khorasan, while Tashtimur must

come here to be punished." The two former chiefs went as they were commanded. Tashtimur was examined by a commission composed of the great amirs and the Vizier. He denied having made a pact with his colleagues, and was then brought face to face with the young officers who had denounced him. He still persisted in his denial. The amirs did not wish to push matters to extremity, being of opinion that the catastrophe that had overwhelmed the family of Choban had already cost too much blood, while the Vizier, who felt the conspiracy was really directed against himself, faithful to his policy of doing good to his enemies, obtained from the Sultan a pardon for Tashtimur, who received orders to go to Khorasan. Other officers implicated in the conspiracy, as Ibrahim Shah, grandson of the amir Sunatai, who was a great favourite of the Sultan, and Haji Togai, son of Sunatai, were pardoned in consideration of the long services devoted by that general to the empire ; but they were ordered to go to Diarbekr, and to reside there with their father, who was governor of that province.*

Tashtimur, *en route* for his place of exile, met, near Ebher, Narin Togai, who was on his way to Sultania without leave. They were both discontented, and exchanged views and discussed plans for overthrowing their enemies, especially the vizier, Ghiath ud din. Narin Togai was the son of Guch-buka, and grandson of the great noyan Kitubuka, who had been killed in Syria, as we described, and had been attached to Abusaid's household when the latter was a child and living in Khorasan. His ambitious and restless character had given umbrage to Dimashk Khoja, who had accordingly sent him to live away, and forbidden him to stay in the ordu. He turned to Choban, who patronised him and reproved his son. Under Choban's ægis he secured a ready access, as before, to the young prince. He soon found how the latter disliked Dimashk, and proposed a plan by which he might rid himself of the favourite, which he was prepared himself to carry out ; and notwithstanding the obligations he was under to Choban, he was one of the chief instruments in overthrowing his family, and shared in the plunder of their property. He thus acquired great wealth and became very powerful. His arrogance was correspondingly augmented, and the Sultan, feeling this, got rid of him by appointing him governor of Khorasan. Foiled in his purpose, which was to secure the post of Amir ul Umera, he found vent for his ambition in trying to enlarge the limits of his government, and thus fell out, as we have seen, with the Malik of Herat. When he heard how the Malik had complained against him, he summoned his son, Shems ud din, and when he refused to go sent troops to Herat to seize him. The young prince fought bravely, and after some sharp combats Narin's troops were obliged to withdraw, and when he marched in person he was not more successful. He revenged himself on

* D'Ohsson, iv. 701-706. Ilkhans, ii. 307-308.

his retreat by plundering the baggage of the Malik, which was *en route* from the ordu, and learning that that chief, when he heard of this, had set out in all haste from Sultania, he posted troops to waylay him, but the Malik gave them the slip by passing through the desert of Tabs.

Meanwhile, Narin Togai was ruining Khorasan by his exactions and tyrannical conduct, and aroused the anger of the Sultan and his council, and it was to try and calm the storm which his conduct had created that he was on the way to the Court, after levying a heavy contribution on the unfortunate province, when he was met by Tashtimur. After consulting with the latter, he sent a secret embassy to Ali Padishah, who was also discontented with his exile and joined in their scheme, as did some other officers. It was decided that Narin Togai should try and seduce the Ilkhan into putting away his present courtiers, and if he failed his supporters were to collect together and seize the reins of authority by main force, after which they could easily obtain Abusaid's sanction. Having formed this plan, the conspirators separated. Tashtimur continued his journey leisurely, awaiting the course of events, and delayed at Kazvin under some pretence. But Narin, when he arrived at Sultania, could not obtain an audience. The Sultan was too irritated at his acts of violence and injustice in Khorasan, while Baghdad Khatun inflamed him further against one whom she deemed the author of the deaths of her father and brothers. Narin then tried to win over some of the officers of the Court, and addressed himself specially to one of his relatives, named Turt, who informed the Vizier. The latter would not credit it. Seeing he made no way, Narin determined to attempt a more daring policy. He went to Ghiath ud din's palace, under pretence of paying him a visit, and accompanied by a number of men-at-arms, whom he planted at the gate of a medresseh or college, past which he expected the Vizier would have to walk. The latter gave him a private audience, but his brother, the amir Ahmed, arrived just at the moment when Narin appeared, followed by some of his people, and told him he had orders to admit no one armed. Having taken Narin's arms off, he passed him in, and stood himself at the door to prevent his suite from following. Narin, seeing that his plan had failed, began to flatter the Vizier, and begged him to speak to the Sultan on his behalf. Ghiath ud din promised to do this, and said further he would go at once and see his master. Narin then posted himself at the gate of the medresseh to await the passage of his enemy, but the Vizier left his palace by another door, and went to the Sultan, to whom he reported that Narin desired to do obeisance to him, and went on to speak on his behalf with his usual kindness. Abusaid, who had already been warned by his confidential people of the designs of the general, demanded from the Vizier with surprise if he was aware what Narin's real intentions were towards him, and at the same time gave orders for his arrest. One of his people having overheard this order, duly warned him, whereupon he returned to his lodgings by a tortuous road, took a horse, arms, and some servants, and withdrew with haste. Khoja Lulu was sent in pursuit on the Khorasan road, but Narin changed his direction, and went towards the mountains, and passing by Ebher reached Rai in four-and-twenty hours, whence he again reached the road to Khorasan, and went to join his military household and his troops. Khoja Lulu having lost trace of him, returned again. Couriers were thereupon dispatched in various directions with orders to arrest Narin.* With his horses worn out, and himself exhausted with hunger and fatigue, he hid away in a little valley near Rai, and sent one of his followers to a neighbouring village for provisions. A Uighur officer, named Haji Uyunmas, who had his yurt in this district, noticing the terrified look of the man, had him arrested and interrogated. He was confused in his answers, and was bastinadoed until he confessed. Thereupon the Uighur chief went with a party of horsemen, using him as his guide, and arrested Narin, who was presently sent in chains to Sultania.

The very day of his flight, Abusaid had sent orders to Tashtimur, who was at Kazvin, to return at once. He was constrained to go, and was also put in chains. Both he and Narin were condemned to death through the influence of Baghdad Khatun, in revenge for the death of her father and brothers. They were executed on the 5th of October, 1327, on the holy day of Kurban Bairam, before the palace of the generalissimo, Sheikh Hassan. Their heads were hung from the walls of Sultania, whence that of Dimashk Khoja was removed. Their houses were pillaged, and commissaries were sent into the provinces to appropriate their property there. Alai ud din Muhammed, who had been implicated in their treachery, was pardoned through the Vizier's influence. Ali Padishah, out of regard for his sister, the Sultan's mother, was merely disgraced. Sheikh Ali, son of Ali Kushji, was appointed governor of Khorasan, with the Khoja Alai ud din Muhammed as vizier ; and they were forbidden to exact from its inhabitants more than the ordinary taxes, for it had been ruined, first by the exactions of Choban, then by those of Narin Togai, and the inhabitants, driven by desperation, had begun to emigrate.†

Ghiath ud din, the Malik of Herat, died there in October, 1329, and was succeeded by his eldest son, Shems ud din, who was handsome, brave, and intelligent, but addicted to wine ; and after the death of his father, who had kept him in restraint, it was said he was only sober for ten days during ten months. He died in 1330, and was succeeded by his brother Hafiz, a timid young man, handsome, and skilled in caligraphy, who was in the hands of some of the grandees. Notwithstanding which, they assassinated him in 1332, and proclaimed Moiz ud din Hussein, who

---

* D'Ohsson, iv. 701-712.   Ilkhans, ii. 307-309.   † D'Ohsson, iv. 712-713.

was still an infant. His nomination was approved by Abusaid, who sent the young prince a robe and a diploma.*

In 1332, Sheikh Hassan was accused of carrying on a secret correspondence with his former wife, Baghdad Khatun, and with having arranged with her a plot to murder the Sultan. Abusaid had him arrested and condemned to death, but he pardoned him at the request of his aunt, who was Sheikh Hassan's mother, but issued orders that he was never again to come into his presence. He was sent to the strong fort of Kémakh, where his mother was also sent. These rumours weakened Abusaid's regard for Baghdad Khatun, but having learnt that the accusation was false, he had those who reported it put to death, and restored her to favour. The following year the government of Rum became vacant by the death of Devlet Shah. It was given to Sheikh Hassan ; and about the same time the Sultan married Dilshad Khatun, daughter of Dimashk Khoja, for whom he conceived a great attachment, and promoted her above his other wives. In 1333-4, Abusaid conferred ' the government of Fars upon the amir Muzaffer Inak, thus supplanting Mahmud Shah Inju, who had held this post for a long time. The latter was very rich, and his domains in Fars produced him a revenue of 100 tumans. He conceived a violent hatred for his supplanter, and formed a league with other amirs, such as Mahmud Issenkutlugh, Sultan Shah (son of Nikruz), Muhammed Bey, Muhammed Pelten, and Muhammed Kushji, who were all jealous of Muzaffer. They went with a party of followers, and made an attack on his palace. Muzaffer got on to the roof, and jumping from one house to another, eventually reached the royal palace, where they followed him as far even as the Sultan's vestibule, whose walls were presently studded with their arrows. They demanded that Muzaffer should be given up to them, and the Sultan, alarmed for his own safety, was about to comply, when Shiburgan, son of Choban (called Siyurgan by Von Hammer), and Khoja Lulu arrived with some troops and drove the rebels away. They were condemned to death, but this sentence was commuted to imprisonment at the request of the benevolent vizier, Ghiath ud din. They were, except Muhammed Shah, imprisoned in various fortresses : Issenkutlug in Khorasan, Sultan Shah (son of Nikruz) in the castle of Sirjan, Muhammed Pelten at Bum, Muhammed Kushji in the castle of Nasir ; Mahmud Shah Inju at Taberek, a fortress of Ispahan ; while Masud Shah, son of Mahmud (D'Ohsson says it was Mahmud himself), was sent to Rum. Except the last-named, they were all confined until Abusaid's death.

In August, 1334, Uzbeg, the Khan of the Golden Horde, was meditating another invasion by way of Derbend. Abusaid made preparations to meet him, but was attacked with illness, and died at Karabagh, in Arran,

on the 30th of November, 1334. There are strong grounds for believing
he died by poison. The Persian writers, such as Mirkhond, suggest it ;
but it is expressly stated by the traveller, Ibn Batuta, who tells us he
was the victim of Baghdad Khatun, who was jealous of the promotion
of the beautiful Dilshad, the daughter of Dimashk Khoja, for whom he
had conceived a great passion. She also naturally did not forgive the
murder of her father and brothers, and is said to have put some poison
into his bath towel. The same author goes on to say that when the
amirs were assured of his death, the Greek eunuch Khoja Lulu went
and found Baghdad Khatun in her bath, and killed her with a blow of
his mace, and her body was exposed for several days with a mere
rag over it.* It must be remembered that she had many enemies,
since she had caused several grandees to be put to death. The author
of the " Shajrat ul Atrak " says that, having stayed at Shirvan for nearly
three months, Abusaid was attacked by a fatal disease, due to the heat
of the weather and the noxious air there. On the fifteenth day after
the attack he went to his bath, and grew so much worse that he died.
He adds that it was reported that Baghdad Khatun had poisoned him
from jealousy of Dilshad Khatun.†

The same work tells us Abusaid was fond of the society of men of
learning, and that he wrote a good hand, for which accomplishment he
was indebted to his tutor, Khoja Abdulla Syrufi.‡ Ibn Tagri Birdi
describes him as an illustrious and brave prince, with an imposing aspect,
generous and gay. He says he wrote well, played the lute, and composed
songs. He had amiable manners, suppressed several taxes, proscribed
strong drinks, demolished churches, and belonged to the Hanéfi sect.§
Ibn Batuta describes him as an excellent and generous ruler, and as
good-looking, without any down on his cheeks. Our traveller says
he once saw him in a boat (*harrákah*) on the Tigris, with Dimashk
Khoja. On either side were two other boats filled with musicians and
singers. He also tells us that he was witness of an act of generosity
on his part when he presented a number of old men with a robe
each, and also appointed each of them a guide.‖ He left Baghdad
with Abusaid's suite, and was thus a witness of the mode in which he
marched and encamped. He tells us the Mongols were in the habit of
starting early in the morning and of halting an hour before the sun
reached his greatest altitude. Each amir arrived with his soldiers, his
kettledrums, and standards, and halted in a certain place which had been
previously assigned him. When all had arrived, and the ranks were
made up, the Sultan mounted his horse, the drums, trumpets, &c.,
sounded, the amirs advanced and saluted their sovereign, and then
returned to their posts. The chamberlains and nakibs then presented

---

* Op. cit., 122-123.   † Op. cit., 309.   ‡ Id., 308.   § D'Ohsson, iv. 717.
‖ Op. cit., ii. 116-117.

themselves. They were followed by the musicians, to the number of about
a hundred, dressed in splendid robes, and mounted. In front of the
musicians went ten horsemen carrying kettledrums hanging from their
necks, and five horsemen bearing flutes. They played the drums and
flutes and then stopped, and were followed by ten musicians singing.
The music having played again, another set of musicians sang, and this
was repeated ten times, the ceremony preceding the pitching of the camp.
While on the march, the chief amirs, to the number of about fifty, rode
on the Sultan's right and left. The standard bearers, kettledrum beaters
the trumpeters, &c., followed the prince; then came the Sultan's slaves,
then the other amirs according to their rank. Each amir had his own
kettledrums and standards. The amir jandar was intrusted with these
arrangements, and was accompanied by a numerous detachment. If any
man lagged behind, his shoe was taken off, filled with sand, and hung
from his neck, and he had to walk to the next station, where he was taken
before the amir, made to lie down on the ground, and had twenty-five
blows of the bastinado on his back, whatever his rank might be. In the
camp the Sultan and his mamluks, or slaves, had a separate quarter.
Each one of the Sultan's khatuns, or wives, also lodged apart, and she
had her special imaum and mueddhins, her Koran readers, and a special
place to do the marketing. The viziers, the kàtibs, and employés also
encamped apart, as did each amir. They attended the Sultan's audience
after " the *Asr*," and returned after the last evening prayer, lanterns being
carried in front of them. When the march began the great kettle was
first sounded, then that of the principal khatun, then that of the Vizier,
and lastly those of the amirs altogether. Thereupon the amir commanding
the advance guard mounted with his men, and was followed by the
khatuns ; then the Sultan's baggage and that of the khatuns ; then
came another amir, with his men, to prevent intruders from approaching the khatuns or the baggage ; then came the rest of the army. Ibn
Batuta says he accompanied the royal cortège for ten days, and then
set out for Tebriz with the amir Alai ud din Muhammed. He tells us he
stayed outside the town, in a place called Sham, where was Ghazan's
tomb, and near it a fine medressch or college and a hermitage, where
travellers received refreshments, consisting of bread, meat, rice and
butter, and sweetmeats. He stayed there himself. The next day he
entered the city, and describes the market-place, called after Ghazan,
as very fine, each trade having its own quarter. The jewels were sold
by beautiful slaves, dressed in splendid costumes, and wearing silken
handkerchiefs for girdles. They sat in front of their masters, the
merchants, and sold the jewels to the wives of the Turks (*i.e.*, of the
Mongols), who rivalled each other in extravagance. He also visited the
quarter where ambergris and musk were sold, and then visited the Jami
mosque, built by the vizier, Ali Shah, and known as Jilan. He tells us

R *

its court was paved with marble, and its walls covered with tiles. It was traversed by a river, and there grew on its banks trees, such as jasmines, vines, &c. In the court of this mosque it was customary to read daily, after the prayer of *Asr*, the *surat ya sin* (*i.e.*, the 36th chapter of the Koran), that named *Victory* (*i.e.*, the 48th), and that called *al-naba* (*i.e.*, the 78th). Our traveller spent only a night at Tebriz, and then returned with Alai ud din to Abusaid, to whom he was presented. The Sultan asked him about his country, and gave him a robe. Hearing that he intended going to the Hijaz, he assigned him a camel, a litter, and provisions, and had letters written for him to the amir of Baghdad.*

Ibn Batuta tells an anecdote of Abusaid. He says that the atabeg Ahmed, of Luristan, went to pay him a visit. Some of his courtiers said to him, "The atabeg came into your presence dressed in a cuirass" (for they fancied the horsehair shirt that Ahmed wore was a cuirass). Abusaid told his courtiers to get to know the fact, whereupon, when the atabeg presented himself one day, Choban, Sunatai, and Sheikh Hassan approached him and felt him with their hands, as if playing with him, and in this way found that he was wearing a horsehair shirt. Thereupon Abusaid bade him sit close by him, and said to him in Turkish, *Sen alayim* (*i.e.*, "Thou art my father"), gave him a present much more valuable than the one he had received, and also a yarligh or diploma exempting him and his family in future from having to give presents to himself or his successors.† The "Yuan shi" records an embassy sent in 1332 by Busai yin, Prince of the Si yu (*i.e.*, western regions), to the Emperor Wen tsung (Tob Timur). The name of the ambassador is given as Ho dji k'ie ma ding, and he took as tribute seven precious stones and other articles. By Busai yin was no doubt meant Abusaid Khan‡.

Abusaid's reign was a very active period of literary culture. Two famous historians flourished at this time, viz., Binaketi and Hamdullah. Ibn Suliman Daud, styled Fakhr Binaketi (*i.e.*, the Glory of Binaket, the place now called Shash, or Tashkend), had already been nominated king of the poets by Ghazan Khan. He composed a poem on the occasion of the marriage of Prince Bestam with Oljai Kutlugh. He also wrote a compendium of universal history, ending in the third year of Abusaid, and which is largely founded on that of Rashid. Hamdullah Mustaufi, of Kazvin, was the author both of a history of the world, entitled "Tarikhi Guzideh," and also of a geographical work, entitled "Nuzhetol Kolub." The former is a famous repository of facts. In the part dealing with the Mongols it is also largely founded on Rashid ud din's great work, but it becomes an independent authority as it nears its own time. I am indebted to Mr. Guy L'Estrange for a MS. translation of the work, which has been of great service to me. Besides these two

---

* Ibn Batuta, ii. 125-131.　　† Op. cit., ii. 33-34.　　‡ Bretschneider, Notices, &c., 104.

historians, there also lived at Abusaid's Court the famous jurist, Nisam ud din, of Herat, and the two sheikhs, Rokn ud din, of Semnan, and Abderrezak, of Kiash, the dervish poet Nasir, of Bukhara, and two other poets, named Khoju Kermani and Mir Kermani, the former the author of the romantic poem "Huma and Humayun."* I may add here some other writers who lived in the previous reigns, and whom I overlooked naming. The great mystical poet of Uljaitu's reign was Iraki ibn Shehriar, of Hamadan, a disciple of the famous mystic Sheikh Shihab ud din Suhrwerdi, who died in 1309, at the age of 82. Two other mystical poets of the time, also scholars of the same master, were Awhad ud din, of Meragha, and Hussein. The former wrote the poem called "Jami Jem," and died at Ispahan in the time of Ghazan. Hussein, of Herat, was the author of over thirty mystical works in prose and verse. These three Sofi poets met together, after travelling in many lands, in the monastery of Ewhadi, and after a quarantine of forty days of fasting, and contemplation, reported to their master their experiences in their several journeys. Their tombs at Damascus, Ispahan, and Herat are still the goal of pilgrimage to the professors of Sofiism. The so-called king of the poets (which does not mean, any more than Laureate with us necessarily means, the best poet) in the earlier part of Abusaid's reign was Ibn Nesuh, the author of the "Deh nameh," which was dedicated to Ghiath ud din, the son of the vizier Rashid ud din. After him Seraj ud din Kumri occupied this post. He is less famous in the history of Persian poetry (says Von Hammer) for his emulation with Obeid Zakyani and Selman Saveji, than for the box on the ears he received for a bad joke from Konkurat, the wife of Abusaid. The pious female hermit Safiyeh, of Abher, who was much esteemed by Abusaid and the ladies of his Court, was one day having a meal with Konkurat when Kumri was also present. Konkurat said, "Give me what is left of what the holy woman has eaten, that I may take it with me into the house." "All that I have eaten, most noble lady, is at your service," said the saucy poet. She thereupon struck him several blows, which made him black and blue, in which state he presented himself before Abusaid, and on his asking the cause, replied, "Formerly a king of the poets was rewarded with a thousand gold pieces. Your wife Konkurat has rewarded me with ten boxes on the ear," at which Abusaid was greatly amused. The most famous poets of this time were Zakyani, Selman, and Hafiz, who, however, flourished somewhat later, and I shall speak of them further on. To revert to Abusaid's own reign, we read of some other poets, as Said Behai Jami, Iz ud din Hamadani, who, like Julahi Abheri, wrote in Pehlevi; two poets named from the town of Feryumed, in Khorasan, where they were born; and Yamin ud din

---

Toghrayi, whose letters, written from Rum to his son Mahmud in Khorasan, are good examples of condensed style in Persian. More famous are the ethical and philosophical verses of the son just named (i.e., of Mahmud ibn Yamin), and known as "Makataati Yamin," and Osman Moti, of Kazvin, who dedicated a collection of satires, under the title of "Razi nameh," to the judge Kazi ud din Razi, who, either from fear of his satirical power, or from regard for his fame, rewarded him with an honorarium of from thirty to forty thousand dinars, which he spent as quickly as he acquired it.*

Coins of Abusaid occur in the British Museum collection representing every year from 717 to 734 HEJ., except the year 718, and one, doubtless a posthumous coin, in the year 737. On one of his gold and fifteen of his silver coins the date, instead of being given in the year of the Hejira, is given in "the Khanian era" (i.e., the era established by Ghazan). These are all of the year 33 (i.e., 733 HEJ.—1332-3). The coins of Abusaid occur in gold, silver, and copper, and the mints on the fine series in the British Museum comprise Jorjan, Erzenjan, Jajerm, Sultania, Kerman, Kazerun, Sivas, Tebriz, Saweh, Shuster, Shiraz, Kashan, Mosul, Tokat, Yezd, Basrah, Baghdad, Hamadan, Baran, Wasith, Baiburt, Sinjar, Irbil, Arivend (i.e., Elvend), Akhlatt, Kaisaryah, Bekkar, Gulistan, Samsun, Meragha, Berdaa, Ispahan, Erivan, Sebzevar, Hisn-Kaifa, Bekbek, Karmin, Hillah, Abusaidin, Mardin, Walashjerd, and Khelat. In the museum are also two coins of Abusaid struck at Halab, or Aleppo, in 732 and 733. Fræhn, in his memoir on the coinage of the Ilkhans, adds a number of other mints—Sheristan Rashidi, Keliber (in Azerbaijan), Arran (i.e., another name for Berdaa), Nakhchivan, Erzerum, Tiflis, Terjan (a town near Erzerum), El Bazar, Aujan, Selivas, Tesui (a town north of Lake Urmia), and a name doubtfully read Hausem. He also says that the style of Alai ud din, given to Abusaid by Khuandemir, nowhere occurs on his coins.† In his supplement Fræhn gives the additional mints of Abiverd and Arran (i.e., Berdaa). De Saulcy publishes a coin struck at Irbil, and another apparently struck at Khotan.‡ On the coins bearing a bilingual inscription the Khan's name is written Bu Said, in Mongol characters.§ Abusaid having made a public profession of the Sunni faith, the names of the orthodox khalifs occur on his coins; yet, strangely enough, two coins at least are extant in which the imaums venerated by the Shias are specially named.||

Note I.—It was during the reign of Abusaid that Friar Odoric of Pordenone traversed Persia, passing on from one Franciscan house to another until he reached India and China. His travels apparently began between 1316 and

---

* Ilkhans, ii. 262-267.　† Op. cit., 518-532.　‡ Journ. Asiat., 3rd ser., xiv. 142 and 145.
§ Fræhn, op. cit., 631.　|| De Saulcy, op. cit., J. A., 3rd ser., xiv. 149.

1318, and he died in 1331.* He went by way of Trebizond to Erzerum, which he calls Arzeron, and which he says in time long past had been a fine and most wealthy city, " and it would have been so unto this day but for the Tartars and Saracens, who have done it much damage. It aboundeth greatly in bread and flesh, and many other kinds of victual, but not in wine or fruits. For the city is mighty cold, and folk say that it is the highest city that is at this day inhabited on the whole face of the earth, but it hath most excellent water."† Tebriz, where the Franciscans had two convents, Odoric describes as " a nobler city, and better for merchandise than any other which at this day existeth in the world," containing great store of all kinds of provisions and goods, well situated, and very rich, the whole world having dealings there. The Christians there declared the Emperor (i.e., Abusaid) had more revenue from it than the King of France from all his kingdom. Near the city was a great mountain of salt, whence anyone might take what he wanted. There were many Christians there.‡ Thence Odoric went to Sultania, which, he tells us, was the Emperor's summer residence, a great and cool city, with good water, and well supplied with costly wares. Dominicans and Franciscans each had a house there. Our traveller passed thence to Kashan, a royal city of great repute which the Tartars had greatly destroyed. It abounded in bread and wine and many other good things.§ Then he went to Yezd, rich in victuals, but especially in figs and green raisins. It was the third city, he tells us, in the realm of the Emperor of the Persians, and the Saracens said no Christian was ever able to live there more than a year.‖ He then went to Comerum (the Camara of Barbaro, and probably the Kinara of Rich), on the site of the ancient Persepolis. He speaks of its walls as fifty miles in circuit; that it contained palaces yet entire, but without inhabitants. Thence he went to Huz, which Colonel Yule identifies with the Hazah of Eastern writers, said by Assemanni to be the same as Adiabena. He speaks of it as beautifully situated and abounding in victuals, with fine mountain pastures all round. Manna was common there, and four partridges could be bought for a Venetian groat. Men were accustomed to knit and spin there, and not women.¶ He then went to Baghdad, which he calls Chaldœa, where the men were comely but the women ill-favoured, the former having rich robes on, and golden fillets with beads on their heads; while the women had only a miserable shift, reaching to the knees, and with sleeves so long and wide that they swept the ground. They had drawers reaching to their feet, which were bare, and their hair was neither plaited nor braided, but dishevelled. It was the custom for women to go first, and not men, as in Europe. He describes seeing a young man who was taking a beautiful woman to be married. She was accompanied by a number of young women, wailing, while he, dressed in handsome clothes, hung his head down. By and bye, the young man mounted his ass, and the bride followed him, barefoot and wretchedly clad, and holding by the ass, and her father went behind, blessing them, until they reached the husband's house. From Baghdad, Odoric went to the district on the Persian Gulf which he calls inland India, and which he

---

* Cathay, and the Way Thither, 6.   † Id., 46.   ‡ Id., 48-49.   § Id., 50-51.
‖ Id., 51-52.   ¶ Id., 52-53.

says the Tartars had greatly wasted. The people there lived mainly on dates, forty-two pounds of which you could buy for a groat. He then went on to Hormuz, a well-fenced city, abounding in costly wares. It was situated on a barren island, without tree or fresh water, but with plenty of fish, flesh, and bread. It was unhealthy, and incredibly hot. The people were tall. When he passed one was just dead, and they got together all the players in the place, and set the dead man on his bed in the middle of the house, while two women danced round him, and the players played on their cymbals and other instruments of music. Then two of the women took hold of the dead man, embraced him, and chanted his praises, and the other women stood up one after another, and took a pipe and piped on it awhile, and when one had done piping she sat down; and so they went on all night, and in the morning they carried him to the tomb.* Thence Odoric set sail for India.

In a tract written about 1330 by the Archbishop of Sultania (probably John of Cora), setting out the estate of the Great Khan, Abusaid is called Boussay, and we are told that, like the ruler of Armalech (i.e., of the Jagatai Khanate) and Uzbeg Khan, he sent yearly live leopards, camels, and gerfalcons, and great store of precious jewels to the Great Khan of Cathay, whom they acknowledged as their lord and suzerain; and to show the power of these chiefs he says that when Uzbeg was at war with Boussaye he brought 707,000 horsemen into the field, without pressing hard on his empire.†

Pegollotti, in his "Notices of the Land Route to Cathay," calls Abusaid Bousaet. His list of the payments exacted upon merchandise from the frontiers of Armenia to Tebriz, shows what a gauntlet of black mail merchants had to traverse. I will quote it here from Colonel Yule's edition.

	Aspers.
At Gandon, where you enter upon the lands of Bousaet, on every load	20
At the same place, for watching	3
At Casena	7
At the Caravanserai of the Admiral (? Karavansarai ul amir)	3
At Gadue	3
At the Caravanserai of Casa Jacomi	3
At the entrance to Salvastro (Sebasti or Sivas) from Aiazzo	1
Inside the city	7
Leaving the city for Tauris	1
At Dudriaga (? Divrik)	3
At Greboco	4
At Mughisar	2½
At ditto as tantaullagio for the watch (i.e., for the Tangauls)	½
At Arzinga (Erzingan), at the entrance to the town	5
Ditto inside the city	9
Ditto for the watchmen, on leaving	3
At the Caravanserai on the Hill	3

* Cathay and the Way Thither, 54-57.　† Id., 238.

Aspers.

At Ligurti		2
At ditto, at the bridge, for tantaullagio		½
At the Caravanserai outside Arzerone (Erzerum)		2
At Arzerone, at the baths		1
Ditto	inside the city	9
Ditto	as a present to the lord	2
Ditto	at the baths towards Tauris	1
At Polorbech		3
At ditto		½
At Sermessacalo (? Hassan Kala), for tantaullagio		½
At Aggia, for the whole journey		½
At the middle of the plain of Aggia, for duty		3
At ditto, for tant		½
At Calacresti (Karakalisa)		½
At the Three Churches (Uchkilesi), for tant		½
Under Noah's Ark (probably Bayezid), for duty		3
Ditto	for tant	½
At Scaracanti,	ditto	½
At Locche	ditto	½
At the plain of the Falconers, for tant (twice altogether)		1
Ditto	ditto for a ticket or permit from the lord	½
At the Camuzoni, for tant		½
At the plains of the Red River (? the Aras), for tant		½
At Condro, for tant		½
At Sandoddi ditto		½
At Tauris ditto		½

And you may reckon that the exactions of the Moccols, or Tartar troopers, along the road, will amount to something like 50 aspers a load. So that the cost on account of a load of merchandise going from Aiazzo of Armenia to Tauris, in Cataria (? Tartaria), will be, as appears by the above details, 209 aspers* a load, and the same back again.†

Note 2.—Heyd has described some traces of the intercourse between the Venetians and the Mongols. The first evidence of dealings between the Republic and the Ilkhans is a document which once existed both in a Tartar original and a Latin translation, but the former recension is lost. This is dated in the beginning of November, 1306 (a mistake for 1305), and begins with the words, "Verbum Zuci Soldani duci Venetiarum," which Heyd has ingeniously explained as the order of the Sultan's yulduchi (or sword bearer) to the Doge of Venice. The Sultan was Uljaitu; the yulduchi was Tomazo Ugi, of Sienna, whose mission to the Pope, &c., we described in a former chapter. The document conferred certain privileges on Venetian traders, and stated that the traders who had frequented the Persian realm for a long time past had had no reason to complain of ill usage.‡ Tomazo also took with him

---

* Really 203 (i.e., about £2. 8s.) † Id., 299-301. ‡ Op. cit., ii. 123-125.

a letter from a certain Khoja Abdallah, in which he assured a Venetian named
Pietro Rudolfo that he would not exact any further satisfaction for some injury
he had done him, nor revenge himself for it upon any other Venetian. This
promise he confirmed by the signature of two witnesses, viz., Tomizo himself
and Balduccio Buffeto (? Buffero). A few years later we find the Venetian
Senate sending an embassy to Persia, to Abusaid Khan, who is called " Imperator
Monsait " in the record of the event. The leader of the embassy was Michele
Dolfino, and it went in 1320. He received *en route* a sum of money from the
Venetian bailo at Trebizond, Giovanni Sanuto. *Inter alia*, he asked for the
restoration of the goods of a Venetian called Francesco de Canale, who
had died at Erzenjan, and which had been seized by a certain Badradin Lulu.
This embassy secured a diploma from Abusaid, allowing the Venetians free
ingress to Persia, and to stay where they liked, and to be allowed to feed their
horses for three days at the post stations. No tribute was to be taken from
them, except the ordinary dues and the payment for the guards. Nor were
these dues to be enlarged or to be charged at other than the recognised stations.
The guards were to convoy them when requested, and if anyone refused he
was to be imprisoned for any damage that followed. If a Venetian was robbed,
the plundered goods were to be made good by the magistrates, watchmen, and
inhabitants of the district. The head man in a district was, on the requisition of the Venetian consul, to afford assistance to the Venetians or their
caravans. Neither their couriers nor sumpter cattle were to be delayed on any
pretence. No Venetian was to be imprisoned on account of the crime or
default of a countryman ; each was to be responsible for his own faults only. The
magistrates were to be helpful to Venetians in disposing of their wares, and to
aid them in recovering their debts. No one was to touch the goods of dead
Venetians except their consul. If a Venetian invoked the aid of a Court, the
highest judge was to pass sentence on him. If disputes arose between Franks
(*i.e.*, Venetians) their consul should decide matters. Lastly, the Latin monks,
who had the cure of souls of the Venetian mercantile community, were given
the valuable privilege of founding places of devotion in any part of the empire,
at their pleasure.*

The Venetians had a consul, probably in several of the larger towns of the
empire. He was styled Maçor. The first of their consuls at Tebriz whose
name is known was Marco de Molino, who, on the 6th of June, 1324, wrote to
the Doge a letter which does not put the Venetian mercantile body in a very
favourable light. He had forbidden the Venetians, in their own interest, having
dealings with a certain Saracen. Notwithstanding this, Francesco Quirini,
with two other Venetians, went to him to buy some spices. This having been
noised abroad in the caravanserai known as Delle Telle, Quirini was set upon
by four other Venetians and badly beaten. He complained to the Khan's
mother, and succeeded in having his assailants seized and imprisoned, and the
consul had to pay 270 bezants to have them released. The commonwealth was
at the same time threatened with a penalty of 5,000 bezants, in consequence of
the ill behaviour of another Venetian, named Marco Davanzo. All this filled

---

the consul with forebodings for the future. He described Tebriz as a dangerous station, which they would have to give up if the Doge did not see his way to some course by which these occurrences could be prevented in future. It was probably in answer to this letter that, in 1329, Marco Cornaro was sent to Tebriz. We are not told what the result of his mission was. In 1332 we have a further notice of the Venetian colony at Tebriz, when we read of a heavy burden being put upon it. Haji Suliman Taibi, of that city, asserted a claim of indemnity for 4,000 bezants, and he was allowed to impose upon all Venetians who went to Tebriz, or left it, a sum of four bezants for each load, until his claim was discharged. The Tebriz merchants had, in addition, to pay three aspers to two Venetians of the house of Sanuto, without any apparent reason. About the same time, the Venetian Senate empowered the bailo of Trebizond to exact an asper for each load of Venetian merchandise, which was to be paid to a dragoman, called Avachi, who was to see after the interests of Venice in that district.*

Note 3.—The author of the work entitled " Divan Alinsha " gives us some curious information about the common form used in writing letters to the Mongol sovereigns of Persia by the Egyptian rulers. He says : " In the letters addressed to the great Khans of the Mongols in the land of Iran, it is customary, according to the author of the ' Tarif,' to write on Baghdad paper after the formula in the name of God and a line of the khutbeh (introduction). Then follows the togra, which is drawn in embossed gold, and which, like all the togras, contains the titles of our sultan. The introduction is then completed ; then follows the titles of the prince to whom the letter is addressed, viz. : ' His noble, lofty majesty, the august Sultan, the king of kings, unique brother like a kân.' The term ' royal ' is excluded, since it is not valued among the Mongols. Then follow some pompous phrases invoking glory for the Sultan and victory for his allies. Good wishes are expressed for the prince, such as length of days, a free display of his banners, a strong army, numerous subjects, &c. Then follow phrases expressing very strongly a constant friendship and sincere veneration, &c., and the motives which have led to the letter being written. The conclusion contains several good wishes in pompous terms, and the good intentions of the sender. In such a letter the introduction, the togra, and the title are written in raised gold. The same kind of material is used in writing important names, such as that of God, of the Prophet, or of other prophets or angels, the name of Islam, those of the Sultans who send and receive the letter, and pronouns referring to either. The rest is written in black ink. Minute instructions on small points are given by the author of the ' Tarif,' and duly set out in Quatremere's translation." *Inter alia*, we are told that the style of the Ilkhan was Abusaid Behadur Khan, and he adds that when peace was made between Sultan Kelavun and Abusaid, the Kadhi Alai ud din Ibn Alathir reflected for a month on the form which the correspondence should take. The titles of " brother " and of " the Mamluk " were both discarded, and it was eventually determined to use the same style as was used in speaking of the Sultan himself.†

---

* *Id.*, 128-129.   † Makrizi, ii. (part 2), 307-315.

CHAPTER XII.

THE LATER ILKHANS AND THE JELAIRIDS.

ARPAGAUN KHAN.

ABUSAID was virtually the last of the Ilkhans. After his death the dynasty collapsed with almost incredible rapidity, and the empire so carefully got together by Khulagu and his successors broke into fragments, which eventually became the prey of the great Timur. It is a curious fact that the year of Abusaid's death was the birth-year of Timur himself, and the Arabs have formed a chronogram from the numerical value represented by the letters corresponding to the figures in this fateful year, 736 HEJ., which well represents its character. The letters so read form the word *Laudh*, meaning refuge, and it was said that men had need of an asylum under such great calamities.* Ghazan had decimated the royal family so persistently that when Abusaid died without male issue nearly all the princes directly descended from Khulagu were living in greater or less obscurity. The turbulence of the great amirs during Abusaid's minority had also loosened the bonds of authority all over the empire, and on his sudden death, to prevent immediate anarchy from supervening, it was necessary to select some one to fill the throne at once—some one, too, who had not aroused jealousies, and was not too vigorous to frighten the late Khan's immediate friends. The vizier, Ghiath ud din Muhammed, accordingly assembled the khatuns, the amirs, and other grandees, and their suffrages fell upon Arpagaun, sometimes corrupted into Arpa Khan, the son of Susu, son of Sinkian, son of Malik Timur, son of Arikbuka, who was the brother of Khulagu, and who had been pointed out by Abusaid as the fittest person to succeed him. He was proclaimed before Abusaid's funeral, after which the body of the latter was duly conveyed to Sultania for burial. We are told that at his inauguration Arpagaun addressed the assembled notables, and said, " I do not, like other Padishahs, need pomp and luxury. Instead of golden girdles woollen ones will suffice for me, and instead of a jewelled crown a cap of felt. From the army I expect obedience, and promise to treat it with kindness." In the Friday prayer he took the title of Muiz ud dunia ved din. Baghdad Khatun did not approve of his election, and was speedily charged, as we have seen, with having poisoned Abusaid,

* De Guignes, iii. 285.

and with having secret communications with Uzbeg Khan of the Golden Horde, who was then threatening an invasion, and was put to death. Her niece, and partially her rival, Dilshad Khatun, who was *enceinte*, deemed it prudent to fly, and went and sought refuge with Abusaid's maternal uncle, Ali Padishah, the governor of Irak Arab. Arpagaun strengthened his position by marrying Satibeg, the widow of Choban and sister of Abusaid, by which he hoped to secure the allegiance of those devoted to Khulagu's family. He then marched against Uzbeg, the Khan of the Golden Horde, who had invaded his borders. This was in the middle of winter. He cleverly planted a force behind the enemy's camp, so as to place him between two fires, a manœuvre which led to Uzbeg's retreat, and also gained for Arpagaun the confidence of his troops. On his return from this expedition he put several chiefs, dangerous from their high birth, rank, or fortune, to death—among others, Sherif ud din Mahmud Shah Inju, the Crœsus of his time, who had administered the finances of Fars, as we have mentioned. He was killed on the pretext that he had secretly nominated a boy of the family of Khulagu to occupy the throne. Inju's sons fled from Tebriz; Masud Shah went to Asia Minor, to Amir Sheikh Hassan, known afterwards as Sheikh Hassan the Greater, the first husband of Baghdad Khatun, who was then governor of Rum. Prince Tukel Kutlugh, of the house of Ogotai, who had sought refuge in Persia, was put to death with his two sons, on the plea that he wished to interfere in the administration of the State. Arpagaun would also have put to death Mahmud Issen Kutlugh, Sultanshah, son of Nikruz, and Muhammed Pilten, who had been imprisoned in various castles by Abusaid, as we have seen, and had been released on his death. Their lives were spared at the solicitation of the vizier, Ghiath ud din.

The Amir Ali Padishah, who was the maternal uncle of Abusaid, and was descended from Tengiz Gurkan, the Uirad, who had been sent by Khubilai against his rebel brother Arikbuka, disapproved of the election of Arpagaun. We are told he assembled the Uirads, and told them Arpagaun had always been the enemy of their tribe. He had been lately joined, as we have seen, by Abusaid's widow, Dilshad. Having assembled the amirs of the Uirads and Arabs, they proceeded to elect a rival Khan, in the person of Musa, the son of Ali, son of Baidu Khan, son of Tarakai, son of Khulagu, and then marched against Arpagaun, after sending seductive messages to his generals whom he knew to be unfavourably disposed towards him and to his vizier, Ghiath ud din. Arpagaun collected forces from various quarters to surround the army of Ali Padishah, but their commanders loitered in the hope that some accommodation would be arrived at. Ali Padishah offered to submit if he were appointed Amir ul Umera. Ghiath ud din advised that this should not be conceded, and also persuaded Arpagaun not to put to death several officers of whose loyalty he was suspicious. A struggle now ensued between the two chiefs at Bagatu (the "Shajrat ul Atrak" says at Chagatu, otherwise Yughtu), not far from Meragha. This was on the 29th of April, 1336. In the midst of the action, the amirs Issen Kutlugh and Sultanshah, who had a violent dislike for the Vizier, deserted, and Arpagaun, notwithstanding his superior force, was obliged to fly. The Vizier and his brother, Pir Sultan, who stood firm for some time, were caught near Meragha and taken before the victor. Ali Padishah wished to spare Ghiath ud din, who was so distinguished for his learning, wit, and generosity, but the other generals insisted on his death, which took place on the 11th Ramazan, 736, and was followed shortly after by that of his brother. His goods and those of his clients at Tebriz were plundered, and the people went so far as to pillage the quarter of the town known by the name of Raba Rashidi, and to carry off an immense booty in coin, objects decorated with precious stones, gold and silver bowls, and precious books. The houses of many people unconnected with the Vizier were plundered at the same time.*

The poets and other learned men, whom the Vizier had liberally patronised, naturally felt his death keenly, and one of the former commemorated it in plaintive verses. An example of his munificence is given by Khuandemir. One day, Sheikh Abu Ishak, who was then Governor of Shiraz, was having a controversy with the Kadhi Alai ud din, and put him the question, whether in the time of Abusaid merit was more rewarded than it was during his own reign. The Kadhi replied, laughing, "See what happened to myself. One day I was seated with Ghiath ud din, who beckoned me to him three times with his hand, and followed this up by a gift of money and lands so extensive that the tithes annually made upon their revenue amounted to 30,000 dinars." Abu Ishak having asked how this came about, Alai ud din replied that every Friday Ghiath ud din used to gather round him a number of doctors and wise men to discuss science and literature. All seated themselves according to their rank. When anyone made a remarkable observation the Vizier used to beckon with his hand for him to approach him. Alai ud din then explained how he had been fortunate enough to signalise himself three times successively at one of these meetings, whereupon Ghiath ud din made him rich presents, as did the Vizier's friends. On hearing this, Abu Ishak released Alai ud din from the annual payment of 30,000 dinars which he had to make to the Treasury.† Alai ud din lived habitually with Ghiath ud din, and the prefaces to his works are filled with the Vizier's praise. He gave one the title of Favaid ghiathiah (i.e., useful observations dedicated to Ghiath ud din). A poet named Ibn Nasuh composed ten pieces addressed to him; and Devlet Shah has preserved a long poem, in which Selman Savaji sang his praises. The famous author of

---

* D'Ohsson, iv. 719-723. Ilkhans, ii. 312-315. Shajrat ul Atrak, 310-311.
† Quatremere, Rashid ud din, lii.-liii.

the "Tarikhi Guzideh" was secretary to Rashid ud din and Ghiath ud din, and that work was dedicated to the latter.* In tracing the end of the Vizier, we had almost overlooked his master Arpagaun. He was speedily captured in the district of Sijas after his defeat, and was taken to Aujan, and handed over to the heirs of Sherif ud din Mahmud Shah Inju, by whom he was put to death.†

Frœhn, in his memoir on the coins of the Ilkhans, describes one of Arpagaun struck at Tiflis in 736 (i.e., 1335-6). Another of his coins has been recently added to the British Museum collection. It was struck in the same year at Shirvan.

MUSA KHAN.

Ali Padishah was now the virtual ruler of the kingdom, the Khan Musa, who mounted the throne at Aujan, being his *protégé* and nominee. He appointed Jemal ud din, the son of Taj ud din Ali, of Shirvan, as Vizier, who exerted himself to fulfil its duties with vigilance. But the calm was of very short duration. The fact was, the empire was breaking asunder, and the amirs in the various districts were becoming more and more independent. Among these, the most important in the West were Sheikh Hassan, the Jelair, the son of Amir Hussein Gurkan by a daughter of Arghun Khan, who had control of Rum ; and Haji Toghai, who was the son of the powerful Amir Sunatai, or Suntai. Sunatai had been appointed governor of Armenia, and also of Diarbekr, at the accession of Abusaid. In this he was apparently succeeded by his son, Haji Toghai, in the year 732 HEJ. The latter was at enmity with Ali Padishah and his people, the Uirads, and he accordingly sent to his neighbour Sheikh Hassan, urging him to seize the reins of power. Sheikh Hassan thereupon nominated a Khan of his own, viz., Muhammed Shah, son of Yul Kutlugh, son of Kuiji, or Kushji (called Ans Timur in the "Shajrat ul Atrak"), son of Anbarji (called Itiarchi in the "Shajrat ul Atrak"), son of Mangu Timur, son of Khulagu. So that there were now two rivals for the throne, both descended from Khulagu—one supported by Ali Padishah, the other by Sheikh Hassan. The latter nominated Shems ud din Muhammed Zakaria, son of Rashid's daughter, as Vizier.‡ Sheikh Hassan left the Amir Irshad, or Yeshed, in charge of Rum, and marched against Ali Padishah, who was at Tebriz with an army composed of the troops of Rum (i.e., of the Turkomans and Georgians). Before coming to arms he proposed to Ali Padishah that they should agree upon some candidate worthy of the throne, and should then withdraw each to his own province. Ali Padishah was willing to do this, but his generals would not consent to abandon the provinces they had so lately won. He accordingly sent word to Sheikh Hassan that they must fight it out, and

* Quatremere, liii.-liv. † D'Ohsson, iv. 723.
‡ Quatremere, lv. Ilkhans, ii. 315. D'Ohsson, iv. 723.

that the fight should take place at Alatagh. The "Shajrat ul Atrak" says the two amirs came together near the town of Naushehr. Before the fight Ali Padishah sent his opponent a message to this effect : "We are both Mussulmans. The princes are about to fight for empire. What need have we to take part in the struggle, and to answer in the next world for the blood about to be shed?" Sheikh Hassan accepted the proposition, and each one of them planted himself on a neighbouring height, Sheikh Hassan having 2,000 men with him. The struggle took place on the 24th of July, 1336, and a plentiful harvest of infidels was reaped, says the "Shajrat ul Atrak." Musa won the battle, and went in pursuit of the fugitives. Meanwhile Ali Padishah descended from his vantage in high glee, and went down towards a fountain to bathe, before returning thanks for the victory in a namaz of two re'kats ; but he was suddenly attacked, when he and his men had discarded their armour, by Sheikh Hassan, who killed him and slaughtered his people. Musa upon this withdrew towards Baghdad. Sheikh Hassan sent Kara Hussein, his naib, and the Amir Akurpukh, the Uighur, in pursuit. After killing many people, he returned to Tebriz with his nominee, Sultan Muhammed, who took up his residence there, and married Dilshad Khatun, the widow of Abusaid. He treated the family of the vizier, Ghiath ud din Muhammed, with great consideration, and divided the vizierate between Masud Shah, son of Mahmud Shah Inju, and Shems ud din Zakaria, nephew of Ghiath ud din, above named. Sultanshah, son of Nikruz, was executed for having caused the death of Baghdad Khatun. He also sent Satibeg, the widow of Choban and of Arpagaun, with her son Siurgan, or Shiburgan, to Mughan.

Amir Akurpukh Uighur, who was sent in pursuit of Musa Khan, quarrelled with Kara Hussein, and went to Sultania to Amir Pir Hussein, the grandson of Choban, who had married his niece, Oghlan Khatun. When he arrived there and related what had happened, they together marched to Khorasan, and in concert with the amirs of that country, and by the united exertions of Amir Sheikh Ali, son of Ali Kushji (who was governor of some cities in Khorasan), Amir Ali Jaafer, Amir Arghun Shah (son of Amir Nuruz, son of the famous Amir Arghun Aka), and the amirs who had fled from Amir Sheikh Hassan, brought Toghai Timur from Mazanderan and installed him as Khan in Khorasan, and read the khutbeh and struck money in his name. The "Shajrat ul Atrak" tell us the grandfather of Toghai Timur Khan, Baba Behadur, had migrated in the year 705 HEJ. with 10,000 families, dependants of Kaidu Khan, into Khorasan, where he entered the service of Uljaitu Khan, and was afterwards put to death. His descendants lived in the district of Jorjan, where Toghai Timur was also brought up. This authority calls Baba Behadur the son of Abukan, son of Alkan, son of Turi, son of Juchi, son of Kibad, son of Yissougai (i.e., it makes him descend from Juchi Khazar, the brother of Jingis Khan). The father of Toghai Timur was called Surikuri. He himself lived, according to the same author, in Mazanderan (doubtless Jorjan is meant). Toghai Timur having been proclaimed Khan, marched towards Tebriz, and on reaching the borders of Azerbaijan, was joined by the fugitive Musa Khan, who was living with the Uirads. An agreement was made that if successful Toghai Timur should retain Khorasan, while Musa should take Irak and Azerbaijan. They met Sheikh Hassan, his nominee, the Khan Muhammed, and his army at Meragha. The "Shajrat ul Atrak" says at Germrud. This was in June, 1337. Toghai Timur Khan fled. Musa Khan held his ground with his Uirads and a body of Khorasan troops, but they were eventually defeated. Musa was captured during the fight and taken to Sheikh Hassan, who had him put to death on the 10th of July, 1337.*

Frahn mentions a copper coin of Musa Khan, with the names of the four orthodox khalifs on it as struck at Tebriz in 736.† There are also two copper coins of his in the British Museum, but apparently without the names of mints on them.

We will now make a short digression to bring up the story of the intercourse of the Persian Mongols with Egypt. In the year 732 Darenda, a town situated some distance west of Malattiya, was surrendered by its Persian governor to the Egyptians, and the traitor was thereupon made an amir by the Sultan Nasir. Abusaid, who had been on friendly terms with the Sultan as we saw, asked for its restitution, as it lay in the midst of his territory. Nasir granted the request, and it again became part of the government of Rum.‡ We must always remember that the Mongol domination over the greater part of Asia Minor was rather in the form of suzerainty than of direct control. It had been so in the days of the Seljuks, and it was still more so now. The greater part of the country west of the salt steppes that stretch from Conia to Angora, was occupied by several Turkoman principalities, which no doubt owed only a slight allegiance to the Mongol governor of Rum. That official held court at Sivas, and, in addition, according to Ibn Batuta, controlled directly the towns of Ak Serai, Nigdeh (or Nicdeh), Cæsarea, Sivas, Amasia, Sunisa, and Kumish Khana. His government bordered immediately on the territory of Armenia, controlled by Haji Toghai. At that time the Mongol governor of Rum was called Alai ud din Artena, who had married Taghy Khatun, a royal princess. She was styled Agha (i.e., great), by which name the relatives of the Ilkhan were known. Ibn Batuta tells us he paid her a visit at Cæsarea. She rose on his arrival, saluted him graciously, and gave him refreshments, and afterwards sent him a horse saddled and bridled, a robe of honour, and money. At Sivas

---

* D'Ohsson, iv. 723-726. Ilkhans, ii. 315-317. Shajrat ul Atrak, 311-317.
† Frahn, de Ilchanorum, &c., numis 534.   ‡ Weil, iv. 347. Note.

our traveller visited Artena himself, who came out to meet him to the
door of his palace. The amir spoke Arabic well, and questioned his
visitor about the two Iraks, Ispahan, Shiraz, Kerman, the atabeg of Lur,
Syria, Egypt, and the Turcoman chiefs.*

　　To revert to Egypt. During the recent troubled times, Nasir, the
Egyptian Sultan, was naturally courted by the various aspirants to the
throne. Thus in June, 1336, directly after the death of Arpagaun, envoys
went to Cairo from Musa and Ali Padishah, conveying their good
wishes, which were reciprocated.† Makrizi seems to say that Arpagaun
himself sought Nasir's help, and promised him in return possession of
Baghdad.‡ In the beginning of 1337 envoys went to Egypt from Sheikh
Hassan and his *protégé*, who were received equally heartily with those of
Musa. The Sultan had promised the latter his help, and had sent a
contingent of troops to the frontier. At the same time he informed
Sheikh Hassan the Great that these troops were ready to help him.
When the Sheikh defeated his rival he sent Nasir his thanks for this
proffered aid, and although he had been in allegiance with Musa, the
Sultan did not hesitate to send the victor his congratulations on his
victory, with costly gifts. Soon after, Nasir courted an alliance with
Artena, the Mongol governor of Asia Minor, already named, who was
threatened on one side by Sheikh Hassan, and on another by Karaja Ibn
Abi Dulkadr, who had appropriated certain districts near Ablestan.
Artena promised to acknowledge Nasir as his suzerain, and to strike coin
and have the Friday prayer said in his name. The Sultan was highly
pleased at this, and instructed the Sheikh Shihab ud din to draw up a
formal diploma, investing Artena as his deputy in Rum. On neither side,
however, was there any intention to keep these easy promises any longer
than was necessary for self-interest. Karaja surprised the fortress of
Darenda, in Asia Minor, which belonged to Artena's government, and
asked the governor of Aleppo to send an Egyptian governor and a
Mussulman garrison there. The treacherous Sultan thereupon,
disregarding his alliance with Artena, at once rewarded his more useful
friend. Artena, on his side, when he no longer feared Sheikh Hassan
the Great, refused to have the coin struck and the prayers said in Nasir's
name. He renewed his promise, however, when troops from Syria, in
alliance with the Turkomans of Karaja, invaded his district.§

MUHAMMED KHAN.

　　The death of Musa still left two contending Khans, viz., Toghai Timur,
who, after the battle above-mentioned, withdrew to Bostam, and continued
to be obeyed in Khorasan and Jorjan; and Muhammed, the *protégé* of
Sheikh Hassan. I described how Timurtash, the son of Choban, who
had been governor of Rum, fled to Egypt, and was there killed. One of

---

* Op. cit., ii. 284-295.　　† Weil, iv. 345-346.　　‡ *Id.*, 346.　　§ *Id.*, 346-347.

his sons was also called Sheikh Hassan, and to distinguish him from the chieftain of the same name, whom we have so often named, was called Kuchuk, or the Little, the other being known as Buzurg, or the Great. After the proscription of his family, he hid away. He now reappeared, and began to collect his father's partisans, and produced a Turkish slave named Karajar (perhaps the Karaja already named), who had been in the service of Haji Hamza, an officer of Timurtash, whom he tried to pass off as his father.* He pretended that Timurtash, having escaped from the prisons of Cairo, had been wandering for some years in distant countries. He married the slave to his own mother, treated him publicly with respect, walking beside his horse ; and many people, misled by this conduct, ranged themselves under the banners of the pretender. His patron, Sheikh Hassan the Little, now sent to announce to his name-sake the return of his father. Haji Hamza, who was there with the latter, went to verify the fact, and having been gained over, attested that ' the impostor was really Timurtash. The partisans of the family of Choban and the Uirads, who did not willingly obey him, now abandoned Sheikh Hassan the Great and went over to the camp of the false Timurtash, to whom they remained faithful, even after his imposture was discovered.

When Nasir, the Egyptian Sultan, heard of the outbreak of the Lesser Hassan and of the reappearance of Timurtash, he was greatly embarrassed, for those who had been commissioned to put the latter to death were all dead, and he was afraid they might have deceived him, and that if Timurtash succeeded in reinstating himself he would become his bitterest enemy. He therefore sent Ahmed Assahbi, one of his most trusted amirs, to Haji Toghai, the Prince of Diarbekr, to form a close alliance with him, which was to be strengthened by the marriage of a son of Toghai with one of his own daughters. Toghai received this envoy well, and sent word back that he did not believe in the pretender, and that he had allied himself with Hassan the Greater against him. The same night, however, the troops of the Lesser Hassan drew near, and Toghai fled to Baghdad. Ahmed, the Sultan's envoy, was plundered, and escaped with difficulty to Aleppo, where he informed his master of what had occurred. Soon after, news came from the Prince of Mardin that the Lesser Hassan had won a victory over his enemy, and that he felt it necessary to submit to him. Nasir again sent Ahmed to Toghai, and renewed his former proposals. Toghai replied that this was not the time to think of marriage. He hoped to be more successful in a second encounter with their enemy, and he begged him to send a few thousand men to Aleppo, who might assist him in case of necessity, and would in any case be a

---

* Shirgai does not call him a Turk, but says that Choban, having married a daughter of the King of Georgia, had a son by her. When he subsequently married Abusaid's sister, he sent the Georgian princess and her son back to her father. Shirgai says the pretender was this son. Weil, iv. 343. Note 1.

S *

moral support to him. We don't know what was Nasir's answer to this, but we know he exerted himself to oppose the Chobanians, and he also succeeded in winning over Hafiz, a brother of Ali Padishah, who was inclined towards them, to join the Greater Hassan, while he, on the other hand, strengthened the alliance between the latter and Toghai, the chief of Diarbekr.

The two Hassans, with their respective armies, at length encountered each other on the 10th July, 1338, at Nakhchivan. Hassan the Great suspected some of his generals, one of whom, Pir Hussein, went over to the other side, while the two armies were ranged opposite to one another. He thereupon fled hastily, and went to Tebriz, where he remained hid for some days, but his *protégé*, the Khan Muhammed, who was still a youth under age, fought bravely with the troops of Khorasan. He, however, fell into the hands of Sheikh Hassan the Little, who put him to death immediately. This was in July, 1338.* The " Shajrat ul Atrak " says the poet Awhad ud din, of Ispahan, flourished during Muhammed's reign. It also says Muhammed was buried at Meragha. In the British Museum there are several coins of Muhammed Khan, both in silver and copper, struck in the years 737-738 of the HEJ. Three of his mint towns were Tebriz, Shiraz, and Hamadan. Mr. Stanley Poole also reads doubtfully the names of Kebir-Sheikh, and Shakan.† Fræhn adds the mint town of Kara Aghach.‡

SATIBEG KHAN.

After the battle, one portion of the army of the victor went to Tebriz and the other to Sultania, and both towns were plundered. The false Timurtash now tried to rid himself of his probably too exacting patron, the Lesser Hassan. He made an ineffectual attempt to assassinate him, and stabbed him with a knife, but the wound was not mortal, and the wounded Sheikh went to Georgia. The false Timurtash having made an attempt to secure Tebriz, and been beaten by the elder Hassan, was joined by the amirs of the Uirads, who had been driven from Irak Ajem, and marched towards Baghdad. Hassan the Little denounced his *protégé*, and went and joined the Princess Satibeg, called Saki Begum in the "Shajrat ul Atrak," the sister of Abusaid, and the widow of his grandfather Choban, who, with her son, Shiburgan Shireh, whom she had had by Choban, had refused to join Hassan the Great. With her he returned to Tebriz, and we read that sixteen of the family of Choban (*i.e.*, of his own relatives) assembled before him and demanded a ruler of the race of Khulagu Khan. The amirs of the Hazaras said that Satibeg, the daughter of Uljaitu Sultan, and sister of Abusaid Khan, was entitled to the throne, since no male of the family of Khulagu remained. She was thereupon

---

* D'Ohsson, iv. 726-728.　Ilkhans, ii. 317.　Weil, iv. 343.　Shajrat ul Atrak, 315-318.
† Catalogue, &c., 94-97.　‡ De Ilk, &c., numis 534-536.

seated on the throne.   This was in 639 HEJ.   Her name was inserted in
the khutbeh or public prayer, and also on the coins.   Having assembled
an army, the Lesser Hassan marched with her to Sultania.   The Greater
Hassan, on hearing of this, went to Kazvin, but presently again advanced
to oppose his rival.   Some negotiations took place between them, and
eventually the latter paid a visit to the Princess Satibeg, acknowledged
her right to the throne, and the two Hassans embraced each other.   It
was arranged that that winter the elder Hassan should remain at Sultania
and the Princess Satibeg at Karabagh, and that in the spring a kuriltai
should be called together to make arrangements for the future.   The
princess, with the Lesser Hassan, thereupon went to Irak and Azerbaijan,
while the elder Hassan went to Sultania.   The former put to death the
pretender, Timurtash, at Aujan.   He had been seized and handed over
by the Uirads.

Although the Greater Hassan had outwardly made peace with Satibeg
and her patron, the Lesser Hassan, he apparently did not really mean all
he said, and invited Toghai Timur Khan to come from Khorasan.   This
was in the year 739 HEJ.   The Khan was accompanied by the Amir
Arghun Shah, and by his vizier, Khoja Alai ud din Muhammed.   But
Hassan speedily repented his complacency, for Toghai Timur and the
amirs of Khorasan followed in everything the advice of the Vizier, who
adopted a parsimonious *régime*, and cut down the salaries, &c., while he
augmented the taxes.   Meanwhile, negotiations were furtively opened
with the Lesser Hassan, to whom Toghai Timur sent Akurpukh, the
Uighur, who had been the atabeg of Abusaid Khan, to say that friendship
had existed between his father and Choban, Hassan's grandfather, and
when an order had been issued to put his father, Surikuri, to death, the
Amir Choban had resisted it, and that he could not forget such an act,
nor could he in consequence permit any injury to be done to any of the
Chobanians, and that it would be well to do away with all cause of
enmity that existed.   Sheikh Hassan the Little was pleased with this
message, and readily said that whatever the Padishah proposed was
undoubtedly for the good of religion, the kingdom, and the people ; and
having made rich presents to Akurpukh and his companions, he
begged him in private to convey his assurance of devotion to Toghai
Timur, to urge upon him that he himself had no real authority in the
kingdom, and that he was suspicious of the motives of Sheikh Hassan the
Great, who was his father-in-law, whom it would be prudent to put aside,
after whose death he would give Dilshad Khatun and Satibeg to Toghai
Timur, and all his family would unite in his service.   When Akurpukh
returned with this message, Toghai Timur, captivated by the hope of
securing Satibeg as his wife, returned a favourable answer, and
suggested that some agreement should be drawn up on these terms
Sheikh Hassan the Little said he was quite firm in his purpose, and that

the best solution would be for Toghai Timur himself to sign a document
embodying the agreement, and calling upon the Chobanians to put to
death Sheikh Hassan the Great and all his dependants, and to make over
the viceroyalty of the kingdom and the tribes to the Chobanians. Toghai
Timur unwittingly fell into the trap which the astute Hassan planted for
him, and duly signed a document embodying the provisions above
mentioned, which he sent to the deceitful Sheikh. The latter, with his
diabolical craft, at once set off, with Pir Hussein and Amir Ali Tulpin, to
the camp of Sheikh Hassan the Great. He summoned one of his
servants to tell him his namesake was waiting outside, and sent the
following message by him : Although Sheikh Hassan the Great had
separated from him, and introduced a stranger to rule the country who had
no claim to the succession, and who was besides the enemy of the house of
Khulagu ; although he had expended treasures on armies to support that
stranger and had hazarded his life and fortune in his defence, notwithstanding that same stranger had written the agreement he now sent him, in
his (Sheikh Hassan's) absence indulging himself in visionary projects, while
professing in public to be his friend. He went on to say that although
he, Sheikh Hassan the Great, deemed him (Sheikh Hassan the Little)
his enemy, yet he could not permit such treachery to be concerted against
him, nor that any hurt should befall him, and he therefore warned him to
take care of himself. Having delivered this artful message he returned
laughing to his camp and again prepared for war, although he told his
officers that all necessity for war was at an end, and that Toghai Timur's
sedition had been quelled, which they would shortly see. When Sheikh
Hassan the Great received the incriminating document, "awaking from
the slumber of negligence," he ordered his men to get under arms, and
then summoned Arghun Shah, who was Toghai Khan's deputy, before
whom he threw the paper, and asked him if it was in the Padishah's handwriting. On his saying yes, he asked, "What evil have I done to him,
that he should have such treacherous intentions towards me and my
family?" Arghun, being ignorant of the whole matter, fell on his knees,
and after praising the kindness and goodwill of Sheikh Hassan, said he
knew nothing of the agreement, but thought it might be traced to the
wiles and fraud of the Lesser Sheikh Hassan, for that Toghai Timur,
although a king, was yet a simple, plain Mongol, and not able to cope
with the insidious arts of the Chobanian ; that if he, Sheikh Hassan,
would grant permission, he and all the troops of Khorasan would either
devote themselves to death before him or root out the family of Choban.
When Sheikh Hassan heard these manly words he said, "May God
forgive the sins of your predecessors," and having embraced him said,
"You are the grandson of the Great Arghun, who by the authority of
Mangu Khakan was chief of all the tribes of Khulagu Khan. It is also
written in the regulations of Khulagu that he was always guided by the

advice of Amír Arghun, and as you also are faithful and trustworthy, carry this writing to the traitor Toghai Timur, and ask why, without any cause of enmity, he entertains such villainous designs against me." He accordingly went to his master, who, "putting the finger of astonishment between the teeth of thought," at last said, "I did not believe the Chobanians capable of such exceeding fraud and villainy, but as it is, having fallen to the ground between two horses, it is not wise to remain here." He was, in fact, so ashamed of having been tricked in this way, that he left the same night for Khorasan, and the various troops who had collected under his banner dispersed. His ready credulity thus lost him the empire.

Hassan the Great now again repaired to Satibeg. He kissed her hand and made excuses for his conduct, dismissed his troops, and left with her for Aujan. They rode side by side, while the Lesser Hassan went on the flank or in front of the cavalcade to preserve order. When they reached Aujan, Hassan the Great encamped two parasangs away, while some of the amirs went to Tebriz. His namesake and rival now became suspicious that he was going to be superseded, and rebelled. He left the camp of Satibeg, which he plundered ; and professing that it was indecent for the kingdom to be ruled by a woman, he caused several of her officers and those of her son Shiburgan to be put to death, and set up Suliman as Khan.*

Several of Satibeg's coins are extant. In the British Museum there are eighteen in silver, struck during the years 739-741 at Khelat, Arbuk, Hisn, Baghdad, and doubtfully at Arzen and Turkan.† Frœhn only names two in his monograph on the coins of the dynasty, one struck at Berdaa and the other at Tebriz, the former in 739.

SHAH JIHAN TIMUR KHAN.

When Hassan the Great heard how his rival had set up Suliman, he also raised a puppet of his own to the throne in the person of Shah Jihan Timur, surnamed Iz ud din, who was the son of Alafrenk, son of Gaikhatu, son of Abaka, and appointed Shems ud din Zakaria as his vizier. At the approach of winter he went to Baghdad, the capital of Irak Arab, of which province he had control, as well as of Khuzistan and Diarbekr. In 1340 he determined to attack his rival and namesake, and advanced from Baghdad with his nominee, Shah Jihan Khan, as far as the river Baghatu (called Chagatu or Yughtu in the "Shajrat ul Atrak"), near which they fought a battle, which he lost. On returning to Baghdad after this reverse he deposed the Khan whom he had created, as an ignorant man, unfit to reign.‡

---

* Shajrat ul Atrak, 319-327. D'Ohsson, iv. 729-732. Ilkhans, ii. 318-319.
† Poole's Catalogue, 103-106. ‡ Shajrat ul Atrak, 327-329. D'Ohsson, iv. 732-733.

There are two silver coins of Jihan Timur in the British Museum, both struck in the year 740 HEJ., one at Arzenjan and the other at Khelat.*

SULIMAN KHAN.

Hassan the Less, as we have seen, had nominated Suliman as ruler of the Khanate. He was the son of Muhammed (Abulghazi leaves out Muhammed, and makes him son of Sanga), son of Sanga, son of Yashmut, son of Khulagu. He married Satibeg,† and acquired authority over Irak Ajem, Azerbaijan, Arran, Mughan, and Georgia. The rest of the empire had been appropriated by different chiefs, who were virtually independent, although all of them did not exercise, so far as we know, the right of coinage. There were Satibeg and the Chobanians in Arran, and Haji Toghai in Diarbekr. Rum, in so far as it was subject to the Mongols, was shared between Artena and Ashraf, son of Timurtash. The sons of Mahmud Shah Inju held Fars. The province of Ispahan obeyed the Seyid Jelal ud din Mirmiran and A'mad ud din Lenbani. Mubariz ud din Muhammed Mozaffer remained master of Yezd ; Malik Kutb ud din Ghuri, of Kerman ; Malik Shuja ud din, of the district of Kum (called Bam by D'Ohsson) ; and Malik Moiz ud din Hussein, of Herat. Toghai Timur Khan reigned in Mazanderan and part of Khorasan. Arghun Shah, son of Nuruz, held the district of Tus, and the Amir Abdullah Mulai, Kuhistan.‡

When, in 740, an embassy came from Sheikh Hassan the Great asking Nasir, the Egyptian Sultan, to send one of his sons with some troops to Irak, and offering to acknowledge him as his suzerain, he replied that his sons were too young, but that he would himself march to Irak with his men if Sheikh Hassan, Hafiz, and Toghai would first swear allegiance to him. In the following year the Chobanians again invaded Irak. Hassan sent once more to Egypt to ask that an army might be sent to him under a competent general, and offering to strike money and have the Friday prayer said in the Sultan's name. The suspicious Sultan asked that these two conditions should first be fulfilled, and then he would send an army. The amir Ahmed was again sent, and Toghai at Diarbekr, and Sheikh Hassan at Baghdad, swore allegiance to the Sultan ; while the latter, in the presence of the envoy, in March and April, 1341, had both money struck and the Friday prayer said in his suzerain's name. Ahmed took some of this money with him back to Egypt, and also Ibrahim Shah, a nephew of Sheikh Hassan, and Berheshin, a son of Toghai, as hostages, together with several kadhis, who carried the documents in which the two chiefs

---

* Poole, op. cit., 102.
† The "Shajrat ul Atrak" says it was Hassan the Less who married the princess by force.
‡ D'Ohsson, 729-730. Ilkhans, ii. 318.

declared themselves Nasir's vassals. They reached Cairo on the 23rd of May, 1341, whereupon the Sultan set about the equipment and provisioning of the army. A few days before it was to march, however, messengers came from the Prince of Mardin to say that the Lesser Hassan, angry at the alliance of his rivals with the Egyptians, had sent envoys to his namesake and rival, and that the latter had made peace with him. Nasir now countermanded his orders to the army, and sent messengers to make some fresh inquiries, but before their return he had died.* This was in the summer of 1341. The same year Hassan the Little defeated Toghai Timur Khan, who had made another attack upon Irak, and drove him away.† He now marched against Haji Toghai, the friend and ally of the Elder Hassan. Toghai sent his nephew and the mollah Suti to take him a proposition for an accommodation, but he himself fled in all haste to Mush in Armenia. He was killed by Ibrahim, brother of Ali Padishah, in 743 or 744.‡ Hassan paid no heed to the communications he had received, but ravaged the district of Mush with fire and sword. The ruler of Mardin went to do homage to him and his nominee Suliman, and was well received and given presents. From Mardin Hassan marched with a large army against his rival of many years' standing, Hassan the Great, who sent an army against him under the amirs Ali Jaafer and Kara Hussein, who defeated him. He thereupon turned his steps towards Rum, plundering the properties of the grandees on the way. Erzerum was laid under contribution, while the mihrab and minbar (i.e., the altar and pulpit) in the mosque in that town, which had been recently rebuilt by Haji Toghai, were burnt, and the grave of Toghai's son was opened and despoiled. In the autumn he went to Tebriz, where Suliman, his nominee, had taken up his residence. In the summer he sent the vizier, Ghiath ud din Muhammed Ali Shahi, to Sultania, for his uncle Shiburgan (called Siyurgan by Von Hammer). After detaining him some time at Tebriz he sent him to Kara Hissar, in Asia Minor. He himself went to Alatagh and Bulak to lay waste the yurt of Haji Toghai and the possessions of the family of Sunatai, and having returned to Tebriz, built a monastery, a medressch or college, and a mosque surpassing all others there in elegance. Meanwhile, his brother Ashraf had gone to Shiraz to drive out Abu Ishak, who had set up authority there, but returned unsuccessful. Ashraf then allied himself with his uncle Yaghi Basti, the son of Choban, against his brother Hassan the Little, and they went over to his rival, the Great Hassan, while several amirs of the Uirads, who had quarrelled with the latter, followed their example. Hassan Kuchuk thereupon killed the remaining amirs of the Uirads who had not been lucky enough to escape. With his usual crafty power of

---

* Weil, op. cit., 348-349.
† D'Ohsson, iv. 732-734. Shajrat ul Atrak, 329-330. ‡ Weil, iv. 344. Note.

aspersion, he succeeded in arousing his rival's suspicions against Yaghi Basti and Shiburgan, who both fled. The Great Hassan now went to Baghdad, while the Little Hassan repaired again to Tebriz. Yaghi Basti and his nephew Ashraf having plundered the camp of Issen Kutlugh, near Rai, marched against Fars, which was then the great *entrepôt* of Persian trade, and, naturally, well worth plundering. On the way they were joined by Pir Hussein, son of Mahmud, son of Choban, and, therefore, Ashraf's cousin ; and they carried off from Ispahan what they could in the shape of banners, drums, trumpets, standards, horses, and camels. Ebrkiuh was burnt, and two thousand of its inhabitants, who had sought shelter in a cave, were suffocated with smoke. The Sheikh Abu Ishak Inju, the ruler of Shiraz, prepared once more to resist them; but when they were within a day's journey of the place, Ashraf heard that his brother Hassan had been assassinated, and accordingly returned.*

The "Shajrat ul Atrak" tells us that in the beginning of the year 744 HEJ. Hassan dispatched his *protégé* Suliman Khan, with Yakub Shah and other amirs, to Rum, but they were defeated and obliged to return. He thereupon imprisoned Yakub for having misbehaved in the battle. One of Yakub's friends, named Arab Mulk, afraid that he might be executed, associated two or three women with him, and on the night of the 27th Rajab of the same year got access to him and killed him, by violently twisting his generative parts, as related in the lines of the poet Khoja Selman Saveji. The next day they secretly left the palace, and the murder was not discovered till the third day, whereupon a search was made for Arab Mulk, who, with his confederates, was seized and put to death.† Another account tells the story differently, viz., that Hassan's wife, Izzet Malik, who had an intrigue with the Amir Hassan Yakub Shah, whom Sheikh Hassan had imprisoned for some grave offence, fancying that her intercourse with that chief was discovered, took advantage of an occasion when Sheikh Hassan was very drunk, and killed him (*lui ôter la vie, en lui pressant les testicules*). This was in 1343. Inasmuch as the strictness and severity of Sheikh Hassan were so well known, no one dared go near the harem, and two days elapsed before his death was known. Izzet Malik, with the women who were her accomplices, had gone out under the pretence of bathing, and had hidden herself. On the third day, the amirs, who had become nervous, sent a woman into the harem, where she found only the corpse of the dead chief, who had expiated a life of violence and chicanery by a corresponding end. Izzet Malik was punished for her crime by a most ignominious death. Her body was cut in pieces and given to the swine to be eaten, and there were people who cut off portions of her flesh and ate them. Among others,

---

* Ilkhans, ii. 330-331.   † Op. cit., 331-332.

the poet Selman, as I have said, the contemporary and rival of Hafiz, wrote an epigram on this tragedy. Suliman Khan, having distributed the considerable treasures of his dead patron among his generals, returned to Karabagh, where he presently imprisoned Kadhi Hussein, the son of the Amir Hassan, and released from prison the Amir Yakub Shah, who had put the murderers of his late patron to death, and nominated him, as the bravest and most prudent of all his amirs, to the post of Amir ul Umera. He was so troubled by the ambitious pretensions of the amirs that he summoned Ashraf and his uncle Yaghi Basti, who accordingly went to Tebriz.

Meanwhile, Shiburgan, who as we have seen had been imprisoned at Kara Hissar, in Asia Minor, killed the commander of that place, seized the treasures which Sheikh Hassan had deposited there, and went and joined Ashraf and Yaghi Basti at Maamurieh, whence they went to Alatagh and the Blue Lake (i.e., that of Erivan), where Choban's head-quarters had formerly been. Their forces rapidly augmented, while those of the titular ruler, Suliman Shah, melted away, and he thought it prudent to fly to Diarbekr. Several of his amirs, such as Haji Hamza, Mulai and his son Muhammed Ali Shah, the vizier, and Bestal the Georgian, joined the three Chobanian chiefs. His vizier, Amad ud din Zeravi, tried to sow discord among the triumvirate, and reported to Ashraf on the one side, and to Yaghi Basti and Shiburgan on the other, that the other side meant that very night to strike a blow. Both parties remained under arms during the night. In the morning messengers passed between them, and the matter having been explained, Amad ud din was put to death. The three Chobanian chiefs now repaired together to Tebriz; but as the citizens paid greater honour to Yaghi Basti and Shiburgan than to Ashraf, the latter became jealous, and under pretence that it was contrary to the yasa of Jingis Khan that Mongols should live in towns, he pointed out to the two amirs that there was good pasturage at the foot of the mountain Sehend. Meanwhile, he heard from the Amir Jelal ud din Amir Kutlugh Shah Ghazani that they intended falling upon him in the morning. Taking his brother, Malik Misr, with him, he repaired to Tebriz, where he learnt that his two associates had gone in the night to Khui. He followed them on foot, and overtook them at Maamurieh. There a struggle took place, in which he won the day. This was apparently in the year 745 HEJ. The "Shajrat ul Atrak" calls the battlefield the field of Aghbabad, and says that after it Ashraf encamped at the town of Babul. Von Hammer calls it Totil.* I do not know what became of Suliman.

There are numerous silver coins of Suliman Khan in the British Museum, struck in the years 740, 741, 742, 743, and 745, and one, possibly

---

* D'Ohsson, iv. 734-735. Ilkhans, ii. 319-320 & 329-333. Shajrat ul Atrak, 333.

a posthumous coin, struck in 751, at Sivas.* The other mints on coins in the Museum are Kashan, Baiburt, Hisn, Nakhchivan, Ḥar'an, Keek (?), Erzenjan, Shirvan, Sivas, Kebir Sheikh, Tebriz, Yezd, Tus, Saveh, Sinjar, and Nubenjan (?) General Houtum-Schindler has published the additional mints of Kerman and Sirjan.† Fræhn also mentions Sultania, Erzerum, and Kazvin as mints on his coins.

ANUSHIRVAN KHAN.

Suliman, shadow that he was, was not the last of the Ilkhans. We must add a notice of another puppet who was set up by Ashraf. After his victory over his relatives at Maamurieh, we are told by Mirkhavend that he raised to the throne a Persian of an ancient stock claiming descent from the famous smith, Giave or Gavian, who lived in the heroic age of Persian history. Khuandemir and Jennaby call him a Kipchak, while Abulghazi makes him descend from Khulagu.‡ Ashraf ordered him to be styled Aadil (i.e., The Just), as his namesake, the Sassanian ruler, had been.

After this, Ashraf went to Kuntsak, in Arran, where Mohai ud din also arrived with propositions of peace from Yaghi Basti and Shiburgan. Ashraf was willing to agree to this, but Shiburgan could not trust him, and went to Diarbekr, where he was at first well received by Ilkan, the son of Sheikh Hassan the Great, who governed there. It was not long, however, before he lost confidence in him and killed him. Yaghi Basti joined Ashraf, his nephew, with whom he went to Tebriz. When they arrived there the Malik made away with his uncle in u, way that no one ever became acquainted with. After this he oppressed the people so much that all who could fled from Iran and Azerbaijan to other countries for safety. At Karabagh he received the allegiance of Kaus or Kavus, the son of Kaikobad, the ruler of Shirvan, to whom he presented a golden cap and jewelled girdle; but the prince of Shirvan was so affected by his truculence (he cut off an amir's head in his presence) that he fled at night, and when Ashraf sent an apologetic embassy with a request for the hand of Kaikobad's daughter in marriage, Kaikobad refused, as he foresaw that this connection would speedily bring him with an army to Shirvan. Having returned to Tebriz, Ashraf shut his brother, Malik Misr, up in prison, and put several begs to death. He imprisoned Ortok, son of the Khoja Jelal ud din, in the Rashidian citadel at Tebriz, and when he left the city in the next winter he intrusted it to the care of a Greek slave, named Muhammed, with a guard of 2,000 men. This slave revolted in his master's absence, and set the two State prisoners free. Ashraf set out in the middle of winter from Karabagh to Tebriz, to put down the revolt. Famine and pestilence

---

* See Poole's Catalogue, p. 4. † Num. Chron., N.S., xx. 320-321.
‡ Fræhn de Ilchanorum, &c., numis 541. Von Hammer, Ilkhans, ii. 333.

meanwhile prevailed, and the tyrant having returned to Karabagh, marched
with an army against Shirvan; but while both armies faced one another
on each side of the frontier river, in full panoply, Ashraf and Kavus, the
Shirvan prince, were reconciled. He then, in the year 748 HEJ., turned
his arms against Sheikh Hassan the Great, at Baghdad, which he laid
siege to. Hassan wished to retire to Kumakh, but his wife, Dilshad,
begged him to stay, and presently Ashraf was constrained to retire by the
bold front of the besieged. Returning to Tebriz, he distributed among
his amirs the various commands of troops in Azerbaijan, Arran, Mughan,
Gurjistan (*i.e.*, Georgia), and Kurdistan, and piled up immense treasures.
His right hand man was his vizier, the Khoja Abdulhaji, whom, however,
notwithstanding his great services, he shut up the next year (*i.e.*, 1349) in
the fortress, with his son-in-law, Masud Dameghani, a famous caligraphist.
They were afterwards sent to Alamut, and intrusted to its guardian, Kia
Irmail, who received the Vizier with great honour, and gave him his
daughter in marriage. Abdulhaji kept up a correspondence with the
ruler of Ghilan, and Ashraf soon repented of having confided his custody
to the governor of Alamut. To make sure of the Vizier, he loaded his
relatives with gifts, and urged him with most flattering messages to
return. He accordingly did so, and his confidence was characteristically
rewarded by his being imprisoned in the castle of Alinju. His son-in-law,
Masud, of Dameghan, was made Vizier in his place. The next year (*i.e.*,
in 1350) the latter, together with the Khoja Muin the Serbedar, were
seized and immured in the castle of Ruyin. Ashraf now set out to attack
Ispahan. Although his army was 50,000 strong, he could only blockade
it on two sides. The siege was protracted for fifty days, and the struggle
was very fierce. At length the citizens sent him word that they were
prepared to resist to the last breath, if necessary, to defend the town; but
that if all he wanted was an acknowledgment of his supremacy, they were
willing to have the khutbeh said and the coin struck in his name. Ashraf
consented, and thereupon sent the mollah Sherif ud din, of Nakhchivan,
into the place, to see these terms carried out, and the Friday prayer was
duly said in the name of Anushirvan, the titular ruler and *protégé* of
Ashraf, while 2,000 dinars were also struck with his name. The town
also paid 100,000 dinars in value of linen and staple.*

Ashraf now raised the siege, and returned to Tebriz. Anushirvan was,
of course, a mere shadow, like the puppets set up by Timur, the real
authority being in the hands of Ashraf. He was of a very avaricious
disposition, and put to death wealthy men, and the governors whom he
himself appointed, in order to secure their wealth. At Aujan he put to
death the judge Shems ud din, the khoja Ghiath ud din Shekerleb
(*i.e.*, Sugar Lips), and Sultan Shah Zeravi. The deposed vizier, Abdulhaji,

---

* Ilkhans, ii. 336-338. D'Ohsson, iv. 736-737. Shajrat ul Atrak, 334.

he sent to a fortress in Kurdistan and placed him under the supervision
of the *kiotval* (*i.e.*, castle governor) Musa, and as he treated him humanely
Ashraf sent him a sharp message to alter this.   He was accordingly
immured in a dungeon, and fed on bread and water until he died.
Feeling that the people of Tebriz hated him, he summoned the mollah
Nisam ud din Ghafiri to use him as a tool in securing their goodwill.
The latter came to the neighbourhood of Tebriz, and Ashraf went out
to meet him.   The mollah told him to his face that no one any longer
believed his word, and Ashraf now went to Karabagh.   When he had
crossed the Araxes he heard at Eskeshehr that Dali Bayezid, whom he
had sent away from Aujan with Abdulhaji, had revolted.   In order to
allay this outbreak he delayed some five days near Mughan, and then
returned to Tebriz, where he distributed largess among the soldiers to make
sure of them.   In the fight which followed, one of the bravest of the
enemy's amirs, Toka Timur, unhorsed Ashraf with a blow from a club,
and dragged him along for some time.   He was only rescued with
difficulty, while his army was beaten.   He now sent Alpi bey against
Dali Bayezid, but soon learnt he had made common cause with the rebel.
Once more did Ashraf open his treasures and distribute largess, and, in
order to save the trouble of counting, had it measured out in sacks.
Luckily for him, Alpi bey and Dali Bayezid disagreed ; the latter committed suicide by running a sword into his own breast, while the former
fled, but was apparently overtaken and killed, for we read that their two
heads were sent to Ashraf, who set them up by the side of his throne, and
afterwards had them perambulated about the country.   After this revenge
on two very dangerous enemies, he chiefly devoted himself to fortifying
Tebriz.   He lived in the Rashidian quarter, around and through which he
cut ditches, in order to parcel it out into sections.   He ordered all who
were able, to build themselves houses to live in.   Those who could not
afford this were to find shelter in the mosques, medresshi, monasteries, or
hospitals.   He himself (says Von Hammer) was like a bat in a dark
room, afraid of everybody.   The birds and sheep which were dressed
for his meals were throttled and cooked in his presence.   He saw the
water which he drank taken from the stream; so greatly did he fear
assassination, and naturally, for there was no one about him who had
not lost some relative through his tyranny.   Outside the door of his
palace was hung a chain, which those who had petitions to present
rattled, and then hung their requests, whence it was known as the Chain
of Justice.   It was an imitation of a similar chain which Anushirvan the
Just had similarly used, truly a parody of justice at the very gate of tyranny.
Ashraf was engaged to be married to the daughter of the ruler of Mardin,
and a year later (in 1252) the wedding took place at Tebriz amidst great
rejoicings.   To escape his tyranny many notable people migrated, among
whom was Mohai ud din, already named, who went to Janibeg, Khan of

the Golden Horde, whom he urged to march against the tyrant. "The emperor," he said, "has it in his power to deliver the people from this tyranny. If he will not do so he will answer for it in the next world." He accordingly entered Azerbaijan, and speedily· dispersed the few troops which Ashraf sent against him. Ashraf himself fled, but was captured between Merend and Khui. The "Habib Alsüar" says he sent his harem and treasures to the mountain. of Muzid, situated where the fountain of Rashid ud din springs. Janibeg wanted to take him with him to Serai, but Kaus, the prince of Shirvan, and the Kadhi Mohai ud din urged that so long as he remained alive, his old subjects would not remain quiet. He was accordingly put to death, and his head was suspended from the door of a mosque ; one account says at Meragha, another at Tebriz. Janibeg took possession of Tebriz, and the Friday after attended the public prayer at the mosque of Ali Shah, after which he subdued Azerbaijan. As Janibeg ruled with justice, and was the patron of literature, Molna Said ud din Tuftazani, in the year 756 HEJ., dedicated the abridgment of the Tulkhiz to him. Having completed the subjugation of the country, Janibeg withdrew again to his own dominions, taking with him Timurtash, the son, and Sultan bakht, the daughter of Ashraf, and leaving his own son Berdibeg as his deputy there. Janibeg died shortly after, and was succeeded as Khan by Berdibeg, who, on leaving Tebriz for the Kipchak, left his vizier Akhijuk there in charge, who followed the oppressive policy of Ashraf.*

I do not know when Anushirvan died. His coins in the Museum are dated in 747, 752, and 754, and doubtfully in 756, and were struck at Berdaa, Nishapur, and Tebriz, and apparently at Shirvan.† One curiously enough has a Chinese inscription in the seal character. Fræhn describes coins of his struck in 743, 746, 747, and 749, and adds the mints of Ani, Van (?), Bazar, Bailekan, and Kazvin.

The last Roman Emperor of the West was called Romulus Augustulus, uniting in his own name those of the founders of the power and of the empire of Rome. The commander of the Persians, in the fatal battle by which the Arabs overthrew the Sassanian empire, bore the famous name of Rustem. The Turkish dynasty of the Seljuks in Persia began and ended with a Tughrul. The Byzantine empire was virtually founded by one Constantine and perished under another. So did the last wearer of the crown of the Ilkhans bear the great name of Anushirvan. Soon after, strangely enough, the famous palace of the Sassanians, called Taki Kezra, whose dome had cracked during the night when Muhammed was born, fell into complete ruin.‡

---

* Ilkhans, ii. 337-340.   Shajrat ul Atrak, 233-234 & 333-335.   D'Ohsson, iv. 735-737 & 740-742. Howorth's Hist. of the Mongols (part ii.), 178-179.
† Poole, Catalogue, 115-117.   ‡ D'Ohsson, iv. 735-736.   Ilkhans, ii. 333-334.

THE ILKANIANS OR JELAIRIDS.

HASSAN BUZURG KHAN.

The most important of the various fragments into which the empire of the Ilkhans broke in pieces, and the only one which the author of the "Shajrat ul Atrak" deems worthy of recording the story of, was the line of princes sprung from Hassan, who, to distinguish him from his rival of the same name, is generally referred to as Hassan Buzurg, or the Great. They were also known as the Jelairs, from the fact that they were derived from the Jelair tribe ; and Ilkanians from Ilka Noyan, who came into Persia with Khulagu, and held a prominent position in his army. He left ten sons, of whom Akbuka was the ninth. He filled a famous *rôle* in the history of Gaikhatu Khan, having, *inter alia*, been the cause of the introduction of the famous paper money which distinguished it, and was put to death in 1295, as we have seen. His eldest son was Hussein Gurkan.* Hussein was styled Gurkan from the fact that he married the daughter of Arghun Khan. He was an important personage during the reign of Abusaid Khan, and was governor of Khorasan. He was the father of the Amir Hassan, styled Hassan the Great, whose life has occupied us so much already, and who as a grandson of the Ilkhan Arghun had the blood of Jingis Khan in his veins. We have seen how, when the various local commanders set up authority in various parts of the empire, he, the son of an Imperial princess, and who married Dilshad Khatun, the widow of Abusaid, got possession of the Arabian Irak, with his capital at Baghdad, and set up one puppet after another. The last of these apparently was Jihan Timur, whom he deposed in 1341, and then put himself on the throne. We have traced his turbulent career in the previous pages. He died at Baghdad in 757 HEJ. (*i.e.*, 1356).† In the British Museum is a silver coin struck by Hassan at Baghdad, in the year 755 HEJ. (*i.e.*, 1354).

SULTAN OWEIS KHAN.

By his wife Dilshad Khatun, the daughter of Dimashk Khoja and grand-daughter of the Amir Choban, Hassan had a son Oweis, who succeeded him at Baghdad. In 759 HEJ. (*i.e.*, in 1358) he marched upon Tebriz, which was then in the hands of Akhijuk, who had been left in charge of Azerbaijan by Berdibeg, the Khan of the Golden Horde, when he withdrew from it as we have described. Akhijuk went to meet him as far as the mountains Sanatai (called Meenai in the "Shajrat ul Atrak").

---

* Erdmann, Voll. Ueb., &c., 26-27.   † D'Ohsson, iv. 742.

The battle was indecisive, and the two armies spent the night in presence of one another. The following morning Akhijuk fled. Oweis pursued him to Tebriz, into which he made a stately entry, while his enemy again fled to Nakhchivan. Oweis, we are told, took possession of the Raba-Rashidi, where he took up his residence, and in the month Ramazan of the same year he put to death forty-seven rebellious chiefs (D'Herbelot says forty) of the party of Malik Ashraf, and in consequence many other amirs and chiefs went and joined Akhijuk at Karabagh. Oweis Sultan sent the Amir Ali Tulpin to attack him, but the latter, on account of some feud he had with the Sultan, allowed himself to be beaten, or rather disgracefully fled to Akhijuk, and his army being dispersed, Oweis returned to Baghdad in the depth of winter, causing much suffering to his men.* The next year Amir Muhammed Muzaffer† advanced from Fars upon Tebriz, which he captured, but in the course of two months, having heard that Sultan Oweis was on the march again from Baghdad, he retired to his capital. When he arrived at Tebriz, the Sultan summoned Akhijuk and treated him kindly, but presently the latter was accused of some treachery, and was duly beheaded. Azerbaijan and Arran were thereby added to the dominions of Sultan Oweis.‡ In the year 765 HEJ. (i.e., 1363) Khoja Merjian, or Murjan, who was governor of Baghdad, rebelled, and the Sultan marched against him. The Khoja was defeated in a battle fought near the city, which was obliged to open its gates. The rebel was pardoned. Having spent eleven months enjoying himself at Baghdad, he went by way of Diarbekr to Mosul, which he captured. He then marched upon Mardin, which he also secured, and thence by way of the desert of Mush and Kara Kalisa to Tebriz, where he spent the winter. He now paid another visit to Baghdad, after which he again returned to Tebriz.§

In 767 an envoy went to Cairo from Murjan, stating that his master had thrown off his allegiance to Oweis, and had acknowledged the Sultan as his suzerain, had had the prayer said and the money coined at Baghdad in his name, had induced the people there to swear allegiance to him, and was ready to march against Oweis. If he won, then he would deem himself as governing Baghdad on behalf of the Sultan, while if he were defeated he would escape to Egypt. This embassy was well received and richly rewarded, but instead of sending troops to help the rebel the astute Egyptian ruler contented himself with providing him merely with two banners, one with the Khalif's arms upon it and the other with that of the Sultan, and sending him a diploma. Meanwhile, another embassy arrived from Sultan Oweis to report Murjan's rebellion, that he was preparing to repress him, and asking that no shelter should be offered to him in Egypt or Syria. The Egyptians were relieved from this

---

* D'Ohsson, iv. 742-743. Shajrat ul Atrak, 335-336.   † Vide next chapter.
‡ D'Ohsson, iv. 745. Shajrat ul Atrak, 336.   § Shajrat ul Atrak, 336-337.

embarrassing position by the fact that the rebel was duly defeated, blinded, and imprisoned, and Baghdad again acknowledged the supremacy of Sultan Oweis.*

Oweis Khan now marched upon Mosul, then governed by the Turkoman chief Bairam Khoja, and having secured it compelled the Prince of Mardin to submit.† He also fought against the Prince of Shirvan.‡ In the year 772 (i.e., 1370) he marched against Vali, the ruler of Jorjan and Kumuz.§ He defeated him near Rai, and pursued him to Simnan. This was apparently the last expedition of Sultan Oweis, who ended his career in the year 776 (i.e., 1374), at Tebriz. Quatremere tells us that Nejib ud din, brother of Shems ud din Zakaria, and grandson of Rashid ud din, was his Vizier.‖ In 775 the amir Wajih ud din Ismael, son of the Vizier Zakaria, was nominated governor of Irak Arab. When Oweis was dying, we are told, his amirs and Kazi Sheikh Ali Kumkhani gathered round him, and asked him to declare his last will, and to nominate his successor, as he had four sons, viz., Hassan, Hussein, Ahmed, and Bayezid. He replied, " Let the sovereignty remain with Hussein, and the government of Baghdad with Sheikh Hassan." The amirs said, "Sheikh Hassan is the elder brother, and will not submit to this arrangement." The Sultan replied, "You know best," and the amirs accepted this as a commission to do as they pleased. They accordingly put Sheikh Hassan in confinement. After this the Sultan lost the power of speech, and died on the night of Jumadi il avul, 776. The same night the amirs put Sheikh Hassan to death in the Dimashk, palace, and the body of Oweis Sultan was taken to Shirvan (?) for burial.¶

In the narrative of Clavigo's embassy we read that on the left-hand side of Tebriz is a high hill, " which they say was once bought by the Genoese from a ruler named Sultan Oweis, for the purpose of building a castle upon it. After it was sold he changed his mind, and when the Genoese wished to build the castle he sent for them and told them that it was not the custom for merchants to build castles in his country. He said they might take their merchandise away, and that if they wanted to build a castle they might move the hill out of his territory. When they answered him, he ordered their heads to be cut off." We read in the same work** that Sultan Oweis had a great house at Tebriz, surrounded by a wall, very beautiful and rich, in which were 20,000 chambers and apartments. This he built with the treasure paid him as tribute by the Sultan of Babylon (i.e., by the Egyptian Sultan, who was generally so styled by Western writers at this time). He called this house Tolbatgana (? Daulat Khana), which means "the House of Fortune."†† There are two letters extant, without date, one sent by Vayscham (i.e., Oweis Khan), the other

---

* Weil, iv. 526-527.   † Id., 527.   ‡ Id., v. 15.   § See next chapter.   ‖ Op. cit., lvi.
¶ Shajrat ul Atrak, 337-338.   D'Herbelot, quoting Khuandemir, 139.   D'Ohsson (who is here much confused), iv. 746.
** Op. cit., ed. Hackluyt, 88-89.   †† Id., 89.

by Sichuarscam, which Heyd treats as a corruption of Sich vais cam (*i.e.*, Shah Oweis' Khan), and sent to the Bailo at Trebizond, and the Venetians there. They contain a pressing invitation for the Venetians to repair to Tebriz, as they had done in the days of Abusaid Khan ; stating that the roads were safe, that it would be well to take advantage of them, and that the merchants would meet with a good reception in Persia, and would not have to pay higher dues than heretofore. In the reply to the first of these letters it is stated that only two years before a large fleet of merchants arrived, intending to open up trade again with Tebriz, but would not undertake the risk until the large customary caravan had arrived from Tebriz with the assurance that the road was safe, when they proposed to return with it. Their fears were justified, for several Venetian merchants who tried to traverse the old route were robbed, and although the Khan punished the robbers and promised to indemnify the merchants, it proved too clearly that the roads were unsafe, and that the strong hand of the Ilkhanid rulers of Persia had become palsied.* The chief poet at the Court of Oweis was Selman Saveji.† The British Museum contains several coins of Sheikh Oweis Khan, struck at Tebriz, Sultania, Baghdad, Irbil, and Shiraz, the dates on which are all partially obliterated.‡ A late acquisition was struck at Halab (Aleppo) ; another one, undated, at Ispahan. Fræhn, in his supplement, gives one struck at Hamadan, in 762 HEJ. He styled himself on his coins Sheikh Oweis Behadur Khan.§

SULTAN HUSSEIN KHAN.

At this time the Turkomans of Armenia were beginning to consolidate a considerable power south of Lake Van. Their leaders were Bairam Khoja and Kara Muhammed.|| Three years after his accession, viz., in the year 779 HEJ. (*i.e.*, 1377), Sultan Hussein marched with an army against the fortress of Akhus, which was in the hands of Kara Muhammed. A peace was arranged between them, and the Turkoman chief visited Hussein, and was well received by him. Weil says he became tributary. Next year some of his amirs, viz., Ismael, Abdul Kadir, Rehman Shah, &c., were driven to rebellion by the tyranny of Adil Agha, and advanced upon Baghdad. Adil Agha, by the Sultan's order, marched from Sultania against them, and came up with them at Aub i Gurkan, and most of the rebels were killed. The same year several amirs of Amir Israel (Quatremere says Ismael), the son of Amir Zakaria, who was governor of Baghdad on behalf of the Sultan, were put to death. Thereupon his nephew, Sheikh Ali, the son of Sheikh Hassan,¶ was raised to the throne there. Ismael had incurred the hatred of Sheikh Ali, who was then ruler of Baghdad, and he induced a number of low people, who

---

* Heyd, op cit., ii. 130-131. † Ilkhans, ii. 265. ‡ Poole, Catalogue, 207-211.
§ Fraehn Recensio, 647. || Weil, v. 16. ¶ Weil, v. 17, calls him his brother.

T *

had received many benefits from Ismael, to ill use their benefactor, and one Friday, in the year 780, one of them, named Mobarek Shah, struck Ismael on the face with his sword, and cut him down. The Amir Masud, brother of Zakaria, rushed to his nephew's succour, but he too was killed. Sheikh Ali asked to see his enemy's head. The assassins accordingly returned and cut it off, and it was hung from a house he was building. It was reported that a few days before, Ismael was watching the workmen busy upon this house, and a carpenter wishing to cut off a projecting beam, Ismael told him he had better let it remain, as it would do to hang a head from ; and, as fortune had it, his own head was the very one with which the experiment was tried.*

Sultan Hussein did not fear any danger from the young pretender at Baghdad. He wrote him a friendly letter, and consented to his being considered as the governor of the city ; but the young prince sent for one of his father's retainers, Pir Ali Baduk, who had transferred his services to Shah Shuja, the ruler of Fars, and was then governing Shuster, and who had apparently begun to encroach upon the Imperial authority, to try and secure Irak Arab for himself. Thereupon Sultan Hussein, with his general, Adil Agha, marched against them with a large army, and not being strong enough to oppose them, Pir Ali returned to Shuster, taking the young prince Sheikh Ali with him. Sultan Hussein then entered Baghdad, and gave himself up to pleasure, and sent Adil Agha in pursuit of Sheikh Ali. He besieged the latter at Shuster, and having got a promise from him not to interfere with Baghdad, but to content himself with his position at Shuster, he retired again to Baghdad, and thence went to Sultania. The young prince and his patron were presently invited to return by some of the Baghdad people, who were probably growing weary of the debaucheries of the Sultan. The latter sent Mahmud Uka and Omar Kipchak against them, but these officers were captured and their army destroyed. Sultan Hussein thereupon fled to Tebriz, which he reached with difficulty.† We also read that Shah Shuja, the ruler of Fars, marched against him from Ispahan with 12,000 picked horsemen, captured *en route* the town of Kazvin, and defeated the Sultan, who had 30,000 men, near Jerbadekan. He compelled him to take shelter in his southern dominions, and occupied Tebriz for some months, after which he retired.‡ This was probably a part of the same campaign. Meanwhile Hussein ordered Adil Agha to attack Rai, and sent all his troops with him. This opportunity was seized by his brother, Sultan Ahmed, who is described as a most bloodthirsty person. He withdrew from Tebriz and went to Irbil, and refused to return again when summoned, but proceeded to collect an army, with which, on the 11th Suffur, 784 (Ibn Arabshah says in Jumadi-l-akhir, 783, *i.e.*, August-September, 1381§), he fell upon

---

Tebriz, "and like a sudden calamity entered the city." Sultan Hussein tried to hide away, but was discovered and put to death. The author of the ."Shajrat ul Atrak" says it was reported he was a very handsome man, but much given to debauchery.* Shems ud din Zakaria, grandson of the great Rashid ud din, was his vizier.†

In the British Museum there are silver coins of Hussein, struck at Basrah, Baghdad, Wasith, Halab, Tebriz, Shamakhi, Hamadan, and Van.‡ Frœhn, in his "Recensio," gives the additional mints of Meragha, Bakuya, and Asterabad.§ On his coins he styles himself Jelal ud din Hussein Khan.

SULTAN AHMED KHAN.

Sultan Ahmed now had himself proclaimed. His brother Bayazid, fearing he might have the same fate as Hussein, fled to Adil Agha, who raised him to the throne. They marched against Ahmed, who fled to Marvand (the "Shajrat ul Atrak" says to Muzid). Adil Agha proceeded to Tebriz, but meanwhile several of his officers mutinied, and he was obliged to return to Sultania. Ahmed now once more returned to Tebriz, but was almost immediately attacked by his nephew, Sheikh Ali, and his patron, Pir Ali Baduk. They met at Heft rud (i.e., the Seven Rivers). Omar Kipchak, who commanded Sultan Ahmed's left wing, deserted, which decided the battle, and the Sultan had to flee by way of Khui to Nakhchivan, where he joined the Turkoman chief, Kara Mahammed, who supplied him with 5,000 troops. In the battle which followed, both Sheikh Ali and Pir Ali Baduk were killed, and the Turkomans returned home with a great spoil. Sultan Ahmed now returned to Tebriz, but Adil Agha, with Bayazid, retained possession of Sultania.‖ D'Ohsson says it was agreed that Ahmed should retain Azerbaijan, and Bayazid Irak Ajem, while Adil Agha should control Baghdad conjointly with an officer of Ahmed.¶

Sultan Ahmed, as we shall see, invited Shah Shuja, the Muzaffarian, to suppress Agha Adil. He marched against him, and relieved Sultania for a while, when his officers were again driven out, and it presently fell into Sultan Ahmed's hands.

But these arrangements were only transitory, for we have reached the time when the great Timur, with his heavy foot, was marching upon Iran, and scattering the various small chiefs who had divided the Empire of the Ilkhans. He had overrun Kumuz in the autumn of 1384, and ordered the amirs Akbuka and Uchara Behadur to winter at Asterabad with the heavy baggage, and having selected three men out of every ten he marched upon Rai. Sultan Ahmed was then at Sultania, which was

---

* Op. cit., 340-341. D'Ohsson, iv. 746-747. † Quatremere, lvi. ‡ Poole, Catalogue, 213-216
§ Op. cit., 185-186. ‖ Shajrat ul Atrak, 341-342. D'Herbelot, 139. ¶ Op. cit., iv. 747.

put in a state of defence by his orders, while its garrison was intrusted to
his son Akbuka and the Amir Sebtani. Sultan Ahmed himself withdrew
to Tebriz. Thereupon Omar Abbas marched, amidst deep snow, with
60,000 men upon Sultania. The garrison, intimidated by the approach
of this force, determined to abandon the city, and withdrew with the
young prince Akbuka. Omar Abbas, having occupied the place, sent
word by Irmakshi to Timur, who was at Rai.* Timur himself arrived at
Sultania in the spring of 1385, and summoned Adil Agha (also called Sarik
Adil), who had gone to live in the dominions of the rulers of Fars. He
set out at once from Shiraz, and was very well received by Timur, who
confided to him the government of the province of Sultania, with its
dependent districts, after having ordered Muhammed, son of Sultan
Khan, to secure them.† Sultan Ahmed, we are told, was so angry with
Sebtani, who had commanded at Sultania, that he had him ignominiously
paraded about Baghdad.‡ After overrunning Mazanderan, Timur
returned once more to Samarkand. Meanwhile, during the winter of
1385, Toktamish advanced from Kipchak, with an army of 90,000 men,
upon Tebriz, which he captured and sacked, as we have described.§
Thomas of Medzoph says that Toktamish, whom he calls chief of the
Uluz and Azakh, and who dwelt at Serai, near the Krim, sent an envoy to
Ahmed, to propose an alliance. This envoy was treated with indignity,
and returning to his master, tore his collar in his presence, meaning that
he had been insulted, and that the insult called for vengeance. Toktamish
thereupon dispatched one of his officers, named Janibeg, with an army,
by way of Derbend. Ahmed, at its approach, fled to Osdan, and thence
to Baghdad. Toktamish captured Tebriz, after an attack of seven days,
killed and plundered many people, and laid waste the whole province.
Thence the invaders passed towards Nakhchivan and the district of
Siunia, where they committed similar ravages and devastated twelve
cantons. It was then winter, and a heavy snow having fallen, many of
the prisoners captured by the northern invaders escaped.||

The disaster which thus befell Tebriz and its neighbourhood seems to
have affected Timur, and to have induced him to once more march
westwards. His tavachis were duly sent out to summon the troops from
various directions, and having nominated the Amir Suliman Shah, son
of Daud, and the Amir Abbas with two others to rule Transoxiana in his
absence, he crossed the Oxus and arrived at Firuz Kuh, a fortress east of
Rai. This was in 1386. On his march he was joined by Seyid Kayas
ud din, son of Seyid Kemal ud din, Prince of Sari, who enrolled himself
among his officers.¶ On the way Timur, detaching a flying column,
marched at its head against the Lurs, whose turbulence and various
robberies had recently culminated in the plunder of a pilgrim caravan on

---

* Sherif ud din. i. 397-398.   † *Id.*, i. 399-400.   ‡ De Guignes, iii. 291.
§ Ante, ii. 235-236.   ‖ Op. cit., 30-31.   ¶ Sherif ud din, i. 405-406.

its way to Mekka. The line of the Atabegs of Little Lur had survived through the Mongol supremacy in Persia, and the present Atabeg was the Maljk Iz ud din. Timur captured Urudgurd, and also the famous stronghold and seat of the robber tribe Kurramabad, and having captured many of the Lurs and thrown them down from the rocks, he returned and joined his army again at Nehavend.* Having heard that Sultan Ahmed had hastily gone to Tebriz from Baghdad with his troops, Timur determined to make a forced march, and to attack him at once, but on hearing of his approach, Ahmed retired again to Baghdad. Thomas of Medzoph says he withdrew by way of Osdan, a town in the district of Reshduni, in the canton of Vasburgan, south of Lake Van, whence he was conducted by the Kurdish chieftain Iz ud din to Baghdad. Seif ud din was sent in pursuit of him, but Ahmed did not wait to meet him, and fled, abandoning his baggage, horses, &c. ; and, says Sherif ud din, " when our troops had pillaged his baggage, they raised the great cry, ' Surun,' and returned."† Meanwhile, Elias Khoja, son of Sheikh Ali Behadur, having passed Nakhchivan with a small body of cavalry, found himself in the salt marshes of Nemekzar, in the presence of Ahmed, who had retired thither, and who had a larger force than his own. A struggle ensued, in which Elias Khoja was wounded and doubtless defeated. Amidst these events great disturbances broke out at Nakhchivan, and, *inter alia*, we are told that Komari Inak, having a grudge against its governor, set fire to the grand dome of the palace of Ziael mulk, and fifteen persons were suffocated by the burning straw which he placed in it.

•* Having subdued Azerbaijan, Timur encamped at Shenob Gazani, where Seyid Razi Khoja, Haji Muhammed Bendghir Kattat, Kadhi Kayas ud din, Kadhi Abdul latif, and other grandees of the country, went and offered their submission. A sum of money was also exacted from Tebriz, as a ransom for its not being pillaged. Timur spent the summer at Tebriz, and while there he had Adil Agha executed, and his house pillaged. The most skilled artisans in various arts were now transported to Samarkand.‡ Timur now proceeded to overrun and lay waste the mountain districts of Armenia, where the Turkomans and Kurds had become practically independent, but he reserved his most cruel attacks for the Christians of Georgia and the mountaineers of the Caucasus. These campaigns form no part of our present subject, nor do his dealings with Shirvan and Kipchak. They will be dealt with in the next volume.

At this time the Turkomans, as we have said, dominated over the mountainous country of Armenia, as far as the Tigris, and began to exercise considerable influence on the fortunes of Western Asia. They were nominally subject to Ahmed Khan, and their redoubted chief Kara Muhammed, who bore on his standards a black sheep, whence his

---

* *Id.*, i., 405-407.   † *Id.*, 408-409. Thomas of Medzoph, 29.   ‡ Sherif ud din, i. 408-411.

followers were known as the Kara Koinlu, or Black Sheep Turkomans, was, according to D'Herbelot, appointed the commander of Ahmed's troops. Timur now turned his arms against these Turkomans, who were charged with continually assailing the pilgrims on their way to Mekka. This struggle will come more naturally in the next volume. The only other incident, in fact, of the protracted campaign which is immediately interesting to us, is the capture of the fort of Alenjik, which, we are told, was held by one of Sultan Ahmed's lieutenants. This was apparently a great stronghold. The lower citadel was captured, but the garrison withdrew to the upper one. Want of water at length compelled it to offer to surrender, and the besiegers accordingly retired a short distance. Meanwhile a big cloud appeared, and the cisterns were speedily filled again, whereupon the garrison refused to surrender. Timur ordered the siege to re-commence.* At length, in the year 1387, he returned once more to Samarkand. It was not for some years, nor till after he ·had crushed almost every opponent within the old dominions of the Ilkhans, that, in 1393, he again had intercourse with Sultan Ahmed. We read that, having passed the sacred season of Bairam at Akbulak, Timur two days after received the Grand Mufti, Nur ud din Abderrahman Esferaini, who was much renowned for his knowledge, and who went to him from Baghdad as an envoy from Ahmed. He received him with the honours he generally paid to learned men and famous doctors. The Mufti reported that his master was willing to submit himself, but that he was much disturbed at the vast armament which Timur had with him, and which he knew he was powerless to resist, but that he was not willing to go in person to him. The envoy presented the customary gift of nine kinds of each object offered. Among the presents were murkens (a large kind of deer), leopards, and Arab horses with golden saddles. These presents were not what Timur wanted ; this was that the Friday prayer should be made and the coin struck there in his name. He therefore did not receive them cordially, although he treated the Mufti himself with distinction, gave him a rich robe, a costly horse, and some silver plate, and sent him back to his master without a definite answer.† This is Sherif ud din's report. Ibn Arabshah says he sent him the head of Shah Mansur, and demanded his submission. Makrizi says the same, adding that to deceive Ahmed he sent him a robe of honour. He also says that Timur expressly promised not to attack Baghdad.‡

Timur had, however, made up his mind to conquer Irak Arab, and having sent his heavy baggage and harems back to Sultania, in charge of Mirza Pir Muhammed Jehanghir, he ordered each soldier to furnish himself with two skins of water, and set out on the 1st of October, 1403, for Baghdad. By forced marches he reached Yanbulak, near Irbil, and

---

* Sherif ud din, i. 430-431.   † Id., ii. 219-221.   ‡ Weil, v. 41.

Kura Kurgah, near Shehrezur, in Kurdistan, and defeated the Turkomans under Kara Muhammed near the latter place, where he halted. Selecting a number of picked men, he advanced by broken paths and defiles. He was carried in a litter himself, while torches lighted the path, which was traversed in all haste, and the main army had difficulty in following him. Having reached Ibrahim Lik, a sacred site some 27 leagues north-east from Baghdad, and asked the inhabitants if they had not informed the people there by pigeon post of his approach, they could not deny that on seeing the dust raised by his troops some pigeons, with notes tied to their wings with silk, were sent to inform the citizens. Timur thereupon obtained another pigeon, and caused the same people to write another letter, stating that the dust was not due to his army, but to a body of Turkomans, who had been mistaken for Timur's people. On the arrival of this note Sultan Ahmed was reassured, but he had already taken the precaution to place his more valuable property in safety on the other side of the Tigris. Having paid his devotions and distributed alms at the tomb of the saint at Ibrahim Lik, Timur set off again, and covered without dismounting the twenty-seven leagues, each of 3,000 paces, separating it from Baghdad, where he arrived on the morning of the 30th of August, 1393.* The citizens, fearing a repetition of the horrors of Khulagu's campaign, opened the gates of the town. Meanwhile Ahmed had withdrawn across the Tigris. He had broken the bridge and sunk the boats there, or conveyed them to the other side, where he intended to remain till Timur's withdrawal, or till he was discovered. Finding his whereabouts was known, he retired towards Hillah. Timur's men, who extended for two leagues on each side of the town, plunged into the river. Ahmed's royal galley, which he had named Shams, or The Sun, was captured, and Timur crossed the Tigris in it. Mirza Miranshah, with his men, swam over opposite Kariatula Kab (*i.e.*, the village of the eagle), situated below the town. "They spread over the land," says Sherif ud din, "like ants or locusts, and the people of Baghdad were as much surprised to see these Jagataians swarming over the river, as their ancestors, the citizens of Babylon, had been with the confusion of tongues, and bit their fingers in wonder." Timur wished to go himself in pursuit of Ahmed, and having passed Serser, three leagues from Baghdad, the first stage made by the pilgrims *en route* thence to Mekka, went as far as Karbaru, where he was entreated by Aibaj Oghlan and the noyans and other chiefs to return to Baghdad and rest himself while they went in pursuit of his enemy, whom they promised to bring back bound hand and foot. He did so, and put up in Sultan Ahmed's palace, many of whose rare treasures had fallen into his hands. His officers marched day and night, and the day but one following reached the Euphrates, and learnt that Ahmed

---

I Sherif ud din, ii, 222-224. Weil, v. 41.

had already crossed it the night before, after breaking the bridge and
sinking the boats, and had gone on towards Damascus by way of Kerbela.
Osman Behadur was for crossing the river there and then in pursuit, but
the other chiefs advised that they should march along the bank until they
found a ford, where there would not be any risk. This advice was
followed, and presently four empty boats were found. The amirs got on
board, while their horses swam over beside the boats. The army imitated
their example, and having remounted on the further side, sped on again.
They overtook the Sultan's baggage and his tents, and secured all the
money and other valuables which he had carried off, and now deemed
it prudent to abandon. The Mirza Miranshah halted at Hillah, and sent
on Aibaj Oghlan (a descendant of Juchi, Khan of Kipchak), Jelalhamid,
Osman Behadur, Sheikh Arslan, Seyid Khoja (son of Sheikh Ali
Behadur), and other amirs—forty-five in number—in pursuit. They were
compelled to advance alone, since the horses of their troopers were
broken down with fatigue, and required rest. They came up with
Ahmed in the plain of Irbil. He had 2,000 men with him, of whom
200 turned to meet the pursuers. The amirs dismounted and fired
showers of arrows, and, having thus compelled them to retire, mounted
again, and again pursued them. Again the amirs were charged, and this
time sheltered under their horses, whence they poured in their arrows.
The third time the manœuvre was tried the amirs had not time to
dismount, and a terrible hand-to-hand scuffle ensued, in which many on
both sides were killed. Osman Abbas fought very bravely, and had his
hand disabled with a sabre cut. Sherif ud din makes out that the
victory rested with the amirs, but tells us they ceased their pursuit. This
struggle, in which a small body of superior officers alone were engaged
on one side, if it really happened, as the panegyrist of Timur just quoted
relates, has a certain special interest, but it sounds like a great exaggeration.
The day of the struggle, he tells us, was very hot, and the plain of Kerbela
being destitute of water, the amirs were afraid of perishing from thirst.
Each one tried to find some. Aibaj Oghlan and Jelalhamid sent people
out, but they only found two pots of water, with which they returned.
The former emptied one without having his thirst quenched, and then
declared he should die of thirst if his companion did not give up his
portion (i.e., the other pot) to him. Jelal replied that a certain Persian
having on one occasion been travelling with an Arab, a similar mischance
befell them in a desert that had now occurred. The Arab still had a
little water, and the Persian said to him : "The generosity of the Arabs
is so famous that it has become proverbial everywhere. It would be a
great proof of this truth if, to save me from certain death, you were to
make over to me the water which you still have." The Arab having
thought a little, said : "If I give you the water it is certain I shall die of
thirst, nevertheless this dire necessity will not justify me in transgressing

the prerogative of the Arabs. I prefer a good reputation to this perishable existence. I prefer to hazard my life, and to sacrifice my life, by giving up this water which I have saved, so that the virtue of the Arabs may still be universally renowned." The Arab accordingly gave the Persian the water. Jelal having told the story, continued : " I wish to imitate the Arab on this occasion, and will give you the water on condition that you make known to the princes of the house of Juchi and their subjects this sacrifice, so that the memory of this deed may always redound to the credit of the descendants of Jagatai Khan, of whom I have the honour to be one, and on condition also that you will tell it when you reach the camp of his highness, so that the event may be recorded in history and be cited as a proof of my courage to all our descendants." Aibaj Having consented to these conditions, Jelal took some bystanders to witness, and surrendered the water, which his companion consumed. He did not die, however, but survived. The amirs having retraced their steps, reached the Euphrates at Makad, where Hussein, son of Ali, was killed. Each of them kissed the portal of the holy place, and went through the ceremonies usual with pilgrims. They took back with them Ala ud daulat, son of Sultan Ahmed, and others of his children, together with his wives and servants. Aibaj and his companion, the amir Jelal, related the story of the water, which pleased Timur greatly, and he poured out his praises, not only on Jelal himself, but on his father Hamid.*

During this campaign the Mirza Muhammed Sultan had a savage campaign among the Kurds, after which he joined Timur at Baghdad, and was appointed governor of Wasith, a town on the Tigris, and of the province dependent on it.† Timur gave orders for the transport of the wives of Sultan Ahmed and his son Ala ud daulat to Samarkand, with all the learned men of Baghdad, and the other proficients in art and science there. This order was duly executed, and the Khoja Abdul Kader, author of a famous work on Oriental music, was among those thus transported. Letters with the news of the capture of Baghdad were sent to Samarkand, Kashgar, Khoten, Khuarezm, Azerbaijan, Persia, Irak, Khorasan, Zabulestan, Mazanderan, Tabaristan, &c. Timur spent two months at Baghdad in various festivities. As a rigid Mussulman he ordered all the wine that could be found in the place to be poured into the Tigris. The people of the city had to pay the usual ransom in consideration of its having been spared from pillage.‡ He confided the city of Baghdad to the Khoja Masaud Sebzevari.§ Having secured Wasith and Hillah, he ordered one of his sons (Sherif ud din says it was Miran Shah) to march on Basrah. Its governor was Salih Ibn Julun. Makrizi says a son of Timur was captured by the garrison, while Ibn Iyas says that the titular Grand Khan of Jagatai, Mahmud Shah, was killed, and a son

---

* Sherif ud din, ii. 221-234.   † Id., 235-236.   ‡ Id., 235-238.   § Id., 259.

of Timur was captured. The latter sent to demand the surrender of his
son. The governor replied that he would not release him until Timur had
released Ahmed's son. Timur thereupon sent fresh troops against the
place, which were again defeated, and he had to postpone its capture till
the winter.* He now proceeded to subdue a number of strong places
in Mesopotamia and Armenia, each governed by its petty ruler. The
details of this campaign, in which Tekrit, Irbil, Mosul, Mardin, Amid,
Mayafarkin, Belbis, &c., were subjected, will more fitly occupy us in the
next volume. He punished the Turkomans of Kara Yusuf for their
depredations, and captured their stronghold of Van.†

When Timur captured Baghdad he sent Sheikh Shah as his envoy to
the Egyptian Sultan, with some Mongol companions and a grand
equipage, reminding him of the long struggle which had been waged
between the Mongols and the Egyptians, which had terminated with the
peace made with Abusaid, and how, after the latter's death, all kinds of
misrule and usurpation had ensued in Persia, where there no longer
reigned a sovereign of the stock of Jingis Khan, but the various governors
had usurped authority, and caused great troubles. To put a stop to this
he had been permitted by Providence to conquer Iran, as far as the
limits of Irak Arab and the borders of Egypt, and in order to complete
the good work he asked that there might be friendly intercourse between
them, and that the merchants of each country might be allowed to
freely traffic in the other.‡ The Sultan put the envoys to death. He
was not wishful for intercourse with such a dangerously treacherous
neighbour, and was afraid they might be spies. He also gave orders to
the governor of Aleppo to welcome the fugitive successor of the Ilkhans,
Ahmed (who, as we have seen, had fled to Syria), as a prince, and to
supply all his needs. He also sent his chief cupbearer, Az Timur, to
conduct him to Egypt. As he approached Cairo the chief amirs went out
to meet him, and conducted him to the Sultan, who rose to receive him,
embraced him, and would not permit him to kiss his hand. He also gave
him some rich robes, and a choice horse with a golden saddle, and rode
by his side through the city. At the citadel he left him to be conducted
by the amirs to the palace set apart for him, where a lordly meal had
been prepared, and he afterwards sent him a costly present of horses,
richly caparisoned, Circassian mamluks and slave girls, costly stuffs, and
5,000 dinars. The Sultan also married his niece, daughter of Hussein
ibn Oweis, but presently put her away, as he heard that she had previously
been courted by her cousin, Jelal ud din Hasan ibn Oweis. These
civilities towards his enemies and truculence towards his envoys was not
likely to be forgiven by Timur, who had already placed his heavy foot on
the Prince of Mardin, and on hearing that his army was *en route* for

---

* Weil, v. 42. Note 1.   † *Id.*, v. 42-46.   ‡ Sherif ud din, ii. 238-241.

Rabah, the Egyptian Sultan in turn set out, and presently repulsed some of Tímur's people near Basrah ; while Ahmed, who had hitherto accompanied the Sultan, invaded Mesopotamia with a suitable force, in conjunction with the Bedouin chief Nuyir, in order to once more occupy Baghdad on behalf of the Egyptian Sultan.*

Meanwhile Timur had been prosecuting his campaign against Tok-tamish, who was in alliance with Barkuk, but, as usual, derived small advantage from his Egyptian friend. In the spring of 1396 he returned from this campaign by way of Derbend, Shirvan, Irbil, and Sultania, and having reached Hamadan, he appointed the Mirza Miran Shah prince of Azerbaijan, Kuhistan, and the Arabian Irak, including all the country from Baku, on the Caspian, to Baghdad, and from Hamadan to the Ottoman frontier, with his capital at Tebriz.† The Bedouin chief Nuyir soon succeeded in occupying Baghdad, and the Friday prayer was said there in the name of the Egyptian Sultan, who duly appointed Ahmed as its governor ; but this arrangement lasted only a short time, since Barkuk died on the 20th June, 1399, and in the confusion which ensued in Egypt Ahmed secured his independence. On another side, the Turkoman Kara Yusuf won several victories over the people of Timur, re-occupied the town of Van, and captured Atilmish, who was a connection of Timur's, having married his niece, and who was sent as a prisoner to the Egyptian Sultan. Mosul, Sinjar, Yezd, and Neharvend also broke away from their allegiance. Meanwhile Miran Shah, who had remained at Tebriz during Timur's campaign in India, having had a serious accident, •became partially demented, and very tyrannical. We are told that for a simple suspicion he would put a man to death, while he dissipated his treasures, destroyed wantonly some of the most famous buildings there, &c.‡ In the early summer of the year 802 (1399) he tried to retake Baghdad. He marched day and night, and made two marches in twenty-four hours, fancying the very mention of his name would strike terror into the Sultan and cause him to abandon the place. When he reached the Dome of Ibrahim Lik, situated twenty-seven leagues from Baghdad, he heard that a conspiracy had broken out among the chief people of Tebriz against him. He nevertheless continued his march. Sultan Ahmed, who knew that the season was unfavourable for siege operations, on account of the drought and extreme heat, resolved to defend the place. Miran Shah had only been two days there when the news from Tebriz became so threatening that he was constrained to return thither post haste, and put to death the principal conspirators, including the Kadhi. He now had to face a revolt of the Georgians. Sultan Ahmed's son Tahir was in command of the fortress of Alenjik, which had been besieged for a long time by Sultan Sanjar, son of Haji Seif ud

---

* Weil, v. 48-55.   † Sherif ud din, ii. 390-391. Weil, v. 56.   ‡ Sherif ud din, iii. 190-191.

din, by order of Timur, who, to make the blockade more complete, had girdled it round with a wall. The Georgians marched to make a diversion in Tahir's favour, and laid waste Azerbaijan. This led to the raising of the siege. They now advanced on Alenjik, and were joined by Tahir, who left the place in charge of Haji Salih and three famous Georgian oznaurs or princes.* These disorders compelled Timur, on his return from his fatiguing campaign in India, to at once face another expedition. He ordered Mirza Rustem, his grandson, with the amirs Sevinjik, his nephew Hassan Jandar, and Hassan Yagadaul, who were at Shiraz, to advance upon Baghdad. They ravaged Luristan *en route*, and reached Mendeli in the month Yumazinlevel (802 HEJ). Mendeli was governed by the Amir Ali Kalander, on behalf of Sultan Ahmed. The place was taken and pillaged.† Mir Ali Kalander escaped, and repaired to his master to report to him what had happened. The latter closed the city gates and broke down the bridge. At this time Shervan, who governed Kuristan on behalf of Timur, having misappropriated its revenues and squeezed large sums of money out of the principal inhabitants there, went to Baghdad with a thousand well-armed followers. He was well received by Sultan Ahmed, but proceeded to corrupt the Sultan's officers with various bribes, from 3,000 to 10,000 dinars of Baghdad. The receipts for some of these payments having fallen into the hands of Kaureh Behadur, one of Ahmed's officers, he reported the matter to his master. This was at the time when he had closed the gates of the place, as we have mentioned. He was greatly disturbed, and himself decapitated Rafeh, one of the men who had been bribed. Shervan, with Kutub Haidari, Mansur, and other amirs, had been dispatched to Uirad. He now sent Yadghiar Ektachi with orders to the amirs to put Shervan away, which was accordingly done, and his head was sent to Baghdad, where the suspicious Sultan in less than eight days put nearly 2,000 of his officers to death. He sent the Khatun Vefa, who had been his foster mother and had brought him up from infancy, to Wasith, where he had her suffocated with a pillow. Having killed with his own hands the other women and officers of the household, and had them thrown into the Tigris, he closed himself within the doors of the seraglio, and would allow no one to enter, and when the bavarchis or butlers brought his food they shouted at the door, where *they deposited the plates, and withdrew without entering. Having passed some days in this fashion, he ordered five or six of his most devoted followers to provide six horses from the stable and to have them conveyed across the Tigris. He himself crossed the river one night in a boat, and having reached his faithful followers who had charge of the horses, fled with them to the district dominated by the Turkoman chief.*

---

* Sherif ud din, iii. 191-194.	† *Id.*, 229-231.

Kara Yusuf. The butlers continued to take the food as usual, and handed it to an officer who was in the secret, without the people outside knowing what had happened. Meanwhile he felt he could not rely on the citizens, and sought out Kara Yusuf, and persuaded him to go and plunder Baghdad, himself acting as guide. When the Turkomans neared the city they were assigned quarters on the other side of the river. The Sultan crossed the river in a boat and entered his palace, and lavishly rewarded them with presents of money, rich stuffs, arms, Arab horses, golden girdles, &c. His real object was, no doubt, to secure protection, and we are told he did not allow his guests to injure the place. At this time Timur was advancing upon Sivas, and Sultan Ahmed, who was well informed of his movements, fearing that if the great Conqueror entered Anatolia and Syria he would cut off his own retreat, determined to leave Baghdad in charge of Farruj, and taking with him his wives and children, and all his treasures, he left Baghdad with Kara Yusuf. They passed the Euphrates and went towards Aleppo. Its governor, Timur-tash, went out to oppose them, but he was defeated, and they thereupon continued their march with their Turkoman followers, but having been defeated by the Bedouin chief, Nuyir, and the governor of Behesna, they turned aside towards Anatolia.* De Guignes reports from the " Tarikhi Bedr ud din " that after defeating the governor of Aleppo they laid waste all the district of Tel bashr and Aintab.† These events took place in the summer of 802 HEJ. Timur being on the Georgian frontier, sent to demand from the Georgian king the surrender of Tahir, son of the Sultan, who, as we have seen, had been rescued by the Georgians at Alenjik. To this message a defiant reply was apparently sent, and the consequence was a cruel invasion of Georgia.‡ When Timur heard that the Sultan and Kara Yusuf had gone to Anatolia he dispatched some cavalry after them, which plundered their baggage and, *inter alia*, captured Sultan Dilshadeh, the elder sister of Kara Yusuf, with his wives and daughter. Kara Yusuf himself escaped, and joined the Ottoman chief Bayazid.§

In 803 HEJ. Timur, having laid waste a good deal of Syria and the country by the Tigris, sent a fresh army to attack Baghdad, where Farruj still commanded on behalf of Sultan Ahmed. The nominal chief of the Jagatai dominions, Mahmud Khan, whom another account makes to perish in an earlier campaign, the Mirza Rustem, and the amir Suliman Shah were selected to lead it. They speedily reached the city and encamped on its southern side. Farruj, inspired by the great number of Turks and Arabs who garrisoned it, had the temerity to march out and offer battle to the invaders. He was assisted by the amirs Ali Kalander, of Mendeli, and Jian Ahmed, of Baku (both towns of Kuristan), who crossed the Tigris at Medaineh. On another side, Farruk Shah, of

* *Id.*, 233-238. Weil, v. 79. † Op. cit., iii. 296. ‡ Sherif ud din, iii. 241. § *Id.*, 268-269.

Hillah, and Mikail, of Sib, set out on the same errand, and reached Serfer with 3,000 well-armed men. These allies were sharply attacked by the Timurid chiefs, and Jian Ahmed, with many of his men, perished in a fight at a building known as that of the Amir Ahmed. Many threw themselves into the water and were drowned, while others escaped with difficulty. Farruj was not discouraged by this defeat. " Sultan Ahmed, my master," he said, "made me swear that if Timur came in person I was to surrender the town to save the people from misfortune ; but if he does not come in person, however great an army assails it, I will defend it courageously against everyone. I cannot disobey my master's orders." He inspired the garrison with his own spirit, and they not only manned the walls, but also poured a shower of weapons upon the enemy from the boats in the river. When Timur learnt of the determination of Farruj, and its pretext, he selected the bravest of his people, left the Empress Chelpan Mulk Aga, with the rest of the army and the baggage, in charge of Shah Rokh, and ordered him to go to Tebriz, while he himself set out for Baghdad by way of Altun Kupruk, a bridge over the river Altun Sui, near Mosul. Having reached the city, he posted his camp opposite the gate of Kariet Ulakab, while his people surrounded the place. He gave orders that the sappers were at once to begin. Farruj, wishing to know if Timur was there in person, sent a confidential person, who knew him, as his envoy. He was well received, and presented with a robe, &c. On his return he reported faithfully what he had seen and heard, but Farruj pretended not to believe him, ill-used him, and for fear he should speak to others in the same strain, imprisoned him. He persuaded the garrison that Timur was not present. Meanwhile the Khojas Masaud Semnani and Mengheli, who were engaged in raising a platform whence to command Baghdad, were both killed by weapons from the place. Timur now sent word to Shah Rokh to join him at Baghdad with the baggage, and on his arrival he held a grand review. Although the city was more than two leagues in circumference, it was beleaguered on all sides, a bridge was made over the Tigris below Kariet Ulakab, while skilled archers were posted in ambush upon it to prevent people escaping. So carefully was guard kept, that although the Tigris flowed through the city and many ships were upon it, no one could escape. At this time the fortress of Alenjik having been captured, its governor, Ahmed Ogulshai, was taken to Timur, who had him put to death. The garrison fought desperately. As soon as the invaders had breached the wall and caused part of it to fall, it was repaired with lime and burnt bricks, and an entrenchment made behind. It was summer, and the heat was so great that the birds fell down dead, "and as the soldiers were encased in their cuirasses," says Sherif ud din, "they might be said to have melted like wax." Presently a great platform was raised, which commanded the city, whence an incessant bombardment with great stones was kept

up. The mirzas and amirs several times asked permission to be allowed
to storm the walls sword in hand, but Timur refused, as he hoped the
garrison would repent and surrender, and thus save the place from
destruction. The siege had lasted forty days, and want began to be felt,
but this did not constrain the defenders. On the 9th of July, 1401, when
the citizens had gone for shelter from the sun's rays into their houses, and
left the walls, having placed their helmets on rods which they had
erected, Timur deemed it a favourable opportunity for a general assault.
Scaling ladders were attached to the walls, and the Sheikh Nur ud din was
the first to mount, and planted there a horsetail standard, surmounted by
a crescent. The big trumpet was sounded and the drums and cymbals
beaten. A general rush was thereupon made, and the walls were pushed
over into the ditch. The troops entered the place sword in hand, while
Timur watched the proceedings from the bridge he had built over the
Tigris. All the means of exit had been closed. Many threw themselves
into the river ; others seized boats, but when they reached the bridge the
archers who were in ambush speedily destroyed them. Farruj and his
daughter managed, nevertheless, in spite of all Timur's efforts, to escape
in a boat. They were pursued along the banks by the soldiers, who
poured in upon them a shower of missiles, and he was at length forced
to throw himself into the water with his daughter, when both of them
were drowned and the boat was sunk. His body was fished up and
thrown on the bank. As the besiegers had lost so many men, each
soldier was ordered to bring the head of a citizen of Baghdad, and neither
the old people of eighty nor children of eight were spared ; quarter was
given neither to rich nor poor, and the slaughter was so great that no one
ever knew the number of victims, although the tavachis were ordered to
furnish an account. "Their heads," says the panegyrist of Timur, "were
built up into towers as a warning to posterity, and so that men should not
raise their feet higher than their capacity." Only a few learned men
secured pardon. They were given robes and escorted to a place of safety.
All the rest of the inhabitants were exterminated. Orders were given
that every house was to be destroyed, but (and it is a curious proof how
religious fanaticism in all climates discriminates on these occasions) the
mosques, colleges, and hospitals were to be spared. In consequence of
this order the markets, bezestins (?), caravanserais, hermitages, cells,
monasteries, palaces, and other buildings were laid low. "It is thus,"
says Sherif ud din, "according to the Koran that the houses of the
wicked are overturned."* It is appalling to think that such destruction
should have overwhelmed for a second time the old capital of the Khalifs,
and this time, not at the hands of an unbeliever, who had learnt his war
policy in the far-off steppes of Mongolia, but of a good Mussulman, who

---

* Op. cit., 361-371.

had been brought up amidst all the amenities of culture. . "After the Tigris was reddened with the blood of the inhabitants of Baghdad," says Sherif ud din ambiguously, "the air began to be tainted with their corpses," and Timur moved away to pay his devotions at the tomb of the Imaum Abu Hanifa, founder of one of the four orthodox Muhammedan sects. He then ordered Sultan Muhammed Khan and Mirza Khalil Sultan to ravage the district round, which they effectually did, including Hillah and Wasith in their raid, and then rejoined him. He thereupon withdrew again.

During this campaign Sultan Ahmed and the Turkoman chief Kara Yusuf were living with the Ottoman Sultan Bayazid, with whom they had found shelter, and whom they persuaded to attack Erzenjan, in revenge for the reverse Bayazid had sustained at Sivas. He captured this place, and brought upon himself ample revenge presently. When Sultan Ahmed learnt that Timur's armies had withdrawn from Baghdad and were being directed towards Turkey, he left Bayazid at Cæsarea of Cappadocia, and returned to Irak Arab by way of Kalat Errum, along the Euphrates. He went to Hit, and thence to Baghdad, where he set about rebuilding the place and collecting such of his people as he could bring together. When Timur heard of this he determined to at once stamp out the process, and ordered four divisions of cavalry to march upon the city by four separate routes. The Mirza Pir Muhammed, with one division, marched by way of Luristan, Kuzistan, and Wasith; the Mirza Abubekr advanced straight upon Baghdad; the Mirzas Sultan Hussein and Khalil Sultan were to occupy certain districts in Chaldea; while the Amir Berendah was told off to attack Jeziret ibn Omar, a town on the Tigris. The troops set out with alacrity, although it was mid-winter. The Mirza Abubekr, having with him the Amir Jihan Shah as his lieutenant, arrived suddenly one evening before Baghdad. Sultan Ahmed was so taken by surprise that he had only time to reach a boat in his shirt, and thus crossed the river with his son, Sultan Tahir, and the officers of his household, with whom he took horse and went towards Hillah. Having rested his troops for the night, Abubekr the following morning dispatched Jihan Shah in pursuit of the Sultan as far as Hillah, where Ahmed broke the bridge and then sought shelter in the islands of Khaled and Malek, on the Lower Euphrates, where Jihan Shah did not care to follow him. Meanwhile, the Mirzas Sultan Hussein and Khalil Sultan marched by way of Shepshemal, and pillaged Mendeli, whence the Amir Ali Kalander, who governed it for Sultan Ahmed, fled, crossed the Tigris with some troops, and made a stand on the other side. Mirza Khalil Sultan dispatched a body of picked men to cross the river by swimming at another point, and to attack Ali Kalander in rear, while he traversed the river opposite to him, and thus attacked the enemy on both sides. This was successfully accomplished. Most of them were killed or captured ; among the latter was Ali Kalander, who was burnt alive " for having had the audacity to defend himself."* Timur having collected his forces for an attack upon Bayezid, Ahmed seized the opportunity, re-occupied Baghdad, and drove out his deputies in Chaldæa. The Turkoman Kara Yusuf also returned to Mesopotamia and collected his friends.

We now read that Sultan Tahir, in conjunction with one of his officers, Agha Firuz, formed a conspiracy against Sultan Ahmed, which included Muhammed Bey, governor of Ormi, the Amir Ali Kalander (who is stated elsewhere by Sherif ud din to have been previously burnt alive), Mikail, and Farrak Shah. They crossed the river together by the bridge during the night, and encamped on the further side. Sultan Ahmed broke the bridge, and dispatched a messenger to summon Kara Yusuf to his assistance. The Turkoman chief accordingly went with some troops, and having united them with those of the Sultan they attacked the rebels together. Tahir was speedily defeated and fled, and in trying to jump his horse across a ravine they fell together to the bottom and he was killed. His people dispersed.† Sultan Ahmed soon quarrelled with his doubtless exacting ally, Kara Yusuf, and repaired to Baghdad. Thither he was pursued by the Turkoman chief, and was obliged to hide himself. Kara Hassan, a faithful dependant, having found him, carried him off on his shoulders, and conveyed him to a spot five leagues off. *En route*, they found a man who had a cow, on which the Sultan mounted and went to Tekrit with Kara Hassan. Sarik Omar Uirad made him a present of forty horses, with all the money, arms, stuffs, and girdles he had. He was then joined by several of his officers, as Sheikh Maksud, Dolat Yar, Adil, and others. From Tekrit he went to Damascus, and Kara Yusuf, his former dependant, became undisputed master of Irak Arab. This was only for a short time, however, for Timur nominated the Mirza Abubekr as governor of that province, as far as Vassit, Basra, Kurdistan, Mardin, Diarbekr, Uirad, and their dependencies, with orders to rebuild Baghdad.‡ In conjunction with the other Timurid princes, he attacked Kara Yusuf on the River Nahreljanam, beyond Hillah (Weil says by Sib, near Hillah), and completely defeated him. All his baggage and flocks were pillaged, his chief wife and the other women of the harem were captured, his brother killed, and he himself compelled to find shelter, like Sultan Ahmed, in Syria, under the ægis of the Egyptian Sultan.§

Abubekr now set to work to rebuild Baghdad, and to restore prosperity to the other towns of Irak Arab.‖ This was during the year 806 HEJ. Soon after, Mir Eluerd, a son of Sultan Ahmed, who was eighteen years old, was removed from Irak Arab and taken to Samarkand.¶ The Egyptian Sultan, afraid to embroil himself with Timur, arrested the two fugitives,

---

* Sherif ud din, iii. 387-391.   † *Id.*, iv. 94-95.   ‡ *Id.*, 93-96.   § *Id.*, 127-129.
‖ *Id.*, 129-130.   ¶ *Id.*, 135.

U *

and sent to ask him what he should do with them. Timur replied in a long letter, written in letters of gold by the Mulana Sheikh Muhammed, famous as a scribe. In this note he asked that Sultan Ahmed might be sent to him bound, while the head of Kara Yusuf should also be sent. With this letter he also forwarded some rich presents.* The Sultan, unwilling to break the laws of hospitality, and wishful at the same time to please Timur, put them under arrest, but not so that they could not communicate together. They thereupon formed a close alliance, promising each other that on regaining their liberty they would help one another. This good fortune came to them on Timur's death, on the 17th of February, 1405. When news of it reached him, the Sultan paid the two some attention, and gave them their liberty. Kara Yusuf having arrived at home, put himself at the head of his Turkomans, and speedily secured the greater part of Chaldæa and Mesopotamia, acting apparently on behalf of Sultan Ahmed, against whom the Egyptian Sultan was much irritated, and to whom he complained bitterly, having expected probably to be treated as suzerain. Not receiving any satisfaction, he withdrew the protection he had given him. Sultan Ahmed did not lose courage, but having dressed a number of his supporters as beggars they entered Baghdad quietly, and excited a seditious movement against the governor, who held the town on behalf of Omar Mirza. He was eventually driven out, whereupon Ahmed was duly proclaimed Sultan by the people. In the latter part of the year 808 HEJ., while Abubekr Mirza, grandson of Timur, was engaged in the siege of Ispahan, the amir Ibrahim went from Shirvan and occupied Tebriz.† Sultan Ahmed thereupon himself advanced upon that city, where the citizens were growing tired of the turbulence of the Timurids. Sheikh Ibrahim, hearing of Ahmed's approach, said : " For a long time we were attached to this august family. As the country was without a master, and the people were tormented by the ambition of those who wished to attack it, we came to its succour ; now that the real ruler is at hand we will return home." Sultan Ahmed reached Tebriz in the middle of Moharram. The chief people of Azerbaijan and the citizens of Tebriz were greatly delighted, and the place was *en fête*. They renewed their oaths of allegiance. Sheikh Ali Uirad and Muhammed Saru Turkoman made him a present of some horses. Khoja Muhammed Kejeyi, the amir Jafar, Khoja Masud Shah, and Zain ud din Kazvini, were appointed heads of the administration. Meanwhile, the Timurid mirzas, Miran Shah and Abubekr, advanced upon Tebriz. Sultan Ahmed, afraid to meet them, went to Aujan, and having consulted with his amirs, determined to return to Baghdad. He accordingly set out, and the troops of Azerbaijan dispersed. Mirza Abubekr speedily reached the city, but would not enter it, as the plague

---

* Sherif ud din, iii. 202-203.   † D'Herbelot, sub voce Avis.

was then raging. He then went to Nakhchivan, and summoned Malik Iz ud din from Kurdistan. They agreed to jointly attack the Turkoman Kara Yusuf. In the fight which took place on the Aras or Araxes, the allies were beaten. Abubekr retreated. On his way his demoralised troops pillaged Tebriz. Abubekr continued his retreat to Sultania, where he passed the winter.* Kara Yusuf now occupied Tebriz, and became the undisputed master of Azerbaijan. Thence he marched on Sultania, which he captured, and carried off its inhabitants to Tebriz, Meragha, and Irbil. He thus added Irak Ajem to his rapidly growing dominions. All this was doubtless very unwelcome to his former suzerain, Sultan Ahmed.

During the year 810 HEJ. Sultan Ahmed devoted himself with great energy to fortifying Baghdad, building up its ramparts and deepening its ditches. Towards the end of the year his son Ala ud daulat Sultan, having been released from the prison where he had been confined at Samarkand, arrived in the neighbourhood of Baghdad. When he saw his son, who was so dear to him, approaching, he descended from his horse, and they embraced. The two went into the city together and spent some days in feasting. Some time after, Ahmed sent him to Hillah, where he might enjoy himself to his heart's content, while he gave himself up to wine and debauchery. On the last day of the month Shaban, Ala ud daulat had a son, to whom he gave the name of Sheikh Hasan. Ahmed went to Hillah to join in the rejoicings at this event, and then returned to Baghdad, where he passed the winter. In the spring he sent for his son, and the two re-commenced their pleasure making.†

We now read how the Timurid prince, Mirza Pir Muhammed, marched to repress a revolt in Khuzistan, and afterwards invaded Beyat, which belonged to Baghdad. He was obliged to withdraw from it on account of the fevers prevailing there.‡ Sultan Ahmed, on hearing of this, himself set out, and suddenly appeared before Havizah. The garrison were stupefied, took to horse, and fled, but they were for the most part overtaken and killed. Pir Haji Kukeltash, hearing of the fate of Havizah, abandoned Dizful and retired towards Shuster, whose governor was so afraid that he even doubted the resources of the citadel of Selasil, and retired hastily upon Ramhormuz. Mirza Pir Muhammed, having heard of this invasion, sent Tulek and Nushirvan Borlas to help the amirs of Khuzistan. They went to the governor of Shuster, who had fled to Ramhormuz. Meanwhile Ahmed, master of Khuzistan, marched upon Ramhormuz. The place resisted for a while, when the garrison fled, and he entered it, killed the Kadhi, Kotb ud din, who had been the soul of the defence, and a great number of foot soldiers, and carried off the rest of the inhabitants to Shuster,

---

after razing the fortress of Ramhormuz. Meanwhile, he learnt that his son, Sultan Ala ud daulat, had revolted, and had marched by way of Hillah and Irbil towards Azerbaijan, which was subject, as we have seen, to the Turkoman Kara Yusuf. When Kara Yusuf heard that he had gone thither without his father's consent, he assigned him a certain moderate sum for his daily expenses, and after the fast of Ramazan ordered some of his people to re-conduct him home again. But when in the neighbourhood of Khoi, the Kurds subject to Malik Iz ud din arrested him, and took him to their master, who detained him for two months, treating him with great consideration. Kara Yusuf wrote to complain of this, and said that he himself having found that the young man was acting without the approval of his father, had expelled him from his dominions, and bade Iz ud din do the same if he did not wish to regret it. Unable to resist, he was obliged to obey this notice, and said goodbye to his guest, who, bereft of resources, went to Irbil, to pass some days with the anchorite, Sherif ud din Ali Sefevi. When he arrived near Tebriz some perverse people fell on him and put him in chains. Kara Yusuf now had him sent to the fortress of Abd Eljuz.*

About the year 812 HEJ. Sultan Ahmed, after having conquered Khuzistan, left Maksud Nizehdar in command there, and returned to Baghdad. Having fallen ill towards the end of winter, he was recommended to remove to his summer quarters, and went on a pilgrimage to the tomb of the Sheikh of Sheikhs, Oweis Karni. His health being completely restored, his amirs pressed him to go to Sultania, urging that Bestam Jaghir, who commanded there, had left the place for Ispahan, leaving his young son Bayazid in charge of it, who would at once abandon the fortress on his approach. Khurrem Shah Derguzmi specially urged this policy. Ahmed accordingly set out. When he reached Hamadan he learnt that Bestam had returned to Ispahan, where having heard of the Sultan's plans, he had put his brother Masum at Sultania, while he himself had gone to Irbil. The Sultan now took up his residence in the kuruk of Sultania. The commanders of the neighbouring districts, and especially Abder Rezak, who had forcibly seized upon Kazvin, submitted to him, and he conferred honours on them worthy of a king. He then sent a *message to Masum, with fair promises, but they were not heeded. Meanwhile, he heard that a man named Oweis, at Baghdad, professing to be his son, had risen in revolt, having been encouraged to mount the throne by those about him. He hastily returned thither. At his approach the conspirators scattered. Oweis and his principal supporter, Mani ibn Shoaib, were captured.

At this time the Turkoman chief, Kara Yusuf, had a successful campaign against Kara Osman, during which he secured Mardin, and afterwards made a raid into Shirvan. Sultan Ahmed sent to ask if he would allow him to have his summer quarters in the district of Hamadan, a request which Kara Yusuf refused. This no doubt irritated Ahmed, who was the heir to the Ilkhans and formerly suzerain of the Turkoman chief, and soon after, while Kara Yusuf was engaged in an attack upon Erzenjan, he determined to march upon his capital, Tebriz.[*] According to Ibn Hajr, he was in alliance with the Timurid chief, Shah Rokh. These events all happened in 813 HEJ. He left Baghdad on the 12th Moharrem, 813 (*i.e.*, 17th May, 1410), *en route* for Azerbaijan, and having reached Hamadan he subdued the Kurds there. Shah Muhammed, son of Kara Yusuf, who was in his yilah or summer quarters at Aujan, set out for Khoi. Ahmed reached Tebriz on the 26th of Rebi the first (29th July). He sent troops in pursuit of Shah Muhammed, which overtook and defeated him near Salmas.[†] At Tebriz Ahmed was visited by the Armenian prince Sempad, son of Ivaneh, grandson of Purthel, to whom he gave as an appanage the town of Arkéghaguth, whose position is unknown.[‡] When Kara Yusuf heard of the doings of his late ally he was busy in settling the affairs of the province of Erzenjan. He set out for Tebriz, and met the Sultan near the town of Asad, two parasangs from Tebriz. In the battle which followed, and which was fought on the 30th of August, 1410, the Sultan fought bravely, but his men were defeated, and he himself was wounded by an arrow in the arm. He sought refuge in a neighbouring garden, but a native of Tebriz, Beha ud din Julah, disclosed his hiding place to the Turkoman chief, who sent some troops to arrest him. He was put in chains and taken before Kara Yusuf, who upbraided him for having been treacherous towards him. He did not deprive him of his life, or even of his title, at once, but appropriated his kingdom, and enjoined him to do nothing against his authority; presently his courtiers suggested that he was a dangerous person, and by their advice he was condemned to death, and strangled. His body was allowed to lie in the dust for two or three days, to prevent turbulent people from setting up a pretender on the ground that he was not really dead. At the request of the people of Tebriz, and according to the wish expressed in Ahmed's will, he was buried in a building called Dimashkuh, near his father and mother. His son, Sultan Ala ud daulat, who had been confined in the fortress of Abd Eljuz, was also put to death and buried in the Dimashkuh. Kaiumars, son of Sheikh Ibrahim, the ruler of Shirvan, arrived to help Ahmed the day after his defeat. He was arrested and imprisoned in the fortress of Arjish, nor would Kara Yusuf accept the money offered by his father for his release. We are told that the Turkoman chief abstained from pillaging Tebriz, although it had sided with his rival.[§]

---

[*] *Id.*, 178-181. Weil, v. 141.   [†] Abdur Rezak, Notices et Extraits, &c., xiv. 193-194.
[‡] Thomas of Medzoph, 95.   [§] Notices et Extraits, &c., xiv. 193.   D'Herbelot, sub voce Avis.

Thus passed away a prince who had probably seen more changes of fortune and more stirring adventures than any of his ancestors. Shah Rokh, son of Timur, when he heard of Ahmed's death, asked Abdal Kadir, a learned man, who was one of his intimate friends, if he had composed nothing on such a tragical event. He thereupon recited four lines with the following meaning : "Tears of blood were shed on this occasion, and on asking wherefore from Fate the only answer was 'Kasd Tebriz' (the expedition to Tebriz)." The two latter words, which only comprise eight letters in Arabic, make the chronogram 813, which was the year of Ahmed's death. Ibn Arabshah cites a witty reply which Ahmed sent to Timur when he was obliged to fly before him, referring to the latter's lameness, &c.: "If I was maimed for the fight, I was not lame for the flight," referring to Timur's double lameness in his leg and hand.[*] In the British Museum there is a large and interesting series of Ahmed's coins. There is a gold coin struck at Baghdad, and silver ones struck at Irbil, Baghdad, Tebriz, Shamakhi, Hamadan, Halab, Arminiyah (?), Vassit, and Mosul. Frœhn, in his supplement, gives the additional mints of Van, Sultania, Khoi, Asterabad, Alemja, and Gushtasf.

SHAH WALAD.

On Sultan Ahmed's death considerable confusion arose at Baghdad, and the authorities differ. The Egyptian writers tell us that Kara Yusuf sent his son, Shah Muhammed, to occupy Baghdad. Nakhsais, a Mamluk of Ahmed's, was in authority there on behalf of one of the latter's nephews, Shah Walad ibn Shahzadeh Sheikh Ali ibn Oweis. Nakhsais defended the place with great bravery, but was killed in a tumult, whereupon his place was taken by another Mamluk, Abdur Rahman Ibn Mullah, who spread the report that Ahmed was still living, and had the khutbeh said in his name. He also was killed in an outbreak, and Shah Walad was again recognised as Sultan. Reports continued to spread that Ahmed was still living, and a fresh conspiracy arose against Sultan Walad, which ended in Baghdad being captured by Shah Muhammed. This was in April, 1411. Presently Walad was driven out by the party which still professed to believe in Ahmed's survival. Kara Yusuf thereupon sent him fresh troops, with which he recaptured the city in August and September of the same year.[†] Other accounts make out that Sultan Walad was put to death with his uncle, Shah Ahmed, and buried in his grave, and Abdur Rezak attributes the events assigned by the Egyptians to him, to his son Mahmud. He tells us how Shah Muhammed, who was then at Irbil, having heard of the turbulence of Baghdad, marched upon the place, and encamped before "the gate

---

of the Sultan's Market." Shah Mahmud had meanwhile secured the supreme authority in the place, and Abd Ibrahim Mallah commanded there in his name; but the latter was murdered by some turbulent people who, in Ahmed's reign, had filled the posts of shahnah and daruga. Great disorders now broke out, and Shah Mahmud, with his brothers, Sultan Muhammed and Sultan Oweis, together with Dendi or Tendu Sultan, his stepmother, left it and fled towards Shuster. Shah Muhammed occupied Baghdad, and also subjected Hibet and a part of Kurdistan, and for some years exercised sovereign rule over those districts.*

SHAH MAHMUD.

After the capture of Baghdad, Tendu, a daughter of Hussein Ibn Oweis, who had been to Cairo with Ahmed, and had married, first Sultan Barkuk, and then her own cousin, Shah Walad, continued to rule at Vassit, Basrah, and Shuster, in the name of her stepson, Mahmud, son of Walad.† In 817 she sent messengers to Shah Rokh with her submission. Shortly after, news arrived from Shuster that she was dead.‡ In the year 819 (i.e., 1416) she was, in fact, put to death, and Mahmud reigned himself.§ He reigned until his death, in 822, when he was followed by his brother,||

SULTAN OWEIS.

He would seem to have held some authority previously, for Abdur Rezak tells us that in 818 Shah Rokh, who was then at Shiraz, sent a courier to Shuster, to Sultan Oweis, with affectionate greeting. Shah Rokh's envoy was received with honour, and Sultan Oweis sent one of his own people back with him.¶ Sultan Oweis was killed by the Turkoman Shah Muhammed in the year 829 HEJ.**

SHAH MUHAMMED.

Sultan Oweis was succeeded at Shuster by his brother, Shah Muhammed, and on his death there succeeded

• HUSSEIN,

son of Ala ud daulat, son of Ahmed Sultan, who conquered the whole of the Arabian Irak, except the capital, Baghdad. Ispahan, the son of Kara Yusuf, had a long struggle with him, and eventually beleagured him at Hillah, which he captured on the 3rd Safai, 825, when Hussein was killed. With him ended the rule of the Jelairids in Irak, which now fell completely into the hands of the Turkomans.††

---

* Notices et Extraits, xiv. 210-211.   † Weil, v. 142. Note.   ‡ Notices et Extraits, xiv. 258-259.
§ Weil, v. 142. Note. | Id.
¶ Notices et Extraits, xiv. 285-286.   ** Weil, v. 142. Note.   †† Id.

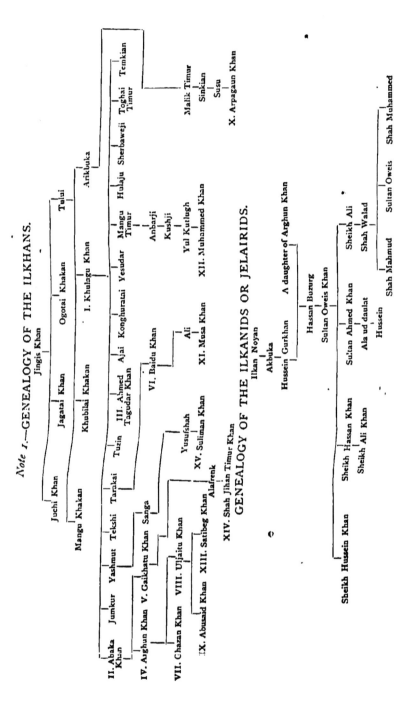

Note 1.—GENEALOGY OF THE ILKHANS.

GENEALOGY OF THE ILKANIDS OR JELAIRIDS.

Note 2.—Little Armenia. The kingdom of Cilicia, or Little Armenia, was so intimately connected in its history with that of the Ilkhans, that I have incorporated short notices of it in my narrative. The last of these was on page 604, where I described the investiture of Leo V. by the Egyptian Sultan. According to Abulfeda, this was during the year 1328-9. M. Dulaurier dates his accession in the year 770 of the Armenian era (*i.e.*, in 1321). Leo reigned till the year 1341, and was the last king of Armenia of the old royal stock of the Rupenians. He was succeeded by John (Juan), son of Amaury de Lusignan, Count of Tyre, brother of Henry III., King of Cyprus,[*] by Zablun, or Zabel, daughter of Leo III., with the style of Constantine III. He was killed, after reigning a year, by his grandees, and was succeeded in 1343 by his brother Guy, who similarly perished after a reign of two years. He was succeeded in 1345 by Constantine IV., a descendant of Leo V. and son of the Baron Baldwin, marshal of the kingdom. He died in 1362, when there was an interregnum of two years, during which there were several competitors, amongst others, Peter I. of Cyprus. In 1365 Leo VI., a supposed son of Constantine IV. by an Armenian mother, mounted the throne. He married Maria, niece of Philip of Tarentum, titular Emperor of Constantinople, and was made prisoner in 1375 by Malik ul Ashraf-Shaaban, Sultan of Egypt, when the kingdom came to an end. Leo was released in 1378 at the request of the kings of Castile and Arragon, and died at Paris on the 29th of November, 1393.[†]

The following table shows the descent of the rulers of Little Armenia :—

```
                    1.  Rupen, called the Great,
                         a relative of Kakeg II.,
                    the last Bagratic ruler of Armenia
                    Established himself in the Taurus
                                  |
                         2.  Constantine I.,
        seized the fortress of Vahga and definitely began the Armenian
                    domination in Cilicia, died in 1100
                    styled Baron by the Crusaders
                                  |
        _____
       |                                                    |
   Thoros I.,                                            Leo I.,
 succeeded as Baron 1100,                            succeeded in 1129,
     died 1129                         made prisoner and taken to Constantinople, 1136
        |                                                    |
   _____            |
  |                       |                      |           |
5.  Thoros. II.,      Sdéphané,        7.  Mleh or Milo,
died in 1167 or 1168  killed in 1164      killed in 1175
  |                       |
6.  Rupen II.,         _____
died in 1170         |                           |
                  8.  Rupen III.,    9.  Leo II. succeeded as Baron, 1187,
                    died in 1187          consecrated King, being the first
                                          King of Little Armenia in 1198,
                                                died in 1219
                                                    |
                                                  Zabel
                                                    |
                                                  Philip,
                                        son of Raymond, Prince of
                                          Antioch, succeeded to
                                        the throne in 1222, but deposed
                                                within a year
```

On the death of Leo, Adam de Gastim became bailly of the kingdom for two years, when he was killed by the Ismaelites, and was replaced by the Grand Baron Constantine.

---

[*] Journ. Asiat., 5th ser., xviii. 293.   [†] Id., xvii. 386.

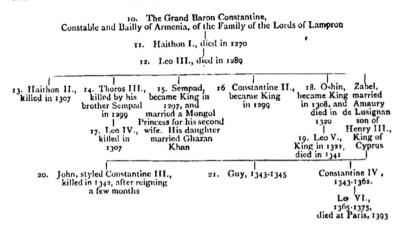

10. The Grand Baron Constantine,
Constable and Bailly of Armenia, of the Family of the Lords of Lampron

11. Haithon I., died in 1270

12. Leo III., died in 1289

13. Haithon II., | 14. Thoros III., | 15. Sempad, | 16 Constantine II., | 18. Oshin, Zabel,
killed in 1307 | killed by his | became King in | became King | became King married
| brother Sempad | 1297, and | in 1299 | in 1308, and Amaury
| in 1299 | married a Mongol | | died in de Lusignan
| | Princess for his second | | 1320 son of
| 17. Leo IV., | wife. His daughter | | Henry III.,
| killed in | married Ghazan | | 19. Leo V., King of
| 1307 | Khan | | King in 1321, Cyprus
| | | | died in 1341

20. John, styled Constantine III., | 21. Guy, 1343-1345 | Constantine IV ,
killed in 1342, after reigning | | 1343-1362.
a few months |

Leo VI.,
1365-1375,
died at Paris, 1393

Note 3.—A second Christian kingdom, whose history was most closely
intertwined with that of the Persian Mongols, and which I have incorporated
largely in my narrative, was that of Georgia. As we have seen, it became
divided into two kingdoms, Karthli and Imerithi. I have carried down its history
to the installation of George V. (styled the Brilliant) as king of Karthli.* This
period of Georgian history, as Brosset says, is singularly obscure. During the
confusion following on the death of Abusaid Khan, Georgi drove out the
Mongols from his country. He then called together a solemn assembly of the
eristhafs and princes of Karthli at Mount Tziv in Hereth, and having put to
death those who would not obey him, set up eristhafs of his own in various
provinces, and was obeyed in all the country as far as Derbend, while the robber
Lesghs paid him tribute. He appointed Sargis Jakel atabeg and generalissimo,
and made a successful raid into Rum and Shirvan.

Meanwhile things had been going on badly in Imerithi. David Narin, as we
have seen, left three sons.† Of these, Constantine, the eldest, died without
issue in 1327, whereupon his brother Michael succeeded to the throne. He
died in 1329, leaving a young son, Bagrat, who could not control the eristhafs,
some of whom would not acknowledge him. Taking advantage of this, George
the Brilliant of Karthli invaded the country and speedily subdued all Imerithi,
appointing governors in various districts, and nominating Bagrat as eristhaf of
Shorapan. This was in 1330. Thereupon the Dadiap Mamia, the Guriel, the
Eristhaf of the Suans, and Sharwashidzéh, eristhaf of Abkhazia, acknowledged
George as king of Imerithi and of all Georgia.‡

The noyan who occupied Azerbaijan (*i.e.*, the Ilkhan for the time being) now
began to invade Somkheth, Ran (*i.e.*, Arran), Mowakan (*i.e.*, Mughan), and
advanced as far as Ganja, or Kantsak, intending to conquer him. He thereupon
fought with him, defeated him, and returned with much booty. He also defeated
the Ottoman Sultan, Orkhan. All Georgia, we are told, was subject to his laws,
and the Caucasians obeyed him from Nicophsia to Derbend, while Ran,
Mowakan, and Shirvan were tributary to him. Thus Georgia, which had been

---

* Ante, 544-587.   † Ante, 357.   ‡ Hist. de la Géorgie, 646-647.

disunited for so long, was once more controlled by one hand. He died at Tiflis in 1346,* and was succeeded by his son, David VII., who had a peaceable and prosperous reign, in which he employed himself in restoring the ruined churches, fortresses, and other buildings in his kingdom. He died in 1360, and was buried at Kelath.†

He was succeeded by his son Bagrat V., in whose reign Timur invaded Georgia. I shall remit an account of him to the next volume, in which the history of Timur will be related. The following table gives the descent of the Georgian kings during the Mongol supremacy in Persia :—

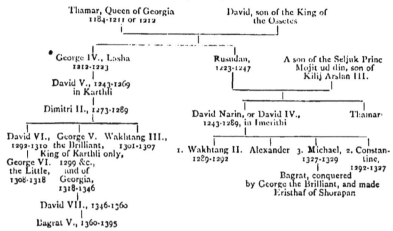

Note 3.—*Mardin*. When the Mongols invaded Western Asia, Mardin was governed by the Ortokid Turkoman dynasty. Its prince resisted them ; but was undone by the treachery of his son, who thus acquired his father's throne, and, as we have seen, it preserved its separate life during the domination of the Ilkhanian dynasty. I have introduced what I could find of its history during the progress of the previous narrative, and will bring together here what is to be said of the later doings of its rulers.

The Malik Salih succeeded his father Mansur in the year 712 HEJ. Von Hammer calls him Salah ud din Yusuf, son of Nejm ud din Mansur.‡ He was the ruler of Mardin when Ibn Batuta visited the town. He tells us he was famous for his acts of generosity. Poets and fakirs visited him, and he gave them splendid presents, thus following in his father's footsteps. He was visited by Abu Abdallah Muhammed, son of Jâbir Alandusi Almeruz, surnamed Alkapf, who praised him greatly, and he gave him 20,000 dirhems. He distributed much alms, and supported colleges and zainahs which supplied strangers with refreshments. His vizier was Jemal ud din Arsinjary. He had made profession at Tebriz, and was on familiar terms with the principal ulemas. His chief kadhi was Borhan ud din Almansily, who was pious, modest, and virtuous, and wore a rough woollen garment not worth ten dirhems. His turban was equally homely. He generally pronounced his judgments in the

---

* *Id.*, 648-649. † *Id.*, 649-650. ‡ Ilkhans, ii. 321.

courtyard of the mosque outside the college, where he performed his devotions, and anyone who did not know him thought him a servant. A woman once passed close by him, when he was away from the mosque, and she did not recognise him. "O, sheikh," she said, "where does the kadhi sit?" "What do you want with him?" "My husband has beaten me," she said; "besides, he has another wife, and does not treat me equally with her in his attentions. I have cited him before the kadhi, but he did not appear, and I am poor and cannot pay the kadhi's people to bring him before the court." "Where does he live?" said the judge. She replied, "In the sailors' village outside the town." "I will go with you thither," he said. "But I have nothing to give you," said the woman. "I would not take anything from you," he replied. He then added, "Go towards the village, and wait for me outside. I will come to you, there." Presently he went as he had promised, and all alone and together they entered her husband's house. The latter said, "Who is this importunate sheikh with you?" "I am what you describe," said the judge, "but you do fairly by the woman." The interview was prolonged, and presently some people came in who knew the judge, and saluted him. The husband was afraid and covered with confusion; but the judge said, "Do not fear, but repair the injustice you have done the woman." The kadhi, having paid what would keep them during the day, left.[*]

Salih reigned for fifty-four years, and died upwards of seventy years old, in the end of HFJ. 765, or beginning of 766. Dr. Rieu has collected some notices of his successors from the chronicles of Ibn Habib (A.H. 648-801) and Ibn Hajar (A.H. 773-850) which he has kindly sent me.

Salih was succeeded by his son Mansur Ahmed, who died in in A.H. 769, at the age of 60; and he by his son, Salih Mahmud, who was deposed, after a reign of a few days, by his uncle, Muzaffer Dåüd, son of Salih, who succeeded him, and died in A.H. 778, when his son Zahir Majd ud din Isa (Ibn Arabshah calls him Al Malik Addahir) succeeded him.

He was ruler of Mardin when Timur invaded the West. When he was on his march from Mosul to Roha or Edessa, Isa sent him his submission, and accordingly at the end of the month Sefer, 796 (i.e., 1404), when the great conqueror neared Mardin, he summoned him to bring his contingent and to march with him against Syria, and then marched on.[†] Isa sent as his envoy Alhaj Muhammed, son of Shasebeki, with presents and excuses for his not going in person, whereupon Timur declared that it was not his policy to leave a rebel behind him. He accordingly returned towards Mardin on the 10th of February, 1404. Isa now went to him with rich presents, horses, mules, &c., nine of each kind, and was presented to Timur two days later. He humbly begged pardon of the great chief, who gave him a robe.[‡] Isa agreed to pay tribute, and to have such taxes imposed as were customary in the case of captured towns. Timur's commissaries went into the place to receive these taxes, and some of his soldiers also trusted themselves carelessly, having gone to make some purchases. They were attacked by a crowd, whereupon Timur summoned Isa

---

once more, and made him kneel down. It then transpired that when he left the city he had told his brother and his people not to part with their arms, and, under no circumstances to surrender the town, and that they must take no heed of any letters he might write to them when under durance bidding them receive a governor or otherwise, as he had determined to sacrifice his life for them and his country. This having been proved against him, he was put in chains; but Timur did not deem it prudent then to besiege Mardin, since the country round was bare of forage.* He accordingly withdrew; but returned again on the 15th of April, and planted his camp about it, and the following day stormed it, the garrison withdrawing to the citadel on the adjoining mountain. Many of them were killed in the flight, while their pursuers captured many youths and girls. They also secured a large booty in horses, mules, and camels, and then proceeded to invest the mountain fortress, which was approachable only by one path, and which was watered by a spring strong enough to turn a mill. Several poets have praised its situation, among others, Ibn Ferai, who called it Kala Shahba (*i.e.*, the White Castle). The besiegers attacked it, and, notwithstanding the stones rolled down upon them, forced their way up to a level with the walls, capturing some prisoners who had not sought shelter in time. According to Sherif ud din, the garrison were cowed by the persistence of the attack, and approached Timur demanding quarter. He thereupon withdrew from the walls into his camp again. The garrison now came out, and gave him numerous presents, some very valuable Turkoman horses, and large sums of money. They swore to be faithful, and undertook to pay a heavy tribute. Timur's heart was softened at this moment by the birth of his grandson, Shah Rukh, which happened just at the time.† Ibn Arabshah makes out he was frustrated in his attack, and revenged himself by destroying the chief buildings in the town ‡ Sherif ud din goes on to say that he pardoned the people of Mardin, and even remitted the taxes he had already received. He gave the principality to Salih, brother of Isa, and gave him patents of office sealed with his own hand.§ Isa was taken to Sultania.‖ Presently, Timur having visited Sultania, took pity on him, after he had been in prison for three years, with his hands and feet chained. He consoled him, gave him a robe, pardoned him, and restored him the principality of Mardin, giving him the necessary patents. He in return swore before Timur's amirs that he would never fail in his obedience to his suzerain, would accompany him in all his wars, &c., and he thereupon returned again to Mardin.¶

Turning once more to the authorities consulted for me by Dr. Rieu, we read how in the year of the HEJ. 809, the amir Jekim, a mamluk of Al Zahir (Barkuk) became Sultan of Aleppo, conquered the neighbouring district, crossed the Euphrates, and marched against the Turkoman amir Muhammed Ibn Karailik, Lord of Amid or Diarbekr. Isa, the chief of Mardin, joined him *en route.* Their united forces were defeated by the Turkomans, and both Jekim and Isa were killed. The battle was fought on the 15th Zulkadah,

---

* Sherif ud din, ii. 274-275. † *Id.*, 280-283. ‡ Op. cit., li. 163.
§ *Id.*, 285. ‖ *Id.*, 312. ¶ *Id.*, 396.

A.H. 809. Thereupon Salih, Isa's brother, who had been nominated chief of Mardin for a short space by Timur, as we have seen, once more mounted the throne; but two years later, being hard pressed by the Turkoman amir, Ibn Karailik, he sought aid from Kara Yusuf, the famous chief of the Black Sheep Turkomans, and proposed to him to give up Mardin to him in exchange for Mosul. His offer was accepted, and he went to Mosul, but died a few days later; and thus ended the dynasty of the Ortokids of Mardin. I append its genealogy as a probably useful table.

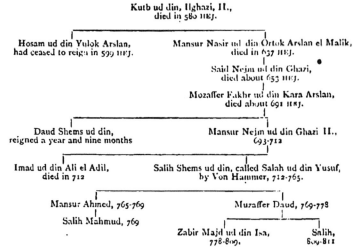

This table is compiled from that given in the first volume of the Eastern section of the work, entitled "Les Historiens des Croisades," xxiv., and from notes supplied me by Dr. Rieu.

Note IV.—GENEALOGY OF THE SELJUKS OF RUM

DURING THE ILKHANIAN RULE.

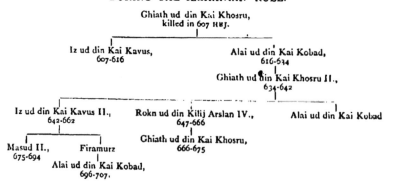

CHAPTER XIII.

THE EASTERN FRAGMENTS OF THE ILKHANIAN EMPIRE.

I.—THE SULTANS OF FARS.

THE PRINCESS KORDOJIN.

FARS had long been the seat of a famous line of princes, the Salghurids. They managed to conciliate the Mongols, and retained their power as subordinates of the Ilkhans until the dynasty came to an end with the Princess Abish, daughter of the last atabeg of Fars, of the Salghurid dynasty, who married Mangu Timur, son of Khulagu.* By Mangu Timur, Abish left a daughter called Kordojin, or Kurdujin, who married Siyurghatmish, the ruler of Kerman, by whom she had a son, Kutb ud din Shah Jihan, who succeeded him, and a daughter, Ismet ud din Aalem Shah, who married the Ilkhan Baidu.† On her husband's death she retained control of Kerman by permission of Baidu Khan. Early in the reign of Abusaid, the Princess Kordojin was nominated by him governor of Fars,‡ and, pleased at her for having sent him a present of standards, &c., on a certain occasion, he gave her the title of Terkhan, which exempted her from the payment of taxes.§ It was probably at this time she also acquired the privilege of coining money, for a coin with her name upon it is known. She was a great builder, and during her rule Shiraz became a magnificent city, the resort of numbers of learned men. Wassaf mentions among the finest buildings of his native city all due to her, (1) the robath, or caravanserai of Sonkor ; (2) that of the Princess Abish, (3) the new mosque, (4) the robath of Shehrolla, (5) the hospital of Mozaffer, (6) the robath of Khair, (7) the robath of Sarbend, bearing an inscription supplied by the Princess Abish, (8) the aqueduct of the old mosque, (9) the khan with two doors, (10) the robath Iddeti, (11) the medresseh Adhadiyeh, which had a revenue of over 200,000 ducats ; (12) the medresseh called after Kordojin herself, in the praise of which Wassaf composed some verses, which were written in gold on a blue ground above the entrance of the medresseh. About this medresseh was planted a garden, traversed by an aqueduct borne on high pillars ; a bath, about

---

* Vide ante, 318-320.   † Ilkhans, ii. 48.   ‡ Id., 270.   § Id., 276. Ante, 593.

which Wassaf wrote some pretty verses, and a fountain, whose good
qualities were contrasted with those of the four rivers of Paradise.   The
students and attendants at the medresseh were well paid, and alms were
also distributed there by the wish of the pious foundress, who, in the
words of Wassaf, combined the virtues of many famous women—the
Balki of Time, the Sarah of the World, the Asia of chastity, the Safura of
purity—nor could the three famous women of Islam—Khadisha, Fatima,
and Aisha—claim to rival her.*   The princess apparently married the
great amir Choban.†   I do not know when she died, the last reference to
her I can find being about the year 1327, when the Malik of Herat asked
for her as a reward for having undone Choban.‡

SHARF UD DIN MAHMUD INJU.

Sharf ud din Mahmud ibn Muhammed, surnamed Inju (i.e., the tax-
gatherer),§ was, according to the "Jihanara" of Ahmed Ibn Muhammed
ibn Abulghaffar al Kazvini, descended from the Khoja Abdallah Ansari ;
others made him descend from Abu Aiyub Ansari.   He was fortunate
enough to secure for himself the lucrative post of tax farmer of Fars,
whence he was promoted to the position of governor of that province by
the influence of Choban.   The date of this I have not been able to trace.
He drew an immense income from the province, and was deemed the
Crœsus of his time, and drew a revenue of 100 tumans from Shiraz and
Shibânkareh, and was thus able to secure himself many powerful friends
among the amirs ; and when Abusaid deposed him in 736 HEJ. (D'Ohsson
says 734), and appointed Muzaffer Inak in his place, some of them, such as
Mahmud Issen Kutlugh, Sultan Shah (son of Nikruz), Muhammed Bey,
Muhammed Pilten, and Muhammed Kushji, gathered round him, and he
tried to seize his rival at his own house.   When he presently fled to the
royal palace, he was pursued thither, and it was only when Shiburgan, the
son of Choban, and the Khoja Lulu happened to come up with assistance,
that the rebels were driven back.   The leaders were now captured, and
would have been put to death but for the intercession of the vizier, Ghiath
ud din.   They were imprisoned in strong fortresses, and Mahmud himself
was locked up in Tabarek, the citadel of Ispahan.   The amirs remained
in custody until Abusaid's death, but Mahmud was released.‖   He
apparently went to live at Tebriz, and was put to death by Arpagaun,
Abusaid's successor, soon after his accession, on the charge that he had
secretly brought up a descendant of Khulagu in his house.¶

---

* Ilkhans, ii. 270-272.   † Id., 298.   ‡ Id., 299.
§ "Jihanara" says the word had this meaning in Mongol. Wassaf says it was taken from the
Khuarezmian dialect.   See Bergmann Munzen der Indschuiden, Numismatische Zeitschrift, iii.
146-147.   Quatremere's Rashid ud din, 130.   Note.
‖ Bergmann, op. cit. 147-149.   D'Ohsson, iv. 715-716.
¶ Mirkhond, in Bergmann, op. cit., 149-150.

MASUD SHAH INJU.

On the murder of their father, Mahmud,[*] his sons fled—Masud, to Rum, to Sheikh Hassan Buzurg, whose representative he became, and who married him to Bakht, the daughter of Dimashk Khoja, son of Choban. His two brothers, Mahmud Shah and Abu Ishak, fled to Ali Padishah, Abusaid's maternal uncle,[†] at Diarbekr, who had set up Musa as a rival khan to Arpagaun. In conjunction with his *protégé*, he defeated the latter in a battle at Karabagh, and Arpagaun having been captured, was handed over to Mahmud's sons, who revenged their father's blood upon him. Presently, Ali Padishah and Musa were driven out by Hassan Buzurg, who set up Muhammed as Khan, and was assisted in his campaign against Ali Padishah by Masud Shah Inju, who, as we have seen, had sought refuge with him. Doubtless, as a reward for his conduct, he was now sent with Pir Hussein, son of Mahmud, son of Choban (the "Jihanara" says with Yaghi Basti, which is doubtless a' mistake), to take charge of Shiraz.[‡]

Ibn Batuta makes out that Abusaid appointed Pir Hussein to succeed Mahmud Shah Inju, that he lived for some time at Shiraz, and then determined to pay a visit to his master in Irak ; but, before doing so, arrested Abu Ishak, son of Mahmud, his two brothers, Rokn ud din and Masud Bek, and his mother, Tash Khatun, under the pretence that he intended to take them to Irak, and to make them disgorge what their father had appropriated. When they arrived in the market-place of Shiraz, Tash Khatun raised her veil, as she did not want them to see her covered, "it not being the custom of the Turkish women to appear veiled in this way." She appealed to the people of Shiraz to help her, saying, "Is a woman such as I am to be carried off from your very midst in this fashion?" Thereupon a carpenter named Pehluvan Mahmud rose and said, "We will not let her leave the place." A tumult arose, several of the soldiers were killed, the treasury was plundered, and the princess and her sons were rescued. The amir Hussein and his adherents fled. He repaired to Abusaid, who gave him an army, and ordered him to occupy Shiraz again. The citizens, feeling that they were not strong enough to resist, sought out the Kadhi Mejd ud din, and begged him to prevent any bloodshed, and to arrange an accommodation. The Kadhi accordingly went to meet Hussein, who dismounted as he approached, and saluted him. Peace was arranged, and the amir encamped outside Shiraz. Next day, the people went out to greet him. They decorated the town and illuminated it with torches, and he entered it in triumph, and behaved very cordially to its citizens.[§] The same author says that

* Ibn Batuta mentions his tomb at Shiraz, and also a post station he built at Yezd Khast.
† Hergu...a, op. cit., 150.   ‡ Bergmann, 151-152.   D'Ohsson, iv. 743.
§ Ibn Batuta, ii. 65-67.

on Abusaid's death, Hussein, afraid that the people of Shiraz would destroy him, left the place.*

Ibn Batuta's narrative is often somewhat suspicious, and these events were probably the same as those elsewhere recorded, where we read that Pir Hussein, in the year 740 (i.e., 1339-40), sought the alliance of Mubariz ud din Muhammed, the governor of Yezd, and having united forces with him at Istakhr, marched upon Shiraz. Masud Shah thereupon fled, and Shiraz presently surrendered.†

Hamdullah Meztufi, writing in 741 HEJ., tells us that Fars and Shebankareh were then subject to Masud.‡ The authorities agree that Masud was put to death in 743 HEJ. (i.e., 1343). The "Jihanara," however, attributes the deed to Yaghi Basti, while elsewhere we read, more probably, that it was Pir Hussein who killed him, through jealousy.§

SHEIKH ABU ISHAK.

The contradictions in the story do not cease with the death of Masud. One account says that Pir Hussein deprived Sultan Shah of the government of Ispahan, and conferred it on Abu Ishak, who was Masud's brother, and that the same year the latter joined his forces with the amir Ashraf, who had advanced into Irak against Pir Hussein. Pir Hussein, not being able to resist the two allies, went to Tebriz, to Sheikh Hassan the Lesser, who had him put to death, merely giving him the option between poison and the sword. He chose poison.‖ In the "Jihanara" we read that Abu Ishak had been sent by Masud against the Shebankarehs. On hearing of his brother's death he returned to Shiraz, whence he drove out Yaghi Basti (? Pir Hussein), and seated himself on the throne, struck money, and had the Friday prayer said in his own name. His power grew greatly, and Ibn Batuta, who visited Shiraz during his reign, tells us his army consisted of 50,000 men (Persians and Turks); that the people of Ispahan, who also obeyed him, were his most faithful subjects, while those of Shiraz were unruly, and were forbidden carrying weapons on pain of death. Our traveller himself saw a man dragged off before the hakim, with a chain about his neck, for having been found with a bow in his possession.¶

Soon after his accession, Abu Ishak had a struggle with Ashraf, the brother of the Lesser Hassan. Ashraf withdrew to Irak, but presently made another attempt, in conjunction with Yaghi Basti, to invade Fars, and plundered and burnt Abrkuh. Abu Ishak marched against them. The allies were only a march from Shiraz, when news arrived of the murder of Hassan. This was in 745 HEJ. They thereupon returned again.** Abu Ishak now began a series of attacks on Muhammed, son

* Ibn Batuta, ii. 68. † Journ. Asiat., 4th ser., iv. 101. ‡ Ilkhans, ii. 321.
Bergmann, op. cit., 151-152. D'Ohsson, iv. 743. ‖ Id., 743-744. ¶ Op. cit., ii. 64.
** Mirkhond, in Bergmann, op. cit., 152-153.

of Muzaffar, ruler of Yezd and Kerman. In one of these he captured
Yezd. Muhammed sought refuge at Mibud, a fortress six miles from
Yezd, surrounded on all sides by sands, where he defended himself
bravely, and made successful sorties, in one of which he killed ten of
Abu Ishak's horses. In another he was caught in an ambuscade, with only
a hundred followers, but cut his way out. At length his brave conduct
attracted the attention and regard of Abu Ishak, and he said : " I wish to
see him, and will then withdraw." He accordingly planted himself near
the fortress, and Muhammed went to the door of the citadel and saluted
him. The Sultan said : " Come down, and trust to my safe conduct."
Muhammed replied : " I have sworn before God not to come to you until
you have entered into the fortress." Abu Ishak consented to do this, and
went with only ten companions. When he reached the gate, Muhammed
dismounted, kissed his stirrup, entered before him, and introduced him
to his house, where his guests took some food. After this, host and guest
returned together on horseback to Abu Ishak's camp. The latter seated
him beside him, gave him his own robes and a present of money, and
they agreed that Muhammed and his father should remain in possession
of the province, while the khutbeh or Friday prayer should be said in
Abu Ishak's name.*

In the year 1350 Abu Ishak again attacked Yezd, and again retired,
and two years later his general, Bikjar, with his nephew, the amir
Kaikobad, son of Kaikhosru, were defeated in the plain of Penj Angusht
(the five fingers), and lost considerable booty. Muhammed, pursuing his
success, laid siege to Shiraz, during which his eldest son, Sherif ud din,
died, which greatly discouraged him. He was helped, however, by the
imprudence of Abu Ishak, who, without cause, put to death the commanders of the quarters of the city called the New Mosque and the New
Garden, while he gave himself up to wine and women. Eventually, the
commander of the quarter called Murdistan, whom he intended putting to
death, surrendered the gate in his charge to Shah Shuja, Muhammed's
son. Abu Ishak thereupon left Shiraz, and withdrew towards Shulistan,
and thence to Ispahan,† where he continued to rule.

During the year 1354 Muhammed attacked Ispahan. In the spring
the governor of that town, despairing of further resistance, agreed to pay
a sum of money to save the place from pillage, whereupon Shah Shuja
who had been left in command of the besieging army by his father,
returned to Shiraz. This was only a short respite, however, and presently
a fresh army from Shiraz beleaguered Ispahan, commanded by Shah
Sultan, nephew of Muhammed, which pressed Abu Ishak sharply. His
troops began to abandon him, and to join the besiegers. The governor
of the fort of Tabrek, close by, treacherously surrendered it, and the

---

¹ Ibn Batuta, ii. 68-70.   † Journ. Asiat., 4th ser., iv. 106-107.

Amir ul-Umera, having said good-bye to his children and servants, left Ispahan with a single attendant. In his distress Abu Ishak took refuge with the Maulana Asil ud din, who was Sheikh-ul-Islam of that district, and who gave him up to Shah Sultan. The latter sent him to Muhammed, who had just returned from pacifying Luristan. He was conducted to the hippodrome of the gate of Istakhr, where Muhammed had assembled the ulemas, the kadhis, and the notables of Fars. Convicted on his own confession of having put to death the amir Haji Dharrah, he was handed over to the younger son of his victim, who decapitated him with a single stroke of his sabre. This happened on a Friday in May, 1357.* Yahia Ibn Abdallatif says he was executed on the Maidan Saadet, which he had himself made. The "Jihanara" speaks of his noble and upright character. Ibn Batuta tells us he was named after the famous Sheikh Abu Ishak al Kaziruni, that he was one of the best sultans of the time, with a good figure and handsome presence. He had a generous soul, and was humble.† He determined to build a splendid building in Shiraz. The foundations were dug out, and the richer inhabitants joined in the labourers' work with the rest, going so far as to make panniers of leather, covered with gold brocaded silk, to carry the earth in, and showing the same luxury in the housings of the beasts of burden they used. They used picks made of silver, and burnt numerous candles during the work. They dressed themselves for their work in their best clothes, and fastened aprons of silk to their girdles. The Sultan watched the work from a belvedere. Ibn Batuta tells us he himself saw the building, when it was three cubits high. When the foundations were finished, the rest of the work was left to the regular workmen, who were paid wages, and it was said that the larger part of the revenues of the place were devoted to it. Jelal ud din ibn Alfeleky Attawrizy superintended the work, and had several thousand workmen under him.‡

Many coins of Abu Ishak are extant. Some of them bearing his name as early as 719 and 724 of the HEJ., are perhaps rather mementoes of his godfather, the holy sheikh, Abu Ishak. Several undoubted coins of his own are in the British Museum, only one, however, with a mint mark, viz., Shiraz. Other mint places of his mentioned by Bergmann are Abrkuh and Shebankareh.§ Abu Ishak does not style himself Sultan on his coins, but simply gives his name. It must be remarked also that while Ibn Batuta calls him Sultan, Mirkhond and the "Jihanara" call him merely Sheikh Amir Abu Ishak.||

---

* Journ. Asiat., 4th ser., iv. 109.   † Op. cit., ii. 64.   ‡ Ibn Batuta, ii. 71-72.
       ¶ Op. cit., 162-163.   § Id., 162.

THE MUZAFFARIANS.

MUZAFFAR.

When the Mongols invaded Khorasan, Ghiath ud din Haji Khorasani, a native of Sejavend, in the district of Khaf, abandoned his country, and removed to Yezd with his three sons, Abubekr, Muhammed, and Mansur. The two former entered the service of Alai ud daulat, atabeg of Yezd, who sent Abubekr with three hundred horsemen to help Khulagu when marching on Baghdad. Abubekr was posted on the Egyptian frontier, and fell in an encounter with the Khafajah Arabs. Mansur devoted himself to his father, after whose death he continued to live near his mausoleum. He left three sons, the amirs Muhammed, Ali, and Muzaffar. Muzaffar was patronised by Yusuf, son of the atabeg Alai ud daulat, and was given command of his troops. Eventually, he joined the service of the Ilkhan Arghun, who appointed him a yassaul, or usher. Ghazan gave him command of a thousand men, with the right to use standards, drums, &c., and also gave him a paizah, or official tablet. Uljaitu Khan gave him charge of the roads and the government of Abrkuh, Luristan, and Mibud. In the year 712 HEJ., or beginning of the next (1312-1313), he received orders to subdue the Shebankarehs, who had revolted. He conquered the rebels, and soon after died. This was in the year 713 (i.e., 1314). His body was transported to Mibud, and there buried in a medresseh, or college, he had built. It is from him the dynasty we are now interested in was named.*

MUBARIZ UD DIN MUHAMMED.

Mubariz ud din Muhammed, son of Muzaffar, was born at Mibud in the year 700 HEJ. (i.e., 1301). On Muzaffar's death he repaired to the court of Uljaitu Khan, who gave him the post held by his father. He remained four years with Uljaitu, on whose death, and the accession of Abusaid, he returned to Mibud. Soon after he allied himself with the amir Kai Khosru, son of Mahmud Shah Inju, against Haji Shah, the atabeg of Yezd, who had killed one of Kai Khosru's officers in order to appropriate one of his slaves. They attacked Haji Shah in the bazaar of Yezd, defeated him, and compelled him to fly with his slaves and wealth This put an end to the dynasty of the atabegs of Yezd, who, according to Mirkhond, had reigned for three centuries.† In 1319, Muhammed was given the government of Yezd by Abusaid. At this time the Nigudarians living in Seistan revolted against the Ilkhan, and commenced to plunder the caravans on the roads. Muhammed, with only sixty men, attacked the Nigudarian chief, named Nuruz, who was at the head of three

---

hundred. He received seventy arrows on his cuirass, was wounded in two places, and had two horses shot under him ; but, some reinforcements coming up, the Nigudarians were defeated, Nuruz was killed, and another of their leaders was captured. Muhammed sent the latter to Uljaitu in an iron cage, with the head of Nuruz fastened round his neck. The Nigudars, determined to avenge their chief, advanced as far as Yezd. Muhammed attacked them again with a smaller force, and again defeated them. The feud continued for some time, and it is said he fought twenty engagements with them in the course of four years. In the year 729 HEJ. (1328-9), he married Khan Kutlugh Makhdum-Shah, daughter of Kutb ud din Shah Jihan, former Sultan of Kerman. In the year 734 HEJ. (1333-4) he went to the court of Abusaid, who gave him a royal robe, a girdle enriched with stones, drums, and a standard, fixed his salary at 100,000 kopeghi dinars, and ordered that in future he should be styled "Amir Zadeh Muhammed."*

After the death of Abusaid, confusion arose in the Persian Irak and Fars. The eldest of the sons of Mahmud Shah Inju, Masud Shah, secured Shiraz and all Fars, and the younger one, Abu Isak, tried to seize Yezd by a ruse ; but Muhammed compelled him to fly, after an engagement which was put an end to by the mediation of the Sheikh Ali Omran. In the year 740 HEJ. (1339-40) the amir Pir Hussein allied himself with Muhammed, as we have seen, and marched upon Shiraz, from which Masud Shah withdrew. Pir Hussein, to reward his ally, gave him the government of Kerman. Muhammed arrived there in July, 1340, and finding great distress, dispersed his soldiers in the different cantons. Malik Kutb ud din Nikruz, who held Kerman for the Mongol rulers, fled to the Prince of Herat, who gave him some troops under Malik Daud. At their approach Muhammed retired, but returned again presently with his son, Shah Sherif ud din Mozaffar. A sharp fight took place at the Gate of the Four Vaults, which was followed by the siege of the town. Presently, the Malik Kutb ud din, feeling discouraged, withdrew toward Khorasan, and Malik Daud was obliged to surrender the town. The capture of Kerman was followed by that of the very strong fort of Bem. Meanwhile, Abu Ishak made himself master of Fars, and, elated by his successes, he, in the spring of 1347, marched against Kerman with a considerable force ; but when he arrived at Behramjerd, fifteen parasangs from that town, he heard that the Afghans, Yermans,† and the nomad Arabs had sided with Muhammed, and he withdrew again to Shiraz. Presently, he again marched on Kerman, ravaging the country *en route*. Muhammed met and defeated him, and compelled him to withdraw. Muhammed now

---

* Journ. Asiat., 4th ser., iv. 98-100.
† These, we are told, were troops sent by the Ilkhan to protect the frontier, at the demand of Jelal ud din Siyurghatmish.

had to repress a revolt of the Yermans and Afghans. He pillaged their homes, and killed seven of their chiefs. The rest fortified themselves in the very strong fortress of Suliman, and presently dared to meet him in the plain of Thavan. They defeated him, and killed 800 of his men, while he received seven wounds, and had his horse killed under him. Although he had previously made peace with Abu Ishak, the latter sent the amir Sultan Shah Jandar, with 2,000 men, to help the rebels, while he himself marched upon Yezd with a considerable army. Muhammed took refuge at Mibud, where, as we have seen, he successfully resisted Abu Ishak's attacks. I have already described the issue of the struggle, which lasted some time.* It ended in Muhammed securing Shiraz and becoming master of Fars. He still had to face some rebellious chiefs, who possessed fortresses of their own. *Inter alios*, Malik Ardeshir, belonging to the royal race of the Shebankarehs, who raised the standard of revolt at Ij, the capital of that principality. Muhammed sent his son, Shah Mahmud, against him, who speedily captured the fortress, and compelled the rebel to fly.†

In 1354 Muhammed marched upon Ispahan, where his enemy, Abu Ishak, still retained authority. He planted his camp at Mardanan, and there he swore allegiance to the Abbasidan Khalif, Mutadhid Billah Abubekr Mustasimi, who lived in Egypt,' and whose name was now inserted in the Friday prayer and on the coins. His example in this respect was presently followed by Abu Ishak himself, as we know from his coins, and subsequently by Muhammed's successor, Shah Shuja. After attacking Ispahan for some time, Muhammed left an army there, under his nephew, Shah Sultan, to press the siege, while he himself marched to Luristan with his two sons, Shah Shuja and Shah Mahmud, to punish Kaiumars, chief of the Lurs, for his continued hostility against his family. On his return from this expedition, he repaired to Shiraz, where he put Abu Ishak to death, Ispahan having in the meantime fallen into his nephew's hands.‡

Master of Kerman, Fars, and Irak Ajem, Muhammed, in the year 1359, determined to march upon Tebriz, the famous capital of the Ilkhans. Ashraf Khan, who had reigned there for some time, had been killed, as we have seen, by Janibeg, the ruler of the Golden Horde, who, claiming the heritage of the Ilkhans, had sent an envoy to Muhammed, confirming him in the humble position of yassaul held by his father. Muhammed sent back the envoy with an equally insulting message. Presently, having heard that Janibeg had withdrawn beyond the Caucasus and left his lieutenant, Akhijuk, in charge at Tebriz, he set out against him with an army of 70,000 horsemen. Akhijuk met him with 30,000 men at Mianeh. Muhammed won the battle, entered Tebriz, and on the Friday following

---

* Ante, 691.   † Journ. Asiat., 4th ser., iv. 108.   ‡ *id.*, 108-109.

mounted the minber, recited the khutbeh, and swore allegiance to the Abbasidan Khalif; then, coming down from the pulpit, he performed the functions of an imaum. Two months later he heard that Sultan Oweis was marching from Baghdad against him, and he speedily withdrew again to Ispahan. Meanwhile, Muhammed had to face troubles in his own family. The partiality he manifested for his grandson, Shah Yahia, son of Shah Muzaffar, who had shown great bravery at Mianeh, aroused the jealousy of his two sons, Shah Shuja and Shah Mahmud; in addition to which he used towards them very foul language, and threatened to blind or put them to death. He also determined to displace them from the throne in favour of their younger brother, Abu Yezid. In order to forestall him, they allied themselves with their cousin, Shah Sukan, and arrested their father in his palace, while engaged in reading the Koran, and while he had no one with him except Maulana Rokn ud din Heravi, afterwards known among the poets as Rokn Sain. Under cover of the night Muhammed was taken to the fortress of Tabrek, and on the night of Friday, the 19th of Ramazan, 759 HEJ., Shah Sultan, no doubt with the concurrence of Shah Shuja, had him blinded with a red hot bodkin. He was then sent to Kalahi Séfid (the White Castle), in Fars. In the course of two months, having won over his guardians, he captured the fortress, and fortified himself there. Shah Shuja sent envoys to him, and it was agreed that he should return to Shiraz, that his name should be replaced in the khutbeh and on the coin, and that matters of state should not be decided without his concurrence. Presently, a plot, having arisen, in which Muhammed was implicated, to put Shah Shuja to death, and to replace him by Abu Yezid, he had the conspirators executed. Muhammed was remitted to prison, and there died on Rebi the 1st, 765 (December, 1363). D'Ohsson says it was in January, 1364.*

Muhammed was undoubtedly a brave prince, and secured a large dominion for himself. He was zealous for religion, and a patron of learned men; but he was rude in manner, and of a hard disposition, perfidious, and bloodthirsty. Maulana Said Lutfallah, who was his constant companion, is reported to have said: "I often saw culprits brought before Muhammed while he was reading the Koran. He used to stop reading, kill them with his own hand, and then resume his pious occupation." Imad ud din Sultan Ahmed said of him: "My elder brother, Shah Shuja, one day asked him, 'Have you killed a thousand men with your own hand?' 'No,' he replied; 'but I think I may have killed eight hundred.'"† He had five sons, Shah Sherif ud din Muzaffar, born in 1325, who died during the siege of Shiraz, as we have seen, in 1353; Shah Shuja, born in 1332, whose mother was Khan Kutlugh Makhdum Shah, daughter of the former Sultan of Kerman, Shah Jihan;

Kutb Shah Mahmud, born in 1337; and two younger sons, named Sultan Imad ud din Ahmed and Abu Yezid.*

.Mirkhond tells us that Muhammed became so zealous and strict that he even exceeded the letter of the law itself in his prohibitions, whence his sons and the wits of Shiraz named him the Mohtésib (i.e., superintendent of police) of Shiraz, and he cites a distich of Hafiz, ending with the words, "Drink your wine in secret, or you will be accused of impiety." Shah Shuja, Muhammed's son, also made verses on the same subject, ending thus : "All the good-for-nothing people have renounced the use of wine, save the mohtésib of the town (i.e., his father, Muhammed), who finds it possible to get drunk without drinking wine."†

In the British Museum are several coins of Muhammed, chiefly with date and mint obliterated. Two of them were struck at Kashan and Yezd respectively.

JELAL UD DIN SHAH SHUJA.

After his father's arrest, Shah Shuja went from Ispahan to Shiraz. He gave the government of Kerman to Sultan Ahmed, that of Abrkuh and Ispahan to Shah Mahmud, and ordered Shah Yahia to be imprisoned at Kohendiz. He soon had to put down a revolt of the Afghans. This was in 1359. It was not long before he also quarrelled with his brother, Shah Mahmud, in consequence of his claiming the revenues of Abrkuh for himself. This was in the year 764 HEJ. (i.e., 1362-3). Shah Mahmud invaded and occupied Yezd, and on his return to Ispahan he suppressed the name of Shah Shuja in the khutbeh, and renounced his allegiance to him. Thereupon the latter beleagured him at Ispahan, but having fallen into an ambuscade, he was captured and condemned to the same punishment he had inflicted on his father. Presently, peace was restored between the two brothers, on condition of Shah Shuja's name being replaced in the khutbeh and on the coins.

Shah Yahia, having gained over his keepers, fortified himself in the fortress of Kohendiz. Feeling too weak to resist his uncle, he submitted, and was pardoned. He was thereupon sent to Yezd with a considerable army, and made his way into the place with a hundred determined followers through a subterranean aqueduct. The commander of the town for Shah Mahmud fled, and he remained master of Yezd, where he rebelled. Shah Shuja sent the vizier Khoja Kavam ud din against him ; but, at the humble solicitation of Shah Yahia and his partisans, he withdrew him. In the year 765 HEJ (1363-4), Shah Mahmud, having secured help from Sultan Oweis, the ruler of Baghdad and Tebriz, marched upon Shiraz. He was joined by the chiefs of the Lesser Lur, of Rum, and Kashan, and won over Shah Yahia by promising him the

---

* Id., 99-100.   † Id., v. 445. Notes.

government of Abrkuh. Shah Shuja abandoned Shiraz and repaired to
Beidha, from whence he marched upon his enemies. He was abandoned
by Sultan Ahmed, who was angry at the little confidence he put in him.
Nevertheless, he joined battle in the plain of Khansar and Seh-Chah
(the Three Wells). The result was indecisive ; but, during the night, he
withdrew to Shiraz, where he fortified himself, and where he was besieged
by the army of Baghdad and Irak.

Meanwhile, Devlet Shah and Malik Muhammed, whom he had sent to
collect the tribute of Kerman, rose in revolt, and captured his son,
Muzaffar ud din Shébéli, and imprisoned him in the green kiosk. Devlet
Shah then recognised the authority of Shah Mahmud, and put his name
in the khutbeh and on the coin. The siege of Shiraz had lasted for eleven
months, and Shah Shuja was being hard pressed, when an accommodation
was arranged between the two brothers, by which he agreed to retire to
Abrkuh, and to remain there for a month, while Shah Mahmud undertook to send back the troops of Sultan Oweis, and to treat with his elder
brother; but, having got possession of Shiraz, he forgot his promise. Shah
Shuja, after staying a few days at Abrkuh, determined to punish Devlet
Shah, and in a battle in which the latter's force was four times as numerous
as his own, he defeated and compelled him to fly to Kerman. Devlet Shah
now submitted ; but, having begun to conspire again, he was put to death,
with his accomplices. The people of Shiraz, weary of the domination of
the troops of Tebriz and Ispahan, sent to invite Shah Shuja to return. He
accordingly did so, and encountered Shah Mahmud near Bésa. The
latter was beaten, and withdrew to Shiraz, which he abandoned in three
or four days, and went to Ispahan. This was the 26th Dulkadeh,
767 (1366).*

In the year 770 (1368-9), Shah Shuja recognised the Khalif Khir
Billah Muhammed, son of Abubekr, and hearing that his brother Mahmud
was again intriguing at Baghdad, he sent an envoy to Sultan Oweis to
ask for the hand of a princess of his house. Shah Mahmud, hearing of
this, also sent an envoy asking for the hand of one of his daughters. The
embarrassed Sultan, on the advice of his amirs, preferred the alliance
with Mahmud. He sent his daughter to Ispahan, and also sent an army
to his assistance. On the arrival of these troops, Shah Mahmud again
advanced on Shiraz, and a battle was fought in the plain of Chasht Khar.
Shah Shuja's right wing defeated the left wing of Mahmud, while his left
wing was defeated by the enemy's right. The indecisive battle was
followed by the flight of a portion of Shah Mahmud's army to Ispahan,
and of that of Sultan Shuja to Shiraz. Ashamed of this flight, the
latter wished to march again against his brother, but was dissuaded
by his nephew, Shah Mansur, and soon after Shah Mahmud returned
again to Ispahan.

---

* Journ. Asiat., 4th ser., v. 437-441.

Towards the end of Ramazan, 776 HEJ. (1375), a messenger arrived at Shiraz with the news of the death of Sultan Oweis, which was followed by that of the death of Shah Mahmud. This last happened on the 13th of March, 1375. The people of Ispahan were not agreed about his successor. The greater number favoured Shah Shuja, but others supported Kutb ud din Oweis, his son. Eventually, the Khoja Bahai ud din Kurchi and Salah ud din, the treasurer, sent people to Shiraz, to announce that the greater part of the amirs, and the chiefs of Shah Mahmud's Court, had recognised Kutb ud din as Sultan. The two chiefs just named had transported Mahmud's treasure to the fortress of Tabrek, where they had sustained two assaults. Thereupon Shah Shuja set out for Ispahan. The chiefs of Irak came to him at each station to offer presents. When Kutb ud din heard of this he took counsel with his partisans, and sent to ask his father's pardon. This was granted, and he went to his camp, where, shortly after, according to Mirkhond, he was poisoned, by his order. When Shah Shuja reached Ispahan, the Malik Fakhr ud din, ruler of Little Luristan, who had been a dependant of the Muzaffarians, and had afterwards gone over to Sultan Oweis of Baghdad, sent him envoys, with a present of racehorses and other precious things. He also inserted his name in the khutbeh and on the coin. The amir Siyurghatmish Arghani also joined him, with 2,000 horsemen. Meanwhile, he heard that the chiefs of Azerbaijan were discontented with Sultan Hussein, son of Sultan Oweis, who passed his time with musicians and singers, and he accordingly determined to capture Tebriz. He therefore marched thither with 12,000 of the men of Irak and Fars. On the way, the people of Kazvin resisted, but the place was speedily captured ; he would not allow it to be pillaged, however, and moved on. He met Sultan Hussein, who was at the head of 30,000 men, at Jerbadekan, defeated him, and captured Abd el Kadir and the Pehluvan Haji Kharbendeh, two of the principal amirs of Azerbaijan, whom he sent in chains to Shiraz. Having announced his victory in various directions, he now approached Tebriz, whence the Seyids, the mollahs, and the grandees of Azerbaijan went to meet him, and were allowed to kiss his hand. He passed the winter there enjoying himself, sent his nephew, Shah Mansur, to Karabagh with 2,000 horsemen, dispatched Ferrukh Agha to Nakhchivan, and ordered the amir Seljuk to occupy Aujan. The latter was not long left alone, but having been joined by another amir named Isfahan Shah with some troops, the two were attacked by two amirs of Azerbaijan, who lived on the rivers Jaghatu and Baghatu. Isfahan Shah was made prisoner, while Seljuk, in attempting to escape, fell from the roof of his house, and broke both his legs. Their troops were either killed or dispersed. News also arrived that Sultan Hussein was on the march. Thereupon Shah Shuja, whose army was dispersed in various places, and who was suffering from gout in the foot, had to retreat. It was the middle of winter, and he was

carried in a litter. When he arrived at Kazvin the people refused to
supply him with food and forage. According to one açcount he did not
punish them, but marched on. According to another, exasperated at
their having killed a messenger of peace sent to them by one of their own
people, the Khoja Mejd ud din Kakum, he ordered the town to be attacked.
It was captured and pillaged. Two months afterwards, Sultan Hussein
having re-entered Tebriz, sent a messenger to Shah Shuja, offering to
exchange Isfahan Shah for the two amirs captured at Jarbadekan. This
was arranged.*

Shah Shuja, having secured the hand of a daughter of Sultan Oweis
for his son, Zain al Abidin, gave the young prince the government of
Ispahan. He now discovered a plot, in which his nephew, Shah
Yahia, was in correspondence with the Pehluvan Akad, son of Toghan
Shah Khorasani, whom he had appointed governor of Kerman, and
was inciting him to rebel. He accordingly sent an army to attack
Yezd, which was Shah Yahia's appanage. The latter fought a fierce
battle outside the town, but was driven inside, and being hard pressed
there, had recourse, we are told, to his usual weapons—perfidy and
trickery. He sent a message out to the besiegers, begging them not to
attack him, as he was about to send Shah Shuja an envoy. Believing in
his word, the soldiers withdrew to their tents, where they were presently
attacked unawares and routed, and a large booty fell into Shah Yahia's
hands. When he heard of this, Shah Shuja determined to march in
person on Yezd. From this he was dissuaded by Shah Mansur, who
offered to go himself and capture the place. Shah Shuja gave him some
troops, with which he proceeded to attack his brother. Almost every day
there was a struggle, in which the people of Yezd had the worst of it.
Eventually, Shah Yahia sent out their common mother to make terms
with his brother. Peace was arranged, and the besieging army returned
home, except Shah Mansur and a few of his friends, who wished to enter
the town. Shah Yahia replied that the place was too small for his needs,
and, in a characteristic way, suggested that Shah Mansur should go to
Asterabad to the amir Vali, and, having obtained an army from him, they
should make common cause against their uncle, Shah Shuja. Shah
Mansur delayed two or three days, but not being able to enter the town,
he set out for Asterabad, cursing in his heart his brother's treachery.
When Shah Shuja heard of all this, he marched himself upon Yezd. Shah
Yahia, thoroughly frightened, sent a deputation to appease his uncle,
which included the latter's son, Sultan Shah, his sister, Khan Zadeh, and
his own son, Sultan Jihanghir, with his other relations. Once more they
succeeded in reconciling Shah Shuja, who again returned to his capital.
Meanwhile, the Pehluvan Akad, of whom we have spoken, was besieged

in Kerman, first by Sultan Ahmed, and then by the Pehluvan Taj ud din Khurrem. He was compelled to surrender, and was put to death, with his principal wife. This was in Ramazan, 776 HEJ. (February, 1375.)*

In the year 781 HEJ. (1379-80), having learnt that Sarik Adil, one of the amirs of Sultan Hussein, was collecting an army at Sultania, with the intention of attacking him, he determined to forestall him, and marched thither. He was sitting at table outside his tent, when a cloud of dust announced the approach of Sarik Adil and his people. He speedily put his men in order of battle, gave the command of the right wing to Sultan Ahmed and Zain al Abidin ; the left wing to Sultan Shébéli and Shah Hussein, brother of Shah Mansur ; and himself retained command of the centre. The enemy numbered 24,000 horsemen, whose charge broke Shah Shuja's ranks. He himself was dismounted and surrounded, and defended himself bravely with his scimitar. Malik Baurji, one of his bravest warriors, mounted him on his horse, on which he would have fled ; but Akhi Kuchuk, one of his chief officers, said to him, " If you fly,* not one of these men will remain alive." While they were discussing, ten or fifteen brave men rallied round them. A cloud of dust arose. Some said, " It is the enemy." Akhi Kuchuk rode on to see, and when he came near the dust he found it was some men of Shah Hussein, who were fleeing with the standard and an assload of kettledrums. He ordered them to sound an announcement of good news. When those who were fleeing heard this, they rallied to the standard. They charged the enemy, who were engaged in pillaging, and put them to flight. Shah Shuja passed the night on the field of battle, and encamped the next day at the gates of Sultania, where Sarik Adil and his amirs fortified themselves. Presently they sued for peace (which was granted them), and sent out some rich presents. Sarik Adil went, unaccompanied, and kissed the hand of Shah Shuja, who presented him with a khelat from his own wardrobe, and a girdle incrusted with precious stones, and then returned home again.

Zain al Abidin, who was very young, misbehaved himself at Ispahan. He was deprived of that government, and imprisoned for some days. His place was given to the amir Pehluvan Khurrem, who kept it till his death. Pir Ali Barik, one of the principal amirs of Azerbaijan, having aroused the jealousy of the notables of that country, fled to Shah Shuja, who presented him with several strings of horses, mules, and camels loaded with rich objects—among other things, kettledrums and a standard. He intrusted him with a considerable army, and ordered him to march upon Shuster. He speedily captured this place, and leaving one of his men, named Islam, took 5,000 men with him, with which he attacked Baghdad. This he also captured, and placed Shah Shuja's name in the

---

* *Id.*, 451-454.

Friday prayer and on the money. When Shah Shuja heard of these successes, he sent Pir Ali a shoulder belt decorated with stones, and an affectionate letter.* From the account in the "Habib ûs siyer," it would seem that Pir Ali was in league with Sheikh Ali, who was then governor of Baghdad on behalf of Hussein.† Hussein, it would seem, marched against Baghdad in 782 HEJ. with Sarik Adil, and drove out his brother and Pir Ali, who retired again to Shuster; but it was not long before an outbreak of the people compelled him to retire, and the two allies again entered the place.

In August or September, 1381, Hussein was murdered by his brother, Sultan Ahmed, who occupied Tebriz.‡ Sheikh Ali, another brother, now marched against him with Pir Ali. A violent struggle ensued, in which they were both killed, and Baghdad thus fell into Sultan Ahmed's hands. Meanwhile, Shah Mansur had been wandering for some time in the district of Mazanderan, and had gone from there to Sultania, where he was imprisoned and bound by Sarik Adil. Escaping, he fled to his uncle, Sultan Ahmed, who received him well. Islam, who commanded at Shuster, as we have seen, sent to inform Shah Shuja of Mansur's return, and for fear of a surprise he sent Pehluvan Ali Shah Mezinani to Islam's help. On his arrival, he tried to murder Islam, but was himself killed in the attempt. After this murder, Shah Mansur occupied Shuster. He made continual attacks upon Luristan, whereupon Shems ud din Pésheng, prince of that country, implored the aid of Shah Shuja, and offered, if assisted, to re-conquer Shuster. Shah Shuja would have marched against Shuster, when he received a message from Sultan Ahmed in these terms: "Sarik Adil has put my young brother, Sultan Bayazid, on the throne of Sultania, and aroused animosity between us. As his majesty Shah Shuja stands in the position of father towards me, I trust he will try to restore peace between two brothers." Shah Shuja sent a messenger back with a favourable reply, while he sent word to Pésheng that he was about to set out for Sultania, but that on his return they should meet at Shuster. The people about Sultan Shébéli incited him against his father, and also reported to Shah Shuja the various bitter things which his son said against him, both in private and public, and fearing that he might have to submit to the same punishment he had inflicted on his own father, he, in July, 1383, ordered Shébéli to be arrested, and to be transported to Kalahi Séfid, in Fars, and thence to the castle of Aklid and Sarmak. Two or three days after, he, when drunk, ordered the amir Ramadhan Akhtaji and the khoja Jauher Kuchuk to blind his son. In vain, at the solicitation of the khoja Turanshah, he tried to recall his order; the courier he had dispatched for the purpose arrived too late.§

---

* Journ. Asiat, 4th ser., v. 454-457.   † Weil, v. 17.   ‡ Id., 17-18.
§ Journ. Asiat, 4th ser., v. 459-461.

When Shah Shuja reached Sultania, Sarik Adil, fearing his junction with Sultan Ahmed, came out of the citadel with Bayazid and went to meet him, and was received with every mark of honour and respect. He then sent an envoy to Sultan Ahmed, and succeeded in making peace between him and Bayazid. After this he went to Shuster by way of the Lesser Lur, and accompanied by Sarik Adil. It was winter, and we are told that during their march it rained five days and nights successively. According to his promise, the atabeg Pesheng went to meet him. The river at Shuster proved impassable, and Shah Mansur being posted on the further side with 7,000 well-armed men, Shah Shuja deemed it prudent to withdraw; but he promised the atabeg to send an army under his brother Abu Yezid, to help him to conquer the place. He returned to Shiraz by way of Kuhi Kiluieh. As he passed through Shulistan, he was taken ill, and halted awhile till he had recovered. When he reached Shiraz he abandoned himself to his love for wine, and as he took little nourishment, he undermined his constitution and lost his appetite. As he saw that he could not live long, he superintended the preparations for his own funeral, and ordered ten hafiz to be continually with him to read the Koran through, every day. The amirs were meanwhile divided into two factions, one of which swore allegiance to his son, Zain al Abidin, and the other to his brother, Imad ud din Ahmed. Shah Shuja having heard of this, summoned his brother. The two when together began to weep, and we are told that sobs stopped their conversation. Ahmed went out, in order that his brother might recover his self-possession. Thereupon, the latter summoned Pir Shah, Ahmed's confidential servant, and bade him tell his master these words: "The world is like the shadow of a cloud, and like a dream; like the former, it is continually moving, and when a man awakes he merely possesses a faint memory of the latter. I foresee much trouble in this town. Our first residence was the city of Kerman. I have never had to complain of thee. Now that I am going to set out on my journey to the other world, if thou beginnest to create strife, God will be dissatisfied as well as I, and it will be necessarily a subject of rejoicing for our enemies. Go at once to Kerman, and leave this city full of trouble." Ahmed consented, and the same day set off thither.*

Shah Shuja now wrote a letter to the Great Timur, who had then advanced into Khorasan, and was menacing Mazanderan. He intrusted this letter to Omar Shah, one of his principal amirs, and in it he commended to the great conqueror his son, Zain al Abidin, as well as his other sons, his brothers, and nephews. A similar letter was addressed to Ahmed, the Sultan of Baghdad. This was in the year 784 HEJ. (1382). Among the presents he sent to Timur are mentioned precious stones and pearls, gold rings and gold coins, rich stuffs, Arab horses, fast mules, with golden saddles ; several teams of six mules each, richly harnessed ; cuirasses decked with silk, rich furniture, a grand scarlet daïs, a royal pavilion, a tent, and a grand umbrella. Omar Shah was duly presented, and delivered the letter to Timur, and the presents to his officers. He was well received, and Timur gave him gold coin, a khelat from his own wardrobe, a girdle, a khanjar, a scimitar decorated with precious stones, and some horses, and sent him back with a favourable answer, and accompanied by one of his officers bearing rich presents for his master. He also sent to ask for the hand of his daughter for his grandson, Mirza Pir Muhammed. According to Mirkhond, the wife of Pir Muhammed was a daughter of Sultan Oweis, son of Shah Shuja. According to Sherif ud din, in the beginning of the year 783 HEJ., Timur sent Unkaitu and Haji Khoja to Shiraz to bring home the princess. The Princess Serai Mulk Khanum, Tuman Aga, and other great ladies went out to meet her, and received her with great ceremony ; scattered precious stones, seed-pearls, and gold dust over her, and prepared the marriage banquet, &c.*

Shah Shuja continued his preparations for his funeral. He chose the linen for his own shroud, and ordered the carpenters in his presence to make the coffin which was to hold his corpse. He nominated a ulema, distinguished for his piety, to wash his body ; and summoned the Amir Ikhtiar ud din Kasam Kurji, from Kerman, in order that he might transport his coffin to Medina. Nor did he forget to leave rich presents for the pious people at Mekka and Medina. At length he died on the 22nd of Shaban, 786 HEJ. (the 9th of October, 1384), at the age of 53 years', and two months. The date 786, according to several writers, was expressed by an Arab phrase, meaning, "It is a loss to Shah Shuja," which chronogram, according to Sherif ud din, was constructed by Hafiz.†

According to Mirkhond, Shah Shuja bore an excellent character. He was learned, liberal, and brave. It is said that, when nine years old, he knew the Koran off by heart, and his memory was so good that he could remember seven or eight Arab verses after hearing them once. He composed a large number of verses in Persian and Arabic, several of which are preserved by Mirkhond, and in the ,Atesh Kédeh of Lutf Ali Beg. Mirkhond justifies his eulogy by quoting, *inter alia*, this anecdote about him : One day he was returning to his palace from the archery ground, when a woman put the following statement in his hand : "I am a woman without means, and have lost my husband ; my two daughters are in pledge to a Jew, who has recently embraced Islam, for the sum of 400 dinars. If the Padishah would deign to rescue these poor girls, God will remember the good act, and I shall be grateful all my life." Shah Shuja, having learnt the contents of the paper, replying, said : "At the day of resurrection, and when everyone shall be rewarded according to his works, if I am asked how it came about that during my reign two young girls, Muhammedans by birth, had been pledged to a converted Mussulman, what shall I say?" Having dismounted and sat down, he said : " Let each one that is devoted to me bring me something, according to his means." The amirs, notables, and even servants thereupon placed on the ground what they could spare in money, precious things, &c., till the whole heap was of the value of 100,000 dinars. He then said to those about him, "Which of you would become my relation ?" A young man, named Adineh, belonging to the company of the amir Isfahan Shah, bending the knee, said : "I am the first to claim this privilege." Shah Shuja asked him what was the value of his annual income. He replied, "One thousand dinars." The king ordered 19,000 dinars to be added to this. Khosru Shah, who served in the company of the amir Alai ud din Inak, having next offered himself, his pay was fixed at 20,000 dinars. Shah Shuja then ordered 400 dinars to be paid to the man who had the two girls in pawn, to redeem them. One of them was then taken to the household of the Princess Durri Mulk, and the other to that of Muhab Shah Khatun, and he sent 30,000 dinars to each princess to prepare a trousseau with. The pile of wealth was made over to the widow ; and when the preparations were ready, he went with all his amirs and the princesses to the marriage of his two *protégés* with the two young men whom he had constituted his relations in this odd fashion.

As I have said, Shah Shuja was an accomplished literary man, and, *inter alia*, wrote verses. It was possibly from a feeling of jealousy or pique that he took a dislike to the famous poet Hafiz, the greatest ornament of his kingdom. Khuandemir tells a singular story in regard to this. It was the fashion in writing a gazel, love poem, &c., to limit it to a single subject, and not to introduce parenthetical matter. Shah Shuja said one day to Hafiz: "The verses of your gazels do not limit themselves to one subject only. On the contrary, three or four verses are devoted to a description of wine, two or three to the doctrine of the Sufis, and one or two to a description of the object of love. This mixture in one and the same gazel is contrary to the practice of the best authors." The Khoja replied : "The sacred words of His Majesty the King are most true ; but, nevertheless, the poems of Hafiz have acquired great fame everywhere, while the verses of other poets are not known outside Shiraz." This repartee was accepted as a personal affront by Shah Shuja, who determined to be avenged. It happened that about this time Hafiz wrote an ode, of which the last verse ran somewhat thus : "Alas, Islamism is the creed of Hafiz, and is it true that a to-morrow (*i.e.*, a day of resurrection) follows upon to-day." Shah Shuja having read this, declared that Hafiz did not believe in a resurrection, and some learned doctors who

x *

were envious of him determined to obtain a *fetva*, or judicial decision, declaring anyone to be an infidel (Kaffir) who should deny the resurrection. Hafiz, greatly troubled at this charge, repaired to the Sheikh Zain ud-din Abubekr Taiabady, who had made a pilgrimage to Mekka and returned to Shiraz, and reported to him what had happened. The latter advised him to interpolate a verse before the incriminated one, putting the latter in some one else's mouth, when he would come under the rule expressed in the proverb: "To quote the opinion of a heretic is not heresy." Thereupon the Khoja composed the following verse: " How much I have been troubled by these words spoken by a Christian one morning at the door of a tavern, to the sound of the drum, flute," &c. Having duly interpolated this, he escaped the penalty.

Khuandemir mentions among the contemporaries of Hafiz a poet named Khoja Imad. He was a jurisconsult of Kerman, and head of a monastery; and Shah Shuja professed a great reverence for him. It was said that whenever he prayed his cat imitated him, which the Sultan looked upon as a miracle. Hafiz being jealous, composed the following gazel : " The Sufi has hung up his nets and opened the cover of his box. He has commenced to practise tricks towards heaven which are fertile in illusions; but the latter will break the eggs in his cap: for he has exhibited his juggleries before those who are in the secret. Approach ! oh cupbearer ! for the elegant mistress of the Sufis has presented herself in all her splendour, and commenced her coquetry. Whence comes this musician, who, having commenced after the manner of Irak, has changed into that of the Hijaz ? Come ! oh my heart ! let us seek refuge with God against those who shorten their sleeves and lengthen their hands (*i.e.*, against the hypocritical Sufi who, with a pretence of austerity, allows injustice to be done). Never employ artifice ; for whoever does not play the game of love with sincerity, will close the gate of reality to himself. To-morrow (*i.e.*, in the day of resurrection), he who has been moved by human motives only shall be covered with shame at the sight of the throne of spiritual doctrine. Oh ! partridge with the graceful step, where goest thou ? Stop ! Be not led away because the cat of this saint has been praying. And you, Hafiz, do not blame the drunkard; since, from all time, God has permitted both devotion and hypocrisy."*

Shah Shuja left four sons, Zain al Abidin, Muzaffer ud din Shébéli, Kutb ud din Oweis, and Moiz ud din Jihanghir. Shah Shuja's various mint towns prove how widely his authority was obeyed. General A. Houtum Schindler enumerates the following : Bender, Lúrdeján, Ráhin, Kerman, Idej, Shiraz, Yezd, Bázuft, Marlahú, Shebánkáreh, Kazirun, Abrkuh, Sirjan, and Beljan (?)† Mr. Poole, in his catalogue of the coins in the British Museum, adds Kashan, Shabiran, and Lár.‡

---

* *Journ. Asiat.*, 5th ser., xi. 408-413.   † *Numismatic Chronicle*, lxxx. 324, &c.
‡ *Op. cit.*, vi. 235-241.

General Houtum Schindler has some valuable notes on the more obscure of these, towns. Sirjan, he says, was formerly the capital of Bardsir. It is now a district of the Kerman province, its principal place being Saidabad, a large village in a fertile plain, five stages from Kerman on the road to Shiraz. Bender, he suggests, was the principal port of Kerman or Fars, perhaps Hormuz or Kish, adding, that the people of Shiraz and Kerman now say shortly Bender, when they speak of the chief ports of their province, Bushin and Bender Abbas.

Idij was the capital of Great Luristan, its ruins being still visible on the Málámir plain, 121 miles from Shuster on the Ispahan road. Bázuft and Lúrdeján were also towns of Luristan, about half-way between Shuster and Ispahan. Two head streams of the river of Shuster have the same names.

Ráhin was probably the modern Rayin, or Rahin, a large village sixty miles south-east of Kerman on the road to Bam. In 1877 it had 2,546 inhabitants. Marlahú is probably the modern Maharlu, about twenty miles south-east of Shiraz.

ZAIN AL ABIDIN, SHAH MANSUR, &c.

On Shah Shuja's death, his empire was divided among his relatives. His eldest son, Zain al Abidin, took Fars, with its capital of Shiraz ; his brother, Sultan Ahmed, became the ruler of Kerman ; his nephew, Shah Yahia, took Yezd ; and another nephew, Shah Mansur, took Ispahan. Zain al Abidin soon found himself at strife with his cousin Yahia and his uncle Abusaid.

A little before Timur reached Ghilan, in his western progress, he sent Merahem to Zain al Abidin to summon him to his presence, as he wanted to show him the goodwill he bore him on account of his father, from whom he had received the letter already referred to. Instead of going he arrested Timur's envoy, and the latter thereupon determined to invade Fars and Irak. He ordered some of his best troops to go on ahead with the baggage by way of Rai, and to winter at Sarik Kamish. With them went the Mírza Miran Shah, the Amir Seif ud din, and the Sheikh Ali Behadur ; the rest of the army, with the imperial standard, advanced by Hamadan and Jerbadeken upon Ispahan. Zain al Abidin's maternal uncle, Sai Muzaffer Shashi, who then governed the town for his nephew, went out to implore Timur's clemency, accompanied by the Khoja Rokn ud din Said, and the grandees, Sherifs, doctors of the law, &c. They were well received, and allowed to kiss the imperial carpet. The troops having seized all the entrances into the town, Timur entered it in triumph, and afterwards retired to the fort of Tabarek. He nominated Aikutmur its governor, appointed guards for the gates, and ordered that all the horses and arms in the town should be surrendered to him, an order which was obeyed. The grandees went to the imperial camp, where the sum to be paid for the ransom of the inhabitants was fixed. For the purpose of collecting this money, the place was divided into different quarters, an amir being assigned to each, and the sum was ordered to be paid to Nur Mulk Berlas and Muhammed Sultan Shah, Timur meantime detaining the grandees as hostages. Unfortunately, a reckless young blacksmith, named Ali Kuchapa, having beaten a drum in the night in the town, this collected a number of vagabonds, who killed all Timur's commissaries, except those who were protected by some of the more prudent inhabitants. A large number of soldiers who had ventured into the town were also murdered, among them being Muhammed, the son of Katei Behadur. More than 3,000 men are thus said to have perished. The mob then seized the gates, and proceeded to fortify them. The next day, Timur having heard what had happened, was in a terrible rage, and ordered the place to be stormed. It was speedily captured, and a general massacre ensued, save of those who had protected the commissaries, and in whose houses a certain number of fugitives found safety. The quarter of the Sherifs, the street of the Turekés, where the doctors of the law lived, and especially the house of the Khoja Imam ud din, although that famous preacher had been dead a year, were also spared. So great was Timur's rage, that he ordered each section of his army to furnish a certain number of heads, and it is said, adds Sherif ud din, that many soldiers, unwilling to slaughter Mussulmans, bought heads from the police authorities. At first a head was sold for twenty dinars kopeghi; but presently, when each soldier had furnished his tale, the price fell to half a dinar, and at last was unsaleable. The soldiers, exasperated by the slaughter of their comrades, followed the wretched victims by their footsteps in the snow. The smallest calculation, based on the registers, puts the number of the slain at 70,000. The heads were placed in heaps outside the walls of Ispahan, and eventually were built up into pyramids. This slaughter took place on the 18th of November, 1387, and Sherif ud din, moralising on the unaccountable ways of the infinite providence of God, concludes that the tragedy was connected with the unfortunate conjunction of the stars, when the two ill-omened planets, Saturn and Mars, were in the sign of the Crab, and the eleventh of the aërial triplets was in the sign of the Twins.*

Having named Haji Bey and Nunan Shah, governors of Ispahan for a year, Timur marched upon Shiraz. Zain al Abidin, on hearing of this, fled to his cousin, Shah Mansur, who was at Shuster, and with whom he was not on good terms, thus imitating, says Timur's historian, the man who threw himself into the fire to avoid the heat of the sun. When he reached the River Dudankeh, Shah Mansur corrupted his troops, who nearly all

---

* Sherif ud din, L 447-454.

went over to him. Zain al Abidin was seized, put in irons, and taken to the castle of Selaseh. Shah Mansur eventually imprisoned the soldiers who had deserted their master, and confiscated their goods. Timur now occupied Shiraz without resistance, and the imperial standard having been planted at a place called Takt Karajeh, outside the town, the various officials of the kingdom went to do homage, and agreed to pay a thousand tumans into Timur's treasury. A grand fête was also held, which was honoured by the presence of Timur himself; the khutbeh was said in his name, and, after prayer and sacrifice, he retired to his camp. In such a case, the sacrifice consisted of a camel. Shah Yahia, prince of Yezd, with his eldest son, Sultan Muhammed, Sultan Ahmed, prince of Kerman, Abu Ishak, a nephew of Shah Shuja, lord of Sirjan, with all the neighbouring rulers, including the atabegs of Lur and Gurghin Lar, said to have been of the race of Gurghin Milad, all made their submission, and were treated with great honour ; and letters were written, describing the recent successes, which were sent in various directions.* Troubles in Transoxiana compelled Timur to return to Samarkand. Before doing so, he gave the government of Shiraz to Shah Yahia, that of Ispahan to the latter's eldest son, Muhammed ; Kerman to Sultan Ahmed, and Sirjan to Abu Ishak. All these princes received letters patent, sealed with the imperial seal, called "altamgha." The Sherif Gerjani, the Amir Aladin Inak, and the principal amirs of Shah Shuja were sent to Samarkand, together with the most skilled handicraftsmen, and set out for his capital. This was Muharrem, 790.† En route he passed through Abrkuh, where the Pehluvan Muhaddeb Khorasani, who had governed the town on behalf of Shah Shuja for a long time, was very submissive, gave the great conqueror a feast, and was duly confirmed in his government.

Of all the Muzaffarian princes, Shah Mansur alone had failed to send in his submission to Timur, and on the latter's withdrawal, he left Shuster and marched upon Shiraz. The Sherifs of the town, with their disciples, opened the gate of Salem to him. As he entered, his brother, Shah Yahia, left the place by the gate of Sadet, and went to Yezd. He thus secured the capital of Fars. Meanwhile, Zain al Abidin, who had been imprisoned, as we have seen, at Kerikerd, four leagues from Shuster, escaped, and went to Malik Az ud din Kerit at Urudgerd. Together they went to Ispahan, whence Sultan Muhammed withdrew into the citadel, and a month later went to join his father at Yezd. Shah Mansur having secured the fortresses of Bid, Sermak, Meruset, and Abrkuh, soon after heard that Zain al Abidin, at the head of an army, was marching against him. The two armies met at the foot of the fortress of Ishtakr, at the end of the new bridge. Shah Mansur's troops swam the river, and attacked their enemy so sharply that Zain al Abidin was compelled to

---

* Id., 454-462.   † Id., 470-471.

retire to Ispahan. The Pehluvan Muhaddeb, fearing that Shah Mansur would conquer the whole kingdom, having entered into a solemn alliance with Shah Yahia, invited him to Abrkuh, and lodged him in a small palace, which he had built and decorated for himself. The prince's officers persuaded him to act treacherously towards his host, who was seized, put in chains, sent to the castle of Melus, on the borders of Yezd, and eventually put to death. Leaving the Amir Muhammed Kurchi with a force to garrison Abrkuh, Shah Yahia returned to Yezd.

Shah Mansur speedily heard of these events, advanced upon Abrkuh, and captured the town, its governor having withdrawn to the citadel. This, however, he surrendered by order of his master, on condition that Shah Mansur did not advance upon Yezd. The latter, therefore, contented himself with ravaging the country on the way to Ispahan, then returned to Shiraz. The next year he repeated his destructive raid in the same district. Zain al Abidin, feeling himself thus hard pressed, appealed to his relatives for help. The princes of Kerman and Sirjan, at the head of their troops, with some other chiefs of the house of Muzaffar, went to his aid, and they advanced upon Shiraz, wasting the country, especially the country of Kerbal. Shah Mansur encountered them at a place called Yuruz, and defeated the allies, who each fled to his own appanage. In the spring of the following year, Shah Mansur again advanced upon Ispahan, which was surrendered to him apparently through the treachery of the Khoja Az ud din. Zain al Abidin fled, apparently intending to join Timur; but he was captured near Rai, between Veramin and Shehriar, by Musa Shuker, who sent him to Shah' Mansur. He was now blinded with a hot iron. Mansur now made two attempts upon Yezd. On the second occasion, Shah Yahia's mother, who was also his mother, came out of the town, and said to him : "Your elder brother, with his children, contenting himself with the town of Yezd, has given up to you the kingdoms of Fars and Irak. If you molest him there, you will be much blamed." She persuaded him to withdraw from Yezd, and to return to Shiraz.

Timur's historian proceeds to say that "during these years there were other struggles between the princes of the house of Muzaffar, whose nature it was to make war upon one another, which caused great disorder in the empire of Iran. The kingdom of Persia was only one, and ten kings claimed to rule it ; but they were, so to say, so many plagues, who maltreated the people and desolated the country."* These events made Shah Mansur master of Fars, Khuzistan, and a part of Irak Ajem. They doubtless also tempted Timur to make another attack upon Persia. He reached Urudgerd on the 26th of February, 1393, a few days later captured Kurram Abad, in Luristan, severely punished the predatory Lurs,

* Sherif ud din, i. 173-183.

and dispatched an army under the Mirza Miran Shah against Kashan, which was surrendered by Meluk Serbedal, who commanded there on behalf of Shah Mansur. Timur himself crossed the River Zal, a tributary of the Kerkhan, which falls into the Tigris, near Kurneh, by a famous bridge, said to have been built by order of the ancient Persian king, Sapor Julektaf. It was made of dressed stone and brick, and consisted of twenty-eight great arches and twenty-seven small ones. He then advanced, by way of Dezful, upon Shuster, and halted by the River Karun, which flows past the town. Ali Kutual and Izfandiar Nami, who commanded there for Shah Mansur, had fled, and the grandees and Sherifs thereupon surrendered it. The surrounding country was, however, pillaged, and a large number of horses and mules were captured and distributed among the soldiers.*

Early in March, Timur crossed the River Karun, and encamped in a grove of palms outside the town. There he was joined by the Mirzas Muhammed Sultan and Pir Muhammed, who had been traversing Kurdistan and Luristan, and punishing their unruly inhabitants.† He also summoned the Mirza Omar Sheikh, and having left the Khoja Masud Sebzevari with the troops of Sebzevar (which he commanded) in charge of Shuster, he set out for Shiraz, crossing the rivers Dudankeh and Shurukan Kendeh. On the 15th of March, 1393, he reached Ram Hormuz, where Pir Muhammed, the prince of Great Luristan, did homage, offered presents, and joined his army. On the 25th of the same month he reached the famous fortress of Kalaa Sefid, where Zain al Abidin was then imprisoned. It was situated on a mountain with scarped sides, and approached only by one small pathway. The top of the mountain was a plateau, a league square, well watered, and well cultivated. It was deemed impregnable, and had never been stormed, and many of the princes had built country houses there. Sherif ud din says that the approach to it was so difficult, that three men could defend the path against 100,000. Its natural defences were supplemented by walls built of great stones, cemented by lime, and as its pastures could support large herds of cattle and much game, it was deemed useless to try and starve the garrison out. Sadet commanded it on behalf of Shah Mansur. Timur having arrived there, and closely inspected the place, determined to try and capture it. Divided into several sections, his soldiers clambered up the rocks, pick in hand, amidst showers of arrows and stones. One of the officers, named Akbugha, scrambled on to a scarped point, deemed unapproachable, and, under cover of his shield, pressed the enemy there so hard that they were discouraged, and ceased fighting. Others followed him, others again made their way up by similar routes, while the Mirza Muhammed Sultan with his men reached the gate of the fortress, raised their horsetail standards, and cried, "Victory!" The fort was stormed, and its garrison thrown over the precipices. The governor Sadet was executed, as a punishment for holding out so stubbornly. Zain al Abidin was captured, and taken before Timur, who gave him a robe, consoled him, and promised to punish Shah Mansur. Timur also released all the women whom the soldiers had captured. Akbugha was rewarded for his bravery, and loaded with money, tents, horses, camels, and mules, and also young girls ; "and this officer, who the previous day had no other property than his horse, could hardly realise that it was not a dream." Leaving Malik Muhammed Aoubehi in charge of the fortress, he set out for Shiraz through the valley of Shaabbevan.

On reaching the gardens outside Shiraz, the amir Osman noticed a party of Shah Mansur's scouts planting some men in ambush. He allowed the scouts to pass, then fell on them from behind, and captured one, who was cross-examined as to ·the position of the enemy. Continuing his march, he encountered three or four thousand of Shah Mansur's horsemen, dressed in mail, with helmets and corslets of velvet, covered with iron, and their horses protected by a kind of cuirass made of quilted silk, with their banners flying. "Shah Mansur advanced at the head of these his picked troops," says Sherif ud din, "like a furious lion," and encountered at a place called Patila 20,000 of Timur's veteran horsemen, overthrew their ranks, cut his way right through them, and reached a most important vantage behind them. Turning round, he charged again, utterly reckless of his life. Timur, seeing him rushing straight at him, wished to seize his lance to oppose him, but found his lance-bearer had been thrust aside in the struggle. Although he had *only fourteen or fifteen men about him, he did not move*, and Shah Mansur struck two heavy blows with his scimitar on the Emperor's well-tempered helmet. They did him no hurt, as the scimitar slipped along his arms. Adel Aktashi held a buckler over his head, and Komari Yasaul advanced in front of him. Timur was wounded in the hand during the fight. Eventually, Shah Mansur was driven away from the neighbourhood of the Emperor, and joined his own infantry. The right wing of this body was broken by the Mirza Muhammed Sultan, with terrible slaughter. The Mirza Pir Muhammed similarly broke the left wing, killing some, and putting the others to flight. Meanwhile, the centre of Timur's army, which had been disintegrated by Shah Mansur's brave charge, was rallied by Shah Rokh, who was then only seventeen years old. Shah Mansur was surrounded, and the young prince cut off his head. Throwing it at the feet of his father, he said : "May the heads of all thine enemies be cast at thy feet like that of the proud Mansur."[*]

---

* Sherif ud din, ii. 192-197. Weil, v. 39-40.

According to Ibn Arabshah, the head was afterwards sent as a warning to Ahmed, the ruler of Baghdad.* The death of their leader discouraged his men, and "the leopards were changed into deer." Those who escaped the sword, fled. Timur mounted a hill, where he embraced his sons and the noyans, and thanked heaven for his victory on his knees. The amirs went to congratulate him, and offered him the golden cup on their knees in the Mongol fashion. Meanwhile, a division of the enemy, well armed, and in battle array, suddenly appeared. Timur and Shah Rokh turned quickly upon them, and routed and drove them as far the mountain of Kalatsurk (i.e., the red fort). The next day, Timur entered Shiraz in triumph, and the imperial standard was unfurled over the gate of Selm, where he lodged. This gate was alone opened, the others being closed; thereupon his great officers entered the place, and obtained inventories of the treasures, furniture, horses, and mules of Shah Mansur and of his courtiers, from the magistrates, &c. These were conveyed outside the town and presented to Timur, who distributed them among his amirs. A sum was also paid as ransom by the citizens. Mirza Muhammed Sultan was sent to Ispahan with orders to establish a garrison there, and to exact a similar ransom. The Mirza Omar Sheikh, who had remained behind in charge of the baggage, had devoted himself to following and punishing the fugitives from Shah Mansur's army, together with the robber tribes of Luristan, the Shuls, and Kurds. He passed by Nubenjian, and having reached Kazirun, he received orders to plant garrisons and appoint officers in the Mongol fashion over that part of the country. He then went to Shiraz to congratulate his father.†

The power of the house of Muzaffar was now broken, and the remaining princes determined to submit. Shah Yahia set out from Yezd, with his sons and Sultan Ahmed, of Kerman. They offered Timur presents of precious stones, horses, mules, tents, &c. Sultan Mehdi, son of Shah Shuja, and Sultan Gadanfer, son of Shah Mansur, were already at Shiraz. Timur spent a month there, feasting. They played there, says Sherif ud din, on the organ and the harp, and the good red wine of Shiraz was presented in golden cups by the most beautiful maidens of the city. Sultan Abu Ishak, grandson of Shah Shuja, also arrived from Sirjan with presents. The great conqueror now proceeded to regulate the affairs of Fars, to remit taxes, and to restore order; and he appointed Mirza Omar Sheikh governor of that province, which, says Sherif ud din, is the heart of the empire, and most full of cities, towns, and villages of any in Asia. The Mirza prepared a grand feast for his father, offered him presents on his knees, and assured him of undying fidelity to him.‡ Timur's historian enlarges on the troubles brought upon Fars by the conflicts of the various members of the family of Muzaffar, each of whom

---

* Weil, v. 41. Note.   † Sherif ud din, ii. 197-201.   ‡ Id., 197-201.

claimed to be sovereign and to strike money in his own name, and among whom father did not spare son, nor son, father. We are told the sheikhs, doctors, imaums, and the people of Fars and Irak presented petitions to Timur in reference to the changes necessary to restore prosperity to the country, which was almost ruined. He determined upon a crucial remedy, and ordered that all the Muzaffarian princes should be seized, put in chains, and that their houses should be pillaged. The amir Osman sent people to Kerman to fetch the treasures of Sultan Ahmed. Mirza Omar Sheikh was, as we have seen, appointed governor of Fars. He was given some trusty amirs as counsellors, and some troops to maintain his authority. Ideku, son of Kaias ud din Berlas, was given the government of Kerman, Temukeh Kuchin that of Yezd, and Lalam Kuchin that of Abrkuh. Guderz, who had governed Sirjan on behalf of Abu Ishak, thinking himself strong enough in that fortress to resist attack, set up as sovereign there.

'Zain al Abidin, who had been blinded by Shah Mansur, and Shébéli, who had been similarly blinded by Shah Shuja, were sent to Samarkand, and Timur assigned for their subsistence some of the best property in that town, so that they might spend, says Sherif ud din, the rest of their lives in some comfort under the shadow of his clemency. In the simplicity of their retreat, he continues, they enjoyed many pleasures unknown to the ambitious. The men of letters and the best artisans of Fars and of Irak abandoned their country, and went to live at Samarkand.*

Leaving Shiraz, Timur went towards Ispahan, enjoying his favourite sport of hunting on the way. At Kumsha he issued, says his historian, the famous order so much desired by the people, and so remarkable in history, for the execution of the princes of the house of Muzaffar. All the males of that house, who were at Yezd and Kerman, were put to death by the governors of those provinces.† Ibn Haji says the number so executed was seventeen. Ibn Arabshah says they were put to death because some of them had determined to kill Timur when with him in a tent.‡ Sirjan was attacked by Timur. It held out for three years, and when it eventually surrendered, we are told that only the usurper Guderz and six other people remained alive out of its garrison and inhabitants. He was put to death.§

I don't know when Zain ud din or Shébéli died. Of Shah Mansur there are three coins in the British Museum, struck respectively at Shiraz, Kazirun, and a place read doubtfully as Suk.||

---

* Sherif ud din, ii. 203-206.   † *Id.*, 207-208.   ‡ Weil, v. 40. Note 1.   § Sherif ud din, ii. 392.
|| Poole Catalogue, vi. 241-242.

SULTAN MOTASSEM.

When the Muzaffarian princes were put to death, as above described, one of them, at least, escaped — a son of Zain al Abidin, by a sister of Sultan Oweis. When his father was taken to Samarkand, he fled to Syria. After the death of Timur, he returned to Irak and Azerbaijan. He was well received by the Turkoman chief, Kara Yusuf, who sent him to Hamadan and Luristan, whence he went to Tebriz. There he spent some days feasting, when, by the advice of the Amir Bestam-Jaghir and the Kadhi Ahmed Saidi, he went to Ispahan, where the Timurid prince Mirza Omar Sheikh was governor. The latter withdrew towards Yezd. The Mirza Izkander, who was then at Koshk-zer, having heard of this, advanced against Sultan Motassem. The two armies met near Atesh-gah. The chief people of Fars deserted, and joined Sultan Motassem. Mirza Izkander, undaunted by this defection, fought bravely, and compelled the enemy to fly. The chief warriors of Irak and Azerbaijan were captured. Sultan Motassem having reached the gates of Ispahan, tried to swim his horse across the river, but being a stout and heavy man, he fell off his horse, and a soldier, who was pursuing him, cut off his head. The Malik Fakhr ud din Muzaffari was also killed in this action.*

Thus ended the dynasty of Muzaffar. During its rule at Shiraz, there dwelt there Hafiz, who, with Saadi and Jami, were the most famous Persian poets. Muhammed Shems ud din, otherwise known as Hafiz (*i.e.*, one who knows the Koran off by heart), was born at Shiraz at the beginning of the fourteenth century, and died, according to some, in 791 HEJ. (1389 A.D.), or, according to others, the year following. It would seem from his writings that he was attached to a monastic life, and was perhaps at the head of a monastery. It is certain that he studied jurisprudence and theology in a college founded by Haji-kivam ud din, whose munificence he often praises, and that eventually he became a professor there. His verses were recited to his scholars, and were collected by them after his death. We have seen that he did not get on very well with Shah Shuja ; but while his own Sovereign neglected him, his fame induced the neighbouring princes to make him very tempting offers to go to their Courts. The poet, however, like Victor Hugo in our own day, was too much attached to his own country to be thus tempted. "The perfume exhaled by the soil of the Mosalla," he says, "will not permit me to travel." "O, cupbearer," he says, in another song, "pour me out the rest of this wine, for you will not see in Paradise the banks of the River Roknabad, nor the rose gardens of the Mosalla." After refusing an invitation from Ahmed, the ruler of Baghdad, he accepted one from Yahia, the Muzaffarian prince of Yezd, whom, however, he did not flatter. In comparing his avarice with the generosity of the King

---

* Abdur Rezak, Notices et Extraits, xiv., part i. 133-174.

of Hormuz, he said : "The King of Hormuz has never seen me, and yet he has granted me a hundred favours ; the King of Yezd has seen me, I have sounded his praises, yet he has given me nothing. Such is the way of kings. As to you, O Hafiz, do not trouble yourself ; the Supreme Judge, who supplies the daily needs of all, will help and assist you." Hafiz was also invited by Mir Faiz Allah Inju, a judicial officer of Mahmud Shah Behmény, the ruler of the Deccan, to go to him, and he sent him a sum of money to pay his expenses on the way. The poet divided a portion of this money among his sister's children and some unmarried women, paid his debts with the rest, and then set out. On arriving at Lur, he met one of his friends, who had been plundered by robbers, and gave him all he had. On reaching Hormuz, he went on board a ship there, but before it started a tremendous storm came on, which disgusted him so much that he landed under pretence of saying farewell to his friend, and did not go on board again, but sent Faiz Allah his excuses in these words : "Even to secure the whole world, it is not worth while to pass a single moment in misery. Sell your coat for a glass of wine, for it is not worth more than that. The wine merchants will not accept it for a cup. O, strange prayer-mat which is not worth even a cup. . . . The sickness caused by the sea seemed formerly a small matter in the presence of the perfume of gold ; but I made a mistake, for a hundred pounds of gold is not sufficient solace for one of these waves. The splendour of the imperial crown is very fascinating, but it is not worth losing your life for it. . . . Hafiz, force yourself to be moderate in your desires, and renounce contemptible riches ; not for the whole world put yourself under obligation to anyone." Faiz Allah, having received this poem, read it to his master, who was much pleased with it, and in consideration of Hafiz having set out to visit him, he ordered a thousand gold pieces to be spent in presents to send to the poet.

When Timur conquered Shiraz the first time, Hafiz composed an ode, in which he said : "If this Turk of Shiraz would accept the homage of my heart, I will give him Samarkand and Bokhara for his black ephelis." Timur being nettled at this, summoned him, and said : "By the blows of my well-tempered sword I have conquered the greater part of the world, and have desolated a thousand places in order to enlarge Samarkand and Bokhara, my capitals and residences, and you, pitiful creature, would sell me both Samarkand and Bokhara for one black ephelis." Hafiz was not disconcerted, but replied : "O ! Sovereign of the world, it is by a similar liberality that I have been reduced to my present condition," a happy repartee which secured the conqueror's favour. Hafiz was buried in the Mosalla of Shiraz, which he so often praised, and a tomb was afterwards built over it by Muhammed Mo'ammáiy, the tutor of the famous Baber, when that prince captured Shiraz in 1451.*

---

* Journ. Asiat., 5th ser., xi. 406-425.

THE PRINCES OF JORJAN.

TOGHAI TIMUR KHAN.

In an earlier page * I have described the origin of Toghai Timur Khan.
I ought to add here that his grandfather Baba Baghatur (or Behadur), who
migrated westward with his ulus, consisting of 10,000 families, in 705, as
I there mentioned, is no doubt the same Baba Behadur who had a
struggle with Uzbeg Khan, and who ravaged Khuarezm in 1315.† He
was put to death, together with his son, by Uljaitu Khan.‡ I have
described how Toghai Timur was first proclaimed Khan in 1337 A.D.§
How, after a defeat by Sheikh Hassan, he withdrew to Khorasan and
Jorjan, where he continued to rule, while Muhammed, the *protégé* of
Sheikh Hassan, controlled the other dominions of the Ilkhans.‖ Some
time afterwards, during the struggle for supremacy between the two
Hassans, Toghai Timur was again invited to occupy the Ilkhanian
throne. I have related the unfortunate result of his accepting this
invitation and his return to Khorasan in an earlier page.¶ This was in
the year 739 HEJ. (*i.e.,* 1346 A.D.) Next year, Toghai Timur made
another invasion of Irak. This was at the instigation of his brother, Ali
Kaun (called Amir Sheikh Ali Kowun, or Gawan, in the "Shajrat ul
Atrak "), to whom he intrusted the command of his army, and who was
assisted by Shiburgan, the son of Choban, and by the Princess Satibeg,
who then governed Irak ; but Hassan the Lesser sent his brother Ashraf
against him, who defeated him near Abher, and drove him out of Irak
Ajem, while his supporter Shiburgan retired into Dilem.**
　Shortly after this, the Serbedarians, a fanatical religious community,
overran Khorasan, as I shall presently describe, with their emissaries, and
speedily ousted the authority of Arghun Shah, Toghai Timur Khan's
deputy there. The Khan marched against them. The army of the Khan
(called the army of Jorjan by Khuandemir) took to flight, and was
pursued by the Serbedarians, and Toghai Timur himself fled to Lar i
Kafran. Thenceforward he entirely lost any control in Khorasan. A
few years later, when the truculent Khoja Yahia was the ruler of the
Serbedarians, Toghai Timur sent him a summons to go and submit to
him. This was written in verse, and in not very courteous terms. It was
replied to by the Khoja Yahia, also in verse. The latter then repaired to
the camp of Toghai Timur, professedly to submit to him. This visit was
in answer to the summons to go and acknowledge his suzerainty. There
were no precautionary guards about, neither were there gatekeepers nor
tent openers in front of the royal tent of Toghai Timur, who only had

* Vide Ante, 638.　† Ante, vol. ii. p. 149.　‡ D'Ohsson, iv. 575.　§ Ante, 639.
‖ Ante, 640.　¶ Ante, 643-645.
** D'Ohsson, iv. 733-734.　Shajrat ul Atrak, 329-330.

with him Khoja Muhammed Bahrabady and two students   A feast, lasting for three days, was held.   At the end of the third day, Hafiz Shekhani, Muhammed Habish (*i.e.*, the Abyssinian), and other Serbedarians, consulted together, saying: "We have so far made no treaty, nor have we pledged our word.   We can make away with the chief (*i.e.*, Toghai Timur) during the banquet."   The latter, it would seem, had also formed the plan of arresting the Serbedarians after the feast.   An opportunity offering, the Khoja Kerrair and Hafiz Shekhani pressed forward close to the Khan, and while they were discussing the affairs of Khorasan, Shekhani struck him on the head with a battle-axe, and he was then killed by the Khoja, while all his companions were also put to death.   The Khan's treasury was now plundered by the Serbedars.   His territory had for some years been limited to the district generally known as Jorjan.

Toghai Timur Khan was assassinated in 1353.*   Fræhn published one of his coins, struck in 738, with the mint mark obliterated,† and others struck at Meshed and Amol.   In the British Museum there are coins struck by him at Amol, Zeidan, Keez (?), Kazvin, Kushan, Basrah, Baghdad, and Jajerm, in the years 738, 739, 740, and 741.   They prove what a wide area obeyed his nominal rule.   The coins struck at Basrah (in 742) and Baghdad are especially interesting in this view.   One coin published by Fræhn in his posthumous work, has the inscription in Mongol characters, the name reading Toghan Timur.‡

AMIR VALI.

The father of Amir Vali, called Sheikh Ali Hindu,§ was one of the grand amirs of Toghai Timur Khan.   When the Khan was killed by the Serbedarians, as we have described, Amir Vali repaired to Nisa with some of his attendants.   This is Mirkhavend's statement, and also that of Sherif ud din.   Khuandemir makes him go to Nishapur, which is improbable.   The governor of that district, Amir Shibly Jany-Kurbany, married his sister, and presently Amir Vali set out, as Khuandemir says, with great hopes and little power, for Jorjan.   On his arrival in Dahistan, he was joined by two hundred horsemen and foot soldiers, dependents of his father, Ali Hindu.   He was speedily attacked by the Serbedarian who had been nominated governor of Asterabad, Hasan Dameghani, who rushed upon him with great energy at the head of five hundred horsemen.   Vali stood firm and defeated his aggressor, killed most of his men, and captured a large number of horses, weapons, &c.   When the news of the defeat of the Serbedarians was spread about the district, many of Toghai Timur's former supporters, who had been hiding away, gathered

---

* D'Ohsson, iv. 738.   † *Recensio*, 645.   ‡ Poole Catalogue, vi. 98-101.
§ *Sherif ud din calls him Sheikh Ali Bisud.* Op. cit., i. 396.

round.   Thereupon Abubekr, who was governor of Shasman, or
Shemasan, on behalf of Hasan Dameghany, attacked Vali with a force
of 2,000 men, but was also beaten, and retired to Khorasan.   Hasan now
gave him 5,000 brave men, and ordered him to return, as he was encamped
at Sultan Dowin.   Meanwhile, Vali had got together a body of brave
people from the district of Jorjan, and in order to make his army look
more numerous and imposing, had dressed some women as soldiers, and
mounted them on horseback.   The Serbedarians were frightened, and
began to break.   Their opponents shouted out, "Tat Kashti" (*i.e.*, the
Tajiks have fled).   They pursued them.   Abubekr, their commander,
jumped into the River Gurgan, but was caught and beheaded.   Many
of his people were killed, while those who escaped fled to Khorasan.
Amir Vali now repaired to Asterabad, whither he summoned Lokman,
the son of Toghai Timur Khan, with the intention of putting him on the
throne ; but, says the Eastern rhetorician, who has recorded the fact, "the
sweets of power overcame the obligations of gratitude for former favours,"
and he thereupon sent him word before he arrived that he must go
elsewhere, and ordered that no one who had been in close relations with
Toghai Timur was to live in his neighbourhood.   He was acknowledged
as master at Asterabad, Bostam, Dameghan, Semnan, and Firuzkuh, and
ruled there until the arrival of the great Timur.*   In 1370, Vali sustained
a defeat near Rai at the hands of Oweis, the ruler of Baghdad, and was
pursued as far as Semnan.†

. It would seem that when the empire of Abusaid broke to pieces,
a small fragment became subject to Arghun Shah, who was the chief of
the district of Yun Garbani (called the province of Yuin by Abulfeda).
He encamped about Kelat, north-west of Sarrakhs, and was probably
partially dependent on Vali.   His son Ali Bey submitted to Timur about
the year 1379, and the latter married his grandson, Mirza Muhammed
Sultan, to his daughter, elsewhere called his sister.   Splendid feasts were
held to celebrate the betrothal ; and, we are told, Timur consulted Ali
Bey on his proposed campaign against Herat, and it was arranged
that when he marched against that city in the spring, Ali Bey should
march to his help.‡   When the latter heard that he was actually setting
out, he dispatched a courier to beg to be allowed to take part in
the campaign.§   Timur accordingly sent a messenger to tell him to
assemble his people and march; but he not only refused to do so, but even
imprisoned Timur's envoy.|| After the capture of Herat, Timur, troubled
by the contumacy of Ali Bey, marched in person towards Kelat and Tus,
and when he reached the tomb of Abu Muslem Meruzi, the famous
general of the founder of the Abassadan line of khalifs, he dismounted

---

* Dorn, Mem. on the Serbedarians, 180-182.   † D'Herbelot quoting Khuandemir, i. 138.
‡ Sherif ud din, i. 310.   § *Id.*, 314. || *Id.*, 317.

and paid it a visit, asking God to aid him. The news of the Emperor's
advance naturally troubled Ali Bey, and he determined to submit, and
accordingly hastened to the camp and did so.*

Timur now marched westwards, secured Nishapur and Sebzevar, and
then advanced upon Isferain, which was subject to Vali. He ordered the
place to be stormed at once. The town was speedily captured, and a
terrible slaughter ensued. "The captors destroyed the place so," says
Sherif ud din, "that only its name now remains." Timur then sent an
envoy to the Amir Vali, promising to reward him if he came at once and
offered his submission ; otherwise, he threatened him with the consequences. Vali kissed Timur's letter on receiving it, and placed it on his
head as a proof of his submission, and promised to go to him shortly,
whereupon Timur returned to Samarkand.† Not long after this, Ali Bey
made a league with Vali, and persuaded him to march upon Sebzevar, where
the Serbedarian, Ali Muayid, was governor. Timur at this time lost his
daughter Akia Beghi, and was greatly depressed in consequence. His
sister, Kutlugh Turkan Aga, in order to rouse him, urged him to march
against Vali and his ally, and to punish the insolent rebels. These words
had their effect on him, and he prepared to set out again westwards.‡
He was joined *en route* by the Malik of Herat and by Miran Shah from
Sarrakhs. On his approach to Kelat, Ali Bey shut himself up with his
people in the fortress. Timur sent a messenger to inquire what was the
cause of his fear, bidding him come to him, and assuring him that no harm
would befall him. He preferred to trust in the strength of his stronghold, which he deemed impregnable. Timur thereupon had recourse to
a ruse. He left the neighbourhood, and set out in the direction of Koran,
a dependency of Abiverd, and gave out that he was going to attack
Mazanderan ; but he merely made a detour. Ali Bey was misled on
hearing that the imperial army had left. He brought his horses, sheep,
&c., out of the fortress, and let them graze in the fields around. Timur
now returned, ravaged the whole district, and built a mound opposite the
gate of Kelat, on which he unfurled the imperial standard. The place
was now closely beleaguered. Miran Shah posted himself opposite the gate
of Jia ; Mirza Ali, son of the Amir Muayid Orlat, near that of Lobra ; Haji
Seif ud din at that of Arghun Shah, and Omar Sheikh before another
gate. Ali Bey now sent a supplicating letter, asking pardon for what he
had done, and offering, if Timur would go to the gate of the town with
only a small escort, that he would go like his slave and throw himself at
his feet. Timur accordingly went, accompanied by only five horsemen.
The walls of Kelat were built on the slope of a high mountain, and a small
path led alongside of the wall among the rocks. This path, which was
closed by a gate of its own, led to the town. Ali Bey noticing that Timur

---

* Sherif ud din, i. 329-330.   † *Id.*, 331-332.   ‡ *Id.*, 331-337.

was ill-attended, planted an ambush, and ordered his men to kill him, whatever the consequences; but the plan failed, and he returned to his camp to receive the congratulations of his generals on his escape. He then ordered the place to be stormed, and the men of Mekrit and Badakhshan, who were skilled crags-men, made their way over the rocks in the night, and reached the gates of the town amidst the clash of cymbals and the bray of trumpets. The enemy was speedily defeated in a struggle outside, and demanded quarter. Ali Bey sent to beg that the conqueror would order the carnage to cease, and promised in a solemn writing to go himself and submit the next day, and as a gage of his sincerity, sent Nikruz and Muhammed Sheikh Haji, the principal amirs of the tribe, Yun Garbani, and also Kand Sultan, who had been betrothed, as we have seen, to Mirza Muhammed Sultan. These people begged him to forgive Ali Bey, which he consented to do. He ordered the slaughter to cease, while Nikruz and Muhammed were told to go to his camp. Next morning at sunrise Timur mounted on horseback and went to the gate of the town, where he was met by the recalcitrant chief, who confessed his faults, and begged that his life might be spared, and that he might have one day's grace only, after which he would go to his camp. He was still treacherous, and devoted the night to fortifying the road of Lohra and other routes by which Timur's people had scaled the mountain, and then sought safety among the crags. Here he for a time escaped. Timur left in a fortnight, and went to the fortress of Kahkaha, between Abiverd and Kelat (still called Kaka), and gave orders that it was to be rebuilt, which his eager soldiers accomplished in two days and two nights. He planted a garrison there under Haji Khoja, and then sent his nominal suzerain, Siyurghatmish Khan, with Mirza Ali and Sheikh Ali, and their division, to blockade the mountain of Kelat, which was most effectually done.*

Timur now proceeded to secure Tershiz, a fortress which was held by some Ghurian dependants of the Malik of Herat, and had refused to submit. Having captured it, he marched onwards against Jorjan. He went by way of Rughi, and passed Kebud Yaemeh and Shamsan. Vali, on hearing of this, sent the Amir Haji with presents of strings of horses, each consisting of nine, with other curiosities, and a humble letter, begging Timur to excuse his going to him on this occasion, asking him to turn his army aside, and offering his submission. Timur reciprocated these messages, and marched upon Kelat. From there news arrived that Sheikh Ali Behadur had secretly during the night clambered up to the fortress with a few followers; that he had been seen by, and had a struggle with some of the garrison; but after fighting for some time, some religious people brought about an interview between him and Ali Bey, and that they

---

* *Id.,* 338-347.

had embraced mutually. The sheikh spent some days with Ali Bey, who tried to secure his good offices with Timur. Meanwhile, the latter returned home by way of Shamlagan and Charmagan, and pitched his camp in the pleasant meadows of Radikan. There he was joined by Sheikh Behadur, who presented Ali Bey, holding a sword, and a winding sheet in his hand, and begged the emperor to pardon him. Timur conferred Radikan upon the sheikh as an hereditary fee. He also gave Sebzevar to Ali Muayid, while the Malik of Herat, with Ali Bey and his people, were conveyed to Samarkand. He then distributed the district of Yun Garbanian among his amirs, but ordered that its inhabitants should be carried off to Transoxiana.* Ali Bey, with the son and brother of the Malik of Herat, were now arrested and sent to Andikan, where they were put in charge of the Mirza Omar Sheikh, while the tribe of Yun Garbanian was planted in the district of Tashkend.† It was not long before an outbreak took place at Herat, and Timur, to remove all danger, had the malik put to death, and with him Ali Bey. This took place in 1382.‡ The following year apparently he again marched westwards, intending to attack Mazanderan. He went by way of Termed, and having crossed the Murghab, pitched his camp there, and received information of two outbreaks, one at Sebzevar (i.e., the Sebzevar south of Herat) and the other in Seistan, and he dispatched Sheik Ali Behadur and Uchara Behadur with a large force to ravage the frontiers of the Amir Vali.§ Having overrun Seistan and Kabulistan, he returned to Samarkand, where he only stayed three months, when he again set out for Jorjan. He crossed the Oxus at Termed, halted some days at Balkh, so that his troops could gather round from various sides, and was speedily joined by 100,000 horsemen in armour. Having reviewed these troops, he crossed the Murghab, took the road to Burkei Tash, and descending to Sarrakhs, he went on to Abiverd and Nissa, where he learnt that the Amir Vali had fortified the citadel of Durun, and had shut himself up there with a body of troops. Sheikh Ali Behadur, Sevinjik Behadur, Mubasher, and other amirs, who commanded the advance guard, had an engagement with the army of the Amir Vali in a place called Ghiaukarsh. The struggle was well maintained. Mubasher was struck by an arrow in the mouth, which came out at his neck ; but he rushed, notwithstanding, upon his opponent, and cut off his head with his sabre. Timur, in consequence of this heroic act, gave him Ghiaukarsh and Hurberi as an hereditary fee. The enemy having been beaten, the victors advanced upon Durun, which was captured, its governor and garrison being put to the sword. They then entered the district of Chilaun, which was full of villages, and having crossed the river at Jorjan, went on to Shasuman. There they proceeded to build bridges over the rivulets, to

---

* Sherif ud din, i. 353-356.   † Id., 357-358.   ‡ Id., 359-362.   § Id., 367.

cut down trees, make roads, and other useful works. The advance guards of either army had a series of sharp encounters during twenty days, and when on the twentieth Timur crossed the bridge of Dervish, Vali advanced against him with the utmost bravery, but was beaten, and many of his people were pursued and killed. To avoid the danger of surprises, each regiment was ordered to entrench itself, then to make a kind of rampart of its shields, and in front of these to plant stakes. All this seems to show either that Vali was a more powerful prince than we suppose, and that Timur was, in fact, hard pressed, or else that the district was a singularly difficult one to fight in. Timur also planted an ambuscade. At length, in the night, Vali with a large force came out of his fortress to attack the camp. They raised a great shout on the right of the army where Miran Shah commanded, and cut down with their swords the palisades and shields which we have mentioned, many falling victims to their courage. Miran Shah now ordered a volley of arrows to be fired, and at the same time the force which had been put in ambush rushed out, sword in hand, and defeated their assailants. Vali had, *inter alia*, dug a number of pitfalls, into which many of his own men fell and perished. The enemy was pursued as far as Asterabad, the neighbourhood of which was laid waste, neither age nor sex being spared. This struggle took place in Shaval, 786 HEJ. (*i.e.*, 1384). Vali fled, with his wives and children and a few soldiers, towards Dameghan, by way of Langaru, and having left his family in the fortress of Girdkuh, famous, as we have seen, as the old stronghold of the Assassins, he went himself towards Rai. Timur sent Khodaidad Husseini, Sheik Ali Behadur, Omar Abbas, Komari Inak, and other chiefs in pursuit of him, and they almost caught him at Rai. He, however, escaped to the forests of Rustemdar, amidst whose recesses and rocks he eventually baffled his pursuers.

LOKMAN PADISHAH.

We have seen how Vali was a usurper, and how he refused to instal Lokman, the son of Toghai Timur, on his father's throne. That young prince, we are told by Sherif ud din, became a wanderer from one land to another until Timur conquered Asterabad, which he made over to him.[*]

In the spring of 1385, Timur again advanced westwards, and secured the district of Kumuz. He then crossed the mountains into Mazanderan proper, when the princes of Rustemdar fled, and his troops overran the country and secured a vast booty. The Amir Vali, who had retired to a place named Yalus, was so terrified that he also took flight, and Timur then marched towards Amol and Sari. Seyid

---

[*] Op. cit., i. 388-397.

Kemal ud din and Seyid Razi ud din, who were the princes of this part of the country, sent their naibs, or lieutenants, with precious stones and gold dust " to scatter over the Emperor's feet," with a considerable tribute. They also struck gold coins with the name and surname of the invincible Timur upon them, and had the khutbeh said in all the mosques of their country, and Timur ended by ordering them to follow the commands of Lokman Padishah, to whom he had given the principality of Asterabad, "hoping thus to restrain them from doing anything but what that prince thought best."* The Amir Vali fled with Mahmud Kalkali to Tebriz, where he helped to defend the place against the attack of Tokhtamish, Khan of Kipchak. When the town was captured by the latter, he, with Mahmud, withdrew to the district of Kalkal.† Presently Vali was seized by his treacherous host, Mahmud Kalkali, and handed over to Komari Inak, who put him to death, and sent his head to Timur.‡ This was apparently in 1386. I don't know when Lokman died, but it was before August, 1402.

PIR PADISHAH.

In August, 1402, Pir Padishah, son of Lokman, on whom Timur had conferred his father's government, feasted the great conqueror at Asterabad.§ A few months later he took part in Timur's campaign in Mazanderan.‖ In the summer of 807 HEJ., Timur being near Bostam, was visited by Pir Padishah, who offered him presents, including nine strings of horses, each consisting of nine animals. Timur gave him a robe, and sent him home again.¶ About the year 807 HEJ., a revolt broke out in the district of Sebzevar, headed by Sultan Ali, who was supported by a number of Serbedarians. The rebel marched against Seyid Khoja, who commanded in that district on behalf of Timur Sultan Ali, and who was being hard pressed when Pir Padishah, who is here called Perek, king of Mazanderan, went to his assistance. Thereupon a struggle ensued outside Sebzevar.** It ended in the defeat of the allies, and Perek, after showing great gallantry, was compelled to retire.†† Some time after, the Mirza Miran Shah having arrived at Kalpush, the late rebel Sultan Ali Sebzevari went to meet him there, and was apparently accompanied by Sultan Hussein, son of Pir Padishah. They were both arrested at the instance of the Mirza's officers, and were sent in chains to Herat.‡‡ Seyid Khoja defeated Perek Padishah, and captured his son,§§ who was called Sultan Ali, like the Serbedarian chief. We next read that Sultan Ali, son of Pir Padishah, fled from Samarkand, where he had been detained, and went to Asterabad. He was, however, arrested by Seyid Khoja, and sent to Shah Rokh, who treated him kindly and sent

---

Sherif ud din, i. 400-401.   † Id., 403.   ‡ Id., 411.   § Id., ii. 143.   ‖ Id., 149.
¶ Id., iv. 170-171.   ** Vide infra, sub voce Serbedarians.
†† Abdur Rezak, Notices et Extraits, xiv. 30.   ‡‡ Id., 54.   §§ Id., 81.

him to his father, together with the latter's wife. He, however, detained Sultan Hassan,* brother of Sultan Ali. Notwithstanding his rebellion, sent word to Pir Padishah that he should retain the government he held in the time of Timur, and bade him have no hesitation therefore in going to him, as he would be received with the greatest affection.†

We now read of one of those numerous revolts which shattered the power of the Timurids. Seyid Khoja, already named, one of Shah Rokh's most distinguished commanders, rebelled, and eventually sought shelter with Pir Padishah in Mazanderan. Shah Rokh sent Mengli Timur Naiman to the latter, to tell him how the rebel, whom he had raised from being a slave to be one of his generals, had revolted, and pillaged his towns and killed their inhabitants, while as soon as he marched against him he had fled like a fox before a lion. He reminded him how able he was to recompense him, and urged that he should surrender the fugitive.‡ Pir Padishah, in reply, sent Mengli Timur and Khoja Mekki from Asterabad, offering, if the province of Asterabad was secured to him, and if some fugitives who had sought shelter with him were pardoned by Shah Rokh, and allowed to return to Herat, he would take care to send the rebels back, together with his son. This letter was deemed impertinent, and Shah Rokh set out for Asterabad, but he could only advance slowly, on account of the heat. He reached Tenaseman, which was held for Pir Padishah by Bayazid Jupan ; but as the city of Asterabad was the real key to the position, he did not deign to attack the smaller place ; "just as falcons," says our author, "deem it a dishonour to kill sparrows, and lions to capture jackals." He accordingly advanced upon Asterabad. A skirmish took place at Feraskhaneh, in which some of Pir Padishah's behadurs were captured and put to death. The imperial army having reached the plain of Asterabad, was engaged in digging a ditch to protect itself from attack, when Pir Padishah, with his forces, emerged from a forest. He commanded the centre; Shems ud din Uj Kara, Shir Ali, and Jafar Sahib the left wing, and Seyid Khoja the right. After a severe engagement, the imperial forces were successful. Pir Padishah, much dejected, fled towards the desert, and thence to Khuarezm.§ Mazanderan, Jorjan, Dahisan, Asterabad, and Dameghan were now confided to the Mirza Omar Behadur.‖

On the fête day following the month of Ramazan in the year 808 HEJ. some people of Irak deserted the Timurid chief, Mirza Khalil Sultan left Samarkand, and went towards Khuarezm. Having arrived there they revolted in favour of Pir Padishah, who had been defeated by Shah Rokh in Mazanderan, as we have seen, and had sought refuge there. They scattered the gold over him which they had received from Mirza Khalil Sultan. He now proposed to himself to conquer

* (?) The same person previously called Sultan Hussein.
† Abdur Rezak, Notices et Extraits, 80.   ‡ Id., 91-92.   § Id., 96-99.   ‖ Id., 100.

Mazanderan.* Elsewhere the same writer tells us that, having collected together the partisans of Jaun Korban and Tavakkul, he entered the province of Mazanderan (really Jorjan) and attacked the fortress of Asterabad, and that Shems ud din Ali and Jemjed Karen, who were its governors on behalf of the Timurid ruler, were in great danger. Shah Rokh having heard of this, set out to repress the revolt. He duly arrived at Shebertu and the town of Kusuiaah, and arranged to attack the enemy by more than one route. While he himself advanced by way of Jam and Meshed, the Amirs Hasan Jandar Firuz Shah, Sheikh Ali Hasanek, and Ajeb Shir went by way of Zavah and Mahulat. When Shah Rokh reached Tarak, he heard that Pir Padishah had been abandoned by the troops of Khorasan, and had fled towards Rustemdar. He thereupon advanced into the plain of Asterabad, and Mazanderan was once more at peace.† Sultan Ali, son of Pir Padishah, now went with his submission, and presented himself to Shah Rokh, at Mehneh Mubarak. He was well received, and Shah Rokh took him with him to Seistan; but after the capture of the fortress of Férah, he suddenly fled, and repaired to Rustemdar, where his father had died.‡

SULTAN ALI.

Sultan Ali received the support of the Amir Kaiumars Rustemdari, while his father's dependants and the people of the surrounding districts joined him. He then determined to attack Asterabad, where Abulleith was governor on behalf of the Timurids. The two came speedily to blows, each at the head of a considerable force. In the struggle, Sultan Ali was so badly wounded that he died. His head was sent to Herat, and the victors secured a large booty.§

THE SERBEDARIANS.
ABDUR REZAK.

The history of this dynasty has been abstracted from Mirkhavend's and Khuandemir's narrations, by Dorn and Von Hammer. They tell us how there lived at the village of Pashtin, or Bashtin, in the district of Baihak, in Khorasan, a man named Shihab ud din Fazlullah, who was descended, on the father's side, from Hussein, the son of Ali, and on the mother's, from the Barmekid Yanhia ibn Khalid. He had five sons, Amir Amin ud din, Amir Abdur Rezak, Amir Wejih ud din Masud, Amir Nasrulla, and Amir Shems ud din. At this time, there was at Abusaid's

Court, a certain Ali Surkh Khuaji, or Surkh juni, surnamed Abu Muslim, who had a great reputation as an athlete and archer. One day, the Sultan remarked, "Is there no one to be found in our empire who can compete with Abu Muslim?" Amin ud din, who was at the head of the Pehluvans, replied that he had a brother in Khorasan, named Abdur Rezak, who was equal to this challenge. Abusaid thereupon sent a messenger to bring him, and in two months he arrived at Sultania, where his appearance and qualities made a great impression on the Sultan. Two or three days later, Abdur Rezak, passing through the market-place of Sultania, noticed that some one had hung up a bow and a bag of gold coin, in an archway, with a notice that anyone who could bend the bow might have the money. He took the bow and bent it, and then distributed the coin. When Abusaid heard of this his interest was still further aroused, and he ordered him to have a trial of skill in archery with Abu Muslim. They accordingly had a struggle, and, as Abdur Rezak's arrow overshot that of his opponent ten paces, the Sultan told the Vizier to give him a profitable post. He was accordingly assigned the collectorship of taxes in Kerman. These amounted to 120,000 gold kopeghi, of which 20,000 was assigned to his own use, and he had to account for the rest. He, however, squandered the whole sum ; and, when the troubles which succeeded the death of Abusaid came, he was in great embarrassment, and returned to Bashtin, where there was a great excitement on account of a demand which had been made upon the brothers Hasan Hamsa and Hussein Hamsa (with whom he lodged, and who were both pious men), not only for men, but also for some boys or maidens on behalf of the government (*i.e.*, the government of Toghai Timur Khan). The brothers said they were prepared to supply the first, but not the second—"On this we stake our heads" (serbedarim). They thereupon killed the messengers. Abdur Rezak, who arrived at this time, strengthened their resolution to resist ; and expelled the fifty men who had been sent by Khoja Alai ud din Muhammed, the Vizier of Khorasan, and who lived at Feryumed, to seize the brothers, and aroused the whole village to resist the scandalous tyranny. "A thousand times better," he said, "that our heads should go to the gibbet rather than suffer such gross tyranny. We lay our heads on that." From this phrase they derived their name of Serbedars (*i.e.*, those who hazard their heads). This took place in the year 737 HEJ. (*i.e.*, 1336).*

The "Chronicle of Herat," which writes in a very hostile spirit to Abdur Rezak, says that, having murdered a distinguished reis, he had surrounded himself with a body of abandoned people and criminals, who had sworn to put away all the agents of the government, or to take their own heads to the gibbet (ser be dar).† Abdur Rezak's repute now grew daily. He

---

* Dorn Mems. St. Peter's Acad., 6th ser., viii. 161-163.   Ilkhans, ii. 324-326.
† Journ. Asiat., 5th ser., xvii. 506.

attacked and defeated the Vizier, who withdrew from Feryumed to Asterabad, where he was pursued and overtaken by Wejih ud din, Abdur Rezak's brother, and was killed at the village of Valayed, near that town. Khuandemir calls the place Sherek i nau. The Vizier's son fled to Sari, while Abdur Rezak returned to Bashtin with the booty he had secured. He now found himself at the head of 700 brave men, and proceeded to occupy Sebzevar, where there was no one to oppose him, and which he now made his head-quarters. The Amir Abdulla Mulai, the ruler of Kuhistan, wished to marry the daughter of the late Vizier, and sent some splendid presents as a marriage gift. Having heard of this, Abdur Rezak sent Muhammed Aitimur to waylay the treasure. The Kuhistanis, however, who had 700 well-armed men, defeated him. Thereupon, Wejih ud din Masud, his brother, came up with 300 men, dispersed the mountaineers, and carried off the presents. Abdur Rezak now determined to secure the Vizier's daughter for himself, but suspecting that he merely wanted to make use of her for a base intrigue with her brother, who was very beautiful, she fled from Sebzevar towards Nishapur.* Von Hammer makes out it was the Vizier's widow whom the Serbedarian chief proposed to marry, and she thought he wanted by this means to secure her beautiful daughter. Abdur Rezak, on hearing of her flight, sent his brother Masud in pursuit, who overtook her at Sengkinder, but allowed her to escape on her urging upon him that, as a good Mussulman, he ought not to take part in such a violation of the law. On his return without her, Abdur Rezak, who was indignant, declared he was no man. He replied sharply, and from a strife of tongues, it came to one of swords. Abdur Rezak sprang through the window, broke his foot, and was put to death by his brother, who succeeded to his authority. This took place, according to the "Chronicle of Herat," 12th Zilhidjeh, 738 (i.e., July, 1338).†

WEJIH UD DIN MASUD.

During the reign of Abusaid, there flourished at Sebzevar a sheikh known as the Sheikh Khalifa, the founder of the order of Dervishes. Sebzevar had already distinguished itself as an abode of heresy. He was apparently a heretic, and aroused great animosity and feeling. He was at first a pupil of the Sheikh Balu Amoly, in Mazanderan; but, some time after that sheikh's doctrines being deemed deficient, he went to Semnan, where he joined the Sheikh Rukn ud din Alai ud daulat Semnany, and spent some time in his monastery, which was a famous seat of learning. He then went to Bahrabad, where he sought out the Khoja Ghiath ud din Hibet ulla Hamevy, and ended by settling at

---

* Dorn, op. cit., 163-164.
† Ilkhans, ii. 324-326. Journ. Asiat., 5th ser., xvii. 506. Dorn, op. cit., 164-165.

Sebzevar, where he entered a monastery, and devoted himself to a close study of the Koran, and to the inculcation of various austerities to a large number of scholars. He seems to have acquired a reputation for working miracles, but his teaching was disapproved of by the ulemas or professors of the law, who forbade him access to the ruler's house. He did not heed them, however. Thereupon his enemies prepared a case for judicial decision, viz., "If a man who establishes himself in a mosque, and discourses on worldly matters, and not only pays no heed to the directions of the learned, but even becomes more stubborn, is it lawful to put him to death?" The greater part of the authorities decided in writing that it was lawful, and sent the decision on to the Court. Abusaid gave the judicious reply that he did not wish to shed the blood of the dervish, and that the officers of the law in Khorasan were to deal with him in the way prescribed by the law. This answer caused the greatest discord among the authorities in such matters at Sebzevar, who had tried so hard to ruin the sheikh; and the discord was at its height when the Sheikh Hasan Juri, a native of Jur, came to Sebzevar, and proclaimed himself a disciple of the unpopular and probably heretical sheikh. This greatly increased his reputation; but his enemies were determined to undo him, and he was found one morning hanging from one of the pillars of the mosque, while a number of stones were at his feet, as if he had hanged himself. His scholars thereupon chose Sheikh Hasan as their leader, and left Sebzevar. He went to Nishapur, Abiverd, Khabushan, and Meshed, spreading the doctrines of his dead master, and when anyone declared himself his scholar he took his name down, but bade him stay at home for the present, and to hold himself in readiness in case he were wanted, when he would be duly summoned. His manner was very persuasive, and he speedily secured an immense following, and the authorities began to be afraid.*

To stop this dangerous excitement the Amir Arghun Shah Juni Kurbani, who commanded in Khorasan on behalf of Toghai Timur Khan, had the sheikh imprisoned in the castle of Tak, or Takh, in the district of Yazer, or Yaser. Thereupon, the Serbedarian chief, Wejid ud din Masud, who doubtless saw in the whole matter a chance of furthering his ambitious schemes, declared himself an adherent of the imprisoned sheikh. Amir Arghun marched against Nishapur with three armies, numbering altogether some 70,000 men. Masud defeated two of these armies, led respectively by the Amir Mahmud of Izferain and Tukal, and compelled them to retreat; and as the third army, under Arghun, arrived too late, and the latter found his other contingents defeated, he withdrew. This religious strife naturally led to the great increase in power of the leader of the Serbedars,

---

* Dorn, op. cit., 166-167.

who now became master of Nishapur and Sebzevar. A disciple of the Sheikh Hasan Juri, called Khoja Asad, now set to work, with seventy companions (Von Hammer says seven), to obtain the release of his master, which he succeeded in doing with the help of the Serbedarian chief, who surprised the citadel of Yazer, released the sheikh, and carried him off in triumph. Meanwhile, Amir Mahmud, the son of Arghun Shah, who, in the absence of his father (who had gone to Irak with Toghai Timur), held his authority, wrote a letter to the sheikh, dissuading him from causing confusion throughout the land by his opinions. The sheikh replied in a long letter (preserved by Mirkhavend), in which he pictured his whole life, and his alliance with Wejih ud din, in the fairest light, and stating that he only aimed at furthering the interests of the Faith.

Wejih ud din now marched upon Herat, which he hoped to conquer. That town and province had recently begun to prosper again, under the beneficent rule of Moiz ud din Hussein Kert, who had succeeded Malik Hafiz about the year 1332. He naturally dreaded the approach of the fanatical Serbedarians, and brought together the contingents of Ghur, Khaisar, the Sinjars from Seistan, and the Nigudars, left Herat, and advanced as far as the neighbourhood of Savah, near Nishapur. There a bloody struggle ensued. It seemed as if the Serbedarians would win the day, for the bravest officers in the Herat army were killed, and their men began to disband, when the Malik, who had planted himself with his guards on an eminence, ordered the charge to be sounded, and the great standard to be unfurled, and putting himself at the head of his wavering men, restored the fight. The struggle began again with terrible energy, but meanwhile Sheikh Hasan was stabbed by one of his own disciples. This news having spread among the Serbedarians, they were seized with sudden panic, and dispersed. Khuandemir says Sheikh Hasan had told his friend that, if he should fall, Masud must withdraw at once, and return to Sebzevar. The Serbedarians were closely pursued, and "a great number of these infidels," says the chronicler, "were killed or made prisoners." Among the latter was the Amir Ibn Yemin, whose life was spared on account of his skill as a poet. After this victory Moiz ud din returned to Herat with a rich booty.*

After his defeat, Wejih ud din Masud returned to Sebzevar, where his fanatical followers do not seem to have lost faith in him; for he was able to make head against Toghai Timur, who was *de jure* ruler of Khorasan, and who, no doubt, deemed it quite time that the turbulent Serbedarians were put down. We have seen how he marched against their chief, and how he was defeated by him. Toghai Timur, by his defeat, lost hold entirely of Khorasan, and his dominions were restricted to Jorjan.

After his victory, Masud went to Asterabad, and issued a decree to the
people of Mazanderan, in which he summoned them to submit to him.
This district Had retained a quasi-independence, under its own princes,
during all the Mongol domination. Kia Jemal ud din Ahmed Jelal, an old
man of experience, " who," says Khuandemir, " had already tested the heat
and cold of time," had recently usurped authority in that province, and
recognised no one as his superior.   Fearing an invasion by the Serbedarian chief and its probable consequences, he arranged with his two
nephews, Kia Tash ud din and Kia Jelal, to pay Masud a visit. They were
received with honour. This strengthened Masud's intention to proceed ;
and he set out, in a peaceful humour apparently, accompanied by these
chiefs. Presently, he sent word to Jelal ud daula Izkander, the governor of
the town of Rustemdar, to fix a place of meeting. The latter consulted with
his brother, Fakhr ud daula Shah Ghazi. They agreed that some sacrifice
must be made, and that it would be well to surrender some districts to the
Serbedarians ; but if the summons meant they were going to enter by
force into Rustemdar and occupy, they must be opposed by arms. Masud
set out in 1342, and went to Amol, pitching his camp, which he inclosed
with palisades, in the district of Buran. The troops of Izkander and Shah
Ghazi mounted, during the night, their Arab horses, made a raid on the
outskirts of this camp, and succeeded in securing some plunder ; their
motive probably being that the erection of the palisading looked like a
permanent occupation of the country. This attack was repeated ; and,
we are told, Kia Jemal ud din sent to tell the inhabitants there and the
other people of Mazanderan that, notwithstanding his being in the
company of Masud, they were to spare no efforts to drive the Serbedarians
out.   The Mazanderan people, therefore, collected in considerable
numbers, made night attacks on the camp, and cried out : "O, men of
Khorasan ! Mazanderan is a wood full of fierce lions. You have with your
own hands opened the gates of misfortune for yourselves, and planted your
feet in the net of calamity and woe. You have fallen together into the
decoy of destruction, and none of you will escape from this dangerous
place." " Masud," says the chronicler, "felt like a fish in a net, in view of
these unseen enemies and unseen threats—hardly knowing which way to
turn."   At length, after waiting for nine days, he determined to go on to
Rustemdar.   When he reached the village of Yasminelateh, he found
the men of Mazanderan barring his path in hostile array. He thereupon
put to death Kia Jemal ud din and his nephews, and then set out on his
retreat by the route of Lavij. He and his men were vigorously pressed,
and some were killed and others captured at almost every step. He
reached the Irlu river with a few followers, but found that his retreat
was cut off in that direction by a force under Sherif ud daula Gustehem
ibn Taj ud daula Siar, and turned aside ; but the troops of that commander
followed him, and captured him in the village of Bandis. They carried

him off to Izkander.   For two days he was kept in custody, and on the
third day was put to death.*   Mirkhavend says the people of Rustemdar
did not want to kill him, but the son of the vizier, Khoja Alai ud din,
whom he had put to death, demanded his blood.   Masud's vizier, the Seyid
Sehir ud din, was also captured, and, when questioned, said that forage
had to be provided every night for 14,000 horses, 600 mules, and 400
camels ; whence the number of the invading force may be guessed.†
Masud was put to death in 1344, and had reigned seven years.

MUHAMMED AI TIMUR.

When Wejih ud din Masud marched against Sheikh Ali Kaʿan, he
left Muhammed Ai Timur, also called Aka Muhammed Timur, one of his
father's servants, and distinguished for his courage and liberality, as his
substitute in Sebzevar.   When he heard of Masud's death, he took upon
himself the government.   We have seen how important the dervishes
and religious devotees were in that town.   Muhammed seems to have
neglected them.   When he had reigned for two years, the Khoja Shems
ud din Ali, who was distinguished alike by his birth and his other good
qualities, gathered round him a body of dervishes and scholars of Sheikh
Hasan, entered Muhammed's room unexpectedly, and addressed him in
the following words : " It is a strange thing that the dervishes have no
longer any power or consideration, although your fortunes and those of your
master were really made by this venerable brotherhood.   You treat them
as if they were mere ragamuffins and a thievish rabble."   As the Khoja's
companions used similar language, Muhammed was confounded.   Being
alone, and having no weapons by him, he addressed them in a kindly
speech.   " Hitherto," he said, " I have not injured a single dervish.   I
have taken the greatest pains and solicitude in the affairs of the state.
Meanwhile, I am ready to do what you think advisable."   Thereupon
they replied, " Get up and enter this room, for we no longer desire
your government."   He now entered the room named, whereupon the
mutineers closed the door behind him, and begged the Khoja to seat
himself on the throne, promising to support him.   Although he
apparently coveted the position, he did not wish to make it appear as if
he had personally profited by Ai Timur's death, so he refused the post for
himself, and suggested that Kalu Izfendiar should be raised to the throne,
while Ai Timur should be put to death as a punishment for his evil
doings.   His death is variously dated in 1346 and 1347.   His reign lasted
two years and a few months.‡

---

* Dorn, op. cit., 170-171.   † Id., 171. Ilkhans, ii. 329.
          ‡ Dorn, op. cit., 171-172. Ilkhans, ii 340.

KALU IZFENDIAR.

Shems ud din, as we have seen, raised Kalu Izfendiar, who was favourably disposed towards him, to the throne. He was not long, however, in falling out with the dervishes, and was accused of tyranny and ill conduct, and was put away, as Ai Timur had been, after reigning only a few months. This is the account given by Khuandemir. Fasih says that Muhammed was immediately succeeded by Lutf Ulla, the son of Masud, who was deposed in ten days, on account of his youth and incapacity. Mirkhavend says Lutf Ulla was merely proposed, but never acknowledged. Jennabi says he was actually acknowledged, but Shems ud din nominated him as regent.

SHEMS UD DIN FADL ALLAH.

The dervishes once more took counsel with their guide, Shems ud din, as to who should be placed on the throne. Some were in favour of Masud's son, but he was deemed too young, and eventually his nephew, Shems ud din Fadl Allah, the son of Abdur Rezak (Fasih calls him Shems ud din, son of Fazl Ulla, brother of Masud), was fixed upon. Devoted to luxury, he neglected taking precautions against the aggression which threatened him from Jorjan, whither Toghai Timur had retired, and left the control of matters in the hands of the Khoja Shems ud din 'Ali, and in the course of seven months he was pointedly urged to resign his position, which he accordingly did. One of his couplets is preserved by Khuandemir :—

The kingdom of poverty and a life of hardship are better than to be a ruler ;
A breath of tranquillity of mind is better than anything a man can wish for.[†]

SHEMS UD DIN ALI.

The Khoja Shems ud din Ali, who had hitherto been a kingmaker, now accepted the responsibility of government himself. We are told he ruled with the greatest severity, to restrain on the one hand the turbulence of the dervishes, and on the other the licentiousness of the Serbedarians. He banished the use of wine from his dominions, and five hundred prostitutes were on one occasion put to death ; Fasih says they were burnt. He used to perambulate the streets at night, to see for himself how order was kept. Meanwhile, his valour, determination, and continual state of preparation for war, deterred his neighbours—Toghai Timur, in Jorjan, and Hussein, the Malik of Herat—from attacking him. He suppressed the outbreaks which took place at Tus and

---

* Dorn, op. cit., 172.   † Id., 171-172. Ilkhans, ii. 340.

Dameghan, but was eventually assassinated by Haidar Kassab, after he had reigned four years and nine months. This Hassan had charge of the Customs. His accounts having been overhauled, he was found to be a considerable sum in arrear. The Khoja thereupon gave orders that what property Haidar had appropriated should be confiscated. As he had nothing left, and the demands kept accumulating upon him, he seized an opportunity and laid his distressed condition before his master. Shems ud din, who was a harsh man, impervious to shame, replied: "Let thy wife enter a brothel, and in this way cancel thy debt to the Divan.' Affronted by this brutal answer, he determined upon revenge, and confided his purpose to the Khoja Yahia Kieravi. At the time of evening prayer, he went up to the fortress when Yahia was in Shems ud din's assembly-room, and, having demanded his compassion for his misfortune, sprang forward and thrust a dagger through him. Hasan Dameghani would have killed the murderer, but was stopped by Khoja Yahia. He replied that he did not wish the deed to appear as if it had Yahia's approval. The date of these events is variously given. Khuandemir apparently puts it in 753 HEJ. (*i.e.*, 1352), Fasih in 755 (*i.e.*, 1354), Mirkhond in 753, and Jennabi in 756 (*i.e.*, 1355), adding that Shems ud din had reigned five and a half years, and was 65 when he died.[*]

YAHIA KIERAVI.

The Khoja Yahia Kieravi, who was privy to the murder of Ali, now became chief of the Serbedarians. He also maintained stringent discipline and strict manners, and devoted the greater portion of his time to profound disputations with the learned. His whole Court was dressed in woollen garments, and only the most learned men were appointed to public offices. In consequence of this he was held in the highest respect by his neighbours. Ghazan Khan, the ruler of Mavera un Nehr, sent him splendid presents, and Toghai Timur, whose realm was now reduced to the district of Jorjan, made a treaty of peace, and interchanged verses with him. Toghai Timur having summoned him to go to him with his submission, he duly went, but took advantage of his visit to assassinate him, as I have described. The Serbedarians now occupied Asterabad, Bostam, Dameghan, Khuar, Semnan, and Taberan, and withdrew to Sebzevar with their booty. When Yahia had reigned four years and eight months, his brother-in-law, Alai ud din (Fasih calls him Alai ud daula, while Mirkhond says it was Yahia's brother Iz ud din), sprang on to the saddle behind him when entering the courtyard of his house (Khuandemir says on his horse, Mirkhavend on his mule), and thrust a dagger into his side. The Khoja himself

---

[*] Dorn, op. cit., 173-174. Ilkhans, ii., 340-341.

seized him. Both fell from the horse together, and as they struggled, the wounded chief gave his antagonist a mortal blow, so that both perished together. Haidar Kassab was at this time at Asterabad with Sultan Maidan. As soon as he heard the news, he went to Sebzevar. Those who had concurred in the murder of Yahia withdrew to the fortress of Shekkan, to which Haidar applied fire for sixteen days, so that it was burnt, together with all who had sought refuge in it. These events, according to Khuandemir, took place in the year 756 HEJ. (*i.e.*, 1355). Fasih and Jennabi say in 759 (*i.e.*, 1357).*

KHOJA DHAHIR UD DIN KERIAVY.

Khuandemir says that, according to the Matla el Sadain, the prince who now succeeded was a nephew of the Khoja Yahia Keriavy, while the author of the "History of the Serbedarians" says he was his brother, as does Fasih. He was appointed chief of the Serbedars, with the concurrence of Haidar Kassab. He was liberal, good natured, and steadfast, and fond of playing at dice and cards. Haidar Kassab had charge of the administration. He deposed the Khoja, according to Mirkhavend and Khuandemir, after a reign of forty days; Fasih says in the year 760 (*i.e.*, 1359), after a reign of eleven months; while Jennabi says after a year and a month.†

HAIDAR KASSAB.

After Haidar had reigned for four months, he was assassinated by the Pehluvan Hasan Dameghani, named Kutluk Bugha, at the instigation of his master. Fasih calls the murderer Kutluk Aga, and says it was in 762 (*i.e.*, 1360). Mirkhavend says Haidar had previously expressed his intention of marching against the Amir Vali in Asterabad, but gave up his plan after he had advanced a day's march. He then went to Izferain.‡

LUTF ULLA.

The Amir Lutf Ulla, son of the Amir Wejih ud din Masud, now succeeded to the chieftainship of the Serbedars, under the tutelage of Hasan Dameghani. After a reign of a year and three months at Sebzevar, discord broke out between him and Hasan, in consequence of which the latter had him imprisoned in the fort of Destjerdan, and gave orders that he was to be put to death. The Serbedars styled him Mirza.§

---

* Dorn, op. cit., 174-177. Ilkhans, ii. 340-342   † Dorn, op. cit., 177-178.
‡ Id., 178.   § Id., 178.

HASAN DAMEGHANI.

The Pehluvan Hasan Dameghani now reigned (*i.e.*, in 1361). During his rule the dervish Azis, a disciple of Sheikh Hasan Juri, acquired great fame for his pious observances in the holy city of Meshed, and collected many people about him. With their aid he acquired some power, and captured the fort of Tus. Hasan, on hearing of this, marched to Tus, which he re-captured, gave the dervish some loads of silk, and bade him not remain long in this district. He accordingly went away to Ispahan, and settled there. Hasan seems at this time, according to Mirkhavend, to have marched against the Amir Vali, at Asterabad, whence, having been defeated, he returned again. Some time after, Khoja Ali Muayid Sebzevary revolted in Dameghan, drove out Nasr Ulla, who commanded there on behalf of Hasan, and ordered Mahmud Riza to go to Ispahan to fetch the dervish Azis. He said he would do so on condition that, if the Khoja became the chief of the Serbedarians, he would nominate him his vizier. He agreed to this, and Mahmud duly brought the dervish, known as "the father of excellence," to Dameghan. Khoja Ali enlisted himself among the disciples of the dervish, an example which was followed by many inhabitants of the place. Presently, a conspiracy broke out in the fortress of Shekkan, or Shegghan, against Hasan. He accordingly left Sebzevar, and marched against and attacked the rebels. When news reached the Khoja Ali and his *protégé*, the dervish Azis, that Hasan had left Sebzevar, they marched thither, occupied it without any trouble, and speedily secured the place and the control of the army. They seized the vizier, Khoja Yunis Semnani, and put him to death, in revenge for that of Amir Lutf Ulla. Hasan saw there was no other course than to submit, and set out for Sebzevar with the intention of becoming a disciple of the dervish Azis, and a subject of the Khoja Ali ; but the latter had issued orders that he was to be put to death. A number of people therefore went out to meet him, held the reins of his horse, and assisted him to dismount, and one of them cut off his head, which they sent to Khoja Ali. According to Fasih, this occurred in the year 763 (*i.e.*, 1362).* Hasan had reigned four years and four months.

KHOJA ALI MUAYID

Jennabi gives Ali Muayid the surname of Nasr Ulla ; while Fræhn, who has published some of his coins, gives his name, in full, as Nasr Allah Nejm ud din Ali ibn el Muayid. As he attained his position through his devotion to religion, he continued to be very attentive to its calls, and spared no pains to give honour to the great Seyids ; while, in the

---

* Dorn, op. cit., 178-179.

expectation of the coming of the Lord of Time, he had a horse prepared
for him each night and morning. He abstained from the use of wine and
hemp juice (? bang).* We are told in the "Chronicle of Herat" that he
gave a great impetus to the spread of the Shia doctrines, and had the names
of the twelve imaums placed on the coins.† When he had been on the
throne nine months, he gave the Dervish Azis the command of an army,
and sent him against the Malik of Herat, Moiz ud din Hussein Kert.
When the dervish reached Nishapur, the Khoja, who had apparently
changed his mind, sent word to the commanders of the army that they
were to abandon Azis and return to him. They viewed this as a piece of
good fortune, and duly returned to Sebzevar; but the dervish, with a body
of his scholars, set out for Irak. Thereupon the Khoja sent some people
in pursuit, who put them all to the sword. This he followed up by
ruining the tombs of the dervish Khalifah and of Sheikh Hasan Juri in the
market-place of Sebzevar, and making them over to the market people as a
latrine. All this was done probably in the interests of orthodoxy. In 777
(*i.e.*, 1375), Ghiath ud din, the Malik of Herat, captured Nishapur, which
was one of the Khoja's possessions, and appointed Izkander Sheikhy, son of
Afrasiab Chelabi, as its amir. In 778 HEJ. (*i.e.*, 1376), the dervish Rukn
ud din, a disciple of Sheikh Hasan Juri and the dervish Azis, went to
Fars, and asked for a contingent of troops from Shah Shuja. When he
presently arrived in Khorasan at the head of this army, Izkander Sheikhy
acknowledged himself as his disciple, and both went together to Sebzevar.
Khoja Ali was not in a position to resist them, so he withdrew towards
Jorjan; while the dervish Rukn ud din planted himself at Sebzevar, and,
in 779 (*i.e.*, 1377), had the khutbeh said and money struck in his own name.
In 780 (*i.e.*, 1378), however, Amir Vali, who, on the death of the Khan
Toghai Timur, had become the ruler of Jorjan, offered the Khoja Ali his
assistance, and they marched together on Sebzevar. Rukn ud din now
fled, and the Khoja once more mounted the throne. In 781 (*i.e.*, 1379),
he visited Asterabad and asked assistance from the amir (*i.e.*, the Amir
Vali) against the great Timur, who was now pursuing his victorious
career; but finding that he could not get any substantial help from him,
he determined to submit, and handed the great conqueror the keys of
his towns. He visited him at Nishapur, and there offered his submission,
was well received by him, and lived peaceably under his protection for the
rest of his life.‡ Sherif ud din says Timur conferred Sebzevar on him.§
He died in 1386, apparently of a wound received in the campaign of
Timur against Luristan.|| Fræhn has published three of his coins, struck
in the years 772, 777, and 780 HEJ., at Sebzevar, Dameghan, and Lais
Abad (? Asterabad) respectively.¶ A coin of his in the British Museum
was struck in 775 HEJ. at Asterabad.

---

* *Id.,* 179.   † Journ. Asiat., 5th ser., xvii. 515.   ‡ Dorn, op. cit., 179-180.
§ Op. cit., i. 356. || *Id.,* 407.   ¶ Recensio, 632-633.

z *

On the death of Ali Muayid, Timur seems to have nominated one of his own people governor of Sebzevar. About the year 791, while he was encamped at Alkushun, news arrived that the Serbedarian princes and Hajibey Yun Garbani had revolted, and had been joined by the garrisons of Kelat and Tus. Timur sent the Mirza Miran Shah against them. He encountered the rebels near Behrabad, a town not far from Tus. They charged with their usual brave recklessness, " but," says Sherif ud din, "they were inclosed by the right and left wings of the Mirza's army like birds in a net." They were cut in pieces. Prince Meluk, half dead, escaped to Persia, with two or three companions. Meanwhile, the Amir Akbuka had marched upon Tus from Herat. He captured that town after Hajibey, who was a younger brother of Ali Bey Yun Garbani, had managed to escape; but he was captured at Semnan by the Sherifs of Hezarégheri, and sent to the Mirza, who had him put to death.*

SULTAN ALI.

About the year 807 HEJ., Sultan Ali, son of the Khoja Masud, forgetting the benefits he had received from Timur, rebelled against his son, Shah Rokh. He put himself at the head of a body of Serbedarians and others, and occupied the districts about Sebzevar. Seyid Khoja sent a body of 600 picked men to attack him, while he opposed them with 200 well-armed horsemen, who made a vigorous charge, and as they consisted of old, well-seasoned men, they cut in pieces a large number of their opponents. Seyid Khoja was much disturbed by this defeat, and went hastily to retrieve it with 2,000 men. On reaching the battle-field, however, he only saw a number of mutilated corpses. Going on to Jajerm, he overtook a party of the rebels, who retained the martial virtues for which the dervishes had been famous. The place was taken, and the people in it slaughtered, but it cost him dear, he himself having been wounded. Seyid Khoja then went on to Firumed, whence the people went for shelter into the citadel. He accordingly proceeded to destroy the gardens and orchards in the vicinity. The Seyids and learned men of the place entreated his clemency, and he agreed to spare it on the payment of a ransom. He then went on to Mezman, which he captured. Thence he passed to Sebzevar, which he ordered to be girdled round with a deep ditch. After an attack of ten days, news arrived that Pir Padishah (or Perek), the king of Mazanderan, was marching to the rescue. Thereupon the siege was raised, and Seyid Khoja went to meet the latter. Meanwhile, Sultan Ali-left the town and joined Perek. The two armies met. As is so often the case, the

* Sherif ud din, ii. 31-34.

right wing of each broke the opposing left wing; but eventually Perek, who commanded the centre, was obliged to give way. Sultan Ali, who was in charge of the right, hearing of this, went after him. A cruel slaughter of the fugitives followed, and Seyid Khoja returned towards Sebzevar with a great booty.[*]

After his defeat, Khoja Sultan Ali Sebzevari repaired to Asterabad. When he heard that the Mirza Miran Shah had arrived at Kalpush, he went to meet him there. The Mirza's officers, Seyid Khoja and Midrab, urged that his recent rebellion should be duly punished. He was accordingly arrested and sent in chains to Herat, together with Sultan Hasan, son of Perek Padishah, while the rest of the conspirators at Sebzevar were put to death.[t] Seyid Khoja afterwards defeated Perek Padishah a second time, captured his son, crushed the power of the Serbedars, and re-organised the province of Khorasan.[‡]

Thus closes the history of a singular episode in Eastern history, in which we have a number of religious devotees and fanatics virtually controlling, for a considerable period, the fortunes of such an important province as Khorasan.

THE MALIKS OF HERAT.

MOIZ UD DIN HUSSEIN KERT.

The history of the Ghurian chiefs descended from Shems ud din Kert, who had authority at Herat from the days of Khulagu to the death of Abusaid, has been incorporated in the previous pages with that of the several Ilkhans, their contemporaries, to whom they were in fact subject and subordinate. When the power of the Ilkhans broke to pieces, like other dependent chiefs elsewhere, the maliks of Herat, who had a more established position than any of them, naturally followed their example, and became independent. We have seen how, in the year 732 HEJ. (i.e., 1331-32), Malik Hafiz was murdered by a number of turbulent chiefs.[§] He was succeeded by his brother, Moiz ud din Hussein, thanks to whose vigour the rebels who had caused confusion during the previous two reigns were subdued, and the people who had been driven from Herat by the civil commotions there, returned to their hearths. This tranquillity was broken by the attack of the Serbedars, which we have described.[||] The result of the war was not only the defeat of the Serbedars, but the Malik of Herat was acknowledged as master in Kuhistan. He then proceeded to subdue the districts of Shiburghan and Andkhud, and surprised the tribes of

---

[*] Abdur Rezak, Notices et Extraits, &c., xiv. 26-30. Ante, 724-725. [t] Id., 53-54.
[‡] Id., 81. [§] Ante, 622-623. [||] Ante, 730.

Arlat and Eperdi (Aibirdi) in Badghiz, dispersed them, and built two columns out of the heads of the prisoners whom he had caused to be decapitated. These two columns were placed one on each side of the Khiaban, near the tomb of Fakhr ud din Razi. "They are still in existence," says the chronicler of Herat.* At this time there was a struggle going on in Transoxiana between the Amir Kazghan and his master, the Jagatai Khan, Kazan Khan, which we shall describe later on, and Moiz ud din deemed the time favourable to declare himself independent. He accordingly did so, and adopted the various signs of royal prerogative, including the five flourishes of trumpets (nôobet), &c.† His enemies, including the chiefs whom he had vanquished at Badghiz, together with the sheikhs of Jani, denounced him to Kazghan, who, in a rage, demanded if the stock of Jingis Khan was extinct, that no account seemed to be made of the royal majesty. "Does not this Ghurian plebeian any longer recognise it? Does he think there is no one greater than he?" Kazghan declared he would ruin his towns and fortresses, and make a river wider than the Oxus out of the blood of his most valiant soldiers.‡ He consequently assembled at Balkh all his forces from Andkhud as far as Kashgar, and was joined by the Khan and the other princes, and marched upon Herat. When the Malik heard of this, he sent an amir with 300 men as far as the Murghab to explore, who speedily returned with the news that Kazghan was on the march, and that the dust raised by the tramp of his men and war machines rose up to heaven, &c. The Malik now summoned his council, and said to them: "An army, so large that the dust it raises obscures the sun, has invaded Persia from Tartary, composed of men who in the assault are as firm as a mountain, and when charging with their heads low are as impetuous as torrents rushing down from the rocks. These heroes only put on their helmets when they have resolved to sacrifice their lives for the sake of victory." Each one present then said his say, and notwithstanding that the invaders were more numerous and more skilled in fighting a pitched battle—the Malik Hussein's forces amounting only to 4,000 cavalry and 15,000 foot soldiers—he determined not to stay in the town, nor the citadel, nor in the gardens and streets of the environs, in order not to create alarm, and so that the enemy might be the more easily surprised. It was determined to go out and meet the foe, and also to make an intrenched camp to the east of the town, and extending from Kehdistan as far as the village of Bui Murgh. The Malik addressed his men, and bade them fight bravely, saying that it was not the strong battalions that always won, but that victory depended more on courage and skill, and if they showed these, the enemy would find the world too small as a place of refuge.§

---

* Journ. Asiat., 5th ser., xvii. 505-508.   † Id., 508-509.   ‡ Sherif ud din, i. 5-9.
§ Journ. Asiat., 5th ser., xvii. 509.

Meanwhile, Kazghan passed Pashnan, and advanced upon Kehdistan, and the next day he mounted on horseback with the Khan Bian Kuli and the Princes Sitilmish, Uljaitu, and 30,000 troops, who occupied the heights surrounding Kazurgah. Kazghan surveyed the Herat army, and remarked that the Malik was ignorant of the art of war, and that the place where he had planted his camp would secure his defeat on two grounds—first, because his people would have to charge uphill to reach their enemy, while the latter would be able to charge down upon them ; and, secondly, that when the sun rose, its light would be in their faces, and they could not see who was approaching. The Amir Kazghan and his companions descended from their vantage already assured of victory. The next day their army was ranged in the form of a crescent, and advanced after the usual patriotic address. Kazghan mounted a height, and ordered the attack to begin. The battle was hotly fought, and the cry of " Sela " arose from both hosts. The word meant that no quarter was to be given ; and soon the ground was strewn with blood, bucklers, cuirasses, and lances, mingled with corpses. Hussein's men were at length defeated, and as he had opened some dykes behind him to prevent the fugitives from retreating, a portion of them perished in the morass thus created, while the rest were pursued by the Mongols, who made a terrible carnage. Hussein retired to Herat with difficulty, accompanied only by his guards, who occupied the gardens and crooked streets in the environs, while Kazghan returned to his camp. The next day he began the siege, and his men fought night and day, at night by the light of fires which were made for the purpose. The siege lasted for forty days, when terms were proposed. Sherif ud din and the chronicler of Herat are at issue as to which side proposed them, but it would seem probable that they came from Hussein, who sent his enemy a present of some splendidly caparisoned horses, curious stuffs, and rich carpets, together with a large sum of money, and promised that if Kazghan would withdraw, he would go the next year in person and make his submission to him. This promise was, as usual, accompanied by a solemn oath. Kazghan, who was of an amiable and generous disposition, readily assented, as he saw the country would be utterly ruined if the war continued, and set out again for Transoxiana, accompanied by the Khan. This was in the year 752 HEJ. (*i.e.*, 1351 A.D.)*

The reverse which he had sustained aroused the animosity of his subjects against the Malik, especially of the turbulent Ghurian chiefs in his service, and his brother, Malik Bakir, was set up as a rival by the rebels. Seeing himself surrounded by a number of Ghurians, whose conduct was menacing, he had the presence of mind to point out some Mongols to them who had come from Badghiz with horses for sale as proper subjects for pillage, and profiting by the disorder which followed, he fled from Herat to Amankuh. There he learnt that his brother was being proclaimed, and determined accordingly to go to Transoxiana, and in effect to keep the promise he had made the year before. When he arrived on the borders of Transoxiana, he met Kazghan with a party of people hunting. He thereupon dismounted and approached him with only two servants, and with every confidence. Kazghan embraced him, saying: "Friend or enemy, you are always a man of spirit." The Malik having told him of what had happened at Herat, he promised to reinstate him there; but the Mongol chiefs were envious of these attentions, and they asked Kazghan to arrest him. As he refused to do this, they formed a plot to put him to death. Kazghan thereupon summoned him, and told him what was going on, adding that it was not in his power to save him, and begging him accordingly to escape on a fleet horse. He did so, and re-entered Herat without being recognised, made his way to the citadel, where his partisans gathered round him, and gave orders for the arrest of his brother. That young prince, who had been a mere plaything in the hands of his officers, was exiled to Fars, and lived there till his death. Directly Moiz ud din re-mounted the throne, he invaded Kuhistan, whose chief, Sitilmish Bey, asked assistance from Muhammed Khajeh, who ruled at Andkhud, Shiburghan, and the country as far as the Oxus. The latter marched with his army to Badghiz, and there joined Sitilmish. The Malik met the confederates at Firamurzan on the route to Sarrakhs. The two allies rushed with impetuous bravery on the troops of Herat, and were both killed. Their men then fled in disorder. After this victory, which was won on the 25th of February, 1358, the Malik returned to Herat, where he had an interview with the Amir Chaku, who had been sent to him as an envoy by the Great Timur, and whom he received with due honour. He promised to go to Sarrakhs to meet Timur, and there make a treaty of peace with him. The latter, who was well aware of the way in which the Kert princes had murdered Nuruz, Danishmend, and Choban, did not place much confidence in these promises; but he deferred any hostile intentions he might have, and sent his son Jihanghir, under the care of Mubarek Shah Sinjari, to Herat. Shortly after, in the year 1369, when Timur's position was more assured by the death of his rival Hussein, Moiz ud din was taken ill with a dangerous sickness. Seeing his end was approaching, he summoned his chief dependants, and made them swear allegiance to his son, Ghiath ud din Pir Ali, whom he nominated his successor. His other son, Malik Muhammed, whose mother belonged to the Arlat tribe, received Sarrakhs as an appanage. The dying chief made Ghiath ud din promise not to molest his brother's possessions, and gave him some wise counsel, reminding him that God and the Prophet had never permitted an unjust and wicked ruler to hold dominion long at Herat. Moiz ud din died in June, 1370, and was buried under the northern dome of the Great Mosque, near the tombs of his father, Ghiath ud din Muhammed, and of Muhammed, son of Sam.*

GHIATH UD DIN PIR ALI.

The new ruler, faithful to his father's advice, used every effort to live at peace with his brother, Malik Muhammed ; but the latter was led astray by perfidious counsels. He suppressed his brother's name from the khutbeh, and put his own on the coins. Thereupon Ghiath ud din marched against Sarrakhs in person ; but the severity of the winter caused an interruption of operations, and he consented to enter into an arrangement. The two brothers had an interview, and parted apparently reconciled.†

In the spring of 773 (i.e., 1371), Ghiath ud din sent Haji Vizier, as his envoy, to Timur, who was then hunting near Karshi, with presents of Arab horses, sumpter mules, mules to be used as palfreys, and a quantity of rich stuffs, girdles, and robes. He was presented to Timur at Kabamiten. Among his presents, we are told, was a famous piebald horse, named Konk Oghlan, bearing a golden saddle. Timur received the envoy courteously, and gave him a robe and other presents ; and then wrote the Malik a letter full of friendly phrases, which he sent him, together with a present of a state robe.‡

Meanwhile, Khoja Ali Muayid, the ruler of the Serbedarians, profited by the power he had acquired in the district of Sebzevar, gave an impetus to the spread of Shia doctrines and caused the twelve imaums to appear on the coins. At the instigation of several ulemas of the Hanefi sect, who represented to him that it was his duty to oppose the progress of this sect, Ghiath ud din made several invasions of the Nishapur district in successive years. That district was governed by the deputies of Muayid. The third of these campaigns was marked by a cruel devastation of the country, in which the canals were laid dry, and the trees uprooted. The chronicler of Herat quotes the following story from the " Matlá es Saadin " : "One day a peasant from the neighbourhood of Nishapur was captured and taken before the Malik, who, wishing to learn his religious views, said : ' Good man, how many fundamental dogmas are there in Islamism ? ' The peasant replied, without hesitation, ' Sire, according to your sect, Islamism reposes upon three dogmas—destroy the crops of the Mussulmans, fill up their canals, and tear up their trees.' " This answer had such an effect on Pir Ali that he returned, with his army, to Herat. But the following year (i.e., 1375), he renewed his invasion, and captured Nishapur, of which he nominated Izkander Sheikhy, son of Afrasiab

* Journ. Asiat., 5th ser., xvii. 510-515.   † Id., 515.   ‡ Sherif ud din, i. 229-231.

Jelali, governor. The next year, Timur sent an envoy to Herat to conclude a treaty of peace. The Malik was much pleased at this, and ordered his son, Pir Muhammed, to repair to his camp; that young prince accordingly set out for Mavera un Nehr. This was in 1377. Timur received him kindly, and affianced him to his niece, Sevend Kütluk Agha, daughter of Shirin Beg; and, having given him many presents, sent him back again, and a few days after the princess was sent after him. Pir Ali prepared a splendid reception for her, and several triumphal arches, richly decorated, were raised between the Jui nu and the round space of the great bazaar. The nuptial festivities lasted several days, and the Malik, to show how much honoured he felt by the alliance, heaped presents upon amir Daud and Muayid Arlas, who had conducted the princess.* The good feeling between the two rulers did not, however, last long. Timur, in 1378, sent an envoy to the Malik to inform him that he proposed holding a kuriltai, or diet, in the spring, where the various princes of the empire would assemble, and that he hoped he would attend it. The Malik replied that if his friend the amir Haji Seif ud din Berlas was sent to escort him—he had much faith in him—that he would at once set out to pay his respects. Timur assented, and the following day sent Seif ud din to Herat. The Malik delayed setting out for a long time, under various pretences—his real object being to provision the town, to supply it well with weapons of defence, and to perfect its fortifications. Only the previous year he had surrounded the place with a new wall two leagues in circuit, which inclosed the faubourgs and gardens outside the old town. The amir having noticed what was going on, returned alone to his master, and reported what he had seen.†

This was a sufficient excuse for Timur to prepare his sword. In the autumn of 1380 he nominated his son, Mirza Miran Shah, governor of Khorasan, although he was only fourteen years old, and appointed as his assistants the amir Jehanghir, brother of the amir Haji Berlas, together with the amirs Haji Seif ud din, Akbuka Osman Abbas, Muhammed Sultan Shah, Komaré (brother of Temuké), Taban Behadur, Uruz Buka (brother of Sarbuka), Pir Hussein Berlas, Hamza (son of the amir Musa), Muhammed Kazghan, Sarik Etekeh, Muzaffar (son of Ushkara), and others. He also gave him fifty companies (? *hazarahs*) of cavalry. They crossed the Oxus by an admirable bridge of boats, and spent the greater part of the ensuing autumn and winter at Balkh and Shiburghan. Towards the end of the winter they occupied Badghiz, where they secured a large booty of horses and other valuables. In the spring, Timur himself collected his forces, and prepared to advance. He crossed the Oxus and proceeded to Andkhud, where he paid a visit to the devotee, Babasenku, one of the class of dervishes who are either half-witted or profess so to

---

be. When Timur approached him, the dervish sent a sheep's breast at his head. Timur accepted this as a good omen, saying, " I am assured that God has granted me the conquest of Khorasan, for this kingdom has always been called the breast, or the middle, of the habitable world."*

Meanwhile, Malik Muhammed, the brother of the ruler of Herat, who, as we have seen, held the appanage of Sarrakhs, went to offer his submission. He was well received, and given some presents. Timur now advanced across the Murghab, and determined to cut the communication between Herat and Nishapur. When he reached Kusuyeh, called Kusupa by Sherif ud din, its governor, Pehluvan Mehdi, submitted, whence the inhabitants of that district were not molested. Timur turned aside to Taibad, and visited the learned doctor, Zain ud din Abubekr Taibadi, who was greatly renowned for his piety and austerities, and with whom he had a long conversation.† Rejoining his army, he went to Fushenj, which threatened resistance. It was duly environed, and four days after an assault was ordered. Timur, we are told, dispensed with his cuirass in order to encourage his men ; and a heavy storm of weapons fell on either side, Timur himself being struck by two arrows. Mirza Shah, son of the amir Muayid Arlas, who resembled Timur, Aiku Timur Belkut, Omar, son of Abbas Mubasher, and others, scaled the ramparts. The Sheikh Ali Behadur and his younger brother, Khosru Buket, with Mirek, son of Elchi, and other warriors traversed the ditch and reached the city gate, which they burst open. Meanwhile, breaches were made in various places, and, after desperate fighting, the fortifications were captured, the garrison was put to the sword, and the town pillaged, while the walls of the citadel were razed. This being the first town of Iran which had fallen, the capture was the subject of great rejoicings. The town of Fushenj, we are told, was famed for its strength, being surrounded by high walls and an excellent rampart. Its citadel was deemed one of the strongest in the world, and was girdled by a deep ditch filled with water. It was also well provisioned, and otherwise supplied.‡ Timur now proceeded to attack Herat. The army laid waste the gardens in the environs, and destroyed their walls ; and then built lines of contravallation about it. Meanwhile, Talek, with a body of Ghurians, deemed the bravest people of Iran, made a sortie, but were repulsed. Sherif ud din says " that the citizens, who knew little of what had passed, preferred the repose of their houses, decorated with the beautiful porcelain of Kashan—a town four days' journey north of Ispahan—to the troubles of war, and thought only of their own safety." Talek determined to arouse them, and sent public criers round, ordering them to repair to the walls and to share their defence among them. A deaf ear was turned to these appeals.§

---

* Id., 314-316.   † Id., 316-318.   ‡ Id., 319-322.   § Id., 323-324.

The "Chronicle of Herat" says that Timur's people made no progress for four days, when some intrepid soldiers advanced towards Kiushk Murghani, and discovered a subterranean conduit, which conveyed water into the city. This they seemed to have followed, and a combat ensued. Pir Ali, who was posted with the *élite* of his men at the gate near Pul Enjil, could not drive them back, and withdrew to the citadel. Khalil Yasaul now stormed the walls, and drove out the defenders. Two thousand Ghurians were made prisoners, and taken before Timur, who greatly praised their prowess, presented each of them with a tunic, gave them their liberty, and bade them tell the citizens that no harm would come to them if they remained quietly in their houses.* Seeing that resistance was hopeless, the Malik determined to send his mother, Sultan Khatun, daughter of Toghai Timur Khan, with his eldest son, Pir Muhammed, accompanied by Izkander Sheikhi, who was said to be of the race of Bijen (a famous princely stock of ancient Persia), to offer his submission. Timur presented the young prince with a magnificent robe, and then sent him back with his mother to bid him come to him in person, while he retained Izkander Sheikhi to inform himself of the affairs of the country, and of what had passed in the town. He then went to a kiosk in the garden of Baghé Zaghan (*i.e.*, the garden of the ravens). The Malik spent the next day in preparations. "The day following," says Sherif ud din, "abandoning his pride, he quitted the city, and went to find Timur, whose carpet he kissed while on his knees, and asked pardon for his fault." He was presented with a robe of honour and a jewelled girdle. The following day the Sherifs, mollahs, imaums, and grandees of the town also went to kiss the conqueror's carpet, and to greet him in the accustomed fashion. The conquest of Herat took place in the month of Moharrem of the year 783 (*i.e.*, March, 1381). Timur now moved his camp to Kehdistan, east of the city, while the treasures amassed by the Ghurian princes were carried away. Among these were money, uncut stones, very rich thrones, crowns of gold, silver bowls, gold and silver brocades, &c.† The old walls of Herat, as well as those recently built, were razed. An impost was levied on the town in lieu of its being pillaged, which was paid in four days. Molla Kutb ud din, son of Molla Nizam ud din, who was the chief imaum and doctor in the country, with 200 old men of consequence, were ordered to go to Shehr i Sebz as hostages, and Timur Tash, nephew of Akbuka, governor of Termed, was ordered to escort them. The gates of the town, which had iron bands of chiselled work upon them containing various inscriptions (the "Chronicle of Herat" says they were inscribed with the names and titles of the various Maliks), were transported to Kesh, "where," says Sherif ud din, "they remain to this day."

Pir Ali had another strong fortress, named Izkelejeh or Amankuh, which he had intrusted to his younger son, Amir Ghuri, who determined to resist. Timur accordingly sent the Malik, his father, to Izkelejeh, to persuade him to surrender, forbidding him, under grave penalties, at the same time to enter the citadel. Pir Ali went to the foot of the walls, and after a parley with his son, persuaded him to be prudent, and returned with him to Timur,* who forgave him, and gave him a royal robe. Timur now sent Jihanshah Yaku to secure Nishapur and Sebzevar, and seems, after making a detour towards Kelat, to have gone in person to Nishapur, and then to have received the submission of the Serbedarian. chief of Sebzevar. His troops then captured Izferain, which belonged to Vali, the ruler of Mazanderan. Having wintered at a country house, called Oghul Yatu Yailak, he made several regulations for the government of Khorasan. He sent Ghiath ud din back to Herat, and the other princes and governors to their several posts, and sent his son, Miran Shah, to Sarrakhs, who arrested Muhammed, brother of Ghiath ud din, and sent him to Samarkand.†

The next year Timur once more crossed the Oxus, and marched westwards. Near Kelat he was joined by Ghiath ud din, with the troops of Herat, while Miran Shah similarly came from Sarrakhs. He first attacked the rebellious ruler of Kelat, and, having left a force to blockade that place, marched upon Tershiz, where Ali Sedid Ghuri held command on behalf of the Malik of Herat. Timur passed by way of Yassi Dapan, and reached Kabushan, whence he sent the princess Dilshadaga, who was not well, back to Samarkand. Tershiz was situated among mountains, and had the reputation of being impregnable, on account of the exceptional height of its walls and the depth and width of its ditches. It was garrisoned by Ghurians, who were a race famous for their prowess, and was well supplied with provisions and munitions of war. Timur complained to Ghiath ud din that these people, who were his subjects, should resist him so stoutly. He said it was due to their ignorance, and he volunteered to go and speak to them. He did so, but they would not heed his orders, and the siege accordingly proceeded. The walls were sapped, and endeavours were made to draw the ditch. The resistance was bravely sustained, but at length the walls and parapets were almost ruined by the stones fired from the catapults, and the garrison, worn out, asked for quarter. Timur treated them well, praised them for their bravery, took them into his service, and appointed their commander governor of the frontier between Turkestan and Kashgar. Tershiz itself was confided to the care of Sarik Etekeh.‡ Timur now marched against Mazanderan, whose ruler submitted to him. He thereupon returned to Samarkand, taking the Malik Ghiath ud din and

---

* Sherif ud din, i. 327-328. Journ. Asiat., 5th ser., xvii. 519.   † Sherif ud din, i. 332-334. ‡ Id., 347-350. Journ. Asiat., 5th ser., xvii. 520.

his children with him.* The Malik's grandson, Amir Ghuri, and his brother, Malik Muhammed, were arrested and placed in confinement at Andikan. He himself, with his eldest son, Pir Muhammed, were also imprisoned.† Herat was confided to Amirgah, son of Timurgah. Meanwhile, a serious outbreak took place in the recently conquered district of Khorasan. Malik Muhammed and Abusaid his brother, the sons of Fakhr ud din, formerly ruler of Herat, had led a life of great poverty during the reigns of the two last maliks. When Timur conquered Herat they represented to him how they had been robbed of their fortunes by Malik Hussein and his son. Timur pitied and released them. He intrusted the government of Ghur to the elder of them, Malik Muhammed, and a famous Ghurian chief, Abusaid Espahbed, who had been kept in prison for ten years by Ghiath ud din, was released.

Meanwhile, the Mirza Miran Shah, with the troops of Herat, had gone in the autumn of 1381, to encamp on the Murghab, in a place called Yendi, which the Persians called Penchdeh (i.e., the five villages). Malik Muhammed deemed it a good opportunity to surprise Herat. He advanced against it with a body of people, and was joined en route by Abusaid Espahbed, with a party of vagabonds. The governor and garrison were obliged to withdraw to the fortress of Ikhtiar ud din, where the governor died. Its gate was burnt by the rebels; whereupon the Turkish garrison, to escape the flames, tried to escape over the walls, but were cut to pieces. Mirza Miran Shah having heard of this, at once sent the amirs Seif ud din and Akbuka towards Herat, and followed with the bulk of his army. The Ghurians marched against them, and a struggle took place in the promenade of Khiaban, in which they were defeated. A terrible massacre of the inhabitants now followed, and a very high tower was built up of their heads. Herat itself was largely reduced to ruins. Ezfizar, whose governor, Ali Daud Khototai, had also rebelled, suffered the same fate. Ali Daud himself was burnt in his house; and two thousand of the citizens were buried alive in holes filled with mud. "It was thus," says the chronicler of Herat, "that an outbreak, fomented by miserable adventurers, spread over Khorasan, and devastated its two most beautiful districts; and the ambition of two fools cost the lives of many innocent people, and converted a land, which was the image of paradise, into an arid desert. We belong to God, and we supplicate Him to prevent the return of such terrible calamities."‡

When Timur heard of the outbreak, he gave orders that the Malik Ghiath ud din, his brother Malik Muhammed, his grandson Amir Ghuri, and Ali Bey Yan Garbani "were to say their prayers and prepare for death" (i.e., he had them executed).§ In the year 791 HEJ. the garrisons of Kelat and Tus having revolted, Timur returned hastily to Samarkand,

---

* Sherif ud din, i. 356. † Id., 357-358.
‡ Journ. Asiat., 5th ser., xvii. 520-522. Sherif ud din, i. 358-360. § Sherif ud din, i. 361.

and put to death Pir Muhammed, son of the Malik Ghiath ud din, and his two sons, Zain al Abidin and Mahmud, and thus put an end to the 'royal stock of Herat.*

SIVAS.

BURHAN UD DIN.

I must conclude my notice of the fragments into which the Ilkhanian Empire broke with a few words about a small and short-lived principality at its western extremity. When the general break up of authority took place in the latter years of the Khanate, which led to the supremacy of the Turkomans in Armenia, we are told that Ahmed Kadhi Burhan ud din, who had been a judge in the service of the Prince of Kaiseriyah, and who combined literary and political abilities, secured the allegiance of various Tartar tribes, which had obeyed the Ilkhans, and which were encamped in the district between Sivas and Kaiseriyah. Elsewhere we read that on the death of the Prince of Kaiseriyah, Ahmed Burhan ud din, with his amirs, divided that principality among them, Haji Geldi took Tokat, Sheikh Mejik took Amasia, and Burhan ud din himself Sivas. Burhan ud din was doubtless supreme, and was able to put 20,000 or 30,000 armed followers in the field. His turbulent followers speedily made attacks on the borders of the Ottoman Sultan Bayazid's territory, and having determined to prevent this new enemy from acquiring too great an authority, he marched against him. Burhan ud din was too prudent to face the Sultan, and retired towards Diarbekr to the mountains of Kharpurt. There he was eventually attacked by Osman beg of the tribe Bayanlu, who then ruled at Diarbekr, and was killed with the greater part of his following. Sherif ud din says it was Bayazid who put him to death, and who also captured his son, Kara Osman.† Another son, Abed Abbas Burhan ud din, we read, attacked his father-in-law, the Prince of Erzenjan, and appropriated his principality, and was afterwards defeated and killed by Kara Yuluk, the Turkoman chief.‡

SEINOL AABIDIN.

On his death, Burhan ud din's people were divided as to what should be done. The greater part of them wished to transfer their allegiance to his son Seinol Aabidin, but a minority deemed it more prudent that their country should accept the powerful Ottoman ruler as its master, and Seinol Aabidin, who was apparently a boy only, was sent to his relative Nasir ud din beg. The Prince of Sulkadr and Bayazid appropriated his dominions, which included the towns of Sivas, Tokat, and Kaiseriyah.§ These events took place in 1392.

* Id., ii. 32.   † Op. cit., iii. 256.   ‡ Von Hammer, Osm. Reiches, i. 153.
§ Zinkeisen Gesch Osm. Reiches, i. 351-352.   Von Hammer Gesch des Osm. Reiches, i. 189.

GENEALOGY OF THE MUZAFFARIANS.

GHIATH UD DIN KHORASANI.

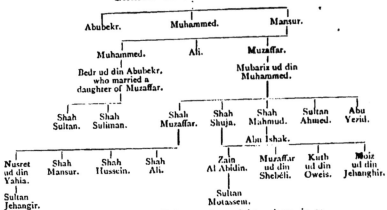

This table is copied from Defremery, Journ. Asiat., 4th ser., iv. 95.

GENEALOGY OF THE MALIKS OF HERAT.

Taj ud din Osman Merghani.

Malik Rokn ud din Abubekr.

Shems ud din (first Malik),
1245-1283.

Rokn ud din (Shems ud din II.),
1278-1283.

Fakhr ud din,   Ghiath ud din,
1285-1307.   1308-1328.

Shems ud din II.,   Malik Hafiz,   Moiz ud din,
1328.   1329-1331.   1333-1370.

Ghiath ud din Pir Ali,
1370-1373.

Note 1—Hosn Kaifa.—A small principality which survived, as a vassal state, throughout the Mongol domination in Persia, was Hosn Kaifa, or Hosnkeif. Quatremere has given a detailed account of it.* I have described its capture by Khulagu Khan.† The place was ruled by princes of the Ayubit family; and I owe to Dr. Rieu some notices of them which may be welcome to some of my readers. It was in the year 629 HEJ. that the Ortokids were dispossessed of Amid and Hosn Kaifa by the Ayubit chief, Malik al Kamil, son of Malik al Adil, who was Sultan of Egypt. Kamil, who died in 635 HEJ., was succeeded as ruler of the Ayubit dominions by his son, Abubekr, surnamed Malik al Adil, and he by his brother, Malik al Salih. During Salih's reign, his son, Moassem Turan Shah, was in possession of Hosn Kaifa. He went to

---

Egypt on his father's death in 647 HEJ., and the next year was assassinated. Turan Shah was the last Ayubit Sultan of Egypt. His son, Muayid, succeeded him at Hosn Kaifa, and, as we have seen,* was ruler there when that place was captured by the Mongols, who put him to death. His full name was Al Muayid Abdallah. · His son was called Kamil Abubekr Shadi, and his son Salih Yusuf. Salih Yusuf paid his respects to the Sultan of Egypt, Malik al Nasir, son of Kelavun, in Ramazan, 726, and was dismissed with marks of favour, and a recommendation addressed to the amir Choban. On his return to Hosn Kaifa, he was put to death by his brother, Adil Mujir ud din Muhammed, who reigned after him.† He was succeeded by his son, Adil Shihab ud din Ghazi, who left two sons, viz., Salih Saif ud din Abubekr and Adil Fakhr ud din Suliman. The first of these succeeded him, and set out, in the year 776 HEJ., for Makka, with the intention of abdicating. He was, however, persuaded to return again; but was soon after superseded by his brother, Suliman, who, in 783, fought with Gharz ud din, chief of the Sulimanis, and the amirs of Bedliz and Diarbekr.‡ In the year 803, he went to Timur, when he was on the way to Mardin, to whom he gave presents and did homage. Timur gave him a robe of gold tissue, a girdle of precious stones, and a sword with a golden scabbard.§ Suliman died in the year 827 HEJ., and was succeeded by his son, Ashraf Ahmed, who was a poet and a patron of learned men. Having encountered the Sultan Barsabai before Amid, he was slain by a body of Turkomans, and was succeeded by his son, Malik Khalil, who compelled Karailik—who had made an inroad into his territory—to come to terms.‖ He died in 832 HEJ , according to the "Sherif Nameh," which says his nephew, Khala ibn Suliman, defeated Hasan beg Ak Kuyunli. He was murdered in his bath by his cousin, and Hosn Kaifa then fell into the hands of the Turkomans.

Note 2—Great Luristan.—I carried the story of this principality down to the death of Afrasiab and the succession of his brother, Nusrat ud din Ahmed, in 696 HEJ.¶ Hamdullah speaks of the latter as an excellent ruler, who made the province rich and prosperous. He made Malik Kutb ud din, son of Imad ud din Pehluvan, lieutenant, and nominated him his successor; and gave the command of the army to Khosru Shah, son of Hosam ud din, who joined their efforts to make the country prosper. Ibn Batuta speaks of the atabeg in the same terms. He says that he built 460 hermitages in his kingdom, of which there were 44 at Idhej, his capital. He divided the taxes of the kingdom into three equal parts. The first was devoted to supporting the hermitages and schools, the second to paying his troops, and the third to the maintenance of his family and servants. He sent a present annually to the King of Irak (*i.e.*, Abusaid Khan), and often paid him a visit. The greater part of his pious foundations were on high mountains (*i.e.*, the Kurdish mountains), through which the roads had to be cut through the solid rock, which was very hard, and yet were so well made that sumpter beasts could traverse them with their loads.

---

* Ante, 161. † Abulfeda, v. 364. Tarikh al Aini, add. 22,360, f. 176.
‡ Inba al Ghumr, add. 7,321 ff. 14-272. Shifa al Kulub, add. 7,311 ff. 126-127.
§ Sherif ud din, iii. 355-357. ‖ Inba, f. 307. Suluk, 164.
¶ Ante, 407. I have erroneously called the latter Nasir ud din, as Von Hammer in one place does. Ilkhans, ii. 321.

Oaks grew there, with the acorns of which they made bread. At each station was a hermitage, at which food for the traveller and forage for his beast were provided, whether he asked for it or no; each person receiving two round loaves, some meat, and sweets made of grape juice mixed with flour and butter, which were duly provided from the legacy left by the atabeg. At each hermitage there was a sheikh, an imaum, a mueddhin, a servant to assist the poor, and servants of both sexes to cook the food. Ahmed was an ascetic, and wore a horsehair shirt close to his skin, which, as we have seen, attracted the attention of Abusaid.[*] According to Defremery, he died in the year 1332.[†] The "Habib Ussiyer" says he died in 733 HEJ. He was succeeded by his son, Yusuf, who, Ibn Batuta says, reigned for ten years; but Defremery makes his successor mount the throne in 1339.[‡] The " Habib Ussiyer " says he died in 740 HEJ. Hamdullah says he followed in his father's footsteps. In the history of Mirza Izkander, abstracted for me by Dr. Rieu, we read that he captured Shuster, Huwaizah, and Basrah, and that he reigned for five years and nine months. Yusuf was succeeded by his brother, Afrasiab, who was reigning when Ibn Batuta visited Idhej, which, with Shuster, were then the chief towns of Great Luristan. He describes Shuster as a beautiful city, with fertile surroundings, and girded about by a river, on which were water wheels. There was living there a famous preacher named Sherif ud din Musa, whose preaching was unapproached elsewhere. He tells us that, after he had finished his discourse, a number of people sent him pieces of paper with critical questions upon them, which he took up and answered in the most finished style. He surrounded himself with learned men and with disciples. The town of Idhej, he tells us, was the residence of the atabeg. Ibn Batuta failed to see him for some time, as he only went out on a Friday, being an habitual drunkard. He had an only son, who fell ill. The authorities fancied that our traveller was a leader of a body of fakirs. They sent him some refreshments and money, and some musicians, with a request that he and his fakirs would dance and pray for the young prince. He protested that he and his companions knew nothing of music or dancing: they offered prayers for the Sultan. The same night, however, the young prince died. The next day, the sheikh bade him accompany the grandees of the town, khadis, fakirs, sherifs, and amirs to present their condolences. He went, and found the hall of audience filled with men, children, slaves, soldiers, &c. They were dressed in rough carpets and horse covers, and had their heads covered with dust and straw; some had also cut off the hair from the front of their heads. They were divided into two bodies, planted at either end of the audience chamber, and advanced towards one another, striking their breasts, and saying Khoudsarima (i.e., " My Lord "). The spectacle disgusted Ibn Batuta. The khadis, khalifs, sherifs, &c., in the audience chamber had covered their ordinary dress with dirty cotton cloths, &c., roughly made, and some of them having pieces of dervishes' habits or black veils on their heads. They leaned against the walls and wept, or pretended to do so, or looked down at the ground. This mourning lasted forty days, and the Sultan sent each of the mourners a robe after it was over.

Ibn Batuta, looking round and not seeing a place to sit, noticed a raised platform, on one corner of which, apart from other people, sat a man, dressed in a woollen garment made of felt, such as was worn by travellers in rainy or snowy weather. To the astonishment of everybody, our traveller went up to this man, mounted the platform, and saluted him. He returned the salute, half rising in order to do so. Ibn Batuta then sat down at the opposite angle, while the audience stared. One of the khadis beckoned him to descend, but he sat on, and he then realised that it was the atabeg himself. Presently, the bier was brought in; orange trees, lemons, and citrons, bearing fruit, were carried on either hand—lanterns, fixed on lances, going in front. A prayer having been said over it, it was conveyed to a place called Hóláfihân, four miles from the town. There was a large college there, traversed by a river, and inclosing a mosque; outside was a bath and a large orchard, where refreshments were given to travellers. Some days after, the atabeg summoned Ibn Batuta. He found him seated on a cushion in a room in which there was no carpet, on account of the mourning, and before him were a pair of covered bowls, one of gold and the other of silver. There was a small green prayer carpet in the room, which was spread for the traveller near the prince. There was no one else present except his chamberlain, the fakir Mahmud, and an attendant. The atabeg inquired from him about the Sultan of Egypt, and about the province of Hijaz. He noticed that the prince was rather drunk; and on his telling him, in Arabic, to speak, he ventured to say: "You are one of the children of the atabeg Ahmed, celebrated for his piety and devotion. There is nothing to reproach you for in your government save that," and he pointed with his finger to the two vessels. He was ashamed at these words, and presently said, "It is a mark of the Divine pity to be allowed to consort with such as you." Presently, seeing that his host was nodding from side to side, and wished to go to sleep, he rose and left. He had left his sandals at the door, but did not find them on leaving. Two fakirs went up and down stairs to find them, and, when found, one of them kissed the sandals and put them on his head in sign of respect, saying, "God bless you; what you have said to our Sultan no one else could say. I hope it will have made an impression on him." On his way from Idhej, Ibn Batuta stayed some days at the place of royal sepulture previously named. The atabeg sent him and his companions a present of coin. On leaving there, he travelled for ten days through the mountains, staying nightly at one of the hospices already named. On the tenth day he reached a station called Giurisva's rokh, which marked the boundary of the atabeg's dominion.* According to the historian of Mirza Izkander, already mentioned, Afrasiab reigned eleven years, when he was succeeded by his son, Nur al Ward, who spent the treasures accumulated by his forefathers in pleasure and in pious foundations, and reigned for thirty-nine years. "The Jhanara" says he was taken and blinded by Muhammed, the Muzaffarian, A.H. 756, and that he was succeeded by his nephew, Pesheng. Mirkhond tells us how, when the Muzaffarian princes, Shah Shuja and Shah Mansur, were at strife, the atabeg Shems ud din Pesheng, who was being hard pressed by the

---

* Ibn Batuta, ii. 30-42.

A 2*.

latter, sent to ask the aid of the former, promising to conquer Shuster for him.
They eventually attacked Shuster together, and captured it, after which
Pesheng set out for Idhej.* Pesheng, according to the "Jihanara," was
succeeded by his son, Ahmed. The historian of Izkander says Nur al Ward
was succeeded by his *son*, Ahmed, who alienated all by his suspicious
disposition, and killed his brother, Hosheng, doubtless the Pesheng of the
other story. He paid homage to Timur, when the great conqueror attacked
Fars in 789 HEJ.† The same writer tells us that on his return to Fars, in the
spring of 795 HEJ., the Atabeg of Great Luristan, whom he calls Pir Muhammed,
went to do him homage at Ram Hormuz, and made him presents, and
was well received.‡ This interview was at the instance of some of Timur's
grandees. In the history of Izkander, we read that on Timur's return to
Samarkand from this campaign (*i.e.*, in the year 798 HEJ.), he brought Ahmed's
two brothers, Afrasiab and Masud Shah, as hostages. He subsequently
released Afrasiab, and divided the country between him and Ahmed. After
Timur's death, Mirza Pir Muhammed seized upon the latter, who remained
four years confined at Kuhendiz. He was released and restored in the year
811°HEJ., and was eventually slain by his own people. His son, Abusaid
after being kept for one or two years by Izkander at his court, was sent to
Luristan as his father's successor. He was still living in 815 HEJ., when the
biographer of Izkander wrote. "The Jihanara" tells us he was succeeded by
his son, Shah Hussein, who was, some time after, killed by his relative, Ghiath
ud din ibn Kavus ibn Hosheng. This was in 827 HEJ. Ghiath ud din,
who was the last of his line, was expelled by the Mirza Ibrahim, son of
Shah Rukh.

Note 3—*Little Luristan.*—I have already related the earlier history of Little' ;
Luristan during the Mongol domination.§ I brought it down to the accession
of Iz ud din Mahmund in 695 HEJ. He is called Iz ud din Amir Muhammed
in Mr. L'Estrange's MS. translation of Hamdullah, which is before me. That
author tells us he was very beautiful when a child. Bedr ud din Masud, the
son of his uncle Hasan, who was his senior, opposed him. In the reign of
Uljaitu orders came that Bedr ud din Masud should be governor of one section
of the country, and take the title of atabeg, and that Iz ud din should be governor
of another.‖ But presently both provinces came into the hands of Iz ud din again,
and he reigned over them until his death, whereupon Daulat Khan became queen
of the country. During her reign various governors were appointed by
the Mongol divan, which, says Hamdullah, is the case to the present day.
Dr. Rieu has abstracted some later notices of Lesser Luristan for me from
a general history written in A.H. 815 for Timur's grandson, Mirza
Izkander. We read there that Iz ud din Hussein, brother of Daulat Khatun,
who had married Yusuf Shah, the ruler of Great Luristan, was appointed
atabeg of Little Lur by Abusaid at the beginning of his reign, and that
he reigned fourteen years. He was succeeded by his son, Shuja ud din
Mahmud, who disgusted his subjects by his tyranny, and became the victim of

---

a general conspiracy. His son, Iz ud din, succeeded him at the age of twelve, A.H. 750, and had a long and prosperous reign. He gave a daughter in marriage to Ahmed ibn Oweis Ilkani, and another to Shah Shuja, the Muzaffarian. Defremery calls him Fakhr ud din, and tells us how, after having been a dependant of Shah Shuja, he afterwards sought refuge with Sultan Oweis. He then sent Shah Shuja a present of horses and precious objects, had the khutbeh said in his name, and also put his name on the coins.* Sherif ud din tells us that in the year 788 HEJ, Timur determined to prosecute a campaign in Iran, and, having crossed the Jihun and reached Firoz Kuh, he made inquiries about the behaviour of various rulers in the West, and learnt of the actions of Malik Iz ud din, prince of Lesser Lur, and of the crimes continually perpetrated by his people, who were great robbers, and had recently plundered a caravan on its way to Mekka. He determined to punish the freebooters, and, having selected two men from every ten, he put himself at the head of a flying column of picked troops and invaded Luristan. He ravaged Urudgurd and its environs, and eventually captured Kurrambad, an almost impregnable fortress, which he razed to the ground, and put to death the greater part of its defenders, who were thrown down from the rocks. Having thus subdued Little Luristan, Timur returned to his camp in the plain of Nehavend.†
It is necessary to remember that at this time there was a second prince of the same name, and who is qualified as Iz ud din Shirin, who ruled over Van and Vastan, and who also felt the weight of Timur's arms.‡ After his various victories, Timur held a reception at Shiraz, which was attended *inter alios* by the princes of Luristan. The biographer of Izkander tells us that Iz ud din fell into Timur's hands, A.H. 790, at the capture of Kalah Rumiyan, and was sent with his son, Sayyidi Ahmed, to Andijan. After three years of captivity, he was restored to his government. During the interval between 790 and 795 we find him assisting the Muzaffarian prince, Zain al Abidin, to escape from confinement, and conducting him to Ispahan, which was soon besieged by Shah Mansur, whereupon Iz ud din visited the latter's camp, and soon after the place surrendered.§ He must afterwards have proved insubordinate, for we read in Sherif ud din how, in the year 795, Timur sent troops to ravage Luristan; while Iz ud din, who was then at Kurrambad, fled, and was pursued by the Mirza Omar Sheikh as far as the fortress of Munkereh, not far from Vasith, on the Tigris, reducing the country *en route*.‖ In this campaign other districts of Kurdistan and Luristan were overrun by the Mirzas, Muhammed Sultan, and Pir Muhammed.¶ The same year Timur made over Little Luristan to Pir Ahmed, the ruler of Great Luristan.** The biographer of Mirza Izkander tells us that Iz ud din accompanied Timur in his Syrian campaign. I am not sure that he has not confused him with a third prince of the same name, who was then ruler of Jeziret. He goes on to say that his son Sayyidi, having failed to pay the tribute, Iz ud din was accused of treachery and flayed alive after a reign of fifty-four years. "The Jihanara," which also mentions this fact, dates it in HEJ. 804. Sherif ud din tells us how, in the year

---

* Journ. Asiat., 4th ser., iv. 447.   † Sherif ud din, i. 406-7.   ‡ *Id.* 437-440.
§ Sherif ud din, ii. 174-181.   ‖ Op. cit., iii. 267-168.   ¶ *Id.* 172.   ** *Id.* 206.

806, there arrived at the Imperial camp a Circassian officer, who had with him the head of Malik Iz ud din of Little Lur, who had revolted. The officer said that he had flayed his victim, stuffed his skin with straw, and exposed it to public view as a terror to marauders.* His son, Sayyidi Ahmed, called Pir Muhammed and Pir Ahmed by Sherif ud din, succeeded him. The latter author tells us how he did homage to Timur in the year 795 HEJ., and, the same year, was confirmed in his principality by the great conqueror, and we are told he returned to his ancient appanage with 2,000 families whom Shah Mansur had pillaged.† The "Jihanara," also abstracted for me by Dr. Rieu, tells us he reigned till 815 HEJ., and was then succeeded by Shah Hussein Abbasi. He captured Hamadan in the time of the Timurid Abusaid ; and was surprised and killed by a Baharlu chief, Kur Pir Ali, in A.H. 871. Our notices are now very broken. We read in the same author of Shah Rustem attending upon Shah Ismail Safavi ; then of Aghur, son of Shah Rustem, accompanying Shah Tahmasp in his campaign in Khorasan in the year 940 HEJ. During his absence the throne was seized by his younger brother, Jehanghir, who slew him on his return. Jehanghir was in turn put to death, by order of Tahmasp, in A.H. 949. He was succeeded by his son, Shah Rustem, who is the last prince of this line of whom I have any notice.

Note 4— Yezd.—The first atabegs of Yezd, according to the "Jihan numa," as reported by Von Hammer, sprang from the Dilemite, Abu Jaafer Muhammed Kakuyeh, which last name the Arabs corrupted into Kakeweih. His son, Abu Mansur Firamur, was appointed governor of Yezd by the Seljuk ruler, Tughrul, in the year 443 HEJ., and was succeeded by his son, the Amir Ali Ibn Firamur, who fell fighting against the Karakhitai, 536 HEJ. Sultan Sinjar gave the government of Yezd to Sain ibn Wirdan, whose mother was a daughter of the amir Ali just named. He built a great mosque at Yezd, and surrounded the grave of Ali at Meshed with a wall ; and was succeeded by his brother, Iz ud din Beshker in 590 HEJ. He was a brave prince, and was intrusted with Shiraz and Ispahan by the Seljuk rulers. He was succeeded in 604 HEJ. by his brother, Wirdansor, and he by his brother, Abu Mansur, surnamed Kutb ud din, the Khaljeh, who died in 616 HEJ., and was succeeded by his son, Mahmud. Mahmud died in 621 HEJ., and was followed by Salghur Shah, the builder of Salghurabad ; and he by Toghan Shah, who was a contemporary of Khulagu Khan,‡ whom he conciliated, and was permitted to continue in his principality. He was apparently the father of Alai ud daulat, whose sister, Turkan Khatun, married the ruler of Fars, and came to a very tragic end, as I have described.§ When Khulagu marched against Baghdad, Alai ud daulat sent the amir Abubekr, with 300 horsemen, to his assistance.‖ He was succeeded by his son, Yusuf Shah, who, in the year 1291, and during the reign of Gaikhatu, the ruler of Yezd, rebelled, and the Ilkhan sent Yisudar with a force to punish him.¶ Ghazan, it would seem, attacked Yezd, and incorporated its revenues in the

* Op. cit., vol. 4, 130.   † Op. cit., iii. 184 and 206.   ‡ Ilkhans, i. 68-69
§ Ante, 202-203.   ‖ Journ. Asiat., 4th ser., iv. 96.   ¶ Ilkhans, i. 40 492-3,

imperial treasury (*i.e.*, displaced its ruling family).* In 1294, the Ilkhan
Baidu appointed Sultan Shah, son of the great amir Nuruz, amir of Yezd, with
a diploma for 10,000 dinars.† About the year 713 HEJ. (1314), the Muzaffarian
prince, Mubariz ud din Muhammed, attacked Haji Shah, the atabeg of Yezd,
fought with him in the bazaar at Yezd, and compelled him to fly with his
servants and riches, thus putting an end to the dynasty which, according to
Mirkhond, had lasted 300 years.‡

Note 5—Shebankareh.—I have already given a short account of the earlier
history of this Kurdish principality.§ Jelal ud din Taib Shah, who mounted the
throne there, as I mentioned, in the year 664 HEJ., was put to death, by order
of Sultan Ahmed Khan, in the year 681 HEJ. He was succeeded by his brother,
Baiji ud din, who reigned prosperously till the year 688 HEJ. Ghiath ud din
and Nizam ud din, the sons of one of these brothers, were reigning at
Shebankareh when Wassaf wrote his history in 706 HEJ. About 755 HEJ., their
successor, Ardeshir, was attacked at Ij, or Idhej, his capital, by Shah Mahmud,
son of the Muzaffarian ruler, Mubariz ud din Muhammed. The place was
captured, and Ardeshir fled by a road behind the citadel. The principality
was then appropriated by the Muzaffarians.‖

---

* *Id.*, '. 69. † *Id.*, ii. 27. ‡ Journ. Asiat., 4th ser., iv. 98. § Ante, 204-205.
‖ The "Jihan numa" quoted by Von Hammer, Ilkhans, ii. 139. Journ. Asiat., 4th ser., iv. 108.

NOTES, CORRECTIONS, AND ADDITIONS.

Page 1 line 13—Insert " boundary of " before Mongol.

" 2 " 1.—I have overlooked describing the end of Rokn ud din. When Muhammed Khuarezm Shah fled from Kazvin to the Caspian his second son went to Kerman, where, reinforced by the troops of Zuzan, the governor of that province, he secured the capital and the treasure, which he distributed among the troops. After a stay of seven months he returned to Irak and prepared to attack Jemal ud din Muhammed, who had seized it. When he neared Rai he heard of the approach of a Mongol force under Taimaz and Tainal, and accordingly shut himself up in the fort of Sutun Abend, near Rai, which was situated on a scarped rock, and deemed impregnable. The Mongols assailed it, and after an attack of six months captured it, and Rokn ud din refusing to do homage was put to death with his people. This apparently happened in the year 619 HEJ. (1222). Zakaria of Kazvin, in his geographical work entitled "Assar-ul-Bilad," under the heading Demavend, states that it was in the latter fortress, situated near Rai, that Rokn ud din Gursaiji locked himself in the year 618, i.e., 1221. Rashid ud din says it was at Firuzkuh. (D'Ohsson, i. 347-48). Jemal ud din having heard of the death of the prince' offered his submission to the Mongols, hoping thus to retain the district of Hamadan. The Mongol generals sent him a robe of honour and invited him to their camp, where they killed him, with his suite. (D'Ohsson, i. 348-49.)

" 2 " 13—For " Hasan Karak," read " Hasan the Karluk." (See Note to page 43.) Hasan the Karluk has left coins struck in 633 and 634 HEJ., but without mint names. His full name was Saifud din al Hasan Karluk. (See Thomas' "Chronicles of the Pathan Kings," 92-98.) Uzbeg also struck coins, on which he styles himself Uzbeg Pai, and minted at Multan. (Id., 98-99.)

" 3 lines 10 and 11—The names " Salgar and Salgarids " are sometimes written " Salghur and Salghurids," and I have spelt them both ways.

Page 8 line 33—Taimaz was probably the Aitmas of Abulghazi. That author says that Ogotai sent Charmaghan and Aitmas with 30,000 men, and that Charmaghan dispatched Aitmas ahead with an advance guard. (Op. cit. 5.)

„ 8 „ 34—Tainal was doubtless the Tainal or Ainal who is named as a noyan in the service of Juchi Khan. (D'Ohsson, i. 223 ; Erdmann, 374.) One of Jingis Khan's cousins, the son of his uncle Daritai, was called Tainal Ineh. (*Id.*, 252.)

„ 11 „ 12—For " Lóró," read " Lôrhó."

„ 11 „ 17—By " Serirs," the well-known " Sirhghers " or " Kubachi," a tribe of armour-makers in the Caucasus, are meant.

„ 15 „ 9—For "the latter," read " Chin Timur."

„ 17 „ 16—For " they," read "the Mongols."

„ 18 and 20 lines 2 and 4—Or Khan, Uz Khan, and Otuz Khan are doubtless three forms of one name. He is called Azar Khan in the " Shajrat ul Atrak," which says he was one of Jelal ud din's confidential servants. When he heard of the advance of the Mongols he went to the bedside of the Sultan, who was asleep, and awoke him. The Sultan being confused from drink, had some cold water thrown on his head, mounted his horse and fled, leaving Azar Khan to resist the enemy, which he did for some time to allow his master to escape. He then followed him. The Mongols mistook him for the Sultan, and pursued him to Rai, but, finding out their mistake, they went after the latter and massacred every follower of his they met. (Op. cit., 185.)

„ 18 „ 30—The " Shajrat ul Atrak " says that Ogotai when he dispatched Taimaz with Charmaghan prophesied that Jelal ud din would fall by his hand, which actually came to pass. (Op. cit., 206-77.)

„ 21 „ 4—As I have described this campaign in so much greater detail in these pages, I think it well to repeat the anecdotes here referred to, which I have already quoted in the first volume. The princes of Diarbekr, Mesopotamia, &c., hid away, and the people were stupefied. " I have been told," says the historian Ibn al Athir, " things which are almost incredible, so great was the terror which seized upon everyone. I heard, for example, of a single Tartar horseman entering a populous village and proceeding to kill the inhabitants one after another, without resistance. I heard of a Tartar who, having no arms with him, and wishing to kill a person whom he had made prisoner, ordered him to lie still till he went to find a sword, which the man actually did, and was put to death accordingly. One person told me : 'I was going along with seventeen people. We met a Tartar, who ordered us to tie ourselves together, with our hands behind us. My companions began to do so, when I said, "This man is alone; let us kill him, and escape." "We are too much afraid," they replied. "But this man will kill you," I answered.' None of them would, however, help, and he had to kill him himself, and they all then escaped." (D'Ohsson, iii. 69 and 70.)

Page 39 line 6—Isferain I have elsewhere written Ezferain.

„ „ „ 15—Nussal was a Kerait of the tribe Tubaut, Erdmann calls him Bisel (Erdmann, 243).

„ „ „ 19—Erdmann says Chin Timur was an Ongut. (Op. cit., 243.)

„ „ „ 32—Ungu Timur is called Atgu Timur by Erdmann. (Op. cit., 243.)

„ 40 „ 1—Instead of " Bahu ud din," read " Bahai ud din."

„ 40 „ 12—On reaching the Imperial Court we are told that Ogotai wished to dine in the tent where Ongu Timur had done homage to him. After the banquet he went out for a short time, when a gust of wind came and blew down the tent. The Emperor ordered it to be cut in pieces. A few days later Kurguz entertained Ogotai in a tent in which all kinds of precious objects, which he thought would please him, were set out, among them a girdle made of certain stones called yarkand (i.e., doubtless of jade). The Emperor was in a good humour, and drank freely. He eventually decided in favour of Kurguz. When the latter's rival, Ongu Timur, was ordered to be handed over to Batu, Chinkai went to Ogotai with a message from him: " The Khakan is Batu's superior. Does it seem right that a dog like I am should cause two such sovereigns to deliberate? The Khakan should decide himself." " You say well," said Ogotai, " for Batu would not in such a case have pardoned his own son." (D'Ohsson, iii. 114.)

„ 40 „ 42—Instead of "Ante 134," read "Ante i., 134."

„ 41 „ 35—After " there," insert " called Sertak Gajan."

„ 42 „ 1—After " Rogdi," read " he was arrested by Tubedai, son of Bisel Noyan (i.e., Nussal)."

„ 43 „ 12—In regard to this obscure campaign, Rashid ud din mentions a General Hukatu who was sent by Ogotai to conquer Kashmir and Hindustan. Stephen the Orpelian calls him Hogata, and says he advanced towards Heutgasdan, by which India is doubtless meant. (St. Martin Memoirs ii., 121 and 269.) Bar Hebræus in his Syriac Chronicle (page 503) says this expedition went towards India, while in his Arab Chronicle (page 306) he says it went towards Tibet. In the " Yuanchao pi shi," published by the Archimandrite Palladius, Ogotai is made to say of the Khalif of Baghdad : " Chormakhan has already been sent against him; I shall now send thither Okhotur with Mungétu to reinforce him." (Mems. Peking Mission, iv. 152.)

Page 59 line 16—This name, Ilchikidai, I have elsewhere also written Ilchikadai.

„ 59 „ 25—*Vide* ante, page 42.

„ 60 „ 23—He is called Kara Buga by Chamchean, see Avdall's tr., p. 241.

„ 62 „ 42—After "Albania," insert "*i.e.*, Arran."

„ 72 „ 36—Instead of "their," read "his."

„ 76 „ 37—After "Vastak," insert "*i.e.*, Vataces."

„ 80 „ 37—Erase the note of interrogation after Ammorico.

„ 81 „ 32—For "dogs," read "a dog."

„ 84 „ 36—On this city of Kurmán, now called Kirmán, situated about 82 miles from Ghazni, in Afghanistan, see Thomas' "Coins of the Pathan Sultans," p. 26, note 2.

„ 85 „ 21—The British Museum has recently acquired some silver coins with the name of Mangu upon them struck at Herat and Nimruz. David the Georgian king struck coins with the name of Mangu upon them. One of them has on one side the inscription, "King David son of George by the aid of God," struck at Tiflis; on the other, "By the power of God and the good pleasure of the padishah Mangu Kaan in the year 650" (*i.e.*, 1252). On a bilingual coin of the same prince we read, "King David the slave of the padishah of the world, of Mangu Kaan." (Melanges Asiatiques, St. Pet. Acad. ii., 105.)

„ 92 lines 29, 32, &c.—Orkhan is otherwise called Uzkhan.

„ 96 line 27—For "Yushmut," read "Yashmut."

„ 97 „ 2—"Tokuz" is also written "Dokuz."

„ 98 „ 14—Erase the words "set out in February, 1254, and "

„ 99 „ 37—After "Journal Asiatique" read "5th series."

„ 100 „ 25—For "Sigistan," read "Sijistan, or Seistan."

„ 101 „ 6—Insert "in the 'Herat Chronicle'" after "event."

„ 102 „ 19—For "Hadari" read "Haidari."

„ 103 „ 34—For "they" read "he."

„ 110 „ 6—For "those," read "their."

„ 110 „ 10—For "Arkauns," read "Arkhauns."

„ 116 „ 41—For "Jehangir" read "Jehanghir."

Pages 132 and 133 lines 37 and 29—For "Arbil," read "Erbil or Irbil."

Page 135 line 7—Instead of "asking that," read "but asked for."

„ 140 „ 16—"Sidak" ought perhaps to be read "Susak."

„ 140 lines 17 and 24—For "Arslan," read "Arghun."

„ 143 line 42—"Maghreb" really means "the West."

„ 153 „ 1—For "Gazan," read "Ghazan."

„ 160 „ 32—For "Loró," read "Lôrhé."

„ 161 „ 4—For "Mowahid," read "Muayid."

„ 162 „ 23—For "Gazan," read "Ghazan."

„ 165 „ 3—Insert "were," after "orders."

„ 170, 171 and 176, lines 15, 29 and 2—For "Ain Julat," read "Ain Jalut."

„ 183 line 29—For "Pershwa," read "Peishwa,"

Page 184 line 8—For " Siraj," read " Saraj."

" 185 " 30—For " Kurt," read " Kert."

" 190 " 21—For " Ain," read " Ani."

" 192 " 8—For " he," read " she."

" 203 " 10—" Hasneviyeh " is otherwise written " Hasuieh " and " Hasnuieh " (See the two next pages. Von Hammer writes it " Fashiyeh." Ilkhans, ii. 139).

" 203 " 24—Coins of this princess, who is also called Ayish—one in gold, struck at Shiraz in A.H. 679, and the other in silver, in 684—are in the British Museum.

" 204 " 24—"Almarz" is called " Mobariz ud din" by Von Hammer. (Ilkhans, i. 69.)

" 204 " 43—For " Mirkhoud," read " Mirkhond."

" 205 " 9—Von Hammer calls " Nusret ud din" " Nasir ud din." (Ilkhans, ii. 139.)

" 207 " 35—For " Khurestan," read " Khuzistan."

" 208 " 1 and 4—For " Gazan," read " Ghazan."

" 209 " 14—Khanikof says he found Khulagu's grave on the right bank of the Jaghatu or Chagatu, not far from its outfall into Lake Urmia. The Nomades call the place Kizil Kurujan, and it seems some remains of the mausoleum erected over him by his sons still exist. (Melanges Asiatiques, St. Pet. Acad., ii. 508.)

" 213 " 13—"Kunkurtai" I have elsewhere written " Konghuratai," following Von Hammer.

" 213 " 40—The British Museum has recently secured a gold coin of Khulagu.

" 215 " 29—Aujan is mentioned by Clavigo under the name of Hujan or Hugan.

" 219 " 26—Mirkhond calls Tudan the son of Sugunjak. (See D'Ohsson, iii., Corrections et Additions.)

" 220 " 19—For " Yesd," read " Yezd."

" 221 " 37 For " Sheikh of Islam," read " Sheikh ul Islam."

" 224 " 1—After " and " insert " he."

" 225 " 19—For " Ablastan" read " Ablestin."

" 225 " 18, 30, and 226, line 6— For " Toros," read " Thoros."

" 225 " 30—In the " Mesalek Alabsar Ji memalek alamsar " of Shihab ud din Abul abbas Ahmed we read that, in regard to the pretensions set up by the princes of Kipchak to the towns of Tebriz and Meragha, our author was told by the Mollah Nidam ud din Abul Fazl Yahia Taiari that Khulagu, after making his conquests, assigned to the contingent sent by the Khan of Kipchak and Khuarezm to help him, Tebriz and Meragha and their revenues. On the death of Khulagu the soldiers of Kipchak assured his son and successor, *Abaka, that Bereke wished to build a mosque at Tebriz. They obtained permission to build one, and put Bereke's*

name on it. They then obtained permission to set up a
manufactory there, where stuffs were made for their own
use and that of Bereke. When the war broke out between
Abaka and Bereke the former destroyed this manufactory.
On the restoration of peace it was rebuilt, and it was agreed
that the subjects of Bereke should continue to receive the
dues from their appanage as before, and also should manufacture what they wished. Presently, the princes of
Kipchak set up claims to the possession of Tebriz and
Meragha, on the ground that Bereke had built the mosque
and manufactory, and sent envoys to Ghazan to assert
their claims, declaring that it was their forefathers who
had conquered these places, and that they consequently
belonged to them by way of inheritance. Ghazan replied
that it was to his sword, and not to inheritance, that he
owed his possessions, among which were Tebriz and
Meragha, which he meant to keep, and the sword should
decide between them. Uzbek, Khan of Kipchak, continually
reasserted these claims. (Notices et Extraits, &c., xiii.
283-284.)

Page 240 line 9—For " him " read " them."
  „   253   „   7—For " Soukor " read " Sonkor."
  „   255   „ 20—For " was," read " had been."
  „   255   „ 28—For " Kai Kobad," read " Kai Kavus."
  „   256   „ 37—For " now," read " then."
  „   258   „ 32, and 267, line 24—For " Kongurtai " read " Konghuratai."
  „   259   „ 41—For " Musud," read " Masud."
  „   260   „ 19—The Ering here named is called "Irenchin" by Eastern writers.
  „   260   „ 30—For " Ibek," read " Eibeg."
  „   266   „ 22—For " they," read " the Egyptians."
  „   269   „ 43—For " Armeine," read " Armenie."
  „   270   „   5—Abulfaraj says that Jelal ud din was the patron of the Papa
                   previously named, and had been put to death with him, and
                   that Masud's death was brought about on the accusation of
                   Jelal ud din's sons.
  „   272   „   6—I owe to the courtesy of Sir Charles Wilson a reference to the
                   following translation of a letter written by a crusader, Sir
                   Joseph de Cancy, a Knight Hospitaller, to King Edward
                   the First of England, giving him an account of these events.
                   This document, as the handiwork of a contemporary who
                   actually shared in the struggle, is most valuable. With
                   the king's answer it is preserved among the Royal Letters
                   in the Record Office, and the translation is from the accom-
                   plished pen of Mr. W. B. Sanders. It will be seen that
                   *the letter somewhat mitigates the accounts of the disastrous
                   rout of the Mongols given by other writers.*

NEWS FROM SYRIA.

"To the Most High and Puissant Lord, my lord Edward, by the Grace of God, most worthy King of England, Lord of Ireland, and Duke of Aquitaine, the least and lowest of his servants, Joseph de Cancy, humble brother of the Holy House of the Hospital of St. John of Jerusalem, dwelling at Acre, kneeling in the service of your Highness, sendeth greeting.

"Forasmuch as your worthy lordship commanded us to continue sending you news of events as they befell in the Holy Land, know ye, sire, that after our Master was come to Tripoli in the close of the month of October, as we have already informed you by our letter written during the passage of the Holy Cross, the hosts of the Tartars and Saracens drew so near as to place the Saracens between our men and the Tartars, so that neither we nor the Prince [of Antioch, Boémond VII.]—the King of Cyprus [Hugh III.] not being yet come up—could join the Tartars, nor they send to us as they had settled to do. Upon this the armies advanced to the close. The Soldan divided his army, which consisted of 50,000 horsemen, into three battalions, and he himself was with that of the centre, which they call the "Heart," after their custom. Sangar Layfscar, Lord of Saone and our marches of Margat, was captain of the left, and the right was commanded by a valiant Turk named Heysedin Laffrain. The Tartars, seeing the array of the Saracens, also formed their people, who numbered 40,000 horsemen, into three battalions, for their Chief had sent the rest of his men to his eldest brother Abagua, who was marching through La Berrie, imagining that Abagua would reach Damascus before him. In one of these three battalions was the King of Armenia with his power and 2,000 Tartars and 1,000 Georgians ; and a Turk named Samagar, who had become Tartar, was also in his company with 3,000 of his countrymen whom he had brought from Turkey, and who called themselves Tartars. The King of Armenia, thus arrayed, threw himself upon the Saracens' left, and so broke and discomfited it that few escaped being put to the sword, and of this left battalion none would have escaped but for the disloyalty of Samagar, who fled with most of his people without either striking or receiving a blow. The right battalion, commanded by Manguodamor, closed with the Soldan's right, in which he had 10,000 Tartars without counting their allies, and put them to rout, but their discomfiture was not nearly so complete as that which had been inflicted upon their comrades in the left. Manguodamor, who is a valiant, bold, and trusty knight, with the remnant of his people, threw himself upon the division in which was the Soldan, and then ensued a great carnage, and the battle raged from before the hour of tierce until sunset. And now, had it not been for the Soldan's gallant bearing, and his prudence and valour, the fate of the left wing would have befallen himself also ; but in the midst of the disasters which surrounded him, seeing his men so evil-handled and killed and some turning in flight, he commanded his trumpets and nakirs to sound, and rally round his person those who survived; without which all would have been destroyed, for of his entire host 600 men alone obeyed the call. The Tartars, imagining that the Saracens were completely defeated, rushed to the pillage, and entirely took the tents of the Soldan and other Saracens, with so great a spoil that no one could with certainty tell us the value thereof. And of the rabble who' followed the camp, who made it like a city full of people, so many were slain that the number could not be known. With which said spoil most of the Tartars returned to their fastnesses, as men who are very covetous, riding on the horses of 'the dead Saracens, which were better than their own, and leaving their sorry beasts behind them. And know ye this, sire, which is considered a great marvel, never was booty taken from one side or the other that could be reckoned, nor could anyone say that anyone had been wounded or afterwards hurt to the death [onqes piles niot trait d'une part ni d'autre qui aconter face ni qe nul puisse dire qe nul fust feri ni nafré de pues à la mort].

" The Soldan, seeing the great cloud of dust raised by those who were thus departing with the spoil, and fancying it was caused by the Tartars, marched towards it. Manguodamor, who was at hand, and had got together a few men amounting to no more than sixty horsemen, advanced to meet him, thinking they were his own people: for the Kings of Armenia and Georgia had gone forward with their following into the country of the Saracens. Now when the Soldan and his people saw Manguodamor, and recognised his companies by their ensigns, they suspected that an ambush was laid for them, and that the display of so small a force was intended to betray them into it. Manguodamor on the other hand, seeing the weakness of his own hand, and the danger of awaiting an attack by the Soldan, fell back and went his way. The Soldan saw this, and imagining him to have done so for the purpose of hastening up his whole army, retired in haste. And so night parted them. So neither the one nor the other held the field; but because the Soldan was the last to retreat men thought the victory ought to belong to him. But well may one say with truth that never since the first conquest of their country have the Saracens received so great a check or been so completely cowed as they were then and are still.

" The King of Armenia, with a great portion of his host, returned to the battle-field, and finding it unoccupied thought to pitch his tents and remain there till the morrow, which, as he was preparing to do, came the traitor Samagar with a part of his men, saying, " Sir King, why dost thou this? Our lord Manguodamor is gone." The King answered that he wished to encamp there for the night, for his men were worn out with fatigue; but Samagar maintained that it would be great treason and disloyalty to do so after their chief had left. So, after many words, the King believed him, and ordering his troops to horse, rode all night till he had passed the place from which the tents had been moved, but found not Manguodamor. The King halted for a short time to rest his horses, but Samagar went his way. Then the King turned towards his own country and passed through the Dry Lands, where there is neither water nor grass, insomuch that many of his horses and companions died of thirst upon the road or perished through the toil they underwent, till he reached his kingdom at last safe and sound, but in evil plight, while many of his followers who had tarried behind came as they best could: for Samagar's people had robbed them by the way, stripping them to the skin and leaving them no horses to ride. The Soldan took counsel with his people by which road he might safest return to

his dominions. Some advised that he should go by the sea-coast into the
country of the Christians, with whom he had truce; others by La Berrie,
where the Tartars should not find him; while others again advised him to
choose the shortest and straightest path. With these he agreed, and so came
to a town which is called Le Lagon, where he had formerly camped on his
advance against the Tartars. The Count of St. Sevrin, bailiff of Acre, sent
messengers and presents to him in order that he might see and ascertain his
condition, which they found poor and little enough and his attendance scanty.
The Soldan, because he would not that the Franks should know his poverty
and misfortune, making courteous reply to the Count, departed by night, and
marched into Babylon. There he tarried some days, and caused a tax to be
levied on all his subjects, taking a third upon those who had 10,000 bezants,
and so from each, rich and poor, according to his condition, whereby his subjects are much discouraged with him, and think themselves doomed to death or
ruin. Then he caused to be proclaimed throughout the land of Egypt that all
those who wished to receive their pay to go to Margath and into Armenia
should come and take it and make ready for the journey. And he caused this
proclamation to be cried once in each week for one month, in spite of which
that most persons say he will not quit Babylon because of his great losses in
men and horses.

" On the other hand, sire, he has put to death fifteen Emirs, as well of those
who deserted him in the field, as of those whom he had left behind in Babylon,
and those whom he had cast into prison, by reason of which things his subjects
are much disheartened and filled with hatred against him. None of his people
for all these threats which he has made, are as yet come to Babylon or Damascus,
at the time of writing these present letters, yet it is true that the Chastelain of
Saphet and his other bailiff on our Marches have made the Bedouins, who were
in the pasturage near us, retire into the mountains, because they say that the
herbage must be kept for the coming of the Soldan. And we suspect them to
give out this that they may make us wish to enter into some evil truce with
them, which may God forbid we should do! Moreover, sire, we understand by
the mouths of good and trustworthy persons lately come from the parts about
Hamous (i.e., Hims) that there is so great a panic there, and in Hallamp and
La Chamelle, that each day men fear a surprise by the Tartars, who have sworn
to come without fail; but this we think cannot be till the setting in of winter.
Wherefore the Soldan of Hamous seeing these things, has sent his wife and
children and most of the treasure of the city into Babylon. On the other side
the men of Baudac understanding by the Soldan's letter that the Tartars had
been defeated, rose in revolt against the rulers whom the Tartars had set over
them. Abagua being then near at hand in La Berrie, hearing this rode thither
and took the city of Baudac, which was subject to him at the time of the
revolt, putting all the men-at-arms to the sword and cutting off the thumbs of
the footmen and . . . . for you know, sire, that they draw with the thumb.

" Other news have we none, at the time of writing these presents to send to
your Highness, save that we have garrisoned our castle with brethren and
men-at-arms, as it behoved us, promptly. Our Master, at the prayer of the

King of Armenia, and considering the evil plight he was in and the ravages committed by the Turcomans in his kingdom since his return by burning and laying waste the city of Lays and other of his towns and villages, has sent him 100 horsemen, 50 brethren well appointed, and 50 Turcoples. But know ye, sire, that never in our remembrance was the Holy Land in such poor estate as it is at this day, wasted by lack of rain, divers pestilences, and the paynim—the greater part of Babylon left unsown for fear of war, and the reason above mentioned; and not only this country but Cyprus and Armenia are in the same condition . . . . the King of Sicily will suffer no provisions to be sent out of his dominions into Syria because of his war with the Greeks, as we understand. Therefore, sire, as we have already written to your Highness, if any of the great lords of your country should come to these parts he would do well to advise the King of Sicily to permit provisions to be carried into Syria as in former times they were wont to be.

"And know, sire, the Holy Land was never so easy of conquest as now, with able generals and store of food; yet never have we seen so few soldiers or so little good counsel in it. May your worthy and royal Majesty flourish for all time by increase of good for better. And would to God, sire, that this might be done by yourself, for it would be accomplished without fail if God would give you the desire of coming here. And this is the belief of all dwellers in the Holy Land, both great and small, that by you with the help of God shall the Holy Land be conquered and brought into the hands of Holy Christendom. These news, sire, are . . . . those which you may believe in spite of any other things that may be told to you. And pardon us, sire, that our letter is so long, for we could not more briefly inform you of these things, the certainty of which your Majesty has left me here to record.

"Written on the last day of May.

"To the most noble, excellent, and puissant King of England."

It was probably Sir Joseph de Cancy's account of Sultan Kelavun's victory that occasioned the following letter from King Edward, the draft of which is still preserved among the Royal Letters in the Record Office, though a good deal damaged by damp and time.

"EDWARD, BY THE GRACE OF GOD, King of England, Lord of Ireland, and Duke of Aquitaine, to his dearest in Christ and faithful secretary, brother Joseph de Chauncy, greeting: For the accounts which you have sent us in your letters from the Holy Land we give you great thanks, because we are made the more joyful the oftener we hear good news of that land and its condition, the which we vehemently wish and desire to hear more frequently. And whereas you desire to hear prosperous reports of our state, we signify unto you, in order to the increase of your comfort, that on the day of the making of these presents, we and our Queen and our children are—blessed be the Most High—flourishing in full health of body; which we would rather know of yourself by true relation than hearsay. For the rest we have received, with cheerful hand, your New Year's gift of jewels which you have sent to us—to wit—two Circassian saddles and two saddle-cloths; and two Gerfalcon's hoods and four Falcon's hoods, for which we return you our abundant thanks. Wishing you to know that we have not considered these presents as small, because we have weighed the goodwill of the giver more than the gifts themselves in this case. Nor indeed do we at present want any more hoods, as by reason of arduous matters of our kingdom which intimately concern us, and do not as yet wish to keep more falcons than we already have. But as regards those stones of rubies which you have sent us . . . . And because we much wish that you should be near us, for our solace and convenience we will and require you that you hasten your arrival in England by the best and quickest means you can—and this as we entirely trust in you—you shall in no case omit . . . . . . of the Hospital in England or the possessions of the same we will have in commendation and uphold them as far as we can by law, as you have requested. Concerning your own estate, which may the Most High prosper, we desire that you certify us thereof by frequent notification. Given at Worcester on the 20th day of May, in the tenth year of our reign." [1282].

Page 278 line 24—The rulers of Hosn Keif were "Ayubits," and not "Ortokids."

,, 282 ,, 3—For " Safr," read " Safi."

,, 283 ,, 4—For " Urmeir," read " Urmevi."

,, 283 ,, 32—Among the recent acquisitions of the British Museum is a gold coin of Abaka, struck at Baghdad, and a silver coin struck at Tebriz. Others struck in the same town are in the Hermitage collection at St. Petersburg. On the word read " darugha " by Schmidt, see page 355.

,, 290 ,, 2—Von Hammer calls him Bulghuvan in the citation given, but on page 368 of his work he names him " Bulughan."

,, 290 ,, 3—For " Toghajar," read " Togachar."

,, 298 ,, 19—For " ask," read " urge upon."

,, 298 ,, 20—Erase " to state."

,, 298 ,, 40—For " Sengi," read " Zengi."

,, 300 ,, 11—For " Bekhr," read " Bekr."

,, 300 ,, 30—For " 80 tumans," read " 8 tumans."

,, 304 ,, 33—For " Abugan," read " Abukian."

,, 307 ,, 19—For " Mosellemin," read " Mosellemi."

,, 310 ,, 33—Mr. S. Poole refers me to a coin in one of the Russian collections, on which is inscribed " Sultan Ahmed," in Mongol characters.

,, 315 ,, 17—For " famous," read " also noted."

,, 326 ,, 5—For " Bekhr,' read " Bekr."

,, 328 ,, 18—Insert " and " before " had."

,, 332 ,, 41—For " they," read " the Mongols."

,, 335 ,, 30—This Choban Sulduz became very famous, and was the founder of the later dynasty of the Chobanids.

,, 338 ,, 39—For " while," read " for."

,, 342 ,, 14—Transfer the words "put to death" to the end of the sentence.

,, 344 ,, 44—For " 5-8," read " 58."

,, 347 ,, 36—Erase " and."

Page 352 lines 7 and 8, &c.—Instead of the sentence "having heard of it he had" read "having heard that some of the principal Saracens had greatly rejoiced," &c.

„ 352 line 9—Erase " and that."

„ 355 „ 21—In his posthumous essays, Fraehn publishes a coin of Arghun struck at Izferain, and Mr. Stanley Poole writes me that there is one in the Hermitage collection struck at Meragha.

„ 357 „ 27—For " Imeuthi," read " Imerithi."

„ 365 „ 42—For " Kabushane," read " Kabushan."

„ 377 „ 34—Fraehn, in his posthumous essays, publishes a coin of Gaikhatu, struck at Hamadan. There is another in the British Museum.

„ 392 „ 45—The Japanese first used paper money, which they called kami zeni, in the years 1319-1331, but in Japan the issue was never so excessive, nor did the value depreciate in the same way. (Klaproth Mems. Relatifs a l'Asie, 388.) In regard to the Persian experiment, Colonel Yule says aptly that block printing was practised, at least for this one purpose, at Tebriz in 1294. (Marco Polo, i. 416.)

„ 394 „ 10—For " in charge of," read " in charge to."

„ 400 „ 27—For " Erzenrum," read " Erzerum."

„ 407 „ 36—For " 669 HEJ," read " 696 HEJ."

„ 407 „ 39—For " Nasir ud din," read " Nuzrat ud din." (See D'Ohsson, iv. 170. Ilkhans, ii. 39. Ibn Batuta, ii. 31.) In Von Hammer's Ilkhans, ii. 321, he is called Nasir ud din.

„ 416 „ 2—For " Chaunters at funerals, as prescribed by the Koran," read " Chaunters of the Koran at funerals."

„ 427 „ 42—Insert " ibn " between Kaikobad and Firamurz.

„ 430 „ 33—For " Thoras," read " Thoros."

„ 435 „ 14—Insert " for " after " and."

„ 435 „ 25—In the " Chronicle of Cyprus," by Florio Bustron, published in the collection of " Documents Inedits etc.," we read that Ghazan sent a messenger to Henry, King of Cyprus, and to the Templars, Hospitallers, and Teutonic knights there, asking them to join in the campaign, promising to make Egypt (here called Babilonia) over to them; saying that he had sent Rabanata as his envoy to " Franza," and now sent Cariedin to the kings of Armenia and Cyprus to bid them march with all their men, and that he himself was setting out with one hundred tumans. The letter was written from Averel on the 21st October, in the year of the *pig*. The envoy duly arrived at Cyprus on the 3rd of November and was received with honour, and was followed by a second messenger, who arrived on the last day of November. (Op. cit., 129-130.) The King of Cyprus sent two galleys and two frigates to Butron under the command of

Polo del Anecto, with orders not to advance beyond that place. He was tempted, however, by the statements of the Christian fugitives who went to him there to make an attempt to capture Tripolis, and set out with a small force, but was attacked about a league from Butron by the Egyptians, who compelled him to retire. He barely escaped with his life, and his companions returned to Cyprus with the galley. (*Id.*, 130-131.) Thereupon the King consulted with the master of the Templars, the " Comendator" of the Hospitallers, and "Signior Chiol," ambassador of Ghazan, and dispatched fifteen galleys under the Captain Bennondo Visconte and the Admiral Baduin de Pinquani, and made a descent upon a place called Rasni, situated on one of the mouths of the Nile. We are told that in the skirmish which ensued Chiol carried a banner of Ghazan. The fleet made descents upon Alexandria, Acre (where the Cypriots landed and defeated the garrison), Tortosa, Maraclea, and returned again to Cyprus by way of Little Armenia. (*Id.*, 131-132.)

Page 440 line 32—Transfer the comma from after Ghazan, to after hyperbolically.

,, 442 ,, 42—For " Christian," read " Christians."

,, 446 ,, 20—For " he," read " the Sultan."

,, 446 ,, 41—The "Chronicle of Cyprus," which calls Kutlugh Shah Catholesso, says it was he who sent on the King of Armenia, and with him Giudo de Iblia, Count of Zaffo, and Geormane, Lord of Siblet, to have an interview with Ghazan. (Op. cit., 132-133.)

,, 450 ,, 36—For " chapter," read " volume."

,, 458 ,, 5—Insert a comma after " before."

,, 477 ,, 25—For " my," read " our."

,, 481 ,, 21—Erase the words "at this time."

,, 486 ,, 36—For " Armiiah," read " Arminiyah."

,, 486 ,, 43—In his posthumous essays Fraehn publishes Amid and Mosul as mint towns of Ghazan. In the Hermitage collection at St. Petersburg, as Mr. S. Poole tells me, there occur the additional mint towns of Khelat and Khartapirt ; in Major Trotter's collection is a coin struck at Amasiya (Numis. Chron. for 1887, page 334) ; while among the recent additions to the British Museum are coins of Ghazan struck at Jajerm, Nishapur, Yezd, Erzenjan, Samsun, and Wasith.

,, 513 ,, 6—For " they," read " and."

,, 530 ,, 20—Insert " that of" before " older."

,, 543 ,, 40—In the 55th chapter of the book of the Histories of Johannes of Dzar there is inserted a notice about Uljaitu, which is apparently taken from some unknown contemporary document. In this we read that during the year 1306 " Karbanda Khan, autocrat of the nation of the Archers, a wicked man,

who hated the Christians, led away by sorcerers and
heretical sheikhs, and inspired by the wicked counsels of
their assistant, Satan, began the struggle against the
invincible rock of Christ. A decree was published in all
the universe, referring to the Christians under his domination, that they should adopt the stupid religion of Mahomet,
or that each person should pay a kharaj tax of eight
dahecans; that they should be smitten in the face, their
beards plucked out, and should have on their right shoulders
a black mark—all on account of his hatred to Christ. The
wicked ones thereupon spread themselves over the towns
and villages, and convents, sowing terror everywhere; for
the fatal decree forbade the performing of mass, the entering
of a church, and the baptism of infants. It was determined
to overwhelm the Christian religion by one blow. Meanwhile, the Christians remained faithful. They paid the
exactions, and bore the torments joyfully. Karbanda Khan,
seeing that these means were insufficient, ordered them all
to be made eunuchs, and to be deprived of an eye, unless
they became Muhammedans. Many perished in this
persecution." (Brosset, Collection d'Historiens Arméniens,
i. 568-569.)

Page 557 line 37—The "Chronicle of Cyprus" in relating these events calls the
town here named "Anazarba," "Navarsan." It says that
Bilarghu demanded that it should be made over to him in
order that he might defend it against the Saracens. Haithon
replied it was not in his power to make it over, saying
he was merely governing the country on behalf of his
nephew Leon, who had been crowned when a child, and
was still an infant, and that he could not dispose of any
part of the kingdom without the consent of the Holy See.
Bilarghu said he did not want the town making over to
him, but only to be allowed to put a garrison there. Haithon
thereupon consulted with his barons, and as the result sent
to his friend Almeric Signor de Sur, governor of Cyprus,
asking him to send him some help, who sent him 300 horse
and 1,000 foot soldiers. Bilarghu having heard of this sent
to the king bidding him and his chief men go to the court
to have a conference. He set out, according to this
account, with his nephew, with the Constable and Marshal
of Armenia and others. When they arrived they were all
decapitated. Haithon, according to this authority, left two
brothers, Chwysia (i.e., Oshin) and Alinach, who were
twins, but as Chwysin had first seen the light, his brother
ceded the throne to him, and he was duly crowned. On
the withdrawal of Bilarghu the Cypriots returned home
again. (Op. cit., 156-157.) The "Chronicle of Cyprus"

agrees with Abulfeda in making Alinach go to. the royal
ordu to complain of this murder of his relatives (*Id.* 161).

Page 560 line 38—For " bull," read " paizah."

„ 562 „ 1—For " the latter," read " Said," and transfer this and the two
following lines to the end of the next paragraph.

„ 566 „ 2—For " filled," read " fitted."

„ 568 „ 31—For " art now," read " at now being."

„ 568 „ 37—For " Yesaul," read " Yassaul."

„ 569 „ 13—For " King," read " Malik."

„ 570 „ 8—Baba was not a prince of the Golden Horde, but a descendant
of Juchi Khasar, and grandfather of Toghai Timur, whose
history occupies a later chapter.

„ 580 „ 22—In his posthumous essays, Fraehn adds the mints of Fenzen
and Hamadan, while among the recent additions to the
British Museum are coins of Uljaitu struck at Izferain,
Asterabad, Jajerm, Nishapur, Amid, Erzerum, El Bazar
Ordu, Kaiseriyah, Kerman, and Med Meruyeh (Isfahan).
In the Hermitage collection Mr. Poole tells me are coins.
struck at Tiflis, Baran, Jezair el ma'mur, and one with the
double mint of Sultania and Tiflis, while in the Asiatic
Museum at Moscow there is a coin of Uljaitu struck at
Amol.

„ 590 „ 40—For " is," read " will be."

„ 592 „ 10—In the year 1318, Jiamaluk, the turbulent chief of the Arab
tribe Kafaja, having encamped between Shiraz and Yezd,
molested the caravans there. He was attacked by Muzaffer,
and his head was taken to the Ilkhan Abusaid. (Journ.
Asiat., 3rd ser., xi. 306-307.)

„ 597 „ 10—For " turtu," read " turtur."

„ 601 „ 5—For " Ismal," read " Ismail."

„ 628 „ 9—M. Khanikof found on the walls of a mosque at Ani an
inscription containing a yarligh or order of Abusaid Khan,
which reads thus : " God, who embraces in his immensity
all his slaves. Abu Said Behadur Khan, who is at this
time the ornament of the capital of the sovereignty of the
world, the grandeur of the earth, and of religion. May his
reign be eternal. Yarligh: As everything from the East to
the West is under the shadow of his clemency and his
justice, may God, the all-powerful, exalt his might and
commandments. His command to us is that the people who
are subject to his orders, or group themselves about the
tribunal of his victorious name, must do neither little nor
much to any of his creatures. They must not receive any
dues, in the form of tamghas, and of Cusoms, and may not
demand from any creature anything else under the form of
kalan or of wine cup, inasmuch as formerly in this town
of Ani, and in other parts of Gurjistan (Georgia), under

pretext of kalan, of wine cup tax, and other illegal taxes, great iniquities have been committed. And inasmuch as, on account of the oppression, the place has begun to be ruined, and its people to disperse, and the assemblies of the people in the town and country are abandoning it because of the kalan ; and they have appealed to me in regard to their movable and immovable goods and their houses ; in order that the Almighty may not remove his shadow from over our heads . . ." The inscription is here broken off, but doubtless continued with the statement of the privileges conferred. Abusaid, as we have seen, took the title of Behadur in the year 719 HEJ. (i.e., 1319-1320), (Melanges Asiatiques, St. Pet. Acad., ii., 63-68.)

ge 628 line 32—In the Melanges Asiatiques of the St. Petersburg Academy, vol. viii., p. 443, etc., is the description of a very fine medallion of Abusaid, struck at Tebriz, in the year 724 HEJ., and found in the bed of the river Ijim, an affluent of the Ausa, in the district of Minussinsk, on the Chinese frontier. It is made of silver, and of the value of 4 roubles 87 kopeks. It contains, inter alia, the names of the four orthodox Khalifs. Mr. Lane Poole tells me that in the Hermitage collection at St. Petersburg are coins of Abusaid struck at Izferain, Kalensuwar, Berdaa, Abulshak, Astera-bad, and Jeziret. In Mr. Leggett's collection is a coin of Abusaid, struck at Arjish, near Khelat. (Numis. Chron., 1886, 230.)

„ 629 „ 23—Dr. Rieu suggests that " Comerum " may be " Kerman."

„ 629 „ 25—He also suggests that Huz is the modern Khuzistan.

„ 639 „ 18—There is a copper coin of Musa struck at Tebriz in 736 HEJ. at the Hermitage, St. Petersburg.

„ 639 „ 38—The amir called Artena by Ibn Batuta is apparently the same person who is called Irshad or Yeshed by other authorities. (See page 637.)

642 „ 22—At the Hermitage there is a coin of Muhammed's struck at Arrachin (?) in 738, and at the Oriental Institute at St. Petersburg is one struck at Sultania in the same year. In Mr. Calvert's collection is a coin of Muhammed, struck at Baghdad in 738 HEJ., and another at Erzenjan in 739. (Numis. Chron., 1885, 235.) Major Trotter has one of his coins struck apparently at Kaiseriya in 728. (Id., 1887, 334.)

„ 645 „ 27—At the Hermitage there is a coin of Satibeg struck at Sultania in 739 HEJ.

„ 653 „ 29—At the Hermitage is a coin of Anushirvan struck at Ganja, i.e., Kentsak, while at the Asiatic Museum, also at St. Petersburg, is one struck at El Kab.

Page 653 line 40—Ibn Batuta tells us the dominions of the Ilkhans were divided among eleven princes: (1) Sheikh Hassan, son of the Sultan's maternal aunt. (2) Ibrahim Shah (really Haji Toghai), son of the Amir Sunitai, who took Mosul and Diar Bekr. (3) The Amir Artena took Rum. (4) Hassan Khoja, son of Timurtash, took Tebriz, Sultania, Hamadan, Kom, Kashan, Rai, Weramin, Farghan (Werkan), and Karaj. (5) Toghai Timur took a part of Khorasan. (6) The Amir Hussein, son of Ghiath ud din, took Herat and the greater part of Khorasan. (7) Malik Dinar took Mekran and Kij. (8) Muhammed Shah, son of Mozaffer, took Yezd, Kerman, and Warku. (9) Malik Kotb ud din Tehemten, took Ormuz, Kish, Kathf, Bahrein, and Kalhat. (10) Abu Ishak took Shiraz, Fars, and Ispahan. (11) Afrasiab the Atabeg took possession of Idhej and other lands. (Op. cit., ii. 124-125.)

„ 655 „ 29—After the death of Abusaid, the town of Hillah was conquered by the Amir Ahmed, son of Romaitha, Prince of Mekka, who held it several years, and was praised by the people of Irak. Eventually he was vanquished by Hassan, ruler of Irak, who put him to the torture, killed him, and secured his treasures, &c. (Ibn Batuta, ii. 99.)

„ 657 „ 9—Insert "had" before "arrived."

„ 657 „ 36—Shems ud din Zakaria was Hussein's vizier. (Quatremere, Hist. des Mongols, lvi.)

„ 664 „ 38—For "that," read "to that which."

„ 667 „ 37—For "there," read "at Baghdad."

„ 672 „ 40—Insert "he" before "crossed."

„ 673 „ 40—For "Mir," read "Nur."

„ 678 „ 25—Erase the comma before "Sheikh."

„ 682 „ 23—For "Imirethi," read "Imerithi."

„ 692 „ 29—It was Abu Ishak's ambition to rival the generosity of the Sultan of Delhi, says Ibn Batuta, "but what distance separates the Pleiads from the Earth." The most generous act recorded of Abu Ishak was when he gave the Sheikh Zâdeh al Khorasani, who came to him as envoy from the Malik of Herat, 70,000 dinars, while the ruler of India had repeated such a gift many times. (Ibn Batuta, ii. 71-73.) Tash Khatun, wife of Abu Ishak, built a great medresseh and hermitage, where travellers found refreshment, and where the Koran was continually read, at Shiraz, near the sacred tomb of Ahmed Ibn Musa, descended from Abu Thatel. The khatun used to go to this sepulchral chapel on the Sunday night, where the khalifs, fakirs, and sherifs also met. The Koran was then recited antiphonally, refreshments were brought, and then the preacher preached. Meanwhile the Khatun occupied an elevated chamber protected from the public gaze by lattice work. Then cymbals were beaten and trumpets were sounded, as was customary at the gates of royal residences. On Thursday and Friday nights she visited the tomb of the saint Abu Abdallah, son of Khafif, when similar services took place. Near this last chapel was the mausoleum of the Amir Muhammed Shah Inju, father of Abu Ishak. (*Id.*, 77-79.) In one of the mosques at Shiraz, Ibn Batuta saw a number of Korans in their silken cases on a kind of stand. On the north side was a hermitage with a lattice opening on to the market place. There sat a sheikh who was well dressed, and was reading the Koran. He told one traveller that he had built and endowed the mosque, while the hermitage was intended for his tomb. He lifted up a carpet and disclosed a grave covered with planks, and close by showed a box containing money which was to be devoted to his funeral expenses and to alms for the poor. (*Id.*, 85-87.) This is very like a picture of a Christian anchorite. At Kazirun, two days' journey from Shiraz, and situated in a district inhabited by Shuls, Ibn Batuta went to see the tomb of the Sheikh Abu Ishak Kaziruni, after whom the Sultan had been named. The Saint was specially venerated by the navigators in the Indian seas, who addressed their prayers to him when bad weather or pirates threatened, and who devoted large sums in offerings. (*Id.*, 89-90.)

Page 692 line 38—Three of his coins have the name of Muhammed Khan upon them, whom he probably recognised as his suzerain. A coin in Mr. Kay's collection, struck in 719, was minted at Kazirun, the birthplace of the saint, his namesake.

„ 695 „ 17—M. Khanikof describes a mosque he saw at Kerman with an inscription saying it was built by order of Muhammed the Muzaffarian in the year 750. (Mémoire sur la partie meridionale de l'Asie Centrale, 195.)

„ 699 „ 6—For "his," read "Shah Shuja's."

„ 712 „ 42—Ibn Arabshah attributes the victory in a considerable measure to the defection of one of Mansur's officers, a native of Khorasan, named Muhammed Ibn Zain ud din. He went over to Timur at the beginning of the fight, and Shah Mansur was left with barely 1,000 men. He also gives a different account of the hero's death, and tells us that when his men were defeated, and he could hardly hold his sword from exhaustion, he buried himself under the slain and wounded, while he discarded his armour, in the hope that he would not be recognised, and might thus escape. Timur, however, had the battle-field searched, and a Mongol stumbled upon him. He disclosed who he was to him, and

offered him many precious stones he had with him if he
assisted him to escape; but he beheaded him, and took his
head to Timur. (Weil v., 39 and 40, notes.) Thomas of
Medsoph tells us a curious anecdote about this struggle.
He says that Shah Mansur had sent an envoy to Timur
with considerable presents, on whose arrival the latter
feigned to be ill. A lamb was killed, of which he drank
the blood; and at the audience with the envoy, his face
seemed like that of a corpse, while he had an iron bowl
brought in, into which he spat out the blood he had taken.
The envoy was delighted at the spectacle, declaring his
end would come that day or the day following. He
mounted a fast horse, and sped to his master. When he
reported the news, general rejoicings broke out throughout
Fars, and those who were reduced to misery set about
providing themselves with arms. Timur now set out by
forced marches towards Shiraz. (Op. cit., 14)

Page 718 line 23—Major Trotter has a coin of Toghai Timur, struck at Amasiya
in 737 HEJ. (Numis. Chron., 1887, 334.)

,, 723 lines 8 and 9—For "we suppose," read "is generally supposed."

,, 725 line 3—Insert "he" before "sent."

,, 734 ,, 2—For "Hassan," read "Haidar."